Ichthyology Handbook

Ichthyology Handbook

B.G. Kapoor □ Bhavna Khanna

Springer-Verlag Berlin Heidelberg GmbH

B.G. Kapoor
Former Professor and Head
Zoology Department
Jodhpur University
Jodhpur, Rajasthan, India

Bhavna Khanna
Kendriya Vidyalaya, Gandhinagar, India

Originally published by Springer-Verlag Berlin Heidelberg New York in 2004
Softcover reprint of the hardcover 1st edition 2004

ISBN 978-3-662-07846-4 ISBN 978-3-662-07844-0 (eBook)
DOI 10.1007/978-3-662-07844-0

To
my parents,
family and teachers

Preface

Fishes and their element move over most of the earth. The diversity of fish sharply reflects their mastery of water. They act as monitors of the aquatic environment, and serve for experimental investigations. The commercial and sport fisheries are of great economic importance. Fish stocks are highly vulnerable to pollution and overfishing. The field of fish biology has experienced exciting growth in recent years and it appears impossible to be conversant with all published literature. For the economy of space, a supplementary list of consulted research articles has been avoided.

This handbook is a culmination of determinate endeavour to cover the curriculum of fish biology in postgraduate class and to show future approaches and directions to researchers. The richly illustrated chapters are centered on two aspects-structure and function, presenting integrated, up-to-date account with relevant references.

Undoubtedly, we would have failed in this attempt without the encouragement and help of many people world-wide; especially thanks are due to Drs. W.S. Hoar, D.J. Randall, A.P. Farrell, R.G. Northcutt, J. Caprio, S. Coombs, G.W. Barlow, T.J. Hara, D.H. Evans, R. Fänge, P. Dejours, J.W.M. Osse, D.E. Kime, C.E. Bond, D.R. Jones, W.K. Milsom, M.H. Horn, K.W. Beyenbach, J.M. Jorgensen, E. Zeiske, M.L. Burleson, N.J. Smatresk, S.H. Weitzman, J. Meseguer, R. Fujii, G.K. Iwama, A.D. Pickering, L.S. Smith, R.M. Santer, K.R. Olson, L. Fishelson, F.De Vree, T.E. Finger, K.B. Døving, B. Pelster, P.S. Enger, M.P. Schreibman, P.J. Walsh, P.K.T. Pang, F.J. Meunier, Y.W. Kunz-Ramsay, R.E. Weber, A.G. Zapata, A.L. Val, J.D. Piganelli, P. Sebert, M.J. Manning, H. Greven, G. von der Emde, M. Wulliman, S.P. Collin, K.A. Jones, J.M. Redding, M.B. Bennett, L.K. Chopin, S.M. Kisia, B.A. Hills, J. Gillotreaux, C.R. Marshall, M. Chardon, F.A. Sibbing, K.L. Martin, M.A. Esteban, G. Zaccone, F. Bretschneider, J.G. Sivak, J. Mogdans, H. Bleckmann, R.C. Peters, M.H. Hofmann, W.G. Raschi, V.B. Meyer-Rochow, G. Tagliafierro, R. Diogo, G. Arratia, M.H. Keenleyside, R.N. Gibson, E.G. Drucker, L.R. Parenti, P.M. Anderson, P.A. Wright, F.A. Huntingford, R.E. Davis, E.L. Brainerd, E.D. Stevens, P.J. Motta, E.H. Hamdani, B. Rüpp C. Lamb, E. Pisano, G. Flik, S. Holmgren, M. Reinecke, K. Reutter, J.D. Crawford, F. Ladich, M.L. Fine, T.J. Pandian, S.S. Guraya, K. Swarup and A.K. Ray.

B.G. Kapoor acknowledges the underlying assistance, influence, tolerance and forebearance of my wife Kamal, my daughters Bhavna (Co-author), Kamna and son Anurag during the preparation of the text.

Bhavna Khanna offers sincere thanks to R.S. Khanna, R.J. Khanna and Swati Kapoor for extending valuable, unflagging cooperation.

B.G. Kapoor expresses profound gratitude to: University Grants Commission, New Delhi; Jiwaji University, Gwalior; Dr. R. Mathur (Prof. & Head) and Colleagues of School of Studies in Zoology, Jiwaji University, Gwalior; University of Jodhpur, Jodhpur; University of Delhi, Delhi; my distinguished pupils, organizations, institutions, the typist and the artist.

Further, we sincerely apologise and feel indebted to authors for taking the liberty of skipping their citations which benefitted us in every respect.

Finally, we are extremely grateful to Dr. Dieter Czeschlik (Editorial Director, Life Sciences), Springer-Verlag, Heidelberg, for the approval of this encyclopaedic "vade-mecum" and for arranging its publication through Mr. N.K. Mehra, (Publisher and Managing Director) Narosa Publishing House, New Delhi, keeping in view the marketing considerations. Thanks are also due to press personnel for painstaking setting and fair, final production.

AUTHORS

Contents

CHAPTER 1
Evolution; Phylogeny; Classification

1. INTRODUCTION

Fish is a popular epithet describing aquatic gill-breathing ectothermic anamnia. In zoological terms fish (= Pisces) comprises three major groups: Agnatha (= Cyclostomata); Chondrichthyes; and Osteichthyes. They constitute an ancient class that outnumbers, in species, the four classes of terrestrial vertebrates put together. 60% of all vertebrate species are fishes.

Fishes are important members of aquatic ecosystems and are of paramount economic importance, both in fisheries and aquaculture. Especially in the last few decades increasing proportions of public funds have been channelled into supporting research on the diseases of fish species used in the production of protein for the worldwide increasing human population. Fish are frequently used as experimental animals in toxicological and immunological research. Fish are suitable for developmental studies. They are also a useful model for the study of the physiological effects of pressure.

In more than 300 million years of evolution, the adaptive radiation of fish has led to the present spectacular diversity of structure and function. In terms of structure, fish can be as large as the whale shark (12.1 m) or as gobies (0.0008 m).

Fishes, the charismatic fauna, havė adapted extremely successfully to the extremes of the aqueous environment. It is difficult to over-emphasize the diversity of animals that exist within these classes, for among them are such contrasting types as sharks, lampreys, tunas, eels and lungfishes. Amongst fishes are pelagic species which must maintain certain swimming speeds in order to be buoyant, species which migrate over thousand of miles, abyssal species living in waters with markedly reduced oxygen content, species living in the subzero waters of the Antarctic, species living in dark caves, species living in torrents, species inhabiting a tropical soda lake and also the obligatory air-breathing species of the tropics, etc. Even in a comparatively benign environment such as relatively shallow waters over continental shelves, the life style of fish species varies greatly, with sedentary benthic and pelagic shoaling species coexisting within a comparatively narrow depth-range. The extent of their radiation in virtually every aquatic habitat documents a particularly successful interplay between physiology and environment.

Given the heterogeneity of fishes and their environments, it is hardly surprising that there is great diversity of physiological mechanisms required for locomotion, circulation and respiration. Parasitic fish funnel metabolic energy primarily into reproduction. Migrating fish command powerful mechanisms of salt and water balance to support life in diverse aquatic environments and no other class of vertebrates commands as many and seemingly bizarre strategies of reproduction as fish.

Commensalism between invertebrates and fishes is extremely rare. It requires sedentary habits and particular features of the anatomy and ecology of both the fish and the attaching invertebrate. Commensalism is apparent between midge larvae (Diptera: Chironomidae) and the catfishes of the families Loricariidae and Astroblepidae.

2. OVERVIEW OF PISCINE EVOLUTION AND PHYLOGENY

To skip a host of references, the reader is referred to the books edited by T. Orvig (1968), P.H. Greenwood, R.S. Miles and C. Patterson (1973), and M.L. Stiassny, L.R. Parenti and G.D. Johnson (1996). The title "Fishes of the World" authored by J.S. Nelson (1984; also refer latest edition of 1994) is especially recommended. Other important references are of J.A. Moy-Thomas and R.S. Miles (1971), Gilbert (1993), Forey and Janvier (1994) and J.A. Long (1995). The two titles: "The biology of hagfishes" edited by J.M. Jorgensen, J.P. Lomholt, R.E. Weber and H. Malte (1998), and "Sharks, skates and rays—the biology of elasmobranch fishes" edited by W.C. Hamlett (1999) are resources of information for respective groups of fishes.

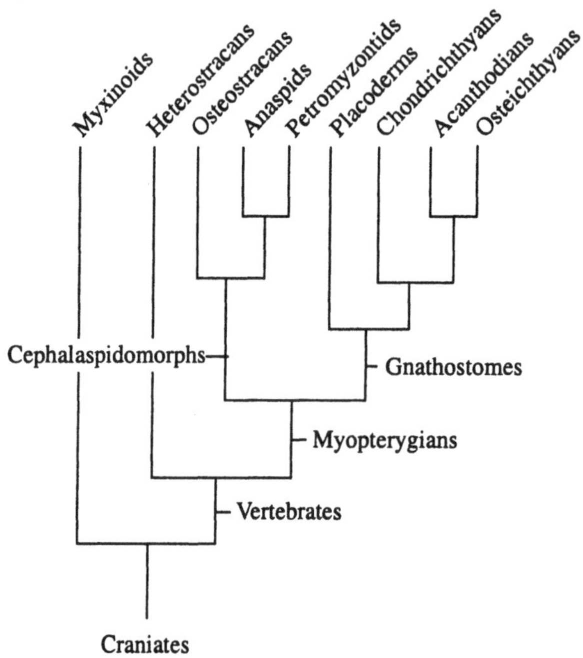

Fig. 1.1 Cladogram of the suspected interrelationships of extinct and living groups of craniates. ***Source:*** **Northcutt (1989b)**

Figures 1.1, 1.2(A)(B) show presumed interrelationship of craniates and of fishes respectively.

Bjerring (1985) presented certain facts and thoughts on piscine phylogeny. Craniate animals, it is estimated, are represented by some 50,000 species, about half of which constitute the piscine world. Their fossil record extends back to upper Cambrian. Although craniate animals have many forms and shapes, they are readily subdivided into ten large groups; osteolepipods, urodelomorphs struniiforms, actinopterygians, brachiopterygians, coelacanthiforms, dipnoans, elasmobranchiomorphs, pteraspidomorphs, and cephalospidomorphs.

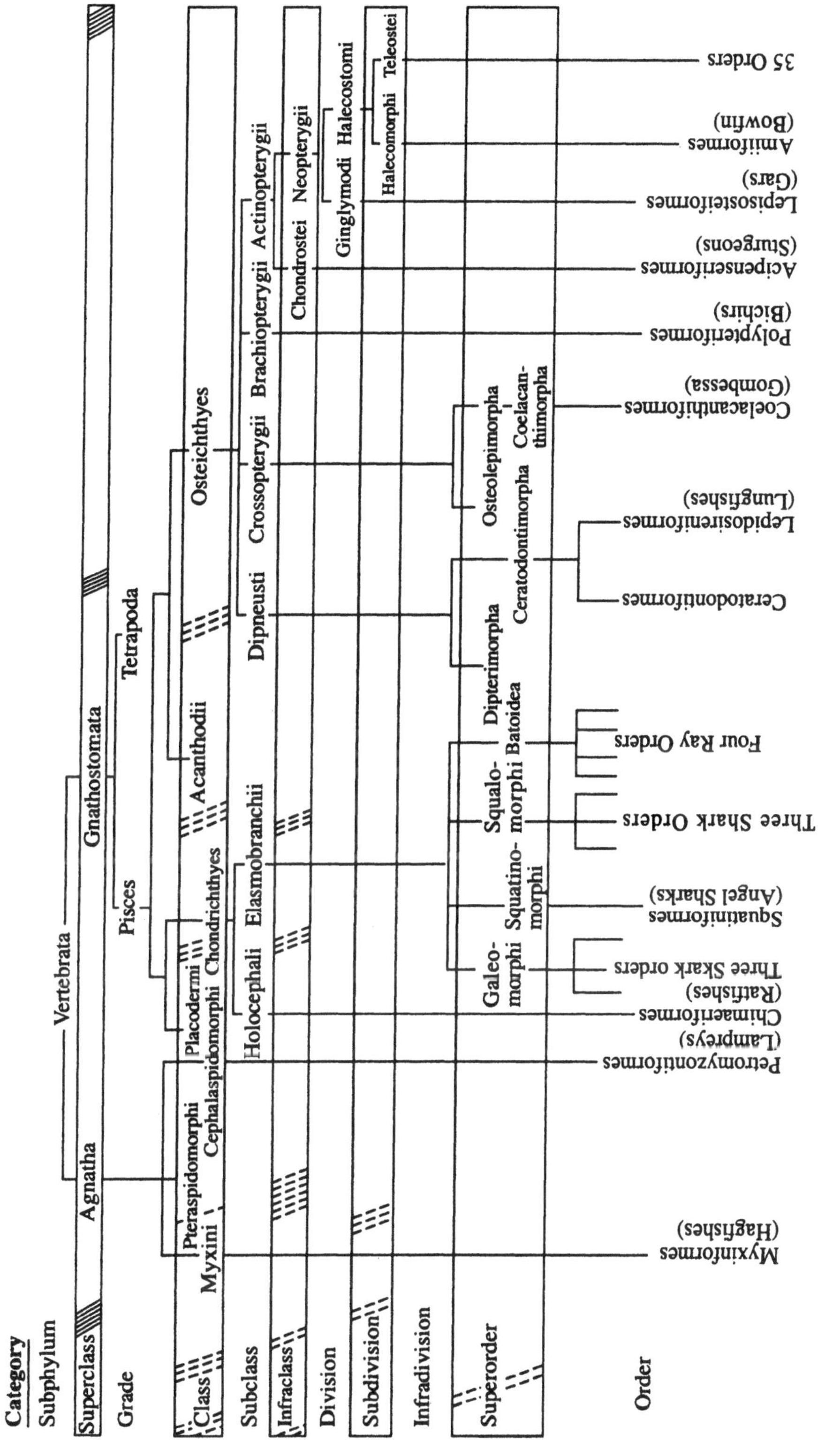

Fig. 1.2(A) **A hierarchical classification of the major groups of fishes, showing presumed evolutionary relationships.** *Source:* **Nelson (1984)**

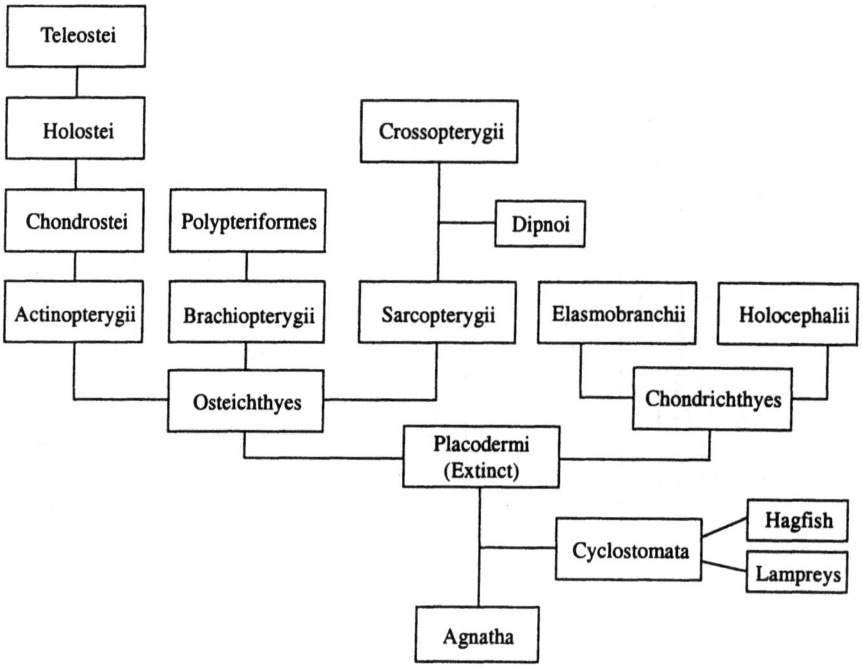

Fig. 1.2(B)
***Source:* Zapata (1996a)**

The cephalospidomorphs have a dorsal nasohypophyseal opening; the pteraspidomorphs have a nasopharyngeal duct. Their living representatives, the lampreys and hagfishes, are collectively known as cyclostomes. Cephalospidomorphs and pteraspidomorphs have entodermal gills directed toward the pharynx; together, therefore, they are termed entobranchiates. The eight other groups of craniate animals have outwardly directed gills of ectodermal derivation; accordingly they are referred to as ectobranchiates. They also have jaws formed from branchial-arch elements. Consequently ectobranchiates are commonly called gnathostomes.

At the very beginning of their known history the ectobranchiates had already subdivided into two major phylogenetic lines, termed plagiostomes and teleostomes. The former are characterized by an utricular diverticulum; the latter, by a dental arcade, i.e, by dentigerous dermal bones around the jaw apparatus.

Plagiostomes comprise the elasmobranchiomorphs, which have open endolymphatic ducts, and the dipnoans, which have pseudotrabecles. The six other ectobranchiate groups constitute the teleostomes. Each of these may be characterized thus: the osteolepipods have an occipital artery; the urodelomorphs have an intercerebral olfactory bulb; the actinopterygians have a swim bladder; the brachiopterygians have a multipinnulate dorsal fin; the coelacanthiforms have an electroreceptive rostral organ; and the struniiforms have a single, long submandibular between the lower jaw and the ipsilateral gular.

2.1 Extinct Forms

The fossil and living forms may be separated from common origin by tens of hundreds of million years (Table 1.1).

Table 1.1 The first appearance of the major groups of fishes and of representative living genera*

	First recorded appearance	
	Geologic period	Age +
Superclass Agnatha		
Class Cephalaspidomorphi		
Subclass Cyclostomata	Pennsylvanian	325
Class Cephalaspidomorphi		
Petromyzon (lamprey)	Recent	
Class Myxini		
Myxine (hagfish)	Recent	
Class Acanthodii	Silurian-Permian	375–275
Class Placodermi	Lower Devonian	400–350
Class Chondrichthyes		
Subclass Elasmobranchii (sharks and rays)	Lower Devonian	375
Heterodontus (Port Jackson shark)	Middle Jurassic	160
Chlamydoselache (frilled shark)	Tertiary	70
Hexanchus (six-gill shark)	Lower Jurassic	175
Rhinobatus (guitarfish)	Upper Jurassic	145
Subclass Holocephali (ratfishes)	Mississippian	350
Chimaera (ratfish)	Upper Cretaceous	100
Class Osteichthyes (higher bonyfishes)	Lower Devonian	400
Subclass Actinopterygii (ray-finned fishes)	Middle Devonian	385
Infraclass Chondrostei	Middle Devonian	385
Acipenser (sturgeon)	Upper Cretaceous	100
Polyodon (paddlefish)	Lower Tertiary	70
Polypterus (bichir)	Lower Tertiary	70
Infraclass Neopterygii	Upper Permian	250
Division Ginglymodi		
Lepisosteus (gar)	Upper Cretaceous	100
Division Halecostomi		
Subdivision Halecomorphi		
Amia	Upper Cretaceous	100
Subdivision Teleostei	Middle Triassic	250
Clupea (herring)	Lower Tertiary	70
Perca (perch)	Lower Tertiary	70
Subclass Crossopterygii (lobe-finned fishes)	Middle Devonian	375
Latimeria (living coelacanth)	Recent	–
Subclass Dipneusti (lungfishes)	Lower Devonian	400
Neoceratodus (Australian lungfish)	Upper Cretaceous	100
Protopterus (African lungfish)	Lower Tertiary	70
Lepidosiren (South American lungfish)	Middle Tertiary	25

*The Acanthodi and Placodermi are extinct.

+ Age is given in millions of years, plus or minus.

2.1.1 Jawless Fishes (Agnathans)

Fossil History: Ostracoderms
The earliest known vertebrates are jawless fishes commonly called ostracoderms (meaning "shell-skin"). The oldest remains are found in Ordivician rocks, and by Silurian, two distinct groups had evolved. In the one that included the osteostracans and the anaspids, the nasohypophyseal duct had a dorsal opening between the eyes. In the other, represented by the heterostracans and the poorly known coelolepids, this duct apparently opened into the roof of the mouth.

The osteostracans have a continuous dorsal head shield composed of dermal bone. The mouth and gill openings are on the ventral surface which indicates that they were bottom dwellers. The body, which is covered with bony plates, is oval or triangular in cross-section, and the tail is heterocercal. Pectoral appendages of limited mobility evolved in Devonian representatives, but were absent in earlier examples. The cartilaginous neurocranium is covered with a thin layer of perichondral bone, which also lines the blood vessels and the nerve canals. Lateral and dorsal sensory fields on the head shield receive a number of canals from the vestibular part of the labyrinth cavity. These canals may have housed lymph ducts or nerves. There is no evidence that these superficial fields contained electric organs.

The anaspids are fusiform fishes with a terminal mouth, lateral gill openings, many small dermal head plates, and body scales. Paired appendages are represented by fin folds and the tail is hypocercal. These presumed bottom feeders of the Silurian and Devonian have no perichondral bone and the internal structure of the head remains unknown.

There appear to be two types of anaspids, both of which are reminiscent of lampreys in having large eyes with the nasohypophyseal and pineal openings on top of the head. One group has enlarged scales along the length of the animals back and immediately behind the gills. The members of the other group completely lack scales, or have poorly developed scales. The best known genus of these "naked" anaspids in *Jamoytius kerwoodi*, which resembles modern lamprey in having a circular or annular cartilage surrounding the subterminal, circular mouth, large eyes and a basket-like skeleton that supported the gills beginning immediately behind the orbit. These similarities have led some workers to suggest that naked anaspids are immediate relatives of lampreys.

The heterostracans (Ordovician-Devonian) range in body form from fusiform to dorsoventrally compressed bottom dwelling types. The front part of the body is encased in bony plates and the caudal half is covered with small scales. The mouth is subterminal and there is a lateral common opening for the gill pouches. A pineal opening is situated between the eyes, but the nasal sacs probably opened into the roof of the buccal cavity. There are no paired appendages and the tail is hypocercal. Although endocranial ossification is absent, the dorsal part of the endocast and the branchial pouches can frequently be observed within the dermal bone cover. In the heterostracans the sensory organs of the head are distributed in canals instead of being confined to "fields" as in the osteostracans. The former have two visible semicircular canals, the latter have three. Illustrated account of fossils of Jawless fishes has been given by Janvier (1993). (For hagfish systematics, read Fernholm, 1998).

2.1.2 Jawed Fishes (Gnathostomes)

Fossil History: Acanthodians and Placoderms
Although evolution of jawed fishes commenced in the Silurian period, it was during the Devonian

that the proliferation of these animals becomes particularly evident in the fossil record, a circumstance that has resulted in this period being called the "Age of Fishes". In addition to hinged jaws and paired fins, these more advanced fishes differed from agnathans in a number of other major features, including presence of paired lateral nares and a three-chambered semicircular canal. Once acquired, these characters were not only retained in the fishes but in all other subsequent vertebrate groups as well.

Acanthodians: The origin of the gnathostomes whose development was a major event in vertebrate evolution, is unfortunately, not documented by the fossil record. Silurian rocks contain dissociated remains of acanthodians, the earliest known jawed fishes. Ranging in shape from fusiform to eel-like, they have small dermal bones covering the head and tiny scales on the body. Their eyes are far forward and the jaws well developed. Two rows of spines (which probably supported finlets) are usually present along the belly. The tail is heterocercal. Unfortunately most specimens are compressed and our knowledge of internal head anatomy is confined to one Permian genus that is commonly preserved in nodules.

Placoderms (armoured fishes): Another early gnathostome group, the placoderms, is known from Devonian rocks, although its ancestry must have extended back into the Silurian. The placoderms include an array of blunt-headed, more or less fusiform predators with well-developed head and shoulder armour, bottom feeders with box-like thoracic dermal bone cover, and several different groups of depressed, ray-like bottom dwellers, with enlarged pectoral fins and reduced dermal ossification. The placoderm visceral skeleton is poorly known, but apparently it was mostly or entirely situated beneath the neurocranium, as in the ratfishes. Shearing or crushing plates of dermal bone are associated with the primary jaws. The body ends in a pointed or weakly heterocercal tail. For further details supported by figures, refer Schultze (1993).

2.2 Extant Forms

Primitive is a controversial concept in Ichthyology as in other branches of Zoology. What is advanced as opposed to 'primitive' is not always clear. Teleosts are considered more advanced than non-teleostean fishes. But among, the latter, the dipnoans (lungfishes), the coelacanth (*Latimeria*) and the brachiopterygians (bichirs: *Polypterus*, ropefishes: *Erpetoichthys*) show relationships to the tetrapods, i.e., the higher vertebrates, and certain sharks have a remarkably large brain volume, a feature usually regarded as 'advanced'.

The myxinoids (hagfishes) and petromyzontids (lampreys) undoubtedly are primitive, but some biologists put these agnathan groups at the side of, not within the systematic category of fishes. The non-teleostean fishes, if, for practical reasons, they are considered as primitive, comprise an extremely heterogenous group.

Although superficially similar, the two main goups of living agnathans, lampreys and hagfishes, are in fact quite different (Table 1.2).

The agnathans are the oldest of the vertebrate radiations. These forms possess neither jaws nor paired fins and kind of prey object and the efficiency of their locomotion are therefore far more restricted than those of the other anamniote radiations. These jawless fishes are recognized not only for their unsavoury feeding habits (lamprey rasp the flesh and suck the blood of other animals, whereas hagfish literally eat their way through dead or dying fishes), but also for their

Table 1.2 Differences between adult lampreys and hagfishes (slime eels) (Cyclostomata or "round-mouth")

Characteristics	Lampreys	Hagfishes
Habitat	Marine; fresh water	Marine
Dorsal fin	One or two	None
Pre-anal fin	Absent	Present
Eyes	Well developed	Rudimentary or vestigial, essentially blind
Oral disc	Present	Absent
Lateral line system	Well developed	Degenerate
Membranous labyrinth	Two semicircular canals	One semicircular* canal
Barbels	Absent	Three pairs, fleshy and tactile
Intestine	Ciliated	Unciliated
Spiral valve intestine	Present	Absent
Heart	Systemic	Systemic; portal, and caudal (accessory hearts)
Buccal glands	Present	Absent
Nostril location	On top of head	In front of head
Nasohypophysial sac	Does not open into pharynx	Opens into pharynx
External gill openings	Seven	One to 14
Internal gill openings	Unite into a single tube connecting to oral cavity	Each enters directly into pharynx
Cranium	Clearly cartilaginous	Poorly developed
Branchial skeleton	Well developed	Rudimentary
Vertebrae (cartilaginous)	Present	Absent
Number of spinal nerve pairs per body segment	Two	One
Ducts of Cuvier to heart	Right only	Left only
Pronephric kidney	Absent	Present
Osmoregulation	Isosmotic	Hyper-or hypo-osmotic
Eggs	Small without hooks	Very large with hooks
Cleavage of embryos	Holoblastic	Meroblastic
Development	Larval stage of development	No metamorphosis; development is direct into a miniature adults at the time of hatching

*Two semicircular canals (one in myxinoids, but some evidence of two sensory ampullae)

primitive appearance. Agnathan larval structure and habitat are reminiscent of protochordates. At metamorphosis agnathans cease life as filter feeders and transform to either parasitic predators or nonfeeders. The living agnathans have lost dermal armour of their ancestors, have elongated the trunk, and have radically reorganized many of the head structures for this special sort of predatory existence. All of these changes indicate that these organisms are not "living fossils", but a very specialized part of the once flourishing agnathan radiation.

There are over 70 living species of lampreys and hagfishes, of which about 60 percent are lampreys. As a group the lampreys are euryhaline, although the majority of species live entirely in fresh water. They are about equally divided between parasitic and non-parasitic forms, of which the latter feed only during the larval (or ammocoete) stage and the adults live only long enough to reproduce. Hagfishes are completely stenohaline and are the only vertebrates in which the body fluids are isosmotic with sea water. Unlike lampreys, the hagfishes lack a larval stage. Lampreys have an antitropical, or bipolar, distribution, being absent from the tropics but displaying a broad distribution in temperate areas of the northern hemisphere (South America, Australia and New Zealand). Hagfishes are distributed in temperate marine waters throughout the world.

Fishes have classically, if imprecisely, been divided into two great "tribes", the "osseous" and the "cartilaginous". Yet, as many mid-nineteenth century naturalists observed, some members of the first "tribe" have a great deal of cartilage, and members of the second have, at any rate, some calcareous matter in their skeleton. To many naturalists of that era, however, modern sharks, rays and holocephalans obviously differed from bony fishes not only in their skeletal composition, but also in their assumed level of organization.

The teleostomes and elasmobranchiomorphs radiations comprise the vast majority of living anamniote vertebrates. Both radiations appear equally old and can be characterized by the development of jaws and paired fins. These acquisitions have allowed entry into new adaptive zones. The origin of jaws has increased the efficiency of feeding and allowed the utilization of new and large prey. Concomitantly the origin of paired fins has increased locomotive stability and manoeuvrability, resulting in new and more active predatory forms. Whereas teleostomes have evolved in both freshwater and marine environments, the elasmobranchiomorphs have evolved primarily in a marine environment.

Chondrichthyans are characterized by a cartilaginous endoskeleton, a unique dental histology and development, the usual presence of urea in the blood and tissues, the inner margins of the pelvic fins in males modified into reproductive structures called "claspers", no diverticulum of the oesophagus that functions as a respiratory organ or buoyancy device, and an unusually well developed electrosensory system. Living forms can be divided into two distinct groups. One is characterized by four pairs of gill arches and a single external gill opening on each side of the head (subclass Holocephali: rattails, ratfishes or chimaerids). The other has five to seven separate gill openings on each side (subclass Elasmobranchii: sharks, skates and rays). In addition to the above, holocephalans differ from elasmobranchs in having a smooth epidermis, a long flexible tail, large eyes, absence of cloaca, absence of ribs, and the upper jaw (palatoquadrate) fused with the braincase (hyostyly). All chimaerids have the teeth consolidated into grinding plates (crushing or grinding dentition), an adaptation for feeding on the molluscs that comprise their chief food (see Section 3).

There are almost 800 living species of chondrichthyans, including about 425 skates and rays, 340 sharks, and 30 chimaerids. All available evidence suggests that the group had a marine origin, and with few exceptions all living species live in a marine environment. The most notable exceptions are the freshwater stingrays of South America (comprising an endemic family and about 14 species), in which the urea concentration of the blood and tissues has been reduced to the point that the group is now incapable of living in salt water.

The elasmobranch fishes include the sharks, skates and stingrays. These fishes comprise a provocative vertebrate group that occupies a pivotal evolutionary position. They diverged some 350 million years ago from their sister group the holocephalans. Contemporary elasmobranchs are usually considered primitive, and this is certainly true of many of their structural characteristics, many of which have been conserved with little change. In other respects, they demonstrate characteristics that are considered advanced. They have large brains, a well-developed nervous system, internal fertilization, viviparity, an endocrine regulatory system similar to higher vertebrates, and physiological processes and mechanisms common to vertebrates in general. Elasmobranchs are among the most adaptable and enduring of animals. They have outwardly changed little during million of years and currently have a cosmopolitan distribution, occupying diverse ecological niches.

The elasmobranchs represent a currently under-exploited model system that may be employed to investigate and elucidate many important biological and evolutionary processes. Apart from their importance as the basic vertebrate used in comparative anatomy classes worldwide, they have special relevance as laboratory animals. Many mechanisms common to all vertebrates are present in elasmobranchs in an elegantly simple model, unencumbered by adaptations to land and homeothermy. There are also many specialized mechanisms unique to them. Numerous important questions remain to be answered about the basic biology of this group, and it is vitally important to stimulate continued research in this direction as well.

Living chondrichthyans form a reasonably coherent group, but the relationships of various taxa within this assemblage (both to each other and to fossils) are far from clear. A three-fold divisions of living chondrichthyans into chimaeroids, sharks and rays is commonly accepted; beyond that there is little assent.

In the last few years, cladistic analysis has been increasingly utilized. Living chondrichthyans are believed to be a monophyletic group based on a number of shared derived characters (synapomorphies): prismatic tissues (tesserae), gill septa extending to the lamellar margin, micromeric scales, and replacement tooth rows. The ratfishes (holocephalimorphs) are generally considered the sister group (Fig. 1.3) of all other living chondrichthyans, which are collectively termed the elasmobranchs, however, this taxon almost certainly does not include all "sharks", as *Cladoselache* and the fossil symmoriids lack the synapomorphies that unite elasmobranchs and they therefore probably comprise the sister groups of a monophyletic group consisting of ratfishes and elasmobranchs. For special interest on chondrichthyan phylogeny, the reader is referred to comprehensive coverages (Maisey 1984a, b, 1986, 1988).

The Osteichthyes, according to the paleontological record, appeared in inland waters of the Middle Devonian, from then on progressively increased in importance, invaded the saltwater seas by the end of the Paleozoic Era, and today include not only fresh-water fishes but the majority of marine as well.

According to one classification, this class falls into two subclasses: Actinopterygii, fishes in which paired fins are supported by dermal rays (ray-finned fishes), and Choanichthyes, fishes having internal nostrils (choanae). The first, rather, rare at the outset, rapidly varied and became the dominant type, their most modern and abundant examplars being the teleosts which comprise 23,637 living species, not to mention a vast multitude of extinct one. The second subclass, the Choanichthyes, contains the aberrant *Dipnoi* (*di* twice; *pnoe*: breath; two breathing apertures, the so-called lungfishes) and the Crossopterygii (*krossi*: fringe; *pterygos*: wing or fin), the stock from which the land vertebrates sprang.

All of the Osteichthyes are characterized by the possession of a specialized or much reduced spiracle (lost in the teleosts), hyoid support for the jaws, gills (never more than five on each side) shielded by an operculum, and some kind of lung or air bladder. In addition to a caudal fin and

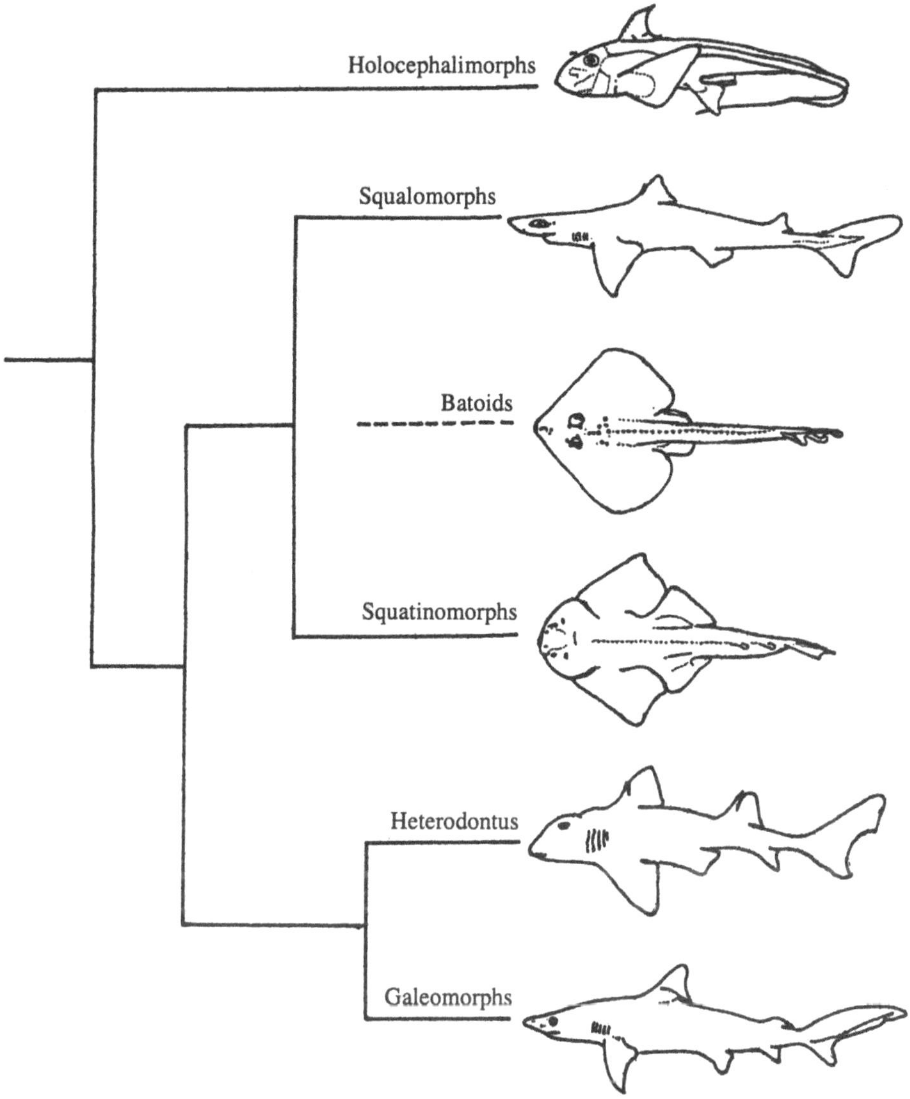

Fig. 1.3 Cladogram of the suspected interrelationships of the living groups of the chondrichthyan fishes. *Source:* Northcutt (1989a)

the paired pectoral and pelvic fins, there was one dorsal fin in the early actinopterygians, two in choanichthyans, and a single anal fin occurs in all. Even in the earliest species known, the bony nature of the skeleton was far advanced, and they also had a dorsal covering of thick scales and plates, attributes which were retained in most of the later members of the class. Only in a few forms-for example, the sturgeons and paddlefishes among the actinopterygians, and the dipnoans among the choanate fishes-there has been a marked retrogression from the highly ossified skeleton.

In all Osteichthyes the head skeleton originates from two sources: internal bones, replacing embryonic cartilage, and external or dermal bones derived from enlarged bony plates or scales. In skull and jaws (and even in the shoulder girdle) the two kinds of bones are combined in a complex composite. Indeed, the skull of the Osteichthyes presents the basic plan on which the skulls of all tetrapod vertebrates are built.

As said before, lung breathing probably had its beginning early in the evolution of the bony fishes. It is thought the living actinopterygian, *Polypterus*—the "bichir" of the Nile, which is adapted to the alternate seasons of drought and flood, has kept the most generalized condition of the lung. This is a simple, bilobed sac opening from the lower side of the pharynx and situated in the same position as the lungs of the terrestrial vertebrates.

Perhaps, a similar lung occurred in extinct crossopterygians, and such a functional organ is found in all existing species of Dipnoi: *Protopterus*, from the swamps of central and western Africa; the eel-shaped *Lepidosiren*. from the swamps of the Amazon and La Platte and their tributaries; and Australian lungfish *Epiceratodus*; the first two can withstand a complete drying up of their locale as they retreat into burrows, and the last, during the periodic drought, comes to the surface in foul and stagnant pools to breathe air.

The ancient Devonian bony fishes, which inhabitated fresh-water streams and lakes, lived under like conditions of seasonal aridity, as evidenced from the sedimentary deposits in which their remains are found. Associated with the apparatus of pulmonary respiration in the Choanichthyes was the presence of internal nostrils which permitted the intake of air without opening the mouth.

To the fishes that dwelt in large bodies of inland waters or that took to the open seas, the lung in its original position in the ventral part of the thorax becomes a disadvantage because it tended to make the animal topheavy. A readaptation took place. Even in the extant "lungfishes" the organ has shifted around towards the top of the body, although its duct may still be attached to the under side of the throat cavity. In all the remaining living bony fishes, the lung or air bladder serves primarily as an hydrostatic organ, which during development grows from the upper side of the pharynx. In some of them even this connection of the swim bladder with gut is lost altogether, and the sac manufactures the gas within its own walls.

The Polypteridae are a monophyletic group of primitive fishes, the exact phylogenetic position of which is still in doubt. They have been classified as crossopterygians but have usually been considered in more recent decade to be primitive actinopterygians. This systematic position has been questioned, however, there is a proposal to include the Polypteridae in an entirely separate group among the teleostomes. [Bjerring (1985) suggested that *Polypterus* is not to be classified as an actinopterygian or a descendent of the paleoniscids. Hence, for the present, *Polypterus* must be placed in a group of teleostomes of its own]. The extant 11 species live in freshwaters of Africa, and all belong to the genera *Polypterus* and *Erpetoichthys* and are all allocated to one family. They appear to be most generalized chondrosteans.

With ray-finned fishes (Actinopterygii), there are five major radiations-Cladistia, Chondrostei, Ginglymodi, Halecomorphi, and Teleostei (Fig. 1.4).

The sturgeons (Acipenseridae) and paddlefishes (Polyodontidae) are the only surviving chondrosteans and their anatomy is more specialized than that of polypterids. The gars (*Lepisosteus*: Ginglymodi) and bowfin (*Amia*: Amiidae), both found in American freshwater systems are the only surviving holeoostean genera. Acipenseridae, Polydontidae, Amiidae and teleost fishes comprise a monophyletic Actinopteri. The surviving teleosts form a group that surpasses all

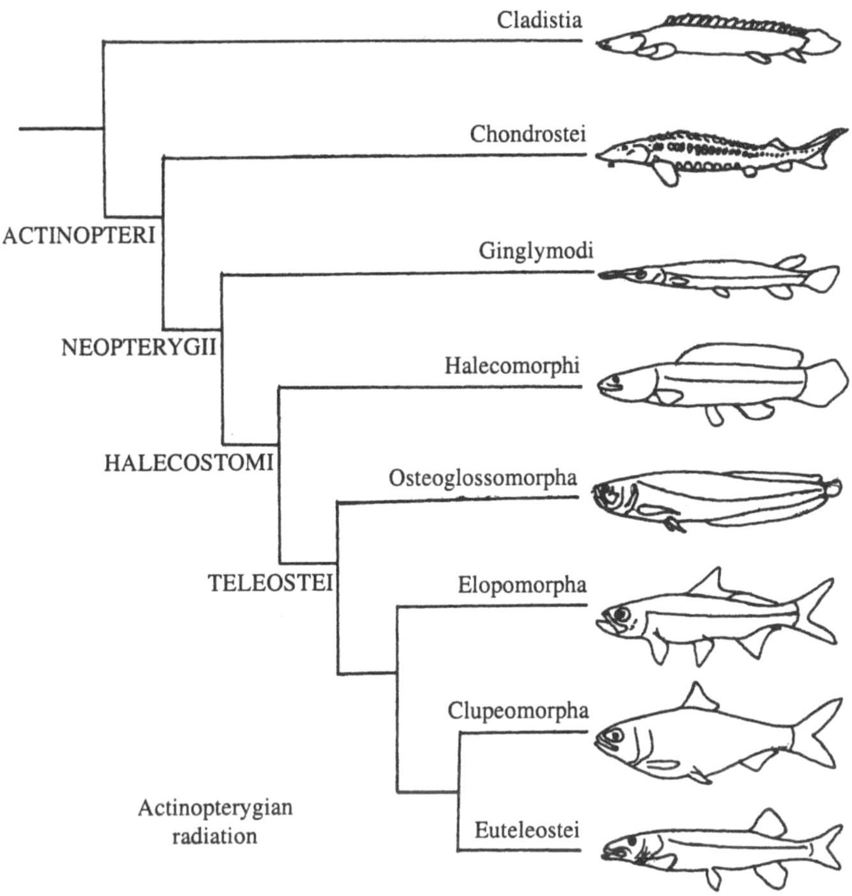

Fig. 1.4 Dendogram of actinopterygian fishes.
Source: **Butler and Northcutt (1993) [After Lauder and Liem 1983; Nelson 1984]**

other living vertebrate groups in number of species and morphological diversity. Within the Teleostei, the nonteleost clade is an extremely diverse assemblage of fishes and contains many monophyletic groups including the Acanthopterygii (spiny-finned fishes).

Paddlefishes appear to be axoltls of the fish world, offering one of the best documented cases of paedomorphosis (or neoteny) known to date (see details, Grande and Bemis 1991). Normal-size, reproductively active adult paddlefishes lack particular ossifications of the scapulocoracoid, vertebral column and neurocranium present in other osteichthyans but these elements eventually ossify in very large, older individuals.

Paddlefishes present excellent examples of two basic types of intraspecific morphological variation: individual and ontogenetic. Individual variation has been documented in other primitive groups of actinopterygians (e.g., Amiidae and Acipenseridae).

Teleosts are the single largest radiation of living vertebrates. The four major groups of teleosts (Fig. 1.5) are the Osteoglossomorpha, Elopomorpha, Clupeomorpha and Euteleostei. The teleosts are the successful fishes of today, the culmination of the series of phylogenetic stages seen in non-teleost bony fishes.

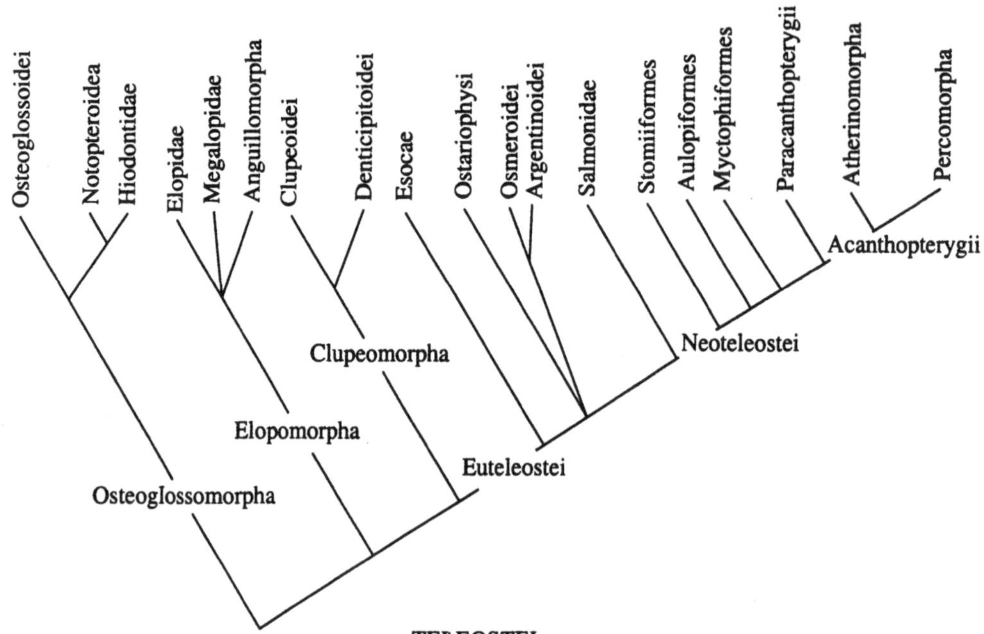

Fig. 1.5 Dendogram of extant teleosts radiations.
Sources: **Butler and Saidel (1991); Butler et al. (1991) and Northcutt and Butler (1991) [Modified from Lauder and Liem 1983]**

The Choanichthyes, numerous at first and the most common inhabitants of freshwaters during the late Paleozoic are now near extinction, there being only four surviving forms (dipnoans and *Latimeria*).

The bony fishes (Osteichthyans) are the sister group of the extinct acanthodians, and cartilaginous fishes and the extinct placoderms, are, in turn, sister groups of the acanthodians and bony fishes, respectively. Following current cladistic interpretations, the osteichthyans are believed to be a monophyletic taxon comprising two sister groups, the ray-finned fishes (actinopterygians) and the lobe-finned fishes and their descendents (sarcopterygians). Further, the living sarcopterygians (the actinistian *Latimeria*, the lungfishes, and the tetrapods) are regarded phylogenetically as an unresolved trichotomy. Uncertainty also exists regarding the phylogenetic relationships of the placoderms. It is very possible that these taxa do not constitute a monophyletic group, but they will be treated as such in this analysis because of our lack of information regarding their phylogenetic relationships.

Finally, the fossil and living agnathans do not constitute a monophyletic group, but the cephalaspidomorphs, comprising of the extinct osteostracans and anaspids, as well as the extant lampreys, are the sister group of jawed vertebrates, and the extinct heterostracans are, in turn, the sister group of all other vertebrates (myopterygians). The hagfishes (myxinoids) are not considered vertebrates as they do not exhibit any shared derived traits (synapomorphies) with cephalaspidomorphs or myopterigians and are thus considered the sister group of the vertebrates.

Heterostracans are the sister group of myopterigians on the basis of four shared derived characters (synapomorphies): two vertical semicircular canals, a well-developed lateral line system, the presence of arcual elements, and a mineralized exoskeleton. These characters are absent in myxinoids, and their absence is assumed to represent the primitive condition. However, if their absence represents an apomorphic condition, the relationship of myxinoids and heterostracans is unresolved.

According to Forey and Janvier (1994), among the living forms, it seems clear that the lampreys are more closely related to gnathostomes than either group is to the hagfishes. The hagfishes are the most primitive vertebrates known, living or extinct. Thus most features of the hagfish that were considered to be degenerate are probably primitively simple, including the lensless eye, the simple hypophysis, the simple ear, the lack of nervous regulation of the heart, the absence of a lateral line system, electroreceptors and a cerebellum, and an inability to regulate internal ionic concentrations.

It also seems to be the case that the majority of ostracoderms are more derived than either lampreys or hagfishes. This would suggest that the lampreys and hagfishes represent a primitive naked state, and that the dermal skeleton arose after these vertebrates had diverged. At first the dermal skeleton consisted of dentine and acellular bone; and only later was it made of cellular bone. A calcified endoskeleton also appeared in the ancestors of gnathostomes after the lampreys and hagfishes diverged.

Perfected paired fins appear to have arisen first in the common ancestor of the osteostracans, the pituriaspids and the gnathostomes. Osteostracans and gnathostomes have pectoral fins and cellular bone. The pituriaspids are the only other group of agnathans that might have had pectoral fins. Lampreys probably never had pectoral fins, although they might have had paired fin folds along part of their body length. Hagfishes and their ancestors never developed paired fins.

Some features of modern hagfishes and lampreys—such as the median nasohypophyseal opening, the prenasal sinus, the pouch-like gills and the complex tongue-belong to the history of gnathostomes. Indeed, these features may have been present in the common ancestor of all known vertebrates but have become modified or lost in gnathostomes. There are other viewpoints on the relationships between modern and fossil agnathans based on other methods. Erik Jarvik of the Swedish Museum of Natural History (Stockholm) considers that hagfishes, lampreys and ostracoderms are more closely related to each other than any are to gnathostomes. He also suggests that agnathans are more derived than gnathostomes. However, this theory assumes a number of parallel evolutionary paths for which the authors see no evidence.

Although many questions in the evolution of vertebrate structure remain unanswered, the ostracoderms provide information that no other source could have given. It seems that much could be gained by further excavations in Early Ordovician and Cambrian deposits where ostracoderms should be present. As the recent history of the field shows, new techniques of preparing and observing the fossils may reveal previously unnoticed details.

3. SPECIAL EMPHASES: HOLOCEPHALI; DIPNOI; COELACANTH

Holocephali: The class Chondrichthyes contains approximately 800 living species and constitute about 3% of the species of living fish. The class is divided into two clearly definable modern subclasses, Selachii (Elasmobranchii) and Holocephali (Chimaeriformes). The Holocephali consists of about 30 species. There are no clear morphological links between the two subclasses.

Derived characters of the class Chondrichthyes include: a cartilaginous endoskeleton, frequently superficially prismatically calcified; a complete exoskeleton of cyclomorial placoid or lepidomorial scales; teeth in numerous series, not fused to jaws and replaced serially through linguo-labial progression of a periodontal ligament; inner ear remains open to the environment through endolymphatic ducts; internal fertilization with copulation; male claspers (copulatory organs) developed from extensions of the pelvic fin axes; anterior dorsal fin only preceded by a superficial spine; and retention of a spiracle.

The modern Holocephali are distinguished by the following derived characters: a hyoidean opercular gill cover; holostyly; neurocranium lacking precerebral fontanelle; postorbital braincase very short; no spiracular opening; dentition of two paired upper and one paired lower, sharp-edge plates growing from below rather than linguolabially, and never replaced; dental plates of osteodentine and/or acellular bone with pleromin tritoral columns; a synarcuum, associated with a mobile first dorsal spine, fin and basal plate; pelvic girdle with high dorsal process; a short, triangular pelvic basipterygium supporting most or all fin radials; male with a medium frontal clasper and paired prepelvic tentaculae articulated to the pelvie girdle.

The holocephalan Chimaeriformes have always occupied an isolated position among the Chondrichthyes, but have been related to the Paleozoic Cochliodontiformes. Analysis of the morphology of all available Holocephalimorpha, including three new cochliodonts, indicates that only two character complexes separate the Chimaeriformes from other. Holocephalimorpha: the basally evergrowing tooth plates with pleromin tritors; and the prepelvic tentaculum, high dorsal pelvic process and short triangular basipterygium. Petalodontiformes and a separate group related to "*Climaxodus*" are contrasted and included in the superorder Paraselachimorpha. There is no factual validity to support the taxon "Bradyodonti" (Lund 1986; read also Orvig 1985).

The "pleromic hard tissue" of tritural columns of the chimaerid tooth plates has been discussed by Orvig (1985). The chimaerid tooth plates show resemblance in mnay ways to those of the ptyctodontids.

Dipnoi: A comprehensive review on the biology of the lobe-finned fishes was published by Thomson (1969). Details of the skull, feeding and respiratory mechanisms, and aestivation (in Dipnoi) have also been given.

Living lungfishes occur in three geographically isolated regions (Table 1.3).

Table 1.3 Extant species of dipnoans+

Species	Family	Continent found
Neoceratodus forsteri	Ceratodontidae	Australia
Lepidosiren paradoxa	Lepidosirenidae	South America
Protopterus aethiopicus	Lepidosirenidae	Africa
Protopterus amphibius	Lepidosirenidae	Africa
Protopterus annectens	Lepidosirenidae	Africa
Protopterus dolloi	Lepidosirenidae	Africa

+ Fossil dipnoans are found on all continents (Marshall 1986a)

The Dipnoi are thought to be a monophyletic group of fishes. They first appeared in the geologic record in the Eary Devonian and persist to the present. Many Paleozoic genera are known from relatively complete cranial material. About 50 extinct genera are represented by some 120 species, although some genera and species may not be valid. Three extant genera contain six species. For details of the phylogeny and parsimony, the reader is referred to Marshall (1986b).

Characterization of the Dipnoi, a monophyletic group

During their long evolutionary history, dipnoans either lost or grossly modified many skeletal characters known to have been present in primitive members of the group. We have no direct knowledge of features of the soft anatomy that left no traces on the skeleton; nor do we have at present any adequate knowledge of the post-cranial skeleton of primitive dipnoan. Soft anatomical characters of the post-cranial skeleton of the advanced genera should not be taken as representative of the Dipnoi in the absence of corroborative evidence.

Fossil and extant dipnoans form a well-defined group of osteichthyans. Tooth plates, a feature in common for extant and the majority of the fossil dipnoans, are not found in all dipnoans. Nevertheless primitive dipnoans can be defined by 21 characters of the head skeleton: bone arrangement in the posterior part of the skull roof, relation between supraorbital sensory line and bone, five extrascapulars, ossified (soft in post-Devonian dipnoans) upper lip, lack of premaxilla and maxilla, number and arrangement of bones in cheek region, lack of coronoids, the presence of an adsymphysial plate, ossified (soft in post-Devonian) lower lip, relationship of oral and mandibular sensory canal to "infradentaries", course of neurovascular system in lower jaw, symphysial tubuli in lower jaw, gular-shaped submandibulars, anterior naris at the edge and posterior naris within the mouth, median contact of pterygoids back to jaw articulation, posterior position of parasphenoid (hucco-hypophysial foramen very anterior in parasphenoid), unpaired "vomer", autostyly, neurocranial support of posterior dermal skull roof, no isolated hypobranchials, and pharyngobranchials reduced or lacking. These 21 features distinguish the dipnoans from all other sarcopterygians. The Lower Devonian genus *Diabolepis*, which is said by some authors to have a closer relationship to dipnoans than to any other sarcopterygian, is considered to be inadequately known at present for definite statements about its relationship (Schulze and Campbell 1986).

Dipnoans as Sarcopterygians

Dipnoans are osteichthyans, and are the sister group of crossopterygians (actinistians, onychodontiforms, porolepiforms, osteolepiforms, and including tetrapods). They share with crossopterygians the following derived features: anocleithrum, connection between preopercular and infraorbital sensory lines, true enamel on teeth, cosmine, sclerotic ring with more than four plates, submandibular series, archipterygium, and process of endochondral bone formation. These features characterize the sarcopterygians (crossopterygians and dipnoans), whereas the intracranial joint, doubleheaded hyomandibula, and three extrascapulae are synapomorphies of crossopterygians. Rhipidistians (onychodontiforms, porolepiforms, osteolepiforms, and including tetrapods) are characterized by two synapomorphies, the presence of an extratemporal and narrow submandibular bone(s). Plicidentine, four infradentaries, three coronoids, and a fenestra ventro-lateralis are synapomorphies of porolepiforms, osteolepiforms, and tetrapods. The tetrapods are most clearly related to panderichthyid osteolepiforms (with which they share labyrinthodont plicidentine, three pairs of median skull roof bones, flat skull with high dorsally situated orbits, and marginal position of external naris). The common ancestor of dipnoans and tetrapods is also the common ancestor of crossopterygians (including tetrapods) and dipnoans; in other words, the hypothetical common ancestor of all sarcopterygians. The dipnoans are not the closest sister group of tetrapods, independently if living forms only are considered or fossil forms included (Schultze 1986).

Inference has been drawn from studies of mitochondrial DNA sequences that lungfishes are more closely related to tetrapods than the coelacanths and it has been further reported that analyses of mitochondrial cytochrome oxidase I gene sequences equally support coelacanth/lungfish and lungfish/tetrapod clades, but rule out a coelacanth/tetrapod clade. It has been concluded that the status of the diponoans as a monophyletic group is undisputed, but their phylogenetic affinity is not resolved (read Nieuwenhuys 1998c).

Coelacanth (*Latimeria chalumnae*): The modern record of the coelacanth starts on December 22nd 1938, when a local fisherman brought to Miss Courteny-Latimer, the Director of the small natural history museum at East London on the east coast of South Africa, a large and strange fish he had caught. Prof. J.L.B. Smith gave the fish the scientific name *Latimeria chalumnae* (*Latimeria* in honour of Miss Latimer; *chalumnae* to note the place, the bay of the Chalumna River, where the fish had been caught.

Very few living organisms have received as much scientific consideration as the living coelacanth, *Latimeria chalumnae.* Since its discovery in 1938, approximately 200 specimens have been caught on Grande Comore and of these only a small number have been carefully investigated morphologically. The fish is to be considered a seriously endangered species. Because of its phylogenetic position with respect to the origin of tetrapods and its proposed status as a "living fossil", the anatomy of *Latimeria* has been studied with a broad spectrum of techniques and approaches.

Various aspects related to the biology and physiology of the living coelacanth were covered in a volume edited by McCosker and Lagios (1979). For the current information, the reader is referred to Musick et al. (1991).

L. chalumnae agrees with most other living fishes in having a pair of nasal sacs, each of which opens to the exterior by an incurrent and excurrent nostril and has lost the choana. However, *L. chalumnae* is unique among living fish in having a rostral organ. This structure, accommodated by a cavity in the ethmoidal region of the endocranium, consists of three pairs of unconnected tubes that open to the exterior and lead to a system of crypts invaginated into a median sac composed of connective tissue. The tubes contain a gelatinous substance and, like the median sac, they are separated from the endocranium by a conspicuous layer of adipose tissue. On the basis of its morphology and the fact that it is innervated by ophthalmic lateralis fibres terminating in the dorsal octavolateral nucleus, the rostral organ has recently been ascribed an electroreceptive function.

L. chalumnae is unique in possessing an intracranial juncture apparatus. The abducens nerve persues a most unusual course. The right and left sacculus are connected by a tubular endolymphatic passage, the saccular communicating duct (see details, Bjerring 1985; refer for literature). Thomson (1970) suggested that anterior movement of the shoulder girdle is an important feature of the intracranial mechanism and that such movement allows the geniocoracoid to contract essentially isometrically when the gape is opened. The reader is also referred to a review (Thomson 1969).

L. chalumnae is the only living vertebrate with paired basicranial muscle spanning an intracranial joint that divides the neurocranium into an anterior (ethmosphenoid) and posterior (oticooccipital) portions. The basicranial muscles (subcephalic muscles of some authors) are large, and the joint has well-developed synovial surfaces. There are also specializations that allow the flexure of the organ systems which cross the region, such as the supraorbital sensory canal. This remarkable case of intracranial kinesis has been the subject of functional anatomical interpretations. According to Bjerring (1993), the coelaconthiform basicranial musculature and the nerve supply which pertains to it are trimetameric in character rather than monometameric, as usually reported.

L. chalumnae, the only living representative of a group of sarcopterygian fishes, was thought to be extinct since the Late Cretaceous, approximately 66 million years ago. The evolution of actinistians (coelacanths) has been cited as one of the classic example of conservative (i.e., having no major phenotypic changes since the Devonian) and bradytelic (i.e., having a slow rate of morphological change through time) evolution.

Evidences in the fossil record indicate that actinistians originated as early as the Middle Devonian (specifically the Givetian); thus, they span a period of time of approximately 380 million years. Although the Actinistia is an ancient group, it is thought by many authors to be an evolutionary conservative group.

The interrelationships of 31 actinistian species (including *L. chalumnae*) are analyzed based on a cladistic analysis of 75 osteological characters. Inference of evolutionary trends (e.g., modification of body shape and skull morphology) from the phylogenetic patterns demonstrates that the morphology of actinistians is less conservative than has been proposed previously. This empirical cladistic approach supports two distinct tempos of evolution during an evolutionary history of 380 million years. Along a phylogenetic pathway originating with a Devonian stem-species and ending with the living *L. chalumnae* (including 101 morphological changes and 18 cladogenetic events), the first tempo occurred during the Devonian-Permian periods as a decreasing rate of morphological changes, which was followed by a stabilizing tempo during Permian-Recent periods. The decreasing tempo is characterized by a sequence of gradual versus quantum temporal changes and low versus faster rates, whereas the stabilizing tempo primarily is gradual and low. In contrast to a common assumption, no significant correlation was found between the rates of morphological evolution and the temporal diversity of species (Cloutier 1991).

Latimeria is the product of a long coelacanth lineage, usually viewed as having changed very little. It has been worked out that throughout coelacanth history there have been changes in the structure of the vertebral column involving an overall increase in the number of vertebral elements, and a consequent crowding of these elements within the abdominal region. These changes may be associated with increasing lobation of the second dorsal and anal fins. In the skull, parameters involving intracranial joint have also changed in such a way that the anterior part of the skull has lengthened in relation to the posterior part and this may be associated with an increase in length of the basicranial muscle (Forey 1991).

Schulze (1991) described the palate of *L. chalumnae* based mainly on three-dimension CT scan reconstruction, and compared with that of other osteichthyans. The palate of *L. chalumnae* compares best with that of rhipidistians; it is more advanced than that of actinopterygians in having fewer bones. This tendency toward bone reduction in the palate is even more pronounced in dipnoans. The interpretation of features of the Early Devonian genus *Diabolepis* determines if authors consider dipnoans or actinistians more closely related to tetrapods. Both groups are only distant relatives of tetrapods.

The phylogenetic relationships of *L. chalumnae* have been among the most controversial issues in vertebrate systematics. Coelacanths are most commonly united with lungfishes and tetrapods in the taxon Sarcopterygii, which forms the sister group to the Actinopterygii within the Osteichthyes. Within the Sarcopterygii, however, relationships among the three taxa are unclear. The 'traditional' view has been that coelacanths are the sister group of tetrapods among extant vertebrates, although a number of authors have suggested that lungfishes are the sister group of tetrapods, with coelacanths forming the sister group of this pair. The final possibility, a sister group relationship between coelacanths and lungfishes has only been proposed recently. This heterogeneity of interpretations reflects a level of uncertainty that has led some authors to represent relationships within the Sarcopterygii as an unresolved trifurcation.

4. DEVELOPMENT AND PHYLOGENY

A highly developed head region, specialized for sensory reception, information integration, protection and prey capture is a major vertebrate characteristic. Specialization of the head, or cephalization, is one of the responses to natural selection which lead to establishment of the vertebrate phylum.

The sequence of developmental events known so far is sufficient to produce a plausible reconstruction of the vertebrate head from the point of view of ontogenetic changes which resulted in increasing cephalization. However, to complete the story, we must bridge a major subdivision within the vertebrate taxon, the division into the Agnatha and Gnathostomata which is based chiefly on a further ontogenetic modification of a neural crest derivative, the branchial arches, into jaws. This last cephalization event, which has been used to subdivide the vertebrates into Agnatha and Gnathostomata, two broad groups at the class or superclass level (depending on the particular scheme used) is acknowledged as having provided a major influence on the subsequent evolution and radiation of the vertebrates.

Since long, presence or absence of jaws has been used as the principal criterion for subdividing the vertebrates into two taxa, the Agnatha and Gnathostomata. This has created a phylogenetic problem as to whether or not any agnathan vertebrate is directly related to the gnathostome vertebrates or whether two groups arise from two different stocks of protochordates, displaying some common characteristics only because of convergence.

In the past, researchers have attempted to resolve this problem of agnathan-gnathostome relationships via systematic cladistic analysis of sets of character traits. This has been done either arguing that one or more groups of agnathans share a number of derived characters with the gnathostomes and are thus monophyletic or that these characters are not shared-derived and therefore that jawed and jawless vertebrates are not sister taxa (monophyletic) sharing a common ancestor. Epitomy of this view runs as: 'it is apparent to us that no known agnathan, fossil or Recent, can be regarded as the sister group of the gnathostomes'.

One major problem with this ontogenetic approach is the dependence on direct developmental data which is obviously non-existent for most if not all fossil vertebrates which must therefore be excluded for such analysis. Developmental evidence does exist for neural crest and epidermal placode derivatives in extant agnathans, but unfortunately this evidence is restricted to the petromyzontiforms (lampreys). Very few myxinoid (hagfish) embryos have been found to date and not enough is known about the tissue primordia to tell what (if anything) is neural-crest derived or placodal in origin in myxinoids.

The evidence for placodal derivative in lampreys is entirely morphological in nature. However, both morphological and experimental evidence are available for identifying structures as derivatives of the cranial crest in lampreys.

The full extent of the neural crest and epidermal placode derivatives in petromyzontiforms is not yet known.

The combined contribution of the neural crest and epidermal placodes to the development of the cranial nerves parallels the situation observed in those gnathostomes which have been studied. Likewise, the neural crest contribution to the anterior portion of the trabeculae and trabecular commissure in lampreys is comparable to the situation in gnathostomes.

However, the most convincing evidence for homologous structures linking the lampreys and gnathostomes, evidence which is both morphological and experimental and which has only been disputed only once, lies in the neural crest derivation of the petromyzontiform branchial arches. The visceral arches (the jaws, hyoid elements and branchial arches) in gnathostomes are all neural crest derivatives, previously postulated as being derived phylogenetically by modifications of the branchial arches of an ancestor that possessed a series of unmodified branchial arches as the sole components of the splanchnocranium. Thus, not only do the branchial arches of lampreys share a common developmental origin with those of the gnathostomes, but they fulfil the unmodified state postulated to be the ancestral condition of the gnathostomes.

On the basis of developmental homology, the branchial arches and cranial nerves can be stated as being shared-derived characters of the petromyzontiforms and gnathostomes that are not shared by the protochordate outgroup. The lampreys can then be placed early in the gnathostome-vertebrate evolution and radiation by virtue of the fact that the fundamental development neural-crest-derived branchial arches has already occurred in lampreys and that all that is 'needed' in the evolution of a 'gnathostome' is a modification of these existing structures, i.e., lampreys possess the developmental precursor of jaws.

The development of jaws which occurred after the lampreys arose was clearly not a trivial event; it enabled the subsequent radiation and evolution of the vertebrates (see Langille and Hall 1989).

5. CLASSIFICATION OF SOME LIVING FISHES (See Further Tables 1.4 and 1.5).

We have freely culled information from many sources:

Superclass	Agnatha (Pisces; Group of jawless fishes)
Class	Myxini
Order	Myxiniformes
	Family Myxinidae (hagfishes: Atlantic hagfish, *Myxine glutinosa*)
Class	Cephalaspidiomorphi
Order	Petromyzontiformes

	Family Petromyzontidae: (lampreys: Sea lamprey *Petromyzon marinus*, Pacific lamprey *Entosphenus tridentata*)
Superclass	Gnathostomata (Pisces; Group of jawed fishes)
Class	Chondrichthyes
Subclass	Holocephali
Order	Chimaeriformes
	Families Callorhynchidae (plownose chimaeras: *Callorhynchys*), Chimaeridae (shortnose chimaeras, ratfishes: *Hydrolagus* sp.), Rhinochimaeridae (longnose chimaeras: *Neoharriota* sp., *Harriota* sp., Rhinochimaera sp.)
Subclass	Elasmobranchii (=Selachii)
Superorder	Galeomorphi
Order	Heterodontiformes
	Family Heterodontidae (horn sharks: *Heterodontus*)
Order	Lamniformes
	Families Odontaspidae (ragged-tooth or sand sharks or sand tigers: *Odontaspis taurus*, Lamnidae: (thresher, basking and mackerel sharks: *Alopias*), Megachasmidae (megamouth shark)
Order	Orectolobiformes
	Families Rhincodontidae (whale sharks: *Rhiniodon*), Orectolobidae (nurse sharks: *Ginglymostoma cirratum*)
Order	Carcharhiniformes
	Families Scyliorhinidae (cat sharks: *Scyliorhinus* sp.), Carcharhinidae (requiem sharks; *Carcharinus* sp.), Sphyrnidae (hammerhead sharks: *Sphyrna* sp.)
Superorder	Squatinomorphi
Order	Squatiniformes
	Family Squatinidae (angel shark, monkfishes: *Squatina californica*)
Superorder	Squalomorphi
Order	Hexanchiformes
	Families Chlamydoselachii (frill sharks: *Chlamydoselachus anguineus*), Hexanchidae (cow sharks, sixgill sharks: *Hexanchus*, "broad headed" sevengill shark: *Notorhynchus*)
Order	Squaliformes
	Family Squalidae (dogfish sharks: *Squalus acanthias*)
Order	Pristiophoriformes
	Family Pristophoridae (saw sharks: *Pristiophorus, Pliotrema*)
Superorder	Batoidea
Order	Rajiformes
	Families Rhinobatidae (guitarfishes: *Rhinobatos* sp.), Rajidae (skates: *Raja* sp.)
Order	Pristiformes
	Family Pristidae (sawfishes: *Pristis perottei*)
Order	Torpediniformes
	Family Torpedinidae (electric rays: *Torpedo nobiliana, Narcine brasiliensis*).
Order	Myliobatiformes
	Families Dasyatidae (sting rays: *Dasyatis* sp.), Potamotrygonidae (river sting rays:

	Potamotrygon), Hexatrygonidae, Myliobatidae (eagle rays: *Myliobatis*), Mobulidae (manta: *Manta birostres* and devil rays)
Subclass	Crossopterygii (Lobefins) (Tasel-finned fishes)
Order	Coelacanthiformes
	Family Latimeriidae (gombessa or coelacanth: *Latimeria chalumnae*)
Subclass	Dipneusti (= Dipnoi) (Lungfishes)
Order	Ceratodontiformes
	Family Ceratodontidae (Australian lungfish: *Neoceratodus forsteri*)
Order	Lepidosireniformes
	Families Lepidosirenidae (South American lungfish: *Lepidosiren paradoxa*), Protopteridae (African lungfish: *Protopterus* sp.)
Subclass	Brachiopterygii (Cladistia) (Arm-finned fishes)
Order	Polypteriformes
	Family Polypteridae (bichirs: *Polypterus* sp., reedfish *Erpetoichthys* (= *Calamoichthys*) *calabaricus*)
Subclass	Actinopterygii (Rayfins; Higher bonyfishes)
Infraclass	Chondrostei
Order	Acipensiformes
	Families Acipenseridae (sturgeons: *Acipenser* sp., *Scaphirhynchus*, Polyodontidae (paddlefishes: *Polyodon spathula*)
Infraclass	Neopterygii
Division	Ginglymodi
Order	Lepisosteidae (gars: *Lepisosteus* sp.)
Division	Halecostomi
Subdivision	Halecomorphi
Order	Amiiformes
	Family Amiidae (bowfin: *Amia calva*)
Subdivision	Teleostei
Infradivision	Clupeomorpha
Order	Clupeiformes
	Families Denticipitidae (denticle herring), Clupeidae (herrings: *Clupea*), Engraulidae (anchovies: *Stolephorus*), Chirocentridae (wolf herring or dorab *Chirocentrus*)
Infradivision	Elopomorpha
Order	Elopiformes
	Families Elopidae (tenpounders: *Elops saurus*), Megalopidae (tarpons: *Megalops atlanticus*), Albulidae (bonefishes: *Albula vulpes*)
Order	Notacanthiformes
	Families Halosauridae (halosaurs), Lipogenyidae, Notacanthidae (spiny eels: *Notacanthus*)
Order	Anguilliformes (Apodes)
	Families Anguillidae (freshwater eels: European eel *Anguilla anguilla*, American eel *Anguilla rostrata*), Heterenchelyidae, Moringuidae (spaghetti eels), Myrocongridae, Muraenidae (moray eels), Nemichthyidae (snipe eels: *Nemichthys scolopaceus*), Cyematidae (bobtail snipe eels), Xenocongridae (false morays),

Synaphobranchidae (cutthroat eels), Macrocephenchelyidae, Colocongridae, Congridae (conger eels: *Heteroconger*), Nettastomidae, (duckbill eels), Serrivomeridae, (sawplate eels), Ophichthidae (snake and worm eels: *Echelus*), Derichthyidae (longneck eels), Saccopharyngidae (swallowers), Eurypharyngidae (gulpers: *Eurypharynx*), Monognathidae

Infradivision Osteoglossomorpha

Order Osteoglossiformes

Families Osteoglossidae (bonytongues: *Heterotis niloticus, Arapaima gigas*), Pantodontidae (butterflyfishes: *Pantodon buchholzi*), Hiodontidae (mooneyes: *Hiodon* sp.), Notopteridae (knifefishes: *Notopterus*), Mormyridae (elephantfishes: *Gnathonemus*, *Mormyrus*), Gymnarchidae (electricfish: *Gymnarchus niloticus*)

Infradivision Euteleostei

Superorder Protacanthopterygii

Order Esociformes

Families Esocidae (pikes: *Esox* sp.), Umbridae (mudminnows: *Umbra* sp., *Dallia pectoralis*), Lepidogalaxiidae: (*Lepidogalaxias salamandroides*)

Order Salmoniformes

Suborder Salmonoidei

Families Salmonidae (salmon: Atlantic salmon *Salmo salar*, chinook salmon *Oncorhynchus tshawytscha*, trout: *Salmo gairdneri*, whitefish: *Coregonus*), Retropinnidae (New Zealand smelts: *Retropinna*), Aplochitonidae (southern smelts: *Lovettia*), Galaxiidae (galaxiids: *Galaxias maculatus*), Osmeridae (smelts: *Hypomesus transpacificus*), Plecoglossidae (ayu: *Plecoglossus altivelis*), Salangidae (oriental icefishes: *Salangichthys*), Sundasalangidae (sundaland noodlefishes)

Suborder Argentinoidei

Families Argentinidae (argentines: *Argentina silus*), Bathylagidae (deepsea smelts), Opisthoproctidae (barreleyes: *Opisthoproctus*), Alepocephalidae (slickheads: *Alepocephalus*), Searsididae (tubeshoulders)

Superorder Ostariophysi

Series Anotophysi

Order Gonorhynchiformes

Families Chanidae (milkfish: *Chanos chanos*), Gonorhynchidae (mousefish: *Gonorhynchus*), Kneriidae (shellear: *Kneria*), Phractolaemidae

Series Otophysi

Order Cypriniformes

Families Cyprinidae (minnows and carps: goldfish *Carassius auratus*, common carp *Cyprinus carpio*, *Catla*, *Labeo*), Psilorhynchidae, Homalopteridae (hillstream loaches), Cobitidae (loaches: *Cobitis*, *Misgurnus*), Gyrinocheilidae (algae eaters, gyrinocheilids), Catostomidae (*Catostomus*)

Order Characiformes

Families Citharinidae (moonfish), Hemiodontidae, Curimatidae, Anostomidae (headstanders), Erythrinidae (trahiras), Lebiasinidae, Gasteropelecidae (freshwater hatchetfishes: *Gasteropelecus*), Ctenoluciidae (pike-characins), Hepsetidae (kaufe pike), Characidae (characins: *Astyanax*).

Order Siluriformes

Families Diplomystidae (diplomystid catfishes), Ictaluridae (North American freshwater catfishes: *Ictalurus*), Bagridae (bagrid catfishes: *Mystus*), Cranoglanididae (armorhead catfishes), Siluridae (Eurasian catfishes, sheatfish catfishes: *Wallago*, *Kryptopterus*), Schilbeidae, Pangasiidae (pangasiid catfishes: *Pangasianodon*), Amblycipitidae (Asiatic torrent catfishes, amblycipitids), Amphiliidae (amphiliids), Akysidae (stream catfishes, akysids), Sisoridae (sisorid catfishes: *Sisor*), Clariidae (airbreathing catfishes: *Clarias*), Heteropneustidae (airsac catfishes, stinging catfishes: *Heteropneustes fossilis*), Chacidae (squarehead catfishes, chacids), Olyridae, Malapteruridae (electric catfishes), Mochokidae (upsidedown catfishes), Ariidae (sea catfishes: *Arius*), Doradidae (thorny catfishes), Auchenipteridae, Aspredinidae (banjo or obstetrical catfishes), Plotosidae (eeltail catfishes), Pimelodidae (long-whiskered catfishes), Ageneiosidae (barbel-less catfishes, ageneiosids), Hypothalmidae (loweye catfishes, hypothalmids), Helogeneidae (helogeneids), Cetopsidae (whale catfishes, cetopsids), Trichomycteridae (parasitic catfishes), Callichthyidae (callichthyid armored catfishes: *Callichthys*), Loricariidae (suckermouth armored catfishes), Astroblepidae (South American torrent catfishes)

Order Gymnotiformes

Families Sternopygidae, Rhamphichthyidae (knifefishes), Hypopomidae, Apteronotidae (freefin knifefishes: *Sternarchus*), Gymnotidae (knife eels, naked-back knifefishes), Electrophoridae (electric eel, electric knifefish *Electrophorus electricus*)

Superorder Stenopterygii

Order Stomiiformes

Families Gonostomatidae (bristlemouths or light fishes: *Cyclothone elongata*), Sternoptychidae (marine hatchetfishes: *Argyropelecus olfersi*), Photichthyidae, Chauliodontidae (viperfishes: *Chauliodus sloani*), Stomiidae (scaly dragonfishes), Astronesthidae (snaggletooths), Melanostomiidae (scaleless black dragonfishes), Malacosteidae (loosejaws: *Malacosteus niger*), Idiacanthidae (black dragonfishes)

Superorder Scopelomorpha

Order Aulopiformes

Families Aulopodidae (aulopus), Chloropthalamidae (greeneyes), Scopelarchidae (pearleyes), Notosudidae (waryfishes), Synodontidae (lizardfishes: *Synodus foetans*), Giganturidae (giganturids), Paralepididae (barracudinas: *Lestidium*), Anotopteridae (daggertooths: *Anotopterus pharao*), Evermannellidae (sabertooth fishes), Omosudidae, Alepsauridae (lancetfishes: *Alepisaurius*), Pseudotrichonotidae

Order Myctophiformes

Families Neoscopelidae, Myctophidae (lanternfishes: *Myctophum*)

Superorder Paracanthopterygii

Order Percopsiformes

Families Percopsidae (trout perches: *Percopsis omiscomaycus*), Aphredoderidae (pirate perch: *Aphredoderus savanus*), Amblyopsidae (cavefishes: *Chologaster agassizii*)

Order Gadiformes

Families Muraenolepidae, Moridae (moridcods, flatnose cods: *Antimora*),

	Melanonidae, Bregmacerotidae (codlets), Gadidae (cods: Atlantic cod *Gadus morhua*), Merlucciidae (merluccid hake, hakes: *Merluccius*), Macrouridae (grenadiers or rattails *Coelorhynchus, Macrourus*)
Order	Ophidiiformes
	Families Ophidiidae (cusk-eels: *Pteridium*), Carapidae (carapids, pearlfishes), Bythitidae (viviparous brotulas), Aphyonidae
Order	Batrachoidiformes
	Family Batrachoididae (toadfishes and midshipmen: midshipmen *Porichthys*, oyster toadfish *Opsanus tau*)
Order	Lophiiformes
	Families Lophiidae (goosefishes: *Lophius americanus*), Brachionichthyidae (warty anglers), Antennariidae (frogfishes, sargassumfish: *Histrio histrio*), Chaunacidae (sea toads), Ogcocephalidae (batfishes: *Ogcocephalus*), Caulophrynidae (fanfins: *Cauliphryne*), Melanocetidae (black devils: *Melanocetus*), Diceratiidae, Himantolophidae (football fishes), Oneirodidae (dreamers), Thaumatichthyidae, Gigantactinidae (whipnoses), Neoceratiidae, Centrophrynidae (deepsea anglerfish), Ceratiidae (sea devils, deepsea anglerfish: *Photocorynus spiniceps*), Linophrynidae (netdevils: *Linophryne*)
Order	Gobiesociformes
	Families Gobiesocidae (clingfishes: *Gobiesox maeandricus*), Alabetidae (singleslits)
Superorder	Acanthopterygii
Series	Atherinomorpha
Order	Cyprinodontiformes
Suborder	Exocoetoidei
	Families Exocoetidae (flying fishes: *Cypselurus*), Hemiramphidae (halfbeaks: *Hemirhampus*), Belonidae (needlefishes: *Belone*), Scomberesocidae (sauries: *Cololabis saira*)
Suborder	Adrianichthyoidei
	Families Oryziidae (ricefishes), Adrianichthyidae: (adrianichthyids), Horaichthyidae (horaichthyids)
Suborder	Cyprinodontoidei
	Families Aplocheilidae (rivulines: *Rivulus marmoratus*), Cyprinodontidae (killifishes and topminnows: *Orestias*), Poeciliidae (livebearers: *Poecilia reticulata*, *Xiphophorus* sp., *Gambusia affinis*), Goodeidae (Mexican livebearers), Anablepidae (four-eyed fishes: *Anableps*), Jenynsiidae (jenynsiids).
Order	Atheriniformes
	Families Atherinidae (silversides: *Leuresthes tenuis*). Isonidae (surf sardines, isonids), Melanotaeniidae (rainbow fishes), Neostethidae (neostethid priapiumfishes), Phallostethidae (phallostethid priapiumfishes)
Superorder	Acanthopterygii
Series	Percomorpha
Order	Lampridiformes
	Families Lampridae (opah), Veliferidae, Lophotidae (crestfishes), Radiicephalidae, Trachipteridae (ribbonfishes: *Regalecus*), Regalecidae (oarfishes), Stylephoridae

(tube-eye: *Stylephorus chordatus*), Ateleopodidae, Mirapinnidae (hairyfish) Eutaeniophoridae (tapetails), Megalomycteridae (largenoses)

Order Beryciformes

Families Stephanoberyeidae (pricklefishes), Melamphaidae (bigscales), Gibberichthyidae (gibberfishes), Trachichthyidae (slimeheads), Diretmidae (spinyfins), Korsogasteridae, Anoplogastridae (fangtooth), Berycidae (alfonsinos), Monocentridae (pinecone fishes: *Monocentris japonicus*), Anomalopidae (lanterneyes: *Anomalops, Photoblepharon*), Holocentridae (squirrelfishes: *Myripristis*), Rondeletiidae (redmouth whalefishes), Barbourisiidae, Cetomimidae (flabbly whalefishes), Polymixiidae (beardfishes: *Polymixia*).

Order Zeiformes

Families Parazenidae (parazen), Macrurocyttidae, Zeidae (dories: *Zeus faber*), Oreosomatidae (oreos), Grammicolepididae, Caproidae (boarfishes: *Antigonia*).

Order Gasterosteiformes

Families Aulorhynchidae (tubesnouts: *Aulorhynchus flavidus*). Hypoptychidae (sandeel), Gasterosteidae (sticklebacks: *Gasterosteus aculeatus, Pungitius*), Indostomidae, Pegasidae (seamoths: *Pegasus umitengu*), Solenostomidae (ghosh pipefishes), Syngnathidae (pipefishes: *Syngnathus* and seahorses: *Hippocampus*) Macrorhamphosidae (snipefishes), Centriscidae (shrimpfishes), Aulostomidae (trumpetfishes), Fistularidae (cornetfishes)

Order Dactylopteriformes

Family Dactylopteridae (flying gurnards: *Dactyloptena*)

Order Scorpaeniformes

Families Scorpaenidae (rockfishes and scorpionfishes: *Sebastes, Pterosis*), Synanceidae (stonefishes: *Synanceia*), Triglidae (searobins: *Trigla*), Caracanthidae (orbicular velvetfishes), Aploactinidae (velvetfishes), Pataecidae (prowfishes), Anoplopomatidae (sablefishes), Hexagrammidae (greenlings), Zaniolepidae (combfishes), Platycephalidae (flatheads), Hoplichthyidae (ghosh flatheads), Congiopodidae (pigfishes), Ereuniidae, Cottidae (sculpins: *Oligocottus*), Cottocomephoridae (Baikal sculpins), Comephoridae (Baikal oilfishes), Normanichthyidae, Cottunculidae, Psychrolutidae, Agonidae (poachers), Cyclopteridae (lumpsuckers or lumpfishes: *Cyclopterus lumpus* and snailfishes: *Liparis*)

Order Perciformes

Suborder Percoidei

Families Centropomidae (snooks), Perichthyidae (temperate basses: *Morone*), Serranidae (sea basses), Grammistidae (soapfishes), Psudochromide (dottybacks), Grammatidae (basslets), Plesiopidae (roundheads), Acanthoclinidae, Glaucosomatidae, Teraponidae (tigerperches), Banjosidae, Kuhliidae (aholeholes), Centrarchidae (sunfishes: *Lepomis, Micropterus*), Priacanthidae (bigeyes), Apogonidae (cardinal fishes), Acropomatidae, Percidae (perches: *Perca, Percina*), Sillaginidae (smelt-whitings), Malacanthidae (tilefishes), Labracoglossidae, Lactariidae (false trevallies), Pomatomidae (bluefishes), Rachycentridae (cobia), Echeneididae (remoras: *Echeneis naucrates*), Carangidae (jacks), Nematistiidae (roosterfish), Coryphaenidae (dolphins), Apolectidae, Menidae (moonfish), Leiognathidae (ponyfishes), Bramidae (pomfrets), Caristiidae (manefishes), Arripidae (Australian salmon), Emmelichthyidae (rovers), Lutjanidae (snappers: *Lutjanus campechanus*), Caesionidae, Nemipteridae (threadfin breams), Lobitidae (triple tails), Gerreidae (mojarras), Haemulidae (grunts), Lethrinidae (emperors), Sparidae (porgies), Centracanthidae, Sciaenidae (croakers and drums), Mullidae (goatfishes), Monodactylidae (moonfishes), Pempheridae (sweepers), Leptobramidae (beach-salmon), Bathyclupeidae, Toxotidae (archerfishes), Coracinidae (galjoens), Kyphosidae (see chubs), Ephippididae (spadefishes), Scatophagidae (scats), Rhinoprenide (threadfin scat), Chaetodontidae (butterflyfishes), Pomacanthidae (anglefishes), Enoplosidae, Pentacerotidae (boarfishes), Nandidae (leaffishes), Oplegnathidae (knifejaws), Embiotocidae (surfperches), Cichlidae (cichlids: *Symphysodon, Pterophyllum*) Pomacentridae (damselfishes: *Pomacentrus*), Gadopsidae (Australian blackfish), Cirrhitidae (hawkfishes), Chironemidae (kelpfishes), Aplodactylidae, Cheilodactylidae (morwongs), Latrididae (trumpeters), Owstoniidae, Cepolidae (bandfishes)

Suborder Mugiloidei
Family Mugilidae (mullets: *Mugil*)

Suborder Sphyraenoidei
Family Sphyraenidae (barracudas: *Sphyraena barracuda*)

Suborder Polynemoidei
Family Polynemidae (threadfins: *Polydactylus*)

Suborder Labroidei
Families Labridae (wrasses), Odacidae, Scaridae (parrotfishes: *Scarus*)

Suborder Zoarcoidei
Families Bathymasteridae (ronquils), Zoarcidae (eelpouts), Stichaeidae (pricklebacks:*Chirolophis*), Cryptacanthodidae (wrymouths), Pholididae (gunnels), Anarchadidae (wolfeels), Ptilichthyidae (quillfish), Zaproridae (prowfish), Scytalinidae (graveldiver)

Suborder Nototheniodei
Families Bovichthyidae, Nototheniidae (cod icefishes, Antarctic blennies), Harpagiferidae (plunder fishes), Bathydraconidae (antarctic dragonfishes), Channichthyidae (crocodile icefishes)

Suborder Trachinoidei
Families Opisthognathidae (jawfishes), Congrogadidae (eelblennies), Chiasmodontidae, Champsodontidae, Notograptidae, Pholidichthyidae (convict blenny), Trichonotidae (sandfishes), Trachinidae (weeverfishes), Uranoscopidae (stargazers), Trichonotidae (sanddivers), Creediidae (sandburrowers), Leptoscopidae, Percophidae (duckbills), Mugiloididae (sandperches), Chemarrhichthyidae

Suborder Blenniodei
Families Tripterygiidae (triplefins, threadfin blennies), Dactyloscopidae (sand stargazers), Labrisomidae (labrisomids), Clinidae (clinids), Chaenopside (pikeblennies and flagblennies), Blennidae (combtooth blennies: *Aspidodontus taeniatus*).

Suborder Icosteoidei
Family Icosteidae (ragfish: *Icosteus aenigmaticus*)

Suborder Schindleriodei
Family Schindleriidae: *Schindleria*

Suborder Ammodytoidei
Family Ammodytidae (sand lances: *Ammodytes*)

Suborder Callionymoidei
Families Callionymidae (dragonets: *Callionymus*), Draconettidae

Suborder Gobioidei
Families Elecotridae (sleepers), Gobiidae (gobies: *Periophthalmus*, *Gillichthys mirabilis*), Rhyacichthyidae (loach gobies), Kraemeriidae (sand gobies), Gobiodidae (eel gobies), Trypauchenidae (burrowing gobies), Microdesmidae (wormfishes)

Suborder Kurtoidei
Family Kurtidae (Nurseryfishes, humpheads, forehead brooders: *Kurtus*)

Suborder Acanthuroidei
Families Acanthuridae (surgeonfishes: *Acanthurus*), Siganidae (rabbitfishes)

Suborder Scombroidei
Families Scombrolabracidae, Gempylidae (snake mackrels, escolors: *Ruvettus pretiosus*), Trichiuridae (cutlassfishes), Scombridae (mackerels: *Scomber* sp. and tunas: *Thunnus thynnus*), Xiphiidae (swordfish: *Xiphias gladius*), Luvaridae (luvar), Istiophoridae (billfishes: *Istiophorus*)

Suborder Stromateoidei
Families Amarsipidae, Centrolophidae (medusafishes), Nomeidae (driftfishes), Ariommatidae, Tetragonuridae (squaretails), Stromateidae (butterfishes)

Suborder Anabantoidei
Families Anabantidae (climbing gouramies: *Anabas testudineus*), Belontiidae (gouramis: *Betta splendens*), Helostomatidae (kissing gourami: *Helostoma temmincki*), Osphronemidae (giant gourami: *Osphronemus goramy*)

Suborder Luciocephaloidei
Family Luciocephalidae (pikehead: *Luciocephalus pulcher*)

Suborder Channoidei
Family Channidae (snakeheads: *Channa* sp.)

Suborder Mastacembeloidei
Families Mastacembelidae (spiny eels: *Mastacembelus*, *Macrognathus*), Chaudhuriidae

Order Synbranchiformes
Family Synbranchidae (swamp eels: *Synbranchus, Monopterus albus*, *Amphipnous cuchia*)

Order Pleuronectiformes
Families Psettodidae, Citharidae, Bothidae (lefteye flounders), Pleuronectidae (righteye flounders: *Hippoglossus*), Soleidae (soles: *Achirus, Solea solea*), Cynoglossidae (tonguefishes: *Symphurus*)

Order Tetraodontiformes

Families Triacanthodidae (spikefishes), Triacanthidae (triplespines: *Tricanthus oxycephalus*), Balistidae (triggerfishes and filefishes: *Balistes*), Ostraciontidae (boxfishes), Triodontidae (threetooth puffers, pursefishes, *Triodon bursarius*), Tetraodontidae (puffers, pufferfishes: *Canthigaster)*, Diodontidae (porcupinefishes: *Diodon*), Molide (molas or ocean sunfishes: *Mola mola*)

Table 1.4 Classification of fossil and extant Chondrichthyes

Subclass Subterbranchialia	Order Symmoriida
Cohort Iniopterygia	Order Eugeneodontida
Cohort Holocephali	Order Orodontida
Order Chondrenchelyidea	Order Petalodontida
Order Cochliodontida	Order Squatinactida
Order Psammodontida	Order Desmiodontida
Order Copodontida	Cohort Euselachii
Order Helodontida	Order Ctenacanthiformes
Order Menaspoidea	Order Hybodontiformes
Order Chimaerida	Order Hexanchiformes
Suborder Chimaerina	Subcohort Neoselachii
Subclass Elasmobranchii	Superorder Heterodontimorphii
Cohert unnamed	Superorder Galeomorphii
Order Xenacanthida	Superorder Squalomorphii
Order Cladoselachida	Superorder Squatinomorphii
Order Coronodontida	Superorder Batomorphii

Source: Schultze (1993)

Table 1.5 Classification of Teleostomi

Class Acanthodi	Subclass Sarcopterygii
Order Climatiida	Infraclass Dipnoi
Order Ischnacanthida	Infraclass Crossopterygii
Order Acanthodida	Actinistia
Class Osteichthyes	Onychodontiformes
Subclass Actinopterygii	Porolepiformes
Cladistia	Osteolepiformes
Palaeonisciformes	Panderichthyiformes
Acipenseriformes	Tetrapoda
Neopterygii	
Ginglymodi	
Halecostomi	
Halecomorphi	
Teleostei	
(Classification and phylogeny of lower Actinopterygii not settled)	

Source: Schultze (1993)

CHAPTER 2

Morphometrics; Taxonomy; Meristic Variation

1. MORPHOMETRICS

1.1 Introduction

While morphometrics can be broadly defined as a body of techniques for describing body form, a somewhat more narrowly defined usage is currently popular. It has been considered as the empirical fusion of geometry and biology. Further, morphometrics has been defined as 'the statistical analysis of biological homology treated as geometrical deformation'. The 'new morphometrics' of the Michigan Morphometrics Study group is not simply about quantitative description of the phenotype, but also about phenotypic change with respect to a homology function in a multivariate setting.

1.2 An Overview

The specific goals of a morphometric approach are paramount to the appropriate choice of method. The fundamental goal is the description of form. Most applications of the method have sought contrasts between classes of objects as a guide to discrimination, be they individual shapes (i.e., organisms or their parts), populations, species, or communities.

Data

The new morphometrics offers no universal protocol for selecting variables to be studied, except that they be numerous and sample the entire form. Morphometric data typically fall into four categories (and often in some combination): distances, angles, co-ordinates, and radial functions of boundary outlines. Traditional data sets from bilaterally symmetric organisms (e.g., fishes) involve distance measures chosen by convention, often because of *a priori* notions of taxonomic value. The use of such data has been criticized on several grounds. First, these distances generally tend to be aligned with only a few axes, thereby biasing the potential contrasts toward longitudinal directions and resulting in an uneven coverage of form. Second, the choice of long distances, as opposed to short segments, can bias the covariance structure of the data. The information obtained in short segments is more localized, with a tighter covariance among variables. Often objections stem from the omission of biological homology. The extent that different intraspecific growth trajectories contribute toward intraspecific shape differences will dictate the degree to which such traditional data fail to describe significant shape contrasts.

Angles are rarely used as primary data. Measurement error for short distances can be propogated when extrapolated to longer distances. Some workers believe the expressing shape contrasts in terms of the angle between two points with reference to a third will remove the confounding effects of size. Departures from isometry in most circumstances violate this assumption and involve statistical problems similar to those caused by using ratios.

Landsmarks and Homology

The Michigan Group has argued for a different convention. They focus instead on homologous landmarks of form. Anatomical landmarks are identified by conventional rules of biological homology (spatial, ontogenetic, and phylogenetic congruence). Positions of the landmarks relative to a particular coordinate system become the raw data for analysis. 'Extremal' variables, representing maximum or minimum distances, are strictly avoided since these rarely provide information on homology. Geometric location is thus wedded to biological homology. The geometric deformation indicated in the statistical analysis can thus be given an evolutionary or ontogenetic interpretation.

The insistence on homologous landmarks is an important contribution of the Michigan Group to the foundation of morphometric method. However, the procedure by which homologous landmarks are identified is not discussed and deserves some comment. Homology in morphometrics does not adequately conform to the concept of biological homology; instead it merges geometric and phylogenetic definitions.

Geometric coordinates (i.e., Cartesian, polar, elliptic, etc.) are not considered to be acceptable data since they fail to incorporate homology information.

The Fourier method attempts to describe form in terms of a radius function: the distance from a central axis to the boundary outline as a function of angle relative to a starting reference ray, integrated over the entire outline.

It has pointed out that neither homologous point measures nor Fourier descriptions capture complete information about shape. The argument is not about which method is better but whether it directly measures those aspects of shape likely to be biologically informative and of specific interest.

The Truss Network

To overcome inherent weaknesses of traditional character sets, the truss network has been proposed: a geometric protocol for character selection. The truss is systematic pattern of measurement intermediate between simple triangulation and a globally redundant design attempts to account for random measurement error by averaging individual mismeasurements over many measures. Given any reference triangle, distances are measured to any three other ones. Note that this set is not totally redundant in that distances between all possible landmark pairs are not measured. In the truss, each quadrilateral shares one edge with the preceding quadrilateral, adding only one extra distance measure over that in the triangular network, while still over-representing the form.

Apart from the advantage of including oblique and vertical distances for a more even coverage of shape, the principal advantage of the truss network includes: (1) the ability to reconstruct the form from the original data (archival), and (2) recognition and compensation for random measurement error. When landmarks are suitably spaced upon the outline, the quadrilaterals will approach an idealized truss. Measurement errors are indicated when the truss cells depart form coplanarity. In a coplanar truss, the sixth distance is uniquely specified by the first five and will vary from this exact value by measurement error (its own plus that present in first five distances). It is suggested that measurement error be removed by 'flattening the truss', An average form is computed by regression of measured distances on a composite measure of body size (i.e., the first principal component in intergroup analysis), followed by reconstruction of the form (archival) using the distances provided by regression functions at the particular body size.

Methods of Data Acquisition

Given a defined set of variables, there are three basic methods of acquiring the data for morphometric analysis: (1) manual distance measurement, (2) manual digitization, and (3) automated video digitization. Several factors can influence the choice of method, but two criteria stand out. First, the method of data acquisition should yield variables that are accurate, precise in repeated measures, and unaffected by specimen orientation or change of operator. Second, ease of operation and speed are desirable.

Manual distance measurement is by far the most widely used method. Dividers, calipers or ocular micrometers (measuring reticles) are used to measure linear distances between landmarks. With even a few variables per specimen, this method can be quite tedious and time consuming, with fatigue reducing data quality. Digital-readout calipers interfaced with the computer can save much time. Users without access to more sophisticated methods will continue to rely on normal measurement.

Digitized data acquisition is increasingly popular. In manual digitization for landmark data, the 'object' is placed on a digitizing tablet and *x*, *y*, co-ordinates are produced for each landmark. The term 'object' refers to one of several possible means of representing the specimen, usually in two dimensions, although data in the third dimension can be obtained by repositioning the specimen and digitizing again. There is another method for morphometrics in three dimensions. The object can be a photograph, radiograph, perimeter tracing, or the specimen itself. For specimens without landmarks, outlines can be digitized in stream mode, yielding a series of *x*, *y* co-ordinates describing the form.

Video digitization offers automated data acquisition. The basic system consists of a video camera, high resolution video monitor, microcomputor, mouse, imaging board (frame grabber) for analogue video-to-digital signal conversion, and software. Several software packages are available and differ in hardware requirements, menu format and flexibility. Some offer data acquisition as both co-ordinates and outlines. Images can also be stored on computor disc for re-analysis.

Comparative studies of the utility of these methods are rare. While the advantages of rapid data collection and absence of transcription errors in automated data acquisition are easily recognized, are there other advantages? Are the methods comparable in variance due to measurement error? What are their limitations?

The problem of measurement is rarely addressed in morphometric studies.

Analysis of Measurement Error

There are quite different potential sources of measurement in the two methods of data acquisition.

Manual measurement error may occur at three steps: (1) reading of caliper dial, (2) transcription onto paper, and (3) keypunch errors. The video system eliminates transcription error, but has other sources not present in manual mode. The accuracy and precision with which one can digitize landmarks depends on the quality and resolution of the video system.

In video mode, the three dimensional object is viewed and digitized in two dimensions. The level of error introduced for a measure is a function of the magnitude of the component in the third dimension, perpendicular to the midsaggital plane.

The manual and digitized data acquisition methods do not differ significantly in precision. There may be real differences in accuracy. Of the potential sources of error, image resolution may be the most critical for caliper measurement. A four-fold decrease in time required per specimen may be achieved when using a video system for measuring many variables on many specimens. However, given the large potential for measurement error observed here for both methods (range 2–17%), caution must be exercised and users should attempt to evaluate and correct for measurement error in morphometric studies.

2. ANALYSIS AND APPLICATIONS

Four multivariate methods most frequently applied to biological problems are discriminant functions (DFA), the related canonical variates (CVA), canonical correlation (CCA), and principal-components analysis (PCA). All are related in using a linear compound of the *n* variables multiplied by their respective coefficients to maximize intragroup variance.

If group structure is known, or assumed in advance, then DFA or CVA are appropriate. DFA attempts to construct the linear compound, or discriminant axis, which maximizes group separation. Specimens of unknown group membership can then be classified by the discriminant function. Except for classifying unknowns, DFA is seldom used alone. CVA is the correlate of PCA when groups are known or assumed. Its main use is in revealing underlying relationships among groups and variables. While in PCA the latent factors are generally orthologous, or uncorrelated with one another, in CVA this is not necessarily so. CCA is useful when one is interested in knowing if (any how) two or more sets of variables covary. If groups are not known in advance, then PCA is to reduce the dimensionality of the data set by representing the observed variables as functions of a smaller number of latent factors which are uncorrelated with one another and ideally can be interpreted. A modification of PCA, known as shared principal component analysis (S-PCA), is a method for discriminating *a priori* groups (recognized by other criteria) on the basis of shape, independent of size.

Principal Components Analysis

PCA is by far the most widely used multivariate method in morphometrics. However, applications typically involve comparisons of *a priori* group (i.e., populations, species, sexes, etc.) and thus depart from generalized PCA. Its use, instead of the more appropriate CVA, is directed toward evaluating hypotheses of *a priori* group structure.

Shared principal components analysis (S-PCA) has received much attention; however most applications have involved fishes. The rational behind S-PCA follows from two problems: (1) most applications involve multi-group comparison, while PCA was intended for single-group analysis; differences in size among groups can confound shape variance and PCs, (2) samples of organisms often incorporate several different ontogenetic stages so morphometric variation incorporates both shape and allometry. In S-PCA, the second and subsequent components are calculated so as to remove intergroup size.

Two comments about the method have been well noted. First, S-PCA does not improve discrimination among groups, but aids in the interpretation of their contrasts, since shape separation is 'sheared' perpendicular to PC2. Second, S-PCA is effecting only when patterns of within-group size are similar among groups. Otherwise, nonsensical components can result.

The morphometric methods have been widely applied in systematic ichthyology: multivariate analysis of hybrids; multivariate morphometrics in the study of the association between morphology and habitat preferences (ecology); morphological discrimination of taxa (systematics) and phylogenetics. There must be caution in extrapolating ecomorphological relationships across species groups. The role of morphometrics in empirical phylogenetics has yet to be precisely defined.

3. BIOCHEMICAL AND SEROLOGICAL TECHNIQUES OF TAXONOMY

The literature on isoenzymes has been quite extensive. The application of their study to fish taxonomy is quite recent and the enzymes present in serum, red blood cells and in a number of tissues have been studied by experimental taxonomists. The techniques used for isoenzymes are almost entirely biochemical although serological methods have been used to clarify certain problems in isoenzyme studies.

Taking isoenzymes, probably more is known about lactic acid dehydrogenase (LDH) than any other enzyme. Other enzymes which have been studied in fishes include among others, estrases of serum and kidney, peroxidase in serum and aspartate aminotransferase (AAT), malate dehydrogenase (MDH), phosphogluconate dehydrogenase, creatine kinase and phosphoglucomutase in tissues.

As in the case of serum proteins, haemoglobins, etc. changes in isoenzymes occur during ontogeny although they have only been extensively studied in LDH, MDH and some esterases. Environmental factors also influence isoenzymes patterns.

Free amino acids have been studied by the chromatographic methods. Fish mucus is species specific. The presence of blood serum antigens in both body mucus and gill mucus has been reported. Serum proteins have been extensively studied. The soluble proteins of eye-lens have great value in taxonomic studies. Water-soluble muscle proteins (myogens), and haemoglobins (multiple haemoglobins; interspecific variation; ontogenetic difference; haemoglobins of hybrid fishes) have significant role in taxonomy.

Serological methods particularly precipitin reactions have proved profitable.

4. MISCELLANEOUS INFORMATION

(1) Genetics of Sex

Flow cytometry: The amount of nuclear DNA, i.e., genome size can be estimated by flow cytometry. Cells are stained with a DNA-specific fluorochrome and are analyzed individually for fluorescence, which is directly proportional to nuclear DNA content. Within the fishes, flow cytometry has been used to measure DNA in a variety of cell types such as erythrocytes and sperm; and in cells from liver, kidney and adipose fin.

Estimation of nuclear DNA is of potential significance in connection with the genetics of sex determination and the study of taxonomic relationships.

(2) Stock Separation

(i) *Phenotype markers*: Vertebral number, fin ray number, and other phenotypic characters have continued to be used by some workers as at least partial markers for races. More cautious views of such phenotypic variation are often expressed by taxonomists who have continued to study the variation in phenotype characters as a topic of research in its own right.

(ii) *Red blood cell antigen frequency markers*: Red blood cells have surface antigens and the blood plasma contains the complementary immune antibodies in solution. These two components give rise to the well-known "blood type" effect.

Different red blood cell antigens are regarded as allelic variants of one or more loci; consequently, red blood cell antigen frequencies can provide the basis for analysis of the genetical structure of populations.

Different views have been expressed. Some geneticists believed that blood group genetic variation had considerable adaptive significance, and generally their point of view has been substantiated. Nonetheless, a very large number of studies were conducted aimed at stock separation.

The use of red blood cell antigen and antibody frequencies to define stocks fell into disuse, but not for the obvious reason that they were adaptive features of blood cells. The main reason was that the proteins involved proved to be systematically altered by various posttranslational modifications. A wide number of "factors", small and large polypeptides, are involved in the immune response so that observed red blood cell antigen frequencies would very often depend on the physical state and disease history of the fish. There are also some experimental difficulties encountered when otherwise accessible red blood surface antigens are "coated" with certain plasma proteins, making it necessary to undertake a variety of cleaning and other preparatory steps. Other red blood cell antigen/antibody studies in fish had already shown very complicated patterns of occurrence that would require lengthy breeding and biochemical analyses to resolve.

(iii) *Protein allele markers*: Electrophoresis of protein allelic variants indicated a system of a few codominant alleles with a readily interpretable genetic basis (allele variants are the proteins coded for by the different alleles occurring at a polymorphic locus). Protein allelic variants rapidly replaced red blood cell antigens and antibodies as genetic markers that could be considered free of any environmental influence.

A comprehensive discussion of all aspects related to stock separation is available (see Gauldie 1991).

Buth et al. (1991) presented information on cyprinid allozymic, isozymic, cytogenetic characteristics and their research applications.

5. ENVIRONMENTAL INFLUENCES ON MORPHOLOGY

Variation in size and shape of fishes will not be treated as fully as meristic variation.

Differences in size and shape are apparent in comparisons between fish from waters of different temperatures. Almost invariably the more northern representatives of a species, or of a genus, are larger than those to the south. The changes in body proportions have not been as thoroughly investigated as those in segmental structures, but some tentative generalizations can be made. Northern, slowly growing races of a species usually have smaller heads, eyes, maxillaries, and fins than do their southern counterparts, although opposite effects are not uncommon.

Growth rate apparently determines body shape by altering the time of transition from one growth stanza to another. Each growth stanza is recognized by a characteristic relative growth of the body part. The transition from one stanza to the next is usually abrupt and produces a new relationship between the growth of the part and the body as a whole. If the logarithms of the part and of the length of the fish are plotted, a straight line is obtained for a given stanza. The transition of the next stanza is revealed by an inflection of the slope of the line. Differences in the relative sizes of body parts between fish of the same species are the results of changes in the timing of the inflections, not of the slopes of the lines. Because of this, growth changes in early life are particularly efficacious in modifying the form of fish. Changes in relative growth sometimes occur at sexual maturity, but seldom after.

Evidently, the environment can modify the shape of a fish acceleration or retardation of the rate of development.

Variation in shape of fishes is poorly understood, but in general retarded growth seems to cause small heads and associated parts, and short fins.

Meristic Variation

Countable structures are of special value systematically because they are relatively easy to record unambiguously and because they are easy to summarize in tabular fashion.

In the meristic characters, slow development is commonly associated with high numbers of elements, fast development with low numbers. Different factors in environment, such as temperature, salinity, and oxygen, may control the role of development. The number of elements is determined during a short "sensitive" period in the embryonic, or larval, fish. The term meristic variation is used in widest sense, referring to all countable characters, specially fin rays and vertebrae.

Plots of temperature during development versus number of elements formed, in precisely conducted experiments, have revealed a family of *V*-, or inverted *V*-shaped curves. However, changing the temperature just before the sensitive period always produced an increase in the counts when the temperature was lowered, and a decrease when the temperature was raised. The rate of differentiation of each meristic character is thought to be more or less independent of the other meristic characters in its relationships to the overall growth of the embryo. Inflection points on the curves indicate reversals in the relationships. Disagreements in findings from the field and from the laboratory may be result of monotonous, uniform conditions in the laboratory, an unnatural and possibly harmful situation. Experiments are needed which employ controlled, fluctuating parameters of the environment.

Genetic determination of variation in meristic characters has been demonstrated through typical breeding experiments, hybridization, and transplantation. It has also been deduced by comparing the nature of variation within a given year-class, between successive year-classes, and between populations.

The variation itself may be a more stable and useful indicator of genetic determinants than the mean.

Regular changes in count, such as geographical clines, probably reflect adaptive changes of a genetic nature.

A few examples of meristic variation are given here. A comparison is made between parasitic lamprey *Mordacia mordax* and the nonparasitic species *M. praecox* based on the counts of myomeres, circumoral cirri and teeth.

Studies of the distribution of mean values in five different meristic characters (vertebral centra, dorsal, anal, and pectoral fin rays, and gill rakers) have demonstrated that at least three populations of northern anchovy (*Engraulis mordax mordax*) inhabit the waters of the Pacific coast of North America.

Meristic characters that have proved valuable systematically in the Sardini include numbers of fin rays (first dorsal spines, second dorsal rays, dorsal finlets, anal finlets, pectoral rays), gill rakers, teeth (especially on the upper and lower jaws), vertebrae, and laminae in the olfactory rosettes.

Within Mormyridae, *Mormyrops curviceps* differ from *M. deliciosus* in seven meristic characters: the number of lateral line scales, caudal-peduncle scales, total vertebrae, caudal vertebrae, anal-fin rays, and teeth on the upper and lower jaws.

A study by Mayden et al. (1991) employd a phylogenetic treatment of morphometric and meristic traits to assist identification and ascertain relationships among extant species of the cyprinid genera *Ptychocheilus* and *Mylopharodon*.

CHAPTER 3

Integument; Dermal Skeleton; Coloration and Pigment Cells; Epidermal Breeding Tubercles and Bony Contact Organs; Biotoxins; Alarm Pheromones

1. INTEGUMENT

1.1 Introduction

The vertebrate integument proves to be one of the main parts by which the mode of adaptive evolution can be continuously outlined.

The original classical studies of the integument indicated that it consisted of two clearly distinguishable layers, an epidermis of ectodermal origin and a mesenchymal dermis. Since then investigations of the fish integument have become increasingly more sophisticated in parallel with the technical advances in mammalian dermatology that have become available.

The outermost component of the fish integument is the cuticle, a delicate extracellular layer. Below it lies the epidermis, an avascular layer of filament containing (Malpighian) cells usually encompassing within it a diversity of other cells types, neural processes, and sensory structures. The dermis consist of two layers, the *stratum spongiosum* and *stratum compactum*, the latter providing the structural strength. Within the upper dermis (*stratum spongiosum*) are found pigment cells, neural processes, and blood vessels. The underlying *stratum compactum* consists primarily of collagen fibres. The integument is attached to the underlying musculature by a layer of areolar tissue, the hypodermis, which provides a degree of flexibility allowing lateral movement between the skin and underlying tissue.

In addition to its anatomical position, several other characteristics of fish skin make its study compelling. First, fish skin varies widely in structure, bearing large ganoid scales in gars (Lepisosteidae) and polypterids (Polypteridae), small scales in sharks (Elasmobranchii) and tunas (Scombridae), and no scales in swordfish (Xiphiidae) and some catfishes. In spite of these differences, fish skin appears to possess an underlying structural ground plan-cross-helical fibres of collagen embedded in dermis. Second, skin has a complex two-dimensional stiffness. Circumferential strain stiffens it axially, giving it the capability to transmit axial forces in lemonsharks *Negaprion brevirostris* and eels *Anguilla rostrata*; however, this capability is lacking in spot *Lepistomus xanthurus* and skipjack tuna *Katsuwonus pelamis*. Third, skin is attached directly to the underlying muscles, putting it in series between muscle and propulsive elements.

The function of the swords and spears of several fast, pelagic, predatory fish species is still enigmatic. The surface structure of the sword of a swordfish *Xiphias gladius* is shown to be both rough and porous. The function of the porous surface structures on the second, and of the porous skins of sharks and of the castor oil fish, will probably be discovered soon using new particle image velocimetry techniques applied under strong magnification to visualize the local behaviour of the flow (Videler 1995). Long et al. (1996) remarked that the functions of fish skin during swimming remain a riddle. In the longnose gar *Lepisosteus osseus*, results of a study demonstrate a casual relationship between skin, the passive flexural stiffness it imparts to the body and the influence of body stiffness on the undulatory wave speed and cycle frequency at which gar choose to swim.

A ciliated embryonic epidermis is found in paddlefish *Polyodon spathula*, the gar *Lepisosteus platyrhincus*, and Dipnoi. The ciliated epidermal cells of young Australian lungfish *Neoceratodus forsteri* are not associated with movements of the embryo within the egg capsule, nor are they a part of a feeding mechanism. They are not related to oxygen uptake. The ciliated epidermis appears to function as a mechanism for clearing the animal of particles and settling organisms before hatching, when the egg membranes have developed holes, and after hatching, when the young fish is living among the submerged rootlets of trees growing on the river bank or in dense strands of aquatic plants (Kemp 1996).

A significant feature of the fish integument, like that of its mammalian counterpart, is the structural modification observed within different regions, this is particularly true of the epidermis, the structure of the integument over appendages usually being very different from that of the trunk. The integument of fish and its products are important in various life supporting functions. Apart from the obvious mechanical protection provided by the dermis, the skin acts as a complex limiting membrane serving to preserve and regulate the homeostasis of the "milieu interieur" by osmoregulation. The differing thicknesses and cytological developments in the skin of marine eels protect these crawling and digging creatures against abrasive interaction with their sea bottom habitat (Fishelson 1996). Certain areas of the body are well endowed with a variety of sensory structure of chemosensory and tactile nature and in some species venomous glands and light organs are prevalent.

In some species, the skin may be an accessory respiratory organ with a rich blood supply and even may have buoyancy and hydrodynamic functions (e.g., castor oil fish *Ruvettus pretiosus* and stromateiod fish *Schedophilus medusophagus*, see Figs. 3.1a, b). Reader can get details of drag reducing mechanisms (Bone 1974; Videler 1995) and of external defence mechanisms (Fletcher 1978). The skin structure and sloughing is a functional adaptation to the mode of life (e.g., stone-fish *Synanceja verrucosa*).

Further, it has been pointed out that fish skin is a living tissue with its lipids in metabolic equilibrium with those of other body tissues, particularly the underlying muscle. In lipid-rich fish, the fat deposits under the skin are usually located in extracellular sinuses, between muscle

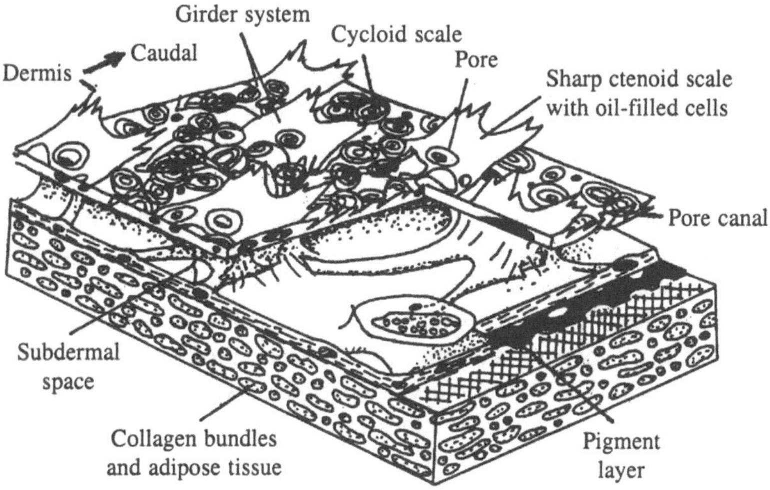

(a) *Ruvettus pretiosus*: Dermis
Source: Bone (1972)

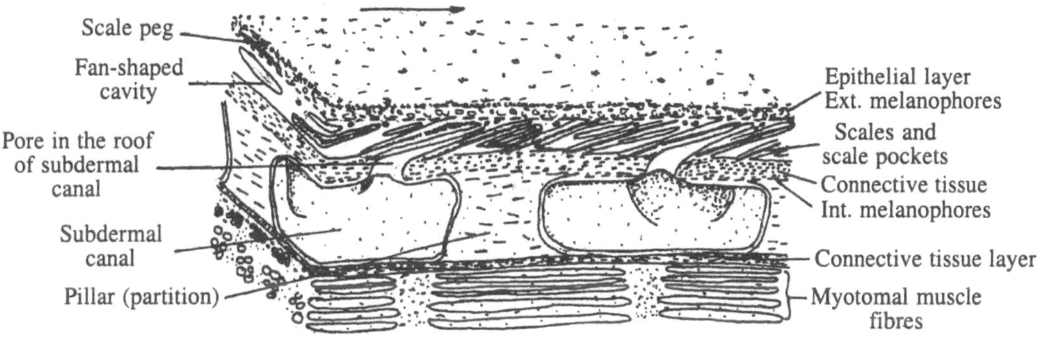

(b) *Schedophilus medusophagus*: Integument.

Fig. 3.1 (a) *Ruvettus pretiosus*. Diagram to show structure of dermis and arrangement of pillars formed by the bases of the ctenoid scales. Only a few cycloid scales are shown. (b) *Schedophilus medusophagus*. Diagram illustrating structure of integument. The arrow points towards the tail of fish.
***Source:* Bone and Brook (1973)**

fibres. Details about lipids and functions have been described in a review on biochemical aspects of buoyancy in fishes (Phleger 1991).

The reader is referred to Whitear (1986a) and Zaccone et al. (2001) for comprehensive accounts of fish skin. Reviews: (Bullock and Roberts 1975; Bullock 1980) are also of interest in this respect. Genten and Danguy (1990) have performed a comparative histochemical analysis of glycoconjugates in secretory cells (mucous and alarm substance cells) of fish epidermis by use of biotinylated lectins. Danguy et al. (1991) initiated a systematic glycohistochemical mapping of endogenous sugar receptor expression, focussing on fish epidermis. Burton (1999) described teleost epidermal plasticity in relation to environment.

1.2 Cuticle (External Mucoid Layer; Mucous Cuticle)

The presence of an outer amorphous layer overlying the epidermis has long been recognized. It is perhaps not surprising that the vast majority of studies on the fish integument appear to ignore the presence of a cuticular layer. Its extreme vulnerability to mechanical damage combined with the apparent ability of some species to divest themselves of this layer explains why it is so rarely noted.

This layer comprises cell protoplasm, sloughed Malpighian cells and some secretion of mucus. It is comprised of a number of protein-polysaccharide complexes. The thickness of the cuticle appears to vary considerably. The consistency of the cuticle varies considerably between species and in one species at least, *Blennius pholis*, the cuticular layer of the leading fin rays of the pectoral fins undoubtedly appears to afford some protection to the fish against abrasion within its rock-pool environment. The *Schedophilus medusophagus* is weakly keratinized. In general, however, the cuticle is a very delicate structure and various suggestions have been presented as to its function as an ion trap which would undoubtedly prove of benefit over chemosensory structures. The mucoid sercretions (cuticle?) of many fishes greatly decreased the friction of turbulent water thereby reducing drag. The protective potential suggested, however, whereby microbial or parasitic infections of the epidermis may be avoided, or at least their incidence diminished, would appear more likely.

1.3 Epidermis: Cell Types

Representations of cell types in the epidermis of fish are illustrated in Fig. 3.2 A-E.

The epidermis forms the major interface between the animals and their surroundings. Many epithelia have a multiplicity of functions, and this is especially true of the skin epithelium, the epidermis, of aquatic animals. Multiple functions can be reflected in morphological complexity. It can also be considered as a compound sensory organ (see Chapter 15). Immunoreactive Tamm-Horsfall protein has been reported in the superficial layers of skin of several fish (also of oral mucosa and gills) (see Howie et al. 1993). It acts as a water-proofing agent.

Fish are covered by a stratified squamous epithelium which, typically, is metabolically active throughout. The basic cellular element contains numerous filaments, notably the tonofilaments associated with desmosomes that knit the tissue into a structural unit. In cyclostomes, undoubtedly homologous cells have been called "mucous" (in other fish that adjective usually implies a goblet cell). The club-shaped "skein cells" is lamprey epidermis showed that twisted components are bundles of tonofilaments (see Zaccone et al. 1995 for cytokeratins). The function of skein cells may be to provide support against distorsion in the soft mucigenic epidermis and to protect the germinative cells. The superficial epithelial cells have characteristic terminal web. Teleosts have microridges or micropapillae at the surface, which is papillate in dipnoans and pitted like a sponge in the primitive actinopterygians and in cyclostomes.

The localization of calmodulin positive immunoreactivity in the superficial epithelial cells of brown trout, *Salmo trutta*, might have a physiological significance in the activation of microvillar skeleton and in the control of permeability of the skin epithelium (Zaccone et al. 1989a).

Exceptionally, the surface layers may be keratinized. The main characteristics of keratinizing epithelia are related to the usual retention of keratin or keratin-like products within the dense layer (proteinaceous envelope) of the hardened cells. The teeth of lampreys are nested cones of strongly keratinized cells, and keratinized oral epithelium also occurs in a number of teleosts.

Cell types in the epidermis of fish

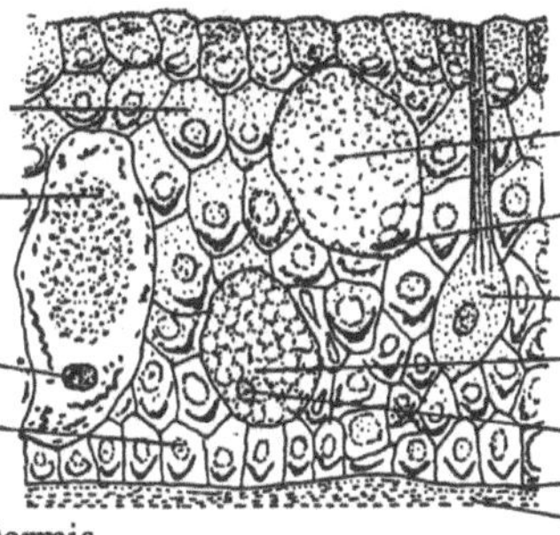

Myxine glutinosa: Skin.

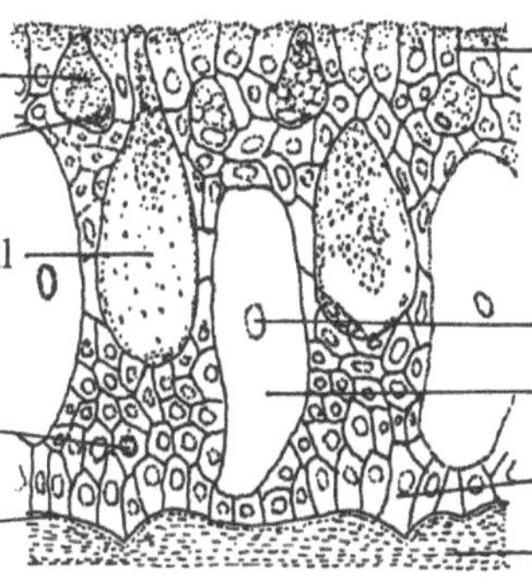

Lampetra spp.: Skin

Fig. 3.2A *Myxine glutinosa*. Mucigenic epithelial cells with a capsule of tonofilaments in the proximal cytoplasm. Thread cell with peripheral thread and central granular mass. Large mucous cells (goblet cells) and a sensory cell.
***Lampetra* spp. Composite section with mucigenic epithelial cells, a granular cell, two basal layer cells modified as skein cells. The Merkel cell, oligovillous cell and multivillous cell are all associated with nerves (not shown), the polyvillous cell is not.**
***Source:* Whitear (1986a)**

Exceptionally, there are keratinized breeding tubercles in males of species of Salmoniformes, Cypriniformes and some other orders, and occasionally similar structures that are not secondary sex characters. There are also attachment organs on the lips or fins of a number of ostariophysan genera; typically the superficial cells bear horny projections, or unculi, of bizarre shapes which aid in adhesion. The Asiatic catfish *Bagarius* has skin roughened by ridged plaques. In mormyrids and gymnotids the middle layer of epidermis form an insulating layer, interrupted by the canals of ampullary organs and cellular plugs of tuberous organs which offer pathways of electrical conduction. Zander (1979) reported chemoreceptor-bearing papillae on the head of the gobiid *Paragobiodon echinocephalus*. These are considered as an adaptation of fish to its hidden way of life between the coral branches in which other sense organs like eyes would be of lesser advantage. Sasse et al. (1970) observed non-sensory "flask-shaped organs" in head epidermis of a cyprinid *Morulius chrysophakedion*. Best and Bone (1970) reported unusual compartmented

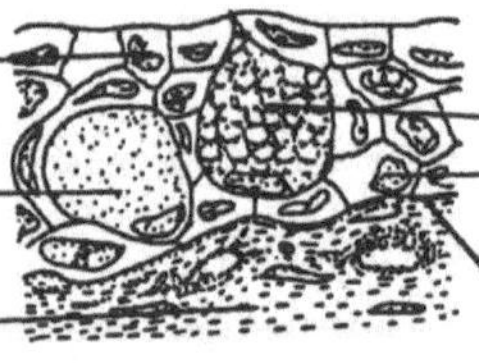

Raja sp.: Skin.

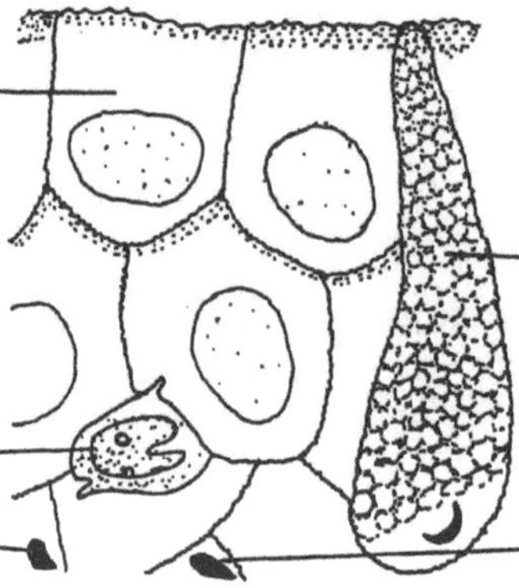

Protopterus spp.: Epidermis.

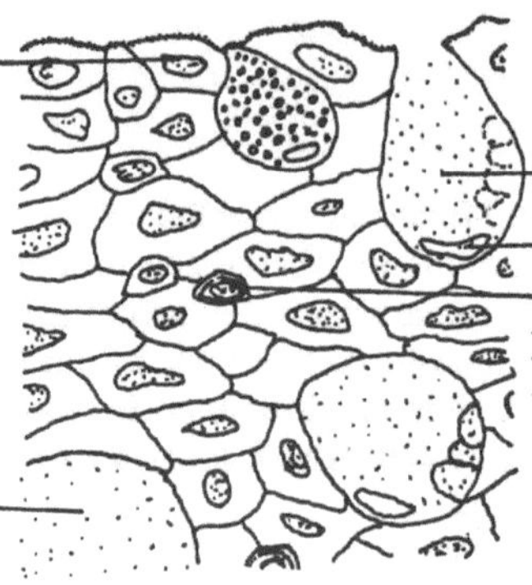

Latimeria chalumnae: Epidermis.

Fig. 3.2B *Raja* sp. Thin epidermis with a goblet cell and a sacciform cell.
***Protopterus* spp. Outer part of epidermis of a pelvic fin; note the large size of cells. There are mucoid vesicles at the distal border of the superficial epithelial cell and the epithelial cells of the second tier. The goblet cells have globules of a varying size and electron-density, the one shown appears to correspond to type B of Kitzan and Sweeny 1968. Merkel cells have an associated nerve, not shown.**
***Latimeria chalumnae*. Outer layers of epidermis near base of pectoral fin. The superficial epithelial cells have an apical palisade of mucoid granules. Mucous goblet cells are numerous. This specimen, from the Comoro Islands expedition 1972, also showed serous goblet cells with electron-dense secretory granules; the cells at the surface to the left is believed to be a developing serous goblet cell. The epidermis is infiltrated by numerous leucocytes.**
***Source:* Whitear (1986a)**

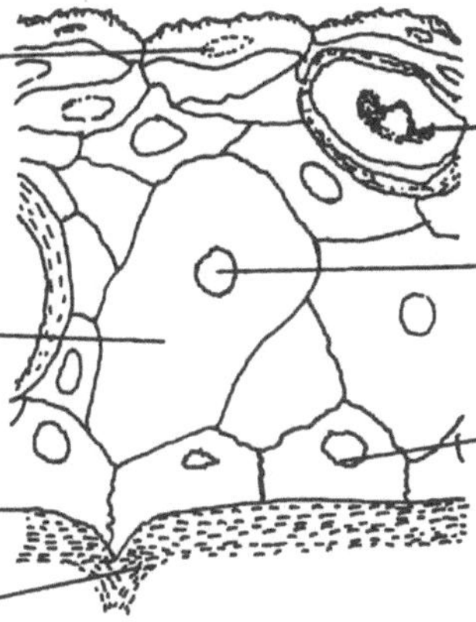

Periophthalmus kohlreuteri: Skin.

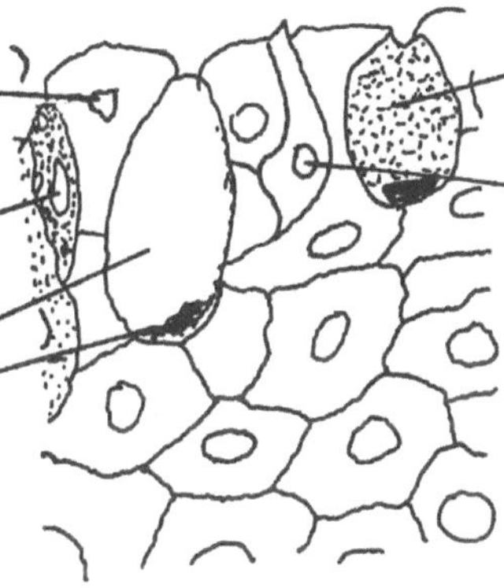

Pomatoschistus spp.: Epidermis.

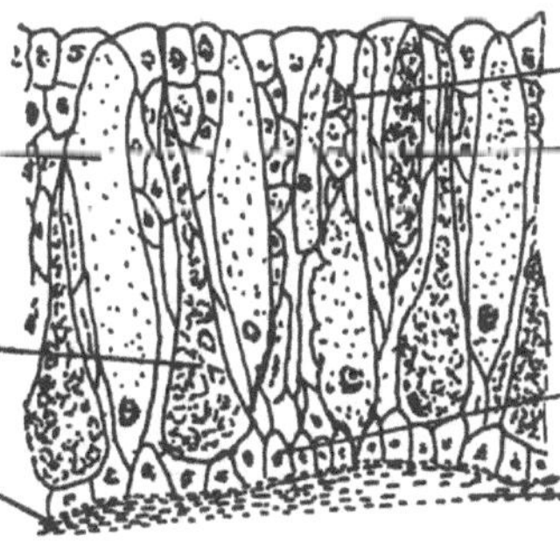

Monopterus cuchia: Skin.

Fig. 3.2C ***Periophthalmus kohlreuteri*. The epidermis of the head penetrated by a blood capillary accompanied by some dermal collagen. The superficial epithelial cells have a well-developed terminal web. The mid-epidermal cells contain few organelles and are swollen by electrolucent contents.**
***Pomatoschistus* spp. Outer layers of the epidermis with a chloride cell and its satellite, a sacciform cell, a sensory cell and a mucous goblet cell.**
***Monopterus cuchia*. Large sacciform cells, two types of mucous goblet cell, large and slender and an ionocyte.**
***Source:* Whitear (1986a)**

Anguilla anguilla: Skin.

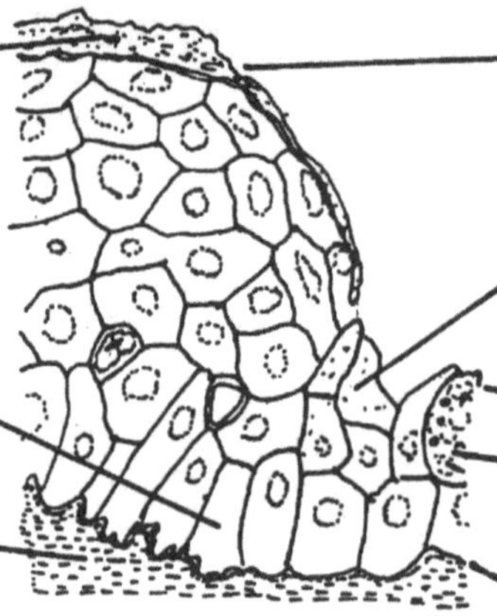

Bagarius sp.: Skin.

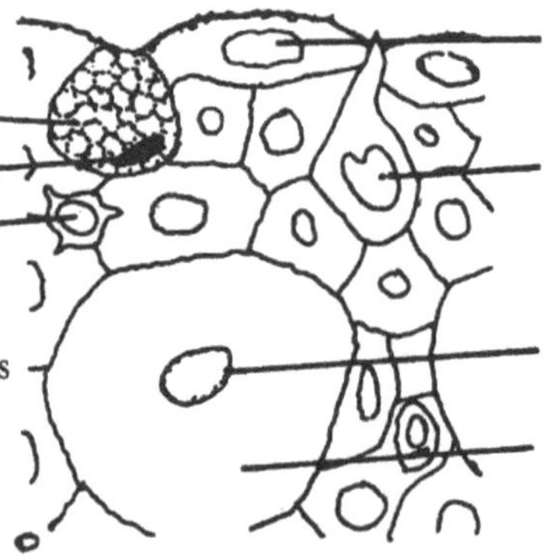

Phoxinus phoxinus: Epidermis.

Fig. 3.2D ***Anguilla anguilla*. small specimen. Large and numerous goblet cells, and club cells with a secretory vacuole.**
***Bagarius* sp. Side of a keratinized plaque. Only the superficial layer of cells is keratinized. Other cell types, such as goblet cells and ionocytes, are confined to the areas between the plaques.**
***Phoxinus phoxinus*. Outer layers of epidermis with a goblet cell, sensory cell, Merkel cell and club cell without secretory vacuole.**
***Source:* Whitear (1986a)**

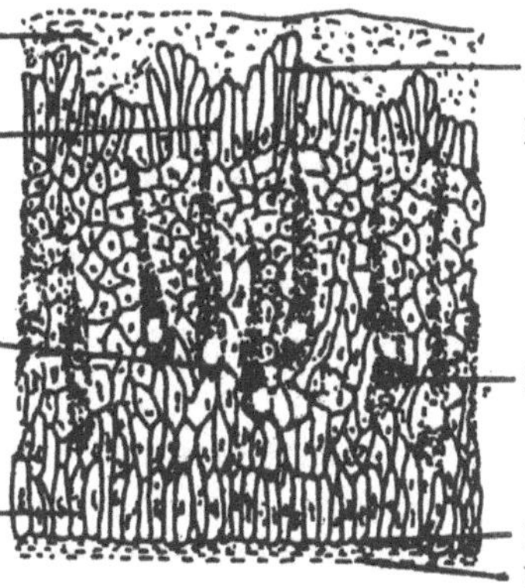

Fig. 3.2E *Trigla* spp. Distal part of a free pectoral fin ray, with a thick cuticle secreted by columnar superficial epithelial cells. Protruding groups of cells contain sensory cells and nerve fibres (not shown). The large elongated cells can be classified as serous goblet cells.
***Source:* Whitear (1986a)**

subdermal spaces filled with a fluid more dilute than sea water in mesoplegic alepocephalid fishes *Xenodermichthys* and *Phytostylus* which contributes to low density of fish. The great number of tonofilaments seen in the basal cell of *Anguilla anguilla* is characteristic of juvenile stages, here in an elver. These are "figures of Eberth". Ballonfish (*Diodon holocanthus*) skin is specialized for inflation.

The thickness of epidermis varies with species, age, position on body, in many cases, stage in the oestrous cycle. In most fish, especially benthic species, the ventral epidermis is thicker. The epidermis is usually thicker in those species which are naked or have vestigial scales. In scaly fish it covers and follows the outline of the scales. The epidermis over fins and tail is often qualitatively different from that of the body apart from variations in thickness, it usually lacks the more specialized cells.

Goblet cells. Goblet or mucous cells are important functional constituents of fish epithelia and are almost universally present in the skin with few exceptions. A goblet cell is considered as an exocrine unicellular gland. The mucous goblet cells are present within the epidermis of teleost fish although the numbers may vary greatly depending upon species, site, and stage of sexual cycle. The cells are usually first evident within the middle layers of the epidermis and during migration they increase in size as they move towards the surface. Biochemical and histochemical analyses of fish mucus present a large degree of variance between species in respect of the nature of the carbohydrate prosthetic component within the cells. *Latimeria* and dipnoans possess goblet cells. Mucous goblet cells normally occur in selachian skin cells. Lamprey epidermis is peculiar in having no goblet cells but the hagfish have a "large mucous" type that qualifies as a goblet. The integument of paddlefish (*Polyodon spathula*) is unusual as a relatively small amount of mucus is produced by epithelial cells that are not modified into regular mucous gland cells. A thick compact epidermis and dermis compensate for the slight amount of mucus secreted.

In general, goblet cells contain rough-surfaced endoplasmic reticulum and prominent Golgi systems, and have fewer desmosomes than epithelial cells. The number and distribution of mucous cells within the epidermis suggests that there is a continuous turnover.

Discrete glandular patches (multicellular glands comprised of goblet cells) situated under the operculum of labrids (e.g., *Labroides phthirophagus*) and scarids (parrot fishes *Scarus dubius*, *S. perspicillatus*) produce transparent mucoid shirt or cocoon that fish wears at night. The envelope functions as a protective device which serves to reduce predation by nocturnal organisms. Multicellular glands with large mucous goblet cells in the tip of the tail produce mucus to line the burrows of the semi-sedentary garden eels (Heterocongridae). Red Sea chingfish *Lepadichthys lineatus* produces large amounts of mucous to protect from contact with the rough surface of the crinoid host.

Also, mucus in fish has elements for defensive purposes, and for feeding-cum-immunity (e.g., *Symphysodon discus*) and as a source of food (e.g., in some cichlids). The distinct odours of certain fishes reside in their mucus. The clownfish *Amphiprion clarkii* is unharmed by its anemone because its mucus does not stimulate the host's nematocysts.

The skin mucous secretion of fishes plays an important role in their behaviour, influencing schooling, recognition of individuals and homing migration. Mucus serves as a messenger of chemical signals. Fish, such as catfish can also show behavioural responses to dissolved mucus.

Protective components, such as proteolytic enzymes (i.e., trypsinogen/trypsin) may be present in fish mucus, and the continuous replacement of mucus, together with antibiotic properties, may effectively impede bacterial infections and ectoparasitic infestation.

An increased production of mucus during the reproductive period has been reported for different species. In *Pimephales promelas* there is an increase in the production of mucus in epidermal mucous cells or goblet cells and thus serves to protect the body regions injured during abrasive behaviour.

The epithelial thickness and degree of mucification are affected by environmental factors and an increased secretion of mucus as a common response to acid stress, to physical handling and to sea water adaptation.

Recent studies indicate that mucification and the composition of mucus may be affected by the nutritional status of the fish. Fig. 3.3 shows various functions of mucus (see also Lebedeva 1999; Zaccone et al. 2001).

Sacciform or Sac Cells. The sacciform cells are unicellular glands of a serous nature. Their contents are complex secretions, usually with a protein moiety, various enzymes, and serotonin immunoreactivity. Sacciform cell secretions consist of high amount of cholesterol and lipids resembling those of granular gland of the amphibian skin (Zaccone et al. 1988).

Gastrin-releasing peptide (GRP)-like and bombesin (BOM)-like immunoreactivities have been localized in exocrine sacciform gland cells in the epidermis of clingfish, *Lepadogaster candollei* (Zaccone et al. 1987). One function of exocrine peptides could be as a response to predators, since the sacciform gland cells may function in the synthesis of toxin principles (crinotoxins). Sacciform cell secretions may serve as intraspecific alarm signals in some cases. The reader is also referred to Zaccone et al. (1994, 1999a, b, 2001). Fasulo et al. (1993) identified by immunohistochemical methods serotonin, and the peptides bombesin and caerulein in the epidermal sacciform cells of *Lepadogaster candollei* and *Mastacembelus erytrotaenia*.

The staining reactions of contents are as varied as their appearance. The most common feature ascribed to these cells is the presence of an acidophilic granular content. Sacciform cells also occur in the polypterids, and in rays, also in *Chimaera*. In African freshwater stingray, *Dasyatis*

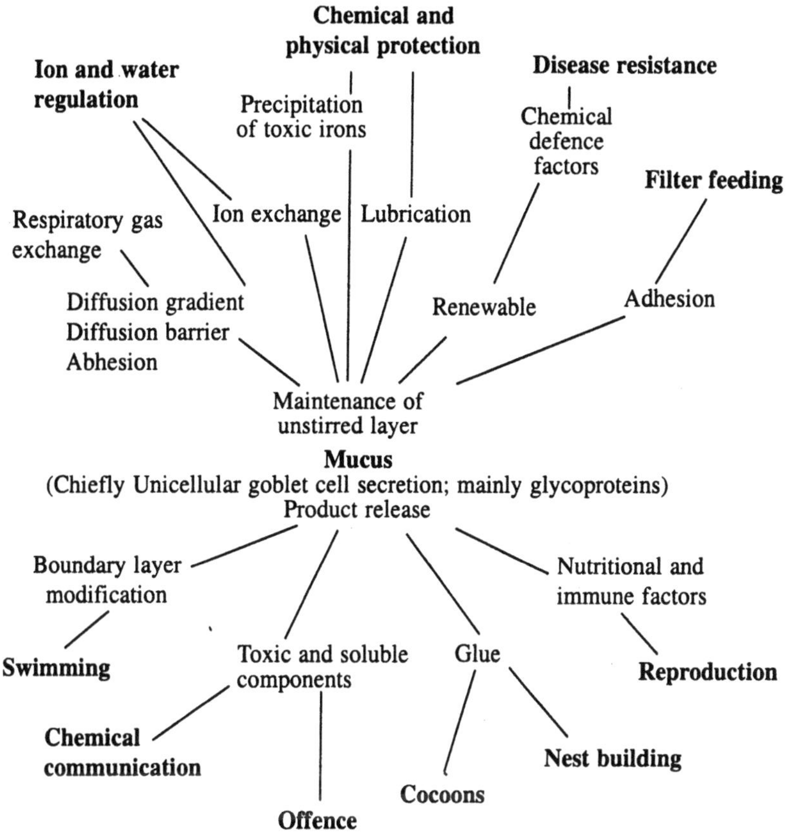

Fig. 3.3 Multiplex functions of mucus.
Sources: **Shephard (1994) and Zaccone et al. (2001)**

garouaensis, integument, interspersed among these sacciform-like cells are large numbers of smaller pseudo-spongiotic cells (i.e., possessing large intracellular spaces surrounding a central nucleus).

Club or *Clavate Cells* (*Giant Cells*). Club cells do not have the universal distribution. Club cells proper are large, sometimes binucleate, serous cells and lack pores. Club cells have been misidentified in lampreys. The club cell of eel is different from that of ostariophysans; only eel club cell has a secretory vacuole. Club cells have a peculiar cytology: most organelles are perinuclear in position while the outer cytoplasm is characterized by helically coiled intermediate filaments associated with desmosomes in place of normal tonofilaments.

In ostariophysans, the club cells are the source of pheromone (alarm substance) releasing a fright reaction only when the skin is physically damaged. Polypterids and eels do not show a fright reaction (see further Section 6). Serotonin (5-HT) and bombesin (BOM) have been demonstrated also in the holocrine club cells of some teleosts. In the oriental catfish *Plotosus lineatus* and *Arius thalassinus* club cells contain toxins while in eels they contain hemaglutinin (read Fasulo et al. 1993; Zaccone et al. 1990, 1994, 1999a, b, 2001). The precise functions of bioactive peptides of amphibian and fish are not known (Zaccone et al. personal communication).

Granular Cells. This category should be limited to those in lamprey epidermis that contain electron-dense membrane-bounded granules and have filament-containing appendages penetrating the deeper epidermis. Granular cells do not open at the surface. The secretion is proteinaceous. The function of appendages is not known.

Thread Cells. The hagfishes, most primitive scaleless vertebrates, have cells producing threads of a unique protein. The nucleus is proximal, the threads mainly peripheral and there is central granular mass. Most of the thick slime discharged by an irritated hag comes from thread cells in glands along the body. The threads are spun from intermediate filaments associated with microtubules and ribosomes (polysomes). The protein nature of the thread has been confirmed. The molecular weights of these proteins range between 14000–18000. The thread is strongly birefringent (refer Terakado et al. 1975; Fernholm 1981).

Venom Cells. Venomous fish have multicellular holocrine glands associated with spines. It has been assumed that venom cells are derived from the epidermal unicellular glands, but in catfish the fine structure of the venom cells is not identical to that of the club cells. Most venoms from not ducted glands are complex secretions, usually with a protein or polypeptide component (see further Section 5).

Ionocytes. Chloride cells are the ionocytes (ion-transporting cells) characteristic of teleosts, found in the gill and oropharyngeal epithelium and in external skin especially of euryhaline species. They have numerous mitochondria and a reticulum of branching tubules opening on the laterobasal membrane. The tubules are associated with enzyme systems. In lamprey epidermis, the polyvillous cells have many of the features of ionocytes but do not have special tubules. Somewhat similar cells have been described from selechian gills but not yet from selachian skin (see also Komnick 1986 and Zaccone et al. 2001).

It should be pointed out that fish have two sites for the uptake of calcium: their gills and their intestine. The branchial epithelium may be considered to be a specialized skin area. In the euryhaline teleosts *Oreochromis mossambicus* and *Oncorhynchus mykiss*, parts of the skin covering the inner operculum have also been shown to transport Ca^{2+}. It has been demonstrated that this transport is probably mediated by the ionocytes found in this part of skin under these conditions. The skin has been shown to contain numerous ionocytes at typical densities of around 2000 mm^{-2}. Extrapolation of positive correlation between skin ionocyte numbers and inward calcium transport rates suggests that skin devoid of ionocytes is essentially impermeable to Ca^{2+}. The contribution of extrabranchial calcium transport via the skin is considered to be small, and this is reflected in the absolute numbers of ionocytes in gills and skin.

Epithelial Sensory Cells and *Supporting Cells.* Supporting cells separate the sensory cells in neuromast and taste buds, and produce secretions covering the surface of these sensory organs (see Chapter 15 and respective figures).

Merkel Cells. Epidermal Merkel cells (one of the components of paraneuronal family) have been found in teleosts, dipnoans and lampreys, but not yet in selachians or hagfish.

Serotonin (5-HT), neuron-specific enolase (NSE), metenkephalin (Met-ENK) and vasoactive intestinal polypeptide (VIP) have been immunocytochemically detected in the Merkel cells of the Conger eel (*Conger conger*), and the catfish *Heteropneustes fossilis*. Mechanoreceptive function, and a trophic function in epidermal growth and differentiation are suggested with unanimous confirmation. (see Hartschuh et al. 1986; Zaccone 1986; Zaccone et al. 1989b, 1994; Whitear, 1993; for innervation, refer Kotrschal et al. 1993) (Fig. 15.69C).

Cement Secreting Glands. Many aquatic vertebrate larvae attach to solid substrates by an adhesive cement. Cements glands have been identified in ganoids, dipnoids, anurans and teleosts. They occur in the larvae of several fish groups, viz., Centrarchidae, Osphonemidae, Characinidae and Cyprinidae. The cement glands are ectodermal derivatives in dipnoids and anurans but they are of endodermal origin in ganoids.

According to Hamlett (1990), the angelfish, *Pteryphyllum scalare*, possess three pairs of transient cephalic cement glands, one pair superior and ventral and the other two pairs superior and dorsal to the eyes. Glands are composed of a single, columnar secretory epithelium with apical secretion granules. Gland cells are distinctly polarized with an oval, basal nucleus. A well-developed rough transitional endoplasmic reticulum and smooth-walled elements occupy a region adjacent to an elaborate Golgi complex. Mature secretory granules dominate the apical region of the cells. When viewed with scanning electron microscope, the cement is initially identifiable as thin filaments. Filaments fuse to form strands which in turn coil with other strands to produce helical cables. Discoidal plaques, possible anchoring sites, occur at the terminal end of the cable.

Intrusive cells. Intrusive cells include chromatophores, neuromast cells, leucocytes and rodlet cells.

1.4 Innervation

Teleost epidermis receives a dual sensory innervation; on the head by trigeminal and facial fibres, and on the trunk by facial (recurrent facial nerve = ramus lateralis accessorius) and spinal fibres. The trigeminal and spinal nerves are presumed to be similar in providing general cutaneous (general somatic afferent) innervation. The facial nerve fibres (special visceral afferent) are presumed to innervate taste buds. Nerve fibres of the different origins are, however, mingled with the dermal plexus. From there bundles of fibres penetrate the basal lamina to ramify into a network of delicate epidermal nerves.

2. DERMIS

The thickness of the dermis, compared with epidermis, varies with species, position on the body and stage of life history. The basement membrane is counted as a dermal element, although epidermal participation is necessary for its formation.

Brainerd (1994) has described pufferfish inflation. Because the collagen fibres in the dermis of unstretched skin are wavy, little force is required to extend the skin up to 40% of its rest length. At high strains the collagen fibres are stressed directly and the skin suddenly becomes stiff. Stiff skin surrounding a ball of incompressible water provides a rigid framework for the support of the erectile spines, and the ballonfish (*Diodon holocanthus*) becomes a formidable, spiny sphere. When *D. holocanthus* is uninflated, the spines (modified scales) lie flat against the fish's body, but when inflated, the spines point at right angles to the body. The spines have triradiate bases: one short, axial process and two long, flattened, lateral processes orthogonal to the axis of the spine. On a resting ballonfish, the spine points caudad, the axial process of the base points cranial, and the lateral processes point around the circumference of the fish. Histological sections through the long axis of a spine show that collagen fibres attach to the axial process. In the uninflated state the collagen fibres originate on the tip of the anterior pointing axial process and curve around toward the inside of the skin. It may be mentioned that ballonfish can triple its body volume by pumping water into its stomach.

2.1 Dermal Skeleton: Scales (see also Chapter 4)

Reader can find a variety of information related to dermal skeleton in a number of publications (Reif 1982; Sire and Geraudie 1983; Sire 1985, 1986, 1990, 1993; Whitear 1986b; Sire et al. 1990, 1997; Sire and Huysseune 1993; Sire and Meunier 1993; Meunier 1994; Meunier and Poplin 1995; Zylberberg and Meunier 1996; Huysseune and Sire 1998). In most bony fish the body is protected by a dermal skeleton of scales that may correspond to the "exoskeleton" that covered the early vertebrates (e.g., Heterostracans, Osteostracans). A clear tendency toward a reduction of the postcranial dermal skeleton is observed in the course of evolution, and only a few tetrapod species (mainly reptiles, some amphibians and mammals) possess dermal sclerifications. In contrast, bony fish (i.e., osteichthyans) show a considerable diversity of their dermal skeleton (various types of scales and bony plates) and of the constituent tissues (various types of bone, of dentine, enamel, enameloid, and highly derived tissues).

It is well admitted now that (i) the dermal skeleton of fishes (and of vertebrates in general) develops from a single integrated modifiable morphogenetic system and (ii) the evolutionary changes are largely the results of developmental changes in timing of appearance and rate of growth of the various tissues.

Possibly the most dramatic morphogenetic shift in the history of osteichthyan squamation was the change from scales with enameloid (or enamel), dentine and bone to scales with weak ossification to epithelium without scales. The ornamentation of the scale surface seems to be related to the distance from the epidermis. The extent of the so-called limiting layer at the external layer surface is directly related, probably, to the epidermal covering, i.e., the limiting layer exists only on the scale surface, that is directly covered by the epidermis or lies at a small distance. Elsewhere, i.e., in the anterior part of scale not covered by epidermis, the external layer proper is only present. The degree of development of the epidermal covering also depends on the position of the scale in the dermis and on the degree of imbrication of the scales, both being related to the ecological adaptation of the species, mainly with regard to a mechanical protection against the substratum

Scales in Primitive Actinopterygian Fish. The ganoid scales of polypterids have a dorsal peg fitting into a socket in the next scale of the geodesic row, and paraxial connective tissue fibres between basal plates. Rows of scales are connected by interacting fibres emerging from the sides of the ridge along the base of the scale and looping across the stratum compactum. The articulating row of bony scales has been regarded as a compression-resisting device, preventing torsion so that the body does not twist as it bends; stress is resisted by the paraxial and intersecting fibres on alternate sides. The basal plate is of lamellar bone, covered by a layer of dentine, topped by highly mineralized ganoine, whether enamel or enameloid is still not clear. The scale is permeated by canal carrying capillary blood vessels to the narrow space between the ganoine and the basement of the epidermis (Figs. 3.4, 3.5).

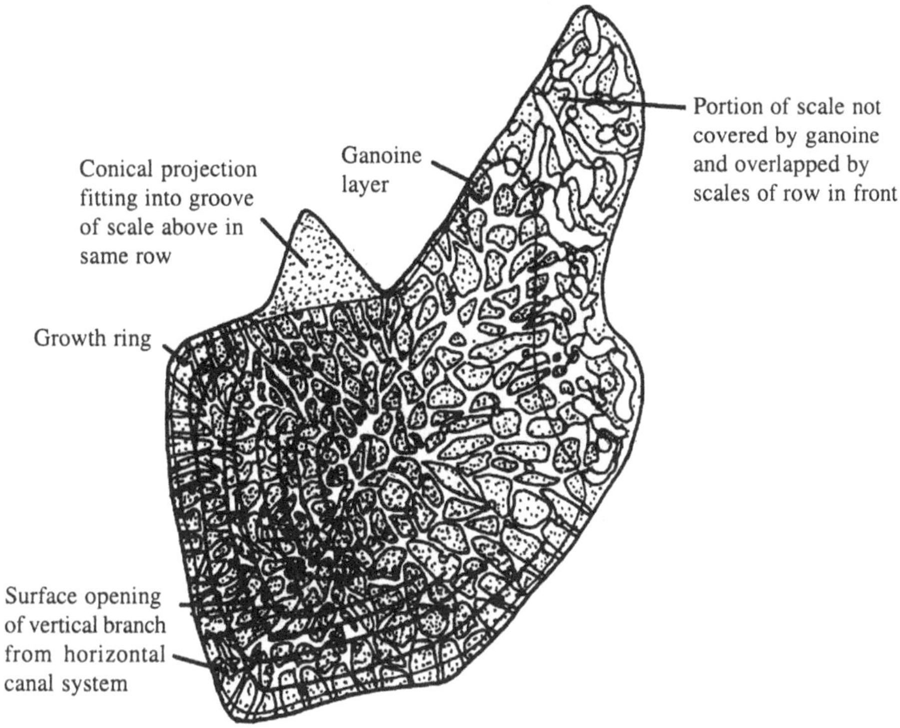

Polypterus: Scale.

Fig. 3.4 ***Polypterus* scale from right side of body, seen as transparency from outer surface. Thicker parts of scale more deeply strippled including sides of groove for projection of scale immediately below in same row. Head end of fish to right.**
***Source:* Kerr (1952)**

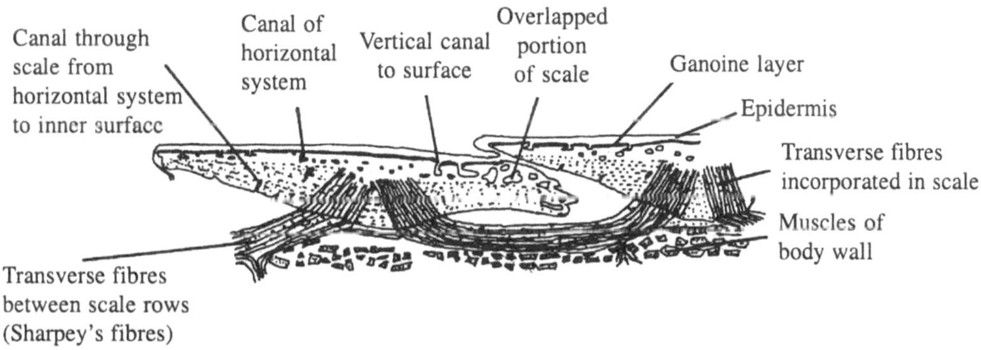

Polypterus: Scale section

Fig. 3.5 ***Polypterus*. Section of scales and soft tissues cut transversely to the scale rows. Head end to the right.**
***Source:* Kerr (1952)**

The *Polypterus* scale does not appear to depart greatly from the generalized paleoniscoid plan. Sire (1989) has reported that the young ganoid scales of *Polypterus senegalus* have a structure similar to that of typical elasmoid scales.

The sturgeons have reduced dermal ossification to bony scutes, small in *Polydon* (Figs. 3.6,

Polyodon spathula: Denticle.

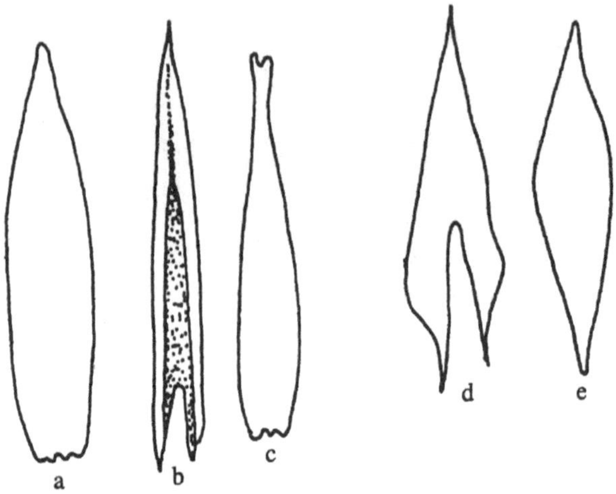

Polyodon spathula: Scales.

Fig. 3.6 Scales of *Polyodon*. Surface view of a horny denticle. a. The first dorsal fulcrum of the caudal fin. b. Ventral view of the ninth dorsal fulcrum. c. The ventral fulcrum of the caudal fin. d. A grooved rhomboidal scale of the caudal lobe. e. A rhomboidal scale of the caudal lobe.
***Source:* Weisel (1975)**

3.7 a, b), large and small in *Acipenser* and *Scaphirhynchus platorynchus* (Fig. 3.8). The scales of *Lepisosteus* still have ganoine, and collagenous fibres between the basal plates, but no dentine layer (Fig. 3.9). The structure of the scale of *Amia* is extremely modified and shows a convergence towards that of *Protopterus*. In *Amia* the scales have become elasmoid, not connected by fibres but set in pockets and over-lapping considerably. The basal plate is composed of sheets of collagen fibres with directions approximately at right angles, set in pairs with the angles of each pair rotated with respect to that above; in *Amia* this rotation is left-handed (Fig. 3.10).

Elasmoid Scales of Teleosts. Data have accumulated on the structure, organization, development and regeneration of fish scales. Among them, the scales of teleosts have been the most thoroughly studied, and due to modern technical approaches, we have now a good knowledge of their structure. The most widespread type of scale is the elasmoid one.

In Osteichthyes there is a fairly sharp distinction between growing dermal bones on the head and growing dermal scales on the trunks. Nongrowing tesserae or nongrowing scales resembling placoid scales have never been found. Much more is known about the histology of scales than of dermal bone in the Osteichthyes. The evolution of the scales in Actinopterygii is very much different from that of the other Osteichthyes.

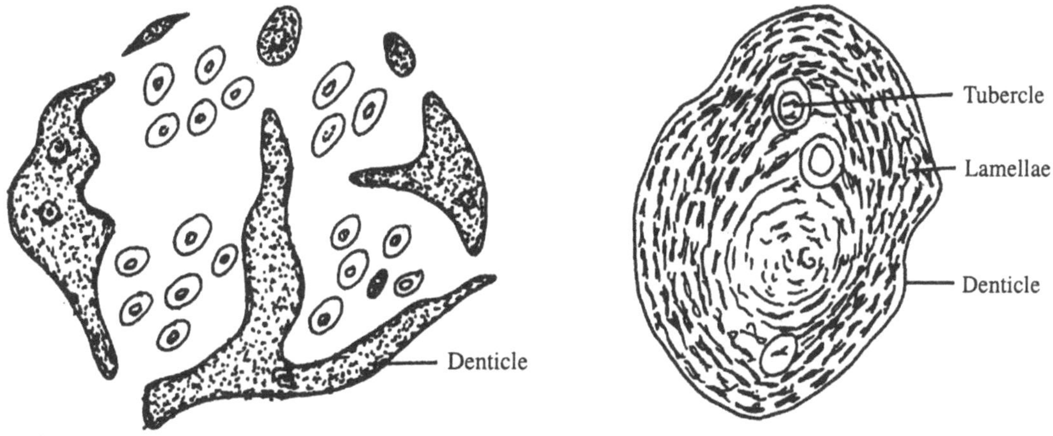

(a) *Polyodon spathula*: Denticle. (b) *Polyodon spathula*: Denticle.

Fig. 3.7 (a) Scales of *Polyodon spathula*. Denticles surrounding openings to surface pits on the rostrum. (b) *Polyodon spathula*. A single denticle showing lamellae, lacunae and three tubercles.
***Source:* Weisel (1975)**

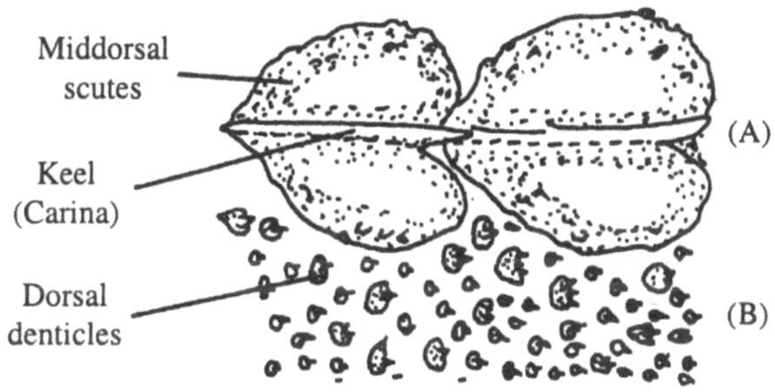

Fig. 3.8 Exoskeleton of *Scaphirhynchus platorynchus*. Scalation from middorsal side of the body.
***Source:* Weisel (1978)**

From a structural point of view, the elasmoid scales of teleosts could be derived from the "odontodal" tissues (i.e., odontocomplexes) that covered the dermal skeleton in ancestral osteichthyan fish. From a developmental point of view, the scales in teleosts are closer to teeth than to dermal bone. Sire et al. (1997) strongly suggest that epidermal products, possibly enamel-like proteins, are deposited at the scale surface and contribute to the thickening of the upper layer in zebrafish (*Danio rerio*) scales.

A scale consists of a plate of collagenous tissue, with superficial mineralization, surrounded by scleroblasts and fibroblasts, set in an oblique pocket lined at least on deep side, by modified fibrocytes with desmosomes and caveolae (for studies on fish scale formation and resorption, see Yamada and Watabe 1979; Onozato and Watabe (1979). The scale is anchored in the pocket by sparse bundles of collagen fibres attached to the calcified layer. In some species the posterior edge of the scale approaches the epidermis which forms a slight tuck beneath it, in others, usually pelagic, the posterior edge of the scale and its epidermal covering protrude, overlapping the epidermis covering the scale behind. The posterior field of the scale is that part not overlapped by other scales, which the lateral and anterior fields usually are. Figures 3.11A, B and 3.12 show the localization and detailed structure of elasmoid scales.

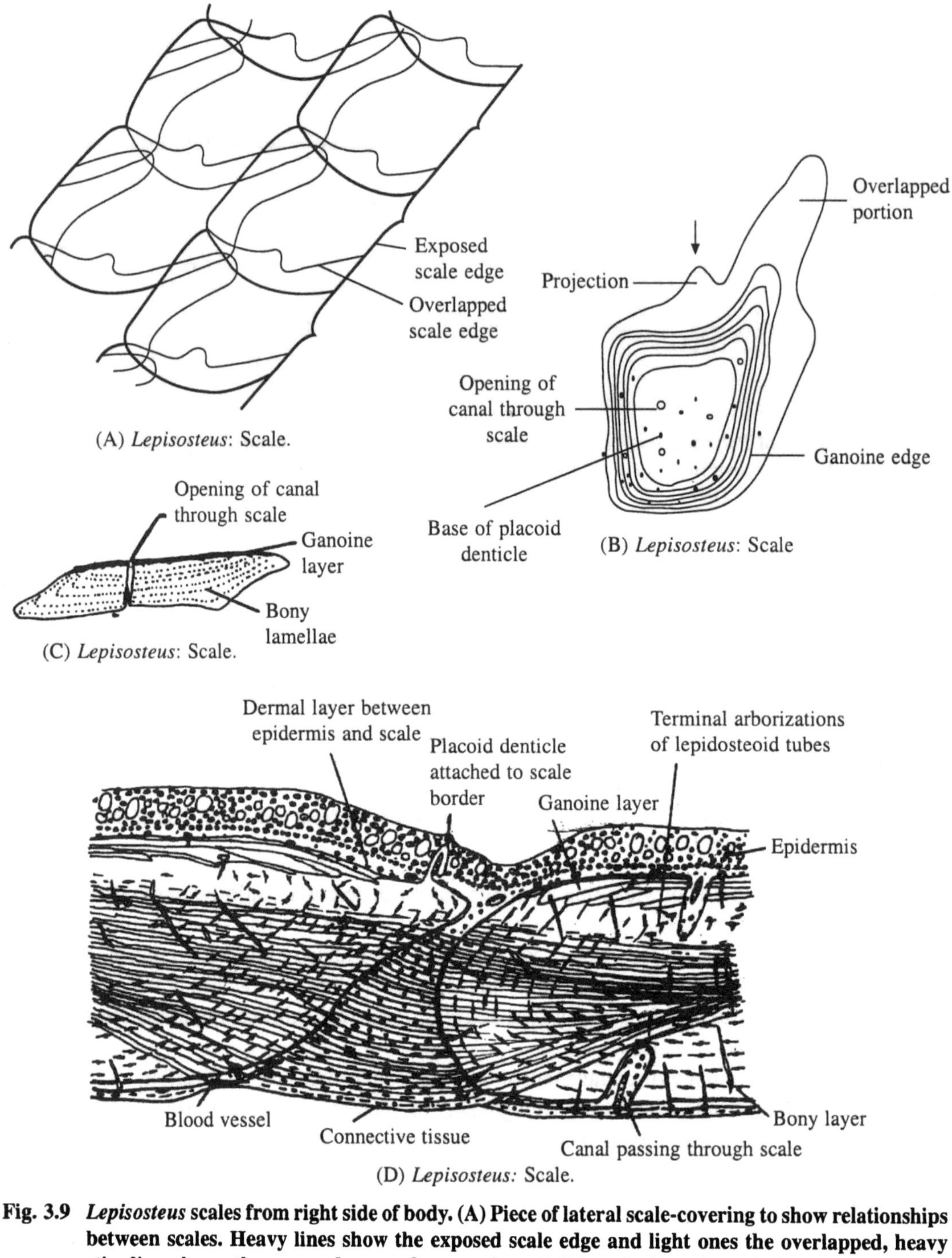

Fig. 3.9 ***Lepisosteus*** **scales from right side of body. (A) Piece of lateral scale-covering to show relationships between scales. Heavy lines show the exposed scale edge and light ones the overlapped, heavy stippling shows the exposed area of one scale and light stippling that of the overlapped. (B)** ***Lepisosteus*** **scale shown as a transparency. In A and B. Head end to right. (C) Vertical section of** ***Lepisosteus*** **scale cut along line of arrow in Fig. (B). (D)** ***Lepisosteus*****, to show junction of two scales. Cut longitudinally to scale row, parallel to Fig. C but passing through scales between conical projection and posterior border. Ventral surface to right.**

Source: **Kerr (1952)**

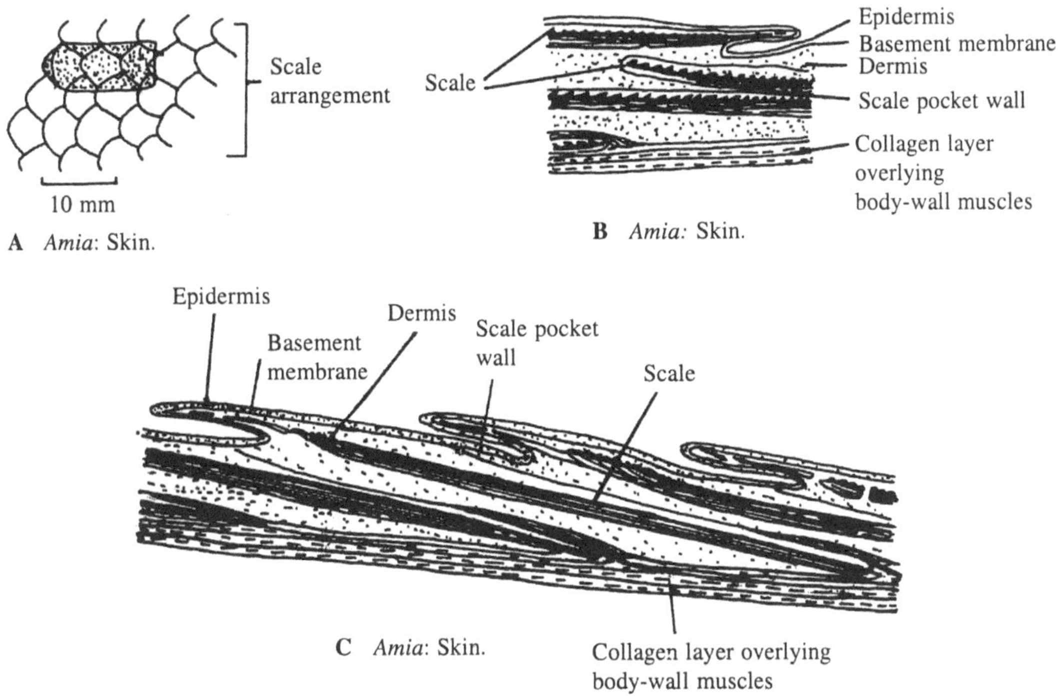

Fig. 3.10 ***Amia*** **scales from right side of body. A. Piece of lateral scale covering seen as transparency from outer side, scales arranged in sloping lines; single scale stippled densely to show exposed portion, and lightly to show overlapped portion. B. Transverse section. C. Section of skin cut longitudinally to body axis. Head end to right in both cases.**
Source: **Kerr (1952)**

Teleost scales are marked by circuli (striae), calcified ridges often with a finely dentate edge. Uncalcified grooves (radii or sulci), which allow flexibility, converge on the scale focus, which is not the geometric centre (Fig. 3.13). The basal plate may or may not be partially calcified and consists of collagen fibres arranged as noted for *Amia*, although the change of angle varies and in some species is such as to result in an almost orthogonal arrangement. Most of the basal plate is infiltrated with a special protein, 'ichthyolepidin', which is absent below the radii. Large hollow scales filled with oil and small superficial scales also occur in the castor oil fish, *Ruvettus pretiosus* (see Fig. 3.1a). The development of elasmoid scale is similar in diverse species. If a scale is removed, the broken epidermis quickly covers the gap and a new scale regenerates in the pocket, probably by proliferation of fibroblasts from the lining.

Not all teleosts have cycloid or ctenoid scales representing the most widespread type of postcranial dermal skeleton; apart from naked species, there are also armoured groups such as Loricariidae, Asonidae and Ostraciidae which are covered by large bony shields and spiny groups such as the Cottidae and Cyclopteridae, whose scales take the form of heavy dermal spines.

Sire (1993) describes the development and fine structure of the bony scutes in callichthyid *Corydoras arcuatus*. The body of adult *C. arcuatus* is covered by two rows of thick scutes, dorsal and ventral, that overlap anteroposteriorly within both rows and also partially between

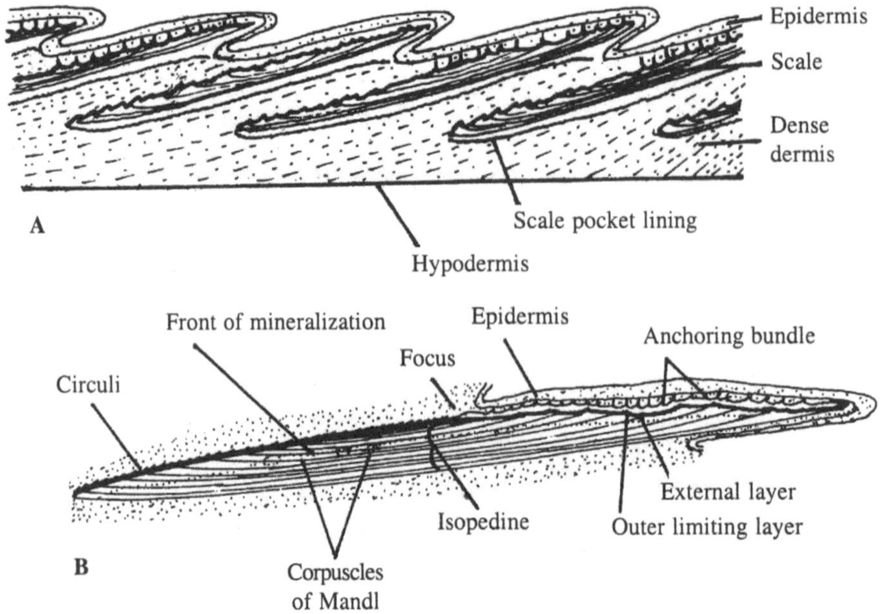

Fig. 3.11 A. Localization of the elasmoid scale in the skin of a teleost fish. B. General structure of a scale. (Drawn from semithin and ground sections of the scales of the cichlid *Hemichromis bimaculatus*). *Source:* Sire (1952)

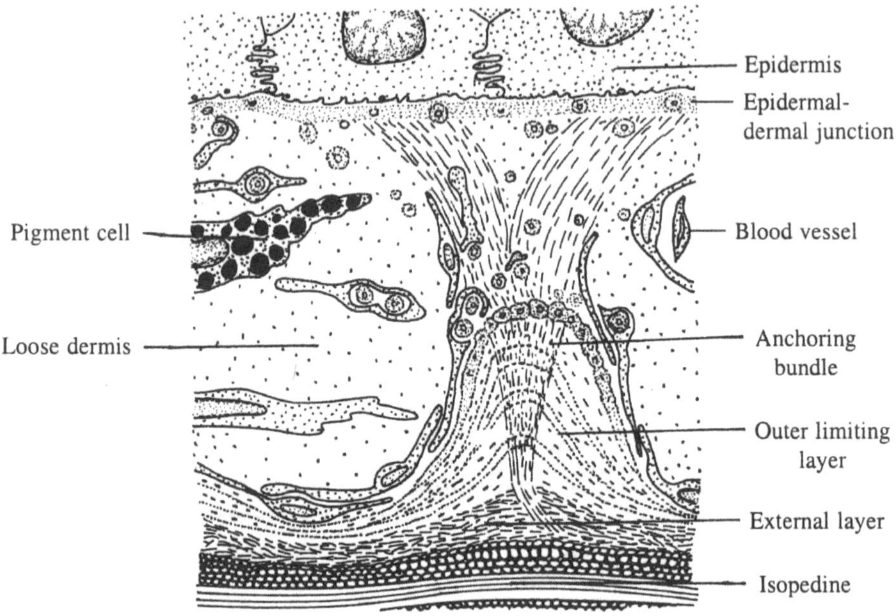

Hemichromis bimaculatus: Structure of a decalcified elasmoid scale.

Fig. 3.12 Detailed structure of a decalcified elasmoid scale of the cichlid (*Hemichromis bimaculatus*) in the posterior region covered by epidermis. The three layers (outer limiting layer, external layer and isopedine) of the elasmoid scale are shown as well as the different localizations of mineralized spherules between the epidermis and the scale. *Source:* Sire (1990)

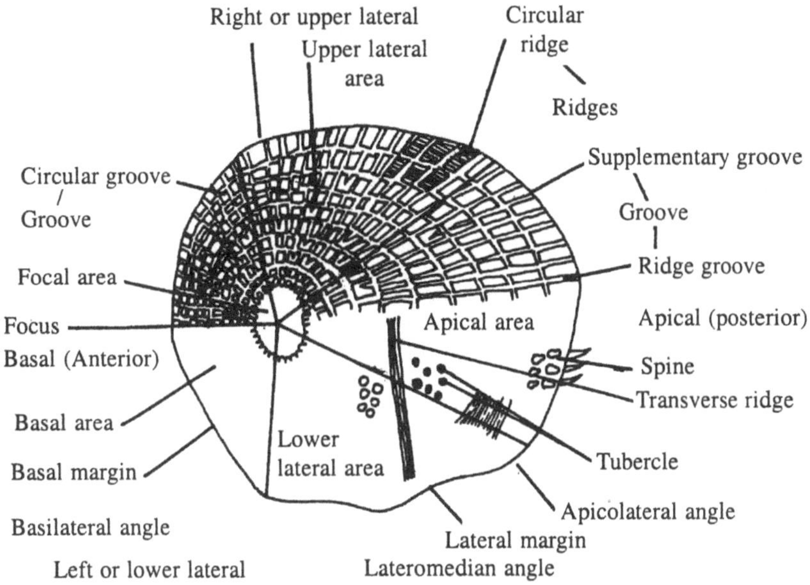

Scale: Technical terms.

Fig. 3.13 Important technical terms of scale.
***Source:* Kobayasi (1953)**

rows at the level of the horizontal septum (close to the lateral line nerve) (Fig. 3.14). The scutes are localized in the dermis, within which they are firmly anchored by large bundles of collagen fibres, the so-called Sharpey's fibres (Fig. 3.15). Such anchoring fibres also attach the scutes to one another.

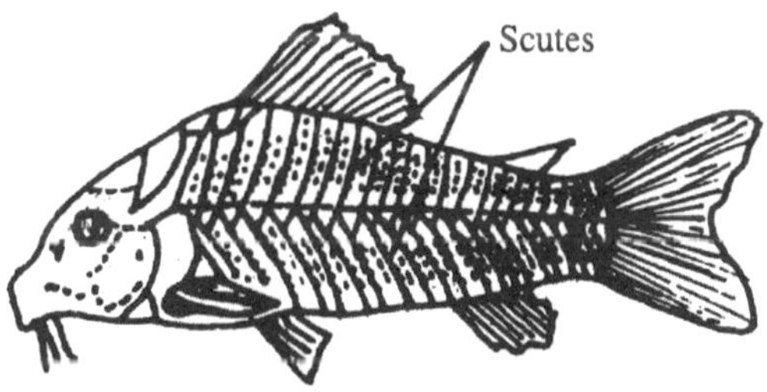

Corydoras arcuatus: Scutes.

Fig. 3.14 *Corydoras arcuatus*. Schematic drawing of the left side of an adult specimen showing the localization of the two rows of scutes.
***Source:* Sire (1993)**

A new term, hyaloine, is introduced for the nonosseous, highly mineralized layer constituting the upper part of the scute. Hyaloine differs from bone in that (1) there are no osteocytes, (2) there are no striated collagen fibrils, (3) it possesses peculiar tinctorial properties, (4) there are thin striations corresponding probably to growth periodicity and (5) it is more mineralized than bone.

The reader is also referred to Sire and Meunier (1993) for the superficial ornamentation and structure of the postcranial osseous dermal plates (scutes) in some armoured siluriforms (Loricariidae, Callichthyidae, Doradidae). In Callichthyidae and Loricariidae, the scute surface

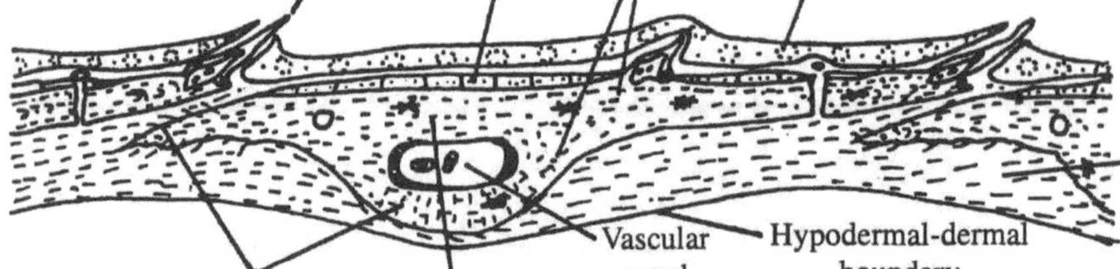

Fig. 3.15 ***Corydoras arcuatus*. Schematic drawing of a longitudinal section in the skin of an adult specimen. *Source:* Sire (1993)**

shows numerous odontodes but they lack in Doradidae the scutes of which are characterized by a large central osseous spine. The odontodes are conical true teeth; they have various species specific shape but they show the same organization: dentin, covered by an hypermineralized substance (enamel or enameloid), surrounding a central pulpar cavity which communicates with the vascular network of the scute. The odontodes are always attached to the scute surface by a ligament. Sire and Huysseune (1996) described the structure and development of the odontodes in *Corydoras aeneus* (Callichthyidae).

The structure of the scutes is compared to that of the ganoid and elasmoid scales and the evolutionary problems raised by the existence of the scutes are discussed.

Fedrigo et al. (1996) described the morphology and histology of scutes in Macroramphosidae (Gasterosteiformes). Scutes are constituted entirely of parallel-fibered bone which is homogeneously mineralized. This bone is avascular and acellular, and its matrix shows clear lines of arrested growth. This structural organization is completely different from that elasmoid scales. The morphology of scutes varies according to location on the body. Characteristics of ornamentation are different from one genus to another.

It has been postulated that the primary function of spines on ctenoid scales is to modify the boundary layer and make it turbulent, thereby to lower the resistance to movement through the water. While this may not be a function of ctenoid scales, it seems unlikely that they originally evolved for this reason. The fastest swimming fishes (Scombridae, Istiophoridae, Xiphiidae) have reduced scales or have lost the scale cover entirely. Further, it has been suggested that ridges and tubercles of some Indian hill stream fishes (Homalopteridae, Sisoridae) might have a hydrodynamic function. There appear to be no fundamental structural differences between cycloid scales with contact organs and ctenoid scales (see Section 4).

Scales are studied as an index of age, for information of the life history, and for taxonomic position of fishes.

In general teleost scales consist of two different layers; the upper calcified layer and the lower fibrillary plate, and represent a massive deposit of calcium. Scales are resorbed under certain physiological conditions such as starvation, sexual maturation, and severe exhaustion due to spawning migration. Rachitogenic diets also result in the resorption of scales. These facts suggest a physiological function of scales as a calcium reservoir.

Scales in Living Fishes other than Actinopterygians. In modern dipnoans the scales are elasmoid, set in pockets, anchored by vertical fibres, and have a superficial layer of plates or squamulae, with spinous sculpture. *Neoceratodus* has large thick scales with an uncalcified basal plate (Figs. 3.16, 3.17). In *Neoceratodus* the scale still shows considerable differentiation, and its parts can be related in some degree to those of the ancestral cosmoid scale. The cosmine layer has disappeared, but the outer calcified layer probably represent the remains of the lacunar layer, and the inner laminated layer may represent the isopedine. In *Protopterus* and *Lepidosiren* the scales are much simplified but their structure can still be understood through that of *Neoceratodus*. The basal plate is not calcified in *Protopterus aethiopicus* but is in *P. annectens* and in *Lepidosiren*.

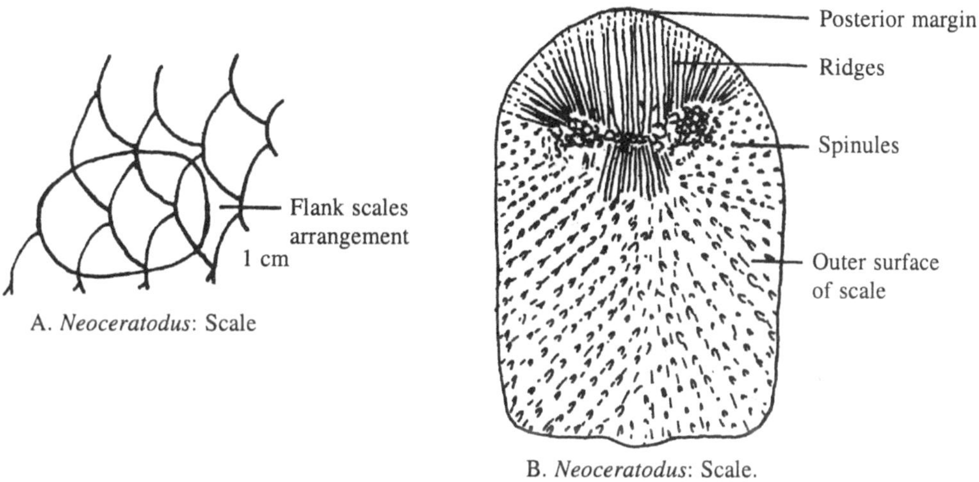

Fig. 3.16 A. Flank scales to show relationship and average degree of overlap. B. Diagram of outer surface of scale of larger fish: *Neoceratodus*. Superficial ridges shown as continuous lines, rods by broken lines, spinules by cones. Posterior end of scale uppermost.
***Source:* Kerr (1955)**

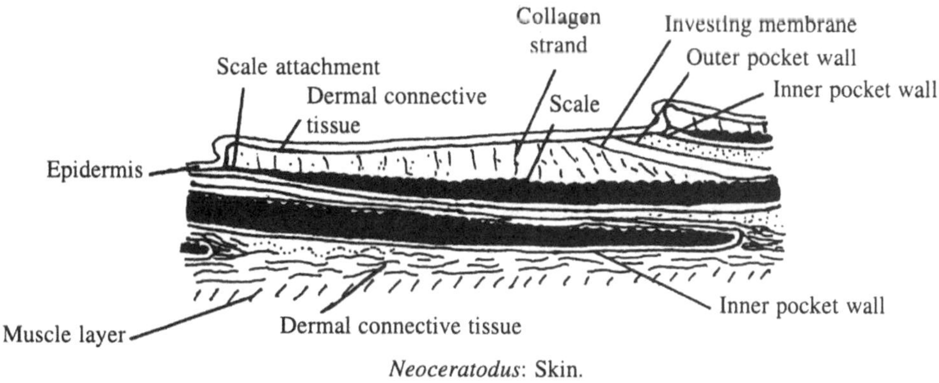

Fig. 3.17 L.S. skin through mid-line of scales of smaller fish *Neoceratodus*.
***Source:* Kerr (1955)**

In all, the basal plate has a twisted plywood arrangement as in *Amia* but with a right-handed twist.

Latimeria has a thick uncalcified basal plate with a right-handed twisted plywood arrangement. Above this is a mineralized laminated layer and a thin homogeneous layer ornamented with ridges. The posterior field carries a number of odontodes which lie on the previously formed ridged surface formerly covered by melanocytes. Odontodes can be shed and reformed; the ridges on the scale plate seem to give no indication of age.

The spinous placoid scales of living chondrichthyans are isolated odontodes with a cap of enameloid covering dentine, a pulp cavity, and a bony base. The odontodes may to a varying degree undergo modifications in external morphology when they are adapted to serve specific functions: (defence against enemies in the form of thorns or spines in skates and rays; protection of free neuromasts in the skin; attachment during mating-then forming hooks or spurs on the genital appendages of the pelvic fins in male sharks and holocephalans and on the frontal clasper and tentacula of the latter, tentacular hooks in male iniopterygians and retractile 'alar' spines on the lateral parts of pectoral fins in male rajids; teeth on the lateral sides of the rostrum, e.g., pristids; read Raschi and Tabit 1992 for an updated review on functional aspects. In *Pristis* rostral teeth that are lost are not replaced and those not severely damaged in use increase in size throughout life.

Recent sharks, skates, rays, and holocephalans have nongrowing scales which cover the body and in many, probably most, species also the integument of the mouth cavity. The scales consist of an enameloid cap, a dentine crown, and a basal plate of acellular bone. Growth of the scales takes place by addition of complete odontodes. In some genera the basal plate of the scale remains fairly thin; in others the base is thickened secondarily. The number of neck canals varies. Nongrowing scales (= placoid scales) of modern sharks have four neck canals. Resorption has never been observed in the dermal skeleton of any recent elasmobranch. Scales and teeth are shed by a resorption of anchoring collagenous fibres. However, as far as is known, this does not affect the hard tissues.

Many fossils and recent sharks and holocephalans have growing dorsal spines. As far as is known, all spines consist of enameloid and dentine. In recent sharks embryos the finspines develop within a single, enlarged papilla. Many finspines have tubercle-shaped, dentinous surface ornamentations. That tubercles represent secondarily fused scales is not correct. Finspines are obviously highly derived organs. Their homology with other elements of the dermal skeleton is difficult to determine.

Poisonous spines of the tail of several batoid families differ considerably from the dorsal spines. They are nongrowing spines, which are replaced. Thus they are clearly homologous to ordinary nongrowing placoid scales. They consist of vascular dentine with an enameloid cap.

Rostral teeth of sawfishes and sawsharks are also derived from normal placoid scales, but intermediate stages between placoid scales and rostral teeth are unknown. In the recent sawfishes (genus *Pristis*) the number of rostral teeth remains constant throughout ontogeny. As far as is known, the rostral teeth are persistently growing by addition of new dentine at the bases. Recent sawsharks (*Pristiophorus* and *Pliotrema*) have nongrowing rostral teeth which are replaced by others of increasing size and number during ontogeny. The condition in the sawfishes is clearly the advanced condition compared to that in the sawsharks (growth and no replacement versus no growth and replacement).

A very specialized type of placoid scale is the gillraker of *Cetorhinus*. These form a fine-meshed sieve for gathering plankton. Since a continuous replacement of the scales would impede the function of the sieving apparatus, the scales are probably replaced as a whole during a feeding pause in winter (read Reif 1982; See also Orvig 1977).

Evolution of Elasmoid Scales. It is now well admitted that the elasmoid scales of the advanced actinopterygian fishes (i.e., teleosts) are derived from the ancestral ganoid scales of "primitive", actinopterygians (i.e., holosteans, palaeoniscids) issued themselves from the rhomboid scales of the "primitive" osteichthyans (see Reif 1982; Maisey 1988; Sire et al. 1997). The structural characteristics of the ancestral ganoid scales have been retained nowadays in the polypterids, and to a lesser extent, in the lepisosteids.

Sire (1990) presents a new phylogenetical interpretation of the routes leading from the ancestral thick ganoid scales to the modern thin elasmoid scales. The different regions constituting the ganoid (ganoine, dentine and osseous basal plate) and elasmoid (limiting layer, external layer and isopedine) scales are compared in view of possible homologies. The outer limiting layer of the elasmoid scale is not ganoine, but some common features indicate that ganoine and outer limiting layer might be closer than generally admitted. On the other hand, the study of the development of the ganoid scales in polypterids has undoubtedly shown that the very young ganoid scales are in fact elasmoid ones. It is concluded that the elasmoid scales of the actinopterygian fishes are derived from the ancestral ganoid scales by a progress of paedomorphosis and that the osseous basal plate, and probably the dentine have not been retained in the elasmoid scales.

The reader is also referred to Sire and Huysseune (1993) for evolutionary considerations.

Development of Squamation in Fishes-Phylogeny and *Systematics*. The literature is poorly documented on the development of squamation in fishes other than teleosts except in the primitive actinopterygian fishes, the holosteans *Lepisosteus* and *Amia*. This lack of information is probably due to the difficulty of breeding dipnoans or polypterids. Studies on the scale development are available on dipnoans and on *Polypterus* but the authors do not comment on the order of scale development. In *Lepisosteus*, the scales appear first along the lateral line in the tail region (= site 1), then the squamation spreads anteriorly. In *Amia calva*, the scales appear at the lateral line just behind the pectoral girdle (= site 3) then the squamation extends posteriorly. In the absence of other data on the development of the squamation in primitive osteichthyan fishes, it is difficult to discuss the phylogenetic implications of the pattern of squamation development in fishes. Nevertheless, in closely related fishes such as *Lepisosteus* and *Amia*, the scales appear first in two opposite regions. Moreover, we can note that the squamation developed in an extinct osteichthyan, *Acanthodes* sp. followed a pattern (= site 1) similar to the majority of teleosts. The squamation development differs within orders and families of teleosts, and it appears similar in families as phylogenetically distant as Callichthyidae and Cyprinidae (Ostariophysi), and Poeciliidae and Cichlidae (Acanthopterygii). Thus, the localization of the first site of scale formation and development of the squamation cannot be linked to primitive or evolved condition. Nevertheless, the overwhelming preponderance of site 1 seems to be some indication that it, if any, is the "primitive" site. Moreover, within a family (Cyprinidae for instance) as well as in holostean fishes, the squamation can develop from site located in different regions of the body. Consequently, the development of squamation may be used with caution to establish phylogenetic relationships.

However, a possible use of squamation development in systematics to discriminate scombrid species has been reported (see Sire and Arnulf 1990).

Proposed Factors Involved in Scale Induction. Sire and Arnulf (1990) have discussed the hypothesis: the lateral line induction. Hypothesis proposed by authors is that scale induction could be related to mechanical constraints imposed on fish skin during swimming. This induction could be related to the shape of the fry before scale formation (elongated or not, flat or high) and to its behaviour (continuous swimming or intermittent swimming, pelagic or benthic, etc). During swimming, the repeated movements of the muscle in a definite region of the body (as for instance in the caudal peduncle or pectoral peduncle), transmitted as tension forces to the skin could induce a previously genetically programmed cell population to enter a differentiation process leading to scale formation. This epigenetic induction by mechanical constraints would explain the main different patterns of squamation development in fishes when the lateral line is not concerned. This hypothesis is still under investigation.

In *Poecilia reticulata*, the first scales appear on the caudal peduncle (site 1) when the fry are still in the belly of the female parent. Consequently, the initiation of the scale development cannot be related to an external factor. The first scales appear in the caudal peduncle, in the lateral line region, favouring the hypothesis of a lateral line induction; but this region of the fry is mobile, as seen in fry freshly removed from the female parent, and that could also be explained by the hypothesis of movement induction. In the case of *P. reticulata* also, the hypotheses of nervous, or mechanical, inductions may not be rejected, and further results on the behaviour of the fry in the mother's belly are needed for better understanding.

Thus, in the actual state of our knowledge of the squamation development in fishes, we can note that the induction by the lateral line, if it exists, is not the only factor influencing the first scale formation.

Epidermal-Dermal and Fibronectin Cell-Interactions during Fish Scale Regeneration. In fishes, there are only a few studies on the reepithelialization of the skin after infliction of experimental or normal wounds. Sire et al. (1990; see literature) have carried out immunofluorescence and TEM studies in this regard in cichlid fish *Hemichromis bimaculatus*. In normal skin, specific localization of fibronectin in the interface between the scale-associated cells and the outer surface of scale-pocket lining cells can be implicated in the attachment between the scale-pocket lining surface favouring a flexibility of scale movement during swimming. In addition, fibronectin could play an important role as a major extracellular matrix protein during wound-healing after scale removal. For regeneration of scales in *H. binaculatus*, see Sire (1982, 1989).

Fin Rays. Fin rays (dermotrichia), which occur in different groups of fishes, are part of the dermal skeleton and according to some authors are probably derived from some type of scale.

Lepidotrichia are segmented and consist of bone; they occur in Actinopterygii and Crossopterygii. Hard rays of Actinopterygii are unsegmented. They can be derived from round scales. *Latimeria* has actinotrichia but dipnoans do not; they have campotrichia, partially mineralized collagenous rods. The ceratotrichia of selachians correspond to the actinotrichia of bony fishes.

3. COLORATION AND PIGMENT CELLS (CHROMATOPHORES)

Features such as striking coloration attract evolutionary biologists because they offer the allure of a simple, perhaps key, trait in communication. Schliwa (1986) has given a review of pigment cells in fish. Fish pigment cells are considered to be derived from neural crest cells. Indeed, pigment cells are known to derive from multipotent stem cell from which mesenchymal lineages could also originate. On the other hand, fish pigment cells are supposed to be able to change their phenotype expression, even after postembryonic development under the influence of extracellular signals. Scale scleroblasts as well as melanophores are neural crest derivatives, a hypothesis recently supported by Zylberberg and Meunier (1996). Demski (1992) has discussed the neural control of chromatophore display in cephalopods and teleosts in the context of convergence of functional-anatomical pathways and mechanisms at several levels of organization. Recent comprehensive reviews by Fujii (1993a, b; see literature; refer also Sugimoto et al. 1994; Fujii and Oshima 1994; Meyer-Rochow 2001) cover a wide spectrum of information on the subject. Obika and Meyer-Rochow (1986, 1990) focused studies on chromatophores of Antarctic teleosts, *Pagothenia borchgreviniki* and *Trematomus bernacchii* respectively. Barlow (1983) has described the behavioural consequences of polychromatism in the midas cichlid, *Cichlasoma citrinellum*. Adult midas cichlid, *Cichlasoma citrinellum*, occur in two color morphs: 'normals' which are able to use melanophores to alter their markings, and amelanics which cannot are mostly 'gold'. The difference is genetic. Coloration has important social consequences because of the effect of gold on communication about aggressive and mating behaviour. Further, Dickman et al. (1988) have dealt with the histological basis of color metamorphosis in *C. citrenellum*. *C. citrenellum* has been the object of further studies (Dickman et al. 1990; Barlow and Siri 1994). The reader is also referred to papers on behavioural studies by Rowland and Towlands (1978) and Slovin and Rowland (1978). Animal coloration is based on three phenomena: (1) general absorption of light, usually be melanin (2) filtering or selective absorption of visible light such as by carotenoids and (3) reflection (should be better termed refraction) of visible light.

Many fishes are capable of changing color in response to environmental changes. Color changes reflecting different phases of behaviour also occur in a number of fishes, such as sunfish (*Lepomis megalotis*), the bitterling (*Rhodeus armatus*) and certain cichlids.

Apart from background conditions, age also seems to play a significant role in the determination of color patterns. Barry and Hawryshyn (1999) described the adaptive significance of reef fish coloration. Color patterns change during the life of fish as well as with the mood of the fish or disease. A specific combination of a number of these patterns produce a color pattern that constitutes a specific signal conveyed to conspecific fish. However, the meaning or significance of particular pattern would differ from species to species. Diurnal visually hunting piscivores more often attack contrastingly coloured fish, resulting in selection for crypsis. A dominant or territory-holding fish is often brightly coloured.

The arrangement of highly specialized pigment cells (chromatophores) in the integument is more varied and less highly organized in fish than in either adult amphibians or reptiles.

The pigmentation of different species of upside-down catfishes of the genus *Synodontis* (a silurid fish) under natural conditions is obviously related to their orientation. Species which habitually remain inverted showed reversed pigmentation, having dark bellies and pale backs.

Larvae or young fish are very commonly pellucid (e.g., young sardines, leptocephalus larvae of Anguilliformes, Elopiformes, Notacanthiformes, and Gonorhynichiformes). During their migration upstream a river, for example, larvae of the eel (e.g., *Anguilla* spp.) take advantage of their transparency to avoid predation. Even among adult forms of teleosts, there exist some transparent ones (e.g., *Kryptopterus bicirrhis, Physailia pellucida, Salangichthys microdon*). Their bodies are compressed although not so much as leptocephalus larvae. Although the population of the chromatophores in the ice goby *Leucopsarion petersii* is much smaller than that of many other fishes these cells may be important in the strategy of survival.

Modes of Color Change. Chromatophores are capable of changing body coloration of the animal by virtue of two different mechanisms. In the first, chromatophores may actively redistribute pigment-containing organelles within the cell boundary in such a way that more or less of pigment organelles are exposed to view. These changes can be rather prompt being mediated, in many cases, through the activity of nerve endings that directly stimulate the chromatophore. If the activities of many cells are spatially and temporally coordinated, the macroscopic effect can be rather dramatic in that it produces changes in overall body coloration of pattern. Examples include the adaptive coloration of fishes in response to background or changing light condition, or the display of color patterns reflecting and signalling mood or physiological state of the animal. These phenomena reflect an alteration in the physiological state of their pigment cells and therefore are referred to as "physiological color change". A different mechanism of changes in body coloration, termed "morphological color change", is based on a much slower, long-lasting process that involves either a decrease or an increase in the total number of pigment cells or the amount of pigment they contain. Such changes are gradual and not easily perceived, but the result may be quite dramatic once the transition from one state to the other, which takes several days or weeks, is completed. As a rule, both morphological and physiological color change occur simultaneously, with the former providing the baseline coloration upon which fine-tuned response of the animal to the internal or external stimuli are superimposed.

Chromatic Elements: Classification of Chromatophores. Chromatic elements are defined as cellular systems that effect coloration of the skin. The chromatophores are neural crest derivatives with pigment-containing intracellular bodies called chromatosomes, as a general type, or by names specific to the different chromatophores are now categorized as "paraneurons".

Melanophores: These are the pigment cells containing black or brown pigments. Melanins (black or brown) are found in many parts of the body. Pigmentary melanin are polymers of 3, 4 dihydroxyphenylalanine Quinone (DOPA-Quinone) synthesized by oxidation of the amino acid tyrosine to DOPA then to DOPA-Quinone by the enzyme tyrosinase. These activities take place on laminar intracellular organelles, the premelanosomes and mature pigment granule, the functional pigmentary organelle known as the melanosome. In many fishes without melanin pigment, such as the orange-red variety of *Oryzias latipes*, melanophores are still present but lack pigment. Such amelatonic melanophores are also found in normal gadoid skin.

Lipophores: Lipophores are those chromatophores containing pigments which are insoluble in water but soluble in organic solvents. They are sub-divided into erythrophores (red or yellowish pigments) and xanthophores (primary yellow pigments). The pigments of lipophores are mainly carotenoids.

Results, mainly from Japan, indicate that tunaxanthin is a rather common pigment in marine fish. This carotenoid is especially abundant in yellow-coloured fishes. Astaxanthin, by contrast, seems to be dominant in red marine fish. Lutein is also widely found in many marine species. Carotenoids commonly found in freshwater fish include beta-carotene, lutein, taraxanthin, astaxanthin, tunaxanthin, alpha-and beta-doradexanthins, and zeaxanthin. Carotenoids cannot be synthesized, de novo within the body and must, of necessity, come from food, although ingested carotenes can be converted to specific fish carotenoids. The absence of carotenoids from the diet eventually results in loss of lipophore color component. Pteridines are associated with red or yellowish pigmentation. Pteridines are compounds with a pteridine nucleus as their basic skeleton. In contrast to carotenoids, pteridines are soluble in water. The cell organelles that contain pteridines are called "pterinosomes" (better term "xanthosomes" or "erythrosomes").

Leucophores and iridophores (colorless pigments primarily guanine): Both of these groups of chromatophores contain purines which are responsible for the white and silver colors. Being barely soluble in water, purines are usually found in crystalline form in iridophores (the light reflecting, rounded type). A wide variety of names, such as antaugonophores, guanophores, iridocytes, iridoleucophores, etc. have been used for these reflecting chromatophores.

Since fish have only four biochromes, the source of the wide range of blues and greens and other hues must depend on these interplays.

The pigments that give these classes of chromatophore their color are contained in membrane organelles named after either the color of the pigment (e.g., melanosomes), or the pigment itself (e.g., pterinosomes, carotenoid droplets-carotenoid appear in lipid droplets not limited by a membrane, or organelles physical properties (e.g., reflecting platelets, refractosomes).

Light reflecting cells in killifish, *Fundulus heteroclitus*, have been differentiated in three groups; one group on the surface of scales reflects white light by a Tyndall light-scattering effect, while two other deeper groups (the melaniridophores and stratum argenteum) are irridescent by multiple thin layer interference system. The shiny surface of a ratfish (*Hydrolagus colliei*) possess characteristics in common with the silvery exterior of teleosts. The metallic sheen is caused by reflection of light from platelets in the skin that are oriented mostly vertically. They lie in a layer of oriented reflecting cells in the dermis (Figs. 3.18, 3.19). Additionally there is a stratum argenteum in the subdermis. Lamprey and juvenile rockling have only a stratum argenteum.

Many fish have silvery reflecting layers in their external surfaces, in their eyes, and in their photophores. The basic reflecting units in fish are small flat crystals, made of a mixture of guanine and hypoxanthine. They are generally found inside living cells; their optical thickness being between about 100 and 200 nm, i.e., quarter wavelengths for the light contained in the visible spectrum. They are arranged in stacks of crystals spaced apart by layers of cytoplasm, also approximately one quarter wavelength in optical thickness. Such a stack (platelet) gives a very high reflectivity with very little material. Platelets are often highly colored, and there is usually a good correlation between the wavebands best reflected and four times the optical thickness of the crystals they contain. If the cells containing the platelets are made to shrink or swell by changing the osmotic strength of the medium in which they are bathed, the spacing of the crystals changes and as a consequence the reflectivity and colour of the platelets also change.

Platelets differ from one another in different parts of the surface of the fish in their size, number per unit area, their orientations with respect to the body surfaces and in their reflectivities, including spectral reflectivities. In some platelets the optical thickness of the crystals and the spaces between them are almost constant and these platelets are highly colored. Typically, such a platelet's spectral reflectivity, and the degree to which it polarizes light, will change with the

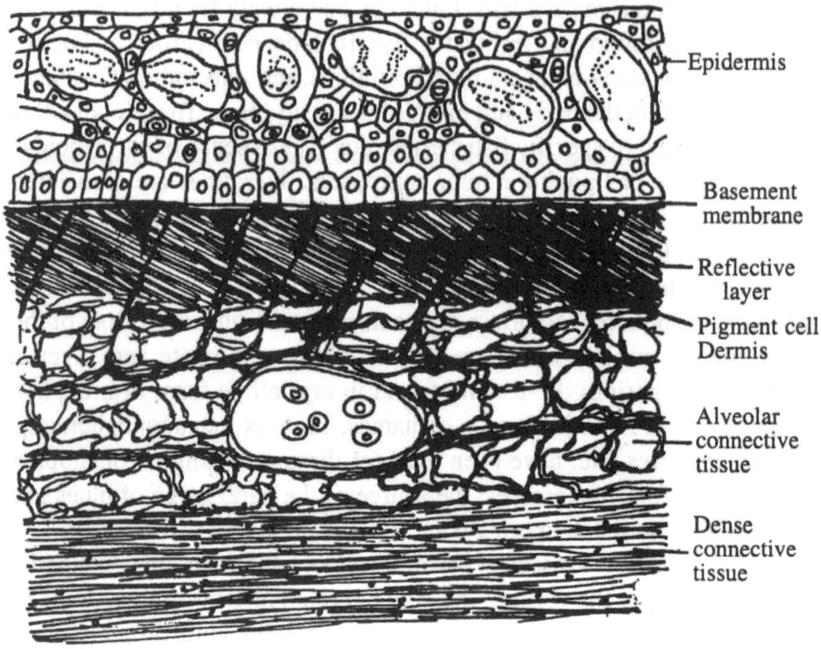

Hydrolagus colliei: Skin.

Fig. 3.18 ***Hydrolagus colliei*. Drawing of a transverse section through the integument of the dorsolateral flank.**
***Source:* Arnott and Nicol (1970)**

angle of the view. The maximum reflectivity moves towards shorter wavelengths as the platelet is viewed more obliquely. Other platelets, for which the number of crystals is higher but whose thickness and spacing are less regular, reflect well throughout the visible spectrum at all angles of incidence. Some reflectors lie on or under the scales other than in the skin. They sometimes overlie white scattering layers; sometimes they are associated with pigmented layers.

Denton and Rowe (1994) have discussed the reflective communication between fish, with special reference to the greater sand eel, *Hyperoplus lanceolatus*. Platelets can also be useful in signalling information about the movements of fish that the fish are making. Such signals are clearly graded. Some, e.g., those that accompany C-starts, will be strong and visible at distances many times the length of the fish. Others, e.g., those given by a spot of silveriness on the tail, will only be visible to near neighbours. Authors have considered the movements of fish in terms of three components, rolling, pitching and yawing and have shown that, by virtue of the organization of the silvery layers, a given movement will produce, to the eye of a neighbour, a characteristic changing pattern of bright and dark surfaces. The changes in brightness accompanying quite small movements are large and to us (authors) and almost certainly to the fish, much more visible than the change in shape or position (the detection of which must also depend on the differences in brightness). Diffusely reflecting surfaces can, to some degree, serve the same purpose but they are less good at doing so. They certainly cannot give a fish the facility of passing quickly from being almost invisible to presenting striking unequivocal signals to their neighbours, as silvery surfaces do. Authors have given an account of the reflecting surfaces of the greater sand eel in relation to this hypothesis.

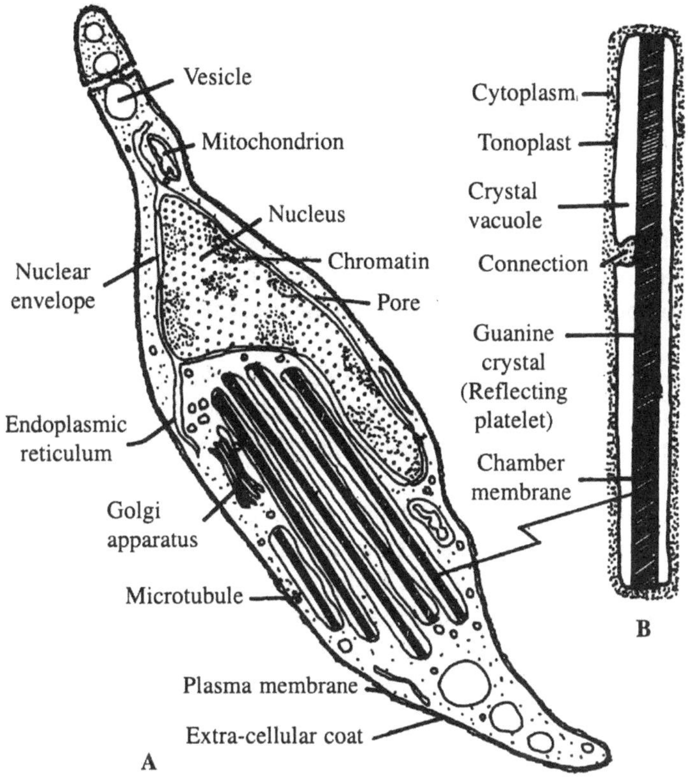

Reflecting cell in *Hydrolagus colliei* skin.

Fig. 3.19 A. Diagrammatic representation of guanine crystal containing cell of the oriented reflecting layer in the skin of the ratfish (*Hydrolagus colliei*). B. Break at left indicates that individual cells may have very long marginal extensions.
***Source:* Arnott and Nicol (1970)**

The reader is also referred to Fujii (1993a, b) for details of optics of iridophores, static iridophores and motile iridophores. Goda et al. (1994) have described the details of blue coloration of common surgeon fish, *Paracanthurus hepatus.* Goda and Fujii (1995), named dendritic blue chromatophores as "cyanophores" and the organelles "cyanosomes".

Migratory fish, such as salmon smolt and freshwater eels, assume the pelagic pattern becoming darker above and more silvery on their flanks, when they forsake the rivers for the sea.

Fine structure. Melanophores: Cells possess flattened nucleus and dendrites running parallel to the surface. Each melanophore contains electron-dense melanosomes, mitochondria, Golgi structures, endoplasmic reticula, free RNP particles, microfilaments, microtubules, pinocytotic vesicles and premelanosomes. In melanophores, the melanosomes are embedded in a three-dimensional system of microtrabecular strands, which interconnects them with microtubules and the plasmalemma and which appears to be a major structural framework controlling pigment granules translocation. Teleost melanophores are sensitive to several hormones: melanocyte stimulating hormone (MSH), melanocyte concentrating hormone (MCH), pineal melatonin and adrenal medullary or chromaffin catecholamines. It should also be noted that certain fish melanophores demonstrate a direct light-sensitivity that involves pigment aggregation.

Erythrophores. Pterinosomes, glycogen particle and vesicular structures probably concerned with carotenoid accumulation are the main components of the cells, besides some other reported in melanophores.

Xanthophores. (i) *Routine fixation observations*: Cytoplasm having endoplasmic reticulum-like structures, Golgi apparatus, ribosome-like particles, mitochondria, pterinosomes.

(ii) Dalton's fixation observations: Numerous vesicular structures. These vesicles were thought to include the endoplasmic reticulum, vesicles of Golgi origin, and vesicles from micropinocytotic vesicles. Lipid soluble particles in which carotenoid pigments were thought to be dissolved.

Iridophores: Their cytoplasmic organelles are reflecting platelets, spherical granules of pleiomorphic internal structure, Golgi apparatus, mitochondria, endoplasmic reticula and ribosomal particles. At the plasma membrane pinocytotic vesicles are observed.

Leucophores. They are light, dendritic; their pigment granules are migratory and they exhibit a dull whitish color; cytoplasmic organelles include pinocytotic vesicles at the plasma membrane, endoplasmic reticula, mitochondria and Golgi apparatus. Leucophores should, however, be distinguished from iridophores because the mechanism of reflection of light by the former is entirely different from that by the latter. The cellular organelles responsible for cellular reflectivity are called leucosomes, and they scatter light in all directions. To date, neither the localization nor the chemical nature of the substance(s) that are responsible for the light-reflecting properties of leucosomes has been determined. Fujii et al. (1997) have described an improved photoelectric method for recording motile responses of individual leucophores.

Position, Structure, Pigment Movement and Control of Chromatophore Activity; Chromatic Units: Multifunctional *Systems*. Fishes may contain both dermal and epidermal chromatophores but the former are more abundant in most species. Dermal melanophores are located just underneath the epidermis in close apposition to the basal lamina. Epidermal pigment cells may be found at various depths of epidermis. While epidermal pigment cells may be of highly variable shape, dermal chromatophores, especially those actively engaged in physiological color change, are more or less flattened cells with a stellate outline. In most chromatophores, cell proceses or arms extend from the cell body (dendritic chromatophores). These processes, may be of variable length and branched in some species, giving the cell an irregular outline, while in others they are of uniform length and unbranched. When pigment granules (organelles, chromatosomes) are more or less evenly distributed through the cell body, the area covered by the granules appears in the color of the pigment. This state of the chromatophore is referred to as the "dispersed" state. Conversely, pigment granules may be induced to pile up in the cell centre, leaving the cell processes pigment-free. This process causes a blanching of the chromatophore-covered area and is referred to as "aggregation". In many cases the transition from one state to another is accompanied by a dramatic change in the thickness of the cell centre as granules are collected in the cell region, while the cell processes may collapse or flatten. The time required to change from one state to the other varies considerably between cell types and between species. While at one or the other extreme state of pigment distribution, granules are by no means at rest and may show constant shuttling back or forth. Since, however, the net result of these independent granule activities is zero, the cell does not change its state of pigment distribution. In most species the duration of pigment aggregation (centripetal migration) and dispersion (centrifugal migration) are not of equal length, the latter being slower by a factor of about 2.

The mechanism of the mass movement of pigment granules, which forms the cellular basis for color change, has been the subject of considerable interest. Basically two models are currently under investigation. The first envisions the presence of a contractile and/or elastic system possibly not unlike the actinomyosin contractile machinery of muscle cells that provides the motive force for granule movements. The second model holds that pigment granules are moved by direct interaction with microtubules mediated by molecules similar to ciliary or flagellar dyneins. Both the "contractile system model" and the "microtubule dynein model" are supported by circumstantial evidence. A solution of this problem has to await further experimentation.

In the dendritic chromatophores (melanophores, xanthophores, erythrophores, leucophores) a number of microtubules have been found emanating from the perikaryon centre into the dendritic processes. Apparently, they function as cytoskeletal elements that maintain the stellate shape of the cells. In addition, recent studies strongly suggest an active role for these microtubules in the centripetal and centrifugal translocation of melanosomes. It has been pointed out that the remarkable degree of architectural order in the lamellar gridwork of collagen fibres is of great importance in the accomplishment of effective and visible chromatic responses.

While the precise molecular mechanism of granule movements in pigment cells is still not understood, the cellular factors that may control these movements are, at least in part, better known. Recent studies demonstrated changes in the level of protein phosphorylation in association with granule movements, even though a direct link to action of cAMP (intracellular cyclic AMP) was not established. The cellular messenger involved in the initiation of pigment aggregation is not known for most chromatophore types, but there are reasons to believe that a rise in the level of cytoplasmic free calcium plays an important role (see details ahead).

In order for adaptive and signaling color changes to be physiology relevant, animals must be able to control such changes at will and without much delay. Two modes of regulations operate either alone or in combination in essentially all fish species: humoral and neural control. Of these humoral control is, of course, not as prompt as that exerted by nerve endings; it requires stimulation of the cells releasing the hormone, its transport through the blood stream, diffusion through the tissue surrounding the effector cells, and binding to appropriate receptors at the chromatophore plasma membrane. Neuronal stimulation via nerve endings directly in contact with the effector cell, on the other hand, provides for an essentially instantaneous response and is therefore well expressed in those species requiring immediate and specific changes in body coloration. It may be pointed out here that chromatophores in teleosts are innervated by post-ganglionic sympathetic nerve fibres that form multiple synaptic endings on the cell body and its processes and thereby affect pigment aggregation or blanching. There is physiological evidence for polyneuronal innervation. Norepinephrine is the post-ganglionic neurotransmitter in most fishes, the exception being catfish in the family Siluridae which utilize acetylcholine as the neurotransmitter. The configuration and density of chromatophore innervation patterns varies regionally as well as with chromatophore type. These silurid melanophores lack alpha-adrenoreceptors, and sympathomimetic agents do not induce the aggregation of pigment. In several catfish species that belong to families other than Siluridae, alpha-adrenoreceptors and muscarinic cholinoreceptors, both of which mediate pigment aggregation, coexist in the melanophores. However, the neural regulation of these cells involves an adrenergic mechanism as it does in many species examined to date. The species of catfishes included species within the genera *Corydoras* (Callichthyidae), *Pimelodella* (Pimelodidae), and *Mystus* (Bagridae). The physiological roles of these cholinoreceptors are still unknown. Recent studies suggest that, although the melanophores of *Zacco* spp. (Cyprinidae) are controlled by an orthodox manner by the sympathetic nervous system, some of them possess extra muscarinic cholinoreceptors that also mediate the aggregation of pigment (Hayashi and Fujii 1993). Further, it has been reported that the receptors that mediated the cholinergically evoked aggregation of melanosomes in *Kryptopterus bicirrhis* and *Corydoras paleatus* were of the M_3 muscarinic subtype (see details: Hayashi and Fujii 1994).

In many fishes, ET (endothelin) may be involved in the subtle and delicate modification of hues and patterns of the integument that are associated with their elaborate and effective strategies for survival (read Fujii et al. 1993 Hayashi et al. 1996; Fujita and Fujii 1997; Fujii 2000; Murata and Fujii 2000; Note: ET secretion has been reported from neurohypophysis, urophysis and adrenal chromaffin cells). Control by hormones may also be regarded as the more basic and evolutionary conserved principle since it is the exclusive mechanism used by Cyclostomata, and the prevailing means of control in elasmobranchs. Among teleosts, both hormonal and neural control mechanisms are expressed to varying degrees. For example, in the eel, hormonal control is dominant, while neural control is of little importance. In more "advanced" teleosts, particularly those showing very rapid or sophisticated changes in body coloration, such as the flatfishes or the squirrelfish, neural control is absolutely dominant. This does not mean that the chromatophores of these latter species are unable to respond to hormones, as demonstrated by the "paling" reaction of frightened fishes that even affects the pigment cells whose nervous connections were cut. Also isolated chromatophores placed in tissue culture will respond to appropriate hormones added to the culture medium, indicating that the receptor machinery is present but is normally overridden by neural control mechanisms. Winter flounder (*Pseudopleuronectes americanus*) chromatophores includes three major melanophore systems with differences in effector distribution and chromatic behaviour. Response characteristics suggest that each of these pattern components represents a different neuroeffector system and that there may be prolonged hormonal control. One pattern component has the capacity for rapid melanosome dispersion and a second has the capacity for rapid melanosome aggregation on different backgrounds. Chromatic behaviour of the third component is not necessarily background related.

Among the agents involved in hormonal control, melanocyte-stimulating hormone (MSH), also called intermedin, is of prime importance not only for fishes but also other vertebrates. The action of MSH is mediated by specific receptors on the outside of the chromatophore membrane. The involvement of the pineal gland as a humoral skin-lightening control centre has not been confirmed. The neural control of teleost chromatophore has been fully established. Adrenergic neurons regulate both aggregate and dispersive melanophore activities, the reaction of the chromatophore depending on the stimulation of either alpha-adrenergic or beta-adrenergic receptors which, in turn, is a function of the monoamine concentration: is it low, beta-receptors are stimulated, resulting on melanosome dispersion, while at higher concentrations, the stimulation of alpha-receptors produces aggregation. Thus, only one fibre system releasing just one neurotransmitter may control physiological color changes in fishes. A brief on the location of chromatomotor neurons and composition of motor units in fishes is covered by Demski (1992). The pathway controlling the eyebar in *Haplochromis* has been studied in detail. The melanocytes are controlled by motor fibres in a branch of the maxillary division of the trigeminal nerve. HRP applied to the nerve is retrogradely transported to cell bodies located in ganglia situated more proximally along the trigeminal complex. The filled cells are most likely part of cranial sympathetic system.

Various factors are known to influence the chromatic state of animals. As might naturally be supposed, the colors of fish are strongly influenced by the photic environment. The information perceived by the lateral eyes and/or the pineal is processed in the central nervous system which transmits signals to chromatophores via hormonal and/or nervous pathways.

Recently, it has been reported that, together with norepinephrine, ATP is liberated from chromatophore nerve terminals. ATP is successively dephosphorylated by exoenzymes to adenosine. By interacting with adenosine receptors on the pigment cell membrane, adenosine acts to disperse pigment. Although liberated concurrently, the true transmitter, norepinephrine, acts first. The amine is quickly removed by reuptake into the nerve terminals, or it is inactivated by catechol *O*-methyltransferase (COMT) or by monoamine oxidase (MAO). After the cessation of sympathetic pulses, adenosine survives for a longer time and disperses pigment, reversing the effect of the true transmitter, norepinephrine. In this way, adenosine, as a functioning cotransmitter, acts to disperse chromatosomes rapidly, thereby enabling fish to change their coloration very rapidly. Thus, the singly innervated chromatophores of teleosts have acquired the ability to react very quickly, resembling in this regard reciprocally innervated effectors.

Receptors for signaling ligands, namely, hormones and neurotransmitters, are found on the cell membrane of chromatophores. Among the principal receptors, the alpha-adrenoreceptors and receptors for MCH and melatonin mediate the aggregation of chromatosomes, while beta-adrenoreceptors and receptors for MSH and adenosine mediate their dispersion. The extracellular signals are translated by the receptor system into an increase or a decrease in the levels of intracellular second messengers. Until recently, three kinds of second messengers were known to take part in the motile responses of chromatophores, namely, cyclic adenosine-3′, 5′-monophosphate (cAMP), Ca^{2+} ions, and inositol 1, 4, 5-trisphosphate (IP_3). Adrenoreceptors of the $alpha_1$ type cause the breakdown of phosphatidyl inositol in the membrane to IP_3, which in turn acts to release Ca^{2+} ions from the sites of intracellular storage. Stimulation of beta-adrenoreceptors, or the receptors for MSH or adenosine activates the adenylate cyclase system to increase intracellular levels of cAMP. Through the stimulation of their specific receptors, melatonin and MCH are thought to inhibit the cyclase system, with resultant aggregation of chromatosomes via decreases in levels of cAMP. Other hormonal substances, autacoids, or local hormones may have some modifying influences on the actions of major ligands. The changes in the levels of second messengers are thought to induce the phosphorylation of certain proteins that are more directly coupled with the motile activities of chromatophores (Fujii 1993b Fig. 3 showing the system for controlling motile activities of light-absorbing chromatophores in teleosts, and Fig. 4 illustrating dynamics of a catecholaminergic true transmitter and a purinergic cotransmitter liberated from a sympathetic postganglionic fibre that controls chromatophores in a teleost; Oshima and Wannititul 1996). For the details of IP_3 function in tilapia (*Oreochromis niloticus*) melanophore, read Fujii et al. (1991).

In most fishes coloration is intended as camouflage and one of the most successful and widely used patterns is the dark above-light below countershading pattern of the pelagic fish. This is, however, a total effect, which does not have the same capacity for rapid change that is manifested with chameleon-like facility by the heterosomated flatfishes and some other marine species.

Information on chromatic arrays in teleosts is available. More details of arrays in a variety of teleosts are needed before any comprehensive model of the morphofunctional systems can be put forth. However, it is apparent that many of the distribution patterns follow outlines related to scale position (see Demski 1992).

The chromatic systems in cichlids not only function as separate units that span many body areas but each appears to develop at different times. Observations on the vertical bands of sunfishes also support the idea. The bands and overall coloration system most likely represent chronological units composed of chromatophore-neural systems of different ages.

The few detailed studies concerning the developmental mechanisms involved in teleost chromatic systems (mostly longitudinal stripes) have been carried out mainly in the zebrafish, *Brachydamio rerio*. The results support the idea that waves of melanophore development result in age classes of melanophores, which populate different embryonic, juvenile and adult patterns. Melanophore death and removal appears to be an important mechanism in some teleosts (see Demski 1992).

Details are available on the neural mechanisms of several of the better-studied chromatic components and patterns and their related behaviours (overall coloration; vertical banding, and specialized features which include spots and eyebars (see Demski 1992).

Circadian changes in pigmentary patterns have been reported. Observations strongly suggest a positive role of melatonin in the diurnal alternation between contrasting dark and light patterns in the pencil fish, *Nannostomus beckfordi*. Circadian changes in color of the blue-green longitudinal stripe of the tetras (characins) are primarily due to high degree of photosensitivity of the iridophores in the stripe. In addition, the hue is influenced by some environmental factors through the activities of the sympathetic nervous system. In tetra fish (*Paracheirodon innesi* and *P. axelrodi*) the pineal hormone, melatonin, plays a definite role in the circadian changes in the coloration of red ventral (abdominal) skin, while other hormonal and nervous factors may modify the distribution of the pigment in the erythrophores (read Hayashi et al. 1993; Masagaki and Fujii 1999) (Fig. 3.20).

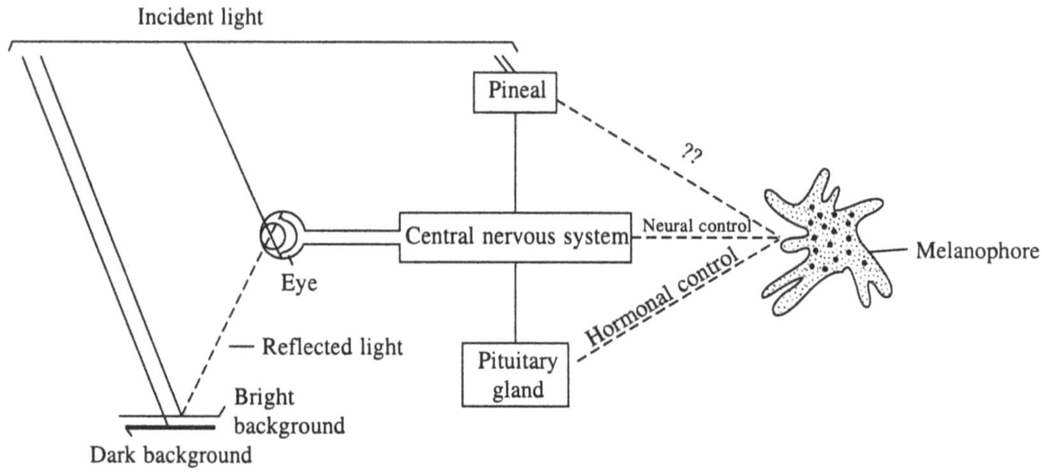

Fig. 3.20 Simple schematic representation of the controlling system of melanophores in typical teleost fishes.
***Source: Fujii* (1973) (Modified)**

Crypsis in the Pelagic Environment. McFall-Ngai (1990) has discussed various aspects of crypticity. Four types of antipredation mechanism have been recognized in marine and freshwater system: (1) time and space (e.g., vertical migration; (2) body-size; (3) visibility; and (4) behaviour (e.g., schooling).

In the homogeneous pelagic aquatic habitat, evolutionary convergence of three main forms of crypsis is evident: (1) transparency; (2) reflection of most, if not all visible wave lengths; and (3) ventral bioluminescence as counterillumination; thus to be cryptic most animals in these habitats use one or a combination of these modalities to variously transmit, reflect or mimic environmental light.

Patterns in the use of these three modes of crypsis in any given animal are related to (1) the biogeography of the species and (2) characteristics of biology of animal, such as size and life history strategy. In clear ocean waters, where the light quality is relatively predictable and stable, certain biogeographic trends in the occurrence of bioluminescence, reflectivity and transparency are apparent (Fig. 3.21A)

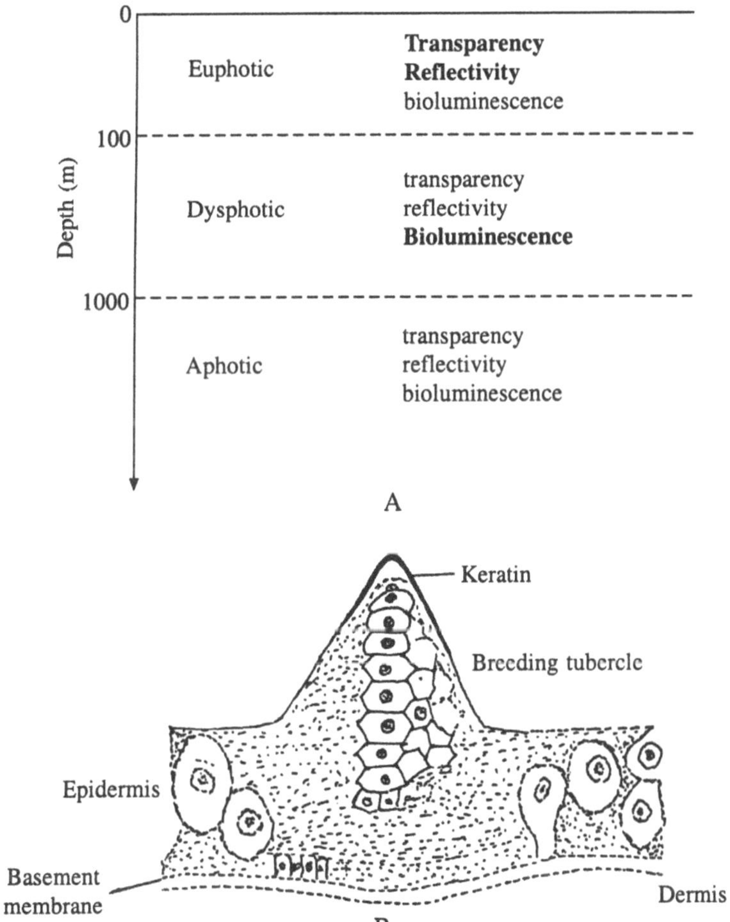

Fig. 3.21 (A) Zones of ocean. (B) Breeding tubercle. Drawing of a section through a small tubercle from the dorsal head surface of male of *Kneria* sp.
***Source:* (A) McFall-Ngai (1990); (B) Wiley and Collette (1970)**

The oceanic euphotic zone: Diurnally in the euphotic zone, transparency is the dominant mechanism and all phyla have transparent representatives. Reflectivity is also a conspicuous mode in the euphotic zone and is best exemplified by schooling fishes. Often these fishes will be countershaded, with the dorsum dark and the ventral light, to camouflage from above and below. Because ambient light levels are so high in this portion of the ocean, the use of bioluminescence is precluded during the day as a form of camouflage.

The oceanic dysphotic zone: The dysphotic zone is characterized by the prevalance of bioluminescence as a cryptic mechanism (see Chapter 18.)

The oceanic aphotic zone: In the aphotic zone, where solar illumination is absent, animals are most often black or red, colors that would most effectively absorb point sources of bioluminescent light produced by potential predators. The cryptic modes of transparency, reflectivity and luminescence, which use or mimic light in the environment, are not adaptive in this environment.

Into this seemingly set of trends is woven a number of other factors, such as diurnal photic cycle and life history strategy of the animal (see Chapter 18).

Nearshore pelagic environment: Transparency and reflectivity are the primary modes of crypsis in these waters, where bioluminescence is rare (see Chapter 18).

Ethological Implications. Fujii (1993b) describes as follows: (1) Cryptic (concealing) coloration: The "cryptic (concealing) colorations" are the body colors and patterns that enable the animal to blend into the background. In this way, a predator can approach its prey without being noticed. The prey, too, can easily escape from predators, with the latter phenomenon being commonly known as "protective coloration". These phenomena usually involve resemblance, not only of body hue but also of color pattern to the background.

When an animal exhibits a color, color pattern, form or posture, such that it resembles another organism, either an animal or a plant, or sometimes a lifeless object, the phenomenon is commonly called "mimicry". When the animal becomes inconspicuous as a result, the phenomenon is called "concealing mimicry", or more appropriately "mimesis". Among the parameters involved in mimesis, color is considered to be of special significance, and this aspect has been designated "chromatic mimesis". A few examples of "chromatic mimesis" are as follows: leaf fish (e.g., *Monocirrhus polyacanthus*) mimic withered leaves; the leafy sea dragon (e.g., *Phycodurus eques*) mimics algae, and anglers (e.g., *Lophius litulon*) mimic rocks.

(2) Aposematic coloration: The purpose of "aposematic coloration" is also to increase the fitness of the fish but, in this case, the increase in fitness results from the possessor advertising its presence. Thus, the term "advertisement coloration" is also frequently used. Among the phenomena included in this category are "warning coloration" and "threatening coloration". Neudecker (1989) concluded that the primary selective force behind chaetodontid coloration, particularly eyespots, has been predation and color patterns have evolved to minimize this threat.

(3) Recognition coloration: Recognition coloration is exclusively for mutual identification among individuals of the same species. The well known "nuptial tubercle" is an example of a display for conspecifics, and it normally appears in the male forms during the breeding season. It functions, of course, to attract females.

Transparency in Predator-Prey Interactions. No organism is completely transparent, and the degree of transparency among pelagic species varies over a wide range. In addition, the extent to which an increase in transparency translates into a significant advantage in predator-prey interactions is difficult to ascertain. However, anything that reduces the contrast and brightness of animal would make it more difficult to see and would in the context of predator-prey interactions, be an advantage.

Morphological correlates for the use of transparency in predator-prey interactions can be found among larval fishes. A wide variety of fishes are transparent during their pelagic larval and juvenile stages. Upon metamorphosis onto the two-dimensional environment of the reef, the fish lose their transparency and assume coloration that render them cryptic against the substratum. One striking example of the phenomena occurs in the post-larval or "acronurus" stage of acanthurid surgeonfishes. During this stage, which immediately precedes settlement on the reef, individuals averaging roughly 20 to 25 mm are virtually transparent except for their silvery gut tract. Upon settlement, they transform to a fully pigmented adult fish within as little as 24 hr. While the biochemical and structural mechanisms underlying this process is not understood, the resulting phenomenon suggests that transparency is a developmentally programmed cryptic adaptation and specifically during the larval and postlarval stage of these fishes. Goda and Fujii (1996) describe the biology of the transparent ice goby *Leucopsarion petersii* (Gobiidae, Perciformes).

The Confusion Effect. The confusion effect is probably widespread in varying degrees where predators seek prey from aggregations. The many animals that tighten their aggregations when threatened place more individuals within the visual field of attacking predators, thus further hindering those predators troubled by multiple targets. In refining this strategy more than most, fishes have emphasized features that make it difficult for predators to concentrate on individuals. All members of a school are about the same size and look much alike (schooling fishes are noted for lacking external differences between the sexes). As noted above, when threatened they increase not only their swimming speed but also the rate at which they dart in and out among one another. Coloration can be important too. Many schoolers are silvery and sunlight reflecting from their bodies at rapidly changing angles during evasive maneuvers presents the attackers with a bewildering shower of brilliant flashes. Many others are barred or striped, and when they swim in schools these patterns blend together as shifting maze of lines that conceal the individuals. Some fishes are barred or striped, only when they school: the goatfish *Mulloidichthys samoensis*, for example, carries a prominent yellow stripe on its side when schooling, but when it forages independently, or in small groups, the stripe is replaced by a solitary black spot. Dissimilar members of a school seem vulnerable.

The confusion effect, being a close-range phenomenon, is particularly suited to the aquatic environment, when visual ranges are short and most large predators myopic. Furthermore, the aquatic medium permits fishes their living integument and thus the highly developed and changeable colorations that so often contribute to masking individuals.

4. EPIDERMAL BREEDING TUBERCLES AND BONY CONTACT ORGANS

Breeding tubercles (pearl organs) (Figs. 3.21B, 3.22) are considered to be local keratinizations distinctly different from normal epidermis. On account of the similarity between the pearl organs

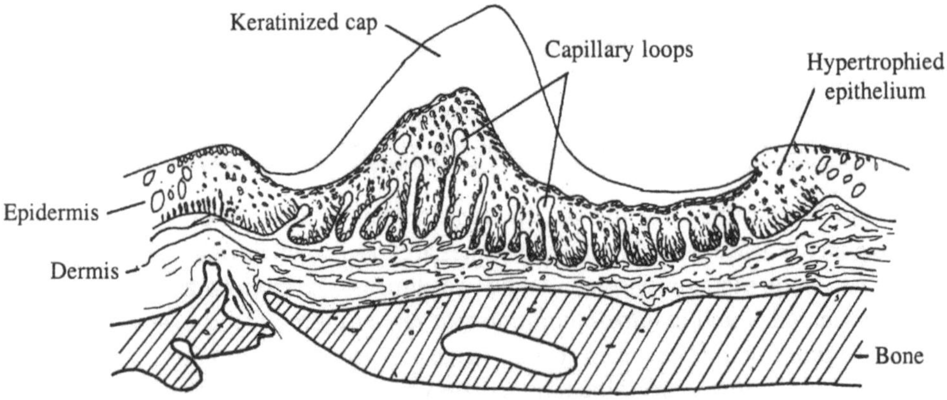

Phractolaemus ansorgei: Postorbital tubercle.

Fig. 3.22 Diagrammatic drawing of a section through a postorbital tubercle of *Phractolaemus ansorgei*. The hypertrophied epithelium supporting the keratinized cap is vascularized by blood vessels from the dermis.
***Source:* Collette (1952)**

and the differentiation of amniote skin the former may possibly be regarded as a primitive phylogenetic type of epidermal derivatives such as reptile scale, feathers and claws.

Breeding tubercles are epidermal structures which function primarily in facilitating contact between individual fishes during spawning. Tubercles are present on some species in at least 17 families of bony fishes in five orders: Salmoniformes (Salmonidae, Plecoglossidae, Osmeridae and Retropinnidae); Gonorhynchiformes (Kneriidae and Phractolaemidae); Cypriniformes (Characidae, Lebiasinidae, Paradontidae, Cyprinidae, Gyrinocheilidae, Psilorhynchidae, Catastomidae, Homalopteridae and Cobitidae); Scorpaeniformes (Cottocomephoridae), and Perciformes (Percidae).

There are three types of roughening structures ("nuptial tubercles"): tubercles consisting of aggregations of non-keratinized epidermal cells; tubercles containing substantial numbers of fully keratinized cells that are organized to form a discrete, usually conical cap; and contact organs which are bony outgrowths from fin rays or scales. In most species, breeding tubercles and contact organs are present only on the male or are much better developed on the male than on the female. Development of these roughening structures is apparently controlled by pituitary and gonadal hormones. They begin to develop before the spawning season, reach their maximum contact just before or during the spawning season, and then regress.

Breeding tubercles or contact organs develop only on freshwater or inshore marine spawning species. These structures probably originally evolved to enable breeding individuals to maintain close contact during spawning to ensure fertilization of eggs, which is particularly important in fishes that spawn in fast-moving water. Of the 24 families with roughening structures, 21 belong to groups of fishes with cycloid scales. Only three families belong to the Acanthoterygii, which usually have ctenoid scales. The structure of contact organs and ctenii is remarkably similar leading to the hypothesis that ctenoid scales may have evolved in higher fishes to permanently replace the contact organs and breeding tubercles found during the breeding season in lower fishes.

Chondrichthyan fishes (sharks, skates and chimaeras) may be able to avoid the problem of needing to maintain contact because they all practice internal fertilization. The pelvic fins of males are modified into intromittent organs-claspers or myxopterygia-for this purpose. Furthermore, both sexes of sharks are covered with rough placoid scales. The placoid scale cover is essentially absent in the Holocephali (chimaeras) but males of a number of species have a club-shaped organ (tentaculum) on the forehead, whose suggested function is to assist holding the female during copulation. Skates and rays have the scale covering greatly reduced or absent. Adult males of many species of Rajidae develop one to five rows of alar spines on the dorsal side of the outer part of each pectoral which may assist in holding the female during copulation. Some species also develop spines in the pelvic region. Pelvic fin denticles present in both sexes of *Raja ocellata* occur only in females of the smaller *R. erinacea*. Both alar and pelvic spines develop at the same time as other sexually dimorphic structure involved in reproduction, such as calcification of clasper cartilages.

Most chondrostean fishes have large rhomboid bony plates which should not be slippery. Even *Polyodon*, the paddlefish, generally considered to be naked, is not slippery but has tiny denticles in the skin.

5. BIOTOXINS: ZOOTOXINS OR ANIMAL POISONS

Marine zootoxins can be classified into 3 major groups: (1) oral biotoxins-those that are poisonous to eat; (2) parenteral biotoxins or venoms-poisons produced by specialized venom glands and injected by means of traumagenic devices (teeth or spines); and (3) crinotoxins-biotoxins that are produced by specialized poison glands. Crinotoxic organisms lack a traumagenic organ; consequently, these biotoxins are usually released directly into the water. The degree to which crinotoxic organisms are involved in human oral intoxications is not known.

The term "poisonous" may be used in the generic sense referring to both oral and parenteral poisons, but it is more commonly used in the specific sense to designate oral poisons. Thus, all venoms are poisons but not all poisons are venoms. Oral marine zootoxins are generally thought to be substances of small molecular size whereas most venoms are generally believed to have a large molecular size-a protein or a toxic substance in close association with a protein. Our knowledge of the chemistry of crinotoxins is too meagre, and the types of poisons too diverse, to permit generalizations to be made at this time.

In most instances, the chemical and pharmacological properties of these poisons are unknown.

Most marine biotoxications of vertebrate origin are caused by fishes.

Poisonous fishes. Defined as fishes, which, when ingested cause a biotoxication in humans due to a toxic substance present in the fish.

(1) *Ichthyosarcotoxic fishes*: Those fishes that contain a poison within the flesh in the broadest sense, i.e., musculature, viscera, skin or slime (mucus), which when ingested by humans will produce a biotoxication. The toxins are oral poisons believed to be substances of small molecular size that are generally not destroyed by heat or gastric juices. The various kinds of biotoxicity recognized among ichthyosarcotoxic fishes are as follows:

(i) Poisonous cyclostomes (lampreys and hagfishes) causing cyclostome poisoning;
(ii) Poisonous elasmobranchs (sharks and rays) causing elasmobranch poisoning;
(iii) Ciguatoxic fishes causing ciguatera poisoning;

(iv) Tetrodotoxic fishes causing puffer-fish poisoning;
(v) Scombrotoxic fishes causing scombroid-fish poisoning;
(vi) Hallucinogenic fishes causing hallucinatory fish-poisoning; and
(vii) Gempylotoxic fishes causing gempylid-fish poisoning.

(2) *Ichthyotoxic fishes*: Those fishes that produce a poison which is generally restricted to the gonads of the fish. The musculature and other parts of the fish are usually edible. There is a definite relationship between gonadal activity and toxin production. Fishes in this group are mainly freshwater species, but a few marine species have been incriminated.

(3) *Ichthyohaemotoxic fishes*: Those fishes having poisonous blood. The poison is usually destroyed by heat and gastric juices.

Ichthyocrinotoxic Fishes: Those fishes that produce a poison by means of glandular structures, independent of a true venom apparatus, i.e., poison glands are present but there is no traumagenic device.

Venomous or Acanthotoxic Fishes: Those fishes that produe poisons by means of glandular structures and are equipped with a traumagenic device to transmit their venoms. The poisons are parenteral toxins, usually large molecules, and are readily destroyed by heat or gastric juices. Venomous fishes are of no direct concern as oral marine intoxicants, the most serious biotoxications of public health importance being produced by ichthyosarcotoxic fishes.

Poisonous fishes number some 1,000 species but only a small number are truely venomous, i.e., have a specific apparatus for production and delivery of toxin. Other poisonous fishes contain poisons within their tissues or secretions.

Although venomous fishes are worldwide in distribution, they occur in greatest concentration and diversification within the torrid zone, and particularly within the Indo-Pacific regions.

Table 3.1 provides a list of representative venomous fishes.

As described earlier, certain fishes secrete noxious substances (crinotoxins) in the unicellular glands associated with spines. The venom glands of teleosts consist of an aggregation of holocrine cells derived from the epidermis but often sunken into the dermis, on either side of a spine, usually of dorsal or pectoral fin. There is no duct, the secretion is released when the spine penetrates the tissues of an assailant. A recent coverage is given by Zaccone et al. (2001).

According to Whitear et al. (1991a), the cytology and histochemistry of pectoral venom glands of *Heteropueustes fossiles* (Indian stinging catfish) are so distinct from those of club cells that the cell types cannot be derived the one from the other. They must be considered as secretory elements separate from the club cells of the epidermis.

The axillary glands are also of epidermal origin with holocrine secretory cells, but typically open to the exterior by a pore situated in catfish (*Ictalurus nebulosus*), between a posterior process of the cleithrum and the base of the pectoral spine. The fine structure differentiates the holocrine cells of the axillary glands from the club cells of the epidermis and from the venom glands associated with the spines of catfishes. The function of the axillary gland secretion remains unknown (Whitear et al. 1991b).

The conclusion based on the comparative study of catfish club cells with axillary glands, and the venom glands is that although they may be classified as holocrine glands, the cytology is

Table 3.1 List of representative venomous fishes (Figs. 3.23–3.29)

Fish		*Body part associated with venomous tissue*			*Mechanism of envenomation*
A.	Class: Chondrichthyes				
	Squalus acanthias (spiny dogfish)	Dorsal fin			Wounds are inflicted by two dorsal stings which are located adjacent to the anterior margin of each of the two dorsal fins
	Dasyatis dipterurus (stingray)	Tail			Venom is secreted and introduced into the body of the victim by the dorsal sting located on the tail
	Hydrolagus colliei (Pacific ratfish)	Dorsal fin			Wounds are inflicted by the single dorsal sting located on the anterior margin of the first dorsal fin
B.	Class: Osteichthyes				
	Catfishes				
	Galeichthyes felis *Clarias batrachus* *Heteropneustes fossilis* *Plotosus lineatus* *Synodontis nigrita*	Dorsal and pectoral fins			Wounds are inflicted by a single sharp, stout spine (in some with short retrorse teeth) immediately in front of the soft-rayed portion of the dorsal and pectoral fins
	Stargazers				
	Uranoscoupus scaber	Shoulder			Intoxication results from contact with the two shoulder spines
	Scorpionfishes	Dorsal	Anal	Pelvic fin spines	
	Pterois volitans	13	3	2	Wounds are inflicted with the use of their dorsal, anal and pelvic stings
	Scorpaena guttata	12	3	2	
	Synanceja horrida	13	3	2	
	Toadfishes				
	Barchatus cirrohus	2 dorsal fin spines 2 opercular spines			Wounds are generally contracted as a result of stepping on toadfishes which are generally buried in sand or mud
	Blenny				
	Meiacanthus nigrolineatus	Dentary, canine fangs			Injected into the victim through specialized fangs; injection usually results in rejection of the prey by its predator

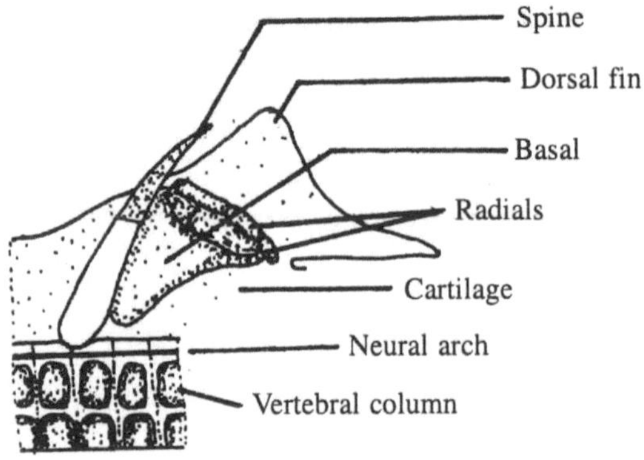

Lateral view.

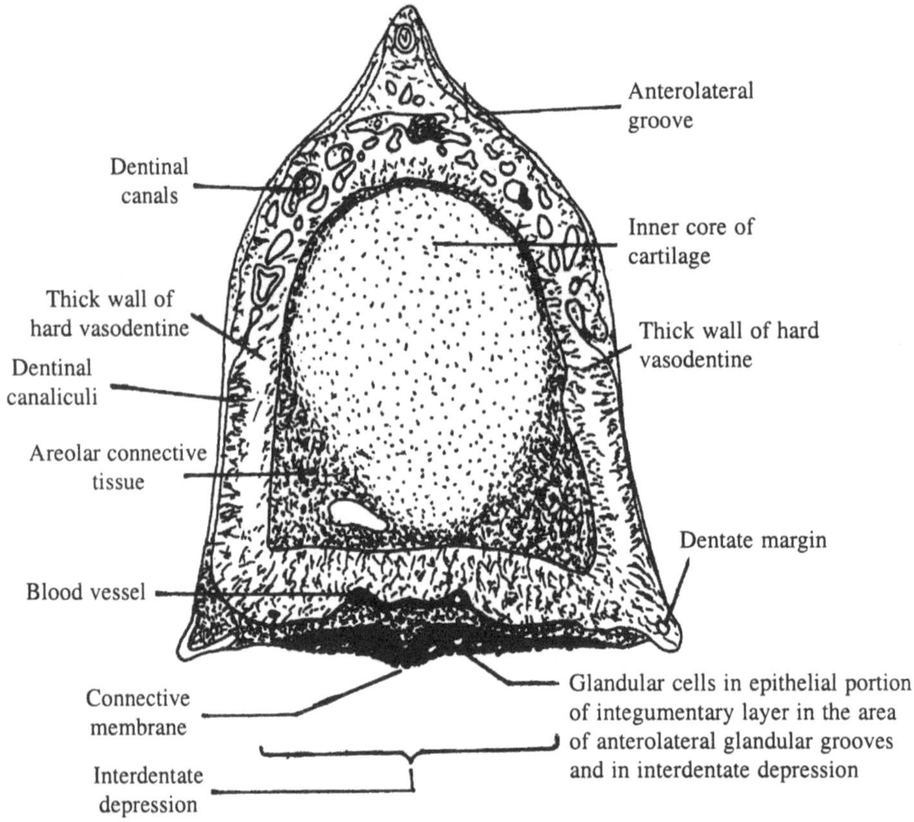

Squalus acanthias: Venom apparatus.

Fig. 3.23 Venom apparatus of *Squalus acanthias*. *Top*: Lateral view of second dorsal fin spine. *Bottom*: Cross section of anterior sting.
***Source:* Halstead (1971)**

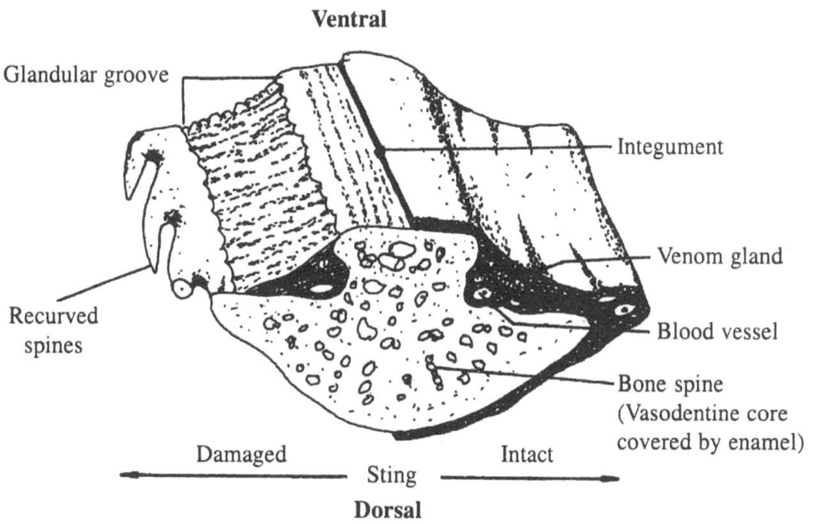

Stingray sting.

Fig. 3.24 Cross section of the venom apparatus of a stingray showing both damaged and intact stings.
***Source:* Halstead (1971)**

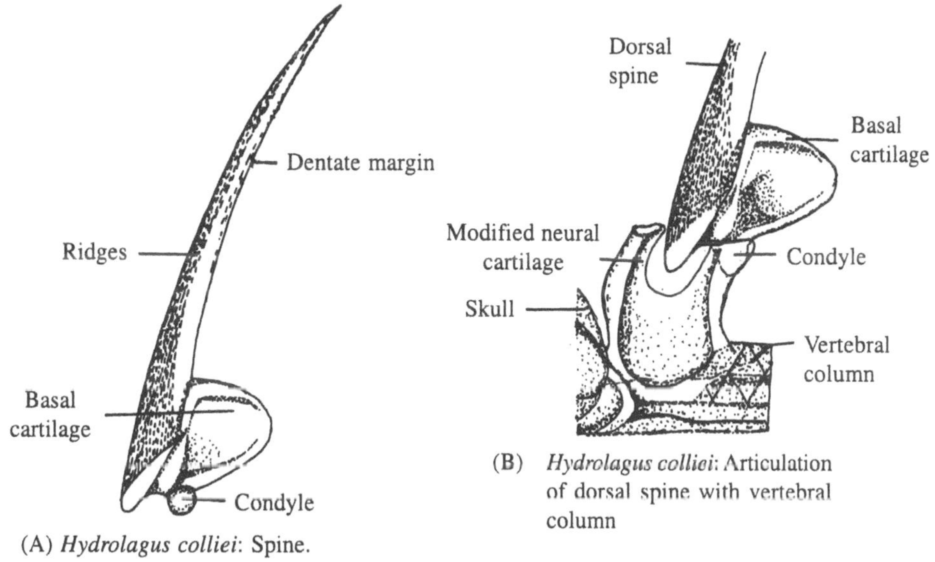

(A) *Hydrolagus colliei*: Spine.

(B) *Hydrolagus colliei*: Articulation of dorsal spine with vertebral column

Fig. 3.25 *Hydrolagus colliei.* (A) Lateral view of spine. (B) Lateral view of articulation of dorsal spine with vertebral column.
***Source:* Halstead and Bunker (1952)**

different in all three, indicating that they possess dissimilar secretory mechanisms and should be considered independent epidermal derivatives. Some toxic components may occur in more than one type (e.g., oriental catfish *Plotosus lineatus*), but this does not indicate homology at the cellular level. The morphology also suggests that the types of gland would be expected to produce different secretions with different properties and functions in the life of fish.

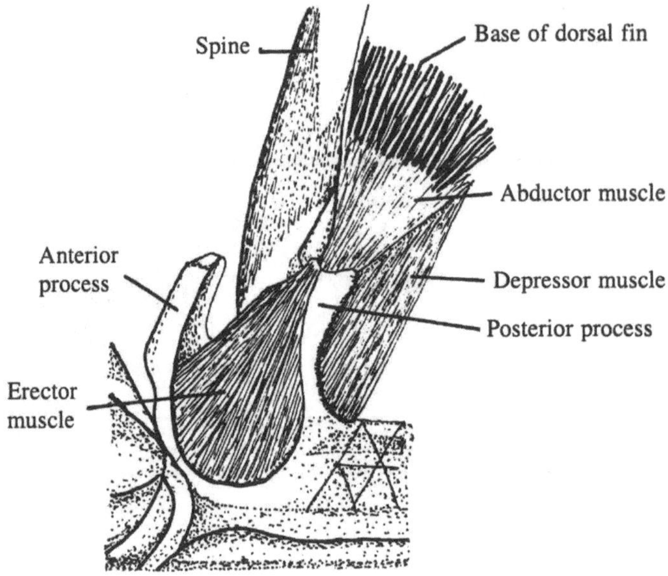

Hydrolagus colliei: Dorsal spine base with muscles.

Fig. 3.26 ***Hydrolagus colliei.*** **Lateral view of base of dorsal spine showing origins and insertions of muscles.**
Source: **Halstead and Bunker (1952)**

The buccal gland of *Meiacanthus,* as in other poison glands of fishes, has a holocrine secretory function.

6. ALARM PHEROMONES (SCHRECKSTOFF SYSTEM)

The chemical signal which transfers information between organisms is called a semiochemical. Semiochemicals which transfer information between member of the same species are called pheromones; those that are interspecific are called allelochemics. Like hormones, both types of semiochemicals are effective in minute amounts.

The club cells are the main source of a specific alarm substance acting as a pheromone ("alarm substance", or "Schreckstoff"), probably hypoxanthine-3 (N)-oxide. Although the alarm substance was first ascribed to some specialized types of epidermal club cells, the so-called "alarm substance cells", recent studies showed that the two cell types are homologous. Such pheromones are characteristic of the superorder Ostariophysi, including the order Cypriniformes and have also been found in the order Perciformes. They are responsible for the fright reaction that may reduce the effectiveness of predators against the alerted individuals, and may also serve as visual signal reading the alarm more rapidly through the social group.

The serotonin contained in the club cells may have the ability to affect the pheromonal or other possible functions of these cells.

The fright reaction represents a communication system. The loss of alarm substance cells during the reproductive period in males of Canadian fathead minnow, *Pimephales promelas*, and some other North American Cyprinidae resulted as an adaptation to the abrasive breeding behaviour. Species without such breeding behaviour retain their alarm substance cells through the year.

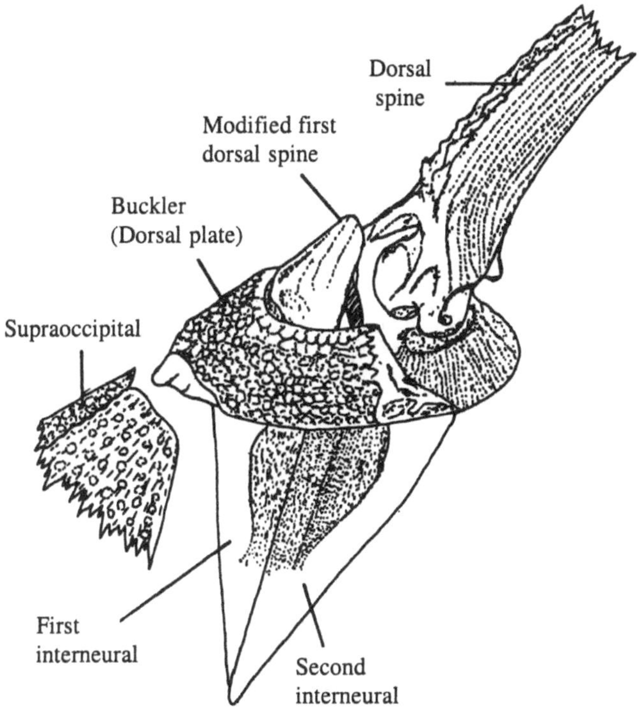

(A) *Galeichthys felis*: Locking device of the dorsal spine

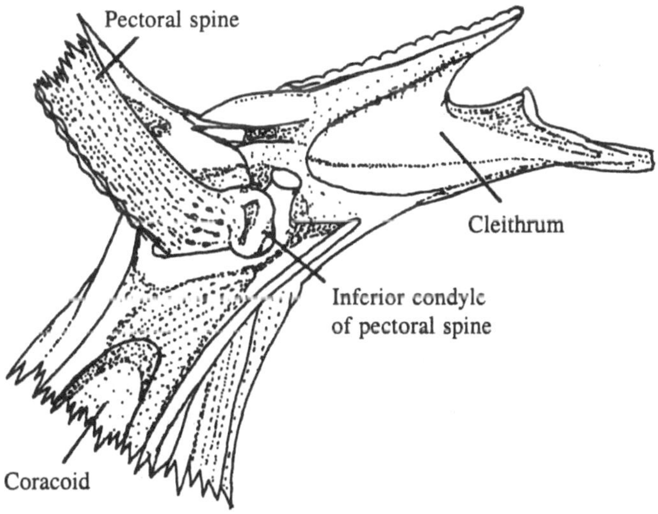

(B)

Fig. 3.27 ***Galeichthys felis*. (A) Anterolateral view of the locking device of dorsal spine showing its relationship to the supraoccipital of the skull. (B) Pectoral spine in articulation with the cleithrum.**
***Source:* Halstead et al. (1953)**

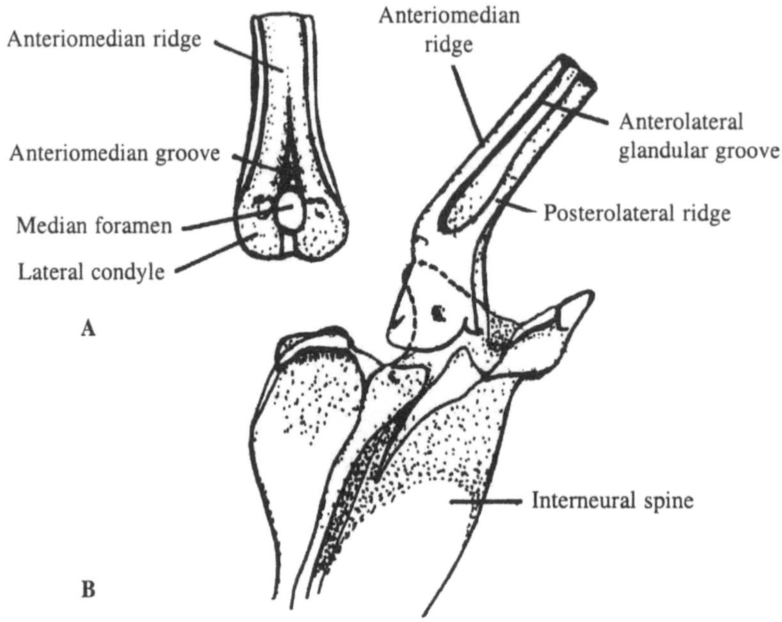

Pterois volitans: Typical dorsal spine.

Fig. 3.28 (A) *Pterois volitans* (Zebrafish). Anterior view of the proximal portion of dorsal spine.
(B) *Pteroid volitans* (Zebrafish). Lateral view showing proximal portion of dorsal spine in articulation with the interneural spine.
***Source:* Halstead et al. (1955)**

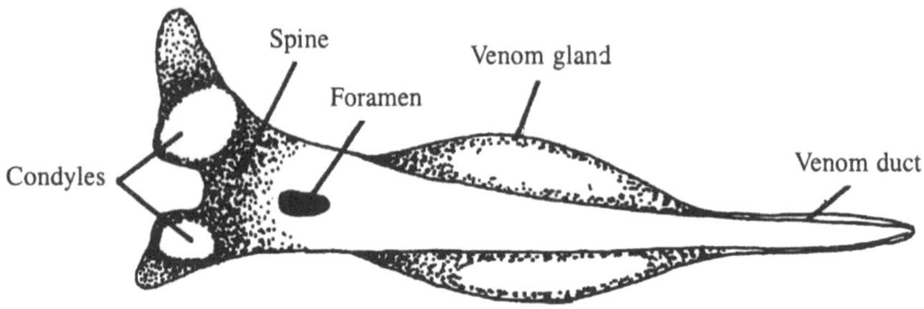

Synanceja sp.: Dorsal spine with venom glands.

Fig. 3.29 *Synanceja trachynis* (stone-fish). Drawing of posterior view of first dorsal spine, showing twin venom glands. The integumentary sheath has been removed.
***Source:* Endean (1961)**

It has been indicated that the mechanism controlling the formation of alarm substance is not uniform in the Ostariophysi and there exists atleast two such mechanisms, one of which is operated by androgens and perhaps other steroid hormones of the gonad.

The fright reaction is mediated by olfaction. Fright reaction vary considerably among species. Responses within a species are highly variable. Alarm substance cells develop before the young fish become responsive to the alarm substance. The reader is referred to: (Smith 1982, 1986, 1992 and Mathis and Smith 1992).

CHAPTER 4

Minerals and Tissues: Bone and Cartilage; An Overview of Endoskeleton

1. INTRODUCTION

The origins of vertebrate calcified tissues have aroused considerable interest. During evolution, skeletons appear with the multiplication of species. Four general evolutionary trends in the vertebrate skeleton have been identified: (i) decrease in the range of skeletal types; (ii) decrease in the amount of ossification within the skeleton; (iii) decrease in the number of bones (i.e., ossification centres), and (iv) decrease in the topographic extent of bones (e.g., dermal armour). None of these trends involves increased specialization of skeletal tissue (see Maisey 1988 for further information).

As its most elemental level, the vertebrate skeleton comprise a superficial dermal component and a deeper lying endoskeleton. The neural crest undoubtedly plays a vital role in forming components of the vertebrate skeleton. As a consequence of various skeletogenetic processes, the vertebrate skeleton is formed at various sites within the body. Ontogenetic growth of the dermal skeleton is accomplished by periodic additions or replacements, fusions, and continuous enlargements (="continuous" growth, which may, however, be periodic). Interstitial expansion of embryonic primary cartilage allows rapid growth of the endoskeleton; this ability becomes somewhat limited with the onset of ossification.

Vertebrate cartilage is endoskeletal, apart from adventitious cartilage in some birds and mammals, and is derived either from mesoderm or ectomesenchyme. Mineralization can occur within and upon endoskeletal matrix. In Chondrichthyes calcification is mainly confined to superficial regions (although it is not strictly perichondral). In osteichthyans, bone deposition may be perichondral or endochondral. Perichondral ossification is more widespread than endochondral among lower vertebrates.

Bone occurs in the endoskeleton (including membrane bones), in ossified tendons, and in the dermal skeleton. It has been characterized on the basis of relative cellularity, collagen fibre orientation, degree of vascularization, texture, and various other criteria such as resorption phenomena or density of matrix. Bone may be cellular or acellular, and there has been much debate over which condition is phylogenetically the more primitive.

Dentines and "enamel" (including lower vertebrate enameloids) are characteristic of the dermal skeleton only. The process of skeletal development and mineralization are evidently interdependent and cannot be treated in isolation. Equally clearly, however, a complex internal skeleton can develop in the absence of calcification (e.g., the cartilaginous endoskeleton of cyclostomes). Among recent vertebrates, calcification of the dermal skeleton is restricted to gnathostomes (note: horny, uncalcified teeth of cyclostomes are part of a dermal skeleton).

Skeletal mineralization of osteichthyan and chondrichthyan fishes is fundamentally similar, and involves the hormonally controlled deposition by special scleroblastic cells of an inorganic calcium-rich matrix upon a fibroblastically generated collagenous framework. Beyond this histochemical similarity, however, there is little resemblance between the mineralization patterns of these groups.

It has been speculated that the evolution of the skeleton was linked to the requirements for a certain number of factors: (i) the ability to float and balance, which relates to the distribution of muscle masses around the centre of gravity; (ii) locomotion; (iii) feeding; and (iv) protection. In all of the evolution there appears to have been an attempt to compromise between the requirement for these different functions.

2. EVOLUTION, ENVIRONMENT, MINERALS AND TISSUES

From the Cambrian (the first Palaeozoic period) onwards, mineralized skeletons have played an important part in evolution (see Table 4.1).

Bone-body fluid continuum: The bone-body fluid continuum is a closed-cycle or *servo* system for retention and rapid turnover of calcium, phosphate, sodium, magnesium, carbonate, hydronium, and other ions. The storage unit is the apatite microcrystal. The transport unit is the extracellular and plasma water. The system as a whole functions to maintain ionic independence for the organism, to sustain the blood concentrations of essential ions, and to conserve or recycle critical elements like phosphorus at a rate sufficient to meet the requirements of fast-moving, large vertebrates. In a boneless vertebrate, the organism is continuously dependent upon the sea for ion supply, transport, and homeostasis as if it were a part of the structure of its own body. A scarce element such as phosphorus exists in sea water in concentrations less than one part in 10 million, yet, indirectly, it determines the availability of nitrate and oxygen to the cells. In order to maintain a blood level of 1.0 mM of inorganic phosphorus per litre, lower vertebrates must turn over enormous volumes of water and ingest plants or other animals. Higher vertebrate having bone tissue in the skeleton store large amounts of phosphorus and other essential elements and utilize the available supply with maximum efficiency. This is an example of the physiological and fundamental function of the skeleton.

Taking first the serum of the lamprey *Petromyzon marinus*, it has been reported that the total ion concentration is significantly low during larval stage. After metamorphosis, Na^+, K^+, and Cl^-, and total serum ion concentrations increased 22 per cent to the levels found in parasitic and full grown spawning adults. The calcium ion concentration during the parasitic phase, when the lamprey is attached to a teleost fish, appears to be lower than in independent free-swimming forms.

The larval lamprey is hypoproteinemic relative to metamorphosed downstream parasitic and migrating adults. Both major components are involved, with the globulins being especially low.

Table 4.1 Occurrence of the skeleton during the course of evolution

Period	Epoch	Distinctive records of life	Skeleton
		CENOZOIC ERA	
Quaternary	Recent	Modern man	
	Pleistocene	Early man	
	Pliocene	Large carnivores	
	Miocene	Whales, grazing forms	Bone cells
Tertiary	Oligocene	Large browsing mammals	
	Eocene		
	Palaeocene	First placental mammals	
		MESOZOIC ERA	
Cretaceous			
Jurassic		First small mammals	Bone cells
Triassic			
		PALAEOZOIC ERA	
Permian		Reptiles evolved	
Carboniferous		Amphibians	Bone cells
Devonian		Teleost fishes	Acellular and cellular bone
Silurian		Early land animals	
		Marine Elasmobranchii	Calcified cartilage
Ordovician		First primitive fishes	
		Protovertebrates	Cartilage
Cambrian		Marine invertebrates	Conective tissues
		PRE-CAMBRIAN TIME	
		Animals with soft tissues	

(Modified from Istin 1975).

Because the lamprey is able to maintain Na^+, K^+, $C1^-$, and total ion and protein concentrations in the fluids of the body in a markedly hypotonic environment, without any calcified tissues in which to store essential inorganic elements, it is assumed that its gill membranes function as an *ion pump* to concentrate and extract ions from large volumes of water. It has been observed that the proportion of extracellular-to-intracellular water is high in the cyclostomes and low in the bony vertebrates.

Elasmobranchs present the most primitive kind of physiological function of skeletal system. The exoskeleton consists of teeth and scales of dentine and enameloid, and the endoskeleton contains large amounts of calcified cartilage. In both exoskeleton and endoskeleton, growth occurs by a one-way process of deposition of new tissue. Remodelling, which involves both resorption of old and deposition of new and which is the basis of the growth process in higher vertebrates, seems not to have evolved in elasmobranchs. A simple physical-chemical equilibrium exists between the apatite of the elasmobranch exoskeleton and endoskeleton and the ionic composition of the fluids of the body. This differs from the complex dynamic equilibrium seen in bony vertebrates which brings osteoclastic and other cellular activity into play to resorb and mobilize the elements stores in mineral and matrix. There are no chondroclasts in elasmobranchs to resorb calcified cartilage; once deposited, cartilage remains for the duration of the life of the individual. Teeth and scales are exfoliated and shed during growth, or when worn out, and immediately are replaced with new structures produced by the integument.

In lower vertebrates having only cartilage and no bone tissue in the skeleton, calcium and phosphorus metabolism is regulated by the skin, gills, membranes, gut, and kidneys. In higher vertebrates, vitamin D-parathyroid hormone complex is added to control the rate of the intracellular chemical processes that are involved.

Herman-Erlee and Flick (1989) have described a comparative account on endocrine involvement in bone metabolism.

Bone is a vertebrate invention. Vertebrate calcified tissues evolved in marine life forms. Seawater is and probably always has been an environment rich in calcium and poor in phosphate. Calcium homeostasis is a prerequisite for life in a marine environment: in seawater calcium ions will enter the body and its cells by diffusion and Ca^{2+} levels may become toxic in the absence of (energized) Ca^{2+} extrusion systems. It is thought that the process of ATP-driven Ca^{2+} extrusion originally yielded calcium phosphate as a by-product; deposition of calcium phosphate in the skin (formation of bone plates) is an effective way to remove Ca^{2+} from the body fluids. At the same time, however, calcium phosphate deposits provide an internal phosphate buffer that makes an animal largely independent of seasonally fluctuating phosphate availability (as is the case in seawater environment).

The earliest fossil vertebrates, the ostracoderms, had skeletons which included the hard tissues: calcified cartilage, bone, dentine and enamel. It seems plausible that the modes of mineralization pioneered in these primitive tissues have persisted throughout vertebrate radiation without major changes. All four types utilize the calcium phosphate mineral apatitite to form inorganic crystallites which develop in close association with matrix proteins. In the collagen-associated tissues-calcified cartilage, bone and dentine-the short, needle-or lath-like crystallites develop along collagen fibrils. Non-collagenous components of the matrix in these tissues include proteoglycans, glycoproteins and matrix vesicles. Special calcium-binding or phosphate-binding constituents may promote crystal nucleation and growth. In enamel (enameloid) large hexagonal crystallites develop in association with glycoproteins amelogenin and enamelin secreted by cell of the inner dental epithelium (ameloblasts). Enamel (enameloid) of the shark tooth resembles mammalian enamel in size and shape of crystallites (Kemp 1984).

The intensity of mineral impregnation of the protein matrix varies greatly from one species to another. Cyclostomes lack a calcified skeleton. (read Wright et al. 1998 for three cartilage types in *Myxine glutinosa*). Chondrichthyans possess skeleton of calcified cartilage. The shark teeth and denticles are covered with a true ectodermal enamel, although the outer layer of these structures is structurally complex, consisting of both calcified ectodermal enamel and uncalcified areas of ectomesenchymal origin. The intradermal base of these structures most probably consists of acellular bone. The structural array of apatite crystallites in the teeth seems strongly correlated with the specific shape and function of individual teeth.

The calcified cartilage of sharks differs significantly from that of other vertebrates, not only in composition, but in the fact that the areas of calcification are composed of many vital and non-hypertrophic cells.

In chondrichthyans calcification is mainly confined to superficial regions (although it is not strictly perichondral). Teleosts have either acellular or cellular bones. In the most highly evolved forms of bony fish, it is probable that a reduction in ossification has occurred in order to adapt to the requirements for floating and rigidity.

The vertebrate skeleton is usually considered to be a living tissue containing a certain number of a well-defined cell types. In the vertebrates, unlike invertebrates, the bone tissue is diffuse. Bony fishes, in contrast to tetrapods, are characterized by a possession of a wide range of skeletal tissues, some of which cannot be readily classified as typical bone or cartilage (chondroid bone). Teleost bone has inspired interest in its role in calcium homeostasis and its possible role as a tool in age or growth studies. Many fishes have bony armour (e.g., boxfishes Ostraciontidae).

The bones of Teleostei may be classified into two groups: cellular and acellular, cellular and acellular bone matrixes are similar (including collagen fibre orientation). The only difference between these groups is the complete absence of enclosed osteocytes in the latter case. This type of cell is generally associated with the destruction of bone tissue in other vertebrates. Acellular tissue is only seen in the most highly evolved forms of Teleostei, which are most numerous, widespread and diversified. Eel, herring and salmonids have cellular while *Fundulus* and *Tilapia* have acellular bones. In the bone-fish (*Albula vulpes*), the skeleton contains both types of bone. However, since scales are generally accepted to represent an acellular type of bone, actually all species with cellular bone possesses both cellular and acellular bone compartments. Fish with acellular and cellular bone are equally distributed in freshwater and seawater environments. This indicates that the type of bone is not related to calcium availability in the aqueous environment. The latter conclusion is further supported by the essentially similar compositions of acellular and cellular bone from either freshwater or seawater fish. Apparently, modern fish succeeded in attaining precise regulation of plasma ion concentrations by endocrine control of ion exchange mechanisms in the branchial (calcium) and intestinal (phosphate) epithelia separating the "milieu interieur" from the "milieu exterieur", thereby realizing constant condition for bone mineral deposition. There is general consensus that the acellular bone of the advanced teleosts has evolved from a cellular type of bone, as found in the extant older order of fish. In teleost fish enlargement of the bony skeletal parts occurs primarily by appositional ossification. Such bones lack marrow cavities. The bones may continue to grow in length and width as long as the fish live.

Cellular bones show numerous histological inter-specific and intra-specific differences. It is believed that cellularity and acellularity can only be correlated at the phylogenetic level.

In osteichthyans, bone deposition may be perichondral (where it can overlie globular calcified cartilage) or endochondral (involving resorption of primary cartilage and its replacement by bone). Perichondral ossification is more widespread than endochondral among lower vertebrates.

Although bones appeared for the first time during the evolution of fish, the physiology of their bone tissue is still but little known. The vertebral column of an eel, *Anguilla anguilla*, for instance, is formed of compact of spongy, lamellar, cellular bone. These animals possess the three categories of cell characteristic of the bones of the upper vertebrates: osteoblasts, osteocytes and osteoclasts. The osteoblasts contribute to the formation of bone tissue, whereas the osteoclasts and osteocytes play a part in its destruction. The eel bone is in many respects very similar to that of mammals.

In fish, calcitonin, hypocalcin, prolactin, estrogen and vitamin D_3 are considered in bone metabolism (for details, see Hermann-Erlee and Flik 1989; read Flik 1990 for calcium homeostasis in fish).

Fish may continue to grow throughout their lives, during periods of increased need for minerals such as during egg production or during periods of restricted mineral availability (migration to soft water), fish may recruit calcium and phosphate from their bones and scales.

Bone resorption in fish may involve two types of demineralization: cell-mediated and non-cell-mediated demineralization. Cell-mediated bone resorption as observed in eels is essentially similar to the process observed in other vertebrates. The presence of multinucleated osteoclasts with clear zones and ruffed borders has been demonstrated in eels and catfish. For *Tilapia macrocephala*, which have acellular bone, as well as for goldfish, which have cellular bone, evidence has been given for mononuclear resorptive cells. However, the contribution of the resorptive cells to bone demineralization in acellular-boned fish seems of minor physiological importance. In both cellular and acellular bone of fish, a phenomenon known as *halastasis* occurs. In this case demineralization of matrix occurs without matrix destruction or removal. This type of demineralization, which may be a reaction to changes in plasma composition, seems of greater importance for acellular-boned fish than is cell-mediated bone resorption. It is indicated that mineral pool in scale is more readily available than the mineral in the other bones. Scale removal may result in a significant decreases in the calcium and phosphate content of skeletal bone, indicating that this bone compartment in fish is not refractory to demineralization but may be mobilized to allow regeneration of scales.

For remodelling and mineralization in fish, the reader can find interesting information in the contribution of Hermann-Erlee and Flik (1989).

It has been pointed out that both in fish with cellular bones and in fish with acellular bones, resorption of skeletal bone occurs after removal of scales (which imposes an extreme need for calcium and phosphate to mineralize newly formed scales), indicating that enclosed osteocytes do not necessarily play a major role in bone resorption. Also, no multinucleated osteoclasts were seen at the resorption sites; the process of bone resorption is attributed to resorptive cells which lack multinucleated and a ruffed border but are able to remove ground substance from the osseous zone of the bone. These cells contain a swollen granular endoplasmic reticulum, indicating that they are actively synthesizing proteins (collagenase?).

Meunier and Huysseune (1992; see literature) have thoroughly discussed the concept of bone tissue in Osteichthyes. Bone typically is made up of three components: (1) bone cells: the osteogenic cells or osteoblasts, the trophic cells or osteocytes, and the clastic cells or osteoclasts; (2) extracellular organic material or organic matrix, i.e., the prominent network of collagenous fibres and proteoglycans; (3) extracellular mineral material essentially crystals of hydroxyapatite. Moreover, bone is frequently vascularized and is submitted to resorption and reconstruction processes, i.e., remodelling (see Witten et al. 2000).

Witten (1997) reported that osteoclasts in the cichlid *Oreochromis niloticus* occur in two forms: mononucleated single cells and cell aggregations. Osteoclasts in *O. niloticus* are not multinucleated; however, during intense bone resorption, they form cell aggregations that resemble multinucleated giant cells in mammals. Conversely, during less intense bone degradation, osteoclasts are flat, have long narrow cytoplasmic processes and resemble the bone-lining cells of mammals. All bone-resorbing cells in *O. niloticus* are mononucleated and lack a ruffled border.

Generally, fish continue growing throughout life and require bone resorption throughout development. Consequently, bone resorption by flat mononucleated osteoclasts presumably also takes place in those fish species in which histological observations have failed to identify multinucleated osteoclasts.

Witten and Villwock (1997) further reported that bone resorption in *O. niloticus* takes places during growth at particular elements of the endoskeleton. Possible reasons have been discussed for the different results in respect to two questions: (a) are osteoclasts in fish with acellular bone mono- or multinucleated and (b) what are the experimental or environmental conditions for the presence of osteoclasts (see Table 4.2A, B). In fish with cellular bone as well as in Chondrichthyes (with bone rudiments), the participation of multinucleated osteoclasts in bone resorption is still a matter of debate (see Sire et al. 1990; Witten et al. 1999a, b, 2000).

Table 4.2A Investigation results with evidence of osteoclast occurrence in teleostean fish with acellular bone. All studies performed with members of the order of Perciformes

	A. Osteoclasts		B. Osteoclasts occur				Methods
Authors	Mono-nucleated	Multi-nucleated	At normal environ-mental conditions	At calcium deficiency conditions	In the case of bone implan-tation and calcium deficiency	In the case of bone implan-tation	
1 Weiss and Watabe (1979)							Light microscopy TEM
2 Glowacki et al. (1986)							Light microscopy, TEM
3 Sire and Huysseune (1993)							Light microscopy, TEM
4 Witten and Villwock, (1997)							Light microscopy, TEM, enzyme histo-chemistry

(Modified from Witten and Villwock 1997; see literature)

In the Osteichthyes, we can find bone with these typical basal components. However, the histological features of the bony tissues can vary according to the species and the bones considered, sometimes according to a particular part of bone.

Although the three basic components of bone tissue can be present (i.e., osteocytes, an organic matrix, and a mineral phase), it has long been known that bony tissue in Osteichthyes can lack trapped cells and/or mineral. This phenomenon has blurred the classical distinction between the generally adopted categories of connective tissues in a way that the osteichthyan skeleton should be described rather in terms of a continuum of structures.

Acellularization process: Osteocytes are the typical bone cells which progressively become entrapped in the bone matrix as the osteoblasts synthesize new bone. Osteocytes are star-shaped, with cytoplasmic processes located within canaliculi in the bone matrix. In Osteichthyes the osteocytes show various features, especially related to the number and the length of the cytoplasmic

Table 4.2B The features of bone-absorbing cells of an advance teleost species, *Oreochromis niloticus* are compared with known attributes of mammalian osteoclasts.

Features of osteoclasts of juvinile *O. niliticus*	Known attributes of mammalian osteoclasts
Histochemical features	
Binding of antibodies directed against V-ATPase	Binding of antibodies directed against V-ATPase
Positive staining for TRAP	Positive staining for TRAP
Positive staining for TraATPase	Positive staining for TraATPase
Negative stainng for ALP	Negative staining for ALP
Histological features	
Cells located inside of Howship's lacunae	Cells located inside of Howship's lacunae
Parimarily mononucleated cells	*Primarily multinuclead giant cells*
Ontogenetic features	
Present at the endosteal bone surfaces of growing bones	Presents at the endosteal bones surfaces of growing bones
Present at sites of tooth eruption	Present at sites of tooth eruption

Features that differ between both groups are in italics
Source: Witten et al. (1999b; see details)

processes. For example, in *Cyprinus carpio* they are numerous, long and regularly distributed on the cell surface whereas in Thunnidae they are scarce and inserted only in diametrically opposite positions. Further, it has been shown that bone lacks osteocytes in numerous fishes (e.g., *Trachurus trachurus* of Carangidae).

The histological features of such bone tissue are very similar to those of cellular bone, except that the osteocytes are missing. Such bone can show primary vascular bone as well as secondary bone, delimited by cement lines, and a varying spatial organization of the collagenous network associated with remodelling processes. In some cases, osteoblastic processes cross the bone matrix.

Bone deprived of osteocytes, including bone with osteoblastic processes, is called acellular bone (anosteocytic bone). The entire skeleton is made up either of cellular or of acellular bone. Within the Teleostei, cellular bone is found in lower groups, whereas acellular bone has developed especially in advanced species. Nevertheless, there are two important exceptions: (1) the Protacanthopterygii (generally considered as relatively primitive fishes) have acellular bone, except the Salmonidae; (2) within the Acanthopterygii (the more evolved teleosts), the Thunnidae have cellular bone. The presence of cellular bone is generally considered the consequence of these animals' high metabolic activity. All the living non-teleost fishes have cellular bone: Coelacanthidae, Dipnoi, Lepisosteidae, Amiidae, Polypteridae and Chondrostei.

Neverthless, this scheme is somewhat oversimplified because it does not take into account the integumental skeleton, i.e., the scales. In teleostean scales, the superficial osseous layer and the deep basal plate are both partly, if not completely, derived from typically bone tissues. A detailed study of the histological feature of the entire skeleton in the various osteichthyan lineages (endoskeleton on the one hand, exoskeleton including scales and fin rays on other hand) indicates that the process of acellularization is a more general feature in scales than in bone *sensu stricto*. In this way, some relatively primitive lineages, such as the Ostariophysi, show a more or less advanced stage of acellularization in the scales, whereas the bone has osteocytes.

As a conclusion it can be stated that acellularization is a common process in the Osteichthyes (Fig. 4.1). Acellularization is a general trend within the various osteichthyan lineages and especially in Teleostei, where it appears to be a heterochronic phenomenon. Within the Teleostei, acellularization is a more general feature in scales than in normal bone: e.g., Salmonidae and Cyprinidae have cellular bone and acellular scales; within the Ostariophysi, Characiformes are known with cells (elasmocytes) included in the basal plate of their scales (Erythrynidae, Anostomidae) whereas Characidae have acellular scales. Moreover, in the Teleostei, the more evolved the taxa are (with the exception of the Thunnidae), the more developed is the acellularization process.

It has been claimed that acellular bone in the Osteichthyes is an apomorphic condition, since it has developed especially in the higher Teleostei; the plesiomorphic condition is cellular bone, which has been maintained in Tetrapods.

Loss of mineralization: In general, the degree of mineralization of the bony tissue in Osteichthyes is similar to that of mammalian bone. Yet the extreme values obtained for Osteichthyes (0.70 to 1.60 g/cm^3) lie wider apart than for mammalian bone (1.15 to 1.50 g/cm^3). The degree of mineralization of bone in the Teleostei may be linked to its histological type (cellular or acellular) and to the nature of the aquatic environment (freshwater or marine) but confirmation is needed.

Some peculiar skeletal tissues that are homologous with true bone have lost the possibility of mineralization, e.g., the posterior area of the campototrichia in Dipnoi or the basal plate of the elasmoid scales. In the latter case, the structure of the tissues shows typical features of what has been described as isopedine. Some of the most important characteristics of isopedine, besides the spatial organization of the collagenous network, are the more or less extensive unmineralized areas and the particular way in which the mineralization front in the basal plate progresses, usually with the presence of Mandl's corpuscles.

Furthermore, in certain areas these unmineralized tissues have lost their osteocytes: in the camptotrichia as well as in the isopedine. The authors end up with cases of skeletal tissues having a similar origin and/or function to bone but lacking both mineral and osteocytes (Fig. 4.1). In view of these findings, it is difficult to adhere to the currently adopted concept of bone tissue as a collection of specialized cells, some of which produce (and then become embedded in) a network of collagenous fibres supporting a solid phase of minerals.

Two further examples illustrate the very close relationship between bone and unmineralized (possibly acellular) isopedine. In the scales of the coelacanth, the mineral component of the isopedine is missing, but in the lateral line scales, the basal plate has vascular canals as in bone and remodelling areas where typical bone lies close to unmineralized isopedine. In the same species, the anterior bony part of the post-temporal merges relatively imperceptibly (without clear limits such as cement lines) with the posterior isopedine part. Moreover, it seems to be the same scleroblast population which gives rise to cellular mineralized bone and the cellular unmineralized isopedine.

Teleost chondroid bone: Another type of skeletal tissue which balances at the limit of classical definitions, and which is frequently found in articular areas in the head of teleosts, is chondroid bone. Mature chondroid bone is a skeletal tissue with large, haphazardly dispersed chondrocyte-like cells embedded in a mineralized bone-like matrix (Fig. 4.2).

Chondroid bone forms *de novo* from osteogenic precursors, not by transformation of an existing tissue, implying that there is no metaplasia involved.

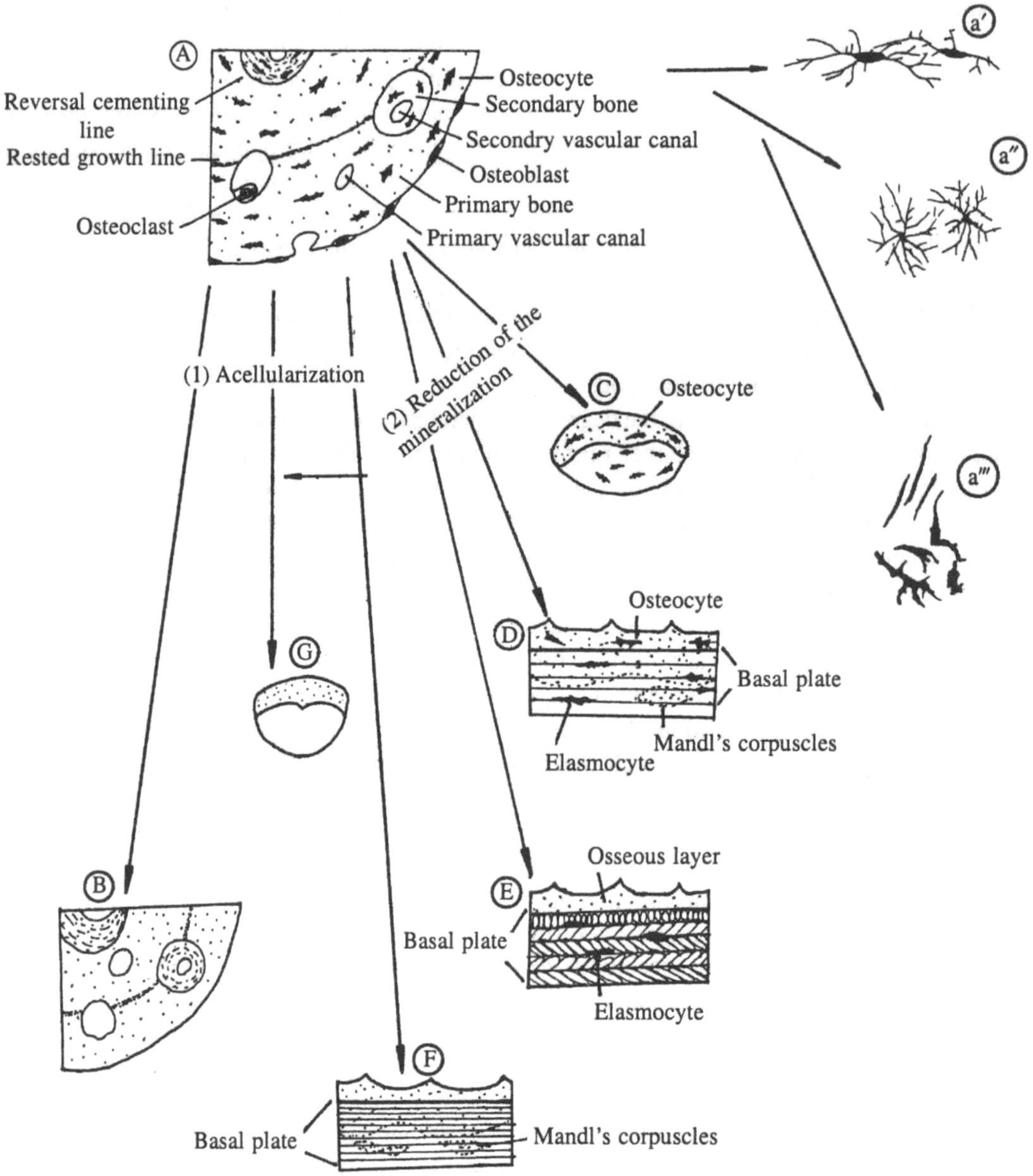

Fig. 4.1 **Recapitulative scheme to show the two main evolutionary trends of bony tissues in Osteichthyes: acellularization (B, E to G) and more or less important lack of mineralization (C to G). A: Compact, cellular, vascularized and pseudo-lamellar bone with localized remodelling: *Latimeria chalumnae* (Crossopterygii) with three types of osteocytes: a′ *Acipenser*, a″ *Anguilla*, a‴ *Thunnus* with a reduced number of cytoplasmic processes; B: Compact, acellular, vascularized with remodelling: *Lethrinus nebulosus* (Teleostei, Acanthopterygii, Perciformes); C: Cellular osseous and cellular unmineralized permanent 'preosseous' tissues; campotrichia of *Neoceratodus forsteri* (Dipnoi, Neoceratodidae): D: Cellular bone and partly unmineralized cellular isopedine: scale of *Amia calva*: E: Acellular bone and unmineralized cellular isopedine: scale of *Latimeria chalumnae* and *Neoceratodus forsteri*: F: Acellular bone and partly unmineralized acellular isopedine: scale of *Hemichromis bimaculatus* (Teleostei, Acanthopterygii, Perciformes); G: Acellular bone and unmineralized acellular 'preosseus' tissue: campotrichia of *Protopterus annectens* (Dipnoi, Lepidosirenidae).**
Source: **Meunier and Huysseune (1992)**

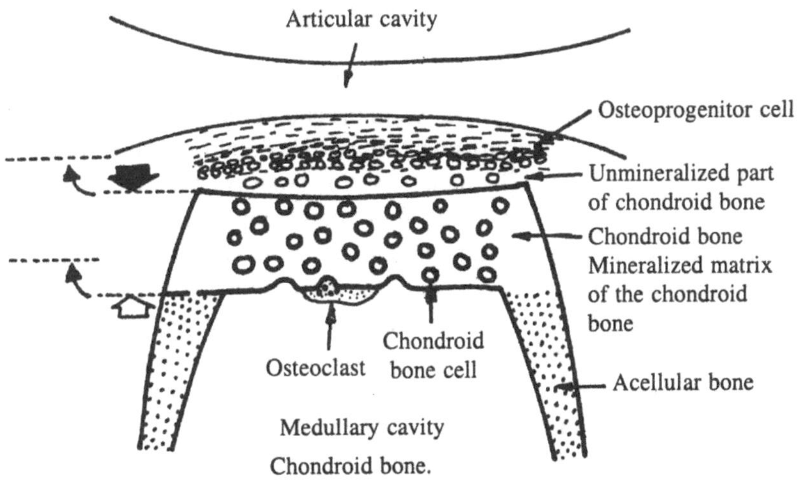

Fig. 4.2 Schematic drawing of the structure and localization of chondroid bone on an acellular bony element. Open and solid arrow respectively indicate the direction and resorption and deposition. Dotted lines and small arrows indicate the shift in position when chondroid bone grows distally while being broken down proximally.
***Source:* Meunier and Huysseune (1992)**

Chondroid bone can be anything between almost acellular (e.g., on the maxillary of rainbow trout) and highly cellular (e.g., on cichlid pharyngeal jaws), and therefore the resemblance to cartilage can be less or more pronounced. In its highly cellular form, chondroid bone resembles (mineralized) secondary cartilage, but is distinguished from it by its fine structure, its histochemical properties, as well as by immunofluorescence and microradiographical data.

Chondroid bone in Teleostei is found in joints and sites subjected to mechanical stress and seems to meet two demands: (1) demands of movement, requiring the adjustment of the shape of the bone (formation of an articular process, apophysis) which probably can be rapidly achieved by the incorporation of cells, and (2) demands of support and resistance (presumably to shear stress); the latter may explain the particular matrix properties and the chondrocytic phenotype of the cells.

Meunier and Huysseune (1992) disagree with the view that chondroid bone forms as the outcome of incomplete differentiation by lack of appropriate stimuli. Rather, their developmental, morphological and histochemical data indicate chondroid bone is a fully differentiated and functional tissue.

Huysseune and Sire (1990) have given transmission electron microscopical observation on the chondroid bone. Chondroid bone, a skeletal tissue morphologically intermediate between cartilage and bone, is composed of a dense mineralized collagenous matrix, resembling that woven-fibred bone, and large chondrocyte-like cells. The latter vary considerably in their morphological features (functional cells, cells containing a large vacuole and degenerating cells). The chondroid bone is mineralized except for its upper layer. Mineralization is initiated in matrix vesicles. Clusters of apatite crystals coalesce at the mineralization front. Distally, the tissue grows by incorporation of cells which exhibit the features of osteoblasts, and which derive from less differentiated fibroblast-like cells located in the outermost layer of the tissue. Proximally, the chondroid bone is subjected to erosion by multinucleated elastic cells. The deposition of bone against the wall of lacunae which have been opened by clastic resorption may suggest a possible active involvement of chondroid bone cells.

The carbohydrate histochemistry of mature chondroid bone in cichlid *Astatotilapia elegans* has been given by Huysseune and Verraes (1990). In the chondroid bone, the narrow unmineralized layer contains sulphated proteoglycans (amongst which are chondroitin-4-sulphate or chondroitin-6-sulphate) and little glycoproteins. The mineralization front is characterized by an enrichment with neutral carbohydrates, probably in the form of glycoproteins. The mineralized matrix contains glycoproteins and a significant amount of carboxylated in addition to sulphated proteoglycans. Possibly a relation exists between the increasing amount of acid mucosubstances during ontogeny and the arrival of unmineralized zone in the mature chondroid bone. Since the studies on the histochemical or biochemical composition of piscine skeletal tissues are quite scarce, further studies are needed to acquire a more complete picture of the constituents that compose the chondroid bone and other piscine skeletal tissues.

The reader is further referred to contributions on chondroid bone (see Huysseune 1986; Huysseune and Verraes 1986; Huysseune et al. 1986 cf. Benjamin 1989).

Sire et al. (1990) have published observations on the osteoclasts which is a common occurrence during normal and experimental bone resorption in a number of teleosts fishes. Osteoclasts acting along the bone surface usually lie in a Howship's lacuna whereas others are wrapped around bone extremities. Electron-microscopical observations reveal that teleost osteoclasts show features similar to those of higher vertebrate osteoclasts, e.g., the presence of a ruffed border and the occurrence of numerous vacuoles, lysosomes and mitochondria. The multinucleated aspect that characterizes osteoclasts in other vertebrate groups is not a distinct feature of teleost osteoclasts since some are possibly mononucleated. Teleost osteoclasts are also able to resorb uncalcified tissues adjoining bone resorption areas, either as a primary process directed toward the tissue (basal plate of elasmoid scale) or as a secondary phenomenon (cartilage).

It may be mentioned that osteoblasts are present in different species of teleosts phylogenetically as distant as *Salmo fario* (Protacanthopterygii), *Eigenmannia virescens* (Ostariophysi), and cichlids (Acanthopterygii). The presence of Howship's lacunae (as evidence of osteoclastic activity) has been reported in osteichthyans, e.g., salmonid *Salmo salar*, cyprinodontid *Orestias albus*, acanthurid *Acanthurus dussumieri*, and dipnoans.

The osteoclasts resorb dermal, perichondral, and endochondral bones. Osteoclasts resorb bone of different structural types (cellular, acellular and chondroid bone).

According to Parenti (1986), cellular bone is typical of lower teleosts, whereas acellular bone is typical of higher teleosts. Ontogenetic data indicate that acellular bone is derived relative to cellular bone. Even though identification of cellular and acellular bone can be made readily with histological preparations, acellular bone has been used infrequently as a character in analyses of teleosts phylogeny. Acellular bone is considered here to be a derived character within teleost fishes. It is found in all Nonteleostei as well as some, but not all Salmoniformes. Acellular bone and teeth of higher teleosts have several properties, including a large fraction of collagen. Teleosts lack a parathyroid gland; bone type is critical to the mechanism of calcium regulation. It is proposed that the character of acellular bone be incorporated into the phylogenetic analyses of teleost fishes by correlating it with derived types of tooth structure.

It has been pointed out that both cellular and acellular bone have osteoblasts or osteocytes in early stages of development. In cellular bone these cells persist, as in salmonids. In contrast, cells of mature acellular bone quickly become incorporated into the surrounding matrix. Cell death is rapid, and cell spaces become filled in; acellular bone has a relatively featureless matrix.

Further discussion in this respect is given by Maisey (1988). Endochondral bone is considered to be an osteochthyan synapomorphy. In recent elasmobranchs the endoskeletal calcification is endochondral. Bone cells have been reported in various regions of the dermal skeleton and endoskeleton of recent and fossil chondrichthyans. The most parsimonius interpretation of these data is that the general absence of cellular or acellular bone is a derived condition for chondrichthyans, because these tissues are widespread among the outgroup (e.g., in osteichthyans and fossil stem gnathostomes). Viewed in this light, osteichthyan endochondral bone represents a cladistically primitive calcification pattern (ossification) occurring in a cladistically primitive location (gnathostome endochondral tissue), rather than a novelty of bony fishes.

Teleost acellular bone develops from cells of the periosteum, within tendons, and also by metaplasia of secondary cartilage cells. Osteocytes are either not enclosed by bone matrix, or else are secondarily replaced by acellular matrix.

The acellular bone is phylogenetically primitive for craniates. Conclusions about the vertebrate skeleton have been drawn from phylogenetic approach, which, according to author, are consequently open to falsification, either directly or indirectly.

3. TELEOSTEAN CARTILAGE

A new classification of teleostean cartilage is presented. Excluded from the scheme are the chordoid and mucochondroid tissues and chondroid bone:

(A) *Cell-rich cartilage* (cells or lacunae occupy 50% of tissue volume)
- (i) Hyaline-cell cartilage—Lips and labial folds of bottom-dwelling cyprinids. Subtypes include (a) fibrohyaline-cell (b) elastichyaline-cell, and (c) lipohyaline-cell.
- (ii) Zellknorpel-Gill filaments of most teleosts.
- (iii) Fibro/cell-rich cartilage-Intra-articular menisci of numerous teleosts.
- (iv) Elastic/cell-rich cartilage—Barbels of numerous catfish.
- (v) Cell-rich hyaline cartilage—Neurocranium and gill arches of many teleosts.

(B) *Matrix-rich cartilage* (cells or lacunae occupy 50% of tissue volume)
- (i) Matrix-rich hyaline cartilage—Neurocranium of many teleosts.
- (ii) Fibrocartilage—Most fibrocartilages in small/young teleosts are cell-rich.
- (iii) Elastic cartilage—Most elastic cartilages in small/young teleosts are cell-rich.

Note: Scleral cartilage in eye consists of cartilage, dense fibrous connective tissue or bone (see Benjamin 1990).

A brief of hyaline-cell cartilage (HCC) (chondroid) is given here. HCC occurs in lips and rostral folds, in pre-palatine and submaxillary menisci, in ligaments, at the anterior end of the basihyal, in the pectoral girdle, in adhesive discs, in gill arches, beneath the basioccipital chewing pad, in barbels, next to the facial nerve, around the olfactory region and in the core of the nasal skin flaps. It is particularly important tissue in cyprinids and related fish, and enormous masses of it are present in the black shark, *Morulius chrysophekadion* and the Hong Kong pleco, *Pseudogastromyzon myersi.* It acts as a damper against the contractions of the heart or the pressure of occluding pharyngeal teeth, and it provides the mouth region of bottom-dwelling, algal eaters with flexible support.

The cells of HCC in numerous teleosts (e.g., *Echeneis naucrates* and *Gyrinocheilus aymonierir*) do not lie in lacunae (Benjamin 1989).

Benjamin and Ralphs (1991) described the extracellular matrix of connective tissues in the heads of teleosts. Tissues related to cartilage are also found in teleost. They include mucochondroid, a mucous connective tissue common beneath the skin (Benjamin 1988), chordoid tissue as represented by the annular ligament of the eye, and the notochord itself. Chondroid bone is also common.

Type II collagen in the bone of certain teleosts (e.g., *Barbus chonchonius, Brachydanio rerio,* sp) has been reported. It is well known that cartilage is the major skeletal tissue in elasmobranchs. Certain teleosts also rely heavily on cartilage for support, notably the lumpfish *Cyclopterus lumpus*. Cartilage collagen often occurs in teleost bone. It could be that there is a range of bone composition in teleosts, ranging from highly cartilaginous fish like the lumpfish, through those that contain characteristic cartilage molecules, to fish that do not contain such molecules in bone. Chondroid bone may contain type II collagen. Type II collagen (a characteristic of mammalian hyaline cartilage) is present in fish hyaline cartilage. Type II collagen is reported in the dense connective tissue of *Pangasius pangasius, Gasterosteus aculeatus* and *Poecilia reticulata.* It was striking that this molecule was either in all dense connective tissues or none. Type II collagen has been reported in the notochordal sheath of *Gyrinocheilus aymonieri.* Type II collagen has been observed in the mucochondroid of *P. pangasius.*

On the derivation of modes of suspension of the first and second splanchnical arches in fishes: An overview

Starting from a hypothetical paleostylic configuration an ancestral autodiastylis has developed in which the processus palatoquadrati are in function and the mandibular and hyoid arches articulate separately with neurocranium.

The autodiastylisis gives rise to two different trends: the "chondrogene" and the "osteogene" amphistylisis. The first trend leads to the euhyostylisis, and the second to the tritohyostylisis. Both have the character of a functional "extremity".

Besides, the methyostylisis-similar in morphology and function develops from a teleohyostylis. The methylostylisis does not show any maxillary mechanism.

Euhyostylisis (Batoidei), tritohyostylisis (*Stylephorus*) and methyostylisis (Acipenseriformes) are probably the "final" stages of different phylogenetical trends. Their morphological and functional congruence must be considered as a convergence.

The klasmatohyostylisis in Balistidae shows a hyostylic character.

In *Lepidosteus* the suspension is recognized as amphihyostylisis because the first and second splanchnical arches articulate without any connection to the neurocranium (Figs. 4.3 and 4.4).

4. ENDOSKELETON

4.1 Axial skeleton: Skull; Vertebral column

4.1.1 Skull: An Overview

The skull of fishes is an extremely complex structure designed to accomplish a variety of functions. Several separate factors are involved in the adaptation of skull morphology (e.g., respiration and the brain and sense organs).

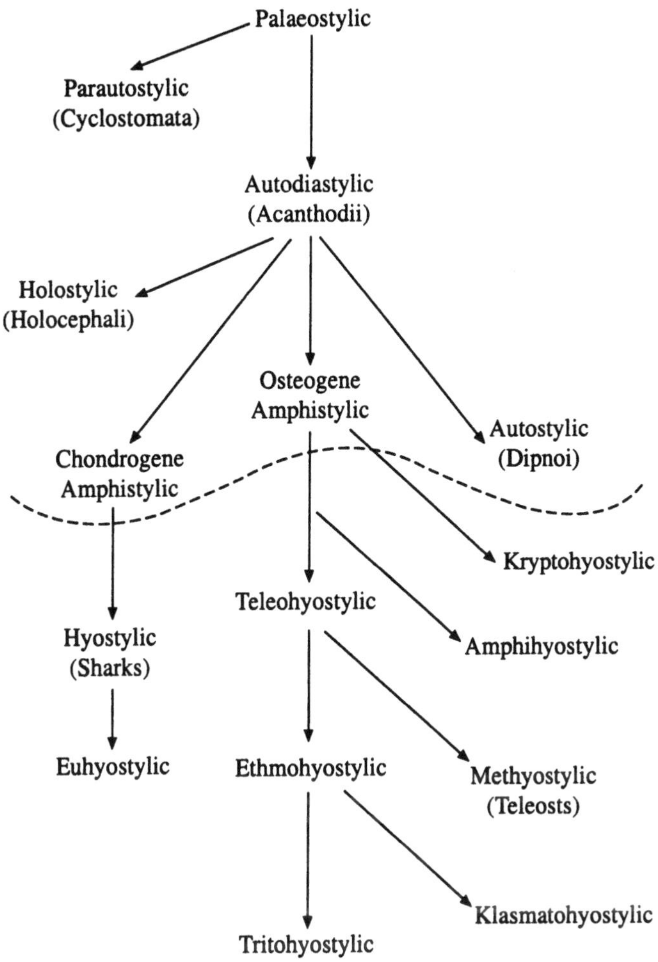

Fig. 4.3 The origin of different types (modes) of suspension of the first two splanchnical arches in fishes. Starting from a hypothetical paleostylic configuration an ancestral autodiastylisis has developed in which the processus palatoquadrati are in function and the mandibular and hyoid arches articulate separately with the neurocranium.
The autodiastylis gives rise to two different trends: the "chondrogene" and the "osteogene" amphistylisis. The first trend leads to the euhyostylisis, the second to the tritohyostylisis. Both have the character of a functional "extremity".
Besides, the methyostylisis-similar in morphology and function-develops from a teleohyostylisis. The methyostylisis does not show any maxillary mechanism. Euhyostylisis (Batoidei), tritohyostylisis (*Stylephorus*) and methyostylisis (Acipenseriformes) are probably the "final" stages of different phylogenetical trends. Their morphological and functional congruences must be considered as a convergence.
The klasmatohyostylisis in Balistidae shows a hyostylic character.
In *Lepidosteus* the suspension is recognized as amphihyostylisis because the first and the second splanchnical arches articulate without any connection to the neurocranium.
It is pointed that further cognizance of types of suspension in fossils may contribute to a completion of the system presented in this paper (Meinel, W. 1968) Über dii Abeltung der Suspensionsmodi des erstenund zweiten splanchnischen Bogens am Neurocranium der Fische Naturwiss. 36: 59–79.

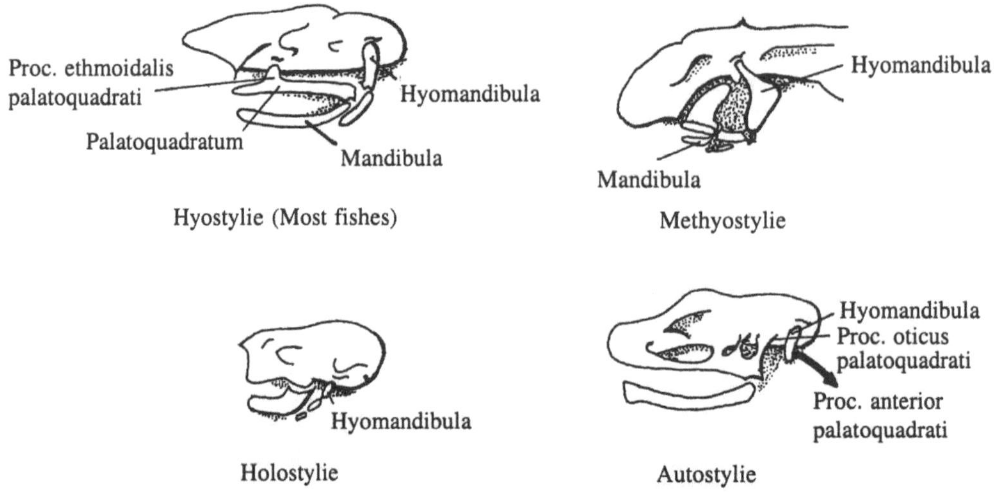

Fig. 4.4 Hyostylie; Methyostylie; Holostylie and Autostylie only drawn.

In describing skulls, no claim is made to have dealt the subject exhaustively. Illustrations are provided in textbooks to give an idea of skull in particular group. In the head skeleton several distinct entities are more or less closely knit together.

Jawless Fishes. Janvier (1993) has given patterns of diversity in the skull of jawless fishes. Atlantic hagfish, *Myxine glutinosa*, has been selected for description. The skull of hagfishes is made up of sinuous cartilaginous arches and plates onto which muscles attach, and which strengthen the wall of ducts. As in lampreys, it acts mainly as an elastic antagonist of muscles. Since the embryology of hagfishes is poorly known and the various skull components fused together, it is difficult to separate the neurocranium from visceral skeleton. Major recognizable units described are: (1) the braincase and nasal basket: the "axocranium"; (2) basicranial series; (3) the hypotic arch complex; (4) the velar membrane; (5) the lingual skeleton; (6) the structure of teeth; (7) the skeleton of the tentacles; and (8) branchial skeleton.

In *Myxine*, there is no skeleton associated with the gill pouches, thus no branchial skeleton.

The skull of lampreys, book lamprey (*Lampetra fluviatilis*) and sea lamprey (*Petromyzon marinus*) has been abundantly described. As in hagfishes, the skull of the adult lampreys consists of sinuous cartilaginous arches and plates. There is a well-defined branchial skeleton which surrounds the series of gill pouches (called branchial basket). Moreover, there is an orbital cavity into which are attached the extrinsic eye muscles. The mouth is surrounded by a large sucker (oral disc) which is strengthened by an annular cartilage, and the snout is armed by large cartilaginous plates which overlap like tiles and give the snout relative mobility. The skull of adult lampreys is generally regarded as an entirely cartilaginous or fibrous but some calcified areas have been reported.

Unlike hagfishes, lampreys pass through a larval stage and metamorphosis, during which the skeleton and the musculature are considerably reworked. Consequently, the nature of the various skeletal elements of the adult skull is obscured and it is sometimes difficult to decide what is neurocranial or visceral in origin. The account of adult lamprey has been given under different heads; (1) the braincase; (2) the prebranchial visceral skeleton; (3) the structure of the teeth, and (4) the branchial skeleton.

It has been pointed out that the question of skull homologies and jawed craniates (gnathostomes) bears the same kind of uncertainties as that of the hagfish-lamprey homologies.

One of the most diversified areas of the skull of primitive craniates is the naso-hypophysial complex (see Chapter 7). In all known gnathostomes, there is no naso-hypophysial complex. Paired and well-separated nasal sacs occur with certainty in the gnathostomes (read Chapter 15).

Jawed Fishes Schultze (1993) deals with the patterns of diversity in the skull of jawed fishes. Jawed fishes possess biting jaws that, together with the presence of trabeculae cranii, external gills, three semicircular canals in the labyrinths, and paired pectoral and pelvic fins supported by endoskeletal girdles, distinguish gnathostomes and agnathans. The jaws are formed by the palatoquadrate dorsally and Meckel's cartilage ventrally. Teeth are located either in connective tissue above and lingual to the palatoquadrate and Meckel's cartilage, or on dermal bones that cover them. If the jaws are ossified, ossification occurs first at the articulation between the upper and lower jaws; thus, the articular ossifies at the posterior end of Meckel's cartilage and the quadrate at the posteroventral end of the palatoquadrates. In this primary jaw articulation the head of the quadrate reaches into the glenoid fossa of the articular.

The palatoquadrate and Meckel's cartilage may be homologous with the first arch of agnathan, which supports the velum. All branchial arches, including the hyoid arch, are composed of paired ventral (hypohyal, ceratohyal/hypobranchial, ceratobranchial) and dorsal (epihyal = hypobranchial, pharyngohyal/epibranchial, pharyngobranchial) elements; the unpaired basihyal and basibranchials unite each side of the branchial skeleton in the midline. Because the gills are located externally on the gill arches, the internal surfaces of the arches may bear teeth or dental plates as does the mandibular arch.

The endocranium of adult chondrichthyans represents a single unit while the teleostome endrocranium is divided by fissures. The tetrapod skull-roof pattern, consisting of paired, medial postparietals, parietals, and frontals, is found only in panderichthyid fishes among osteichthyans. The postparietals cover only the posterior part of the otic region of actinopterygians; the anterior part of the otic region is covered by the posterior parts of the parietals. In actinopterygians, the parietal covers the anterior part of the otic region and the orbitotemporal region including the pineal organ. The external opening of the pineal organ lies between the parietals in all the osteichthyans.

The composition of the dermal cover of the palatoquadrate and Meckel's cartilage in actinopterygians is consistent with that in crossopterygians and tetrapods, except that the dentary (dentalosplenial) includes the splenial and postsplenial with the anterior portion of the mandibular canal in actinopterygians. Dipnoans differ greatly from osteichthyans. They lack a premaxilla and maxilla, and the homologies of the palatal bones are uncertain. Except for the prearticular medial to Meckel's cartilage, the dermal cover of the lower jaw is unique.

Chondrichthyes. Recent Chondrichthyes, or cartilaginous fishes, are represented by elasmobranchs and holocephalans which share one unique, derived character—presence of prismatically calcified cartilage in the endocranium, the visceral skeleton, and the skeleton of the fins.

Chondrichthyans lack an actual exocranium. Their heads may be covered by scales, which may be enlarged somewhat, and are less antero-posteriorly directed than the body scales.

It is generally accepted that the chondrichthyan endocranium is a single unit, and that the occurrence of an otico-occipital fissure is a synapomorphy for xenacanth elasmobranchs and convergent to the structure of the same name in teleostomes. The presence of an otico-occipital fissure may be primitive among chondrichthyans and, therefore, a symplesiomorphy of chondrichthyans and, teleostomes. Similarly, the presence of a post-orbital articulation with the otic process of the palatoquadrate may be a symplesiomorphy of the two groups.

The lack of chondrification of the anterior wall of the cranial cavity in front of the pineal foramen (precerebral fontanelle) is a unique feature of the chondrichthyan endocranium. Chondrichthyans also lack a nasal capsule floor. The orbital is surrounded by a thickened ring which forms preorbital and prominent postorbital processes where the palatoquadrate attaches with the ethmoidal and postorbital articulations. An eyestalk is present in neoselachians. An open endolymphatic fossa occurs in the dorsal midline of the endocranium of elasmobranchs, but is absent in holocephalans. Lateral otic processes occur in elasmobranchs but not in holocephalans.

The chondrichthyan endocranium is limited in its diversity, being a single unit. The otico-occipital region can change in length in relation to the length of the whole endocranium from long (over 50%) to short (to much less than 50%); it is short in neoselachians except in Heterodontimorphi (50%). The nasal capsules and the orbits are always quite large, although relative size differences can be observed.

Variations occur in the formation of processes, rostral, pre-and postorbital processes, and the size of the precerebral fontanelle. Pre-and postorbital processes are small in most elasmobranchs; they can extend laterally, and even far, as in the hammerhead sharks, the Syphyrinidae and mobulid rays. In Sphyrinidae, the pre-and postorbital processes form narrow stripes which connect distally; the preorbital process is joint with the distally extended nasal capsule. The nasal capsule is ventrally closed. In mobulid rays, the lateral extension is formed by a compact structure. In most batomorphs the distal part of the preorbital process is separated from the nasal capsule; it forms a separate cartilage. In Torpedinoidea, the preorbital process is enlarged and branched into many separate processes which bridge the area between the reduced rostral process and the preorbital process.

The rostral process is a structure of the Neoselachii excluding the Heterodontimorphi. It is formed either by a median ventral process, by two lateral processes combined with a median ventral one (all three are fused anteriorly), or by a troughlike structure. Elongation of the rostral process occurs within different neoselachian groups, in Mitsukurinidae within Galeomorphi, in Pristiophoriformes within Squalimorphi, and in Batomorphii. The sawsharks, the Pristiophoriformes, have a solid, heavy calified rostral process with lateral teeth; the rostral process is much longer than the neurocranium proper. In rhinobatid and rajid batomorphs, the length of the rostral process exceeds also the length of the neurocranium; here the precerebral fontanelle is elongated as in other batomorphs.

The mandibular arch of elasmobranchs is formed by two free cartilaginous elements, the palatoquadrate and Meckel's cartilage. The palatoquadrate is fused to the endocranium in holocephalans (holostylic). The articulation of the palatoquadrate with the endocranium is variable. It has been noted that rostral, ethmoidal, orbital, palatobasal, and postorbital articulations, not all of which occur in a single specimen. Because the epihyal articulates with the palatoquadrate in all elasmobranchs to some degree, all are hyostylic. Amphistyly is a special case of hyostyly in which the palatoquadrate is more firmly attached to the braincase; in advanced sharks, the epihyal supports the more freely movable palatoquadrate ("typical" hyostyly).

The palatoquadrate and Meckel's cartilage form a double joint with an articular knob at the posterior end and a joint pan on the posterior medial side of the palatoquadrate. Meckel's cartilage forms the cotyle to accept the quadrate head, and the dorsal process to articulate with the joint pan on the medial side of the palatoquadrate. Lateral to the palatoquadrate and Meckel's cartilage, there are a number of labial cartilages to which some superficial jaw muscles attach. The jaw articulation may be placed below (e.g., Scyliorhinidae) or behind (e.g., Sphyrhinidae) the otico-occipital region.

The hyoid arch consists of epi-, cerato- and median basihyal in elasmobranchs; a hypohyal is absent in chondrichthyans. Holocephalans are the only gnathostomes with a pharyngohyal. There is a half gill slit, a spiracle, between epihyal and palatoquadrates in chondrichthyans (closed in adult holocephalans). The gill arches posterior to the spiracle have a complete set of posteriorly directed pharyngo-, epi-, cerato-, and hypobranchials, and a median basibranchial. Except for holocephalans, there is a gap between the basihyal and basibranchials corresponding to the anteriorly directed ceratohyal and posteriorly directed hypobranchials. The posterior gill arches are modified in some extant sharks; thus, the pharyngo-and epibranchial or the cerato- and hypobranchial may fuse within one arch or between the two last arches. Extrabranchials are present outside the gill arches, but they are rarely illustrated.

The gill arches are narrowly spaced below the posterior endocranium in Subterbranchialia, but widely spaced in elasmobranchs so that the shoulder girdle (scapulocoracoid) is widely separated from the skull. Subterbranchialia have an opercular flap supported by hyoidean rays.

Osteichthyes. Each osteichthyan group has distinct exocranial and endocranial patterns. A mosaic pattern of bones on the snout seems to be primitive for all osteichthyans, even though it is unknown among primitive actinopterygians. Consolidation of paired bones along the midline starts posteriorly with two bones infront of the extrascapulars. This is true of primitive dipnoans too; however, in most dipnoans, an unpaired median bone lies in the midline. Extrascapulars with the occipital commissure of the lateral line system lie behind the endocranium and are not solidly enclosed in the skull roof. Although the presence of a tooth-bearing outer dental arcade (premaxilla, maxilla, dentary) is cited as a common character of Osteichthyes, dipnoans lack all three bones. On the other hand, all osteichthyans possess a tooth-bearing prearticular. The cheek bones are without direct contact with each other in the coelacanth *Latimeria.*

The endocranium of osteichthyans ossifies perichondrally and endochondrally. Ossification proceeds in a posterior-to-anterior direction, and perichondral ossification precedes endochondral ossification. Thus, in actinopterygians, the anterior ethmoidal region is unossified or bears only weak perichondral ossification. The endocranium is partly divided to regions by fissures. The ventral fissure in actinopterygians and dipnoans is homologous with the ventral part of intracranial joint of crossopterygians, and separates the otic and orbitotemporal regions. In advanced actinopterygians and dipnoans, the ventral fissure is posterior in position. The otico-occipital region may be single as one unit or be divided into many ossification centres. Primitive actinopterygians and dipnoans lack separate ossification (except *Dipnorhynchus* with three ossification centres in the otic region), whereas advanced actinopterygians and crossopterygians have many ossification centres. In the occipital region, a supraoccipital is known only in teleostean actinopterygians, some crossopterygians, and tetrapods. Primitively, there seem to be three segments in osteichthyans, but there may be more or fewer among extant actinopterygians. In extant dipnoans, there are three segments in *Neoceratodus* and one in *Lepidosiren*, and there is only one in the extant actinistian *Latimeria*.

Primitively in osteichthyans, a massive lateral commissure penetrated by the jugular canal forms the lateral wall of the trigeminofacialis chamber in the otic region. The facet for the epihyal lies dorsal to the jugular canal in actinopterygians and straddles it in sarcopterygians. A spiracular groove is developed in primitive actinopterygians together with an open spiracle on the skull roof, but a spiracular groove is rudimentary or missing in sarcopterygians. Except for dipnoans, the endolymphatic ducts are closed in osteichthyans. Paired fossae bridgei are not a common feature of osteichthyans, occurring only in actinopterygians. Crossopterygians have paired posttemporal fossae for axial musculature, and dipnoans lack both fossa bridgei and posttemporal fossa. Although a posterior myodome is absent in the otic region of primitive osteichthyans, it occurs in actinopterygians. Two pairs of anterior myodomes are found in some primitive actinopterygians, actinistians, and few rhipidistians. Primitively, these structures may be absent; based on their variable positions in different osteichthyans, they seem to have developed independently in various taxa. Endochondral ossification invades the nasal capsule of crossopterygians, but not that of actinopterygians.

The mandibular and hyoid arches are covered laterally by dermal bones. The dermal covers of the mandibular arches of actinopterygians, dipnoans, and crossopterygians are each distinctive but the endoskeletal bauplan is similar in each group. The dermal hyoid cover (operculogular series) extends posteriorly to cover posterior gill arches laterally. An opercle and subopercle occur in all osteichthyans, but there are differences in the bones that lie between the subopercle and gular. The prearticular on the medial surface of Meckel's cartilage and the pterygoid (= endopterygoid) on the medial side of the palatoquadrate (upper jaw) are present in all osteichthyans, whereas the occurrence of ossifications (i.e., dermal plates with teeth) on the gill arches varies within groups. An unpaired median parasphenoid covers the roof of the mouth.

Actinopterygii. Taking the exocranium, the largest part of the skull-roof of actinopterygians is occupied by two pairs of bones, which are usually called "parietal" and "frontal". An unpaired median bone, the rostral, lies in front of the parietals and between the anterior nasal openings in primitive actinopterygians. The posterior margin of the skull is formed by an even number of extrascapulars (two in primitive and four in advanced actinopterygians). The jugal and lacrimal are divided into smaller bones known as infraorbitals (usually 1–5) in advanced actinopterygians. In advanced actinopterygians, a small bone, the antorbital, occupies the area between the posterior nasal opening and orbit. Dorsal to the orbit are additional bones, the supraorbitals. The orbit is large, and the eyeball is surrounded by four sclerotic plates (two or none in advanced actinopterygians). The cheek region is occupied by a dorsal extension of the posterior maxilla, the preopercle, and the dermal dermohyal, which is attached to the epihyal. In some advanced actinopterygians, the preopercle is separated from the posterior infraorbitals by additional suborbital bones. The cheek region is rigid in primitive actinopterygians. In advanced actinopterygians, the cheek bones are separated from each other, the maxilla and premaxilla become movable (protrusible mouth), and additional bones, supramaxilla (e) (two are characteristic for teleostean actinopterygians), appear above the maxilla.

Variation of exocranium is extreme within actinopterygians. The exocranium varies in proportions, in relation of bones to each other, in number of bones, and shape and thickness of bones. In *Polypterus*, the most primitive extant actinopterygian, accessory bones occur lateral to the medial pairs of extrascapulars, supratemporotabulars, and parietals, which are numerous and variable in number. There is a very long cheek region with a broad preopercle and two bones below it and posterior to the maxilla, of which the posterior one can be homologized with the quadratojugal, whereas the anterior one is an additional bone. The Acipensiformes have numerous additional bones in the snout region which vary intraspecifically in shape and number; the snout is elongated to a long rostrum formed by many independent ossifications in Polyodontidae (Grande and Bemis 1991). The Ginglymodi (*Lepisosteus*) develop a bone mosaic in the cheek region, which occupies the place of the large preopercle of *Polypterus*. The anteriorly elongated upper jaw of *Lepisosteus* is bordered by many toothed small bones, lacrimals or infraorbitals, which fuse with the maxilla. Within teleosts, the number of bones is reduced, with the exception of some Scorpaeniformes, but even in this group and others like the Dactylopteriformes and Siluriformes the bony cover is formed by enlarged bones only. Change in size and proportions of these bones results in the great variability of the teleostean skull. Commonly, the cheek region between the orbit and opercle is short; nevertheless it can be very elongated as in some osteoglossomorphs, such as *Arapaima*, *Notopterus*, and *Mormyrops* or in synbranchid eels. In *Arapaima*, the elongated cheek region is covered by elongated infraorbitals, whereas the lateral side of the palatal bones is merely covered by skin in the other genera.

In different teleostean groups, the snout is elongated as in Ginglymodi or other lower actinopterygians (e.g., saurichthyids, aspidorhynchids). The elongation can be formed by premaxilla and dentalosplenial (e.g., in the Belonidae, needlefishes). In teleosts with small premaxillae, maxillae, and dentalosplenial, the parietal with the mesethmoid in the upper side of the snout and articular (Mormyridae, elephantfishes), or preopercle (Acanthuridae, surgeonfishes; Balistidae, leatherjackets, and others) or preopercle and quadrate (Syngnathiformes, trumpetfishes, cornetfishes, snipefishes, pipefishes, and seahorses., and Aulorhynchidae, tubesnouts) in the lower side of snout are elongated.

As the other extreme, the skull can be very deep, as in Zeiformes and in some Perciformes; here the median supraoccipital reaches between the postparietals to form a deep crest. In flatfishes, the Pleuronectiformes, ventral and dorsal margins of head become the lateral margins. The skull is asymmetrical by ontogenetic movement of one eye to the upper side; the skull roof lies on one side, but the other bones keep their positions. Deep sea fishes reduce the skull bones to narrow stripes with the extreme in some Lophiiformes, the anglerfishes. As a whole the teleosts have greater freedom in the variability of the skull bones than lower actinopterygians. The skull bones are tightly sutured to a boxlike skull in lower actinopterygians, whereas many bones in the heads of teleosts are not in close contact with their neighbours. That gives great possibility for differential growth and thus for great diversity in skull shapes of teleosts.

Typically, the main body of the actinopterygian endocranium is formed by the otic and orbitotemporal regions, with the occipital and ethmoidal regions being short additions to the posterior and anterior ends, respectively. The occipital region is limited anteriorly by occipital fissure which is completely uninterrupted in primitive actinopterygians. The fissure is confluent dorsally with the posterior dorsal fontanelle, and extends latero-ventrally into the vestibular fontanelle and the ventral otic fissure. Thus, the dorsal extent of the occipital region is abbreviated but its ventral component is long.

The otico-orbitotemporal region is formed by a single ossification in primitive actinopterygians but by many separate ossifications in advanced actinopterygians, expecially teleosts. The epihyal (hyomandibular) articulates in an oblique facet anterior and dorsal to the jugular groove in primitive actinopterygians, whereas its position is horizontal in advanced actinopterygians. The spirtacle runs in a groove from the orbitosphenoid fissure to the cranial roof in front of the lateral commissure and opens on the surface in primitive actinopterygians, whereas it is enclosed in a canal and ends blind in advanced actinopterygians; it is bounded anteriorly by the postorbital process. The left and right orbits are separated by a narrow median wall.

Dorsally, an anterior dorsal fontanelle may be present as an extension of the pineal foramen. The endolymphatic duct has a blind end in the otic region. Fossae bridgei-characteristic for advanced actinopterygians are absent in primitive actinopterygians. In advanced actinopterygians there is a posttemporal fossae that is separated from, or confluent, with fossa bridgei.

The basisphenoid surrounds the pituitary fossa laterally and ventrally. The pituitary fossa opens via a large foramen in the parasphenoid in primitive actinopterygians. A basipterygoid process that articulates with the palatoquadrate is present.

The ethmoid region is perichondrally ossified in primitive actinopterygians. The nasal capsule is surrounded by a thin perichondral ossification in actinopterygians, whereas additional endoskeletal ossifications occur in teleosts.

The endoskeletal mandibular arch is formed primitively by two elements, the palatoquadrate (= upper jaw) and Meckel's cartilage (= lower jaw). The palatoquadrate bears two articulations with the neurocranium: (1) an ethmoidal articulation (between autopalatine and lateral ethmoid), which becomes double in higher actinopterygians, and (2) a palatobasal articulation (between the basal process of metapterygoid and basipterygoid process), which is lost in advanced actinopterygians. The palatoquadrate of advanced actinopterygians ossifies from three centres-the halecostomian autopalatine, metapterygoid, and quadrate. It is invested by two series of dermal bones; the outer series is composed of the vomer, dermopalatines, and ectopterygoid, and the inner series is constituted by the endopterygoid (= pterygoid) and tooth plates fused to the metapterygoid, the so-called dermometapterygoid. The lateral dermal cover is composed of the premaxilla, maxilla, infraorbitals, preopercular, dermohyal, and quadratojugal. In advanced actinopterygians, the bones of the cheek region are reduced such that only soft tissue covers the space between the infraorbitals and preopercle, which is associated with the opercular series. The premaxilla and maxilla are mobile.

The condyle of the quadrate articulates in the glenoid fossa of the posterior end (articular) of Meckel's cartilage. In addition to the two Meckelian ossifications, there is a third ossification (the retroarticular) in advanced actinopterygians; in the latter, the retroarticular and angular can take part in the formation of the glenoid fossa. The tooth-bearing bone of actinopterygians, the dentalosplenial, represents the dentary + splenial + postsplenial of crossopterygians. The angular and surangular (lost in advanced actinopterygians) lie posterior to the dentalosplenial. The surangular is not a canal-bearing bone as in crossopterygians. The number of coronoids dorsal to Meckel's cartilage varies from many in primitive actinopterygians to none in advanced taxa.

The jaw articulation is primitively below the posterior part of the skull. The jaws are shortened and placed behind the orbit in Acipensiformes, resulting in a ventral mouth as in sharks. In Ginglymodi, the jaw articulation lies in front of the orbit, and the dentalosplenial is differently elongated depending on the species. Elongation of the premaxilla, sometimes elongation of maxilla, and dentalosplenial can be found in some lower actinopterygians (e.g., *Saurichthys*, aspidorhynchids) and in different teleosts (e.g., Belonidae, needlefishes). In some cases the upper jaw is more elongated than the lower jaw (e.g., Xiphidae, swordfishes; Istiophoridae, billfishes), whereas in others only the lower jaw is elongated (Hemiramphidae, halfbeaks). In very few forms is the jaw articulation placed far behind the head so that the mouth becomes very large (Saccopharyngidae, swallowers) or enormous (Eurypharyngidae, gulpers); in the latter case the jaws are six to eight times the length of the neurocranium. In contrast, the jaw can be small in many teleosts, as in Syngnathidae where small premaxillae, maxillae, and dentalosplenials lie far anteriorly at the end of an elongated snout. The maxillae are free from the cheek bones in halecostomes so that the jaws could differentiate widely, especially in teleosts. This is one reason, if not the main reason for the explosive radiation of teleosts. In many ostariophysans and nonteleosts, the premaxillae become movable and protrusible, giving these teleosts further feeding advantages.

The hyoid arch and four branchial arches articulate with a long, unpaired basibranchial. The fifth branchial arch is reduced in primitive actinopterygians; in advanced actinopterygians, the branchial arches are reduced further in posterior-to-anterior direction. The epihyal (= hyomandibula) and interhyal form the dorsal portion of the hyoid arch; the symplectic connects the epihyal with the quadrate in advanced actinopterygians. The epihyal articulates with the endocranium. Of the posterior branchial arches, the first infrapharyngobranchial articulates with the neurocranium, whereas the other pharyngobranchials (supra- and infrapharyngobranchials) are connected to the neurocranium, via connective tissue only. Tooth plates on the branchial arches are often more important for mastication than the teeth on the jaws in advanced actinopterygians. The urohyal, a tendon bone, is known only in advanced actinopterygians.

The operculogular series forms a continuous series from the dorsolateral opercle to the subopercle, branchiostegals, and the medio-ventral gular. Gulars are lost in advanced actinopterygians. The development of bony skull is quite variable in teleosts. The first bones always seem to be of dermal origin. Ossification of the branchial basket first affects the centres of the ceratobranchials, then the centres of the epibranchials, and finally the centres of the pharynobranchials. The hypobranchials and axes of the basibranchials ossify last. The first pair of ceratobranchials and three pairs of pharyngobranchials bear upper and lower pharyngeal toothed plates even before they ossify (read Wagemans et al. 1998).

Sarcopterygii. Dipnoans and crossopterygians have distinct exocranial patterns. In both dipnoans and crossopterygians, the number of lateral skull-roof bones is greater in primitive taxa than in advanced taxa. Dipnoans have more (10 or more) bones in a short cheek region than do crossopterygians (8 or fewer). The maxilla and premaxilla are absent in dipnoans; among crossopterygians, the maxilla is absent in actinistians. The pattern of the skull-roof, cheek, and jaw bones evolved independently in dipnoans and crossopterygians from a mosaic of bones so that comparison between the pattern of advanced forms in both groups is not possible.

There are more differences than similarities in the endocrania of dipnoans and crossopterygians. The dipnoan endocranium is a single unit to which the palatoquadrate is fused; in crossopterygians, the ventral fissure extends dorsally to form an intracranial joint, thereby separating the endocranium into two separate units. The otico-occipital fissure is closed perichondrally in dipnoans, partially closed in rhipidistians, and is represented as a suture in actinistians. A supraoccipital is present in crossopterygians, but not in dipnoans.

On the postero-dorsal surface of the otic region there are paired posttemporal fossae with axial musculature in crossopterygians; structures in the same region in dipnoans carry adductor muscles and are not homologous to the fossae in crossopterygians.

The upper part of the mandibular arch, the palatoquadrate, is fused to the endocranium in dipnoans, but free in crossopterygians. Thus, only the quadrate can be identified in dipnoans. In contrast, three palatoquadrate ossification centres are identifiable in some crossopterygians (actinistians), but primitively, there is only one large ossification. In crossopterygians, the palatoquadrate bears four endocranial articulations: (1) an ethmoidal articulation between autopalatine and ethmoid (may become divided into two articulations); (2) the palatobasal articulation between the basal and basipterygoid processes; (3) the postorbital articulation between the ascending and suprapterygoid processes; and (4) the paratemporal articulation behind the intracranial joint. Additional articulations occur in some crossopterygians.

The lingual and buccal dermal covers of the palatoquadrate differ in dipnoans and crossopterygians; the only palatal similarity shared is the presence of a parasphenoid and pterygoids in both groups.

Meckel's cartilage ossifies as a single unit in dipnoans, whereas it arises from two ossification centres that give rise to the anterior mentomeckelian bone and posterior articular bone in crossopterygians. Both groups posess a prearticular on the medial surface of Meckel's cartilage, but other mandibular bones are difficult to compare. Coronoids and dentary occur in crossopterygians, but not in dipnoans.

The branchial skeleton is reduced in extant sarcopterygians. The hyoid arch is formed by the hypohyal, ceratohyal, and epihyal in dipnoans, whereas crossopterygians possess an additional interhyal. In the following arches, hypo- and pharyngobranchials occur only in crossopterygians. Presence of a "urohyal" preformed in cartilage is common in both groups.

The operculogular series varies in size but not composition. The submandibulars are unknown in actinistians.

Dipnoi. In contrast to Devonian dipnoans, the three extant dipnoan genera have reduced exocrania and cartilaginous endocrania. Possession of median, unpaired bones that are flanked by two series of paired bones is a characteristic feature of dipnoans. The relation between skull-roof bones and lateral line canals is lost completely in the extant African, *Protopterus*, and South American lungfish, *Lepidosiren*. The infraorbital bones are developed in *Neoceratodus*, they do not suture with each other. They are missing in *Protopterus* and *Lepidosiren*. A maxilla and a separate premaxilla are not developed.

The endocranium and the palatoquadrate form a single unit. Because the palatoquadrate is fused to the neurocranium, only the quadrate can be identified with certainty. The medial and lateral dermal covers of the palatoquadrate differ from those of other osteichthyans except for the pterygoid and parasphenoid, which are the same as in other osteichthyans.

4.1.2 Vertebral Column

The vertebral column in fishes ranges in development from a cartilage sheath around a notochord (hagfishes), through a cartilaginous vertebral column (vertebrae without centra in lampreys but with centra in elasmobranchs-with cyclospondylus/tectospondylus/asterospondylus conditions), to one of partially ossified cartilage (ratfishes), to one of solid bone (teleosts). Although the centra (round centres of the vertebrae) are aligned in a series in fishes, the vertebrae lack elaborate interlocking processes, which are necessary in terrestrial vertebrates to counteract gravitational forces. The gars (Lepisosteidae) are an exception to this rule; they have interlocking vertebrae that resemble those of reptiles. This is presumably related to heavy load of scaled armour they carry around.

Despite their apparent simplicity in structure compared with the vertebrae of terrestrial vertebrates, the vertebrae of fish are structurally specialized. The anteriormost vertebrae (the atlas and the axis) are structured for articulation with the skull. At the rear of the spinal column, the posteriormost vertebrae are modified into series of flattened elements (penultimate vertebrae, hypurals, epurals, and urostyle), which articulate with the rays of the caudal fin. Between these two sets of highly modified vertebrae are the trunk vertebrae, most of which bear ventral (pleural) ribs, which usually attach to the basapophyses (parapophyses). In addition to the ventral ribs, most bony fishes also have dorsal ribs (intermuscular bones) that extend into horizontal skeletogenous septum, and usually just loosely associated with the vertebral column. Other intermuscular bones are named according to structures that bear them; for example, those borne on the neural arch are called epineural, and those on the ribs are called epipleurals. Intermuscular bones are exceedingly numerous and complex in some flounders (Pleuronectiformes); they may occur in four rows; each bone may be brushlike at both ends. Ribs and intermuscular bones appear to have developed in response to stresses and strains set up by the musculature. They do not impede contraction but serve somewhat the same function as tendons. They seem to lend strength without detracting from lateral flexibility.

Both teleost and elasmobranch trunk vertebrae also feature neural processes which form an arch to accommodate the spinal cord. Through the length of these fish, the neural arches together compose the neural canal. The fusion of transverse processes produces a haemal canal ventral to the centra in the caudal vertebrae, which may bear a haemal spine.

Vertical flexibility of the axial skeleton does occur in some fish at the junction between the skull and the vertebral column. The girdle is connected to the skull via a two-plane hinge system (involving the forked posttemporal and the supracleithrum) that allows some head movement without dislocating the lateral line sensory system where it passes forward onto the skull. In many teleosts the whole cranium tilts where the mouth is opened during feeding. Some myctophids can bend their neck upward. The clingfish *Lepadichthys lineatus* has a dorsal gap between the first two cervical vertebrae and can bend its head up while attached to the substrate by its ventral sucker (Fig. 4.5). A modification in some Asian cyprinids involves a curious "neck bending" which is probably associated with escape by skittering along the surface. *Chela maasi* can snap its head back so that the dorsal surface is almost at right angles to the contour of the back. The action swings the large pectoral fins downward, so as to thrust the whole fish upward. The pivot for this action is between the first vertebra and the skull and on either side between the upper end of the cleithrum and a large rounded expansion of the transverse processes of the first two vertebrae. A tough fibrous sheath protects the nerve cord at the site of bending. The large, elongate, predatory *Rhaphiodon* (family Cynodontidae) has a sharp-edged thorox and the ability to bend neck upwards, perhaps coupled with the downward thrusts of the strong pectoral fins. The coelacanth *Latimeria* has a jointed cranium, the anterior part of which tilts up when the mouth is opened. Some dorsoventral flexion of the vertebral column may be possible in sticklebacks and in tunas. But there are limited abilities; the skeletal architecture of the fish generally suits them for lateral, not vertical, undulation.

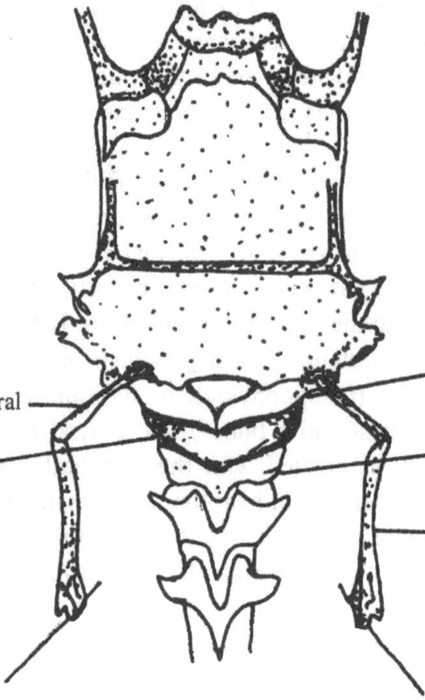

Fig. 4.5 The skeleton of the head and anterior vertebrae of *Lepadichthys lineatus*. ***Source:* Fishelson (1968)**

Flexibility may vary markedly along the vertebral column, controlled in part by the degree that the zygapophyses of one vertebra overlap the next. At each of the prepeduncular and postpeduncular joints of some thunniform swimmers the zygapophyses are reduced and the adjacent vertebral faces form a lateral hinge. Between the two joints, the zygapophyses greatly overlap so as to lock several vertebrae into a rigid unit, further strengthened by a wide bony lateral keel.

Structural details of the notochord and elastic longitudinal ligaments (dorsal and ventral) of fish are presented with discussions on their possible contribution to the speed, power and modes of swimming by conferring an automatic spring-like resilience to the vertebral axis as a whole. The notochord is believed to function as a series of ring-like hinges placed intervertebrally which dictates that the centra must be concave (amphicoelous) to support and house them. The dorsal logitudinal ligament is always present in most fish, the ventral is found in primitive teleosts only. The dorsal and ventral ligaments are suitably situated to assist the circulation of lymph (see Chapter 8 for attention and comment) and blood respectively in the small lateral vessels associated with the main longitudinal one of the appropriate system.

The otophysic connection involving or not involving the intercalation of bony elements are reported in many teleostean fishes.

Though there are notable exceptions, movements of the head relative to the body is rarely very extensive in modern fishes. In all ostariophysine fishes the presence of Weberian apparatus seems to impose even greater than usual restriction on the flexibility of the head and anterior vertebrae. Within this group potential head movement, to judge from the amount of fusion in the anterior vertebrae, is greatest in characins and least in catfishes, though some variation occurs within cyprinids and more within the catfishes. Loss of flexibility in the post-temporal part of the pectoral girdle parallels that in the vertebral column.

The second, third, and fourth vertebrae of the Weberian apparatus in Siluriformes are fused to a single ossification called a "complex vertebra". Additional posterior vertebral centra are frequently fused to or united by exceedingly tight joints with the complex vertebra.

The number of vertebrae varies in fish. While many selective factors must converge to determine the optimum number for each species, locomotory ability is likely to be prominent. There is a definite inverse correlation between vertebral number and robustness of the body. The generalization is less valid between widely differing groups of fishes. There is suggestion of a phyletic decrease in vertebral number in the course of evolution. No primitive fish (cyclostomes, elasmobranchs, lower teleosts) have low counts. Strikingly low counts occur irregularly among some but not all the higher teleosts. The forms retaining the most cartilage have many vertebrae and high flexibility. A widespread tendency for fish species with larger adult size to have more vertebrae may have locomotory significance. The ubiquity of pleomerism is surprising. If, as is suggested, there is some functional correlation between the body size and segment number, it may well be related to locomotion. Pleomerism seems to be so general that its explanation, if forthcoming, is likely to shed some light on the functional significance of segment number in fish. Numbers of vertebrae (and of fin rays and other meristic series) are subject to several other puzzling sorts of variation which may have locomotory implications. Details regarding pleomerism can be obtained from the article of Lindsey (1975).

4.2 Appendicular Skeleton (Fins and Fin Girdles)

The appendicular skeleton, compared with skull, is relatively simple consisting of the internal supports for various fins—most prominently the supports (girdles) for pectoral and pelvic fins.

The lampreys and hagfishes lack not only paired fins, but also the girdles that support them.

4.2.1 Pectoral Fin and Pectoral Girdle

Paired Structures. The pectoral fins are less mobile in sharks than in most teleosts, but are more mobile than the dorsal or anal. The low and horizontal positions of the pectorals allows them to act as hydroplanes, which counteract the negative buoyancy and the pitch produced by the heterocercal tail. In a few mesopelagic sharks which have achieved nearly neutral buoyancy the pectorals have shifted upward and rotated, coming to resemble the paddlelike pectorals of higher teleosts. Except perhaps in these aberrant forms, sharks do not use their pectorals as brakes the way teleosts do, and sharks are incapable of making sudden stops. In skates the pectorals are highly modified and become the principal locomotor organ. The pectorals of more primitive teleosts are low on the body and largely horizontal, like those of sharks. In higher teleosts the pectorals have moved up on the sides, and their bases have changed to an essentially vertical alignment.

In the Chondrichthyes the support for pectoral fins (the pectoral girdle) consists of an upper, or scapular, section and a lower coracoid piece. A small suprascapular cartilage may be present. In sharks the two halves of the girdle are separate from each other dorsally and do not attach to the vertebral column. In rays the two halves join each other or the vertebral column. The pectoral fin articulates with the coracoid. There are three basal cartilages—the propterygium anteriorly, a middle mesopterygium, and in posterior metapterygian. Series of jointed radials, which attach to the basal cartilages, bear the fin rays (ceratotrichia) at their distal ends.

In Osteichthyes, the pectoral girdle is composed of paired coracoids and scapulae, four pairs of radials (actinosts), the posttemporals, supracleithra, cleithra and postcleithra. The upper branch to the posttemporal attaches to the epiotic and the lower branch to the pterotic or opisthotic (intercalary).

In the course of teleostean evolution the pectoral fin moves from its essentially ventral position on the body in such "holosteans" as *Amia* to a location well upon the sides in higher teleosts. Accompanying this shift in position the part of the scapulocoracoid border with which the pectoral fin support (actinosts) articulate changes from nearly horizontal in *Amia*, through more or less L-shaped in lower teleosts, to approximately vertical in higher teleosts. In the course of this change two series of structural developments occur, each of which seems to correspond to a shift in the nature of pectoral fin movements. One of these has to do with the anteriormost pectoral ray and differentiates modern lower teleosts from their "holostean" ancestors. The other is a series of individually minor changes associated with the loss of the mesocoracoid arch in the iniomous fishes and higher teleosts.

The postcleithra of osteoglossiform fishes is the single anocleithrum inherited from ancestral forms. Other modern teleosts may have two additional postcleithra which seem to be scale derivatives. The iniomous (myctophiform) genus *Aulopus* appears to have three postcleithra, as do various modern lower teleosts, but more advanced forms have a mixture of two. In each of the major groups of modern teleostean fishes the postcleithra seem to have undergone loss. They are absent in, for example, the mormyrids among osteoglossimorph series, the eels among elopomorphs, the engraulids among clupeomorphs, and the catfishes among ostariophysines. Postcleithra are absent in all modern gonorynchiform fishes. In the majority of teleosts, however, the postcleithra are not completely lost but undergo reduction in number, with specialization in the remaining units.

In bony fishes the vertebral column, with a horizontal axis, and the pectoral girdle with a basically vertical axis, form the only skeletal links between the head the body. The individual bones of the dermal girdle are considered as supporting units joined by a series of articulations that permit differential movement between adjacent bones. The movements mediated by this linkage system are: lateral swinging of the head relative to the body, expansion of the distance between the central areas of two pectoral girdles to permit passage of large food items, and fore-and-aft movements of the anteroventral ends of the cleithra relative to the skull. Among other factors affecting the structure of the dermal pectoral girdle are the provision for the support of the pectoral fin base and the requirement for the effective operation of a sleeve valve between the girdle and the opercular cover.

The various parts of pectoral girdle generally serve a number of separate functions. The posttemporal and supracleithrum carry the lateral line from the head on to the body. The posteroventral surfaces of the cleithrum provide a support for the structures of the pectoral fin base, i.e., the primary pectoral girdle. The anteroventral ends of the cleithra lie external to, and presumably protect, the heart; they also provide a surface for the attachment of muscles forward to the urohyal and for others passing upward to the bases of the gill arches. Within teleostean group of bony fishes two additional roles have evolved. A lower postcleithral element develops into anterior component of the rib case. In higher teleosts, the cleithra serve as support for the pelvic girdles. The pectoral girdle has many peculiarities directly related to the enlargement of the branchial chamber.

4.2.2 Pelvic Fin and Pelvic Girdle

Paired Structures. The contribution of the pelvic fins to locomotion is usually minimal. They seldom contribute to the forward propulsion but may serve as stabilizers and manoeuvering vanes. They are frequently modified for nonlocomotory purposes, and become copulatory organs, defensive spines, or adhesive discs. In quite a few groups the pelvic fins are absent.

The pelvic girdle of elasmobranchs is generally a single transverse cartilaginous bar. It bears backwardly directed basipterygium on either side. The lateral margin of the basipterygium carries a row of triserial radial rods, and these in turn support the closely spaced ceratotrichia. The girdle is imbedded in the ventral myotomes.

In lower teleosts the two halves of the pelvic girdle are fused in the midline, and there are a few skeletal nodules, equivalent to the radials in the pectoral base, along the posterior margins where the rays insert. The innermost is the largest. This allows some rotation of the fin about the outermost ray as an axis. In higher teleosts the girdle halves are usually separate, there are no radials, and the fins can be extended or retracted along an essentially single plane. There is a concomitant reduction in the number of pelvic fin rays. In higher fish where the pectoral fins have moved upward the pelvics often move forward, and the pelvic girdle becomes strongly attached by ligaments, to the cleithra of the pectoral girdle. Forms having such a firm pelvic base may develop strong pelvic spines, which may articulate with the girdle through a joint-allowing movement only in one plane. In the anterior position, the locomotory role of pelvic fins is largely to provide braking or turning, and to counteract the pitch produced when the pectoral fins are extended.

4.2.3 Functional Value of Paired Fins

The pectoral fins of teleost fishes serve a great variety of functions (Table 4.3). A primary function can be made between functions associated with locomotion, i.e. swimming and other functions.

Table 4.3 Functions of pectoral fins in Teleostei

- Pectoral fin functions
  - Aquatic locomotion
    - Unidirectional locomotion
      - Straightforward swimming
        - Paddle/ hydrofoil (labriform)
        - Undulation (diodontiform)
      - Steering / braking
        - relatively simple movements of the pectoral fins
    - Multidirectional locomotion (manoeuvering)
      - rhythmic movements (undulation)
      - non-rhythmic movements
  - Other functions (Benthic and terrestrial locomotion, flying, attachment to the substratum, digging, defence, camouflage, copulation)

From: Geerlink (1989)

Flight has been reported for several families of teleost fish; there are no such claims for elasmobranch fish. In the case of the 'flying gurnards' (Dactylopteridae) there is a considerable controversy about their ability to fly at all, despite the greatly enlarged pectoral fins of the group. A freshwater African osteoglossid, the butterfly fish *Pantodon* and the gastropelicid flying hatchet fish *Gasteropelecus* undoubtedly leave the water on short flights that invlove flapping of pectoral fins. *Gasteropelecus* has extremely well-developed pectoral muscles that make up 25% of the body weight, more than in most birds. The flying hatchet fish beat the fins both during taxiing and in flight. One or two hemirhamphid species posess enlarged pectoral fins and show true gliding flight rather than simple leaps.

A review by Davenport (1994) concentrates upon exocoetid flying fish (Exocetidae; Beloniformes; examples: *Cypsilurus, Exocoetus*). Adult flying fish may be broadly divided into two categories; 'two wingers' (e.g., *Exocoetus*) in which the enlarged pectoral fins make up most of the lifting surfaces, and 'fourwingers' (e.g., *Cypsilurus*) in which both pectoral and pelvic fins are hypetrophied. The head and body are cigar-shaped, with a pronounced squared-off cross section for much of the body length. In some genera the cross section is virtually rectangular (e.g., *Cypsilurus*), in others more polygonal. The flat-bottomed nature of the body has been implicated in yielding extra lift. The presence of large swim bladders in flying fish is probably a consequence of the heavily muscular nature of the fish and the size of dense structures such as the pectoral girdle and fins; a large gas space will be needed to achieve neutral buoyancy. In all flying fish there is distinctly asymmetrical, hypocercal tail, the lower (hypocaudal lobe) of the caudal fin being elongate and stiffened.

The pectoral girdle of *Exocoetus* is considerably enlarged by comparison with most teleosts, the coracoid and scapula being particularly hypertrophied. The pectoral fin rays are considerably modified from the normal teleost pattern.

A continuing controversy concerns the question of whether flying fish do or do not have control over the direction of flights. Control of flight direction would seem to be feasible, but as well as requiring appropriate muscular function, such control would also imply to see well in air, to permit monitoring of sea surface. Flying fish have large eyes, and triangular section of the head means that they have a good downwards field of view (also read Chapter 15). Details of centre of gravity (CG) are given. All flying fish are 'shoulder-wing' monoplanes, in which the CG is appreciably below the centre of pressure of the expanded and highly cambered pectoral fins. Flying fish probably fly mainly to excape from predators.

Pectoral fins are used in terrestrial locomotion (e.g., *Clarias*, snakeheads, mudskippers). In Lophiiformes the pectoral base is thick and muscular, and the rays are handlike. The frogfish *Histrio* climbs among sargassum weeds, moving each pectoral and pelvic fin independently, and actually clasping the branches with its prehensile fingerlike dermal rays. The goby *Gobius* and scorpionfish *Scorpaena* have less highly modified pectorals, but can move along the bottom either by drawing the whole fin in toward the body, or passing a wave of successive contractions along the adductors of the fin rays.

In *Pomatoschistus lozani* (a benthic goby fish), pectoral fin adduction is of great importance in generating a forward propulsion. The pectoral fins of gobies seem better adapted to powerful adduction than a generalised teleost. The proximal radials form a large rigid shoulder plate with a long distal margin on which a high pectoral fin articulates. The fin muscles are strongly developed and assure, together with the large pectoral fin powerful-dragbased-pectoral propulsion.

The morphological adaptations for powerful adduction, however, are at the cost of the manoeuvering abilities of the pectoral fins (Adriaens et al. 1993).

Morphological adaptations to a semi-terrestrial existence stem mainly from the differences in the densities of air and water. The much lower density of air means that locomotion on land requires greater muscular strength and a modification of the locomotor organs. In the amphibious blenny, *Alticus kirkii*, the pectoral fin radials are movable, the pelvis is strongly developed, the fish has short strong pelvic fin rays and the muscles are modified strongly. There are hooks at the ends of anal, pectoral, and pelvic fin rays. All these features are considered to air climbing on steep rocks while out of water. Similar modifications are reported in *Mnierpes macrocephalus* in which the anal fin has an incised membrane between the rays. The ventral aspects of the body, anal, pectoral and pelvic fins are padded.

In the family Synancejidae, the goblinfish *Choridactylus multibaris* has the three lower rays of each pectoral fin detached from the web, protected by conical horny caps, and capable of independent movement. The searobins, family Triglidae, also have the lower rays of the large fanlike pectorals separate and capable of independent movement as the fish "feels" its way forward over the bottom. The hillstream loaches, family Homalopteridae, adhere to the rocks with large sucking disc formed by the pectoral and pelvic fins; they can creep using the anterior fin rays. The lungfish *Protopterus* often rests on the bottom supported on the pectoral and pelvic fins.

The reader is referred to Geerlink (1983, 1987, 1989) for further details of pectoral fin function in locomotion.

4.3 Median Fins

The median fins of most fishes are large, conspicuous structures whose evolutionary origin within the chordate clade probably preceded that of appendicular structures, and yet we presently know remarkably little about how the median fins function during fish locomotion.

In elasmobranchs the internal supports to the dorsal or to the anal fin is a flat mosaic of cartilaginous plates called radials. The outer of the two or three rows of radials (which are joined by ligaments) projects out into the flat fin base, sometimes extending almost to the fin margin. The tightly packed ceratotrichia greatly outnumber the underlying radials, which they grasp. The fin of the shark cannot be collapsed, nor apparently can its area be altered.

In the Chondrostei, the most primitive living teleosts, including *Acipenser* and *Polypterus*, the median fin rays also outnumber their internal supports, but in all other teleosts the two correspond serially except, at the fin extremities. The rays or spines are more widely spaced, and so are their internal supports. Each of these median supports, the pterygiophores, has two or three elements: a large proximal dagger-shaped element called a basal (whose inner end may overlap the tips of the neural or haemal spines), a small mesial element which slopes backward to form a "spacer" between the successive pterygiophores and which during ontogeny usually becomes fused as part of the basal, and a distal radial element which may be grasped by the ray base.

Proterocercal (diphycercal) caudal fin probable original type with body axis continued to support caudal fin as if it were divided equally into dorsal and ventral parts. Examples are lampreys and hagfishes.

Heterocercal caudal fins, found in most elasmobranchs and in chondrosteans (sturgeons and their relatives), have the posterior section of the vertebral column bent upwards and continued almost to the tip of the fin. The haemal processes of the vertebrae are elongated, and project down and back to provide a stiff base to the expanded lower (or hypochordal) lobe of the fin. Some anterior haemal processes involved in the fin base may bear detached radial elements. The neural processes are shorter than the haemal, and carry an outer row of more numerous radials. These radials support a tightly packed row of fin rays (ceratotrichia in elasmobranchs, lepidotrichia in sturgeons) which occupy the long low ridge which is the upper (or epichordal) fin lobe. The hypochordal lobe is much deeper, and is strengthened by longer rays which grasp the expanded haemal processes. It is often indented on its trailing edge to form a ventral hypochordal lobe with long rays, and a longitudinal hypochondral lobe with shorter rays. In more swift pelagic sharks the vertebral column is bent more sharply, the ventral lobe is extended, and the resultant high aspect-ratio outline resembles the tail of a tuna.

In higher teleosts, the caudal fin is superficially symmetrical top and bottom (homocercal). Internally, the vertebral column turns up sharply, but does not reach the hind edge of the fin as it does in elasmobranchs. On the dorsal side of the upturned vertebrae are a few epural elements probably derived from the neural processes. None of the major caudal fin rays arises here, and only minor rays above and ahead of the main caudal fin. Almost all the caudal fin arises from the lower surface of the upturned column, which may retain several centra or which may be reduced to a single rod, the urostyle. The haemal processes of the last few centra are greatly expanded, and together form a fanlike plate which retains several distinguishable elements (hypurals) in more primitive fish but which may become almost wholly fused to each other and to the urostyle in higher fish. On the dorsal side of the upturned terminal vertebrae are a few epural elements probably derived from the neural process. Anteriorly, the caudal skeleton grades almost imperceptably into the backward sloping haemal and neural processes from the vertebrae of the caudal peduncle. In those teleosts which swim primarily with pectoral, dorsal, or anal fins, the propulsive importance of the caudal fin is often reflected in the number of caudal fin rays. These are typically 19 in soft-rayed fishes and 17 in spiny-rayed groups, but in groups relying on the dorsal, anal, or pectoral fins as principal propellers the caudal rays may be reduced to 12–15 (Zeiformes, Labridae, Scaridae), 9–12 (most Tetraodontiformes) or even none in Molidae.

The caudal skeleton has received a great deal of recent attention, for it has proved to be of considerable value in interpreting teleostean relationships. "Pleurostylar" and "stegural" types of caudal skeleton are reported in lower teleostean fish groups.

In salmoniform, cypriniform, and a few other teleost groups there is an adipose fin lying between the dorsal rayed fin and the tail. It is often merely a small tab in adult fish, but in some catfishes the adipose fin is both long and high. The adipose fin almost always lacks supporting structures other then fin actinotrichia in the membrane. Muscles are absent, or rarely form inclinators inserted on a median membrane. The adipose fin may be significant chiefly for hydrodynamic reasons during the juvenile stages.

5. ANATOMY OF RAYS AND SPINES OF FINS

Fishes rely largely on their fins for swimming. Whether the locomotory action consists of movements of the various fins or of wave-like lateral flexures of the trunk, fins of fishes serve as the major planes against which the propulsive force is generated. In locomotion by lateral flexures of the trunk, the caudal fin is the main plane producing the propulsive thrust. On the other hand, in the type of swimming utilizing fin movements only, the pectoral, dorsal, and anal fins, used singly or in combination, provide the principal thrust. In addition to their propulsive function, paired and unpaired fins also serve as hydrofoils, stabilizers, rudders, and brakes. Fins also serve other functions, such as in reproduction, protection, and adhesion. To perform these numerous functions, fins must not only be moveable and flexible but also be fairly rigid at other times to effect counter forces. In teleostean fishes, the narrow attachment of the fin and the unique muscular and endoskeletal arrangements of the fins permit considerable mobility. Similarly, fin rays provide a compromise between flexibility and support.

The mechanical plan of the fin rays and fin spines is easier to comprehend then that of the myomeres. The flexible web of each fin is a double sheet supported by stiff rods like spokes of a fan. Each rod is attached at its base through a hinge or swivel joint to internal supports buried in the body. Each rod is operated by sets of muscles which can swing it in one plane or more planes. The whole complex of web, rods, internal supports, and muscles which make up a fin is usually inserted into the body as a largely self-contained unit without serial correspondence to the adjacent myomeres.

Teleost fin rays and spines arise embryologically as thickenings in the basement membrane of the dermal fin fold. The opposed pairs become fused along most of their length to form the shaft of the ray or spine. The halves remain separated at the base, where they are expanded to form a pair of tongs which in fin rays grasp the outer edge of their internal support. Fin rays may branch fore-and-aft several times as they approach the outer edge of the web, and their shafts are divided into segments by regularly spaced rings which probably contribute flexibility. Fin spines, in contrast, come to a point at their tips, rather than branching; the halves of the shaft are fused almost to the base to form a rigid, unsegmented spear, and the base typically forms a strong hinge on the internal support which can swing in the median plane but not laterally.

In adult teleosts the spines, rays, and internal supports are ossified; the bony rays are called lepidotrichia, and are thought to be homologous with the scales. The fins of elasmobranchs are supported by ceratotrichia composed of collagen; they are more closely packed than teleost lepidotrichia but are also paired structures, each grasping a median cartilaginous support. Very fine horny filaments, the actinotrichia also occur in the fin fold of larval teleosts, and may persist in the outer edge of the adult fin membrane.

Rays in the dorsal and anal fins of most teleosts correspond one-to-one with their internal supports, and can be erected, depressed, or inclined freely on a universal joint. In the other teleost fins, and in elasmobranchs, the rays are more numerous then their supports and are usually more restricted in their movements.

The dermal fibres (ceratotrichia) are homologous with fibrous rays (actinotrichia), which are localized at the distal ends of bony fins in teleosts and also within the adipose fins of salmonids, siluroids, characins, synodontids and certain other fish. It is equally clear that ceratotrichia and actinotrichia are members of the collagen class of scleroproteins.

Most catfish have large spines in their dorsal and pectoral fins. Similar spines occur in quite a variety of other fish, for instance *Gasterosteus* (Gasterosteiformes) and *Tricanthus* (Perciformes) (in both these spines are in the dorsal and pelvic fins). The principal function of these spines is protective. Some catfish have evolved poison glands associated with their spines. The dorsal pterygiophores, anterior vertebrae and pectoral girdles of catfish are modified to form a strong and rigid framework to support the dorsal and pectoral spines. There is some variation in the strength of the framework and this can to a large extent be correlated with variation in the size of the spines.

Nauen and Lauder (2000) have focused on the importance of finlets in the locomotion of *Scomber japonicus*.

5.1 Structure of Fin Rays

Here, the structure of fin rays of *Tilapia mossambica* has been described. The fin rays consist of a number of lepidotrichial segments, each segment consisting of a pair of plate-like elements separated by a thin band of tissue. These elements are called hemisegments. The area in between each two hemisegments is called intrasegmental region. The intersegmental region, on the other hand, separates the adjacent segments from one another.

The mode of development of the hemisegments appears in several ways similar to that of fish scales. Each hemisegment is contained by an envelope of scleroblasts which secrete collagen fibres in a unipolar fashion. Calcification takes place as a result of deposition of hydroxyapatite-like crystals between the collagen fibres. However, the orientation of these fibres is not as regular as that of fibres occurring in scales.

The epidermal cells of the finfold or fin membrane are similar to those in other parts of the body. The collagen layer system of the fin membrane closely follows the wavy profile shown by the proximal epidermal cells. The bulk of fin tissue is formed by the dermal constituents including many fibroblasts interspersed with occasional capillaries, nerve fibres, melanophores and erythrophores.

The development of the peculiar folding pattern of the basal epidermal cells and the underlying collagenous tissue is limited to the pliable fin membranes. The skin of the body does not show the extensive folding. It may have to do with the stresses occurring in the fin membranes when the fin rays are raised or lowered during swimming or during a change in the behaviour pattern. This would apply for the unpaired as well as the paired fins. At the same time, much of the intersegmental tissue consists of loosely connected components because of the presence of massive aggregations of collagen fibres.

The evolution of hard spines out of soft fin rays is one of the main criteria separating the more modern Acanthopterygii from the Malacopterygii.

Details on the relation between structure and bending properties of teleost fin rays are given by Geerlink and Videler (1987).

6. COLLATERAL INFORMATION TO SKELETON

Reduction in bone ossification in deep-sea fishes is accompanied by protein reduction. This is an obvious advantage to the maintenance of neutral buoyancy, because skeletal tissue may have a density as high as 2–3 and protein 1.33 g/ml. The loss of protein occurs mostly in the trunk and caudal musculature of bathypelagic fishes, where it includes skeletal reduction.

Although the liver of sharks is probably the most important buoyancy organ since it may contain a good deal of low density squalene, density regulation also results from skeletal tissues as well as white muscle and skin. In certain teleosts, such as the lumpsucker (*Cyclopterus lumpus*), low density is derived from a cartilaginous skeleton. The skull, vertebral column and branchial skeleton are almost completely uncalcified, with the average density 1.04 g/ml. Much persistent cartilage and weak skeletal ossification as well as incomplete mineralization of scales also characterizes the swim bladderless notothenoid (Antarctic toothfish *Dissostichus mawsoni*), which is neutral buoyant. Cartilage (density of 1.1) is substituted for bone (density 2.0) in the skull, pectoral girdle and caudal skeleton of *D. mawsoni*.

A limited number of studies indicate that lipid storage in fish is fairly common in the ocean. The castor-oil fish *Ruvettus pretiosus*, stores 30% lipid in frontal bone. Oil-filled bones in the cirrhitid, *Cirrhitus pinulatus* and the labrid, *Cheilinus rhodochrous*, have 52–90% lipid. 66–80% lipid in bones of a chaetodontid, *Chaetodon ornatissimus*, is present. Some species of fish including the labrid *Semicossyphus pulcher* and the sablefish *Anoplopoma fimbria*, store the majority of their body lipids in bones (25–93%). Most Jamaican reef fishes stored more lipid in the neurocranium than the vertebral centra. Triglyceride is the major lipid class in most of these fishes with oily bones (74.1–93.7% as percent lipid); cholesterol and phospholipid were two other lipid classes in the bones.

Skeletal fatty acid composition of the cardinal snapper, *Pristopomoides macropthalmus*, is probably controlled by its diet consisting of larval forms from deeper colder water.

The neurocranium is a major lipid storage tissue in the ocean surgeon, *Acanthurus bahianus*. The neurocranium in these tropical herbivorous marine fishes provides a readily available pool of energy-rich lipid for use during short term starvation stress, which might be incurred during intense tropical storms in the Caribbean. The neurocranium of the uieja, *Pimelometopon darwini*, and the cojinoba, *Seriolella violacea*, contained 29.5 and 26% lipid (as percent dry weight), mostly as triacylglycerol.

The reader is referred to Phleger (1987 1988a, b, 1991), Phleger and Laub (1989) and Phleger and Wambeke (1994). For the significance of bone lipids, and reduction of skeleton density in buoyancy at depth, also read Pelster (1997). (see Tables 4.4 and 4.5).

Table 4.4 Bone lipids of some species of marine fishes

Tissue	Tissue lipid (mg)	lipid (% dry wt.)	Triglyceride (%)	Sterol (%)	Phospholipid (%)
Skull	65.4	5.5	76.5	15.3	8.2
Spine	104.7	5.0	57.8	30.0	11.9
Bone lipids of *Alectis ciliaris*					
Skull	14,041.7	23.9	96.8	0.9	2.3
Spine	2000.5	4.5	94.4	4.0	1.6
Bone lipids of the giant hawkfish, *Cirritus rivulatus*					
Skull	29.0	0.8	88.8	15.4	5.8
Spine	76.4	2.0	34.6	25.2	9.4
Bone lipids of the spotted-faced rock cod, *Cephalopholis pachycentron*					
Skull	12.1	0.2	45.9	51.0	3.1
Spine	27.9	0.4	31.1	65.8	3.1
Bone lipids of the male parrot fish, *Scarus sordidus*					

Source: Phleger (1987)

Table 4.5 Bone lipids of some species of marine fish

Tissue	Lipid (mg)	Lipid (%)	Triglyceride (percent of lipid)	Cholesterol (percent of lipid)	Phospholipid (per cent of lipid)
Lipids of *Acanthurus bahianus*					
Vertebral centra	17.0	6.9	84.9	5.5	9.6
Neurocranium	600.0	29.7	89.7	5.9	4.4
Lipids of *Priacanthus arenatus*					
Vertebral centra	136.0	10.1	90.7	3.9	5.4
Neurocranium	163.0	14.7	93.7	4.1	2.2
Lipids of *Synodus intermedius*					
Vertebral centra	2.6	0.4	27.6	3.4	69.0
Neurocranium	6.8	2.5	71.0	6.4	22.6
Lipids of *Tylosurus crocodilus*					
Vertebral centra	134.5	6.5	89.6	5.5	4.9
Neurocranium	20.4	1.7	76.0	2.1	21.9
Lipids of *Urolophus jamaicensis*					
Vertebral column	12.6	0.6	72.9	1.9	25.2
Chondrocranium	0.8	0.1	35.7	7.1	57.1

Source: Phleger (1988a)

CHAPTER 5

Swimming Muscles: Myotomal Muscle Fibres

1. INTRODUCTION

Many of the most influential studies in fish locomotion during the recent past have involved a fruitful combination of hydrodynamic modelling, kinematics, and analyses of external body shape. It has been clarified how hydrodynamic performance is related to the distribution of external body surface area in different species. In contrast, scant information presently relates interspecific variation in external body shape to variation in the internal axial morphology. Furthermore, longitudinal variation in the axial musculature and skeleton within the body of individual fish generally has not been quantified despite the fact the such variation has significant consequences for understanding how locomotor forces are generated and transmitted. Some studies of longitudinal and interspecific variation in the axial morphology and function of fishes have examined the importance of myotomal fibre trajectories, biochemical properties such as stiffness, and muscle strain. Only recently, studies have been initiated to quantify both intraspecific longitudinal variation and interspecific variation in the axial structures that are relevant to understanding locomotor function. Especially lacking are quantitative studies of the overall shape of myotomes and their relationship to the axial skeleton.

The description of axial morphology is critical to understanding locomotor mechanics for at least three reasons. First, a single myotome may span many vertebrae and potentially induce bending at many segments simultaneously. Second, the fish myotome changes shape considerably along the body, and these changes in shape will affect the transmission of force to the axial skeleton. However, there are no published measurements of longitudinal shape changes in the myotomes of fishes. Third, the relationship between the axial skeleton (neural and haemal spines and vertebral centra) and myotomes may also change shape along the body, yet there is little information available on skeletal-myotomal relationships.

Previous structural research on myotomes has focused on considering fibre types and their innervation and spatial distribution, and on general description of the position of one or a few myotomes within the body. In addition, the phylogenetic distribution of the taxa has varied widely. Recently, a study has been aimed to quantify myomeric and skeletal morphology in a few fish taxa that vary significantly in body shape within a relatively small monophyletic clade (e.g., Jayne and Lauder 1994b).

Cyclostomes. Muscles comprise the major type of tissues in the lamprey body. They are important both for the mechanical functions of swimming, feeding and ventilation and for storage of lipid and glycogen. The metabolic stores in body muscles are particularly important during their upstream migration, sexual maturation and spawning. Likewise, lipid stores in adipose tissues of the ammocoete support the metabolism of the lamprey of the animal during metamorphosis.

Striated muscles of the lamprey are classified as somatic or visceral, according to their embryological origins. The somatic muscles, including the extraocular, myotomal and fin muscles, are derived from the myomeres, whereas the visceral muscles, which are used for feeding and ventilation, arise in the lateral plates of the visceral arches in the head and gill regions. The visceral and somatic muscles are innervated by separate motor systems and remain anatomically distinct at all stages in the life cycle of lamprey.

The organization of visceral and somatic muscles in cyclostomes is phylogenetically significant because it is, in some respects, intermediate between those of acraniate chordates and the jawed vertebrates. In particular, the myotomal subunit in the lamprey may be a special transitional stage between the myotomal lamellae of *Branchiostoma* and the functionally separate muscle fibres of higher fish. One of the most striking examples of divergence between the jawless and jawed vertebrates is the organization of the visceral muscles of the head.

Gnathostomatous Fish. A great variety of body shapes and modes of locomotion are found among fishes. Locomotion is achieved using a combination of the paired and unpaired fins and segmentally arranged myotomal muscles. Adaptations for aquatic locomotion account for many of the specialized features of fish muscle. The energy expenditure of locomotion in water is not linearly related to speed as it is for many terrestrial vertebrates. In general, the power requirements of swimming are thought to arise as a function of body size and velocity. This reflects the way drag forces on body increase with speed. Some species have sacrificed speed for manoeuverability using combinations of their paired and unpaired fins as the prime means of locomotion. In others, body shape and the arrangement of fins is important in providing an effective body-camouflage.

2. SOMATIC MUSCLES: ORGANIZATION OF MUSCLE FIBRES

2.1 Fins

Few studies have examined the organization of muscle fibres in the fins. The muscles of the dorsal fins in lamprey have a subunit organization. These muscles consist of bundles of fibres which are situated at the base of the fins parallel to the fin rays. Their contraction bends the fin to the same side. In small adult *Petromyzon marinus* the subunit in the fin muscle consists of about 10 central fibres and an outer layer of parietal fibres. In some species with enlarged pectoral fins (e.g., notothenoids and *Chimaera*) individual fibres often run in a parallel fashion within the muscles, in other fish the fin muscles are pinnate. Different fibre types are often mixed together, as in the limb muscles of terrestrial animals rather than anatomically separated, as in the myotomal muscles (see Johnston 1983). Not surprisingly the anatomy of the skeletal elements and muscles in the caudal part of fish varies enormously with swimming style, particularly with the power needed to be transferred to the tailfin (Johnston and Altringham 1991, see literature; see Chapter 6).

2.2 Myotomal Muscles: Axial Segmental Body Propulsive Muscles

A wide variety of body forms and modes of swimming have been described among the fishes. The overwhelming majority of papers published on fish muscle have dealt exclusively with the segmental myotomal musculature. In agnathans the myotomes are of shallow W-shape. In teleosts and elasmobranchs the myotomes have a complex W-shape and form a series of overlapping cones (Fig. 5.1). The muscle fibres in each myotome inserts into a connective tissue sheath or myocommota. Functional aspects of the shape and arrangement of the myotomes has recently been reviewed both in different groups of fishes and in development. The orientation of fibres within the myotome is complex and varies along the body and with distance from the vertebral column. In general, superficial fibres (dark or red or slow) run parallel to the surface whereas deeper fibres (white or light or fast) make angles upto 40° with the long axis of the body of gnathostome fishes. A comparison of some principal characteristics of red and white muscles of fishes has been given by Webb (1993a; Table 2). The significance of this arrangement is thought to be to allow similar degrees sarcomere shortening at different body flexures. Thus optimal overlap between thick and thin filaments and hence maximum tension generation is achieved at all depths within the myotome (Figs. 5.2–5.5) (Bone 1978; Johnston 1981; Johnston and Altringham 1991). The reader is also referred to a recent contribution by Webb (1993a). All myotomal red fibres are approximately equidistant from, and paralled to, the median axis of body bending. All sarcomeres shorten by the same amount in the length range where force generation is > 96% of the maximum. In contrast, myotomal white fibres vary in distance from the median plane, and subtend angles 10 to nearly 40° to the body axis.

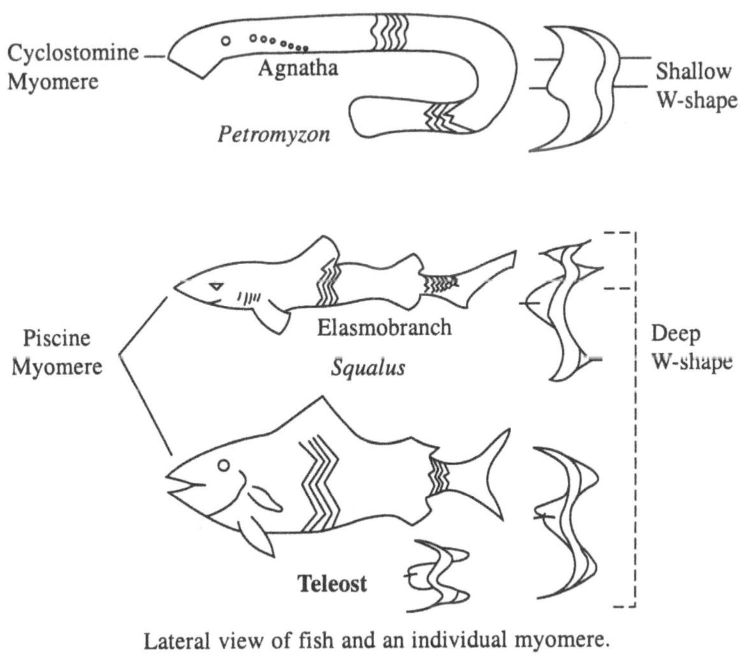

Lateral view of fish and an individual myomere.

Fig. 5.1 Simplified diagrams of myotomal shape in various fish groups. An enlarged view of a midtrunk myotome in each is shown at right; a teleostean caudal, myotome at bottom.
***Source:* Bone (1978)**

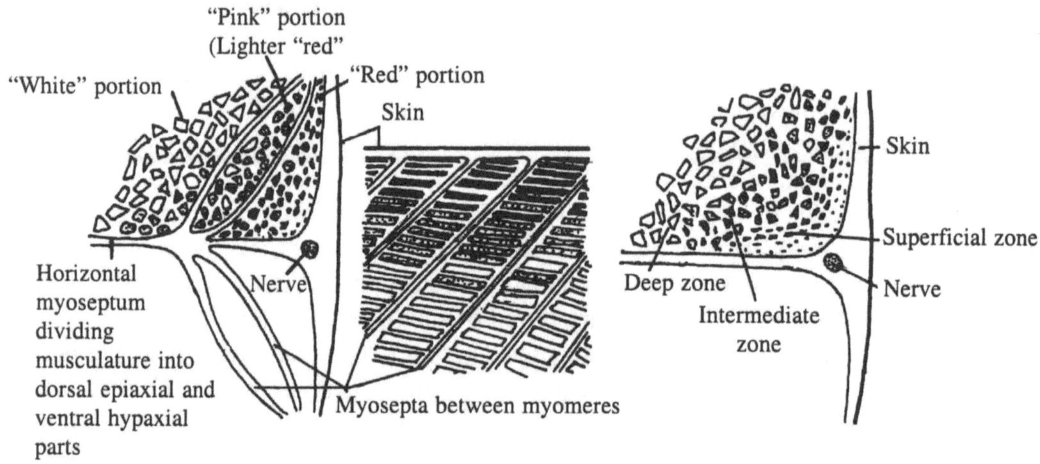

Gadus morhua: Trunk musculature

Fig. 5.2 A–C. Arrangement of trunk musculature in cod (*Gadus morhua*). A. Shows a typical transverse section through the trunk musculature in the lateral line region. Connective tissue septa appear to divide the muscle into portion with different macrocopical colour. B. Shows the location of muscle fibres with different colour relative to myomeres in a horizontal section. No connective tissue septa separate groups of muscle fibres within the myomeres. C. Shows a transversal section within the confines of one myomere (compare with Fig. 5.2A). No septa separate groups of fibres. Fibres with different macroscopical colour are instead arranged into three zones with intermingling of fibre types in the transitional regions. Macroscopically "red fibres" are drawn black; "pink" ones stippled, and "white" fibres are left white.
***Source:* Korneliussen et al. (1978)**

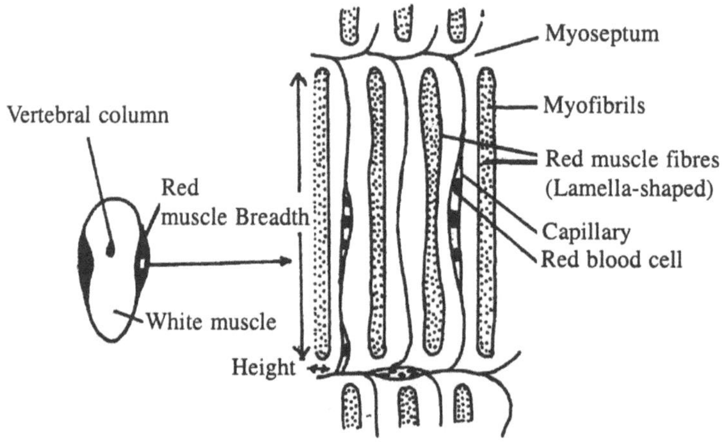

Fig. 5.3 A diagrammatic representation of the shape and orientation of anchovy (*Engraulis* sp.) red muscle fibres in a position one third of the length from the tail.
***Source:* Greer-Walker et al. (1980)**

The gear ratio of white fibres is four times higher than for red fibres, which, combined with higher Vmax' results in white fibres producing movements that are an order of magnitude faster than those due to red fibres.

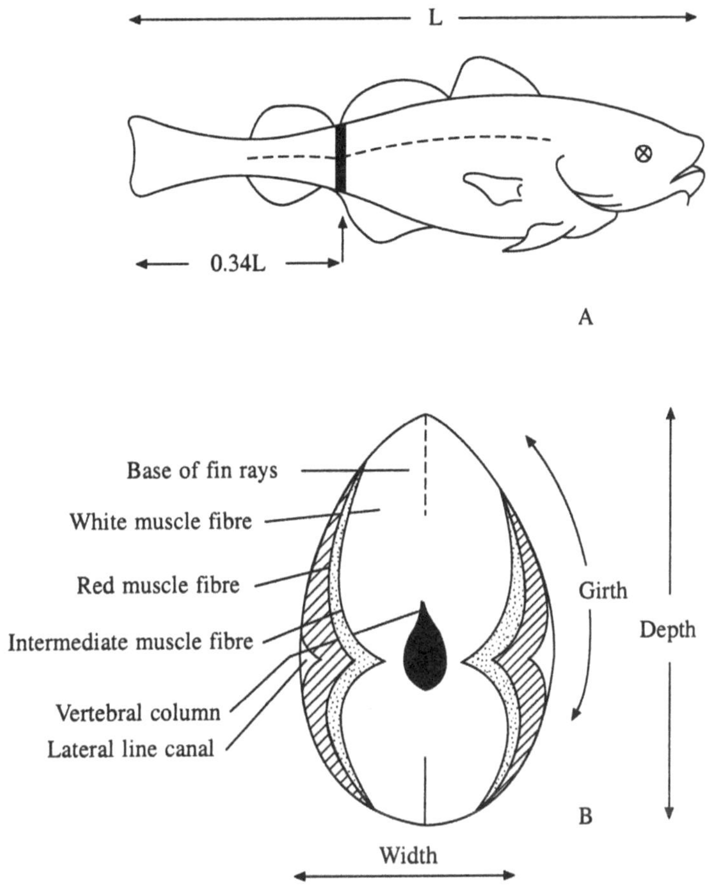

Gadus morhua:
Maximum flexure during sustained swimming.

Fig. 5.4 (A) Arrows indicate the sample area and the point at which maximum flexure occurs during sustained swimming. (B) Transverse section at the point of maximum flexture.
***Source:* Greer-Walker (1970)**

Fibres are also classified in this order: slow superficial red fibres, fast red fibres between slow and white fibres, and fast white fibres. The number of distinct fibres types varies from two to five depending on species (Johnston 1981, 1983b). Johnston and Altringham (1991) had divided muscle fibres into two main types, tonic and twitch. Tonic fibres occur as a thin superficial layer in some bottom living sharks (e.g., dogfish) and in certain teleosts (e.g., *Noemacheilus barbatulus*) and are thought to have a postural role in bending the trunk. The white, and red fibres are the twitch fibres. The reader is also referred to Kilarski and Kozlowska (1987).

The subunit organization of fast and slow muscle fibres in the myotomes is a unique characteristic of lampreys. Each subunit spans the horizontal dimensions of the myotome and consists of several inner layers of central fibres of the slow type. The central fibres are unusual in that they are electronically coupled to one another and only the most medial fibres receive direct motor innervation. In other respects, the cellular properties in the somatic muscle fibres in the lamprey are similar to those of fast and slow fibres in other vertebrates.

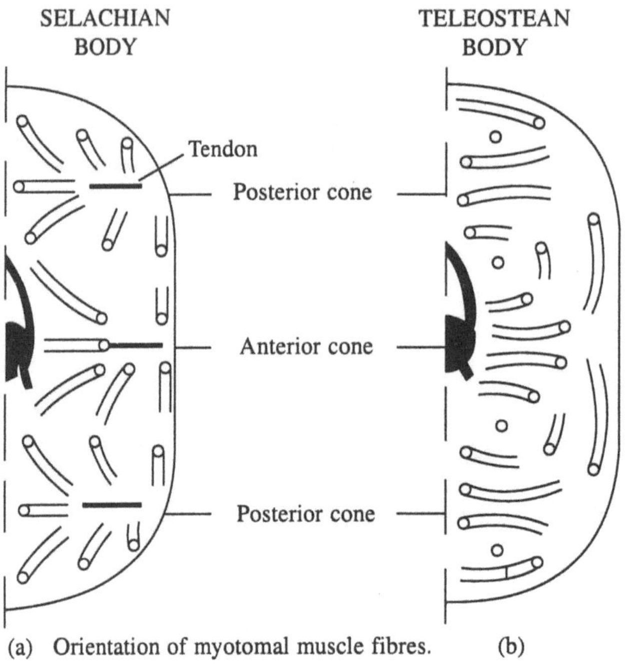

Fig. 5.5 Thick transverse slices across (a) selachian, and (b) teleostean body showing orientation of myotomal muscle fibres as viewed from behind.
***Source:* Bone (1978)**

The relationship between classifications of fish skeletal muscle based on colour and their physiological and metabolic characteristics together with a suggested terminology is given in Table 5.1

Fig. 5.6 shows the distribution of fibres in the segmental muscles of dogfish (*Scyliorhinus canicula*) and carp (*Carassius carassius*) based on histochemical criteria. Meyer-Rochow et al. (1994) presented cytochemical and histological details of muscle fibres in the Southern smelt (*Retropinna retropinna*).

Raamsdonk et al. (1978) discussed the differentiation of muscle fibre types in *Brachydanio rerio*. They rejected the possibility that neuronal innervation is involved in the early stages of the differentiation of muscle fibre types. A mechanical influence can not be excluded as a determining factor in the differentiation of muscle fibres. It appears that the contractile properties of the muscle fibres imply the development of a distinct type of energy metabolism. Of the muscle fibre types in the adult musculature, the so-called white fast fatigue fibres appear first, the next the show fatigue-resistant red fibres. There are no grounds to suppose that the different fibre types arise from distinct populations of myoblasts. The adult fibre types develop from a common embryonic fibre type, first by a differentiation in the myofibrillar properties and in the later stage by a differentiation in the metabolic properties.

2.3 Ultrastructural Characteristics of Myotomal Muscle Fibres

The parietal and central muscle fibres in lamprey myotomes have the same general ultrastructural appearance as in other vertebrate striated muscles. Nevertheless, the two types of fibres exhibit

Table 5.1 Relationship between classifications of fish muscle fibre types based on colour and physiological characters

Colour Classification	Physiological Characters
Dogfish (*Scyliorhinus canicula*)	
Superficial (S)	Tonic
Outer red (OR)	Slow, highly aerobic non-action potential generating fibres
Inner red (IR)	OR fibres distinguished histochemically by their higher staining for succinic dehydrogenase activity and less alkaline-stable myofibrillar ATPase
Outer white (OW)	Fast twitch, intermediate aerobic and highly developed glycolytic capacity
Inner white (IW)	Fast twitch glycolytic, low aerobic capacity
Common carp (*Cyprinus carpio*)	
Red (R)	Slow, highly aerobic, non-action potential generating fibres utilized for sustained swimming
Pink (P)	Fast twitch highly aerobic and glycolytic capacities Fatigue resistant, utilized in steady performance, pink and white fibres are involved in addition to red muscle particularly at speeds approaching *U*-crit*
White (W)	Fast twitch glycolytic, low/intermediate aerobic capacity Although carp pink and white fibres are twitch fibres they differ from the corresponding fibres of dogfish in that they are multiply innervated Fast fatiguing, utilized for "burst swimming"

* Fatigue speed = Critical velocity, *U* crit
(Modified after Johnson 1983a)

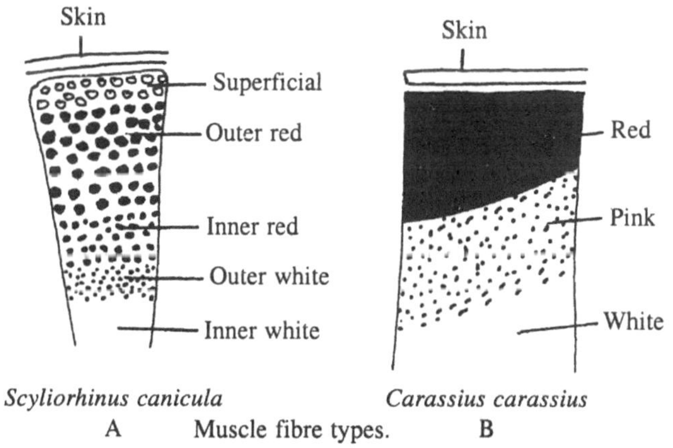

Fig. 5.6 Diagrammatic representation of the distribution of muscle fibres types in (A) dogfish (*Scyliorhinus canicula*) and (B) crucian carp (*Carassius carassius*). *Source:* Johnston (1983a)

a number of quantitative differences (e.g., in mitochondria, lipid droplets, myofibrils, sarcoplasmic reticulum plus T-tubules) which are consistent with their slow and fast physiological properties. Myofibrils in parietal and central fibres have a typical banding pattern, with A, I, M and Z lines.

A striking characteristic of central fibres is the pronounced development of transverse (T) tubular system and of the sarcoplasmic reticulum. The T-tubules and sarcoplasmic reticulum are present but more sparse in the parietal fibres. Mitochondria and lipid probably contribute to the pink or brownish colour of parietal fibres in lamprey myotomes, while their infrequent occurrence leaves the central fibres relatively clear.

The myotomes of the hagfish also contain red and white muscle fibres. They do not form tight subunits characteristic of lamprey. Whereas T-tubules are situated along the Z-line of myofibrils in lamprey myotomal fibres, they occur adjacent to the A–I bands in myofibrils of the hagfish.

A comprehensive account in this respect is given by Johnston (1983a; see also Johnston and Altringham 1991). Table 5.2 summarizes a selection of available data on the ultrastructure of fish slow and fast glycolytic fibres.

Table 5.2 Ultrastructure of fish myotomal muscles

Parameter	Muscle fibre type	
	Slow fibres (S)	Fast fibres (FG)
Z-disc	'thick'	'thin'
M-line	present	present
Myofibrillar volume (%)	60.1 Eel (*Anguilla anguilla*)	80.4 Eel
	60.3 Perch (*Perca fluvitilis*)	78.0 Perch
	43.1 Tench (*Tinca tinca*)	73.1 Tench
Mitochondria fraction (%)	21.4 Eel	1.2 Eel
	22.0 Perch	6.0 Perch
	22.9 Tench	4.5 Tench
Sarcoplasmic reticulum percent muscle fibre volume	2.1 Eel	6.0 Eel
	4.4 Tench	3.2 Tench
	5.1 Rainbow trout (*Salmo gairdneri*)	13.7 Rainbow trout
T-system percent myofibrillar volume	0.31 Eel	0.38 Eel
	0.18 Perch	0.27 Perch
	0.29 Tench	0.40 Tench
	0.10 Rainbow trout	0.40 Rainbow trout

(Modified after Johnson 1983 a)

In general, the characteristics of fast aerobic fibres are intermediate between those of S and FG fibres. Myosin filament lengths (~1.6 μ) are similar in all fibres; however, actin filament lengths increase in order: slow > fast aerobic > fast glycolytic. This reflects the relative contraction velocities of these fibre types because a larger actin filament means fewer sarcomeres in series per unit fibre length and a slower speed of shortening.

For most fish, the vascularization and mitochondrial volume of slow fibres correlates well with their predicted scopes for aerobic activity. For example, mitochondria occupy 18% to 24% of slow fibre volume in *Scyliorhinus canicula* and 34% in *Etmopterus spinax*, sedentary bottom-living and active mid-water elasmobranchs, respectively.

An interesting exception to this general correlation is that of skipjack tuna (*Katsuwonus pelamis*). This species maintains elevated brain and slow muscle temperatures by means of a countercurrent vascular heat exchanger. Both field observations of swimming behaviour and morphologic physiological data on the respiratory and cardiovascular systems of this species point to one of the highest aerobic scopes for activity of any fish; however, slow muscles of the skipjack tuna have a mitochondrial fraction (MF) that is only half (~16%) that of another scombroid, the Atlantic mackerel (*Scomber scomber*), which does not operate muscle temperatures much above ambient temperature. It seems possible that an elevated body temperature allows mitochondria to operate somewhat more efficiently such that a higher scope for aerobic activity can be achieved at the same time as reducing the proportion of muscle fibre volume given over to mitochondria.

The volume density of mitochondria is usually 10–30 times higher in slow than fast myotomal fibres. Mitochondria occupy around 46% of the cell volume of slow fibres of the European anchovy (*Engraulis encrasicolus*). Anchovies are active pelagic species that during filter feeding expand the gape of their mouths, increasing the body cross-section by a factor of around four times. The continuous activity and high drag forces associated with this method of feeding require a high and sustained power output from the slow muscle. Mitochondria in anchovy slow fibres are not only numerous but also have a highly developed cristae structure. About 50% of the external surface of the fibre is vascularized. In contrast, mitochondrial volume in the slow fibres of juvenile eel is only 20%, and around 66% of fibres have no direct capillary contact. Compared to anchovy, eel muscle mitochondria also have a relatively simple cristae structure. These observations are compatible with a limited and rather inflexible aerobic scope for activity in the eel.

The volume density of mitochondria is not always a good descriptor of aerobic capacity, due to variations in cristae density and aerobic enzyme activities. For example, slow muscles in certain sluggish Antarctic fish living at – 1°C contain almost 60% mitochondria by volume. In this case the amplication of mitochondria is thought to represent an adaptation to overcome catalytic and diffusional limitations at low temperature. A similar proliferation of mitochondria occurs with seasonal temperature acclimation in teleosts. For example, in striped bass (*Morone saxatilis*) the fraction of cell volume occupied by mitochondria increases 0.29 to 0.45 between 25°C and 5°C. The mean intermitochondrial spacing decreased from 2.6 to 1.4 μm over this temperature range.

A large number of factors other than capillary density affect tissue perfusion and oxygenation, including blood flow, perfusion distribution, myoglobin concentrations, and factors affecting the haemoglobin-oxygen equilibrium. Nevertheless, there is good correlation between certain parameters of vascularization and mitochondria volume. In case where mitochondria volume is subjected to modification in relation to temperature or oxygen availability, changes in capillary supply occur in parallel, suggesting that the area available for diffusion of oxygen or nutrients is a limiting factor in fixing the maximum aerobic capacity of the muscle.

Compared to slow fibres, there is even greater interspecific variation in the capillary supply and mitochondrial volume of fast glycolytic fibres. It seems likely that this reflects the different extents to which FG fibres are involved in sustained activity. Low mitochondrial densities are likely to be sufficient only for resting metabolism. Somewhat higher volumes have been found in teleosts that recruit their FG fibres for sustained activity.

Myofibril packing is more regular and dense in fast glycolytic fibres than slow fibres, reflecting the requirement of this fibre type to develop tension extremely rapidly during burst swimming.

Tonic fibres are densely packed (76%) with myofibrils which are large in diameter (2–4 μm) and irregular in cross section. Twitch fibres contain smaller diameter myofibrils (0.5–1.0 μm) which occupy from 40 to 65% of fibre volume in slow fibres and from 70 to 90% in fast fibres.

In general, the volume of sarcoplasmic reticulum (SR) is somewhat higher in fish S and FG fibres that it is in analogous amphibian and mammalian muscles. This is in part related to the higher frequency of contraction-relaxation cycles in fish compared to most types of terrestrial locomotion, necessitating a high rate of Ca^{2+} uptake from myoplasm. In most myotomal muscle fibres, the T-tubules (sarcoplasmic transverse tubular system) are located at the Z-disc. Location of T-tubules at A/I band junction is found in certain fast-acting, sound-producing muscles, suggesting that the location of the T-system is not particularly correlated with either actin filament length or muscle speed.

The structure of the A-band in teleost muscle is different from that of other vertebrates, including other fish (see details, Johnston and Altringham 1991; read also Kisia 1996).

Pool et al. (1976) studied the sera raised against actin and myosin, extracted from white muscle of the tench (*Tinca tinca*) and compared the classification of muscle fibre types in fish skeletal muscle as found with immunohistochemical techniques, with the classification as found with enzyme histochemical reactions for succinate-dehydrogenase (SDH), lactate dehydrogenase (LDH) and myofibrillar ATP-ase. It appeared that both the actin and the myosin serum were specific for white muscle fibres in fish. Further it was found that in both the A- and the I-band of the sarcomeres, fibre type specific high histochemical myofibrillar ATP-ase activity.

2.4 Distribution of Capillaries in the Somatic Musculature

Blood supply to a working muscle is of vital importance to the transport to and from the cell. Absolute blood supply is determined by two factors, vascularization as measured by capillary densities, and the rate of blood flow through the capillaries.

Teleost white muscle appears to be mostly sparsely vascularized skeletal muscle yet reported having a capillary fibre (C:F) ratio between 0.2 and 0.3 and between 0.6 and 1.0 peripheral capillaries around a muscle fibre (data obtained) for grass whiting *Haletta semifasciata*, leather jacket *Acanthaluteres* sp., flathead *Platycephalus bassensis*.

Red muscle (in flathead) is heavily vascularized having a C:F ratio between 1.9 and 2.5 and between 5.3 and 6.6 peripheral capillaries depending on the method used.

Vascularization of flathead pink fibres is intermediate between the red and white fibres having around 3.0 peripheral capillaries per muscle fibre.

The heavy vascularization reported here, has the potential to ensure adequate transport of substances to and from the continuously active muscle fibres. In contrast the reliance of white muscle on anaerobic metabolism seems a necessary consequence of the sparse vascular supply. Consequently the vascularization results reported here fit the generally accepted concepts of the role of the red and white muscle in fish. For further information, see Kisia (1996, 1999). Mathieu-Costello et al. (1996) provided morphometric data on fibre vascularization and aerobic capacity in red muscle of tuna (*Katsuwonus pelamis*).

2.5 Innervation

Agnathous fish. Parietal and central muscle fibres in lamprey myotomes receive separate systems of motor innervation. In the hagfish the red muscle fibres receive multiple innervation distributed over their surfaces, whereas each white muscle fibre receives a focal, basket-shaped synapse at one end.

2.5.1 Focally Innervated Fast Muscles

Gnathostomatous fish. Despite a wide variety of body forms and locomotor habits, the basic arrangement of muscle fibre types in the myotomes of different fishes is remarkably constant; however, this similarity belies major differences in the pattern of motor innervation between the fast muscles of different fish groups. In elasmobranchs, chondrosteans, dipnoans, and a few teleosts, fast fibres are focally innervated at one myoseptal end. Single motor end-plates represent the usual pattern of innervation found among vertebrate twitch fibres; however, elasmobranch fast fibres show an interesting deviation from this plan. In *Scyliorhinus*, *Squalus*, *Raja*, *Torpedo*, and *Rhina*, each fibre is innervated by two separate axons of different sizes that fuse to form a single end-plate. Furthermore, there are two types of motor terminal, each containing vesicles of different size ranges. Because only acetylcholinesterase has so far been demonstrated in the subsynaptic folds of the end-plates, the function of this "dual innervation" remains a mystery. It is also unclear whether a similar arrangement is found among other groups that have focally innervated fast fibres. For example, fast fibres of the European eel *Anguilla anguilla* appear to be innervated only by a single motor axon (Johnston 1983a; Johnston and Altringham 1991). (Fig. 5.7 A–D).

2.5.2 Multiply Innervated Fast Muscles

Among the teleost fishes only those considered to have relatively primitive taxonomic features have fast fibres with single end-plates. In the majority of teleosts, including all the acanthopterygian orders, the fast fibres are multiply innervated. This type of fast muscle innervation is not common among vertebrates with exception. In the case of myotomal muscles, each fibre receives a complex distributed innervation. Typically there are a large number of end-plates and several innervating motor axons.

It seems highly likely that within the basic framework of multiply innervated, fast motor systems, there will exist many individual difference between different species both in relation to mode of life and phylogeny (Johnston 1983a).

The evolution of multiply innervated fast muscles may be related to the need to achieve rapid tail-beat frequencies. There is evidence that multiply innervation has arisen independently on several occasions during the adaptive radiation of teleosts. Indeed in some orders both focal and multiply pattern co-exist. In general, the multiply innervation of pattern is associated with a greater capacity for aerobic metabolism than is found in focally innervated muscles (Johnston 1983b; Johnston and Altringham 1991; also refer Altringham and Johnston 1988; Archer et al. 1990).

2.5.3 Slow Muscles

In contrast to fast muscles, it appears that slow fibres receive a broadly similar type of innervation in all fishes. Typically, each fibre receives a number of terminations of *engrappe* type and may be innervated by one or more different axons (Johnston 1983a).

2.5.4 Muscle Recruitment and Innervation Patterns

Recruitment patterns correlate with innervation patterns. Red muscle is comprised of tonic fibres with distributed multi-terminal innervation. Power requirements are probably matched to swimming resistance by graded contractions and by varying the number of active motor units. White fibres of elasmobranchs, sarcopterygians, and primitive actinopterygians (Halecostomes, Ginglymodi, and basal elopomorph and clupeomorph teleosts) are focally innervated. These fish use red muscle at low aerobic swimming speeds, and recruit white muscle only for sprints and rapid

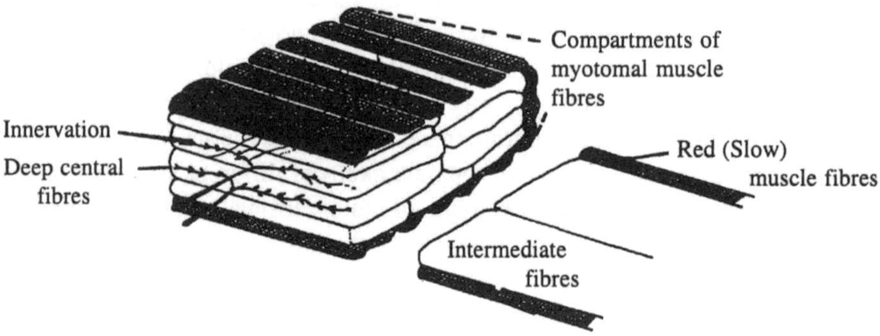

(A) Lamprey: Myotome

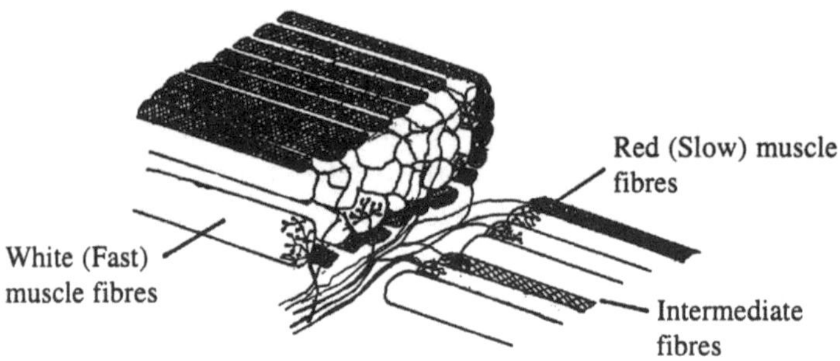

(B) Hagfish: Myotome

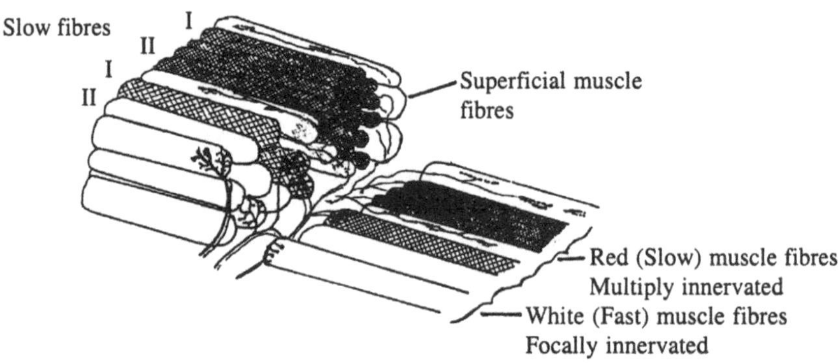

(C) *Scyliorhinus*: Myotome

Fig. 5.7 *(Contd)*

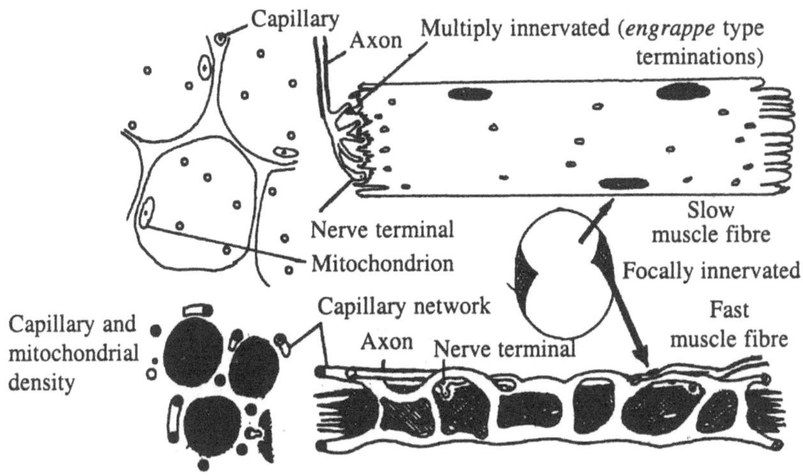

(D) Capillary and mitochondrial density

Fig. 5.7 (A) Diagrammatic view of lamprey myotome from internal aspect. Note absence of innervation of intermediate fibres; innervation of deep central fibres; and multiple innervation of slow fibres bordering muscle unit. (B) Hagfish myotome viewed from an internal aspect. Fast fibres terminally innervated, slow fibres innervated by axons passing onto the fibres from both myoseptal ends. Occasional intermediate fibres from both myoseptal ends. Occasional intermediate fibres scattered around borders of muscle unit next to slow fibres; these are apparently terminally innervated. (C) *Scyliorhinus* myotome from an internal aspect. Superficial muscle fibres overlie slow fibres (I and II); fast muscle fibres are terminally innervated and mainly of Type I. Note the dual innervation of fast muscle fibres. (D) Schematic comparison between slow fibre (above) and fast fibre. The fast fibres of some teleosts are multiply innervated. On the left, a transverse section showing comparative capillary and mitochondrial density.
***Source:* Bone (1978)**

acceleration. White muscle fibres of more derived teleosts are unique in having multiterminal innervation like red muscle fibres. This white muscle is also believed to show graded contractions permitting white muscle recruitment at relative lower speeds when it can support red muscle in sustainable swimming (Webb 1993a; see for literature; read also Section 6, and Coughlin and Rome 1996).

Several new findings of the axial undulatory locomotion of bluegill sunfish (*Lepomis macrochirus*) suggest that fishes have a much greater diversity of motor output than has been previously described. The timing of white and red muscle activity at a given longitudinal position need not be synchronous, and this was apparent near the threshold levels for recruitment of white muscle. During moderate speed unsteady swimming, durations of white muscle electromyograms (emgs) were consistently less than those recorded from the more superficial red locations. Furthermore, the onset of white activity consistently lagged behind that of red while the offsets of red and white activity were effectively synchronous. During the most rapid burst and glide swimming there was a pronounced decrease in the intensity of red recruitment, and this finding differs from previous additive models of fibre type recruitment.

It is important to emphasize that in this study of *Lepomis*, muscle recruitment over a continuum of locomotor speeds and behaviours that excluded the startle response, was examined. The startle response involves specialized neurocircuitory that produces an initial motor output with no detectable (< 1 ms) longitudinal lags in the activation of axial muscles of *Lepomis* (see details: Jayne and Lauder 1994a; see also Jayne and Lauder 1993).

3. MECHANICAL PROPERTIES OF MUSCLES

On the basis of their mechanical properties the superficial red fibres may be classified as slow. Maximum rates of contraction for cold-water species are in the range 0.4 to 0.7 muscle length/s or around 25–50% of that for white fibres. Slow fibres in fish are multiply innervated by one or more axons. During sustained swimming they are probably activated by junction potentials.

The concentrations of myoglobin in slow muscles are in the range 50 to 350 n mol mg^{-1} wet weight. Interesting exceptions are the Channichthyidae found in Antarctic waters in which myoglobin gene is not expressed. Fish slow fibres generally have higher volume fractions of mitochondria and are better supplied with capillaries than comparable locomotory muscles in terrestrial animals. This is probably related to the very high proportion of cardiac output this tissue receives during swimming. Red slow muscle mitochondria are able to utilize both carbohydrates and lipids as fuels and possible also certain amino acids. The main substrates for lipid metabolism are triacylglycerol and fatty acids in teleosts and ketone bodies in elasmobranchs.

In contrast to slow muscles in amphibia, those in fish have a conspicuous M-line and a well developed sarcoplasmic reticulum (Johnston 1983b).

4. REGULATION OF MUSCLE CONTRACTILITY

Some features of the regulation of contractility in fish skeletal muscle are illustrated in Figs. 5.8, 5.9. Tension development rises as a function of cytoplasmic free calcium concentration within the range 0.1 to 10 μM. In multiply innervated fibres the release of calcium from SR is dependent on the degree of depolarization of muscle membrane which is in turn determined by the number of motor neurons activated and their firing frequency. In fish white but not red fibres cytoplasmic calcium binding proteins (parvalbumins) play an important role in regulating free calcium concentrations. Parvalbumins are found in all vertebrate fast muscles but occur in particularly high concentrations in fish white fibres. Calcium bound to parvalbumins can be exchanged and accumulated by SR vesicles. The current view is that parvalbumins function in aiding rapid relaxation. Relaxation occurs before all the Ca^{2+} is resequestered within the SR.

Activation of myosin crossbridges in fish muscle is achieved through direct effects of Ca^{2+}–binding to troponin C as in other vertebrates. In addition, phosphorylation of myosin light-chain kinase may lead to a modulation of force production (Johnston 1981; read also Johnston and Altringham 1991).

It is noteworthy that fish myosins exhibits a subunit composition similar to that of endothermal vertebrates. The white (fast-twitch) muscle myosin displays three distinct light chains while red (slow-twitch) and cardiac muscle myosins possess only two chains. As all these light chains are different in electric charge and/or molecular weight, they constitute a very good criterion for the characterization of fibre types and fish species. Fish sarcoplasm includes a relatively light content in low molecular weight Ca^{2+} binding proteins, the parvalbumins. These proteins present several isotypes whose amount and distribution vary according to the fibre type. This subject has been studied by Huriaux et al. (1990, 1997; see literature).

(For relation with development, see Huriaux et al. 1996; Focaut et al. 1996) on barbel (*Barbus barbus*) populations.

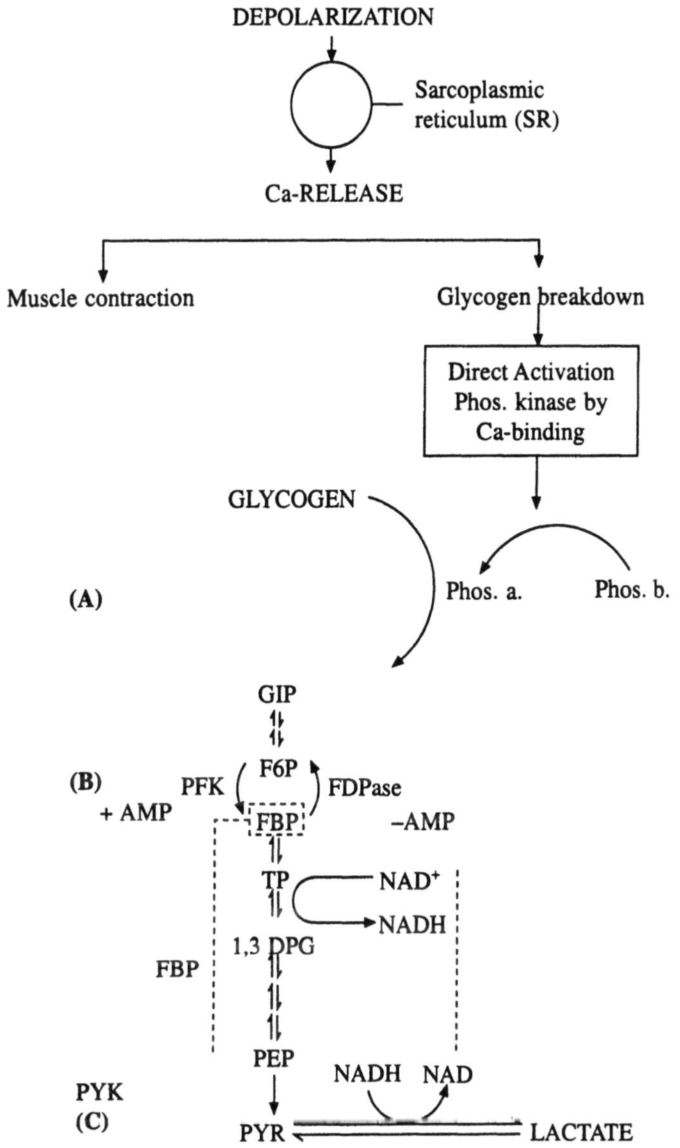

Fig. 5.8 Control of anaerobic glycogenolysis in fish white muscle. Regulatory steps include (A) Glycogen phosphorylase, (B) Phosphofructokinase and (C) Pyruvate kinase. Abbreviations: F6P-fructose-6-phosphate; Phos, phosphorylase; GIP, glucose-1-phosphate; PFK, phosphofructokinase; FBP, fructose 1, 6-biphosphate; AMP, adenosine 5′ monophosphate; TP, triosephosphate; NAD, β-nicotinamide adenine dinucleotide; NADH, β-nicotinamide adenine dinucleotide reduced form; 1, 3 DPG, 1, 3 diphosphoglyceric acid; PEP, phosph (enol) pyruvate; PYK, pyruvate kinase; PYR, pyruvate.
Source: **Johnston (1981)**

5. METABOLISM

Typically while muscles have a higher glycolytic potential (phosphorylase, PFK activities) and lower oxidative capacity (citrate synthetase) than red muscles. It has been demonstrated for a

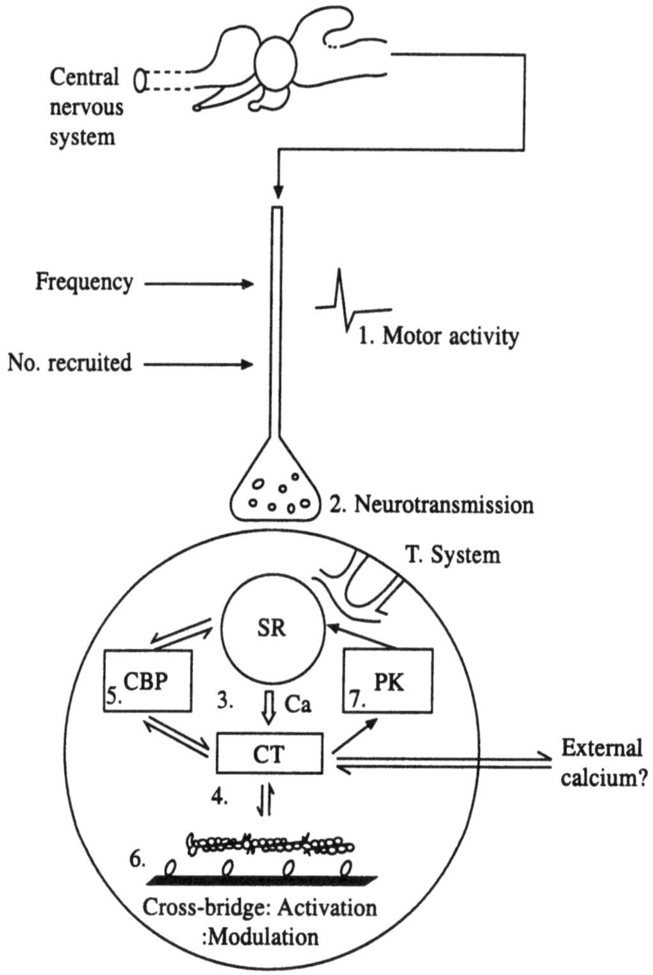

Fig. 5.9 The control of contractility in fish muscle. Abbreviations: SR, sarcoplasmic reticulum; CBP, cytoplasmic calcium binding proteins; CT, calcium transient; PK, Ca^{2+}-dependent protein kinase. Factors modulating force production include (1) The number and frequency of motor units activated. (2) Neurotransmission. (3) The muscle action potential (or distributed depolarization) is transmitted to the SR via the T-tubule system causing release of bound calcium. (4) This results in a transient increase in free calcium from 0.1 to 10 uM. (5) The duration of the calcium transient is thought to be reduced in fast muscles by the presence of high concentrations of cytoplasmic calcium binding proteins (parvalbumins). (6) Calcium ions activate (a) muscle contraction and (b) glycogen breakdown by direct binding effects to troponin C and phosphorylase kinase activity. There is also evidence that force production can be modulated by myosin phosphorylation through a specific Ca^{2+}-activated myosin light-chain kinase. (7) In slow mammalian muscles calcium uptake into the SR is modulated by a cytoplasmic factor thought to be a Ca^{2+}-dependent protein kinase. Slow muscle of salmonid fishes do not appear to have such a mechanism. The extent to which this is true of other species is unknwon. Incubation of *Tilapia* red and white muscles in ringers containing zero calcium or verapamil ($5 \times 10M^{-5}$) leads to a rapid and reversible loss of tension. It appears, therefore, that fish muscles have a requirement for extracellular calcium. Whether Ca^{2+} crosses the sarcolemma or remains bound to sites on the outside is unknown.
Source: Johnston (1981)

wide range of vertebrate muscles that myofibrillar ATPase activities parallel the unloaded speeds of shortening. On this basis, white muscles of trout and other species studied (dogfish, carp, tuna) are around two to five times as fast as red muscles. Activities of enzymes responsible for maintaining ATP supply during contraction such as creatine kinase, adenylate kinase and 5′ AMP amino-hydrolase parallel those of myofibrillar ATPase. 5′ AMP amino-hydrolase is particularly important in fast muscles since it catalyses the deamination of AMP to IMP thus shifting the equilibrium of adenylate kinase in favour of ATP production. It has been calculated that in fatigued muscles upto half of the ATP for contraction is supplied by this pathway.

It has been found that muscles with the lowest ATP/AMP ratio usually have the lowest rate of ATP utilization and the least variation in energy requirement. The ATP/AMP ratio in the red and white muscle of brook trout *Salvelinus fontinalis* (R, 18.6; W, 38.3) and tuna (R, 26.3; W, 61.6) reflect the relative activity levels of these muscles and lend support to the concept.

Both metabolic concentrations and enzyme activities can be modified by environmental factors such as temperature, activity and nutritional status.

Burst swimming is accompanied by an extremely rapid activation of glycolysis in white muscle. The regulation of glycogenolysis in fish white muscle is illustrated in Fig. 5.8.

Non-equilibrium or regulatory steps in anaerobic glycolysis include phosphorylase, phosphofructokinase and pyruvate kinase.

The metabolism of muscles during steady swimming is much less well understood. In general, fish can increase their oxygen uptake by a factor of 10–15 times between rest and maximal activity. It seems likely that red muscle receives a significant proportion of the cardiac output during steady swimming. This is particularly true for species in which red muscles alone support sustained activity (e.g., elasmobranchs, dipnoans, primitive teleosts). Since red fibres constitute a small fraction of the trunk musculature (5–10%) they undoubtedly possess a highly active aerobic metabolism. Little is known about the regulation of the tricarboxylic acid cycle in fish. Lipid is known to be an important fuel during steady swimming particularly in those species which undertake a spawning migration. There are also differences in the sites of storage and metabolism of lipid between different fish groups. It seems likely that most important fat fuels are ketone bodies in elasmobranchs and fatty acids in teleosts. It seems likely that oxidation of fats as fuel is much more important in red than in white muscles even when the latter are recruited for sustained activity. The reader is further referred to Coughlin and Rome (1999) and Coughlin (2000).

Numerous studies have shown glycogen depletion during sustained swimming in both the red and white muscles of teleosts. However, the pathways of carbohydrate metabolism by multiply innervated white fibres at sustained speeds are not known for any species. Two basic possibilities exist: ATP generation could be aerobic using local glycogen stores or blood glucose, or largely anaerobic resulting in the production of lactic acid. For the latter case, a number of schemes have been proposed involving lactate exchange with other tissues to keep the fish in overall aerobic balance.

The importance of aerobic glucose utilization is likely to vary considerably between species for both myotomal muscle types. In trout, and possibly tuna, the aerobic capacity of white muscle is probably sufficient to support sustained activity and blood glucose may be an important fuel. However, in many species it would appear that anaerobic pathways are important in providing ATP for sustained activity. Locomotor muscle, of high-performance fishes, tuna described as "energy speculators" has been studied in detail (Brill 1996; Dickson 1996b; Korsmeyer et al. 1996). (Figs. 5.10–5.12).

The metabolic fate of lactate produced during sustained and following burst swimming, has been the subject of a large number of papers. It is likely that major part of the lactate produced during exercise is oxidized to pyruvate. Fish may be able to maintain overall aerobic balance by transferring lactate produced by anaerobic glycogenolysis in the white muscle to the gills and other peripheral sites for subsequent oxidation. The extent to which fish liver can utilize lactate as a glyconeogenic substance is unclear. On the basis of indirect evidence it has been proposed that red muscles does have a significant role in synthesizing glucose from circulating lactate. The lactate turnover rate of teleosts is one or two orders of magnitude lower than in mammals of equivalent size, but lactate has the same importance as a fuel in both vertebrate groups. However, lactate turnover rate and oxidation rate do not scale with body mass in the same fashion as does metabolic rate. For details of this metabolic fuel for endurance locomotion, the reader can see Weber (1991).

Many species of fish undergo periods of seasonal starvation linked to fluctuations in food supply. In fish that do not store large quantities of lipid, white muscle myofibrillar proteins serve as an important source of metabolic precursors during starvation. It seems likely that amino acids released from muscles during starvation, are converted to glucose via gluconeogenic pathways in the liver and hence constitute an important fuel for muscular activity under these conditions.

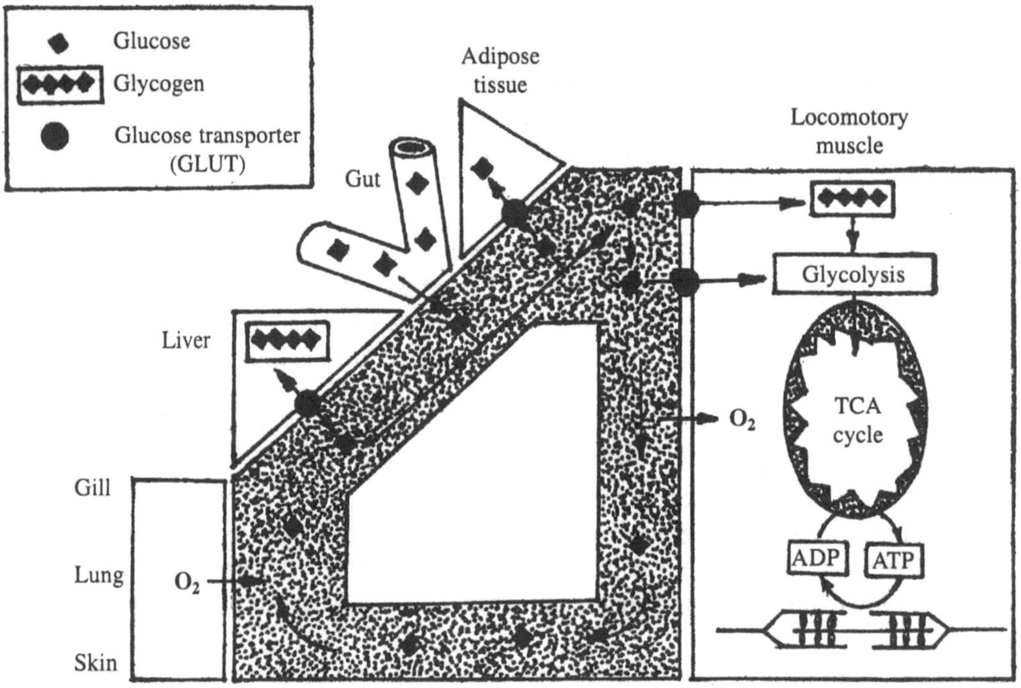

Fig. 5.10 Pathway for carbohydrate supply to locomotory muscles during sustained exercise. Simple carbohydrate subunits are derived from intestinal absorption or hepatic gluconeogenesis. They can be stored as liver glycogen, muscle glycogen or converted to triacylglycerol in adipose tissues. Glucose exchange between liver, adipose tissue, blood and muscle is controlled by tissue-specific glucose transporters (GLUT) in cell membranes.
***Source:* Weber and Haman (1996)**

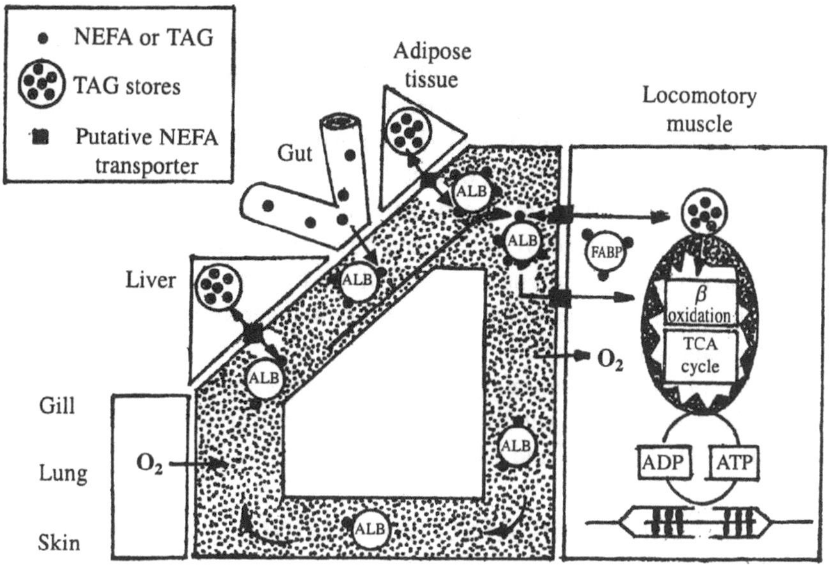

Fig. 5.11 Pathway for lipid supply to locomotory muscles during sustained exercise. Nonesterified fatty acids (NEFA) and triacylglycerol (TAG) of intestinal origin can be stored as liver, adipose tissue or muscle TAG reserves. Lipids are transported through plasma bound to an albumin-like protein and through the cytosol bound to fatty acid binding proteins (FABP). NEFA exchange between liver, adipose tissue, plasma and muscle may be mediated in part by presumed NEFA transporters in cell membranes.
***Source:* Weber and Haman (1996)**

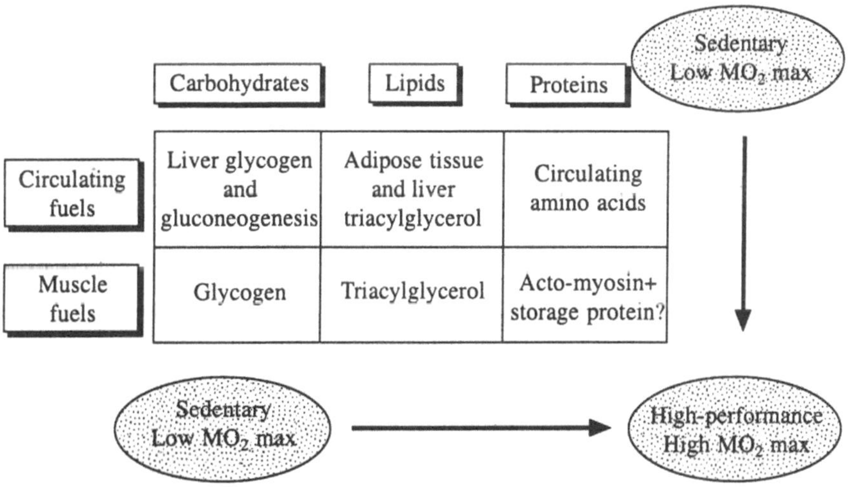

Fig. 5.12 Kinds of oxidative fuels and their storage sites. Arrows indicate hypothetical changes in fuel perference when going from a low-aerobic, sedentary fish to a high-performance fish. Highly aerobic species will tend to use proportionately more proteins and lipids, and less carbohydrates. Also, endurance-adapted species will obtain a much larger fraction of their energy from intramuscular substrate sources and a smaller fraction from circulatory sources than low-aerobic species.
***Source:* Weber and Haman (1996)**

'Sit-and-wait' predators generally have low endurance at high speed since predatory manoeuvres occur over short distances. Typically, the fast muscle in species using this feeding strategy, such as the Antarctic fish *Notothenia neglecta* have high activities of creatine phosphokinase, adenylate kinase and AMP deaminase and low activities of glycolytic enzymes and modest glycogen stores. *N. neglecta* swum to exhaustion (1–3 min) show a respiratory but not a metabolic acidosis with no significant accumulation of lactate in the blood or muscles. In contrast, species with greater endurance at maximum speed, such as skipjack tuna, have muscle glycogen concentrations of 100 μ mol glucosyl unit/g, and have extremely high activities of glycolytic enzymes. Muscle lactate concentrations can reach 40 mM following 15 min strenous activity in salmonids. Thus the relative proportions of the different anaerobic energy supplying pathways are matched to swimming behaviour.

Finally, a number of authors have suggested that fish muscles are a major site of amino acid oxidation. There is, however, as yet little direct evidence for amino acid oxidation in fish muscles (Johnston 1981; Johnston and Altringham 1991).

6 FIBRE RECRUITMENT DURING SWIMMING

The relative activities of twitch and slow motor units during swimming in the lamprey are not well known. It is not known whether the central twitch fibres in the lamprey myotome are more specialized for glycolytic metabolism than the slow parietal fibres. Lampreys do not swim well over long periods, probably due to fatigue of muscles. The mechanisms of fatigue and differences in susceptibility between slow and twitch fibres are unknown in lampreys. The arrangement of two types of fibres in unique myotomal subunits suggests that their respective types of metabolism may be different and that metabolic interactions may occur.

Locomotion in fish is achieved using a combination of the paired and unpaired fins and segmentally arranged myotomal muscles. Red muscle fibres constitute a major portion of the fin muscles and a relative minor component of the trunk muscles. The proportion of red fibres in the myotomes of different species is correlated positively with sustained swimming performance and negatively with the involvement of pectoral fins during swimming at low and intermediate speeds. For example, red fibres constitute around 29.8 of the trunk muscle of *Scomber colias* (active pelagic swimmer), 7.2% in the angler fish *Lophius piscatorius* (sedentary bottom dweller) and 5.5% in *Notothenia rossii* which primarily uses enlarged pectoral fins for sustained swimming (Johnston 1983a).

The contractile and metabolic characters of red muscle are adapted for sustained effort. In general, active continuous swimmers such as anchovy have the greatest amounts of red muscle. Electromyographical studies have shown that in elasmobranchs and some primitive teleosts only the red fibres are active during sustained swimming. In order to swim at higher speeds the fish must contract its white muscle which is specialized for anaerobic ATP generation (Johnston 1983b).

The simple division of labour between red and white muscle is not found for the majority of fishes. The advanced teleost orders show a more complex behaviour. In many cases, both red and white muscle are recruited at low speeds. White fibres in these fish are more aerobic in character and are able to utilize a variety of different substrates as fuels. This more complex division of labour between muscle types appears to be correlated with developments in the nervous system, particularly with respect to the innervation of the white muscle (Johnston 1983b).

Slow muscle fibres are arranged in parallel to the longitudinal body axis whereas fast fibres make varying angles of upto 40°. This pattern of organization enables fast fibres to produce large bending movements in the trunk for almost isometric contractions (7–9% resting lengths for elasmobranchs and 2–3% resting length for teleosts). There is an orderly recruitments of slow red > fast red > fast white fibres with increasing swimming speed. The tail-beat frequency required to produce maximum speed varies inversely with body-length. Larval and small fish have tail frequencies upto 60 Hz. Adaptations for rapid cycles of contraction in fast muscles include an extensive sarcoplasmic reticulum and high concentrations (~1 mM) of cytoplasmic Ca^{2+}-binding proteins (Johnston 1983b).

Johnston (1989) reviewed the structure and function of swimming muscles in Antarctic fish, especially of Notothenoidei. Adult stages of the vast majority of species swim at low speeds using large pectoral (labriform locomotion). This is supplemented with sub-carangiform swimming in pelagic juvenile stages and in the adult stages of some other species. The thrust for sustained activity is provided by the recruitment of slow muscle fibres. Slow muscle myofibrils typically occur in columns one fibril thick entirely surrounded by mitochondria. The resulting high volume density of mitochondria (30–60%), and short inter-mitochondrial spacing, is thought to represent an adaptation which serves to compensate for the detrimental effects of low temperature on enzyme reaction and diffusion rates. Sub-carangiform swimming is used to achieve burst speeds associated with prey capture and/or predator avoidance. Burst speeds require the recruitment of fast twitch fibres in the myotomes. In many demersal species the energy supply for burst swimming comes from phosphogen hydrolysis, and the capacity of fast twitch fibres of anaerobic glycogenolysis is severely reduced. Antarctic fish are characterized by delayed maturation, slow growth and low metabolic rates. The fast myotomal muscles of adult stages often contain few fibres less than 80 μm diameter, fibres 200–450 μm diameter forming the major size class in numerous species. It is therefore likely that hyperplasia ceases at a relatively early stage in development and that subsequent muscle growth involves hypertrophy of existing fibres. Studies of contractile properties of isolated muscle fibres suggest that this is most noticeable for juvenile stages, which have higher maximum tail-beat frequencies than adults. For special attention on specializations of swimming muscles in the pelagic Antarctic fish *Pleurogramma antarcticum*, consult: (Johnston et al. 1988; see also Johnston and Dunn 1987 and Fleming et al. 1990) for temperature acclimation. Archer and Johnston (1991) focused on the density of cristae and distribution of mitochondria in the slow muscle fibres of Antarctic fish. Slow muscle fibres are not diffusion limited for O_2 in species lacking haemoglobin and myoglobin. The high and homogeneous distribution of mitochondria in Antarctic fish results in low values for the mean free spacing of mitochondria (1.04–2.96 μm) relative to temperate species. This may represent an adaptation to overcome the effects of low temperature on the diffusion of metabolites between capillaries and mitochondria and between mitochondria and myofibrils.

Altringham and Johnston (1986) reported in the Antarctic species, both tension and ATPase activity of the fast fibres were higher than those of slow fibres, and the slow fibres were 30% more economical than fast fibres. After correction for differences in myofibrillar density, tensions were very similar.

Johnson and Johnston (1991) investigated the contractile properties of swimming muscles in some marine teleosts from Antarctic, temperate, and tropical latitudes. The thermal range over which live fibres produced maximum tension was highly correlated with habitat temperature.

Fast muscle fibres from Antarctic, temperate, and tropical fish produce similar absolute tensions at their normal body temperatures. Although the contractile proteins in polar fish show substantial resistance type adaptations which enable contraction to proceed at very low temperatures, muscle power output remains significantly lower than in warm-water species. Temperature compensation of muscle power output would appear to largely involve adaptations in maximum force generation, with relatively minor contributions from time-dependent contractile properties. Meyer-Rochow and Devine (1986) reported ultrastructural observations of red and white muscle from Antarctic fish (*Dissostichus mawsoni* and *Pagothenia borchgrevinki*). Aerobic metabolism is dominant and these species would not use white trunk muscle for sustained or slow swimming. Rapid contractions of white trunk muscle as in prey capture or predator evasion are more likely.

Webb (1993a) has discussed the muscle performance in relation to temperature. While temperature affects muscle power and V_{max}, kinematics vary little with temperature. As a result, the ability of muscle to meet power requirements to swim at a given speed decreases at lower temperatures.

Bunshnell et al. (1994) have tested the hypotheses of metabolic cold adaptations in two species of cod (*Gadus morhua* and *G. ogac*) in arctic waters.

Fig. 5.13 shows the inter-relationships of some of the elements which influence muscle

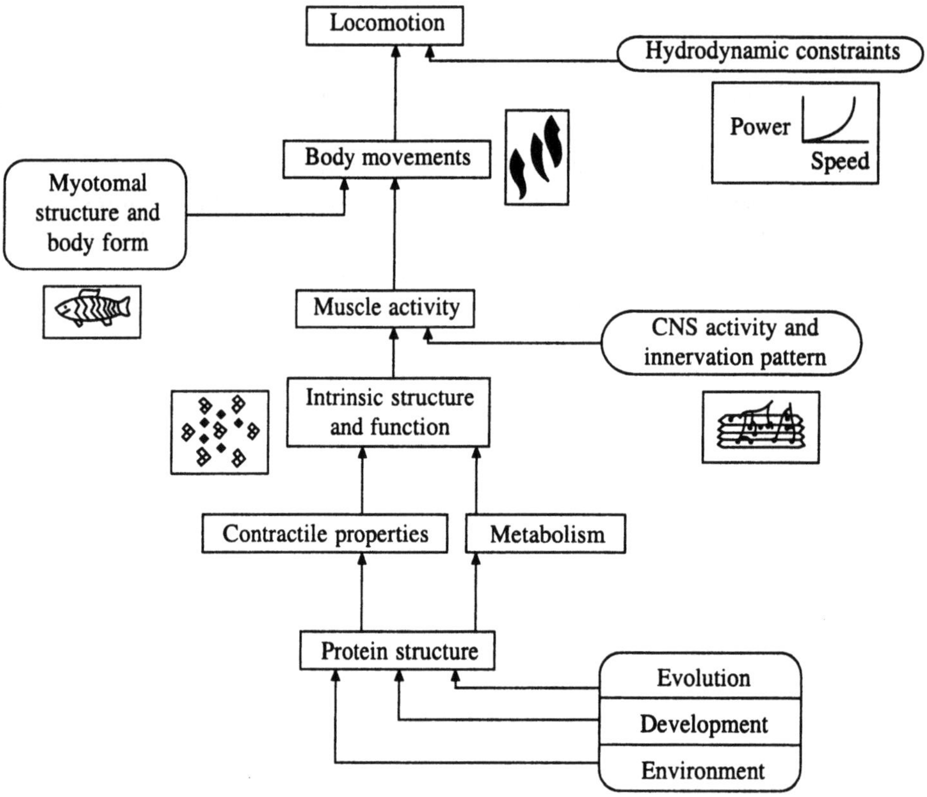

Fig. 5.13 Factors influencing muscle structure and function, and locomotion in fish. A number of elements are highlighted, which are either of particular importance to swimming fish. (the arrangement of the body musculature into myotomes, undulatory body movements, and the high power demands of fast swimming), are of uncertain significance (the simple lattice structure of myosin and the presence of polyneuronal innervation in the fast fibres of many teleosts).
***Source:* Johnston and Altringham (1991)**

biochemistry and ultimately locomotion, and highlights a number of features either of particular importance to swimming fish, or of uncertain significance (Johnston and Altringham 1991).

In brief, classification of muscle fibres is generally based on criteria correlated with their speed of contraction and on criteria correlated with their resistance to fatigue.

Speed of contraction is correlated with the level and pH stability of myofibrillar ATPase activity. The occurrence of enzymes of aerobic metabolism is correlated with the resistance to fatigue.

Red muscle has a relatively low maximum velocity of shortening (*V*max) and slow activation-relaxation kinetics, while white muscle has a high *V*max and fast activation-relaxation kinetics. It has been shown that pink muscle is recruited at intermediate swimming speeds relative to red and white muscle. Pink muscle may play a cruicial role in during sustainable swimming.

Red muscle has been implicated in slow, continuous, non-fatiguing swimming, whilst white muscle has been implicated in fast, rapidly fatiguing activity. Coughlin et al. (1996) reported that pink muscle in the scup (*Stenotomus chrysops*) is found in anatomical positions in which red muscle produces very little power during swimming; the anterior region of the fish, which undergoes the lowest strain during swimming. Pink muscle supplements the relatively low power generated by red muscle in anterior regional during swimming.

Fish show a wide range of habitat and resource utilization which impose different demands on swimming ability and consequently the underlying propulsive musculature.

Adaptation to a particular mode of life is also reflected in muscle biochemistry, and relations between biochemical parameters and activity patterns have now been demonstrated.

Blood supply to a muscle is an important determinant of muscle function; absolute blood supply being determined by capillary density and rate of blood flow through the capillaries.

Phleger (1991) in his comprehensive article on biochemical aspects of buoyancy in fishes, has covered information on muscles. Reduction of musculature, more dense than sea water, is fairly common in deep sea fishes and help them maintain neutral buoyancy. Red muscle is reduced in some midwater fish that lack a swim bladder.

Oily muscles are also considered to be an important buoyancy mechanism in marine animals. Wax esters are more commonly present in muscle of deep and midwater fishes than epipelagic fishes where they are usually associated with roe. Some families of teleost fishes which have more than 10% wax ester specifically in muscle are Trachichthydae, Paralepididae, Latimeridae, Moridae, Myctophidae, Gempylidae, and Zeiformes.

Lumpsucker (*Cyclopterus lumpus*) muscle has a high lipid content. The high fat plus low osmolarity of the muscle give it a density of 1.01 g ml^{-1}. The dorsal muscle located above the abdomen may be a specialized buoyancy organ, not important in locomotion.

Fresh water fishes have more than 10% muscle lipid (as percent wet weight of the fillet). These include *Cyprinus carpio*, *Coregnus clupeaformis*, *Anguilla japonica* and *Pimelodus claria*.

An interesting article by Fetcho (1992) on the spinal motor system in early vertebrates and some of its evolutionary changes has provided additional information related to locomotion.

6.1 Aerobic and Anaerobic Scaling in Fish

These aspects have received the attention of Goolish (1991); some of the points are given here:

The maximum metabolic rate of fish, measured during aerobic swimming, scales according to mass exponent ≥ 1.0. This is not in accord with the negative allometry of aerobic metabolism predicted by the scaling of oxygen uptake and delivery.

An analysis of the tissue-specific distribution of aerobic capacity in fish indicates that only a small proportion, perhaps 10–20%, of total capacity occurs in the red muscle tissue used during aerobic swimming. This is in sharp contrast to the situation in mammals where the majority of the whole-body aerobic capacity is found in the skeletal muscles.

Because of its large mass, the characteristically anaerobic white muscle tissue of fish contains most of the whole-body aerobic capacity. This aerobic potential may be responsible for oxygen consumption rate following exercise (oxygen debt) which are higher than during aerobic swimming.

Because different physiological activities are responsible for maximum metabolic rate in different sized fish, no single 'use-based' theory (e.g., locomotor muscle power) should be able to account for the scaling of aerobic metabolism.

Enzymatic and metabolite evidence indicates positive allometry in the anaerobic potential of fish white muscle. This appears to be the situation for both intraspecific and interspecific scaling. The allometry of anaerobic demand is influenced by (1) higher weight-specific power requirements in larger fish during burst and sprint swimming and (2) decreasing weight-specific aerobic power and sprint swimming and (3) decreasing weight-specific aerobic power production by the muscle of larger fish.

The swimming of very small (larval) fish appears to be powered primarily by aerobic metabolism even at burst and sprint speeds. This conclusion is based on (1) the scaling of performance at burst and aerobic swimming, (2) comparisons of power requirements and red muscle power output, and (3) the enzyme profiles of larval fish. Increased dependence on anaerobic energy production seems to occur at approximately 1–3 cm.

Upper limits to anaerobic metabolism in large fish are constrained by the scaling of glycogen reserves ($= L^3$), which do not increase as fast as the power requirements at burst speeds ($= L^{4.4}$). More importantly, decreasing aerobic capacity at large sizes results in estimated glycogen restoration times which are prohibitively long (> 10 days). Estimates of daily weight-specific anaerobic energy production, based on glycogen restoration times, decrease by approximately two orders of magnitude with increased size. This expression of anaerobic metabolism indicates, not positive, but rather negative allometry.

The scaling of anaerobic metabolism displays characteristics of both positive and negative allometry, which means that it may not be possible to accurately describe it with any general mass exponent. These considerations also suggest that there is an optimal size for the practical exploitation of anaerobic metabolism during locomotion.

Jones (1982) has discussed the anaerobic exercise in teleost fish. In fish, as in all vertebrates, the highest levels of exercise performance are achieved anaerobically. This "burst-type" activity can only be maintained for brief periods (upto 30 s), being terminated by exhaustion of intracellular energy supplies or by accumulation of waste products. On the other hand fish are also capable of "marathon-type" performance in which activity can be sustained for hours or days. Sustained performance is aerobic and metabolic demand matches supply while waste production is balanced by disposal. Between these two extremes is a zone of steady performance in which exercise is kept up for periods from 2 to 200 min, terminated by fatigue. The degree to which anaerobic energy contributes to swimming performance in the steady exercise zone is unknown. The relationship between oxygen consumption or scope for activity and swimming speed (U), the repayment of an oxygen debt during recovery from fatigue, and the reduction in maximum sustained speed (critical velocity, U crit) by reduced environmental oxygen or lowered blood haemoglobin suggest that steady performance in salmonids is aerobic. In salmonids, sufficient power is provided from the oxygen consumed to swim steadily at least at 80% of *U* crit. On the other hand depending on the temperature, aerobic power in cyprinids is only sufficient to reach 30–50% *U* crit. Aerobic relationships holding for salmonids do not apply to cyprinids. In fact, high respiratory quotients early in exercise along with glycogen depletion and lactate accumulation rates in muscle during vigorous swimming suggest that steady performance in cyprinids is worthy of further examination.

Randall et al. (1987) investigated the effects of burst swimming on aerobic swimming in chinook salmon (*Oncorhynchus tshawytscha*). Burst swimming in fish results in a marked metabolic acidosis. *O. tshawytscha* blood was shown to have a marked Root shift, such that burst swimming and subsequent metabolic acidosis should impair oxygen delivery to the tissues and, therefore, aerobic swimming capacity. Burst swimming was found to have no effect on aerobic swimming capacity in chinook salmon and it is concluded that any effects on aerobic swimming, of the induced metabolic acidosis following burst swimming, was offset by the release of catecholamines.

Videler and Nolet (1990) reported that fishes and turtles use less energy for swimming at optimum velocity (*Uopt*) than the other submerged swimmers. Fishes and turtles use approximately 5.5 times less energy than the submerged swimmers. Bernier et al. (1996a) investigated the role of adenosine on anaerbic metabolism during hypoxia.

7. FISH MUSCLE GROWTH

In fish, muscular tissue forms a large part of the body mass: some 40–60% of the total body mass in most fish is locomotor muscle. Growth of fish muscle is different from that of mammals and birds. In growing fish, in addition to hypertrophy, hyperplasia can play an important role in muscle growth even after the juvenile stage. However, the relative importance of hyperplasia is maximal in small (young) fish and decreases with increasing growth.

Another aspect of fish (muscle) growth is the large variation of growth-rate that can occur depending on the circumstances as the availability of food and differences in temperature. Therefore, age is not a good parameter in the study of fish muscle growth. Growth related events as the cessation of hyperplasia and fine diameter distribution show a much better relationship with fish length. Comprehensive accounts on muscle development and growth can be obtained from different articles published in the volume edited by I.A. Johnston (2000). (See also Weatherley and Gill 1987).

7.1 Regeneration

Regeneration is a special form of muscle growth after injury or transplantation. The extent and success of regeneration varies with the nature of injury, but in all situations the process involves: revascularization, cellular infiltration, phagocytosis of necrotic damaged muscle, proliferation of muscle precursor cells fusing either into multi-nucleated myotubes or with the ends of damaged muscle fibres and finally reinnervation.

7.2 Myosatellite Cells

Myosatellite cells are small spindle-shaped cells having a heterochromatic nucleus. They are situated between the sarcolemma and the basal lamina of fully differentiated skeletal muscle fibres. During gowth and regeneration of muscle fibres, these cells are the source of additional myonuclei. Myosatellite cells are present in the striated muscle of all investigated vertebrate species. Although the presence of myosatellite cells has been demonstrated in the skeletal muscle of some species of fish (shark, eel, carp, trout), out knowledge of the role of myosatellite cells in the growth of fish is still limited (see details: Koumans et al. 1990, 1991, 1993a, b; see literature). The reader is also referred to papers on muscle growth by Kundu and Mansuri (1990, 1992), Kundu et al. (1990) and Stoiber and Sänger (1996).

Meyer-Rochow and Ingram (1993) reported the red-white muscle distribution and fibre growth dynamics: a comparison between lacustrine and river populations of the Southern smelt *Retropinna retropinna*. The results convincingly demonstrated the usefulness of the fish's muscle plasticity under natural conditions in relation to the environment and behaviour of the fish.

7.3 Rapid Somatic Growth and Muscle Damage

Physical exercise is known to stimulate protein turnover, muscle fibre hypertrophy, somatic growth, and improve general welfare in salmonid fish, and GMD (growth-associated muscle damage) was observed, predominantly, in the exercised fish.

Muscle damage in fish has often been linked to malnutrition (e.g., vitamin E-selenium deficiency) and exposure to toxic agents but whilst it is frequently reported that exercise leads to muscle fibre hypertrophy in fish, swimming activity is not known to induce muscle damage. Pathologically induced muscle damage (PMD) is characterized by an irregular, indistinctive and multifocal distribution of damaged areas within the myotomes.

Christiansen et al. (1992) have described a novel category of extensive and distinct muscle damage within the white swimming muscle of Arctic charr (*Salvelinus alpinus*, Salmonidae). Based upon myomorphological characters and somatic growth patterns displayed by individual fish, it appears that this type of muscle is associated with healthy and rapidly growing fish.

Unlike PMD, that seen in healthy and rapidly growing fish is not scattered within the myotomes but relates to the myosepta in a well-defined and systematic pattern.

CHAPTER 6

Locomotion: Propulsion System

1. INTRODUCTION

Evolution leads towards an optimization of all processes underlying movement, bound by constraints of life in water. The biochemical consequences of this evolutionary process can only by understood within the context of the hydrodynamic and biomechanical constraints operating on a swimming fish. These constraints influence the structural design of fish, and the properties of their neuromuscular system. The hydrodynamic constraints are determined by the physical properties of water and the speed, size and shape of fish moving through the water (Johnston and Altringham 1991). Kotrschal and Essler (1995) have discussed the goals and approaches in the analysis of locomotion in fish.

Animals are denser than either fresh water or sea water, and therefore tend to sink, unless they have adaptations that give buoyancy. Very small organisms sink slowly, reproduce rapidly and can be kept suspended by natural turbulence: individuals lost by sinking are replaced by reproduction. Large animals will sink unless they swim or evolve buoyancy organs (Alexander 1990).

Swimming is basic to many activities, including agonistic and reproductive behaviour, foraging and predator avoidance, and numerous mechanisms are recognized for effective swimming (Webb 1988; Videler and Wardle 1991). The recent book on fish swimming by J. Videler (1993) presents an account of all aspects of fish locomotion. Webb (1994) focused on different features of exercise performance of fish. Exercise is the topic of a review by Butler (1986). A paper regarding benthic living, density and station-holding by Webb (1990) is quite informative. Many studies on fish locomotion have concentrated on bony fish, especially teleosts. As a result, modern ideas on fish locomotor functional-morphology are dominated by knowledge of only one of the major groups of fish. However, there are many unique features among cartilaginous fish that suggest they have exploited some different biomechanical properties. Sharks appear to swim like elongate teleosts, but in contrast they have discrete, often widely spaced median fin more typical of fusiform teleosts. They are a few teleosts which depart from the "classical" ichthyan body form to such an extent as the seahorse *Hippocampus.*

Much of the diversity in the locomotion of fishes has been categorized by the longitudinal extent of the undulating axial structures, and it would be useful if the amplitude of the lateral undulation had a clear skeletal or myological correlate. However, the angles of the muscle fibres even within a single myomere differ considerably and this complicates prediction of the functional consequences of variability in myomeric shape. However, the myosepta are the primary site for attachment of the axial muscle fibres, and hence must be an important mechanical link for force transmission.

Perhaps a grossly symmetric myomeric morphology facilitates lateral bending while minimizing torsion and/or dorsal-ventral flexion of the vertebral column. Current literature has examined neither the within-myomere patterns of muscle activation nor the mechanical behaviours of the axial skeleton that are relevant to the issue of the potential torsional forces that may occur during the swimming of the fish.

Although the caudal region of many fish taxa is fundamentally important in the transmission of locomotor forces, muscle activity is not limited to the caudal region of most undulatory swimmers. As muscle activity propagates from head to tail during undulatory swimming, the extent to which all of the contractile tissue within a myomere is activated synchronously is little known. Some evidence suggests that it is indeed possible to not have coactivation of all of the contractile tissue comprising an axial muscle segment.

The pattern of muscle activation within a myomere as well as the arrangement of myomeres relative to the axial skeleton suggests the following question: What are the fundamental units of muscle recruitment (parts of myomeres, entire myomeres, or multiple portions of different myomeres) that are involved in the axial undulations of fishes? The pattern of innervation within a myomere is complex and does not itself suggest a clear a priori hypothesis of myomeric function.

Current neuroanatomical and neurophysiological information is not sufficient to predict whether an entire myomere can or must be activated synchronously and how many spinal segments innervate a single myomere.

In the light of tremendous complexity in the neuroanatomy and axial morphology of fishes, it is extremely difficult at the present time to make informed predictions about axial function during undulatory swimming. Systematic comparative electromyographic studies of muscle activity during swimming should help to reduce the number of possibilities for neural and mechanical mechanisms underlying swimming. By studying behaviours over a wide range of speeds that require a greater portions of the total musculature to be recruited, insights should be gained into the extent to which entire myomeres are utilized.

Several species of genus *Synodontis* (Siluroidei) regularly invert and swim on their backs. It is thought that the habit has probably evolved more than once in response to ecological pressures.

Swordfishes (Xiphiidae) and billfishes (Istiophoridae) show bodily features no other fish have. Reference is related to the extensions of the upper jaws, the swords, and to the shape of the head. Sword bearing fishes swim at speeds exceeding those of all other nektonic animals reaching values of well over 100 km h^{-1}. All the swords have rough surface. There is considerable controversy about the function of these extensions of the head. The arguments in favour of the drag reducing function are stronger than the observations supporting the killing device purpose.

Exercise is the stress which animals most frequently experience and may be defined as work performed on the environment by locomotory muscles. Exercise is accompanied by an increase in the rate of energy conversion from the resting state. This increase provides for the energy requirement of locomotor muscles as well as for extra work performed by the heart and respiratory muscles in supplying oxygen demanded in exercise (Jones and Randall 1978; see for details).

The aerobic and anaerobic scaling in fish is currently a subject of interest (for details, see Goolish 1991).

2. DETERMINANTS OF LOCOMOTION

Webb (1988) discusses this subject comprehensively. Swimming is a common and important component of vertebrate behaviour. Therefore the study of swimming form and function is essential to understanding vertebrate biology. An interactive process between biologists and hydrodynamicists has proven extremely successful in providing such understanding. This process starts with a defination of forces that may act on an animal; viscous and pressure drag, acceleration reaction, lift and body inertia, and ground reaction. Dominant force components are selected on the basis of ground contact, Reynolds number, reduced frequency and shape. Conservation of momentum requires forces and their moments be in balance leading to models describing swimming motions and associated forces. These provide the bridge between hydrodynamic studies and biological applications. Life history patterns are also affected by locomotor capabilities of various morphologies. Scale effects limit the distribution of propulsors (see also Webb 1978). Various facets of swimming have been covered in a number of publications (Webb 1991, 1992, 1993b, c, 1998; Webb et al. 1991, 1992).

Lauder (1989) has focused on the historical, and functional analyses of the caudal fin locomotion in ray-finned fishes in order to understand the design of the locomotor apparatus in vertebrates.

Fish may propel themselves by successively generating and destroying tail tip eddies of alternating direction. Each eddy is produced by a sideways tail flip, and is destroyed by the return stroke during which a forward force $F \approx 0.6 \text{ Kph } A^3 f^2$ is generated, where p is the density of water, h the height of the tail fin, A the amplitude of maximum excursion from the centre line and f, the tail frequency. A dimensionless constant with a typical value K = 40 describes the entrainment of water into the tail tip vortex. The average forward power of this motion is $P = 2.4 \text{ Kph } A^4 f^3$. This model accounts for both starts and steady swimming (Ahlborn et al. 1991).

Under the title "Fish Biomechanics: purpose or means?", Aerts (1992) provides useful information on different facets of fish biomechanics.

The reader is referred to Fig. 6.1 and Tables 6.1 to 6.5 for information on allied aspects.

3. BODY FORM, LOCOMOTION AND FORAGING IN AQUATIC VERTEBRATES

Webb (1984) has defined four functional categories to embrace the range of locomotor diversity of aquatic vertebrates: (1) body/caudal fin (BCF) periodic propulsion where locomotor movements repeat, as occurs in cruising and sprint swimming; (2) BCF transient propulsion where kinematics are brief and non-cyclic, as occurs in fast-starts and powered turns; (3) median and paired fin (MPF) propulsion, with very diverse fin kinematics, used in slow swimming and precise manoeuver, (4) occasional propulsion or "non-swimming". Specialization in any one of these categories compromises performance in one of more of the others, thereby reducing locomotor diversity and hence behavioural options (Table 6.1). Food characteristics influencing the role of locomotion in search and capture are: (1) distribution in space and/or time and (2) evasive capabilities. BCF periodic swimmers take food that is widely dispersed in space/time; BCF transient swimmers consume locally abundant evasive items and MPF swimmers consume non-evasive food in structurally complex habitats. Locomotor specialists under-utilize smaller food items in exposed habitats. This resource is exploited by smaller fish, which are locomotor generalists, because of predation pressures. For such locomotor generalists, locomotor adaptations for food capture are

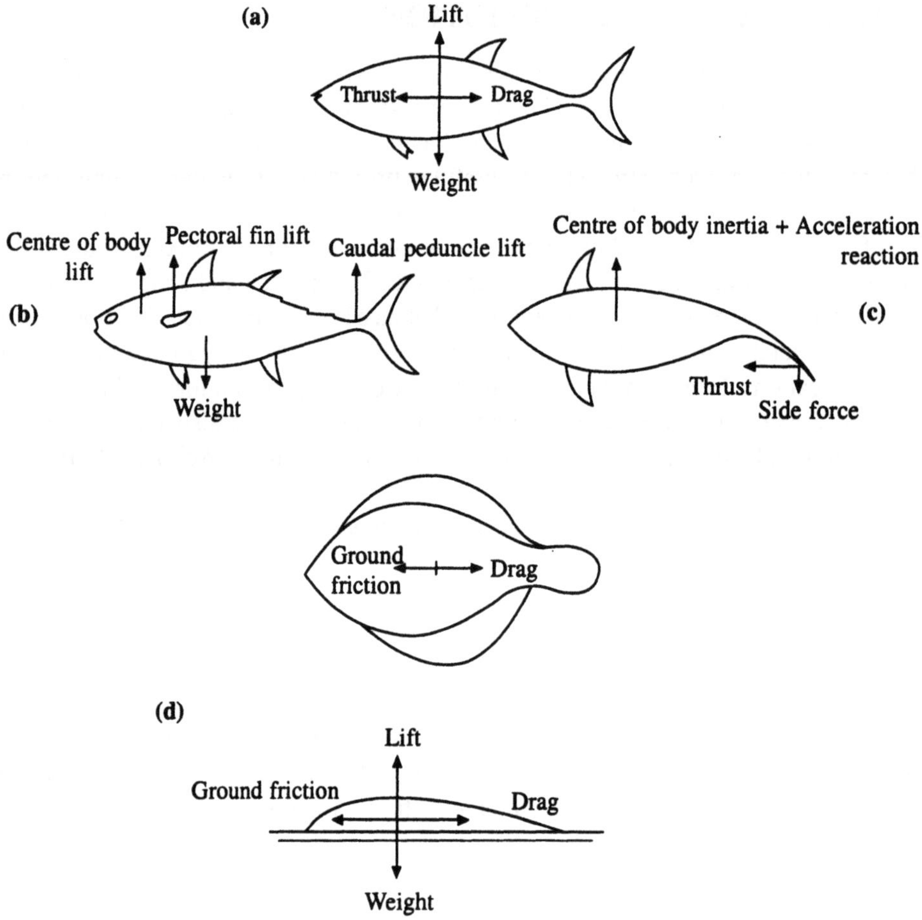

Fig. 6.1 Forces and moment balances for swimming vertebrates. (a) Force balance for rectilinear swimming. (b) Moment balance for forces in the horizontal plane. (c) Recoil moment balance. (d) Plan and side views with force balance for a benthic fish holding station on the bottom. (a–c) show tuna and (d) plaice, *Pleuronectes platessa*.
***Source:* Webb (1988).**

of diminished importance and other adaptations such as suction and protrusible jaws in fish are common (Fig. 6.2). For coverage of body shape and feeding behaviour, see Horn (1998).

4. BODY AND FIN FORM

Recent theoretical and experimental studies and various field observations on juvenile and adult fish suggest that the functional locomotor morphology of most bony fish can be understood in terms of three activity patterns. These are (1) *transient swimming* during acceleration and powered turns, (2) *periodic* or *cyclic* swimming during cruising and sprinting at relatively cruising speed, and (3) *slow swimming* and manoeuvering.

It has become traditional to use exemplary species to name morphologies associated with various locomotor patterns, and those relevant are included in Table 6.2. Many characters could

Table 6.1 A functional classification of aquatic locomotor propulsion mechanisms

	Characters	Optimal design features	Functions
(A)	*Body/caudal fin (BCF) periodic propulsion* Cyclically repeating kinematics. Relatively high power. Small linear and angular accelerations.	High aspect ratio lunate tail. Narrow caudal peduncle. Relatively stiff, streamlined anterior body. Large anterior depth/mass. Endothermy?	Swimming sustained for several seconds to several weeks in cruising, prolonged and sprint swimming (steady and two phase patterns), during chases, patrolling, station holding, migration, etc.
(B)	*Body/caudal fin (BCF) transient propulsion* Brief non-cyclic kinematics. High power. Linear and/or angular acceleration usually large.	Large body depth and area, especially caudally. Flexible body. Large muscle mass relative to body mass.	Fast-starts and powered turns lasting tens of milliseconds and used in prey capture, predator evasion, etc.
(C)	*Median and paired fin (MPF) propulsion* Variable kinematics involving discrete fins. Low power. Low speed, low acceleration rates.	Lateral insertion pectoral fins. Anterior ventro-lateral insertion of pelvic fins. Extended anal and dorsal fins. Deep, laterally flattened body.	Slow swimming and precise manoeuver in searching, stalking, feeding, hiding, etc.
(D)	*Occasional and non-swimmers* Whole body movements relatively unimportant.	—	—

Source: Webb (1984a)

be used to quantify the functional morphology of the fish (Table 6.2). The reader can find interesting details in the paper by Webb and Weihs (1986).

4.1 Performance Criteria

Animal motions can be described in terms of speed, acceleration, and turning radius. These three variables are the most suitable kinematic measures of performance. In addition, animals must accumulate some surplus energy for growth and reproduction, and cost and efficiency of locomotion are criteria that reflect the consequences of movement to the overall energy budget of an animal (Daniel and Webb 1987).

Progress of fish manoeuvering in water, the density of which is approximately equal to that of animal itself, can be classified into three major categories: sustained, prolonged, and burst swimming speeds. The biological constraints on performance listed are: size (length, weight), sex, and disease. The environment constraints include temperature, oxygen, carbon dioxide, salinity and pollution (Beamish 1978; read Webb 1993a for details).

Tale 6.2 Body form characters of teleostean locomotor specialists for three activity patterns Characters are the main entries; a method of quantifying each is shown in parentheses[a]

Character (Index)[a]	Thunniform (cruiser)	Esociform (accelerator)	Chaetodontiform (manoeuverer)
Body shape (shape factor)	Streamlined (intermediate)	Esongate (large)	Gibbose (small)
Caudal peduncle (*dcp/db*)	Narrow (small)	Deep (large)	Intermediate (intermediate)
Caudal fin (aspect ratio)	Lunate (large)	Rounded or square-shape; indented or square trailing edge (small)	Rounded or square-shape; indented or square trailing edge
Effective area of posterior of body (transient swimming thrust factor)	Small (small)	Large (large)	Intermediate (intermediate)
Muscle mass (percent muscle mass)	Large (large)	Large (large)	Small (small)
Position of dorsal fin (s) (*La/Lt*) and (*Lp/Lt*)	Anterior (small) (small)	Posterior (large) (large)	Anterior to posterior (small) (large)
Location of pectoral fins (*Lpec/Lt*)	Various (various)	Various (various)	Lateral (small)
Location of pelvic fins (*Lpel/Lt*) and (*dpel/db*)	Various (various) (various)	Various (various) (various)	Anterior and ventrolateral (small) (small)

a Key: *dcp* depth of caudal peduncle.
db maximum depth of body.
dpel distance of pelvic fins from ventral surface.
La distance from nose to anterior edge of dorsal fin.
Lp distance from nose to posterior edge of dorsal fin.
Lpec distance from nose to pectoral fin base.
Lpel distance from nose to pelvic fin base.
Lt total length.

Source: Webb and Weihs (1986).

Jensen et al. (1983) reported that short-term severe exercise resulted in a drastic decrease in arterial blood pH which is attributed to a mixed respiratory and metabolic acidosis. Concomitantly arterial Po_2 rose in apparent compensation for the detrimental effects of acidosis on O_2 transport by the blood.

Butler et al. (1992), in case of adult brown trout (*Salmo trutta*), stated that seasonal temperature had significant but small effects on resting Mo_2 (oxygen uptake) and *U*crit.

There is much debate as to the possible causes of fatigue during exercise in vertebrates and one possible cause is a reduction in intracellular pH (pHi). Butler and Day (1993) have reported in brown trout (*S. trutta*) that *U*crit (fatigue) coincides with a particular pHi of the red muscles (read for possible mechanisms).

Webb (1993a) has described the muscle systems of fins. Separate muscle systems move the fin blade of the median caudal, anal, and dorsal fins and the paired pectoral and pelvic fins, while additional muscle systems and spring-like fin rays control the shape of the fin blade.

Table 6.3 Shape factors maximizing thrust and reducing resistance and energy wastage for propulsors

Propulsor and kinematic pattern	Dominant thrust force component	Thrust maximization	Resistance or source of energy loss	Resistance minimization
Undulation of the body/caudal fin	Viscous and pressure drag and acceleration reaction	Large span increasing caudally correlating with velocity and increasing in the same direction	Energy viscous drag ("boundary layer thinning")	Reduce body area anterior to caudal fin
			Lateral recoil	Reduce body area anterior to the caudal fin
Undulation of appendage	Drag acceleration reaction	Large span	Enchanced viscous drag	No special mechanisms recognized
Oscillation of caudal fin or tail	Lift	High aspect ratio	Lateral recoil	Narrow necking; tail attached to body by streamlined caudal peduncle
			Induced drag	Large aspect ratio tail
Oscillation of appendage	Drag and acceleration reaction	Triangular, oscillating about apex	Drag during recovery stroke	Feather paddle-blade and/or slow speed recovery stroke
			Interference drag	Triangular fin oscillating about apex
	Lift	High aspect ratio	Induced drag	High aspect ratio
			Acceleration reaction	Distal tapering (high aspect ratio)
			Interference drag	Narrow tapering body connection

Source: Webb (1988)

Table 6.4 Predicted optimal morphologies for various behaviour patterns in different habitats

	Cruising and sprinting	Cruising	Acceleration	Station holding	Low speed manoeuvering
Typical habitat	pelagic	pelagic	various	benthic	Structurally complex
Typology	thunniform	carchariniform	none	pleuronectiform	chaetodontiform
Vertebrate groups	thunnids, lamnids	many selachians	esocids cottids	pleuronectiformes, rays	acanthopterygian fishes, e.g., chaetodontids, centrarchids
Morphological characteristics					
Thrust maximization					
Caudal fin	high AR lunate	moderate to high AR	small AR large	–	small AR
Body cross section	–	variable along length	elliptical	–	elliptical
Body flexibility	–	large	large	–	variable
Anterior median fins	slot effect possible	large fins spaced to enhance thrust	large along body length	–	large in caudal region
Paired fins	median hydrofoils	ventral hydrofoils	variable	variable	anterior and ventrolateral
Resistance minimization					
Body shape	streamlined rigid	fusiform, flexible	elongate, flexible	streamlined, gibbose	streamlined, gibbose
Fitness ratio (L/H)	4 – 5	4 – 5	< 4	12 – 14	1 – 2
L_H/L	0.5 – 0.7	0.3 – 0.5	0.3 – 0.4	0.2 – 0.3	0.3 – 0.5
Anterior median fins	large over centre of mass	large fins widely spaced	–	large along body length for fanning	–
Caudal area	small	small	–	–	–
Recoil minimization					
Caudal peduncle	narrow	narrow	–	–	–
Anterior median fins	large over centre of mass	large anteriorly	–	–	large anteriorly

Key: AR = aspect ratio; H = maximum depth, or height from the ground in benthic species; L = total length; L_H = distance of H from nose.
Source: Webb (1988)

Table 6.5 A generalized summary of recruitment patterns for muscles, propulsors and swimming behaviour which define various gaits as performance level increases

Propulsor recruitment sequence	Muscle recruitment sequence	Behaviour recruitment sequence
Low performance		
MPF undulatory propulsion	MPF red fibres	
MPF undulatory/ oscillatory propulsion	MPF red fibres	
MPF plus BCF propulsors	MPF and BCF red fibres	
BCF propulsor	BCF red fibres	Burst-and coast swimming, drag enhancement
BCF propulsor	BCF red fibres	Steady swimming
BCF propulsor	BCF red and white fibres	Steady swimming
BCF propulsor	BCF red and white fibres	Burst-and-coast swimming
BCF propulsion	BCF white fibres	Steady swimming (sprints)
BCF propulsion	BCF white fibres	Fast-starts

Source: Webb (1993a)

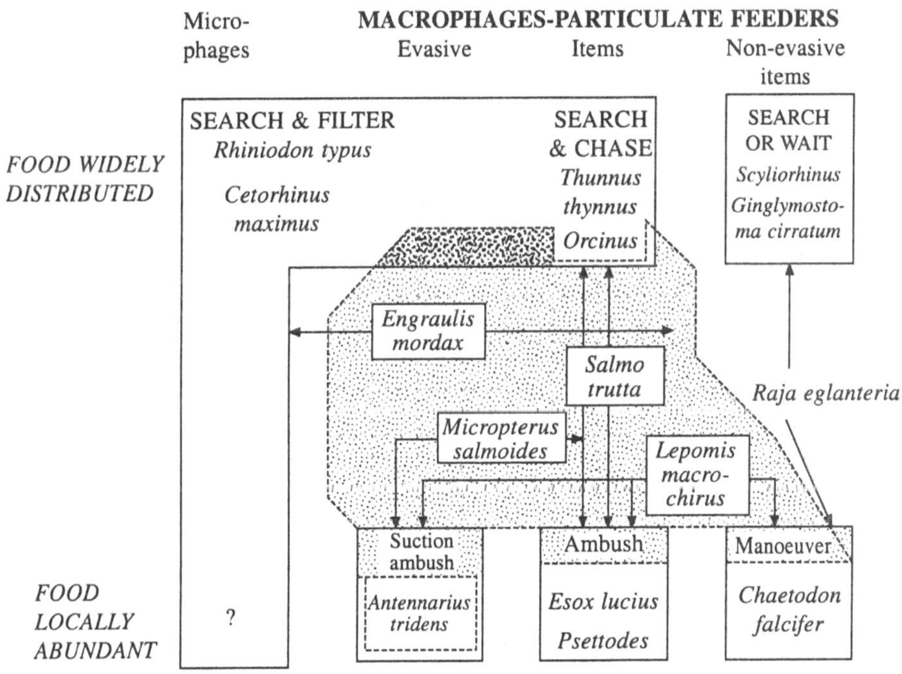

Fig. 6.2 Aquatic locomotor forms and food properties.
Source: Webb (1984a)

The principal muscles in the pectoral fins are deep and superficial adductor and abductor muscle blocks inserting on expanded T-shaped bases of the paired fin rays. The first ray, which would be subject to the greatest fluid forces, also has arrector and flexor muscles. Similarly, erector and depressor muscles of the median fins link lepidotrichia or fin rays to the pterygiophores.

Various oblique muscles control fin area, for example the inclinator muscles of the dorsal and anal fins and arrector muscles of the pectoral fins.

Actinopterygian fin muscle bundles tend to be small and fusiform or moderately pinnate while those of batoidimorph fish tend to be more extensive and organized in sheets. Superficial red fibres in the pectoral fin rays are parallel to and insert about halfway along the ceratotrichia. White muscle fibres are arranged obliquely to the ceratotrichia with a pinnate-like arrangement, which probably increases power output.

The anatomy and fibre arrangements of myotomal muscle are different from those of the median fins, and the way the muscle, myosepta, and skin drive the tail is not understood. Segmental myomeres in the least derived chordates and larval fishes are folded into V-shapes, with the apex pointing rostrally. The flexures become increasingly W-shaped during development as well as in more derived species. Myosepta between the myomeres are flexible in bending but stiff under tension and undoubtedly are key to force transmission from the muscle to the body. Myosepta are often stiffened in teleosts, probably to control torsional moments due to muscle contraction.

The majority of vertebrate swimmers behave like 'hybrid oscillators' which flex from side to side, 'resistance-dominated' posteriorly and 'stiffness-dominated' anteriorly.

It has been emphasized that pectoral fin swimming is of the most important modes of locomotion for ray-finned fishes. In addition to being the primary mode of locomotion for several large families of teleost fishes, it is used by most ray-finned species for manoeuvering and swimming at low speeds. Despite the critical importance of this method of locomotion to fishes, most research on fish locomotion has focused on caudal fin and body swimming. For details of pectoral fin and locomction, see Wilga and Lauder (1999, 2000) and Drucker and Lauder (2000, 2001), and for caudal fin and locomotion, refer to Lauder (2000).

There is a suggestion that oscillating locomotor movements of fish pectoral fins involve either a drag-based or a lift-based mechanism to produce thrust. During drag-based propulsion, the fin is pulled posteriorly through the water like a paddle, with its surface perpendicular to the direction of movement. Lift-based pectoral fin propulsors may use their fins as wings with predominantly dorsoventral movements and a fin surface oriented obliquely with respect to movement. In addition to lift-and drag-based mechanisms, the acceleration reaction is another mechanism which may contribute to forward thrust during pectoral fin locomotion. The acceleration reaction is force generated by resistance of the fluid surrounding the fin in response to changes in velocity of the fin. The acceleration reaction is especially important during periods of unsteady flow.

Much of the work examining the mechanisms of thrust production during pectoral fin locomotion has been theoretical. The effects of speed on pectoral fin movements have also received little attention. Most tetrapods change limb kinematics when changing from a walk to run. Similarly, fish that swim using their pectoral fins may change the pattern of fin movement as speed increases. Fish in the family Embiotocidae may use a different fin movement at different swimming speeds, but no data are available for more generalized teleosts.

Gibb et al. (1994) reported kinematics of pectoral fin locomotion in the bluegill sunfish *Lepomis macrochirus*. The more ventral and proximal portions of the fin edge usually had smaller amplitudes of movement than did the more dorsal and distal locations. With increased swimming speed, the amplitudes of the lateral and longitudinal fin movements generally decreased.

They used two distal markers and basal reference point to determine the orientation of various planar fin elements. During early adduction and most of abduction, these planar fin elements usually had positive angles of attack. Because of fin rotation, angles of attack calculated from three-dimensional data differed considerably from those estimated from a simple lateral projection. As swimming speed increased, the angles of attack of the planar fin elements with respect to the overall direction of swimming approached zero. The oscillatory movements of the pectoral fin of bluegill suggest that both lift-and drag-based propulsive mechanisms are used to generate foreward thrust. In addition, the reduced frequency parameter calculated for the pectoral fin of *Lepomis* ($0 \approx 0.85$) and the Reynold number of 5×10^3 indicate that the acceleration reaction forces may contribute significantly to thrust production and to the total force balance on the fin. Further information on pectoral fin swimming has recently been described in surfperches (Embiotocidae) (Drucker and Jensen 1996a, b, 1997).

Webb (1993a) has described the interactions between the muscles and propulsors. The mechanisms translating muscle contraction into propulsive movements are poorly understood. The presence of discrete muscle bundless rotating fin rays, clearly definable force trajectories, and the ability for manipulation should make studies on MPF systems comparatively simple. Instead, research has focused on the more complex BCF system. Each different combination of propulsor and muscle type defines a gait (Fig. 6.3 read details:Webb 1994, 1998; Drucker 1996). The contribution from Webb (1998) has important Fig. 1 on a hierarchical classification of fish propulsors, MPF and BCF gaits, and Fig. 3 on the pattern of gait recruitment with performance level.

The simplest situation for BCF swimming is alternate and complete muscle contractions on opposite sides of the body. This is probably the situation in prehatching and early larvae, involving only the anterior myomeres and in high-speed sprints.

Swimming speed limits and endurance are directly related to food capture, escape from predators and reproduction. Therefore they are subjected to strong selection pressures and serve the evolutionary fitness of the species (Videler and Wardle 1991). Swimming behaviour is strongly dominated by the viscosity of the water, and the general principles found for adult fish, being strongly determined by inertial forces, will not apply to larvae. Morphological adaptations to a semiterrestrial existence stem mainly from the differences in the densities of air and water. The much lower density of air means that locomotion on land requires greater muscular strength and a modification of locomotor organs. Intertidal fishes are generally poor swimmers and keep close to the bottom (see details: Gibson 1982).

Bone (1974) discussed various drag reducing mechanisms especially related to body form, opercular slots, scales and mucus. For further details on the drag, reader can refer to Webb (1978), Blake (1983) and Yates (1983).

4.1.1 Modes of Swimming

There is spectacular variation in the number, shape, size and position of fins. All these have functional consequences.

Figure 6.4 shows the involvement of fins and body in propulsion (Lindsey 1978; read Blake 1983; Webb 1984, 1993a, 1994, 1998; Hebrank et al. 1990; Coughlin et al. 1992; Long 1992, 1995; Long et al. 1994; McHenry et al. 1995; Coughlin and Rome 1996; Dickinson 1996a, b; Drucker 1996; Drucker and Jensen 1996a, b, 1997; Gillis 1996; Jayne and Lauder 1996; Jayne et al. 1996; Long and Nipper 1996; Long et al. 1996). Modifications of the locomotor pattern are not apparent in Dipnoi as they are in other fishes. Many teleosts can move backwards by means of fin movement. Cyclostomes and eels seem to move backwards by means of reversed waves. The entire biology of scombroids is tightly linked to swimming as a way of life. All life's activities occur in motion (Magnuson 1978). The reader is also referred to Tables 6.3 and 6.4.

During continuous swimming by larvae and elongate fishes, the head and anterior of the body make large lateral movements. These movements not only waste energy but also reduce the

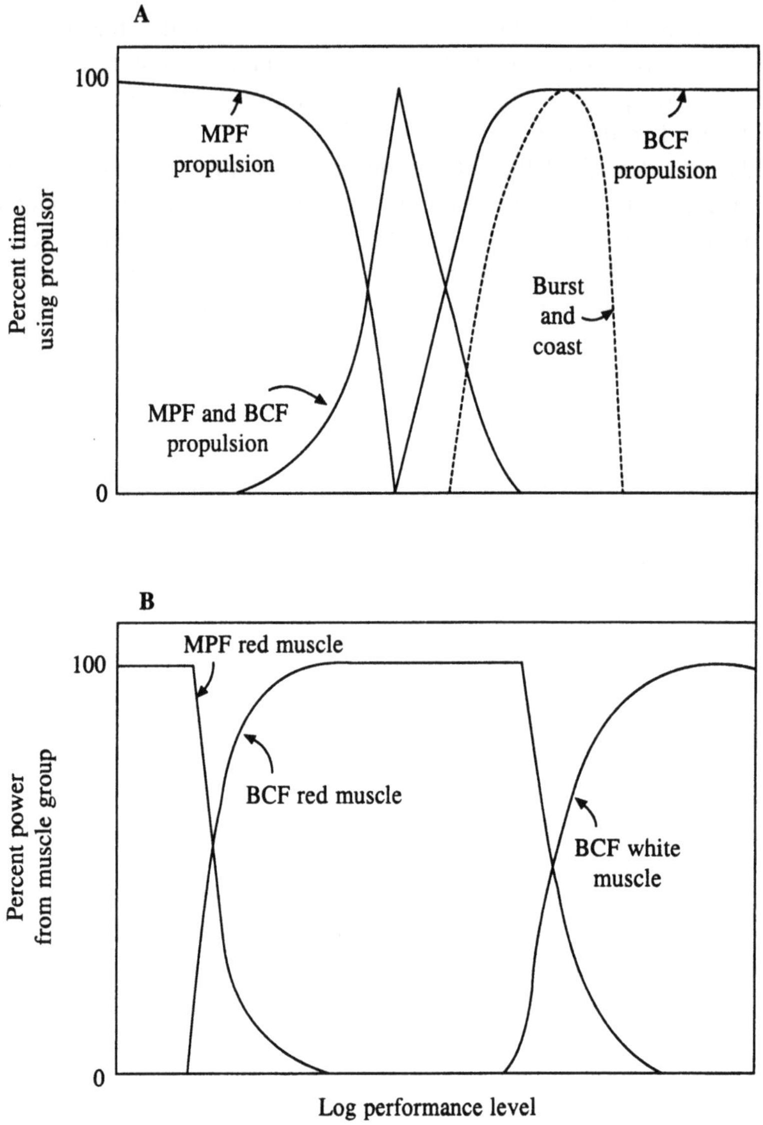

Fig. 6.3 Diagrammatic illustration of fractionation of the speed and acceleration performance range by (A) propulsors and (B) muscle type. Each different combination of propulsor and muscle type defines a gait. Note that performance is shown on a logarithmic scale to expand the lower performance range where gaits are more numerous.
Source: **Webb (1994)**

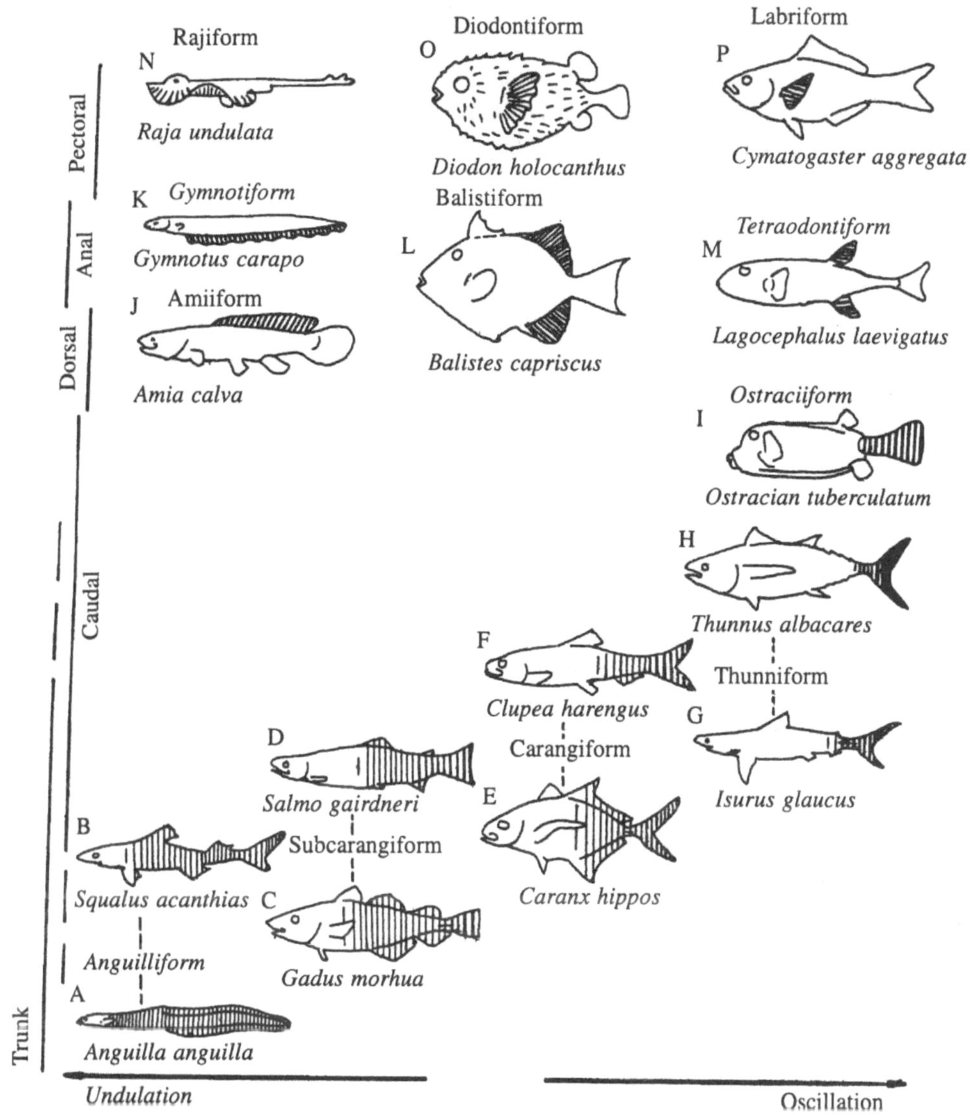

Fig. 6.4 Modes of forward swimming in fish, arranged along the vertical axis according to the propulsive contributions of body and fins (indicated by density of shading), and along the horizontal axis according to a scale running from serpentine undulation (more than one wavelength present) to oscillation (a rigid wigwag or fanline motion), with examples of species.
***Source:* Lindsey (1978)**

effectiveness of sensory organs concentrated rostrally. During ontogeny, and with the onset of steady swimming after a fast-start, the head of fusiform fishes is stabilized relative to tail. A feature of this transition appears to be the faster caudal propagation of waves of muscular activity compared with the speed of the propulsive wave. This has three effects: (1) the body is stiffened rostrally, (2) bending moments are generated at the tail, and (3) the posterior body and tail are stiffened such that the curvature induced by resistive forces at various speeds results in appropriate angles to orient momentum changes close to the axis of progression of the body.

According to Webb (1993a), although fish swimming patterns are very variable, swimming gaits are readily defined in terms of muscle fibre type, propulsor type, and propulsor kinematics (Table 6.5). These gaits divide a large performance range into small parts in which thrust power and muscle power are generated efficiently. Maximum swimming speeds achieved in different gaits vary considerably among species, reflecting interspecific differences in muscle and propulsor characteristics.

Tilting behaviour. At low swimming speeds, negatively buoyant fish 'tilt', with the body axis subtending an angle to the axis of progression. Tilting compensates for reduced control forces that occur at low speeds. According to Webb (1993b), tilting is a general phenomenon of low speed swimming at which stabilizers lose their effectiveness. Tilting is interpreted as an active compensatory mechanism associated with increased drag and concomitant increased propulsor velocities to provide better stabilizing forces.

Trimming mechanisms. There are two classes of *passive* and *active* trimming mechanisms used by fish for stability regulation at low swimming speeds. Passive mechanisms increase the area and/or angle of attack of control surfaces. For example, negatively buoyant elasmobranchs, tuna and mackerel increase pectoral fin extension as speed increases, and fin surface area is larger in species that routinely swim at low speeds. Neutrally buoyant fish also show greater fin extension at low speeds. Tilting by negatively buoyant fish increases the angle of attack of paired fin lifting surfaces while increasing the total area generating lift by recruiting body area. Active mechanisms use propulsor motions to generate stabilizing forces, involving gait changes and tilting (read Webb 1993b).

4.2 NON-SWIMMING LOCOMOTION

4.2.1 Jet Propulsion

It is now generally believed that in most fish the small volume of water expelled during respiration makes the thrust which it could generate negligible. In fish it may be important most often in providing impetus during acceleration particularly from a starting point (e.g., when a ray takes off from the bottom). The slight forward thrust resulting from water ejected through the gill slits during normal respiration requires some locomotory compensation in a fish holding position in midwater. Exhaled water may have another effect in fish locomotion. It has been suggested that water from gill slits may keep the boundary-layer fluid surrounding the body charged with enough kinetic energy to avoid separation. Exhalation has been shown to be coordinated with swimming movements (Lindsey 1978).

4.2.2 Terrestrial Locomotion

Various fish (e.g., *Clarias*, *Anabas*, *Periophthalmus*, *Mnierpes macrocephalus*) can move on land by different modes where fins also have a role. Suctorial discs formed by the mouth or the paired fins may also be used in combination with inching movements of the body in Petromyzontidae, Loricariidae and Gobiidae (Lindsey 1978).

In the amphibious blenny, *Aticus kirkii*, the pectoral fin radials are moveable, the pelvis is strongly developed, the fish has short strong pelvic fin rays and the muscles are modified accordingly. There are also hooks at the ends of the anal, pectoral and pelvic fin rays. All these features are considered to air climbing on steep rocks while out of water. Similar modifications are reported in *M. macrocephalus* in which the anal fin has an incised membrane between the rays. The ventral aspects of the body, anal and pelvic fins are padded. Land locomotion in this species is performed by alternate flexures of the tail very similar to that in *Lophalticus (Aticus) kiekii*, *A. saliens*, and *Gobionellus sagittula*. This method differs from those used by mudskippers which include "crutching", skipping and climbing on vegetation by means of the pectoral and fused pelvic fins (read Gibson 1982).

4.2.3 Moving on the Bottom and Burrowing

Many fish which are heavier than water habitually rest on the bottom, but move forward by convential body and tail movements that carry them on slight excursions away from the surface. The pectorals, generally modified serve in such movements (e.g., *Lophius*, *Ogcocephalus*, *Scorpaena*). Hillstream fish use paired fins for creeping. Burrowing into soft bottom may be achieved by the same anguilliform mode used in swimming (e.g., *Myxine*, Synbranchidae). Non-anguilliform fish (e.g., mudminnow *Umbra*) can "swim" into the soft bottom. Rays and flatfish cover themselves by throwing up sand with undulatory fin movements (Lindsey 1978).

4.2.4 Jumping, Gliding and Flying

A fish usually jumps into the air by swimming rapidly upward through the surface of the water; its momentum alone carries it forward after the tail has left the water entirely. Examples are tarpon, salmon and *Mugil*. Some fish skim or slip along the surface in a series of small leaps or a hydroplanelike precursor to gliding flight (e.g., *Mugil*, *Hemiramphus*). The most proficient aerial gliders are the flying fishes (Exocoetidae) with specialized pectorals and pelvic fins. The Asian cyprinid *Chela maasi* elude capture by skittering away over the surface in what may be a rapid sequency of a curious "neck bending". *Rhaphiodon* (Cynodontidae) (Fig. 6.5) also has a different neck-bending type of locomotion. In both fishes the pectoral fins have a role. The South American hatchetfishes (Gasteropelecidae), unlike flying fish, apparently beat their wings (pectorals) in flight emitting an audible buzz. The osteoglossid butterfly fish *Pantodon buchholzi* can also flap its wings (pectorals) (Lindsey 1978; see review by Davenport 1994).

Pectoral fin adduction during locomotion (hops and darts) of goby *Pomatoschistus lozanoi* is of great importance in generating a forward propulsion (Adriaens et al. 1993).

Many post-larval benthic fish live in nutrient-poor habitat where food is rare or episodic carcass falls provide unpredictable inputs. Such fish often have reduced axial skeletal and muscle systems inhibiting locomotion and probably metabolic expenditures between feeding (Webb 1984).

5. PROPULSIVE ANATOMY

Lindsey (1978) has discussed the following aspects:

5.1 Trunk

Segment number: A fish is a chain of metameres with a head on one end and a tail on the other. More than in most other vertebrates, locomotion in fish depends on the propagation of lateral propulsive waves by the sequential action of basically similar units arranged in a series. Each unit has a set of muscles, a set of nerves, and a set of arteries and veins.

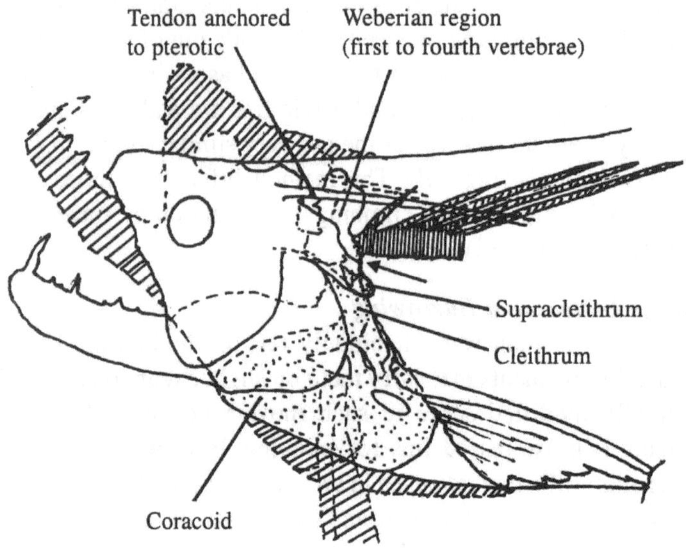

Rhaphiodon: Head

Fig. 6.5 ***Rhaphiodon:*** **Head with head horizontal and pectorals folded against body (heavy solid outline) and with head inflected and pectorals projecting down (hatched broken outline). The pectoral girdle in horizontal head position is stippled. Hyoid apparatus is omitted. Arrow and dot indicate hinge between fourth and fifth vertebrae.**
Source: **Lesiuk and Lindsey (1978)**

The number of vertebrae (and hence of metameres) in living fish species varies. While many selective factors must converge to determine the optimum number for each species, locomotory ability is likely to be prominent. The greater the number of myotomes, the most flexible the body is likely to be. There does not seem to be a simple correlation between speed and number. There is a definite inverse correlation between vertebral number and robustness of body. The thickness of the body rather than its segment number may be dominant in determining flexibility.

There is a view of phyletic decrease in vertebral number in the course of evolution. No primitive fish (cyclostomes, elasmobranchs, lower teleosts) have low counts. Strikingly low counts occur irregularly among some but not all the higher teleosts. The forms retaining the most cartilage have many vertebrae and high flexibility. Ontogenetically, the young of all species have high flexibility quite uniformly distributed along the body; rigidity of the anterior may set in with ossification coupled with increased body width.

A widespread tendence for fish species with larger adult size to have more vertebrae than smaller related species may have locomotory significance. The ubiquity of this phenomenon "pleomerism" is surprising, probably has locomotory importance. Number of vertebrae (and of fin rays and other meristic series) may have locomotory implications.

Skeleton: The axial skeletons of fishes vary in structure from unrestricted notochords to fully ossified vertebral columns. Adult Acipenseriformes, the sturgeons and paddlefishes, are one of the few groups to retain a notochord, which lacks intervertebral joints while possessing a continuous internal lumen filled with vacuolated cells. This persistent notochord is correlated with swimming motions that differ from those of vertebrated species. Lake sturgeon, *Acipenser fulvescens*, modulate three kinematic variables with increasing speed: tailbeat frequency, tailbeat amplitude, and propulsive wavelength, which is the length of the body's undulatory wave as it appears from superimposed midline images of one tail beat cycle. On the other hand, many teleost species modulate only one or two variables: tailbeat frequency and sometimes tailbeat amplitude. According to Long (1995), in the precaudal notochord of white sturgeon, *Acipenser transmontanus*, the axial pattern of maximal stiffness moments was congruent with the pattern of maximal notochordal curvature in the precaudal region, but in the caudal notochord maximal angular stiffness was located cranial to maximal curvature. One interpretation of this pattern is that the precaudal notochord resists bending moments generated by the muscles and that the caudal notochord resists bending moments generated by hydrodynamic forces acting on the tail. The vertebral column, the vertical flexibility at the junction of the skull and the vertebral column, and fluid-filled cushions between adjacent vertebrae, pleural ribs and intermuscular bones, and other bones unattached to the vertebral column have a functional interpretation in respect of locomotion. Symmons (1979) reported the role of notochord and elastic longitudinal ligaments (dorsal and ventral) in swimming by conferring an automatic spring-like resilience to the vertebral axis as a whole. Forms like mackerel (*Scomber seombrus*) and tuna fish, where longer centra (for a given length of fish) mean fewer joints, which are interlocked by apophyses, swim with a stiff body and vibrating tail fin. The dorsal ligament is a large and strong one, making this type of vertebral column exceptionally resilient. This must help to set the rigid high aspect caudal fin into the characteristically fast vibrations. The long body of an eel (*Anguilla anguilla*) has numerous vertebrae with very flexible joints. Nevertheless, the dorsal ligament is well-developed. In consequence, movement is very sinuous with several curves at any one moment along the body. Resilience contributed by ligament is segmentally confined from vertebra to vertebra as such moves on its joint, extending to each body wave. The joint between the head and first vertebra of herring (*Clupea harengus*) and similar fish is quite flexible; the dorsal ligament between the two is well-developed. For ventral ligament details, see the cited paper.

The equation of motion, which describes the relationship between force and motion, has been used to determine the mechanical properties that are important to the dynamic functions of intervertebral joints. This approach has yielded several insights. First, the mechanical properties of angular stiffness and damping coefficient vary with the motion of the joint. Second, there is a regional variation in angular stiffness and on damping coefficient; this variation is correlated with structural variation. Third, the mechanics of the intervertebral joints are dominated by stiffness moments (read details: Long 1992).

All istiophorids have a backbone morphology that is unique among teleosts. Over most of its length, the backbone of the blue marlin *Makaira nigricans*, the billfish with spectacular high performance, like that of all other istiophorids, contains enlarged and flattened neural and haemal spines and zygapophyses, all of which span the intervertebral joints. The plates of bone restrict dorsoventral bending of the backbone but their arrangement permits a high degree of lateral flexion. The spines and zygapophyses also appear to be important in stabilizing the relatively large intervertebral joints against axial compression and lateral shearing during bending. Although bone is an elastic material, these overlapping structures are not arranged so as to contribute to elastic recoil of the backbone during normal swimming movements.

These sprinters are capable of remarkable transoceanic migrations (Hebrank et al. 1990).

Another aspect of flexural stiffness and steady swimming in respect of functions of skin has been a subject of interest. The skin is in a position, by means of its mechanical properties, much of the body's undulatory motion. In addition to its anatomical position, several other characteristics of fish skin make its study compelling. First, fish skin varies widely in structure, bearing large ganoid scales in gars and polypterids, small scales in shark and tuna and no scales in swordfish and some catfishes. In spite of these differences, fish skin appears to possess an underlying structural ground plan-cross-helical fibres of collagen embedded in the dermis. Second, skin has a complex two-dimensional stiffness. Circumferential strain stiffens it axially, giving it the capability to transmit axial forces in lemon sharks *Negaprion brevirostris*, and eels *Anguilla rostrata*; however, this capability is lacking in spot *Lepistomus xanthurus* and skipjack tuna *Katsuwonus pelamis*. Third, skin is attached directly to the underlying muscles, putting it in series between muscle and propulsive elements.

Considering the evolutionary sceneries for the reduction of armoured skin, it has been reported that the first vertebrates, which appeared some 450 million years ago, are characterized by the presence of external dermal bone. An extensive exoskeleton is clearly seen in the oldest complete fossil fish, *Sacabambaspis*, with articulated plates oriented in obliquely inclined scale rows, covering the body. While the first robust exoskeletons are often interpreted as adaptations against predation, the presence of ganoid scales in living gar and polypterids has other mechanical consequences, including enhanced station-holding and passive lung inflation. The armoured skin in longnose gar *Lepisosteus asseus* can play a central role in the control of body stiffness and undulatory wave motion during steady swimming. The armoured skin may represent an *alternative* rather than a maladaptive solution to the evolutionary challenges of aquatic locomotion.

The gar's axial musculoskeletal system is unusual in its lateral flexibility. When the dermis is cut and caudal scale row is removed, the remaining muscle and backbone still show a neutral zone of curvature, a region of bending without detectable stiffness. This is a surprising result, given that the backbones of other fishes show no such neutral zone. Gars are the only fish with opisthocoelaus joints, with the centrum convex anteriorly and concave posteriorly. In a manner yet to be determined, this unusual intervertebral joint design may play a role in creating a neutral zone. Neutral zones are also present in the intervertebral joints of mammals, which possess an intervertebral joint with a flat articular surfaces on the centra (Long et al. 1996; read also Long and Nipper 1996).

Myomeres: The setting of myomeres, the presence of tendons, horizontal septum of connective tissue between dorsal epiaxial and ventral hypaxial portions of myomere, longitudinal ligaments, and the fibre-types in various fish and inter-relationships with bones have significance in locomotion.

The lateral musculature provides almost all the propulsive force. Other muscle systems or modifications in trunk particularly in relation to fins are of no less importance.

5.2 Innervation and Co-ordination of Trunk Muscles

The reader is referred to Chapter 5 for the required information. Receptors (specially of octavolateralis system, Chapter 15) and Mauthnerian system (Chapter 14) are useful in locomotory coordination.

5.3 Metabolism

Bilinski (1974), Driedzic and Hochachka (1978) and recent coverage in Chapter 5 should be consulted.

6. ESCAPE RESPONSE

The escape response of fishes is one of the most rapid locomotor behaviours of vertebrates, and as such it is of great interest to such diverse areas as predator-prey ecology, functional morphology, muscle physiology and neurophysiology.

Fish employ fast-starts for escapes and prey capture in their environment. Fish lacking a fast-start response possess other structural or behavioural antipredator adaptations (e.g., toxins, spines, burrowing habits).

Escape response in fish have been studied from biomechanical, physiological and behavioural perspectives. The ability of fish to escape predators may depend upon linear performance (velocity, acceleration), accurate timing and turning capabilities.

Escape response consist of three kinematic stages: a preparatory stage (stage 1) involving a strong unilateral muscular contraction bending the fish into a C-shape, a propulsive stage (stage 2) in which the tail bends in the opposite direction to that in stage 1, and 3 a variable stage involving coasting, steady swimming, or subsequent fast-start.

Morphology predicted to be optimal for maximizing fast-start performance includes a flexible body, a large muscle mass relative to total body mass, and a large depth of section caudally; previous studies have shown that the caudal fin plays an important role in thrust production.

The initial contraction during stage 1 is usually initiated by a single pair of reticulospinal neurons (Mauthner cells), although alternative circuits may exist. A parallel network of neurons interacting with the Mauthner cells controls the extent of stage 1 contraction.

Escape paths tend to be away from the stimulus, since the axial musculature contralateral to the stimulus contracts during stage 1, forming the C-bend. It has been suggested that turning angles are programmed at the onset of the escape response in fish and other animals. Recently, it has been shown that escape trajectories are linearly related to stimulus direction. However, they vary considerably after the initial turn away from the stimulus. Domenici and Blake (1993) in the study of escape trajectories in anglefish (*Pterophyllum eimekei*), define escape trajectories as a circular variable with 0″ as stimulus direction. Authors suggest that the presence of multiple preferred trajectories may be adaptive in preventing predators from learning any single fixed pattern of response and compensating for it. The reader can get comprehensive information from recent papers in this subject (Webb 1986; Domenici and Blake 1991; Harper and Blake 1991; Kasapi et al. 1993; Jayne and Lauder 1996).

The amplitude and timing of lateral bending during escape behaviour of fishes are poorly understood and these quantities have not been quantified with electromyograms. However, during escapes, the fish's centre of mass is known to remain stationary during the initial synchronous ipsilateral activation of muscle along the entire length of the fish. It is far from obvious from prior studies whether or not other kinematic and mechanical events occur in standing or travelling patterns during the different stages of the escape response. The extent of lateral bending is also of central importance to all undulatory locomotor behaviours including the escape response, because this kinematic variable is proportional to the magnitude of the strain undergone by the axial musculature.

Jayne and Lauder (1993; see discussion and literature) quantified midline kinematics with synchronized electromyograms (emgs) from the red and white muscles on both sides of bluegill sunfish (*Lepomis macrochirus*) during escape behaviours which were elicited from fish both at a standstill and during steady speed swimming. Analyses of variance determined whether or not kinematic and emg variables differed significantly between muscle fibre types, along longitudinal positions, and between swimming versus standstill trials.

Jayne and Lauder (1996) reported that the functional units for axial locomotion of fishes do not correspond simply to the anatomical units of individual myomeres. Rather the *in vovo* motor pattern is a primary determinant of the functional units involved in swimming. Results were based on comparative analyses on two taxa within a single monophyletic clade, the sunfish family Centrarchidae.

CHAPTER 7

Digestive system; Growth; Gastrointestinal Hormones; Age Determination

1. INTRODUCTION

The most primitive vertebrates, the fish evolved in the ancestral oceanic environment, whereas the most advanced vertebrates, the mammals, attained dominance on dry land. For both, success depends on adjusting the essential problems of survival: feeding, escape from predators and reproduction.

Fish larvae are the smallest independently functioning vertebrates. Within the given limits of organism size (Kotrschal et al. 1990; Osse 1990), larvae have to effect simultaneously drastic growth rates, progressive differentiation of both transitory larval and permanent juvenile characters, and complete organ and systemic functioning (Blaxter 1988). To meet these challenges, larvae must be equipped with efficient mechanisms for the uptake [(i.e., search and capture (e.g., Blaxter 1988) and processing (i.e., digestion, absorption and metabolism (e.g., Dabrowski 1984; Govoni et al. 1986)] of food. Segner et al. (1994) have given details on the ontogeny of digestive and metabolic functions in larval turbot, *Scophthalmus maximus*. Turbot larvae at first feeding possess: (1) a functional intestine; (2) a functional pancreas; and (3) a functional liver and gall bladder. The larvae lack: (1) Gill respiration. Instead of branchial respiration the larvae rely on cutaneous gas exchange. (2) Anaerobic muscle energy metabolism. This implies that larvae do not use carbohydrates, via glycolysis, for swimming energy metabolism, but rely on the aerobic breakdown of amino acids and/or fatty acids. (3) A functional stomach. The absence of structures such as stomach and gills in fish larvae may be explained by size problems. Segner et al. (1995) described the organogenesis in larval turbot (*S. maximus*).

Aquatic vertebrates, particularly fishes, incorporate some of the most diverse and versatile feeding morphologies and behaviours within the Vertebrata. Studies on their feeding apparatus have attempted to describe the mechanisms and versatility of feeding in these vertebrates to gain a better understanding of their functional morphology and the evolution of the vertebrate feeding mechanism. Consequently, functional studies on the feeding of bony fishes have centered on the mechanics of prey capture, prey processing, and feeding performance. To better understand the evolution of feeding mechanism, phylogenetic interpretations of bony fish feeding structures have been forwarded. However, an understanding of the mechanics and evolution of aquatic feeding suffers from a profound lack of information on elasmobranch feeding.

The elasmobranchs incorporate a diverse group of aquatic feeders that have radiated from a grasping ancestor to include cutting, suction-crushing, gouging, crushing, and filter-feeding forms. Many studies provide valuable data on the anatomy of the head and jaws of sharks, and include descriptive studies of the feeding structures. There have been few synthetic studies on shark feeding mechanisms (see Motta et al., 1997 for a carcharhinid shark; Wilga and Motta 1998 for a squaliform shark). Durophagy (feeding on hard prey) is known in Holocephali, Heterodontiformes, Orectolobiformes, Carchariniformes, Rhinobatoidei, Rajoidea and Myliobatoida (Wilga and Motta 2000; see Fig. 9. Schematic diagram of the functional components involved in jaw protrusion and retraction in *Sphyrna tiburo*).

Morphology is generally believed to be a major determinant of feeding habits in fishes. Dietary differences among species and among individuals within species are often claimed to be due to differential capture abilities inferred from morphological variation. Mouth size, mouth position, jaw length, dentition, gill raker spacing, eye size, skull shape, and body form are some factors frequently assumed to be indicative of fish feeding abilities. However, only rarely are the inferences related to morphological differences to functional abilities supported with direct empirical data. Mouth width has been shown to limit maximum prey size in several fish taxa, and gill raker spacing can set a lower limit on prey size. In young and small fish, eye size can limit visual acuity because larger eyes can detect smaller prey. In spite of these and other efforts, the feeding constraints imposed by morphology are poorly known for most fish taxa.

Two principal modes of aquatic feeding in lower vertebrates have received attention: (1) suspension feeding and (2) suction feeding (also called inertial suction feeding or gape-and-suck feeding). In sharp contrast to suspension feeding, which may be modelled by steady-state hydrodynamic equations, the salient feature of suction feeding is its nonsteady, dynamic nature. Prey capture by suction feeding is primitive for the Osteichthyes and is common in all the major lineages of lower vertebrates, including ray-finned fishes (Actinopterygii), coelacanths, and lungfishes. The major feature of the evolution of the actinopterygian feeding mechanism is the increase in structural complexity in both the pharyngeal and front jaws. Structural diversification is a function of the number of independent biomechanical pathways governing movement. Two major trends that have been noted during the evolution of feeding systems in teleost fishes are: (1) the great majority of fishes retain suction feeding, the phylogenetic primitive means of prey capture in vertebrates, as the principal means of capturing and ingesting food and (2) associated with this emphasis on suction feeding, many lineages of teleosts possess a highly mobile and protrusible upper jaw. A reversal of both of these trends has been recorded in percomorph teleost fishes (Tetraodontiformes, e.g., queen triggerfish *Balistes vetula*).

The study of feeding has undergone a renaissance in recent years (Elshoud-Oldenhave and Kershaw 1976; Osse 1976; Anker 1978; van Hasselt 1978, 1979; Liem 1978, 1979, 1980, 1993; Dutta 1979; Elshoud-Oldenhave 1979; Gosline 1980; Lauder 1982, 1983a, b, 1985a, b; Motta 1982, 1984, 1985, 1988, 1989; Muller et al. 1982, 1985; Sibbing 1982, 1988, 1991a, b; Leeuwen and Muller 1983, 1984; Lauder and Clark 1984; Muller and Osse 1984; Muller et al. 1985; Osse 1985; Sibbing and Uribe 1985; Drost and Boogart 1986; Lauder et al. 1986; Sibbing et al. 1986; Aerts et al. 1987; Wainwright 1987, 1988, 1989a, b, 1992, 1993, 1996, 1999; Benmouna et al. 1989; Hourigan et al. 1989; Sanford and Lauder 1989; Wainwright et al. 1989; Westneat and Wainwright 1989; Aerts 1990, 1991; Hoogenboezm et al. 1990, 1991, 1992, 1993a, b; Claes and De Vree 1991a, b; Claes et al. 1991; Drucker and Jensen 1991; Lammens and Hoogenboezm

1991; Harper and Blake 1991; Motta et al. 1991; Smith and Reay, 1991; Berg et al. 1992; Lauder and Wainwright 1992; Aerts and De Vree 1993; Turingen 1993; Turingen and Wainwright 1993; Wainwright and Turingen 1993; Frazetta 1994; Huysseune et al. 1994; Ringler 1994; Sanderson et al. 1994; Vandewalle et al. 1994, 1995, Richard and Wainwright 1995; Wainwright and Richard 1995; Gartner et al. 1997; Grubich and Wainwright 1997; Ralston and Wainwright 1997; Clifton and Motta 1998; Gluckmann and Vandewalle 1998; Wilga and Motta 1998, 2000; Huet et al. 1999; Motta and Wilga 1999, 2001; Viladiu et al. 1999; Diogo and Chardon 2000; Wilga et al. 2000; Edmonds et al. 2001). A number of aspects (e.g., morphological constraints, population consequences in respect of foraging have been described by Persson et al. (1997).

Aquatic food is diverse in size, shape, location, motility, chemical and mechanical properties as are adaptations to fish to capture and process food. The wide variety of aquatic food is considered to be instrumental for the diversification in fish species. A variety of information regarding diet selection (e.g., hunger state, learning and memory) is provided by Hughes (1997).

Diversity in regional morphology of fish digestive tracts is evident. This variation could be due to differences in phylogeny, feeding biology, nutrition, age, physiological conditions and the special functions which the gut may perform.

Kapoor et al. (1975b; see literature), Fänge and Grove (1979), Smith (1989), Sinha (1992, 1993), Kapoor and Khanna (1993, 1994) and Horn (1998) have comprehensively covered several aspects of the subject.

2. RESOURCE USE AND FEEDING: THE OVERALL PICTURE OF EXPLOITATION OF ENVIRONMENT

Foraging behaviour is constrained by the physiology and morphology of fish. The behaviour of the individual will be influenced by the presence of conspecifics and by predators.

Different aspects of foraging have been discussed by Hyatt (1979), Webb (1984a, 1986b), Crowder (1985), Hart (1986), Milinski (1986), Pitcher (1986), Magurran and Pitcher (1987), Pitcher and House (1987), Hart and Hamarin (1988), Gotceitas and Colgan (1989), Cailliet and Ebeling (1990), Domenici and Blake (1991), Harper and Blake (1991), Van Duren and Glass (1992), Hughes (1997), Overholtzer and Motta (1999) and Muñoz and Motta (2000).

Fishes use several sensory systems to detect food. The sensory systems include primarily taste, touch, vision and olfaction, and possibly audition, lateral line, and electroreception. For detecting food at some distance from the stimulus source, however, olfaction, taste, vision (depending upon the species and water clarity), and probably audition predominate. Of these systems, vision and olfaction are similar across the vertebrate lineage in terms of the general distribution of the receptors involved; however, in catfish, taste buds are found not only within the mouth, as is the rule for tetrapods, but are located across the entire external body surface. The feeding and escape behaviours in channel catfish, *Ictalurus punctatus*, have been illustrated as an example (Fig. 7.1) (see Valentinčič and Caprio 1994).

A summary of possible controlling factors is given (Fig. 7.2). This suggest mechanisms by which feeding behaviour is switched on and off, but does not explain how feeding is integrated with other behaviours into a fitness-maximising sequence. This is because the model in Figure 7.2 does not account for the dynamic interactions that will occur between activities (Hart 1986).

For details of bodyform, locomotion and foraging in aquatic vertebrates (see Chapter 6).

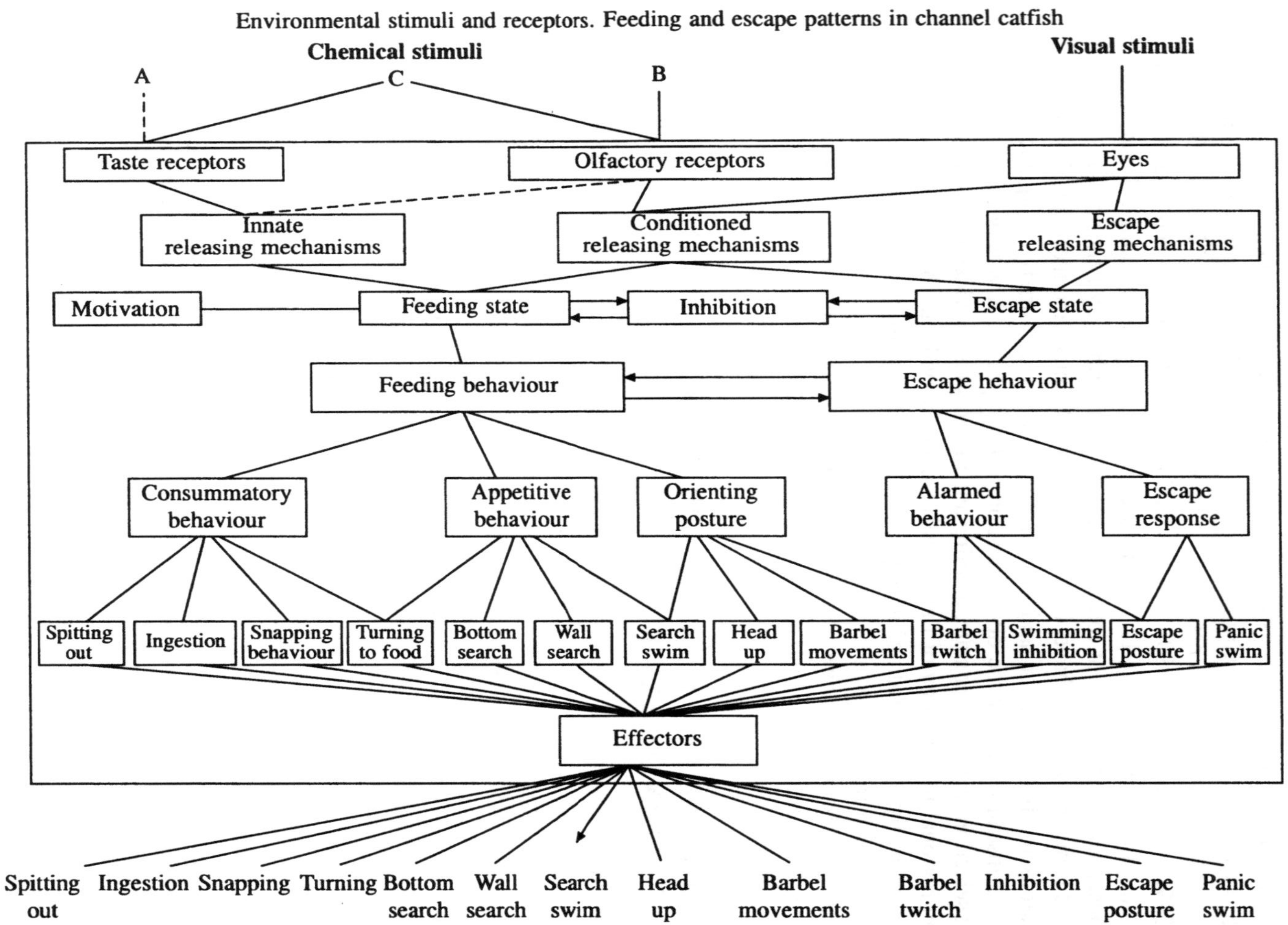

Fig. 7.1 **Hierarchical organization of the patterns of feeding and escape behaviours. Environmental stimuli and receptors are presented at the top and effectors and consummatory behaviour patterns are presented at the bottom of the figure. Feeding and escape behaviours are released by innate releasing mechanisms, whereas the conditioned releasing mechanism influence the duration and speed of the searching swims. Small fonts indicate individual behaviour patterns and their relationships at a lower level of hierarchical organization; larger fonts indicate complex behaviour patterns and control mechanisms that release these behaviours. Arrows within the box indicate mutual inhibition of escape and feeding behaviours. The large filled arrow pointing to search swim and the thick line connecting this arrow with the conditioned releasing mechanism of box indicate events released by an amino acid stimulus that were reinforced by a food reward.**

Source: **Valentinčič and Caprio (1994)**

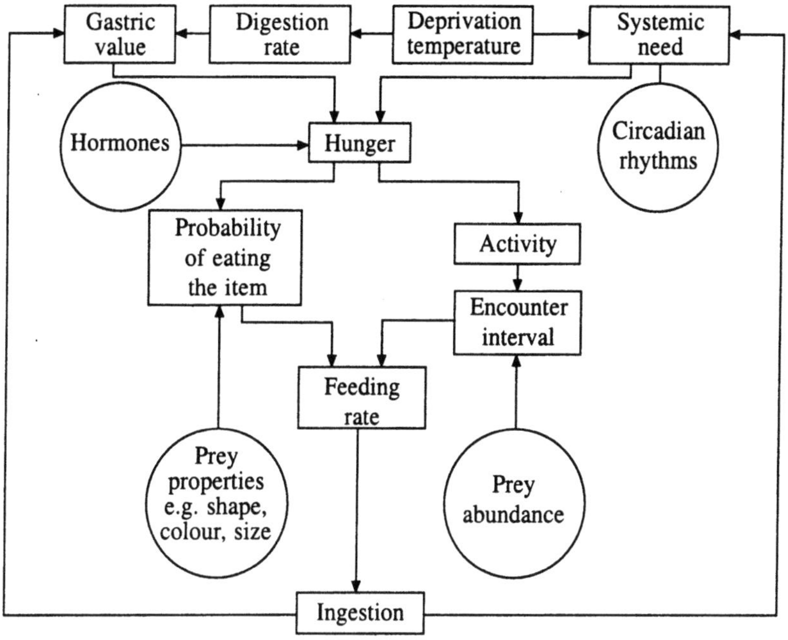

Feeding behaviour in teleost fish.

Fig. 7.2 A model of feeding behaviour in fish, illustrating how internal and external factors influence the need to feed (hunger).
***Source:* Hart (1986)**

In order to feed, fishes have to generate a flow of food-laden water through their bucco-pharyngeal cavity. This can be achieved in two ways: (1) by translational movements of the fish with both the mouth and the opercula wide open, leaving the water and prey relatively unaffected (imparting minimal momentum to the water and prey), or (2) by active suction, resulting from fast expansion of the mouth cavity (inertial suction). In fact, both techniques are combined to a greater or lesser extent. Suction always involves displacements of the fish as it sucks itself forwards or because it is assisted by active swimming. Moreover, advanced fishes protrude the border of their mouth towards the prey during suction, thus improving the flow relative to the mouth (e.g., *Epibulus insidator*).

Particularly in cases where transitional movements of the fish are minimized, the subsequent expansion of the mouth and opercular cavities for suction proceeds very fast.

It is a common belief that expansion of the mouth for suction occurs as a result of a sequence of muscle activities governed by a central pattern generator that is not subjected to peripheral feedback modifications subject to the onset of a specific motor programme. The selection and assemblage of the functional components of the motor programme depend entirely upon peripheral *pre-strike* information (mainly visual) on the type, size, position and behaviour of the prey. Once started, the motor pattern of the strike cannot be modulated by further feedback control and the suction activity progresses in a programmed fashion.

Analysis of the feeding kinematics of *Astatotilapia* (a small cichlid fish) suggests the presence of peripheral feedback modulation of a motor pattern, allowing the act of suction to be abbreviated. In this way, the effort spent in suction is minimized. The biological significance of the development of such a modulating feedback system is not immediately obvious from a 'classical ecological' point of view. It is postulated that the muscular metabolism itself might constrain the short, transient and strenous motor output typical of suction feeding. Thus, reducing the suction effort makes the sense when successive strenous headpart movements are immediately required for additional suction, buccal transport or spitting (see details: Aerts and DeVree 1993).

Teleostean piscivores often do not chase prey that make escape attempts and those chases that occur are not very successful. Fish piscivores are normally 3 to 4 times larger than their prey. The apparent disadvantage of larger size on manoeuverability is small because fish can turn in very small radius circles while advantages of speed and power for large fish are substantial. Therefore, chases would appear to favour predator success. However, predator performance advantages could be eroded if predator response latencies are large. Webb (1984a) reported that the relatively long chase response latency, with unpredictable prey manoeuverability and usually proximity to cover, appear to favour prey escape in chases and may explain not only the apparent reluctance of predators to chase but also reluctance to persue at high speed. Weihs and Webb (1984) presented a kinematic analysis of optimal avoidance and evasion techniques for prey.

There is now extensive body of literature on different aspects of antipredator responses (Table 7.1). For detailed information, the reader is referred to Csányi and Dóka (1993), Grant (1993), Huntingford and Wright (1993), Magnhagen (1993) Krause and Godin (1995) and Smith (1997) (see Figs. 7.3–7.7).

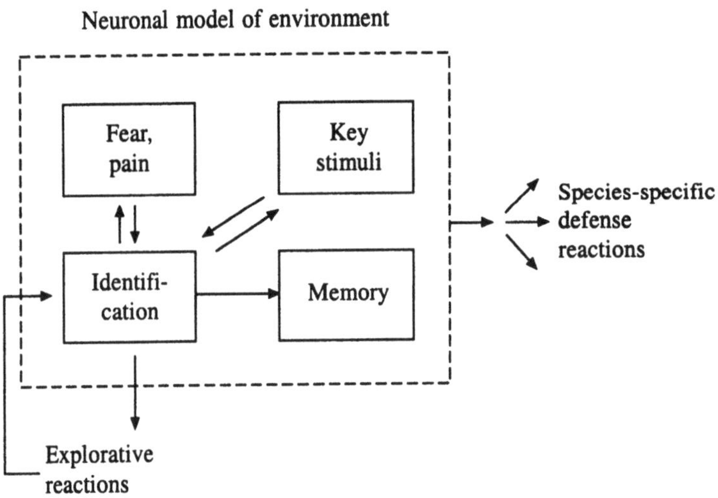

Fig. 7.3 Sketch of the interactive learning hypothesis.
***Source:* Csányi and Dóka (1993)**

Findings by Richard and Wainwright (1995) have demonstrated that body size can have major effects on feeding kinematics, as recorded in large mouth bass (*Micropterus salmoides*). Further, Wainwright and Richard (1995) have reported that scaling effects in *M. salmonides* may include changes in the relative resistance of the jaw and head structures to movement through water and changes in the intrinsic contractile properties of the muscles of the feeding apparatus.

Predator-prey interactions invariably involve locomotion. Locomotor capabilities are determined by physical laws, which in turn limit locomotor behaviour (Webb 1986b). Predator-prey relations have important consequences for the structure of many fish communities, although competition is also an important factor (see Hart 1989 for predicting resource utilization: the utility of

Table 7.1 Adaptive variation in behaviour as a consequence of increased predation risk

Behaviour effect of increased predation pressure	Species
Schooling	
larger and more cohesive school	*Poecilia reticulata*
	Phoxinus phoxinus
Evasion tactics	
more effectively integrated in	*P. reticulata*
high-risk populations	*Gasterosteus aculeatus*
	P. phoxinus
Inspection and predator assignment	
increase in inspection frequency	*G. aculeatus*
	P. phoxinus
increase in inspection groupsize	*P. phoxinus*
predator-specific attack cone avoidance	*P. reticulata*
more efficient assessment of predator motivation	*P. reticulata*
more likely to employ tit for tat	*G. aculeatus*
	P. reticulata
Response to overhead threat	
greatest in high-risk population	*P. reticulata*
	G. aculeatus
Response to alarm pheromone	
more pronounced behavioural change	*P. phoxinus*
following exposure to Schreckstoff	
Acquisition of defence skills	
predisposition to respond to experience	*G. aculeatus*
greater in high-risk populations	*P. phoxinus*
Habituation of antipredator responses	
slower in high-risk populations	*G. aculeatus*
Habitat selection	
remain near surface and seek cover	*P. reticulata*
at edge of river	
Foraging	
increased foraging tenacity	*P. reticulata*
Female choice	
preference for less brightly	*P. reticulata*
coloured males	
avoidance of sneaky mating attempts	*P. reticulata*
compromised by predator avoidance	
Male mating tactics	
increased use of sneaky mating	*P. reticulata*
tactics in high risk predation	
Aggression	
individual aggression reduced	*G. aculeatus*
	P. reticulata

Source: Magurran et al. (1993)

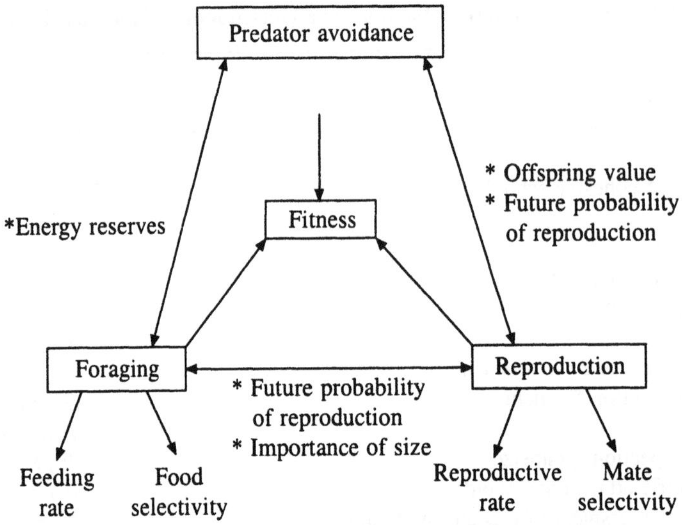

Fig. 7.4 **An animal's fitness depends on several conflicting demands. For example, predation risk can influence fitness both directly and through influencing foraging and reproductive behaviour in different ways. Trade-offs between conflicting demands (↔) are influenced by several factors (examples marked by*).**
Source: **Magnhagen (1993)**

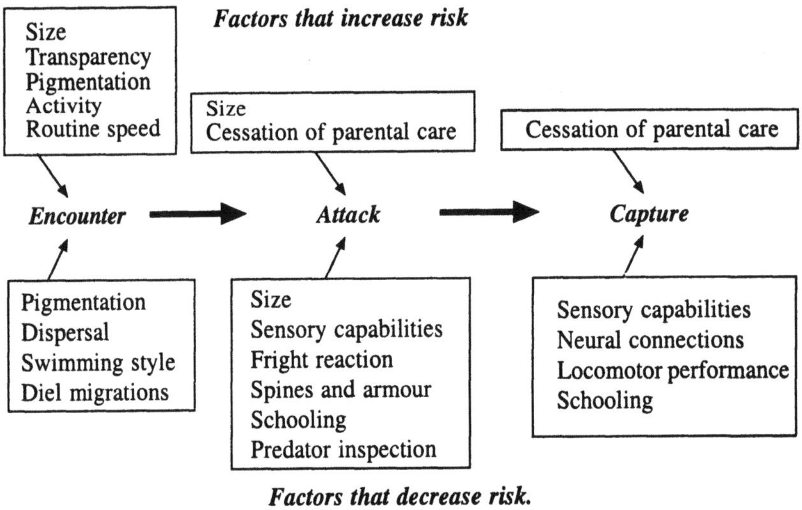

Fig. 7.5 **Summary of some effects of development on the predator sequence. Listed item undergo predictable changes during development, which influence one or more steps in the sequence. For example, size increases during development and transparency decreases. This renders a larva a larger, more conspicuous target for a predator, thereby increasing encounter rate. Larger size also increases the profitability of the larva to the predator, making an attack more likely. Yet with continued growth, the prey becomes too large for the predator to consume, lowering the risk of attack. The less predictable effects of experience, natural selection, and environment are not included here.**
Source: **Fuiman and Magurran (1994)**

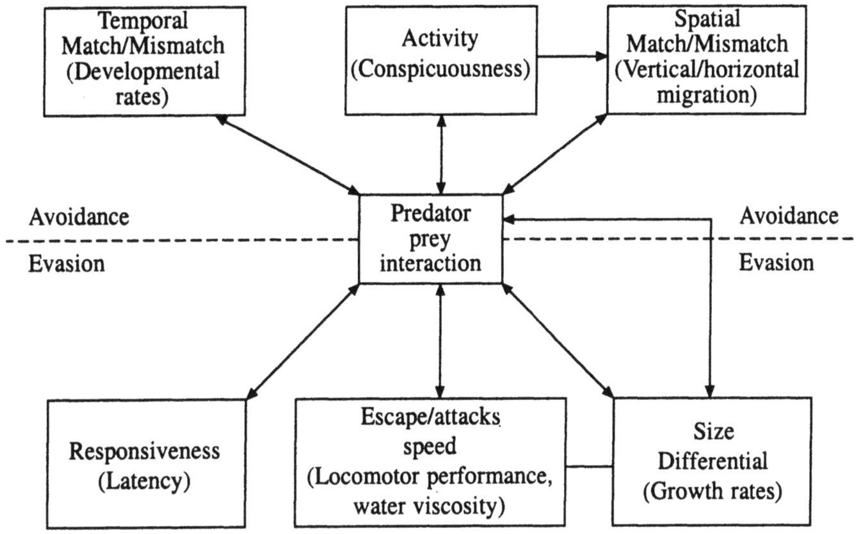

Fig. 7.6 Box digram showing predatory-prey interrelations with temperature-dependent processes. *Source:* Blaxter (1992)

optimal foraging models). The capture efficiency of predators is often inversely correlated with habitat complexity. Hart and Hamarin (1988) suggested that the prey size eaten was determined by an interaction between jaw morphology and the relative swimming speeds of pike (*Esox lucius*) and rudd (*Scordinius erythrophthalmus*), rather than by a behavioural choice.

Gotceitas and Colgan (1989) supported the hypothesis that the relationship between foraging success and habitat complexity is non-linear, with some threshold level of complexity necessary to reduce predator success significantly. A similar, but directionally opposite, relationship also exists between habitat complexity and prey choice when avoiding predation.

Many species of fish forage in groups. Such groupings may result from fish independently aggregating at preferred feeding sites or from social attraction between conspecifics. When grouping is the result of social attraction commonly referred to as schooling or shoaling, it is often assumed that the group confers an advantage over being solitary by increasing foraging efficiency, decreasing predation risk or both, with regard to foraging, being in a group may increase encounter rates with clumped or patchily distributed prey, socially facilitate feeding behaviour speed the recognition of new prey types and increase success in exploiting prey that are difficult to capture.

In fish shoals, forage area copying has a number of well-established benefits. For example, fish locate food more rapidly in larger social groups through this effect (Pitcher and House 1987).

Krause (1992) reported that neither shoal size nor food delivery rate alone can play a role in the initial choice among patches. Rather an individual's initial choice of patch appears to be influenced by the excitement that conspecifics show to food, which in nature may lead to selection of the more profitable patch.

Several studies of shoaling fish using either pelagic or non-pelagic species have emphasized the effect of hunger on the structure of the shoal. Robinson and Pitcher (1989) suggested that hunger motivation at lower supplies of food induces a behaviour where co-ordinated swimming

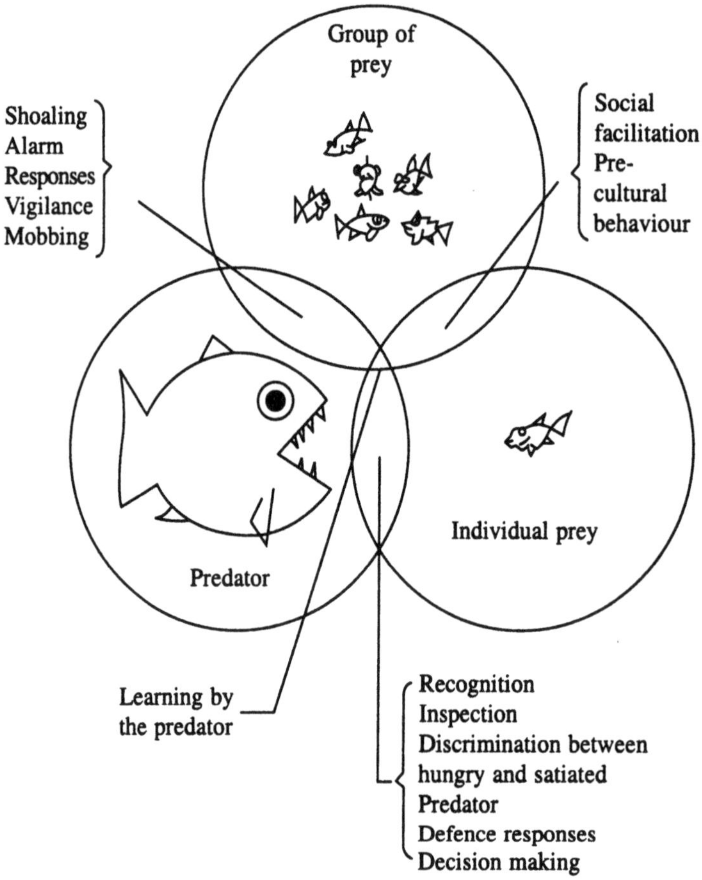

Predator-prey interactions.

Fig. 7.7 Different classes of predator-prey interactions.
***Source:* Csányi and Dóka (1993)**

is changed in favour of more individualistic and independent search for food. Reduction of polarization increases the total area searched by members of the shoal and could increase the probability of finding food through the shoal enhancement effect. Furthermore, a separate search for food could reduce the competition between shoalers. However, security is sacrificed, since co-ordination and synchrony are bases of many predator evasions presented by shoaling fish.

Krause (1994) studied how food-competition and predation risk affect size-assortative shoaling and what the ecological consequences are, in respect of chub (*Leuciscus cephalus*). Results indicated that predation plays a more important role than food competition for size-assortative shoaling in chub. Krause and Godin (1994) reported in banded killifish (*Fundulus*) that body size plays a key role in shoal choice, overriding the effects of shoal size and species preference. Shoal choice was affected by predator threat only when differences between shoal size or body size of stimulus fish were large.

Size-assortative shoaling is usually confined in two different forms: large and small fish separate completely and form two distinct shoals or they remain in the same shoal but they segregate horizontally or vertically. The underlying reasons for this behaviour are believed to be twofold. Large fish are more competitive and therefore gain a larger share of the food resources which puts small fish at a disadvantage in a mixed shoal of small and large fish. Small fish should therefore try to separate themselves from large fish. On the other hand single large fish in shoals of small fish are conspicuous to predators and might suffer from increased predation (oddity effect). It is suggested that the latter could be reason which prevents large fish from invading shoals of small fish.

Shoaling fish take account of the foraging success of others in the group but also maintain vigilance against predators. Although more vigilant fish larger groups allow a stalking predator to approacher closer before halting feeding to flee, implying that fish in a shoal may trade-off predation risk against feeding rewards (see Pitcher et al. 1988). Ryer and Olla (1995) indicated that social aspects of foraging strategies are flexible, but are also influenced by prior experience with resource distribution. Chivers et al. (1995) suggested that preferential shoaling with familiar conspecifics leads to an increase in cooperative antipredator behaviour and may thereby lower a minnow's risk of predation (read for details on predator inspection behaviour).

Approaching predators is particularly ubiquitous in fishes and is generally referred to as predator inspection behaviour. As with other vertebrates, prey fish typically approach a potential predator from a distance in a tentative, saltatory manner. The approach is characterized by a series of moves toward the predator, interrupted by short stationary pauses, during which the prey is visually fixated on the predator. An approach is terminated by movement away from the predator. Fish inspect potential predators most commonly alone or in pairs and less often in larger groups. However, considerable inter-individual variability exists within fish populations in the tendency to inspect potential predators and in the minimum approach distance. Such variability is likely based on differences between individuals in fitness benefits and costs associated with their inspecting a potential predator during a given encounter. The benefits and costs of any individual decision will depend on several factors, including the animals' current internal state (e.g., body size, hunger level, reproductive status), its morphology (e.g., body armour) and local ecological factors (e.g., risk of predation, food availability). Godin et al. (1994) reported in threespine stickleback (*Gasterosteus aculeatus*) the hungrier fish in a shoal are more willing to take greater risks to inspect a potential threat at a distance, compared with their well-fed shoal mates and suggest that they may gain a foraging benefit in doing so. If marked asymmetries in hunger state exist among members of fish shoals, then mutual cooperation during predator inspection visits may be difficult to achieve because well-fed individuals are not as likely to initiate or participate in inspection visits as are hungry individuals. Godin and Davis (1995) presented strong evidence for a predator pursuit-deterrence function of predator inspection behaviour in the guppy (*Poecilia reticulata*), and also have implications for the conditions necessary for the evolution of predator approach behaviour as a cooperative strategy in prey. Krause and Godin (1996) indicated that the ability to respond to approaching predators and the risk of mortality to predation in the guppy is strongly influenced by their foraging activity, and in particular their foraging posture, and that cichlid (*Aequideus pulcher*) predators preferentially select less wary and more vulnerable guppies.

Turingan et al. (1995), on the basis of relationship between prey utilization and jaw biomechanics in two Carribean populations of triggerfishes, hypothesized that the inter-population differences in morphology are induced by the activities of feeding on different prey and enhance the feeding abilities of fishes for locally dominant prey. Plasticity of the feeding mechanism may be a widespread attribute of fish feeding systems that promotes the ability of species to occupy multiple habitat types successfully. Trophic breadth may be correlated with behavioural and functional versatility or flexibility, such that species with limited diets would be predicted to be less versatile or flexible in their feeding capabilities than would trophic generalists. In pufferfishes, (Ralston and Wainwright, 1997) the results support a direct relationship between trophic breadth of a species and its functional versatility for buccal manipulation. In contrast, prey capture was a relatively stereotyped behaviour for pufferfishes (*Chilomycterus schoepfi*—a specialist on hard-shelled prey and *Sphoeroides nephalus*—a generalist predator) as few prey-type effects were found (Fig. 7.8).

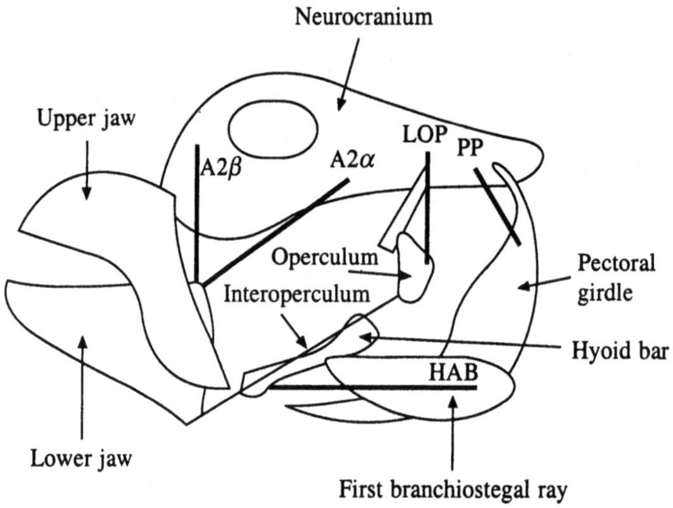

Fig. 7.8 Schematic diagram of the skull of *Chilomycterus schoepfi* with their suspensorium removed. Illustrating the functional units of the feeding mechanism. Thick lines indicate muscles and their attachments. Most of the suspensorium and branchial apparatus are not shown in this figure. Abbreviations: A2 α section 2α of the adductor-mandibulae; A2 β section 2β of the adductor mandibulae; HAB, hyohyoideus abductor; LOP, levator operculi; PP, protractor pectoralis. *Source:* Ralston and Wainwright (1997)

Alarm signals, as indicators and modifiers of predation threat, may alter predator-prey interactions, foraging tactics and physiological condition, and may affect morphology and life history (Smith 1992). Alarm signals that are released only following capture by predators may be functionally distinct from other alarm signals and may have evolved through direct benefits to the sender (Mathis et. al. 1995). Predator recognition has been indicated in fathead minnows (*Pimephales promelas*) (Chivers and Smith 1995).

Fishes can be characterized by their lipid stores as starvation tolerant or starvation sensitive. An example of starvation sensitive fish is the plaice, *Pleuronectes platessa*, which can have upto 7% lipid by wet weight in the muscle, but depend primarily on visceral fat stores for short term food deprivations. Both salmon and trout can store large quantities of lipid, and are capable of tolerating starvation for upto six months. The most starvation resistant fish is the eel *Anguilla rostrata*. *A. rostrata* adults have been known to tolerate upto four years of food deprivation. *A. rostrata* deposits very little visceral lipid, and are thought to utilize primary free amino acids in the muscle as an energy store during starvation. Starvation tolerance is enlarged by the low temperature and the inactivity of dormant state in many species during winter months. Lauff and

Wood (1996) reported rise in protein use, fall in lipid metabolism replaced by rise in carbohydrate metabolism during starvation in juvenile rainbow trout (*Oncorchynchus mykiss*).

Overwintering survival of freshwater fishes is determined by a combination of temperature, oxygen availability and energy stores (either in the form of body lipids or availability of prey during winter). Several species of fish undergo prolonged periods of fasting in their natural environment, frequently associated both with migration and with spawning (Sullivan 1986).

Size-selective predation has been widely documented in aquatic ecosystems and is often assumed to be an important component of individual fitness. Stream salmonids are size selective predators under both laboratory and natural stream conditions (read Grant and Noakes 1986 for a test of a size-selective predation model with juvenile brook charr *Salvelinus fontinalis*).

In some circumstances, juvenile salmonids in streams defend feeding territories. Their primary food is invertebrate drift carried by water currents, so one might expect these currents to have an effect on the size and shape of the defended and foraging areas. Grant et al. (1989) covered many aspects related to this topic. They suggested that in brook charr (*S. fontinalis*) aggressive radius may increase at a faster rate than foraging radius with increasing body size.

Pleuronectiform fishes (flatfishes) hatch as bilaterally symmetrical larvae, but during metamorphosis one eye migrates to the opposite side of the head and the fish come to rest on their blind (or eyeless) side. As a consequence of eye migration, all pleuronectiform fish possess some degree of cephalic skeletal and muscular bilateral assymetry. Gibb (1995) reported on the kinematics of prey capture in a flatfish, *Pleuronichthys verticalis*. They are morphologically asymmetrical teleosts with substantial bilateral asymmetry in the neurocranium, suspensorium and anterior jaws. Prey capture in *P. verticalis* contains significant differences in the timing of kinematic events from that described for other suction feeders. Perhaps, more remarkably, these fish display an extreme degree of bilateral functional asymmetry in the movements of upper and lower jaws which allows them to suck up prey from the substratum. Morphological asymmetry present in the skull is concentrated in the regions involved in opening the mouth and extending and protruding the jaws. Thus, functional asymmetry in feeding behaviour is related to, and perhaps, caused by morphological asymmetry in the jaw structures (Fig. 7.9). Berget and Wainwright (1997) reported the mechanism of prey in two syngnathid fishes, the lined seahorse *Hippocampus erectus* and the dusky pipefish *Syngnathus floridae*. The dominant features of the morphology of Syngnathidae include extreme elongation of the suspensorium and neurocranium with a small mouth located at the anterior tip of the head (Figs. 7.10–7.13). These species exhibit the generalized kinematic pattern of prey capture in bony fishes. Prey are captured during a sudden upward swing of the head, which brings the mouth to the prey. Suction is used to draw the prey into the buccal cavity.

In deep-see environment, differences in parasite patterns between two groups of fishes (midwater and benthic) are correlated with differences in food and migratory habits, life histories, body size, length of life, physical environment features and energy relationships. Gartner et al. (1997) reviewed the current state of knowledge regarding feeding in deep-ocean fishes. They focused on the patterns in the diets, sources of food, and deep-sea energetics related to feeding.

In the natural environment fish lie near the apex of the food pyramid, consequently almost all of them have evolved as carnivores; omnivorous or herbivorous habits are developed in very few species. This is especially true of the marine environment, where omnivorous fish are conspicuous by their absence.

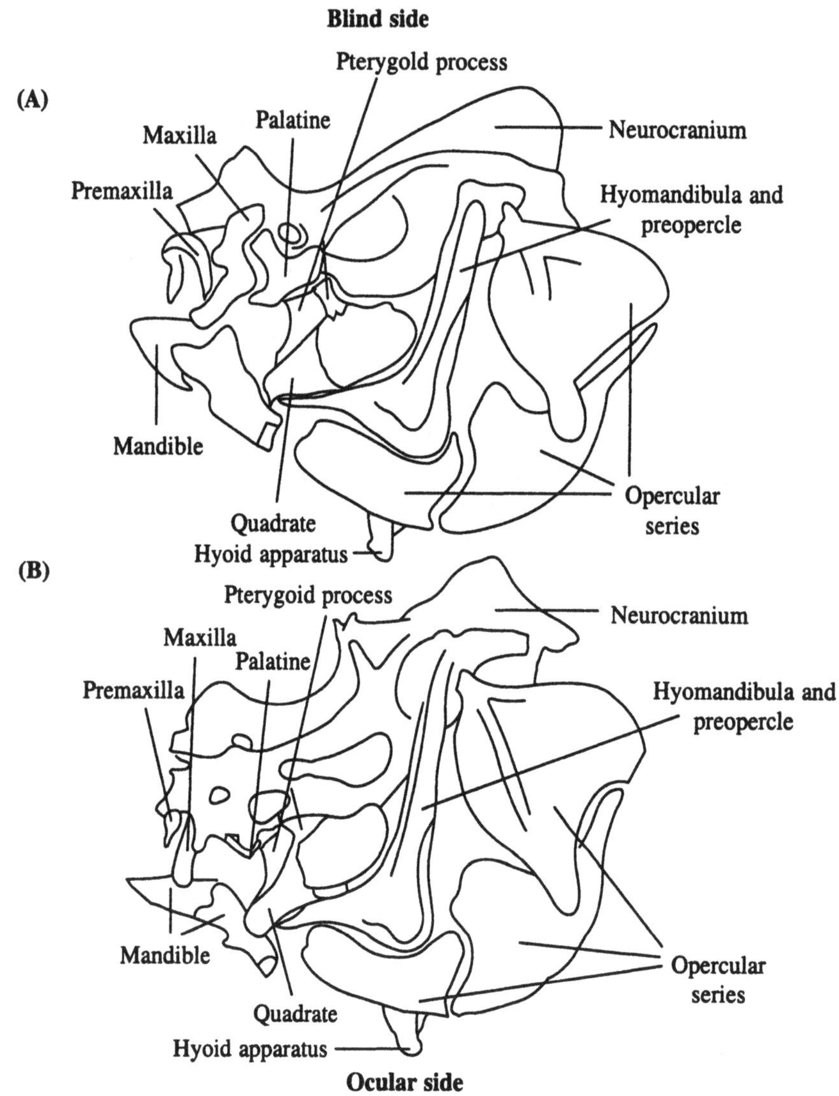

Pleuronichthys verticalis: Skull.

Fig. 7.9 Camera lucida drawings of a cleared and stained skull of *Pleuronichthys verticalis*. (A). The blind side of the head; (B). The ocular side. Note that the ocular side has been reversed horizontally in order to facilitate comparison with the blind side.
***Source:* Gibb (1995)**

Herbivory is recognized as a common dietary mode among fishes of inshore habitats of tropical seas. In contrast, this mode of feeding is considered to be uncommon among shallow-water fishes of temperate latitudes. In temperate rocky intertidal communities, fishes exhibit a wide variety of feeding types, ranging from carnivory to omnivory to strict herbivory. This broad range of feeding habits is attributed to richness and abundance of food available in the intertidal zone (Horn 1982). Fig. 1 given by Horn (1998) on predators and prey is quite informative.

Coral reef fishes have drawn a lot of attention (Hobson 1974, 1994; Myrberg and Thresher

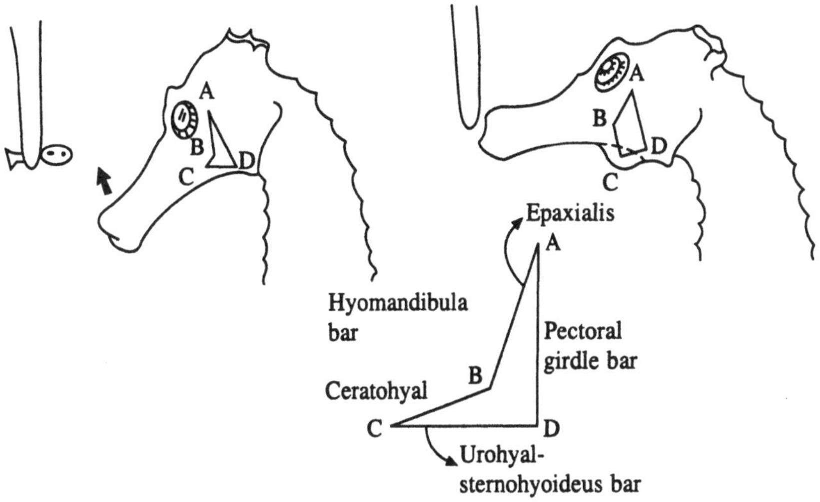

Feeding mechanism in *Hippocampus erectus.*

Fig. 7.10 ***Hippocampus erectus*****. Proposed four-bar linkage model for cranial elevation and hyoid depression in the seahorse feeding mechanism. The pectoral bar (D-A) remains in a fixed position while the hyomandibula bar (A-B) moves anteriad and dorsally with head elevation because of contraction of the epiaxialis muscle, resulting in the passive movement of the ceratohyal (B-C) anteriad and ventrally. Slight shortening of the urohyal-sternohyoideus bar (C-D) by contraction of the sternohyoideus muscle causes the release of the four-bar system from the initially stressed and rigid conditions.**
Source: **Bergert and Wainwright (1997)**

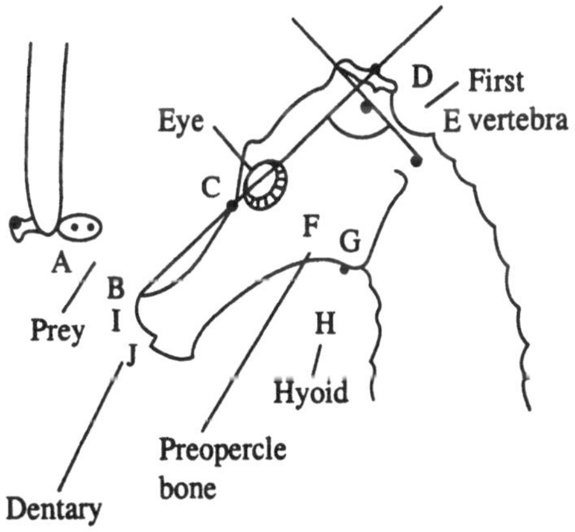

Feeding mechanism in *Hippocampus erectus.*

Fig. 7.11 ***Hippocampus erectus*****. Variables measured for kinematic study. Distance to prey was the distance in millimeters from the tip of the premaxillae to the nearest point on the prey item (A-B). Angle of the head elevation was the angle formed by one line connecting the tip of the premaxilla to a spot on the neurocranium just anterior to the eye (B-C) and a second line connecting the posterior tip of the crown with the position of the first vertebra (D-E). Hyoid depression was the distance in millimeters of the distal tip of the hyoid (H) from a horizontal line drawn along the ventral arm of the preopercle bone (F-G). Gape distance was the distance in millimeters between the anterior tip of the premaxilla and the anterior tip of the dentary (I-J).**
Source: **Bergert and Wainwright (1997)**

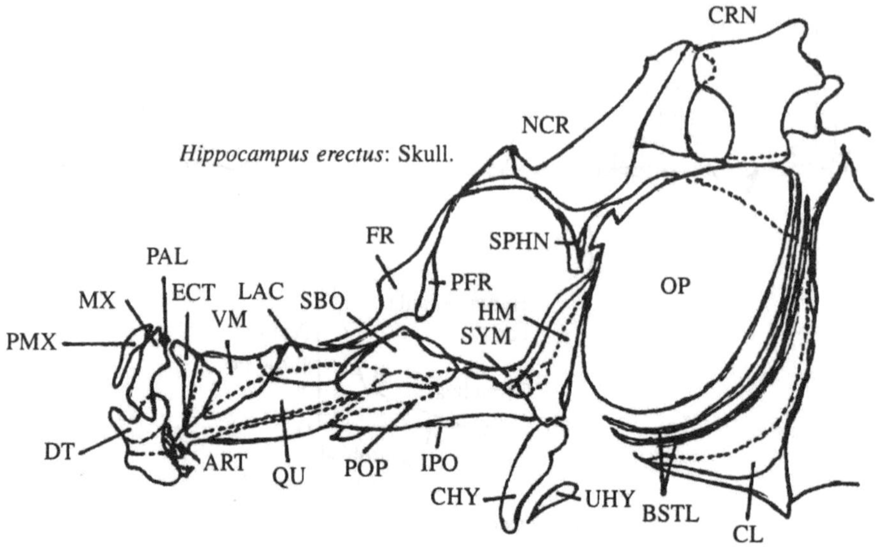

Fig. 7.12 ***Hippocampus erectus*. Diagram of the seahorse skull. ART, Articular; BSTL, Branchiostegal rays; CHY, Ceratohyal; CL, Cleithrum; CRN, Crown; DETH, Dermethmoid; DT, Dentary; ECT, Ectopterygoid; ENT, Entopterygoid; FR, Frontal; HM, Hyomandibula; IOP; Interopercle; LAC, Lacrimal; MX, Maxilla; NCR, Neurocranium; OP, Opercle; PAL, Palatine; PFR, Prefrontal; PMX, Premaxilla; POP, Preopercle; QU, Quadrate; SBO, Suborbital; SPHN, Sphenotic; SYM, Symplectic; UHY, Urohyal; VM, Vomer.**
***Source:* Bergert and Wainwright (1997)**

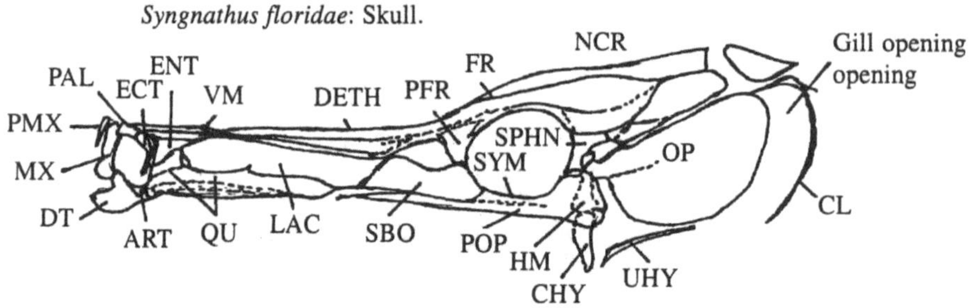

Fig. 7.13 **Diagram of the skull of the dusky pipefish, *Syngnathus floridae*. Abbreviations as in Fig. 7.12.**
***Source:* Bergert and Wainwright (1997)**

1974; Neudecker 1979; Birkeland and Neudecker 1981; Hourigan et al. 1988; Motta 1988; Harmelin-Vivien 1989; Horn 1989; Tricas 1989a, b; Choat 1991; Choat and Bellwood 1991; Hay 1991; Leis 1991; Reese 1991, 1993; Van Rooij et al. 1995, 1996; Fishelson and Sharon 1997).

Obligate corallivores (e. g., butterflyfishes, Chaetodontidae) could serve as indicator organisms. Butterflyfishes are divided into coral-feeding specialists and generalists. Many butterfly fishes exhibit food preferences among their coral food resources. Coral-feeding specialists may respond differently to changes in corals than do species with more generalized diets. Most coral-feeding species are territorial, with pairs defending areas where all feeding and sheltering takes place. Obligate coral-feeding butterflyfishes are exclusively associated with living coral, a necessary requirement for their use as indicator species. Positive correlation between coral cover and butterflyfish numbers is common. The distribution and abundance of the fishes closely follows the distribution of their preferred food corals. The distribution and abundance of coral-feeding generalists in relation to food-coral distributions differs from that of specialists. The size of foraging areas varies as a function of coral cover, especially preferred coral cover. It has been noted that angelfishes (Pomacanthidae, e.g., *Holocanthus, Pomacanthus*) feed extensively on sponges.

Feeding by herbivores on coral reefs is considered to have important consequences for reef plants. A number of fish species defend areas of reef substratum against a range of other species. Resident species may return substantial amount of nutrients and other organic materials to the defended site by excretion and defecation. Access to defended areas involves complex patterns of antagonistic and synergistic interactions among a diverse range of herbivorous fishes. Grazing of the reef substratum is a kind of bioerosion. Scarids and acanthurids have been identified as major bioeroders and sediment producers on reefs.

Barlow (1993b) has given a fine account of marine reef fishes related to feeding territories and mating systems. Feeding territories are well known among marine fishes that are associated with substrate, such as herbivores and corallivores. Although less immediately apparent, feeding territories also occur among marine carnivores. Territoriality probably seldom occurs among freshwater fishes because it is not economic. This may be exacerbated by the unstable distribution of food in space and time. In comparison, coral reefs are more stable, though at first glance they might seem not.

In order to satisfy their requirements for an adequate food supply many intertidal fish species have to make extensive foraging excursions away from their low tide locations. In the majority of cases feeding excursions are made at high tide when conditions are most favourable. Although feeding may be possible when immersed, only those forms which have evolved a semi-amphibious mode of life can do so efficiently. The main defence of intertidal fish against predators seems to be their agility, cryptic coloration, visual alertness and intimate knowledge of their immediate surroundings. This knowledge enables them to gain refuge rapidly when danger threatens (Gibson 1982).

Fish-food interactions largely depend how efficient each fish is at foraging on particularly food items; efficiency is ultimately determined by food's abundance, size, escape ability and other characteristics, as well as by the feeding apparatus, how it functions and its versatility.

Food choice in animals is a complex and poorly understood subject. Results of investigation have partially explained food choice in herbivorous fish *Cebidichthys violaceus* and helped explain why fucoid algae are not eaten by north temperate herbivorous fishes.

The reader is referred to a contribution by Neighbors and Horn (1991) for nutritional quality of macrophytes eaten and not eaten by two temperate-zone herbivorous fishes-*C. violaceus* and *Xiphister mucosus*.

By minimizing the time spent feeding, a fish will have more time, to carry out other activities; alternatively, by minimizing the energy spent to capture its prey, a fish will have more energy left for metabolism and growth. So, time minimization or energy maximization are the two ways in which a fish can be efficient as a forager (Hart 1986). Strategies to maximize the profit-prey selection by coral-feeding butterfly fish are aspects well discussed by Tricas (1989). The reader is referred to Reese (1989) for information on coral feeding guild, habit formation and spatial learning, and foraging paths.

Feeding success and growth are determined by differing efficiencies of foraging on particular items, rather than by major differences in physiological conversion. Feeding efficiency is usually taken as a ratio between energy gain and cost per unit of time, and should be conceived as the product of the efficiencies for each separate part of the feeding process. Energy cost of feeding is particularly hard to quantify and in practice energy gain merely refers to caloric value of food.

Parasites can have profound effects on the behaviour of their hosts. Milinski (1993) suggested that parasitized fish give more importance to food acquisition and less importance to the avoidance of predators than unparasitized fish do, because the nourishment of parasites and higher locomotory costs increase their energetic requirements. Todate, behavioural studies on parasitism in fishes have mainly focused on how parasites affect the fish's trade-off between foraging and predation risk. If parasitized fish actually have high energetic costs than unparasitized conspecifics, they should spend more time in shoal positions that offer high feeding rates and reduce intraspecific competition, such as peripheral positions. Parasitized fish might also leave the shoal and forage solitarily to avoid competition for food with conspecifics. There is anecdotal evidence that parasitized fish are less often found in shoals. However, by occupying peripheral positions in shoals or by foraging solitarily, parasitized fish may incur a greater risk of predation (Magurran 1990; Pitcher and Parrish 1993).

Most fish species forage primarily during day or night, with a few species primarily active during crepuscular periods of dawn and dusk. Several similarities in these diel patterns of activity occur in very different habitat types. In coral reef, temperate lake and temperate marine kelp-bed assemblages, diurnal fishes include herbivores, zooplanktivores, invertebrate feeders, piscivores and cleaners (Helfman 1986). Horn et al. (1990) discussed the diurnal feeding periodicity of an herbivorous blenniid fish, *Parablennius (=Blennius) sanguinolentus*, in the western Mediterranean. The afternoon feeding peak occurs during the period of greatest nutritional content of the fish's algal food resource.

Predators are said to gain an advantage over their prey if they strike from below under low light levels. As light levels rise above a threshold, an image of maximum contrast, such as a silhoutte seen against the downwelling light from the surface, is the first to become visible. Retinal rods, upon which low-light vision depends are the best at detecting such silhouttes, as they first appear during dawn. The twilight phenomenon may have shaped not only the crepuscular diurnal behaviour of prey fish and search behaviour of predatory fish, but also the physiology of predator's retina and various means of minimizing silhouttes in prey fish. Pitcher and Turner (1986) supported the 'twilight hypothesis' that under low light levels, stalking predators have an advance over shoaling prey.

The best strategy for an animal competing intraspecifically for food depends on its relative competitive ability, its needs, and on the strategies its competitors are using (Milinski 1986). Crowder (1985) suggested that feeding mode shifts seems to depend on the relative profitability of each mode. Microhabitat selection by juvenile sticklebacks (*Gasterosteus aculeatus* and *G. wheatlandi*) may serve to reduce the risk of parasitism (Poulin and FitzGerald 1989). The reader is also referred to Overholtzer and Motta (1999) and Muñoz and Motta (2000).

Recent work has highlighted the important advantages that individual animals can derive from membership of a social group. In fish shoals there is now a substantial evidence for twofold benefits. First, fish in larger shoals forage more efficiently because they locate food with greater speed, allocate more time to feeding and are better at adjusting their foraging in an unpredictable environment. Secondly, shoals enjoy a number of antipredator advantages not shared by solitary fish. Large shoals detect predators at a greater distance and, as a consequence, achieve a better compromise between vigilance and foraging. Once spotted by a predator, they are less likely to be attacked and if attacked have a small chance of being caught.

Based on the prey processing studies in haemulid fishes, Wainwright (1989) has supported a growing data base on aquatic feeding that: (1) motor patterns tend to be highly conserved among closely related taxa; (2) the ability to modulate motor patterns in response to different prey types appears to be a general property of fish feeding mechanisms; and (3) variations in experimental EMG data is ubiquitous and, when unaccounted for, confounds comparisons among treatment groups.

Generally, the diet of a certain species is likely to vary in time and in space, independently of the alterations linked to the age of individuals. Strictly specialized diets, consisting of one certain kind of diet, are rare, and they constitute a heavy handicap for the survival of the species, these being threatened by the smallest change in environmental conditions. Most species show a certain natural disposition to modify their feeding habits to whatever is available at the moment. Numerous researchers have classified fish feeding habits into guilds such as herbivory, carnivory, omnivory and the like. Views have been expressed on this issue. For example, how and where a fish feeds are more important than what it eats. Fish are notably opportunistic in their feeding, and the general impression is that how they feed tells much more than what they eat. It has been pointed out that the classification of fishes into trophic group is not consistent. It can be based on the systematic position of food item (molluscivores), on the size (ontogenetic stage) of the food (paedophages) or on the location of food (periphyton versus phytoplankton). It has been noted that morphologically, most trophic groups are recognized by a particular facies. A facia and its constituent structure is related more to the way food is collected than to the food itself, as with oral shelling or pharyngeal crushing of molluscs. No simple relation between food and feeding apparatus has been found in cichlid fishes. This is the result of at least two factors: the feeding function is not necessarily the only demand determining the structure of the feeding apparatus and the functional significance of the feeding apparatus is to be understood in mechanical terms and the food categories are subordinate to that. The validity of exclusively classifying fishes by trophic guilds has been questioned on the belief that much about the organization of the fish assemblages can be understood by focussing on morphological and behavioural features in relation to the microtopographic features of the environment.

Certain carnivorous fishes feed on prey that are permanently accessible and that possess passive protective devices: spicules, spines, shells and so on. Some fishes, such as angelfishes (Pomacanthidae) are thus specialized in the predation of sponges, and others, as some Balistidae, feed more particularly on urchins or on gastropods, as do some large wrasses (Labridae). Other species, thanks to the shape of their body and their snout, are able to reach prey hidden in reef crevices during the day. Thus, the Labridae *Gomophosus*, with its long and tubular snout, can catch small crustaceans hidden between the coral branches. Some species of goatfish (Mullidae) have peculiar foraging organs, the barbels, with which they detect invertebrates living in the sediment. Lizard fishes (Synodontidae) and flatheads (Platycephalidae) remain motionless on the bottom or on the corals, waiting for a prey to pass within their reach. They seize it with a rapid lunge. Such fishes are often referred to as ambush predators.

In the marine environment, among the invertebrate-feeders, there exists a very important group of fishes with a particular diet; these are planktophagous fishes. The marine environment is characterized by the presence of plankton. Plankton represents an important food resource, not only for the pelagic fishes but also for the littoral and reef fishes swimming above the bottom, as indicated by the important number of planktophagous fishes encountered on the reef. Certain planktophagous fishes, such as certain Pomacentridae and Nasidae, feed during the day. Others have nocturnal behaviour (Holocentridae, Apogonidae, Atherinidae) and catch organisms that become planktonic at night, in addition to those which are always part of the plankton.

Lying in wait is common among certain reef fishes, either diurnal (Platycephalidae) or nocturnal (Scorpaenidae). These fishes are generally camouflaged by their shape and colour and have strong resemblance with the bottom where they hide. These escape notice and suddenly dash at their prey when the latter passes within their reach.

On the reefs, hunting method is used by trumpetfishes (Aulostomidae) and cornetfishes (Fistularidae). These species slowly approach another fish, and when they are at a suitable distance, they dart upon the prey, open mouthed, to snatch it. (Harmelin-Vivien 1978).

Desert fishes are opportunistic in their feeding, as are most fish species. Many pupfishes appear to feed almost exclusively on bottom sediments, at least at certain sizes and at certain times of a year. Some sediments of springs are copropelic and may be quite high in various organic nutrients. One of the most extreme, and unusual, feeding habits of a desert fish involves an endemic catfish, *Clarias cavernicola*, which lives in a deep, pitlike spring in the Kalahari Desert, and appears to subsist mostly on the excrement of a local colony of baboons.

Foods used by desert fishes provide a pattern of increasing restriction in variety in the diet as icthyofaunal diversity increases. Tolerant forms living alone usually are ubiquitous in feeding. Kinds living in faunistically more saturated habitats, increase specialization of feeding niches.

High water temperatures in deserts may create exceptional nutritional problems for fishes. Metabolic demands at high ambient temperatures must be intense, and a number of authors have noted a maximum in feeding activities at higher environmental temperatures, so long as the temperatures do not rise high enough to suppress activity or to initiate an avoidance reaction (see Deacon and Minckley 1974).

High-latitude environments are characterized by substantial seasonal variations in climatic conditions and food availability. Anadromy may have evolved in high-latitude fishes as a means to take advantage of the favourable feeding conditions in the ocean when conditions for feeding and growth in fresh water are poor.

Teleost fishes less than 10 cm in maximum length comprise at least 10% of this group. Adaptiveness of small size in fishes relates to efficiency of feeding on small prey and possibilities for exploitation of spatially restricted habitats. Three basic types of ecotype for small fishes- nektonic, epibenthic and cryptobenthic- are described. Greater susceptibility to predation is consequence of small size in at least nektonic and epibenthic habitats. Some clingfish (*Diademichthys*, *Arcos*, *Cellichthys*) and gobies (*Ginsburgellus*) have become vertebrate ectoparasites, feeding on the tube feet of urchin. The famous "Candiru" (*Vandellia*) of Amozonia, small pygidiid catfish, enter the gill chambers of larger fish to browse on the gill filaments (Miller 1979).

Evidence of consistent endosymbiosis in the gut of a herbivorous marine fish (from Red Sea) has been reported. Surgeonfish (*Acanthurus* species) harbour gut symbionts that include bacteria, trichomonadid flagellates, and a peculiar putative protist that attains densities of 20,000 to 100,000 cells per millilitres of gut contents (Fishelson et al. 1985).

Cannibalism has been recorded in 36 out of 410 teleost families. The families that have provided the most information include the Engraulididae, Esocidae, Poeciliidae, Gasterosteidae, Percidae and Cichlidae. Cannibalism has been classified into seven types, depending on life-history stage, age difference between cannibal and prey, and whether or not they are related. A number of factors potentially influence the extent and rate of cannibalism in fish; the factors may be interspecific, relating life history, or intraspecific, in which case both endogenous and exogenous factors are important. The main proximate advantage conferred by cannibalism is assumed to be nutritional. In an ultimate sense the behaviour may have evolved as a particularly effective competitive strategy (Smith and Reay 1991). It has been reported that cannibalism can form the principal food for adult pike where other species are scarce. Giles et al. (1986) supported the view that intraspecific predation can potentially be of major importance in regulating the number of pike recruits to a year class. A comprehensive account of cannibalism and prey specialization in Arctic charr, *Salvelinus alpinus* is given by Amundsen et al. (1995).

Filial cannibalism is especially prevalent in fishes with parental care. Recent research supports a central tenet of parental-investment theory; that is, parents consume their offsprings when it maximizes their lifetime reproductive success (FitzGerald 1992; see also Belles-Isles and FitzGerald 1991). On the question of egg cannibalism by sticklebacks; spite or selfishness, FitzGerald (1992a) presented data suggesting that female threespine sticklebacks (*Gasterosteus aculeatus*) show spiteful behaviour.

Valerio and Barlow (1985) described the occurrence of filial cannibalism and the ontogeny of agonistic behaviour in the Midas cichlid, *Cichlasma citrinellum*, during the first three months of life. The fry (young fish) grew at uneven rates producing an increasing disparity of sizes. This was paralleled by increasing cannibalism which eliminated most of the small siblings; that in turn, resulted in a decline in cannibalism.

Relative size differences had a profound effect on the expression of agonistic behaviour throughout the study. Large and medium fry were similar, but small ones expressed little aggressive behaviour and avoided the other fry.

The confusion effect is probably widespread in varying degree where predators seek prey from aggregations. The many animals that tighten their aggregations when threatened place more individuals within the visual field of attacking predators, thus further hindering those predators troubled by multiple targets. In refining this strategy more than most, fishes have emphasized features that make it difficult for predators to concentrate on individuals. All members of a school are about the same size and look much alike (schooling fishes are noted for lacking external differences between the sexes). As noted above, when threatened they increase not only their swimming speed but also the rate at which they dart in an out among one another. Coloration can be important too. Many schoolers are silvery, and sunlight reflecting from their bodies at rapidly changing angles during evasive manoeuvers presents the attacker with a bewildering shower of brilliant flashes. Many others are barred or striped, and when they swim in schools, these patterns blend together as a shifting maze of lines that conceal the individuals. Some fishes are barred or striped only when they school: the goatfish *Mulloidichthys samoensis*, for example, carries a prominent yellow stripe on its side when schooling, but when it forages independently, or in small groups, the stripe is replaced by a solitary black spot (Hobson 1978).

Mosquito fish (*Gambusia affinis*) are widely used for biological control of mosquito larvae in California rice fields but the presence of different types of alternate arthropod prey affects the performance of the fish (Linden and Cech 1990).

Aspidontus taeniatus (Blenniidae) is an aggressive mimic of the cleaner fish, *Labroides dimidiatus*. Its strong behavioural and morphological resemblance to *L. dimidiatus* appear to facilitate its predatory habit of biting off proportions of the fins of other fishes. *L. dimidiatus* feeds on the surface of host fishes and removes ectoparasites, scales and other material. Host fishes frequently solicit these cleaning services by assuming an unusual posture or posing while the cleaner inspects the host; but the cleaner may also chase its host in order to feed. Cleaning symbiosis appears to be a mutualistic relationship in that it contributes to the survival of both cleaners and hosts but it has been suggested that this relationship could, at times, exist as a parasitic symbiosis (read Losey 1974). The reader can get allied information from the contribution of Losey (1993).

A deeper understanding of the factors initiating feeding behaviour requires knowledge of the gustatory physiology. At present, knowledge of physiological factors controlling appetite is poor. A fish in the reproductive state is most likely to stop feeding.

There are few studies dealing with the neural control of feeding in teleosts. Hypothalamus has a role in organization and control of feeding behaviour.

3. GUT

In some fish the gut is a straight tube from mouth to anus but usually an oesophagus, stomach, and intestine can be identified, as well as subdivisions apparently related to diet. In Agnatha, the morphological subdivisions of the gut are poorly defined: the oesophagus is narrow, followed by a straight intestine leading to the anus. There is no defined stomach. In fishes having no stomachs, no acid phase of digestion occurs.

Carnivores have short guts in relation to their body length, with herbivores having the longest. Omnivores are supposed to have intermediate gut lengths.

Pigmentation of the gut walls is not unusual in deep-sea fishes, to hide any light produced by luminescent prey, which might expose the predator, from shining through the gut walls (Fishelson 1994; see also Fishelson et al. 1997). Black gut phenomenon is also reported in Antarctic nototheniid fishes (Eastman and De Vries 1997).

3.1 Mouth: Emphases-Predigestive Mechanisms, Dentition, Filter Equipment and Aids

Feeding mechanisms in Cyclostomata are different from those of other vertebrates. The head of *Petromyzon* is without jaws, the mouth being a suckerlike adaptation for attaching to suitable prey from which food is scraped by means of horny "teeth" situated on the surface of the funnel-shaped mouth and on the centrally placed tongue (Fig. 7.14).

There may be species differences in the feeding habits of adult parasitic lampreys. These may vary with the stage in the life cycle and the food available. It has been suggested that the term "predatory" is much more appropriate than "parasitic". Lampreys often feed in deep water, rely primarily on olfaction. Anticoagulant secretions allow tissue fluids and blood to be ingested and passed directly to the intestine, since the stomach is lacking (Bertmar 1985). Cytokeratins have been reported in the horny teeth of the adult sea lamprey (*P. marinus*) (Zaccone et al. 1995).

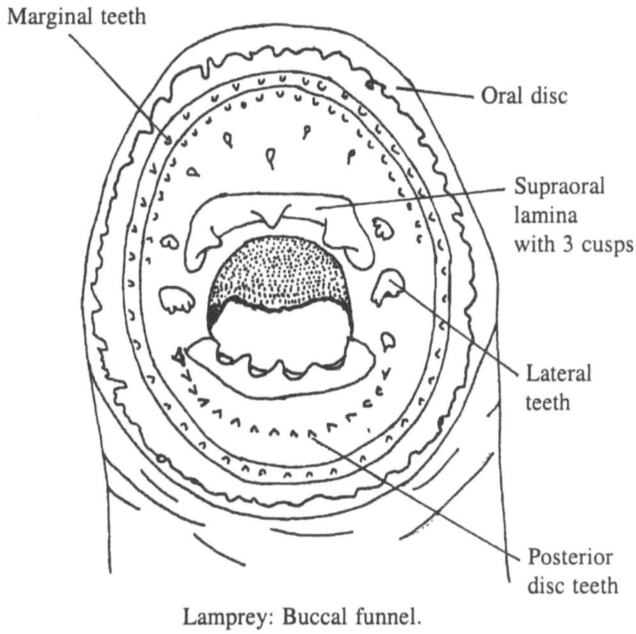

Fig. 7.14 Oral disc or buccal funnel of lamprey showing positions of teeth.

Adult *P. marinus* possess a prominent pair of sac-like buccal glands which produce a secretion known to have anticoagulating, haemolytic, and cytolytic properties. Each gland is situated on the ventral side of the head at the level of the eyes, imbedded in the basilaris muscle, and is drained by a duct which opens into the mouth just inside the infraoral lamina. Buccal glands are not present in larval lampreys, but develop during the metamorphosis of the ammocoete into the adult. The secretory epithelium of the glands of the actively feeding adult is typically pseudostratified. Each cell contains an elongate nucleus, scattered lipid droplets, small apical PAS-positive secretory granules, abundant granular mitochondria, and prominent basal chromideal substance.

The epithelium of the glands undergoes a marked regression when the lamprey ceases to feed and migrates upstream to spawn. The cytological evidence indicates that the buccal glands function only during feeding.

Larval lampreys (ammocoetes) generally feed on the microflora and microfauna in the regions surrounding their burrows. The branched cirri at the posterior end of the oral hood act as a sieve through which water must pass before entering the pharynx. The larger particles accumulate in the oral hood where they become enmeshed in mucus and are periodically expelled by a kind of "coughing action". The sensory basis for the selective feeding is largely unknown. During metamorphosis, the animal is generally not feeding (Bertmar 1985).

In spite of the fact that keratinization is most common in terrestrial vertebrates, cyclostomes, including *Myxine* and *Lampetra*, have keratinized teeth. Myxine teeth are more complex than those of *Lampetra* where the simple conical teeth are distributed over the circumoral sucker and each one is underlain by one or more similar cones which are known to replace the functional tooth at intervals. The horny teeth of hagfishes have long been an enigma as to their mode of growth or replacement. Contrary to those of lampreys, no replacement tooth has been observed

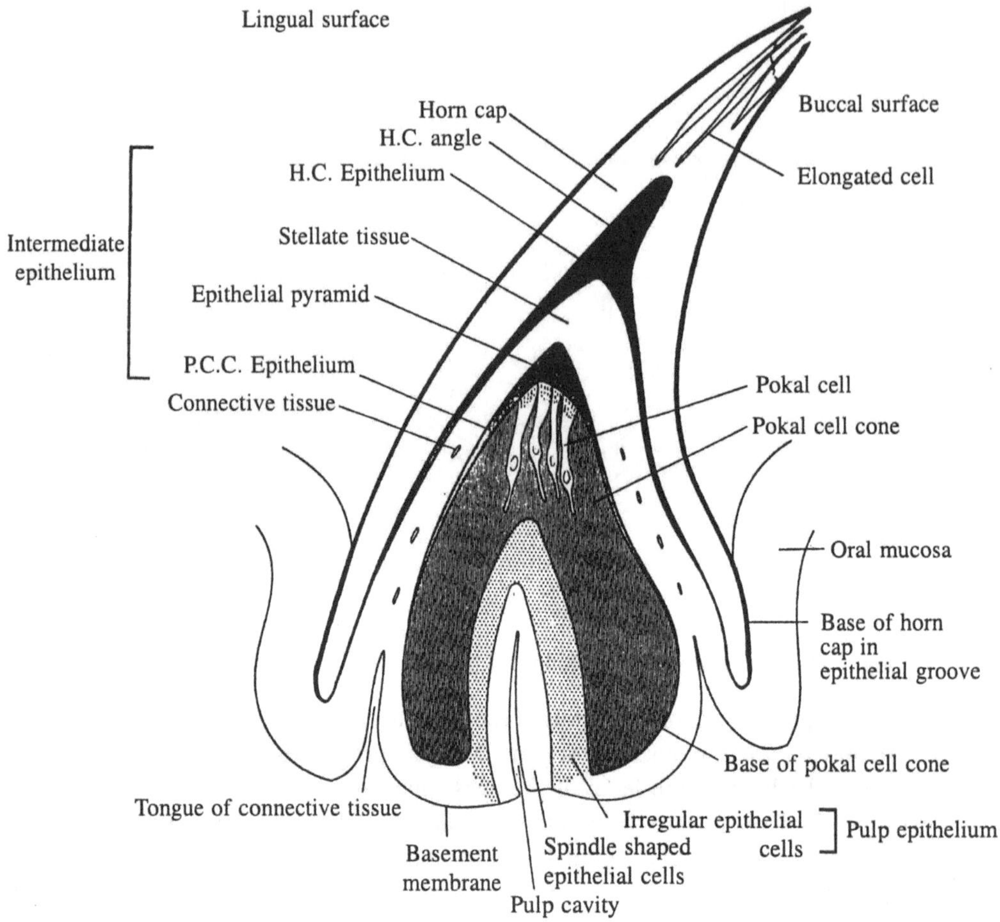

Fig. 7.15 Diagram of a vertical section through a single tooth cusp, showing the different regions of the tooth in *Myxine glutinosa*.
***Source:* Dawson (1969)**

in hagfishes until recently (read Janvier 1993). The teeth of myxinoids seem to exhibit a replacement, yet probably slower than lampreys.

In *Myxine*, all but the single palatal tooth are arranged in four longitudinal rows on a moveable lower "jaw", those in each row are joined at the sides of their bases by hinge tissue to form a tooth comb. Beneath the outer cone shaped horn cap (HC) is another cone, the pokal cell cone (PCC), which in contrast to the replacement tooth of *Lampetra* has several features by which it may be distinguished from the functional horn cap (Fig. 7.15). It has been shown that *Myxine* are burrow dwelling animals, emerging mainly for food capture. Animals in the process of tooth replacement might remain in their burrows and escape capture. The arrangement of efferent branchial system in *Myxine* is an adaptation to a burrowing habit. Hags do not suck or rasp, but bite their food by everting and retracting the dental plates on the floor of the buccal cavity. Hags locate food by olfaction, swimming against the current towards the source of stimulation. Hags are not parasitic, but are unselective scavengers of benthic invertebrates and dead, moribund, or captive fish. Decayed fish are eaten in preference to fresh fish. Hags penetrate the skin of fishes whereever they are able, and do not necessarily bore into the body cavity. It is advantageous for a burrowing aquatic animal to have its exhalent respiratory or feeding current directed away from the inhalent current so as to minimize recirculation, and this is provided for in *Myxine*. *Myxine* can survive for months without feeding. This is presumably related to their lethargic habits and to the extensive adipose tissue which is characteristic of most freshly caught individuals.

Hags are less efficient than most fishes in locating food. They are blind, their capacity for locomotion and precise orientation is poor and they must be in contact with their food before they can bite.

Although many sharks have a rather general vertebrate body plan, they display a number of specializations for feeding that belie the notion they are "primitive". These specializations include a battery of highly developed exteroreceptive systems such as vision, olfaction, acoustico-lateralis sense and electroreception; and a cranial morphology that has been moulded into a number of functionally adaptive forms. These forms result in grasping, sucking, crushing, gouging, cutting and filtering systems of feeding. With relatively few exceptions, elasmobranch feeding mechanisms share such features as subterminal or inferior mouths, a dynamic tooth replacement system, hyostylic jaw suspension, and a kinetic, protractile upper jaw. The evolution of the high diversity of mechanical feeding systems in such a small group of vertebrates has probably been facilitated by the morphological simplicity of basic feeding mechanisms. This rediation was accomplished by modifications in jaw length, the length and supporting angle of the hyomandibula, the size of the gape, dentition, and changes in the relative size of cranial musculature. The evolutionary pattern of shark feeding is complex, there being several examples of both parallelism and convergence. A long-jawed, grasping form (similar to, but not identical with *Chamydoselachus*) is considered primitive. From a subsequent benthic sucking and grasping ancestor, similar in many respects to some living batoids, radiated crushing, ray-like forms; cutting squaloid forms; and gouging, lamniform and carcharhiniform types. From the latter developed sucking and grasping, or crushing forms such as modern orectolobiforms, triakids and heterodontiforms sharks. From several levels (primary crushing, secondary crushing and gouging) there emerged filter-feeding forms represented today by mobulids, rhinodontids and cetorhinids (Moss 1977; for specific attention on sharks (Carcharhinidae), see Moss 1972; Tricas 1979, 1982, 1985; Tricas et al. 1981; Tricas and McCosker 1984; Motta et al. 1997). Frazetta (1994) has published a detailed account of feeding mechanisms in sharks and other elasmobranchs. Throughout the elasmobranchs, the protractable jaws appear to be a major motif in the design of the feeding mechanism. While the components of this apparatus are constructed of calcified cartilage, not bone, elasmobranchs have been variously adapted to subdue and reduce prey that may be active, or possess durable body coverings. However, the significance of certain aspects of the feeding mechanisms, such as the loosely affixed teeth in modern sharks, is not understood, despite their importance in any analysis of jaw mechanics. Recent contributions in this regard are from Wilga and Motta (1998, 2000), Motta and Wilga (1999, 2001) Wilga et al. (2000) and Edmonds et al. (2001).

Compared to the relatively numerous studies of bony-fish jaw systems, investigations of elasmobranch cranial function are sparse.

Motta et al. (1991) provided the first quantitative electromyographic analysis of biting in lemon shark, *Negaprion brevirostris*. Wu (1994) has emphasized that elasmobranchs are able to protrude their jaws (see teleost description). Importance of labial cartilage chain, palatoquadrate and Meckel's cartilage in generating suction has been mentioned. The wobbegong sharks (Orectolobiae) are specialized for jaw protrusion (see also Motta and Wilga 2001).

Reif (1982) has given a comprehensive account of evolution of dentition in vertebrates. The dentition of all Elasmobranchii is organized into discrete tooth families, each of which consists of a functional tooth and all its successors. Growth of the whole dentition is accomplished either (1) by regular increase in size of successive teeth within the tooth families, (2) by splitting of tooth families, or (3) by insertion of new tooth families. A fusion of neighbouring tooth families has never been observed in any recent shark. A reduced tooth number occurs in two families of rays, Myliobatidae and Rhinopteridae. No information is available how the reduced tooth number of Myliobatidae and Rhinopteridae evolved. For further details, see Huysseune and Sire (1998).

Elasmobranch teeth consist largely of dentine and thus can be regarded as advanced odontodes. They are anchored by collagen fibres which originate in the bony base of the teeth (read details ahead). The hypermineralized cap is enameloid. The skates and rays (batoids) have a secondarily reduced enameloid layer. In Neoselachii the enameloid is covered with a "shiny layer" probably ectodermal enamel.

In all elasmobranchs, teeth are restricted to the upper and lower jaw. In the rest of the oral cavity, scales occur in numerous genera, but never teeth. Tooth replacement is related to body growth in sharks. This fact makes possible the estimation of body growth rates from tooth replacement measurements. Rapid tooth replacement in crustacean-eating specialists is a necessary adaptation to insure the maintenance of an adequate dentition. The dentition in the durophagus *Sphyrna tiburo* have molariform teeth; the anterior teeth are cuspidate and used for grasping, and the posterior teeth are molaroform and used for crushing. Crushing and grinding teeth are typically associated with sharks and rays that feed on hard prey, such as crustaceans and molluscs.

Chimaeroids (Holocephali) have tooth plates which grow throughout ontogeny and are never replaced. Orvig (1985) has described the histology and growth of tooth plates in various fossil and recent chimaerids, with special reference to the structure of their vascular and compact pleromin (characteristic columns of pleromic hard tissue). It is concluded that statodont elements of this kind cannot have had connection with the dentition of iniopterygians and cannot either have evolved from lyodont dental units of such bradyodonts as, for example, the cochliodontids. As compared with statodont elements in other fishes, the chimaerid tooth plates show resemblance in many ways to those of the ptyctodontids and may well have developed by certain structural changes from tooth plates like those of the latter. For related information, the reader is referred to an earlier contribution by Orvig (1980).

The dentition is made up of two pairs of tooth plates in the upper jaw, and one pair in the lower. In certain forms (*Elasmodectes, Rhinochimaera*), the plates are comparatively thin, but generally assume more massive proportions. They may be fairly sharp-edged, but are often rather blunt, which might indicate an adaptation to crushing instead of shearing. However, a clear correlation between tooth plate shape and dietary habits has yet to be demonstrated.

It has been demonstrated in recent form that chimaerid tooth plates may be subjected to a fairly wide range of variation. They may thus be to some extent differently developed in different specimens of the same species, and even undergo marked changes with advancing age (e.g., *Harriota, Hydrolagus, Chimaera*).

Chimaerid tooth plates are clearly statodont dental elements, which in the living organism grow in their basal part at the same time as they are subjected to superficial abrasion.

Almost all members of the classes Chondrichthyes, Osteichthyes, Amphibia, and Reptilia lack the cementum seen in the class Mammalia and a few species of the class Reptilia. Instead, they have various other tooth-attachment mechanisms: fibrous attachment, hinged attachment, pedicellate attachment, ankylosis, or thecodont attachment. Pedicellate attachment is tooth-support by a dental pedicle interposed between the tooth and the jaw bone; it occurs in many species of teleosts. The dental pedicles are derived, at least in part, from the dental papilla and are calcified. Whenever the teeth are replaced, their pedicles are absorbed and then newly formed on the base of the renewed teeth. The pedicles in teleosts are usually fused to the bone; however, the spongy pedicle in adult dog salmon (*Oncorhynchys keta*) is easily distinguishable from dentin and jaw bone, but fused with the jaw bone; and in ayu sweetfish (*Plecoglossus altivelis*) the pedicles are interposed between a layer of gelatinous connective tissue and the jaw bone and terminate on the periosteum. *Zacco temmincki* and *Z. platypus* (Cyprinidae) have pharyngeal teeth which are directly attached to the pharyngeal bone without a pedicle.

The dentitions of fish present a tremendous variety (see Figs. 7.16–7.23). Unlike the elasmobranchs where teeth occur only on the jaws, the teeth in bony fishes may be present on the tongue (e.g., *Salmo*), roof of the mouth or in the throat. In fishes, amphibians and many reptiles teeth may be borne on a number of bones of the palatal complex, the pterygoids, palatines, vomer and ectopterygoids. In size and form the teeth of sharks exhibit great diversity, varying from long, slender, awl-like structures to large flat triangular teeth. The main spine ("cusp") is often accompanied by subsidiary "cusps". So characteristic is the form of teeth that it is often possible to identify a species simply by the characters of one or two teeth. Sometimes, the teeth of the two jaws are substantially different. In some sharks the teeth are small and flattened in a horizontal plane and in the Port Jackson shark (*Heterodontus*) there is an interesting example of a dentition consisting of different forms. The teeth at the front of the jaws are small and conical, changing on passing backwards into teeth having the form of pads or nodules of varying size (Figs. 7.16, and 7.17).

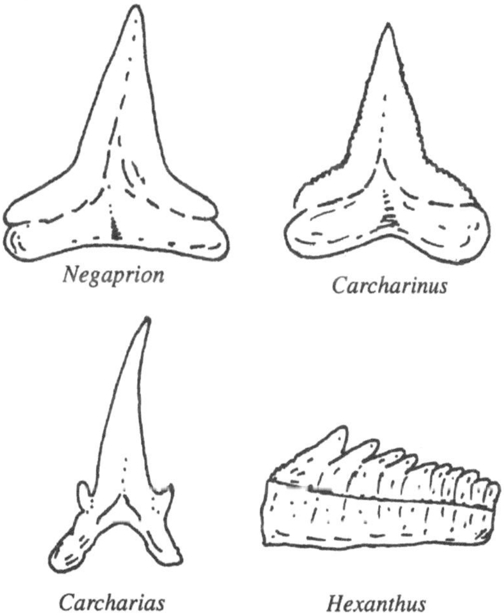

Placoid teeth in sharks.

Fig. 7.16 Variations in the form of placoid teeth in sharks.
***Source:* Miles and Poole (1967)**

In skates and rays, which have dorsoventrally flattened bodies, the jaws are characteristically equipped with a mosaic pavement of flattened crushing teeth for dealing with molluscs and crustaceans collected on the sea bed (Fig. 7.18).

In bony fishes the distribution of teeth may be quite irregular or they may be placed in one or more regular rows parallel with the edges of the jaws or disposed over a large area as broad band or patches. As a rule the teeth are more rigidly attached than those of elasmobranchs although in some bony fishes they are freely moveable. In the eels, for example, the teeth are connected to pedestals by means of an annular ligament; in other species the ligament is so modified on the outer and inner aspects of the tooth that it bends to allow food material to pass, becoming erect again and rigidly resistant to movement in opposite direction so effectively preventing the escape of prey. Such "hinged" teeth occur in fish such as the cod (*Gadus*), hake (*Merluccuis*, Fig. 7.19) and pike (*Esox*). In yet other bony fishes the tooth and bone of attachment are continuous and there is no intervention of soft tissue. This type of attachment, which also occurs in some tetrapods, is known as ankylosis. Whether fixed or moveable, teeth are generally recurved backwards towards the throat so that the movement of prey away from pharynx is hindered or prevented. (Fig. 7.20) (Table 7.2). Parmentier et al. (1998) focused on the morphology of the buccal apparatus in respect of feeding in carapids.

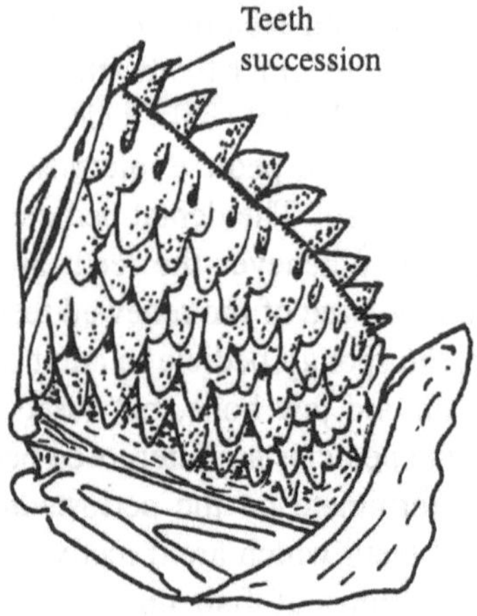

Scymnus lichia: Lower jaw.

Fig. 7.17 Right half of the lower jaw of a shark, *Scymnus lichia*, showing the continuous tooth band. As functional teeth are shed, the next row of successional teeth is pulled around the edge of the jaw and erected into the functional position.
***Source:* Miles and Poole (1967)**

In some bony fish teeth (e.g., cyprinids) are present on the inner margins of pharyngeal bones, (Fig. 7.21). External features of the pharyngeal teeth have often been studied as a criterion for the classification of cyprinids. The replacement pattern of pharyngeal dentition in the Japanese cyprinid fish, *Gnathopogon coeruleoceus*, is similar to that of the jaw tooth replacement in many reptiles (Nakajima 1979). The development

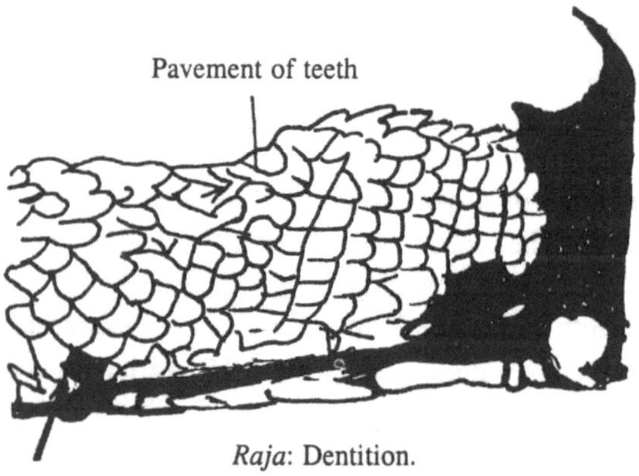

Raja: Dentition.

Fig. 7.18 Inner aspect of the lower jaw of *Raja* showing continuous "pavement" of teeth passing gradually upwards over the jaw from the tooth band below. The pointer indicates some dried membrane, the operculum, which partly covers the developing teeth. The teeth interdigitate and lie in rows in which a spiral arrangement can be discerned. "Rows which spiral around the jaw".
***Source:* Miles and Poole (1967)**

of the pharyngeal dentitions often seeming to be inversely related to the development of the jaw dentition; that is when one is pronounced, the other is feeble. Pharyngeal dimorphism and considerable plasticity in the oral dentition have been observed in cichlids. The potentially molluscivorous cichlid *Astatoreschromis alluadi* is known to exhibit phenotypic plasticity in its pharyngeal jaw apparatus (see Huysseune et al. 1994). Parenti and Thomas (1998) discussed pharyngeal jaw morphology in gobioid systematics.

The pharyngeal jaw apparatus of cichlids is characterized by two upper pharyngeal jaws, which articulate with the neurocranial base, and a single lower pharyngeal jaw. Cichlids use their pharyngeal jaws for mastication, transport and swallowing of the food. Detailed cineradiographic studies of *Orechromis niloticus* (Claes and De Vree 1991a) and *Haplochromis burtoni* (Claes and De Vree 1991b) revealed that food processing involves cyclic movements of the pharyngeal jaws, and that each cycle contains a transport, reduction, and swallowing component. Food processing in *O. niloticus* is generated by a characteristic contraction pattern of pharyngeal jaw musculature (see details: Claes and De Vree 1992). Vandewalle et al. (1994) remarked that, during evolution, the extent of intra-oral dentition is generally reduced and the

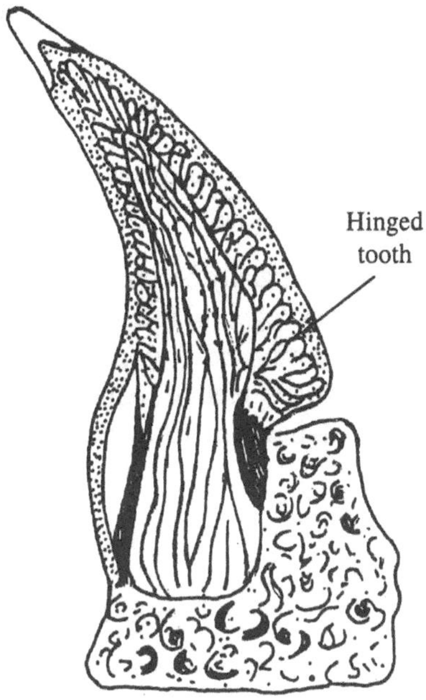

Merluccius: Hinged tooth of a bony fish.

Fig. 7.19 Hinged tooth of a bony fish such as the hake, *Merluccius*. The teeth is recurved backwards and, on the anterior side, its base is attached to the bone of the jaw by elastic ligaments which stretch and allow the tooth to bend backwards as food is swallowed.
***Source:* Miles and Poole (1967)**

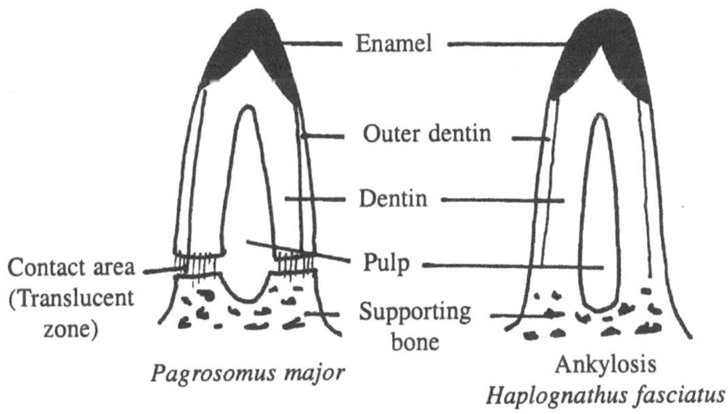

Tooth attachment.

Fig. 7.20 Diagrammatic illustration of two types of the attachment found between a tooth and its supporting structure.
***Source:* Isokawa et al. (1970)**

Table 7.2 Manner of the supporting bone to jaw teeth

Fish	Relationship of the supporting bone
Pagrosomus major	Connected via translucent zone (contact area) existing between the supporting bone and teeth
Semicossphus reticulatus	Teeth joined to the supporting bone by ankylosis
Haplognathus fasciatus	Teeth joined to the supporting bone by ankylosis

Source: Isokawa et al. (1970)

pharyngeal jaws gradually become the most important food-processing apparatus. The contribution covered (i) morphology, (ii) role of the pharyngeal apparatus in food processing (iii) development of pharyngeal jaws, and (iv) polymorphism and plasticity of the pharyngeal apparatus in teleosts. For jaw and teeth of American cichlids, see Casciotta and Arratia (1993). The aspect that "function influences form during ontogeny" was discussed by Galis (1993) citing the

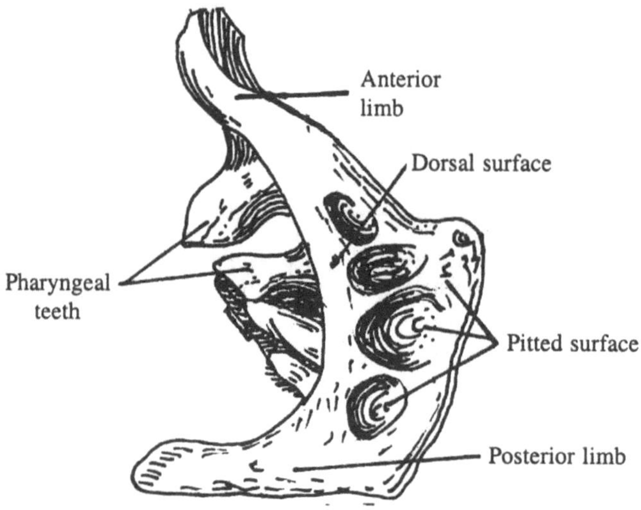

Megalobrama terminalis: Pharyngeal teeth.

Fig. 7.21 Pharyngeal teeth.

example of polymorphism in the pharyngeal jaw apparatus in South American cichlids. Distinctly heterodont dentitions are found in certain bony fish, the sheep's head fish, *Sargus*, being a good example (Fig. 7.22).

The porgy's (*Pagrosomus major*, *Evynnis* and *Sparus swinhonis*) pharyngeal bones consist of two parts in upper and lower regions, and the pharynx is surrounded with these bones. Upper bones are made of three ossicles firmly bound with connective tissue. These bones form a dental plate and these attach to pharyngeal regions exclusively with soft tissues. These bones are composed of anterior, medial and posterior bones (Fig. 7.23).

The sparid, *Diplodus sargus*, with heterodont dentition (incisiform, molariform) of buccal jaws and frail pharyngeal jaws, displays an original pattern of specialization toward a wide diet specturm including shells (Vandewalle et al. 1995).

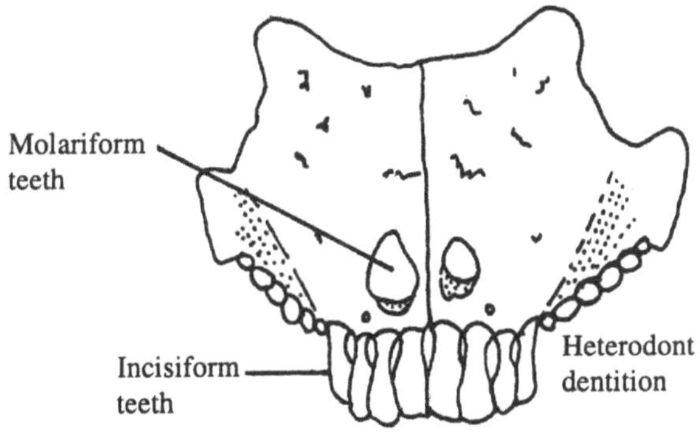

Sargus ovis: Premaxillae.

Fig. 7.22 Premaxillae of the sheep's head fish, *Sargus ovis*. Anterior teeth are incisiform, posterior teeth are molariform.
***Source:* Miles and Poole (1967)**

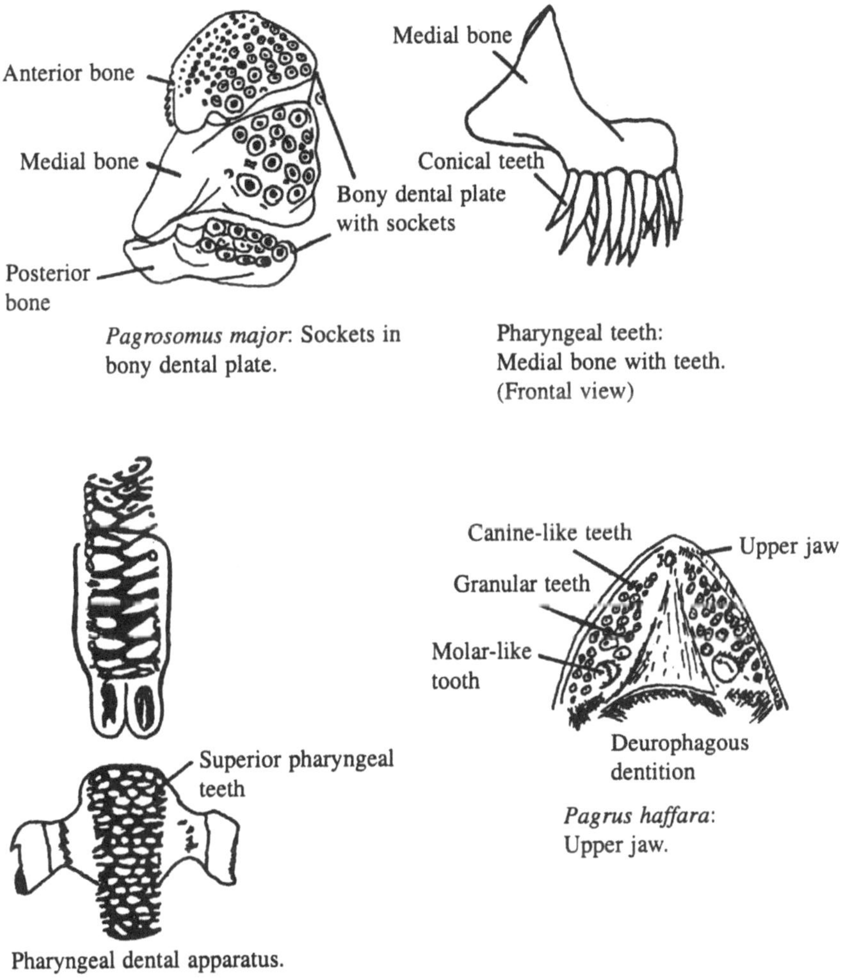

Pagrosomus major: Sockets in bony dental plate.

Pharyngeal teeth: Medial bone with teeth. (Frontal view)

Pharyngeal dental apparatus.
Scarus sordidus (Coral-feeding fish)

Pagrus haffara: Upper jaw.

Fig. 7.23 Teeth in *Pagrosomus major, Scarus sordidus* and *Pagrus haffara*.

The oral jaws of scarids are highly modified for a scraping mode of feeding (read Choat 1991) (Fig. 7.23).

To accommodate for strong wear, the teeth of some advanced families (Scaridae, Diodontidae, Triodontidae, Tetraodontiade, and Molidae) have evolved "beaks". Each quadrant of a beak contains numerous tooth families in Scaridae, Triodontidae, and Diodontidae. In Tetraodontiade each quadrant contains only one very long and slender tooth and all its successors. In four families tooth replacement is automatic and is accomplished by abrasion of the previous tooth. In summary, all fish beaks consist in one way or another of dentine teeth with a thick enameloid cap.

Summarily, in actinopterygians, two forms of tooth attachment are found based on the degree of mineralization of the collagen between the tooth and the base and the bone of attachment. While primitive actinopterygians possess teeth ankylosed to the bone (Type 1), the predominant attachment of teeth in teleosts is through a thin ring of unmineralized collagen between the tooth and bone (Type 2) allowing some tooth movement (Fink 1981).

As mentioned, fishes exhibit a peculiar diversity of dentition. At the most fundamental level, it is generally true that the more active predators have strong jaws with sharp teeth, and that on the edge of the jaws serve to bite and catch prey while those on the walls of the pharynx either prevent escape or aid in additional processing during food ingestion. In the snapper family (Lutjanidae) the piscivores have well-developed canine dentition, mollusc and crustacean feeders possess short, heavy dentition, and zooplankton feeders have the least developed dentition.

Uehara et al. (1983) described the fine structure of the horny teeth of the lamprey, *Entosphenus japonicus*. Most of the horny teeth consisted of two horny and two nonhorny layers. The primary horny layer was well keratinized. The secondary horny layer displayed various stages of keratinization. The appearance of the nonhorny layers varied according to the degree of keratinization of the horny layers beneath them.

Cyclostome teeth are indeed part of a dermal skeleton. Recent immunochemical and biochemical studies of cyclostome teeth reveal the presence of distinctive enamel-like antigens and proteins with molecular weights as high as those occurring in gnathostome tooth enamel (see Maisey 1988).

The structure of teeth has been described in detail in some bony fishes. For example, in the butterflyfish (Perciformes, Chaetodontidae, coral feeders), basically they are comprised of a form of dentine partly covered by an enameloid substance. Each tooth has a shaft with a pulp cavity, and a tooth cap capped with iron. The functional teeth are all attached to bony pedestals. Relatively simple changes in dentition, accompanied by other changes in a few key elements of the head such as premaxilla and dentary have resulted in a diversity of feeding types among butterflyfishes. The iron layer, which has been shown by others to be harder than underlying enameloid, probably represents a specialization to harden the teeth, resisting abrasion and cracking (Motta 1987, 1989; for electron microscopic picture, see Sparks et al. 1990- three mineralized (hydroxyapatite) layer). Uehara and Miyoshi (1987) examined the structure of upper teeth of the filefish, *Stephanolepis cirrhifer* (Monacanthidae) by light and scanning electron microscopy. The point of discussion has been the tubular enamel of pleurodont teeth which are tightly fixed to the jaw bone by bundles of fibrils.

Uehara and Miyoshi (1993) have described the structure and tooth attachment of the comblike teeth and denticles of the aye sweetfish *Plecoglossus altevelis* by light and scanning electron microscopy. Constituents of the denticle are crown with a spine pointed anteriorly, a triangular plate in the cervical region and a root that curves laterally and tapers off to a point. The root apex is fused to a long thin pedicle. Enameloid, fibrous dentine, dentinal tubules are reported.

An integration of the gustatory and filtering structures is typical in the pharyngobranchial region and is reported in numerous groups of bony fishes. Complex filtering structures are found in several species of planktonphage teleosts such as Coregonidae and Clupeidae.

In spite of a lack of critical evidence many authors continue to conclude that there will be a precise relation between gill-raker numbers, length, spacing, food type, and food size. The failure to find a precise relationship between gill-raker dimensions and size may be because the rakers function as rather "leaky sieves" or it may be that the "selection" of prey is based on perceptual processes during search, detection and attack.

The capacity of a filter (volume per unit time) is an important property determining (together with the retention ability) the amount of prey that can be retained per unit time. The capacity of a filter is determined by the velocity of water in the filter and the area through which water flows in the filter (the filtering area). Details regarding branchial sieve can be found in a number of recent publications (Hoogenboezem et al. 1990, 1991, 1992, 1993; Lammens and Hoogenboezem 1991; Eggold and Motta 1992; van den Berg et al. 1992, 1993, 1994; Hoogenboezem and Boogart 1993, Osse et al. 1997). Gill apparatus of ray-finned fishes, long supposed to have a role in regulating water flow during respiration, plays an important role in the nature of fluid flow through the mouth during feeding (read Lauder et al. 1986). Vandewalle et al. (1998) presented findings on branchial basket in relation to feeding in four carapids.

During the catching of prey, some labrid fishes or wrasses (Perciformes) show rapid and wide jaw-movements with about equal amounts of rotation by both jaws. During the opening of the mouth there is a remarkable adduction of the suspensorium and the jaw apparatus with a prominent protrusion of the premaxilla. This seems to contrast with the opinion generally held, for other fishes, that mouth opening and protrusion coincide with abduction.

Food collection is performed entirely by oropharyngeal apparatus. In some, mainly predatory, fish the mouth is specialized for fast and voluminous suction. In others, mainly planktivorous fish, prey capture proceeds by over-swimming and suction plays a minor role. The complex of muscles, bones and ligaments composing the suction-pressure pump of the respiratory apparatus eventually performs a wide range of diversified feeding actions. By accurate timing of volume changes in the oral, buccal, pharyngeal and opercular cavities and by differential operation of the oral and opercular valves, carp can perform as many as ten various stereotyped pattern of movements, each serving a particular role in food processing (Fig. 7.24 A, B). In addition, size-selective retention of small food particles is effected by the branchial sieve, if such particles are abundant. Whereas size, location, and motility of food are crucial factors for capture strategies, taste, degree of mixture with non-food and size are determinants in sorting out from non-food.

The ability of animals to utilize ingested food resources depends on their ability to commute them, increasing the surface area for digestive enzymes, or merely breaking indigestible capsules. Such plant materials impose extraordinary demands on mechanical and chemical breakdown by fish, it is not surprising that, even in tropical zones, with continuous crops, very few fish are true herbivores.

Mastication is not common among fish. Oral jaws serve to seize and bite prey and, primitively buccopharyngeal teeth serve transport and swallowing. True mastication is bound to the evolutionary development of posterior branchial arches into pharyngeal jaws as in some lower (e.g., cyprinid)

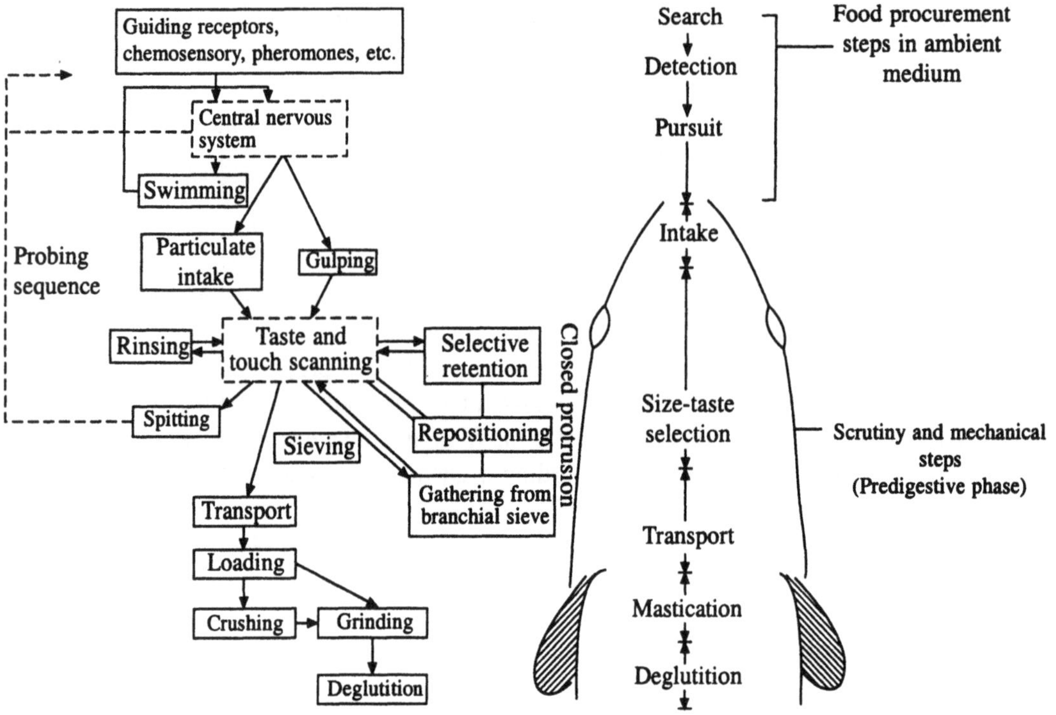

Fig. 7.24(A) Interrelations between distinct components of food handling in carp. Their actual site of action in the fish is indicated at the right. These stereotyped patterns are variously integrated into effective feeding sequences, adjusted to the type of food. "Probing" refers to repetitive intake and spitting of bottom debris while evaluating its profitability.
***Source:* Sibbing (1991b) (Modified)**

and many higher (e.g., cichlid) teleosts. They are freely suspended in muscular slings from the skull. Cyprinid fish lack oral teeth, a stomach and cellulases. High demands are imposed on pharyngeal mastication for increasing the surface area of food for effective digestion or for breaking indigestible capsules (Sibbing 1991b) (read also Osse et al. 1997).

Durophagy has been reported in queen triggerfish, *Balistes vetula* (Tetraodontiformes). This fish combines primitive behaviours and motor patterns (suction feeding and buccal manipulation) with specialized morphology (strong teeth, robust jaws, and hypertrophied adductor muscles) and a novel behaviour (blowing) to exploit armoured prey such as sea urchins, molluscs, and crabs. A blowing behaviour was periodically employed during prey handling, as prey were forcefully "spit out" from the mouth, either to reposition them or to separate unwarranted material from flesh (Turingan and Wainwright 1993).

The development of premaxillary protrusion has been one of the major themes of teleostean structural evolution. Most teleost use some combination of two basically different feeding mechanisms: one is jaw bite and the other is the development of negative pressure in the oral and opercular cavities as the mouth is opened. Whether the fish emphasizes the bite or suction depends in part on the size of items eaten. The lip-like structure appears to be a constant feature of cyprinoids.

Protrusion of upper jaw is a general occurrence in advanced teleosts (e.g., percomorph fishes (Fig. 7.25), but among lower teleosts it has uniquely and independently evolved in cyprinids.

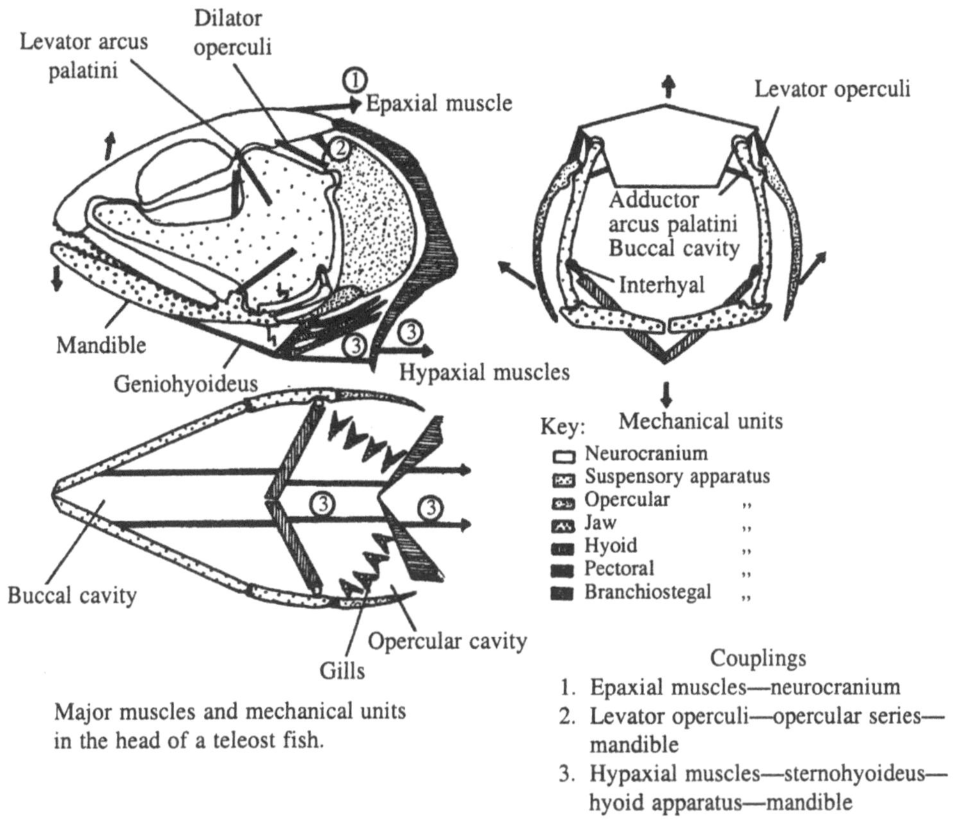

Fig. 7.24(B) The major muscle and mechanical units in the head of a teleost fish. Three major musculoskeletal couplings are involved in the expansive phase of prey capture: (1) the epiaxial muscles coupling, which causes cranial elevation; (2) The levator operculi, which mediates mandibular depression via the opercular apparatus and interoperculomandibular ligament (ligament 1) and (3) the hyoid coupling which also governs mandibular depression via the mandibulohyoid ligament (12).
***Source:* Lauder (1985a)**

Among its many proposed feeding functions, upper jaw protrusion has been suggested to increase the energetic efficiency of suction feeding by increasing prey capture success and by decreasing energy expended in ingesting food. Analysis of winnowing by *Embiotoca jacksoni* and by *Cyprinus carpio* indicates that, in addition, premaxillary protrusion is of critical importance to hydrodynamic prey processing. Similarly the hyoid apparatus, which may contribute to buccal cavity pressure reduction during suction, should be viewed as an agent of hydraulic prey capture. Dorsoventral movements of the hyoid during winnowing position prey precisely within the oropharyngeal cavity (see Motta 1982, 1984, 1985; Osse 1985; for winnowing see Drucker and Jensen 1991).

Pharyngeal jaws in cyprinids bear upto three rows of teeth. They are so varied in shape, size, number and arrangement that they are commonly used as keys for species identification. This wide variation is generally accepted as reflecting their role in adaptive radiation and trophic segregation.

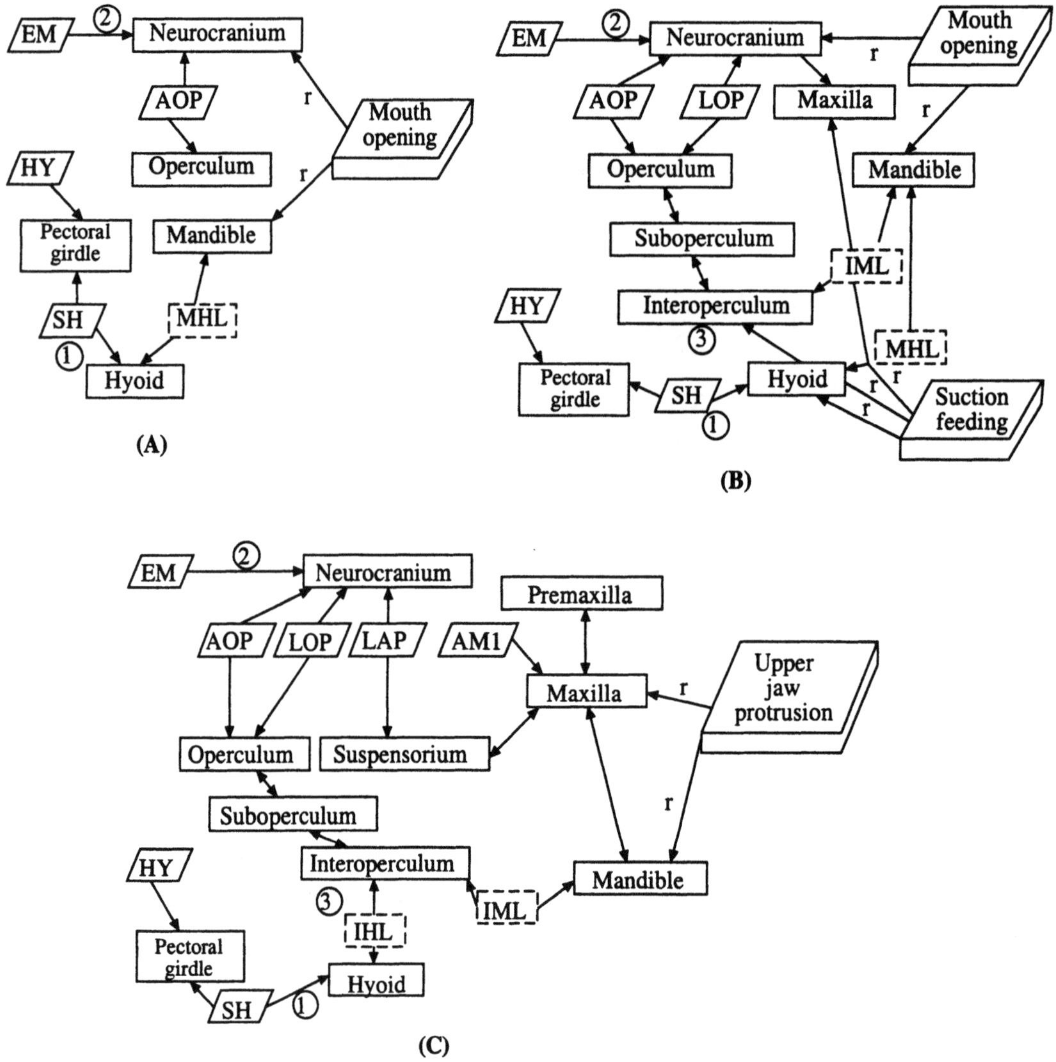

Fig. 7.25 Structural network in the head of a primitive actinopterygian (A), a primitive halecostome (B) and a percomorph (C) to show the biomechanical pathways governing mouth opening, suction feeding, and jaw protrusion functions. Homologous biomechanical pathways are similarly numbered. Note the increase in complexity of the structural network in actinopterygian evolution. Only the function of jaw protrusion is shown in (C); the primitive functions of mouth opening and suction feeding are omitted for clarity. Solid rectangles-bony elements; dashed rectangles-ligaments; parallelograms = muscles. Arrows run from the muscle to the bone of insertion double-headed arrows indicate ligamentous connections between bony elements. Three dimensional rectangles indicate major functions which are realized. (r, arrows) by the biomechanical couplings indicated. Abbreviations: AM1, division of A1 of the adductor mandibulae; AOP, adductor operculi muscle; EM, epiaxial muscles; HY, hypaxial (obliquus inferioris) musculature; IHL, interoperculohyoid ligament; IML, interoperculomandibular ligament; LAP, levator arcus palatini muscle; LOP, levator operculi muscle; MHL, mandibulohyoid ligament; SH, sternohyoideus muscle.
Source: Lauder (1982)

Osteoglossomorph fishes (commonly known as bony-tongued fishes) have a characteristic tongue-bite apparatus for intraoral prey processing to shread and disable prey. This apparatus is considered to be a synapomorphy for osteoglossomorph fishes (read Sanford and Lauder 1989 for *Notopterus*).

In shovelnose sturgeons, *Scaphirhynchus platorynchus* (Chondrostei) the mouth is highly protractile with the upper jaw connected to a peculiar palatine apparatus, that is, in effect, a tongue on the roof of the mouth. The fish is well adapted for feeding and maintaining its position in fast flowing water on river bottoms (Fig. 7.26).

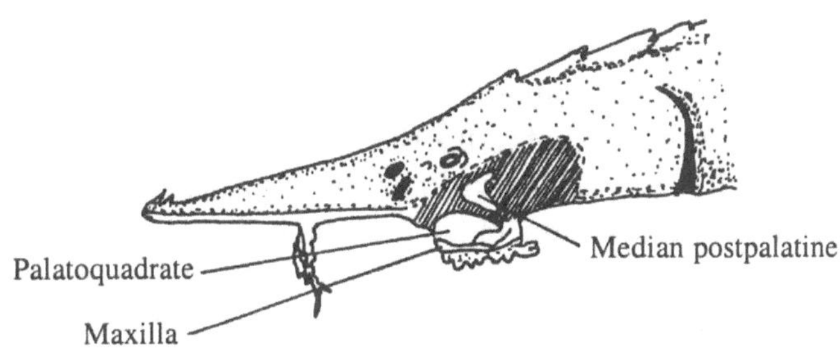

Fig. 7.26 Head of *Scaphirhynchus* partially dissected to reveal the palatine apparatus. *Source:* Weisel (1979)

In *Acipenser transmontanus* (Chondrostei) the jaws are highly degenerate and the oral apparatus is highly modified for feeding: extreme ventral position of the highly modified suctorial mouth and foraging nature of feeding in an often silty substrate of river and lake bottoms. Sanderson et al. (1994) reported that both swimming velocity and buccal flow velocity were significantly higher during suspension feeding than during ram ventilation. As the paddlefish (*Polyodon spathula*) shifted from ventilation to feeding, buccal flow velocity increased to approximately 60% of the swimming velocity. During prey processing, buccal flow velocity was significantly higher than the swimming velocity, indicating that prey processing involves the generation of suction.

It may be mentioned here that suspension-feeding fish filter enormous quantities of water to extract minute prey that are too small to be sensed and engulfed as individual particles. Such species have been reported in at least 16 families belonging to 11 orders (Sanderson and Wassersug 1993). These fish are either intermittent suction feeders ('pump' suspension feeders) that generate repetitive pulses of negative pressure to suck parcels of water containing prey into the buccal cavity, or continuous ram feeders ('two-net' suspension feeders) that use forward body velocity to transport water through the mouth.

The recent *Latimeria* (Actinistia) has a well-developed dentition which is restricted to the jaw bones. The teeth consist of an enamel cap, dentine, and basal bone and are ankylosed to dermal bones. They are formed in a fairly superficial region, although deeper than placoid scales. A permanent dental lamina is probably not developed. Nothing is known about the general organization and the replacement patterns of the dentition. Coelacanth is a noctural, piscivorous drift-hunter that moves very slowly in up- and downwelling currents, using its paired fin to stabilize the drift motion. The fish feeds on a variety of epibenthic and benthic fishes associated with rocky reefs at moderate depths. Squids have also been reported in the diet. The data available suggest, however, that the coelacanth is euryphagous and may switch its feeding to whichever prey is most abundant.

Early (Devonian) Dipnoi, on the other hand, had three dental structures: "buccal denticles", tooth ridges, and tooth plates. "Buccal denticles" were replaced. Recent dipnoan genera have large tooth plates. There is convincing evidence that the 'tooth ridges' were periodically shed or resorbed and replaced as the fish grew. "Buccal denticles" and tooth ridges seem to be primitive features within group compared with the rest of the vertebrates. Tooth plates are an advanced feature which evolved only once within the Dipnoi. The plates can be compared to a polyphyodont dentition. Each plate consists of radiating ridges. Each ridge results from a fusion of numerous cusps, and is homologous to one tooth family. At the side of the plate to which the ridges diverge new material is added throughout morphogenesis. At the side where the ridges converge material is resorbed. This process is equivalent to the shedding of individual teeth. The morphogenetic analysis of dipnoan tooth plates has proved that the tooth plates of dipnoans evolved from a well-organized polyphyodont dentition, rather than the fusion of scattered denticles which were enlarged and arranged in rows. Tooth plates, in which the individual plate elements are not completely fused, are more primitive than plates that have continuous ridges. For details of lungfish (*Protopterus* and *Lepidosiren*) tooth plates, consultation of the paper by Bemis (1984) is advised.

The tooth plates of *Neoceratodus* are considerably broader than those of *Protopterus* and *Lepidosiren*, which are reduced to a series of 'blades'. The diet of *Neoceratodus* is reported to include vegetable and non-vegetable stuff. Kemp (1996) discussed the phylogenetic importance of marginal tooth-bearing bones in *Neoceratodus*. *Lepidosiren* is reported as feeding on a variety of animal and vegetable. The food of *Protopterus* seems to be molluscs, small fishes, and crabs but vegetable material is also found in the stomach.

The manner of feeding in all three seems to be similar. The food is detected almost exclusively by olfaction, with sensory assistance from the pectoral appendages. The food is sucked into the mouth in a rapid movement and a lengthy chewing process commences. Chewed food is pushed in and out of the mouth and outside is suspended in a cloud of mucus.

Bemis and Lauder (1986) described the feeding mechanism in *Lepidosiren*. The function of hydraulic transport is analogous to that of the tongue in many tetrapods.

It should be pointed out that the Dipnoi have an akinetic skull and lack a marginal dentition on the maxillae, premaxillae and dentaries. The crossopterygian skull on the other hand is characteristically a highly kinetic system featuring a unique "intracranial joint" that divides the braincase into ethmosphenoid and oticooccipital portions (Fig. 7.27). In Rhipidistia, a full marginal dentition with dentigerous premaxillae, maxillae, and dentries is present, while in Coelacanthini the dentaries alone are absent.

Various teleosts have organ-aids in oropharynx: lamellar organ of the palate (e.g., *Cirrhina mrigala*, Fig. 7.28); tongue lappet, unlike in others, in adult *Plecoglossus altevelis*; cushiony and papillated palatal organ (e.g., in some cyprinids); pharyngeal glands (in mouth breeding forms); pharyngeal valve hung in pharyngeal roof (e.g., coral feeder *Scarus sordidus*, see Fig. 7.23); epibranchial organs (e.g., *Chanos*, Figs. 7.29, 7.30. see Bertmar et al. 1969; Kuwatani and Kafuku 1978); pharyngobranchial (pharyngeal) organs (in Mugiliidae, Fig. 7.31a, b); spiny pad on the roof and floor of pharynx (some gobies); and pharyngeal or oesophageal sacs (e.g., *Nomeus albula*, *Pampus argenteus*, *Psenopsis anomala* (Fig. 7.32)(read Kapoor et al. 1975b).

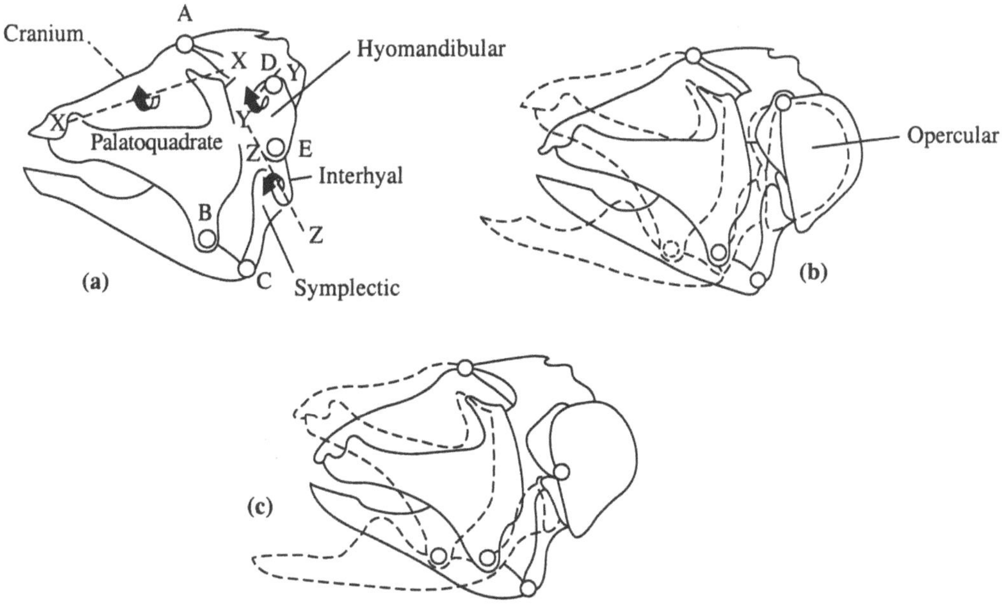

Latimeria chalumnae: Skull.

Fig. 7.27 Outlines of the skull of *Latimeria* (a) shows axes of joints. Small circles represent transverse axes and broken lines represent axes roughly parallel to the median plane. The arrows show movements which occur when the orobranchial chamber if enlarged: each indicates movement of the distal skeletal element relative to the proximal one. The operculum is omitted, (b) shows the mechanism of mouth opening postulated earlier and (c) shows the recently postulated mechanism.
***Source:* Alexander (1973)**

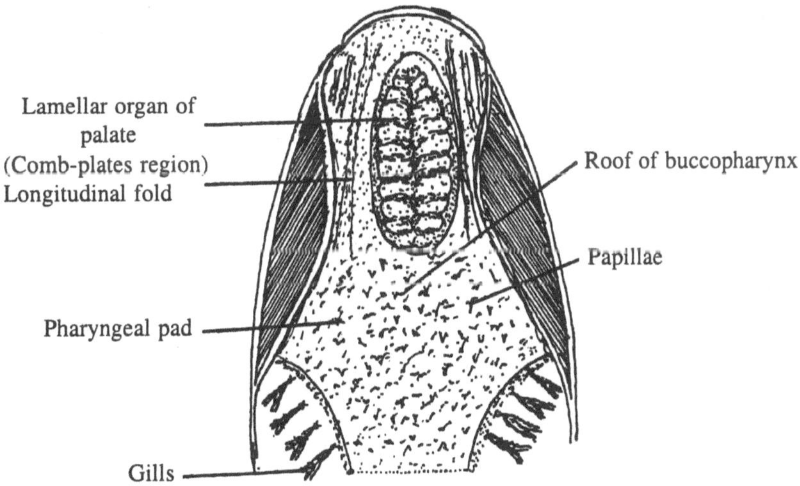

Cirrhina mrigala: Oropharynx.

Fig. 7.28 Oropharynx of *Cirrhina mrigala* showing lamellar organ.
***Source:* Majumdar (1952)**

Further information on oesophageal sacs is given in Tables 7.3 and 7.4.

A buccal, dental toxic gland is reported in *Meiacanthus nigrolineatus* (Blennidae).

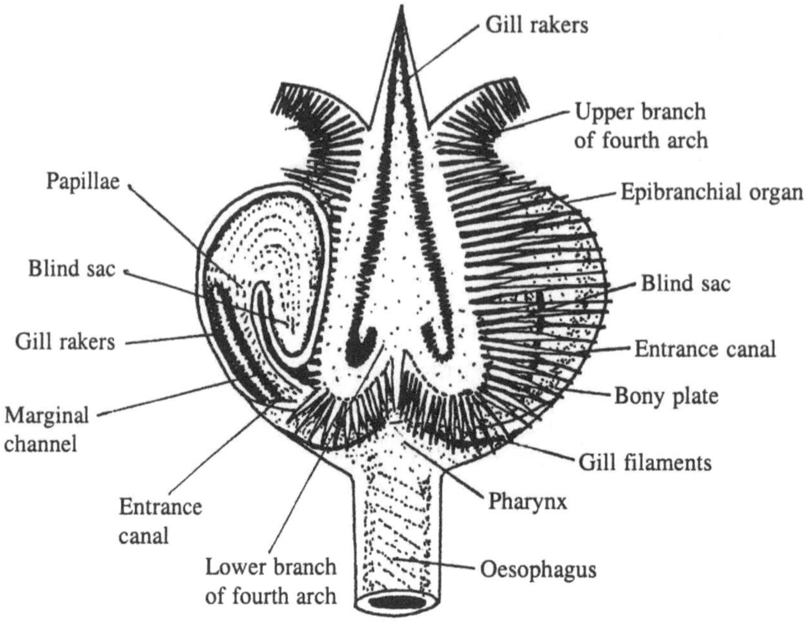

Chanos chanos: Epibranchial organs.

Fig. 7.29 Diagram of ventral view of entrance canal-blind sac epibranchial organs of milkfish *Chanos chanos* (Type 5), right organ cut horizontally to show position of marginal channel and gillrakers along the entrance canal.
***Source:* Bertmar et al. (1969)**

Generally, teleost brains reflect life styles. Habitat requirements and food preference can readily be associated with brain shapes, specially the brain stem.

3.2 Post-Branchial Gut: Orientation, Structure and Function

In both lampreys and hagfishes the gut is a simple tube (Figs. 7.33–7.35). Behind the pharynx the short oesophagus leads into the intestine at the junction of the bile duct with the gut. The oesophagus is lined with ciliated columnar epithelium and mucous cells. In some species, the anterior intestine is expanded into one or two blind-ending diverticula. The internal surface of the intestine is increased by a spiral fold (typhlosole) which extends along the length of the intestine and in the larva contains submucosal haemopoietic tissue. In addition, the wall of the anterior intestine is folded longitudinally. The wall of anterior part of the intestine contains mainly absorptive cells with microvilli at the luminal surface, ciliated cells, enzyme-secreting zymogen cells and endocrine cells. The posterior intestine is lined mainly with absorptive cells, mucous cells and endocrine cells. The zymogen cells are concentrated in the foremost part of the intestine, and in some species are confined to the diverticula. They have been equated with the exocrine pancreas of vertebrates, in which this organ is a separate entity. Larval and adult cyclostomes feed in different ways-the larvae are microphagous filter feeders while the adults are non-feeding parasitic, predatory or scavengers according to species. In lampreys, at metamorphosis the gall bladder and bile duct degenerate, while they are retained in hagfishes. As they become sexually mature, adult lampreys cease to feed and the gut gradually degenerates.

For specific attention on the digestive system, the reader is referred to Youson (1985). Although

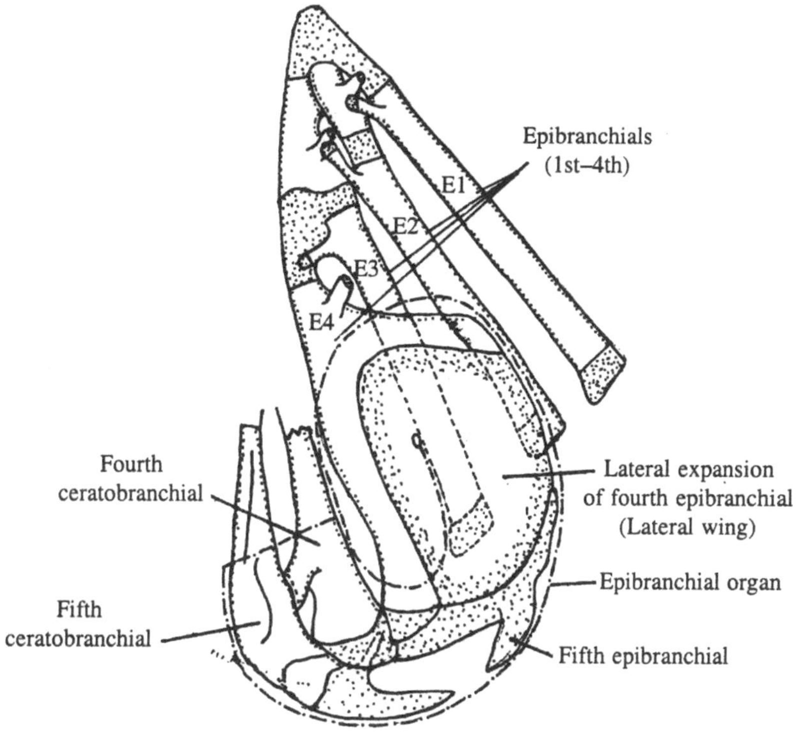

Chanos chanos: Epibranchial organ.

Fig. 7.30 Epibranchial organ of milkfish *Chanos chanos*: supporting elements of the organ: dotted area representing cartilaginous structure, and dots-and-bars line indicating outline of the organ (Dorsal view).
***Source:* Kuwatani and Kafuku (1978)**

isolated from the alimentary canal through the absence of bile ducts, the liver of adult lampreys cannot be considered primitive. The existence of bile ducts in larvae and their programmed atresia at metamorphosis is an indication that a liver without bile ducts may have evolved in lampreys as either a specialization in function (e.g., storage), a requirement of ontogenesis (caudal endocrine pancreas), or due to a lack of need (gut excretion of bile).

The histological organization of the gastrointestinal tract of dogfish (*Scyliorhinus stellaris*) is very similar to that of higher vertebrates. Some points of difference concern the oesophagus, which is lined by stratified epithelium provided with mucin-secreting cells and has haemopoietic tissue (the organ of Leydig) between the mucosal and muscle layers of its ventral and dorsal walls.

The tubular gastric glands are of two types. In the proximal part, they are lined by cupuliform cells secreting pepsin and hydrochloric acid, whereas in the distal or pyloric part, they are lined by columnar, apparently non-secreting cells.

Smolka et al. (1994) showed structural and immunocytochemical characteristics of acid secretory cells of gastric glands of Atlantic stingray *Dasyatis sabina*. These were of two types, oxyntic and oxynticopeptic. Oxyntic cells were found immediately adjacent to gastric pits in all gastric glands, whereas oxynticopeptic cells were found adjacent to oxyntic cells in proximal and midregion glands only. The first evidence for an H, K-ATPase in specific cell types of the elasmobranch stomach has been provided.

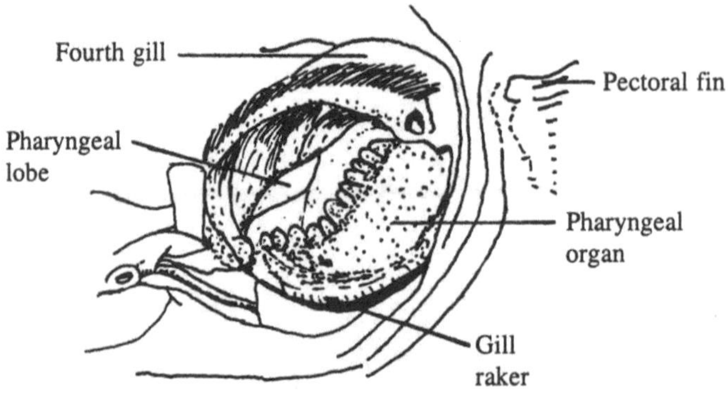

(a) *Mugil cephalus*: Pharyngeal organ.

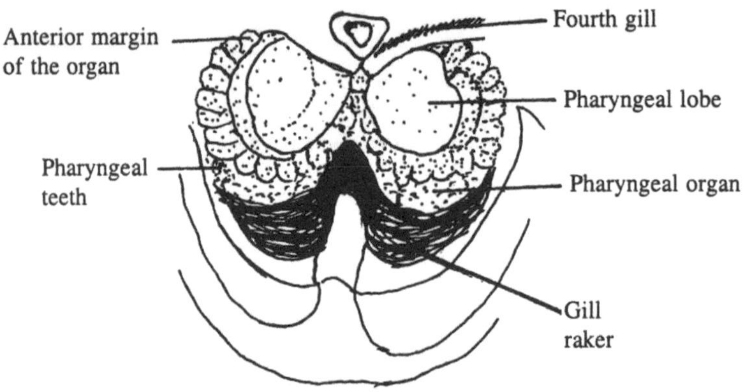

(b) *M. cephalus*: Pharyngeal organ.

Fig. 7.31 (a) *Mugil cephalus* showing left pharyngeal organ after removing operculum and gills. (b) *M. cephalus* showing anterior side of the pharyngeal organ. *Source:* Kawamoto and Higashi (1965)

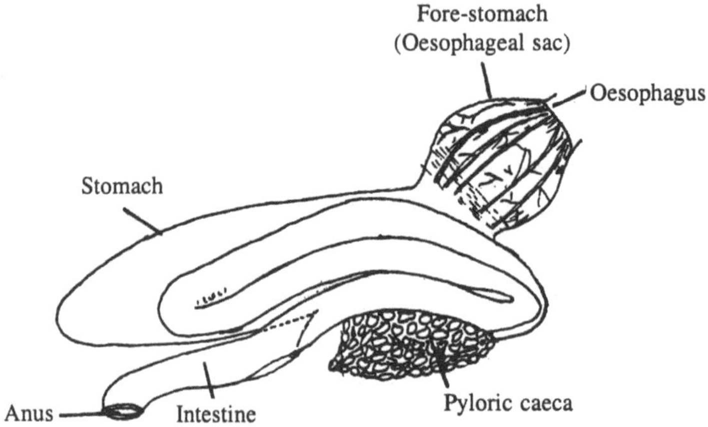

Psenopsis anomala: Post-pharyngeal gut.

Fig. 7.32 Post-pharyngeal gut of *Psenopsis anomala* showing oesophageal sac.

Table 7.3 Classification of oesophageal sacs in three points of view

Fish	Shape	Mucosa	Teeth
Ocycrius japonicus	Kidney	Wrinkled	Teeth
Psenopsis anomala	"	"	"
Ictius pellucidus	"	"	"
Ariomma lurida	"	"	"
Nomeus albula	"	Polypoid	"
Pampus argenteus	Elliptical	"	"
Pampus echinogaster	"	"	"
Tetragonurus cuvieri	"	"	No teeth
Tetragonurus atlanticus	"	"	"

Source: Isokawa et al. (1965)

Table 7.4 Classification of polypoid process

Fish	Supporting tissues	Form of basal bones	Ramification
Nomeus albula	Bone	Radial	–
Pampus argenteus	"	"	–
Pampus echinogaster	"	"	–
Ariomma lurida	"	Scaly	–
Tetragonurus cuvieri	Cartilage	–	Ramifiform
Tetragonurus atlanticus	"	–	"

Source: Isokawa et al. (1965)

The spiral valve, which represents the small intestine in most elasmobranch and teleost fishes, is lined by very high columnar epithelium. The proximal tract is often described as equivalent to the duodenum in higher vertebrates. The hind-most tract of the gut (rectum) receives a duct from the rectal gland that is thought to regulate the volume of the body fluids under the influence of atrial natriuretic factor (Fig. 7.36).

The alimentary canal of the dogfish is provided with parasympathetic and sympathetic innervation. The oesophagus and stomach, including the pylorus, receives pneumogastric fibres, whereas the nerves of the intestine have a spinal sympathetic origin. Visceral ganglia are present within the myenteric and submucosal plexuses of gut wall (Cimini et al. 1989).

Figures (7.37 and 7.38) show the organization of post-pharyngeal guts in fishes. In the oesophagus of an African freshwater fish *Phractolaemus ansorgei*, numerous folds suggest the spiral valve, similar to that reported in elasmobranchs, lungfishes and some primitive teleosteans (Danguy et al. 1985). The function of giant "barrel shaped cell" (not mucoid) reported in buccal cavity and oesophagus of some catfishes remains conjectural (Danguy et al. 1986).

Yamamoto and Hirano (1978) reported morphological changes in the oesophageal epithelium of the eel, *Anguilla japonica*, during adaptation to seawater. In freshwater-adapted eels, the stratified epithelium is provided with many mucous cells. In seawater-adapted eel, the stratified epithelium is extensively replaced by a simple columnar epithelium free of mucous cells.

Amongst the teleost the gastric types possess a distensible stomach, even gizzard-like, with gastric glands composed of only one type of secretory cells (oxynticopeptic cells) and a short intestine. The oxynticopeptic cells contain two immunochemically distinct types of pepsinogens (PGI and PGII).

Gastric gland cells comprising two cell types designated Types I and II have been indentified in freshwater bream *Oreochromis mossambicus*. Type I cells, the chief component of the glands are large cells characterized by tubulovesicles in the apical cytoplasm. Type II cells are identified by the characters of their small dense granules in the cytoplasm (Coetzee et al. 1990). The stomachless types possess an intestine recognized into unconvoluted bulb and long, coiled part, devoid of cilia and multicellular glands. According to Elbal and Agulliero (1986), the light and dark cells in the gastric glands of *Sparus auratus* can be related to an alternative morphological expression of a single cell type, producing mainly pepsinogen (light cells) or acid (dark cells), the first being characterized by the granular reticulum, a well-developed Golgi apparatus and zymogen-like granules and the latter by the more numerous mitochondria and by a tubulo-vesicular system. In the perch (*Perca fluviatilis*, fish with stomach and pyloric caeca), the enterocytes of anterior segment of intestine are involved in fat absorption, and those of middle intestine in ingestion of proteins by pinocytosis. Plasma cells, packed with a well developed rough endoplasmic reticulum and Golgi apparatus and many free ribosomes occur and may represent a local immune system. Pear-shaped cells with a thick border of microfibrils, are found (Noaillac-Depeyre and Gas 1979).

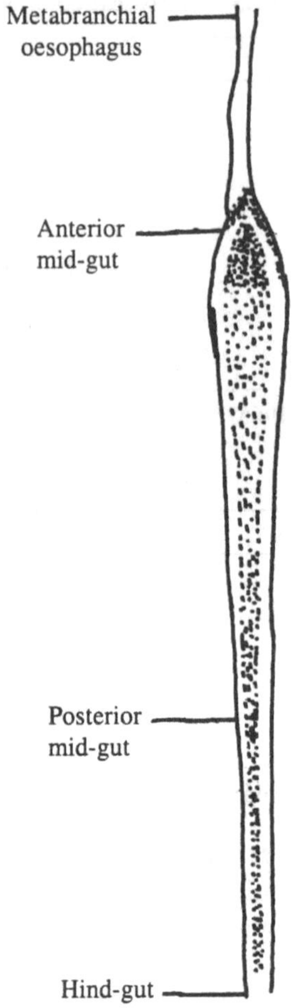

Petromyzon: Intraperitoneal alimentary canal.

Fig. 7.33 Alimentary canal of *Petromyzon*.

The intestinal caeca (pyloric caeca) wherever present, augment alimentary function (Fig. 7.38). Pyloric caeca are lacking in sharks and rays (Chondrichthyes), stomachless teleosts (carps and sucker: Cyprinidae and Catostomidae) and some teleosts with stomach (e.g., pike: Esocidae). Within certain families of fish with stomachs, such as blennies (Blennidae), caeca occur in some species but not in others. While the presence or absence of caeca in fish is not clearly related to diet or gut length, caeca when present tend to be better developed in carnivores than herbivores, especially in carnivores with relatively short guts. The number of pyloric caeca ranges from 1 to over 1,000, is species-specific, and is often utilized as a taxonomic character for identifying fish specimens to species.

The functions of pyloric caeca have been disputed. Fish caeca are entirely different from the distally located caeca of birds and mammals, which have fermentation functions. That the functions of the pyloric caeca include enzymatic and absorption have received support (Buddington and Diamond 1987; read also Hossain and Dutta 1995). For pattern of caecae distribution in teleost fish, the reader is referred to Dutta and Hossain (1993).

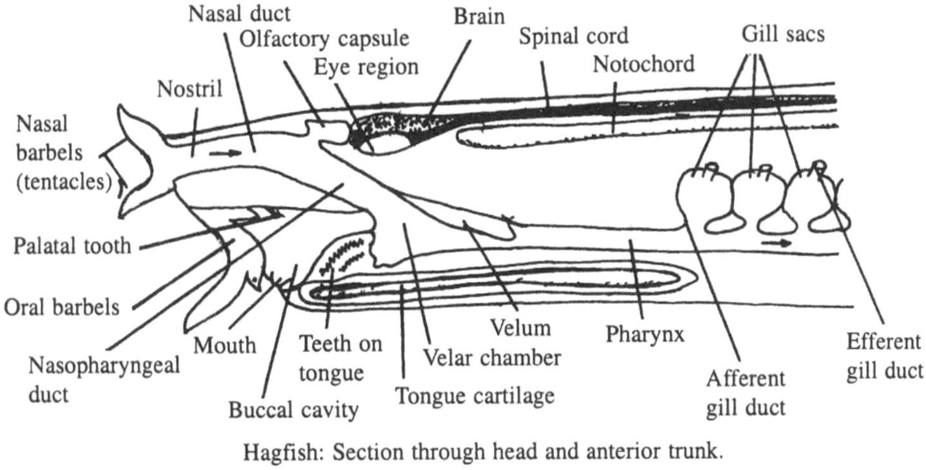

Fig. 7.34 Diagram showing sagittal section through the head and anterior trunk of *Myxine*. Arrows indicate direction of water flow.
***Sources:* Brodal, A., Fänge, R. (eds.) (1963); Burggren et al. (1985)**

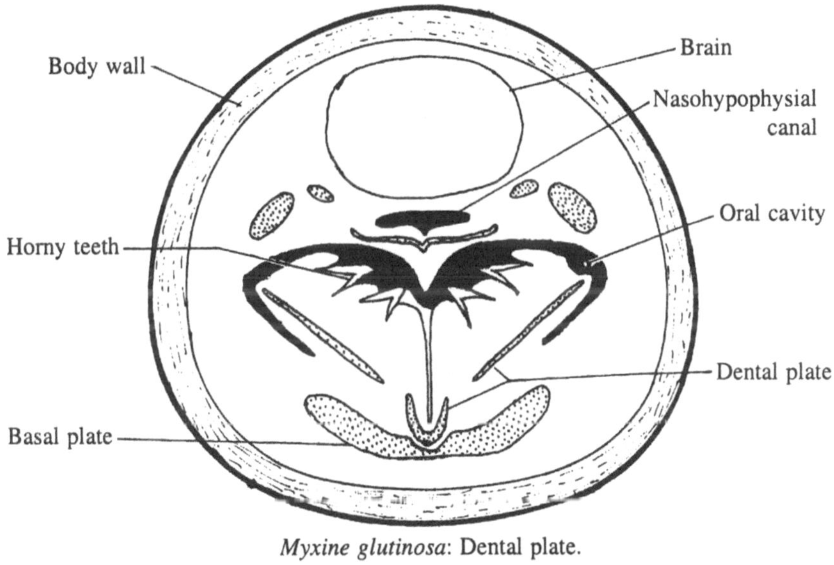

Fig. 7.35 Diagram showing a transverse section through the region of dental plate showing cartilages and teeth in *Myxine glutinosa*.
***Source:* Brodal, A., Fänge, R. (eds.) (1963)**

Only in *Coccorella*, evermannellid genus, pyloric caecum extends into the head; the appearance of this unique neomorphous feature still awaits satisfactory explanation.

For information on SEM pictures of gut, the reader is referred to Sinha and Chakrabarti (1989) and Chakrabarti et al. (1992).

Difference in gut morphology are also apparent among different fish species with similar diets. Many studies of fish have concluded that herbivores tend to have longer intestines than

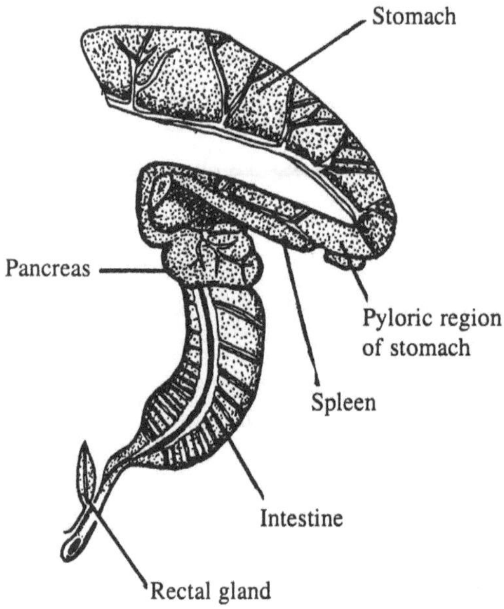

Notorynchus maculatum: Post-pharyngeal gut.

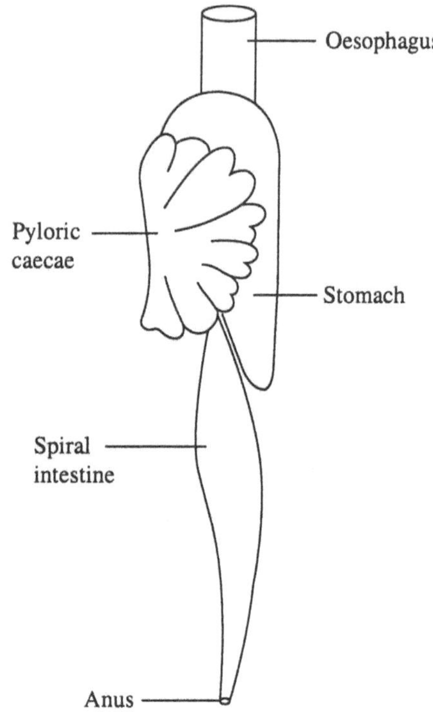

Polydon folium: Post-pharyngeal gut.

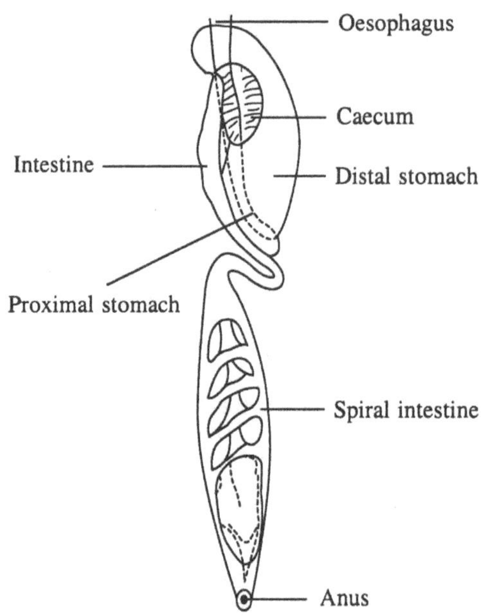

Scaphirhynchus: Post-pharyngeal gut.

Fig. 7.36 Post-pharyngeal guts in different fish.

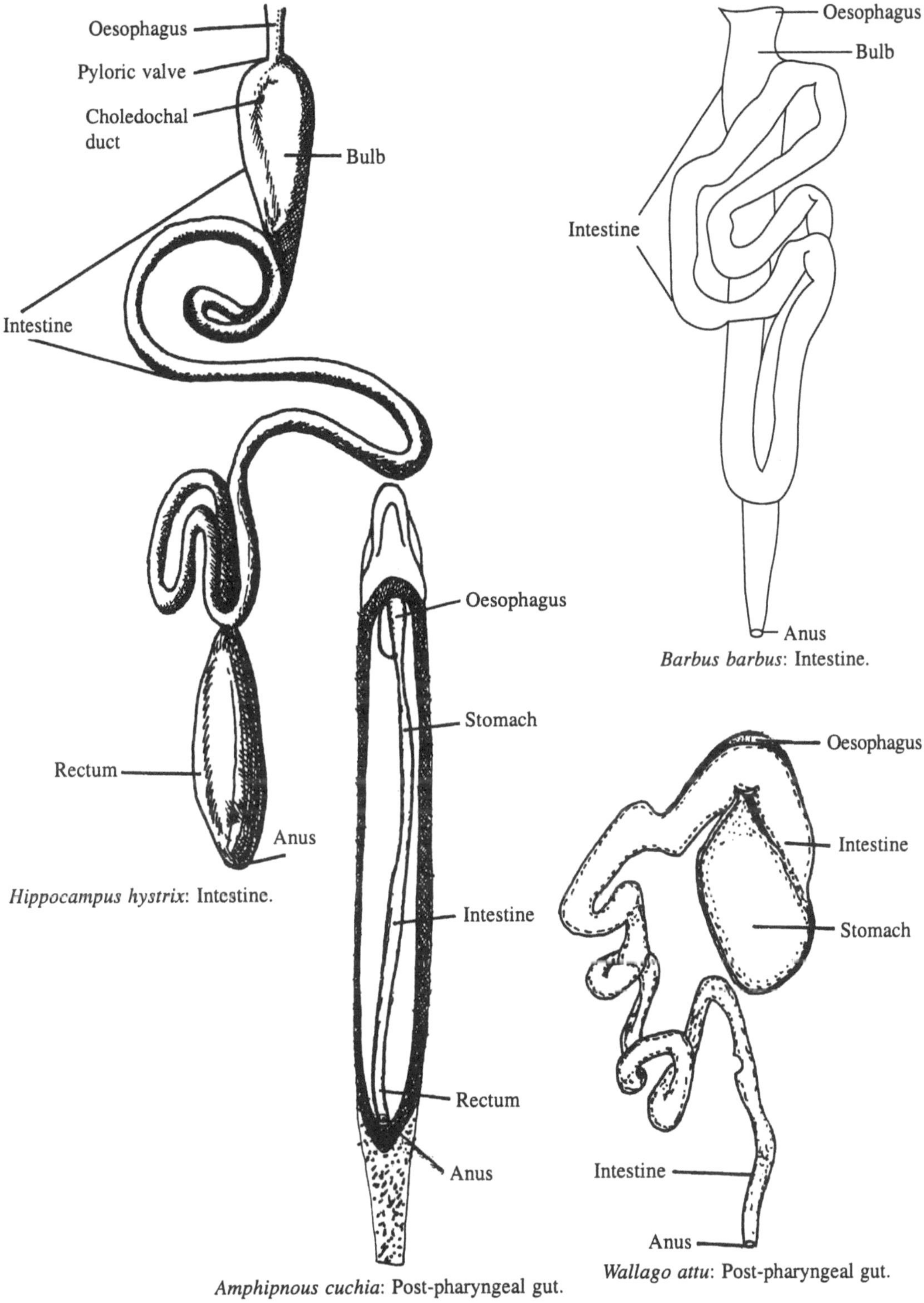

Hippocampus hystrix: Intestine.

Amphipnous cuchia: Post-pharyngeal gut.

Barbus barbus: Intestine.

Wallago attu: Post-pharyngeal gut.

Fig. 7.37 Post-pharyngeal guts in different fish.

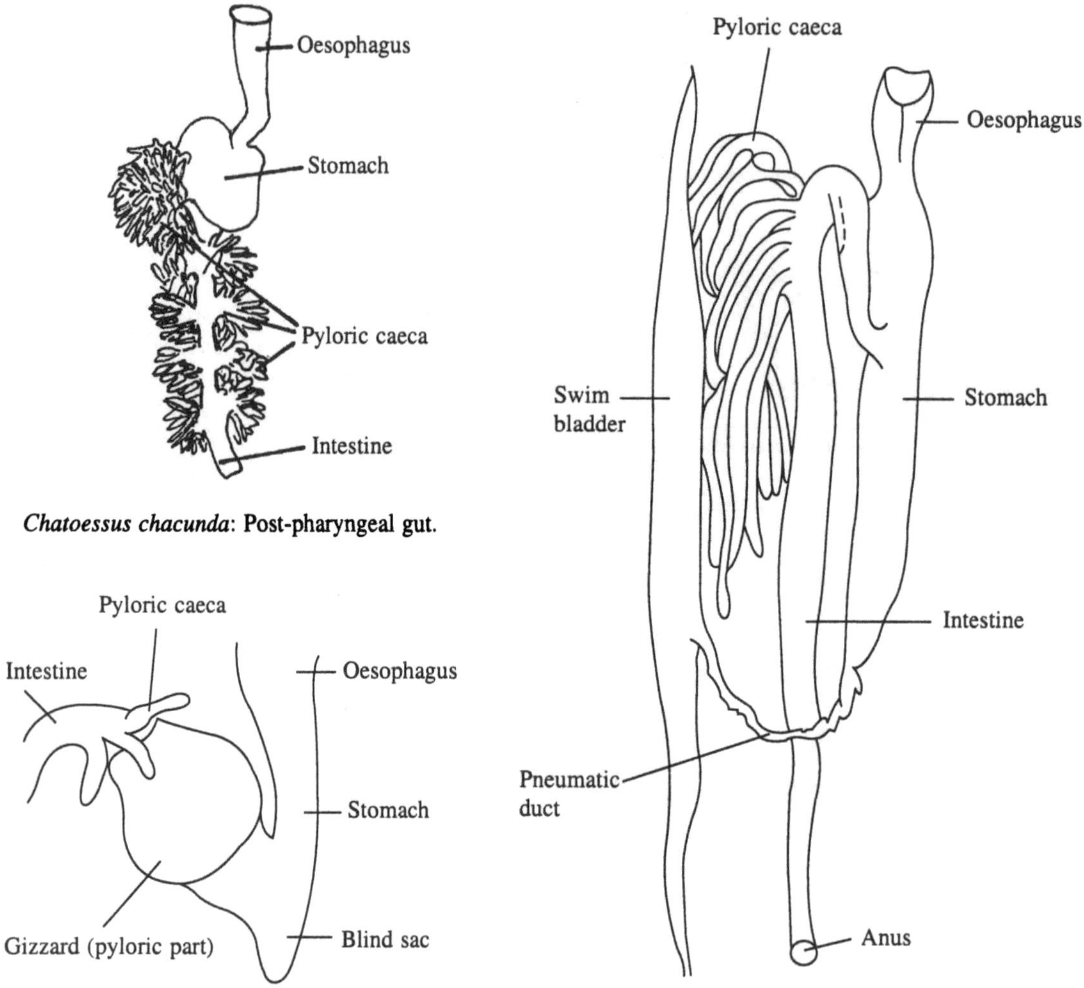

Chatoessus chacunda: Post-pharyngeal gut.

Mugil cephalus: Post-pharyngeal gut.

Clupea harengus: Post-pharyngeal gut.

Fig. 7.38 Post-pharyngeal guts in different fish.

carnivores and that intestinal length in a given species increases with dietary fibre content. It is difficult to generalize on the relationships of complexity of gut and feeding habits of fishes.

The intestinal regions, easily distinguishable in higher vertebrates, are not present in teleosts. It is generally accepted that only the gastric glands in Mammalia have distinct acid-producing parietal cells and zymogenic chief cells. In bony fishes, amphibians and birds, both hydrochloric acid and pepsinogen are assumed to be secreted by one cell type.

The "inflatable sac", a ventral diverticulum of the stomach occurs in certain members of Diodontidae and Tetraodontidae (Fig. 7.39). The stomach of the ballonfish, *Diodon holocanthus* has lost its digestive function. There is only transitional epithelium. It lacks tubular gastric glands and seems to have no digestive function. However, the stomach may sometimes serve for food storage. Abundant epithelial mucous cells, also common in the gastrointestinal tract of

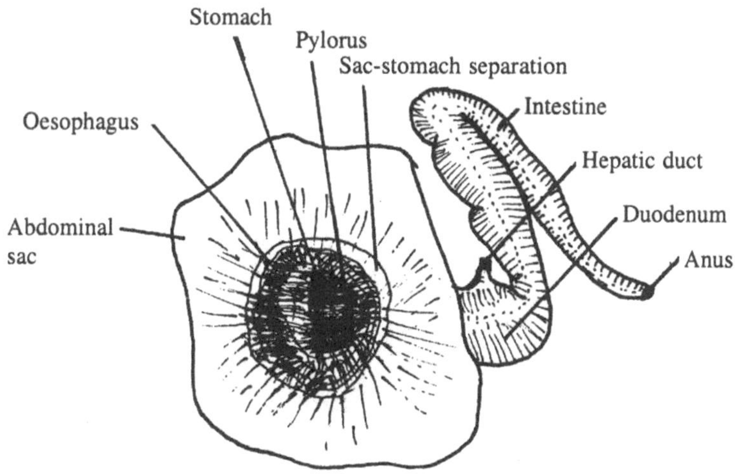

Spheroides maculatus:
Post-pharyngeal gut.

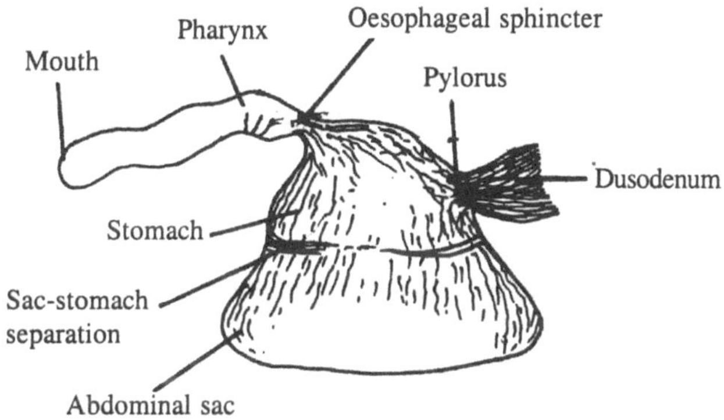

Spheroides maculatus: Inflatable sac.

Fig. 7.39 Inflatable sac in *Spheroides maculatus*.

other fishes, most likely serve to lubricate the unfolding of the multiply folded stomach wall (Brainerd 1994). Ballonfish can triple its body volume by pumping water into its stomach. Details of inflation (mechanism of the buccal pump) have been described for the striped burrfish, *Chilomycterus schoeffi* (Diodontidae) (Wainwright et al. 1995). Evolution of pufferfish inflation behaviour has been given by Wainwright and Turingan (1997).

Spiral valves are found in the intestine of elasmobranchs, holocephalans, polypterids, holosteans, acipenseroids, and in the coelacanth, *Latimeria chalumnae*. Teleost fish have a gut which is typical of the higher vertebrates in many respects, although the midgut villi (absorptive papilli) of mammals are absent in fish. A cloaca never occurs in teleost fish, except the Dipnoi, although it is universal in sharks and rays. For innervation of gut and adjacent structures, (see Figs. 7.40–7.42).

Reviews on digestive physiology (Kapoor et al. 1975b; Fänge and Grove 1979; Vigna 1983; Vonk and Western 1984; Grove 1986; Smith 1989; Hofer 1991) have provided information in

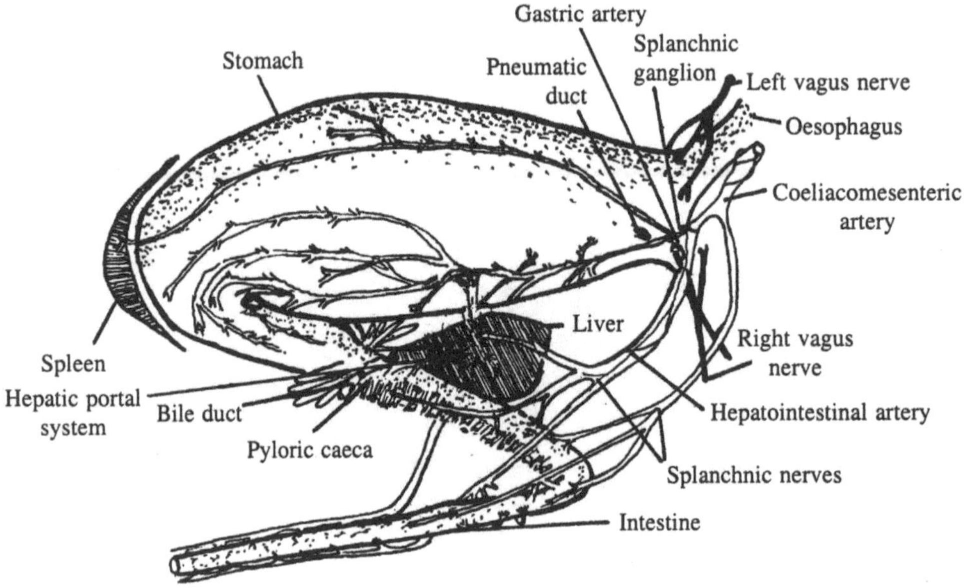

Salmo trutta: Splanchnic nerve path.

Fig. 7.40 Innervation of the brown trout (*Salmo trutta*) rectum by posterior autonomic nerves. *Source:* Burnstock (1959)

different fish taxa. Recent contributions, especially on herbivorous fishes (Horn 1989, 1992; Horn and Messer 1992; Clements et al. 1994; Clements and Choat 1995) have augmented our knowledge. Articles by Jeuniaux (1993) and Jobling (1993) are also significant.

The cell proliferation in the digestive mucosa of fishes has been studied by Danguy et al. (1988). Mucus-secreting cells seemed unable to proliferate. The oesophagus and the intestine are endowed with the highest proliferative rate and the stomach with the lowest.

Proteolytic enzymes predominate in the typical elasmobranch or primitive bony fish, if it is carnivore, not an herbivore. Hofer (1991) describes the digestive physiology with emphasis on cyprinids. The most remarkable feature of the digestive system of cyprinids (and some other teleost families, i.e., Cyprinodontidae, Labridae, Blennidae, and Gobiidae) is the lack of a true stomach. In consequences, cyprinids have dispensed with the secretion of hydrochloric acid and of pepsin (pepsinogen), a proteolytic enzyme which requires an acid environment for its activation and activity. In other specialized teleosts, e.g., cichlids, high concentrations of hydrochloric acid (pH 1.5–2.0) effect the denaturation and subsequent lyses of bacterial and plant cells, thus significantly increasing the utilization of plant, algal and detrital matter.

The stomach also plays an important role as a storage organ, particularly for carnivorous species which swallow large food items. The lack of this storage capacity in cyprinids is partly made up for by an extension of the foregut, the intestinal bulb, where most of important digestive processes take place.

Stroband et al. (1979) rejected the hypothesis that the uptake of protein macromolecules is to be attributed to the absence of a stomach and therefore to an inefficient digestion of proteins (read also Stroband and Kroon 1981).

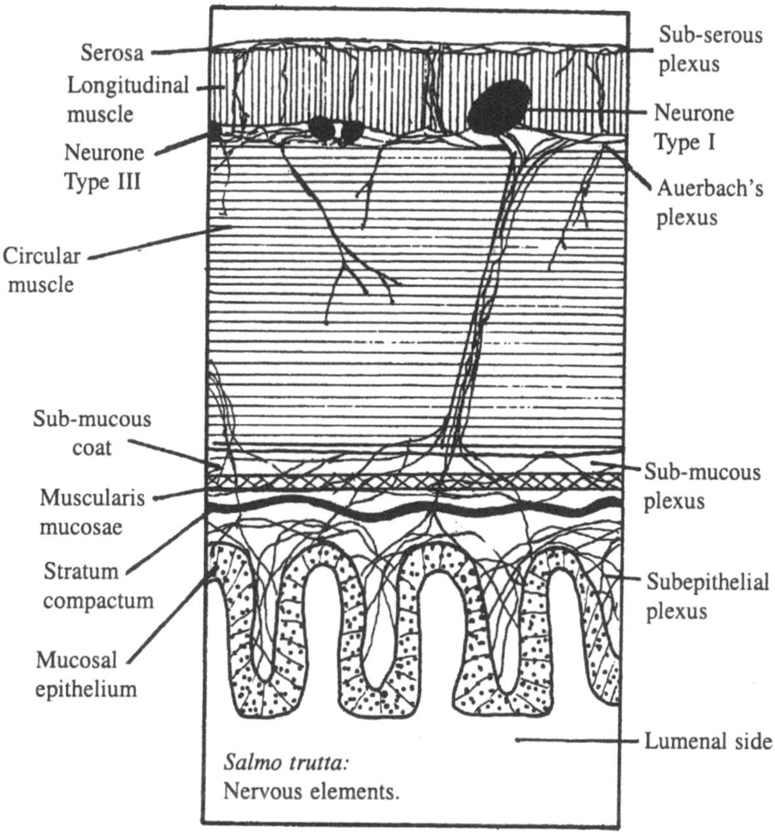

Fig. 7.41 Diagrammatic transverse section through the wall of the brown trout (*Salmo trutta*) stomach to show the general arrangement of nervous elements.
***Sources:* Burnstock (1959); Kapoor et al. (1975b)**

The intestinal epithelium represents a cell renewal system (e.g., in goldfish *Carassius auratus*; carp *Cyprinus carpio*). Details of this in juvenile grasscarp, *Ctenopharyngodon idella* are given by Stroband and Debets (1978).

The digestive juice secreted by exocrine pancreatic tissue into the intestinal bulb contains enzymes for processing proteins, carbohydrates and lipids, as well as bicarbonates.

Specific enzymes for resistant polysaccharides, like chitin or laminarin (a component of algal cell walls) have been found in the digestive tract of cyprinid and non-cyprinid fish.

Elongation of intestinal tract is a common adaptation to vegetable diets, is also found in cyprinids. The intestinal elongation is not necessarily accompanied by an increase of the total mucosal surface. The intestinal elongation in macroherbivorous cyprinids results in a proportional increase in gut passage time, thus promoting digestive processes and probably to a lesser extent microbial degradation of the more resistant diet.

The positive correlation between the amount of protein or starch in the diet of fish and the activity of specific enzymes in the digestive tract is well documented, although this correlation may be the result of different degrees of dilution of digestive juices. In fact, with a plant diet (of low protein content), the specific proteolytic activity of digestive juice is lower, but the daily production of proteolytic enzymes is even higher, than with an animal diet. The large amount of

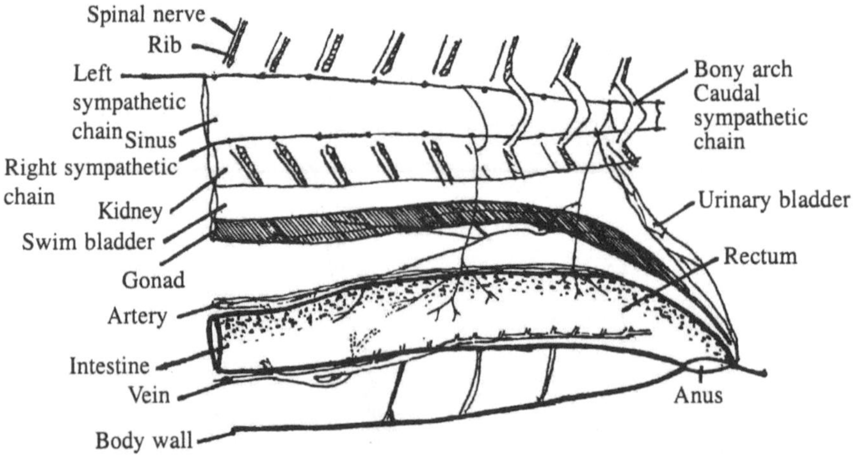

Salmo trutta: Innervation of rectum by autonomic nerves.

Fig. 7.42 Path of splanchnic nerve to the brown trout (*Salmo trutta*) gut. Fine splanchnic nerves follow all the major branches of the coeliacomesenteric artery. The stomach is shown from the dorsal surface. *Source:* Burnstock (1959)

ingested plant material not only dilutes the digestive juice but also accelerates passage through the gut, so that a higher enzyme production is required.

Proteins ingested by teleost fish can be transferred to the circulatory system, intact or partially hydrolyzed. Studies in rainbow trout (*Salmo gairdneri*) have shown that the transfer of ingested proteins into the plasma occurs in the posterior gut via enterocytes, which are structurally differentiated for absorbing proteins and carrying out intracellular digestion. Numerous macrophages and lymphoid cells are infiltrated between the epithelial cells in teleost fish. Le Bail et al. (1989) reported interference of granule cells in the intestinal transfer of growth hormone into the circulatory system of rainbow trout. The oral administration of GH can be potentially used in fish culture.

Since readily digestible polysaccharides (starch, glycogen) are present to any significant degree in plants, the activity of amylases is more specifically adapted to the diet of a given species than other enzymes. Therefore, it is not surprising that in cyprinids, which generally tend to omnivory, and particularly in carp, amylase activity is higher than in carnivorous families of fish.

Jeuniaux (1993) has reviewed the chitinolytic systems in the digestive tract of vertebrates. Chitinases have been found in the digestive tract and in glandular secretions of numerous fish species. The only exceptions seems to be *Lophius piscatorius, Abramis brama* and *Protopterus aethiopicus*. Chitinase secretion has been observed in gastric mucosa, intestinal mucosa, pyloric caeca and/or pancreas.

The digestive tracts of marine herbivorous fishes show a wide degree and range of specializations for handling an alga or seagrass diet. These fishes browse or graze (if sediment is ingested) on their plant foods with varying selectivity, using closely spaced cropping or scraping teeth. With certain exceptions, herbivorous species have longer guts and shorter gut-transit times than carnivorous or omnivorous species, and the food is attacked with an apparently convential set of vertebrate enzymes once the contents of the plant cell are released. Many browsers appear to have least-specialized guts among herbivorous fishes, and to require only a stomach pH slightly more acidic than that of carnivore in order to leach enough nutrients from algal cells without actually breaking down the cell wall. Other browsers have acidic stomach as well as apparently fermentative microbial digestion in the hindgut. Grazers often have either a pharyngeal or gastric grinding mechanism to break down the food into small particles, and to break open plant-cell walls. These species usually have non-acidic stomachs, or lack stomachs entirely (Horn 1989, 1992; Horn and Messer 1992). Edwards and Horn (1982) reported a temperate zone intertidal fish, *Cebidichthys violaceus*, can digest (assimilate) microalgal tissue. Other than a digestive tract that increases in relative length with size, *C. violaceus* lacks the morphological features associated with herbivorous fishes, including grinding teeth, pharyngeal mill or muscular stomach. This fish does possess, however, acidic gastric environment, compared with pH values near neutrality for its intestine. Acidic gastric secretions in other fishes effectively break down algal tissue and bacterial cells. In *C. violaceus*, the low pH of the stomach coupled with selective feeding pattern results in the relatively efficient digestion of microalgal material.

Ingestion, digestion and conversion of food are highly temperature dependent processes (Smit 1980). The majority of digestion and all the nutrient uptake occurs in the intestine. It is of interest, therefore, to examine temperature effects on intestinal enzymes. No quantitative changes occurred with temperature acclimation for two major proteases, trypsin and chymotrypsin, from rainbow trout (*Salmo gairdneri*). However, the eel (*Anguilla japonica*), *Salmo trutta* and *Tinca tinca* possess trypsin isozymes, and possibility exists that changes in isozymes or in the ratio of isozymes may allow thermal compensation in rainhow trout. Similarly, genetic differences may optimize trypsin for the temperature encountered by a population. According to McLeese and Stevens (1986), two differently reared groups of trout display different intestinal and pancreatic characteristics, due to environmental and/or genetic factors. The significance and extent of these differences requires further study, both in other systems within the fish and in other groups of trout.

There are three obvious variables to be accounted in studying gastric evacuation time (GET) and gastric evacuation rate (GER) size of the meal, size of the fish, and temperature. Differing food quality may or may not affect GET (Smith 1989).

Horn and Gibson (1990) reported that food consumption increased and gut transit times decreased with increase in temperature in all three temperate species of sea-weed-eating fishes *Linophrys pholis*, *Parablennius sanguinolentus* and *Gobius cobitis*. Assimilation efficiencies for algal carbon and nitrogen did not vary with temperature in *G. cobitis*. Carbon but not nitrogen assimilation efficiency varied with temperature in *P. sanguinolentus*, and both carbon and nitrogen assimilation efficiencies changes broadly with temperature in *L. pholis*, the most omnivorous of the three species. Assimilation efficiencies were generally within the previously reported ranges for herbivorous marine fishes. These species are thus clearly capable of extracting energy from algal food sources.

Jobling (1993) discussed various factors: (1) influence of body size on ingestion, (2) temperature effects on ingestion, and (3) other factors known to influence ingestion (e.g., dissolved oxygen levels, seasonal differences, stages in the reproductive cycle). Factors influencing faecal losses were considered (e.g., absorption efficiency, assimilation efficiency). Ingestion of food will lead to an increase in the rate of nitrogenous excretion. Factors influencing metabolism were considered (for further information, see Section 6). The reader is also referred to Houlihan et al. (1993).

Lipids have probably been the most intensively studied of the major biochemical constituents of aquatic organisms. Of all the animals currently farmed, fish are the richest source by far of n-3, polyunsaturated fatty acids (PUFA), which has fundamental implications for fish nutrition and, in the final analysis, for human nutrition (see details of digestion, absorption and deposition: Sargent et al. 1989). The reader is also referred to Bilinski (1974), Sire and Vernier (1981), Sire et al. (1981), Lie and Lambertsen (1985), Babin and Vernier (1989), Vernier (1990) and Phleger et al. (1997). The key steps in lipogenesis and their control by insulin are well illustrated (Mommsen 1998; Fig. 1).

In teleost fish, intestinal absorption of fats and secretion of lipoproteins into the lymph by the intestine are basically comparable to the mammalian process, even though assimilation is slower in fish (cf. Steffensen and Lomholt 1992: Secondary vascular system, Chapter 8). Dietary lipids are absorbed in the anterior intestine and the pyloric caeca by species possessing them, similarly to the first third of the mammalian intestine. Triglycerides and wax esters are the two main forms of neutral lipids available to fish in their natural environment. In contrast to mammals, free fatty acids and glycerol are the major products of the luminal hydrolysis of dietary triglycerides in fish resulting from the nonspecificity of pancreatic lipase or the additional action of a β-monoglyceride lipase. Fatty acids are then re-esterified in intestinal epithelial cells, primarily as triglycerides. Enzymatic hydrolysis and absorption of wax esters also occurs in the pyloric caeca and the intestine of the various species studied (e.g., anchovy, cod, rainbow trout) but the process is slower than that for triglycerides. Most of the fatty alcohols produced by the hydrolysis of wax esters are oxidized to their corresponding fatty acids in the intestinal tissue. They are then re-esterified there into acyl lipids, mainly triglycerides and phospholipids as shown in the gourami (*Trichogaster cosby*), carp (*Cyprinus carpio*), and rainbow trout (*Salmo gairdneri*). Dietary lipids are present in two forms in intestinal epithelial cells, either in the form of large droplets or exported as lipoproteins.

Export of particles apparently occur only via the portal route in the carp, only via the lymph in the trout, or via both in the tench (*Tinca tinca*) and the perch (*Perca fluviatilis*). In spite of the fact that nearly all fatty acids are re-esterified in intestinal cells and are incorporated into newly formed lipoproteins, the direct passage of free fatty acids from the intestinal lumen into the circulatory system has been described. This process appears to occur during early absorption and under special alimentary conditions. Free fatty acids represent only a small portion of total fatty acids carried by blood and their level is comparable to that observed in mammals and birds. They are probably combined in the blood with plasma fatty acid-binding proteins comparable to their binding to albumin in mammals.

In fish, lipids are integrated into lipoproteins in the proportion of at least 85% in the spiny dogfish (*Squalus acanthias*) and 93% and 95% in mature male rainbow trout and channel catfish (*Ictalurus punctatus*) respectively.

In fish, lipids are absorbed as either fatty acids or triacylglycerols aggregated into chylomicra particles. The interorgan transport of lipid in fish is generally similar to that of mammals. Fish possess both an exogenous and endogenous transport system (Sheridan 1988; read details) (Fig. 7.43). The lipoprotein complexes of the endogenous transport system resemble those of other animals; certain apoprotein constituents are similar to human apoproteins. The interconversion of lipoprotein in fish is not well-defined as is the mechanism of LDL-cholesterol uptake. Lipids in fish are stored in several depots organs (liver, muscle, mesenteric fat) primarily as triacylglycerol

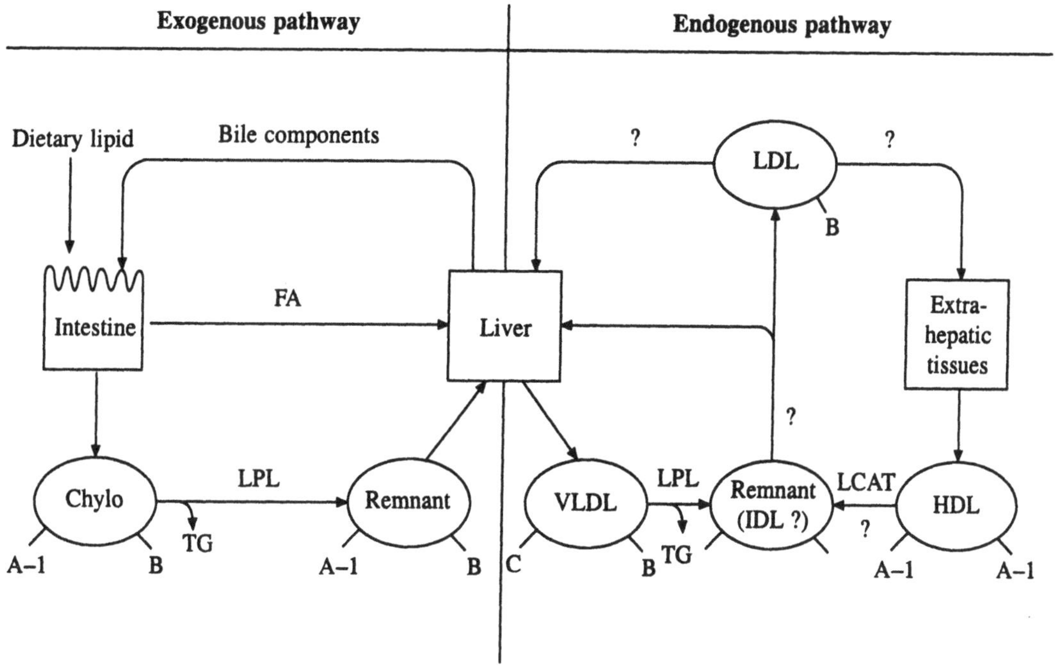

Abbreviations: Chylo: chylomicra; FA, fatty acids,
FFA, free fatty acids; HDL, high density lipoprotein;
IDL, intermediate density lipoprotein;
LDL, low density lipoprotein: LPL, lipoprotein lipase;
TG, triacylglycerals; VLDL, very ow density lipoprotein.

Fig. 7.43 Exogenous and endogenous lipid transport pathways in fish.
***Source:* Sheridan (1988)**

(Fig. 7.44). Mobilization of lipids proceeds from the activation of lipolytic enzyme activity (triacylglycerol lipase) and results in the hydrolysis of stored TG and subsequent release of FA. Fatty acids are carried in the plasma of fish by one, or perhaps more, albumin-like binding proteins. Lipolytic activity is hormonally modulated. A variety of slow acting (thyroxin, cortisol, growth hormone, prolactin) and fast-acting (epinephrine, norepinephrine, somatostatin) agents have been shown to stimulate lipase activity; perhaps indicating multiple mechanisms of hormone action (Sheridan 1988) (Fig. 7.45). According to Jobling and Miglavs (1993), food appetite of Arctic charr, *Salvetinus alpinus*, was greatest in the fish with reduced lipid depots. It is suggested that appetite declines as lipid accumulates and energy reserves become repleted, such that energy reserves are then maintained close to some set-point. McClelland et al. (1995) found that two fishes, the European seabass (*Dicentrarchus labrax*) and gilt-head sea bream (*Chrysophyrys auratus*) with similar lifestyle store lipid in the same fashion and that dietary FA composition affects the FA composition of TL and TAG fractions but not PL (phospholopids) in tissue and plasma.

Jobling et al. (1998) reported changes in lipid class composition of tissues of anadromous Arctic charr. Females lost opprox. 80% of their body lipid during spawning and overwintering, the majority of this being TAG. Males also lost large amounts of TAG, during the winter, but in contrast to the females lipid depletion amounted to only 50–55% of that present at the time of ascent into the fresh water. For lipid biosynthesis, refer Mommsen (1998).

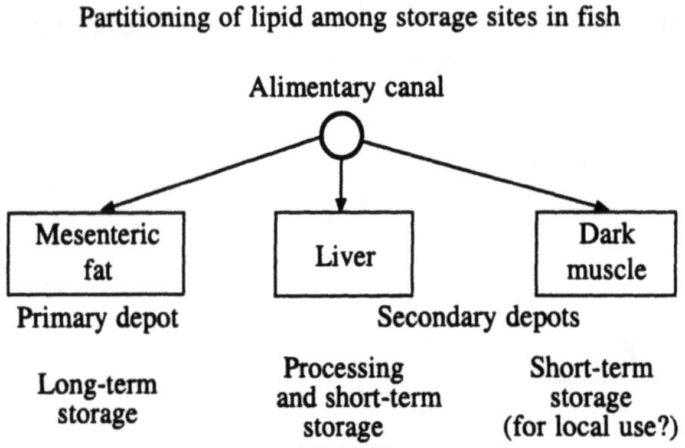

Fig. 7.44 Partitioning of lipids among depot organs in fish. Lipid is absorbed from the alimentary canal and transported in the blood as lipoprotein (chylomicra). Lipids are removed from the blood by the action of lipoprotein lipase and deposited among the various storage sites.
***Source:* Sheridan (1994)**

Fish have been shown to absorb short-chain fatty acids (SCFAs) from the gut by membrane carrier-mediated transport and passive diffusion. SCFAs may contribute to the nutrition of some tropical and temperate species of marine herbivorous fish. The high concentrations of SCFAs in

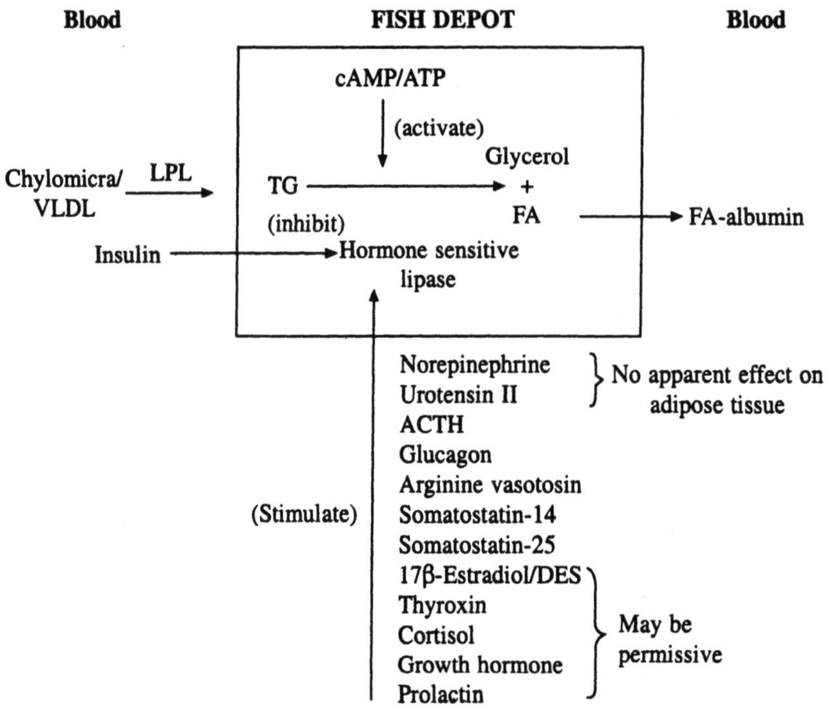

Fig. 7.45 Summary of the hormonal regulation of lipid mobilization in depot organs of fish. Abbreviations: FA, fatty acids; LPL, lipoprotein lipase; TG, triacylglycerols; VLDL, very low density lipoprotein.
***Source:* Sheridan (1994)**

the intestine of some species suggest that gastrointestinal microorganisms contribute to the nutrition of many of the species examined (see details: Clements and Choat 1995; refer also Clements et al. 1994). Abundant populations of microorganisms occur in alimentary tracts of many herbivorous marine fishes, including acanthurids, girellids, kyphosids, odacids and aplodactylids. SCFAs produced by endosymbionts in the posterior intestine are used as a blood fuel either for energy purposes or lipid synthesis by the host fish (Clements et al. 1994).

In recent years, many ecologists have attempted to estimate food consumption by fish using direct methods on stomach contents analysis and rates of gastric emptying. Two of the most important factors governing gastric emptying are meal volume and food composition (refer Saether and Jobling 1997 on gastrointestinal evacuation). These factors have also been demonstrated to influence the secretion of gastric acid, digestive enzymes and gut hormones (see for details: Grove 1986; Jobling 1986). Elucidation of factors controlling feeding behaviour is only in its infancy (read gustatory stimulants, and neuropeptides: Mommsen 1998).

The least digestible, hard skeletal elements of prey are often the last part of a meal to be emptied from the stomach but little is known about the mechanisms controlling gastric retention and evacuation of large indigestible food fragments in fish species. Carnivorous fishes such as cod, *Gadus morhua*, may require several days to digest and evacuate large prey and will usually consume additional food before all remnants of the previous meal have been evacuated.

In cod, the distal stomach and pylorus are highly muscular and are richly invested with endocrine cells showing gastrin/CCK-like immunoreactivity, suggesting that the pyloric region may be of key importance in the regulation of gastric emptying in this species. It often appears that there is some form of 'meal stacking' such that digestion and evacuation of one meal is completed before digestion of meals ingested later commences. Both particle size and density may influence clearance from the stomach. Clearance of indigestible material from the stomach of mammals also appears to be affected by density, with dense particles being cleared more slowly than lighter material. The consumption of additional meals seem to retard, rather than hasten emptying of large indigestible particles from the stomach of cod. Whether the vomiting reflex is the mechanism involved in the expulsion of large indigestible remains from the stomach is questionable. It has been hypothesized that the intake of further meals, leading to an increased digestion within the stomach, might considerably shorten the gastric life of these fragments. Thus, neither IMMC (interdigestive migrating myoelectric complex) phase III contraction nor a vomiting reflex may be essential mechanisms involved in the emptying of refractory solids from the cod stomach, but such fragments are, nevertheless, cleared effectively (dos Santos and Jobling 1991). Olsson and Holmgren (2001) reviewed the control of gut motility.

In teleost fish, an entire intestinal segment, designated as the middle intestine, posterior intestine or rectum, depending on the species and/or the authors, is characterized by the presence of an endocytosis complex and a highly developed vacuolar system in the epithelial cells.

At all stages of the development of teleost fish, from the first feeding by larvae or juveniles until the adult stage, ingested protein is segregated-regardless of molecular weight-in the supranuclear vacuolar system after its transit via the tubulo-vesicular system of the apical cytoplasm.

Acid phosphatase activity has been demonstrated at the ultrastructural level in the vacuolar system of the rainbow trout and goldfish.

Lysosomal proteases, cathepsins D and B, have the same localization. Cathepsin D is an endopeptidase whose action is comparable to that of pepsin and which plays an essential role in the breakdown of exogenous proteins. Cathpsin B is a protease comparable to papain. In addition to acting on natural substrates, it also acts on synthetic substrates containing arginine residues. The enzyme is detected in the vacuolar system of the apical cytoplasm.

Acid phosphatase is visualized at the level of the vacuoles containing exogenous proteins, Cathepsins B and D, the essential lysosomal proteolytic enzymes, are also localized at this level.

In conclusion, the epithelial cells of the posterior intestine are characterized by a heterophagous process for dietary proteins. This process, in both larvae or juveniles and adults, begins immediately after feeding and lasts throughout the phase of alimentary bolus transit. The reality of plasma of ingested proteins is established.

The calcium transport phenomena in intestine has been addressed by Flik and Verbost (1993).

In conclusion, a regional differentiation of the intestine of teleost fish gives rise to a posterior intestine where proteins are absorbed as macromolecules. There is a permanent double function of this segment: nutritional and immune (Vernier 1990; also see Georgopoulou et al. 1985, 1988).

The transport of various substances across brush border of intestine has been focused in a series of publications (see Bell et al. 1987; Cassano et al. 1988, 1990; Reshkin et al. 1988, 1989; Vilella et al. 1988, 1989a, b, 1990; Romano et al. 1989; Storelli et al. 1989; Ahearn et al. 1992; Verri et al. 1992; Maffia et al. 1993). A few results are given here. (1) In eel intestinal brush-border membrane (BBM) vesicles, there is the presence of at least two distinct agencies for Na^+-dependent proline and alanine translocation. The first system is specific for alanine and short-chain neutral amino acids; the second system, specific for amino acids and the N-methylated analogues, is regulated by alanine concentration (2) Data has suggested the presence of an electrical-potential-dependent H^+/glycyl-glycine cotransport system in the eel intestinal (BBM) vesicles (3) Results have indicated that vitamin L-ascorbate is transported across brush-border membranes by an electrogenic Na^+-dependent process in conjunction with more than one sodium ion (4) Results have suggested that the intestinal border may have a pathway for myoinositol (a sugar-related cyclohexanol vitamin) transport entirely separate from that for D-glucose but inhibited by D-glucose via binding to a regulator site on the myoinositol transporter. Markedly dessimilar influx kinetic constants suggest possible differences in myoinositol needs by carnivorous and herbivorous fish. It is well documented that fish and other vertebrates show definite transport patterns related to genetically determined dietary preferences. Sugar absorption rates are higher in herbivorous/omnivorous species than in carnivores, while carnivores preferentially absorb amino acids over sugars regardless of the dietary content of protein and carbohydrate. Further, it is reported that herbivores/omnivores are able to regulate the activity of intestinal sugar carriers with variations in dietary carbohydrate levels, while carnivores are apparently unable to do this (5) Genetic dietary transport adaptation (inositol) occurs on the brush-border membrane while no such adaptation is present at the epithelial basolateral pole. (6) Organization of intestinal brush-border amino acid transport system is comparable with that found in mammals, but with small differences that may partly relate to the metabolic requirements of these poikilothermic vertebrates, as well as to possible specific dietary needs that they may illustrate.

According to Dabrowski and Kock (1989), the sites of absorption of ascorbic acid (AA) in the rainbow trout (*Oncorhynchus mykiss*) gut were located in the stomach (20.7%), pyloric caecal region (23.4%), middle intestine (21.9%) and posterior intestine (20.1%).

For the digestive enzymes in the gut of fishes, their sources, sites of action, their substrates, and the products resulting from their digestive actions, see Horn (1998; Table 1).

3.3 Associated Glands

3.3.1 Liver

The liver of Atlantic hagfish, *Myxine glutinosa*, is of a laminar structure and resembles the mammalian, sacculosinusoidal type. The liver of young ammocoetes larvae is like that of hagfish but in the course of its growth the laminar structure is lost. It assumes the character of a tubular gland. It is of sacculosinusoidal type in sturgeon *Acipenser ruthenus* (Elias and Bengelsdorf 1952).

The liver is an integrator of physiological and biochemical functions. It is noteworthy that the structure of fish liver varies between species. Obvious differences are stated to exist even within the same family, and there is therefore definite interest in studying hepatic structure for a particular species. There are, however, interorder structural similarities. Approximately 85% of teleost liver volume is occupied by hepatocytes, the most numerous cell type. (Hinton and Laurén 1990; see Fig. 2 for structural stratification of the teleost liver).

The concept of a double-layered muralium (i.e., 2-cell thick interconnecting laminae) for the arrangement of hepatic parenchyma has been a well entrenched fact for several decades. Notwithstanding this, some recent reports clearly defended for all the cartilaginous and bony fishes the parenchyma array (e.g., in brown trout *Salmo trutta fario*), i.e. tubules in which the lumen is a canaliculus, a preductule or an initial ductule, the tubular muralium.

The ultrastructural aspects of the hepatic cell show endoplasmic reticulum, lipid droplets, Golgi apparatus, electron-dense bodies in supranuclear position, coated vesicles, microtubules and multivesicular bodies. Gap japs join the hepatocytes laterally. In the hepatic cell/capillary interface, there is a classical space of Disse.

The biliary tree comprises canaliculi, preductules, ductules and ducts. The canaliculi are formed by the apposition of hepatocytes; in rare instances the canaliculi are intracellular. Canaliculi without microvilli are also reported in fishes. Microvilli project from the hepatocytes into the lumen. Macrophages, melanomacrophages, Ito cells (fat-storing cells) in the space of Disse (perisinusoidal space) and endothelial cells are found.

Sexual differences in structure can be observed. Seasonal/nutritional factors are also important in relation to cell contents,

In the rainbow trout (*Salmo gairdneri*), the connection between bile ductules and the initial portion of the bile duct has been considered as being analogous to the mammalian canals of Hering; the latter are connecting channels between canaliculi (in lobules) and bile ducts (in portal tracts). The canal of Hering has two portions: an initial portion, in which the wall is formed by epithelial cells and hepatocytes and a subsequent formed solely by 2–4 epithelial cells. The initial portion of the canal of Hering is thus very similar to the preductule of salmonids and its terminal segment is identical to the salmonid ductule. This being so, the preductules and ductules form a segment connecting canaliculi to ducts; consequently, that segment has been considered as the homologue of the canal of Hering (Rocha et al. 1994; see literature; refer Gilloteaux et al. 1996 for biliary tract and gall bladder; also a review of fish liver by Rocha and Monteiro (1999).

Rocha et al. (1996) discussed that the pale-grey inter-hepatocytic cells in *S. trutta* may represent either a distinct cell population (pit cells?) or a subpopulation of resting macrophages. For stereological data of liver, the reader is referred to Rocha et al. (1997). Truslé and Anadon

(1995) contributed on the structure and function of fish liver, for instance for digestion, immune defence, detoxification, nitrogen catabolism, blood turn-over, reproduction, and storage (Fig. 7.46), For stressor-associated alterations of hepatocytes, information is available from the contribution of Hinton and Laurén (1990).

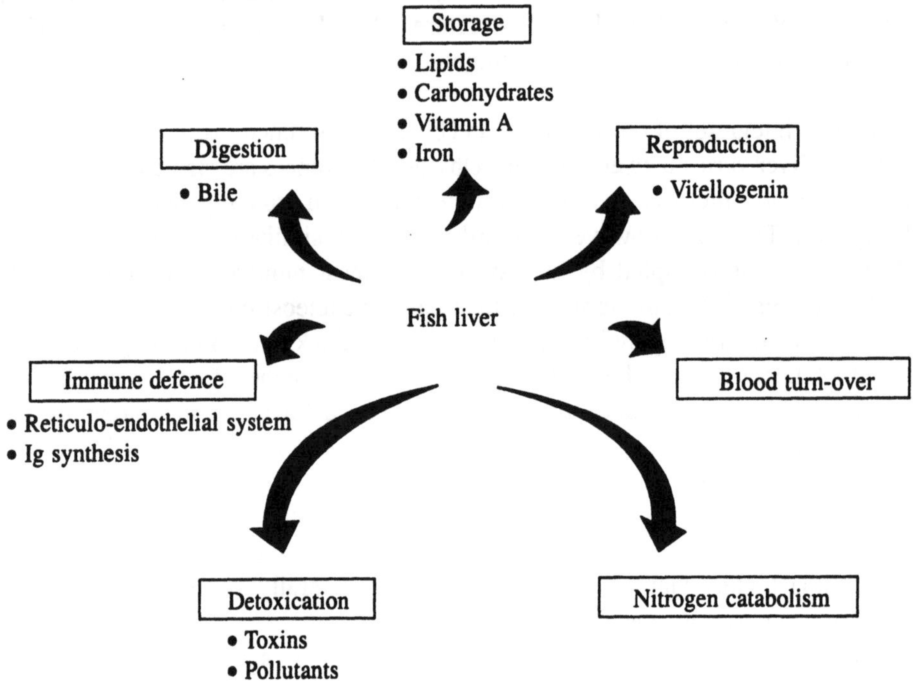

Fig. 7.46 Prominent functions of the fish liver.
***Source:* Buslé and Anandon (1996)**

Despite the absence of lobulation, the liver of the brown trout has the stromal elements spatially organized as venous-biliary-arteriolar tracts (VBAT), venous arteriolar tracts (VAT), biliary-arteriolar tracts (BAT), venous-biliary tracts (VBT), biliary tracts (BT), arteriolar tracts (AT), and isolated veins. These components are not two- but three dimensional entities, and the anatomical interrelations among all entities are displayed. The VBAT, VAT and VBT are considered portal tracts; the adjacent parenchymal zones are viewed as periportal areas. The veins emerging from those tracts are regarded as afferent, and related with periportal zones. The veins that do not communicate with VBAT, VAT or VBT are viewed as efferent. Isolated veins occupy around 60% of the stromal areas. Nevertheless, VBAT, VAT and BT are also considerably important, occupying varying proportions of the stromal areas (8–12%). The VBT and BAT are less important in quantitative terms. No sexual differences appear in either qualitative or quantitative terms. There is no structural support for an eventual macroorganization of hepatic tissues (Rocha et al. 1995). Rocha and Monteiro (1994) reported the presence of rodlet cells in the intrahepatic biliary ducts of the brown trout.

Nopanitaya et al. (1979) has reported new type of perisinusoidal cells containing numerous microfilaments, situated in the space of Disse and adhering firmly through desmosomes both to sinusoidal endothelial cells and to hepatocytes. This perisinusoidal cell in fish is morphological distinct from other sinusoid cells including endothelial cells, Kupffer cells, fat-storing cells, and "pit cells". They may provide a supportive framework within the liver.

Hepatic ultrastructure of Antarctic fishes have a unique type of perisinusoid (Ito) cells, indicating a role in protein synthesis. Perisinusoidal cells contain numerous lipid droplets.

Compared to the liver of higher vertebrates, most teleost species (e.g., *Haplochromis burtoni, Barbus (Puntius) tetrazona, Carassius auratus, Xiphophorus helleri, Corydoras melanistius*) possess a liver poor in connective tissue. In contrast the liver of *Tetradon leiurus* is characterized by an abundance of connective tissue. This phenomenon can be attributed to the high proportion of sinusoids in the parenchyma, which confers a structure similar to that of mammalian liver. The resulting loss of mechanical stability is obviously compensated for by the enlarged contribution of the connective tissue to the structure of parenchyma (Langer 1979a).

Tetradon leiurus, lacks true Kupffer cells and Ito-cells, in the sinusoidal region. Nevertheless, the close relation of granulocytes to the sinusoidal epithelium suggests that phagocytotic processes cannot be excluded for the sinusoidal region. Exceptionally, in *Hemihaplochromis multicolor*, there were signs of possible phagocytosis by sinusoidal endothelial cells. Defence-functions are undertaken by the region of venae hepaticae (with granulocytes and melanocytes in wall) in the liver of *Haplochromis burtoni*, which is free of Kupffer cells (Langer 1979b).

All fish livers produce bile which is stored in the gall bladder and drained via the bile duct into the anterior intestine or the pyloric caecum. Fish bile is alkaline, can contain digestive enzymes, and has the same function of emulsifying lipids as in all vertebrates. In general, factors controlling the production and secretion of bile in fish are not understood.

The fine structural details of shark and skate liver are virtually identical (Fig. 7.47). Shark liver, however, does possess proportionately more lipid than the skate. The amount of lipid present is related to seasonal changes and in the case of females to their reproductive condition. Hepatocytes are large polyhedral cells characterized by numerous, large, round lipid deposits. No Kupffer cells and Ito-cells are reported. Studies have not confirmed the ubiquity of Ito-cells in either shark or skate liver. A capsule invests the elasmobranch liver. It is a typical mesothelium equivalent to Glisson's capsule (Hamlett and Stoner 1987; Hamlett 1988).

3.3.2 Pancreas

Some vertebrates (adults, but not larval lamprey, pigeon, and rat) lack a gall bladder. Cyclostomes have no discrete pancreas. In the lungfish, the pancreas is embedded in the wall of the intestine instead of being suspended in the peritoneal cavity.

In sharks and rays, the pancreas is discrete and lies along the intestine near the spleen, and some of the primitive bony fishes such as sturgeon, bowfin, and paddlefish still show this characteristic. NPY-like peptide has been isolated from the pancreas of the dogfish and skate.

The reader is referred to Chapter 13, Section 9 for allied details (see Figures). The pancreatic anatomy of bony fishes is variable. Most often there is no discrete pancreas—it is diffuse. Pancreatic nodules can be scattered in the mesenteries, in the liver, clustered around the bile duct or in combination of sites. NPY-like peptide has been isolated from the pancreas of bony fishes. Two subsets of NPY-related peptides have been reported in the pancreas of anglerfish (*Lophius americanus*).

The normal exocrine pancreas (compact) of *Pimelodus maculatus* has the acinar cells, the argentaffin cells, the intermediate cells and the centroacinar cells. The most prominent cytoplasmic

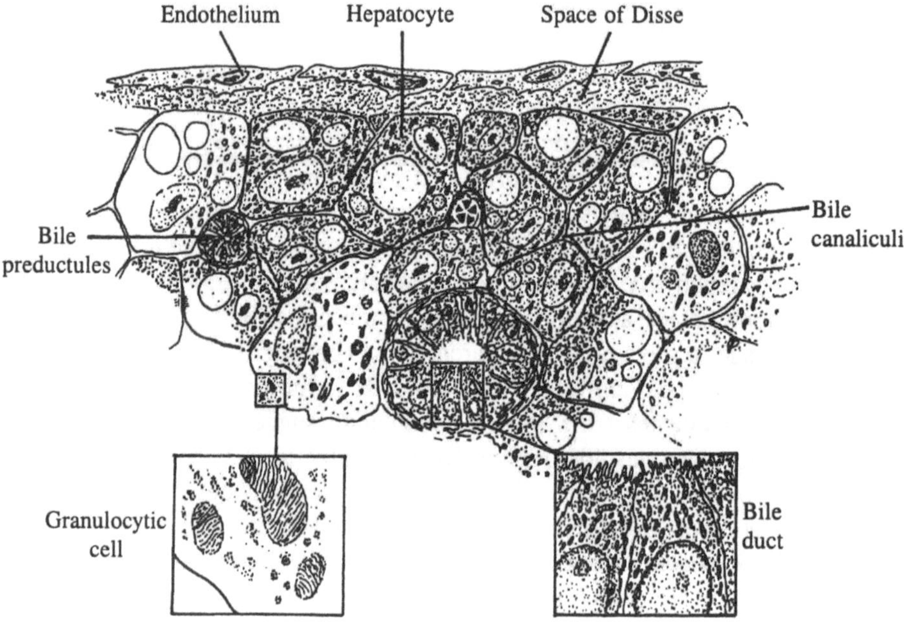

Elasmobranch: Liver ultrastructure.

Fig. 7.47 A composite line drawing of the ultrastructural details of the elasmobranch liver. The sinusoidal endothelium is separated from the liver parenchyma by the space of Disse. This space is rich in collagen. Hepatocytes dominate the liver parenchyma. Granulocytic cells are frequently seen. Bile canaliculi, bile preductules and bile ducts are easily discernable. The cells of the bile ducts are columnar with a rich array of apical mitochondria and a Golgi apparatus (inset). The apex of the cells have microvilli and the lateral boundaries form elaborate interdigitations.
***Sources:* Hamlett and Stoner (1987) and Hamlett (1988)**

feature of the acinar cells is the well developed rough endoplasmic reticulum, which appears predominantly under the vesicular form. The argentaffin cells are found lodged between the acinar cells and duct cells, in the connective tissue they are isolated principally that surrounds the ducts. The typical granules are the cytoplasmic component which characterize the argentaffin cells. The intermediate cells are characterized by the presence of two distinct granule types: one resembling that found in the endocrine cells and the other resembling the granules of the acinar cells. The centroacinar cells resemble that found in other species (Stipp et al. 1980).

The study of lipid digestion has been hampered by the lack of discrete pancreas in most teleosts.

4. ONTOGENETIC ASPECTS OF NUTRITIONAL REQUIREMENTS

Dabrowski (1986) has covered this aspect. Nutritional requirements of aquatic poikilotherms appear to differ from those of terrestrial homeotherms.

Digestive tract morphology changes during ontogenesis and aspects of fish metabolism, protein synthesis and body growth rate are interconnected. Protein and amino acid requirements given the body mass perspective and the generalization of the protein maintenance requirement in protein requirement for maximum growth was estimated to amount to 5–20%. Several cases of amino acid deficiency symptoms showed strong dependence on fish weight (age), but even

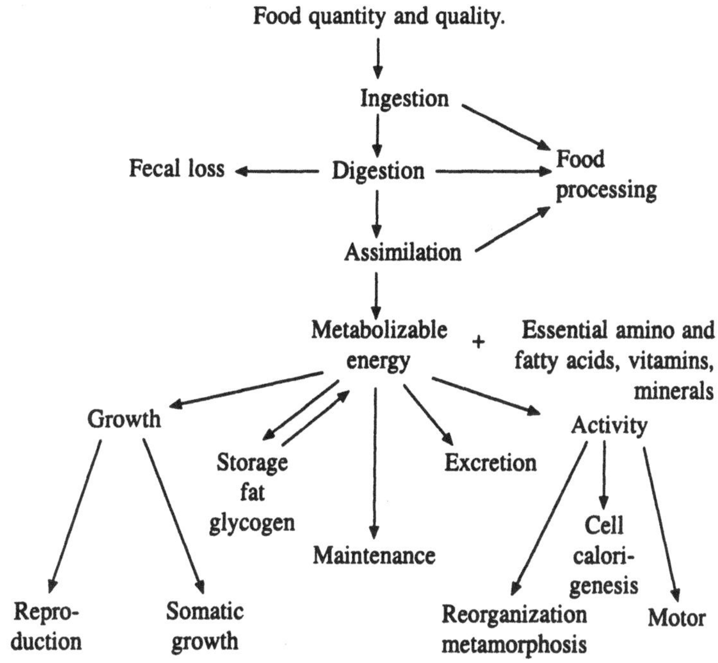

Fig. 7.48 Generalized pathways for use of food and metabolizable energy in animals. ***Source:*** **Eales (1988)**

most numerous studies on salmonids are lacking complete research throughout the life history of one species in defined nutritional and environmental conditions. Deficiency symptoms for amino acid reported are eye cataracts, caudal fin erosion, scoliosis and lordosis at different timings.

Larval and juvenile fish have reduced capacity of catabolic adaptability and this fact links them to strictly carnivorous animals.

Young fish are much more susceptible to lack of vitamins in food than are juveniles or subadults. This appears to be the case for vitamins B6, E, C and A. There are virtually no studies on vitamin requirements in larval warm-water fishes and very few on first feeding salmonids. The same applies to vitamin need in reproductive fish.

Fatty acid deficiencies manifest themselves faster in juvenile fish, but larval fish might require separate classes of lipids, phosphatidylcholine, in the diet to develop and grow at all. Growth depression and mortality is caused by fatty acids deficiency.

The deficiency of minerals (e.g., P, Zn, Mg, Se) leads to poor growth (dwarfism), and even lens cataracts. On comparing the deficiency time for the same minerals, it seems that the warmwater catfish is much more susceptable than salmonids. However, carp seems to be more resistant to Zn deficiency than Mg, when the inverse seems to be true in salmonids. For attention on Mg, read van der Velden (1991) and van der Velden et al. (1991).

4.1 Vitamin D

Teleosts are known to require vitamin D but the physiological function of this vitamin in teleost calcium metabolism is but poorly understood. One reason for this lack of understanding is that teleosts store large quantities of this vitamin and as a result it has not been possible to obtain teleosts which could be considered as vitamin D deficient. There are, however, sufficient data available to indicate that vitamin D is involved in calcium metabolism in both fresh water and marine teleosts. Investigators have reported hypercalcemic effects of various vitamin D compounds in different fresh water teleosts. There are less data on the physiological role of these seco-steroids in sea water teleosts in spite of the fact that the highest concentration of vitamin D_3 occurs in the liver of these fish and that a number of seco-steroids were identified in the plasma. Recently, a report has appeared on the presence of both vitamin D_3 and 1, 25-(OH_2)-D_3 together with specific plasma binding proteins for these seco-steroids in the Atlantic cod (*Gadus morhua*). Also are found specific high-affinity receptors for these seco-steroids in the intestine and the gills, which are well known to perform important functions in calcium regulation of fishes.

Fenwick et al. (1994; see references) reported that both D_3 and 1, 25- $(OH)_2$-D_3 can have calcitropic effects in the marine Antarctic teleost, *Pagothenia bernacchii*, and these two forms of vitamin D can exert different effects within the same species. The Notothenioidei apparently have poorly calcified bone, a weight reducing adaptation thought to be associated wtih the development of neutral buoyancy in these swim bladderless fish.

5. NUTRITIONAL ENERGETICS

The generalized pathways for use of food and metabolizable energy in animals is given in Fig. 7.48. The topic has received a critical attention from Smith RR (1989). Fish, like other animals, require energy to sustain life. This energy is derived from that stored in the chemical bonds of the feed they eat and it is released when these bonds are broken by oxidative reactions. The energy available to the fish is dependent on how well they digest the variety of ingredients presented to them. To the fish, waste heat has little use and is rapidly lost to the surrounding water. Some fish, like the tunas, may use this waste heat to elevate the digestive tract temperature to accelerate digestive processes.

Fish are among the most efficient of animals in converting feed to body tissue. Several factors contribute to high energetic efficiency of fishes and some other aquatic organisms. Fish have an efficient mode of transportation. Their bodies are supported by the water and they do not need the antigravitational muscles of land animals. A streamlined body moving through the water is the most efficient form of transportation.

There are species differences in the energy used for travel. Some fish move no more than is necessary to find feed. Others are actively swimming all the time. Adult salmon may travel hundreds of miles upstream in their spawning runs. These fish do not feed after leaving salt water. All the energy for travel and gonad maturation must come from body stores.

As far as is known, freshwater fish do not attempt to maintain a body temperature, which is different from the water temperature. Fish excrete about 85% of their waste nitrogen as ammonia down a concentration gradient through the gills into the water with little or no energy cost. The energy cost of detoxification, concentration, and excretion of waste products appears as part of the heat increment. Fish require little energy per unit of protein than do mammals or birds.

The demands for energy for maintenance and physical activity must be satisfied before any energy is available for growth. Excitement and associated activity caused by external disturbances will waste energy. Temperature probably has the greatest effect of any of the factors on energy requirements. The metabolic rate of fish is reduced when the water and consequently the body temperature of fish is reduced. This reduction continues until the lower lethal limit is reached and the fish dies. This lower limit differs with species. For some species it may be as high as 10°C and for others it is below the freezing point of fresh water. This adaptation to cold makes it possible for some fish to survive in waters which may be covered with ice nine months out of the year. There is little food available and oxygen is low, but metabolic rate is reduced to the point where little is required.

On the high end of the temperature scale the body temperature of the fish rises with the water temperature. Metabolic rate increases, appetite increases, and growth rate increases until the upper critical temperature is reached, at which it can no longer function properly. Food intake is reduced, growth rate is reduced, and metabolic rate remains high. As the temperature is further elevated, death occurs, usually as a result of hypoxia or enzyme failure. Again this lethal temperature varies with the species. With some species it may be as high as 40°C.

Although no widely accepted value has been established, it is clear that small fish have a higher energy requirement per unit weight than do large fish of the same species. It is difficult to make comparisons if metabolic rate and hence maintenance energy requirements are given only as kcal/kg and the size of the fish is not considered.

Stress, whatever the source, will increase the energy requirement and reduce the growth rate. Any energy used to combat stress will not be available for growth. Physical and chemical stress can come from accumulation of waste products, low oxygen, crowding, handling, external disturbances, water pollution, poor-quality feed, and many other sources.

Energy is required to synthesize body protein and fat from the available amino and fatty acids. A rough generalization can be made that about one-third of the energy of the feed will be lost as combustible waste. This will be uneaten feed, faeces, and urine and gill excretions. One-third will be oxidized to supply the energy needs of the fish, and one-third will be returned as fish growth.

Fats, carbohydrates, and proteins can all be used by the fish as a source of energy. Protein is usually the most expensive, and nonprotein sources of energy should supply most of the energy needs. When protein is catabolized for energy, the amino acids are deaminated and the ammonia is excreted. The excretion of excess ammonia poses no problem for fish because of the efficient nitrogen excretion ability. The question of whether to supply energy in the form of protein or nonprotein is strictly economic, and it is sometimes less expensive to add extra protein to obtain the essential amino acids and energy than it is to supplement diets with these amino acids. The excess protein is used efficiently for energy.

Fats contain more energy per unit weight than any other biological compound. Fats are used efficiently to supply energy. The diet of carnivorous fish in natural conditions on the dry basis is about 50% protein and about 50% fat. Fat increases the palatability of feeds. Fat also reduces dust problems and is an aid in pelleting upto 11% of the diet. If more than this fat is desired it is applied to the feed after pelleting.

There is very little carbohydrate in natural diets of carnivorous species. The herbivorous species consume much plant material with a high content of carbohydrate. Carbohydrate molecules range from simple sugars which are readily digestible by all species to the complex carbohydrates, cellulose and lignin which can be digested only by bacteria. Carbohydrate is the least expensive source of energy. It is therefore economically advantageous to use as much carbohydrate as possible in fish diets. There is some controversy of how much carbohydrate fish can effectively use.

Detailing further, it is noteworthy that food categories vary in their contents of essential amino acids. Omnivorous fishes utilize animal food for growth and plant food for metabolic energy. The grass carp (*Ctenopharyngodon idella*) grows ten times faster on animal food, although its diet in the natural habitat consists mostly of plant material. For many herbivorous fishes, plant food supplies sufficient metabolic energy but traces of animal material are required for growth. Fishes that are herbivorous as adults are often carnivorous in the juvenile stages of their life history; this is probably because amino acids and other essential nutrients are necessary for early stages when development and growth are most important. As the fish matures and its growth rate slows, the fish can maintain itself on plant food. However, many of these herbivorous fishes will readily accept animal food if made available.

Coral tissue is abundantly available throughout the year on coral reefs, and the process of using it as a major source of food energy provides additional feeding niches, thereby contributing to the potential diversity of coral reef fishes and the carrying capacity for fishes by the coral reef habitats. Coral tissue appears to be an adequate source of food energy by its abundance, but chaetodontids tend to feed on a variety of scleractinians and must still occasionally obtain a few worms, crustaceans or fish eggs to augment this diet with essential nutrients. Although *Chaetodon capistratus* and *C. aculeatus* obtain most of their food energy from different sources, both may be limited in growth or reproduction by the same essential nutrients (Birkeland and Neudecker 1981).

Fris and Horn (1993; see literature) provided the first evidence that a temperate zone fish (e.g., stichaeid *Cebidichthys violaceus*) can grow on a strictly seaweed diet. Further, it has been unequivocally shown that a marine fish, from either temperate or tropical waters, can grow on an all-plant diet. Results showed that gut retention times of *C. violaceus* are directly related to dietary protein content. Results contradicted statements that plant-eating fishes must consume animal material to obtain enough protein for growth.

Further, it should be noted that food conversion efficiency and energy conversion efficiency in fishes has been shown to increase with an increase in protein content of the diet.

Calculations of net protein utilization for a variety of fishes and other vertebrates show that the values decrease significantly with increasing dietary protein.

Many authors have claimed that fishes have higher protein requirements than terrestrial vertebrates. This difference is based on evidence that the proportion of protein in diets that elicit maximum growth in fish is higher than that providing maximum growth in terrestrial vertebrates.

The dictum of higher protein requirements for fishes needs to be re-evaluated, at least with respect to herbivores.

That the nitrogen excretion rate in *C. violaceous* did not increase with an increase in dietary protein content may be explained at least in part by the marked decrease in food consumption rates with increase in dietary protein levels (see Horn et al. 1995).

6. GROWTH

Starting with the fertilized egg, the growth of a fish can be divided into a series of stanzas, marked off by developmental or ecological crises such as hatching, transformation from larva to fingerling, migration from salt to fresh water or vice versa, or maturation. Ricker (1979) discussed the growth rates and models. The reader is also referred to Ursin (1979) for principles of growth in fishes and to Brett and Groves (1979) for physiological energetics (read also Jobling 1999).

Osse and van den Boogart (1995) described the development and also metric growth of fish larvae. It has been concluded that the pattern of the development and growth shows in many cases a close parallel with the successive functional requirements (see Table 7.5).

Table 7.5 Relations between egg diameter and (total) length of larvae at hatching, first branchial ventilation, first feeding and reaching the juvenile stage arranged according to egg diameter.

		Total length (mm of larva at			
	Egg diameter (mm)	Hatching	Branchial ventilation	First feeding	Reaching the Juvenile stage
Tinea tinca	1.2	4.8	5.6	7.2	16.0
Cyprinus carpio	1.9	4.5	5.5	5.8	21.0
Hucho hucho	4.5	13.7	15.6	25.0	39.0
Salmo nykiss	5.9	10.4	29.0	26.0	30.0
Oncorhynchus keta	7.0–7.3	20.0	19.5	40.0	40.0

Source: Osse and van den Boogart (1995)

In order for the commercial production of fish to be economically viable the fish should have high survival rates, grow rapidly and make efficient use of food provided. From the marketing viewpoint, it is also desirable that the harvested fish are of uniform size, are of attractive appearance and reach high standards of nutritional quality. Numerous environmental factors are known to influence feeding and growth (Brett 1979; Jobling 1993), but in order to provide the fish with rearing conditions that ensure optimal growth and effective production, it is essential to have information about the ways in which the feeding and growth performance of individual fish are influenced by changes in various biotic and abiotic factors. Meunier et al. (1994) carried out skeletolochronological studies of various species of *Arius couma* living in French Guiana. They all seem to show, every year, two fast growth periods alternating with two slow growth periods that correspond respectively to rainy and dry seasons. It is possible the amount of food, abundant during the rainy season, that regulates growth performance. Van Rooij et al. (1995) worked on the growth of different life phases and social categories of the protogynous parrotfish *Sparisoma viride* and suggested that 'bachelors' trade growth against current reproduction and thereby enhance their chance to acquire the status of a territorial male with high reproductive success. Shearer et al. (1997) found that food intake, and possibly growth could be retarded if a feeding regime results in high body fat levels in juvenile chinook salmon (*Oncorhynchus tshawytscha*).

Growth has been explained in different ways:

(i) If growth, food consumption and the losses associated with maintenance are all measured in energy units then growth is defined as:

$$\text{Growth} = \text{In} - \text{Out}$$

Growth of a fish can be conceived as primarily a problem of internal control at the cellular level or at the organ level. It can also be conceived as primarily a problem of interaction with the environment.

(ii) Growth can be defined as the change in size (length, weight) over time or, energetically, as the change in calories stored as somatic and reproductive tissue. The energetic definition is particularly useful for understanding the factors that affect growth in fishes because ingested food energy (*I*) measured in calories, must emerge either as energy expended for metabolism(*M*) or growth (*G*), or energy excreted (*E*). This can be simply expressed in the equation.

$$I = M + G + E$$

Metabolic energy expenditures can include calories expended for body maintenance and repair, for digesting food, and for movement. Excreted energy can take the form of faeces, ammonia and urea, and the small quantities of mucus and sloughed epidermal cells of the skin. The remaining factor in the energy equation is growth.

(iii) Partioning of biological energy has been described by Smith (1989). The energy intake of an animal is divided among many functions within the animal body. There are many places where energy is lost between intake energy and a useful product. Losses occur as combustible products and as heat. The goal of the fish producer is to keep losses as low as possible and thereby maximum returns of a useful product.

Energy equations

The first law of thermodynamics, conservation of energy required that

$$\text{IE} = \text{FE} + \text{GE} + \text{UE} + \text{ZE} = \text{SE} + \text{HE} + \text{RE}$$

(IE = Intake of food energy; FE = Faecal energy)

Energy flow in animal system

The idealized flow of energy through animal system is given in Fig. 7.48. There may be quantitative differences because of the ability of some species to consume and digest feeds high in fibre. The magnitude of the losses as faecal energy, urinary and gill waste energy, and heat production are dependent on the diet and the level of feeding. The inclusion of fibrous materials which are poorly digested by fish will increase the loss through faecal energy. The energy lost as combustible gas, which is large in some animals, is very small in fish and can usually be neglected in calculations. The waste energy excreted through the urine or gills represents absorbed nutrients that cannot be metabolized by the fish. Recovered energy is that as growth of the fish or sexual products.

The demands for energy for basal metabolism and voluntary activity have priority and retained energy may be a positive or negative quantity. If intake energy is not sufficient to fill these needs, body stores are used and the animal loses weight.

Basal metabolism: It reflects the need to sustain life processes of an animal in the fasting and resting state. This energy is used to maintain vital cellular activity, respiration, and blood circulation and is referred to as basal metabolic rate (BMR).

Standard metabolism: It is the minimum rate observed for well-acclimated unfed fish in the midpoint of their temperature range. This is roughly equivalent to basic metabolism. Standard metabolism has also been estimated by measuring metabolic rate at two levels of activity and extrapolating to zero activity.

Energy utilization: The energy intake of an animal is divided among many energy-requiring processes. The magnitude of each fraction is dependent on the quantity of energy intake and the ability of the fish to digest and utilize that energy. The requirement for basal or standard metabolism is fairly constant and is a function of species and temperature. Voluntary activity would probably be greater in fish which were well fed than in starved fish. The heat of nutrient metabolism is roughly proportional to feeding rate as also are urinary and gill excretions.

Effect of body size on energy requirement

The relationship of body weight to metabolic rate in animals can be described by the general equation:

$Y = aW^b$, where Y is the metabolic rate, W is the body weight, a is a constant which is dependent on species and temperature. The exponent b describes the slope of the regression line of the log of weight to the log of the metabolic rate. In mature mammals $Y = 70.5\ W^{0.75}$, where Y is kcal/day and W is in kg. The value of the exponent for fish has been described as ranging from 0.50 to 1.00.

Although no widely accepted value has been established, it is clear that small fish have a higher energy requirement per unit weight than do large fish of the same species. It is difficult to make comparisons if metabolic rate and hence maintenance energy requirements are given only kcal/kg and the size of the fish is not considered.

*Equations and definations used in energy distribution scheme.

$$\text{Digestible Energy (DE)} = \text{IE} - \text{FE}$$

$$\text{Metabolizable Energy (ME)} = \text{IE} - (\text{FE} + \text{UE} + \text{ZE})$$

$$\text{RE} = \text{ME} - \text{HE}$$

Apparently digestible energy (DE) is energy in food consumed minus energy in faeces:

$$\text{DE} = \text{IE} - \text{FE}$$

Metabolizable energy (ME) is the energy in the food less energy lost in faeces, urine, and combustible gas:

$$\text{ME} = \text{IE} - (\text{FE} + \text{UE} + \text{GE}), \text{ For fish,}$$

ME = IE – (FE + UE + ZE). Gaseous energy need not be considered in fish measurements.

Gill excretion energy (ZE) is the gross energy in the combustible compounds excreted through the gills of aquatic animals.

Recovered energy (RE), commonly called energy balance, is that portion of the feed energy retained as part of the body or voided as a useful product.

Total heat production (HE) is the energy lost from an animal system in a form other than as a combustible compound.

Articles by Jobling (1983a, b), Boeuf and Le Bail (1999) and Boeuf et al. (1999) deserve special mention for the understanding of different aspects of growth studies. The scientists at the Norwegian Institute of Fisheries and Aquaculture, Tromso and at Norwegian College of Fishery Science, University of Tromso, Tromso, have in recent years, focused on a broad spectrum of growth (see Jorgensen and Jobling 1992, 1993a, b; 1994; Arnesen et al. 1993a, b, 1994; Dos Santos et al. 1993; Jobling et al. 1993a, b, c, 1994; Jobling and Baardvik 1994; Jobling and Koskela 1996; Koskela et al. 1997, 1998; Johansen and Jobling 1998). The reader is also referred to Lemm et al. (1993).

Salmonids in general, and Atlantic salmon (*Salmo salar*) in particular, are considered to be visual feeders, feeding predominantly during the hours of daylight. Studies carried out on rainbow trout (*Oncorhynchus mykiss*) and Arctic charr (*Salvelinus alpinus*) have demonstrated that substantial proportions of daily food consumption may be eaten during the hours of darkness, indicating that there may be interspecific differences in feeding patterns displayed by salmonid fishes.

A number of studies have demonstrated that there is a positive correlation between the daylength and growth in Atlantic salmon, possibly as a result of increased food intake as days become longer. The relationship between photoperiod and growth may depend upon a series of physiological interactions, entrained by photostimulated neuroendocrine activity, but growth of a visual feeder may also be limited by the number of hours when food is available. If this is the case, the parr-smolt transformation may also be impaired by limiting food availability, since both small body and depressed growth rates have been shown to delay smolting in salmonids.

Amongst fish species, compensatory growth has been reported for salmonids, cyprinids and pleuronectids. It has also been reported that Atlantic cod may display a pronounced compensatory growth following the return to good feeding conditions after a period of nutritional deficiency.

Gonadosomatic indices tended to be larger in cod, especially the males that had experienced 'nutritional stress'. Whilst increased investment in reproduction by fish subjected to poor feeding conditions has been found in some studies, other data tend to equivocal, and there is a lack of consensus..

Sustained exercise can result in a number of physiological and behavioural effects in salmonid fish. For example, improved growth has been reported in juvenile brown trout, post-smolt and adult Atlantic salmon, juvenile rainbow trout, juvenile brook trout, and fry and juveniles of Arctic charr, subjected to intermediate velocities at flow rates between the critical swimming speed of the fish. Improved growth appears to be associated with red and white muscle hypertrophy caused by the physical stimulation of the muscles.

There are some indications that exercising salmonids display better food conversion than fish performing routine activity and in recent experiments with Arctic charr, in which food intake was measured with great accuracy, this has been confirmed.

It has been suggested that fish which are exposed to moderate water currents exhibit schooling behaviour, display less agonistic encounters and have a reduced tendency to form hierarchies. The development of such hierarchies may affect the pattern of growth, leading to an increased variation in fish size within the group, elevated metabolic rates and stress.

Exercise in juvenile Atlantic salmon affected the behaviour of the fish, appeared to result in reduced levels of agonistic activity, and the rate of fin wound healing seemed to increase with increasing swimming speed. Exercise did not, however, lead to any improvement of the osmoregulatory capacity of the fish undergoing parr-smolt transformation.

Growth performance of Arctic charr, *Salvelinus alpinus*, is depressed at low (20 kg m^{-3}) stocking densities and growth has been reported to reach a maximum when fish are stocked at densities of 50–70 kg m^{-3}. Growth rates may, however, be high when fish are stocked at densities in excess of 100 kg m^{-3}, provided that good water quality can be maintained. Findings show that Arctic charr may feed and grow well in complete darkness, and also provide evidence that feeding in darkness eliminates the negative effects of low stocking density on growth. Seasonal variations in feeding and growth have been reported for a number of fish species including

Arctic charr *Salvelinus alpinus*. There is generally a marked increase in growth in spring and a depression during the autumn. Among salmonids this growth pattern may be influenced by the annual cycles of parr-smolt transformation and reproduction. The annual cycles are thought to reflect endogenous circannual rhythms which are controlled by the seasonal photoperiod cycle.

While protoperiod appears to be most important factor governing the timing of parr-smolt transformation, other factors may also affect the performance of salmonid fish following seawater transfer. Chronic stress induced by crowding, or high levels of intraspecific social interactions may interfere with the parr-smolt transformation, and an abrupt transfer from fresh to seawater may act as a severe stressor. Both appetite and growth were suppressed for more than 30 days following the transfer of Atlantic salmon smolts from fresh to seawater.

Taking salmonids, some studies on the effects of salinity on growth and metabolic rate appear to support the hypothesis that the energetic costs of osmoregulation are lowest in an isotonic medium (10 ppt) and that the energy savings are sufficient to result in increased growth relative to that seen in both hypo-and hyperosmotic media. Other studies have failed to reveal significant differences in growth and metabolism in fresh- and iso-osmotic water. It has been reported that growth was highest in freshwater which declined with increasing salinity. Metabolic rates increased with salinity and, thus, were inversely correlated with growth rates. Thus, the effects of salinity on growth responses seem to vary with salmonid species, ontogenetic stage of the fish, body size, season and the transfer and acclimation procedures. The results of the study with Arctic charr do not suggest that any metabolic costs associated with adaptation to different salinities were sufficiently large to cause differences in growth rates between the different groups of fish.

It has been demonstrated that body size affected osmo- and iono-regulation in Arctic charr both in freshwater and seawater. This effect was manifested as a positive correlation between fish length and plasma ion levels in freshwater and a negative correlation between fish length and plasma ion levels in seawater. It is also well established that hypo-osmoregulatory performance increases with body size in other salmonids. It has been suggested that the Arctic charr displays an endogenous circannual rhythm in seawater tolerance which is synchronized by changes in photoperiod.

It has been indicated that feed-growth relations for Arctic charr were affected by both salinity and season. Gross feed conversion efficiency in Arctic charr was lower in seawater than in freshwater. For further information on this fish, the reader can refer to Saether et al. (1996).

Fatty acid requirements of freshwater and marine fishes are not the same. Dietary sources of n-3 HUFAs with more than 20 carbons play an essential role in the nutrition of marine fishes. The theory that a marine environment results in a critical requirement of n-3 long-chain fatty acids for fishes has some support.

Protein and energy requirements of fish decrease with increasing fish size and increase with increasing water temperature.

Jobling et al. (1995a) discussed monitoring voluntary food intake under practical conditions, methods and applications (refer the Coverage of food intake and growth performance of juvenile Atlantic salmon and Arctic charr). The reader is also referred to Jobling et al. (1995b) on this topic. For special attention on charr, see Jobling (1995). A title 'Fish Bioenergetics' from M. Jobling (1994) deserves a close reading. Comprehensive account of growth and metabolism has been given by Mommsen (1998).

Considerable progress has been made towards increasing the growth rate and food conversion efficiency of a variety of fish species reared under aquaculture conditions. Essentially, two approaches have been adopted, the traditional approach of selective breeding from desirable phenotypes and, recently, the application of biochemical/molecular technical such as genetic engineering. In all situations, however, the performance of the fish is governed not only by its genetic potential but also by its immediate environmental conditions. Certain environmental factors such as water temperature and food supply have an obvious and major influence on growth rate and food conversion efficiency but are not necessarily stressful in themselves. Other factors such as poor water quality, physical disturbance and social domination of one fish by another can act as powerful environmental stresses, thereby reducing the performance of the fish. It has been demonstrated that a variety of hormonal pathways are responsible for stressed-induced growth suppression in teleost fish. Any attempts to improve the potential for growth by selective breading or by direct gene manipulation must, therefore, take into consideration the fish's environment if the returns from such an approach are to be maximized. Scope still exists for the genetic selection of fish with a reduced stress response (i.e., acceleration of the rate of domestication) and progress along these lines ought to result in improved disease resistance and reproductive performance as well as increased growth rates.

6.1 Scaling of Metabolism

Jobling and Davis (1980) and Jobling (1981) have assessed the influence of feeding on the metabolic rate of fishes. Following feeding, the rate of oxygen consumption of fishes increases to a peak and then slowly declines to the pre-feeding level. Many factors have been shown to influence the magnitude and duration of post-prandial oxygen consumption and these include quantity and/or quality of ration and environmental temperature.

Studies on the scaling of active metabolism in fish have usually involved intraspecific comparisons. Therefore, not only is size a variable, but individual fish are compared at different stages in ontogeny when they have vastly different energy distribution patterns. In particular, small fish must allocate a greater proportion of energy to processing food and allow for a much large scope for growth if they are to grow rapidly.

Total viscera aerobic capacity is also higher than red muscle capacity for small, i.e., fast growing, individuals. This is aerobic capacity associated with food processing, and its negative allometry can account for the observation that postprandial elevations in oxygen consumption can exceed the active metabolic rate for small fish but not for large ones. The use of postprandial metabolic rate as the maximum metabolic rate for small fish should result in a scaling function for maximum metabolic rate which displays the expected negative allometry.

Because different physiological activities are responsible for maximum metabolic rate in different sized fish, no single 'used-based' theory (e.g., locomotor muscle power) should be able to account for scaling of aerobic metabolism (see details: Goolish 1989).

Jobling (1993) discussed comprehensively factors influencing metabolism. Losses dut to metabolic demands usually constitute a large proportion of the energy budget of a fish. These energy costs can be partitioned into the minimal costs required for maintaining basic bodily functions, those associated with activity and those related to the digestion, absorption and processing of food. While this partitioning may be relatively easy to grasp from a theoretical standpoint, it is far from easy to quantify the different components experimentally. This contribution (Jobling 1993) focused further on the factors affecting growth and reproduction.

Houlihan et al. (1993) reviewed the biochemical correlates of growth rate in fish. Food consumption may regulate protein growth in fish through a stimulation of protein synthesis. Protein degradation rates also appear to increase in fish as protein intake increases. Within a species, environmental factors or changes in metabolism may significantly alter food/growth relationships. Little is known about stress and alterations in food/protein synthesis relationships and/or growth/synthesis relationships. Protein synthesis and oxygen consumption are closely linked. It is likely that the free amino acids resulting from digestion will play a key role in the regulation of protein synthesis rates although it is by no means clear how such regulation may be achieved. It is an unsettle point whether all tissues demonstrate the same stimulation in protein synthesis. Certainly food reduction reduces protein synthesis rates in fish tissues. The white muscle emerges as the strongest candidate tissue which could act as a sensitive marker for the growth rate of the whole fish. Protein synthesis rates in the white muscle have been shown to correlate very well with whole-fish growth rates. RNA concentrations have frequently been used as measures of growth rate.

A variety of other correlates of growth rate have been used by various authors. Among them are scale protein synthesis rates and enzyme activities.

Temperature changes affect, among other processes, food intake, maintenance requirements, metabolic rates, enzymic processes, diffusion of small molecules, membrane functions and protein synthesis rates.

7. GASTROINTESTINAL HORMONES (GASTROINTESTINAL ENDOCRINE SYSTEM; ENDOCRINE CELLS; NON-ABSORPTIVE CELLS)

Vigna (1986) has given a fine account of this system. The dual distribution of bioactive peptides in endocrine cells and nerves of gastrointestinal tract of mammals is now well established and emerging knowledge indicates that the digestive apparatus of lower vertebrates is just as complex.

The gastrointestinal endocrine system is unlike most other vertebrate endocrine system, because it is diffuse (a diffuse neuroendocrine system- DNES).

Gastrointestinal hormones play important roles in regulating the chemical traffic in the vertebrate digestive tract. The gut hormones in fishes are given in Table 7.6.

In the dogfish gut, it seem possible that three different and unrelated peptides (bombesin-like, substance P- like and PH1-like) may be present in the same cells.

Co-localization of different peptides seems also to be present in nerve fibres. The recent contribution on this aspect is of Reinecke et al. (1997).

In *Barbus conchonius* gut, coexistence of glucagon- and PP- like peptides is found. The coexistence of PP- and glucagon-like substances appears to be very common through the vertebrate series. Whether the coexistence is caused by the presence of a common precursor for PP- and glucagon-like peptides or that separate precursors, encoded by two different genes, for both peptides are present, remains to be investigated.

In contrast to the relatively well-studied Amphibia, there is scant information available about the effects of CCK/gastrin peptides on osteichthyan gastric acid secretion. The situation in Chondrichthyes is unclear, due to scarce information avilable. The few data support the concept that pattern of regulation of stomach secretion in dogfish, and perhaps the other chondrichthyan species possessing a stomach, resembles the tetrapod more closely than the osteichthyan pattern, further emphasizing the high degree of specialization of osteichthyan endocrine system.

Table 7.6 Occurrence of immunoreactive endocrine cells and nerves in the guts of fishes*

Taxon	Stomach	Intestine	Nerves
Cyclostomes (lampreys)		Glucagon like; Gastrin/CCK; Peptide yy (Pancreatic polypeptide/Neuropeptide); Substance P Vasoactive intestinal peptide; Somatostatin;	Bombesin (gastrin-releasing hormone) Calcitonin, gene-related peptide; Serotonin
Chondrichthyes (dogfish *Scyliorhinus stellaris*)*	Gastrin/CCK; PHI/PHM; Glucagon; Substance P; Bombesin/GRP; PP; PYY; NPY; FMRF-NH_2; Somatostatin; Met-enkephalin	Gastrin/CCK; PHI/PHM; VIP; Glucagon; Substance P; Bombesin/GRP; Neurotensin; PP; PYY; NPY; FMRF-NH_2; Somatostatin; Met-enkephalin;	Gastrin/CCK; PHI/PHM; VIP; Substance P; Bombesin/GRP; FMRF/NH_2; Somatostatin; Met-enkephalin; Galanin
Osteichthyes (teleosts)	Serotonin; Bombesin; Somatostatin; Gastrin; Met-enkephalin; Substance P; Glucagon; PP; Secretin	Glucagon-like; PP; VIP; Neurotensin; Gastrin/CCK; Met-enkephalin; Substance P; GLP; GIP	Bombesin; Enkephalin; Neurotensin; Substance P; VIP

* Specific locations in stomach, intestine and nerves have been skipped off (see sources). For taxa in teleosts, see references.

Legend: GIP (Gastric inhibitory polypeptide); GLP (Glucagon-like peptide); NPY (Neuropeptide tyrosine); PHI (Peptide histidine isoleucine); PHM (Peptide histidine methionine); PP (Pancreatic polypeptide); VIP (Vasoactive intestinal polypeptide; PYY (Peptide tyrosine tyrosine)

Sources: Noaillac-Depyre and Holland (1981), Reifel et al. (1983), Tagliafierro and Faraldi (1984), Cimini et al. (1985, 1989), Noaillac-Depeyre (1985), Tagliafierro et al. (1985, 1989a, b), Vigna (1985), Elbal and Agulleiro (1986), Holmgren et al. (1986), Rombout et al. (1986a), Abad et al. (1987), Elbal et al. (1988, 1991), Nozaki (1988), Boon et al. (1990) and Van Noorden (1990). See for ontogeny, distribution, and differentiation (Faraldi et al. 1988; Chiba et al. 1995).

Little is known about the function of VIP in fishes; in *Gadus morhua* VIP inhibits the gastric acid secretion while in *Salmo gairdneri* it exhibits a dual action on the smooth muscle of the stomach wall; in a cartilaginous fish *Squalus acanthias*, VIP clearly inhibits gut motility. VIP has no effect on rectal gland secretion in cartilaginous fishes. The regulative role in growth and differentiation of intestinal portion and in absorption and secretion has been suggested in *Scyliorhinus stellaris* (read Tagliafierro et al. 1988).

The agnathans (hagfishes and lampreys) have no acid-secreting stomach and this is thought to be a primitive feature of this group. In contrast, the lack of stomach in some other vertebrates (e.g., ratfish, lungfish, some teleosts such as the Cyprinidae) represent a secondary loss.

There is some evidence for the endocrine control of gall bladder smooth muscle contraction in lower vertebrates (see Rajjo et al. 1988).

There are no studies of the endocrine control of pancreatic enzyme secretion of osteichthyans, probably because the gland does not exist as a discrete, compact organ in most bony fish species. Instead, the pancreas is usually diffuse; glandular tissue can often be found along the portal veins, penetrating the liver, suspended in the mesenteries between the intestine and liver, associated with principal islet of endocrine tissue, and dispersed among the pyloric caeca.

The living Agnatha do not possess an exocrine pancreas, and thus seem to represent a stage of vertebrate evolution before the appearance of this gland as a separate organ. However, the modern Agnatha, the cyclostomes, do have zymogen cells scattered in the intestinal epithelium which are considered to be homologues of the pancreatic acinar cells of higher vertebrates.

The actions of hormones of cholecystokinin/gastrin family of peptides have been widely studied. It has been concluded that regulation of pancreatic enzyme secretion is a primitive action of cholecystokinin and that the sensitivity of gall bladder muscle and gastric acid-secreting cells to these peptides evolved later, possibly in the ancestral lineage that led to the gnathostomes. The need for increased digestive efficiency to support the higher levels of activity made possible by the evolution of jaws is suggested as a strong selection pressure leading to this pattern of evolution of endocrine regulation of gastrointestinal functions. It may be mentioned here that enteroendocrine cells have been reported in some fishes. Argyrophil cells have been reported in several species. Serotonin- containing or argentaffin cells have not been reported in teleosts or cyclostomes with the exception of *Anguilla occidentalis australis, Salmo trutta S. irideus*, and *Mugil auratus*. Several enterohormones (secretin, cholecystokin, pancreozymin) have been found in the intestine of teleosts (see details: Rombout 1977). The role of peptides in the control of feeding has been discussed by Mommsen (1998). The recent contribution in this direction is from Tagliafierro (1999). To get details on the neuroendocrine system in the intestinal tract and pancreas in Antarctic fish, read Tagliafierro et al. (1998).

8. SUPPLEMENTAL FUNCTIONS OF GUT

Suction, taste, respiration, buoyancy, sound generation, bioluminescence, ionic regulation, endocrine regulation, immunity, defence, parental care are some other functions associated with gut (see details: Kapoor and Khanna 1993).

9. AGE DETERMINATION

9.1 Elasmobranchs

Little is known about the age, growth, and reproduction of elasmobranchs because many species are difficult to sample, are of relatively large size, are highly mobile, exhibit seasonality, and are of minor commercial value. In addition, many of the convential age determination methods used for bony fishes are not applicable to elasmobranchs because elasmobranchs lack calcareous otoliths and other skeletal hard parts.

Centrum edge analysis

Elasmobranchs have been aged using growth zones within the vertebral centra by several methods. These growth zones are comprised of opaque bands that have cells with high concentrations of calcium and phosphorus and translucent bands that are less mineralized (Fig. 7.49).

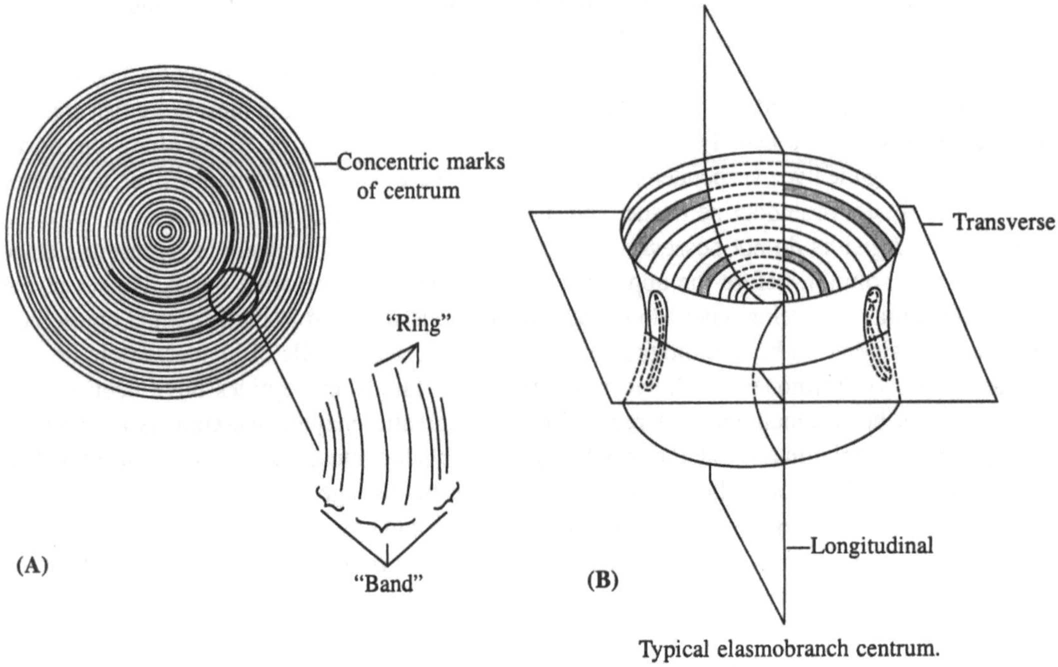

Fig. 7.49 Diagram of typical elasmobranch centrum showing (A) bands mode of fine "rings"; and (B) the two sectioning planes used.
***Source:* Cailliet et al. (1983a)**

Centrum edge characterization is an inexpensive and useful means of age verification but can only be used if the centrum edges can be elucidated and read. In cases where centrum edges are difficult to read, other methods are available for verifying age estimates and ascertaining the temporal periodicity of band deposition.

The most successful technique was X-radiography for study of bands.

Growth curves are fitted independently for females and males of sharks *Mustelus californicus* and *M. henli* using *von Bertalanffy growth equation*:

$$Lt = L\propto (1-e^{-k(t-to)})$$

where: -

Lt = total length at time t

L ∝ = asymptotic length

k = growth constant

to = theoretical age at zero length

Factors that mediate the temporal periodicity of calcium deposition in elasmobranch centra are not known. Little attention has been given to how external factors such as day length, temperature, salinity, diet and/or metabolism mediate opaque and translucent band formation in elasmobranch fishes. Virtually no studies have addressed the hormonal regulation of calcium deposition into the vertebrae.

Radiometric dating

This technique uses relative radioactivity levels of naturally occurring isotopes from different parts of a vertebral centrum to estimate age.

Laboratory and field growth with tetracycline

The combination of known growth with tetracycline injection is essential to any age validation study. This approach provides somatic growth information along with a discrete mark in the calcified structure, which enables a direct comparison of growth zone deposition with time at liberty. The direct nature of this approach has made it popular, despite the problems associated with laboratory and low probability of recapturing organisms that have been in the wild a sufficient amount of time (see Caillet et al. 1983a; read also 1983b; Martin and Cailliet 1988; Yudin and Caillet 1990; Caillet 1990).

New methods of age determination and verification have been described by Caillet and Tanaka (1990).

Image analysis and computer-aided ageing

With the advent of personal computors, microscopy has been coupled with computor and video monitor technology to aid researches in analyzing growth information from calcified structures. These systems generally use image analysis to increase objectivity and speed in estimating growth. This assumes, of course, that objective criteria for growth zones can be established at the start of the project to insure interinvestigator consistency. Also, this approach requires calibration and standardization procedures for each species and structure studied.

Electron microprobe and elemental microanalysis

The technology to analyze concentrations of elements in calcified structures recently has been developed and is an additional method of verification of growth zone periodicity. This technique can be used to assess chemical differences that may be correlated with visually identified growth zones in the calcified structure. While this approach has been applied to bony fishes, it has rarely been used to evaluate elasmobranch fishes (see also Cailliet and Radtke 1987).

Stable isotopes

To assess the growth history of individual organisms and the potential influence of environmental factors on growth, stable isotopes have been analyzed in calcified structures of shellfish and bony fish. This approach has not been used for elasmobranchs, but certainly has promise.

Tissue protein concentrations

Certain proteins are hypothesized to accumulate in the tissue of organisms. One of these, lipofuschin, has been proposed as an independent index of age because it appears to accumulate at a predictable rate and is not resorbed over the life of an organism. While experiencing methodical difficulties in its initial use as an age-determination technique, this approach has distinct promise for ageing large, fast-moving and relatively rare organisms, such as elasmobranchs and billfishes.

Precision and accuracy

Several mathematic expressions have been created to evaluate the precision of age estimates. These include the average percent error (APE) and the index of precision (D); both have been used in studies of bony fish growth, but seldom have been applied to elasmobranch studies.

Tagging studies

One of the most successful age validation methods is tag and recapture analysis. This enables one to measure the specimen, remeasure it once it is recaptured, and thus calculate growth over a known time period. Additionally, marking the calcified structures of the organism with oxytetracycline (OTC) allows an evaluation of growth zone deposition periodicity. Although this approach has been successfully used with many fishes, its use with elasmobranchs has been minimal.

9.2 Teleosts

The usual means of age determination in bony fishes, by examining scales, otoliths, or bones are not applicable to elasmobranchs. However, the evidence that does exist indicates that growth rates in sharks may be slow, compared with many teleosts, with size at first sexual maturity estimated at approximately 60 to 90% of the asymptotic length. Because of these growth and reproductive traits, intensive fisheries directed toward subadults could deleteriously affect the total population size of sharks very quickly, assuming a close relation between stock and recruitment. Without information on size and age at which reproduction first occurs, effective management measures, such as setting size restrictions, would be difficult to implement.

Edge characters in scales and otoliths

The use of scales and otoliths for age determination presupposes that annual changes are reflected in the edge structure. As fish grows, the sclerites (circuli) along the scale edge are widely spaced and can be traced around the entire, or nearly the entire, anterior part of scale (open scale edge). When fish growth is retarded or terminated, the sclerites formed are norrowly spaced and usually have a dead end (dense scale edge) (Jonsson 1976).

Teleosts generally have three pairs of otoliths, large sagittae, and smaller lapilli and asterici, located in the inner ear and composed of calcium carbonate (usually in the aragonite morph) deposited on an organic matrix composed largely of keratinlike protein, termed otolin, which is rich in aspartate and glutamate residues. Biological information is stored in various aspects of the ultrastructure and chemistry of otoliths. For example, the shapes of otolith, particularly the sagittae, have been used to identify species; the chemistry of otoliths has been used to recover information about the fish environment; but most frequently otoliths are used to determine the age of fishes.

Three kinds of periodic structures are present in otoliths, all of which have been used to estimate fish age: opaque and hyaline zones, daily growth increments, and check rings. A convergent series of opaque and hyaline zones proceeding from the nucleus to the edge of otoliths is often observed at low magnification by either transmitted or reflected light (and often after heating for optical enhancement). The patterns formed by such series of zones are often interpreted to give the age of the fish. Daily growth increments have been used to estimate the age of juveniles and some adult fish. Unvalidated microscopic growth increments, assumed to be daily, have been used as a basis for age estimation in other species of fish. Daily growth increments appear to form as a consequence of diurnal rhythms in the mobilization of calcium and neuropeptide production. The widespread occurrence of daily increments in teleost otoliths has greatly benefited studies of growth in a number of short-lived species, by enabling the accurate determination of both age and the timing of life-history events. However, validation of daily increment periodicity is necessary for each new species and life-history stage studied, since increments may not be deposited daily in all species or life-history stages. Further, increments may not be easily discernable or formed daily in slow-growing individuals. Interruptions in increment deposition may also lead to inaccuracies in the growth record. The presence of prominent sub-daily increments too also make it difficult to discern the fundamental rhythmicity of increment formation. The check rings of otoliths have a more complicated origin.

A variety of techniques have been developed to reveal increments and interpret growth patterns in otoliths. These include the "crack and char" method, staining, thin section and the use of multiple thin sections embedded in pigmentd resin. Methods have been developed to examine scales and use semi-automated techniques for interpretation of scales and otoliths. X-ray techniques and chemical analytical procedures have been used to examine the internal structure of calcareous skeletal structures and otoliths. The presence of daily increments in otoliths provides a time-consuming but potentially precise method for validating gross structures such as seasonal growth increments and age determination, particularly where the seasonal increments are difficult to interpret. The examination of daily increments also offers the opportunity to quantify short-term changes in growth patterns and is of value when investigating larval development. Chemical markers such as tetracycline or biological stains have been used to time-mark seasonal growth patterns and acetazolamide can be used to introduce checks among daily increments. Thus far it has not been possible to determine which are the most effective methods for determining the age of individual species, because few validation experiments have been undertaken to confirm and compare procedures. In addition, the requirement for precision will be different depending on whether the data is to be used in a scientific experiment with a small number of fish, or used for a gross estimation of population age structure in a commercial fisheries survey. In the former case, otoliths are the preferred structure to be examined with validation undertaken by the use of micro-increment analysis. In the latter, estimates of age taken by evaluating scales might be more practical because a large numbers of fish could be sampled rapidly. However, estimates of age based on interpretation of scales must be viewed with caution because these underestimate the true age.

Although the check rings of fish otoliths play a major role in the age determination of many species of fish, and their subsequent management, there is very little information published on either the composition or ultrastructure of otolith check rings. One view based on staining techniques was that check rings were simply very narrow, protein-rich opaque zones, and that perhaps through a very slow growth of the otoliths, annual opaque zone became compressed to check rings. Recently, it has been shown that check rings are due to changes in crystalline structure, but there are no accessory check rings (i.e., check rings that do not fit in with the overall zonation pattern) (Gauldie 1990, 1991; Gauldie and Nelson 1990; Gauldie et al. 1990). For growth rates and model, the reader can consult Ricker (1979). Recent studies related to otoliths are of Wright et al. (1992) and Wright and Huntingford (1993). For age determination in Antarctic fish, see White (1991).

Age is in essential parameter in the study of population dynamics of commercial aquatic species. In fish, age (in years) has been determined by examining hard tissue such as vertebrae, operculae and otoliths. Age determination of larvae and juvenile stages in the wild can facilitate the study of the influence of biotic and abiotic factors on growth and survival. Thus ageing of early stages of fish with an accurate daily temporal scale can produce precise estimates of these parameters. However, it is important that due attention should be paid, firstly, to validation of the increments according to a definite time scale, and secondly to factors which might affect the production of the increments, e.g., temperature, photoperiod and feeding periodicity. There is no evidence that temperature is an environmental modifier of otolith increments: deposition has been found not to vary between different temperatures which fall within the range of the metabolic activity of the fish. There are contradictory reports about the effects of photoperiod. Little information is available on the relation between body growth, ring deposition rate and otolith growth. Study by Al Hossaini and Pitcher (1988) confirmed that feeding and photoperiod regimes could affect the daily ring deposition in sagittal otoliths of plaice (*Pleuronectes platessa*) juveniles.

Sagittae, vertebrae, and anal and dorsal fin spines have been utilized for age estimation of Pacific blue marlin, *Makaira nigricans* (Istiophoridae). Although more difficult to collect and interpret, sagittae provide more detailed age information (Hill et al. 1989).

Assessment analyses of exploited fish stock commonly require age and size information. Many widely-used stock assessment methods such as virtual population analysis depend upon ageing. Determination of the ages of the individuals is, however, often difficult, tedious and expensive, in contrast to length information which is easier to collect. Therefore, ageing length-frequency data using statistical methods can have great utility in applying age-based stock assessment methods (see Liu et al. 1989).

9.3 Sturgeons

Growth patterns of clavicles, cleithra, opercles, medial nuchals, dorsal scutes, and pectoral fin ray sections have been compared in white sturgeon *Acipenser transmontanus*. All calcified structures contained concenteric growth zones that increased in number with the size of the fish and were interpreted as annual events. There was a direct linear relationship between size of structures and size of the fish. Pectoral fin sections were the most practical ageing structure in terms of ease of collection, processing, legibility, and precision of interpretation. Age estimates from other structures resulted in poor precision between readers but relatively good intrareader precision. This suggests possible corroborative use with the development of better interpretive criteria and elucidation techniques (Brennan and Cailliet 1989).

CHAPTER 8
Cardiovascular System

1. INTRODUCTION

The structural and functional transformations of the cardiovascular systems in the phylogenetic development of vertebrates bear a close relationship to altered requirements for circulatory transport services. Among these the variable requirements for efficiency in gas exchange and respiratory gas transport have undoubtedly exerted the greatest selection pressures on the phylogenetic development of the cardiovascular system. The most momentous transformation of the heart and major central vessels is closely associated with the transition of vertebrate life from aquatic to aerial breathing.

The cardiovascular system undergoes radical change from the invertebrate to the vertebrate animal. The open-type cardiovascular system commonly found in invertebrates is converted to a closed system. In the evolutionary process the transition from an open to a closed vascular system has, however, been a gradual one.

Fishes are the most successful vertebrate group in terms of biomass and number of species. Widely varying physiological demands are made on fish species occupying different environments and exhibiting different life styles, and the successful provision of an adequate oxygen supply to the tissues is therefore of paramount importance to fish. Studies of the haematology and blood chemistry in different species are of a comparative physiological interest. They contribute to a greater understanding of the variation of blood characteristics in relation to such factors as phylogenetic position, ecological habitat, food selection and mode of life.

The circulatory system has provided important taxonomic characters in the Scombroidei. The arterial branching patterns in the head are relatively constant within the major elasmobranch groups and that some phylogenetic hypotheses can be made from their analysis (Muñoz-Chápuli and Garcia-Garrido 1986).

Circulations of most fishes follow a typical vertebrate pattern: blood is driven by a central pump around a closed vascular system in which exchange takes place in microscopic capillary beds. Nonetheless, the organization of these circulatory systems displays uniquely piscine features in that blood from an undivided heart perfuses the gas exchanger (gill) and systemic circulation *in series* with no intervening pump. Thus the entire cardiac outflow enters the ventral aorta, which divides into arterial arches supplying the gills, and efferent vessels from the gills fuse to supply the systemic circulation, largely by means of a dorsal aorta. Although series coupling of distinct vascular beds (e.g., portal systems) are found throughout the vertebrates. two important features distinguish the coupling of the gill and systemic beds in fishes and these are of major importance when considering circulatory dynamics. First, all of the cardiac output usually perfuses both system in a single circulation; second, the vascular flow resistances of two beds are comparable and a large pressure drop occurs across both systems. The result is a system in which each of the two beds influences the perfusion and haemodynamics of the other and influences the nature of cardiovascular coupling.

The cardiovascular system of fish has been the subject of a number of reviews (Randall 1970; Yamauchi 1980; Farrell 1984, 1993; Santer 1985). The reader is further referred to books: Physiology and Form of Fish Circulation (G.H. Satchell 1991) and Volume 12 of Fish Physiology (edited by W.S. Hoar, D.J. Randall and A.P. Farrell 1992). Two other important contribution are of Satchell (1978) and Butler and Metcalfe (1988). Coucelo et al. (1996) focused on ultrasonography characterization of heart morphology and blood flow of lower vertebrates. "Circulation of body fluids" is the theme of a contribution by Farrell (1991b). The recent overview of cardiovascular system is given by Olson (1998).

2. GROSS MORPHOLOGY, ANATOMY, FINE STRUCTURE, INNERVATION OF HEART

The fish hearts and heart-like organs include the branchial (or systemic) hearts in different fish groups, and the hepatic portal vein heart and the caudal heart in myxinoids. The typical fish heart is often called a 'venous heart' because it only pumps venous blood; the concept of piscine heart as a typical "venous" type is an over-simplified generalization, at least on morphological grounds. It represents the main propulsive organ in fish circulatory system and at the same time is homologous to the pulmonary or main heart in higher vertebrates.

The fish heart is positioned farther anterior than the tetrapod heart and is closely associated with the primary branchial arches. It is always posterior to the gills. Cyclostome gills are further posterior than in other fish and the heart is in a correspondingly more posterior position.

Agnathous fish: The branchial heart lies in a cartilaginous pericardium which communicates with the peritoneal cavity in hagfish such as *Myxine* via a pericardioperitoneal canal, but is completely closed in lampetroid (petromyzontid) cyclostomes (Fig. 8.1). Due to the fact that only a single duct of Cuvier persists into the adulthood, the sinus venosus (which contains smooth muscles in wall) lies in a dorso-ventral plane in the midline. In hagfish it is the left duct of Cuvier that persists, whilst in lampreys the right duct remains. The atrium is made up of a thin network of trabeculae composed of myocardial cells. The ventricle is composed of an entirely trabecular myocardium and is made up of endothelial cells, myocardial cells and interstitial cells. In cyclostomes the "conus" has no cardiac muscle in its wall and is not rhythmically contractile. The hagfish conus is much thinner than that in lamprey, which is probably on account of much lower blood pressures in hagfishes than in lampreys. Specialized pacemaker cells have been reported in the atrium of Pacific hagfish (*Eptatretus stouti*). The heart index (heart weight/body weight × 100) of lampetroid cyclostomes is high, e.g., 0.59% in *Petromyzon marinus*, but in myxinoids 0.16 in *Myxine glutinosa*; it is comparable with that of sedentary teleosts. The high heart index in lampetroids is similar to that of mammals and higher than in any other poikilothermic vertebrates. It is suggested that this reflects a high oxygen utilization by *P. marinus*. The distinctive feature of cyclostome heart is the occurrence of many cells containing

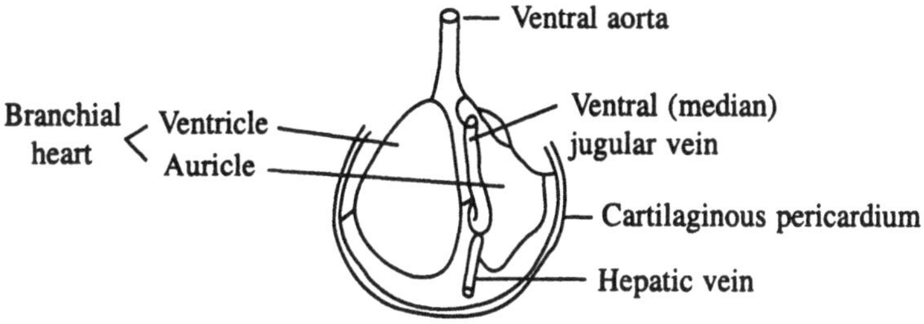

(A) Branchial heart of *Lampetra fluviatilis* (adult).

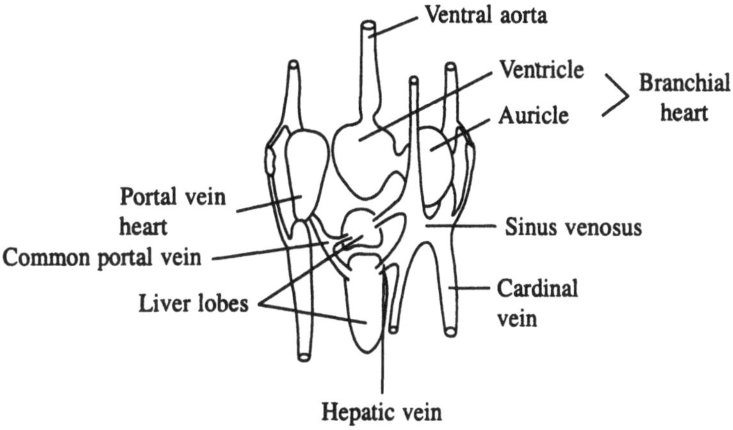

(B) Branchial heart of *Myxine glutinosa*.

Fig. 8.1 (A) Diagram of the branchial heart of adult *Lampetra fluviatilis* (ventral view). (B) Diagram of the branchial heart and adjacent structures of heart of *Myxine glutinosa*. *Source:* Augustinsson et al. (1955)

high concentrations of catecholamines and in consequence heart catecholamine contents are very high.

The systemic heart of myxinoid cyclostomes is aneural, but the sinus venosus of lampetroid cyclostomes receives an excitatory parasympathetic nerve supply.

The hepatic portal vein heart occurs only in the hagfishes as an expansion developed along the course of the hepatic portal vein. It pumps the blood from the gut and the gonad, as well as from the anterior cardinal vein, through the portal vein into the liver. Although it has been customary to regard the portal vein heart as a unichambered organ, stress has been given on the importance of the terminal enlarged portion of one of the afferent vessels to the chamber, the sinus supraintestinalis, which is known to pulsate as a pacemaker for the portal vein heart. The time relation that exists between the sinus supraintestinalis, and the portal vein heart is the same as that existing between the contraction of the sinus venosus and the atrium of the branchial heart. It has been pointed out that portal vein heart is an organ that shares many attributes of the branchial heart, being only smaller and simpler in structure and beating at faster rate.

Another heart-like organ in hagfishes is the caudal heart, which is composed of paired expansions (caudal heart sacs), one on each side of a median cartilage plate. These sacs collect the sinus fluid from subcutaneous sinuses in the tail region and pump the fluid into the efferent vessels, which unite to form a median caudal vein. The inlets and outlets of the caudal heart sacs are guarded by valves. However, these sacs are not contractile in themselves, but are filled and emptied alternately by the action of two extrinsic, somatic muscles (*musculi cordis caudalis*) that run on the outside of each and attach to the cartilaginous plate. A purely neurogenic contraction of the muscles on the side causes a bending of the plate toward that side, which results in a filling of the sac on the same side and a simultaneous emptying of the sac on the opposite side (Figs. 8.2, 8.3) (see Yamauchi 1980; Santer 1985; Satchell 1992).

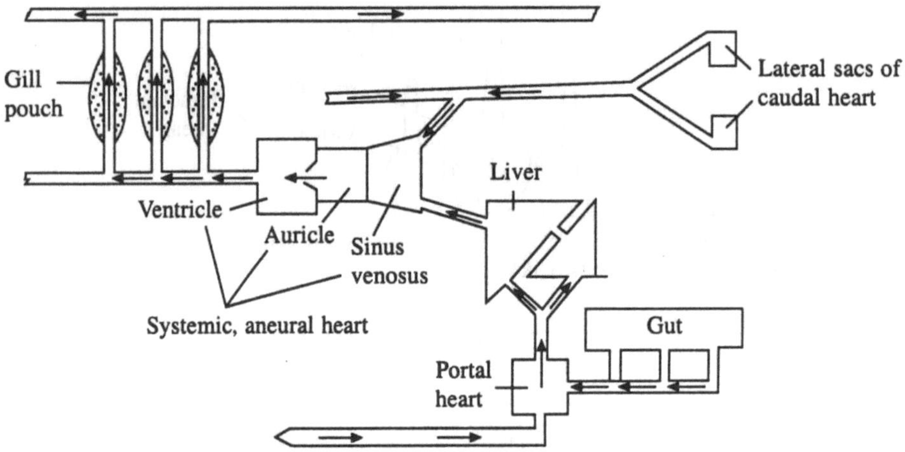

Fig. 8.2 Highly schematic illustration showing the relationship of the systemic, portal and caudal hearts of the hagfish. The paired lateral sacs of the caudal heart are represented as separate structures. Only three of the 12 pairs of gill pouches characteristic of *Eptatretus stoutii* are shown.
***Source:* Jensen (1965)**

The scheme of blood circulation in the adult lamprey is given in Fig. 8.4.

Gnathostomatous fish: The elasmobranch heart presents a fairly uniform gross morphological and histological form. The piscine, myogenic heart consists of four principal chambers arranged in series, in tandem, the sinus venosus, atrium, ventricle, and bulbus cordis (conus arteriosus) in elasmobranchs and dipnoans and in teleosts, the bulbus arteriosus (Figs. 8.5–8.7). Undetermined tachykinin is secreted in sinus venosus of the dogfish. (*Scyliorhinus canicula*) (Muñoz-Chápuli et al. 1994). The cardiac chambers are surrounded by a pericardium of variable size and rigidity. The sinus venosus is a thin-walled chamber and its volume is similar to that of the atrium in teleosts, but less than that of the atrium in elasmobranchs. The sinus venosus receives venous blood via paired Cuverian ducts, hepatic veins, and anterior jugular veins. The sino-atrial ostium is guarded by a large valve, and the openings of the hepatic veins into the sinus venosus are guarded by muscular sphincters. However, the Cuvierian ducts are not valved; they freely communicate with the major systemic veins, which are particularly capacious in elasmobranchs. The major functional role of the sinus venosus is related to the initiation and control of the heart beat. The sinus venosus is the site of specialized cardiac pacemaker tissue in many fish. A ring of specialized myocardial cells (nodal tissue) is usually located at the base of the sinoatrial

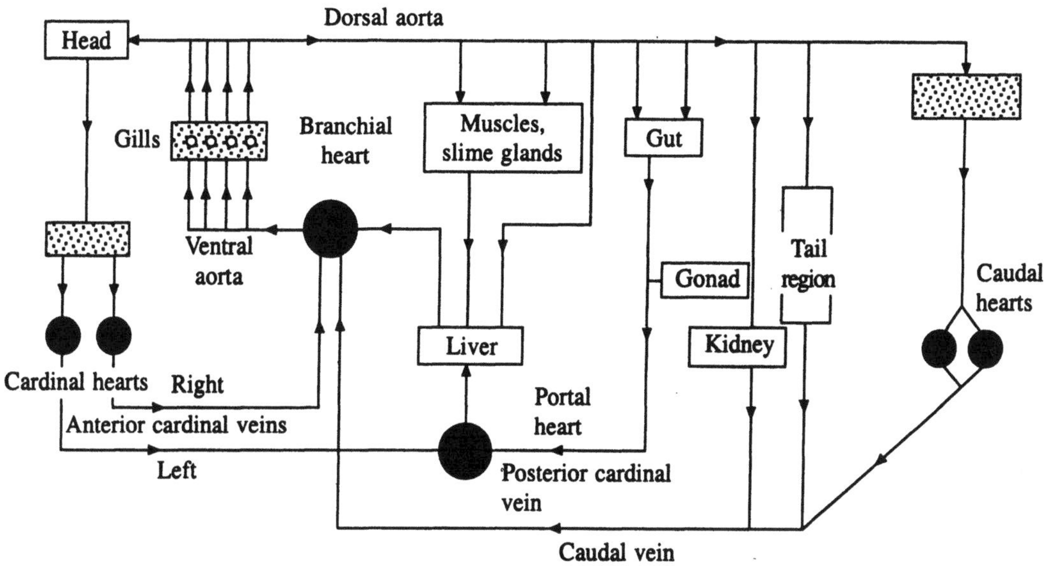

Fig. 8.3 Diagrammatic representation of the cardiovascular system of a hagfish. Black circles represent contractile structures ("hearts") while sinuses are represented by shaded areas.
***Source:* Burggren et al. (1985)**

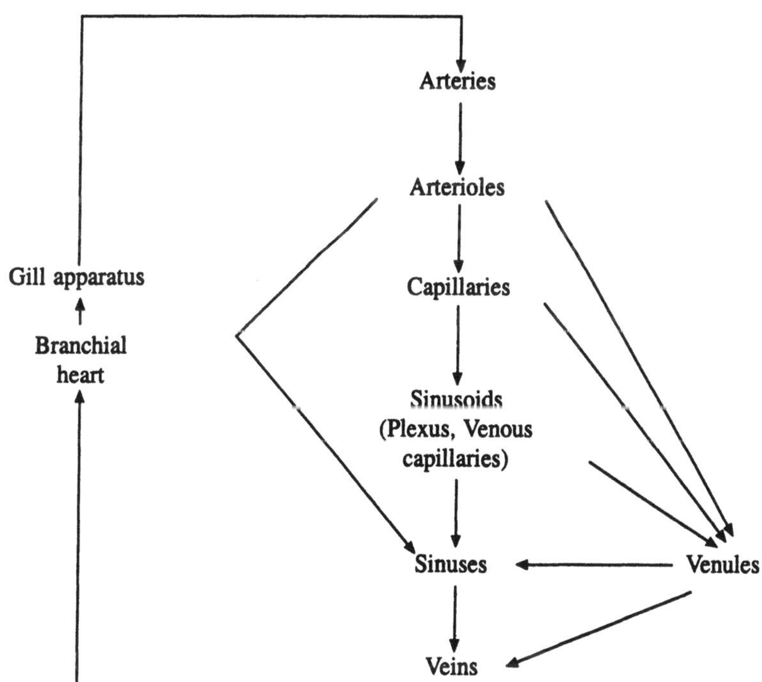

Fig. 8.4 Scheme of blood circulation in the adult lamprey.
***Source:* Fänge (1972)**

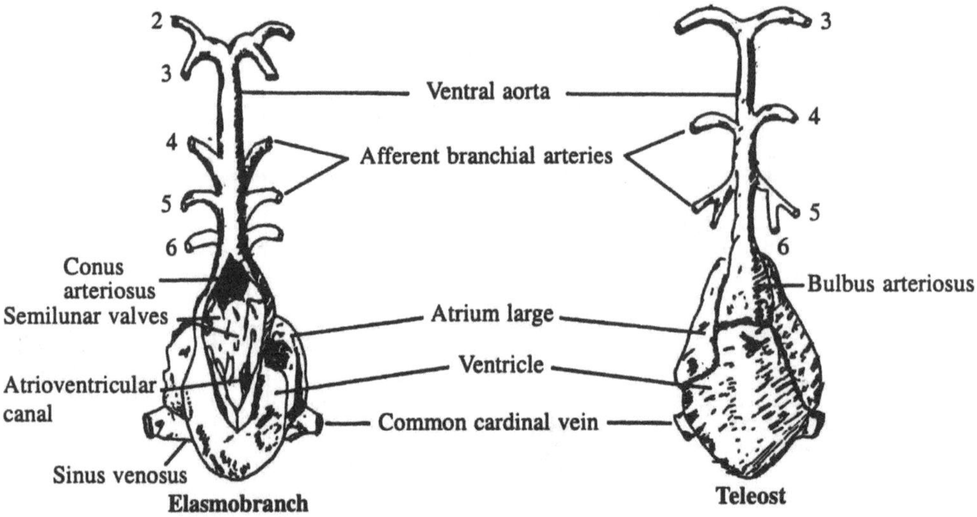

Fig. 8.5 The hearts and major afferent arteries of an elasmobranch (left) and a teleost (right). ***Source:*** **Langille et al. (1983)**

ostium and connects with atrial myocardium. Nodal tissue can have other locations. In the eel (*Anguilla vulgaris*) and lungfish (*Protopterus aethiopicus*), the pacemaker is located at the junction of the sinus venosus and Cuverian ducts. Specialized pacemaker cells have also been reported in the atrium of plaice, brown bullhead (*Ictalurus nebulosus*), as well as in the atrioventricular region. The thin-walled sinus venous is sparsely equipped with cardiac muscle, while the atrium is more richly supplied. Most importantly, the size of the atrium when fully distended is as large or larger than the maximally filled single ventricle into which it communicates mediodorsally through the atrioventricular opening, also guarded by valves. The atrium in elasmobranchs has never been observed to be in any extensive contact with the conus arteriosus, in contrast to the situation found in some teleosts. The atrium is an irregularly shaped chamber with thin trabecular walls; the size may vary, for example, exceptionally large atria are reported for tunas and red-blooded Antarctic fishes, *Pagothenia* species. The single ventricle is a thick-walled chamber of tubular, saclike and pyramidal shapes (teleosts) or normally rounded, sac-like (elasmobranchs) contours. Cardiac ligaments have only been observed in one species of elasmobranch, namely *Raja clavata.* A distinct feature of the piscine ventricle is that its wall is composed of a spongy trabeculate myocardium projecting into and nearly filling the entire ventricular cavity. This spongy myocardium is invested in a compact but often very thin layer of densely arranged muscle fibres, the cortical, denser myocardium. Generally there is a

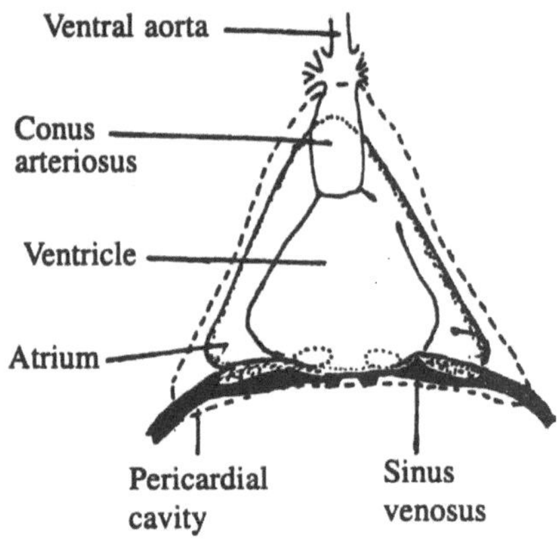

Fig. 8.6 Elasmobranch heart.

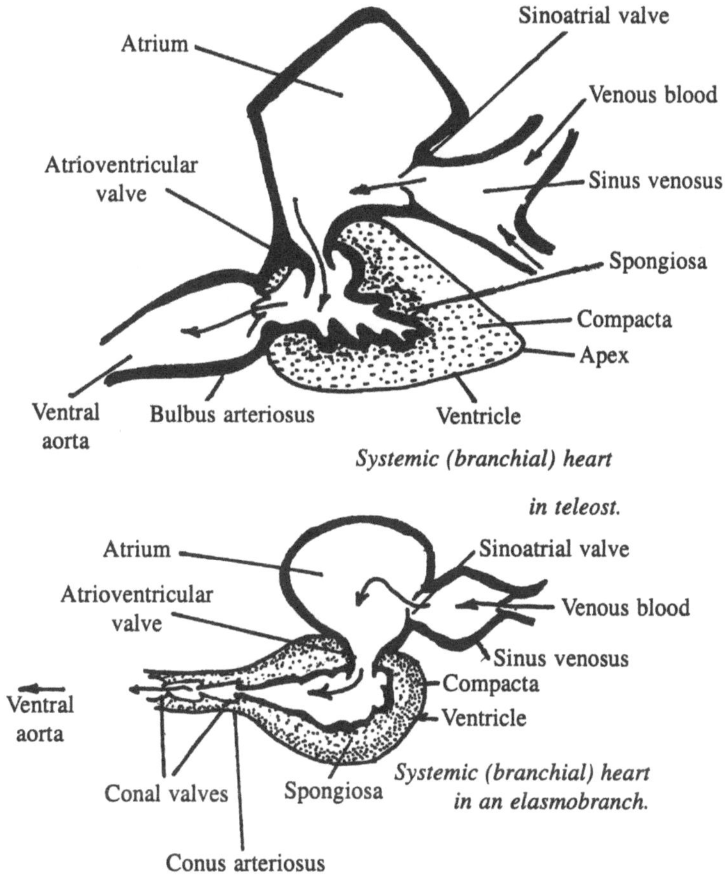

Fig. 8.7 Diagrams of the heart in a teleost (trout) and an elasmobranch (shark).
***Sources:* Randall (1970) and other articles**

coronary vasculature with some differences in origin (cephalad, in some an additional caudal). Multiple coronary arteries are found on the conus arteriosus of elasmobranchs (details of the coronary system in elasmobranchs are reported by Andrés et al. 1990, 1992, 1993; Muñoz-Chapuli et al. 1994b). A coronary circulation is not found in all fishes, but when present it supplies arterial blood to the outer compact myocardium of the ventricle (see details: Davie and Farrell 1991). The remainder of ventricle or the entire ventricle in fish lacking a coronary circulation, is composed of spongy myocardium which drives oxygen from the venous blood pumped through the ventricle. The relative proportion of the compact myocardium is directly related to the activity level of fish species. The coronary circulation is important for maximum aerobic swimming and implicit in this conclusion is that maximum cardiac function is probably necessary for maximum aerobic swimming performance. The functional significance of ventricle shapes has been interpreted in different ways. Four ventricle types have been reported (Figs. 8.8, 8.9; Table 8.1A, B, C). The trabecular arrangement of spongiosa accounts for the greater proportion of ventricular mass in almost all fishes. Active species have larger ventricles. Endothermic sharks have larger ventricles than ectothermic sharks. The fourth cardiac chamber, the bulbus arteriosus in teleosts, and the bulbus cordis (chondrichthyan fishes) represents the ventricular

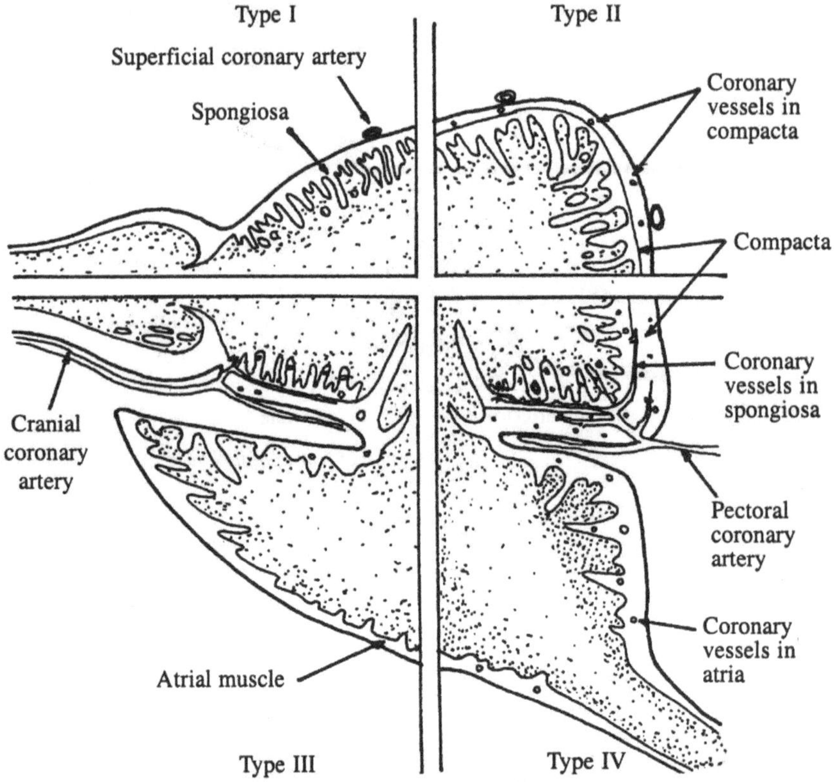

Ventricle types in fish.

Fig. 8.8 A schematic illustration of the main features that distinguish the four ventricle types in fishes. The basic anatomical design of the fish heart has been divided into quadrants with each quadrant illustrating particular features of a given ventricle type. Type I is characterized by having a single myocardial type (spongiosa) and no capillaries in the ventricular muscle. Most type I hearts have no coronary vessels whatsoever. Type II is characterized by two muscle layers in the ventricle (inner spongiosa and outer compacta), a coronary circulation, and capillaries only in the outer compacta. Type III is similar to Type II, but capillaries are found in both the spongiosa and compacta, and to a limited extent in the atrium. Type IV has a larger percentage of the ventricle as compacta (> 30 %) and a more extensive capillarization of the atrium. The coronary circulation, regardless of the type, is derived from a cranial supply, as depicted in the lower left quadrant, and an additional pectoral supply in some fishes, as depicted in the lower right quadrant.
Source: **Farrell and Jones (1992)**

outflow tract, but have in addition important auxiliary functions. In teleosts, the bulbus consists of vascular smooth muscle and elastic tissue and, in most teleosts is separated from the ventricle by a single set of valves. The bulbus cordis (conus) of chondrichthyan, holostean, and dipnoan fishes by distinction consists of cardiac muscle and thus serves a cardiac chamber adding to and extending ventricular outflow. Outflow through the bulbus cordis occurs past several sets of valves guarding ventral aortic backflow. Sans-Coma et al. (1995) have described the microscopic architecture of the elasmobranch conal valves, specifically of adult dogfish (*Scyliorhinus canicula*). Hamlett et al. (1996) reported on the anatomy, histology, and development of the three sets of cardiac valves in embryonic and adult elasmobranch fishes.

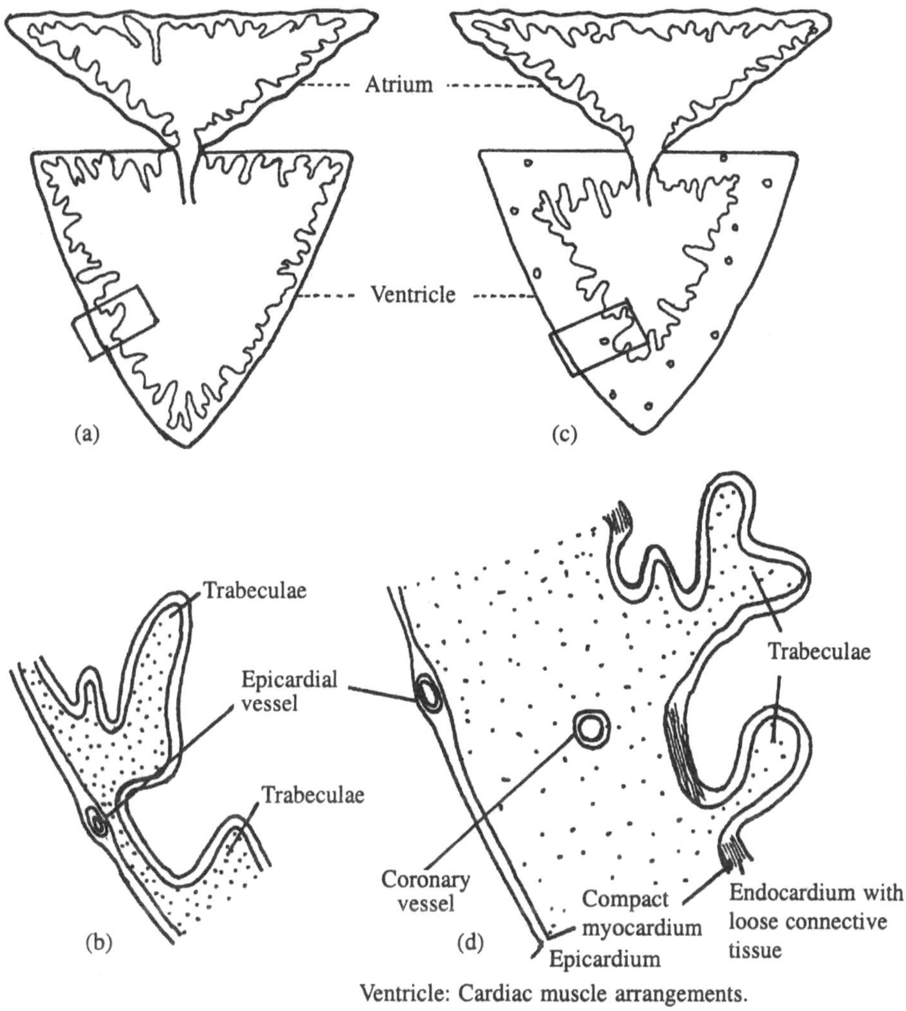

Fig. 8.9 Diagrams of transverse section through fish hearts at the level of the atrio-ventricular orifice. (a) represents the commoner type, lacking a compact layer, a portion of which is enlarged in (b). (c) represents a heart with a compact layer of myocardium supplied by a coronary vasculature, a portion of which is enlarged in (d).
***Source:* Santer and Green Walker (1980)**

In chondrichthyan fishes the four cardiac chambers and a short segment of the ventral aorta are invested in a rigid or semirigid pericardium. The outside pericardial wall (parietal pericardium) is adherent to surrounding cartilages and muscles making it into a rigid boxlike structure. The posterior pericardial wall makes up a septum toward the peritoneal cavity. The pericardium communicates with the primary perivisceral coelom via a pericardio-peritoneal canal, but this canal is flaccid and is not potent unless the pericardial pressure exceeds that in the peritoneal cavity. In teleost fishes the pericardium is generally less fibrous and rigid and has less contact and anchorage with external structures. Teleosts lack a pericardio-peritoneal canal.

The pericardium of all fishes is filled with a lymphlike fluid, but having a lower tonocity than regular lymph. (see Section 10. Secondary vascular system).

Table 8.1A Types of ventricular organization in fish*

Type	I (spongy)	II (mixed)	III (mixed)	IV (mixed)
Ventricular myoarchitecture	Spongiosa	Spongiosa compacta	Spongiosa compacta (< 30%)	Spongiosa compacta (> 30%)
Blood supply	Lacunary (venous blood)	Lacunary (venous blood) + vascular (oxygenated blood)	Lacunary (venous blood) + vascular (oxygenated blood)	Lacunary (venous blood) + vascular (oxygenated blood)
Degree of vascularization	Absent or limited to the epicardium	Coronary vessels limited to the compacta	Vascularization of compacta and spongiosa Thebesian-like shunts	Vascularization of compacta and spongiosa Thebesian-like shunts
Interface	Endocardium	Endocardium + endothelium	Endocardium + endothelium	Endocardium + endothelium
Representative species	*Scorpaena*, *Pleuronectes* icefish	*Oncorhynchus mykiss*	*Scyliorhinus* Torpedo	Sharks Tuna

Source: Tota and Gattuso (1996)

* Most of the spongy hearts (type I) are typical "venous and avascular" hearts (subtype a). Other subtypes are related to the presence of epicardial vascularization. The types II-IV are all mixed types of ventricle. The differences are in the amount of compacta and the degree of vascularization. The term *interface* is used to indicate the internal boundary mediating the interaction between myocardial cells and blood. In the spongy heart, the endocardium is the only barrier between the superfusing blood and the myocardium, while in the higher types there is, in addition, the vascular endothelium. Whether and how this different endocardial/endothelial composition might exert a different control of cardiac performance is completely unknown. In most fish, the coronary artery(ies) originate cranially from the hypobranchial artery. In some species (eel and swordfish) a caudal coronary artery originating from a branch of dorsal aorta is also present.

Vagal cardiac innervation is found in all fishes except the hagfishes. In all fishes the parasympathetic innervation is cholinergic and inhibitory whilst it is most unlikely that the elasmobranch heart is innervated by the sympathetic nervous system (for specific information in *Carcharhinus melanopterus*, see Chopin and Bennett 1995); there is now unequivocal evidence of an adrenergic sympathetic innervation of the heart in holosteans and in several teleost genera. However, there is no evidence at all of the presence of adrenergic nerves in pleuronectid teleosts. There is no evidence of a sympathetic innervation of the dipnoan heart. Further details can be obtained from the review of Nilsson (1997).

Butler and Metcalfe (1988) described the neural control of the elasmobranch heart. Branchial cardiac vagal (bcv) motoneurons in the medulla are distributed in the vagal motor column over about 1 mm rostral to obex (Fig. 8.10). Like other vagal motoneurons, the majority of the bcv motoneurons are located medially, close to the wall of the fourth ventricle, while some bcv motoneurons are located in a ventrolateral group. Further details on the activity are given of both groups of neurons.

Table 8.1B Cardiac morphometrics in selected fishes

	Relative ventricular mass (%)	Compacta (%)	Relative atrial mass (%)
Teleosts wth compacta			
Katsuwoncis pelamis (skipjack tuna)	0.38	65.6	0.061
Makaira nigricans (Pacific bluemarlin)	0.087	48.0	0.013
Salmo gairdneri (rainbow trout)	0.08–0.13	30–39	0.017
Anguilla dieffenbachii (longfinned eel)	0.03–0.10	40.9	0.007–0.013
Elasmobranchs			
Isurus oxyrinchus (shortfin makoshark)	0.14	41.5	0.028
Squalus acanthias (dogfish)	0.086	22.1	0.017
Teleosts with only spongiosa			
Chaenocephalus aceratus (haemoglobin-free icefish)	0.30	none	—
Pagothenia borchgrevinki (red-blooded Antarctic fish)	0.16	none	0.040

Source: Farrell and Jones (1992)

Table 8.1C Levels of cardiac organization in fishes

	Type I		Type II	Type III	Type IV
Representative species	Subtype(a) *Scorpaena*	Subtype(b) *Pleuronectes*	*Conger*	*Scyllium*, *Torpedo*	Sharks, xiphoids scombroids
Myoarchitecture					
Spongiosa	+	+	+	+	+
Compacta	–	–	+	+	++
Compacta (%)			19–21	22–24	31–42
Circulation					
Venous lacunary	+	+	+	+	+
Vascularization of epicardium	–	+	+	+	+
Capillary vascularization of the compacta	–	–	+	++	++
Capillary vascularization of the spongiosa	–	–	–	+	+
Shunts	–	–	–	+	+

Symbols: – absent; + present; ++ highly developed.
Sources: Tota (1983) and Tota et al. (1983)

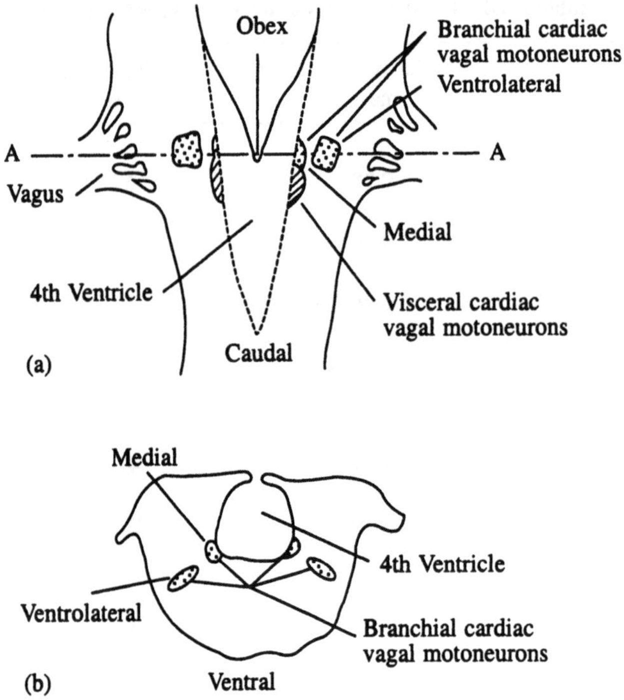

Fig. 8.10 A diagrammatic illustration of the location of cardiac vagal motoneurons in the hindbrain of the dogfish *Scyliorhinus canicula.* (a) A "longitudinal" section through the hindbrain showing the location of the visceral cardiac vagal motoneurons, together with the medial and ventro-lateral branchial cardiac vagal motoneurons relative to the obex. (b) A "transverse" section through the hindbrain at the level *A* showing the position of the medial and ventrolateral branchial cardiac vagal motoneurons.
***Source:* Butler and Metcalfe (1988)**

Cardiac growth and Myocytes
Considerable interspecific variation exists in heart mass relative to body mass in fish. Values range from 0.04 to 0.2%, with larger heart masses generally found in fish with a higher sustained swimming ability. However, interspecific comparisons tend to hide the fact that relative heart mass can vary as a function of life stage and in response to external perturbations such as water temperature. For example, relative ventricular mass increases with acclimation to cold in catfish (*Ictalurus punctatus*), goldfish (*Carassius auratus*), trout (*Salmo gairdneri*) and carp (*Cyprinus carpio*). This increase in ventricular mass apparently compensates for the direct temperature effect on inotropy by maintaining cardiac stroke work. Similarly, swimming stimulates protein synthesis in the trout heart, as do elevated work loads in perfused hearts. Such effects may help explain the observation that heart growth is slower than body growth of Atlantic salmon (*S. salar*) during the freshwater phases of its life history, but faster than body growth in the seawater phase when they undergo an extensive migration. In contrast, heart growth in tuna, *Thunnus thynnus*, is relatively slower in fish larger than 20 kg. For cardiac growth in rainbow trout (*S. gairdneri*), the reader is referred to Farrell et al. (1988). Although cardiac growth is largely tied to the growth of the animal, cardiac development is also influenced significantly by environmental factors. Cardiac growth in O+ to 3 year class rainbow trout is characterized by a mixture of

myocyte hyperplasia and hypertrophy. The reader is referred to Muñoz-Chápuli et al. (1994c) for the cardiac development in the dogfish *Scyliorhinus canicula*.

A number of features distinguish fish cardiac muscle cells from those in mammals. These include: (1) small size compared with that in mammals, (2) limited anatomical development of sarcoplasmic reticulum (SR), (3) peripheral arrangement of myofibrils, and (4) variable amount of intracellular myoglobin. The SR is sparse in fish hearts and the sarcolemma (SL) lacks the T-tubule system typical of all skeletal muscle and mammalian cardiac muscle. Elasmobranch myocytes have less SR compared with teleosts. There is no ultrastructural evidence of specialized conducting fibres in the fish heart.

Tables 8.2 and 8.3 provide information regarding myocytes.

Table 8.2 Quantitative ultrastructural characteristics of teleost ventricular myocytes

Species	Habitat & activity pattern	Myoglobin (n mol g⁻ wet wt.)	V_v(mt.f)	V_v(my.f)	$\frac{V_v\text{(mt.f)}}{V_v\text{(my.f)}}$	Myocyte diameter (μm)
I Myoglobin-poor hearts						
Chaenocephalus aceratus	Polar, marine Sedentary, benthic	1.02– 1.51	± 0.43 0.04	± 0.31	1.39	+ 5.95 1.92
Cyclopterus lumpus Lumpsucker	Temperate, marine Sedentary, benthic	nd	0.27 ± 0.01	0.42 ± 0.02	0.64	2.17 ± 1.00
II Myoglobin-rich hearts						
Notothenia neglecta	Polar, marine largely benthic	19.6 ± 0.1	0.25 ± 0.01	0.43 ± 0.02	0.58	5.29 ± 1.51
Myoxocephalus scorpius *Cottus*	Temperate, marine Sedentary, benthic	20.7 ± 0.4	0.17 ± 0.06	0.51 ± 0.15	0.33	3.59 ± 2.59
Thunnus thynnus	Marine Active pelagic	581– 1175	0.39	0.47	0.62	–Compacta
	Tropical/ temperate	581– 1175	0.32	0.49	0.65	–Spongiosa

Abbreviations: V_v(mt.f) = Mitochondrial volume density
V_v(my.f) = Myofibrilar volume density
nd = Not detected
(Abridgement: Tota et al. 1991a)

It has been suggested that the sparsity of the sarcoplasmic reticulum and the absence of T-tubules in the hearts of lower vertebrates had dictated that the myocyte diameters be small in order to maintain appropriate diffusional distances for Ca^{2+} to the myofilaments.

Table 8.3 Ultrastructural characteristics of myofibrils from fish myocardial cells

Species	Recognizable bands, discs or lines
Myxine glutinosa	Z, I, A, H, M
Petromyzon marinus	Z, I, A, H
Chimaera monstrosa	Z, I, A, H, M, N
Perca fluviatilis	Z, I, A, H, M
Acipenser stellatus	Z, I, A, H, M

Source: Santer (1985)

For details and historical background on various aspects, the reader is further referred to a number of publications (Randall 1970; Johansen and Burggren 1980; Yamauchi 1980; Tota 1983; Tota et al. 1983; Santer 1985; Nel et al. 1991; Burleson et al. 1992; Farrell and Jones 1992; Tibbits et al. 1992; Agnisola et al. 1994; Driedzic and Gesser 1994; Chopin and Bennett 1995).

2.1 Multiple Ventral Aortae

The presence of multiple ventral aortae, which arise directly from the bulbus arteriosus in facultative air-breathing catfish *Heteropneustes fossilis* is a feature also reported in air-breathing climbing perch *Anabas testudineus* and in snakeheads *Channa maculata* and *C. argus*. However, in the latter three species the ventral aortae are believed to minimize mixing of oxygenated air-breathing organ venous blood and deoxygenated systemic venous blood and thereby assist in separating the respiratory and systemic circulation. The function of multiple ventral aortae in *H. fossilis*, in which peripheral blood separation does not appear possible, is problematic (Olson et al. 1990).

2.2 Carotid Labyrinth

Among teleosts, the catfishes (Siluroidei) are the sole possessor of the enigmatic structure, the carotid labyrinth (see details: Srivastava and Singh 1980, Olson 1990, Burleson et al. 1992, Srivastava et al. 1988, 1994; Srivastava, 1996).

2.3 Dipnoan Heart

Three genera of dipnoans (lungfish) exist: *Protopterus* (Africa), *Lepidosiren* (South America) and *Neoceratodus* (Australia). Of the three, only *Neoceratodus* is an obligate water-breather, restorting only to air-breathing when the water becomes hypoxic, whilst *Protopterus* and *Lepidosiren* alternate freely between gill and lung breathing and rely entirely on lung breathing during aestivation.

The lungfish represent the first phylogenetically important step in the transition from a single to double circulation in the vertebrates.

The hearts and vascular arrangements in lungfish are quite different from the usual piscine pattern. Notable differences are related to the partial structural division of the heart. The development of divided hearts in lungfish is associated with the special pulmonary vascular circuit having a separate return to the left side of the heart but coupled in parallel with the systemic or internal gas exchange circuit on the arterial side.

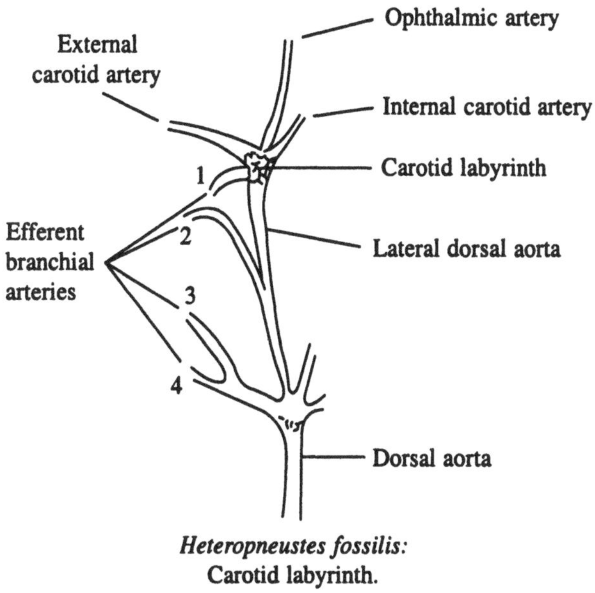

Fig. 8.11 Carotid labyrinth in *Heteropneustes fossilis*.
***Source:* Srivastava and Singh (1980)**

Among lungfishes the extent of cardiac division is related to the prevalence of air-breathing. Cardiac division is least complete in *Neoceratodus* and best developed in *Lepidosiren*. The heart in *Protopterus* resembles that of *Lepidosiren* and both are similar in appearance to hearts in amphibians, while the cardiac morphology of *Neoceratodus* has retained some typical fishlike (elasmobranch) features.

Special features of the dipnoan hearts compared with those of other fishes are well represented in the heart of *Protopterus aethiopicus*. These special features include the location of heart behind the pectoral girdle, the relatively anterior position of the sinus venosus and atrium with respect to the ventricle, and the dipnoan atrioventricular plug (atrioventricular "cushion") rather than atrioventricular valves, and the inner subdivision of heart by partial and ventricular septa and the spiral folds in the bulbus cordis (conus arteriosus). In addition to the cardiac features, the dipnoan vascular pattern is unusual among the air-breathing fishes in that blood from the lungs passes directly to the left side of the atrium and the ventral aorta is very short or non-existent, thus the afferent branchial arteries originate from the anterior end of sharply twisted bulbus cordis (or truncus) as do the homologous vessels in the amphibians (Figs. 8.12–8.14).

The return of blood to heart depends on the driving, or *vis-a-tergo* force and the attractive *vis-a-fronte* force. The driving force may result from the energy remaining from the cardiac populsion of the blood or from body movement such as those involved in swimming. Suctional attraction depends on the lowering of the intravascular pressure at the cardiac end of the venous circulation. The reduced pressure is created inside the pericardium due to the reduction in volume of the heart with each cardiac systole.

In parallel vascular circuits, the distribution of blood will vary with the relative resistance in various circuits. It has been shown that pulmonary vascular resistance is lower than the systemic resistance both in lungfishes and amphibians. The difference of the resistances is most evident as a lower pressure in the central systemic veins than in the pulmonary veins.

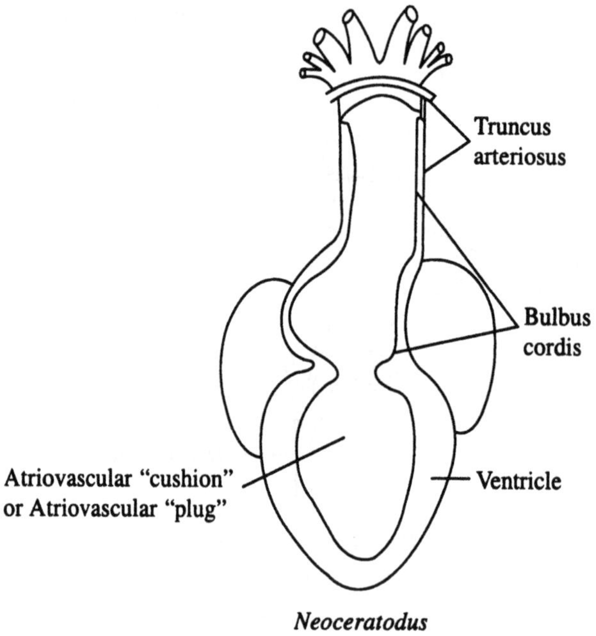

Fig. 8.12 Schematic saggital section through the heart of *Neoceratodus*. Heart is shown from a ventral aspect.

As in elasmobranch fishes, the cardiac suctional attraction depends on the heart rate. An important extracardiac mechanism which increases systemic venosus return is the contraction of skeletal musculature.

The dipnoan ventricular septa may be of little consequence in this comparison since trabeculae in both lungfish and amphibian hearts may be primarily responsible for the maintenance of functional separation of blood in this chamber. It may be that the ventricular trabeculation in lungfish hearts is less marked and of lesser importance in the functional separation than the prominent trabecular subdivision of the ventricle in Amphibia.

As in cyclostomes, large numbers of catecholamine-containing cells occur in heart of dipnoan.

The cardiac innervation of the lungfish is poorly understood but it is known that cardiac branches of the vagus reach the heart and can be observed on the wall of the sinus venosus. Despite the lack of recognition of nodal tissue or conduction pathways, the distribution of ganglion cells corresponds fairly accurately with the electrophysiologically defined pacemaker regions, and it is therefore likely that there is a cholinergic vagal influence on the pacemaker activity of dipnoans. There is no evidence for a sympathetic innervation of the dipnoan heart.

In all lungfishes the heart is surrounded by a tough, semirigid pericardium through which the central veins enter and the bulbus cordis exists. Although teleost fishes lack a stiff pericardium, it is a highly characteristic feature of all elasmobranchs, and its integrity is essential for the normal cardiovascular function in sharks. The condition of the pericardium in most actinopterygians is unknown (Johansen and Hanson 1968; Johansen and Burggren 1980; Santer 1985; Burggren and Johansen 1986; Farrell and Jones 1992; Chopin and Bennett 1996b).

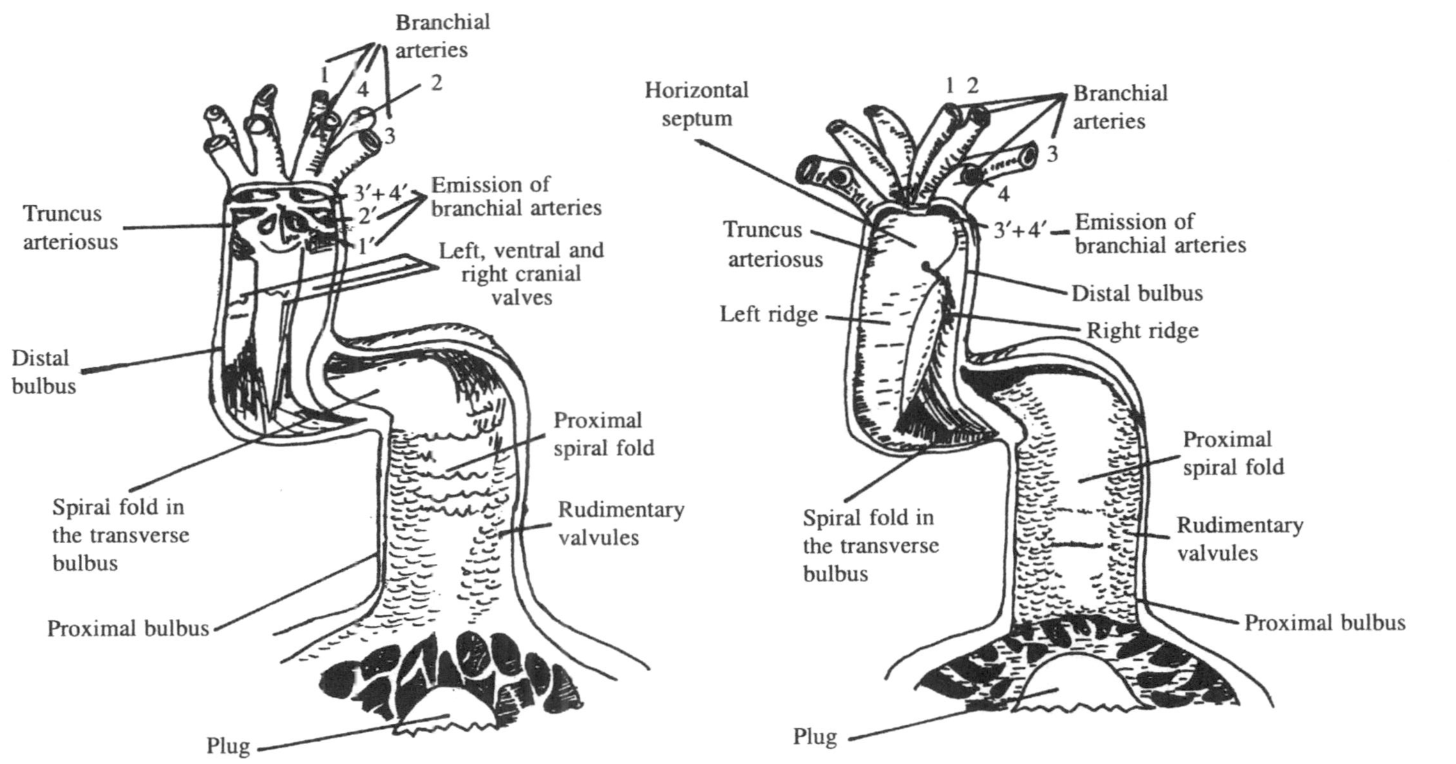

Fig. 8.13 Diagrams showing the bulbus cordis and truncus arteriosus of *Neoceratodus forsteri* and *Lepidosiren paradoxa* respectively. The dorsal roof of the bulbus and truncus is opened and partly removed. 1′–4′ emission of branchial arteries, 1–4 seen from the lumen of the truncus. Note the common emission of No. 3 and 4 in both species.
***Source:* Klitgaard (1978)**

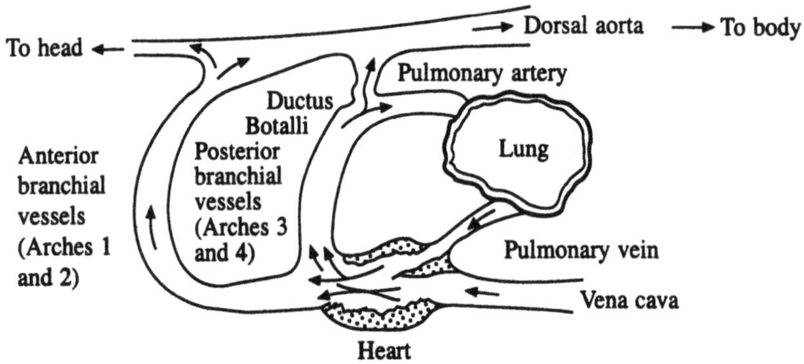

Fig. 8.14 Highly schematic diagram of the anterior circulation of the lungfish. ***Source:*** **Burggren et al. (1985)**

3. CARDIAC PHYSIOLOGY

3.1 Background

Effective gas exchange in either gills, skin or air-breathing organ requires equally effective cardiovascular function if gas exchange at the level of the tissues is to be maximized.

The cardiovascular gross morphology of most phyletically ancient fishes–Chondrostei, Chondrichthyes, Holostei, Coelacanthini—is qualitatively similar to the general pattern of more advanced bony fishes. The heart ejects blood into a single ventral aorta. This vessel conveys blood in a closed circulation to the capillary bed of the gills, where gas exchange occurs. The dorsal aorta forms from the confluence of the branchial efferent vessels, and carries blood directly on to the systemic vascular bed. Deoxygenated blood draining the tissues return via veins directly to the sinus venosus.

This basic arrangement of the systemic tissues in series with, and downstream from, the gills imposes specific haemodynamic constraints and demands on the circulation, regardless of phylogeny. Both branchial and the systemic vascular resistances can be neurally and hormonally regulated, and in concert with adjustments in heart rate and stroke volume, fishes exhibit a wide range of cardiovascular performance suitable for rest, exercise or hypoxia.

The hagfishes, lampreys, and the lungfishes represent striking exception to this general piscine circulatory pattern.

In cyclostomes we find a semi-closed vascular system where extensive sinuses of lacunar spaces are in functional connection with other vascular channels.

The circulation in *Myxine* differs from circulation in vertebrate generally by a partial shifting of the blood-moving power from the systemic heart to the gills and the portal, cardinal and caudal auxiliary hearts. The cardioregulation in the aneural branchial heart of the hagfish is accomplished by two major factors: tension by venous return and a biochemical factor produced in the heart. The hagfish's branchial heart is considered to be homologous to all other vertebrate hearts, but has the most primitive features. The heart of the hagfish has no coronary circulation nor any innervation, unlike the heart of all gnathostome fishes. Arterial blood pressure in hagfish is generally very low.

The cardiovascular morphology of the lamprey appears intermediate between that of the hagfish and more advanced fishes.

3.2 Overview of Cardiovascular and Haemodynamic Energetics with Emphasis to Teleost Fish: The Heart as a Pump

With the possible exception of the hagfish heart, fish hearts are primarily aerobic tissues. The ability of a fish heart to tolerate anoxia, i.e., to support its demand for energy through anaerobic metabolism, seems to be largely related to the resting power output of the heart. Consequently, the hagfish heart can meet at least 70% of its power output (<< 1 m W/g) through anaerobic metabolism. About 25% of the resting cardiac power output of the sea raven (*Hemitripterus americanus*) can be supported by anaerobic metabolism. Elasmobranch hearts can withstand anoxia for upto 140 min. However, perfused trout hearts fail rapidly with moderate hypoxia.

The teleost heart has a single ventricle, which pumps blood into the ventral aorta, a single strium and a sinus venosus, which acts as a reservoir for venous blood. The cardiac pacemaker is located at the sino-atrial junction.

Power output of the heart is determined primarily by the product of cardiac output ($\dot{V}b$) and ventral aortic pressure (Pva). Power output is a measure of the heart's ability to work as a pump, that is, it's ability to provide sufficient blood flow (cardiac output) to meet the metabolic needs of the tissues, and to generate sufficient pressure to overcome vascular resistance. Power output is about 1 mW g^{-1} ventricle wet weight in benthic fish such as the sea raven (*H. americanus*) and lingcod (*Ophiodon elongatus*) and about 1.7 mW g^{-1} in active fish such as the trout (*Salmo gairdneri*). Power output increases to about 7.4 mW g^{-1} at critical swimming speed in trout and in experimentally restrained tuna it is about 12 mW g^{-1}. For the cardiovascular data measured in fish at rest and during either exercise or hypoxia, see Farrell (1993) (Table 3). Also refer to Table 2 given by Olson (1998).

The ability of the teleost heart to vary its power output is dependent upon a number of intrinsic and extrinsic mechanisms which ultimately affect Pva and $\dot{V}b$. The factor which affect Pva and $\dot{V}b$ are discussed.

3.2.1 *Blood Pressure*

Blood pressure is determined by the product of blood flow and vascular resistance. In teleosts, the heart pumps venous blood through the gills before oxygenated blood is distributed to various segments of the systemic circuit. Vascular resistance is, therefore, readily subdivided into gill and systemic resistances (Rg and Rs). Thus, for a given $\dot{V}b$, Pva is set by the sum of Rg and Rs, which are in series, and Pva is indicative of the pressure work done by the heart. The dorsal aortic pressure, Pda, is set by Rs and is indicative of the blood pressure needed to distribute Vb through the various, in-parallel systemic circulations. The mean Pva is normally of the order of 35 to 65 cm H_2O, with the corresponding Pda being about 8 to 20 cm H_2O lower. This indicates that Rs is about three times greater than Rg.

Both Rs and Rg are regulated by neural and humoral influences, which produce important changes in Pva. The teleost heart can cope with physiological changes in Pva through homeometric regulation. This intrinsic mechanism enables the heart to vary pressure development to accommodate changes in Rg and Rs without changing $\dot{V}b$.

Few studies have measured the central venous blood pressure (Pcv) in teleosts. All indications are that Pcv is normally less than 2 cm H_2O above ambient pressure. The Pcv represents the filling pressure of the heart.

3.2.2 *Cardiac Output*

The majority of direct measurements of $\dot{V}b$ range from 6 to 30 ml min^{-1} kg^{-1} for teleosts in temperate water. $\dot{V}b$ increases several-fold during sustained swimming in trout and in experimentally restrained tuna $\dot{V}b$ is about 90 ml min^{-1} kg^{-1} at 20°C. Cardiac output is the product of heart rate (fH) and stroke volume (SV_H), both of which are regulated by independent mechanisms (Fig. 8.15). Figure 3 given be Olson (1998) illustrates the determinants of the cardiac output. (See also Table 1 for cardiovascular variables).

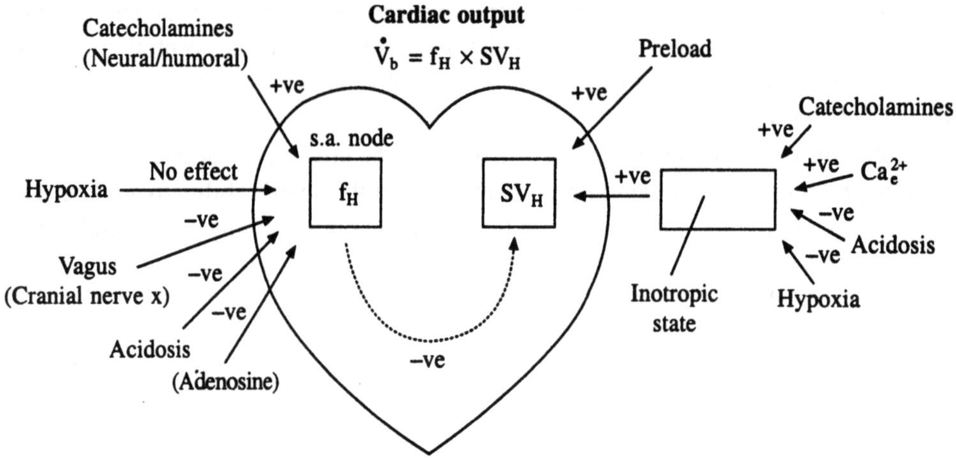

Fig. 8.15 A summary of the factors which affect cardiac output ($\dot{V}_b$) in teleost fish.
***Source:* Farrell (1985)**

Heart Rate. The intrinsic f_H is set by the discharge frequency of the sinoatrial pacemaker. The intrinsic f_H is in turn modulated by several neural and humoral mechanisms. A primary control mechanism is the inhibitory vagal tone which causes bradycardia. In resting fish, f_H is somewhat lower than the pacemaker frequency because of a resting vagal tone to the heart. Thus, the resting fish can increase f_H by removing this vagal tone. Tachycardia is also produced by circulating catecholamines and/or adrenergic innervation when present. Hypercapnic acidosis and extracellular adenosine both decrease f_H. A moderate level of hypoxia ($Po_2 > 40$ mm Hg) does not affect the intrinsic f_H, even though environmental hypoxia causes a reflex bradycardia through an increase in vagal tone.

There are many physiological factors which set the upper limit for f_H in fish. One factor that may be particularly important is the movement (diffusion and transport) of calcium for excitation-contraction (EC) coupling.

It has been reported in *Conger conger* heart that prostacyclin (PGI_2) has a positive inotropic effect on the atrium ventricle, but it does not affect the heart rate.

Stroke Volume. Studies with perfused hearts have clearly established the importance of preload in determining SV_H through the Starling mechanism. Work with trout, salmon, sea raven and ocean pout (*Macrozoarces americanus*) hearts has shown that small changes in preload (< 2 cm H_2O) produce formidable changes in $\dot{V}b$. This response occurs without a major change in f_H and at a constant diastolic afterload.

In addition to the preload, the inotropic state of the myocardium affects SV_H. Catecholamines and extracellular Ca^{2+} improve the hearts's inotropic state. Acidosis and hypoxia have negative inotropic effects. However, a marked change in the inotropic state of myocardium does not necessarily alter the resting performance of the heart considerably. Instead, the change in inotropic state is manifest as a modification of the Starling response. For example, severe extracellular acidosis (pH 7.4) reduces the resting $\dot{V}b$ by only 15% to 20%, but the Starling response to preload is reduced over three-fold. A similar situation exists when the heart is exposed to hypoxia. Such observations indicate that the teleost heart has an intrinsic reserve for maintaining its *resting* work level.

SV_H is also dependent on f_H because SV_H is in part determined the time available for ventricular filling ($1/f_H$). The importance of this is that whenever f_H changes the counteracting change in SV_H will tend to regulate $\dot{V}b$ at a given level. This type of regulation has been observed in perfused hearts and probably occurs *in vivo* during environmental hypoxia when there is a marked bradycardia and little in change $\dot{V}b$.

Maximum stroke volume is limited by anatomical constraints among other factors. The ventricle must be larger to accommodate a larger maximum stroke volume especially since end-systolic volumes in fish hearts are quite small. Modulation of stroke volume is important in increasing internal O_2 convection during exercise, whereas maximum stroke volume appears to be much less plastic between species compared with f_H. Antarctic fishes are exceptional in having relatively large stroke volumes (2–10 ml/kg body mass) and ventricles masses but very low arterial blood pressures. The large variability in ventricle mass among fishes undoubtedly reflects adaptations for generating higher blood pressures, for functioning at cold temperatures, and for accommodating larger stroke volumes.

Tunas and Antarctic fishes are two exceptions to the generalization that fish primarily vary stroke volume. Both fish groups appear to increase f_H by two-fold during exercise, while keeping stroke volume relatively fixed. Ventricles producing mainly pressure work and those producing mainly volume work are distinguished. Most species including elasmobranchs, are located between these two extremes. Indicative examples, such as tuna and icefish, show the existence of a definite relationship between the myoarchitecture of ventricle and the mechanical behaviour of the whole heart expressed in terms of pressure generation and volume movement (Tota and Gattuse 1996; see Table 8.4; for tuna refer reviews by Farrell 1996; Tibbits 1996).

3.2.3 Myocardial Oxygen Uptake

The only studies of myocardial O_2 consumption (Mo_2) are with teleosts without a separate coronary circulation, i.e., they rely on the venous blood being pumped through the heart for their O_2 supply. For such fish the following general points have been established: (1) The mechanical efficiency of the heart is about 15%; an efficiency comparable to the mammalian myocardium. (2) The anaerobic contribution to the metabolism is < 5%. (3) Calculations indicate that at rest, less than 4% of the available O_2 is extracted from the venous blood during its passage through the heart and so venous blood provides an adequate O_2 supply. (4) The O_2 requirement of the heart in the resting fish is less than 1% of the standard O_2 uptake of the fish.

It is clear from above that the heart can do extra work by increasing $\dot{V}b$ and Pva.

Myocardial oxygen consumption appears to be independent of the heart's anatomical type (refer Davie and Farrell 1991).

Table 8.4 Ecophysiological and cardiocirculatory characteristics of tunas and icefish

Tunas	Icefish
Ecophysiology	
Continuously swimming	Sedentary
Very high oxygen demands	Low oxygen demands
Tropical/temperate; warm-bodied	Polar; Temperature range from – 1.9 to 6.0°C
Circulation	
Blood volume similar to that of other teleosts	High blood volume
High haematocrit	Very low haematocrit
Very high Hb concentration	Hb absent
High systemic vascular resistance	Low systemic vascular resistance
Countercurrent retial system	Gill blood vessels + capillaries with large diameter
High blood pressures	Relatively low blood pressures
High cardiac output	High cardiac output
Ventricular enlargement (increased compacta)	Cardiomegaly of the trabeculate type of heart.

Source: Tota and Gattuso (1996)

3.2.4 Temperature Effects

A change in environmental temperature has an immediate and substantial effect upon heart performance in fish. A decrease in temperature results in a decrease in resting heart rate and subsequent output in many species. At least in rainbow trout, low temperature also results in a lowering of maximum heart rate induced by swimming activity. Lowering of temperature leads to a decreased frequency of contraction by isolated heart preparations and to a lowering of stroke volume by isolated sea raven hearts perfused with Ringer's solution containing physiological levels of calcium and paced at rates of 12–48 contraction min^{-1}. The cumulative evidence shows that in the absence of intervening factors an acute change in temperature has a direct impact on both the chronotropic and inotropic responsiveness of the fish heart. Chronic exposure to low temperature may invoke alterations in heart performance.

Temperature effects on the neural control of f_H are established. Studies with perfused heart have demonstrated positive chronotropic and inotropic responses to temperature. This means that as the metabolic needs of the tissues increase with temperature, $\dot{V}b$ can increase intrinsically to meet their needs.

Perhaps more unusual is the observation of a seasonal adaptation to temperature. Both the intrinsic f_H and the inotropic capabilities of the sea raven adapted to cold conditions (5°C) were significantly lower than those of warm-adapted fish (10°C) that were acutely cooled to a lower temperature (3°C). Conversely, the inotropic capabilities of fish adapted to warm conditions were greater than those of cold-adapted fish that were acutely warmed to a higher temperature (13°C). The significance of this inverse compensation of the heart's intrinsic capabilities with temperature may be related to the quiescent state of these fish during winter when the water temperature is near 0°C.

Understanding of the response of cardiac muscle to low temperature acclimation of non-polar teleost fish is still poorly understood (Driedzic et al. 1996). In rainbow trout (*Oncorhynchus mykiss*) acclimation to low temperature is associated with an increase in the use of fatty acids to support energy metabolism and heart growth. Trafficking of fatty acids is increased following acclimation to low temperature. Increases in the oxidation of exogenous fatty acids are associated with elevations in the level of mitochondrial-based fattly acyl-CoA synthetase and carnitine acyl-CoA transferase I (Patey and Driedzic 1997).

3.2.5 The Role of Catecholamines

It is clear from the above that catecholamines provide inotropic and chronotropic stimulation of the heart. Until recently, it was unclear under what conditions adrenergic stimulation of the heart was important. This was specially true in view of the equivocal results obtained with adrenergic drug infusions into intact fish, and the finding that adrenergic stimulation did not increase the *resting* power output of perfused hearts.

Present data indicate that catecholamines play an important role in protecting the integrity of the myocardium under adverse and stressful circumstances. Bernier et al. (1996b) suggested the role of adenosine in modulating the circulating level of catecholamines in hypoxic conditions.

The first point is that circulating catecholamines increase during stress. Stressful exercise also results in as much as 0.5 pH unit decrease in blood pH and a decrease in venous Po_2. Both the acidosis and hypoxia impair the inotropic capabilities of heart based on perfused heart and isolated ventricular strip studies. However, following strenous exercise *in vovo* when there is a respiratory acidosis, cardiac performance is enhanced rather than diminished. The explanation for this is probably adrenergic stimulation of the myocardium during acidosis since a physiological level of adrenaline (0.1 u*M*) restores and even improves the Starling response of the perfused heart exposed to severe (pH 7.4) hypercapnic acidosis. The mechanism(s) underlying the protective effect of adrenaline on the teleost myocardium is (are) unclear at this time.

Nilsson (1994) suggested that the adrenergic vasomotor innervation is a major nervous regulator of blood pressure in Atlantic cod *Gadus morhua* during rest, exercise, and hypoxia. Reports are that there appears to be no adrenergic control of blood pressure in the eel (*Anguilla australis*) and an adrenergic innervation of the heart is lacking in pleuronectids.

The potential control of the cardiovascular system by circulating catecholamines is of particular interest in elasmobranchs since these fish lack any direct sympathetic innervation of the heart and possess no motor innervation of their gill blood vessels; consequently, it would seem that humoral control of the heart and the branchial circulation is the only type of extrinsic control in these animals. Furthermore, when their environmental oxygen level is reduced, dogfish release comparatively large quantities of adrenaline and noradrenaline into their bloodstream, suggesting that these hormones may play some role in their cardiovascular response to hypoxia.

Although the elasmobranch heart lacks an adrenergic innervation but is amply supplied with adrenergic receptors.

Recently, Xiang et al. (1994) reported that NPY (neuropeptide Y) could potentiate the effect of NA (noradrenaline) in the dogfish (*Squalus acanthias*) hearts. The changes in inositol 1, 4, 5-triphosphate (Ins(1, 4, 5)P_3) may be implicated in the cardiac effects of NA and NPY in these hearts.

The reader is referred to reviews on cardiac performance in teleost fish (Farrell 1985, also see Farrell 1983, Nilsson 1997). For information related to vascular actions of prostaglandins (PGs) in fish, refer Piomelli et al. (1985), Acierno et al. (1990) and Mustafa et al. (1992).

Farrell (1991a) gave a comparative view of cardiac physiology in fish to provide insights into cardiac control mechanisms and the degree of cardiac plasticity between species. Between the extremes represented by skipjack tuna *Katsuwonus pelamis* and hagfish *Eptatretus cirrhatus*, there is a 15-fold difference in cardiac output that is largely accounted for by species differences in heart rate rather than stroke volume. The relative importance of heart rate and stroke volume is reversed in many fish in terms of explaining the up to three-fold increase in cardiac output observed during swimming. A conspicuous and unexplained observation is that all fish studied so far, with the exception of tuna, have heart rates lower than 120 bpm. In many fish luminal O_2 in venous blood is the sole myocardial O_2 supply, and venous Po_2 threshold may ultimately limit cardiac performance. More active fish have bigger ventricle that generally beat faster and generate higher blood pressures and flows. The bigger ventricle in active fish, while improving convection to working skeletal muscle, requires the development of a coronary circulation to supplement to varying degree the luminal myocardial O_2 supply.

The review provided certain important information. Cholinergic inhibitory control of heart rate (via the vagus nerve) is universal to all elasmobranchs, teleosts, amphibians, reptiles, and mammals so far examined (hagfish and lamprey excluded). Adrenergic influences on heart rate are not fully understood. Cardiotonic effects of catecholamines are present in fish without adrenergic innervation of the heart. Adrenergic cardiac innervation is absent in hagfish, elasmobranchs, and relatively inactive teleosts. Therefore, the range for heart rate during aerobic exercise in these fishes will be determined by cholinergic nerves. Other teleosts have adrenergic cardiac innervation, and positive chronotropic effects could be mediated through these nerves.

Cardiac anatomy is also more complex in active fish. Whereas the ventricle is entirely trabecular in sluggish fish, only the inner 30%–80% of the ventricle is trabecular in more active fish. The outer layer of their ventricle is more compact and has a coronary circulation. The myoarchitecture of the compact layer is more elaborate in highly active fish, increasing from an arrangement of one to three muscle fibres. Moreover, there is a shift from a saclike ventricle to a more pyramidal shape and a concentric arrangement of the compact fibres around the apices of the ventricle. These anatomical specializations presumably relate to advantages in terms of pressure development.

The role of a haematocrit on cardiac performance has not been fully examined in fish.

Myocardial $\dot{V}o_2$ in fish is directly related to cardiac power output. Myocardial O_2 supply is derived from one of two sources. Sluggish fish lack a coronary circulation and rely entirely on the O_2 contained in the venous blood (the blood bathing the cardiac tissue and it is pumped throughout cardiac chambers). More active fish, with higher $\dot{Q}$'s and higher blood pressures, have a higher myocardial O_2 demand and have a coronary circulation.

The anaerobic potential of the heart as measured by biochemical indices does not vary greatly between fish species. Consequently, anaerobic ATP production by the heart will be negligible in very active fish with a high ATP requirements (Fig. 8.16) but can be an important component during hypoxic exposure in the less active fish such as the sea raven, in which cardiac work is much lower. Hagfish have the lowest cardiac ATP requirement, and most, if not all, of the ATP requirement for resting cardiac work can be met by anaerobic metabolism.

Further details on cardiac physiology can be obtained in the contributions by Farrell and Jones (1992) and Driedzic and Gesser (1994). The reader should also refer to contributions by Butler and Metcalfe (1988), Tibbits et al. (1992) and Coucelo et al. (1996). The ECG (electrocardiogram) terminology is the same in fishes as in mammals. The P wave represents

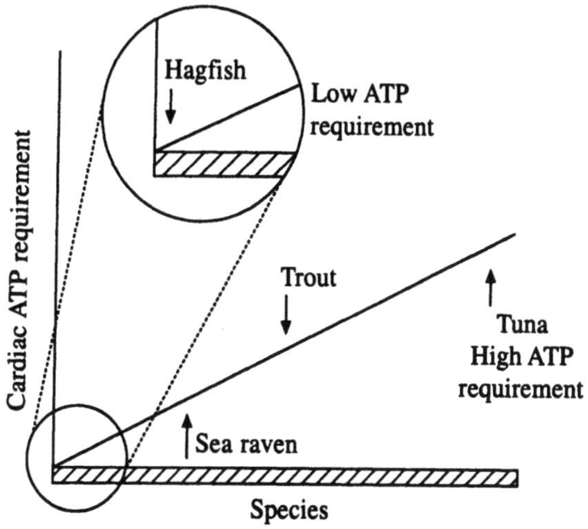

Fig. 8.16 A diagrammatic analysis of the relationship between the myocardial ATP requirement of the heart (line) and its anaerobic ATP-generating potential (hatched area). It is assumed that the anaerobic potentials in fish heart are relatively similar in sluggish and active fish. The low ATP requirement of the hagfish heart means that the majority of this need can be met through anaerobiosis. The high ATP requirement of the tuna heart means that ATP generation through anaerobiosis is insignificant. Other fish lie somewhere within this range, depending upon myocardial power output.
***Source:* Farrell (1991a)**

atrial contraction, the QRS complex, ventricular contraction and the T wave, ventricular relaxation. However, additional waves are present in fish compared with mammalian ECGs. Contraction of conal cardiac muscle in elasmobranchs produces a B wave during the S-T interval (Fig. 8.17). Contraction of cardiac muscle in sinus venosus appears as a wave during the P-T interval and is prominent in *Anguilla anguilla* and hagfishes.

The propagation of the action potential from the atrioventricular node to the various parts of the ventricle is coordinated and synchronized despite a lack of histological evidence for fast-conducting fibres in fish ventricles. Whether the various layers of fibre bundles of different myocyte diameters in the ventricle are important in the spread of excitation is unknown. At the level of individual myocytes, current flow between cells is affected by the degree of folding and number of gap junctions in the intercalated discs. It has been noted that teleosts have intercalated discs that are less convoluted and have few gap junctions than those of higher vertebrates. Gap junctions may be absent or few in number in hagfish myocytes.

4. ROLES OF PERICARDIUM, CONUS AND BULBOUS ARTERIOSUS

4.1 Role of Pericardium in Cardiac Performance and Venous Return

The tough, fibrous pericardial membrane found in sharks has a limited compliance, and intrapericardial pressure is subambient. As a result, contraction of the ventricle reduces intrapericardial pressure and assists atrial filling. The concept of *vis-a-fronte* filling (suction filling) of the atrium in sharks is also supported by the fact that venous pressure and diastolic blood pressures in the atrium and ventricle are subambient. Furthermore, when the pericardium

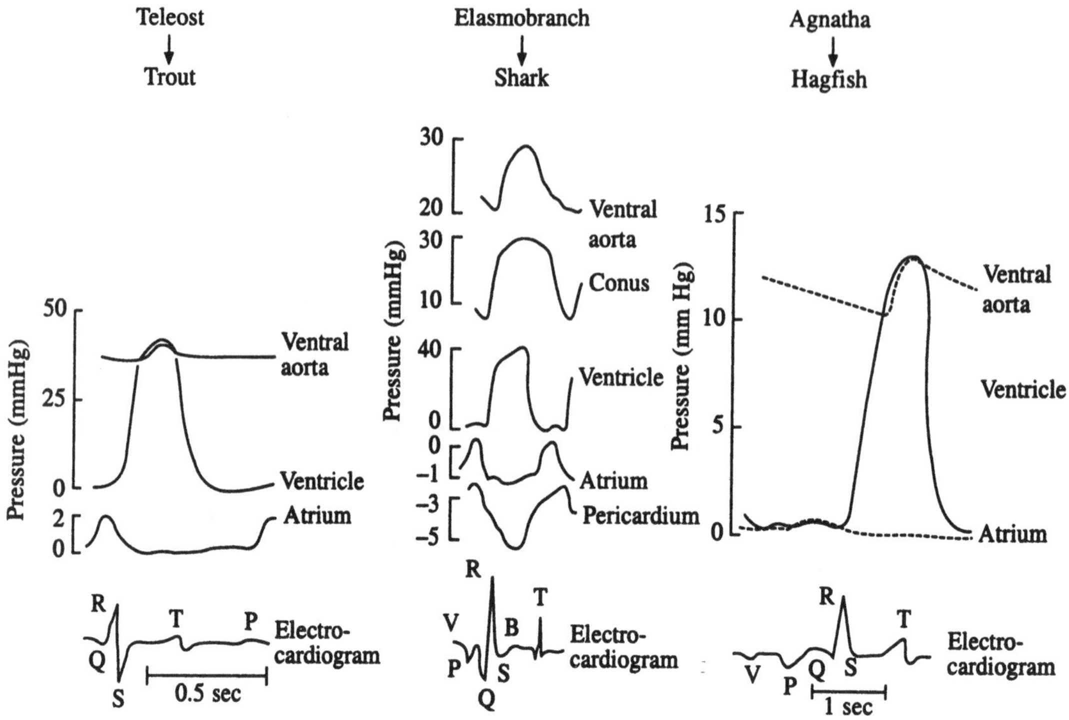

Fig. 8.17 Electrocardiograms.
***Source:* Farrell and Jones (1992)**

is cut, ventricular systolic pressure and cardiac output are reduced and relaxation time of ventricle is increased. Clearly the pericardium has an integral role in cardiac performance in elasmobranchs.

It has been suggested that *vis-a-fronte* filling, as found in elasmobranchs, was advantageous because of the evolutionary development of the venous system. This generalization could certainly be true for cyclostome and elasmobranchs where arterial pressures are generally low, and regulation of venous pressure may be less precise. However, because trout are capable of generating high arterial pressure, and in trout, at least, venous pressure can be increased, this generalization needs refining.

In this regard, it may be considered that ventilation-perfusion synchrony is more common in elasmobranchs than in teleosts. Furthermore, teleosts such as trout and tuna stop ventilatory movements and ram ventilate as swimming speed increases, which is consistent with the observation that *vis-a-fronte* filling becomes less important at higher levels of cardiac performance.

One disadvantage of *vis-a-fronte* filling and a stiff pericardium is that increases in stroke volume may be compromised by the finite volume of the pericardium. Sharks apparently overcome this problem by means of a pericardioperitoneal canal which allows fluid shifts between the pericardium and coelom when stroke volume increases. Whether teleost fish are also capable of adjusting the volume of pericardial fluid or their pericardium is sufficiently compliant to allow large increases in stroke volume is not clear.

It has been suggested that *vis-a-fronte* of the heart is very important under resting conditions in trout (*Salmo gairdneri*) and that *vis-a-tergo* filling of heart becomes progressively more important as maximum cardiac output is reached (Farrell 1988; see also Farrell 1991a; Farrell and Jones 1992).

4.2 Roles of Conus Arteriosus and Bulbus Arterious

At present, the role of the elasmobranch conus arteriosus is obscure. Nevertheless many aspects of its function are clear (see Bushnell et al. 1992).

Depulsation and prolongation of aortic flow during diastole are undoubted functions of the teleostean bulbus. The functional significance of differing bulbar shapes is unknown (read Bushnell et al. 1992; for discussion see also Johansen and Burggren 1980 for 'windkessel effect').

Details regarding pressure and flow relationships are available in Bushnell et al. (1992). Information regarding physical factor affecting blood flow can also be obtained from Bushnell et al. (1992). Jones (1999) reviewed form and functions of bulbus arteriosus.

4.3 Venous Return

The importance of the *vis-a-fronte* or aspiring effect of the heart in fishes, and in elasmobranchs in particular, takes on special significance, since their very large and capacious central venous reservoirs (in particular the cardinal sinuses of elasmobranchs) are so thin-walled that no functionally important vascular smooth muscle in their walls can control the capacitance of the reservoir function. A controlled *vis-a-fronte* aspiration becomes essential for venous return and increased cardiac output.

The apparent lack of smooth muscle venomotor activity in fishes, and in elasmobranchs in particular, takes on special significance by the presence of numerous sphincters on small arteries and veins of skates, and above all the prominent sphincters on the hepatic veins of dogfish and skates. The diffusely distributed small sphincters may be essential to peripheral blood flow distribution. The prominent hepatic vein sphincters situated right where these large veins enter sinus venosus are likely to be of utmost significance for both cardiac output adjustment and liver function. Venous valves have also been described for teleost fishes. No study has reported on their significance but these valves will certainly exert an influence on venosus return during swimming (Johansen and Burggren 1980; Satchell 1992). (Fig. 8.18). Zhang et al. (1998) presented the first demonstration of tonic regulation of vascular capacitance in any fish. The sympathetic nervous system is an active effector of venous tone and compliance in rainbow trout.

4.4 Atrial Contraction: Essential for Ventricular Filling

Venous return, ventricular distensibility, and the resulting end-diastolic volume of the ventricle are well known determinants of the ventricular output in all vertebrates. After pointing out factors influencing venous return special to fishes, we should address the role of atrial filling and contraction for ventricular-end-diastolic volume in fishes. The presence of sinus venosus as a contractile cardiac chamber together with a *vis-a-fronte* aspirating effect of the subambient intrapericardial pressure are the responsible filling mechanisms for the atrium. The atrium itself by virtue of its large capacity and, most importantly, by the presence of the sinoatrial valves becomes an important determinant of ventricular filling. The large capacity atrium and its great compliance together with its compliment of cardiac muscle fibres and the sinoatrial valves make atrial contraction a principal mechanism for ventricular filling. In fish, especially the ancient elasmobranchs, the atrium and ventricle are crucially depending on each other's activity for the filling and hence the output of two chambers. The measure of reciprocity that occurs in fish is due to the fact that the two chambers are enclosed within a semi-rigid pericardium (Johansen and Burggren 1980).

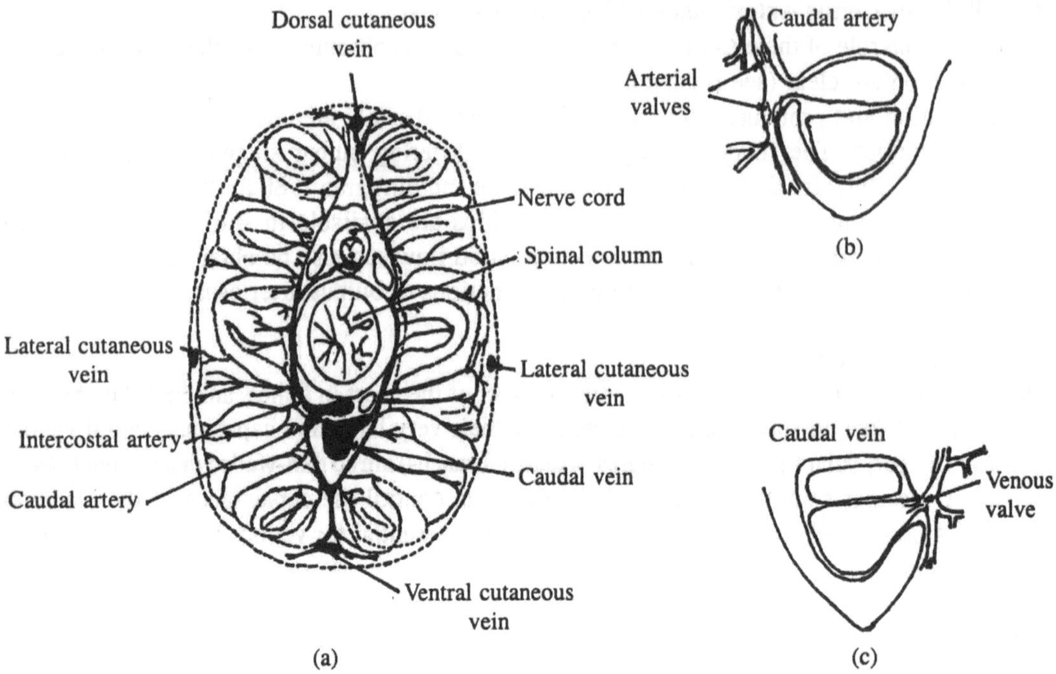

Fig. 8.18 (a) Diagrammatic cross-section of the trunk near the base of the tail of Port Jackson shark, *Heterodontus portusjacksoni*. The intercostal artery and intercostal vein have their origins half a segment apart; they are here shown at the same level. (b) The location of the arterial valves. (c) The location of the venous valves.
***Source:* Jones and Randall (1978)**

4.5 Ventricular Function in Fishes

The spongy region may act as many auxilliary pumps to the large pump of compact myocardium enveloping the entire ventricular lumen. In the dipnoan heart the trabeculation of ventricular myocardium plays an additional significant functional role in the prevention of mixing inside the ventricle between oxygen rich, pulmonary venous blood and oxygen poor systemic venous blood (Johansen and Burggren 1980).

5. BLOOD DISTRIBUTION

There is an increase in blood flow to the working muscles, but a decrease in gut blood flow during exercise in fish. There is a contraction of spleen during both exercise and hypoxia, increasing the number of circulating erythrocytes. There is some evidence for a reduction of blood volume, due to both water loss and transfer to other tissues, during exercise which also raises haematocrit. The peripheral circulation has α-adrenergic vasoconstrictor and a β-adrenergic dilator response. In general, however, little is known of the control of the distribution of blood flow in the peripheral circulation of fish, especially in terms of the regulation of gas transfer to the tissues (Randall 1990; see also Bushnell et al. 1992). Egginton and Rankin (1998) covered this aspect in Antarctic icefish (channichthyids).

Cardiac output increases with exercise, due to an increase in heart rate and stroke volume, the exact contribution of each varying with the species and conditions. This is achieved by a reduction in cholinergic and an increase in adrenergic activity in nerves innervating the heart. There is an increase in venous return and an elevation in circulating catecholamines that probably also contribute to an increase in cardiac output, as well as ameliorating any adverse effects of an acidotic state on cardiac contraction.

Hypoxia results in a marked bradycardia and sinus arrhythmia, both due to increased cholinergic inhibitory activity in the vagus innervating the heart. The afferent arm of this reflex is from external chemoreceptors in the gills detecting a reduction in oxygen in the water flowing over the gills. The bradycardia is offset by an increase in stroke volume in trout, such that there is little change in cardiac output until oxygen levels are low, when cardiac output is reduced. The changes in cardiac function with hypoxia alter the pattern of filling of the gills and the relationship between water and blood flow. These changes presumably enhance gill diffusing capacity and, therefore, oxygen transfer.

Stimulation of nociceptors on the gills cause a bradycardia and a reduction in blood pressure in fish (Randall 1990; see also Bushnell et al. 1992).

It is generally accepted that swimming in teleost fish is accompanied by increases in cardiac output ($\dot{V}b$) and systemic vascular conductance such that there are much smaller increases in arterial blood pressure relative to change in $\dot{V}b$. The increase in systemic vascular conductance is attributed to several events including the opening of resistance vessels by loss of excitatory cholinergic and/or α-adrenergic tone, passive distension through the increase in blood pressure, changes in vascular compliance and possible autoregulation of blood flow. However, the vascular compression, which must accompany undulatory swimming, will compromise any increase in systemic vascular conductance.

Vascular compression, if profound enough can have important cardiovascular consequences. It can elicit a barostatic reflex bradycardia which decreases $\dot{V}b$, reduces blood flow to the muscle, and thereby may prevent the heart from exceeding its intrinsic capacity for pressure work. The accommodation of the barostatic reflex with time, plus a modest, passive vasodilation as $\dot{V}b$ is restored, ensures that the reduced blood flow is only temporary. The set point for the barostatic reflex appears to be temperature dependent (Farrell 1986; see also Butler and Metcalfe 1986).

Driedzic and Gesser (1985) reported that an increase in contraction frequency over the physiological range by teleost hearts results in a decrease in tension development but that elevated Ca_0(extracellular calcium concentration) may reduce or eliminate the negative effect. This relationship could have relevance to the maintenance of cardiac performance during/or following intense swimming when both heart rate and plasma calcium may be elevated.

An extracellular acidosis of both respiratory and metabolic origin normally occurs after exhaustive exercise in teleosts. Typically, the respiratory component predominates during the initial phase of recovery (0–2 h), producing about a 0.5 pH unit reduction in extracellular pH (pHe). *In vitro* experiments clearly demonstrate than an extracellular acidosis of this magnitude depresses cardiac performance by reducing both cardiac contractility and intrinsic heart rate (f_H). Respiratory acidosis decreases myocardial intracellular pH (pHi) in isolated ventricle strips from trout, and the lower pHi probably reduces cardiac contractility as is the case in the mammalian myocardium. However, *in vivo* myocardial pHi does not fall in trout and sea raven despite the pronounced extracellular acidosis after exercise; in fact, the myocardium tends to become alkalotic. Since a physiological concentration of adrenaline protects the normal range of *in vitro* cardiac performance during extracellular acidosis, it is possible, therefore, that the increased level of circulating catecholamines associated with exercise stress promotes some form of active regulation of myocardial pHi, and hence prevents the deleterious effect of acidosis on contractility.

Catecholamines restored contractility in the rainbow trout (*Salmo gairdneri*) heart, in the absence of pHi regulation under perfusion conditions. The mechanism for this recovery may be related to adrenergic modulation of sarcolemmal calcium channels. Further studies on these exchanges are clearly warranted (see Farrell and Milligan 1986).

5.1 Overview of Skeletal Muscle Circulation

There are two types of muscle fibres in the myotome of fishes (red and white). These types form discrete motor systems, receiving separate nervous innervation and having different enzyme distributions, mitochondrial content, and fibre diameter. White muscle represents a power reserve, is used in burst swimming of short duration for escape and attack, and energy is supplied by anaerobic glycolysis. During burst swimming there is rapid depletion of muscle glycogen and a large lactate production. White muscle fatigues rapidly. The sparse vascularization limits the oxygen supply and during activity the muscle must operate anaerobically. The poor vascularization will also contribute to the slow release of lactate from the muscle and minimize the effects of a massive lactate production on the rest of the body.

Red myotomal muscle is used in sustained swimming and is almost impossible to fatigue. There are no changes in red muscle glycogen, and energy is supplied by oxidation of fats. The blood volume and number of capillaries per unit weight are three times that in white muscle.

The muscles of fish are cooled by blood, loss of heat through the general body surface is insignificant. Metabolic heat, which warms the blood, is quickly lost at the gill surface, where blood comes in close proximity to water.

5.2 Circulation in Air-Breathers

The development of air-breathing organs in bimodally breathing fish has necessitated a degree of vascular remodelling in order to enhance gas exchange and support other homeostatic activities. Macrocirculatory changes include several plumbing schemes that allow perfusion of gills, air-breathing organ and systemic circulations in a variety of in-parallel and in-series arrangements. The incorporation of structural adaptations designed to minimize admixture of oxygenated and deoxygenated blood in transit through the heart as well as vascular shunts further increases the efficiency of the gas exchange process. A number of anatomical modifications in capillary architecture and endothelial cell structure are found in air-breathing fish and appear to be unique to these vertebrates. The physiological significance of the microcircular adaptations remains, to a large extent, speculative (Olson 1994; see also Randall, 1994). Of note are publications on *Heteropneustes fossilis* (Olson et al. 1990), *Channa* species (Munshi et al. 1994; Olson et al. 1994). Other significant articles are by Satchell (1976). Burggren and Johansen (1986) and Laurent (1996) (Figs. 8.19–8.24).

5.3 Circulatory Adaptations in Haemoglobinless Icefish

The adaptations in icefish, Antarctic teleosts (Channichthydae) include: (1) increases in blood volume, from two to four times larger than that in many teleosts; (2) increases in blood flow associated with lowered oxygen demands; (3) corresponding increase in cardiac output by

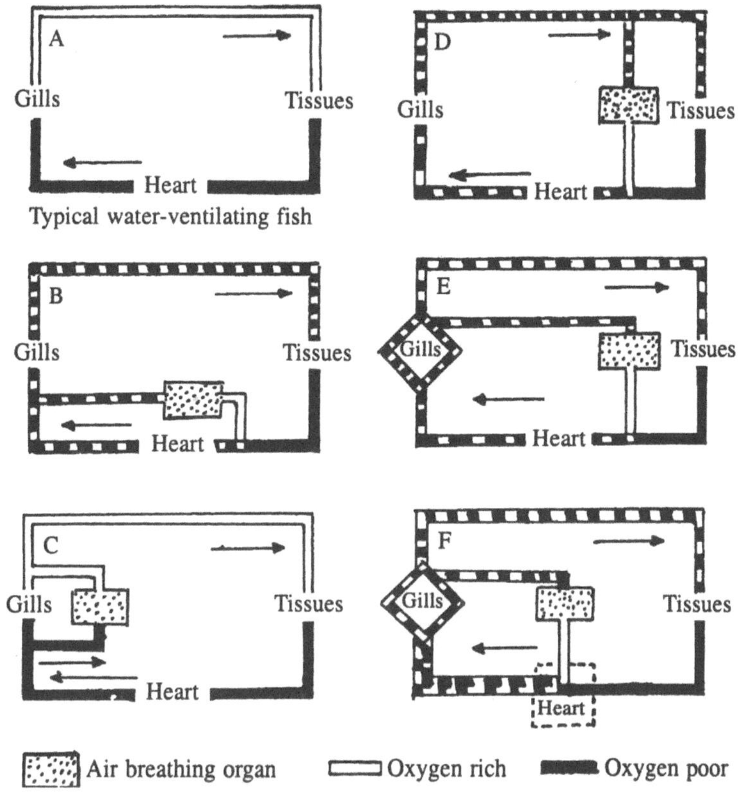

Fig. 8.19 Schematic representation of the flow of oxygenated (white) and deoxygenated (black) blood through the circulation of various air-breathing fishes. A. General arrangement of typical strictly aquatic fish breathing with gills. B. Air-breathing organ (ABO) derived from pharyngeal or opercular mucosa—e.g., *Monopterus, Ophiocephalus* (= *Channa*), *Electrophorus, Amphipnous, Periophthalmus, Anabas*. C. Gills, buccal mucosa, or chambers extending from the opercular cavity serving as ABO—e.g., *Clarias, Saccobranchus* (=*Heteropneustes*). D. Intestinal tract used as ABO—e.g., *Hoplosternum, Plecostomus, Ancistrus, Misgurnus*. E. Air bladder serving as ABO—e.g., *Polypterus, Calamoichthys, Amia, Lepisosteus* F. A lung homologous to that of tetrapods serving as ABO—e.g., *Neoceratodus, Lepidosiren, Protopterus*.
***Source:* Burggren and Johansen (1986)**

means of very large stroke volume (6–15 times greater than in other teleosts) at low heart rate and at relatively low ventral aortic pressure; (4) gill blood vessels and muscle capillaries which have larger diameters compared to other teleosts; (5) increased vascularity in the skin and in the fins; (6) impressive cardiac enlargement resulting in a relative heart weight similar to that of small mammals. The interrelationships between these compensatory adaptations are illustrated in Fig. 8.25 (Tota et al. 1991a, b). Further details can be obtained in the review by Driedzic and Gesser (1994).

1. Life without haemoglobin has been made feasible by three differently directed developments. The metabolic requirements have been suppressed to the maximum extent compatible with a competitive existence. This has been accomplished by (1) selecting a low temperature habitat; (2) eliminating all but the absolutely essential muscle activities; and (3) reducing protein synthesis and other biochemical processes at least as a consequence of reduced activity, and possibly, slow growth.

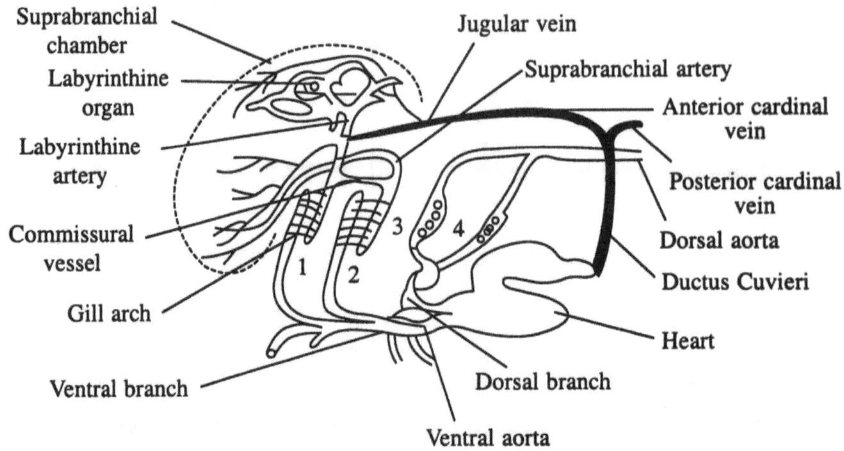

Anabas testudineus: Vascular supply to accessory respiratory organ.

Fig. 8.20 Diagram of major vessels connecting heart, gills, and accessory respiratory organs in *Anabas testudineus*. The ventral aorta divides into a dorsal and a ventral branch: the latter supplies gill arches 1 and 2, the former supplies arches 3 and 4. The efferent branchial artery of arch 1 becomes the labyrinthine artery and supplies the labyrinthine organ. The efferent branchial artery of arch 2 becomes the suprabranchial artery supplying the suprabranchial chamber. A commissural vessel connects efferent branchial arteries of arches 1 and 2. The labyrinthine and suprabranchial chamber are drained through the jugular vein into the anterior cardinal vein. The ductus Cuvieri connects the anterior cardinal vein and posterior cardinal vein with the heart. Blood from gill arches 3 and 4 passes into the dorsal aorta.
***Source:* Munshi et al. (1986)**

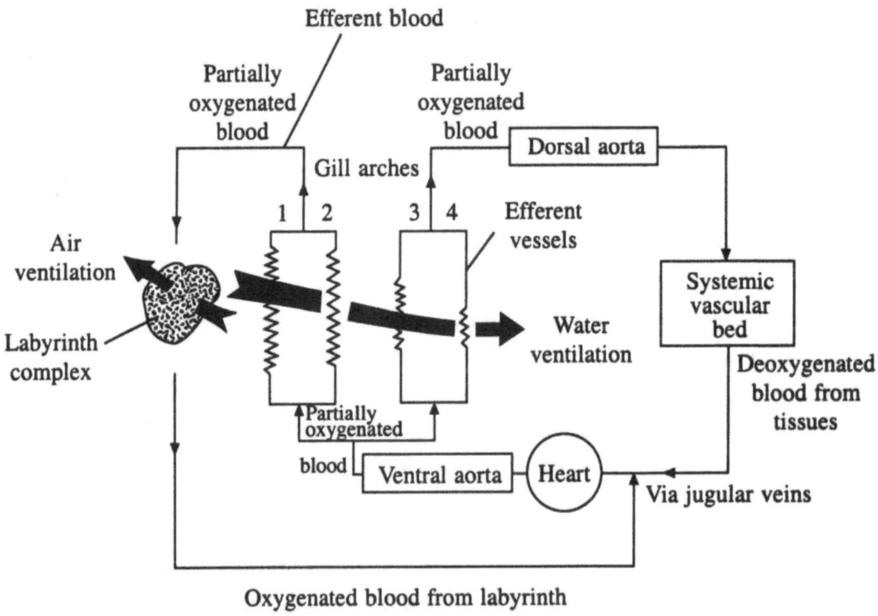

Trichogaster trichopterus: Vascular supply of labyrinth.

Fig. 8.21 Diagrammatic representation of the circulation of the air-breathing fish *Trichogaster trichopterus*.
***Source:* Burggren (1979)**

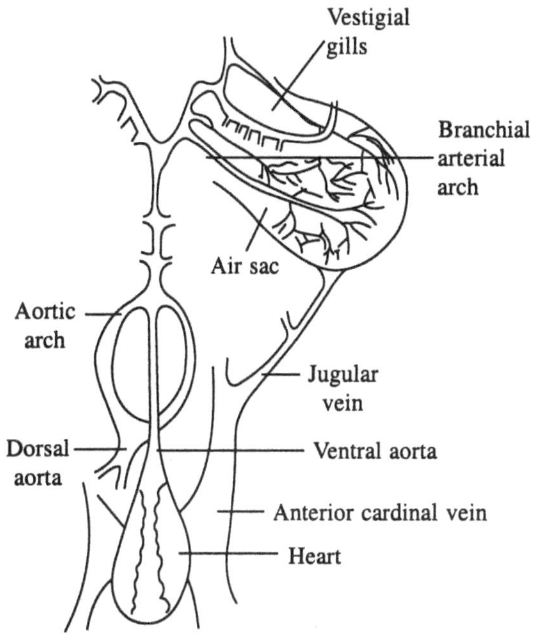

Fig. 8.22 Vascular supply of air sac in *Monopterus* (= *Amphipnous*) *cuchia*.
***Source:* Lomholt and Johansen (1976)**

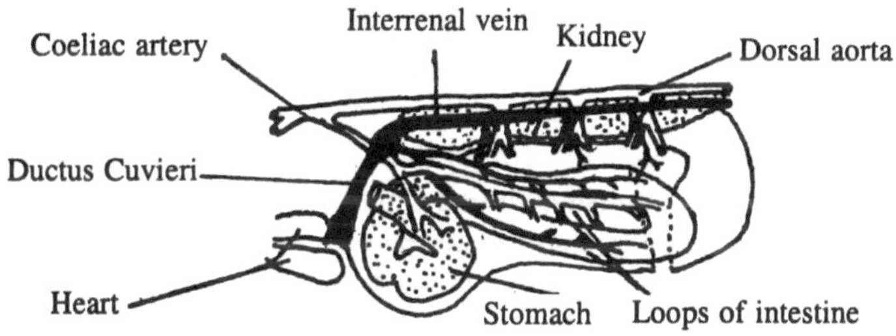

Hoplosternum: Return of oxygenated blood from air-breathing organ to a central vein.

Fig. 8.23 Vascular supply of air-breathing organ in *Hoplosternum*.
***Source:* Satchell (1976)**

2. Features that provide for better gas exchange have been maximized, most notably the vascularization of cutaneous tissues and fins. The added capabilities, combined with the lowered metabolic needs for oxygen, have aided to reduce substantially the demands placed on the circulatory system.

3. Most prominently, the circulatory volume and the rate of turnover have been greatly increased. The many different elements embodied in this development make it difficult to suggest an evolutionary sequence of events (read details: Hemmingsen 1991).

6. HEAT-EXCHANGE RETIAL SYSTEMS

Vascular heat exchangers, consisting of small interdigitating arteries and veins, are associated with swimming muscles in all true tuna (Scombridae), lamnid sharks, and swordfish (refer also

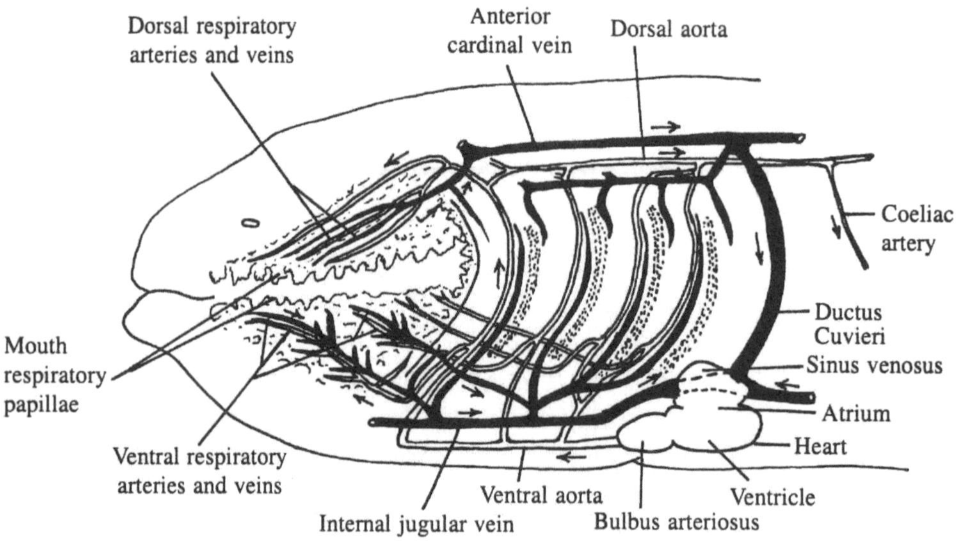

Electrophorus: Return of oxygenated blood from air-breathing organ to a central vein.

Fig. 8.24 Vascular supply of air-breathing organ in *Electrophorus*.
***Source:* Satchell (1976)**

Chapter 15 for heat exchange ratio). The heat exchanger's effectiveness in trapping heat is determined, in part, by three factors; the shunt fraction (how much blood bypasses the exchanger), the surface area of the rete available for heat exchange, and rate of blood flow through the rete. It has been suggested that physiological thermoregulation occurs in endothermic fish. This , presumably, is a result of changes in the heat exchange efficiency of the retial systems resulting from alterations in blood flow through or around the retia. Unfortunately, there is no direct evidence for neuronal and humoral control of blood flow in retia. Future research will make an important contribution towards our understanding of thermoregulation effected through changes in blood flow (read details: Bushnell et al. 1992; see Butler and Metcalfe 1988 and Muñoz-Chápuli and De Andrés 1995 for specific information on elasmobranchs).

7. CARDIAC PHYSIOLOGY IN TUNAS

Farrell et al. (1992) described the cardiac physiology in tunas. Tunas are amongst the most atheletic of fishes. Their metabolic rates are almost an order of magnitude greater than those of rainbow trout (*Oncorhynchus mykiss*), approaching those of similarly sized mammals. Many tuna species have body temperatures higher than the water in water as a consequence of specialized vascular countercurrent systems for conserving metabolic heat. Tunas generate higher arterial blood pressures than other fishes to drive the blood through the relatively high vascular resistance of the branchial circulation. Correspondingly, the relative heart mass is greater than in other fishes. The ventricle has a thick layer of compact myocardium (> 40% by mass), and the coronary circulation supplies both the outer compact and inner spongy layers of myocardium. The metabolic zonation between the compact and spongy layers has been characterized. The vascular heat exchanger of skipjack tuna (*Katsuwonis pelamis*) has been illustrated (Figs. 8.26, 8.27).

In summary, tunas are very active fishes and need high rates of oxygen delivery to their

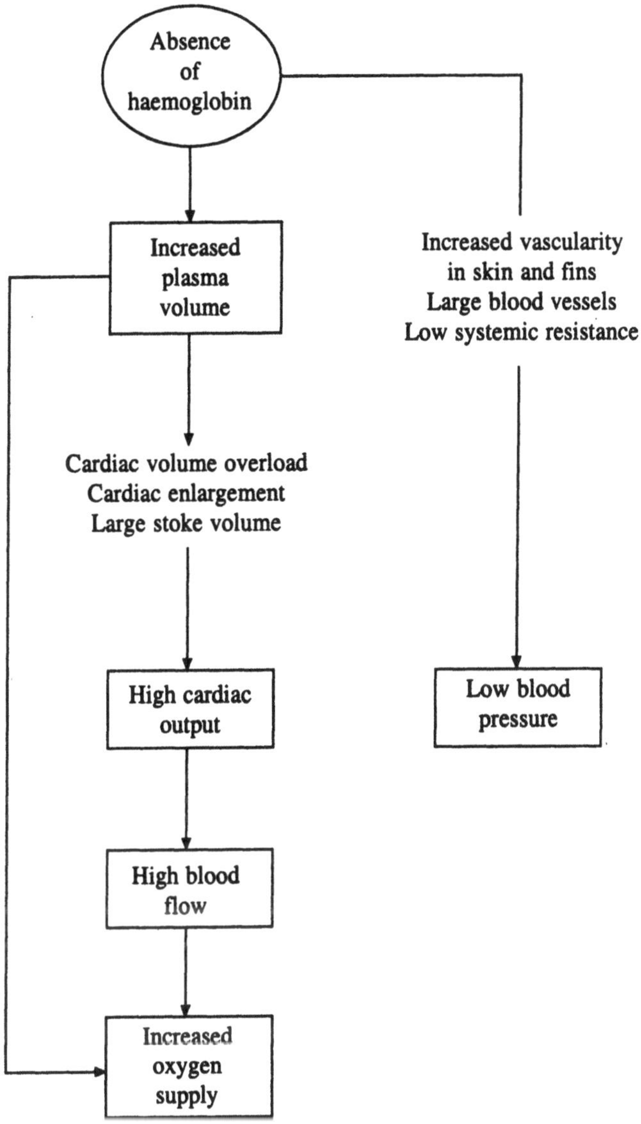

Fig. 8.25 Scheme of circulatory adaptations in icefish *Chionodraco hamatus*.
***Source:* Tota et al. (1991b)**

tissues. High levels of blood flow, ventral aortic blood pressure, and blood oxygen carrying capacity are needed to sustain this activity. High flow and pressure are generated by a large heart which has a relative mass comparable with that of mammalian hearts. Because of the ventricle's high myocardial oxygen demand and large diffusion distance from the lumen, venous blood is unlikely to ever satisfy the normal myocardial oxygen demand. Thus, tunas differ from other fishes examined so far in that the venous blood apparently cannot support resting cardiac performance; a coronary circulation is essential. Furthermore, the very high body mass specific cardiac output in skipjack tuna (*K. pelamis*) appears to be matched by a

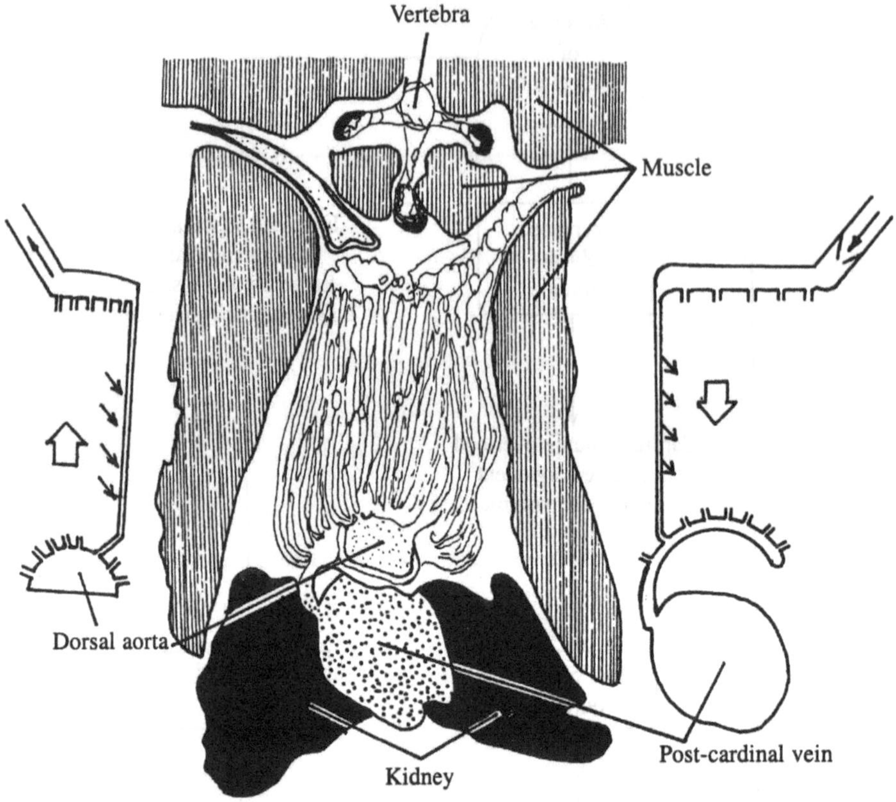

Fig. 8.26 Schematic drawing that illustrates the major features of the vascular heat exchanger of skipjack tuna. Left inset shows pattern of arterial blood flow, and small arrows indicate heat transfer from venules to arterioles in the exchanger.
***Source:* Stevens et al. (1974)**

very high body mass specific coronary blood flow rate. The outer compact myocardium and its coronary oxygen supply appear to be more important for pressure development by the heart than for flow output. Consequently, if dorsal aortic blood pressure is reduced, either through impaired cardiac performance or elevated gill resistance, then coronary flow will likely decrease. The effect of interruption of coronary flow in tunas are probably similar to those in mammals, where ischemia depresses cardiac performance which, in turn, further impairs coronary flow and infraction ensues.

Jones et al. (1993) concluded that 'Windkessel' dominates central cardiovascular dynamics of tuna despite heart rates and blood pressures that fall in the mammalian range.

Details of cardiovascular performance of tuna heart are given by Farrell (1996) and Tibbits (1996).

8. RELATIONSHIP BETWEEN CARDIAC ENERGY METABOLISM AND CARDIAC WORK DEMAND IN FISHES

Sidell and Driedzic (1985) addressed these aspects and summarized as (1) Resting power develop-

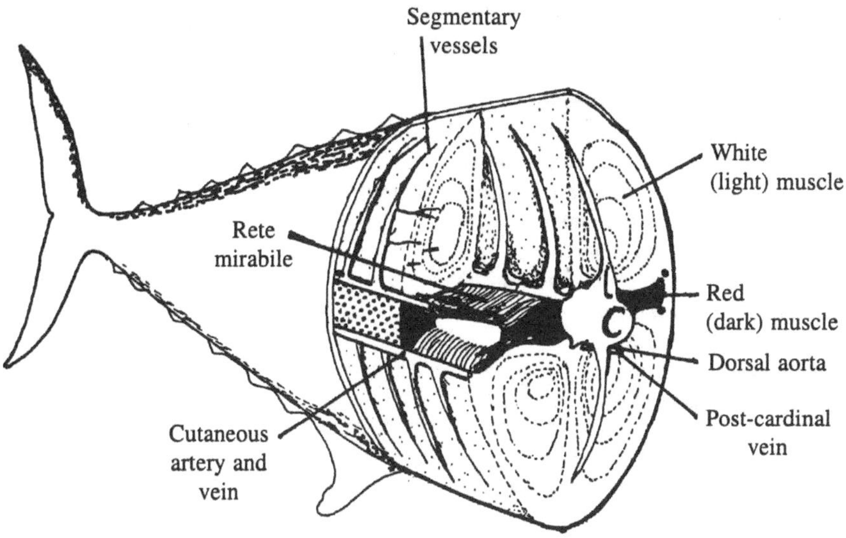

Fig. 8.27 A schematic diagram of the blood supply to muscle of bluefin tuna. Four pairs of cutaneous vessels branch into many smaller vessels and form the countercurrent heat exchange retia above and below the red swimming (dark) muscle. The white (light) muscle is supplied by bands of alternating arteries and veins that pass through the muscle from the segmental vessels, thus acting as small exchangers. Other non-heat-exchanging arteries and veins (not shown) run out from dorsal aorta and post-cardinal vein.
***Source:* Bushnell et al. (1992)**

ment of fish heart varies over a considerable range, from 0.1 m Watt kg^{-1} body wt. to > 3 m Watt kg^{-1} body wt. However, compared to endothermic vertebrates (e.g., rat: 45 m Watt kg^{-1} body wt.) the fish myocardium is a low-work system. (2) Cardiac power development increases from cyclostomes to elasmobranchs and increases further from sluggish to active teleosts. (3) Capacity for aerobic energy metabolism generally expands through the fishes as cardiac power development increases. (4) Anaerobic tolerance of fish hearts is related more to species' ecology and life history than phyletic position or cardiac workload. (5) Physiological results indicate that resting heart work may be supported by metabolism of carbohydrate in cyclostomes, by either carbohydrate or ketone bodies in elasmobranchs and by either carbohydrate or fatty acids in teleosts. Fuel preferenda appear more related to phylogenetic position than cardiac workload. (6) Biochemical indices suggest a significantly greater capacity for oxidation of carbohydrate than fatty acids in teleost hearts, but may overestimate the true differential. (7) The range of cardiac workloads among fishes is apparently not sufficient to have caused the marked predominance of fatty acid utilization characteristic of myocardial energy metabolism in endothermic vertebrates.

The contribution by Driedzic (1988) on the matching of cardiac oxygen delivery and fuel supply to energy demand in teleosts and cephalopods-representing two diverse groups of animals which have evolved quite independently but exhibiting similar life-styles, competing for the same ecological niche-should prove informative to readers.

The intracellular protein myoglobin occurs in high concentrations in vertebrate red skeletal muscle and in the hearts of most vertebrate animals. Myoglobin content in fish hearts is extremely variable (Table 8.5).

Elevated levels appear to occur in two distinct groups of fishes, those that exhibit sustainable

Table 8.5 Heart myoglobin content in selected teleosts (nmol/g wet weight)

Slow moving species, normoxic environment	
Lumpfish (*Cyclopterus lumpus*)	< 1
Ocean pout (*Macrozoarces americanus*)	5.2 ± 1.8
Sea raven (*Hemitripterus americanus*)	63.8 ± 2.1
Three-bearded rockling (*Gairdropsarus vulgaris*)	108 ± 15
Sustainable high speed swimmers	
Chain pickerel (*Esox niger*)	285 ± 12
Atlantic mackerel (*Scomber scombrus*)	332 ± 24
Bluefin tuna (*Thunnus thynnus*)	> 600
Hypoxia-tolerant species	
Eel (*Anguilla rostrata*)	325
Carp (*Cyprinus carpio*)	488 ± 23

Source: Driedzic (1988)
Note: Values are expressed as means ± SE.

high-speed swimming and those that are especially tolerant of environmental hypoxia. In both cases, venous blood reaching the spongy layer has a particularly low P_{O_2}. In contrast, animals that are sedentary and (or) live in constantly oxygen-rich marine environments have very low myocardial myoglobin concentrations. In Atlantic salmon (*Salmo salar*), myoglobin content in spongy layer is more than twice as high as that in the cortical layer.

It has been shown that the function of myoglobin in fish hearts is similar to that in red vertebrate muscle, although the precise window with respect to oxygen availability over which the molecule operates may differ.

Further information regarding correlations between metabolic fuel supply and ATP demand is given. The importance of myoglobin has been detailed further by Driedzic and Gesser (1994).

Bailey and Driedzic (1988) suggested that myoglobin renders oxygen extraction by fish hearts independent of the rate of perfusion. Canty and Driedzic (1987) confirmed the role of myoglobin in protecting heart performance during hypoxia. Further, it has been suggested that myoglobin may not be essential in meeting the demands of maximal respiration, at least under normoxia. The role of myoglobin appears to be confined to situations of low extracellular oxygen availability. Bailey et al. (1990) concluded that the presence of myoglobin in heart allows for the maintenance of respiration, and secondarily contractility, over a wider range of extracellular O_2 levels than is possible in myoglobin-poor heart. The impact of myoglobin in intact heart is apparent at relatively high extracellular P_{O_2}s (40–80 mm Hg) in this model system.

Driedzic (1982) covered various facets of cardiac energy metabolism. Notable points are: (1) The use of glucose as a fuel for aerobic metabolism in all fish hearts examined to date with the possible exception of the freshwater elasmobranch, *Potamotrygon magdalenae.* (2) Lactate appears to be an excellent fuel of metabolism, especially for the teleost heart. (3) The use of fatty acids as metabolic fuels is more diverse. Hagfish hearts appears to have a limited capability to utilize this fuel. (4) Elasmobranch hearts do not utilize exogenous free fatty acids as metabolic fuels. (5) Fatty acids are an excellent metabolic fuel for teleost heart. (6) The use of ketones as metabolic fuels is well developed in elasmobranch hearts. Fig. 8.28 illustrates the major soluble and protein-bound circulatory fuels in fish. Fig. 8.29 illustrates lipid fluxes.

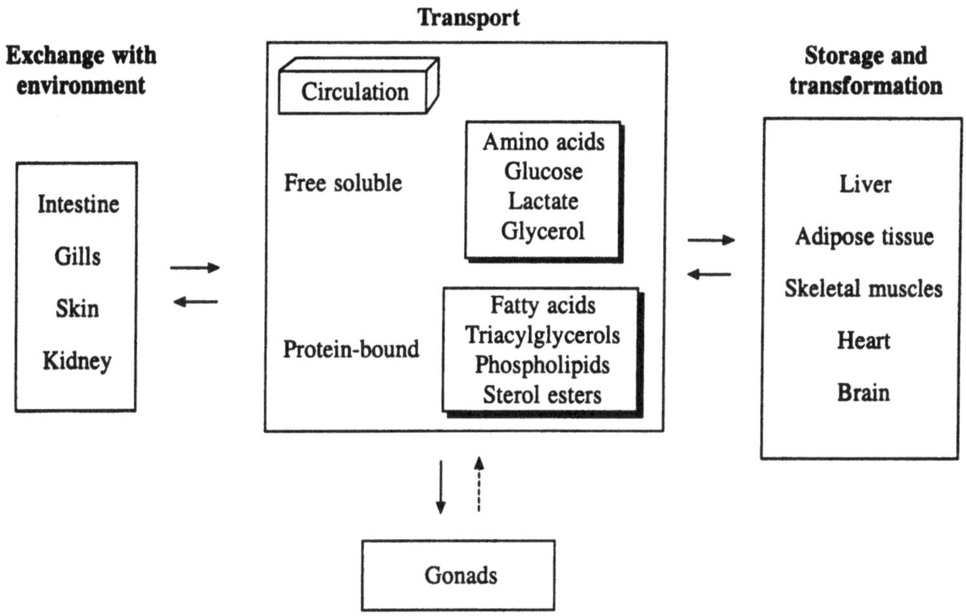

Fig. 8.28 Major soluble and protein-bound circulatory fuels in fish: sources and destinations.
***Source:* Weber and Zwingelstein (1995)**

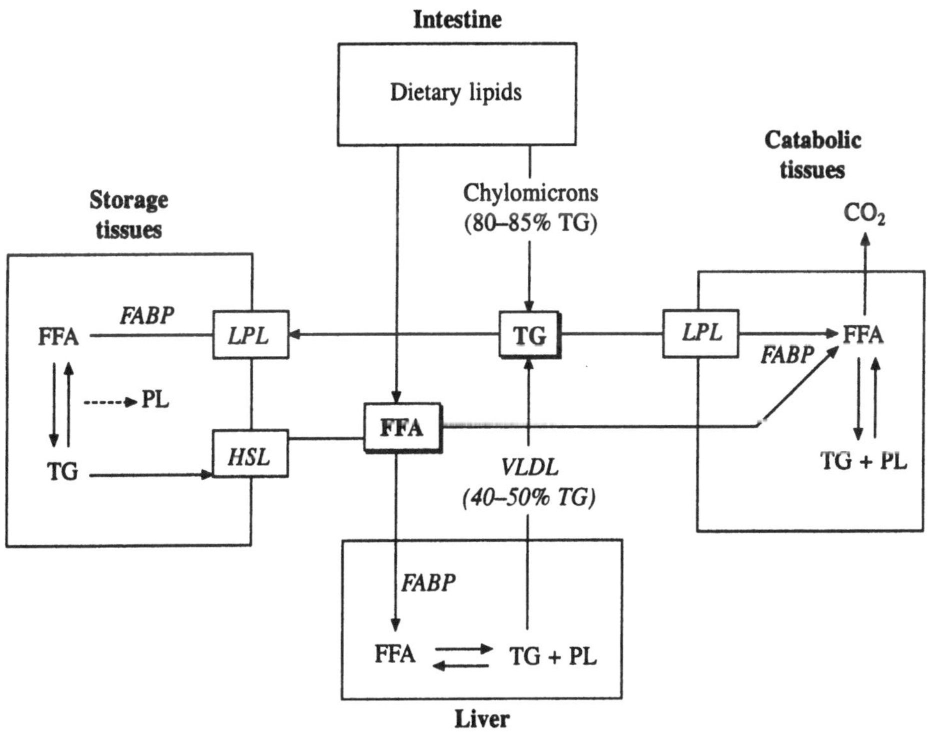

Fig. 8.29 Source, destination and composition of plasma lipids. FABP-fatty acid binding protein; FFA-free fatty acids; HSL-hormone sensitive lipase; LPL-lipoprotein lipase; PL-phospholipids; TG-triacylglycerols; VLDL-very low density lipoproteins.
***Source:* Weber and Zwingelstein (1995)**

On the metabolism in epicardial and endocardial layers, it has been summarized that it is not possible to generate a comprehensive and uniform picture of metabolic differences between endo-and epicardium. Overall the differences in metabolic characteristics between the compact and spongy layers seem small relative to differences in oxygen delivery.

Metabolism under dysoxic condition has been discussed.

Driedzic and Almeida-Val (1996) remarked that the inability of muscle of marine elasmobranchs to utilize fatty acids as an energy source may not be a common feature of all elasmobranchs.

9. AFFERENT INPUTS ASSOCIATED WITH CARDIOVASCULAR CONTROL OF FISH

Mechanoreceptors associated with respiratory passages have been implicated in the control of respiratory pattern and cardiovascular coupling in fish although the specific reflex roles have not been worked out, as yet, for any receptor group. Some fishes may possess intracardiac receptors homologous to mammalian atrial and ventricular stretch receptors. Very few studies, however, have examined cardioventilatory reflexes arising from receptors located in the heart of fishes. The importance of chemoreceptors in cardiorespiratory responses of fishes to hypoxia, hypoxemia and elevated metabolic demands is now being recognized. The mechanism of O_2 chemotransduction in fish chemoreceptors has not been studied.

The anatomical location, afferent pathways, and discharge characteristics of the receptors that mediate cardioventilatory reflex responses to arterial acidosis in fishes, however, remain unknown. The sensitivity to hypercapnia and acidosis in terrestrial vertebrates arises from the combined responses of peripheral and central chemoreceptors. It is likely that responses to hypercapnic acidosis in fishes are also mediated by peripheral, central, or both chemoreceptors, but there have been no concerted efforts to identify whether the characterized branchial chemoreceptors of fishes are Co_2/pH sensitive, and there is compelling evidence for central Co_2/pH chemosensitivity in fishes. Given that systemic arterial chemoreceptors in all animals examined respond to both O_2 and CO_2, it seems probable that the branchial chemoreceptors of fishes mediate reflex responses to hypercapnic acidosis as well as to hypoxemia. The responses are likely to be weak at the relatively low temperature and low Pco_2 in blood or water, but it could be predicted that at elevated temperatures the responses to Pco_2 and acidosis should be more pronounced. The respiratory responses to Co_2 at different temperatures have not, however, been systematically investigated in fishes.

There is ample evidence that fish possess nociceptors, but few studies have described the location, afferent discharge patterns in response to specific stimulants, or specific cardiorespiratory reflexes of these "defense" receptors.

Data regarding central sensory areas controlling cardioventilatory control in fishes are sparse. (Burleson et al. 1992; Milson et al. 1999 refer also Chapters 9, 10, 14, 15).

10. SECONDARY VASCULAR SYSTEM

Recent evidence suggests that the fish cardiovascular system may be unique among vertebrates in that it lacks a true lymphatic system (Vogel 1985; Steffensen and Lomholt 1992; Ishimatsu ct al. 1995; Chopin and Bennett 1996a; Olson 1996b). The vasculature previously thought to be lymphatic in nature, and now termed the secondary circulation, differs from the lymphatic system of higher vertebrates in two important ways: (1) a direct arterio-arterial vascular connection exists between the primary and secondary circulation in fish, and (2) the secondary circulation forms its own system of capillary beds in the gills, fins, mouth, skin, and peritoneum (Vogel 1985; Steffensen and Lomholt 1992; Ishimatsu et al. 1995; Olson 1996b). The secondary circulation is formed from near-capillary sized arteriorles that arise directly from the primary arterial circulation and subsequently reanastomose into an independent secondary arterial system in parallel with the primary system. Downstream of the capillary beds, the secondary veins join the primary circulation at the caudal vein and near the heart (Vogel 1985; Steffensen and Lomholt; 1992). Another secondary circulation pattern has been observed in dorsal musculature of *Blennius pavo* and *Zosterissesor ophiocephalus* where the secondary vessels associated with the segmental arteries anastomosed with the primary red muscle capillaries and drained directly into the primary cutaneous venous circulation. The exact function of the teleost secondary circulation is unclear. Although the close association of the secondary capillary beds with internal and external surfaces suggest an osmoregulatory-homeostatic function, this remains to be demonstrated.

The many descriptions of lymphatic vessels in the anatomical literatures are in most cases venous vessels of the secondary system (Fig. 8.30). The number and diversity of species in which the secondary vascular system has been observed seem to be sufficiently large to justify the statement that the system is of general occurrence among the teleosts. In Fig. 8.31 a schematic diagram of the secondary vascular circulation of a teleost is shown. Many details of its distribution remains to be worked out, and most likely many differences among species will appear.

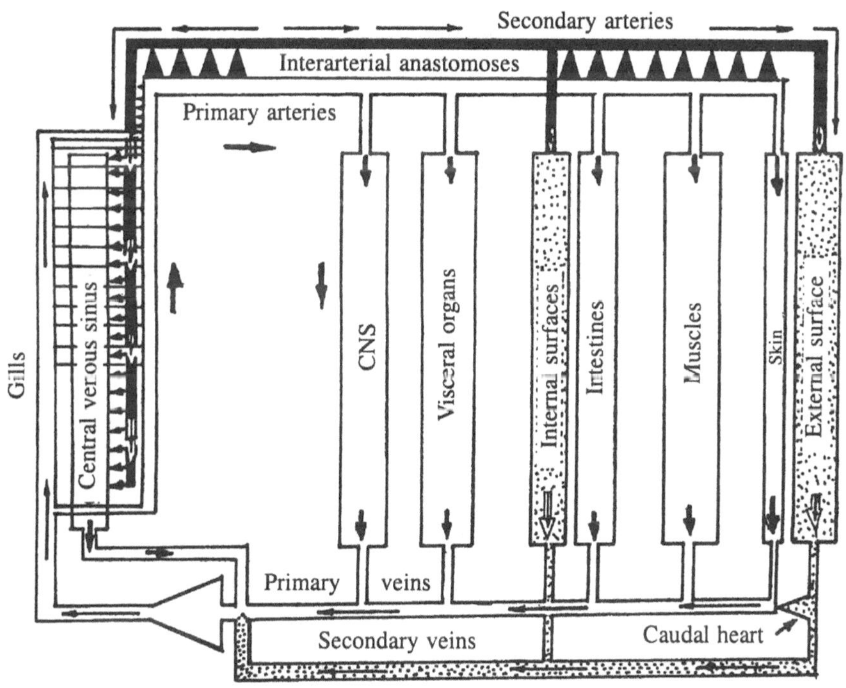

Fig. 8.30 Vogel's concept of the primary and secondary circulatory systems in a typical teleost. *Sources:* Vogel (1985), and Steffensen and Lomholt (1992)

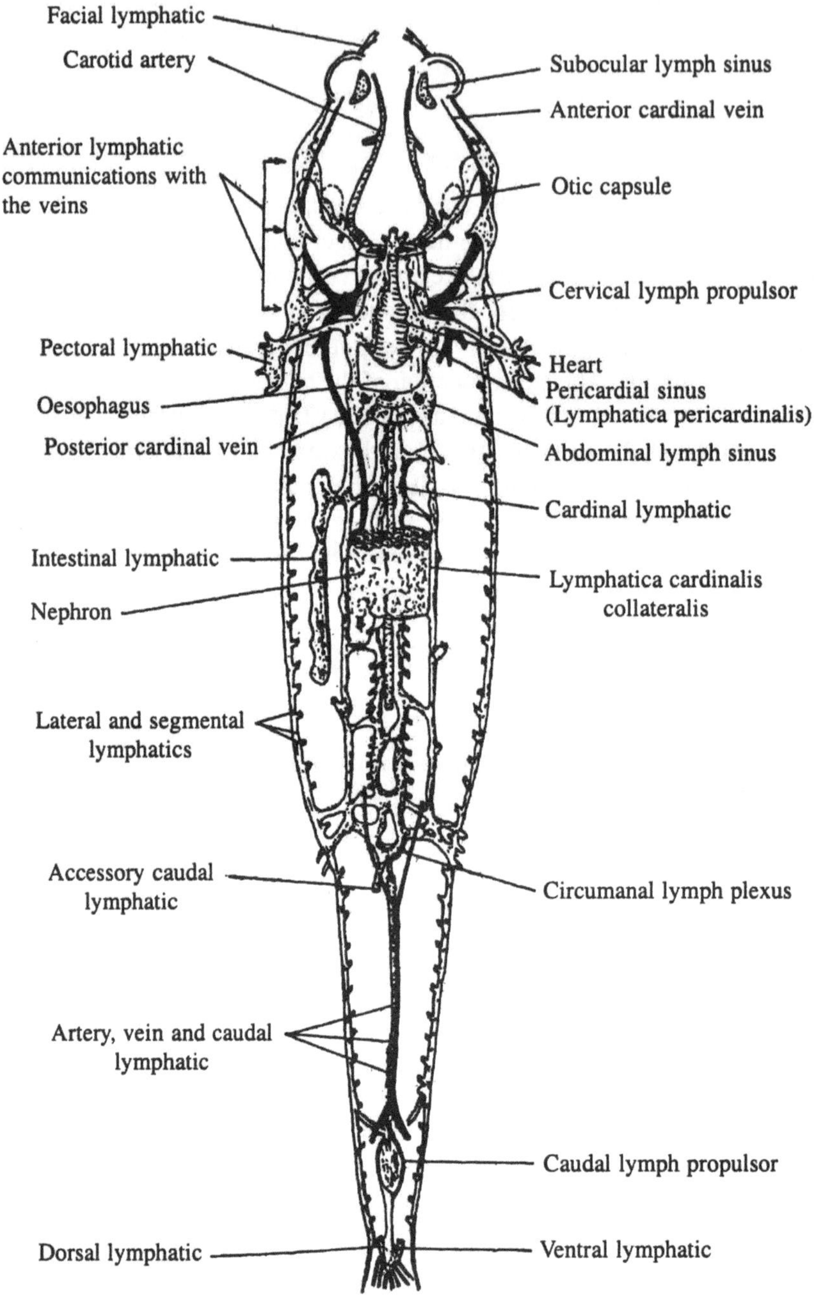

Secondary vascular system.

Fig. 8.31 Secondary vascular system. Composite scheme of systemic lymphatics in teleosts; Veins, black; arteries, cross-lined. The caudal lymph propulsor is shown at its best development, as seen, for example, in the eels.
***Sources:* Kampmeier (1969), and Steffensen and Lomholt (1992)**

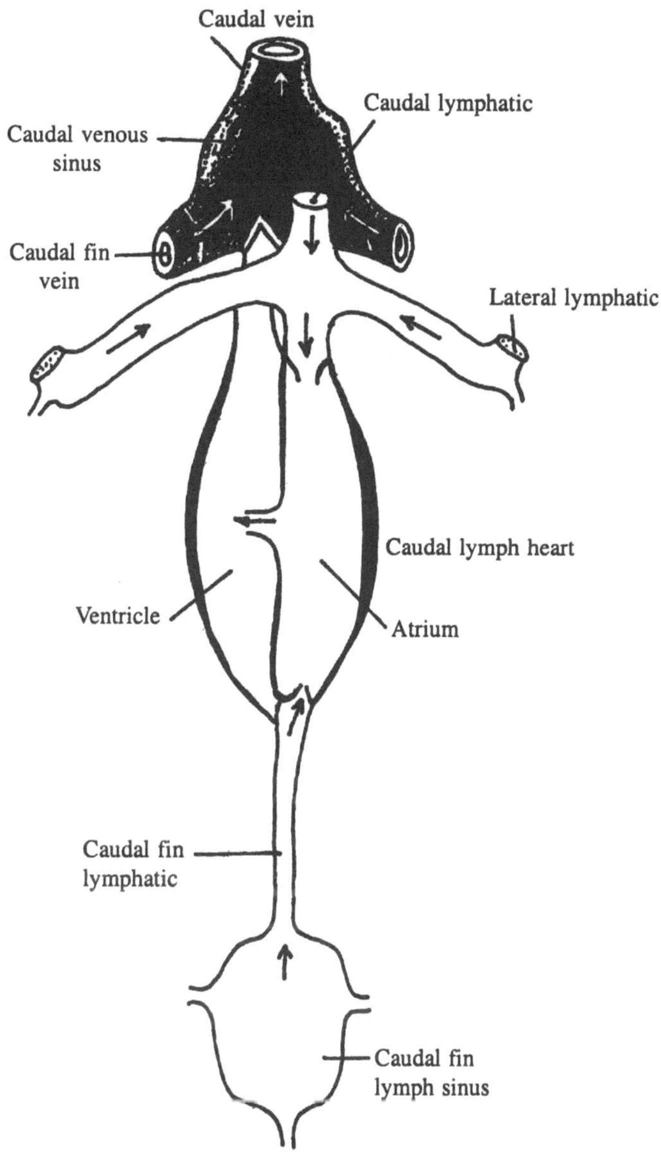

Secondary vascular circulation.

Teleosts: Lymphatic venous relations in the tail.

Fig. 8.32 Scheme of the best-developed pulsatile mechanism (caudal lymph propulsor and its parts) in the tail of teleosts.
***Source:* Kampmeier (1969)**

Taking the elasmobranchs, the anatomical literature reflects the same difficulties in defining lymphatic vessels as in the case of cyclostomes; the problems have been detailed out.

Practically nothing is known about control of flow through the secondary system. Observations on the glass catfish *Kryptopterus bicirrhis* have shown that flow may vary considerably between neighbouring segments. Whether these variations reflect resistance changes in the anastomoses

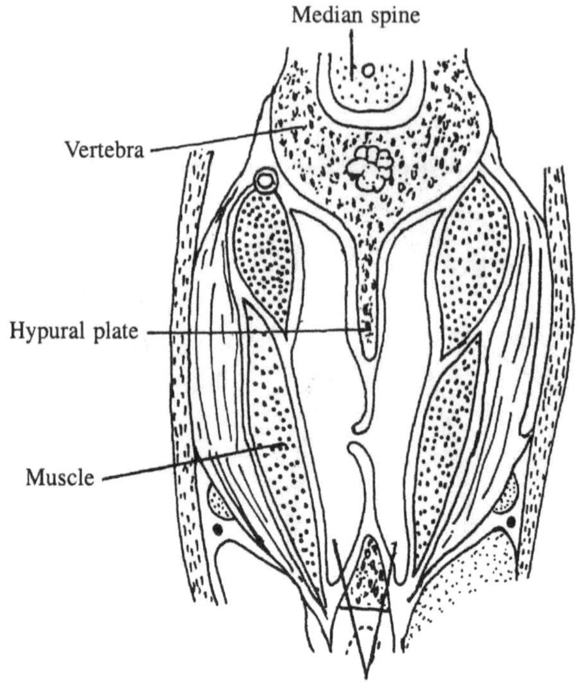

Fig. 8.33 Transection of the caudal lymph propulsor of the common eel. *Anguilla vulgaris*. Note hypural plate, extending down from vertebra, and its foramen.
***Source:* Kampmeier (1969)**

or in the peripheral secondary vessels is not known. At the capillary level local variations in flow have been observed. The secondary capillaries are equipped with a precapillary sphincter (read also Ishimatsu et al. 1995).

The secondary venous system is connected to the primary system as well as in the tail via a caudal heart (see details: Ishimatsu et al. 1995) (Figs. 8.32, 8.33).

The functional significance of the flow of plasma between the primary and secondary system is not known, but mast depend on the time required to exchange plasma from one system to the other. The pressure in the secondary system is not known, but it must be lower than that in the primary system. The presence of a caudal heart suggests that the pressure in the secondary system afferent to the caudal heart is lower than that in the primary venous system.

According to Wardle (1971), the pressure in the neural lymph duct, which is a secondary vein, is negative due to a pumping action generated by respiratory movements. The neural lymph duct now must be considered to belong to the secondary vascular system (Fig. 8.34).

The respiratory gas exchange of the skin of fishes has been shown to take place directly with the surrounding water. An obvious implication is that secondary capillary networks in the skin may supply nutrients, particularly to the mucus-producing cells. There is evidence that the secondary circulation of the gill plays this role in relation to the salt-secreting chloride cells.

Based on the observation that secondary capillary networks are restricted to surfaces, that is,

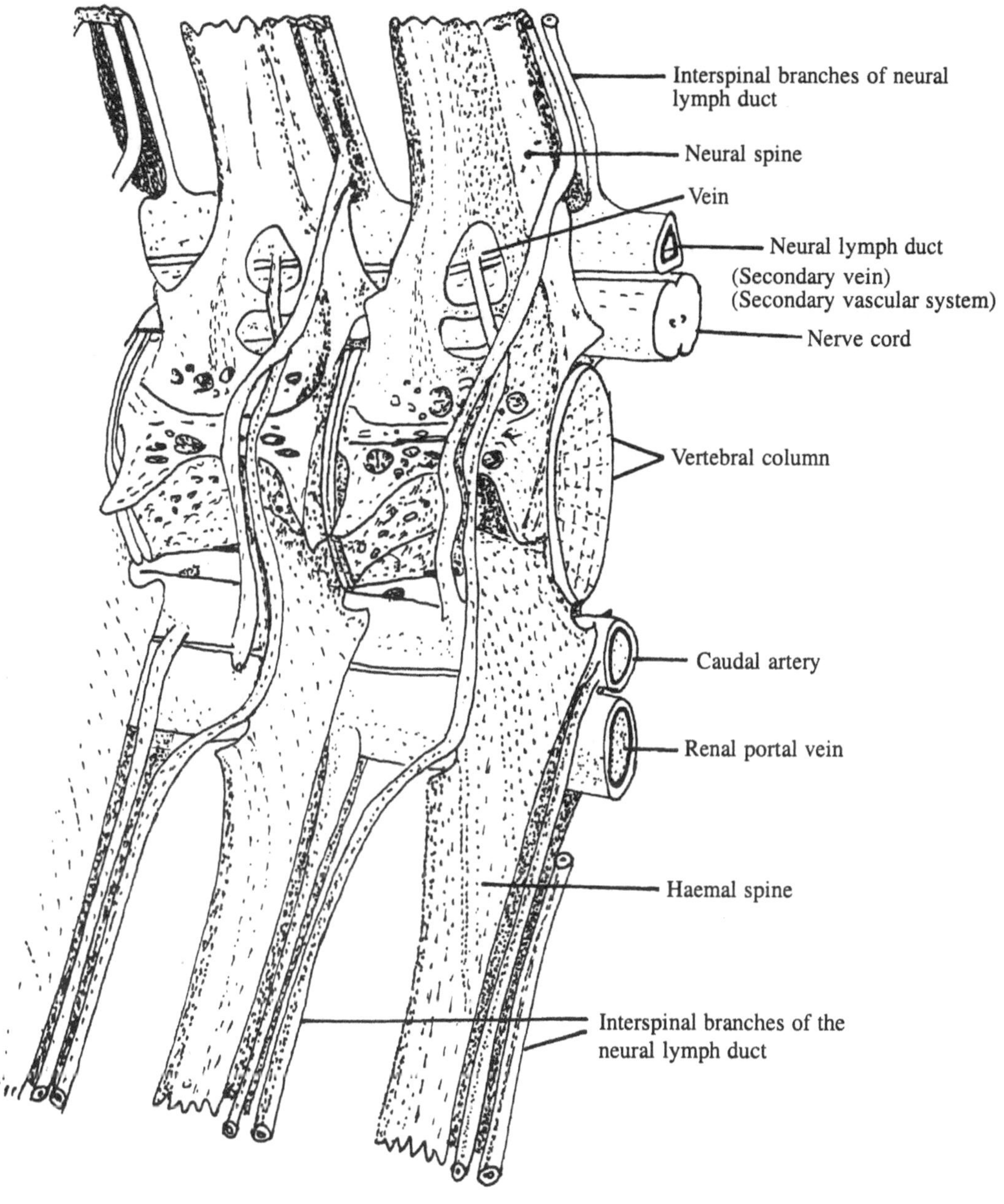

Fig. 8.34 Diagrammatic illustration showing relation of the neural lymph duct to the interspinal ducts and indicating the position relative to the vertebral column and major blood vessels.
***Source:* Wardle (1971)**

to potential sites of infection, it has been proposed that the secondary vascular system may be of importance to the distribution of white blood cells. There is evidence that the process of plasma skimming favours the entrance of white cells into the secondary vascular system. A comparison with mammalian lymphatic system opens up a number of problems which deserve further attention.

Secondary circulation of the vascular heat exchanger in skipjack tuna, *Katsuwonus pelamis* has been recently described (Dewar et al. 1994). Small bore vessels, with characteristics typical of those associated with the secondary circulation, are abundant in the central and lateral heat exchangers. These vessels arise from large-bore retial arterioles and, after a tortuous path, anastomose repeatedly to form secondary collecting vessels which ultimately exit from either pole of the rete. Chopin and Bennett (1996a) reported the ultrastructural features of the secondary vessel system of the blue catfish *Arius graeffi*. Immunohistochemistry revealed the presence of tyrosine-hydroxylase, an enzyme involved in catecholamine synthesis in nerve varicosities close to secondary vessels. This vessel system could therefore be regulated by adrenergic nerves.

True "tetrapod type" lymphatics are present in the South American lungfish, *Lepidosiren paradoxa*, and in the Australian lungfish, *Neoceratodus*. It thus seems that true lymphatics appear for the first time in the evolutionary lineage leading to the earliest tetrapods, a lineage to which the lungfishes are closely related.

The idea of a continuous evolution of the lymphatic system from cyclostomes to mammals has not received favour. (cf. Lomholt and Franko-Dossar 1998).

11. FISH BLOOD CELLS

Unlike the well understood haemopoiesis of mammals, the origin of the different blood cells in fish is still undetermined. Physiology of haemopoiesis has been the subject of a review by Fänge (1986b).

Contribution by Fänge (1992; see for literature) is the main basis of the following details on blood cells. Figures (8.35, 8.36) illustrate different blood cells in shark and lungfish respectively.

Other recent papers on blood cells of gillhead seabream (*Sparus aurata*) (López-Ruiz et al. 1992), melano-macrophages of sea bass (*Dicentrarchus labrax*) and *S. aurata* (Meseguer et al. 1994a), and leucocytes of *S. aurata* (Meseguer et al. 1994b) have added further information. Recent contribution on blood cells of *D. labrax* is from Esteban et al. (2000). Details of fish leucocytes in different fish groups can be obtained from Zapata et al. (1996a). Secombes (1996) discussed various aspects of fish leucocytes in nonspecific cellular defences.

Findings have been reported on head-kidney of the sea bass (*D. labrax*) erythropoiesis and thrombopoiesis (Esteban et al. 1989); granulopoiesis (Meseguer et al. 1990), and stromal cells, macrophages and lymphoid cells (Meseguer et al. 1991). It has been mentioned that like mammals, stroma cell types here include endothelial and adventitial cells comprising the sinusoidal wall, fibroblast-like reticular cells related to scarce reticular fibres, and macrophage-type reticulum cells, the last probably corresponding to the resident macrophage population of higher vertebrates. Their possible role in the haemopoietic microenvironment has been considered. Studies on phagocytic defense mechanism in sea bass (Esteban and Meseguer 1994) and reticulo-endothelial stroma of the head-kidney of sea bream (Meseguer et al. 1995) are significant additions.

Further, the value of fish haematology in establishing accurate diagnostics and the main role of teleost head-kidney in both humoral and cell mediated immunities has been demonstrated. Ultrastructural findings have been reported in the head-kidney of the sea bass (*D. labrax*): erythropoiesis and thrombopoiesis (Esteban et al. 1989); granulopoiesis (Meseguer et al. 1990), and stromal cells, macrophages and lymphoid cells (Meseguer et al. 1991).

Red cells: The concentration of erythrocytes in the blood can be expressed as haematocrit or as the number of red cells per volume blood. The haematocrit values of fish blood range from

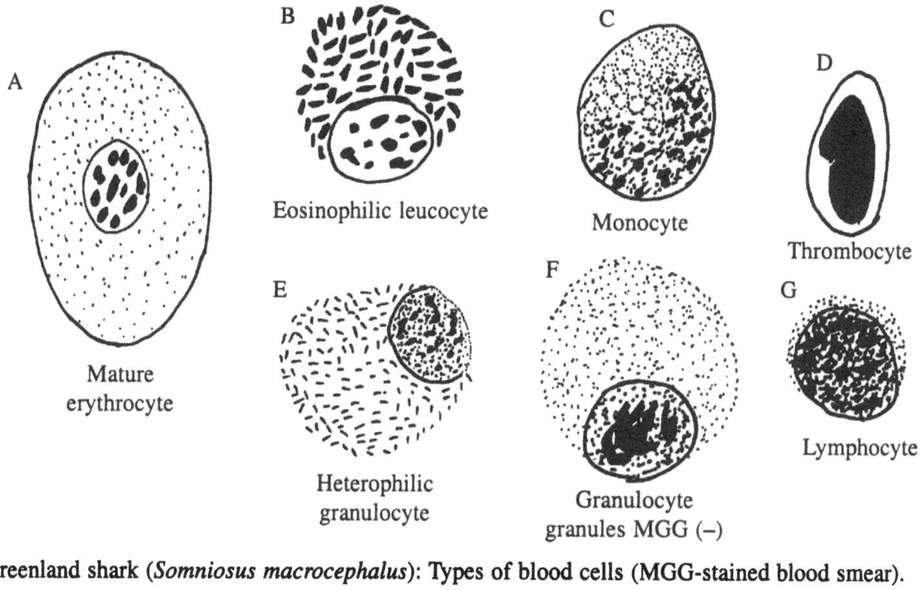

Fig. 8.35 Greenland shark (*Somniosus microcephalus*). Types of cells in a MGG-stained blood smear. (A) mature erythrocyte. (B) Eosinophilic leucocyte (G1). (C) Monocyte: basophilic vacuolated cytoplasm, bean-shaped nucleus. (D) Spindle-shaped thrombocyte. (E) Heterophilic granulocyte (G2); small rod-shaped weakly eosinophilic granules. (F) Granulocyte with granules that do not stain with MGG (G3): nucleus eccenteric, pycnotic. (G) Lymphocyte: basophilic cytoplasm forms a rim around the nucleus.
***Source:* Fänge (1987)**

almost 0 to more than 50% in actively swimming, surface-feeding species. In most teleosts the haematocrit is between 20 and 40%, but some members of the family Channichthyidae (Antarctic icefishes) has colourless blood containing extremely few erythrocytes which are fragile and lack haemoglobin (see details: Wells et al. 1990; Table 1). Sharks, rays and *Chimaera* show not only lower haematocrits but also lower haemoglobin values than the teleosts. The haematocrit, haemoglobin concentration, and other haematological parameters are highly sensitive to

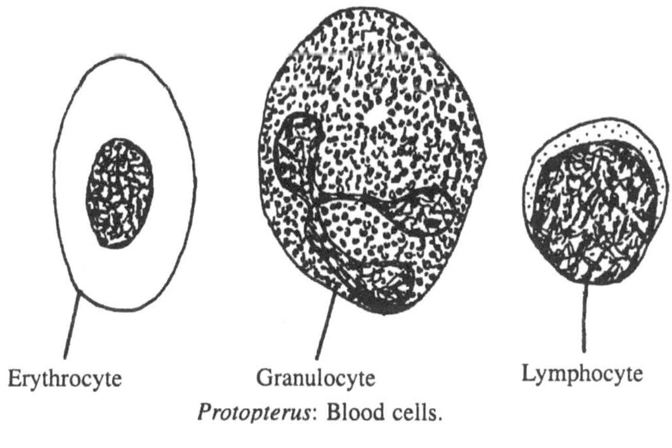

Fig. 8.36 Erythrocyte, granulocyte and lymphocyte from a dipnoan *Protopterus*.
***Source:* Fänge (1985, 1992)**

Table 8.6 Sizes of blood cells (length or diameter in μm) and amounts of nuclear DNA (pg/N) in fishes

Fishes	Erythrocytes	Granulocytes	DNA
Eptatretus	18	14–15	5.6
Squalus	21	15	14.6
Protopterus	40	44	160–284
Teleosts (several)	10–15	-	0.8–6.6

Source: Fänge (1985) (Abridged)

physiological changes, for instances those occurring due to stress. Contraction of splenic vessels may force stored red cells into the general circulation.

Mature red cells usually are oval and disc-shaped with a compact nucleus. The erythrocytes of lamprey, *Lampetra fluviatilis*, are bioconcave with an almost circular outline, rather similar to mammalian erythrocytes. Round disc-shaped red cells occur in some teleosts (e.g., *Siphonostoma typhle*). The average red cell size differs between systematic groups of fishes; active species have slightly smaller erythrocytes than in nonactive, and exceptionally large red cells are reported in dipnoans. Elasmobranchs, holocephalans, and myxinoids also possess relatively large blood cells. Nonnucleated red cells are found in some marine deep water teleosts (e.g., *Maurolicus mulleri*). The sizes of blood and other cells are correlated to the content of DNA (Table 8.6).

Erythrocyte nuclear measurements have been used to distinguish between diploid and triploid individuals of fish species.

The cytoplasm of mature red cells is reported to contain mitochondria, Golgi complex, so-called segregation apparatus, pinocytotic vacuoles and microtubules (marginal bands).

The amount of erythrocytes and the total haemoglobin concentration in the blood vary in accordance to life habits. Fast swimming species of fishes on average have more erythrocytes, larger haematocrits, and more haemoglobin than less mobile forms. Under normal conditions fishes usually are able to keep the concentration of red cells in the blood relatively constant. Such a red cell homeostasis results from a dynamic equilibrium between new formation (erythropoiesis) and destruction of erythrocytes. New erythrocytes are continuously entering the circulation, and effete erythrocytes are destroyed at the same time. Destruction and elimination of aged or damaged erythrocytes are brought about by macrophages in the kidney and spleen. The life span in tench, *Tinca tinca*, is more than 150 days.

The metabolism of vertebrate red cells generates energy for the maintenance of cell shape and for the transport of substances across the cell membrane. The sodium pump and phosphorylation processes consume about 50% of the total energy. Fish erythrocytes are metabolically more active than mammalian erythrocytes. The haemoglobins of vertebrates with the exception of cyclostomes are tetrameric, built of four peptide chains, each with one heme and a molecular weight of about 17,000. Monomeric and dimeric haemoglobins are found in the cyclostome red cells. The haemoglobin of the coelacanth, *Latimeria*, shows features of fish as well as tetrapod relationships. Sequence analyses of the haemoglobin indicate that *Latimeria* is more closely related to tetrapods than are dipnoans. Multiple haemoglobin systems in many fishes may be important in physiological adaptations to variable environments.

There are reports about the presence of other proteins, phosphates, and nitrogen metabolites in fish red cells.

Red cell membranes consist of a bimolecular lipid layer associated with carbohydrate-containing antigens intermingled with proteins, among which are contractile proteins and enzymes. Spectrin, actin, and other proteins form an intricate membrane skeleton interacting with haemoglobin, membrane transport proteins, and tubulin of the marginal bands. Spectrin and actin occur in the red cell membrane of elasmobranchs but in cyclostomes spectrin is lacking. A few studies have been made in order to find blood group systems in fish. Somewhat less than half of the erythrocyte membrane consists of lipids, mainly phospholipids. The function of the membrane lipid components is influenced by environmental factors such as diet and temperature. The red cell membrane contains transport proteins for both anions and cations. The exchange of cations may be brought about by a sodium pump. Little is known about molecules responsible for cation transport across the erythrocyte membrane in fishes. Hagfish RBCs lack anion exchanger (Dohn and Malte 1998).

Many vertebrate erythrocyte membranes possess β-adrenergic receptors for catecholamines. The erythrocytes of elasmobranchs and cyclostomes, in contrast to those of other vertebrates, are not influenced by catecholamines, and adrenergic receptors may not occur in the erythrocyte membrane of cyclostomes.

The osmotic fragility of the erythrocytes differs between species. The erythrocytes of euryhaline species (e.g., rainbow trout, gars) show higher tolerance against hypotonic salt solutions than those of exclusively marine teleosts. The erythrocytes of the hagfish (*Myxine glutinosa*) seem more resistant against hypotonic salt solutions than those from an elasmobranch. Taurine and amino acids are important in the intracellular osmotic volume regulation of the erythrocytes of teleosts and elasmobranchs.

The oxygen binding capacity of the blood is dependent on the concentration and properties of the haemoglobin containing red cells. Fish in different ecological environments need different functioning haemoglobins. The oxygen binding capacity of the haemoglobin molecule varies with the chemical nature of the protein moiety of the molecule and with the intraerythrocytic content of certain substances such as phosphate, which modulate the properties of the haemoglobin molecule. The oxygen binding properties of the blood are controlled by changes in the physiochemical environment of the erythrocytes. The oxygen affinity of the blood of the water breathing fishes generally is higher than in air-breathing fishes due to different intrinsic properties of the haemoglobins. The biochemistry of oxygen transport in red-blooded Antarctic fish has been discussed by diPrisco et al. (1991). For linkage between O_2 and CO_2 transport, read Brauner and Randall (1998).

The adaptive regulation of intraerythrocytic levels of nucleoside triphosphates (NTPs) has been extensively reported in many fish species. Adenosine triphosphate (ATP) and guanosine triphosphate (GTP) have been described as the main NTPs in fish erythrocytes. They are negatively charged phosphates that bind to deoxyhaemoglobin but not to oxyhaemoglobin and both act as negative modulators of haemoglobin oxygen-affinity. Nonetheless, the mechanism(s) by which oxygen tension affects NTP levels is unknown (see further Val et al. 1995). Val et al. (1991) reported that the adoption of an air-breathing habit in *Arapaima gigas* (an Amazonan fish, the pirarucu) is accompanied by significant changes in haemoglobin allosteric cofactors, both metabolically labile ones (ATP and GTP) and IPP (inositol pentaphosphate), which are appropriate for a transition from gill ventilation with hypoxic water to air breathing. Contribution from Nikinmaa and Boutilier (1995) may be consulted for details of production of ATP and GTP. Feuerlin and Weber (1994) reported a method for rapid and simultaneous assay of ATP and GTP, and of anodic and cathodic Hbs, in minute quantities of fish blood. For recent review of red blood cell metabolism, see Walsh et al. (1998).

Oxygen consumption by RBCs of fish is a well documented phenomenon. However, metabolic fuel used by fish erythrocytes is poorly understood. *In situ*, RBCs are bathed in plasma rich in glucose, lactate, fatty acids and amino acids, all potential metabolic fuels. It is probable that some fish erythrocytes are fuelled primary by exogenous glucose. This contention is based upon the presence of a number of the prerequisite glycolytic enzymes in yellow perch *Perca flavens* and rainbow trout *Oncorhynchus mykiss* erythrocytes, the ability of lungfish, electric eel and *Anguilla japonica* RBCs to take up glucose and the production of radioactive CO_2 from ^{14}C-labelled glucose by lungfish, nurse shark and rainbow trout erythrocytes. Sea raven (*Hemitripterus americanus*) erythrocytes have an aerobic metabolism that is fuelled by exogenous glucose (see Sephton et al. 1991; read also Sephton and Driedzic 1994a, b).

A rise in haematocrit is associated with an exponential rise in blood viscosity, and consequently in the work that has to be performed by the heart. Fish biologists have documented a number of factors triggering haematocrit increase, both quantitatively through a rise in the number of cells per unit blood volume, and qualitatively through erythrocyte swelling. There must therefore be a balance between haematocrit and viscosity, with additional regulation of oxygen transport provided by changes in blood-oxygen affinity (read Wells and Weber 1991).

The efficiency of transport of carbon dioxide in vertebrate circulation depends on a rapid exchange of chloride and bicarbonate through the erythrocyte cell membrane. The enzyme carbonic anhydrase facilitates the anion exchange through the erythrocytic membrane. However, in fish the anion permeability of the red cells varies. In cyclostomes the capacity for CO_2 excretion probably is lower than any other vertebrates. For special interest on carbonic anhydrase, see Lessard et al. (1995), Randall and Val (1995) and Randall and Brauner (1998, 1999).

Wood (1987) compared the oxygen transport in aquatic and terrestrial vertebrates. Many aspects of the external milieu impact upon the respiratory properties of blood. Water is often polluted or otherwise deficient in oxygen, making adaptations to hypoxia necessary. Air breathing ectotherms must adapt to hypoxia from an internal mechanism-intracardial shunt. The principle of symmorphies applies to red cells in that their number and qualities are sufficient for the required O_2 transport. Blood gas transport is focused in a review on gas change (Perry and McDonald 1993; see also Gilmour 1998).

It has been suggested that the role of the red cell in metabolic homeostasis is less important in ectotherms.

The position of the O_2 dissociate curve is much more labile in ectotherms than in homeotherms due to large daily fluctuations of temperature and pH.

Burggren et al. (1985) covered certain aspects of blood respiratory properties. Interesting information is provided on haemoglobins of hagfish, lampreys and marked changes in haemoglobin-O_2 binding associated with aestivation in lungfishes. Also, changes in respiratory properties of blood associated with the evolution of air breathing are discussed. Wells and Weber (1983) investigated the efficiency of the internal oxygen transporting system of the dogfish *Squalus acanthias*.

Weber (1992) discussed molecular adaptations to hypoxia in fish. Fish Hbs exhibit large variations in intrinsic O_2 affinities and cofactor sensitivities, both interspecifically and among multiple Hbs. Whereas relatively little is known about the molecular basis for variations in intrinsic O_2 affinities, plausible explanations exist for differences in ligant sensitivities. Attention has been paid on (1) allosteric effects, (2) functional heterogeneity, and (3) the Root effect. In an earlier article, Weber (1990) described the incidence and classification, physiological importance, ontogenetic changes, molecular explanations, and acclimatory and nongenetic multiplicity under the purview of multiple Hb systems in fish. For functional adaptations in haemoglobins, see Weber and Jensen (1988); Weber and Wells (1989); Nikkinmaa and Weber (1993), Fago and Weber (1995, 1996, 1998), Brauner and Randall (1998), Jensen et al (1998), Brauner and Jenson (1999), and Val et al. (1999). Physiological adaptation of Hbs of marine and freshwater fish are described by di Prisco and Tamburrini (1992) (see also Perry and McDonald 1993; Gilmour 1998).

Scholnick and Mangum (1991) discussed the sensitivity of Hbs to intracellular effectors. It has been suggested that ATP lowers O_2 affinity of the shark but not ray Hbs, which may reflect the origin of intracellular modulation within the elasmobranchs.

Protons are usually passively distributed across the red blood cell membranes of both freshwater and marine teleost fishes, as shown for carp, rainbow trout and Atlantic cod. However, in the red blood cells of the rainbow trout, catecholamines activate a sodium/proton exchanger. The resulting net proton efflux shifts the equilibrium between carbonic acid and bicarbonate inside the red blood cell towards the latter, bicarbonate leaving the red blood cell in exchange for chloride. As a result of intracellular sodium and chloride accumulation, the red blood cells swell due to osmotically induced water uptake. The activation of the sodium/proton exchanger has important physiological consequences. Facing extracellular acidosis, the catecholamine-induced proton efflux and the resulting intracellular alkalization will increase haemoglobin O_2-affinity and offset the potentially detrimental Bohr and Root effects associated with decreased pHe. Catecholamine-induced changes in red blood cell ion and water content as well as changes in pHi, pHe and P_{50} have been reported in many teleosts other than rainbow trout. Gilmour (1998) has given Fig. 3 on a schematic model outlining the effects of catecholamines on the teleost red blood cell. Recently, Berendrink and Bridges (1994a) reported a sodium dependent pHi regulatory mechanism in the red blood cells of the marine Atlantic cod. Further, Berendrink and Bridges (1994b; see literature) suggested the activation of a sodium/proton exchanger by catecholamines in the red blood cells of the Atlantic cod. The red blood receptor affinity for adrenaline was three times higher than that for noradrenaline. Comparison with data in the literature for *in vivo* catecholamine concentrations indicated that adrenaline was more effective than noradrenaline in activating the red blood cell sodium/proton exchanger in the Atlantic cod *in vivo*.

The effects of beta-adrenergic stimulation of teleost erythrocytes have been well documented. Addition of beta-adrenergic agonists to teleost erythrocytes *in vitro* results in an increase in erythrocyte pH and water content. It has also been demonstrated *in vivo* that an increase in circulating catecholamine levels results in an increase in erythrocyte pH and water content.

Nikinmaa and Boutlier (1995) described in detail stimulation of the sodium/proton exchanger by catecholamines and its effect on red cell pH. Sorensen and Weber (1995) reported the combined effects of adrenaline and cortisol and of oxygen state on trout blood viscosity. Viscosity changes continuously in parallel with arterio-venous changes in oxygenation and red cell volume.

Randall (1990), in his contribution on control and coordination of gas exchange in water breathers, described the transfer of oxygen and carbon dioxide alongwith other aspects. Further details can be obtained from the contribution of Randall and Brauner (1998, 1999).

Red cells and spleen: The teleost spleen needs special attention in this context. It is a discrete organ containing, in addition to erythrocytes sequestered from circulation, erythropoietic tissue involved in the synthesis of new erythrocytes. In this division into supply and synthesis, there appear to be some species differences. The European eel (*Anguilla anguilla*) spleen is an erythrocyte reservoir but is not thought to be a major erythropoietic organ. The splenic mass in goldfish (*Carassius auratus*) does not change with induced anaemia. The rainbow trout (*Salmo gairdneri*) spleen, in contrast, plays an erythropoietic role during anaemia, in addition to serving as a store of erythrocytes. Exercise in rainbow trout is accompanied by haemoconcentration and the spleen appears to contribute to elevated circulating haemoglobin levels. Capture stress induced a 25% reduction in spleen haemoglobin concentration in the marine teleost *Girella tricuspidata*, and splenic histology revealed melanomacrophagic centres and erythropoietic tissue indicative of a major role in erythrocyte destruction and synthesis. Neither role can be played by the spleen of icefish, which lacks haemoglobin.

Although storage and synthesis of erythrocytes in the spleen have been examined in few fish, several intraspecific factors affect the size of the spleen. These include reproductive status and body weight, exercise and hypoxia. Active fish appear to have highly contractile spleens that assist in the expulsion of erythrocytes (refer Wells and Weber 1990 for the spleen in hypoxic and exercised rainbow trout) (see Chapter 19 for Lymphomyeloid system).

White cells: Except for a few hours during which they are transported by the blood, vertebrate leucocytes function outside the circulation. In fishes large amounts of leucocytes, in all phases of development, occur in specific lymphomyeloid tissues and organs and infiltrate the skin, mucosal membranes, and connective tissue areas all over the organism. Fish blood is remarkably rich in leucocytes. Recently fish leucocytes have been described by Zapata et al. (1996).

Fish leucocytes may be divided into granulocytes (neutrophil, eosinophil, and basophil) and nongranulocytes (lymphocytes, monocytes, and thrombocytes).

The majority of granulocytes are mobile, phagocytically active cells. The cytoplasm contains lysosomal granules, vacuoles, mitochondria, and other organelles or particles. Contractile vacuoles occur in elasmobranch granulocytes. The properties of granulocytes vary extremely, especially between systematic groups of fishes. In teleosts, granulocytes constitute 4.5–18% or more of the leucocytes in the blood. Neutrophils (the name given to those leucocytes that possess granules which do not stain markedly with either acid or basic dyes at neutral pH) (heterophils), eosinophils (acidophils) (cytoplasmic granules stain with acid dyes) and basophils (granules stain by basic dyes) have been distinguished. The scanty information available on leucocytes of bony fishes is mainly based on a few species of cyprinids and salmonids. The basic haematological pattern of vertebrates is probably conserved in fish, although some significant differences have been detected. The nomenclature and classification of teleost leucocytes needs to be clarified. Furthermore, the morphofunctional significance of teleost blood cells is not well understood. Studies on the ontogenetic development of fish leucocytes, particulary in fish of evolutionary and/or economic importance, are needed.

General agreement exists concerning the presence in fish of heterophilic, acidophilic and basophilic series although more than three granulocyte types have been described and one or two granulopoietic series have frequently been found. Ultrastructural descriptions of granulopoietic stages in fish are scarce, their nomenclature and characterization being unclear. Granulopoietic stages have been characterized by the size, structure and staining properties in some teleost species, whilst some morphological similarities between mature and immature cells have been considered in others.

Some authors have reported on the granulocyte function in the non-specific defenses of fish.

Fish heterophilic granulocytes have been described as active amoeboid and phagocytic cells; the influence of stress on the number of circulating heterophils has also been observed. Likewise, the degranulation during the inflammatory response and the amoeboid capability of fish acidophils have also been considered. However, the roles of fish granulocytes, mainly acidophils and basophils remain unresolved and "what is unclear is the functional heterogeneity in fish where several morphologically distinct granulocytes are present".

Heterophil, neutrophil and polymorphonuclear leucocytes designate the same cell type. However, in the peripheral blood of carp both heterophil and neutrophil cells are described.

Acidophilic granulocytes may be present or absent in fish peripheral blood, the mature acidophils being the least frequent and the only granulocyte type in circulating blood of *Fundulus heteroclitus* and *Cyprinodon variegatus*. Two types of acidophilic granulocytes have been observed in some fish. Furthermore, numerous mature acidophils are present in the inflammatory response.

Basophilic leucocytes have not been described for some fish species, although observed in others. They have been named "blood mast cells" and functionally equated to mast cells. In the electron microscope, the neutrophils show Golgi apparatus, mitochondria, ribosomes, endoplasmic reticulum, vacuoles, glycogen particles, and specific granules. The blood of chondrosteans contains neutrophils and eosinophils. The elasmobranch blood is exceptionally rich in granulocytes. There is little agreement on the number of existing types of elasmobranch granulocytes. Eosinophils, heterophils, or neutrophils are reported. At present there is no generally accepted nomenclature for elasmobranch granulocytes. Dipnoan white blood cells are exceptionally abundant and prominent. White blood cell pattern of lungfishes is relatively complex. Heterophils, small and large eosinophils, and neutrophils are found. Granulocytes constitute the majority of leucocytes in the blood of *Latimeria chalumnae*. Granulocytes constitute about 50% of the total leucocytes in the blood of the hagfish, *Myxine glutinosa*. Only one granulocyte type is recognized. The nuclei are polymorph with one to three segments. The granulocytes of lamprey (*Lampetra fluviatilis*) resemble those of the hagfish but are smaller. As in the hagfish, they constitute about 50% of the leucocytes and have nonsegmented or 2–3 lobed nuclei. The granulocytes of the sea lamprey (*Petromyzon marinus*) resemble teleostean neutrophils ultrastructurally.

The presence of mast cells, in fish, similar to those found in mammalian connective tissue, has been debatable. PAS-positive granulocytes have been reported as mast cell-like cells but there is no direct evidence that they contain histamine and heparin.

Lymphocytes of different sizes with a round or oval nucleus are found in fish. They are mobile but usually nonphagocytic. The cytoplasm is nongranulated or contains few minute granules and usually stained blue in routinely stained smears. The cytoplasm shows mitochondria, rough and small endoplasmic reticulum, ribosomes and a Golgi system.

Functional heterogeneity in fish lymphocytes has been suggested by different methods, such as mitogenicity response, but the presence of different lymphoid cell populations and the existence of T and B cell analogues have not been convincingly demonstrated. Using morphological criteria, fish lymphoid cells have, at present, been subdivided into lymphoblasts, lymphocytes, and plasma cells, or lymphoblasts, prolymphoblasts, prolymphocytes and lymphocytes, or into mature and immature lymphocytes.

The origin of the head-kidney lymphocytes is still unresolved.

Plasma cells have a basophil, usually nongranulated cytoplasm and an eccentric spherical nucleus. The cytoplasm exhibits numerous ribosomes in a rough endoplasmic reticulum that forms flat or irregular cisternae. Plasma cells in lymphomyeloid tissues of elasmobranchs, teleosts, lampreys, and hagfishes have been identified electron microscopically. The plasma cells are considered as the main producers of immunoglobulins, but lymphocytes and other cells probably also produce antibodies.

The structure and function of teleost head-kidney plasma cells have been poorly studied. Plasma cells can be abundant or scarce in fish haemopoietic tissues. This cell type was not always recognized in the blood, it being absent or in very small numbers. The few circulating plasma cells reflect their limited release from lymphoid or lymphomyeloid tissues.

Monocytes are mobile, phagocytic cells. Macrophages are also reported in different positions in fish. The head-kidney seems to be the main source of fish macrophages. Unlike other vertebrates, fish macrophages have been described in both haematogenous and histogenous forms. According to the localization, ultrastructural features or differing glass adherent properties, several types of macrophages have been recognized in the same teleost species. Ultrastructurally, monocytes closely resemble heterophilic granulocytes and blast cells. Macrophages have many of the features seen in monocytes in addition to also having an increased number of primary and secondary lysosomes. Fish macrophages operate as the first defence line and some functional abilities of their have recently been reported. They remove senile or dead cells as well as antibody-antigen complexes. Extensive literature is available on fish melanomacrophages (MMØ), these being defined as cells with ultrastructural features similar to those of macrophages but containing a considerable amounts of pigments. Incidentally the origin and development of melanin-containing cells is poorly understood. Although tissue catabolism seems to be the main process for forming pigments within melanomacrophages. It is still a matter of debate as to whether they are melanogenic or melanin-phagocytic cells.

Isolated MMØ have been reported as exclusive to primitive fish including lampreys and elasmobranchs, the small "melanomacrophage centres" (MMCs) observed in some teleosts being considered early stages of their final encapsulated stage. However, there is very little information on MMC ultrastructure, and although there is evidence which points to an evolution from a random distribution of MMC to the highly organized MMCs of advanced bony fishes, the presence of free MMØ in teleosts has also been suggested.

Histochemical studies of MMØ have revealed contradictory results. Many aspects of MMØ biology remain unresolved. Thrombocytes, or spindle cells, usually are oval or spindle-shaped. Contrary to the mammalian platelets (or thrombocytes), they are nucleated and occasionally consist of almost naked nuclei. The number of thrombocytes varies. Ultrastructurally the thrombocyte shows mitochondria, ribosomes, glycogen granules and bundles of microtubules and well-developed surface connected canalicular system (SCCS).

Following the erythrocyte, the thrombocyte is the most numerous cell type in fish blood; however, its morphology and function have been little studied compared to erythrocytes and leucocytes. The structure and origin of teleost thrombocytes and mammalian blood platelets are quite different and thus may reflect evolutionary differences. It has been suggested that fish thrombocytes have a function similar to mammalian platelets but the participation of fish thrombocyte in blood coagulation has not been definitively established. On the other hand, several studies on the possible phagocytic function of thrombocytes have been carried out, although some evidence of their non-phagocytic capabilities has been observed and lysosomal enzymes such as peroxidase or acid phosphatase have not been found.

Blast cells represent early stages of red or white blood cells that arise by transformation of haemopoietic stem cells. Blast cells have increased nuclear volume, and the cytoplasm is intensely basophilic due to numerous ribosomes. Blast cells have the capacity to divide and form new cells. Blast cells are regularly present in small numbers in fish circulation. In teleosts the blast cells, to a great extent, may remain in the haemopoietic tissues, but in the blood of cyclostome, hagfish (*Myxine glutinosa*) erythroblasts, and leucocytoblasts are abundant.

The frequency of leucocytes in the blood of fishes is influenced by physiological and environmental conditions (see Moura et al. 1994 for effects of temperature). The number of leucocytes is affected by hormones in similar ways as in other vertebrates. The amount of circulating white blood cells increases with infection and parasitic infestation. In higher vertebrates the granulocytes are relatively short-lived and must be replaced by continuous granulocytopoiesis. Conditions in fishes are unknown, but the granulocytes probably have a short life span in fishes too.

Phagocytically active cells in fish blood are mainly neutrophil (heterophil) granulocytes and monocytes (mononuclear phagocytes). Macrophages are the main cells responsible for phagocytosis in the peritoneum, spleen, kidney, liver, gills, and other tissues. The phagocytes destroy ingested microorganisms by chemical mechanisms. Deficiency of phagocytic killing mechanisms may explain which microorganisms and parasites cause disease in fish and other animals. Phagocytic uptake of antigens is an important step in the initiation of humoral immune response. Special antigen-trapping or antigen-presenting cell systems may be distinguished. Phagocytes are affected by nerve transmitters and stress hormones.

Phagocytically active granulocytes destroy bacteria and fungal parasites by using enzymes localized in the cytoplasmic granules. Several antibacterial peptides are present in fish granulocytes. Hydrolytic enzymes and oxygen-dependent bactericidal systems are reported.

Immunocompetent lymphocytes constitute the basis of immune reactions. Fishes like all other vertebrates, show both cell mediated and humoral immune responses. Nonspecific cytotoxic cells or natural killer cells (NK cells) are also produced in the lymphomyeloid tissues of fresh water teleosts (for further information, see Chapter 19 and Meseguer et al. 1999).

In cyclostomes and elasmobranchs blood clotting is relatively slow, in teleosts, and lungfishes very fast. Haemostatic responses of fishes have not been much investigated, but evidences are that similar mechanisms are at work as in higher vertebrates. The detailed knowledge on the roles played by thrombocytes and other cells in fish blood coagulation is lacking. Many lacunae in the knowledge related to blood cells still exist.

12. CHEMICAL PROPERTIES OF BLOOD

McDonald and Milligan (1992) focussed on the blood variables such as hormones, metabolites, proteins, lipids and electrocytes as well as variations resulting from such factors as temperature, hypoxia, exercise, salinity, feeding, development, and reproductive state.

Irrespective of the endogenous or exogenous factor of interest, attention must first be paid to two factors that have profound influence on blood chemistry: the method employed to obtain the blood sample for analysis and the time of day the samples are collected.

Diel cycling of hormones is reported. Diel fluctuations often appear in circulating hormones and plasma electrocytes (see Chapter 13).

Taking up the metabolites, glucose is one of the most frequently measured blood metabolites and is, perhaps, the most variable. The main factors are activity, stress, nutritional status, and reproductive cycle. Although blood glucose levels vary considerably among species, within a species, and over time within an individual, severe hypoglycemia (i.e., > 50% decline in blood glucose) does not occur. Lactate is the major end product of anaerobic metabolism in vertebrates and is routinely found in low levels, < 1 mM, in fish blood. Its accumulation in blood is influenced by oxygen limitation, life-style and habitat. In a few species, ethanol also has been reported under anaerobic conditions. Level of ketone bodies mainly in elasmobranch blood is influenced by both starvation and reproductive status. In most fish, the bile pigments are removed from the body when food moves along the gut, and thus, plasma levels of these pigments are quite low. The lamprey (*Lampetra lamottenii*), however, is a notable exception. Upon metamorphosis from the ammocoete to adult form, the liver undergoes extensive reorganization; the bile ducts degenerate and the gall bladder is lost.

As regards the nonprotein nitrogenous components, the distribution of free amino acids between erythrocytes and plasma differs among species, and there are no obvious trends. Furthermore, in some instances, the presence of red cell free amino acids is independent of that of the plasma. Although the data are sparse, it appears that plasma amino acid levels are higher in marine elasmobranchs than in either marine or freshwater teleosts or marine agnathids. This is most likely a consequence of the high extracellular osmotic pressure maintained by organic osmolytes in elasmobranchs. Whole blood and plasma levels of essential amino acids (EAA) are mainly a function of dietary protein intake and are positively correlated with their levels in the diet. Nonessential aminoacids, in contrast, are independent of dietary levels. Ammonia, urea and uric acids, trimethylamine oxide, creatine and creatinine are also considered (see Chapter 11).

Very little information is available on plasma proteins. Reduction in plasma (protein) are less common, and are generally associated with prolonged starvation and severe stress. In fish, albumin-like proteins have been identified in lamprey, hagfish, and several teleost species. There are, however, some questions as to the presence of albumin in elasmobranchs. Immunoglobulin levels are quite variable in fish. Steroid hormone-binding in fish tend to have a higher affinity and higher capacity than those in mammals. Little is known about the thyroid-binding hormones in fish. A number of marine teleosts inhabiting polar and subpolar regions produce antifreeze proteins (AFP) or antifreeze glycoproteins (AFGP). A few studies have attempted to use plasma enzyme values for the assessment of general fish health.

Why is there no carbonic anhydrase (CA) activity available to fish plasma has been explained by Lessard et al. (1995).

In fish, triglycerides and phospholipids are the most abundant lipid classes. Total plasma lipid levels are influenced by many factors, including diet, stress, and reproductive state. Lipoproteins, cholesterol, nonesterified fatty acid (NEFA; most metabolically active form of lipids in the blood, and affected by starvation, migration and spawning), triglycerides (TG) (the level of plasma TG varies throughout the life cycle of fish and is affected by such factors as sexual maturation, smoltification in the salmonids, spawning, and nutritional status) are also considered. Generally, elasmobranchs tend to have lower plasma TG levels than do teleosts and are less dependent on the physiological and nutritional state of the animal.

The major trends in electrolyte/osmolyte composition in fish from different groups is well known. Na^+ and Cl^- are the major ions in the blood of all fishes. The concentration varies in fish.

Plasma Na^+ and Cl^- levels are sensitive to a wide variety of environmental and endogenous influences. Salinity, pollutants, and stress have effect on plasma electrolytes. In freshwater fishes total plasma Ca^{2+} levels are quite uniform, falling within the narrow range of 2–3 m*M*, except for the euryhaline elasmobranchs where plasma Ca^{2+} is 3.9 m*M*. In marine fishes, Ca^{2+} levels are higher. A variety of normal stresses have either no effect on plasma (Ca^{2+}) or can cause a slight increase. Circadian fluctuations in plasma (Ca^{2+}) appear to be negligible. A fraction of total plasma calcium is always protein bound. Plasma Mg^{2+} levels are lower than Ca^{2+} levels in all freshwater fishes (except *Potamotrygon* sp.) and in marine elasmobranchs. In the other marine fish, Mg^{2+} levels exceed Ca^{2+} levels with the highest Mg^{2+} levels being found in myxinoids. Like Ca^{2+}, Mg^{2+} is bound by plasma proteins. Less than 2% of the total body K^+ (potassium) is contained in extracellular fluids. Strenous exercise causes an increase in plasma (K^+). Data for phosphate and sulphate anions are sparse relative to other plasma ions.

13. BLOOD AND EXTRACELLULAR FLUID VOLUME REGULATION: ROLE OF RENIN-ANGIOTENSIN SYSTEM, KALLIKREIN-KINN SYSTEM, AND ATRIAL NATRIURETIC PEPTIDES

Organization of the various fluid compartments has been described by Holmes and Donaldson (1969). Olson (1992) comprehensively reviewed aspects concerning blood and extracellular fluid volume regulation. Total body water (TBW) is the amount of water in an intact fish. Total body water reflects the net water balance between a fish and its environment and is determined by osmotic and hydraulic components. Osmotic forces are affected by active and passive exchange of electrolytes across the gills, skin, gastrointestinal and renal epithelia, and nonionic osmolytes generated through metabolic activities. Net water balance is also affected by mechanical compliance of the body (total body fluid volume divided by total body fluid pressure). Total body compliance determines the relationship between total fluid volume and average internal hydraulic pressure. Whole body compliance has not been measured in fish.

The total body water makes up roughly 70% of the body weight (or volume) of fishes and other vertebrates. Rather higher values are found for cyclostomes than for fishes generally (75.6 for a lamprey, *Petromyzon*, and 70.8% for a marine teleost).

Control of intracellular fluid volume (ICFV) and composition enables solvation and movement of appropriate molecules upon which cellular function are achieved. Fluid movement in and out of the intracellular compartment appears to be an obligatory response to changes in intracellular and extracellular osmolarity, and it is probable that the regulatory responses of individual cells are designed to protect intracellular volume rather than to control extracellular fluids.

Extracellular water is the sum of interstitial, plasma, lymph, and transcellular fluids. Regulation of extracellular fluid volume (ECFV) and composition is necessary to achieve a stable, or at least manageable, environment around the cells and to serve as an avenue of molecular commerce between cells and tissues. Extracellular fluid is interposed between the intracellular compartment and the environment and is thereby directly subjugated to hydraulic and osmotic forces at both the cellular level in the tissues and across the various body epithelia. Extracellular fluids volume is largely governed by transepithelial exchange processes. Tissues responsible for these activities include the gill, skin, gastrointestinal tract, and the kidney.

Few factors pose a greater immediate threat to fish volume regulation than the osmolarity of the environment. Extracellular fluids appear to be more sensitive than intracellular fluids to an initial change in transepithelial osmolarity gradients. When ECFV is evaluated in the context of the change in TBW, both organismal and cellular regulatory systems can be appreciated. Besides the salinity, exercise, environment (temperature, photopteriod, pH, and Po_2), and other stress affect ECFV.

Total blood volume is influenced by the sum of factors that govern the volume occupied by red cells and the plasma volume. It is assumed that regulation of total blood volume is achieved primarily, if not exclusively, through adjustments in plasma volume, as it is in mammals. Blood volume is regulated by factors between intravascular and interstitial compartments.

The largest blood volumes (ml/kg) are found in the hagfish followed by lampreys and skates (*Eptatretus stouti*-206; *Petromyzon marinus*-85; *Raja binoculata*-80). In the hagfish, *Eptatretus cirrhatus*, 30% of the blood volume is located in a subcutaneous sinus that extends over most of the body. Haematocrit is lower in the subcutaneous sinus than in the remaining vasculature (4.3 versus 13.5), and the time necessary for equilibrium between the two compartment is in excess of 8h. The blood volume accounts for around 50% of the extracellular space in myxinoids. Osteichthyes collectively have the lowest blood volume of any vertebrate. The tunas appear to be an exception (BV = 132 ml/kg), perhaps due to their extensive heat-exchanging rate (Table 8.7).

Table 8.7 Volumes of extracellular fluid compartments in some fish groups in percent of body weight/volume

Fish group	Extracellular fluid	Plasma volume	Blood volume
Cyclostomes			
Lamprey	23.9	5.5	8.5
Hagfish	25.9	13.8	18.7
Elasmobranch	21.2	5.4	6.6
Chondrostean	18.4	2.5	3.4
Holostean	16.0	2.1	3.6
Teleosts (several)	15.4	1.9	2.9

Source: Fänge (1985)

Tissue fluid compartments have seldom been determined in fish, although they have been perhaps most extensively studied in salmonids.

Exhaustive exercise is perhaps the best known effector of apparent fluid flux from the vascular compartments of fish.

The relationships between blood pressure and blood volume and the effects of environmental salinity on these cardiovascular parameters have received surprisingly little attention.

How the RAS (renin-angiotensin system), KKS (kallikrein-kinin system) (Fig. 8.37) and ANP (artial natriuretic peptides) may contribute to the blood volume homeostasis has been described. RAS and ANP have been shown to affect a variety of vascular and extravascular effectors and these systems may play an important role in volume homeostasis. The kallikrein-kinin system may also prove to have important systemic as well as intraorgan cardiovascular functions. Undoubtedly, other central and peripheral systems are also involved (see also Fänge 1985; Nilsson and Holmgren 1992; Platzack et al. 1993, and Chapter 13). A heterogenous distribution

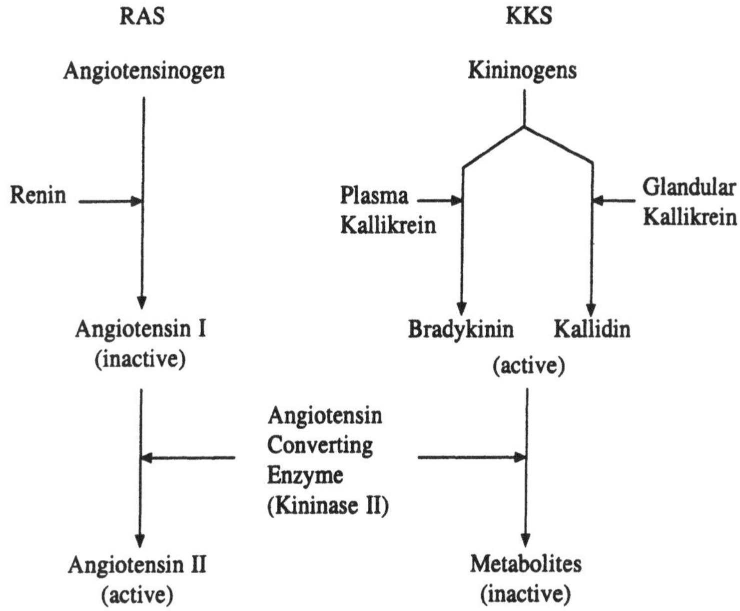

Fig. 8.37 Activation steps in the renin-angiotensin (RAS) and kallikrein-kinin (KKS) systems. ***Source:*** **Olson (1992)**

of ^{125}I-ANF binding exists in the different cardiac regions of *Conger conger*. Some insights concerning ANF receptor binding activity and its putative regulatory role of different cardiac functions are discussed (Cerra et al. 1992).

Recent information of cardiovascular effects of bradykinin (Olson et al. 1997a), ANP (Duff et al. 1997), AVT (Conklin et al. 1997) and ANG (Olson et al. 1997b) is available. A reappraisal of the volume of the secondary circulation has been given by Bushnell et al. (1998).

Reverting to natriuretic peptides (NPs), it may be pointed out that in several species of teleosts at least three types of NPs have been isolated and the sequence determined. Two of them, A-type NP (ANP), and C-type NP (CNP), are structurally very similar to the NPs identified in other vertebrates namely ANP, B-type NP (BNP) and CNP. The third type of NP found in fish is mainly synthesized by ventricle and is called ventricular NP (VNP). It is longer than the other natriuretic hormones but it shares a high sequency homology with all other vertebrate NPs. It remains unclear whether this peptide is an additional type of NP or it belongs to the BNP family.

The identification of three types of NP receptors (NPRs) in organs such as the gills, kidney, and intestine provided a molecular target for the osmoregulatory functions of NPs in fish. Cerra et al. (1998) have given details of cardiac NR receptor heterogeneity in Antarctic teleosts.

Antarctic teleosts are very specialized fishes and ANPs may be essential for their survival in their extreme environment. Further work is necessary to characterize its molecular sequence of the peptides, and the affinity and structure of the receptor subtypes. The administration of homologous ANPs may clarify the physiological role by this family of peptides in the different environmental and haemodynamic conditions encountered by the Antarctic teleosts.

CHAPTER 9

Respiratory System

1. INTRODUCTION

Respiration is basic for the main sustenance of life and it seems to be comparable in most fishes. There is no spatial separation of the functional components for feeding, coughing and respiration, which makes a separate analysis of the basic respiration patterns highly relevant, before the study of the integration of these functional components can begin. Internal respiratory status in fishes is determined by the combined process of branchial gas transfer and blood gas transport.

Several physical characteristics appear to render water much less suited as a respiratory medium than air. O_2 solubility of water is some 30 times less than that of air, and declines with increasing temperature, when the metabolic demand of most fish increases. Viscosity and density of water are about 60 and 800 times higher than that in air.

Considerable information regarding respiration in water and air (see Dejours 1988; also literature) and gas exchange (Piiper and Scheid 1989-theory, model, and experimental data) should prove stimulating and interesting to physiologists.

Vertebrates evolved primarily in water and thus they needed to adapt to water breathing. The breathing of air to supplement aquatic gas exchange by gills and skin has arisen independently in several different groups of fishes. Because of this diverse ancestry, modern air-breathing fishes show remarkable diversity in air-breathing organs, with different species using various combinations of gills, skin, buccal and pharyngeal cavities, as well as the swim bladder and digestive tract.

The gills are the most important site of contact between fish and water, covering a surface area 10–60 larger than the area of skin covering the body surface.

Considering its modest size and fragile structure, the fish gill is charged with impressive functions. Gills are not solely respiratory organs but also serve in osmo-and ionoregulation, nitrogen excretion, and acid-base regulation. Furthermore, the gills play a role during feeding. The gills form a sieve-like structure which serves to filter food particles. Adjustment of gill and position can be expected to control the mesh width of the gill sieve. In addition, it is also apparent that gills actively participate in calcium homeostasis and they are considered, at least in freshwater fish, to be the most important calcium-regulating organs. The gills obtain this status by virtue of their ability to transport calcium directly from the environmental water into the blood. Indeed, some authors have considered that the calcium dissolved in the environment and the calcium in the body fluids of fish are part of the same system. Certainly, fish are known to absorb calcium directly from their environment and may even rely entirely on environmental calcium during periods of dietary calcium deprivation. Bones and scales, crucial elements in vertebrate life, derive their sturdiness from deposited calcium phosphate minerals. Most fish grow continuously and, to do this, accumulate calcium in their body throughout their life. This is realized through a continuous uptake of Ca^{2+} via their gills (for details of calcium transport in fish gill, see Flik and Verbost, 1993).

Phospholipids are important gill membrane structural components through which transport of ions and gases must occur.

To meet the constraints placed on the respiratory system for gas exchange, gas exchange organs must be designed to match diffusing capacity to metabolic demands. Metabolic demands, on the other hand, vary greatly among vertebrate groups, particularly between the ectotherms and endotherms, as well as within any given group, as a function of both life-style and body size.

The diffusing capacity of the exchange organ will be most clearly reflected in its anatomical diffusion factor (ADF). This variable represents the anatomically measurable component of the oxygen-diffusing capacity of the tissue and is equal to the ratio of respiratory surface area to the harmonic mean thickness of the tissue barrier between the blood and the respiratory medium.

The anatomical trends associated with the reductions in ADF that accompany increasing metabolic demands in vertebrates include the development of a large surface area (through compartmentalization), a rich vascular supply, and a reduced barrier thickness. With the increase in complexity of the gas-exchange surfaces, particularly with invagination of these surfaces to form protected sites inside the body, different demands are placed on the mechanisms for ventilating these surfaces. General body movements are replaced by specific respiratory pumps which must be designed to provide sufficient ventilation to match metabolic demands. The design of the pumps, however, is heavily influenced by other physiological, environmental, and behavioural constraints. They therefore represent the results of many experiments in design, each with its own unique features (Milsom 1989).

Cardiac and respiratory rhythms interact in various ways in vertebrates. Teleost fish, which leave the aquatic environment for short periods and are exposed to air, show a bradycardia. Bradycardia has also been observed in teleosts during periods of naturally occurring apnoea. Cycle by cycle synchrony between the respiratory and cardiac rhythms occurs in a number of teleosts and elasmobranch fish. The synchrony is, however, seldom perfect, usually quite labile and dependent on a variety of environmental conditions.

Cardio-respiratory synchrony would be expected primarily to influence the efficacy of gas transfer at the secondary lamellae. Cardio-respiratory synchrony in some teleosts may enhance gas transfer in stress: hypoxia and heat stress. The functional significance of cardio-respiratory synchrony may vary between species and a common mechanism may control the coupling of heart beat both with individual breaths and the cycles of longer period seen in arrhythmic breathing.

A review entitled, "Environmental perturbations of oxygen transport in teleost fishes: causes, consequences and compensation", encompasses varied information particularly related to hypoxia, combined hypoxia-hypercapnia, temperature, and salinity change (Jensen et al. 1993). Different aspects of fish gills are published in other reviews (Potts and McWilliams 1989; Laurent and Perry 1991; McDonald and Wood 1993). The recent contributions on gas exchange are by Perry and McDonald (1993) Gilmour (1998) and Randall and Brauner (1998).

According to Olson (1998), the gill has several anatomical attributes that indicate its potential as a highly efficient endocrine regulator and perhaps even endocrine organ.

2. GENERAL STRUCTURE AND ORGANIZATION: AN OVERVIEW

The gills of vertebrates are formed as a series of paired pouches in the lateral walls of pharynx which open externally by gill slits. Adjacent pouches are separated by interbranchial septa containing skeletal elements that support the gills.

The gills of the Agnatha lie inside the pharyngeal basket whereas those of the Osteichthys and Chondrichthyes are found outside the branchial skeleton. Branchial gross structure in the agnathan fishes contrasts markedly with the independently-supported branchial arches of gnathostomes. The cartilaginous 'branchial skeleton' is quite extensive in the lampreys, but less well defined in myxinoids (Bartels 1998).

The detailed form and arrangement of gills varies amongst the fishes. The degree of development of interbranchial septa is a major difference between the bony and cartilaginous fishes and there exists a morphological series between elasmobranchs and teleosts in the degree of reduction of the gill septum (Figs. 9.1, 9.2). Gonzalez et al. (1996) described the early gill development in the rainbow trout, *Oncorhynchus mykiss*.

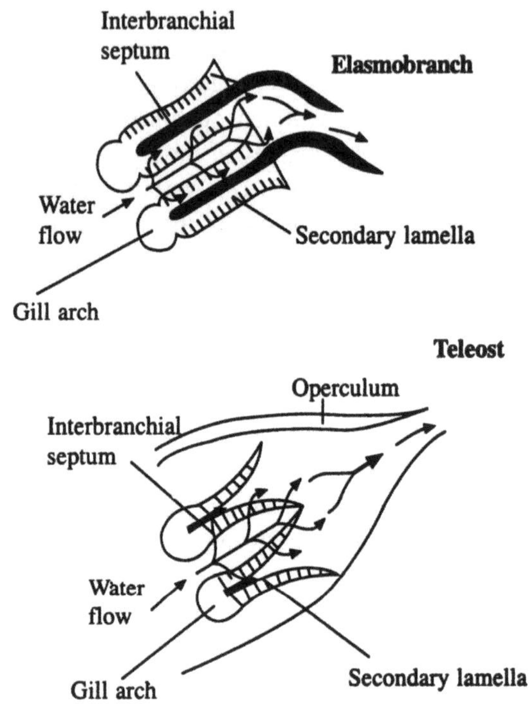

Fig. 9.1 Pattern of water flow over fish gills.
***Source:* Randall and Wright (1989)**

On the long gill filaments (primary lamellae) which extend outwards into the gill chamber (i.e., external, opercular side) an actual gas exchanger is mounted in the form of densely spaced leaf-like secondary lamellae; transverse folds on both surfaces of the filaments. There does not seem to be any fundamental difference in the basic structure of the secondary lamellae of teleosts and elasmobranch fishes. The tips of adjacent gill filament rows are in contact for nearly all the time during a normal respiratory cycle, thus forming a continuous 'gill curtain' (Figs. 9.3, 9.4).

Two distinct gill modifications occur among oceanic fishes. The filamentar fusion ("reticulate"), unlike the lamellar fusion, is not restricted to the tunas. Modifications are associated with continuous passive ventilation at moderately high rates of swimming.

The respiratory system in fish acts as a double pump, consisting of a buccal force pump separated from paired opercular suction pumps-powered by respiratory pump muscles-by a continuous mesh of gills. Buccal and opercular valves ensure unidirectional water flow from the mouth towards the opercula, both during inspiration and expiration (Fig. 9.5). The gills are endowed with goblet (mucous) cells, chloride cells, acidophil cells, and receptors including for taste (Fig. 9.6).

Each filament has an afferent and efferent vascular supply derived from branchial arteries in the gill arches. In the capillaries of the secondary lamellae blood flows in the opposite direction

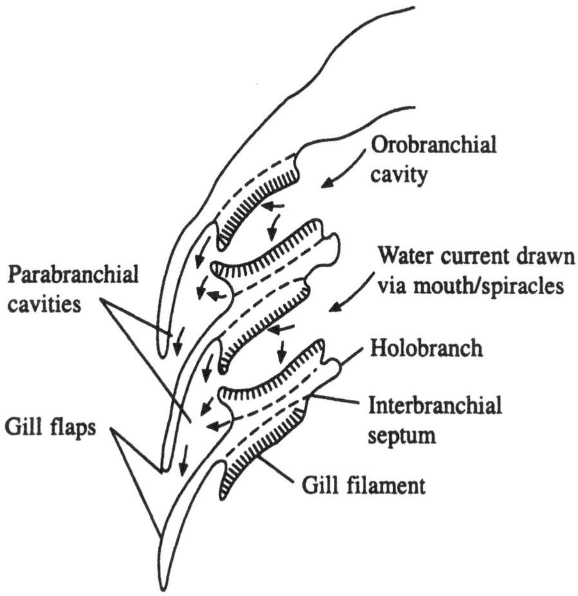

Elasmobranch gills.

Fig. 9.2 A diagrammatic representation of the gills of a dogfish showing the water movements (*arrows*) from the orobranchial cavity into the first two parabranchial cavities on the *left side*.
***Source:* Butler and Metcalfe (1988)**

to the external respiratory water current. This countercurrent design increases the efficiency of gills as gas exchangers so that the low O_2 content of water can be better exploited.

It is important to note that in the fish, the heart pumps only deoxygenated blood. The gill vasculature is inserted between ventral and dorsal aorta; it thus is in series with the tissue vessels which it precedes, and is therefore exposed to high blood pressures. This will have important effects on the construction of the secondary lamellae which must resist this high pressure and still must serve gas exchange efficiently.

Gill structures are, therefore, evaginated to allow smooth uninhibited ventilation by water which needs to flow through at high rate and meet a low resistance, whereas blood perfusion at high pressure is easy. This design eliminates pockets of "residual water" whose PO_2 would fall rapidly because of slow diffusion of O_2 in water. In skipjack tuna (*Katsuwonux pelamis*), the water flow through the gills is continuous; the oxygen uptake ratio are the extreme case (i.e., highest values) for water-breathers.

Almost nothing is known of the respiratory physiology of the Holocephali.

Cutaneous gas exchange, typically associated with branchial or pulmonary gas exchange, also contributes significantly to tissue respiration, mostly CO_2 elimination, in various fishes.

Gill rakers on the buccal (i.e., internal) side of the gill arches serve to remove large particles from the water current before it passes over the gills. The form of gill rakers varies according to the feeding habits of the fish.

The appearance of the aerial respiration was a pivotal development in the evolution of terrestrial vertebrates. Air breathing in all marine fishes seems to be an integral part of their natural history from early ontogeny. Air breathing enables a number of marine species to exploit habitats that

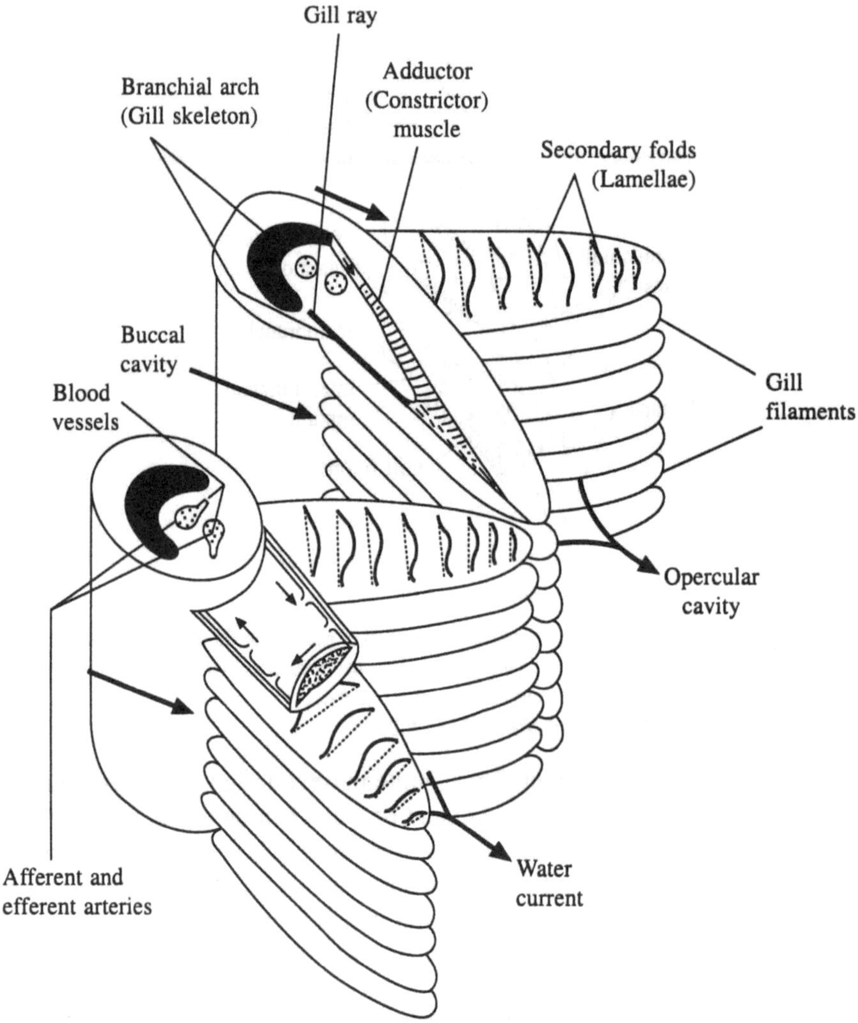

Fig. 9.3 Gills arches and blood flow.
***Sources:* Randall (1990) and other related articles.**

are not saturated with competitors and predators for food and shelter. Most freshwater air breathers are not amphibious and breathe air in response to seasonally recurring conditions of stagnation and hypoxia. The complex array of aerial respiratory organs commonly seen in freshwater air breathers is not found in marine species. In general, air breathing adaptations in two groups do not seem to directly relate to the different physical properties of salt and fresh water.

Among air-breathing fish, in addition to reduction in number of arches, there may also be a reduction in the number and lengths of the gill filaments. A similar reduction also occurs in some deep-sea fishes.

Further, the gills are of great importance in CO_2 excretion whilst the accessory respiratory organs which may or may not be modifications of the gill region are mainly concerned with O_2 uptake (see Wilson et al. 1999 for fine structure of gill epithelium of mudskipper *Periophalmodon sehlosseri*).

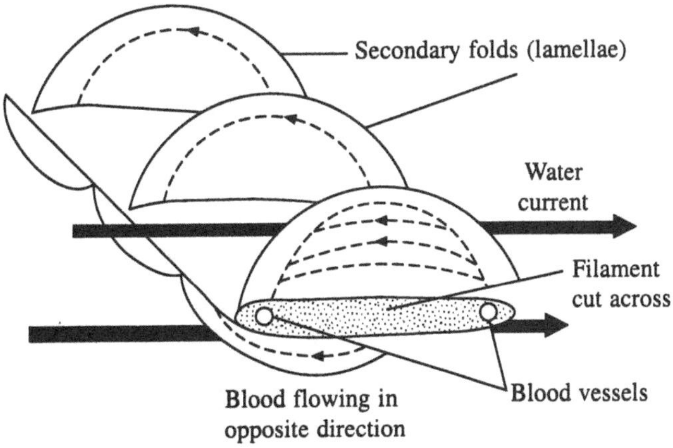

Fig. 9.4 Part of a single filament with three secondary folds on each side. The flow of blood is in the opposite direction to the water.

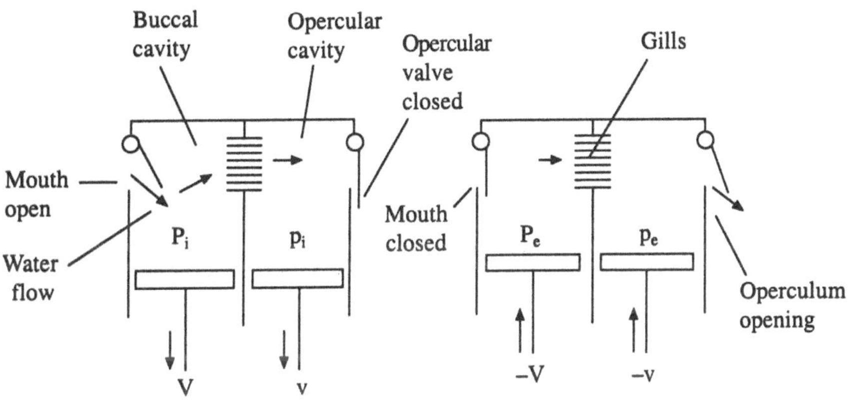

Fig. 9.5 Model of ventilating system, showing buccal and opercular pumps operating during (a) inspiration and (b) expiration.
***Sources:* Shelton (1970), Milsom (1989) and Randall (1990)**

Most species of air-breathing fishes rely on a variety of modifications of the buccal cavity and pharynx, on the walls of the stomach and intestine, on air sacs or on lungs for aerial exchange. The mechanisms used to ventilate these organs are equally diverse as, undoubtedly, are the control mechanisms involved.

Various cardiovascular and respiratory reflexes have been described which are associated with inflation and/or deflation of these organs. Furthermore, changes in volume of gas exchange organs have been implicated in adjusting breathing frequency to match metabolic rate and buoyancy state.

Specialization in gill structures, the use of accessory respiratory organs, and the behavioural and motor capacity to avoid stress would seem of primary importance in the successful utilization of amphibious life (see also Wilson et al. 1999).

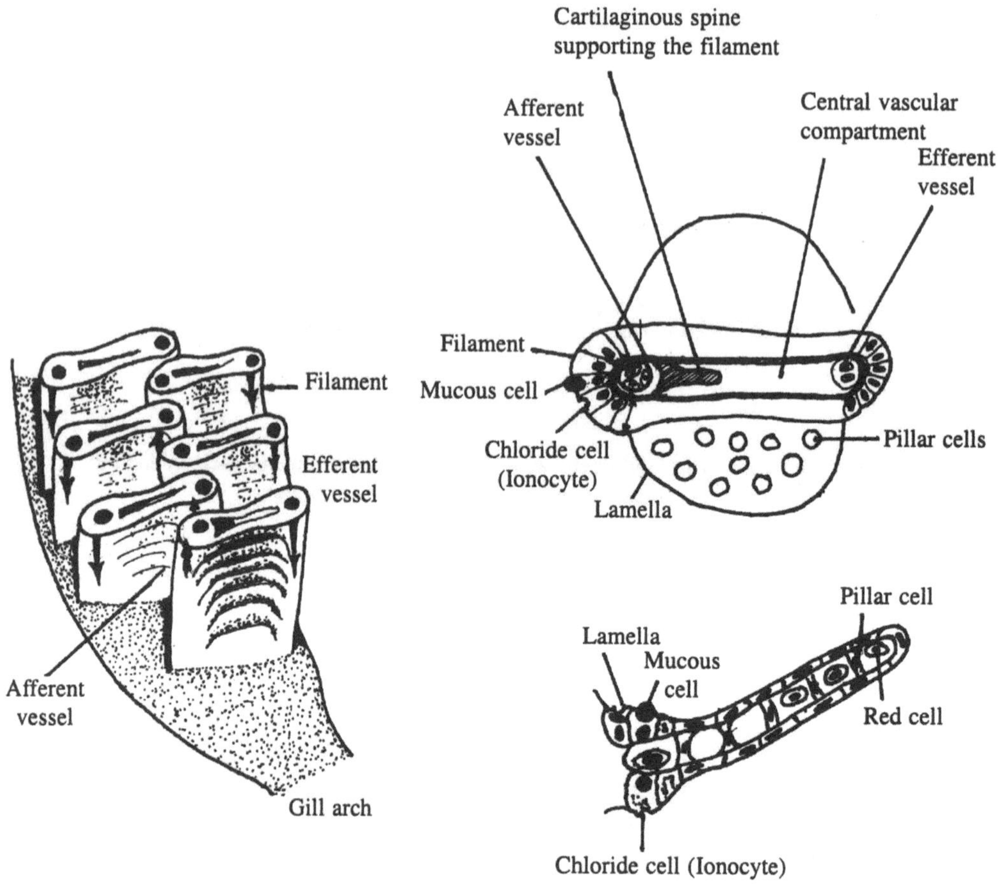

Fig. 9.6 Simplified anatomical and histological features of the teleostean gill. ***Source:*** **Maetz (1971)**

3. BIOCHEMICAL ASPECTS OF GILLS

Gills are composed mostly of biomembranes consisting of such molecules as protein, phospholipids, and cholesterol. Thus, they do not contribute to the overall buoyancy of the fish with respect to density. However, gill chloride secretory cells pump out monovalent ions (Cl^-, Na^+, K^+) to osmoregulate and maintain hypotonicity, which is important in buoyancy (see also Wilson et al. 1999).

There are few published studies on gill lipids of marine fish. Phosphatidylcholine (PC), phosphatidylethanolamine (PE), phosphatidylserine, phosphatidylinositol and sphingomyelin have been reported. No major differences in component phospholipids were observed in pre-and post-spawning salmon (*Oncorhynchus keta*), fishes from shallow water, fishes from great depths, some air breathers and a non-air breathers of Amazon river. Environmental temperature does influence phospholipid composition. Acidity also affects the PC/PE ratio in fish gills (see Phleger 1991).

Lin et al. (1994) have reported the immunolocalization of H^+-ATPase in the gill epithelia of freshwater rainbow trout (*Oncorhynchus mykiss*). H^+-ATPase is located in the apex of both chloride cells and epithelial cells of fish. Findings have supported the proposed role of a proton pump in sodium uptake in freshwater fish. The reader is also referred to Randall and Lin (1993).

4. CONSTRUCTION OF GILL LAMELLA

The gill lamella must resist the shear forces of water flow. They are accordingly lined by a relatively thick two layered epithelium reinforced by a felt of collagen fibrils (Fig. 9.7). In order to resist the high vascular pressure the "capillaries" of the lamella are spanned and lined by pillar cells; these are transformed fibroblasts and are associated with collagen columns which brace the two strong basement membranes. In addition, pillar cells contain cytoplasmic bundles of actin filaments and thus have contractile properties similar to smooth muscle cells. This construction of gill lamella results in a thick tissue barrier, but this may not be a major disadvantage because the countercurrent flow of blood and water allows for efficient gas exchange. The high blood pressure prevailing in the gill vasculature is also responsible for maintaining the gills turgid, and thus allowing them to stand erect and resist the shear forces of water (Bicudo and Weibel 1987; see also Ojha 1993).

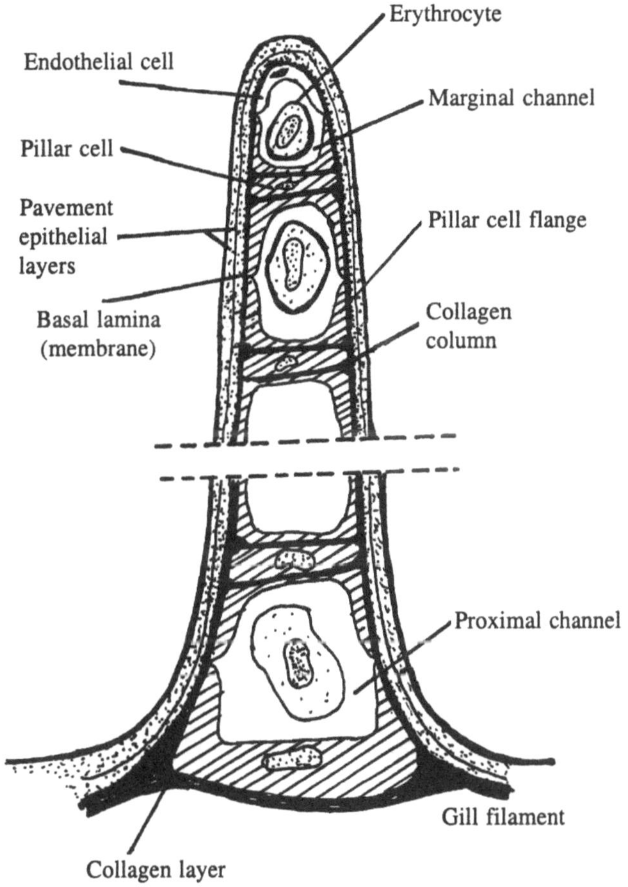

Fig. 9.7 Diagram of a transverse section through the secondary lamella of a fish gill. Note the connections between the collagen layers across the base of the lamella and by columns within the pillar cells. The two other layers of the basement membrane (homogenous and fine fibrous layers) are also indicated.
***Source:* Hughes (1977b)**

Zaccone et al. (1994) discuss about paraneurons in the gills of fishes. Serotonin has been detected in the cells of branchial epithelium in both teleost and selachian species. More recent ultrastructural and immunohistochemical evidence led to the discovery of receptosecretory cells within both the filamental and lamellar epithelium. In all the species examined the neuroendocrine cells occur singly or in small groups; they are more numerous in trout, eel and *Heteropneustes fossilis* than in *Blennius sanguinolentus* or the dipnoan *Protopterus annectens*. Recently, neuroendocrine cells have been found in the gill filamental and lamellar epithelium of *Amia calva*; they resemble the neuroendocrine cells of *Amia* lung, but have larger dense cored vesicles.

The gill is a rich source of calcitonin gene-related peptide (CGRP). In the catfish *H. fossilis*, open neuroendocrine cells in both lamellar and filamental epithelia of the gills react to antibodies against Calbindin DK 28 and calmodulin. It is speculated that in fish gills calcium signals could be transmitted to chloride cells concerned in the uptake of calcium in fresh water.

The neuroendocrine cells of the gills of *H. fossilis* also show cytokeratin immunoreactivities of AE1 to KL1 antibodies and current investigations have shown the occurrence of S-100 protein in these cells. This is one of the calcium-binding proteins; its function is still unknown, but may, like calbindin, be concerned with regulation of intracellular calcium levels and calcium-modulated cellular processes.

The role that these substances play in the gills is still at the level of hypothesis. In general, the morphology and content of bioactive substances of the neuroendocrine cells suggest they have a receptosecretory function in the gills. This is particularly likely that in the open type of neuroendocrine cells, reacting for enkephalins and some calcium-binding proteins, that have the apex at the epithelial surface in contact with water in the respiratory current and base turned towards the vasculature. Probably the two types of neuroendocrine cells (open and closed) play different roles, that of the serotonin-containing closed type is enigmatic, but they could monitor the gaseous exchange across the filamental epithelium.

Neuroendocrine cells have also been detected in the gill epithelium of *Lampetra japonica*. They are differentiated from the other cells of the epithelium by positive immunoreaction for leuenkaphalin, and show a negative reaction for serotonin and bombesin. Secretion of serotonin by neuroendocrine cells of lung epithelium in *Polypterus* has been confirmed. Neuroendocrine cells in the pneumatic duct region of *Protopterus* are present. Related references (Zaccone et al. 1989a, b, 1992, 1995, 1997) also deserve attention.

Pillar Cells. Pillar cells are characteristic of secondary lamellae. Each pillar cell has a large nucleus and the dense cytoplasm contains many small osmiophilic granules which are often grouped in rosettes, and there are numerous mitochondria. Other characteristics have already been described. According to Olson (1991), the pillar cells may have substantial metabolic effects on circulating hormones.

In the gills of tunas and certain air-breathing fish, the pillar cells form definite rows with adjacent cells in contact and sometimes sharing common columns. In most secondary lamellae, however, the blood pathways are less definite except for the marginal channel. The general arrangement of pillar cells suggested a preferred pathway for blood along this channel. Blood mainly flows around the secondary lamellar margin but under conditions of stress there is an increased flow to other parts of the secondary lamellae.

The junctions between the flanges of adjacent pillar cells are often overlapping or abutting and in certain cases fairly tight junctions (desmosomes) are seen (Hughes 1977b).

The distribution of pillar cells in gill lamellae of lampreys is different from that of higher fishes and it may be inefficient for respiratory exchange.

The marginal channel. This channel along the free edge of the secondary lamellae forms a distinctive pathway. The cells lining the outer surface of this channel differ from pillar cells both in their development and ultrastructure. Collagen columns are absent but they contain oval osmiophilic bodies characteristic of endothelial cells as are their many micropinocytotic vesicles. Junctions between these cells and flanges of the outermost row of pillar cells are similar to those between individual pillar cell flanges. Fine filaments are generally lacking from the marginal channel which appears to be less capable of changing its diameter. It seems reasonable to conclude that these are true endothelial cells which line much of the marginal cells (Hughes 1977b).

The fine structure and function (s) of non-respiratory pathways are largely unknown. Clearly the ultrastructure, function and origin of these endothelial cells as well as pillar cells remains to be determined (Olson 1991).

The marginal channel is particularly well-developed in the gills of the haemoglobin-lacking icefish *Chaenocephalus aceratus*. In most fish, blood probably flows along the length of the secondary lamella. In the gills of tuna, blood flows across the width of the lamella, from its base to the marginal channel. This shortened route reduces the drop of blood pressure across the gill circulation.

Sculpture of Respiratory Surface. All gradations in surface sculpturing, from microvilli to microridges, have been observed on the gills of rainbow trout, *Salmo gairdneri*. The surface of the gill filament is generally made of microridges whereas the secondary lamellar surfaces have both microvilli and microridges which vary in their length. It is concluded that in view of dimensions of interlamellar spaces the flow across the epithelial surface will be mainly laminar. The microridges probably serve to anchor the thin films of mucous covering, which may serve other functions not directly related to that of gas transfer. Kendall and Dale (1979) reported SEM and TEM observation of *S. gairdneri* gill. Ojha and Singh (1986) have described SEM of the gills of a hill stream fish *Danio dangilla*. SEM of the fish gill has recently been described by Olson (1996a). Hills (1996) has described the effects of gill dimensions on respiration.

5. VASCULARIZATION OF RESPIRATORY SURFACES

The vascularization of various respiratory surfaces may be defined by two indices: (1) the density index of vascularization for a definite unit of the organ surface and (2) the vascularization index calculated for a weight unit of the animal body. Both these indices can be expressed by the length, surface area, and capacity of subepithelial capillaries, but the first of them refers to the vessels, say, of 1 sq. mm. of the surface of an organ and the other to all the vessels of the organ (or organs) calculated for 1 gm of body weight. The index of vascularization per unit of body weight determines the magnitude of the share of the organ examined in the global gaseous exchange, or it indicates the respiratory value of the organ for the organism. Conclusions based on morphological data are approximate and need confrontation with the results of physiological and ecological studies, in which such factors as the Hb level in the blood, its affinity to oxygen, the ways of living of a species, etc. are taken into account.

The size of the respiratory surface of gills depends on such factors as the length of gill arches, the number and total length of lamellae, the density of plates on the lamellae, the size of the plates themselves, and the density of the vascular network in them.

Gill Vasculature. The afferent branchial arteries of both hagfish and lampreys divide to supply the anterior and posterior hemibranchs of adjacent gill pouches. As in the majority of more advanced fishes, the bulk flows of water and of blood through the gill pouches of hagfish are countercurrent, an anatomical arrangement maximizing the potential for gas exchange (Malte and Lomholt 1998). Countercurrent flow has not been confirmed for the lamprey.

In the lamprey gill filament, arterio-venous anastomoses were found between afferent filament artery or the cavernous body and filament vein, and between efferent filament artery or the collecting arteries of the efferent branchial artery and the peribranchial venous sinus (Figs. 9.8–9.10).

Randall (1985) has given an account of shunts in fish gills. The gill vasculature is complex and varies between species, but at the level of gas exchange unit, the gill lamellae the microvascular structure is more uniform. The entire cardiac output flows first through the gills before passing into systemic circulation. Blood from the heart passes into the ventral aorta, to the filament afferent arteries, through the lamellae and then is collected into the filament efferent arteries passing into the dorsal aorta and the systemic circulation. The lamellae are the site of gas transfer and are the respiratory unit of the gills. Regulation is exerted both over the degree of flow in each gill arch, the number of lamellae perfused and the pattern of perfusion of each lamella. Within the gills the circulation can be divided into 2 interconnected systems. The 1st is the lamellar circulation which, in some species, receives all the cardiac output and in others, most of the cardiac output. The 2nd one is the circulation to the body of the gill filaments.

The Filament Circulation. The filament circulation consists of a large central venous sinus situated within the body of the filament, supplied by arterial-venous anastomoses (AVa's) arising from the efferent and/or afferent side of the lamellar circulation (Fig. 9.11), and draining via branchial veins back to the heart. The size and shape of the lamellae, the extent of the central venous sinus and the nature and number of input capillaries to the filament circulation, whether they arise from the efferent or from both the afferent and efferent sides of the gill lamellar circulation varies between species.

Some fish species have only efferent anastomoses connecting the lamellar circuit to the central venus sinus (Fig. 9.12A), whereas other may have both efferent and afferent connection (Fig. 9.12B). The arterio-venous connections may be long and fairly tortuous, presumably serving a nutritive function before entering the central venous sinus, or they may be short and numerous, as in the eel, forming a direct connection between the pre-and post-lamellar circuit and the central venous system. The actual pathways may be complicated in the trout, *Salmo gairdneri*, there are arterio-venous connections between the efferent filament artery and the central venous sinus (Fig. 9.13), but in addition Fromm's artery arises from the efferent branchial artery, runs parallel to the afferent filament artery and gives off a series of arteriovenous anastomoses to the central venous sinus. There is infact great variability in the type and number of connections between species within each group. From a functional point of view, those fish with afferent arterio-venous connections can shunt a portion of the cardiac output away from the lamellar

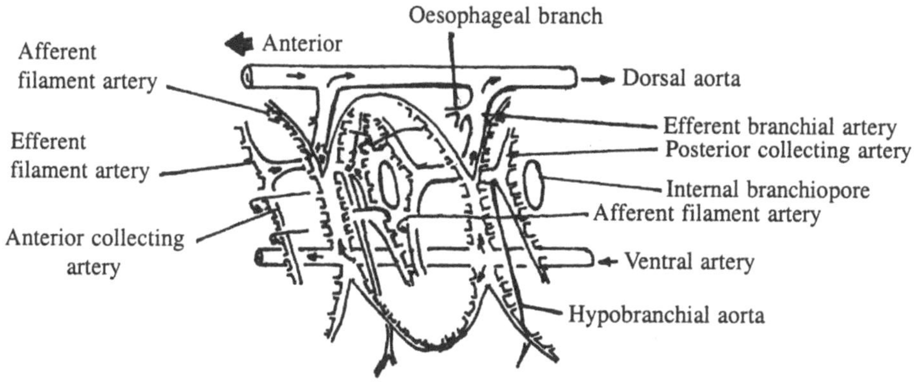

Fig. 9.8 ***Lampetra japonica*: Gill arteries which supply gill pouches on the left side. A lateral view after resection of lateral halves of gill pouches. Arrows indicate directions of blood flow.**
***Source:* Nakao and Uchinomiya (1978)**

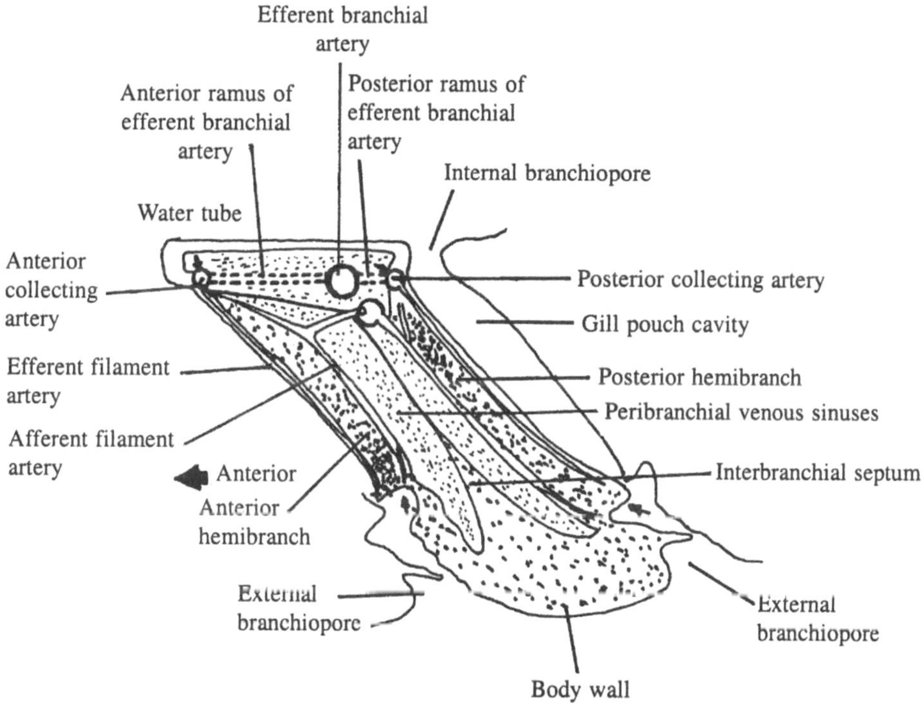

Fig. 9.9 ***Lampetra japonica*: Gill arteries, afferent and efferent filament arteries.**
***Lampetra japonica*: Diagram illustrating gill arteries which supply afferent and efferent filament arteries to each gill filament. The anterior and the posterior hemibranchs are separated from the interbranchial septum by extensive peribranchial venous sinuses (dotted area). Note that the interbranchial septum extends anteromedially as a thin septum which passes in front of the efferent branchial artery and ends on the lateral side of the anterior collecting artery of the efferent branchial artery. Two stacks of axial plates form the anterior and the posterior hemibranch, respectively, and the axial plate in each hemibranch is represented by a striped area in this figure.**
***Source:* Nakao and Uchinomiya (1978)**

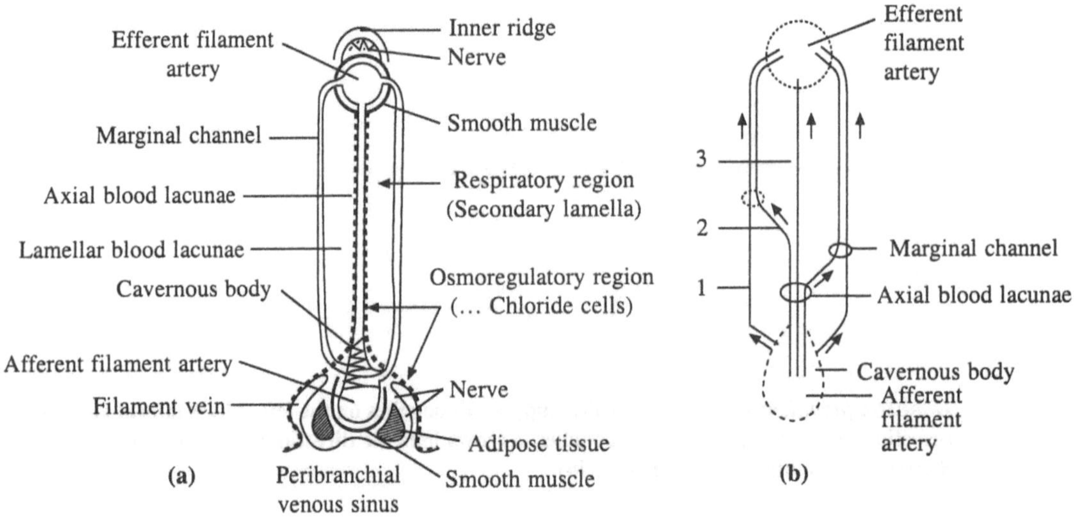

Lampetra japonica: Gill filament blood vessels.

Fig. 9.10 ***Lampetra japonica*. Fig. (a) shows the organization of blood vessels within a cross-sectioned gill filament. Note that the surface of the gill filament is subdivided into the respiratory region and the osmoregulatory region. The former is represented by secondary lamellae covered by respiratory epithelial cells. The latter is the axial plate and the interfilament regions covered by chloride cell-containing epithelium (dotted lines); (b) exhibits different courses of blood streams within a gill filament, which flow from the afferent filament artery finally to the afferent filament artery. A part of the blood in the afferent filament artery flows, via non-respiratory shunts, to the peribranchial venous sinus, although it is not shown in this figure.**
***Source:* Nakao and Uchinomiya (1978)**

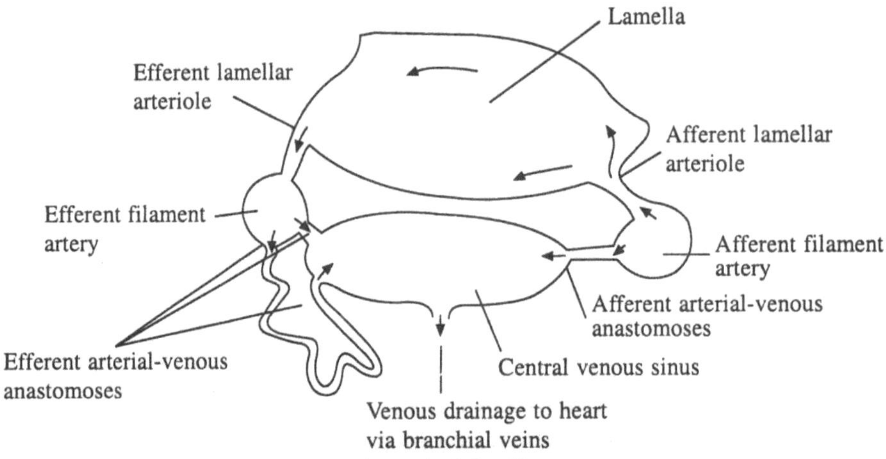

Shunts in fish gills.

Fig. 9.11 Schematic diagram of the lamellar and central venous sinus circulations in fish gills indicating several possible interconnections between the 2 systems.
***Source:* Randall (1985)**

circulation and this deoxygenated blood will pass into the central venous sinus and back to the heart. This blood potentially has a very short recirculation time. Those fish with efferent arterio-venous connections will supply the filament with oxygenated blood which presumably serves a nutritive function.

The central venous sinus is extensive and not usually a simple sac-like structure, but is a complex of interconnected veins, often extending as finger-like processes across the filament and connected to lateral veins surrounding the afferent and efferent filament arteries. The central venous sinus drains into the branchial veins and back to the heart. Valves have been observed in the surface of the central venous sinus and fluid can pass from the interstitial spaces in the gill into the sinus. In the African lungfish there are small micropumps designed to move fluid from the interstitium into the veins within the gills. Thus, fluid in the central venous sinus may

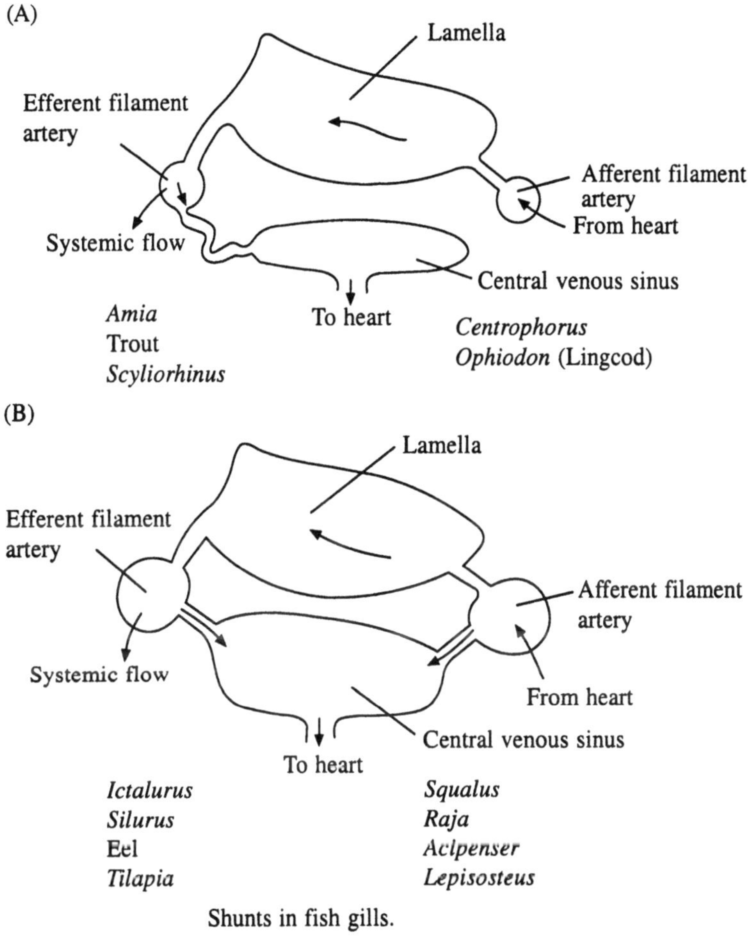

Fig. 9.12 The fish in group (A) have nearly all efferent arterial-venous anastomoses, whereas both afferent and efferent anastomoses are found in group (B).
***Source:* Randall (1985)**

originate from the efferent and/or afferent gill circulation, and, in addition, may drain the interstitial spaces of the gills. The central sinus is often referred as "venolymphatic", but no doubt veins are not lymphatic and the spaces they drain are not lymph spaces; no unequivocal evidence for the presence of any lymphatics has been reported.

There are different reports on plasma skimming, and central venous sinus pressure.

The function of the filament circulation are not clear, but several have been suggested, not all

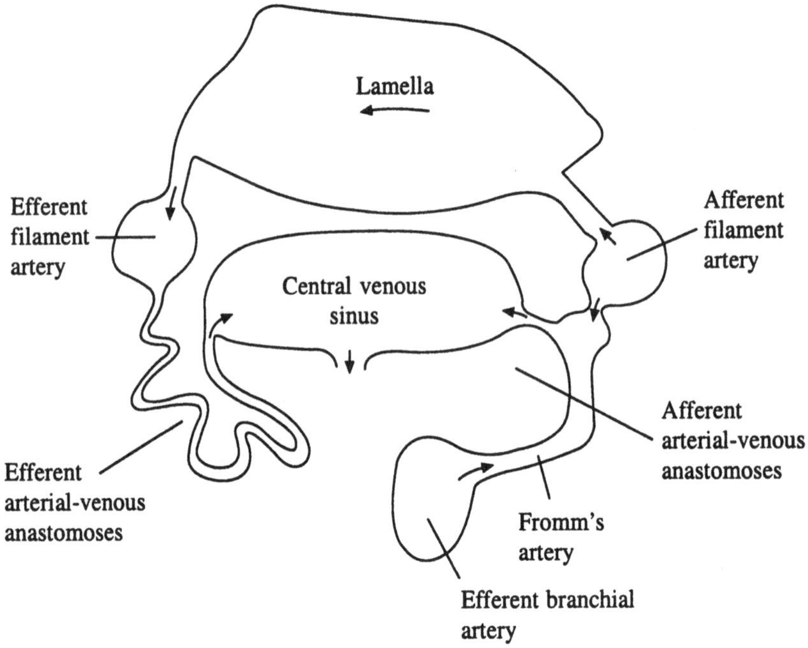

Salmo gairdneri: Shunt in fish gills.

Fig. 9.13 A schematic diagram of the lamellar and central venous circulations and their interconnections in rainbow trout *Salmo gairdneri*.
***Source:* Randall (1985)**

mutually elusive. The central venous sinus could act as a plasma reservoir, its volume representing about 5% of blood volume; could play role in osmoregulation, and hormone degradation. Special attention on arterio-venous anastomoses (AVa's) of a euryhaline teleost, *Tilapia mossambica*, has been given by Vogel et al. (1973, 1974). Vogel (1978; see details) reported AVa's in the afferent region of trout (*Salmo gairdneri*) gill filaments originating from two small filament arteries (Fromm's arteries), which parallel the main afferent filament vessel on either side. The review (Ishimatsu et al. 1995) is highly informative in respect of central venous sinus. Metcalfe and Butler (1986) have described the vascular anatomy of the gills of the dogfish (*Scyliorhinus canicula*). Two distinct pathways are present: a 'respiratory' blood pathway which conveys deoxygenated blood from the heart to the gas exchange surface of the gills and then carries oxygenated blood away to the systemic circulation, and a 'non-respiratory' blood pathway which interconnects with the respiratory blood pathway and which diverts blood into the venous drainage from the gills. As in teleost fishes, elasmobranchs possess a central venous sinus within the core of the filament tissue between the corpus cavernosum (afferent filament in teleosts) and the efferent filament artery. In both groups of fishes, the central venous sinus is connected to afferent and efferent companion vessels which run close to the afferent and efferent filament arteries, respectively. Efferent arterio-venous anastomoses have been reported in elasmobranchs. In addition, afferent arterio-venous anastomoses have also been observed. Osmotic function and a nutritive supply to the gill tissue in respect of 'non-respiratory' blood pathway are discussed. The role of the corpus cavernosum in elasmobranch gills is as yet unresolved. It may act as hydraulic skeleton. The reader is also referred to Glass and Soncini (1999).

6. INNERVATION PATTERN OF GILLS OF A TELEOST

The gill arches and the adjacent parts of the roof and the floor of the buccal cavity of teleost are innervated by branches of the glossopharyngeal (IX) and vagal (X) nerves. These branchial nerves are of a mixed characters, conveying both sensory and motor information.

In carp (*Cyprinus carpio*), each branchial nerve comprises an epibranchial ganglion from which three nerves emerge: (i) the pretrematic branch, (ii) the dorsal pharyngeal branch, (iii) the posttrematic branch (read De Graaf 1990 for detailed course and branching pattern). The control of gill blood flow has been in-depth discussed by Nilsson (1986).

7. BREATHING

Various facets on breathing have received coverage in the articles of Lauder (1984), Burggren et al. (1985), Shelton et al. (1986), Scheid (1987), Butler and Metcalfe (1988), Milsom (1989), Randall (1990), Burleson et al. (1992), Perry and McDonald (1993), Van Rooij and Videler (1996) and Gilmour (1998).

One of the largest constraints placed on the design of ventilatory pumps in aquatic organisms arises from the characteristics of H_2O as a respiratory medium. Both the low oxygen content and high density of this medium are limiting factors for the performance of these pumps.

In general, aquatic vertebrates breath continuously, maintaining a unidirectional flow of water over the gills to ensure adequate rates of oxygen, carbon dioxide and ammonia transfer. Certain exceptions can be mentioned (read following account: lamprey and sturgeon).

Breathing in Agnathans. Tidal breathing of adult lampreys is produced by a single phase of excitation that compresses the branchial basket for exhalation. Inhalation is due to passive recoil. The frequency and intensity of breathing in intact lampreys increases with temperature. The ventilatory behaviour of intact lampreys is modified by various sensory inputs. The neural control of respiration in lampreys is not yet well characterized. Lampreys exhibit tidal ventilation of their gill pouches, an adaptation associated with the fact that their mouth is often buried in the flesh of other animals.

The course of respiratory current in *Myxine* is unique. Water is taken in through the terminal nostril and, after passing through a very long pharynx, enters the afferent branchial ducts, passes through the gill pouches, and emerges from the efferent branchial ducts, by a common aperture on each side of the body. The respiratory current is maintained, as in the ammocoete, by a pulsating velum.

The pharyngeo-cutaneous duct serves only for expulsion of accumulated particles from the pharynx. *Myxine* can apparently obtain enough oxygen through its skin to sustain minor activity. It appears that *Myxine*, when feeding, has no respiratory current. It is suggested that the posterior portion of the branchial region is due to the displacement by the tongue musculature and that the backward direction of gill ducts is an adaptation to the burrowing habit. It is likely that the velum has originated independently in the petromyzonts and myxinoids. Thus, the unidirectional ventilation of the hagfish is produced by a velum of more complicated structure and by muscular activity of the gill sacs.

The particular water-pumping device, the velum, in hagfishes consists of a trifid and posteriorly directed lamina arising from the roof of pharynx, just behind the posterior end of the nasopharyngeal duct. The velum possesses a complex skeleton, which consists of thin cartilaginous rods and arches (read details: Janvier 1993). This contribution is also referred to for information on the skeleton and movements of lingual apparatus by muscles.

It may be mentioned that the lamprey tends to inhabit more highly-oxygenated waters that does the hagfish, and feeding probably does not introduce a functional reduction in gill ventilation as it does in hagfishes.

Tidal ventilation of a gas exchange organ with a dense medium such as water is generally considered to be prohibitive in terms of metabolic cost.

Gill Ventilation and Locomotion: Ram Ventilation in Gnathostomous Fish. The specific details of the respiratory pumps associated with gill ventilation in elasmobranchs and teleosts vary greatly from species to species, although the basic mechanics of operation is the same in all. Water is usually taken through the mouth, flows through a buccal cavity, is passed over the gills into an opercular cavity, and leaves through common opercular openings on either side of the body. There are specific differences in the degree of branchiostegal apparatus that supports the gills. There are differences in the number of inhalent openings (mouth and (or) spiracles) and exhalent openings (common operculum or separate branchial openings), the relative sizes of inhalent and exhalent openings (and consequently the timing distribution of inspiration and expiration), as well as relative contributions of opercular and buccal pumps to the net movement of water. Many of these variations are associated with differences in habitat. Thus, plaice, a benthic fish which often lies partially buried on the ocean bottom, has a well developed branchiostegal apparatus, a dominant opercular pump, and a reduction in the size of exhalent openings. These features tend to produce a slow, deep inspiration and a short, rapid expiration of relatively high velocity, which are presumably of adaptive value for fishes buried in silt or sand.

Fish, which are predominantly swimmers, have the buccal pump better developed, although in some cases neither pump operates during the swimming movements of the fish.

There is also a wide variation in specific muscles involved in powering ventilation, although interestingly, the same breathing pattern is maintained regardless of which sets of muscles are involved. Fig. 9.14 illustrates a general scheme of the respiratory muscles involved in powering ventilation in teleosts.

It is evident that a muscle can serve more than one behavioural function. Respiration and feeding in teleost fishes are controlled by muscles of the opercula (gill-covers) and buccal cavity. Some of these muscles participate in both behaviours. In many fish species, display behaviours used during aggression and courtship also involve movement of opercula. Thus, muscles used in respiration and feeding can also serve a communicative function. The neuromotor control of respiration in teleosts has been studied primarily is species that do not perform opercular display, such as carp and trout. Reader interested in motor innervation of respiratory muscles and an opercular display muscle in Siamese fighting fish *Betta splendens* can refer to Gorlick (1989).

Work of breathing

The respiratory fish acts as a double pump. Fig. 9.5 illustrates a model of the ventilating system based on dual mechanism (an opercular suction pump and a buccal force pump) operating with four phases. During the first phase, an opercular suction pump draws water in through the

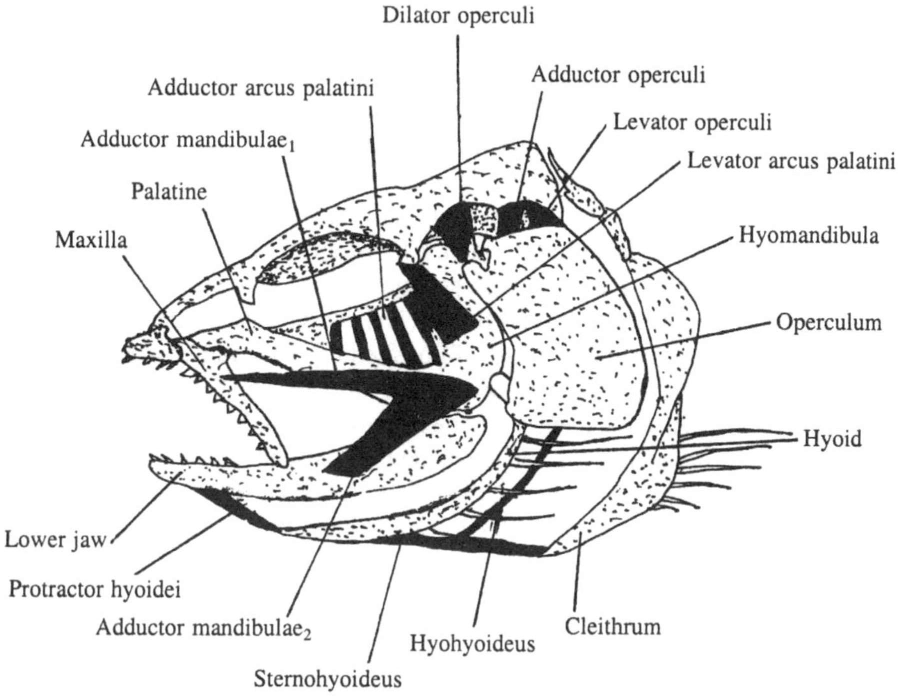

Fig. 9.14 Lateral view of the skull and important respiratory muscles in a typical teleost. *Sources:* Shelton (1970) and Milsom (1989)

mouth, and through the spiracle in those fish that passes one. The water enters into buccal cavity (volume increases by V and the pressure decreases by P_1) and some passes over the gills into the opercular cavities, whose valves are closed (volume decreases by v and the pressure decreases by p_1). The second phase is a period of transition, when the pressure differential between the buccal and opercular cavities is reduced. During the third phase, the mouth and buccal valves are closed and water passes over the gills from the contracting buccal cavity (volume decreases by V and pressure increases by Pe) and leaves the system along with the water in the opercular cavities (volume decreases by v and the pressure increases by pe) through the common opercular (or separate branchial) openings on either side. Phase 4 is another transitional period during which differential pressure may be reversed in the system.

Although this model serves as a simple description of the functional events associated with the ventilation cycle, there is mechanical interaction throughout the system and thus, anatomically, two systems (opercular suction and branchial force) are not really distinct. No muscles are specifically involved in generating opercular suction or buccal force and consequently, that the two mechanisms incorporated into such models are not distinct.

Regardless of these variations, the work that must be performed by these pumps, compared with those of air-breathing vertebrates is great. This is because of constraints imposed by water as a respiratory medium. In contrast to air breathing, water-breathing eliminates surface forces and thus reduces elastic work. Flow resistive and inertial forces are expected to be much higher in water-breathers than in air-breathers because of the higher viscosity and density of water compared with air.

The basic requirement to enable fish to extract water across the gill is that the respiratory surface be irrigated. So far we have assumed that the fish dwells in still water, and that the ventilatory water flow is produced by the action of the respiratory pump. However, many fish irrigate their gills by simply opening their mouth when swimming or remaining stationery in flowing water, and this mode has been termed ram ventilation.

Ram ventilation has been observed in a number of near-shore, midwater and pelagic fish whose habit is to cruise (Table 9.1). These fish convert from pump to ram ventilation when reaching a critical velocity relative to the water. Almost all our knowledge about the energetics of ram ventilation comes from studies of such pelagic species. However, ram gill ventilation also occurs in ram filter-feeding fishes such as anchovies (*Engraulis capensis*) and menhaden (*Brevoortia tyranns*), in which it is an obvious correlate of this feeding mode rather simply a way to reduce drag during swimming (although this may also occur). From the perspective of energy acquisition, filter feeding is an energetically expensive mode of prey capture, requiring continual swimming movements instead of the sporadic brief movements of sit-and-wait predator. The expense of continuous swimming for filter feeding could be partially offset by reduced metabolic costs for gill ventilation if filter feeding is coupled with ram ventilation.

Table 9.1 Examples of ram ventilating fish

Species and systematic position	Habitat
Chondrichthyes	
Odontaspididae	
Odontaspis taurus	subtidal, beaches and bays
Carcharinidae	
Triakis semifasciata	subtidal, beaches and bays
Osteichthyes	
Salmonidae	
Oncorhynchus nerka	pelagic and midwater, anadromous
Echeneidae	
Remora remora	open ocean, surface to midwater
Carangidae	
Caranx crysos	near shore and bays, pelagic to midwater
Scombridae	
Katsuwonus pelamis	pelagic, open ocean
Scomber japonicus	near shore, pelagic

Source: Roberts (1975a)

Burggren and Bemis (1992) studied metabolism and ram gill ventilation in juvenile North American paddlefish, *Polyodon spathula. Polyodon* lacks the oral and opercular valves that are typical for fishes that rely on a buccal pump system to ventilate the gills, and the jaw opening system of *Polyodon* is poorly suited for regular pumping movements. *Polyodon* is essentially an obligate ram ventilator under normal conditions. During steady swimming in normoxia, paddlefishes normally maintain 70%–80% of the maximum sustainable speed. This results in a normal minimum metabolic rate that is about twice that of the minimum (resting) rate of other acipenseriform fishes. From a phylogenetic standpoint, other acipenseriform fishes also use ram ventilation, leading to the hypothesis that the evolutionary origin of a reliance on ram ventilation in *Polyodon* probably predates the origin of the filter feeding habit. Constant swimming may be metabolically expensive, but it would appear to allow some energy to be conserved by ram ventilation. This may be particularly advantageous for species such as *P. spathula* that combine filter feeding and ram ventilation.

It has been pointed out that buccal and opercular valves are also absent in the other living polyodontid, the Chinese paddlefish *Psephurus gladius*. These valves are present in such outgroup forms as *Polypterus* and are quite likely plesiomorphic for actinopterygians so that the authors regard their absence in polyodontid as a derived condition. These features warrant a broader phylogenetic survey among actinopterygians.

Another point concerns the size and function of the spiracle. The spiracle, although well developed in *Polypterus* appears to play a minor role in respiration. Sturgeons of the subfamily Acipenserinae also have large spiracles and in these fish the spiracle has a demonstrated respiratory function (e.g., *Acipenser transmontanus*). Although scaphirhynchine sturgeons lack the spiracle as adults, this is clearly a derived character based on recent ontogenetic information. Both living genera of paddlefishes (*Psephurus* and *Polyodon*) have minute spiracles; thus, it is probable that spiracle reduction is a derived character at some level within Polyodontidae. The spiracle is an insignificant site for either intake or outflow of water in Polyodontids. Thus, water taken into the oral cavity has only one exit over the gill epithelium.

Two major costs of gill ventilation are the energy loss due to resistance of the gills to water flow and the energy loss from cyclic acceleration and deceleration of water as it is pumped through the branchial chambers. The former will occur in both typical respiratory pumping and ram ventilation while the latter is completely eliminated during ram ventilation, in which mouth is held open during swimming and the forward velocity of the fish generates water flow over the gills. Ram ventilation essentially transfers the work of respiration from the buccal and opercular muscles to the swimming muscles of the trunk. The transition to ram ventilation somewhat reduces the profile drag caused by swimming at high velocities. Although the profile drag reduction is very species-dependent, in the bluefish (*Pomatomus saltatrix*) and striped bass (*Morone saxatilis*) the transition to ram gill ventilation may allow as much as 50% increase in swimming speed without an increase in metabolic rate. Ram ventilation must be regarded as an extreme example for the potential interaction between locomotor and respiratory activity, in which the locomotor muscles perform the entire work of breathing. In contrast, during pump ventilation locomotor energy is obtained from the respiratory muscles when expired water is expelled from the opercular cavity. With ram ventilation, the muscle mass available for respiration is enormous and will not be limiting for O_2 uptake.

The work of breathing can be expected to be smaller with ram ventilation since much of the inertial work is eliminated with the steady flow of water.

Interestingly the sharksucker *Echeneis naucrates*, who attaches to sharks and perform ram ventilation on the expense of the locomotor energy of the host, the only respiratory energy being spent by the sucking muscle.

Extraction Efficiency and Gill Diffusing Capacity. Gill systems of fish usually have higher extraction efficiencies than lung systems of other vertebrates at rest. For man, the figure is about 25% at rest, but it improves with exercise as the ventilation-perfusion ratio sharply rises. The gill system of fish contrasts dramatically, in that relatively high percent utilizations, ranging from 40 to 80% have been determined for resting fish, but the figures fall rapidly as activity levels increase. As activity increases gill ventilation imposes an increasingly heavy metabolic load upon the life of a gill breather.

Piiper (1989) has compared the gas-exchange efficiency of fish gills and bird lungs. The gas exchange efficiency of the countercurrent water-blood flow system is considerably reduced by diffusion resistance, a great part of which resides in the water between the secondary gill lamellae. Gas exchange in birds can be analyzed using the serial multicapillary or cross current model in which a number of blood capillaries distributed over the length of the parabranchi have gas-exchange contact with parabranchial gas. The efficacy of the system is further enhanced for CO_2 due to Haldane effect.

The secondary lamellae greatly increase the surface area of the gills in comparison with the volume they occupy. However, the surface area and the mean distance between blood and water (diffusion distance) do vary between fishes. For example in the bottom-dwelling, scavenging dogfish (*Scyliorhinus canicula*), gill area is 210 $mm^2 g^{-1}$ and diffusion distance is 11.3 μm, whereas in the very active skipjack tuna (*Katsuwonus pelamis*) the values are 1350 $mm^2 g^{-1}$ and 0.6 μm respectively. As might be expected, To_2 is an order of magnitude greater in the continuously swimming tuna (read Butler and Metcalfe 1983); (see Farrell 1996; Table 1 for total, branchial and systemic vascular resistances in selected species of fishes, and Table 2 for interspecific differences in respiratory variables in selected fishes).

Gill diffusing capacity probably increases in response to hypoxia and exercise. The gills are innervated by adrenergic and cholinergic fibres, and respond with β-adrenergic vasodilation and α-adrenergic and cholinergic vasoconstriction.

Gill blood volume, is somewhat larger than heart stroke volume, so each stroke volume remains in the gills for one heartbeat. Blood residence time in the gills is about 3*s* at rest, decreasing 1*s* during exercise. At low blood flows, oxygen uptake is largely perfusion limited, but at higher rates of oxygen uptake there are both diffusion and perfusion limitations. Increases in diffusing capacity probably enhance oxygen uptake during exercise and hypoxia. Many factors, including changes in both water and blood flow, increases in gill epithelial permeability and a thinning of epithelium, probably contribute to this increase in diffusing capacity.

The Work Load of Gill Ventilation. Estimations of how large a work load gill ventilation adds to the metabolic budget of a fish have been subject to much disagreement. No one disputes the view that the work load of ventilation increases out of proportion with the energy load of swimming. The low resting estimate for fish is about 0.5% and almost certainly, it is too low to be given serious consideration. Other estimates range from 18 to 43%, and from 5 to 15%. The fact is that the gill breather has a very efficient respiratory mechanism just as long as activity is kept minimal.

Compensation for Ventilation Work Load and Habitat. If the activity behaviour of marine fish is examined alongwith the design and function of the branchial apparatus, the contention that niche selection has been influenced by the high metabolic cost of gill ventilation gains additional support. Benthic fish typically depend largely on the opercular phase of ventilatory cycles. In animals, like the sculpin *Myoxocephalus octodecimspinosus* or the toadfish *Opsanus tau*, the opercular aspiration phases may approach 90% or better of simple cycles when the fish are quiet and left undisturbed. Ventilation in pelagic fish contrasts markedly for they either depend largely on the buccal force pump to overcome gill resistance to flow or, like most midwater fish, they rely on the balanced action of both pumps. The aspiration function of the opercular pump in benthic species is greatly facilitated by extensive development of the branchiostegal apparatus that ballons out during opercular aspiration to allow the peribranchial spaces to expand many times over their volume at the beginning of their opercular phase. Only remnants of the branchiostegal rays remain in association with the opercular doors of fast midwater and pelagic species. Resting respiratory cycles range in frequency as might be expected for fish in different environments: the longest cycles are found in benthic species and the shortest in midwater and pelagic fishes. Cardiac rates show the same trend but, proportionately, the difference in the range of cardiac frequencies is not as large so that the heart rates of benthic species at rest generally are faster than their opercular rates.

8. CONTROL SYSTEMS

The Central Pattern Generator for Respiration in Fish. In fish, neurons situated in the medulla are of fundamental importance in the generation of breathing patterns and the execution of breathing movements. It is also recognized that the water flow through the gills of fish results from breathing movements that are caused by complex mechanical interactions of the muscles and skeleton of the head. It has been indicated that rhythmic sensory input from the periphery is not necessary for the maintenance of respiratory rhythmicity and that a central pattern generator for respiration exists in fish.

These pumps are controlled by motoneurones with cell bodies in the medulla, as in other vertebrates.

It has been estimated that there are atleast 2000 neurons in the medulla of tench that show rhythmic activity in phase with breathing and that between 5% and 40% are branchial motoneurons. Activity in these neurons maintains the sequential contraction of a series of muscles controlling the buccal and opercular pumps. In some instances neuronal activity is confined to a discrete phase of the breathing cycle, but, in other cases, a neuron may be active for most of the cycle. This, to some extent, reflects the timing of muscular activity; in some muscles like the adductor mandibulae show strong phase activity, whereas the sternohyoideus can be active throughout most of the breathing cycle. The motoneurons extend dorso-laterally throughout the medulla and comprise the trigeminal Vth, facial VIIth, glossopharyngeal IXth, the vagal Xth motor nuclei. These motonuclei and the descending trigeminal nucleus and the reticular formation constitute the medullary respiratory centre (Fig. 9.15). It has been reported that the respiratory motoneurons of the dogfish fire in order of the sequential rostro-caudal distribution of their motonuclei in the brainstem. In fact, there are extensive interconnections between all respiratory motonuclei and the reticular formation, presumably to maintain the complex sequencing of activation of various skeletal muscles involved in gill ventilation.

The respiratory motor-nuclei receive inputs via the facial nucleus from vagal afferents innervating gill mechanoreceptors (Fig. 9.16). In addition, areas of mesencephalic tegmentum in the midbrain, with afferent and efferent connections to the reticular formation, initiate periodic breathing seen in some fish that are either inactive or exposed to hyperoxic conditions. Fish change breathing pattern in response to exercise, hypoxia, feeding and probably many other conditions, so there are many other inputs to respiratory neurons, presumably including descending inputs from the forebrain and midbrain and peripheral chemoreceptors.

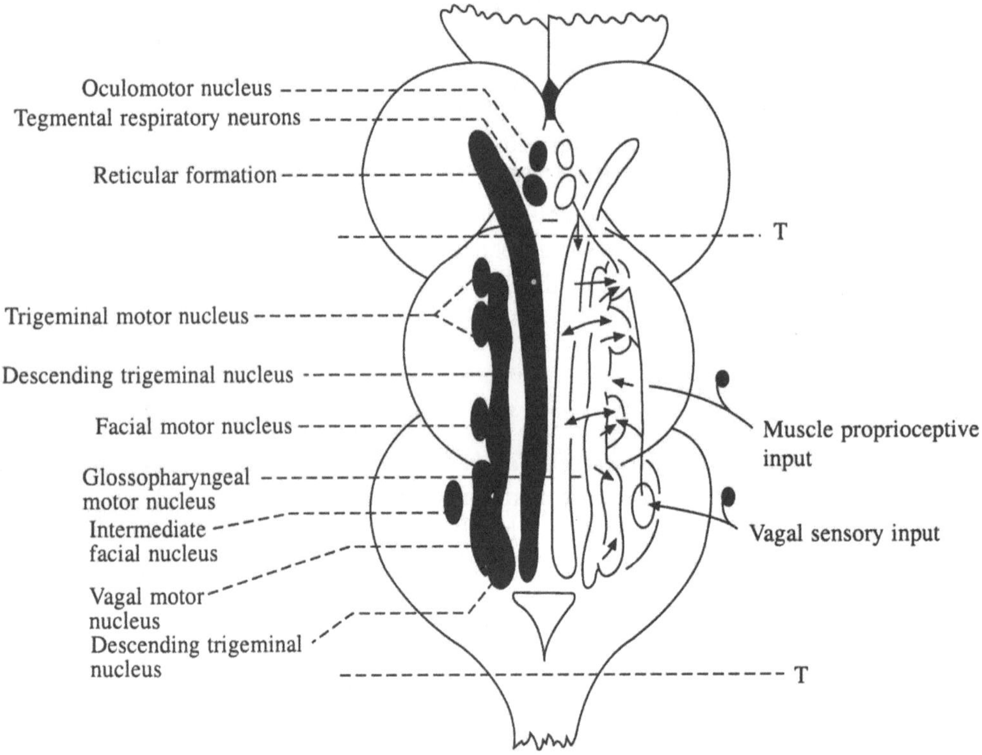

Fig. 9.15 Projection onto dorsal surface of location of respiratory neurons within brain stem of teleosts. *Left*: main sites (in black) located below paired optic lobes and in medulla below cerebellum and facial lobe. *Right*: main interconnections (arrows). *Dashed lines* (T), area defined as essential for normal rhythmic breathing by transection experiments.
***Source:* Shelton et al. (1986)**

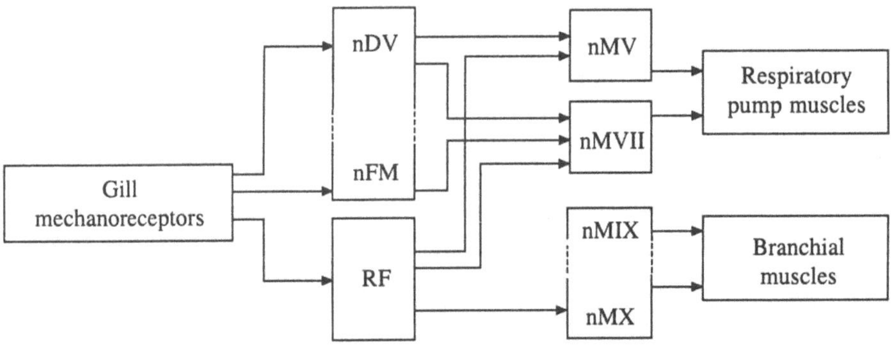

Fig. 9.16 Diagram of possible pathways of sensory signals from gill mechanoreceptors to respiratory motor nuclei (pathways via the gustatory lobes L-IX and L-X are excluded). The descending trigeminal nucleus (n DV) and the median funicular nucleus (n FM) are presented as a structure, because the largest part of trigeminal root fibres end in the funicular region.
Abbreviations:
nMV-trigeminal motor nucleus, NMVII-facial motor nucleus, nMIX-glossopharyngeal motor nucleus, nMX-vagal motor nucleus, RF-reticular formation,
***Source:* De Graaf (1989)**

During ram ventilation there must be some afferent input, sensitive to swimming velocity, that inhibits the output of the rhythmic medullary respiratory neurons. The source of the input may be mechanoreceptors on the gill arches, flow-sensitive receptors of the acousticolateralis system, or probably stretch-sensitive receptors in the swimming musculature or skin.

Modulation of Rhythmic Respiratory Movements. Chemoreceptors (*Oxygen-sensitive receptors*): Matching oxygen uptake to metabolic demands requires that animals sense and respond to respiratory and blood oxygen concentrations. In fishes, the need for fine control over branchial gas exchange is exacerbated by the low O_2 capacitance of water and the tremendous spatial and temporal variability in oxygen availability in the aquatic environment.

That sensory input may change the respiratory pattern is to be expected, since the respiratory movements need to be adapted to momentary demands. Several types of receptors have been demonstrated in the gill region.

Environmental hyperoxia occurs in natural waters and depresses gill ventilation in cyclostomes, elasmobranchs, and a wide range of teleosts including icefishes.

Environmental hypoxia has been shown to elicit both respiratory and cardiovascular responses, of varying magnitude, in all fish that have been studied to date. Whereas the ventilatory response in teleost fish is strong, consisting of increase in respiratory frequency and tidal volume, very little response is shown by elasmobranch fish. Although there is some evidence to implicate the pseudobranch of teleost fish as a possible receptor site involved in ventilatory response, most data implicates putative arterial chemoreceptors, possibly located within the central nervous system.

The cardiac response to environmental hypoxia, in both elasmobranchs and teleosts, consist of an immediate, intense bradycardia accompanied by an increase in cardiac stroke volume. Often in teleosts, these changes are evenly matched and thus produce little change in cardiac output. In elasmobranchs the receptors involved in producing this response are diffusedly spread throughout the orobranchial and parabranchial cavities and are innervated by cranial nerves V, VII, IX and X (Milsom et al. 1999 see Sundin 1999).

Although the general location and innervation of the receptors involved in these cardiac responses have been described in elasmobranchs and teleosts, nothing is known of their exact location and response characteristics. The arteries supplying the first two branchial arches in these fishes are homologous with the carotid and aortic arches of higher vertebrates and it has been suggested that the phylogenetic fate of the O_2 sensitive receptors in the branchial arches and gills, and innervated by the IXth and Xth cranial nerves in fish may be in the discrete carotid and aortic bodies of terrestrial vertebrates. If these were the case, the O_2 receptors located in the first gills might be expected to exhibit similar properties to those of carotid and aortic bodies (Fig. 9.17).

Fish gills are known to show a high degree of intrinsic vasomotion and thus a variety of stimuli associated with smooth muscle contraction/relaxation, changes in vessel flow or local changes in blood gas or metabolic concentration remain as possible sources of rhythmic stimuli.

Indirect evidence (based on O_2 sensitive receptors in the first gill arch of yellowfin tuna, *Thunnus albacares*, suggests that these receptors are located near the gill vasculature and are more sensitive to changes in arterial P_{O2} than water P_{O2}. Taken together with the phylogenetic evidence indicating that the branchial artery supplying the first gill arch in teleosts gives rise to

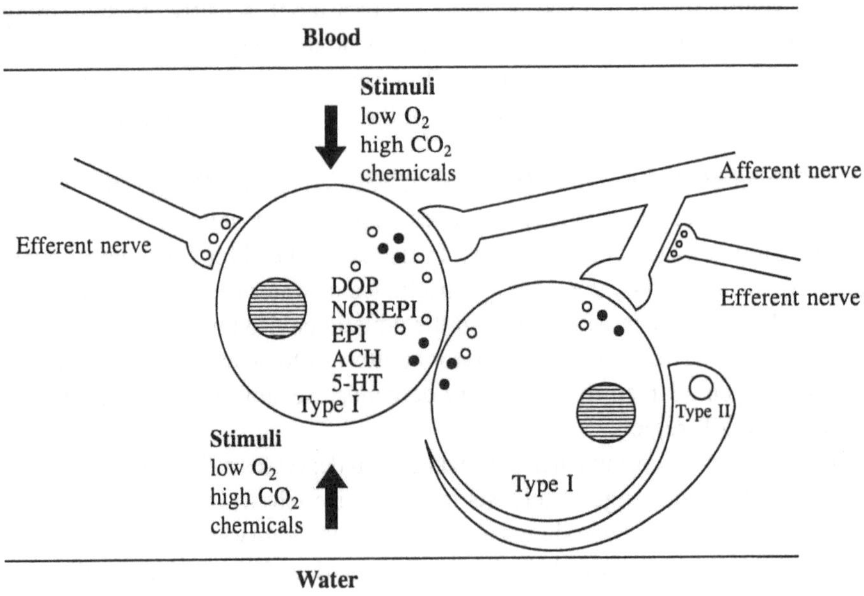

Fig. 9.17 Schematic diagram of the primary elements of a putative branchial O_2 chemoreceptive unit in a fish gill. Stimuli (hypoxia, circulating (neuro) chemicals, hypercapnia?) may reach Type I cells from either the external environment (water) or internal environment (blood). Type I cells are thought to be the primary O_2 transducer and release neurochemical (s) that trigger action potentials in the afferent neurons. Type II cells are glia-type cells that provide cellular support for Type I cells but appear to serve no active function in O_2 chemotransduction.
***Source:* Burleson (1995)**

carotid artery of mammals, this evidence lends strong support to the hypothesis that the O_2 sensitive chemoreceptors located in the first gill arch of teleosts are homologous to carotid body chemoreceptors of higher vertebrates.

The finding of a widespread system of neuroepithelial cells in the primary lamellae of all gill arches of teleosts, however, suggests that O_2 sensors may be located more widely than just the first gill arch (Milsom and Brill 1986; see also Shelton et al. 1986; Perry and Wood 1989) (Figs. 9.18, 9.19).

Both internal and external oxygen-sensitive receptors are located in the gills, the external receptors facing the water flow, the internal receptors monitoring blood oxygen and flow. These oxygen chemoreceptors cause changes in breathing and/or heart rate via neuronal pathways because the responses are rapid, occurring within seconds of the appearance of hypoxia. Chemoreceptors may not be involved in the hypoxic stimulation of ventilation in elasmobranchs. The gill blood oxygen content receptors, if present in elasmobranchs, may be involved in catecholamine release rather than having any direction on the respiratory centre. The external group of gill oxygen receptors presumably exists in elasmobranchs, however, because hypoxia is associated with a marked bradycardia with a rapid onset (read Randall 1990). Recent studies indicate that fish have at least two functional groups of O_2-sensitive chemoreceptors. Reflex hypoxic bradycardia is mediated exclusively by externally (water) oriented receptors while ventilatory reflexes are mediated by externally and internally (blood/tissue) oriented receptors (read Burleson and Milsom 1993; Fig. 9.17, read Burleson 1995; Milsom et al. 1999). For arterial chemoreceptors in vertebrates, the reader is referred to Milsom (1993).

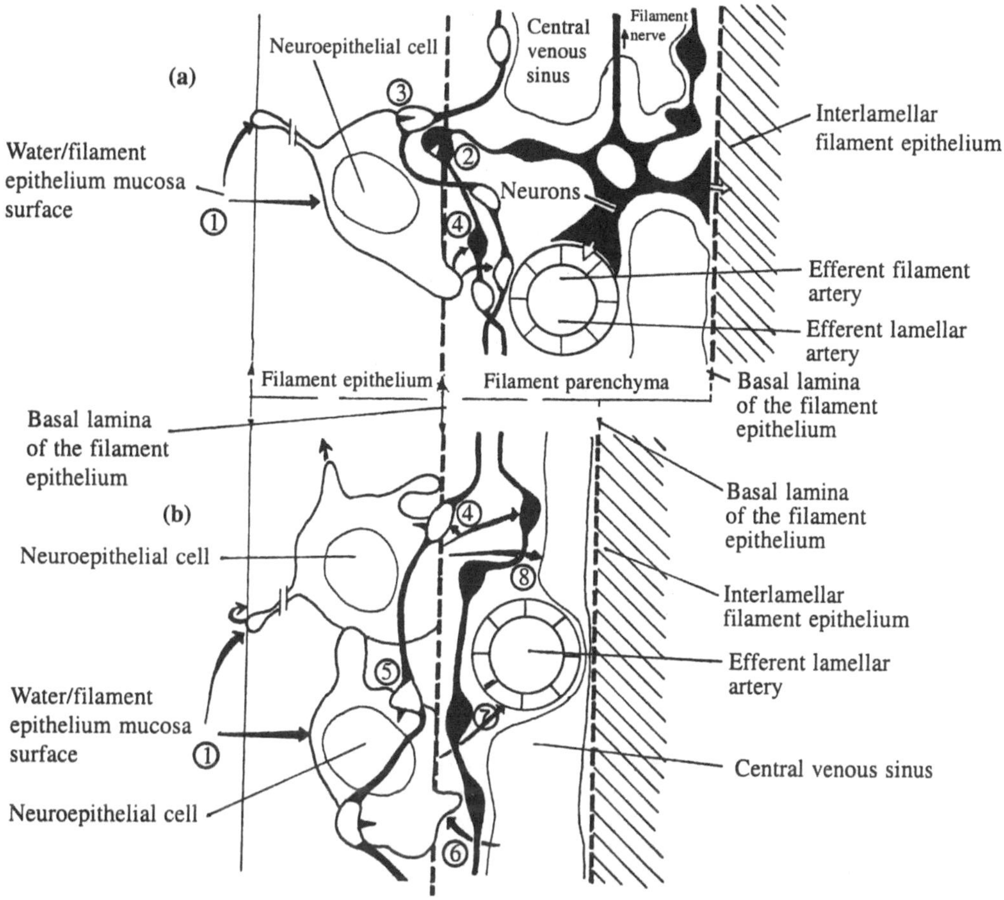

Fig. 9.18 Illustration of possible interactions between the neuroepithelial cells (NECs) and their environment and innervation in basal and apical areas of teleost gill filament. (a): Chemoreceptive function and innervation of the neuroepithelial cells in the basal area of the filament. (b): Secretory function, targets, and innervation of the neuroepithelial cells in the apical area of the filament. (1) Environmental factors act differently on the NECs or through the filament epithelium (i.e., O_2, CO_2, ions, pH, temperature, intercellular messengers). (2) Chemosensory pathway: afferent synapses on the 5-HT filament neurons connect to the filament nerve. (3) Afferent synapses on the catecholaminergic sympathetic nerve (functioning). (4) Substances (i.e., 5 HT and peptides) diffuse from the neuroepithelial cells and act on the subepithelial nerve plexus (development, functioning). (5) Influence of the catecholaminergic sympathetic nerves system on the functioning of neuroepithelial cells via efferent synaptoid contacts. (6) Active substances (i.e., hormones and ions) diffuse from the arterio-venous blood of the central venous system upto neuroepithelial cells. (7) Local control of respiration through an intrinsic sensorimotor way: 5 HT vasomotor control of efferent lamellar arteries by the neuroepithelial cells. (8) 5 HT control of central venous sinus by the neuroepithelial cells (permeability and extent).
***Source:* Bailly et al. (1992)**

Hedrick and Jones (1993) investigated the mechanisms and physiological control of air-breathing in an extant halecomorph fish, the bowfin (*Amia calva*). The bowfin is a primitive (non-teleost) actinopterygian fish which uses both gills and a vascularized gas bladder for

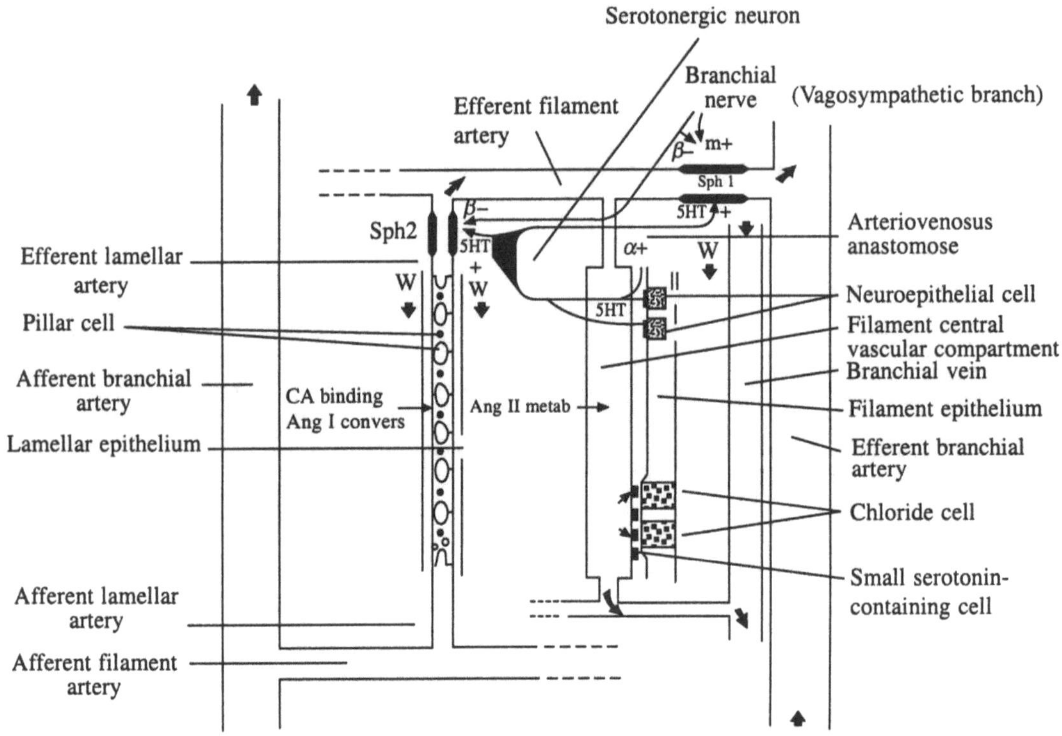

Fig. 9.19 **Schematic representation of some aspects of gill control machinery as suggested by recent findings from ultrastructural, histochemical and immunocytochemical studies. These new results mainly relate to the discovery of a rich intramural serotonergic nervous system. This system relates the serotonin-containing neuroepithelial cells, serotonergic large neurons and strategically located vascular sphincters. Large, short processes of serotonergic neurons, filled with 5 HT, surround the sphincters tunica media, whereas another type of process (dendrites) run close to the neuroepithelial cells. Environmental hypoxia has been shown to "degranulate" the neuroepithelial cells. Thus, this organization suggests that vasoconstriction of sph 1 and sph 2 is mediated by 5 HT (a potent vasoconstrictor in fish) release from serotonergic neurons, the functioning of which depends, in turn, on transduction of environmental or tissular factors.**

Another system consists of small cells containing serotonin located within the wall of the central vascular compartment of the filament beneath its epithelium and close to chloride cells. This organization raises the hypothesis of a functional link between 5 HT and epithelial ion transport by chloride cells by adenylate cyclase activation in a way similar to that shown recently in the mussel. Finally, the following are also presented: catecholamine binding to lamellae pillar cells, angiotensin converting activity (Ang I-II) of the lamellar tissues, and angiotensin II the metabolism in the arteriovenous bed.

Abbreviations: α: α-adrenoceptor (probably noninnervated in the gill) mediating arteriovenous anastomose. β^-: β-adrenoceptor (presumably innervated) mediating Sph1 and Sph2 dilation m^+: muscarmic (innervated) receptor mediating: Sph cholinergic contraction. $5HT^+$: serotonin receptor (presumably $5HT_2$) mediating Sph 1 and Sph 2 constriction.
Source: **Laurent (1989)**

respiratory exchange. Recent studies have confirmed the observation that exhalation precedes inhalation during aerial ventilation. The exhalation-inhalation sequence described for *A. calva* is similar to that in gar (*Lepisosteus osseus*) and several teleost species. The gas bladder of *A. calva* and other primitive fishes, such as gar, lungfish and polypterids function both as a gas exchanger and a buoyancy organ. There is some evidence to indicate that both O_2-sensitive chemoreceptors and volume-sensitive gas bladder stretch receptors are involved in the control of aerial ventilation. For example, studies have shown that *A. calva*, like many other air-breathing fish, increase air-breathing during periods of hypoxic stress or thermal stress, and that air-breathing frequency also increases after gas deflation.

The chemoreceptor sites that mediate the oxygen-related reflexes have not been established, but there is evidence from gar that both externally and internally oriented O_2 receptors may be involved.

Air flow measurements in *A. calva* revealed two different types of ventilatory patterns: type I air-breaths were characterized by exhalation followed by inhalation; type II air-breaths consisted of single inhalations with no respiratory phase.

Type I breaths appear to be under feedback control from O_2-sensitive chemoreceptors since they were stimulated by aquatic or aerial hypoxia and were nearly abolished by aerial hypoxia. These results also indicate that *A. calva* respond to changes in intravascular Po_2; however, externally facing chemoreceptors that stimulate air-breathing in aquatic hypoxia cannot be discounted. Type II air-breaths, which occurred in aerial hyperoxia despite aquatic hypoxia, appear to be stimulated by reduction gas bladder volume (VB), suggesting that type II breaths are controlled by gas bladder stretch receptors. Type II breaths are likely to have a buoyancy-regulating function.

Randall et al. (1992) have shown that diet enriched in Omega-3 polyunsaturated fatty acid (PUFA) and/or vitamin E contribute significantly to regulation of metabolism in hypoxia in sturgeon (*Acipenser naccarii*).

Catecholamines: Catecholamines stimulate ventilation in fish but the response is seasonal. Catecholamines are released into the circulation of fish in response to a wide variety of stresses including hypoxia, hypercapnia, anaemia, blood acidosis, air exposure, and both exhausting and violent exercise. Butler et al. (1986) considered levels of plasma catecholamines during different levels of exercise in fishes.

Noradrenaline is known to cross the blood-brain barrier of fish so elevations in blood catecholamine levels will lead to hyperventilation by direct stimulation of respiratory drive in the medulla. As outlined above, many stresses cause the release of catecholamines into the blood, and this release could be an important factor in mediating and associated hyperventilation. In the case of hypoxia, the increase in ventilation in the trout is far too rapid to be solely due to central effect of circulating catecholamines. Stimulation of gill oxygen chemoreceptors presumably drive the initial increase in ventilation. The subsequent rise in catecholamines during severe hypoxia, however, could play an important role in the stimulation of ventilation during maintained hypoxia. It is possible that, in elasmobranchs, the increased rate of catecholamines into the circulation may account for all of the increasing in breathing (Randall 1990; read also Perry and Wood 1989; Ishimatsu et al. 1995 for discussion).

According to Metcalfe and Butler (1988), there is no substantial evidence that the increase in circulating levels of catecholamines plays any crucial role in the cardiovascular or respiratory responses of the dogfish (*Scyliorhinus canicula*) to hypoxia.

CO_2/H^+ *Receptors and Gill Mechanoreceptors*: Piscine gill mechanoreceptors, like other vertebrate mechanoreceptors, appear to be simple free nerve endings situated in muscle or connective tissue. It is thought that the anatomical location of a mechanoreceptor determines its sensory modality and response characteristics.

The ability to sense and respond to changes in CO_2/pH represents an important stage in the evolution of terrestriality and also is an important component of the responses of fishes to environmental hypercapnia, acidosis and exercise. The anatomical location, afferent pathways, and discharge characteristics of the receptors that mediate cardioventilatory reflex responses to arterial acidosis in fishes, however, remain unknown. Are branchial chemoreceptor CO_2/pH sensitive? Is there a clear-cut evidence for central CO_2/pH chemosensitivity? Both the questions need a solution. The role of CO_2/pH chemoreceptors in the regulation of respiration in lower vertebrates has been reviewed (Milsom 1995).

Hypercapnic and acidotic states cause an increase in ventilation in teleosts and elasmobranchs. Chemoreceptors highly sensitive to carbon dioxide are found in the palatal organ in the roof of the buccal cavity of carp. Probably, some sort of central H^+ receptor exists in fish. Gill ventilation, however, is not well correlated with cerebral spinal fluid pH.

In addition, hyperoxic hypercapnia, although causing an acidosis, can result in hypoventilation or hyperventilation.

The gill region of both elasmobranchs and teleosts contain large numbers of mechanoreceptors, innervated by IXth and Xth cranial nerves and stimulated movements of the branchial apparatus. These receptors modulate the breathing pattern and adjust the geometry of the gill apparatus to the water flow. The mechanoreceptors show a wide variety of activity patterns, both tonic and phasic. The intermediate facial nucleus, which receives afferent information from gill mechanoreceptors, projects to the respiratory motonuclei. The probable overall effect of mechanoreceptor input is to adjust the relationship between breathing rate and amplitude, inhibiting breathing frequency in elasmobranchs but increasing rate while reducing amplitude in teleosts.

The functions of mechanoreceptors suggested are (i) they play a role in the control of respiration, e.g., by maintaining a closed gill curtain, (ii) a protective function against mechanical hazards. They may also induce cleaning or expulsion reflexes and coughs (see De Graaf 1989).

Further studies are required to assign specific role for orobranchial mechanoreceptors. The mechanoreceptors associated with respiratory passages may regulate breathing pattern and cardiorespiratory synchronization in fish; though the procedure is still unclear. Details of mechanoreceptors associated with respiratory passage can be obtained in a comprehensive review (Burleson et al. 1992).

The switch from rhythmic to ram ventilation occurs at a specific water velocity, depending on the fish, temperature and oxygen content of the water. Mechanoreceptive input, located in the gills, acoustico-lateralis system, or skin and detecting water velocity may trigger the switch. These inputs may operate through the mesencephalic tegmentum in the midbrain. Oxygen levels modify the water velocity at which the switch occurs. In continuously ram-ventilating tuna, aquatic hypoxia results in a larger mouth gape. Thus, chemoreceptors appear to play, rather initiating role in ram ventilation.

Nocireceptors have been located in dogfish (*Squalus acanthias*) gills that show similarities to type J receptors in mammalian lungs. They reduce the possibilities of gill edema. These are labelled as "defense" receptors implicated even in coughing (read Burleson et al. 1992).

Proprioreceptors in gills play a part in relating the rate of respiration to its amplitude in dogfish. Proprioreceptors associated with displacement of gill filaments and gill rakers are found in teleosts (e.g., *Salmo salar*). Although the physiological characteristics of proprioreceptors mentioned above have been well described, the role of these receptors in respiratory control is unknown, since there is lack of knowledge about mechanoreceptor activity under normal conditions in an actively breathing fish.

Overview of gill control machinery in teleosts. Fig. 9.19 summarizes the findings of gill control machinery for teleosts (see Laurent 1989). New results mainly relate to the discovery of a rich intramural serotonergic nervous system. This system relates the serotonin-containing neuroepithelial cells (NEC), serotonergic large neurons (SN) and strategically located vascular sphincters. Large, short processes of serotonergic neuron, filled up with 5 HT, surround the sphincters tunica media, whereas another type of process (dendrites) run close to the NEC. Environmental hypoxia has been shown to "degranulate" the NEC. Thus, this organization suggests that vasoconstriction of sphincter 1 and sphincter 2 is mediated by 5 HT (a potent vasoconstrictor in fish) release from SN, the functioning of which depends, in turn, on transduction of environmental or tissular factors.

Another system consists of small cells containing serotonin (SCC) located within the wall of the central vascular compartment of the filament (CVS) beneath its epithelium and close to chloride cells (CC). This organization raises the hypothesis of a functional link between 5 HT and epithelial ion transport by chloride cells by adenylate cyclase activation.

Finally, the following are also presented: catecholamine binding to lamellae pillar cells, angiotensin converting activity (Ang I-II) of the lamellar tissues and angiotensin II metabolism in the arteriovenous bed. It has been supposed that serotonin vasoconstriction corresponds with the nonadrenergic, noncholinergic (NANC) effects.

9. DIPNOI-MECHANICS OF GILL AND LUNG VENTILATION

Inspite of some structural differences (e.g., reduction of opercular and hyoid bones in *Protopterus* relative to *Neoceratodus*), the mechanics of both gill and lung ventilation are qualitatively the same in *Lepidosiren*, *Neoceratodus*, and *Protopterus*. Ventilation of the gills with water occurs through the action of a positive pressure buccal pump anterior to the gills and an opercular suction pump posterior to the gills. These pumps operate in concert to generate a nearly continuous water pressure gradient favouring water flow in the mouth, through the gills and out the opercular opening. Such a ventilatory system is widely shared amongst aquatic vertebrates, occurring in phyletically ancient fishes, highly derived teleosts, and even preserved somewhat in larval and adult amphibians.

Ventilation of the lungs with air in *Protopterus* is achieved by action of the same musculoskeletal elements involved in aquatic ventilation. In essence, a single aquatic ventilatory cycle irrigating the branchial chamber is replaced by a modified cycle that generates a single air breath. The single-breathing cycle occurs immediately after the snout has been thrust above the water surface into the air. The glottis, which has been tightly closed during water-breathing cycles, is opened, and the pressure in the lungs decreases as expiration occurs. There follows a particularly tight closure of the opercular flaps, brought about by extended activity of constrictor hyoideus muscle, an exaggerated depression of the buccal floor drawing a large volume of fresh air through the opened mouth into the buccal cavity. The mouth is then closed, and the head reimmersed in water. A subsequent strong elevation of the buccal floor increases buccal pressure to approximately 20 mm Hg, which is much higher than during a normal aquatic ventilatory cycle. The elevated buccal pressure drives inspired air past the open glottis into the lungs, completing air-breathing cycle. The aquatic ventilatory cycles both preceding and following the cycle in which the lungs are ventilated show different patterns of muscle activation and corresponding motor responses. Lungfishes usually ventilate their lungs with a single air breath followed by a period of apnea of variable length, though under some experimental condition tightly-grouped multiple breaths may occur. Estivation also cause changes in patterns and mechanics of lung ventilation.

Lung ventilation in the Dipnoi thus incorporates a buccal force pump mechanism. Although several workers have suggested that aspiration breathing involving rib movement for generation of subambient pleural pressures may occur in lungfish and other air-breathing fishes, these claims have not been substantiated. It is most likely that lung ventilation mechanism seen in extant Dipnoi is similar to that of earliest tetrapods, and that aspiration ventilation of lung replaced buccal force ventilation only later as ribs took on increasing structural importance in fully terrestrial vertebrates.

10. VENTILATION IN *POLYPTERUS, ACIPENSER* AND *AMIA*

Polypterus: Spiracles are used to inhale air. Uptake of air via the mouth is also observed. When again under the water surface, air may leave the animal through the gills behind the opercula and also through the spiracles (bubble formation). Such air bubbles appear to be an indication of an exhalation process. The mechanism of inhaling through the spiracles is difficult to understand, as it remains unclear by which mechanism the air is drawn into the lungs. Inhaling by mouth, like in dipnoans, seems most plausible, because muscles of mouth and pharynx may provide the power for pushing air into the lungs. Some teleosts fill their swim bladder by the same type of oropharyngeal contraction.

Acipenser: The sturgeon *Acipenser*, strictly aquatic water breather, retains many characteristics of their palaeoniscoid ancesters. Associated with the sedate form of feeding (soft suctorial mouth for benthic feeding) and low activity levels in the sturgeon is a reduced metabolic rate and a unusual ventilatory mechanism whereby the gills can be effectively ventilated in a quasitidal fashion strictly through the opercular slits when the mouth is otherwise engaged in feeding.

The pharyngeal jaw apparatus in the "primitive" living ray-finned fishes *Polypterus*, *Lepisosteus* and *Amia* has two important biological roles: swallowing of prey and air ventilation (see details: Lauder and Wainwright 1992).

The four-stroke lung ventilation pattern is evident in *Polypterus senegalus*, *Lepisosteus oculatus* and *Amia calva* (Brainerd 1994).

According to Smatresk (1994), studies on extant biomodally breathing vertebrates offers us a chance to gain insight into the changes in respiratory control during the evolutionary transition from water to air breathing. In primitive actinopterygian air-breathing fishes (*Lepisosteus* and *Amia*), gill ventilation is driven by an endogenously active central rhythm generator that is powerfully modulated by afferent input from internally and externally oriented branchial chemoreceptors, as it is in water-breathing actinopterygians. The effects of internal or external chemoreceptor stimulation on water and air breathing vary substantially in these aquatic air breathers, suggesting that their roles are evolutionary malleable. Air breathing in these bimodal breathers usually occurs as single breaths taken at irregular intervals and is an on-demand phenomenon activated primarily by afferent input from the branchial chemoreceptors. There is no evidence for central CO_2/pH sensitive chemoreceptors and air-breathing organ mechanoreceptors have little influence over branchial-or air- breathing patterns in actinopterygian air breathers. In the sarcopterygian lungfish *Lepidosiren* and *Protopterus*, ventilation of the highly reduced gills is relatively unresponsive to chemoreceptor or mechanoreceptor input. The branchial chemoreceptors of the anterior arches appear to monitor arterialized blood, while chemoreceptors in the posterior arches may monitor venous blood. Lungfish respond vigorously to hypercapnia, but it is not known whether these responses are mediated by central or peripheral chemoreceptors. A major difference between the sarcopterygian and actinopterygian bimodal breathers is that lungfish can inflate their lungs using rhythmic route of air breathing, and lung mechanoreceptors influence the onset and termination of these lung inflation cycles. The control of breathing in amphibians appears similar to that of lungfish. There are other recent papers of interest in this regard (Smatresk and Azizi 1987; Smatresk 1988; Hedrick et al. 1991; Burleson et al. 1992).

A comparative analysis of actinopterygian aerial buccal pumps indicates that the primitive patterns of air transfer differs fundamentally between these two clades. Actinopterygian fishes ventilate their lungs with a four-stroke buccal pump: the buccal cavity expands and fills with expired air, compresses to expel expired air, expands again to take in fresh air, and then compresses again to pump fresh air into lungs. Lungfishes, caudates, and anurans expand and compress the buccal cavity only once for expiratory-inspiratory cycle, and thus use a two-stroke pump. Both of these bidirectional, aerial buccal pumps evolved from unidirectional, aquatic buccal pumps. The two-stroke aerial pump and the primitive aquatic pump used for gill irrigation share slow movements and may both be triggered by the same central rhythm generator. These similarities suggest that the two-stroke buccal pump evolved from the gill irrigation pump. Similarly, the four-stroke pump shares rapid movement and afferent triggering with aquatic suction feeding and coughing, suggesting that the four-stroke pump may have evolved from a combination of two suction feeding or coughing movements. Thus the differences between the actinopterygian and sarcopterygian aerial buccal pumps may be due to their independent evolution from different aquatic buccal pumps, rather than due to divergence from a single aerial buccal pump (Brainerd 1994).

11. SECONDARY AIR BREATHING IN FISHES

11.1 An Overview

Transition to air breathing has occurred many times in natural selection during vertebrate evolution. One of the most interesting groups of animals with evolution to air breathing are the fishes. Most air-breathing fishes are teleosts, but dipnoans, chondrostean, and holostean fishes are well represented by air-breathing species. The reader is referred to Chapter 8 for cardio-respiratory modelling with air-breathing. (For further details, see Graham 1997, 1999).

Shortage of O_2 and high cost of extracting it from water compelled some fish to select means of absorbing O_2 directly from the atmosphere. Seasonal drought further aggravated the conditions for respiratory exchange in some water breathers. Environmental conditions, thus, called for direct absorption of atmospheric O_2 in some fish, and fossil records and extant fish suggest that diverse solutions for obtaining O_2 directly from the atmosphere did occur and that air breathing in fish evolved repeatedly in many different ways. Thus, the development of aerial respiration by fish seems to have evolved in a niche selection as a means to escape oxygen lack in poor quality respiratory environments and also to avoid dessication (read Fernandes et al. 1999).

Reduction in the exchange surface area of gills is an identifiable trend in the air-breathing fish as well. On the other hand, expansion of the available exchange surface seems to have followed with the evolution of fishes adapted to high-speed swimming. In fact, the progression of increase in gill surface per unit body weight directly relates to the metabolic demands of activity in different habitats, from lowest to highest in the order: benthic, midwater, pelagic. Morphologically, the surface area is a function of gill filament and lamellar spacing as well as lengths of individual lamellae, and the total surface is inversely related to the body size within a species.

Synchronous air breathing appears to occur widely in conjuction with aerial respiration. It is suggested that synchronous air breathing has anti-predator functions analogous to schooling, but with clustering in the temporal rather than spatial dimension.

The diversity of air-breathing habits in fish is great. Among teleosts, however, most species are facultative air breathers dependent on air breathing only when conditions for aquatic exchange become severely limited or O_2 requirements are increased during activity. Other species practice air breathing routinely as a supplement to aquatic gas exchange. Very few species of teleosts are obligate air breathers at ambient conditions similar to those prevailing in their natural habits.

Characteristically, air breathing fishes utilize two modes of gas exchange by use of gills and skin for aquatic breathing and a highly vascularized receptacle for air breathing. These fishes utilize aquatic and aerial breathing to various degrees depending upon environmental factors and activity. The relative importance of the modes of breathing is reflected structurally. In obligate air breathing fishes the gills tend to show reduced surface area and a coarse, thick epithelium while the well vascularized air breathing organ has a large surface area compatible with efficient gas exchange. A converse situation prevails in the facultative air breathers which rely predominantly on aquatic breathing. In very few species both the aquatic and aerial gas exchange organs are well developed.

The usage of the phrase ‘amphibious fishes’ covers the physiology of terrestrial adaptations in those amphibious fishes lacking specialized accessory respiratory organs.

Sayer and Davenport (1991) have summarized the possible casual factors which may elicit an emergence response in Fig. 9.20 (A) and (B). Amphibious fish inhabiting closed systems, as typified by freshwater or intertidal pools, may leave water for any of the reasons detailed in Fig. 9.20(A). The relative importance of any one stimulus is likely to vary between different species. However, it is possible that in closed systems, adverse fluctuations in physico-chemical parameters will have a more important effect in eliciting amphibious behaviour than will biotic factors. In open systems, such as coastal waters or large freshwater bodies, effectively two routes of escape from adverse aquatic conditions are available to amphibious fish. They may move onto land, or alternatively they may move underwater to find better conditions. In such a system, where physico-chemical parameters remain relatively constant, abiotic factors are unlikely to have a significant influence on amphibious behaviour. The dominant stimulus in open systems is possibly the three-way interaction between predation, competition, and short-or long-term food availability (Fig. 9.20(B)).

It is unlikely that any one of the factors will act alone in causing amphibious behaviour, and

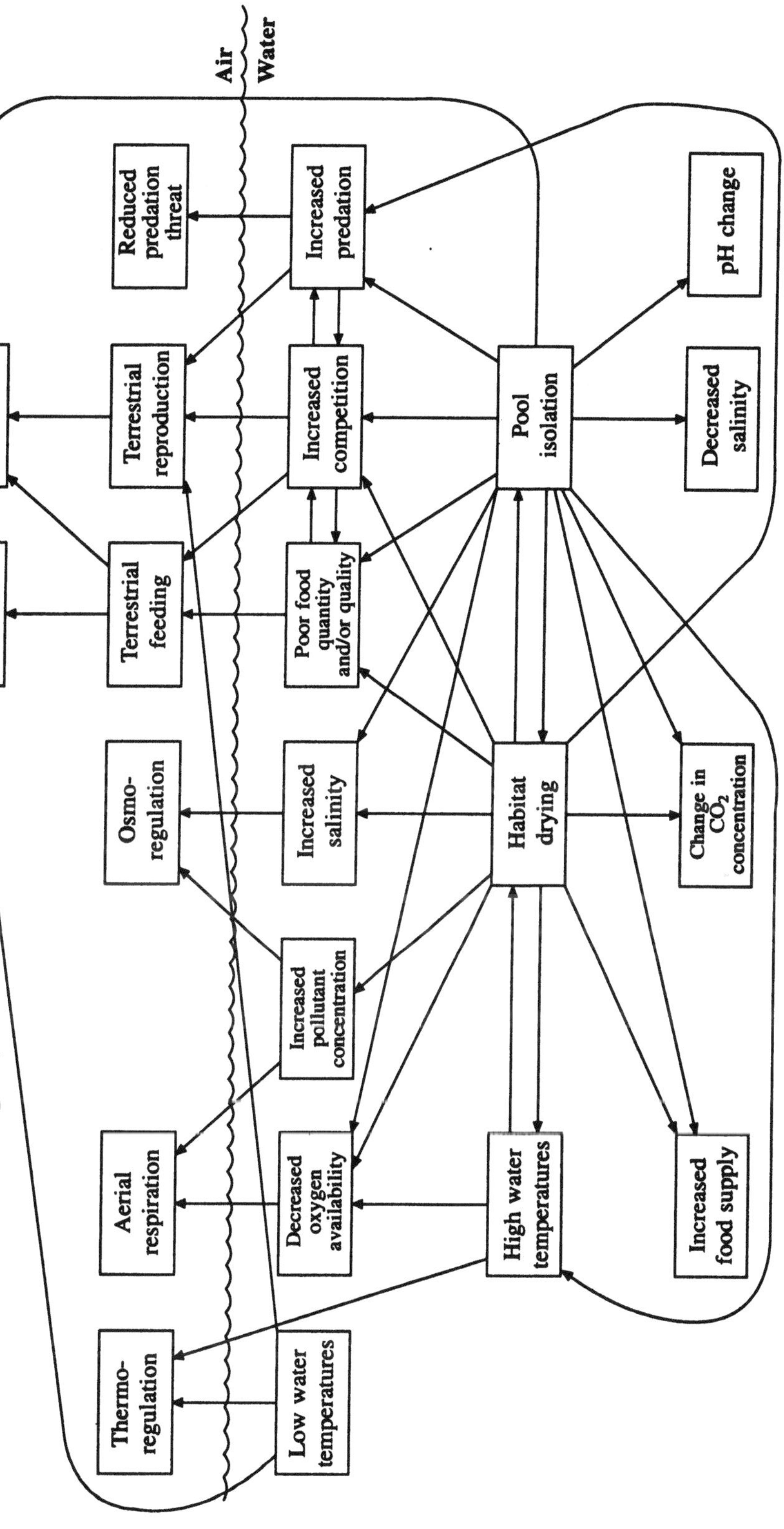

(A) Emergence behaviour by amphibious fish.
Biotic and abiotic influences.
Closed systems.
(Freshwater or intertidal pools)

(B) Emergence behaviour by amphibious fish. Biotic and abiotic influences. Open systems (Coastal waters or large freshwater bodies).

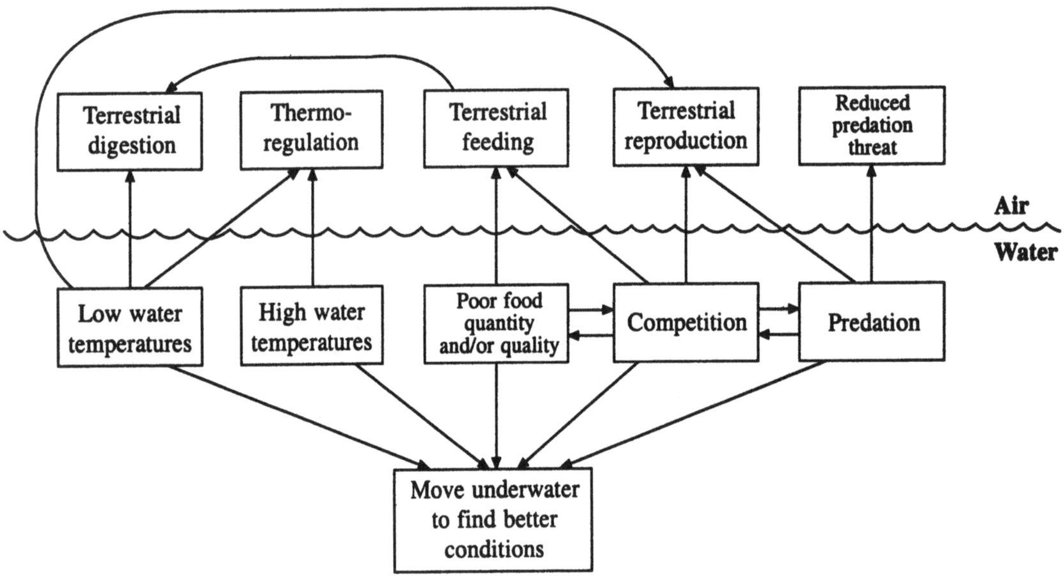

(B)

Fig. 9.20 Summary of the major biotic and abiotic influences eliciting emergence behaviour by amphibious fish. (A) Closed systems (i.e., freshwater or intertidal pools); (B) Open systems (i.e., coastal waters or large freshwater bodies).
***Source:* Sayer and Davenport (1991)**

in this respect the available literature on fish leaving water is lacking. Much of its fragmentary and partly anecdotal, and the limited amount of experimental work tends to concentrate on individual causative factors. There is evidently scope for detailed examination of emersion in a number of amphibious fishes, testing a matrix of environmental and biotic stimuli, in an attempt to determine in more detail the reasons for such behaviour.

Dejours (1994) concluded that the problem of amphibious breathing cannot be viewed separately from all functions which are influenced by the nature of the milieu where animals live. To succeed in the colonization of land, it is necessary that all functions are successfully adapted to their milieu. Such statements are very very vague, and raise the problems of the mechanisms of adaptation and evolution. But since aquatic life, amphibious life, and terrestrial life are provided with similar adaptations related to the milieu (respiration, water economy, ionic secretion, nitrogen end product, locomotion, reproduction, communication, etc.), it is obvious that onto the phylogenetic division of the animal kingdom must be superimposed a transphyletic division which shows clearly primordial aquatic life, the transitional amphibious life and the terrestrial life.

Morris and Bridges (1994) have discussed the properties of respiratory pigments in bimodal breathing animals: air and water breathing by fish and crustaceans. Oxygen is nearly 30 times more available in air compared with water, whereas the carbon dioxide capacity of water ~28.5-fold greater than water, presenting bimodal breathing species with two different respiratory milieus. The respiratory pigment plays a variable role in animals switching between the two media. In vertebrates the transition to air breathing involves two main strategies: a decrease in oxygen affinity and changes in other haematological parameters such as haematocrit. When appropriately analyzed, data reveal a decrease in blood oxygen affinity during transition to air. This may arise via differences in the intrinsic affinity as occurs in some amphibians, or be due to increases in the organic phosphate: haemoglobin ratio when acclimating to air breathing. Adopting air breathing often promotes increased haematocrit.

11.2 Designs of Air Breathing Organs

Among teleosts, several families of fishes have developed varied types of accessory respiratory organs *viz*: (a) specialization and modifications of the buccopharynx in the genera *Monopterus* and *Synbranchus* (Synbranchidae); *Channa* (Channidae) and *Electrophorus* (Electrophoridae); (b) branchial chambers including either suprabranchial labyrinthifom organs found in Anabantoidei (e.g., *Anabas, Ctenopoma, Sandelia, Macropodus, Trichogaster*, or dendritic organs such as those found in the family Clariidae (*Clarias, Clariallabes, Dolichallabes, Channallabes*, or extension as air sac in Heteropneustidae (e.g. *Heteropneustes*); (c) opercular chambers: certain species of the families Gobiidae (e.g. , *Apocryptes, Boleophthalmus, Gobius, Periophthalmus, Pseudopocryptes*, and Blenniidae (e.g., *Andamia*) have their opercular chambers modified for air-breathing purposes; (d) special parts of the post-pharyngeal gut are modified and adapted to aerial respiration: *Brochis, Corydoras, Hoplosternum* (Callichthyidae); *Leptocephalus* and *Misgurnus* (*Cobitidae*) use intestines for the aerial breathing whereas *Ancistrus, Loricaria, Plecostomus, Pterygoplichthys* (Loricariidae) and *Pygidium* (Trichomycteridae) use their stomachs for this purpose; oesophagus is modified as an accessory respiratory organ in Alaska blackfish, *Dallia pectoralis*; (e) finally the swim bladder functions as an accessory respiratory organ in *Arapaima* (Osteoglossidae), *Erythrinus* and *Hoploerythrinus* (Erythrinidae), *Gymnarchus* (Gymnarchidae), *Gymnotus* (Gymnotidae), *Notopterus* (Notopteridae), *Phractolaemus* (Phractolaemidae), *Kneria* (Kneriidae), *Piabucina* (Lebiasinidae; for special attention see Graham et al. 1978), *Umbra* (Umbridae) and *Pantodon* (Pantodontidae).

A few selected examples will suffice to demonstrate the different designs of the air breathing organs. The vast majority of air-breathing fishes are tropical freshwater teleosts. Marine species in general differ from freshwater species in that accessory respiratory organs are mainly lacking. A coverge on the structure, function and evolution of the accessory respiratory organs of air-breathing fishes of India (Munshi 1985) is of special interest. For morphometrics, refer Roy and Munshi (1995).

The air breathing organ in *Anabas testudineus* consists of a suprabranchial chamber enclosing within it a labyrinthine organ borne by the epibranch of the first gill arch on either side of the pharynx (Fig. 9.21). Details of the respiratory islets are given by Munshi and Hughes (1991).

The accessory respiratory organs in catfish *Clarias lazera and C. batrachus* comprise (i) suprabranchial chambers situated dorsally to the gill cavities and lined by respiratory membrane; (ii) fans (gill-plates) borne on each gill arch and (iii) the respiratory tree or dendritic (arborescent) organs borne by the second and fourth gill arches (Figs. 9.22, 9.23a, b). SEM study is strongly indicative of their common origin from the gills. Vandewalle and Chardon (1991) presented a new hypothesis on the air breathing in *C. gariepinus*. (1) At rest on the bottom, the suprabranchial cavity is filled with air. (2) While coming upto the surface, a part of the air of the suprabranchial cavity passes to the opercular cavities, especially in their anterior part, through slits between the fan-like valves borne on the epibranchials. (3) Somewhat later, opercular adduction results in

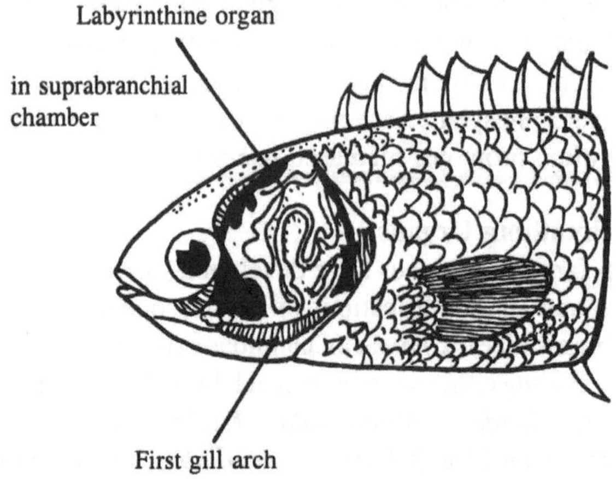

Anabas testudineus: Air-breathing organ.

Fig. 9.21 ***Anabas testudineus*. Lateral view of the head showing the disposition of the labyrinthine organ and first gill arch.**
***Source:* Munshi (1976)**

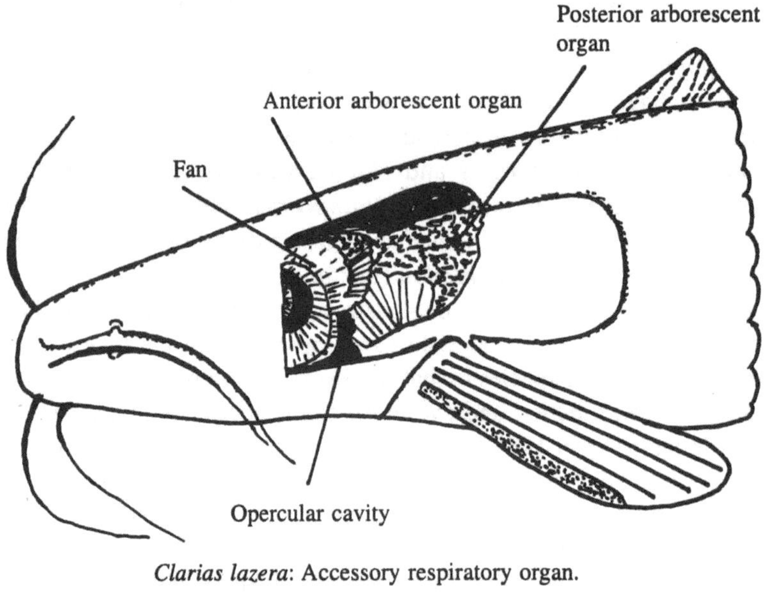

Clarias lazera: Accessory respiratory organ.

Fig. 9.22 **Lateral view showing the gross structure and relative position of arborescent organs in *Clarias lazera*.**

expulsion of bubbles through the opercular slits. (4) When the snout reaches the surface, the mouth is wide open. Depression of the hypid bars and opercular abduction produce aspiration of air into the buccopharyngeal cavity. (5) Air is forced into the suprabranchial cavity, through the third branchial slit by mouth closure and buccopharyngeal constriction. Hyoid bars elevation plays a prominent role in that constriction.

The air-breathing organ of *Heteropneustes* (*Saccobranchus*) *fossilis* has been derived by the

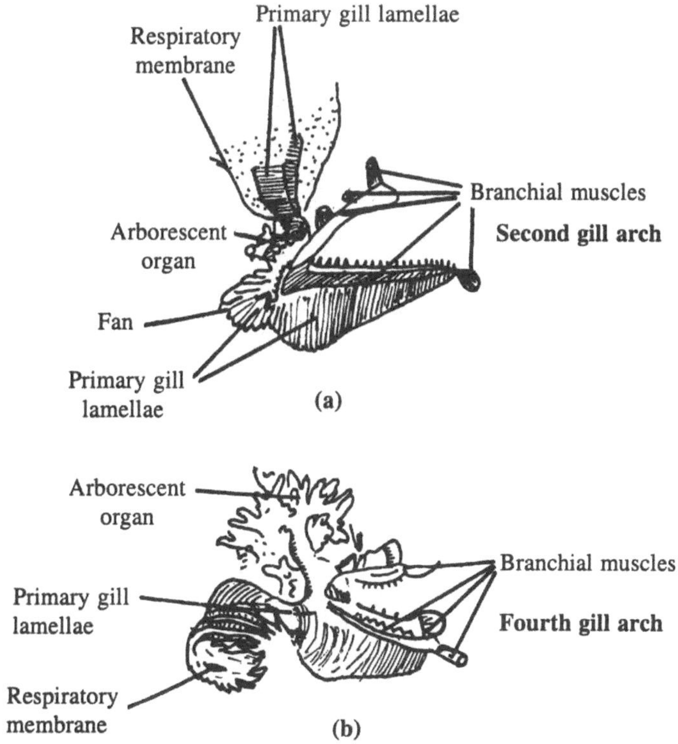

Clarias batrachus: Accessory respiratory organ.

Fig. 9.23 ***Clarias batrachus*. Lateral views of (a) second and (b) the fourth gill arches. Arborescent organs are present on the second and fourth gill arches.**
***Source:* Munshi (1976)**

modification of the gills (last branchial arch). The suprabranchial chambers extend backward into the trunk region in the form of a tube deeply imbedded in the trunk myotomes (Figs. 9.24, 9.25). Air enters first the suprabranchial chamber through the inhalent aperture guarded by the second and third gill fans and then into the posterior tubular part. The respiratory surface is derived from transformed secondary lamellae which lie flat on the surface of the respiratory sac, having two distinct regions; the non-respiratory basal and the respiratory distal parts. The lamellae retain their relative position in relation to the afferent and efferent primary filamental arteries. Pillar cells in the columns containing collagen fibrils, as found in the gills of fish, are also found in the lamellae-like structures of the respiratory islets of the air sacs, in both the respiratory and non-respiratory regions, which suggest that they are structures homologous to gill secondary lamellae (for various facets, see Munshi 1993). Investigations (immunohistochemical) have shown that more immunoreactive neuroendocrine cells are present in the gills than in the air sac epithelium (read Zaccone et al. 1995).

The air breathing organ in *Channa* sp. consist of a suprabranchial chamber located in the pharyngeal roof and a pair of dentritic plates borne one each by the hyomandibula and the epibranch of the first gill arch on either side of pharynx (Fig. 9.26).

The gills of *Channa* and *Heteropneustes* are quite capable of meeting the increase in oxygen demand with increasing body mass although the gill area per gram decreases rapidly in these

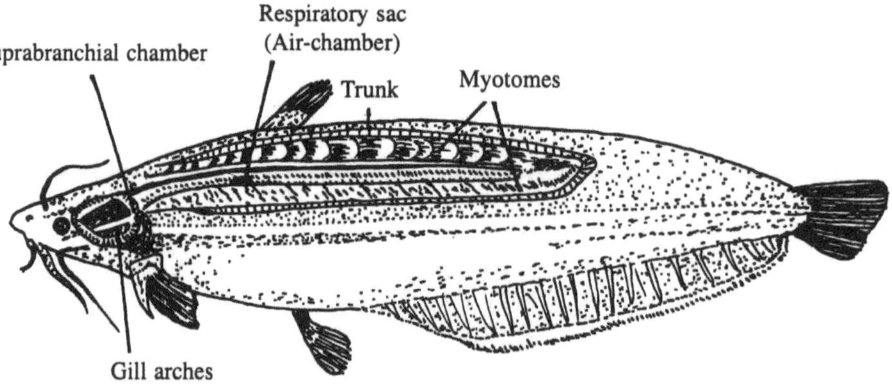

Heteropneustes fossilis: Accessory respiratory organ.

Fig. 9.24 Dissection of the respiratory sac of *Heteropneustes fossilis*.
***Source:* Munshi (1962)**

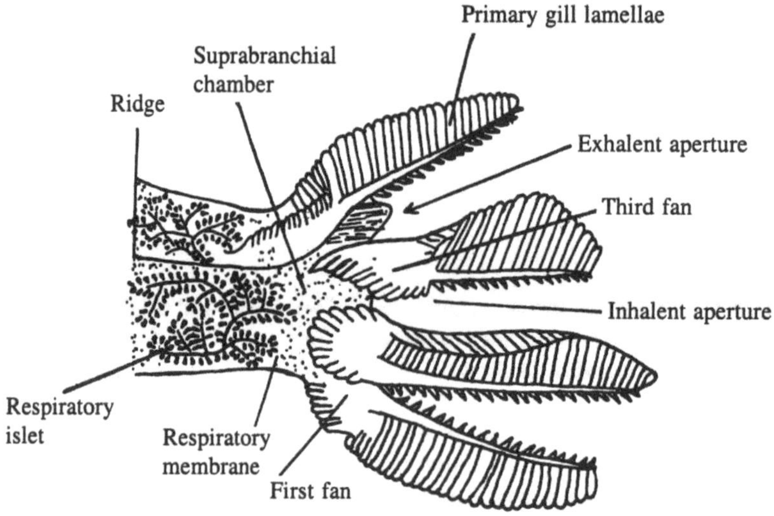

Heteropneustes fossilis: Accessory respiratory organ.

Fig. 9.25 *Heteropneustes fossilis*. The respiratory sac opened out by means of a mid-ventral incision to show the relation between the "fans", the inhalent and the exhalent apertures, the gill arches and the suprabranchial chamber and respiratory sac.
***Source:* Munshi (1976)**

species. In consequence it has been suggested that the suprabranchial chamber may be capable of gaseous exchange even when it remains under water.

The mud eel *Monopterus cuchia* is an obligatory air breather, highly terrestrial and can survive indefinitely on land if it is kept moist. Air breathing is mostly accomplished with a pair of pharyngeal inflatable sacs, which receive their blood supply from the heart via afferent branchial vessels having lost nearly all functional gill filaments (Figs. 9.27, 9.28). The air sacs are posterodorsal extensions of the pharynx and they are lined by a membrane having vascular respiratory islets and non-vascular areas. Also the pharynx, the hypopharynx and the pharyngeal region of the branchial arches participate in air breathing, and are lined by the same type of respiratory islets. The entire respiratory membrane is covered by a protective mucous layer

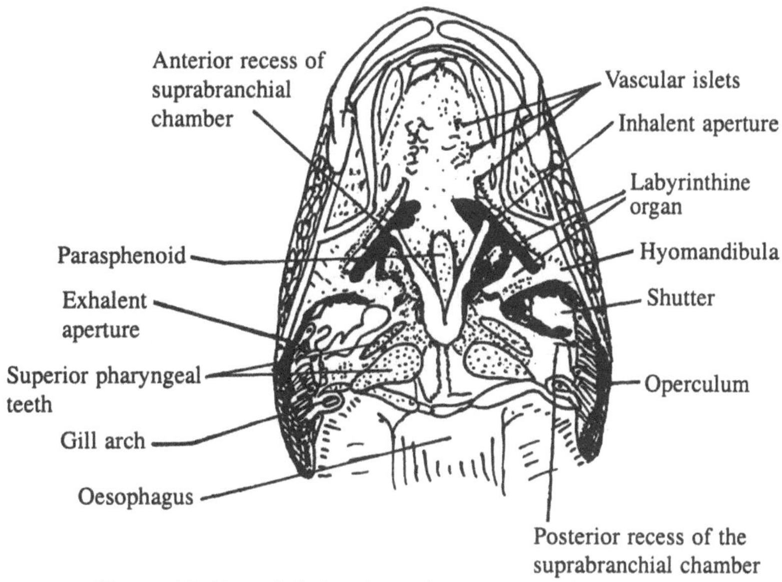

Channa (Ophiocephalus) striatus: Accessory respiratory organs.

Fig. 9.26 ***Channa (Ophiocephalus) striatus*. A view of fish dissected from the ventral side to show the anterior and posterior recesses of the suprabranchial chamber, inhalent and exhalent apertures, transverse process of the hyomandibula bearing the nodular labyrinthine organ, 'shutter', gill arches and vascular "islets" on the margin of the anterior recess and on the roof of the posterior section of the pharynx.**
***Source:* Munshi (1976)**

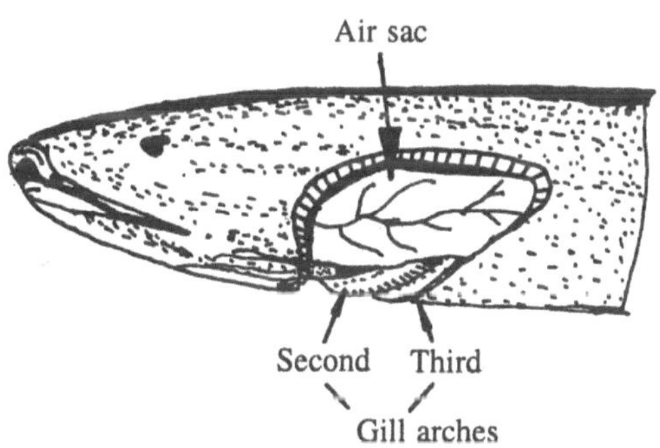

Fig. 9.27 **Diagram of the head *Monopterus cuchia* showing relative position of air sac and second and third gill arches.**
***Source:* Munshi et al. (1990)**

which is anchored by microridges on the outer epithelial surface. The respiratory islets consist of vascular papillae, which are specialized capillaries which take the form of spirals. This provides an efficient mechanism to bring every individual red blood cell into intimate contact with respiratory gas exchange. It also provides enough resistance to increase their residence time and ensure proper oxygenation.

In *Periophthalmus vulgaris* (mud-skipper; an amphibious gobiid fish), the opercular chambers function as lungs for aerial respiration. The chambers become extensive as they extend forward

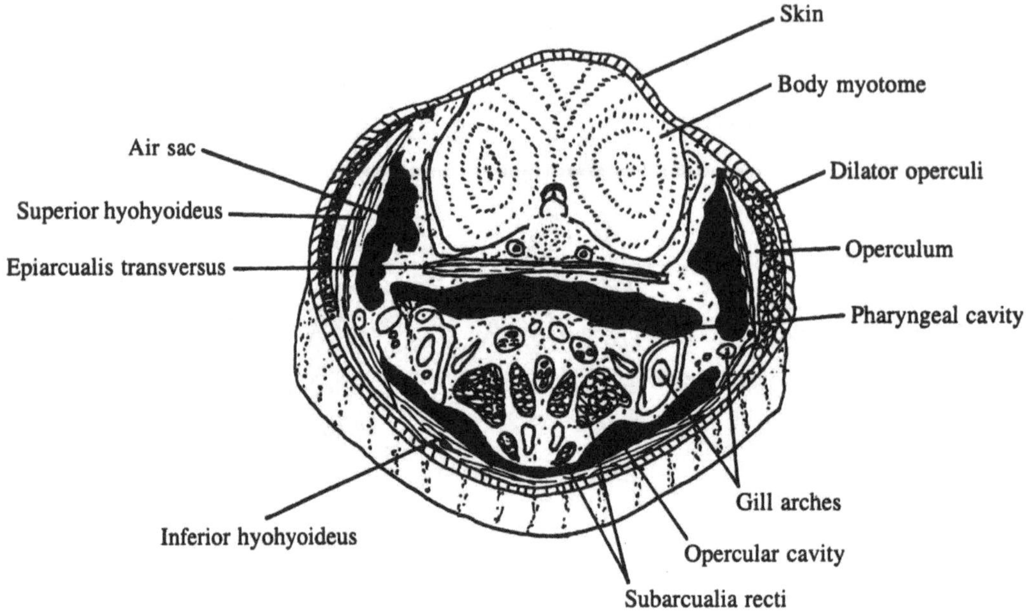

Fig. 9.28 ***Monopterus cuchia.*** **A diagram of a T.S. of the head passing through the middle part of the air sac showing the opercular cavity, branchiostegal membrane and pharyngeal cavity.**
Source: **Munshi (1976)**

below the buccal cavity as the infrabranchial recesses and over the gills as suprabranchial recesses (Fig. 9.29). The epithelium of the opercular chamber is thrown into a number of folds that form air-pockets or the alveoli. The respiratory epithelium lining these alveoli is highly vascular and is very thin.

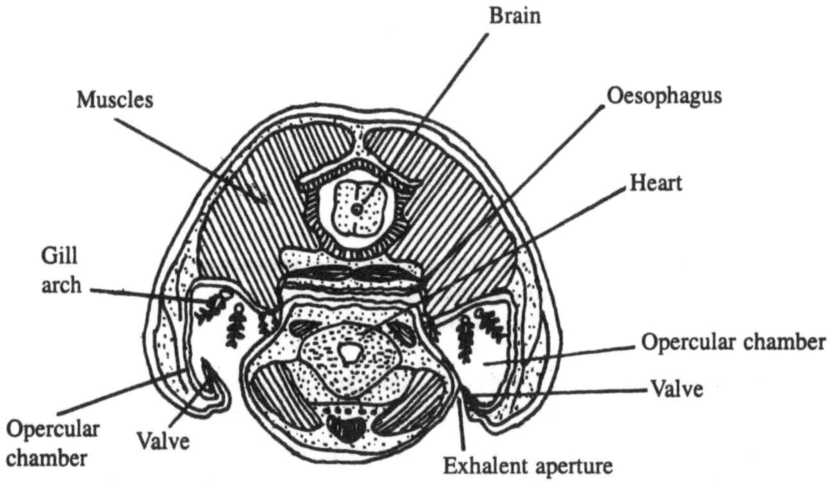

Fig. 9.29 ***Periophthalmus vulgaris.*** **Diagrammatic sketch of a T.S. passing through the posterior part of the pharynx showing disposition of four gill arches, the respiratory opercular chambers, valves, the exhalent aperture, branchial muscles and the heart.**
Source: **Munshi (1976)**

The branchiostegal apparatus in mud-skipper appears to have a special role in respiration as the epithelium lining it is highly vascularized and bears a characteristic valve to regulate air ventilation.

The Japanese weatherloach (*Misgurnus anguillicaudatus*) can exchange gases with water, via gills and skin, and with air, via the posterior region of the alimentary canal with air taken in at the mouth and simultaneous expulsion of intestinal gas from the vent. Although the weatherloach is not an obligate air-breather, aerial gas exchange normally occurs even at 10°C in air-saturated water.

The alimentary canal of *Misgurnus anguillicaudatus* is divided into three zones, glandular portion separated by a spiral section from the posterior, respiratory zone which has the greatest capillary densities and shortest gas diffusion distances. At rest (20°C), the intestine takes up about 20% of the total O_2 but accounts for less than 3% of total CO_2 elimination (gas exchange ratio = 0.08 for intestine). O_2 extraction averages 50%. Increasing temperature causes only slight increases in total metabolic rate, but highly significant increases in intestinal gas exchange develop as temperature rises. Intestinal gas exchange also rises with decreasing O_2 availability. A strong hypoxic drive and weak hypercapnic drive exist for aerial ventilation of the intestine, but are reduced or absent for aquatic ventilation of the gills. In spite of having to function in respiration, absorption, secretion and buoyancy regulation, the potential effectiveness of intestinal gas exchange is shown to be similar to that of other structures used for aerial gas exchange in facultative air-breathing fish (McMahon and Burggren 1987).

Gas exchange in the gourami, *Trichogaster trichopterus*, an obligate air breather, is achieved by branchial exchange with water and aerial exchange via labyrinth organs lying within the suprabranchial chamber. *Trichogaster* is able to maintain a respiratory homeostasis under varying environmental condition by exploiting whichever respiratory medium at a particular time is most effective for O_2 uptake and CO_2 elimination.

Fig. 9.30A,B illustrates the lung structure in Dipnoi. The African lungfish *Protopterus aethiopicus* is an obligate air breather provided with both lungs and gills for aerial exchange. The sac-like lungs are internally divided by trabeculae which largely consist of smooth muscle bundles. *P. aethiopicus* has paired, symmetrical sac-like lungs attached to the dorsal body wall. Anteriorly, they are fused over a short distance, forming a common anterior chamber which is connected to the alimentary canal by a short pneumatic duct that whirls around the oesophagus and opens on the ventral side of the latter through a glottis (Fig. 9.31). The cavity of the pneumatic duct, which is not reinforced with cartilage, and the adjacent parts of the single anterior lung chamber are lined by three different types of cells: pneumatocytes, ciliated cells and mucous cells. Neuroepithelial endocrine (NEE) cells are only found to occur solitarily in the basal part of the cilio-mucous epithelium which is restricted to the pneumatic duct and adjacent parts of the common anterior chamber. NEE cells may perceive changes in the airway gases whereupon they could respond by releasing a chemical modulator, influencing contacting afferent nerve terminals or nearby smooth muscle bundles. Furthermore, nerve fibres or NEE cells might be stretch sensitive (Adriaensen et al. 1990).

The neuroepithelial cells (NECs) are present in the gill filament epithelium of teleosts and non-teleosts such as *Scyliorhinus canicula* and *Acipenser baeri*. These cells share common characteristics with the NECs described as isolated or clustered in neuroepithelial bodies in the respiratory mucosa of mammals, and submammalian vertebrates including air-breathing fishes.

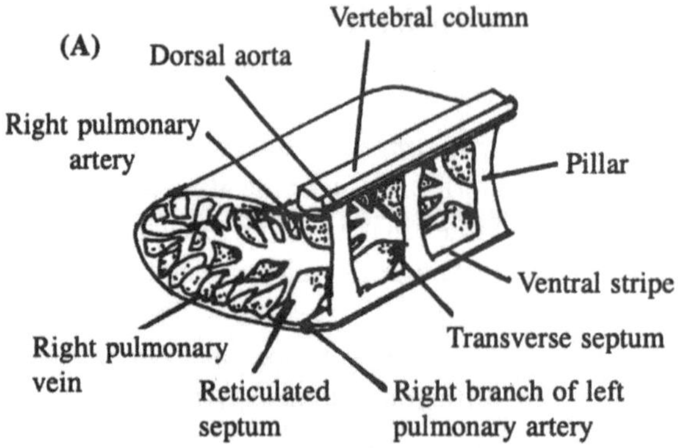

Neoceratodus forsteri: Lung.

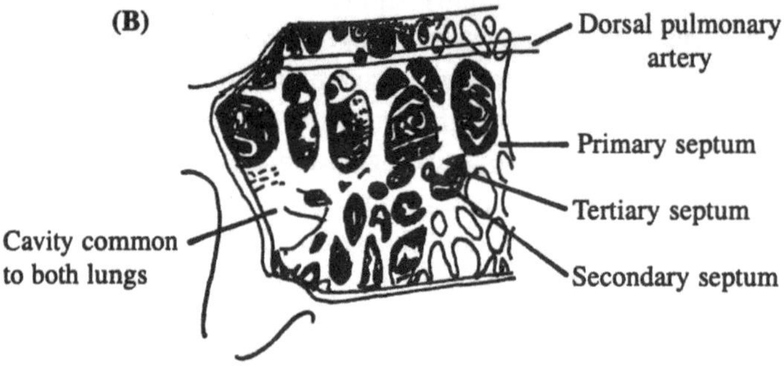

Protopterus: Lung.

Fig. 9.30 (A) (B) Lung structure in Dipnoi.
***Source:* Burggren and Johansen (1986)**

The serotonin-like immunoreactivity of the NEC's is evident. The secretory function of NECs is discussed with reference to the probable involvement of serotonin in the modulation of fish gill function. In addition, their connection with central and branchial nervous systems suggest a possible chemoreceptive role (Bailley et al. 1992) (see Fig. 9.18).

Zaccone et al. (1992) identified neuroendocrine (NE) cells in the gills of fish by serotonin and peptide immunohistochemistry. They occur either as single cells or as cell clusters within the filamental epithelium. Two populations of NE cells were identified that exhibit an immunoreactivity either for serotonin along or for met-enkephalin and leu-enkephalin co-distributed in the same cells. Authors reported the occurrence of open-type endocrine cells in the fish gill epithelium and confirmed the presence of only serotonin-immunopositive cells of the closed type. The NE cells of the species studied did not react positively to other neuroendocrine markers occurring in the mammalian lung. Great differences probably exist in the expression of these epitopes on antigenic structures among various animal species. The physiological role of neuropeptides and serotonin in the fish gill NE cell is unknown. Zaccone et al. (1994) further mentioned that in fish gills, a closed type neuroendocrine cell reacts for serotonin, an open type for enkephalins and some calcium-binding proteins (calbindin, calmodulin and S-100 protein). The functions of neuroendocrine cells in fishes await investigation, but it is assumed that are regulatory.

It has been also reported that the dipnoan *Protopterus* has innervated pulmonary neuro-

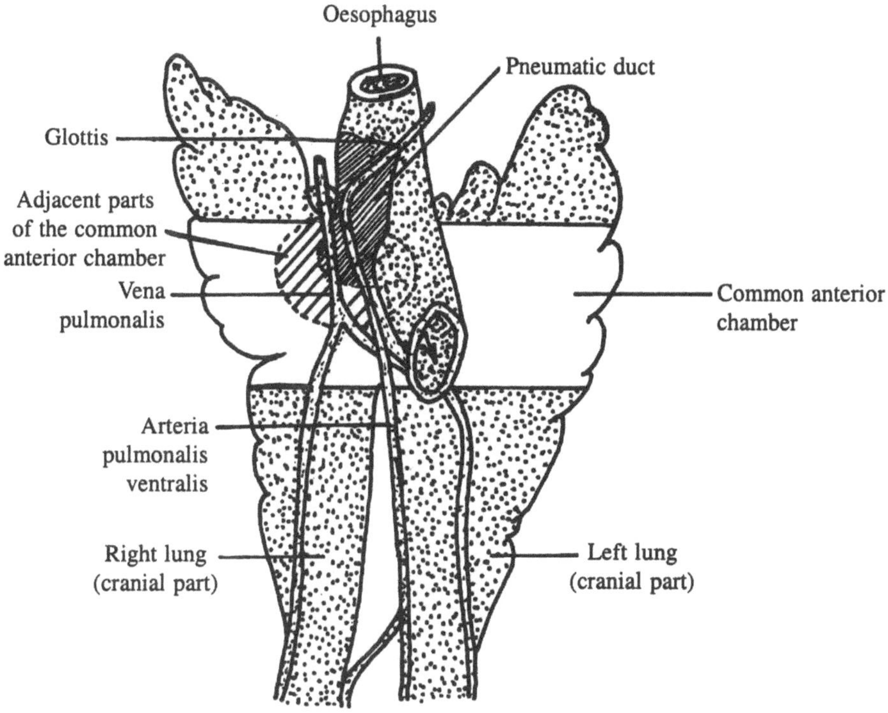

African lungfish *Protopterus aethiopicus*: Lungs.

Fig. 9.31 Macroscopic organization of the cranial part of the lungs of *Protopterus aethiopicus* as viewed from the ventral side. The unshaded area represents the 'common anterior chamber' whereas the narrow-spaced hatching marks the 'pneumatic duct' and the wide-spaced hatching an area termed 'adjacent parts of the common anterior chamber'.
***Source:* Adriaensen et al. (1990)**

endocrine cells in the pneumatic duct region with dense-cored vesicles. In *Polypterus* and *Amia* the lungs have serotonin-positive neuroendocrine cells that are apparently not innervated (see Zaccone et al. 1992 for localization of calbindin D28K-like immunoreactivity in fish gill).

In the South American lungfish, *Lepidosiren paradoxa*, the air breathing organ consists of two long dorsal tubes, originating from the foregut. The main airway which opens into the posterior region of the pharynx, continues through the whole length of the lung. The internal lung surface is much folded, especially in the dorsal regions. Folding takes the form of septa and trabeculae. In some ways the lungs of *Lepidosiren* resemble the lungs of amphibians. The surface of the lung is provided with a lining film which contains phospholipids, similar to the surfactant in mammalian and amphibian lungs except for the absence of tubular myelin figures characteristic of true lung surfactant. Also the capillaries do not continue at the edges of some of the septa, and consequently non-respiratory surfaces are present. The septa are relatively thick, containing much connective tissue and smooth muscle, and there are capillaries on both sides as in the African lungfish and amphibian lung septa.

The Polypteridae possess unchambered sac-like lungs—a small left one and a large right one-which lie dorsal to the intestine in the body cavity and are ventrally connected with the oesophagus via a "ductus pneumaticus" (Fig. 9.32). In addition to the lungs, these animals have well-developed and richly vascularized gills; thus, the animals are not exclusively dependent on air-breathing as is, for example, the lungfish *Lepidosiren paradoxa*, which have reduced gills. An air-breathing function is based on cells composing the lining epithelium of the lungs that represent the typical basic types of a vertebrate lung. The pneumatocytes I, together with the closely apposed capillary endothelia, form gas-exchanging area. The ciliated and mucous cells, representing the epithelia of airways, are also arranged longitudinally. Cells corresponding morphologically to pneumatocytes II of higher vertebrates are found in gas-exchange area of the lungs. The lungs of *Polypterus* are covered with phospholipids and glycoproteins as in mammals. The presence of lamellar bodies in pneumatocytes II point to the strong possibility that *Polypterus* produces surfactant. Further, the thin blood-air barrier suggests that lungs of *Polypterus* are obviously very efficient gas-exchanging organs. In addition to respiratory function, the lungs seem also to have hydrostatic tasks (Lechleuthner et al. 1989).

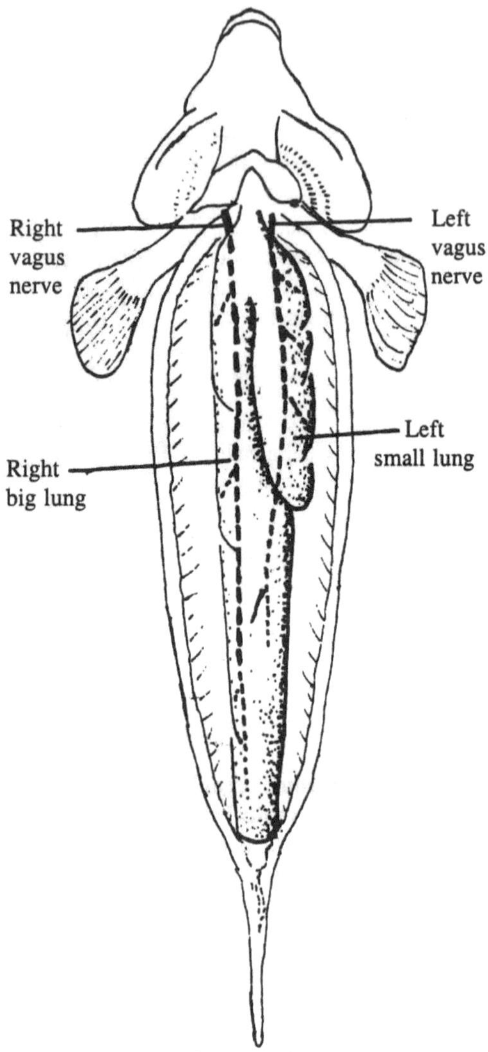

Polypterus: Lungs.

Fig. 9.32 ***Polypterus*. Schematic drawing of a ventrally opened specimen. The intestine is removed so that both lungs are visible. The dorsal location of the vagi nerves is indicated. *Source:* Lechleuthner et al. (1989)**

Despite the preponderance of buccal pumping in aerial ventilation of both fishes and amphibians, buccal filling and compression has several disadvantages and has been replaced in higher vertebrates by systems which use an aspiratory or suction powered pump. An aspiratory suction mechanism is seen first in a 'primitive' air-breathing teleost, the osteoglossid pirarucu (*Arapaima gigas*). The swim bladder and associated structures are shown in Fig. 9.33. This view, however, has recently been strongly criticized. A sub-ambient pressure in the air-sac (if actually occurring) could not supplement the volume pumped by the buccal force pump and it is argued that there is yet no evidence for aspiratory air ventilation of the lung or gas bladder in fishes.

11.3 Development, Origin and Evolution of Neo-morphic Air Breathing Organs in Teleosts

Singh (1993) hypothesized the involvement of gill materials derived from the gill arches-initiated

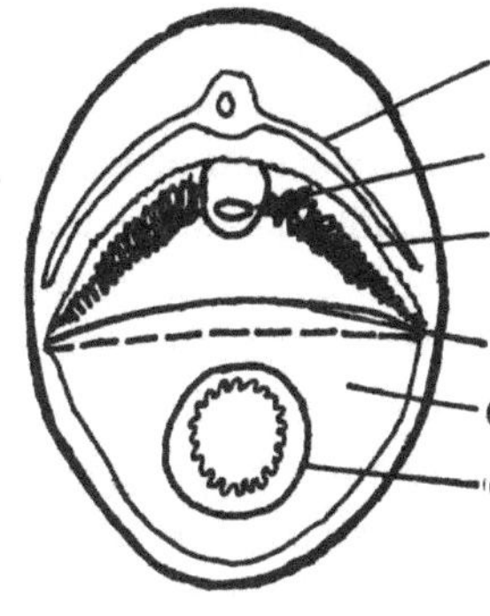

Fig. 9.33 Diagrammatic cross section of the pirarucu (*Arapaima gigas*) midway along the body showing the typical arrangement of swim bladder and associated structures. During inhalation the diaphragm probably moves downwards, as indicated by the broken line, thereby creating aspiration.
***Source:* Farrell and Randall (1978)**

by the fifth one in the evolution of these organs. Further, the implication of adjoining ectoderm is significant in the culumination of this phenomenon. Quite interesting is the transformation of aquatic into aerial respiratory structure, as evidenced in Siluroidei and Gobilidae, and this hints the dual potency of the incipient embryonic material.

12. PERIODIC BREATHING IN BIMODAL EXCHANGE

Breathing patterns in air-breathing fish and in amphibians, both of which show well-developed biomodal exchange, are enormously varied. There have been few exhaustive investigations of ventilation mechanisms, but in all cases the active pumping systems rely extensively on the buccal apparatus for moving both water and air. The buccal pump is used in a more complex series of movements in animals with an air bladder or lung.

Most air-breathing fish visit the surface briefly to execute just one cycle of ventilation movements, after which they quickly submerge. The intervals between such visits are frequently longer than 1 h in facultative air breathers, such as *Neoceratodus*, *Amia calva*, *Synbranchus marmoratus*, *Heteropneustes fossilis*, *Clarias batrachus* and *Hoploerythrinus*, particularly when they are in well-aerated water at low temperatures. Many of these animals do not surface at all when the water conditions are good. Obligate air breathers such as *Arapaima*, *Protopterus* and *Lepidosiren* ventilate the air exchanger at much shorter intervals, which vary between 0.5 and 15 min under normal resting conditions. Some animals, such as *Electrophorus*, are so dependent on air breathing that they succumb in a very short time if access to the surface is prevented.

In air-breathing ectotherms there are three major sources of peripheral input that may interact with nervous elements within the medulla to produce breathing pattern. There are mechanoreceptors associated with the lungs, chemoreceptors associated with the lungs, and intravascular chemoreceptors. All three are certainly involved in regulating the duration of lung ventilating and breath holding periods.

All lung mechanoreceptors respond to the degree of lung inflation (static change), the rate of change in lung volume (dynamic change), or both. In the fishes *Amia*, *Lepidosiren* and *Protopterus*, most receptors demonstrate both static and dynamic responses. In *Lepidosiren* and *Protopterus*, discharge correlates best with transpulmonary pressure.

In *Amia*, lung inflation is signalled by a rate-sensitive inhibition of mechanoreceptor discharge. *Lepidosiren* and *Protopterus* have both rapidly and slow adapting mechanoreceptors, and most rapidly adapting receptors show a brief bout of discharge during lung inflation.

The pulmonary mechanoreceptors of some lunged fish also show varying degrees of sensitivity to CO_2. On average, receptor discharged is halved by levels of inspired CO_2 of 6% in *Protopterus* and *Lepidosiren*. Pulmonary receptors in *Amia* do not appear to exhibit any sensitivity to CO_2.

Air-breathing organ mechanoreceptors have been detailed (see Burleson et al. 1992). It seems likely that the types of mechanoreceptors already described for the gills and orobranchial cavity would also be adequate for providing proprioreceptive feedback for the control of air breathing, or cardiorespiratory modulation, when air breathing was confined to the mouth, gills, or pharynx. Afferent activity arising from these receptors correlated to inflation or deflation of these areas during air breathing, however, has not yet been recorded although reflex tachycardia following inflation of the buccal cavity in the Panamanian swamp eel (*Synbranchus marmoratus*) and the electric eel (*Electrophorus electricus*) provide evidence for volume mechanoreceptors in teleost air breathers.

When air breathing involves swim bladder "lungs", however, proprioreceptive feedback for the control of air breathing and cardiorespiratory modulation would be expected to arise from these structures and, indeed, mechanoreceptors arising from swim bladder lungs have been described in detail in lungfish (*Protopterus*), bowfin (*Amia calva*), and spotted gar (*Lepisosteus oculatus*). These receptors are all innervated by a branch of the ramus intestinalis of the vagus and including slowly and rapidly adapting receptors. The slowly adapting receptors display both dynamic and tonic changes in activity on inflation or deflation, indicating that they are able to provide information about the rate of change, as well as the absolute change in the volume of the air-breathing organ. The rapidly adapting receptors respond to inflation, deflation, or both. Both groups of receptors in lungfish and gar are inhibited by 6–10% inspired CO_2.

In the lunged fishes *Amia*, *Protopterus*, and *L. oculatus*, lung deflation stimulates breathing whereas inflation inhibits.

The persistent selection pressure directed towards evolution of the air-breathing habit in certain environments had led to a great proliferation of structure and physiology in many unrelated species. The balance between gas exchange in water and in air can obviously be set at many different levels and can be changed in many different ways, by organismal and environmental factors, while still satisfying the animals overall requirements. Because of diversity of response that is inevitable in such varied developments, broad generalizations are difficult to make.

Oxygen: The influence of the aquatic medium on bimodal breathers will vary with the degree of development of aquatic exchange organs, whether they are ventilated or not, and the feasibility of simultaneous aerial and aquatic breathing. In most facultative and many "regular but not obligate" air breathers, air and water ventilation occur together. Aquatic ventilation may be continuous, often increasing in frequency with time after an air breath (e.g., *Amia*) or discontinuous, starting just before and stopping at the air breath (e.g., *Piabucina*) or shortly afterwards (e.g., *Anabas*). In *Synbranchus*, gill ventilation does not start until after an air breath and only continues for 2–3s, although long periods without air breathing may be interrupted by short episodes of gill ventilation. In *Monopterus* (*Amphipnous*) *cuchia* air breathing or water breathing occurs exclusively; there is no short-term alteration from mode to the other.

In most air-breathing fishes the highest gill ventilation rates occur at some stage of aquatic hypoxia (e.g., in *Clarias*, *Hoplosternum thoracatum*, *Brochis splendens*, *Umbra limi*, *Heteropneustes fossilis*, *juvenile Protopterus*, *Synbranchus*, *Polypterus senegalus*). In ancient teleosts *Amia* and *Lepisosteus*, the highest gill ventilation frequencies occur in normoxia and rate falls at all levels of aquatic hypoxia. According to Burleson et al. (1998), air breathing in *L. oculatus* allows gas to exercise to the same extent in normoxic and hypoxic water and enables them to re-establish blood gas and acid-base balance after exhaustive activity even in hypoxic water. In *Anabas*, however, gill ventilation rate is higher in aquatic hyperoxia than in normoxia or hypoxia, perhaps because of the cessation of air breathing in hyperoxia. This last response is quite unusual for gill ventilation volume/or rate fall during aquatic hyperoxia in *Amia* and *Lepisosteus*, in the air-breathing teleost *Synbranchus*, in the lungfish *Neoceratodus*, juvenile *Protopterus aethiopicus* and *P. dolloi*. The decline in gill ventilation in hyperoxia implies that there must be an O_2 stimulus to gill ventilation even when these animals are under normoxic conditions.

In the most air-breathing fish, aquatic hypoxia increases and hyperoxia decreases air-breathing frequency. In many of these animals a threshold level of hypoxia must be attained before air breathing starts or its frequency is significantly increased. The P_{O2} at which this occurs varies between species and ranges from 24 mm Hg in *Synbranchus* to value as high as 100 mm Hg in *Heteropneustes*. Increases in activity and temperature affect the threshold P_{O2} level because of their effects on O_2 demand. For example, *Amia* significantly increases air-breathing frequency in response to environmental hypoxia only at temperatures of 20°C–30°C.

Several obligate air breathers (adult lungfish *Protopterus* and *Lepidosiren*, electric eel *Electrophorus*, and gourami *Trichogaster trichopterus*) are claimed to be insensitive to water O_2 content, although it has been claimed that aquatic hyperoxia reduces air-breathing frequency in *Electrophorus*.

In some air-breathing fishes the air in the air-exchange organ is completely changed with each breath; thus changes in air-breathing frequency are usually a good guide to changes in minute ventilation. The obligate air-breathing teleosts *Trichogaster* and *Monopterus* (*Amphipnous*) in which the air-breathing organ is derived from orobranchial cavity, seem able to vary V_T (tidal volume). Aquatic hypoxia has no effect on V_T in *Trichogaster*, whereas it increases V_T in *Monopterus*. Aquatic hyperoxia reduces V_T in both species.

Ventilation volume, or air-breathing frequency seems to be increased by aerial hypoxia and decreased by hyperoxia in all obligate air breathers despite the fact that Pa_{CO_2} falls in hypoxia and rises in hyperoxia. In most nonobligate air breathers, ventilation is increased by aerial hypoxia, though *Anabas* and *Brochis* respond by increasing gill ventilation rather than air breathing. In addition, gill ventilation in *Brochis* decreases during exposure to aerial hyperoxia.

Carbon dioxide: In several bimodal breathers, toleration of wide swings in blood P_{CO_2} and pH is the strategy employed to cope with both chronic and acute hypercapnia.

In some air-breathing fishes, low levels of aquatic hypercapnia stimulate gill ventilation, but progressive inhibition almost always occurs as CO_2 concentrations increase.

The aquatic CO_2 level required to stimulate air breathing ranges widely. Some obligate or regular air breathers, such as *Electrophorus* and *Monopterus* seem not to be affected by hypercapnic environments. Almost 1% CO_2 is necessary to stimulate air breathing in *Anabas, Protopterus, Piabucina, Hypostomus, Plecostomus* and *Amia*, and levels of 7% to 10% are required in the case of *Saccobranchus*.

High levels of O_2 in air or water have varying effects on the responses to aquatic hypercapnia. In some species, air breathing is curtailed (e.g., *Trichogaster*); in others it is scarcely affected (e.g., *Anabas* and *Electrophorus*).

There are some interactions in the effect of aquatic hypoxia and hypercapnia. There is some evidence to suggest that hypoxia is more important than hypercapnia in stimulating air breathing. In most bimodal breathers the gills and/or skin are the major avenues of CO_2 elimination, so an increase in gill ventilation will do nothing to alleviate the consequence of aquatic hypercapnia. As in the case of O_2, the effects of changing CO_2 concentration in inspired air depends enormously on the efficiency of branchial and cutaneous routes for CO_2 elimination. Breathing is reduced in chronic hypercapnia, so ventilation changes would appear to be independent of blood pH changes.

Temperature: Increase in temperature causes the metabolic rate to increase in all bimodal breathers and, in facultative air breathers, the increased aerobic requirement is met by an increase in the importance of the air-breathing organ in O_2 exchange, compared with skin and gills. In some animals, its role in CO_2 elimination also increases, which contrasts with the role of the skin, where cutaneous CO_2 conductance is little influenced by alterations in body temperature (e.g., *Amia*).

In most bimodal breathers, blood pH and Pco_2 change with temperature, but it is not clear whether these changes are the consequence of other regulatory processes or have a major role in determining the relationships between temperature, metabolism, and ventilation. In general, total CO_2 concentrations in bimodal breathers are substantially unaffected by changes in temperature (*Synbranchus* is an exception).

In many bimodal breathers, temperature interacts with O_2 in establishing or stimulating a particular mode of breathing but the relation between the water Po_2 and temperature in determining or affecting the primary mode of breathing is complex. The effect of changes in water P_{O2} appear to predominate in *Umbra, Ancistrus*, and *Hypostomus*; thus increases in temperature only promote air breathing in hypoxic water. In normoxic water an increase in temperature only stimulates branchial ventilation. However, higher temperatures increase air-breathing frequency in *Amia* and *Lepisosteus*, even in normoxic water, and branchial ventilation is unaffected. Ventilation of the air-breathing organ is stimulated by increased temperature in obligate air-breathing fish.

13. AMPHIBIOUS FISHES

Coming again to amphibious fishes, it is noted that the independent evolution of amphibious fishes in several families of marine teleosts has resulted in a variety of specialization for terrestrial life and in varying degrees of terrestriality among individual species.

Gobius is entirely aquatic. *Boleophthalmus* is terrestrial but less so than *Periophthalmus*. The reader is also referred to Table 9.2.

The ecological significance of air breathing in intertidal fishes seems to be that it enables them to exploit a habitat relatively free from competitors and predators. Their ability to remain out of water at low tide also means that they avoid the necessity of making energetically costly migrations with each tidal cycle (Gibson 1982, 1986).

Table 9.2 Ratios of gill and body surface areas in gobies in relation to their degree of terrestriality

Species	Length (cm)	Body surface area (cm^2)	Gill surface area (cm^2)	Gill area/ Body area
Gobius caninus	10.0	34.6	48.6	1.40
Boleophthalmus boddaerti	12.6	40.0	33.6	0.83
Periophthalmus koelreuteri	10.7	44.6	21.6	0.46

Intertidal rockpools provide a challenging environment for rockpool fish with rapid changes taking place in many environmental parameters over a tidal cycle. Intertidal fish exhibit a number of behavioural adaptations such as the avoidance of hypoxic situations or remaining inactive during aerial "stranding", other species are, however, well adapted to breathe air and exhibit morphological adaptations such as smaller gill areas, specialized buccopharyngeal epithelia and a proliferation of cutaneous blood vessels in the skin. Oxygen consumption in rockpool fish is comparable to non-intertidal fish and responds in a similar manner to temperature changes. The ability to regulate oxygen consumption down to oxygen tensions below 40 Torr is, however, marked in rockpool species. Aerial and aquatic rates of respiration are similar in those species which are able to breathe air and the Respiratory Quotient normally remains between 0.7 and 0.9. A number of intertidal fishes are well adapted for cutaneous respiration, satisfying half of their oxygen and carbon dioxide exchange through the skin. Ventilatory responses to increased temperature, hyperoxia and hypoxia are similar to those of other fish but cardiac responses may differ in that no change in heart rate is seen under hypoxia or hyperoxia. Ventilatory and cardiac responses to aerial respiration are well adapted in some species maintaining ventilation and perfusion during aerial exposure. A marked Bohr effect, low temperature sensitivity and a temperature dependent Haldane effect have been measured in the haemoglobin of some intertidal fish. These properties may assist oxygen transport and carbon dioxide exchange during cyclical changes in environmental parameter within an intertidal pool (Bridges 1985, 1988, 1999; Fig. 9.34).

Amphibious marine air-breathers do not have highly specialized respiratory structures but have developed the capacity to release CO_2 in air. On the other hand, modern air breathers with lungs or special lung-like structures are primarily aquatic and remain bound to water for the release of CO_2.

In general, tropical 'skippers' have a Respiratory Exchange Ratio (RER, the ratio of CO_2 released at the respiratory surfaces to O_2 consumed) of approximately 0.7 to 1.0 in air (Table 9.3). This is similar to the Respiratory Quotient (RQ) which is the ratio of the CO_2 produced to the O_2 consumed during tissue metabolism. Because some CO_2 can be stored in the body, an animal's short-term RER may not always equal the RQ; a lower RER indicates CO_2 is accumulating in the body, leading to respiratory acidosis. The RER of many freshwater air-breathing fishes is low, near or below 0.4 (Table 9.4), for the air-breathing organ, but the sum of RERs of both aerial and aquatic respiratory organs totals approximately 0.7–1.0.

The aerial RER of two low-tide emerged, intertidal fishes have been reported as 0.46 for the cottid *Clinocottus recalvus* and 0.34 for the stichaeid *Xiphister atropurpureus*. However, in two

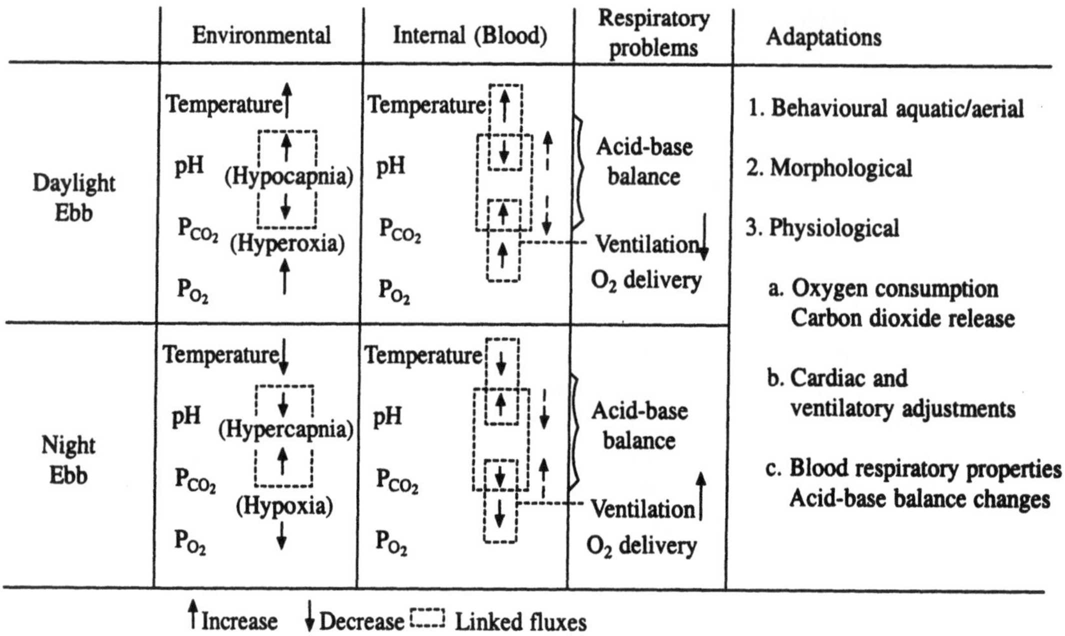

Fig. 9.34 Intertidal rockpools-problems and solutions. Summary of the changes in environmental parameters and their influence on the "internal milieu". The areas in which respiratory problems may arise are indicated together with possible adaptations.
***Source:* Bridges (1988)**

other species, the RER was higher, more similar to that seen in the tropical 'skipper' fishes: 0.80 for *Blennius pholis* and 0.73 for *Clinocottus analis.*

Response to emersion seems to be fundamentally different in marine and freshwater air-breathers. These fish demonstrate an unusual combination of metabolic and cardiovascular adjustments to emersion. Emersion produces a marked shift toward urotelism in waste nitrogen production. There appears to be no systemic accumulation of ammonia during emersion.

Table 9.3 Aerial Respiratory Exchange Ratio (RER) for marine amphibious fishes of the skipper type, fully emerged from water.

Mean = 0.75 ± 0.03

Family; species	RER
Blennidae	
Alticus kirki	0.73
Gobiidae	
Periophthalmus dipus	0.77
Periophthalmus schlosseri	0.72
Periophthalmus vulgaris	0.78
Labrisomidae	
Mnierpes macrocephalus	0.73

Source: Martin (1993) (see references).

Table 9.4 Respiratory Exchange Ratio (RER) for freshwater fishes that breath bimodally, in water and air simultaneously. The first column is the RER of the air breathing organ only, the second is the combined RER of both the air breathing organ and the gills in water. Mean Aerial RER = 0.28 ± 0.22; mean combined RER = 0.92 ± 0.17.

Family; species	Aerial RER	Combined RER
Ceratodontidae		
Neoceratodus forsteri	0.00	1.24
Lepidosirenidae		
Lepidosiren paradoxa (Juvenile)	0.73	1.12
Lepidosiren paradoxa (Adult)	0.44	0.73
Protopteridae		
Protopterus aethiopicus	0.26	0.73
Protopterus aethiopicus	0.27	0.76
Amiidae		
Amia calva	0.61	1.05
Amia calva	0.21	0.91
Belontiidae		
Trichogaster trichopterus	0.25	0.79
Trichogaster trichopterus	0.18	0.88
Cobitidae		
Misgurnus anguillicadatus	0.08	0.90
Lepisosteidae		
Lepisosteus oculatus	0.10	1.05

Source: Martin (1993) (see references).

A majority of marine species maintain the same metabolic rates when taken from water and placed in air while most of the freshwater fishes reduce metabolism in air (Table 9.5).

Amphibious fishes may thermoregulate while out of water, by altering the colour of skin to regulate heat absorption.

A few amphibious fishes are described here. The rockskipper *Mnierpes macrocephalus* has loosely fitting scales, the distal ends of which form pockets that hold water on the body of the fish when it is out of water. The water in these scale pockets presumably keeps the body moist,

Table 9.5

Species	Habitat
I. Aerial metabolic rate ⩾ Aquatic	
Periophthalmus sobrinus	marine
Blennius pholis	marine
Gillichthys mirabilis	marine
II. Aerial Metabolic Rate < Aquatic	
Channa striatus	freshwater
Anabas testudineus	freshwater
Clarias batrachus	freshwater

Source: Graham (1976)

and would facilitate aerial gas exchange. The metabolism of rockskippers is high compared to non-amphibious littoral species, and reflects the rockskippers very active life style. Combination of low gill area and close spacing of secondary lamellae could be a specialization that prevents the gills from desiccating while the fish is out of water. The close spacing would also add to the rigidity of the lamellae. The combination of low total gill area with the close spacing is also present in the climbing perch *Anabas testudineus* and this may be a general pattern among amphibious fishes; however, more data are needed.

Sicyases sanguineus in nature remains closer to water than the mudskipper fish, *Periophthalmus sobrinus*. However, it can survive periods of total emersion equally well. The mudskipper makes the greater physiological adjustment, however, since it lives at environmental temperatures about 20° above those in the clingfish's environment. The occurrence of periods without oxygen uptake is circumstantial evidence that larger *S. sanguineus* do not respire continuously through the skin. Occasional cutaneous gas exchange may occur across areas of thinner skin on the ventral surface, just anterior to the attachment disc. This region occasionally becomes reddish in colour, due to suffusion with blood, at times when the fish actively raises its head from attachment surfaces.

S. sanguineus can survive more than $1\frac{1}{2}$ days out of water, if not exposed to severe dehydration or thermal stresses. Rates of evaporative loss while out of water are low. Upper lethal temperatures are also low, reflecting the uniformly cool water temperatures of their environment. In nature the fish rarely leave areas in which are frequently washed over by waves.

For details on the morphological and physiological adaptations to intertidal life, the reader is referred to Gibson (1986) and further to an edited volume by M.H. Horn, K. Martin and M. Chotkowski (1998). Information on this aspect in the amphibious Red sea rockskipper fish *Alticus kirki* (Martin and Lighton 1989; Brown et al. 1992), rocky intertidal fishes of California and Oregon (Yoshiyama and Cech 1994) and cottid fishes (Martin 1991, 1996) is available.

14. ADAPTATIONS TO RESIST DESICCATION

Fish in air lose water from their gills and skin, which impairs the function of these tissues in aerial gas exchange and limits terrestrial activity. To minimize water loss in air, amphibious marine fish regulate the amount of time they stay out of water and avoid prolonged exposure to sunlight. Additionally some species such as *Periophthalmus* and *Sicyases* can change skin-pigment concentration to regulate heat absorption.

According to Horn and Riegle (1981), in general, adaptive relationships were found between body size, surface area, tolerance of water loss and intertidal distribution pattern in amphibious stichaeoid fishes. No information is available on the morphological adaptations by which the stichaeoids might reduce water loss.

Rolling, loosening of body scales, reduction in frequency of opercular pumping, mucus production around air-breathing surfaces are means to resist or slow down desiccation in various marine air-breathers.

Air breathing clariid catfishes may survive whilst buried in moist sand, or in burrows with water-air interface, but survival is unlikely in dried mud or sand. *Clarias* species show marked physiological adaptation to aestivation. In *C. mossambicus* oxygen uptake by the haemoglobin is highly efficient, and this species is remarkably resistant to high carbon dioxide levels in the blood. *C. batrachus* has a high haemoglobin content and a high oxygen capacity in the blood. *Clarias* species can become quiescent for long periods. *C. batrachus* can lower their activity and metabolic rate in response to respiratory stress.

Although clariids secrete copious mucus there is sound evidence that they can form a *Protopterus*-like cocoon, which would be essential for aestivation in a completely dry substrate.

15. AESTIVATION AND SUCTIONAL BREATHING IN LUNGFISH *PROTOPTERUS AETHIOPICUS*

The aestivating lungfish *Protopterus aethiopicus* relies on suctional breathing. The mechanism of the breathing process and the activity of the associated muscles and skeletal parts are unknown.

When the lungfish is in water the lung is ventilated by essentially aquatic breathing movements involving a buccal, positive-pressure swallowing of air, whereas expiration is largely effected by passive compression of the lung from the external water as the head is raised to the surface and the glottis opened. This basic mechanism of inspiration was passed on in only slightly modified form to the lunged amphibians. When the lungfish aestivates, however, in a soil-encased cocoon, conditions for breathing are quite different. Air breathing alone can support its aerobic metabolism when, at the start of aestivation, the lungfish becomes enveloped by a hardened mucous secretion. This cocoon covers the entire surface and even enters the mouth to form a hollow, non-collapsible breathing tube connecting the buccopharyngeal cavity with the air channel in the soil leading to the outside air. This prevents the movements of the buccal floor, such as practised in the free swimming lungfish. More importantly, it prevents the complete closure of the mouth and thus compels the animal to suspend positive-pressure buccal pumping for filling its lung.

During aestivation, the epithelial layer of the gill lamellae becomes thinner. The entire gill vasculature, including the capillaries and afferent-efferent shunts on arches IV-VI, are very dilated which presumably promotes blood flow through these gill arches to the lungs.

The activity of fishes ranges widely but few if any undergo such profound and dramatic changes in metabolism and its supporting organ functions as the lungfishes when they enter aestivation.

Aestivation does not involve major changes in temperature, yet the metabolism of aestivating fish may drop to 10% of the awake rate while excretion and elimination of metabolic end products except CO_2 are suspended. Species of the African genus of lungfishes, *Protopterus*, experience this remarkable state of suspended animation for recorded periods upto 4 years (predictable 7 years).

During aestivation conditions for breathing and gas exchange in the lung are markedly changed, whereas in the awake fish the convective forces of ventilation cause a near complete tidal exchange of gases in the lung, the aestivating lungfish has very small tidal volume; and a large residual lung volume making diffusion the principal means for renewal of gas inside the lung.

The striking increase in O_2-Hb affinity during aestivation is regarded as an adaptation to a reduced alveolar O_2 availability associated with aestivation. Altered mechanics of pulmonary ventilation and reduced ambient O_2 availability caused by subterranean habitat may have been selection pressure for the development of the high O_2 affinity.

A fresh analysis of observations and interpretations is given by Lomholt (1993) (see further Harder et al. 1999).

16. VENTILATION OF EXTERNAL RESPIRATORY SITES

Skin respiration. Cutaneous gas exchange may have played a very important respiratory role in fish ancestors and many early fishes. This may apply in particular to larval stages which had to survive in hypoxic water during the development of more complete adult ventilatory mechanisms. Cutaneous respiration is often viewed as a passive complement to locomotor or other movements.

Interestingly, at least 80% of O_2 uptake by *Myxine glutinosa* can occur across the skin. Certainly, the very thin epidermis and dense sub-epidermal capillary network of the hagfish will facilitate cutaneous gas exchange. A wide variety of fishes, both primitive and advanced, depend heavily upon cutaneous gas exchange. The respiratory rate of the skin of adult lampreys has been estimated to be only 20% of total O_2 uptake. Cutaneous gas exchange declines in importance as metabolic rates rise in a wide variety of vertebrates, perhaps explaining its diminished importance in adult lampreys relative to hagfish.

The capillary network in the skin of some fish is surprisingly rich and may play a significant part in gas exchange. In all fish with scales, the capillary network is located between the scale and the epidermis. Many fish have skins that are just as vascular as those of amphibians, where cutaneous gas exchange is acknowledged to be important. Jakubowski (1989) reported a comparative data on skin vascularization of fish and amphibians. The intensity of skin vascularization does not reflect the real degree of skin participation in respiratory gas exchange. The sub-epidermal capillary network in some species may perform the nutritive function only while in others both the nutritive and respiratory functions jointly.

The appearance and position of cutaneous respiratory blood vessels varies in different species: *Boleophthalmus* (tail); *Periophthalmus* and *Periophthalmadon* (between scales), blennies (anterior part of body) and *Dormitator* (frontal part of head).

Since cutaneous respiration is responsible for a considerable portion of aerial gas exchange in intertidal fish, the skin in many species is highly specialized. The presence of numerous cutaneous blood vessels in the blenniids from the Red Sea increase with the degree of terrestrialism. The presence of minor ridges and grooves in the skin of *Andamia* together with the ramification of superficial blood vessels on operculum are specializations for cutaneous respiration.

In fishes, findings are very disparate; for example in a single species, the European eel, the cutaneous oxygen uptake is reported as 10–61% of the total oxygen consumption by different authors. This variation may reflect different adaptations to survival in hypoxia water or in air where conjugating of gill leaflets results in deficient function. In fishes, the oxygen obtained by cutaneous respiration may complete an insufficient gill respiration or at least allow for a reduced ventilatory activity and consequently for an economy of energy.

The great diversity of oxygen exchanges and oxygen consumption in the skin of teleosts in sea water cannot yet be explained and emphasized the present inadequacy of knowledge of skin physiology of aquatic vertebrates.

The functions of the skin, particularly mucus production, are not the same in fresh-water and sea-water fishes.

In some fishes, all the oxygen absorbed by the skin is consumed by the skin tissue itself, whereas in other species there is a net inward transcutaneous oxygen flux. The skin on the blind side of the sole (*Solea solea*) is more efficient in oxygen uptake than the frog skin, the usual reference for cutaneous respiration.

External gill filaments in elasmobranchs. Embryos of elasmobranchs, viz., sharks, skates and rays possess transient external gill filaments. These originate on the posterior surface of the gill arch and emerge through the gill slits. Each filament has been reported to be a single capillary loop that connects to both the afferent and efferent branchial arteries. During later embryonic life, the filaments are retracted and resorbed as definitive gill lamellae develop at the base of the filaments.

The external gill filaments of sharks and rays have been generally considered a respiratory structure although they are suspected also as sites of nutrient absorption.

External gills in Dipnoi and Polypterini. In addition to the four branchial arches confined within the branchial cavity, all larval lepidosirenids possess external gills (Fig. 9.35). *Neoceratodus* never has external gills as larva or adult. Some species of *Protopterus* retain external gills throughout the life cycle. These external gills are located downstream and in series with the three posterior gill arches.

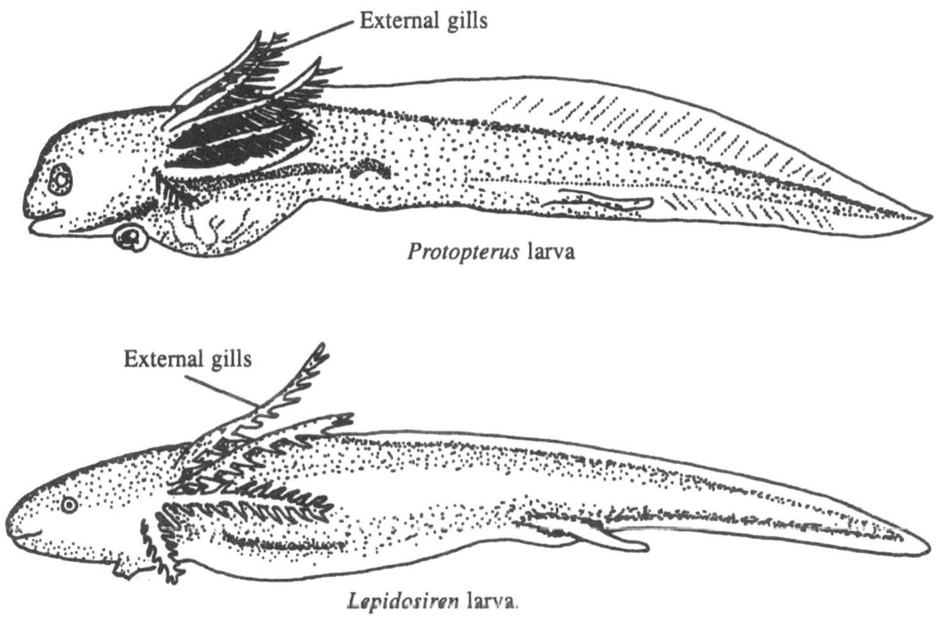

Fig. 9.35 External gills in Dipnoan larvae.

External gills occur in the larvae of the bichir (*Polypterus*) (Fig. 9.36).

17. COUGH (EXPULSION REFLEX)

Usually, fish breathe continuously. However, carp (*Cyprinus carpio*) may exhibit an intermittent breathing pattern ('bout respiration') at high oxygen levels, when series of three to ten breathing cycles are separated by pauses of several seconds.

Further characteristic of the breathing pattern of many fishes is a recurrent respiratory movement differing from the normal, the so called "cough". The water is forcibly ejected from the respiratory openings: the mouth, spiracles, gill slits or opercular openings.

A non-tidal respiratory system in fish is vulnerable to danger that particulate matter suspended in inspired water may clog the entrances to the finest channels, i.e., the openings of the interlamellar spaces in the orobranchial cavity.

Mechanoreceptors in the gill of lamprey can initiate "cough" responses and may modulate normal breathing.

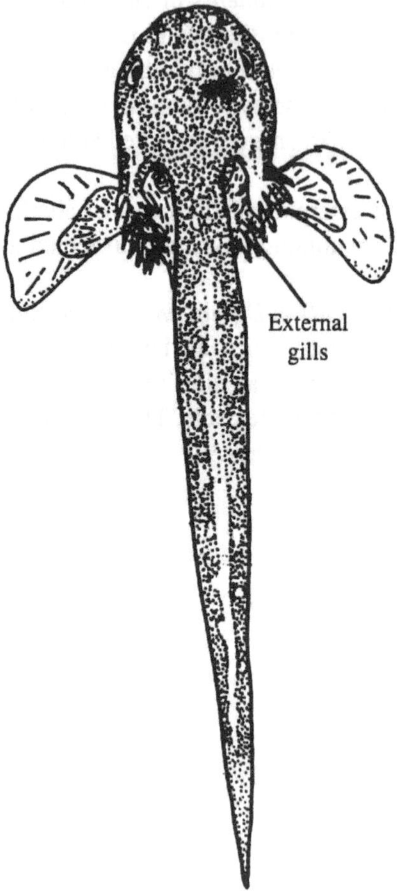

Polypterus larva.

Fig. 9.36 External gills in *Polypterus senegalus* larva.

The cough in the carp (*Cyprinus carpio*) is primarily achieved by muscles concerned in filling the parabranchial cavities. In elasmobranch fish (*Heterodontus portusjacksoni*) the parabranchial cough is primarily achieved by muscles such as constrictor branchiales concerned with emptying the parabranchial cavities. In elasmobranch fish there is always a rise in pressure in the parabranchial cavities and water is ejected from the mouth. In the carp there is a fall in pressure in the parabranchial cavities and water is aspirated into them whilst the mouth is closed. These differences doubtless relate to the presence of the bony operculum which, with its dilator and levator muscles, is able to effect an active vigorous abduction impossible with the membraneous investment of the elasmobranch gill pouch. Moreover, the retention in the elasmobranch fish of the interbranchial septum and its associated septal canals has increased the complexity of the water pathway and may require a back-flush generated at higher and positive pressure for its efficient cleaning. The parabranchial cough in *H. portusjacksoni* back-flushed the secondary lamellae by a vigorous constriction of the pharyngeal constrictor muscles against the gill bars braced by the contraction of the constrictor muscles was inhibited. The parabranchial cough may serve to clear the respiratory channels of debris and parasite and the orobranchial cough denies access of noxious substances to the secondary gill lamellae. Often the cough appears at regular intervals of a few respiratory cycles.

18. MODIFICATIONS OF GILLS IN RELATION TO ENVIRONMENT

The variety of environments occupied by fish and their many different modes of life is associated with adaptations of their respiratory system. The thickness of the water-blood barrier also differs with the mode of life.

The number of filaments also varies with age. As the surface of the fish increases so does the number of filaments and, consequently, the total respiratory surface.

Teleosts normally pump water over the gills by alternately expanding and contracting the buccal and opercular cavities. Fish that live in sandy substrates probably are all endowed with respiratory specializations which enable them to respire while buried. Normal breathing movements in wet sand probably require considerable effort, not only for expansion of the buccal and opercular cavities (voids produced by these breathing movements would soon be filled by sand) but also for the expulsion of water into the sand. The flatfish (Pleuronectiformes) bury themselves in sand except for mouth, eyes, and upper operculum. These fish beat both opercles, and water is passed over both sets of gills but they possess a channel between the two opercular cavities which allows water to be passed only out of the upper operculum (Fig. 9.37).

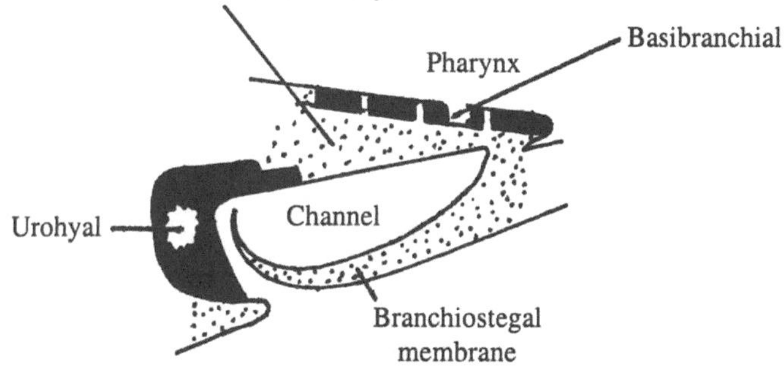

Fig. 9.37 Diagrammatic median section showing the structures surrounding the channel which connects the opercular cavities of a flatfish (Pleuronectiformes). The right side of the figure is anterior. *Source:* Yazdani and Alexander (1967)

In the benthic sand stargazers (Teleostei; Dactyloscopidae) the branchiostegal rays, muscles and membranes have evolved a substitute for the opercular pump. Opercles are immobile and function to screen substrate particles out of the branchiostegal pump. Branchiostegal and oxygen consumption rates compare similarly with other species using an opercular pump. The power of branchiostegal pump is probably closely related with substrate porocity (Todd 1973).

The ability of the sturgeon *Acipenser transmontanus* to alternate between buccal and opercular water intake for the purpose of gill ventilation without compromising O_2 uptake represents a highly effective and flexible adaptation for an aquatic life style involving bottom dwelling and feeding. The reduced spiracles play little or no role in gill ventilation.

The deep-sea fish generally have gill dimensions far less than those of rapid swimmers of the ocean, such as tunas and that some of them have far less gill than such a sluggish littoral species as *Opsanus tau*. Nevertheless it is also clear that some deep-water fish have gill dimensions which range between those of the sluggish forms and the large heterogeneous group known as "Gray's Intermediates".

The poor development of the gills of the coelacanth *Latimeria* is probably related to its deep water habitat rather than to its particular position.

Air-breathers have evolved toward minimizing their total amount of gill tissue, but compensate this loss of gill area by adding amounts of tissue that function more efficiently in aerial respiration. Freshwater air-breathers, including some species that never leave water, also have low total gill area. Lower gill area reduces the energetic cost of branchial ventilation in water and species that can breathe aerially may be opting toward this means of saving energy. Another factor favouring reduced total gill area in aquatic air-breathers is the possible loss of O_2 from the gills as the blood is circulated from accessory air-breathing organs through the gills to body.

The length, height, spacing and total number of secondary lamellae are characteristic of each species and correlate fairly well with their mode of life. In active pelagic forms there are large numbers of closely spaced, short secondary lamellae whereas in more sluggish species they are more widely spaced and higher, and the total gill area is smaller.

In air-breathing fishes the gills are of great importance in CO_2 excretion whilst the accessory respiratory organs, when may or may not be the modifications of the gill region, are mainly concerned with O_2 uptake.

For the habitats and gill surface areas of various fish (see Table 9.6).

Table 9.6

Fishes	Gill area ($cm^2.g^{-1}$)	Habits	Respiratory modifications
Euthynnus pelamis (Scombridae) skipjack tuna	13.20	Aquatic, open ocean, very active, pelagic	Ram ventilator and obligate swimmer
Stenotomus crysos (Sparidae) scup	5.06	Aquatic, midwater, open ocean, active	Ram ventilator
Callionomus lyra (Callionymidae) dragonet	2.92	Aquatic, benthic, open ocean, moderately active	Opercular tube, directs excurrent respiratory stream upward
Opsanus tau (Batrachoididae) oyster toadfish	2.14	Aquatic, benthic, in oyster beds, sluggish	Withstands low Po_2 habitats at low tide exposure, respiratory conformer
Anabas testudineus (Anabantidae) climbing perch	1.50	Amphibious, wet season migratory, moderately active	Air labyrinth, gulps air, CO_2 eliminated via gills

Source: Roberts (1975b)

Niche selection by gill breathing animals has resulted in a very few different redesigns of gill structure and functional modes of mechanical ventilation to minimize the metabolic load of ventilation, and to allow (1) occupancy of habitats requiring high activity levels for survival, as exemplified for life on beaches and mudflats, for terrestrial life in general, and for life as midwater and pelagic predators; and (2) partial escape from O_2 deficient waters by resorting to aerial respiration in, for example, eutrophic estuaries, marshes, swamps and seasonal ponds.

Some fish have managed to strengthen the gill structure with a parallel loss in gill area, but with the addition of accessory respiratory surfaces to facilitate amphibious life.

19. GILL MORPHOMETRY AND SCALING

From the respiratory point of view, the most important measurements of gills are the surface area (*A*) for gas exchange and the water-blood distance (*t*). The relation between these measurements is given by the following equation:

$$Vo_2 = (KA\Delta Po_2)/t,$$

where $Vo_2 = O_2$ uptake in ml O_2/min; *A* is the area in cm^2 for gaseous exchange, ΔPo_2 is the mean differences between the oxygen tensions of water and blood; *K* is the permeation coefficient (ml O_2/*u*m/cm^2/mm Hg/Min). Other measurements relevant to gill function are the dimensions of the sieve through which water is pumped. Further information can be obtained from the articles of Munshi (1985) and Roy and Munshi (1995).

Various methods have been employed and hence the differences in results obtained. A survey of these different components supports the view that they are related to activity, at least where this can be clearly defined.

Hughes (1977a) has discussed scaling of important parameters of the respiratory system of fish in relation to body mass. Particular attention is given to the dimensions of the gas exchange surface, and relationships between these parameters and the amount of oxygen transferred. It is shown that, although the exponent relating to standard oxygen consumption to body mass is usually about 0.75, this may vary from 0.5 to more than 1.0 for individual species. There are also variations in the corresponding exponents for respiratory area. Ojha and Mishra (1993) have reported data on the dimensions and scaling of gills in relation to body weight of the Gangetic mullets, *Rhinomugil corsula* and *Sicamugil cascasia* (also see Jha and Haque 1999). The pertinent article by Roy and Munshi (1995) is quite informative.

In some organisms the ratio between oxygen consumption and respiratory area increases with body mass and consequently the respiratory surface may be limiting. In others there is an increase in surface area relative to the oxygen demand and consequently a possible increase in scope for activity with body size.

In general, while such studies involving a range of species with different body size provide important guidelines, intraspecific studies may not necessarily show the same trend.

Studies of scaling in juvenile and adult fish have shown that the gill surface area and aerobic capacity expand at roughly the same rate as fish grow. This is usually interpreted as indicating that aerobic capacity is effectively limited by gill surface area. Recent ablation studies tend to support this hypothesis.

Relatively few studies have examined scaling during early life but what information is available suggests that gill area and aerobic capacity are not closely correlated in very young fish. In particular, the gill area of fish larvae appears to expand at a much more rapid rate in relation to body mass than does oxygen consumption. Allometric mass exponents for gill area typically seem to be in the range of 1.5–3.5, although values as high as 7.1 have been reported. Mass exponents for metabolic rate, in contrast, are usually close to unity ($b \approx 1.0$). Recently, it has been suggested that the reason for the poor correlation between the gill area and aerobic capacity in young fish is that other surfaces, in particular the skin, play a much more important role in gas exchange during early life that do in older fish. Further, it has been speculated that, because of this, the appropriate indicator of larval exchange capacity is total surface area (gills + skin), not gill area (read Rombough and Moroz 1990).

20. BLOOD RESPIRATORY ADAPTATIONS

O_2 binding properties of blood (e.g., haemoglobin concentration, O_2-H_b affinity, Bohr factor, heme cooperactivity) are major factors influencing both the movement of gases across the respiratory membrane and their transport in the blood. The respiratory properties of the blood of phyletically ancient fishes reflect both phylogenetic and environmental influences. Val (1993) has given an account of adaptations of fishes to extreme conditions in fresh water (read information on blood).

Fish blood commonly has an oxygen capacity of 7–10 volume % and ventilation: perfusion ratios of 15:1–19:1 are common.

A feature that appears common to all these accessory respiratory organs is that the gas exchange ratio (R) for CO_2 excretion: oxygen uptake is often of the order of 0.2 and seldom exceeds 0.5. Thus these accessory breathing organs are utilized to obtain oxygen from the air but most of the carbon dioxide is excreted across the gills or skin into the water.

All available data on blood respiratory properties of air-breathing fishes support the contention that increased dependence on air breathing is associated with higher P_{50} values or an O_2 transport system favouring O_2 unloading at higher tissue O_2 tensions. This tendency can be regarded as adaptive to fishes able to utilize atmospheric O_2 for which O_2 loading presents no problem. P_{50} value of *Protopterus amphibious* is by far the highest recorded. The relative importance of air breathing for O_2 uptake is also high amounting to more than 90% of total even when the fish is in well-aerated water. Conversely it is demonstrable that fishes confined to water and gill breathing have higher O_2 affinities the more O_2 deficient their habitat.

Physicochemical parameters of intertidal rockpools have been shown to change rapidly over a daily tidal cycle. Fishes living in these habitat may be exposed to hypoxic and hyperoxic conditions, together with short-term changes in temperature. These extreme environmental conditions may therefore influence the respiratory homoeostasis of fish and create problems for gas transport (see Fig. 9.34).

The 'typical response' to hypoxia in fish is an increase in ventilation volume and a reflex bradycardia. The mechanism by which an increase in ventilation volume is achieved varies among fish species. The bradycardia is offset by an increase in stroke volume, thereby maintaining perfusion of the respiratory surfaces at a high level. The bradycardia response together with an increased stroke volume are thought to promote lamellar recruitment.

Sublittoral soft sediments are colonized by some burrowing teleosts, for example, the red band-fish *Cepola rubescens* and the snake blenny *Lumpenus lampretaformis*. The burrows of these animals can be considered as a microbiotope, since the physicochemical parameters within the burrow water may differ from those in the surface water. The gas-transporting properties of blood of both burrowing species seems to be similar, although there are differences in the magnitude of these properties. The correlation of these properties with the O_2 supply in the environment seems to be stronger than the correlation with burrowing lifestyle.

Although burrow construction and structure may differ in the two species, both are exposed to environmental hypoxia and hypercapnia. However, both appear to be well adapted to their fossorial mode of life (see details: Atkinson et al. 1987). The reader is referred to Pelster et al. (1988) for aerobic and anerobic metabolism in relation to environmental oxygen tension in *L. lampretaformis*. 'Icefishes' (Chaenichthyidae) are congenitally haemoglobin-free fishes. Studies of gills in *Chaenocephalus aceratus* (sluggish bottom dweller) and *Champosocephalus esox* (active, small fish which undertake vertical migration) indicated that *C. aceratus* had unusually coarse gills with few and thick lamellae and a long distance between blood and water. *C. esox* had gills with dimensions similar to those of normal fishes. The gill area of *C. aceratus* was normal, that of *C. esox* relatively large when compared to normal fishes of similar sizes. It is reported that there is no parameter in the gill structure which parallels lack of haemoglobin.

Ultsch (1989) discusses various aspects concerning overwintering in freshwater fishes. Adaptations for coping with hypoxia and anoxia among overwintering freshwater fishes may include metabolic depression, a decrease in blood O_2 affinity, microhabitat selection, air breathing, short-distance migration, biochemical modifications aimed at adjusting glycolytirates, and alcoholic fermentation.

21. PSEUDOBRANCH

Laurent and Dunel-Erb (1984) have reviewed the structure and function of the pseudobranch.

In the teleosts the pseudobranch is located in the cranial part of the subopercular cavity to which it is attached by one of its sides. The pseudobranch is basically a hemibranch, in other words, a gill arch supporting only one row of filaments. Each of them bears a series of lamellae as in the gill. The lack of central venous sinuses in the pseudobranch or at least their extreme reduction is one of the characteristic differences of the gill filament.

Different types of pseudobranch organization have been observed according to fresh-or saltwater origin of fish, but it has been noticed that forms, sizes, and external structures vary widely within

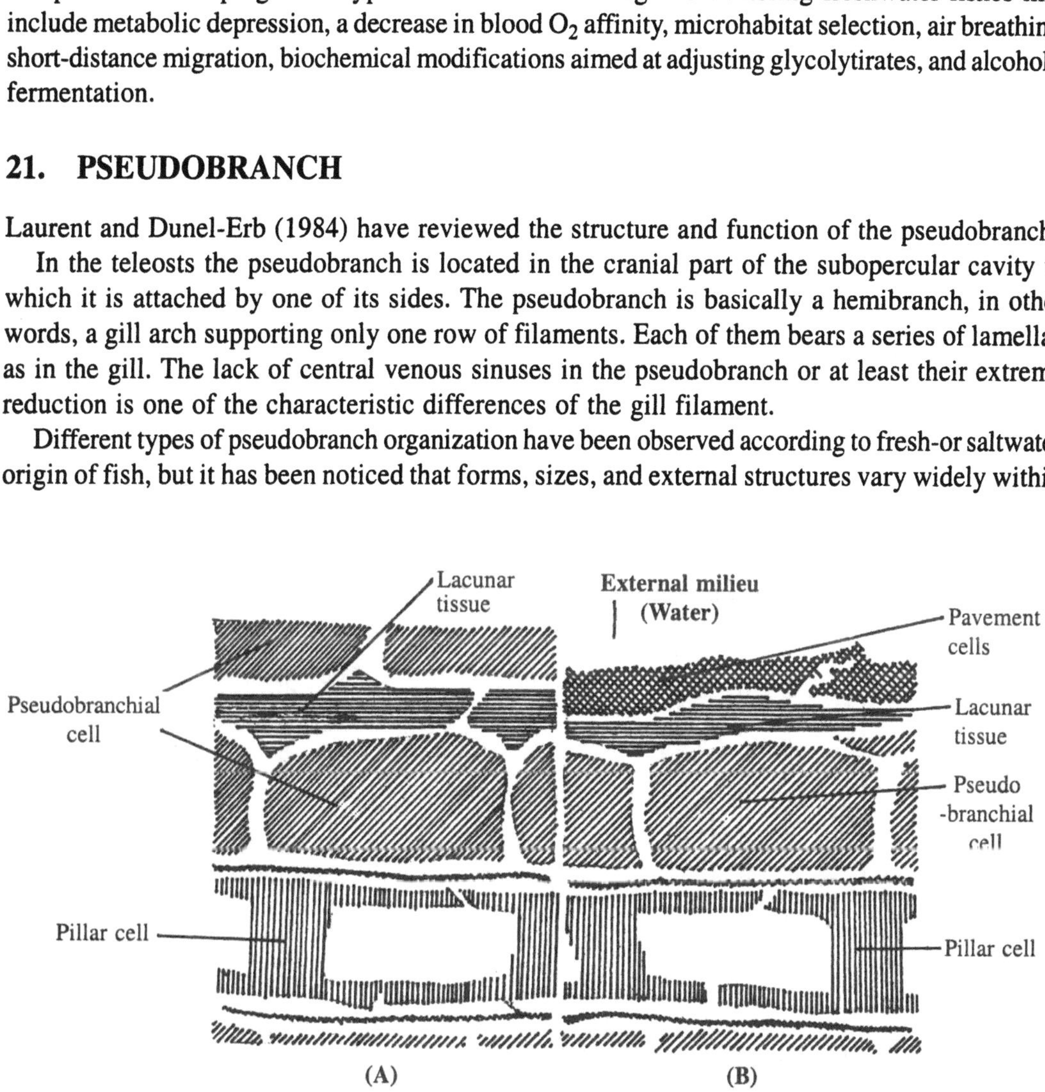

Fig. 9.38 Schematic representation of the pseudobranchial cell environment. (A) In the case of fused lamellae, a layer of lacunar tissue common to two lamellae is interposed between adjacent pseudobranchial cells. (B) In the case of free lamellae, a layer of lacunar tissue is still interposed between the pseudobranchial cells and the pavement cells, which cover the free lamellae. In each case the pseudobranchial cells line the pillar capillary.
Source: **Laurent and Dunel-Erb (1984)**

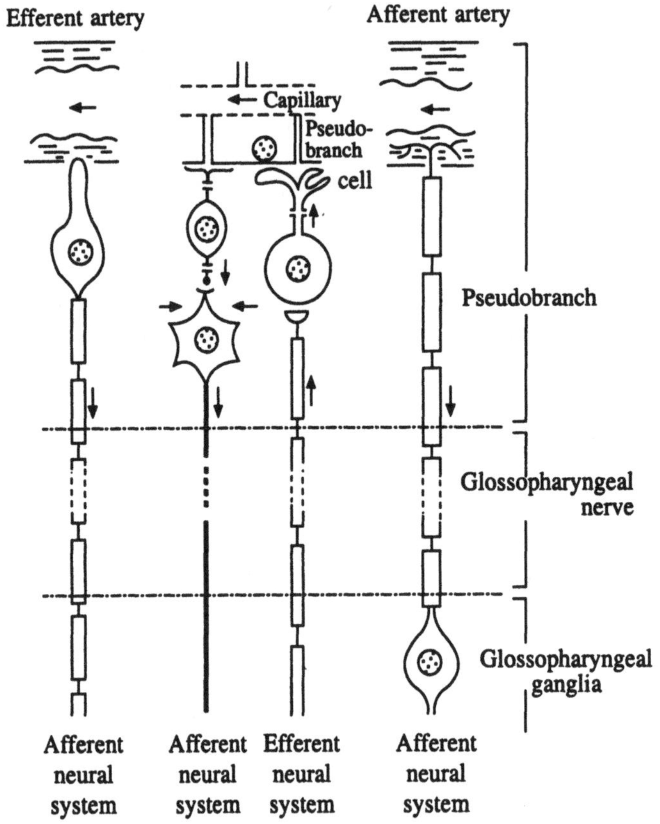

Fig. 9.39 Schematic diagram of the pseudobranch innervation (teleosts).
***Source:* Laurent and Rouzeau (1972)**

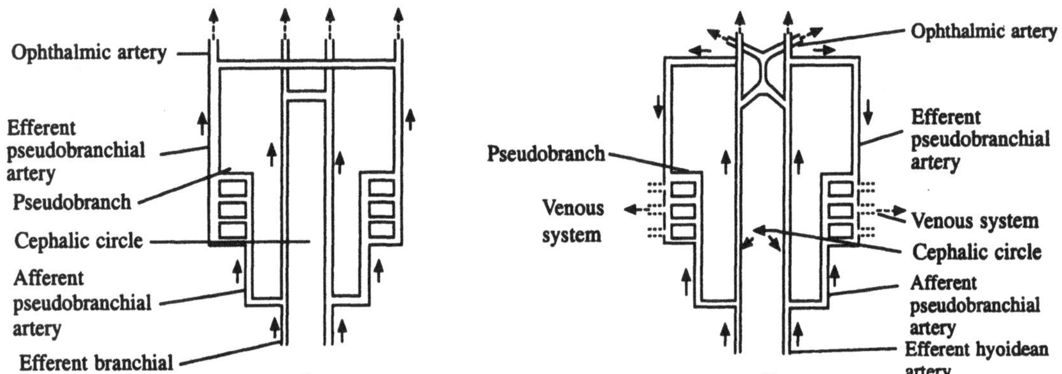

Fig. 9.40 Diagram showing differences between teleost and elasmobranch pseudobranch vascular organization. (A) Teleost pseudobranch receives blood from efferent branchial artery via the afferent pseudobranchial artery. Efferent pseudobranchial artery drains off blood toward the ophthalmic artery. Cephalic and ophthalmic circles are separated. (B) In elasmobranchs the cephalic and ophthalmic circles are interconnected, so afferent and efferent pseudobranchial arteries, both contribute to the blood supply of the pseudobranch. Blood is finally drained off by a venous system.
***Source:* Laurent and Dunel-Erb (1984)**

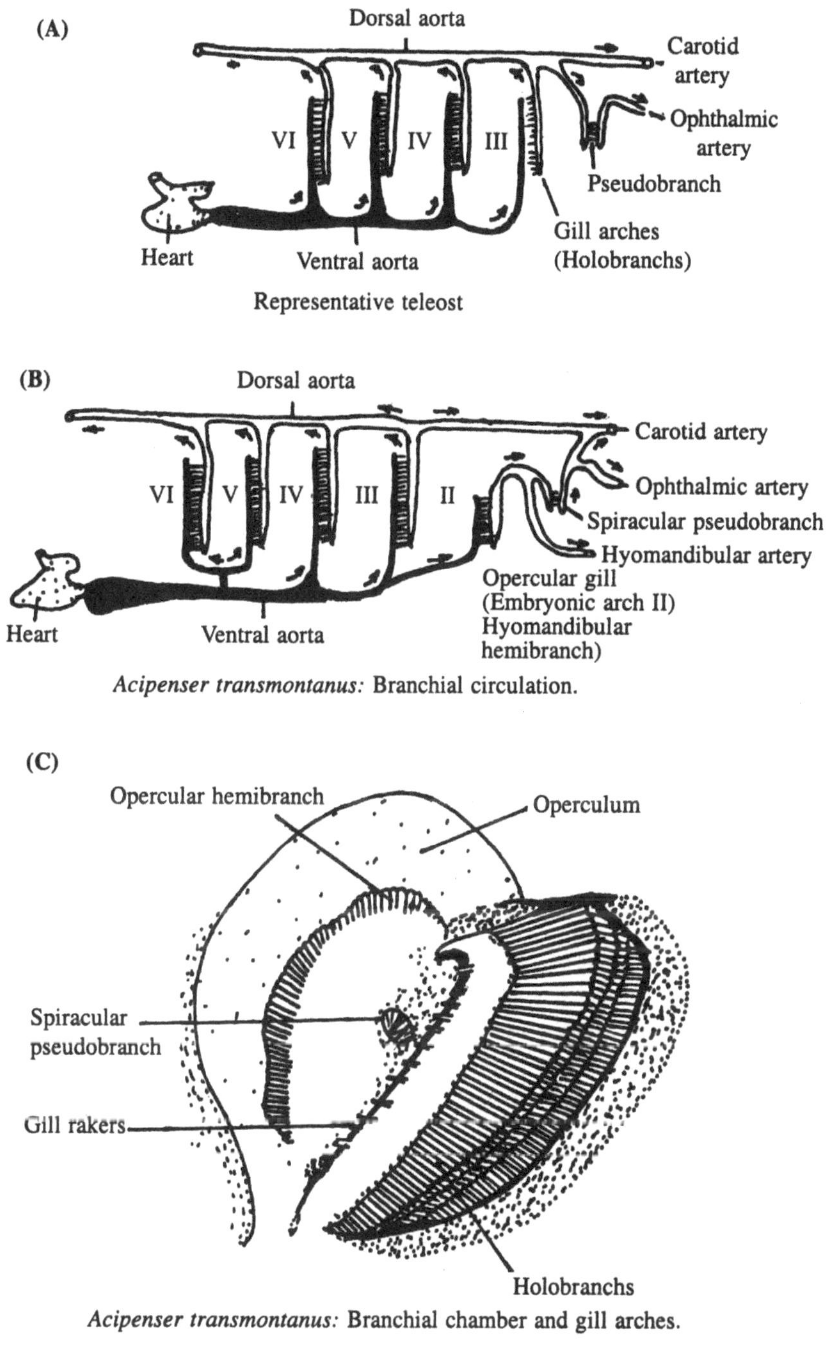

Fig. 9.41 Diagrammatic illustrations of the branchial circulation of (A) a representative teleost and (B) the sturgeon *Acipenser transmontaules* and (C) depicts the gross morphology of the branchial chamber and gill arches of the sturgeon, revealed by extreme abduction of the operculum. *Source:* Burggren et al. (1979)

orders, families, and species without any obvious reason, whereas vascular organization and innervation keep a similar organization throughout.

Pseudobranch is a much impressive organ in Percoidei and in the great majority of marine fish. In Cyprinoidei and Salmonidei, which include a large number of freshwater fish, the pseudobranch is of smaller size. For specific details on the pseudobranch of paradise fish *Macropodus opercularis*, see Ladich (1985, 1987).

A useful classification of the pseudobranchs is based on whether or not they are covered invisible (e.g., cyprinids) covered by opercular epithelium, or completely free (e.g., *Gobius niger*). The various patterns have important consequences concerning structure and cellular components of the pseudobranchial epithelium. This distinguishes two kinds of epithelia, with respect to their situation regarding the external medium.

Lamellae are vascularized by pillar capillaries similar to those of the gill. Almost all pseudobranchs have typical pseudobranchial cells; similar cells are never found in gills (Fig. 9.38). In the pseudobranch, chloride cells exhibit the characteristic differences of fresh-and saltwater species. This is in contrast with pseudobranchial cells, which have morphological characteristics remarkably constant regardless of the medium the fish inhabits. Differences between the pseudobranchial and chloride cells are related to mitochondria, tubules, apical vesicles, apical pit, innervation and vascular compartment.

The teleost pseudobranch may be innervated by a branch from facial, or glossopharyngeal or from both (Fig. 9.39).

The pseudobranch is a very minute organ in elasmobranchs, located on the mandibular arch as in teleosts, but shifted forward within the spiracle (spiracular pseudobranch). The spiracle is a channel connecting the pharyngeal cavity to the external medium. The spiracular pseudobranch lies on the anterior spiracular wall and consists of seven to eight filaments oriented parallel to the water medium. Above the pseudobranch an erectile organ has the capability to close the spiracle. The vascularization of elasmobranch and teleost pseudobranch has been given in Fig. 9.40.

The outer stratified layer of cells of epithelium showing desmosomes, terminal web, exocytotic vesicles is backed by long columnar cells resting on basal lamina. The columnar cells possess glycogen granules, osmiophilic lysozomes, apical microvilli and a characteristic luminal pit. The palatine branch of the facial nerve innervates the epithelium.

The pseudobranch in elasmobranchs and particularly in *Torpedo marmorata* is mostly a sensory organ that has no other significant structure from a functional point of view than the parenchymal epitheloid cells and neuroepithelial cells.

The pseudobranch has also been found in certain chondrosteans and holosteans (Fig. 9.41).

The function of the pseudobranch is still debatable. It has been considered a salt-regulating, a respiratory, a glandular, or a sensory organ. Besides, a view that pseudobranch acts as a vascular rete mirabile functionally associated with vision has been advanced.

The reader can find information regarding the distribution of visceral and branchial arches in phyletically ancient fishes (Table 1; see Burggren et al. 1985). The pseudobranch (external) is reported in *Scyliorhinus, Hexanchs, Heptanchus, Raia, Acipenser Huso, Lepisosteus, Amia;* spiracle is found in *Scyliorhinus, Hexanchus, Heptanchus, Raia, Acipenser,* and *Polypterus*.

CHAPTER **10**

Swim Bladder and Weberian Apparatus

1. INTRODUCTION

The early vertebrates had oxygen resorbing equipment—the lung-like structures and it is generally accepted that vertebrates acquired functional lungs prior to a locomotor apparatus for invasion of a terrestrial environment. Lungs supposedly evolved in paleozoic freshwater vertebrates as an adaptation to hypoxic stress.

Certain fishes are able to concentrate gases from the blood stream in a special organ, the hollow swim bladder (also termed gas bladder, air bladder, an enclose of free gases or gas-filled sac). Thus, the swim bladder is not necessary for life, as it is missing in many fishes.

Both swim bladders and lungs emerge embryologically from a single or paired pouch of the cranial part of the digestive tube. As a consequence, the swim bladder and to some degree the lungs, contain the same principal tissue elements as the gut wall.

Lungs probably originated as ventral diverticula of the gut, so the dorsal position of most fish lungs and swim bladders is secondary.

In a majority of fishes living now, the swim bladder is a hydrostatic organ with no respiratory function, but some primitive fishes and a few teleosts breath air by a lung or swim bladder. Swim bladders and lungs are homologous.

The regulation of swim bladder volume holds interest for special physiology and behaviour because buoyancy compares to gravity on land in its postural significance and adaptive influence. This regulation also has a comparative interest because of a phylogenetic lineage in a bladder and lung physiology and in the neural processes of their control.

The contributions of Fänge (1983), Danguy and Lenglet (1988), Scheid et al. (1990), Pelster and Weber (1991), Phleger (1991), Pelster and Scheid (1992, 1993a, b, 1994); Pelster (1994 1995, 1997, 1998); Pelster et al. (1994) Maina et al. (1995) and Scheid and Pelster (1995) cover comprehensive accounts of various aspects of swim bladder.

2. POSITION AND OCCURRENCE OF THE SWIM BLADDER

The swim bladder lies usually in the body cavity immediately beneath the vertebral column in a retroperitoneal space above the abdominal viscera and provided by "service tissues" (blood vessels and nerves) (Fig. 10.1). In this peritoneal space the swim bladder is surrounded by loose connective tissue and fat. It may lie relatively free or be fixed dorsally to parapophyses. In many Cypriniformes, the swim bladder lies immediately under the skin, the lateral body wall muscle to the swim bladder are absent, which augments the function of Weberian apparatus.

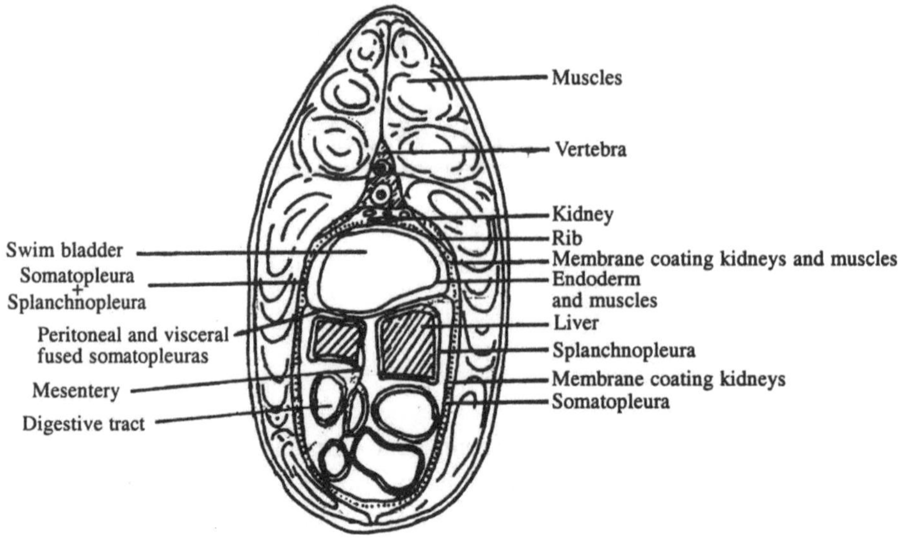

Fig. 10.1 Schematic transversal section through the anterior part of the body of an adult *Leuciscus rutilus*. *Source:* Chardon and Vandewalle (1989)

A swim bladder is not found in all fishes (Table 10.1), but some teleosts which lack a swim bladder in adult stage possess a swim bladder in the juvenile stage: pleuronectids (flounders), the angler (*Lophius*), and stromateoid fishes.

Table 10.1 Occurrence and type of swim bladder in fish groups

Division	Swim bladder
Cyclostomata	Absent
Chondrichthyes	Absent
Crossopterygii (coelacanth *Latimeria*)	Regressed
Dipnoi	Respiratory
Polypteriformes	Respiratory
Chondrostei	Nonrespiratory
Holostei	Respiratory
Teleostei	Respiratory (very few) Nonrespiratory or absent

Source: Fänge (1983)

Among marine teleosts a swim bladder occurs in the majority of epi-, meso-, and benthopelagic species, but is lacking or regressed in bathypelagic ones. The benthopelagic fishes have a depth range of 200–7000 m. They comprise relatively large fishes with a well-developed swim bladder (e.g., macrourids, morids). Although thousands of fishes have a swim bladder, this organ has been examined more closely only in relatively few species.

3. STRUCTURE: AN OVERVIEW

The varying shapes of the swim bladder closely reflect functional diversities. Its general anatomy in adult fish is higly variable; in some, "physostomatous" species, the connection to the alimentary canal (the pneumatic duct) persists while in other "physoclistous" ones it is lost (Fig. 10.2A). The shape of the swim bladder in teleosts varies markedly in different species, although it is most frequently cylindrical or ovoid. In some species the swim bladder is a single chamber, whereas in others it is divided into two or more interconnected ones. The two chambers may be lateral to one another as in *Prionotus*, but in ostariophysans and many nonostariophysans they are generally rostrocaudal to one another. In some species the swim bladder is divided into two or more compartments by septa or diaphragms, with openings to allow movement of gas (e.g., *Opsanus*). The swim bladder in many species (*Myripristis*, *Notopterus*), also have anterior projections which increase its proximity to the ear and are presumably an adaptation for improved audition. The two-chambered swim bladder of serrasalmid fishes has an extension of the posterior end to varying degrees. The swim bladder in herring *Clupea harengus* and sprat *Sprattus sprattus* is connected with narrow ducts to both the gut and cloaca. Air is swallowed and enters the swim bladder through the pneumatic duct and the gas is released through anal opening, the anal duct

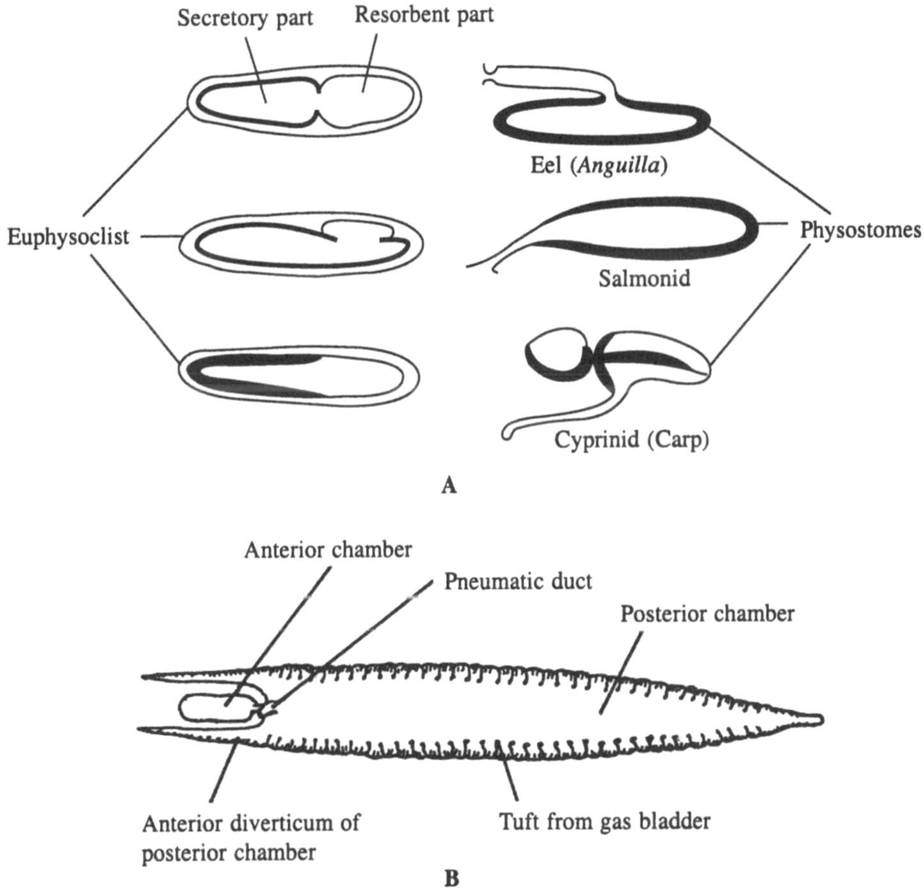

Fig. 10.2A Diagram of a few types of teleostean swim bladders. *Thick lines* indicates smooth muscle layers with a-adrenergic responses to catecholamines. These muscles act in gas exchange reflexes.
***Source:* Fänge (1983)**
B. *Rhapiodon:* Swim (gas) bladder.
***Source:* Lesiuk and Lindsey (1978)**

seems to have a sphincter. Swim bladders have in some cases been reduced in compensation for air carried in aerial respiratory organs. The swim bladder is also reduced in a number of surface-living marine fish. The size of the swim bladder may vary markedly within a genus (e.g., *Rhinichthys*, *Gobius*). *Plecostomus* has a perforated bony sheet on the sides of the swim bladder, the pores covered by epidermal and pigment cells. In *Rhaphiodon*, the diverticulae of the posterior sac reach just below skin (Fig. 10.2B).

Fishes with well-developed muscles and skeleton (organs which tend to increase body density above water density) usually have a large swim bladder. Also for buoyancy reason the swim bladder is usually larger in freshwater than in marine forms.

Within the Catastomidae the swim bladder has long been recognized as being a useful taxonomic tool because of its basic variation among the genus-groups.

Extinct lobe-finned fishes are assumed to have possessed a primitive lung, the only representative living now, the coelacanth (*Latimeria chalumnae*) has a regressed swim bladder replaced by a mass of wax esters. The African (*Protopterus*) and the South American (*Lepidosiren*) lungfishes have paired lungs connected with the gut via a glottis. The Australian lungfish (*Neoceratodus*) has a single lung. The African freshwater polypterids (*Polypterus*, *Calamoichthys*) possess paired lunglike swim bladders. The holosteans (*Lepisosteus*, *Amia*) have swim bladder resembling a primitive lung.

Compared with the greater number of modern bony fish species, extremely few teleosts have a swim bladder adapted for oxygen uptake from air. Teleostean respiratory swim bladders have an alveolated or spongy richly vascularized inner lining and a short, wide pneumatic duct. Trabeculae (septa or honey-comb pattern), with well vascularized thin epithelium is reported in the gas bladder of *Pangasius sutchi* (Danguy and Lenglet 1988). Both the degree and type of partitioning of gas bladder vary extensively among species using the gas bladder as an accessory "lung". In *Phractolaemus ansorgei*, intra-epithelial capillaries have been demonstrated in gas bladder. *Pantodon* (Pantodontidae) has air-filled processes from the swim bladder which penetrate into the vertebrae, forming avian like "pneumatic bones".

The chondrosteans (*Acipenser*, *Polydon*) are primitive physostomes with poorly vascularized nonrespiratory swim bladders. The clupeids, salmonids, esocids, siluroids, cyprinids, and anguillids are teleostean physostomes with nonrespiratory swim bladders. A great number of teleosts are physoclists (mainly euphysoclists; members of the families Gadidae, Macrouridae, Triglidae, Cichlidae, Sciaenidae, Labridae and Percidae) with nonrespiratory swim bladders. In physoclists gas exchange takes place between the swim bladder and the circulating blood through specialized parts of the swim bladder wall. The pneumatic duct is absent (Fig. 10.3).

Brainerd (1994) discussed the evolution of lung-gill bimodal breathing and the homology of vertebrate respiratory pumps. The four-stroke lung ventilation pattern is evident in *Polypterus*, *Lepisosteus* and *Amia*. Among teleost fishes, the osteoglossomorphs *Arapaima gigas*, *Gymnarchus niloticus*, and *Notopterus chitala*, and the catfish *Pangasius sutchi* all use four-stroke ventilation. *G. niloticus* adds an oesophageal pump to the system but the basic pattern of two buccal cavity expansions is retained. The exceptions to this pattern are the lung ventilation mechanisms of two ostariophysan fishes, *Gymnotus carapo* and *Hoploerythrinus unitaeniatus*. These fishes reverse the order of expiration and inspiration: fresh air is first pumped into the oesophagus and from there into the lung. Air is then expired from the lung into the oesophagus and out the mouth in *G. carapo* or out the opercular valves in *H. unitaeniatus*. In contrast to four-stroke buccal pump,

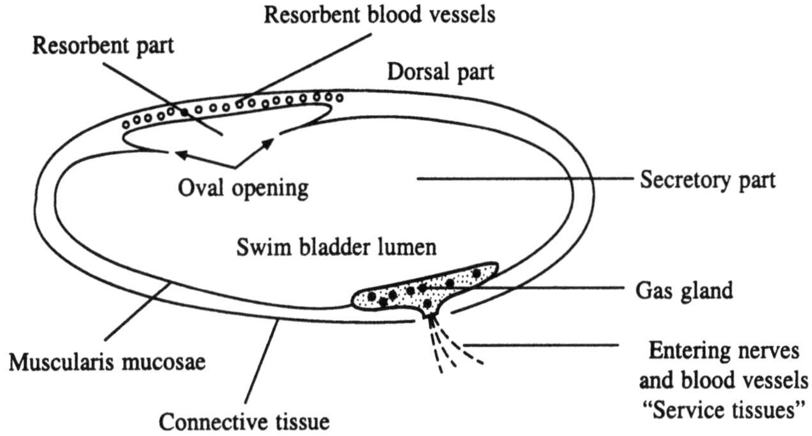

Fig. 10.3 ***Gadus morhua*. Diagrammatic figure showing the main organization of the cod swim bladder.**
Source: **Nilsson (1971)**

lungfishes and amphibians use a two-stroke pump in which both expiration and inspiration occur within one buccal expansion-compression cycle. First, the mouth or nares open and the buccal cavity expands to draw in fresh air. While the buccal cavity is expanding, air is exhaled from the lungs into and through the fresh-air filled buccal cavity. Second, the mouth or nares close, the buccal cavity compresses, and a mixture of fresh and exhaled air is pumped into the lungs. Two stroke lung ventilation pattern is evident in sarcopterygians: *Lepidosiren paradoxa*, *Protopterus aethiopicus* and *Necturus maculosus*.

Several animals also pump one or more mouthfuls of fresh air into their lungs after the first two-stroke, expiration-inspiration cycle. *Lepidosiren* normally uses one or more accessory inspirations, and *Protopterus*, when under duress, is also capable of performing accessory inspirations (also read Chapter 9).

Vascular designs The primitive pulmonary circulation of lungfishes is kept partly separate from the systemic circulation by incomplete septa in the heart. Fishes have a simple circulation, and oxygenated blood from a respiratory swim bladder becomes mixed with venous blood from different parts of the body when passing the heart (except in dipnoans).

Taking the nonrespiratory swim bladders, the swim bladder of physostomes are supplied with blood from the coeliacomesenteric artery, and from the dorsal aorta. In most physostomes countercurrent vascular bundles in the swim bladder are relatively weakly developed, but retia mirabilia or red bodies of considerable size are found in *Esox*, *Anguilla*, and *Conger*. A double circulation is found in the swim bladder of the euphysoclists (and *Anguilla*, *Conger*). The secretory part of the swim bladder, usually supplied from the coeliacomesenteric artery, has well-developed countercurrent vascular bundles (retia mirabilia). All the blood entering and leaving the gas gland area and the secretory mucosa passes the countercurrent vessels. The veins leaving the countercurrent bundles join the hepatic portal system. The resorbent part of the swim bladder is supplied from the dorsal aorta, or branch of the coeliacomesentery artery that bypasses the countercurrent systems. The veins from the resorbent capillary plexus join the posterior cardinal vein system. In eels (*Anguilla*, *Conger*) two well-developed rete mirabile portions receive blood from the swim bladder artery and supply the secretory part of the swim bladder.

The rete vessels deliver blood to the hepatic portal vein. The pneumatic duct of eels, which serves as a resorbent area, is vascularized by a pulmonarylike capillary network. From the resorbent capillary plexus of the pneumatic duct a vein carries blood to the ductus Cuvieri, or directly to the atrium of the heart (Figs. 10.4–10.6).

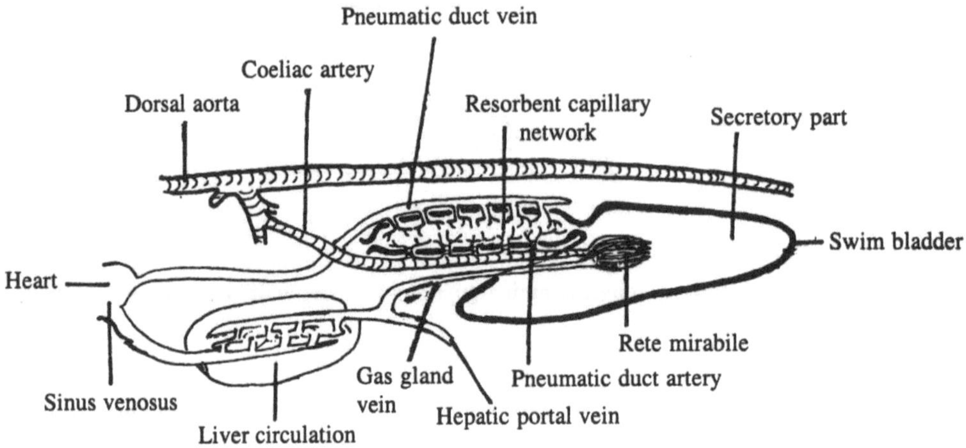

Fig. 10.4 Diagram of the vascular supply of the swim bladder. Physostome type: *Anguilla*-type
***Source:* Fänge (1953)**

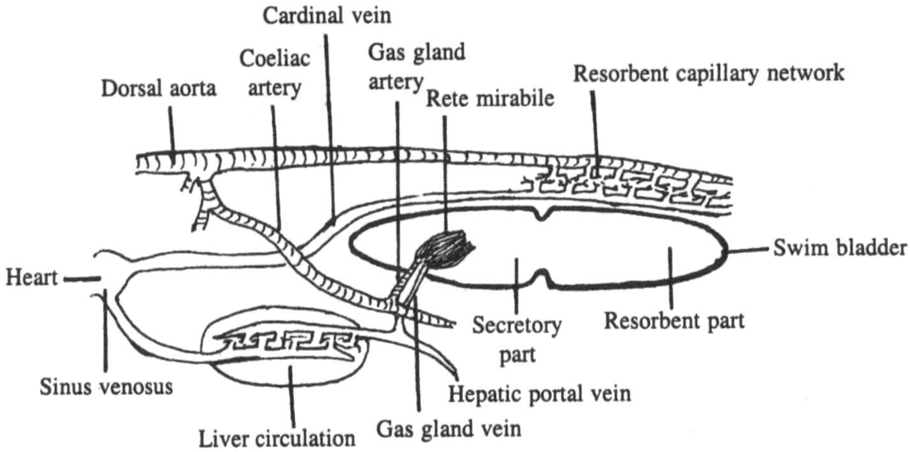

Fig. 10.5 Diagram of the vascular supply of the swim bladder: Euphysoclist type.
***Source:* Fänge (1953)**

Microscopic observations The lungs and respiratory swim bladders of nonteleosts have folded lining with a rich vascular supply meant for gas exchange. No countercurrent systems occur. Similar to lungs of higher vertebrates, lamellar bodies (intracellular osmiophilic bodies) are reported in *Polypterus* and dipnoan lungs. Smooth or striated muscles are reported in the swim bladders or lungs in several nonteleostean fishes.

In a physostomatous fish *Carassius auratus*, the lining cells of anterior chamber possess short microvilli, extracellular osmiophilic layer, free ribosomes, mitochondria, Golgi body, lamellar

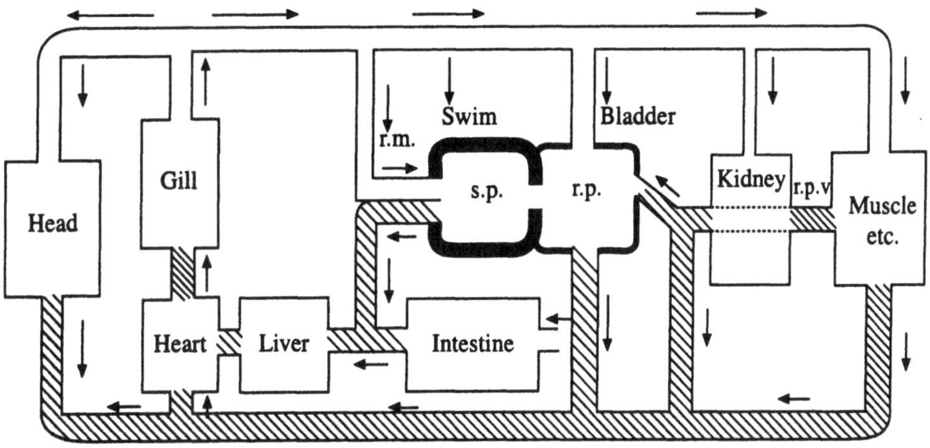

Fig. 10.6 Blood circulation in the euphysoclist type of swim bladder in relation to the body circulation. r.m., rete mirabile; r.p., resorbent part; r.p.v., renal portal vessels, s.p., secretory part, Shaded area, venous blood.
***Source:* Fänge (1983)**

bodies (osmiophilic bodies), and endoplasmic reticulum. Desmosomes and interdigitations are seen. In its posterior chamber, there are no lamellar bodies. Smooth muscles are found in most physostome swim bladders.

The euphysoclist (more highly organized physoclists) swim bladder has tunica externa and tunica interna. The tunica externa is a dense connective tissue capsule. It is composed of collagen fibrils but may contain some smooth or striated muscles, the latter are sound producing. The tunica interna is made up of submucosa and mucosa. The submucosa consists of a loose connective tissue, often jelly-like in consistency and contains nerves and vessels (such as retia mirabilia). The mucosa is comprised of an epithelium which is associated with blood capillaries, a delicate connective tissue membrane (lamina, propria), and a smooth muscle membrane or muscularis mucosae. Ostariophysans have a slit on the anterior-dorsal side of tunica externa, to which the most posterior part of the Weberian ossicles, the tripus is connected.

The division of the euphysoclist swim bladder into a "secretory" and a "resorbent part" usually concerns the submucosa and the mucosa. The boundary between the two regions is formed by a mucosal fold called the "diaphragm" or "oval" (Fig. 10.7). Circular and radial smooth muscles are arranged in a sphincterlike way in this fold. The secretory mucosa-submucosa is much thicker than the resorbent mucosa-submucosa. The muscularis mucosae of the secretory part of the swim bladder is strongly developed, but it is very thin in the resorbent part. The submucosa of the secretory part is impregnated with guanine crystals, lipid membranes, or platelets of unknown nature. These materials are thought to make the secretory submucosa impermeable to gases. The resorbent submucosa-mucosa is extremely thin and closely associated with resorbent capillary network.

Gas gland. The gas gland consists of gland-like cells supplied with blood from the countercurrent vascular bundles. The gas gland cells typically derive from thinner epithelium, but in certain physostomes supposed gas gland cells look like modified connective tissue cells, and their epithelial nature is uncertain (e.g., cyprinids, and *Argentina*—a deep water physoclist).

In physoclists the gas glands are generally more highly developed than in physostomes. The composition and form may vary. Histochemically, the presence of carbonic anhydrase and glycogen has been demonstrated. The fine structure shows microvilli, endoplasmic reticulum (SER, RER),

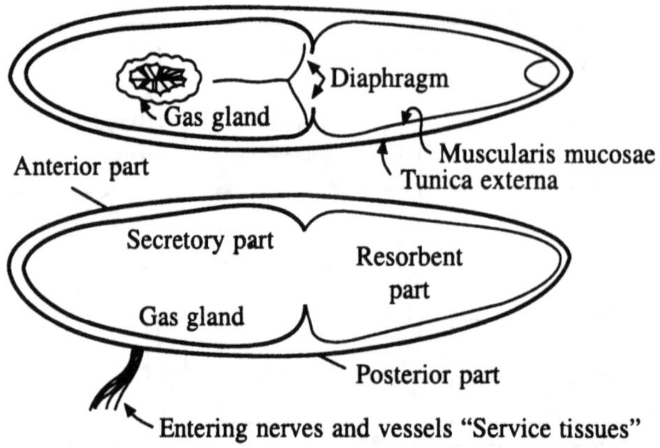

Fig. 10.7 Semi-diagrammatic drawing of the swim bladder of the goldshinny wrasse (*Ctenolabrus rupestris*); dorsal view (upper figure) and lateral view (lower figure). The anterior (secretory) part and the posterior (resorbent) part are separated by a transverse diaphragm wtih a central hole. The muscularis mucosae moves inside the tunica externa, with the loose connective tissue of the submucosa serving as a "lubricant". The entering blood vessels branch to form the parallel bundles of capillaries of the rete mirabile, the countercurrent blood supply to the gas gland tissue. Blood vessels, together with a nerve trunk, also run along the ventral surface of the muscularis to the diaphragm and posterior part.
Source: **Fänge et al. (1976)**

glycogen particles, a Golgi complex mitochondria, lipid-like spherical formations (multilamellar or dark bodies), and desmosomes. The osmiophilic bodies resemble multilamellar bodies in vertebrate lung. The basal perivascular cell base of gas gland cells slows a complex folding. The basal infoldings lack mitochondria. The gas gland capillaries have very thin endothelial lining with large mitochondria.

Chemical nature of the lining layer of swim bladders. Very little information is available about the chemical nature of the lining layer in swim bladders. Phosphatidylethanolamine and phosphatidylserine have been reported in the extracellular linings of the swim bladders of deep-sea fishes. An extracellular layer has been found in the toadfish (*Opsanus tau*) swim bladder; its chemical interpretation is difficult.

There is a high percentage of cholesterol in the extracellular lining of the swim bladders of deep-sea fish. The crystalline needles that are found in the extracellular lining of toadfish swim bladder has been interpreted as an indication of cholesterol.

Nerve supply. Nerve cells and vagal nerve fibres have been observed in lungs and swim bladders of nonteleostean fishes. Adrenergic fibres occur in *Lepisosteus* swim bladder wall, but not so in the case of *Protopterus* lungs.

The teleostean swim bladder has dual innervation—vagal and sympathetic origin. The main nervous supply is from the ramus intestinalis of the vagus. This ramus is in reality a vago-sympathetic trunk containing sympathetic components which join the vagus close to the brain. Sympathetic fibres, partly anastomising with the vagus, also reach the swim bladder from the coeliac ganglion (Fig. 10.8). A ganglionic layer in the isthmus that connects the two chambers

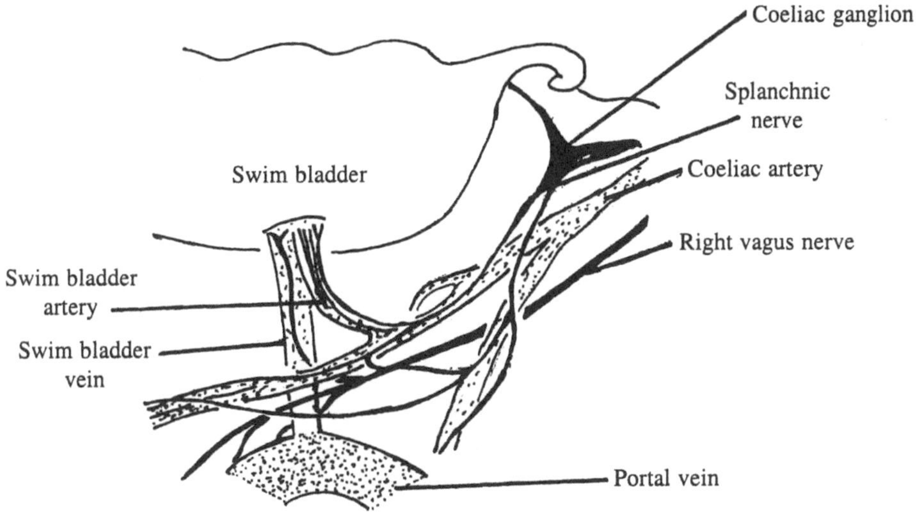

Fig. 10.8 ***Gadus morhua*: Nerves and blood vessels outside the swim bladder on the right side. The head kidney has been removed to expose the coeliac ganglion.**
***Source:* Nilsson (1972)**

of the cyprinid swim bladder; large nerve cells in the pneumatic duct of *Anguilla anguilla*; a "gas gland ganglion" in euphysoclists (e.g., *Gadus morhua, Perca fluviatilis*) have been described. The gas gland ganglion shows strong acetylcholine esterase activity but no catecholamine fluorescence. High activity of acetylcholine synthesizing enzyme choline acetyltransferase is found in the gas gland and the secretory mucosa. Fibres with strong acetylcholine esterase activity, probably originating in the "gas gland ganglion" are distributed at the base of the gas gland cells. The "swim bladder nerve ganglion" is positive for catecholamines and acetylcholine esterase. The large ganglion cells may constitute adrenergic sympathetic postganglionic neurons.

The muscularis mucosae of the secretory part of the swim bladder of *Ctenolabrus* and *Gadus* contain considerable amounts of noradrenaline and small quantities of adrenaline.

Depth adaptations of the swim bladder structures. Anatomical modifications of the skeleton supposed to increase the strength of the swim bladder have been found in certain deep-water gadids.

The amount of guanine in the swim bladder wall increases with depth. The swim bladder of the conger eel (*Conger*), which lives deeper than the common eel (*Anguilla*) contains more guanine and is about ten times less permeable to O_2 and CO_2 than the swim bladder of the latter. However, when the common eel metamorphoses to the silver eel stage before migrating to spawning depths, the guanine content of the swim bladder increases drastically.

A direct relationship exists between the degree of development of the gas gland-rete mirabile complex and the depth range. In deep-living fishes the numbers and sizes of the retia and the surface of the gas gland increase with depth. Whereas the retia of surface-living fishes are less than 1 mm long, those of deep-sea fishes measure 25–65 mm in length. In the common eel (*Anguilla*) the two retia increase considerably in size during migration.

The large size of the countercurrent vascular system (rete) in deep-sea fishes is important for deposition of gases, and by serving as a lock against diffusion loss of gases. The rediffusion of oxygen across the rete capillaries permits deep-sea fishes to keep an oxygen tension in the swim bladder vessels, which is perhaps 1000–3000 times larger than that in the systemic blood vessels (Figs. 10.9, 10.10).

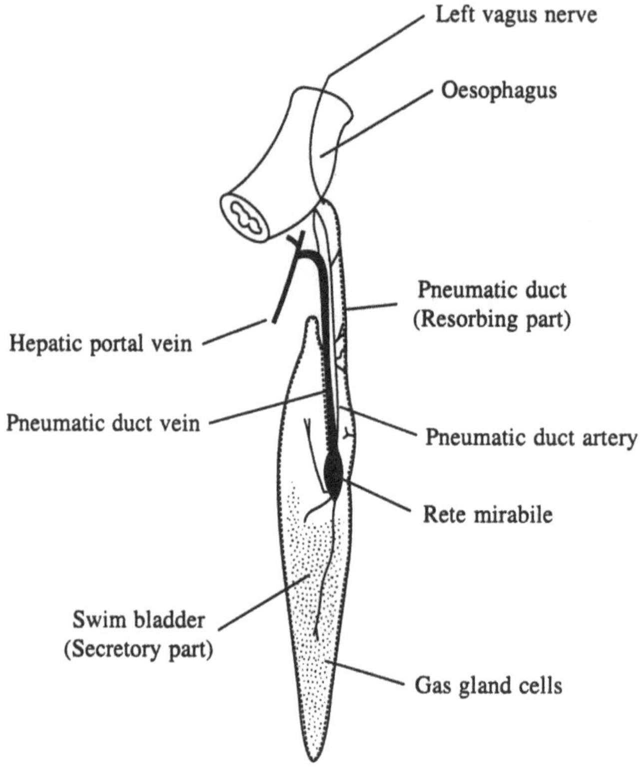

Fig. 10.9 ***Anguilla vulgaris*, anaesthetized specimen. Normal appearance of the swim bladder.** ***Source:*** **Fänge (1953) (Modified)**

4. COMPOSITION OF SWIM BLADDER GASES

The total pressure of the swim bladder gas at any depth is in equilibrium with the hydrostatic pressure. The swim bladder contains the same gases that occur in the air, the ambient water, and the body fluids. In the living fish the swim bladder volume is normally kept constant, but gases are continuously entering and leaving the swim bladder at the same rate (steady state).

Steady state composition. In shallow-living fishes the oxygen concentration in the swim bladder is usually near that of the air but in *Anguilla* and *Opsanus* high percentages of oxygen may occur. The percentage of oxygen in the swim bladder increases with depth. The buoyancy effect of the swim bladder of deep-sea fishes must be rather low because of the increased densities of the compressed gases. The values of the swim bladder as an oxygen store for respiration ought to be higher in these fishes than in the shallow-living ones, but it is not known whether this is of biological importance.

The concentration of carbon dioxide in the swim bladder is normally low. About 99% nitrogen

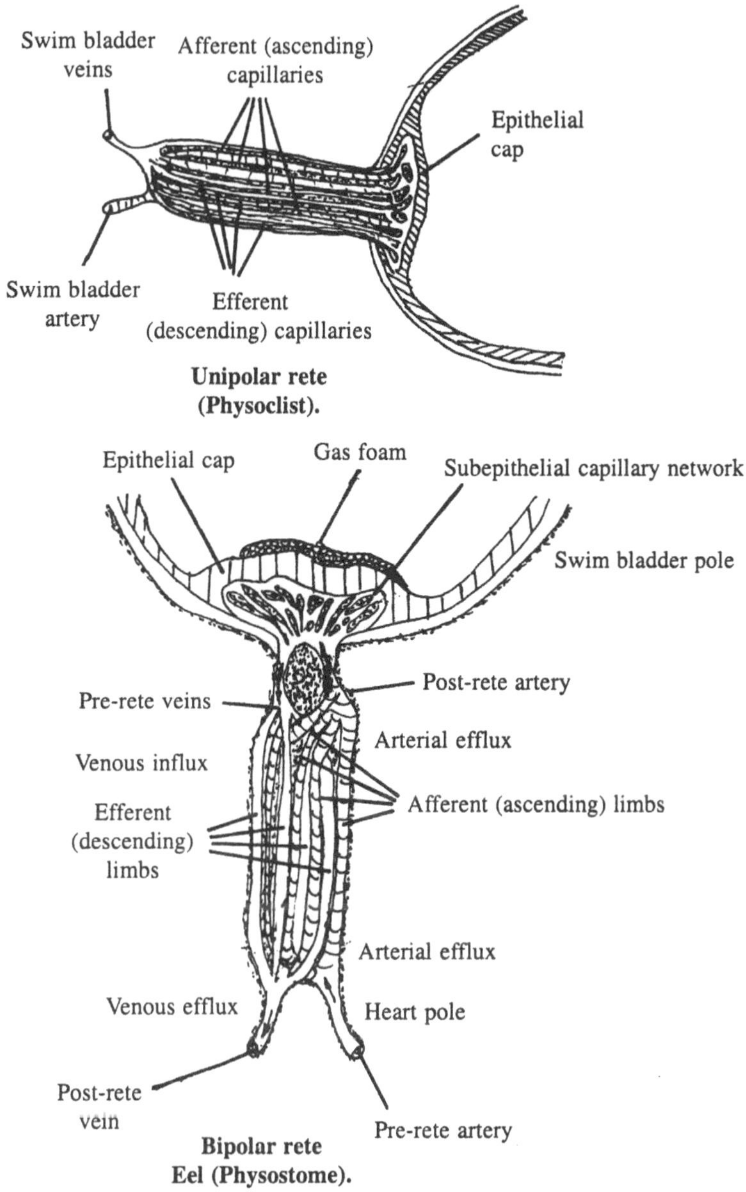

Fig. 10.10 Types of rete mirabile: Unipolar (physoclist) (upper) and Bipolar (physostome)(lower). ***Source:*** **Blaxter and Tyler (1978)**

is found in the swim bladders of certain species of physostomes living at 60–70 m depth. In deep-sea fishes the percentage of nitrogen is low. Argon, like nitrogen, is enriched in the swim bladder of deep-water fishes. Because argon is inert, it does not pass into the swim bladder as a result of chemical reaction, but some "physical mechanism" must exist for the transport of argon and other inert gases. Probably all gases dissolved in the surrounding water are deposited into the swim bladder.

Newly secreted gases. The gas mixture introduced into the swim bladder by secretion differs in composition from the steady state composition. Oxygen is the main gas deposited. Carbon dioxide also reaches high values. The highest values of CO_2 occur at the peak of gas secretion, when resorption of gases is minimized due to expansion of gas-impermeable secretory mucosa-submucosa (= oval closure). Concentrations of CO_2 and O_2 deposited by the slow functioning gas secretory systems of physostomes are of similar magnitudes to those in physoclists.

Three important factors which keep the gases inside the swim bladder must be emphasized here.

(1) The entrance to the pneumatic duct may be closed off by muscles in the wall.

(2) The swim bladder wall is covered by a silvery layer that is impregnated with guanine and some hypoxanthine, constituting an impermeable coat. Removing this silvery layer increases the gas permeability of the swim bladder wall some 100 times.

(3) The rete countercurrent system acts as a barrier in that gas is shunted from the rete venous capillaries to the rete arterial capillaries. This is analogous to the diffusion shunt that has been assumed for muscle tissue in mammals. The rete, thus, does not play a role in gas deposition but also in its maintenance. It should, however, be noted that complete cessation of blood flow to the swim bladder epithelium would constitute an even more effective way of preventing gas loss.

The reader is referred to cardiovascular system (Chapter 8) for gas transport properties of fish blood.

5. GAS EXCHANGE THROUGH THE PNEUMATIC DUCT—THE CONDUIT

Lungs and respiratory swim bladders

(i) Nonteleosts

Protopterus and *Lepidosiren* are obligate air-breathers with degenerated gills. *Neoceratodus*, a facultative air-breather, uses its lung during nocturnal active periods as an accessory structure for oxygen uptake. In dipnoans inspiration is brought about by a buccal pumping mechanism. Nerve-induced opening and closing of glottis are involved in the chain of events of pulmonary ventilation. Expiration is largely passive, but smooth muscles of lung wall may have some importance. These muscles are innervated by the vagus nerve. The motor nerve fibres may be cholinergic. Catecholamines contract or relax the muscles, but the effects are weak.

Buccal pump mechanisms are also responsible for inspiration in *Polypterus*, *Calamoichthys*, *Lepisosteus*, and *Amia calva*. Gars (*Lepisosteus*) inhale through the mouth and exhale through the opercular openings. At low water temperatures (20–25°C) the gars obtain 70%–80% of their O_2 requirements from the swim bladder. Elimination of CO_2 proceeds entirely through the gills at low temperature, but at warm temperature upto 8% of the respiratory CO_2 is excreted through the swim bladder. The gar swim bladder is ventilated several times per hour. Expiration is thought to be passive, brought about by the hydrostatic pressure. However, the physiological importance of the striated and smooth muscles in the wall of *Lepisosteus*, *Amia*, *Polypterus*, and *Calamoichthys* has not been investigated.

The blood streaming to the lung/respiratory swim bladder comes from the efferent gill arteries.

If, as may often happen with air-breathing fishes, the ambient water is poor in oxygen, oxygen may diffuse into the water when the blood passes the gills. In *Neoceratodus* oxygen loss from the gills is reduced by a high oxygen affinity of the blood. In other dipnoans reduction of gill function protects against diffusion loss of O_2 when the fishes are surrounded by stagnant water. In *Lepisosteus* O_2 loss from the gills is probably minimized by shunts allowing some blood to bypass the gills when the fish is breathing air. Such shunting of parts of the ventral aortic blood past the gas exchange surface of the gills is supposed to take place in several air-breathing fishes. However, no shunts in *Amia* are found.

The neuromuscular control of the ventilating movements in air-breathing nonteleostean fishes is imperfectly known. The brain centre, which produces rhythmic signals to the respiratory muscles, seems to be more sensitive to lack of O_2 than to excess of CO_2.

(ii) Teleosts

Quite a number of teleosts of different families possess anatomic structures used for the uptake of atmospheric O_2. Organs occasionally adapted for aerial respiration include the skin, gills, mouth and buccal cavity and special structures within these, intestine, and swim bladder. Only a handful of species use the swim bladder for O_2 uptake.

Most air-breathing teleosts live in stagnant tropical fresh waters. As a rule they are facultative air-breathers, using the swim bladder as an accessory respiratory organ when the O_2 content of water is low. A minority are obligate air-breathers (e.g., *Arapaima, Pantodon, Gymnarchus*).

Arapaima gigas has reduced gills and depends on the lung-like swim bladder for its normal respiration. The mode of ventilation is unusual in that it inspires by aspiration, probably using the diaphragm-like ventral wall of the swim bladder to produce suction (see Fig. 9.33). The swim bladder gets its blood supply from the dorsal aorta. Because of gill reduction this carries blood with an unusually low pO_2.

Facultative air-breathing teleosts with a respiratory swim bladder use this almost exclusively for O_2 uptake and depend on gills and the body surface for the excretion of CO_2. The obligatory air-breathing *Arapaima*, unlike other teleosts, has a relatively high excretion of CO_2 via the swim bladder. Due to reduced gill function this fish excretes insufficient amounts of CO_2 via the gills.

The ventilation of the teleostean respiratory swim bladders is based on nervous reflexes similar to those controlling gas exchange through the pneumatic duct of other physostomes. Nervous control of the pneumatic sphincter and smooth muscles in the wall of the swim bladder is probably important in the ventilatory activities.

Non-respiratory physostome swim bladders

Physostomes change their swim bladder volume both by gas secretion and resorption, and by intake and extrusion of gases via the pneumatic duct.

If the swim bladder is emptied by pressure decrease or puncture, most physostomes are able to fill it again by gulping air at the surface. The air is pumped into the swim bladder by buccal muscles. Intake of air through a pneumatic duct is essential for the initial inflation of the embryonic swim bladder in certain physoclists.

The term "gasspuckreflex", gas spitting reflex means a release of gases via the pneumatic duct. A release of gas may also be initiated by various disturbing factors. The gasspuckreflex involves opening of pneumatic sphincter, contraction of smooth muscles of the swim bladder wall, and expulsion of gas bubbles from the mouth. A reflex centre may be situated within the diencephalon of the brain. Excess internal pressure in the swim bladder, contraction of the body wall muscles and external hydrostatic pressure contribute to the expulsion of gases through the pneumatic duct.

(iii) Nervous control of gas expulsion through the pneumatic duct

A principal similarity between the gasspuckreflex of physostomes, and the deflatory reflex of physoclists has been pointed out. Both responses involve opening (relaxation) of a sphincter area, and simultaneous contraction of smooth muscles of swim bladder wall. The sphincter or the whole pneumatic duct of physostomes contracts by acetylcholine and relaxes by catecholamines, the latter response being mediated by beta-adrenergic receptors. The smooth muscles of the physostome swim bladder wall contract by catecholamines, probably an alpha-adrenergic effect. Adrenergic influences provoke or facilitate gas release from the swim bladder in *Oncorhynchus* and *Clupea*.

Information is scattered and incomplete, but indicates that adrenergic influences decrease swim bladder volume in physostomes by opening the pneumatic sphincter and contracting the swim bladder muscles. Cholinergic stimuli may close the sphincter more firmly (e.g., *Oncorhynchus*). However, great species differences probably exist.

6. COUNTERCURRENT CONCENTRATION AND GAS SECRETION IN SWIM BLADDER

Many fish possess a gas-filled swim bladder, and the gas molecules enter the bladder by diffusion from its capillaries and epithelium. The high gas partial pressures necessary to drive diffusion are produced in two steps. First, in the blood perfusing the swim bladder capillaries the physical solubility of gases is reduced via the salting-out effect, and gases are released from a chemical binding site via blood acidification (e.g., Root effect for O_2; conversion of HCO_3^- to CO_2). These effects result in an increase in gas partial pressure in blood. In a second step, this initial increase in partial pressure (the "single concentrating effect") is multiplied by back diffusion of gas molecules in the countercurrent system of the rete mirabile. For inert gases, the salting-out effect, induced by the release of lactic acid from the gas gland cells in the swim bladder tissue, results in a probably small reduction of solubility, while the acid-induced decrease in haemoglobin O_2-carrying capacity (Root effect) and the release of CO_2 from HCO_3^-, combined with release of anaerobically produced CO_2 from gas gland cells, results in a significant concentrating effect for O_2 and CO_2. The relatively high permeability of the rete capillaries for gases allows back diffusion, and thus for countercurrent concentration of gases. The rete mirabile is, however, permeable not only for gases but also for metabolites and salt molecules, and the countercurrent concentration of solutes enhances the countercurrent concentration of gas molecules. This is particularly important for inert gases because their single concentrating effect is small. Back diffusion of CO_2 in the rete mirabile not only results in a countercurrent concentration of CO_2 but also enhances the partial pressure increase of O_2 by decreasing the haemoglobin O_2-carrying capacity within the arterial capillaries of the rete. This interdependence of O_2 and CO_2 may apply to countercurrent systems in other tissues as well, for example, to the skeletal muscle. Hence, the swim bladder rete mirabile may serve as a model tissue for studying of gas exchange and countercurrent blood flow in other organs. Furthermore, the regulation of countercurrent efficiency and gas secretion promises to be an interesting area of research, including also the regulation of metabolic activity and acid production in the gas gland cells (Pelster and Scheid 1992; see Figs. 10.11, 10.12).

Most teleostean physostomes are able to fill the swim bladder by gas secretion. In clupeids (herring) the situation is unclear. The herring lacks obvious gas depositing structures (rete mirabile, gas gland), but has relatively high O_2 values in the swim bladder and observations of release of gas bubbles from vertically migrating herrings indicate that some gas secretion may occur. Gas secretion is known from the swim bladder of esocids, umbrids, cyprinids, siluriforms, and eels. In the goldfish, *Carassius auratus*, gas secretion is linear in time and independent of swim bladder volume. Many physostomes require a considerable time to complete gas secretion, but in *Anguilla* gas secretion is as rapid as in physoclists. In physoclists gas secretion is generally fast. The swim bladder of deep-sea fishes contain a "swim bladder foam" formed by expansion of gases within the swim bladder wall when fishes are hauled to the surface from deep water. The foam of bilayered membranes has neutral fat, phospholipids, cholesterol, and lipoproteins.

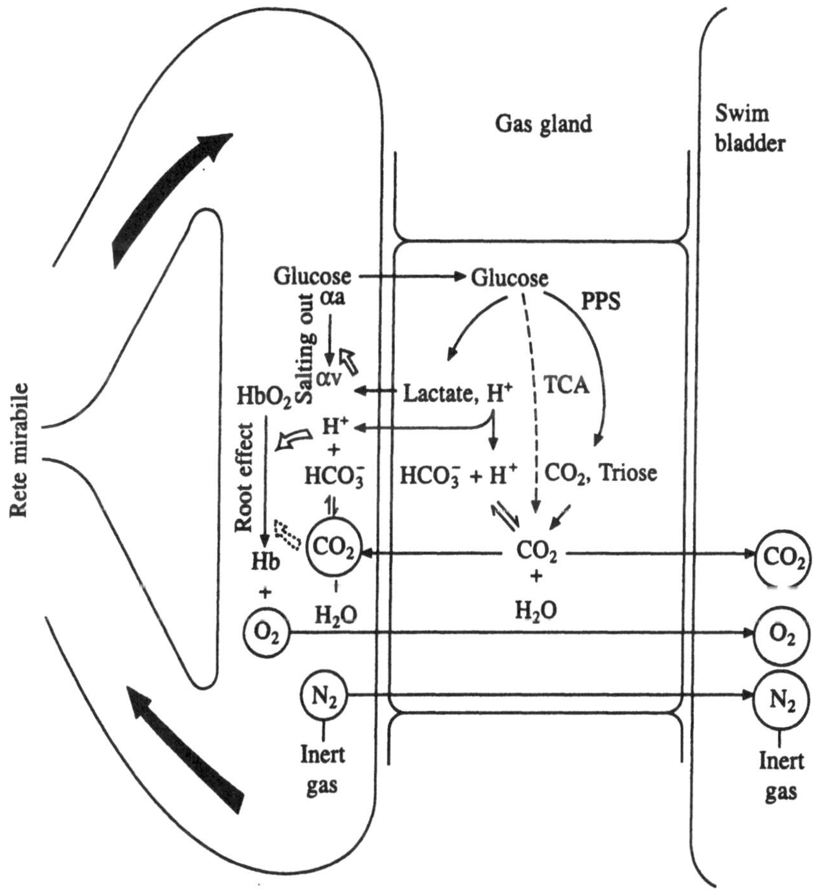

Fig. 10.11 Present concept of metabolism of the gas gland cells and its influence on the physical solubility of inert gases or the release of gas from a chemical binding site in the blood; TCA, tricarboxylic acid cycle; PPS, pentose phosphate shunt.
Source: Pelster and Scheid (1972)

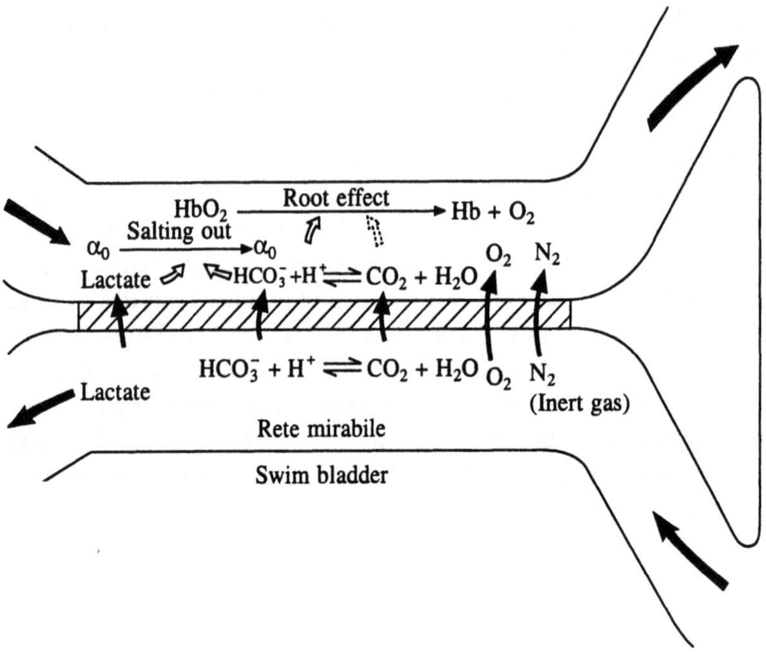

Fig. 10.12 Present concept of movements of gases and solute in the rete mirabile and the resulting countercurrent concentration of gases and solutes. Fluxes have not been determined and it cannot be excluded that further ion movements are involved, e.g., for osmotical or electrical reasons. Open arrows indicate influence or possible influence on the inert gas solubility and haemoglobin O_2 binding characteristics. Closed thick arrow indicate movement of substance. ***Source:*** **Pelster and Scheid (1992)**

Metabolism of the swim bladder epithelium and the single concentrating effect.
Figure 10.13 shows the general model proposed by Kuhn et al. (1963) for the generation of high gas partial pressures by means of countercurrent multiplication, that is applicable to any gas transported in the blood.

A summary of the main catabolic pathways of gas gland cells is in Figs. 10.14, 10.15. The main fuel for this metabolism is blood glucose, internal glycogen stores usually being too small to be of significance. In the active, gas depositing swim bladder of the European eel (*Anguilla anguilla*) the main fraction of glucose (about 75–80%) is fermented to lactate in the anaerobic glycolysis. Aerobic metabolism with tricarboxylic acid cycle and the respiratory chain account for only 1% of the glucose degradation. More important is the pentose phosphate shunt, probably responsible for about 20–25% of glucose catabolism, in which glucose is decarboxylated. Part of the CO_2 is released into the bladder where it takes part in acidifying the blood (see details: Pelster and Scheid 1993; also refer Pelster and Scheid 1992a, b; Pelster 1995; Pelster and Pott 1996; Pelster et al. 1997). The reader is also referred to the coverage on swim bladder gas and oxygen secretion by Phleger (1991). Attention should be given to the contribution of Pelster et al. (1990) with focus on reduction of gas solubility in the swim bladder and of Scheid et al. (1990) on the gas exchange in this organ. The physiology of the Root effect has been comprehensively reviewed by Pelster and Weber (1991; read also Randall and Brauner 1998). Recent review on the metabolism of the swim bladder tissue is from Pelster (1995).

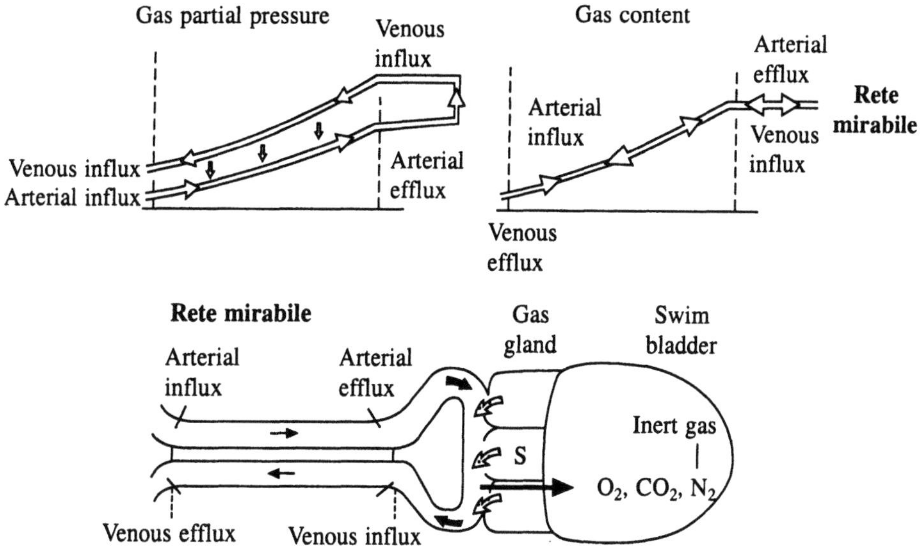

Fig. 10.13 Scheme of the basic principle for the generation of high gas partial pressures in a countercurrent system as proposed by Kuhn et al. (1963). A substance (S) is released from the gas gland cells and reduces the physical solubility of gases in blood. The gas partial pressure increases and along the partial pressure gradient gas molecules diffuse back to the arterial side in a countercurrent system. Arterial influx and efflux of the rete mirabile; venous influx and efflux of the rete mirabile are shown.
***Source:* Pelster and Scheid (1993a)**

The mechanism of lactate extrusion from gas gland cells is unknown (Pelster and Scheid 1993b).

7. GAS RESORPTION

Rates of resorption. In air-breathing physostomes, the swim bladder functions like a primitive lung. But in the majority of physostomes, which do not use the swim bladder for respiratory uptake of O_2, the absorption of O_2 is slow, because poor vascularization of the swim bladder wall prevents rapid diffusion. Some species of physostomes have an efficient gas secretory mechanism but no extensive resorption.

The physoclist swim bladder represent a mixture of gases enclosed within tissues. Due to continuous O_2 consumption by the tissues, the total pressure of gases within the swim bladder slightly exceeds that of tissues. This pressure gradient forces gases to disappear by diffusion. The individual gases differ in their diffusion rates through the tissues.

Carbon dioxide diffuses so easily that the concentration within the swim bladder is usually close to zero. Oxygen is the main gas which is exchanged during volume control of the swim bladder, while N_2 is least movable. The rate at which O_2 is resorbed depends upon how fast it can be carried away by the blood. In deep water resorption is much faster than secretion (although the rates of both processes are influenced by nerve reflexes).

Some species of fish probably maintain neutral buoyancy only when actively swimming. During resting periods they may stay at the bottom with a half-filled swim bladder.

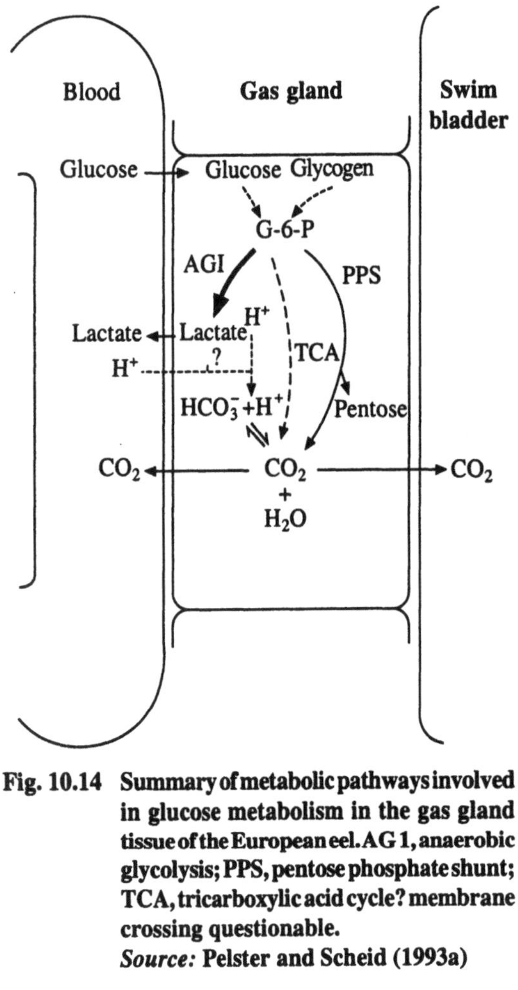

Fig. 10.14 Summary of metabolic pathways involved in glucose metabolism in the gas gland tissue of the European eel. AG 1, anaerobic glycolysis; PPS, pentose phosphate shunt; TCA, tricarboxylic acid cycle? membrane crossing questionable.
Source: **Pelster and Scheid (1993a)**

8. FUNCTIONS

The functions of the swim bladder have been well discussed by Alexander (1966), Blaxter and Tyler (1978), Harden Jones and Scholes (1985) and Pelster (1997, 1998).

Approximately half of the 23,637 known species of teleosts possess a swim bladder at some stage of their lives. Its function vary with the life style of the species.

8.1 Audition

The swim bladder has numerous functions including that of a hydrostatic organ to adjust the density of the animal to changes in depth. The swim bladder is also involved in both sound production and sound detection, and perhaps an air-storage device for respiration in some fishes. The specific role in sound detection has not been intensively studied, but that it does have a role is clear from a small number of behavioural and physiological studies.

Enhancement of auditory sensitivity, by proximity of swim bladder to the inner ear, has also been shown by behavioural studies showing that squirrelfish (*Adioryx xantherythus*) is sensitive only to sounds to about 1,000 Hz, whereas a related species, *Myripristis kuntee* can detect sounds to at least 3,000 Hz with sensitivity exceeding that of several Ostariophysi. Significantly, the swim bladder is much closer to the auditory bulla than in *Adioryx*; thus less acoustic information is lost in body tissue between the swim bladder and the ear.

Similarly, removal of the small air bubbles near the inner ear in anabantids and mormyrids significantly decreases hearing sensitivity. It has been shown that a species of tuna with a swim bladder (*Thunnus albacares*) has about a 20–dB better sensitivity over much of its hearing range than a species without a swim bladder (*Euthynnus affinis*), although the range of sensitivity is similar in both species (50–1,000 Hz).

The swim bladder contributes to sound detection by providing a medium of different density than that through which the stimulus is travelling. Many investigators approximate the function of the swim bladder by comparing it to a free bubble of air in water. It is then possible to infer the swim bladder's functional features from equations for the response of such an air bubble in water. This model would be heavily damped, have a narrow resonance frequency, and respond sharply to signal onset and offset. Thus, the swim bladder should respond best to particular frequencies and essentially "amplify" signals and not pass others outside the region of resonance. However, recent workers suggest that the swim bladder is not as sharply tuned as a free bubble,

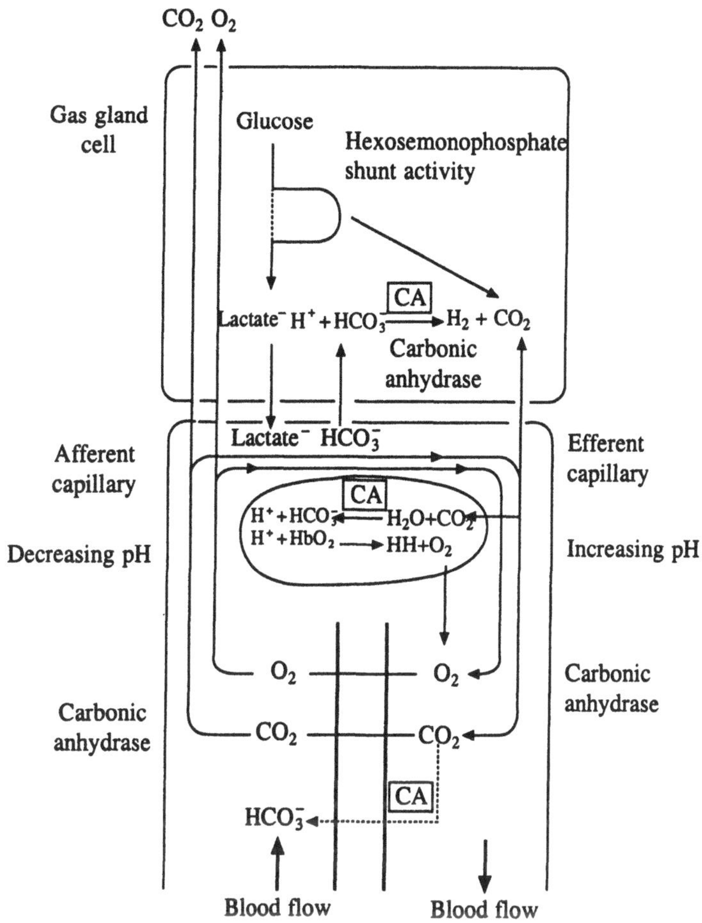

Fig. 10.15 Possible mechanisms responsible for deposition of O_2 and CO_2 in the swim bladder and the role of carbonic anhydrase in gas secretion. In the gas gland cells carbonic anhydrase causes dehydration of HCO_3^-, whereas in the red cells and in the rete cells the same enzyme may serve to transform CO_2 into HCO_3^-. Most of the CO_2 released during gas secretion comes from bicarbonate of the blood plasma, but some is reduced by the hexosemonophosphate shunt activity in the gas gland cells.
***Source:* Fänge (1983)**

so there is a fairly wide frequency range over which the swim bladder should respond. The decrease in tuning sharpness results from the swim bladder's being within the body cavity of the fish, where the body wall affect damping, and from the shape of the swim bladder, which differs from that of a typical air bubble. Data are limited, but studies on the intact swim bladder of *Carassius auratus* and *Gadus morhua* demonstrate that the swim bladder's response characteristics are relatively flat over the range hearing sensitivity for both species.

Several additional experiments indicate that the response of the swim bladder is not as simple as that of a broad-band sound radiator. Further, it has been suggested that the overall flat response of the swim bladder in several species may not reflect the total picture of the swim bladder response.

There may be considerable interspecific variation in the way sound gets to the ear. It has been concluded that the swim bladder is involved in pressure detection but does not act as a release system for the ear in Ostariophysi, as the displacement response of the ear does not change with removal of swim bladder.

The swim bladder is also a pressure detector in the cod (*Gadus morhua*), a nonostariophysan. The swim bladder is involved in audition of signals about 1,000 Hz, whereas another mechanism, presumably direct displacement detection, is operative at lower frequencies.

The swim bladder is clearly a highly dynamic organ and does not seem suitable either as a proprioreceptor or as a structure for giving information on absolute pressure level if fish are constantly changing depth.

There is increasing evidence, however, that the swim bladder is not always acting as a compliant structure "obeying" Boyles law, but can sometime be maintained by action of its own or surrounding muscles with a higher internal pressure than the outside.

There is some experimental evidence that vertical migration may not impair pressure sensitivity. Pressure sensitivity is restored after a depth change much more quickly than gas could be secreted or resorbed.

Acoustic coupling between the inner ear and the swim bladder can be altered by changing the physical relationship between the two systems. The best known mechanism to enhance coupling is in the superorder Ostariophysi, where the swim bladder and inner ear are mechanically coupled by the Weberian ossicles, a series of bones derived from vertebrae. In these species, movement of the swim bladder walls during acoustic stimulation causes anterior and posterior movements of the Weberian ossicles. The anterior-most ossicle, the scaphium makes up lateral and posterior walls of the sinus impar, a fluid-filled space that connects to the transverse canal and then to both sacculi. Ossicular movement causes fluid movements in the sacculi, which "catch" flutes on the saccular otoliths, causing them to "rock", thus stimulating the sensory hair cells on the maculae.

In nonostariophysans (a nontaxonomic term for all fishes without the Weberian ossicles) the sound reradiated from the swim bladder must transverse body tissue to get to the inner ear. In some species this distance is minimized by the swim bladder's having anterior diverticula that contact the auditory region of the cranium. Such diverticula have been shown to enhance auditory sensitivity in a number of species, but data are not sufficient to indicate a quantitative relationship between auditory sensitivity and swim bladder—inner ear distance.

It has been indicated that species with swim bladders relatively distant from the inner ears would be expected to have poorer acoustic sensitivity than species with closer coupling between the organs (we have no idea as to the effects of intervening body tissue, such as muscle and bone, on attentuation characteristics of the reradiated sounds).

The second mechanism for sound detection involves direct acoustic stimulation of the ear without involvement of a swim bladder, even if one is present. In nonostariophysan, the cichlid *Tilapia macrocephala*, it has been suggested that the swim bladder plays little or no auditory role and that the ear in cichlids, and perhaps many nonostariophysans, detect sound only through relative movement of the fish's body and the otoliths. Similar data have been reported for several species of flatfish that do not have swim bladders. Data for goldfish and cod indicate that the swim bladder plays little or no role in detection of lower frequencies and that the inner ear may be stimulated directly (for details: read Popper 1983).

8.2 Buoyancy

The locomotion of fish in the aquatic environment presents problems which are not encountered by terrestrial animals. Since two main body constituents, the proteins and the skeletal material, are denser than water, fish either have to evolve devices to lower their body density or have to provide an additional muscular effort to overcome sinking.

It is important here to define a few terms—The sinking factor of an aquatic animal is 1000 times the ratio of its density to that of the water in which it lives. When the sinking factor is greater than 1000 the animal has negative buoyancy and tends to sink, when it is less than 1000 the animal has positive buoyancy and tends to rise, and when it is 1000 then animal has neutral buoyancy. The early Osteichthyes probably had negative buoyancy in spite of the presence of a lung.

The specific gravity of fresh water is of course 1, and that of sea water is about 1.026. Fish without swim bladders therefore usually have sinking factors between 1030 and 1090. Teleosts with swim bladders usually have sinking factors between 990 and 1010 at the depth to which they are adapted.

Neutral buoyancy is of great advantage to any object in water, a submarine like an animal, since it will help to save energy to avoid sinking or rising. Depending on their composition, fish are usually somewhat denser than the surrounding water because of the skeletal and protein masses (densities around 2–3 and 1.3 g/cm^3, respectively) which overrides the "floating" mass of fat (density 0.9). Thus their overall density is some 5% larger than that of water, and this results in a sinking force of about 5% of their weight, which to overcome in water can be shown to be a formidable energy expenditure. There are several ways in which animals have solved this problem to become neutrally buoyant, and the swim bladder is one such floating device.

An air-filled swim bladder reduces the density of fish without adding much weight, since the density of air is only 0.00125 g/cm^3 at 1 atm, and still only 0.7 g/cm^3 at 700 atm, a pressure experienced in the sea at a depth of 7000 m below the surface. There are, however, problems with such a floating device. (For densities of various fishes, see Alexander 1993; Table 1).

The swim bladder is a primary effector for the regulation of buoyancy in fishes. This buoyancy is in turn a direct gas volume correlate. Buoyancy determines locomotor reflex pattern in aquatic animals.

The swim bladder acts a hydrostatic organ by replacing part of the fish, which is heavier than water, with gas. The specific weight of the fish as a whole is thereby brought toward that of ambient water, and the fish becomes neutrally buoyant.

The function of the swim bladder rests on its ability to maintain a gas space inside a fish and to vary this amount of gas according to changing hydrostatic demands. A fish with a certain amount of gas in the bladder is in buoyancy equilibrium at one depth only. The swim bladder saves locomotory energy for the fish.

Gas secretion implies transport of gases against pressure gradients. The gases are mainly oxygen and carbon dioxide, but inert gases (nitrogen, argon) are involved too. Gas secretion is best developed in closed swim bladders. It may exist in respiratory swim bladders but has not been found in lungs, although it seems to be based on common physiological principles realized in many systems (countercurrent circulation, glycolytic activity, etc.). The volume homeostasis of the swim bladder permits the fish to keep buoyancy, saving energy by not having to swim continuously to avoid sinking. This function probably is of enough biological importance why, during phylogenesis, primitive lung structures of the ancestors to modern fishes evolved into highly specialized swim bladders.

In the living physoclists, gases are continuously lost by diffusion in resorbent mucosa, but this is counterbalanced by gas secretion. The constant volume in a living fish is in a steady state controlled by autonomic nerve reflexes.

If a fish with neutral buoyancy is to be in stable equilibrium when motionless in mid-water, horizontal and the right way up, its centre of gravity should lie directly ventral to its centre of buoyancy. This is seldom or never achieved in practice. Though the swim bladder usually lies immediately ventral to the kidneys, the centre of gravity is nearly always slightly dorsal to the centre of buoyancy.

For a swimming fish to be in equilibrium, that is, to swim at constant speed and depth, certain forces acting on the fish must be equal (see Fig. 6.1). The buoyancy of the fish plus the lift produced by fish's hydrofoils must equal the weight of the fish. Likewise, thrust produced by the fish must equal to drag that results from the forward movement of the fish through the water. In addition moments of forces acting on the fish must be balanced so that it does not roll, pitch or yaw.

Many species of fish use the swim bladder as a hydrostatic organ, thereby attaining neutral buoyancy. Demersal and bathybenthic species possessing swim bladders have been observed to be neutrally buoyant at depth and it is known that their swim bladder is mainly oxygen.

The buoyancy of a fish with a swim bladder changes with depth, and the fish can only maintain its station by the action of its fins within a certain range of depths near the depth at which it has neutral buoyancy. The physical properties of the swim bladder of Cypriniformes reduce the rate of change of buoyancy with depth.

The body of a fish is generally denser than the environment. Fat is one component, aside from gas, which has appreciably lower density than water.

Although several species of elasmobranchs gain most of their buoyancy from fatty livers, in many other fish in this group a significant amount of lift is provided by other tissues other than liver. The electric rays, *Torpedo nobiliana*, are almost neutrally buoyant due to the presence of considerable amounts of electrical tissue of comparatively low density occurring concurrently with a fatty liver.

Although the fat-filled swim bladder of *Latimeria* does not allow wide changes in the relative density of the fish, it nonetheless seems to be efficient in providing neutral buoyancy within a limited depth range beyond which the fish would be able to move only by actively swimming upward or downwards. In this connection, it is noteworthy that the fats present in the swim bladder and tissues of *Latimeria* are mainly lipids and wax esters of low relative density.

When the teleostean swim bladder functions as a hydrostatic organ it usually occupies from 7 to 8% of the volume of a freshwater fish and from 4 to 5% of the volume of a marine fish. If the swim bladder is closed, volume is maintained by the addition or removal of gas-usually mainly oxygen-through the gas gland and the oval, respectively. As gas secretion and resorption proceed relatively slowly, it follows that there must be some restriction on the extent and speed of the changes in depth that a fish can make while still enjoying the advantages of neutral buoyancy.

The cod *Gadus morhua* has a closed, compliant swim bladder which occupies 5% of its volume. The rate of gas resorption was much faster than that of gas secretion, and the difference between two rates increased with pressure. It has been suggested that when a closed swim bladder has a hydrostatic function it may be advantageous if neutral buoyancy is maintained only at upper limits of the diurnal vertical range.

The role of the swim bladder as a buoyancy organ has been questioned but evidence has suggested that some fish which undertake extensive vertical migration have neutral buoyancy only at the top of their vertical range. Swordfish (*Xiphias gladius*) appear to maintain neutral buoyancy only at the top of the vertical range.

The suggestion that fish maintain neutral buoyancy for only part of the day would appear to be compatible with many of the known vertical migrations undertaken by physoclists in shelf seas, and with those of species which are found occasionally (cod) or more frequently (redfish, *Sebastes* sp., blue whiting, *Micromestius poutassou*) in mid-water over great depths. Further, there should be a saving of energy for fish which maintain neutral buoyancy at the top of their vertical range and rest on the bottom for part or all of the remainder of the day.

Various factors are important. The rate of gas secretion, an active biochemical process, increases with temperature in the bluegill sunfish *Lepomis macrochirus*, *Tilapia mossambica* and cod. Gas resorption is primarily dependent on diffusion and a direct effect of temperature would be expected to be small; in cod there was no significant increase in the rate of gas absorption with temperature.

Migrations subject fishes to altered temperature, density and hydrostatic pressure. Effects on the swim bladder are reflexly compensated. Thus, fishes moving from salt into fresh water have to secrete more gas in order to maintaining buoyancy. In addition to reflex mechanisms, endocrine factors may influence swim bladder functions.

The coelacanth, *Latimeria chalumnae*, has a spine which is a notochord made of tough, thick cartilage filled with liquid. The name 'coelacanth' means 'hollow spine'. Perhaps this liquid is hypotonic and aids in buoyancy. In certain teleosts, such as the lumpsucker (*Cyclopterus lumpus*), low density is derived from a cartilaginous skeleton. Neutral buoyancy in the castor-oil fish, *Ruvettus pretiosus*, which lacks a swim bladder, is attained by storage of large amounts of wax esters in the flesh, skeleton and skin. Reduced skin mass in piscivorous fish improves buoyancy and acceleration performance.

Some pelagic deep-sea fishes attain or approach neutral buoyancy by having reduced heavy tissues (e.g., muscle and bone). This strategy is particularly common among fishes of deep (usually > 1000 m) oceanic midwaters (bathypelagic zone) where food supply is sparse. Energetically expensive tissues such as a swim bladder or compounds such as lipids are usually reduced or absent in bathypelagic fishes. In benthopelagic fishes, however, which swim slowly or hover just above deep sea floor and have an access to an increased food supply compared to fishes of deep midwaters, a swim bladder is frequently present as are other energetically costly materials (e.g., lipids).

Acanthonus armatus (Ophidiidae) is a slightly negatively buoyant benthopelagic without a swim bladder that has reduced tissues and components (muscle, bone, brain, gills, lipids) and a massive head. The enlarged cranial cavity filled with dilute (low density) fluid is an adaptation to approach neutral buoyancy and is supported by information on body morphology and behaviour and ecology of fish. The buoyancy adaptation evolved is unlike that known for other benthopelagic fishes. The large head contains most of the heavy components of the body. The strategy of compensating these elements amounts to a localization of positively buoyant material in the cranium which allows the fish to swim or hover in a horizontal position with greater energetic efficiency (see Pelster 1997: Watery tissues).

Notothenioids, large and heavily muscled endemic swim bladderless Antarctic bottom fishes are heavier than sea water. They attain neutral buoyancy on account of specializations of skeletal, integumentary, muscular and digestive systems.

Some scombroids fish permanently inhabit the pelagic environment despite the fact that they are denser than seawater. They overcome their negative buoyancy through continuous swimming with pectoral fins extended as lifting hydrofoils and thus maintain hydrostatic equilibrium. Their swim bladder is often reduced or absent (see details: Horn 1970, 1975).

The forces involved in the buoyancy of the genus *Synodontis* are usually large owing to the well-developed cephalo-nuchal shield possessed by all members of the genus, and the large swim bladder and high levels of body fat needed to counteract this. Negative correlations were found between the total quantities of fat present and the volumes of the swim bladder for several species. Several species of this genus regularly invert and swim on their backs. The position of the food source was found to be the most important factor controlling the orientation of the species.

The accessory respiratory organ in catfishes *Hoplosternum thoracatum* and *Brochis splendens* provides about 75% of the lift required to attain neutral buoyancy whereas the swim bladder provides less than 5%. Both are negatively buoyant, and are continuous but not obligatory air-breathers. The respiratory and hydrostatic functions of the accessory respiratory organ are compatible.

Considering the centre of gravity, in pelagic fish *Cryptoterus*, the centre of gravity lies well posterior to the centre of buoyancy. The seahorse, *Hippocampus brevirostris*, has a swim bladder divided into anterior and posterior sacs. It adjusts the relative sizes of the sacs, and so the relative positions of the centres of gravity and buoyancy, to suit its position. It keeps its body almost horizontal in fast horizontal swimming, and almost vertical (head-up) while resting.

The swim bladder compensates for downdrift caused by the high densities of the skeleton and muscles. In lipid-rich fishes, such as the herring (*Clupea*) the swim bladder volume is correlated with the fat content of the body. In *Tilapia mossambica*, the ability to compensate for pressure changes decreases with age, and presumably with an increase in length and weight.

Gases in the swim bladder reduce the density of fish significantly even at the greatest depths at which fish with swim bladders have been caught.

Negative buoyancy is advantageous in providing frictional forces when a fish rests on the bottom, and has often been secondarily acquired by reduction of the swim bladder, especially in fish exposed to currents.

Control of swim bladder volume to regulate buoyancy, and thus compensate for changes in water velocity, is common in North American freshwater fishes. In this response, buoyancy is usually lowered as water velocity is increased, and is usally raised when still water is encountered. Changes in swim bladder volume by physostomes and physoclists can be considerable and may require upto 8 days to complete.

Lipids assume the primary buoyancy function as the gas-filled swim bladder regresses with age (e.g., in some myctophids). This change may eliminate the physiological constraints imposed by a gas-filled swim bladder and permit the more extensive diel vertical migration of adults.

It has been estimated that the presence of a swim bladder, giving a fish neutral buoyancy, results in a substantial saving of energy at low and moderate swimming speeds, but cannot alter appreciably the maximum speed for short bursts.

The hypothesis that conserving energy, discouraging predation and obtaining energy are the important selective factors affecting the evolution of deep-sea fishes and that these factors have greatly influenced the chemical composition of fish.

To some extent the concept of a swim bladder as a buoyancy organ must be down-graded. The swim bladder has perhaps evolved from an original buoyancy function (though this is by no means sure) to take a number of roles: the enhancement of hearing, sound production, as an oxygen source or as a reservoir of gas for an ancillary system such as the otic bulla. Most of its functions depends to some extent on the maintenance of volume near some optimum level. The loss of the swim bladder in bottom living species is an obvious adaptation to the absence of need for buoyancy; but in these species other functions of the swim bladder are also lost.

Many deep sea fish are found to have other buoyancy mechanisms such as oil-filled swim bladders or they have a high water content and degeneration of the skeleton.

Theoretical considerations show that it is more economical metabolically to maintain buoyancy by a gas-filled swim bladder near the surface rather than to make compensating swimming movements or store lipid. For a vertically migrant lipid storage must be least costly and the safest way of permitting large scale vertical migration. The physostome, though entirely safe from swim bladder bursting, has an insuperable problem to give buoyancy at great depths. It may be that the greater flexibility with an open swim bladder for vertical movement in the short term outweights the adaptive value of an efficient buoyancy organ (at least at a fixed depth) in the life of physoclists.

For details on buoyancy adaptations, important contributions are of Alexander (1990, 1993, see for various calculations), and Phleger (1991, 1998—very comprehensive accounts). Relevant articles (Pelster 1997, 1998) are very significant and informative. The reader is also referred to Chapter 9.

8.3 THE CONNECTION BETWEEN THE SWIM BLADDER AND THE EAR: WEBERIAN APPARATUS

For any fish with a swim bladder, sound could reach the ears by both direct and indirect pathways. The 'indirect' path involves the response of the swim bladder to the sound pressure waveform. As the swim bladder extends and contracts in rarefaction and compression phases, the movement of its wall may be transmitted to the ears via specialized routes (e.g., the Weberian apparatus, rostral swim bladder projections to the ear, etc.) in some species and not so specialized routes in others. Whether or not a given otolith organ is stimulated by the sound pressure waveform depends upon the efficiency of the sound conducting pathway to that organ from the swim bladder. It has been indicated that this mode of hearing could not be involved in sound source localization since the source of the sound, as far as the animal is concerned, is always the swim bladder, thereby stimulating the two ears directly and equally.

Generally, greatest sensitivity is associated with special adaptations for conducting the pressure-dependent movements of the swim bladder wall to the ear (usually the saccule). Poorest sound pressure sensitivity is associated with lack of a swim bladder, or the lack of specialized conduction pathways to the ear.

The best known sound conduction specializations are the Weberian ossicles of the Ostariophysi. These bones link the anterior end of the swim bladder complex to the saccules of both ears. Other well-known adaptations generally include arrangements to bring anterior projections of the swim bladder into the close proximity with the ear, as in the mormyrids where a small bubble of air is associated with the ear itself, and in clupeids, where there is an extension of the swim bladder to the utricle. While it can be assumed that these structural modifications may increase sound pressure sensitivity over their absence, no biomechanical data yet explain the differences in sensitivity among those species having the various types of specialization for pressure detection.

The otolith organs and more peripheral sound-conducting structures have apparently evolved together to determine the frequency range of hearing for a particular species.

The Weberian apparatus in Ostariophysi consists of four or more supporting anterior vertebrae (pars sustentaculum) and three, in some cases four, small moveable, bony parts (ossicles) connected to the internal ear (pars auditum). The Weberian ossicles pivot on the anterior vertebrae (Fig. 10.16).

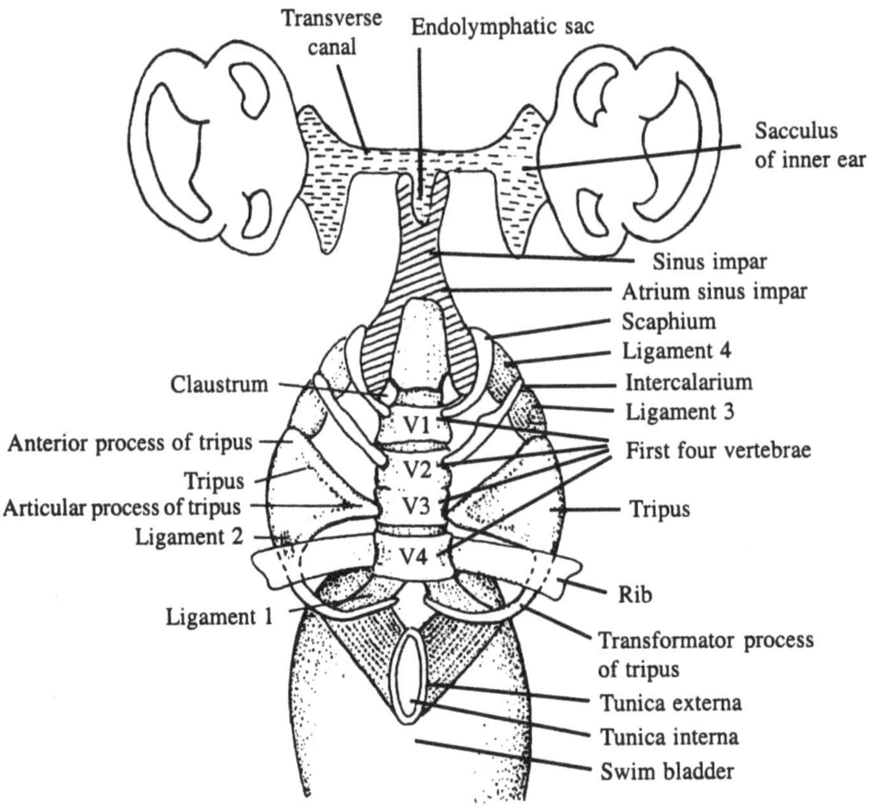

Fig. 10.16 Generalized dorsal view of the peripheral auditory system in the Ostariophysi, showng the swim bladder and its connection to the inner ear via the Weberian ossicles.
***Source:* Popper (1983) See also Chardon and Vandewalle (1991)**

Rademaker et al. (1989) have described the ontogeny of the Weberian apparatus of *Clarias gariepinus*. The morphological and ecological diversity of the Ostariophysi supports the assumption that possession of a Weberian apparatus confers a significant advantage.

According to Rademaker et al. (1989), the Weberian apparatus consists of two symmetrical groups of tiny bones that represent modifications of the lateral processes of the four anterior vertebrae. These ossicles connect the anterior part of the swim bladder to the perilymphatic spaces adjacent to the sensory epithelium of the sacculus. The anterior part of the swim bladder, the camera area Weberiana, has a thick and fibrous tunica externa, which is interrupted by a longitudinal slit along the dorsal midline.

Each chain is composed of four ossicles (from anterior to posterior: scaphium, intercalarium, tripus, and os suspensor) which are generally moveably attached to the vertebral centra. The posterior end of the tripus tapers into a hook-shaped transformator, which is attached to the tunica externa, at the edge of the slit. The scaphium has articular and ascending processes and an anterior spoon-shaped part, the concha. Tripus, intercalarium and scaphium are connected by an interossicular ligament. The part of intercalarium attached to the ligament is called the manubrium. The fourth ossicle, the os suspensor, is connected to the transformator of the tripus by a ligament. A separate supradorsal of the first vertebra, the claustrum, is independent from other ossicles. In catfishes the second, third, and fourth vertebral centra are believed to be completely fused to form the "complex vertebra". However, the fifth vertebral centrum, which is also fused to the complex vertebra, remains recognizable.

The lagena is larger than the sacculus. The left and right labyrinths communicate ventrally by the sinus endolymphaticus weberianus, the posterior end of which projects out into a median, unpaired sinus impair perilymphaticus. The concha of the scaphium forms the posterolateral wall of the sinus impar perilymphaticus.

Pressure fluctuations cause volume changes in the elastic camera area Weberiana of the swim bladder. The volume changes are accommodated by widening and narrowing of the dorsal slit in the tunica externa. Consequently, the lips of the slit slide laterally and medially, carrying along the posterior end of the tripus. The vibrations are transferred to the concha of the scaphium through the chain of ossicles and then to the perilymphatic and endolymphatic fluids.

In many catfishes the Weberian apparatus is modified in response to functional necessities such as increasing density in benthic families or improvement of sound transmission. The reduced swim bladder of clariids is enclosed in a bony capsule, which is formed by the fusion of the enlarged parapophyses of the fourth and fifth vertebrae and by newly formed ventral processes. The capsules are open laterally so that the wall of the swim bladder lies just beneath the skin at the level of the "lateral cutaneous area". The transformator enters the capsule through in the bony wall. The interossicular ligament is short and the intercalarium is either reduced to the small manubrium, or lacking.

Development of the Weberian apparatus of Clarias gariepinus (as given by Rademaker et al. 1989). It is clear that vertebral elements participate in the formation of the skeletal elements. Some particular aspects of development, however, deserve mention. Comparison has been made with data reported for other fishes (*Scardinius erythrophthalmus*, *Galeichthyes felis*, *Carassius auratus*, *Labeo rohita*).

Scaphium: The scaphium, homologous to the basidorsal 1, is in one piece; the concha appears soon, and from the beginning is continuous with the rest of anlage. Others report that it appears to be a separate mesenchymatous ossification, which is independent from the basidorsal.

Claustrum: The claustrum appears late in the one piece. The present data do not support any previous statements (i) homologous to supradorsal 1, (ii) mesenchymatous ossification, and (iii) supradorsal 1 + mesenchymatous ossification.

Tripus: The tripus is homologous to the basiventral of the third vertebra. At no time in their development is it possible to discern a rib distinct from the basidorsal. A distal costal anlage is described (in fry) and even (in an adult) in different fish.

Os suspensor: The os suspensor (very early stage) is clearly homologous to the basiventral of the fourth vertebra and may include its rib, although the latter is never separated from the main ossicle as observed in other fish fry. Later on, the os suspensor resembles the adult catfish-like structures. The laterodorsal part of the basiventral is called in the adult catfishes a "transverse process" or "parapophysis" and appears later. The dorsal wall of the capsule of the swim bladder is made of these "fourth parapophyses" associated with the fifth ones. The "ventral process" develops late, without any conspicuous anlage.

The most important problem in the development of the Weberian apparatus of *Clarias gariepinus* is the apparent absence of an intercalarium.

Further development has been interpreted as: (i) The second basidorsal never becomes an intercalarium, the so-called intercalarium of *Clarias* is a part of the tripus; (ii) the manbrium is an ossification of the interossicular ligament as in goldfish and the major carp; (iii) the manubrium is a reduced second basidorsal partially fused with the tripus (this hypothesis was not applicable).

It has been concluded that the claustrum, scaphium, tripus, and os suspensor appear to form one single piece. The scaphium is a basidorsal of the first vertebra, the tripus is a basiventral of the third vertebra, and the os suspensor is a basiventral of the fourth vertebra. The intercalarium may not be a transformed second basidorsal, but an anterior part of the tripus.

Vandewalle et al. (1990) reported that the Weberian ossicles of *Barbus barbus* develop from the beginning in one single piece, contrary to several other fishes where several distinct primordia contribute to their formation.

It consists not only of two symmetrical chains of four ossicles (scaphium surmounted by a clustrum, intercalarium, tripus and os suspensor linked by inter ossicular ligament) but also of modified surrounding organs; the anterior chamber of swim bladder (camera area Weberiana), anterior vertebrae, the labyrinths. Modifications of the latters are particularly important.

The ontogeny of the Weberian apparatus in different Cyprinidae shows that they are homologous to basidorsal and basiventrals of the first four vertebrae, that is consistent with the observations on catfishes and on the Characidae *Brycon* sp. Recently Chardon and Vandewalle (1997) presented a synthesis of anatomical, paleontological and ontogenetic knowledge of the Weberian apparatus.

According to Popper (1983), although data are limited, the Ostariophysi, as a group, appear to have considerable better auditory sensitivity than many nonostariophysan species without Weberian ossicles. It is fairly certain that Weberian ossicles function in audition, but there are virtually no data to indicate whether they passively and precisely transfer energy from the swim bladder to the ear, or whether they alter the acoustic signal in some way. It has been suggested that the tunica externa could provide some high-pass filtering to incoming signals, and to some degree, affect sound in the same way as do the middle ear bones of mammals. Further it has been pointed out that there is a potential lever action in the ossicles, particularly in the tripus, with the posterior arm of the bone farther from the fulcrum (the point of attachment to the third vertebrae) than is the anterior arm. Consequently, the ossicles could provide some signal amplication. Other questions centre on the damping of the ossicles and their acting as filters for sound. Perhaps the most intriguing questions relate to interspecific variation in the structure of the ossicles and the effect of this variation on auditory characteristics of different species.

The claustrum and intercalarium tend to be lost from the pars auditum independently in members of several families, and there is a distinct tendency in bottom-dwelling forms for the normally two-chambered swim bladder to be reduced and enclosed in a bony capsule. There is also a tendency for centra of the Weberian vertebrae to ankylose and for additional vertebrae to be incorporated into the apparatus.

The Weberian apparatus of *Astroblepus pholeter* (Astroblepidae; cave-dwelling and microphalm silurid) sensu lato differs from *A. grixalvii*, the type species of the genus, and from all anatomically known catfishes, by one peculiarity of great interest. The two half bladders (formed by a reduction of a normal swim bladder) are, as for *A. grixalvii*, separated and enclosed in a bony capsule built by the transverse processes of the complex vertebra. But, in *A. pholeter* the side walls of the bladder undergo a dome-shaped ossification, but keep free the bony capsules. The ossified parts of the swim bladder act as a mammalian tympanum and greatly improve the reception of vibrations by the Weberian apparatus, by eliminating the elastic compliance normally introduced by the elastic wall of the unossified swim bladder.

The Weberian apparatus of Cypriniformes incorporates a mechanism which tends to prevent its being out of action by depth changes. In many Cypriniformes, an area of the body lateral to swim bladder lacks muscle. This probably increases the sensitivity of the Weberian apparatus to sounds near the resonant frequency of the swim bladder. Many Cypriniformes have enlarged parapophyses which keep the surrounding tissues clear of the parts of the swim bladder which move with the greatest amplitudes under the influence of the sound.

8.4 Sound Production

Musculature, extrinsic or intrinsic to the swim bladder wall, is reported in fish which are known to make sounds. Presumably the swim bladder serves as a resonator. The role of resonance is far from clear. Interference with swim bladder by damage, deflation, or filling the water reduces the sound output (see Chapter 16).

9. CONTROL OF SWIM BLADDER HYDROSTATIC FUNCTION

As given in preceding account, many fish possess a gas-filled swim bladder as a hydrostatic organ. To retain neutral buoyancy during vertical migration in the face of varying hydrostatic pressure, the swim bladder volume must be kept constant by depositing gas when swimming down and by resorbing gas when ascending. Thus, to maintain neutral buoyancy, a sophisticated control system is necessary. Stretch receptors provide afferent information from the bladder wall. The efferent loop consists of at least two separate elements: control of the connection between the resorbing and secretory part of the bladder, i.e., control of gas resorption, and control of secretory activity of the bladder.

Opening and closing of the connection between the secretory and the resorbing part of the bladder appear to be under nervous control. In *Gadus morhua* and *Pollachius virens*, vagal stimulation caused opening of the oval, probably due to β-adrenergic relaxation of circular muscles and α-adrenergic contraction of radial muscles. Application of acetylcholine resulted in a contraction of the circular muscles, but this appears not to be mediated by cholinergic nerves; it has been suggested that closure of the oval may occur because of lack of opening impulses.

The secretory activity of the bladder can be modified by a change in acid production and acid release by gas gland cells. The acid is responsible for the initiation of the so-called single concentrating effect, the first step in gas deposition. A decrease in the single concentrating effect will reduce the gas partial pressure increase and thus the partial pressure gradient towards the swim bladder lumen, reducing the rate of gas deposition. The rate of gas deposition may also depend on swim bladder blood perfusion. Recent studies have revealed a significant correlation between the rate of gas deposition into the swim bladder and the blood perfusion of the bladder. Furthermore, under hypoxic conditions both were reduced, indicating that the regulation of the swim bladder perfusion is involved in the regulation of the rate of gas deposition.

Catecholamines may be important for the control of swim bladder perfusion, as a vasoconstriction in swim bladder vessels has been observed during stimulation of the splanchnic and vagus nerves, mediated by α-adrenergic receptors. Injection of catecholamines, however, did not affect the rate of gas deposition. Vasodilatory nerves have not been described in the swim bladder.

Recent studies on the adrenergic control of swim bladder perfusion in the European eel, *Anguilla anguilla*, have indicated that α-and β-adrenergically controlled resistance vessels are located close to the rete mirabile, probably at the arterial entrance into the rete. These vessels control perfusion of the rete mirabile and of the swim bladder and are thus involved in the control of gas deposition into the swim bladder.

The physiological significance of adrenergic control has been further discussed (see Pelster 1994).

Ewart and Driedzic (1990) reported that lactate production by the cod gas gland may be enhanced by both adrenaline and cholinergic stimulation, a finding that is consistent with the innervation pattern of the tissue. The tissue displays very high levels of anaerobic relative to aerobic enzymes, but this profile is not consistent with the actual rate of lactate production, which is quite low.

10. SWIM BLADDER AND HYDROSTATIC EQUILIBRIUM

Fish are believed to be hydrostatically unstable, with the centre of buoyancy both below and anterior or posterior to the centre of mass. A fish balances these destabilizing hydrostatic moments with forces generated by propulsors. In fish with swim bladders, the requisite thrust forces are commonly produced using the median and paired fins rather than body and caudal fin undulation. The median and paired fins are also the major propulsors for routine swimming for many fishes, dominating time-activity budgets. However, the functional characteristics of median and paired fin propulsion are poorly known (also see Chapter 6).

Webb and Weihs (1994) discussed this subject in four species of teleostean fishes with swim bladders: elongated-bodied European eel (*Anguilla anguilla*), the fusiform yellow perch (*Perca flavescens*), and two gibbose species, rock bass (*Ambloplites rupestris*) and bluegill (*Lepomis macrochirus*). Based on the findings, in general, it has been suggested that fish with swim bladders can be at or close to neutral hydrostatic equilibrium but any disturbance will be destabilizing. This probably explains continuous fin-beating of fish in apparently still water. In this context, the reader is referred to Alexander (1967, 1993), Blake (1983), Videler (1993) and Bone et al. (1995).

CHAPTER 11

Excretory System; Acid-Base Regulation; Ion Exchange

1. INTRODUCTION

The occurrence of osmolytes (osmotic solutes) in the intracellular fluids of diverse groups of organisms indicates an independent, evolutionary "discovery" of an effective way of dealing with the problems of water stress that has been faced numerous times, by diverse organisms.

The distribution of osmolyte systems in reference to fishes has been discussed in various reviews. In the animal kingdom, the convergent evolutionary trends in the occurrence of the organic osmolytes suggest an important point: the bases for selecting these osmolytes are certain ubiquitous features of the chemistry of water-solute-macromolecule interactions that apply in all living organisms (see Somero 1987).

Regulation of acid-base status is an imperative task for maintenance of homeostasis in living animals. The mechanisms involved in acid-base regulation do not differ among and within vertebrate classes, but the extent to which certain processes are, or can be, utilized may be vastly different.

It is clear from the extant literature that various fish groups face chronic osmoregulatory problems that depend on the surrounding salinity. Freshwater teleosts face chronic, potential hypervolemia and hyponatremia (hypochloremia). Significant numbers of species of teleosts are euryhaline. Sharks, skates and rays are slightly hyperosmotic to sea water. Elasmobranchs face chronic hypernatremia (hyperchloremia) (read Evans 1990).

The osmoregulatory apparatus of advanced aquatic animals may be regarded as a concentric, three compartment systems; the outer ring comprising the environment, the middle the extra-cellular fluid, and the inner sphere the tissues. The organism must first of all be capable of regulating the composition of the extracellular fluid different from that of the external environment by using specialized osmoregulatory organs such as the gills and kidneys. Secondly, the membranes of tissue cells must maintain as intracellular fluid composition that is radically different from that of the extracellular fluid.

The evolutionary shift from ammonia excretion in primary aquatic vertebrates (fishes and some Amphibia) to urea excretion, and subsequently uric acid is a classic scenario in the history of the vertebrates, despite the well-known exceptions such as uricotelic amphibians and ammonotelic crocodilians.

The vertebrate kidney plays an important role in maintaining homeostasis, regulating body fluid volumes and salt concentration, and eliminating nitrogenous wastes. The ultrafiltration of the plasma in the glomerulus is the first step in the formation of urine by almost all kidneys. The almost universal presence and great elaboration of the glomerulus in vertebrate kidneys reflects its value in renal function. Glomerular filtration is therefore a prominent and vital process and has been the focus of many reviews. The magnitude of glomerular filtration and variability in its rate may have significant influences on the transport functions of the rest of the nephron and therefore may greatly affect the composition as well as the volume of the final urine. Despite controversy concerning the environment in which the glomerulus originated (whether in fresh or salt water), it is usually considered to have evolved largely for the purpose of volume regulation although an opposing view also prevails. Indeed, filtration of plasma is a mechanism admirably suited for eliminating volume loads quickly and efficiently. And, since the filtration of plasma relies on the hydrostatic pressure in the arterial system which is maintained for other purposes, producing the filtrate is not energetically expensive. However, if the solutes associated with the filtered water need to be conserved then there may be significant costs associated with the reabsorption of these substances. Nevertheless, the expense of reabsorbing solutes from a filtrate would be considerably less than the cost of forming the primary urine by active solute-linked fluid secretion and subsequently reabsorbing solutes. Moreover, filtration of the plasma may be rapid and may be quickly altered, providing not only a large capacity for eliminating volume loads quickly, but also growing for the precise regulation of excretory water loss.

Glomerular filtration is also an important mechanism for the clearance of many endogenous substances from the plasma (e.g., urea) and for the nonspecific elimination of many potentially toxic exogenous substances for which no transport mechanism exists. Even for substances which are secreted by specific transport mechanisms in the renal tubules, the quantity filtered by the glomerulus may contribute a significant fraction to the total amount excreted (Yokota et al. 1985).

Various facets of the subject under purview are covered in a number of publications (for example: Hickman and Trump 1969; Watts and Watts 1974; Hentschel 1977, 1979; Evans 1980, 1986, 1987, 1990, 1993, 1995; Dantzler 1982, 1985; Heisler 1984, 1986, 1988, 1990; Henderson et al. 1985; King and Goldstein 1985; Braun and Dantzler 1987; Brown and Brown 1987; Larsen and Bern 1987; Nishimura 1987; Shuttleworth 1987; Randall and Wright 1989; Wright et al. 1989; Avella and Bornacin 1990; Perry and Laurent 1990; Wood and Gross 1990; Potts and Hedges 1991; Vilella et al. 1991; Walsh and Henry 1991; Wood 1991, 1993; Wright 1991; Kormanik 1992; Mommsen and Walsh 1992; Takei 1993; Zono et al. 1993; Anderson and Walsh 1995; Claiborne 1998; Karnaky 1998; Grosell and Jensen 1999; Fines et al. 2001). Dantzler (1995) focused on renal transport of organic acids and bases in nonmammalian vertebrates.

Different effector organs and tissues, such as gills, kidneys, gastrointestinal tract and skin contribute to the process of adapting to the environment.

Fish gills are multifunctional organs which are involved in acid-base regulation, nitrogen excretion, respiratory gas exchange, and monovalent ion regulations. The fish gill can be considered as the sum of two epithelia, the respiratory and nonrespiratory.

2. KIDNEY: STRUCTURE AND FUNCTION

Considerable attention has been paid to kidney structure, especially its development, in considering vertebrate relationships. In fact, few vertebrate organs have so extensive a developmental and morphological data base. Those data have been treated in terms of physiological and ecological correlates, and in terms of evolutionary relationships, including the origins of major vertebrate groups.

Excretory system in the genus *Notopterus* is quite peculiar. The reduction in size of the coelom has resulted in certain interesting modifications of the alimentary canal, swim bladder, kidneys and gonads. An anomalous feature has also been observed in the organization of the kidney ducts and urinary bladder. The ureters lie deep within the mesonephros (the posterior kidney). The common ureter, runs forward superficially over the entire length of the left mesonephros, enters the swim bladder and after piercing the wall opens into a long, curved, tubular urinary bladder which extends as far as the middle of the body cavity.

Comparison of vertebrate nephrons reveals a basic sequence of tubular segments: glomerulus → neck segment → proximal tubular—intermediate segment → distal tubule → collecting tubule and duct system. Evolution of vertebrate renal anatomy appears quite conservative when compared, for example, to evolution of respiratory and cardiovascular systems in vertebrates. Major anatomical changes in vertebrates kidneys separate those of birds and mammals from kidneys of lower vertebrates. General increase in animal size from fish to mammals is reflected by an increase in total number of nephrons per kidney, rather than by constant change in tubular dimensions.

Comparative renal physiology has a rich history with a record of many major advances attributed to work on non-mammalian species. There are a number of distinguishing features of kidney morphology among vertebrate groups. These are: (i) glomerular size and complexity varies markedly between groups, correlated at least in part with the habitat of the organism; (ii) the segmental morphology of the nephron varies primarily with regard to the presence or absence of a mammalian-like loop of Henle-collecting duct parallel tubule arrangement, and (iii) presence or absence of a renal portal system which can perfuse tubule elements of the kidney with a venous flow from the animal's hindquarters (see Laverty and Clark 1989) (Fig. 11.1).

2.1 Agnatha

In adult myxinoids the pronephros is a paired organ, situated close to the atrium of the branchial heart on the left side of the body, and the portal heart proper on the right side. Tubules, central mass, and glomus are distinguished (Fig. 11.2). The pronephros of *Myxine* is not excretory in the usual sense of word; it is probably a lymphoid tissue. The mesonephros (Fig. 11.3) is a typical vertebrate kidney, one of the simplest urine-forming system but remarkable for the large size of the Malpighian corpuscles and the lack of nephric tubules. The long ureters (archinephric ducts) may have taken over some, but not all, of the functions, attributed to the kidney tubules of other vertebrates; they also serve as urinary bladder, being able to distend with urine. Each archinephric duct drains 30–40 large glomeruli. Glomeruli and archinephric ducts are connected by very short neck segments. However, the morphology of the archinephric duct epithelium suggests similarity with the proximal convoluted tubule of mammals (Raguse-Degener et al. 1980; see also Fänge 1963; Rankin 1997; Karnaky 1998). *Myxine* probably belongs to the ureothelic vertebrates.

Youson (1985) described the excretory system in lamprey (pronephros, larval opisthonephros and adult opisthonephros) (see Fig. 11.1). The adult lamprey has a highly evolved, unique and specialized kidney which is characterized by the presence of single renal corpuscle, regional

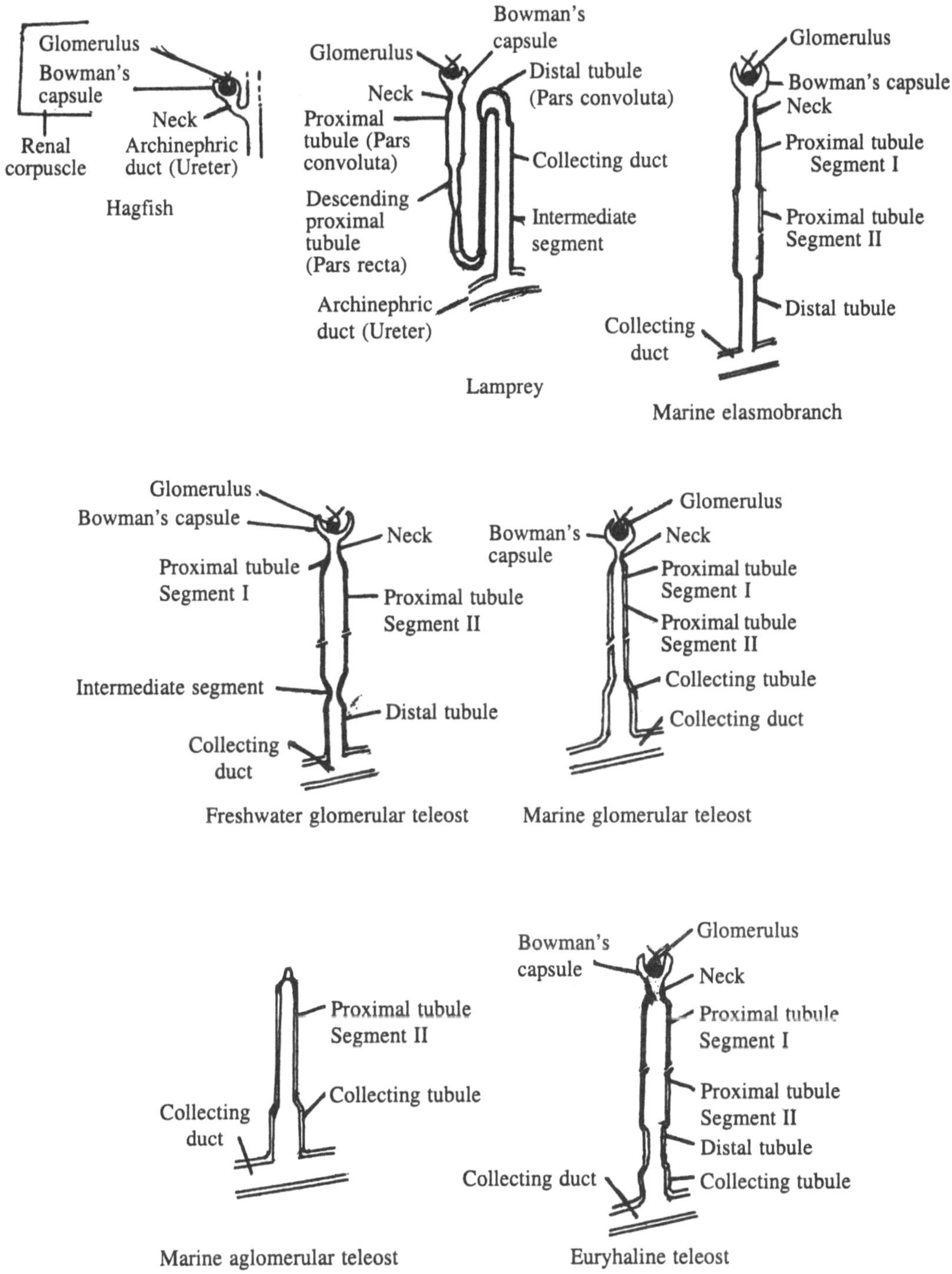

Fig. 11.1 Schematic drawings of the nephrons from various fish groups. The various tubule segments for each type are labelled but folding patterns (except for loop structure present in lampreys) and actual sizes of individual nephrons are not represented. Breaks in nephron segments indicate relatively greater segment lengths.
***Sources:* Dantzler (1985) and Laverty and Clark (1989)**

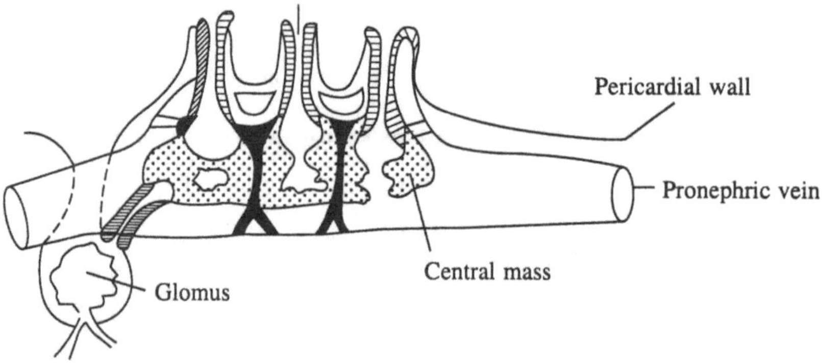

Fig. 11.2 ***Myxine*: Diagram of the structure of the pronephros.**
***Source:* Fänge (1963)**

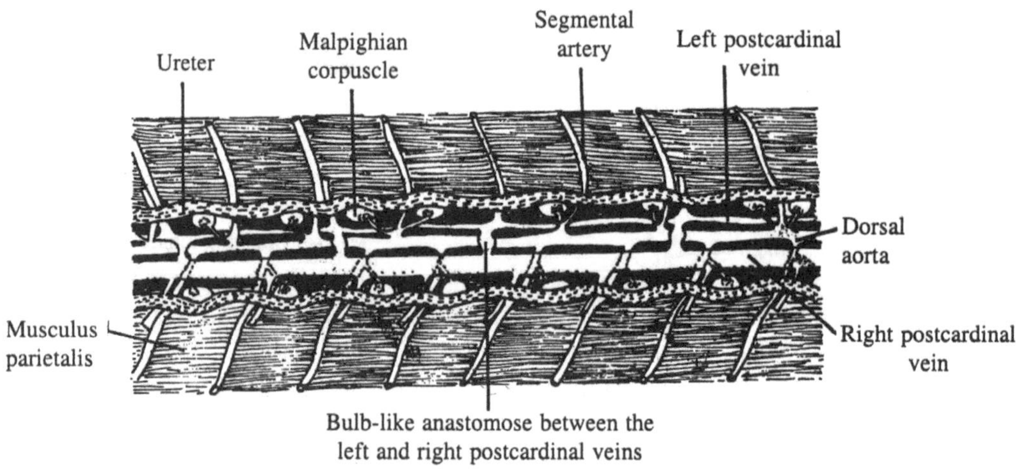

Fig. 11.3 ***Myxine*: Mesonephros as seen during dissection and after removal of the intestine (schematic drawing of living specimen; the mesenteric vessels are not shown).**
***Source:* Fänge (1963)**

differentiation of the tubules, and the absence of a renal portal system. The degree of elaboration of the kidney is variable among lamprey species. In general, definitive kindey of nonparasitic species show similarities to the ammocoete kidney. On the other hand, the kidneys of anadromous species show a high degree of specialization which has been equated with their ability to osmoregulate in salt -and freshwater. Features of the kidney which are important to this mechanism are the chloride-like cells of the distal tubule, the extensive vascular surface of the glomus, and perhaps, the parallel arrangement and looping of the tubules in the ventral part of the kidney. Unlike the opisthonephroi of larval lampreys and many gnathostome fishes, the opisthonephroi are not a storage site for haemopoietic tissue. However, cells lining the sinusoids play a major role in erthyrophagocytosis. The reader is also referred to Rankin (1997) and Karnaky (1998).

2.2 Chondrichthyes

The elasmobranch kidney is significant not only for its unique structure among vertebrates but also for its function of reabsorbing nearly all urea (and trimethylamine oxide) filtered through the glomerulus. Urea is maintained in high plasma and tissue concentrations and is primarily responsible for the elevated osmolarity of body fluids over that of sea water. Although it has been postulated that active urea transport from tubular fluid takes place in the elasmobranch kidney, recently it has been proposed that passive reabsorption of urea may occur what appears to be a unique arrangement of nephron segments (read Smith and Wright 1999). Further, numerous comparative observations have reported the absence of a renin-angiotensin system. These functional peculiarities have their morphological counterpart in the complex arrangement of the nephrons forming the elasmobranch kidney, as established so far by light microscopic investigations, in the absence of granular juxtaglomerular cells, which form the morphological basis for renin synthesis, storage and secretion. For urinary system, see Lacy and Reale (1999).

The kidney construction in the elasmobranch (e.g., *Raja*) is described here:

(i) Configuration of elasmobranch nephron—Each nephron is composed of a large renal corpuscle and of an extremely long highly convoluted tubule emptying into the collecting duct system. The segments of tubule and their sequence, beginning with the renal corpuscle are (1) the neck segment with ciliated cells; (2) the proximal segment showing epithelial cells with a brush border; (3) the intermediate segment; (4) the distal segment emptying into (5) the collecting duct. The path of the tubule has been schematically shown in Figs. 11.4 and 11.5.

The presence of a renal countercurrent system in elasmobranchs has been demonstrated. The tubules and capillaries inside the bundles form renal countercurrent system in these fish. The possible participation of this part of the tubular bundle in the urea reabsorption has been indicated.

(ii) The vascular pole of the elasmobranch corpuscle—In mammals, the justaglomerular apparatus (JGA) consists of afferent glomerular vessels, of the macula densa of the distal tubule and of the cells of extracellular mesangium (Goormaghitigh cells). Usually, the smooth muscle cells of the afferent arteriole contain granules containing renin and angiotensin II as well as other components of the renin-angiotensin system.

Neither a macula densa nor smooth muscle cells with granules could be recognized by light microscopy. Further, an extraglomerular mesangium should occur only in mammals.

The absence of a JGA in the elasmobranch kidney seemed to be in agreement with the lack of renin activity from these fishes. Recently, experiments have indicated that in an elasmobranch and a cyclostome "renin-like materials" are present. In addition, it has been demonstrated that at the vascular pole of the renal corpuscles, structures can be identified which are morphologically comparable to those composing the vertebrate JGA. Thus, the assertions that the elasmobranch kidney lacks a renin-angiotensin system and a JGA are no longer tenable. Even a conclusion has been drawn that "renin activity occurs in all classes of vertebrates". (See Lacy and Reale, 1999).

There are certain recent findings. The flagellar cells are also reported in the renal tubules. Results suggest a vary efficient, undulating beat of flagellar ribbons which promotes the flow of glomerular filterate down the long tubule of the elasmobranch kidney. This mechanism should be added to those already proposed to explain the movement of fluid along the renal tubule of the elasmobranchs: the low residual filteration pressure and the contraction of the epithelial cells both in Bowman's capsule and in the tubule. For special interest on flagellar ribbon beat cycle, see Rivers et al. (1995).

Granular epithelial cells at the vascular pole of the renal corpuscle, peripolar cells, have been found in the kidneys of some marine elasmobranch fish species. Nothing is known about the function of the peripolar cells although some hypotheses have been advanced.

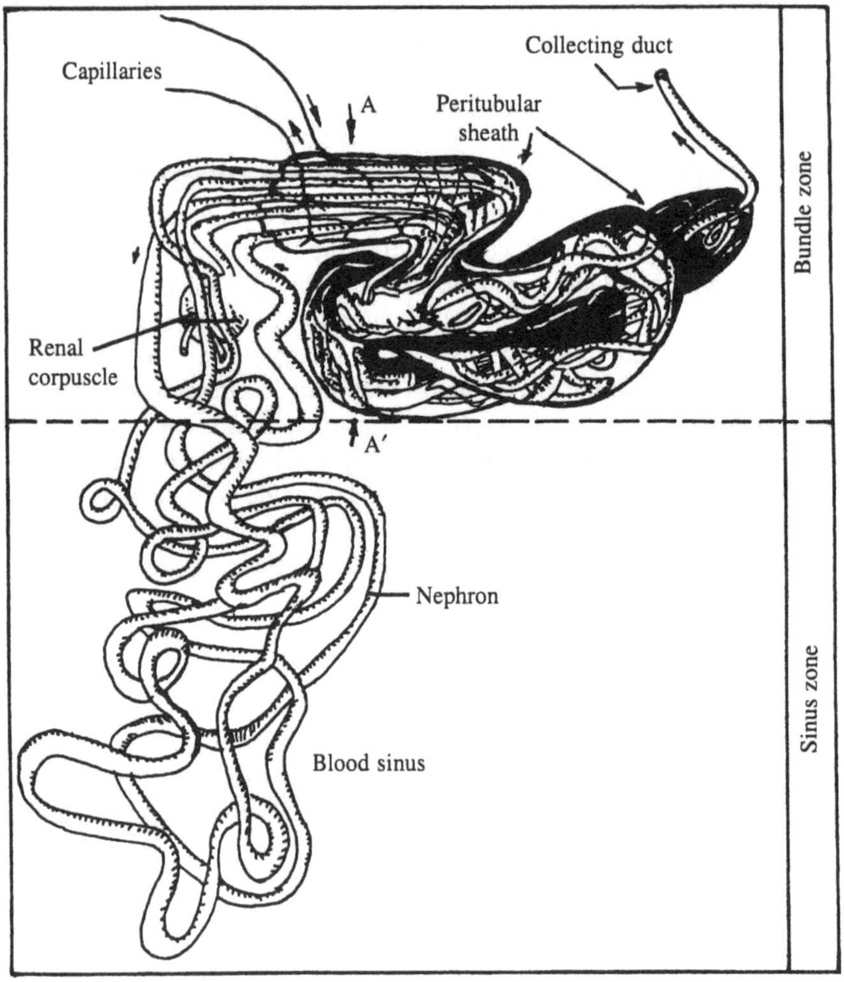

Fig. 11.4 Drawing of the nephron configuration in the bundle and sinus zones of the skate kidney. The drawing is based on serial section reconstruction but is highly simplified. Arrows next to renal corpuscle indicate the direction of the urine flow in different segments of the tubule and in the late distal tubule. The peritubular sheath and enclosed nephron segments have an initial straight portion near renal corpuscle followed by a convoluted portion. Five tubules (each limb of two loops, loop I and III, and the distal tubule) enter and exit the sheath close to the renal corpuscle. One tubule (late distal) exits the sheath at the end of the convoluted portion of the bundle. Capillaries also enter and exit (arrows) the peritubular sheath near the renal corpuscle and form an anastomotic network around the tubules along the entire length of the bundle. A-A' defines a cross section plane through the tubular bundle. In the sinus zone the tubules of two ventrally located loops (Loop II and IV) are highly intermingled.
Source: **Lacy and Reale (1985b)**

Findings on the elasmobranch kidney support the possibility of a close relationship between kallikrein-kinin and renin-angiotensin systems. To date, demonstration to reveal any renin or kallikrein activity in the elasmobranch kidney by use of immunohistochemistry using mammalian antibodies has failed.

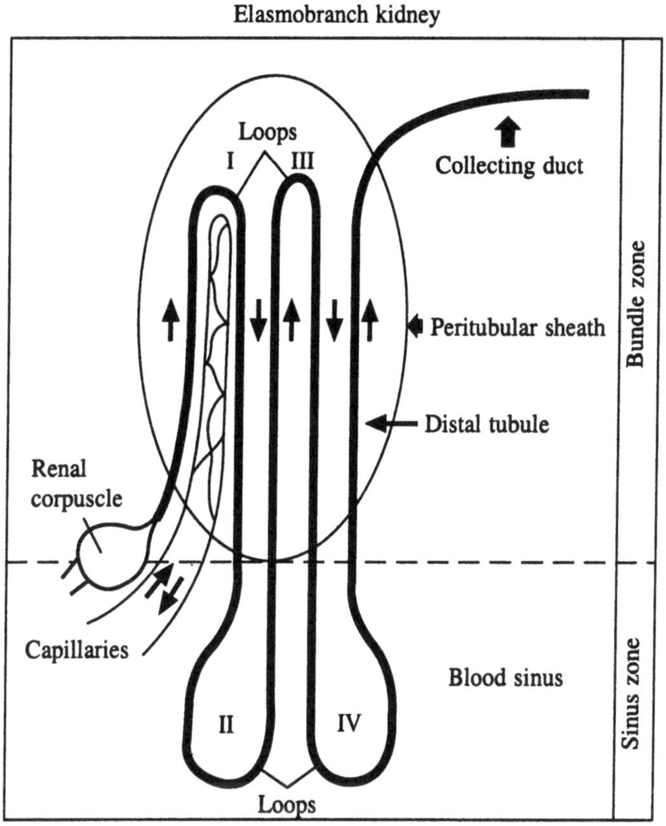

Fig. 11.5 Path of a tubule in the kidney of the *Raja clavata*. I-IV, the four loops of the nephrons. Loops II, III and the distal tubule, as well as some blood capillaries, form a bundle which is surrounded by the peritubular sheath. All the bundles lie in the bundle zone. Loops II and IV, within the blood sinuses, form the sinus zone. *Source:* Reale and Lacy (1987)

The potential importance of peripolar cells to renal physiology is not clear.

Certain features associated with kidney of elasmobranchs are of interest. In both suborders of elasmobranchs (Batoidea: skates and rays, Selachii: sharks), there are marked sexual differences in the kidney. In males, the anterior part of the kidney comes into service of the genital system while in the females the anterior part is either reduced or consists of rudimentary nephrons. The kidneys of both sharks and skates lie in a retroperitoneal position. Blood supply to the kidneys of both skate and shark is provided by branches (renals) of dorsal aorta and a venous portal system from the tail and body wall.

Great structural variations are present in the kidney depending on the species, sex, and age as well as individual variation. Species differences are based on renal corpuscles and tubular bundles; mainly position of urinary pole to the renal corpuscle. Another species difference concerns the kidney. Unrelated to the species differences between the kidneys of the skate and the shark, the compartmentalization of the parenchyma and the path of the nephrons are the same in these two elasmobranchs. For alkaline glands, see Grabowski et al. (1999).

The elasmobranch kidney is unique in several respects relative to that of other lower aquatic vertebrates. The Agnatha (*Myxine*) have renal corpuscles connected to a long archinephric duct by very short tubules. In the sea lamprey (Cyclostomata) the nephron is significantly more complex than in the hagfish (*Myxine*) and consists of 5 segments; neck, proximal, intermediate, distal, and collecting duct. There is, however, no compartmentalization in those kidney, as observed in those of *Raja* and *Squalus*.

Studies of teleost nephrons also have not reported a strict compartmentalization of the kidney. The elasmobranch thus appears to be unique among lower marine vertebrates in that the kidney is divided into anatomically distinct zones in which there is strict segregation of nephron segments. The "special segment" of the nephron for nearly complete reabsorption of urea does not exist. Rather, "special devices" are present: "the tubular bundles" which probably contain the countercurrent multiplier which may facilitate urea reabsorption and the peritubular sheath which probably forms a selective barrier to solute and water movement and thus segregates the microenvironment bathing the tubules and capillaries of the bundles from that outside the peritubular sheath (see Lacy and Reale 1985, 1989, 1995; Reale and Lacy 1987; Lacy et al. 1989, Hazon et al. 1997). Hentschel (1991) has provided conclusive evidence for morphological homology of elasmobranch nephron segments and collecting tubule-collecting duct system with those of other vertebrates. The origin and nature of the central vessel and the bundle sheath is clarified. Those specific structures of marine elasmobranch kidney supposedly are of great functional relevance for the renal countercurrent system that in turn is essential for ion-and osmoregulation.

2.3 Osteichthyes

The kidney/urinary bladder (renal complex) of teleosts has major hydromineral regulatory duties. The urinary bladder of teleosts is not homologous to the urinary bladder of higher vertebrates.

It is remarkable in teleosts that the difference among species shows not only the difference of the internal structure of the kidney but also the extreme difference of the external shape (Fig. 11.6). The mesonephros is the kidney which has chiefly renal functions in the adult. The head kidney has no renal function. The structure of the head-kidney in teleosts consists entirely of lymphoid tissue and has no nephron. The interrenal tissue is located on the wall of the cardinal vein passing through the head-kidney. In *Fugo chrysops*, the distinction between the kidney and the head-kidney is not clear. A pair of ureters leave the posterior end, and they join in front of the small urinary bladder. Corpuscles of Stannius lie on the dorsal side of the junction of ureters. In typical fresh-water fishes, the heterotopic thyroid follicles are often found in the kidney or the head-kidney. In general, the active and fast swimming fishes have more numerous and better developed glomeruli than the sluggish fish have.

Beyenbach and Baustian (1989), Beyenbach and Dantzler (1990), Baustian and Beyenbach (1993), Beyenbach (1995) and Beyenbach and Liu (1996) have provided a wealth of information on the comparative physiology of the proximal tubule. As the first major tubule along the length of the nephron, the proximal tubule is invariably present in all vertebrate kidney. It performs most renal transport functions in glomerular kidneys, and virtually all transport functions in aglomerular kidneys. Because of its ubiquitous presence in vertebrate kidney and its central role in extracellular fluid homeostasis of glomerular vertebrates, the proximal tubule has received considerable attention from biologists and physiologists.

Present-day studies of nonmammalian kidneys have a peripheral status in renal physiology, and nonmammalian renal systems are typically interpreted in terms of mammalian physiology.

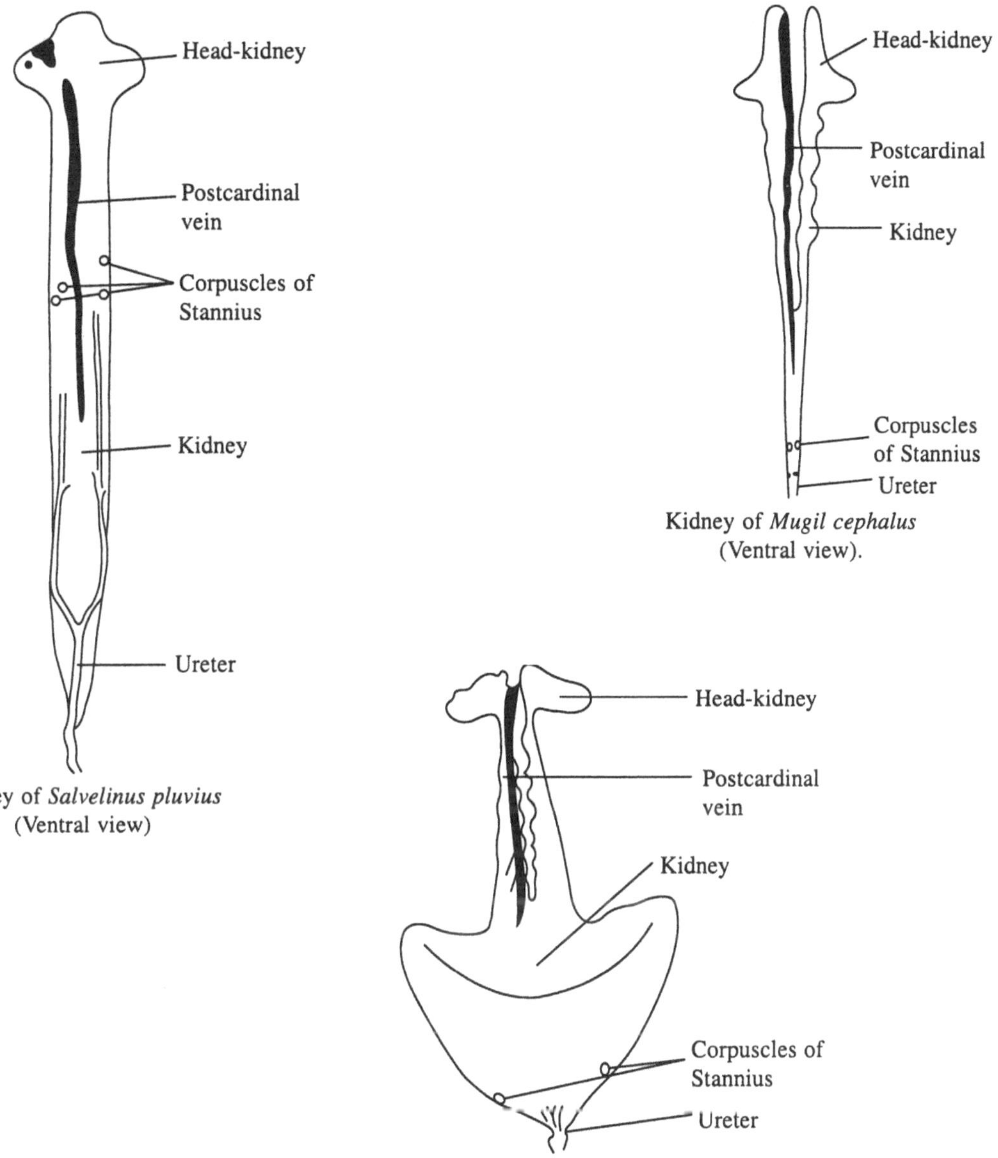

Kidney of *Salvelinus pluvius* (Ventral view)

Kidney of *Mugil cephalus* (Ventral view).

Kidney of *Parasilurus asotus* (Ventral view).

Fig. 11.6 Diagram of the kidney shape from ventral view in the teleostean fishes: *Salvelinus pluvius* (freshwater fish) (Type I). *Parasilurus asotus* (freshwater fish) (Type II), *Mugil cephalus* (marine fish) (Type III). Types after M. Ogawa (1961), Comparative study of the external shape of the teleostean kidney with relation to phylogeny. Sc. Rep. T.K.D. Sect. B., Vol. 10, No. 149, pp. 61–88.

2.3.1 Aglomerular Urine Formation in Fish (Overview)

Of the 45,000 vertebrate species existing today, there is a vanishingly small group, some 30 species, which are unique in that they produce urine without the aid of glomerular filteration. These species are anatomically aglomerular, i.e., lacking glomeruli altogether. These anatomically aglomerular vertebrates are all teleost fish. They are randomly distributed among unrelated phyletic groups in 6 families and 13 genera. Most are marine, but there are also freshwater aglomerular fish. In addition to the aglomerular nephron, these fish are among the most exotic, highly specialized, and at the same time the least active fish known (Fig. 11.7).

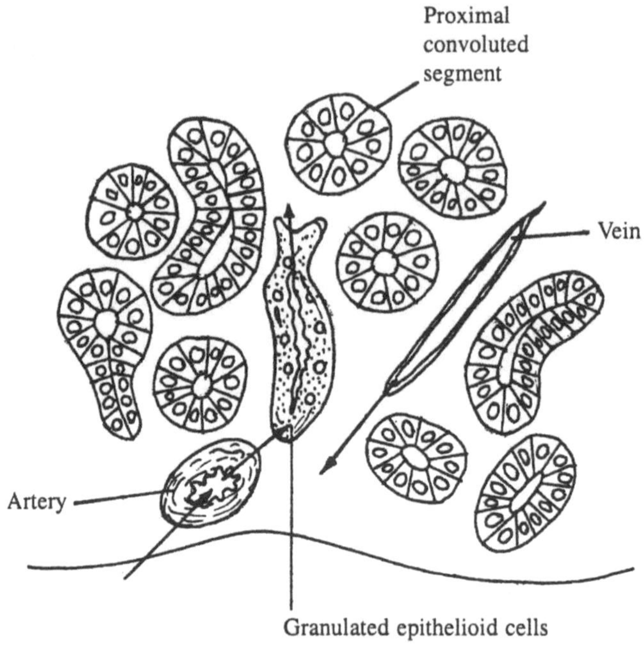

Fig. 11.7 Schematic illustration of aglomerular kidney of the toadfish. Arrows indicate direction of blood flow.

The emphasis on anatomically aglomerular renal tubules is made here to distinguish it from functionally aglomerular renal tubules. Functionally aglomerular renal tubules do have viable glomeruli, but they function intermittently according to physiological needs. Like anatomically aglomerular species, functional aglomerular species are also teleost fish.

The presence of anatomically aglomerular fish in both seawater and fresh water, and the tolerance of dilute media by normally marine aglomerular fish indicate that glomeruli and glomerular filteration are not anatomical or physiological prerequisites for life in either concentrating or diluting habitats. In the case of fish, this is not surprising since there are extrarenal as well as renal mechanisms involved in salt and water balance. Among the extrarenal mechanisms are transport across the gill and the gut. Other anatomical and functional properties of fish, which have not received as much attention, e.g., the permeability of body surface, can play a critical role as well. A low permeability to salt and water would be advantageous in either sea water or fresh water. Indeed, the skin of aglomerular toadfish *Opsanus tau* is considered to be the least permeable to Na^+ and water of all marine fish, and it becomes even less permeable upon transfer to fresh water. Clearly, a low permeability to salt and water reduces any environmental osmotic and ionic load and consequently reduces demand for osmotic and ionic regulation. A recent paper in this direction is of Baustian et al. (1997).

The essential regulatory activities pertaining to salt and water balance in sea water or fresh water are carried out by only two tubule segments in the aglomerular kidney. These are the proximal tubule and the collecting duct system including the urinary bladder. Of these, the proximal tubule is anatomically and functionally dominant. It is brush-bordered, as is every other vertebrate renal proximal tubule regardless of net secretory or reabsorptive functions. The proximal tubule in aglomerular fish is similar to proximal tubule II of glomerular fish. Apparently, the first segment of proximal tubule has been lost in aglomerular kidneys as the lack of glomerular filtration has made the reabsorption of filtered solutes superfluous. Likewise, in the absence of glomerular filtration, the arterial supply to the kidney has degenerated in aglomerular species. Aglomerular kidneys receive blood exclusively from venous drainages, the caudal vein, the caudal accessory vein, the posterior caudal vein and other veins.

Overall, epithelial transport in aglomerular kidneys is secretory. In general, net secretion takes place across the proximal tubule; however, this does not preclude the capacity for reabsorptive transport in this segment as well as in the collecting ducts and urinary bladder. Osmotically driven water secretion is considered secondary to primary solute secretion. Secretion rates can be impressive, particularly for divalent ions (Mg^{2+} and Ca^{2+}) and organic solutes (ammonia, urea, trimethylamine oxide TMAO), but also for water. Secretion pressures in the renal tubule of toadfish exceed the dorsal aortic pressure. Urine flow rates from aglomerular kidneys, reflecting net secretory activities of renal tubules, respond to changes in venous perfusion pressure in a manner qualitatively similar to glomerular fish when arterial perfusion is changed.

The aglomerular kidney like the glomerular kidney contributes more to extracellular homeostasis than the secretion of divalent ions. In this function, the proximal tubule of aglomerular kidney is the primary site of bulk transport as it is in the glomerular kidney.

It has been suggested, that the kidney of aglomerular fish responds directly to divalent ions without involvement of a systemic hormone. The physiology of aglomerular kidneys in permanent denizens of freshwater habitat, where divalent ion loading does not occur, has not been studied at all. Further, it is of interest that urea is a diuretic in aglomerular kidneys. Less diffusible organic solutes which are not readily excreted from the gill are secreted by the aglomerular kidney. The aglomerular kidney also excretes amino acids. Proteins also appear in the ureteral urine of aglomerular fish. The proximal tubules of aglomerular kidneys have the capacity for reabsorptive glucose transport despite the absence of a filtered load.

In mammals, the renin-angiotensin system is an endocrine system with broad effects, all targeted toward restoring extra-cellular volume after volume loss. The studies of endocrine system in aglomerular fish are too few and contradictory to make conclusive statements.

According to Nishimura and Madey (1989), results suggest that calcium influx may mediate inhibitory messages for renin secretion, while the β-adrenoreceptor mediated activation of granulated cells appears absent in toadfish.

2.3.2 Glomerular Urine Formation in Fish (Overview)

The nephrons of glomerular fish lack a distal segment, as do aglomerular nephrons. Thus, with the exception of glomeruli and a proximal segment I, the glomerular nephron of marine fish is quite similar to the aglomerular nephron. This may also be true for the nephrons of most euryhaline species which resemble those of stenohaline marine fish. The anatomical similarities of marine and euryhaline kidneys are not surprising since most euryhaline fish (of at least North and Middle America) are members of marine families.

All degrees of glomerular degeneration are present in marine fish, both between and within species. Among the species, glomerular development ranges from well-vascularized glomeruli to glomeruli of small size and number. Even closely related species may be either functionally glomerular (longhorn sculpin) or virtually aglomerular (daddy sculpin). Within species, glomerular degeneration may not be a pathology. Instead, it may reflect a physiological response under regulatory control. This is best observed in the reduction of filtering nephrons in euryhaline species making the transition from fresh water to seawater.

The blood flow to the glomerular kidney of marine fish is predominantly venous by way of the caudal and other veins. Even in marine fish with well-vascularized glomeruli, filtration of the arterial supply contributes little to urine flow. For example, in the longhorn sculpin, renal plasma flow is 37 times greater than glomerulation filtration, while in the nearly aglomerular daddy sculpin renal plasma flow is more than 1,300 times greater than glomerulation filtration.

Like aglomerular kidney, glomerular kidneys excrete a urine with high concentrations of Mg^{2+} and SO_4^{2+}. Urine flow rates are low, with low concentration of Na^+ and Cl^-, quite comparable to urine output in aglomerular fish. Even though the rates of glomerular filteration generally exceed urine flow rates, filteration rates are low and tubular transport activities are high, such that epithelial transport, in particular secretion, influences urine flow and urine composition for more than glomerular filteration. Urine flow rates that are significantly greater than glomerular filtration rates, indicating net tubular secretion, are observed often in glomerular seawater fish and occasionally also in a glomerular freshwater fish.

With these broad perspective of the anatomical and physiological similarities of glomerular and aglomerular urine formation, similarities in the details of epithelial transport across the proximal tubule can be expected. So close are the functional similarities between glomerular and aglomerular proximal tubules of marine fish that it is difficult to distinguish between the two. The quantitative differences that do exist pertain mostly to reabsorptive capacities which are greater in glomerular proximal tubules.

Like aglomerular proximal tubules, glomerular tubules of several marine fish secrete taurine. As in aglomerular fish, the renal secretion of protein is low in glomerular fish. The mechanisms involved in protein secretion are unclear.

The actions of the renin-angiotensin system in glomerular fish also support extracellular fluid balance. As in other vertebrates, this system responds to extracellular volume depletion. Whereas angiotensin has purely tubular effect in the aglomerular kidney, its effects in glomerular kidneys appear to be mediated primarily via changes in glomerular filtration.

The occurrence of NaCl and fluid secretion in glomerular proximal tubules of teleosts, elasmobranchs, and reptiles and in mammalian renal tissue cultures suggests that the genetic potential for NaCl secretion is present in every vertebrate kidney (Beyenbach 1986). Beyenbach et al. (1993) attempted to understand the mechanisms of transepithelial Mg transport in the kidneys of fish. Further advances in this respect have been made by Freire et al. (1996) and Chandra et al. (1997).

2.3.3 Transepithelial Potentials and Kidney Types

Potts and Hedges (1991) have given the results of their survey of transepithelial potentials and kidney types (Table 11.1).

Taking the evolutionary aspect, Potts and Hedges (1991) stated that although the aglomerular

Table 11.1 Distribution of transepithelial potentials and kidney types among some marine teleosts of different taxonomic position.
Potential-Potential of peritoneal fluid in relation to seawater

Fish	Potential	Kidney
Salmon, herring	High + ve	Glomerular
Carp-like fishes	High + ve	Glomerular
Eel	High + ve	Glomerular
Pipefishes, seahorses	Low – ve	Aglomerular
Cod, etc.	High + ve	Glomerular
Perch, bass, catfish	High + ve	Glomerular and Aglomerular
Gurnards, bullheads, lumpsuckers	High + ve	Glomerular and Aglomerular
Flatfish	High + ve	Glomerular
	Low – ve	
Sunfish and triggerfish	?	Glomerular and Aglomerular
Toadfishes	Low – ve	Aglomerular
Anglerfishes	High + ve	Aglomerular

Data source: Potts and Hedges (1991)

kidney probably evolved in sea water, some glomerular fishes are tolerant of diluted sea water. Filtration is low in marine teleosts and in euryhaline marine teleosts a proportion of the glomeruli may not function in dilute media. Furthermore, the aglomerular kidney secretes magnesium and sulphate ions, an obvious adaptation to marine conditions. Neverthless, some aglomerular fish can now tolerate dilute brackish water or even fresh water. Similarly, *Opsanus* is now a euryhaline fish although the change from chloride secretion to sodium and chloride secretion is probably an adaptation to seawater life. The pipefishes and toadfishes are also both morphologically eccentric teleosts, which diverged far from the streamline, fast swimming ancestral stock.

2.3.4 Glomerular Filteration

For glomerular nephrons, the rate of formation of ultrafilterate at each glomerulus determines the rate at which fluid is delivered to the lumen of the corresponding proximal tubule and therefore is the initial determinant of the volume and composition of the final urine.

Some examples of direct measurements of filtration rates for individual nephrons in some fishes are shown (Table 11.2).

Clearance measurements of whole-kidney GFR have been made for a number of species of fishes (Table 11.3).

For comparative aspects of glomerular filtration, the reader is referred to Yokota et al. (1985). Many phylogenetic differences in morphological, physiological and metabolic design have a decisive impact on the magnitude of the GFR. The endothermic classes, with more numerous glomeruli, high metabolic rates, and high ultrafiltration pressures, have proportionately higher rates of glomerular filtration than the ectothermic groups. As a group, the reptiles, with presumably the lowest rates of water influx, exhibit the lowest GFRs. Within each class, there are trends toward species with greater access to free water having GFRs (e.g., fresh water vs. marine; mesic vs. xeric). The clearest examples exists for teleosts, with marine forms having lower GFRs than their fresh water relatives.

Table 11.2 Examples of single-nephron filtrate rates (SNGFR)*

Species, environment, and mode of existence	Condition	SNGFR (ni-min^{-1})
Atlantic hagfish (*Myxine glutinosa*), marine	Sea water	24.2 ± 6.8 (6)
Pacific hagfish (*Eptatretus stouti*), marine	Sea water	20.3 ± 2.13 (71)
River lamprey (*Lampetra fluviatilis*), freshwater	Fresh water	7.02 ± 0.027 (89)
Rainbow trout (*Salmo gairdneri*), euryhaline	Fresh water Sea water	1.31 + 0.20 (5); 45% filtering 3.74 ± 1.12 (3); 5% filtering

* Values are means ± SE. Figures in parentheses indicate numbers of determinations, except in the case of the trout, where they indicate number of animals.
Data source: Dantzler (1985)

Table 11.3 Examples of changes in whole-kidney GFR in fishes*

Species, environment, and mode of existence	Condition	GFR (ml-kg^{-1} hr^{-1})
European eel (*Anguilla anguilla*), euryhaline	Adapted to fresh water Adapted to sea water	4.6 ± 0.54 1.0 ± 0.22
Rainbow trout (*Salmo gairdneri*), euryhaline	Adapted to fresh water Adapted to sea water	8.6 ± 0.95 1.2 ± 0.05
Plains killifish (*Fundulus kansae*), euryhaline	Adapted to fresh water Adapted to sea water	25 1.4

* Values are means for ± SE. All means with SE are for more than four values.
Data source: Dantzler (1985)

Single nephron glomerular filtration rate (SNGFR) is subject to a moderate dependency on body mass in the mammals, reptiles and amphibians, whereas in birds and fishes the SNGFR is essentially mass dependent.

In addition to the relatively long-term adaptation of renal function to the environment of a species, animals may also modify renal function acutely to respond appropriately to physiological perturbations.

Many studies have been performed on anadromous and catadromous fishes, investigating changes in renal function as a result of acclimation to salt water or fresh water. The similar findings of these studies are that these fish (e.g., *Salmo gairdneri*, *Fundulus kansae*; *Anguilla rostrata*) have higher GFRs in fresh water than in salt water. Thus, the changes in GFR in euryhaline fishes with acute transfer to fresh water or salt water environments mirrors the differences in GFR between stenohaline fresh water and salt water fishes. Changes in the fractional reabsorption of the glomerular filtrate may also occur with acclimation to water of different salinities. For example, the European eel (*Anguilla anguilla*) eliminates about 90% of the filtered volume in fresh water and 65% of the filtered volume in salt water. The river lamprey (*Lampetra fluviatilis*) displays a fivefold increase in GFR upon acclimation to fresh water from 50% sea water.

2.3.5 Tubular Transport

Tubular Transport of Inorganic Ions:

Filtered sodium and chloride is absorbed by the nephrons in glomerular teleost, lungfish, and freshwater lampreys. Even a smaller fraction may be absorbed by the nephrons of marine and, possibly, freshwater elasmobranchs. The primitive marine hagfishes (for example, *Myxine glutinosa*) conform to their environment, and no filtered sodium is absorbed along the archinephric ducts. In marine teleosts, whose nephrons lack distal tubules, most of the sodium and chloride appears to be absorbed in the proximal tubules. However, recent studies on flounder (*Pseudopleuronectes americanus*) tubules indicate that sodium and chloride may actually be secreted in the proximal tubule and that this may play a role in fluid secretion. Presumably, much of the sodium is absorbed farther along the nephron and in the bladder. In freshwater teleosts, significant absorption occurs in the proximal tubules, but a large additional component apparently occurs in the distal tubules, where dilution take place. The situation is probably similar in the elasmobranchs, although the bulk of the absorption appears to occur in the second segment of the proximal tubule. In the freshwater lampreys, however, approximately 80% of the filtered sodium and chloride appears to be absorbed in the distal tubules, whereas in freshwater teleosts, dilution takes place.

Even less is known about the tubular transport of potassium than of sodium. However, both net tubular absorption and net tubular secretion can occur in representatives of freshwater teleosts, and possibly marine elasmobranchs. In euryhaline teleosts, net tubular absorption generally occurs following adaptation to fresh water. Only net tubular absorption is observed in freshwater lampreys, whereas only net tubular secretion is observed in hagfishes. There is little or no potassium absorption along the proximal tubules of elasmobranch fishes (little skate, *Raja erinacea*) whereas 25 to 35% of filtered potassium is absorbed along the proximal tubules in freshwater lampreys (*Lampetra fluviatilis*).

In elasmobranchs, marine teleosts and euryhaline teleosts adapted to sea water, net tubular secretion of calcium occurs. In the little skate (*R. erinacea*), this secretion appears to occur in the second part of the proximal tubule. The secretory mechanism responds directly to the concentration of calcium in the peritubular blood. Uptake of calcium across the peritubular surface of isolated renal tubules from seawater-acclimated winter flounders (*P. americanus*) involves a fast and a slow component. The slow component, which may be most important for net secretion, is saturable, ATP-dependent, sodium-sensitive, and partially inhibited by magnesium.

As regard phosphate, both net tubular absorption and net tubular secretion are observed in fishes. Among the fishes, the hagfishes, lampreys, and marine elasmobranchs always exhibit net tubular secretion (the major site of secretion in little skate, *R. erinacea*, is the second segment of the proximal tubule). In true stenohaline freshwater teleosts, net absorption always appears to occur, although net secretion can occur in the euryhaline European eel (*Anguilla anguilla*) adapted to fresh water. Net secretion generally occurs in glomerular marine teleosts and apparently can occur in aglomerular and pauciglomerular marine teleosts. In general, it appears that euryhaline teleosts exhibit net absorption when adapted to fresh water and net secretion when adapted to sea water, although the mechanism controlling the change is not clear.

In respect of magnesium and sulfate transport among the fishes, both net tubular absorption and net tubular secretion are observed. Net tubular absorption occurs in freshwater lampreys and probably in all true stenohaline freshwater teleosts. However, net secretion does occur in some euryhaline teleosts adapted to fresh water (e.g., the American eel, *Anguilla rostrata*), but even in these animals, it is less marked during adaptation to fresh water than during adaptation to sea water.

Net secretion occurs in all marine fishes examined: hagfishes, elasmobranchs, and teleosts. This appears to be major adaptation to eliminate these divalent ions ingested with sea water. Such net secretion is striking in pauciglomerular, as well as glomerular, marine teleosts; very little excreted magnesium or sulfate derives from filtered ions. In general, net secretion ceases or, at least is marked reduced with adaptation to euryhaline teleosts to fresh water. Although chloride always undergoes net tubular absorption, a large fraction appears in the urine when the need to excrete magnesium exceeds the need to excrete sulfate. In *R. erinacea*, the primary site of magnesium and sulfate secretion is the second segment of the proximal tubule and this is assumed to be the case in marine teleosts as well.

As noted earlier, magnesium tends to inhibit the major slow component of calcium transport across the peritubular side of flounder tubules and may be transported by a similar mechanism. Studies of sulfate transport with tubules and membrane vesicles from two species of flounder (*Pseudopleuronectes americanus* and *Paralichthys lethostigma*) reveal a fast and a slow component for transport across the peritubular membrane. The slow component, which, as in the case of calcium, is probably the major one involved in secretion, is carrier-mediated and sodium-dependent and is inhibited by antimycin A and the inhibitor of anion transport 4-acetamido-4′-isothiocyano-2, 2′-disulfonic stilbene. The transport of magnesium and sulfate is directly influenced by the concentration of ions in the peritubular blood *in vovo*, but there are no known local or systemic humoral factors that regulate the transport.

Tubular Transport of Fluid:

Fluid absorption is quite variable among the fishes. For example, the marine cyclostomes, such as the Atlantic hagfish (*Myxine glutinosa*), conform to their environment, and no fluid is absorbed along the archinephric ducts. Approximately 40% of the filtered water is absorbed by the nephrons of freshwater lampreys (*Lampetra fluviatilis*), but only approximately 10% is absorbed along the proximal tubule. In marine elasmobranchs, 70–85% of the filtered water is normally absorbed in the renal tubules, although this may vary from a low of 9% to a high of 93%. Among the freshwater teleosts, as in the freshwater lamprey, an average of approximately 40% of the filtered water is absorbed, but this can vary from 25% to 75%. In marine teleosts, or euryhaline teleosts adapted to sea water, over 90% of the filtered water apparently may be absorbed, but actual absorption of filtered water is difficult to determine because of tubular fluid secretion. The major sites of tubular fluid absorption have not been determined in elasmobranchs or teleosts.

In the marine teleosts, the initial urine may be produced by ultrafiltration of plasma in glomerular species, by tubular secretion of fluid in aglomerular species, or by both mechanisms in some glomerular forms. Apparently, tubular secretion of fluid accounts for initial urine formation in aglomerular freshwater fish as well, and it may even account for a portion of urine formation in euryhaline fish adapted to fresh water. In euryhaline eels adapted to fresh water it is not possible to implicate the tubular secretion of any single inorganic ion in the process of fluid secretion. However, the possibility of potassium, calcium, or magnesium secretion in a proximal segment followed by absorption in a more distal segment cannot be ruled out. Some similar mechanisms must function in the aglomerular freshwater teleosts.

Numerous studies have shown a strong correlation between magnesium and, to a little extent, sulfate and calcium excretion and urine flow rate in aglomerular and glomerular marine teleosts. In the seawater-adapted winter flounder (*P. americanus*) more than half of the urine volume is secreted fluid, despite a relatively high GFR. The dominant divalent cation in the urine is magnesium, and the fluid secretion rate shows a highly significant linear correlation with the magnesium secretion rate, strongly suggesting that magnesium secretion is the driving force for fluid secretion. Studies have demonstrated that the primary ions secreted are sodium and chloride and that fluid secretion depends on the secretion of these ions.

Tubular Transport of Organic Substances:

Taking first the amino acids, tubular secretion of taurine occurs in flounder (*P. americanus*), and both net tubular secretion and net tubular absorption of taurine occur in dogfish (*Squalus acanthias*) and little skate (*Raja erinacea*).

Net tubular secretion of urea can occur in hagfish (*Myxine glutinosa*). Net passive absorption by the renal tubules plays an important role in urea excretion. Net tubular absorption of urea that may be active has been described in marine elasmobranchs, where high concentrations in body fluids help to maintain a favourable osmotic balance. This absorption in the dogfish (*S. acanthias*) occurs between the end of the proximal tubule and the beginning of the collecting duct, apparently in the distal tubule.

Glucose, which is freely filtered, is normally completely absorbed by the renal tubules of all vertebrate species studied, except those of an aglomerular teleost (*Lophius americanus*), in which a very small net secretory flux can occur.

Integrative Summary of Some Glomerular and Tubular Functions:

(1) Hagfishes: These animals, which are entirely marine, are in osmotic equilibrium with the surrounding sea water, the osmolarity of the body fluids being determined primarily by the concentrations of sodium and chloride. Glomerular filtrate empties directly into primitive archinephric ducts along which glucose is absorbed and urea, potassium, phosphate, magnesium and sulphate are secreted, but essentially no sodium or fluid is absorbed. The final urine is isosmotic with the plasma and sea water.

(2) Lampreys: The kidneys of freshwater lampreys have large fused glomeruli and nephrons with proximal and distal tubules and collecting ducts. They have high filtration rates and produce large quantities of dilute urine. Less than half of the filtered fluid is absorbed, most of it in the distal tubules and collecting ducts. However, most of the filtered sodium, chloride, and even potassium is absorbed, also primarily all along the distal tubules. And this absorption of ions, far in excess of the absorption of water, leads to the production of a urine only approximately one-tenth the osmolarity of the plasma, thus helping the animals to maintain a plasma osmolarity well above that of the surrounding fresh water.

(3) Elasmobranchs: The marine elasmobranchs, like the hagfishes, conform to the osmolarity of their environment, but unlike the situation for hagfishes, much of their osmolarity of their extracellular fluid is determined by the concentrations of urea and TMAO. Their nephrons have very large glomeruli and long, complex proximal and distal tubules. The filtrate rate is higher

than that for marine teleosts, and much of the filtered water is normally absorbed, although great variation can occur. About two-thirds of the filtered sodium chloride is absorbed, and some dilution of tubular fluid takes place, apparently primarily along the collecting ducts, but the final urine is close to isomotic with the plasma and sea water. Much of the sodium chloride absorbed by the renal tubules may be secreted by the rectal gland, a structure specialized to secrete these ions in excess of water. Net tubular secretion of magnesium, sulfate, calcium, and phosphate occurs in the second segment of the proximal tubule, and these are important constituents of the final urine. Urea, a major determinant of the plasma osmolarity, is absorbed along the distal tubules by what may be an active process.

(4) Marine teleosts: The marine teleosts maintain the osmolarity of their body fluids well below that of the surrounding sea water. They ingest sea water and must excrete the absorbed divalent ions (e.g., magnesium and sulfate) and monovalent ions (e.g., sodium and chloride). Their kidneys have small, few, or no glomeruli, and the nephrons lack distal tubules. The filtration rate for species with glomeruli is low, and the urine flow rate is low for all species. Most of the filtered water is apparently absorbed secondary to the absorption of sodium chloride. Net secretion of magnesium, sulfate, calcium and frequently phosphate occurs, apparently primarily along the second segment of the proximal tubule. This is the major route for the excretion of these ions. Net tubular secretion of fluid, apparently dependent on the secretion of sodium chloride or possibly magnesium, determines the actual urine flow rate in glomerular and aglomerular species. The final urine is close to isosmotic with the plasma, but the concentrations of sodium and chloride are usually low, and the urine may be slightly hypoosmotic. The excess sodium chloride absorbed by the nephrons and gastrointestinal tract is secreted by the gills.

(5) Freshwater teleosts: The extracellular osmolarity of true stenohaline freshwater teleosts is well above the osmolarity of the surrounding water. The kidneys of most species have larger and more numerous glomeruli than marine forms, and the nephrons contain well-developed distal tubules. The filtrate rate also is much higher than in marine forms, and the urine flow is copious and dilute. Net tubular secretion of divalent ions is absent. In fact, net tubular absorption of those filtered occurs. Most of the filtered sodium and chloride is absorbed, and this absorption in the distal tubules, which are essentially impermeable to water, produces a urine as dilute as one-tenth the osmolarity of the plasma. The few aglomerular freshwater teleosts apparently produce urine by net tubular secretion of ions (probably sodium and chloride, but possibly magnesium or calcium) and water in one nephron (possibly the second segment of the proximal tubule) and absorption of ions without water in a more distal nephron or even in the bladder.

(6) Euryhaline teleosts: These fishes can maintain the osmolarity of their body fluids above that of fresh water, when adapted to fresh water, and below that of sea water when adapted to sea water. The kidneys, like those of stenohaline freshwater teleosts, tend to have numerous and fairly large glomeruli, and the nephrons have distal tubules. When these fishes have adapted to sea water, the number of nephrons filtering and, as a result, the whole-kidney GFR are low, possibly from the presence of AVT or angiotensin and the absence of prolactin. At this time, marked net tubular secretion of divalent ions (magnesium, sulfate, and calcium) and fluid occurs. The nephrons are permeable to water, but this does not appear to be produced by the presence of AVT. Most of the filtered sodium chloride is absorbed, and there can be some dilution of tubular fluid. However, the final urine tends to be close to isomotic with the plasma, with high concentrations of divalent ion and low concentrations of sodium and chloride.

When these animals are adapted to fresh water, the number of nephrons filtering and, consequently, the whole-kidney GFR are high, possibly from the absence of AVT or angiotensin or the presence of prolactin. Net tubular secretion of divalent ions and fluid tends to cease (although some may continue, especially in aglomerular species). The distal tubules become largely impermeable to water, but sodium chloride absorption continues in the distal tubules, and a dilute urine is produced.

As these renal changes occur with adaptation to different aqueous environments, there are comparable appropriate changes in ion and water movements across the gills and the bladder. It must be kept in mind that regulation of the various renal functions in all the fishes is coordinated with regulation of ion and water movements across the gills, integument, and bladder and specialized ion-transporting structures, such as the elasmobranch rectal gland, in the adaptation of the animals to their environment.

The reader can get related information in the papers by Long and Giebisch (1979) and Dantzler (1982) (Figs. 11.8 and 11.9).

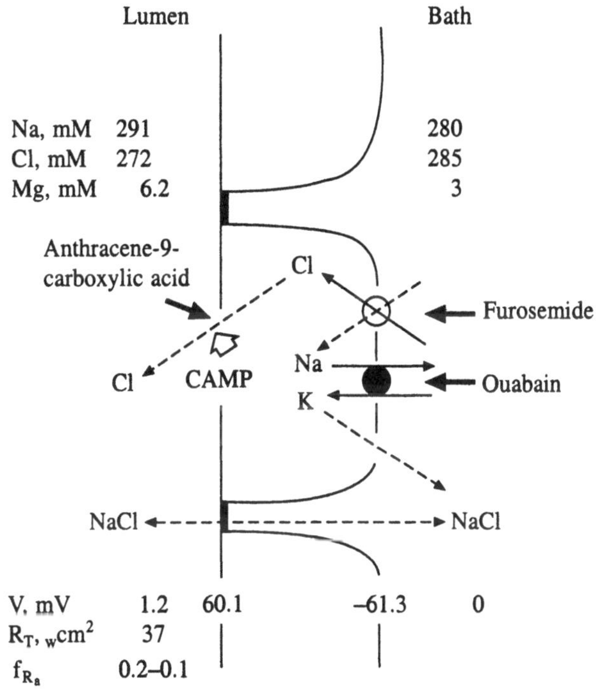

Fig. 11.8 Mechanism of NaCl secretion in shark renal proximal tubules. ***Solid arrows*****, movements against electrochemical gradient;** ***broken arrow*****, movement down electrochemical gradient; V, voltage, Rr, transepithelial resistance; fRa, fractional resistance of apical membrane.**
Source: **Beyenbach (1986) (See also Baustian and Beyenbach 1993)**

2.3.6 Endocrine Control of Renal Function

The structure and function of kidneys differ among vertebrates in different taxonomic groups and habitats. Hence, hormones that control renal function differ. Furthermore, the chemistry and function of native hormones and the properties of hormones have changed during vertebrate evolution. There are several unique aspects of control of renal function in nonmammalian

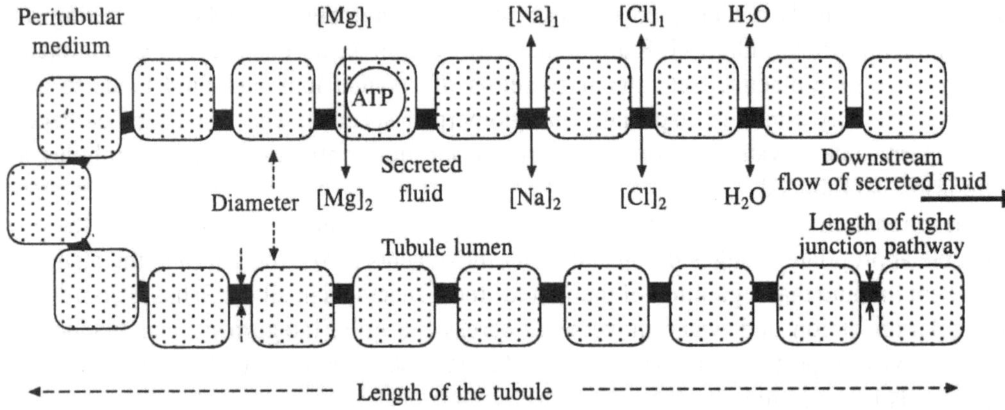

Fig. 11.9 Model of transepithelial fluid secretion in aglomerular and glomerular fish proximal tubules. Only transepithelial Mg secretion is active; transport of Na and Cl is passive.
***Source:* Beyenbach and Liu (1996)**

vertebrates. First, in both aquatic and terrestrial animals volume and osmoregulation are attributed, in addition to the kidney, to organs such as the gill, skin, and gut. Moreover, there are the rectal, salt, and nasal glands of elasmobranchs, reptiles and birds, respectively. Thus, endocrine control of renal function is a part of the integrative control of osmoregulation, and the relative contribution of the kidney varies depending on the species, and their habitats. Second, the response to the same endogenous or exogenous hormone will vary in the same species depending on the environment, their hydromineral status, sexual maturity, and life span. Third, all vertebrate kidneys, except those of cyclostomes, stenohaline freshwater teleosts, and mammals, receive dual blood supplies; one from the renal arterial system and the other from the renal portal vein. The relative contribution of the two systems toward renal blood flow varies among species and conditions; hormones control renal function via both routes. Fourth, glomerular filtration may operate intermittently, and glomerular filtration rate (GFR) is often a major determinant for urine flow and urinary electrolyte excretion. Fifth, heterogeneity of nephron structure is large, and the hormonal response of the whole kidney and of the individual nephrons often differs. Kidneys of nonmammalian vertebrates are often autoregulated poorly or not at all. Hormones and humoral substances which influence systemic blood pressure or intrarenal haemodynamics readily influence salt and water excretion.

Hormones influencing renal water and solute excretion can be classified as extrarenal or intrarenal hormones. Extrarenal hormones are transported to the kidney through the systemic circulation; intrarenal hormones, such as the renal prostaglandins, kinins and possibly renin-angiotensin system, are synthesized in the kidney and act locally. Further, hormones alter the renal handing of solutes and water by their action on (1) the systemic cardiovascular system, (2) the intrarenal vasculature, and (3) the tubular transport system.

Let us first recall that condition of osmoregulation in lower vertebrates differ greatly with each other.

Among Agnatha, hagfishes display atypical osmoregulatory characteristics and their endocrine repertoire is poorly understood. Hormonal actions are unclear although the kidney appears to act as a regulator of extracellular fluid volume. Lampreys (Petromyzontidae) show many similarities with amphihaline potamotocous, such as salmon and the shad, teleost fish, with respect to osmoregulation, but again their endocrine system requires further defination.

There are many similarities in the endocrine repertoire of fishes (adrenocortical homologue, hypothalmohypophysial system, thyroid tissue, etc.) and the osmoregulatory effector systems (gills, gut, kidney for example) seem to be uniformly present. Superimposed upon these homologies, there are, however, unique endocrine glands (e.g., corpuscles of Stannius, urohypophysis) and special devices (e.g., rectal gland of elasmobranch fishes) with intriguing biochemical and physiological variations.

The Myxinoidea, lowest in the zoological scale and somewhat aberrant among contemporary vertebrates, are poecilosmotic, like most invertebrates, i.e., they maintain their blood plasma at or about the same osmolarity as their environmental seawater, and there are, with one or two exceptions, equal ionic activities in the two solutions. The functional kidney consists of a small number of relatively large glomeruli sited segmentally along the archinephric ducts. The process of ultrafiltration is sensitively dependent upon arterial pressure. The short tubular segments are somewhat equivocally concerned with reabsorption and/or secretion of ions such as sodium, magnesium and calcium and possibly with filtered macromolecules. There appears to be little or no water absorption along the nephron and it has been concluded that in hagfishes the kidney is primarily a volume-regulatory device.

The general endocrinology of hagfishes is poorly understood. The basic anatomy of the myxinoid kidney suggests that perhaps renal function amounts to volume regulation, and being pressure dependent than vasoactive hormones are responsible for its functional integrity. The short tubular segments may simply act as conduits for the volume-regulating activities of the glomeruli. Clearly the basic endocrinology and renal physiology demands extensive examination before firm statements can be made.

The lampreys (Petromyzonoidea), more advanced Cyclostomata, superficially at least, present many similarities with the teleost fishes in terms of their general renal function. Indeed, based on their general osmoregulatory characteristics, it has been affirmed persuasively that there is a close parallel evolution between lampreys and Osteichthyes; such parallelism is not so readily identified when endocrine system is considered, however. In this latter context the relative dissociation of the hypothalamohypophysial components, the somewhat equivocal nature of the form and secretion (s) of the adrenocortical homologue, and the possible absence of a renin-angiotensin system are of special note.

Under particular circumstances, most notably when undergoing reproductive migrations, lampreys can adapt to both freshwater and seawater environments, and there is every reason to suppose that the underlying mechanisms permitting the adaptations are akin to those employed by euryhaline teleosts.

In summary, the lampreys possess kidneys that functionally resemble those of teleost fishes to a very considerable extent, and in many respects the euryhalinity of these fish depends upon appropriate and equivalent renal responses. The renal function would again appear to centre on volume regulation, although the processes involved in regulating the fractional excretion of glomerular ultrafiltrate and its contained solutes are not known. These latter processes are likely, but as yet undemonstrated, sites for hormonal action.

Elasmobranch fishes are predominantly marine and maintain their plasma osmolalities at or slightly above that of the environmental seawater as a result of retention of urea and trimethylamine oxide. There is at the same time careful homeostasis of body fluid volume and individual solute composition. The general processes responsible for fluid and electrolyte regulation are dependent on the conjoint actions of the gills, rectal gland and kidney. Endocrine impingement upon these structures, more especially upon the kidney is however poorly understood and this is in part because elasmobranch fishes without digressing significantly from the general vertebrate endocrine system possesses a number of unique hormones.

A primary function of the elasmobranch kidney would appear to be the retention of nitrogenous waste products and the elimination of divalent cations. It probably also plays a role in volume regulation, although, since the net flux of water across the gills will be minimal in the presence of almost no osmotic gradient, this function may not be significant. Typically the elasmobranch kidney possesses extremely long nephrons with very well-developed glomeruli. Although there are brisk rates of glomerular filtration, urine production is relatively sparse and voided urine is approximately iso-osmotic with plasma. Urine is virtually urea free, and the process of urea reabsorption may be passive or active sodium-linked process. The iso-osmotic character of urine with respect to plasma largely results from the intense secretory activity for divalent cations and some attempts have been made to allocate particularly secretory and/or resorptive functions to many identifiable nephron segments.

This is a certain amount of admittedly circumstantial evidence that both adrenocortical steroids (or more precisely 1-α-hydroxycorticosterone) and at least one of the neurohypophysial hormones act upon the kidney. There are indications that the elasmobranch, unlike teleost fishes for example, adjusts its urine production more by altering renal tubular fluid reabsorption than by changing rates of glomerular filtration. Thus antidiuretic hormones come into play in these fishes to allow both glomerular and renal tubular mechanisms to temper overall urine output.

Before discussing the renal control of urea retention, it is important to place the phenomenon of the 'physiological uraemia' into proper perspective, since this substance plays such a crucial role in elasmobranch osmoregulation and yet it is for other animals a relatively toxic material, being a potent protein-denaturing agent. Certainly the concentrations of urea that exists normally in elasmobranch plasma interfere with the structure and function of non-elasmobranch proteins and enzymes. While some elasmobranch proteins such as haemoglobin display urea tolerance, rather curiously others do not. The problem of how they function *in vovo* was unsolved until the methylamine compounds, in particular trimethylamine oxide were found to act as 'biochemical antagonists' to disruptive actions of urea. *In vitro* there appears to be an optimal trimethylamine oxide to urea ratio of 1 : 2 to achieve protection of proteins against denaturation by urea. Such a ratio has been observed in studies of intracellular environment of some elasmobranch species.

Thus the regulation of urea production and maintenance of appropriate amounts of trimethylamine oxide within the body fluids of elasmobranchs are of central importance to osmoregulation, particularly when the animals are exposed to either experimentally or in nature to changing environmental salinities. Of course most elasmobranchs are unable to withstand abrupt transfer to significantly more dilute or concentrated environments, but a number of studies show that stepwise and gradual adaptation allows good survival of a number of purely marine species in both dilute and more concentrated seawaters. On adaptation of elasmobranchs to more dilute environments there is a reduction in the total plasma osmolality with concomitant reductions in plasma sodium, chloride and urea concentrations; significantly there is usually greater reductions in urea than in other solutes. The reduced urea concentrations may result from an increased urea excretion and/or reduced hepatic synthesis depending on the species. Adaptations of fishes to more concentrated seawater produces broadly opposite effects in terms of solute composition.

There is a tentative indication that the corticosteroid may moderate the production rates of urea and its renal excretion, while at the same time it participates in sodium regulation at reduced environmental salinities.

Clearly if corticosteroids are involved in osmoregulation in the suggested manner, then, as in other vertebrates, feedback control of their secretory patterns must be present. Certain features provide circumstantial evidence for ultimately a brain-interrenal feedback system, but there are unsolved aspects that require clarification. Besides, certain of pre-requisites suggestive of a renin-angiotensin system are present in certain elasmobranch species.

It is clear that conclusions regarding hormonal control of renal function of elasmobranch fishes must be at present be very tentative. At the same time recent observations on both the basic physiological regulatory mechanisms with special regard to the metabolism of nitrogenous waste, the 'osmotic ballast' so essential to elasmobranch osmoregulation, and the application of renal tubular micropuncture alongside sensitive assessment of endocrine status render this area of comparative physiology an extremely fertile one.

Among the osteichthyans with focus on teleosts three general types of renal osmoregulatory mechanisms may be identified: stenohaline-marine; stenohaline-freshwater; and euryhaline. In freshwater, teleosts encounter a large osmotic influx of water and incipient depletion of ions, and to counter these, the kidney eliminates large volume of osmotically free water and the gills offset ion depletion by active uptake from the environment. When in the sea, osmotically free water is at a premium for the fish, since there is osmotic loss across permeable surfaces and is a passive gain of sodium chloride; to gain water, marine teleosts actively imbibe the environmental water, but at the same time they obligatorily gain solute which must be eliminated. Again conjoint renal and extrarenal (predominantly gills) systems are responsible for maintaining homeostasis. Characteristically marine teleosts produce sparse volumes of urine rich in divalent ions, while the gained monovalent ions are extruded by the gills.

For ionic and osmotic regulation during migration of fish, the reader is referred to Fontaine (1975). Anadromous salmonids (amphihaline migrators) usually undergo a parr-smolt transformation (smoltification) involving several behavioural, morphological, physiological and biochemical changes before they enter seawater (See reviews: Hoar 1988; Boeuf 1993). These changes are commonly regarded as being prospective preparations for saltwater life. Another view expressed is that these changes represent a reemergence of ancestral marine adaptations that compromise the continuation of freshwater residence. For the salmonid species that normally go through a parr-smolt transformation this process is regarded as being critical to successful growth in full seawater.

Anadromous salmonids display seasonal variations in seawater tolerance, and one of the major effects of physiological changes that occurs during smoltification is that fish become capable of restoring and regulating the ionic composition of their body fluids at approximately 300–340 m Osm/kg, within a few days after entering seawater. Therefore, the temporal changes in blood plasma electrolyte levels, after direct transfer from freshwater to seawater, will express the rate at which seawater acclimation is taking place, and will thereby give valuable information about the development status of the fish at the time of transfer.

The timing of the parr-smolt transformation appears to be controlled by an endogenous rhythm synchronized by external factors. Photoperiod seems to be the most important "Zeitgeber" whilst temperature acts mainly through its effect on developmental rate. The influences on environmental factors are mediated through the nervous and endocrine systems, and during parr-smolt transformation substantial changes occur in the dynamics of several hormones such as thyroxine (T_4), triiodothyronine (T_3), prolactin, growth hormone, cortisol and insulin.

Halvorsen et al. (1993) reported that the hypoosmoregulatory capacity of seaward-migrating Arctic charr (*Savelinus alpinus*) is well developed before they enter the sea, and may be comparable to that seen in salmonids which undergo parr-smolt transformation. There was a negative correlation between blood plasma electrolyte levels and both body length and weight, indicating that hyposmoregulatory capacity was affected by the size of the fish. The reader can get comprehensive information on the seawater adaptation in Arctic charr, *S. alpinus* dealing specifically aspects of osmoregulation, feed intake and growth (Arnesen 1994).

It is apparent that intrarenal blood flow with respect to glomerular perfusion, rates of filtration in single nephrons, degree of renal portal venous perfusion, filtration pressures and glomerular intermittency, is a central factor in renal adaptive responses to teleost fishes to changing environmental salinity. The control mechanisms within the kidney involve arginine vasotocin, prolactin, adrenocortical steroids, and the renin-angiotensin system. A generally accepted view, if somewhat tentative view places prolactin as a key factor of freshwater adaptation, while cortisol and angiotensin are principally concerned with renal performance in seawater teleosts. The nature of actions of arginine vasotocin remain obscure but it is becoming increasingly clear the interplay between this peptide and the substances mentioned above is of key importance.

It is now established that both angiotensin and arginine vasotocin are significantly altered by environmental salinity of the teleost; the major influence would not seem to be plasma volume in terms of regulating their secretion, rather plasma sodium concentrations appear to be the best correlate so far examined. There are of course other likely regulatory influences, arterial pressure itself; it has been congently argued atleast for the renin-angiotensin system, a primary role in regulating blood pressure.

The actions of angiotensin II and arginine vasotocin cannot, however, be viewed in isolation, and although they demonstrably affect each other's secretory patterns, they both have primary and secondary actions on other hormonal systems and osmoregulatory structures including gills, gastrointestine and kidney. Moreover the dipsogenic actions of angiotensin II and the actions of the renin-angiotensin system in perhaps maintaining increased adrenocortical activity in seawater fish must be set against synergistic and perhaps antagonistic actions of other hormones such as prolactin and possibly arginine vasotocin under varying regimes of salinity. Moreover in teleostean renal function and hormones, there is impingement of the physiological actions of the secretions of the corpuscles of Stannius, the urophysis and pars intermedia materials. In addition the possible osmoregulatory function of the pancreas as well as establishing a role for vitamin D_3 endocrine system require further analysis and interpolated into a general scheme to describe the endocrine control of the kidney in both marine and freshwater teleosts.

Lungfishes impose upon themselves a severe osmotic stress when aestivating effectively converting themselves from ammonotelic to ureotelic organisms. When ureotelic there is a condition of anuria, but on emergence into freshwaters ammonotelism develops and the animals undergo an enormous diuresis to remove retained urea. The diuretic and sometimes natiuretic phases appear to be pressure dependent and it has been persuasively argued that arginine vasotocin is a key hormone in the response. Virtually nothing is known about the delicate interplay that must be necessary for the responses, with other hormones in particular the renin-angiotensin system and the adrenocortical steroids, both of which have been delineated in lungfishes (Dantzler 1985; read Henderson et al. 1985; Nishimura 1985; Braun and Dantzler 1987).

Nishimura (1985) discussed endocrine control of renal handling of solutes and water in vertebrates, and concluded that hormones influence renal function by both extrarenal and intrarenal mechanisms. Extrarenal mechanisms include the effects through systemic haemodynamic and neural pathways, whereas interrenal mechanisms can be largely divided into effects on intrarenal haemodynamics and those on tubular transport epithelia. Neurohypophysial hormones and renin-angiotensin system appear to act primarily on systemic and preglomerular vasculature in primitive vertebrates, while direct tubular action appears to have evolved at a later stage of phylogeny. Although aldosterone is an essential hormone for fluid mineral balance in mammals, the action of mineralocorticoids on tubular Na transport has not been established in nonmammalian tetrapods. In bony fishes in hyperosmotic environments, cortisol accelerates active Na extrusion from the gill. In contrast, prolactin is important for maintaining low osmotic water permeability of the transport epithelia in fishes in hypoosmotic media. Thus, both function and site of hormone action appear to have changed during the evolution of vertebrates interacting with changing environments, and in response to the demands from other bodily functions. Furthermore, evolution of interactions, at the cellular level, between systemic and locally formed hormones such as prostaglandin, kinins, and perhaps angiotensin may have developed more elaborate controlling systems of renal handling of solutes and water (Fig. 11.10). For further details on the functions of endocrine organs, the reader is referred to Dantzler (1985), Brown and Brown (1987), Grau (1987), Larsen and Bern (1987), Nishiniura (1987) and Hirano (1989). There is evidence in several salmonid species that growth (GH) may facilitate seawater adaptation (see Brown and Brown 1987).

Autonomic innervation of the teleost kidney of fishes is not well understood. In general, cholinergic innervation of blood vessels appears to be dominant in primitive animals, whereas adrenergic innervation increases and may be dominant in more advanced vertebrates (Nishimura 1985; read also Braun and Dantzler 1987).

Dantzler (1985) provided comprehensive account of renal function. The physiological role of adrenocorticosteroids in regulating sodium balance is not clear (Braun and Dantzler 1987).

Laverty and Clark (1989) reviewed the renal regulation of calcium and phosphate in fish. Calcium and phosphate regulatory systems appear to be more complex than those of tetrapods, since these organisms inhabit aquatic environments of differing salt and mineral content and have several unique systems that appear to play a role in calcium homeostasis.

In fishes, calcium is mainly taken up via the gills, and gut plays a minor role. For phosphate, the gut seems to be only uptake route, and sodium-dependent phosphate uptake has been demonstrated in carp intestine (Wendelaar Bonga and Pang 1989).

Fishes have no observable parathyroid glands. Pituitary gland, vitamin D, ultimobranchial glands and Stannius corpuscles regulate calcium levels. Calcitonin may also affect kidney function in fish. Calcitonin increased blood pressure, urine volume and glomerular filtration rate but did not change the clearance of sodium, potassium, chloride or calcium in freshwater eels. Calcitonin receptors do not occur in fish kidneys. Any effects of the hormone on renal excretion patterns may result from changes in blood pressure or from changes in plasma concentrations of electrolytes, in turn altering the amount of substance delivered to the kidneys. Calcitonin may not be the primary controller of blood calcium levels in fish. There is good evidence that the corpuscles of

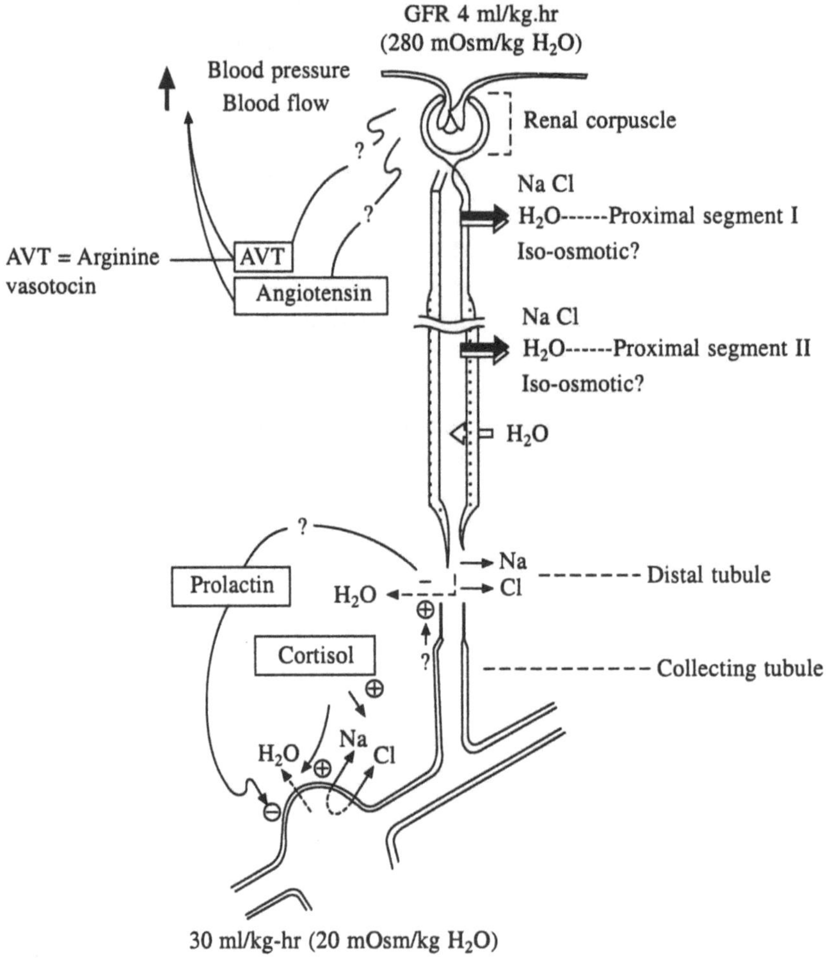

Fig. 11.10 Schematic illustration of the renal handling of water and NaCl and effects of hormones in the freshwater teleost. Solid and open arrows indicate active and passive movements, respectively, of electrolytes and water. Brokwn arrows show low permeability to water. + = Stimulation; – = inhibition.
***Source:* Nishimura (1985)**

Stannius, located on or embedded in the kidneys of teleosts, cause hypocalcemia in fish. Unfortunately, there are yet no published accounts of renal responses to PTH in fish. PTH may stimulate phosphate transport in fish via an intracellular pathway (diacylglycerol and protein kinase C) that is also found in the tetrapod kidney. Whether there are endocrine systems present in fish that regulate phosphate transport through this mechanism is unknown (read Laverty and Clark 1989; see also Robertson 1989).

Currently a wealth of evidence points to an osmoregulatory function for active urophysial principles (see Larson and Bern 1987). Evans (1990) has given an analysis of the hormonal control of fish osmoregulation. Evidence for a role of atriopeptin in fish osmoregulation has been presented. The hormonal control of clasmobranch osmoregulation is poorly understood, with the major exception of the rectal gland. Salt excretion by the gland is controlled by a variety of hormones and neurotransmitters. Rather surprisingly, it appears that arginine vasotocin and prolactin are important modulators of osmoregulation in fresh water, where fishes are volume-loaded and salt-depleted. Other hormones that inhibit salt excretion across the fish gill epithelium and therefore important in freshwater osmoregulation include: α-adrenergic agonists, somatostatin, and urotensin II. There has been clear cut demonstration of hormonal control of active salt uptake by gill epithelium of freshwater fishes, but there is some indication that stimulation of β-adrenoreception in the gill epithelium may enhance Na^+ uptake. In seawater teleosts, it is clear that more data are needed before the role of the RAS in fish osmoregulation is fully understood. The role of aldosterone in osmoregulation is unknown. The importance of cortisol has been expressed. Other neurohumoral agents, which have been found to stimulate salt excretion by the gill epithelium and therefore may play roles in teleost osmoregulation in seawater, include: β-adrenergic agonists, vasoactive intestinal peptide (VIP) and glucagon.

3. DIPNOI

Wake (1986) surveyed and reviewed the urogenital morphology of dipnoans, with comparisons to other fishes and to amphibians. Lungfish kidneys are paired, somewhat lobed, elongate, highly pigmented retroperitoneal structures. Two distinct portions, the genital or vesicular posterior and the anterior urinary, are distinguished. The nephron in each portion has a typical corpuscle and glomerulus, a ciliated neck segment, a proximal tubule of two components, a ciliated intermediate segment, a distal segment, and a collecting duct system. The cytological features of the nephron of lungfishes are similar in all aspects to those of a number of "typical" freshwater teleosts. The posterior nephrons are capable of urinary function. It is not known specifically whether they are functional at any time, nor is it known whether lungfish urine would inhibit sperm in any way.

The modifications of the kidney for sperm transport is distinctly different in dipnoans when compared with amphibians. In dipnoans the posterior kidney is involved in sperm transport, and the nephrons are essentially unmodified from typical urinary structures. In amphibians the anterior mesenephros is utilized for sperm transport if the kidney is used reproductively. In gymnophiones the nephrons are unmodified; in anurans and urodeles the nephrons are altered for sperm transport, usually lacking a capsule and glomerulus.

The chondrosteans, holostean neopterygians, amphibians, dipnoans, elasmobranchs, and even cyclostomes share a primitive pattern of pronephric development and structure. Reduction in number of pronephric units is a general derived state in each lineage. Teleosts show a similar pattern but with a greater diversity of derived states. In addition, mesonephric development shows considerable diversity among cyclostomes, elasmobranchs, primitive osteichthyeans, dipnoans, teleosts, and amphibians. Therefore, there are no shared derived states that support the hypothesis of a sister-group relationship between amphibians and dipnoans, unless one were to assume that gymnophiones are ancestral to sharks, osteichthyans, and other amphibians, and are a sister group to cyclostomes. The pattern is one of convergence in reduction of pronephric units in all vertebrate lineages.

Wake (1986) pointed out that analysis of comparative biology and evolution of the dipnoan urinogenital system is incomplete for several reasons (see details in paper) (refer also Chapter 12).

4. EXCRETION OF NITROGENOUS COMPOUNDS

The renal excretion of nitrogenous compounds plays an important role not only in the elimination of nitrogen end products but also in regulating the acid-base and osmotic balance of body fluids. The major nitrogenous compounds excreted by the vertebrate kidney are ammonia, urea and uric acid. An edited volume by P.A. Wright and P.M. Anderson (2001) has provided details of nitrogen excretion in fish.

What makes the fishes unique and of particular interest from a comparative standpoint is the wide variety of nitrogenous end products found in these species and their adaptive significance with respect to varying degrees of water availability and embryonic habitat.

A review by Mommsen and Walsh (1992) has presented the biochemical and environmental perspectives on nitrogen metabolism in fishes-specifically the teleosts. Catabolism of proteins and amino acids make a major contribution to the total energy production of fishes.

The general features of ammonotelism, ureotelism and uricotelism as they relate to evolution of vertebrates have been widely documented. Wide differences in patterns of nitrogen excretion exist in different kinds of fishes such as cyclostomes, cartilaginous fishes, teleosts and lungfish and within similar groups of fishes occupying different ecological niches. Thus, the cyclostomes and teleosts are ammonotelic and the cartilaginous fishes are ureotelic. One type of lungfish, *Neoceratodus*, which does not leave the water is ammonotelic whereas *Protopterus* is primarily ammonotelic in water but becomes entirely ureotelic during aestivation. It is clear, therefore, that among the fishes the evolution of the part of metabolism concerned with the formation of nitrogenous end products has been greatly affected by the pressures of the environment. Related information is obtainable in the contribution by Karnaky (1998).

Gulf toadfish (*Opsanus beta*) has been subjected to a variety of experiments related to the subject (Barber and Walsh 1993; Walsh et al. 1994a, b. 1999; Anderson and Walsh 1995; Walsh and Milligan 1995; Wood et al. 1995; Hopkins et al. 1999; Part et al. 1999; Kong et al. 2000).

Ammonia: Amonia is the end product of protein catabolism and is stored in the body of fish in high concentrations relative to basal excretion rates. Ammonia, if allowed to accumulate, is toxic and is converted to less toxic compounds or excreted.

In fishes, ammonia and urea are the major nitrogenous end products with ammonia comprising at least 80% of nitrogen excretion in most teleosts. The major pathway for the production of ammonia is through the transamination of various amino acids. The primary site for ammonia production is probably the liver, but the necessary enzymes have also been located in the kidneys, gills, and skeletal muscle tissue. Figs. 11.11–11.13 illustrate the sources and fate of ammonia in fishes.

The gills are the major site of ammonia excretion in fishes, but smaller quantities of ammonia may also be eliminated by the kidney and skin. Although the majority of branchial ammonia excretion represents clearance from the blood, gill metabolism may contribute between 20% and 5–8% to net ammonia excretion. The excretion of ammonia by fishes is variable, depending on the state of the animal, the environmental conditions and the species (Randall and Wright 1987; see also Evans 1987). The mechanism (s) of branchial ammonia excretion remain controversial (read Wood 1993). A reading of contribution by Watts and Watts (1974) should convey preliminary information on metabolic origin and fate of ammonia.

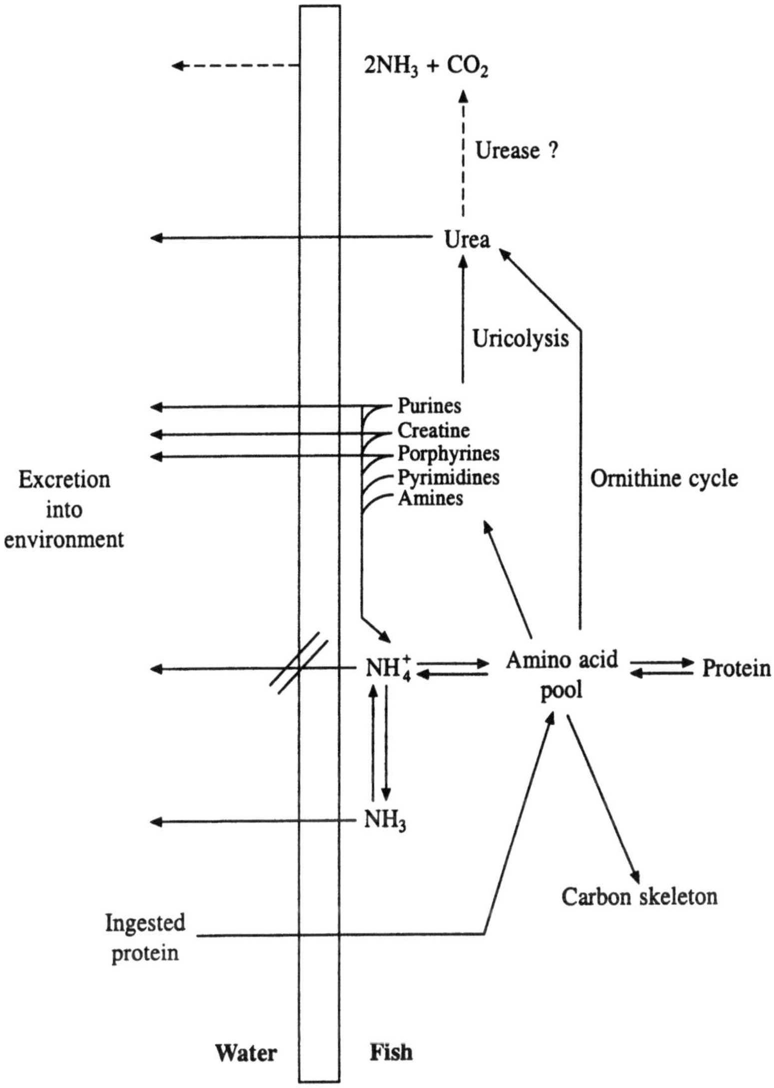

Fig. 11.11 Metabolic pathways of ammonia production and utilization of fish. It is not clear whether urease is present in fish, but some ammonia may be formed via this pathway.
***Source:* Randall and Wright (1987)**

Ammonia as the main nitrogenous metabolic end product of ammonotelic organisms is produced at a comparatively high rate of up to 15% of oxygen consumption. As opposed to urea, the nitrogenous end product ammonia is advantageous with respect to the amount of energy produced during protein catabolism, but provides problems by its pronounced toxicity. According to its high pK′ value of about 9.6, it is immediately ionized upon production, resulting in a net alkalinization. This alkalinizing effect is reversed upon elimination from the body fluids, as long as NH_3 is released by non-ionic diffusion leaving behind the H^+ ions originally buffered at the site of production. Release as NH_4^+ by ionic diffusion or by active carrier-mediated ion exchange with the associated permanent removal of H^+ ions may be advantageous during acidic acid-base disturbances, but will result during steady-state conditions in a progressive alkalinization of the

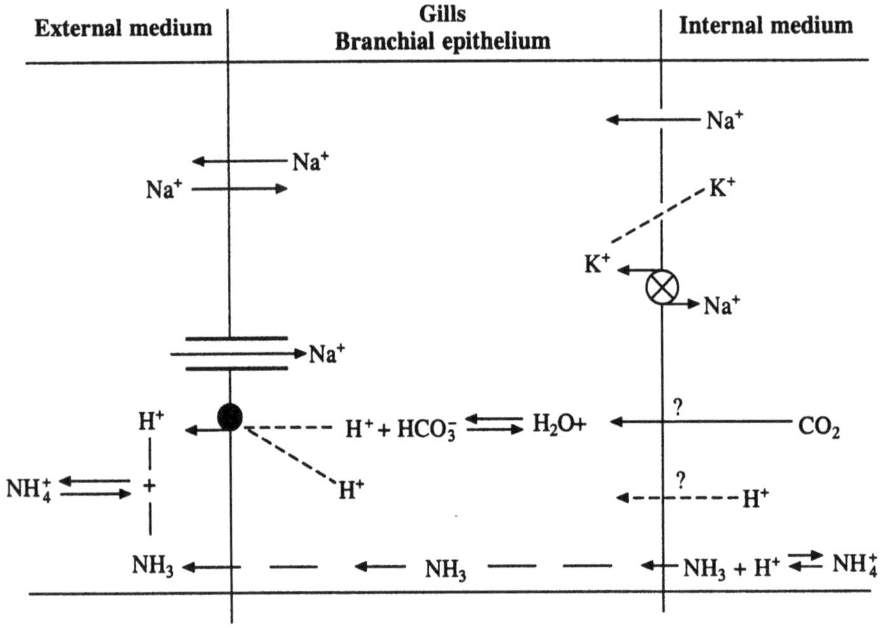

Fig. 11.12 **Schematic representation of sodium, ammonia and proton transport through the gill epithelium of freshwater rainbow trout (*Salmo* gairdneri). Active pumps are denoted by circles → and → are diffusion processes; ? indicates diffusion when electrochemical conditions are favourable. CO_2 enters the branchial cell, where it is hydrated by carbonic anhydrase to produce bicarbonate and protons. The latter are actively excreted electrogenically (together with internal and intracellular protons?) and could be responsible for the passive entry of Na^+ through an apical channel. Then, intracellular sodium is actively transported across the basal membrane by Na^+-K^+/ATPase. Ammonia crosses the serosa and mucosa very quickly as the NH_3, form, the rate being a function of Δ pNH_3 between the internal and external media. In the external medium, it equilibrates with NH_4^+ according to the environmental pH and proton availability.** ***Source:*** **Avella and Bornancin (1990)**

body fluids if this load is not corrected for by acid-base regulatory mechanisms. Depending on the mode of transfer through various body fluid compartments, ammonia may accordingly provide a more or less significant stress for pH regulation.

The following conclusion has been drawn after a close analysis of the contribution of mechanisms involved in ammonia transfer along the elimination pathway through the body fluids.

The similarity of the relevant diffusion coefficient of NH_3 and NH_4^+, together with the much higher diffusion gradients for NH_4^+, suggests that the ammonia transfer along aqueous diffusion pathways takes place mainly by ionic diffusion, facilitated by simultaneous transfer of HCO_3^- originating from metabolic CO_2. The evidence with respect to the form in which ammonia is preferentially transferred across cell membranes is still controversial. Large ratios of $D_{NH_3}/D_{NH_4^+}$ determined in tight epithelia and the extremely unlikely implications of ionic movement clearly favour non-ionic diffusion as the key mechanism. Elimination from body fluids of ammonia takes place mainly at the gill epithelium. At low, natural environmental concentrations translocation occurs mainly be passive diffusion, non-ionic in freshwater, both ionic and non-ionic in marine species. Carrier-mediated ionic transfer processes for NH_4^+ are utilized as major elimination mechanisms only when the environmental conditions do not allow establishment of sufficient

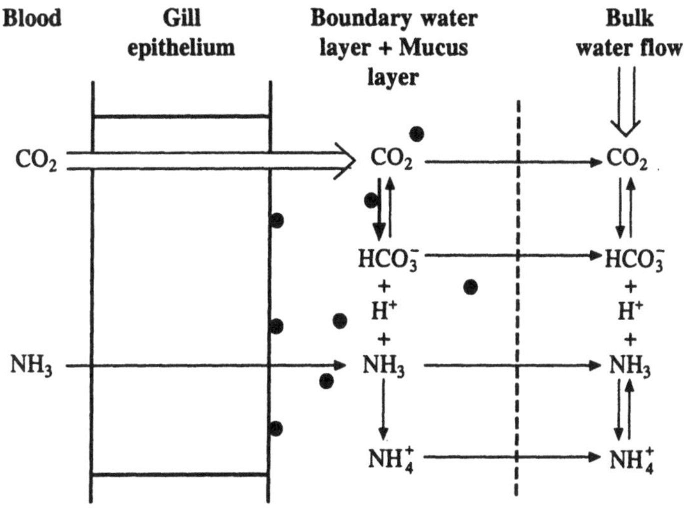

Fig. 11.13 **A simplified cross section through the gill epithelium showing the bulk water flow, the boundary water layer and associated mucus layer containing carbonic anhydrase activity (represented by filled circles). The catalysed hydration of CO_2 in the boundary layers to HCO_3^- and H^+ facilitates ammonia excretion by promoting the interconversion of NH_3 to NH_4^+, thereby preventing reductions in the transepithelial NH_3 diffusion gradient due to accumulation of NH_3.**
Source: **Perry and Laurent (1990)**

P_{NH_3} gradients across epithelium at normal plasma ammonia levels. At high water pH an unstirred gill water layer combined with rapid ionization of CO_2 by carbonic anhydrase at the mucosal epithelial surface is suitable to provide a favourable microenvironment of relatively low ammonia partial pressures and pH at the epithelial surface (Heisler 1990; Walsh 1998).

As evident, in fish, acid-base balance is normally attributed to ion exchange mechanisms operating at the gill. There is, however, growing evidence that renal ammonia excretion, albeit small, may play an important role in the acid-base regulation of these animals, especially for those fish species confronting an acid environment. At low environmental pH, ionic regulation at the gill is severely impaired, resulting in a net influx of acid at this epithelium and leaving renal acid excretion as the only means for compensation to acid-base stress. The tolerance of fish species to acid environments could be related to physiological and biochemical capacities for acid-base regulation including the potential for renal ammoniagenesis. Recent studies have shown that in at least three freshwater species (rainbow trout, catfish and goldfish) acidosis provokes a rise in renal ammonia and titratable acid excretion. A similar response to acidosis has been identified in elasmobranchs, an older species in terms of evolution, suggesting the presence of renal ammoniagenic response as an adaptation for acid-base regulation early in vertebrate history. It has been shown that the rise in ammonia excretion is due to an increase in renal ammonia production rather than changes that may occur in urine flow, urine pH, or plasma ammonia concentration, but on the biochemical capacity of the renal cells to produce NH_3 from a number of amino acids available in the kidney including glutamine, glutamate, and aspartate. At the goldfish gill, on the contrary, acidosis did not stimulate ammonia excretion indicating that the ammoniagenic response to acidosis is specific for the kidney. In addition, it should be noted that the form in which ammonia excretion across the gill occurs (via the diffusion of the NH_3 or NH_4^+/Na^+ ion exchange) determines the effect that this excretion has on acid-base balance (see King and Goldstein 1985).

Although, generally, the renal excretion of ammonia is secondary to branchial ammonia release, it is possible that under specific environmental conditions, renal excretion of ammonia prevails. Such conditions may exist in *Chalcalburnus tarichi*, a cyprinid living in a low-soda, alkaline lake in Turkey, which on average excretes about 63% of its nitrogenous waste as ammonia with urea-nitrogen making up the remainder. Although the environmental conditions are likely to impair, if not prevent, branchial release of ammonia, it is possible that instead of detoxifying ammonia into urea prior to excretion-*de novo* synthesis of urea is absent in the tissues of *C. tarichi*—these fish release ammonia directly via the kidneys or utilize an as yet unknown mechanism to release ammonia across the skin.

In spite of the fact that actual rates of ammonia release through renal pathways may be small compared with total release of ammonia across the gill or, as in the case of marine fishes, across the skin, renal production and release of ammonia may play an important role in acid-base regulation. Teleosts can and do use renal ammonia production, mainly from amino acid amides, to remove and excrete (read Mommsen and Walsh 1992).

Further information is available (see Ye et al. 1991). Fish exposed to acid water become acidotic and lose body electrolytes, particularly Na^+ and Cl^-, due to a decrease in Na^+ influx and an increase in Na^+ and Cl^- efflux across the gills. The sodium chloride loss during acid exposure is associated with a reversal of the gill transepithelial potential resulting in a decrease in the Cl^- electrochemical gradient and an increase in the sodium electrochemical potential.

Acid water causes an extracellular acidosis in fish which reduces blood oxygen content via the Root effect (see also Wood 1989).

Urea: Although the vast majority of teleost fish excrete their waste nitrogen principally as ammonia (ammoniotely), recent studies have expanded the list of teleosts which excrete the bulk of their waste nitrogen as urea (ureotely). Ureotely in teleosts can be obligate or facultative. Elasmobranch fishes and lake Magadi tilapia (*Oreochromis alcalicus grahami*, read further) are obligate ureoteles. Among facultatively ureotelic teleosts, the gulf toadfish (*Opsanus beta*) have received recent attention.

At present, no simple method of detoxyfying ammonia is known, only by complex indirect pathways can ammonia be converted to be less toxic nitrogenous end product such as urea. Urea is a polar compound. Despite the similarity in diffusion coefficients, ammonia passes through many biological membranes much more readily than urea. The reason is thought to be high lipid solubility of the NH_3° form of ammonia and the low lipid solubility of urea.

Urea is a major end product for the excretion of waste nitrogen in mammals, marine elasmobranchs, besides many species of amphibians and chelonian reptiles. For marine elasmobranchs ('ureosmotic'), the function of urea also includes an important osmotic role; it can be accummulated in the whole animal acting as an osmotically active particle to regulate water balance. Urea is retained in body fluids of marine elasmobranchs upto levels of 300 mosm in order to help balance the high osmolarity of seawater environment and thereby maintain water balance. Urea is reabsorbed by marine elasmobranch kidney. In the dogfish (*Squalus acanthias*), approximately 95% of the urea filterated at the glomerulus is returned to the blood. The link between the renal reabsorption of urea and urea retention in elasmobranchs is demonstrated in the freshwater elasmobranch, *Potamotrygon*, which has low concentrations of urea in its body fluids and does not display renal reabsorption of this solute.

The mechanism of urea transport in the marine elasmobranch is unclear. The rate of urea reabsorption has been shown to be directly proportional to plasma concentrations of urea over a wide physiological range. Urea reabsorption takes place somewhere along the distal tubule in dogfish kidney, but the exact site is unknown.

In at least two situations the urea end product has been found advantageous. First, cartilaginous fishes, elasmobranchs and holocephalans since early in their evolution have employed internal fertilization and the sequestration of the young either within a uterus or leathery egg case. Under these conditions where ammonia could reach toxic levels, incorporation of this compound into urea would be advantageous for the embryo. Furthermore, retention of urea has provided both adult and embryonic cartilaginous fishes with a method of maintaining osmotic balance with the otherwise hypertonic seawater environment. Second, the ability of certain lungfishes to convert from ammonotelism to ureotelism during water deprivation has enabled these fishes to aestivate for long periods of time while the relatively inert urea accumulates to high levels (see Goldstein and Forster 1970; Watts and Watts 1974; King and Goldstein 1985).

A high tissue urea concentration (1000–2000 mg%) is characteristic of marine elasmobranchs. A study of *Raja binoculata*, an oviparous form, suggests that this high urea level is established very early in embryonic development and remains constant through to adult life. Only the coelacanth, *Latimeria chalumnae*, is known to have a tissue urea level matching that of marine elasmobranchs, the reported values being 2130 mg% in the blood and 1700 mg% in the liver. Urea in the African lungfish *Protopterus* builds up to a high level during aestivation, when excretion is prevented, but the normal level is quite low.

The Holocephali also possess the ability to retain urea in their body fluids. 244–287 mmoles per litre of urea in the blood of *Chimaera monstrosa* was recorded. This was compared with 349–356 mmoles per litre for the hagfish, *Myxine glutinosa*. It was concluded that the blood of *Chimaera* was probably iso-osmotic with seawater, the balance of osmolarity being made up by sodium chloride.

In freshwater elasmobranchs, the urea concentration ranges around 200–300 mg% but in a single genus, *Potamotrygon*, with a long history of living in freshwater, the concentration is only 2–3 mg%. This value for *Potamotrygon* is considerably lower than those reported for freshwater teleosts of 4–6.7 mg% for goldfish and various species of trout.

The metabolic routes potentially leading to the production of urea in fish tissues are given in Figs. 11.14–11.16 (see Mommsen and Walsh 1992, 1998 for details; refer also Wood 1993). A review has been addressed to ornithine-urea cycle (= urea cycle) (see Anderson (1995). The urea cycle in *Squalus acanthias* is shown in Fig. 11.17 (see abbreviations in Figs. 11.18, 11.19). Walsh et al. (1994) covered the topic of urea transport by hepatocytes and red blood cells in fish. (Fig. 11.18).

The site of urea release varies among species and depends on such factors as environmental salinity, branchial urea permeability and urine production rates. Potentially, it can be voided through gills, skin, urine and faeces, or any combination of these four sites. Although largely branchial excretion was identified for freshwater species, a more recent study of five marine teleosts-in which urea accounted for less than 20% of nitrogenous waste showed that release

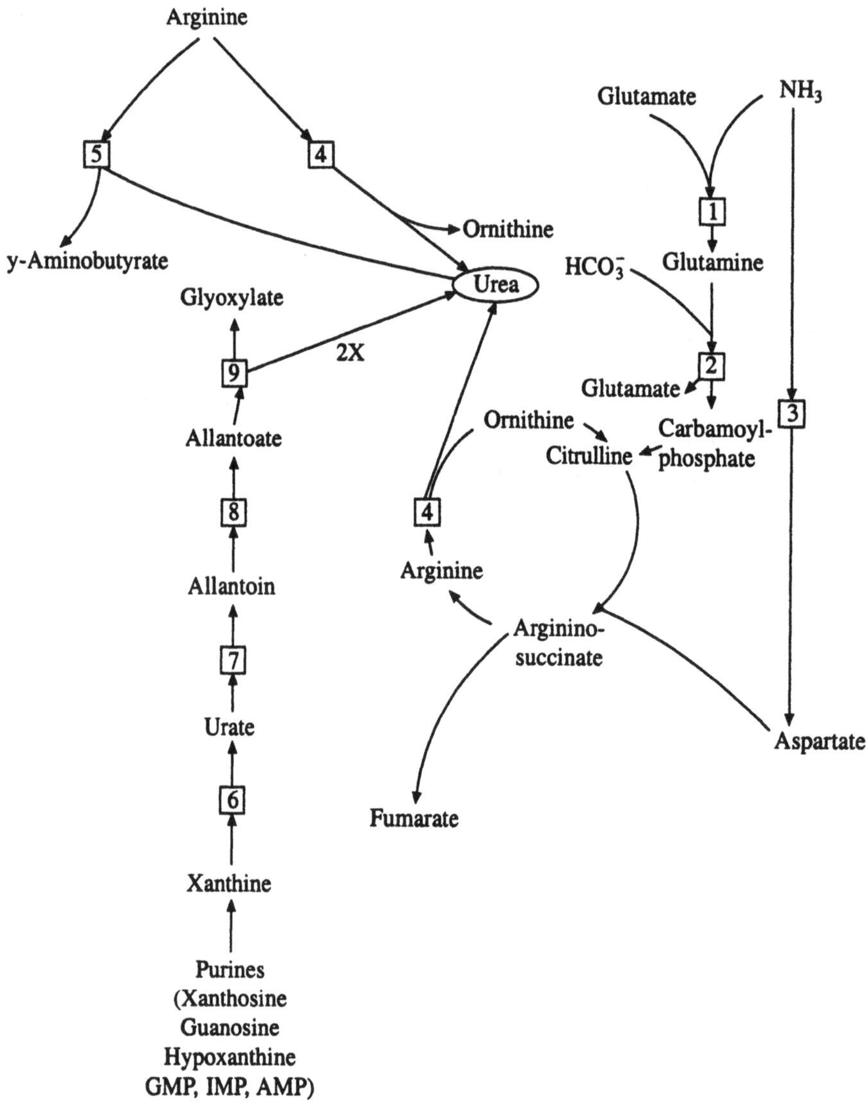

Fig. 11.14 Metabolic routes potentially leading to the production of urea in fish tissues. Three different pathways are presented: Arginine breakdown, purine degradation and synthesis by the ornithine-urea cycle. Key enzymes involved are: 1. Glutamine synthetase; 2. Carbamoyl phosphate synthetase (CPS III); 3. Glutamate dehydrogenase followed aspartate aminotransferase; 4. Arginase; 5. The y-guanidino urea hydrolase pathway; 6. Xanthine oxidase; 7. Urate oxidase; 8. Allantoinase; 9. Allantoicase. ATP-consuming reaction in the ornithine-urea are: synthesis of glutamine from glutamate and ammonia (1 ATP), synthesis of carbamoyl-phosphate from bicarbonate and glutamine (2 ATP), and synthesis of argininosuccinate from citrulline and aspartate (2 ATP).
***Source:* Mommsen and Walsh (1992)**

across the skin or via the urine was two-to three-fold higher than branchial excretion. The potential separation between sites for ammonia and urea excretion is exceptionally well illustrated in an air-breathing catfish (*Heteropneustes fossilis*), which releases more than 99% of its waste

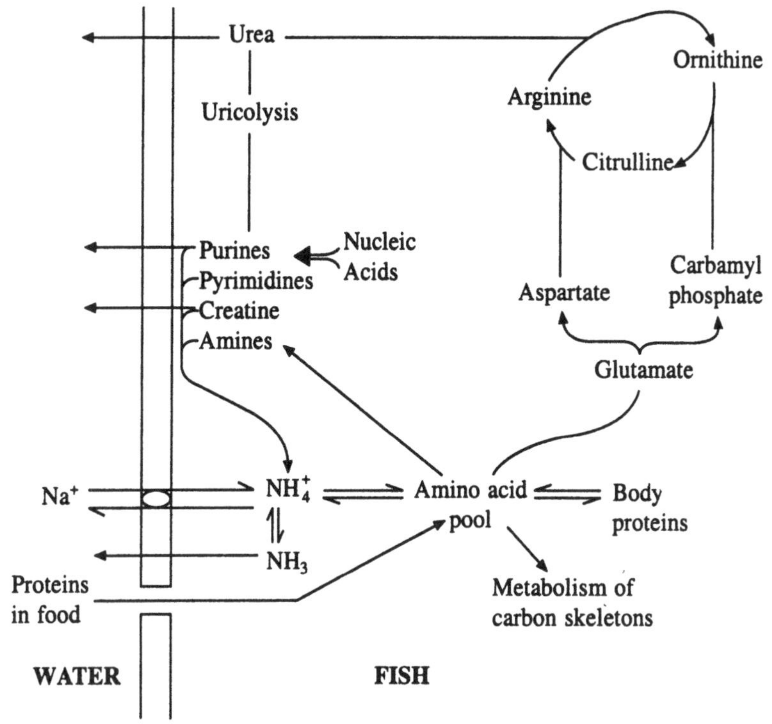

Fig. 11.15 Pathways for production and excretion of nitrogenous waste in fish. Vertical bar indicates the water-fish barrier, and the ion exchange of Na^+ for NH_4^+ is also shown. In the majority of teleosts, urea is synthesized by uricolysis since some key enzymes of the ornithine-urea cycle are absent. Urea synthesis in cartilaginous fish occurs primarily through the ornithine-urea cycle. ***Source:*** **Jobling (1993)**

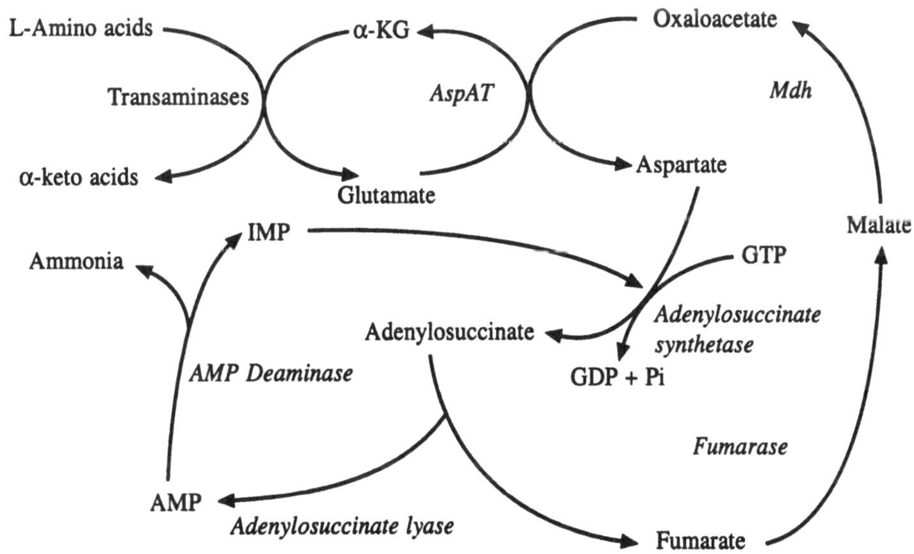

Fig. 11.16 Purine nucleotide cycle which operates extensively in fish muscle postexercise to scavenge AMP (produced by ATP hydrolysis). Ammonia production is believed to promote muscle intracellular pH adjustment by subsequent consumption of protons to yield NH_4^+ -AspAT = aspartate amino transferase; Mdh = malate dehydrogenase; other abbreviations as in Figure 11.18. ***Source:*** **Walsh (1998)**

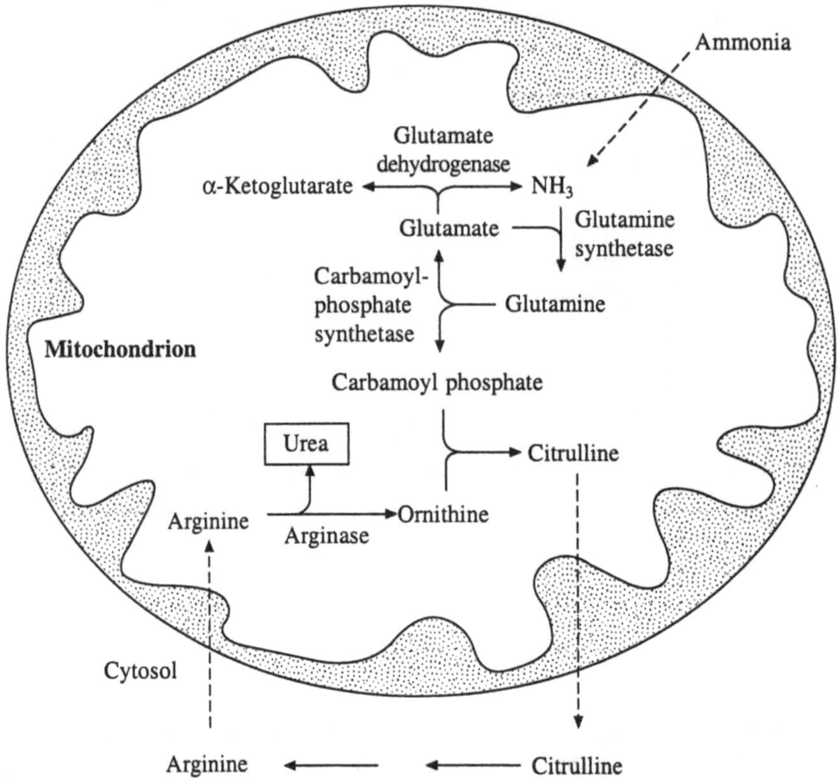

Fig. 11.17 Illustration of glutamate-dependent urea cycle in elasmobranchs.
***Source:* Anderson (1995)**
***Source:* Anderson (1995)**

ammonia across the gill, while renal and urinary release accounts for more than 25% of total urea discharge. Urea normally makes up less than 40% of the nitrogenous waste and may build up in fish plasma and tissues to much higher concentrations (1–10 mM) than the toxic ammonia. This partial accumulation may indicate that voiding the system is not quite as straightforward as expected, and ultimately, it is possible that the molecule functions as a minor osmolyte in teleostean fishes. In contrast to continuous release of ammonia, the excretion of urea may be intermittent. Ureogenic toadfish (*Opsanus beta*), for instance, temporarily store urea in the bladder before pulsatile release into the surrounding water. The pulse event is not triggered by a specific plasma urea threshold, but may involve the hypothalamus-interrenal axis (Wood et al. 1997). For further reading, refer Walsh (1997, 1998) and Part et al. (1999).

Wood et al. (1995, 1997) focused on the rate and routes of urea-N and ammonia-N excretion in the ureagenic toadfish (*O. beta*) and on the possibility that urea-N excretion occurs in pulses. They speculated that the normal adaptive significance is that ureotelism facilitates cryptic behaviour, allowing the toadfish to virtually eliminate N-waste excretion during long periods while it remains sheltered in burrows. However, during severe stress, the effects of high cortisol levels overwhelm the ammonia and urea retention mechanisms, and both substances leak across the general body surface. (Fig. 11.19).

Urea is quite evenly distributed throughout the fish's body, since it readily crosses internal

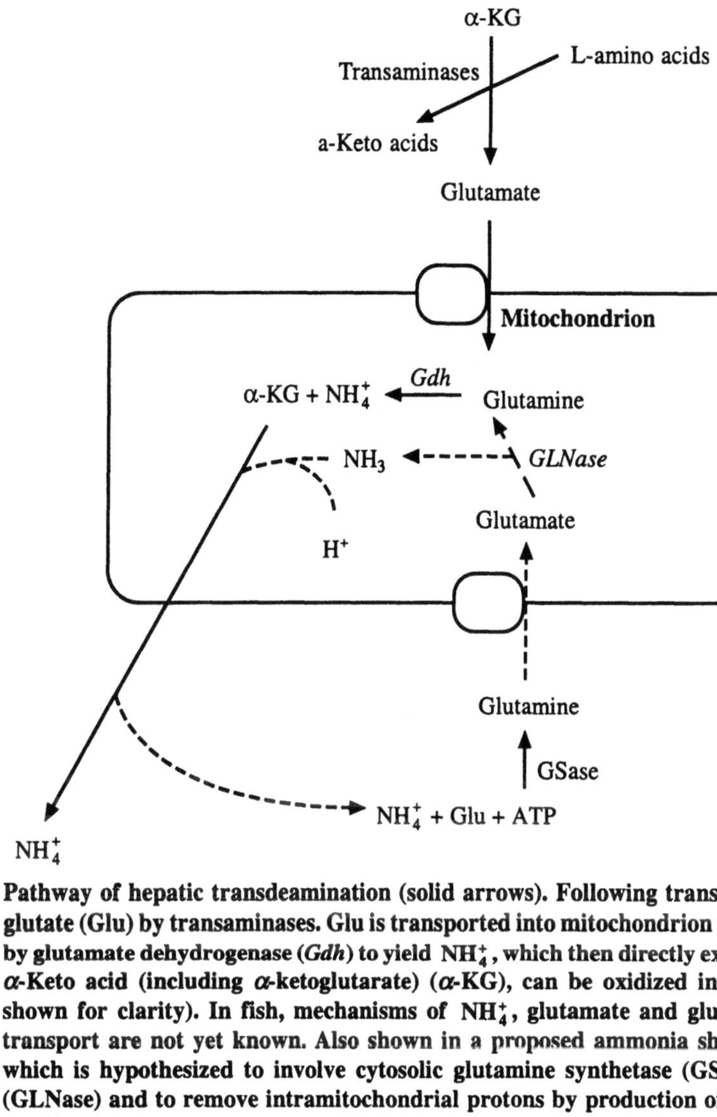

Fig. 11.18 Pathway of hepatic transdeamination (solid arrows). Following transfer of aminogroups to glutate (Glu) by transaminases. Glu is transported into mitochondrion where it is deaminated by glutamate dehydrogenase (*Gdh*) to yield NH_4^+, which then directly exits the mitochondrion. α-Keto acid (including α-ketoglutarate) (α-KG), can be oxidized in the Krebs cycle (not shown for clarity). In fish, mechanisms of NH_4^+, glutamate and glutamine mitochondrial transport are not yet known. Also shown in a proposed ammonia shuttle (dashed arrows) which is hypothesized to involve cytosolic glutamine synthetase (GSase) and glutaminase (GLNase) and to remove intramitochondrial protons by production of NH_3 and subsequent reaction to NH_4^+.
***Source:* Walsh (1998)**

membranes. There are some restrictions to the free movement of urea, however, mainly pertaining to the gill and membranes associated with the kidney. Even though elasmobranch fishes utilize urea as the important intracellular osmolyte, the elasmobranch kidney excretes urea. Nonetheless, a certain portion of urea is subsequently reabsorbed by the kidney. Nothing is known about tubular transport of urea in teleosts, but the phenomenon has been extensively studied in marine elasmobranchs. The proximate controls on urea reabsorption remain unclear (Walsh et al. 1994b; also read Walsh 1998).

It is possible that the mere measurements of urea excretion in fishes will conceal the true rates of internal urea production. In marine elasmobranchs, which possess internal levels of urea in excess of 200 mM, a certain degree of urea destruction may occur internally. In sharks, this

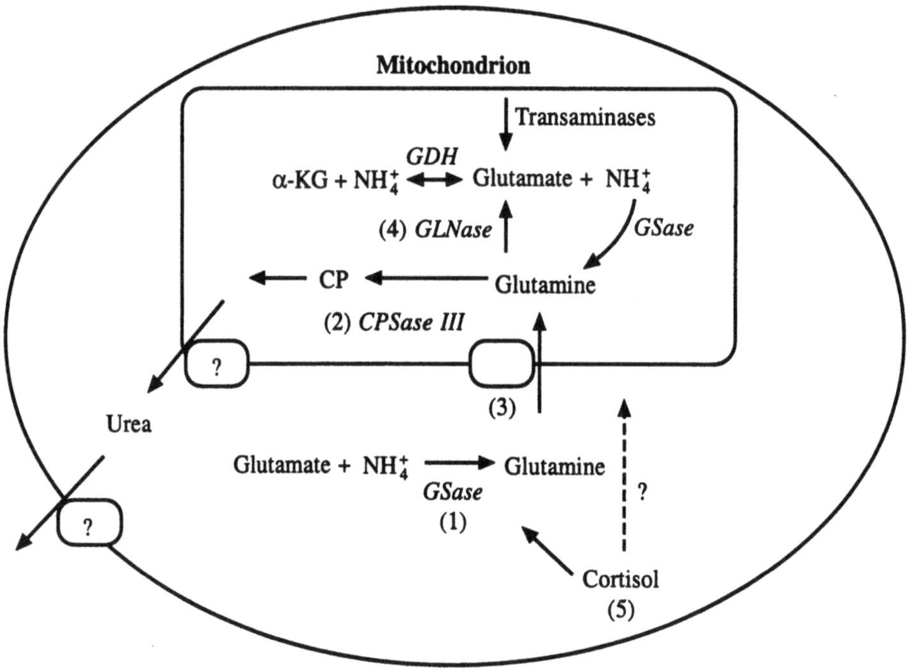

Fig. 11.19 Schematic of proposed regulatory points for upregulation of ureogenesis of in gulf toadfish *Opsanus beta* liver. Abbreviations: KG,-ketoglurate; CP, carbamoyl phosphate; CPSase III, carbamoylphosphate synthetase III; GSase, glutamine synthetase; GDH, glutamate dehydrogenase; GLNase, glutaminase, Downstream region of O-UC and Krebs' cycle omitted for clarity. Also depicted are putative urea transporter.
***Source:* Walsh (1997)**

process is mediated by microbial processes in several tissues as well as in the lumen of gastrointestinal tract. In the teleost toadfish, which may achieve plasma urea concentration in the 10 mM range and bladder concentrations approaching 20 mM, similar urea destruction occurs only in the intestine. (see for details: Mommsen and Walsh 1992).

Mommsen and Walsh (1989) described the evolution of urea synthesis in vertebrates, with emphasis on piscine connection. Elasmobranch fishes, the coelacanth, aestivating lungfish, amphibians, and mammals synthesize urea by the ornithine-urea cycle; by comparison, urea synthetic activity is generally insignificant in teleostean fishes. Isolated liver cells of two teleost toadfishes, *Opsanus beta* and *O. tau* synthesize urea by the ornithine-urea cycle at substantial rates. Because toadfish excrete ammonia, do not use urea as an osmolyte, and have substantial levels of urease in their digestive systems, urea may serve as a transient nitrogen store, forming the basis of a nitrogen conservation shuttle system between liver and gut as in ruminants and hibernators. Toadfish synthesize urea using enzyme and subcellular distributions similar to those of elasmobranchs: glutamine-dependent carbamoyl phosphate synthethase (CPS III) and mitochondrial arginase. In contrast, mammals have CPS I (ammonia-dependent) and cytosolic arginase. Data on CPS and arginases in other fishes, including lungfishes and the coelacanth, support the hypothesis that the ornithine-urea cycle, a monophyletic trait in the vertebrates, underwent two key changes before the evolution of the extant lungfishes: a switch from CPS III to CPS I and replacement of mitochondrial arginase by a cytosolic equivalent. The reader is also referred to Watts and Watts (1974), Brown and Brown (1985) and Anderson (1995). For the effects of confinement/crowding on ureogenesis in *Opsanus beta*, see Walsh et al. (1994a). Anderson and Walsh (1995) supported the hypothesis that glutamine synthetase may be an important regulatory control point in determining rates of ureogenesis in *O. beta*. Other contributions of interest are from Walsh and Milligan (1995), Walsh (1997, 1998), Chadwick and Wright (1999), Lindley et al. (1999), Terjesen et al. (2000) and Todgham et al. (2001).

Griffith (1991) analyzed a scenario for the evolution of urea retention in fishes, with special attention on *Latimeria*. *Latimeria* is a classic marine ureosmotic regulator in which urea is used as an osmolyte that allows osmotic equilibrium with sea water while maintaining low ion levels. It is suggested that ureosmotic regulators evolved from freshwater or anadromous ancestors that entered the marine habitat. Once early ureosmotic regulators were established in the sea there would have been strong selection for internal fertilization and development, as is seen in *Latimeria* and many elasmobranchs. It is suggested that ureosmotic regulation was a common strategy in Paleozoic marine gnathostomes.

From the foregone account, it is evident that ammonia is the major end product of N-metabolism in almost all Osteichthyes, agnathans, and freshwater Chondrichthyes, but less important than urea in marine Chondrichthyes, dipnoans during aestivation, and a very few teleosts living in extreme environments (Tables 11.4–11.6).

Table 11.4 Rates (μ mol-N-kg^{-1}.h^{-1}) and relative percentages of N-excretion as ammonia-N and urea-N through gills and kidney

Species	Medium	Gills		Kidney	
		Ammonia-N	Urea-N	Ammonia-N	Urea-N
Agnatha					
Entosphenus tridentatus	Fresh water	139 (95%)	0 (%)	6 (4%)	1 (1%)
Osteichthyes					
Cyprinus carpio[a]	Fresh water	315 (88%)	27 (7%)	15 (4%)	3 (1%)
Cyprinus carpio[b]	Fresh water	115 (82%)	11 (8%)	15 (10%)	0 (0%)
Heteropneustes fossilis[a]	Fresh water	236 (85%)	30 (11%)	1 (0%)	11 (4%)
Blennius pholis[b]	Seawater	51 (35%)	26 (18%)	58 (39%)	12 (8%)
Chondrichthyes					
Pristis microdon[a]	Fresh water	250 (18%)	759 (55%)	27 (2%)	336 (25%)
Squalus acanthias[a]	Seawater	13 (2%)	519 (91%)	0 (0%)	37 (7%)

Note: The percentages assume that the total amount of ammonia-N plus urea-N is 100%.

(a) Kidney excretion measured by urinary catheter. Excretion through skin and gastrointestinal tract is therefore included in the gill component.

(b) Kidney excretion measured by divided chamber. Excretion through skin and gastrointestinal tract is therefore largely included in the kidney component.

Data source: Wood (1993). Abridged.

Table 11.5 Concentrations of (mmol-N.1^{-1}) Ammonia-N in the blood plasma

Species	Medium	Ammonia-N	Urea-N
Agnatha			
Lampetra fluviatilis[a]	Fresh water	0.38	"0"
Myxine glutinosa[a]	Seawater	"0"	5.60
Osteichthyes			
Cyprinus carpio[a]	Fresh water	0.29	1.71
Heteropneustes fossilis[a]	Fresh water	0.47	1.52
Amia calva[b]	Fresh water	0.31	0.22
Oncorhynchus clarki henshawi[b]	10% Seawater[c]	0.17	8.15
Oreochromis alcalicus grahami[a]	50% Seawater[d]	0.77	10.52
Chalcalburnus tarichi[a]	63% Seawater[e]	1.30	36.18
Opsanus beta[b]	Seawater	0.26	16.10
Protopterus aethiopicus[a]	Fresh water	"0.71"	1.21
Protopterus aethiopicus[b]	Aestivation (13 months)	–	406
Latimeria chalumnae[a]	Seawater	–	754
Chondrichthyes			
Potamotrygon sp.[a]	Fresh water	"0"	2.20
Squalus acanthias[a]	Seawater	0.33	672
Scyliorhinus canicula[a]	Seawater	0.66	814
Chimaera monstrosa[a]	Seawater	0.31	664

(a) Sampled by caudal or cardiac puncture.
(b) Sampled by in-dwelling catheters.
(c) Pyramid Lake, pH 9.4.
(d) Lake Magadi, pH 10.
(e) Lake Van, pH 9.8.
Data source: Wood (1993). Abridged.

Uric acid: In fish, the liver appears readily able to convert uric acid to urea which, in teleosts, explains the presence of significant quantities of urea in the blood despite the absence of complete hepatic urea cycle in these fishes.

Taurine: Amino acids are normally avidly conserved by the renal tubule. Since these solutes are readily filtered at the glomerulus, conservation is dependent on renal tubular transport mechanisms for reabsorption. Net reabsorption is characteristic of the renal handling of most amino acids for the kidneys of all the vertebrate classes, including fish.

Although net renal secretion of amino acids is rarely observed, recent studies of three marine fish species have shown that taurine, a sulfonated amino acid, is secreted by the renal tubules of these fish. Taurine is one of the most commonly occurring amino acids in the urine of vertebrates ranging from marine fish to mammalian species. The renal clearance of taurine has been found to be modulated in response to particular stresses, including changes in environmental salinity for the elasmobranch. The regulatory mechanism controlling secretion is unknown.

Trimethylamine oxide (TMAO): Another nitrogenous compound of general occurrence among the fishes is trimethylamine oxide. In at least one group of fishes, the elasmobranchs, it occurs

Table 11.6 Whole-body rates (μ mol-N.kg^{-1}.h^{-1}) of ammonia-N and urea-N excretion in aquatic Osteichthyes, Osteichthyes of alkaline lakes, and air-breathing Osteichthyes

Species	Medium	Ammonia	Urea-N	OUC
Aquatic Osteichthyes				
Cyprinus carpio	Fresh water	330 (92%)	30	
Oreochromis nilotica	Fresh water	763 (85%)	138	
Engraulis mordax	Seawater	293 (80%)	71	
Agonus cataphractus	Seawater	645 (84%)	127	
Limanda limanda	Seawater	490 (93%)	36	
Osteichthyes of Alkaline Lakes				
Orechromis alcalicus grahami	Lake Magadi pH = 10, ~50 seawater	0 (0%)	7,600	Y
Oncorhynchus clarki henshawii	Pyramid Lake pH = 9.4, ~10% seawater	123 (66)	63	N
Chalcalburnus tarichi	Lake Van, pH = 9.8, ~63% seawater	1,041 (63%)	607	N
Air Breathing Osteichthyes				
Clarias batrachus	Fresh water	421 (92%)	37	Y
Heteropneustes fossilis	Fresh water	254 (91%)	24	Y
Anabas testudineus	Fresh water	417 (96%)	16	Y
Channa punctatus	Fresh water	292 (97%)	9	Y
Opsanus beta	Seawater	75 (77%)	22	Y
Periophthalmus expeditionium	Seawater	771 (67%)	383	N
Periophthalmus gracilis	Seawater	548 (86%)	92	N

Note: The percentage (%) occurring as ammonia-N assumes that the total of ammonia-N plus urea-N is 100%. The presence (Y) or absence (N) of significant ornithine urea cycle activity (OUC) in the liver is noted for those species for which the information is available.
Data source: Wood (1993), Abridged.

in high concentrations (approximately 80 mM). It is actively retained and contributes about one-third to the total organic osmolarity of body fluids.

In *Squalus acanthias*, the muscle level is much lower and ranges around 200 mg%. In teleosts, the concentration range is lower but equally variable, 15–250 mg% for marine teleosts and generally less in freshwater teleosts, 25–90 mg%. Euryhaline species captured from seawater were found to have a higher TMAO concentration than when taken from freshwater. The red muscle is generally considered to be richer in TMAO than the white muscle. TMAO may be of dietary origin or perhaps as a detoxification product in choline metabolism.

In contrast to the elasmobranchs where TMAO is actively reabsorbed by the kidney, in certain teleosts such as *Lophius americanus* it is actively secreted by an organic base secretory mechanism of general occurrence throughout the vertebrates. Furthermore, TMAO may play a role in regulation of intracellular osmotic balance. Creatine, creatinine and a variety of other nitrogenous end products are found in the urine of many fishes (King and Goldstein 1985; see also Goldstein and Forster 1970; Watts and Watts 1994).

5. UREA SYNTHESIS AND STRATEGIES FOR SURVIVAL

The change from an aquatic existence to terrestrial life demands radical alteration in nitrogen metabolism. An immediate need is for the detoxification of ammonia, the normal end product of nitrogen metabolism in the aquatic phase, to a less toxic end product. The change from ammonia to urea demands either suppressing production of ammonia or incorporating it into urea.

In lungfish (*Protopterus dolloi*), the production of urea is lower during aestivation. The rate of metabolism is reduced by two-thirds during aestivation and thus the load of degraded protein nitrogen is reduced. In the aquatic phase some 50% of the energy in *Protopterus* comes from protein, in aestivation approximately 100% of the energy comes from protein. Therefore, urea synthesis operating at its usual level of activity could handle the reduced rate of non-protein nitrogen. No other route of disposal would be required. Ammonia production *per se* could be suppressed during aestivation by some sort of feedback control. Ammonia could block its own synthesis from glutamate by some type of feedback inhibition.

It is interesting to note that the lungfish, *Neoceratodus forsteri*, has either not acquired or has lost the capability of living on land and uses its lung, only as accessory respiratory organs during periods of excessive activity. These fish produce and excrete mainly ammonia and very little urea. The need to detoxify ammonia does not exist in these fish (Goldstein and Forster 1970).

An entirely different approach to longer-term ammonia detoxification or to the production of nitrogenous wastes is observed in some adapted teleostean fishes. This strategy is found in species that may routinely encounter conditions of erratic water availability, when dilution of toxic ammonia into the surrounding water is clearly finite, or when the external conditions negatively impact ammonia excretion. At present, this growing list includes representative species of clingfishes, toadfishes, catfishes, gouramis, mudskippers, snakeheads, frogfishes (?), a tilapia and several air-breathers. The African lungfishes also belong into this heterogenous group. All these animals may be able to switch to preferential production of urea during periods of water limitation or they may be permanently ureogenic. However, instead of using urea liberated from the catabolism of purines or endogenous arginine, these fishes synthesize urea *de novo*. They do so by use of ornithine-urea cycle. Because of the substantial cost involved-about 5 high-energy phosphates are expended for two nitrogens fixed—it is usually assumed that aquatic fishes resort to this metabolic expenditure only under extreme physiological conditions. Thus, these species avoid the long-term buildup of toxic NH_3, while as usual, the brain is protected from short-term impacts by the deoxification mechanisms outlined above (read details: Mommsen and Walsh 1992; see also Evans 1987).

Interesting details regarding urea excretion as a strategy for survival in a very alkaline environment are available in recent papers on tilapia fish *Oreochronus alcalicus grahami*, the only cichlid fish living in Lake Magadi (Kenya), an alkaline 'soda' lake, fed by hot springs and almost covered by solid crust crystalline 'trona' (upto 5 m in depth) and composed of Na_2CO_3 and $NaHCO_3$. Temperatures may reach more than 40°C, osmolarity more than 600 m Osm/kg, and pHs more than 10.

The Lake Magadi tilapia possesses a complete set of ornithine-urea cycle (OUC) enzymes in the liver and appears to produce urea mainly be OUC (Walsh 1997). However, the sites, mechanisms and all of the possible functions of urea excretion remain unknown. Complete reliance on ureotelism makes this fish species unique; most teleosts are ammoniotelic and have negligible hepatic OUC activity. Indeed, only a handful are capable of OUC-based ureogenesis, and these are generally ammoniotelic under normal conditions [e.g., the toadfish (Mommsen and Walsh 1989) and certain air-breathing fish of India (Saha and Ratha 1987, 1989)]. It has been argued that this fish excretes urea-N because it would be difficult or impossible to excrete ammonia across the gills into a highly buffered water where the pH (10) is so much greater than pK (9.1) of the NH_3/NH_4^+ reaction. Urea production in the Lake Magadi tilapia does not appear to remove exogenous HCO_3^- or to play a role in normal acid-base regulation.

Further, it has been predicted that control of ureogenesis in the tilapia is probably regulated substantially at the substrate level, namely in the supply to the liver of ammonia and amino acids from the diet, and bicarbonate from the environment, rather than by large changes in enzyme concentrations *per se*. This strategy may be contrasted with the response of the normally ammoniotelic, but potentially ureogenic, teleost *Heteropneustes fossilis*, an air-breathing fish with hepatic OUC capacity (Saha and Ratha 1986). See further: (Randall et al. 1989; Wood et al. 1989; Walsh et al. 1993, 1994b, Anderson 1995). The reader is also referred to Walsh and Henry (1991) for carbon dioxide and ammonia metabolism and exchange. For comprehensive information on the end products of amino acid metabolism, see Cowey and Walton (1989).

6. ENERGETICS OF OSMOTIC REGULATION IN UREOTELIC AND HYPOOSMOTIC FISHES

Kirschner (1993) reviewed the major mechanisms of urea turnover and of active ion transport in osmoregulatory organs. The energetic cost of osmotic regulation in hyporegulating marine teleosts and in ureotelic elasmobranchs was estimated. Both groups of fish used 10–15% of their standard metabolism for osmotic regulation. There are a number of uncertainties that may require modification of these values. It is not known whether TMAO lost by ureotelic regulators is replaced by ingestion or by synthesis. If by synthesis, the cost of regulation would be higher than calculated. Even more serious is the uncertainty regarding the use of organic osmolytes to maintain equilibrium between cells and blood. This strategy almost certainly requires some energy input, but there is insufficient information to permit its quantitative estimation, so the factor was omitted from the calculation. Among hyporegulators the sole unknown is whether the oesophagus may absorb some of the ingested salt by diffusion. If so, it would decrease the energy expended in the intestinal tract and lower the total cost of regulation. The conclusion, based on known factors for the species examined (flounder *Platichthys flesus*, rainbow trout *Salmo gairdneri*, dogfish *Scyliorhynus canicula* and skate *Raja erinacea*), is that both modes are about equally efficient. If it becomes possible to factor in the unknowns it may be that hyporegulation is slightly more economical than ureosmotic.

7. ACID-BASE REGULATION

Claiborne (1998) has dealt with this aspect in fish. pH is a central parameter for the function of metabolism and, accordingly, for the production of energy, which is essential for maintenance of life. Certain key enzymes (such as phosphofructokinase, PFK) exhibit pronounced pH active optima, such that deviations by more than a few tenths of a pH unit may considerably reduce metabolic flux, or even completely inhibit energy production. Regulation of pH within narrow limits is accordingly an indispensable prerequisite for maintenance of life. Heisler (1993; see literature) has described characteristics of fish acid-base regulation in response to exogenous and endogenous disturbances and delineated features of principal mechanisms available to fish for maintenance of homeostasis. Metabolism provides a continuous load of CO_2 and nonvolatile acid-base relevant end products, which have to be eliminated in order not to accumulate. The rate of endogenous production of acid-base relevant ions may rise tremendously during extreme muscular activity and during extreme hypoxia, when lactic acid is released by anaerobic metabolism. Exogenous disturbances of acid-base regulation are induced by changes in environmental temperature, oxygen, and CO_2 concentrations, as well as by low water pH and associated effects of toxic substances on ion transfer epithelia.

Butler and Day (1993) supported the observations of other workers that pHi (intracellular pH) of some tissues can be regulated independently of extracellular pH. At least at the lower end of temperature range, the pHi of some tissues, especially the muscles, may be completely independent of temperature. Whether there is a seasonal component to this phenomenon remains to be seen, as does its biological relevances, particularly in an animal that is capable of high levels of swimming performance under winter conditions.

Generally, three possible sites are involved in transepithelial ionic transfer: the skin, the gills, and the kidneys. Chloride cells are one of the important participants in transepithelial ionic transfer. The involvement of secondary circulatory system in acid-base regulation is also indicated. The kidney has a minor contribution. The predominant role of the gills is well documented (see account of chloride cells). For detailed information, the reader is referred to Ishimatsu et al. (1995).

Ionic transfer in fish is generally considered to be best described as active electroneuronal ion exchange processes of H^+ or NH_4^+, and of HCO_3^- against Cl^-. As an additional mechanism, an electrogenic H^+ pump in the apical membrane of branchial chloride cells has been postulated recently, being linked only indirectly to movement of Na^+ or other cations.

Carbon dioxide and ammonia are both gases that are unique in physiological and biochemical systems because of their high solubilities in water; the solubility, or capacitance of carbon dioxide in water is roughly 28 times that for O_2 and for ammonia it is 23,000-fold higher. The total capacities of water for carbon dioxide and ammonia are even greater as both CO_2 and NH_3 undergo chemical reactions with water and thus exist as both gaseous and ionic species. Other gases are highly soluble in water (e.g., nitrogen) but are chemically inert and thus have less biochemical consequence. CO_2 in water forms an acid, and because the pK′ for the carbon dioxide system is about 6.1, at many environmental pHs (e.g., seawater = 8.0) and at *resting* physiological pH in fish, over 95% of the total carbon dioxide is in the form of HCO_3^-. Conversely, NH_3 in water acts as a base, and since the pK′ for ammonia is high (pK′ = 9 – 10) at typical environmental and physiological pHs about 99% of ammonia is in the protonated form, NH_4^+.

The aqueous solubility and pK′ of both carbondioxide and ammonia are dependent on physical factors such as temperature, salinity and pH. Furthermore, the total amounts of both substances are influenced by biological activity such as photosynthesis and denitrification (see Walsh and Henry 1991).

According to Heisler (1986), the normal acid-base status is the result of a balance between uptake and of acid-base relevant ions (H^+, OH^- or HCO_3^-), and the elimination of those ions from the organism. Maintenance of this balance is complicated by the ammoniotelism of most fish species as compared to terrestrial animals. In addition to the quantitatively most important acid-base relevant metabolic end product, CO_2, which is produced depending on the type of metabolic substrate in an amount of 0.7–1.0 (respiratory ratio) of the oxygen consumption (variable with body temperature), ammonia is produced in teleost fish species in a relative amount of about 10% of the oxygen consumption which in turn represents 60–90% of their nitrogenous waste. Even in the ureotelic elasmobranch which produce more than 70–90% of their nitrogenous waste as urea and utilize this relatively non-toxic and acid-base irrelevant substance as significant fraction of their osmotic plasma activity (~ one half) the ammonia production is higher than the net amount of H^+ ions originated from metabolic production of fixed acids and bases (net ΔH^+ excr). This difference between non-volatile acid production from break-down of sulfur-, phosphorous-and chlorine-containing organic compounds and from incomplete oxidation of fatty acids, amino acids or carbohydrates, and the production of non-volatile bases from oxidation of carbohydrate alkali metal salts or incomplete oxidation of organic bases, is in fish comparable to that in mammals and is eliminated by transepithelial ion exchange mechanisms of H^+ against Na^+, or HCO_3^- against Cl^- in opposite direction.

CO_2 is eliminated from fish mainly in molecular form by passive diffusion through the branchial epithelium, and in small specimens and fish with large relative surface area (e.g., eels) also through the skin.

Much emphasis has now been given on fish gill boundary layer: a site of linkage between carbon dioxide and ammonia excretion (read details: Wright and Randall 1987; Wright et al. 1989; Lin and Randall 1990; Perry and Laurent 1990; Randall 1990; Walsh and Henry 1991, Walsh 1998).

Fish excrete molecular CO_2 and NH_3 into the water passing over the gills. The excretion rates are an order of magnitude greater for CO_2 than for NH_3. Some carbon dioxide is excreted as bicarbonate in exchange for chloride and some ammonia as ammonium ion in exchange for sodium. The former represents about 10% of the total resting carbon dioxide excretion and the latter upto 50% of the total ammonia excretion. Because molecular CO_2 is the dominant excretory product in the gills, the water should be acidified because of CO_2 hydration catalyzed by carbonic anhydrase in the gill mucus and water boundary layer and on the apical surface of the gill epithelium. In neutral and alkaline waters most of the excreted CO_2 should be converted to bicarbonate, but under acid conditions only a small fraction should be hydrated. Thus, the extent of acidification of water as it passes over the gills should decrease with water pH. The opposite should be true for the NH_3/NH_4^+ reaction. Under acid conditions almost all the NH_3 should be converted to NH_4^+ but under alkaline conditions only a fraction of the excreted NH_3 should form NH_4^+ in the gill water.

Carbon dioxide excretion and the subsequent catalyzed hydration reaction acidifies the water boundary layer at the gill surface (see Fig. 11.13). The reader is referred to Gilmour (1998) for Fig. 2 showing a schematic representation of the CO_2 excretion pathway in most fishes and to Fig. 5 showing a schematic model of the linkage between CO_2 and NH_3 excretion in the boundary layer of water immediately adjacent to the gill surface.

Wood (1991) has comprehensively discussed the branchial ion and acid-base transfer in freshwater teleost fish, taking specifically environmental hyperoxia as a probe. Relevant theory, methodology and a synthesis of current ideas are presented based on experiments with hypercapnia, low pH, acid and base infusion, temperature change, enforced exercise, toxicants, and other challenges. Use of this protocol has demonstrated that (1) respiratory acidosis is largely compensated by adjustment of Cl^- versus base exchange, while both Cl^- versus base and Na^+ versus acid exchanges contribute during metabolic alkalosis; (2) these dynamic adjustments in Na^+ and Cl^- influx are achieved by complex changes in both substrate affinity and maximum transport capacity (*Jmax*); Na^+ and Cl^- affinities appear to be maximal under control conditions and can only be decreased but *Jmax* values can be either increased or decreased depending on the internal acid-base status; (3) differential diffusive efflux of Na^+ relative to Cl^- provides an additional mechanism of net acid flux during alkalosis; (4) large decreases in branchial intracellular fluid volume, Na^+, and Cl^- levels occur during respiratory acidosis and are corrected during normoxic recovery; (5) mean gill intracellular pH remains extremely stable. The importance of the internal acid-base status of the transport cells, and a possible regulatory role for extracellular pH, are discussed.

The reader is also referred to Wood and Goss (1990) for kinetic analysis of the relationships between ion exchange and acid-base regulation at the gills of freshwater fish. The agreement of the intracellular and extracellular analyses suggested that the acid-base status of the transporting cells may be linked directly to the acid-base status of the blood.

Based on the measurements of transepithelial potentials of a wide variety of species of marine fishes, Potts and Hedges (1991; see details) reported that most lie within the range 20–24 mV positive indicating that chloride is actively excreted while sodium is close to passive equilibrium. In most species the permeability to potassium is greater than to sodium. In marked contrast to most other families, members of the Syngnathidae showed low of negative potentials. It has been indicated that sodium as well as chloride ions must be actively extruded in this species.

Barber and Walsh (1993) indicated that, in the ureogenic teleost toadfish (*Opsanus beta*), the primary driving force for urea synthesis and excretion appears to be the need to detoxify ammonia, and that acid-base consequences, if any, are probably secondary.

Heisler (1982) discussed acid-base regulation in South-American fresh-water teleost *Synbranchus marmoratus*, a fish capable of switching from exclusive water breathing to exclusive air breathing. Unlike other air breathing fish, *Synbranchus* is not equipped with any specialized gas exchange structures such as lungs, modified swim bladders or extensively enlarged surface areas in the buccopharyngeal chambers, and performs its respiratory gas exchange in both modes with the same structures, mainly the gills and, probably to a smaller extent, the skin.

The acid-base balance tends to differ between obligate water breathers such as cyclostomes, elasmobranchs and teleosts on the one hand, and obligate air breathers such as reptiles on the other. Amphibians and facultative air-breathing fishes, for which gas exchange may take place in both water and air, are in intermediate position (read details: Driedzic and Gesser 1994).

The strategy of utilization of a limited amount of bicarbonate for the protection of the intracellular pH appears to be the most efficient way to provide a constant milieu for the intracellular structures, and exemplifies the general priority of intracellular over extracellular acid-base regulation in vertebrates.

The contributions related to this field which deserve attention from readers are of Albers (1970), Maetz (1971), Pang et al. (1977), Evans (1979, 1980, 1986, 1987, 1993), Cameron and Heisler (1983, 1985), Isaia (1984), Heisler (1984, 1988, 1993, 1999), Bridges (1988, 1993),

Table 11.17 Overview of general acid-base responses in fish to external and internal experimental challenges

Perturbation	Acid-Base relevant internal change	Extracellular Acid-Base effect	Compensatory response in ECF (concentrations in extracellular spaces)	Degree of compensation	Fish↔Water Compensatory transfers
Water hypercapnia	Hypercapnia	pH decrease	Increase plasma [HCO_3^-]	Good	Net H^+ excretion
Water hyperoxia	Hypercapnia	pH decrease	Increase plasma [HCO_3^-]	Excellent	Net H^+ excretion
Water hypoxia	Hypercapnia, metabolic acidosis	pH decrease, plasma lactate increase	Reduce P_{CO_2}, increase plasma [HCO_3^-], metabolize lactate	Excellent	Initial net H^+ excretion, later HCO_3^- excretion
Water acidification	Metabolic acidosis, gill permeability increase	pH decrease, ion loss	Decreased gill permeability (some species)	Poor to excellent	Little
Water alkalinization	Hypercapnia, metabolic alkalosis	pH increase, plasma P_{NH_3}/NH_4^+ increase	Decrease plasma [HCO_3^-] increase lactic acid	Fair to excellent	Increased ammonia urea excretion, H^+ uptake
Switch from water to air breathing	Hypercapnia	pH decrease	Increase plasma [HCO_3^-]	Poor	–
Water temperature increase	Temperature increase	pH decrease, P_{CO_2} increase, plasma [HCO_3^-] decrease	–	Little	Net HCO_3^- excretion
Exercise	Hypercapnia, metabolic acidosis	pH decrease, plasma lactate increase	Reduce P_{CO_2}, increase plasma (HCO_3^-), metabolic lactate	Excellent	Initial net H^+ excretion, later HCO_3^- excretion
Acid infusion	Metabolic acidosis	pH decrease	Recovery of plasma [HCO_3^-]	Excellent	Net H^+ excretion
Base infusion	Metabolic alkalosis	pH increase	Reduction of plasma [HCO_3^-]	Excellent	Net HCO_3^- excretion

Source: Claiborne (1998)

Potts and McWilliams (1989), Wood (1989), Avella and Bornacin (1990), Wright (1991), Claiborne et al. (1997), and Claiborne (1998). Table 11.7 gives an overview of physiological adjustments to acid-base imbalance.

8. ION EXCHANGE

Hagfish: Myxine glutinosa is a non-osmoregulator. Blood and urine are isotonic or only to a small degree hypertonic in comparison with the environment or the medium.

Observations about the ionic composition in serum and urine of the hagfish have shown that the concentrations of sodium and chloride are similar in serum and urine, while the concentrations of potassium, magnesium, calcium, sulfate and phosphate are elevated. With fluid absorption zero as revealed by the urine/plasma ratio of insulin, these latter ions are secreted. Glucose concentration is lower in urine than in the serum, which reveals reabsorption in the archinephric duct.

The process of urea handling in *M. glutinosa* needs further elucidation. Probably the excretion of urea is effected by the liver-gall bladder system as well as by the kidney. However, it has as yet not been determined whether *Myxine* manages its waste nitrogen excretion exclusively with the aid of the ornithine cycle (Raguse-Degener et al. 1980) (refer also Rankin 1997; Karnaky 1998).

In lampreys, in general ionic and osmoregulatory mechanisms are strikingly similar to those of teleosts (Karnaky 1998). Plasma ionic concentrations of various fishes is given by Evans (1993; see Table 1)

Marine Elasmobranshs: Marine elasmobranchs are hypoionic regulators, maintaining blood concentrations of the major monovalent ions, Na and Cl, at levels below that of surrounding seawater. The specifics of salt and water balance in the adult elasmobranchs have been extensively reviewed (Shuttleworth 1988; Evans 1990; Hazon et al. 1997). The potential effectors for salt and water balance are the kidneys, rectal gland, gut and gills.

The kidney is a net excretor of salts and water and produces a urine that is isoionic and slightly hypoosmotic compared to blood. It is of lesser importance in solute regulation. The kidney reabsorbs nearly all the urea and TMAO filtered, which helps to maintain high blood levels of

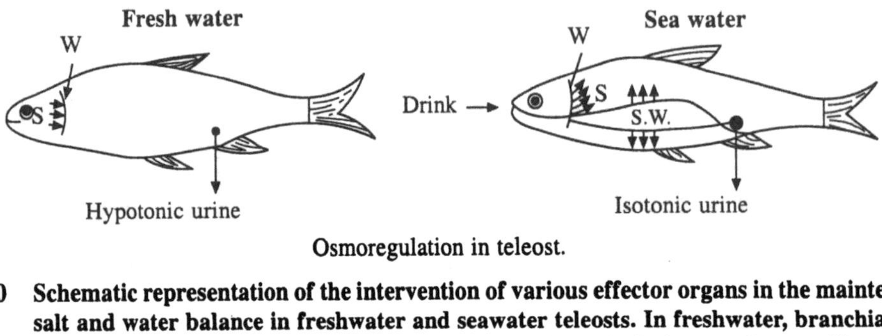

Osmoregulation in teleost.

Fig. 11.20 Schematic representation of the intervention of various effector organs in the maintenance of salt and water balance in freshwater and seawater teleosts. In freshwater, branchial gain of water and loss of NaCl is balanced by excretion of relatively large volumes of hypoosmotic urine and active branchial extraction of NaCl from the medium. In seawater, branchial loss of water and gain of NaCl is balanced by oral ingestion of the medium, excretion of nearly iso-osmotic urine, and active branchial extrusion of NaCl.
***Source:* Maetz (1971)**

these osmolytes. The gut eliminates some mono-and divalent ions but is of little use since elasmobranchs drink very little seawater (about 1% of that seen in marine teleosts). Urine flow is low, and combined with the rectal output is about equal to the net osmotic flow across the gills.

The digitiform rectal gland (see further) which drains into the cloaca, can secrete a concentrated solution of NaCl that is hyperionic and isosmotic to the blood. Since the discovery of substantial amounts of the transporting enzymes Na-K-ATPase in the rectal gland, this tissue has been used extensively as a model for salt transport. The rectal gland plays a major role in the elimination of excess salts from the body.

Salt excretion by rectal gland is controlled by a variety of hormones and neurotransmitter. The primary stimulant appears to be either VIP or another peptide from the gut termed rectin.

According to Kormanik (1992), embryos of more advanced viviparous elasmobranchs develop in a solution that is similar to maternal plasma. Iono-and osmoregulation by these embryos would appear to be minimal. It is clear that in the oviparous elasmobranchs, the ability of the egg/embryo to maintain salts and urea at appropriate levels is present at the earliest stage of development. The ability of prenatal elasmobranch embryos to ions-and osmoregulate would allow the evolution of a diverse array of reproductive strategies in the elasmobranchs.

The role of gills in ion regulation of seawater teleosts is well known, however, the role of the gills in elasmobranchs has been the subject of some debate. The amplified surface area of the gills is a site for the influx and efflux of Na and Cl. Debate centers on the role of the gills as excreters of salts. While the levels of transporting enzymes like Na-K-ATPase are low compared to teleosts, both gill permeability and the corresponding flux of ions are low. The gills of elasmobranchs possess mitochondria rich "chloride" or "chloride-like" cells as least as indicated by their morphology and are therefore likely contributors to salt secretion.

The reader can find interesting information on the role of RAS in respect of osmoregulation (see Chapter 13). Figure 11.20 shows the features of osmoregulation in teleosts.

Chondrostean: White sturgeon *Acipenser transmontanus:* Although there have been many investigations of osmoregulation in adult teleosts, there have been fewer on the more 'primitive' fish groups. This is especially true of the Chondrostei, as a result of their large size and the difficulty of obtaining them. Many of the living sturgeon species are put into three categories: anadromous, semi-anadromous, and landlocked. *A. transmontanus* is a semi-anadromous fish, which spends most of its adult life in sea water, but close to the shore, and migrates into fresh water to spawn.

White sturgeons, like anadromous teleosts, are hypoosmotic in seawater and hyperosmotic in fresh water. Regulation of plasma electrocytes appears slower than in anadromous teleosts. Also, comparisons of urine electrolyte concentrations indicate fundamental differences in ionic regulation between North American and Eurasian sturgeon species.

The ability of juveniles to regulate plasma osmotality is limited. The young adult fish are able to tolerate higher salinities than juvenile sturgeon but probably are also characterized by low activity of the necessary ion exchange mechanisms in the gills which permit rapid adjustment of blood electrolytes with graduate change in external salinity (McEnroe and Cech 1985, 1987).

Coelacanth Latimeria maintains a blood osmolarity near that of its sea water environment by retaining urea and trimethylamine oxide in concentrations similar to those of chondrichthyans. Ion levels are fairly low, as in teleosts, and evidence is presented suggesting that the rectal gland is involved as a sodium secreting organ. Bladder urine of *Latimeria* is isouremic with the serum, contrasting with the elasmobranchs which reabsorb most of the filtered urea. It has been suggested that urea retention in *Latimeria* evolved independently of its occurrence in the cartilaginous fishes (Griffith and Pang 1979).

A survey of osmoregulation in the aquatic vertebrates revealed four basically different patterns into which all known fishes could be accommodated (read Pang et al. 1997).

9. AUXILIARY ORGANS OF OSMOREGULATION

Chloride cells (Ionocytes). Chloride cells (ionocytes) and salt glands are auxiliary organs of osmoregulation. By the absorption and excretion of ions, they support the limited potency of excretory organs to achieve homeostasis under extreme conditions of hydration and dehydration.

Chloride cells occur in aquatic, branchiate vertebrate and are endowed with a functional dualism. They exert an ion-excretory function in marine fish under the dehydrating condition of seawater, while serve an ion-absorptive function in freshwater fish which are exposed to permanent hydration in their habitat. In anadromous and catadromous species they temporarily change between the excretory and absorptive functions.

Chloride cells usually occur as unicellular or paucicellular elements interspersed in the stratified epithelium of the gill filaments, opercular skin facing the gill chamber of an euryhaline killifish, and on the secondary gill lamellae of air-breathing labyrinth fish and lungfish. They are also encountered outside the branchial chamber in the normal epidermis, and in the renal collecting ducts of marine catfish that are capable of producing hypertonic urine. They are not restricted to teleosts, but occur in the gills of elasmobranchs and cyclostomes.

Hagfish possess ionocytes in their gills that have been suggested to be the site of calcium ion exchanges and these ionocytes resemble the chloride cells of teleost fish, which are thought to be the principal site of calcium uptake. It may be mentioned that calcium is an extracellular ion that hagfish regulate to concentrations markedly different from that of the medium in which they live. Otherwise they have ionic concentrations in their extracellular fluids similar to those in seawater. Ultrastructural evidence suggested that the gills of hagfish are a tight barrier to diffusion. Experiments indicate that gills of hagfish are, like those in teleosts, potentially important sites for calcium uptake (see Forster and Fenwick 1994).

Chloride cells conspicuously exceed in size the adjacent epithelial cells. They often extend through the entire height of the stratified epithelium with their apex in direct contact to the external medium and their base resting on the basal lamina. Their apical plasma membrane is locally indented to form a small crypt filled with mucus and subdivided by interconnected septa or by microvilli. A dense population of mitochondria and an elaborate system of ramifying and anastomosing tubules constitute the prevailing cytoplasmic structures. The complex system of basolateral plasma membranes is rich in Na^+/K^+-ATPase (Komnick 1986) (Fig. 11.21). Wendelaar Bonga et al. (1990) described the ultrastructure of the chloride cells on the gills of the euryhaline teleost *Oreochromis mossambicus* during exposure to acidified water. Function of chloride cells has been recently reviewed (Jurss and Bastrop 1995; McCormick 1995; Zaccone et al. 2001; see also Isaia 1984; Fig. 5; Kultz et al. 1995; Karnaky 1998; Fig. 4(A) current model for ion transport mechanisms in seawater teleost chloride cell and (B) an older model for ion uptake mechanisms in freshwater teleost gill epithelium). A TEM study of chloride cells in the channel

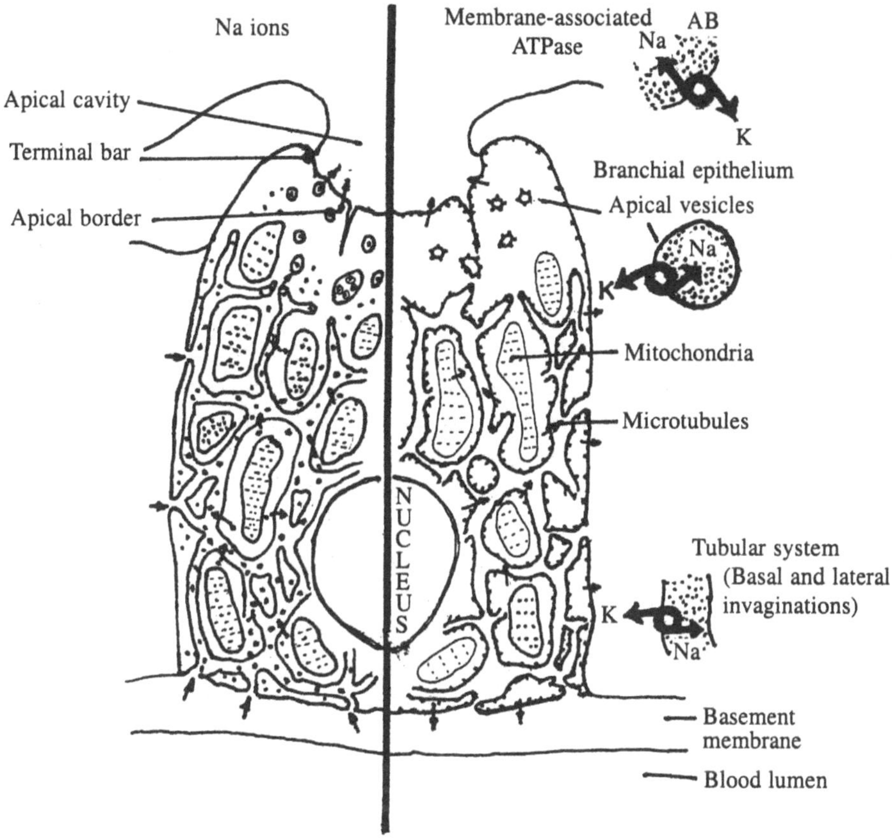

Fig. 11.21 A model for the functioning of the teleostean chloride cell in seawater. On the left, distribution and pathway of Na^+ ions. On the right side, tentative localization of the membrane-associated ATPase and the pathway of active Na^+ transport. Inserts: Na^+/K^+ exchange effected by the transport of ATPase in the tubular system, in the apical vesicles and the apical border. ***Source:*** **Maetz (1974)**

catfish *Ictalurus punctatus* has been made to find the relationship of the tubular system with other organelles (Henderson 1995). Silvering in the eel, *Anguilla anguilla*, produces an increase in the number of chloride cells, as well as the development of their tubular system and mitochondria (Fontaine et al. 1995). Details of ion transport by chloride cells and urinary bladder are given in the contribution by Marshall (1995). The function of chloride cells (MR cells) remains highly controversial. Recent data suggest that MR cells in freshwater gills are involved in Cl^- uptake, but not in Na^+ uptake. MR cells in freshwater trout contributes to Ca^{2+} but not to NaCl ionic balance (read Marshall et al. 1992, 1995; refer also Marshall et al. 1997, Hazon et al. 1997). The reader is also referred to a concise review on the elimination of excess salt in nonmammalian vertebrates (Shuttleworth 1987).

Salt glands in Elasmobranchs: The functional relations between the rectal gland and the branchial chloride cells both jointly occurring in elasmobranchs await future elucidation.

The rectal gland, a major desalinizing organ, of sharks and rays is usually a single, digitiform, occasionally lobate appendage in the mesentery of the hind-gut, while in chimaeroid fishes it is represented by a number of small intramural glands. Sharks rectal gland (SRG) has been a subject of intensive investigation (Beyenbach and Dantzler 1990; Ecay and Valentich 1991; Karnaky et al. 1991; Valentich et al. 1995; Valentich, in press; also see Hazon et al. 1997 and Karnaky 1998). For the current model for NaCl secretion by SRG, refer Evans (1993; Fig. 3). Freshwater elasmobranchs do not need to excrete excess salt; the rectal gland is reduced in size, and presumably in function.

The extramural rectal gland is a highly vascularised organ which consists of tightly packed, branched, blind tubules radiating from a central collecting duct that communicates via a short extra-glandular duct with the rectal lumen. The tubule cells have sparse and short microvilli, mitochondria, and richness in Na/K-ATPase. The central collecting duct shows mitochondria-rich cells, cells of presumably endocrine function, flask cells of unknown functions, and goblet cells (mucocytes). Rectal gland epithelial cells transport chloride ions from the blood to the lumen of secretory tubules against an electrochemical gradient by the process of secondary active chloride secretion. The ion transport moieties present in the apical and basolateral membranes of rectal gland cells have been identified and characterized.

The chloride secretory activity of the spiny dogfish (*Squalus acanthias*) rectal gland is under endocrine (atriopeptin AP), neuronal vasoactive intestinal peptide (VIP) and somatostatin, and autocrine (adenosine) control. Evidence supporting a role for elevated intracellular calcium in the activation of rectal gland chloride secretion by some of these regulatory inputs has been presented (see coverage: Valentich et al. 1995; Hazon et al. 1997; Fig. 5.4 a schematic diagram showing the currently accepted pathways and membrane protein involved in the active secretion of NaCl from rectal gland epithelium). Table 11.8 provides information of the extrabranchial sites in elasmobranch acid-base regulation*.

Table 11.8 Extrabranchial sites in elasmobranch acid-base regulation*

Species	Stress factor	Kidneys + urinary bladder	Skin	Rectal gland + abdominal pores
Squalus acanthias	Hypercapnia	< 1	—	—
	HCO_3^- infusion	< 1	—	—
	HCl infusion	< 1	—	—
Scyliorhinus stellarus	Hypercapnia	< 1	< 1	< 1
	Exercise	< 1	—	< 1
	HCL infusion	< 1	< 1	< 1
	HCO3 infusion	< 1	< 1	< 1
	Temperature changes	3	—	—

*Contributions given as percentage of total.
Source: Heisler (1989; see details).

The occurrence of salt glands in teleosts is rather exception and so far only reported for marine catfish of the family *Plotosidae*; a dendritic appendage behind the anal opening and the urogenital papilla. The dendritic appendage is a highly vascularized integumental gland consisting of acini and short tubules whose lumina are in open communication with the seawater. The acinar and tubular epithelium comprises also two types of mitochondria-rich glandular cells named principal and clear cells.

It may well be that the dendritic appendage represents a local, organogenetic accumulation of cutaneous chloride cells.

The occurrence of this organ appears somehow related to the unusual invasion of marine habitat by these siluroids, the unusual production of hypertonic urine, and the unusual equipment of their renal tubules with chloride cells (Komnick 1986; see also Isaia 1984).

A cloacal cation-secreting gland in *Protopterus*, functionally, similar to rectal glands of elasmobranchs, holocephalans, and *Latimeria* is present though the anatomic relations differs from that of the latter groups (see Wake 1986).

CHAPTER **12**

Reproductive system; Rhythms; Physiological Aging (Senescence) in Fish; Development; Phenotypic Variability in Fish Populations

1. INTRODUCTION

Reproduction ensures the production of gametes, embryos, and young so that, given certain external conditions, the survival of the species is secured; therefore, reproduction is the key phenomenon for biological evolution such as we know it. On the other hand, reproductive mechanisms are not necessary for the individual and are not permanent functions.

Reproduction comprises a number of steps, preceding (gametogenesis, gamete release) and following (protection or nutrition of egg, embryo, and young) the main event, which is fertilization. This pattern shows great diversity in its modalities and in its temporal organization. Indeed, vertebrates made the conquest of nearly all the biosphere (of very different external media) with an astonishing number of reproductive adaptations (e.g., various kinds of oviparity, ovoviviparity, viviparity). Much diversity is also seen in the articulation between life span and the sex span (age at puberty, pattern of reproductive cycles, number of breedings, etc.). Subtle interactions between sex hormones and developmental hormones are certainly involved in the determination of these patterns. These differences often have little phylogenetic meaning; many analogous adaptations occurred in very different classes, such as viviparity in cartilaginous fish and mammals, and quite different adaptations may be realized within a given group (e.g., teleosts).

Reproduction of any vertebrate species cannot succeed unless the sexual behaviours of the individual are exhibited at the correct time, i.e., correctly synchronized with its own gonadal development and with the activities of potential males. But within and between individuals, chemical signals play important role in ensuring that the correct timing of sexual behaviour is achieved. Within the individuals of many species, hormones are released into the blood in amounts related to gonadal condition; they also function as a common internal chemical signal which stimulates sexual behaviours at the appropriate stage of gonadal development. Reproductive pheromones perform functions analogous to those of hormones but transmit chemical information between individuals to alter reproductive behaviour or physiology.

Reproduction among fish at high latitudes is usually an annual event with the seasonal timing of gonadal maturation being under the influence of an endogenous cycle of gonad development that is synchronized with environmental cues. Amongst salmonids, individuals within a population will not all mature in any given year, and not all mature individuals will reproduce each year. Nutritional status appears to have an important influence upon whether or not a given individual will reproduce within any given year and factors related to nutritional status and rates of deposition of energy reserves also appear to be of major importance in influencing the timing of maturation. More specifically, maturing fish are reported to have higher growth rates and/or display more rapid rates of energy deposition during the spring months than immature conspecifics. These differences in growth and energy acquisition between maturing and immature individuals are, however, of limited duration because maturing fish tend to become anorexic during the course of summer, there is a decline in growth rate and condition tends to become poorer. This aspect has been analyzed in Arctic charr (*Salvelinus alpinus*) by Tveiten et al. (1996).

The teleosts presents a range of behavioural repertoires. At one extreme are species whose sexual behaviours are restricted to a brief and evidently promiscous shedding of the gametes, while at the other extreme are species that form pair bonds, prepare and defend a spawning site, and care for the young. Finally, from an evolutionary perspective, teleosts provide an instructive example of how a major change in reproductive strategy—the shift from external ("donator of oocytes") to internal ("recipient of sperm") fertilization has been accompanied by an equally dramatic change in the regulation of female reproductive behaviour.

Although internal fertilization is relatively uncommon among fishes, it is the sole mode of reproduction for all extant elasmobranchs. Because most species are wide ranging and difficult to observe, field documentation of their mating behaviour is extremely rare, generally fortuitous, and particularly inconclusive.

2. GONADS; FERTILIZATION; SOCIAL CONTROL OF SEXUAL MATURATION; ALLIED ASPECTS

This topic has been reviewed by Hoar (1969). Representative types of urogenital systems are illustrated (Figs. 12.1 and 12.2). A brief account of testis and ovary of fish has been given by Redding and Patino (1993) in the chapter entitled 'Reproductive Physiology'. Another title "Fish reproduction: strategies and tactics" encompassing many significant contributions has been edited by G.W. Potts and R.J. Wooton (1984). A recent article on reproduction is from Kraak et al. (1998)

A synthesis of the extensive literature on testis in vertebrate is available (see Nagahama 1986; Grier 1992). The reader is also referred to a review related to gestation, with emphasis on corpus luteum biology, placentation, and parturition (Jones and Baxter 1991).

Lampreys become sexually mature and carry out gametogenesis once and die soon after spawning. In contrast, it is commonly assumed that most hagfishes are repetitively cyclic breeders without seasonally defined reproductive cycles, (Figs, 12.3 and 12.4) The cyclostomes have a single, medially placed gonad. They lack Mullerian ducts, and ovulated eggs are released into the body cavity and spawned through an external genital pore. Fertilization occurs externally. The male gametes of the cyclostomes develop in cysts. The cysts gradually transform into spermatozoan masses and eventually rupture, shedding spermatozoa directly into the body cavity.

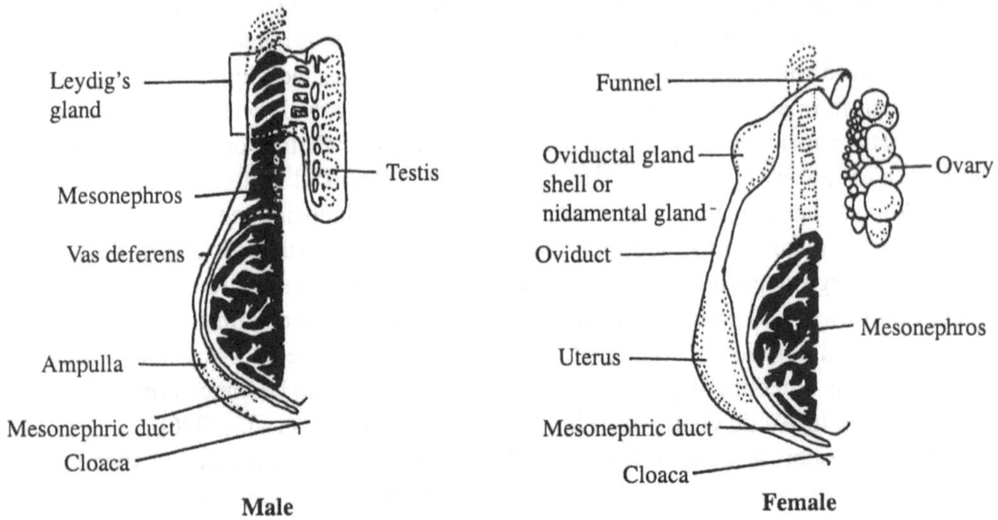

Fig. 12.1 Selachian urogenital systems: Male and Female.
***Source:* Hoar (1969)**

The fine structure of Sertoli cells of the hagfish, *Eptatretus stouti*, indicates a supporting function for the germ cells rather than a secretory role. In the period immediately proceding spermiation, the Sertoli cells in river lamprey *Lampetra fluviatilis* show some characteristics of steroidogenic tissues. Very few observations are available to suggest the presence of interstitial cells in hagfish testis. In lampreys, the presence of 3 β-HSD has been histochemically reported in interstitial cells. Ultrastructurally, the interstitial cells of lamprey show all of the features characteristic of steroidogenic tissue.

Testicular structure in agnathans and in elasmobranchs is polyspermatocystic (Grier 1992; Parsons and Grier 1992). For gonads and reproduction in hagfishes, read Patzner (1998).

The urogenital morphology of the chimaeroid fish *Hydrolagus collei* corresponds well with that of elasmobranch fishes.

Paired testes, equivalent in size and comprising 1–5% of body weight in adult elasmobranchs, are located anteriorly in the body cavity and suspended from the dorsal body wall by double mesorchia. Each is embedded in a long irregular epigonal organ which has no known reproductive function but is lymphomyeloid in nature and a major source of leucocytes; elasmobranchs having no bone marrow or lymph nodes.

Elasmobranch testis shows a distinct zonation. The unit is the closed spheroidal ampullae, each containing approximately 500 so-called spermatocysts arranged around a central cavity. Each spermatocyst consists of a single Sertoli cell in which one to 64 germ cells are embedded. Sertoli cells (also referred to as nurse cells, sustentacular cells, and supporting cells), the only non-germinal elements within the seminiferous tubule, yield a positive histochemical reaction for 3 β-HSD activity. The ampullogenic zone runs the length of the gonad and propagates germinal cysts that mature as they move in dorso-lateral direction. All the spermatocysts within a single ampulla are, approximately, at the same stage of development. In transverse section, the testis shows a distinct concentric zonation of successively more advanced germinal stages. Moving from the peripheral wall to the hilus, three distinct zones are distinguishable: (1) a

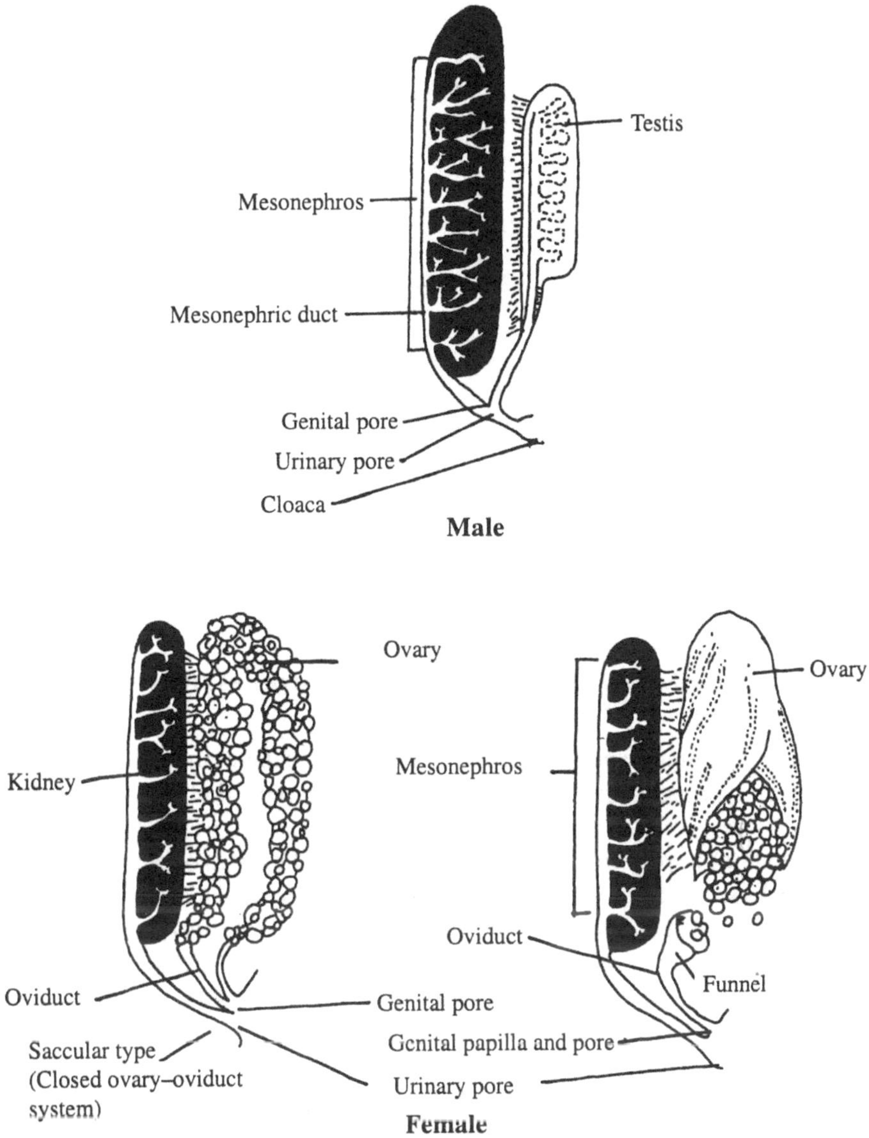

Fig. 12.2 Teleost urogenital system: Male, Female.
***Source:* Hoar (1969)**

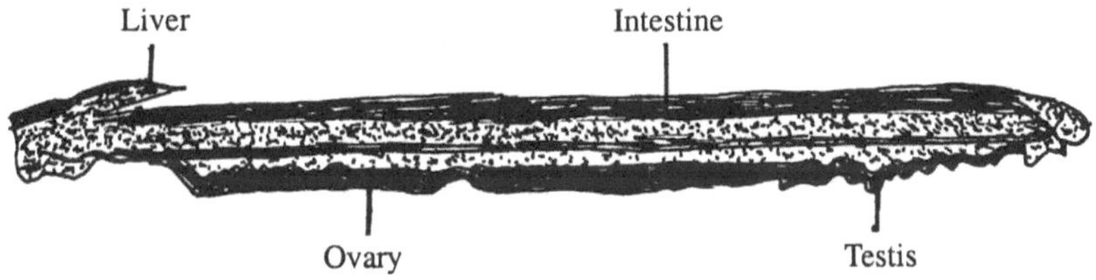

Fig. 12.3 Hermaphrodite gonad of *Myxine* ♀
***Source:* Brodal, A., Fänge, R. (eds.) (1969)**

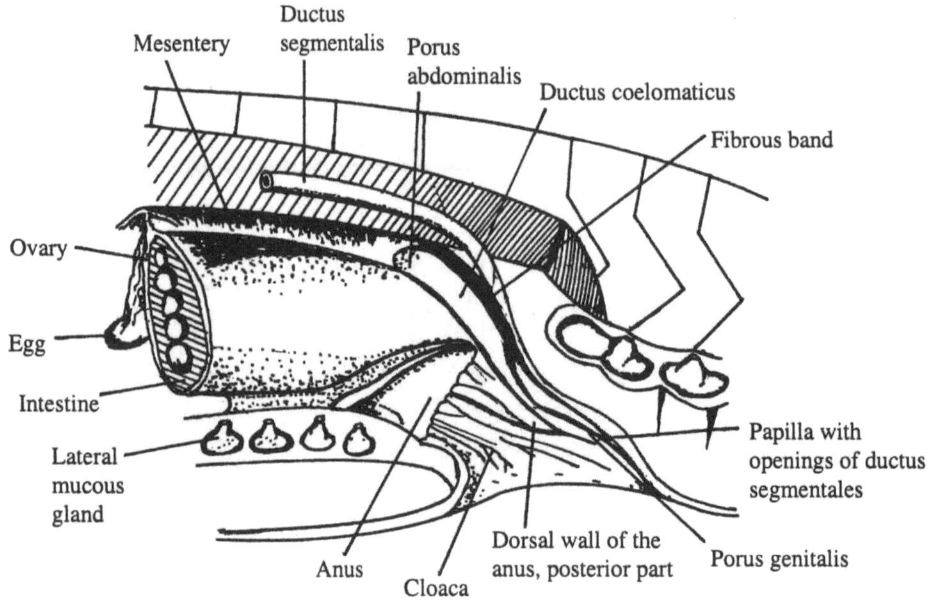

Myxine glutinosa: Cloacal region.

Fig. 12.4 Schematic longitudinal section through the cloacal region of *Myxine glutinosa*. Left view. *Source:* Brodal, A., Fänge, R. (eds.) (1963)

tubulogenic zone, containing spermatogonia; (2) an intermediate zone, containing tubules at different stages of spermatogenesis; and (3) a zone containing both mature or degenerating Sertoli cells.

Although interstitial Leydig cells have been described in the testes of representatives of every extant vertebrate class, in elasmobranchs there is controversy as to the identity of typical vertebrate Leydig cells (see Nagahama 1986; Fasano et al. 1989; Callard 1991a, b). For overview of spermatogenesis, the reader is referred to Nagahama (1986) and Callard (1991a, b).

The organization of the ovary is highly variable in elasmobranchs because of species diversity and a wide range of reproductive patterns. The appearance of the ovaries of different elasmobranchs varies according to the stage of the reproductive cycle. Ovarian development is characterized by a proliferation of cells from the surface epithelium. In some species (*Torpedo marmorata*, *T. ocellata*, *Chimaera monstrosa*), oocytes with a diameter larger than 0.1 mm are covered by a multilayed, heteromorphic epithelium that is composed of two cell types: (1) large cells, with reticular nucleus and abundant cytoplasm, sometimes full of globules similar to yolk platelets; and (2) small cells interspersed with large cells.

At the end of vitellogenesis, the granulosa becomes flattened and single-layered and the large cells become sparser and finally disappear. In other elasmobranchs (*Scyliorhinus canicula*, *S. stellaris*, *Mustelus canis*), the follicular epithelium is multilayed but monomorphic. In all cases, the follicular epithelium is surrounded by a theca interna and theca externa with a layer of cuboidal columnar cells between.

The formation of corpora lutea from the postovulatory and atretic follicles has been described in numerous species of elasmobranchs. In *Squalus acanthias*, the corpus luteum develops from postovulatory follicles and by hypertrophy of granulosa cells, surrounded by a separate theca sheet. In *Scyliorhinus stellaris* and several species of *Raja*, it is formed by the hypertrophy of both thecal and granulosa cells. The preovulatory corpora lutea (atretic follicles) in *Torpedo* spp. develop from the ovarian follicle, usually during advanced vitellogenesis, after the phagocytosis of yolk by the granulosa cells. At full development, they are characterized by hypertrophy of the granulosa cells, whereas the theca interna forms the connective stromal tissue (Fasano et al. 1989; Jones and Baxter 1991; refer also Nagahama 1986).

Elasmobranch oviducts are of pronephric (Mullerian) derivation. Eggs are shed by the ovary into the body cavity, and are carried to the opening of the oviduct by ciliary action. The oviduct can be subdivided into several specialized regions. The most anterior region is the oviducal shell or nidamental gland, which is best developed in oviparous species. The shell gland in viviparous species is smaller, and in *Squalus*, for example, the shell is reduced to a membrane in which several eggs may be packaged to form the "candle" seen in Stage A pregnancies. In viviparous Japanese smooth dogfish, the shell membrane persists through development and may form part of the placenta. Estradiol is implicated in the development and maintenance of elasmobranch reproductive tracts, as shown by the effects of injections of the hormone and the correlation between plasma estradiol levels, ovarian follicular and shell gland size. The action of relaxin has also been pointed out (Callard and Klosterman 1988).

In most teleosts, the testes are elongated paired organs attached to the dorsal body wall; in some species (e.g., poeciliids) they are combined into a sac. Phallostethid testes and ovaries are single, not paired. Testicular structure in teleosts varies from species to species, although two basic types, lobular and tubular, can be identified according to the differentiation of the germinal tissue (Figs. 12.5 and 12.6). Grier (1992) proposed the "Extracellular Matrix Hypothesis" to explain how changes in the chronological formation of basement membranes resulted in the

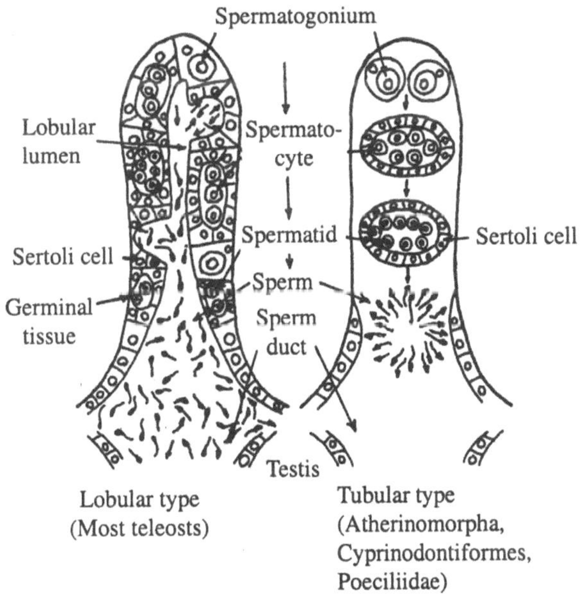

Fig. 12.5 Diagrammatic representation of testis structure in teleosts: Lobular type; tubular type.
***Source:* Nagahama (1983, 1986)**

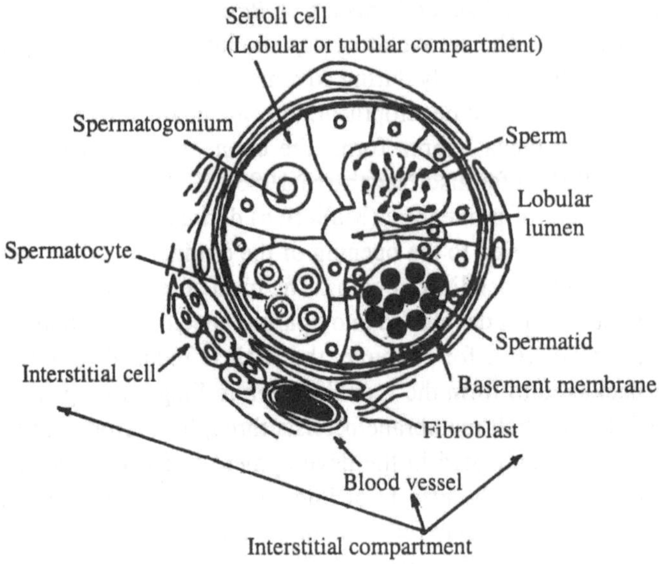

Fig. 12.6 Diagrammatic representation of a portion of testes of the Tubular type demonstrating a close relationship between germ cells at various stages of development and Sertoli cells. Interstitial cells are located in the interstitium. *Source:* Nagahama (1983)

evolution of lobular and tubular testes from spermatocystic testes. The testis of the lobular type, which is typical of most teleosts, is composed of numerous separate lobules joined by a thin layer of fibrous connective tissue; the arrangement of the lobules varies considerably. Within the lobules, primary spermatogonia undergo numerous mitotic divisions, producing cysts containing several spermatogonial cells. Duration maturation, all of the germ cells within each cyst are at approximately the same stage of development. As spermatogenesis and then spermiogenesis proceed, the cysts expand and eventually rupture, liberating sperm into the lobular lumen which is continuous with the sperm duct. In salmonids, the cells in most cysts composing the same testes are synchronized and contain germ cells at the same spermatogenic stage.

The alternative testicular structure, the tubular type is restricted to the atheriniform species, such as the guppy, *Poecilia reticulata*. In this situation, the tubules are regularly oriented between the external tunica propria (blind end) and a central cavity into which spermatozoa are released. Primary spermatogonia are located only at the blind end of each tubule. As spermatogenesis and spermiogenesis proceed, the germinal cyst moves centrally within the testis toward the vas efferens (efferent duct); there is no structure corresponding to the lobular lumen.

Grier et al. (1987) described the mechanism of tubule elongation during testicular recrudescence in the redfish, *Sciaenops ocellatus* (Perciformes). The mechanism of spermatogonial depletion from the testis in this seasonally breeding teleost is not understood but is basic to the understanding of reproductive processes in fish.

Based on the distribution of spermatogonia, the testicular structure in four orders of teleosts (Salmoniformes, Perciformes, Cypriniformes and Atheriniformes) has been divided into two basic tubular designs-unrestricted spermatogonial testes and restricted spermatogonial testis. The former type is common to most teleosts, and the latter type (in which spermatogonia are totally restricted to the distal terminus of the tubule immediately beneath the tunica albuginea) is characteristic of the Atheriniformes. Grier (1992, 1993) recognized the three following testes' types in teleosts: anastomising tubular, lobular with an unrestricted distribution of spermatogonia, or lobular with a restricted distribution of spermatogonia, so far known in Atherinomorpha (cf. Nagahama 1986) (Fig. 12.7).

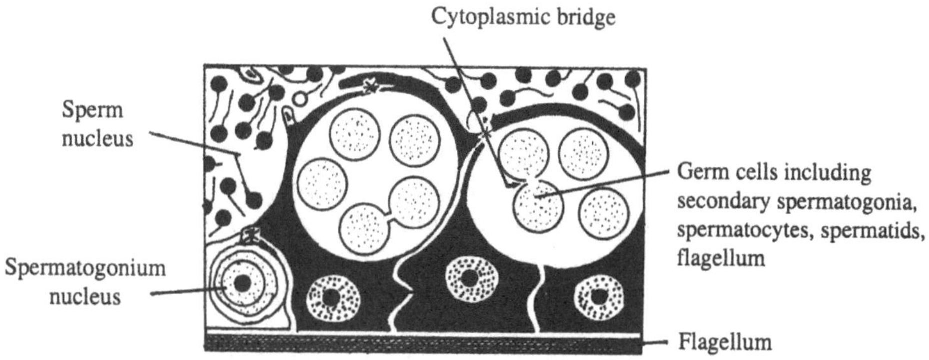

Fig. 12.7 The continuous spermatogonial germinal epithelium of a teleost. Spermatogonia, its nucleus being labeled, occur along the lengths of lobules. Germ cells reside within spermatocysts, but Sertoli cell processes are entirely restricted to the periphery of the spermatocyst. At spermiation, the spermatocyst opens (arrows), and sperm are shed into the lumen of the germinal epithelium. *Source:* Grier (1993)

In certain species of teleosts, numerous lipid droplets are contained in the cytoplasm of the Sertoli cells forming the wall of testicular cysts. In teleosts, it is still uncertain whether Sertoli cells can produce steroids. Histochemical studies of these Sertoli cells in some fish (e.g., *Fundulus heteroclitus*, *Tilapia mossambica*, *Salmo gairdneri*) reveal the presence of 3 β-HSD activity.

Sertoli cells can phagocytize residual bodies and degenerating germ cells and are involved in the transport of metabolites. In some teleosts, Sertoli cells are transformed into efferent duct cells following spermiation (see details: Grier 1993).

Leydig cells (interstitial cells) in the teleost testis are usually distributed singly or in small groups in the connective tissue matrix between lobules of germinal elements. In *Poecilia latipinna*, there are interstitial cells around the efferent duct and at the periphery of the testis, but not between testicular lobules. The testis of some gobiid species has two parts; the seminiferous, and the glandular. The glandular portion (so called "mesorchial gland") shows steroidogenesis. Various enzymes involved in steroid hormone synthesis have been demonstrated by histochemistry in the interstitial cells of the testes of a number of teleosts. The electron microscopial observations have supported the steroidogenic features. For a discussion on lobule boundary cells, the reader is referred to Nagahama (1983, 1986), Grier et al. (1989) and Grier (1993).

In the protogynous, rice field eel, *Monopterus albus*, extensive development of the interstitial Leydig cells precedes the formation of testicular lobules during natural sex reversal. Proliferation of Leydig cells is also observed in parallel with the progress of sex reversal from the functional female phase to the functional male phase in two species of wrasses, *Halichoeres tenuispinus* and *Thalassoma duperrey*.

During the course of testicular morphogenesis, teleost sperm ducts (vas deferens) are formed by somatic cells derived from the coelomic wall. These structures, unlike those of most vertebrates, are in no way a part of the nephric duct or Wolffian duct. The sperm duct system of species with tubule type testes consists of two parts, efferent duct (vasa efferentia) and the main sperm duct (vasa deferentia).

Seminal vesicles ("accessory glands", "seminal glands"), not homologous to the mammalian accessory glands of the testis, are reported in some teleost families (Blenniidae, Gobiidae, Clariidae, Bagridae, Tripterygiidae) (Fig. 12.8). These vesicles, attached to the sperm ducts, in gobiid fishes, are lined with an excretory epithelium composed of three types of cells: (1) columnar cells with giant Golgi cisterns in the form of large rings apical to the nucleus; (2) excretory cells with remarkably interdigitative basal lamina, that resemble sperm duct cells, and (3) groups of interstitial, Leydig-type cells that possibly from a part of the mesorchial gland. Morphologically the vesicles appear to be taxa specific, the simplest one in the form of a few tubules on the sperm duct, and the most elaborate one forming large, wing-like structures, with an elaborate mass of passages. In many species the final form of the vesicles develops during the fish ontogenesis; in other, the form of this organ does not change during the life time of fish. With the onset of reproduction, secretion and sperm accumulate in the tubule of the seminal vesicle and are expelled from there onto the eggs during fertilization (see Fishelson 1991).

Hurk et al. (1987) reported that the seminal vesicle of the African catfish, *Clarias gariepinus*, consists of 36–44 finger-like lobes built up of tubules in which a fluid is secreted containing acid

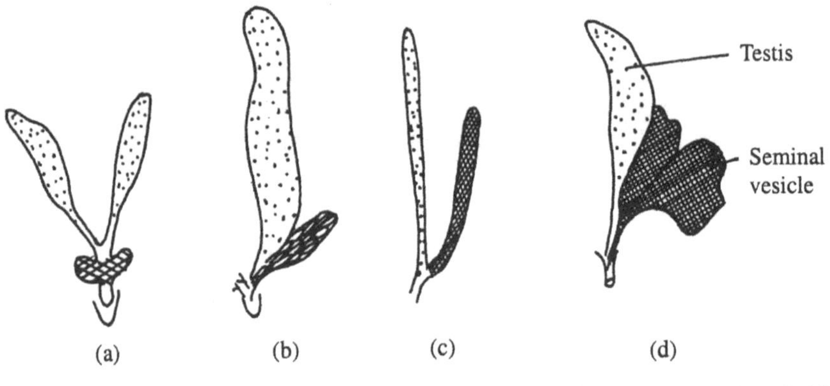

Fig. 12.8 Seminal vesicles in male gobiid fishes Gobiidae, Teleostei. Types: (a) *Knipowitchia caucasica*; (b) *Periophthalmus sobrinus*; (c) *Fussigobius* sp. and (d) *Goblus guttatus*.
***Source:* Fishelson (1991)**

polysaccharides, acid-, neutral- and basic proteins, and phospholipids. In this fluid sperm cells are stored. After ejaculation, it prolongs the period of sperm activity. The seminal vesicle fluid is secreted by the epithelium lining the tubules (simple cylindrical to simple squamous). These epithelial cells contain enzymes involved in energy-liberating processes. Interstitial cells between the tubules have enzyme-histochemical and ultrastructural features indicative of steroid biosynthesis. Similar characteristics are found in testicular interstitial cells. In blennid fishes, these glands function in the differentiation and nutrition of germ cells, in the secretion of sialomucins, in phagocytosis of germ cells, and in lipid storage (Lahnsteiner and Patzner 1990; Lahnsteiner et al. 1990).

The female reproductive system of teleosts, unlike that of mammals, is highly variable, reflecting the wide range of reproductive patterns, including viviparity. In most teleosts, the ovary is a hollow paired organ; however, in some species, paired structures become fused into one solid, single organ during their early development. The ovary consists of oogonia, oocytes and their surrounding follicle cells, supporting tissue or stroma and vascular and nervous tissue (Fig. 12.9) (see Van der Kraak et al. 1998; Fig. 3 illustrating the endocrine signals identified as regulators of ovarian follicular development in teleosts). Most teleosts are cyclical breeders and the ovary varies greatly in appearance at different times in the cycle. Three ovarian types have been classified according to the pattern of oocyte development. The synchronous ("synchronisme total") ovary contains oocytes all at the same stage of development; this type is found in teleosts which spawn only once and then die, such as anadromous *Oncorhynchus* species or catadromous eels. The group synchronous ("synchronisme par groups") ovary consists of at least two populations of oocytes at different developmental stages; teleosts with this type of ovary, such as the flounder, *Liposetta abscura*, and the rainbow trout, *Salmo gairdneri*, generally spawn once a year and have a relatively short breeding season. The asynchronous ("metachrone") ovary contains oocytes at all stages of development; this type occurs in those species (e.g., the medaka, *Oryzias latipes* and the goldfish, *Carassius auratus*) which spawn many times during a prolonged breeding season.

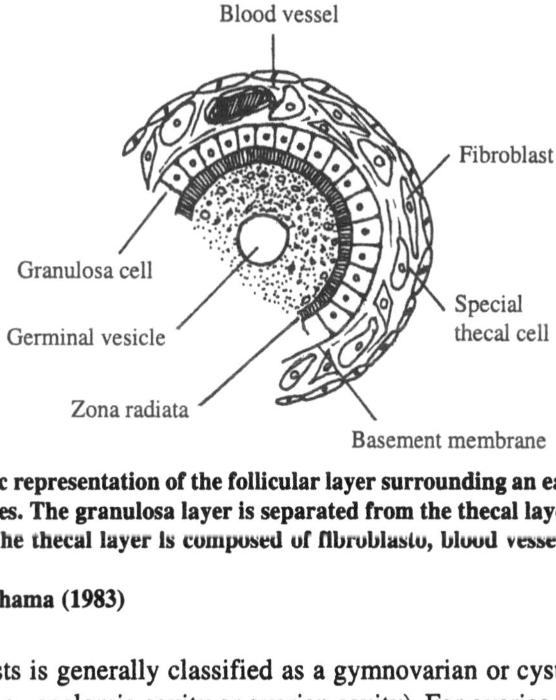

Fig. 12.9 Diagrammatic representation of the follicular layer surrounding an early vitellogenic oocyte of salmonid fishes. The granulosa layer is separated from the thecal layer by a distinct basement membrane. The thecal layer is composed of fibroblasts, blood vessels, etc., and large special thecal cells.
***Source:* Nagahama (1983)**

The ovary of teleosts is generally classified as a gymnovarian or cystovarian type based on the site of ovulation (i.e., coelomic cavity or ovarian cavity). For ovaries of both types, ovulated eggs are immersed in a fluid, namely, coelomic fluid in fish with ovaries of the gymnovarian type or ovarian fluid in those with ovaries of the cystovarian type. These fluids have various functions, such as preservation of fertile eggs, promotion of sperm motility, lubricating at spawning, and formation of spawned egg mass. In masu salmon, *Oncorhynchus masou*, and chum salmon, *O. keta*, which have ovaries of the gymnovarian type, the epithelial covering the dorsolateral coelomic wall and mesovarium can synthesize and secrete coelomic fluid (Fig. 12.10). In ovaries of the cystovarian type, it was observed that the ovarian fluid partly originated from the inner epithelial lining of the ovarian wall. Seasonal changes in secretory activity of the epithelium have only been reported for one species, the viviparous teleost *Cymatogaster aggregata*. Elkhorn sculpin, *Alcichthys alcornis*, a marine cottid, has a viscous ovarian fluid in the ovarian cavity, which plays an important role in its unique mode of reproduction called internal gamete association (i.e., internal insemination and sperm-egg association but a delay in the physiological fertilization until spawning). Seasonal changes in fine structure of the inner epithelial lining and capillary endothelium of the ovary revealed that ovarian fluid originated as a result of the secretory activity of the tissues (see Koya et al. 1995; for internal gamete association, also refer Munehara et al. 1989, 1991, 1994; Koya et al. 1993a, b; cf. internal fertilization in certain phallostethid fishes (Grier and Parenti 1994 and in Glandulcaudinae-characid fishes (Burns et al. 1995). For cheirodontines (Characidae), reference of Burns et al. (1997) can be made.

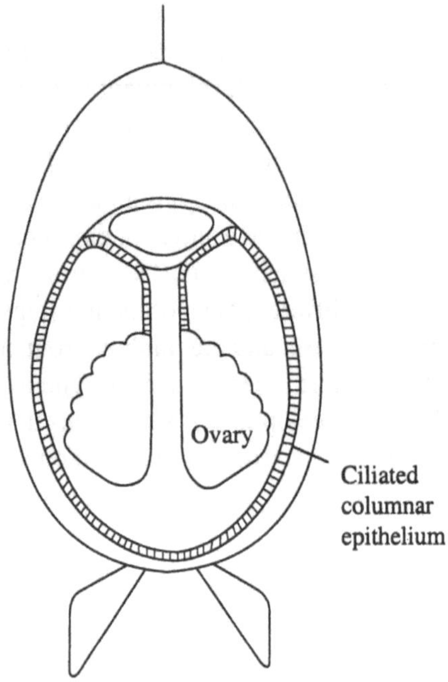

Fig. 12.10 A schematic drawing illustrating the distribution of ciliated columnar epithelium lining the coelonic cavity in mature female *Oncorhynchus keta*.
***Source:* Matsubara et al. (1993).**

Koya et al. (1993c) reported that the components of the jelly-like ovarian fluid present in the ovarian lumen during the spawning period of female masked greenling, *Hexagrammos octogrammus*, are synthesized and secreted not only from ovarian wall epithelium, but also from ovigerous lamella epithelium, the modes of secretion from these epithelia comprising microapocrine, macroapocrine and exocytotic mechanisms. Fishelson (1977) reported, contrary to general picture in teleosts, that ovary epithelium in pteroid fish *Dendrochirus brachypterus* does not possess a germinal function and acts a secretory organ.

A coelomic fluid-specific protein (CFSP) has been detected in the coelomic epithelium and mesovarium of mature female chum salmon *Oncorhynchus keta*. CFSP is synthesized and accumulated in the epithelial cells, and then secreted into coelom as a component of coelomic fluid (Matsubara et al. 1993). For transport of circulating protein component into the coelomic cavity during the coelomic fluid production in female masu salmon *O. masou* (see Matsubara et al. 1994).

The protein of coelomic fluid in female salmonid fish originates from two different organs, blood and the inner epithelium of coelom. The production mechanism of this fluid including the endocrine control remains unclear (Matsubara et al. 1995).

The teleost ovary generally has an ovarian cavity continuous with the oviduct; mature oocytes are discharged into the ovarian cavity and move to the oviduct. The development of ovarian cavity varies from species to species.

For an overview of oogenesis scenerio: (1) oogenesis, (2) vitellogenesis, (3) atretic oocytes (preovulatory corpora atretica), (4) oocyte maturation, and ovulation, the reader is referred to Nagahama (1983; see review, Nagahama et al. 1994). Reviews on atretic oocytes (Saidapur 1978; Guraya 1993; Shanbhag and Saidapur 1996, vitellogenesis (Mommsen and Walsh 1988; Ho 1991; Nath 1999), reproductive cycles (Malhotra et al. 1989; Neelakantan et al. 1989; Saidapur 1989), oocyte assembly (Mommsen and Walsh 1991), gonadal development (Nayak and Nadkarni 1993, Guraya 1999), and nuclear extrusions and their functional significance in the fish oogenesis (Guraya 1994) should infuse interest to readers. Coverage of these aspects is also given by Jones and Baxter (1991) and Redding and Patino (1993). Jackson et al. (1994) have reported the oocyte maturation triggered by the presence of male in the blue gourami, *Trichogaster trichopterus*. Jalabert et al. (1991) presented a comprehensive account of oocyte maturation in vertebrates. Flores and Burns (1993) presented ultrastructural observations of embryonic and early adult germ cells and their support cells in both sexes of *Xiphophorus*.

An oil droplet is very important in the buoyancy of freshwater eggs, whereas a dilute aqueous fluid is more important in the buoyancy of marine fish eggs. The buoyant pelagic eggs of marine fish have the highest water content of any known vertebrate eggs; in eggs of freshwater and other vertebrates water is usually 50–70%. Neutral buoyancy of eggs and larvae of the sablefish, *Anoplopoma fimbria*, changes markedly during development. The vertical distribution of eggs of Atlantic halibut (*Hippoglossus hippoglossus*) closely correlates with occean salinity which affects egg buoyancy. Biochemical changes which occur during fish egg ripening cause a massive influx of water after vitellogenesis but before ovulation (for further information of egg in relation to buoyancy, see Phleger 1991).

Miller (1979b) discussed the adaptive value of various reproductive mechanisms, from highly fecund oviparity, via internal fertilization and simple viviparity, to superfoetation, along with proximal estimates of reproductive effort made for several species contrasting in offspring size and number. Author focused on the adaptiveness and implications of small size in teleosts.

To cover steroidogenic aspects related to: (1) interstitial cells or Leydig cells, (2) Sertoli cells or lobule boundary cells, (3) preovulatory follicles-granulosa cells, special thecal cells, (4) postovulatory follicles, (5) atretic follicles (corpora atretica or preovulatory corpora lutea, (6) 17 β-estradiol and maturation-inducing steroid (17 α, 20 β -dihydroxy-4-pregen-3-one), see Nagahama (1983; for special attention on MIH, read Nagahama et al. 1994). Nagahama and Yamashita (1989) briefly reviewed the current status of investigations, mainly based on the amago salmon, *Oncorhynchus rhodurus*, on the mechanisms of synthesis and action of 17 α, 20 β -dihydroxy-4-pnegnen-3-one (17 α, 20 β -diOH prog). The reader is also referred to Redding and Patino (1993) for a concise account of gonadal hormones. Saidapur (1982) described the structure and function of postovulatory follicles (corpora lutea) in the ovaries of nonmammalian vertebrates. Details of maturation-promoting factor (MPF)—a tertiary mediator of oocyte maturation are given by Nagahama et al. (1994) (Figs. 12.11 and 12.12).

In general, vitellogenesis and its hormonal control in the elasmobranch fishes have received much less attention than in teleost fishes (see Mommsen and Walsh 1988).

The synthesis of vitellogenin, or vitellogenesis, is under the control of ovarian estrogen. Vitellogenin—a large molecular, lipo-phosphoglycoprotein is the yolk precursor. The actual purification of fish vitellogenins from plasma has proven difficult. In chum salmon, rainbow trout, the Japanese eel, and medaka, vitellogenin exhibits unusual iron-binding capacity, a property

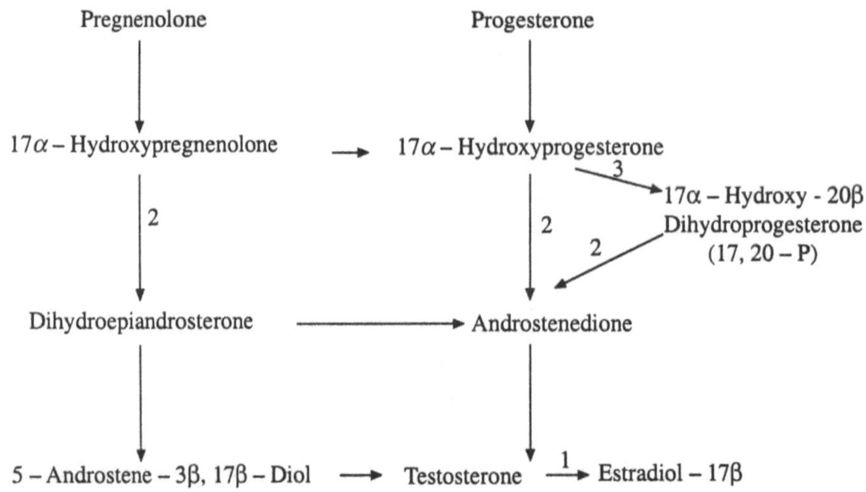

Fig. 12.11 **Biosynthetic pathways linking production of 17-hydroxylated progesterone, estradiol-17 *β*, and testosterone in teleost ovarian follicles. 1. aromatase; 2. desmolase; 3. 20*β*-hydroxysteroid oxireductase.**
***Source:* Jones and Baxter (1991)**

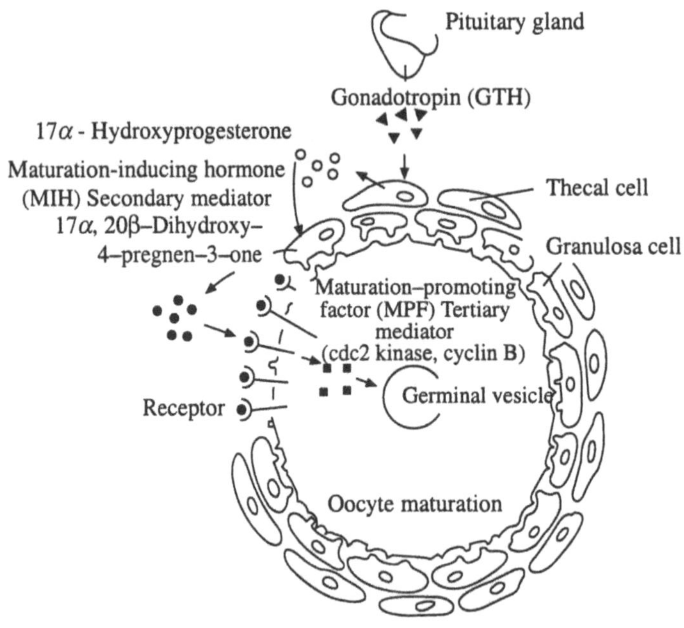

Fig. 12.12 **Hormonal regulation of oocyte maturation in teleosts. Three major mediators, gonadotropin, maturation-inducing hormone, and maturation-promoting factor, are involved.**
***Source:* Nagahama et al. (1994)**

seldom found in vitellogenin of other vertebrates. Information on piscine vitellogenesis is especially sensitive to environmental influence, temperature being considered a key regulator. Photoperiod and circadian rhythm are other regulating factors of vitellogenesis (read Ho 1991). Female-specific serum proteins (FSSPs), collectively called vitellogenin, appear during the normal breeding season. In addition, to vitellogenin, an FSSP that is hard to define as a true vitellogenin has been recently immunologically identified in medaka, *Oryzias latipes*. This FSSP is closely associated with the egg envelope and is produced in hepatocytes under the influence of estrogen. An FSSP (FS3), clearly distinct from that in medaka, has been reported in *Sebastes taczanowskii*, as a major constituent of ovarian fluid (see Takemura et al. 1991). The appearance of a novel sex-limited protein (FS3) is correlated with the active reproductive periods of the female fish (viviparous rockfish *S. taczanowskii*) and may be related to the peculiar reproductive systems of this species such as deposition of spermatozoa and/or developing embryos in the ovary (Takemura et al. 1993).

One of the most striking features observed during teleost oogenesis is the formation of a thick, highly differentiated zone (egg membrane, vitelline membrane, zona radiata, zona pellucida). The egg membrane of many teleosts is structurally modified to form a small opening, the micropyle, through which the sperm gains access to the enclosed egg. The micropyle is occupied by a highly specialized micropylar cells until mature oocytes are discharged from their enclosing follicles at the time of ovulation. Hamazaki et al. (1989) reported that the "spawning female-specific (SF) substance" from the liver is a precursor substance of the major component, ZI-3, of the inner layer of egg envelope in the spawning female fish, *Oryzias latipes*. This (SF) substance is distinct from vitellogenin but closely resembles a major glycoprotein component, ZI-3, of the inner layer (zona radiata interna) of the ovarian egg envelope with respect to some biochemical and immunochemical characteristics. The reader is referred to for specific details of a jelly coat in oocyte envelope (Abraham et al. 1993), surface structure of egg (Britz et al. 1995) and micropyle (Riehl and Gotting 1974; Riehl 1993, 1996; Riehl and Kokoscha 1993; Riehl and Meinel 1994; Guraya 1996). Maternal influences on the variation in egg sizes in temperate marine fishes are discussed by Chambers and Leggett (1996).

Diversity in the structure of spermatozoa is more apparent in poikilotherms, especially in fishes, in which forms with round or elongated sperm-heads such as blennies or sperm with or without acrosomes, or even with two flagella are often encountered. One of the modifications observed in the sperm-tail is the enlargement of the flagellar plasma membrane, which forms lateral protruding extensions in some species (e.g., *Oreochromis macrochir*). Concentrations of polysaccharide in the lateral expansions of the *Ophidion* sp. sperm-tail have been reported. The fin-type flagella has been regarded as an aplesimorphic primitive character. It would thus appear that additional energy supplied is deposited in such finlike structures that power sperm motility. Fishelson (1995) reported unilateral winged flagellum of sperm in *Badis badis*. Afzelius (1978) reported a close correspondence between the teleost and holostean spermatozoa (see Table 12.1).

The posterior region of the testis is "aspermatogenic" and serves as a storage zone for mature spermatozoa (e.g., glandulocaudine species) (see Burns et al. 1995). Extensive sperm storage areas are also present in the testes of a number of internally fertilized teleosts distantly related to the glandulocaudines. The sperm ducts store spermatozoa or sperm packets in the Poeciliidae, Anablepidae, Hemiramphidae, Goodeidae, Phallostethidae and Embiotocidae. In Lepidogalaxidae, the majority of testis functions as a sperm storage area. In some blennies the seminal vesicles store spermatozoa. In some cyprinids the testicular main and spermatic ducts function as sperm reservoirs. Details of the mechanism of packet formation in the form of unencapsulated sperm bundles (referred to as spermatozeugmata) can be obtained from the paper of Downing and Burns (1995) on Hemiramphidae. The reader is also referred to Grier et al. (1981).

Table 12.1 Characteristics of fish spermatozoa

Fish	Sperm features
Squalus	Acrosome, long nucleus, long midpiece, tail with lateral columns
Raja	Acrosome, long nucleus, tail with lateral columns
Neoceratodus	Acrosome, long nucleus, short midpiece with many mitochondria, tail with lateral fins
Protopterus	Same, but two tails with lateral fins
Polypterus	Acrosome, long nucleus, short midpiece, two tails (sperm axis hairpinbent)
Acipenser	Acrosome, short nucleus and midpiece, tail with sidefins
Amia	No acrosome, short nucleus and midpiece
Lepisosteus	No acrosome, short nucleus and midpiece, tail with lateral fins
Salmo and most lower teleostean species	No acrosome, short nucleus and midpiece, tail with lateral fins
Many other teleostean species	No acrosome, modified midpiece
Latimeria	Acrosome, long nucleus, long midpiece, tail with sidefin

Source: Afzelius (1978)

Elongation of the sperm nucleus may correlate with sperm transport and storage in internally fertilizing species. The formation of distinct sperm packets has been reported for many internally fertilizing teleosts, and appears to increase the chances of effective transfer of sperm to the female by maintaining high sperm densities during transport. Among the teleosts, most externally fertilizing species produce typical "aquasperm" having spherical nuclei. Grier and Fishelson (1995) reported colloidal sperm-packaging in mouthbrooding tilapline fishes.

The mature sperm with acrosome in hagfish has a long middle piece. A long middle piece is also reported in the sperm of lamprey (Jespersen 1975). Sperm morphology appears to reflect the mode of fertilization.

There is a large intergroup diversity in the number of spermatozoa produced during spermatogenesis and in the yield of spermiation in teleost fish. There is also a large diversity of endocrine regulation patterns for both spermatogenesis and spermiation. GTH usually relayed by steroids is necessary for multiplication of B type spermatogonia, meiosis, and in some cases spermiogenesis. The yield of spermiation appears directly related to the circulating level of hormones. There is also evidence that hormones influence the physiology and quality of spermatozoa. Sperm production can be related to the mode of reproduction. Large amounts of readily available sperm are required in species reproducing in open water; high primary sperm production and high hormonal stimulation are the conditions for the release of adequate amounts of spermatozoa. Complex social interactions mediated by pheromones also play a role. Males stimulate ovulation/oviposition in females and females stimulate spermiation and sperm release in males. An interesting dual stimulation occurs in goldfish. The male is stimulated by the female at 2 successive stages, preovulatory (increase of volume of milt available) and post-ovulatory (stimulation of courtship behaviour and most likely of the volume of milt released). It is interesting to note the economy of the process; mobilization of sperm required, and double function of 17–20 P (steroid) and prostaglandins involved as "hormones" for ovulation and pheromones for male stimulation. In species with a low GSI (testis weight at the end of spermatogenesis-expressed in % of the male body weight = GSI), some economy of sperm occurs and results from sophisticated spawning behaviour with probably also a complex hormonal regulation which remains to be investigated

(Billard et al. 1990). Males and females of pelagically spawning marine fishes face different problems in maximizing reproduces success. Females must ensure successful fertilization of their egg clutches. A study has shown that overall, the sperm allocation pattern can be explained both as a form of sperm economy in which males release the minimum number of sperm needed to fertilize most of the eggs in clutches of different size, and as a response to an increased probability of sperm competition with females of increasing size (Marconato and Shapiro 1996).

Billard and Cosson (1992) concluded that detailed analysis of the mechanisms involved in sperm motility are still limited. The main problem limiting the development of research is the short duration of motility, its rapid changes and the difficulty in obtaining homogeneous mixing of the semen. A two-step procedure, with a high final dilution, eliminates this problem and clearly demonstrates the rapid decline in the flagellar beat frequencies and the sperm velocities. After the first brief phase of motility involving all spermatozoa, a second phase can be identified with only a few spermatozoa showing some flagellar activity, without noticeable displacement, over a long period of time (upto 30 min in carp). In the literature, these two phases have often been confused, and this confusion, together with an inadequate dilution rate, explains the large variation in the duration of the motility reported previously.

It is known that the sperm of various teleosts of a true oviparous type shows high motility in response to hypotonic freshwater or hypertonic sea water. In fish with internal gametic association, the environment for sperm movement is the ovarian fluid which may be isotonic to body fluids of the fish, and is expected to be different in its physicochemical aspects from that for the sperm of true oviparous fish. Koya et al. (1993b) reported that the spermatozoa of the marine sculpins with the internal gametic association show their motility in environmental conditions appropriate to respective reproductive modes. Marshall et al. (1989) indicated that gonadotropin may directly regulate ion transport functions of the blood-testis barrier of vertebrates and in this way regulate seminal plasma ionic composition. For additional account, see Marshall et al. (1993).

The fertilization cone is a temporary structure formed on the egg surface just after gamete membrane fusion. The formation of a fertilization cone at the sperm entry site (SES) has been reported in a number of fishes. With progress of sperm penetration the SES of rose bitterling (*Rhodeus ocellatus ocellatus*) eggs transform into a swollen mass (SM) with a striking resemblance to the fertilization cone reported in carp eggs. Eggs in which the micropyle has been plugged by the SM do not accept spermatozoa following the fertilizing spermatozoon. Accordingly, the SM seems to function as a micropylar plug that prevents penetration of excess spermatozoa. Similarly it has been reported that the fertilization cone of carp eggs functions as a micropylar plug. Although the SM of the rose bitterling eggs formed swiftly after gamete membrane fusion, it did not disappear afterwards. In contrast, the fertilization cone of the carp eggs and other fish eggs recede quickly.

There are few reports on fertilization cone formation on egg surfaces at sites other than the SES just beneath the micropyle. Formation and recession of fertilization cones other than the SES is observed in the rose bitterling. The fertilization cones of polyspermically fertilized eggs differs from the SM of monospermically fertilized eggs in the rose bitterling. Although the differences can not be wholly explained, chorion with a micropyle and differences in components of the cortical cytoplasm beneath the SES and other areas may play significant roles (read Ohta 1985, 1991; Ohta et al. 1990; Ohta and Nashirozawa 1996).

A better understanding of the biology of sperm will be useful for management in fish farming. Improvements are needed particularly in cryopreservation, which has not been very successful and which has found no practical application so far.

Calcium ions play an important role in the chain reaction of fertilization in fish eggs, i.e., as stimulants, reception of stimulation, evocation of excitation (fertilization-wave or impulse), activation of enzymatic dissolution of the alveolar envelope, release of the alveolar contents, and separation of chorion by colloidal osmotic pressure (Yamamoto 1961). Calcium ions are necessary not only for the primary phase (reception of stimulation) but also for the later phases (Yamamoto 1961), according to experimental results with medaka (*Oryzias latipes*) eggs. Later researches reported that the fertilization wave in the medaka egg is propagated by calcium-stimulated calcium release, primarily from some internal sources other than the large cortical vesicles. Although it is not yet clear by which step fertilization of elkhorn sculpin (*Alcichthys alcicornis*) eggs is impeded in the ovarian cavity, it is clear that the chain reaction of fertilization in this species is triggered by a rise in calcium concentration of the surrounding water after egg deposition. It has been pointed out that low calcium concentration in the ovarian fluid may be essential for 'internal gametic association type' which is unique reproductive mode for teleosts (Munehara et al. 1994).

The egg fertilization rate or development decreases with time after ovulation. Such a phenomenon is called 'over-ripening', and is well known among many teleost species. In the turbot (*Scophthalmus maximus*) fertility of eggs is lost 2 to 3 days after ovulation, and the hatching rate decreases to 0% just one day after ovulation. For the Pacific herring, delayed spawning results in a progressive loss of larval viability, although fertilization rates remain high. For the barfin flounder (*Verasper moseri*), the fertilization rate during mid-spawning season remain at about 70% for 2 days after ovulation, but the viability of embryos might decrease faster than the fertilization rate.

The fertilization rate of eggs just after ovulation during the late spawning season was slightly lower than during mid-spawning season, and the decline in the fertilization rate with time after ovulation was also faster during the late spawning season. For the Japanese sea bream *Pagrus major*, the buoyant egg and hatching rates are lower during the early and late spawning season than during the mid-spawning season. The phenomenon of deterioration in egg quality during the late spawning season may be widespread among multiple spawners.

For multiple spawning fish with a definite cycle of ovulation and spawning during a single spawning season, in some species the interval of ovulation is known to be affected by water temperature (Koya et al. 1994).

The prominent features in which differences between the pelagic eggs, and demersal eggs are as follows: (i) *pelagic eggs*—subdermal space large; embryonal fins large and inflated; no vitelline circulation; eggs hatch an early embryo, (ii) demersal eggs—subdermal space small; embryonal fins small and little developed; prominent vitelline circulation; eggs hatch a postlarva. Largely transparent osteichthid eggs and larvae are mostly associated with highly illuminated places and are of most common occurrence as oceanic pelagic eggs. Hatching occurs at what is evidently the earliest possible time. Often present are many slow hatching, heavily pigmented eggs, usually adhesive and mostly placed in shaded locations.

Transparency and opacity are both recognized as protective specializations, with few exceptions. Fishes which produce these hyaline, pelagic eggs usually have well exposed pineal organs, and generally exercise no parental core, while those which produce heavily pigmented demersal eggs have well covered pineal organs which often do exercise parental care. Intermediates often have pigmentary contral of chromatophores over the pineal area and exercise it to occlude the pineal in the dark but expose it in the light.

The association of pigmented eggs laid in dark places with covered pineal areas and of hyaline eggs laid in brightly lighted places with exposed pineal areas suggest that there is here a primary condition governed by events at the enzymatic or molecular level. At the organismic level the role transparency plays in protecting against predation would seem to be of a secondary value especially as much predation on planktion is carried out on an essentially nonvisual basis.

The demersal eggs of teleosts comprise three functional types: non-adhesive, adhesive and twine eggs. The contact mechanisms of adhesive eggs vary according to the species. For example, in Plecoglossidae and Osmeridae, an adherent membrane covering the animal hemisphere of the egg is turned inversely, giving rise to an adherent stalk. Eggs, which attach to substrates by threads or filaments on the chorion have been reported in Pomacentridae, *Oryzias latipes, Fundulus heteroclitus* and *Gobiidae*. On the other hand, fish eggs with an adhesive material on the surface of the chorion are found in many taxonomic groups, i.e., Clupeiformes, Cypriniformes, Gasterosteiformes, Percoidei, Scopaeniformes and Pleuronectiformes. The manner of adhesive materials is considered to vary according to species. The vitelline envelope of a teleost egg can be divided into three compartments according to the origin; primary (of oocyte origin), secondary (of follicle cell origin) and tertiary (of oviduct origin). Details on the formation of the vitelline envelope can be obtained from the paper of Koya et al. (1995).

Recently there is a growing interest in the study of egg ultrastructure. In Cichlids, eggs attachment systems from phylogenetic angle have been studied. Cichlids are often divided into two polyphyletic behavioural groups: substrate-spawners and mouthbrooders. Substrate-spawning is the plesiomorphic cichlid condition and mouthbrooding, a derived condition, has arisen independently at least three times (and probably many more times) within the Afro-Neotropical cichlid radiation. The presence of a distinct mucus-filament attachment system is unique among percomorph taxa examined to date and is interpreted as a further support for the monophyly of Afro-Neotropical Cichlidae (see details: Stiassny and Mezey 1993).

With one known exception, all coral reef fishes produce pelagic eggs or larvae that spend variable periods of times in the water column prior to metamorphosing and settling onto shallow reefs.

The fish egg contains levels of nutrient that must meet demands for both energy and growth of the embryo and of the resultant fry upto the stage of yolk sac absorption and first feeding. Triacylglycerols commonly serve as a major energy reservoir. Details about lipid contents of eggs can be obtained from the articles of Sargent et al. (1989) and Phleger (1991). According to Jobling et al. (1995), thermal environment influenced the deposition of lipids during oocyte development in female broodstock Arctic charr, *Salvelinus alpinus*.

The most obvious phenological modification distinguishing annual killifishes—the egg-laying toothcarps (Cyprinodontidae) from other short-lived teleosts are found in relation to individual survival over the dry season, from fertilization of the egg in the flooded basin of one year to sexual maturation of the resulting offspring in the reflooded basin of the next wet seasons. Aestivation takes place, not in a cocoon, like lungfish, but in drought-resistant egg, which is buried in the mud bottom of a temporary pool. Physiologically, the state of the prehatching embryo may be very similar to that an aestivating lungfish. Eggs have developed thickened chorionic membrane and the chorionic filaments so characteristic of other rivulin eggs are either absent or modified into stiff, short structures, all features contributing to a hard, incompressible "shell" which will withstand severe mechanical stress and dessication. Not only are the eggs protected in this way, but they show also unique and incredible modifications. Embryonic diapause and hatching response to rainfall are reported (see Simpson 1979).

Aspects of spawning and fertilization have also been covered by Redding and Patino (1993). The discrete timing of spawning and hatching in many temperate fishes is believed to be adaptive, reflecting the phase of the mean seasonal cycle of environmental conditions most favourable for egg and larval survival. Deviations in either the phase of seasonal cycle or the timing of spawning and hatching are assumed to have consequences for the survival of the annual cohort of pre-recruits. Hypotheses proposed to explain the need for precise timing of life-history events have emphasised the relative importance of temporal variability in either food-chain processes, predation pressure, physical dispersive patterns of an area or the interaction between physical forcing and conditions favourable for larval growth.

Asynchrony between the time when larvae occur and favourable environmental conditions would be expected to lead to differences in the temporal distribution of hatch dates between surviving recruits and newly hatched larvae. Such a comparison of hatch-date compositions is possible through the analysis of daily otolith increments. Water flow and spawning time in a coral reef fish *Thalassoma bifasciatum* has been discussed by Shapiro et al. (1997). Some elementary processes of hatching are shown in Fig. 12.13 (see details: Inohaya et al. 1995).

The blood-testis barrier (BTB) has been demonstrated in guppy and the cyprinodont *Aphanius dispar* (Abraham et al. 1980; Nagahama 1983). The tight junctions between the Sertoli cells provide a permeability barrier between the vascular spaces of the stroma and the lumina of ripe cysts.

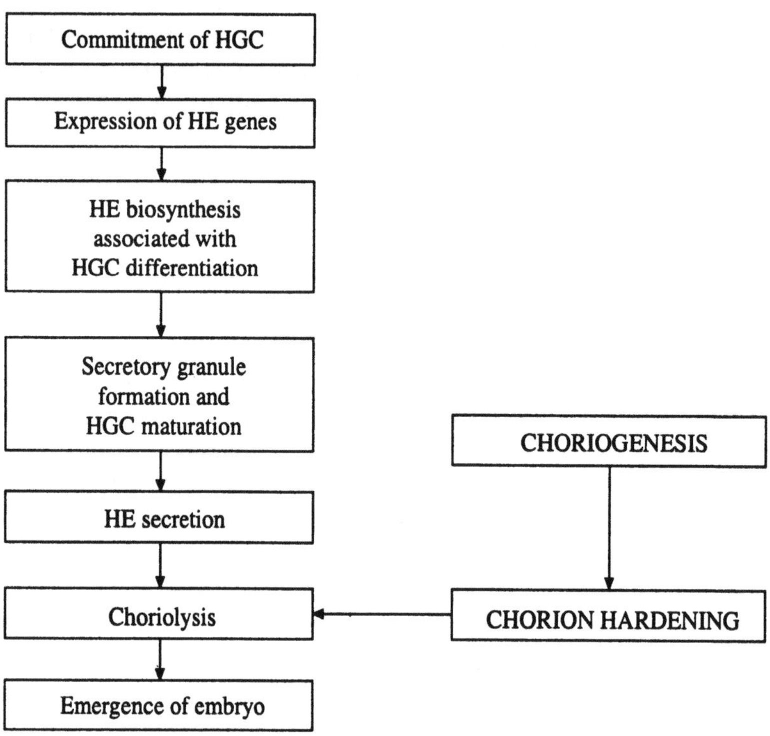

Fig. 12.13 **Some elementary processes directly or indirectly related to the enzymatic hatching in fish. He, hatching enzyme. HGC, hatching gland cell.**
***Source:* Yamagami (1996)**

A phenomenon exists in some fishes in which their sexual maturation is under social control (e.g., *Xiphophorus variatus*, *Poecilia gillii*, *Gambusia affinis*, *Pseudolabrus celidotus*, *Thalasoma duperrey*, *Sarotherodon mossambicus*). Recently, Bushmann and Burns (1994) reported the existence of a social maturational control in the neotropical swordtail characin *Corynopoma riisei*. The presence of adult male *C. riisei* was found to inhibit the maturation of juvenile males.

Seasonal cycles of gonadal development have been described for many species of teleosts. Such cycles are most readily observed in fishes of temperate regions, where substantial variations in temperature and day length may occur over the course of a year.

In many fishes gonadal recrudescence appears to depend upon the attainment of both a certain temperature and day length. Included among fishes with this type of gonadal regulation are *Phoxinus laevis*, *Oryzias latipes*, *Gasterosteus aculeatus*, *Culaea inconstans*, *Lepomis cyanellus* and female *Enneacanthus obesus*.

Such synchronizers may not play major roles in the regulation of tropical fish reproduction. The environmental factor in tropical habitats that shows the most marked variation seasonally is the alternation of dry and rainy seasons. Some tropical species, such as the weakly electric gymnotids reproduce exclusively during the rainy periods, while others show peaks of reproductive activity during that time (e.g., *Anableps dowi*, Anablepidae).

The deleterious effects of inbreeding are well recognized, but often the physiological mechanisms that lower fitnesses are poorly understood. Burns and Kallman (1985) reported that the highly inbred Coatzacoalcos (Cp) strain of the platyfish, *Xiphophorus maculatus* was noted for a high percentage of infertile females (XX). The ovaries of approximately one-quarter of all females regress. The time of gonadal atrophy varied before sexual maturation upto 11 months of age. It is suggested that the physiological lesion responsible for the ovarian regression syndrome involves the processes that control the determination and differentiation of the germ cells similar to those found in human 46, XY gonadal dysgenesis.

Various external stimuli influence the development of characters associated with reproduction. Some teleosts, for example, various anabantids change colour rapidly by means of chromatophore pigment dispersal or concentration. These changes are often associated with the fish's social life, particularly during the reproductive periods, and have assumed a signal function.

The male three-spined stickleback *Gasterosteus aculeatus* displays nuptial coloration as a result of erythrophore proliferation. The visual pathway mediates most of the responses elicited by social stimuli.

Dipnoan testes are elongate structures (Fig. 12.14), bound to kidney and dorsal body wall by mesenteries, and overlain by fat, especially before the dry season. The testes are stout medially, tapering laterally. In *Protopterus* and *Lepidosiren* the posterior part of the testis is not spermatogenic, but is "vesicular". The vesicular testis invades kidney tissue. *Neoceratodus* lacks a vesicular posterior region. The vesicular region is spongy, trabecular, and lacking spermatogenic tubules.

During the breeding season the spermatogenic part of the testis is filled with tubules containing mature sperm. The tubules are enlarged and have nests of primary spermatogonia peripherally, secondary spermatogonia somewhat more medially and spermatids in the lumen. Tubules open into a longitudinal testis duct that extends posteriorly to a series of lateral ducts, the "vasa efferentia", which carry sperm to the nephrons.

The regressed testis has much stromatous tissue. Tubules are reduced.

All three genera of dipnoan have spermatogonial distribution identical to the salmoniform,

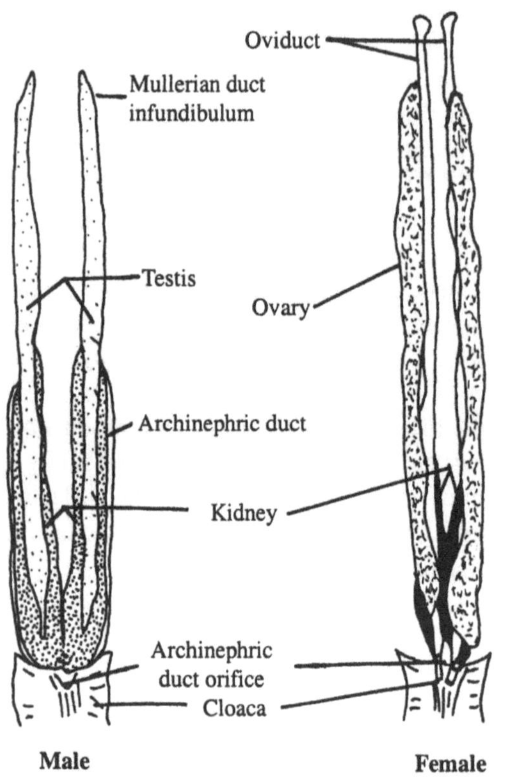

Fig. 12.14 Urogenital morphology of *Protopterus aethiopicus*: male and female.
***Source:* Wake (1986)**

perciform, and cyprinform type. Lungfish therefore share the primitive pattern with many teleosts and are unlike amphibians.

Sperm of *Protopterus* have a long tapering head, a short middle piece, and two flagella each about twice the length of the head. The head piece of *Neoceratodus* is similar to that of *Protopterus* but sperm of *Neoceratodus* have a single flagellum about three times the length of the head. The monoflagellate state is primitive.

Very little attention has been paid to the morphology of female urogenital systems, perhaps because it presents a "typical vertebrate" pattern. The ovaries are large, elongate structures that extend from near the heart to the posterior end of the kidneys. They are paired, and held to the dorsal body wall by a stout mesovarium. The ovaries are also tightly bound to the oviducts by a mesentery. The ovarian structure and pattern of ovum development in dipnoans shed no light on dipnoan relationships. The pattern is a primitive one, shared with most osteichthyans and many amphibians, particularly frogs.

The oviducts are Mullerian duct derivatives that extend from near the heart to their entrance into the cloaca dorsal to the urinary caecum.

The primitive vertebrate state has the embryonic Mullerian ducts becoming the adult oviducts. Among osteichthyans, this pattern is retained by dipnoans, actinistians, chondrosteans, and some holostean neopterygians. The condition in which most neopterygians do not develop Mullerian ducts (their "oviducts" are therefore not homologous to those of other vertebrates) is a derived state. The presence of oviducts of dipnoans therefore gives no positive support to hypotheses of relationship using presence of Mullerian duct derivatives as a character. The abdominal pores of *Protopterus* opening by a single aperture was likened to the condition in "primitive osteichthyan", but this conclusion seems very tenuous.

Males of all three genera of lungfish retain Mullerian ducts, at least to some extent. Those of *Neoceratodus* extend from the anterior end of the testis to the cloaca, where they fuse. They do not open to the cloaca. In *Protopterus*, adult males retain vestiges of the infundibulum at the anterior end of the kidney and a few millimeters of each duct posteriorly. These ducts unite in the wall of the cloaca. The ducts are even more reduced in males of *Lepidosiren*. Only the anterior infundibula are present.

The abdominal pores are probably not associated with ovum transport. They may be vestigial or involved in other fluid flow.

All three genera of dipnoans retain a simple cloaca with the urinary ducts opening by a single aperture. Presence of cloaca is of course the primitive vertebrate pattern, and all bony fish that lack a cloaca are derived (read Wake 1986).

3. MATERNAL-FOETAL RELATIONSHIPS

There has long been interest in the evolution of viviparity among groups of vertebrates. Many correlates to live-bearing have been observed, such as internal fertilization, reduced numbers of ova, and prolonged developmental periods. Yet, there is still not real agreement on the vocabulary of viviparity, on the modes of selection and costs and benefits of the phenomenon (Packard et al. 1989). Viviparity is generally construed as any mechanism for live-bearing, or maintenance through part or all of development, by either maternal (usually) or paternal parent in or any part of the body (e.g., ovary, follicle, oviduct or uterus, skin, vocal sacs, stomach (see Wake 1989). Other terminological schemes recognize categories of live-bearing, such as ovoviviparity (maintenance of developing young through part or all of development with *no* provision of nutrition beyond that of yolk) and viviparity (provision of parental nutrition *after* the yolk is exhausted, and categories of nutrient provision (e.g., lecithotrophy; matrotrophy). Fig.12.15

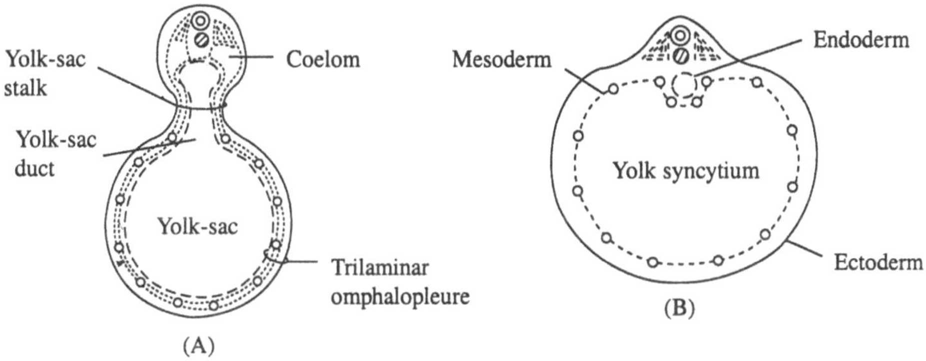

Fig. 12.15 (A) Arrangement of the extraembryonic membranes in sharks and (B) embryonic membranes in teleosts. Open circles indicate vascular mesoderm.
***Source:* Jones and Baxter (1991)**

illustrates the arrangement of the extraembryonic membranes in sharks and embryonic membranes in teleosts. Wake (1992) stated "Viviparity has evolved many times in many lineages of vertebrates. Many evolutionary scenarios for evolution of viviparity, or structures associated with it, have been proposed. Many of these are testable using methodologies developed recently. Some structures that facilitate viviparity (e.g., corpora lutea and oviducts) are homologous, for they are inherited from an ancestor in a lineage. More such structures (e.g., placentas and placental analogues) are convergences, for they appear in diverse lineages but are not traceable to a common ancestor. Conservatism and constraint characterize, but not all, such convergences. A phylogenate hypothesis facilitates identification of homologies and convergences and the testing of evolutionary scenarios".

Callard et al. (1992) suggested that ovarian progesterone (follicular or luteal in origin) has a dual role in the evolution of viviparity: (1) to inhibit myometrical contractions, thus providing a primary condition for egg retention and viviparity, and (2) to inhibit estrogen-induced hepatic vitellogenin synthesis as part of both normal oviparous cycles and as a concomitant of placental evolution.

Viviparity makes its first appearance in the vertebrates among fishes, but has not been indentified in the cyclostomes (hagfishes and lampreys). Viviparity has been reported in the living coelacanth, *Latimeria.*

Wourms et al. (1988) contributed a detailed account of maternal-embryonic relationship in viviparous fishes. Jones and Baxter (1991) reviewed information about the patterns and endocrine control of gestation and parturition (or spawning, oviposition) in vertebrates. A brief account on maternal-embryonic relationship is also given by Redding and Patino (1993) in the chapter on reproductive physiology related to fishes.

The adaptive value of viviparity in general (including ovoviviparity) lies in the protection it affords to developing offspring during the most vulnerable stage in their life histories, while at the same time accommodating all metabolic needs. Vulnerability of eggs to disease and predation is avoided and embryos can be transported to favourable temperatures for development. Viviparity also conveys certain advantages in dispersal; a single pregnant female is capable of colonizing a new environment.

The shark yolk sac represents the most primitive typical vertebrate foetal membrane. Although homology is not direct with the placentae of mammals, the same basic endocrine hormones govern the placental development in both groups. Further, the yolk sac and placenta of sharks may be regarded as the "common denominator" vertebrate fetal membrane.

Viviparity appears to have evolved enumerable times in ectothermic vertebrates, and often more than once within the same genus.

Chondrichthyan fish

Korsgaard and Weber (1989) reviewed the different patterns of maternal-foetal trophic and respiratory exchange, its hormonal control, the anatomical structures involved, and the adaptations to environmental conditions and metabolic requirements in viviparous ectothermic vertebrates. Prenatal nutrient absorptive structures in selachians (Hamlett 1986), and evolution and morphogenesis of the placenta in sharks (Hamlett 1989), are allied coverages. An overview of gestation can be obtained from Jones and Baxter (1991). For special attention on uterogestation and placentation in elasmobranchs, the reader is referred to Hamlett et al. (1993).

The reproductive biology of the chondrochthyan fishes is remarkably sophisticated. Using both oviparous and viviparous reproductive modes, the group in general has adapted the strategy of giving birth to relatively few young at one time, each representing the investment of a great deal of maternal energy. The oviparous species foreshadow the situation common in oviparous reptiles and universal in birds. On the other hand, viviparous species range from internal incubators, in which large yolked eggs are retained, to other species in which the complexity of placentation and egg yolk reduction approaches the eutherian condition. Further, in certain viviparous elasmobranchs, the phenomenon of histiotrophic nutrition attains an importance and complexity not seen in any vertebrate group. Associated with the elasmobranch priority on these female reproductive adaptations, internal fertilization, amniote patterns of reproductive tract development, sex differentiation and a typically vertebrate of reproductive endocrinology are seen for the first time in the vertebrates (Callard and Klosterman 1988).

Viviparous reproduction is widespread among the chondrichthyan fish. Among the elasmobranchs, skates are primarily oviparous, and all rays are viviparous. Sharks exhibit a range of parity types (253 of 367 extant species are viviparous). Elasmobranch fishes produce small quantities of large eggs, ranging in size from 1 mm in *Scoliodon* to 100 mm in *Ginglymostoma* and *Chlamydoselachus*. Each size is a reflection of reproductive strategy (oviparous or viviparous) with small eggs being correlated with an enhanced maternal nutritive role during gestation. In contrast, large eggs are the rule when the embryo is exclusively dependent on the yolk reserve or maternal histiotroph for development. Oviparous elasmobranchs lay eggs surrounded by leathery egg cases; the latter are secreted by glands of the oviduct(s). The oviducts are true derivatives of the embryonic Mullerian ducts. All extant holocephalans are oviparous. The gross reproductive anatomy of the viviparous species is fairly similar to that of oviparous ones, and internal fertilization, a prerequisite for viviparous reproduction, typically occurs in both reproductive forms. In viviparous species, however, part of the oviduct is modified to form an uterus, the structure which varies with species and with various adaptations facilitating the maternal-foetal exchange of the different materials.

Oviparous species normally lose 20–30% of the original weight of the egg during development. The cat shark, *Scyliorhinus*, shows a 21% loss of organic material during development. This loss represents energy used for tissue growth and maintenance. The egg case-fluids are nearly identical with seawater.

Viviparity in chondrichthyans may be grouped into nutritional categories: lecithotrophy in which the embryos rely on their yolk sac as the sole nutritional source during development; and matrotrophy, which is found in species with a placental analog or a true yolk sac placenta. Oophagy, defined as consumption of unfertilized eggs by the growing embryos, or adelophagy, which is embryonic cannibalism in utero, are less common in these fish.

Interestingly, the nutritional profit in terms of increased embryonic organic weight does not appear to be related to the intimacy of maternal and foetal structures. In species with a placental analog nutritional transfer to the embryos primarily results from a secretory function of the modified uterine epithelium. Although secretions from uterus may be abundant, the actual amount of nutrients may be small as indicated by temporary weight losses of about 35% in some species of *Torpedo*.

In placental sharks embryonic dry weight may increase eight-to-ten fold compared to that of the egg. The true placental forms, which have yolk sac placentas are limited to two families of sharks, the Carcharinidae (requiem sharks) and Sphyrnidae (hammerhead sharks) (Figs. 12.16–12.20). However, individual genera of these two families may show markedly different nutritional forms. Thus, *Mustelus manzano* lacks a placenta and the embryo is typically lecithotrophic, whereas *Mustelus griseus* embryos becomes increasingly dependent on the placenta as the yolk mass becomes reduced with growth. The placenta consists of two parts, the maternal uterine wall and the embryonic yolk sac wall, which show increasing interdigitation with embryonic growth. Exchanges between maternal organism and the embryos occur across an embryonic membrane, which is found between maternal and embryonic portions of the placenta, and is considered to play an important role in the nutrition, respiration and excretion of the embryos of *M. griseus*.

The ultrastructure of the yolk sac placenta of different viviparous sharks has recently been described in relation to its function in gas exchange and haematotrophic nutrient transport. The preimplantation yolk sac is believed to secrete proteins and act as a respiratory organ and thus to be implicated in synthetic as well as in transport processes. It has been hypothesised that the maternal shark liver continues to synthesize vitellogenins during gestation and that these precursor proteins are transferred to and taken up by the foetal yolk sac epithelium.

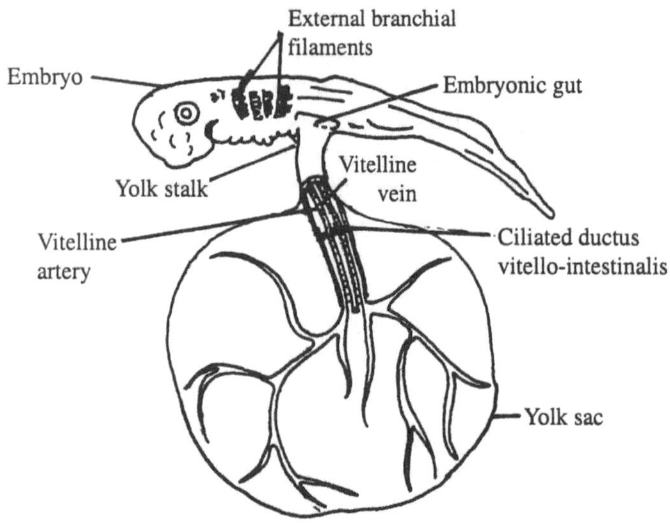

Fig. 12.16 Preimplantation stage carcharhinid embryo with external branchial filaments. The yolk stalk contains a vitelline artery, vein and ciliated ductus vitellointestinalis, confluent with the embryo gut.
***Source:* Hamlett (1986)**

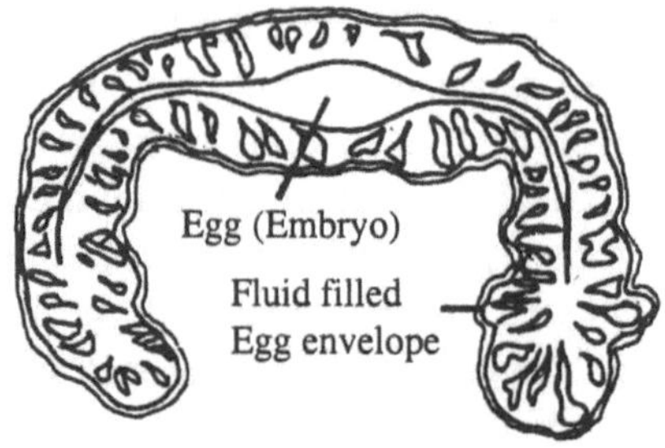

Fig. 12.17 Diagrammatic representation of partially unfolded egg envelope with fertilized egg.
***Source:* Hamlett (1986)**

Stress should be given on paraplacental nutritive absorptive structures: (1) external gill filaments of selachians restricted to early embryonic period. They contain a single vascular sinusoidal loop that is continuous with the afferent and efferent branchial arteries. External gill filaments are present prior to the establishment of a functional placenta and may function in respiration and as a paraplacental nutritive membrane prior to the establishment of a functional placenta (see Hamlett et al. 1985e for EM observations), and (2) appendiculae. After a functional placental attachment is effected with the maternal uterine lining, the yolk stalk may be correctly termed as umbilical cord. Most species of *Carcharhinus* possess a smooth umbilical cord. Other species display umbilical cords festooned by arborescent appendage (appendiculae). Genera possessing, appendiculae include *Scoliodon*, *Rhizoprionodon* (Fig. 12.21), *Sphyrna* and *Paragaleus*. The morphology of appendiculae varies with species. Appendiculae may be persistent or transient and even absent in certain forms.

Complex ontogenetic changes as regards the source of nutrition have recently been described in the sand tiger shark, *Odontaspis taurus*, where the embryos initially are lecithotrophic, but progressively they change to adelphophagic and oophagic nutrition.

In the viviparous spiny dogfish, *Squalus acanthias*, the embryos are retained in the uterus for a gestation period of about 22 months. This species is lecithotrophic and shows no signs of nutritive secretory activity of the uterine epithelium. A considerable weight loss of upto 55% during embryonic development is accompanied by ultrastructural changes that convert the uterine mucosa into a respiratory membrane, which may also participate in transport of water and electrolytes. During the first 6 months of gestation when the embryos are enclosed in an elongated egg case, the embryonic kidneys are still immature, and the urea concentration of the uterine

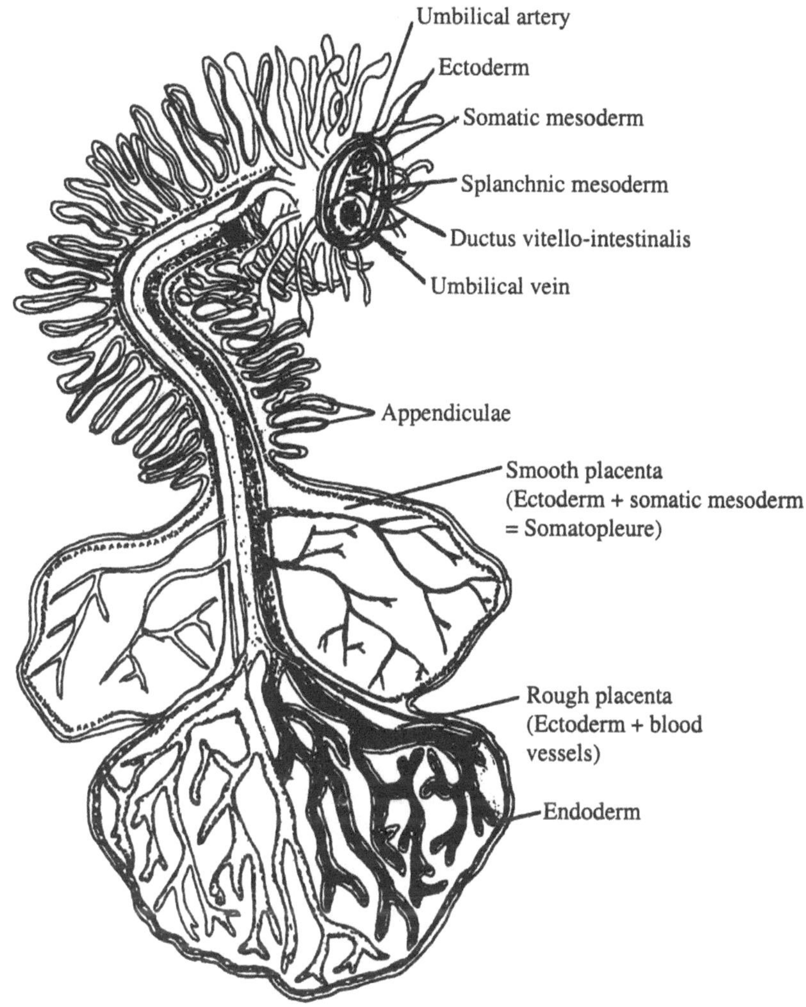

Fig. 12.18 Stereodiagram of the germ layer contributions to the shark yolk sac placenta. The appendiculae are vascularized outpocketings of ectoderm that are filled with loose connective tissue. The smooth placenta has contributions from ectoderm and somatic mesoderm. The rough placenta is richly supplied by a mesodermally-derived vasculature. Endodermal cells coat the blood vessels and the inner aspect of the ectoderm.
***Source:* Hamlett (1986)**

fluid is at the same level as in the maternal blood. After 6 months the embryos hatch and are able to maintain their own uremia. Even enclosed in the presumed egg case ("candle") and incubated in sea water, the embryos are capable of regulating their plasma electrolyte and urea concentrations. Older hatched embryo in utero are capable of osmoregulation against sea water and display a reduced ionic permeability and a functional rectal gland, as well as other pathways for extrusion of NaCl.

In some sharks gas and metabolite exchanges occur across placentae which are modifications of the distal portions of the yolk sac that are in contact or interdigitate with vascularized areas

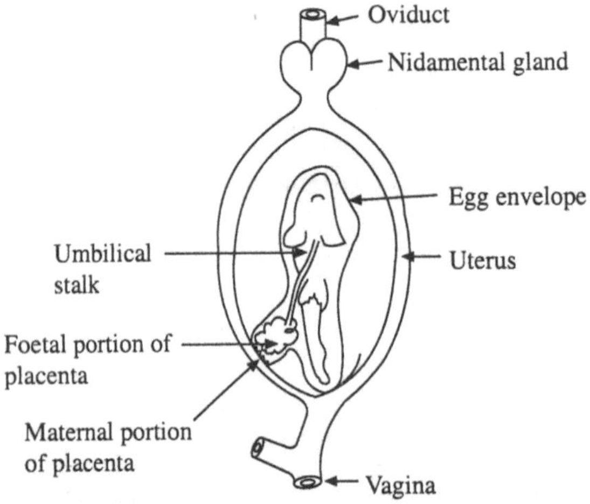

Fig. 12.19 Diagram of a near term carcharhinid embryo within the maternal uterus. The now empty yolk sac has become implanted in the maternal uterine wall. The egg envelope intervenes between embryonic and maternal tissues. Concomitant with growth and elongation of the embryo, the egg envelope reservoir has been obliterated and the embryo has enlarged to fill the egg envelope.
***Source:* Hamlett (1986)**

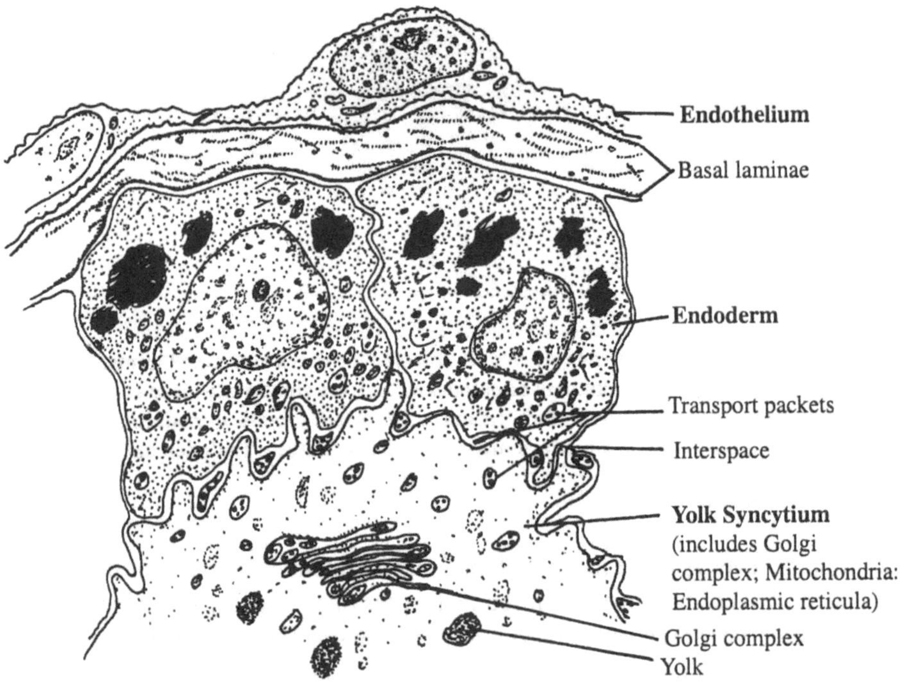

Yolk-endoderm-endothelial complex in a preimplantation yolk sac.

Fig. 12.20 Yolk platelets are partially digested within the yolk syncytium. The products are modified within the yolk syncytical Golgi complex and transported to the endoderm in membrane-limited packets. Metabolites are then passed to the endothelium.
***Source:* Hamlett (1986)**

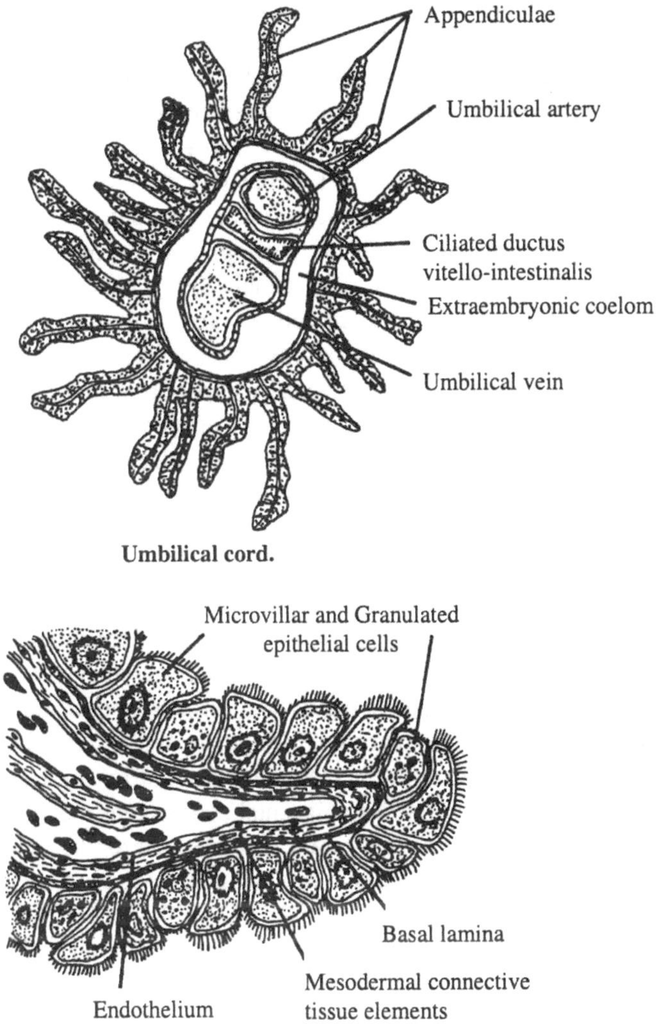

Fig. 12.21 A transversely sectioned umbilical cord illustrates the umbilical artery and vein, ciliated ductus vitello-intestinalis, extraembryonic coelom and appendiculae. An enlargement demonstrates the histological features of an appendicula, viz. epithelium with microvillar and granulated cells, basal laminae, mesodermal connective tissue elements and endothelium.
***Source:* Hamlett (1986)**

of uterus. In skates and rays, long thin villi (trophonemata) extend from the uterine wall. and at least in one species (*Pteroplatea micrura*) these enter the foetal oesophagus via the spiracles, actually a kind of suckling. The foetuses are not otherwise attached to the mother. In *Squalus acanthias*, which lacks a placental analog, a vascular plexus covers the foetal body in the initial phases of development, and the yolk sac throughout most of intrauterine existence, while external gill filaments, followed by internal gills, are found in the prenatal period.

There is very little evidence for hormonal control of viviparity of elasmobranchs beyond the process of vitellogenesis and ovulation.

A summary of evolution and morphogenesis of the placenta in sharks is that placental sharks sequentially utilize three modes of embryonic/foetal nutrition to nourish their developing young-viz., vitellogenesis, histotroph secretion, and haematotrophic placentation. These are the principal modes employed by vertebrates in general. The progressive shift from reliance on yolk, to histotroph, then to the placenta mirrors the probable stages in the evolution of placentae in sharks. Preimplantation shark embryos begin their ontogenetic development by reliance on yolk. In vitellogenesis, energy-rich precursors of hepatic origin are stored in oocytes prior to ovulation. Following fertilization, embryogenesis and foetal development proceed at the expense of yolk stores sequestered in the yolk sac. Yolk is made available to the young by two means. Yolk is partially solubilized in the yolk syncytium of the yolk sac and subsequently modulated through the yolk sac endoderm. Yolk metabolites are then transported across the vitelline endothelium to reach the foetal circulation. Yolk granules are physically transported to the foetal gut by ciliated cells lining the ductus vitellointestinalis. The ductus is an endodermally lined conduit that extends from the yolk sac to connect to the foetal gut where digestion and absorption occur.

Following depletion of the yolk stores and prior to the development of the placenta, nutrient substances may be supplied to the foetus by the secretory activity of the uterus in the form of histotroph. Histotroph may then be ingested/or absorbed.

In the transition from the yolk-reliant and histotrophic phases to a placental situation, there is a coincident shift in the function of the maternal uterus to one of nutrient-producing or transporting organ. Essential adaptations of the uterus include: enhanced secretory function, expansion to accommodate the embryos, respiration, osmoregulation, waste disposal, and protection of the young. As the yolk sac differentiates into placenta, focal areas of the uterus become modified as uterine attachment sites. It is at the utero-placental complex that nutrient and metabolite exchange between mother and foetus will be effected. Paraplacental uterine sites may also continue to play a role in the elaboration of histotroph. The placenta is specialized into two segments: a proximal, smooth portion that has characteristics of a steroid-producing tissue and a more richly vascularized distal portion that is nutrient-transport tissue.

Evolution of placentation involves modifications of existing maternal and foetal membranes. The ontogenetic transformation of the shark oviduct into a functional uterus for the retention and metabolic support of developing young certainly is modulated by endocrine regulation. In the developing foetus, the yolk sac and stalk display terminal differentiation into a regionally specialized placenta and an umbilical cord respectively. Accessory vascular structures, of the umbilical cord, termed appendiculae, are present in some species.

Kormanik (1992) discussed the ion and osmoregulation in prenatal elasmobranchs, with focus on evolutionary implications (see Chapter 11).

Details of ultrastructure of the full-term yolk sac placenta have been given by Hamlett (1985a, b, c). Hamlett et al. (1985d) described the structure of trophonemata in cownose ray, *Rhinoptera bonasus*. Hamlett (1987) reviewed the morphological development and function of the shark yolk sac placenta. For subcellular organization of the yolk synctial-endoderm complex in the preimplantation yolk sac of the shark, *Rhizoprionodon terraenovae*, see Hamlett et al. (1987). The reader is also referred to Callard and Klosterman (1988) for account of chondrichthyan viviparity.

Bony fish

In bony fish reproduction is typically oviparous with external fertilization. Of the 425 families of extant osteichthyan fishes, 14 families contain viviparous species. Only 510 out of approximately 23,637 species of osteichthyan fishes are viviparous. The lungfish order Dipnoi (subclass Sarcopterygii) contains 6 extant species, all of which are oviparous. Within the order Crossopterygii (Sarcopterygii), the extant coelacanth *Latimeria chalumnae* is viviparous. Fourteen of the fifteen families of bony fish containing viviparous species are in the subclass Actinopterygii. Most viviparous species are in the superorder Teleostei, orders Cyprinodontiformes (families Goodeidae, Anablepidae, and Poeciliidae) and Perciformes (families Embiotocidae, Clinidae and Labrisomidae).

Unlike other vertebrates, teleosts lack oviducts of Mullerian duct derivation. Both the lungfishes and *Latimeria*, however, possess oviducts that are homologous to those of other vertebrates. In teleosts, neither males nor females develop Mullerian ducts; however, ducts (gonoducts) do develop from either the peritoneum overlying the ovaries or from ovarian tissues. In contrast to other vertebrates, the teleost gonoduct is continuous with the lining of the saccular ovary. Ova are ovulated into the ovarian cavity, which connects with the gonoduct. However, in some teleosts portions of gonoducts proximal to the ovary may degenerate, and ovulated ova exit the ovarian cavity and enter the coelomic cavity rather than the gonoduct; ova exit through either the remaining portion of the gonoduct or through pores (e.g., salmonids, osteoglossids).

Internal fertilization, which is obligate for the evolution of teleostean viviparity, is accomplished by either an intromittent organ (derived from anal fin rays) or cloacal apposition. Viable sperm may be stored within the ovary for at least 10 months. Four distinct processes are involved in teleostean reproduction, and variation in the sequence of these processes is found in viviparous teleosts. The four processes are: (1) ovulation (0), (2) fertilization (F): (3) hatching (H) from egg envelopes; and (4) parturition (P); exit from the female's body through the gonoduct or pore. The sequence of these events in the typical ovuliparous teleost with external fertilization is O-P-F-H. In viviparous teleosts, the sequence of these events changes because of internal fertilization and the potential for intraovarian fertilization and gestation. The embryos of viviparous teleost fish are gestated within the ovary, a feature which is unique among vertebrates. Intraovarian gestation is either "follicular", with the eggs fertilized and developing within the ovarian follicles, or "ovarian", in which case the eggs, fertilized or unfertilized, or developing embryos are released from the follicles into the ovarian cavity for further development. The nutritional relationships may be defined as either mainly lecithotrophic or as matrotrophic. A number of morphological specializations have developed in embryonic as well as in maternal tissues to facilitate nutrient transfer to matrotrophic developing embryos. Thus, the embryonic epidermis, fin, epithelium, gut, or even yolk sac may be specialized for nutrient absorption, while the maternal ovarian epithelium may similarly be differentiated to increase its secretory function. It is generally considered that, with respect to viviparity, the lecithotrophic fishes are more primitive and conversely, the matrotrophic ones are more specialized.

Placentation can be divided into: (1) nonplacental matrotrophy (e.g., sharks and (2) teleostean placentation, which is further categorized into (i) branchial placentation (e.g., in embiotocid and jenynsid fishes); (ii) trophotaenial placentation (e.g., in two ophidioids, one parabrotulid, and all goodeids except one species; and (iii) pericardial-sac placentation (e.g., in families Poeciliidae, Clinidae and Anablepidae) (see following account).

Regarding the nonplacental matrotrophy, as in sharks, oophagia and adelophagia is reported in some ophidioid fishes with ovarian gestation. The viviparous coelacanth *Latimeria chalumnae* is probably oophagous (see Wourms et al. 1991). The coelacanth embryonic dry weight increases by as much as 23% during gestation; thus, oophagic nutrient supplementation may be quite important. Incubation of embryos in an abdominal integumentary pouch of male seahorses offers the rare possibility of a placental relationship in a male vertebrate. During gestation of *Hippocampus guttulatus*, which lasts 21 days, the embryos lie in honeycomb-like chambers in the pouch lining and are bathed in "marsupial fluid" (not of serum origin). This fluid contains yolk protein, which decreases in concentration during gestation. The yolk is derived from lysis of some of the eggs originally deposited in the pouch. This lysis is caused by a protease secreted by the pouch epithelial lining, and protease secretion is controlled by pituitary prolactin. It is not known if the embryos use the yolk for nourishment (i.e., oophagia). It has been proposed that pouch of *Hippocampus erectus* acts as an osmotic chamber; sodium ion concentration in the pouch fluid increases during incubation, whereas calcium ion concentration decreases. Radioactive calcium injected in the fluid is taken up by the embryos. The internal pouch lining is very thin and highly vascularized and there is exchange of nitrogenous wastes and gases between embryos and the male.

During late gestation in *Ogilbia cayorum*, a viviparous brotulid fish, ovigerous bulb form in the ovarian cavity, and these extend through the mouth into the oesophagus of some, but not all, embryos. The bulbs are fluid-filled and composed of vascular ovarian epithelium. These ovigerous bulbs appear secretory based on their histology, and they may function in embryonic nutrition. Ovigerous bulbs have also been described in the Cuban blindfish, *Lucifuga*. The apposition between ovigerous bulbs and pharyngeal epithelium may constitute a branchial placenta; however, there may be no physiological exchange between these tissues, since pharyngeal epithelium in most vertebrates is stratified and non-absorptive. Therefore, this relationship may simply enhance the delivery of ovarian secretions into the embryonic gastrointestinal tract. Until further investigation determine that intimate maternal-embryonic tissue apposition occurs, it cannot be classified as placenta.

Schindler and Hamlett (1993) discussed the maternal-embryonic relations in viviparous teleosts.

Unlike elasmobranchs and amniotes, teleosts have no extra-embryonic membranes, and the ectoderm and mesoderm that surround the yolk are not discarded at hatching. An exception to this is seen in embryonic goodeids (*Ameca splendens*), which lose trophotaeniae following parturition. Therefore, placental conditions in teleosts are not homologous to placentae seen in elasmobranchs or amniotes.

A few examples of teleostean placentation are given here.

Viviparous Teleosts with Intrafollicular Gestation

(1) Family Anablepidae: The viviparous, four-eyed fish *Anableps* displays follicular gestation with two marked morphological specializations, the follicular placenta and spectacular enlargement of the gut (Figs. 12.22 and 12.23). Development is typically matrotrophic. Embryonic postfertilization weight increases about 3000 to 8400 times, reflecting a massive transfer of maternal metabolites across follicular placenta.

Jenynsia, viviparous cyprinodonts from southern South America have recently been lumped together with *Anableps* and *Oxyzygonectus* to form the expanded family Anablepidae. This family includes a rather heterogeneous group of fishes, each genus having developed its own strategies for providing nutrition during embryogeny. The monotypic *Oxyzgonectus* is oviparous, and the free viviparous species of *Anableps* pass through prolonged follicular gestation; however, the embryos of *Jenynsia* are released into the ovarian cavity at rather immature stages of development.

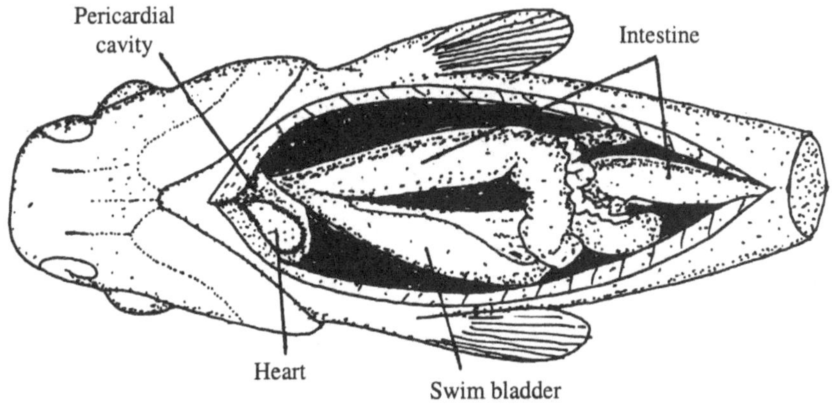

Fig. 12.22 Dissection of embryo of *Anableps anableps* at the end of the period of gestation. The enlarged belly sac has been absorbed and the differentiation of the belly wall is complete. The expanded intestine has undergone involution and is greatly decreased in length.

The embryos of *J. lineata* absorb their yolk early in gestation and thereafter depend on maternal nutrition until term. *J. lineata* retains its embryos within the ovarian cavity for a prolonged gestation. In the absence of egg envelopes, maternal-embryonic transfer occurs through ovarian fluid across apposed epithelia, relatively lining the ovarian lumen and the surface of the embryos. Outgrowths of the inner lining of the ovary, villi ovariales, enter the pharyngeal cavity of the embryos via an opercular cleft remaining from early stages of development, after depletion of yolk reserves, until shortly before term.

In *Jenynsia*, fecundity is much greater than the birth rate which may reflect the failure of some embryos to catch a villus ovarialis; embryos that do not obtain connection to the parental supply system may die (see details: Schindler and de Vries 1988).

(2) Family Poeciliidae: Among the variable reproductive adaptations in teleosts, a remarkable array is found within a single family of viviparous fishes, the Poeciliidae. A fairly wide range of productivity levels persists within the family. In addition to their diverse mechanism of viviparity, poeciliids also utilize a range of adaptations characteristic of oviparous poikilothermic vertebrates. Within the Poeciliidae, positive correlations of increasing fecundity with increasing female body size also appear to be the rule, but here, too, exceptions occur.

Reproduction in the Poeciliidae evolved along two major pathways: (1) single brooded ovoviviparity, and (ii) superfetation.

Most species of poeciliid fish display follicular gestation, with a typical gestation period of approximately 1 month. Superfoetation (simultaneously occurring multiple stages of embryos) is common. The expanded pericardial sac constitutes the principal foetal morphological adaptation (Fig. 12.24). Nutritionally they are lecithotrophic as well as matrotrophic, the former often resulting in dry weight loss during embryonic development, and the latter in constant or increasing embryonic dry weight during gestation. Thus, the maternal-foetal trophic relationship may be quantified by a low rate of weight loss in the embryo. In *Poecilia reticulata* which appears to be minimally dependent on maternal nourishment, the intrafollicular exchange seems to be limited to minerals and respiratory gases.

Following high salinity adaptation of the pregnant female guppy *P. reticulata*, increases in

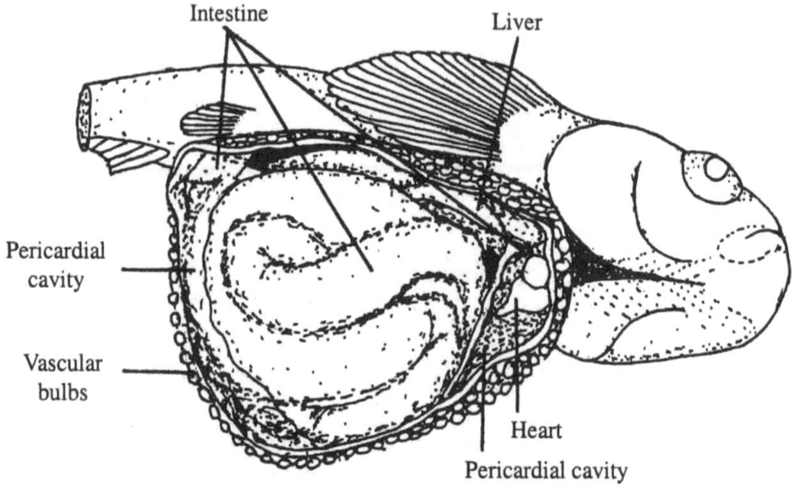

(A) *Anableps anableps*: Embryo.

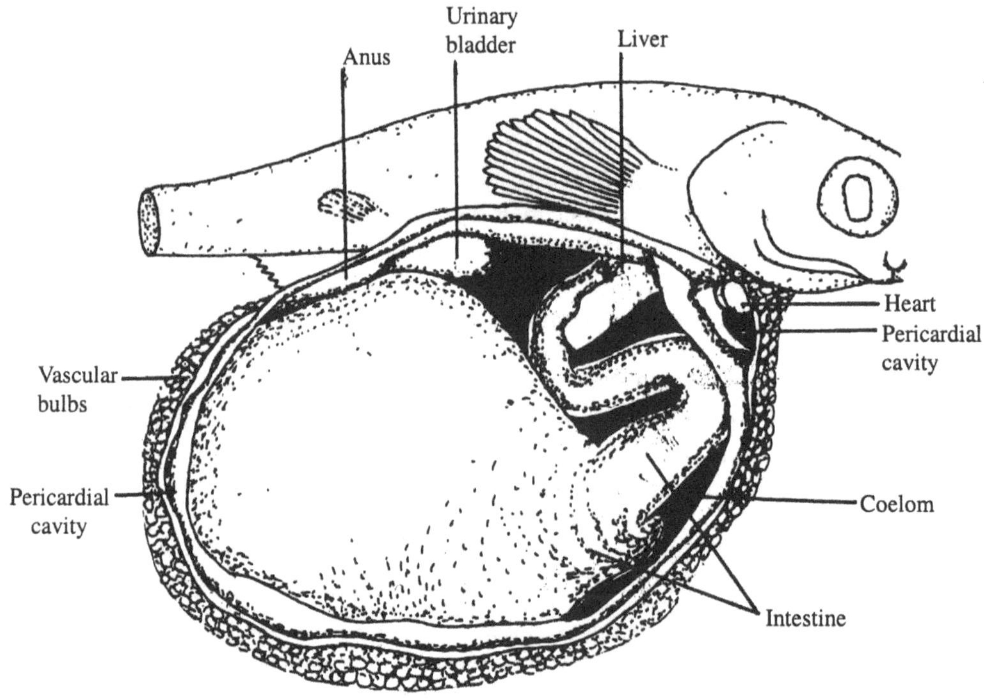

(B) *Anableps dowei*: Embryo.

Fig. 12.23 (A) Dissection of 21 mm. embryo of *Anableps anableps*. Pericardial sac is in maximal stage of development. Expanded middle intestine is in maximal stage. (B) Dissection of 24 mm. embryo of *Anableps dowei* to show the relations of the pericardial cavity, coelomic cavity and expanded posterior intestine.

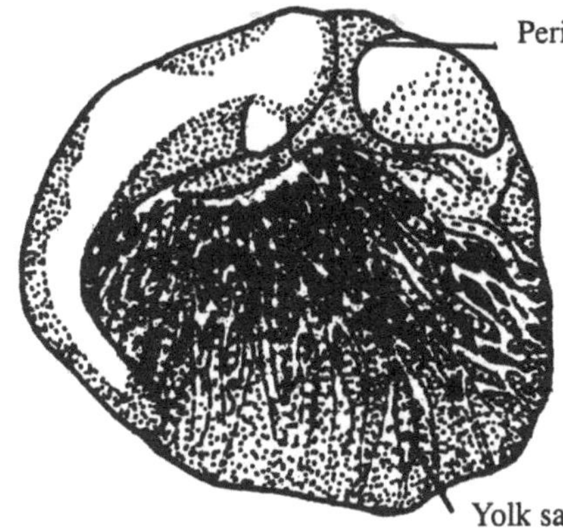

Fig. 12.24 Embryo of poeciliid fish with fairly large yolk sac. A part of the pericardial sac is drawn over the head in the form of a hood. The yolk sac and a part of the sac containing the pericardial cavity are covered by a portal network.

embryonic electrolyte, urea, and amino acid concentrations have been demonstrated. The contribution of urea to embryonic osmoregulation seems to dominate in early development, whereas the amino acid pool appears to be more important during the later stages, indicating an ontogenetic exchange in embryonic nitrogen metabolism. However, the mechanisms underlying nitrogen accumulation in the embryos, including adaptive changes in enzymatic activities of an ornithine-urea cycle, has not yet been specified, although there is evidence for a urea cycle in oocytes and developing embryos of the guppy.

Pituitary hormones, thyroid hormones and 17, 20 progesterone have been indicated to play a role during various steps of development.

Views have been presented on superfoetation in fishes and the cost of reproduction. In extending a hypothetical life table to enable calculation of *r* (the intrinsic rate of increase), fishes without superfoetation have a slight advantage. Regardless of female survivorship, the ability to produce a large number of young in the initial reproductive period may be adaptive in transient environments where *r* selective is operative.

Superfoetation must convey some evolutionary advantages to viviparous fishes; however, the causative selective force, still remains an enigma (read Downhower and Brown 1975 and refer Table 1). The reader is referred to contributions on follicular placenta of viviparous poeciliid fish, *Heterandia formosa* (a matrotrophic)- ultrastructure and development of the embryonic absorptive surface (Grove and Wourms 1991) and of the follicular epithelium (Grove and Wourms 1994). The presence of cells in the surface epithelium of *H. formosa* embryos that morphologically resemble teleost chloride cells indicates that the embryonic surface in *H. formosa* may serve an osmoregulatory role as well as an absorptive function.

As mentioned earlier, embryonic nutrition in poeciliids ranges from lecithotrophy to matrotrophy. In *P. reticulata* and in *Gambusia affinis*, embryos are reported to display anywhere from 0% to 35% change in dry weight, indicating that there is little or no maternal-embryonic nutrient transfer. On the other hand, embryonic dry weight increases of 1,800% and 3,900% in *Poeciliopsis turneri* and *H. formosa*, respectively, indicate substantial embryonic absorption of exogenous nutrients.

It has been found, for a variety of organisms, including fish, that the size of offspring can influence offspring fitness. In fish, the aspects of fitness most often studied are early survival in the presence or absence of food, egg development rate, and early growth. There is also evidence that the mother's environment influences offspring size and composition. The environment experienced by mother thus potentially influences offspring fitness, mediated through this maternal effect. Reznick et al. (1996) studied this aspect in three species of poeciliid fishes, taking food availability as the environmental factor, and proposed that production of large young in the face of low food availability might represent adaptive plasticity; matrotrophy might represent a constraint that prevents such an adaptive response. Superfoetation had no impact on this maternal effect.

(3) Family Clinidae: The embryos of *Clinus superciliosus* and *C. dorsalis* which develop intrafollicularly and are supplied with a small amount of yolk, absorb nutrients through various absorptive areas of the body. Nutritional transfer in *C. superciliosus* follows from a foetal weight increase of 34%. The follicular fluid is rich in lipids and amino acids, which are believed to be actively secreted by hypertrophied portion of the follicular epithelium.

Viviparous Teleosts with Intraluminal Gestation

(1) Family Zoarcidae: Within the Zoarcidae, viviparity appears to have evolved only in three species of the genus *Zoarces*. *Z. elongatus* is a viviparous, teleost, which inhabits the Pacific coast of northern Japan. Embryos of *Z. elongatus* are retained in the ovarian cavity for more than five months after fertilization. Ripe eggs of *Z. elongatus* are very large, about 4.4 mm in diameter, indicating that embryos have access to abundant yolk reserves during early development. In addition, the embryos of this species grow from 29.1 mm at hatching to 61.0 mm by late gestation. It has been demonstrated that embryos of *Z. elongatus* increased from 7 mg to 65 mg in dry weight during gestation. Such an increase indicates that *Z. elongatus* is a matrotrophic species in which embryos depend on a continual supply of maternal nutrients. Hind-gut is the main site of nutrient absorption in *Z. elongatus* (Koya et al. 1994).

In the blenny *Z. viviparous* the eggs leave the ovarian follicles before fertilization. Gestation is thus ovarian and lasts approximately 5 months. After hatching the embryos develop in the ovarian fluid secreted by the ovarian epithelium. During the period between fertilization and hatching, dry weight of eggs/embryos remain stable, indicating that the embryos are mainly nourished on the yolk. After hatching, while the embryos carry an external yolk sac, embryonic dry weights increase considerably and oxygen consumption increases correspondingly. The voluminous ovarian fluid is rich in glucose, free fatty acids, and amino acids but no proteins. The rate of clearance of intraovarian glucose or amino acid loads shows that the ovarian fluid is not a static pool. The observed growth to respiration ratio of 0.91 indicated a relative high efficiency for converting food to growth.

The impact of pregnancy on the maternal organism is reflected in considerable exchanges in biochemical events. Thus, during the latter part of pregnancy a marked decrease is observed in maternal liver weight, liver lipids, and glycogen, and lipid simultaneously accumulates in the maternal blood. The activity levels of glutamate dehydrogenase and two transaminases of maternal liver increased significantly during midterm pregnancy suggesting changes in overall protein metabolism. These metabolic changes are consistent with increases in maternal-foetal energy transfer.

The osmolarity and the concentration of inorganic ions in the maternal blood and the ovarian fluid of *Z. viviparus* are approximately equal, but very different from that of the surrounding brackish water. The newborn fish adapt well to ambient osmotic stress indicating development of an osmoregulatory capacity already before birth.

The mechanism underlying maternal-foetal oxygen transfer in viviparous teleosts is known only in a few species. The hormonal control of gestation in *Z. viviparus* is still unexplored.

Korsgaard (1994a) studied the calcium metabolism in relation to ovarian functions and embryonic development during early and late pregnancy in *Z. viviparus*. Results indicated that an embryonic net uptake of calcium from ambient ovarian fluid does occur during development and that the maternal organism contributes by an increased calcium turnover in the ovary as indicated by the tracer experiments performed *in vovo*. The embryonic metabolism of the assimilated calcium in ovario needs to be elucidated further. Korsgaard (1994b) reported that the capacity for urea excretion by post-yolk sac embryos of *Z. viviparus* may be of adaptive significance for their prolonged stay in ovario. The capacity of excretion of urea seems to decrease after acclimation to sea water.

(2) Family Goodeidae: Prolonged gestation of goodeid embryos within the ovarian cavity is associated with maternal-embryonic nutrient transfer.

As a structural character that is related to viviparity, the goodeid embryos, except those of *Ataeniobius toweri*, have developed perianal processes. These exceptional appendages, occurring also in some species of the Parabrotulidae and the Bythitidae originate from the proctodaeum and, in accordance with their assumed functions, have been named "trophotaeniae" that physiologically substitute for a placental exchange surface.

Embryos of the Goodeidae develop within the ovarian cavity changing from lecithotrophy through progressive matrotrophy via the so-called trophotaeniae placenta (Figs. 12.25 and 12.26). In *Ameca splendens*, the yolk reserves are sparse and consumed in early embryonic development. The placenta consists of two different components, the extraembryonic membranes or trophotaeniae, and the maternal component, the internal ovarian epithelium. It is suggested that maternal nutrients are released by the internal ovarian epithelium into the ovarian cavity and taken up via the embryonic trophotaeniae, which increase dry weight by about 99% in the fully developed embryo. It has been demonstrated that trophotaenial cells are specialized for endocytosis of exogenous proteins dissolved in the ovarian fluid and their subsequent enzymatic degradation. In another, goodeid species *Xenoophorus captivus*, the trophotaenial structure seems also to function in gas exchange and osmoregulation.

Trophotaenial placenta has been considered to be equivalent to the inverted yolk-sac placenta found in some rodents and lagomorph mammals. The highly vascular trophotaeniae are longer than the embryo, and they increase the embryonic surface area approximately four-fold. Trophotaenial epithelial cells contain apical alkaline phosphatase and moderate amounts of acid phosphatase, indicating the absorptive capabilities of the epithelium and the presence of lysosomes capable of HRP proteolysis.

Details of goodeid teleosts are given in a number of publications, for instance: *Ameca splendens*, *Chapalichthys pardalis*, *Girardinichthys viviparous*, *Ilyodon furcidens*, *Xenotaenia resolanae* and *Xenotoca eiseni* (Schindler and de Vries 1986a), *Girardinichthys viviparus* (Schindler and de Vries 1986b; Schindler 1990); *Xenoophorus captivus* (Schindler and de Vreis 1987a; 1988a,

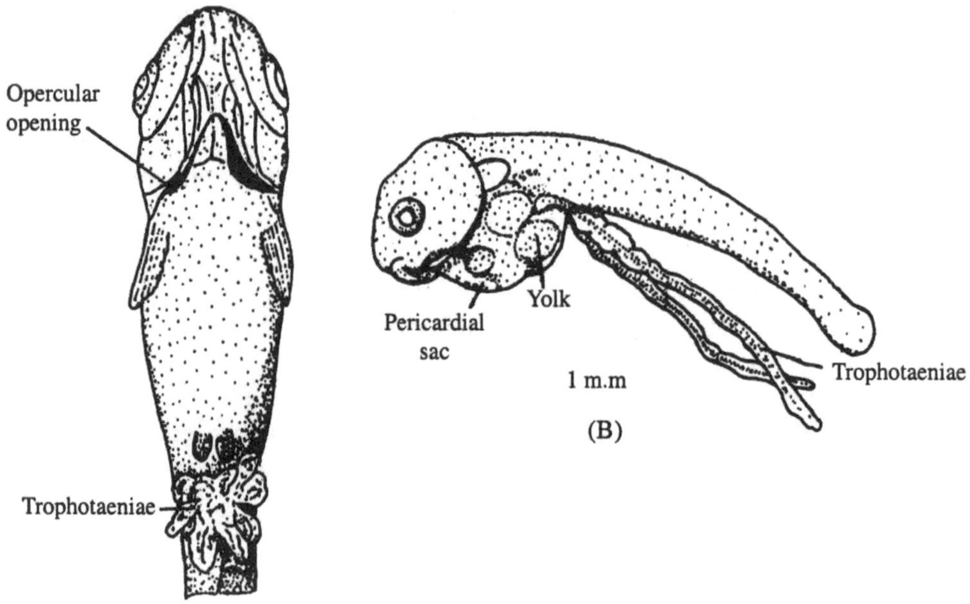

Fig. 12.25 **(A) Embryo of *Goodea luitpoldii* in late gestation; to illustrate wide opercular openings and trophotaeniae. (B) Early embryo of *Characodon lateralis*; yolk sac is receding pericardial sac is at its maximal stage. Trophotaeniae are growing rapidly and becoming vascularized. The extensive portal network on the pericardial sac is not illustrated.**

b; Schindler et al. 1988), *Xenotoca eiseni* (Schindler and de Vries 1987b, 1990), *Skiffia bilineata* (Schindler and Kujat 1990), *Allodontichthys* (Schindler and Radda 1990) and *Ameca splendens* (Schindler and Greven 1992).

(3) Family Scorpaenidae: The scorpaenid genus *Sebastes* exhibits predominantly lecithotrophic viviparity. Recent work on black rockfish, *Sebastes melanops*, however, demonstrates the occurrence of matrotrophism with a maternal contribution of approximately 70% of the catabolic energy used during embryonic development estimated on the basis of nitrogen and protein budgets and calorimetrically determined embryonic energy contents. Efficiency of yolk untilization in approximately 37 days that on average elapse between fertilization to parturition was estimated to be approx. 36%. As 81% of the initial energy investment is calculated still to be present at parturition, the difference appears to be supplied by maternal nutritional contribution. Oxygen consumption per embryo increases with development. The oxygen tension inside the ovary is unknown, but *Sebastes* have a specialized dual arterial system supplying blood to both ovaries. The embryonic weight and caloric and protein contents gradually decreased through development, but total nitrogen remained nearly constant indicating that nitrogenous energy sources are involved in exogenous nutrition of the embryos. This is also evident from investigation of kurosoi, *S. schlegeli*. Uptake of maternal nutrients was histologically demonstrated to occur by ingestion and absorption of ovarian fluid by the hind-gut of the embryos. These studies demonstrate a similar pattern of matrotrophic embryonic energetics in two different *Sebastes* species. Details on embryonic energetics are given by Boehlert and Voklavich (1984), Boehlert and Yamada (1986) and Boehlert et al. (1986, 1991).

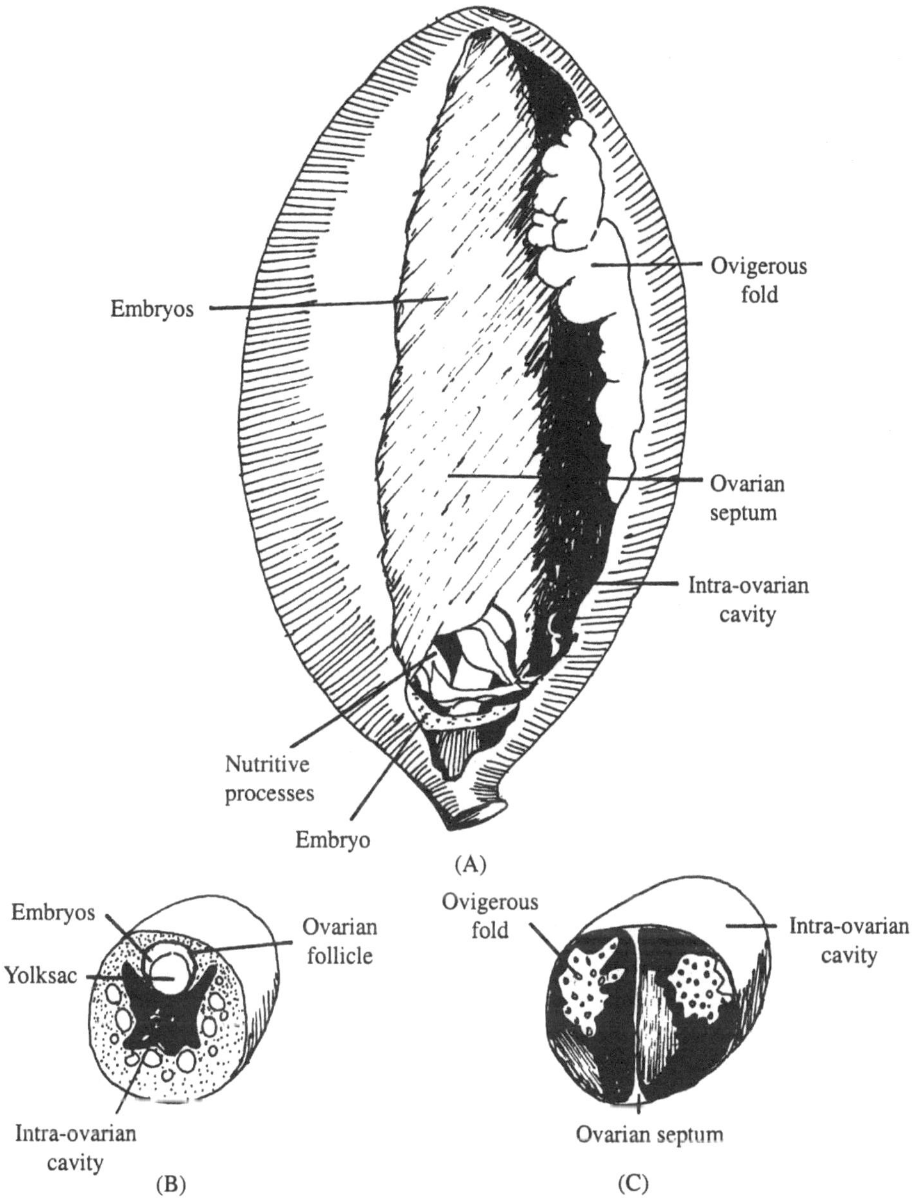

Fig. 12.26 **(A) Dissociation of ovisac of *Goodea bilioneata* with left part of ventral wall removed. Embryos on one side of median septum removed to expose ovigerous folds. Embryos in undissected half of ovisac visible through transparent septum. (B) Stereogram to represent ovary and intra-ovarian cavity in poeciliid type of top minnow. Yolk sac is large and yolk furnishes only means of subsistence for embryo till birth. Embryo remains in ovarian follicle until birth, and is not retained in intra-ovarian cavity. (C) Stereogram to represent condition of ovisac in *Goodea bilineata*. Yolk sac very small (0.4 mm. in diameter) at its maximum. Embryos evacuated from ovarian follicle at very young stage and develop thereafter in the intra-ovarian cavity. Embryo have proctodoeal process for nutrition and respiration. Ovary becomes nutritive organ while young are in ovary. Ovigerous portion confined to ovigerous folds.**

Conte et al. (1991) supported the hypothesis that the intestinal tract is functional prior to gestation and could be transporting important nutritive material found in maternal ovarian fluid in *Sebastes*.

Takemura et al. (1995) reported in *S. taczanowskii* that ingestion of intact macromolecules occurs during the late embryonic stages and that the yolk-like substances detected within the hind-gut epithelium of such embryos may be derived from the bathing ovarian fluid.

Shimizu et al. (1991) suggested that the embryonic epidermis in *Sebastes* is structurally loose and takes up low weight molecules, while rectal cells, after the opening of the mouth, actively ingest exogenous, high weight molecules. Yoklavich and Boehlert (1991) reported that oldest embryos of *Sebastes* have fully developed mouths and guts, and require additional nutrition at this stage of development.

(4) Family Embiotocidae: In *Cymatogaster aggregata* and *Embiotoca lateralis*, the eggs that are ovulated in early segmentation stages occur in the cavity of the ovary as free floating blastulae surrounded by a thin proteinaceous ovarian fluid. Pregnancy usually lasts several months. During early and midgestation the ovarian epithelium becomes hypertrophic and secretes ovarian fluid. The amino acids and lipids found in the ovarian fluid correspond to those of the maternal serum, but occur at lower levels. It has been indicated that novel peptides distinct from those of maternal serum are synthesized by the ovarian epithelium. Through gestation ovarian fluid contains large quantities of urea, whereas urea could not be detected in the maternal serum. The capacity to concentrate urea thus appears to be lost or reduced in the adult fish. The high level of glutamic acid observed in the ovarian fluid might reflect pronounced amino acid catabolism by the embryos, as glutamate might serve as the amino group donor in the urea production. Endocrine support of gestation appears to be of limited significance since estradiol and progesterone occur at low levels throughout gestation.

Striped sea perch, *E. lateralis*, foetuses are characterized by extensive, richly capillarized fin webs that appear to be involved in gas and ion exchange. In pile perch, *Rhaccochilis vacca*, the ovarian fluid oxygen tension falls markedly during gestation, whereas the oxygen content of the ovarian fluid initially decreases and subsequently increases as fluid volume rises towards term. Ovarian tissue intrudes the foetal gill cavities (branchial placentation) in pile perch and several other embotocids, reducing the diffusion distance and increasing the structural transfer capacity. Birth evidently is a traumatic event requiring a rapid shift from fin-dermal to gill breathing, and corresponding shunting of blood circulation.

In the sea perch, blood oxygen affinity is higher in midgestational than in late gestational stages, and lowest in adult.

Considered in conjunction with the decrease in blood oxygen affinity during mid- to late gestational development in sea perch, the fall in ovarian fluid oxygen tension during gestational development and the increased weight-specific foetal oxygen consumption observed in pile perch, present perflexing questions concerning the regulatory mechanisms underlying foetal oxygen uptake in fish.

The compendium on the maternal-embryonic relations in viviparous teleosts given by Schindler and Hamlett (1993) is as follows. Teleosts do not possess uteri and ovulation occurs into the ovarian lumen. Hence, osteichthyan viviparity is characterized by intrafollicular gestation or intraluminal gestation (within the ovarian cavity), and thus, the ovary functions both in egg production and maintenance of the developing embryos. Because there is no attachment between embryonic and maternal tissues, maternal-embryonic exchange of gas, electrolyte, nutrients, metabolic wastes, and possibly effector molecules and immunoglobulins is mediated by embryotrophe. In several instances it has been demonstrated that embryotrophic proteins originate from maternal serum rather than from elaboration within ovarian tissues. In intraluminal gestation, the ovarian wall becomes hypervascularized and the inner ovarian epithelium lining the ovarian cavity secretes embryotrophe. Trophotaeniae provide extraembryonic endodermal exchange surfaces in many species undergoing intraluminal gestation. There is a strong indication of a trophotaenial membrane binding system for proteins containing appreciable amounts of lysine. In intrafollicular gestation, yolk, pericardial and coelomic sacs play a major role in physiological exchange. In most instances, intrafollicular gestation is associated with extraembryonic fine structural modifications consistent with respiratory and osmoregulatory functions. The postulated role in nutrient transfer requires experimental evidence. In matrotrophic viviparous teleosts, the gut may provide an effective absorptive surface. On the other hand, the skin and perhaps the gills and buccopharyngeal tissue, serve as respiratory surfaces. As yet, no experimental evidence suggests exchange of metabolites at these sites.

Weber (1994) discussed haemoglobin-based O_2 transfer in viviparous animals. Although the available information remains fragmentary, fish generally appear to have foetal Hb's with distinct electrophoretic properties and oxygenation characteristics, as witnessed by the barndoor skate *Raja binoculata*, the spiny dogfish *Squalus acanthias*, and the ovary-breeding teleosts, seaperch *Embiotoca lateralis*, and blenny *Zoarces viviparus*. Adult red cells of *S. acanthias* and *E. lateralis* additionally have higher nucleotide triphosphate (NTP) levels than foetal cells. These data reveal exploitation of strategies I and II (see this paper). While ATP is the only measurable phosphate in early embryos of the smooth dogfish *Mustelus canis*, GTP (the more potent allosteric effector) dominates in adults, indicating an adaptive shift in cofactor synthesis. Foetal *E. lateralis*, moreover, have lower erythrocytic Hb and NTP concentrations than adults, which may further increase the affinity shift. In *Z. viviparus*, where adult as well as foetal Hb's are heterogenous (different components occurring in the same individuals and polymorphic (different heterogeneity patterns occurring in different genetic strains), the O_2 tensions of the ovarian fluid (38 mm Hg), predict about 90% saturation of foetal blood at 10°C and pH 7.5, and O_2 transfer may be enhanced by lower heme-heme cooperativity and a greater Bohr effect in the adult Hb.

4. INTROMITTENT ORGAN AND OTHER SPECIALIZED STRUCTURES IN MALES

In viviparous fish, sperm are transferred to the female, at least in some species, with the aid of an intromittent organ. Male elasmobranchs have modified pelvic fins which are used as "claspers" during copulation. The further modification of the clasper as a phallus-like organ for internal fertilization has been described in *Scyllium*, *Acanthias*, *Raja* and *Cetorhinus*. The specialization of this organ in the frilled shark, *Chlamydoselachus anguineus*, is remarkable. The male discharges into a groove on the medial side of the clasper a mixture of sperm from the urogenital papilla and sea water from the siphon sacs. At copulation the mixture is ejected into the oviducts of the female.

The anal fin of species of *Xiphophorus*, *Molliensia*, *Gambusia*, and *Fundulus* becomes specialized after puberty into a gonopodium which is used as an intromittent organ. Before the onset of sexual maturity, the anal fin, almost identical in structure in males and females of poeciliid fishes, is a generalized organ with ten bony rays. The approach of sexual maturity in the males of *Gambusia affinis* is marked by a rapid metamorphosis of the fin, in which there is a characteristic acceleration of growth in rays 3, 4 and 5 and a subordination of growth in the other rays. The anal fins may give rise to true tubular "penises" (e.g., jenynsiids). The males of cottids which are known to be copulating species are generally similar in so far as they have a huge penis compared with their body size: *Orthonopias triacis*, *Clinocottus recalvus*, *Oligocottus snyderi*, *Synchirus gilli*, *Clinocottus analis*, *Taurulus bubalis*, *Pseudoblennius cottoides*, *P. percoides*, *Furcina ishikawae*, *Artedius harringtoni*, and *A. alcicornis*. In *Blepsias cirrhosus* (the little dragon sculpin another case of internal gametic association and external fertilization), on the contrary, the penis is too diminutive to be noticed. An intromittent organ often occurs in internally fertilizing groups, such as the gonopodium of the Auckenipteridae, Poeciliidae, Anablepidae, Adrianichthyidae and andropodium of the Hemiramphidae. The gonopodium has been shown to develop under the influence of testicular hormones in the poeciliid fishes.

In some poeciliids this fin, although elongated, is not used for intromission, but instead is employed to "fan" the ejected sperm toward eggs which also have just been discharged (see Godin 1995 for mating tactics in male *Poecilia reticulata*). Phallostethids have an intromittent organ, the priapium, that is formed from the pelvic fin and parts of the pectoral fin, rather than, from the anal fin (see Grier and Parenti 1994). Burns (1991) has described the morphological transformation of the anal fin into the intromittent gonopodium in *Anableps dowi*.

The males anal is employed for internal fertilization by at least one oviparous fish, *Apogon imberbis*. The fertilized eggs are then deposited in a clump which the male picks up, carrying the mass in his mouth during most of their incubation. In the atherinid *Labidesthes sicculus*, the genital duct extends as a short tube through which free sperm are transferred to the female reproductive tract.

Examples of other specialized structures: Late in his life the male pink salmon, *Oncorhynchus gorbuscha* grows a huge hump on his back, and his head becomes large and bizzare in appearance. The "sword", a greatly elongated portion of the caudal fin which gives *Xiphophorus* its name, is another conspicuous male secondary sex character. The novel gill-derived gland in swordtail characin, *Corynopoma riisei* appears to be a male secondary sex character that develops as sexual maturity is attained. Based on histological characteristics of the gland, its location, and the complex courtship behaviour of this species, it has been hypothesized that the gill gland is producing some type of female attractant (pheromone) (Burns and Weitzman 1996).

An endocrine control of nuptial coloration has been reported. The dramatic changes in coloration and body shape characteristic of spawning male Pacific salmon appear to be governed by androgens. In addition to their effects upon coloration, androgens may also affect the development and maintenance of diverse morphological structures (e.g., pearl organs and breeding tubercles), and fin modifications such as simple elongation in the dorsal fins of blue gourami *Trichogaster trichopterus*, and the gobiid fish, *Pterogobius zonoleucus*, extension of the caudal fin into a sword into the swordtail, *Xiphophorus helleri*, and the development of the gonopodia from the anal fin in poeciliids. In the medaka, distinctive structures of the male anal fin, teeth, and body shape all appear to be under androgen control (see Liley and Stacey 1983).

5. PARENTAL CARE BEHAVIOUR AND ALLIED ASPECTS: REPRODUCTIVE GUILDS

Behavioural ecology, with special emphasis on mating systems, parental care and sexual plasticity, has been discussed in recent contributions from Barlow (1986, 1989, 1993a, b; read also Bakker and Milinski, 1993; Dugatkin and Godin, 1993; Godin, 1995; Godin and Dugatkin, 1995, 1996; Godin and Briggs 1996; Van Rooij et al 1996; Rosenqvist and Houde 1997). Noakes and Godin (1988) covered many facets of social responses in teleost fishes- genetic kin recognition, parental care, and development. Allopaternal care has also been reported (see details: Santos 1995). The cost of parental care has been discussed in teleost fishes (Smith and Wooton 1995) (Fig. 12.27A).

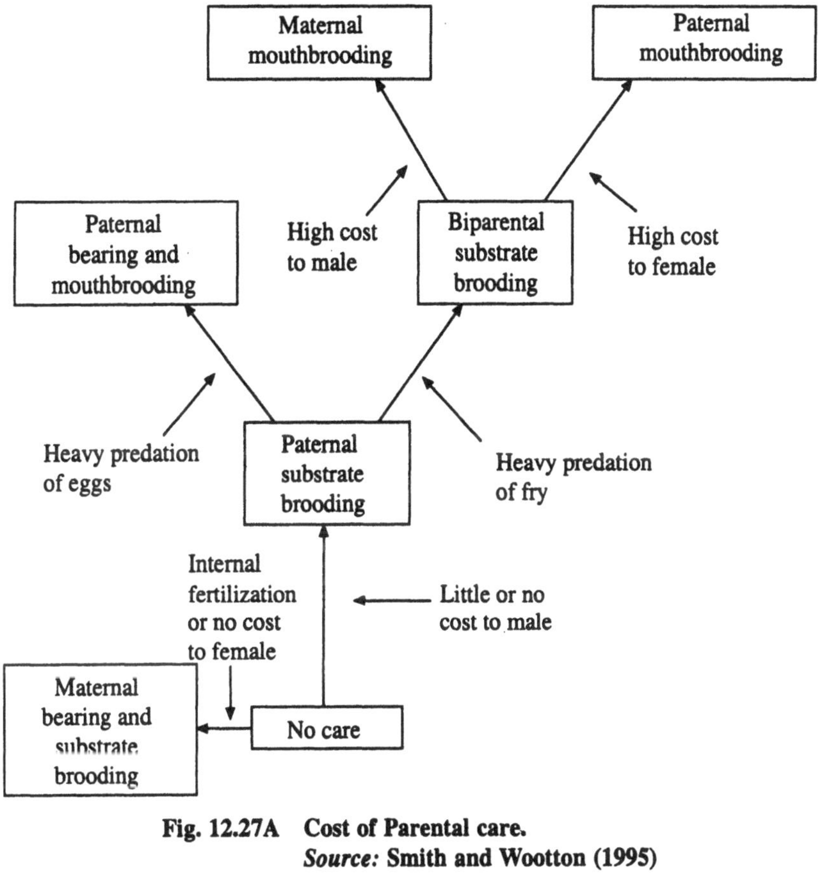

Fig. 12.27A Cost of Parental care.
***Source:* Smith and Wootton (1995)**

A long-term pair bond is believed to exist in most species of chaetodontid fishes. The general pattern is that most individuals of many species are found in pairs (see Lobel 1989 for *Chaetodon multicinctus*).

The operational sex ratio (OSR): The relative number of males and females available to mate is believed to be the primary determinant of sex roles in mating competition. Sex differences in potential reproductive rate are the best empirical predictor of the OSR, incorporating elements such as sex differences in parental investment, risk of predation and anatomic constraints. However, the OSR also depends fundamentally on the primary sex ratio and on sex differences in reproductive life span, mortality, and time of arrival to breed. For OSR in seahorses, see Vincent (1994) and for sand gobies *Pomatoschistus minutus*, refer Kvarnemo et al. (1995).

Parental care behaviour is highly specialized in many actinopterygians and chemical stimuli may regulate various aspects of parent's responses to egg and fry. Parental behaviour consists of various motor patterns, such as fanning of eggs, preparing and maintaining bubble or substrate nests and mouth brooding.

Parental investment (PI) is any behaviour performed by a parent that increases the probability that the young under care will survive while decreasing the parents' ability to produce future offspring. The key assumption of PI theory is that at any instant, the level of PI should be a trade-off between the reproductive value of the current brood (number and age of young) and the reproductive value of the parent (its chances of additional offspring) as determined by age of parent. For discussion of various predictions of PI theory, the reader is referred to Lachance and Fitzgerald (1992; see also Coleman et al. 1985; Coleman and Gross 1991).

The quality of parental care appears to correlate positively with egg size, both *among* and *within* species of fishes (Sargent et al. 1987). Several studies have shown the importance of eggs in a nest for increasing the probability of obtaining more eggs in fish with male parental care (e.g., *Gasterosteus aculeatus, Cottus gobio, Padogobius martensi*).

Parental care in fishes appears in most instance to have been derived directly from the defence of a territory for reproductive purposes. It provides a suitable environment for the development of the zygotes, for their defence, and, less commonly, for the mobile fry from predators. The eggs and newly hatched fry are fanned and/or mouthed while predators are driven away by the guardian's attacks. In most teleosts the period of parental care is brief, extending only to the eggs and larvae. In a few instances where parental care of the mobile fry is practised, a behaviour whose occurrence is limited almost entirely to freshwater fishes, such a commitment rarely lasts more than six weeks.

Two allied aspects deserve attention: lekking and monogamy. Lekking is defined as arena behaviour. It is a temporary aggregation of sexually active males for reproduction. The aggregation of males on the lek is visited by female single or *enmasse*, who select the males with whom they mate. Once mating is accomplished, the females leave the lek. The monogamy has been described elsewhere.

Parental care from different angles has been well described (see Keenleyside 1972, 1978; Rowland 1974; Timms and Keenleyside 1975; Keenleyside and Prince 1976; Smith-Grayton and Keenleyside 1978; Blumer 1979; Perrone and Zaret 1979; Gross and Sargent 1985; Barlow 1987, 1993a, b; FitzGerald et al. 1989; Stiassny and Gerstnev 1992; also refer to Magnhagen 1992; Magnhagen and Vestergaard 1993; Torricelli et al. 1993 for freshwater goby; Almada and Santos 1995; Smith and Wooton; 1995).

Lekking has been the focus of paper by Barlow (1978). It has become increasingly apparent that lekking behaviour is common among teleost fishes, for example in Cyprinidae, Catastomidae (Cypriniformes), Atherinidae, Melanotaenidae, Cyprinodontidae, Poeciliidae, (Atherinomorpha), Gasterosteidae (Gasterosteiformes), Centrachidae, Percidae, Spariade, Cichlidae, Pomacentridae, Scaridae, Callionymidae, Belontiidae (Perciformes).

With the exception of the viviparous surfperches (Embiotocidae) for which there is only circumstantial evidence of lekking, and a single viviparous poeciliid, *Poeciliopsis occidentalis*, all lek fishes so far known are viviparous.

Among lekking teleosts, four possibilities exist with regard to parental care: (1) Noncustodial lekking (e.g., most cyprinodonts, Melanotaenidae and Atherinidae); (2) Lekking with maternal care of spawn (e.g., poeciliid *Poeciliopsis occidentalis* and in the maternal mouth-brooding species of the family Cichlidae (see Barlow 1974 for hezagonal territorles); (3) Protocustodial lekking (e.g., many cyprinids, *Cyprinodon*, Etheostomatinae), and (4) Paternal-custodial lekking (e.g., some cyprinids, many darter perches, centrarchids, sticklebacks, some wrasses, Sparidae, Pomacentridae).

Lekking teleosts are characterized by polygamy. Sequential polygyny is found in all lekking species, and sporadic instances of polygyny have been reported among the Centrarchidae and the Labridae. Polyandry has been reported among cyprinodonts.

There is little information on the mechanisms determining access to favoured territories. There are only fragmentary indications that position effects characterize lekking in teleosts. Intermittent lekking is also practised by some teleosts.

The dependence of many teleosts on a hygienic location for their relatively vulnerable eggs means that environmental factors directly determine the location of lek sites in species practicing noncustodial, protocustodial, and paternal-custodial lekking. The bluehead wrasse *Thalassoma bifasciatum* provides an example of coral reef fish with traditional lek sites. Substratum-independent or "free water" spawning behaviour has been described for the maternal mouth-brooding Tanganykan cichlid *Tropheus moorii* and *Limnochromis microlepidotus*. Regardless of their pattern brood care cichlids, gouramies, and other teleosts are limited in their selection of spawning sites by predation on both the breeding adults and their spawn. In some instances, the siting of leks is constrained by adverse physiographic factors.

The evolution of lekking in teleosts is linked to dependence upon suitable spawning sites, by the localized substrata for the deposition of demersal eggs on points assuring a favourable launch window for pelagic eggs. While synchrony of reproductive activity is imposed by environmental factors, proximity is determined by the availability of suitable spawning sites. Where these spawning are limited in either occurrence or extensiveness, their sequestration, either totally or in part, by an individual may be a positive adaptation. Males would be in a better position to implement such an adaptation because of the greater energetic demand that the maturation of eggs places on females.

Lekking arises automatically in teleost fishes when the size of a discrete spawning territory sequestered by a male is small relative to extent of the available site. As the number of males entering the area and setting up territories increases, the size of each territory will contract to the smallest area required by each male in order to reproduce successfully. Under most circumstances, and for most animals, the resulting geometry will converge upon but seldom achieve, an hexagonal array of territories.

It is suspected that intermittent lekking is the most primitive manifestation of this reproductive strategy in teleosts. The defense of a spawning territory, even if only from conspecifics, confers serendipitously a degree of protection from predation to the eggs deposited there in. Among teleosts that practice parental care, the primitive condition is for the male to establish a territory and for the female to visit him there. Regardless of whether they practice parental care or not, lekking teleosts have the potential advantage of swamping spawn predators with the sheer number of eggs and fry produced within a limited area, depending on the number of eggs and fry, and on predators and their capacity to devour eggs and fry (read Loiselle and Barlow 1978).

Keenleyside (1978) stated that the parent-offspring links in most of the fish species showing any parental behaviour at all are generally looser and briefer than those of most birds. Four major aspects of parental behaviour can be emphasized here.

(1) Postzygotic investment and nest construction: Many fishes, especially freshwater species practice parental care. The caretaker is almost always the male. When the female is involved, the course of evolution appears to have gone from exclusively male care to joint care, then to exclusively female care. That the male is usual caretaker in fishes stands in contrast to the situation within the Vertebrata, with the exception of Amphibia.

Social systems of coral reef fishes have been thorougly studied (Shapiro and Boulon 1987; Shapiro 1991). Barlow (1981) commented that parental care among coral reef fishes cannot be discussed without regard to the size of the fish. Caretaking predominates among male species and is virtually nonexistent among large ones. But in all of them, postzygotic investment is minimal compared to freshwater species because of the pervasive need to disperse as eggs or young. Almada and Santos (1995) reported prevalence of parental care in rocky intertidal fishes. In these fishes parental care is much more frequent than for marine fishes as a whole. Guarding was already present in the ancestors of the resident species, at least in the majority of cases—a hypotheses proposed.

Coral reef fishes live in a patchy environment, and adults cannot migrate between patches. Guarded demersal eggs, requiring a greater individual investment by the small short-lived species, further reduces their individual fecundity. These costs to fecundity and the reduction in agile young appear to account for the lack of postzygotic investment in larger long-lived species.

Fitzerald and Caza (1993) reported that males of a fresh-water population of threespine sticklebacks (*Gasterosteus aculeatus*) invested more in older/larger broods whereas environmental factors played no role in PI decisions. In contrast, males of an anadromous population adjusted levels of PI in response to environmental factors rather than brood characteristics. Males increased their level of aggressive defence in response to a rise in water level, indicating that they can adjust their level of investment in response to changes in environmental conditions that affect brood survival.

The extent and variation in nest building is very much restricted in fishes. First of all, external fertilization, with simultaneous release of male and female gametes into the water, is the rule among fishes and many species produce buoyant, pelagic eggs that are carried by the prevailing currents; those species do not make nests or provide any form of parental care. Among those with demersal eggs, some species (e.g., salmonids, lampreys, some cyprinids) dig excavations in gravel substrate, lay their eggs in the pits, cover them with gravel, and desert them. Other species (e.g., centrarchids, blennids, some cichlids and pomacentrids) spawn in excavations they have made in the substrate, often under stones, sunken logs and debris, or inside empty mollusc shells. At least the eggs, and sometimes also the newly hatched fish are guarded in these locations.

Several schemes have been devised to group fishes according to their choice of spawning site and degree of any post-spawning care. Most European cyprinids do not guard their eggs and are open substratum spawners. Further grouping rests on the type of substratum over which the adhesive eggs are scattered: (1) Lithophils deposit eggs on the rock or gravel bottom of rivers (e.g., *Leuciscus leuciscus*); (2) Psammophils scatter their eggs on sand or near the roots of the plants (e.g., *Gobio gobio*); (3) Phytophils spawn on aquatic or flood terrestrial vegetation (e.g., carp and *Tinca*); (4) Pelagophils spawn in rivers where the turbulence and flow conditions ensure the suspension of the semipelagic eggs (e.g., *Hypophthalmichthys molitrix*; (5) Phytolithophils are adaptable species. A good example of a European species with a more specialist mode is the bitterling, *Rhodeus amarus*. Although with several related species, this fish is a brood hider which deposits its eggs into the siphon of freshwater mussels. *Leucaspius dilineatus* provides a rare example of a European cyprinid that belongs to second major grouping, the 'egg guarders'. The female spawns its eggs in a ribbon over aquatic macrophytes and the male fans and guards them (see Mills 1991).

The most common form of parental care is that directed to hidden clutch of adhesive eggs. Caretaking ceases at hatching, when the young ascend into the plankton.

Nest building and parental care is almost absent from pelagic marine fishes but is a common feature of coastal marine fish. In some species these behaviours are relatively simple but in some, the behaviour reaches an elaborate degree of sophistication (e.g., *Crenilabrus melops*, a labrid). Only a few fishes construct nests. The best-known of these nest-building fish are the sticklebacks (Gasterosteidae). The male builds a spherical or elongate nest by collecting plant fragments, rootlets, and the like, and then building them together with adhesive kidney secretions. The various probing, boring, sucking, and "glueing" activities result in the formation of a compact nest with an internal chamber to receive the eggs. The most elaborate nest is made by *Apeltes quadracus*. With this species, a cup-shaped nest is attached to rooted plants close to the bottom. After a clutch of eggs is laid, the male builds an extension of the nest up and over the eggs, with a concave upper surface to the extension. A second clutch of eggs is laid on the new nest floor, and this procedure may be repeated several times, until the male has several clutches of eggs stacked vertically within a single multitiered nest. Jamieson and Colgan (1989) reported that the presence of eggs had a general priming effect on the courtship behaviour of *Gasterosteus aculeatus* males, but evidence for active female choice was lacking.

A number of other species build less complex nests of collected materials. These are less tightly bound together than the stickleback nests and are usually placed under rocks or in dense vegetation. Presumably, in these locations they are protected against turbulent wave action, and the capacities of the parents to guard against predation are enhanced.

Potts (1985) described the nest structure of the corkwing wrasse, *Crenilabrus melops* (a gonochorist with discrete sexes; Labridae). Large and complex algae nests are built in the littoral and sublittoral regions and have been shown to predominantly occupy north facing crevices. The shaded aspect of the nest crevice and the bulk of algae that it consists of provide protection against dessication, temperature extremes and the prevailing turbulence of the sea. The nests comprise many different species of algae and it has been shown that their composition is to some extent influenced by algae availability. The nest is built along a crevice so enabling the male fish to spawn with several females during the spawning season. The crevice situation also enables the male fish to protect the nest more effectively and increases the probability of paternity by the nesting male over secondary males that intrude during spawning. The physiological mechanisms controlling nest building in *C. melops* are likely to be similar to that found in sticklebacks and other freshwater species in which hormone levels are set early in the reproductive season leaving the fish ready to respond to environmental triggers. Such a system would have considerable advantages in an environment as variable as a tidal coastal zone.

Guarding eggs or fry may also restrict feeding to the extent that the parent must remain at the nest. In some cases the parental individual may guard the nest continuously with little or no feeding, and in others it leaves the nests at intervals to feed. For example, in the family Pomacentridae the males of *Chromis dispilus*, *C. punctipinnis* and *Hypsypops rubicunda* remain at the nest guarding continuously, but males and females of *Amphiprion chrysopterus* leave the nest at frequent intervals to feed. The distribution of food items in time and space also alters the costs of guarding in relation to feeding. Findings on dynamics of nest defense by male centrarchid fish indicate species-specific differences in nest defense and males did discriminate intruders, possibly on their potential threat to the brood. The defense of the nest was related to nest stage, a finding which supports current theory of parental care (Colgan and Brown 1988). In the marine fish, *Pomatoschistus microps*, the common goby, both nest owners and sneakers modify their behaviour in order to decrease the risks of being eaten (Magnhagen 1995).

African lungfishes are known to deposit their eggs in prepared, mostly U-shaped, tunnel-like nests, and to guard their young. The breeding nests of *Protopterus dolloi* are the same structures used by adult animals as dry-season refugees. The Australian lungfishes do not make a nest, nor do they guard their young.

(2) Incubation of eggs: Fish eggs are not maintained at temperatures above ambient, either by the parents or by surrounding substrate or nest materials, and hence no true incubation occurs. Even though some species maintain contact with their eggs by oral brooding (e.g., some cichlids), by wrapping the body around the eggs (some stichaeids and pholids), and of course by carrying them internally after copulation and internal fertilization (many cyprinodonts), there is no evidence that these egg tending behaviours influence the temperature of the eggs. Remaining coiled around the eggs (by females and males) appears to keep the eggs moist and reduces their exposure to the air, just as splashing keeps eggs (by males) moist in similar situations.

(3) Feeding the young: Two activities can be considered. First, the "fin-digging". In this behaviour the adult fish settles to the bottom and, with vigorous undulations of the body and fins, stirs up loose substrate materials that stimulate the fry to congregate around the parent and feed on loosened particles. Male and female parents are equally aggressive in defense of their broods (e.g., cichlid *Cichlasoma nigrofasciatum*). Second, this is a form of plunge-feeding found among *Geophagus* species, in which the fish thrusts its snout vigorously into the substrate, withdraws it and expels the mouth contents. Loose particles may then be ingested, and if an adult *Geophagus* is foraging in this manner with a school of its progeny, the young fish also gather and feed on the expelled material.

Another form of "feeding the young" also occurs among cichlids (*Symphysodon discus*, *Etroplus maculatus*, *Cichlasoma citrinellum*). As the young reach the free-swimming stage of development, they begin to graze on the mucus on their parents side. Mucus-producing cells in the adults' epidermis are most numerous at this stage of the reproductive cycle, and it is clear that mucus is being ingested by the young fish. In fact, feeding from the parent's body can be induced in a number of cichlid species by depriving the fry of other food sources. Similar parent-contacting behaviour by juveniles occurs in the coral reef pomacentrid *Acanthochromis polyacanthus*, although no direct proof of mucus ingestion has been obtained.

Noakes (1979) analyzed the functions of glancing (maintenance of group cohesion) and micronipping (trophic function). Parent-touching behaviour by young fishes occurs in a number of species. Most of the reports have come from the family Cichlidae. The behaviour appears to serve a trophic function in many species, but may also serve to maintain cohesion of family groups, to keep adults in a parental state, or to communicate the motivational state, of the young (e.g., fear, hunger). It has been suggested that prolactin may regulate the behaviour and epidermal mucus condition in parental fish involved in such behaviour, but the evidence is not conclusive. There appear to be similarities between cases of parent touching as trophic behaviour, and mucus- and/scale-feeding by cleaning symbionts or predators.

(4) Protecting of the young: Internal gestation by female parallels the oral brooding and brood-pouch egg carrying performed by males. Oral brooding appears to inhibit feeding. Probably the most efficient protective device used by fishes while guarding their young is oral brooding. This is common among cichlids, especially species of the African Great Lakes, but it also occurs in other groups. The only way a predator can capture young fish that are being carried in their parent's mouth is to injure or kill the parent, force it to jettison its brood, or wait until the parent releases its young voluntarily, then dash in and capture them. The best available evidence for this type of predation comes from several species of *Haplochromis* in Lake Victoria, which feed either exclusively or principally on the eggs and yolk-sac larvae of other cichlid fishes. The method these fish use to capture their food is uncertain because of the difficulty of observing them in the turbid waters of Lake Victoria.

Much has been reported about the pharyngeal mucus gland and parental activity, for example bacteriocidal activity in mucus, adjusting the pH in the buccal cavity of the incubating parent. Mucus from the parent has been suggested as protective agent in species.

Both sexes of a cardinal fish, *Apogon semilineatus* (Apogonidae) are reported to mouth brood eggs but not fry although all other apogonids are male mouth brooders. It is proposed that in these marine fishes both parents are needed to protect eggs from predators. Almost all other cases of biparental tending (Cichlidae, Ictaluridae, Ophiocephalidae, Pomacentridae) involve both eggs and fry as expected. Two exceptions are the pomacentrid anemonefish *Amphiprion* and the blind goby *Typhlogobius californiensis*. In both cases adults live in male-female pairs and guard eggs which produce pelagic larvae. However, anemonefish form pairs upon settling in a sea anemone, and remain in the anemone permanently. In most species males do nearly all of the egg-tending alone. Hence, it seems that the fish stay paired to keep possession of the anaemone. *Typhlogobius* pairs stay permanently in a burrow shared with a shrimp.

The Syngnathidae (pipefishes and seahorses) are small reef fishes in which the males bear the embryos in a ventral brood pouch (see details on sex roles; Vincent et al. 1992, Vincent 1994). In the closely related monotypic family, the Solenostomidae, the female retains the fertilized eggs in a pouch formed by her pelvic fins. After hatching, the young are released into plankton.

(5) More specialized features of parental behaviour: A kind of brood-parasitism is reported in some fishes. The best-known examples occur among the North American freshwater Cyprinidae, where some species frequently spawn in gravel mound nests built by other cyprinids or by centrarchids. Since some of the "host" species practise parental care of the eggs, whereas the "brood-parasites" do not, the latter presumably benefit from this association.

Multiparental brood care has been reported in two cichlid genera, *Apistogramma* and *Lamprologus* and less certainly in a third, *Nannacara*. All three are "dwarf" of very small cichlids and nest in cavities among rocks and tree roots. In these the females tend the eggs. Since very small species should be most vulnerable to predation, multiparental care should be looked for in the smallest egg-tending species in a community.

Certain conclusions regarding parental care in fishes are drawn. Among fishes, embryonic developmental rates depend on ambient temperatures, and clutches usually consist of large numbers of small eggs. Most species have no parental care at all, and among those that care for the eggs only, the young after hatching either join the plankton and drift away from the hatching site, or settle close to the substrate and remain near hiding places. In either case the parent-offspring association is quickly lost. Most fishes with posthatching parental care breed in shallow, fresh, or brackish waters, where turbulence and rapid change in water levels are common. Associated with these conditions one finds the furthest development of parental care, including what must surely be the most efficient and safest way to move freely swimming young away from danger—that is, in the parent's mouth.

Gross and Sargent (1985) pointed out some facts regarding parental care in fishes: (1) Parental care exists in 21% of the families of bony fishes, and is widely distributed phylogenetically; (2) Guarding is predominant care behaviour. Guarding typically involves display or active chasing of egg or fry predators which include both conspecifics and heterospecifics. These predators rarely threaten the adult fish directly; (3) Uniparental care is most common; (4) Male parental care is more common than female parental care; (5) Neither sex is physiologically or physically preadapted to give care. An exception may be internal fertilization for live bearing or gestate the young (e.g., *Syngnathus*, *Hippocampus*); (6) Male care is significantly correlated with external fertilization. Female care is significantly correlated with internal fertilization. Male care with internal fertilization is known in only three families (Apogonidae, Cottidae and Pantodontidae). Live bearing has evolved from internal fertilization independently several times within fishes; (7) Male parental care is associated with male territorial behaviour and polygyny; (8) Parental care is more common among freshwater than marine families. In general, pelagic eggs do not receive care. Rather, parental care with external fertilization is usually found when eggs are attached to substrate. It has been suggested that more variable nature of the freshwater and the relative availability of the substrate as compared with the marine environment, favours egg clumping. This clumping is probably a preadaptation to the evolution of parental care and may explain the predominance of parental care in freshwater fishes; (9) Uniparental male care has evolved from a state of no care; biparental care has evolved from a state of male care.

Further conclusions derived by Gross and Sargent (1985) are: (i) Parental care has only one benefit, the increased survivership of young. It may, however, have three costs: a "mating cost", an "adult survivorship cost", and a "future fertility cost". (ii) On average, male and female will derive the same benefit from care. They probably also pay the same adult survivorship cost. However, their mating cost and future fertility costs may differ. (iii) A mating cost usually applies to males. However, this cost may be reduced by male territoriality and, in some situations, be entirely removed. Under this condition, natural selection on present reproductive success is equivalent for males and females. (iv) When fecundity accelerates with body size in females, while male mating success follows a linear relationship with body size, future fertility costs of parental care are greater for females than males. The predominance of male parental care in fishes is not explained by males deriving greater benefits from care, but by males paying smaller future costs. Males thus accrue a greater net fitness advantage from parental care. (v) The evolution of biparental care from uniparental male care may occur because male care selects for larger egg sizes and increased embryo investment by females. This increases the benefit to the female parental care. (vi) By contrast, uniparental female care may originate from biparental care when males are selected to desert. This occurs when females care creates a mating cost to females. In some cases male desertion may "lock" females into uniparental care. However, in many other cases females may be selected to desert, giving rise to "no care". (vii) The origin of uniparental female care from no care is rare in externally fertilizing fishes. This is because the benefits of care rarely outweight a female's future fertility costs. For internally fertilizing species, however, the benefit of care is high whereas the cost is probably low. Most of these species have evolved embryo retention. (viii) When parental care begins with male care and moves to biparental care, it has been suggested that care evolution will include cyclical dynamics. Parental care in some fishes may thus be seen as transitional and changing through evolutionary time rather than as an evolutionary stable state. In theory, "no care" may be a phylogenetically advanced state.

The fitness cost of care-giving in teleost has been shown to be related to the sex of the parent and to their mating system. The cost to a solitary male can be reduced survival and in some cases reduced breeding frequency. The cost to a solitary female parent is often one of reduced breeding frequency and future fecundity. Biparental males have reduced spawning frequency, unless they desert their males or show bigamy. In externally fertilizing species, parental care has evolved principally in male but also in females under conditions where there is little or no fitness cost to the parent (Smith and Wooton 1995) (see Fig. 12.27A).

According to Kindler et al. (1991), the hormonal mechanisms underlying parental care behaviour in fish are poorly understood. On the basis of investigations on parental care behaviours of nesting male bluegills (*Lepomis macrochirus*) they suggested that prolactin, mediated through the action of dopamine, is involved in the expression of parental care behaviour in this species. For *Reproductive Guilds*, the reader is referred to Table 12.2.

Longitudinal clines: The causes of variation in egg number have long interested evolutionary ecologists. The latitudinal variation in clutch size found in many animal species, including Pacific salmon, has been an enigmatic problem in ecology. Fleming and Gross (1990) analyzed the egg number and egg size of 17 populations of coho salmon (*Oncorhynchus kisutch*) distributed over a latitudinal gradient in North America. These populations have a significant latitudinal increase in their egg number. But this increase is accompanied by a significant latitudinal decrease in their egg size. The total biomass of eggs produced also declines with latitude. Thus, the positive latitudinal trend in egg number cannot be explained by latitudinal variation in the total investment in eggs. This suggests that local optima in egg size may result in latitudinal clines in egg number. This reasoning, that egg number evolves around selection for egg size, is in fact predicted by life history theory and may explain the clutch size patterns observed in many organisms.

Courtship and reproductive success: Courtship is a common tactic used to attract mates, but it is no means universal. In many fishes of the live-bearing subfamily Poeciliinae, males use a mixture of courtship displays and direct thrusts to mate with females. Courtship is usually some variant of the 'sigmoid' display, in which males flex their bodies in front of females, giving females the opportunity to choose their mates. When a female is receptive to these displays, the male briefly inserts his gonopodium (modified anal fin) into the female's genital pore for internal fertilization. In contrast, a thrusting male approaches the female from behind and jabs his gonopodium at the urogenital pore without prior display. This behavioural alternative may circumvent female choice, but it carries a low success rate for fertilization owing to female evasiveness and the likelihood of the sperm being displaced by a subsequent full copulation.

In guppies, *Poecilia reticulata* (Poeciliidae), courtship by males encourages female receptivity. However, in the presence of predators or at high light levels, when predation risk is greatest, courtship decreases. Thrusting may increase in frequency or as a proportion of mating behaviour because it is safer for males or because it circumvents female reluctance to mate under the threat of predation. Thus, courtship and gonopodial thrusting appear to have evolved under opposite selection pressures: courtship may attract both females and predators, whereas thrusting may minimize both female choice and predation threat.

Reynolds et al. (1993) tested for coadaptation between morphology and mating behaviour of individual male guppies from a single wild population in Trinidad. They concluded that covariation between alternative mating behaviour and male morphology is mediated by changing environmental conditions influencing the strengths of sexual and natural selection.

Studies on breeding competition have been carried out on female coho salmon (*Oncorhynchus kisutch*). Results indicate that the morphology and life history of adult female salmon respond evolutionarily to local selection pressures, including both the biotic demands of breeding competition and the abiotic demands of migration (Fleming and Gross 1989; see also van den Berghe 1989).

Reynolds and Gross (1992) reported that females may prefer males that provide heritable benefits ('good genes') for offspring development or survival in *P. reticulata.* Female mate preferences may have important effects on the inheritance of life history traits by offspring. The results are consistent with the 'good genes' theory of sexual selection but they also illustrate some of the pitfalls inherent among alternative theories.

A study by Gross (1991) provides empirical evidence in a wild population for frequency-dependent sexual selection between alternative male reproductive strategies. The bluegill sunfish (*Lepomis macrochirus*) has two males reproductive strategies, cuckolder or parental, used by different males to compete in fertilizing the same eggs. As the density of cuckolders in colonies of parental males increases, the average mating success of cuckolders initially peaks but then declines. The cuckolder density at which their success peaks is determined by ecological characteristics of each colony. A theoretical analysis assuming random and omniscient cuckolder distributions among ecologically different colonies shows that cuckolders will fertilize decreasing proportions of eggs, relative to parental males, as cuckolders increase in frequency in the population. This supports evolutionary models that assume negative frequency-dependent selection between the competing strategies. Cuckolder and parental strategies may therefore have evolved as an Evolutionary Stable State. The reader is also referred to Gross (1982).

Findings by Kindler et al. (1989) suggested that the elevated levels of 11-ketotesterone (11 KT) are associated with the behaviours displayed by spawning parental males. The levels of testosterone (T), however, seem to be associated with the occurrence of a phenomenon common to both parental and cuckolder males, such as development of gonads and/or spermiation.

Another interesting aspect is the relationship between parental care and anti-predator behaviour. In the bullhead, *Cottus gobio*, parental care involves the guarding of multiple clutches of eggs stuck to the underside of flat stones. This species has excellent cryptic camouflage and is largely

Table 12.2 Reproductive guilds of teleost fishes

I. Nonguarders of eggs and young
   A. Open substrate spawners
      1. Pelagic spawners, e.g., *Mola mola*
      2. Benthic spawners
         (a) Spawners on coarse bottoms (rocks, gravels etc.)
            (1) Pelagic free embryo and larvae, e.g., *Morone saxatilis*
            (2) Benthic free embryo and larvae, e.g., *Phoxinus phoxinus*
         (b) Spawners on plants
            (1) Nonobligatory, e.g., *Rutilus rutilus*
            (2) Obligatory, e.g., *Exos lucius*
         (c) Spawners on sandy bottoms, e.g., *Gobio gobio*
   B. Brood hiders
      1. Benthic spawners, e.g., *Oncorhynchus nerka*
      2. Cave spawners, e.g., *Anoptichthys jordani*
      3. Spawners on invertebrate, e.g., *Rhodeus amarus*
      4. Beach spawners, e.g., *Leuresthes tenuis*
      5. Annual fishes, e.g., *Nothobranchius guentheri*

II. Guarders
   A. Substrate choosers
      1. Rock spawners, e.g., *Chromis chromis*
      2. Plant spawners, e.g., *Pomoxis annularis*
      3. Terrestrial spawners, e.g., *Copeina arnoldi*
      4. Pelagic spawners, e.g., *Channa spp.*
   B. Nest spawners
      1. Rock and gravel nesters, e.g., *Amblolites rupestris*
      2. Sand nesters, e.g., *Cichlasoma nicaraguense*
      3. Plant material nesters
         (a) Gluemakers, e.g., *Gasterosteus aculeatus*
         (b) Nongluemakers, e.g., *Micropterus salmoides*
      4. Bubble nesters, e.g., *Betta splendens*
      5. Hole nesters, e.g., *Cottus aleuticus*
      6. Miscellaneous material nesters, e.g., *Lepomis macrochirus*
      7. Anemone nesters, e.g., *Amphiprion* spp.

III. Bearers
   A. External bearers
      1. Transfer brooders, e.g., *Oryzias latipes*
      2. Forehead brooders, e.g., *Kurtus gulliveri*
      3. Mouth brooders, e.g., *Oreochromis mossambicus*
      4. Gill-chamber brooders, e.g., *Typhlichthys subterraneus*
      5. Skin brooders, e.g., *Bunocephalus*
      6. Pouch brooders, e.g., *Syngnathus abaster*
   B. Internal bearers
      1. Ovi-ovoviviparous, e.g., *Glandulocauda inequalis*
      2. Ovoviviparous, e.g., *Sebastes marinus*
      3. Viviparous, e.g., *Poecilia reticulata*

Sources: Balon (1975a), Moyle and Cech (1988), Wooton (1990)

inactive during the day, defending exclusive small reproductive territories beneath individual stones. Conspicuousness is minimized in two ways, first by having no reproductive display save for a rapid head-bite administered by the male, and secondly by ritualizing territorial disputes into sequences of head darkening, opercular raising, fin spreading and head-nodding (grunting) displays. Males are markedly more active than females in the breeding season *C. gobio* therefore minimizes predation risk from birds by remaining out of sight. Parental care in three-spined stickleback, *Gasterosteus aculeatus*, is undertaken exclusively by the male. Eggs and fry are at risk both from predators and from nest-raiding rival male sticklebacks which commonly steal eggs. Female *Gasterosteus*, playing no part in parental care must produce and spawn as many batches of eggs as possible. Both sexes must evade predators. Giles (1984) proposed that the basis of the difference between sexes in anti-predator behaviour lies in the necessity of nest-tending and territorial defence by the male. Females apparently lose nothing by being very cautious in the presence of predators whilst males may lose the entire contents of their nest if they take too long to recover from an encounter with a predator.

Krack and Videler (1991) investigated the criteria for female mate choice in a natural population of Mediterranean blenny, *Aidablennius sphynx*. Male display did not increase the probability that females spawned. Male size did not correlate with the numbers of eggs received. Female preference for large broods might be adaptive because the number of eggs present in the nest is probably a reliable predictor of the survival chances of the eggs.

Variability in timing and location of spawning and reproductive behaviour is common in sympatric or parapatric morphs of fish. Corresponding genetic differences have been shown in some cases (e.g., brown trout, *Salmo trutta*; chum salmon, *Oncorhynchus keta*).

Monogamy: Mating systems are distinguished by the composition of the group at the time of reproduction. In the most common situation, polygyny, one male fertilizes the gametes of several females, though the details may differ greatly, as in harems and leks. In nonpromiscuous polyandry, a rare situation, several males mate with one female. In monogamy, the effective sex ratio is 1:1.

Monogamy is defined in two ways: (1) a pair remains together until the end of a reproductive cycle that entails biparental care of the young; (2) a pair spawns at short intervals with the same partner over a long period of time, with no parental care. Contributions (Barlow 1984, 1986, 1988, 1991, 1993a, b; Tricas and Hiramoto 1989) are related to various aspects of this topic.

Six families are known to display monogamy. They are the bony-tongues (Osteoglossidae), the North American catfishes (Ictaluridae), the Asian catfishes (Heteropneustidae), the African catfishes (Bagridae), the snakeheads (Channidae) and the cichlids (Cichlidae).

These families share a number of adaptations. The most remarkable of these is that all are biparental and that caretaking extends to protecting free-swimming young not just to eggs. In addition, all possess breeding territories, and the eggs are deposited on the substrate. They tend to be seasonal breeders. Generally they are large fish. Another important ingredient is the ability to disperse at any age because of the continuity of the environment. In most of these families, there are but a few species that are monogamous, except for the Cichlidae. Cichlids are truely exceptional in that monogamy is ancestral. Mouth brooding usually, but not always, results in polygyny, typically lekking. But the prominent theme among cichlids is biparental care of eggs, which is relatively common among fishes (see details of mating systems in cichlid fishes, Barlow 1991).

Midas cichlid (*Cichlasoma citrenellum*) has been subjected to a number of studies from different angles (see Barlow et al. 1990; Holder et al. 1991; Rogers and Barlow 1991).

The instability of riverine environments may select against feeding territories that promote nonparental monogamy. Male care of eggs plus distinct seasonality make it more profitable for a large female to remain with the male and assist in caretaking.

Monogamy among coral reef fishes presents a completely different picture. Only one species is known to show biparental care, the plankton-feeding damsel *Acanthochromis polyacanthus*. All other examples are of exclusive monogamy without biparental care. Parental care of young is selected against because coral reef fishes live in a hierarchically structured patchy environment. They are so closely adapted to their reef habitats that they are ill-equipped to move from one patch to another. Most coral reef fishes spawn abundant planktonic eggs at short intervals. In many of the small species the eggs are demersal; they develop into planktonic larvae.

So far, thirteen families of reef fishes are counted in which there is reasonable evidence of monogamy: ghost pipefishes (Solenostomidae), pipefishes (Syngnathidae), sea basses (Serranidae), tilefishes (Branchiostegoidae), butterfly fishes (Chaetodontidae), angelfishes (Pomacanthidae), damsels (Pomacentridae), wrasses (Labridae), gobies (Gobiidae), surgeonfishes (Acanthuridae), triggerfishes (Ballistidae), filefishes (Monacanthidae) and sharp nose puffers (Canthigasteridae). Detailed information is available on syngnathid fishes (pipefishes and seahorses) (Vincent et al. 1992; Vincent 1995; Vincent and Sadler (in press); Vincent et al. (in press).

Monogamous coral reef fishes differ from their freshwater counterparts in several fundamental ways. Many of them are strongly territorial over food or over a place of refuge. They spawn repeatedly over long periods of time, often daily. Spawning within a territory, as in the damsels, or the fish may ascend out of their territory, as in the surgeonfishes and butterfly fishes. They tend to be relatively small species, though the triggerfishes are large. With the one exception mentioned, there is no biparental care of free swimming one, though male care of eggs occurs in some families. It is possible, in some instances, to relate this type of monogamy to the distribution of resources. Information for chaetodontids can be obtained from Neudecker and Lobel (1982).

6. DIVERSITY IN REPRODUCTIVE PATTERNS AND BEHAVIOUR

Alternative reproductive behaviour is performed by males that have a low competitive ability regarding territory defence and/or mate attraction. In many fish species, small males can reproduce successfully only by trying to fertilize eggs of females spawning with larger, territorial males. In genera such as, for example, Pacific salmon (*Oncorhynchus*) and North American sunfishes (*Lepomis*), this reproductive behaviour is part of a separate life-history strategy with early maturing males acting as sneak maters, in contrast to males that grow larger and become territory holders. In other species, individuals may switch between different reproductive tactics depending on circumstances, such as size, ontogenetic stage, or intensity of competition.

In Atlantic salmon, *Salmo salar*, and chinook salmon, *Oncorhynchus tshawytscha*, male parr that reproduce as sneakers suffer a high risk of mortality from injuries caused by the larger, territorial males. In these species, with discrete life histories, male parr have to take this risk in order to reproduce. Sneaking behaviour in fish usually elicits aggressive behaviour in the territorial males and sometimes also in the spawning females and this conspicuous behaviour might attract the attention of other predatory fish species.

The risk of being eaten can affect the behaviour of an animal in many ways. Both foraging behaviour and mating have been found to change under predation risk in a variety of species. Usually there is a trade-off between the benefits accrued and the risk of mortality when performing a certain behaviour. The optimal choice of action depend on, for example, the individual's metabolic state, during both foraging and parental care. During the different phases of reproduction, prospects of future reproductive success should also be an important factor for decisions about risk-taking. In several fish species, for example, in the common goby, *Pomatoschistus microps*, the three-spined stickleback, *Gasterosteus aculeatus*, smallmouth bass, *Micropterus dolomieui* and bluegill sunfish, *Lepomis macrochirus*, nest-guarding males increase risk-taking as the reproductive value of the brood increases. Prespawning behaviour is also modified by predation risk in the sand goby, *Pomatoschistus minutus*, probably to decrease detection.

The teleosts exhibit the greatest diversity in reproductive patterns among vertebrates. The patterns range from (1) gonochoristic species having testis-bearing (male) and ovary-bearing (female) individuals which form typical monogamous pairs; (2) deep-sea anglerfish in which male attaches himself to the female, possibly relinquishing control over his own testis; (3) sequentially hermaphroditic species (protogynic and protandric forms) in which individuals change one sex to another; (4) simultaneously hermaphroditic species in which individuals trade eggs and sperms during mating, frequently reversing their sexual roles; and (5) simultaneously hermaphroditic species in which individuals self-fertilize under natural conditions.

Protogyny (first female, then male) is favoured where dominance of males is a major factor. This reproductive pattern is observed in species where a male guards a number of females (harem) or protects a territory (lek) which female enter to mate.

Protandry (first male, then female) occurs in smaller stable groups with essentially a monogamous relationship between that reproduces. The largest males change sex, become females; large females produce the greatest number of eggs. Protandry is common in instances where small males can fertilize the eggs of the larger females, and for one reason or the other, large males cannot monopolize groups of females.

Rhinomuraena spp.—The first vertebrate genus with postanally situated urogenital organs - are specialized, marine sand-dwelling ribbon-like eels, protandric in their sexual activity. With maturation they develop black, blue-yellow, and yellow colour morphs, according to stages of sexual change (Fishelson 1990).

For the details of gonadal structure in hermaphroditic fish species, the reader is referred to Sadovy and Shapiro (1987) and Shapiro (1992).

6.1 Sex Change as an Alternative Life-History Style

One of the most dramatic and all-encompassing alternatives that is known in the animal kingdom involves adult sex change. Individuals begin life as one sex and subsequently switch to the opposite sex.

A diverse array of fishes, mainly shallow-water marine species, change sex. Although the diagnosis of sex change requires care, sex change has been well documented in 20 families (for example in Salmoniformes, Synbranchiformes, Scorpaeniformes, Perciformes). Individuals mature as females and latter change to males (protogyny), or mature as males and later change to females (protandry).

Most protogynous are monandric, i.e., there is only one way to produce males, by sex change of a female. Other protogynous species are diandric: a portion of the population matures initially as 'primary' males which do not change sex, while other 'secondary' males are produced by sex change in females. In some protandric species individuals may be capable of differentiating either as a female or as a male that can later change sex. If so, then these species would be digynic.

Both types of sex change are initiated, as far as is known, by behavioural or demographic alternations within a fish's social system. Removal of a male from social groups of protogynous fish induces a female to change sex. Removal of a female from protandric groups induces a male to change to female. The stimulating cues do not appear to be simple visual, chemical, or acoustic signals communicated over long distances among group members but rather disruptions in ongoing patterns of close-contact behavioural interactions. Behaviours involved include aggressive chases, fin displays, and elaborate movements such as U-swims, which males sweep rapidly past females along U-shaped pathways.

In protogynous species, sex change may also be triggered by events other than male loss. Gradual reduction in the frequency of a female interactions with the male, migration of females from a bisexual group, and entry into a group of more than a threshold number of new adult females are all thought to induce a female to change sex in some species. It has been suggested, for one species, *Anthias squamipinnis* that females may be able to influence the onset of their own sex change by altering the way they use space within their social group.

In several protogynous and protandric species, more than one individual of the initial sex must be present in order for loss of an individual of the second sex to induce complete sex change. The cues indicating the presence of several members of the initial sex are not necessarily the same for all species. In the wrasse *Thalassoma duperrey*, cues are visual. In the goby *Coryphopterus glaucofraenum*, cues are chemical.

Sex change in either direction represents a thorough alteration in the physiology and behaviour of the individual. The changes appear in the morphology of the gonad, in the levels of circulating steroid hormones, often in external coloration, in social mating behaviour, in H-Y antigen levels, and possibly in the quantity of Bkm (banded krait minor) satellite DNA in the chromosomes. The changes in H-Y antigen levels and in Bkm DNA are particularly impressive because they were found in a species in which sex change is known to be induced behaviourally. The results imply that behavioural changes within the social system produce alterations in the switching mechanism controlling H-Y antigen expression and in nuclear DNA in the sex-changing individual.

Details of the proximate mechanisms controlling sex change suggest that these systems are designed to respond to alterations in the demographic environment of the individual. More concretely, particular demographic alteration within the specific social unit occupied by an individual cause the individual to change sex. If social units vary in size, sex ratio, or internal structure within a population, then the onset of sex change, with respect to the life history of an individual, will also vary among social units (Shapiro 1979, 1980, 1981, 1984, 1987a, b, 1988a,b, 1989; Shapiro and Lubbock 1980; Sadovy and Shapiro 1987; Wachtel et al. 1991a, b). Vincent (1992) discussed the prospects for sex reversal in teleost fishes. Coral reef fishes have been further subjected to allied studies (Shapiro 1986a, b) (read also Guraya 1999).

In pipefishes and seahorses (Syngnathidae), the male receives eggs from the female and then broods the embryos on the ventral body surface or in a brood pouch. Subsequently he releases live offsprings. Obviously there is a physical limit to the number of eggs a male can accommodate, and males therefore to some extent may be limiting sex in the above sense. Accordingly, sex role reversal has been recorded for a number of species in this family. The degree of sex role reversal differs among pipefish species, however,

The pipefish *Nerophis ophidion* has a complicated courtship dance during which the female transfers her eggs to the ventral part of a male's body. The male provides the embryos with nutrition and oxygen. Mate choice by males and competition between females can be expected in *N. ophidion* because male limits the reproductive success of the female: the female can remate faster than the male can. In this species, the 'normal' sex roles are reversed in that (1) the female transfers her eggs to the male who broods them, (2) the male faces higher reproductive costs than the female, (3) the female is larger, (4) she has secondary sexual characters, lacking in males, such as a blue sexual coloration and a ventral skin fold which makes her appear larger, and (5) she is more active during courtship than the male. The females's skin fold develops during the breeding season and disappears afterwards. The blue sexual colouring consists of several deep blue stripes along the anterior sides of the body. The coloration is permanent, but becomes pale during the winter. The size of the coloured part is positively correlated with body size and fecundity. The sex ratio is equal. Females with a large amount of blue colour are known to be preferred by males independent of female body length. The strong dimorphism in *N. ophiodion* is likely to have evolved in response to some constraint other than energy investment imposed by males on the female's reproductive success (see Berglund et al. 1986a, b; Rosenqvist 1990, 1993). Both male mate choice based upon female secondary sexual characters and female-female interactions determine the non-random mating pattern.

Noakes et al. (1989) considered the widespread occurrence of local phenotypic differentiation within salmonine species, especially arctic charr, *Salvelinus alpinus*, from the perspective of the evolution of life-history styles. Heterochronic shifts in ontogeny in relation to niche utilization are discussed as a possible proximate mechanism to account for this phenomenon. The relatively high degree of philopatry, especially homing to particular spawning locations, combined with localized differentiation of physical conditions on the spawning grounds predisposes the salmonines to the evolution of localized, differentiated spawning stocks. This differentiation ranges from relatively minor phenotypic distinctions, such as colour variants, through more distinct trophic morphs to sympatric sub-species or possibly even distinct species.

6.2 Hermaphroditism

The number of known hermaphroditic fish species has multiplied enormously during the past two decades. There has been an outburst of interest on behavioural, ecological, and evolutionary aspects of hermaphroditism. Sadovy and Shapiro (1987) enumerated and assessed the criteria that may be used to determine whether a fish species is hermaphroditic. The variations in configuration of germinal tissue in hermaphroditic fishes has been illustrated in Fig. 12.27B. It has been pointed out that bimodal age-frequency distributions have fewer problems associated with their interpretation than do bimodal size-frequency distributions.

Functional hermaphroditism is a normal mode of reproduction in teleost fishes and is common in certain families. Acquired hermaphroditism occurs in deep-sea certatoid anglers (e.g., *Edriolychnus schmidti*, Linophrynidae) in which males become parasites attached to females, which appear to gain control of their testis. The fish demonstrate extreme sexual dimorphism, perhaps the greatest among all vertebrates. Sequential sex change is common in wrasses (Labridae) and seabass (Serranidae), the coral reef fishes. In the former, the most complex systems have two types of males (diandry) and two different colour phases. Sex-changed females (secondary males) are produced in response to the particular needs of their social group. Absence of a male

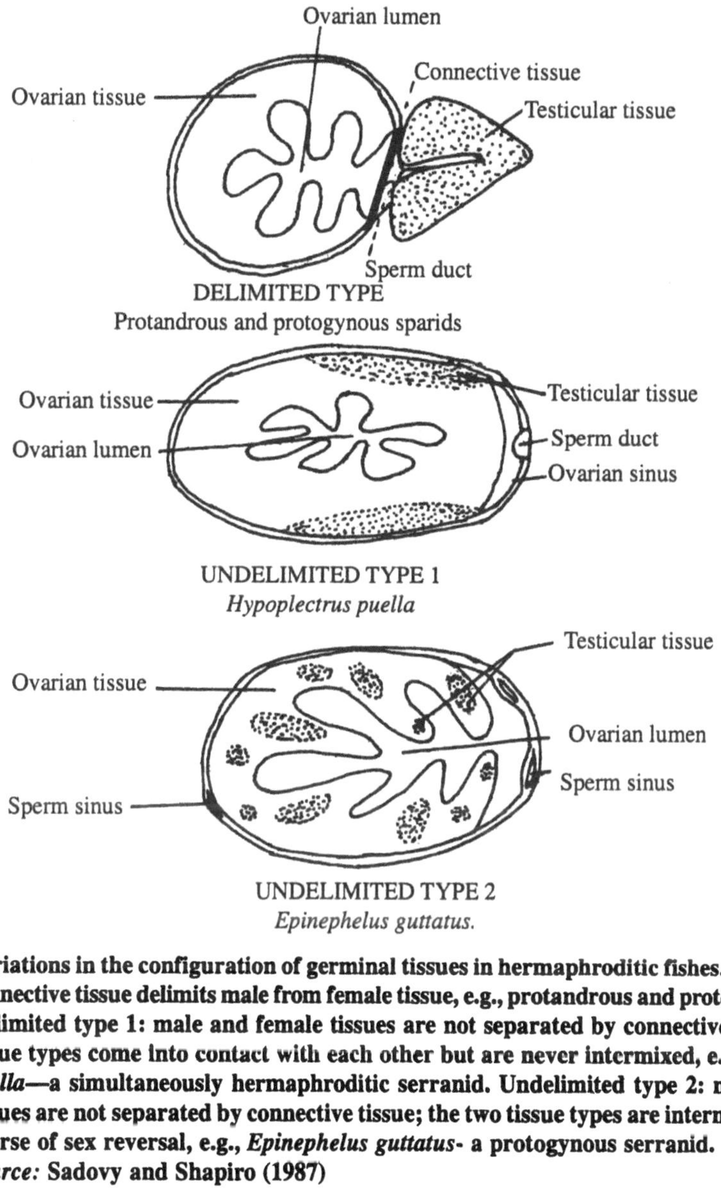

Fig. 12.27B Variations in the configuration of germinal tissues in hermaphroditic fishes. Delimited type: connective tissue delimits male from female tissue, e.g., protandrous and protogynous sparids. Unlimited type 1: male and female tissues are not separated by connective tissue; the two tissue types come into contact with each other but are never intermixed, e.g., *Hypoplectrus puella*—a simultaneously hermaphroditic serranid. Undelimited type 2: male and female tissues are not separated by connective tissue; the two tissue types are intermixed during the course of sex reversal, e.g., *Epinephelus guttatus*- a protogynous serranid.
***Source:* Sadovy and Shapiro (1987)**

and the presence of a smaller fish are the critical cues that trigger sex reversal in female wrasses of the genus *Thalassoma*. In the small anthiine sea bass *Anthias squamipinnis* of Pacific coral reefs, changes in complex behavioural interactions within a social group following removal of male result in male-typical behaviour in the largest female. Later she develops a male colour pattern and functional testes.

Most simultaneous hermaphrodites (Fig. 12.28) are the sea bass in the subfamily Serraninae (genera *Serranus* and *Hypoplectrus*). Individuals have functional ovotestes and typically occur in pairs that mate several times in succession with alternate reversal of sex role. The behaviour is called "egg trading" (reciprocal parcellation of small units of eggs) and presumably helps to prevent male cheating or nonreciprocation from the partner in the male phase. Hermaphroditic sea bass expend more reproductive effort in female functions. Some of the species have developed visual signals for sex-role identification. The ultimate in hermaphroditism is a small self-fertilizing killifish. This species normally produces homozygous, isogenetic clones. Males can develop under specific environmental conditions, and pairing of males and hermaphrodites may serve to limited the genetic uniformity.

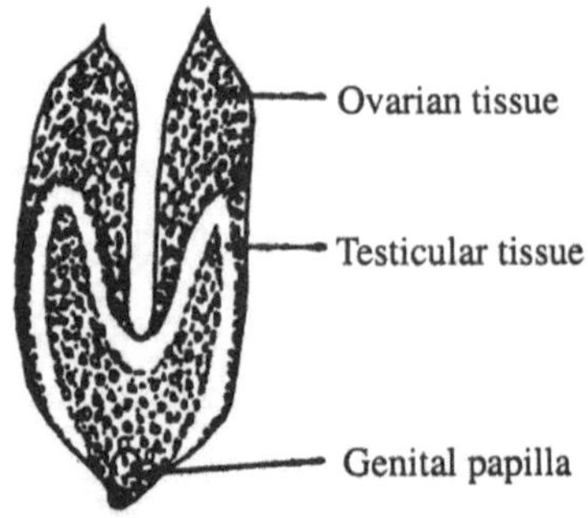

Fig. 12.28 Drawing of the ventral aspect of the gonad of *Serranus tigrinus*. Note that the testicular tissue forms a continuous band on the underside of the ovarian tissue, fusing anteriorly to form a loop on either side. The two types of tissue have separate openings in the region of the genital mound or papilla.
***Source:* Demski (1987)**

Further observations of deep-sea anglers indicate that the female controls the male's testes through hormones circulated in a common vascular system. This control may entail secretion of GnRH (gonadotropin-releasing hormone) by the females, with subsequent release of gonadotropins by either one or both of the animals. Sperm and egg release may be synchronized by female hormones and/or male detection of cues from the female. In sequential sex-changers, testosterone administration can trigger sex-colour, behaviour, and gonadal reversal in female wrasses. Androgens can also initiate sex change in the protogynous sea bass, *A. squamipinnis*, while estrogens appear to at least temporarily inhibit the response.

Considerable information is available on gonadal development and function in "egg trading sea bass" but little is known of their pyschobiology. Telencephalic regions and diencephalon are implicated. Unfortunately, almost nothing is known concerning the mechanisms of behaviour in the self-fertilizing killifish *Rivulus marmoratus*. The killifish demonstrate dramatic reproductive adaptations to environmental factors. The natural self-fertilization of the hermaphrodite is most likely a response of alternate drying and flooding. Studies have demonstrated that *R. marmoratus* has at least the capability under certain conditions to produce sexual phenotypes similar to non-self-fertilizing hermaphrodites (Demski 1987).

Cochran and Grier (1991) discussed the possibility that sexual succession results from inhibition of ovarian tissue response to steroidogenic gonadotropin in protogynous hermaphrodite black sea bass, *Centropristis striatus*.

6.3 Social Control of Primary Sex Differentiation

Sexual differentiation in teleost fishes is characteristically labile. The most dramatic form of sexual lability is postmaturational sex change, which is common among teleosts, although rare or absent in other vertebrate taxa. In many cases this process is regulated by social cues, particularly dominance interactions. In Midas cichlid, *Cichlasoma citrinellum*, these same sorts of social interactions affect much earlier stages of sexual differentiation. In this species, males are much larger than females. This sex-based size difference does not arise from endogenous factors associated with sex. Rather, sex is determined by relative size as a juvenile. This mode of sex determination, which may be common among teleosts, is a heterochronic variant of postmaturational sex change, one in which some individuals are deflected from a default female trajectory before maturation, as a result of social signals. The size-advantage model, which specifies the optimal size for sex change in hermaphroditic species, can be extended to account for the decision whether to mature as a male or a female in Midas cichlid (Francis and Barlow 1993).

7. ROLE OF HORMONES AND SEX PHEROMONES

7.1 Endocrine Control of Reproduction

Considerable information on reproductive physiology of fish is currently available (see Peter 1983; Belle et al. 1988; Chiba and Honma 1989; 4th Int. Symp. ed. Scott et al. 1991; Dulka et al. 1992; Liley et al. 1993; Olsen and Liley 1993; Abdullah and Kime 1994; Magliulo-Cepriano et al. 1994; Liley and Kroon 1995; Pankhurst 1998; Van Der Kraak et al. 1998).

Steroid hormones: From cyclostomes to mammals, the localization, cytological aspects, and the enzymatic equipment of steroidogenic cells did not undergo changes in their general outlines. Major sex steroids released by gonads in blood are androgens, estrogens, and progestagens in proportion fluctuating with sex, age, and sexual stage. No compound is exclusively produced in one sex or the other. Major pathways of steroidogenesis are not the same in all vertebrates. In teleosts, a 20β – hydroxysteroid dehydrogenase allows the production of a powerful progestagen, 17α – hydroxy, 20β-dihydroprogesterone, which plays an important role in the last steps of male or female gametogenesis; a uridine diphosphate glucurosyl transferase allows the gonadal release of conjugated androgens and estrogens, which act as pheromones; and high $11\ \beta$ – hydroxylase and $11\ \beta$ – hydroxy-steroid dehydrogenase are implied in the synthesis of a very powerful androgen, 11- ketotesterone.

Gonadal synthesis of partical compounds, especially in teleosts (11-ketotesterone, $17\ \alpha$ - hydroxy $20\ \beta$ -dihydroprogesterone, glucuroconjugated derivatives) show a taxonomic diversification.

Peptide hormones: In several species of teleosts, a single GTH was purified and shown to stimulate various aspects of gonadal development; also a sequence data demonstrated that its β -subunit is homologous to the mammalian GTH β -subunit. Thus, for a while, we have thought only two lineages (TSH) and one GTH) existed in teleosts. However, it was recently shown that salmonids possess two GTHs homologous to mammalian hormones. Therefore, "single GTH" situations have to be looked for in lampreys and hagfish.

Messenger RNAs for GTH α and GTH β were detected in teleosts. Comparative studies are required on the number and the expression of GTH β-subunit genes in lower vertebrates. Several GnRHs have been characterized in lower vertebrates.

Gonadal secretion: Sex steroids are released in blood, some of them after they are conjugated in the gonads. Plasma levels are very different depending on class and even species. In a given species, they vary qualitatively and quantitatively, largely in relation to the sex cycle.

Circulating steroids can be bound to specific plasma proteins, characterized by a high affinity and low capacity, proteins which underwent an important evolution. A binding system with a high affinity does not seem to exist in cyclostomes. A binding protein with a broad specificity is present in elasmobranchs: androgens, estrogens, and progestagen (and corticosteroids) are competitive for the same site. They may represent a primitive situation from which the binding systems of more recent vertebrates evolved. In other ectotherms (e.g., teleosts), the binding system is made of two components, the specifications of which are different, one binds only androgens and estrogens, the other binds progestagen (and corticosteroids).

In teleosts the recent data in salmonids indicate that the presence of two pituitary LH- and FSH-like GTHs; one of them (now called GTH II) has been extensively studied and appears to stimulate specific phases of steroidogenesis. Quantitative changes in plasma GTH II levels are important factors for determining large quantitative and qualitative sex steroid variations, during teleostean sex cycles. As far as GTH I is concerned, it appears to differ from GTH II, showing a lower specificity for 17 α -hydroxy, 20 β -dihydroprogesterone production during oocyte maturation, whereas both hormones have a similar effect on estradiol production by midvittelogenic ovarian follicles from salmonids. Radioimmunoassays of GTH I and II suggest that synthesis and secretion of GTH I are predominant during prematurational stages of spermatogenesis and vittellogenesis. Thus, physiologically, GTH I could be important at these stages and GTH II more important later on.

In teleosts, other mechanisms certainly exists. Experiments have indicated that extrapituitary, nervous, or paracrine factors may stimulate ovarian aromatase activity. In lampreys, the interrenal and meta-adenohypophysis appear to be involved in sex steroid synthesis and its control.

Central nervous system-pituitary axis: In female trout (*Salmo gairdneri*), GTH pulses, the frequency and the amplitude of which may vary during sexual cycle, were observed. GTH appearance is also rapid in teleosts. The metabolic clearance of GTHs depends on their structures.

In primitive vertebrates, such as lampreys, chondrichthyans, and even coelacanths, there is no direct connection between hypothalamus and pituitary. GnRH should reach GTH cells by general circulation or by diffusion. Such a pattern appears compatible with a tonic rather than with a precisely timed pulsatile GTH stimulation.

The primitive situation evolved in two very different ways, with the establishment of either a venous portal system, (in tetrapods) or a direct innervation of the pituitary (in teleosts). In some teleosts, GnRH neurons end in the digitations of the neurohypophysis within the adenohypophysis (as in eels and salmonids); in others they go across the basal membrane of the neurohypophysis and come into direct contact with gonadotropic cells (as in cyprinids, poeciliids, or gobiids); they may even form synaptic contacts with these cells.

In teleosts a dopaminergic (DA) inhibition of the action of GnRH on GTH cells was demonstrated, which may be mediated, at least in part, through a down-regulation of the pituitary GnRH receptors. Moreover, not only GnRH and DA neurons but also other neurons (e.g., GABA-containing neurons were shown to end on GTH cells in cyprinids; also in these fish, serotonin has a stimulatory effect on GTH release. Thus, in teleosts, GTH cell is an important site of integration for neuroendocrine signals.

In all gnathostomes, sex steroids were shown to control the gonadotropic pituitary-hypothalamus axis. In fish, beside the classical negative feedback that exists in adults, a peculiar effect of sex steroid was demonstrated in juveniles of some species (eels, salmonids) (i.e., an intense stimulation of pituitary GTH II synthesis). A positive effect of sex steroids was also demonstrated on GnRH synthesis in the eel and GnRH release in juvenile trout.

Hormonal receptors: With regard to the specificity of H-R (hormone-receptor) interaction, there are indications of an increase from lower to higher vertebrates, for both GTH-TSH and Gn RH families.

Teleost TSH receptors recognize all TSH and GTHs, whereas no such heterothyrotrophic effects are seen in mammals. Also in teleosts, preliminary data on the action of GTH I and GTH II on steroidogenesis suggest that their receptors could exhibit less specificity than LH and FSH receptors in mammals.

In the eel, experiments have suggested the existence of ovarian "isoreceptors" for GTH-like molecules, which differ in their specificity pattern.

Receptors for sex steroids appear to exist in all vertebrates, in nuclear and sometimes cytosolic fractions.

The expression (or nonexpression) of hormonal receptors in various tissues is a very important evolutionary mechanism. The most obvious example is the variety of steroid-controlled reproductive organs, secondary sexual characters, or sexual behaviours, exhibited in the animal kingdom; all of these characters result from the expression of steroid receptors in a select variety of different targets.

Thus, estradiol and progesterone receptors exist in female organs, allowing the steroid controlled maintenance and development of oocytes in all oviparous vertebrates.

With regard to secondary sexual characters, androgen receptors have been detected in the skin of several vertebrates, including teleosts. The reader is referred to the contribution by Fontaine et al. (1991). Reviews (King and Millar 1991; Sathyanesan 1994; Singh and Lal 1994) have covered allied aspects.

Takano et al. (1991) indicated that estradiol-17 β (E_2) is involved in vitellogenesis, and 17 α, 20 β -dihydroxy-4- pregnen-3-one (17 α, 20 β -di OH prog; (17, 20 *P*); (17, 20 β-*P*) may have some important roles in gestation in the multiple spawner *Sebastiscus marmoratus*. The reader is also referred to Nagahama et al. (1991) for *Sebastes taczanowskii* and *S. schlegili*.

Schreibman et al. (1990) described the life cycle of the brain-pituitary-gonad axis in teleosts. In teleosts as in other vertebrates, investigations into the developmental events in the brain-pituitary-gonad (BPG) axis have centered about specific peroids in the life of the organism-largely puberty (sexual maturation), to a lesser extent the neonatal period, and least to senescence. The scarcity of longitudinal (birth to death) studies is reflected in the information concerning the sequence of events leading upto, and falling between, these two key periods.

One vertebrate that has been studied extensively from a longitudinal point of view is the platyfish, *Xiphophorus maculatus*. The characteristics of the reproductive system of this small freshwater teleost are remarkably similar to those of higher vertebrates, including mammals. Their studies in platyfish indicate that there is a complex system of interactive events and control mechanisms involving brain, pituitary gland, and gonad that characterize specific periods in the life of an organism (Fig. 12.29)(see also Chapter 13, Fig. 13.10 and Chapter 14).

Peter et al. (1990) discussed action of gonadotropin-releasing hormone (GnRH) in the goldfish. GnRH receptors are located on both gonadotrophs and somatotrophs, and that the two cell types do not have communication. The differences in responsiveness of GTH and GH release to sGnRH and cGnHR-II suggests that there may be some difference in the GnRH receptors on these two pars distalis cell types. Results may also indicate the presence of two separate high affinity GnRH receptors, perhaps one originating from gonadotrophs and other from somatotrophs.

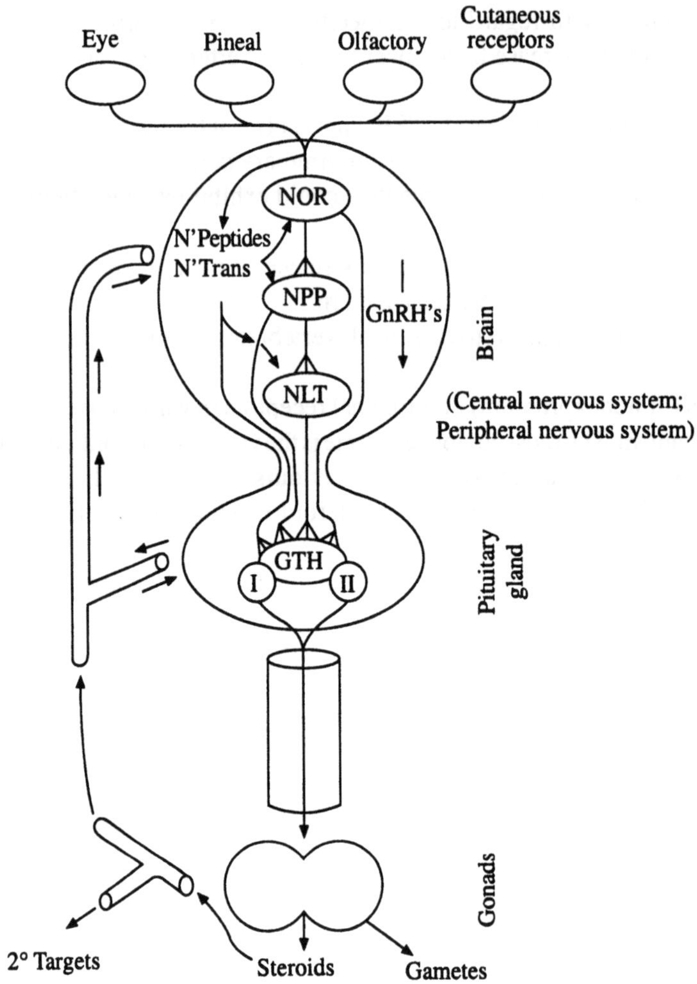

Fig. 12.29 A schematic representation of the BPG axis depicting its relationship with receptor organs of environmental cues. Hormones involved in the axis are indicated within their appropriate organ; arrows indicate paths of movement of humoral agents and feedback controls. 2°, secondary; N' trans, neurotransmitters; N'peptide, neuropeptides
***Source:* Schreibman et al. (1994)**

Although there is some differential activity of sGnRH and cGnRH-II in stimulating GH and GTH release, and in pituitary receptor binding in the goldfish, whether the observed differences have physiological significance remains open. The reader is also referred to Peter et al. (1987) on vertebrate gonadotropin-releasing hormones: phylogeny and structure-function relationships.

Throughout vertebrates, the gonadotropin releasing hormone (GnRH) decapeptide family is known to be the primary group of neuropeptides regulating reproduction. Multiple forms of GnRH are found within several species of teleost fishes. GnRH neurons in the preoptic area of teleosts exert direct control over pituitary/gonadal function. It is likely that the spatial and temporal events determining the development and adult expression of forebrain GnRH cell populations is associated with the expression of sexual plasticity (see Grober et al. 1995).

7.2 Sex Pheromones

Fishes have well-developed chemical senses, a predictable evolutionary consequence of life in a medium where chemical cues abound and visual cues are often limited. Hormonal pheromones of fish appear to be among the best characterized vertebrate sex pheromones. Sex pheromones have been shown to regulate spawning behaviour and endocrine events related to reproduction in a variety of teleost species. The concept of a distinction between primar and releaser pheromones, the former instilling slow, developmental processes and the latter evoking rapid behavioural responses, also applied to these situations. Although perception of sex pheromones by taste has been suggested by some authors, in most cases, olfaction appears to play a dominant role in mediating the response to water-borne pheromones. All teleost sex pheromones identified to date are sex hormones or derivative (Table 12.3) whereas non-sex pheromones are of a different nature (e.g., proteinaceous).

Table 12.3 Identified reproductive pheromones in the gonads of teleosts, their source and function

Species	Compound (source)	Function
Brachydanio rerio (zebrafish)	Steroid glucoronides (testes)	Induction of ovulation
Brachydanio rerio (zebrafish)	Steroid glucoronide mixture containing estradiol-17 β (ovary)	Attraction of male
Carassius auratus (goldfish)	17 α, 20 β -Dihydroxy-4-pregnen-3-one (ovary)	Raising milt volume in male
Carassius auratus (goldfish)	Prostaglandins (ovary)	Attraction of male
Clarias gariepinus (African catfish)	Steroid glucoronides (seminal vesicle)	Attraction of ovulated female
Gobius jozo (black goby)	Etiocholanolone glucuronide (mesorchial gland)	Attraction of gravid female
Poecilia reticulata (guppies)	Free and glucuronidated estradiol-17 β	Attraction of male

(Modified after Van Weerd and Richter 1991; Stacey et al. 1991)

Pheromonal steroid glucoronides act upon olfactory system. The existence of an olfactory-brain-pituitary axis, has been demonstrated in *C. auratus* and *C. gariepinus*. Little is known about pheromonal stimulation in early phases of the reproductive cycle. In teleost fish, sex pheromones are involved in advanced phases of reproduction, priming actions influencing oocyte maturation and ovulation in females, and spermiation in males; releasing actions involving attraction of perspective mates and triggering spawning behaviour and release of gametes. In the African catfish *C. gariepinus*, sex pheromones were recently demonstrated to also influence earlier phases of the gonadal cycle, i.e., ovarian development in females.

Ovarian development is under the control of the endocrine system, through the hypothalamic-pituitary-gonad axis. The endocrine events encompass the release of GTH, which stimulates ovarian follicles to produce the female sex hormone estradiol-17 β (E2). E2 induces hepatocytes to synthesize the female-specific plasma protein, vitellogenin (VTG). VTG is actively sequestered from the maternal blood stream by developing oocytes in the ovary and, after processing, is deposited as yolk material. Through accumulation of yolk, the oocytes increase enormously in size and cause the ovary as a whole to grow. The whole process is known as vitellogenesis and

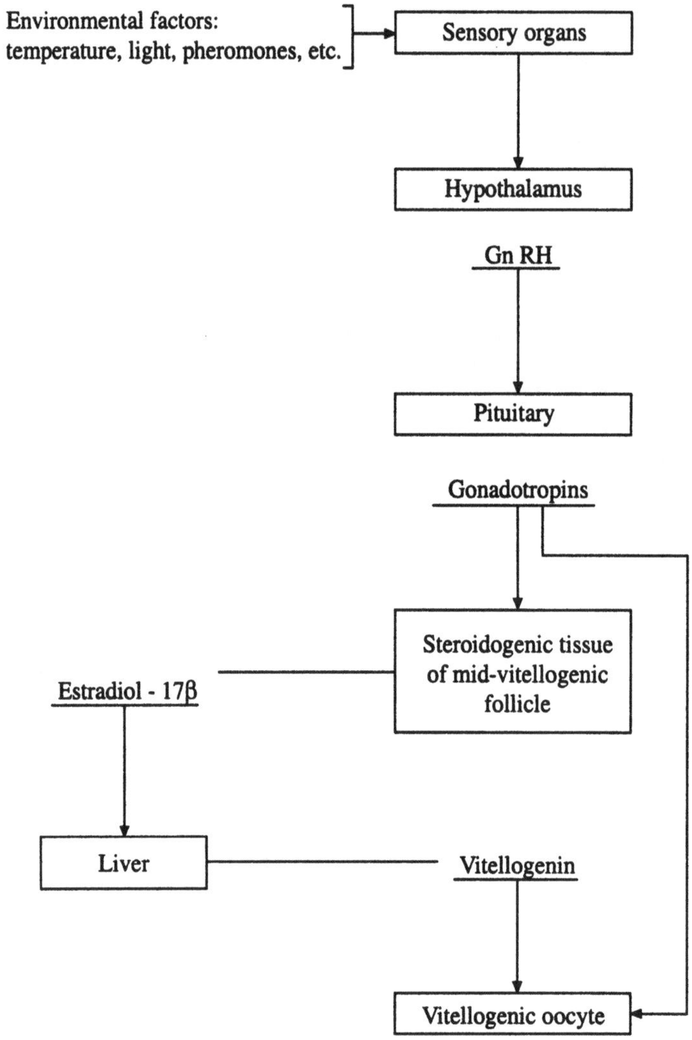

Fig. 12.30 Schematic representation of the hypothalamus-pituitary-gonadal-liver axis in a female teleost showing organs and pathways involved in ovarian development. ***Source:*** **Van Weerd and Richter (1991)**

is depicted in Fig. 12.30. For details of vitellogenesis and oocyte assembly, see Mommesen and Walsh (1988) and Ho (1991) (see also Dréan et al. 1994; Fig. 1: The main estradiol actions in rainbow trout during the reproductive cycle).

In teleosts, there is interaction between sexes in order to synchronize gamete maturation and facilitate sexual encounters and in some species, pheromones are part of this interaction. Experiments have also indicated that female tactile stimuli tended to curb male gonadal development and the emission of 'male-directed' male olfactory stimuli suggesting inhibitory female cues. General social inhibition of maturation (gonadal development) was described for *Xiphophorus* spp., *Poecilia reticulata*, *Haplochromis burtoni*, but gender-related inhibition of teleost gonadal development has been recently described.

Being involved in reproduction, a gonadal origin of sex pheromones seem obvious (refer Table 12.3). Female sex pheromones are produced in the ovary, and the male sex pheromones in the testes. Males of several teleost species possess specialized structures where sex pheromones are produced. *G. Jozo* has a mesorchial gland, with an abundance of Leydig cells, where sex attractants are produced. Evolutionary advanced catfish species, including Clariidae, possess a seminal vesicle, in the lobes of which fluid is produced containing proteins, polysaccharides and phospholipids. Apart from possible functions related to sperm storage and facilitating egg fertilization, the content of the seminal vesicle is responsible for attracting ovulated females. However, the skin of *C. gariepinus* may also produce pheromones. Urine may be the vehicle for gonadally produced pheromones as in the plaice *Pleuronectes platessa.*

For pheromones to be species-specific, the basically limited number of sex steroids in teleosts must be differentially used in the various species, especially those with overlapping biotopes. Several explanations have been offered in this respect.

In the specific cases of hormonal pheromones, this evolution likely involves three distinct stages (Fig. 12.31). There is evidence both for and against the proposal that changes in responses to pheromones are due to changes in olfactory receptor function.

In the osteichthyan fishes so far examined, EOG (electro-olfactogram) responses to water-borne hormones and hormone metabolites have not been found throughout the higher taxa, but only within the salmonid (order Salmoniformes) and ostariophysine fishes (the related order Cypriniformes, Characiformes, Siluriformes and Gymmotiformes).

Recent contributions on these reproductive chemosignals have covered the dramatic progress (see Liley and Stacey 1983; Van den Hurk and Lambert 1983; Stacey and Sorensen 1986, 1991; Stacey et al. 1986, 1991, 1993; Dulka et al. 1987; Resink et al. 1989; Anderson and Døving 1991; Dulka and Stacey 1991; Van Weerd and Richter 1991; Sorensen 1992, 1996; Van den Hurk and Resink 1992; Defraint and Sorensen 1993; Irrine and Sorensen 1993; Sorensen and Scott 1994). The reader is also referred to an overview (Redding and Patino 1993). Section 6 on pheromones and behaviour in Proc. 4th Int Symp. ed. Scott et al. 1991) deals with papers related to this aspect.

Prostaglandins (PGs) are involved in a number of physiological processes that govern reproduction; they have been shown to activate or facilitate sexual receptivity in diverse vertebrates. PGs induce spawning behaviour in female goldfish, a paradise fish, *Macropodus opercularis*, a cichlid, *Cichlasoma bimaculatum* and *Puntius gonionotus*. PGs of the F series also are released as pheromone by spawning female goldfish and induce spawning behaviour in male goldfish through an olfactory mechanism. In the fathead minnow, *Pimephales promelas*, PGF_2 stimulates the release of a semiochemical by females that induces male courtship behaviour (see Whittier 1991; read also Sorensen et al. 1988; refer Chapter 13).

8. REPRODUCTIVE EFFORT; FECUNDITY; RHYTHMIC BEHAVIOUR

Reproductive effort: The process of reproduction in the adult fish requires an allocation of material and energy from that in transit or already stored within the body (Fig. 12.32). The proportion of physiologically useful resources devoted to reproduction, subject to both proximate and ultimate controls, has been defined as "reproductive effort". Within the energy budget of the fish, expenditure on reproduction may be divided into (1) a "primary" sector, involving prezygotic

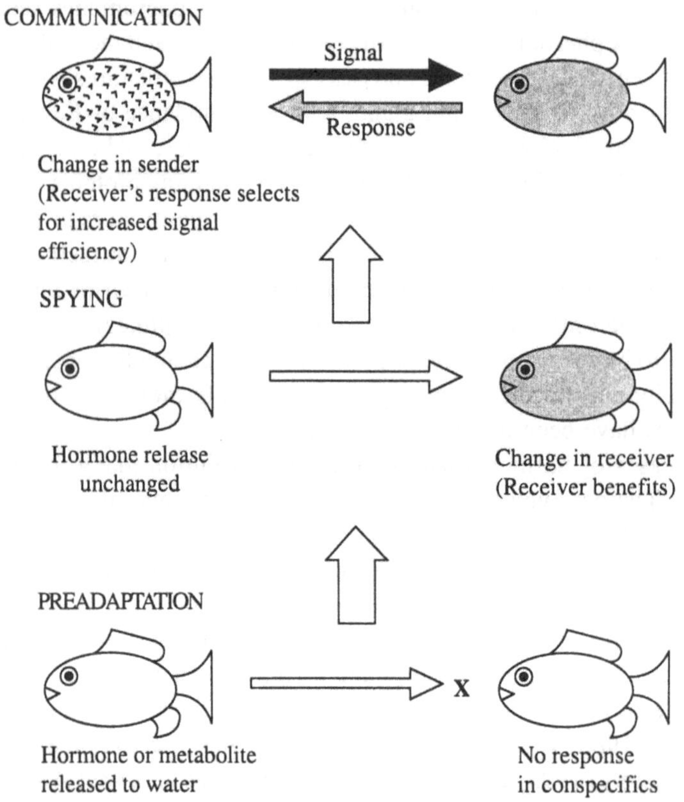

Fig. 12.31 Proposed stages in the evolution of hormonal pheromones. Hormones are preadapted for pheromonal function because they are released to the water in synchrony with specific reproductive events. The first stage in evolution pheromonal function likely involves the acquisition of responsiveness to a released hormone or metabolite. The result may simply be a unilateral benefit to the receiver (chemical spying). If the receiver's response can benefit the individual releasing the hormonal pheromone, chemical communication may evolve.
***Source:* Stacey et al. (1991)**

maturation biosyntheses and storage within the gametes, notably expressed in oocyte size and number, with postzygotic contributions in viviparous species and (2) a "secondary" sector, itself divisible into (a) the anabolism of secondary sexual characters, not only the material of more obvious sexual dimorphism, as shown in coloration and finnage, but even pheromones known to feature in teleost reproduction and, rarely, material for nest building or mucophagy by young and (b) the catabolism of reproductive behaviour, expended, in vary proportions on the activities of migration, territoriality, nidification, courtship, and, in oviparous species, brood care. Of these, gonadal maturation, and mating behaviour immediately prior to fertilization, are obviously reproductive necessities. Among extremes in secondary expenditure must be the anadromous behaviour of salmonids. The spawning migrations of female sockeye salmon entail loss of over 90% of body fats and 50–60% of total body protein, admittedly without an in-built cost ceiling on reproductive effort. It is also important to remember that reproductive effort varies in kind between the sexes, most obviously in relation to yolk deposition in the eggs. Among teleosts in general, when the sexes differ in mortality rate, it is usually the males which have a shorter

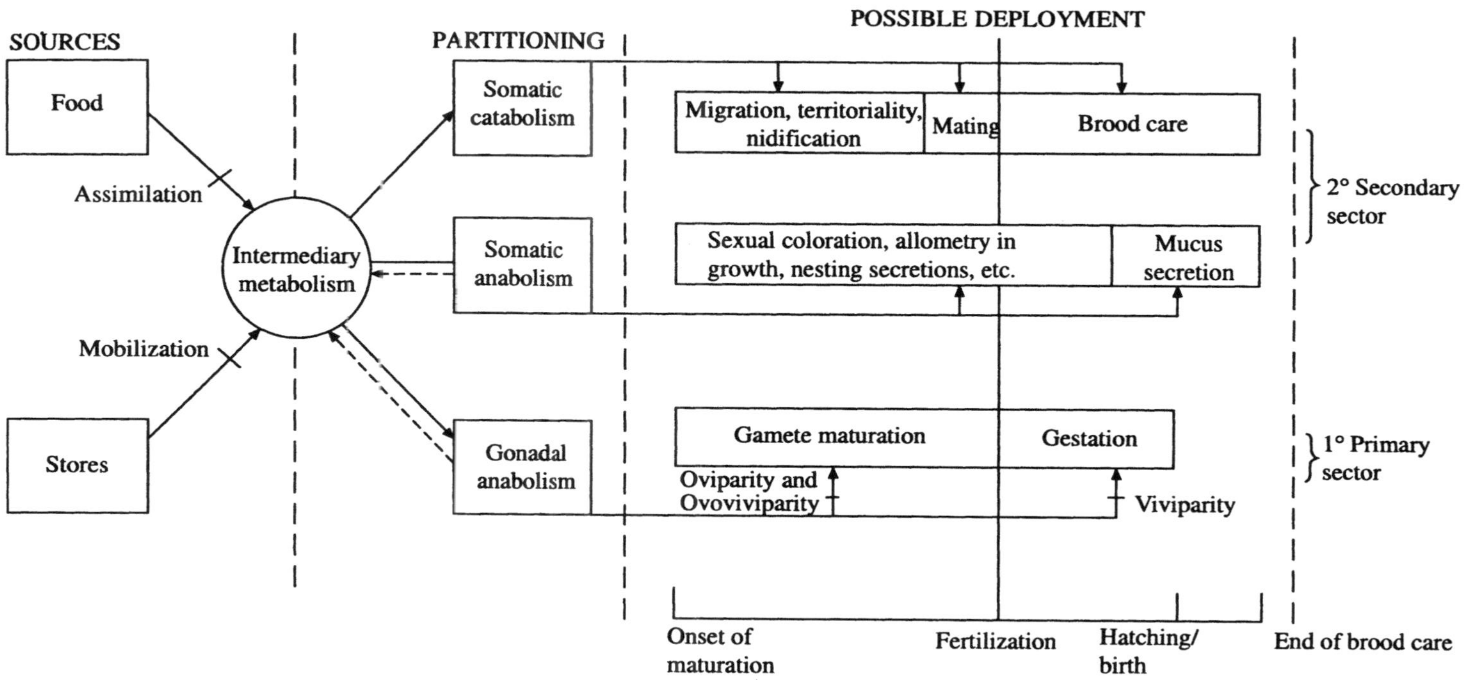

Fig. 12.32 **The sources, major partitioning and possible deployment of energy and material as reproductive efforts in teleosts, essential requirements with horizontal edging; reclaim possibility indicated by broken lines.**
Source: **Miller (1979a)**

lifespan. Perhaps the "risk" element of increased activity, with greater exposure to predation, contributes towards this phenomenon.

Closely tied with reproduction is the anabolism of storage within the fish system. In teleosts, depot substances, typically triglyceride lipids, may reach as high as 30% of body weight at the end of a feeding season. Phenology of storage is most familiarly associated with temperature or low Arctic marine species in a markedly seasonal but fairly predictable ecosystem. The final magnitude attained by the fish-maximum size- and the temporal sequence of intermediate stages- the growth pattern-may be expected to have adaptive significance relative to survival, feeding and reproductive opportunities during individual lifespan in a particular ecotope. An equally important area of research has been the study of environmental influences on fish growth, whose susceptibility is shown by widespread seasonality. Photoperiod and salinity are two other easily adjusted environmental features which are known to influence growth, besides a numerous and potentially variable range of factor believed to act on intrinsic control of growth. More subtle factors, such as scent, sight and proximity of fellow fishes must interact to exert a complex sociobiotic effect, quite apart from the possibility of more obvious resource limitations (read Miller 1979a).

The annual timing of gonad production in fish is usually controlled by endogenous factors sensitive to exogenous indicator variables such as photoperiod (Bye 1984). This timing should be adapted to the seasonal cycle in production and availability of food for adults (Iles 1984) and the availability of suitable environments for incubation and for the first-feeding young. There is a strong relationship among timing of first feeding, survival and growth rate. Timing of spawning can be of ultimate importance in this context. The reader can obtain information regarding the relationships among time of spawning, incubation temperature, timing of first feeding and early growth in four sympatric morphs of Arctic charr, *Salvelinus alpinus*, in Thinguallavatn, Iceland (Skúlason et al. 1989).

Most reproductively competent female fish exhibit annual cycles of ovarian development (Lam 1983; Bye 1984). The cycle begins with ovarian recrudescence, continues through a breeding period, and ends with ovarian regression. The recrudescence (which primarily involves vitellogenesis) is mostly under the influences of seasonal temperature and photoperiod. Later, during the breeding period, one or more clutches of ovarian follicles develop progressively, often in synchrony with a combination of transient changes in temperature, illumination, spawning substrate availability, rainfall, moon phase, and tide, etc. (Lam 1983; Bye 1984; Stacey 1984). Ultimately, mature follicles ovulate and fish spawn at a specific time, a particular site, or both to maximize the immediate survival of eggs and hatchlings. Afterwards, a reproductive regression sets into ensure that no hatchlings emerge and subsequently face a harsh environment inimical to their survival: follicles cease their growth and the ovary remains regressed until the next seasonal recrudescence. The seemingly direct responses of reproductive cycles to environmental influences do not necessarily exclude the involvement of endogenous organismal rhythms (Lam 1983; Bye 1984; Stacey 1984). Such self-sustained rhythms schedule reproductive preparation, time receptivities for external cues, and, most important of all, can be entrained to follow repetitive environmental parallels.

Comparisons between field and laboratory data have revealed mechanisms underlying the annual and semilunar reproductive cycles in female *Fundulus heteroclitus* (Hsiao et al. 1994).

Empirical studies have shown that in teleosts there is a significant correlation between the life history parameters, age at first reproduction, natural mortality, and growth rate. It has been hypothesized that these correlations are the result of evolutionary adjustments due to the trade-off between reproduction, growth, and survival (see Roff 1984).

Wooton (1992) provided a preliminary review of some of constraints that are likely to operate on life history traits. The review addressed to: (1) A hierarchy of constraints; (2) Size as a physical constraint (body size, total egg volume, fecundity, egg size); (3) Allometric constraints; (4) Physiological constraints; (5) Demographic constraints, and (6) Genetic constraints.

Different temporal allocation of finite energy resources to alternative components such as reproduction and somatic growth can be used to characterize the life-history variation in fishes.

Salmonid fishes show extensive life-history variation among (e.g., iteroparus versus semelparous), and within species (e.g., sympatric trophics morphs).

RNA/DNA ratios were used to estimate recent somatic allocation of spawning arctic charr, *Salvelinus alpinus*, and pink salmon, *Oncorhynchus gorbuscha.* Smaller Arctic charr morphs had lower ratios than larger morphs. Larger male pink salmon had lower ratios than smaller males. Thus recent somatic allocation (anabolic metabolic activity and growth) is coupled with cumulative lifetime growth in iteroparous Arctic charr, but is uncoupled in semelparous pink salmon. These data suggest somatic investment in these species is being differentially effected by the energetic costs of reproduction (Ferguson et al. 1990). For salmon life history, the reader is referred to Gross (1985, 1991).

Fecundity: Reliable estimates of fecundity are essential for studies of population dynamics and life histories of fish. For fish whose eggs mature all at once during a relatively short annual breeding season, fecundity estimates are simply made by counting the number of eggs per female. However, many marine and freshwater fish from both temperate and tropical zones spawn several times over a protracted breeding season. In these species oocytes develop in several distinct batches and are laid successively during the breeding season. The factors determining fecundity patterns (clutch size, inter-spawning intervals, total number of clutches) for these species are largely unknown. Photoperiod and food have been implicated as important determinants in laboratory studies, but little is known about egg laying patterns of multiple spawning fish in nature. Food is unlikely to directly limit egg production of fish in salt marsh tide pools because of abundant food supply. Temperature may adversely affect reproductively active females by acting on locomotor and feeding activity and on respiratory behaviour. High water temperatures also adversely affect oogenesis. For control of the timing of ovulation by exogenous factors (see Stacey 1984). The biochemical embryology of fishes is known in a few fishes. It is chiefly concerned with molecular biology (DNA-RNA protein) and energy metabolism (glycolysis, respiration).

Various chapters in "Fish Reproduction: strategies and tactics" (edited by G.W. Potts and R.J. Wooton 1984) provide interesting details on the related aspects. Magnhagen (1991) discusses predation risk as a cost of reproduction in animals. Kime (1995) reviewed comprehensively the effects of pollution on reproduction in fish.

Coral reef fishes, regardless of pattern of parental investment, are characterized by great fecundity and a virtually continuous breeding season. In general eggs are small, which increases the number of eggs per clutch, even in species that invest postgygotically. Selection for high fecundity probably also explains why in some monogamous species of coral reef fishes the female is larger than the male, which contrasts sharply with comparable freshwater fishes. In the anemonefishes the female is the largest individual of a reproducing unit and the male *Amphiprion* does most of the caretaking, which allows the female to continue feeding and thereby develop more eggs. In the peculiar little *Solenstomus cyanopterus*, however, it is the female who carries the embryos, and she is huge compared to her mate. The female is much the larger sex in monogamous pairs of the surgeonfish *Acanthurus leucosternon*; a pair is better able to defend a feeding territory, and paired females have more fat reserves than do solitary ones.

Broadcasting pelagic eggs does not typify all reef species. Yet species that broadcast comprise upto half those of the reef (Barlow 1981).

Rhythmic behaviour: Studies have indicated a fine temporal pattern of anabolic activity, derived ultimately from the earth's rotation and juxtaposition rather than from its evolution. This is biorhythmicity of a circadian or circatidal nature. Such rhythms have been amply demonstrated in various behavioural and other catabolic functions studied in teleosts, and are typically entrained by photoperiod or tidal oscillation. These catabolic changes may be expected to impose effects of comparable periodicity on daily scope for anabolism within the same integrated metabolic system, perhaps by affecting food intake, growth efficiency or degree of muscular activity. There is now, in fact, good evidence for circadian rhythms in various aspects of teleost anabolism. For example, within the endocrine systems influencing growth, storage and maturation, daily cycles have been found in plasma levels of growth hormone and prolactin, and in pituitary gonadotrophin content. The timing of gonad maturation in a number of teleosts has also been shown to depend on a circadian rhythm in sensitivity to the inductive effects of light. Among anabolic end-products in teleosts, a daily fluctuation in hepatic lipogenesis is noted, while, for somatic growth, a rhythm of similar period in glycine incorporation by scale-pocket cells is described. Long-term running of a circadian growth cycle is suggested by the preservation of daily growth layers in teleost otoliths. Experiments with the cichlid, *Tilapia mossambica*, suggest that formation of these daily marks require entrainment of an internal clock by a 25 h light/dark regime. A monthly and bimonthly pattern is also apparent in the growth history of otoliths stuided. Annual changes in otolith structure are often used for age-determination of individual fish when manifest in relation to marked circannual fluctuation in growth and reproduction. Discovery of a circadian rhythm in otolith deposition now holds great promise as suggesting a means for age determination in young fishes or in species, especially many of the tropics, which have not experienced seasonal growth checks or limited breeding seasons, a useful practical outcome from a somewhat academic field of fish phenology.

At present, it is debatable whether the persistent circadian rhythms of teleost metabolism are basically controlled by some internal timing pulse or by responsiveness to regular environmental variation, including geophysical factors imperceptible to human consciousness. The study of endogenous rhythms in fish anabolism has never excluded such subtle influences. The teleost otoliths, whose growth pattern may show seasonal change under constant temperature and salinity, might even qualify as suitable receptors for variation in, say, electromagnetic fields, since a piezoelectric function has been attributed to these structures. Alternatively, one theory of internal timing does depend on a cyclical anabolic phenomenon of intrinsic production, envisaged as transcription from DNA "chronons". From whatever causation, cellular reactions and enzyme activity would seem to present a picture of routine fluctuation (read Miller 1979a).

Gibson (1982) discussed the rhythmic behaviour of intertidal fishes. Three main functions have been ascribed to these intertidal movements: feeding, avoidance of predation, and reproduction. Few intertidal animals which live in an environment where regular and predictable changes in conditions are the rule, synchronization of activity with such changes is clearly of importance. Synchronization may be brought about by a direct response to changing conditions such as the flooding of a pool and subsequent alterations in temperature, salinity or light intensity but numerous recent studies have shown that many species also possess endogenous tide-related rhythms of behaviour. The endogenous nature of this rhythmic behaviour has been demonstrated in the laboratory where the fish show alternating bouts of rests and activity with the periods of activity occurring at or near the time of high tide. Such tidal rhythms seem to be of common occurrence because they have been recorded in representatives of many different families (Blenniidae, Tripterygiidae, Cottidae, Gobiesociadae, Gobiidae, Periophthalmidae, Pleuronectidae, Bothidae, Soleidae).

Endogenous rhythms operating independently of the environment are usually considered to enable an animal to be prepared for predictable change. Such preparation may be particularly important for intertidal species whose opportunities for feeding or mating may be limited to a few hours each tidal cycle or even each 24 h if they are strongly diurnal. On occasions, however, environmental conditions may be unsuitable for normal activity and the underlying inherent rhythm may be overridden by a direct response to prevailing conditions.

Specific behaviour patterns or local tidal conditions may also have an effect on the phase relationship of rhythmic activity. In those rocky shore species which show rhythmicity the activity peaks are usually centered on the predicted time of high tide whereas in species from sandy shores the peaks occur on the ebb tide. Although the adaptive significance of such a phase difference has not been investigated experimentally it may be related to the different movement patterns of the two groups.

In areas where the tidal range is small and fluctuations in water level are often more dependent upon meteorological conditions fish seem to regulate their activity according to light: dark cycle rather than to the tides.

Even in areas where tidal fluctuations may be considerable, however, several field studies have demonstrated that many species are strongly diurnal, drastically reducing or ceasing activity at night. Such diurnal changes in behaviour have been recorded in mudskippers, clingfishes, blennies, clinid *Mnierpes* and pholid *Apodichthys flavidus*.

The presence of diurnalism in the great majority of intertidal residents is probably related to their dependence on sight for carrying out most of their activities, particularly the detection of predators and for feeding.

A special note can be given here on the reproductive behaviour and tides. During the reproductive season many marine fishes synchronise their spawning with the tidal cycle. In most species spawning takes place at, or near, high water of spring tides. As a consequence, the release of gamates follows a semi-lunar or lunar cycle. Most coral reef fishes whose spawning is related to tides spawn at high tide or when it is ebbing. Not all pelagic spawners have a tidal component to their reproductive behaviour, however, nor is tidally-related spawning absent from benthic spawners; some pomacentrids and blennies, for example, show lunar or semi-lunar rhythms in their spawning patterns (read Gibson 1992; see literature).

Details of rhythmic activity were also covered in "Rhythmic activity of fishes" edited by J.E. Thorpe (1978). Also, Schwassmann (1971) focussed on biological rhythms. A recent title "Rhythms in fishes" edited by M.A. Ali (1992) is an interesting collection of papers. A chapter by Reebs (1992) is an interesting description on sleep, inactivity and circadian rhythms. Whether or not can fish sleep depends on the criteria we use to define sleep. Sleep behaviour has been recorded in *Tinca tinca.*

9. PHYSIOLOGICAL AGING (SENESCENCE) IN FISH

The study of senescence is, essentially, a science in its infancy. The process of aging begins at birth (if not earlier) and differs from, but includes, senescence, the processes of growing old.

Attempts to discuss the phenomenon of aging or senescence in fish were made by Simpson (1979), Woodhead (1979) and Craig (1984). Schreibman (1987) dealt specifically the aging of the endocrine system. For aging of lymphoid tissue, read Tatner (1986).

Aging is the increase in the rate of mortality and decrease in fecundity that occurs after maturity and at advanced age in a population. At present there appears to be two main theories of aging: (i) aging by wear and tear and (ii) aging genetically determined and controlled by a biological clock. There is more evidence supporting theory of genetic aging controlled by a biological clock than aging by a random damage.

Growth and natural mortality are closely correlated between and within species and populations. Fast growth and early maturity curtail the life-span; slow growth and late maturity have the reverse effect. There appears to be a trade-off between reproductive effort and longevity to ensure the production of the maximum number of progeny over the life-span of mature adults. The trade-off is influenced by a feedback system dependent on reproductive success. Genetic control of growth and thus mortality is not well understood and is obscured by environmental factors. Empirical data have shown that populations of fish from one gene pool can respond rapidly to environmental conditions by altering their growth, reproduction, and life-span. The extent of this response depends on the gene mixing within the population.

The genetics of life-history traits are little understood. Each species has a characteristic maximum life-span (T_{max}) which is genetically determined, but the mean life-span comes under environmental influence and will vary widely. There is evidence for the importance of "optimum condition" (i.e., relationship of metabolic rate, temperature, growth, reproduction). Crowding of fish can result in reduced growth rates, late maturity, decreased fecundity, and increased longevity. This may be the result of food shortage but other factors such as pheromones have been suggested. Modern theories about the immune response and control by environmental factors support original "rate of living" hypothesis.

A hypothesis states that living organisms undergo bioscenescent and antibioscenescent processes. The life-span of cell components depends on the balance between these processes with biosenescent factors always predominating. Thus living systems can never be immortal, only the continuity of genetic material is immortal and this is made possible by the separation of soma from the germ cell. The evidence for immortality is sparse. It is claimed in flatworms. The death of somatic cells can be prevented only by their transformation into malignant cells. Malignant tumors, are in general, more frequent in aged individuals and these have been noted in fish species.

A number of studies have been made on the degenerative changes in the soma and the gonad, changes in the endocrine glands, and the reduction in regenerative processes with age. These include histological evidence of deterioration in the pituitary, thyroid, thymus, stomach, pancreas, liver, spleen, kidney, skin, and cardiovascular system. Further evidence of aging by a biological clock rather than wear and tear comes from the studies at the cellular level.

Changes in growth and natural mortality have an important influence on yield and recruitment in fisheries. The understanding of aging process lies in an improved knowledge of bioenergetic processes and the environmental factors that control them (see Craig 1985 for literature and mathematical details).

Schreibman (1987) covered the changes that occur in the brain, pituitary and gonad from the onset of sexual maturation into the senescence of platyfish (*Xiphophorus maculatus*; Poeciliidae), a small viviparous freshwater teleost. Sufficient information has been amassed on the structure and function of its reproductive system, and the hereditary mechanisms that control its age at sexual maturation, to make the platyfish invaluable for studies involving the mechanism of aging of the brain-pituitary-gonad (BPG) axis. The information presented represented the most complete "longitudinal" study available for any ectotherm. Yet, it is clear that there are still wide abysses in our knowledge even for the platyfish. The trends observed were that there are distinct changes in the distribution and relative quality of neuropeptides, neurotransmitter and pituitary and gonadal hormones, as well as in the structure of the cells that manufacture, transport, or are target organs for them, during the life-span of platyfish. The interdigitating cyclical events which occur during platyfish aging suggest the exchange of information among components of the BPG axis. Studies indicated that each ir-GnRH-containing region of the brain plays a significant, and perhaps unique role during specific phases of the life-span of platyfish.

Further, it has been pointed that the process of senescence exhibits sexual dimorphism. Age-relative differences are quantitative. Genetic factors determine the timing and sequence of events in the brain and pituitary gland that culminate in puberty. Genetic regulation in platyfish of the structure and function of the BPG axis is carried into senescence. It has been suggested that further studies concentrating on the development and interactivity of neurotransmitters and neuropeptides in animals of different genotypes will be rewarding in our general comprehension of aging. The basic question that needs to be resolved, of course, is where and how genes exert their control over the development and aging of the BPG axis. We need to know more about the relationship between development, aging and longevity and how these processes are influenced by genotype.

According to Simpson (1979), senescence in annual killifishes is indicated by changes in soluble to insoluble collagen ratios in skin, but adult life in captivity is long in relation to early maturation and attainment of limiting size.

10. TERMINOLOGY OF INTERVALS IN FISH DEVELOPMENT

An illustrated and well-discussed account on terminology of intervals in fish development has been given by Balon (1975b). The entire life of a fish, from the time development.commences until death consists of five periods: embryonic, larval, juvenile, adult and senescent.

Each period is divided into phases as a convenient means of morphophysiological identification of developmental levels. Initially, the rapid developmental processes are reflected in rapid morphological and physiological changes; however, in each successive period, the developmental speed decreases until a complete standstill marks the end of the senescent period and death of the organism.

The embryonic period can be divided into three phases: cleavage, embryonic, and eleutheroembryonic. The larval period of most species can be divided into two phases-protopterygiolarval and pterygiolarval. The *protopterygiolarval* phase encompasses the interval between the transition to exogenous feeding and the commencement of differentiation of the embryonic median fin fold-the first appearance of mesenchymatic buds of lepidotrichia and dorsal and anal fin outlines within the fin fold. In true annuals this phase is missing and well-developed pterygiolarvae hatch at the end of diapause III (Balon 1975a). The *pterygiolarval phase* lasts from the beginning of the differentiation of the unpaired fins until the embryonic median fin fold is entirely differentiated or no longer apparent.

The salmonids and mouthbrooding cichlids, for example, have no larval period but instead the juveniles go through a special alevin phase. In eels, the phases of toothed and toothless leptocephalid larvae replace the former pterygio or fin fold phases. Notwithstanding the extremely long larval period, a special metamorphic elver phase at the beginning of the juvenile period parallels the alevin phase of salmonids. The unusual ontogeny of eels is marked not only by the long larval period (1–3 years) but by size of the larvae before metamorphosis and the reduction in size after. Most often the leptocephalid larva reaches 50–100 mm, but once a 1.8 m larva was recorded off southern Africa. Thus, in some instances the larva can be larger than the adult fish.

The *alevin phase* exists in species that do not have a larval period and lasts until the development of scalation or ossified spines. It occurs mainly in salmonids, but is synonymous with the elver phase in anguillids, the metamorphic phase in elopids, the thiolichthys phase in chaetodontids, the acronurus phase in acanthurids, and in brotulids with the exterilium phase. These latter names can be used to identify the phases of fishes with more unusual larval periods.

In salmonids the alevin phase can be succeeded by the smolt or parr phase. This phase encompasses the interval of downstream migration following the alevin phase and preceding the juvenile phase. The juvenile period of salmonids would, in this case, be divided into three phases: alevin, smolt, and juvenile. The age designation could then be accordingly modified.

So far no phases have been identified for the juvenile period of other species nor for the adult and senescent periods of any species. Fishes at any given level of development should bear names of corresponding phases or periods. Only in this manner can definitive and clearly characterized names correspond to an exact level of development. Vernacular or popular names, like green egg, eyed egg, fry, young spawn, hatchling, wriggler, and all the previously rejected prelarvae, postlarvae, and other similar illogical names, should be avoided, at least in scientific literature.

Actually, the periods and phases are a qualitative framework that groups smaller developmental intervals into a kind of model for most fishes while steps are the smaller, more species-specific intervals.

Chambers and Leggett (1992; see literature) synthesized observations on metamorphic ages and sizes in flatfishes (Pleuronectiformes) within a hierarchial framework which ascends from individual to population to species levels. Flatfish metamorphosis is abrupt, systemic, and apparent, even at casual observation. At the species level, size at metamorphosis ranges from ≈ 4 to 120 mm and is long—normally distributed. Age at metamorphosis is more plastic than size in its response to different temperatures; as temperatures decrease the mean metamorphic age increases as much as twofold. Metamorphic size is expected to increase with decreasing temperatures although the effect is not as strong and the data are less convincing. At the individual level, data are available for starry flounder, *Platichthys stellatus*, and winter flounder, *Pseudopleuronectes americanus*. Bivariate distribution of metamorphic ages and sizes for these flounders, schematized as metamorphic envelopes, have different locations and orientations, but similar shapes. Metamorphic age has greater within-population variation and is less influenced by parentage than is metamorphic length. Furthermore, maternal contributions to metamorphic traits exceed paternal ones. In a benign environment lacking mortality bias, the effect of metamorphic age on juvenile size variation was ≈ sixfold greater than the effect of metamorphic size. The impact of metamorphic age on juvenile size variation lessened with extension of spawning interval and with increased duration of juvenile life prior to capture.

A distinguishing characteristic of elopomorph fishes (orders Elopiformes, Albuliformes, Notacanthiformes, Anguilliformes, and Saccopharyngiformes) is that they have a leptocephalus larval stage. In addition to a distinctive morphology (a laterally compressed body composed mainly of acellular gelatinous material), premetamorphic leptocephali (Phase I) are unique among teleost larvae because although a considerable increase in size may take place during this period, ingestion of prey or other macroscopic particles has not been demonstrated. There is little data on the trophic relationships of leptocephali. Direct absorption of dissolved organic material (DOM) has been suggested and more recently it has been shown that particulate organic matter (POM) may also be an important food source.

Hydromineral studies on premetamorphic leptocephali indicate a high water and salt content conditions that may be associated with synthesis of gelatinous matrix.

Specific glycosaminoglycan (GAG) components of the gelatinous matrix has been identified for several species. Lipid and carbohydrate stores generated during Phase I are utilized during metamorphosis in bonefish (*Albula*).

According to Donnelly et al. (1995), trends for biochemical components and nucleic acids suggest that growth of Phase I leptocephali occurs in two subphases (Ia and Ib). Phase Ia is characterized by cellular proliferation, preferential synthesis of protein and carbohydrate relative to lipid, and growth manifested more as increased length rather than increased mass. Phase Ia extends from yolk-sac absorption to ≈ 90 mm TL. In Phase Ib, nucleic acid content level off, lipid deposition increases , and mass increases exponentially.

11. PHENOTYPIC VARIABILITY IN FISH POPULATIONS

Each fish has a unique history. In theory, this history can be described at any moment by an attribute vector,

$$a = (p_1, p_2, p_3, \ldots, p_i, x, y, z, t)$$

Here, the p_i's are the measures of an individual's status, such as its age, size, condition, life stage, gender, and behaviour elements x, y, z are spatial coordinates; and t is time. Taken together, the attribute vector describes what, where, and when. Repeated observations on individuals can be differentiated to estimate rate variables, such as growth and movement rates, and integrated to yield dynamic performances, such as lifetime reproductive effort.

The phenotype is a manifestation of the underlying genotype as expressed in the local environment. The interaction of the phenotype and the environment results in an individual's experiences, such as its encounters with prey, predators, and mates. The attribute vector is a snapshot view of the individual fish, so these past experiences and events are implicit. An individual's current condition, however, can be viewed as a "zoometer" of the quality of the environment the individual has experienced. The resolution and scope of the zoometer is characteristic of the index used to measure condition (e.g., length-weight, lipid-lean, and RNA-DNA ratios). Only the exceptional technique would allow an explicit reconstruction of a longitudinal record for individual fish.

The uniqueness of individuals implies variability in at least some life history attributes. Anticipation of consequential variability in individual-level attributes has motivated the individual-base approach. Variability in trait has origins or sources that can be divided into genetic and environmental categories (Table 12.4). The genetic components contribute to the resemblance between parents and offspring, the response of a trait to selection, and the genetic correlations between two or more traits.

Table 12.4 **Potential sources of phenotypic variability in fishes. Some sources are likely to be interrelated, for example, the amount of prey eaten could vary with temperature and turbidity, or the range of phenotypes produced by a genotypes produced by a genotype could vary with local environmental conditions. Parents make genetic and nongenetic contributions. The nongenetic components are considered as one class of environmental sources of variability.**

	Environmental		
Genetic	Nongenetic parental	Biotic	Abiotic
Additive component	Spawning site and time	Prey abundance and quality	Temperature
Epistasis	Egg size and quality	Crowding	Duration of growing season
Allelic dominance	Brooding	Competitors	Salinity
Correlation with other traits	Nest guarding	Predators	Turbidity
		Diseases	Turbulence
		Parasites	Oxygen
		Sublethal toxins	Sublethal toxins

Source: Chambers (1993)

The environmental contributions to the phenotypic variance are all nongenetic sources. These sources are many, vary with species, location, and time, and often are interrelated. For example, the effect of temperature on the duration of embryonic period varies with the prevailing salinity. The degree of influence of an environment on a trait can depend on the underlying genotype of the individual.

Nongenetic parental contributions are considered as a type of environmental influence. They are usually maternal in origin and act early in life. For most marine fishes this contribution is in the form of yolk and it ends with yolk absorption. The size and quality of eggs and their site and season of release are several of the variables by which a female may influence the status and prospects of her offspring. For species, that spawn multiple batches, variability can be further attributed to a within-female, among-batch component. Maternal influences may represent an effect of a female's age and condition rather than a genetic basis per se, hence they are treated as an environmental source of variation in Table 12.4. Despite their limited temporal scope, maternal influences may have a particularly large effect on offspring fitness by increasing the chances of survival of some advantaged individuals through the period of extraordinarily high mortality characteristics of early larval life.

The variability in many life history traits is the result of numerous genetic and environmental factors, each contributing small amounts. Traits often have a normal statistical distribution (or one that is transformable to a normal distribution) because of the multifarious sources contributing to their variability. This feature allows analysis of the trait distribution by the parametric statistical methods of quantitative genetics.

Results of a study by Hutchings (1996; see literature), in the case of brook trout (*Salvelinus fontinalis*) illustrated how environmental variation can favour the evolution of adaptive phenotypic plasticity in life history through growth-dependent effects on fitness. Details of behavioural implications of intraspecific life history variation are covered by Hutchings (1993).

CHAPTER 13
Endocrine System

1. INTRODUCTION

We owe much to our current appreciation of the endocrine repertoire to comparative studies on lower vertebrates, which have shown that while identical or highly homologous hormones occur in all vertebrate classes, their physiological roles may differ. A hormone may be involved with several diverse functions even within one animal, but the emphasis on a particular function can shift during the course of evolution, so that homologous hormones in fishes and mammals may come to serve different roles in two classes.

The fundamental biochemical and physical processes which are concerned in the development, growth, metabolism, and reproduction in fishes are integrated and controlled to a greater or less extent by hormonal means. In common with other vertebrates, fishes possess, therefore, a well-developed endocrine system and most, if not all, of the mammalian endocrine glands have their piscine counterparts.

It must be admitted that a large part of work has been restricted to freshwater fish. Moreover, there is astonishing diversity of hormones and endocrine mechanisms in fish. Studies hold an inherent interest for fish physiologists and general biologists. Hormones and endocrine mechanisms are important when one considers the relation between animals and their external medium.

Information on endocrine system or hormones covering different aspects of fish life is contained in a number of reviews (Barr 1965; Matty 1966; Chester Jones 1974; Woodhead 1975; Fontaine 1975) and books, viz., Fish endocrinology by A.J. Matty (1985), Vertebrate endocrinology: Fundamentals and biomedical implications edited by P.K.T. Pang and M.P. Schreibman (1986, 1987, 1989, 1990), and Progress in comparative endocrinology edited by A. Epple, G.G. Scanes and M.H. Stetson (1990). Of specific interest are recent reviews dealing with cardiovascular system and hormones interaction in fish (McDonald and Milligan 1992; Olson 1992). Scott et al. (1991) edited proceedings of a symposium on reproductive physiology of fish. Wendelaar Bonga (1993) gave a comprehensive coverage of endocrinology with a focus on specific aspects. Hazon and Balment (1998) published a contribution on endocrinology of fish. Olsson (1990) has attempted to reveal the evolution of some of the complicated chordate endocrine organs based on the knowledge about disparate "protochordate" (procraniate) survivors. A very stimulating discussion of hormonal peptide evolution has been presented by Fontaine (1985). Nozaki (1985) discussed the perplexing brain and pituitary relationships in the cyclostomes. It is possible that a number of peptides in hagfish may have been eliminated during evolution, or more precisely, regression, the author opines. Henderson et al. (1985) have discussed renal osmoregulatory mechanisms in the context of hormones in fishes. Takei (1993) focused on the role of peptide hormones in fish osmoregulation. Holmgren and Jensen (2001) discussed the evolution of vertebrate neuropeptides.

Fig. 13.1(A) shows, in schematic form, the position of endocrine glands or hormone-producing tissues in a bony fish. Fig. 13(B) illustrates other aspects of the endocrine system.

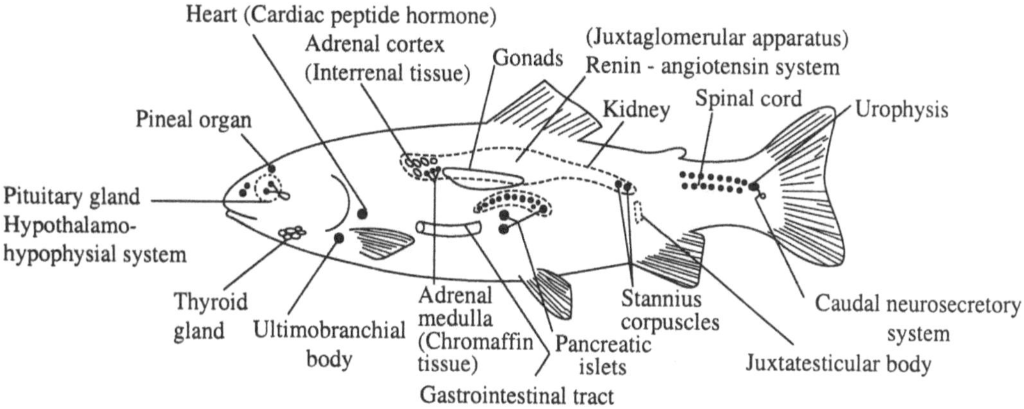

Fig. 13.1A Endocrine glands or hormone-producing tissues of bony fish.

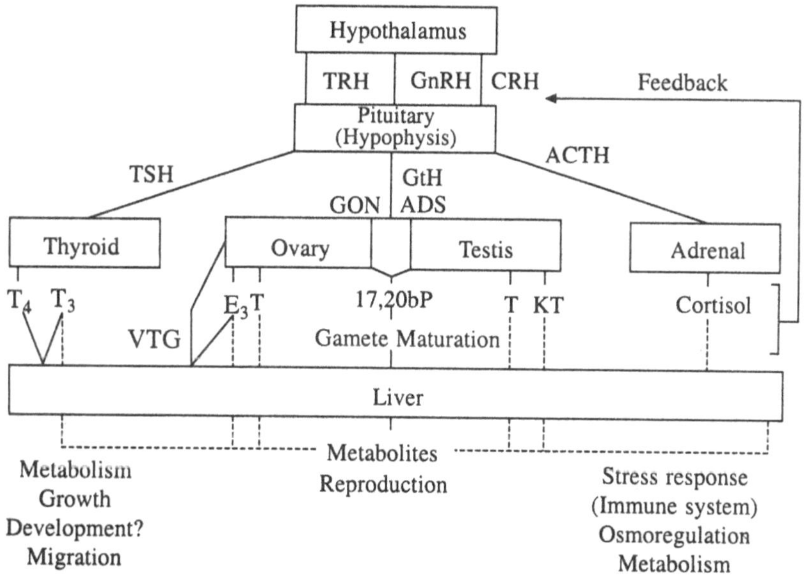

TRH = thyrotrophin releasing hormone;
GnRH = gonadotrophin releasing hormone;
CRH = corticotrophin releasing hormone;
TSH = thyroid stimulating hormone;
GtH = gonadotrophins I and II;
ACTH = adrenocorticotrophic hormone
T_4 = thyroxine;
T_3 = triiodothyronine;
E_2 = estradiol;
T = testosterone;
$17,20\beta P$ = $17,20\beta$-dihydroxy-4-pregnen-3-one;
KT = 11-ketotestosterone;
VTG = vitellogenin.

Fig. 13.1B A schematic diagram of the endocrine system of fish.
***Source:* Kime (1999)**

2. PITUITARY GLAND (HYPOPHYSIS)

All vertebrates have a well-developed pituitary gland consisting of a neural component, the neurohypophysis, and an ectodermal component, the adenohyposis. The structural intimacy of neurohypophysis and adenohyposis that is established early during the embryology of the pituitary gland reflects the direct functional interaction between the central nervous and endocrine systems. The extent of the anatomical intimacy between neuropituitary and adenopituitary components ranges considerably among the vertebrate classes from no apparent contact in cyclostomes to intimate interdigitation in teleosts. The reader is referred to a review by Schreibman (1986; see literature) on pituitary gland in vertebrates. A brief coverage of the neuroendocrine control of pituitary gland in teleosts is given by Prasada Rao (1994).

The great disparity between the way hagfishes and lampreys regulate reproductive functions may be reflected, in part, by variations in structure and function of their pituitary glands. The hagfish pituitary gland appears to be the most primitive of the two. Its neurohypophysis is hereby developed, flattened, saclike and contains many aldehyde-fuchsin-positive (AF^+) neurosecretory endings. The hypophysial-portal system of hagfish is usually markedly reduced but may show species variation. The adenohypophysis is separated from the neurohypophysis and is not differentiated into zones except, perhaps, in the posterior of older and larger animals (*Myxine*) (Fig. 13.2). Generally, islets of cell are embedded in a poorly vascularized, loose connective tissue which is continuous with the layer separating the neurohypophysis from the adenohypophysis. Some cells form follicles and contain periodic acid Schiff positive (PAS^+) colloid.

The lamprey neurohypophysis consists of a thin portion and a thickened posterior part, which, because it possesses neurosecretory neurons ending in a neurohaemal structure, has been termed a "pars nervosa". The adenohypophysis is compact and is differentiated into two regions, a pars distalis and a pars intermedia. A well-developed pars intermedia forms a neurointermediate lobe with the pars nervosa; however, there are no direct nervous or vascular connections between the two components. The lamprey pars distalis is separated from the infundibulum by connective tissue, and its cells display a regional distribution, at least according to their stain affinity, ultrastructure, and purported function. There is little evidence to suggest that hypothalamus exerts any influence on the pars distalis. Lamprey represents the most primitive of hypothalamus-pituitary associations (Fig. 13.2).

In elasmobranch fishes, the neurohypophysis consists of a thin-walled portion that is a true median eminence, which is connected to the pars distalis by a hypothalamo-hypophysial portal system. The median eminence is divided into anterior and posterior portions in at least some species. The posterior region receives both hypothalamic aminergic and peptidergic axons and is linked by the portal system to the CPD (caudal pars distalis) and perhaps indirectly to the ventral lobe; the anterior region contains less neurosecretory material and is linked to the RPD (rostral pars distalis). It has often been suggested that this arrangement could reflect a more rapid and efficient method for controlling different pituitary cell types. Since, in contrast to teleosts, there is no direct innervation of the pars distalis, the portal system is the only route for hypothalamic control. There is also a portal supply to the intermediate lobe. A saccus vasculosus apears for the first time in elasmobranchs. It is not, however, quite as developed as its homologue in teleosts. It develops from the hypothalamus as a ballon-like structure just above the neurointermediate lobe, but since it is considered to be sensory and nonendocrine, it will not be discussed further.

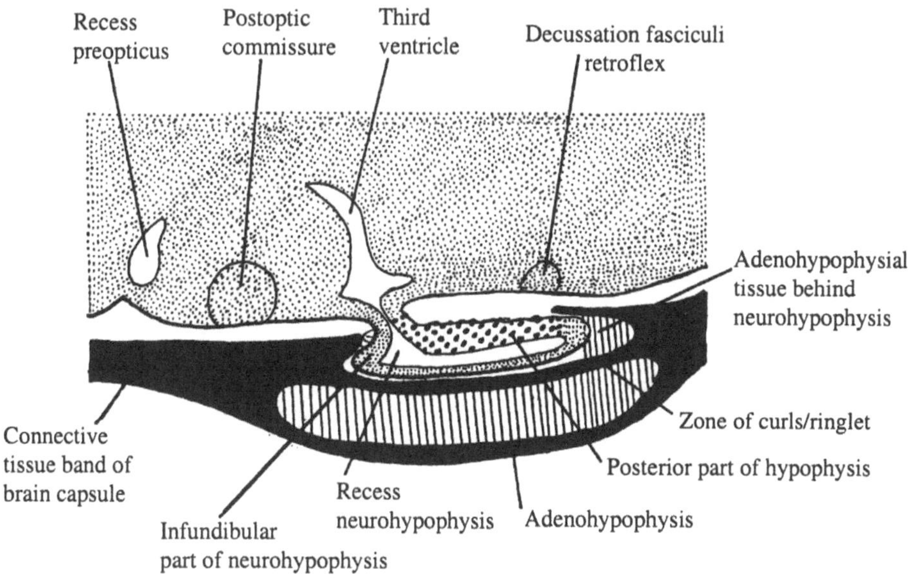

(A) *Myxine glutinosa*: Hypophysis.

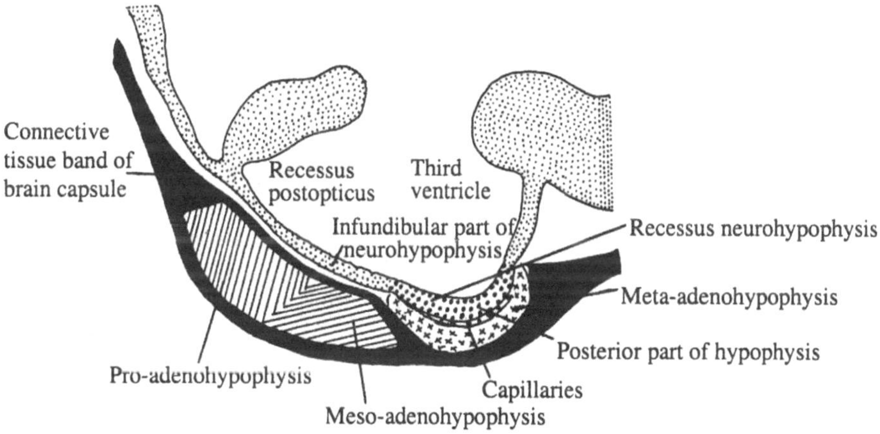

(B) *Lampetra fluv.*: Hypophysis

Fig. 13.2 (A) Section through hypophysis of *Myxine*. (B) Semidiagrammatic median section through the hypophysial-hypothalamus region of *Lampetra fluviatilis*.
***Source:* Adam (1959)**

The adenohypophysis is divided into four areas. There is an elongated pars distalis (the "dorsal lobe"), with its characteristic RPD and CPD; a ventral lobe attached by a stalk to the CPD; and a large pars intermedia that is heavily penetrated by neurohypophysial tissue to form a typical neurointermediate lobe (Figs. 13.3, 13.4). A pars tuberalis is not seen.

The holocephalan neurohypophysis (e.g., ratfish) has a pars nervosa and a prominent median eminence that sends many short blood vessels into RPD and CPD. A typical intermediate lobe is seen. In holocephalans, the entire pars distalis is hollow and is not as clearly divisible into rostral and caudal regions as in elasmobranchs. A ventral lobe, characteristic of elasmobranchs,

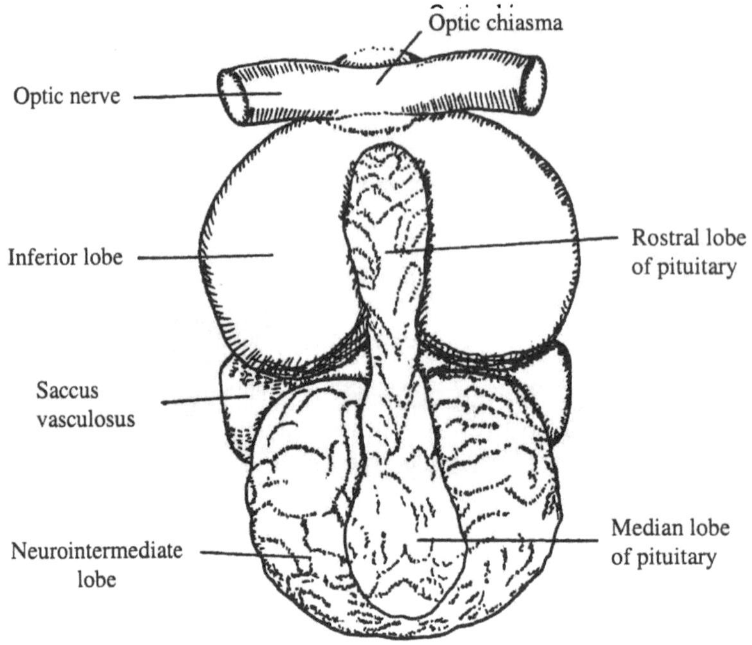

Scyliorhinus canicula: Pituitary gland.

Fig. 13.3 Ventral view of the hypothalamus and pituitary of the dogfish, *Scyliorhinus canicula*, excluding the ventral lobe.
***Source:* Knowles et al. (1975)**

is not present. In adult holocephalans, however, there is a large compact follicular structure called the Rachendachhypophyse (the pharyngeal lobe). It lies outside the cranium in the roof of the buccal cavity and has an independent blood supply. It is derived from embryonic pars distalis rudiment and is ultimately separated from it when a connecting stalk disappears.

The pituitaries of bony fish are shown in Fig. 13.5. The pituitary glands of *Polypterus* and *Calambichthys* are apparently similar in nature. Type A fibres (peptidergic and AF^+) originate in the preoptic nucleus and end in the pars nervosa. The origin of the type B fibres (aminergic and AF^-) is unknown (some suspect it is also the preoptic nucleus), but they terminate in the median eminence. A typical neurointermediate lobe is present. *Calamoichthys* and *Polypterus* both possess a typical median eminence and a portal system and, in this respect, are different from the phylogenetically more recent teleosts. There is no direct innervation of the pars distalis cells, a feature common to teleosts. The most remarkable feature of the Chondrostei is the presence of a duct or canal that connects a cavity in the ventral region of the RPD to the roof of the mouth (the buccohypophysial or orophysial duct). The gland subdivisions are easily discernable and resemble those seen in teleosts.

In Acipensiformes, the most notable anatomic feature is a large hypophysial cavity with many tubular extensions unlike the palaeoniscoids, however, there is no hypophysial duct. A thin-walled neural lobe penetrates the deepest regions of a large pars intermedia. There is no direct innervation of the pars distalis, but there is a well-developed median eminence and portal system.

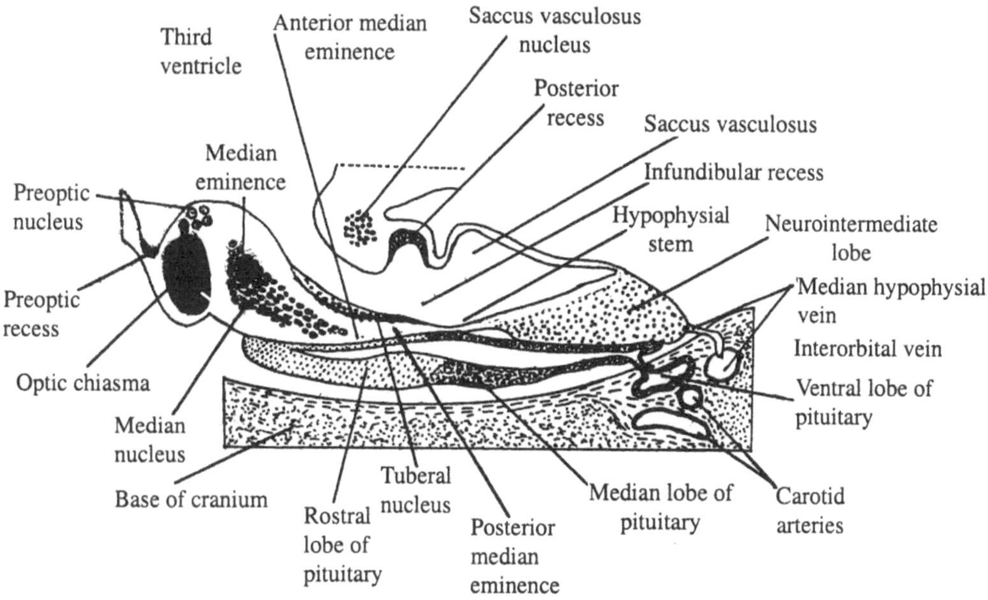

Fig. 13.4 Schematic sagittal section through the hypothalamus, pituitary and cranial base of the dogfish, *Scyliorhinus canicula*.
***Source:* Knowles et al. (1975)**

In Holostei, neither *Amia* nor *Lepisosteus* has a hypophysial cavity or duct as adult. The gland is attached along most of its length to the infundibular floor, and a vascular connective tissue is found in between. There is a large saccus vasculosus. In contrast to teleosts, the pars distalis is not penetrated by large strands of neurohypophysial tissue. A few fibres with neurosecretory materials (alcian blue-positive) extend in CPD. Innervation of pars intermedia of *Amia* is similar to that of other ganoids and primitive teleosts, in that A fibres contact the basement membrane of the intervascular space that separates the neural processes from endocrine cells. Holosteans also maintain a median eminence that is somewhat better developed than their phylogenetic ancestor, the acipenseroids.

Teleost pituitaries in relation to midbrain are shown in Fig. 13.6. The adenohypophysis is clearly divisible into three regions, the RPD, CPD, and pars intermedia (Fig. 13.7). (There is no pars tuberalis). The clarity of this separation is essentially due to a restriction of specific cell types to distinct regions of the gland. In teleosts, there is a structural intimacy of the neurohypophysis with both RPD and CPD as well as with par intermedia, and a unique feature is that the adenohypophysial cells are innervated directly. A few exceptions to the general rule have been noted, and in these forms an elaborate system of extravascular circulation serves to convey neurosecretions to pituitary cells (e.g., *Salmo salar*, *Mugil cephalus*). There is generally no true eminence or portal system present in teleosts.

Variations in the structure of the pituitary gland appear in the axis formed between the pituitary gland and the brain, and thus, in the orientation of the various adenohypophysial regions. Specific cell types may not be found in the same region of the gland in all fishes (e.g., thyrotropes). The arrangement of cells may also vary. For example, the follicles that are common in the Salmonidae, Clupeidae and Anguillidae generally are not found in other fish. A persistent

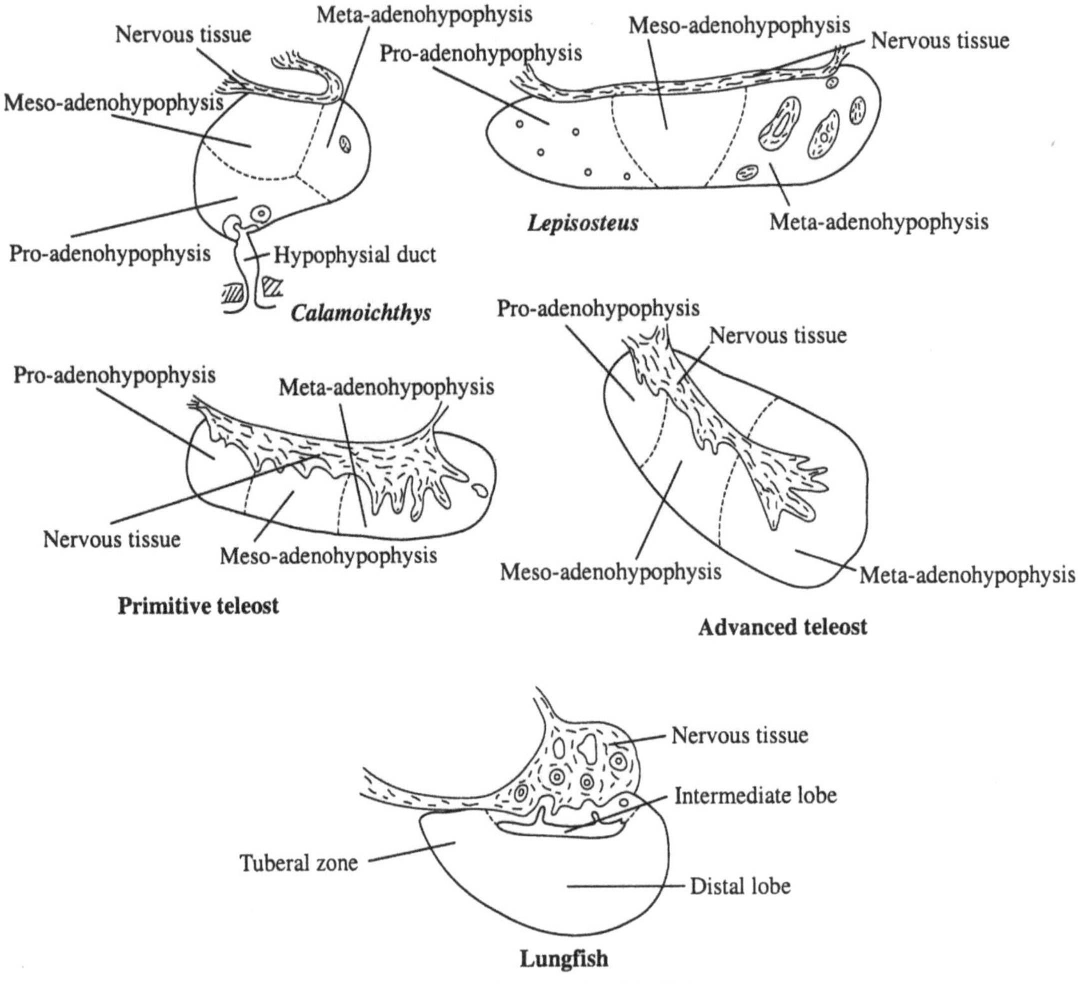

Pituitary gland in Fish.

Fig. 13.5 Pituitaries of fishes in sagittal section. Head end to the left, nervous tissue indicated by short wavy lines. In *Calamoichthys* there is a persistent hypophysial duct passing through the parasphenoid bone and associated with vesicles in the pro-adenohypophysis; the meta-adenohypophysis receives nervous processes from the brain floor but only to either side of the mid-line; the meso-adenohypophysis receives no processes. In the holostean *Lepisosteus* nervous penetration is again restricted to lateral processes into the meta-adehypophysis. In the primitive teleost the three glandular regions still follow one behind the other from infront but in the advanced they tend to enwrap the nervous lobe. In the lungfish the large nervous lobe contains tubes from the infundibulum and diverticula of the hypophysial cavity.
***Source:* Dodd and Kerr (1963)**

orophysial duct may be seen in adult *Chanos chanos* and in some Clupeidae. Different hypothalmic control mechanisms also have been noted among the teleosts.

The neurohypophysis receives neuronal endings from hypothalamic and extrahypothalamic nuclei. The neurohypophysis typically can be divided into two parts, an anterior region with AF^-, type B fibres (and a modicum of pituicytes) in contact with pars distalis, and a posterior

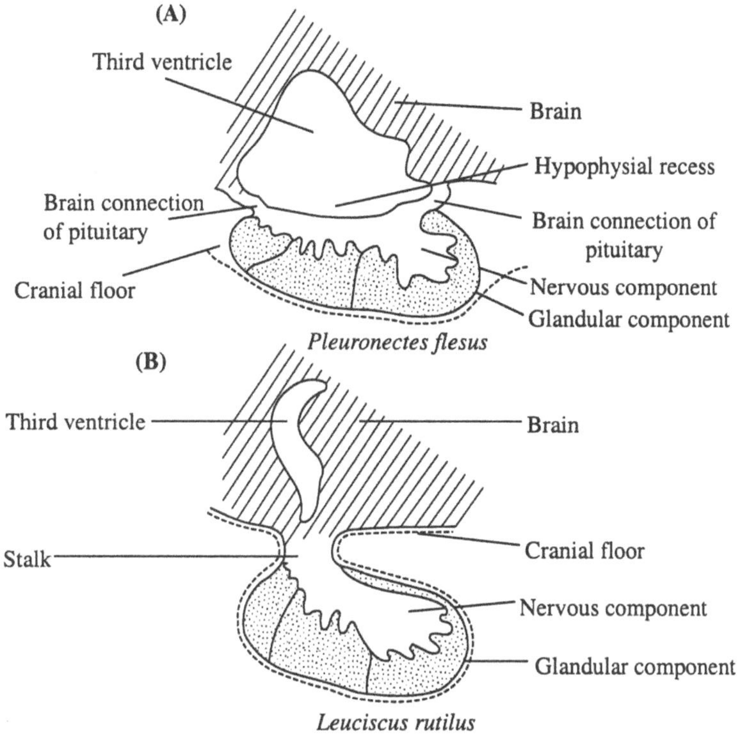

Pituitary gland in Teleosts.

Fig. 13.6 Diagrammatic L.S. of teleost pituitaries showing (A) type with no stalk and open hypophysial recess, and (B) type with stalk and obliterated recess.
***Source:* Kerr (1942)**

region composed mainly of AF^+, type A fibres (but with many pituicytes) in direct association with the pars intermedia. It has been suggested that these two regions correspond to the median eminence and the neural lobe, respectively. However, the existence of a true median eminence in teleosts, as seen in other groups, is a question mark. Although there is the possibility that the neurosecretions could be released into the blood vessels in the neurohypophysis that course through the pars distalis and thus reach pituitary secretory cells, there is no vascular pattern seen that is comparable to that of other vertebrates. A true median eminence is composed of a primary capillary plexus, which then drains to a secondary plexus by way of a portal vein. From a functional point of view, it would be a redundant, unnecessary system, since pituitary cells are directly (or almost directly) innervated by these nerve fibres.

In general, hormone-producing cells of a particular type are restricted to specific regions of the adenohyposis (Fig. 13.7). In platyfish, *Xiphophorus maculatus*, the age at which the gonadotropic zone develops is under genetic control and is dependent on the prior maturation of specific LHRH system in the brain. The number of gonadotropins produced by teleosts appear to be an irreconcilable problem; however, based on immunochemistry, one gonadotrope cell type cell is present in the pars distalis. The pars intermedia generally contains two cell types (PbH^+ type presumed to be the source of MSH; PAS^+ cells, function obscure). In platyfish PAS^+ pars intermedia cells are a second type of gonadotrope called GTH^p cells, which serve to regulate

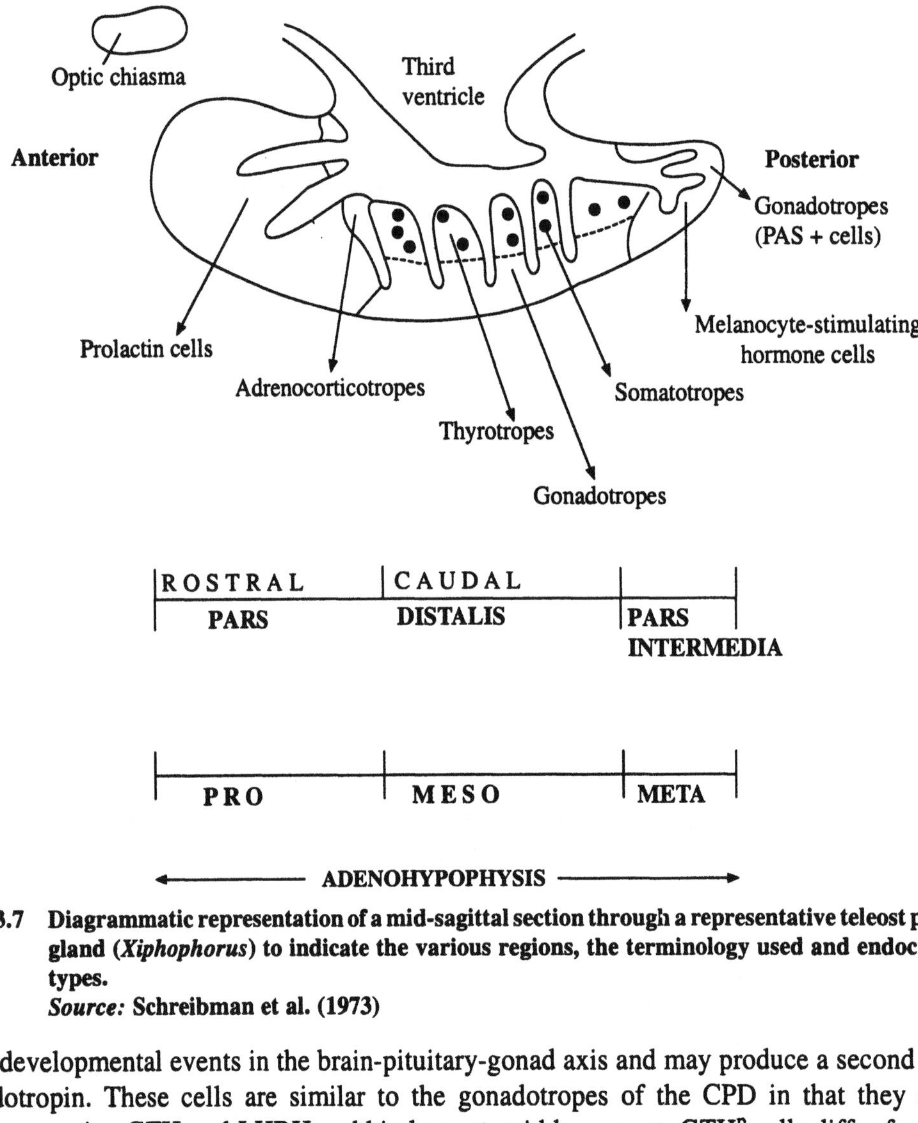

Fig. 13.7 Diagrammatic representation of a mid-sagittal section through a representative teleost pituitary gland (*Xiphophorus*) to indicate the various regions, the terminology used and endocrine cell types.
***Source:* Schreibman et al. (1973)**

early developmental events in the brain-pituitary-gonad axis and may produce a second type of gonadotropin. These cells are similar to the gonadotropes of the CPD in that they contain immunoreactive GTH and LHRH and bind sex steroid hormones. GTH^p cells differ from CPD gonadotropes in their morphology, temporal pattern of proliferation and immunoreactive response to serotonin and tyrosine hydroxylase, as well as to GTH subunit.

In Dipnoi, the neurohypophysis is divisible into a median eminence, infundibular stalk, and neural lobe. The preoptic nucleus is the major hypothalamic nucleus projecting to the neurohypophysis. The projecting neurons are peptidergic and terminate around the portal vessels, the plexus intermedius, the neural lobe, and in the pars intermedia. In *Lepidosiren* many small aminergic neurons whose perikarya lie just below the ependyma of the median eminence and whose axons end in the pars distalis, pars intermedia and the neural lobe close to the pars intermedia have been described.

The adenohypophysis of lungfish is more similar in structure to that of amphibians (urodeles specially) that it is to other fishes. Among the amphibian characteristic of lungfish are a well-formed neural lobe, a prominent median eminence, the absence of a saccus vasculosus, and a less obvious regional distribution of cell types. Nevertheless, certain fishlike characteristics persist. Most notable is the direct aminergic innervation of the pars distalis cells, the interdigitation of the neurohypophysis and pars intermedia, the prominent hypophysial cleft between the pars distalis and the pars intermedia, the absence of a pars tuberalis, and the presence of follicles, especially in *Neoceratodus*. The size of the pars intermedia, the arrangement of its cells, and its structural association with the neural lobe processes varies among the lungfish; however, in all cases it is separated from the pars distalis by the hypophysial cleft.

There is still to be learned about the lungfish pituitary, in the three genera (*Protopterus*, *Lepidosiren*, *Neoceratodus*).

The orthodox features of the coelacanth *Latimeria chalumnae* include a neurointermediate lobe (typically fishlike) ventral to the brain, direct contact between pars distalis and pars intermedia, follicles and tubules containing glycoprotein colloid seen in many fish, a well-developed hypothalamo-hypophysial portal system, and a probable median eminence. Other features that are more "fishlike" include a saccus vasculosus, penetration of neurohypophysial tissue into pars distalis, and regional distribution of the gland based on cell type.

Among the exceptional features is a greatly extended pars distalis that is tripartite in nature. Another orthodox proximal portion lies close to (even interdigitating with) a neurointermediate lobe and can be separated on the basis of vascular supply and cell type into dorsal (rostral) and posterior (caudal) regions. The rostral portion contains orangeophils and erythrosinophils, and orangeophils and basophils populate the caudal region. An elongate extension of the pars distalis comprises its most rostral lobe (buccal portion). What is unique is a long, cylindrical, tubular hypophyseal cavity that extends from the rostral lobe of the pars distalis and contains vascularized masses of adenohypophysial cells, the so-called rostral islets or pars buccalis. These structures are comparable with the Racendachhypophyse of holocephalans rather than with the ventral lobe of elasmobranchs in so far as their anatomical connection is concerned; however, cell types of the islets are similar to those in the ventral lobe.

Thus, the pituitary gland of *Latimeria* shows many specializations and peculiarities. It possesses many features of the teleost and elasmobranchiomorph gland rather than those of dipnoans and amphibians.

The hypothalamic-hypophysial relationships in different fish groups have been illustrated in Fig. 13.8. Chiba and Honma (1989) described the innervation and vascularization of hypothalamo-hypophysial system of the Japanese charr (*Salvelinus leucomaenis*) in relation to its reproduction. Polenov et al. (1997) studied the regeneration of the hypothalamo-hypophyseal system in the sterlet (*Acipenser ruthenus*) after hypophysectomy. Meurling et al. (1996) reported that in cartilaginous fishes, the hypothalamic neurophysin neurons, in addition to ending on systemic capillaries of the neural lobe, could participate in the control of the three lobes of the adenohypophysis (rostral, median and intermediate) by a dual mechanism, i.e., a neurovascular pathway and a direct neural input. The extent of the latter varies considerably between the species investigated.

2.1 Pituitary Hormones

Hormones from pars distalis and cellular types are listed in Table 13.1.

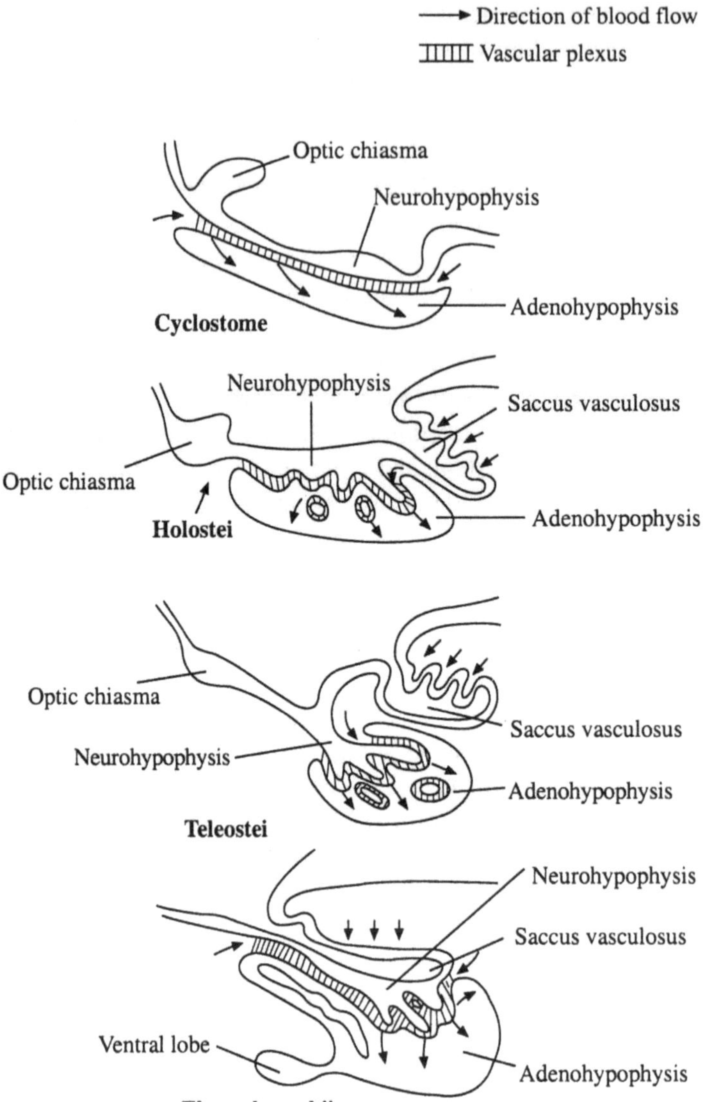

Fig. 13.8 Diagrams of sagittal section through the hypothalamic-hypophysial region of various fish. ***Source:*** **Jorgensen and Larsen (1960)**

TSH: (The Gonadotropin Family)

Peter and Fryer (1983) have discussed data on regulation of thyrotropin. The involvement of nucleus lateral tuberis (NLT) in the thyrotropin inhibitory hormone (TIH) in goldfish *Carassius auratus* has been indicated. It was not possible, however, to correlate the distribution of a particular cell type in NLT of *C. auratus* with the functional region for control of thyroid activity.

The TSH cells in teleosts may be innervated directly by neurosecretory endings, the endings may be adjacent to the cells without synaptoid contact, or the endings may be separated from the TSH cells by a base membrane.

Table 13.1 Cytology of the pars distalis from representatives of different fish groups

Hormone	Class : Group			Cellular names or types
1. Thyroid-stimulating hormone (TSH) Thyrotropin	Chondrichthyes	:	Selachii	Type 1. basophil
	Osteichthyes	:	Holostei	Type 1. basophil
	Osteichthyes	:	Teleostei	Type 1. basophil, delta
2. Gonadotropin (GTH)	Chondrichthyes	:	Selachii	Type 2. basophil
	Osteichthyes	:	Holostei	Type 2. basophil
	Osteichthyes	:	Teleostei	Type 2. basophil, beta, gamma
3. Adrenocorticotropin (Adrenocorticotropic hormone) (Corticotropin) (ACTH)	Chondrichthyes	:	Selachii	Type 3. basophil
	Osteichthyes	:	Holostei	acidophil?
	Osteichthyes	:	Teleostei	Type 3. basophil, epsilon
4. Prolactin (PRL)	Chondrichthyes	:	Selachii	Type 1. acidophil
	Osteichthyes	:	Holostei	Type 1. acidophil
	Osteichthyes	:	Teleostei	Type 1. acidophil epsilon
5. Somatotropin (GH)	Chondrichthyes	:	Selachii	Type 2. acidophil
	Osteichthyes	:	Holostei	Type 2. acidophil
	Osteichthyes	:	Teleostei	Type 2. acidophil, alpha

Note: Acidophils : Type 1 (stain with azocarmine, acid fuchsin, and erythrosin)
Type 2 (stain with orange G)
Basophils : Type 1 (stain with PAS, AF, alcian blue, and aniline blue)
Type 2 (stain with PAS, alcian blue, aniline blue; some classes AF, sometimes orange G granules)
Type 3 (almost chromophobic and often PbH^-)

Abbreviations:
Periodic acid Schiff (PAS)
Alcian blue (AB)
Aldehyde fuchsin (AF)
Lead haemotoxylin (PbH)

Modified after Schreibman (1986)

Although TRH (thyrotropin releasing hormone, the tripeptide pyroglutamyl histidyl proline amide of mammals) is apparently present in teleosts, data regarding its role in regulation of TSH secretion are contradictory.

An annual cycle of thyroid gland activity has been described for some teleosts, for instance, *Salvelinus fontinalis*, implying that the secretion of TSH and the neuronhormones regulating its secretion fluctuate annually. The relationship between seasonal changes in external environmental stimuli and TSH secretion, and the neural mechanisms that regulate TSH secretion need to be investigated.

Thyroid hormone exerts a negative feedback via the hypothalamus and the pituitary.

GTH (or GtH) (The Gonadotropin family)

Peter and Fryer (1983) have described regulation of GtH cells of teleosts. The brain normally has a stimulatory influence on GtH cells, presumably via a gonadotropin-releasing hormone (GRH).

Nucleus lateral tuberis (NLT) pars posterior and NLT pars anterior in the pituitary stalk region are involved in regulation of GtH secretion for gonadal recrudescence. The involvement of the NLT in GRH secretion is also suggested. The role of preoptic region is undecided. In *Carassius auratus*, innervation of GtH cells by both types A and B neurosecretory endings, assumed to originate from the NPO (nucleus preopticus, preoptic nucleus) and NLT, respectively. Another report mention of three cell types in NLT of *C. auratus*, all of which could innervate the pituitary. Cell type III in NLT seems to correspond to the functional area for regulation of gonadal activity; however, the possible relationship of this NLT cell type to regulation of GtH cell activity remains to be more directly investigated. Innervation of GtH cells by type A endings in *C. auratus* suggest NPO involvement in regulation of GtH cell activity. The specific function of each type of neurosecretory ending on GtH cells in *C. auratus* is not known; there is no direct evidence for either GRH or GRIH in these neurosecretory fibres, but the presence of two types of endings does imply a dual neurohormonal control system.

Other studies have also demonstrated direct innervation of GtH cells, but there is large interspecific variation among teleosts in this regard.

Characteristically, teleosts have annual reproductive cycles, spawning at one time or during a certain period of the year. Photoperiod and temperature seem to be primary environmental factors cuing teleost reproduction (see Bhat and Gopal Dutt 1990 for *Puntius sarana*). A neural efferent pathway originating in the preoptic region has been demonstrated.

Gonadotropin secretion is also influenced by the pineal, prostaglandins, and negative and possible feedback effects of steroid hormones.

Ishii (1991) has given an account of fish gonadotropins. Findings of different groups of investigators vary, for example, (i) single type gonadotropin stimulates various aspects of the gonadal function such as follicular growth, steroidogenesis, vitellogenesis, oocyte maturation, ovulation in females, and spermatogenesis and androgen production in males, (ii) two types of gonadotropin, named maturional gonadotropin (or carbohydrate-rich gonadotropin), and vitellogenic gonadotropin (or carbohydrate-poor gonadotropin) according to their biological and chemical properties, (iii) two types of gonadotropins, GTH I and GTH II, from salmon pituitary glands, have partially different amino acid sequence. GTH I and GTH are contained in different cells in the adenohypophysis of the salmon, (iv) tuna (*Thunnus albaceus*) has two types of gonadotropin-like glycoproteins. First is active on the testis of homologous species and inactive on the goby testis, and the second is inactive on the tuna testis but active on the goby testis. The second, or heterologous, type gonadotropin seems to occur in many other teleost species. For further information and discussions, see also Nozaki et al. (1990), Schreibman et al. (1990), Fontaine et al. (1991), Kawaguchi et al. (1991), Swanson (1991), McDonald and Milligan (1992), and Nagahama et al. (1994). Illustrations in this respect are given by Van der Kraak et al. (1998): Fig. 1 shows the neuroendocrine pathways involved in the control of pituitary GTH-II secretion in teleosts, and Fig. 2 shows the intracellular signaling pathways involved in the regulation of GTH-II release in teleost pituitary gonadotrophs.

There has been only a single original paper on the purification of gonadotropin in elasmobranchs and none on cyclostomes.

King and Millar (1991) have described gonadotropin releasing hormones (peptides). cGnRH-II almost universally present in gnathostome fish, and sGnRH also occurs in most species investigated. Multiple forms of GnRH were described in various species of bony fish, on the basis of immunological and chromatographic criteria. In most teleost fish studied, sGnRH occurs together with cGnRH-II. mGnRH and a peptide similar to cGnRH-I are present in several species. Among primitive species of bony fish, mGnRH occurs together with cGnRH-II. There is also evidence for novel forms of GnRH in species of Osteichthyes.

In Chondrichthyes, cGnRH-II and sGnRH have been detected as well as novel forms that differ from the known GnRHs. GnRH is present in cyclostomes.

The involvement of calcium in the mediation of GnRH action has been established in several species of fish including tilapia and African catfish (*Clarias gariepinus*).

In goldfish, both cGnRH-II and sGnRH are released from the preoptic area and the pituitary. Thus, in fish two forms of GnRH may regulate pituitary hormone secretion. Whether several forms of GnRH are involved in the physiological control of pituitary gonadotropic function is equivocal. Indeed, GnRH has been shown to stimulate not only gonadotropin secretion but also the release of growth hormone in goldfish.

Pituitary GnRH receptors differ among the different vertebrate classes. In all species, the receptor has a specific requirement for the NH_2– and COOH-terminal sequences of GnRH.

Peter et al. (1990) have discussed the actions of GnRH in the goldfish. GnRH receptors are located on both gonadotrophs and somatotrophs, and that the two cell types do not have communication. The differences in responsiveness of GtH and GH release to sGnRH and cGnRH-II suggests that there may be some differences in the GnRH receptors on these two pars distalis cell types. Further, results indicate the presence of two separate high affinity GnRH receptors, perhaps one originating from gonadotrophs and one from somatotrophs. A number of questions remain for investigation with this unique neuropeptide system. Habibi and Peter (1991) have advanced our understanding on GnRH receptors in teleosts. Peter et al. (1991) reported the actions of catecholamines and sex steroids in regulation of gonadotropin-II (GtH-II) in the goldfish (Fig. 13.9).

Sower (1990) has reviewed recent data on GnRH in relation to reproduction function in the primitive groups of agnathan and chondrichthyan fishes. GnRH as a stimulator of the pituitary-gonadal axis appears to be conserved through vertebrate evolution. In most vertebrates, GnRH released by the hypothalamus reaches the anterior pituitary by a hypophysial portal system and in teleost fishes by direct innervation. Unlike most vertebrates there is the absence of a hypothalamo-hypophysial link of either a hypophysial portal system or nervous connections between pituitary (gonadotropic region) and hypothalamus in agnathan and chondrichthyans. This suggests that GnRH reaches the pituitary either by a systemic route or by simple diffusion. Hypothalamus plays little or no role in reproductive control in hagfish. In lampreys, unlike the hagfish, there is immunocytochemical evidence for hypothalamic influence on the pituitary by diffusion of GnRH from the neurohypophysis to the pars distalis of the pituitary. GnRH-like neurons identified by immunocytochemistry project their fibres primarily into the neurohypophysis from the preoptic region. There is evidence of normal occurrence of GnRH in a part of lamprey brain homologous with that brain region in higher vertebrates in which GnRH localization forms part of a neuroendocrine mechanism for gonadotropin secretion.

In chondrichthyans, there is a hypophysial portal system between the hypothalamus and pituitary except to the lobe of pituitary that is considered to be the gonadotropic region; the ventral lobe (in elasmobranchs) and the buccal lobe (in holocephalans) are separated from the rest of the pituitary. It has ben reported that GnRH must reach these lobes by peripheral circulation.

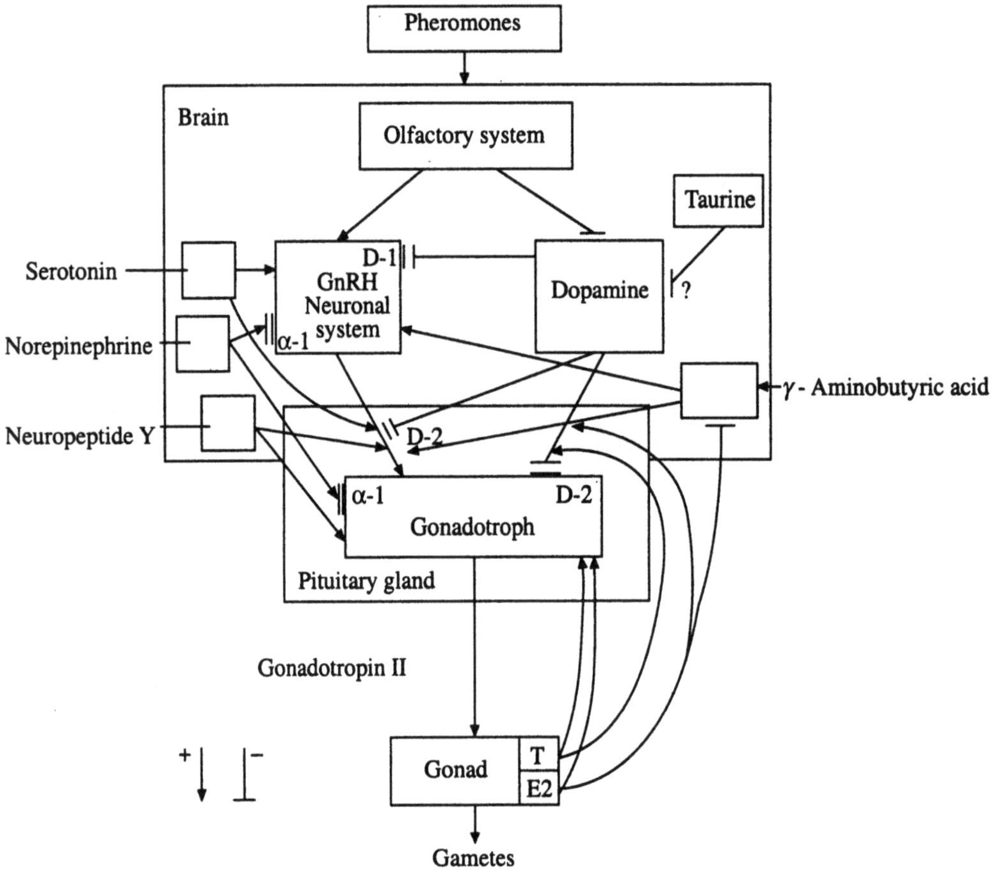

Fig. 13.9 Model for the neuroendocrine regulation of gonadotropin-II release in teleosts. A line with an arrow indicates a stimulatory effect; a line with a bar indicates an inhibitory effect. Abbreviations: α1-noradrenergic receptor, α1; estradiol, E2; Testosterone, T; type-1 dopamine receptor, D-1; type-2 dopamine receptor, D-2.
***Source:* Peter et al. (1991)**

There has been only one reported immunocytochemical study of GnRH distribution in the brain in chondrichthyans.

With the exception of hagfish, evidence has been presented for the existence of a gonadotropic hormone in the pituitary gland of species of agnathans and chondrochthyans. In hagfish, there is no evidence as yet for a role of GnRH in stimulating reproductive activity. In lampreys, there is evidence for the regulatory influence of the hypothalamus on the pituitary-gonadal axis. In chondrichthyans, there are very few studies on the biological activity of GnRH. The role of GnRH in these primitive fishes is far from understood. Fontaine et al. (1991) have discussed the anatomy of hypothalamus-pituitary connection, with emphasis on GnRH, and multiendocrine cerebral control. In teleosts, GTH cell is an important site of integration for neuroendocrine signals. The steroid control has also been addressed.

The reader interested in details of phylogeny and structure-function relationship of vertebrate gonadotropin-releasing hormones may refer to Peter et al. (1987). Contributions (Omeljaniuk et al. 1989; Anderson et al. 1991; Sherwood and Coe 1991) have advanced understanding on GnRH.

Schreibman et al. (1990) have discussed the brain-pituitary-gonad axis in teleosts, especially the platyfish, *Xiphophorus maculatus*. There is complex system of interactive events and control mechanisms involving brain, pituitary gland, and gonad that characterized specific periods in the life of an organism (Fig. 13.10). A variety of signals from external and internal milieus, of both short- and long-term duration, are channeled directly or indirectly into the GnRH containing centres of the brain.

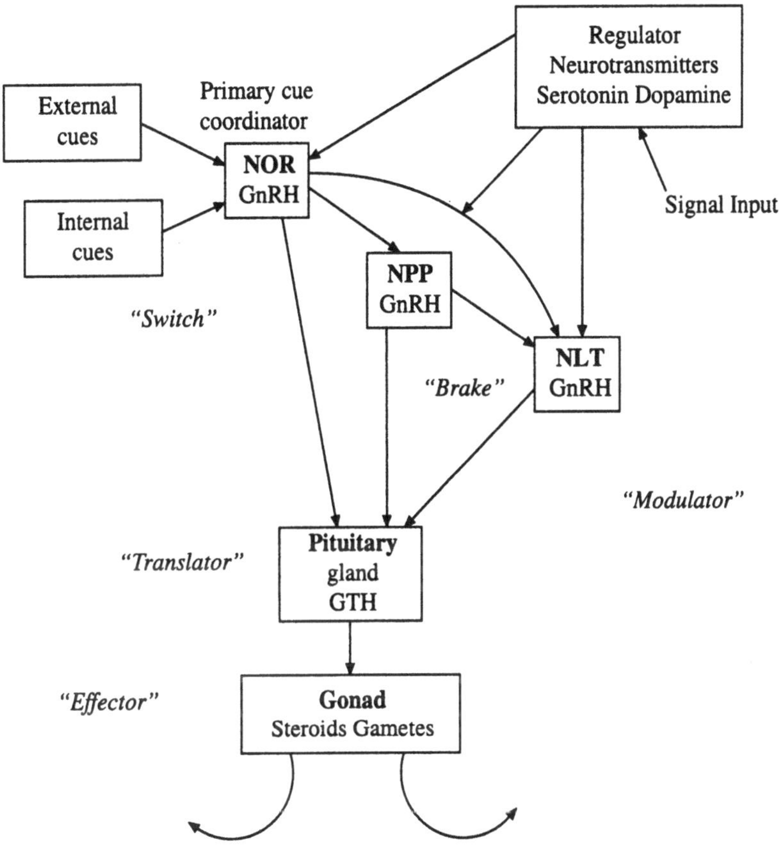

Fig. 13.10 This diagram suggests a structural and functional relationship among the various components of BPG axis, and the select sites where a variety of cues may affect this axis in a teleost. The black arrowheads point to suspected primary, preferred pathways of interaction.

Maturation of the reproductive system is characterized by a sequential appearance of three ir-GnRH containing regions of the brain (i.e., the nucleus olfactoretinalis (NOR) first, followed at puberty by the nucleus preopticus periventricularis (NPP) and then the nucleus lateralis tuberis (NLT).

***Source:* Schreibman et al. (1990)**

ACTH: The POMC (proopiomelanocortin)-Gene Derived peptide

The secretion of ACTH from the pituitary is subject to stimulatory influences of the hypothalamus. The presence for a teleost corticotropin-releasing factor (CRF) has been reported. The nature of this teleostean CRF is not known. It has been suggested that NLTa (nucleus lateral tuberis pars anterioris) and rostral NLTp (nucleus lateral tuberis pars posterioris) is the source of CRF.

NLT cell type involved in regulation of ACTH has not been defined. Neurohypophysial peptides may play a role in regulating ACTH secretion in teleosts. An inhibitory involvement of the epithalamus in the regulation of ACTH secretion has been suggested. It has been demonstrated in *C. auratus* that corticosteroids exert feedback effects on brain to suppress ACTH secretion. Corticosteroids act on the NLT to inhibit the synthesis and/or release of CRF. In teleosts, hypothalamic afferent pathways which may influence ACTH secretion have not been investigated. There is little information concerning neural control of daily rhythms of adrenocorticoids in teleosts. Activation of the pituitary-interrenal axis of teleosts has been observed during spawning migrations, during various disease conditions, and osmotic challenges (Peter and Fryer 1983).

PRL: (The Prolactin family)

The PRLs purified to date have been found to have an apparent molecular weight of 22–23 kD, and isoelectric points between 7.3–10.3. An exception to this is the 20 kD tilapia PRL variant which has a pI of 6.7 (see Rand-Weaver and Kawauchi 1993; read also for structure).

Prolactin (PRL) and growth hormone (GH) are simple protein hormones synthesized and secreted from the adenohypophysis of the pituitary in most vertebrates. The considerable structural homology between PRL and GH has led to the suggestion that they evolved from a single ancestral gene by gene duplication. PRL has been well known as the most versatile of all the pituitary hormones in its action in vertebrates, whereas GH has been considered to be almost exclusively concerned with stimulation of animal growth. Recently, PRLs have been isolated and characterized from pituitaries of chum and chinook salmon (see Hirano et al. 1987). No studies have been conducted on the actions of PRL in seawater fish, although it is clear that PRL cells are present as shown by immunocytochemistry (see Rand-Weaver and Kawauchi 1993).

As shown in Table 13.2 the effects of PRL on teleost osmoregulation are also manifold, reflecting its versatile nature in physiological processes.

Prunet and Auperin (1994) state that the first event in PRL action is its binding to a specific receptor located on the cell membrane. Multiplicity of PRL action should be reflected at the receptor level. This chapter on PRL details the biochemical characteristics of fish PRL, the characterization and regulation of PRL receptors in tilapia, and finally the biological interpretations of these new data on PRL receptors.

Of the many hormones produced by the pituitary gland in fish, three, in particular have been considered as candidates for the hypercalcemic action of this gland. PRL and ACTH, produced in the cells concentrated in the rostral pars distalis of the pituitary, and the putative hormone produced by the PAS-positive cells of the pars intermedia (DIPAS cells) (see Wendelaar Bonga and Pang 1989). The reader is also referred to a review by Wendelaar Bonga and Pang (1991) on control of calcium regulating hormones in vertebrates.

These are evidences for the importance of PRL in maintaining hydromineral balance in fish. Gills, skin, opercular membrane, kidney, urinary bladder and intestine are PRL targets in teleosts. A common action of PRL in teleost osmoregulation is that of reducing epithelial permeability. Supporting this view are facts that (a) PRL decreases osmotic water influx across the intestines of seawater eels and trout, and across the urinary bladder of seawater goby, and (b) PRL decreases passive Na^+ loss and Cl^- extrusion from the gills, and NaCl absorption by the intestine, in the eel and trout. By contrast, PRL, possibly in conjunction with corticosteroids, stimulates Na^+ transport in the flounder and goby urinary bladder and stimulates Na^+ and water absorption by the seawater tilapia intestine.

Table 13.2 Actions of prolactin in teleosts

Osmoregulation	
	Survival of hypophysectomized fishes in fresh water
	Decreased gill Na efflux
	Decreased gill osmotic water permeability
	Restoration of plasma Na in hypophysectomized eels when given with cortisol
	Inhibition of gill Na, K-ATPase and stimulation in kidney
	Increased skin and gill mucus secretion
	Decreased ion and water absorption from intestine
	Diuresis and increased glomerular size
	Decreased water and increased ion absorption from bladder
	Hypercalcemia
Reproduction	
	Parental behaviour (nest building, fanning, buccal egg incubation)
	Seminal vesicle secretion
	Maintenance of brood pouch in male seahorse
	Gonadotrophic or antigonadal effects
Growth	
	Proliferation of mucus cells and melanocytes
	Seminal vesicle growth
	Renal glomerular growth
	Stimulation and proliferation of renal tubules
Metabolism	
	Increased muscle free fatty acid
	Increased or decreased body fat stores
Steroid synergism	
	No retention by gills, kideny and urinary bladder (corticosteroids)
	Decreased ion and water absorption from intestine (corticosteroids)
	Parental behaviours (sex steroids)
	Maintenance of brood pouch in seahorse (corticosteroids)
	Stimulation of seminal vesicles (androgen)
	Fat metabolism (corticosteroids)

Source: Hirano (1986)

Besides, (1) it increases skin and gill mucus secretion, (2) it increases glomerular and renal tubule growth and (3) it maintains whole body TEP (transepithelial potential) in *Gillichthys* (Brown and Brown 1987).

The major stimulus to prolactin release appears to be a reduction of plasma osmolarity, and/or specific ions such as Ca^{2+}, or both. Nevertheless, PRL levels can change independently of changes in plasma osmolarity or electrocytes. Prolactic levels rise in response to stress in either fresh water or seawater and decline with smoltification in fresh water. The high variation in plasma PRL among individual fish (all collected at similar times of the day) suggests pulsatile release of the hormone. There is some evidence of a marked difference between the sexes of PRL release. Plasma PRL levels rose about 4-fold higher in mature female chum salmon compared to males when both were transferred from seawater to freshwater (see McDonald and Milligan 1992; see also Rand-Weaver and Kawauchi 1993).

Investigation on prolactin release in the tilapia (*Oreochromis mossambicus*), has indicated that Ca^{2+} loss may not always reflect intracellular events that lead to PRL release (Richman et al. 1991). Mudminnow *Umbra pygmaea* is extremely acid tolerant, having an optimum water pH range of about 3.5 to 6.0, which is consistent with ecological observations. Prolactin secretion is at a minimum in this range. Exposure to neutral water represents osmoregulatory stress for the mudminnow, in contrast to all other teleost species examined so far (Wendelaar Bonga et al. 1990).

Prolactin secretion is regulated mainly by a release-inhibitory hormone, possible dopamine, from the nucleus lateral tuberis, and by the effects directly on the prolactin cells of changes in osmolality of the medium (Peter and Fryer 1983).

GH: (The Prolactin family)
The isolated proteins have been characterized by gel electrophoresis to have a molecular weight of 21–22 kD and isoelectric points between pH 5.6 and 7.1. An exception to this is Atlantic salmon GH which has been shown to be phosphorylated and so exhibits an isoelectric point between pH 6.5 and 8.2, the latter being the dephosphorylated form. Although a consensus sequence for N-linked glycosylation has been found in fish GHs there is presently no evidence that the hormones contain carbohydrate.

A characteristic of the members of the prolactin gene family is their pleitropic nature. It is a protein of less than 200 amino acids which contain four half-cysteine residues (see Rand-Weaver and Kawauchi 1993). In fish growth hormone (GH) acts as a growth-promoting hormone, but it also appears to be involved in osmoregulation, stress adaptation and reproduction. Growth, osmoregulation, stress adaptation and reproduction are strongly interdependent in fish, and all these processes depend on a well-regulated calcium balance; Ca^{2+} uptake via the gills is pivotal in calcium handling by fish. The interdependence of these processes and their dependence on GH may underlie the pleiotropic characters of the effects of GH in fish. A pivotal target for GH is the liver, where it stimulates receptors with nanomolar affinity. Liver cells activated by GH produce a fish insulin-like growth factor (IGF, probably IGF-I) that may subsequently stimulate growth. IGF also has osmoregulatory actions, as has recently been detected in rainbow trout (*Oncorhynchus mykiss*). The demonstration of GH receptors in rainbow trout gills and of an osmoregulatory role of IGF suggests that an analogous 'GH-IGF system' is present in gills and liver. The presence of GH receptors on ionocytes is in good agreement with the consensus that GH exerts osmoregulatory effects in seawater-adapting and seawater-adapted salmonids, and this action could be independent of IGF production by the liver (Fig. 13.11). Further, a comparison of the roles of GH and prolactin in calcium regulation of the tilapia has shown that GH has specific calcitropic effects on freshwater tilapia that differ from those of prolactin (see Flik et al. 1993). Details on the structure and evolution of GH can by obtained from a comprehensive review from Chen et al. (1994). The reader is also referred to Takei (1993) on the role of peptide hormones in fish osmoregulation (Fig. 13.12).

McDonald and Milligan (1992) have briefly dealt with levels of growth hormone (GH) in blood of teleosts in different stages: resting, exercise, seawater transfer, generally showing elevation trend. There is also a consistent relationship between GH levels and the age (size) of the fish, with small fish invariably showing much higher GH circulating levels. In addition to daily variations, there are variations associated with changing season and daylength. In goldfish,

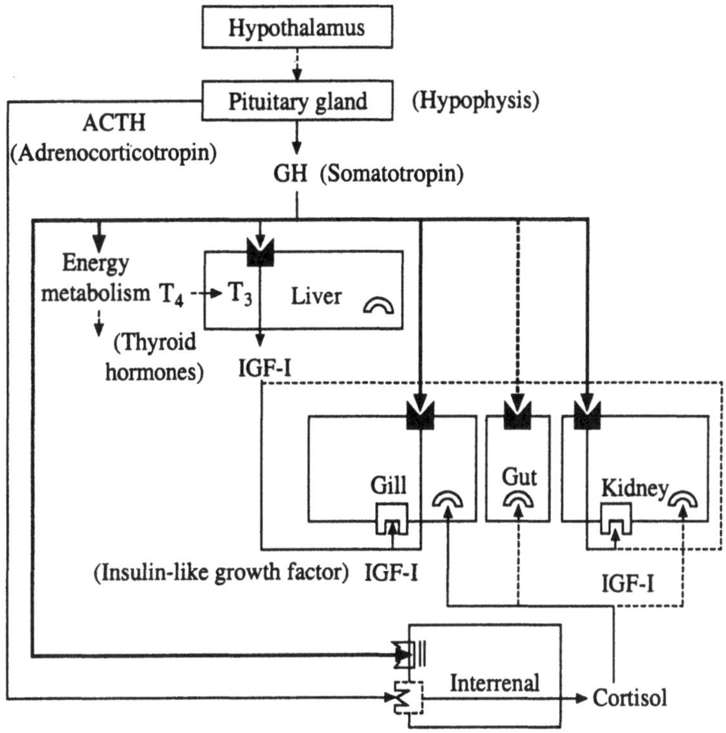

Fig. 13.11 Summary representation of GH action during seawater acclimation of salmonid fish. Continuous lines (arrows) denote confirmed pathways and broken lines possible pathways. Cortisol receptors are present in the gill, liver, intestine, kidney and muscle of brook trout and Atlantic salmon (Chakraborti *et al.* 1987; Chakraborti and Weisbart 1987; Weisbart *et al.* 1989). IGF-I receptors have yet to be identified. Response in the gill is reflected in stimulation of Na^+, K^+ -ATPase activity and chloride cell differentiation.
***Source:* Sakamoto et al. (1993)**

Carassius auratus, the seasonal peak in plasma GH levels occurs in spring to late summer where levels of GH are about 2-fold higher than seasonal minima in November.

There is evidence in several salmonid species that GH may facilitate seawater adaptation. In salmonids GH may have independent actions on growth and plasma electrolytes (Brown and Brown 1987). GH also has been implicated in the inhibitory control of antifreeze protein production in winter flounder and in gonadal steroid synthesis (see Wendelar Bonga 1993).

Peter and Fryer (1983) reported that there is little known about the control of GH in teleosts. A role for the hypothalamus in the control of GH secretion has frequently been suggested, but has not been defined. Little is known about afferent pathways to the teleost hypothalamus, which may be involved in controlling GH secretion, but evidence suggests a variety of inputs. A neural input is suggested by the diurnal variation in plasma GH in *Oncorhynchus nerka*. In addition, several species of teleosts exhibit seasonal growth patterns which are sensitive to photoperiod manipulations. A correlation exists between seasonal growth with changes in the pituitary cytology of GH cells in *Salmo salar*. The GH cells exhibited hyperplasia and hypertrophy during the rapid growth phase of salmon parr, which occurs under long-day conditions. In *Anguilla anguilla* and *Cyprinus carpio* the GH cells become highly active during periods of starvation. Also, there is a reduced number of GH cells in female *A. anguilla* at sexual maturity. For each of these

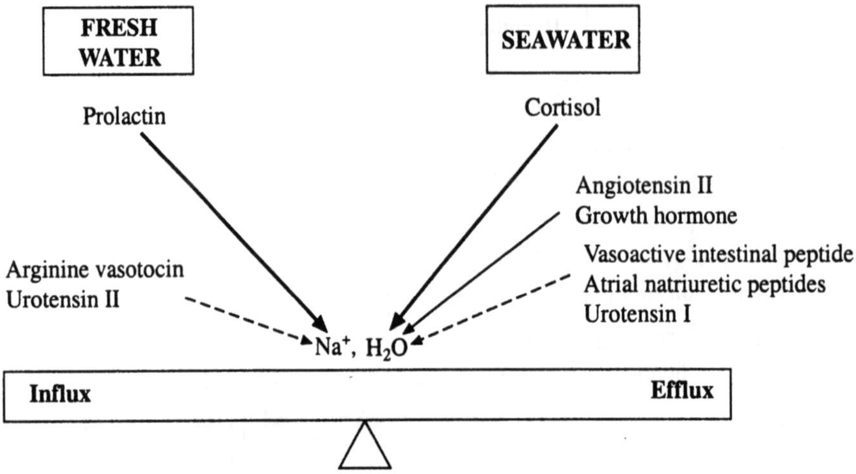

Fig. 13.12 Diagram to show how hormones interact to adjust influx and efflux of Na^+ and water to enable fishes to adapt in either fresh water or seawater. The relative contribution of each hormone to the adaptation, which can be assessed from the data accumulated thus far, is expressed by the thickness of the line and broken lines indicate that the role of the hormone is far from being established.
***Source:* Takei (1993)**

observations, however, the pathways and mechanisms that may regulate GH secretion have not been defined. Details of growth hormone and IGFs can be read in Mommsen (1998) and Eliès et al. (1999).

Somatolactin *SL: (The Prolactin family)*
Somatolactin, a hormone of the somatolactin cells of the pars intermedia cells, has structural homology with prolactin and growth hormone. Its function is unclear. The somatolactin cells of teleosts respond to a variety of stimuli, such as the calcium concentration, pH, salinity, change in background color (read Rand-Weaver and Kawauchi 1993; Wendelaar Bonga 1993). The primary structures of SL from three teleosts, flounder, chum salmon and cod, have been determined. The hormone consists of 207–209 residues; flounder SL being two amino acids shorter than cod and chum salmon SLs. In contrast to GH and PRL, the SLs can be aligned without the introduction of gaps (read details: Rand-Weaver and Kawauchi 1993). SL belongs to GH/PRL family. Presently GH genes from rainbow trout, Atlantic salmon and carp, the PRL gene from carp and the SL gene from chum salmon have been analyzed (see Rand-Weaver and Kawauchi 1993; refer Ono and Kawauchi 1994 for details of SL gene).

MCH:
Baker (1991) has reviewed data on melanin-concentrating hormone (MCH)—a neuropeptide. In teleost fish, most of the MCH activity is found in the neurohypophysial lobe where the peptide is stored, ready for release into the blood. The relative abundance of MCH in the brain and pituitary depends on a number of factors including the method used to measure the hormone. MSH (melanocyte stimulating hormone) secreted by intermediate lobe of pituitary is the functional antagonist of MCH on melanophores. Age and background color are important factors.

The MCH neurons appear to be bi-or multipolar cells, sometimes with branching fibres. Most of the axons entering the pituitary gland terminate in the neural lobe but a few penetrate the pars distalis, coursing near the corticotropes and other cells. No pituitary portal system exists in teleosts but pituitary cells are controlled by hypothalamic neurons which terminate on or near the secretory cells. The presence of MCH fibres in the pars distalis could indicate a potential regulatory effect on one or more pituitary cell types. MCH fibres in the neurohypophysial lobe concentrate near major blood vessels and also in the case of eel, terminate on the vascular basement membrane which separates the neural and intermediate lobes. This location permits MCH access to the general circulation and also to cells of the pars intermedia, which accords with the observation that MCH can exert an inhibitory effect on the melanotrope cells. In some species (e.g., carp, *Poecilia*), in which a basement membrane is less developed, or in the trout in which fibres can penetrate the base membrane, the MCH nerve terminals make direct contact with the pars intermedia cells.

In all nonteleostean vertebrates that have been studied, the MCH perikarya are located more centrally within the hypothalamus and it seems likely that the ventral situation of these neurons in teleosts is associated with their strong axonal projection to the pituitary gland. The hypothalamo-hypophysial component is usually much less well developed in nonteleostean vertebrates, while the central projections may be very extensive.

In lampreys, MCH neurons can be detected even in young ammocoetes (larvae) and become more apparent with increasing age. The cell bodies are located in a single group near the third ventricle of the hypothalamus, level with pituitary intermediate lobe. Many neurons are strikingly bipolar, one axon making contact with the ventricular cavity while the other is directed laterally into the brain. The most prominent axonal tracts are seen in the lateral hypothalamus, projecting backward down the spinal cord and forward toward the optic chiasma and olfactory lobes, and some apparently crossing from side to side above the optic chiasma. Less prominent axonal tracts are directed dorsally toward the region of the habenular nuclei. Immunoreactive MCH fibres are evident in the neurohypophysial lobe of the ammocoete but are less obvious in this location after metamorphosis. Immunoreactive MCH cell bodies in elasmobranchs (dogfish *Scyliorhinus canicula*) are found in the nucleus saccus vasculosus, nucleus tuberculus posterioris, and nucleus lateralis tuberis. No MCH fibres have been detected in the pituitary neural lobe, but they project throughout the posterior hypothalamus, and are found also in the dorsal thalamus and central grey of the mesencephalon.

Terlou et al. (1985) reported in the rainbow trout, *Salmo gairdneri*, perikarya of the lateral part of the nucleus lateralis tuberis (NLT-pl), located in the ventral hypothalamus, are connected via their axons with the neurointermediate lobe of the pituitary. Immunoreactive material produced in the NLT-pl cells represents the melanin-concentrating hormone.

Considering the physiology of MCH (as a color-change hormone in fish), it is already known that several types of chromatophores are present in the skin, each distinguished by the pigment they contain (see Chapter 3). The pigmentary effects of MCH are not confined to the melanophores; the hormone will also cause aggregation of the red and yellow pigment in the erythrophores and xanthophores, but because of the greater difficulty in observing these cells, they have received relatively less attention.

In all teleosts examined, MCH and MSH have mutually antagonistic effects on melanin dispersion, the end result depending on the concentration of either hormone. Fish may differ in their use of two hormones. Apart from its direct effect on the melanophores, MCH also influences pigmentation directly, by depressing the release of MSH from the pars intermedia of the pituitary gland.

The only other vertebrates as yet found capable of responding to MCH by melanin concentration are holostean fish, which currently are considered to be closely related to the teleosts.

MCH plays a modulatory role in the stress response of fish. Secretion of MCH can be enhanced by repeated stress. The concentration of MCH available for release in the brain can be influenced by factors such as background color and stress, which regulate the release of MCH from the pituitary. While this does not mean that neuromodulatory MCH is necessarily released within the brain in parallel with hormones release from the pituitary (although this might be the case), it tends to suggest that the neuromodulatory MCH system and the neurohypophysial system are closely connected. The results could support the idea that MCH fibres projecting to the neurohypophysis and to the brain may arise from the same perikarya.

MSH: The POMC (proopiomelanocortin)-Gene Derived peptide

The teleost pars intermedia contains two types of secretory cells: (1) PAS+ve and PbH-ve and (2) PbH+ve and PAS-ve. Not all teleosts display these typical cells types, however, as salmonids appear to have only PbH+ve cells. Evidence from several sources indicate that the PbH+ve cells are the source of MSH, which induces melanin dispersion in the melanocytes and darkening of cells. Immunofluorescence studies have demonstrated the presence of MSH and/or its precursor ACTH in the PbH+ve cells of several species of teleosts (e.g., *Gasterosteus aculeatus, Salmo gairdneri*). Hypothalamic involvement in control of two cell types of the pars intermedia has been suggested. Melanocyte-stimulating hormone-releasing hormone (MRH) or a melanocyte-stimulating hormone release-inhibiting hormone (MIH) are discussed.

The neuroendocrine control of teleost pars intermedia shows a considerable anatomical variation. For example, in *Anguilla anguilla, Perca fluviatilis*, and *S. gairdneri*, neurosecretory axons do not penetrate the pars intermedia but terminate in extravascular channels bordering the pars intermedia, or terminate at the basement membrane and its network in the neurohypophysis. In other teleosts the pars intermedia cells may be directly innervated by the neurosecretory axons. In *Carassius auratus, Hippocampus* sp., MSH-secreting cells are doubly innervated by both type A and type B fibres.

A variety of evidence suggests that MSH secretion in teleosts may be inhibited by a catecholaminergic mechanism. It is suggested that a catecholaminergic mechanism inhibits the release of MSH within the neural lobe. This action may be direct, in which case a catecholamine may be MIH. Alternatively, the catecholamine may influence MSH secretion indirectly, by promoting the release of an MIH, or by inhibiting the relese of MRH from nerve terminals within the pars intermedia. The involvement of hypothalamus, and nucleus preopticus and adrenocortical hormones has been suggested. Several histological studies have noted a stimulation of the pars intermedia associated with reproduction. Although these observations suggest that the pars intermedia plays a role in reproduction, the functional significance of this activity and the pathways that initiate the activity are not known (see Peter and Fryer 1983).

Figure 13.13 shows the target organs and functionally linked sequence of events related to pituitary hormones.

3. NEUROHYPOPHYSIS

Goossens et al. (1977), Braun and Dantzler (1987), Perks (1987), Guibbolini et al. (1988) and Hazon and Balment (1998) have discussed various aspects of neurohypophysial hormones (see also Urano et al. 1994; Fig. 1).

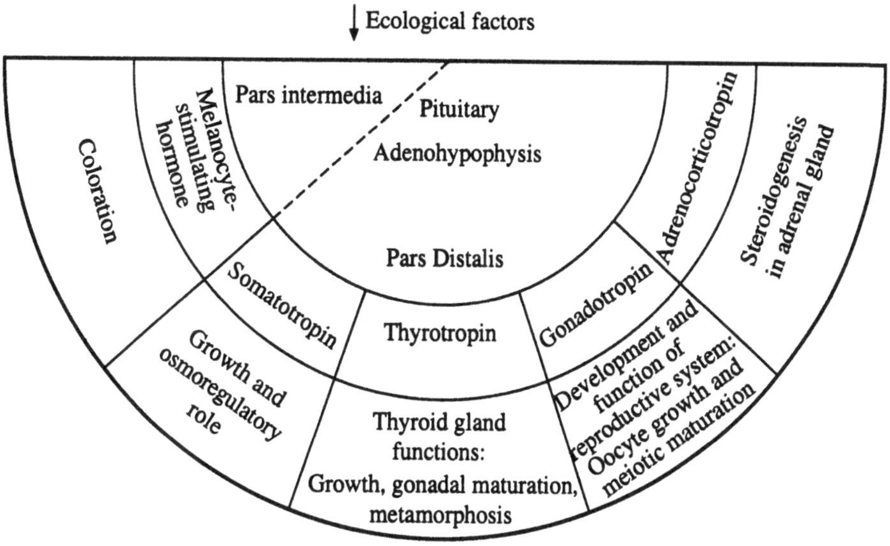

Fig. 13.13 The regions of the pituitary gland, hormones and target organs or processes in teleosts. Each radial segment represents a functionally linked sequence of events.

The teleost fish neurohypophysial peptide hormones, arginine vasotocin (AVT) and isotocin (IT) (nonapeptides) are synthesized by separate hypothamic neurosecretory neurons (Fig. 13.14). The axons of these neurons terminate in the neurohypophysis, where AVT and IT are released into circulation. AVT the non-mammalian counterpart of the well-known vasopressin, is believed to play a role in the maintenance of salt and fluid balance, cardiovascular activity, endocrine secretion, reproduction, and probably in neurotransmission and neuromodulation processes in the central nervous system in teleosts. While AVT displays antidiuretic activity in birds and Amphibia, the results for fish are inconsistent. Arginine vasotocin was reported to be diuretic or antidiuretic depending on the dose. Although it is known that synthesis and secretion of AVT is sensitive to environmental stimuli, the precise role of AVT in fish osmotic adaptation mechanisms remains unclear. There are a few studies reporting variations in plasma AVT levels with respect to salinity. Moreover, changes in the pituitary AVT content during osmotic stress has been

Arginine vasotocin (AVT)

Cys – Tyr – Ile – Gln – Asn – Cys – Pro – Arg – Gly – NH_2

Isotocin (IT)

Cys – Tyr – Ile – Ser – Asn – Cys – Pro – Ile – Gly – NH_2

Fig. 13.14 Amino acid sequences of the natural neurohypophysial hormones in teleost fish. ***Source:*** **Kulczykowska (1995b)**

described. Much less is known about the physiological significance of IT. To date measurements of plasma isotocin concentrations are limited.

AVT is found in all vertebrates, but IT is restricted to teleosts. It has been suggested that AVT is probably the ancestral neurohypophysial peptide. According to Suzuki et al. (1994), the lamprey VT precursor is more closely related to gnathostome VT and VP precursors than to the hagfish VT precursors.

The synthesis and/or release of AVT and IT are controlled independently, so that these nonapeptides have different physiological roles in teleost fish. AVT might participate in circadian time-keeping system in fish (Kulczykowska and Stolarski 1996; read Kulczykowska 1997).

To know about the precursors of these hormones, the reader is referred to Heirhorst et al. (1989, 1990) and Buchholz et al. (1995). Van Kesteren et al. (1996) provided a theory of the molecular co-evolution of the functionally distinct ligand-receptor sites of the vasopressin/oxytocin superfamily of bioactive peptides.

4. THYROID GLAND

The thyroid gland has been a focal point of investigation by comparative endocrinologists for several decades. The exhaustive review of thyroid functions in cyclostomes and fishes by Eales (1979; see for literature) is a pioneer effort from which a reader can benefit. Donaldson et al. (1979) have contributed considerable information related to hormonal enhancement of growth and have analyzed the role of thyroid hormones. Contribution from Leatherland (1994) on thyroidology of fishes deserves a thorough reading. All juvenile and adult vertebrates possess a distinctive thyroid gland (tissue), comprising thyroid follicle units.

There is no definitive thyroid gland in adult cyclostomes, but thyroid follicles are found in connective tissue around the ventral aorta. It is not uncommon for the non-encapsulated follicles to migrate to a variety of ectopic (i.e., non-pharyngeal) locations, most notably to the well-vascularized head-kidney, providing a means of increasing functional thyroid mass under conditions of iodine deficiency. In some exceptional species (Bermuda parrotfish, swordfish, tuna), the follicles are massed together in single (or tandemly arranged, double) median lobes (Fig. 13.15). In general, the thyroid glands of teleosts exhibit great heterogeneity with respect to follicular cells.

The ontogeny of the thyroid tissue in lampreys probably provides the best indication of the evolutionary origin of the thyroid gland in fishes and other vertebrates. In the ammocoete, there are no thyroid follicles. Iodine is actively absorbed from the ambient water by specific ciliated epithelial cells of the endostyle (subpharyngeal gland), and these cells synthesize and secrete iodinate amino compounds.

In fishes, iodide is taken up from environmental sources via the gills and gut, and is transported in the blood either as free iodide, or in some species bound to plasma proteins (Fig. 13.16).

Detailed histological studies representative of the major groups of fish indicate a well-vascularized thyroid. In contrast, follicles are scantily vascularized in lampreys. In hagfish, the follicles are embedded in fat, and although peculiar thin-walled and blind-ended spatulate vessels have been shown to come into intimate contact with at least some follicles, secretion is presumably made into perifollicular spaces some distance from the nearest blood vessels.

In both the diffuse thyroid and compact thyroid of teleosts, the oxygenated blood supply is supplied by branches of efferent branchial arteries, while the venous return is via the inferior

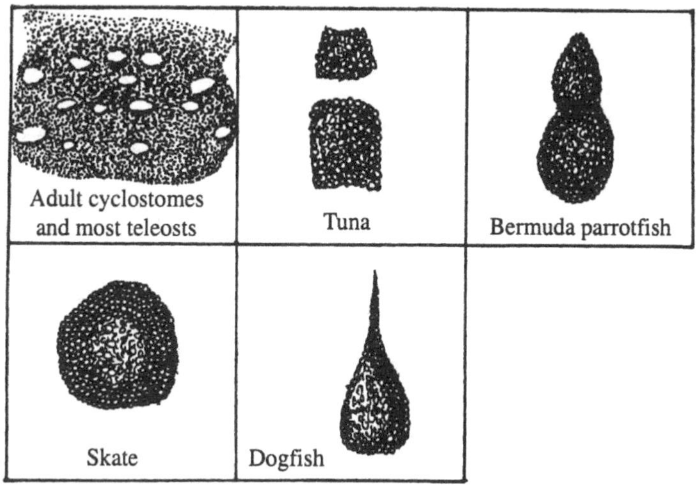

Fig. 13.15 Outlines of the shapes of thyroid glands in fishes.
***Source:* Dent (1986)**

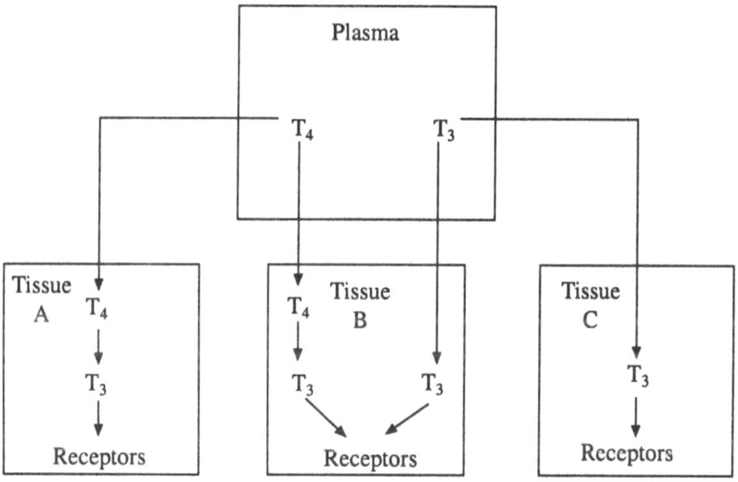

Fig. 13.16 Hypothetical sources of T_3 to different tissues. (A) All T_3 is derived intracellularly from T_4. (B) T_3 is derived intracellularly and from plasma T_3. (C) T_3 is derived solely from plasma T_3.
***Source:* Eales and Brown (1993)**

jugular vein (see Eales 1979; Dent 1986). For evolutionary aspects, the reader is also referred to above cited articles.

Fig. 13.17 gives an idea of peripheral thyriod model (see details: Eales 1985). Fig. 13.18 shows a schematic flow chart showing the hypothalamus-pituitary gland-thyroid gland axis in teleost fish (see details: Leatherland 1994).

Eales (1990) has reviewed the thyroid function in poikilotherms. The poikilotherm thyroid system is usually assumed to conform to the homeotherm plan. The thyroid secretes primarily L-thyroxine (T_4) in response to thyroid stimulating hormone (TSH), whose pituitary release is regulated by hypothalamic hormones, including thyrotropin releasing hormone (TRH). Over

99% of T_4 binds to plasma proteins but free T_4 enters cells and is enzymatically deiodinated to several iodothyronines including 3, 5, 3′-triiodo-L-thyroine (T_3). T_3 is considered the active thyroid hormone (TH) since it binds avidly to nuclear receptors promoting mRNA and protein synthesis, orchestrating changes in metabolism and growth (Fig. 13.19).

In salmonid fish and parrotfish, TSH stimulates acute release of T_4 but not T_3 implying that plasma T_3 is not generated from thyroid sources. For other poikilotherms studied to date, T_4 is probably the primary thyroidal secreting but the extent of T_3 release is unknown. With increasing emphasis on regulation of thyroidal status through extrathyroidal T_4 to T_3 conversion, it is important to ascertain the extent of thyroid T_3 release in poikilotherms and its contribution to the plasma T_3 pool.

Extrathyroidal T_4 to T_3 conversion depends on deiodinase activity evident in the microsomes of various tissues. The deiodinase properties and tissue locations vary between species. In salmonid fish and eels a high proportion (> 65%) of circulating T_4 may be converted to T_3. In trout deiodination may be solely 5′ (T_3 but not rT_3) occurring mainly in liver. In salmonids the *Km* of the main deiodinase falls in the physiologic range of 1–2 nm, but systems with a higher *Km* may exist.

T_4 to T_3 conversion represents an activation and a potential regulator of peripheral thyroidal status. In support of this view, the V*max* (level of functional enzyme) changes rapidly and drammatically in response to physiological state. Increase in V*max* is induced by testosterone and growth hormone and decrease is induced by estradiol and starvation. Employing other indirect indices of deiodinase activity several workers have shown temperature, salinity, cortisol, and caloric intake are also influential. Clearly the amount and activity of T_4 5′-monodeiodinase is of considerable importance in the regulation of thyroidal status and is finely controlled by a broad spctrum of variables.

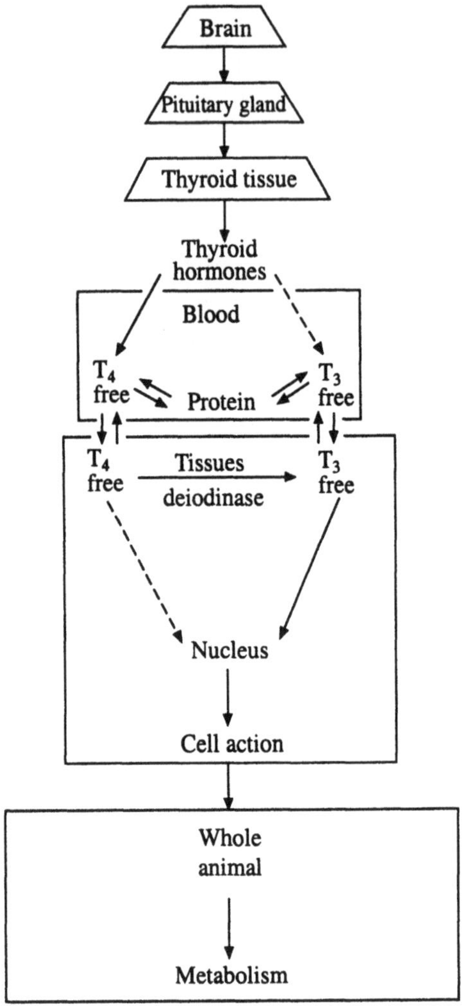

Fig. 13.17 The peripheral thyroid model in which thyroidal status is determined in any given cell by the number of nuclear receptor sites occupied by thyroid hormone (mainly T_3). In the extreme form of the model, the brain-pituitary axis may do little more than regulate the level of the inactive prohormone (T_4), the predominant regulation of thyroidal status depending on local or systemic deiodinase activity responsible for generating the active hormone T_3. *Source:* Eales (1985)

Putative high affinity (K*d* 10^{-10}M), low capacity ≤ (1000 sites/cell) T_3-binding sites have

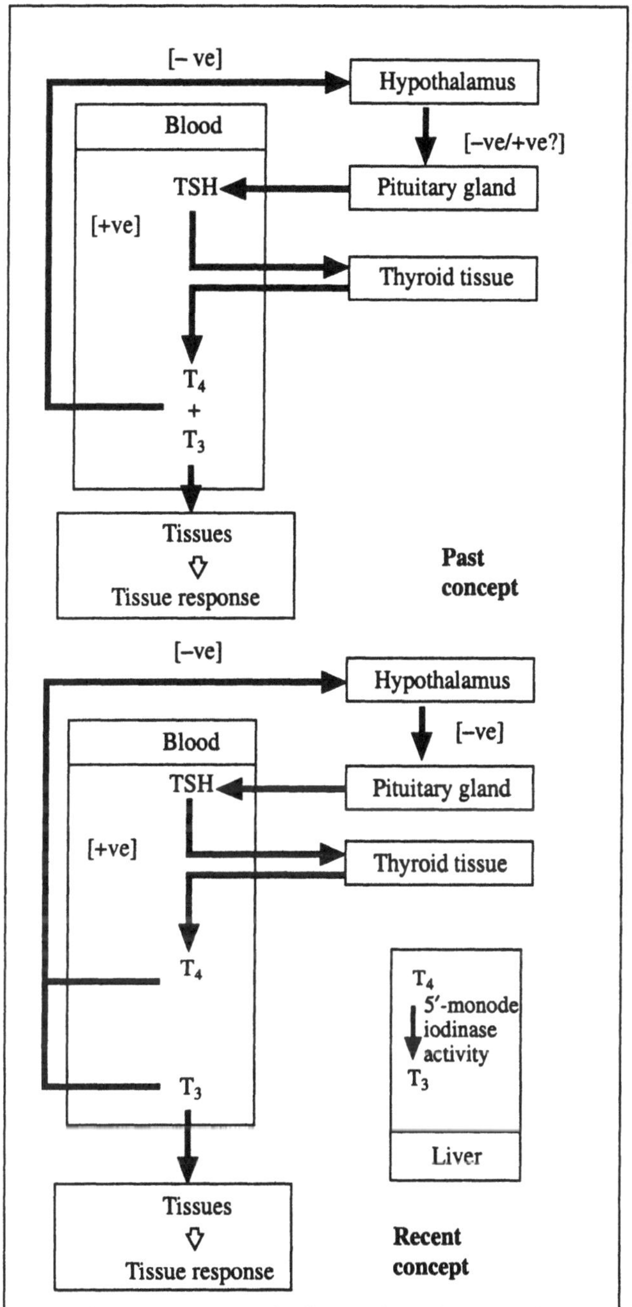

Hypothalamus-pituitary gland-thyroid gland axis in teleosts.

Fig. 13.18 **Shows a schematic flow chart showing the hypothalamus-pituitary gland-thyroid gland axis in teleost fish (Past) and (Present) concepts. In past, the nature of the hypothalamic control over the pituitary gland secretion of thyrotropic hormone (TSH) was unclear, and the thyroid tissue was thought to be the sole source of both thyroxine (T_4) and triiodothyronine (T_3). In the present concept, the hypothalamic control of secretion appears to be inhibitory (although the nature of hypothalamic factors responsible is not clear), and second, the primary source of circulating T_3 was argued to be peripheral tissues (such as the liver and kidney) and not the thyroid tissue.** ***Source:*** **Leatherland (1994)**

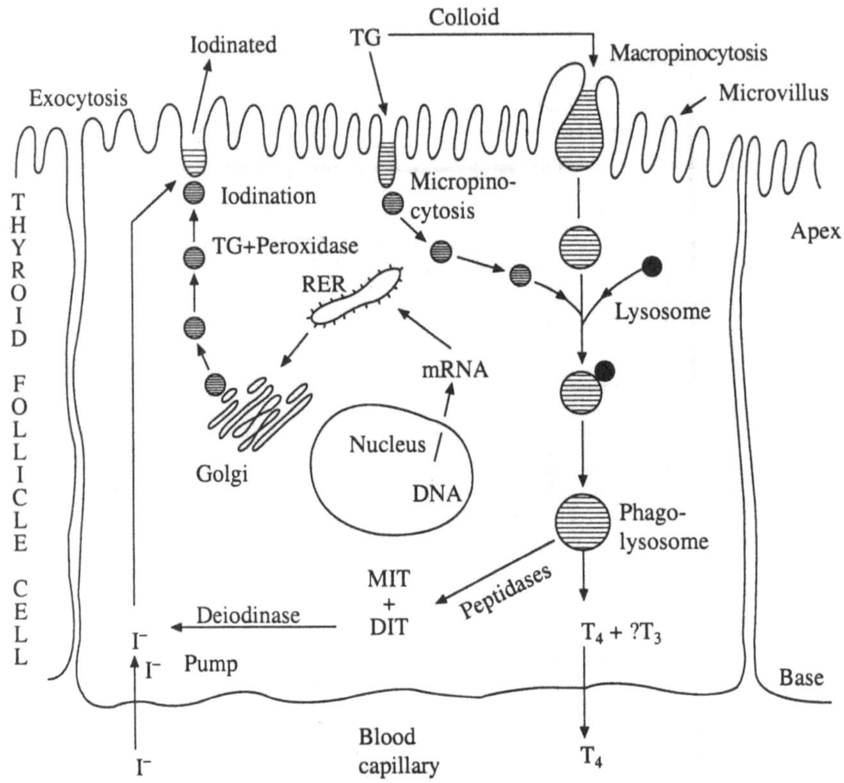

Fig. 13.19 A generalized thyroid follicle cell, showing the main stages in thyroid hormone biosynthesis and secretion. The initial steps involve I^- pumping into the cell and synthesis of peroxidase and the large protein precursor, thyroglobulin (TG). At the apical membrane, peroxidase facilitates iodination of tyrosyl moieties of TG, and the iodinated TG is stored extracellularly in the colloid. Colloid is ingested by micropinocytosis or by macropinocytosis under strong TSH stimulation. The ingested colloid vacuoles fuse with lysosomes, and peptidase activity in the resulting phagolysosome digests TG, liberating amino acids including T_4, possible small amounts of T_3 and their precursors monoiodotyrosine (MIT) and diodotyrosine (DIT). The T_4 (and any T_3) is secreted to blood but MIT and DIT are deiodinated to release iodide, which may be recycled within the cell to synthesize TH.
Source: **Eales and Brown (1993)**

been partially characterized from the nuclei of liver and other tissues for cyclostomes and salmonid teleosts. Observations led to conclusion that during growth suppression in fish some regulation of thyroid status may occur through change in receptor abundance but that relative to 5′-monodeidinase regulation, receptor regulation is not a key determinant of peripheral thyroidal status.

Documented amongst poikilotherms mainly in teleost fish, enhanced growth represents one of the most predictable responses to chronic physiologic TH administration. T_4 is less potent than T_3 which in turn is more potent in combination with other anabolic hormones. Both increased appetite and food conversion efficiency contribute to the growth response to T_3 (Fig. 13.20).

Regarding gonadal maturation, most studies have focussed on ovarian development. There is an established role of TH in ovarian maturation in elasmobranchs and a comparable situation

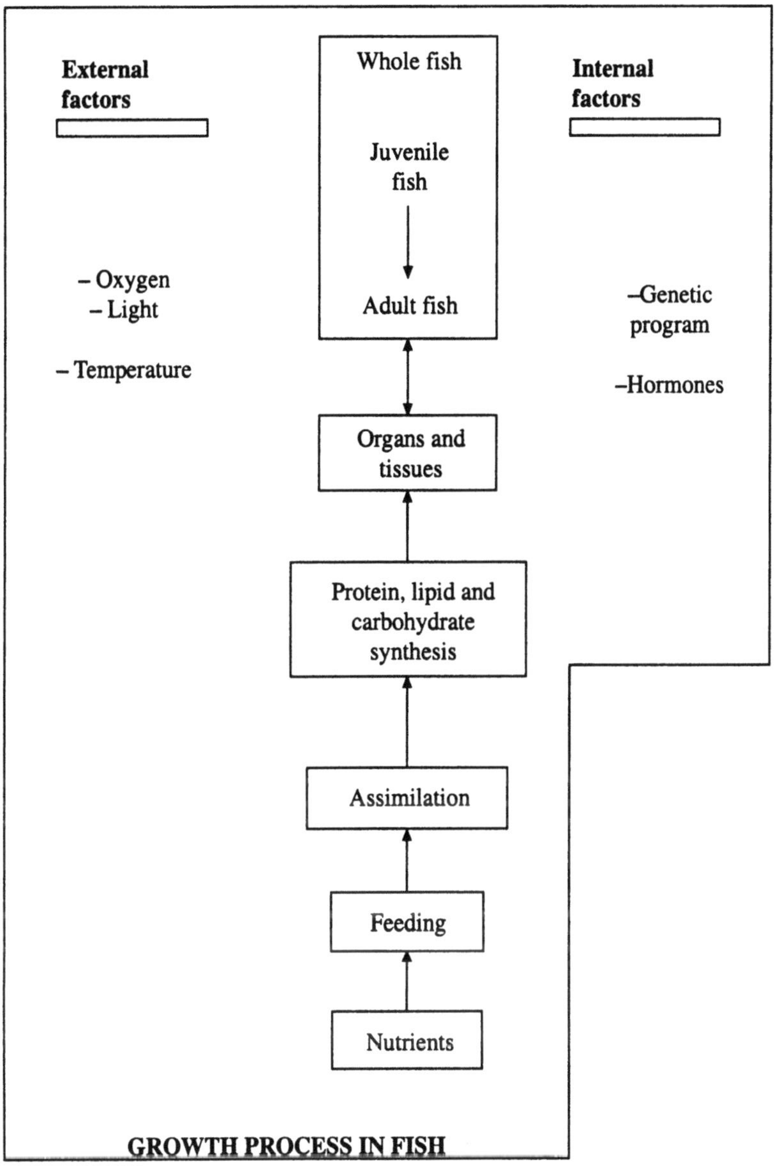

Fig. 13.20 **The schematic flow chart illustrates, in a simple fashion, the interactive nature of major factors involved in the "growth" process of fishes. The central column illustrates the basic components of energy uptake, via food, production of macromolecules and the incorporation of these molecules in tissues in the form of metabolite reserves, or for somatic growth. Even somatic growth processes represent a form of metabolite reserve, since they can be mobilized as required (hence the two way arrow connecting the "whole fish" and "organs and tissues" operational boxes in the figure). The nutritional needs for growth change with ontogeny, represented by the references to juvenile and adult stages within the "whole fish" box. Growth rates are heritable and the processes represented the central column are significantly induced by genetic program; moreover, most of the processes are coordinated by hormones. However, this internal regulation is greatly influenced by environmental factors.**
Source: **Leatherland (1994)**

may exist in teleosts. TH enhances the effect of gonadotropin on early ovarian function and later ovarian maturation.

The role, if any, of the TH during salmonid smoltification is obscure (for details, see Boeuf 1993). Thyroid secretory activity is markedly stimulated prior to seaward migration. The thyroid probably facilitates actions of other hormones (e.g., GH, cortisol) with more visible effects. In flounders (*Paralichthys olivaceous*), a TH role in promoting metamorphosis has been demonstrated more convincingly. TH are also present in teleost eggs and a prehatching role is possible. Such a TH role is suggested by the finding that immersion of developing medaka eggs and early embryos in T_4 decreases the meristic counts (numbers of fin rays and vertebrae).

Conclusions can be given here. Although a thyroidal contribution of T_3 to the plasma T_3 pool cannot be ruled out for most poikilotherms, it seems likely that, as in mammals, most T_3 is generated by deiodinase systems in peripheral tissues. Deiodinase activity and level changes during ontogeny and in response to physiologic/environmental perturbations, providing extrathyroidal control of T_3 production. Although T_3 receptor properties can change, regulation at this level is less evident than at the level of deiodinase. Direct actions of T_4 cannot be excluded but most TH action is probably exerted through T_3. TH (T_3) regulates metabolically demanding events, the emphasis of which depends on the animal and its physiological state.

Eales (1988) has given details (Figs. 13.21, 13.22) which are related with the influence of nutritional state on thyroid function. The ectothermic teleost fish is profoundly sensitive at several levels to acute and chronic alterations in the quality and quantity of ingested food. Predominant amongst the thyroid changes is a consistent alteration in peripheral T_3 production due to regulation of hepatic T_4 5′-monodeiodinase activity, which causes decreased T_3 blood levels in caloric- (mainly carbohydrate) deficiency states.

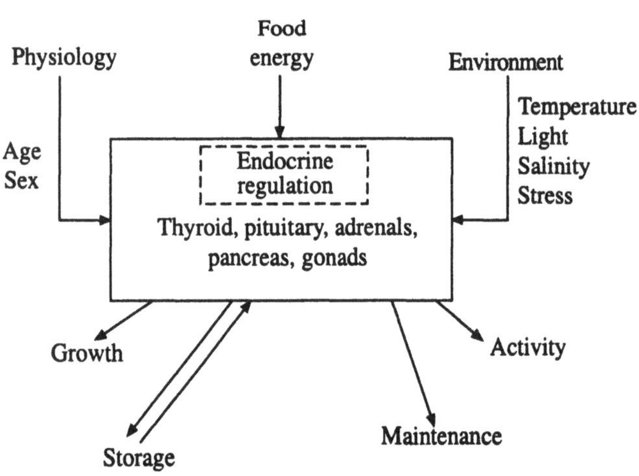

Fig. 13.21 Role of endocrine system in regulating use of metabolizable energy derived from food in relation to physiologic state and environmental variation. ***Source:*** **Eales (1988)**

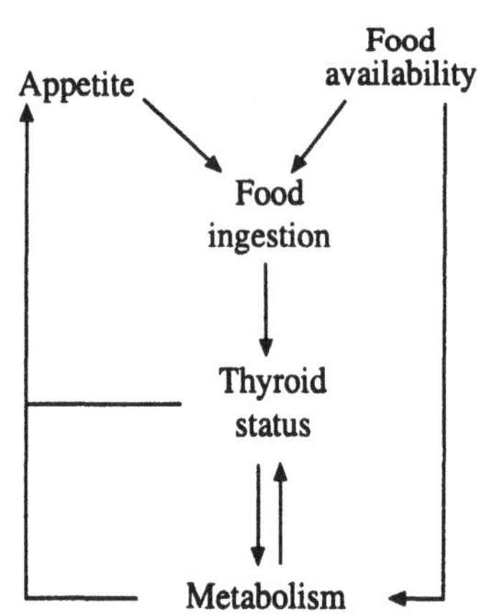

Fig. 13.22 Pathways whereby food might influence thyroid function. ***Source:*** **Eales (1988)**

In teleosts, studies of diel cycles in plasma T_4 and T_3 also indicate that food intake may have major acute effects on the thyroid. The diel cycle is usually well defined for T_4; for T_3 the cycle is less pronounced and in some instance may not be significant statistically.

The theme of the paper by Grau (1988) is related to environmental influences on thyroid function in teleost fish. Evidences indicate that the 'thyroidal system' can be exquisitely sensitive to such environmental changes and perturbations as day length, lunar cycle events, temperature, pH, salinity, and nutrition. It has been pointed out that thyroid status appears to respond to environmental cues in a predictive and mediated fashion. The thyroid system, including relevant brain, pituitary and peripheral components, in addition to the thyroid itself, may act in anticipation of developmental and/or environmental requirements. Leatherland (1994) has given details of iodine deficiency and goitres.

McDonald and Milligan (1992) have given a brief account of thyroid hormones, along with other hormones present in blood of fish. Plisetskya (1985) has discussed the role of thyroid hormones in intermediary metabolism in agnathans. It has been speculated that the slow development of the T_3 lipolytic effect is the resultant of a combined lipolytic and lipogenic effect of thyroid hormones. A lipolytic effect of T_3 in lamprey development has been tentatively indicated.

The role of thyroid hormones on fish Ca and bone metabolism remains little understood (Dacke 1989).

5. ULTIMOBRANCHIAL BODY

As the name implies, the ultimobranchial bodies are embryologically derived from the last pharyngeal branchial pouch. Dacke (1979) discussed calcium regulation in fish. Open and closed calcium systems of Ca regulation are focused (Fig. 13.23; see Table XV of Dacke's article for concentrations of Ca and total ions in the plasma of various groups of fish). Robertson (1986) has presented a comprehensive account of the ultimobranchial body in vertebrates.

The ultimobranchial body is a discrete endocrine gland in fish. The phylogeny and development of the ultimobranchial body throughout the vertebrates, illustrates a functional separation of two components, the granulated endocrine "C cell" hypocalcemic hormone (calcitonin= CT- containing cells) and the "U cell", involved in a holocrine secretion.

The ultimobranchial body, as a discrete organ, is not present in the Agnatha, but is only found in jawed vertebrates. In sharks, the ultimobranchial gland exhibits an assymetry, it being larger on the left than the right. It is derived directly from the last pharyngeal pouch or from the floor of thc pharynx at the level of the last pouch regardless of the number of pouches.

The parenchyma consists of granulated and non-granulated cell types in fish. A tall columnar type, with light-staining cytoplasm, a supranuclear Golgi apparatus, and a well-developed infranuclear rough endoplasmic reticulum (ER) may contain dense osmiophilic inclusion in the apical cytoplasm that are similar to other cell debris in the lumen. In granulated cells, the ER is less conspicuous; however, the Golgi apparatus has numerous dense vesicles, and the infranuclear cytoplasm is filled with dense-cored membrane-bound granules. A clear distinction of these cells shows the non-granulated cell, with an abundance of microfilaments, as a supporting cell which extends to the apical surface with extensive interdigitations between like cells and situated between granulated cells. The apical surface has long microvilli encircling the edge of the cell and commonly possess a single cilium. Only the granulated cell is immunoreactive (Fig. 13.24).

Myelinated and unmyelinated nerve bundles are commonly seen in the underlying connective tissue of fish ultimobranchial glands but direct synaptic contacts have not been observed. In fish, the polarity of granulated ultimobranchial cells is directed toward the vascular bed, as opposed to the apical luminal border.

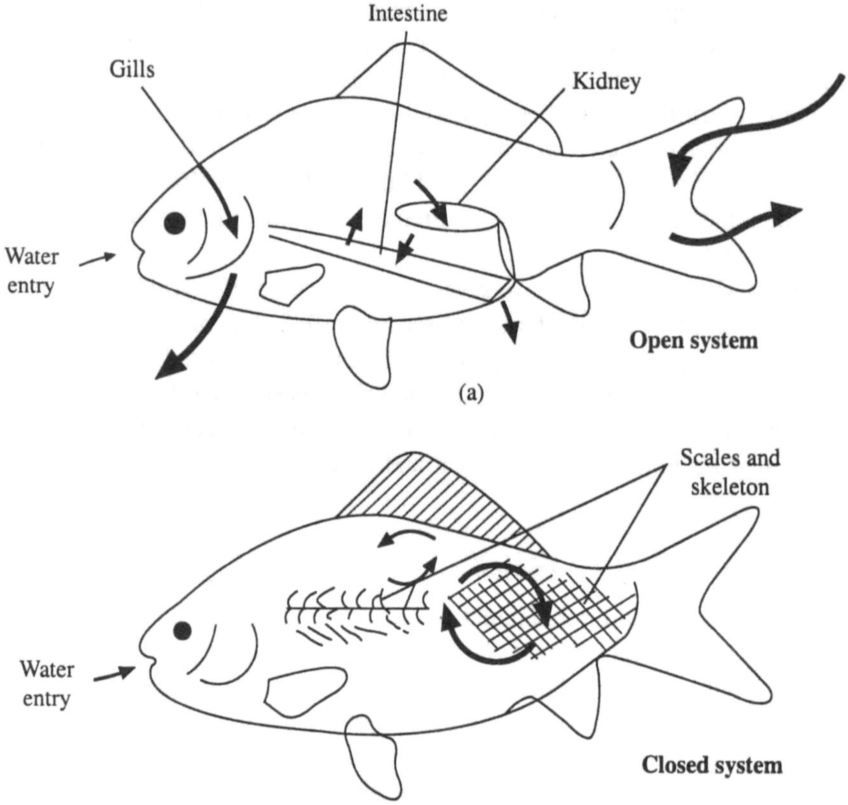

Calcium regulation.

(b)

Fig. 13.23 **(a) A fish considered as an "open system". Ca mainly enters and leaves via the fin and gills. A small amout of Ca is absorbed via the intestine and the kidneys exert some control over Ca loss. In its most extreme form this concept considers that the surronding water acts as an extension of the body fluids. (b) A fish considered as a "closed system". Ca is stored in scales and skeleton and continually recycled in and out of the structures so as to maintain a steady-state in the animal. (a) and (b) represent extreme forms and most teleost probably have both systems which are clearly not mutually exclusive.**
***Source:* Dacke (1979)**

In the salmon *Salmo salar* and the zebra fish *Brachydanio rerio*, the ultimobranchial glands are relatively inactive in immature fish but become active and degranulated during spawning phase.

The physiological function of CT in fish is still under debate. It does not appear to play a dominant role in the homeostatic control of extracellular calcium, although some effect cannot be excluded under extreme conditions. CT may play a role in Ca^{2+} homeostasis by protecting the skeleton during periods of the increased Ca^{2+} demand of oogenesis. Available data indicate that the gills rather than the bone tissue are the target for CT in fish. Reports on the effects of CT on fish bone are scarce but some data suggest that CT may effect bone cells in fish (Braun and Dantzler 1987; Fenwick 1989; Hermann-Erlee and Flik 1989; Wendelaar Bonga and Pang 1991; McDonald and Milligan 1992). Calcitonin secretion is stimulated during gonadal maturation in female fishes. A specific role for calcitonin in gonadal maturation has also been suggested, but

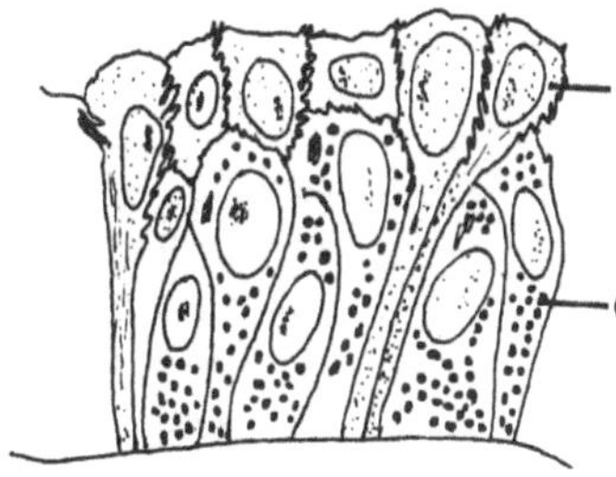

Ultimobranchial gland of *Salmo gairdneri.*

Fig. 13.24 Diagrammatic representation of cell types in Ultimobranchial gland of trout. *Source:* **Robertson (1986)**

direct evidence is lacking. Calcitonin, like other hormones, shows peaks in plasma levels during ovarian maturation, penetrates the eggs and becomes involved in early development of the embryos. The factors controlling calcitonin, with exception of oestrogens, have not been identified in fishes. Immunoreactivity to calcitonin- and to calcitonin gene-related peptide, originating from the calcitonin gene by alternative RNA splicing has been found in the nervous system. The peptides may function as neurotransmitter and neuromodulators, rather than hormones (read Wendelaar Bonga 1993).

In fish there is no evidence that calcitonin affects tubular transport. In fresh-water-adapted European eels, calcitonin increases blood pressure. Associated with this is an increase in urine flow, GFR, and electrolyte excretion. These effects of calcitonin are not seen in salt-water adapted eels.

For an evolutionary and overall consideration of hormones and calcium regulation in vertebrates, the reader is referred to Pang et al. (1989).

6. STANNIUS CORPUSCLES (CORPUSCLES OF STANNIUS) (CS)

The corpuscles of Stannius (CS) are small, spherical, white or yellowish pink, unique encapsulated endocrine glands found on or in the mesonephric kidney of holostean and teleostean fishes (Figs. 13.25, 13.26). They produce the predominant hypocalcemic calcitropic hormone in fish, stanniocalcin (STC), previously referred to as hypocalcin (HC) or teleocalcin (TC) or the corpuscle of Stannius protein. The number of CS varies; mostly higher in the holosteans. The position varies; even the number may vary in different sexes; and within one species, individual differences may occur that are related to age or sex.

The embryological origin makes the CS unique among the endocrine glands of the vertebrates (may be outgrowths of pronephric/mesonephric/opisthonephric ducts).

The outlines and arrangement of CS may vary. Data on vascular supply are limited. It has been indicated that the products released by CS first pass through kidney before they enter the general body circulation. CS have a rich autonomic innervation. The differences in embryological and morphological characteristics make it rather unlikely that CS represent the fish (parathyroid) glands. There is a possibility that the two endocrine tissues may produce different hormones of the same family. In addition that the CS contains a pressor substance similar to renin is a distinct possibility.

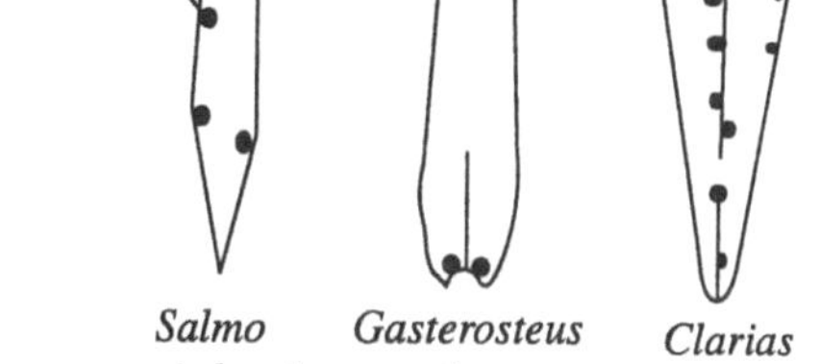

Fig. 13.25 Location of Stannius corpuscles in fish kidneys. *Source:* **Wendelaar Bonga and Pang (1986)**

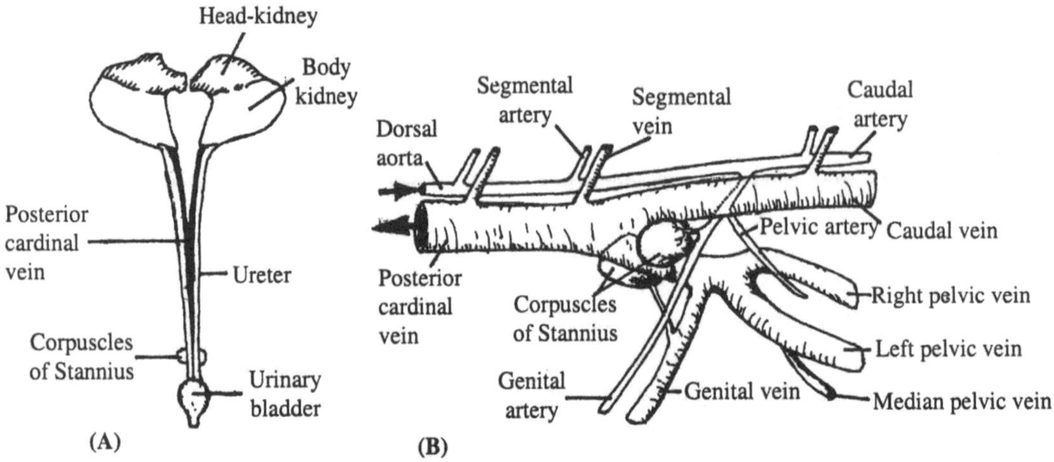

Fig. 13.26 (A) Diagrammatic ventral view of the urinary organs of the Kusafugu puffer, *Takifugu niphobles*, showing the location of the corpuscles of Stannius. (B) Diagrammatic lateral view (*anterior at the left*) of the vascularization of the corpuscles of Stannius showing arterial supply (small arrow) and venous drainage (large arrow).
***Source:* Chiba (1990)**

Stanniocalcin (STC) is a glycosylated homodimeric disulfide-linked glycoprotein with a molecular weight ranging from 39 kDa in sockeye salmon to about 60 kDa in rainbow trout. Neither parathyroid glands nor true PTH (–like) molecules have been demonstrated in fish; true parathyroid glands occur only in terrestrial tetrapods. The view that PTH of higher vertebrates and HC of fish are homologous hormones has not received support. There are no clear indications that PTH or HC affects bone cells in fish.

The presence of two cell types (1 and 2) has been shown in the CS of various species. Type-1 cells are more abundant, rounded and contain extensive granular endoplasmic reticulum and large secretory granules, whereas type-2 cells are slender, often with cytoplasmic extensions and characterized by the presence of small granules. The number of granules in type-2 cells is much less than that in type-1 cells. Recent work on immunocytochemical localization of hypocalcin in the endocrine cells of CS suggests that type-1 and type-2 cells represent structurally different forms of the same functional cell-type.

The gills are likely to be the main target tissue for the active CS principle which acts by reducing the entry of calcium ions from the environment. However, it is also possible that hypocalcin affects metabolism in the kidney, intestine, bone, and scales. Only more recently have the intestinal transport of calcium and the renal handling of phosphate been identified under the influence STC (read Wagner 1994; for phosphate specifically, see Graham et al. 1994).

In trout, in the basolateral plasma membranes of branchial epithelium (and also in intestinal and renal membrane preparations), a hypocalcin-sensitive adenylate cyclase activity could be demonstrated. Hypocalcin inhibits this adenylate cyclase. These results strongly suggest that these tissues contain hypocalcin receptors, which exert inhibitory control over adenylate cyclase.

For describing mechanisms of action of hypocalcin on fish gill, studies of the effect of this hormone on the high-affinity Ca^{2+} -ATPase must be carried out. Hypocalcin may affect the permeability of the chloride cells, and this may be the key control mechanism of branchial calcium uptake.

Wagner (1994) reviewed the molecular biology of CS and regulation of STC gene expression. STC has been cloned and fully characterized in only two species of fish (eel and salmon). Calcium regulation of STC m RNA levels has been described. Contributions (Wagner et al. 1992, 1993; Sterba et al. 1993; Wagner and Jaworski 1994; Ellis and Wagner 1995) deal with special aspects of STC.

Hypocalcin, rather than calcitonin, is a dominant hypocalcemic hormone in fishes. It is a fast-acting antihypercalcemic hormone preventing Ca^{2+} influx via the gills.

STC seems to be, evolutionarily, the first and physiologically the most important calcitropic factor in fish.

Plasma calcium levels control storage and synthesis rates of hypocalcin in the CS. Although nervous control of secretion of the CS cannot be excluded, most authors consider it likely that the nerves are mainly controlling the blood flow through the glands. Only a few studies have dealt with extracellular signal transduction in the CS cells. A recent experiment has shown that in the American eel (*Anguilla rostrata*) it appears that there is a potential for cholinergic control of STC release but that offer factors such as the local plasma concentration may also be involved, at least in response to severe acute hypercalcemia.

Several hormones may influence the secretory activity of the CS, although the effects may be mainly indirect. The evidence is exclusively based on histological observations.

The reader can find interesting information regarding CS in the following publications (Wendelaar Bonga and Pang 1986; Urasa and Wendelaar Bonga, 1987; Wagner et al. 1988; Fenwick 1989; Flik et al. 1989; Herman-Erlee and Flik 1989; Hirano 1989; Kaneko et al. 1989; Wendelaar Bonga et al. 1989; Chiba 1990; Flik 1990; Hanssen et al. 1991; McDonald and Milligan 1992; Fenwick and Verbost 1993; Verbost et al. 1993; Cano et al. 1994; Butler and Cadinouche 1995).

One important question remains to be answered, and this is how the chondrichthyans and all other fishes that have no CS are able to maintain their plasma calcium levels, in particular in seawater.

7. INTERRENAL TISSUE

Chester Jones and Phillips (1989) have given a comparative account of the adrenal and interrenal glands of members in Agnatha and Gnathostomata. There is another informative contribution on adrenal complex in vertebrates from Hanke and Kloas (1995).

There is little doubt that the agnathan "presumptive" adrenocortical homologue (AH) represents the most primitive vertebrate condition. The tendency among vertebrates for a more anterior concentration of this tissue within the coelomic cavity or within kidneys has long been reported. Among the more primitive fishes, the distribution of AH reflects the position of the animal on the taxonomic scale (Fig. 13.27). In Agnatha, the Petromyzontidae and Myxinidae present difficulties in identification of interrenal tissue which, if it exists, may or may not secrete known interrenal/adrenocortical steroid. For the Petromyzontidae, presumptive interrenal tissue may be concentrated within the pronephric region on the dorsal side of the pericardium and along the walls of the cardinal veins (Fig. 13.28). Lampreys may closely resemble an original vertebrate condition which was believed to be the coexistence of AH cells with kidney tubules. Even though the intimacy of AH cells and the arterial circulation is unique, the association of these cells to the posterior cardinal veins is a feature consistent with the Osteichthyes (see Youson 1985). In *Lampetra planeri*, cells with organelles characteristic of steroid production have been

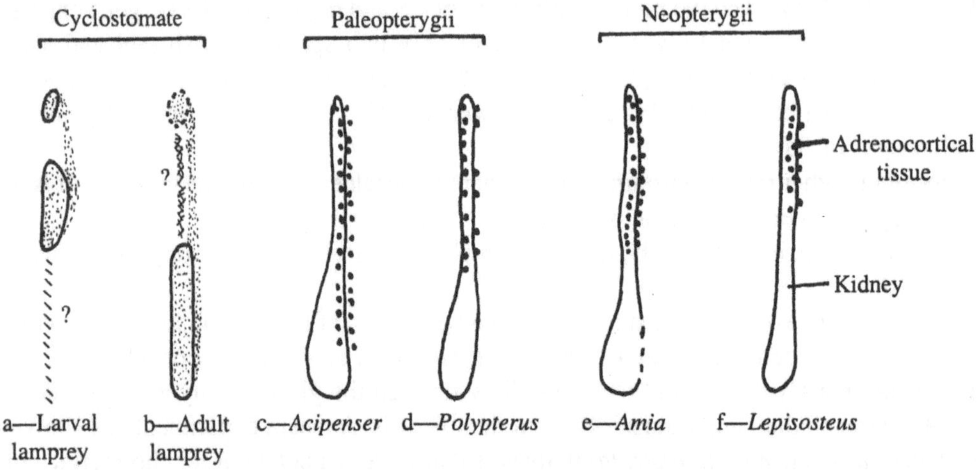

Fig. 13.27 Distribution of adrenocortical tissue (interrenal gland) as diffuaely scattered islets (dots) in the kidneys and around dorsal vessels in larval and adult lampreys is compared with adrenocortical tissue in corpuscles (dark circles) in *Acipenser*, *Polypterus*, *Amia* and *Lepisosteus*. Note the progressive anterior accumulation from *Acipenser* to *Lepisosteus*.
***Source:* Youson (1985)**

reported. The definitive enzyme 5-ene-3 β – hydroxysteroid dehydrogenase is not readily demonstrated but it may occur at certain stages of lamprey life cycle, for example, during migration. The presumptive interrenal tissue in the Myxinidae may lie in the region of the pronephros, but we are here on even weaker ground than in the discussion about the lampreys. There are no certain results.

In Chondrichthyes, it is important to note that the interrenal gland is completely separated from the chromaffin cells, which lie in paired groups segmentally arranged. The interrenal itself

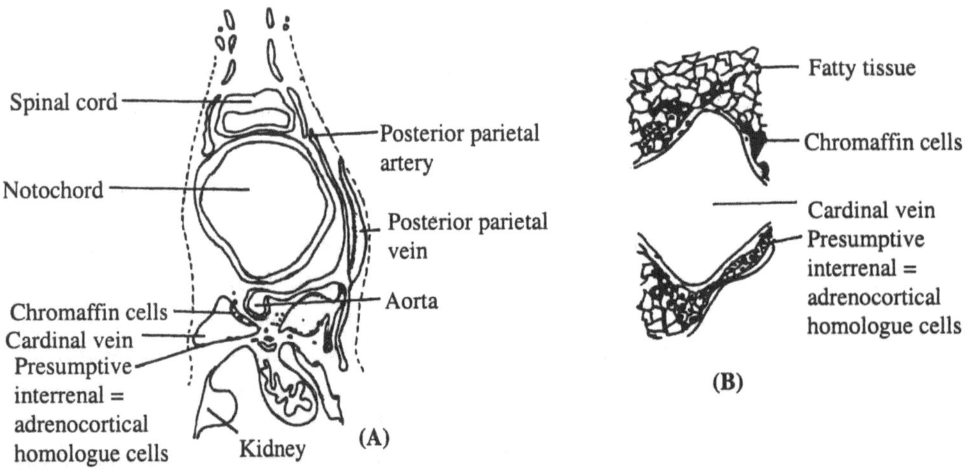

Fig. 13.28 (A) A transverse section of *Petromyzon* which shows the presumptive interrenal = adrenocortical homologue cells arranged about the cardinal veins. (B) Presumptive interrenal cells of *Lampetra* occurring about a cardinal vein with adjacent chromaffin cells with fatty tissue.
***Source:* Chester Jones and Phillips (1986)**

lies in the region of the posterior kidney as blocks of cells, horseshoe-shaped (ray *Raja* spp.) or rod-shaped (sharks *Scyllium canicula*), or is concentrated in one oval form (*Torpedo* spp.) (Fig. 13.29). It is generally found that internenal cells are homogeneous, all grouped in cords, and are round or polygonal, with a large prominent nucleus containing basophilic chromatin. Cells have lipid droplets, cholesterol, and 5-ene-3 β- hydroxysteroid dehydrogenase. Ultrastructurally, two types of cell, dark and light, are indicated in immature *Raja clavata*. Liposomes, mitochondria, smooth endoplasmic reticulum, Golgi material, and polyribosomes are evident. There are suggestions of interrenal cell changes with sexual activity and viviparity, though further investigation is needed. Zonation within interrenal mass of cells may well exist, though it is not usually obvious.

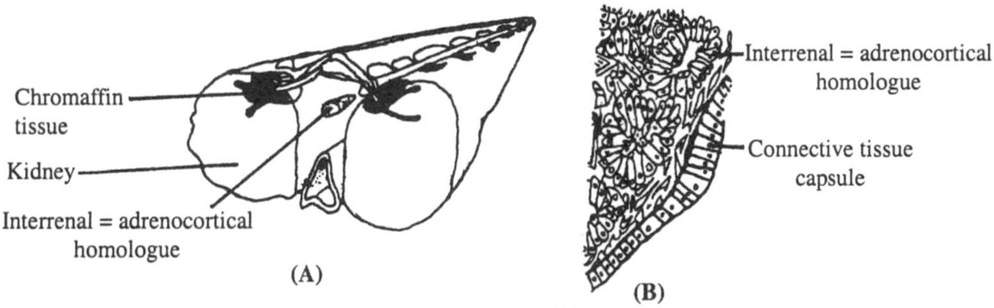

Interrenal tissue of cartilaginous fish.

Fig. 13.29 Interrenal tissue of Elasmobranchs. (A) The main point is that the interrenal = adrenocortical homologue is completely separated from chromaffin tissue, which is segmentally arranged. In the dogfish type the interrenal is rod-shaped and is found midway between the mesonephric groups. (B) The interrenal = adrenocortical homologue is cast into lobules and cords of cells. It is a discrete gland bounded by a connective tissue capsule.
***Source:* Chester Jones and Phillips (1986)**

In general, the teleost interrenal is rather inaccessible and diffuse, and consists of groups, cords, or strands of cells set around the posterior cardinal veins and their branches, vascularized by twigs of the dorsal aorta (Figs. 13.30, 13.31). The interrenal is usually confined to the anterior part of the kidney ("head-kidney"). In some species, e.g., *Perca fluviatilis*, cortical and chromaffin tissue are closely intermingled. The histology of the interrenal cell conforms to the tetrapod pattern, though with less display of lipids droplets. The presence of cholesterol, ascorbic acid, glucose-6-phosphate dehydrogenase, 5-ene-3 β-01 steroid dehydrogenase has been demonstrated in the few species examined.

There is clear evidence for pituitary control by some type of fish ACTH. The major hormonal steroid is cortisol; additionally aldosterone debately occurs. Indeed there is a view that aldosterone, at least as dominant mineralocorticoid, arose with the emergence of the Tetrapoda. Hanke and Kloas (1995) reported aldosterone in Amphibia onwards.

Apparent stressful reactions may be involved in such natural processes as migration and spawning. The interrenal in teleosts undergoes seasonal variations, with enlargement at the reproductive period. The salmonids are prime examples. Hanke and Kloas (1995) have discussed the physiological effects of corticosteroids in osmomineral stress. For details, Wendelaar Bonga's contribution (1993) should be referred to.

The many actions of cortisol are reflected by the complexity of the mechanisms that control its secretion. In addition to ACTH, the major secretagogue, and -MSH, hormones such as growth

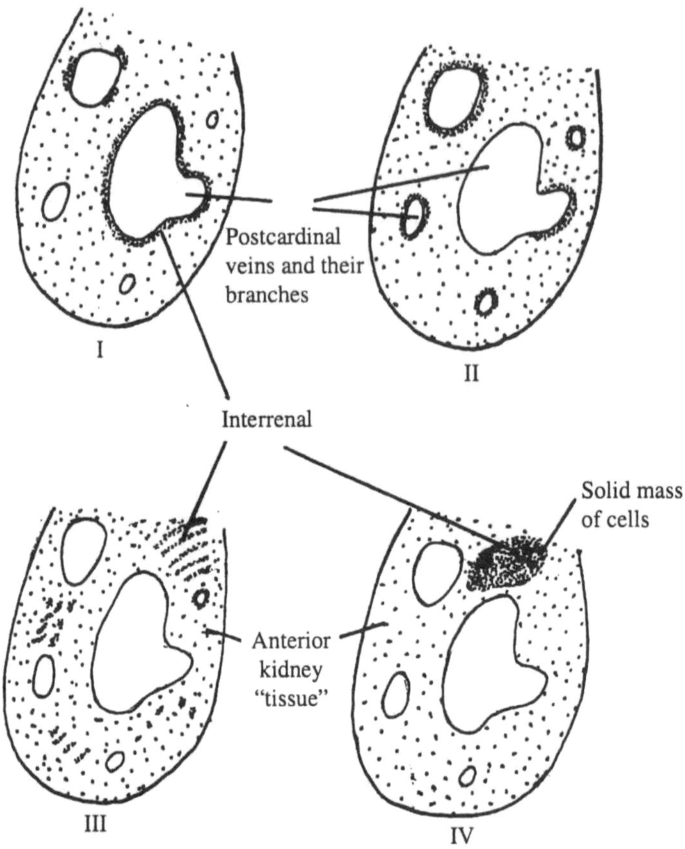

Fig. 13.30 Classification of types of interrenal distribution. Open spaces represent postcardinal distribution. Open spaces represent postcardinal veins and their branches; light stippling indicates anterior kidney "tissue".
***Source:* Nandi (1962)**

hormone, thyroxin, angiotensin II, arginine vasotocin, and catecholamines may have corticotropic activities. Factors such as the N-terminal peptide of the POMC (proopiomelanocortin) and factors produced by immune system may modulate cortisol secretion directly or by affecting other regulatory components of hypothalamo-pituitary-interrenal axis (read Wendelaar Bonga 1993).

In Dipnoi, the presumptive interrenal consists of small group of cells closely associated with the postcardinal veins and their tributaries where the vessels pass through the kidneys. The cells contain lipid, cholesterol and 5-ene-3 -hydroxysteroid dehydrogenase. The morphological site is more amphibian than teleostean, akin to the interrenal tissue of the urodele *Pleurodeles waltii* and tadpole of *Rana*.

In cartilaginous fish, 11-deoxycorticosterone appears to be the predominant mineralocorticoid, whereas in teleosts both aldosterone and 11-deoxycorticosterone appear to be present. However, the physiological role of these hormones in regulating sodium balance is not clear. In freshwater eels, the most complete adrenalectomy that can be achieved apparently leads to an increase in fractional sodium excretion, and in rainbow trout the excretion of sodium apparently decreases after injections of 11-deoxycorticosterone (Braun and Dantzler 1987; see also Nishimura 1985).

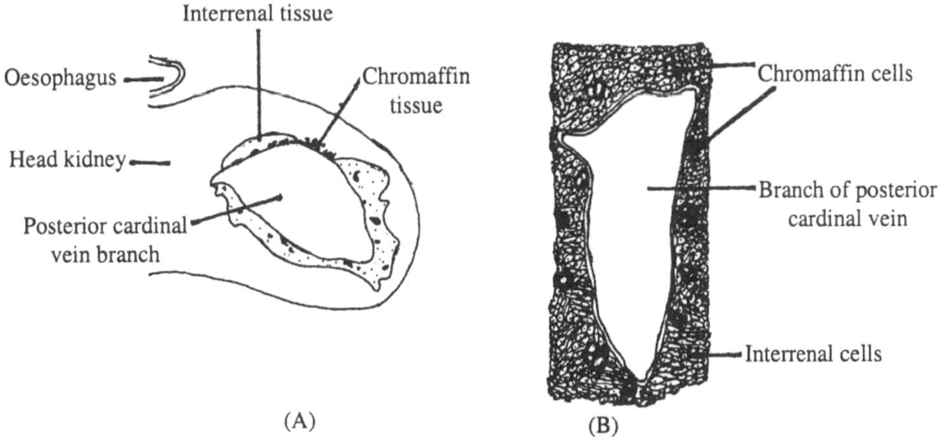

Interrenal tissue of bony fish.

Fig. 13.31 Interrenal tissue of the Teleostei (bony fish). (A) The interrenal tissue lies anteriorly in the head kidney, which is the non-functional part of the mesonephros, rarely displaying tubules, and most lymphocytes. The interrenal = adrenal homologue is disposed about branches of the posterior cardinal veins, and the diagram shows the fish's left side. The oesophagus is central. Chromaffin tissue (black dots) is scattered throughout the interrenal tissue in many species, though the majority of the teleostean species has not been examined. (B) A typical set of interrenal cells set about a branch of the posterior cardinal vein with intermingling of chromaffin cells. ***Source:*** **Chester Jones and Phillips (1986)**

O'Toole et al. (1990) reported the principal corticosteroid, a novel hormone α-hydroxycorticosterone (1 α-OH-B). A definitive role for 1 α-OH-B in the control of the electrolyte economy of the dogfish is far from clear. The reader is also referred to Hazon and Henderson (1985) for its secretory dynamics. Recently Sampath-Kumar (1994) reported the presence of adrenodoxin-like peptide from the interrenal cells of Asian sea bass (*Lates calcarifer*) and the tilapia (*Oreochromis mossambicus*); its physiological role is to investigated.

An important role for cortisol in regulation of Ca fluxes in fish changing between different environmental levels of Ca, which is analogous to its effect on osmoregulation, has been suggested (see Dacke 1989).

The most usual reason for an increase in plasma cortisol in teleosts is stress. Indeed, cortisol methods are the methods of choice for quantifying stress in fishes. Stressful circumstances include handling, close confinement, transport or other physical disturbances, and rapid changes in water quality (e.g., pH, salinity, or temperature). The response to stress is mediated by ACTH (pituitary-adrenal cortex axis) and can be very rapid. Generally speaking, there appears to be an approximate correlation between metabolic scope and cortisol elevation in response to stress. Article from Brown (1993) has provided extensive information, mainly dealing with endocrine responses to pollutants. Schreck (1996) discussed endocrine mediation in respect of stress effects.

Other than strictly stressful conditions, there are other circumstances under which cortisol rises. These include diel cycles, smoltification in anadromous salmonids, feeding, and sexual maturation (McDonald and Milligan 1992). Little is known of the control of elasmobranch interrenal function, although components of a pituitary interrenal axis have been identified (read O'Toole et al. 1990).

Cortisol, corticosterone, and 11-deoxycortisol have been identified in the blood of the most primitive living vertebrates, such as the Atlantic hagfish, *Myxine glutinosa.*

8. CHROMAFFIN TISSUE

The chromaffin tissue of all groups of vertebrates releases the same types of catecholamines into the circulation. These are dopamine, noradrenaline and adrenaline. The amount and function of dopamine as a circulating hormone is not clear.

The chromaffin tissue in lamprey and ammocoete larvae is found in every segment from the second branchial segment to near the end of the postanal region. These chromaffin cells are found in association with the cardinal veins and the dorsal and ventral parietal veins. "Chromaffin cells" also occur in the heart of *Myxine glutinosa.* In *Lampetra,* the sinus venosus contains high amounts of adrenaline and noradrenaline, while the atrium and the ventricle contains mainly adrenaline. In *Myxine,* adrenaline predominates in the ventricle, noradrenaline in the atrium and the portal vein heart. The reason for these differences in regional distribution is unknown (Fänge 1972). Exhaustive account of chromaffin cells in cyclostome hearts has been given by Santer (1985).

The chromaffin tissue of the elasmobranchs is found as distinct bodies lying segmentally, but not always regularly, above the dorsal wall of the posterior cardinal veins and sometimes posteriorly in the kidney. The anatomy of the sympathetic system and the associated chromaffin tissue in *Heterodontus portusjacksoni* (Fig. 13.32) agrees in most respects with that of *Squalus acanthias.* The so-called axillary bodies, i.e., the paired masses of sympathetic ganglion cells and chromaffin

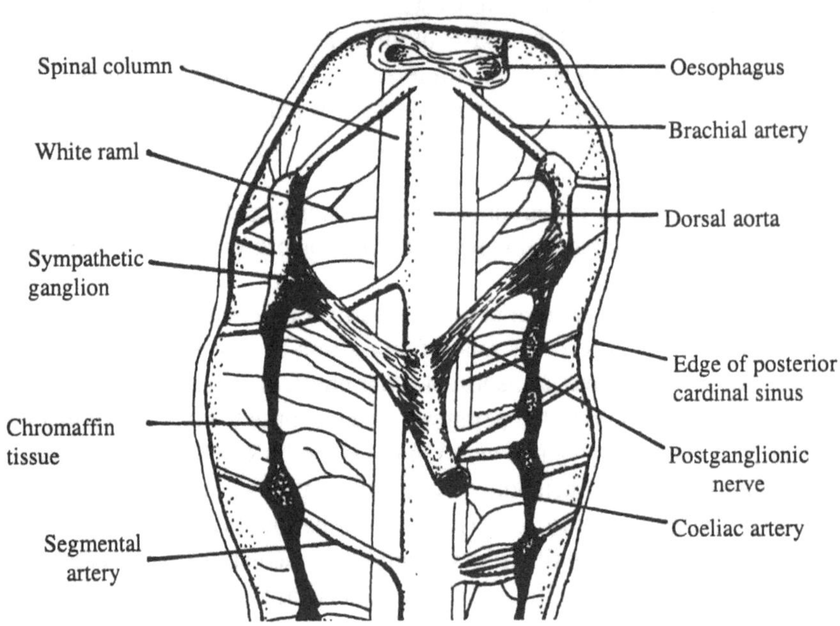

Fig. 13.32 Chromaffin tissue and sympathetic ganglia in the posterior cardinal sinus of *Heterodontus portusjacksoni.*
***Source:* Gannon et al. (1972)**

tissue, which are such a conspicuous feature in the posterior cardinal sinus of other elasmobranch fish, are not so obvious in *Heterodontus*. The chromaffin tissue lies within the capacious blood-filled space of the posterior cardinal sinus. The detailed arrangement of the blood vessels within each mass is unknown but it is likely that the tissue liberates catecholamines directly into the posterior cardinal sinus. The blood from the sinus is aspirated, during each systole directly into the heart and passes from there to the gills. The chromaffin cells have a fine structure comparable to that of chromaffin cells in other vertebrates.

The chromaffin tissue in teleosts is always closely associated with the postcardinal veins in the region of the so-called head-kidney. Chromaffin tissue has been classified in five categories as shown in Fig. 13.33. The chromaffin cells are usually located between the lumen of the veins and the interrenal cells, but in Scaridae and Labridae the position is reversed. Adrenal-cortical and chromaffin tissue are sometimes intermingled. Chromophobic as well as basophilic "chromaffin" cells have been observed in various species, and these may be of various sizes relative to the interrenal cells (see Nandi 1962; Matty 1966; Gannon et al. 1972).

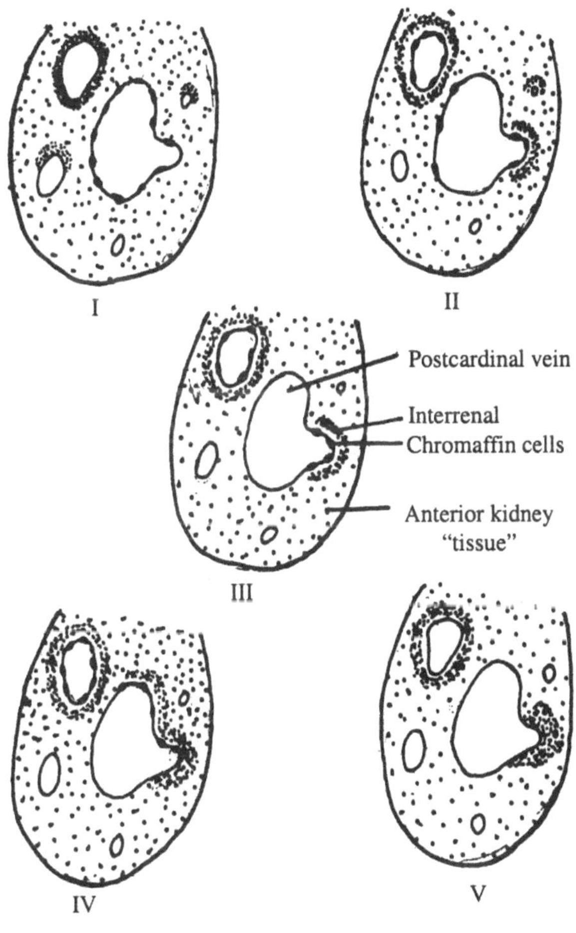

Fig. 13.33 Classification of types of chromaffin cell distribution. Open spaces represent postcardinal veins and their branches; light stippling indicates anterior kidney "tissue".
***Source:* Nandi (1962)**

The catecholamine storage and release system pertaining to the heart in *Heterodontus* is different from both the cyclostomes, in which the chromaffin tissue is located in the walls of the heart chambers and the teleosts, in which an extensive cardioregulatory adrenergic innervation of sinus venous, atrium and ventricle is found. Thus at least three different modes of storing catecholamines for cardiac regulation are present in fish (Gannon and Campbell, 1972).

Plisetskaya (1985) has dealt with the catecholamine levels and functions in cyclostomes. Catecholamines may be assumed to play some role in enhancing gluconeogenesis. According to author, must we agree with a suggestion that the action of catecholamines in agnathans is restricted to cardiovascular, respiratory and osmoregulatory, but not metabolic effects?

Catecholamine release from the chromaffin tissue into the venous blood in teleost fish is under sympathetic control (Fig. 13.34) (see Reid 1999).

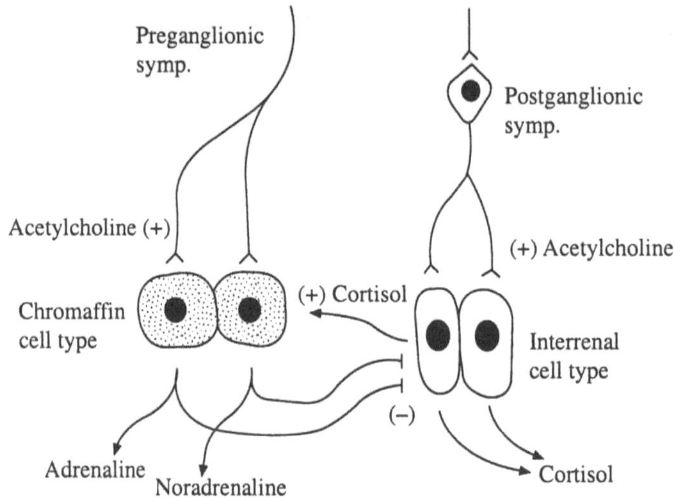

Fig. 13.34 Chromaffin and interrenal connections in carp tissue. Acetylcholine stimulates both chromaffin and interrenal cells. Differences may exist, that preganglionic fibres innervate chromaffin cells while postganglionic fibres run to the interrenal cells. A functional interference exists between catecholamines and cortisol. The acetylcholine receptors respond as well to nicotine as to muscarine. ***Source:*** **Hanke and Kloas (1995)**

The main circulating catecholamines in teleosts are adrenaline and noradrenaline. Plasma levels have been extensively measured but only in a few species. Although only a limited number of teleosts have been examined there does appear to be marked species differences in catacholamine mobilization that may in part be related to the activity level of species. In addition to marked species differences there are seasonal and water temperature related variations in catecholamine release (McDonald and Milligan 1992).

Catecholamines are released in the blood of fish by stress, including hypoxia, hypercapnia, anaemia, blood acidosis, and both exhaustive or violent exercise. However, the proximate cause for release appears to be hypoxia, perhaps modulated by the acid-base status of the animal. Catecholamines can exert a wide variety of effects on the physiology of fish, for example, on the distribution of red blood cells, on the permeability of the gill epithelium, on oxygen transport, on red blood cell volume and intracellular pH and on the branchial vascular resistance to flow. Catecholamines stimulate ventilation in fish but the response is seasonal (see Aota et al., 1990; Randall 1990; McDonald and Milligan 1992; also see Chapter 8). The reader is also referred to (Butler and Metcalfe 1983, 1988; Farrell 1984; Bushnell et al. 1992; Farrell and Jones 1992). Catecholamines are involved in the regulation of hepatic carbohydrate metabolism (read Danulat and Mommsen 1990).

9. PANCREATIC ISLETS (ENDOCRINE PANCREAS; ISLETS OF LANGERHANS)

The comparative endocrinology of the pancreas has not commanded as much attention as that of the thyroid or anterior pituitary though after the finding of principal islets in fish it has received constant, if not prolific attention.

Various aspects of progress in the field of fish pancreatic hormones have been reviewed recently (Plisetskaya 1989, 1990, 1995; Duguay and Mommsen 1994; Plisetskaya and Mommsen 1996). The comparative islet anatomy has been fully described by Epple and Brinn (1986). A. Epple and J.E. Brinn's (1987) book "The Comparative Physiology of the Pancreatic islets", an encyclopedic source of information, deals with most aspects of the structure of the vertebrate pancreas and the functions of its active peptides. A brief review by Gill (1994) has enhanced related information.

Figures 13.35, 13.36 illustrate the position of endocrine pancreas in fishes. It is now generally accepted that the pancreatic islets of the higher vertebrates (gnathostomes) contain four major cell types, which produce glucagon (A cells), insulin (B cells), somatostatin (D cells) and pancreatic polypeptide (F cells). The current situation is summarized in Fig. 13.37.

In lampreys and hagfishes, the islet organ is separated from the equivalents of the exocrine pancreas. The endocrine pancreas consist of an islet organ which develops from the anterior part of the gut in larval lampreys, at the junction of the oesophagus, bile duct and intestine. In hagfishes, it is derived mainly from the wall of the bile duct. In lampreys, the islet tissue may be present in two or three parts, cranial accumulation round the junction of the bile duct and intestine, a caudal or hepatic area between the intestine and the liver, often almost surrounded by hepatic tissue, at the level of hepatic portal vein, and cords of islet cells in the intestinal wall between the two main accumulations. Some of the interspecific morphological variations have recently been described: the location and arrangement of pancreatic endocrine tissue in larval and adult Southern Hemisphere lamprey *Geotria australis* differs markedly from those exhibited by comparable stages of Northern Hemisphere lampreys, (see Hillard et al. 1985).

Lamprey islets consists of groups of cells in lobules, the cells of one lobule secreting either insulin or somatostatin, and the two types of lobule being present in approximately equal numbers. In hagfishes, the islet peptide is predominantly insulin with rather few scattered somatostatin containing cells, at least in *Myxine glutinosa*. There may be a higher proportion of somatostatin cells in *Eptatretus stouti*. Thus, the hagfish islet parenchyma is basically a two-hormone endocrine gland.

The most recent advances as regards the hagfish islet parenchyma is that in some of the insulin cells a co-localization has been found immunohistochemically with an IGF-I-like peptide. It has a molecular weight of 2–4 kD and seems biochemically to be different from both hagfish

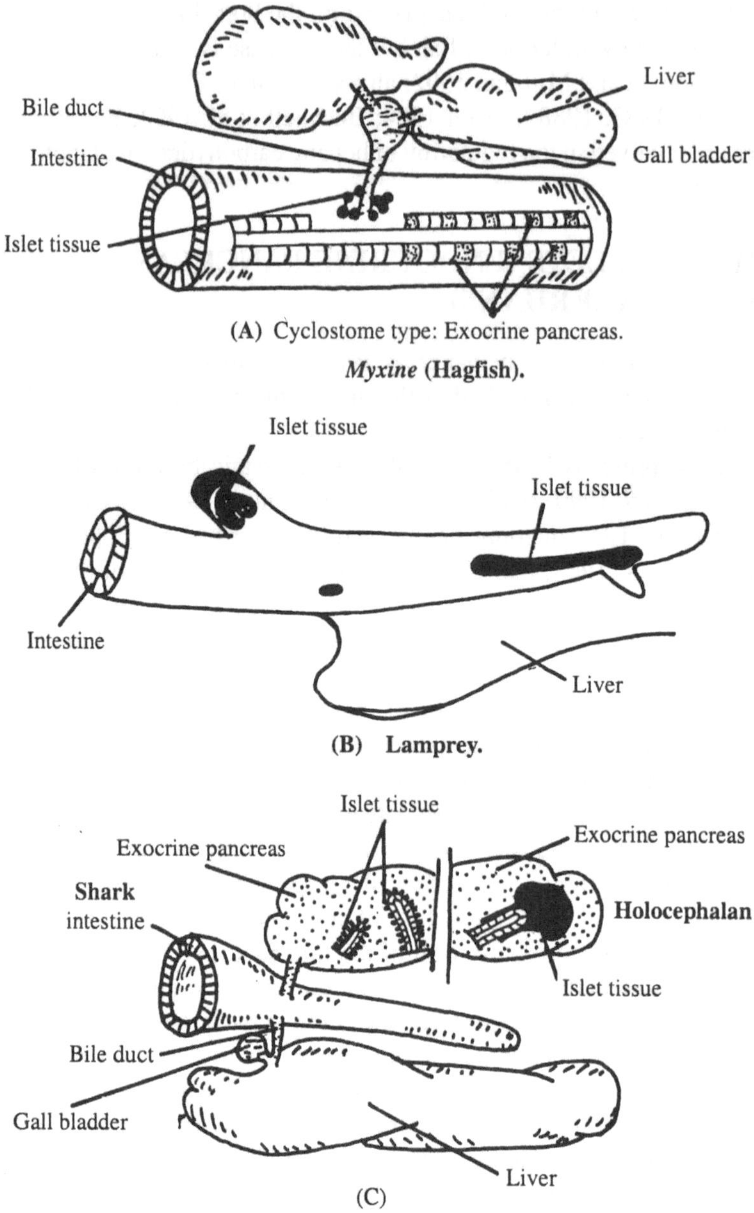

(A) Cyclostome type: Exocrine pancreas.
***Myxine* (Hagfish).**

(B) Lamprey.

(C)

Fig. 13.35 The pancreas types of fishes: (A) *Myxine.* Ringlike arrangement of islet cell follicles near the distal end of the bile duct. The exocrine pancreas is represented by specialized cells in the midgut epithelium. (B) Islet tissue in lamprey. (C) Primitive gnathostome type. Many sharks have a compact pancreas with islet cells occurring as an outer layer of small ducts (left). the endocrine pancreas of the holocephalans (right) exemplifies an intermediate stage to the tetrapodlike type.
***Sources:* Epple and Brinn (1986) and other related articles.**

and mammalian insulin, as well as from mammalian IGF-I. Here, also, the sequence of a preproinsulin-like growth factor cDNA has recently been established (Falkmer 1995).

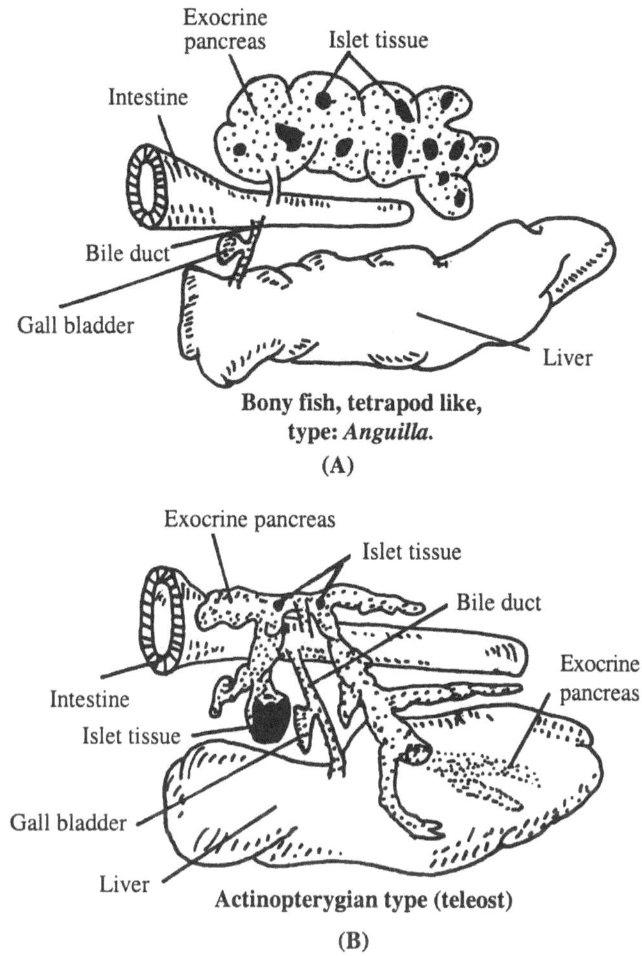

Fig. 13.36 **(A) Tetrapod type. A compact pancreas with scattered endocrine islets, similar to that of the tetrapods, is found in a few teleosts, e.g., *Anguilla*. The pancreas of the lungfishes is basically of the tetrapod type, but surrounded by folds of the stomach and intestine. (B) The majority of the islet tissue stems from a compact endocrine portion of the dorsal anlage. If this endocrine portion is not split up by invading exocrine tissue during ontogenesis, it remains as the compact endocrine mass of the Brockmann bodies. Actinopterygian type (teleost). In most actinopterygians the exocrine pancreas tends to split into small strands of tissue which sometimes invade other organs.**
Sources: **Epple and Brinn (1986) and other related articles.**

A rare third type of islet cell has been identified in the sea lamprey, *Petromyzon marinus*, and there is now some evidence that this cell contains a peptide of the pancreatic polypeptide family, thus bringing the cyclostomes a little closer to their vertebrate relatives (Van Noorden 1990; see also Epple 1969; Youson 1985; Epple and Brinn 1986).

Both elasmobranchs and holocephalans have a distinct extramural pancreas. The structure of the islet organ varies greatly between both groups (Fig. 13.35). Immunocytochemically, A, B, D, F, and GIP (gastric inhibitory peptide) cells have been indentified in the islet organs of

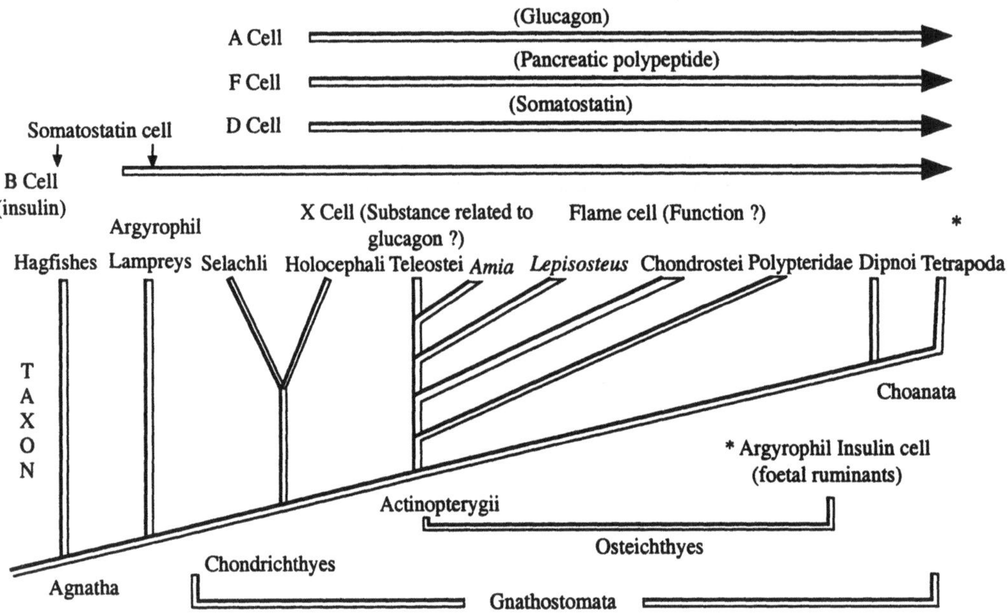

Fig. 13.37 The phylogenetic distribution of the major types of islet cells. *Latimeria* has been omitted, because of the limited information on its islet cytology.
***Source:* Epple and Brinn (1986)**

elasmobranchs. The islet organ of holocephalan shows the presence of A, B, D, F cells besides X cell. In addition, small numbers of enkephalin, GIP, and serotonin cells have been found. In contrast to the elasmobranchs, the islet cells do not seem to be innervated. Information of considerable interest has appeared concerning the immunocytochemical and structural studies of pancreatic hormones (see Plisetskaya 1990). The islet parenchyma in sharks and rays show a co-localization of insulin and IGF-I-like high-molecular peptide. The IGF-I immunoreactive peptide is a high-molecular compound, probably representing a pro-IGF-peptide. IGF-I receptors have also been detected (Falkmer 1995).

In actinopterygians, the topographic relations between the islet organ, exocrine pancreas, and liver show a variety unsurpassed by other vertebrates. Nevertheless, two typical trends are obvious: (1) to concentrate the islet tissue in one or a few large accumulations and (2) to mingle the exocrine pancreas and liver tissue. These two trends are very pronounced in Polyteridae as well as in many of the most advanced teleosts. The large accumulations of islet tissue are referred to as "principal islets", and since these are almost always in contact with some exocrine tissue, principle islets together with the closely associated exocrine pancreas are termed "Brockmann bodies".

In Polypteridae, a large principal islet with a complete rim of exocrine pancreas is buried deep in the liver. A few small islets occur in nearby exocrine regions. All four major types of islet cell appear to be present, and the islets are well innervated. In Chondrostei, the pancreas is rather diffuse and is partly included in the liver; the islet organ seems to consist of scattered inconspicuous islets of unknown cytology. In *Lepisosteus*, the exocrine pancreas mingles with the liver to an even greater degree than in Polypteridae and extends along the bile duct. Here, it surrounds large accumulations of islet cells, forming an aggregation of Brockmann bodies. The islet cytology of garpikes is puzzling, since in addition to the four main cell types, there is a very frequent, fifth, flame-shaped cell of unknown function. The islets of *Lepisosteus* are highly innervated by fibres resembling the classical cholinergic type. In *Amia calva*, the pancreas is rather diffuse and partly invade the liver; the islet tissue is scattered and does not form a distinct accumulation. *Amia* has the four major types cells, and as in *Lepisosteus*, there is a well-developed innervation which ultrastructurally appears to be "cholinergic".

In the teleost *Takifigu niphobles*, Endo et al. (1991) have reported that some 'neurotransmitters' are released non-synaptically in the endocrine pancreas and that same 'neurotransmitters' may enter the blood stream as 'neurosecretory hormones'.

Teleostean species which have principal islets with little associated exocrine tissue in particular have contributed important data in islet research. Very large and almost purely endocrine Brockmann bodies in *Lophius*, *Cottus* and *Ictalurus* are probably the most widely studied teleost islets. Brockmann bodies have been used extensively in studies of insulin biosynthesis. The "reason" for the extremely strong peptidergic islet innervation in teleosts is unknown, but it may play an important role in the integration of osmoregulatory and metabolic signals. The shape of the teleost pancreas varies from compact with scattered (e.g., eel, pike) to extremely diffuse with Brockmann bodies. In the salmonids, there is a Brockmann body near the gall bladder, and considerable quantities of exocrine and endocrine pancreas occur between the numerous pyloric caeca. As in many tetrapods, the teleost islets tend to have a core of B and D cells and a rim of A and F cells. According to Falkner (1995), in the principal islets of the cyprinids, there is obviously no co-localization of the insulin and the IGF-I (insulin-like growth factor) immunoreactivities, whereas in *Cottus* Brockmann bodies the co-localization is still there.

In Dipnoi, there is an almost complete "wrapping" of the pancreas and adjacent spleen region by parts of the stomach and the intestine. Interestingly, the islet organ of *Protopterus* consists of a few well-encapsulated "principal islets" while in *Neoceratodus* the islets are scattered in the mammalian fashion. So far, only A , B, and D cells have been described (Epple and Brinn 1986).

According to Conlon et al. (1988), onto-and phylogenetical studies of the evolution of cells, producing regulatory peptides, belong to the "hormone families" of insulin, somatostatin, glucagon, and PP (the pancreatic polypeptide), have shown that the islets of Langerhans in vertebrates form a substantial part of the large neuroendocrine system (NES). The NES consist of three major parts, viz., (i) neuronal cells of the central and peripheral nervous systems, (ii) disseminated cells in the mucosa of the alimentary tract (and that of other hollow organs), and (iii) the parenchymal cells of the classical endocrine glands.

A separate islet organ does not occur in the NES until the appearance of the first vertebrates, viz., the Agnatha. Here a grossly visible islet organ exists, free from exocrine, acinar pancreatic parenchyma. It is a two—hormone organ with insulin and somatostatin cells only. The first pancreatic gland in evolution appears in holocephalans; it is equipped with large islets of Langerhans, containing—in addition to insulin and somatostatin—also glucagon-/glicentin-producing cells. The final evolutionary stage—the four-hormone islet parenchyma of mammalian type—is, however, reached already in sharks and rays.

Studies by means of analyses of the amino-acid sequences have revealed interesting features in the molecular evolution of the some of the members of the four islet hormone families. In the vertebrates, it is clearly established from the recent reports on the molecular structure of insulins of *Myxine*, *Hydrolagus*, *Squalus* and *Cottus*, that the well-known basic structure of the monomer and the dimer has been conserved throughout the whole of vertebrate evolution. Thus, the A-and B-chains with three cystine-SS-bridges are present in all the insulins sequenced (about 40), so far. Also the biosynthesis of the insulin molecule to the two-chain hormone, is a processing mechanism which is common to all the insulin molecules studied in this respect. When *Myxine* and human insulins are compared, it is found that the number of radical substitutions in the insulin monomer affect less than 25% of the molecule and occurs in those parts of A- and B-chains that are known to be highly variable.

In this way, and at least in vertebrates, insulin of GEP origins represents a highly conserved molecules and thus by association, a ligand of considerable adaptive value. Duguay and Mommsen (1994) reviewed the current knowledge of the molecular biology of insulin, glucagon, glucagon like peptide, somatostatin and pancreatic peptide in piscine systems.

Plisetskaya (1985) has described the metabolic events and their hormonal regulation in agnathans. Many questions dealing with the role of insulin in agnathans are still to be addressed. One of them is the question of the role of this hormone in larval lampreys and during metamorphosis when the larvae stop feeding and many important metabolic changes in the liver take place.

The primary hormones stimulating gluconeogenesis in vertebrates are the glucocorticoids and "the hormone of hunger", glucagon. These hormones are either absent, or present in extremely low quantities, in agnathans. According to Duguay and Mommsen (1994), the best known and principal function of glucagon in fishes is strong glycogenolytic action on liver resulting in hyperglycemia.

Glucagon-like immunoreactivity has been found in the extracts of agnathan intestine and the so-called "cranial" pancreatic islets, though it is possible that the latter activity was due to contamination by gut tissue.

Ince (1983) has discussed the pancreatic control of metabolism (and other processes), and the factors which influence islet activity in teleosts.

Points of summary are:
(1) Insulin enhances proteogenesis and lipogenesis in teleosts, whilst its role in carbohydrate metabolism seems to be directed more towards glucose oxidation than glycogenesis. Growth and osmoregulation are two further processes in which insulin and/or islet hormones have been implicated. (2) Glucagon stimulates glycogenolysis and gluconeogenesis, but there is no convincing evidence of a lipolytic role in teleosts. (3) Plasma insulin levels are dependent on dietary protein level, nutritional state, and season. Amino acids provide a greater stimulus for insulin release than glucose, although intracellular mechanisms involved are poorly understood. Circulating catecholamines both inhibit and stimulate insulin release. Somatostatin exerts a strong inhibitory influence. (4) The effects of fatty acids, gastrointestinal hormones and neural influences on insulin release, and of altered hormone dynamics on steady-state insulin levels, are unknown. (5) The factors regulating glucagon and somatostatin levels and release in teleosts, remain to be determined. (see also Duguay and Mommsen 1994). According to Chan and Steiner (1994), the relative expression of insulinlike growth factors (IGF-I and II) in teleosts during embryogenesis and their actions in specific tissues need to be clarified. It would also be interesting to determine whether hagfish and lamprey contain only a single IGF gene. IGF-I has been the subject of a number of recent investigations (Plisetskaya and Duan 1994; Berwert et al. 1995; Duan et al. 1995, Gutierrez et al. 1995, Moriyama et al. 1995, 1997, Parrizas et al. 1995; Moon et al. 1996; read also Reinecke 1995; Reinecke et al. 1997).

Epple (1987) has focussed on osmoregulatory islet functions in vertebrates. In Agnatha, it appears that the possible osmoregulatory role of the islet organ has been studied only in the myxinid *Eptatretus stouti.* Very little seems to be known concerning the osmoregulatory role(s) of islet hormones in Chondrichthyes. In teleosts, an osmoregulatory role of the islet organ has been clearly established. The osmoregulatory interactions between the islet organ and adenohypophysis of the teleosts have not yet been studied extensively. An interpretation of the osmoregulatory role(s) of the islet hormones requires consideration of their impact on intermediary metabolism.

Feeding typically increases insulin secretion. Amino acids are the most powerful stimulator of insulin secretion in fish with arginine being the most potent of the amino acids tested. Nonetheless, high carbohydrate diets are more effective at stimulating insulin release than low carbohydrate diets. Insulin levels also increase with the parr-smolt transition in salmonids, decrease during spawning and with a drop in water temperature, and show seasonal fluctuations. Starvation may be responsible for some of the drops in insulin levels. Glucagon and glucagon-like peptide levels fluctuate to a smaller extent than insulin. Glucagon and glucagon-like peptide levels decrease with starvation but the ratio of glucagon (and glucagon-like peptide) to insulin increases; thus change is consistent with activation of gluconeogenesis (McDonald and Milligan 1992; for details regarding pancreatic hormones under various physiological conditions-smoltification, spawning migration, and experimental feeding *versus* feeding in teleost fishes, read Plisetskaya 1990).

Regulation of secretion of somatostatin from fish islets has not been studied extensively. Most of the physiological studies of Brockmann body somatostatins have focussed on SST-25 in coho salmon, *Oncorhynchus kitsutch* (read Duguay and Mommsen 1994). The physiological function of pancreatic polyptide (PP) in fish is unknwon (Duguay and Mommsen 1994).

10. CAUDAL NEUROSECRETORY SYSTEM: NEUROHORMONES FROM FISH TAILS

In addition to the cranial (hypothalamo-neurohypophysial) neurosecretory system found in all vertebrates, most fishes have a caudal neurosecretory system composed of spinal cord neurons whose axons project to a neurohaemal area. In teleosts the axonal terminals are aggregated into an organ, the urophysis, represented by a ventral enlargement near the posterior end of the spinal cord (Fig. 13.38). The teleosts urophysis is generally a distinct, easily recognizable, lobate structure of variable external form. Significant morphological variations lie in the organization of the neurosecretory fibres in relation to the vascular bed and in the degree of penetration of the meninx by the neurosecretory fibres to form an organ external to the spinal cord proper.

A urophysis as such does not exist in fish groups other than teleosts and holosteans, but ganoid and elasmobranch fishes possess diffuse neurohaemal areas on the ventral surface of the posterior spinal cord to which large caudal neurosecretory neurons project (Fig. 13.38). The elasmobranch caudal neurosecretory system may represent the more primitive condition, and stages in the evolution of the advanced urophysial types can be visualized. The peculiar histology shown by caudal neurosecretory system appears to have taxonomic significance.

For the holocephalan *Hydrolagus colliei*, the caudal neurosecretory elements might not be present. In cyclostomes, no elements of caudal neurosecretory system were identified. Lower

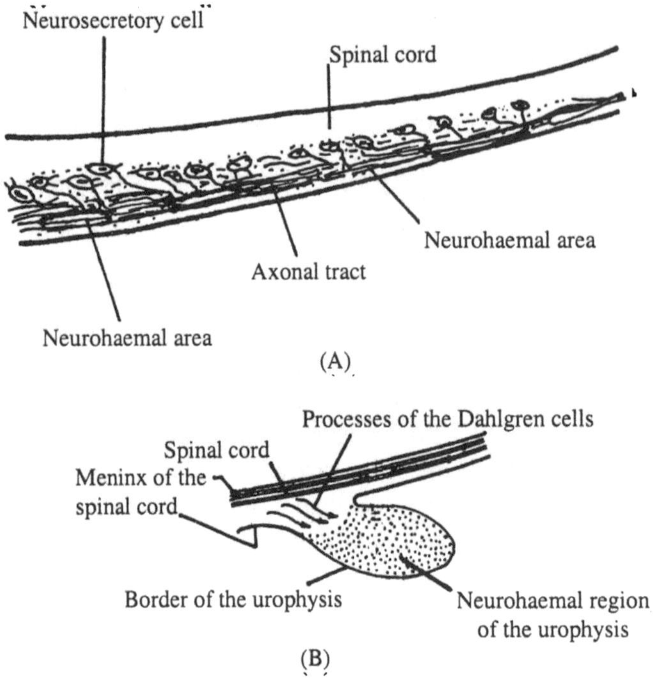

Fig. 13.38 **(A) Diagram of caudal region of the spinal cord indicating neurosecretory cells, axonal tract and neurohaemal areas in the banded dogfish, *Triakis scyllia*. (B) Sagittal section of teleost *Lebistes reticulatus* urophysis**
***Sources:* (A) Kobayashi et al. (1986). (B) Fridberg and Bern (1968)**

actinopterygians, *Acipenser medirostris* and *A. guldenstadti*, possess a caudal neurosecretory system quite similar to that of the elasmobranchs. On the other hand, in other members of primitive bony fishes, such as *Polypterus* sp. and the garpike, *Lepisosteus osseus*, it is not certain whether these forms possess a urophysis. (see for urophysis in *Onchorhynchus*, Meek and Nieuwenhuys, 1998; Fig. 15.57).

In teleosts, the perikarya of the neurosecretory cells are usually located ventrolateral to the central canal and are distributed anteriorly as far as the fifth to eighth preterminal vertebrae. The cell somata are often distributed caudally in the filum terminale (Fig. 13.39). In elasmobranchs, the neurosecretory cell perikarya are huge and occur extensively along the spinal cord.

Caudal neurosecretory cells are usually large and polymorphic, and their nuclei are often lobated. Although caudal neurosecretory cells may generally be stained with acidic dyes, a small number of basophilic cells and chromophobe cells are encountered.

On the basis of cell size, there seems to be at least two or three types of caudal neurosecretory cells in teleosts and elasmobranchs. It is not known whether cells of different sizes have different functions. There is immunocytochemical evidence, however, that they produce the same hormones, irrespective of differing size.

The caudal neurosecretory system develops at the late larval stage and later than the diencephalic neurosecretory system. After total extirpation of the system, the ependymal cells apparently transform into neurosecretory cells, and the system regenerates (e.g., *Tilapia mossambica*, *Clarias batrachus*). However, there are species in which the system does not regenerate (e.g., *Misgurnus*

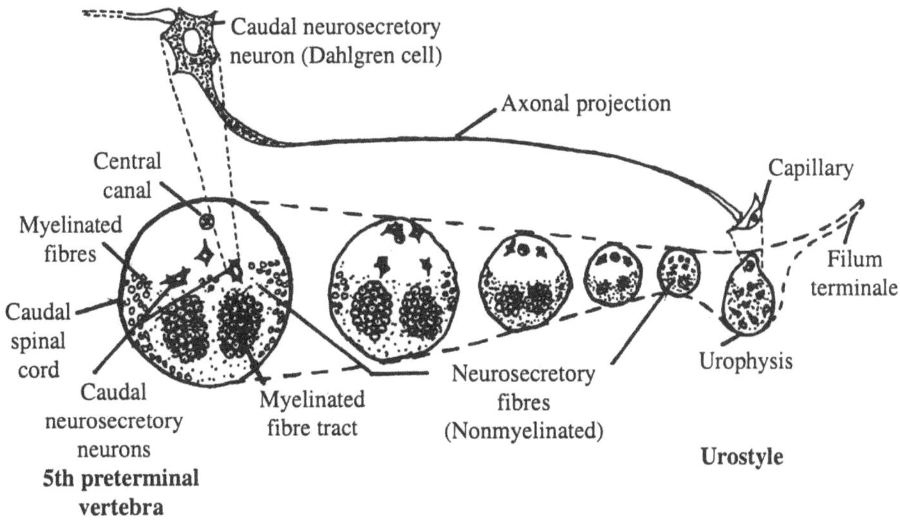

Fig. 13.39 Diagram of caudal spinal cord and urophysis of the gobiid teleost *Gillichthys nobilis* (long-jawed mudsucker), showing several transverse sections. The two types of caudal neurosecretory neurons (formerly called Dahlgren cells) are indicated (stippled, CRF/U_I reactive cells; solid, CRF/U_I and U_{II} reactive cells). Above the spinal cord is the diagram of a single caudal neurosecretory neuron showing its innervation from the brain and its axonal projection to a urophysial capillary.
***Source:* Bern et al. (1985)**

anguillicaudatus, *Oryzias latipes*). The discrepancy between the data may be due to age or species differences in the fishes used. Oka et al. (1993) have studied the early development of the caudal neurosecretory system of the chum salmon (*Oncorhynchus keta*) as revealed by immunohistochemistry for urotensins I and II.

Electron microscopy reveals that axon endings terminating at capillaries in the urophysis contain large neurosecretory granules and small vesicles, at least partly, are the carriers of acetylcholine.

Projections from four brain centres to the caudal neurosecretory perikarya have been demonstrated: (1) cholinergic neurons in the ventromedial division of the reticular formation, (2) small catecholaminergic neurons forming a rostrocaudal longitudinal column in dorsolateral aspects of the medulla, (3) peptidergic neurons in the dorsal tegmental magnocellular nucleus, and (4) small cholinergic neurons in the nucleus of the medial longitudinal fascicle. In addition, the caudal neurosecretory cells seems to be innervated by collaterals of caudal neurosecretory nerves and by catecholaminergic neurons located in the posterior spinal cord.

The urophysis contains monoaminergic fibres in some fishes but not in others. It also shows a high concentration of acetylcholine, suggesting the presence of cholinergic fibres. It is not yet fully understood whether neural mechanisms control hormone release at the level of the urophysis.

Urotensins I (UI) and II (UII) have been characterized from teleost species. There is no *a priori* reason why only two neurohormones should be secreted from the caudal neurosecretory system. UI has homologous amino acid sequence with corticotropin-releasing factor (CRF) and sauvagine, whereas UII is partially homologous with somatostatin. The search for other peptides may prove rewarding. Whether amines and other small molecules of mediator significance are released into the circulation needs further examination in additional species. The presence of large quantities of acetylcholine in the urophysis, has so far not resulted in further investigation, either as to source or to function. In addition, met-enkephalin occurs in carp neurosecretory neurons and the urophysis. Arginine vasotocin is present in the urophysis of a few teleosts. It has already been suggested that the N-terminal portion of U II precursor may be biologically active.

The bulk of the immunocytochemical information presently available indicates that U I and U II may exist in the same neurosecretory neurons in several fish species. It has been reported that urotensins I and II are derived from separate precursors as a result of expression of two different genetic determinants. In view of the fact that these two peptides have often opposing effects, the physiological significance of co-localization is puzzling as well as the manner in which differential synthesis and release are regulated.

The rich and varied innervation of the caudal neurons makes it apparent that the caudal neurosecretory system does not operate autonomously. In addition to catecholaminergic and peptidergic innervation, there may also serotoninergic input in non-teleostean and teleostean fishes. The effect of neurally transmitted information from the anterior central nervous system on electrophysiological activity of caudal neurons was already established. The system could also be affected by blood-borne agents other than ions or osmotic pressure.

Myotropic effects of urotensins have been recorded. The receptor situation in regard to both urotensin I and urotensin II invites further clarification. It is not surprising that urotensin I binds to CRH (corticotropin-releasing hormone) receptors, but it may also bind to VIP (vasoactive intestinal polypeptide) receptors. Urotensin II may bind to somatostatin receptors in some cases. The question of specific receptors for U I and U II remains open in fishes.

Despite recognition of the many actions of urotensins, the physiological role of the caudal neurosecretory system continues to be elusive. Why the caudal end of the central nervous system in most fishes should produce two peptides functionally parallel totally (urotensin I) or partially (urotensin II) to factors produced in the hypothalamus at the cranial end remains puzzling. Urotensin I can mimic the action of CRH, and urotensin II can mimic partially the effects of somatostatin-14 (and *vice versa*). In addition, the bioactivity of urotensin II in fishes has been related to osmoregulation (Fig. 13.40), cardiovascular regulation, reproduction and metabolism, and urotensin I has osmoregulatory effects directly by its action on transepithelial transport, as does urotensin II.

Attention has been drawn to the fact that the urophysis secretes its hormones into the veins joining with the renal portal system, suggesting that likely targets may be in the renal field (Fig. 13.41). However, there are branches from this system that could allow urotensins to reach such organs as the urinary bladder and the intestine by retrograde flow and the liver by connection to the hepatic portal vein. Effects on more distal organs (gills and opercular membrane, pituitary, aorta) may prove to be only of pharmacological significance. The discovery of urotensin II-immunoreactive neurons anterior to the caudal neurosecretory system raises the possibility of other routes by which urotensins can present themselves to various regions of the organism through the systemic circulation as well as the cerebrospinal fluid.

The physiology of the caudal neurosecretory system continues to defy final delineation. For coverage, see (Fridberg and Bern 1967; Bern 1969, 1990; Chester Jones et al. 1974; Fontaine 1975; Bern et al. 1985; Kobayashi et al. 1986; Larson and Bern 1987; McDonald and Milligan 1992; Oka et al. 1993, 1995; Wendelaar Bonga 1993).

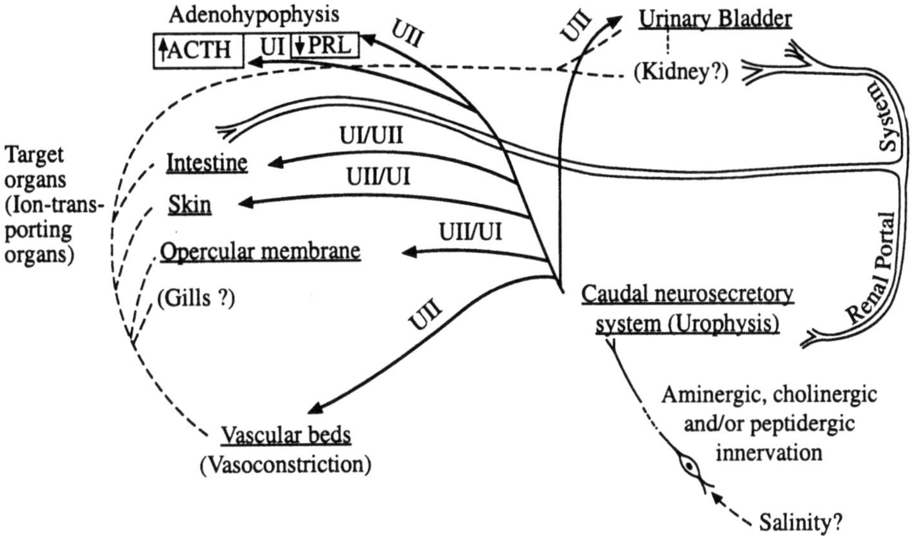

Fig. 13.40 Possible pathways of urotensin action on teleost osmoregulation. Direct (→) and indirect (- - - →) routes of action are shown.
***Source:* Larson and Bern (1987)**

11. GONADS

The principal circulating sex steroids in male teleosts are the androgens, 11-ketotestosterone (KT) and testosterone (T) (in order of importance), while in female teleosts it is 17β estradiol (E2). Final maturation in males and females in many teleosts is brought about by the progestogen 17α, 20β -dihydroxyprogesterone (P). Significant circulating levels of T are also found in females because it is the immediate precursor to E2.

In female teleosts plasma E2 and T gradually increase during vitellogenesis, followed in many cases, by a decrease in E2 before and during oocyte maturation. There is also a temporary increase in T before P increases, which induces oocyte maturation. In males, KT and T rise during spermatogenesis with KT reaching a maximum just prior to the start of spermiation. The concentration of both androgens decrease during spermiation, whereas P shows an important rise related to its putative role in regulation of spermiation and control of male spawning behaviour. However, P is not the final maturation steroid in all teleosts, and absolute plasma steroid concentrations vary dramatically among families (see McDonald and Milligan 1992). According to Brantley et al. (1993), a pattern of 11-ketotestosterone elevation in the courting morph exists across six species with male dimorphism, including *Porichthys notatus* (list included *Salmo salar*, *Lepomis macrochirus*, *Sparisoma viride*, *Thalassoma duperrey*, *Coris julis*). At this point, evidence is strong for the role of this steroid in the induction or maintenance of courtship display structures. Clearly, further study of the relationship of morph-specific behaviour to particular androgens is needed.

Selcer and Leavitt (1991) have given a comprehensive account of estrogens and progestins. It is unknown if progesterone (progestogen) has a physiological function in cyclostomes or elasmobranchs. In teleosts, gonadotropins stimulate the production of a steroid that initiates oocyte maturation. In a variety of teleosts, progesterone, 17α–hydroxyprogesterone, 17α, 20β–

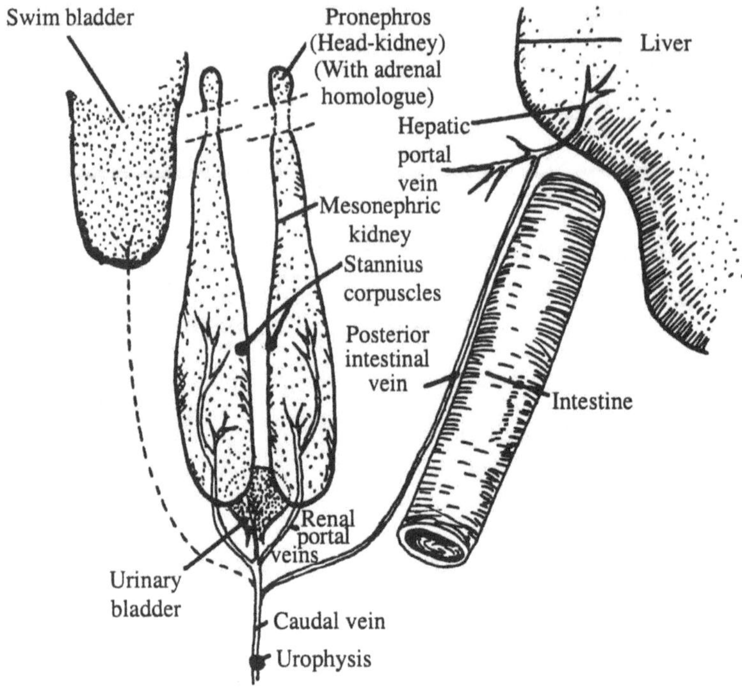

Fig. 13.41 Diagram of some vascular connections in teleost fishes. The renal portal veins carry urotensins to the "renal field" which includes the kidneys and the dorsum of the urinary bladder and possibly other tissues embedded in the kidney such as the calcium-regulating Stannius corpuscles and the juxtaglomerular cells. However, some branches of the caudal vein are known to bypass the kidneys and connect with the liver and probably the urinary bladder. Haemodynamic status of the posterior intestinal vein is such as to allow possible backflow of blood from the caudal vein to the intestinal wall. Hence, "direct" targets of urotensins could include the intestine, the liver, and the swim bladder, in addition to structures in the renal field.
***Source:* Bern et al. (1985)**

dihydroxyprogesterone, cortisol, and deoxycorticosterone have been shown to be effective in the initiation of oocyte maturation. Of these, 17α, 20 β –dihydroxyprogesterone is generally the most potent. This hormone is thought to be produced by the teleost ovary by a two-step mechanism whereby the thecal cells produce 17α–hydroxyprogesterone, which is transferred to the granulosa cells for conversion to 17α–20 β dihydroxyprogesterone. The site of action of this steroid is on the oocyte surface.

Although estradiol –17β is the primary estrogen that promotes vitellogenesis, the hormonal specificity of the vitellogenic response differs among vertebrate groups. In the Pacific hagfish (*Eptatretus stouti*), esterone is as effective as estradiol in stimulating vitellogenin production, and progesterone is also able to elicit the response, although to a lesser degree than estrogens. In some teleost fishes, androgens can induce vitellogenesis; however, the doses required are quite high. This induction may be a result of androgen conversion to estrogens, except possibly in the case of goldfish, where a purported nonaromatizable androgen (11α –ketotestosterone) was shown to induce vitellogenesis. Estrone may also be important in regulating vitellogenesis of some teleosts, and estriol has been demonstrated to be most effective estrogen in eliciting a vitellogenic response in one teleost, *Tilapia aurea*.

In lampreys, estrogens do not appear to influence gonadal differentiation. Among elasmobranchs, estradiol-17β, as well as progesterone, testosterone propionate and deoxycorticosterone, has a feminizing effect on sexual differentiation of the gonads; however, gonadal transplants did not reveal any effect of steroids on gonadal differentiation.

For teleosts, there is a controversy as to the role of sex steroids in gonadal differentiation. One widely accepted hypothesis, based on experiments employing exogenous steroid treatment, is that androgens are inducers of male gonads and estrogens are inducers of female gonads. There is ample evidence that estrogens and androgens given at the appropriate time will result in a functional female or male, respectively, regardless of genetic sex. However, it has yet to be demonstrated conclusively that the sex steroids are available to the gonads at the appropriate time for gonadal differentiation. Sex steroids in teleosts are also important in directing differentiation of the gonadal ducts.

Sex steroids may be a factor in the development of secondary sex characteristics of lampreys. Similar information is not available for hagfish, partly because they exhibit little sexual dimorphism other than gonadal.

It has been reported that sex steroids can cause sex reversals in teleosts but it is yet to be established if the central nervous system is involved in these changes.

In teleosts, estrogen appears to have negative feedback effects on the hypothalamus, inhibiting gonadotropin release. In contrast, for cyclostomes there is no evidence to support a gonadotropin gonadal feedback relationship, despite several attempts to elucidate the control of reproductive cycles in hagfish.

Fontaine et al. (1991) have discussed evolutionary aspects related to endocrine control of reproduction. From cyclostomes to mammals, the localization, cytological aspect, and enzymatic equipment of steroidogenic cells did not undergo changes in their general outlines. Major sex steroids released by gonads in blood are androgens (C19-steroids; e.g., testosterone, androstenedione), estrogens (C18-steroids; e.g., estradiol, estrone) and progestagens (C21-steroids; e.g., progesterone) in proportions fluctuating with sex, age and sexual stage. No compound is exclusively produced in one sex or the other.

Metabolic pathways of steroidogenesis are not the same in all vertebrates. In teleosts, a 20β-hydroxysteroid dehydrogenase allows the production of a powerful progestagen, 17α-hydroxy, 20β-dihydroprogesterone, which plays an important role in the last steps of male or female gametogenesis; a uridine diphosphate glucurosyl transferase allows the gonadal release of conjugated androgens and estrogens, which act as pheromones; and a high 11β-hydroxylase and 11β-hydroxysteroid dehydrogenase are implied in the synthesis of a very powerful androgen, 11-ketotestosterone.

Gonadal synthesis of particular compounds, especially in teleosts (11-ketotestosterone, 17α-hydroxy, 20β-dihydroprogesterone, glucuroconjugated derivatives) show a taxonomic diversification, the meaning of which deserves attention.

Sex steroids are released in blood, some of them after they are conjugated in the gonads. Plasma levels are very different depending on the class and even species. In a given species, they vary qualitatively and quantitatively largely in relation to sex cycle.

Circulating steroids can be bound to specific plasma proteins, characterized by a high affinity (KD 4°C $< 10^{-7}$ M) and low capacity ($<$ uM), proteins which underwent an important evolution. A binding system with a high affinity does not seem to exist in cyclostomes. A binding protein with a broad specificity is present in elasmobranchs: androgens, estrogens, and progestagens (and corticosteroids) are competitive for the same site. That may represent a primitive situation from which the binding systems of more recent vertebrates evolved. In other ectotherms (e.g., teleosts), the binding system is made of two components, the specificities of which are different; one binds only androgens and estrogens, the other binds progestagens (and corticosteroids).

In teleosts, recent data in salmonids indicate the presence of two pituitary LH- and FSH-like GTHs; one of them (now called GTH II) has been exclusively studied and appears to stimulate specific phases of steroidogenesis. During the first steps of the sex cycle, GTH II increases androgen production by a stimulation of "3β–hydroxysteroid dehydrogenase – Δ_5 – Δ – isomerase" complex in Leydig and thecal cells; in females, androgens are further metabolized into estrogens by granulosa cells. At the end of the sex cycle (spermiation in males and oocyte maturation in females), high circulating levels of GTH II are able to induce the production of 17α–hydroxy, $\beta20$–dihydroprogesterone; this involves a decrease in enzymatic activities implied in the cleavage of side-chain of C_{21}-steroids in thecal and Leydig cells. Additionally, a stimulation by GTH of 20β–hydroxysteroid dehydrogenase in granulosa cells was demonstrated. Therefore, quantitative changes in plasma GTH levels are important factors for determining very large quantitative and qualitative plasma sex steroid variations during teleostean sex cycles. As far as GTH is concerned, it appears to differ from GTH II, showing a lower specificity for 17α–hydroxy, 20β-dihydroxyprogesterone production, during oocyte maturation, whereas both hormones have a similar effect on estradiol production by midvittelogenic ovarian follicle from salmonids. Radioimmunoassays of GTH I and GTH II suggest that the synthesis and secretion of GTH I are important during prematurational stages of spermatogenesis and vitellogenesis. Thus, physiologically, GTH I could be more important at these stages and GTH II more important later on.

In teleosts, other mechanisms certainly exist. In lampreys, the interrenal and meta-adenohypophysis appears to be involved in sex steroid synthesis and its control.

In all gnathostomes, sex steroids were shown to control the gonadotropic pituitary-hypothalamus axis. In fish, beside the classical negative feedback that exists in adults, a peculiar effect of sex steroids was demonstrated in juveniles of some species (eels, salmonids) (i.e., an intense stimulation of pituitary GTH II synthesis). A positive effect of sex steroids was also demonstrated on GnRH synthesis in the eel and GnRH release in juvenile trout.

Receptors for sex steroid appear to exist in all vertebrates in nuclear and sometimes also cytosolic fractions. Different receptors for C_{18}–, C_{19}–, or C_{21}– steroids were generally described. However, their specificity in teleostean tissues appears broader than in those of higher vertebrates. It is interesting to note that specific receptors for the main circulating androgen in teleosts (11-ketotestosterone) has, as yet, not been characterized.

The expression (or nonexpression) of hormonal receptors in various tissues is a very important evolutionary mechanism. The most obvious example is the variety of steroid-controlled reproductive organs, secondary sexual characters, or sexual behaviours, exhibited in the animal kingdom, all of these characters result from the expression of steroid receptors in a select variety of different targets.

Thus, estradiol and progesterone receptors exist in female organs, allowing for the steroid-controlled maintenance and development of oocytes in all oviparous vertebrates. For the structure and regulation of genes for estrogen receptors, contribution from Drean et al. (1994) is quite informative.

With regard to secondary sexual characters, androgen receptors have been detected in the skin of several vertebrates, including teleosts. Also, one should recall that a number of nervous and myogenic structures, which are targets for sex steroids, are implied in a variety of sexual behaviour in lower vertebrates (e.g., electric organ of a mormyrid fish).

Gonadal PRL receptors were characterized in several classes including teleosts. Estradiol stimulates bone resorption in teleosts. For the rôle of gonadal steroids, in the turnover of bone and scales in Osteichthyes, see Dacke (1989).

Stacey (1983) discussed the role of hormones in reproductive behaviour of teleosts. There appears to be no clear evidence that hormonal control of male sexual behaviour in teleosts differs substantially from what has been demonstrated in other vertebrate groups, i.e., that the onset and maintenance of male sexual activity occurs in response to increased release of steroids from the maturing testes.

Sex steroids evidently play important roles in regulating female sexual behaviour in the guppy (*Poecilia reticulata*), and male sexual behaviour in a number of species. Specific brain areas which concentrate sex steroids have been identified in several teleosts, although stimulation of spawning or other reproductive behaviours in teleosts by a direction of steroids on the brain has yet to be demonstrated.

The reader is referred to Selcer and Leavitt (1991) for other specific information concerning estrogens and progestins. For sex steroids, useful earlier information can be obtained from Matty (1966), Liley (1969), Chester Jones et al. (1974), Liley and Stacey (1983; see specially secondary sexual characters), Liley and Rouger (1990), Borg (1991), Canario (1991) and Kime (1991, 1993). For seasonal endocrine cycles in marine teleosts, see Pankhurst and Carragher (1991).

The information in connection with (i) mechanisms of estrogen action in liver and (ii) endocrine and nonendocrine factors controlling vitellogenesis has been covered by Ho (1991). Estrogen receptors have only been identified in the livers of a few fishes. Among the estrogens, estradiol is the most potent for the induction of vitellogenesis in fishes. The potencies of other two common estrogens, estrone and estriol, relative to that of estradiol, vary according to species. Certain androgens are able to induce hepatic vitellogenesis, for example, in *Carassius auratus* and *Gobius niger*, as exceptional cases.

The field of behavioural endocrinology in teleosts is in its infancy. Cardwell and Liley (1990) have taken up this line of investigation in stoplight parrotfish. *Sparisoma viride*, a protogynous coral reef fish. High androgen levels during periods of intense aggressive encounters induce increased aggression and territory vigilance, promoting the chances of successful defence against intrusion.

12. KIDNEY: RENIN-ANGIOTENSIN SYSTEM (RAS)

Renin is a proteolytic enzyme that acts on renin substrate to form decapeptide angiotensin I (ANG I), which is subsequently converted to octapeptide angiotensin II (ANG II) by converting enzyme (Figs. 13.42 and 13.43). ANG II possesses a broad spectrum of biological activities, and this entire cascade is known as the renin-angiotensin system.

The major site of synthesis and secretion of renin is juxtaglomerular cells (JG cells) of the afferent arterioles of the kidney. Specific components of the mammalian juxtaglomerular apparatus (JGA) are juxtaglomerular cells containing renin-forming granules, the macula densa (MD) and extraglomerular mesangium (EGM).

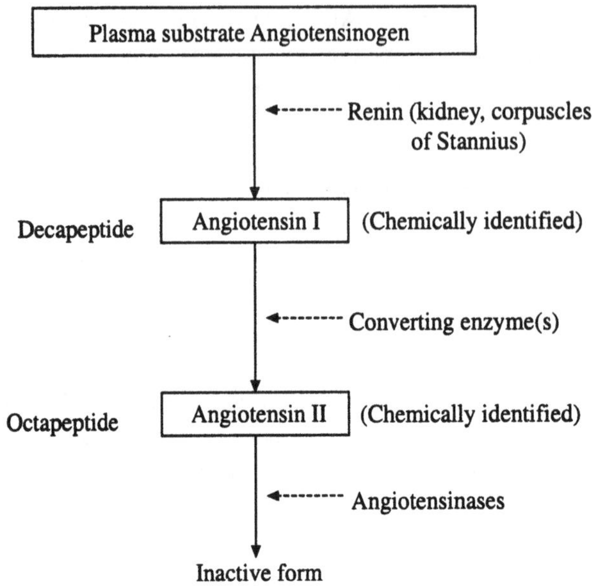

Fig. 13.42 Cascade of the renin-angiotensin system.
***Sources:* Nishimura and Ogawa (1973) and Nishimura (1983)**

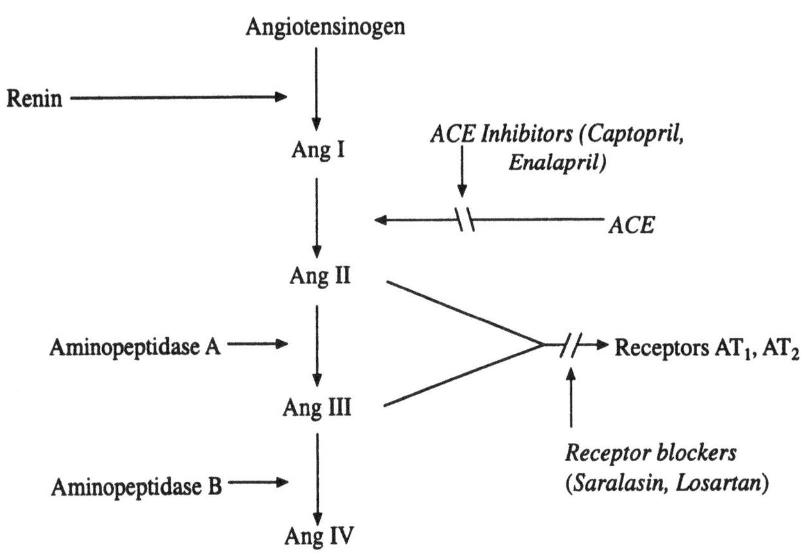

Fig. 13.43 The renin-angiotensin system: enzymes, peptides, inhibitors and receptors.
***Source:* Fuentes and Eddy (1997)**

No other vertebrate group demonstrates the degree of development of juxtaglomerular apparatus seen in mammals. Primitive bony fishes (holosteans), teleosts, lungfishes and *Latimeria* have granulated epitheloid cells similar to the mammalian JG cells located in the media of small arteries and arterioles of the kidney. Furthermore, the kidneys of aglomerular teleosts also appear to have granulated cells. Granulated cells have not been found in the more primitive vertebrates (cyclostomes and elasmobranchs), but holocephalans, do have granulated cells

(Figs. 13.44 and 13.45). No anatomical proximity is found between the JG cells and renal tubules in teleostean kidneys; thus a tubular component of the JG apparatus is absent.

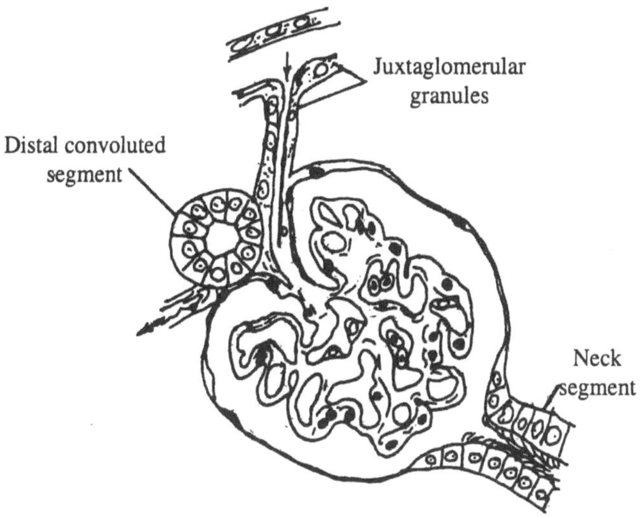

(A) JGA of *Hydrolagus colliei.*

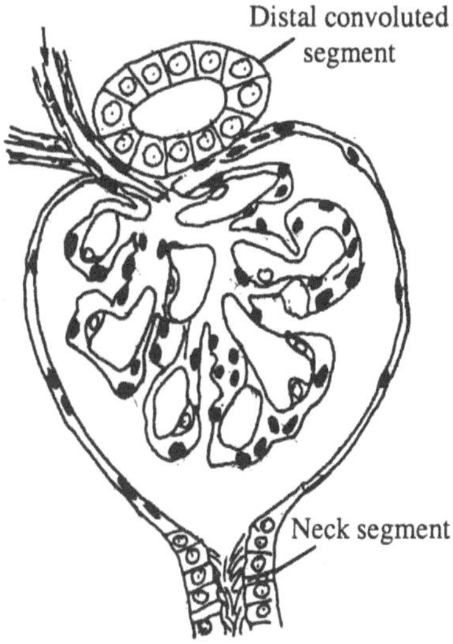

(B) A typical JGA site of elasmobranchs: *Dasyatis akajei.*

Fig. 13.44 (A) Schematic illustration of the juxtaglomerular area of the ratfish, *Hydolagus colliei.* Cells which contain granules are not enlarged or thick as is observed in teleosts or tetrapods. The renal distal tubule attaches to the afferent arteriole near the vascular pole, but cells showing characteristics of the macula densa are not recognized. (B) A typical site of JGA of elasmobranchs, *Dasyatis akajei.* In elasmobranchs JG granules were not recognized. The distal tubule contacts the vascular pole of the glomerulus, but no MD or EGM is observed. The JGA is absent in elasmobranchs.
***Source:* Sokabe and Ogawa (1974)**

Stannius corpuscles contain renin -like activity and immunoreactive ANG II. The amount of renin in Stannius corpuscles is less than 1% of that found in the kidney and it is probable that corpuscular renin-angiotensin has a paracrine function. However, Stannius corpuscles may contribute to the regulation of systemic blood pressure or volume as stanniectomy in the eel lowers blood pressure by 33%.

To date, neither angiotensinogen nor renin from fish has been purified and sequenced. Angiotensins from several teleosts have been sequenced and synthesized.

The amino acid sequence of decapeptide ANG I has been described in four species of teleost fishes: Japanese goose fish (*Lophius litulon*), chum salmon (*Oncorhynchus keta*), Japanese eel (*Anguilla japonica*) and American eel (*Anguilla rostrata*), in which the amino acids in positions 1 and 9 are different from these of more advanced vertebrates. Valine is in position 5 in all nonmammalian vertebrates examined, while isoleucine is found in that position in most mammalian angiotensins.

Angiotensin converting enzyme activity is found in tissues of nearly all teleosts, although enzymatic activity is generally lower in teleosts than in other vertebrates, including subteleostean species.

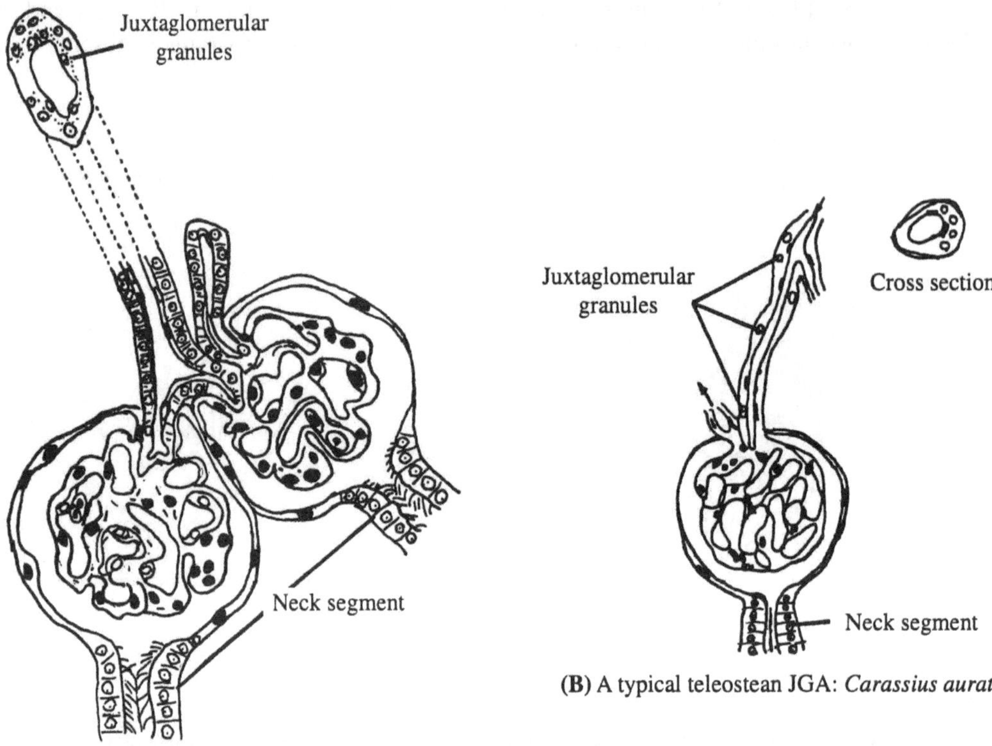

(A) JGA of *Protopterus aethiopicus.*

(B) A typical teleostean JGA: *Carassius auratus.*

Fig. 13.45 (A) JGA of *Protopterus*. (B) Schematic illustration of the juxtaglomerular area of the goldfish, *Carassius auratus*. Granules are disposed along arterioles and small arteries. No anatomical relation is observed between granulated cells and distal tubules.
***Source:* Sokabe and Ogawa (1974)**

Hypotension, hypovolemia, and, to a lesser extent osmotic perturbation are the best known activators of systemic RAS response in fish although there are considerable species and experimentally induced variations.

The RAS appears to serve primarily as an antidop effector of arterial blood pressure, fluid volume, or both in fish. The effects of angiotensins in chondrosteans or holosteans have not been reported. It has been suggested that the RAS evolved as a modulator of the sympathetic nervous system or as an intrarenal hormone to control nephron function. Findings have indicated that the RAS is an important tonic pressor system in many teleosts and that ANG II has direct effects on the cardiovascular system. There is relatively little information on the site or mode of ANG II action in teleosts.

Results of experiments on rainbow trout (*Oncorhynchus mykiss*) indicate that the systemic microcirculation is the major cardiovascular effector of angiotensin-mediated pressor responses. In addition, the RAS has little direct effect on non-vascular smooth muscle or the heart. From an evolutionary perspective, the initial site of direct systemic RAS action appears to be the vascular microcirculation (Nishimura and Ogawa 1973; Chester Jones et al. 1974; Sokabe and Ogawa 1974; Henderson et al. 1985; Nishimura 1985a, b, 1987; Braun and Dantzler 1987; Olson 1992; Olson et al. 1994).

13. KALLIKREIN-KININ SYSTEM

The kallikrein-kinin system is the physiological enantiomer of the RAS (see Fig. 8.37).

Olson (1992) has given evidence for the presence of this system in fish (rainbow trout *Salmo gairdneri*). This system may have a role in cardiovascular homeostasis.

14. NATRIURETIC PEPTIDE HORMONES

Recently, there has been intense interest in what is now known to be a family of natriuretic peptide hormones. These contain atrial natriuretic peptide (ANP), brain natriuretic peptide (BNP), C-type natriuretic peptide (CNP) and the recently described ventricular natriuretic peptide, named, with the exception of CNP, after the respective sites of synthesis: atrium, brain, and ventricle. Interestingly, it is now clear that the atrium is the major site of BNP synthesis, and CNP appears to be the major peptide in the brain, although a substantial amount is apparently present in the dogfish shark heart, suggesting that CNP may be circulating hormone in fish.

The presence of ANP has been shown in atrial and often ventricular cardiocytes of fishes so far examined. Two systems, the renin-angiotensin system and atrial natriuretic peptide have been shown to affect a variety of vascular and extravascular effectors and these system may play an important role in volume (fluid) homeostasis (osmoregulation) (Evans 1990, 1993, 1995; Evans and Takei 1992; Olson 1992). Surprisingly, the bulk of current evidence support a role in salt rather than volume regulation.

Evans et al. (1993) reported that C-type natriuretic peptide (CNP) is released from the dogfish shark (*Squalus acanthias*) heart and is a circulating hormone with potent vasodilatory effects, in sharp contrast to apparent role of CNP predominantly as a brain neuropeptide in mammals. It is also clear that the structure and function of the members of ANP family of hormones in sharks, and other fish groups, deserve further study. The reader is also referred to Evans (1991). Fish NPs have been isolated and sequenced. The intestine, gills, kidney and cardiovascular system have been identified as the main targets the fish (Evans 1993, 1995). Kloos et al. (1994) reported that in teleostean fish (e.g., *Cyprinus carpio*), ANP has stimulatory effects on adrenal cortisol and catecholamine secretion, whereas in all other vertebrate classes investigated inhibitory effects of ANP mediated by guanylate cyclase have been detected.

15 PROSTAGLANDINS (PGs)

Prostaglandins modulate sexual receptivity in diverse vertebrates. Prostaglandins are important biological regulatory compounds derived from 20 carbon fatty acids. Physiological effects are attributed to monoenoic and dienoic prostaglandins in fish. These include effects on ovulation and spawning, cortisol production in eels and hagfish, and melanosome dispersion in carp. Prostaglandins induce spawning behaviour in several species of teleost fish and have been recently implicated in pheromonal communication during spawning in fish. Prostaglandins regulate cardiovascular system in fish. Prostaglandins also affected the renal system of fish.

Female sex behaviour in externally fertilizing vertebrates (fishes and Amphibia) are regulated not by ovarian steroids but by prostaglandins synthesized when ovulated oocytes in the reproductive tract.

Environmental temperature may effect prostaglandin production and function (Liley and

Stacey 1983; Stacey 1983; Herman 1990; Kobayashi and Stacey 1993). The reader is also referred to Dacke (1989) for information on eicosanoids.

16. PINEAL ORGAN (THE PINEAL AND MELATONIN: REGULATORS OF CIRCADIAN FUNCTION) (INTERNAL "ZEITGEBER")

Chapter 15 provides details for the structure of the pineal organ. Organisms, from unicellulars to vertebrates, are structured in time as well as in space. Many, if not most, biochemical, physiological and behavioural parameters exhibited by organisms show daily fluctuations. Significantly, most of these daily rhythms will persist under constant conditions, with period near 24 h in length, showing that they are driven by an internal daily or 'circadian clock'. Under natural conditions these rhythms are 'entrained' by environmental stimuli (such as light and temperature) so that the 'right' events occur at the 'right time of the day'.

Ultradian rhythms are periodic phenomena with periods shorter than 24 hours. At the cellular level they are related to biochemical and neurogenic oscillations. At supracellular levels of organization they appear as rhythmic changes of respiration, heart beat, and locomotor activity. Ultradian rhythms occur both in early ontogenesis and in adult life. Stimulus deprivation, i.e., absence of modulation of rhythmic activity by nonrhythmic ecofactors, seems to be a condition for the occurrence of ultradian rhythms (Peters and Veeneklas 1991).

The use of the term 'circadian clock' does not mean that a single discrete 'clock' is responsible for driving all of an organism's daily rhythms. In recent years it has become appreciated that multicellular organisms are multi-oscillator in nature; that is, more than one circadian clock may exist within a single organism. Normally, however, all of an organism's multiple circadian rhythms exhibit fixed phase relationships with each other under both entrained and free-running conditions. Accordingly, when multiple clocks exist with an individual organism, they must be coupled in some manner.

Several areas have been implicated as being important to circadian organization in vertebrates; these include the pineal organ, the suprachiasmatic nuclei (SCN) of the hypothalamus and the lateral eyes. Interesting similarities, as well as significant differences seem to exist between species on the relative roles that these areas play within an animal's circadian system. Holmqvist et al. (1992) summarized the neuromorphological, cyto-and chemoarchitectonial indications of the presence of a master clock in the preoptic/suprachiasmatic area of teleosts.

Regarding the multiple photic inputs to the circadian system, it has been pointed out that removal of the eyes does not prevent entrainment (synchronization) of the circadian system by 24 h light-dark cycles in any of the nonmammalian vertebrates examined so far. Extraretinal photoreceptors, therefore, exist in fish, amphibians, reptiles and birds which are fully capable of mediating entrainment. The photosensory capacities of the pineal systems of lower vertebrates have made them obvious candidates as the extraretinal photoreceptors (Table 13.3). A major role for the pineal within the circadian systems of lower vertebrates is revealed by observing the effects of pinealectomy on the free-running activity rhythms. Pinealectomy can cause: (1) marked changes in the free-running periods of fish (e.g., *Lota lota*, *Catostomus commersoni*), (2) splitting of activity rhythms into two major circadian components in a fish (*C. commersoni*) or (3) arrhythmicity in fish (e.g., *Heteropneustes fossilis*). Note that a given species may exhibit more than one kind of response to pinealectomy. The pineal serves as a (1) pacemaker and (2)

Table 13.3 Comparison between structure and physiological properties of pineal and retina

	Pineal	Retina
Position	Intracranial, but some species possess an extracranial parietal eye or frontal organ.	Superficial
Optic media	Opaque to transparent.	Clear
Photoreceptor	Cone-like (some rod-like) with small or rudimentary outer segments.	Cones and rods with well developed segments.
Outer segments	< 50 disks	> 1000 disks (*rods*)
Second-order neurons	Projecting neurons (ganglion cells) and possibly a small population of interneurons.	Horizontal, bipolar, amacrine and ganglion cells with several cell populations.
Cellular organization	No apparent stratification	Organization into 3 nuclear and 2 plexiform layers.
Photoreceptor potentials	Hyperpolarization, some with depolarization.	Hyperpolarization.
Time course	Low temporal resolution, slow time course.	High temporal resolution, fast response, short integration times (*cones*) Low temporal resolution, slow response, longer integration times (*rods*).
Intensity-amplitude relation	Similar to retinal receptors, but covering a wide range than in retinal photoreceptors	Follows Naka-Rushton function.
Sensitivity	High sensitivity	High sensitivity (*rods*) Lower sensitivity (*cones*)
Photopigments	One or more photopigments, dependent on species.	Rhodopsin and several cone photopigments, dependent on species.
Adaptation	Adapts over a wide range of intensities, similar to retinal receptors	Adaptation dependent on receptor type.
Spatial resolution	No, because of the absence of a dioptic apparatus.	Good (*cones*)
Color vision	No, but some species show a color-dependent chromatic response	Yes
Endocrine function	Yes	Yes (?)

Source: Meissl and Yáñez (1994)

a photothermoendocrine transducer (read Meissl and Ekstrom 1993). In some species (e.g., trout) pinealectomy does not completely eliminate blood-borne melatonin. Clearly, in those species in which melatonin is important to circadian organization, the potential exists for extrapineal melatonin-producing sites to play a role within the circadian system (Underwood 1990; see also Underwood and Groos 1982). Meissl and Brandstatter (1992) have reviewed the phenomenon (Fig. 13.46). The interconnection of electrophysiological and endocrine aspects of phototransduction are, at present, not well understood.

In recent years, there has been an explosion of interest in the role of melatonin in vertebrate systems. Melatonin (MLT; 5-methoxy N-acetyltryptamine) is an indoleamine secreted by the pineal gland. The most prominent feature of plasma MLT in teleost is marked day-night fluctuations. Studies indicate that MLT secretion is light sensitive as well as having a circadian or endogenous

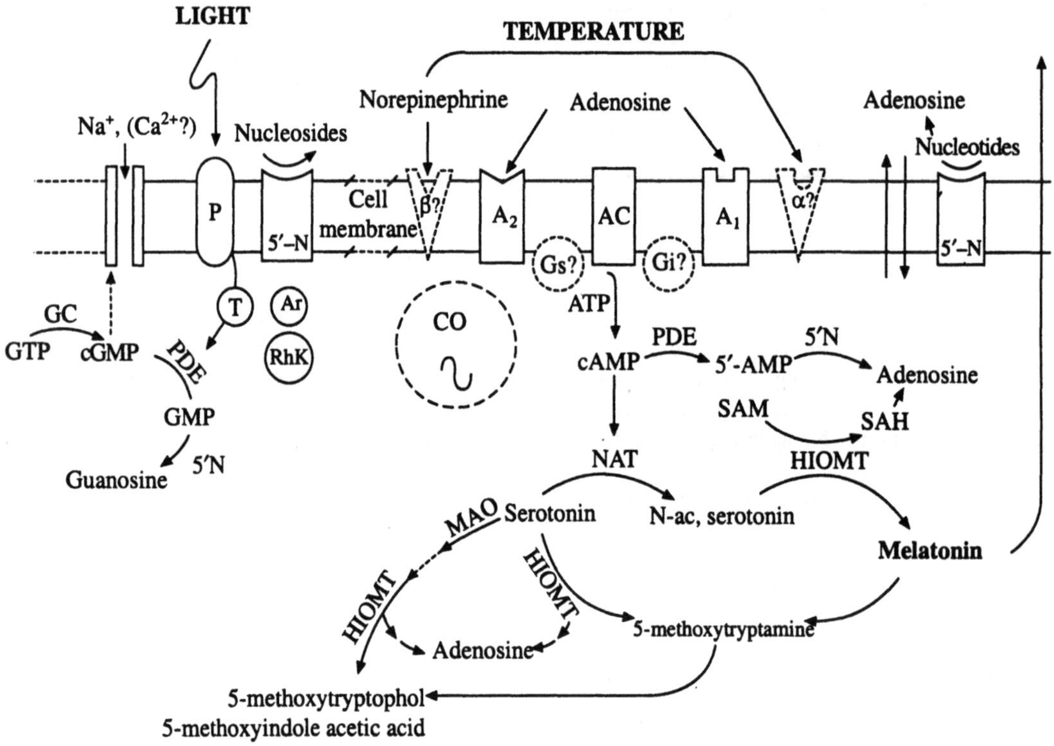

Fig. 13.46 Environmental and chemical factors controlling melatonin production in fish pineal photoreceptor cells,: *α*: *α*-adrenoreceptor; *β*: *β*-adrenoreceptor, A_1, A_2: –adenosine receptors; AC: adenylyl cyclase; Ar; arrestin; cAMP: cyclic AMP; cGMP: cyclic GMP; CO: circadian oscillator, GC: gyanylyl cyclase; Gi: inhibitory guanine nucleotide binding protein; Gs: stimulatory guanine nucleotide binding protein; HIOMT: hydroxyyindole-O-methyltransferase; MAO: monoamine oxidase; 5'–N: 5'–nucleotidase; N-ac. serotonin: N-acetylserotonin; NAT: serotonin N-acetyltransferase; PDE: cAMP and cGMP phosphodiestrases; RhK: rhodopsine kinase; SAH: S-adenosylhomocysteine; SAM: S-adenosylmethionine; T: transducin.
***Source:* Falcón et al. (1992)**

rhythm. Other than strictly day night fluctuations, plasma MLT levels increase with seawater adaptation in salmonids and cyclic MLT levels in plasma have been implicated in sexual maturation by altering daily cycles in plasma Gt H concentrations (McDonald and Milligan 1992).

Randall et al. (1994) have reported that seasonal changes in daylength are reflected in the seasonally-changing pattern of melatonin secretion which potentially provides Atlantic salmon with accurate information of both daily and calendar time. However, it is not yet known whether this information is responsible for the photoperiodic entrainment of seasonal events such as smoltification.

Melatonin is considered to be the most physiologically active indole derivative having actions on the reproductive system, pigmentation, as well as locomotor activity. It has been concluded that melatonin was synthesized and probably stored in the sensory cells during early scotophase and eventually passed through the supportive cells and secreted in the percapillary space. That the pineal organ of fishes contributes to elevated circulating levels of melatonin during the scotophase was demonstrated in trout where pinealectomy resulted in a significant lowering of plasma melatonin at night. The reader is also referred to Garg (1988) in this aspect.

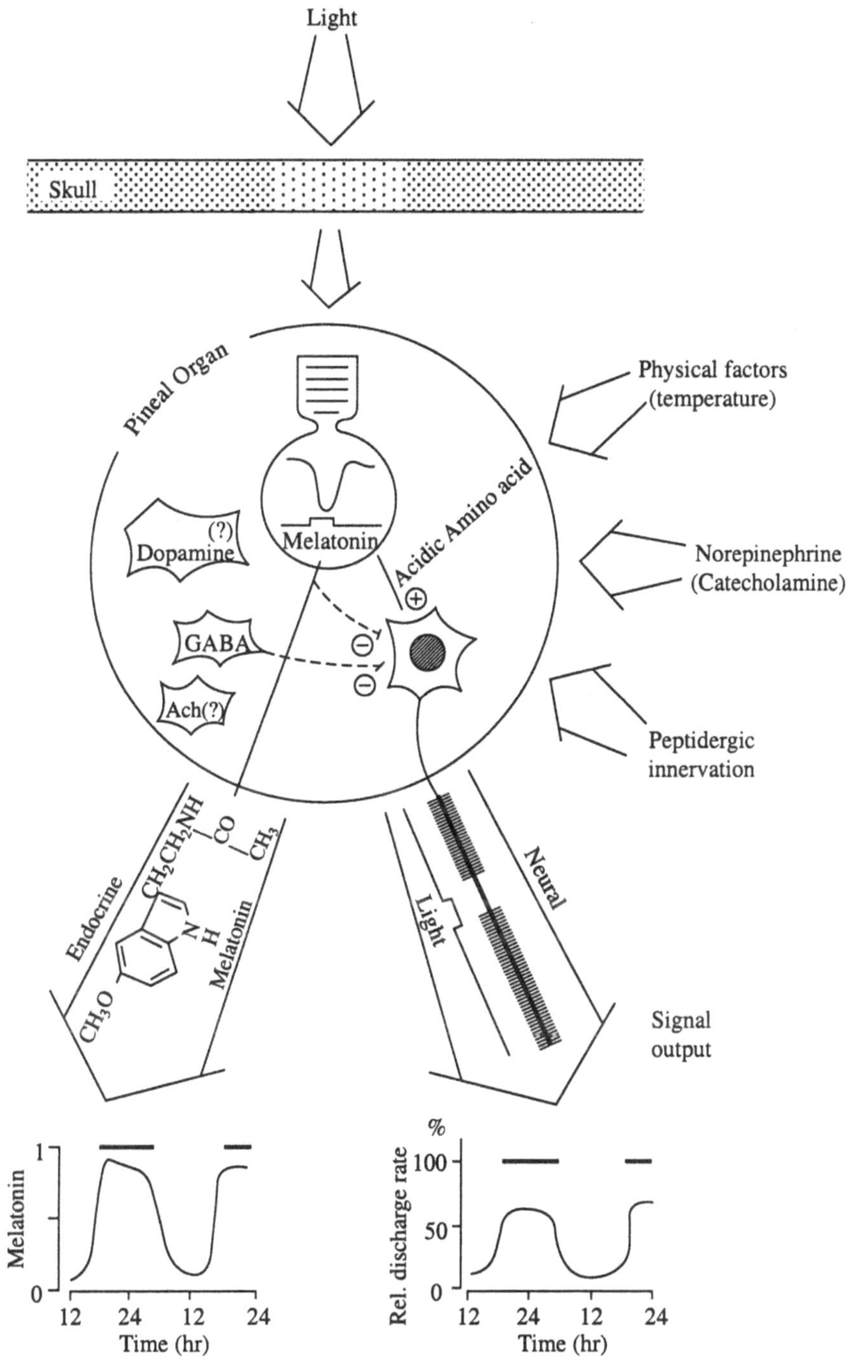

Fig. 13.47 Schematic diagram of a pineal organ of a teleost showing that the photosensitive organ tranduces photic information directly into neuronal and endocrine signals. The diagram also indicates that the output signals are modulated by an interneuronal network and by extrinsic factors such as temperature or noradrenergic and peptidergic innervation.
***Source:* Meissl and Brandstätter (1992)**

Further evidence in support of the possibility that melatonin is synthesized in the pineal organ of lower forms is the presence of the enzyme hydroxindole-O-methyltransferase (HIOMT), which catalyzes the final step in the biosynthesis of the indole. According to Falcon et al. (1992), phototransduction mechanisms in pineal photoreceptors might be very similar to those found in retinal photoreceptors (Fig. 13.47). Together with cAMP (cAMP i), Ca^{2+} ions (Ca^{2+}) i) play an important role in pineal signal transduction and are involved in stimulation of the melatonin biosynthesis in all vertebrate species investigated thus for. The mechanisms which regulate

Arginine vasopressin

Cys – Tyr – Phe – Gln – Asn – Cys – Pro – Arg – Gly – NH_2

Arginine vasotocin

Cys – Tyr – Ile – Gin – Asn – Cys – Pro – Arg – Gly – NH_2

Isotocin

Cys – Tyr – Ile – Ser – Asn – Cys – Pro – Ile – Gyl – NH_2

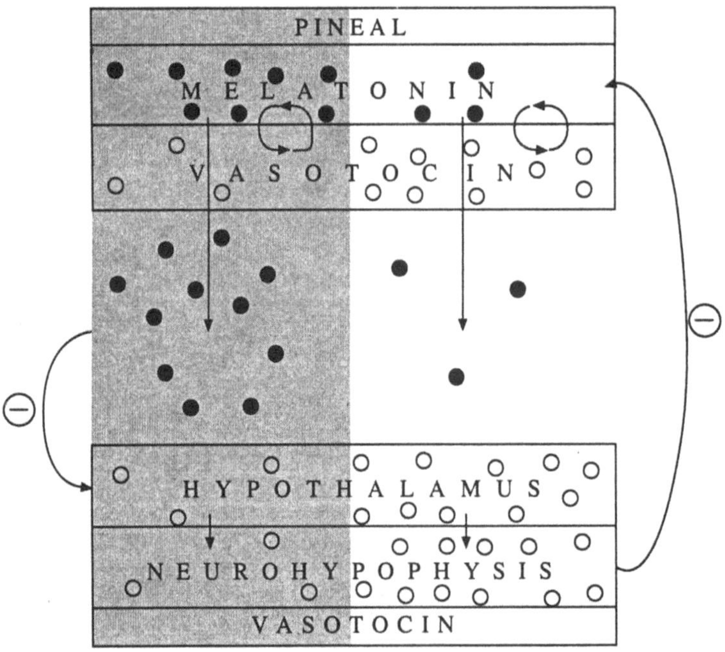

Fig. 13.48 (Upper) Amino acid sequence of selected natural neurohypophysial hormones. (Lower) Proposed model for AVT-melatonin relationships in fish: melatonin (•) synthesized in the pineal during darkness (shaded area inhibits vasotocin synthesis in hypothalmic neurons and/or in the pineal; vastocin (O) synthesized according to osmotic stimuli in hypothalmic neurons (released in neurohypophysis) and/or in the pineal inhibits melatonin synthesis in the pineal.
***Source:* Kulczykowska (1995a)**

[(Ca^{2+})i] in pinealocytes are complex and differ considerably between poikilothermic and mammalian species (see Korf et al. 1997; Kroeber et al. 1997). Details on fish pineal can be obtained from a chain of recent articles (Thibault et al. 1993a, b; Henderson et al. 1994; Meissl and Yáñez 1994; Sriwastawa and Srivastav 1994). It has been pointed out that pineal photoreceptors are related through phylogeny and that their well conserved melatoninergic function appears early in the course of evolution. Sex hormones seem to exert a control on the pineal. Pineal photoreceptors appear as multieffector cells which transduce information from both physical (photoperoid) and internal (chemical) factors.

Table 13.4 Survey of hormones of teleost fishes

Source	Hormone	Nature	Functional Involvement
Pituitary			
Pars distalis	Gonadotropin(s)	Glycoprotein	Egg and sperm production.
	Growth hormone	Protein	Growth and metabolism. Seawater adaptation in several salmonid fishes. Gonadal steroidogenesis.
	Prolactin	Protein	Osmoregulation, reproduction metabolism, behaviour and mucus production.
	Adrenocorticotropic hormone	Peptide	Promotes steroidogenesis in adrenal gland.
	Thyrotropic hormone	Glycoprotein	Thyroxine production
Pars intermedia	Melanocyte stimulating hormone	Peptide	Coloration.
	Somatolactin	Protein	Calcium regulation, acid-base balance, background adaptation, reproduction.
Pars nervosa (Neurohypophysis)	Arginine vasotocin	Octapeptide	Cardiovascular. Osmoregulation activity. Endocrine secretion. Reproductive processes.
	Isotocin	Neurotransmitter/or neuromodulator in CNS	Function less known.
Pineal organ	Melatonin	Indoleamine	Photoendocrine.
Thyroid	Triiodothyronine	Iodinated	Enhancement of growth.
	Thyroxine	Tyrosine derivatives	Gonadal maturation. In promoting metamorphosis.
Ultimobranchial body	Calcitonin	Peptide	Calcium metabolism. Gonadal maturation; early development of embryos.
Interrenal tissue	Cortisol	C_{21} steroid	Potential hypercalcemic action. Dominant osmoregulatory hormone in marine fishes. Direct effect on the immune

(Contd)

Table 13.4 (*Contd*)

Source	Hormone	Nature	Functional Involvement
			system and disease resistance. Can effect distribution of leucocytes, result in lymphocytopenia, and suppress the growth of lymphocytes.
			Stressful conditions;
Chromaffin Tissue	Epinephrine (Adrenaline)	Catecholamine	Hypoxia, hypercapnia, exhaustive exercise, acidosis, social stress, etc.
	Norepinephrine (Noradrenaline)	Catecholamine	
Pancreatic islets	Insulin	Peptide	Metabolism (carbohydrate, protein, lipid).
	Glucagon	Peptide	
	Glucagon-like peptide	Peptide	
	Somatostatin	Peptide	
Stannius corpuscles	Stanniocalcin	Glycoprotein	Calcium metabolism. Sexual maturation. Migration (smoltification).
Gonads			
Ovary	Estrogen (17 β estradiol)	C_{19} steroid	Agonistic behaviour. Secondary sexual characteristics.
Testes	Androgens (11-ketotestosterone)	C_{19} steroid	Events leading to spawning.
Caudal urophysis	Urotensin I	41 aas	May play a "Fine tuning" role in osmoregulation.
	Urotensin II	12 aas	Function remains problematic.
Kidney			
Renin-Angiotensin system	Angiotensin	Peptide	Antidrop effector of blood pressure.
Heart cardiac tissue	Natriuretic peptide hormones	Peptide	Cardiovascular role. Volume homeostasis.
Gastrointestinal tract	Stomach: Serotonin bombesin, gastrin, met-enkephalin, Substance P Intestine: glucagon-like PP, VIP, neurotensin gastrin/CCK, met-enkephalin substance	Peptide	Regulation of chemical traffic.

Among other compounds which have been implicated as active pineal factors is the peptide, arginine vasotocin (AVT). Whether the pineal organ synthesizes this peptides is not known (NcNulty 1984; see also Fenwick 1970; Oguri and Omura 1973; Chester Jones et al. 1975). AVT presence in fish pineal is now known. (Fig. 13.48; see also Fig. 13.14). AVT-melatonin interactions may be of great importance, especially for migrating fish, in which mechanisms of adaptation to both light and salinity play a key role (Kulczykowska 1995a). A new, highly sensitive, rapid method for the determination of low levels of endogenous melatonin has been reported by Kulczykowska and Luvone (1998).

17. GASTROINTESTINAL TRACT

Chapter 7 includes an overview of hormones from gastrointestinal tract. In Table 13.4 the hormones of teleost fishes are listed (see gastrointestinal tract).

18. JUXTATESTICULAR BODY (JTB)

An unusual male organ, the JTB appears to be a consistent feature of the urogenital apparatus of the genus *Opisthognathus*. It is located outside the coelomic cavity, in the posterior part of the trunk, ventral to the mesonephric ducts and dorsal to the urinary bladder and testes. A nephrogenic origin for this organ has been proposed. Strong similarities with the thyroid follicular cells suggest a similar pattern of synthesis and secretion. JTB follicular cells perform two major types of secretory activities: the secretion of a glycoprotein from the apical part of the cells into the follicular lumen and the endocrine, or paracrine, secretion of a still unknown substance(s) from the basal part of the cell either into the extrafollicular space or the blood or both. The role in male reproduction is not yet understood (Rasotto and Manni 1995).

19. PSUDOBRANCHIAL NEUROSECRETORY SYSTEM

Neurosecretory cells of the paraneuron type of unclear function characterize this system reported mainly in Siluriformes, besides Perciformes, Clupeiformes, Channiformes and Atheriformes (Srivastava and Gopesh 1987; Gopesh and Srivastava 1997; see also Srivastava et al. 1981 Fig. 1; 1988; Figs. 1–5).

CHAPTER 14

Nervous System

1. INTRODUCTION

Behaviour is to a great extent a function of an intricably organized nervous system, and that studies in behaviour cannot for long ignore this neurological substrate. Considerable information is available on the neurobiology of fishes (see C.U. Ariëns Kappers, G.C. Huber and E.C. Crosby 1936; R.G. Northcutt and R.E. Davis (editors) (1983a); R.E. Davis and R.G. Northcutt (editors) (1983); R. Niewenhuys, H.J. Donkelaar and C. Nicholson (1998). The reader is referred to Aronsow (1963) and Northcutt and Davis (1983b). The reader is also referred to Aronson (1963), Bernstein (1970), Ebbesson and Northcutt (1976), Northcutt (1977a, 1978, 1989) and Northcutt and Davis (1983b). The contribution from Wullimann (1998) provided considerable details on the functional anatomy of teleostean brain: (1) sensory systems (olfaction, vision, mechanoreception, electroreception, audition, vestibular sense, gustation, general visceral sense and somatosensation, (2) motor and premotor systems (motor nuclei of cranial nerves, descending spinal projections, reticular formation, and (3) integrative centres (cerebellum, tectum opticum and telencephalam).

In all craniates the brain arises embryologically from a neural plate that transforms into a hollow tube by either invagination or delamination. Similarly, in all living craniates the neural tube subsequently divides into a hind-(rhombencephalon), mid-(mesencephalon) and forebrain (prosencephalon) by differential growth of the tube. The similarities in embryology and cytology of these subdivisions support the view that the brains of the earliest craniates must have been similarly divided.

In gnathostomes further development of the brain is characterized by rapid growth of the side-walls and rostral roof (rhombic lip) of the hindbrain that results in the medulla oblongata and cerebellum, respectively. In contrast, a large portion of the hindbrain roof does not thicken but forms the thin roof (tela choroidea) of the fourth ventricle.

In all craniates the midbrain comprises a roof, or optic tectum, and a floor, or tegmentum. The optic tectum is usually bilobed and is major target of the optic tracts.

The forebrain in all living craniates is subdivided into a caudal diencephalon and a more rostral telencephalon. The telencephalon forms telencephalic hemispheres by evagination, except in ray-finned fishes where the hemispheres form by eversion. In all craniates the hemispheres form rostral evaginations termed the olfactory bulbs. These similarities suggest that the earliest craniates also possessed paired hemispheres and olfactory bulbs that formed by evagination (Northcutt 1985).

2. VARIATION IN ANAMNIOTIC NERVOUS SYSTEMS

Even a casual gross examination of anamniote brains reveals marked differences in their total size and differential development (Fig. 14.1). Ebbesson and Northcutt (1976) considered three mechanisms which have operated and are operating to mould brain evolution: (1) the number of units can increase (or decrease) without a change in the capacity of existing units, (2) the information-carrying capacity or "sophistication" of the units can change, and (3) the wiring or circuitry can change.

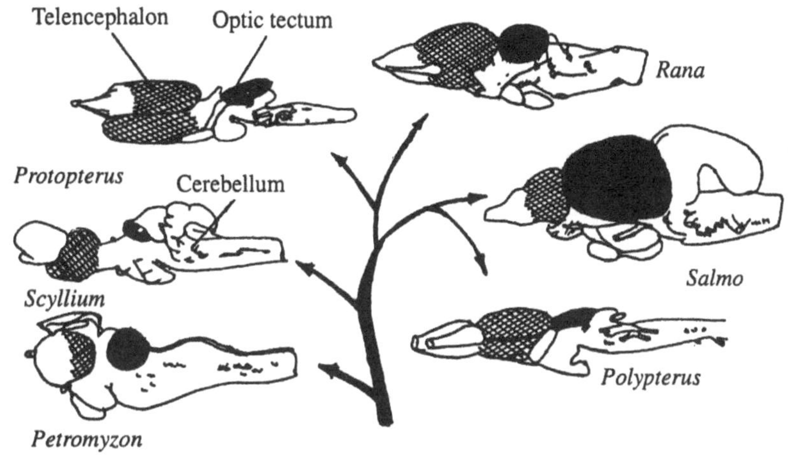

Fig. 14.1 Phenetogram illustrating some of the gross variation that exists in anamniotic brains. The extent of the telencephalon is indicated by the hatching, the optic tectum by solid black, and the cerebellum by stippling. These brains have not been drawn to scale.
***Source:* Ebbesson and Northcutt (1976)**

Brain-body weight ratios: Studies of changes in brain-body ratios probably reflect changes in the number of units. Data suggest that at least three levels of gross brain-body ratios exist in anamniotes. The first level may characterize agnathans, with a ratio of at least 500% less than other anamniotes; a second level may characterize the bony fishes and amphibians; and a third level may characterize elasmobranchs, with brain-body ratios as much as 400% higher than all other anamniotes.

The reader interested in brain-body data can refer Ebbesson and Northcutt (1976) and Northcutt (1977, 1978, 1989).

Ridet and Bauchot (1993) reported encephalization indices for 1184 species of teleosts and emphasized that indices are more correlated with biological adaptations (locomotor, food habits, protection against predators and habitat) of the species than with their phylogenetic relationships. Analysis of major subdivisions of 114 representative teleost species shows that there is, in general, (1) a negative correlation between olfactory and visual sense organs, (2) a positive correlation between cerebellar size and level of muscular activity, and (3) a positive correlation between increase in telencephalic and diencephalic centres and more derived fishes.

3. AN OVERVIEW OF BRAIN ORGANIZATION IN CYCLOSTOMATA

In lampreys the hindbrain exhibits an expansive fourth ventricle but has a poorly developed rhombic lip that has traditionally been identified as a cerebellum. This "cerebellum", however, is not organized into the same layers in lampreys, nor does it exhibit the same classes of neurons as does the cerebellum of gnathostomes. The brain division in lampreys may be comparable only to the eminentia granularis, a closely related cellular population that also arises from the rhombic lip in gnathostomes. This hypothesis can be tested by examining the efferent pathways of this cell group in lampreys, as the eminentia granularis and cerebellum in gnathostomes possess very different efferent connections. For details of CNS of lamprey, read Nieuwenhuys and Nicholson (1998).

In myxinoids, the hindbrain is foreshortened in comparison to that of other craniates, and the fourth ventricle is extremely small. The reduced ventricular system is known to be derived, as in early embryonic stages the brain of *Myxine* exhibits a well-developed ventricular system. The absence of much of the ventricular system and the rather uniform development of the more rostral lobes of the brain has complicated the recognition of homologies with brain divisions in other craniates. The lobes immediately rostral to the fourth ventricle have been interpreted as a cerebellum and as the midbrain roof. Recent experimental studies of *Eptatretus* clearly demonstrate that this division is homologous to the midbrain roof and that a cerebellum does not exist in myxinoids. The question of whether or not the myxinoids possess a cerebellum has received widely different answers in the literature. Choroid plexuses are entirely lacking; thus, the production of the cerebrospinal fluid may occur elsewhere, most likely in the brain itself (read Wicht and Nieuwenhuys 1998).

The optic tectum of lampreys, unlike that of gnathostomes, is characterized by a tela choroidea. Myxinoids do not possess much a midbrain tela and have only a very reduced midbrain ventricle.

The forebrain of post-embryonic myxinoids exhibits only traces of the ventricular system. Recent studies have demonstrated that the brain of myxinoids is several times larger than that of lampreys of comparable body weight. Not only are the most brain regions larger in hagfishes than that in lampreys of comparable size, many of these neural regions also possess more migrated and differentiated cell populations.

Many, if not most, of the cell populations identified in the brains of gnathostomes can be recognized in hagfishes and lampreys (Northcutt 1985) (Figs. 14.2, 14.3).

Chiba and Honma (1986) studied the comparative anatomy of the brain-ventricular system of the lampreys (*Lampetra japonica* and *L. mitsukurii*) and a hagfish (*Paramyxine atami*). The lampreys had a series of broad cerebral ventricles associated with several circumventricular organs such as pineal and parapineal organs, dorsal sac, choroid plexus, subcommissural organ and neurohypophysis, whereas the hagfish possessed small and narrow ventricles solely with subcommissural organ and neurohypophysis (Fig. 14.4). Electron-microscopically, the ependymal layer of the infundibular floor of the lampreys exhibited a distinct regional differentiation in contrast to a simpler structure of the hagfish. However, the frequent occurrence and elaboration of cerebrospinal fluid (CSF)-containing neurons were outstanding in the lampreys than in the hagfish.

Certain conclusions were drawn. No choroid plexus in the hagfish brain may be related to the vestigeal or greatly reduced ventricular cavity. Although the lack of pineal complex in the hagfish is difficult to explain with respect to the grade of development of the ventricular cavity, it seems to be related to burrowing life in deep sea.

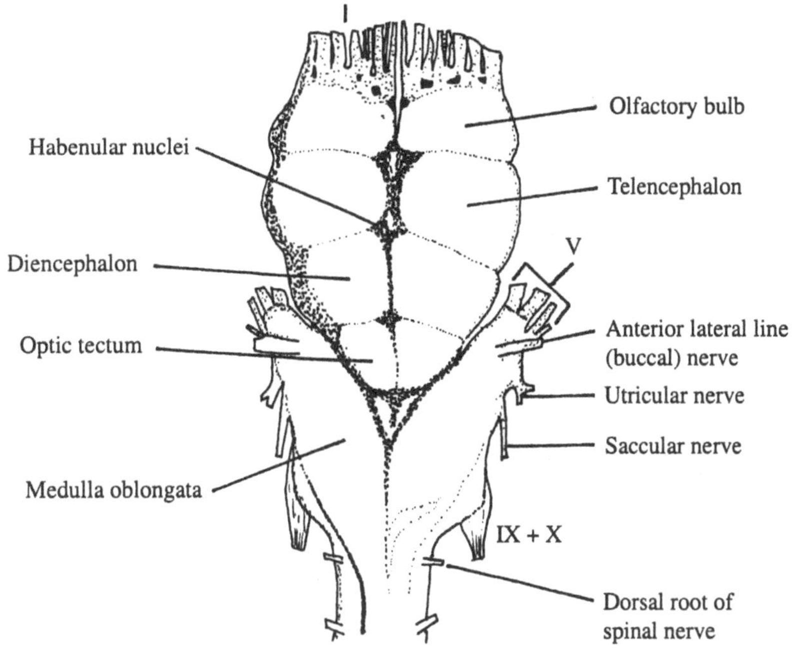

Brain of *Eptatretus stouti.*

Fig. 14.2 Dorsal view of the brain of the Pacific hagfish, *Eptatretus stouti* I, V, IX, X cranial nerves. *Source:* Northcutt (1985) For CNS of *Myxine*, see Brodal, A., Fänge, R. (eds.) (1963)

4. CENTRAL NERVOUS SYSTEM IN GNATHOSTOMOUS FISH: ACTINOPTERYGIANS

4.1 Brain Stem

The brain stem (truncus cerebri) comprises the rhombencephalon and the mesencephalon. Rostrally, it borders the diencephalon, and caudally, it grades into the spinal cord. In all vertebrates this "intermediate" part of the central nervous system harbours the chief centres of origin and termination of all the cranial nerves, except I. In addition, the brain stem contains a fairly well-developed reticular formation and a number of relay centres with their associated ascending and descending connections. The cerebellum, although ontogenetically and phylogenetically a derivative of the rostral part of the rhombencephalon, is not considered to be a part of the brain stem. It is, however, noteworthy that the cerebellum and the brain stem are strongly interconnected. Many of the cerebral afferent systems originate in this brain part and all of the cerebral efferent fibres terminate in it.

The brains of actinopterygian fishes show considerable variation in size and shape (Fig. 14.5). The astonishing structural diversity of teleost brain has been the subject of numerous studies. The teleost rhombencephalon is shorter and more compact than in chondrosteans (*Acipenser*, *Scaphirhynchus*, *Polyodon*) and holosteans (*Amia*, *Lepisosteus*). The ventricular ridges found in chondrosteans and holosteans can also be discerned in most teleosts. Some groups of teleosts show strong development of a particular sense system with corresponding enlargement of its recipient centre in the rhombencephalon. Thus, in mormyrids the lateral line system is strongly

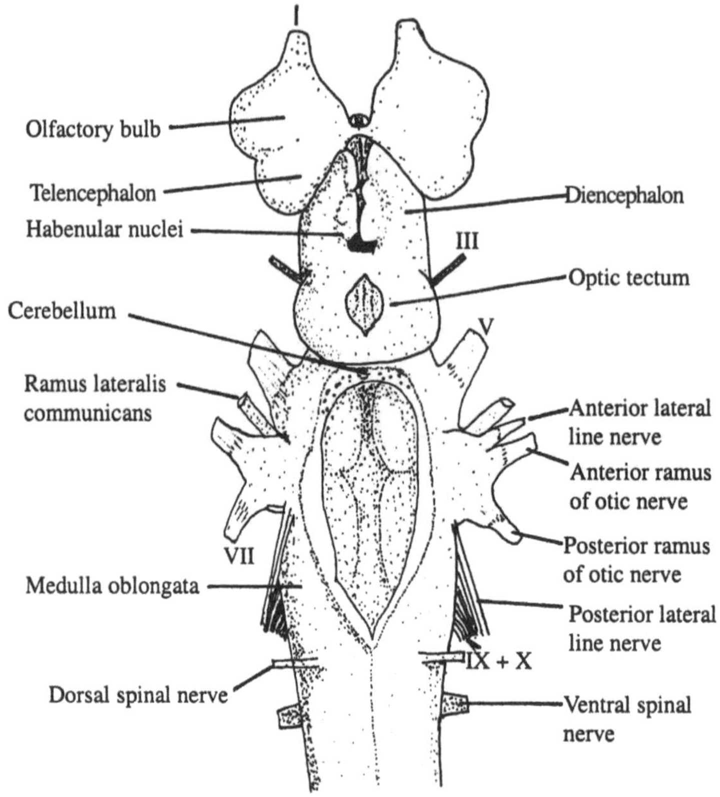

Brain of *Ichthyomyzon unicuspis.*

Fig. 14.3 Dorsal view of the brain of the silver lamprey, *Ichthyomyzon unicuspis*. I, III, V, VII, IX, X cranial nerves.
***Source:* Northcutt (1985)**

developed and its sense organs have become extensively transformed into the electroreceptor organs. Consequently, the lateral line nerves are very large and their centre of termination in the rhombencephalic alar plate is an enormous lobe. In many of the Ostariophysi (cyprinids and silurids) the taste buds are particularly numerous and are not confined to the mouth but spread over the head and body. As a result, the sensory components of VIIth, IXth, and Xth cranial nerves and their centres of termination are very large.

The teleostean cerebellum comprises three main parts: the corpus, the valvula, and the lobus vestibulolateralis. The corpus cerebelli varies in size and is said to be related to the complexity of locomotor performance. In most teleosts the cerebelli arches over the fourth ventricle, but in silurids it extends forward over the tectum. In doctor fish *Acanthurus* and some other species (e.g., sergeant major, *Abudefduf*, kingfish, *Menticirrhus*) a particularly well-developed corpus cerebelli which, projecting rostrally, covers not only tectum, but also a part of the telencephalon. In *Corvina nigra*, transverse furrows divide the corpus into anterior, medial and posterior lobes. However, the corpus cerebelli is best developed in mormyrids, where it is convoluted and differentiated into several lobes. The valvula cerebelli also varies considerably in size and shape. In *Gobius* and anglerfish *Lophius* it is represented by a small, subtectal fold, but in most teleosts

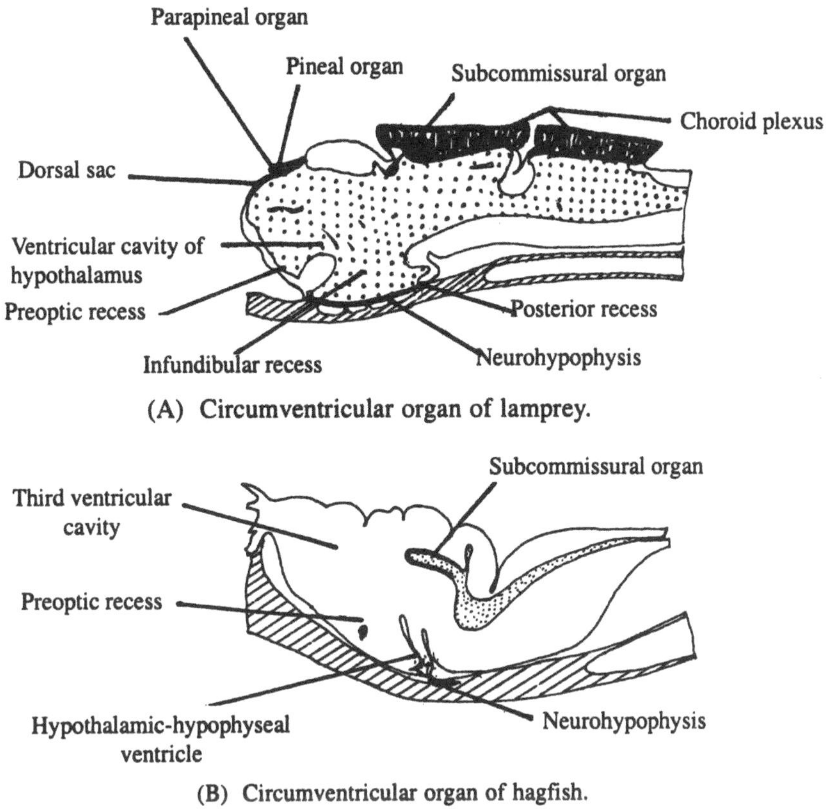

Fig. 14.4 Circumventricular organ of (A) lamprey and (B) hagfish.
***Source:* Chiba and Honma (1986)**

the valvula occupies a considerable part of the optic ventricle. In silurids and cyprinids the valvula is so strongly developed that if effects a lateral displacement of the tectum mesencephali. Mormyrids have an extremely large valvula. In this group the valvula grows out of the midbrain ventricle and becomes a superficially situated structure covering all other parts of the brain (Figs. 14.5, 14.6). The lobus vestibulolateralis of the teleost cerebellum is a solid structure, whose rostrolateral parts consist of large masses of granule cells known as eminentiae granulares. The lateral recesses of the fourth ventricle border the vestibulolateral lobe caudolaterally. These recesses are relatively small and do not extend into lateral diverticula of the cerebellum. Small masses of granular cells surround the teleost lateral recesses and have been designated as auricles by some authors. However, more likely it is considered that the entire lateral part of the vestibulolateral lobe constitutes the homologue of the large auriculae of the chondrosteans.

It is noteworthy that holosteans are intermediate between chondrosteans and teleosts in the development of the lateral parts of the cerebellum. In the midbrain, the size of the tectum clearly varies with the size of the eyes. The most medial parts of both tectal halves extend to form bilateral elongated eminences known as tori longitudinales. Corresponding structures are found in the holosteans and in the chondrostean *Polyodon*. Distinct semicirculares are present in all teleosts, and their size varies with the development of the octavus and the lateral line systems.

These tegmental elevations attain their greatest size in mormyrid fishes. In all actinopterygians the lateral portions of the valvula cerebelli are continuous with the tegmental region lying caudomedial to the tori semicirculares. In teleosts at least, this connection, which is usually a fusion, represents a primary continuity.

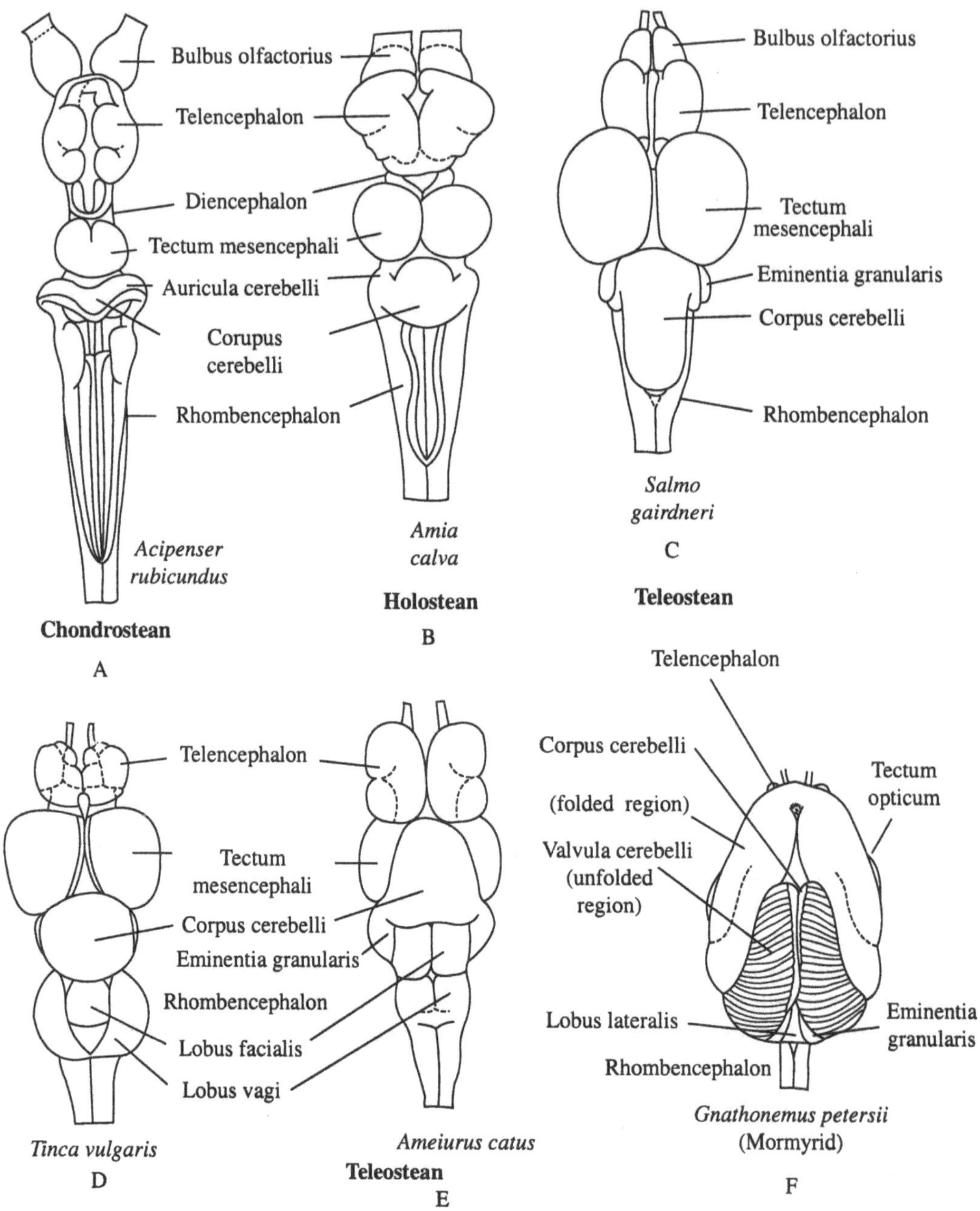

Fig. 14.5 Brains of actinopterygian fishes (Dorsal view).
***Source:* Nieuwenhys and Pouwels (1983) F: (Modified)**

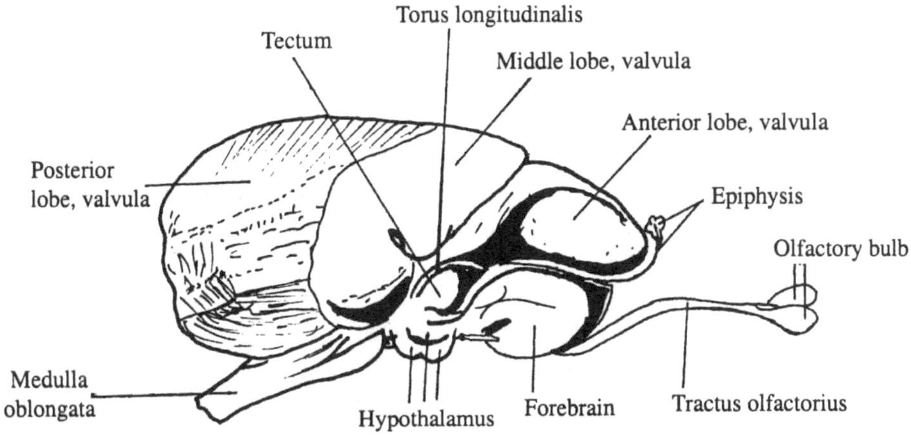

Fig. 14.6 Lateral view of brain of *Mormyrus kanume*, showing enormous expansion of valvula cerebelli. *Source:* Aronson (1963)

Detailed account of (1) nerve components and longitudinal zones; (2) nuclei and fibre tracts, etc., in the brain stem of actinopterygian fishes is given by Nieuwenhuys and Pouwels (1983). The reader is also referred to Aronson (1963) and Ebbesson and Northcutt (1976). Alibardi and Meyer-Rochow (1994) described the ultrastructure of caudal spinal cord in an Antartic fish (*Pagothenia* (= *Trematomus*) *borchgrevinko*. The study suggested that in this cold-adapted species of fish the nervous system is active in producing the observed richness in membrane lipid materials and that lipid materials and the lipids are particularly fluid because of high percentage of unsaturated faty acids, helping to isolate the axoplasm from the surrounding environment of extremely low temperature. For details of CNS of holostean and teleostean fishes, see Meek and Niewenhuys (1998). Literature on the CNS in the brachiopterygians is rather scanty. Physiological studies on CNS have not been reported so far (Nieuwenhuys 1998a). For the coverage of CNS in chondrostean fishes, lungfishes, and coelacanth, read Nieuwenhuys (1998b, c, d) respectively.

Mauthner cells: The Mauthner cells (M-cells) are a pair of neurons found in the hindbrain of most fishes and amphibians. These exceptionally large cells are identifiable by their distinctive morphology, their position within the medulla oblongata, and their electrophysiological response characteristics. The M-cell receives input from a variety of sensory sources such as the auditory, vestibular, visual and lateral line systems (Fig. 14.7).

The M-cell initiates the first stage of a rapid startle response which is thought to be involved in escape from predators (see details of M-cells: Zottoli et al. 1984, 1987; Zottoli and Danielson 1989).

Fetcho (1991, 1992) pointed out that the most prominent feature of this escape behavior is a rapid, forceful bend of body and tail which lead to a characteristic C-bend (Stage 1) early in the escape.

The circuitry by which the M-cell generates this C-bend has been studied in some detail (Fig. 14.8). The axon of each M-cell forms contacts with ipsilateral motoneurons and premotor interneurons along the length of the spinal cord. Each M-cell monosynaptically and chemically excites the very large motoneurons, which are in turn electrotonically coupled to the large motoneurons and contact other smaller motoneurons in the motor column. The mono-and disynaptic excitation of the largest motoneurons leads to a massive excitation of fast white muscle fibres on

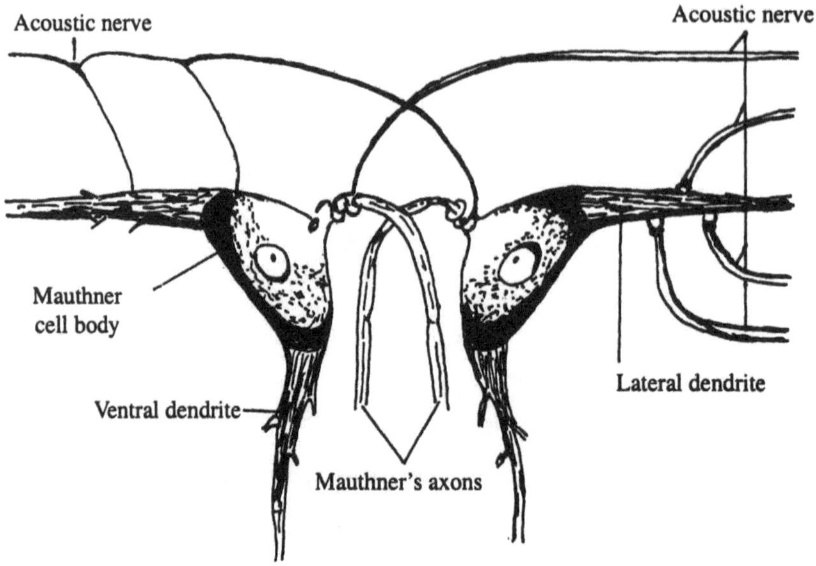

Fig. 14.7 Schematic representation of Mauthner cells in teleosts. On the right, unbranched VIIIth nerve fibres terminate on lateral dendrite of homolateral cell and on axon pole of contralateral cell. On the left, VIIIth nerve collaterals terminate on homolateral dendrite and cell body and on contralateral axon pole.
***Source:* Aronson (1963)**

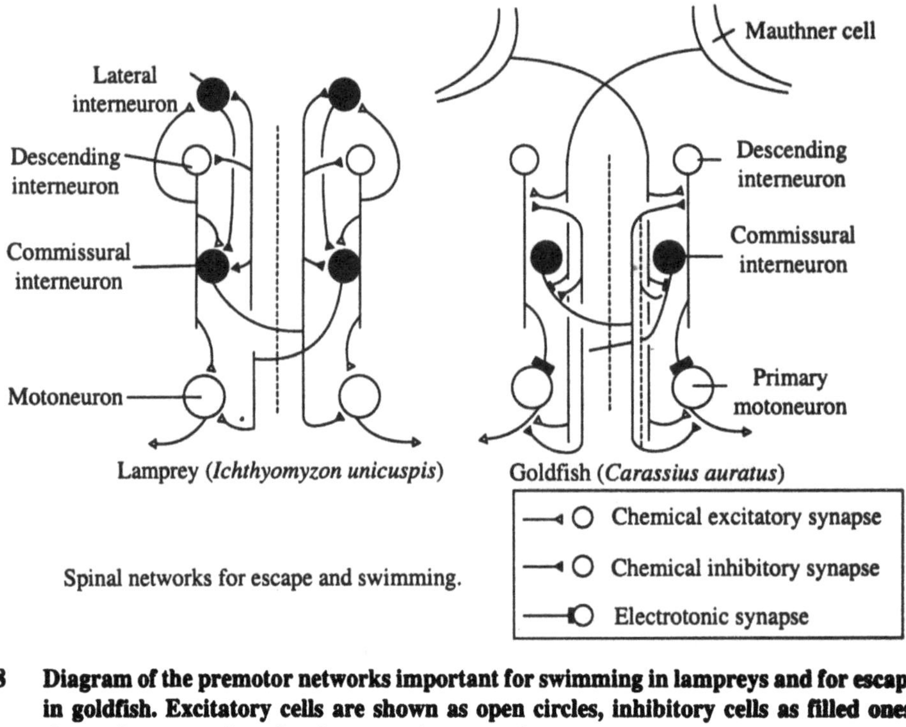

Fig. 14.8 Diagram of the premotor networks important for swimming in lampreys and for escape turns in goldfish. Excitatory cells are shown as open circles, inhibitory cells as filled ones. Inset shows the conventions for the synaptic connections.
***Source:* Fetcho (1992)**

each side of the active M-axon. The M-axon is also electrotonically coupled to commissural interneurons that cross the spinal cord to monosynaptically inhibit motoneurons and interneurons on the opposite side. These glycinergic commissural interneurons inhibit the large motoneurons, the descending interneurons, and the commissural interneurons on the opposite side. They serve to block contralateral motor output and prevent inhibition from controlateral commissural interneurons. The consequence of this network is a massive excitation of ipsilateral motoneurons and a blockade of contralateral activity, leading to the very fast and forceful C-bend in the early part of the movement.

The network shows obvious parallels with the spinal networks for swimming in lampreys and embryonic frogs as shown in Fig. 14.8. These similarities reveal important, primitive features of axial motor networks among vertebrates.

Aronson (1963) described that Mauthner apparatus is a motor association mechanism which contributes to effective swimming and equilibratory movements of the trunk, fins, and tail. It is most elaborate in fishes in which side-to-side tail swimming (carangiform mode) is most highly developed. Conversely, it is poorly developed or absent in tailless fish, in species living at the bottom, and in those showing eel-like (anguilliform) movements. Locomotion in the lungfish *Protopterus*, is typically anguilliform. However, a pair of giant fibres in the spinal cord have been identified as Mauthner neurons, thus providing an exception to this generalization.

Further, most sharks are constantly in motion. In fact, this is necessary in many sharks as an aid to respiration. Sudden spurts in forward locomotion are rarely seen. The absence of the Mauthner apparatus may be the explanation for their inability to accelerate rapidly.

According to Pfeiffer et al. (1986), the fright reaction of most ostariophysan species reacting to alarm substance begins with the "Mauthner reflex" (Mauthner-initiated startle response). The behaviour controlled by the brain and sensory organs (goal oriented fleeing, bottom seeking, forming a dense school, searching for a hiding place) begins to function after the unidirected jumps to the side caused by the Mauthner neurons, which are time gaining. The reader is also referred to Niewenhuys and Pouwels (1983) for the Mauthner cells.

In contrast to the primitive condition represented in cyclostomes, the spinal cord in elasmobranchs has the same basic structural organization as in all higher vertebrates. The fish spinal cord occupies the entire vertebral column and in the actinopterygians ends in a specialization, the caudal neurosecretory system. In cyclostomes the spinal cord is flattened. The spinal cord of the Chondrichthyes is more differentiated than that of cyclostomes. The spinal cord of the Osteichthyes shows a general resemblance in organization to that of the Chondrichthyes. In anglerfish (*Lophius*) and gurnards or sea robins (*Trigla, Prionotus*), there are enlargements (swellings) on the dorsal surface of the anterior end of spinal cord. One of the conspicuous elements of the spinal cord of the bony fish is the large Mauthner fibres which give rise to many short collaterals within the spinal cord. These short collaterals have extensive connections with motor horn cells in the ventral horn. The massiveness of the grey matter increases and the connections which are made by the interstitial cells in the spinal cord become more complex as the evolutionary scale is advanced within the fish from the cyclostomes through the Osteichthyes.

4.2 Cerebellum

In the preceeding account, the morphology of the cerebellum has been described. The organization of the teleost cerebellum has been fully described by Finger (1983). For SEM-observations of teleost fish cerebellum, the reader is referred to Castejon and Caraballo (1980). For most teleosts, the bulk of the cerebellum lies in the corpus. The shape of the corpus may relate, less to function than to morphology of the skull. Cell types in the teleost cerebellum include those found in the cerebellum of other vertebrates. The cerebellar inputs and outputs are well discussed. The role of the cerebellum in motor and sensory systems is not completely understood. The corpus cerebelli may be most closely related to motor control, whereas the vestibulolateral lobe appears to be more involved in processing lateral line sensory input.

Aronson (1963) noted comparisons of central nervous system between sharks and bony fishes. For example, (1) specializations, in relation to the cranial nerves, are much more pronounced in most teleosts; (2) elasmobranchs do not have a valvula; and (3) afferent connections to cerebellum are different. The afferent connections of the acoustico-lateralis system with the cerebellar auricles and derivatives are more specialized in teleosts, but the most significant difference is the more elaborate connection of the midbrain and the diencephalon with the cerebellum in bony fishes (read further for the function of the valvula). Hofmann (1999) described the nervous system of elasmobranchs.

4.3 Optic Tectum and Tegmentum

The tectum opticum, tectum mesencephali, or mesencephalic roof is a bilateral, shell-like structure which rests on the mesencephalic tegmentum. It receives a voluminous input from the contralateral retina, as well as other, lesser inputs from various sources. The tectal output is directed toward the pretectum and thalamus, the torus longitudinalis, the contralateral tectum, and several sites in the brain stem tegmentum.

The tectum and tegmentum of teleosts (with a few possible exceptions in blind species Fig. 14.9; read Riedel 1997 for *Astyanax hubbsi*) form larger, more intricate, and more specialized structures in elasmobranchs. The size and complexity of these structures are more variable among teleost species than in elasmobranchs, depending on the degree of development of the several exteroreceptive sensory systems. Also the tectum and tegmentum are more distinctly separated, indicating the development of specialized functional properties.

The extent of the ventricle is greatly reduced, since much of this space is occupied by the valvula of the cerebellum, by the torus semicircularis, and by the torus longitudinalis.

The tectum in teleosts is composed of a large number of discrete layers than in sharks and rays. There are differences in afferent and efferent tracts (see Aronson 1963 for details).

The optic tectum has been examined in detail in bony fishes by Northcutt (1983), Vanegas (1983) and Butler (1992). The opticum tectum of actinopterygian fishes forms the bulk of midbrain roof. Four zones from superficial to deep have generally been recognized. These are the superficial white and gray zones, the central zone, the deep white zone, and periventicular gray zone (see Northcutt 1983). Various subdivisions of these zones have been recognized. For example, Vanegas (1983) divided the superficial zone into a stratum marginale, stratum opticum, and stratum fibrosum et griseum superficiale. The stratum marginale is composed of unmyelinated axons that originate in the torus longitudinalis (Fig. 14.10).

Retinal fibres projecting to the superficial zone of the tectum form two fascicles in the majority of species termed T1 and T2 by Northcutt (1983), and corresponding, respectively to the stratum opticum and stratum fibrosum et griseum superficiale of Vanegas (1983). Additional fascicles of retinal fibres have also been found in the central and deep white zones of the tectum. Variations in presence and position of the retinal fascicles in teleost fish have been surveyed (see

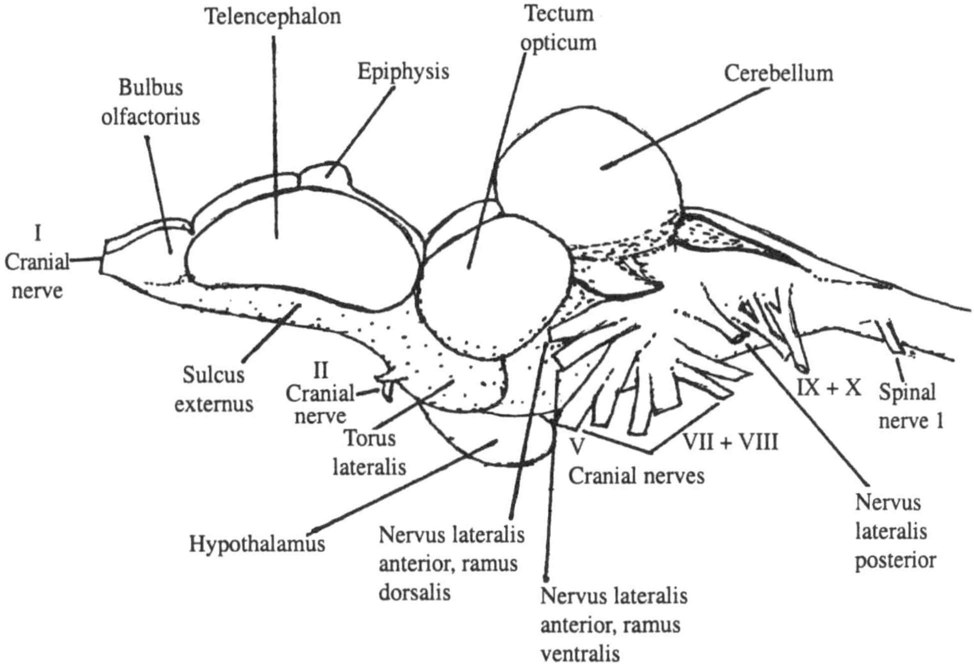

Fig. 14.9 Lateral view of the brain of the blind cave fish *Astyanax hubbsi*. Both the optic nerves (II) and optic lobes are considerably smaller than those in the sighted river species.
***Source:* Riedel (1997)**

literature in Butler 1992). Recently, particular attention has been focussed on the organization of the periventrical gray zone and of afferent and efferent fibre systems within or adjacent to it. It is hypothesized that the dorsal part of the gray, as can be recognized cytoarchitectonically, is in register with the projection of the dorsal optic tract and that the lateral and ventral optic tract (see Butler 1992 for detailed coverage).

Vanegas (1983) described the behavioural aspects of tectal function, related to eye and body movements. Movements elicited from the tectum are not single, discrete, and instantaneous. They are complex, coordinated motor acts which build up after some latency and go on for sometime after the termination of the stimulus.

The structure and function of the retina of teleosts has been extensively investigated (see Powers and Easter 1983). Friedlander (1983) considered the anatomy and physiology of the regions of the forebrain that may play a role in visual processing. Retinal projections to several diencephalic-mesencephalic targets have been demonstrated in several actinopterygians. Anatomical and physiological investigations reveal that the forebrain of teleosts receives visual information. The teleost visual diencephalic is a heterogeneous region. The receptive fields of the neurons in the diencephalic regions have a wide range of properties. Small stationary stimuli are usually adequate to drive the cells. These regions may be involved in pattern discrimination, such as conspecific recognition. The retinodiencephalic and retinotectal systems of teleosts are probably not purely separate parallel processing channels. The localization of function in teleost forebrain has been suggested.

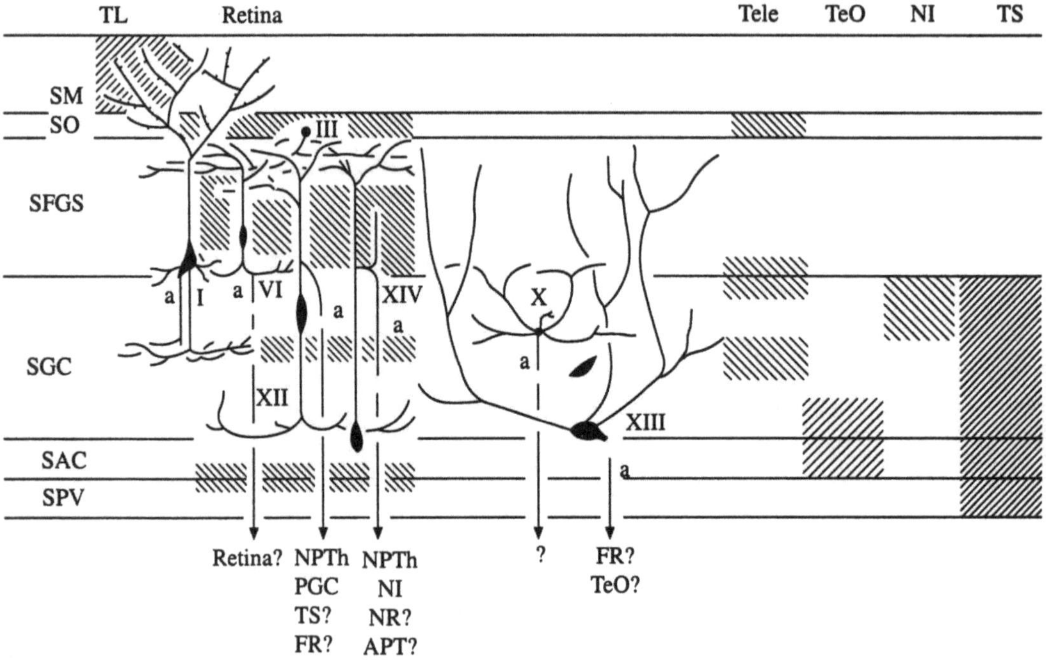

Fig. 14.10 **Idealized diagram of the main neuron types in the optic tectum of teleosts. I, Pyramidal neuron; III, horizontal neuron of SO and SFGS; VI, bipolar neuron of SFGS; X, small multipolar neuron; XII, fusiform neuron; XIII large multipolar neuron; XIV, large pyriform neuron; a, axons. Some sources of afferents are indicated at the top, with their layers of termination shown correspondingly by shading. Neurons contacted by afferents from TL and retina are shown within the respective shaded areas. Efferent tectal neurons have axons leading to downword arrows, and the targets of such axons are correspondingly shown at the bottom.**
APT: Area pretectalis; FR: Formatio reticularis; NI: Nucleus isthmi; NPTH: Nucleus prethalamicus; NR: Nucleus rotundus; PGC: Primary gustatory centres; Tele: Telencephalon; TeO: Tectum opticum; SM: Stratum marginale; SO: Stratum opticum, partes superficialis, intermedia, and profunda; SFGS: Stratum fibrosum et griseum superficiale; SGC: Stratum griseum centrale, inner plexiform layer, inner gray layer; SAC: Stratum album centrale; SPV: Stratum periventriculare; TL: Torus longitudinalis; TS: Torus semicircularis.
***Source:* Vanegas and Ito (1983)**

The cells of origin of the descending telencephalic projection to tectum and visual diencephalon may be receiving inputs from other modalities. This would allow for gating or modulating visual information before the visuomotor transform with information from other systems. The convergent nature of inputs to the diencephalon in at least one teleost, *Micropterus*, suggests a highly integrative function for this brain region. Since this part of the teleost brain has a strong visual input, as well as both ascending spinal and tectal and descending telencephalic inputs, it is in a strategic position to monitor information flow (Fig. 14.11).

Recent advances have been made on visual pretectal nuclei (see literature in Butler et al. 1991) and retinal projections (see recent references in Butler and Saidel 1991; Saidel and Butler 1991; Butler and Northcutt 1992; Northcutt and Butler 1993). The reader is also referred to Bradford and Northcutt (1983) for a comprehensive account on the organization of the diencephalon and pretectum of the ray-finned fishes.

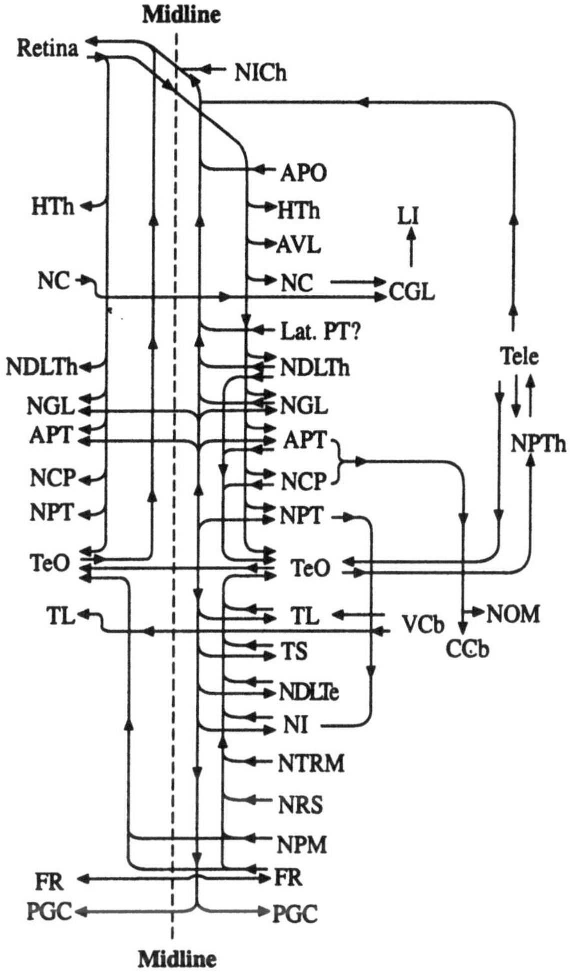

Fig. 14.11 **Connections within the teleostean visual system. APO, area preoptica; APT, area pretectalis; AVL, area ventralis, lateralis; CCb, corpus cerebelli; CG1, corpus glomerulosum; FR, formatio reticularis; HTh, hypothalamus; Lat PT, lateral pretectum; LI, lobus inferior; NC, nucleus corticalis; NCP, nucleus commissurae posterioris; NI, nucleus isthmi; NDL Te, nuclei dorsolateralis tegmenti; NDL Th, nucleus dorsalateralis thalami; NGL, nucleus geniculatus lateralis; NI, nucleus isthmi; NICh, nucleus intrachiasmaticus; NOM, nuclei oculomotores; NPM, nucleus profundus mesencephali; NPT, nucleus pretectalis; NPTh, nucleus prethalamicus; NRS, nucleus reticularis superior; NTRM, nucleus tegmenti rostralis mesencephali; PGC, primary gustatory centres; VCb, valvula cerebelli; Tele, telencephalon; TeO, tectum opticum TL, torus longitudinalis TS, torus smicirculars.**
Source: **Vanegas and Ito (1983)**

Marked variation occurs in the organization of the pretectum among teleost fishes. Northcutt and Wullimann (1988) found strikingly different patterns of organization in the presence or absence of certain nuclei or in their degree of development in representatives of the various teleost radiations. One example of this diversity is the large, complexly organized nucleus glomerulus which has been described in nonteleosts. Although this nucleus is an obvious and striking feature of the pretectum in percomorph teleosts, a homologue is not readily identifiable in species belonging to a number of other teleost lineages.

It has been hypothesized that the posterior pretectal nucleus is homologous to both the pars intermedius and nucleus glomerulus. The presence of a posterior pretectal nucleus is plesiomorphic for teleosts. The presence of two distinct nuclei, the pars intermedius and nucleus glomerulus, is apomorphic in acanthopterygians, and perhaps even in nonteleosts. A number of features of the posterior pretectal nucleus can be recognized as plesiomorphic and used in characterizing and identifying the nucleus in various extant species. Nucleus corticalis, which projects to the posterior pretectal nucleus and to nucleus glomerulus, is also plesiomorphic for teleosts (see details: Butler et al. 1991).

A number of studies of retinofugal projections to the diencephalon and optic tectum have been carried out. The distribution of retinofugal fibres to nuclei in the diencephalon in *Clupea* generally corresponds to that seen in other actinopterygians. The reader is referred to Northcutt and Butler (1993) for details and literature. Retinofugal pathways have been examined in longnosegar *Lepisosteus osseus* (Northcutt and Butler 1976), in lepidosirenid lungfishes (Northcutt 1977), in channel catfish *Ictalurus punctatus* (Prasada Rao and Sharma 1982) and in green sunfish *Lepomis cyanellus* (Northcutt and Butler 1991).

Voneida and Fish (1984) have compared the tectal anatomy and physiology of the blind cave characin, *Astyanax hubbsi* with that of it sighted ancestor *A. mexicanus* (the river fish) and with goldfish. Studies have revealed that, with the exception of a greatly reduced retinotectal projection, connectivity and structure of cave fish tecta are similar to those described in slighted species. It appears that the rudimentary retinotectal input is nonfunctional.

The retinofugal projections in the bowfin, a nonteleost actinopterygian, have been worked out by Butler and Northcutt (1992). Nuclei have been identified in the thalamus, the periventricular portion of the posterior tuberculum, synencephalon, and pretectum which are homologous to like-named nuclei in teleosts and other non-teleost actinopterygian fishes. Of particular note, a posterior pretectal nucleus and, possibly, a homologue of nucleus corticalis have been found to be present in the pretectum. These nuclei have previously been identified only in teleosts. The posterior pretectal nucleus is relatively small in the bowfin, and the distribution of a small, versus a large, posterior pretectal nuclei in Teleostei and Halecomorphi suggests that this nucleus was small plesiomorphologically. The pattern of retinofugal projections in the bowfin is similar to that in other non-teleost actinopterygian fishes and in teleosts in most regards. Contralaterally, the retina projects to nuclei in the dorsal and ventral thalamus, superficial and central pretectum, dorsal and ventral accessory nuclei, and to the optic tectum. Additionally, there are sparse projections to the suprachiasmatic nucleus in the preoptic area, the periventricular nucleus of the posterior tuberculum, and the dorsal and ventral periventricular pretectal nuclei. Ipsilateral projections are sparse and are derived from fibres which do not decussate in the optic chiasma. Undecussated ipsilateral retinal projections, as present in the bowfin, are a widely distributed character in vertebrates and appear to be plesiomorphic for vertebrates.

Among teleosts, undecussated ipsilateral retinal projections have been found in two osteoglossomorphs, and, among euteleosts, in some ostariophysans. Undecussated ipsilateral projections have been found to be absent, however, in an elopomorph and in most euteleosts studied.

With the present data, it is evident that undecussated ipsilateral retinal projections are present in members of all four non-teleost actinopterygian clades, as well as in some teleosts, supporting the hypothesis that these projections are plesiomorphic for actinopterygian fishes. Similar projections have been found in a member of the sister group to Actinopterygii, the Australian lungfish, and in elasmobranchs. Among agnathans, the presence of these projections in *Petromyzon* has been confirmed, and they have also been found in *Lampetra*. Such projections were identified in *Eptatretus stouti*.

The presence of retinopetal cells in the mesencephalon has been reported in the lamprey *Lampetra fluiviatis*, the bichir *Polypterus* and in a euteleost, the pike *Esox lucius*. Mesencephalic retinopetal cells are not widely distributed within actinopterygian fishes, however.

Forebrain retinopetal cells, distributed throughout the ventral telencephalon and the diencephalon, have been found in a variety of percomorph and atherinomorph fishes (acanthopterygians). They do not appear to be present caudal to the immediate region of the olfactory bulb in osteoglossomorph or elopomorph fishes and may thus be apomorphic for acanthopterygians. As *Amia* comprises the outgroup to Teleostei, the present finding in *Amia* of no retrogradely labeled, retinopetal cells supports this hypothesis (see Butler and Northcutt 1992).

African and South American lungfish *Protopterus* and *Lepidosiren*, lack ipsilateral retinal projections. The overall pattern of the primary retinofugal projections is markedly similar to that of amphibians (Northcutt 1977).

4.4 Diencephalon and Pretectum

The diencephalon is a complex brain region lying between the telencephalon, rostrally, and the mesencephalon, caudally. The precise rostral and caudal boundaries of the diencephalon are not easily determined and have been variously drawn by earlier workers.

Braford and Northcutt (1983; see for background information) divided the diencephalon and pretectum into the following regions: Preoptic area; hypothalamus; three thalamic zones: posterior tuberculum, ventral thalamus, and dorsal thalamus; epithalamus (habenula + habenular commissure); synencephalon, including part of the pretectum; and the remainder of the pretectum.

The reader is referred to Table 2 given in the contribution by Bradford and Northcutt (1983) suggesting the nomenclature for cell masses in *Carassius*.

Aronson (1963) reported that the teleost diencephalon is larger, more variable, and contains many more discrete nuclei and fibre bundles than that of elasmobranchs. Epithalamus contributes to the control of chromatophore and negative phototactic responses. Ventral thalamus and the hypothalamus together act as a major centre of correlations for olfactory and gustatory sensations, with the less direct involvement of other sensory modalities. This portion of fish brain is a major area for the organization of feeding responses and projects directly on efferent centres of the branchial motor nerves. It is also an important autonomic centre, as shown by its close relation to the pituitary gland. The highly specialized and variable features demonstrated for this region in teleosts correlate well with the great variety of rather precise feeding habits that have developed in many teleost species. These stand in contrast to the rather indiscriminate mass-feeding habits characteristics of many sharks. The major function of the epithalamus and dorsal thalamus is the correlation of extero-receptive impulses, particularly in relation to the sensitivity to light.

A large nucleus with distinctive cytoarchitecture, named nucleus rostrolateralis has been identified in the rostral part of the diencephalon in the freshwater butterfly fish, *Pantodon buchholzi* (Osteoglossompha). Nucleus rostrolateralis has been found in one other osteoglossomorph fish, *Osteoglossum bicirrhosum*. Most previous studies of central pathways and diencephalic organization in ray-finned fishes have been carried out on species of euteleosts. On cytoarchitectural grounds, no nucleus comparable to nucleus rostrolateralis has been found in any of these fishes. Nucleus rostrolateralis also appears to be absent in species of clupeomorphs and elopomorphs as well as in another osteoglossomorph *Gnathonemus petersii.* Likewise, in a number of non-teleost ray-finned fishes, no such nucleus has been identified. A nucleus rostrolateralis has been also identified on cytoarchitectonic grounds in longnose gar *Lepisosteus osseus* (read Butler and Saidel 1992 for details of structure).

Neuroendocrine interlock

Light-brain-pituitary (-endocrine)/extrahypothalamic axis. An Overview

In teleosts, as in other vertebrates, light and light periods influence different physiological and behavioural events. Circadian rhythms that may be expressed in locomotor activity, feeding, behaviour, thermoregulation and colour changes are entrained from changes in the daily light period. Circadian rhythms can also be transformed into longer rhythms for control of yearly cycles such as reproductive and migratory cycles. The lateral eyes and pineal organ are light receptive organs that convey light/photic information to central brain areas via neuronal projections. In the brain, light recipient neurons relay visual/photic information to other central brain areas and certain brain nuclei may entrain circadian and circannual rhythms.

The light-brain-pituitary/extrahypothalamic axis comprises complex neuronal circuitries serving as interfaces between functional nuclei and subnuclei in which a variety of neurotransmitter/hormone systems may act as mediators of photo-controlled light information.

The hypophysiotrophic system: In the brain of teleosts, the neurosecretory preoptic nucleus (NPO) and different nuclei in the ventral hypothalamus, such as the nucleus lateralis tuberis (NLT), possess neurons that both directly and indirectly innervate the pituitary. These nuclei are considered to comprise major hypophysiotrophic systems in teleosts. In most teleosts, pituitary innervation occurs in the neurohypophysis, which via a short capillary system connects to the pars distalis, and as a direct neuronal innervation of the pars distalis. In teleosts, the NPO also innervates pituitary regulatory neurons in the ventral hypothalamus. Several neurohormones found in the hypophysiotrophic nuclei are known to affect a diversity of endocrine functions by both direct and indirect influence on synthesis and release of hormones synthesized in the pituitary.

Several neurohormones are known in teleosts of which those best characterized are dopamine, the nonapeptides (arginine vasotocin and isotocin), somatostatin, growth hormone releasing factor, corticotropin releasing factor and gonadotropin releasing hormone. Other suggested neurohormone systems in teleosts have been immunocytochemically detected in neurosecretory neurons of teleosts. These systems include neurons synthesizing putative cholecytokinin, enkephalin, β-endorphin, neurotensin, angiotensin II, substance P, vasoactive intestinal polypeptide, FMRF amide, gastrin and glucagon. Hypophysiotropic properties by serotonin, noradrenaline and acetylcholine have also been emphasized. Recently the newly discovered 29 amino acid peptide named galanin has begun to be considered a major hypophysiotrophic system in teleosts. Recent suggestions of additional pituitary projecting neurons outside the "classically" defined neurosecretory nuclei may indicate the presence of additional neurohormone systems.

Extrahypothalamic pathways have been suggested in the whole vertebrate phylogeny, including teleosts. Neurosecretory NPO neurons in teleosts have been indicated to possess extrahypothalamic pathways to a diversity of brain areas and a few immunocytochemical and receptor autoradiographic studies may suggest the presence of actual functional target sites in extrahypothalamic brain areas for nonapeptidergic neurosecretory nuclei in teleosts.

Functional speculations of neurohormones. Action by dopamine on pituitary functions in teleosts includes inhibitory control of pituitary prolactin, thyrotropin, gonadotropin and melanocyte stimulating hormones. Preoptic dopaminergic neurons are suggested to be the major source of dopamine inhibition on both spontaneous and gonadotropin-releasing hormone (GnRH) stimulated gonadotropin hormone release. Recent investigations in teleosts have indicated a role of hypothalamic dopamine in photoperiodically dependent, seasonal cyclic events. In addition, multiple modulatory actions of dopamine in the retina of teleosts and other vertebrates are well established.

In teleosts, the nonapeptide arginine vasotocin (AVT) participates in osmoregulation and control of reproduction. AVT is synthesized in NPO neurons that project to the neurohypophysis, where it is released into central circulation, effecting cells in the kidney and gills, and project to the ventral hypothalamus, influencing actions of pituitary regulating neurons (read Nürnberger and Korf 1996).

So far little is known about the functional properties of galanin-like peptides. Galanin may influence pituitary prolactin and growth hormone secretion. In the eel, galanin synthesis has been suggested to be influenced by sex steroids, indicating involvement of galanin in sexual behaviour and in reproduction in teleosts (see also Magliulo-Cepriano et al. 1993).

In teleosts, indications of an endogenous circadian rhythm generator comprise free-running circadian rhythms in locomotor activity, feeding, behavioural thermoregulation, colour change and extraretinal non-pineal photosensitivity. In some teleost species, neurons in the preoptic nucleus have been shown to express a rhythmic activity, photoperiod dependent preoptic AVT fluctuations have been indicated, melatonin binding has been demonstrated in preoptic areas of goldfish and within the retinorecipient preoptic/hypothalamic nucleus (HON) of the Atlantic salmon.

The extensive extrahypothalamic pathways from the "classical" neurosecretory nuclei indicate that they play multiple roles in a diversity of brain functions in teleosts. In the Atlantic salmon, extrahypothalamic pathways from preoptic galaninergic and nonapeptidergic neurons to different central optic nuclei and to optic tectum indicate a modulatory role of these hypophysiotrophic peptidergic nuclei in visual processing. In Atlantic salmon, the bi/multipolar AVT neurons have the property to exert their central osmoregulatory control synergistically with the neurosecretory preoptico-hypophyseal axis.

No central actions by dopamine have been shown in teleosts. However, the wide distribution in the brain of teleosts indicates an involvement of central dopamine in a variety of extrahypothalamic brain functions.

On the basis of the distribution of galaninergic neurons and their target sites in extrahypothalamic areas, galanin is suggested to act as a neurotransmitter/modulator with involvement in somatosensory, central gustatory, olfactory and visual functions.

For hypophysiotrophic system and related aspects, Atlantic salmon has been extensively studied (see Holmqvist 1993; Holmqvist et al. 1994; Holmqvist and Ekstrom 1995). Information can be obtained on brain peptides in lower vertebrates (Crim and Vigna 1983), FMRF amide-like substance in the brain of cloudy dogfish, *Scyliorhinus torazame* (Chiba et al. 1991), *Xiphophorus maculatus* (Magliulo-Cepriano et al. 1993), and hypophiotrophic neurons in the hypothalamus of the catfish *Clarias batrachus* (Prasad Rao et al. 1993). In dogfish an FMRF amide-like substance participates in the regulation of adenohypophysial function. This molecule may have a role as a neurotransmitter and/or neuromodulator in the central nervous system.

Neuropeptide Y (NPY), a member of the pancreatic polypeptide family, is widely distributed in the central nervous system of various vertebrates, including teleosts, elasmobranchs and cyclostomes.

In lower vertebrates, particularly in teleosts, NPY has been shown experimentally to be involved in regulation of the adenohypophysial function, i.e., stimulation of gonadotropin and growth hormone release. On the other hand, mapping studies of NPY-immunoreactive structures in the piscine brain have been performed in a few species including the goldfish, platyfish (*Xiphophorus maculatus*), scyliorhinid dogfishes (*Scyliorhinus torazame*), brown hagfish (*Paramyxine atami*), and recently in white sturgeon (*Acipenser transmontanus*). In white sturgeon, NPY or a related molecule is involved in the hypothalamic neuroendocrine mechanisms of this primitive bony fish (read Chiba and Honma 1992, 1994; for *X. maculatus*, see Magliulo-Cepriano and Schreibman 1993).

In teleost fish, the nervous terminalis or nucleus olfactoretinalis (NOR) system is a gonadotropin-releasing-hormone (GnRH)—containing neuronal system generally associated with the olfactory system. While evidence suggests that the NOR functions as a modulator of environmental cues (retina, olfactory system, pineal), pheromone receptor, modulator of gonadotropin secretion, reproductive system development and physiology, and reproductive behaviour, the functional significance of this system is still to be definatively determined. In teleosts, immunohistochemical studies on the (GnRH) neuron system have been conducted in various species including salmonids. For ontogenetic development of GnRH neurons in chum salmon brain, the reader is referred to Chiba et al. (1994).

Magliulo-Cepriano et al. (1993) reported the distribution of FMRF-amide, galanin (GAL), and neurotensin (NT) in the brain and pituitary gland of *X. maculatus* and suggested that these peptides are involved in the neuroendocrine regulation of growth and reproduction in this genus.

Holmqvist et al. (1994) reported that nitric oxide (NO), synthesized by NO synthase (NOS)-like enzymes, may play an important role in a diversity of cellular mechanisms in the brain of Atlantic salmon (*Salmo salar*), including in neural systems related to olfactory, visual, hypophysiotrophic, viscero-sensoric and motor functions.

Saccus vasculosus (SV): The saccus vasculosus, reported predominantly in marine fishes, is a highly vasculator protrusion from the ventral wall of diencephalon (a highly vascularized choroidal organ). The lining constituents are mainly the coronet (crown cells) and supporting cells (glial cells); (also named 'Zwischenzellen' or 'Stützellen'); the functions of the former, sensory or secretory or transport of low-molecular substance have been suggested. The question regarding their distribution and role in fish species still remains enigmatic (see Singh 1994).

Chiba et al. (1989) reported the presence of the S-100 protein (an acidic, Ca^{2+}-binding protein) in the hypophysis and saccus vasculosus of *Mustelus manazo* (the seven gummy shark). *Mustelus* SV contains two different types of glial cells; the supporting elements in the epithelium and the astrocytes in the subepithelial nerve tract. S-100 protein immunoreactivity is not detected in chief cells of the SV, the coronet cells or supporting cells. The functional significance of S-100 protein-like substance in the hypophysis and SV of the elasmobranchs is still obscure, but studies on the distribution of this substance in lower vertebrates are of interest from phylogenetic aspects. Chiba et al. (1996b) studied the distribution of perikarya and fibres containing neuropeptide (NPY)-like immunoreactvity in the SV of cartilagincus and bony fishes. Results suggested an involvement of NPY or related substance in the SV function and its regulation.

Subcommissural Organ (SCO): The subcommissural gland (SCO) is an ancient and persistent organ of the brain of vertebrates. SCO belongs to the group of periventricular brain structures known as circumventricular organs. It can be regarded as a truely glandular structure. It is located in the dorso-caudal region of the third ventricle, at the entrance of the Sylvian aqueduct. It consists of a secretory ependyma with one or two layers of columnar cells lining the posterior commissure. SCO is a structure highly specialized in the secretion of glycoproteins. The secretory material is released into the cerebrospinal fluid (CSF) of the third ventricle and condenses to form a thread-like structure, the Reissner's fibre (RF). RF extend along the aqueduct, fourth ventricle and central canal of the spinal cord (see Grondona et al. 1994 for the analysis of secretory glycoproteins of SCO of the dogfish *Scyliorhinus canicula*).

4.5 TELENCEPHALON

The forebrain of sharks shows considerable variability in size and configurations, but this seems to be related for the most part with the degree of development of the olfactory mechanism. The forebrain of bony fishes is even more variable but these elaborations seems to take a rather different direction. It is clear that forebrain of elasmobranchs serves primarily the organization of a very sensitive and effective olfactory system. It may also serve to correlate olfactory data with those of other systems through the epithalamus and hypothalamus. The same appears true for those species of teleosts in which olfaction is dominant. In other teleosts in which olfaction is not dominant, particularly must clearly have another function (Aronson 1965).

In appearance, the telencephalon of ray-finned fishes is strikingly different from that of other fishes and land vertebrates. Although ray-finned fishes possess rostrally located olfactory bulbs and more caudal cerebral hemispheres, like other vertebrates, the hemispheres comprise two solid masses flanking a median ventricular space. The cerebral hemispheres of ray-finned fishes included both the pallium and subpallium (see discussion on eversion hypothesis, Northcutt and Davis 1983b; also refer Northcutt and Bradford 1980).

Studies have recognized both pallial and subpallial areas within the telencephalon of ray-finned fishes, but there is no agreement concerning the boundary between the pallium and sub-pallium or the number of groups comprising these fields. Problems of interpretation are further compounded by the fact that ray-finned fishes reveal four different grades of telencephalic organization. The polypteriforms (*Polypterus* and *Calamoichthyes*) possess the least differentiated telencephalon, with the number of telencephalic cell groups increasing through chondrostean, holostean and teleostean grades.

Details of cytoarchitecture, telencephalic afferents (olfactory afferents, nonolfactory afferents), telencephalic efferents can be obtained from contribution of Northcutt and Davis (1983b). Although numerous descriptive and experimental studies have increased our understanding of telencephalic organization, there is still little agreement whether several nuclei are pallial or subpallial in position and what their homologies are in other vertebrates. This information is not merely of interest in establishing hypotheses regarding the origin and subsequent distribution of a given telencephalic nucleus, it is critical to our ability to recognize how telencephalic functions have evolved, to our perspective on telencephalic organization of ray-finned fishes compared to that of other vertebrates.

At present, there is general agreement on the number of telencephalic nuclei in ray-finned fishes, and considerable progress has been made in establishing a common nomenclature, but it will be impossible to recognize similarities and differences in the functions of individual telencephalic nuclei of ray-finned fishes and other vertebrates until the pallial or subpallial origins and homologies of these nuclei are determined. For a discussion on pallio-subpallial boundary, see Northcutt and Davis (1983b). Table 2 (Riedel 1997) is a literature survey of identified telencephalic elements in the forebrain of actinopterygian fishes.

The behavioural functions of teleostean telencephalon have been covered comprehensively (see Davis and Kassel 1983; Hofmann 2001); Telencephalon is not exclusively an olfactory brain area, but the localization of nonolfactory functions, for that matter, of olfactory functions, is in a very preliminary state of investigation. The inference that the telencephalon serves to activate, arouse, or facilitate behaviours is a generally satisfactory one in view of the limited data. Elucidation of how the telencephalon regulates behaviour will require experiments at the physiological and biochemical level. Investigators have differed in the details of their concept of the arousal functions of the telencephalon.

It has been proposed that the telencephalon specifically modulates reproductive behaviours and defensive reactions to strong or unfamiliar stimulation. The modulation of reproductive behaviour and of defensiveness may be independent telencephalic functions, since defensiveness but not reproductive behaviour is increased by unilateral hemispherectomy. Male responses to female sexual stimuli, nest-building activities, and possibly egg care are disrupted following bilateral hemispherectomy. Male nonsexual social behaviours, which occur early in a spawning trial prior to the onset of explicit sexual responses, are either unaffected or slightly decreased. These conclusions are based on work with a few teleosts, mainly oviparous acanthopterygians. Other teleosts, and representative holosteans and chondrosteans should be examined.

The role of teleostean telencephalon in learning is another aspect (see Hollis and Overmier 1978; Overmier and Hollis 1983) and Demski (1983) reviewed the physiological and behavioural effects of electrical brain stimulation.

4.6 Brain Organization in the Cartilaginous Fishes—A Note

Most of the preceding coverage is related with the bony fishes. The brain organization in the cartilaginous fishes has been comprehensively covered by Northcutt (1977, 1978, 1989) (Fig. 14.12). Examination of shark brain: body ratios reveals that these taxa possess relative brain volumes in a range overlapping those of bony fish as well as birds and mammals. Much of the variation is due to relative development of the telencephalon and cerebellum. Telencephalon weights vary from 24% in *Squalus* to 52% in *Sphyrna*. Analysis of the cytoarchitectonics of the shark brains reveals at least two patterns of development. Squalimorph sharks possess low brain: body ratios, and the telencephalon of these taxa possess well developed lateral ventricles and

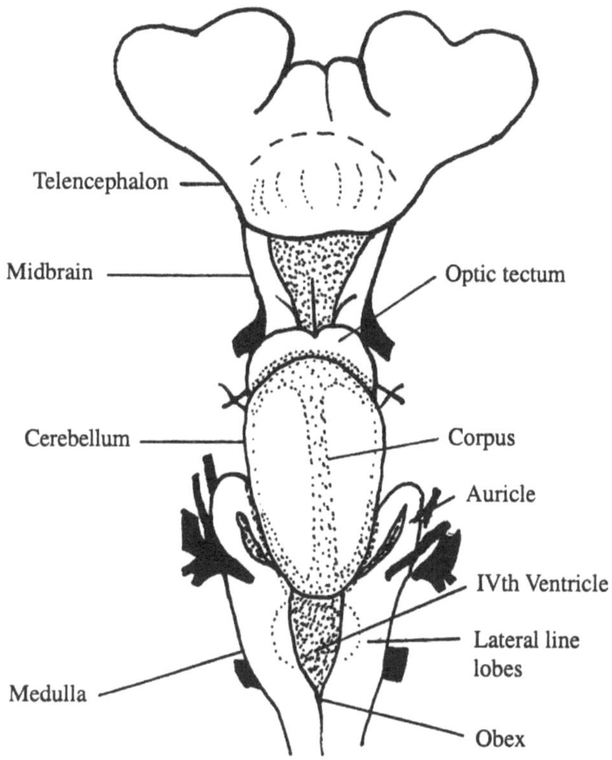

Fig. 14.12 Dorsal view of the brain of *Scyliorhinus*.
***Source:* Roberts (1978)**

poorly developed pallial areas. The diencephalon is characterized by prominent periventricular laminae, and the cerebellum lacks foliation. The lamniform and carcharhiniform sharks are characterized by high brain: body ratios, and there is marked hypertrophy of the telencephalon. The roof (pallial) regions, as well as the diencephalon, are characterized by extensive cellular migrations. The cerebella of these forms possess extensive complex foliation.

In elasmobranch fishes, Northcutt (1989) discussed relative brain size, variations of the octavolateralis column, cerebellum, optic tectum and diencephalon. The biological significance of most chondrichthyan neural trends is unknown.

Aronson (1963) pointed out several major differences between the brains of elasmobranchs and teleosts: (1) Relative to body size, the brains of sharks and rays are large and contain greatly expanded ventricular spaces, whereas the brains of teleosts are small and compact with markedly reduced ventricular spaces. This major difference in structure between the elasmobranch and teleost brains may be attributed to the relative size of the head and the growth processes during early ontogenetic stage. There are no indications that this structural difference in size and compactness accounts for any of the behavioural differences between the two groups of fishes. (2) The teleostean brain contains many more discrete tracts and well circumscribed compact nuclei, whereas the tracts and nuclei in the shark's brain are much more diffuse. (3) The teleostean brain is highly variable. Pronounced structural differences are seen even among some species of a single genus. The elasmobranch brain, on the other hand, is relatively uniform. Although some variability is seen in the forebrain and cerebellum, this never reaches the proportions found among teleosts. This variability in brain structure is correlated with diverse habits in many families of bony fishes. Included here are variations in locomotion, orientation, social and sexual behaviour, and above all, in feeding patterns. Neural adaptability has made it possible for teleosts to enter almost all of the available aquatic ecological niches, whereas elasmobranchs, with their uniform, less adaptable, and more diffuse nervous systems, are confined for the most part to the relatively uniform conditions of more or less open waters. (4) Many species of teleosts develop hypertrophied primary or secondary sensory areas in response to the elaboration of special sensory systems in different parts of the body. In some species they reach huge proportion. Nothing comparable to the expanded neural structures has been observed in elasmobranchs (e.g., the facial and vagal lobes of the medulla in the carp).

Barry (1987b) investigated the central connections of the IXth and Xth cranial nerves in the clearnose skate, *Raja eglanteria*. The most caudal vagal rootlet is the accessory nerve of the skate.

4.7 Brain Organization of the Coelacanth *Latimeria chalumnae:* General Observations

The brain of *Latimeria* (Fig. 14.13) is considerably elongate with little trace of the cephalic flexures seen in most bony fishes except dipnoans (Fig. 14.14). The brain is housed in an expansive neurocranium. The olfactory peduncles are extremely long, and the olfactory bulbs are directly apposed to the nasal cavity. The swelling at the rostral poles of the telencephalic hemispheres do not represent the olfactory bulbs. In this respect *Latimeria* clearly differs from the lepidosirenid lungfishes, in which the olfactory bulbs arise from, and are in direct continuity with, the telencephalon. Elongated peduncles occur in some actinopterygians and in most cartilaginous fishes, as well as in *Neoceratodus*.

The telencephalon of *Latimeria* is relatively longer than that of actinopterygian fishes with the exception of *Polypterus* and its shape is more similar to the telencephalon in dipnoans. The general similarity is further reinforced by an obvious lateral swelling of the ventral telencephalon which is even more pronounced in the Dipnoi.

The pituitary of *Latimeria* appears unique among that of vertebrates as it extends rostrally rather than caudally from the infundibulum. This condition has been interpreted as a consequence of skull growth rapidly out-distancing brain growth and the rostral shifting of the hypophyseal fenestra.

The optic nerves are large and appear relatively larger than do those of most bony fishes. They are far larger than those of any of the living dipnoans. The optic tectum is clearly paired and possess slightly developed lateral recesses and well developed cellular and fibrous laminae. The internal organization of the tectum of *Latimeria* reveals more extensive differentiation than the tecta of dipnoans and salamanders.

The large cerebellum is the most striking feature of the general morphology of the *Latimeria* brain. The corpus is a large spherical dome surrounded laterally and caudally by pronounced auricles and the cerebellum is larger than in living amphibians.

Latimeria has all three somatic motor cranial nerves innervating the extrinsic eye muscles; and they arise from the same brain stem areas as in other vertebrates. The abducent nerve is unusual as it arises from multiple roots, and the most caudal ramus is considerably removed from the other rami.

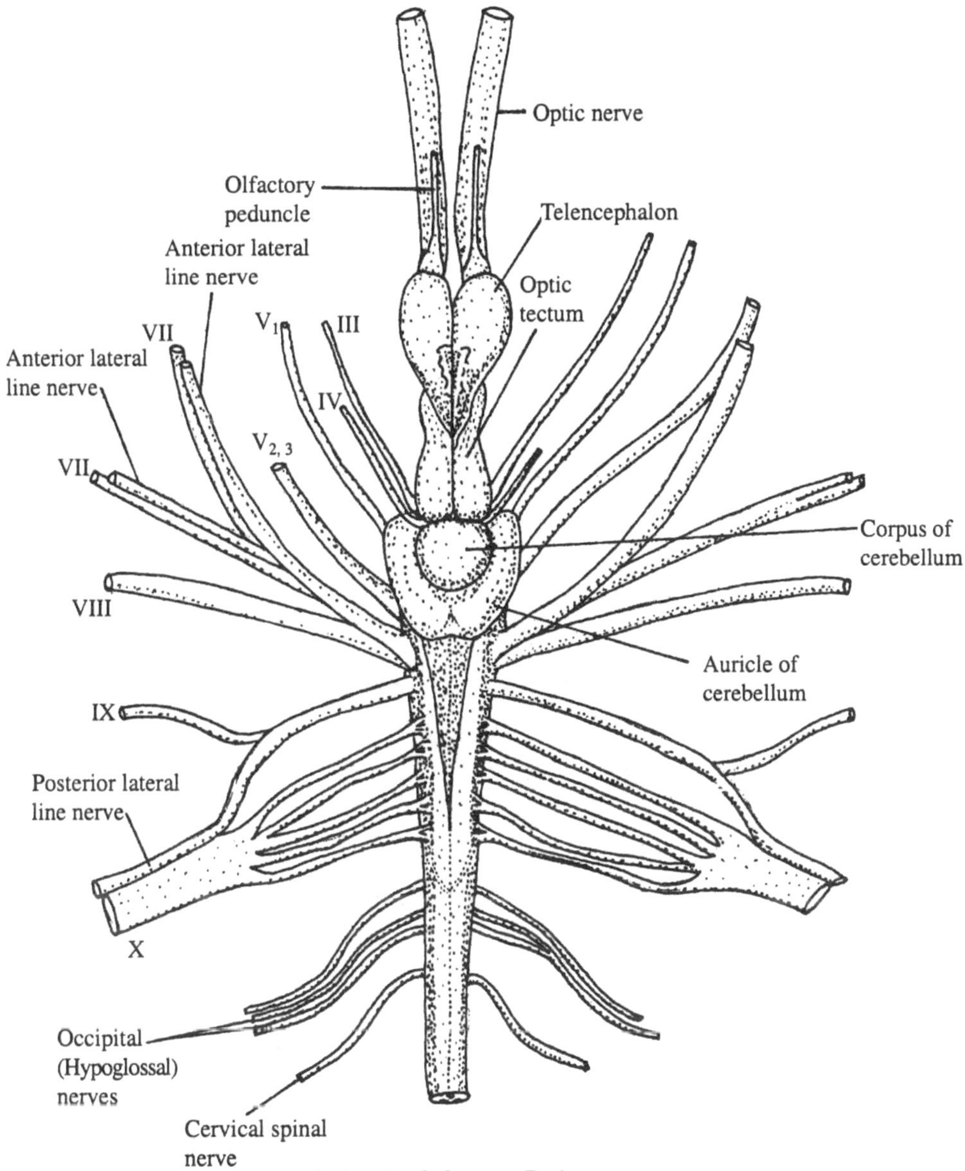

Latimeria chalumnae: Brain.

Fig. 14.13 Dorsal view of brain of *Latimeria*. Olfactory bulbs of the telencephalon not pictured. *Source:* Northcutt et al. (1978)

Latimeria, like other bony fishes, possesses a distinct profundus branch (V_1) of the trigeminal complex. However, V_1 separates very proximally from the mandibular and maxillary branches ($V_{2,3}$). The facial cranial nerves are obvious and divide into rostral and caudal branches. Both branches are closely associated with branches of the anterior lateral line nerves in their course toward neurocranial exit. However, the anterior lateral line nerves are distinct cranial nerves possessing separate entry into the medulla dorsal to the facial rami.

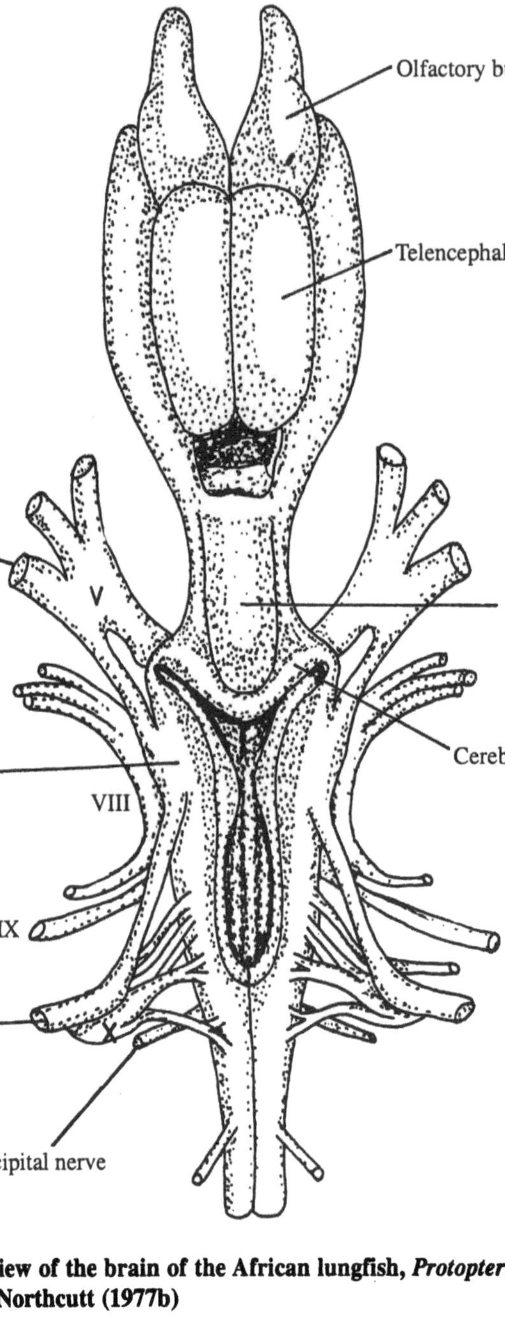

Fig. 14.14 Dorsal view of the brain of the African lungfish, *Protopterus annectens*. *Source:* Northcutt (1977b)

The eighth cranial nerves consist of a single ramus with entry caudal to the anterior lateral line nerves. *Latimeria* is similar in this respect to most other bony fishes.

The posterior lateral line nerves are well developed and are first seen in close association with the vagal nerves as they enter the neurocranium. However, the posterior lateral line nerves rapidly arc rostrally and run with the glossopharyngeal nerves. In *Latimeria* there can be no question that the anterior and posterior lateral line nerves are distinct and separate cranial nerves.

These lateral line nerves are clearly not components of the facial and vagal nerves, even though they have frequently been said to be components of these branchiomeric nerves.

The vagal nerves are well developed and, as in other vertebrates, arise from a series of rootlets. *Latimeria*, like other bony fishes, possesses a number of spino-occipital nerves which innervate the hypobranchial musculature and are probably homologous to the hypoglossal nerves of land vertebrates.

An analysis indicates that the brain of *Latimeria* possesses a number of characters in common with those of *Squalus* and *Scaphirhynchus*. The latter taxa both exhibit relative brain proportions that are most similar to *Latimeria*. *Scaphirhynchus* deviates most by conspicuous hypertrophy of the olfactory apparatus and relative reduction in the visual system. *Latimeria* and squaliform sharks possess well developed visual systems, lateral line organs, and moderate cerebellar development. Contrary to popular belief, sharks (with the exception of sphyrnids) are not characterized by hypertrophy of the olfactory system. Similarly, olfaction in *Latimeria* does not appear as well developed as vision, based on the relative size of the olfactory and optic tracts.

The following neural features characterize *Latimeria*; and they are probably primitive gnathostome features based on their wide distribution in squaliform sharks and non-teleost actinopterygian fishes: (1) moderate to poorly developed olfactory organs; (2) well developed visual system; (3) bilobed mesencephalic tectum with well developed internal lamination; (4) small mesencephalic acoustico-lateralis centre (torus semicircularis); (5) moderate to well developed cerebellum; (6) well-developed lateral line organs supplied by two distinct and separate pairs of cranial nerves; and (7) an evaginated telencephalon. The last feature is not present in actinopterygian fishes, but is present in all other bony fishes, sharks and agnathans; thus an evaginated telencephalon is most likely the primitive gnathostome condition.

Brain component sizes in *Latimeria* are more similar to those extrapolated for amphibian brains than to those for actinopterygians. However, the cerebellum of *Latimeria* is considerably larger than that of amphibian (Northcutt et al. 1978).

5. MONOAMINE-CONTAINING NEURONAL SYSTEMS IN THE TELEOSTEAN BRAIN

Among the various biogenic amines known to be present in the central nervous system of vertebrates, the primary catecholamines, norepinephrine and its immediate metabolic precursor dopamine, and the indolamine known as serotonin, or 5-hydroxytryptamine are the most extensively investigated.

Parent (1983) reviewed the monoamine-containing neuronal systems in teleostean brains, based on his studies in pumpkinseed sunfish, *Lepomis gibbosus*, and the yellow perch, *Perca flavescens*. In addition to teleosts, the distribution of monoamine neurons has also been investigated in the cyclostome, *Lampetra fluviatilis*. In striking contrast with the findings in teleosts, no catecholamine neuronal somata were visualized in the brain stem proper in *Lampetra*. No information is available concerning the distribution and organization of monoamine neurons in the brain stem of elasmobranchs. However, appreciable concentrations of serotonin, dopamine and norepinephrine have been detected biochemically in the brain of *Acanthias vulgaris*. The distribution of monoamine neurons in the brains of chondrosteans and holosteans has apparently not been investigated.

The distribution of monoamine neuronal somata in the brain stem of teleosts resembles that seen in other vertebrates. In fishes the projections of the catecholamine cells of the central medullary groups are still unknown. One of the most impressive monoamine cell groups of the brain stem of teleosts is the small collection of large, intensely fluorescent neuronal somata identified here as the isthmal group. Very little is known of the exact projection of the isthmal cell group in fishes. The apparent absence of catecholamine neuronal somata in the midbrain of fishes is at variance with the situation in reptiles, birds, and mammals, where a large catecholamine cell population occurs within the midbrain tegmentum. Although no catecholamine neurons were visualized in the midbrain proper of teleosts, a small group of medium-sized to large catecholamine cells occurs near the mesodiencephalic junction level in the nucleus posterior tuberis in *Lepomis*. Catecholamine cells of various sizes also occur near mesodiencephalic junction in the so-called tuberculum posterior in the cyclostome *L. fluviatilis*. The small medullospinal catecholamine cell group has also been described in teleosts. Further studies are needed to understand more clearly the exact relationship between the catecholamine cells of the medullospinal group and the sensory intramedullary cells. The catecholamine medullospinal group could also be homologous to the most caudally located cells of the medulla of land vertebrates, such as the so-called A1 and A2 catecholamine cell groups and the catecholamine neurons of the area postrema.

As in teleosts, a multitude of monoamine CSF-contacting cells are found along the ependymal wall of third ventricle in elasmobranchs and cyclostomes. Most of the CSF-contacting cells in the hypothalamus are of the catecholamine type, although some serotonin-type cells are also present along the posterior and lateral recesses.

In the hypothalamus of cyclostomes, CSF-contacting monoamine cells displaying the same morphological features as those in teleosts and sharks are distributed abundantly along the ependymal wall of the preoptic recess, and especially along the infundibular and posterior recesses of third ventricle. In addition to teleosts, the distribution of monoamine axonal varicosities in the telencephalon has also been studied in a cyclostome—the lamprey, *Lampetra fluviatilis*. The telencephalon of *Lampetra* receives an extensive serotoninergic innervation, whereas the catecholaminergic innervation is mainly confined to the so-called "corpus striatum". In *Carassiws auratus*, the telencephalon, after hypothalamus, contains the highest level of primary catecholamines, as also reported in *Lepomis*. The pattern of monoaminergic innervation in the telencephalon of teleosts is strikingly complex. As a whole, the telencephalic catecholamine afferents appear more widely distributed in the teleosts than in land vertebrates. The number and distribution of catecholamine axonal varicosities in the telencephalon is much greater in teleosts than in cyclostomes, where they are confined to the so-called "corpus striatum". Correspondingly, cyclostomes lack an isthmal catecholamine cell group, which occurs in teleosts and is assumed to be homologous to the locus coeruleus of land vertebrates.

6. SPINAL AND CRANIAL NERVES

Northcutt (1985) reviewed the experimental neuroanatomical and embryological studies of these nerves.

Spinal nerves: Gnathostomes are characterized by a single pair of spinal nerves per body segment. Each spinal nerve comprises a dorsal spinal root, bearing a sensory ganglion, and a ventral spinal root. These roots fuse peripherally and reissue as a number of somatic and visceral rami. Traditionally, the spinal nerves of amniotes are believed to possess dorsal spinal roots composed solely of entering sensory neurites, whereas the ventral roots are believed to be composed solely of existing motor axons. Lampreys differ from gnathostomes in possessing two pairs of spinal nerves per body segment. Dorsal and ventral nerves alternate and do not unite but course peripherally as totally distinct rami. The dorsal spinal nerve, like that in gnathostomes, possesses a sensory ganglion. Traditionally, the dorsal spinal nerve in lampreys is believed to comprise somatic and visceral sensory fibres, as well as all visceral motor axons, whereas the ventral spinal nerves are said to consist only of somatic motor axons. Visceral fibres have been reported in the ventral spinal nerves. Myxinoids possess a single pair of spinal nerves per segment. Although symmetry of the spinal nerves is not precise in myxinoids, in that dorsal and ventral spinal roots alternate, these roots fuse peripherally to form a single nerve. Although the components of the spinal nerves in myxinoids have not been determined experimentally, it is possible that the dorsal and ventral roots consist predominantly of sensory and motor fibres, respectively.

Given the difference in pattern of dorsal and ventral spinal root fusion in myxinoids and gnathostomes, it is probable that the earliest craniates possessed separate dorsal and ventral spinal nerves, as do living lampreys.

Cranial nerves: The cranial, or cephalic nerves, of craniates have traditionally been grouped into three phylogenetic categories: (1) ventral cranial nerves (III, IV, VI and XII) that are serially homologous to the ventral spinal nerves of lampreys; (2) dorsal cranial nerves (V_1, V_2, VII, IX and X) that are serially homologous to the dorsal spinal nerves in lampreys; and (3) special sensory cranial nerves (0, I, II and VIII and the lateral line nerves) that arise embryologically from neurogenic placoidal tissue. The optic "nerve" is usually removed from consideration, as it is not a true nerve but a neural tract formed by evagination of a part of the wall of the neural tube. For specific information on the cranial nerves of *Polypterus*, see Piotrowski and Northcutt (1996). It may be pointed out that parietal decussation of the optic nerves, in which bundles of fibres from each nerve interdigitate, appears to have arisen with the origin of craniates, as this pattern occurs in *Eptatretus* as well as most members of all vertebrate groups. Teleost fishes appear to be a notable exception, as their optic nerves cross over each other, with each nerve retaining its integerity, or one nerve penetrates the other. The absence of intedigitated optic nerve decussation thus appears to be uniquely derived character of teleost fishes.

The reader is further referred to a critical re-evaluation of views and conclusions derived by Northcutt (1985) regarding supply of nerves to muscles, especially of head and eye.

7. AUTONOMIC NERVES

A general account of fish autonomic nervous system (ANS) has been given by Campbell (1970). Review by Nilsson and Holmgren (1993; see literature) is the main source of the following account. A comparative description of this system of the dog and dogfish has been provided by Nilsson (1997). The reader is also referred to Goehler and Finger (1996). Donald (1998) has covered various aspects in detail of the autonomic nervous system (see Tables 1–3). The ANS is defined as that component of the nervous system with a ganglionic synapse outside the central nervous system (CNS).

7.1 Agnathous Fish

The anatomy of the autonomic nervous system in fish shows great variation, from the very simple arrangement in cyclostomes to the elaborate and well-developed system in teleosts. In the cyclostomes, particularly in the myxinoids (hagfish), the autonomic nervous system is poorly developed, and it is often impossible to decide whether peripheral neurons are autonomic or sensory. There are no sympathetic chains, but nerve branches that can be regarded as corresponding to a spinal autonomic outflow leave in the ventral (myxinoids) or dorsal and ventral (lampetroids; lampreys) spinal roots. There are ganglionic clusters along the dorsal aorta and elsewhere and fibres are also found in the blood vessels. The nature and function of a spinal autonomic innervation of the viscera in cyclostomes need further attention.

Cranial autonomic fibres are present in the vagus (X) in both cyclostomes groups, and in the lampetroids autonomic pathways may also run in the facial (VII) and glossopharyngeal (IX) nerves. The facial and glossopharyngeal pathways probably innervate the gills. Cardiac vagal fibres occur in lampetroids, but there is no extrinsic innervation of the heart in myxinoids. However, a vagal innervation of the gall bladder and the hepatic ducts in the vicinity of the endocrine pancreas has been shown in *Myxine glutinosa* (read Nilsson and Holmgren 1998).

Ganglionated nerves form a plexus in the cyclostome (*Lampetra* and *Myxine*) gut, but the nature and function of these nerves are not well understood.

7.2 Gnathostomatous Fish

Elasmobranchs: Sympathetic chains of the type found in the teleost (see later) and tetrapods are absent in elasmobranchs, but the segmental paravertebral ganglia are irregularly connected longitudinally and by transverse commissures. The ganglia receives their input from the spinal nerves via white *rami communicantes*, but recurrent grey rami are absent in elasmobranchs.

The most anterior pair of paravertebral ganglia, and a segmentally arranged series or clusters of chromaffin cells (often known as the suprarenal tissue), form the axillary bodies (Figs. 14.15 and 14.16). Posterior to these, the paravertebral ganglia are smaller and lie dorsally in the posterior cardinal sinus (e.g., Port Jackson shark *Heterodontus portusjacksoni*). Also, these ganglia are often found in association with the suprarenal tissue. Nerves from the paravertebral ganglia, the splanchnic nerves, are made up of postganglionic fibres that run to the viscera (see Fig. 13.32).

Cranial autonomic fibres have been postulated in the oculomotor (III), facial (VII), glossopharyngeal (IX), and vagus (X) nerves. A ciliary ganglion is present, and fibres from this ganglion innervate the eye. The function of the facial and glossopharyngeal fibres is not clear, and a vasomotor innervation of the gill blood vessels has been denied. Vagal fibres innervate the heart, where a ganglion is found in the wall of the sinus venosus. The fibres extend posteriorly as far as the pylorus and anterior part of the spiral intestine.

Myenteric and submucous plexuses are present in the elasmobranch gut.

Teleosts: The sympathetic chains in teleost and ganoid fish are generally well developed, and the arrangement resembles that found in the tetrapods. A unique feature in teleosts and holosteans is the cephalic portions of the sympathetic chains: the chains continue into the head, bearing ganglia in connection with the trigeminal/facial, glossopharyngeal, and vagus nerves

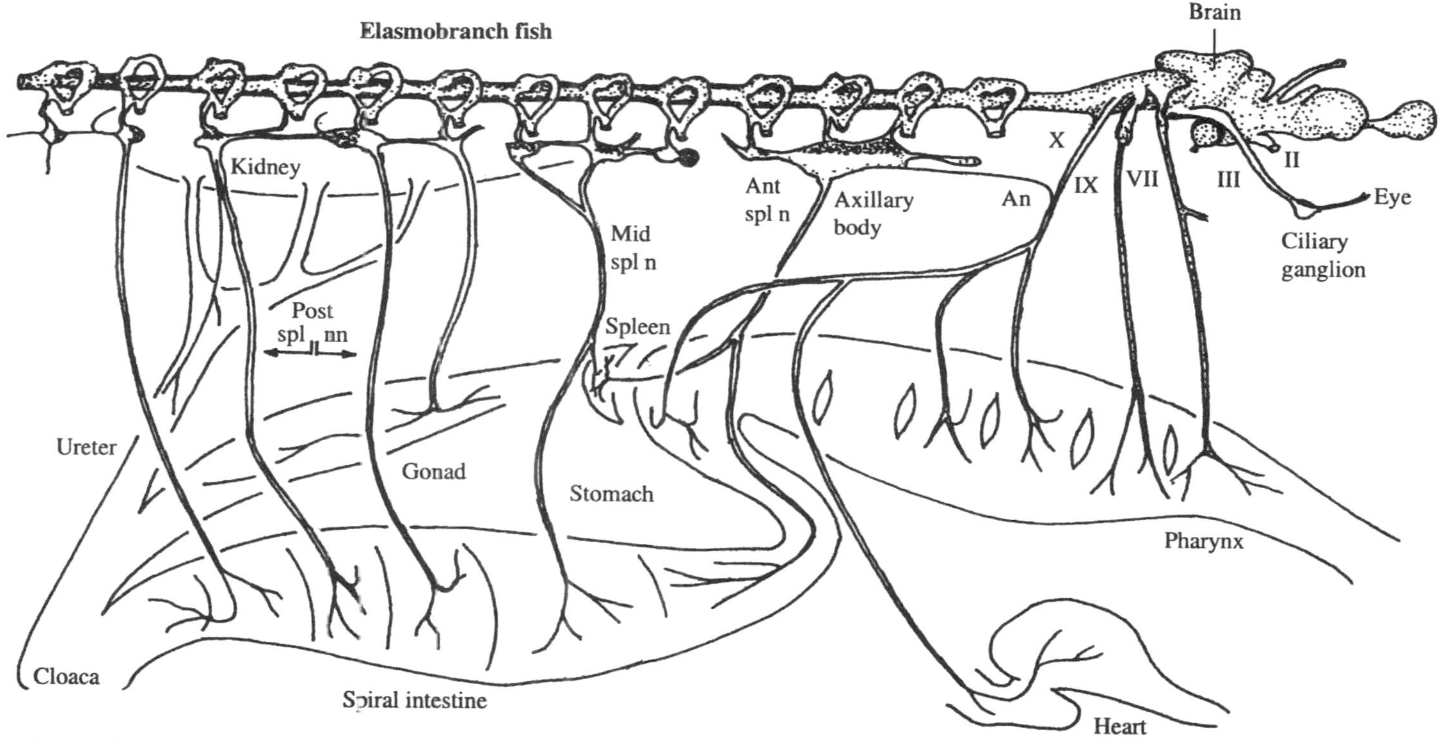

Fig. 14.15 **Generalized view of the arrangement of the autonomic nervous system in elasmobranchs; Note the chromaffin tissue which is associated with the paravertebral ganglia (e.g., the axillary body).**
An, anastomosis connecting the anterior paravertebral ganglia to the vagus nerve;
Ant spl n, Anterior splanchmic nerve;
Mid spl n, Middle splanchnic nerve;
Post spl nn, posterior splanic nerves,
***Source:* Nilsson and Holmgren (1993)**

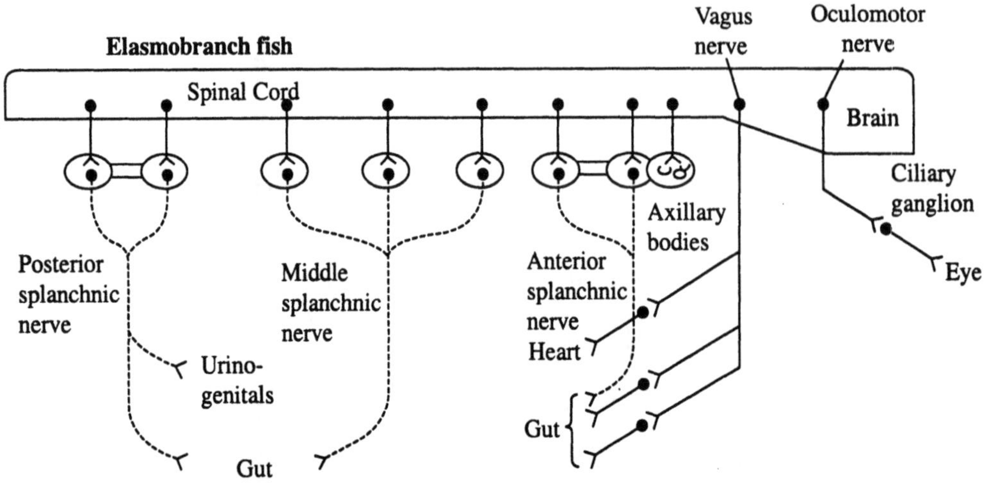

Fig. 14.16 Summary of the main features in the organization of the autonomic nervous system in the elasmobranchs.
***Source:* Nilsson (1997)**

(Figs. 14.17 and 14.18). The connections between these ganglia and the cranial nerves are composed solely of postganglionic fibres (i.e., grey *rami communicantes*), and the preganglionic fibres of these pathways are all of spinal autonomic origin. The presence of adrenergic ganglion cells is the ciliary ganglion may be regarded as an indication that this ganglion, in part, represents the most anterior end of the teleost sympathetic chains.

At the level of the third to fourth spinal nerves on the right side, there is a substantial sympathetic chain ganglion. From this ganglion (the coeliac ganglion) the anterior splanchnic nerves run along the coeliac and mesenteric arteries to the gut, swim bladder, spleen, and liver. The coeliac ganglion in teleosts is part of the right sympathetic chain and not, as in tetrapods, a prevertebral or collateral ganglion situated away from the sympathetic chains. Thus, as in elasmobranchs, the splanchnic nerves are composed largely of postganglionic fibres. However, in some species preganglionic neurons that innervate small groups of ganglion cells are also present in the splanchnic nerve.

The sympathetic chains fuse to a single strand in the posterior part of the abdomen. In this region a posterior splanchnic nerve (vesicular nerve) leaves the chains to innervate the posterior part of the gut and the urinogenital organs. In *Gadus morhua*, *Uranoscopus scaber* and some other teleost species, a substantial ganglion is present in the nerve running along the ureters, which suggest an arrangement of the posterior spinal autonomic ganglia similar to the "sacral parasympathetic" system.

The presence of cranial autonomic ("parasympathetic") fibres has been shown unequivocally only in the oculomotor (III), and vagus (X) nerves of teleosts. In view of the extent of the cephalic sympathetic chains, innervation of the vasculature of the head and gills is probably of spinal autonomic origin. A pair of large vagus nerves run to the gut and swim bladder. There are substantial contributions of spinal autonomic neurons to the vagi, which may therefore be best described as "vago-sympathetic nerve trunks".

As in elasmobranchs, both myenteric and submucous plexuses are present in the enteric nervous system of teleosts.

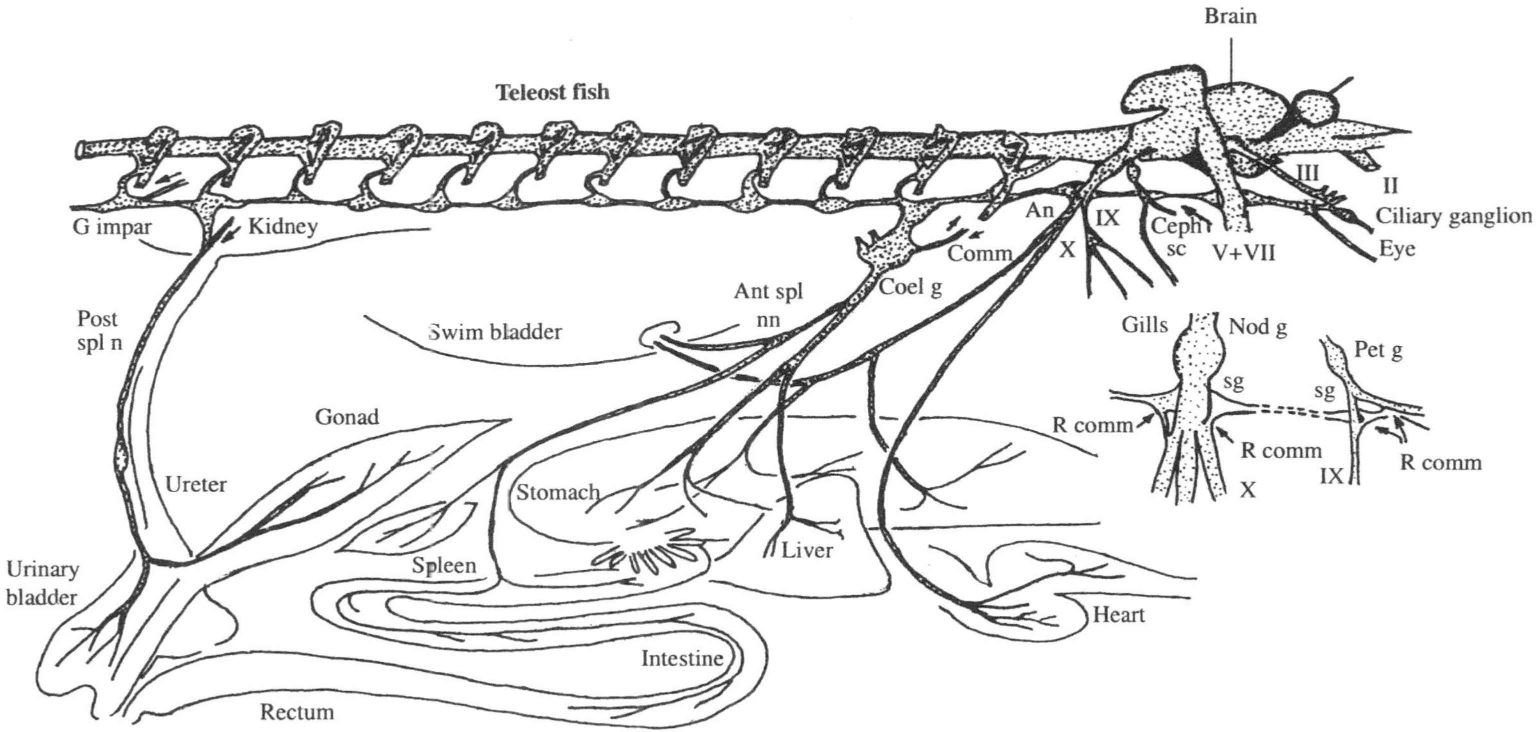

Fig. 14.17 **Generalized view of the arrangement of the autonomic nervous system in teleosts. Note the cephalic sympathetic chain which carries ganglia in contact with the cranial nerves.**
Abbreviations. An, Anastomosis connecting the anterior paravertebral ganglia to the vagus nerve; Ant spl nn, Anterior splanchic nerves, Ceph sc, Cephalic sympathetic chain; Coel g, Coeliac ganglion; Comm, Commissure between left and right sympathetic chain; G impar, Ganglion impar (= posterior fusion of left and right sympathetic chains); Nod g, Nodose ganglion; Pet g, Petrous ganglion; R comm, ramus communicans; Sg, Sympathetic ganglion.
Source: **Nilsson and Holmgren (1993)**

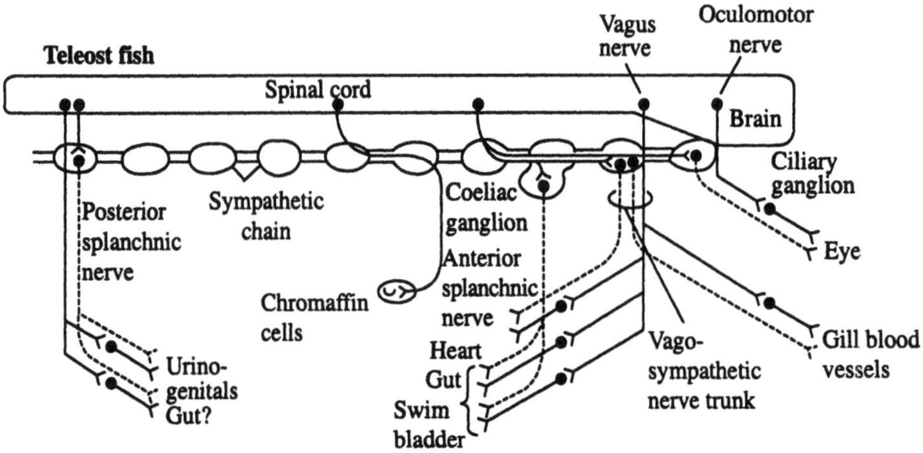

Fig. 14.18 Summary of the main features in the organization of the autonomic nervous system in the teleosts.
Source: **Nilsson (1997)**

Dipnoans: The sympathetic chains in lungfish are extremely fine strands. The strands form loops (annuli) around the segmental (intercostral) arteries, and fibres run to the chromaffin tissue that is embedded in the wall of these arteries. The sympathetic chain ganglia are minute, and adrenergic cell bodies were not observed in these ganglia in *Protopterus aethiopicus* or *Lepidosiren paradoxa*. The nature and function of the spinal autonomic pathways (apart from a probable control of chromaffin tissue) remain to be clarified.

Little is known about the autonomic component in the cranial nerves of lungfish, but a cardiac, pulmonary and probably gastrointestinal innervation by vagal fibres has been postulated. In addition, autonomic fibres to the eye may run in the oculomotor nerve of *Neoceratodus fosteri*.

Control of Iris in Fish Eye

In the elasmobranchs and to some extent, in the teleost *Uranoscopus scaber*, the oculomotor pathway control the iris dilator muscle. In addition, there is a "sympathetic" cholinergic innervation of the iris sphincter in *Uranoscopus* but, in this species, no apparent direct light effect on the sphincter.

In the Atlantic cod (*Gadus morhua*), however, the pattern is completely different. There appears to be no "parasympathetic" (oculomotor) pathway to the iris, and the "sympathetic" innervation (which is adrenergic) contracts the iris sphincter (see Nilsson 1997).

CHAPTER 15
Receptors

1. INTRODUCTION

Animal life is strongly dependent on the functioning of sensory systems. Generally speaking, strong relationships exist between, on the one hand environmental factors and behavioural patterns, and on the other hand the capabilities of exteroreceptive sensory systems. A well-developed sensory organ is accompanied by the presence of large recipient brain structures. In other words, we can say that receptors are essential for the interplay between the organism and its environment. The research on chemoreception in fishes has been growing at a remarkable pace over the last few decades. In recent times, there is a shift towards ecological trend in chemosensory research in all vertebrates, termed as Sensory Ecology.

Figure 15.1 is designed to show the sensory ecology, i.e., sensory perspective of the organism

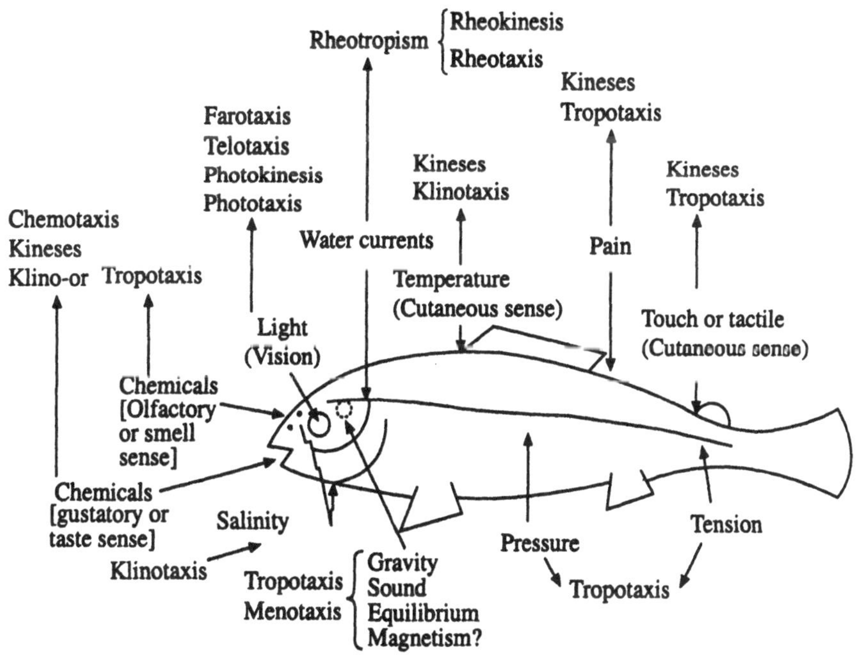

Fig. 15.1 Sensory ecology gives a sensory perspective of the organism in relation to the environment. Different external stimuli are the basis for various direct or indirect orientation movements. ***Source:*** **Bertmar (1985) (Modified)**

in relation to the environment. Table 15.1 gives an overview of sense organs and their function in orientation and orientation behaviour (see Bertmar 1985).

Table 15.1 Sense organs and their functions in orientation and behaviour

Main group of orienting mechanisms	Orienting receptor organs	Orientation behaviour
Orientation in fixed-space reference system: stimulus sources provide the coordinates and/or serve to maintain the normal position, and they are not the goal of the orientation.	Gravity and proprioceptive organs	Maintenance of physical balance. Control of position and direction of action in space.
	Eyes	Orientation related to light. Horizontal near-orientation to landmarks.
	Latero-sensory organs	Tactile, kinesthetic registration and repetition of own movements. Orientation in currents.
Stabilization of posture and movement (locomotion) independent of locality.	Semicircular canals Utriculus-Sacculus (Inner ear)	Orientation related to equilibrium.
	Eyes	Optomotoric (kinetic) control
Object orientation: stimulus sources are the objects of orientation; frequently locating processes (goal and direction is considered).	Eyes	Optical target orientation
	Lagena (Inner ear)	Acoustical orientation
	Skin receptors	Tactile orientation to other fish, etc.
	Gustatory receptors	Chemosensory near-orientation to food, etc.
	Olfactory organs	Chemosensory near-orientation to different water qualities in feeding areas. Homing orientation.

Source: Bertmar (1985).

All living craniates possess optic, otic, lateralis, olfactory and gustatory sense organs that arise in development primarily or in part from neurogenic ectodermal placodes, whereas many of these special sense organs are absent in cephalochordates and urochordates, the obvious outgroups of craniates, or they are distinctly different in their anatomy and embryology. Thus it is possible that the earliest craniates were characterized by all of the special organs found in living craniates and these organs arose with the origin of craniates. Peters (1993) compared the sensory capacities of catfish, *Ictalurus nebulosus* with those of dogfish, *Scyliorhinus canicula*, in an orientational perspective. Both fish are bottom dwellers which implies living at low light levels and feeding on benthic prey. Since the fish orients towards incident light or parallel to the directivity of the light field, the visual system may be considered as a bi-sensor system. The sense of smell is a bi-sensor system in respect of sources of stimuli. The sense of taste operates according to other references in the array mode. For special senses in elasmobranchs, read Bleckmann and Hofmann (1999).

2. EYE (OPTIC ORGAN)

2.1 Overview

Diversity in the photic environment and the modes of life has probably been a major evolutionary factor in the spectacular development of the ocular adaptations in the animal world. (see Fig. 15.2; Table 15.2).

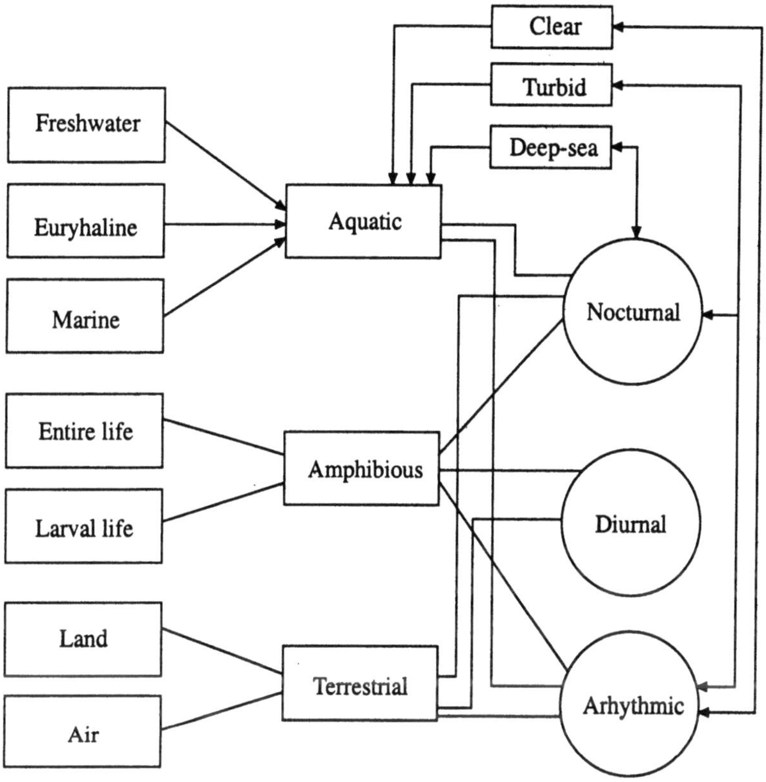

Photoreception: Adaptive radiation.

Fig. 15.2 Diagram showing the relationship among habitats, modes of life and photic environmental conditions. It is noteworthy that a purely diurnal mode of life is not to be found in the aquatic environment. This may be due to greater variation in photic quality that one encounters there. The deep-sea and turbid environments have often been compared, but the only similarity is the paucity or attenuation of light.
***Source:* Ali et al. (1978)**

Light may be perceived by fish via the eye, the pineal and sometimes the skin. With the characteristics of direction, intensity, wavelength and period, light may be involved in many functions of life in fish (Fig. 15.3). Direction is probably the most useful attribute since it enables animals to perceive very precisely the direction of food, predators or other members of the species.

The amount of information that the eye can gain from a light source ultimately depends upon the number of photons that the light carries and in dim light it is a famine of photons rather than

Table 15.2 Summary of visual adaptations to photic environments and habits

		Scotopic	
Nocturnal Crepuscular	Deep-Sea	Turbid Waters	Caves Oceanic Abyss
Pure-rod retinas	Pure-rod retinas		Pure-rod retinas
Rod-rich retinas	Rod rich retinas	Rod-rich retinas	Eyes degenerated
Tapeta lucida	Tapeta lucida	Tapeta lucida	Eyes absent
Optical guides (crystalline cones)	Optical guides (rod myoids)	Optical guides (cone myoids)	
Superimposed layers of rhabdomeres	Superimposed layers of rods		
Closed-type rhabdoms	Large mass of rod outer segments		
Superposition eyes	Grouped rods	Grouped rods and cones	
	Tubular eyes		
Colour vision	Colour vision	Colour vision	Colour vision
Coloured intraocular filters			Coloured intraocular filters
Cone-rich retinas	Cone-rich retinas	Cone rich retinas	Cone rich retinas
Pure-cone retinas	Colourless cone droplets	Polarotaxis	Polarotaxis
Apposition eyes	Double optical axes		
Open-type rhabdoms			
Diurnal	Water surface	Clear shallow water	Terrestrial aerial
		Photopic	

Source: Ali et al. (1978)

inadequate sensitivity of the eye that limits the ability of the eye to detect fine detail, rapid movement, colour and contrast. The absorption of a single photon by a visual pigment molecule is sufficient to isomerise the chromophoric group of the molecule and thus to set in train the events that ultimately lead to the sensation of vision.

In striking contrast to the enormous variety of eye types evolved among invertebrate species, there exists essentially a single type of image-forming vertebrate eye. Because the vertebrate eye first evolved in water it is sensitive to only a narrow band of wavelengths of electromagnetic radiation, which are transmitted through water without significant attenuation.

The exploitation of light for vision has proven a powerful selective force in all organisms. Light travels in straight lines, can be reflected, and varies in both wavelength (colour) and intensity. To inform their owners about the visual world, image forming eyes must capture light and resolve it into images. These two requirements for vision have shaped the evolution of eyes throughout the animal kingdom.

The sensitivity of the eye depends on the brightness of image reaching the retina. The retinal illuminance is directly proportional to the intensity of the source and inversely proportional to

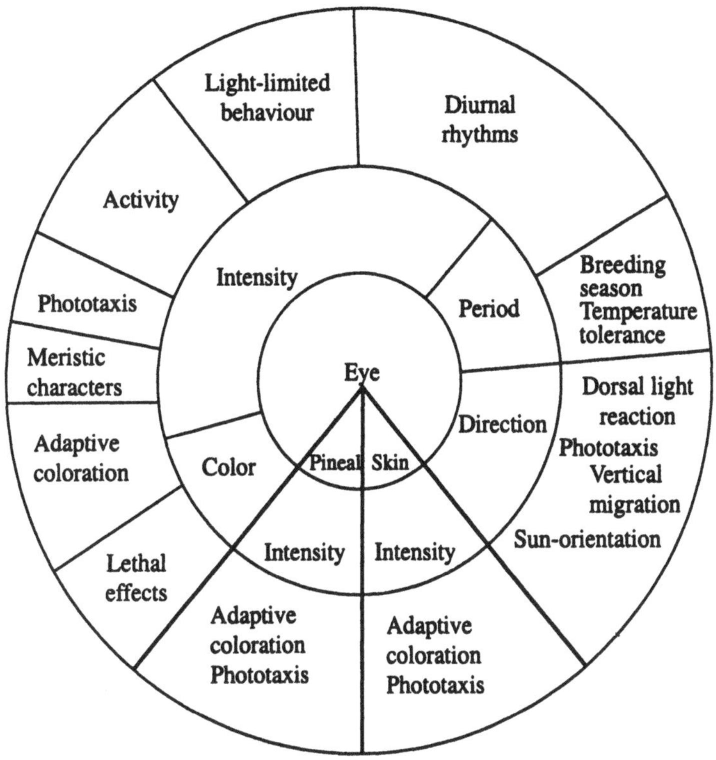

Fig. 15.3 Diagram showing functioning of the eye in fish.
***Source:* Blaxter (1975)**

square of Matthiessen's ratio for the species in question (For details of Matthiessen's ratio, see Fernald 1988; Sivak 1990).

Thus, when viewing an extended source larger fish eyes do not have a brighter retinal image resulting from a proportionally larger pupillary aperture. This result is particularly interesting because most fish eyes do grow larger as the fish grows, suggesting that, for fish living in well-illuminated habitats, increasing the image brightness was not an important selective force for growing a larger eye.

There are other adaptations to the geometry of the eye and head that increase the brightness of the retinal image without inordinately increasing the size of the eye. Many dim-light animals have a tapetum behind the retina that reflects the light that has already passed unabsorbed through the retina, back through it a second time. A rostral aphakic space coupled with sighting grooves in the snout along the direction of the view increases the effective aperture of the eye in the forward direction. The retina of deep-living or slightly nocturnal fishes are usually rod-dominated and the rods may be arranged in layers to catch the greatest possible number of photons. Photon capture is further made efficient in deep sea fishes by the presence of rhodopsins in the rods that have their wavelength of maximum absorption (λ max) at the 470–480 nm region of the spectrum and are thus tuned to absorb most efficiently those wavelengths of light to which the deep oceans are most transparent.

As the conditions for vision become more and more difficult, the visual information that can be got from ever large and more specialized eyes is no longer worth the effort in obtaining it. At this point other sensory modalities, such as various lateral line senses, are developed instead and the eye is reduced in size and complexity and, presumably, in importance.

Only in the extreme cases of cave-dwelling species such as the Iranian blind fish, *Noemacheilus smithi*, is the eye completely absent.

The design for an aquatic visual system must meet a stiff set of specifications, stiffer than for most aerial visual systems because of the photic conditions peculiar to the underwater scene. Light is colored by the selective filtering action of water and dissolved substances, it is scattered and refracted by suspended particles, making distinct objects appear faint and blurred, and more often than not, it is scarce. During evolution, these factors must have shaped the colour vision, pattern vision, and visual sensitivity of fishes, and presumably shaped them as multifariously as the light environments they inhabit.

Visual acuity is the ability to see clearly the fine details of objects, especially as these objects become smaller and closer together. There are four types of visual acuity; detection, recognition, localization, and resolution. The role of visual acuity in the life of pelagic fishes may involve, in addition to the detection of prey and predators, the recognition of transient and permanent body marks. Skipjack tuna and other pelagic fishes exhibit transient color markings at certain times, such as during feeding or during courtship. These markings may convey information about the presence of food to other fish in the school or signal aggressive intentions. Visual recognition of certain permanent body marks may aid a fish to distinguish its own kind from others closely resembling it.

According to Ulinski (1980), the available evidence is consistent with the hypothesis that the visual system evolved in accordance with the synthetic theory. It remains, however, to establish more clearly the relationships between the visual system structure and environmental demands.

2.2 Eye in Fish Groups

The lateral eyes of lampreys are similar to those of gnathostomes in most respects, however, they differ in a number of features that are more similar to the eyes of embryonic or larval gnathostomes and are likely plesiomorphic features common to myopterygians. In lampreys, the cornea consists of an outer dermal segment, devoid of pigmentation, glands and blood vessels, and an inner mesodermal segment which is a continuation of the sclera. The two corneal layers are closely associated but separated by a capillary space. The nuclei of the cells composing the lens are diffuse, unlike the nuclei of the lens cells in gnathostomes, and there are no ciliary muscles or a muscular iris. The egg-shaped lens (in silver lamprey, *Ichthyomyzon unicuspis*), which contains horizontal stutures on both anterior and posterior surfaces, is also asymmetric in the nasotemporal axis. The pupil is elongated in the horizontal plane, with the lens filling the dorsal margin of the eye. The iris consists of a single layer of epithelium, and accommodation occurs by changes of the scleral cornea affected by extrinsic eye muscles. Six extraocular muscles are found but no extraocular cornealis muscle is observed inserting the transparent dermal cornea. The retina in lampreys consists of pigmented and neural layers, as in other myopterygians, and the neural retina appears to possess classes of receptor cells comparable to the rods and cones of other retinas. The usual cellular layers of the neural retina can be recognized, except that a distinct ganglion cell, along with bodies of the bipolar and amacrine cells, form the internal nuclear layer (Northcutt 1985; see also Collin and Fritsch 1993). Considering the primitive phylogenetic position of this species of vertebrate, the retinal pigment epithelium is remarkably similar in most respects to that of avascular retinas of other lower vertebrates. Myeloid bodies are a common feature of this and some other avascular retinas. The lamprey pigment epithelium also contains many phagosomes and secondary lysomes or residual bodies (Dickson and Graves 1979). Dickson and Graves (1981) reported that in young parasitic anadromous sea lamprey (*Petromyzon marinus*) eye, the giant choroidal cells contain large quantities of glycogen and lipid, while in mature spawning animals, the presence of these substance is minimal. The giant glycogen-filled cells of the lamprey choroid and meninges represents a unique adaptation to nourish vital nervous tissues during the lengthy fasting period prior to spawning of these animals.

Collin et al. (1999) presented details of the ocular morphology of the lamprey *Geotria australis*. Collin and Potter (2000) published the first report of a tapetum in cyclostomes (in a lamprey *Mordacia mordax*). In hagfishes, the lateral eyes are poorly developed and are most similar to the eyes of gnathostomes that are highly degenerate. Eptatretids exhibit prominent eyespots of skin with reduced pigmentation over the eyes, whereas myxinoids do not possess recognizable eyespots. Similarly, the eyes of eptatretids and myxinoids lack a cornea, lens and iris, whereas the eyes of eptatretids do possess a central vitreous body which appears to have been lost in myxinoids. The neural retinal of eptatretids is differentiated into an outer receptor layer and inner and outer fibrous layers separated by an inner cell layer. The receptor cells contain outer segments with well organized discs and basal synaptic bodies similar to those of the hair cells of the lateral line organs in gnathostomes. The neural retina of myxinids consist of the same layers, but the outer segments of the receptor cells consist of disordered membranous whorls, and these cells do not appear to possess synaptic ribbons. Similarly, extrinsic eye muscles and their accompanying cranial nerves are absent in all hagfishes (Northcutt 1985; see also Ross 1963) (Fig. 15.4).

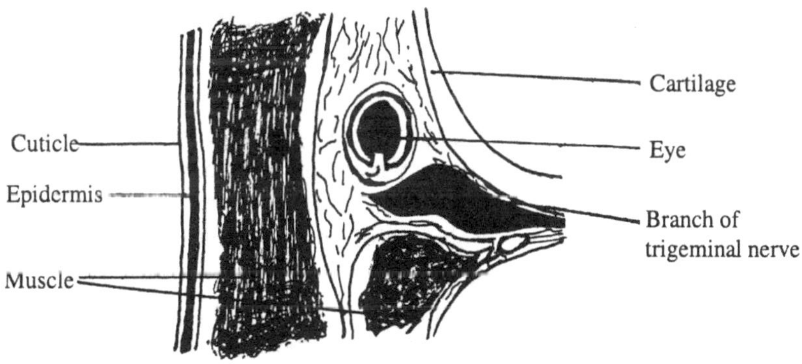

Fig. 15.4 Diagram of section through eye of *Myxine*.
***Source:* Ross (1963)**

The eyes of hagfishes may play an important role in the maintenance of circadian and circannual rhythms.

For comprehensive information of vision in fishes, the reader is referred to "Vision in Fishes: new approaches in research" edited by M.A. Ali (1975). J.A.C. Nicol (1989) published a book entitled "The eyes of fishes".

Gruber (1977) has described the visual system of sharks. Sharks have been evolving as an independent group for several hundred million years and presently occupy the major habitats of the sea, with a few species even restricted to fresh waters. Thus, it is not surprising that their ocular specializations are as uniquely characteristic of sharks as are their naked gill slits, placoid scales, and cartilaginous skeletons. In fact, the eye of elasmobranchs differs in several major ways from its teleostean counterpart; for example, in mode of accommodation, pupilary mobility, ocular adenexa (lids), and tapetal structure. Most astonishing is that many of the adaptations which characterize the visual systems of sharks, such as mobile eyelids, ciliary folds, and a flattened lens drawn back from an arched cornea, are extremely reminescent of the aerial eye.

Elasmobranchs possess the most highly evolved tapetum known in the animal kingdom (Figs. 15.5 and 15.6).

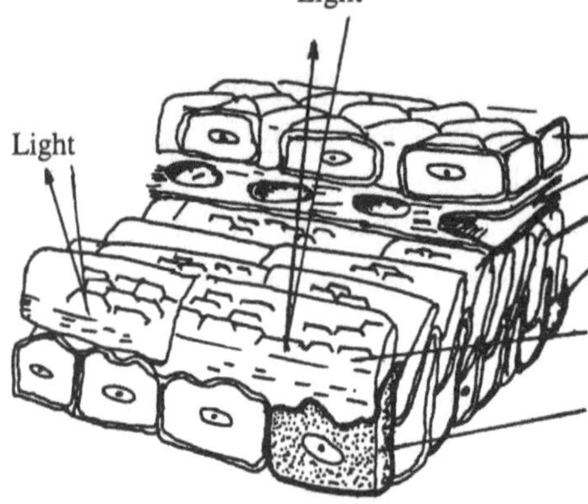

Fig. 15.5 Perspective diagram of the elasmobranch tapetum. Unabsorbed light enters and passes through the eye, retina and retinal epithelium and choriocapillaries to impinge upon the tapetal plates. If the melanin pigment is withdrawn into the pigment cell light will reflect from the hexagonal guanine crystals back through the retina.
***Source:* Gruber (1977)**

Additionally within the elasmobranchs, the visual system is by no means homogeneous. Although many ocular features are shared, many more are unique to the various taxa within the elasmobranchs, ranging from the species to ordinal level. For example, some sharks have fixed eyelids while others have a completely mobile third lid or nictitating membrane. Variability of pupil types and shapes characterize the elasmobranchs. The pupil size of some species is fixed, while others can stop the pupil down to one or more stenopaic or pinhole apertures. Many sharks have a partially or totally occlusible tapetum, while the tapetum of catshark (Scyliorhinidae) is fixed. And so the list could be continued. Thus, it is difficult to describe the "average" elasmobranch eye in a meaningful way.

Elasmobranchs, as all aquatic vertebrates, possess a voluminous lens. Refraction and accommodation are primarily functions of the lens and associated ciliary structures. The elasmobranch has variously been labelled as myopic or hypermetropic. The structural organization of the elasmobranch retina conforms to the usual vertebrate plan, i.e., three cellular and two synaptic layers. However, it is unusual in several ways. Along with cyclostomes, adult elasmobranchs are the only vertebrates whose retina is entirely devoid of blood vessels. Retinal nutrition is apparently mediated by the so-called pigment epithelium, which is actually unpigmented in

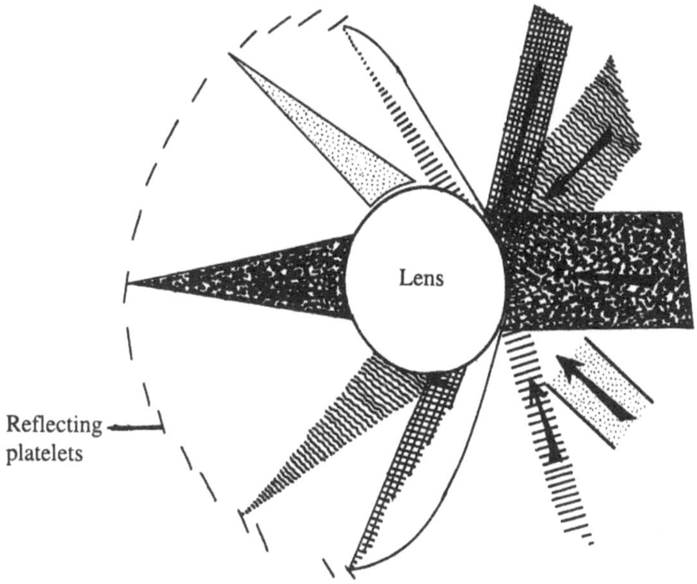

Fig. 15.6 Orientation of reflecting platelets in the eye of the blue shark, and the direction of light rays which can reach them. Vertical section.
***Source:* Nicol (1975)**

sharks. The only light-screening material in the elasmobranch eye resides in the choroid behind the retina. Most notable is the enormous size of several retinal elements such as horizontal cells, amacrine-processes, and some ganglion cells. The visual pigments of deep-sea sharks are environmentally tuned to restricted wavelengths of light which occur at great depths.

Basic data on color vision in elasmobranchs (in sharks) are inconclusive, while the entire areas such as resolution and spatial vision are unstudied. Probably the greatest mystery is the relation between the apparently exquisitely visual system and the natural behaviour of sharks.

Two things can be said with some assurance: (1) these unique ocular features reflect the long and independent phylogenetic history of the cartilaginous fishes and (2) the elasmobranch eye cannot be considered as "primitive". Rather, the shark's visual system is housed in a specialized and apparently well adapted vertebrate eye (Gruber and Cohen 1978).

Sivak (1978) gave details of the refraction and accommodation of the elasmobranch eye. Preliminary measurements of intraocular dimensions and refractive indices of the ocular media indicate that the optical components of the elasmobranch eye are quite different than those of the teleost.

The elasmobranch lens is not spherical. Refractive indices of elasmobranch lenses appear to be significantly lower than the equivalent value of 1.69 reported for goldfish.

Maddock and Nicol (1978) described the structure of eyes of *Hydrolagus affinis* (deep-sea species) and *H. colliei* (coastal species) (Chimaeridae). The eye is nearly globular. The cornea is thin and iridial curtains leave a large pupillary aperture. Eyelids extend well over the upper and lower surfaces of the eyeball and front of the iris. They have a tapetum lucidum located in the choroid. They have pure rod retinae. The ratfish shows eyeshine, and its skin also is very lustrous containing a layer of oriented reflecting cells much like those of the tapetum lucidum. Neither species has a mechanism for regulating the light entering the eye and affecting the visual cells. The pupils are immobile, and no migration of pigment take place across the tapetal surface. *H. affinis* has a golden rhodopsin, λ max 484 nm, similar to that reported for *H. Colliei*. Several features have been related to photic conditions in environment.

Interesting information is now known on different topics: morphological aspects of the teleostean visual system (optic nerve; connections of the retina; connections of the tectum opticum; some pathways, circuits and functional considerations (Vanegas and Ito 1983; Wullimann 1998); cornea (Collin and Collin 1995, 1996, 1998, 2000, 2001; Pettigrew and Collin 1995); retina (Arnot et al. 1970; Zynar and Ali 1975; Ali and Wagner 1976; Best and Nicol 1978, 1979; Meyer-Rochow and Tiäng 1978; Brackvelt 1980 a, b; Meyer-Rochow and Pyle 1980; Meyer-Rochow and Klyne 1982; Powers and Ester 1983; Collin 1988, 1999; Collin and Collin 1988; Collin and Pettigrew 1988a, b, Fritsch and Collin 1990; Wagner et al. 1993, 1998; Collin and Ali 1994; Chiu et al. 1995; Collin et al. 1996a, b, 2000; Shand 1997; Meyer-Rochow et al. 1999; Shand et al. 1999, 2000a, b; Bozzano and Collin 2000; vision in air and water (Land 1987); ontogenetic shifts in visual system of teleosts (Kotrschal et al. 1990); aquatic adaptations in fish eyes, especially in Teleostei (Fernald 1988; Collin and Portridge 1996, Collin 1997; Collin et al.

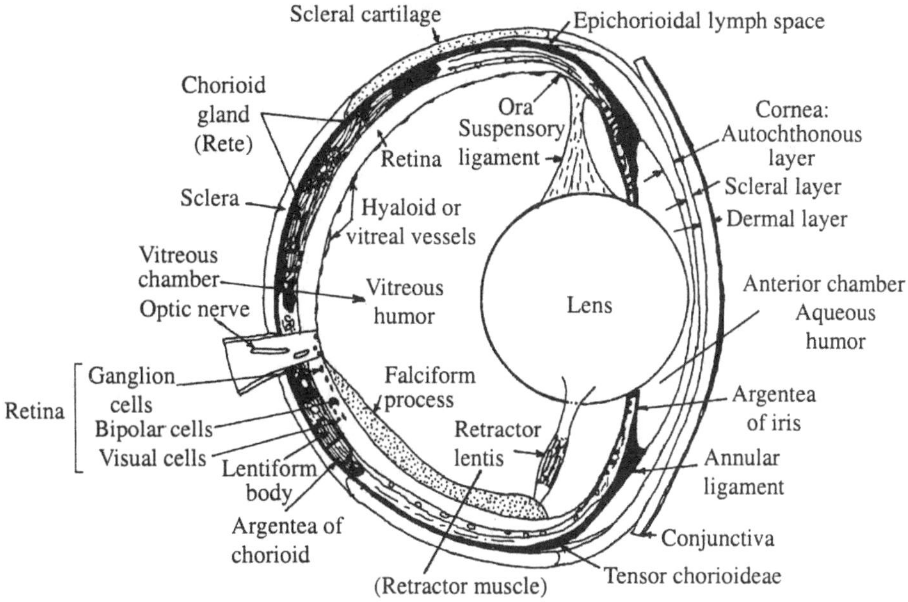

Fig. 15.7 Diagrammatic vertical section of a generalized teleost eye. Some structures shown may not be present in some eyes. For example, either lentiform body and falciform process or hyaloid vessels are present. Retinal constituents are shown.
***Source:* Fernald (1988) (Modified)**

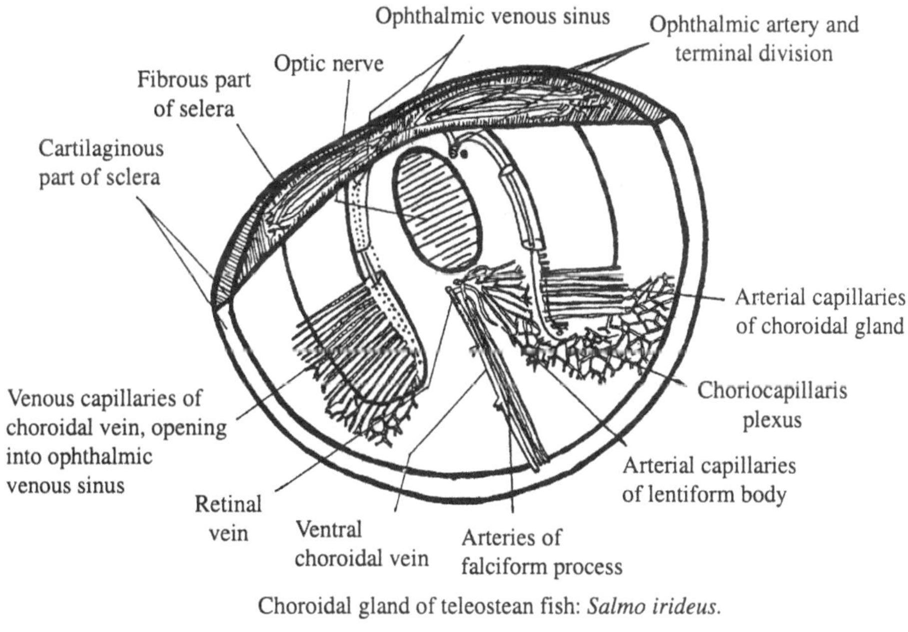

Choroidal gland of teleostean fish: *Salmo irideus.*

Fig. 15.8 The vascular pattern of the choroidal gland and lentiform body (rainbow trout, right eye). The upper part of the eyeball has been cut away to show the limbs of the choroidal gland in cross-section.
***Source:* Barnett (1951)**

living in air in one major aspect: underwater there is no air-cornea interface. The interface is an element in the optics of land-living vertebrates. The air-cornea interface serves as an important refractory element of the eye. Underwater eyes have no optical benefit from the cornea because its refractive index is nearly identical to that of water. Thus, in all underwater eyes, the refractive power of the cornea is neutralized and the lens, which provides the only dioptic power for image formation, needs to be stronger. In addition, lenses of fish need a short focal length to minimize eye size. The exceptions to this rule include skates and rays, which have proportionately larger eyes and presumably, therefore, aspherical lenses, for which no very compelling reasons are apparent as well as a few species that live at the air-water interface, including the "four-eyed" minnow, *Anableps* sp. and the amphibious fish *Dialommus fuscus*. The lens of *Pantodon buccholzi* (Osteoglossiformes) is not spherical. As in *Anableps* this lens may be an adaptation to simultaneous air/water vision. Two mesopelagic teleosts, *Trachipterus trachypterus* and *T. arcticus* (the ribbon or dealfish) have relatively small oval lenses, in which lens thickness (axial diameter) is greater than equatorial diameter, i.e., the long axis of the ovoid coincides with the equatorial axis rather than pupil axis of the eye. These lenses are multifocal devices designed for the abnormally shallow retinal cup located in the head of a laterally compressed fish. Flattened (non-spherical) lenses with equatorial diameters greater than thickness (axial diameter) in several species of pelagic deep-sea teleosts have been reported (read Sivak 1990; Sivak et al. 1999).

It is obvious that not all aquatic vertebrates possess spherical lenses. Nonspherical lens are reported in the sandlance, *Limnichthyes fasciatus*, *Trachipterus trachypterus*, *T. arcticus*, several species of deep-sea anglerfishes, the gulper eel, *Anableps anableps*, and intertidal mudskipper, *Periophthalmus*, elasmobranchs (excepting *Mustelus canis* and *Squalus acanthias* with almost spherical lens). Holosteans also possess spherical lenses.

Although the cornea does not contribute to the optical power of the aquatic eye, shallow-living diurnal fish often have yellow cornea and occasionally yellow lenses. These optical filters are thought to serve at least two purposes. First, in deeper-living animals, the yellow filter serves to counter camouflage coloration of animals swimming above. Second, the yellow filter reduces the amount of shortwave light, which is most scattered within the eye. Degradation of the image due to this scattering of shorter wave lengths is thus reduced. Since any filter necessarily reduces the sensitivity of the eye, some fish species have a differential distribution of corneal pigment, and others have corneal chromatophores, providing a filter in the day, but becoming virtually transparent at night. Exceptions to the general description given here are the eyes of deep-sea fishes (read description ahead). Pettigrew et al. (2000) reported the importance of cornea in prey capture and accommodation in *L. fasciatus*.

The annular ligament or loose meshwork of cellulofibrous tissue is a regular feature of many teleost corneae, including that of the goldfish, *Carassius auratus*, and the trout, *Salmo fario*, but is also found in nonteleost species such as the Florida garfish, *Lepisosteus platyrhincus* and the lamprey, *Lampetra* sp. However, it is either reduced or absent in the pipefish, *Corythoichthyes paxtoni* (Teleostei), the spiny dogfish *Squalus acanthias* (Elasmobranchii), the lungfish *Protopterus* sp. (Dipnoi) and the coelacanth, *Latimeria chalumnae* (Crossopterygii). The function of the annular ligament is unknown; may have a refractive, supporting, and energy providing function in some cases.

The eye lens pigmentation in deep-sea hatchetfish, *Argyropelecus affinis*, and probably other members of this genus is characterized by: (1) an abrupt onset and steady increase in concentration during the life history of the fish, (2) restriction to the outer layers of the lenses of adult fishes, (3) a complex absorption profile in the near UV and blue-violet wavelength, and (4) a chromophore that is tightly bound to a single, specific soluble lens protein, alpha crystatin. No other visual systems with lens pigmentation yet described have these four characteristics. The pigment is a caratonoprotein complex, the carotenoid-like chromophore being strongly attached with a specific soluble lens protein, alpha crystatin. Data on spectral absorbance of the lens and visual pigment of these fishes suggest that the lens pigmentation acts as a short-wave filter to improve acuity of the visual system (see McFall-Ngai et al. 1986 for discussion on function of pigment in the lenses).

Lens pigmentation has risen a number of times in mesopelagic fishes. However, the biological function of such an adaptation remains unclear. Lens pigmentation does not correlate with the presence or absence of tubular eyes, i.e., not all tubular eyes have yellow lenses, while some yellow lenses occur in fishes with non-tubular eyes. In addition, while the perception of intra- or inter-specifically produced luminescence may be enhanced by the presence of yellow lenses in some species, intraspecific communication as a function cannot be deduced from the occurrence of lens pigmentation in the mesopelagic fishes studied thus far. Furthermore, depending on the species, the lens may or may not change the peak wavelength of visual pigment sensitivity. The only conclusion that can be made from all pigmented lenses of mesopelagic fishes is that the pigmented lens would filter out a portion of the visual curve (McFall-Ngai et al. 1988).

Atlantic flying fish (*Cypselurus heterurus*) show an interesting variation of the flat cornea theme. The cornea consists of three almost flat triangular facets, forming a tent with its peak in the centre. Refractive measurements through the facets showed emmetropia in air and slight hypermetropia in water. Paired flat facets have also been found in clinid fish *Mnierpes macrocephalus* which use their eyes in and out of water. A flat cornea has some disadvantages: in air the periphery of the field is compressed and distorted, and in water the field of the view is reduced because the lens cannot protrude as it does in a normal fish eye. To some extent the latter problem is overcome by the facetted structure in flying fish, but at the cost of a disjointed image.

How good is the quality of the image viewed through a spherican lens? Teleost lenses deliver surprising good images right upto their edges. In most fish, the lens protrudes through the pupil, so that the iris cannot limit images to central area of the lens. Thus, to be useful, the image must be of high quality all across the lens. This is in contrast to the lenses found in most vertebrate eye, which evolved for aerial vision, in which a pupil restricts vision to the central portion of the lens. It is important to note that the problem of producing good images with a spherically fish lens is made much more complicated by the fact that fishes grow throughout life requiring that some mechanism must exist to modify any refractive index gradient as eye enlarges.

The optical resolution of fish lens has been measured less often than chromatic aberration. In humans, the optical resolution of the lens approximately matches that of the retina. The fish lens, in contrast, has an angular resolution about 10 times better than that of the retina. Only a very small lens sizes (i.e., the size of the lens when the animal first becomes free swimming) does the theoretical limits of lens resolution match retinal resolution. This suggests that, for fish that do not live in the deep sea, growth of the eyes serves primarily to increase the optical resolution and perhaps to improve contrast detection.

It is worth noting that some fish species do not have spherical lenses. In the elasmobranchs, for example, the lens is subspherical. In two other genera of fish, specialized optical systems reflect selection for good vision at the interface of air and water. The amphibious fish *Dialommus fuscus* is probably unique in that its eyes have flat corneal windows that form a prism enabling it see sharply in both air and water. According to an analysis of the optics of the *D.fuscus* eye, while the animal is out of water the prismatic cornea probably produces a double image of the world in a single eye. The speculation that binocular information might exist within a single eye is extremely interesting and should be persued.

One of the most remarkable structural aberrations among the eyes of teleost fishes is seen in *Anableps anableps*, with a truely amphibious eye. *A. anableps* (a cyprinodontid) called the "four-eyes minnows" from South America cruises with half of its eye above the surface, and half below. This is probably the best known example of a vertebrate eye with an amphibious visual capability. It has two pupils, (a pupil in the shape figure 8) one looking into each medium, and lens whose shape "combine an aquatic optical system harmoniously with an aerial one, in a perfectly static situation". The compromise is achieved by the ovoid (egg-shaped) shape of the lens, with its long axis in the direction which looks down into the water. Images arriving through the air are refracted by both the cornea and lens and pass through the latter along its flatter side. In contrast, rays arriving through the water are imaged only by the lens, and these pass through the point of the lens, where the radius of the curvature is very small and lens essentially spherical. *Anableps* appear to have evolved its optical system in response to an increasingly murky habitat (Fig. 15.9).

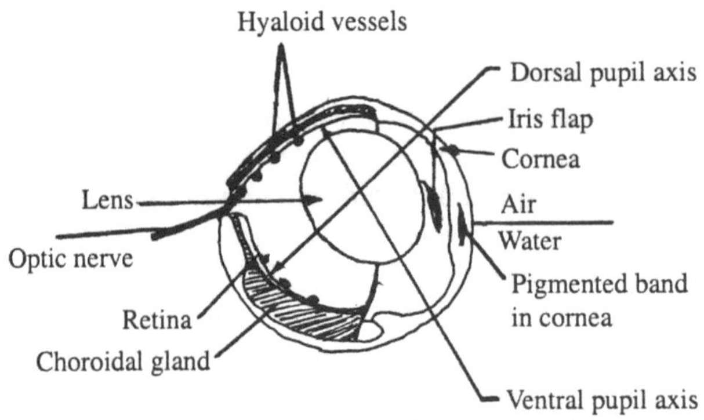

Fig. 15.9 Drawing of the eye of *Anableps anableps* indicating the directions of aerial and aquatic optic paths
***Source:* Sivak (1978)**

Another example indicates that lens shape is in fact determined by the overall refractive needs of the species in question. The mudskipper (*Periophthalmus*) clings to the seashore rocks or tidal flats with its dorsally positioned eye exposed to air. The eyes of this fish are in high turrets and move as though provided with universal joints. The eyes rotate under secondary spectacles or covering which apparently protect the corneas from desicatting. The eyes are adapted for aerial vision. For example, the lens is elliptical, not spherical in shape. However, since no evidence exists concerning an ability to compensate for the refractive change associated with immersion of the eyes in water, acute amphibious vision cannot definately said to exist, i.e., vision may be poor in water. Fresh-water mullet *Rhinomugil corsula* move with their snout in level with water and eyes protruding above. The retina is divided into a superior and an inferior hemispheres and probably adapted for aquatic and aerial vision.

But there are other teleost fishes with constantly submerged eyes, except for occasional leaps out of water, which have excellent out-of-water vision. One of these is the archer-fish (*Toxotes*) which spits a tiny stream of water at an overhead insect, knocking it down to the surface film of the water, thus rending it incapable of escape.

In the eye of the aruana *Osteoglossum bicirrhosum* the division of the retina into upper and lower parts by the septum may be of use in rapid adjustments to markedly dissimilar light levels, with the lower retinal section always bright-adapted and the upper dark-adapted (Fig. 15.10). A septum is also present in *Pantodon buchholzi* eye. The eye of the parasitic fish, *Carapus mourlani* inhabiting sea-cucumbers or large starfish, is an interesting sense organ, because it combines features of degenerated photoreceptors characteristic of cave organisms (keratosis, small number of nerve fibres, no screening pigment between the rods) with adaptations which are commonly met within predominantly nocturnal and deep-sea forms (e.g., pure rod retina, pupillary contraction, receptor-bipolar-ganglion cell ratio). Presumably, the eye is fully functional in the larval and juvenile stage of *C. mourlani* but degenerates with increasing age. Age-dependent retinal degeneration has been reported in the eye of cave fish *Anoptichthys*. The age-dependent deterioration is likely to be connected with the "pseudocave" conditions which must prevail inside the dark coelom of fish's echinoderm host.

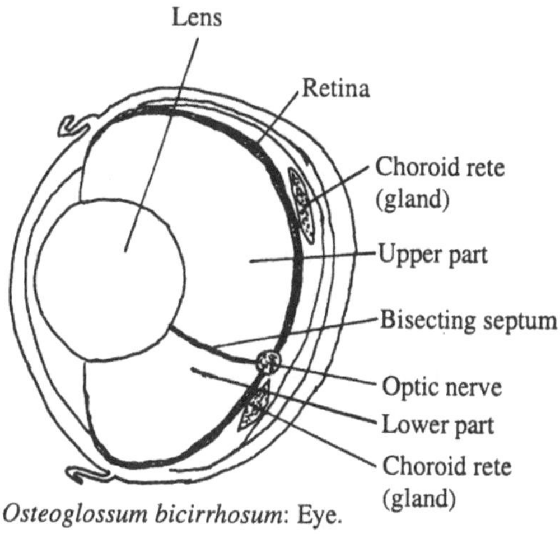

Fig. 15.10 The eye of *Osteoglossum bicirrhosum*, semi-diagrammatic camera-lucida drawing of a preparation ca. 16 mm in diameter.
***Source:* Haedrich et al. (1973)**

The visual image available to an animal depends not only on the quality of the lens, but also on the state of the focus provided by the accommodative apparatus. In contrast to land vertebrates, which focus their eyes by adjusting the refractive strength of the lens by modifying its shape, accommodation in fish, when it occurs, results from the movement of the spherically shaped lens within the globe. In most fish examined, e.g., African cichlid fish (*Haplochromis burtoni*) the lens moves approximately parallel to the plane of the pupil along an axis of pupillary eccentricity. This means that different parts of the eye are focussed at different distances from the eye on different parts of the visual field simultaneously. For example, in the relaxed accommodative state in fish, the lens lies nearer to the nasal pole of the eye. Thus the temporal part of the retina (nasal visual field) is focussed for near vision and the nasal pole retina (temporal visual field) for far vision.

Sivak (1990) described the optical variability of the fish lens: lens shape, relative focal length (Matthiessen's ratio), spherical aberration, chromatic aberration and functional significance of fish lens quality.

The fish eye is slightly myopic since underwater visibility is limited by backscatter and turbidity, however, myopic vision may be quite appropriate. It has been found in *H. burtoni* eyes that as the animal gets larger, its accommodative mechanism is appropriately scaled to body size.

Since the teleost lens is nearly incompressible sphere that protrudes through the pupil, it allows the fish periscopic vision, so that each eye has a field of view larger than 180 degrees. In many fish species, the pupillary opening is slightly eccentric, with a small aphakic area nasal to the spherical lens. The major axis of this eccentricity is colinear with the axis of the lens movement and has been hypothesized to enlarge the field of binocular vision.

Fish have a gelatinous, thin, transparent retinal structure characteristic of all vertebrates. Fish retinae differ fundamentally from those of other vertebrates, however, because they continue to grow throughout life of the animal, both by adding neurons and by stretching existing retinal tissue. In fish, the rate of eye growth relative to body growth depends on the importance of vision to the animal, and in some species, the individual growth rate may be under social control. The proportion of retinal area that increases owing to cell division, as compared with that due to retinal stretching, varies greatly among species. In teleost retina, new retinal tissue is added all round the iris margin at the ora serrata. In addition, the existing retina is being stretched, accommodating the intercalation of newly generated rods exclusively. These two processes preserve visual sensitivity by maintaining constant rod density while adjusting eye size. As a consequence of these essentially competing processes, there is a critical regulatory problem at the margin of the eye. Since the density of rods is constant of all sizes of fish, while the density of all other cell types decreases, the ratio of the number of rods to the number of other cells is increasing.

In the eyes of surface-feeding fish (life close to the interface) *Aplocheilus lineatus* and *Epiplatys grahami* there are two horizontal band-shaped area in retina, with increased densities of both receptors and ganglion cells. The central area is used for lateral vision, below the water surface, but the ventral area is directed upwards at the angle of 47° to the vertical that is, above the surface, just inside the margin of Snell's window. Thus, both area look almost parallel to the water surface, but below and above it respectively. Interestingly, the intermediate zone, which corresponds only to the very close objects in the surface, is poorly developed and even shows evidence of degeneration.

The benthic *Trachinus vipera* occurs at a depth of 50 m and migrates in shore during summer. It lies buried in the sand with the top of its head exposed. The benthic habitat and predatory behaviour, tuned to function in twilight, are reflected in the structure and ultrastructure of the eye (see details: Kunz et al. 1985).

The eyes of Antarctic fish are generally similar to those of coastal fish elsewhere. The position and orientation of eyes varies somewhat within the notothenoids and appears to be correlated with feeding vector. There is a well-developed spectacle over the surface of the eye, which in conjunction with the cornea prevents ocular freezing. The ocular fluids themselves are slightly hypotonic with respect to plasma and contain very low levels of serum proteins, they are not protected from freezing by glycoprotein antifreeze. The ocular fluids must remain permanently supercooled and can only do so because the spectacle and cornea provide a suitable barrier to provide ice propagation into the eye. Both the spectacle and cornea are avascular, and rely on oxygen and nutrient diffusion from the ocular margin (see details of eye movements and retinal structure: Macdonald and Montgomery 1991).

Color vision may have evolved in fish over 400 million years ago. Detecting different colors has been achieved in most species through the evolution of retinal photoreceptors that are maximally sensitive to different wavelengths of light. In a few cases, however, filters positioned in front of photopigments offer another solution. The brain is informed about detected chromatic differences via color-opponent neurons.

Photoreceptors in fish, as in other vertebrates, are elongated structures lying along the axis parallel to that of incident light and subdivisible into obvious regions. Most sclerad is the outer segment, while the light-sensitive photopigments are located in discs lying perpendicular to the axis of incident light. Cones have outer segments shaped like truncated cones, with a slight taper while rods are long, slender cylinders. Vitread lies the barrel-shaped inner segment connected to the outer segment by a thin cilium. The scleral end of the inner segment called the ellipsoid, contains a collection of mitochondria, while the vitread end, called the myoid, contains the machinery responsible for moving the photoreceptors during retinomotor movements. The nuclei lie vitread, with the sparser, lighter staining cone nuclei in a row along the outer nuclear layer and the more numerous, darkly staining, smaller rod nuclei in several rows more vitread. These nuclei connect via thin fibres to their synaptical terminals, cone pedicles and rod spherules. These terminals contain synaptic vesicles and synaptic ribbon(s) and are responsible for connecting with bipolar and horizontal cells and other photoreceptors. The discs containing photopigment are constantly shed and renewed. Visual pigment proteins are synthesized in the myoid, travel through the cilium, and are incorporated into the outer segments. In rods, the incorporation proceeds in an orderly fashion from the base of the outer segment, whereas, in cones, new material is incorporated in a diffuse pattern (Fig. 15.11). (see also Table 15.3).

The circadian clock for many biological rhythms in animals has been assigned to the hypothalamus and the pineal gland. This endogenous clock is synchronized to the external light/dark cycle by a 'Zeitgeber' (timer). The eye is considered to be the principal timer. However, it has emerged that the eye itself contains an endogenous circadian clock. This intraocular clock may interact with, or be dependent on, the other endogenous clocks in the central nervous systems.

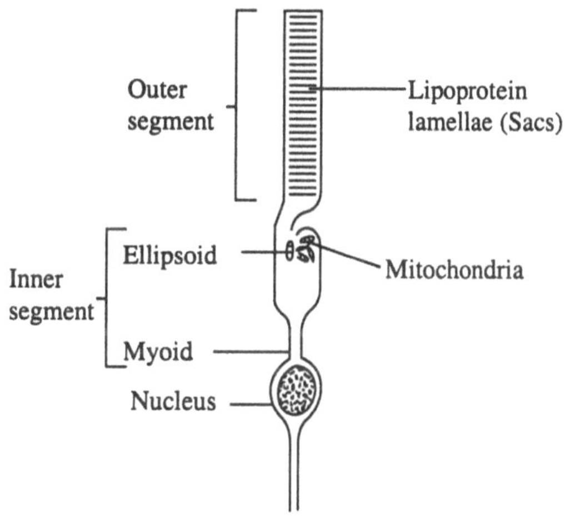

Fig. 15.11 Diagram of generalized photoreceptor. ***Source:*** **Kunz (1971)**

Lacking pupillary movements in response to light and darkness, fish and other lower vertebrates exhibit retinomotor movements (RMM). RMM refer to positional changes of photoreceptors (rods and cones) and melanin granules within the retinal pigment epithelium (RPE). In bright light, cones contract, rods extend and RPE pigment granules move vitreally. In darkness, the reverse movements occur. RMM are regulated by light and by an endogenous rhythm, and are thought to increase the sensitivity of the eyes in dim light and to improve visual acuity in bright light.

Shedding of the tips of photoreceptor outer segments (OS) and phagocytosis of the shed discs, now called phagosomes, by RPE have been observed for some time. Investigations have revealed that cyclic light plays a role in the regulation of discs loss. Twice each day the RPE phagocytizes and destroys packets of shed discs. The peak early in the morning is due to shedding from rods, the peak early at night to shedding from cones. Shedding is driven by light and/or is under endogenous circadian control.

Table 15.3 Differences between rods and cones

Rods	Cones
High sensitivity, nocturnal	Lower sensitivity, diurnal vision
More photopigment, greater photon capture	Less photopigment
High amplication per photon absorption	Less amplication
Single photon detection	
Saturation at low levels of brightness	Saturate only in intense light
Low temporal resolution	High temporal resolution
Slow to respond with a long integration time	Fast response kinetics, short integration time
More sensitive to scattered light	Most sensitive to direct rays emanating from a target
Low acuity: highly convergent retinal pathways, with no foveal or specialized retinal areas	High acuity: less convergent retinal pathways, concentrated in foveal areas and/or specialized areas of the retina
Achromatic: only one photoreceptor and thus no color vision	Chromatic: several types of cones with differential sensitivity and hence the capability for color vision.

Source: Hawryshyn (1998)

The endogenous control of both RMM and OS shedding is thought to be intraocular rather than by remote tissues such as the pineal gland. While there is evidence for the involvement of both dopamine and melatonin as intraocular neuromodulators, the role of cAMP has also been investigated, and it seems that this substance is engaged in retinal melatonin/dopamine interaction. Dopamine has been shown to prevent cAMP production in the retina via a D_2 receptor inhibition of adenylate cyclase activity. Melatonin has a stimulatory effect on cAMP in the retina. Experimental elevation of cAMP results in dark-adaptive RMM and in inhibition of cone shedding at night. Melatonin levels are reported to be high at night and the concentration of dopamine and cAMP to be high during the day.

Attempts have been made to study the changes of cAMP in the eye during development (see McCormick et al. 1989 for brown trout *Salmo truta*; refer literature). cAMP plays a role in RMM.

Very bright light can cause glare effects, especially in fishes living in the superficial layer of water column. Many reputed surface fishes, such as guppy, the four-eyed fish, the killifish and flying fishes, all have colourless "oil droplets" of mitochondrial origin in their cones. It is presumed that glare effects are attenuated by the probable light-collimating properties of these droplets, thus directing light preferentially toward the photosensitive outer segments of optically isolated cones. In some fishes, reduction of infra-ocular glare is apparently not accomplished at the level of the photoreceptor cells, as shown above, but rather by irridescent multilayers in the cornea.

The light-adapted eye of the toadfish *Opsanus* exhibits eye shine. The choroid is yellow and has a yellow stratum argenteum. There is no melanocyte layer in the choroid.

Unlike deep-sea fishes most fish have both rods and cones. Cones are single in many fishes. Many fish species have double or twin cones, which may or may not both have the same photopigment. Such directly opposed photoreceptors are widespread in the animal kingdom, but there is little agreement about their adaptive value. In addition, as noted above, some benthic fish have multiple tiered banks of rod outer segments most likely to optimize photon capture. It may be mentioned that some fish species have more numerous cones and the 14-spined stickleback (*Spinachia spinachia*), the roach (*Rutilus rutilus*), the Japanese dace (*Tribolodon hakonensis*), the shanny (*Blennius pholis*) and others are known to have four types of cone. Some guppies (*Poecilia reticulata*) may have upto six types of cones in their retinae. How these animals incorporate this number of cell types into a colour vision mechanism is a matter for speculation.

Cone photoreceptors in most teleosts retinae are arranged in a highly regular mosaic pattern. Two basic types of patterns have been described: a row pattern and a square pattern. In both, a row of double or twin cones alternates with a row of single cones. The patterns differ in that in one case the twin or double cones are arranged so that lines through the centres of each pair are at right angles with respect to one another (the square pattern), whereas in other case the axes of the double cones are parallel to one another (the row pattern). The square mosaic is generally found in fish species that are highly dependent upon vision. In some species, both patterns exist in different regions of the adult retina or occur at different times during development. In every case, rods when present are interspersed among the cone array throughout the extent of the retina, as if perfectly ordered bowling pins (cones) sat in a field of alfalfa (rods). Reader can find comprehensive, interesting information on the constructional morphology of photoreceptor patterns in percomorph fish (van der Meer 1992) (Fig. 15.12).

Vertebrate visual cells—like other cells of the nervous system—once differentiated do not divide any more. As a consequence, no replacement of damaged cells can take place. However, their photoreceptive outer segments are regenerated. Even under normal conditions they are continually renewed by a constant supply of new proteins. The rod outer segments are renewed by discarding 'packages' of discs at the apex and synthesizing new discs at the base. Shedding is observed in single or twin cones and occurs at any level of the outer segment. The discs are not discarded in packages or as single discs, but are shed in small vesicular portions. This mode of 'disc shedding' may explain why in cone outer segment radioactively labelled protein is diffusely distributed (see Yacob and Kunz 1977). For cones in clupeid fishes, see Novales Flamarique and Hawryshyn (1998).

Taking the vision in the ultraviolet light, sensitivity to UV-A (320–390 nm) is well documented in both vertebrates and invertebrates and a number of fishes have retinal cones sensitive to ultraviolet light including the roach (*Rutilus rutilus*), the Japanese dace (*Tribolodon hakonensis*) and the goldfish (*Carassius auratus*). All these fishes live in shallow fresh water and ultraviolet receptors have not yet been identified in marine fishes. The possession of UV vision is necessarily correlated with the possession of corneas and lenses that are transparent to ultraviolet light (for discussion and importance, read Partridge 1990; Jacobs 1992).

A number of fishes are known to respond to polarized light, and in most cases it is the cone photoreceptors that seem to be involved in polarization detection. Indeed, in the goldfish (*C. auratus*), the UV-, green- and red-absorbing cones are all polarization-sensitive and only the blue cones are insensitive to e-vector orientation. The mechanism by which fish detect light polarization is not yet understood. Further, there is little information about the way in which fishes use this visual input (Partridge 1990; refer also Hawryshyn and McFarland 1987, Douglas and Hawryshyn 1990; Hawryshyn 1992; Parkyn and Hawryshyn 2000). Kunz and Callaghan (1989) described embryonic fissures in teleost eyes and their possible role in detection of polarized lights.

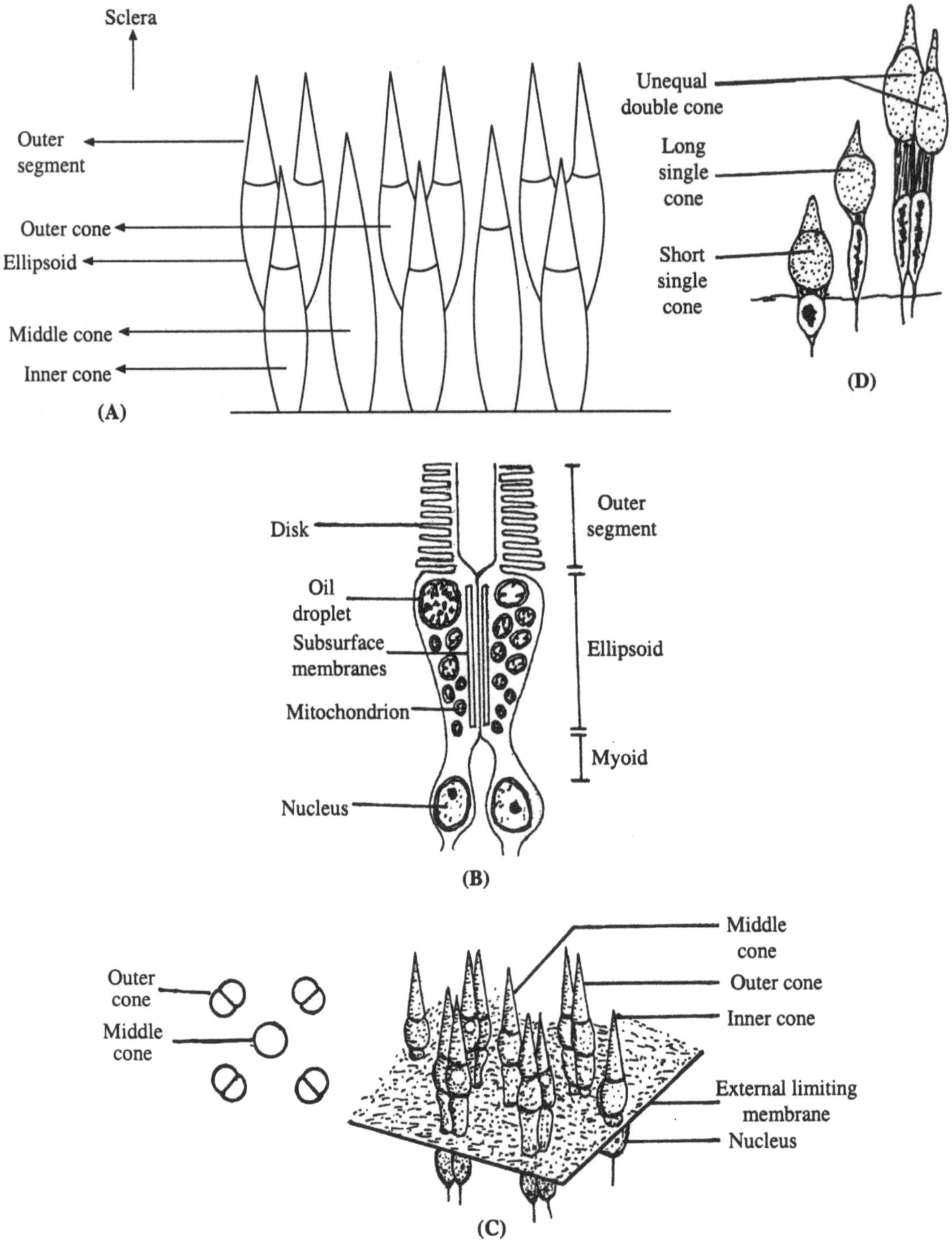

Fig. 15.12 **(A) Schematic drawing showing the tiered arrangement of *Lebistes* cones (central region). (B) Diagram of twin-cones; Note: ellipsoid with oil-droplet, mitochondrial size-gradient and subsurface membranes. (C) Mosaic unit; The rods occupy the spaces between the cones and are not shown. Left: cross section through ellipsoids. Right: three dimensional representation. (D) Diagram of cone types of the roach *Leuciscus rutilus*. Short single; long single; unequal double cone.**
Sources: **Fig. 15.12(A–C) Kunz and Wise (1973), Fig. 15.12(D) Nicol (1975)**

Schellart (1992) tabulated sensory indices, habitat and behavioural feature of a large number fish species. In general, diurnal fishes have well-developed cone systems, in contrast to nocturnal species. Large eyes are found in species with typical diurnal behaviour (e.g., mackerels), but also crepuscular species (e.g., gadids) and sometimes even in nocturnal fish living in water with good visibility (e.g., holocentrids). Perciforms, generally, are expected to have good visual capabilities. This is confirmed by the finding that perciforms have a relatively thick stratum opticum plus stratum fibrosum et griseum superficial and that many of them have dense receptor mosaics and high ganglion cell densities. These fishes are often diurnal, pelagic marine predators, living in water with good visibility. Perciforms are vision specialists, often with well-developed CN (canal neuromast) system whereas otophysans, are hearing specialists with a well-developed SN (superficial neuromast) system.

Cones with different photopigments must be distributed systematically across the retina to be effective as contrast or color detectors. Cones of certain fishes (e.g., *Fundulus heteroclitus*, *Poecilia reticulata*) contain ellipsosomes which may serve as intracellular color filters (MacNichol et al. 1978).

Trichromatic photoreceptor systems and opponent color-coding mechanisms appeared to have evolved in ancient fish before teleosts arose, supporting the view that the vertebrate color vision originated in an ancient fish.

Fish retinas are structurally similar to other vertebrate retinas (Fig. 15.13), and in some respects they are the best retinas for study of certain visual problems. Aquatic retinas are relatively coarse grained, in the sense that cones are large and not tightly packed. This is probably adapted to the low quality of the aquatic image (Powers and Easter 1983). Some degrees of degeneration in the retinas of cave fishes have been reported. Disoriented outer segments and short inner segments are features of the photoreceptor in *Astyanax hubbusi*. In the pigment epithelium, myeloid bodies are present (Yew and Yoshihara 1977). Primitive osteoglossiform fishes, the mooneyes *Hiodon* have tapeta lucida (ocular reflectors) in the pigment epithelium. Reflecting material contains uric acid. The compounds that form the tapetal materials of fishes include guanine, uric acid, 7,8-dihydroxanthopterin, glyceryl tridocosahexaenoate, a tetramer of 5,6-dihydroxyindole-2-carboxylic acid with decarboxylated 5-adenosylmethionine, and 3-(2,5-5-5-dicysteinyl-3,4-dihydroxyphenyl)-alanine (Zyznar et al. 1978).

Retinal structure of some Brazilian tide pool fishes has been grouped into three categories, viz., 1. visually dependent (good acuity and sensitivity), 2. specialized for dim environments (high sensitivity at the cost of acuity), 3. poor visual ability (Wagner et al. 1976).

In most lower vertebrates, the retinal photoreceptors and retinal pigmented epithelium move radially in response to changes in ambient light. In some animals, these changes, known as retinomotor or photomechanical movements, exhibit an endogenous circadian rhythm (see details: Wagner et al. 1992). In the light, myoids on cone photoreceptors contrast, moving the cones so that the inner (vitreous) end of the inner segments lie along the external limiting membrane, while myoids of the rods elongate, moving the rods away from the outer limiting membrane. Pigmented granules of the retinal pigmented epithelium disperse so that rod outer segments are completely immersed in the dark pigmented epithelium. In the dark, these movements reverse. Pigment aggregates in the pigmented epithelium to expose rod outer segments, and rods move into position adjacent to the external limiting membrane. Cones myoids lengthen and cones move out away from the external limiting membrane. Retinomotor movements are maximal in

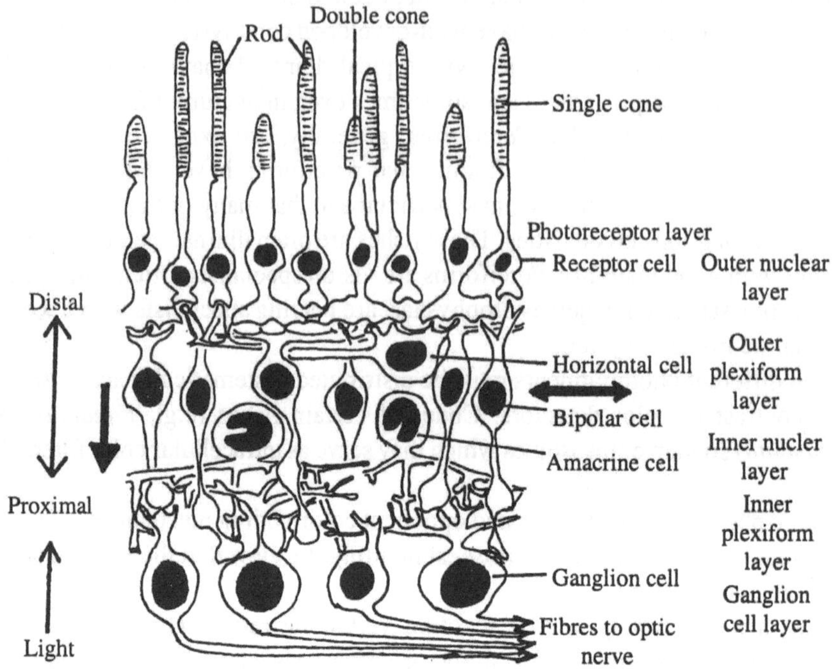

Fig. 15.13 **Retina. The retina has five major classes of neurons arranged into three nuclear layers; photoreceptors (rods and cones), bipolar cells, horizontal cells, amacrine cells, and ganglion cells. Photoreceptors, bipolar and horizontal cells make synaptic connections with each other in the outer plexiform layer. The bipolar, amacrine and ganglion cells make contact in the inner plexiform layer. Bipolar cells bridge the two layers. Information (short bold arrows) flows vertically from photoreceptors to bipolar cells to ganglion cells. Information also flow laterally, mediated by horizontal cells in the outer plexiform layer and amacrine cells in the inner plexiform layer.**
Sources: **Kotrschal et al. (1992) and Hawryshyn (1998)**

teleost with duplex retinae. Coincident with this retinomotor activity at dawn and dusk is a quiescent period in the behaviour of the animal, suggesting that the photoreceptors may not be useful during their movement. The magnitude of endogenous rhythms exhibited by retinomotor movements varied greatly among the species studied (catfish) and depends on both the activity patterns in the natural life of the animal and the immediately prior maintenance conditions.

Munk (1968) discussed the basic structural features of the holostein eye from a phylogenetic point of view and listed structural features which are of particular importance for comparison in Table 15.4 (see paper for discussion). The eyes of *Protopterus* and *Lepidosiren* are built on the same pattern as that of the urodeles.

2.3 Photovisual Pigments and Visual Sensitivity

Light has so widespread an effect upon living organism that it is quite the exception to find a species which is not equipped with some mechanism of response. The importance of the visual pigments lies in the fact that they and they alone, catch the light quanta which form the starting point of the visual process, and colours which they will not absorb will not be seen.

The photopigments of fish photoreceptors are made up of a protein moiety (opsin) and a

Table 15.4

	Brachipterygii *Calamoichthyes calabaricus*	Holostei *Lepisosteus productus*	*Amia calva*	Teleostei
Typical annular ligament	–	–	+	+
Musculus tensor choriodeae	–	–	+	+
Choroid gland	–	–	+	+
Lens muscle	+	+	+	+
Anterior suspensory lens ligament	+	–	+	–
Posterior suspensory lens ligament	–	+	+	+
Linear optic papilla	+	+	+	+
Arteria hyaloidea	+	+	+	+
Paraboloid in visual cells	+	+	+	–
Oil droplets in cones	+	–	–	+
Displaced bipolars	+	+	+	–
Fascicles of optic nerve fibres in ganglion cell layer	+	+	+	+

– = absent; + = present or, in the last column (Teleostei), present in at least some species.
Source: Munk (1968)

chromophore, either vitamin A1 or A2. Pigments based on A1 are called rhodopsin; those based on A2, porphyropsin. Given the same opsin, conjugation with vitamin A1 yields a photopigment which absorbs at shorter wavelengths than if A2 were used. Different opsins yield different wavelengths at maximum absorption (λ max), but the absorption spectra of all rhodopsins have similar shapes, as do porphyropsins. More recently it has been shown that the templete curves are valid only for λ max in the midspectral region (see Powers and Easter 1983).

The mechanism by which chromophoric absorption is modified by a protein is not yet understood; however, several hypotheses have been proposed (see McFall-Ngai 1991 for properties of the visual pigments).

A number of fishes are known to possess paired pigment systems consisting of mixture of rhodopsin and poryphyropsin and in some species the relative proportions of two pigment types change with environmental factors such as light intensity, photoperiod, temperature, and season.

The wide range of rhodopsin λ max (wavelength of maximum absorption) in fishes is the result of selective presence related to the diversity of photic habitats in the aquatic environments. This evolutive process has probably been effected mostly through polymorphism in genes coding for opsin.

In fishes undertaking spawning migrations from one aquatic habitat to another, the visual pigment mixture is converted to pure rhodopsin (eel) or to a mixture largely dominated by porphyropsin (salmonids). However, these changes are occurring in anticipation of the migratory journeys and are not directly influenced by immediate environmental factors. The controlling anticipatory agents are endocrine in nature, involving both the pituitary and thyroid glands. Thus, the endogenous factors regulating sexual maturity and metamorphic events may also influence directly the conversion of one chromophore group to another (for details, read Browman and Hawryshyn 1992). In non-migratory fishes, changes in the proportions of rhodopsin and porphyropsin may occur, and are usually related to seasonal or geographic variations in the colour and intensity of transmitted light.

The visual pigments of deep-sea fishes have been much studied (see Partridge et al. 1988, 1989, 1992; Partridge 1990). The eye of the parasitic fish *Carapus mourlani* combines features of degenerated photoreceptors, characteristics of cave-organisms, with adaptations which are commonly found in nocturnal and deep-sea forms (Meyer-Rochow and Tiang 1978).

Fishes living in surface waters of the sea generally have rhodopsin visual pigments in their rods with λ max values close to 500 nm. In contrast, most deep-sea fishes have blue-sensitive visual pigments with λ max values between 470 and 490 nm.

Deep-sea fishes do not seem to have increased their overall visual sensitivities by developing pigments with higher than normal photosensitivities. Instead they appear to have increased the sensitivity of their photoreceptors by maximizing their photon catches. This is done both by having very long rod outer segments and by having visual pigments apparently at very high concentrations. The maximum specific absorbances of visual pigments in photoreceptors of shallow water living fishes are typically 0.01–0.015 um^{-1}. In contrast, the maximum specific absorbances of visual pigments in deep-sea fishes range from 0.009 to 0.031 um^{-1} with a mean of 0.021 um^{-1}.

The regeneration of deep-sea fish visual pigments has been insufficiently investigated to provide much insight into their regeneration kinetics.

Spectral clustering of λ max values is also found in deep-sea fish visual pigments. There is a high degree of cluster-point conservatism within genera. The general conservatism may be indicative of the way in which the evolution of visual pigments in the deep sea is determined: not by ecological pressures but by phylogenetic affinities.

Although most deep-sea fishes have but one visual pigment, various species are known to have more than one. Of these species most, if not all, have short-wavelength absorbing filters in their lenses or retinae. The group of species known to have more than one visual pigment in their retinae consists of some 14 species of teleost and one elasmobranch. In the case of five of these teleosts, *Pachystomias microdon*, *Aristostomias grimaldii*, *A. scintillans*, *Malacosteus danae* and *M. niger*, their unusual retinal spectral sensitivities are explicable in terms of their equally unusual bioluminescence. These species have red-sensitive retinae and photophores emitting in the far red, thus giving them access to a private waveband with which to see and not be seen, except by conspecifics. In all other multipigment species, however, there is no obvious correlation with bioluminescent emission, whether their own or that of other species.

Little is known about the turn-over of visual pigment in deep-sea fish retinae. Many fishes are known to change their visual pigments in the course of their lives by chromophore substitution, resulting in a shift from rhodopsin to porphyropsin pigments, or *vice versa*. However, few animals are thought to change their visual pigments by opsin substitution, the eel *Anguilla* spp. being a notable exception.

The finding of rods containing two types of visual pigment is an indication of a photoreceptor in the process of changing its spectral sensitivity by opsin change. Unlike mammals, fish have the potential to insert new rods within the photoreceptor mosaic of their retinae and do so as a mechanism for retinal growth.

Like many deep-sea fishes, *Scopelarchus analis* has a larval stage that lives in surface waters (< 100 m depth) and individuals only assume life at greater depths (in excess of 500 m in daytime) as adults. Adults of this species have vertically directed eye and it seems clear that their visual preoccupations are with looking upwards. In addition their eyes possess heavily pigmented yellow lenses with only 50% transmission at *ca* 465 nm which, as argued, may be a device to 'break' the ventral camouflage of animals seen in silhoutte against the down-welling ambient light.

2.4 Bioluminescence and Vision (Visual Adaptations to Dim Light and Bioluminescence)

It is paradoxial that deep-sea fish, some of the most inaccessible animals in the world, have been a focus of attention for vision research for several decades. In large measure this intensity of interest has been prompted by the realization either in a dim homochromatic blue world, or, at greater depths and at night, a virtually lightless environment broken only by bioluminescent emissions.

The topic assessed by Nicol (1978) has provided interesting information. The unique nature of visual environment in deep oceans places obvious constraints on the vision of deep-sea fish. Unlike cave-dwelling species, deep-sea fishes have not, in general, abandoned the visual sense and have instead evolved many remarkable ocular specializations ranging from gross anatomy to retinal organization as supported by following examples.

Dependent upon several factors, it is assumed that eyes of deep-sea fishes have enhanced sensitivity. Among these special features are the great transparency of the lens, large ocular size, large pupils, aphakic apertures, and modification of retina. The retinae of deep-sea fishes are noted for both high amounts of visual pigment and for large sizes of retinal photoreceptor outer segments.

Changes of size and structure of eyes of fishes with depth have been related to the photic environment. Most mesopelagic fishes possess large eyes with wide pupils and correspondingly enlarged lenses; often there is an aphakic space, e.g., lantern fishes, hatchet fishes. Such eyes have potentially good resolving power and, compared with others of the same size with smaller pupils, permit more light to enter per unit of retina. A few species, such as *Cyclothone*, have small eyes. At deeper levels, below 100 m, there is tendency for eyes to be smaller, relative to the size of the head, but still to have greatly enlarged pupils. Eyes of three species of *Gonostoma* exemplify this trend. The majority of these fishes are luminescent and are sensing their own lights or those of others.

Below 1000 and 3000 m many pelagic species have small or degenerate eyes. The tiny eyes have small resolving power and are probably used for detecting or distinguishing objects in the near vicinity. Benthopelagic and benthic fishes extending down to 1000 m, such as macrourids, and brotulids, have large well-formed eyes, but the eyes become smaller and regression occurs in fishes living at greater depths. No deep-sea fish with degenerate eyes is known to be luminescent.

A fairly large number of mesopelagic and bathypelagic fishes have large tubular eyes which are directed forwards, e.g., *Winteria*, or upwards, e.g., *Argyropelecus*. These eyes are in parallel and permit binocular vision. They have a main retina which lies at the bottom of the tube and which continues over the median wall as the accessory retina (e.g., *Ichthyococcus*, Fig. 15.14). The main retina has longer rods, is more sensitive and receives a focussed image. The accessory retina is too close to the lens to receive a focussed image, and presumably detects movements of well-lit objects in near vicinity.

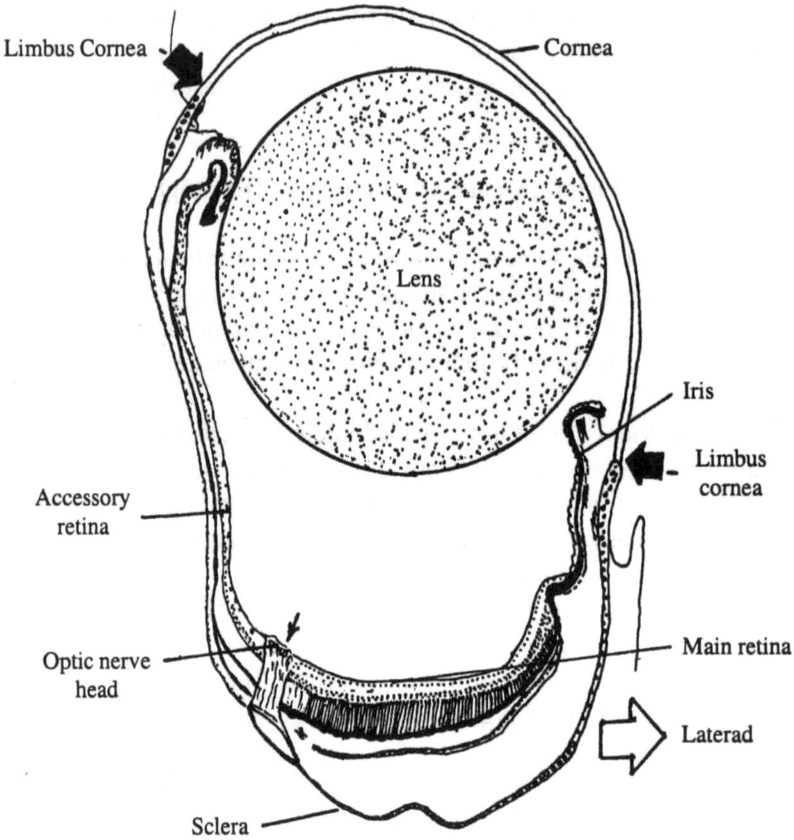

Fig. 15.14 Dorsally directed tubular eye of *Ichthyococcus ovatus*. Thick arrows point at limbus cornea, thin arrow points at optic nerve head.
***Source:* Munk (1966)**

The eye of *Directmus argenteus*, a mesopelagic deep-sea teleost has several unusual structural features, among others an extraordinarily thick ventral multiple bank retina used for binocular vision in the dorsal direction. The ventral retina, which is the thickest part of the retina, contains an inner (vitread) bank of rods with exceptionally long outer segments and an outer 4–7 banks of rods with short outer segments; it also contains a region with cones that are located in the innermost part of the visual cell layer. The ventral retina possesses a retinal tapetum lucidum, i.e., a tapetum where the reflective elements are located within the retinal pigment epithelial cells. The presence of a retinal epithelium and a very thick cell layer in the ventral part of the retina probably means that this portion of the eye has an extraordinarily high sensitivity. It is suggested that the combination of yellow eye lenses and a supposedly ventral sensitive retina in *D. argenteus* may be used to break the camouflage of mesopelagic animals that use their ventral photophores for countershading (Munk 1980).

A peculiar feature of some tubular eyes is a diverticulum in the form of an accessory organ or secondary globe (e.g., *Bathylychnops*, Fig. 15.15). The window of the diverticulum faces

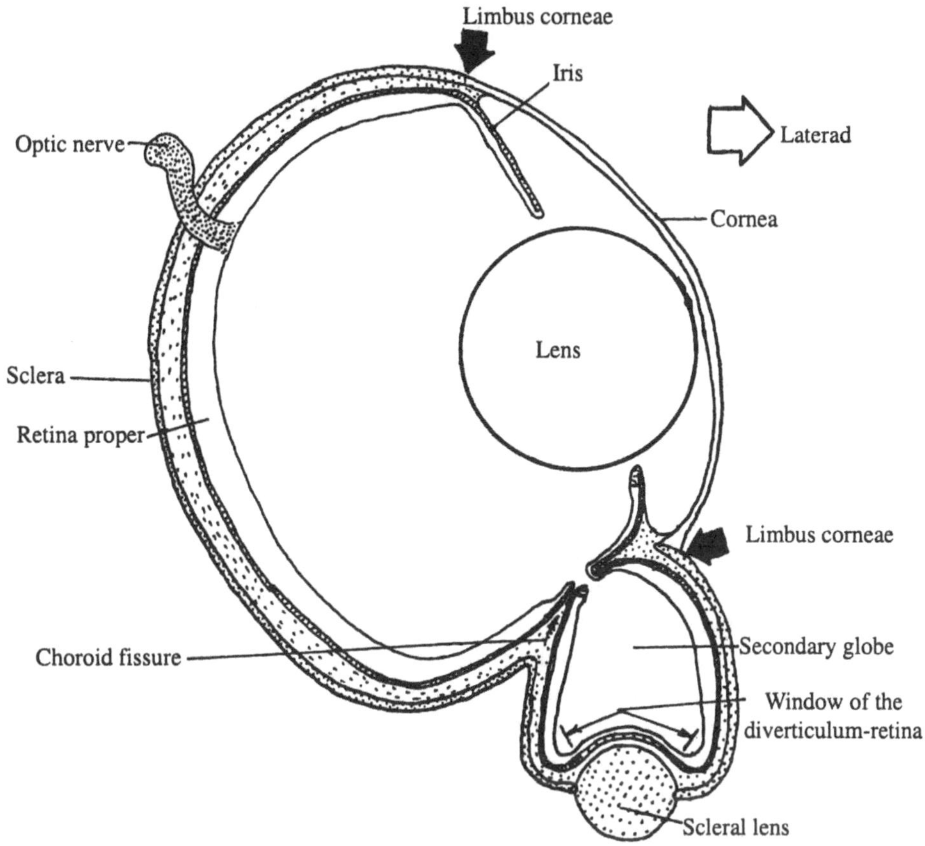

Fig. 15.15 Diagram of the eye of *Bathylychnops exilis* as shown in vertical section at right angles to equator of eyeball. Thick arrows point at limbus corneae. Thin arrow points at choroid fissure. The lumen of the secondary globe is marked.
***Source:* Munk (1966)**

downwards in some species with dorsally tubular eyes, such as *Opisthoproctus*, and temporally in species with rostro-dorsally directed ones, such as *Winteria*. Because of position, light sometimes reaches the retina of the diverticulum indirectly, through the mouth of *Gigantura*, for example. Retinal diverticula appear to have the function of perceiving objects below or behind the fish, objects which because of shape and optic axes, could not otherwise be perceived. In *Bathylychnops* the retinal diverticulum is in contact with a lens-like thickening of the sclera. In this fish, also the ventral field is restricted and the secondary globe compensates by its ability to perceive light and movement in a ventral direction. Some of these fishes are luminescent; in all of them the diverticula could serve to detect luminescence outside the visual fields of the main and accessory retinae.

Aphakic apertures (in which the lens does not entirely fill the pupil) are a common feature of deep-sea fishes. It has been suggested that their basic function is to increase the relative retinal illumination within a localized zone or over the whole retina. This is accomplished by enlarging the part of the lens aperture for rays which are oblique to the axis passing through the centre of the lens and the centre of the fundus. Large circumlental aphakic apertures increase the effective sensitivity of the whole eye by augmenting the overall retinal illumination. An increase of the pupil-to-lens ratio from 1.0 to 1.5 doubles the relative illumination of the peripheral retina. It has been however, noted that a rostral aphakic space can permit movement of the lens during accommodation.

A crescent-shaped rostral aperture, found in species with rostral binocular vision, is often associated with the presence of a caudal (temporal) zone of the enhanced resolving power, i.e., an area on fovea in the retina (e.g., in *Alepocephalidae*). Rods are densely packed in this region, rod outer segments are greatly elongated and there is sometimes a foveal depression. Such areas and foveae are found in mesopelagic and.bathypelagic species such as *Nansenia, Bathylagus*, etc. and some of these fishes are luminescent. Cones as well as rods are found in several deep-sea fishes, and *Scopelsauras* has a pure cone fovea in the temporal region. These adaptations are believed to assist movement detection and to be important in binocular fixation, and are advantageous to predators hunting small objects by sight. Furthermore, binocular vision itself confers greater light sensitivity. For details of main and accessory retinae of *Scopelarchus analis*, the reader is referred to Partridge and Archer (1992).

Collin and Partridge (1996) described the retinal specializations in the eyes of deep-sea teleosts. It was found that the retinal ganglion cell topography changes markedly across the retina with a density range of 6.3–50.6 × 10^3 cells mm^{-2} in a pattern unique to each species. A number of mesopelagic species including the lanternfishes *Lampanyctus macdonaldi* and *Myctophum punctatum* possess a concentric increase in cell density towards the retinal margins, putatively enhancing peripheral vision. Three tubular-eyed species including *Scopelarchus michaelsarsi* possess an area centralis in the centro-lateral region of the main retina supporting the premise that this specialization receives a focused image. Some benthic species, such as the smoothhead *Rouleina attrita* and the searsid *Searsia koefoedi* also boast a structural specialization or fovea in temporal retina with centro-peripheral cell gradients exceeding 30 : 1. Benthic species such as the tripodfish *Bathypterois dubius* possess two regional increases in ganglion cell density or areae centrales, one nasal and the other temporal, thereby increasing spatial resolving power in the caudal and rostral visual fields, respectively. These quantitative analyses suggest that deep-sea fishes, like their shallow-water counterparts, also use a specific region of their visual field of acute vision. This may be an advantage for the detection and localization of bioluminescent light sources.

Readers can get a wealth of information in recent contributions (Collin 1997, Collin et al. 1997).

Trachipterus arcticus, with a rotating eye, has well-developed space on the both side of the lens. The rotation of the eye through 90° from the vertical to the horizontal may be related to the orientation of the fish in the water when the fish swims in the "normal" fashion, with its body parallel to the surface, the pupil would be horizontal when the fish swims upright, its body perpendicular to the surface, the pupil would be vertical.

Ability to resolve photophore patterns or distinguish one among several small luminescent animals depends upon visual contrast and acuity. Fishes seem to have visual acuities much like those of man, and it may be expected that resolution will be much reduced at low luminescence levels when rod vision is operative.

2.5 Visual Alarm Signals

This is an interesting field in fish behaviour. Vision should be a good channel for alarm signalling because transmission is rapid and directional. Some of the aspects covered are goby bobbing; mobbing as a potential visual alarm signal; alarm signals between parental fish and their young; and aposematism as a potential alarm signal (refer Smith 1992; see also Mathis et al. 1993).

3. INNER EAR (OTIC ORGAN)

3.1 Introduction

Hearing is but one member of a large family of modalities that we call, collectively, the octavolateralis system, a better term than the older acoustico-lateralis system.

The structural plan of the inner ear is essentially constant throughout the vertebrates. The membranous labyrinth of vertebrates is a complex of mechanosensory structures in the inner ear. Specialized end organs mediate the reception of gravity, linear and angular accelerations, and provide some signal that is important in regulating muscle tone. These structures and their innervation by the eighth cranial nerve form the peripheral vestibular system.

The membranous labyrinth is surrounded by a cartilaginous or bony capsule filled with perilymphatic fluid and suspensory connective tissue acting as shock absorbers. The labyrinth consists of a membranous sac filled with endolymph. The sensory nerve leaves this capsule through foramina. In vertebrates with a bony skull the labyrinth is enclosed in a bony labyrinth capsule replicating the general shape of the labyrinth, especially that of the semicircular canals. Bony tubes filled with perilymph surround the membranous endolymph-filled canals.

There are several important contributions on the subject (Lowenstein 1971, 1974; Popper and Fay 1973, 1977, 1984, 1993, 1999; Wersall and Bagger-Sjoback 1974, Baird 1974a, b; Fay 1978, 1985; Roberts 1978; Barber and Emerson 1980; Enger 1981; Popper et al. 1982; Fay and Popper 1983; Platt 1983; Popper 1983; Fay and Ream 1986, 1993; Kalmijn 1988; Rogers et al. 1988; Coombs and Fay 1989; McCormick 1992; Schellart 1992; Hawkins 1993; Lu and Fay 1993, 1995; Popper and Platt 1993; Coombs 1994; McCormick and Bradford 1994; Schellart and Wubbels 1998). For general interest on inner ear development, read Corwin (1997).

Chapters in "Hearing and sound communication in fishes" (edited by W.N. Tavolga, A.N. Popper, and R.R. Fay 1981) also focus on various aspects related to the system.

Inner ear is distinguished into: (1) Fish-like form; the otic labyrinth is suspended in the otic capsule by periotic reticulum continuous with perimeningeal tissue, and no specific periotic

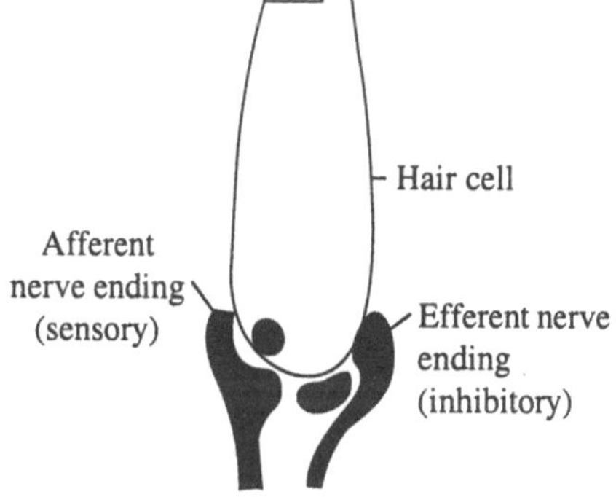

Fig. 15.16 Vertebrate hair cell.
***Source:* Bleckmann (1986)**

channels are delineated and (2) Reptilelike form; with differentiation of the inferior division of the otic labyrinth, a periotic cistern and periotic sac form links in a vibratory transmission pathway extending from the stapes to the region of metotic fissure.

The basic parts of the otic labyrinth (three semicircular ducts with ampullae, utricle, saccule, and lagena) are well known, are varyingly evident in all lower vertebrates except cyclostomes, and, except for the acquisition of a definitive cochlear duct during the transition from amphibians to reptiles, the general morphology of the labyrinth has shown only limited changes throughout its history. Ultrastructural studies in animals ranging from cyclostomes to birds have shown a basic similarity in the architecture of neuroepithelia of the labyrinthine mechanoreceptors. Two major cell types are uniformly present, hair (sensory) cells (Fig. 15.16) and supporting (sustentacular) cells. (read details: Baird 1974a).

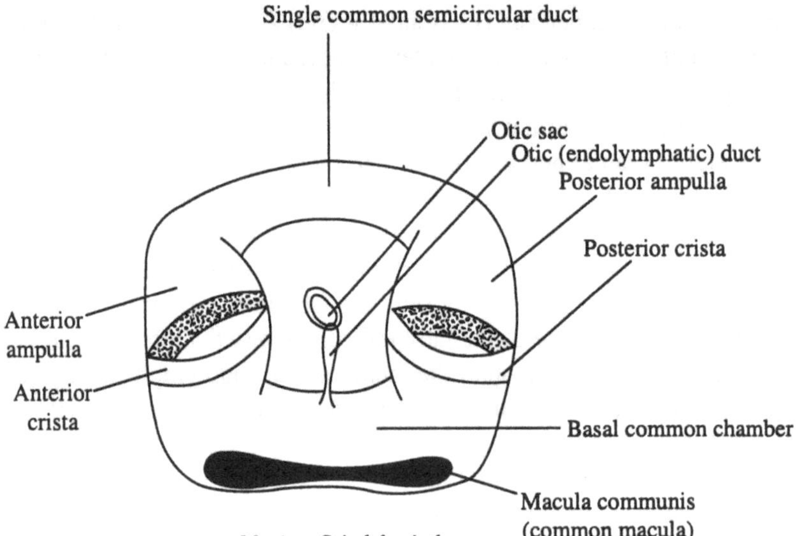

Myxine: Otic labyrinth.
(A)

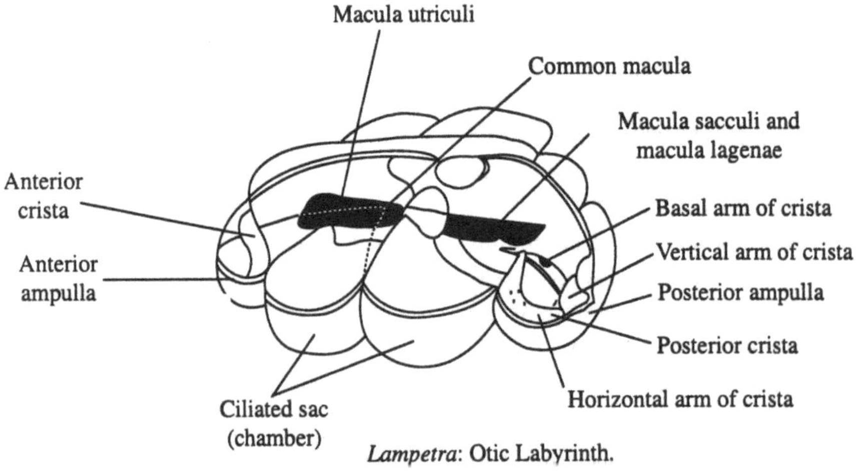

Lampetra: Otic Labyrinth.
(B)

(Contd)

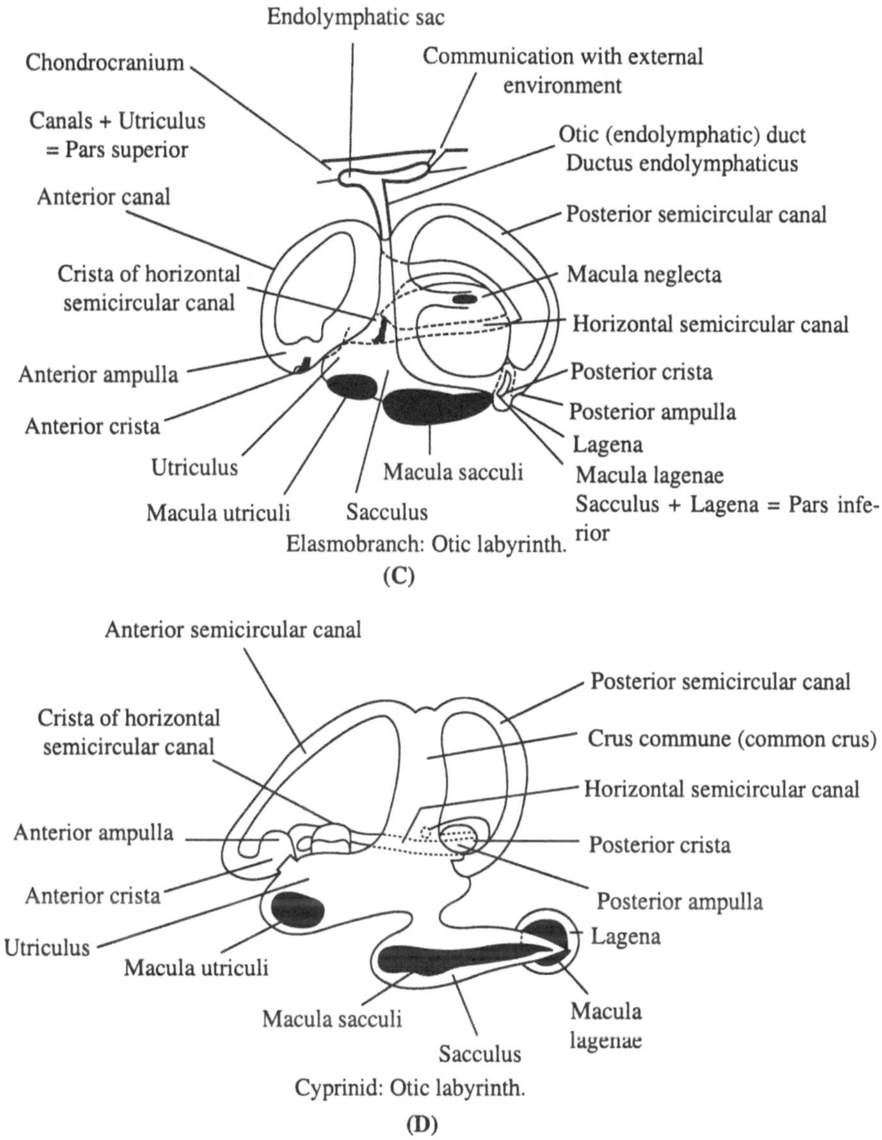

Fig. 15.17 The four chief of morphological types (semidiagrammatic representations of the medial aspect) of labyrinth in the agnathous and gnathostome fishes. A. *Myxine*; B. *Lampetra*; C. Elasmobranch; D. Cyprinid. Endolymphatic duct (omitted in B and D).
***Sources:* Lowenstein (1971, 1974) and Baird (1974)**

3.2 Inner Ear in Fish Groups

Labyrinths without horizontal canals are found in cyclostomes. The membranous (otic) labyrinth of cyclostomes is considerably different from the labyrinth of higher vertebrates. The labyrinth of *Myxine* is formed by an oblique sagittally positioned ring. The median part of the ring contains a macula communis. Anterior and posterior to this are slightly widened ampules, each containing a crista ampullaris. The ampules open into two canals which fuse dorsally to complete the ring (Fig. 15.17A) (read Jorgensen et al. 1998).

The otic labyrinth has an irregular toroidal form, and is oriented obliquely to the major axes of the head. It is enclosed by a kidney-shaped capsule consisting of cartilage except for its medial aspect, which is formed by a membrane of connective tissue. The capsule is penetrated by the branches of the eighth nerve supplying the sensory areas located on the macula communis and the two cristal ampullares.

The labyrinth of *Myxine* differs from those of gnathostomes in that only two ampullar cristae are present. Each crista forms an annular ridge around the inner wall of the ampule and is innervated by a branch of the eighth cranial nerve. A cupula proper is apparently not present. The macula communis forms a continuous sheet covering the median part of the ventral wall of the labyrinth ring. The entire macula is covered by a mass of statoconia embedded in a fibrous matrix (statolithic or statoconic membrane). The macula communis and the two cristae are covered by a sensory epithelium containing sensory cells and supporting cells. A short otic (endolymphatic) duct extends from the dorsomedial aspect of the common chamber, transverses the capsular foramen and terminates blindly into a small otic sac within the cranial cavity. Very little is known about the function of the ear in hagfishes. As compared with that of other vertebrates, the sensitivity of the inner ear of hagfishes is rather low.

The otic capsule of *Petromyzon* resembles that of the hagfish except that it lacks an internal pillar and the medial aperture is subdivided to provide a small dorsal foramen for the otic duct. The periotic labyrinth is limited to a narrow cleft between the capsular walls and a compact, bullous otic labyrinth. The membranous labyrinth of *Petromyzon* of unusual form is considerably more complicated than that of *Myxine*. Two semicircular canals are present, the anterior and posterior canals. Each canal widens at one end into an ampule, which in turn communicates with the elongated sac forming the median part of the labyrinth. It also communicates with one of the two broad ciliated chambers which are unique structures of unknown function appearing only in *Petromyzon* labyrinth (Fig. 15.17B: *Lampetra*),

A short otic duct extends dorsomedially from the common chamber, traverses the dorsal foramen in the otic capsule, and terminates in an otic sac within the cranial cavity. A small "ventral duct" is also described.

The ciliated chambers form two parts of the central cavity of the labyrinth which is partially divided by an incomplete ciliated chamber. The ciliated chambers are lined with ciliated cells except for their central wall. The powerful cilia of these cells keep the endolymph in the chambers in constant movement. The ciliated chambers are divided into two compartments by an invagination of the wall of the chamber, and the two compartments have opposite rotations of the endolymphatic movements. However, the ciliated chambers have an extensive area of communication with the semicircular canals and the sac of the labyrinth. The endolymph in these compartments is undisturbed by the movements.

Three arms, forming the cristae ampullares of the two ampullae, project from the ampullary wall into the lumen of the ampulla and covered with sensory epithelium: one basal arm along the bottom of the ampulla, one vertical and one horizontal arm. The three arms run in planes roughly perpendicular to each other.

Except for the macula neglecta, the maculae of the lamprey labyrinth form a macula communis occupying the base of the labyrinth. Although this macula is covered by an uninterrupted layer of otoconia embedded in a fibrous matrix, it is possible to distinguish an anterior horizontal macula, the macula utriculi, a vertical macula, the macula sacculi, and a posterior horizontal macula, the macula lagenae. The macula dorsalis or macula neglecta lies in a separate recess on the dorsomedian face of the labyrinth.

As already mentioned, the labyrinth of *Myxine* forms a circle where each of the two ampullae is in contact with one half of the united semicircular canal. In *Petromyzon* two canals can be distinguished, which end with a crus in the macula carrying the otosaccus.

Observation on the swimming behaviour of these animals reveals that their movements in the horizontal plane are well controlled in a manner normal for a fish of their body shape, even though optical orientation is either absent (*Myxine*) or indifferent (*Lampetra*).

There are no behavioural data that demonstrate hearing in any agnathan. However, eighth nerve responses to vibratory stimuli in the range of 80 Hz have been recorded in *Lampetra* but not in *Myxine*. Based on circumstantial evidence it was suggested that the presumed macula neglecta and saccule of lampreys are vibratory receptors. Clearly, more data are needed before we can conclude whether inner ear acoustic reception is present in extant agnathans.

In elasmobranchs (sharks, skates and rays), enclosure of otic labyrinth tends to be complete, with only foramina for the eighth nerve and the otic duct actually perforating the capsule. A third membrane-covered opening or thinned area of cartilage is also present in these animals. Originally described as a "fenestra ovalis" ("fenestra of Scarpa"), the opening lies dorsally, adjacent to the foramen for the otic duct; it may have a role in the conduction of sound. The capsule in *Chimaera* is less complete in that the otic duct, utricle, and immediately adjacent parts of the otic labyrinth are exposed at the median aperture and lie enmeshed in periotic reticulum between the primitive menix and the capsule. The intracapsular periotic labyrinth differs considerably among representative cartilaginous fishes, and its dimensions vary regionally within given forms, but definitive periotic channels apparently have not been identified in the group.

As already mentioned, the labyrinth of *Myxine* forms a circle where each of the two ampullae is in contact with one half of the united semicircular canal. In *Petromyzon* two canals can be distinguished, which end with a common crus in the macula carrying the otosaccus. In all chondrichthyans examined so far, however, three semicircular canals and a sacculi are present and varying degrees of utricular and lagenal differentiation are evidenced. Communication of the otic labyrinth with the external environment is characteristic of the group, (i.e., the endolymphatic duct is not a blind pouch, but instead opens as an endolymphatic pore on the dorsal surface of the head; the functional significance of this is unknown), and the organization of the parts of labyrinth is subject to considerable variation (read details: Myrberg 2001).

Certain psammophilous elasmobranchs lack the ability to produce calcium carbonate otoliths and compensate for this by uptake of sandgrains through their endolymphatic ducts. The penetration of sand grains into inner ear may occur when the animals are burrowing in sandy bottoms. A sieve-like function of the duct openings could explain selective uptake of mineral particles of suitable shapes and sizes.

Inspite of the communication between the endolymphatic system of elasmobranchs and the sea water, the endolymph differs considerably in its inorganic composition from the sea water (read Fänge 1982).

There are forms with or without a common crus. In elasmobranchs the semicircular canals are more or less separate structures connected with the utricle only by thin ducts, and the utricle and sacculi with their diverticula form a widely communicating sac, the otosaccus, which contains the various sensory areas of the macular type (Fig. 15.17C) Corwin (1978) reported in some elasmobranchs that the macula neglecta seems to be elaborated in proportion to each species' dependence on raptorial predation on fish. The involvement of inner ear organs in hearing, however, is not altogether clear.

The otic capsule in Osteichthyes varies considerably in its developmental patterns and incorporation of the usual vertebrate centres of ossification, and in the extent to which it encloses the membranous labyrinth. In general, the median aspect of the capsule is widely open, and parts of the otic labyrinth occupy fossae rather than bony canals or compartments, but cartilaginous struts or partitions may contribute to the definition of semicircular canals and/or narrowing of the medial aperture. In some forms, however, (e.g., anabantids and most ostariophysians Fig. 15.17D) bony compartments posteroventral to the main part of the capsule enclose the saccule and lagena; and dipnoan (e.g., *Neoceratodus* and *Protopterus*) exhibit a capsule that despite a large median aperture, is relatively well separated from the cranial cavity and is divided internally. Conversely, an otic capsule is essentially lacking in the seahorse (*Hippocampus*) and the pipefish (*Syngnathus*), the otic labyrinth is suspended within the cranial cavity. The otic capsule is closely associated with foramina for the glossopharyngeal and vagus nerves, derivatives of the embryonic metotic fissure, and extensions of the swim bladder may establish relationships with these or other specialized openings or thinned area in the capsular wall.

Although the known otic labyrinths of bony fishes can be divided into two general morphological categories ("typical" and ostariophysine), the superior part of the labyrinth (or pars superior, canals and utriculus) shows an essentially constant pattern. The relationships between the superior and inferior divisions (pars inferior, sacculus and lagena) of the labyrinth are variable. Dipnoans (e.g., *Neoceratodus*, *Protopterus*) are an exception in that the utricular recess is remarkably large. Topographically, the saccule and lagena are usually juxtaposed to the ventromedial aspect of the utricle, but the inferior division may be large and lie lateral to the utricle (e.g., *Gobius*, *Periophthalmus*), or it may be significantly separated from the superior part (e.g., ostariophysians, *Ostracion*, *Chauliodus*). More notable, however, is the variance in communication between the superior and inferior divisions. There are reports on the presence of a utriculo-saccular canal in a chondrostean (*Acipenser*) and in dipnoans (*Neoceratodus*, *Protptoerus*) and its absence in holosteans (*Lepisosteus*, *Amia*) and some teleosts.

The inferior division of the "typical labyrinth (e.g., *Acipenser*, *Neoceratodus*, *Perca*) reflects varying levels of differentiation of the lagena as a posterior appendage of the saccule. In some bony fishes (e.g., *Acipenser*, *Amia*, *Lepisosteus*, *Neoceratodus*, *Protopterus*, *Belone*, *Coregonus*, *Exox*), the saccular and lagenar maculae occupy, respectively, anterior and posterior positions, within a common chamber. Other non-ostariophysine teleosts show distinct lagenae ranging from shallow evaginations of the posterior saccular wall (e.g., *Solea*) to polypoid posterior appendages that communicate with the saccule by way of a narrow canal (e.g., *Lophius*, *Raniceps*). The topographical relationships between the two parts are variable, depending largely upon their relative sizes, but the basic pattern is apparently present in all non-ostariophysians, including those with air chambers directly related to the labyrinth (i.e., clupeids, mormyrids, anabantids) (Fig. 15.18).

Classical description of conditions in carps, characins, and catfishes suggest a similar constancy in the ostariophysine labyrinth. The inferior division, connected to the utricle by an elongate utriculo-saccular canal, occupies a paramedian fossa in the cranial floor. A relatively slender and a large rounded lagena appear as distinct chambers; a relatively slender sacculo-lagenar foramen

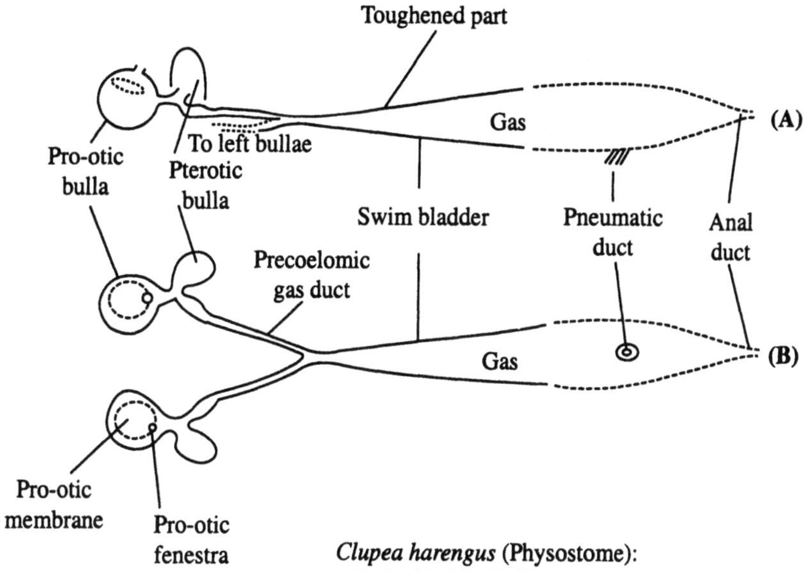

Fig. 15.18 Swim bladder-bulla system from left side (A) and from above (B).
***Source:* Allen et al. (1976)**

unites the two near the mid-point of the lateral saccular wall, and the posterior part of the saccule, largely surrounded by dense periotic tissue, lies medial to the lagena. Medially, just anterior to the level of the sacculo-lagenar communication, the saccule is connected to its adjacent counterpart by a transverse intersaccular canal which abuts posteriorly against the unpaired periotic sinus of the Weberian apparatus. In some forms (e.g., *Wallago*, *Selenaspis*), the saccule lies entirely anterior to the lagena, and numerous variations are noted in the form and relationships of the transverse intersaccular communication, and the otoliths. In both "typical" and ostariophysine labyrinths, the otic duct is frequently absent; when present, it extends from the dorsomedial aspect of the saccule and terminates blindly without notable enlargement.

3.2.1 General Overview of the Auditory System in Bony Fish

The peripheral vestibular system is part of the inner ear, and the sensory organs are the bilaterally paired labyrinths within the cranium, located posterior to each eye. In bony fish each labyrinth consist of dorsal set of three looping tubes, and underneath these, a set of three pouches. Fishes lack a cochlea, the specialized auditory organ of mammals, but otherwise the general structure of the labyrinth has shown remarkably little change during the vertebrates' evolutionary radiation from aquatic to terrestrial and even aerial environments.

The semicircular canals are three hollow fluid-filled loops, named for their form, arching out and returning to a common base. The three canals can be considered as three mutually orthogonal, connected toroids. Two canals, the anterior vertical and posterior vertical lie in a nearly orthogonal vertical planes, each oriented roughly 45° outward from the sagittal body plane; they share a common medial vertical portion the crus. The horizontal canal lies roughly in the horizontal plane in the standard level swimming posture. Each canal shares fluid communication with the other two, and are filled with endolymph, a fluid noted for its relatively high potassium-and low sodium-ion concentrations.

The ears of most species of fish have seven sensory end organs. Four of the end organs, the cristae of semicircular canals and (perhaps) the macula neglecta, are involved in detection of angular acceleration. The three other receptors are the otolith organs, the utricle, saccule, and lagena. The primary acoustic receptors are thought to be saccule and lagena, while the utricle is considered a gravistic receptor in most fishes. While a functional dichotomy of the three otolith organs is implied in the literature, vestibular and acoustic senses may represent ends of a sensory continuum that is responsive to widely varying frequencies. There is evidence that some otolith organs concurrently perform both vestibular and acoustic roles.

Each otolith organ in teleost fishes is a saclike structure containing a single, dense calcareous structure, the otolith. On the wall of the chamber is a sensory epithelium (or macula). Macula contains large number of sensory and supporting cells. The otolith and the sensory epithelium are separated by a thin, otolith membrane relative to one another. The chamber space not filled with otolith is filled with fluid (Fig. 15.19).

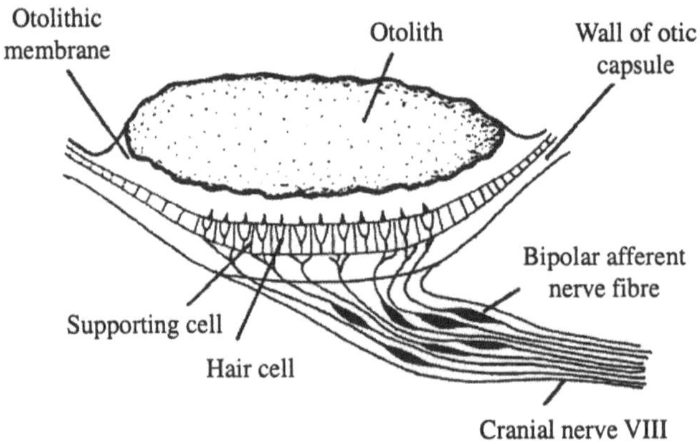

Gadus morhua: Otolith organ.

Fig. 15.19 Schematic cross section through the utriculus, showing the otolith mounted on the hair cells but separated from the cilia of the cells by an otolithic membrane. Bipolar afferent fibres and finer efferent fibres synapse with the hair cells.
***Source:* Hawkins (1986)**

Teleosts generally have three pairs of otoliths, large sagittate and smaller lapilli and asterici, located in the inner ear and composed of calcium carbonate (usually in the aragonite morph) deposited on an organic matrix. These crystalline structures are mechanical components of sound transduction responsible for hearing in fishes. The shapes of otoliths are a determinant of the particular sound frequencies for which the otolith acts as a transducer. Otolith shape can be complex and is of species-specific. All three otoliths often show growth-related rings, in addition to their highly specific variability of shape, making them extremely useful in ecological and systematic studies. For the fine structure of sagitta and otolithic membrane in catfishes, refer Jenkins (1979).

The sensory epithelia of the fish ear contain only vertebrate type II (vestibular) sensory hair cells, each of which has an apical ciliary bundle containing a single, eccentrically positioned, kinocilium and more than 40 stereocilia. The cylindrical form of the cell is innervated by both afferent and efferent synaptic terminal buttons, unlike the flask-shaped type I (found together with type II) cells in birds and mammals that is innervated by an afferent nerve terminal forming a chalice completely enclosing the base of the receptor cells. The sensory epithelia of the otolith end organs may contain thousands, tens of thousands, or even hundreds of thousands sensory hair cells organized into "orientation groups". All the hair cells in each orientation group are oriented with their kinocilium on the same side of the ciliary bundle. The saccular macula of a majority of teleost fishes have four such orientation groups, while the lagenar and utricular epithelia more often have two groups. In contrast to the humans, hair cells are produced throughout life in the ears of fish, amphibians and birds (Corwin and Oberholtzer 1997). For special interest on fine structure of saccular macula in catfishes, see Jenkins (1979).

The morphological polarization of the hair cells is extremely significant for sound detection by all vertebrates, including fishes. Physiological experiments have shown that each hair cell is directionally responsive and generates a maximum depolarizing receptor potential when the direction of bending of the ciliary bundle is towards the kinocilium from the stereocilia. A hyperpolarizing receptor potential occurs when bending is in the opposite direction. Bending in any other direction gives a receptor potential to the cosine of the direction of stimulation. Consequently, each hair cell is, in effect, a directionally sensitive receptor.

3.2.1.1 Stimulation of the Ear

Experimental evidence indicates that sound reaches the ear by atleast two "pathways". One pathway involves detection of the particle motion of an acoustic disturbance through a direct impingment upon the inner ear. This is referred to as the "direct" method of the inner ear stimulation. The "direct" method depends on the similar densities of water and a fish's body and their dissimilarity to the density of an otolith. Acoustically driven motion of a fish and the water mass containing the fish will be the same because of their similar densities. Since the otolith is three times denser than water and the body of the fish, it will move with a smaller amplitude and at different phase from the rest of the body (including the sensory epithelium). Furthermore, the otolith and sensory epithelium are connected to one another by the otolith membrane, so they move relative to one another. Because the cilia of the sensory hair cells are coupled to both the otolith and sensory epithelium (either directly by contacting the otolith or via the otolith membrane), they bend during any particle displacement that produces a physiological response from the sensory hair cells.

A second, "indirect" method for sound stimulation involves detection of the pressure component of the sound field. This mechanism involves the swim bladder, a gas-filled chamber in the abdominal cavity of many fish species. Sound pressure sets the wall of the swim bladder into the motion due to relatively high compressibility of the gas. The motion of the swim bladder reradiates sound energy in the form of particle displacement. This second source of particle displacement also causes the otolith to move relative to the sensory hair cells, thereby producing a shearing of the ciliary bundles (Fig. 15.20).

The "direct" path is thought to function in all fishes, whereas the "indirect" path is found in some, but not necessarily all, species with a swim bladder. The two pathways appear to be complementary in that the "direct" path is much effective below a few Hz, while the "indirect" path seems most effective for signals above a few hundred Hz.

In teleost audiology, three groups of species are distinguished. Group (I) includes most Otophysi, all of which have Weberian ossicles, which have further enhanced the sensitivity for

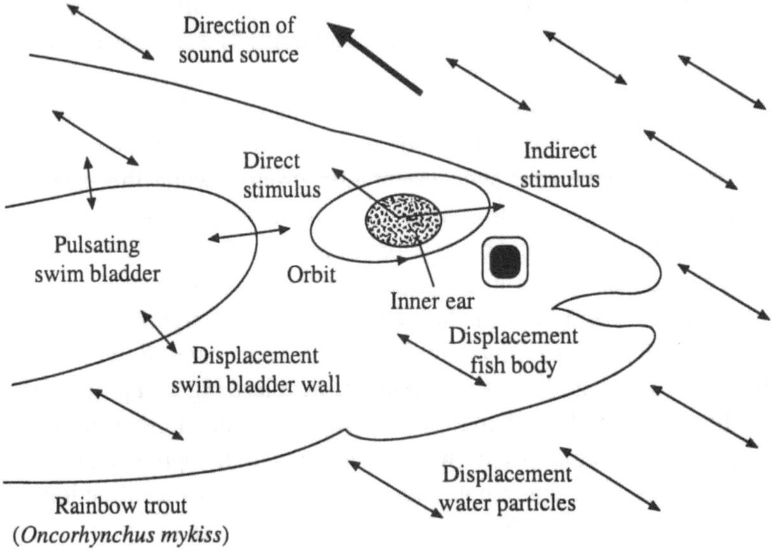

Fig. 15.20 Stimulation of inner ear of swim bladder-bearing fish. Motion component of sound field causes displacement of directionally sensitive hair cells relative to otolith, which lags surrounding tissue because of its higher inertia. Gas-filled swim bladder functions as a sound pressure to displacement transducer, stimulating inner ear indirectly from a body-fixed angle. Vectorial addition of direct displacement and swim bladder mediated displacement yields an elliptical displacement orbit.
***Source:* Wubbels and Schellart (1997)**

sound pressure by effectively transmitting the movements of the swim bladder directly to the ear. Group (II) includes species from several, unrelated taxonomic orders. These species have specializations thought to enhance hearing, by bringing the pressure-to-displacement transducer (swim bladder or other air cavity) closer to the ear than species without such specializations. The adaptations of both groups, whose members are the 'hearing specialists', extend the auditory 'scope' in the space and frequency domains. The group (III) of fish are also drawn from different taxa. None of these species, the 'non-specialists', has structures known to enhance hearing. For further information, see Ladich (1999, 2000) and Wysocki and Ladich (2001).

3.2.1.2 Signal Analysis

Behavioural experiments have demonstrated a considerable interspecific variability in the range of detectable frequencies (bandwidth) and sound pressure levels. For example, the most widely studied teleost species, the goldfish (*Carassius auratus*) can detect sounds from below 50 Hz to about 300 Hz with best sensitivity (-40 to 50 dB *re* : 1 μ bar) at 400 to 1000 Hz. In contrast, another widely studied species, the cod (*Gadus morhua*) can detect sounds from below 50 Hz to only about 500 Hz with best sensitivity (-20 dB *re* : 1 μ bar) at 60 to 300 Hz.

The differences in bandwidth and sensitivity between these two species illustrate a major principle of fish audition. Fishes, such as the goldfish, have auxiliary structures to improve acoustic coupling between the swim bladder and inner ear (the Weberian ossicles), often have a wider bandwidth and better sensitivity than fishes without such coupling. Fishes lacking a swim bladder (such as some flatfish), or fish in which the swim bladder does not function as a pressure detector to subserve the "indirect pathway" (as some cichlids including *Tilapia*), hear even more poorly than the cod.

In addition to detecting pure tones, behavioural experiments have shown that many fish species can perform frequency and intensity discrimination, and detect the pressure of a signal when masked by background noise. For example, the goldfish can discriminate small differences in pulse repetition rate and detect amplitude modulation with extremely small modulation indices.

For further information on hearing capabilities, the reader is referred to Fay (1992, 1994), Schellart (1992) and Schellart and Wubbels (1998).

3.2.1.3 Sound Localization

Considerable historical controversy prevails as to whether fish can determine the direction of a sound source. It was correctly argued that the mechanism(s) used by terrestrial vertebrates for doing sound source localization could not work with fishes, because the greater speed of sound in water, coupled with proximity of ears to the single swim bladder, would not present interaural differences in signals nearly as large as those used by terrestrial vertebrates. More recent studies have begun to address the mechanism(s) employed by fishes for sound source localization and found that, while some fishes can indeed localize sounds, the mechanisms employed are quite unique among vertebrates (see further Chapter 16).

3.2.1.4 Ear and Postures

Control of a variety of postures, at least, can occur without much change in the structural organization of pars superior receptors in teleosts. As a flatfish, the flounder (*Platichthys stellatus*) is unusual in lying on one side as an adult, functionally rotating its vestibular organs by 90° relative to gravity. The upside-down catfish (*Synodontis nigriventris*) is unusual in swimming belly upwards in open water, functionally rotating its vestibular organs 180° relative to gravity.

Gravistic responses in fishes depend on the magnitude of the lateral force applied to the otoliths rather than on pressure against the macula.

In the end organs chiefly utilized for postural and locomotor coordination, the semicircular canals and the utricle, the orientation patterns of hair cell are remarkably consistent for a wide range of fishes representing a variety of swimming behaviours. These patterns seen in teleosts appear common throughout the tetrapods as well.

A few unusual postural behaviours in fishes can be explained by peripheral changes in the vestibular system, but most are thought to result from central functional controls. Both visual and tactile components may contribute to postural regulation; conversely, vestibular input is a major factor in the control of eye movements. The phenomenon of compensation to vestibular lesions is one example of functional plasticity under central control. Demonstration of direct efferent regulation of peripheral gain or bias, although elusive, provides yet another example of central control.

3.2.1.5 Audition and the Central Nervous System

Anamniotic vertebrates are characterized by at least three distinct patterns of medullar octaval organization: agnathan, gnathostome fish, and amphibian patterns. Lampreys possess a single ventral octaval nucleus, within which three distinct aggregates of large octavomotor cells are recognized. The octavomotor neurons form vestibulospinal and vestibulo-oculomotor connections. Further information is awaited in lamprey. Information is further lacking whether this pattern is also characteristic of hagfishes. The ventral octaval nucleus of lampreys probably represents the primitive condition for vertebrates and is probably a field homologue of all the octaval nuclei in other vertebrates.

All gnathostome fishes examined possess at least four primary octaval nuclei: anterior, magnocellular, descending and posterior octaval nuclei. Preliminary data suggest that the anterior octaval is likely an auditory nucleus and that it projects bilaterally to a midbrain auditory centre, the torus semicircularis; the magnocellular and, at least, the rostral part of the descending octaval nuclei form descending vestibulospinal pathways. Teleost fishes possess a fifth octaval nucleus, the tangential nucleus, that has likely evolved from the magnocellular nucleus in other fishes; like the magnocellular nucleus, it projects to the spinal cord. The torus semicircularis forms bilateral ascending projections to the diencephalon, and telencephalic auditory areas are established in some sharks and teleosts, but the exact location and connections of these telencephalic centres must still be determined.

All available evidence strongly suggests that fishes are characterized by most of the same auditory pathways and centres seen in land vertebrates (for details and comparison with amphibians, see Northcutt 1981; for cytoarchitectural analysis of octavolateralis area of the goldfish, *Carassius auratus*, see McCormick and Bradford 1994). Related study in skate, *Raja eglanteria* (Barry 1987a) and in the otophysan *Ictalurus punctatus* (McCormick and Bradford 1993) deserve special reading (Figs. 15.21 and 15.22).

McCormick (1982) discussed the organization of the octavolateralis area in actinopterygian fishes. The octavolateralis area can be divided into a dorsal lateralis area composed of first-order

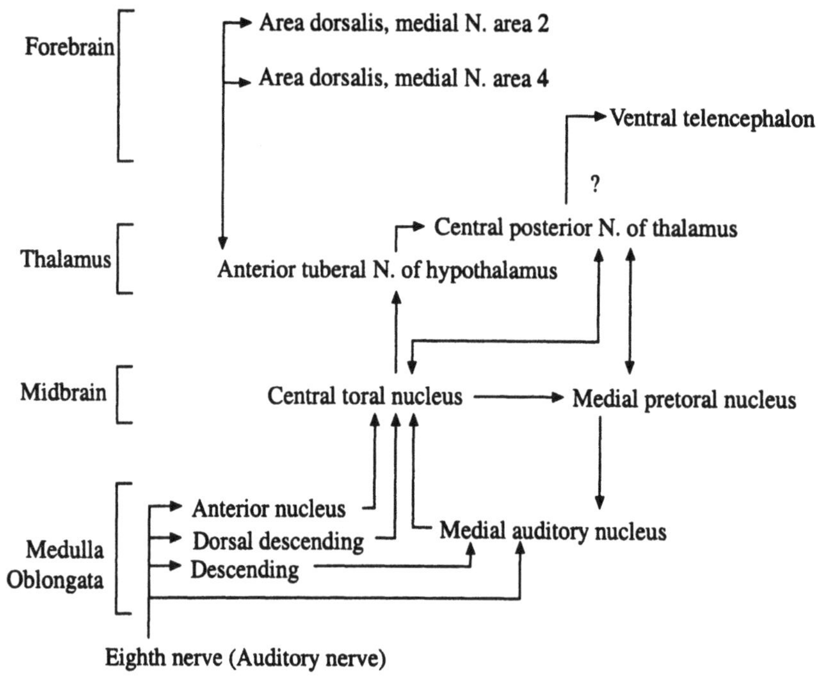

Fig. 15.21 Schematic diagram of the auditory portion of the brain of the channel catfish (*Ictalurus punctatus*). At present, we assume that this is representative of the Otophysi and perhaps, of most fish species as well.
***Source:* Popper and Fay (1993)**

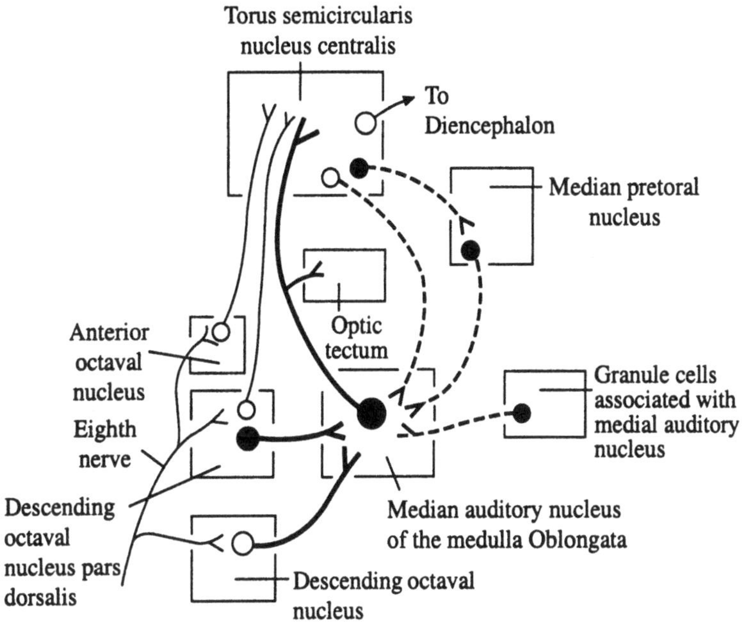

Fig. 15.22 Summary diagram of ascending (solid line) and descending (broken line) auditory/acoustic connections. Note that the bulk of ascending input (heavy solid line) is relayed to midbrain (torus semicircularis nucleus centralis) via a second-order brainstem nucleus (median auditory nucleus of the medulla). ***Source:*** **Finger and Tong (1984)**

lateral line nuclei, and a ventral octavus area composed of nuclei receiving first-order input from the eighth nerve. Three patterns of organization of the lateralis are recognized. The organization of this area in polypteriforms and chondrosteans is similar to that in chondrichthyans. The lateralis area of holosteans and nonelectroreceptive teleosts exhibits a second organizational pattern that is hypothesized to reflect the loss of the electroreceptive portion of the lateral line system. Each group of electroreceptive teleosts is believed to have evolved electroreception independently, a situation that is reflected centrally by a third organizational pattern within the lateralis area, which is distinctly different from that of early radiations of electroreceptive fishes.

The octavus area of actinopterygians exhibits two patterns of organization-that of polypteryforms, chondrosteans, and holosteans, and that of teleosts. The functional significance of these patterns has yet to be elucidated.

Finger and Tong (1984) discussed the central organization of eighth nerve and mechanosensory lateral line systems in the brainstem of ictalurid catfish. Auditory signals reach the midbrain via an entirely separate route. The octaval nerve terminates in the column of five medullary nuclei. Of these, only the anterior and descending octaval nuclei maintain a direct but sparse projection to the auditory nucleus of the midbrain. The bulk of the auditory input to the midbrain involves a newly described medullary nucleus, the median auditory nucleus of the medulla. This nucleus receives input from the descending octaval nucleus and projects bilaterally to the auditory nucleus of the torus semicircularis. It is suggested that the medial auditory nucleus of the medulla is homologous to portions of the superior olivary complex of other vertebrates (see Fig. 15.23).

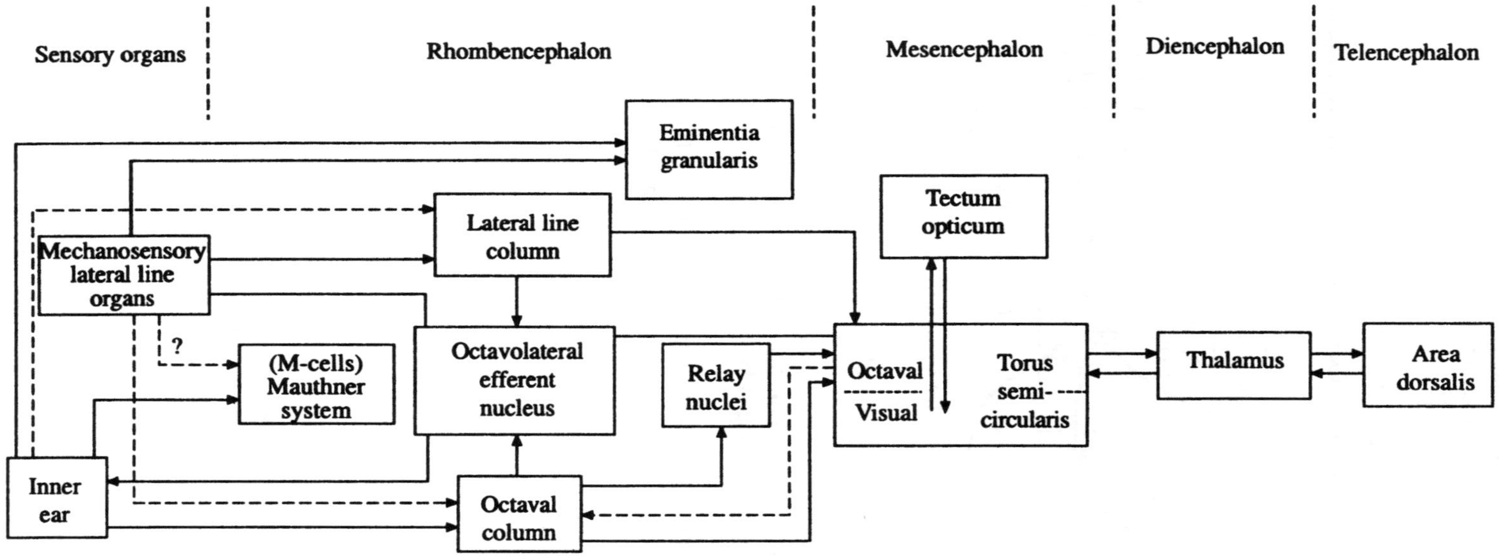

Fig. 15.23 Diagram of the auditory and lateral line pathway. The figure only shows the principal auditory and lateral line structures. The major afferent and efferent connections are indicated by solid lines, minor connections by dashed lines.
Source: Schellart and Wubbels (1998)

Details of the central auditory pathways in anamniotes are given by McCormick (1992; for clupeid *Dorosoma cepedianum*, see McCormick (1997). For neuronal encoding of sound direction in the auditory midbrain in rainbow trout (*Oncorhynchus mykiss*, see Wubbels and Schellart (1997). For an overview, read Wullimann (1998).

3.2.1.6 Auditory Alarm Signals

There are a number of candidates for auditory alarm signalling in fishes. Many of the reported cases of sound production in fishes are sounds produced in response to being held, prodded or hooked. There are reports of fish producing sounds specifically in response to the presence of a predator. Longspine squirrelfish (*Holocentrus rufus*) produce both alarm and distress calls.

At least one acoustic alarm signal system, the staccato call of the squirrelfish, meets the basic criteria, although it deserves more detailed analysis. The gadids and serranids also appear likely candidates. There are many examples of sounds given by fish when they are captured-distress calls. Their functional significance is ripe for investigation (see Smith 1992).

4. MECHANOSENSORY LATERAL LINE

4.1 Introduction

Hair-cell sensory systems are present in all extant vertebrates. In terrestrial vertebrates, hair-cell transducers are utilized by two sensory modalities of the inner ear—the auditory system and the vestibular system. In fish, too, hair-cell receptors for the vestibular and auditory senses are located in the inner ear and are innervated by different branches of the eighth cranial nerve. But fishes possess an additional hair-cell modality, the lateral line system, which is innervated by two cranial nerves unique to aquatic vertebrates—the anterior and posterior lateral line nerves. Despite these differences in innervation, the hair-cell receptors of the lateral line system are quite similar to hair-cell receptors of the otic end organs.

Comprehensive accounts of the lateral line system are available in "The Mechanosensory Lateral Line: Neurobiology and Evolution" (edited by S. Coombs P. Görner and H. Münz 1989). Recent contributions are by Akoev and Andrianov (1989) Webb (1989a, b, c, 1990), Coombs and Janssen (1990), Vischer (1990), Schellart (1992), Popper and Platt (1993), Coombs and Montgomery (1994, 1999), Montgomery et al. (1994, 1995), Coombs (1995), Schellart and Wubbels (1998), Bretschneider et al. (2001) and Mogdans and Bleckmann (2001). The reader is also referred to Zaccone et al. (2001).

The sensory lateral line system in gnathostomes arises from a series of ectodermal placodes and is a primitive vertebrate sensory system. It is present in all fishes and in most larval and aquatic adult amphibians, but it is absent in amniotic vertebrates. In fishes, the lateral line system is primitively composed of three subsystems based on at least three receptor types; mechanoreceptive neuromasts, mechanoreceptive spiracular organs and electroreceptive ampullary organs. Although electroreceptor and spiracular organs are limited in their distribution, the primitive neuromast-based mechanoreceptive lateral line system is present in all fishes.

4.2 Lateral Line System in Different Fish

4.2.1 Basic Morphology and Development

Figure 15.24 is the schematic illustration of the lateral line sense organs of several fishes.

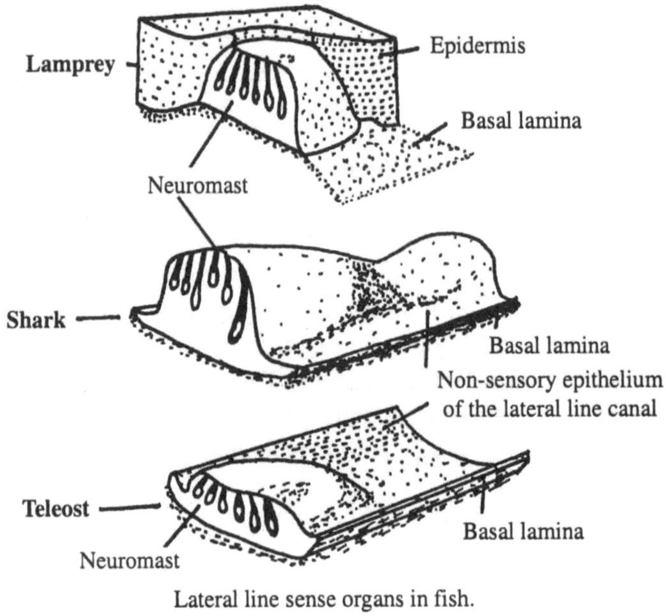

Fig. 15.24 Schematic drawing of the lateral line sense organs of several fish. The free neuromast of the lamprey is embedded in the epidermis. The neuromasts of the lateral line canal organ of the shark are elongate and almost continuous whereas those of teleost fishes are apparently isolated from neighbouring organs. *Source:* Hama and Yamada (1977)

The distribution of neuromasts and their innervation has been examined in lampreys *Lampetra* and *Petromyzon fluviatilis*. In these taxa the neuromasts are located in canals but occur as free neuromasts arranged as various lines on the skin surface. Lateral line canals in Pacific hagfish *Eptatretus stouti* reveal no true canal organization (cells in open grooves are modified mucous cells). They do not contain receptor cells or any innervation, so they probably represent a degenerate system (Fernholm 1985; Northcutt 1985, 1989; see also Jorgensen 1989, Braun and Northcutt 1997, 1998).

Roberts (1978) published a comprehensive description of the mechanoreceptors and behaviour of elasmobranchs with special reference to octavolateralis system. The lateral canal, supraorbital canal, infraorbital canal and hyomandibular canal have been distinguished. Variations are found, for example, the arrangement in rays differs considerably from that of sharks. In other species, like *Pristiurus* with its elongate snout or *Sphyrna* with its bizzare head, the normal pattern is of course distorted, while in bottom-dwelling species like *Raja* the ventral canals have few openings or are absent (*Torpedo*); these canals are well formed and have extensively dichotomizing tubules in open-ocean rays such as *Myliobatis*. Contribution by Boord and Campbell (1977) is specific for sharks. Pit organs or free neuromasts are situated between pairs of modified scales. The distribution of pit organs varies with the species, but they occur along the dorsolateral surface of the body dorsal to the lateral line or, in some species, both dorsal and ventral to the lateral line. Pit organs of the head are confined, in *Squalus acanthias* and *Mustelus canis*, to a pair in front of each endolymphatic pore and a mandibular row back of the lower jaw. For structural and functional features, refer Table 15.5 (Fig. 15.25).

Table 15.5 Some structural and functional features of the lateral line system of sharks

Sense organ	ampullae of Lorenzini	neuromasts
Distribution	head	head, trunk, tail
Receptor	modified hair cell (cilium; no stereocilia)	hair cell (kinocilium; stereocilia)
Innervation	anterior lateral line nerve (dorsal root)	anterior lateral line nerve (ventral root); posterior lateral line; nerve IX
Peripheral termination	afferent	afferent and efferent
Central terminal	anterior lateral line lobe	posterior lateral line lobe, eminentia granularis, and auricle of cerebellum
Function	passive electroreception	mechanoreception
Stimulus	DC and low frequency electric fields	water movements (displacements)
Role	electrolocation	orientation; coordination of swimming movements

Source: Boord and Campbell (1977)

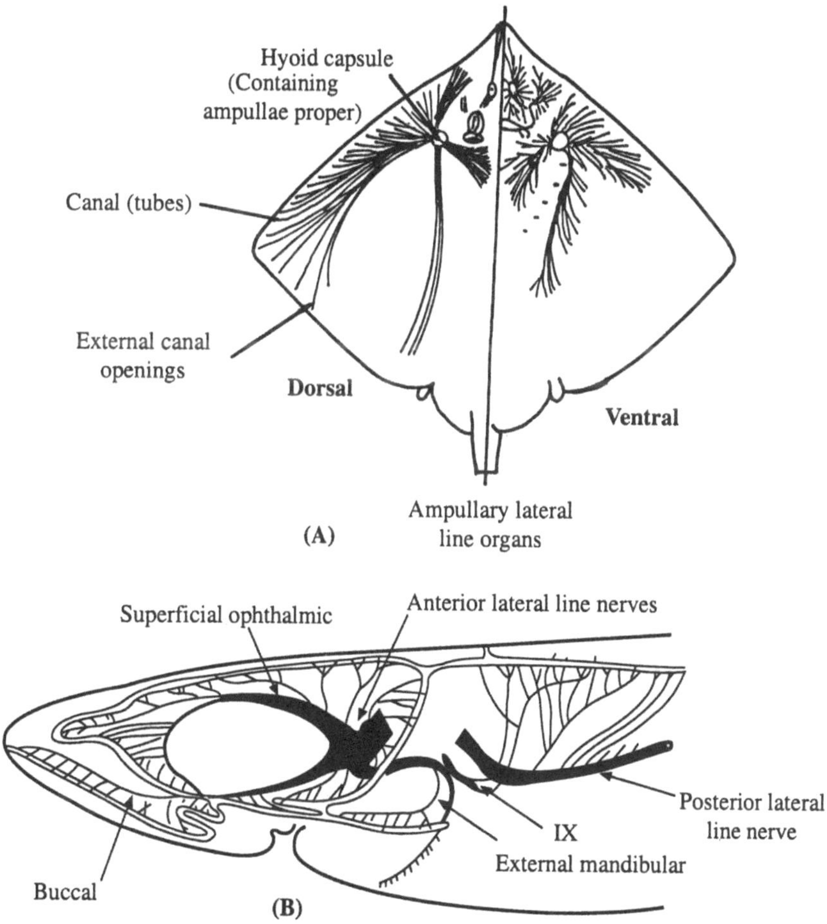

Fig. 15.25 (A) Dorsal and ventral views of the canal system in *Raja Clavata.* (B) Innervation of the head canals of *Squalus*.
***Sources:* (A) Dijkgraaf (1963) and Murray (1974); (B) Roberts (1978)**

Teleosts have received the maximum attention and the details of elasmobranch lateral line system are not therefore covered here. Descriptions of the lateral line system morphology can be found throughout the systematic ichthyological literature but, in this context, morphological variation is most often described without reference to function.

Mechanoreceptive lateral line system is typically composed of a series of neuromasts in pored canals contained in the dermal bones of the head. There are also neuromasts in one or more canals on the body which are generally contained in scales which cover the trunk. The neuromast consist of a hair cells epithelium surrounded by supporting cells (Figs. 15.26, 15.27).

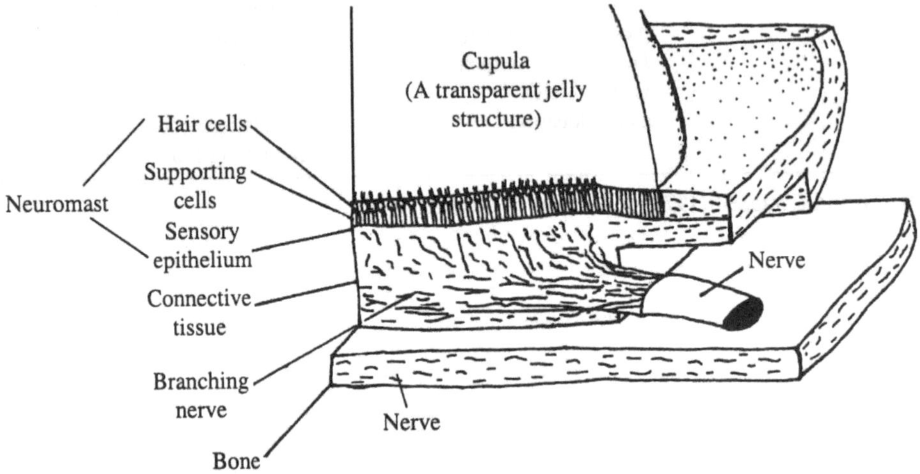

Fig. 15.26 Schematic drawing of a cross section of a neuromast in the lateral line canal of the ruff (*Acerina cernua*). Vascularization is not shown.
***Sources:* Kroese and Netten (1987) and Netten and Kroese (1989)**

In addition, superficial neuromasts ("pit organs"), free neuromasts and sensory hillocks of various authors are generally found on the epithelium of head, trunk and tail (Fig. 15.28, 15.29). In *Esox americanus* and *E. niger* (Esocidae), two types of surface pits, sensory and nonsensory, are reported. Both types of pits are morphologically very similar and are associated with modified (cardioid) scales. The pits are recognized by the presence and absence respectively of neuromasts. The nonsensory pits have central cells; the function and significance is unknown. The nonsensory pit and central cells represent an aplastic condition of the lateral line system. Modified scales and patterns of pit organ distribution have been well described in elasmobranchs.

Teleosts typically have one pair of trunk canal and seven pored, paired and differently innervated canals on the head, the supraorbital, infraorbital, preopercular and mandibular canals, the contiguous otic and postotic canal and the temporal portion of the main canal. There are also one or more commissures (e.g., rostral, supratemporal or occipital) or opposing canal tubules (e.g., the coronal pore in cichlids, which, if complete, link the canals of the two sides of the head) (Fig. 15.30).

On the basis of the relative development of the canals and the orientation of the neuromasts receptors contained within them, four major head canal types have been identified; simple (= narrow); branched; widened, and reduced. The branched, widened and reduced head canal types have arisen convergently a number of times in the teleosts, and adaptive explanations have been suggested for the observed morphological variation based on correlations with ecological

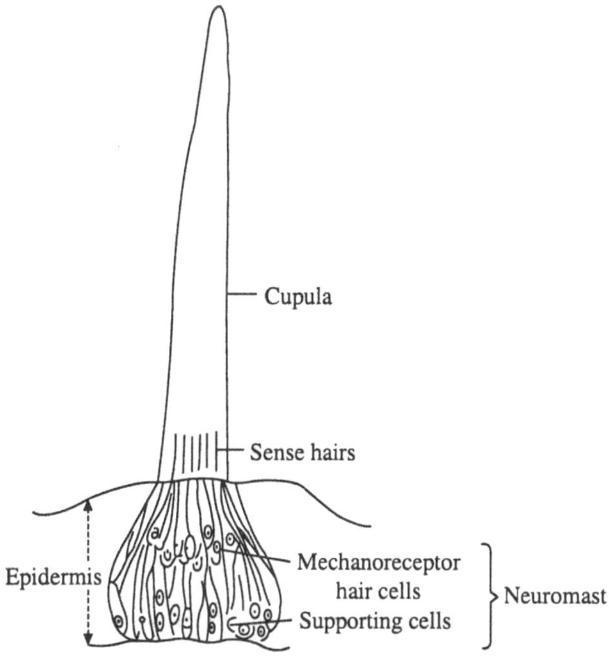

Fig. 15.27 Schematic diagram of one neuromast and cupula, modified to show the tapered cupula typical of *Menidia* fry. In the fry the neuromast bulge out beyond the epidermis.
***Source:* Cahn and Shaw (1965)**

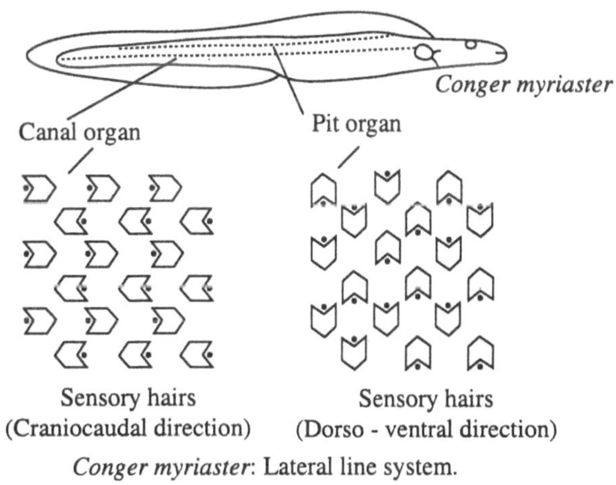

Fig. 15.28 The pig organs of the sea eel, *Conger myriaster* are arranged in a row along the side of body dorsal to the canal organ. The sensory hairs in the pit organs are arranged in a dorso-ventral direction whereas those in the canal organ are arranged in a craniocaudal direction.
***Source:* Hama (1978)**

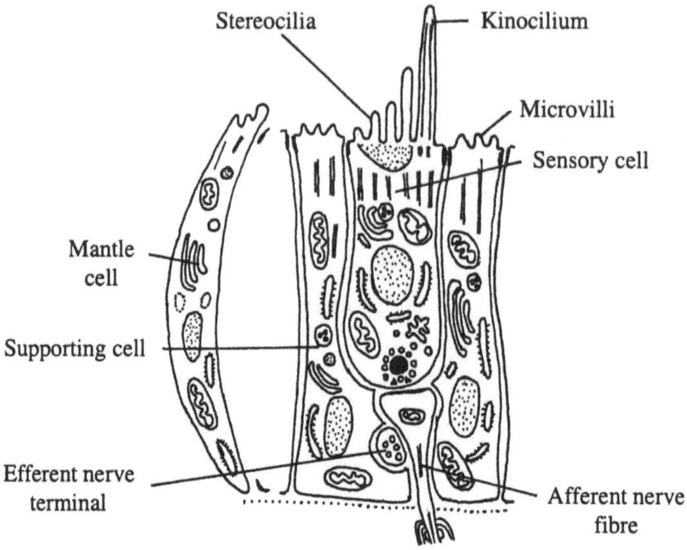

Conger myriaster: Fine structure of pit organ.

Fig. 15.29 Scheme of a vertical section through the sensory epithelium of the lateral line pit organ of the sea eel, *Conger myriaster*.
***Source:* Hama (1978)**

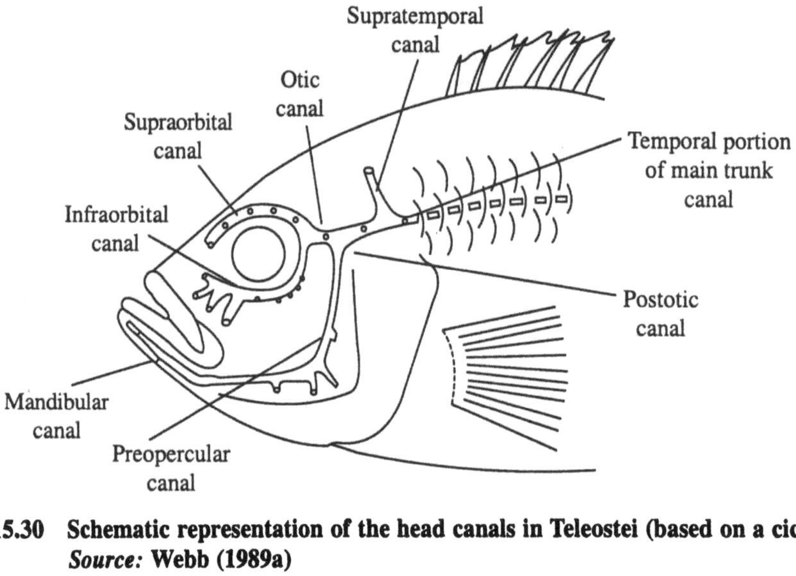

Fig. 15.30 Schematic representation of the head canals in Teleostei (based on a cichlid).
***Source:* Webb (1989a)**

habitat including hydrodynamic environment (schooling fishes tend to have branched head canals, deep-water fishes tend to have widened head canals and benthic fishes tend to have reduced head canals). Data are available to support the hypothesis that these are alternate strategies for changing the signal-to-noise processing capabilities of the system by altering the frequency response properties of the head canal system. There are indications that the trunk canal may have a hydrodynamic monitoring function for the detection of water flow patterns in actively swimming fishes, but in benthic or mesopelagic and bathypelagic fishes its function appears to shift, and morphological specializations like those of the head canal system are common.

Besides, certain interesting observations are on record. Although superficial neuromasts occur in almost all fishes and have been documented in many teleosts, extensive superficial neuromast proliferation, referred to as a 'mixed system', has been specifically documented in a number of species. Such systems are often reported in taxa from light-limited habitats such as the amblyopsid and characid blind cave fishes (Fig. 15.31) and fishes of benthopelagic assemblage, including the eyeless ophidiodid *Typhliasina pearsi* and the bathypteroid *Ipnops murrayi*. The extensive proliferation of superficial neuromasts in the goldfish, *Carassius auratus*, for example, cannot be explained using this line of reasoning.

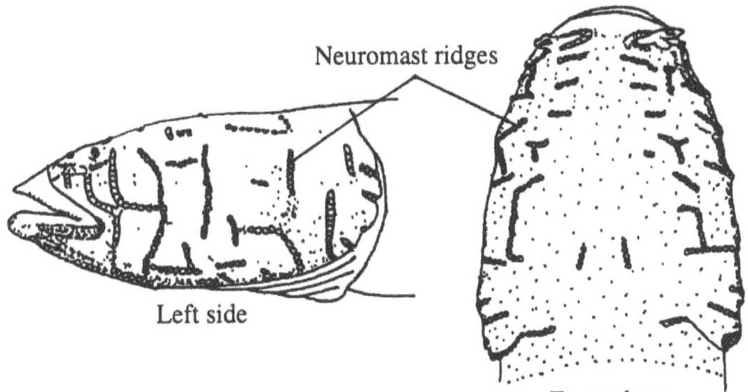

Fig. 15.31 Head of blind cave fish (*Typhlichthys osborni*), seen from left side and from above, with neuromast ridges. Note perpendicular arrangement of rows. *Source:* Dijkgraaf (1963)

The head canal morphology has been reviewed in four major teleost clades (Osteoglossomorpha, Elopomorpha, Clupeomorpha and Euteleostei). It has been suggested that a branched tubule system associated with the head canals is a primitive teleost character, while widened and reduced head canal patterns are convergent functional specialization in many taxa.

Eight trunk canal patterns are described in teleosts: (1) Complete (straight), (2) Complete (arched), (3) Complete (dorsal displacement), (4) Complete (ventral displacement), (5) Disjunct, (6) Incomplete, (7) Multiple, and (8) Absent. An examination of the diversity of trunk canal patterns reveals a nonrandom systematic distribution of trunk canal patterns within teleost orders and families, and an increase in pattern diversity coincident with the diversification of body form, locomotory fin use patterns and ecology in the "higher teleosts".

The nonrandom systematic distribution of these morphological types, the presence of 'morphological gaps', and the hypothesized relationships between neuromast ontogeny and bone ossification indicate the importance of phylogenetic and developmental constraints on the evolution of the system. Patterns of lateral line ontogeny and the systematic distribution of the morphological types of the lateral lines defined here suggest that heterochrony is a major mechanism of morphological diversification, but this has yet to be specifically tested (Webb 1989a, ; see also Coombs et al. 1988).

It appears that the trunk canal pattern placement, and length are largely a product of the combined limitations imposed by myomere geometry and body shape. There are, however, aspects of trunk canal morphology for which the relationship to myomere geometry is unknown. The relationship between the position of the trunk canal in teleosts with arched, multiple, dorsally displaced, ventrally displaced, zigzag (in some tetraodontiforms) and decurved (in some cyprinids and characoids) trunk canal patterns and their respective myomere configurations need to be examined. In addition, it has been suggested that a functional adaptation in response to a change in the hydrodynamic environment on the fish surface due to shift of the pectoral fin on the flank, of the diversification of fin use patterns in swimming, or respiratory flow emanating from operculum may be responsible for the course of trunk canal. The extent to which these potential functional demands have modified the placement of trunk canal is not known (Fig. 15.32). The hair cell orientation is fixed and does not change during neuromast ontogeny in fishes. Canal neuromasts have physiological polarizations parallel to the axis of the canal in which they lie. Superficial neuromasts accessory to canals tend to have physiological polarizations either parallel or perpendicular to the long axis of the animal. Distinct changes in neuromast topography take place during neuromast ontogeny and lateral line canal formation. Interspecific and intraspecific variation in neuromasts in adult fishes may represent various stages in neuromast development. The shape of the sensory strip (distribution of hair cells) and cupula vary ontogenetically and interspecifically depending on the type of lateral line canal pattern possessed by the fish. Superficial neuromasts form lines or stitches on the head and trunk, but their ontogeny has not been studied in detail. The order of formation of the trunk canal appears to be more conservative (Webb 1989).

Further study of the comparative morphology and evolution of the lateral line system in the Labridae has been made to test hypotheses of functional adaptation and morphological constraints by Webb (1990). The reader is further referred to Vischer (1990). Vischer (1989) reported that

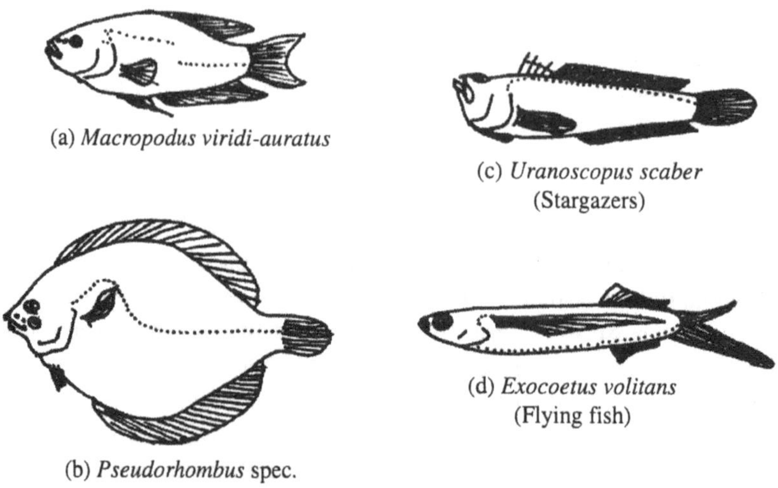

(a) *Macropodus viridi-auratus*

(c) *Uranoscopus scaber* (Stargazers)

(b) *Pseudorhombus* spec.

(d) *Exocoetus volitans* (Flying fish)

Lateral line in fish.

Fig. 15.32 Secondary displacement of lateral line on trunk in connection with dorsal shift of pectoral fin: (a) *Macropodus viride-auratus*; (b) *Pseudorhombus spec*; or as an adaptation to special modes of living, (c) *Uranoscopus scaber*; (d) *Exocoetus volitans*. *Source:* Dijkgraaf (1963)

the mechanosensory system develops before the electrosensory system in the South American weakly electric fish of the genus *Eigenmannia*.

The neuromasts consist of sensory, supporting and basal cells, surrounded by mantle calls. The neuromasts are covered by a gelatinous cupula (probably not in lampreys), which reaches out into the surrounding water or in the canal fluid. The receptor cells occupy the apical surface, but the basal cells are found only in the lower part of the epithelium. The receptor cells are the secondary sensory cells, which usually form synaptic contacts with nerve endings. An individual hair cell itself lacks dendrites or an axon but is connected to the central nervous system by an afferent (sensory) and efferent (inhibitory) nerve fibre.

The neuromast sensory cells correspond to those of the inner ear which are counted as paraneurons but there has been little investigation of this aspect. In free neuromasts of the blind cave fish *Astyanax jordani*, substance P and leu-enkephalin immunoreactivity was found in the cytoplasm of both sensory and accessory cells, but not in the nerve terminal (see Zaccone et al. 1994).

Significant morphological relationship exists between lateral line nerve and electric tissues. In those electric fishes which are purely shocking type (e.g., *Malapterurus* and *Astroscopus*) the lateral line nerve retains the usual teleost pattern having the nerve passing the length of the body situated in the superficial lateral body-wall musculature. In *Astroscopus* the electric organs are far anterior while the body is usually almost completely buried in the sand. In *Malapterurus* the loose connective tissue and the fatty layer intervene between the electric organs and the nerve. In both cases even if the lateral nerve were to be self-stimulated during a discharge it would only be occasionally and of little significance. In the purely pulsating field-field type of electric fishes (e.g., gymnotids and mormyrids) the discharge of the electric organs would produce a more-or-less continuous condition of self-stimulation for the lateral line nerve. In this situation the lateral line system would become useless unless some sort of special modification were developed. This modification appears to be to "bury" the nerve deeply within the body. In the case of *Electrophorus* the electric activity produced is of both types. Here the lateral line is not only buried deeply within the body but in addition, it is surrounded by a sheath of insulating fatty tissue.

4.2.2 Fine Structure and Function

Figure 15.33 gives details of the ultrastructure of the lateral line canal organ of *Lota vulgaris* (read details: Flock 1965). Each neuromast is composed of hair cells which are physiologically polarized as a result of the spatial relationship of the single kinocilium and the group of stereocilia (better called stereovilli) which project from the cell surface of each hair cell. In each neuromast the hair cells are oriented at an angle of 180° to one another, thereby determining the single axis of best physiological sensitivity of the neuromast.

Scale formation and lateral line tube formation are independent, and this suggests that tubed lateral line scales are composite structures. It has been suggested that the dermal bones of the head that contain lateral line canals are made up of two components: a scale and a lateral line ossicle (canal). Although the ossification of the lateral line canal is initiated around a neuromast, formation of scales occurs first as a separate process. Both timing and ultrastructure of lateral line scale formation remain to be investigated (Webb 1989c).

Details of sensory transduction in lateral line hair cells have been given by Kroese and van

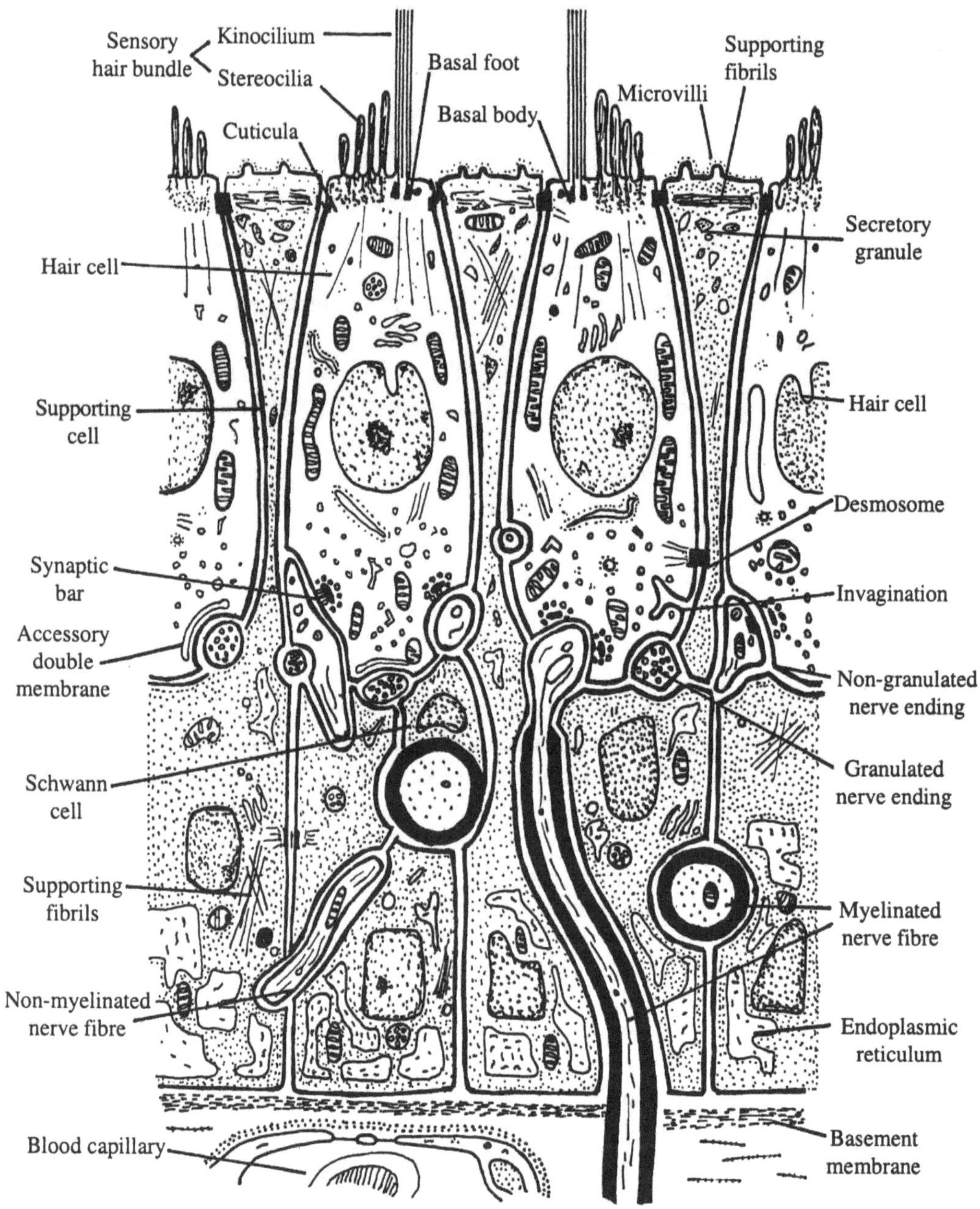

Fig. 15.33 Schematic drawing showing the ultrastructure of the sensory epithelium of the lateral line canal organ of *Lota vulgaris*.
***Sources:* Flock (1965) and Popper (1983)**

Netten (1989) and van Netten and Kroese (1989). Experiments have demonstrated frequency selectivity in the lateral line is mainly controlled by hydrodynamic and cupular mechanics, and not by specific tunic mechanisms in the hair cells. Yet hair cells are in another way strongly involved in frequency selectivity of the organ, since cupular mechanics depends on the hair bundle stiffness coupling of the cupula to the sensory epithelium. In this respect, the lateral line shows a remarkable parallel with the mammalian cochlea, in which hair cells probably make an important contribution to the mechanics of the cochlear partition and thereby to frequency selectivity. Further, the mechanical sensitivity of the lateral line hair cells is comparable to that of cochlear hair cells.

Most likely, hydrodynamic conditions and cupular mechanics play a predominant role in bringing about the response of superficial neuromasts to water velocity and of the canal neuromasts to water acceleration. The degree to which the remarkable difference in the number of hair cells between superficial and canal neuromasts contributes to the different response characteristics of both types of neuromasts remains to be elucidated (Fig. 15.34).

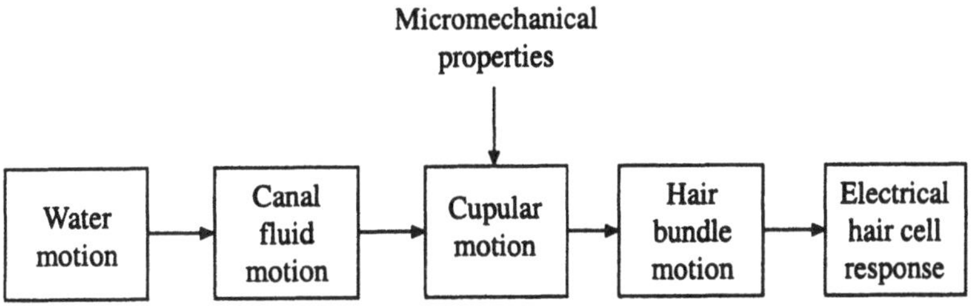

Fig. 15.34 Schematic drawing of the peripheral mechanical part of lateral line transduction. The step from canal fluid to cupular motion and the subsequent transmission to the sensory hair bundles are indicated.
***Source:* Netten and Kroese (1989)**

Regional differences in the architecture and size of lateral line canals and neuromasts were measured in an Antarctic fish, *Trematomus bernacchii*, and the data were used in models of canal and cupular mechanics to predict the frequency of these two peripheral structures. Despite large variations in the peripheral morphology of the head and trunk canals in fish of different sizes, lateral line fibres were relatively homogeneous in their frequency response properties (Coombs and Montgomery 1992; read also Montgomery et al. 1994). In Antarctic notothenioids, some of the lateral line morphological variation seen across different species may very well be explained by pressures favouring the retention of larval pelagicism, rather than any direct pressures on the function of the lateral line (Coombs and Montgomery 1994).

Münz (1989) focussed on the differences between the two types of sense organs, superficial and canal neuromasts, and the implication of these differences for the function of the system. As recognized now, the classification of a neuromast as either a superficial or canal neuromast is usually based on its position. Superficial neuromasts are situated in the epidermis. Their apical structures protrude from the body surface into the surrounding water. Canal neuromasts are completely recessed in the skin or in the bony tissue of the skull or scales. Their apical structures extend into a tubular, fluid-filled chamber which in majority of species communicates with the surrounding water via pores of various shape. Canal neuromasts are formed only in cartilaginous and bony fish. Amphibians do not have canallike structures in their lateral line.

Conclusion has been drawn that the peripheral organization provides the lateral line centres in the brain with information on the hydrodynamic conditions by measuring amplitude, time, and direction of water movements at numerous points on the body surface. Variations in the morphological, physiological and spatial features are presumably adaptations to accomplish this task.

Account by Bodznick (1989) covers a comparative picture of the electrosensory and mechanosensory lateral line systems (for electroreceptors, see Chapter 17). Fundamental distinction between the senses is the absence on all electroreceptors of efferent innervation, which is found in nearly all lateral line neuromasts.

Although the exact relationship between the lateral line and electrosensory systems is uncertain, numerous shared features (Table 15.6), some unique to these systems, are evidence of a close evolutionary link.

Table 15.6 Similarities between lateral line and electrosensory systems

1. Dermal receptor organs distributed over the head (and body) innervated by lateral line nerves.
2. Receptor cells (hair cells) in a support cell epithelium make occluding junctions with support cells and have presynaptic dense bodies and apical microvilli and/or a kinocilium.
3. Receptors and their innervation arise ontogenetically from dorsolateral epidermal placodes.
4. Primary afferents project to alar medullary nuclei with cerebellarlike organization.
5. Medullary nuclei have commissural connections and lemniscal projections to midbrain.
   The receptor cells of electroreceptors and neuromasts are structurally similar (Fig. 15.35).

Source: Bodznick (1989)

It has already been mentioned that lateral line systems and neuromasts are found in virtually all aquatic anamniotes with the apparent exception of hagfishes (read Braun and Northcutt 1997, 1998). Distribution and numerous characters shared by all of the systems indicate that they are homologous and that all have passed down from the same early vertebrate ancestor. The heterostracans, which are the earliest vertebrates to appear in the fossil record, already had well-developed lateral line systems. In contrast, the distribution of electroreception indicates a polyphyletic origin for the sense, once in nonteleosts and two or more times in teleosts. A comparison of the ontogeny of neuromasts and electroreceptors is of interest because a common embryonic origin is evidence for the common evolutionary origin of receptors. The natural stimuli for these two modalities appear to be distinct and overlapping. Perhaps not coincidentally, the central projections of the lateral line and electrosense afferents in the medulla are adjacent but also completely nonoverlapping. The same cannot be said about lateral line and eighth nerve fibres where several instances of overlap of central projections are known that may reflect overlap in their stimulus ranges.

The lateral line system of fishes functions in chemoreception (free neuromasts; salt detection); prey localization; assists in the maintenance of position in a school, and mediates the detection of nearby obstacles through the self-generation of a flow field and then detection of deformations of that flow field (see Katsuki and Onada 1973; Bleckmann et al. 1989; Coombs and Janssen 1990; Webb 1990; Coombs 1999; Coombs and Montgomery 1999).

Kalmijn (1989) elucidated the functional evolution of lateral line and inner ear sensory systems. The two sensory systems respond in principle to acceleration or a combination of acceleration and velocity. The lateral line of free-moving fish detects the spatial differences in imposed flow fields, whereas the inner ear detects the flow field per se. A reevaluation of the earlier data in terms of the proper stimulus parameter confirms the expected low-pass frequency characteristics of the lateral line and inner ear.

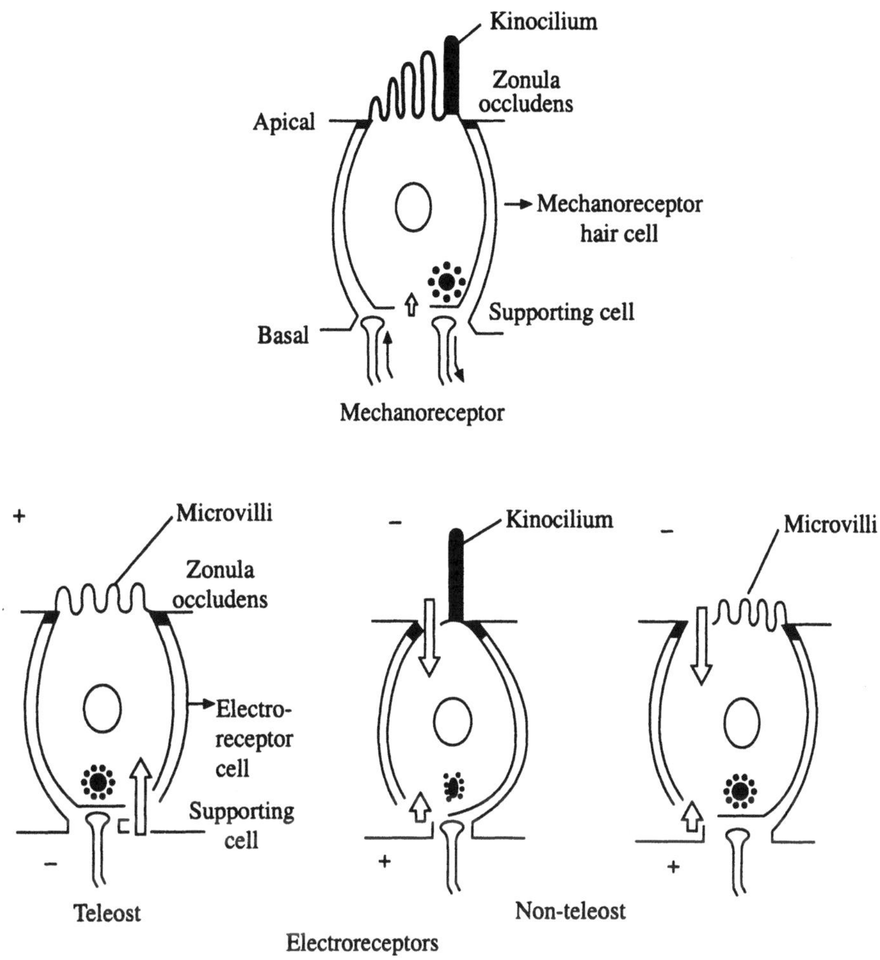

Fig. 15.35 Diagrammatic representation of neuromast mechanoreceptor hair cell and typical teleost and nonteleost electroreceptor cells. Stimulus polarity for excitatory response of electroreceptors is indicated. Main inward Ca currents (demonstrated or presumed) responsible for voltage sensitivity are indicated by large arrows. Although teleost receptors typically have only, apical microvilli and nonteleosts a kinocilium or microvilli, a kinocilium and microvilli occur on electroreceptors of some teleosts and nonteleosts. Only mechanoreceptor hair cell has efferent innervation.
***Source:* Bodznick (1989)**

The lateral line is considered a hydrodynamic detector, attending to the local-flow field in the vicinity of the source. The inner ear is thought to cover the remainder of the local-flow field and, in many fishes to detect the propagating sound wave as well. It is inferred that the acoustic function originated in the near field.

Coombs and Janssen (1990) reviewed the topic of water flow detection by the mechanosensory lateral line. Key factors in the discrimination between hearing and "distance touch" are signal attenuation with distance, signal frequency, and relative motion between the animal and the surrounding water (Table 15.7). Interesting information on lateral line excitation pattern is given by Coombs et al. (1996) and Coombs and Conley (1997).

Table 15.7 A comparison of lateral line and auditory function in fish*

	Lateral line canal	Auditory system
Stimulus	Difference between acceleration of animal and surrounding water Local incompressible flow	Acceleration of whole animal with respect to a single point Local, intermediate, and distant flows, propagated pressure waves
Working distance	Inner region of near field	Near to far field
Single attenuation	$\frac{1}{y^4}, \frac{1}{y^3}$	$\frac{1}{\gamma^3}, \frac{1}{y^2}, \frac{1}{y}$
Frequency range	0–500 Hz	0–2000 Hz

Source: Coombs and Janssen (1990) (*also see details: Coombs and Montgomery 1999; Table 8.3).

The lateral line system is thought to respond primarily to incompressible flow restricted to the innermost regions of the near field of an underwater dipolar source. In contrast, the response of the fish auditory system extends further into the near field and also into the far field. The ear can detect accelerations of the fish body or, in some fishes with specialized auditory systems, fluctuations in ambient pressure. The near field extent of the lateral line response will vary, depending on the source amplitude and frequency. The lateral line system is essentially a low-frequency system.

Although the way in which the lateral line and inner ear respond to hydrodynamic phenomena and the distance range over which they respond may differ, the two systems are surprisingly similar in terms of their sensory capabilities—at least to some extent they can be compared. Absolute sensitivity and frequency discrimination capabilities are quite similar for fish ear and lateral line. The ability to localize with the lateral line, while difficult to compare with auditory localization abilities, appears to be quite accurate and for surface-feeding animals may even depend upon some of the same mechanisms used by auditory system of terrestrial vertebrates, for example, time of arrival differences between receptors.

Denton and Gray (1993) reported that there are major and systematic differences in the patterns of receptor organ stimulation of the acoustico-lateralis system of clupeid fish between those expected from external sources and from a fish's own movements.

The mechanosensory lateral line may serve several different functions both within a single animal and between different species. While the full extent of these functions is yet unknown, the lateral line has been shown to be inolved, as described earlier in school behaviour (e.g., saith, *Pollachius vireus*), feeding behaviour (e.g., *Aplocheilus lineatus*, *Pantodon buchholzi*, mottled sculpin *Cottus bairdi*, and probably the piper *Hyporhamphus ihi*, Antarctic fish *Pagothenia borchgrevinki*; for special interest, see Montgomery and Macdonald 1987; Macdonald et al. 1988; Montgomery et al. 1988; Montgomery 1989; Macdonald and Montgomery 1991) and hydrodynamic "imaging" of the immediate surroundings. Not surprisingly the lateral line system seems to be well adapted for animals that must function in total darkness or very low light levels, such as the naturally blind cavefish (e.g., *Anoptichthys jordani*) or the nocturnally feeding mottled sculpin. However, both the lateral line and the visual system seem to be important in schooling behaviour and bluegills (*Lepomis macrochirus*), which normally rely on vision for day time feeding, are capable of using the lateral line for detecting and capturing prey when vision is prevented. Thus, it is apparent that the lateral line system, while it may have evolved specializations in dark-adapted animals, can and does have normal functions in animals for which vision is not severely restricted.

Schellart (1992) reported that the index of the lateral line canal system, LI, is mainly based on anatomic characteristics. Species with superficial neuromasts generally have low LIs and appear to prefer quiet waters. Well-developed canal lateral line organs seldom go together with very good hearing, but the high hearing index, HI, is often accompanied by large numbers of superficial neuromasts.

4.2.3 Lateral Line and Brain

Roberts (1978) described the innervation of lateral line system of elasmobranchs. Fig. 15.25 shows the innervation of head canals of *Squalus*. The region of the brain that receives the lateral line nerves is dominated by the cerebellum, which in these fishes consist of three portions: the unpaired corpus, the paired auricles, and paired lateral line lobes (Fig. 15.36). The corpus is mostly concerned with spinal pathways, whereas the other two centres are specifically related to the octavolateralis system.

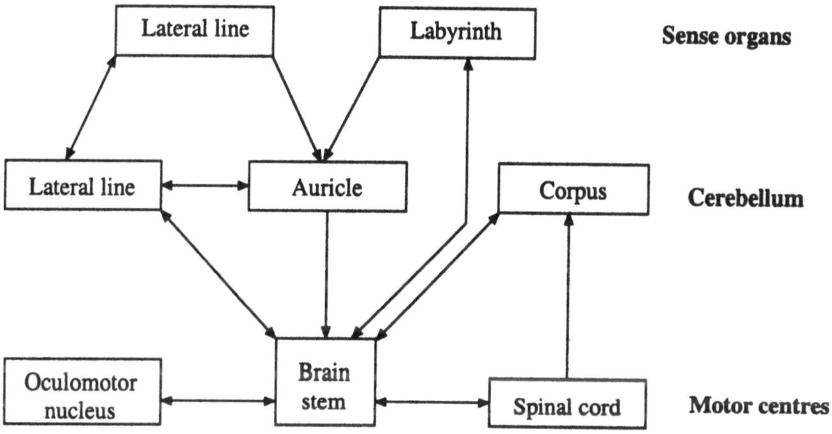

Fig. 15.36 Schematic drawing showing the relationship of the acoustico-lateralis sense organs, the components of the cerebellum, and the motor centres of the selachian brain.
***Source:* Roberts (1978)**

As is revealed in Fig. 15.36 the lateral line lobes, the auricles, and the cerebellar corpus are superimposed on, and have neural circuits that are in parallel, with the hindbrain centres.

Northcutt (1989) reviewed the phylogenetic distribution and innervation of craniate mechanoreceptive lateral lines. It has been emphasized that although numerous behavioural, ecological and physiological factors correlate with various patterns of lateral line distribution and their accessory structures (grooves and canals), the relative roles of different evolutionary constraints (architectural, developmental, historical, and selectional) as possible causative factors are unclear. Thus many, if not most, aspects of the distribution, function, history, innervation, and organization of lateral lines of craniates remain to be examined.

McCormick (1989) reviewed the central lateral line mechanosensory pathways in bony fish. Bony fish have separate lemniscal pathways for electroreception and mechanoreception that extend at least to the level of the diencephalon and probably to the telencephalon. It is also reasonably evident that separate lemniscal pathways for hearing and mechanoreception extend to the level of torus semicircularis, although there are known (e.g., magnocellular octaval nucleus; Mauthner cell) or possible (e.g., superior olive; perilemniscal cells) sites in which information from these modalities converges. The identity of the acoustic and lemniscal pathways beyond the level of the midbrain is not yet agreed upon (see also McCormick 1983 for details of octavolateralis area in fishes; read for further information, Wullimann 1998).

Finger and Tong (1984) contributed on central organization of these modalities (auditory and mechanosensory) in ictalurid catfish. The primary mechanosensory nucleus of the medulla, n. medialis, projects directly to the optic tectum and to the mechanosensory nucleus of the torus semicircularis. Nucleus medialis receives input from primary lateral line fibres as well as from a number of sites within the CNS: n. praeemintialis pars ventralis, and the eminentia granularis and a lobus caudalis of the cerebellum. The n. praeeminentialis itself receives a descending input from the mechanosensory nucleus of the torus semicircularis. These mechanosensory lateral line pathways are parallel to, but distinct from, those of the electrosensory lateral line system (Fig. 15.37). The reader is also referred to Finger and Bullock (1982), Tong (1982), and McCormick (1997).

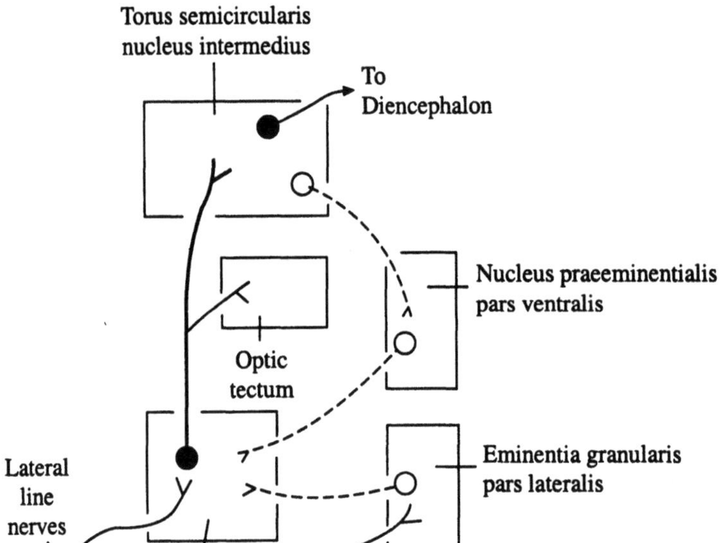

Fig. 15.37 Summary diagram of ascending (solid line) and descending (broken line) connections of the mechanosensory lateral line system. In contrast to the auditory/acoustico system, the ascending input passes directly from the primary sensory nucleus (nucleus medialis) to the midbrain (torus semicircularis nucleus intermedius).
***Source:* Finger and Tong (1984)**

Boord and Montgomery (1989) discussed the substantial differences in the central lateral line organization in primitive versus advanced elasmobranchs and pointed out the need to study more representatives of the various radiations in order to generate a more accurate picture of how central pathways have evolved within this single vertebrate class. The organization of the lateral mesencephalic nucleus (the probable homologue of the torus semicircularis) is a case in point. In the advanced elasmobranch *Raja*, it is well differentiated complex with distinct termination zones for three of the octavolateralis modalities (electrosensory, mechanosensory, and presumed auditory), whereas in two squalomorphs (*Squalus* and *Cephaloscyllium*), it is a poorly differentiated zone in which extensive overlap between the three modalities occurs. Interestingly, the torus semicircularis of osteichthyans is also variable in its degree of differentiation. Whereas the torus semicircularis of electroreceptive teleost is well defined cytoarchitonically, its subdivisions in *Amia* and various nonelectroreceptive teleosts are much less distinct. Unlike elasmobranchs, however, all bony fishes studied thus far appear to have largely separate toral areas for the octavolateralis modalities.

Recent anatomical and physiological studies indicate that the mechanoreceptive lateral line of teleost fishes and aquatic amphibians is represented at all levels of the neuraxis-i.e., from medulla to telencephalon. The same holds true for cartilaginous fishes (see review: Bleckmann and Bullock 1989).

Roberts and Meredith (1989) reviewed the efferent system. An efferent innervation of octavolateralis sense organs, although widespread, is not universally present, nor is it always developed to the same degree. An efferent innervation is completely absent from the inner ear and lateral line of the hagfish *Eptatretus burgeri*, and the lateral line of lampreys.

An obvious view of efferent function is that it provides a feedback inhibitory regulation of hair cell sensitivity; in this circuit (Fig. 15.38) strong stimulation of the hair cells will provoke efferent activity and so reduce hair cell output.

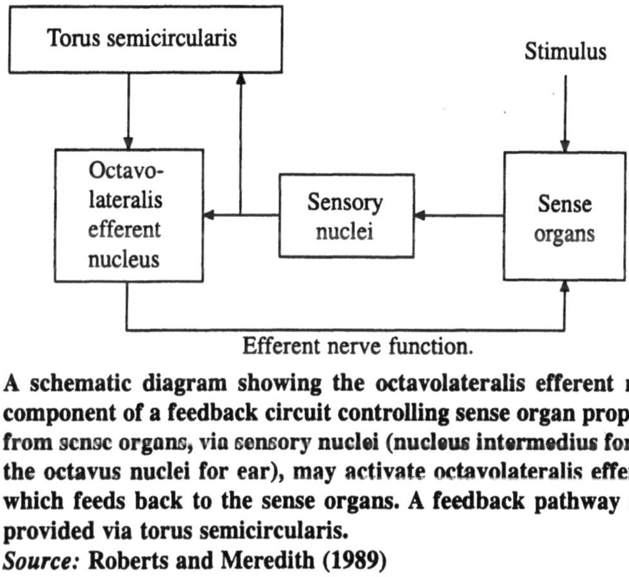

Fig. 15.38 A schematic diagram showing the octavolateralis efferent nucleus as a component of a feedback circuit controlling sense organ properties. Input from sense organs, via sensory nuclei (nucleus intermedius for lateral line, the octavus nuclei for ear), may activate octavolateralis efferent nucleus which feeds back to the sense organs. A feedback pathway may also be provided via torus semicircularis.
***Source:* Roberts and Meredith (1989)**

Stimulation of sensory receptors other than those involved in the ear and lateral line can activate the efferent system, for example tactile and visual stimuli. Although these sensory modalities may directly drive OEN (octavolateralis efferent nucleus) neurons, it is more likely that central circuits, integrating a variety of sensory modalities, are involved, because effective stimuli are known to be "arousing" and to provoke movement.

The efferent neurons can be speculated as a component within some general activating circuit (Fig. 15.39), with the function of providing feed-forward control of sense organ activity. (see also Figs. 15.40, 15.41)

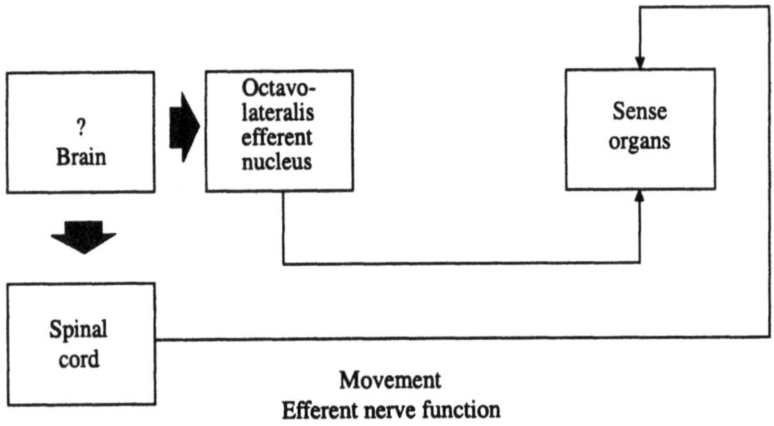

Fig. 15.39 A schematic diagram showing the octavolateralis efferent nucleus as a component of a feed-forward circuit that controls sense organ properties. Postulated, and as yet unknown centres (?) in the brain, involved in arousal and movement control, activate octavolateralis nucleus and locomotor centres. The resulting movement will stimulate the sense organs.
Source: **Roberts and Meredith (1989)**

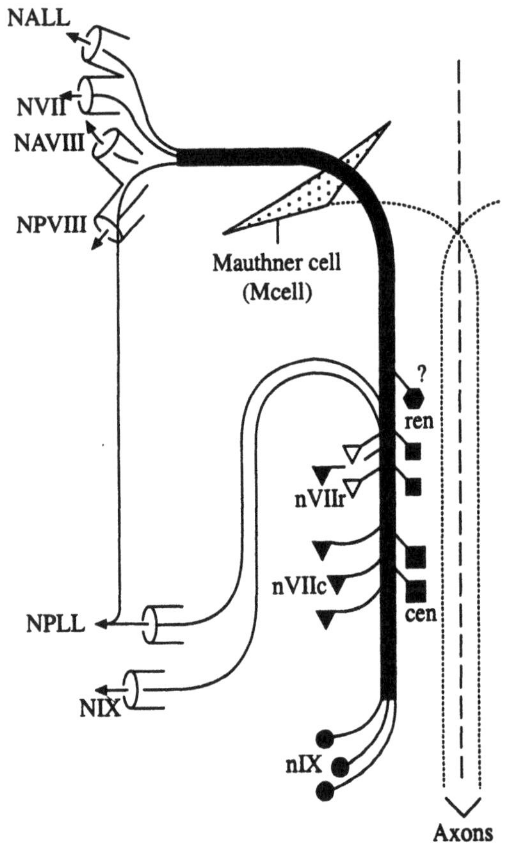

Fig. 15.40 A ChAT-IR pathway in the goldfish central nervous system which includes octavolateralis efferent axons. A horizontal view of the left side of the hindbrain. The midline is represented by the dashed line, the left M (Mauthner) cell by stippling, and both M axons by dotted lines (note how the M axons decussate). The ninth (N IX), posterior lateral line (NPLL), anterior (NA VIII) and posterior (NP VIII) branches of the eighth nerve, and facial (N VII) and anterior lateral line nerve (NALL) roots are on the left side of this schematic diagram. The ChAT-IR pathway in transverse section is located dorsolateral to the film just beneath the fourth ventricle. The ChAT-IT pathway (shown here as a thick, black line) originates caudally (bottom of drawing) with axons of the glossopharyngeal motor neurons. More rostrally are axons of facial motoneurons (both the caudal and rostral divisions), axons of as yet unidentified neurons (?), and axons of octavolateralis efferent neurons join. The ninth motor axons exit near the rostral division of the facial nucleus, as do some of the lateralis efferent axons. The rest of the ChAT-IR axons extend rostrally to the level of the M axon, where they travel laterally toward the eighth cranial nerve root. At this point, the axons diverge and extend to the appropriate cranial nerves.
Source: **Zottoli and Danielson (1989)**

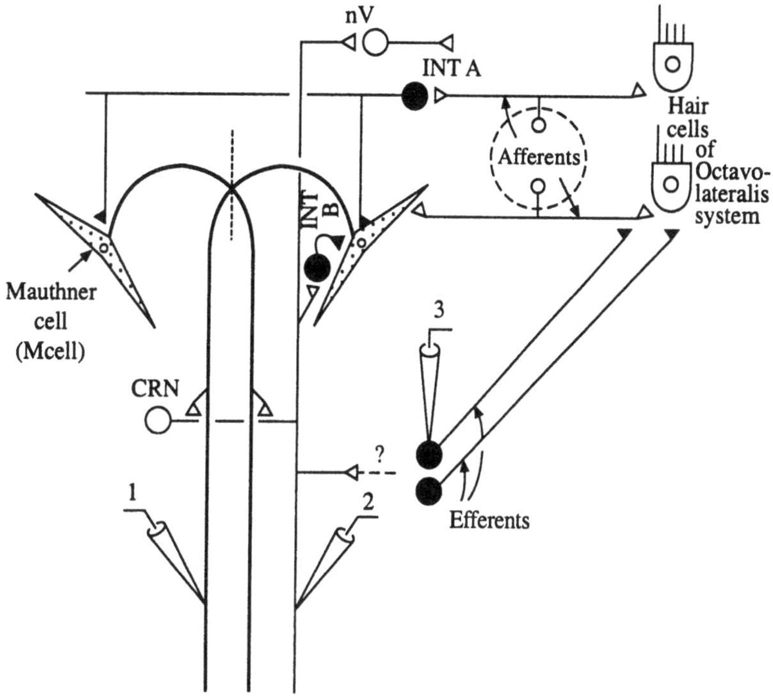

Fig. 15.41 **Schematic diagram of part of the M (Mauthner) cell's feedback inhibitory network and its relation to excitatory input from hair cells of the octavolateralis system. Darkened cell bodies and terminals indicate inhibitory action. Excitation of either M (Mauthner) cell (stippled) results in activation of cranial relay neuron, which excites a number of cells including, at least (1) cranial motoneurons (the fifth motor nucleus, nV): (2) interneurons inhibitory to the M (Mauthner) cell (INT B, PHP neurons, which are responsible for the collateral (IPSP); and (3) efferent neurons. The excitation of efferent neurons presumably contains a di-or polysynaptic pathway, the elements of which are not known (dashed linewith question mark). Efferent neurons are known to inhibit both hair cells and afferent fibres in the goldfish inner ear. Since afferent fibres are known to excite another class of neurons inhibitory to the M cell (INT A, another class of PHPP neurons), it is entirely possible that efferent neurons may synapse on and influence the excitability of hair cells and/or afferent projecting to these interneurons (connections not shown). Electrodes 1, 2, 3 indicate a recording paradigm that might help elucidate the central connectivity of efferent neurons in relation to the M (Mauthner) cell. All the neurons in this diagram except the efferents have been identified and recorded from in the goldfish.**
Source: **Zottoli and Danielson (1989)**

Tricas and Highstein (1991) presented the hypothesis that the octavolateralis efferent system in free-swimming toadfish, *Opsanus tau*, functions in the peripheral processing of information carried by the lateral line in natural settings.

4.2.4 Specialized Lateral Line Receptor Systems: Spiracular Organs and Vesicles of Savi

Spiracular Organs. The spiracular organs occur in most gnathostome fish taxa including elasmobranchs and nonteleost bony fishes. They are not found in *Eptatretus*, *Petromyzon* and *Polypterus*. The presence or absence of a spiracular organ in *Chimaera* and *Latimeria* is uncertain; they occur in lungfishes. The phylogenetic distribution suggests that the presence of a spiracular organ is a primitive character of gnathostomes that has been independently lost in *Polypterus*, amphibians and teleosts, and perhaps in holocephalans and *Latimeria*. A single spiracular organ is associated with the first (spiracular) gill cleft bilaterally. There is considerable interspecific variation in the morphology of the spiracular organ. In all known cases, the spiracular organ is a pouch or tube whose lumen is lined with neuromasts containing typical vertebrate sensory hair cells. A gelatinous cupula completely fills the lumen. In elasmobranchs, the lumen of the spiracular organ opens into a spiracular cleft or diverticulum of the cleft via a pore of variable dimensions, but the pore is apparently of little functional importance. In lungfishes, this opening is lacking. Despite differences in morphology, in both the little skate, *Raja erinacea*, and the smooth dogfish, *Mustelus canis*, (Figs. 15.42, 15,43), the spiracular organ appears to function as a proprioreceptor sensitive to movements about the hyomandibula-cranial joint. The biological role of the spiracular organ in the animal's behaviour is not clear.

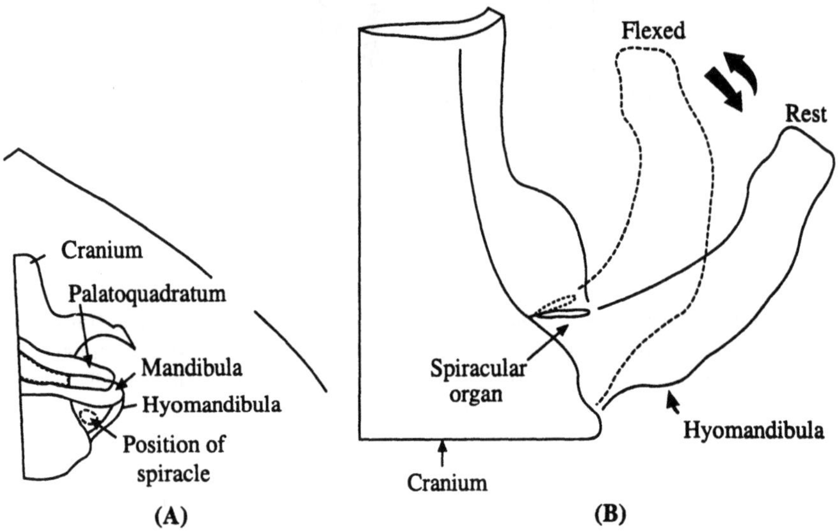

Fig. 15.42 (A) Spiracular organ in little skate *Raja erinacea*. Ventral view of chondrocranium, in relation to the rostral border of the fish. The spiracle is indicated by a dotted line. (B) Ventral view of an isolated preparation consisting of part of cranium, the hyomandibula, and the spiracular organ from the left side. The positions of the hyomandibula and spiracular organ are shown at rest and during flexion.
***Source:* Barry and Bennett (1989)**

Vesicles of Savi (Savi's vesicles). Among elasmobranchs, and in particular among those that are dorsoventrally flattened, the mechanoreceptive lateral line system shows extensive morphological elaboration and interspecific variation on both the dorsal and ventral surfaces. In members of the Torpediniformes, or electric rays, the canal system is lacking ventrally and has been replaced by the vesicles of Savi. Similar lateral line receptors have been described in various species of Myliobatiformes including *Urolophus*. Sac-like vesicles have been described in various dasyatid rays but not in *Urolophus*. Too little is known about these other vesicles to comment about their possible homology to vesicles of Savi in the electric rays. Lateral line canal organs with few or no pores have evolved numerous times among fishes.

Each vesicle of Savi is a closed off pouch approximately 2 to 3 mm in diameter. Vesicles are

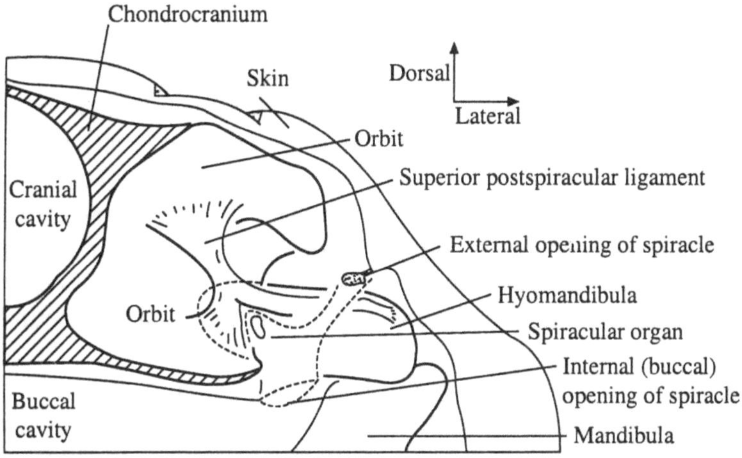

Fig. 15.43 Spiracular organ of *Mustelus canis*.
***Source:* Barry and Bennett (1989)**

located largely on the ventral surface, rostral to the mouth and along the rostral edge of the pectoral fin.

Three neuromasts (one central and two slightly smaller peripheral, whose centres fall on a straight line) and each having own cupula, are found in each vesicle of *Torpedo*. Mechanoreceptive hair cells are polarized bidirectionally in parallel with the line determined by the neuromast centres. The hair cells have both afferent and efferent synapses. Vesicles in *Torpedo* are innervated by the superficial ophthalmic, buccal, and external mandibular division of the anterior lateral line nerve.

The vesicles of Savi might be classified as proprioreceptors, since movements of the underlying cartilages should excite the hair cells. However, their sensitivity to vibrations, as well as their rostral and largely ventral location, is consistent with a role of the vesicles of Savi as exteroreceptors particularly sensitive to substrate-borne vibrations and to deformations resulting from contact of the animal with objects in the substratum. Vibrations and deformations could be transmitted via the skin or the cartilages.

Both the spiracular organ and the vesicles of Savi are mechanoreceptors derived from the lateral line system. Their unique characteristics are due in part to the location and morphology of the organ, but also the response characteristics of the hair cells and afferent fibres. Both the spiracular organ and the vesicles of Savi are effectively isolated from water movements and have specialized attachment to the skeleton. Other specializations may be largely related to differences in vibration sensitivity. The spiracular organ and vesicles of Savi represent extremes of lateral line mechanoreceptor vibration sensitivity.

The spiracular organ is organized for efficient transduction of gross deformations, because the cupula fills the lumen and the receptors are well coupled to the cupula by long kinocilia. The peripheral neuromasts of the vesicles of Savi may function in a similar way.

The spiracular organs and vesicles of Savi exemplify the diversity of the lateral line system in terms of both its morphology and its function (Barry and Bennett 1989; see also Jorgensen 1989; Zaccone et al. 2001).

5. CHEMORECEPTORS (CHEMICAL SENSES)

5.1 Introduction

Sensory receptors are the immediate detectors of environmental stimuli, which bombard the animal continuously. These may be chemical (olfaction, taste, etc.), physical (photoreception, thermoreception, etc.) or mechanical (position, motion, etc.). The behavioural significance of the signals arriving at the brain through the different channels varies greatly from one species to another. For many animal groups vision is most important while in others information about the chemical nature of the external environment is most essential. For fishes whose life is entirely restricted to the aqueous environment, chemoreceptors play a major and sometimes decisive role in behaviour.

The chemical senses in all vertebrates can be separated according to end organ, mode of innervation, and primary terminus within the central nervous system. According to these criteria, three exteroreceptive chemical senses can be distinguished: olfaction, gustation, and the common chemical sense. The olfactory sense is mediated via olfactory receptors, themselves bipolar neurons with axon (olfactory nerve, the cranial nerve 1) extending into the central nervous system and terminating in the olfactory bulb, the rostralmost extension of the forebrain. In contrast, gustation is mediated through specialized axonless neuroepithelial receptors called taste buds. The chemosensory elements of the taste buds synapse on the distal processes of ganglionic cells of either the facial (VII), glossopharyngeal (IX), or vagal (X) cranial nerves, which terminate centrally in the special visceral sensory column of medulla. The common chemical sense, the nonolfactory-nongustatory chemosense, is the least characterized, but appears to be either isolated chemosensory cells innervated by the somatic sensory (trigeminal (V) and spinal nerves or by free endings of the same nerves. In most, and especially aquatic vertebrates, these chemical senses mediate different sorts of behaviour. The olfactory sense is associated with functions such as intraspecific social interactions or being alerted to the presence of food, whereas gustation is used in food localization and determination of palatability. In general, stimulation of the common chemical sense results in avoidance, but in at least some specialized species, such as searobins (*Prionotus carolinus*), this sense is used for food localization.

Receptor cells of both olfactory and gustatory systems must discriminate relevant chemical stimuli from background chemical "noise" that exists in the immediate environment of all vertebrates. This ability of chemoreceptors to act as "peripheral filters" by detecting and preferentially passing information concerning biologically important chemical stimuli occurs through the evolutionary development of different receptor molecules with different chemospecificities built into the membranes of the receptor cells.

The distinction between two sensory modalities, olfactory and gustatory, in fish is not always as clear as in terrestrial vertebrates, mainly because in fish both olfaction and taste are mediated by dilute aqueous solutions. Practically, in terrestrial air breathers, those receptors which have high sensitivity and specificity, which respond to volatile materials, and which are distance chemical receptors are distinguished as olfactory. Those receptors of moderate sensitivity, which are stimulated by dilute solutions, and which are usually associated with feeding are considered as gustatory or contact chemical receptors. However, in fish differences between olfactory and gustatory senses are not as clearcut.

Despite the clearcut separability according to end organ and pattern of innervation, smell and taste are intermingled sensations for air-breathing vertebrates in which the nasal cavity communicates with the oral cavity. Chemical-sensing systems in both tetrapods and fishes are not distinguished by the stimulus vehicle, but rather by the structure and innervation of sensory receptors. This distinction is further reflected in the behaviour mediated by each system.

5.2 Olfactory System

5.2.1 Apertures of Craniate Olfactory Organs

The olfactory organs of craniates appear to derive from an anterior pair of erstwhile branchiothyria and their adjacent neurogenic placodes. The organ is termed *saccus nasalis* if its source is the ectobranchial invagination of the branchiothyrium, and the *saccus rhinalis* if its source is the entobranchial evagination of the branchiothyrium. Nasal sacs occur in gnathostomes and lampreys and rhinal sacs in hagfish. It is difficult, therefore, to avoid the conclusion that the olfactory organ has evolved more than once. The rhinal sac has only one aperture, the postica, which represents an internal branchiothyric orifice. Partial fusion of the right and left sac led to the development of a postica communis. The nasal sac has either one, two or three apertures. Altogether there are nine different apertural sacci nasalis, viz. the naris, the foris, the tremiscus, the two nariculae, the rimilla, the portula, the opiscus and the janua. The janua represents an external branchiothyric orifice and occurs in lampreys. The naris and foris arose by subdivision of the orifice from which the janua emanated; they are found in osteolepipods. The naris is also found in urodeles, porolepiforms and some teleosts. The incurrent and excurrent naricules resulted from bipartition of a naris and are present in actinopterygians, coelacanthiforms, brachiopterygians, dipnoans and elasmobranchiomorphs. Of the four remaining openings of the nasal sacs the rimilla results from an inpushing of the epidermis, the portula from an outpocketing of the nasal sac, and the tremiscus from a combination of two such movements, the one inward and the other outward. Rimillae occur in cyprinodonts, anurans, urodeles, caecilians and porolopiforms, portulae in uranoscopids and bathydraconids, and tremisci in urodeles and porolepiforms. The opiscus, finally, is the opening through which the vomeronasal organ communicates with the oral cavity. It occurs in most lizards and snakes and probably also in some mammals.

Detailing further, living gnathostome fish generally have two pairs of nasal apertures, both of which open to the exterior. Three exceptions to this rule are: (1) a few uranoscopid and bathydraconid teleosts, in addition to two pairs of external nasal aperture, have a third pair of apertures so that each nasal sac communicate with the buccal cavity; (2) some teleosts, such as the sculpins and the sticklebacks, have merely a single nasal aperture on each side of the head, their olfactory organs being alternately filled and emptied by the respiratory movements; (3) holocephalans and dipnoans have one pair of external nasal apertures and another pair of oral nasal apertures.

The oral nasal apertures of living dipnoans have been regarded as choanae by many anatomists. Choanateness has been ascribed also to the holocephalans as well as to the species of the uranoscopid genus *Astroscopus*. Moreover, it has been suggested that the choana is homologous to one of the two external nasal apertures seen in most actinopterygians and chondrichthyans and that the naris is homologous to the other of these apertures. Regretfully there is no general agreement as to any of these apertural homologies.

Like the sculpins and the sticklebacks, living lampreys and hagfish, jointly termed cyclostomes, have merely a single aperture through which water alternately enters and leaves their contiguous olfactory organs. This nasal communication, however, is indirect in cyclostomes, unlike the teleosts just mentioned in which it is direct. The apertures of the olfactory organs of the two cyclostomate groups are generally regarded as homologous structures.

Some teleosts, such as sculpins, the sticklebacks and the needlefish lack a nasal bridge and are therefore nariate. The cyprinodonts, too, are nariate, although each of their nasal sacs presents two external nasal apertures. The explanation for this is that the anterior of these apertures descends directly from the opening of the nasal fovea, whereas the posterior one is a secondary perforation between the back part of the nasal sac and the epidermis. This pseudonaricule is termed the rimilla. In a few nariculate teleosts, among the benthic uranascopids and bathydraconids (Antarctic fish *Gymnodraco acuticeps*), each nasal sac opens into the oral cavity via a diverticular outgrowth. Apparently, this nasal opening is a secondary development as is the rimilla. It is referred as the portula (Figs. 15.44, 15.45).

The nasal sac is either situated well in front of the mouth as is *Squalus acanthias*, or as close

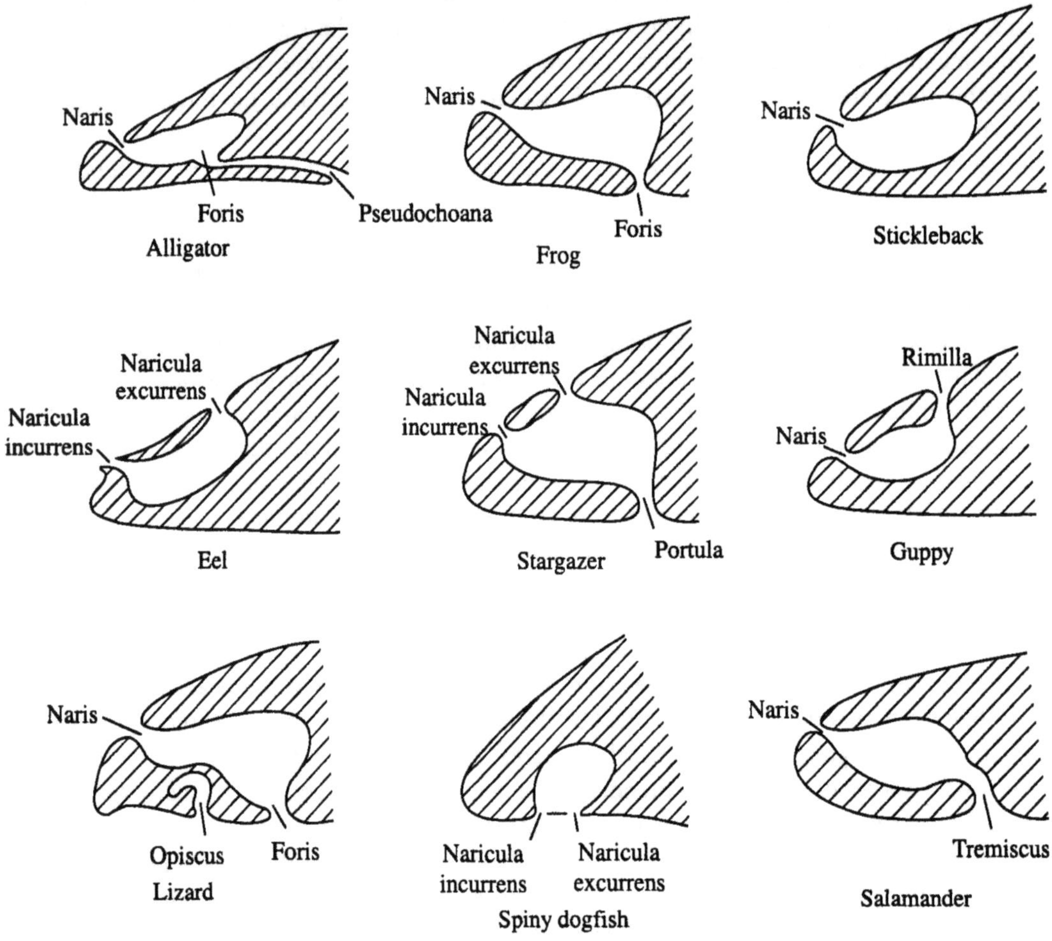

Fig. 15.44 Schematic sinistrial view of apertures of the nasal sac in various gnathostomes. Note that the pseudochoana is not a nasal sac aperture. The precursive nasolachrymal duct (rimillar tube) is omitted in frog and salamander.
***Source:* Bjerring (1989)**

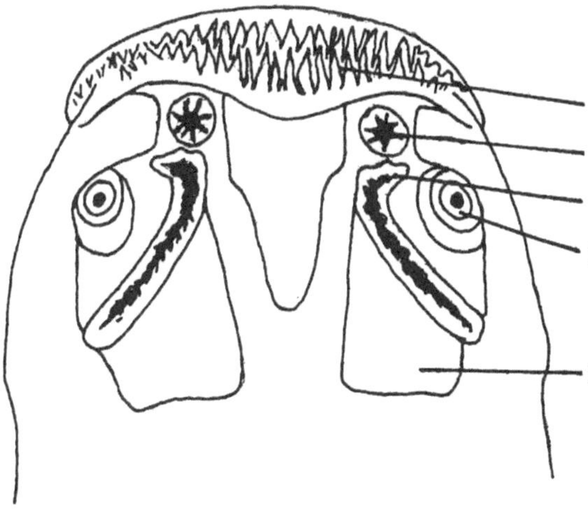

Fig. 15.45 Sketch of dorsal surface of the head of stargazer. *Astroscopus*, showing internal nares.
***Source:* Atz (1952)**

to the mouth as in *Scyliorhinus canicula.* It has been concluded that sharks and rays like most other living gnathostome fish, are nariculate.

The choana very often, but wrongly, has been numbered among the apertures of the olfactory organ. Evidently, it is an opening that owes its occurrence to the development of a secondary palate. Only mammals have choanae (see Bjerring 1989 for discussion and references).

5.2.2 Morphology of the Olfactory Organ

Edited books by T.J. Hara (1982, 1992) contain contributions by various authors which can serve as guide for literature. The salient difference between the fish nose and the nose of higher vertebrates is that it is separate from the respiratory channel. Vertebrates perceive odours through one or two sensory sacs which develop on the forepart of the head. The size of the olfactory system in different fish varies a great deal according to habitat and lineage. The primary plan of fish nose is still that of a covered pit, not connected with the mouth. The olfactory system may roughly be divided into three parts: (1) a peripheral part represented by the olfactory organ which houses the neuroepithelium and the olfactory bulb; (2) an intermediate part, the anterior olfactory nucleus; and (3) a central part mainly in the paleocortical area.

5.2.2.1 Peripheral Olfactory Organ

The cyclostomes (hagfish and lampreys) are unique in possessing only a single olfactory (monorhinal) organ with a single nostril. It ends either in a blind sac (lampreys) or connects with the pharynx (hagfishes: Døving 1998). Lampreys develop their nasal apparatus only when they are transformed into adults. The olfactory organs of fishes are diversely developed. At one extreme they are well developed (macrosmatic forms) such as in elasmobranchs (Fig. 15.46) and most eels (Apodes), and at the other they are poorly developed (microsmatic forms) such as in pike, flying fish (Exocoetidae), stickleback, pipefish and anglerfish (Lophiidae). Macrosmatic fish depend largely upon the acute sense of smell while microsmatic fish rely on senses other than olfaction. In sharks and rays, where the chemical senses are highly important, the paired

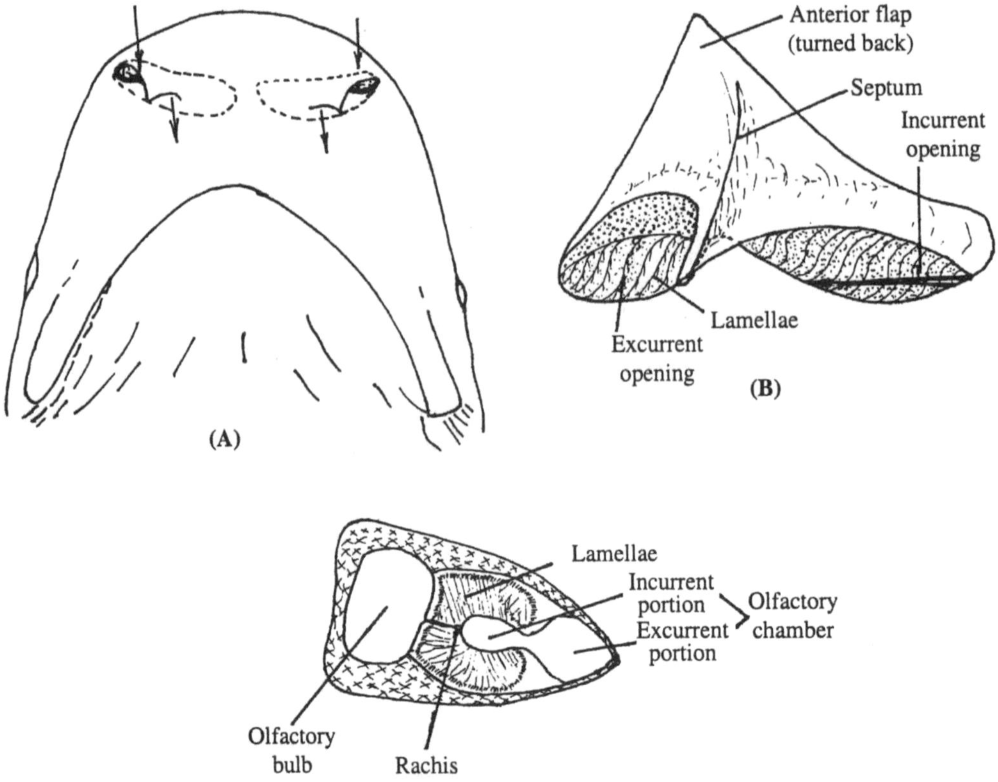

Olfactory organ of *Galeocerdo cuvieri.*

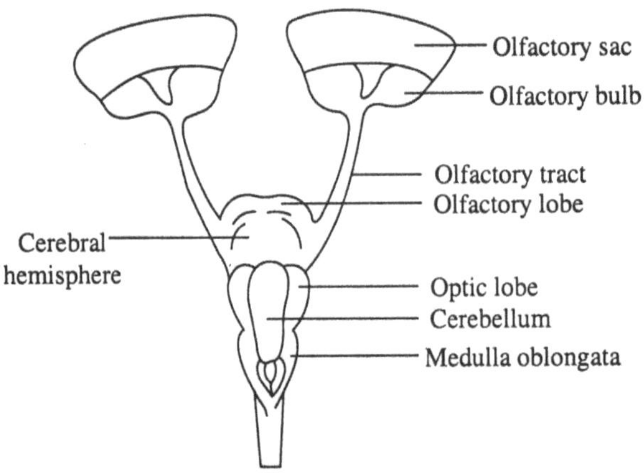

(D) *Galeocerdo cuvieri:* Olfactory organ and brain.

Fig. 15.46 Nostrils. (A) Ventral view of head showing position of nostrils and indicating position of olfactory sacs; (B) Left-hand nostril with anterior flap turned back showing interdigitation of septum with posterior flap; (C) Cross section of olfactory capsules showing olfactory bulbs, lamellae, 'rachis' incurrent and excurrent portion of the chamber. (D) The olfactory organ and brain.
Source: **Tester (1963)**

blind olfactory pits (nasal cavity or olfactory chamber or sac) are usually situated on the ventral side of the snout. Wide variations in the location, size and shape of the nostrils have been described in elasmobranchs. Folds in the nasal pits of chondrichthyans lie predominantly at right angles to the longitudinal axis of the pit; these folds inturn are thrown into secondary folds which carry the olfactory epithelium. The narial pits are covered by variously folded and moulded skin flaps so that water courses through them from a lateral entrance to a medial exit; some pits may be almost divided into two chambers. Ciliary movements within the pits also serve to channel the water. The hammerhead (*Sphyrna lewini*) has a peculiar head shape; the entrances to the nose lie laterally at the head's leading edge where a pronounced groove deflects water outwardly, directly into the large nasal pits, the water then leaves the nose ventrally.

Bichirs (*Polypterus*) appear to possess a well-developed olfactory system, based on the size and complexity of their olfactory organs, olfactory nerves, and bulbs. Not surprisingly, the importance of this system is also reflected in extensive secondary olfactory projections to multiple telencephalic and diencephalic centres. Olfaction is the primary sensory modality in polypterids; nocturnal bottom-dwelling predators that feed primarily on other fish. In bottom-dwelling sturgeons, the eyes are small but their olfactory organs are well developed.

The nostrils of the olfactory organ in *Latimeria* are widely separated. The incurrent nostril lies just above the mouth and the excurrent nostril is a narrow slit close to the eye. No choanae are present.

In the teleost fishes there are considerable variations in size, structure and degree of development of the olfactory organs. Although there is no extensive review of the nasal anatomy of all species, several studies describe or review numerous species, or consider one species in detail. The gross morphology of the olfactory organ of the burbot (*Lota lota*), a freshwater species of the family Gadidae is illustrated in Fig. 15.47.

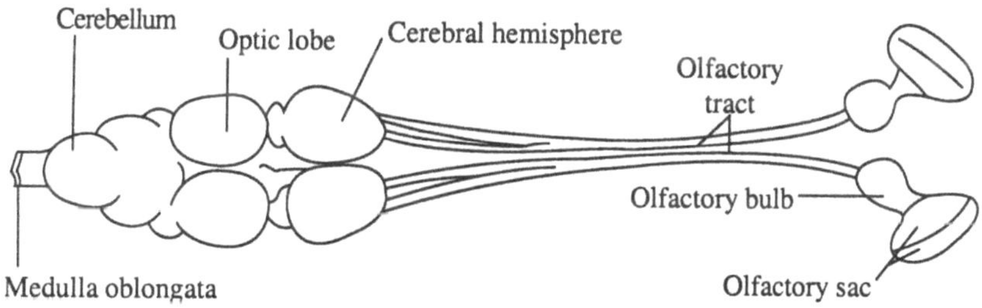

Fig. 15.47 Brain and olfactory organ of burbot (*Lota lota*), dorsal view.
***Source:* Døving and Gemne (1965)**

Only a few of the typical patterns will be described below. The paired olfactory pits are usually located on the dorsal side of the head. The eels and morays (Anguilliformes), believed to have the most acute sense of smell, have large and elongate olfactory pits extending from the tip of the snout to the orbit of the eye. In contrast, certain puffers (Tetraodontidae) which are highly visually oriented reef fishes, have completely lost the nasal sac; there remains only a minute epidermal patch from which the nearly invisible olfactory nerve originates. These puffers also have no olfaction bulb. There are also great differences in the olfactory organs even between different sexes of the same species. The most successful groups of bathypelagic fishes, ceratoid anglerfishes (Lophiiformes) and *Cyclothone* spp. (Myctophiformes) have evolved distinct sexual dimorphism in olfactory organization and size (Fig. 15.48). The males have large olfactory

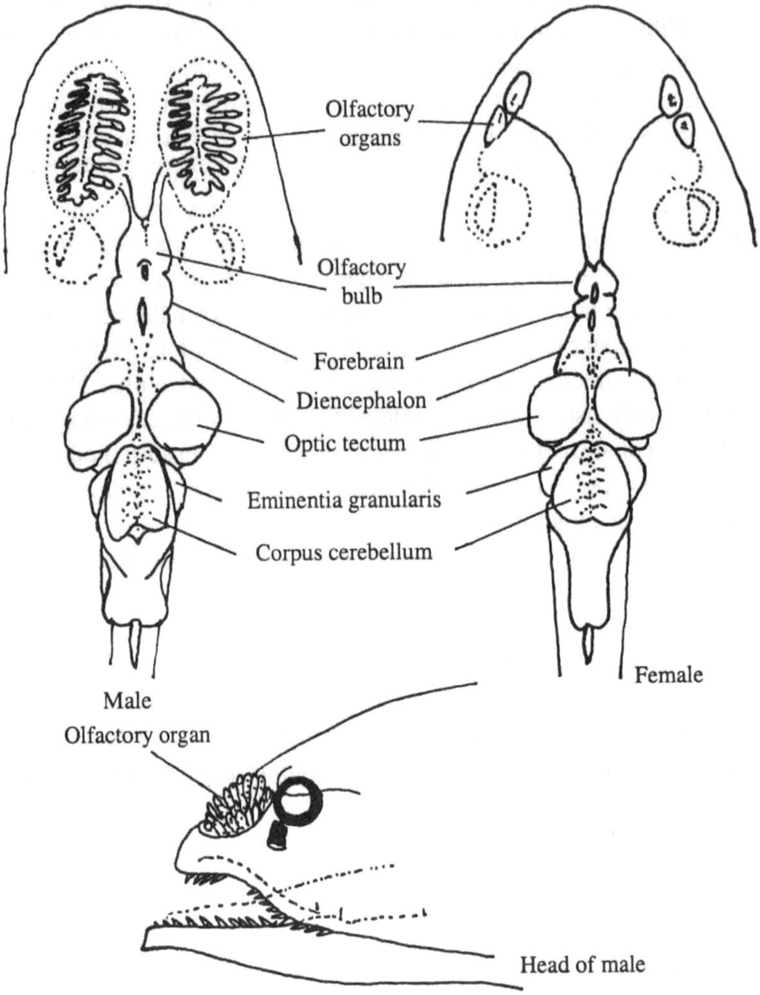

Fig. 15.48 Olfactory organs and brain of *Cyclothone microdon*. Top left, male; top right, female; bottom, head of male.
***Source:* Marshall (1967)**

organs, while these are small or regressed in the females of all species. A sexual dimorphism of olfactory organs is also reported in the deep-sea eel family Monognathidae and gulper eel (Saccopharyngidae). The olfactory organs are weakly developed in females, whereas they are very large in males. In other bathypelagic groups such as the Serrivomeridae, Nemichthydae (Anguilliformes) and Macrouridae (Gadiformes) the olfactory organ is poorly developed in both sexes. It has been considered that bathypelagic fishes generally use senses other than olfaction for procurement of food and that the well-developed olfactory organ found in the males of the Myctophiformes and the Lophiformes may be useful only in searching for females.

Each nasal pit generally has two openings or orifices to the exterior, anterior inlet and posterior outlet (nares) which are separated by a nasal flap or bridge. The patterns of olfactory flap vary greatly from species to species.

Condition of nares has been described in many teleostean genera. There are cases in which only one opening is present: in *Zoarces*, *Centronotus* and *Liparis*, the posterior nostril is secondarily lost, but in *Gastrosteus*, it represents a conjunction of the anterior and posterior nares. Anglerfishes have a tentacular nose.

Mechanisms employed for water sampling by olfactory organs of fish is quite interesting. In one group of teleost fishes, isomates, the olfactory cavity is divided in two compartments, a vestibule and a gallery. The connection between the two compartments is formed by the corridors of the olfactory rosette. Water currents are created by ciliary beats on the walls of the corridors. To illustrate the construction of the olfactory organ in isosmates and the direction of the water flow a schematic diagram of a cross section of the olfactory organ is shown in Fig. 15.49A. In another group, cyclosmates, the accessory olfactory sacs can be compressed and expanded in connection with respiratory and/or mouth movements, these volume variations force water to pass over the olfactory epithelium (Fig. 15.49B).

The study of the water transport through the olfactory organ of *Protopterus annectens* has pointed out an original autonomous cyclic water flow. The nomenclature based on water flow, has been modified: *isomates*, fishes that use the ciliary action to transport water through the

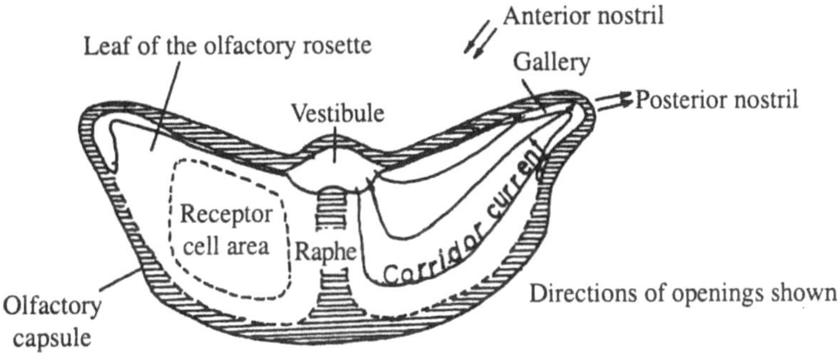

Isosmate

(A) *Anguilla anguilla*: Olfactory organ.

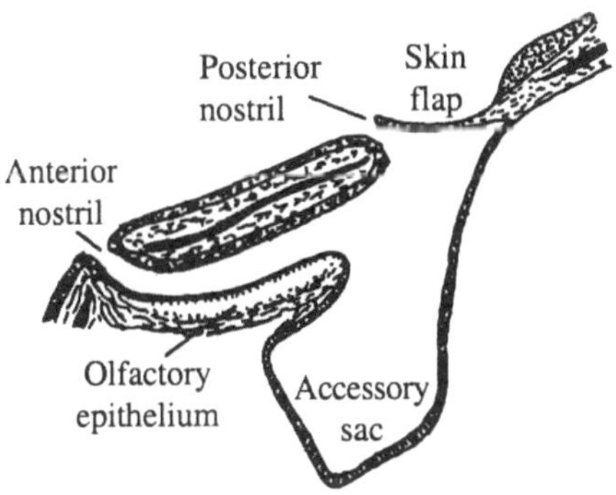

(B) **Cyclosmate** oviparous cyprinodont: Olfactory organ.

Fig. 15.49 (A) Schematic cross-section of the olfactory organ of eel, *Anguilla anguilla*, showing the principal water currents from the vestibule to the gallery via the corridors. (B) The olfactory organ of a cyclosmate. Sagittal section through the olfactory cavity of an oviparous cyprinodont. *Source:* Døving and Thommeson (1977)

olfactory organ; ***heterocyclosmates***, a new denomination for the fishes and the aerial choanates which have an olfactory cycle dependent or linked with the respiratory cycle; ***autocyclosmates***, a new denomination for the vertebrates, lungfish as example, where the olfactory cycle is autonomous. The question is to know if this sampling mechanism is of a peculiar efficiency for the animal living in its natural environment.

Ventilation of the olfactory organs is created by a pumping mechanism associated with the respiratory movements in lampreys. In hagfish the water current is produced by the movements of a muscular velum in the pharynx (Fig. 15.50). The water enters through the nostril, passes to the olfactory organ and leaves through the gill opening (Fig. 15.51). In lampreys the pharyngeal pouch is activated by the pressure changes in the respiratory tube, which create a water current in and out of nostril. In both cases the valve in front of the olfactory chamber directs the incoming water into the spaces between olfactory lamellae. In elasmobranchs the ventilatory flow of water through the olfactory organ is created by pressure differential between incurrent and excurrent nostrils. Forward motion and respiratory activity may assist in the process. In lungfish in water, ventilation of the olfactory organs is provided by pumping mechanisms. In air, the olfactory organs are believed not to be used. During dry periods both *Protopterus* and *Lepidosiren* may aestivate, *Protopterus* makes a cocoon of mud with a tube for breathing, but the nostrils become closed by mucus.

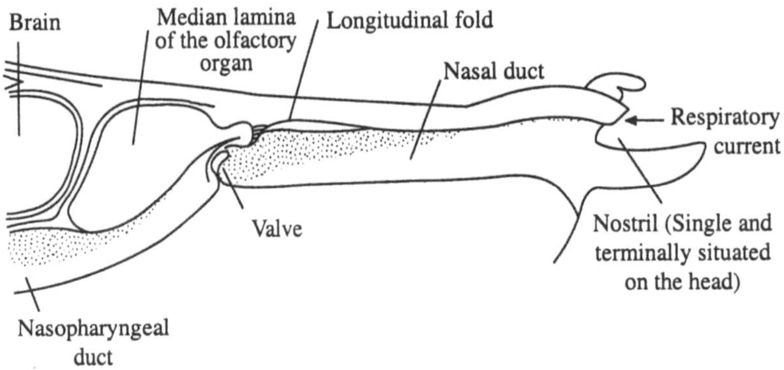

Fig. 15.50 Sagittal section of the anterodorsal part of the head of *Myxine glutinosa* showing the median lamina of the olfactory organ, the nasal duct with the valve and the nasopharyngeal duct.
***Source:* Theisen (1973)**

The olfactory rosette in *Latimeria* appears as five conical sections with many ramifications lodging the olfactory lamellae. Cilia provide the water currents through the organ.

In the Table 15.8 is indicated the presence or absence of accessory olfactory sacs and of ciliary activity.

The olfactory chamber of the deep-sea fish *Aphanopus carbo* (Trichiuridae) is enlarged by three accessory sacs: two ethmoidal sacs and one lacrymal sac (consisting of two parts). These sacs serve as ventilation apparatus which causes a permanent water current within the olfactory chamber and between the olfactory folds. The action of the olfactory sacs is induced by the splanchnokinetic. The forebrain of *A. Carbo* is well developed; its size ranges between that of forebrains in microsmatic and macrosmatic teleost species. In addition to the eye-the olfactory organ is well developed (relative to that in other teleosts). This fact suggests that this fish is

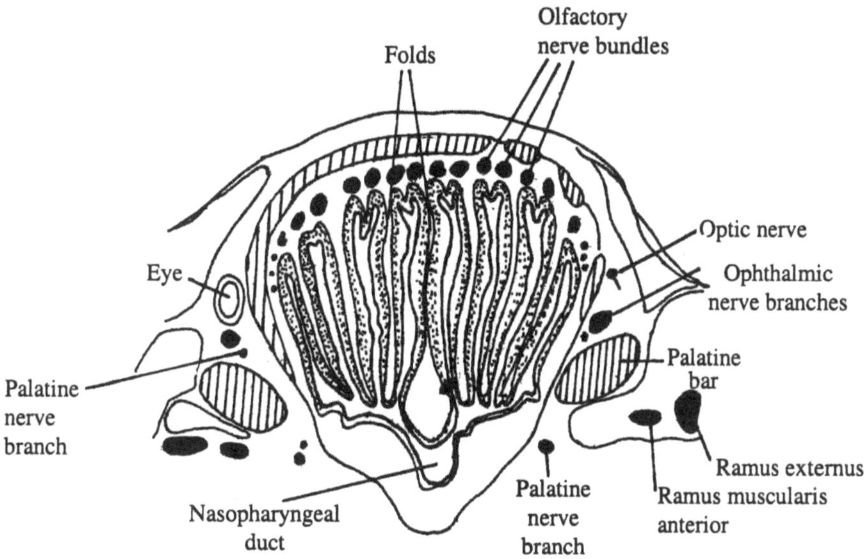

Fig. 15.51 Cross section through the olfactory apparatus of *Myxine* to show the distribution of olfactory nerve bundles.
***Source:* Brodal, A., Fänge, R. (eds.) (1963)**

related to a group of teleost species characterized by optical and olfactory mechanisms of high performance (Holl and Meinel 1968).

The nasal sac is linked with the olfactory epithelium, which is generally raised from the floor of the organ into a complicated series of folds or lamellae to make a rosette.

Considerable variation is apparent in the arrangement of the olfactory lamellae. In hagfish the olfactory rosette is composed of about seven free lamellae arranged longitudinally and parallel to one another, whereas in lampreys it consists of a number of longitudinally arranged lamellae radiating from the wall of the olfactory chamber. In the majority of teleosts, the lamellae radiate

Table 15.8 The presence (P) or absence (A) of ciliary activity and accessory olfactory sacs. (W) Weak ciliary activity

Fish	Ciliary activity	Accessory sacs
Dogfish *(Scylorhinus canicula)*	P	A
Pike *(Esox lucius)*	W	A
Carp *(Cyprinus carpio)*	P	A
Catfish *(Silurus glanis)*	P	P
Perch *(Perca fluviatilis)*	W	P
Scorpion fish *(Scorpaena porcus)*	W	P

from a central ridge (raphe) arising rostrocaudally from the floor of the nasal cavity, to make a rosette-like pattern. In other teleosts, a series of parallel lamellae occur in the absence of a raphe. Various types of arrangement of lamellae have been described and classified. Eight types of arrangement of lamellae in the olfactory organ of teleosts has been shown schematically in

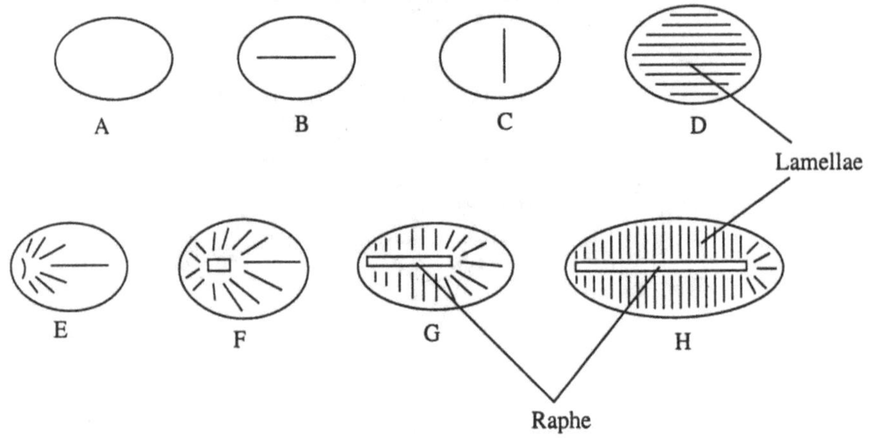

Fig. 15.52 Eight types of arrangement of lamellae in the olfactory organ of teleosts. An olfactory organ has: (A) no lamella; (B) one longitudinal lamella; (C) one transverse lamella; (D) longitudinal lamellae arranged in parallel with one another; (E) lamellae arranged in a fan-shape; (F) lamellae radiating in all directions from a raphe; (G) lamellae arranged transversely or obliquely to an elongate raphe; (H) numerous lamellae mostly arranged transversely to a very long raphe.
***Source:* Yamamoto (1982)**

Fig. 15.52. In Australian lungfish *Neoceratodus forsteri*, the olfactory rosette is composed of a long raphe, on each side of which is situated a row of laminae or folds. Each olfactory fold is provided with secondary folds (Fig. 15.53). The arrangement of the olfactory lamellae has been employed as a diagnostic character in taxonomy of some teleostean groups. The secondary folding of olfactory organ in young and adult sea trout (*Salmo trutta trutta*) has been described

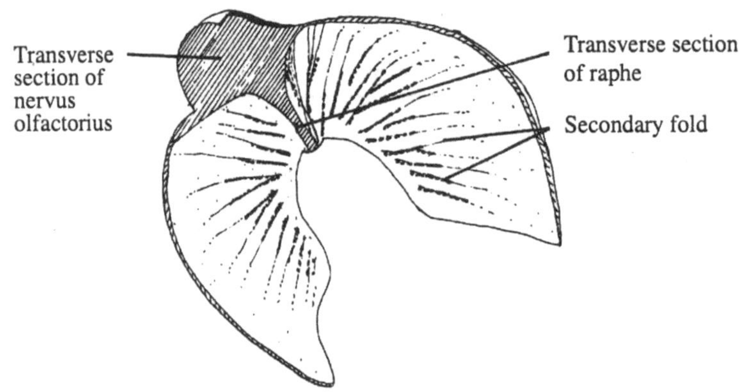

Olfactory organ of *Neoceratodus forsterl.*

Fig. 15.53 A pair of folds from the left olfactory organ of *Neoceratodus forsteri*, in anterior view.
***Source:* Theisen (1972)**

by Bertmar (1972).

The number of olfactory lamellae varies greatly among species. Olfactory lamellae are lacking in small cyprinodonts (nine subfamilies of Cyprinodontoidea). Eels and morays are known to have a large number of lamellae, but the largest number (230 in one nasal cavity) has been reported in *Hoplopagrus guentheri* (Perciformes). The Polypteridae of Africa have an extremely developed nose, in which numerous folds extend from five raphe disposed in parallel, radial symmetry around the olfactory nerve trunk. It differs fundamentally from the actinopterygian olfactory organ and resembles that of the living crossopterygian *Latimeria*. Water circulation is by means of cilia.

The presence of an extensively folded olfactory epithelium in bichirs (*Polypterus*) and the other anamniotic taxa indicates that this character represents the ancestral condition for craniates. Although all these taxa do exhibit folding of the olfactory epithelium, this folding appears to involve different portion of the olfactory organ. In *Eptatretus*, only the roof of the olfactory organ is thrown into folds, whereas in *Petromyzon* the lateral walls of olfactory organ are foliated. Among gnathostomes, the roof (*Protopterus*, *Squalus*), floor (*Amia* and many teleosts), caudal pole (*Chimaera* and *Polypterus*), or one or both walls (*Acipenser*, *Lepisosteus*, *Latimeria*) of the olfactory organ may be folded. Much of this variation is probably due to the final topographical orientation of the olfactory organ differences in the development of its incurrent and excurrent ducts, rather than to differentiation of a sensory epithelium from different portions of the olfactory vesicles. For this reason, the actual pattern of foliation is likely to have no phylogenetic significance. Although similarities in the foliation of the olfactory epithelium of *Latimeria* and *Polypterus* have been claimed to suggest a close phylogenetic relationship, these similarities, however are at best superficial and do not result in the formation of the same number of lobes.

The number of lamellae increases to some extent with growth of individuals but remains constant after a fish reaches a certain stage in growth.

Both faces of the olfactory lamellae are usually flat. Secondary foldings of the lamellar face have been found in many species of the Salmoniformes, the burbot, *Lota lota* and two species of the Perciformes, *Therapon jarbua* and *Prionurus microlepidotus*.

Different views have been expressed on the correlation between the olfactory organ (olfactory sense acuity) and food search. More quantitative studies on the olfactory acuity in many species are necessary to determine the relevances of such correlations.

Elements of the Olfactory Mucosa. The sensory epithelium—a columnar pseudostratified type-is disposed continuously, or in islets distributed uniformly, in large fields on opposing faces of the olfactory lamellae. The non-sensory epithelium, also referred to as indifferent and as nonciliated or ciliated indifferent epithelium, is analogous to the respiratory epithelium of other vertebrates. As the terms indicate, the numbers and ciliated cells in question vary greatly in species (Fig. 15.54) (Table 15.9A, B).

Vertebrate olfactory cells exhibit characteristics which are at the same neuronal and paraneuronal. Olfactory receptor cells are bipolar neurons, either ciliated or microvillous (Fig. 15.55) as demonstrated by means of degeneration and regeneration techniques. Microvillous receptor cells have been considered precursors of ciliated receptor type. However, ontogenetic investigations, as well as studies on degeneration and regeneration, have shown that both cell types represent

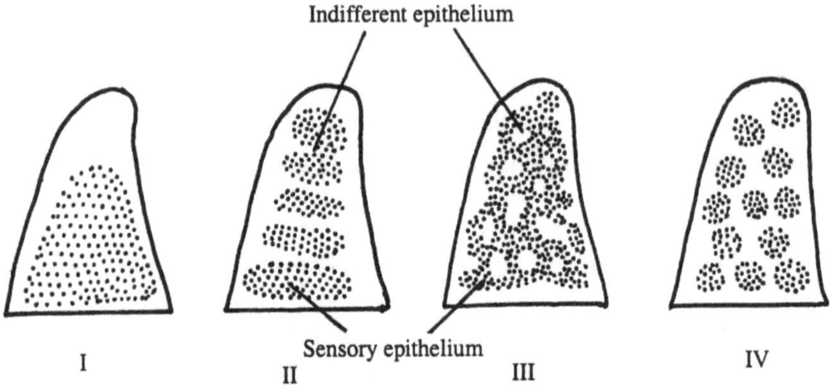

Fig. 15.54 Four types of distribution patterns of the sensory epithelium (shown by dotted areas). The sensory epithelium is: (I) continuous except for the margin of the lamella; (II) separated into some large areas by the indifferent epithelium; (III) mixed irregularly wih the indifferent epithelium; (IV) dispersed in discrete islets.
***Source:* Yamamoto (1982)**

independent entities. Ciliated and microvillous olfactory neurons are present in almost all teleosts. Elasmobranchs exhibit the microvillous type only. While both receptor types have been recognized in hagfish and the African lungfish, *Protopterus annectens*, the river lamprey, *Lampetra fluviatilis*, shows only the ciliated type, and the Australian lungfish, *Neoceratodus forsteri*, only the microvillous receptor type. Olfactory receptor density is quite variable in fishes and has been reported in different species to be between 46000 and 500000 mm^{-2}.

Both cell types have many cytological features in common: a granular endoplasmic reticulum, a Golgi complex, and free ribosomes. The axon contains smooth endoplasmic reticulum and microtubules. Characteristics of the dendrite are longitudinary orientated microtubules and long mitochondria. At the level of the terminal web in the surrounding supporting cells, the dendrite forms a terminal swelling, the olfactory knob, protruding slightly above the epithelial surface.

Ciliated receptor cells dominate in the olfactory epithelium of migratory sturgeons, whereas in stationary species microvillous receptor cells are relatively abundant. Supporting cells are generally restricted to the sensory epithelium. At their free surface, supporting cells give off microvilli or microvillous-like protrusions. In the river lamprey and the reedfish, *Erpetoichthys calabaricus*, supporting cells are found bearing cilia in addition to microvilli. Characteristic internal features of the supporting cells are bundles of tonofilaments generally longitudinally arranged. Adjacent to Golgi complex, profiles of granular endoplasmic reticulum, free ribosomes, and mitochondria are recognizable. Several functions of the supporting cells have been suggested: secretory, absorptive and glia function. Basal cells, goblet cells and ciliated nonsensory cells are also reported.

A unique feature of the olfactory system is the regenerative capacity of the receptor neurons. The receptor cells are continually renewed in normal animals. This renewal process of olfactory receptor cells may be considered to be an adaptation of the system to impairment caused by environmental hazards during the normal life of the animal.

No morphological difference of the olfactory organ between species (in Characoidea) with a fright reaction (e.g., *Gymnocorymbus*, *Copeina*) and species without a fright reaction (e.g., *Anoptichthys*, *Rooseveltia*, *Poecilobrycon*, *Chilodus*) was found.

Table 15.9A Morphological properties of the peripheral olfactory organs of some teleosts

Order and species	Arrangement of lamellae *1	Number of lamellae per organ	Distribution of sensory epithelium*2	Distribution of non-sensory cilia*3
Anguilliformes				
Conger myriaster	H	120	I	+ + + +
Clupeiformes				
Clupandon punctatus	G	26	III	+
Engraulis japonica	G	28	I	–
Salmoniformes				
Oncorhynchus keta	F	18	II	+ + +
Mcytophiformes				
Trachinocephalus myops	E	10	I	+
Cypriniformes				
Cyprinus carpio	G	31	I	+ +
Siluriformes				
Parasilurus asotus	H	68	I	+ + + +
Plotosus anguillaris	D	7	IV	–
Gobiesociformes				
Conidens laticephalus	A	0	I	+ +
Atheriniformes				
Hemiramphus sajori	C	I	IV	–
Oryzias latipes	A	0	IV	–
Atherion elymus	D	3	IV	–
Gasterosteiformes				
Hippocampus coronatus	A	0	I	+
Pleuronectiformes				
Paralichthys olivaceus	G	22, 19*4	I	+ + +
Tetraodontiformes				
Rudaris arcodes	A	0	I	–
Navodon modestus	B	1	I	+ +

*1 Letters (A-H) refer to the types represented schematically in Fig. 15.52.
*2 Roman numerals (I-IV) refer to the types represented schematically in Fig. 15.54.
*3 – absent, + sparse, + + dense, + + + + extremely dense.
*4 Left figures, ocular side; right figures, blind side.
Source: Yamamoto (1982)

SEM study of the olfactory epithelium of a few coldwater hillstream teleosts has indicated a possible correlation in the presence of particular receptor cells with specific niche and feeding nature (Singh 1994). Recent reports on the fine structure of olfactory organs are from Zeiske et al. (1994), Singh et al. (1995), Fishelson (1997), Fishelson and Baranes (1997), Sorenson and Caprio (1998) and Hansen et al. (1999).

Mechanism of Olfaction. The olfactory neuron of fish exhibits membrane characteristics similar to those determined in other vertebrates. Unlike other excitable cells, the ionic concentrations across the olfactory epithelium are not determining factors of the resting membrane potential of the fish olfactory neurons. The injection of depolarizing electric current or the application of

Table 15.9B A summary of the differences between teleost microvillar and olfactory receptor neurons (ORNs)

Microvillar ORNs	Ciliated ORNs
Microvilli extend from apical surface.	Cilia extend from apical surface.
The dendrite is short and the nucleus is located in the upper third of the olfactory epithelium.	The dendrite is long and the nucleus is located in the the lower third of the olfactory epithelium.
Differentiate after ciliated ORNs during ontogeny and during axotomy induced synchrnous ORN differentiation (rainbow trout, goldfish).	Differentiate before microvillar ORNs during ontogeny and during axotomy induced synchronous ORN differentiation (rainbow trout, goldfish).
Axons project to the caudal-dorsal. olfactory bulb (catfish).	Axons project to the ventral olfactory bulb (catfish).
Respond to amino acids (salmonids).	Respond to bile acids (salmonids).
	Respond to amino acids and bile acid (rainbow trout).
	(Respond to amino acids, not bile acids, (goldfish).
Contain V2R homologue that binds L-arginine *(goldfish).*	

Source: Zielinski and Hara (2001; see references)

odorants to the olfactory neurons leads to firing of spike potential. The firing frequency increases linearly with an increase in the current intensity or stimulant concentration. A hypothetical model for the olfactory transduction proposes that specific receptors in the receptive membrane zone, assumed to be located in the olfactory knob and/or cilia, bind to stimulus molecules. This results in the conductance change in the electrogenic zone, leading to a production of a generator potential.

The entire process of fish olfaction takes place in water, therefore olfactory receptors are normally stimulated by water-soluble compounds. Four major chemical groups have been identified as potent odorants for fishes, amino acids, bile acids, gonadal steroids, and prostaglandins. (see Sorensen and Caprio 1998; Table 2; also refer to Hara 1993; Table 1). Amino acids, appear to be detected by atleast three receptor sites, or transduction mechanisms, with varying affinity and specificity (generalist type), while the three other chemical groups are detected by respective receptors with high specificity (specialist type). Furthermore, multiple receptor subtypes with distinct binding characteristics exist within each chemical group. Structural and functional separation of the olfactory system exists in fishes. Information associated with reproductive behaviour is mediated through the medial olfactory tract, and stimulation by amino acids is mediated through the lateral olfactory tract (Hara 1992). Details of physiology of olfactory receptors are now known (see Caprio et al. 1989; Kang and Caprio 1991, 1995; Ivanova and Caprio (1993). For further information, read Devitsina and Chernova (1994), Sorenson and Caprio (1998). Kasumyan et al. (1999) and Zielinski and Hara (2001). Hara and Zhang (1996) reported that in salmonid fishes olfactory neurons responsive to amino acids and taurocholic acid are randomly distributed throughout the olfactory epithelium, and yet project to spatially segregated regions and thereby generated signals are encoded independently in the olfactory bulb (see Fig. 8 of the article).

Hara and Zhang (1998; see Fig. 8) further reported that in salmonid fishes that olfactory

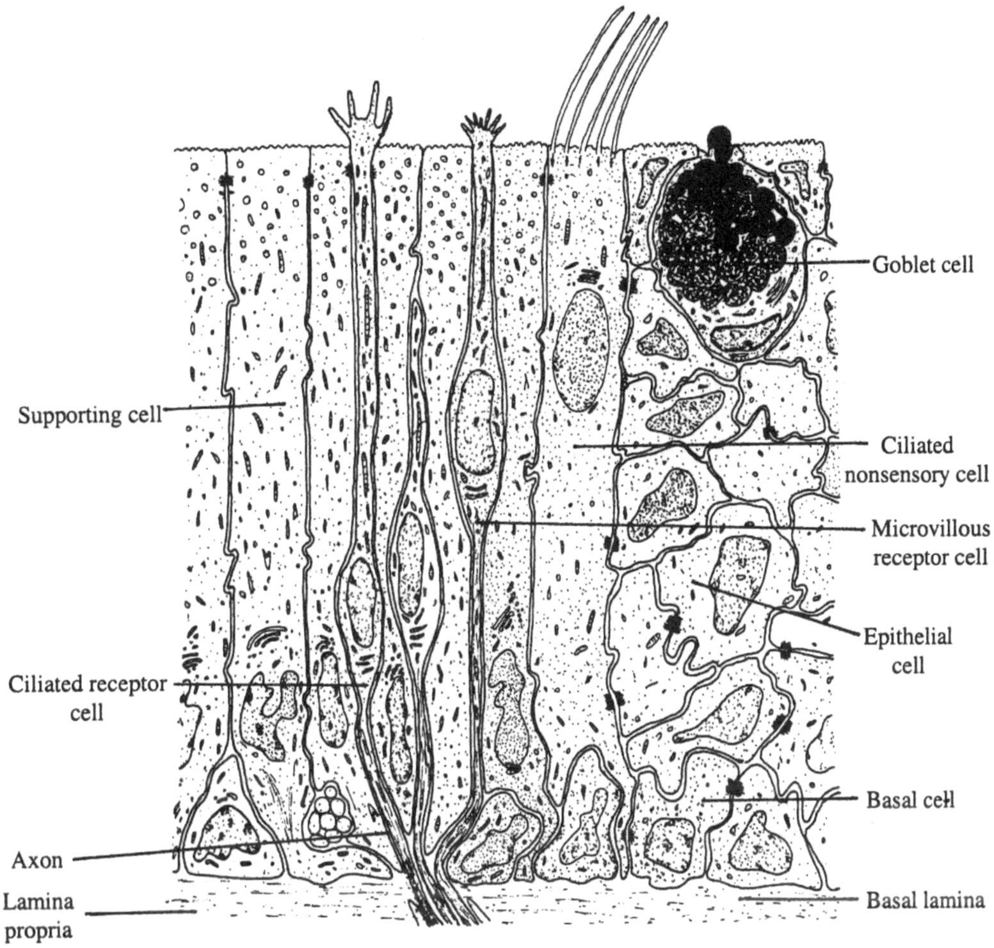

Fig. 15.55 Simplified diagram of the olfactory epithelium of teleosts, showing sensory and nonsensory epithelium.
***Source:* Zeiske et al. (1992)**

signals due to hormonal pheromones are processed in a manner distinct from those for amino acids and bile acids and may possibly be mediated by extrabulbar primary olfactory fibres bypassing the bulb.

Further details on molecular mechanisms of chemosensory transduction: gustation and olfaction are available. It appears as though multiple mechanisms exist for the transduction of chemosensory information in teleosts, and that in some cases multiple transduction pathways can be stimulated by a single receptor class. Possible receptor and transduction for taste and olfaction have been suggested.

Recent results suggest that different conductances modulated by either IP_3 (inositol-1, 4, 5-triphosphate) or cAMP (adenosine cyclic 3′, 5′-monophosphate) are involved in mediating olfactory transduction in channel catfish (*Ictalurus punctatus*). Olfactory receptor neurons and that Ca^{2+}-activated K^+ channels contribute to the termination of the IP_3 and cAMP responses

(Miyamoto et al. 1992; see also Bruch and Teeter 1990; Restrepo et al. 1990; Restrepo and Boyle 1991). For molecular mechanisms of chemosensory transduction, the reader is referred to Brand and Bruch (1992). Other important references on structure and function of olfactory organs are of Marshall (1967), Hara (1971, 1975, 1992, 1993a, b, 1994), Kleerekoper (1978), Yamamoto (1982), Little (1983), Tucker (1983), Caprio (1984, 1988), Zeiske et al. (1992), Devitsina and Chernova (1994), Fishelson (1995, 1997). Kasumyan et al. (1999) and Zielinski and Hara (2001). For the ontogeny of olfactory organ, reader can obtain information from the contributions by Hansen and Zeiske (1993, 1995) on zebrafish, *Brachydanio rerio* and Fishelson (1997) in cichlids. Regeneration in the peripheral olfactory system has been described by Zippel (1993).

Behavioural Function of Olfaction and Taste. Details on gustatory system will follow (see subsection 5.3). Olfactory acuity varies greatly among species. The olfactory acuity of fish also shows seasonal fluctuations. Sexual hormones may be responsible for these fluctuations. Increased sensitivity to food stimuli has been found to occur in male *Carassius* during spawning season.

The fish olfactory and taste systems are highly sensitive to a variety of pure chemical and naturally occurring chemical stimuli. Different species exhibit variable sensitivity, and olfactory sensitivity may differ from those of taste in some species. Certain marine and freshwater species are particularly sensitive to single amino acids, through both smell and taste.

Fish can discriminate among the scents of different species as well as among different conspecific individuals. Attractive scents facilitate the schooling response in some species and enable predatory species to locate prey. Amino acids and natural derivatives may be important components of some of these chemical signals.

In many species, chemical stimuli are important in various aspects of social behaviour, such as initiation of courtship responses and the attraction of males to sexually responsive females (and vice versa). Reproductive behaviour is often elicited by a combination of visual and chemical stimuli. In several species ovarian fluids are an important source of female sexual stimuli. Some males also excrete stimuli through glandular tissue that appears during the breeding season. Besides amino acids, bile acids (especially taurine and glycine conjugates), and two reproductive pheromones, sex steroids and prostaglandins are potent olfactory stimulants.

Chemical signals also function in parent-offspring interactions in facilitating the care of eggs and allowing parental discrimination of their own fry from those of other species as well as those of other conspecific broods. The chemical stimuli of fry initiates searching behaviour in their parents, and in some species the fry are attracted by their parents' scent.

A chemical substance that inhibits growth, fry survival, ovarian development, and aggression has been shown in several species. This substance results from increased population and is species-specific.

Ostariophysian fishes bear epidermal cells containing an alarm substance or Schreckstoff. The substance is released by injury to the skin as during a predator attack, and it elicits fright reactions in other conspecific individuals and in other ostariophysian fishes. The chemical composition of the alarm substance may be a pterine or similar double-ringed compound; amino acids may also be components of the alarm substance. Chemical alarm signals have been reported in a number of fish taxa.

The recognition of prey through chemoreception occurs in a variety of vertebrates. For many prey species (e.g., fathead minnow *Pimephales promelas*), the ability to chemically recognize a predator is an important component of an individual's antipredator arsenal, because it may result in a greater number of ecological situations in which predator recognition is possible. For example, prey fishes with chemical predator recognition abilities may be able to utilize habitats with poor optic properties. As well, nonvisual detection of cryptic ambush predators, such as the northern pike (*Esox lucius*) may be critical to the survival of prey fishes.

In fishes, predator recognition through chemoreception could result from olfaction, gustation, or single-called chemoreception. Solitary chemosensory cells, which are secondary epidermal sensory cells (see Chapter 3), may be used in predator recognition and avoidance, as they respond to mucus of heterospecific fishes. Little information exists on the role of gustation in predator recognition. However, fishes possess numerous taste receptors that could potentially be used in predator recognition. Olfaction has been demonstrated to be responsible for mediating predator recognition in coho salmon (*Oncorhynchus kisutch*), brook trout (*Salvelinus fontinalis*) and the bitterling (*Rhodeus sericesu amarus*). Olfaction is based on nerve cell receptor within the olfactory pits. As these receptors are highly localized in the nares, this sensory system, unlike the gustatory or single-celled chemosensory system can be blocked.

Recently, it has been demonstrated that the olfactory system is necessary for the ability of fathead minnows to recognize northern pike as a predator and that the gustatory and single-celled chemosensory systems are not sufficient for this recognition in the absence of olfactory input. Olfactory impairment was behaviourally confirmed by exposing minnows to alarm substance.

Olfaction is important in the migratory response of some fish. Adult Pacific salmon return to spawn in streams where they had developed as fry. This recognition is based upon an odor unique to the stream, which the salmon learned as fry and remembered into adulthood. Fish possess the ability to discriminate among the odors of different streams. Young salmon fry are imprinted on the odor of the stream in which they develop and later discriminate between the odor of the natal stream and others during the spawning migration. When the fry are exposed in water scented with an artificial chemical, they are attracted to this scent when they return as adults.

Salmon recognize natural waters on the basis of the characteristic combination of odorants that are present, and that calcium is one of this bouquet of odorants that characterizes individual natural waters.

Pheromones released by populations of salmon may scent the spawning area and attract species of *Salmo* during the spawning migration. Pheromones also induce the entrance of young eels into freshwater streams.

Olfaction and taste are essential in some species for detecting the presence of food substances and initiating mechanisms for locating the food. Some species orient against a water current or swim upstream as long as they perceive the chemical stimulus. Other species locate the food by making spatial or temporal comparisons of the chemical stimulus intensity as they approach the food. The taste system predominates in directing consumption after the food has been located. Anatomical lesions and taste-aversion studies have differentiated the roles of external and internal taste in the sensory control of consumption.

Although specific substances are naturally attractive to certain species, learning also plays a major role in the development of food preferences. Extracts of natural food stimuli have been used to identify the nature of the food stimulus. Among a variety of chemical stimuli substances, amino acids are frequently in the fraction that is most effective behaviourally. Extracts of prey species differ in their chemical composition. The attractiveness of most extracts appears to be based on the mixture of component substances. Predatory species differ in their responsiveness to these components (see for details of above coverage: Hara 1971, 1993a, b, 1994a, b; Bardach and Villars 1974; Bodznik 1978; Kleerekoper 1978; Døving et al. 1980; Selset 1980; Selset and Døving 1980; Caprio 1982, 1984; Little 1983; Tucker 1983; Huque et al. 1992; Jones, 1992; Mathis and Smith 1992, 1993a, b; Smith 1992; Stabell 1992; Brown et al. 1995; Chivers et al. 1995, Wisenden et al. 1995; Kang and Caprio 1997; Zippel et al. 1997; Sorenson and Caprio 1998).

5.2.3 Olfactory System within the Brain of Fish

5.2.3.1 Olfactory Nerve

The unmyelinated axons of the olfactory nerve emerge from the base of the receptor neurons to gather into fascicles that penetrate the bone underlying the olfactory mucosa. The nerve runs caudally to terminate in the superficial layers of the olfactory bulb. In some fish, such as *Carassius* and *Ictalurus*, the olfactory nerve is quite short, the olfactory bulb being close to the epithelium, but relatively remote from the brain. In other fish, such as *Salmo*, the nerve is comparatively long with the olfactory bulb directly opposed to the rostral end of the telencephalon. Some olfactory fibres of *Gasterosteus aculeatus* collect into fascicles that course through the bulbs into the telencephalon without forming the glomeruli. The axons within the olfactory nerve are arranged into several groups of bundles. In the carp, *Cyprinus carpio*, for example, the olfactory nerve consists of two main bundles, medial and lateral. The former is derived from the more rostral lamellae, and the latter from the more caudal. The fibres do not branch until they terminate.

5.2.3.2 Olfactory Bulb

The olfactory nerve fibres contact the olfactory bulb at its rostral end and spread across the surface of the bulb to terminate in the superficial glomerular layer. Various types of small juxtaglomerular cells are located in the vicinity of glomeruli, while most of the large efferent cells lie deep to the glomerular layer. Deeper still is the granule cell core of the olfactory bulb.

The olfactory bulb in teleosts appear to have essentially the same organization as that in other vertebrates. The basic organization is illustrated semischematically in Fig. 15.56. The olfactory nerve fibres terminate on the dendrites of juxtaglomerular cells and efferent bulbar cell types. The deep granule cells are predominantly axonless, with apical dendrites extending outwards to form dendrodendrific synapses with some of the efferent cell populations. In the goldfish, *Carassius auratus*, several types of afferent cells are recognizable.One class of such cells include typical mitral/tufted types; another class is the ruffed cell, which may be a specialized form of the mitral/tufted cell population.

In all species of fish examined to date, the olfactory bulb projects to a limited area of the telencephalon. Even higher order olfactory connections appear to involve a limited area of the telencephalon. Thus the concept of the telencephalon in fishes as being entirely devoted to the olfactory sense is outdated and should be abandoned.

The axons of the mitral cells form the majority of the centripetal fibres of the olfactory tract

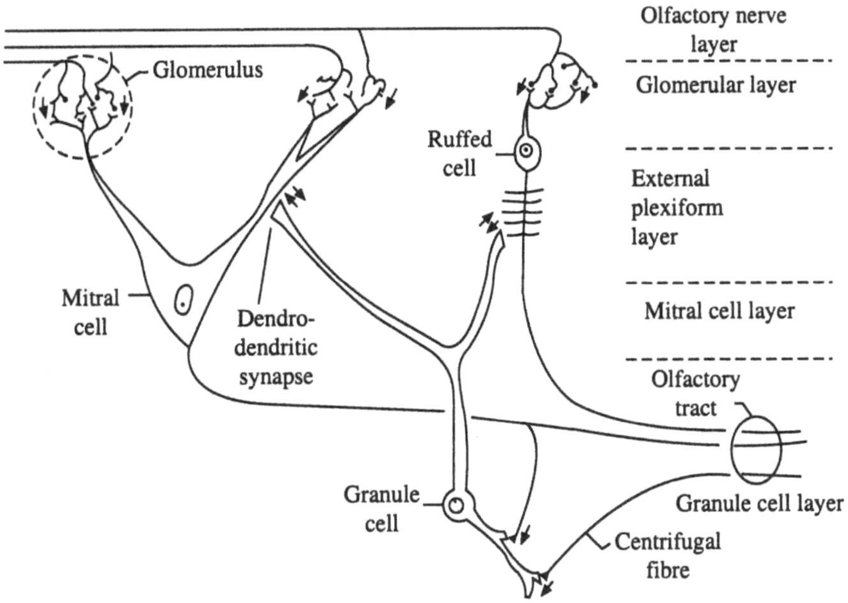

Fig. 15.56 Semischematic diagram of the basic organization of olfactory bulb in teleosts.
***Source:* Finger (1988)**

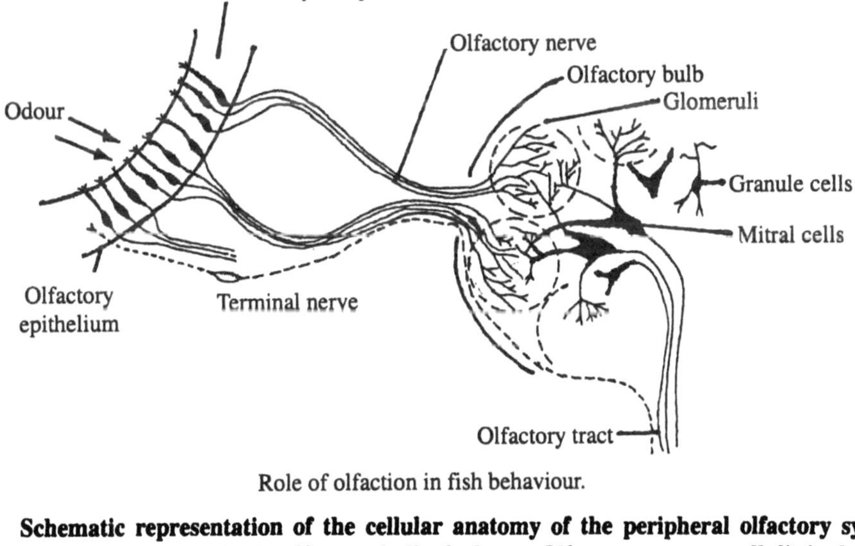

Role of olfaction in fish behaviour.

Fig. 15.57 Schematic representation of the cellular anatomy of the peripheral olfactory system and neural organization in the olfactory bulb of teleosts. Olfactory receptor cells lie in the olfactory epithelium. Bundles of receptor cell-axons form the fila olfactoria, which coalesce as the olfactory nerve. Axon terminals branch at the olfactory bulb in the form of glomeruli, where they synapse with processes of second-order neurons, mitral cells. Mitral-cell axons project centrally as the olfactory tract.
***Source:* Hara (1993a)**

through which information is conveyed to the telencephalic hemispheres (Fig. 15.57). The olfactory tract consists of two main bundles, lateral and medial. Both bundles are further subdivided into several small bundles. Some fibres run directly to the hypothalamus, and some cross in the anterior commissure. The centrifugal fibres originating in the telencephalon run backwards through the medial portion of the tract, terminate in the granule-cell layer, and form synapses with granule-cell dendrites of the bulb. The mitral cells are thus under the influence of higher centres through the centrifugal fibres.

The olfactory tract fibres appear to terminate bilaterally in the following telencephalic hemispheres: (1) medial terminal field in the area ventralis telencephali, (2) lateral terminal field in the ventrolateral part of the area dorsalis telencephali, and (3) posterior terminal field in the central part of the area dorsalis telencephali. Recent studies support the view that the olfactory system is composed of two separate systems: (1) the lateral olfactory system, in which the lateral part of the olfactory bulb receives input mainly from the lateral bundle of the olfactory nerve and sends outputs to the lateral olfactory tract, and (2) the medial olfactory system, in which the medial part of the olfactory bulb receives inputs mainly from the medial bundle of the olfactory nerve and sends outputs to the medial olfactory tract. The segregation of the teleost olfactory system may be homologous to two functionally distinct subsystems, i.e., the main and accessory olfactory systems. The reader is also referred to Wullimann (1998) and Hamdani et al. (2000; see Table 3).

5.2.4 Olfaction in Flatfish: Lopsided Smelling

In all fishes except the flatfish species, the olfactory organs of the two sides of the animal are equal in size, and the central connections of the olfactory bulbs are symmetrical. In the flatfish (e.g., winter flounder *Pleuronectes americanus*) the olfactory organ on the upward-facing side is considerably larger than the one on the downward-facing side. The olfactory centres in the central nervous system are similarly asymmetrical with the olfactory bulbs, and even the telencephalon of the upward side being larger than the contralateral counterparts (Prasad Rao and Finger 1984). Despite the marked asymmetry in size of the olfactory bulbs, the bulb on each side maintains a projection to the contralateral telencephalon.

5.2.5 Terminal Nerve

In all cases, including bony fishes, a separate group of fibres emerges from the telencephalon or olfactory bulb to join the olfactory nerve in entering the olfactory capsule. This distinct nerve has been called the *terminal nerve* (nervus terminalis) and is usually considered a separate cranial nerve, distinct from the olfactory nerve. The axons of the terminal nerve course centrally into the telencephalon in association with the medial olfactory tract in a wide range of vertebrates including teleosts. Investigations indicated that adult bichirs retain a terminal nerve, but the location of its ganglionic cells remains enigmatic. In contrast, terminal nerve does not appear to exist in *Eptatretus*, and its occurrence in lampreys is uncertain. In goldfish (*Carassius auratus*), large neuronal cell bodies (somata) occur within the median half of the olfactory nerves and the rostral part of the olfactory bulb. The peripheral process of some ganglion cells of the terminal nerve end in the olfactory epithelium, and the central processes project to various areas in the telencephalon, diencephalon and mesencephalon as well as to the retina. Those terminal nerve ganglion cells which project to the retina contain LHRH (luteinizing hormone-releasing hormone) and has been named the nucleus *olfactoretinalis* (read also Chiba et al. 1996a).

In goldfish, the large ganglion cells of the terminal nerve receive direct somatic input from a small number of catecholaminergic axons arising from the locus coeruleus. This connection may indicate that this population of terminal nerve cell is involved in a behavioural or attentional, rather than sensory, process. One hypothesized role for the terminal nerve is in control of some aspects of mating behaviour (Figs. 15.58, 15.59). (see Demski and Northcutt 1983; Finger 1988; Hara 1993; Wullimann 1998; read also Kleerekoper 1978).

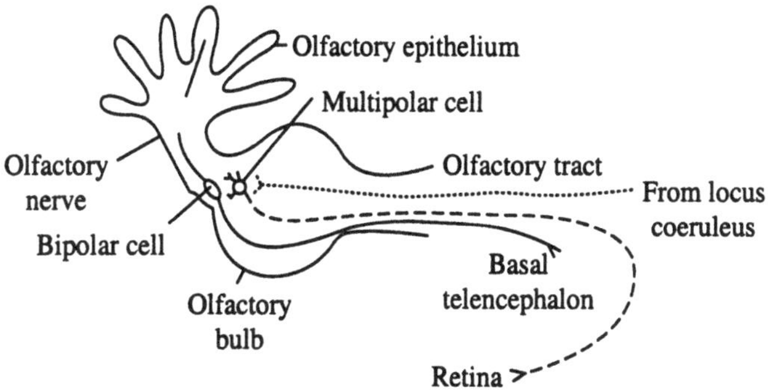

Organization and position of terminal nerve ganglion in *Carassius.*

Fig. 15.58 Semischematic diagram of the organization and position of terminal nerve ganglion in *Carassius*. Two types of cells are present (1) bipolar cell, which has peripherally directed process and centrally directed process that reaches only as far as the basal telencephalon, and (2) multipolar type, with short dendrites and which projects to the retina, as well as to basal telencephalon. Multipolar cell is immunoreactive for LHRH (luteinizing hormone-releasing hormone). At least some multipolar terminal nerve ganglion cells receive centrifugal input from the vicinity of locus coeruleus.
***Source:* Finger (1988)**

5.3 Gustatory System

Various sources have informed us on taste in fish (Atema 1971; Hara 1971, 1993a, b, 1994a, b; Bardach and Villars 1974; Kapoor et al. 1975a; Pevzner 1976; Reutter 1978, 1982, 1986, 1992; Caprio 1982, 1984, 1988; Little 1983; Tucker 1983; Jakubowski and Whitear 1986, 1990; Kanwal and Caprio 1987, 1988; Kanwal et al. 1988; Roper 1989; Finger et al. 1991; Kortschal et al. 1991; Gomahr et al. 1992; Kanwal and Finger 1992; Marui and Caprio 1992; Caprio et al. 1993; Reutter and Witt 1993; Whitear 1993; Zaccone et al. 1994; Finger 1997; Sorenson and Caprio 1998; Jakubowski and Zuwala 2000, and Tagliafierro and Zaccone 2001).

Taste buds are considered as one of the components of the paraneuronal family. Our knowledge of the distribution of taste buds in different systematic groups of fishes is very poor. Their presence has been reported in hagfishes (note Schreiner Organs represent a unique agglomerulation of SSC not homologous to taste buds) (read further), lampreys, *Trygon*, *Raja*, *Lepidosiren*, *Protopterus*, *Polypterus*, *Calamoichthyes*, *Acipenser*, and *Amia*. The position of taste buds differs. No report is yet available on *Latimeria* and *Lepisosteus*. Most of the work in fish taste system is concerned with one of numerous teleost species. (For Schreiner organs, see Braun and Northcutt 1998; Braun 1998).

The taste buds (peripheral gustatory organs, terminal buds) are compound epidermal chemosensory and tectile organs. Besides all the differences between teleostean species there

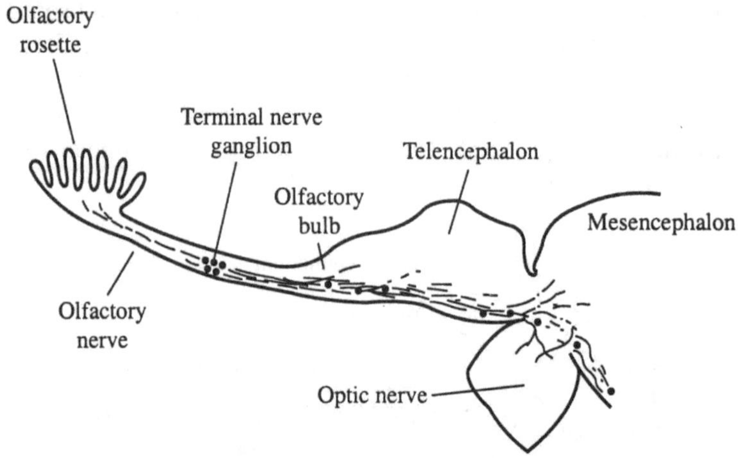

Fig. 15.59 Diagram of the forebrain of the ayu *Plecoglossus altivelis* showing gonadotropin releasing-hormone (GnRH) immunoreactive terminal nerve and hypothalamic neuron systems. Large dots represent large polygonal neurons that belong to the terminal nerve, whereas small dots show small fusiform cells that comprise the hypothalamic neuron system. Lines are GnRH-positive nerve fibres that belong mainly to the terminal nerve. *Source:* Chiba et al. (1996a)

seems to exist one basic type of taste buds in bony fishes. Such taste buds are typically ovoid or pear-shaped, their longitudinal axis is oriented vertically to the surface of the epidermis of the outer skin or the epithelium of the lip, the oral cavity including tongue, the pharynx, the oesophagus, on the gill arches or gill rakers, and gill chambers of most teleosts. From various observations on taste buds occurring on gill arches and gill rakers, certain significant points were put forth: (1) no marked differences in number and arrangement of taste buds were found in plankton feeder and predatory fishes; (2) fresh water fishes had a denser arrangement of taste buds in their branchial regions in comparison to that of marine fishes, and (3) the taste buds of marine fishes were extremely small and not well developed.

A correlation has been found between taste bud development and the variety of food types present in the environment. Fishes living in dark muddy or silty water or even in caves have more taste buds than the cognates living in clear waters of the region. With regard to the visual sense this might be a sign of organ compensation. Taste buds of elasmobranchs are restricted to the mouth and pharynx. Normally taste bud-bearing epithelia are stratified and squamous and not keratinized on the apical surface. The taste bud elements are supporting or sustentacular cells, and receptor cells (gustatory cells). Gustatory cells are also counted as paraneurons. The receptor and sustentacular cells possess surface hair (light microscopy) or microvilli (electron microscopy) of two types. Sustentacular cells (dark cells) have several short microvilli, their basal parts are divided into lobed processes. Receptor cells (light cells) bear a single atypical, often large microvillous extending into the cell as a rod surrounded by vesicles, although some bony fishes are reported to have two microvilli (Fig. 15.60).

The cytoplasm of light cell comprises many organelles: Golgi systems, vesicular structures, endoplasmic reticula (RER, SER), free ribosomes, mitochondria, tonofilaments, microtubules, and occasionally lysosomes, multivesicular bodies and glycogen granules. Dark cells are relatively

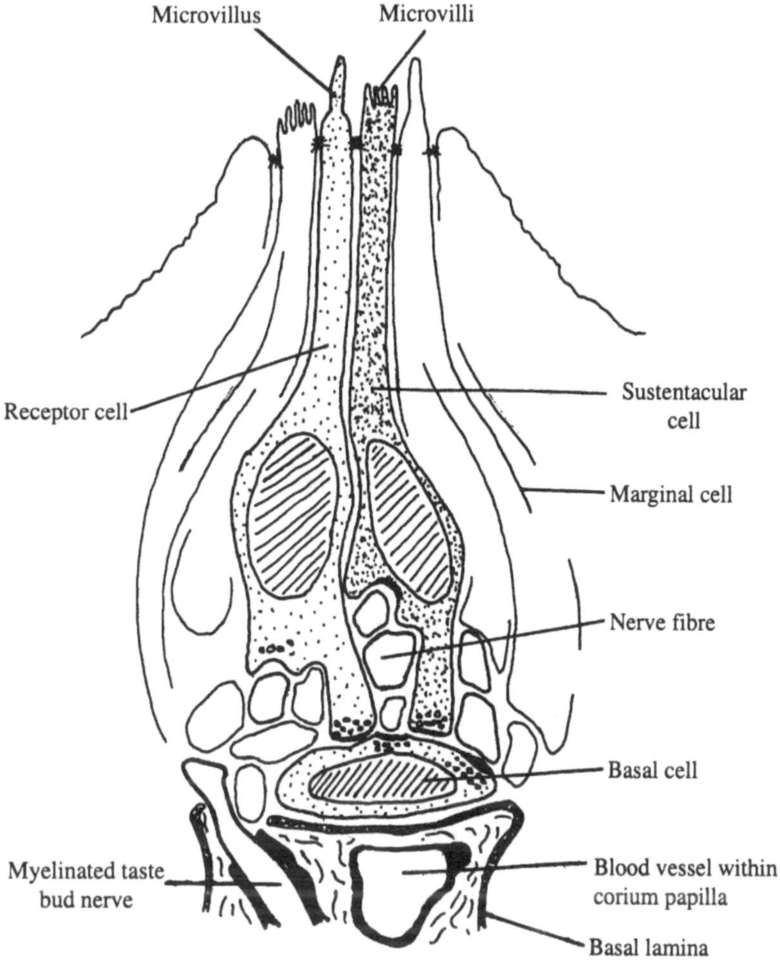

Fig. 15.60 Schematic drawing of the ultrastructure of *Ictalurus* taste bud. Note the sites of synaptic contacts.
***Sources:* Kapoor et al. (1975a) and Reutter (1986)(Modified)**

more electron-dense and rich in organelles compared with the light cells. In nucleus the hetero- and euchromatin are distinct areas and paracrystalline inclusion may also be found.

Elasmobranch taste receptors have 3 to 12 microvilli. Both cell types may synapse with basal cells, and nerve fibres synapse with basal and atleast one of the other cell types. The cytoplasm of basal cells has mitochondria, endoplasmic reticula (RER, SER), Golgi systems, free ribosomes, glycogen granules, intermediate filaments, microtubules, multivesicular bodies, and a centriole. The basal cells are rich in serotonin and neuron-specific enclose (NSE) and possibly related to Merkel cells, which belong to the class of paraneurons. The function of the basal cells is still unclear; may serve different functions. Mucus covering the receptive surfaces of gustatory and olfactory sensory cells probably does not materially impede access by small molecules and ions. Taste bud cells turn over by replacement with cells from the surrounding epithelium. Taste buds are maintained by a trophic influence of nerves. For views on light and dark cells, the reader is referred to Reutter (1986, 1992, 1993); Jakubowski and Whitear (1990) and Zaccone et al. (1994).

For lecting binding sites in the taste buds of fish, the reader is referred to Witt and Reutter (1988, 1990). Lectin-carbohydrate interactions may play a role in recognition phenomena on the plasmalemmata of taste bud sensory cells.

As mentioned the taste buds are not confined to the oropharyngeal cavity, but also occur at the body surface. These external taste buds are particularly abundant in the ostariophysian fishes, like siluroids and cyprinids. They show a polar arrangement at the body surface, with highest densities at the lips, the gular region, barbels (if present) and along other suitable appendages, like pectoral and pelvic fins.

Quantitative studies on densities and distribution of external taste buds are rare. Even less abundant are comparative investigations. A recent study is on the density and distribution of external taste buds in some European cyprinid fishes. Although some relation between degree of benthivory and density of external taste buds exists, a purely adaptive interpretation of taste bud distribution patterns seems questionable when a finer spatial and structural scale is considered. It is known that taste buds only survive under the influence of an intact nerve supply. Taste bud formation is induced by the re-innervating nerve fibres. Thus the information on taste bud density may mainly reside in the innervating nerves, which are of neural crest origin.

Taking the ecomorphological view, at a coarse scale, density and arrangement of the external taste system can be predicted from the species ecology.

Benthivory and a turbid habitat often occur together. Thus in species and environments where vision is only of limited use, the external taste system may increase in importance for detecting food. External taste may serve for gradient search, as in catfish, but generally elicits ingestion upon contact. Touch seems as important for triggering this feeding reflex as chemostimulation. Indeed, all taste systems investigated are also touch responsive. Taste buds on hillocks may even be superior for the perception of tactile stimuli. Taste buds thus integrate information on the chemical quality with exact spatio-temporal location of prey.

It is suggested that there is indeed a compensation by taste when vision is impeded by low light levels or high turbidity. In discussing this subject, one meets the most intriguing question as to how vision and taste "communicate"? How could a deterministic adjustment of taste bud density possibly be triggered by deficient or suboptimal visual input? What causes the changes in numbers and/or size of the neural crest derived sensory ganglion cells which induce taste bud formation.

5.3.1 Organization of Taste System within the Brain of Bony Fishes

Gustatory information is transmitted to the brain of vertebrates by three pairs of cranial nerves: the facial (VII), the glossopharyngeal (IX), and the vagus (X). In ictalurid catfish these make up at least two gustatory subsystems. Facial nerve branches, which innervate primarily the taste buds located on external body surface, together with their central projections constitute the extraoral gustatory system. This system is implicated in the detection and localization of a food source. The glossopharyngeal and vagal nerve branches innervate only those buds located in the oropharyngeal region and thus constitute the oropharyngeal taste system, which is important for selective food ingestion (Fig. 15.61).

The peripheral organization of glossopharyngeal and vagal nerves has been drawn in Fig. 15.61. Pattern of projection of IX-X nerve roots in the brain stem of fishes has been under intensive investigations. The new findings suggest interesting aspects of neural organization and information processing in the teleostean brainstem (Table 15.10).

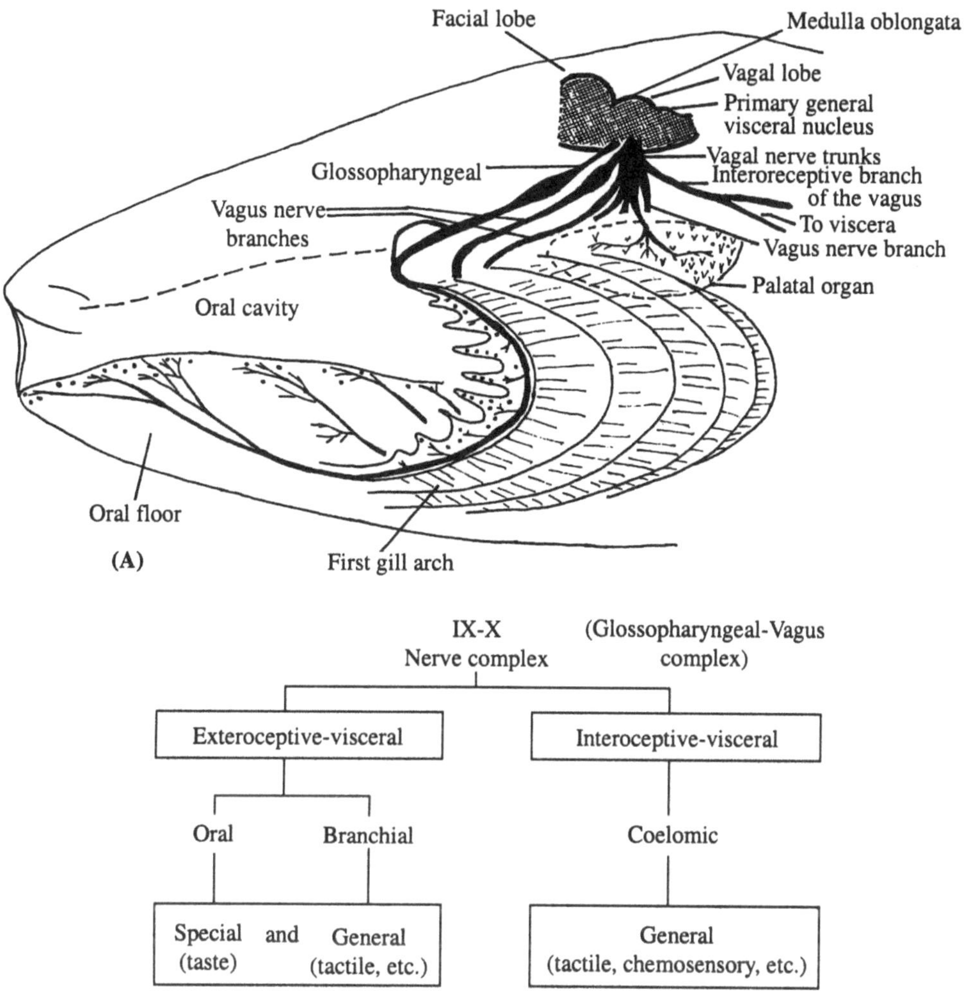

(B) Major components of the Glossopharyngeal-vagal (IX-X) nerve complex in ictalurid catfish.

Fig. 15.61 (A) A diagrammatic sagittal view of the peripheral pattern of innervation of the glossopharyngeal (IX) and exteroreceptive (branchial) and interoreceptive (branchial) and interoceptive visceral branches of the vagus (X) in the oropharyngeal region of the channel catfish. (B) Two functional-anatomical subdivisions of nerve fibres in the glossopharyngeal-vagal complex. The glossopharyngeal nerve and the branchial branches of the vagus together constitute the exteroreceptive-visceral subdivision, which is distinct from the group of exteroceptive-visceral fibres of the vagus innervating coelomic viscera.
***Source:* Kanwal and Caprio (1987)**

The nucleus intermedius of the facial lobe, the dorsal cap of the vagal lobe, and the nucleus intermedius of vagal lobe are implicated as sites for visceral interactions related to feeding. The exteroreceptive-visceral nerve branches remain distinct, peripherally and centrally, from the interoreceptive-visceral branch of the vagus.

Brain patterns allow prediction of sensory types in fishes. Facial and vagal taste inputs

Table 15.10 Correlation of visceral information in the brainstem of ictalurid catfish

Brainstem regions	Converging visceral inputs
Nucleus intermedius of the facial lobe (nIF)	Extraoral and oral gustovisceral inputs
Dorsal cap nucleus (DC)	Spatially segregated oropharyngeal visceral inputs
Nucleus intermedius of vagal lobe (nIV)	Exteroreceptive and interoreceptive visceral inputs.

Source: Kanwal and Caprio (1987)

terminate in the facial and vagal lobes respectively, in the medulla of teleosts (Figs. 15.62, 15.63). In the majority of teleostean species, neither the facial lobe nor the vagal lobe is well developed, and the two may even be continuous and grossly indistinguishable from each other. This general pattern of organization of primary gustatory nuclei applies to elasmobranch as well.

In several species of teleosts the gustatory system has evolved to facilitate feeding by discriminating and localizing food stimuli in the environment (via the facial subsystem) or in the oropharynx (via the vagal subsystem). In general, the primary gustatory centres exhibit similar types of anatomical modifications, such as lobular or layered morphology (exhibiting highly organized units of structure) and topographic patterns of peripheral input for the representation of sensory information. In anyone species, the facial, glossopharyngeal, and chemosensory inputs are encoded and distributed in parallel, at least upto the level of primary sensory centre.

The gustatory tracts from the facial and vagal lobes ascend and descend within the brain. Fibres within the descending tract terminate in the funicular nuclei. In addition, direct (faciospinal) and indirect (reticulospinal) pathways transmit facial taste information to the spinal cord, while vagal taste projects directly to the nucleus ambigous. The supramedullary gustatory pathways may belong to either of two categories. One category of reciprocal connections constitutes a loop between the diencephalic and the brainstem nuclei. The other is a 'lemniscal-type' 'labelled-line' sensory pathway which ascends from the medulla to the telencephalon. However, these ascending pathways are not necessarily homologous to similar pathways (limbic and lemniscal systems) described in mammals. Finally, several descending projections also originate from the telencephalic taste centre. These descend ipsilaterally via the medial and the lateral forebrain bundles and terminate in the vicinity of the preglomerular complex, the nucleus paracommissuralis in the habenula and several other sites within the ventral diencephalon.

Figure 15.64 is the schematic diagram indicating connections of gustatory lobes in teleosts.

Diagram 15.65 shows the basic neuronal circuitory from the palatal sorting system in the common goldfish (see also Fig. 15.66) (see for details: Finger 1983, 1988a, b, 1997; Kanwal and Caprio 1987; Kanwal et al. 1988; Hayama and Caprio 1990; Finger and Kanwal 1992; Goehler and Finger 1992; Kanwal and Finger 1992, 1997; Lamb and Caprio 1992, 1993; Fink and Wullimann 1998). The reader is also referred to Wullimann (1998).

5.3.2 Taste Transduction

Gustatory perception occurs within a few hundred milliseconds following the interaction of stimulus molecules or ions with the exposed, apical membranes of taste receptor cells. This interaction initiates a sequence of events leading to membrane depolarization and release of an unidentified neurotransmitter (s) at synapses with nerve fibres of cranial nerves VII, IX or X.

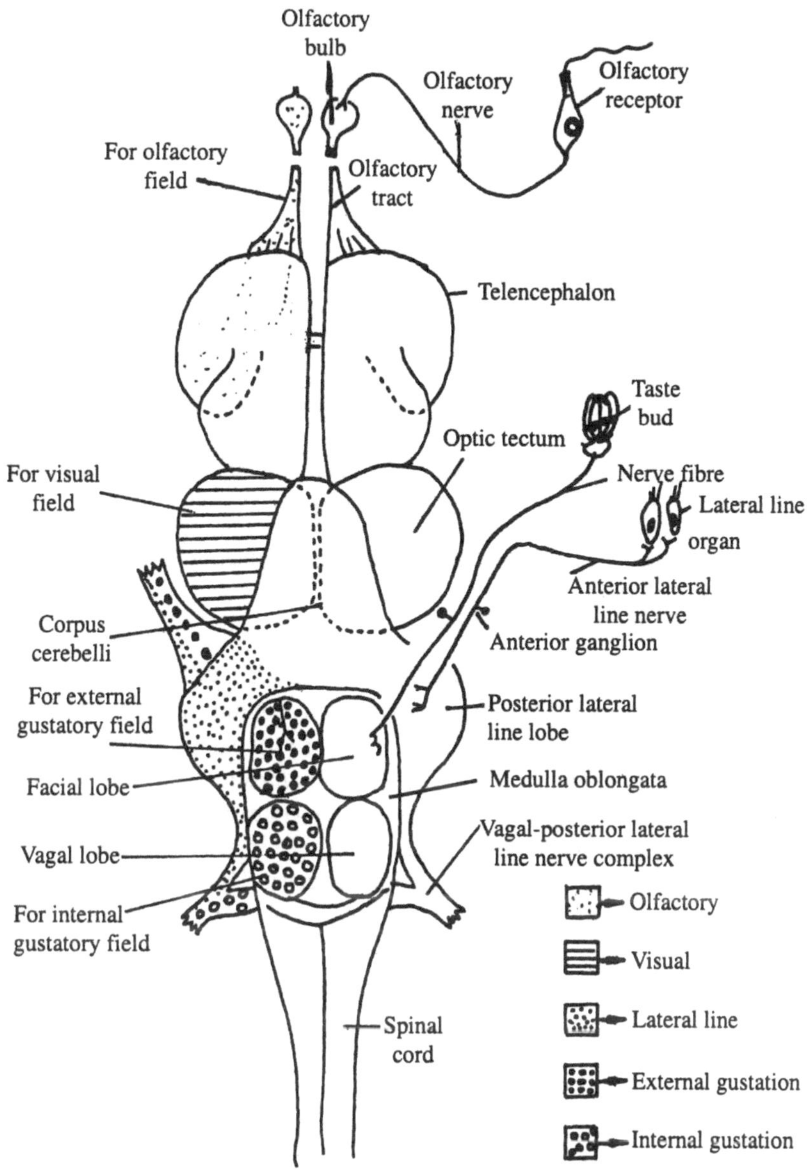

Fig. 15.62 Schematic dorsal view of the catfish brain, showing receptor morphology and primary central terminal fields for various cranial nerves.
***Source:* Finger (1983)**

The resulting increase in impulses to the taste area of the medulla and higher neural centres ultimately results in taste perception.

Different classes of amino acid receptors identified in catfish (*Ictalurus punctatus*) taste cells appear to be coupled to taste receptor cell depolarization and subsequent neural and behavioural events in one of two general ways. The receptor for L-alanine and some other neutral amino acids is G-protein linked to formation of second messengers and may belong to the known large

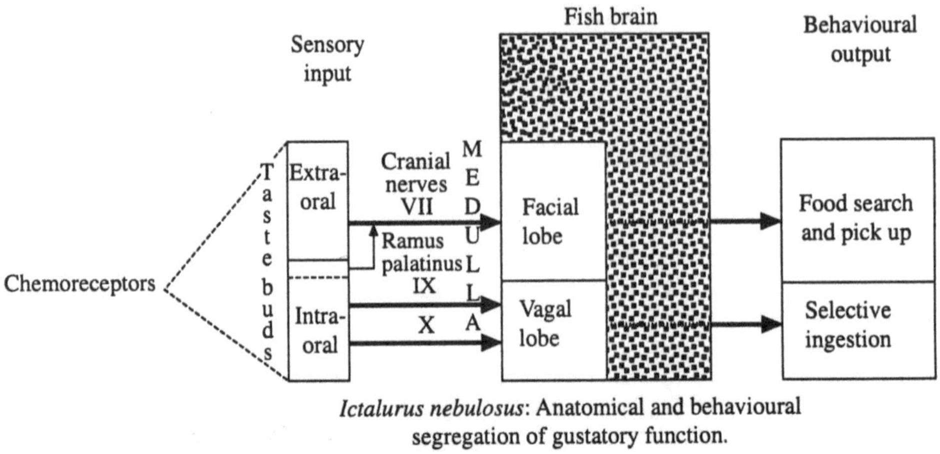

Fig. 15.63 Anatomical and behavioural segregation of gustatory function in the bullhead catfish, *Ictalurus nebulosus*. The extraoral taste subsystem is anatomically distinct from the oropharyngeal subsystem in its innervation and central representation. The two subsystems are also behaviourally distinct in that they coordinate different phases of feeding. rp. ramus palatinus-a branch of the facial innervates the anterior palate.
***Source:* Kanwal and Finger (1992)**

family of G-protein-coupled receptors. In contrast, the high-affinity taste receptor for L-arginine and the low-affinity taste receptor for L-proline appear to contain integral cation channels that are directly gated by stimulus binding and may belong to the superfamily of ligand-gated ion channels. The distinction between the analine and arginine/proline receptor mechanisms is also maintained at the neural level where two major types of gustatory fibres, the alanine and arginine fibres, transmit amino acid information to taste nuclei within the central nervous system that control food search and consummatory behaviour patterns. The reader can obtain details of gustation from Sorenson and Caprio (1998; Table 1) and also from Hara (1993; Table 2).

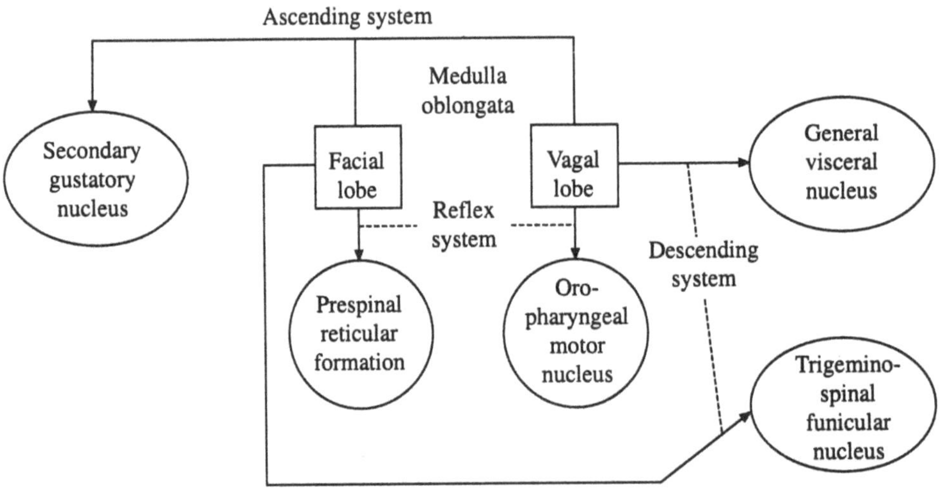

Fig. 15.64 Schematic diagram indicating connections of gustatory lobes in teleosts.
***Source:* Finger (1988a)**

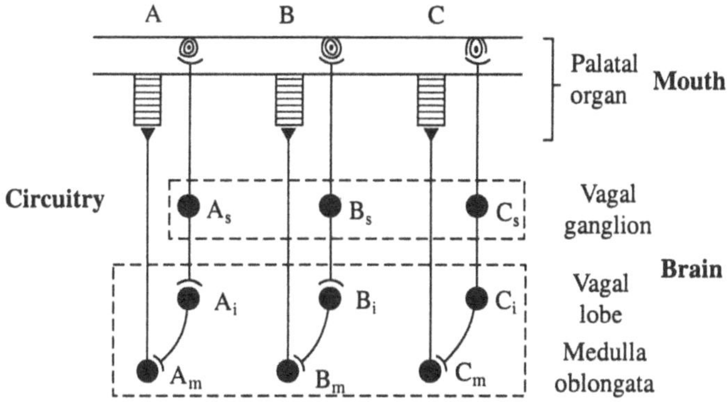

Fig. 15.65 Schematic drawing showing the minimal circuitry involved in the intraoral food-sorting system in goldfish (*Carassius auratus*). At least one interneuron (A_i, B_i, C_i) is interposed between the primary sensory input and the motor output.
***Source:* Finger (1988)**

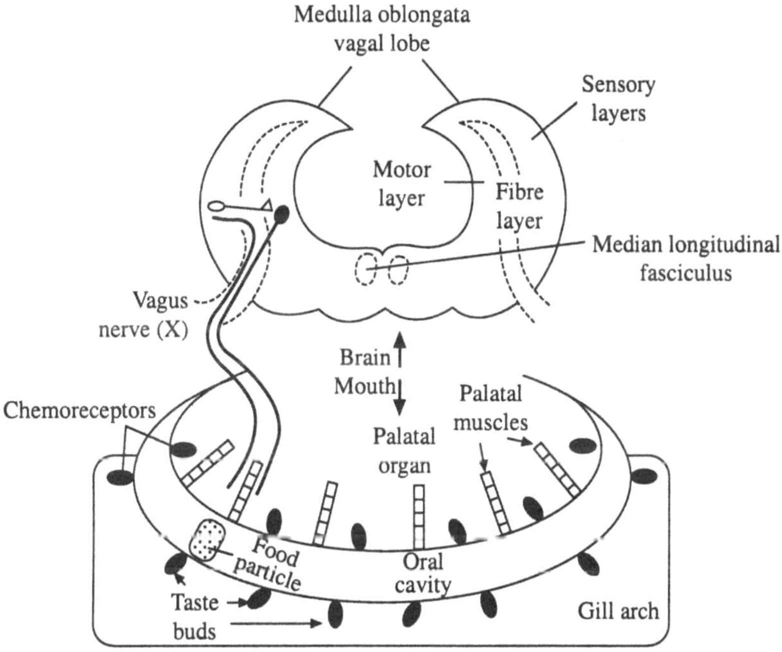

Fig. 15.66 Intraoral gustatory system in cyprinid *Carassius auratus*.
***Source:* Kanwal and Finger (1992)**

Both oral and extra-oral systems of the channel catfish are generally poorly sensitive to the classical vertebrate taste stimuli (salts, sugars and acids; however, a primary exception is the relatively high sensitivity of the oropharyngeal taste system to quinine hydrochloride, a compound that is perceived as bitter by humans (read details: Kanwal and Caprio 1983; Tucker 1983; Caprio 1984; Kanwal et al. 1987; Bryant et al. 1989, 1993; Kalinoski et al. 1989a, b, 1992;

Michel and Caprio 1991; Wegert and Caprio 1991; Huque et al. 1992; Kohbara et al. 1992; Caprio et al. 1993; Michel et al. 1993; Valentinčič 1994; see also Marui and Caprio 1992; Hara 1993b, 1994; Devitsina and Chervova 1994); Kohbara and Caprio 1996; Valentinčič and Caprio 1997; Ogawa et al. 1997a, b; Sorenson and Caprio 1998).

5.4 Solitary Chemosensory Cells

Solitary chemosensory cells (SCC) belong to the paraneuronal family. SSC of unknown function are vertebrate secondary epidermal cells which resemble taste bud receptors and regarded as a major actinopterygian chemosensory system. These cells occur widespread in lower aquatic vertebrates (e.g., hagfish as Schreiner Organs on tentacles and nasopharynx, lampreys, selachians, rocklings, cyprinids, searobins). In lampreys oligovillous cells (solitary chemosensory cells) responded best to mucoid substances, whereas in the searobins (*Prionotus carolinus*) finger-like free pectoral fin rays, supposed solitary chemosensory cell units were amino-acid positive. In the searobin the fin innervation seems entirely spinal (Fig. 15.67) (see Fig. 15.68 for hake) their central sensory representations cause distinct swellings within the rostrodorsal spinal cord. By these rays, devoid of taste buds, these benthic fishes probe the substrate for food. In the presently best known solitary chemosensory system, the anterior dorsal fin of rocklings (e.g., *Gaidropsaurus* and *Ciliata*, Gadidae), the sensory nerve synapsing with anterior dorsal fin solitary chemosensory cells are facial (recurrent facial nerve). They carry no facial tactile component and are furthermore narrowly tuned, responding mainly to dilutions of body mucus of heterospecifics. High convergence of hundreds of receptor cells to a single facial nerve fibre, glomerular (and thus complex structural) organization of the dorsal facial lobe and anterior dorsal fin sampling favour the view of low-threshold discrimination properties of anterior dorsal fin solitary chemosensory system. The cells are flask-or spindle shaped with their nuclei in the basal part of the epidermis. The cells

Fig. 15.67 ***Prionotus carolinus*** **(Searobin). Semischematic drawing of somatotropic representation of pectoral fin and fin rays in rostral spinal cord.** ***Left:*** **Lateral view of fish to show the position of fin rays and pectoral fin.** ***Right:*** **Dorsal view of rostral part of spinal cord, indicating which appendage is represented in which accessory spinal lobe. Note counterintuitive arrangement; i.e., ventral fin ray which is anatomically most anterior, is represented in most posterior accessory lobe; dorsal fin ray or its spinal innervation; middle fin ray or its spinal representation; ventral fin ray or its spinal innervation.**
Source: **Finger (1988)**

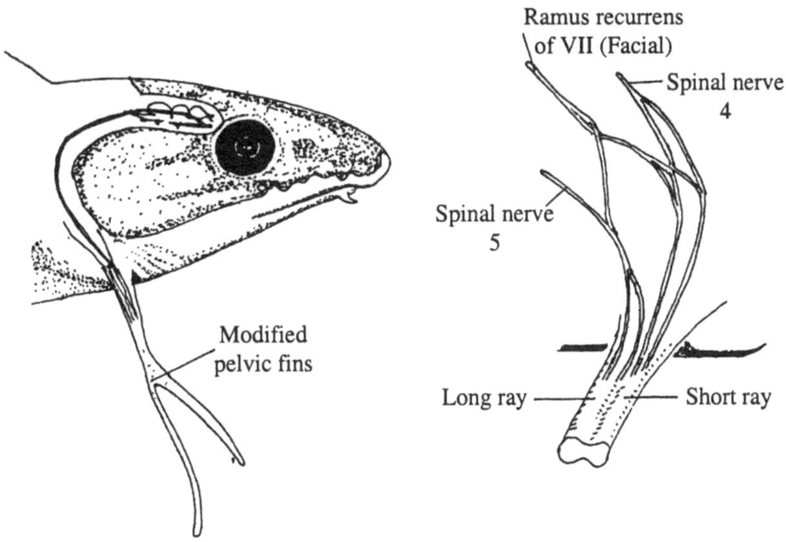

Fig. 15.68 Innervation of the external taste buds on the modified pelvic fins of a hake (*Urophycis*).
***Source:* Bardach and Atema (1971)**

are bipolar, with an apical process, or processes, at the surface of the epidermis, and the proximal region associated with a neurite profile.

There is some variation of solitary chemosensory cells apex structure over the vertebrate pedigree. This suggests a phylogenetic change from a few microvilli at the agnathan level of organization, which eventually merge or were replaced by a single villus at the agnathan-gnathostome transition. From the lower actinopterygians into the teleosts, this single villus may be branched and form a multiple tip. In the most modern, perciform teleosts, as well as in rocklings, the receptive villus is unbranched and shows a single tip only.

They may be found in the external skin and oropharyngeal epithelium, including the gills, in various families of teleost fishes. In lampreys, the corresponding cells were called 'oligovillus' to distinguish them from other bipolar cells present in the epidermis (Fig. 15.69). The cytology is peculiar but sufficiently similar to that in the teleosts to warrant comparison. Oligovillus cells are numerous in the oral region, around dorsal fin, the gill vents and at the tip of male genital papilla. They are present everywhere in the lamprey's skin, but more scattered. Lamprey oligovillus cells have microvillar cores which may extend throughout the neck of cell. Numerous mitochondria, Golgi systems, endoplasmic reticulum, multivesicular bodies, lysosomes, and well-developed vertically set microtubules characteristically surround the core of actin filaments subtending the apical process, or processes; this description applies to the solitary chemosensory cells of other groups besides lampreys (for details, the reader is referred to Whitear 1992). The transmitter involved in the solitary chemosensory cells is unknown.

Solitary chemosensory cells on the head and oropharyngeal epithelium could share innervation with the taste buds in the same area from the facial or vagal special sensory component. In species with a recurrent facial nerve passing to taste buds on the body, this can also carry nerve fibres to solitary chemosensory cells as in the case of the anterior dorsal fin of rocklings. In the triglids, the sensory innervation of modified pectoral fin rays is undoubtedly from spinal nerves.

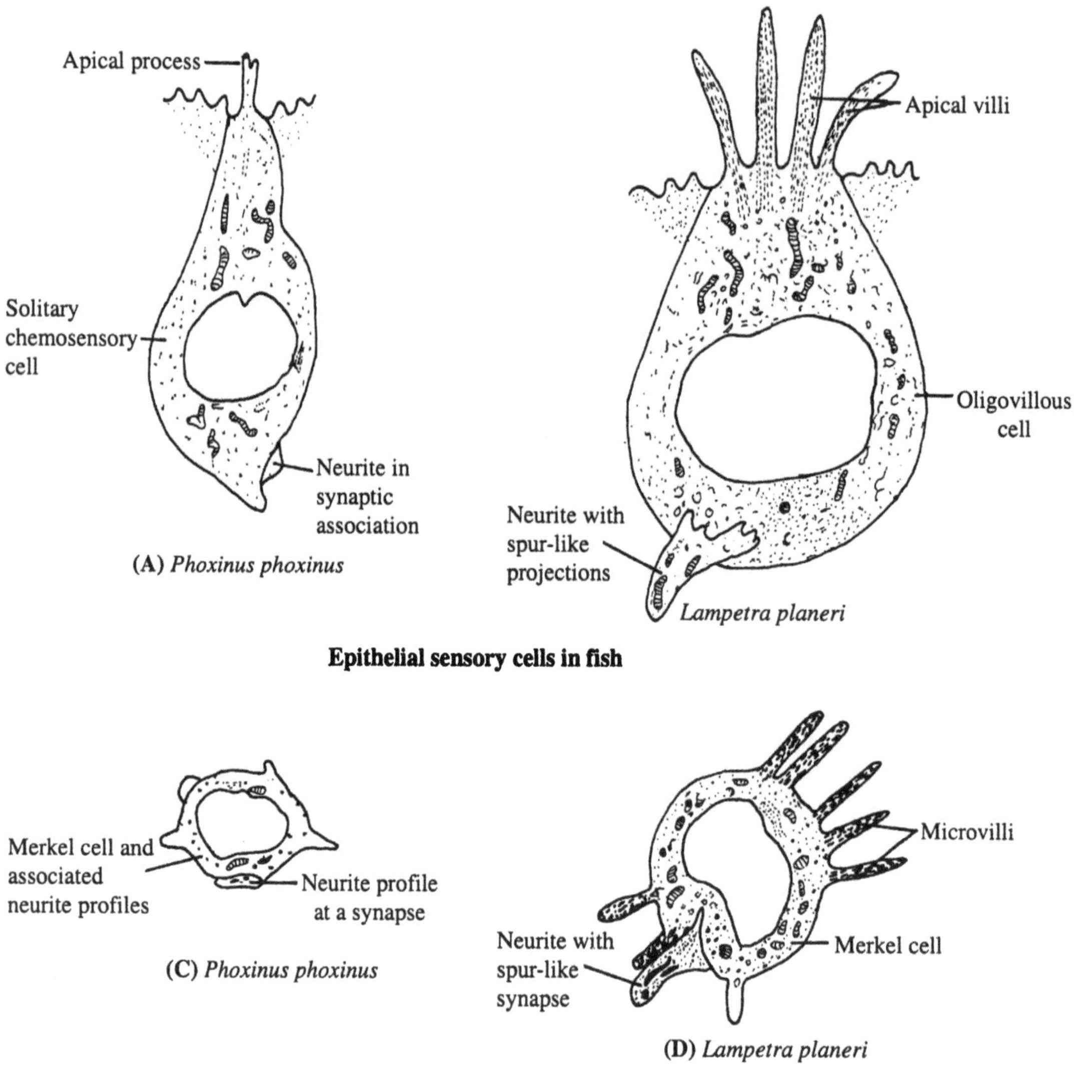

Fig. 15.69 (A) Solitary chemosensory cell. (B) Oligovillous cell. (C) A Merkel cell with associated neurite profiles. (D) A Merkel cell with microvilli.
***Source:* Whitear (1993)**

The nerve endings concerned lie in small projections of the epidermis, the claviform bodies, each of which contains a sensory cell and a loop of free endings that is presumably tactile in functions.

These chemosensors have been the object of recent studies (Finger 1982, 1997; Silver and Finger 1984; Reutter 1986; Kotrschal et al. 1984, 1989, 1990a, 1993; Kortschal 1991, 1992, 1995, 1996; Whitear 1992, 1993; Peters et al. 1989, 1991; Zaccone et al. 1994; 2001; also read Sorenson and Caprio 1998).

Kotrschal (1992) reported that (1) in teleosts chemosensory cells may be very abundant and (2) in interspecific comparison neither solitary chemosensory cell distribution patterns nor their abundance covary with taste bud system. These results support the view that solitary chemosensory cells have functions and biological roles different from taste bud system.

Different structural organization of taste buds and solitary chemosensory cells indicates different transmission properties. Fish taste buds, like other vertebrate taste buds are compound organs, which contain three major specialized cell types, chemosensory, supporting and basal cells. Although there is considerable disagreement on taste bud synaptic organization and terminology in fishes, it seems clear that taste buds respond both to a relatively wide range of chemicals and also to tactile stimuli. This allows them to combine the benefits of touch and chemosense: exact spatio-temporal tactile location can be matched with information on local chemical quality. As also indicated by their polar distribution over the body, taste buds may be characterized as chemosensory touch corpuscle. As such, external taste buds localize food and are involved in oropharyngeal sorting process. In contrast, solitary chemosensory cells, embedded between epidermal cells, are relatively evenly spaced over the body surface.

No matter what function and biological role will be ultimately associated with solitary chemosensory cells in cyprinids and other teleosts, and whether or not solitary chemosensory cells are the structural substrate for Parker's common chemical sense, the quantitative results emphasize their importance and independence from the taste bud system.

Table 15.11 summarizes a few characteristics of chemoreceptors.

Fig. 15.70 is a comparison of three major chemoreceptive systems in fish.

Table 15.11 Summary of proposed sensory input channels in teleost fish, listing their morphological substrate (= sensory cell), cell type (primary or secondary), central connections (termination areas of sensory afferents in the central nervous system), functions (neuronal responses to stimuli) and biological roles (behavioural responses and ecological context). There is no direct evidence for a sensory function of the nervous terminalis system (Demski and Schwanzel-Fukuda 1987; cited from Kotrschal (1991)

Chemosensory input channel	Morphological substrate	Sensory cell type	Central connections	Functions	Biological roles
1. Olfactory system	Olfactory neurons	Primary	Olfactory bulb	Low-threshold, wide spectrum discrimination	Food search, chemo-learning and social behaviour
2. Nervous terminalis system	Olfacto-retinalis neurons	Primary	Retina, hypothalamus	(?)	Reproduction (?)
3. Free nerve endings ('Common chemical sense')	Trigeminal or spinal afferent neurons	Primary	Brainstem, spinal cord	High-threshold discrimination of aversive stimuli	Avoidance behaviour
4. Taste	Taste buds	Secondary	Brainstem viscero-sensory lobe	Close-range (food) discrimination	Feeding
5. Solitary chemosensory cell chemosense	Solitary chemosensory cells	Secondary	Brainstem	Low-threshold mucus discri-mination	(1) Predator or general avoidance; (2) Feeding (?)

Source: Kotrschal (1991)

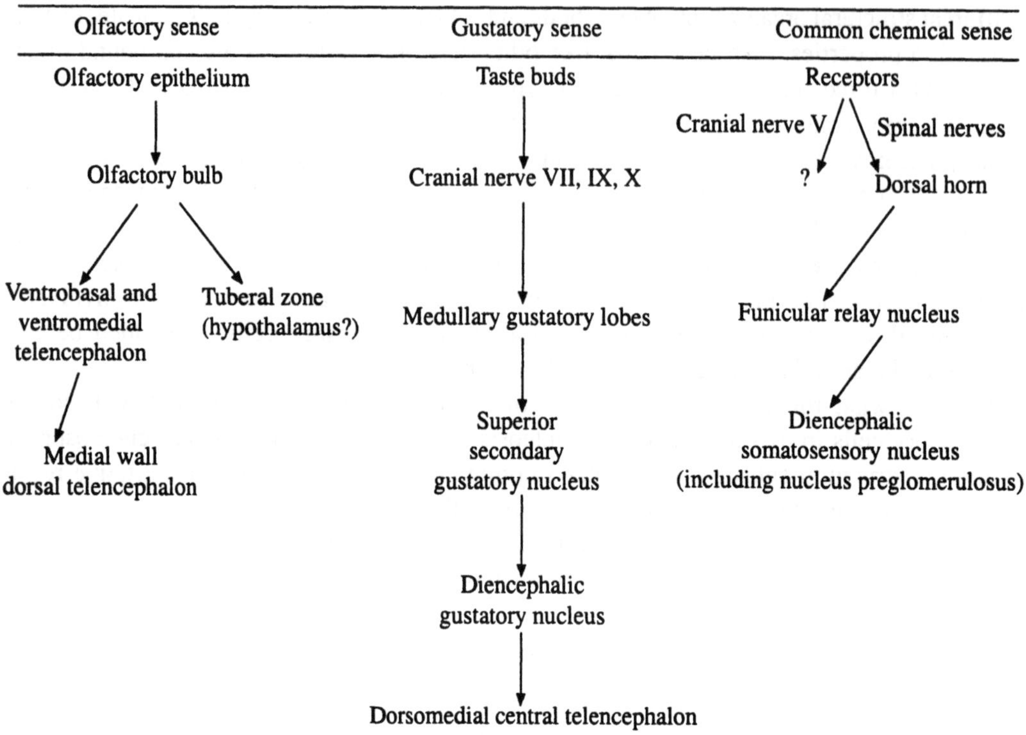

Fig. 15.70 Comparison of three major chemoreceptive systems in fish. *Note:* Little overlap exists for the three systems at any level of the neuraxis studied to date. One might expect, however, on the basis of behaviour, that the chemoreceptor systems ultimately converge on some central structure. Cranial nerves V (trigeminal), VII (facial), IX (glossopharyngeal) and X (vagal). *Source:* Finger (1988)

6 PINEAL COMPLEX

6.1 Introduction

The pineal complex develops from the diencephalic roof in a circumscribed area between the habenular and posterior commissures. The cellular elements of the pineal complex are derivatives of the primitive neuroepithelium. In several species of recent vertebrates the pineal complex is divided into distinct components.

In lampreys and most gnathostome fish the pineal complex consists of a pineal organ proper (epiphysis cerebri) and a parapineal organ, both located inside the cranium (Fig. 15.71). In several species of fish the tissue overlying the pineal complex displays distinct specializations forming a pineal window or fontanelle. The ventral aspect of the epiphysis cerebri is bordered by the dorsal sac. The parapineal organ is located to the left of the pineal stalk. The pineal organ proper is in anatomical continuity with the habenular and posterior commissures; the parapineal organ is connected to the left habenular ganglion. The lumen of the parapineal (very prominent in lampreys) is reduced to a capillary space in most other fishes, and the cytoarchitecture of the parapineal organ closely resembles that of the habenular nuclei. The pineal organ proper is a hollow structure in most species of fish, and its lumen is in open communication with the third ventricle.

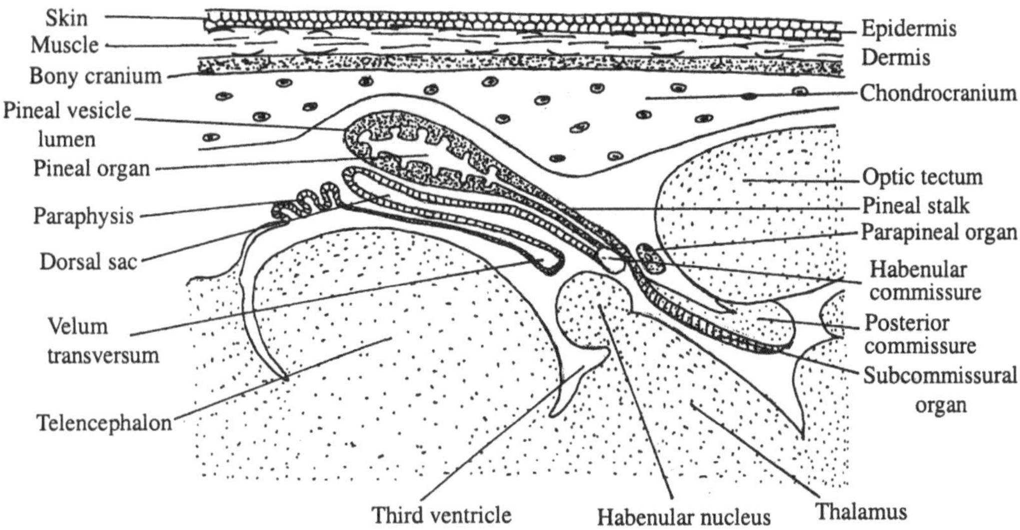

Fig. 15.71 Diagrammatic representation of a sagittal section through the telencephalic and diencephalic roof of the rainbow trout.
***Sources:* Oguri and Omura (1973) and Omura et al. (1990)**

In adult Myxinidae both the pineal and parapineal organs are absent. The anlage of the pineal organ, occurring in early ontogenetic stages, disappears during later development.

The electric ray family (Torpedinidae) possess no pineal organ. In *Narke japonica* a knob-like pineal organ with a central lumen has been demonstrated just in front of the subcommissural organ.

6.1.1 Pineal Organ

Pineal organ has been the subject of numerous contributions (Fenwick 1970; Oguri and Omura 1973; Omura and Ali 1980; McNulty 1978, 1979, 1984; Korf and Oksche 1986; Omura et al. 1990; Underwood 1990; Meissl and Brandstatter 1992; Meissl and Yáñez 1994; Sriwastwa and Srivastav 1994, Shedpure and Pati 1995-in vertebrates; read also Chapter 13). A Comprehensive account on the innervation of pineal gland, with focus on mammals, has recently been published (Korf 1996). Meyer-Rochow et al. (1999) determined immunohistochemically that in the end vesicles and the pineal stalks of Antarctic teleosts (*Pagothenia borchgrevinki* and *Trematomus bernacchii*) at least two population of cells occurred that displayed reactions indicative of typical vertebrate photoreceptors. In the well-developed pineal organ of teleosts, the epithelial parenchyma is composed of (i) typical and modified receptor cells, (ii) uni-bi-and multipolar nerve cells, and (iii) cerebrospinal fluid-exposed supporting cells. The pineal nerve projects via the stalk and the posterior commissure to various brain centres. Accumulating evidence suggests that the pineal nerve is composed not only of the axons of intrapineal nerve cells but also of the axonal processes of pineal photoreceptors and/or cerebrospinal fluid-contacting neurons (Fig. 15.72). The reader is referred to informative reviews from Ekstrom and Meissl (1997), and Shand and Foster (1999).

Photoreception. Photoreceptor cells, the outer segment of which ultrastructurally resembles that of retinal cones, were demonstrated in pineal sense organs of petromyzontids and fishes. The

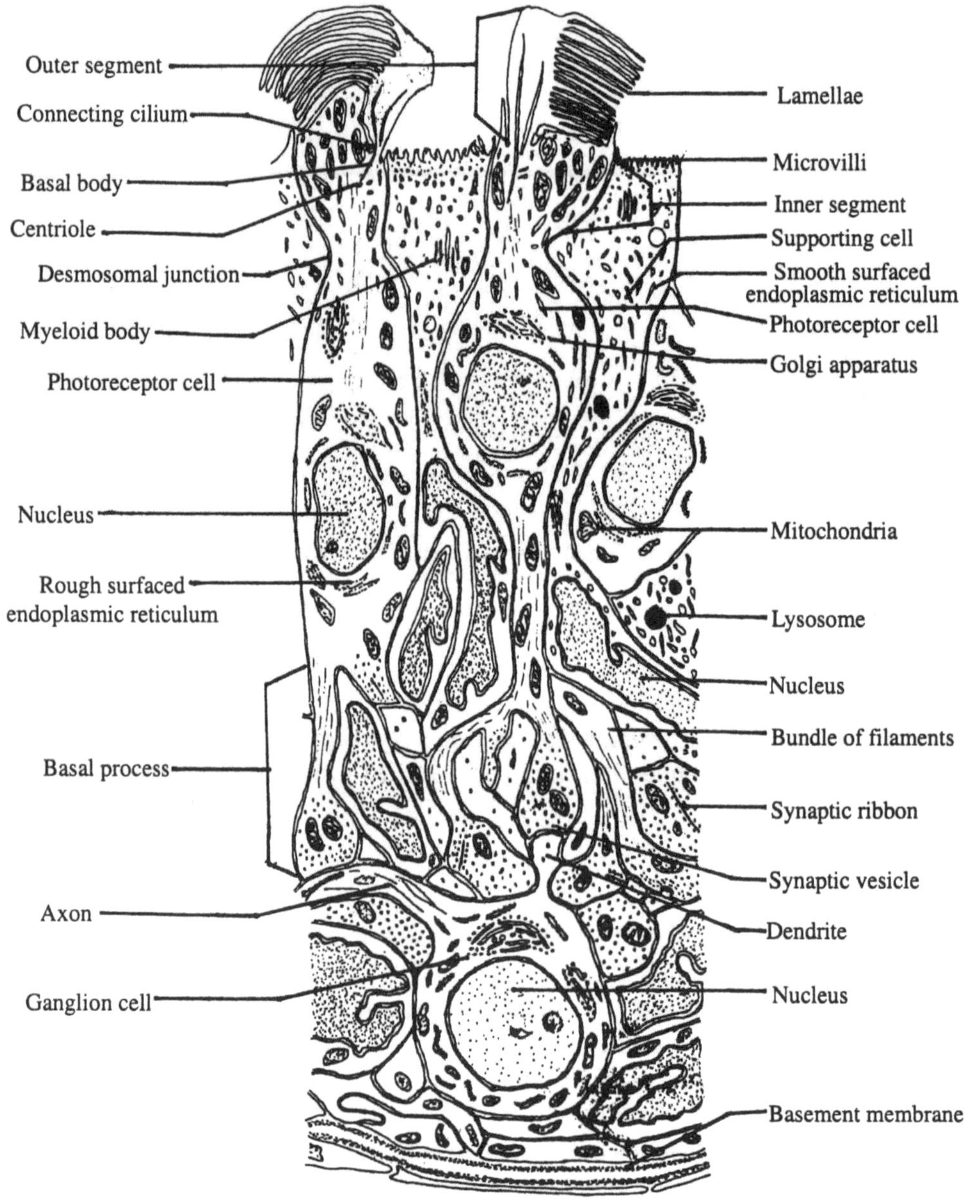

Fig. 15.72 Diagrammatic ultrastructure of the pineal epithelium of teleost.
***Source:* Oguri and Omura (1973)**

outer segment protrudes into the pineal lumen and consists of numerous discs produced by successive basoapical invaginations of the plasma membrane. The discs comprise membrane-limited spaces separated by thin layers of the outer segment cytoplasm and marginally remaining in communication with the extracellular space (i.e., the lumen of the pineal organ). The number of discs in teleost species is small, averaging between 20 and 60. Ontogenetic changes have been reported in outer segment structure. The extent of outer segment degeneration in fishes which live in caves or burrows is variable and appears to depend on the degree to which they are restricted to their underground zones. Comparative studies of the pineal organ from deep-sea fishes indicated a trend for an increased number of discs in the outer segments of deeper-dwelling forms. Long outer segments are also a common feature of retinal photoreceptor cells in many mesopelagic fishes. In troglophilic fish, *Chologaster agassizi*, specializations of the organelles in the pineal cells are similar to those observed in other vertebrates living in environment of low light levels. Receptor cells in the pineal organ of sharks, lungfishes and the coelacanth differ from those of teleostean species in that the outer segments are better developed, with the number of discs ranging upto 150 in the shark and 275 in the coelacanth. The protein opsin, a component of retinal photopigments, has been identified in the outer segment of pineal receptor cells in lampreys, elasmobranchs and bony fishes.

The outer segment of pineal photoreceptor cells is connected to the inner segment via connecting piece containing a cilium of the $9 + 0 \times 2$ type. An accumulation of mitochondria forms the "ellipsoid" of the inner segment which is supposed to be a source of energy required for the transduction processes in the outer segments. The synaptic pedicle (with characteristic synaptic ribbons) has been reported. There appears to be a tendency for the pineal organ of fishes with fewer synaptic ribbons to have lower light-intensity thresholds. It is becoming increasingly clear that synaptic ribbons are dynamic organelles which fluctuate in both their size and number depending upon environmental lightening conditions (Fig. 15.73).

Electrophysiological investigations have shown that pineal photoreceptors are capable of transformation of photic stimuli into an electrical type of response that is transmitted synaptically to intrapineal second-order neurons.

It is still open to question whether pineal photoreceptors may convert photic information

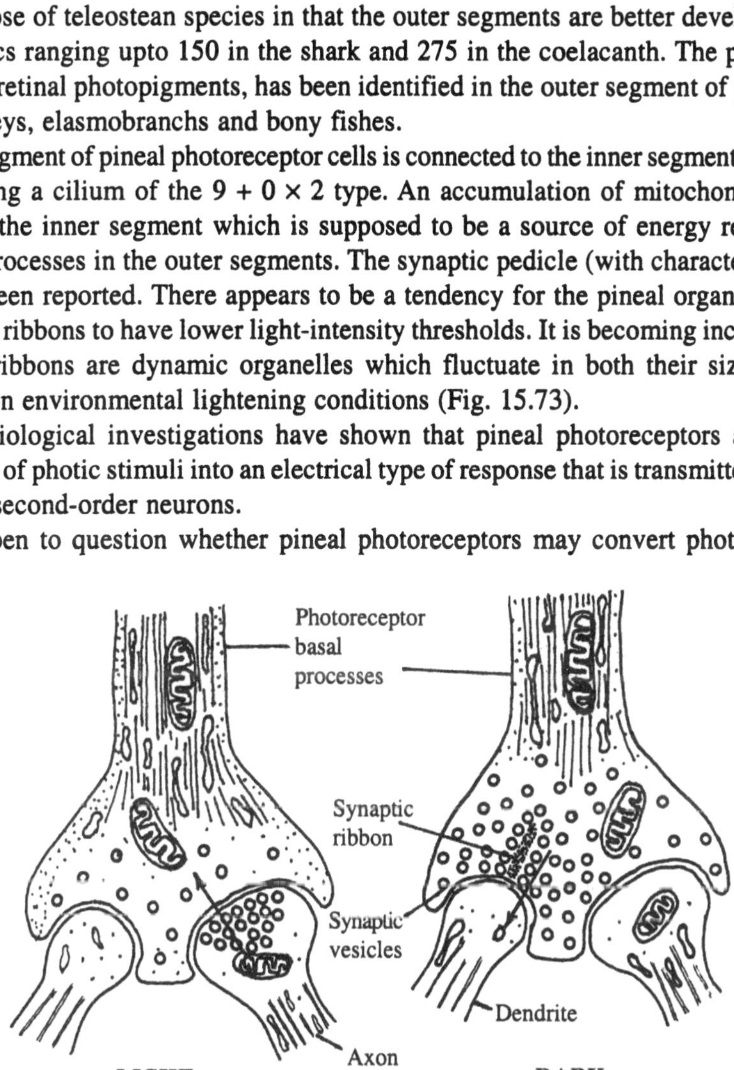

Fig. 15.73 Diagrammatic representation of the synaptic relation in fish pineal organ in the light-and dark adapted status. During the light phase, the inhibitory (?) axo-axonic synapse appears to predominate, while in the dark, the sensory axo-dendritic synapse increases greatly in number. *Arrows* indicate the direction of signal transmission.
***Source:* Omura and Ali (1980)**

directly into an endocrine (hormonal) output. Serotonin, one of the precursors of melatonin has been demonstrated in the pineal parenchyma (see also Meissl 1997).

Innervation. Two types of ganglion cells are found in pineal complex: type I, (usually multipolar or bipolar and type II pseudounipolar. Both populations of nerve cells display a positive acetylcholinesterase reaction. The topographical distribution of AchE-positive nerve cells per unit area increases in a rostrocaudal direction, indicating that the convergence ratio of photoreceptor cells to secondary neurons is high in the end vesicle and lower in the pineal stalk. There is considerable variation in the number of nerve fibres descending the pineal stalk.

The central projections of nerve fibres has been examined in numerous species. Generally, the nerve fibres from the pineal organ enter the area of the posterior commissure. Depending upon the species, branches have also been described to the habenular commissure and habenular nuclei, the subcommissural organ, and the optic tectum. According to Korf (1996), in poikilothermic vertebrates the autonomic (sympathetic) innervation is either missing or rather inconspicuous and does not seem to play an important functional role for the control of pineal functions. Instead, intrapineal photoreception and phototransduction appear as the prevailing regulatory mechanism in these species. Out of all vertebrate classes, mammals possess the densest sympathetic (noradrenergic) innervation of the pineal.

Innervation as well as structure and function of the pineal organ have undergone conspicuous transformation in the course of vertebrate evolution. The sympathetic innervation of the pineal organ appears to develop progressively during phylogeny; this progressive development may be casually related to the changes of the pineal photocyte. In anamniotes, the pineal complex comprises true photoreceptor cells synaptically connected to second-order neurons which give rise to pinealofugal neuronal pathways (central pinealofugal innervation). These convey signals from the pineal photoreceptors to various nuclei in the di- and mesencephalon. Pineal photoreceptors, intrapineal neurons and central pinealofugal pathways are gradually reduced during phylogeny; as a consequence, photoreceptors are lacking, in the mammalian pineal.

The innervation of the parapineal organ in these aquatic vertebrate has been studied less extensively.

Supportive (Interstitial) Cell. Supportive cells are the second predominant cell type in the pineal complex of fishes. In hollow pineal organs these cells extend from the base of the pineal parenchyma to the lumen, and they ensheath pineal sensory cells and neurophil formations. Supportive cells are endowed with numerous microvilli and a few cilia of the 9 + 2 × 2 type protruding into the pineal lumen. The basal portions of these cells may form several interdigitating end feet or a single, broad end foot contacting the basal lamina. The cytoplasm of these cells contains smooth endoplasmic reticulum. In some species, supportive cells contain a considerable amount of rough endoplasmic reticulum with dilated cisternae encompassing granular material. Dense-core vesicle (70–170 nm in diameter) are observed adjacent to the stacks of Golgi cisternae and, as accumulations, in the luminal and basal portions of the cells.

In lampreys and fishes, the supportive cells display flattened stacks or packed arrays of membranes resembling myeloid bodies of the pigment epithelium of the retina. It has been shown that supportive cells incorporate 3H-labelled vitamin A; the most heavily labelled organelles are the myeloid bodies. This finding might point toward an involvement of these cells in the metabolism of photopigments. Furthermore, supportive cells may serve the phagocytosis of outer segments during the action of renewal mechanisms.

Vascular Supply. The vascular relationships between the pineal organ and dorsal sac appear to be very complex. According to a recent study on the pineal vascular system of the rainbow trout, there are two different circulatory pathways: (1) blood vessels ascending and descending via the pineal stalk; (2) vascular supply emanating from both left and right lateral aspects and joining to the former circuit in the pineal end-vesicle. It has been speculated that in the rainbow trout the pineal secretory products are transported via a local vascular pathway into the dorsal sac, and thus gain access to the ventricular cerebrospinal fluid and the brain tissue.

Vascular Permeability (Blood-Brain Barrier) in the Pineal Organ

The diencephalic roof of teleosts is predominantly occupied by the photoneuroendocrine pineal organ, the choroidal dorsal sac, and the functionally still enigmatic parapineal organ. The former two circumventricular organs are in open communication with the third ventricle; furthermore, they are well vascularized by means of numerous capillaries endowed with fenestrated endothelial cells.

In general, the blood-brain barrier is essential for the maintenance of the homeostasis of the cerebrospinal fluid, and to prevent the entrance of noxious agents from the blood into the brain. Accordingly, it might be expected that the blood-brain barrier is a general feature of the central nervous system in all classes of vertebrates. However, there are some contradictory reports showing that the blood-brain barrier in some species of cyclostomes and elasmobranchs differs in its structure and function from the respective barrier in mammals.

In teleosts, Omura et al. (1990) investigated the permeability of the blood-brain barrier of the circumventricular organs and concluded that the capillaries of most of these organs, except the subcommissural organ, are endowed with fenestrated endothelia; such areas are exposed to the haemal environment. Therefore, it is generally accepted that these circumventricular organs are devoid of a blood-brain barrier. Based on studies and results on the blood-brain barrier in the structural differentiations of the diencephalic roof of the rainbow trout, they emphasised that in spite of the leaky capillaries an effective barrier system may exist between the haemal environment and the ventricular cerebrospinal fluid compartment in this area.

This blood-cerebrospinal fluid barrier may involve the following components: (1) a sieve or trap system of perivascular 'stromal' connective-tissue elements (collagen fibrils, macrophages, fibroblasts); (2) a system of circumventricular tight junctions adjoining epithelial cells; (3) a heterolytic system of epithelial endocytosis followed by lysosomal sequestration, and (4) a system of macrophages inhabiting the luminal surface and ventricular compartment. In contrast to the pineal organ, the parapineal organ of teleosts may be located 'inside' the blood-brain barrier.

Samejima et al. (1994) investigated the involvement of multiple receptors in modulating the function of the pineal gland by searching for dopaminergic and cholinergic receptors in rainbow trout (*Salmo gairdneri*) pineal gland. Distribution of dopamine D_1 and D_2 receptors; and muscarinic and nicotinic cholinergic receptors was studied. The results of the study indicated that the pineal gland should not be viewed as a homogenous tissue possessing identical density of various receptors. Furthermore, it has been suggested that different regions of the pineal gland may indeed possess unique functions. The precise function (s) of cholinergic neurons in trout pineal gland remains to be delineated.

6.1.2 Parapineal Organ

The parapineal organ in *Petromyzon* is represented by a vesicle consisting of two "retinae", a dorsal inverted one whose ganglion cells are mainly localized in its rostral part, and a ventral direct one in which only one or two ganglion cells could be localized. Other ganglion cells constitute a parapineal ganglion. This ganglion parapineale is connected to the parapineal organ by a short tract and to the left habenular ganglion by a nervous layer. The ganglion parapineale and the nervous layer represent the rostral part of the left habenular ganglion.

A comparison of the innervation of the parapineal organ with that of the pineal organ and the observation of double parapineal organs suggest that the parapineal organ orginates from a bilateral structure, as is true for the pineal organ.

There is evidence for a generalized structural regression of the parapineal in two species of cyclostomes, *Petromyzon marinus* and *Lampetra planeri.* Photoreceptor cells are less abundant; basal synaptic processes less ramified; smaller synaptic ribbons; decrease in the number of ganglion cells, and a poorly developed smooth endoplasmic reticulum in them are characteristic features. It has been postulated that the reduced photosensitivity of the parapineal may be ralated to decreased photic stimulation as a result of its ventral position beneath the pineal organ.

In spite of the earlier assumption that the parapineal organ is regressed in adult teleosts, this tiny and enigmatic organ has been found in more than 40 species of teleosts. The parapineal organ of teleosts has been considered to be homologous to the parietal eye of lizards and to the parapineal organ of petromyzontids.

The parapineal organ is variable in size and location; in some cases it is difficult to distinguish this organ from the tissue of adjacent habenular nucleus. The parapineal organ of the rainbow trout is located to the left of the pineal stalk and dorso-posterior to the habenular commissure, through which the parapineal tract penetrates into the left habenular nucleus. The neural projections of the parapineal organ is still unproven, although a close relation to the pineal tract has been indicated.

The function of the parapineal organ is, however, still open to discussion, although a limited number of outer segments of photoreceptor cells was demonstrated in this vesicular body by electron microscopy and immunohistochemical use of antibodies against rod-opsin.

Further studies are required to elucidate the interrelationships between the parapineal organ and the habenular nucleus in teleosts. The functional significance of the parapineal organ still remains enigmatic.

7. TEMPERATURE (THERMOBIOLOGY)

There are as yet no specific thermoreceptors known in fish. Murray (1971) discussed the temperature perception and the sensory systems implicated, the lateral line organs and the general, cutaneous, segmental innervation.

Contributions related to temperature and fish life are of Fry (1967, 1971), Brett (1970), Fry and Hochachka (1970), Robert (1974), Crashaw (1979), Graham (1983) and Hazel (1993).

The vast majority of fishes are ectotherms, and their body temperature (*T*b) approximates ambient water temperature (*T*a). Fishes, isothermal with their environment, cope with the problems of *poikilothermy* (variable body temperature) by a diversity of mechanisms. They are not able to produce sufficient amounts of metabolic heat to elevate their *T*b above *T*a or to regulate *T*b as *T*a changes; because they are immersed in water, fish are unable to use mechanisms for evaporative and radiative energy exchange, which are integral strategies in the thermoregulatory repertoires of terrestrial ectotherms. The ability of some fish species to detect thermal gradient and to behaviourally select a preferred *T*a in a heterothermal environment has been well documented. This thermoregulatory option, however, cannot be exercised in all habitats or by all (or even most) species. Physical restrictions (i.e., optimal *T*a may be located beyond the range or it may be part of a large thermal front and drift faster than the swimming capacity of a fish) and ecologic factors (needs for shelter and food) very likely place real limitations on the extent to which behavioural thermoregulation can be used by many fish species. For these reasons virtually all aspects of the life histories of fishes (and many other aquatic ectotherms) are governed by ambient water temperature, which has been described as the *master ecological* factor. The reader is referred to Graham (1983) who considered the heat transfer principles, heat transfer mechanisms, and empirical analyses of heat transfer (Fig. 15.74; see also Fig. 15.75).

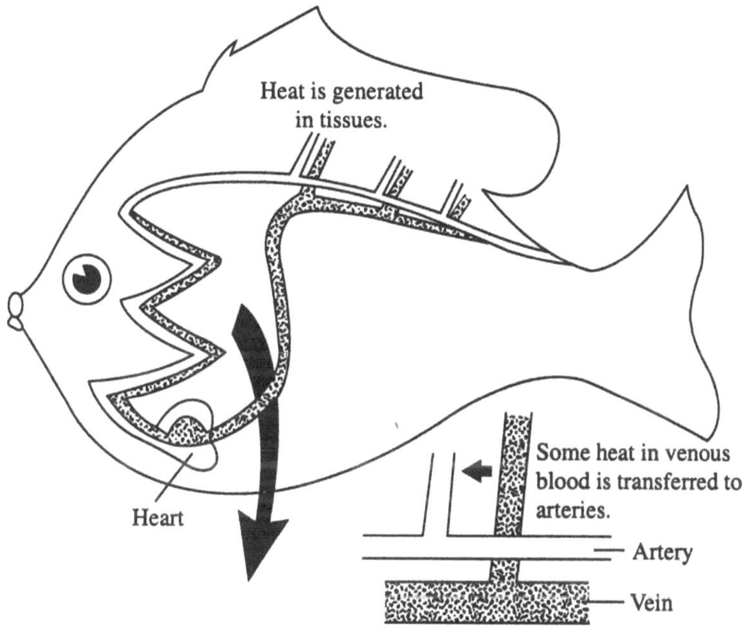

Fig. 15.74 Pattern of heat transfer in a typical fish; metabolic heat is lost to the water as blood circulates through the gills.
***Source:* Stevens et al. (1974)**

7.1 Endothermy

Endothermy, the ability to maintain elevated body temperature by metabolic means, has been documented only within one major assemblage of large oceanic teleosts, the Scombroidei. Sharks of the family Lamnidae and Alopidae have convergently evolved endothermy. The transition from ectothermy to endothermy requires the elevation of aerobic capacity (and hence heat production) and the reduction of the rate of heat loss. An inability to limit heat loss while respiring through gills prevents most of the fish from being warmer than the water in which they swim. Except for the difference in the mode of respiration and the resultant heat loss, the physiological requirements for the evolution of endothermy in fish should parallel those in terrestrial amniotes.

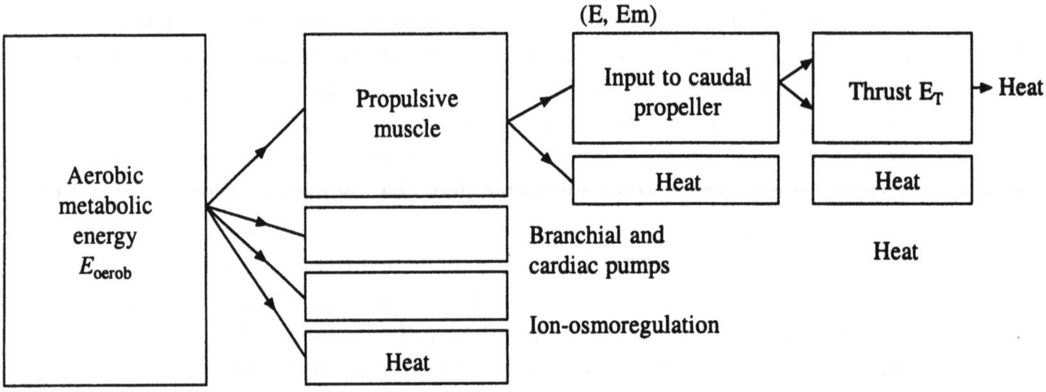

Fig. 15.75 Sources of metabolic heat production in a swimming fish. Only thrust heat would not affect T_a. *Source:* Graham (1983) (read details)

Two distinct strategies for elevating tissue temperatures have evolved in the Scombroidei. Tunas (Scombridae: *Thunnus*, *Katsuwonus*, *Euthynnus*, and *Auxis*) achieve the endothermic condition in a manner similar to birds and mammals. They have exceptionally high metabolic rates and have reduced whole body thermal conductance. In contrast, billfishes (Xiphiidae: *Xiphias*, and Istiophoridae: *Istiophorus*, *Makaira* and *Tetrapturus*) use cranial endothermy and warm only the brain and eyes. The billfishes accomplish this more limited form of endothermy with the use of a thermogenic organ composed of highly modified extraocular muscle fibres. In billfishes, expression of the heater phenotype (thermogenic muscle cells, that produce heat without force generation) is associated with the superior rectus eye muscle (Fig. 15.76). A countercurrent heat exchanger retains the heat beneath the brain, and distinct arterial supply directs warm blood to the retina. One other scombroid, the butterfly mackerel, *Gasterochiasma melampus*, also uses cranial endothermy, but the thermogenic tissue in this species is derived from the lateral rectus eye muscle. The lateral and superior rectus muscles derive from different somites during ontogeny implying separate evolutionary origins of the thermogenic cell type in *Gasterochiasma* and billfishes. However, heater cells of *Gasterochiasma* and billfishes are structurally identical and have similar biochemical properties, suggesting they are homologous. A close phylogenetic relationship between these taxa would support this latter possibility, particularly if billfishes and *Gasterochiasma* share a common ancestor to the exclusion of other species lacking the thermogenic tissue type.

Tunas possess no tissue specialization for heat production and instead retain metabolic heat by way of vascular countercurrent heat exchangers located in the brain, muscle, and viscera (see Fig. 8.26). Red aerobic muscle, which powers sustained swimming, contributes a major fraction of this metabolically derived heat. In most teleosts, the red muscle is located laterally, just beneath the skin. In tunas, this muscle occupies an atypical centralized location, and is closely apposed to the axial column. The central positioning of red muscle in tunas has received much attention as a result of its role in endothermy; however, it may also reflect the unique locomotor style of tunas, thunniform swimming. These fish generate thrust exclusively via the tail and restrict their propulsive movements to the caudal region. Evolution of the locomotor apparatus and endothermy may be tightly coupled in this lineage; thunniform swimming arises concomitantly with a movement of the aerobic red muscle internally, and this central aerobic muscle mass is the major source of metabolic heat in tunas. An edited volume on tuna by B.A.

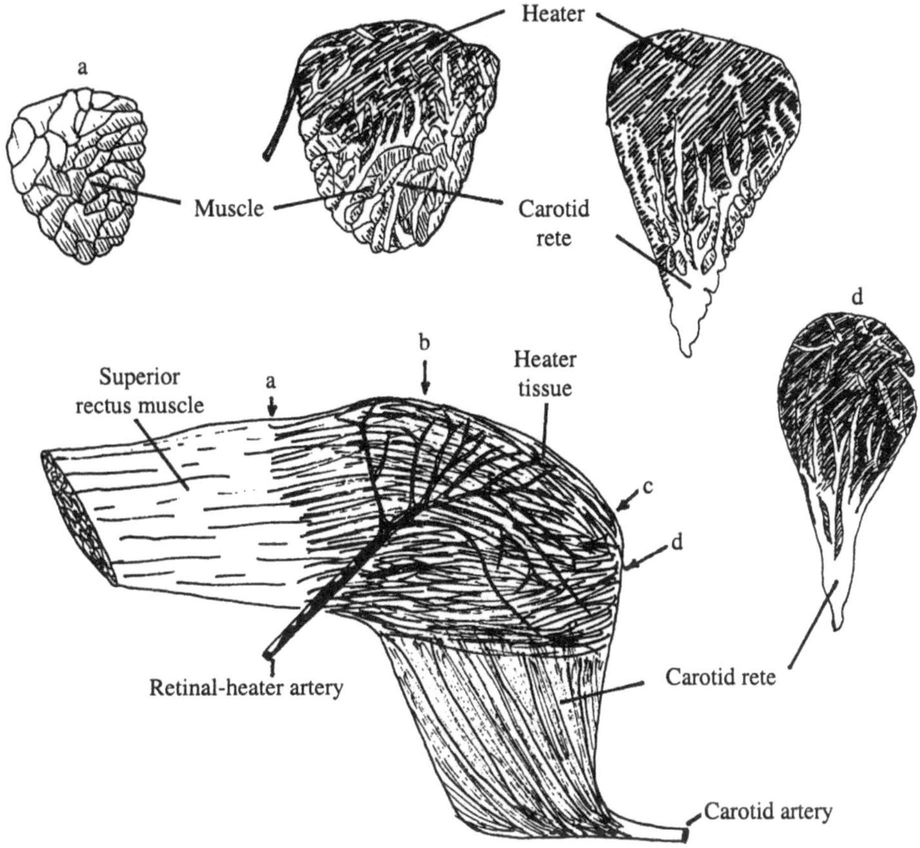

Temperature regulation in *Makaira Indica*.

Fig. 15.76 Diagram of the superior rectus muscle and the associated heat-producing tissue and carotid rete from a black marlin, *Makaira indica*. Cross-section a-d indicate the change in morphology that occours as the eye muscle courses into the orbit toward the base of the braincase. This is a lateral view and the arrows with the letters indicate the position from which the sections were taken. The carotid rete is a countercurrent heat exchanger and prevents the convective dissipation of the heat being produced in the heater tissue. The retinal-heater artery is the major outlet for blood leaving the heater tissue. This artery supplies warm blood from the heater tissue to the eye.
***Source:* Block (1987)**

Block and F.D. Stevens (2001) has provided interesting information covering various facets of life.

An attempt to trace the evolution of endothermy has been made in mackerels, tunas and billfishes (suborder Scrombroidei and Teleostei). Multiple origins and diverse strategies for endothermy exist among these fish. It has been suggested that the evolution of cranial endothermy in fish was selected in order to permit thermal niche expansion and not selected for increased aerobic capacity (Block et al. 1993; see also Block 1987).

For related information the reader is directed to Stevens (1973), Stevens et al. (1974), Sharp and Dixon (1978), Stevens and Neill (1978), and Smit (1980).

Blaxter (1992) described the influence of temperature on fish eggs and larvae from an autecological viewpoint. Temperature may have lethal or sub-lethal effects as well as pacing metabolism, other physiological responses and behaviour. It influences body size, growth, differentiation of muscle and meristic characters. The controlling effects of temperature at different hierarchial levels from the chemical to the ecological are set out in a simplified diagram (Fig. 15.77). The complexity of the interaction between temperature and fish eggs and larvae is summarised in Fig. 15.77 in which it is assumed that the core of the relationship is between predators and prey.

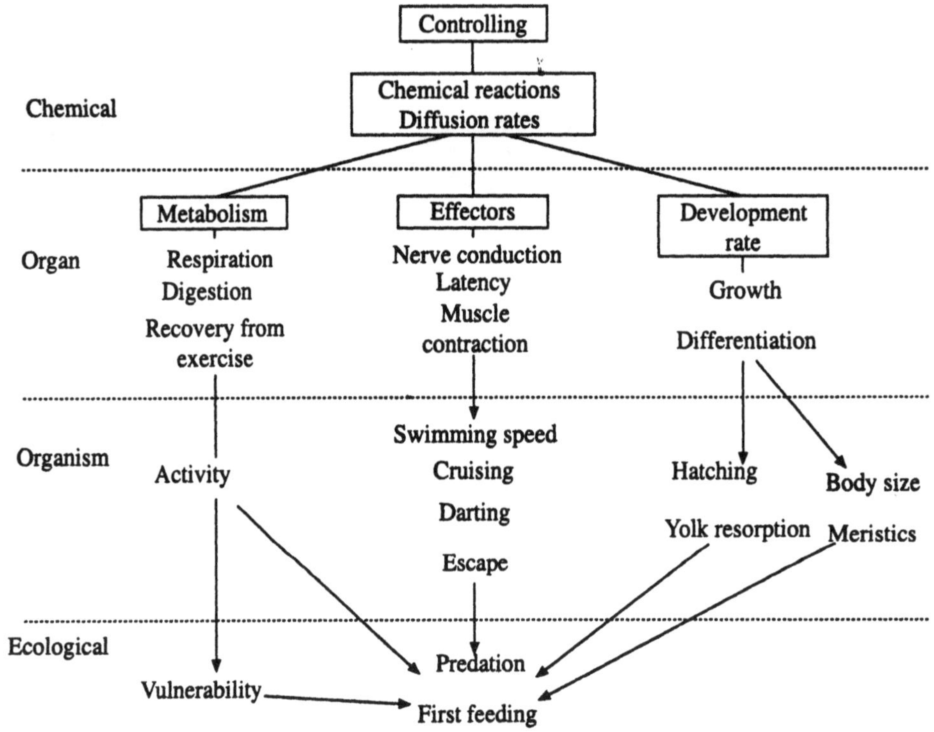

Fig. 15.77 Diagram showing the way in which temperature controls many life processes at different hierarchical levels.
***Source:* Blaxter (1992)**

Fish, like all other vertebrates, initiate thermoregulatory behaviour by responding to input from both peripheral and central thermoreceptors. Processing and integration of this information is accomplished in the preoptic anterior hypothalamus (see details: Hazel 1993).

The reader is referred to Chapter 20 for freezing resistance and Chapter 21 for overwintering.

An extensive (although probably not exhaustive) literature search for documentation of temperature changes experienced by free-living fishes produced the examples shown in Table (15.12).

8. ELECTRORECEPTORS (read Chapter 17)

9. OVERVIEW OF SENSORY ADAPTATIONS OF DEEP-SEA AND ANTARCTIC FISHES (see Table 15.13)

Table 15.12 Documented examples of acute temperature change in fishes

Species	dT	Time
Inundation		
Oligocottus in tide pools	13–23	min
38 species in tropical tidal swamp	25–38	∠ 24 h
Vertical migration over a thermocline		
Chromis sp., feeding station to refuge site	14–21	30 min
3 species fish larvae	∠ 6	min
Many species deepsea fish	10–20	30 min
Other		
Larval cod, mixing by storm	7–10	3 day
White sucker (*Catostomus commersoni*)		
Natural habitat	12–14	8 min
Thermal plume	8–13	7 min

dT, change in temperature, °C
Data from Montgomery and Macdonald (1990)

Table 15.13 Summary of the comparison of the sensory adaptations of deep-sea and Antarctic fishes

	Deep-sea fishes	Antarctic fishes
Olfaction chemosense	Sexual dimorphism	Not so far as we know
	Macrosmatism	
Vision	Dorsal visual axis	*Trematomus hansoni* and *T. nicolai*
	Sensitivity adaptatons	
	Large eyes	No
	Tubular eyes	No
	Aphakic space	No
	Pure rod retina	No
	Long rod outer segments	*Pleurogramma antarcticum*
	400–600 μm	44 μm
	High convergence ratios	*Dissotichus mawsoni*
	600 : 1	58.1
	Multiple rod banks	No
	Tapeta lucida	No
	Matched visual pigment	Yes in spring
	Photophores	No
Octavolateralis system	Hearing	No information
	Gravitation receptors	No information
	Large semicircular canals	No
	Lateral line	
	Papillate neuromasts	No
	Superficcal neuromasts	Yes for *Rheotoxis*
	Canal neuromasts	Standard
	Membranous canals	Partial
	Mechanosensory camouflage	Labriform probably not
Touch	Fin rays	Yes
	Mental barbels	Yes

Source: Montgomery and Macdonald (1998).

CHAPTER **16**

Sound Production, Detection and Communication

1. INTRODUCTION

Sound in air and in water have basically the same characteristics in that the sound is a pressure wave with accompanying particle motion. In water, however, the particle motion component of the sound is substantially smaller than that which would accompany a pressure wave in air with the same amplitude. Whereas few, if any, terrestrial vertebrates use the relatively large, direct particle motion component of the sound field, particle motion is a major stimulus for fishes which is very much involved in many aspects of sound processing by fishes. The pressure wave is a scalar quantity, and thus has no directional components. In contrast, particle displacement is a vectorial quantity and, therefore, has directional components that potentially can be exploited to provide information about the direction of the sound source relative to receiver (Rogers et al. 1988).

A large and growing literature exists on the mechanisms of sound production (see Tavolga 1971; Hawkins and Myrberg 1983; Hawkins 1993) and audition (see Enger 1981; Popper 1983; Popper and Fay 1984, 1993; Kalmijn 1988; Rogers et al. 1988). Behavioural work and biophysical studies form the cornerstones of this subject. Much of the historical analysis of hearing in fishes has been based on the performance of members of the superorder Ostariophysi (or Cypriniformes), a major freshwater group of teleost fishes.

Behavioural investigations of auditory capabilities of fishes have demonstrated that some species are capable of many of the same auditory functions as mammals- including frequency discrimination, detection of signals in the presence of a masker, and sound source localization. In fish there is no equivalent to the mammalian cochlea.

2. SONIC MECHANISMS

Many species of fish are known to produce sounds underwater. The sonic equipment of fishes is as varied as are the fishes themselves. There appears to be little consistency in the type of sonic devices in relation to general classification of evolutionary relationships among many families. Broadly, these devices can be grouped into three categories: stridulatory mechanisms; hydrodynamic phenomena: and muscular contractions involving the swim bladder. Biological relevance is better documented for stridulatory and swim bladder sounds.

Among the vertebrates the fishes have no vocal organs and make sounds with the help of other anatomical elements. In the Dipnoi, the larynx has no skeletal parts; the sphincter of larynx is made by smooth muscle fibres. Orally from the aditus laryngis a strong plate is formed containing elastic elements; opposite this plate a voluminous fibrous cushion is located. Both are under control of constrictors and dilators. But sounds in fishes, as shown are produced in different other ways; they do not originate in the larynx and consequently are not activated by respiratory air current.

Fish sounds vary in structure, depending on the mechanism used to produce them. Low frequency sounds have been recorded for a fairly large number of marine and freshwater fishes and these are usually attributed to mechanisms associated with the swim bladder. Higher frequency sounds, often described as creaks or rasps, are thought to be produced by stridulation or the grinding together of hard surfaces such as finray joints or tooth-bearing bones.

A few examples can be given here. Members of the grunt family, Pomadasyidae e.g., (*Haemulon*) produce a sharp, vibrant call by grating a dorsal patch of pharyngeal denticles against smaller ventral patches, and in the trigger fish *Balistes* (family Balistidae) the fused anterior spines of the dorsal fin produce a grating sound when moved against their socket. Some catfish of the family Siluridae (sea catfish *Galeichthys felis*, gaff-top-sail-catfish *Bagre marinus*, and *Heteroneustes fossilis*) produce a squeak when the enlarged pectoral spines are moved (Fig. 16.1). The sound producing mechanism in *Tilapia mossambica* consists of a single ventral and two dorsal pharyngeals located in the buccal cavity and provided with numerous small, specially modified chisellike teeth (Fig. 16.2). Sound production may take place by moving either one or both of the dorsal pharyngeals against the ventral pharyngeal, the required force being created by a combined effect of the mandibular musculature and the special muscle bundles (sonic muscles) inserting onto the pharyngeals. Stridulation by patches of pharyngeal teeth also occurs in *Heteroneustes fossilis* and *Macrones vittatus*. Sound production by means of modified molariform teeth is common in puffers (order Tetraodontiformes) and parrotfishes (family Scaridae). Sound is produced by friction of adjacent bones, probably the skull and vertebral elements in clown-fish *Amphiprion*

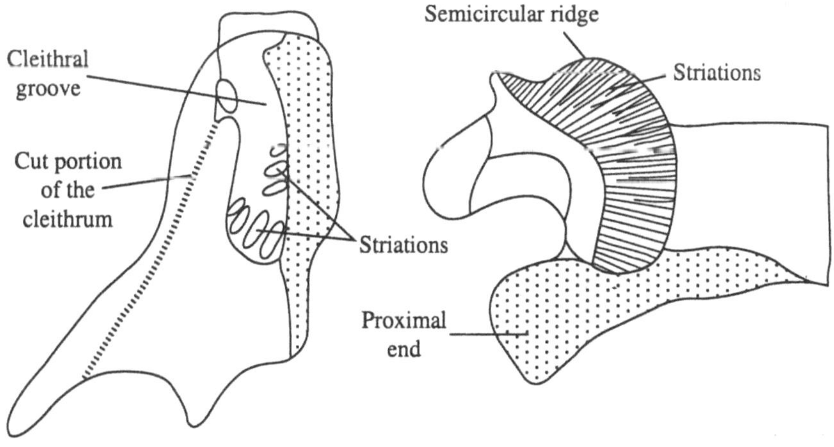

Sound production by pectoral spine in *Heteropneustes fossilis.*

Fig. 16.1 Sound-producing organs of catfish ((Heteropneustes fossilis) Proximal end of the pectoral spine to show the semicircular ridge and its striation. Cleithral groove to show its striations.

and sea horse *Hippocampus*. Stridulation involving skeleton element of dorsal fin and anterior vertebral column is reported in Indian catfish *Sisor rhabdophorus*. Other fish clap or thump different parts of the body together, like the black grouper (*Mycteroperca bonaci*), which bangs the opercula or gill covers against the body to produce a low-pitched thump.

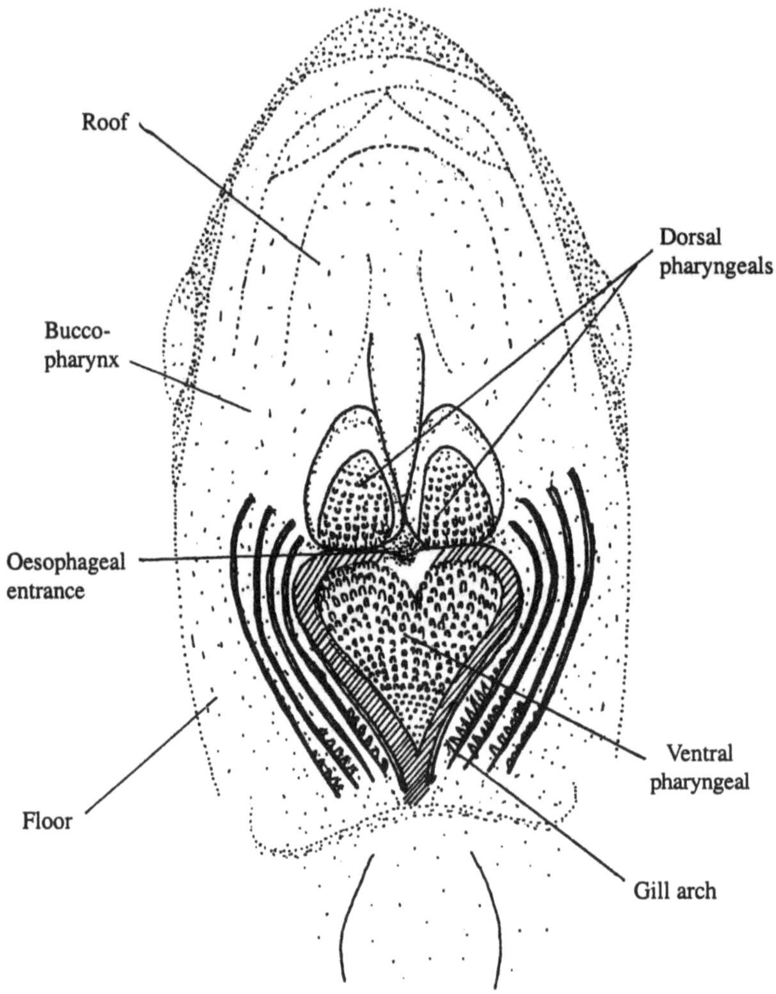

Sound production in *Tilapia mossambica*.

Fig. 16.2 Location and form of dorsal and ventral pharyngeals in the buccal cavity of *Tilapia mossambica*.
***Source:* Lanzing (1974)**

The sound producing organ of the croaking gourami *Trichopsis vittatus* consists of a very enlarged part of the M. adductor superfacialis, found at the inside of the cleithrum, which is attached to the fourth and fifth pectoral fin ray by two tendons, thickened to elastic pads. These pads are situated above an elevation formed by the base of the second and third fin rays. Double-pulsed tone bursts are generated by the pads snapping over this elevation during the contraction of the sonic muscle. Sexual dimorphism of the sound producing organ in the pygmy gourami (*T. pumilus*) is found. The muscles and tendons are very reduced in females. Therefore, probably females are not able to produce sounds.

Sudden expulsion of gas from suprabranchial diverticulae also produces sound in *H. fossilis.*

Hydrodynamic sounds are produced by an animal's moving through the water and are intensified by rapid change in the animal's direction or speed. There are few data on the functional significance of these low-frequency sounds and their biological role, but it has been suggested that they are involved in maintaining an animal's position in school by enabling them to detect neighbours.

With all these mechanisms the gas-filled swim bladder may play a subsidiary part, and may impart a hollow resonant quality to the sound. If the swim bladder of a sound-producing white grunt, *Haemulon plumieri*, is deflated, the sound generated by stridulation of the pharyngeal teeth is less loud, and loses its grunt-like quality. The swim bladder is more directly involved in the production of sounds by some physostomatus fish (with open swim bladders), such as the Atlantic eel (*Anguilla anguilla*) where gas is released from the swim bladder into the oesophagus, giving a sharp pop or squeak.

Sound is produced with the swim bladder in several ways. One such mechanism involves the swim bladder's acting to reradiate sound produced by movement of bony elements. Another type of swim bladder sound involves contraction of the extrinsic muscles (e.g., those originating in the occipital region of the skull) which insert on ribs or connective tissue near the swim bladder (e.g., brotulid *Monomitpus*) (Fig. 16.3). Swim bladder sounds are also produced by two intrinsic muscles of the swim bladder wall. The best known example of this sound is the oyster toadfish (*Opsanus tau*), where the lateral walls of the heart-shaped swim bladder have extensive intrinsic musculature (Fig. 16.4b). In midshipman (*Porichthys*), a genus of fish closely related to toadfishes, the sonic mechanism appears to be virtually identical. Another example of swim bladder with drumming muscles is drum-fish (*Bairdiella chrysura*) (Fig. 16.4a). In some triglids (e.g., *Bellator* and *Prionotus* species) both intrinsic and extrinsic sonic muscles are found (Figs. 16.5 and 16.6).

Various catfish use the swim bladder for the production of sound. The simplest mechanism is found in various Pimelodidae. In *Pimelodus clarias* a large muscle originates on the thickened anterior edges of the parapophysis of the fourth vertebra and inserts on the ventral surface of the swim bladder on each side. The swim bladders of many other fish possess similar extrinsic sound producing muscles (e.g., characins *Serrasalmus*). Another skeletal basis of sound producing mechanism, known as the elastic spring or "springfederapparat" mechanism is found in some Ariidae, and in the Doradidae (except for *Centromochlus=Tatia*), the Mochokidae, *Malapterurus* and *Pangasius*. In these catfish the anterior part of the parapophysis of the fourth vertebra ("Mullerian ramus"), which lies as usual against the lateral part of the anterior surface of the swim bladder, is flexible, and lacks the usual ligamentous connection to the posttemporal. It is often tipped by a plate of thicker bone which is firmly attached to the tunica externa. A muscle runs from it to the occiput and/or ventral surface of the nuchal field. When this muscle contracts it pulls the anterior process of parapophysis (the elastic spring) forwards, enlarging the swim bladder: when it relaxes, it allows the spring to recoil. The swim bladder is thus caused to pulsate, emitting sound (Fig. 16.7)(For details in catfish, see Ladich (2001).

In perhaps the most specialised and characteristic sound producing mechanism, paired striated drumming muscles from the body wall compress the swim bladder. These muscles are derived from the trunk musculature, and may simply overlie the swim bladder, as in the drums (Sciaenidae,

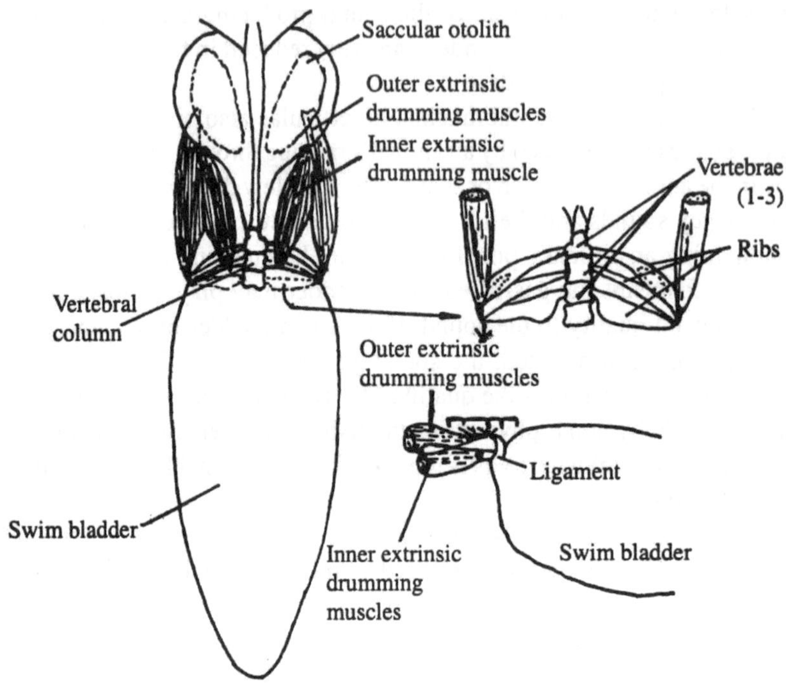

Monomitpus: Swim bladder and presumed sonic mechanism.

Fig. 16.3 Swim bladder and presumed sonic mechanism of a brotulid, *Monomitpus*. Left: The complete mechanism showing two pairs of muscles originating on the rear of the skull and inserting on the ribs; upper right: insertion points of the sonic muscles on the ends of the ribs; lower right; side view attachment of sonic muscles to the ribs and swim bladder.
***Source:* Tavolga (1971)**

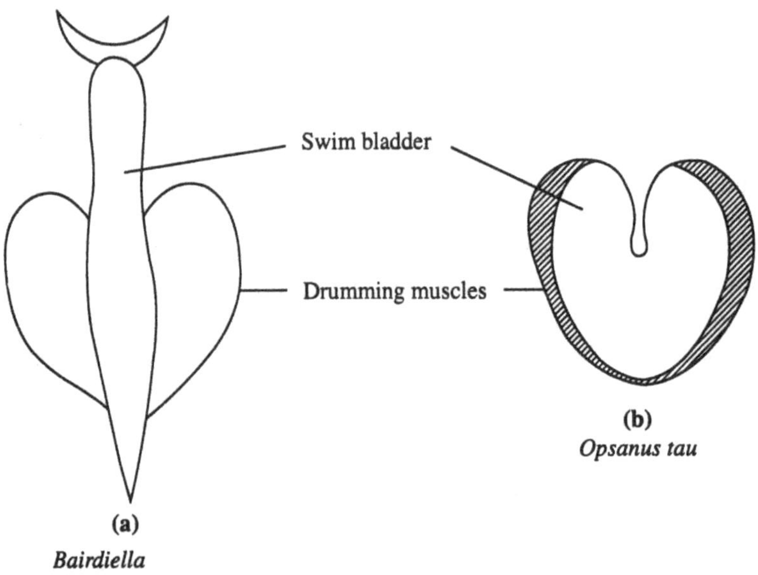

Swim bladder with intrinsic drumming muscles.

Fig. 16.4 Intrinsic drumming muscles of (a) *Bairdiella chrysura* and (b) *Opsanus tau*.

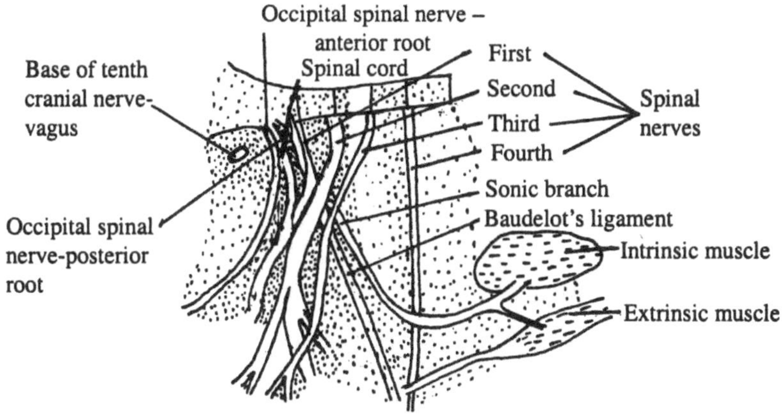

Bellator egretta: Innervation of intrinsic and extrinsic muscles.

Fig. 16.5 Ventral view of the right side showing innervation of intrinsic and extrinsic sonic muscles in *Bellator egretta*.
***Source:* Evans (1973)**

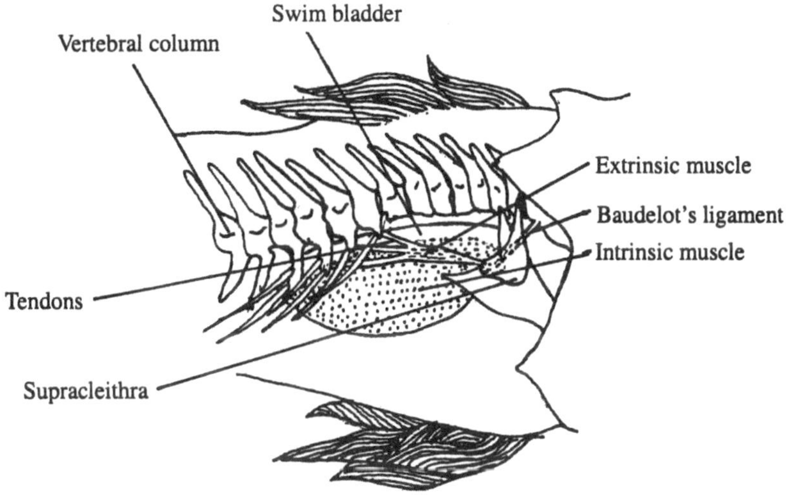

Fig. 16.6 Right lateral view of *Prionotus carolinus* showing swim bladder and associated sonic structures.
***Source:* Evans (1973)**

the largest family of sonic fishes) (e.g., male freshwater drum *Aplodinotus grunniens*; Fig. 16.8), or they may be attached partly or wholly to the organ. Within a family, such as the gurnards (Triglidae) or codfishes (Gadidae), there can be substantial differences between the conformation of the muscles in the individual species (Fig. 16.9) and even between fish and different sexes, the muscles often being highly developed in the male. The muscle fibres themselves are specialised. They are often red or yellow in colour, with a high myoglobin content and a rich blood supply. Their diameter is thin, they contain a well-developed sarcoplasmic reticulum (e.g., oyster toadfish. *O. tau*) and they may show innervation by a large number of nerve fibres, with many nerve terminals along their length (polyaxonal innervation, e.g., toadfish and squirrelfish, *Holocentrus*

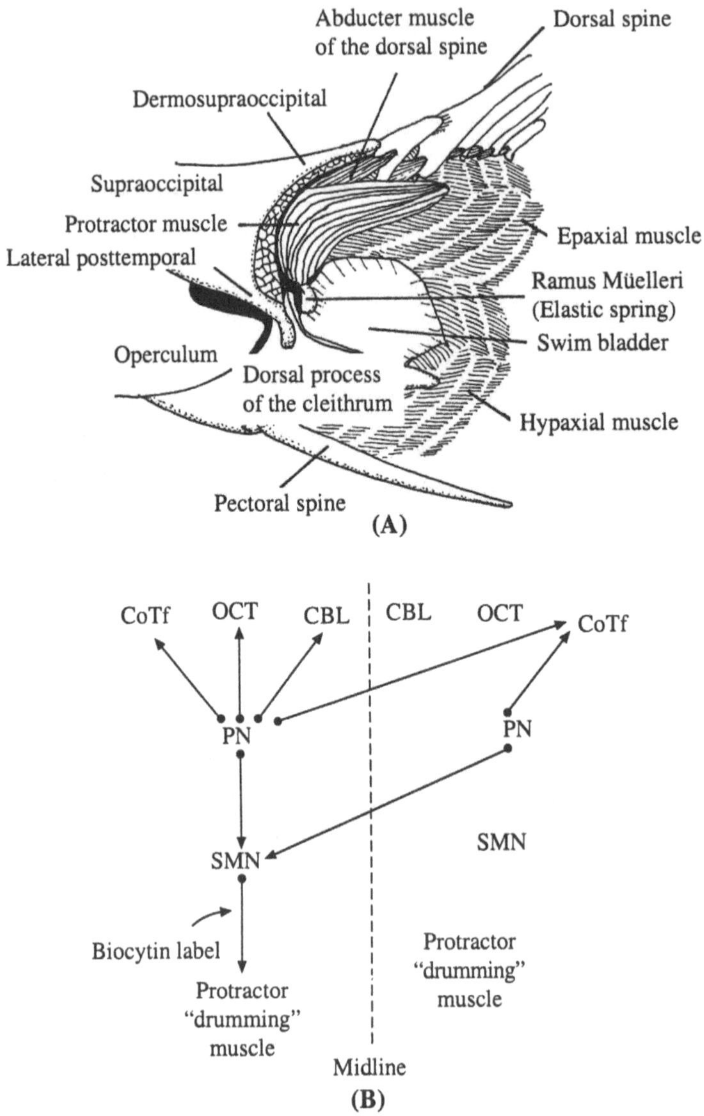

Fig. 16.7 **(A) Line drawing of a dorsolateral view of the postcranial region in *Synodontis nigromaculatus* showing the elastic spring mechanism and swim bladder protractor muscle. Supraoccipital, dermosupraoccipital, lateral posttemporal bones, and dorsal process of the cleithrum have been partially removed to reveal the protractor muscle adjacent to the swim bladder and elastic spring (ramus Muelleri).**
(B) Summary diagram of proposed pattern of organization for the sonic/vocal-acousticolateralis pathways in mochokid catfish. Biocytin application to a single sonic nerve that innervates the ipsilateral protractor muscle results in biocytin-filled sonic motor (SMN, ipsilateral only) and premotor neurons (PN, bilateral), Biocytinlabeled terminal fields mainly occur (1) bilaterally in a commissural region at the level of the caudal pole of the descending octaval nucleus (CoTf), (2) ipsilaterally in octavolateralis-related (OCT) nuclei (octavolateralis efferent nucleus, granule cell division of the medial auditory nucleus, nucleus praeeminentialis), and (3) ipsilaterally in the eminentia granularis and granule cell layer of the corpus of the cerebellum (CBL).
***Source:* Ladich and Bass (1996)**

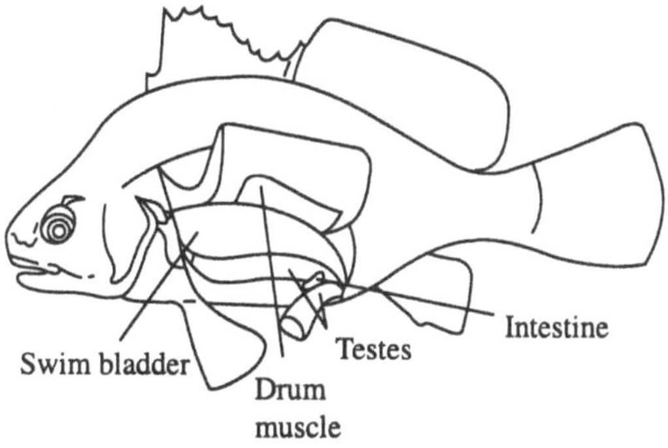

Aplodinotus grunniens ♂:

Swim bladder with extrinsic sonic muscles.
(a)

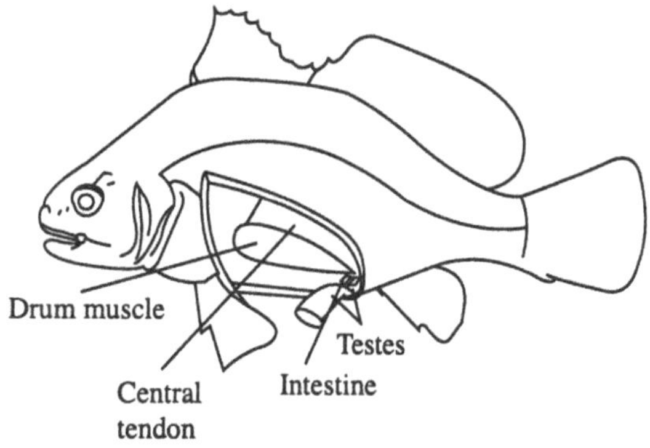

Aplodinotus grunniens ♂:

Swim bladder with extrinsic sonic muscles.
(b)

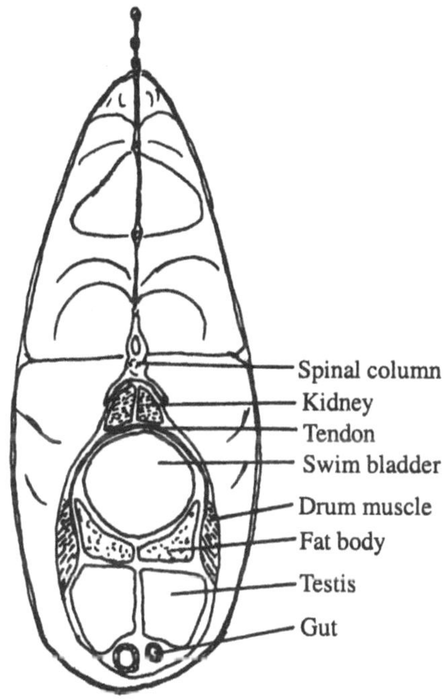

Aplodinotus grunniens ♂:
Sound producing apparatus
(c)

Fig. 16.8 (a), (b) and (c) The swim bladder in the drumfish, *Aplodinotus*, is vibrated by a pair of sonic muscles. *Source:* Schneider and Hasler (1960)

rufus). They contract very rapidly, (in some cases upto 300 to 400/sec) with a high degree of synchrony, and can be stimulated to contract repeatedly at a very fast rate without going into a sustained state of contraction or tetany. The functional morphology of sonic muscle fibres in the oyster toadfish (*O. tau*) has been examined by Fine et al. (1993). For review, the reader is referred to Walsh et al. (1995). For development of sonic neuromuscular system in this fish, see Fine (1989).

The sound producing muscles in many diverse species are innervated by branches of the so-called occipital nerves (e.g., marine catfish *Arius felis* and *Bagre marinus*, *Holocentrus ascenionis*, oyster toadfish *O. tau*). These nerves, which are often double, exit between the vagus and first spinal nerves. The sonic muscles in all these species may be homologous based on this common nerve supply and that the occipital nerves are probably homologous to the hypoglossal nerves of tetrapods. One group of fish, the croakers and drums (Sciaenidae) form a notable exception to this innervation pattern. In these animals the sonic muscles are derived from the lateral body wall and are thought to be innervated by the spinal nerves. The nerves to the sonic muscle in toadfish have been characterized as containing uniform fibres with a diameter of approximately 10 μm and conduction rates of about 25 to 30 m sec at 21°C. In addition, the fibres are known to be cholinergic and to transport certain glycoproteins at the rapid rate of 96–120 mm/day.

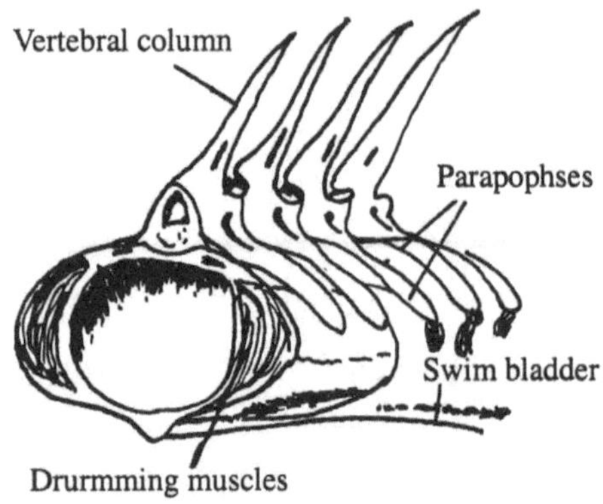

Fig. 16.9 The sound-producing in gadoid fish. The drumming muscles in the haddock (*Melanogrammus aeglefinus*) are dorsally attached to the wall of the swim bladder overlying strong lateral wings (parapophyses) extending from the anterior vertebrae. Ventrally the muscle fibres insert upon the tough outer tunic of the gas-filled swim bladder.
***Source:* Hawkins (1986)**

Swim bladder sounds consist of short pulses containing a pure tone and several harmonics of fundamental frequencies. The fundamental frequency is directly related to the contraction rate of the sonic muscles, but the harmonic content of the signal depends on the animal's substrate. Swim bladder sounds are mostly short duration pulses with a fundamental frequency of 75 to 100 Hz.

The swim bladder appears to act as an impedance-matching device rather than enhancing specific resonance properties of sound signals. The swim bladder provides broad-band amplification of signals produced by extrinsic or intrinsic sonic mechanisms.

3. MECHANISMS OF FISH AUDITION (ACOUSTIC-FIELD DETECTION AND PROCESSING)

Hearing is but one member of large family of modalities that we call, collectively, the octavolateralis system, a better term than the older acoustico-lateralis system (see Popper and Fay 1993 for discussion). Fishes have an impressive complement of hydrodynamic and acoustic sensors, commonly referred to as the lateral line and inner ear sense organs. Sound detection in fishes involves the lateral line as well as the inner ear and its associated structures (read Popper and Fay 1999).

While the otophysans have the best known adaptation for hearing, a number of other species in widely diverse taxa also have specializations that probably enhance hearing. Fishes (including otophysans) having specializations that enhance hearing have been referred to as hearing 'specialists', whereas that do not have such specializations are 'nonspecialists' or'generalists'. Hearing specialists tend to have a wider hearing bandwidth and greater sensitivity then nonspecialists. The limited behavioural data suggest that frequency and intensity discrimination may not be as acute in nonspecialists as in specialists.

Some of the better-known hearing specialists are found among the taxonomic groups Osteoglossomorpha, Perciformes (especially the Anabantidae), Beryciformes (especially the Holocentridae) and Clupeiformes. Despite the taxonomic diversity among these (and other) specialists, it is often the case that these groups have similar specializations for hearing. For example, otophysans and morymyrids (Osteoglossiformes) have similar specialized hair cell orientation patterns in the saccule, gas bladders in close mechanical contact with the ears, and similar hearing capabilities. Yet the species are not closely related. Thus, these specializations have probably arisen independently.

Even though there are hearing specialists among each of these taxonomic groups, most of these groups also contain numerous species that are nonspecialists. Thus, among the family Holocentridae (Beryciformes) there is a genus of hearing specialists, *Myripristis*, all of which have anterior extensions of the swim bladder that abut the ear (Fig. 16.10), and a genus of nonspecialists, *Adioryx*, in which swim bladder terminates far from the ear. Among the Siluriformes, at least one marine species, *Arius felis*, has a greatly enlarged utricle that appears to be used for detection of low frequency sounds (100–200 Hz), while other silurids (and other otophysans) do not have hypertrophied utricle or the excellent low frequency hearing. Similar differences can be found within the anabantids (Perciformes) and osteoglossids (Osteoglossiformes) where some species are hearing specialists and others are not. Popper and Fay (1993) predict that such diversity would be found among other taxa if studied sufficiently. At the same, it should be noted that within the genus *Pomacentrus* (damselfish, Perciformes) all species have very similar audiograms.

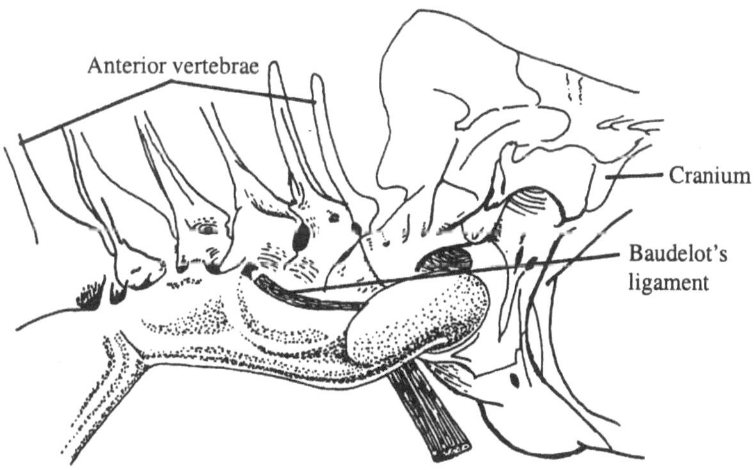

Fig. 16.10 Swim bladder-auditory bulla region of *Myripristis argyromus*. (Lateral view)
***Source:* Nelson (1955)**

Auditory equipment: Experimental evidence indicates that sound reaches the ear by at least two pathways in teleost fish. The "direct" path is thought to function in all fishes, whereas the "indirect" path is found in some, but not necessarily all, species with a swim bladder, serving as an accessory hearing organ. As the swim bladder expands and contracts in rarefaction and compression phases, the movement of its wall may be transmitted to the ear via specialized routes (e.g., the Weberian ossicles, rostral swim bladder projections to the ear, etc.) and fishes referred to as Ostariophysi (e.g., goldfish along with carp, catfish, minnows). Weberian ossicles, a chain of modified vertebrae, link the anterior swim bladder with the fluid system of the labyrinth. The two pathways appear to be complementary, in that the "direct" path is most effective below a few hundred Hz, while the "indirect" path seems most effective for signals above a few hundred Hz.

Behavioural experiments have demonstrated a considerable interspecific variability in the range of detectable frequencies (bandwidth) and sound pressure levels. For example, the most widely studied teleost species, the goldfish (*Carassius auratus*) can detect sounds from below 50 Hz to about 3000 Hz with best sensitivity at 400 to 1000 Hz. In contrast, another widely studied species, the cod (*Gadus morhua*) can detect sounds from below 50 Hz to only about 500 Hz with best sensitivity at 60 to 300 Hz (Table 16.1). The difference in bandwidth and sensitivity between the two most widely studied teleost species (goldfish and the cod) illustrate a major principle of fish audition. Fishes having auxilliary structures to improve acoustic coupling between the swim bladder and inner ear (e.g., goldfish) often have wider bandwidth and better sensitivity than fishes without such coupling. Fishes lacking a swim bladder (such as some flatfish) or fish in which the swim bladder does not function as a pressure detector to subserve the "indirect pathway" (e.g., *Tilapia*) hear even more poorly than cod (Table 16.1).

Table 16.1 Hearing range and pitch discrimination in some teleosts

Species	Upper frequency limit (Hz)	Frequency discrimination at given pitch (various sources)
Ostariophysi		
Phoxinus laevis	5000–7000	19% at 1000 Hz
		6% at 1000
		3% at 400–800
Carassius auratus	3480	3.5% at 500
		3.5% at 400
		4.4% at 1000
		7% at 50
Nonostariophysine		
Gobius niger	800	9% at 150
		12% at 300
Corvina nigra	1000	12% at 385
Sargus annularis	1250	< 9% at 150
		15% at 300–450
		> 30% at 600
Cottus scorpius	300	19% at 60
		8% at 95
		16% at 140

(Modified after Enger 1981). (See Ladich (2000; Table 7) for allied information.

In addition to detecting pure tones, we know from behavioural experiments that many fish species can perform frequency and intensity discrimination, and detect the presence of a signal even when masked by background noise. For example, the goldfish can discriminate small differences in pulse repetition rate and detect amplitude modulation with extremely small modulation indices.

Some fishes can indeed localize sounds, the mechanisms employed are quite unique among vertebrates.

Thus, the best known mechanisms of sound detection by fishes involves the swim bladder. Acoustic coupling between the inner ear and the swim bladder can be altered by changing the physical relationships between the two systems.

From experiments, a hypothesis has emerged that fishes with a swim bladder relate the acoustic pressure to particle motions, and fishes without a swim bladder relate the vertical to the horizontal particle motions, to remove the 180-degree ambiguity in the particle-velocity field. Unfortunately, it was not stated explicitly that the theories applied only to the far field or to the fields of monopole sources. In the instance of a dipole field, the mechanisms would suffer from an unsurmountable three-dimensional ambiguity at only slightly lower frequencies or slightly shorter distances from the source than those chosen from most tests. How might the animals have solved the problem, assuming they did? (see Kalmijn 1988 for discussion; for a comparative study of electric and near-field acoustic detections, read Kalmijn 1997).

It has been shown that, except for the monopole swim bladder fields of more vocal fishes, most biological objects produce predominantly dipole fields, like all animals during swimming. The monopole moment is often negligibly small, while the dipole moment tends to override all higher-order moments. The lateral line responds to the local-flow field only within a few body lengths from the source, whereas the inner ear detects the remainder of the near field, as well as the particle motions or, via the swim bladder, the acoustic pressure of the propagating sound wave.

In order to determine unambiguously the bearing of sound source, a fish must be able to resolve acoustic pressure and the components of the acoustic displacement vector from the signals detected by the otolithic organs. It has been demonstrated in bony fish that much of the processing required to do this may be implicit in the structure of ear and its associated neural innervation. The central nervous system might use to further process the derived information to localize a sound source and discriminate frequency and range.

To add further, it is possible that the substantial interspecific variation in the structure of ear among the 25000 or more extant species represents the variants on information processing for the central nervous system. Only a few basic morphologic and ultrastructural "patterns" are found among fishes, and these patterns recur in taxonomically diverse fish groups. It has been suggested that the structural variations in the fish ear may be related to some aspect (s) of sound processing.

Extensive field data reveal that free-ranging sharks are most readily attracted by sounds of very low frequency, down to 10 Hz, if not less. Sharks and kindred skates and rays are devoid of air cavities and are in their evolutionary development closer to the original function of underwater hearing than are the teleost fishes with swim bladders. Nevertheless, even though the latter may respond to sounds of several hundred, or even a few thousand, hertz, they also show clear reactions to the lower frequencies, down to a few hertz.

Certainly there is need to investigate the hearing abilities of still more species. Next to nothing is known of hearing in cephalochordates, such as *Amphioxus*, and agnathans like the lamprey *Petromyzon* and hagfish *Myxine*. Detailed information on the hearing abilities of sharks and rays is still largely lacking. Certain teleost groups also need closer investigation (e.g., Clupeidae, Engraulidae and Scombridae). More information is required on the ontogeny of sound reception in fish. The majority of behavioural and psychophysical studies have been of adult fish, and the development of hearing has hardly been touched.

4. NEURAL CONTROL OF TELEOST SOUND PRODUCTION

Sonic motor cells in most of the fishes studied are located in nuclei in the caudal medulla and their axons exit the brain as the occipital nerves which innervate the sonic muscle. The main exception to this pattern is found in croakers and drums (Sciaenidae) which have sonic motor cells presumed to be located in the spinal cord with axons running in the spinal nerves. With the great variety of teleosts available it would not be surprising to eventually find more exceptions. As far as is known, and this relates mostly to the toadfish, pattern generators are located in the tegmentum, somewhere between the caudal midbrain and level of the obex. The brain stimulation data suggest that reticular systems are involved, and results indicate that the median longitudinal fasciculus may be a possible output pathway for these mechanisms. The sonic midbrain area appears to have a direct influence on lower sonic motor systems. This area probably receives acoustic inputs from nearby auditory regions in the torus semicirularis and perhaps visual and other sensory inputs as well. The preoptic area and the inferior lobe which are involved in complex sexual and aggressive responses may influence sound production by connections with the sonic midbrain area. In addition, the preoptic area, at the least, and possibly several other regions involved in the central control of sound production in fishes may be influenced by changes in the circulating androgens via local hormone-concentrating neurons (Demski 1981; Fig. 16.11).

Fine and Perini (1994) demonstrated by electrical stimulation that forebrain centres are involved in sonic courtship and agonistic behaviour in oyster toadfish (*O. tau*).

This subject has been further covered in recent publications and synopses are given here. Sound production during social behaviour is known in representative of 40–50 families of teleost fishes. Many sonic organs employ drumming muscles attached to the swim bladder; contractions cause swim bladder to vibrate and emit low frequency sounds (toadfishes, cods, searobins, piranhas, croakers, mormyrids, catfishes). Other morphological structures are also specialized for sound production. Many of these are stridulatory and utilize rubbing of bony elements such as hypertrophied pectoral fin tendons, pharyngeal teeth, or neck vertebrate (as described earlier in Section 2). Several catfish families including pimelodids, doradids, mochokids, callichthyids, loricarids and bunocephalids and ariids have evolved two sonic organs: a swim bladder with extrinsic drumming muscles and a pectoral fin stridulatory apparatus.

The presence of independently derived sonic organs raises interesting questions about their motoneuron pools and innervation patterns. In toadfishes, searobins, sculpins and mormyrids, sonic muscles are innervated by a special sonic motor nucleus (SMN) having no other known function. In croaking gouramis, however, a ventrolateral spinal motor column contains pectoral motoneurons which are responsible for movement as well as sound production.

SMN axons exit in ventral roots of the occipital nerves in toadfishes, searobins and sculpins, representatives of three different orders. In sciaenids drumming muscles are innervated segmentally

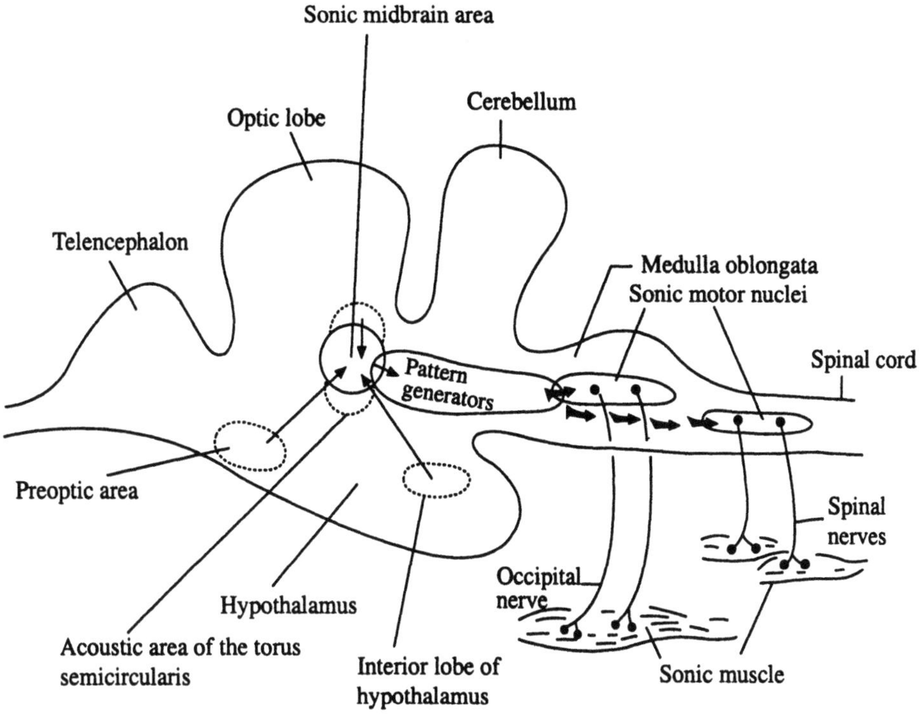

Fig. 16.11 Summary model of neural systems involved in the control of sound production in teleosts. The systems are plotted on a hypothetical sagittal section of fish brain. The occipital nerve system is characteristic of most fishes studied while the spinal sonic motor apparatus is thought to be present in croakers and drums (Sciaenidae).
***Source:* Demski (1981)**

by true spinal nerves, whereas mormyrid drumming muscles have vagal innervation. Thus, fishes have evolved nonhomologous types of swim bladder muscle innervation at least three times. In ariid catfishes drumming muscles are innervated by the dorsal root of occipital nerve whereas the pectoral spine muscles are innervated by the anterior part of the brachial plexus, or 'ramus cervicalis', which is composed of occipital and spinal nerves.

Ladich and Fine (1994; also see literature) localized the motoneurons and occipital and true spinal innervation of sound producing organ in pimelodid catfish. Pimelodids have a stridulatory organ composed of the pectoral girdle and the first pectoral fin ray, a swim bladder with extrinsic muscles to produce drumming sounds, and a tensor tripodis (TT) muscle that inserts onto the swim bladder. Sonic muscles are innervated by three branches (rostral, dorsal and caudal) of the occipital nerve (Oc) and the first true spinal nerves (S1 and 2): pectoral spine muscles (abductor, adductor and ventral rotator) by the rostral branch of Oc and S1 and 2, drumming muscle by the caudal branch of Oc and twigs of the S1 and 2, and TT by the dorsal branch of Oc (Fig. 16.12). Sonic nuclei form ipsilateral medial, intermediate and ventrolateral columns in the caudal medulla and spinal cord. Pectoral neurons form a ventrolateral motor column, and neurons for the first spine occupy the rostral part of the column. The median division of the swim bladder drumming motor nucleus (DMm) is situated on the midline between the central canal and the medial longitudinal fasciculus. The rostral pole of the DM nucleus expands ventrolaterally to include a population of neurons of intermediate position (DMi). The TT nucleus also assumes an intermediate position ventrolateral to DMm. The pectoral, TT, and DMi have a restricted rostrocaudal extent, whereas DMm extends further caudally. These data demonstrate that fish can evolve multiple sonic motor nuclei and that sound producing organs can be innervated in parallel by occipital and spinal nerves. The reader is referred to Ladich and Fine (1992) for details on the localization of pectoral fin motoneurons (sonic and hovering) in the croaking gourami, *Trichopsis vittatus*. Mochokid catfish (e.g., *Synodontis nigromaculatus*) have (1) a pectoral spine stridulatory apparatus and (2) swim bladder 'drumming' muscles. Pectoral spine-related motoneurons are within the ventral motor column whereas swim bladder motoneurons lie just ventral to the central canal or ventral ventricle (Ladich and Bass 1996) (Fig. 16.7).

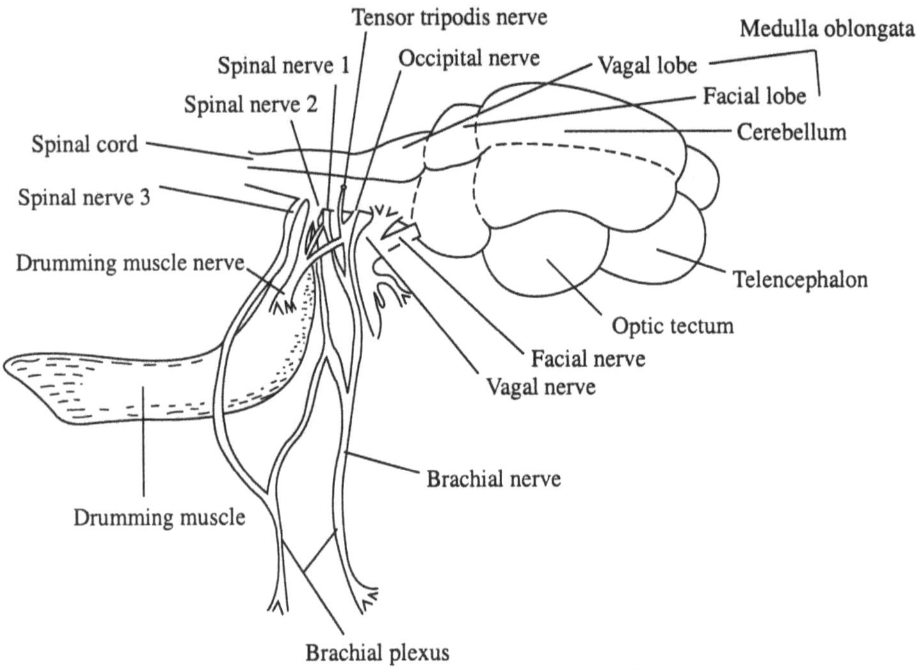

Pimelodid catfish *Pimelodus*: Nerve supply to drumming muscle.

Fig. 16.12 Drawing of occipital and first two spinal nerves from dissections and whole mounts of catfish stained with Sudan Black B. Dorsal nerve roots are not included.
***Source:* Ladich and Fine (1994)**

Recently, plainfin midshipman (*Porichthys notatus*) has been extensively studied (Bass and Marchaterre 1989a, b; Bass 1990; Bass and Baker 1990; Bass and Andersen 1991; Brantley et al. 1993a, b, c; Bass et al. 1994; Walsh et al. 1995—a review).

The sonic motor nucleus (SMN) of *P. notatus* is a midline nucleus located at the junction of the caudal medulla and rostral spinal cord. Its motoneurons innervate sonic "drumming" muscles that are attached to the lateral walls of the swim bladder. There are two classes of sexually mature males referred to as Type I and Type II. The type I males are larger and generate sounds during the breeding season. The type II males are smaller and like adult females have not been shown to generate sounds.

The average soma diameter of motoneurons of Type I males is about 50% larger than that of Type II and females. There is also a small but significant difference in soma diameter between Type II males and females; they are smaller in the former class.

There were *no* sex differences in the presence of distribution of different classes of axosomatic and axodendritic terminal buttons, which include: (1) active zones with either clear, round, or pleomorphic vesicles, (2) active zones with *both* clear, round vesicles, and larger dense core vesicles, and (3) "mixed synapses" with gap junctions and active zones usually associated with pleomorphic vesicles.

Sex differences in soma diameter correlate with a number of sex differences in the gross and ultrastructural features that distinguish the sonic muscles of Type I males from those of Type II males and females, which are similar to each other. The absence of qualitative sex differences in synaptic morphology suggest that the central neuronal circuitory of the sonic motor system is similar among all three adult morphs.

The peripheral sonic motor system of Type I males differs from Type II males and females (which resemble each other) in a number of ways: (1) the mass of their swim bladder and associated sonic muscles is 50% greater, (2) their muscle fibres are several times larger and have a characteristically large volume of sarcoplasm that surrounds the myofibrils and is densely filled with mitochondria, (3) the length of Z-lines of their myofibrils is about 20–fold greater, and (4) their sarcoplasmic reticulum is more highly branched. The ultrastructure of the myofibrils of Type II males and females resembles that found in the sonic muscle of males and females in other related species. The larger mass and specializations of the sonic muscle in Type I are considered to be adaptations related to their known role in sound production and the unique long duration "humming" sounds that they generate during the breeding season. The similarity in the sonic motor system between females and Type II males is considered to be related to the utilization of an "alternative mating strategy" by Type II males. This is probably the first documentation of a sex difference or for that matter, a sexual polymorphism in ultrastructural features of a vertebrate myofibril.

As mentioned, the SMN has a midline position at the caudal end of the brainstem and innervates the ipsilateral muscle via occipital nerve roots. Individual pacemaker neurons, which lie adjacent to the SMN, have been identified as the only source of direct synaptic input onto sonic motoneurons. Pacemaker neurons set the discharge frequency of the motoneurons, which then determines the contraction rate of the muscles and, in turn, the fundamental frequencies of vocalizations. There is direct relationship between the rhythmic, patterned output of a brainstem pacemaker circuit and the physical attributes of species- and sex- typical vocalizations.

Results of vocal-acoustic network investigations strongly imply that links between vocal and acoustic brain regions are traits common to endothermic vertebrates like mammals and ectothermic vertebrates like teleost fishes that vocalize.

Also it has been suggested that (1) sonic muscle in *P. notatus* is an androgen target tissue, (2) fibre structure and fibre number are androgen-sensitive features, and (3) there exist sex- and morph-specific patterns of sonic muscle responsiveness to androgen implants. A discussion on hormonal regulation is given by Walsh et al. (1995).

5. UTILIZATION OF ACOUSTICAL CUES

Numerous reviews have attested to the fully adequate hearing abilities of many fishes. Our present lack of knowledge may be attributed to a lack of appreciation of the details that must be in any ethological study of communication which involves experimental manipulation.

Recent findings have shown that numerous species, both vertebrate and invertebrate, use mate assessment mechanisms to increase fitness and, in the majority of cases, such assessment is carried by females. Although the sensory bases underlying such mechanisms have remained largely ignored, evidence suggests that mate selection depends, at least partly, on vocal assessment among members of several vertebrate classes.

Detailed discussions on the mechanisms can be found in several sources (Myrberg 1981; Hawkins and Myrberg 1983; Bass 1990; Smith 1992; Hawkins 1993; see Table 16.2).

Many species of fish use sound in a wide range of behavioural situations. Sound production capabilities and acoustic repertoire of fishes are not nearly as elaborate as terrestrial vertebrates but a variety of acoustic signals are used in reproductive (courtship), territorial, and agonistic behaviour. In fact, sound is often the best communicative channel for aquatic vertebrates as well as for territorial ones, since it is only highly directional, has a low rate of attenuation, and is not affected by obstacles such as rocky projections and reefs. Thus, sound is a suitable channel for communication over long distance, in murky water, or in areas where visual and chemical signals are disrupted by barriers or spurious currents (read Hawkins and Amorim, 2000).

It is evident that the great majority of sounds emitted by fish are produced in a social context, and involve interaction between individuals. There may be some exceptions. In general, however, calls from fish have been recorded during an encounter between the vocalist and another fish of the same or a different species, and it is usually tactically assumed that the call involved communication between the animals. Both sender and receiver may obtain mutual benefit. Calls do serve in driving away predators, or in reducing the likelyhood of a successful attack. The same calls may alert other fish, and help them to escape predators. The advantages to the sender in these circumstances are not clear, unless the fish alerted are kin-related. An attentive listner reacts to the call of another to its own advantage (interception).

One of the common contexts of sound production is during reproductive activity, where the calls may directly influence the behaviour of prospective mates. Sometimes the sounds accompany visual displays and it is possible that they form only part of a more complex signalling system. Often the calls are produced by the male fish, which in many cases shows territorial behaviour. From the contexts in which these reproductive calls occur, it seems likely that they serve to advertise the presence and reproductive readiness of the sender to the females, and may even arouse reproductive activity in the latter. The calls may provide the bases for mate selection by the female. It has been pointed out that the preponderance of territorial species producing the calls suggests that this kind of behaviour is characteristic of species whose sexes may be separated by considerable distance, or living in habitats where visual and chemical signals are inadequate.

The swim bladder and intrinsic sonic muscles of oyster toadfish (*O. tau*) grow for life and are larger in males who produce a courtship boatwhistle call than in females who merely grunt.

Sound is only a part of a more complex assemblage of signals emanating from the fish. A possibility that sound production by the males may assist in stimulating reproductive maturation in the female is indicated by mouth-breeding cichlid, *Oreochromis mossambicus*.

There are several recorded instances of sound production by one male stimulating others to be vocal. Thus, in the haddock, another male may approach a sound-producing male and may engage it in an aggressive display, both fish emitting sounds. Among colonial males of the

Table 16.2 Application of the framework to sound communication in fishes: some examples

Signaling Predators Families of Fishes whose members produce "Startle" type sounds		Signaling Prospective Mates: Species whose members produce sounds while courting mates	Signaling companions. Species whose members produce sounds or probable near-field effects while interacting in coordination or apparent cooperation	Signaling competitors for mates: Species whose members produce sounds while interacting aggressively with competitors during active courtship	Signaling competitors for resources other than mates: Species whose members produce sounds while showing moderate to high levels of aggression during interactions. In most cases, opponents move away or avoid the sender; often the sender enhances visual display
Internally produced sounds	Sudden swimming sounds				
Ariidae	Clupeidae	Anabantidae: *Trichopsis vittatus*	Gadidae: *Pollachius virens* (blinded individuals)	Pomacentridae: *Eupomacentrus partitus*	Anabantidae: *Trichopsis vittatus*
Balistidae	Gadidae	Batrachoididae: *Opsanus tau*			
Gadidae	Scaridae			Sciaenidae: *Corvina nigra*	Ariidae: *Arius felis*
Sciaenidae	Labriade	Centrarchidae-: *Lepomis macrochirus*	Holocentridae-: *Holocentrus rufus* *Myripristis berndti*		
Diodontidae					Cichlidae: *Tilapia mossambica* Cyprinidae: *Notropis analostanus* Gadidae: *Gadus morhua* Holocentridae: *Holocentrus rufus* Theraponidae: *Therapon jarbua*
Holocentridae	Sciaenidae	Cichlidae: *Tilapia mossambica*			
Scaridae	Serranidae	Cyprinidae: *Notropis analostanus* Gadidae: *Gadus morhua*			

Source: Myrberg (1981)

various damselfish (*Eupomacentus* spp.) courtship chirping by an individual male may initiate chirping among neighbouring males on the reef, and some may move towards the territory of the initiator. These are almost certainly examples of the interception of a call. Since an initiator may lead an approaching female away from the initiator, they point to the adaptive significance of intercepting the calls of others.

Field evidence has demonstrated that free-ranging females of the bicolour coral reef damselfish *Pomacentrus partitus* use the courtship of conspecific males to locate male nest sites during the spawning period. Evidence is also provided that females can distinguish, not only between such sounds and similar sounds on the reef, but also between courtship sounds of different conspecific males, the latter indicating a role in mate assessment by acoustical modality.

Aggressive fish are often vocal. Male croaking gouramis (*Trichopsis vittatus*) may participate in prolonged bouts, with butting, chasing and lateral displays by one male to another. However, there are numerous examples of fish producing sounds for other resources, for example food or space. Female haddock, though silent during courtship, will readily produce sounds during competitive feeding outside the breeding season. Sound production can provide significant benefits in such circumstances (read Henglmüller and Ladich 1999).

It has been suggested that social aggregation of fish may be facilitated by sounds. So far, however, few sounds have been recorded from schooling species. Indeed, within a family like the Gadidae, sound production is often absent from the more actively schooling forms like saite (*Pollachius virens*). One disadvantage in using sound to promote cohesion of the school once assembled is that predators may intercept the sounds, perhaps eliminating any anti-predator advantages provided by the schooling habit. Various studies have shown that predatory fish, and especially sharks, may home in on the incidental sounds produced by struggling or injured prey.

Studies on four marine species: the plainfish midshipman (*Porichthys notatus*), the oyster toadfish (*Opsanus tau*), the sculpin (*Myoxocephalus scorpius*) and the common searobin (*Prionotus carolinus*) have clearly shown that there are neurobiological correlates of dimorphisms in reproductive and vocalizing behaviours (for details, see Bass 1990).

CHAPTER 17

Electric Organs and Electroreceptors

1. ELECTRIC ORGANS

1.1 Introductory Overview

The specialized electric organs (EOs) which provide this stimulus energy are found in several groups of distinct and unrelated fishes. They are thought to have evolved independently in these groups because of their derivation from different skeletal muscles (muscle precursors), or even nerves, in various body regions. It also appears as if natural selection has shaped the detailed arrangement of the modified electrogenic units into a variety of parallel and serial patterns which are adapted to specific resistivities of freshwater and ocean water, a process that could be considered an example of convergent parallel evolution.

Members of the actinopterygian and chondrichthyan radiations have evolved electric organs. The "electric" actinopterygians include members of the osteoglossomorphs (two families of mormyriforms) and euteleosts (six families of gymnotiforms); the electric catfish, Malapteruridae; and the stargazer, *Astroscopus*. The electric chondrichthyans include the torpedinoids (electric rays) and rajoids (skates). The reader interested in phylogenetic analysis of the South American electric fishes (order Gymnotiformes) and the evolution of their electrogenic system can find useful information provided by Alves-Gomes et al. (1995). Alves-Gomes and Hopkins (1997) reported the molecular insights into the phylogeny of mormyriform fishes and the evolution of their electric organs.

Electric fish are often classified as strongly or weakly electric. The majority of species are weakly electric. Some have both strongly and weakly electric organs (Fig. 17.1). A number of weakly electric fishes have small or accessory electric organs in the head region (Fig. 17.2; Table 17.1). Major discoveries related to the electric organ structure and function have focused on weakly electric teleosts, mormyriforms, and gymnotiforms. The anatomy of the electrocyte is complex and varies from species to species (Table 17.2).

Electric fish vary the structure of the pulse sequences they emit. Some strong electric organs discharges (EODs) together to make a wave which is remarkable for its constancy of frequency (wave species). Others insert long pauses between pulses (pulse species). The individual EOD is usually so stereotyped and so constant for the species that it is like a characteristic signature. Morphologically, some species of mormyrids are so similar that they can not be distinguished but they can now be recognized by their electric signatures. Several species show sex differences in EODs. EOD differences within genera are as great or greater as differences between genera.

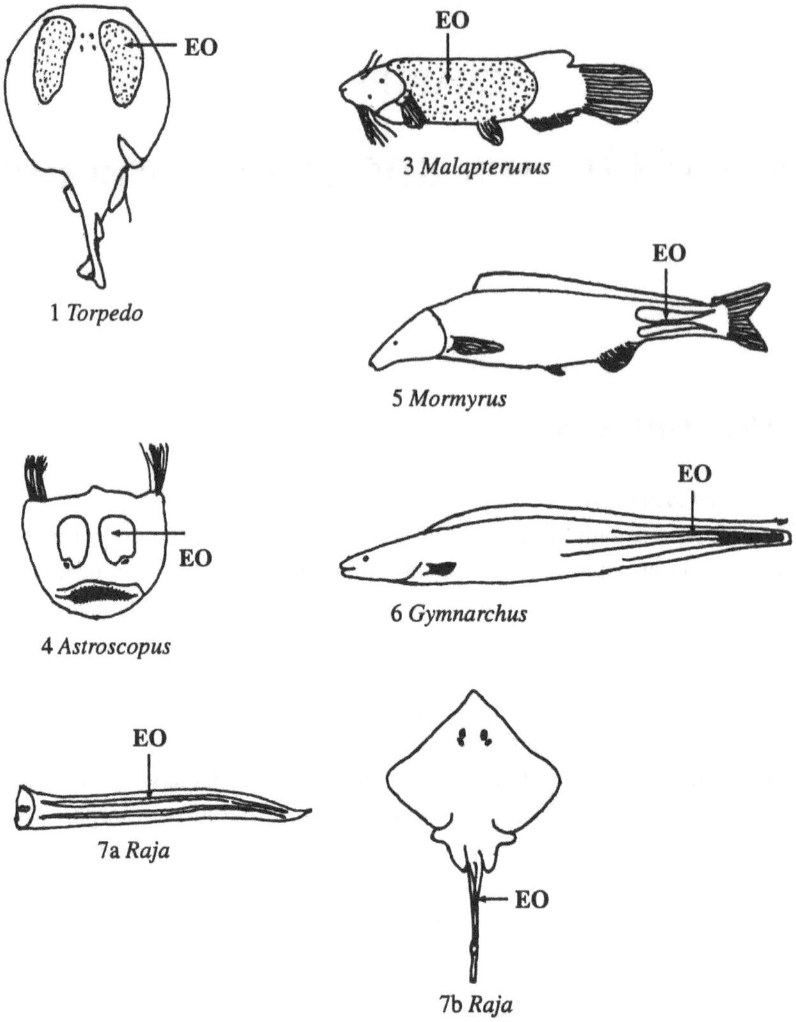

Fig. 17.1 **The electric organs of representatives of the several families of electric fishes. For 2a, b *Electrophorus* (see Fig. 17.2). EO = Electric organ.**
Source: **Bennett (1970, 1971a)**

For information regarding individual and sex specificity in EODs of breeding mormyrid fish (*Pollimyrus isidori*), see Crawford (1992).

The comparative approach to studies of electric organs and EOD's continues to offer a rich source of information on the morphological and physiological bases of behavioural adaptation. It has now been conclusively proven that in addition to its electrolocation role, the electric organ is a specialized vertebrate communication organ, much like the syrinx of songbirds or the larynx of anuran amphibians and mammals (a few of which echolocate).

Electric organs function on the following principle:

Serially stacked electrocytes are synchronously excited by spinal nerve signals to generate small voltage gradients in the direction of their linear arrangement. Since each stack is tightly

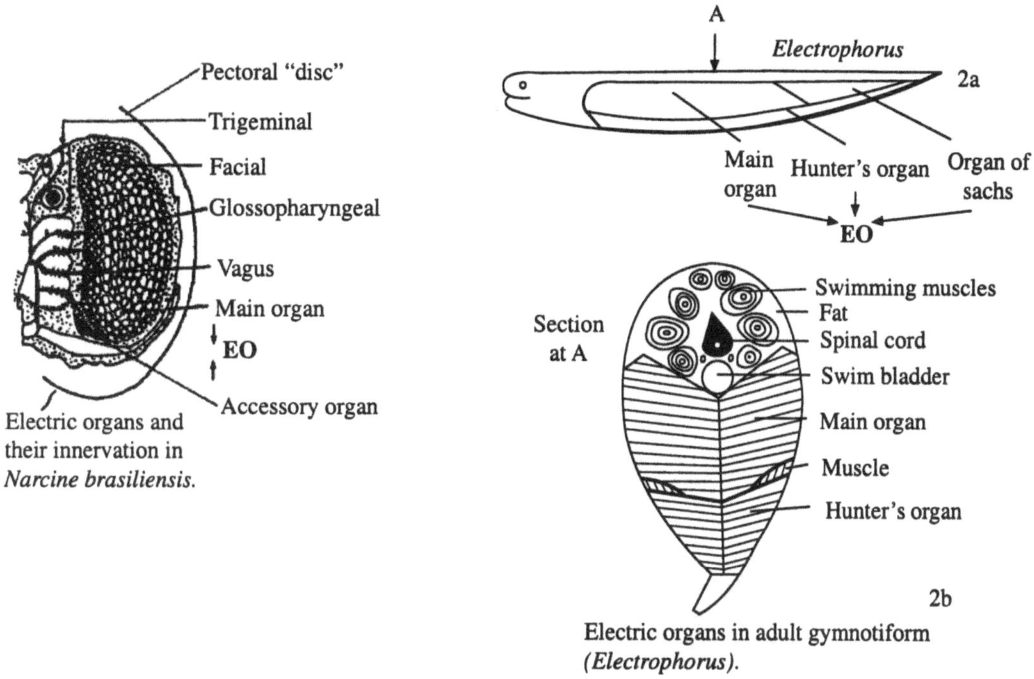

Fig. 17.2 **Accessory electric organs in adult gymnotiform (*Electrophorus*) and a torpedinoid (*Narcine*). *Sources:* Grundfest (1967) and Bass (1986) EO = Electric organ.**

Table 17.1 **Systematic position of certain electrogenic and electroreceptive fishes, electric organ discharge (EOD), origin and distribution**

Position	EOD	Origin	Distribution
Order Rajiformes Family Rajidae *Raja*, many species (electric skates, ordinary rays)	Weak	Myogenic	Marine, Cosmopolitan
Order Torpediniformes Family Torpedinidae A number of genera, many species (electric rays and torpedos)	Strong, upto 60V or/ 1 kW, some perhaps weak Torpedinid *Narcine* has both strongly and weakly electric organs	Myogenic	Marine, Atlantic, Indian and Pacific Oceans
Order Osteoglossiformes Family Mormyridae (elephantnose fishes, mormyrs) About 10 genera, comprised of over 200 species. *Mormyrus* (P), *Gnathonemus* (P) *Mormyrops* (P)	Weak, best known electroreceptive	Myogenic	Freshwater, Africa
Family Gymnarchidae (gymnarchids) Monospecific genus	Weak	Myogenic	

Position	EOD	Origin	Distribution
Gymnarchus niloticus (W, 250–500 Hz)			
Order Gymnotiformes			
Has 55 species of gymnotoids	Best known	Myogenic	Freshwater, central
Family Sternophygidae	electroreceptive		and South America
Sternopygus (W, 50–150 Hz)	Weak	Myogenic	
Eigenmannia (W, 250–600 Hz)			
Family Rhamphichthyidae (knifefishes)			
Rhamphichthys (P)	Weak	Myogenic	
Gymnorhamphichthys (P, has accessory electric organ in head region)			
Family Hypopomidae	Weak	Myogenic	
Hypopomus (P)			
Family Apteronotidae	Weak	Neurogenic	
Apteronotus (W, 750–1200 Hz)			
Sternarchorhamphus (W, 750–1000 Hz)			
Adontosternarchus (W, 950–1250 Hz)			
Sternarchorhynchus (W, 1500–1800 Hz)			
Family Gymnotidae	Weak	Myogenic	
Monospecific genus *Gymnotus carapo* (P) (knife-eels or nackedback fishes)			
Has accessory electric organs in head region			
Family Electrophoridae	Strong, more then 500V	Myogenic	
Monospecific genus			
Electrophorus electricus (P)			
Has both strongly and weakly electric organs.			
Order Siluriformes+			
Family Malapteruridae	Strong, 350V	Myogenic	
Malapterurus electricus, *M. microstoma* (electric catfishes)			
Order Perciformes			
Family Uranoscopidae	Strong? about 5V	Myogenic	Marine, Atlantic
one electric genus	in air from small		Indian and Pacific
Astroscopus; several species (stargazers)	animals		Oceans
Sand burrowing form, eyes on dorsal surface of head look virtually straight upward			

P = Pulse species or pulsers (buzzers): those which discharge brief pulses, separated by long and variable intervals.
W = Wave species or wavers (tone species) (hummers); those which produce long pulses, separated by short and highly stable intervals.
+ = Several species of synodontid catfish-a new group of weakly electric fishes.

Table 17.2 Anatomical and physiological classification of some electric fishes with myogenic electric organs

Different Varieties of Electrocytes

<table>
<tr><td colspan="3">Marine Electric Fishes
Electrically inexcitable</td><td colspan="9">Freshwater Electric Fishes
Also electrically excitable membrane</td></tr>
<tr><td colspan="8">One Surface Innervated</td><td colspan="4">Stalk Innervation</td></tr>
<tr><td colspan="6">One surface electrogenic</td><td colspan="6">Both surfaces electrogenic</td></tr>
<tr><td>Rectification</td><td colspan="3">No rectification</td><td>Discontinuous discharges</td><td colspan="3">Continuous discharges</td><td>Simple stalk system</td><td colspan="3">Complex stalk system</td></tr>
<tr><td>Single organ</td><td>Double organ</td><td colspan="3">Single organ</td><td colspan="3"></td><td>Single innervation</td><td>Multiple innervation</td><td colspan="2">Single innervation</td></tr>
<tr><td rowspan="2"></td><td colspan="4" rowspan="2">Monophasic responses
(Electric Organ discharge EOD; pulses)</td><td rowspan="2"></td><td rowspan="2">Diphasic responses</td><td></td><td colspan="3">Non-penetrating</td><td>Penetrating</td></tr>
<tr><td colspan="3">Monophasic responses</td><td>Diphasic responses</td><td>Triphasic responses</td></tr>
<tr><td>Raja</td><td>Narcine</td><td>Torpedo</td><td>Astroscopus</td><td>Electrophous</td><td>Eigenmannia</td><td>Gymnotus</td><td>Gymnarchus</td><td>Malapterurus</td><td>Mormyrus</td><td colspan="2">Gnathonemus</td></tr>
<tr><td>Rajiformes</td><td colspan="2">Torpediniformes</td><td>Perciformes</td><td colspan="3">Gymnotiformes</td><td>Osteo-glossiformes</td><td>Siluriformes</td><td colspan="3">Osteoglossiformes</td></tr>
<tr><td colspan="3">Chondrichthyes</td><td colspan="9">Osteichthyes</td></tr>
</table>

Shape and Face Innervation:
Raja: cup-and disc shaped electrocytes, anterior face innervated.
Torpedo: disc-like electrocytes, ventral surface innervated.
Astroscopus: flattened electrocytes, dorsal face innervated.
Electrophorus: flattened electrocytes, posterior face innervated.
Gymnarchus: electrocytes flattened cylinders, posterior faces innervated.
Malapterurus: electrocytes like lily pads, main part disc-shaped, candal face innervated.

surrounded by insulating tissues, the voltage contributions of its electrocytes add linearly, as with a series of small batteries (Fig. 17.3).

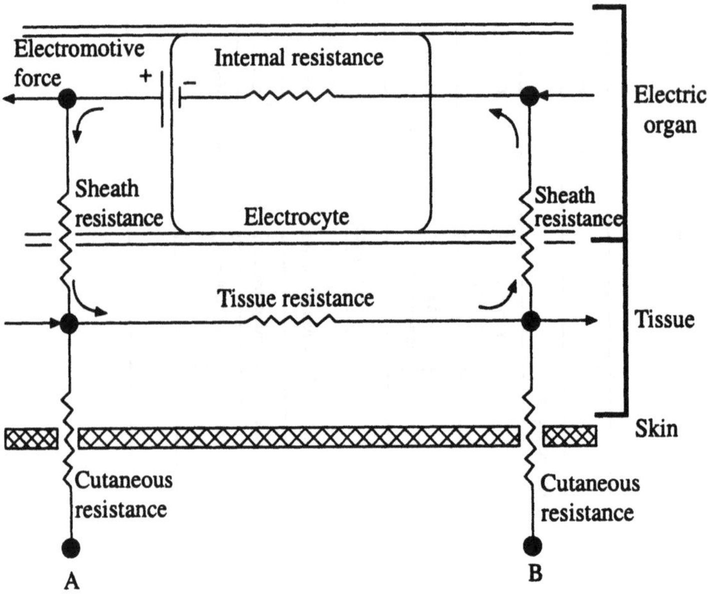

Fig. 17.3 Theoretical representation of the electrocytic generator in its environment. *Arrows*, possible current paths through different resistances. Voltage drop was measured between A and B. (see details: Caputi et al. 1993).

The electric organ is driven by a command centre in the medulla. Diameters and lengths of spinal nerves supplying the electric organ are adapted to ensure synchronous arrival of the pacemaker signal at the electrocytes of the organ.

The similarities and diversities in the properties and in structure lend piquancy as well as theoretical importance to the comparative study. The significance pertains not only to the fields of comparative physiology and morphology but also to the domain of evolutionary theory.

By comparison with electrogenesis, electroreception is far more common among fishes, although most groups of electroreceptive fishes possess only one type of electroreceptor, known as ampullary receptors. The remarkable adaptations of certain fishes to produce and sense electric currents has provided physiologists and behaviour biologists not only fascination with a new sensory modality, but also with a number of model systems in neuroethology, aiding in understanding the control of behaviour, and the evolution of communication signals and sensory modalities (see Hopkins 1983; Bass 1986; Bodznick and Boord 1986; Raschi and Machanos 1987; Zakon 1988; Emde 1998b).

1.2 Comparative Morphology of Electric Organs

In the Torpedinidae, the organs derive from branchial musculature. The electrocytes of Torpedinidae are stacked one above the other in a column of series array. The columns of electrocytes are closely packed in a parallel array as seen in the dorsal view of the main organ of *Narcine brasiliensis*. This species also has an accessory organ which differs in morphology as well as electrophysiological properties from the main. Recent embryological work has indicated that

Torpedo also has an accessory electric organ that degenerates at the 30-mm stage of development. So, two electric organs appear to characterize torpedinoids.

In Astroscopidae, the electric organ replaces eye muscles (see also Fig. 15.45). In the gymnotiforms, the body is extremely short, and particularly so in *Electrophorus*, the electric organ forms most of the soft tissue of tail and differentiated into Organ of Sachs, main organ and Hunter's organ. In *Malapterurus*, the two electric organs surrounding the fish body like a mantle, lying just under the skin, have each evolved from a single motor unit of the shoulder girdle. In the Rajidae, Mormyridae, and *Gymnarchus*, the electric organ comprises a small part of tail musculature. In rajids, the electric organs are bilateral and spindle-shaped; in Mormyridae, the two columns of electric organ on each side lie above and below the lateral line, and in *Gymnarchus*, the electric organ on each side is comprised of four columns of different extent.

As a vestige of their muscular origin, electrocytes in Rajidae retain striations. Similar patterns are also present in electrocytes of Uranoscopidae and Mormyridae, but in none do the cells have contractile capacity. The cell membrane of electrocytes shows some degree of differentiation in that the membrane of the uninnervated major surface has many invaginations. In the electrocytes of the Torpedinidae, the invaginations are extremely numerous and develop a dense network of canaliculi. The canalicular structure might be vestige of the transverse tubular system of muscle fibres which is formed from invagination of the cell membrane and which in the muscle fibres is involved in excitation-contraction coupling.

The canalicular system cannot, of course, function in the same manner. However, the great increase in the surface of the membrane by the proliferations of the invaginations may contribute to the decrease in the resistance of the uninnervated surface of many types of electrocytes. In comparison, the innervated surface is relatively smooth except for short inpocketings. The membrane is very densely innervated in the elasmobranch electrocytes as it also is in electrocytes of *Astroscopus*, *Gymnarchus*, *Electrophorus*, and some other gymnotoids. In others, however, the innervation is restricted to a stalk (Figs. 17.4 and 17.5). In all cases studied, synaptic transmission from the electromotor nerve fibres to the electrocytes is cholinergic, and the sites of innervation are also sites of histochemical cholinesterase localization. In one group of gymnotoid fish (Apteronotidae) the electric organs are derived from nerve fibres.

1.3 Electrical Function of Electric Organs

As electrogenic systems, electric organs are unique in that the output voltage of the organ during activity is considerably higher than the output of any one electrocyte. This result is achieved by series summation of the EMF's of the individual batteries, and the electric organs are uniquely structured so as to permit this summation. The series summation arise from the stacking of electrocytes, either in a dorsoventral direction as in Torpedinidae and Uranoscopidae or in a rostrocaudal direction as in all other electric fishes. That only the innervated surface of electrocytes is electrogenic has been proved to be correct for many types of electrocytes, including those of Torpedinidae, *Astroscopus*, *Gymnarchus* and some of the Gymnotiformes including *Electrophorus*, but in other Gymnotiformes, all the Mormyridae, and *Malapterurus*, both major surfaces are electrogenic and contribute current to the organ though in opposite directions of flow. The electrocytes of the Rajidae form a special category.

The three strongly electric forms (Torpedinidae, *Electrophorus*, and *Malapterurus*) owe the power of their discharges not only to the fact that the number of electrocytes in series is very

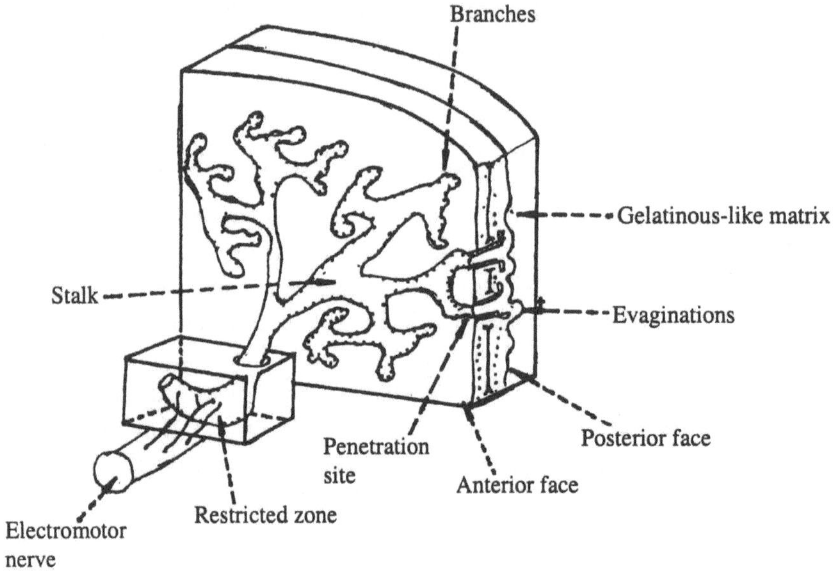

Fig. 17.4 Line drawing showing a cutaway view of one-half of an electrocyte with a penetrating-stalk system. The three fundamental elements of an electrocyte are indicated: anterior and posterior faces and a stalk. The entire cell is enclosed in a compartment containing a gelatinous-like matrix. The finger-like evaginations of the posterior face join into branches that penetrate the main electrocyte body to the anterior side at penetration sites. The stalk's branches fuse into a trunk that is innervated in a restricted zone by axons of the electromotor nerve.
***Source:* Bass (1989)**

high (thus creating a high voltage) but also to the large cross-sectional surface of the parallel arrays of electrocytes.

In many weakly electric fishes the properties of the electrocytes as electrogenic systems are much more complex than those of the electrocytes in the strongly electric fishes. The low output of the organs relates solely to the fact that the number of cells in both the series and parallel arrays is smaller in the weakly electric fishes.

1.4 PHYSIOLOGY OF THE GENERATING CELLS, ELECTROCYTES

Every known type of electric organ has been studied sufficiently to explain its voltage output in terms of membrane potentials of the individual electrocytes. Not only the form but the amplitude of the organ discharges can be accounted for in terms of the responses of the single cells. Although a variety of pulse shapes are produced, the basic operation is always similar to that of the Bernstein mechanism, that is the cell has what may be considered as two faces whose membranes develop different potentials to generate external currents. The individual faces can have a number of combinations of different types of membrane:

(i) spike-generating membrane alone, presumably involving Na and K activation and anomalous rectification; (ii) spike-generating membrane mixed with PSP (post-synaptic potential)-generating (end plate) membrane, the spike-generating membrane some-time without delayed rectification but with anomalous rectification; (iii) PSP-generating membrane only; (iv) membrane exhibiting

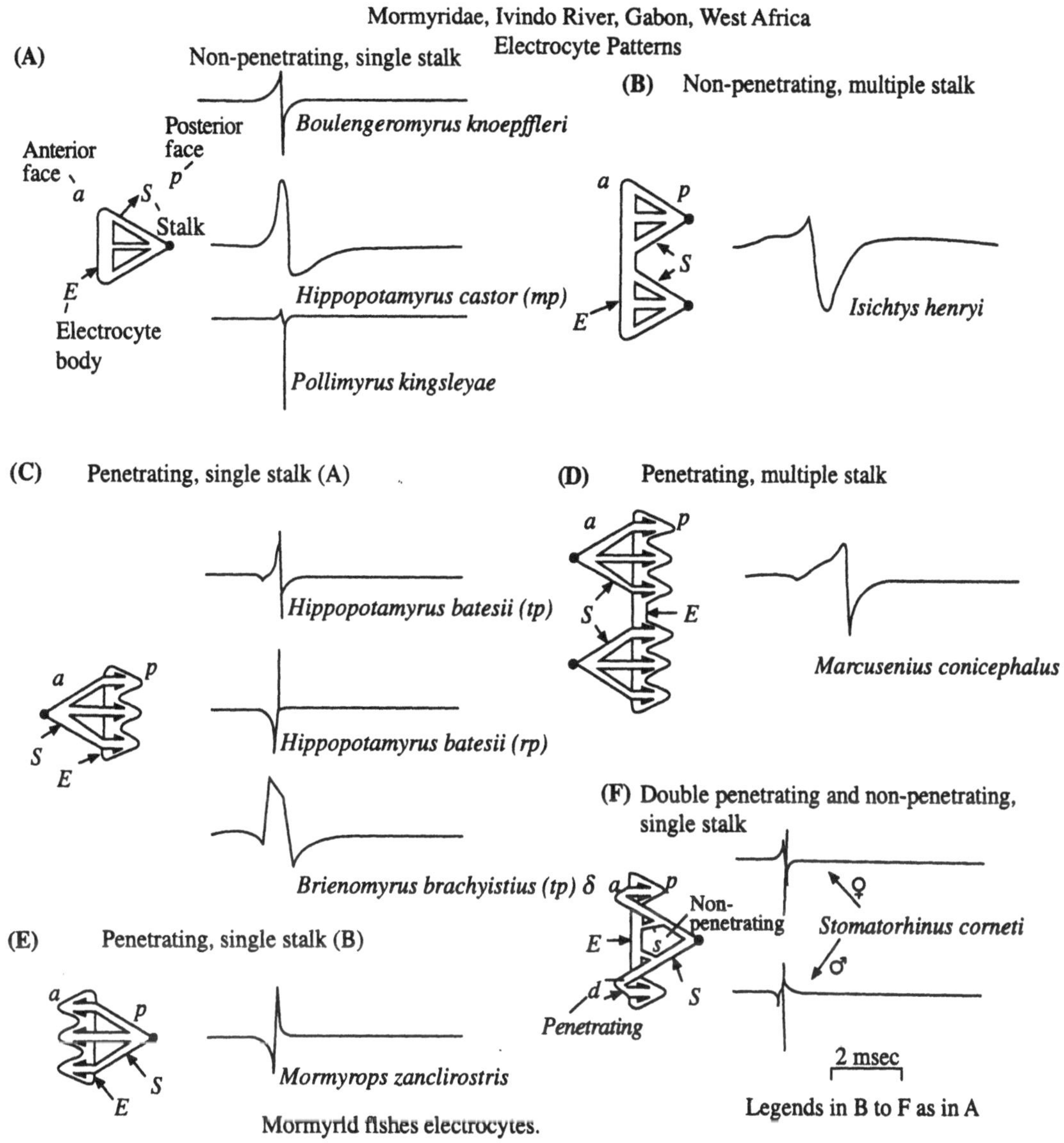

Fig. 17.5 Patterns of spatial organization of electrocytes for species of mormyrid fishes. An electrocyte has two principal components: an electrocyte body (E) and a stalk (S) which is an extension of the electrocyte body and is the site where spinal electromotoneurons synapse (dot at tip of stalk). Each electrocyte body has an anterior (a) and a posterior (p) face which are aligned along the longitudinal body axis. Some representative EOD waveform pulses associated with each anatomical pattern are shown. The EOD waveform pulse describes the appearance of each individual pulse generated by the electrocytes and can be modeled as having two or three phases. Species with one (A) or more (B) innervated stalks that extend away from the posterior face have a biphasic pulse with an initial head positivity. Species with one (C) or more (D) stalks that penetrate the electrocyte body and are innervated on the anterior side have triphasic EODs with an initial negative phase. One species (E) has a penetrating stalk that arises on the anterior side; its EOD is also triphasic, but the initial phase is positive and only seen at high gain. Another species (F) has a stalk with both penetrating (d) and nonpenetrating (s) elements and a polyphasic EOD of very brief duration. Some species, such as *Stomatorhinus corneti* (F), have a sex difference in the EOD, although the electrocyte geometry is monomorphic; there are sex differences in the electrocyte membranes themselves.
Source: **Bass (1986, 1989)**

delayed rectification only; (v) PSP-generating membrane with membrane exhibiting delayed rectification; (vi) membrane having a high capacity and high resistance so that most of the current through it is capacitative; (vii) membrane exhibiting what is apparently a polarization capacity, possibly due to accumulation of potassium in the extracellular space; and (viii) low resistance and inexcitable membrane, that is membrane having a fixed (low) resistance and capacitance. These different kinds of faces are combined in a number of ways.

It has been seen that the same form of organ discharge can be achieved in a number of ways by various types of electrocytes. The significance of the particular specializations is often unknown, although sometimes the properties of the organs or the individual cells can be understood in terms of maximizing the efficiency of power generation.

Another generalization concerns the presence of a dc component in the electric organ discharge. All strongly electric fish emit dc impulses. A few weakly electric fish that fire at low frequencies also emit pulses that are dc or have a large dc component, but in most weakly electric fish including all high-frequency species, the organ discharge has no dc component.

A long recognized difference between the electric eel and *Torpedo* is now known to be universal: electrocytes of freshwater fish have spike-generating membrane; those of marine fish do not. The significance would seem to be related to the need to deliver the maximum power to environments of high vs low resistivity. This difference between marine and freshwater forms is a puzzle.

The values of membrane resistance in electrocytes are far below the values for muscle and correspondingly the current outputs are much greater. Weakly electric organs tend to have somewhat higher membrane resistances.

The property of electrocytes that is most outstanding is frequency of firing. The electric organ of sternarchids discharge constantly at frequency from 700–1800/sec, *Eigenmannia* operates at 200–600/sec, and *Gymnarchus* 250–500/sec.

1.5 Electric Organ Discharge (EOD)

The electric organ of fishes produces a readily observable electric signal, the electric organ discharge or EOD (Table 17.3). In addition to its importance in social behaviours, the electric organ discharge is the carrier signal for active electroreception.

The electric organ discharge of an electric fish is controlled by anatomical organization of electrogenic cells, and by the physiology of electrocyte membranes. The EOD is fixed and stereotyped for a given fish. The electrocytes are exited simultaneously and their action potentials seem to produce the EOD. The discharge frequency is determined by the firing frequency of medullary pacemaker nucleus (PMN) which can be transiently modified for communication or preserve electrolocation abilities (the jamming avoidance response (JAR) (Figs. 17.6–17.8).

The EOD has two elements referred to as the EOD waveform pulse and the EOD rhythm. The EOD waveform describes the shape and duration of each individual pulse which is determined by individual electrocytes of the electric organ. In contrast, the EOD rhythm describes the firing rate or frequency of the electric organ and is determined by the central motor pathway. In considering the diversity of electric communication systems, the design of the peripheral electric organ and its modified muscle cells, the electrocytes, comes to the forefront.

Freshwater fishes with weakly electric organs are divisible into two classes, depending on the appearance of their EOD rhythm: (1) wave species (hummers, tone species) with a constant,

Table 17.3 Diversity of EODs and electric organs in some electric fishes

Group	Number of phases	Polarity	Amplitude (V)	Stalk
Uranscopidae	1	Dorsal, negative	5 (air)	Absent
Malpteruridae	1	Head, positive	350	Present
Rajoidei	1	Head, negative	0.25–1	Absent
Torpedinoidei				
main organ	1	Dorsal, positive	50 (air)	Absent
Accessory organ (*Narcine* only)	1	Dorsal, negative	0.1–1	Absent
Mormyriformes				
Gymnarchidae	1	Head, positive	1–2	Absent
Mormyridae	1–4	Head positive or negative	2–5	Present
Gymnotiformes				
Electrophoridae Main, Hunter's, Sachs' organs	1	Head, positive	10 (weak) 500 (strong)	Absent
Sternopygidae Main organ	1	Head, positive	2	Absent
Apteronotidae Main organ	1, 2	Head, positive or negative	1–3	Absent
Accessory organ	1	Head, positive	25 mV	Absent

Source: Bass (1986) (Abridged)

quasisinusoidal EOD waveform having a high stable frequency. Most 'wave' fish produce a simple, monophasic pulse each time the electric organ fires, and (2) pulse species (buzzers) with an EOD waveform consisting of single EODs that are widely spaced temporally and which occur with a variably frequency. The pulse waveform may be mono-, bi-, and triphasic, and so may be complex in the time domain. Discharge patterns of weakly electric organs in marine fish are intermittent.

Several genera of electric fish (e.g., *Steatogenys, Hypopygus*) possess more than one electric organ. Even though all organs are fired synchronously, they still differ in spectral composition of their discharge. As a consequence, the shape and spectral composition of the overall EOD may vary strongly if recorded at different points of the animals' body surface. Moderate local variations in EOD shape may also be found in species with a single electric organ. Relative current contributions from single electrocytes may vary along the animals' body surface.

Sexual dimorphisms are seen in the EOD in both wave and pulse fish. Male *Eigenmannia* and *Sternopygus* discharge at lower EOD frequencies, whereas male *Apteronotus* discharge at higher EOD frequencies than the females of those species. In sexually dimorphic pulse fish, the waveforms of two sexes may differ in the number, amplitude, and duration of various phases and, usually, in the total duration of the pulse. In all cases observed todate, the EOD pulse of females and juveniles is of shorter duration than that of mature males. These differences are likely due to the effects of steroids on the tissues involved in producing the EOD. It is currently thought than in wave fish, sex steroids affect the PMN, thereby changing EOD frequency, whereas in pulse fish these hormones affect the electric organ, where they influence pulse waveform properties. It has been demonstrated that androgens also effect the EOD pulse duration in the wave fish *Sternopygus*,

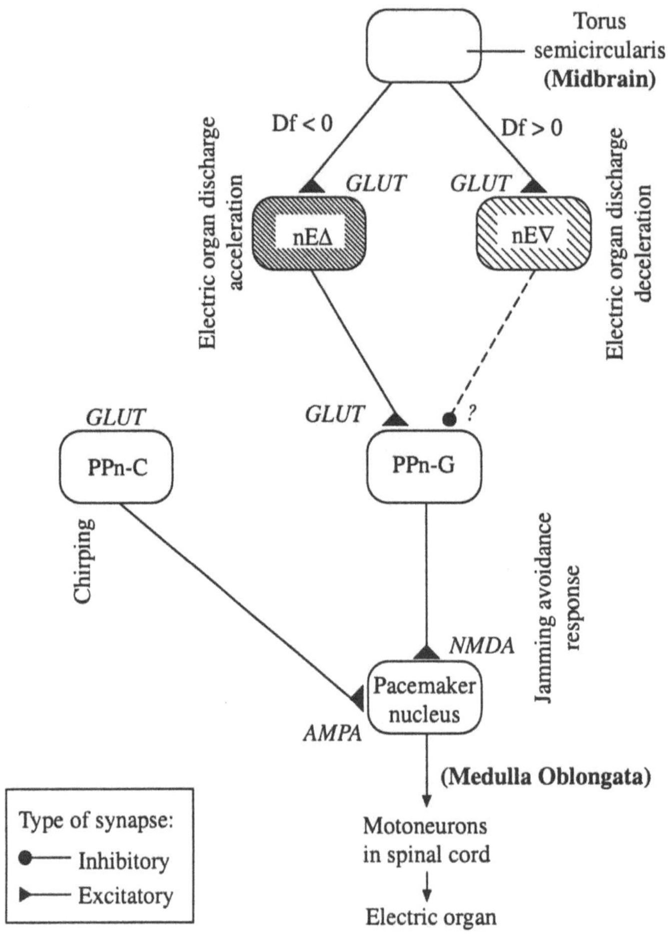

Fig. 17.6 Original hypothesis of the motor control of the jamming avoidance response in *Eigenmannia*. The torus semicircularis contains two classes of 'sign-selective' cells that are excited either by positive or by negative frequency differences (Df) between the jamming signal and the fish's own EOD). The dorsal portion of the nucleus electrosensorius (nE↑) contains higher order sign-selective neurons, excited by negative frequency differences. Its stimulation accelerates the pacemaker. Higher-order sign-selective cells that are positive frequency differences are located in the ventral portion of the nucleus electrosensorius (nE↓). Its stimulation lowers the EOD frequency. The PPn-G represents the top of a neuronal hierarchy controlling the EOD frequency. Its neurons are excited by negative frequency differences and inhibited by positive frequency differences. Their excitation raises and their inhibition lowers the frequency of the pacemaker nucleus at the pacemaker nucleus, which triggers each discharge cycle of the electric organ by a single command pulse. Chirp-like frequency modulations are generated by the PPn-G and mediated by AMPA receptors. The effect of the PPn-G/CP was assumed to be mediated by NMDA-type receptors. Compare with Figure 17.7 which incorporates the new findings about the motor control of jamming avoidance response. GLUT indicates where electrophoretic injection of L. glutamate elicits effects.
Source: **Metzner (1993)**

and that this effect is likely due to direct actions of the hormone on the spiking membrane of the electrocytes. Detailed analysis of *Pollimyrus isidori* (mormyrid) EOD in social communication: the EOD as a sex recognition cue and the EOD as an individual recognition cue, has been provided by Crawford (1992). Using anatomical, biochemical, and physiological methods, we can now characterize the electric organ as an androgen target tissue.

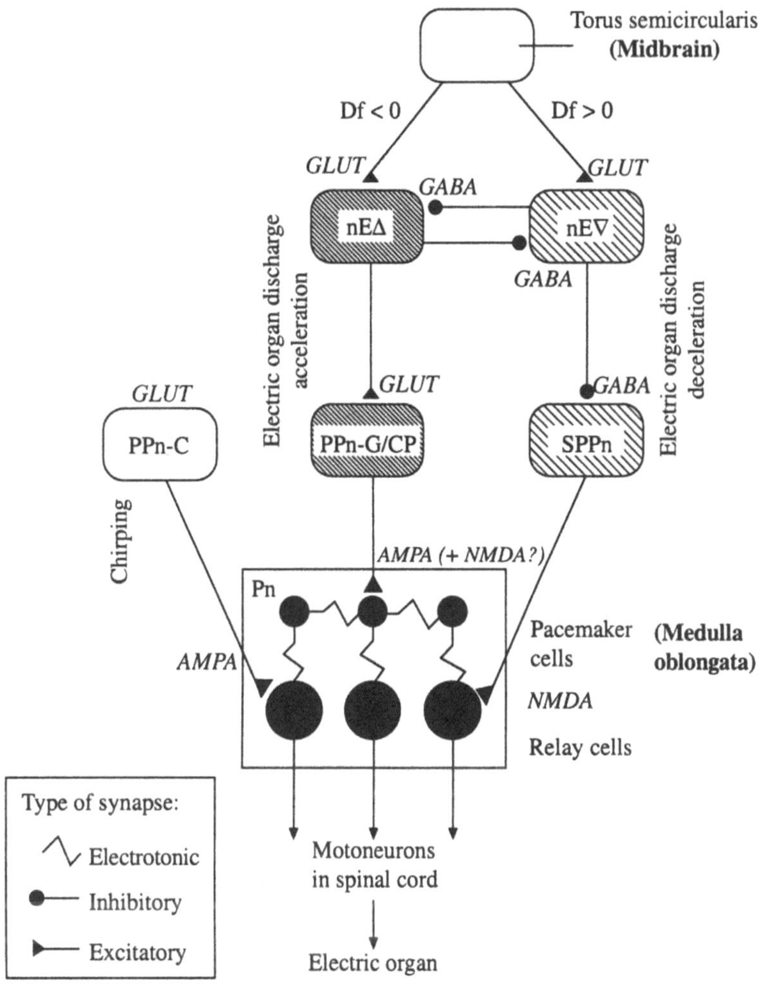

Fig. 17.7 **Present hypothesis of the motor control of the JAR in *Eigenmannia*. Compare with Figure 17.6 which shows the original hypothesis. The response to jamming stimuli with negative frequency difference (Df), that is, EOD accelerations, is controlled by a purely excitatory pathway (darkly stippled structures) that originates in the dorsal nucleus electrosensorius (nE↑) passes via the PPn-G/CP, and is assumed to terminate on pacemaker (Pn) cells where the synaptic transmission is mediated primarily by AMPA-type receptors. The pathway that controls the response to positive frequency differences, that is, EOD decelerations (hatched structures) begins in the ventral nucleus electrosensorius (nE↓) and continues via GABAergic connections within the SPPn. The SPPn projects to relay cells, where NMDA-type receptors mediate synaptic transmission. The second pathway also controls the resting frequency by the tonic activity of the SPPn. Communication signals (chirps) are generated by the PPn-C and its connection with relay cells in the pacemaker nucleus via AMPA-type receptors (lightly stippled). The reciprocal GABA ergic connection between the dorsal (nE↑) and ventral nucleus electrosensorius (nE↓) has been demonstrated GLUT indicates where electrophoretic injection of L. glutamate elicit effects.**
Source: **Metzner (1993)**

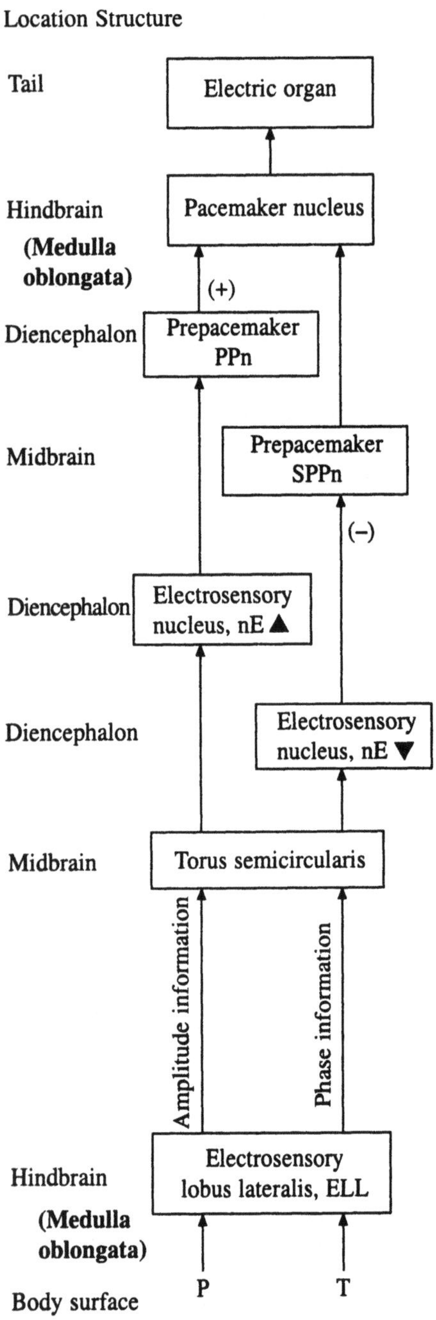

Function

Electric organ discharges at the frequency of the pacemaker nucleus.

Pacemaker nucleus triggers each discharge of the electric organ with a single command pulse.

Neurons of the diencephalic prepacemaker nucleus (PPn) are excited by a negative Df and raise (+) the frequency of the pacemaker nucleus.

Neurons of the sublemniscal prepacemaker nucleus (SPPn) excite the pacemaker nucleus tonically. They are inhibited by a positive Df and, as a consequence, the pacemaker frequency drops (–).

The diencephalic electrosensory nucleus (nE▲) is excited by a negative Df and raises the frequency of the PPn.

The diencephalic electrosensory nucleus (nE▼) is excited by a positive Df and inhibits the SPPn. As a consequence, the pacemaker frequency drops.

Neurons in lamina 6 of the Torus Semicircularis determine phase differences between signals arriving from different points on the body surface. Within deeper laminae, this information is combined with local amplitude information to compute the sign of the frequency difference, Df Neurons that discriminate the sign of Df appear to innervate the nE▲ and nE▼ and to excite these two nuclei in accordance with the sign of Df.

The ELL codes the phase and amplitude of the signal for many points on the body surface in separate, somatotopically ordered layers of neurons and relays this information to different laminae of the Torus Semicircularis.

P-and T-receptors code amplitude and phase locally.

Fig. 17.8 The neuronal flow of information in the Jamming Avoidance Response (JAR) of *Eigenmannia*. The fish decodes the sign of the frequency difference, Df, between a neighbour's EOD and its own by evaluating modulations in the phase and in the amplitude of the interference signal. Phase and amplitude are coded and processed separately up to the level of the torus. A convergence of phase and amplitude information within deeper layers of the torus yields neurons that discriminate the sign of Df. It is not yet known in sufficient detail how these neurons innervate and activate the diencephalic electrosensory nuclei, nE▼ and nE▲.
***Source:* Heiligenberg (1994)**

There is a degree of co-ordination between the PMN and the electric organ in weakly electric wave fish. It is likely that androgens play a role in this process. If, as proposed, the electrocyte membrane properties determine the characteristics of waveform, then the androgen must act on both the PMN and the electrocytes to keep EOD frequency and pulse duration in register.

In *Electrophorus*, discharges of high-voltage and low-voltage pulses are produced by electric organs; the former are generated by the main and Hunter's organ and the latter by the Sachs organ.

1.6 Fine Structure of Electrocytes

The major fine structural differences in the physiologically different electrocytes of freshwater weakly electric teleosts relate to specializations of the cell surface, distribution of mitochondria, and variations in the filamentous inclusions. The degree of surface proliferation of the parallel faces of electrocytes produced by tubular invaginations of these faces is consistent with the physiological desirability for the electrocytes to match impedances across these faces during spike generation.

The distribution of mitochondria shows a less complicated relation to the physiological responsiveness of cells than does the degree of surface proliferation. The mitochondria tend to be packed more densely and closer to innervated face if only that face is active, i.e., in *Gymnarchus*, *Hypopomus*. There is not much difference between faces if both become active, i.e., in *Gnathonemus*, *Gymnotus*. Presumably sodium is pumped back out mostly where it enters during activity, and the proximity of mitochondria to these faces would increase the availability of ATP. The high density of mitochondria is not universal among electrocytes. The exceptions so far noted (*Electrophorus*, *Malapterurus*, *Narcine*, *Raja* and *Torpedo*) occur among the strongly electric fish (with the exception of *Raja*). In these species the electric organs are used rather infrequently and become fatigued relatively quickly. Slower restoration of ionic concentrations is presumably compatible with their intermittent operation.

The surface proliferations in electrocytes may be homologous to junctional folds or transverse tubular system. The latter seems a bit more likely in view of the tubular or canalicular form of the invaginations and the absence of acetylcholinesterase except at the actual sites of innervation. In respect to the origin of the proliferations, it is interesting that the form appears quite different in two non-myogenic cell types that show highly proliferated membrane associated with increased capacitance. In the neurogenic electrocytes of *Apteronotus*, there are polypoid processes that appear to have formed by protrusion from the cell surface. The particular form taken by the proliferations in the electrocytes thus seems to reflect their myogenic origin.

The major feature of the interiors of the electrocytes is the presence of filaments randomly arrayed in all weakly electric species and organized into some more or less orderly pattern in the mormyrids. The distribution of organized filament systems in the electrocytes of weakly electric teleosts together with earlier embryological and morphological studies of these and other species seems to indicate a correlation between filament organization and level of development at the onset of differentiation of the electrocytes. Filaments about 80 Å in diameter are omnipresent and correspond to secondary (thin) filaments of muscle particularly where Z lines are present (*Gymnarchus* and rajids). Well organized myofibrillarlike system with thicker filaments are homologus with primary filaments of muscle identified as myosin.

It has been recently suggested that intermediate filaments are anchored to the postsynaptic membrane of *Torpedo* electrocyte via a 43-kDa protein-ACh-R linked. However, desmin is a minor component of the postsynaptic membrane in the electromotor and is concentrated among and around the ends of the junctional folds in the neuromuscular junction. We do not know whether desmin of *Electrophorus* is involved in the same type association with the membrane via 43-kDa protein, as postulated for *Torpedo*, nor do we know whether it interacts with the putatative F-actin of electrocytes of *Electrophorus* (read Cordeiro et al. 1996).

In *Gnathonemus*, organized fibrils possess all the band patterns (A, I, M and Z bands) of normal striated muscle. Transverse fibres of intermediate diameters are present at the Z band.

The A and Z bands in *Mormyrus* are birefringent. From the ordering of filaments in *Gnathonemus*, it seems likely that there too they are birefringent although some cross bridges normally associated with primary filaments are obviously missing. It has been reported that the loss of birefringence was the first sign of electrocyte differentiation in *Raja*. Fox et al. (1990) reported neuronal cell death and synaptic terminal degeneration in the adult electric organs of the skate (*Raja undulata*, *R. erinacea*).

The significance of the absence of primary filaments and cross bridges or the presence of thin filaments in all the electrocytes, beyond their reflection of myogenic origin (although no electrocyte can move) remains to be determined.

The major portion of the central zone in the electrocytes of all gymnotids contains ribosomes, glycogen particles, elements of rough and smooth ER and occasional thin twisting tubules (which may be ER).

The syncytial surface of the electrocyte in strongly electric fish *Electrophorus* presents tubular invaginations of the cell membrane and that a fibrous coat (fuzzy coat, glycocalyx) covers the outer surface of the membrane penetrating the lumen of the invaginations. The cytoplasm of the electrocytes has few organelles: mitochondria, glycogen particles, dense bodies, Golgi complex, and a network of microfilaments and microtubules. It is possible that in the electrocyte the microtubules are involved, in association with the network of filament, in the maintenance of the form of the electrocyte. The invaginations enlarge the area of contact between the electrocyte and connective tissue, and so must favour the low electrical resistance detected at the anterior face. They should also increase ion transport. The structure of synapses in the electrocytes is similar in general respects to that described in many neuromuscular junctions and in other electric organs. The sparse, large, densely innervated parallel of the posterior face remind one of those electrocytes that have one very large and prominent stalk, the only place where nerve endings abut.

The electrocytes are distributed along the rostral-caudal axis of the electric eel. The rostral face of each electrocyte is not innervated and is enriched for Na^+, K^+, and ATPase, whereas the caudal face is innervated and has numerous acetylcholine receptors (ACh-R). It is possible that this cellular asymmetry of the electrocyte might be established and/or maintained by cytoskeletal proteins. Desmin protein is an abundant constituent of the intermediate filaments in the electrocytes. The intermediate filaments play a role in the maintenance of the morphology of electrocytes (Cordeiro et al. 1995, 1996; Chagas and Neto 1989; see Lilian Ayres et al. 1991 for actin in electric organ).

Although tubulin and actin are present in the electrocytes of electric eel, the study of desmin-main intermediate filament protein is of interest (Costa et al. 1988).

Mormyrids (Fig. 17.9) and gymnotiforms have larval and adult electric organs. Light and electron microscopic observations indicate that the larval organ has ultrastructural features similar to those of adult organs. In *Apternotus*, the only known electric fish whose adult organ is neurogenic, there is also a larval myogenic organ. The adult neurogenic organ lies dorsal to the larval organ within the hypaxial muscle. The behavioural significance of the larval EOD is not clear.

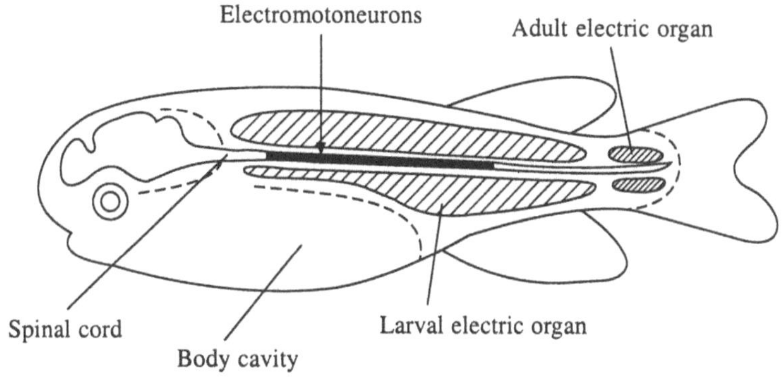

Fig. 17.9 ***Pollimyrus isodori*; Larval and adult electric organ. Schematic representation of the maximal extent (dorsal-ventral, rostral-caudal) of larval electric organ. The electrocytes of this organ are found over the whole length of the medial part of the deep lateral muscle, except in the caudal peduncle where the adult electric organ develops. The body cavity considerably limits the ventral part of the larval electric organ. The extension of the electromotoneurons in the spinal cord is indicated as well.**
Source: **Kirschbaum (1980)**

1.7 Neural Control of Electric Organs

The peripheral electric organ together with the central pathway makes up a special neuroeffector system, the electromotor system. The two elasmobranch forms exemplify the different varieties of neural control of the organ. The branchial origin of torpedine electric organs calls for innervation from the cranial nerves (in *Narcine* the facial, glossopharyngeal and vagus nerves supply the electric organs; the fifth cranial nerve does not enter the electric organ, although in *Torpedo* it is reported to participate in the innervation). The electromotor neurons are grouped into a large electric lobe, and they are under control of a smaller number of neurons in a command centre. Only in the marine teleost *Astroscopus* is the electric organ also innervated by a cranial nerve. In this case, the electrocytes derive from ocular muscles. In the Rajidae, as in Mormyridae, *Gymnarchus* and Gymnotidae, central control mechanisms in the brain send commands to the spinal electromotoneurons which have a segmental distribution. In *Malapterurus*, each organ is innervated by only a single electromotor neuron. The pair of electromotoneurons is located in the cervical spinal cord between the second and third ventral roots and can be identified by their giant somata (50–200) μm. Their axons leave the spinal cord within the third ventral roots and then form bilateral electric nerves containing a single axon each. Both somata and proximal dendrites are densely innervated. Several transmitters contribute to this innervation. The chemical characteristics of the electromotoneuron innervations resemble those of other spinal motor neurons rather than spinal electromotoneurons of other electric fish. The pattern of electromotoneuron innervations in *Malapterurus* reveals little specialization. The capacity for information processing required for the control of the electric organ discharge appears to be achieved by the increased integrational capacity of the newly evolved multiple dendrites and not by an additional parallel channel specific for electromotor system (see Schikorski et al. (1994a).

Schikorski et al. (1994b) concluded that the evolution of the electric organ from muscle tissue and the development of a central control system of the electromotor response in *Malapterurus* involved a minimum of alterations in central nervous circuitory. In contrast to many other electric fishes the electromotor control is mainly accomplished at the level of the electromotor neurons.

The basic picture of neural control is this: A small group of neurons in the higher spinal cord, medulla, or perhaps midbrain constitute the "command nucleus" which takes the "decision" to discharge the organ. On firing of the cells of the command nucleus synchronously, the "command signal" to discharge is initiated. This activity is thereafter passed on to the electrocytes either directly or through one or more neural relays. The firing of individual electrocytes must be synchronous. Synchrony is necessary for largest output (Fig. 17.10). The only exception to the central control is the chin organ of *Adontosternarchus* where all the electrocytes are in parallel and the activity of each tends to excite the others.

The command neurons are probably autoactive in the continually discharging forms; they are pacemaker neurons. The electric fishes that discharge only intermittently, the command neurons decide to call for an organ to discharge. The coupling by means of "electronic synapses" causes highly synchronous firing that is found in the command neurons of electric organ systems. Synchronization in most of the electric organ system is so precise that the negligible delay of electrically mediated transmission is required; chemically mediated transmission would be too slow, there would be too great delay.

The medullary pacemaker nucleus of the gymnotiform electric fish, *Hypopomus*, triggers each EOD by a single command pulse. It consists of electrotonically coupled 'pacemaker' cells, which generate the rhythm, and 'relay' cells which follow the pacemaker cells and excite the spinal motoneurons of the electric organ. The pacemaker cells receive two inputs from the complex of the diencephalic prepacemaker nucleus (PPn), a GABA-ergic inhibition and a glutamatergic excitation. Relay cells, on the other hand, receive two of the PPn, the PPn-C, and a second from a subnucleus sublemniscal prepacemaker nucleus (SPPn) (Fig. 17.11) (see ultrastructural evidence: Kennedy and Heiligenberg 1994; see also Heiligenberg 1994).

1.7.1 Pathways and Patterns of Neural Activity

The circuitary of electric organ systems is now known in many electric fishes. In *Torpedo*, there is a command nucleus on each side lying deep in the medulla. These nuclei activate the very large electromotor nuclei present on the dorsal surface of the medulla. It is widely accepted that neuronal synchronization is realized by electrotonic coupling in electric fish. Electrotonic coupling is lacking in the electromotor system of *Torpedo*; chemically mediated positive feedback might be of importance in the bulbar command nucleus and in the electric lobes to produce responses. Gap junctions are lacking in both the command nucleus and the electric lobe of *Torpedo*. Thus, previous assessment that electrotonic coupling is necessary to produce neuronal synchronization seems not be valid in the case of *Torpedo*.

In the skate (rajid), the electric organ is probably under the control of midline nucleus in medulla, the nature (relay or pacemaker, electrotonically coupling) is not well understood.

In the mormyrids, the control of organ is complex. The entire controlling channel involves a

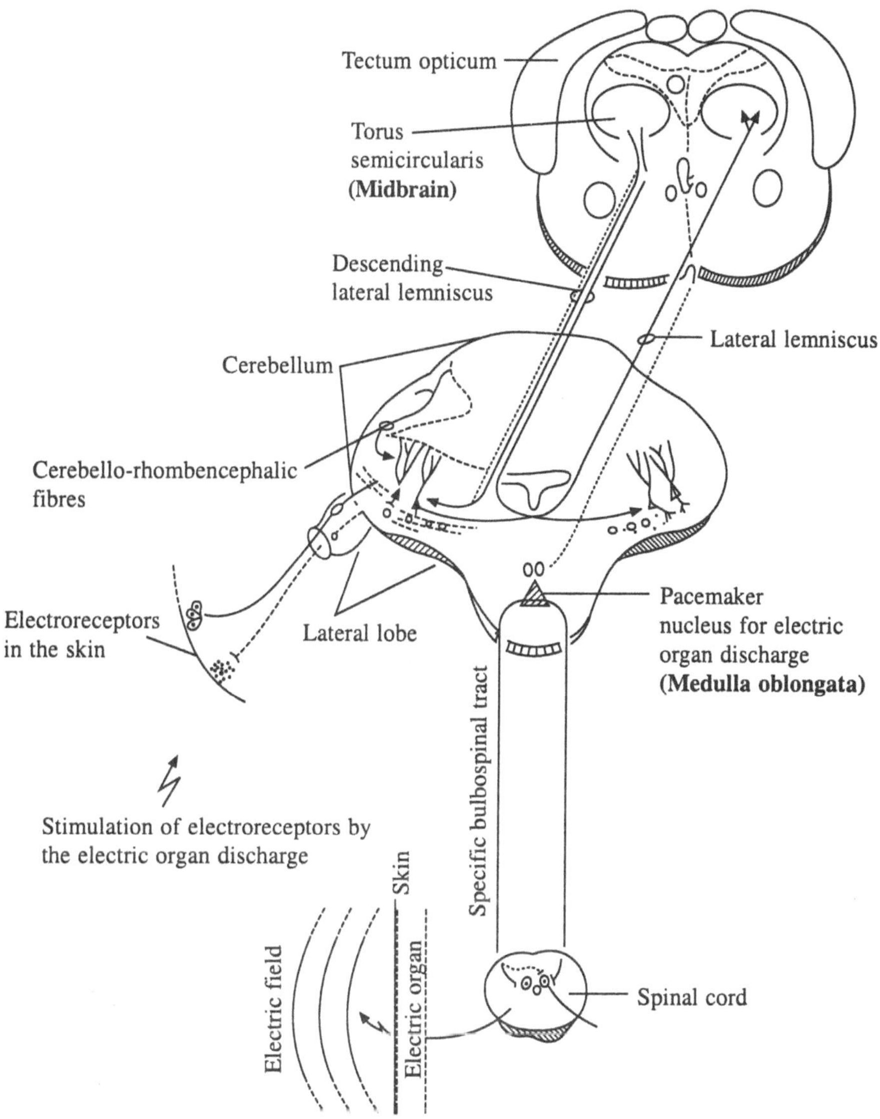

Fig. 17.10 Schematic drawing of the known electrosensory pathways in gymnotid fishes. The fast-conducting pathway (RT-system) is shown by dotted lines; the slow system (AM system) is shown by thick lines.
***Sources:* Szabo and Fessard (1974), Bell (1979) and Hopkins (1983)**

bilateral pacemaker nucleus, a midline nucleus in the medulla, and spinal neurons. Both medullary and spinal neurons are closely coupled electrotonically.

In the gymnotids, more than one neural level is involved in the command system. It includes pacemaker nucleus in the midline of medulla, a relay nucleus also in the midline of medulla, which in turn activates the spinal neurons—in each case all the neurons are electrotonically coupled (see Fig. 17.10)

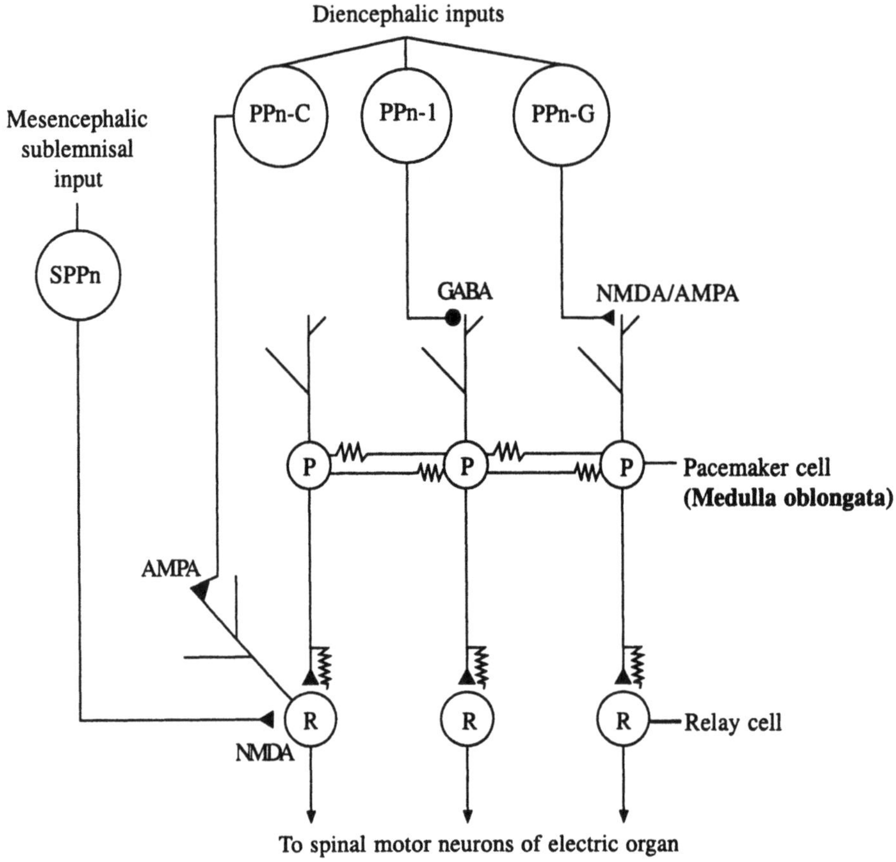

Control of the pacemaker in *Hypopomus.*

Fig. 17.11 The rhythm of the medullary pacemaker nucleus of gymnotiform electric fish, *Hypopomus*, can be modulated by 3 diencephalic inputs (PPn-G, PPn-1, PPn-C) and one mesencephalic sublemniscal input (SPPn). These effects are mediated respectively by GABA receptors or two glutamate receptor subtypes, NMDA and AMPA. The pacemaker nucleus consists of electrotonically linked pacemaker cells (P) and relay cells (R). Activation of the PPn-G gradually accelerates the pacemaker, while activation of the PPn-1 slows it down. Activation of the PPn-C induces rapid bursting, or 'chirping' of the relay cells, while activation of the SPPn leads to a sustained depolarization of these cells and thus shuts off the regular electric-organ discharge. The relay cells fire approximately 1 ms later than the pacemaker cells, which indicates that gap junctions that are visible in the mixed-type synapses between pacemaker cells and relay cells do not provide electronic coupling sufficient for synchronous firing. A minor role of AMPA receptors in the effect of PPn-G has been discovered only recently.
***Source:* (Kennedy and Heiligenberg (1994)**

Recent studies of electric organ (EO) in *Gymnotus carapo* have shown that the innervation pattern and electrophysiological properties of electrocytes are not homogenous along the EO. The existence of groups of electrocytes with different properties has a major influence on the EOD characteristics in *G. carapo*. According to Lorenzo et al. (1993), the pacemaker nucleus of *G. carapo* contains two types of neurons: pacemaker cells which set up the frequency of the EOD and relay cells which convey command signal to the spinal cord. Direct activation of a single relay cell provides enough excitation to discharge a pool of spinal electromotor neurons and electrocytes, generating a small EOD (unit EOD). Different relay cells generate unit EODs of variable size and waveform, indicating the involvement of different groups of electrocytes. Three types of relay cells could be identified: type I commanding the rostral 10% of the EO, type II which distribute their command all along the EO and type III driving the caudal 30%. Waveform analysis of unit EOD indicates that doubly innervated electrocytes which are the most relevant for attaining the specific EOD waveform, receive a favoured command from the pacemaker nucleus.

Further, Caputi et al. (1993) presented data which indicate that the EOD may be considered as the result of the sequential activation of 3 defined portions of the EO: the abdominal portion (included in the rostral 25% of the fish body), the central portion (comprising the intermediate 50% of the fish body) and the tail portion (the caudal 25% of the fish body). The EOD generated at each portion is characterized by: (1) timing respect to the pacemaker nucleus discharge, (2) speed of the progression within the region, (3) waveform, and (4) magnitude.

Spinal sections demonstrated that EMNs (electromotor neurons) serving relatively small portions of the EO are widely distributed (convergence) and that surgical exclusion of relatively small portions of the spinal cord diminishes the amplitude of the EOD along an extended portion of EO (divergence).

Electric catfish presents a simplest control system. Two neurons lie in the first spinal segment, one on either side. Each neuron innervates all the electrocytes on its side and each impulse fires the electrocytes once. The two neurons are closely coupled electrotonically, and an impulse initiated in one cell propagates into the other. Excitatory inputs gradedly depolarize the cells and when one cell is excited the other must fire also.

A different control mechanism exists in electric eel. There are two electric organs to be discharged, the weakly electric (Sachs') organ and the main organ. Hunter's organ has two parts; the caudal part resembles Sachs' organ and the rostral part resembles the main organ. All the electrocytes fire together, but only Sachs' organ and posterior part of Hunter's organ show activity during weak emissions. The organs are under the control of a single midline relay nucleus in the medulla. Each command impulse from the medulla is relayed at every spinal electromotor neuron and transmitted out to each electrocyte. The pacemaker nucleus has not yet been localized in the electric eel.

In the stargazer *Astroscopus*, the command signal is initiated in a midline nucleus in medulla and is relayed in the very large oculomotor nucleus. Neurons are electrotonically coupled in command as well as in relay nucleus. Truely, the oculomotor neurons not only relay but also decide to initiate a discharge.

1.7.2 Synchronization of Electrocyte Activity

In most electric fish there is synchronization of electrocyte activity; the firing of electrocytes is synchronous. Two mechanisms contribute to synchronizations: (i) equalization of path length (of nerve fibre) and thereby conduction time and (ii) lower conduction velocity of shortest paths to nearer parts of the organ tending to equalize conduction time. Probably, most systems use a combination of two mechanisms.

2. ELECTRORECEPTORS (ELECTROSENSORY SYSTEM)

2.1 Introduction

The electrosensory system has proven to be most interesting and useful system for studying a variety of basic neurobiological problems. (See also Chapter 15; Section 4).

A large review of the field was edited by T.H. Bullock and W. Heiligenberg (1986). Since then, however, significant progress has been made in several areas (see Zakon 1987; Bodznick 1989; Bastian 1990; Heiligenberg 1990, 1993; Kramer, 1990, 1996, Moller 1995).

The electrosensory system, as a means of obtaining information about the environment, is one of the outstanding feats of evolution. Electroreceptors are known only in vertebrates and are part of octavolateralis system. They are a heterogenous group of receptors, a new category of sensory transducers. The mechanosensory lateral line as well as electrosensory systems are considered to be of equal age and probably as old as the vertebrate radiation. Although electroreception is rare in teleosts, all nonteleost fishes, except for holostean fishes and hagfishes, and many amphibians outside of anurans are electroreceptive (Fig. 17.12). Receptor organ density, location, and patterning on the body surface all vary widely among species.

Bodznick (1989) lists the following similarities between lateral line and electrosensory systems:

1. Dermal receptor organs distributed over the head (and body) innervated by lateral line nerves.
2. Receptor cells (hair cells) in a support cell epithelium make occluding junctions with support cells and have presynaptic dense bodies, and apical microvilli and/or a kinocilium.
3. Receptors and their innervation arise ontogenetically from dorsolateral epidermal placodes.
4. Primary afferents to alar medullary nuclei with cerebellar-like organization.
5. Medullary nuclei have commissural connections and lemniscal projections to midbrain.

One fundamental distinction between the senses is the absence in all electroreceptors of efferent innervation, which is found in nearly all lateral line neuromasts (see Fig. 15.35).

We see two broad classes of electroreceptors, the ampullary organs and tuberous organs (see Tables 17.4–17.6; Figs. 17.13–17.15). Ampullary organs in their Lorenzinian form are a general possession of almost all the nonteleost fishes. Most of the orders of teleosts apparently lack them. Presumably, ampullary organs have reappeared atleast twice. Tuberous organs, which have evolved independently in gymnotiforms and osteoglossiforms, share certain functional similarities, but otherwise are quite different in two orders. The tuberous organs may be regarded as evolved from ampullary organs, which are "arrested" in their development. Ampullary receptor organs of weakly electric fish are also added with growth. In fact, they are added at about twice the rate of tuberous organs. Addition of ampullary organs is also observed in closely related catfishes (see details: Emde 1998a).

Developmental studies have explored the significance of steroid hormones for the tuning of electroreceptors to the frequency of an endogenous neuronal oscillator which drives the electric current pulses necessary for their stimulation. Embryological studies have revealed that the development of mechanoreceptors and electroreceptors in the fish's skin is induced by the innervation of primary afferent nerve fibres which are specific with regard to their central connections as well as with regard to the type of receptor induced in the periphery (Heiligenberg 1990).

Zakon (1991) surmised that the nerve is necessary for continued development of the electroreceptor organs and that the postsynaptic vesicles may provide a trophic factor for their growth, but some challenging questions in this regard need answers.

The sensory cell of ampullary organ has a round nucleus in the apical cytoplasm, mitochondria, Golgi apparatus, endoplasmic reticulum, free ribosomes, polysomes, inclusion bodies and free

Table 17.4 Summary of octavolateralis sense organs for small electrical currents: Electroreceptors diversity of types and subtypes, taxonomic distribution, structure, and function. Dichotomy between Reception of Signals

I Ampullary Systems
Ampullary organs; low frequency sensitive receptors (*Tonic receptors*) (< 50 Hz) to DC electric fields primarily of biological origin (subsurface ampullae with patent transepidermal mucopolysaccharide jelly filled canal (well-insulated core conductor) to the surface; receptor cells separated by supporting cells, contrary to sensory cells, reach from basal lamina to the luminal surface, involved in the secretion of gelatinous content of ampulla or pouch; number of sensory cells in epithelial lining varies, exceptionally large in *Polydon, Plotosus* ca 50, *Amiurus* 4; pass band ca 0.1–50 Hz; afferent nerve fibres are frequently coders; in electric teleosts, usually not synchronized with EOD, except when that has DC components; electric receptors sensitive to opposite polarities; functions in locating other animals possibly objects, geomagnetic field for orientation purposes at least in elasmobranchs, possibly electrochemical fields).

A. *Lorenzinian ampullae = Ampullae of Lorenzini* (the afferent nerve fibres may be called "L" units) in majority of non-teleost fishes (Petromyzontiformes, Elasmobranchii, Dipneusti, Crossopterygii, Polypteriformes, Chondrostei); Holostei have apparently lost ampullary organs; lamprey receptors are morphologically unlike all other ampullary organs, more properly called end buds, without canals, sense cells possess microvilli but no kinocilium; hair cell receptors with apical kinocilium but no microvilli; sensory cells with microvilli around cilium in *Polypterus, Protopterus, Neoceratodus*; ciliated cells with still uncertain sensory nature in the rostral organ of *Latimeria* excited when ampulla goes negative depolarizing apical membrane (polarity: cathode excites) long canal in marine species, an adaptation to the high conductivity of marine environment; short duct microampullae ("mini-ampullae") in freshwater species, e.g., South American *Potamatrygon*, African *Dasyatis garouaensis*; chemical synapse, neural control: medulla, and in midbrain centre-torus semicircularis.

B. *Teleost ampullae = ampullae of teleosts* ("F" units) may be heterogeneous; a homoplaseous analog in minority of teleost fishes, most of the 30 or so orders of teleosts lack tham (world-wide Siluriformes, mostly nonelectric except for *Malapterurus*, the electric catfish of Nile, South American Gymnotiformes, African Mormyridae, *Xenomystus* of Osteoglossiformes excited when ampulla goes positive, depolarizing basal membrane (polarity: anode excites); hair cell receptors without kinocilium (excepting *Xenomystus* with a short cilium) but with microvilli long canal in marine species (sometimes, inappropriately called Lorenzinian), "small pit organs", with short duct in freshwater species; organs of *Kyptopterus* have no canals at all) they are located in a partially covered pit; organs of marine catfish *Plotosus* are considered as a type intermediate between Lorenzinian ampullae of elasmobranchs and ampullary organs of freshwater teleosts. chemical synapse; neural control: medulla, and midbrain centre-torus semicircularis

II Tuberous Systems Exclusively in electric fishes Gymnotiformes: Sternopygidae, Rhamphichthyidae, Hypopomadie, Apteronotidae, Gymnotidae, Electrophoridae Osteoglossiformes: Mormyridae, Gymnarchidae
Tuberous organs; high frequency sensitive receptors (Phasic receptors) (characteristically situated in a closed cavity; structural variation is great; cavity occupied by sense cells projecting from basal attachment, usually without patent duct; specialized accessory cells-external or apical cells loosely coupled forming a porous "plug" over receptor cells and supporting basal cells; sense cells with microvilli protruding or lining an invagination, or without them; excited by inward current; polarity: anode excites; passband ca. 50–20,000 Hz; usually tuned to a best frequency; series capacity and active filtering contribute to tuning afferent fibres synchronized with EOD and encoding intensity in various ways not appropriately called frequency coders; neural control: medulla, and midbrain centre-torus semicircularis; functions in location and communication)

A. *Burst duration coders* ("B" units) in gymnotiform pulse species (*Hypopomus* spp., *Gymnotus* spp., *Rhamphichthys, Hypopygus*) small axon, unmyelinated preterminals, many sense cells per organ; burst of spikes follows EOD, graded in number at fixed interval, little latency shift; receptor potential an active oscillation, weakly damped subdivided into alpha, beta, gamma and delta types, with some subtypes AM units; chemical synapse; function: communication, electrolocation

B. Pulse marker units ("M" units) in gymnotiform pulse species (*Hypopomus* spp., *Gymnotus* spp., *Rhamphichthys, Hypopygus*)

Large axon; myelinated preterminal branches high threshold in some species; low in others single spike follows EOD with fixed latency; RT units, synapse probably electrical; funtion?

A′. Phase coders ("T" units) in gymnotiform wave species (*Sternopygus, Eigenmannia, Apteronotus*): medium sixed axon; myelinated preterminals single spike follows each EOD in physiologically intensity range; shift latency with intensity with small jitter; receptor potential is rapidly damped, active tuned oscillation more sensitive (absolute) than P units; some are sensitive to the difference between sinusoidal and clipped, EOD-like waveform, some are insensitive and others are intermediate; RT units; synapse electrical or mixed in *Apteronotus albifrons*; function: communication, electrolocation

B′. *Probability coders* ("P" units) in gymnotiform wave species (*Sternopygus macrurus, Eigenmannia virescens, Apteronotus albifrons*): morphology may be similar to preceding; single spike follows some EODs; probability graded with intensity; smaller, more jittery latency shift and more phasic than "T" units in some fish; difference from T in receptor potential is unknown; they may be similar some are sensitive to the difference between sinusoidal and clipped, EOD-like wave form, some are insensitive and others are intermediate; AM units; type of synapse?; function: communication, electrolocation

A″. Knollenorgans ("K" units) in mormyrid pulse species (*Gnathonemus petersii, Marcusenius* spp.): "large pore" organs or large receptors (60 sensory cells are known in *Petrocephalus*), each receptor cell is encase in its own capsule; myelinated preterminals, project to PLLL (posterior lateral line lobe) nucleus; differentiated into KI, most common, being distributed on the head and dorsal and ventral portion of the flanks and the larger KII, few and always located around the operculum or eyes in *Petrocephalus*; more sensitive (absolute) than D units spontaneous receptor potential in most species; rhythmic, usually spikelike, from a few hundred Hz upto several k Hz under artificial conditions; RT units; electrical synapse; function: electrocommunication

B″ *Mormyromasts*; burst duration coders ("D" units) of mormyrids: "medium-pore" organs or medium receptors; unmyelinated preterminals; project to PLLL cortex two froms of sense cells (with or without microvilli, each within its own encapsulated chamber and each with its own source of innervation, complex innervation; threshold can be high, upto 10 mV/cm; receptor potential not spontaneous; damped ringing burst of spikes graded in latency, number and spacing (likely to be a heterogeneous category); too insensitive for sensing distant fish; burst duration code for close fish; AM units; chemical synapse; function: electrolocation

A‴ *Gymnarchomasts type I* ("S" units) in wave species gymnarchid *Gymnarchus niloticus* like preceding but single nerve fibre, with unmyelinated preterminals pair of unequal sense cells, or more like "T" units but high spontaneous firing rate, less tendency to burst below 1:1 range, small dynamic range of phase coding; RT system; chemical synapse?; function?

B‴ *Gymnarchomasts type II* ("O" units) in wave species gymnarchid *Gymnarchus niloticus* single nerve fibre with unmyelinated preterminals; 12 or more large and similar sensory units per organ; sense cells with long microvilli at bottom of invagination; like "P" units but spontaneous with more difficulty following 1 : 1, small dynamic range of phase part of probability coding; high intensity suppression, AM capabilities; chemical synapse?; function?

Tuberous receptors classified according to physiological functions as "rapid timing" units (RT) and "amplitude-modulated" units (AM); RT units appear to be adapted for rapid relay of information to CNS about the timing of external events, usually independent of their amplitude; AM units are specialized for encoding the amplitude of external stimuli; Synapse type: transmission between receptor cell and afferent neuron.

Table 17.5 Tuberous systems: EOD and receptor type in weakly electric freshwater fish

Family	EOD	Receptor types	Family	EOD	Receptor type
Sternopygidae	W	P, T	Mormyridae	P	K, D
Rhamphichthyidae	P	B, M	Gymnarchidae	W	O, S
Hypopomidae	P	B, M			
Apteronotidae	W	P, T			
Gymnotidae	P	B, M			
Electrophoridae	P	?			

EOD type : W = Wave type; P = Pulse type
B units = Burst duration coders
M units = Pulse marker units
T units = Phase or timing coders
P units = Probability coders
K units = Knollenorgans
D units = Mormyromasts
S units = Gymnarchomasts type I
O units = Gymnarchomasts type II
Source: Zakon (1988)

Table 17.6 List of fishes with electrosensory ability (fishes that utilize electrical stimulus energy and possess specialized electroreceptors)

(a) non-electrogenic ("passive" species; no electric organs with ampullary electroreceptors)
- (a_1) marine: many sharks and rays
- (a_2) freshwater: several catfishes, freshwater rays, lungfishes

(b) electrogenic ("active" species; with electric organs, tuberous and ampullary receptors)
- (b_1) wave species ("active" weakly electric freshwater fishes)
- (b_2) pulse species ("active" weakly electric freshwater fishes)
  Some Rajidae (marine skates) with very irregular and weak discharges can be included in this group.
- (b_3) pulses; also additional powerful shocks:
  Electrophorus, Torpedo, Narcine
- (b_4) powerful shocks only; electroreceptors present: *Malapterurus*
- (b_5) powerful shocks only; electroreceptors not known: *Astroscopus*

Source: Schwassmann (1978)

vesicles. For the ultrastructure of the ampullary electroreceptive organs in *Silurus glanis*, see Jakubowski (1987). Bodznick et al. (1993) have discussed accommodation to maintain stimuli in the ampullae of Lorenzini. The tuberous organ sensory cell (e.g., in mormyrid) has a round nucleus in apical cytoplasm, Golgi complex, endoplasmic reticulum, microtubules, vesicles but neither cisternae nor inclusion bodies.

Species, for example *Anguilla* (Anguilliformes) and *Astroscopus* (Perciformes) may make use of electric detection even though they are not convincingly spoken of as having electroreceptors. Skin papillae as possible electroreceptors have been reported in deep-sea eel *Cyema atrum* (Cyemidae: Anguilloidei; no electric organ is present).

There have been a few cytochemical studies on electroreceptors, a member of the paraneuronal family. Adenosine triphosphatases have been identified in ampullary organs of the mormyrid *Gnathonemus petersii*; there is a high concentration of Mg^{2+} (Na^+K^+)-dependent ATP-A and $Mg^{2+}Ca^{2+}$-dependent ATP-A in the microvilli of the sensory cell of the tuberous organ. This is

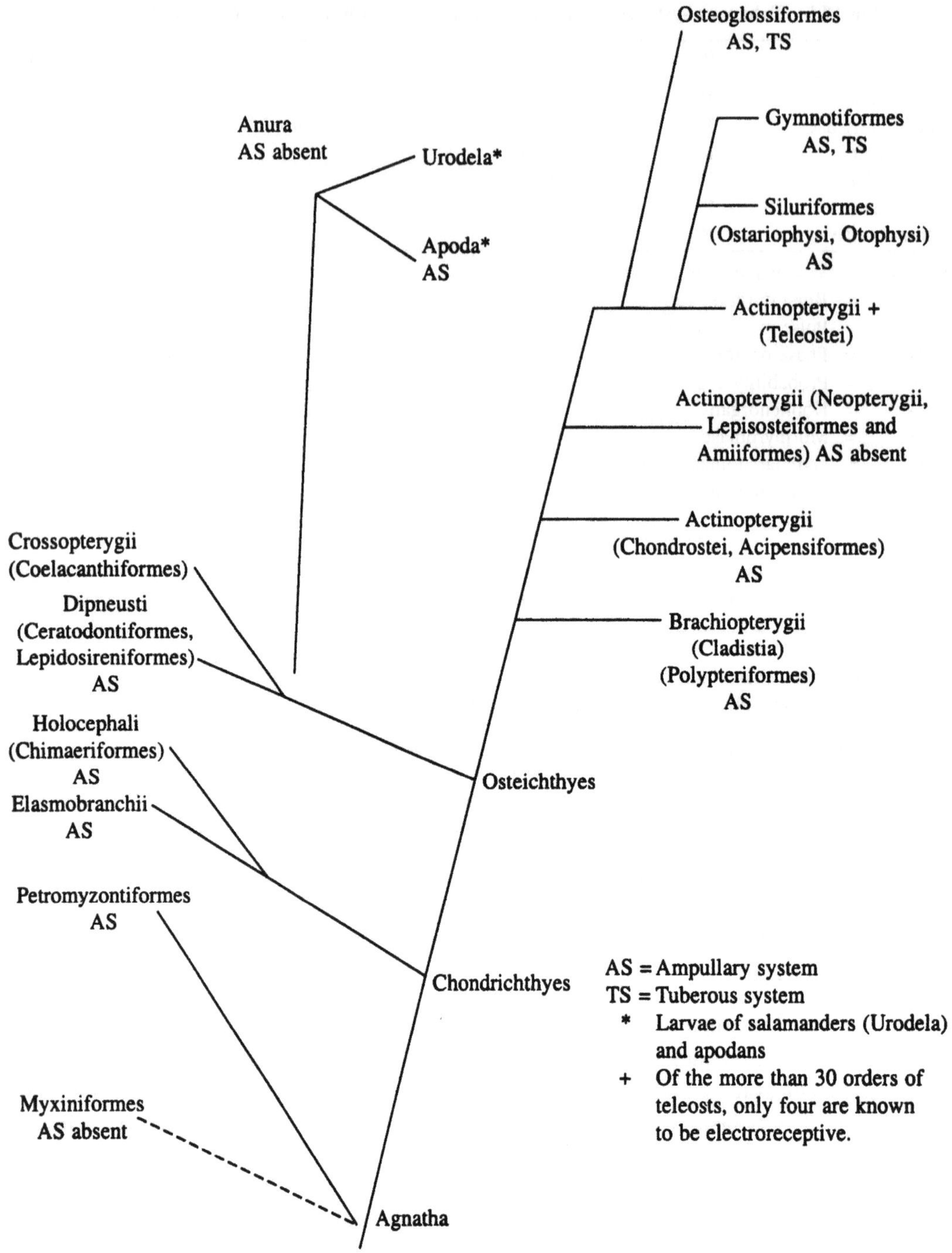

Fig. 17.12 Tree of electrosensory systems in Anamniota.
***Source:* Bodznick (1989) (Modified)**

of interest in view of electrophysiological data suggesting that the brief transitory potentials here are due to the Ca^{2+} crossing the plasma membrane.

Electrophysiological studies in several species have suggested that glutamate could be responsible for neurotransmission at the first afferent synapse of lateralis receptors. The distribution of this amino acid has been studied with a monoclonal antibody in the gymnotid *Apteronotus leptorhynchus*; a high concentration of glutamate was found in the afferent fibres of both ampullary and tuberous organs and in the sensory cells of the tuberous organ. Immunohistochemistry demonstrated substance P in the sensory cells of the tuberous organs in the gymnotids *Sternopygus* sp. and *A. leptorhynchus*. This provides another example of the association of an amino acid and a neuropeptide. In contrast, substance P is present in the cytoplasm of the platform accessory cells of the tuberous organ of *Eigenmannia virescens* but not in that of sensory cells. Recent data show that these sensory cells can differentiate from the platform accessory cells.

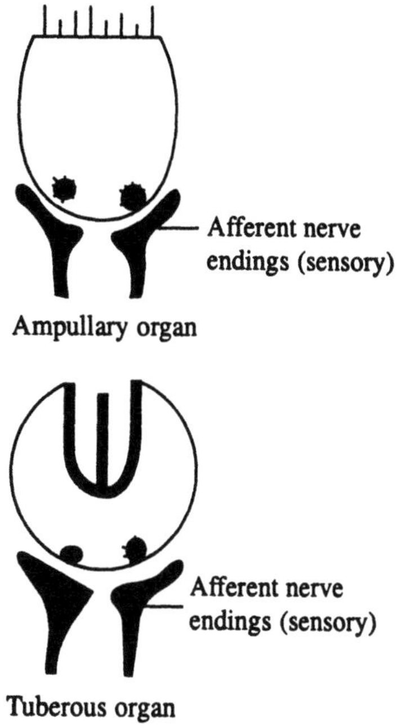

Fig. 17.13 Sensory cells in specialised (electric) lateral line organs of teleost fishes. *Source:* Bleckmann (1986)

In selachian fishes, the ampullae of Lorenzini are electroreceptors; the sensory cells differ from those of bony fish by bearing kinocilia, but immunohistochemical data on the sensory epithelium are not available (see Zaccone et al. 1994) (Fig. 17.16).

Most electroreceptors appear to transmit chemically, and glutamate is said to be a likely candidate transmitter (read Andrianov et al. 1994; Heijmen et al. 1994). Electric transmission is also found in certain electroreceptors. Efferent fibres to electroreceptors seem to be altogether absent. There is ample evidence that transmission between receptor cell and primary afferents of ampullary electroreceptor organs of freshwater catfish *Ictalurus* is glutamatergic and that both NMDA (N-methyl-D-aspartate) and non-NMDA receptors are involved. A resting level of transmitter release is set by a bias voltage of yet unidentified origin, and by the ion composition of the ambient water. Transmitter release is modulated by electric potential differences of both internal and external sources and also by the variation in the ionic contents of the aquatic milieu (Andrianov et al. 1992, 1996,1997). There is the presence of AMPA receptors at the synaptic membrane of ampullary electroreceptor organs in the catfish (Andrianov at al. 1994).

2.2 Somatic Distribution of Electroreceptors

Zakon (1984) has given details on the addition of receptor organs with age in a weakly electric fish *Sternopygus dariensis*.

Ampullary organs occur in many electric and non-electric fish, whereas tuberous organs are

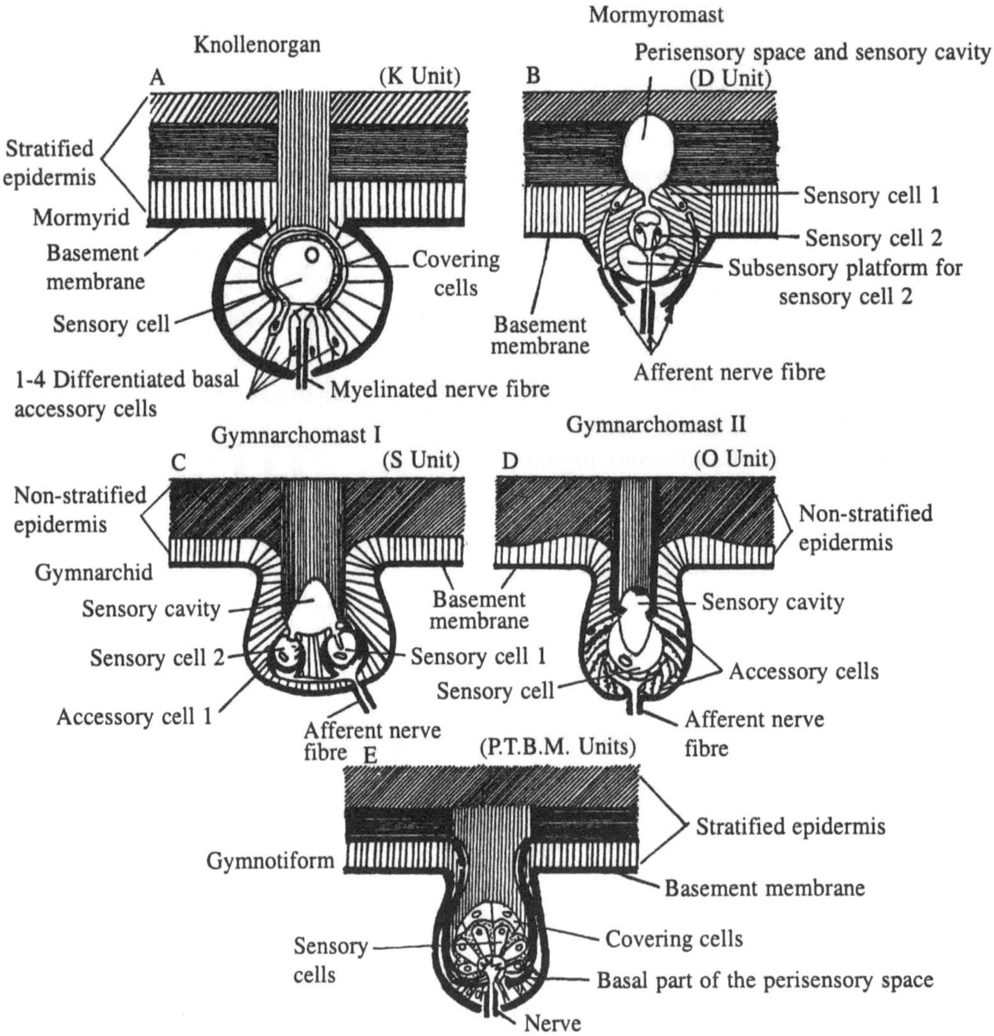

Fig. 17.14 Electroreceptors: Types of tuberous organs.
***Source:* Zakon (1988) (See also Szabo 1974)**

specific to weakly electric fish. The ampullary organs occur singly (probably the primitive pattern) or in clusters as tube-like invagination of the epidermis. They are restricted to the head, except in lungfishes, (Fig. 17.17) where they also occur on the trunk and tail but not on the paired pectoral and pelvic fins. The ampullary organs in *Polypterus* are restricted solely to the head. They occur singly and exhibit density on the snout and around eyes. Ampullary organs in *Polyodon* are restricted to the bill and head, as in sturgeons, and that they are also grouped into the clusters. The living coelacanth *Latimeria* possess a pair of a well-developed deep head organ (snout or rostral organ) which appears to be unique among bony fishes and is believed a collection of ampullary organs.

The long-duct ampullary organs of marine Elasmobranchii have a particular and well-known somatic configuration (Fig. 17.18). The ampullae are grouped in the head in several clusters surrounded by capsules from where the canals (or ducts) radiate with their external openings spread over the head and pectoral fins (Fig. 17.19). Raschi (1986), Raschi and Mackanos (1987, 1989), Raschi and Adams (1988) and Raschi et al. 2001) have substantially added to the studies on ampullae of Lorenzini. In chondrichthyans, the total numbers of ampullary organs of Lorenzini are: *Squalus acanthias* (1200), *Galeocerdo cuvieri* (760), *Carcharhinus plumbeus* (2328), *Sphyrna levini* (2824), *Torpedo marmorata* (162), *Raja laevis* (1480), and *Hydrolagus colliei* (468). Similar is the distribution in marine catfish *Plotosus*, but unlike marine elasmobranchs, *Plotosus* displays in addition ampullary organs with long duct disposed along the base of the anal and dorsal fins (Figs. 17.18 and 17.20). The ampullae are grouped according to myomeres at the base of each of the fins.

In *Ictalurus*, similar to *Sorubim*, the ampullary organs are distributed all over the body surface but are particularly concentrated on the head. In general they are single organs, in *Sorubim* they are sometimes paired. Also in *Kyptopterus* the majority of organs lies in the head region; 5–10 organs are grouped together and they are arranged in 6 rows on each side of the body. African knifefish *Xenomystus nigri* (Notopteridae) has ampullary organs on the head and body. Each organ contains four to eight sensory cells separated from each other by supporting cells. Both sensory and supporting cells are surrounded by mantle cells. No such ampullary organs are found in the skin of the Asian knife- fish, *Notopterus chitala* (Jorgensen and Bullock 1987). In the gymnotid *Eigenmannia* 20 organs can constitute a group which is distributed in a characteristic manner in a dorso-lateral strips close to the lateral line. There are only a few on the head and on the dorsal and ventral skin surface. In the mormyrid *Gnathonemus* the ampullary organs are distributed on the ventral and dorsal trunk surface but the majority is found on the head (Fig. 17.18). They are never found to be grouped. In *Electrophorus* and in *Gymnarchus* the distribution pattern follows the general distribution of other specific lateral line organs.

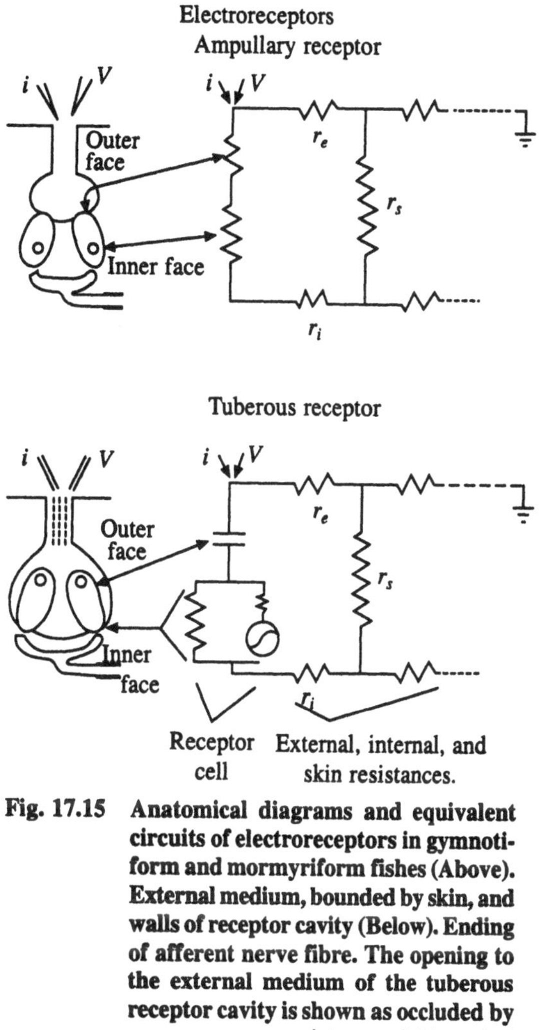

Fig. 17.15 Anatomical diagrams and equivalent circuits of electroreceptors in gymnotiform and mormyriform fishes (Above). External medium, bounded by skin, and walls of receptor cavity (Below). Ending of afferent nerve fibre. The opening to the external medium of the tuberous receptor cavity is shown as occluded by a porous mass; resistance of this and of ampullary canal and of hair cells cytoplasm are assumed to be small and are omitted. Electrodes for stimulating and recording with respect to a distant electrode are shown at the sense organ openings. Note that generally the hair cells of ampullary organs are largely embedded in the wall of the cavity, whereas those of tuberous organs protrude into the cavity.
***Source:* Bullock (1981)**

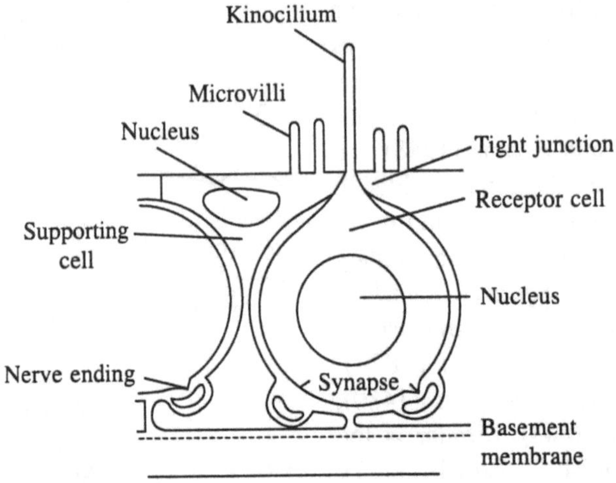

Fig. 17.16 Ampullae of Lorenzini. Section through the sensory epithelium, showing the receptor and supporting cells.
***Source:* Murray (1974)**

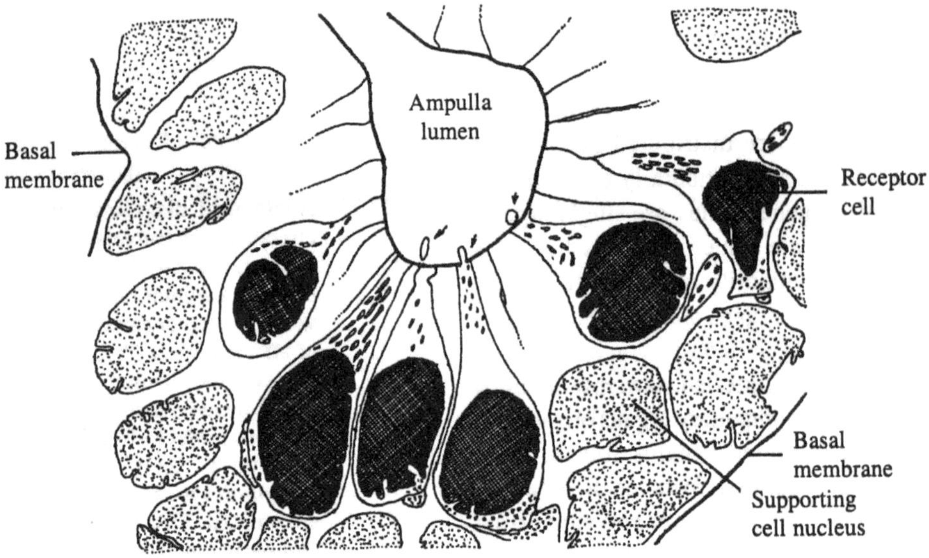

Fig. 17.17 Cross section through an ampullary electroreceptor of *Protopterus*. Receptor cells are in contact with the ampulla lumen.
***Source:* Roth and Tscharntke (1976)**

The mormyromasts are distributed in the same way as the ampullary organs; they follow the extension of the specific epidermis which covers the fish body differently in various species.

In *Gymnarchus niloticus* the gymnarchomasts are found everywhere in skin but show a higher density on the head and the anterior trunk.

In mormyrids, tuberous organs are distributed in the same fashion as the ampullary organs and mormyromasts. They show a striking accumulation in the gill region and on the dorso-lateral sides of the body.

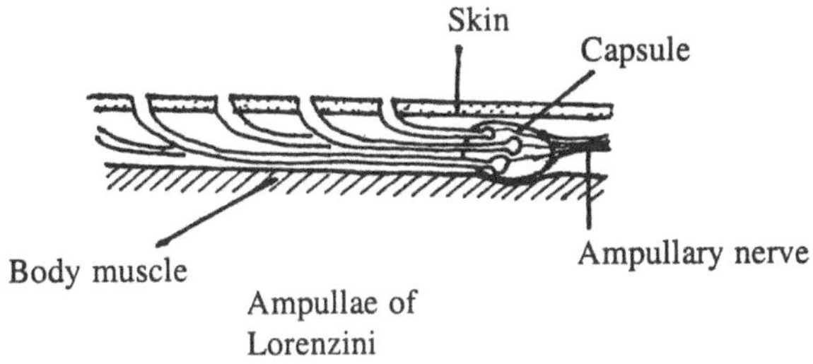

(A) Section through skin showing canals in ampullae within a capsule.
Source: Murray (1974)

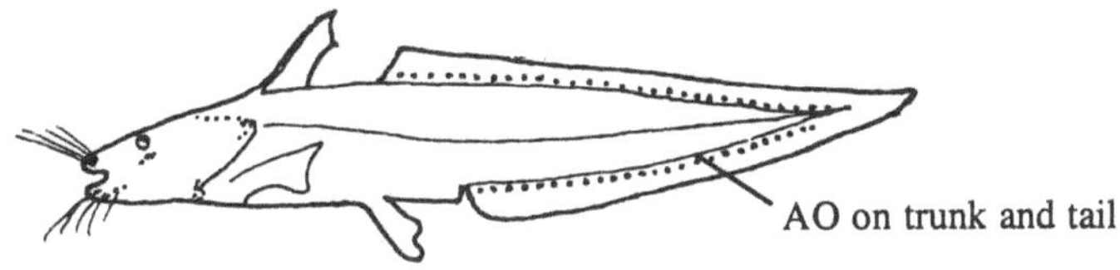

(B) Ampullary organs in *Plotosus*

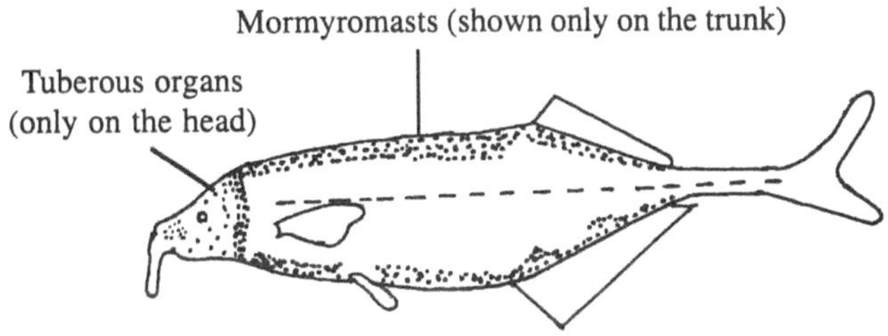

(C) Electroreceptors of *Gnathonemus petersii*
Specialized lateral line organs of electroreception

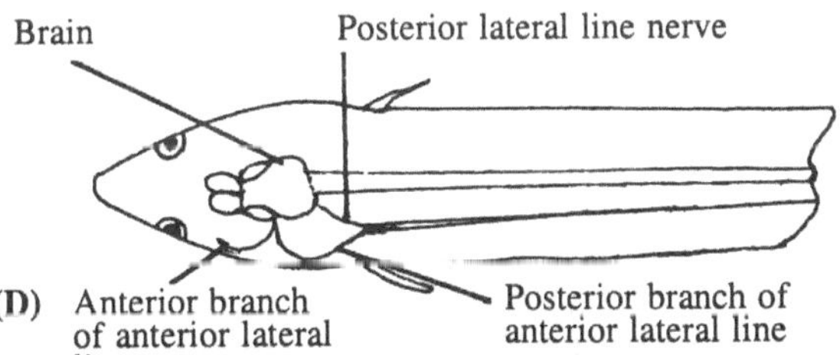

(D)

Fig. 17.18 (A) Diagram to illustrate the anatomy of ampulla of Lorenzini. Section through skin showing canals in ampullae within a capsule. (B) Ampullary organs in *Plotosus*. (C) Electroreceptors of *Gnathonemus*. (D) Course of lateral line nerve in gymnotid fish.
AO = Ampullary organ (read Bennett (1971b)
***Source:* Fig. (B, C, D). Szabo (1974)**

2.3 Relation of Electroreceptors to the Lateral Line Nerve

The course of the lateral line nerves and thus their innervation fields are not the same in different fish. (Fig. 17.18). For details of development of the electrosensory system in gymnotiform form fish *Eigenmannia*, read Vischer et al. (1989).

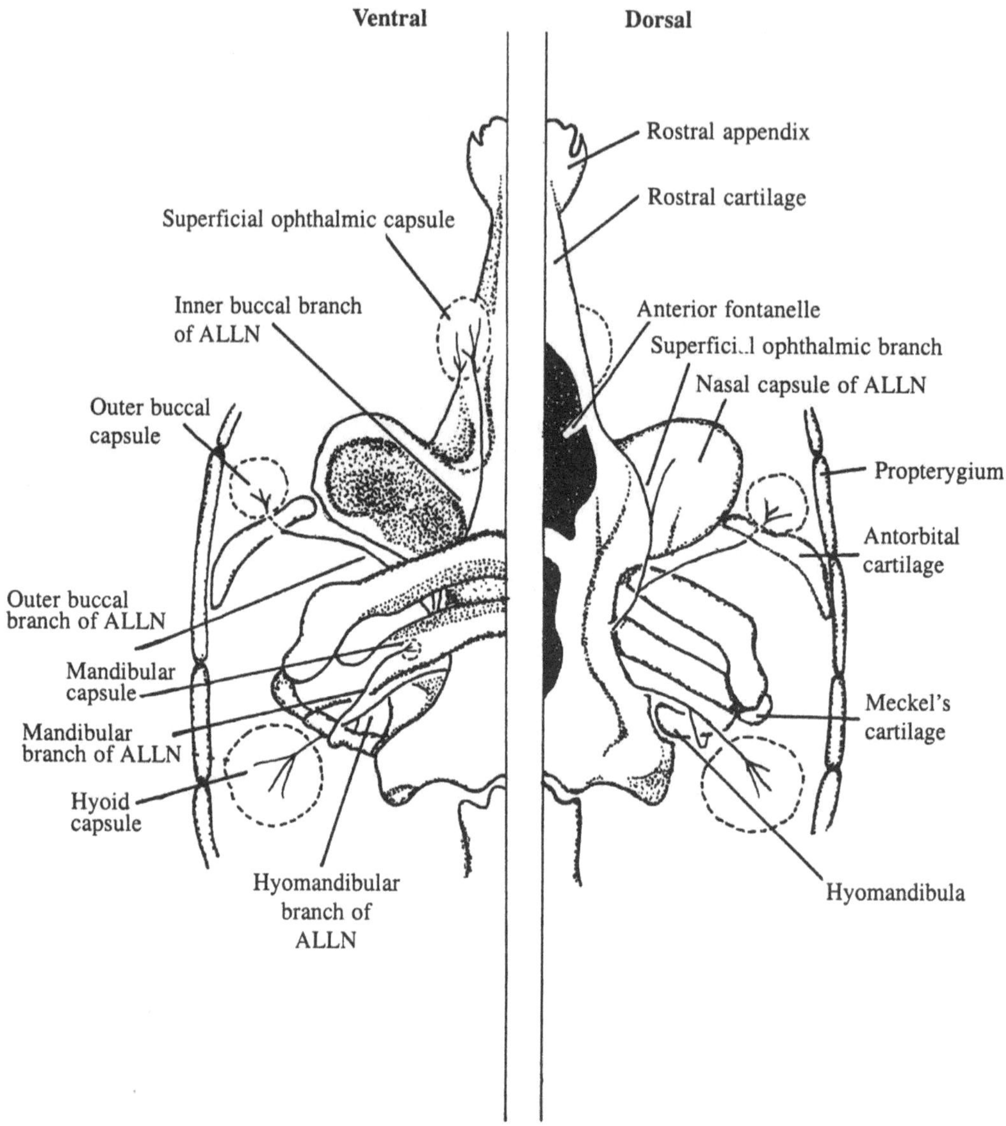

Fig. 17.19 Position of innervation of the ampullary capsules in *Raja eglanteria*. Abbreviation: ALLN = Anterior lateral line nerve.
***Source:* Raschi (1986)**

All ampullary organs in nonneopterygian bony fishes are innervated by ganglion cells of the anterior lateral line nerve, which divides into the following rami; a superficial ophthalmic ramus that innervates ampullary organs on the rostral snout; a buccal ramus that innervates ampullary organs below the eye and on the upper jaw; an otic ramus that innervates ampullary organs on the dorsal head caudal to the eye; and a hyomandibular trunk that innervates ampullary organs on the lower jaw and on the lateral and ventral surfaces of the head. In addition, dipnoans possess ampullary organs on the trunk that are innervated by a recurrent ramus of the anterior

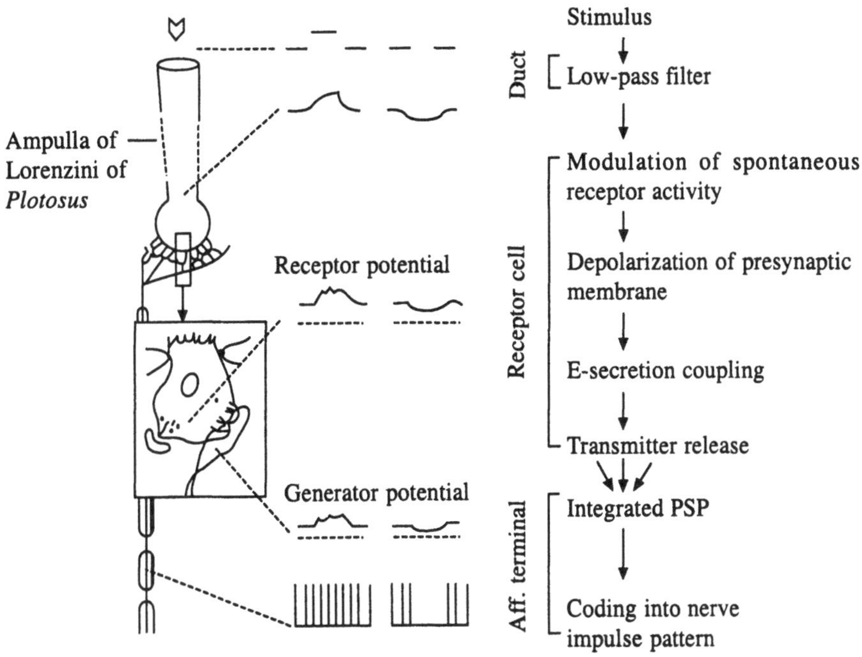

Fig. 17.20 The mode of operation of ampulla of Lorenzini of *Plotosus*. *Left*, anatomical diagram of a single ampulla. Convergence of many receptor cells onto a single afferent nerve is depicted. In the thick square is shown a single receptor cell, with apical microvilli with afferent synapses on the basal surface. *Center*, schematic illustration of potential changes at different places of the ampullary system, shown to the left. Note that the receptor potential must be superimposed on the electrotonic potential and the generator potential has the afferent spikes riding on it, but that the different components are shown separately for the purpose of schematic presentation. Positivity is shown as upward directions. The broken lines under receptor and generator potentials indicate their fully suppressed levels. At rest, both potentials are shifted to more depolarized levels, because of the spontaneous receptor activity. *Right*, casual sequence of intermediate steps in electroreception.
***Source:* Obara (1976)**

lateral line nerve whose fibres course in the trunk as dorsal, main, and ventral lateral line nerves in conjunction with similar branches of the posterior lateral line nerve. All fibres that innervate ampullary organs collect to form a dorsal root of the anterior lateral line nerve. These dorsal root fibres terminate in the ipsilateral dorsal octavolateralis nucleus and in a dorsal component of the eminentia granularis. The dorsal octavolateralis nucleus gives rise to a bilaterally ascending lemniscus that terminates in a lateral subdivision of torus semicircularis and in the central zone of optic tectum. Further connections to the forebrain are currently unknown.

Wide differences are found amongst weakly electric fish which have also been examined most carefully. In mormyrids and *Gymnarchus* the posterior lateralis nerve (*nervus lateralis metaoticus*) innervates "specialized" and "ordinary" lateral line organs in the trunk. In mormyrids a dorsal branch splits off from it and innervates predominantly the "specific lateral line organs" in the dorsal skin. A dorsal branch of this type is also found in *Malapterurus* but, in contrast to mormyrids, it arises from the anterior lateralis nerve. This recurrent posterior branch of the anterior lateralis nerve (*ramus recurrens nervi faciales*) represents a large nerve trunk in gymnotids

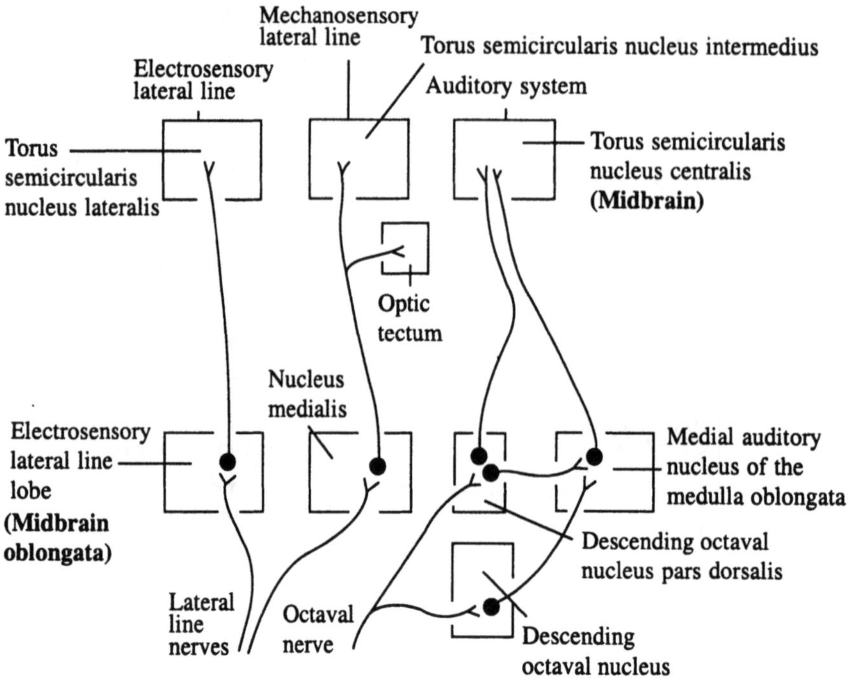

Fig. 17.21 Overall summary diagram showing the principal connections of electrosensory lateral line, mechanosensory lateral line and auditory/acoustic systems.
***Source:* Finger and Tong (1984)**

(and *Electrophorus*). After emerging from the skull it leaves the *nervus lateralis anterior*, and runs in a caudal direction (*recurrens*), supplying all the "specialized" lateral line organs of the trunk. In the gill region the posterior branch of *nervus lateralis anterior* joins the *nervus lateralis posterior* which innervates the "ordinary" lateral line organs. The recurrent posterior branch displays one or several dorsal branches in *Hypopomus* and in *Sternarchus*.

Both the "ordinary" and "specialized" lateral line organs on the head are innervated by the common trunk of *nervus lateralis anterior* (*n. lat. prooticus*).

Connecting the lateral line organs with the brain, the lateral line nerves end in lateral lobe of rhombencephalon. Fish provided with "specific" lateral line organs have an enlarged lateral lobe. All weakly electric fish (mormyrids and gymnotids) have few neuromasts in the lateral line system.

2.4 Electroreception in the Central Nervous System

The information on central processing has snowballed rapidly in recent years. The highlights from the information available are:

(i) *Medulla*—the *dorsal nucleus* of the octavolateral region of medulla in elasmobranchs, chondrosteans, dipnoans, and polypteriforms can be identified with electroreceptor input; apparently all the ampullary afferents end here and essentially all the lateral line and VIIIth nerve afferents end in other nuclei. The dorsal nucleus seems to be diagnostic of the possession of a specialized electric sense. In the electroreceptive teleosts, however, there is never a dorsal nucleus and the ampullary and tuberous afferents terminate in a special portion of the *median nucleus*, whose other portions receive common lateral line afferents. That special portion is often called the electrosensory lateral line lobe (ELL); formerly known as the posterior lateral line lobe (PLLL);

(ii) these nuclei give the appearance of being more than simple relays; a diversity of cell types, complex connections in laminated cortical organizations with a layer of granule cells and input from other brain centres all suggest *functional complexity* and differentiation of the transections performed. The PLLL of mormyrids has a *cortex* that receives axon terminals from ampullary and mormyromast afferents in different zones and, more centrally, a *nucleus* that receives the terminals from knollenorgans. These are quite different from the targets of mechanoreceptive lateral line afferents, as well as VIIIth nerve afferents. Within the cortex of PLLL there are three separate *maps* of the body surface, with curious breaks in continuity of the cell layers between them. One map is for ampullary afferents. Possibly the other two are for two types of afferent fibres from mormyromasts. The cortex is quite cerebellar-like, with a molecular and ganglion cell layer. A fourth map is present in the PLLL nucleus, receiving from knollenorgans though it is less precise. Atleast in sharks, catfish and mormyrids, if not more widely, there are presumably electric, common lateral line, and acoustic nuclei side by side in the octavolateral area of medulla but these projections have not yet been fully delineated from each other. One can expect, moreover, that there will be found somatotopic maps in the dorsal nucleus of elasmobranchs and their allies and in the median nucleus of siluriforms and gymnotiforms;

(iii) the principal midbrain centre for electroreception is the torus semicirularis (TS) and is shared, with acoustic and lateral line mechanoreceptive input, in segregated divisions (Fig. 17.21);

(iv) the cerebellum is generally large in electroreceptive species and achieves in some an unprecedented enormity and histological specializations of Purkinje units (termed "gigantocerebellum" in mormyrids); and

(v) the telencephalon shows both electric and acoustic influence, so far studied with evoked potentials.

Figure 17.10 is a schematic drawing of the known electrosensory pathways in gymnotoid fishes. Details of multiple electrosensory maps in the medulla of weakly electric gymnotiform fish *Eigenmannia* are given by Shumway (1989). For further information, the reader is referred to Tong and Bullock (1982), Tong and Finger (1983), Grant et al. (1996), Amagai (1998), Amagai et al. (1998), Friedman and Hopkins (1998), Emde (1998b) and Wullimann (1998).

2.5 BEHAVIOURAL FUNCTIONS OF ELECTRORECEPTION

There has been an avalanche of information on the electric fishes of Africa (the Mormyridae, *Gymnarchus*, and *Malapterurus*) and South America (the Gymnotoidei). According to Hopkins (1995), new and old-world topical electric fish lack a common electrical ancestor, suggesting that the mechanisms of signal generation and recognition evolved independently in the two groups.

A variety of studies have demonstrated in particular the diversity of EODs found among teleosts—the African mormyriforms and South and Central American gymnotiforms. Some species even have a sex difference in the EOD wave form pulse or rhythm that can be induced by gonadal steroid hormones. These findings have nurtured a renewed interest in the structure and function of electric organs, especially in teleosts that utilize their EOD for electrolocation and communication functions (read also Lorenzo et al. 2001).

Electroreceptors are specialized for different tasks:

Electrolocation: *Active electrolocation* occurs when an individual senses an electric field, which it produces itself, as a current flowing through electroreceptors embedded in the skin, and further, senses the presence of objects in the environment (inhomogeneities in the electrical impedance) by detecting perturbations in the normal local current flowing through the skin. For information in mormyrid *Brienomyrus*, read Friedman and Hopkins (1996) specifically on the daily movement patterns. The details of the function of A- and B- electroreceptor cells of mormyromast and Knollenorgans in electrolocation are provided by Emde and Bleckmann (1997). Both gymnotiform wave-type fish and mormyrid pulse type fish are able to perform a similar capacitance detection task. Their sensory capabilities appear to be analogous in that both fish groups detect capacitive values of a similar range, which includes capacitive values of living natural objects. Moreover, both groups can identify capacitances as such, leading to the ability to discriminate unambiguously between animate and inanimate objects. These sensory capabilities evolved independently in both groups of fish. There is a great selective advantage for this ability in a nocturnal aquatic environment (read Emde and Bleckmann 1997, Emde 1998b; see Fig. 1 for schematic distribution of electroreception among fishes).

Passive electrolocation is defined when electroreceptive fish may exploit weak electrical "clues" emanating from prey organisms in their search for hidden food. Clearly of biological importance, passive electrolocation permits accurate detection of prey, and perhaps predators, by cetain fishes which, because of their murky habitat and nocturnal behaviour, may be totally dependent upon nonvisual mechanisms. The object-detecting function shows great similarities to the echo-locating orientating mechanism of bats.

Passive electrolocation differs from *active electrolocation* in that the fish relies entirely upon exafferent rather than reafferent electrosensory information from its own electric discharge. By comparison with other sensory modalities, the physical cues available for passive electrolocation are limited. For example, there is no velocity cue that can be used for localization because electric signals are electrostatic fields, not propagating waves. In addition, most electric current sources in the fish's body environment are dipole-like so that the electric field vector is curved and does not point directly at the source. Even the magnitude of an electric field varies spatially in complex ways that make independent assessment of source distance and direction ambiguous. Sheih et al. (1996) conducted an experimental study of passive electrolocation in *Gymnotus carapo* and *Brachyhypopomus diazi*. Passive electrolocation is stimulus-bound and that these fish find the electrical sources by tracking instantaneous local electric current vectors.

There is no evidence that any chondrichthyan fishes are capable of active electrolocation, as found in weakly electric teleosts. Electroreceptors of Rajidae and Topedinidae are not specialized to detect electric organ discharge. Rather, the ampullae of Lorenzini respond to incidental electric fields of biological, geophysical, and electrochemical origin, fields that can guide prey localization and orientation within the environment.

The response of the receptor in object detection is a function of several parameters: the impedance of the object, the size of the object, its distance from receptor, and the location of the receptor on the fish's body. It is difficult to make generalization about the magnitude of electrolocation, even for a single species. However, electrolocation of inanimate object is only effective when the object is in close proximity of the fish (see Emde and Schwarz 2001).

Electroorientation: Electroreceptive fish might use electric fields induced through motion in the earth's magnetic field as an orientation mechanism.

Electrical communication: Electric communication occurs when one individual's electric signal (stimulus) evokes a behavioural change (response) in another in a mutually adaptive way. Jamming avoidance response (JAR) is not considered as an adaptation for social communication in the strictest sense but rather for electrolocation.

JAR has been studied most closely in the gymnotiform genus *Eigenmannia*. The neuronal substrate of the JAR is shown in the form of a flow diagram in Fig. 17.8 (Heiligenberg 1994). For details on JAR in electric fish, the reader is referred to: Heiligenberg and Rose (1985), Heiligenberg (1991, 1997), Viete and Heiligenberg (1991), Kawasaki (1993), Metzner (1993), and Kramer (1999) (see Figs. 17.6–17.8).

Electric communication is a surprisingly rich modality for social communication. Signals serve to threaten attack, beg submission, facilitate retreat, embellish courtship, attract mates, or simply identify the signaller's sex, species or size. Some electrical emissions are stereotyped patterns of discharges like burst of pulses, sudden cessation of all discharge, slight modification in discharge frequency—and can be called displays. In many cases the signature structure is clearly adapted to its function. Kramer and Teubl (1993) reported that a true waveform sensitivity for continuous signals should be extremely useful in discriminating between different individuals, conspecific or not, of electric wave fish in these nocturnal creatures. The spectral phase sensitivity of these fish (knifefish *Eigenmannia*) provides an additional parameter for the evolution of signal variety, as clearly seen in the richness of waveforms of electric organ discharges in a sympatric community of Amazonian fish.

The round stingray, *Urolophus halleri*, which lacks an active electric organ, produces an electric field with a complex geometry that is modulated rhythmically by movements of the spiracles and gill slits during ventilation. This weak stimulus is used in the field by reproductively active male stingrays to locate mates, and also by female rays to locate buried consexuals. Electrosensory primary afferent neurons are most sensitive to stimuli that vary sinusoidally at the same frequency as the natural respiratory movements. The match between primary afferent frequency and the ventilatory phasic signals produced by conspecifics indicates that the electrosensory system serves an important biological function in the social behaviour of elasmobranchs (Tricas et al. 1996).

Eigenmannia are seasonal breeders that spawn only during the tropical rainy season. Both sexes modulate their otherwise constant wave-like EODs in the context of courtship and aggressive behaviour by "chirps", abrupt frequency modulations with brief interruptions. Chirp-like EOD modulations can be elicited from a subnucleus of the PPn, the PPn-C. Reconstruction of individual PPn-C neurons showed maturity-dependent changes in the pattern of dendritic proliferation in females (Zupanc and Heiligenberg 1989).

Hypopomus occidentalis, a weakly electric gymnotiform fish with a pulse-type discharge, has a sexually dimorphic EOD. Gravid females could readily discriminate male-like from female-like EOD waveshapes. This feature of the electric signal is sufficient for sex recognition. *H. occidentalis* has the capability to discriminate male from female EOD pulse shape using both temporal and spectral cues, but that frequency discrimination based on spectral tuning of electroreceptors is the most likely neuronal mechanism (Shumway and Zelick 1988). *H. pinnicaudatus* has a marked sexual dimorphism in its tail filament. Sexually mature males have long, 'feathered' tails as compared with females. Mature males have larger electrocytes which are more widely spaced and more numerous than those in mature females. The biphasic, EOD is longer in males than in females. The peak-to-peak amplitude of the males EOD is weaker than a female's of the same total length. It is suggested that the reduction in EOD amplitude is a consequence of the increase in EOD duration among males. Further, female choice probably played a role in the evolution of long duration EODs among males, and males may have secondarily grown long tails to compensate for the loss in active space that would otherwise accompany a weaker EOD (Hopkins et al. 1990). For a detailed account on electrocommunication, the reader is referred to Kramer (1990).

Correlation between the properties of an electric signature and ecological and social conditions of mormyrids in their natural environment suggest that their signature properties are under natural selection, much as are social communication displays. For further information on mormyrids specifically, the reader can refer to recent papers (Moller et al. 1989; Serrier and Moller 1989; Moller and Brown 1990; Crawford 1992; Kramer and Kuhn 1994; Kramer 1994, 1997; Szabo and Denizot 1994).

According to Montogomery (1991), in low-light environments the lateral line and electrosensory systems of fishes can replace vision as the major sensory modality. These systems provide insight into sensory processing for orientation, object detection, and noise suppression.

Sex steroid hormones have the capacity to influence and coordinate the electrical activity of three target tissues (electric organ, medulla, tuberous receptors) that constitute the electrosensory system.

Attention is focused on marine skates and the freshwater mormyriforms, and gymnotiforms. Equal attention is not paid for example, to *Astroscopus*, *Malapterurus*, or torpedinoids, in which little is known about electroreception.

Heiligenberg (1994) discussed the detection and generation of electric communication signals in gymnotiform fish. In the context of their social communication, electric fish generate specific modulations of their EOD rhythm. Afferent pathways involved in the detection of a conspecific's EODs and their modulations lead to diencephalic structures that are tuned individually to a specific type of signal. These structures, in turn, are also linked to nuclei that initiate specific modulations of the fish's own electric organ pacemaker or may affect the fish's motivational state by influencing the pituitary. A comparison of neuronal substrates across related species suggests ways in which nature may have moulded different behaviours by gradual alterations of a network shared by all species (read Kramer 2001).

Androgens produce effects both on the EOD generating circuitary, probably at the level of the pacemaker nucleus, and on electroreceptors, probably, ultimately on receptor cell membrane conductances. These effects occur in parallel allowing the two parameters to remain well matched. In contrast to former predictions, exogenous electric fields alone appear unable to shift receptor tuning (Keller et al. 1986; read also Meyer and Zakon 1982; Bass and Hopkins 1983, 1985; Mills and Zakon 1987). Bass (1989) has shown how gonadal steroid hormones exert their effects directly on the properties of the action potentials and the electrically excitable membranes of the electric organ that generate a spike-like EOD. Ferrari and Zakon (1989) reported that the pacemaker nucleus is unnecessary for the maintenance, development, or hormone-mediated shift of receptor tuning in weakly electric gymnotiform fish *Sternopygus macrurus*.

3. EVOLUTION

Electric fish provide remarkable examples of convergent evolution. Not merely the production of electricity, but many of the properties of individual electrocytes and of their activity patterns show convergences. There are also equally dramatic convergences in the properties of electroreceptors.

Development of electroreceptors and of electric organs did not always occur side by side (Table 17.6). Some fish are electrosensory but lack electric organs. In the two large groups of "active" electric fishes (wave species and pulse species) differentiation of electrosensitive "tonic" receptors could have preceded the development of electric organs, but the appearance of "phasic" electroreceptors must have paralleled by the modification of certain muscle groups, or spinal nerves in the Apteronotidae, into the electric organs. Increasingly high water turbidity, precluding effective use of vision must have been the underlying selective pressure in this development. Possession of an effective electrosensory system for orientation puposes then made possible shifting of the time of activity into the night. The trend towards a nocturnal activity pattern could be a considered a latter adaptation at a time when many waters, with the exception of major rivers, became clear to the advantage of day-active visually orienting predators.

Wave species may have evolved from pulse species, in response to selection for efficient receptor mechanisms, for efficient communication, and for needs for increased temporal resolution in electrolocation. Other factors, such as the presence of electroreceptive predators or prey, or biogeographic considerations would locally influence the pattern of evolution. These factors may be important in explaining present day differences between the types of EODs found in Gymnotoidei of South America.

The electrogenic properties of two large freshwater groups of "weakly" electric fishes (Gymnotiforms of South America and Osteoglossiforms of Africa) are of special importance since they form an essential part of highly active electrosensory system for electrolocation, orientation, and social communication. These groups are phylogenetically far apart and any resemblance between *Gymnarchus* and the gymnotoids may be due to parallel adaptations to a new life style utilizing the electrogenic and electrosensory abilities. Gymnotoids are considered the most highly modified Characoidei of ostariophysan stock, while Mormyridae are not related to any African characins but are placed with the Osteoglossiforms.

A comparison of EOD behaviour among four unrelated groups of strong electric fish (Malapteruridae, Torpedinidae, Electrophoridae, and the stargazer *Astroscopus*) presents a good example of parallel evolution. While there is well-documented similarity in structure and general time course of the EOD volley of all four groups of strong electric fish, characteristic differences concern the fish's other mechanisms involved in predatory behaviour. Prey detection in *Astroscopus* is primarily visual. In the other three groups, visual prey detection plays a minor role, if one at all. *Torpedo* can perceive its prey in the absence of visual cues and seems to detect it chemically and/or mechanically. *Malapterurus* and *Electrophorus* discharge in response to chemical and mechanical stimulation. Both fish can emit a low-frequency-type EOD activity that may be involved in prey detection. The energy of low-frequency pulses in *Electrophorus* is smaller in comparison with the fish's volley pulses (both types are generated by different electric organs); no such difference between pre-volley and volley pulses (generated by the same organ) exists in *Malapterurus*.

Electroreception is not everywhere homologous but has been independently evolved atleast twice (nonteleosts and teleosts). It is hard to believe that within the teleosts there has been less than two independent inventions of the whole electric apparatus, once in ostariophysines and once in osteoglossiforms.

It should, however, be noted that electroreception and electrogenesis may not be as recent an innovation as is suggested by its presence in the rather modern groups. The occurrence of ampullary-like canals (cosmine pore canal, electroreceptor) in the dermal plates of Devonian fishes ancestral to both agnathan and gnathostome lines (Cephalaspidoformes, and Acanthodii), the apparent presence of electroreceptors in the coelacanth *Latimeria*, and their wide occurrence among the more primitive extant fishes indicate an ancient lineage. In as much as the first fishes appear to have had both a mechanosensory lateral line, as well as an electrosensory system, both systems may be considered to be of equal age and probably as old as the vertebrate radiation.

Electrogenesis has evolved at least six times (in the torpedinoids, rajoids, mormyriforms, gymnotiforms, siluriforms, and perciforms) and electroreception at least three times (in the nonteleosts, osteoglossomorphs, and ostariophysans) independently among fishes (see discussion by Bass 1986, 1989).

Vertebrate sensory systems often represent a serious challenge to the synthetic theory of evolution because of the complexity that arises from multiple interrelated functional components. This is somewhat alleviated by considering the comparatively simple chondrichthyan electroreceptive system. Estimates of structural variation in the peripheral component, i.e., the ampullae of Lorenzini, at each of the hierarchal levels suggest possible correlation with morphological and natural history parameters, including dorso-ventral flattening, eye and mouth position, feeding habits, prey availability, and migratory behaviour (Raschi and Mackanos 1987). Electroreception in dogfish, *Scyliorhinus canicula*, shows the characteristics of an array sensor system: dispersed receptor organs designed for the detection of non-uniform electric fields, representing nearby stimulus sources. In catfish, *Ictalurus nebulosus*, there is evidence that the electric sense can be used both in the array mode, when nearby electric sources with non-uniform fields have to be detected, and in the bi-sensor mode, which allows the detection of left and right in uniform fields both dc and ac. (Peters 1993), read Peters et al. (2001). Kalmijn (1997) give details of electric sense in relation to prey detection in elasmobranchs.

4. CONCLUSIONS

Most electric organs are derived from muscle, some from neural tissue; the anatomy of electrocyte is complex and varies from species to species; some have accessory electric organs. A few species can produce strong electric discharges. However, the majority of species is limited to weak electrical discharges.

The electric organ is driven by a command centre in the medulla. Diameter and lengths of spinal nerves supplying the electric organ are adapted to insure synchronous arrival of pacemaker signal at the electrocytes of the organ.

In the freshwater electric fishes, the "pulse species" which are in majority differ markedly from a smaller number of "wave species". Regardless of whether a species is a pulse species or a wave species, all species appear to have a unique EOD wavefrom. The uniqueness derives from a combination of frequency components of the EOD cycle and the repetition frequency of the EODs. From species to species, there is a broad range of species specific EOD waveforms. The EOD may additionally show age-or sex dependent differences that subserve intraspecific communication. Transient stereotyped modulation of EOD (in wave-type fish) or pulse pattern (in pulse-type fish) may also serve as social signals during courtship or aggression.

All mormyrids are pulsers, the gymnarchid *Gymnarchus niloticus* is a waver, and gymnotoids may be either pulsers or wavers.

The electrosense system is a second order sensory system with non-neuronal receptor cells; it is considered to be a part of octavolateralis system, and to have evolved from the ordinary lateral line system.

Electroreceptors are found in almost all ancestral types of fish and aquatic forms of amphibians and they may be as old as the origin of vertebrates. Most curiously, electrosensation was lost with the evolution of holostean fish, but then reappeared in some isolated groups of more modern teleost fishes. Efferent control appears to be absent in all classes of electrical receptors. Peripheral interaction between receptors such as lateral influence is slight or absent in electroreceptors. Both chemical and electrical transmission are found. Glutamate is implicated. By comparison with electrogenesis, electroreception is far more common among fishes.

Electroreceptors, embedded in skin, are small organs composed of receptor cells, supporting cells, connective tissue and afferent nerve terminals. Electroreceptors are separable morphologically and physiologically into two major classes: the ampullary electroreceptors and the tuberous electroreceptors. Morphologically, they differ in regard to the amount of the receptor cell membrane in contact with the water surrounding the fish, the specialization of the apical portion of the receptor cell, and the presence or absence of an accessory plug in the neck of the receptor organ pore. Physiologically, they have different voltage sensitivities, and different ways of encoding electric field intensity.

Whereas ampullary receptors sense low-frequency electric signals of not more than several tens of Hz, tuberous receptors are tuned to the dominant spectral frequencies of the fish's own electric organ discharges (EODs), in the range from tens to thousands of Hz. Ampullary receptors have become famous for their extreme sensitivity, and behavioural thresholds can be as low as several nV/cm.

Electroreceptors are distributed on the fish's body most densely at the rostral end. Many groups of electroreceptive fishes possess ampullary receptors. Ampullary receptors are so different in the nonteleosts and teleosts that they hardly deserve the same name and should be sharply distinguished as the Lorenzinian and teleost ampullary organs, respectively. Lamprey ampullary organs are morphologically unlike all other ampullary organs. Tuberous organs in gymnotiforms and osteoglossiforms have some resemblances by convergent evolution but are unrelated and in many respects different.

The distribution of electroreception indicates a polyphyletic origin for the sense, once in nonteleosts and two or more times in teleosts. It is also presumed that ampullary electroreceptors arose first in teleosts and tuberous electroreceptors were secondarily derived in weakly electric teleosts from prexistent ampullary organs.

The function of electric organs which are mostly, but not always, encountered with specialized electroreception can be listed as:

(1) *electroparalysis*—for defense or feeding purposes;

(2) *electrolocation*—for near-field orientation, escape, feeding purposes (this mode is found in all weakly electric fish, with *Electrophorus* being capable of delivering strong electric shock. Most of the nonneopterygian bony fishes are nocturnal or crepuscular and likely use electroreception in both prey detection and predator avoidance. It is possible that this sense might also aid long-distance migrations in sturgeons, since it is suspected to serve this function in elasmobranchs.

Electric fish have evolved jamming avoidance responses which serve to minimize the detrimental effects of signal interference on their ability to electrolocate objects. Species with wave-type electric organ discharges shift their frequency away from similar EOD frequencies of their neighbours. Species with pulse-type EODs time their pulses in such a manner as to minimize the probability of coincidences with pulses of their neighbours, and different strategies have been adopted by the unrelated African mormyrids and South American gymnotiforms.

(3) *electrocommunication*—for purposes of social interactions within populations, especially during territorial and mating behaviour, perhaps also interspecifically.

Electroreception in active, electrogenic freshwater groups must be recognized as an adaptation to the conditions which precluded vision as an orientation mode, and, secondarily perhaps, as an adaptation for a successful night-active life style. Intra-and interspecific social communication would be other adaptive features. The latter communication function could also include possible synchronization effects, mainly of circadian nature, within population of these fish.

The dual function of electric sense in the electric fishes has proved especially interesting when comparing the demands and specializations of receptors for each function. In the mormyrid fishes, segregation of inputs occurs quite early, and one special class of electroreceptors appears to be designated to function in communication, another in electrolocation. Segregation of function is less clear in gymnotids, as the central nervous connections and physiology.

Unusual central nervous system structures in electric fish include: the cerebellum-like posterior lateral line lobes which are so much larger and more highly differentiated in electric fish than in nonelectric ones; the torus semicircularis of gymnotoids in which different laminae receive input from different receptor classes and in which the somatotopic maps in each layer are superimposed; the large cerebellums in electric fish and in particular that of mormyrids. The mormyrid valvula cerebelli is extraordinary not only in size but also in the organization and regularity of its histological structure.

Separate brain regions appear to be associated with each type of electroreceptor and thus with each of the presumed roles of the electrosensory systems. It has been hypothesized that (1) all animals with a dorsal nucleus, including lampreys are electroreceptive; (2) holosteans and most teleosts, which have only a median nucleus, lack electroreception; and (3) a few teleosts with no dorsal nucleus but rather an electrosensory lateral line lobe (ELL), which is topologically distinct from the dorsal nucleus, had reinvented electroreception. The electrosensory lateral line lobe of the hindbrain contains four electrosensory maps of the body surface. Whereas the most medial map receives input from ampullary electroreceptors, the three remaining maps-lateral, centro-lateral, and centro-medial-receive identical input from tuberous electroreceptors via collaterals of their primary afferents. The three tuberous maps of the ELL differ in their specializations, with the lateral map achieving highest temporal and lowest spatial resolution.

In addition to a dorsal nucleus, all nonteleost electroreceptive anamniotes have tonic electroreceptors that are activated by the cathodal pole of electric field stimulus (teleost electroreceptors are activated by the anode) and they are innervated by only the anterior lateral line nerve. These characters, and the phylogenetic distribution, indicate that nonteleost electrosensory systems are homologous with each other but not with teleosts. The electrosensory systems in lampreys and primitive jawed fishes and amphibians apparently stemmed from the same early vertebrate that was ancestral to both petromyzontid agnathans and gnathostomes. Thus, nonteleost electroreceptors may be as old as or older than lateral line neuromasts, and it is equally likely that neuromasts evolved from preexisting electroreceptors, that electroreceptors evolved from preexisting neuromasts, or that two evolved separately from a third receptor such as the ciliated receptors of photochordates. The phylogenetic distributions of electroreceptors and neuromasts do not favour on one of these possibilities over the others. Hagfishes, which may be considered a sister group of lampreys, and hagfishes appear to lack electroreception as well as well neuromasts and thus contribute little to clarifying this issue.

Some interesting problems remain:

(1) Whether the relevant part of the medial nucleus looks or works differently in the three teleost orders (gymnotiforms, siluriforms, and osteoglossiforms)?
(2) What are the details about the mechanisms by which steroids alter electrical properties of excitable cells?
(3) How changes in electrical activity are coordinated across several steroid sensitive sites?
(4) What are the details of the anatomy and function of the central pathways and centres subserving electroreception in non-neopterygian bony fishes?
(5) Although laboratory studies show that electrolocation is used in a variety of short-range discrimination, little is known about the use of electrolocation in the natural environment.

CHAPTER 18
Bioluminescence

1. INTRODUCTION

Bioluminescence is chemical luminescence or chemiluminescence. One of the most striking features of bioluminescence is the sheer diversity of organisms that have developed the ability to emit light. The light field vertically divides the open ocean into the euphotic (= epipelagic), dysphotic (= mesopelagic) and aphotic zones (= bathypelagic). The luminescence of animals is a source of light in dark environments, indeed it is the only kind of light in the deeper parts of the ocean. The anatomical and physiologícal expression of luminescent abilities reaches its zenith in the fishes.

As an emission of radiant energy, frequently modulated, it is used in a variety of ways. As a stimulus luminescence is of particular interest in relation to its sensitivity to photoreceptors and the visual capacity of animals. Many animals respond in various ways to luminescence and depend on 'living light" for essential behavioural activities.

Luminescence is incoherent light, it radiates into space, the intensity is reduced during transmission through the animal's tissues and the external medium, that is by differential absorption and scattering. The energy of luminescence is very low but the sensitivity of the eye is extraordinarily high.

The sea acts as monochromator: all the ultra-violet is absorbed within a few cm from the surface, practically all the infra-red and much of the visible red are absorbed within 1m, but clear water is relatively transparent to light in the range violet to blue green.

Animals from epipelagic and coastal waters have light ranging from blue to green (465 to 517 nm). In pelagic animals from deep waters the light is generally blue, and the maximum emission lies between about 455 nm to 507 nm. A few meso- and bathypelagic fishes emit light of several colours, green and red. The light from the bacterial light organs of fishes is blue-green or green (> max 484 to 500 nm). Fluorescence is a common feature of many light organs of fish.

During the course of the last few decades bioluminescence has been the subject of numerous reviews (see literature: Herring 1982; McFall-Ngai 1991; and contributions in "Bioluminescence in action", edited by P.J. Herring 1978) (also read Herring 2002).

2. DISTRIBUTION OF LUMINOUS FISH

The known and putative luminous families and genera make up respectively 10% to 5% of the total recent fish families and genera. There are no known luminous freshwater fishes. All luminous fish are marine and the great majority of them are oceanic. Some inhabit shallow or coastal waters. Luminescent species are abundant in mesopelagic waters, i.e., those depths at which ambient light is dim but not imperceptible, and they ascend to the surface at night, e.g., a few deep-water squaloid sharks and numerous macrourids.

Generally a trend in the distribution of light organs in fish can be found. They are numerous and often elaborate in upper and lower mesopelagic species and they are very variable in the bathypelagic realm.

The question arises: why should structurally similar photophores, predominantly ventrally located, evolve independently in squids, crustaceans and fish? The convergence suggests a common environmental factor or set of factors acting to produce similar structural modifications in these organisms. The hypothesis advanced here offers a possible explanation for this convergence. Also, it would interrelate in a logical way three phenomena limited principally to the mesopelagic zone. They are the deep scattering layers, diurnal migrations and bioluminescence. Most of the evidence so far available suggests that the deep scattering layers as well as the organisms performing extensive vertical migrations are limited to the upper 500–800 m of the ocean. This is also the region of the ocean where bioluminescence is most highly developed in a wide variety of organisms. The hypothesis propounded here suggests that there is a definite relationship between the light environment of the organisms of the scattering layers and their bioluminescence. The movements of deep scattering layers and the organisms performing diurnal migrations appear to be most closely correlated with fluctuations in the amount of solar light reaching depth.

3(A). ANATOMICAL DISTRIBUTION OF LIGHT ORGANS

"Photophore" means a luminous organ with structures accessory to photogenic tissue. The position of luminous organs and tissues are documented for all bioluminescent families of fishes. The majority of families, species and individuals possess photophores only or mainly on the ventral half (as opposed to the dorsal half) of the body. The position and in some cases also the photophore structure tend to direct the emitted light downwards.

Large, ventral photophores characterize the mesopelagic fauna. In the bathypelagic realm reductions in the serial ventral photophores are often correlated with an increase in luminous tissue or organs elsewhere.

Internal bacterial organs are commonest among coastal fishes but are also known in the mesopelagic fishes. Bacterial light organs are also found in some epibenthic fishes. In the bathypelagic realm and below, the array of light organs in fishes becomes very variable in distribution and complexity.

Light organs of pelagic animals are most diversely situated and arranged. In deep-sea fishes they usually form specific patterns, especially on the lower surfaces. In addition to ventral photophores, some species also have a multitude of minute photophores scattered over the whole body surface. Still others have internal light organs which lie in the ventral body wall or are attached to the gut. Often lens and reflector are present, sometimes a light guide; these have the effect of channelling and focussing the light. When the photogenic centre is small, accessory structures may serve to enlarge the light-emitting area so that the whole lower surface appears luminous.

Of deep-sea fishes having ventral photophores, descriptions abound of how the light is directed downwards and is chiefly or only visible from below. It has been pointed out that the ventral photophores of fishes lie outside the animal's visual field. This arrangement could offer only one explanation for ventral sitting. It has been concluded that there is less probability of a fish being seen by predators when its bioluminescence is directed downwards.

There are some noteworthy differences in the distribution of light organs in fishes. One feature of many fishes is the sexual dimorphism of light organs. Developmental changes occur in the light organ patterns of fish and, as might be expected, additional photophores being added as the animal matures. Regression of light organs is a rare phenomenon. Most examples of serial ventral light organs in fishes glow steadily whereas single organs flash rapidly.

3(B). TAXONOMIC DISTRIBUTION OF LUMINESCENCE

Luminescence may be produced either by symbiotic luminous bacteria in special organs or by intrinsic chemical systems. Representative examples are described and illustrated here.

While most luminous fishes evolved mechanisms of nonbacterial luminescence, more is known about the biochemistry and molecular biology of bacterially produced symbiotic luminescence.

Three species of culturable marine luminous bacteria have been identified from the light organs of fishes: *Photobacterium leiognathi*, *P. phosphoreum* and *Vibrio fischeri*. As in the light organs of marine squids the symbioses exhibt unilaterál species specificity between hosts and bacteria, i.e., a given fish species will always harbour a given luminous bacterial species, but a given symbiotic luminous bacterial species may occur in more than one fish species or family. Three biological characteristics of the light organs and their bacteria have been invoked to explain the species specific nature of the occurrence of these bacteria with different species of host fishes: (a) the relationship between the geographic range of the host and the different growth temperature optima of the bacterial species, (b) the anatomical position of the light organ, and (c) the phylogenetic position of the host (McFall-Ngai and Toller 1991). According to authors how and why certain bacterial species associate with particular fish families is likely to remain an unresolved question until the mechanism of initiation and development of the symbioses are made clear.

Shallow-water tropical luminous fishes are grouped as those having: (1) cephalic bacterial light organs (anomalopids and monocentrids); (2) ventrally oriented, gut-associated bacterial or non-bacterial light organs (apogonids, leiognathids, and pempherids); or (3) ventrally oriented non-bacterial photophores (batrachoidids, sciaenids, and engraulids). *Gazza minuta*, a leiognathid fish, spans the first two categories, and so shares attributes of both the anomalopid-monocentrid group and the apogonid-leiognathid-pempherid group. The details are given in forthcoming description.

4. BACTERIAL BIOLUMINESCENCE (EXTRINSIC LUMINESCENCE)

This is one mode of light that cuts across phyletic boundaries: the extraordinary culture chambers for symbiotic luminous bacteria. Bacterial symbionts are known to be the source of luminescence in species from atleast five different orders of fishes (Beryciformes, Perciformes, Gadiformes, Salmoniformes and Lophiiformes).

These light organs, which vary from simple sacs to compound photophores with reflecting, diffusing and control structures, occur on body surface and as internal organs. Within the constraints of the overall organization, there are, however, considerable differences. The bacterial light organs are often located in gut diverticula deep within the body, whose external surface is usually heavily pigmented by the slow aggregation or dispersion of melanophores. Fish organs of this type may in consequence have very complex light guides and/or diffusing systems, reaching peaks of elaboration in some cases (refer *Opisthoproctus* and *Leiognathus*). Such light organs are generally described as the "indirect" emission type. These light organs are analogous to intrinsic photophores. In certain fishes luminous material can be voided to the outside and it is thought that the bacteria in the gland are in a steady state of growth.

The bacterial species have been isolated and characterized, identified or unidentified. The distribution patterns of free-living marine luminous bacteria are believed to be controlled not only by temperature but also by salt tolerance, resistance to photoxidation and the ability to grow in nutrient-poor conditions.

Two to five percent of soluble bacterial protein is luciferase. Luciferase catalyzes the reaction between bacterial luciferin (which is $FMNH_2$) and a long-chain aldehyde and molecular oxygen. During the course of the reaction a large amount of energy is liberated; it seems likely that luminescence is produced by the excited enzyme molecule. *In vitro* the reaction slows as $FMNH_2$ is oxidized to FMN. *In vivo* $FMNH_2$ is continuously regenerated to FMN by a further reaction involving NADH. For the molecular biology, biochemistry and physiology of fish bacterial light organ symbioses, the reader is referred to McFall-Ngai and Toller (1991). Coelenterazine, a luciferine, and luciferase activity specific to coelenterazine were detected and assayed in various tissues (e.g., digestive systems and photophores) of mesopelagic fishes from the Strait of Messina. Dietary acquisition of coelenterazine was suggested (Mallefet and Shimomura 1995).

4.1 Taxonomic Distribution of Bacterial Bioluminescence, Microanatomy and Operation of Luminous Organs

4.1.1 Beryciformes

Anomalopidae (flashlight fishes)

This family is widely distributed; *Anomalops* and *Photoblepharon* occur in the Indo-Pacific while *Kryptophanaron* is a new world genus reported from Baja California and Caribbean region. These small dark, tropical, surface-water fishes have a prominent suborbital photophore filled with bacteria (Fig. 18.1). The photophore of *Photoblepharon* is relatively larger than that of any other fish.

The fine structure in *Anomalops* and *Photoblepharon* has been described and that of *Kryptophanaron* is probably similar. The bacteria are contained within a mass of parallel tubules aligned at right angles to the surface of the organ. Each tubule has a thin epithelial wall and the tubules are arranged in rosettes of five or six around a capillary. Near the surface of the organ the tubules open into collecting channels which coalesce and open to the exterior through numerous pores. The base of each tubule closely abuts on to a reflector which in *Anomalops* is composed of two parts, the main interior reflector covers most of the inner surface of the organ and a smaller external reflector along the lower margin of the photophore, both composed of guanine crystals (Fig. 18.2).

The light is controlled either by drawing a pigmented curtain or screen up from the lower

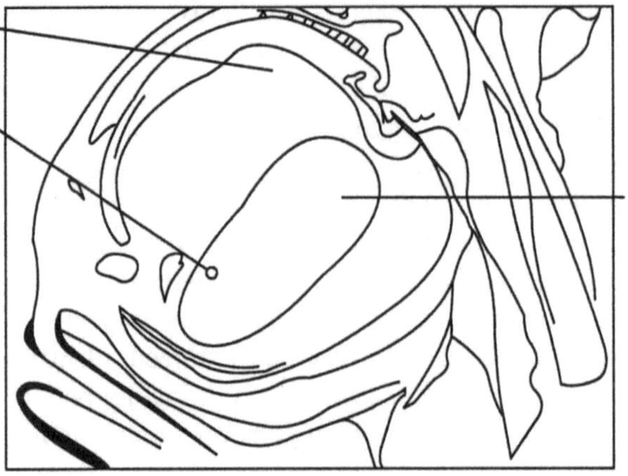

Fig. 18.1 The eye of luminous fish *Anomalops*, showing the elliptical white luminous organ with its opening.
***Source:* Haneda (1955)**

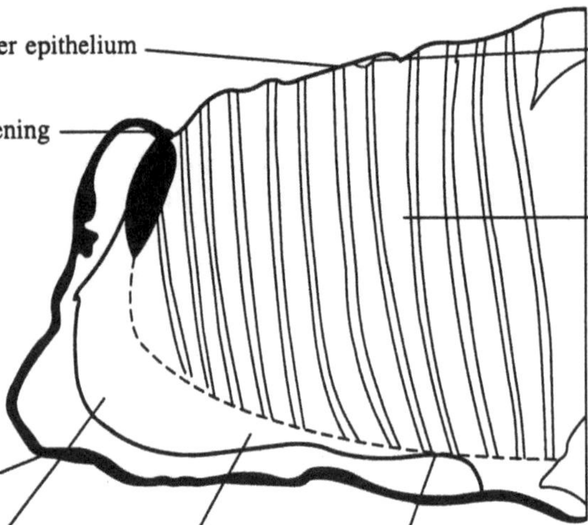

Fig. 18.2 A longitudinal section of the luminous organ of *Anomalops*.
***Source:* Haneda (1955)**

border of the socket and over its emitting surface (*Photoblepharon*), by rotating the whole organ downwards into a dark suborbital pocket (*Anomalops*) or, in *Kryptophanaron*, by a combination of both mechanisms. The musculature for both systems is derived from the levator maxilla. As the bacteria are continuously luminous these movements have the effect of producing a 'blink' in the animals's light. The movements are rapid, upto 100 blinks per minute in *Anomalops* and *Photoblepharon*.

The results of measurements of angular distribution (in air) of light emitted by photophores of freshly killed specimens of *Anomalops* and *Photoblepharon* show marked differences between two species particularly in the dorso-ventral and horizontal planes. *Photoblepharon* with its larger more concex photophore has a considerably broader distribution in both planes. Indeed the field of illumination cast by its photophores is remarkably uniform round much of the head. The external reflector in *Anomalops* may perhaps partially compensate for the very limited dorsal illumination otherwise cast by its photophore. *Anomalops*, however, has some potential for altering the direction of the emitted beam but it can 'squint' its photophores to a limited degree.

The nature of the bacteria in anomalopid photophores has not yet been established. Despite many attempts luminous *in vitro* cultures have not yet been achieved.

Monocentridae (pinecone or knightfishes)
The light organs of *Monocentris japonicus* take the form of two small protuberances on the lower jaw and lack both a reflector and a lens. The surface of the organ is covered with epidermal papillae; ducts running through four larger papillae open to the exterior. The bacteria are present in numerous tubules which empty into a central canal connected to the exterior through the ducts. The bacteria are closely associated with the surfaces of the cells making up the tubule wall and occasional intracellular bacteria have been observed. The whole organ is surrounded by chromatophores which apparently control light emission. The light is produced by *Vibrio fischeri*. Bacteria have been cultured.

4.1.2 Perciformes

Apogonidae (cardinal fishes)
Although many apogonids, from both coastal and oceanic environments have luminous organs, only one genus *Siphamia*, is known to harbour luminous bacteria. The light organ is similar in all the species of the genus and forms a small glandular structure situated dorsal to the ventral musculature just anterior to the pelvic girdle. A number of ducts run from the gland to open into the intestine. Light is diffused ventrally by the translucent muscle bands (Fig. 18.3). The bacteria are probably a strain of *P. leiognathi*.

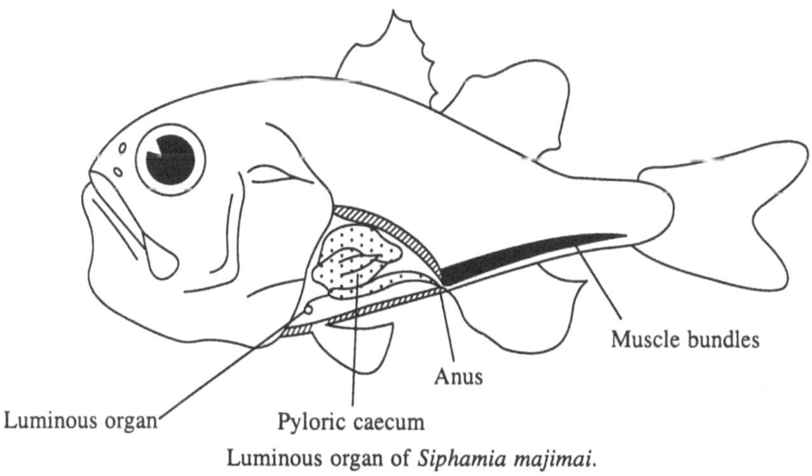

Luminous organ of *Siphamia majimai.*

Fig. 18.3 Lateral aspect of *Siphamia majimai* showing the position of the luminous organ. Pectoral fin, liver and a part of the body wall are removed.
***Source:* Iwai (1960)**

Acropomatidae

There are only two known species, *Acropoma hanedai* and *A. japonicum*. The light organ is a U-shaped body straddling the anus, the loop facing in opposite directions in the two species. Numerous ducts from the organ coalesce into a canal which opens externally just anterior to the anus. Translucent muscles act as light diffusers and a dorsal reflector and numerous chromatophores serve to control the emission (Fig. 18.4).

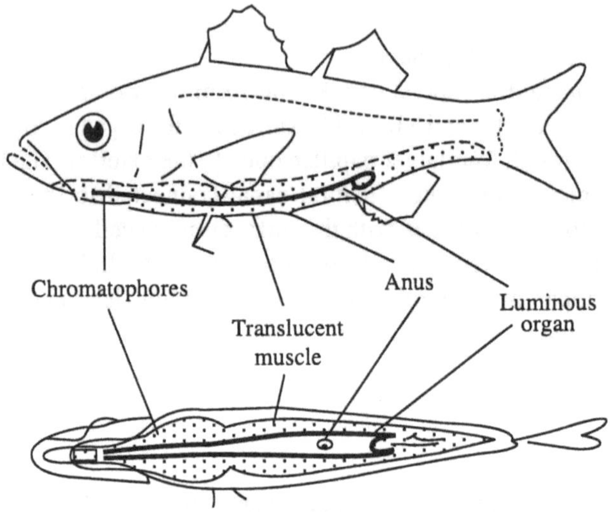

Acropoma handedai: Photogenic system.

Fig. 18.4 ***Acropoma japonicum*. Diagram of the luminous organ system.**
***Source:* Haneda and Johnson (1962)**

Leiognathidae (ponyfishes or slipmouths)

These are commonly found in shallow coastal waters of the tropical and subtropical Indo-Pacific. External and internal sexual dimorphism has been reported in certain leiognathid fishes. The externally dimorphic skin patches are closely associated with the internally dimorphic light organ system in some forms. The luminescent system of ponyfishes differ morphologically from the systems of all other known fishes and especially from those emitting light ventrally. All 20 species of three ponyfish genera, *Gazza*, *Secutor*, and *Leiognathus* have a doughnut-shaped light organ which opens into the oesophagus. In some species the organ encircles the oesophagus while in others it is displaced or enlarged dorsally with clear indications of a paired structure. A series of abdominal accessory structures positioned in tandem are present for controlling light intensity and for directing and dispersing the light. Luminous bacteria are enclosed within tubules lined with a glandular epithelium. The light from the bacteria shines through a chromatophore occluded dorsal window into the tissue-paper thin physoclistous swim bladder. This is silvered dorsally and half-silvered ventrally, acting as a parabolic mirror to direct the light via ventral window into the ventral, abdominal, translucent musculature through which it diffuses over the ventral surface of the fish (Figs. 18.5 and 18.6).

A physoclistous swim bladder of leiognathid fishes functions not only in buoyancy but also in reflection of bioluminescent light from the circumoesophageal light organ. Further, the enhancement of purine in the reflective portion (dorsum) of the swim bladder and the correlation of the differential distribution of purines with the path of light indicate that the common leiognathid, *L. equalus* swim bladder is exquisetely adapted to function as a reflector of bioluminescent light.

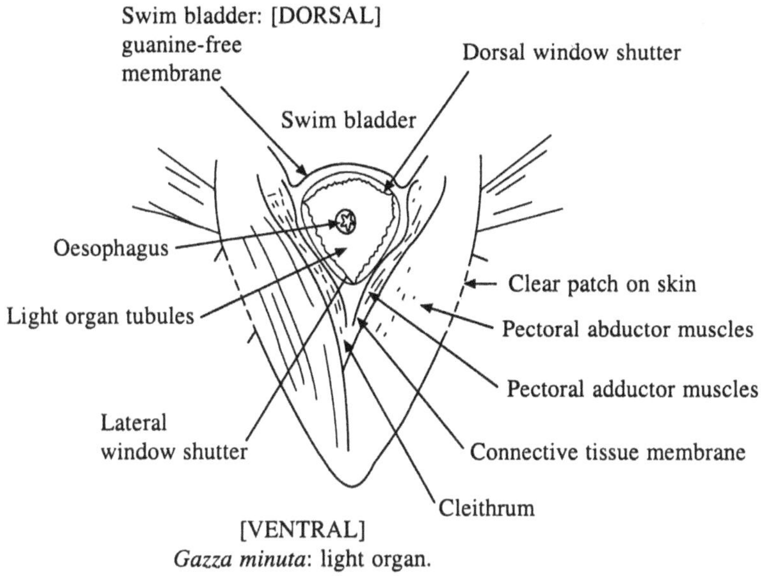

Fig. 18.5 ***Gazza minuta*. Cross section through the light-oesophageal region, showing anatomical relationships of tissue involved in DPL (discrete projected luminescence) expression. *Source:* McFall-Ngai and Dunlap (1983)**

The tissues that intervene between the internal light organ and the external environment are responsible for the observed mottled pattern of the ventral luminescence. Furthermore, these tissues, which have been incorporated into the light organ system, are involved in the control of the intensity, spectral quality and angular distribution of the fish's luminescence.

Three modes of luminescence are described for *G. minuta*: discrete projected luminescence (DPL), ventral body flash, and buccal luminescence. The structural correlate for is a small clear patch of skin lying at the posterior margin of each opercular cavity. These modes of luminescences indicate a much greater than expected diversity of luminous behaviours in leiognaths, perhaps greater than that of any other organism yet studied.

Studies on the fine structure have shown that the bacteria are invariably extracellular. The epithelial lining of tubules is rich in endoplasmic reticulum, has a microvillous border, desmosomes, nucleus, mitochondria and Golgi body.

The light organs of *L. elongatus* and *L. rivulatus* show marked sexual dimorphism.

The leiognathids can produce both short bright flashes and steady ventral luminescence and it is assumed that muscular control of aperture of the light organ is involved in the production of flashes while chromatophores exert a long term intensity control.

The luminescent signalling capability of leiognathids appears versatile and complex. The potential for luminescent behavioural versatility in the Leiognathidae stems from the internal location of the light organ and the concomitant modifications of various tissues (e.g., swim bladder lining, skin of opercular margin, skin of flank) to permit unique and diverse forms of light emission.

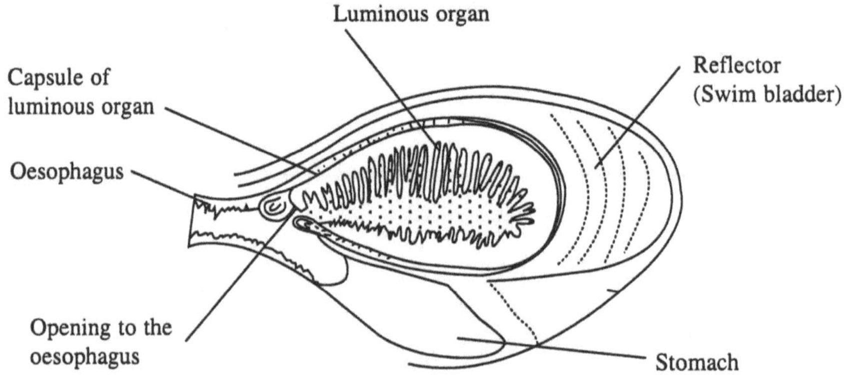

(A) Diagram of L.S. of luminous organ of male *Leiognathus elongatus.*

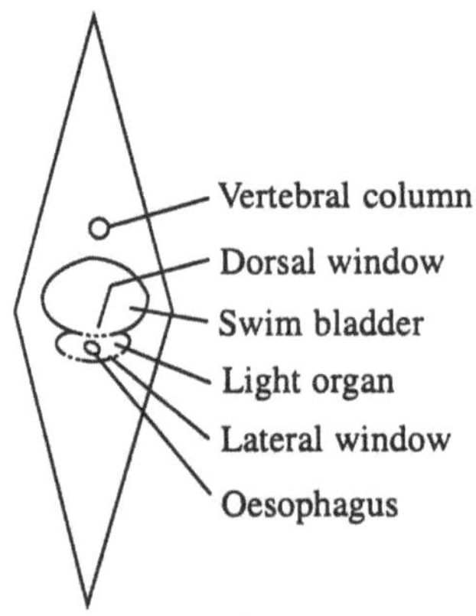

(B) Position of the light organ in a leiognathid fish.

Fig. 18.6 (A) Diagram of longitudinal section of the luminescent system of pony fish *Leiognathus elongatus.* (B) Cross section illustrating the position of the light organ and the relationship of the light organ windows to the swim bladder and lateroventral musculature. In the majority of the analyzed systems, the light organ has three muscular shutters: a dorsal shutter, which permits light to exit into the swim bladder space, and a pair of lateral shutters, which open into translucent muscles anteroventrally. The light entering the swim bladder is reflected outwards through a posterior, transparent area on the anal fin pterygiophore into the translucent posteroventral musculature.
***Source:* Fig. 18.6(A) Yata and Tsuji (1976), Fig. 18.6(B) McFall-Ngai and Morin (1991)**

The luminous bacteria in the gland are easily cultured. In all three genera the bacteria are pure cultures of *P. leiognathi.*

For special focus regarding bioluminescence in leiognathids, the reader is referred to McFall Ngai (1983), McFall-Ngai and Dunlop (1983, 1984) and McFall-Ngai and Morin (1991) and Table 18.1.

4.1.3 Gadiformes

Luminous gadoids are "predominantly slope dwelling fishes of mid-depths ranging from 100–1000 m depth" and no abyssal gadoids are luminous. While most gadoids are benthopelagic in habit, there are a number of mesopelagic macrourids including both luminous and non-luminous

Table 18.1 Summary of modes of luminescent behaviour in leiognathids+

Behaviour	General description	Means of control	Other parts of anatomy used	Proposed function
Counter-illumination	Diffuse glow over entire ventrum	Dorsal and lateral light organ shutters and chromatophores Chromatophores of muscle and skin	Swim bladder, transparent anal fin pterygiophore hypaxial muscles and skin	Antipredation and/or predation (silhoutte camouflage)
Discrete, projected luminescence (DPL)	Antero-ventrally directed bright spot of light from posterior opercular cavity	Branchiostegal, lateral shutters and chromatophores of light organ skin peripheral to patch	Translucent anterohypaxial muscles, transparent cleithrum, and transparent skin	Predation, antipredation, and/or communication
Ventral body flash	Single, intense flash over entire lower two-thirds of fish	Dorsal and lateral shutters and chromatophores of light organ	Same as in counter-illumination	Antipredation (such as startle display or alarm signal)
Opercular flash*	Flashes of variable duration and brightness over entire opercular area	Lateral shutters and chromatophores of light organ	Translucent muscle and transparent bone of the opercle skin	Antipredation
Buccal luminescence*	Glow of variable brightness into buccal cavity	Shutters and chromatophores of light organ	Clear membrane at back of throat	Predation

* Not observed in swimming fish.
\+ Modified after McFall-Ngai and Dunlap (1983)

species. The luminous species do not have depth ranges clearly distinguishable from non-luminous ones and the role played by luminescence in both these and benthopelagic gadoids remain unknown.

Macrouridae (grenadiers or rat-tails)
More than half of the approximately 300 species of macrourids have bulbous ventral light organs which are situated anterior to anus and from which a duct opens into the perianal groove or the rectum (Fig. 18.7). The morphology of the organ varies considerably between genera and species. Some are single bulbs, others double structures joined by a long tubule. Lens and reflector systems may be present and the light may be diffused throughout specialized regions of ventral

musculature. It is likely that the luminescence in some species is brightest in juveniles or even confined to the young stages in which the light organ is relatively large. Under normal conditions the light is probably regulated by chromatophores.

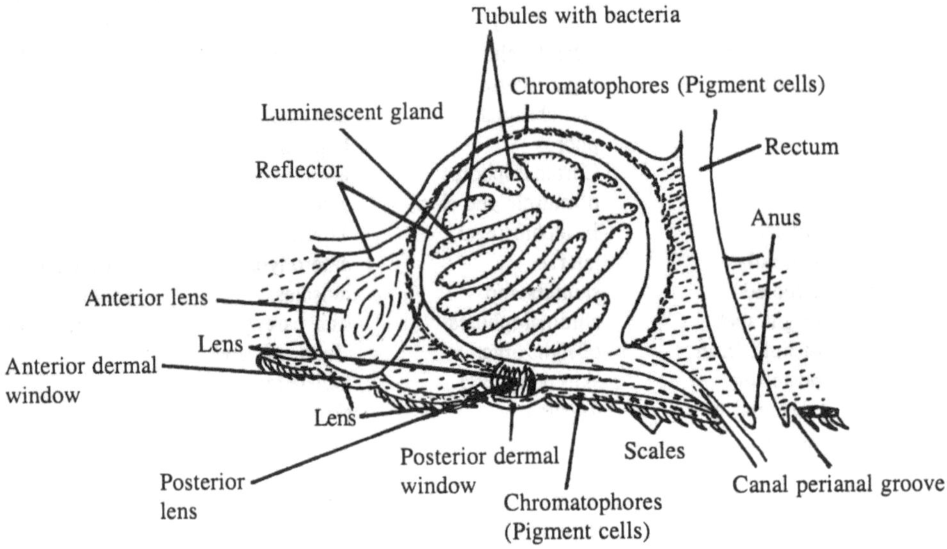

Fig. 18.7 Diagrammatic sagittal section through the bacterial light organ of *Malacocephalus laevis*.
***Source:* Herring, P.J. (ed.) (1978) Bioluminesce in action. Academic Press, London.**

The bulbs of the light organ harbour luminous bacteria which can readily be cultured *in vitro*. The bacteria have been identified as *P. phosphoreum*.

Moridae (morid cods)
These fishes are mainly benthopelagic forms found along the continental slopes. Luminescence in morids has been observed only in species of *Physiculus* but light organs are also present in species of some other genera. The organs are similar to those of macrourids; a bulbous gland containing bacteria opens into the rectum. The bacteria from *Physiculus rastrelliger* have been cultured and identified as *P. phosphoreum*.

4.1.4 Salmoniformes

Bathylagidae
The presence of rectal light organs has been reported from *Opisthoproctus*, *Winteria* and *Rhynchohyalus*. The light organs are rectal diverticula containing clonal populations of *P. phosphoreum*. The light organ in mesopelagic argentinoid *Opisthoproctus* is directed into a hyaline light guide region which diffuses it over the flattened ventral sole of the fish (Fig. 18.8). A similar but less elaborate system obtains in *Winteria* (with tubular eyes facing anteriorly), but the anal light organ of *Rhynchohyalus* probably emits light directly to the exterior as there are no accessory optical structures present.

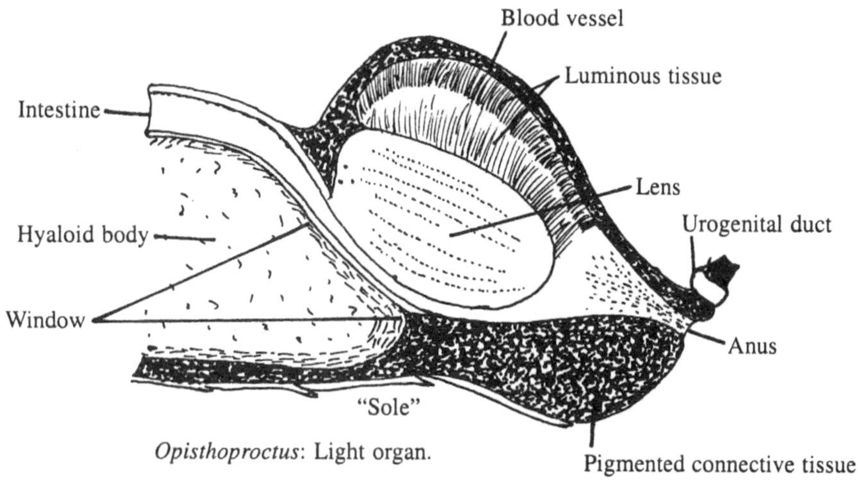

Fig. 18.8 Median section through the light organ of *Opisthoproctus soleatus*.
***Source:* Nicol (1967)**

4.1.5 Lophiiformes

Ceratioidei

The order Lophiiformes contains most of the families with dorsal photophores. Most adult ceratioids are found at about 2000 m. A characteristic feature of most bathypelagic anglerfishes is the presence in the female of a modified first dorsal finray on the head, a mobile illicium ("fishing rod") whose bulbous, fleshy, black-pigmented tip, the esca, is modified to form a light organ or luminescent bait (lure) (Fig. 18.9). The esca bears a dorsal or anterodorsal aperture which can be varied, though the mechanism is not clear. The esca also bears a variety of elaborate optical structures including tubular light guides and lenses, so that the light may emerge from a number of separate and widely spaced apertures, some on elongated appendages.

The escal luminescence has been confirmed in *Melanocetus*, *Oneirodes* and *Linophyrne*. *Linophyrne* has an unique dual bioluminescent systems. The hyoid barbels light organs do not contain bacteria but complex paracrystalline photogenic granules. The barbel organs control may well be invested in the blood supply. The virtual absence of major nerve fibres is unusual in that control of luminescence is believed in most species to be nervous. It is unique that the systems involve two entirely separate systems of light production, one being an open organ harbouring luminous bacteria while the other is a closed organ with intrinsic, intracellular luminescence. In addition, they appear to be derived from different germinal layers, the esca being clearly ectodermal and the barbel organs are probably mesodermally derived. Other sites of luminescence are known in a few species of anglerfish. In *Ceratias* and *Cryptosaras* other modified dorsal rays form caruncles containing bacteria. The caruncles degenerate in adult *Ceratias* but persist in mature *Cryptosaras*.

Male anglers are small and many species are parasitic on females. The swimming power of females is poor and it seems likely that they lure prey into their mouths by means of luminous esca. The mobile illicium is brought down to the mouth in the last phase of luring prey into its jaws. Hence, although it is anatomically dorsal, functionally it may be brought down to mid-lateral or sometimes ventral position.

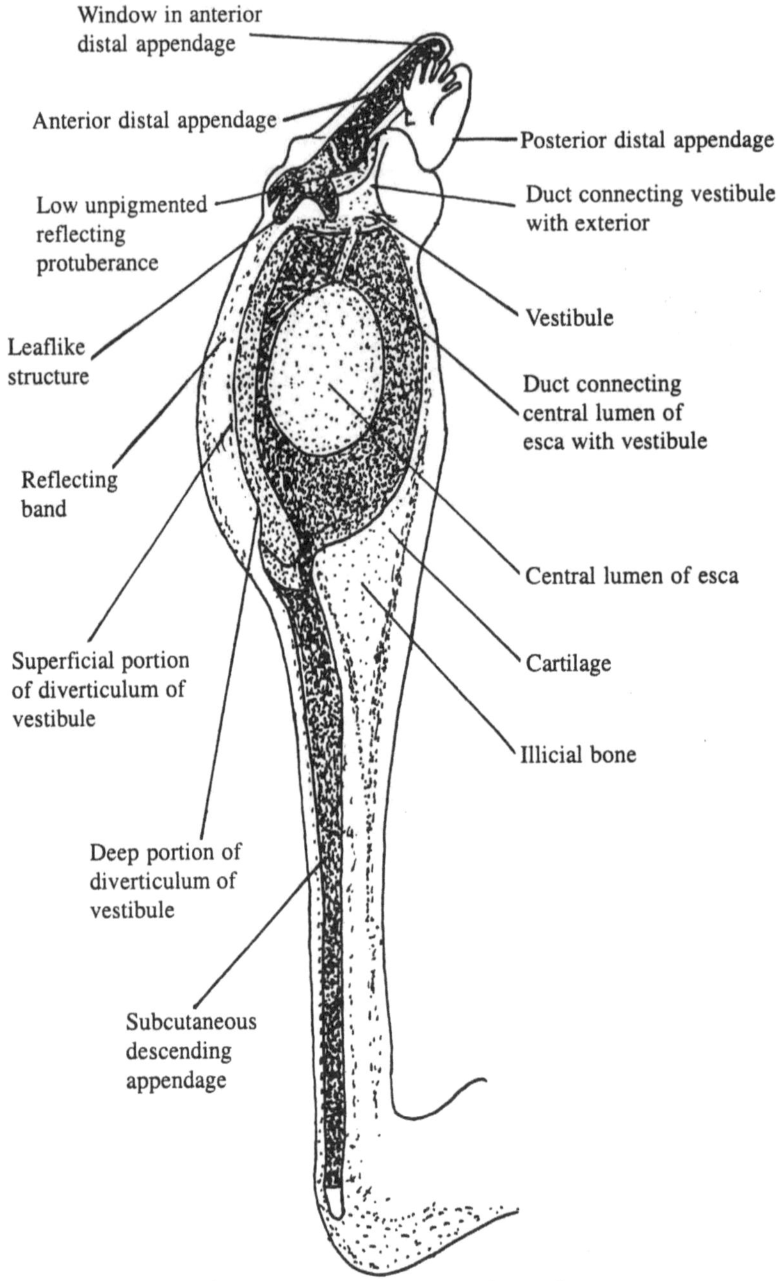

Illicium and esca of *Chaenophryne draco*.

Fig. 18.9 Rostro-lateral view of illicium and esca in female *Chaenophryne draco*.
***Source:* Munk and Bartelsen (1980)**

No luminescent bacterial cultures have been obtained so far. Nevertheless the presence of bacterial luciferase in the escas has been demonstrated. The luciferase has fast kinetics implying that the symbionts may be a species of *Photobacterium*.

5. NON-BACTERIAL BIOLUMINESCENCE (Intracellular Luminescence; Intrinsic Luminescence)

Most luminescent systems in fishes are not bacterial but intrinsic. The organs vary very greatly in their organization and in the complexity of the associated accessory optical structures. Simple light organs are very common either with or without a pigment cup. Compound light organs, with accessory lenses, reflectors and pigments or chromatophore "filters" have attained a very high degree of development and considerable morphological variation is to be found. Multiple organs in which a photogenic mass illuminates several adjacent serial organs occur in some luminous fishes.

5.1 Taxonomic Distribution of Intracellular Luminescence, Microanatomy and Operation of Luminous Organs

5.1.1 Squaliformes

A number of mid-water, pelagic sharks are luminous and have particularly large ventral surfaces. In *Euprotomicrus* (Squalidae, Dalatiinae) the photophores are primarily ventrally distributed, in crescents around each scale, and are tiny compact units. Each photophore has a single photogenic cell in a pigment cup with an overlying lens. In *Etmopterus* (Squalidae, Squalinae) the photophores are a little larger and more irregularly distributed on the ventral and lateral surfaces of the body. They contain 6 to 13 photogenic cells and an iris-like structure as well as lens and pigment. Reflectors are absent as in *Euprotomicrus*. Light intensity may perhaps be controlled by the iris chromatophores.

5.1.2 Salmoniformes

Stomiatoidei

The stomiatoids are typically small, dark brown or black fish (with the exception of silvery-sided hatchetfish: Sternopthychidae) found in the top 1000 m where they make considerable diurnal vertical migrations. They are amongst the most abundant bathypelagic fish.

Gonostomatidae

Numerous ventral and ventrolateral photophores are found in gonostomatid fishes. In some "maurolicid" genera, some photophores are grouped into compound organs in which the light from the common photogenic mass is passed through the apertures of individual units. The photophore contains "A" and "B" cell types. "A" cells lie deep and possess a highly organized endoplasmic reticulum forming a granular product (glycoprotein). "B" cells occur more distally and contain glycogen, glycoprotein, plenty of ribonucleic acid, and abundant ribosomes as membrane bound vesicles with finely granular product close to Golgi zones. Intermediate cell types may be found. "B" cells make up the "filter tissue" present in many photophores, usually purple in colour.

Three photophore types have been recognized. In the alpha type (e.g., *Maurolicus*) the "A" cells are set perpendicular to the long axis of the photophore and the secretory cycle of each cell takes place along the axis of the photophore. In the beta type the "A" cells are arranged round a central lumen or duct which may open outside (e.g., *Gonostoma*) or be closed (e.g., *Cyclothone*). The secretion is found in the apical zones of each cell. In the gamma type (e.g., *Vinciguerria*) the duct is lacking and "A" cells are radially arranged, with a marked polarity of cell products. In all three types "A" cells may be converted into "B" cells, but the precise homologies are not yet clear and the role of "B" cells, in particular remains obscure.

The large ventral photophores always possess a crystalline reflector and a surrounding pigment layer. A clear lens-like area may be present (e.g., *Maurolicus*) and be surrounded by a "lamellar ring".

Sternoptychidae (hatchet fishes)

Argyropelecus (Fig. 18.10) and *Sternoptyx* have compound photophores predominantly ventral in distribution and direction . They are the alpha photophores with polyhedral "A" cells and no duct. The "B" cells are pigmented, with a transmission maximum around 480 nm. The very large tubular reflector channel associated with most of the photophores have been shown to determine the angular distribution of emitted light which very closely matches that of ambient downwelling light. The preorbital organ in *Argyropelecus* shines into the eye and may act as a reference standard for comparison with ambient light. *Sternoptyx* spp. apparently lack this organ in its normal position but have two large organs, lacking reflector and filter, situated in the roof of the mouth. Rees et al. (1992) reported that the substrate for *Argyropelecus* is most probably coelenterazine. The luminous system in *Argyropelecus* photophores is closely related to that of bioluminescent coelenterates.

Studies with isolated photophores have also shown that the bioluminescence control mechanism in *Argyropelecus* is adrenergic in nature (Mallefet and Baguet 1985).

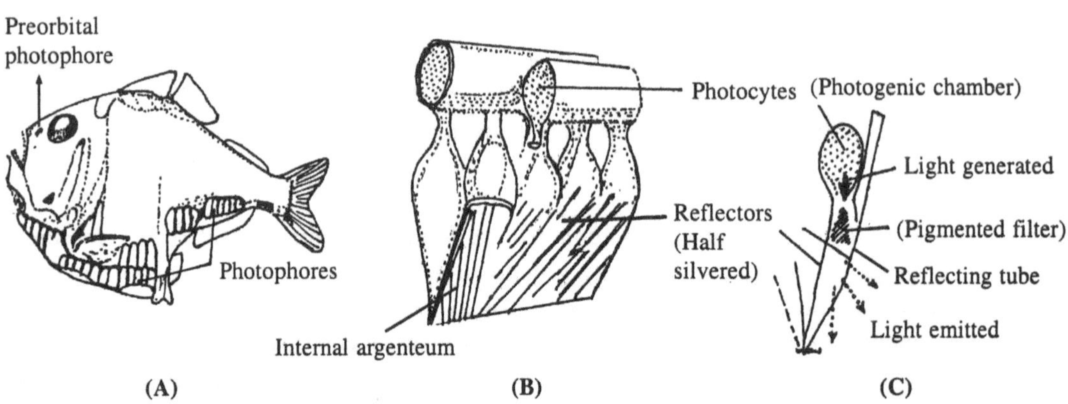

Argyropelecus aculeatus: Light organ.

Fig. 18.10 The hatchetfish (*Argyropelecus aculeatus*. (A) Distribution of photophores over the body. (Diagrammatic side view) (B) Organization of ventral photophore; a common photocyte chamber on each side emits light through separate tubular photophores, each of which has an inner silvered argenteum and outer half-silvered surface. (C) Diagram of a vertical section through a photophore. The filter is positioned in the aperture of the photocyte chamber.
***Source:* Herring (1982)**

Astronesthidae, Chauliodontidae, Stomiatidae, Idiacanthidae, Malanostomiatidae, Malacosteidae

All possess serial ventral photophores but most also bear tiny simple pigmented photophores (Fig. 18.11), usually lacking reflectors; orbital organs, many are rotatable or can be covered by a screen; and glandular patches of luminous tissue elsewhere on body. The photophore pattern is an important taxonomic character, but does exhibit considerable meristic variation. The photophores of *Chauliodus* and *Stomias* are of gamma type. *Pachystomias*, *Aristostomias* and *Malacosteus* emit red light, as well as blue, from separate organs in contrast to blue or blue-green emission so characteristic of almost all marine organisms.

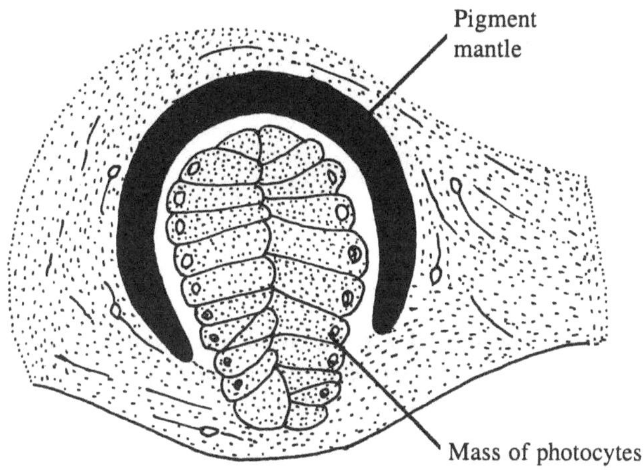

Fig. 18.11 A small pigmented light organ on the head of *Stomias valdiviae*.
***Source:* Nicol (1967)**

A single simple photophore at the tip of elongate dorsal fin ray in *Chauliodus* may serve as a lure. Barbel luminescence has been reported in *Leptostomias*, *Chirostomias*, *Eustomias* and *Astronesthes*. The biological function of presumably luminous chin barbel and the first pectoral finray and its filaments in melanostomatid *Thysanactis dentex* is unknown. The function of limbus photophores of eyeball in *T. dentex* remains an enigma. Such luminous structures are generally thought to function as lures attracting prey.

5.1.3 Myctophiformes

Myctophidae (lanternfishes)

Myctophids include 32 genera. About 300 species of lanternfishes are amongst the most abundant and certainly most accessible of oceanic fishes, typically found from 200 to 600 m in the daytime but migrating closer to surface at night. All species (except *Taaningichthys paurolychnus*) bear serial photophores and other luminous tissue. The serial photophores have been further distinguished into (i) primary, mainly ventral and having lens, and (ii) secondary, spread all over the body and lacking lens.

The primary photophore lies within a scale cup and has a small mass of photogenic tissue, lens formed from overlying scale, reflector, pigment backing and elaborate vascular network. The luminous tissue occurs most commonly on the caudal peduncle, as the caudal organs. Sexual dimorphism occurs in many genera; commonly supracaudal organ in males and infracaudal organs in females. Caudal organs and tissue patches luminescence much faster and brighter than the serial photophores.

The photogenic tissue of all species is fundamentally similar, both in serial photophores and other luminous tissue. The tissue has stacks of thin lamellar cells. The arrangement of the lamellae in caudal organs differs substantially in different species. The photophores have an elaborate associated nerve supply (Fig. 18.12).

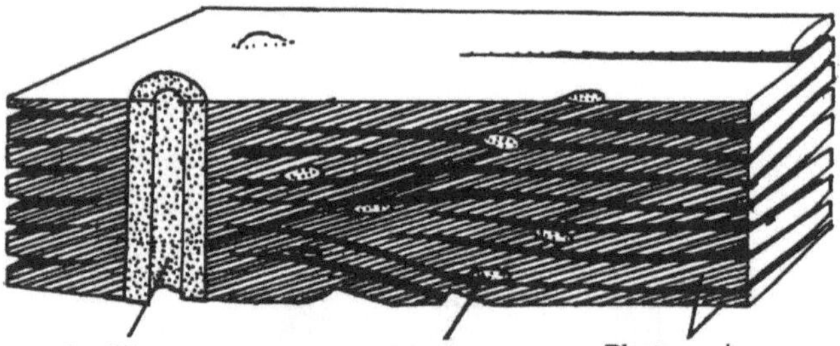

Fig. 18.12 Schematic section of a generalized myctophid caudal organ showing arrangement of lamellae relative to a capillary.
***Source:* Edwards and Herring (1977)**

According to Foran (1991), the myctophids and stomiiforms represent two of the most common groups of luminous fishes, yet the nature of their luminescence has remained a mystery. Data and discussion provided the first indication that luminous bacterial symbionts exist in myctophids and stomiiforms and that these symbionts account for luminescence in these fishes. What form they take and how they might come to inhabit the fish light organs remains to be determined.

5.1.4 Batrachoidiformes

Batrachoididae (toadfishes)
Within this world-wide shallow water tropical and temperate family only *Porichthys* bear photophores. It is a nocturnal animal, submerging itself in mud during the day. The genus, containing several species (three well known are *P. notatus*, *P. myriaster*, and *P. porosissimus*) is found off the coasts of North, Central and South America. In California, these fishes are called "midshipman" because the shining photophores are supposed to resemble the buttons on a midshipman's uniform. They are called "singing fish" or "canary bird fish" because of a curious humming noise they make with their swim bladder. There is a courting behaviour and the male guards a nest.

Many hundred photophores (800) occupy linear tracts in neretic *Porichthys*. These photophores can be separated into a cephalic and trunk series. The cephalic series includes the branchiostegal, mandibular, and opercular rows. The trunk series comprises the lateral, dorsal pleural, anal, gastral, ventral and gular rows (Fig. 18.13).

Porichthys photophore has been described ultrastructurally. The 1 to 2 mm diameter photophore is roughly spherical, containing a dermal lens, a layer of photocytes and supportive cells, a guanine parabolic reflector and sheathing layer (Fig. 18.14). The photocytes containing Golgi

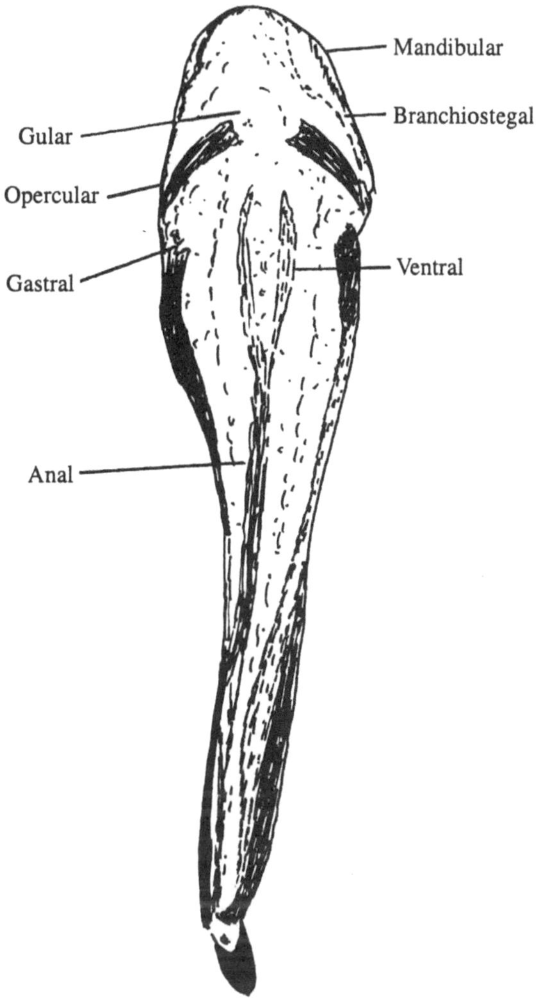

Fig. 18.13 Diagram of *Porichthys myriaster* showing the disposition of the photophores on the ventral side.
***Source:* Baguet (1975)**

regions, mitochondria, ribonucleoprotein particles-have microvillous borders and highly vesicular cytoplasm which communicates with deep infolding of the cell membrane by means of pores. The reflector and sheath are mesodermal while the remainder of the organ is ectodermal. Each photophore has a blood supply and a network of nerve terminals with bead-like varicosities. Nerve terminals do not contact the photocyte and indeed, do not appear to extend beyond a basal lamina which separates the photocytes and an extracellular space or channel from the reflector and sheathing layer. The nerves are probably catecholaminergic fibres from the sympathetic system. There are however isolated desmosomal contact points between supportive and photogenic cells (Fig. 18.15) (read Zaccone et al. 2001).

The very large populations of cytoplasmic vesicles and microvillous borders of the photocytes require special mention. Neither have been observed in luminescent cells of other fish.

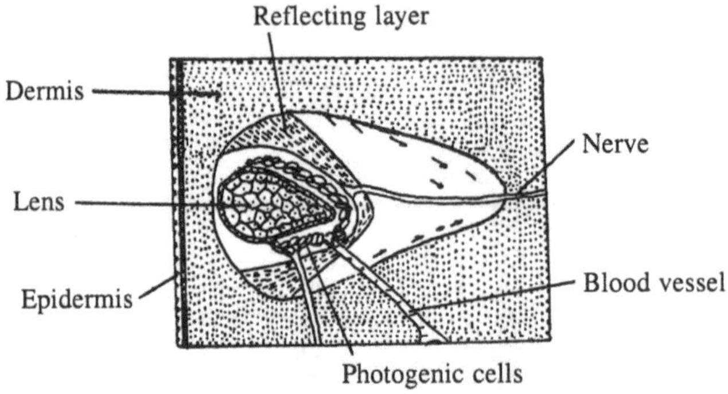

Fig. 18.14 Diagram of a *Porichthys* photophore showing the relative positions of different elements.
***Source:* Herring (1982)**

Perhaps the most probable explanation of their presence in *Porichthys* is that they are the specializations for uptake and storage of luciferin obtained through diet.

5.1.5 Perciformes

Apogonidae (cardinal fishes)

This large family has small nocturnal tropical shallow water fishes. The luminescence in certain apogonids is from endogenous *Cypridina*-type system. These fishes possess glandular light organs associated with gut in three structural arrangements:

(i) In *Apogon ellioti*, there is a small, bright yellow anterior (thoracic) light organ which is a pouch-like anterior protrusion off the second bend of the intestine, and pair of dark bean-shaped posterior (anal) light organs connected to the rectum. The anterior organ, covered by chromatophores, lies in a translucent keel muscle which extends from the isthmus to the pelvic fins. This muscle functions as a light diffuser and there is a collagenous reflector above it. A membranous reflector also lies above, and a lenticular translucent muscle lies below each anal organ;

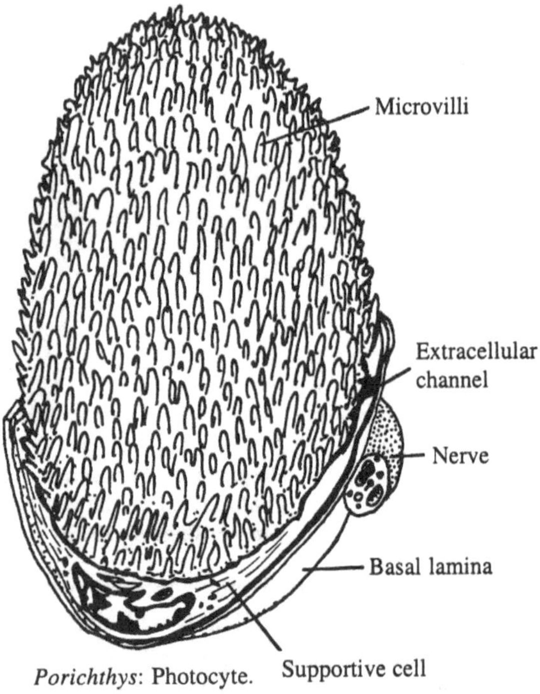

Fig. 18.15 Diagram of a *Porichthys* photocyte with its associated supportive cell and nerve.
***Source:* Herring (1982)**

(ii) In *Archamia* species, the light organ is formed by the second loop of the intestine plus numerous pyloric caeca, both situated in the translucent keel muscle, and

(iii) In *Rhabdamia*, the light organ is formed only at the distal end of a pair of pyloric caeca, one on each side of the body just below the opercular flap. Each caecum abuts on to a small subsurface lens-like structure which is surrounded with black pigment. The light is emitted ventrolaterally from two small spots infront of the pelvic fin (Fig. 18.16).

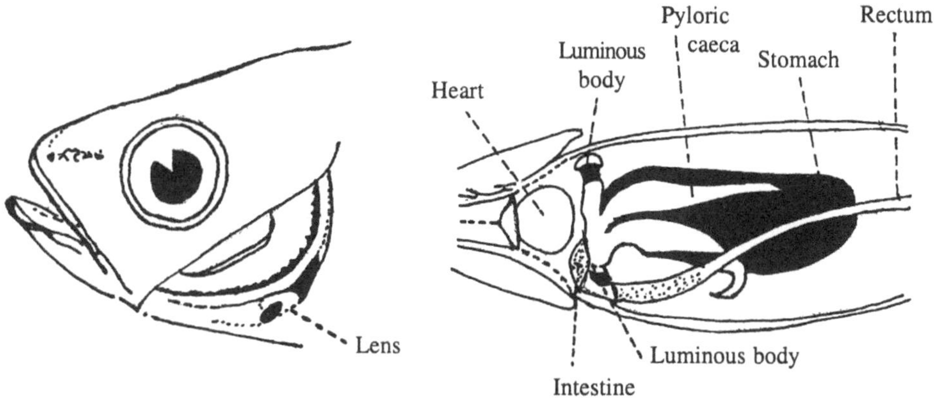

Fig. 18.16 Diagram of the luminescent organ system in *Rhabdmia cypselura* from the left lateral and ventral abdominal aspects.
***Source:* Haneda et al. (1969)**

Pemphridae (sweepers)
Sweepers are small nocturnal fishes which are found in shallow rocky and coral reef habitats of the tropics and sub-tropics. The morphology of luminescent organ system in *Parapriacanthus beryciformes* has a system of indirect illumination. There are two visceral light organs: (i) an anterior (thoracic) "Y" shaped luminous body connected to the most anterior pair (out of 5 pairs) of phyloric caeca and embedded in translucent, ventral thoracic musculature serving as a lens; an opaque white reflector, and chromatophores which function to control the brightness of emitted light, and (ii) a small rod-like mid-ventral, and linear posterior (anal) light organ is attached to the rectum with no structural communication and has a thin ventral lens. It opens outside through a small pore adjacent to the anus (Fig. 18.17).

Parapriacanthus is the only known example where pyloric caeca constitute an integral part of the organ system. The caeca seem to act as luciferin stores. Light is produced by luciferin-luciferase reaction, and extracts cross-react with those of ostracod *Cypridina*. *Cypridina* spp. have occasionally been found in the gut of various of these perciform fish, and it is possible that fish luciferin might be derived from this source.

Table 18.2 is the listing of photophore positions and type of luminescence in some bioluminescent fishes.

6. EVOLUTION OF PHOTOGENIC CONTROL

Most luminous animals, even those with bacterial photophores, are able to modulate their light emission to some extent. The fact in itself is enough to indicate the evolutionary importance of control, and the remarkable variety of devices adopted implies a wide spectrum of selective pressures. The principal adaptive values may be: limiting display time to reduce targetting by predators, conserving energy, and improving communicative efficiency and specificity.

In spite of the manifold functions apparently served by bioluminescence and the diversity of

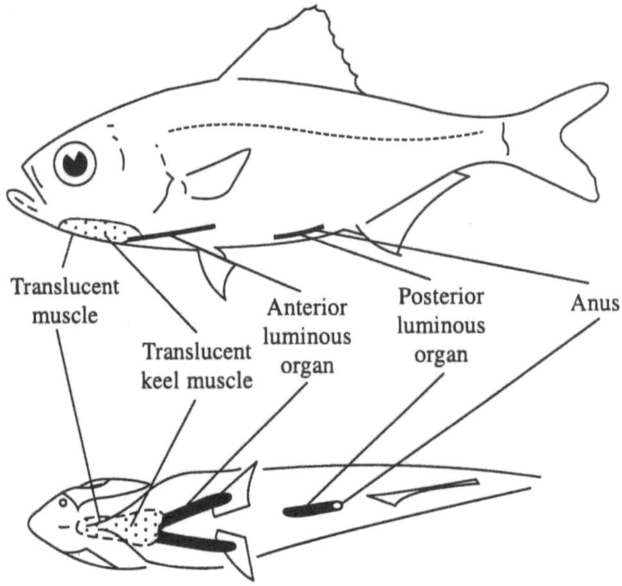

Fig. 18.17 Diagram of the luminescent organ system in *Parapriacanthus beryciformes*. *Source:* Haneda and Johnson (1962)

the emitting structures, it is often possible to recognize or infer major control agencies, neural, hormonal or chromatophoric.

In fishes, they range from the muscular sphincter occluding the gut-associated organs of the leiognathids; through the various shutter or eyelid structures of certain anomalopids and stomiatids; to chromatophore masks of the shark *Etmopterus*, and certain teleost fish. Other interesting example of muscular action opening or closing the clear fenestrae has been reported in *Opisthoproctus*. The light from the rectal bulb escapes though a ventral pigmented area (the "sole"). A hyaline light channel is provided with a system of muscles by which the sole can be lengthened or shortened. When the sole is lengthened, the clear area at the bases of scales become exposed, allowing light to pass through. Thus there is much variety in even these relatively unsophisticated control methods and they are no more uniform in animals with the steadily luminous bacterial photophores than in those with intrinsic luminescence. The orbital organs in anomalopid fishes infact illustrate quite different systems in three closely related genera. In *Anomalops* the whole organ pivots on a hinge and rolls into a pigment lined socket. In *Photoblepharon* a black membranous shade or shutter is pulled up across the photophore. *Kryptophanaron*, which has not been seen alive, appears to have atleast the anlagen of both control devices and is regarded as ancestral to the other two genera. It is puzzling enough that an ancestral form should persist along with its descendents but seems particularly curious that two distinct devices, which can hardly be equally efficient in controlling light emission, have developed and persisted in fishes with very similar habits and habitats.

The different behavioural patterns of *Anomalops* and *Photoblepharon*, with shorter illumination periods of the former, are thought to have been derived from a primitive *Kryptophanaron*-type form adapted to a relatively deep habitat. Invasion of shallower habitats has evolved modifications of the light organ; the shutter mechanism of *Photoblepharon* has allowed the photophore to

Table 18.2 Photophore positions and type of luminescence in some bioluminescent fishes

Taxa	Luminous Tissue or Organ	Position	Type
Chordata			
Sub-Phylum-Craniata (Vertebrata)			
Class- Chondrichthyes			
Subclass-Elasmobranchii			
Order- Squaliformes			
Family- Squalidae			
Isitus	Entire lower trunk, paired and caudal fins	Mainly ventral	Intracellular
Etmopterus	Whole underside and near mouth	Ventral	Intracellular
Class-Osteichthys			
Subclass- Actinopterygii			
Infraclass- Teleostei			
Order- Salmoniformes			
Family- Bathylagidae			
Opisthoproctus	Rectal bulb reflecting ventrally	Ventral	Bacterial
Family- Gonostomatidae			
Maurolicus,	Body organs	Ventral	Intracellular
Gonostoma.			
Cyclothone			
Family- Sternoptychidae			
Argyropelecus	Body organs	Ventral	Intracellular
Family- Stomiatidae			
Stomias	Body and head organs	Ventral	Intracellular
Family- Astronesthidae			
Astronesthes	Body, head organs, barbel, finrays	Mainly ventral	Intracellular
Order- Myctophiformes			
Family- Myctophidae			
(many genera)	Body photophores	Ventral or mainly ventral	Intracellular
	Caudal organs	Dorsal and/or ventral	
Order- Batrachoidiformes			
Family- Batrachoididae			
Porichthys	Rows of photophores on body and head (larger and more numerous ventrally)	Mainly ventral	Intracellular
Order- Lophiiformes			
Family-Melanocetidae			
Melanocetus	Esca in females	Dorsal	Bacterial
Family- Oneirodidae			
Oneirodes	Esca in females	Dorsal	Bacterial

Taxa	Luminous Tissue or Organ	Position	Type
Family-Linophrynidac			
Linophryne	Esca and hyoid barbel organs in females	Dorsal Ventral	Bacterial
Order- Gadiformes			
Family- Moridae			
Physiculus	Anal gland	Ventral	Bacterial
Family- Macrouridae			
Malacocephalus	Anal gland	Ventral	Bacterial
Order- Beryciformes			
Family- Monocentridae			
Monocentris	Submandibular organ	Ventral	Bacterial
Family- Anomalopidae			
Anomalops	Suborbital organ	Midlateral	Bacterial
Photoblepharon			
Order- Perciformes			
Family- Apogonidae			
Apogon	Anal and thoracic glands, intestinal diverticula, translucent thorax	Ventral	Intracellular
Family- Acropomatidae			
Acropoma	Thoracic gland, long duct ending near anus, ventral body wall translucent	Ventral	Bacterial
Family- Leiognathidae			
Leiognathus	Circumoesophageal gland at stomach, pectoral and thoracic region translucent	Ventral	Bacterial
Family- Pempheridae			
Parapriacanthus	Foregut, preanal duct, thoracic and abdominal organs, translucent thorax	Ventral	Intracellular

Source: McAllister (1967)

enlarge and to be occluded in short rapid blinks. Its habitat close to coral substratum renders it difficult for a nocturnal predator to attack despite its bright illumination. *Anomalops*, on the other hand, having become more of a pelagic planktivore, is more vulnerable to predation but rapid continuous blinking of individuals in large shoals makes it harder for a predator to focus upon a single individual.

Among photophores and photocytes with emission control rather than masking there are often such profound structural differences as to leave little doubt that the photophores have evolved independently. It has been indicated that in the evolution of bioluminescence the effector and control aspects may develop independently. Emitting structures, biochemical systems and subcellular reaction frameworks often appear to illustrate *de novo* origin and adaptive radiation, whereas emission is often controlled by mechanisms, such as excitable cell membranes and neurons, which are of great antiquity and are wide-spread in the animal kingdom.

Although bioluminescent elasmobranchs are known, their luminescent systems have not been studied physiologically or biochemically. Among marine teleosts there are many luminescent forms including a few with relatively well-studied neural control mechanisms. The most extensively studied of these is the batrachoid *Porichthys*. Other fish with probable neural control of luminescence include hatchet-fish and stomiatoids. The head light organs receive innervation by branches of the facial nerve. The trunk photophores in various fish are innervated by the branches of spinal nerves. In *Porichthys* and other teleosts the serial photophores are innervated by the sympathetic nervous system (Fig. 18.18). The respective roles of the sympathetic and spinal contribution to the neural control of caudal organ luminescence of lantern fishes are yet unclear. Adrenergic neurotransmission as well as electronic spread of excitation may be involved in functioning of caudal luminous organs.

Photophores of *Porichthys* can be isolated and remain functional for a few hours when immersed in saline. Their luminescence can be triggered by either electrical or chemical stimulation. The external application of adrenaline or noradrenaline on isolated photophores leads to a biphasic light emission. Recent data suggest that purine compounds might exert a very important control on the luminescence of photogenic cells in *Porichthys* photophore. It is concluded that P2-purinergic receptors might be present on photocytes and exert a negative control on the luminescent system (see details: Christophe and Baguet 1983, 1985; Rees et al. 1991). Mallefet and Baguet (1993) suggested that the oxygen consumed during the fast phase of adrenaline luminescence could represent the activity of an extra mitochondrial oxidative pathway involved in the light reaction. The reader is also referred to Rees and Baguet (1988) and Mallefet and Baguet (1989).

Mallefet and Baguet (1991) discussed the hypothesis of a hypometabolic state of the isolated photophores of *Argyropelecus hemigymnus* during light emission. The mechanism of light production by isolated photophores of the epipelagic fish *Porichthys* and the mesopelagic fish *Argyropelecus* are different (Mallefet and Baguet 1985).

While superficially similar to the photophores of *Porichthys*, in myctophids containing a central photocyte mass backed by a reflector and capped by a lens, in this instance a modified scale, the organization of myctophid photocyte and its innervation is radically different from that in *Porichthys*. In contrast with the irregular array of ovoid photocytes with sparse innervation found in *Porichthys*, the myctophid photocyte is flattened and occurs in stacks interspersed with extensive ramifications of nerve terminals. The large light organs of caudal peduncle are highly similar in organization.

Data concerning the control of myctophid luminescence are still very limited. The presence in the photophores of both myelinated and unmyelinated nerve fibres, probably of spinal and sympathetic origin neural control is more finely tuned and complex than the simple excitatory pathway. The question of neurotransmitters or electronic transmission (or a combination of two) remains unsolved.

7. FUNCTIONS OF LUMINESCENCE

Many functions have been ascribed to luminescence and the more obvious come under intraspecific

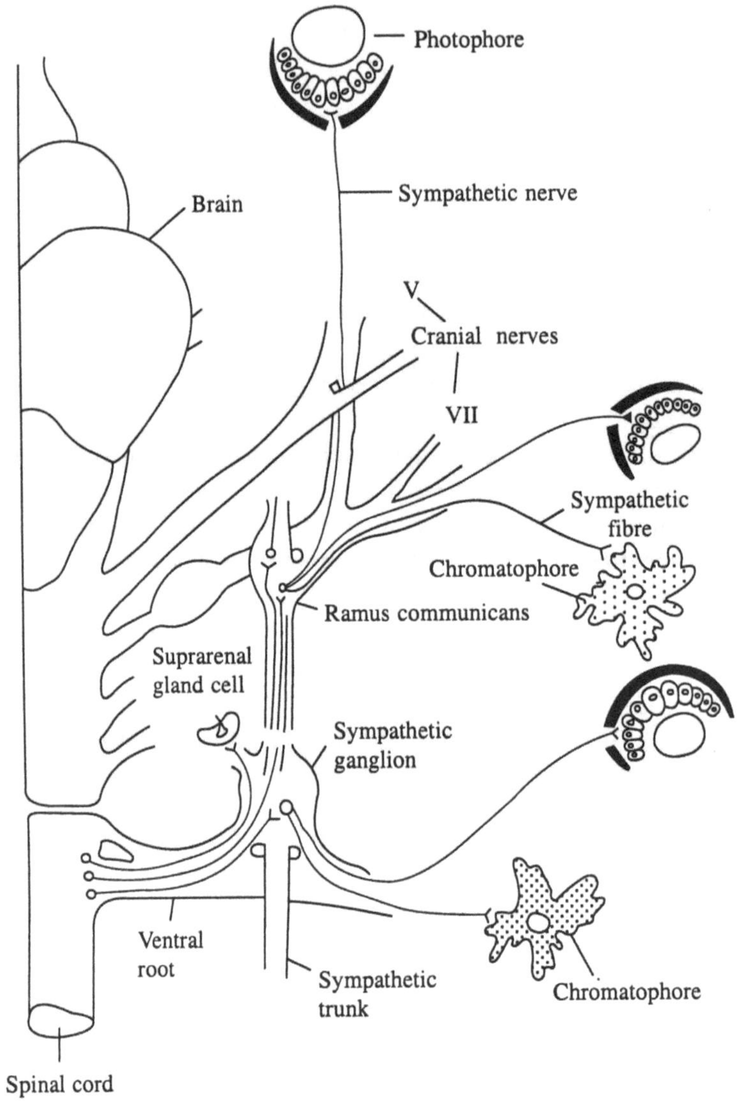

Fig. 18.18 Diagrammatic representation of the sympathetic outflow of a teleost fish including suggested innervation of the integumental photophore and chromatophores.
***Sources:* Nicol (1967) and Herring (1982)**

and interspecific heaings. Herring (1990) gave an overview of bioluminescent communication in the sea. Article by Widder (1999) provided useful information.

7.1 Intraspecific Communication

(i) Courtship: Marine photic courtship has been reported in *Porichthys notatus*.

(ii) Recognition of species and sex: The reasonable use of photophores among myctophids for species recognition has been suggested. Stomiatoid species differ in their photophore arrangement and the luminescence may function for recognition as well as illumination. The luminescent barbels of melanostomiatids vary greatly from species to species. This may be useful in recognition or merely an indication that each melanostomiatid is adapted to attracting a different sort of prey with its lures. The function of esca in anglerfishes is generally believed to be a means of a species recognition. In Elasmobranchii, since the pattern of photophores is different in each species, luminescence may function as a recognition device.

Where there is sexual dimorphism or unisexual distribution of photophores in various fishes, there is reason to suspect a sexual function of luminescence, but has not yet been observed directly in dimorphic animals. A bioluminescent signalling function by males has been suggested for sexual dimorphism in leiognathids, which may play an important role in the schooling behaviour as well as in species and sexual recognition of these coastal fishes.

The shallow, reef-dwelling anomalopids *Photoblepharon palpebratus* and *Anomalops katoptron* exhibit sexual differences in blinking that appear to function in pair formation and territorial defense.

(iii) Congregation: Mass congregations have long been recognized as among the most spectacular manifestations of bioluminescence. They are doubtless rivalled by the striking non-synchronized schools and shoals of various bathypelagic luminous fishes. It has generally been assumed that darkness and practically infinite extent of deep sea water exacerbate not only the uncertainties of finding individual mates but of social assembly for the benefits commonly ascribed to schooling (mutual promotion of feeding, defence, territoriality, mating, etc.). Consequently, it is widely presumed, though without direct evidence, that such functions are aided by intraspecific communication. The suggestion that the schooling of *Photoblepharon* and *Anomalops* is mediated by the blinking of their subocular photophores is reasonable but does not exclude mechanisms that serve in non-luminous species, such as lateral line system Luminescence has been supposed to enable various bathypelagic fish to stay together during their diurnal vertical migrations. Bioluminescence has also been proposed as aiding territorial spacing and defence. The assumption that species recognition for either courtship or congregation could be effected via the characteristic spatial patterns of body photophores has been discussed for myctophid fishes with the conclusion that further modulation of colour, light intensity, or number of photophores active, would be necessary. Pattern recognition should be well within the capabilities of fishes has been suggested.

It has been hypothesized, however, that luminescence may be used by fish to attract potential mates from a distance, so increasing the otherwise small chances of reproductive encounter in the deep sea. Male deep sea anglerfishes do not possess an illicium, and it is possible that the luminous esca of the female is just as attractive to males as it is to the female's prey. Males of the stomiatoid *Idiacanthus* are much smaller than the females, but have well developed cheek light organs; those of the females are minute. Females may make use of flashes from these male organs to track down the relatively feebly swimming males.

7.2 Interspecific Functions

(i) Food-getting: Circumstantial indications of photic predation are offered by the females of deep sea anglerfishes (Ceratioidei) in which the angling apparatus, the luminous lure, dangles or waves near the mouth. Use of light to attract prey seems probable also in fish where there are photophores within the mouth (e.g., stomatoid *Chauliodus*). Photic predation has been suggested for the ventral luminescence of the shark *Etmopterus*. Luminescence is used for feeding purpose by *Anomalops*. Photophores of *Kryptophanaron alfredi* are used to search out prey on the reef face and that the animals normally spend the day in deep water.

The deep water, slow moving gapers (Saccopharyngiformes) may use the luminous caudal organ at the end of their whip-tail as a lure. Most fish of this sort are poor swimmers. The advantages of luring prey that it conserves energy and does not require the development of elaborate distant sensory organs such as eyes or lateral line system. This is especially advantageous in deep sea where prey are few and far between. The chief groups with luminescent lures (ceratoids and gapers) are found deeper (at 1500–3000 m) than other luminous fish, which are most typical of depth between 100 and 500 m. *Chauliodus* also ranges deep. The chief difficulty with regard to such a function for luminescent lures is that they are usually rather dim; in case of *Leptostomias* (Melanostomiatidae), a maximum visible distance of 5 m has been suggested.

(ii) Defence:

(a) Camouflage: Faint but highly directional daylight in midwaters of the open ocean silhouettes an opaque animal which can be visible, therefore, to predators below. Many animals, however, can use their own bioluminescence to eliminate their silhouettes. The idea that photophores are used for counterillumination (bioluminescent countershading) has been suggested by a number of biologists.

The size and distribution of photophores are closely related with the depth and quality of ambient light. This concealment strategy is limited to depths greater than 350–400 m, largely because of high visual acuity of predators and high cost of producing countershading at lesser depths. There is also a changeover in midwater fishes with increasing depth from predominantly reflective species with larger ventral photophores to predominantly non-reflective species with small or no ventral photophores at a depth of 650–700 m. Perhaps silvery fishes rely totally on reflected light for lateral countershading while black fishes (at depth where lateral countershading is necessary) rely to varying degree on bioluminescence. While the latter strategy is energetically more expensive, it has the advantage of simultaneously camouflaging the animal against downwelling light and the flashes or searchlights of nearby animals.

Angular distribution of bioluminescence produced by midwater fishes *Argyropelecus* and *Chauliodus* closely matched the radiance field of daylight in the midwaters. The colour filters in the photophores of *Argyropelecus* and *Sternoptyx* passed only blue light at about 480 nm. This wave length is close to transmission peak of sunlight in oceanic water.

The sub-orbital organs in fishes are not generally countershading devices, the array on the lower jaw and isthmus serving this purpose.

Dangers have been pointed out to a predator of its prey's luminescing while in the gut and the heavily pigmented guts and peritoneal linings found in deep sea fish are means of preventing light from passing out of the body. Presumably the black pigmentation of the skin in some fish is insufficient by itself to prevent escape of internal luminescence. It has been hypothesized that limits on concealment from predation through counterillumination determine the upper limit of mesopelagic fauna during the day. At night near full moon, animals may be found at light levels higher than those at which counterillumination seems to be an effective strategy.

There is evidence that fish eyes are amply sensitive to discriminate backlit from frontlit surfaces at depth, and that the spectral emission of photophores matches the dim blue rays of residual sunlight at depth and that the photophore intensity is in the right range to match the ambient light at particular layer depths.

To counterilluminate properly, an animal must be able to determine the intensity of the downwelling light and must have some means to ensure that its photophores respond appropriately. It has been suggested that eyes of myctophid fishes detect downwelling light. Midwater fishes, however, have a prominent extraocular photoreceptor, the pineal organ, often lying beneath an unpigmented cutaneous window, which could be important in detecting downwelling light. In various midwater fishes, it has been suggested that photophores directed into the eyes feedback information to regulate luminescence. The pineal organ in myctophid fishes lies on the dorsal midline of the head, between the eyes. The ventral surface of *Myctophum spinosum* is covered with many large but rather broadly spaced photophores. The skin above the pineal organ lacks chromatophores, allowing easy location of the organ and accurate positioning of the shield. The organ lies over a heavily pigmented epithelium which provides a light shield form below. Apparently, both the eyes and extraocular photoreceptors are involved in detecting downwelling light for counterillumination. Indeed, the evidence in support of the theory of bioluminescent countershading in the midwaters of the open ocean is now substantial. The only major evidence lacking is the experimental demonstration of reduced predation on countershading animals and field observations of the phenomenon. The apparent importance of bioluminescent countershading cannot be overestimated. It has been suggested that the upper limit of midwater fauna is a result, to a large degree, of the severe limitations placed on bioluminescent countershading at this level. We envision a change over in the macrofauna at this depth from a deeper component in which opaque animals are able to hide, to an upper component in which opaque animals must rely more on speed, size, weapons, etc., than on hiding to avoid predation.

(b) The photophore pattern of *Porichthys porosissimus* "mimics" the appearance of a ctenophore *Mnemiopsis* and that supposedly luminous pelvic organ of bathypelagic *Kasidoron* (Family Kasidorioidae) mimics a colonial siphonophore. These supposed resemblances are just to mislead the predators (concealment).

(c) Warning signals; confusion: The luminescence of *P. porosissimus* serves to warn potential predators of its toxic opercular spine but the experimental evidence is flawed. The luminous escal response in *Linophyrne* on manipulation indicates that the esca may serve a function in confusing or dazzling a potential aggressor.

(d) Diversion of predator: The caudal light organs (sternchasers) of myctophids serve as devices to deflect predators, and in sexual dimorphic species serve for sexual recognition and courtship. In the myctophid *Tarletobenia crenulans*, only the male possesses caudal organs and there is evidence to support the hypothesis that the females evade predation by passivity and reliance on darkness whilst males flash and swim off and in doing so they divert the predator from the reproductively more valuable females, while exposing themselves to greater risk. It is presumed that a single male can fertilize a number of females.

(iii) Illumination: Illumintation is usually an implicit factor in the adaptive value of luminescence. A novel predatory use has been suggested for certain red photophores of a few species of mesopelagic fish that are also capable of seeing red light. Since many midwater fish should be colour-blind judging from their all rod retinas, an animal that can both emit and see red light might have a tremendous "snooperscope" advantge in being able to illuminate prey organisms by light invisible to them (providing also the emitter simultaneously turned off its normal blue-emitting photophores).

(iv) Self stimulation: It has been suggested that orbital photophores produce a relatively feeble glow, thus adapting the eye to light before the main photophores are turned on, which should prevent a fish using a bright flash to startle a predator before being itself dazzled. It is also possible that orbital photophores provide an intensity against which to judge ambient light levels.

(v) Ventral luminescence has also been thought to provide habitat illumination for benthal and abyssal fish.

So much of the information on luminescence in fish is still purely anatomical, and so bewildering in its variety, that functional interpretations almost certainly err on the side of conservatism, an inevitable consequence of our restricted concepts of the normal lives of these remarkable animals. Any purpose that is fulfilled by colour or pattern in the illuminated terrestrial or coastal environment can also be achieved by luminescence in the dark of the deep sea.

8. ORIGIN AND EVOLUTION OF BACTERIAL PHOTOPHORES

The origins of bacterial organs not associated with gut, such as the lure photophore of anglerfishes, the jaw organ of *Monocentris*, and the orbital organs of anomalopid fishes, are still obscure. Consideration of the detailed relation of the light producing organs to the alimentary canal indicates that such systems of indirect illumination must have had a multiphyletic origin. Another puzzling question why bacterial photophores exist only in squids and fishes. Still another is why bacterial photophores almost never occur together with intrinsic luminescence in the same organism. The anglerfish *Linophryne* is an exception to this rule in that it has intrinsic photophores in addition to its bacterial escal light.

The evolution of bacterial light organs appears to have occurred relatively often in shallow water coastal fishes whereas in oceanic fishes intrinsic light organs predominate. It may be that this is related in some way to the relative abundance of luminous bacteria in coastal waters, particularly in the tropics and subtropics.

Much work is needed before the significance and genealogy of bacterial photophores is understood.

9. EVOLUTIONARY ESTABLISHMENT OF BIOLUMINESCENCE- GENERAL

All particulars of present day light emission-chemilluminescence, photophore structure, control mechanisms and behavioural uses are presumably the finely-tuned products of untold millenia of random mutation and intense natural selection. It has been assumed that bioluminescence played a dominant role in the colonization of deep sea habitat. Photogeny probably is a key factor in the success of many present-day bathyal species but it is not so clear that it is a prerequisite for invasion. Two directions of evolution can be predicted because of the likely advantages, on the one hand, of increasing the reach of light, and, on the other, of limiting emission time to reduce predation, conserve energy and increase communicative specificity. Elaborations of photophores and multiple uses of light emission can also be viewed as secondary consequences of such multifactorial selection and of competitive and predatory pressures from co-evolving species.

10. CONCLUSIONS

Bioluminescence is ubiquitous in the oceans, and must have considerable ecological and behavioural significance. It seems clear that the more complex bioluminescence of fish is just one facet of the behaviour of pelagic organisms which is related to the marine photic environment. Luminescent ability seems closely interrelated both with what little is known about the eyes and visual mechanisms of deep sea fish, and with pigmentation, transparency and reflectivity. The greatest development of luminescent organs and the greatest proportion of luminescent forms are found in mesopelagic regions, i.e., those depths at which ambient light is dim but not imperceptible.

There are a variety of light producing mechanisms. Fishes have photophores with (i) highly evolved reflectors and lenses and/or (ii) mechanisms for producing diffuse general luminescence, usually illuminating the ventral surface. It appears usually to serve in predator evasion. Diffuse illumination of the ventral surface is most common in mesopelagic teleosts with internal luminescent organs.

Diverse systems point to diverse selective pressures resulting in the multiple independent origin of luminescent systems. As a result of evolution the mechanisms have converged to serve a variety of functions. Most of the hypotheses necessarily still concern capabilities and physiological mechanisms rather than functions.

Almost nothing is known of the luminescent behaviour of most fish. What is known has been either deducted from the morphology of a particular species or from studies of individuals under atypical environmental conditions; too often, observations have been made on dying or moribund specimens. Better techniques are needed for maintaining bathypelatic species in the laboratory; observations must be made *in situ*. The behavioural importance of coloration of terrestrial organisms is well recognized and luminescence seems to play just as important a role in the marine environment.

CHAPTER **19**

Lymphomyeloid (Lymphohaemopoietic) System; Immune Defence Mechanisms; Fish Pathology; Teratology; Regeneration; Wound Healing

1. INTRODUCTION

The vertebrate lymphomyeloid system produces and stores lymphocytes and other blood cells. This system or complex is most highly developed in mammals. It consists of two categories of tissues, lymphoid and myeloid. Lymphoid tissues contain predominantly lymphocytes and form the structural basis of the immune system. Myeloid tissues are characterised by granulocytes and other blood cells. In mammals lymphoid tissues (thymus, lymph nodes, tonsils, solitary follicles, Peyer's plaques, spleen) and myeloid tissue (bone marrow) are anatomically separate, but exchange of free cells takes place between the tissues. Mammalian bone marrow, although predominantly myeloid, contains some lymphocytes. Thus in higher vertebrates, primary and secondary lymphoid organs are morpho-physiologically well defined.

Adaptive immunity has been reported in fish in connection with the capacity of vertebrates to show immune response products of lymphocytes and lymphoid organs. Despite pioneer works in comparative immunology and modern text-books of Comparative Immunology, studies of lymphoid organs of lower vertebrates still remain scarce.

In fishes the lymphomyeloid tissues are considered to be less specialized than in higher vertebrates and consists of intermingled lymphoid and myeloid elements. A clear separation is not possible because of the lymphohaemopoietic condition of their lymphoid tissues. Instead of bone marrow and lymph nodes, fishes possess other types of lymphohaemopoietic tissues. New blood cells originate within these tissues and to some extent in the circulatory blood. For phylogenetic analogies between fish head-kidney and mammalian bone marrow, see Meseguer et al. (1995).

Haemopoiesis relies on multipotent stem cells, supposedly lymphocyte-like, which are transported by the blood. Stem cells from the blood settle at different sites in the organism and start to proliferate, forming lymphohaemopoietic tissues or organs. Probably under the influence of microenvironmental factors (carbon dioxide, oxygen, pH, hormones), they transform into the blast cells which divide and differentiate into mature blood cells: erythrocytes, lymphocytes, granulocytes and so on.

Lymphohaemopoietic tissues, such as the epigonal and Leydig organs of elasmobranch fishes, are characterized by a continuous flux of cells. They may be considered as aggregations of dividing blood cells, cell cultures, rather than well defined organs. However, because the dividing blood cells have a predilection for certain sites in organism, the lymphohaemopoietic tissues may appear as structurally rather well defined organs. The localization of the sites for lymphohaemopoietic activity probably is phylogenetically anchored.

The lymphohaemopoietic tissues usually contain large numbers of lymphocytes and may well serve functions similar to those of mammalian lymph nodes. Plasma cells, regarded as the main source of immunoglobulins, occur in lymphomyeloid tissues of many groups of fishes.

Although there is still considerable debate about the lymphatic drainage in fish, fish do not require a subsidiary system of true lymphatic vessels to drain fluid from interstial spaces. Therefore secondary lymphoid organs (lymph nodes) associated with the lymphatic system of mammal's are absent in fish.

2. LYMPHOMYELOID TISSUES IN VARIOUS GROUPS OF FISHES

Information on fish lymphomyeloid tissues can be obtained from contributions of Davina et al. (1980), Zapata (1980, 1981), Zapata et al. (1984, 1996a, b), Fänge (1982, 1985, 1986a, b, 1987, 1992), Mattisson and Fänge (1986), Chiba et al. (1988), Mattisson et al. (1990) Tatner (1996), Bernstein et al. (1998), and Manning (1998). The lymphomyeloid tissues in different fishes are given in Fig. 19.1. The Table 19.1(A, B) gives the distribution of lymphohaemopoietic tissues and bone marrow equivalents in various groups of fishes. In total, the lymphomyeloid tissues of fishes compose 0.5–1.5% of the body weight or about half the percentual weight of analogous tissues in mammals.

Agnatha: The intestinal submucosa, the pronephros and the peripheral blood are the main sites for production of new blood cells in myxinoids (hagfishes). Lymphomyeloid cell islands or tissue, mainly granulopoietic, dispersed in the intestinal submucosa, is often regarded as a primitive spleen but probably more resembles bone marrow.

In petromyzonid (lampreys) the main locus for haemopoiesis moves from the intestine (typhlosole, spiral valve) to the supraneural organ. The typhlosole is regarded as a primitive spleen. The supraneural myeloid organ is a yellowish-red column dorsal to the spinal cord. It contains granulocytes, immature red cells and macrophages and serves as a primitive bone marrow.

Neither myxinoids nor petromyzonids have a thymus, but this probably occurs in all other fishes. Diffuse accumulations of lymphocytes in the gill region of larval lampreys are supposed to represent a "prothymus".

Chondrichthyes: Chondrichthyes, including Holocephali and Elasmobranchii, may be the most primitive vertebrates containing histologically distinguishable lymphoid organs, mainly thymus and spleen. Furthermore, especially interesting is the presence in both subclasses of lymphohaemopoietic tissues in areas unlike other vertebrates. In elasmobranchs, the Leydig's and epigonal organs are the principal lymphohaemopoietic tissues. The holocephalans (*Chimaera*,

Table 19.1 (A) Distribution of lymphohaemopoietic tissues in various groups of fishes

Fish group	Thymus	Oesophagus (Leydig organ)	Intestine (GALT)	Spleen	Heart	Meninges/ neural arch	Skeleton head and shoulder	Associated urogenital Kidney (KALT)	With system gonad associated epigonal organ
Cyclostomes									
Lamprey*	0	0	(+)	0	0	+	0	(+)	0
Hagfish	0	0	+	0	0	0	0	(+)	0
Holocephalans	+	0	(+)	+	0	0	+	0	0
Elasmobranchs	+	+/0	(+)	+	0	(+)	0	0 (+)	+/0
Coelacanth	+	0	(+)	+	0	0	0	+	0
Dipnoans	+	0	+	+	0	0	0	+	0
Brachiopterygians	+	0	(+)	+	0	+	0	?	0
Chondrosteans	+	0	(+)	+	+ (sturgeons)	+	0	+	0
Holosteans	+	0	(+)	+	0	+	0	+	0
Teleosts	+	(+)/0	(+)	+	0	0	0	+	0

+ = present; (+) = present in small amounts, or diffusely distributed; *Note Table may need revision*
0 = not present. GALT (Gut-associated lymphoid tissue). KALT (Kidney-associated lymphoid tissue)
* = During metamorphosis of the lamprey larvae the major sites of haemopoiesis change from the kidney and intestinal typhlosole to the supraneural organ (fat column). Lymphomyeloid tissue of intestine of cyclostomes is often regarded as representing a primitive spleen.
Source: Fänge (1985) and other related articles.

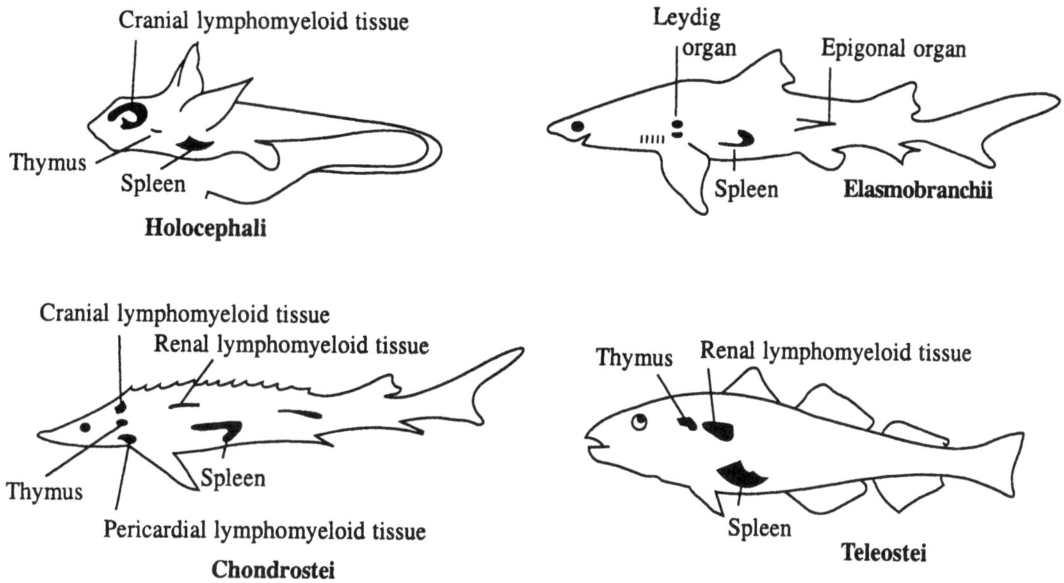

Fig. 19.1 Lymphomyeloid tissues in different types of fishes.
***Source:* Fänge (1982a)**

Table 19.1 (B) Bone marrow equivalents in gnathostomate fish

Chondrichthyes	
Holocephali	Orbit and cranium
Elasmobranchii	Epigonal organ, Leydig's organ, meninges
Osteichthyes	
Actinopterygii	
Chondrostei	Cranium, heart, kidney, gonads
Holostei	Cranium, heart, kidney
Teleostei	Kidney
Brachiopterygii	
Polypteriformes	Olfactory sac, meninges, kidney
Sarcopterygii	
Dipnoi	Spiral valve, kidney, gonads
Crossopterygii	To be studied

Source: Zapata et al. (1996a)

Hydrolagus) contain a granulopoietic and lymphopoietic tissue in the depressions and cavities of the cartilaginous skeleton. Recently, lymphohaemopoietic masses have been found in the "meninx primitiva" of the stingray *Dasyatis akajei*. Lymphohaemopoietic tissue may also occur in the kidney, e.g., in the frilled shark, *Chlamydoselachus anquineus* and in the spring dogfish, *Squalus acanthias*.

The main lymphohaemopoietic masses of holocephalans are found in the cranium [in the orbit, in the preorbital (ethmoid canal) and suprapalatally], and in the coracoid area. They were histologically identified as lymphoid and myeloid tissues. Two granulocytic lines, one related to heterophils and the other to eosinophils have been identified in the ultrastructural study of the cranial lymphohaemopoietic tissue in *Chimaera monstrosa*. No basophils were found. The available information is still confusing and contradictory in Holocephali and Elasmobranchii. The presence of macrophages, lymphocytes and developing and mature plasma cells suggests an important role in eliciting immune responses. A phylogenetic relationship between this tissue and the higher vertebrate bone marrow is proposed on the basis of histological similarities between the cell micro-environments governing haemopoietic differentiation in these organs.

Furthermore, reciprocal relationship could exist among various lymphohaemopoietic organs of primitive fish as proposed in elasmobranchs. Chondrichthyes species with no or poorly developed Leydig's organ possess very voluminous epigonal tissue (*Ginglymostoma cirratum*) but those with a large Leydig's organ (*Squatina squatina*) often lack the epigonal tissue. No other blood-forming loci, except spleen and cranial lymphohaemopoietic tissue exist in chimaeroids. Neither Leydig's and epigonal organs nor lymphohaemopoietic tissue associated with urogenital system occur in holocephalans.

In elasmobranchs the lymphomyeloid tissues constitute 0.4–2.3% of body weight, in a holocephalan (*Chimaera montrosa*) about 1.7% of the body weight but in mammals the lymphomyeloid system weights 2.3–3.0% of the body weight. Higher efficiency of the blood gas transport and the immune functions necessitate a large volume of blood forming and lymphocytes producing tissues.

Fänge (1987) and Zapata et al. (1996b) focused on the lymphomyeloid system in elasmobranchs. *Thymus:* The thymus of elasmobranch is lobated and clearly divided into cortex and medulla. The thymus of the dogfish (*Scyliorhinus canicula*) shows large numbers of different sized lymphocytes within an epithelial reticular network. Lymphoblasts, and macrophages occur, but no Hassall's corpuscles, myoid cells, mucous cells or plasma cells. In some species the thymus involutes with age. It is not known if distinct T and B lymphocytes do exists in elasmobranchs.

The structure of the elasmobranch thymus (as reported in *Raja clavata* and *Torpedo marmorata*) is similar to that of higher vertebrates except for the absence of Hassall's corpuscles.

The kidney and spleen are the main secondary lymphoid organs of teleost fish.

Spleen: The spleen in Pisces is present within the dorsal mesentery in the form of an independent, circumscribed organ outside of the intestinal tract. Among elasmobranchs it is multilobulated or bean-shaped and lies near the caudal part of stomach. It is very well developed, particularly with respect to lymphoid tissue. The relative weight of the spleen, expressed as percentage of body weight, is quite variable. It is lobated and shows ellipsoids which open into red pulp, periarterial lymphoid sheaths (PALS) forming white pulp, and plasma cells responsible for production of immunoglobulins. Trapping of antibody-antigen complex may be one function of ellipsoids. Ellipsoids are not phagocytic but translocate material at the cell surface by a 'capping' mechanism. For specific attention on fish spleen, the reader is referred to Fänge and Nilsson (1985; see Fig. 19.2), Quesada et al. (1990) and Zapata et al. (1996b).

Apart from their particular locations, Leydig and epigonal organs (ovary and testis) are histologically similar.

Leydig organ: The lymphogranulopoietic tissue in the oesophagus is divided into a series of anastomising lobes which consist of a reticulum packed with leucocytes. Blast cells, and plasma cells are reported. Lymphocytes occasionally form follicle-like aggregations.

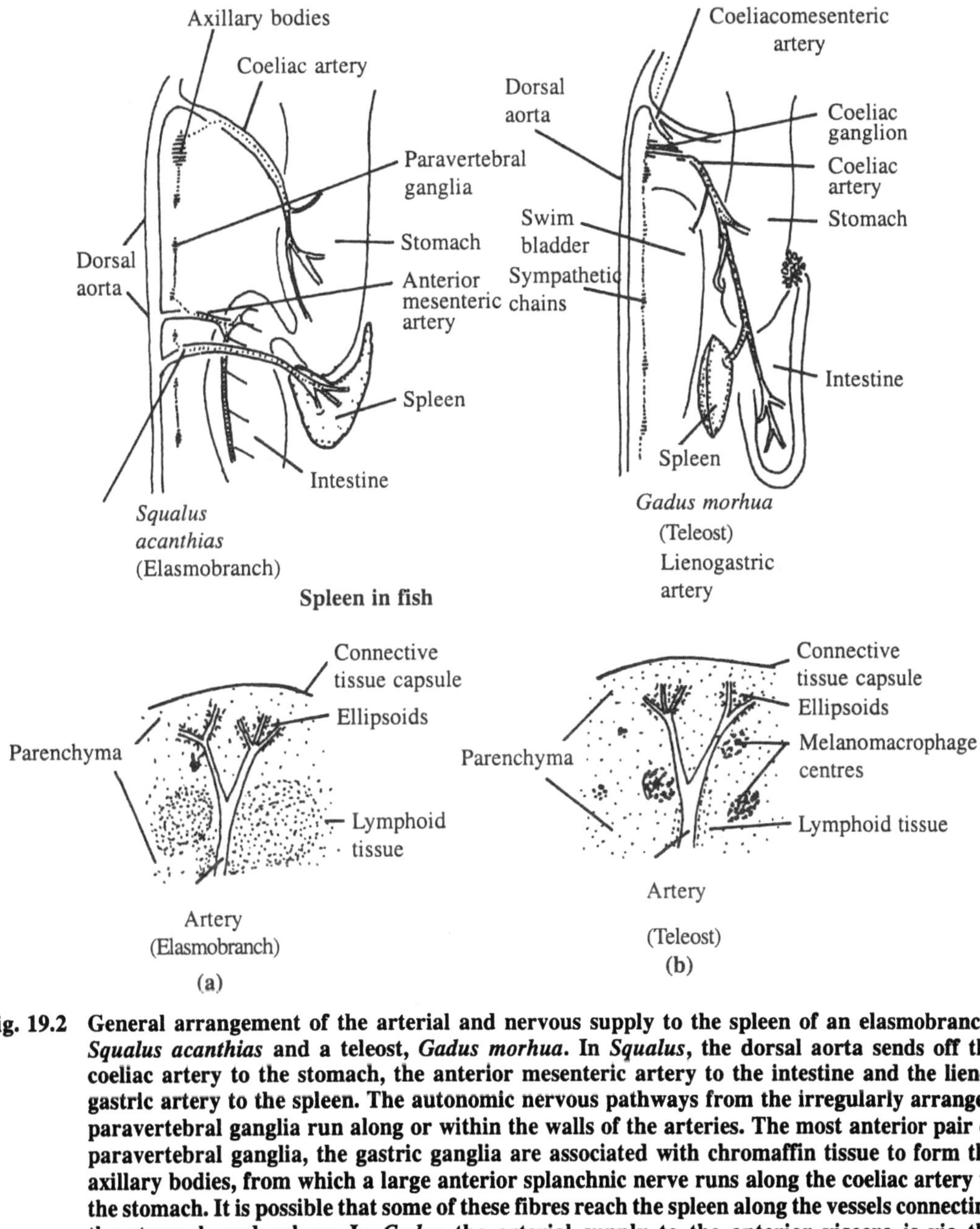

Fig. 19.2 General arrangement of the arterial and nervous supply to the spleen of an elasmobranch, *Squalus acanthias* and a teleost, *Gadus morhua*. In *Squalus*, the dorsal aorta sends off the coeliac artery to the stomach, the anterior mesenteric artery to the intestine and the lienogastric artery to the spleen. The autonomic nervous pathways from the irregularly arranged paravertebral ganglia run along or within the walls of the arteries. The most anterior pair of paravertebral ganglia, the gastric ganglia are associated with chromaffin tissue to form the axillary bodies, from which a large anterior splanchnic nerve runs along the coeliac artery to the stomach. It is possible that some of these fibres reach the spleen along the vessels connecting the stomach and spleen. In *Gadus* the arterial supply to the anterior viscera is via the coeliacomesenteric artery, which branches soon after the exit from the dorsal aorta. Branches reach the swim bladder, stomach, spleen and intestine. The autonomic nervous pathways are collected in the anterior splanchnic nerve, which runs along the arteries from the coeliac ganglion. The paravertebral ganglia of teleost are longitudinally connected to sympathetic chains of the type found also in the tetrapods.
Diagram of the microscopic structure of fish spleens. (a) Elasmobranch type. Lymphoid tissue occurs around blood vessels and lymphocytes also are found elsewhere in the parenchyma. Melanomacrophages are distributed within the whole tissue. (b) Teleost type, Lymphocytes are widely scattered and show only a slight tendency to accumulate around blood vessels and melanomacrophage centres. Germinal centres are not found in fish splenic tissue, or in other lymphoid tissues in fishes (veins and sinuses are not shown in the drawings).
***Source:* Fänge and Nilsson (1985)**

Leydig organ contains similar type of granulocytes as the circulating blood but the cells in Leydig organ generally are less mature than in the blood.

Epigonal organ: The lymphogranulopoietic organ is a paired structure, usually of whitish light colour, may have a dark colour where considerable erythropoiesis is taking place. The histological structure is similar to that of the Leydig organ.

Gut-associated and kidney-associated lymphoid tissue also occurs in elasmobranchs.

Osteichthyes: In *Lepidosiren* (Dipnoi) the thymus persists in the adult. The spleen forms a compact brown organ within the wall of stomach and the anterior part of the intestine. It is erythropoietic. Caudally to the spleen a granulopoietic tissue is developed within the spiral intestine, and similar tissue covers the kidney and gonads.

The lymphomyeloid tissues in Crossopterygii and Polypteriformes have been incompletely investigated.

The meninges of the brain and the spinal cord are an important site of leucocytopoiesis in chondrosteans and holosteans, and probably to some extent also in brachiopterygians and certain elasmobranchs. In chondrosteans and holosteans the heavily leucocyte-infiltrated meninges form a distinct 'meningeal' or 'epimyelencephalic' organ. Certain depressions or canals in the cranium of extinct Devonian fishes indicate that those had a well-developed 'epimyelencephalic organ'. The microscopic structure of the meningeal lymphohaemopoietic tissue is strikingly similar to that of tetrapod haemopoietic bone marrow.

Chondrosteans possess a well-developed thymus and a spleen containing red and white pulp.

In both chondrosteans and holosteans the kidney is an important site of haemopoiesis. In *Polydon* numerous Peyer's plaques occur in the intestine. The size and shape of the pericardial lymphomyeloid tissue is highly variable in *Acipenser*. The highly diversiform and well-developed lymphoid tissues of sturgeons may serve as basis of efficient immune system. The biological significance of the unique lymphoid tissue of the pericardium of sturgeon is unknown but the tissue probably functions in immune process and perhaps in lymphopoiesis.

In teleosts, no uniformity can be found in the histology of lymphoid organs. Indeed, in light of the enormous number of living species, the morphological variations suggest secondary adaptive processes rather than specific specializations of a morphofunctional significance. Teleosts have thymus, pronephros and spleen (for attention on spleen, see Fänge and Nilsson 1985; Quesada et al. 1990, 1994) as their major lymphomyeloid organs. A remarkable feature of the teleostean thymus is its location near the branchial cavity and its permanent continuity with the pharyngeal epithelium Moreover, the existence or not a clear separation between thymic cortex and medulla is moot. An unclear demarcation has been claimed for various teleosts, while other authors have reported three different zones. The thymic ultrastructure of teleosts is generally similar to that described for other fishes and ectothermic vertebrates. Thymocytes appear to be arranged within a network of epithelial-reticular cells. Epithelial cysts are constantly present, especially in older fishes, but Hassall's corpuscles are lacking. Myoid cells with an ultrastructure like that of mammalian cells are also a constant fact of teleostean thymus. Little information is available on thymic vascularization in fish. Few data are currently available on the innevation of fish thymus (see details: Zapata et al. 1996a). Fishelson (1995) has described the cytological and morphological ontogenesis and involution of the thymus in cichlid fishes. Features of three types of cells that produce cytoplasmic extensions, and fate of Hassall's corpuscles are well described. Studies in thymology indicate that the thymic involution of some of the bony fishes takes place in parallel with sexual maturation. With increasing age, on the other hand, diminution of the thymus also occurs in turn in the case of clariid catfish, and even in the immature eel (read also Becker et al. 2001). Tamura and Honma (1973) reported a drastic enlargement in the parenchymatous tissue during the breeding season, while regressive changes took place with a gradual increase in the amount of stromatous tissue in flathead goby, *Luciogobius guttatus*. In any case, including holocephalan fish, the possible relationship between thymic involution and the life-span peculiar to every species of fish is suggested.

Functions of teleostean thymus and spleenic lymphoid tissue are controversial. The pronephros in addition to its lymphomyeloid activity has endocrine functions (chromaffin cells producing catecholamines). An ultrastructural study of pronephros has confirmed its bone-marrow like nature.

Extensive lymphomyeloid masses are found in the viscera of the coelacanth, *Latimeria chalumnae*, but complete anatomical data are missing.

3. SOME PROPERTIES OF LYMPHOMYELOID TISSUES CONTENT OF ENZYMES

The lymphomyeloid tissues, especially the granulocyte-containing Leydig and epigonal organs, are very rich in hydrolytic enzymes such as proteases, acid phosphatase and glucosidases. The hydrolytic enzymes may be released from leucocytic granules. Probably the hydrolytic enzymes contribute in defence against microbes and parasites. The bacteriolytic enzyme lysozyme occurs at high activity in lymphomyeloid tissues of some species and chitinolytic activity commonly occurs. Peroxidases are of little importance in elasmobranch blood cells, but peroxidase positive cells have been observed in the epigonal organ of the nurse shark. The elasmobranch may possess bacteria killing systems different from those of mammalian leucocytes, although so far no information is available on the subject.

An overview of the advances so far made on this topic conveys that fishes of different systematic groups possess a rich variety of lymphomyeloid structures. Characteristic of most of these tissues is an abundance of vascular sinuses. Probably blood, more rarely body cavity fluid or lymph, is filtered through masses of proliferating blood cells into the venous sinuses. Such an activity may resemble the filtering of lymph in mammalian lymph nodes.

The biological importance of the array of hydrolytic enzymes in the lymphomyeloid tissues is not known, but several enzymes may function in connection with phagocytosis. Enzymes like lysozymes and peroxidases may take part in defence against microbes and parasites.

A plasma cell system occurs in elasmobranch, chondrosteans, holosteans and teleosts. Antibody-producing cells, probably partly identical to plasma cell, have been demonstrated by immunological techniques in lymphomyeloid organs of teleosts. Little is known of immunoglobulin synthesis in other types of fishes. The great structural variety of fish lymphomyeloid tissues raises many problems and provides research opportunities for developmental and functional studies of vertebrate immune and defense systems.

Summarily, fish possess a rich variety of lymphomyeloid tissues which resemble lymphoid tissue and bone marrow of higher vertebrates. Predominantly myeloid tissues occur in the oesophagus and the gonads of elasmobranchs and in the head and shoulder region of holocephalans. In chondrosteans granulopoietic tissue occurs in the meninges, erythropoietic tissue in the kidney and lymphoid tissue in the pericardium. The head-kidney is the major lymphomyeloid organ in teleosts. Zapata et al. (1996b) listed further directions for the fish immune system research; (1) Analysis of the role played by distinct types of granulocytes in the defence mechanism of fish. (2) Clarification of the true immunological capacities of both hagfish and lampreys. (3) Analysis of the immunological role of the different bone marrow-equivalent lymphoid organs that appear in primitive fish. (4) A better characterization of the cell components, functional capabilities, and ontogenetic appearance of the mucosal-associated lymphoid tissue. (5) Study of the ontogenetical correlation between the appearance of fish lymphoid organs and maturation of immune capacities.

IMMUNE DEFENSE MECHANISMS

4. INTRODUCTION

Within recent years considerable progress has been made in understanding the immune response in fish, as reviewed by Corbel (1975), Zapata (1983), Rijkers (1985), Evans and Gratzek (1989) and Bernstein et al. (1998). The recent volume on fish immune system edited by G. Iwama and T. Nakanishi (1996) is highly informative. Fänge (1992) covered this aspect in the chapter on fish blood cells. Van Muiswinkel (1992) published a paper on fish immunology and fish health. Evelyn (1996) attempted to explain how microbes that have achieved the attributes necessary to warrant the label "fish pathogen " manage to attach to and penetrate the fish's epithelial barriers, and how they manage to persist and multiply within fish despite the antimicrobial substances and processes available to the fish.

In order to survive, each organism has to maintain its internal integrity, which means that invading micro-organisms have to be successfully combatted. During evolution a number of ways to achieve this goal have evolved, culminating in a highly complex and integrated immune system in mammals. Phylogenetic studies have shown that non-specific defence mechanisms, like phagocytosis are already present in invertebrates. Specific immunity is only observed in vertebrates. In this respect, fishes have a special position because they are the first group of animals in which an immune system, characterized by the presence of specific antibodies, occurs. The reader is referred to Raison and dos Remedios (1998) for hagfish immune system.

The ability of an animal to respond to a particular antigen, and the complexity of immune response, depend upon the developmental status of the lymphoid system. The immunocompetent tissues and organs have already been described.

5. NON-SPECIFIC HUMORAL DEFENCE

In higher vertebrates, a number of components have been described with a potent anti-bacterial and/or antiviral activity which function in a non-specific way. Although some components act totally independently of the immune system, others are activated by antibody (e.g., complement).

Soluble mediators of resistance (non-specific immunity) may be found in the serum, or mucus membranes, or as secretory products of activated cells. Certain of these soluble factors that have been described previously are: components of a system which appear to have complement-like activity, lysozyme, a low molecular weight component which lyses certain microorganisms; C-reactive protein, a normal component of serum that may act as a bacteriolytic substance; transferrin, an iron-binding glycoprotein that may play an important role in non-specific resistance; and some have suggested that interferon-like molecules may be effective in several antiviral responses in fish. The reader is referred to Yano (1996) for better insights of this topic.

Natural agglutinins are 60–300 kD glycoproteins with an activity directed against carbohydrate moieties on surface membranes of xenogeneic erythrocytes. These natural agglutinins have been found in representatives from all classes of fishes. Their role in defence systems is suggested by their opsonic activity.

Complement has been demonstrated by functional criteria in all classes of fish. Activation can occur via either classical (antigen-antibody complexes) or alternative (bacterial cell wall components) pathways. The structural and biochemical studies performed up to now indicate that both complexity and regulation of complement activation are similar in fish and mammals.

The complement cascade can only be activated when antibody molecules (in antigen-antibody complexes) and complement components are from the same or a closely related species. The reader is referred to Manning (1998) for detailed coverage. (See Yano 1996; Fig. 1).

6. NON-SPECIFIC CELLULAR DEFENCE

Non-specific cellular immunity in fish may be expressed as either phagocytes or cytotoxic cells. Differing reports have been published regarding the phagocytic ability of fish reticuloendothelial cells. Macrophages and their roles as phagocytes in inflammation and infectious disease have only been partially identified. Perhaps better characterized is the system of cytolytically active cells referred to as non-specific cytotoxic cells (NCC). These in addition to other non-antigen specific cells may be the fish equivalent of mammalian NK cells. Information on morphology and isolation of macrophages, granulocytes and NCC can be obtained from the contribution of Secombes (1996) (read also Manning and Nakanishi 1996; Manning 1998 and Meseguer et al. 1999).

7. HUMORAL IMMUNITY: ANTIGEN-SPECIFIC IMMUNE RESPONSES

There is an overwhelming amount of evidence demonstrating the capacity of all classes of fishes to mount a specific immune response as measured by the appearance of specific antibodies in serum following proper antigenic stimulation. In Agnatha, generally poor antibody responses can be elicited towards a limited number of antigens. Chondrichthyes are capable of producing antibody against a variety of soluble, particulate and bacterial antigens. The capacity to develop immunological memory (i.e., heightened and accelerated secondary antibody response) is questionable in this class of animals.

Kaatari and Piganelli (1996) provided background on the function of the humoral immune system and the application of this knowledge to the immunoprophylaxis of fish. Coverage included: (1) Antibody molecule. (2) Antibody effector mechanisms. (3) Induction of B cells. (4) Memory. (5) Mucosal immunity. (6) Vaccination and the antibody response.

Teleost fish respond to antigenic stimulation by the production of specific antibodies. Both primary and secondary responses have been observed. The immunological memory and the secondary response itself are specific for the primary antigen.

For a number of species, the antibody-producing cells have been localized. Kidney (pronephros and mesonephros) appears to be the main site of antibody-producing cells, while spleen is of minor importance, at least in some species.

It has been concluded that, especially in bony fish, the capacity to mount a humoral immune response is well developed in terms of specificity and memory.

All vertebrates so far examined for the presence of surface membrane immunoglobulin bearing cells have been found to have this type of cell. Fish immunoglobulins are analogous to mammalian IgM (see Kaattari and Piganelli 1996 for details of Igs). Elasmobranch immunoglobulins are pentameric or dimeric, those of teleosts tetrameric (for structural and functional differences between mucus and serum Ig in carp *Cyprinus carpio*, see Rombout et al. 1993a, b; for special attention on B cell and immunoglobulin heterogeneity in *C.carpio*, see Koumans van Diepen et al. 1995). Channel catfish (*Ictalurus punctatus*) leucocytes have been examined for surface antigen phenotype using monoclonal antibodies. Cytofluorographic examination of these labeled cells revealed that 40% of catfish peripheral blood leucocytes have surface immunoglobulin. Other studies have shown that peripheral blood from fish can be divided into 3 functionally different subpopulations based on the presence of surface immunoglobulin. Functional heterogeneity of fish cells was also shown using monoclonals generated against sIgM. In this study, $sIgM^{+}$ cells were responsive to LPS (lypopolysaccharide) but not Con-A (Concanavalin A), indicating that in fish, B-cells were present and they exhibited surface immunoglobulin. Immunoglobulins (IgM?) are also found in the mucus in plaice and channel catfish. Coscia and Oreste (1998) presented data of Antarctic fish immunoglobulins.

Others have described the presence of T-cell responses in teleost fishes. A monoclonal antibody designated 13C10 detected a T-helper subset in catfish, and additional evidence for T-lymphocyte heterogeneity was provided by experiments where mixed leucocyte responses were generated in rainbow trout.

These studies have shown that, when considered as a group, teleost fish have progressed to an immunologic stage of development where antigen specific antibody and T-cell mediated responses are functional. In addition, considerable evidence has shown that non-specific mediators (including phagocytic and cytotoxic cells) should be considered as principal mediators of parasite resistance.

8. ANTIPARASITIC IMMUNITY

Recently, immune responses to protozoan and helminth infections in teleost fish, have been described (Table 19.2). Types of immunity generated against protozoan parasites (haemoflagellates and ciliated protozoans) consist of both non-specific and antigen specific responses.

Humoral antibody responses appear to be the only mechanism yet described for helminth mediated immunity. These responses can be collectively characterized as consisting of IgM-like antibodies (sensitive to 2-ME treatment) that elicit either a complement dependent or indirect opsomin-like (agglutinin) activity. These antibody responses also varied depending on the ambient temperature at which the fish were maintained post-infection.

In conclusion, both humoral and cellular immune responses are operative in all groups of

Table 19.2 Summarized list of different types of responses to protozoan and helminth infections

Parasite	Response
Trypanosoma	Complement dependent "plasma incubation test" Primary and secondary antibody responses. Phagocytosis
Icthyophthirius	Primary and secondary antibody responses Parasite immobilisation Non-specific cytotoxic cell lysis
Acanthoborthrium	Complement dependent lysis
Gyrodactylus	Generalized "resistance"
Rhipidocotyle	Antibody (IgM)
Pomphorhynchus	Antibody

Source: Evans and Gratzek (1989)

fishes. Moreover, primitive classes also use a number of defense mechanisms which are otherwise mainly observed in invertebrates. Advanced classes such as teleosts, have a more developed immune system which, although less refined in its effectiveness is comparable with mammals.

Cyclostomes have a weak antibody production, but elasmobranchs and teleosts are good producers of immunoglobulins.

Information is lacking on immune responses in non-teleost groups except elasmobranchs and cyclostomes. Probably chondrosteans and dipnoans, etc., have well functioning immune systems.

The habitat may influence the degree of development of immune apparatus. There is also evidence that, in some species at least, nutritional factors and behaviour patterns may influence the immune response. Secombes and Chappell (1996) discussed the selected features of the responses by fish to helminth parasites and compared where appropriate with mammals.

9. PERITONEAL EXUDATE CELLS

Fish peritoneal exudate cells (PECs) play important roles in both non-specific and specific immune mechanisms.

Meseguer et al. (1993) reported the results of ultrastructural and quantitative studies of PECs of two seawater teleosts seabream (*Sparus aurata*) and sea bass (*Dicentrarchus labrax*). They compared the morphological features of the various cell-types with those of ontogenic and circulating forms of the same fish species, and with those of other fish.

The peritoneal exudates of seabream and sea bass consist of granulocytes, lymphocytes and macrophages. These cells show conspicuous ultrastructural differences from the same cell-types of blood and head-kidney. Peritoneal exudate granulocytes differ from their corresponding circulating or head-kidney forms in the following way: (a) they are larger in size, and (b) their abundant cytoplasmic granules have some new ultrastructural features, and a new granule population might also be present. Likewise, lymphocytes also show a noticeable difference; they contain a sparse population of small dense cytoplasmic granules. Monocytes, macrophages, and transitional forms between these two cell-types are also found. The percentage of peritoneal exudate cell-types is different in seabream and sea bass. Macrophages in sea bass represent the most abundant peritoneal exudate cell-type. However, seabream shows lower percentage of macrophages than granulocytes.

10. CELL LYSIS

Cell lysis plays an important role in the vertebrate immune system that enables the organism to eliminate unwanted target cells. Cell-mediated cytotoxicity has recently been analyzed in lower vertebrates. Some cytotoxic assays have been carried out in a few species of fish, and most demonstrate the cytolytic activity of leucocytes against established mammalian cell lines. The fish lymphoid cells, which mediate rapid and unrestricted lysis of targets have been called nonspecific or natural cytotoxic cells (NCC), and it has been postulated that they may be evolutionary progenitors of mammalian NK cells. Recent studies demonstrate that fish cytotoxic effector cells form a cell population of granular lymphocytes, although monocyte-macrophage-like cells and granulocytes also be involved. (read review by Meseguer et al. 1999).

Meseguer et al. (1994a; see for details) concluded that seabream (*Sparus aurata*) and carp (*Cyprinus carpio*) have a leucocyte cell population with ultrastructural features of either monocytes or lymphocytes showing nonspecific cytotoxic ability. In both species, monocyte-like cells usually appeared flattened against the targets and seemed to enclose fragments of the target to form cytoplasmic vesicles and the content of their scarce cytoplasmic granules seemed to be delivered into these vesicles. In the seabream lymphocyte-like cells, dense cytoplasmic granules occurred only occasionally, and neither microvilli nor cell processes were present at the contact areas with the targets. In the carp, the contacts were more numerous and formed regularly interdigitating contact areas and the lymphocytes showed granules with characteristic dense and fibrillar contents.

The reader is also referred to Fänge (1992) on nonspecific cytotoxic lymphocytes. Secombes (1996) comprehensively described the (i) the phagocytes migration, (ii) phagocytosis, (iii) phagocyte killing mechanisms, (iv) phagocytes as accessory cells, and (v) nonspecific cytotoxicity. Many factors have been shown to influence NCC activity, e.g., diet, temperature, stress, strain, age. Recognition and binding to target cells by teleost NCCs appears to be receptor mediated. Little is known about the mechanisms by which NCCs lyse target cells. Reader can find interesting information on enhancement of fish phagocyte activity in the article by Secombes (1994). Table 19.3 provides a general overview of lymphoid cells, organs and fundamental characteristics of the immune responses of fishes.

FISH PATHOLOGY

11. INTRODUCTION

Disease is a constant concomitant of life for any species, normally removing individuals from the population continuously. Fishes are, of course, subject to a wide spectrum of diseases of infectious or noninfectious etiology ("disease" can be defined in the broad sense as "any departure from normal structure or function of an animal" or as "the end result of interaction between a noxious stimulus and a biological system"). Summarily, the phenomenon of disease appears to be as old as life itself. In fact, disease is a major motor and denominator of evolution.

Disfunction and death due to activity of infectious agents constitute the narrower, but often predominant concept of disease. *Infectious diseases*-caused by viruses, bacteria, fungi, protozoa, and other organisms are prime suspects in searches for causes of mortalities often to the exclusion of other possible causes. *Non-infectious* diseases include such phenomena as environmentally

Table 19.3 Overview of the elements of vertebrate immunity

	Mammals	Agnathans	Chondrichthyes	Teleosts
Lymphocytes	T, B	+ ; plasma cells	+ ; plasma cells	T, B
Organs	Thymus, spleen, bone marrow, lymph nodes, GALT	GALT?	Thymus, spleen GALT, Leydig organ, epigonal organ	Thymus, spleen GALT, pronephros
Germinal centres	+	–	–	–
Affinity maturation	+	–	–	?
Somatic maturation/ hypermutation	+	–	+	+
Ig isotypes	IgM, IgG, IgD, IgE, IgA	–	IgM, IgW, IgR NAR[a]	IgM
TCR (T-Cell Receptors)	α/β; γ/δ	–	β; δ	α/β
MHC		?		
I	+		– ?	+
II	+		+	+
Complement	+	C3C4 ancester	+	+
RAG genes (Recombinase activating genes)	+	–	+	+
Graft rejection	+	+	+	+
B cell	+	–	+	+
T cells	+	–	+ ?	+ ?

[a]NAR has been reported in nurse shark. It is not clear whether IgW and IgR are in all Chondrichthyes.
Source: Bernstein et al. (1998)

induced skeletal anomalies, genetic abnormalities, physiological malfunctions caused by chemical environmental factors, metabolic disorders resulting from nutritional deficiencies, many forms of neoplasia, and a host of others. In many instances, it is probably the combination of an infectious agent and environmental stress that eventually causes mortality.

The distinction between "infection" and "diseases" must be kept in mind. Most organisms are constantly hosts to potentially pathogenic microorganisms, but disease results from imbalance of the interactive system which includes virulence of the pathogen, resistance of the host, and effects of environmental stresses.

Infectious disease usually exists in an epizootic form, weakening or disabling individuals and rendering them more susceptible to predators or other environmental stresses. Occasionally, epizootics and mortalities may sweep through animal populations. These epizootics are triggered by a complex interplay of pathogen, environment, and host population. Considering only the environmental aspects of such outbreaks, any departure from normal conditions produces a degree of stress on the population, and may contribute to an increase in prevalence of a pathogen, or of facultative invaders. Some of these environmental factors are drastic changes in temperature, lack of adequate food, or overcrowding. Resistance of the host animal to disease is, of course, intimately related to these stresses.

Environmental stresses have been implicated in a number of fish diseases, but are difficult to quantify. Even a defination of stress can be elusive. Stress has been defined as (i) the sum of all physiological responses by which an animal tries to maintain or reestablish a normal metabolism in the face of a physical or chemical force; (ii) as a state produced by an environmental or other factor which extends the adaptive responses of an animal beyond the normal range, or which disturbs the normal functioning to such as extent that, in either case, the chances of survival are significantly reduced, and (iii) as the product and not the cause of homeostatic changes. Stress is the effect of any force which tends to extend any homeostatic or stabilizing process beyond its normal limits, at any level of biological organization. The stress responses of farmed fish has been described by Pickering (1998).

Human activity has introduced or has increased environmental stresses for fish. We have, for instance, added pesticides and other synthetic chemicals which can, even in low concentrations, drastically affect the physiology of fish, and with which the species may have had no previous evolutionary experience. We have added heavy organic loads, in the form of sewage sludge and effluents, which can produce anaerobic or low oxygen environment and which are often accompanied by other contaminants such as heavy metals, that can interfere with enzymes of the fish and the food organisms they consume. Moller (1985) reviewed the role of pollution as a cause of fish diseases. Fish diseases and abnormalities discussed are: chromosomal abnormalities, skeletal deformities, hepatoma, mesenchymal tumours, epidermal papillomas, skin ulcers, fin rot, oral papillomas, cauliflower tumour and various diseases. Parameters that influence disease rates are (1) environmental parameters: low selection pressure by competition and predation; nutritional condition; adverse environmental conditions during the larval phase; salinity and adult fish; temperature and adult fish; parasites as disease vectors, pollutants, and (2) stock-specific parameters: spawning habits, population density, fish migration, and fish length.

In the contemporary approach toward understanding communicable diseases, epidemiology or epizootiology plays an important role in attempt to explain relationships between the hosts, pathogens, environments, and outbreaks of diseases. It can be represented in the form of an equation:

$$H + P + S^2 = D$$

In this equation, H represents the species and strain of the host, its age and inherited susceptibility to any particular disease; P represents the pathogen causing the disease with all its variability; and S represents stress of the environment. The square of S is used because the stress caused by environment increased in geometrical progression when the conditions are approaching the limits of tolerance by host. The D represents the disease which results if the three components listed on the left side of the equation are in proper qualitative and quantitative proportions.

12. DISEASES—AN OVERVIEW

Unfortunately, the overwhelming amount of knowledge of fish diseases is restricted to a very few species. The host-agent relations and disease manifestation may be different in culture and in nature. Changes in balance between agent and host are induced through genetic, environmental or nutritional factors. A major shortcoming in fish disease research is the frequently confusing taxonomic status of the organisms involved, especially among the protozoan agents, but also in taxa of multicellular parasites.

Environmental stresses may not only be the immediate cause of a variety of diseases, but can also significantly modify the ecological potentials of both agent and host. Hence, they may play an important role in the destabilization of equilibria between disease-causing and disease-counteracting forces. No disease can be explored adequately without taking environmental stressors, present and past, and nutritional disorders into account.

There is much need for broadening the basis of disease research. During the past several years there have been signs of increasing interest in various aspects of fish diseases. Several conferences, reviews and books have provided voluminous information, a representative fraction is given here (see H.H. Reichenbach-Klinke 1966; Sindermann 1966, 1979; Snieszko 1972, 1973; Mawdesley-Thomas 1975; W.E. Ribdin and G. Migaki 1975; Roberts and Bullock 1976, 1989; Ahne 1977, 1979; O. Kinne 1984). Evelyn (1996) focused on "Infection and Disease". Aspects covered were: (1) Colonization and entry into the host (penetration via gills, gastrointestinal tract, skin and miscellaneous routes. (2) Postentry events (spread). (3) Mechanisms of survival in the fish host (survival in the face of humoral factors; survival in phagocytes and (4) Virulence factors permitting growth within the host. The bacterial fish pathogens selected for the topic were: *Vibrio anguillaruni*, *Aeromonas salmonicida* (typical strain), ***Renibacterium salmoninarium***, *Yersina ruckeri*, *Edwardsiella ictalure*, and *Pasteurella piscicida*. Although there is much to learn about the mechanisms employed by fish pathogens to avoid the host's defenses, a considerable body of knowledge in this aspect is beginning to accumulate. What is particularly remarkable about this is that this informaton, prompted by an interest in obtaining vaccines for controlling fish diseases, has essentially been obtained in the last few decades. A major thrust in fish immunology these days is in the direction of cellular immunity, particularly on ways of "up-regulating" bacterial killing by phagocytes.

12.1 Diseases caused by Microorganisms

Microorganisms constitute a most important source of disease in fish. In fact, over millions of years microorganisms have achieved mastership in gaining access to, and in the utilization of, organic materials synthesized and stored in the bodies of living organisms. While the latter fail to prevent or to counteract excessive microbial attack, disease prevails. The disease condition may range from chronic to acute, with varying degrees of host response. Viral and bacterial infections are major hazards confronting the commercial acquaculturist.

12.1.1 Diseases Due to Viral Agents

There can hardly be any doubt: while our present knowledge on fish diseases caused by Virales is still quite limited, important breakthroughs can be expected in the foreseeable future. In fact, viruses may be shown not only to be the primary cause of numerous diseases not yet fully explored or even known, but also to function as pacemakers for numerous other secondarily effective disease agents. For these reasons, viruses require special attention in future research.

Viral agents exhibit the same basic characteristics as the Virales of homeotherms and they comprise the same virus groupings, e.g., herpesviruses, iridoviruses, rhabdoviruses and reoviruses. Additional fish-disease agents are considered to be adenovirus, coronavirus, calicivirus and leucovirus. Still others await investigation. It is predicted that also some of the major virus groups not yet isolated from fish but known to exist in homeotherms will eventually be found to inhabit fish hosts. The overwhelming number of reports on virus-caused fish diseases stem from captive fish, i.e., hosts kept under environmental conditions which deviate from those to which the fish has adapted during its evolution. Some 87% of the viruses detected in fish tissues inhabit hosts (mainly fish raised in fresh-water aquaculture farms) utilized as human food. The remaining 13% have been reported from tropical fishes, mostly pet fish (ornamental fish) kept in home aquaria.

Viruses are best known in marine fishes as suspected or known etiological agents of several neoplastic, hyperplastic diseases. Neoplastic or hyperplastic diseases include dermal and epidermal papillomas of many flatfish species (Fig. 19.3). A remarkable tumorous growth of eels labelled as "cauliflower disease" (Fig. 19.3). is a common chronic fibro-epithelial tumor, often of drammatic proportion, occurs principally in the head region of European eels, *Anguilla anguilla.* Eels with progressive tumors become emaciated and die. A number of highly pathogenic viruses of freshwater fishes do not cause tumor. The diseases produced included viral haemorrhagic septicemia of salmonids in Europe, and infectious pancreatic necrosis of trout, *Salvelinus fontinalis* and *Salmo gairdneri.*

The external lesions of lymphocystis are raised growth of generally granular, wart-like or nodular tissue which is composed of enlarged host cells and their covering membrane (Fig. 19.3).

Thus far, viruses inhabiting fish tissues have not been shown to be infectious to man or other homeotherms. However, the effects viruses exhibit in fishes and the general patterns of agent-host relations are similar in both cases. Two fish-inhabiting viruses, a leucovirus-like agent of esocids and a herpesvirus of *Oncorhynchus masou*, are oncogenic and other viruses, observed in neoplasma, might also turn out to be oncogenic (i.e., tend to induce tumors).

Apart from purely academic interest in piscine viruses, the chief importance of virus diseases is an economic one. About 33% of the total new population of economically useful fishes perish each year through disease; about 60% of these losses are due to microorganisms causing infection, and viruses that are pathogenic for fishes play a very important role. Thus, for example, the Anguillidae are threatened by rhabdoviruses (EVA, EVEX) and probably by a papovirus (Stopa); *Rhabdovirus carpio* often causes death in vast numbers among the Cyprinidae (*Cyprinus carpio*); in Esocidae fry (*Esox lucius*), the pikefry rhabdovirus often causes upto 100% losses; the population of the Ictaluridae is drastically reduced by a herpesvirus (CCV); and considerable losses are caused in the Salmonidae by rhabdoviruses (HN, Egtved), a virus that is presumed to be a reovirus (IPN), and herpesviruses. Various saltwater fishes are threatened by an iridovirus infection (PEN).

12.1.2 Diseases Due to Bacteria

Present in each drop of water, each cubic millimeter of sediment, attached to suspended particles and living in large numbers on and in all other organisms bacteria are of basic importance for the functioning of natural aquatic and terrestrial ecosystems. While most bacteria known so far are benign, a large number have been shown to be the most immediate cause of disease or to be secondarily involved in the development of disease phenomena.

Reports of bacterial epizootics in marine fishes are surprisingly infrequent, and in fact relatively few bacterial pathogens have been recorded from natural populations of marine fishes. Of all the known bacterial diseases of marine fishes, none has a longer or more fascinating history than the "red disease" of eels, caused by *Vibrio anguillarum.*

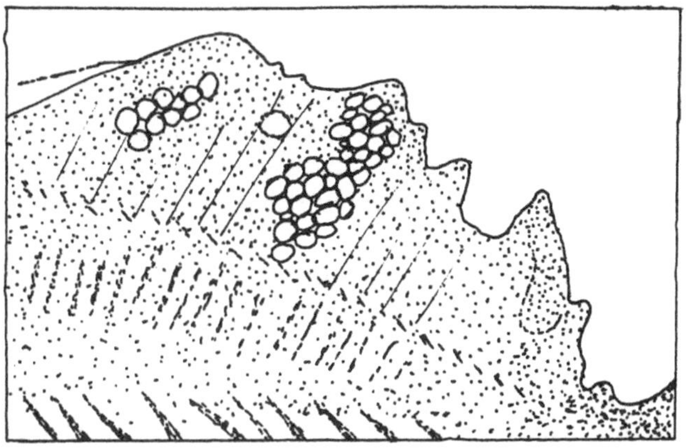

(A) Lymphocystis.

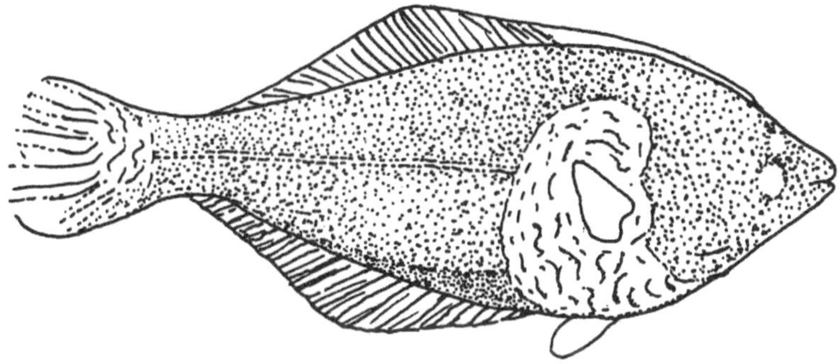

(B) Papilloma of flounder (Virus disease).

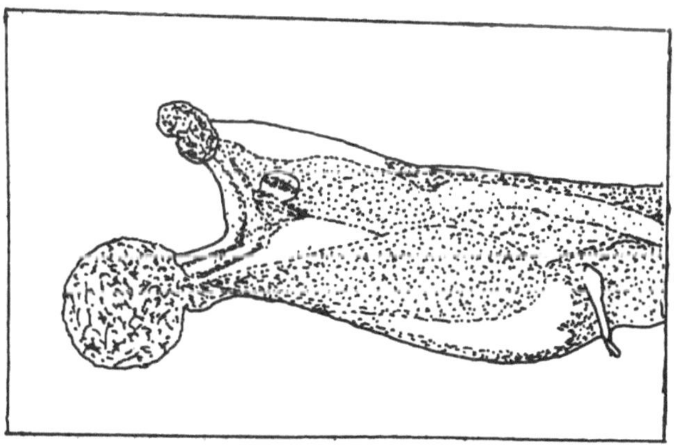

(C) Cauliflower disease of eel (Virus disease).

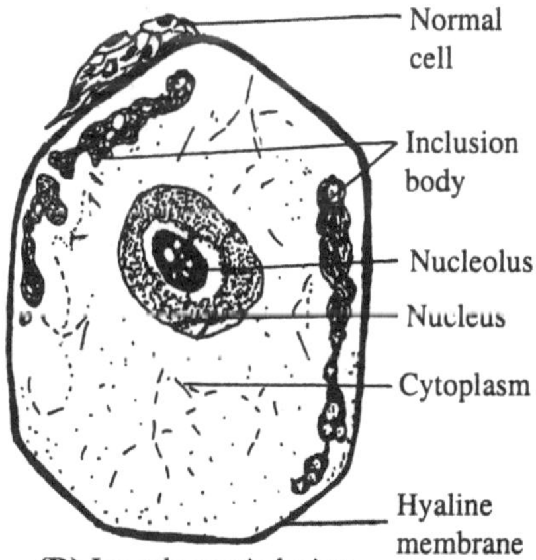

(D) Lymphocystis lesion.

Fig. 19.3 (A) Lymphocystis disease of the dorsal fin. (B) Papilloma of flounder (C) Cauliflower disease of eel. (D) Drawing of a section through part of a lymphocystis lesion. One of the tremendously enlarged infected cells is shown in detail, and adjoining normal sized cells are shown for comparison. The infected cell has a length of about 140 μ.
***Source:* For (A–C) Sindermann (1966)**

Over the last few decades, research on bacteria caused diseases of fishes has received unique impetus from man's rapidly increasing efforts to provide additional food through aquaculture. Intensive fish culture thrives under conditions which tend to change dramatically the natural agent-host balances, typically, in favour of the former. In addition to viral agents, this is particularly so for bacteria. The future of commercial fish culture depends decisively on the development of handy methods for diagnosis, prevention, control and treatment of fish diseases due to scores of bacterial agents.

What means do we have for control of bacterial diseases of aquatic animals? The first one is avoidance of exposure to pathogens. This is possible in relatively few fish cultural facilities—only those having clean water.

Many pondfish culture and marine culture of fish takes place in open waters where fish pathogens are present. Under such conditions, avoidance is not possible, and use of chemotherapeutic drugs is expensive and transitory. Immunization is a recent and growing control measure that is proving highly promising. Although immunization of humans is the most effective means of disease control, the immune response in fishes is dependent on temperature.

Immunization of cold-blooded vertebrates was started many years ago, but little progress was made because the response was negligible at low temperature and the necessity of handling each animal made it impractical on a large scale. Immunization of fish by oral administration of antigen and by immersion of fish, with or without preconditioning, is becoming so promising that commercial laboratories are now manufacturing fish vaccines.

12.1.3 Diseases Due to Fungi

Of the numerous marine fungi only a few associate closely with marine fishes. *Ichthyophonus* (*hoferi*) is the only marine fungus that has been shown to cause diseases in fishes. *I. hoferi* invades its host via their alimentary tract by thick-walled, multinucleate 'resting spores' liberated from disintegrating, infected tissues. Transmission occurs from fish to fish. Ichthyophoniasis is a systemic infection (similar to tuberculosis or mycobacteriosis), characterized by granulomas in the affected organs. It can develop into a severe disease resulting in heavy mortalities in acutely and chronically infected fish. In *Clupea harengus*, external signs of disease include a 'sand paper appearance' of the skin due to minute reddish pustules (pathogen cysts in lateral muscles). In addition, white necrotic areas may form on the skin, as well as ulcerations from which spores are released into the ambient water. *Ichthyophonus* disease can also manifest itself in sex reversal, exophthalmia, hyperaemia and other eye infections, etc. From the marine environment, reports on fish invading fungi other than *Ichthyophonus* sp. are rare. *Ichthyophonus* is known as a menace to aquarium fishes in Europe.

In contrast, numerous instances of fish diseases caused by fungi have been reported from fresh-water habitats. Members of Oomycetes, Deuteromycetes and Ascomycetes can cause severe diseases and heavy losses among fresh-water-living fishes, especially those reared and raised in aquaculture farms.

Species of the genus *Saprolegnia* are usually implicated in fungus diseases of freshwater fish and fish eggs. The presence of fungus is indicated by a white cottony growth which consists of a mass (mycelium) of nonseptate filaments (hyphae). Saprolegnia are characterized by having club-shaped sporangia containing zoospores and round oogonia (sexual reproductive structures) (Fig. 19.4).

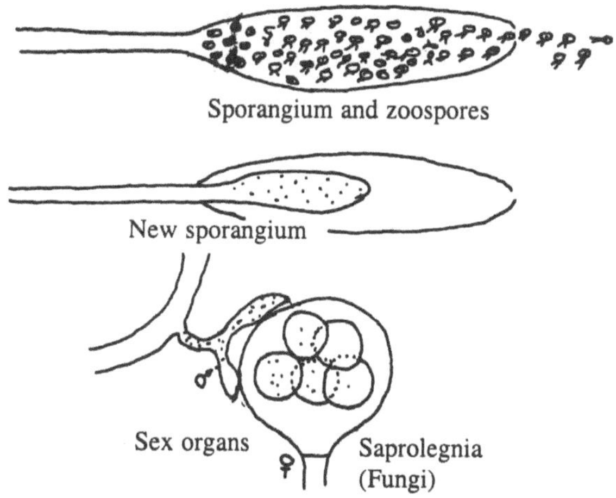

Fig. 19.4 ***Saprolegnia*. Top-sporangium with many flagellated zoospores. Middle-An old sporangium with a new one proliferating into it, a characteristic of the genus, Bottom-Sex organs-antheridium on left, oogonium on right; these are usually found only on special media.**

As with other diseases, prevention and treatment of fungus diseases is possible only under controlled conditions in captive water bodies. Prophylactic measures such as water-quality control, sanitary care, selection of uninfected food material and a rigid quarintine regime seem the primary methods of choice.

12.2 Diseases Caused by Protistans

Most of the protozoan parasites have received attention not primarily under etiological aspects or as disease-causing agents, but with an eye on their taxonomy, morphology or life-cycle dynamics. In fact, although seemingly of paramount importance as potential causes of fish diseases, parasitic protozoans have been studied less extensively than helminths or crustaceans.

Protozoan invasions of marine fish are dependent on ecological conditions of the host. These conditions act upon the fine balance between the protective mechanisms of the fish and the virulence of the protozoa, influencing it either directly or through the host's organism. When adverse environmental conditions prevail against the host, chronic infections may flare up or new one may be acquired. Protozoan invaders or their propagative stages are ever present around their potential hosts. Stress conditions and crowding support protozoan invasions.

There is a large body of information concerned with the disease-causing potential of flagellates, amoebic and opalinid forms, Apicomplexa, Microspora, Myxosporea and Ciliata. Details are known on nutrition, proliferation, life cycles, transmission, host relations, host defense and ecological aspects of protozoan infections. Special attention is paid to diseases in cultivated fishes.

As in the case with other biotic disease agents, protozoan species benign under undisturbed, natural conditions may succeed in obtaining considerable benefits (energy, matter, space) from fish hosts under stress (e.g., pollution, crowding in intensive culture), and consequently in inflicting severe damage. In nature serious lesions have been reported to be due to tissue infecting forms. *Ichthyophthirius multifilis* (Fig. 19.5) cause the dreaded "Ich" disease and can be lethal to catfish of all size. In advanced cases the parasites are evident as small white pustules over bodies and fins of infected fish. It is highly contagious.

The Microsporida are obligatory cytozoic parasite which reproduce by asexual division (schizogony) and fusion (sporogony) within individual host cells sometimes producing cell hypertrophy tumor called xenomas (Fig. 19.6). Ectoparasitic protozoans may be particularly harmful due to rapid flare, no need for intermediate hosts and little or no host specificity. The pathogenicity of protozoans (Flagellata, Amoebae) infecting the intestinal lumen of fishes tends to be quite limited. However, more experiments must be conducted in order to examine whether or not coccidians which are common in fish intestines, but have not yet been shown to cause serious diseases can in fact exert pathogenic effects in their respective hosts.

Fig. 19.5 ***Ichthyophthirius multifilis*** **(Protozoa) "Ich" disease. These spherical granules cause the dreaded "Ich" disease and can be lethal to catfishes of all sizes. A horseshoe-shaped nucleus is visible even in live specimens and positively indentifies the parasite. In advanced cases, the parasites are evident as small white postules over bodies and fins of infected fish. Highly contagious. Single species.**

A widely distributed disease of clupeoid fishes is that caused by *Elimeria sardinae* (Coccidia). The parasite localizes, often in massive numbers, in the testes, where it may reduce productive capacity or cause complete sterility of the host. *Eimeria* may occur in other organs (liver, swim bladder). Members of another coccidian group, the Haemogregarinidae, have long been recognized as of widespread occurrence in blood cells of fishes. Diseases caused by Myxosporidea are common in marine as well as fresh-water fishes. The myxosporideans inhabiting organs such as gall bladder are most frequently studied and described, but those localizing in somatic muscles must be considered the most destructive and of greatest economic importance. Extensive necrosis of the flesh in different fishes causes conditions known as "wormy", "mushy", "jellied", "milkiness". Gill-invading Myxosporidea are not usually of serious consequence to marine fishes. Several species of Myxosporidea invade the cranial cartilages of fishes. *Myxosoma*, a highly infectious disease agent acquired early in life in fresh water, causes "Whirling disease" of juvenile salmon and trout, as well as gross skeletal abnormalities. Several serious diseases of marine fishes result from microsporidean invasion (e.g., to prevent reproduction by mechanical occlusion of the vent; to destroy areas of digestive tract and gonad, and impair other metabolic functions).

12.3 Diseases Caused by Protophytans

To all we know, Protophyta (algae) do not qualify as important disease agents. Only the dinoflagellates have evolved representatives—members of the genera *Amyloodinium*, *Crepidoodinium* and *Ichthyodinium* which parasitize fishes. Thus, *A. occellatum* is the agent of 'velvet disease' known from temperate and warm-water, marine fish species.

Non-parasitic, free living, red-tide causing dinoflagellate such as *Gymnodinium breve* when present in large populations with high individual densities ('bloom'), may cause large-scale mortalities in fish and other taxa by releasing toxic substance in the ambient water.

Microsporidan (*Plistophera cepedianae*) cyst on a gizzard shad (*Dorosoma cepedianum*).

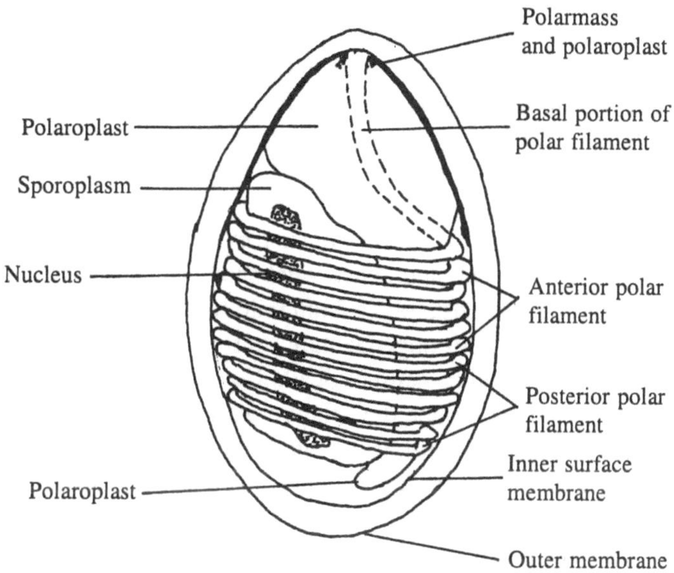

Microsporidan spore (*Thelohania californica*).

Fig. 19.6 Microsporidian (*Plistophora copedianae*) cyst on a gizzard shad (*Dorosoma cepedianum*).

12.4 Diseases Caused by Cnidarians

Intimate associations between cnidarians and fishes are rare. They range from commensalism to parasitism. Several species of the athecate genus *Hydrichthys* have been shown to live parasitically on fishes (mainly juvenile or small-sized species), primarily affecting the host's body surface (epidermis, fins, opercula, gills), but some also invade fish body via wounds caused by fish-associated copepods and exploit deeper tissues and blood vessels.

12.5 Diseases Caused by Helminths

The heterogenous group of Helminthes constitutes the largest, most numerous and most diversified assembly of parasites associating with marine fishes. However, only a few helminth species have been studied in depth and little is known about agent-host interactions. Parasite effects at the individual or population levels, agent, virulence, host defense, host range, host specificity, etiology, as well as disease prevention and therapy avail thorough attention from pathologists and ecologists alike.

The helminths-trematodes, cestodes, nematodes and acanthocephala are common parasites of marine fishes. Usually it is as larvae that the worms are of greatest significance. Adults occur in the digestive tract of marine fishes, but are usually not a cause of serious disease. Adult cestodes are common and occasionally harmful digestive tract parasites of fishes, but as with trematodes, larval stages are of greatest concern to man. Larval tapeworms occur frequently in the viscera and flesh of marine and estuarine fishes. The nematode parasites of marine fishes have received attention primarily because certain larvae that infest the flesh and viscera reduce the commercial value of the host. The "spiny headed" worms (Acanthocephala) are represented as adults and as larvae in marine fishes.

Commercial production of catfishes is a rapidly growing industry. Catfishes may carry a number of minute parasitic worms on their body, fins and gills.

Most experimental studies conducted on helminth biology are concerned with non-marine helminths. The taxonomy and life cycles of many species remain to be analyzed. Pathogenicity and detailed agent effects have been explored in only a few selected species, and practically nothing is known about helminth-fish relations in the unconstrained environment. That reliable reports on mass mortalities in natural fish populations due to helminth parasites are very rare, need not necessarily be indicative of the absence of such events, but may be due to the difficulties involved in the sampling of diseased, and hence weakened, as well as dead hosts.

Helminth-caused effects on fish hosts range from mild to severe to lethal. Several Trematoda may cause popeye (exophthalmia). Monogenea may exert severe effects on gills and skin; they can cause destruction of the eye, and sometimes death. Cestodes may damage the intestinal wall, and larvae of anisakid nematodes reduce the relative weight of the liver as well as condition factor. Parasitic Acanthocephala can perforate the intestinal wall. Added economic effects include discarding of otherwise edible fish products, delay in processing operations and loss in oil yields.

Helminth-caused diseases are more obvious, easier to study and often more frequent in captive fishes (research cultivation and commercial cultivation), then in their free-living counterparts. Interestingly, cultured fish have been shown to lose many of their helminth parasites, while other parasites affect them in increased numbers. Such shifts in agent-host relations are indicative of differential environmental requirements of agent and host or lack of intermediate hosts under culture conditions (Fig. 19.7).

12.6 Disease Caused by Crustaceans

As is the case with other biotic disease agents of fishes, more information is available on freshwater species—both agents and hosts—than on their marine relatives. In addition, few investigators have explored crustaceans with a view on their role as fish symbionts and as potential causes of fish diseases.

Amongst the crustacean groups, copepods display the widest range of adaptation to life on or in fishes. Marine fishes are parasitized by variety of copepods of which several members of the families Lernacoceridae, Pennellidae and Sphyriidae are particularly injurious to the host. Usually the adult females become highly modified and penetrate the flesh, often causing extensive ulceration. As in other crustacean fish parasites, small and mobile ectoparasites do not tend to be highly pathogenic; the intensity of their impact depends primarily on parasite intensity and host distribution. While larger, mobile ectoparasites may cause lesions, except for species of *Argulus*

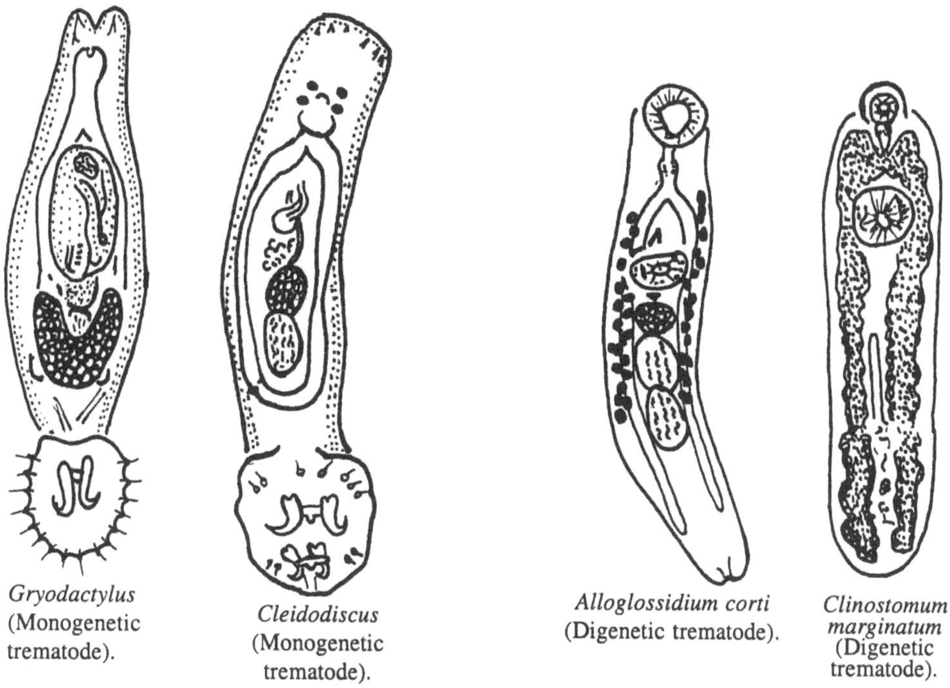

Fig. 19.7 ***Gyrodactylus*** **(Monogenetic trematode). Generally found on the body and fins but may occur on gills. It can be identified by its lack of eye-spots, single pair of large hooks, and by the fact it gives birth to its young. Unborn individuals can be seen within living worms.**
Cleidodiscus. **(Monogenetic trematode). Found only on gills and is distinguished by two pairs of large hooks, each equipped with cuticular barbs, by eyespots, and by the fact that it lays eggs. Frequently unlaid egg can be seen within the worm.**
Alloglossidium corti **(Digenetic trematode). Catfishes are hosts for adult parasites and frequently are intermediate hosts in parasitic life cycles. Many flukes attain adulthood in the intestines, swim bladder, or urinary bladder of fishes. Such organisms generally has a flattened, wormlike form. Stained preparations examined under a microscope will reveal an oral sucker at the anterior end, a ventral sucker near the mid third of the body, and well-developed internal reproductive organs.** ***Alloglossidium corti*** **occurs in the intestine of catfishes.**
Clinostonum marginatum **(Digenetic trematode). Flukes which use fish as intermediate hosts generally penetrate the body and encyst in the flesh, fat bodies, or mesenteries. Most of these flukes are larval stages of worms which are adult parasites in birds. These encysted worms give a "grubby" appearance to fish flesh if abundant.** ***C. marginatum*** **has a characteristic yellow colour. Hence, the encysted worms are often referred to as "the yellow grub of fishes". In most cases, these worms reach adulthood only if eaten by the proper bird host.**

("fish lice" Fig. 19.8) the effects inflicted are usually not severe. In general, crustacean gill parasites are capable of causing extensive damage to their hosts. However, from the marine environment no evidence is available to date for severe respiratory impairments. Serious host damage may be caused by mesoparasitic copepods. In fact, their association with fishes is believed to often result directly in fish mortality. Crustaceans living endoparasitically in fishes have been shown to cause a variety of effects, depending primarily on body size and taxonomic

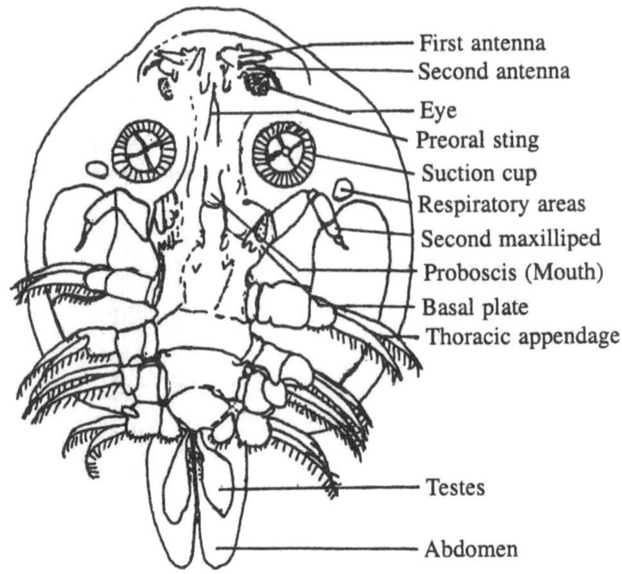

"Fish lice"

Fig. 19.8 Adult *Argulus* "Fish lice" (Parasitic copepod).

identity of agent and host, site and duration of the infestation and, of course, the balance between agent virulence and host defense.

Details are available on a variety of local parasite effects (e.g., on gills, skin, muscle, skeleton, sense organs, heart, liver, kidney, gonads, alimentary canal) and general effects (e.g., on fish weight and chemical composition, growth, metabolism, blood, reproduction, behaviour). Secondary infections facilitated by crustacean parasites are insufficiently investigated. This holds for their importance relative to cases of intimate coexistence between fish hosts and additional group of parasitic symbionts, as well as for defining cause-effects relation. There is a brief consideration of impacts on fishes due to free-living crustaceans and of the economic consequences of detrimental parasite effects in fishery and aquaculture. While attempts have been made to estimate the losses in quality and number of fishes suffered by commercial fishery and aquaculture ventures, and while the losses may be considerable, sound assessments require more solid data than are presently available.

12.7 Diseases caused by Neoplasia ("Neoplasms or Tumors")

Definations: "A disturbance of growth characterized primarily by an unceasing, abnormal, and excessive proliferation of cells", or "That form of hyperplasia which is caused, at least in part, by an intrinsically heritable abnormality in the involved cells". Thus, neoplasia (new growth or new formations) comprise local swellings (tumors) due to excessively growing tissues, benign or malignant. In recent years the study of neoplasms in aquatic organisms has been greatly intensified. An increasing number of organisms have been shown to carry tumors, especially those maintained in captive bodies of water (i.e., under modified environmental conditions) and those living in natural bodies of water subject to severe man-made pollution.

Neoplasms are classified according to the type of tumor origin (epithelial, mesenchymal, pigment-cell, neural tumors) and type of tumors formed (e.g., papilloma, fibroma, lymphoma, melanoma, neurilemmona, etc.).

Neoplasia of fishes are similar to those reported from higher vertebrates. Tumor are widespread among different fish taxa, ranging from Cyclostomata to Chondrichthyes to Teleostei. Many of the neoplasms have been reported from bottom-feeding fish. Almost all organs may be affected. In some cases, tumorous fish diseases have attained epidemic proportions and thus economic importance. Most of our present knowledge of fish neoplasia stems from investigations on freshwater forms. Of the different types of tumors reported, several have not yet been reported in marine fishes. Certain kinds of tumors can be produced experimentally via crossbreeding: they develop regularly in hybrids of different tumor-free parents.

As other disease phenomena, tumor formations are subject to natural selection. Since most fishes reproduce throughout their life, even tumors forming in old individuals partcipate in affecting the evolution of counteractive mechanism (e.g., DNA repair, immunological defense). Tumor diseases may assist in programming the gene pool of a population towards a high level of resistance against tumor formation.

While the occasional occurrence of neoplasia appears to be a natural phenomenon, rates of tumor diseases exceeding 1% of a given population require explanation. They may be attributable to: (i) interbreeding resulting in genotypes with reduced tumor resistance; (ii) infections (virus) on non-adapted (non-immunized) fish; (iii) sudden environmental change (man-made water pollution; invasion of new habitats).

As with other diseases, it is usually the combination of different causes acting in concert that induces tumor growth. Like organisms themselves, tumors respond to the total accumulative impact of their environment.

Table 19.4 is a list of important diseases and related details.

12.8 Diseases Caused by Environmental Stressors

The term 'stress', in its general connotation, refers to 'strain', 'pressure' or 'tension' exerted upon a body. In biology the term 'stress' to many investigators implies forces (e.g., sub-or supranormal, injury, disease-agent impact, social dominance) acting upon an organism or its part, significant enough to interfere with normal functions. Such forces may produce non-specific or specific responses. Non-specific responses are-regardless of causative agent-similar in quality (general adaptation syndrome); specific responses more or less differ from each other, depending on the types of stress concerned. In practice, such differentiation may be difficult to make because a given organism always responds to the summative, concomitant impact of all stress-causing forces encountered. Unfortunately, the terminology employed for describing organismic responses to environmental stress-including that due to coexisting, foreign, free-living or parasitic organisms is inconsistent.

The term 'stress' is employed to designate the response of an organism to an impact; the term 'stressor' to identify the causative agent. While extreme intensities of stressors, acute or chronic, tend to be lethal, sublethal stressor intensities can induce a variety of negative deviations in the normal functions and structures of living organisms. Examples of sublethal stress effects are: increased susceptibility to disease, reduced tolerance to other stressors, diminished rates of growth and reproduction, and a decreased capacity for intra-and interspecific competition. Effects

Table 19.4 List of important diseases and related details

	Disease	Causative agent	Etiologic and pathological features	Control
I	*Viral Diseases**			
(1)	Lymphocystis disease (LCV) Host: Various freshwater and marine fishes during summer months. (*Lepomis macrochirus*, *Pleuronectes* sp.).	Lymphocystis virus (Poxvirus group) Iridovirus	Formation of single or multiple wartlike skin proliferations. Neoplastic lesion is due to proliferation of individuals cells (cell hypertrophy).	Avoidance of infection.
(2)	Fish Pox Mild, chronic disease giving fish an unappetizing appearance.	Fish Poxvirus	Proliferation of the epithelium with production of glossy opaque papillae. On recovery the scales are discoloured. Affected fish eat poorly and grow slower than healthy fish.	Avoidance. Disinfection of eggs with Betadine.
(3)	Epithelioma of Eels (eel papilloma; cauliflower disease; stomatopapilloma) Host: *Anguilla anguilla*	Polyhedral virus Papova?	Cauliflower-like, whitish-red or light brown tumor on the head (excrescences proliferating with tumors). Pathogenicity unknown.	Avoidance. Disinfection of eggs with Betadine.
(4)	Infectious Pancreatic Necrosis (IPN), Pancreatic necrosis virus Host: *Salvelinus* species, *Salmo* species.	IPN virus Reovirus	Horizontal spasmodic spiral swimming leading to exhaustion and sinking to the bottom to die. IPN virus infects fry and fingerlings. Mortality is generally heavy during epizootic. Acute affection; death in fry and young fish. Exophthalmia; darkened colour; faecal pseudocasts; mucoid accumulations in otherwise vacuous alimentary tract; necrosis of pancreas, in some cases slight renal or hepatic necrosis.	Avoidance of infection. Sanitary measures. Fish older than 6 months are resistant.
(5)	Viral Haemorrhagic Scepticemia (VHS) Host: rainbow trout (*Salmo gairdneri*) in Europe. Fry and fingerlings.	Egtved virus Rhabdovirus	Dark colour; exophthalmos; bleeding in gills and fins. Haemorrhages in musculature and perivisceral tissue; bleeding in the air bladder; liver discolouration; kidney swellings; enteritis; necrosis of liver, kidneys, spleen and pancreas. Disturbance of osmoregulation with resultant ascites and severe anaemia. Acute affection and death at all ages.	Avoidance of exposure/infection. Vaccine technology has only recently been developed (read *Kaatari* and *Piganelli* 1996)

(Contd)

	Disease	Causative Agent	Etiologic and Pathological Features	Control
(6)	Spring Viremia of Carp (SVC) Host: carp in central Europe.	Rhabdovirus carpio	Pathogenesis is not clearly understood. Dark colour; exophthalmos; petechial bleeding in skin; prolapsed inflamed anus; distended body; serosanguineous exuduation in body cavity; enteritis; intestine filled with yellow mucus; petechial bleeding in heart, kidney, intestine and skeletal muscles; haemorrhage in air bladder.	Avoidance of infection.
(7)	Channel Catfish Virus Disease (CCVD) Host: Channel catfish (*Ictalurus punctatus*).	Channel catfish virus (herpesvirus)	Dropsical condition; exophthalmos; haemorrhages and erratic swimming movements (uncoordinated swimming in spiral paths); acute affection; death in fry.	Avoidance. Disinfection of eggs with Betadine.
II	*Bacterial Diseases*			
(8)	Fish Furunculosis Worldwide except Australia and New Zealand. Host: Salmonids.	*Aeromonas salmonicida*	Diagnosis can be based on signs or the establishment of the presence of gram-negative bacteria in internal organs or lesions ("Furuncles").	Sulfamerazine. Oxatetracycline. Furoxone. Nifurprazine in water or with food. Chloramphenicol. Disinfection of eggs with iodophors. Sulfisoxsole in brown trout. Oral immunization (only under experimental conditions)

* Characteristics of Viruses

Papova viruses: non-enveloped, isometric virion measuring 45–55 nm; 72 capsomeres; double-stranded DNA; replication in nucleus.
Iridoviruses: icosahedron of varying size (190–300 nm) with about 1,500 capsomeres); double-stranded DNA; replication in cytoplasm.
Herpesviruses: enveloped virion about 150 nm in diameter; icosahedral capsid with 162 hollow capsomeres; double-stranded DNA; virus DNA replication in nucleus; virus protein synthesis in cytoplasm; nucleocaspid formation in nucleus; positive Feulgen intranuclear inclusions (Type A).
Reoviruses: non-enveloped virion; 55–75 nm; icosahedral symmetry; 92 capsomeres; double-stranded RNA; RNA-dependent RNA polymerase; replication in cytoplasm.
Rhabdoviruses: enveloped; rod-shaped virus particle of various sizes (70 × 175 nm); single-stranded RNA, RNA-dependent RNA polymerase; replication in cytoplasm; proliferation from the plasma membrane of the cell. (For details, see Ahne 1979).

(Contd)

	Disease	Causative agent	Etiologic and pathological features	Control
(9)	Bacterial Hemorrhagic Septicemia (Infectious Dropsy) Worldwide occurrence. Host: carp.	*Aeromonas liquefaciens** (= *A. hydrophil*)	Two forms of disease: (i) Dropsical, which is acute. (ii) Ulcerous, which is milder and chronic.	Avoidance of infection. Oxytetracycline or Chloramphenicol orally or intraperitoneally. Furnance. Egg disinfection with iodophors.
(10)	Vibriosis (a Gram-negative disease) (red disease) of eels. Worldwide, mostly in marine and estuarine environment; also aquarium fishes. Usually associated with crowding and increased temperatures. Infection may occur through gills or digestive tract.	*Vibrio anguillarum* (Truely marine pathogen)	Signs and lesions caused by vibriosis include petechiae around mouth, opercula and ventral surface of the body. Abcesses and boils are common. Progressive reddening of fins and skin, visceral haemorrhages, reduced activity and death-often preceded by loosening and fraying of the skin. Pathogenesis is still unknown.	Oxytetracycline, nitrofurazone and sulfamerazine orally. Furnance in bath externally. Nifurprazine as a bath. Sulfamethazine. Oral immunization. Avoid infection. *Vibrio* vaccines are one of the most effective aquaculture vaccines to date (Kaatari and Piganelli 1996) Methods used are: injection, oral, and immersion.
(11)	Mycobacteriosis Universal. Common in aquaria.	*Mycobacterium marinum* *M. fortuitum* and related (Myxobacterioses)	Fish affected with mycobacterioses become emaciated and anoretic; some had deformed jaws, vertebrae; exophthalmos, and changes in body colouration; some had ulcers and damaged fins. Grayish white miliary lesions sometime occurred in internal organs.	Avoid infection. Kanamycin with food.
(12)	Columnaris disease ("Cotton wool" or "mouth fungus") Worldwide occurrence.	*Chondrococcus columnaris* (Myxobacterioses)	Characterized by the appearance of shallow, diffuse necrotic lesions of the body surface.	Avoid overcrowding, rough handling, and temperature above optimum for species. Treatment as cold-water disease.

(Contd)

Disease	Causative agent	Etiologic and pathological features	Control
			Oral immunization very promising.
(13) Bacterial gill disease (BGD) Mainly in hatchery. Host: Salmonids.	Unidentified myxobacteria. When fish kept in crowded conditions (Myxobacterioses).	In typical BGD, characteristic lesions consist of proliferation of gill epithelium leading to clubbing and fusing of gill lamellae and filaments. Thickening of the gill epithelium and loss of functional surface tissue due to fusing lead to respiratory insufficiency and eventually death.	Avoidance of crowding. Furnance added to water at 1 ppm for 1 hr. Roccal, Hyamine and other quaternary ammonia disinfectants. Diquat.
III *Disease Caused by Parasite*			
(15) Ichthyophthiriasis ("Ich")	*Ichthyophthiricus* (Intradermal protozoan)	White pustules over body.	Avoidance. Killing of free-swimming forms with: formalin. Malachite geen, formalin with Malachite green, Copper sulfate.
(16) Whirling disease Host: Salmonids.	*Myxosoma cerebralis* (Protozoan parasite) Myxosporidea		Avoidance of exposure to infection. Chemotherapy is not sufficiently advanced but promising. Disinfection of pond bottoms with quicklime or calcium cyanamide.

* Bergey's manual does not recognize such well-known names as *A. liquefaciens*, or fish pathogenic myxobacteria except *Condrococcus columnaris* whose name has been changed to *Flexibacter*.

** Users are advised to follow instructions available with the drugs, and to take under consideration the concentration of the drugs present in trade-name formulations.

Note In control of fish diseases one should avoid ideas based too much on analogies with disease of humans and domestic animals. Fish and shellfish are aquatic and poikilothermic, and these are important differences.

Disease	Causative agent	Etiologic and pathological features	Control
			Note: The parasite is seldom, if ever transmitted directly from fish to fish. The parasite can be transmitted by feeding infected fishes, by other parasites, or requires aging of spores under natural condition.

of stressors at the community level are difficult to establish. They manifest themselves through changes experienced at the individual level and may be marked by a variety of compensatory mechanisms effective at supra-individual levels of integration.

Fishes challenged by stressors respond by compensative measures at the molecular, subcellular, cellular, endocrine, blood, tissue and organ levels. While such measures tend to be similar in different species, species-specific differences have been reported. Typically, a stressor induces the central nervous system of the responding fish to release 'stress hormones' which, in turn, initiate the compensatory measures available and necessary.

Three levels of stress responses are distinguished: the primary level involving the endocrine system; the secondary level comprising blood and tissue chemistry; and the third level manifesting itself in individuals and populations. Reliable assessments of population level responses are difficult to make, except in cases of severe impact and damage. Considering the potentially pathogenic dimensions of stressor effects, it has been suggested that stressful environmental impacts (including those due to pollution) may predispose fishes to biotic and non-biotic diseases (Table 19.5). Also behavioural factors can affect the fishes'disease predisposition, e.g., through excessive intensities of social dominance or of crowding. In aquaculture, vibriosis, viral erythrocytic necrosis, myxobacterial gill disease and bacterial hemorrhagic septicemia have become classic examples of stress-mediated infectious diseases.

Table 19.5 Chemicals which can be considered as potential endocrine disruptors in fish

Heavy metals:	Arsenic, cadmium, chromium, copper, iron, lead, mercury, tin, zinc
Organochlorine pesticide:	Aldrin, chlordecone (Kepone), 2, 4-D, DDT and metabolites, dieldrin, endosulfan, endrin, β-and γ- HCH (lindone), linuron, methoxychlor, mirex
Orgarophosphate pesticide:	Chlorfenvinphos, cythion, diazinon, elsan, fenitrothion, malathion, methyl parathion, mevinphos, monocrytophos, parathion, quinalphos, temphos, TEPA, tetrachlorvinphos
Other pesticides:	Atrazine, carbofuran, carbaryl, esfenvalerate, imidiazole fungicides
Industrial pollutants:	Alkylphenols, ammonia, asbestos, chlorinated paraffins, 4-chloroaniline, cyanide, detergents, di-*n*-butyl phthalate, polyaromatic hydrocarbons, polychlorinated dioxins, polybrominated naphthalenes, β-sitosterol, sulfide, thiourea, urea, acid water, coal dust. Effluents of oil refineries, textile mills, power stations sewage treatment work, vegetable oil factories

Source: Kimc (1999)

Stress-mediated fish diseases may serve as indicators of environmental deformations, including those due to man's activities (i.e., biological monitoring). Infectious diseases due to bacteria (aeromonads, pseudomonads, Myxobacteria) appear to be particularly well-suited because these pathogens are widely distributed in marine and inland waters. Non-infectious fish diseases that may hold promise as indicators of environmental deformations include chromosomal and structural abnormalities of eggs and larvae, skeletal anomalies and neoplasms.

Much has been documented and discussed on host-pathogen-environmental relations, mechanisms of infections and non-infectious diseases, as well as the effect of fish diseases at the population level. It has been concluded that lack of basic ecological data is hindering the definition of management alternatives for solving many acute fishery resource problems. Schreck

(1996) concentrated only on fish health as related to pathogenic organisms and the immune system as influenced by stress. Stressful situations result in a cascade of events that are transduced centrally and communicated via the nervous and endocrine systems and predispose fish to disease (Fig. 19.9).

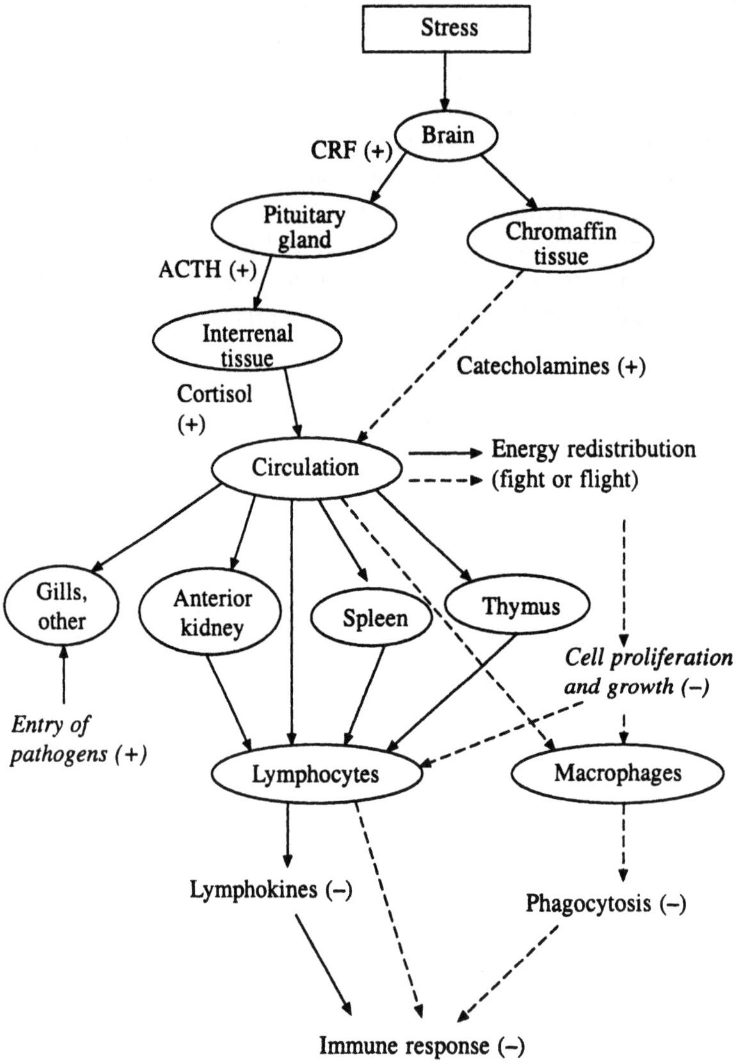

Fig. 19.9 Schematic drawing of the physiological stress response and mechanisms whereby the immune system is affected. Pathways denoted by solid arrows are those mediated by the hypothalamic-hypophyseal-interrenal axis, and those with dashed arrows are mediated by catecholamines. Italicized words denote modes of action that are inferred but not based on direct data (+) and (–) indicate the direction of the response.
***Source:* Schreck (1996)**

Summarily, stress can have marked effects on the health of fish. Prolonged exposure to stress is particularly detrimental to the health of fish. However, even very brief stressful experiences can depress certain of the cellular and humoral immune systems, and consequently lower pathogen-resistance capacity. The physiology of disease prevention and resistance is mediated by the endocrine system, particularly the steroid cortisol. Once a stressor is perceived, a cascade of neuroendocrine events ensues that generally leads to elevation of this steroid hormone in circulation; it then affects lymphocytes and antibody production. Action of this hormone and others appear to operate by activation of specific receptors in the lymphocyte. Stress can also depress macrophage activity and distribution of leucocytes into various body compartments.

The physiological stress response and the immune system are under genetic influence. In addition, the fish's past and present rearing environment affects both the physiological response to stress and the ability to resist pathogens. Different ontological stages have differing magnitudes in the physiological response to similar stressors and differing pathogen tolerances. In addition, the effect of physiological stress factors such as cortisol on disease defense mechanisms change through ontogeny (read Pickering and Pottinger 1995; Pickering 1997)(see Tables 19.6–19.11).

Other points of interest should be briefly mentioned here. Stress affects bacteriocidal activity of leucocytes. Little work has been done to assess the effects of stress on phagocytes in fish.

Table 19.6 Biochemical changes associated with the neuroendocrine response of fish to stress

Determinant	Tissue	Direction of change	Nature of stress
Epinephrine	Plasma	↑	Physical, respiratory, fright
Norepinephrine	Plasma	↑	Physical, respiratory, fright
	Brain	↓	Social (dominant)
Dopamine	Plasma	↑	Physical, respiratory, fright
	Brain	↑	Social (dominant)
ACTH	Plasma	↑	Physical, behavioural, chemical
Cortisol	Plasma	↑	Physical, behavioural, chemical
Cortisol receptors	Liver	↓	Chronic stress
	Gill	↓	Chronic stress
Thyroxine	Plasma	↑	Acute physical
		↓	Chronic stress, starvation
Triiodothyronine	Plasma	↓	Acute physical
Growth hormone	Plasma	↓	Acute physical
	Plasma	↑	Chronic stress, starvation
Prolactin	Plasma	↑	Acute physical
	Plasma	↓	Acute physical
Somatolactin	Plasma	↑	Acute physical
α-MSH	Plasma	↑	Physical
β-Endorphin	Plasma	↑	Physical, thermal shock
MCH	Plasma	↑	Physical
Insulin	Plasma	↓	Toxicants
Renin	Plasma	↑	Ammonia exposure
Testosterone	Plasma	↓	Physical, toxicants
II-Ketotestosterone	Plasma	↓	Physical, toxicants
Estradiol	Plasma	↓	Physical, toxicants
Urotensin I	Caudal neurosecretory system	↑	Low pH
Arginine vasotocin	Brain, pituitary	↑	Low pH

Source: Pickering and Pottinger (1995)

Table 19.7 Effects of stress on endocrine systems of fish other than the sympathetio—chromaffin and hypothalamic-pituitary-interrenal axes

Endocrine gland	Hormone	Nature of stress	Stimulation or suppression (+/–)
Thyroid	Thyroxine (T_4)	Acute physical	+
		Chronic starvation	–
	Triiodothyronine (T_3)	Acute physical	–
Pituitary	Growth hormone	Acute physical	–
		Chronic starvation	+
	Prolactin	Acute physical	+
		Acute physical	–
	Somatolactin	Acute physical	+
	α-Melanocyte stimulating hormone	Physical	+
	β-Endorphin	Physical/thermal shock	+
	Gonadotropin	Chronic cortisol	–
		Acute physical	+
Hypothalamus	Melanin concentrating hormone	Physical	+
Gonad	Testosterone	Chronic physical	–
		Chronic cortisol	–
	II-Ketotestosterone	Chronic physical	–
	Oestradiol	Chronic physical	–
		Chronic cortisol	–
		Acute cortisol	–
Pancreas	Insulin	Toxicants	–
Kidney	Renin	Ammonia	+

Source: Pickering (1997)

Essentially nothing is known about the adrenergic control systems in fishes relative to mediation on physiological factors responsible for health during stress.

Little work has been done to elucidate potential relations between the physical (and perhaps mental) condition of fish and their ability to resist stress, or to determine how this would affect their general health and disease resistance.

Temperature appears to be a major factor that can modify the response to stress. A fish's social environment may also affect its health and general well-being (read Iwama et al. 1999).

12.9 Nutritional Fish Diseases

Nutritional fish diseases are defined as those which can be attributed to deficiency, excess, or improper balance of components present in the food available. Such diseases usually have a gradual onset because symptoms do not appear until one or more of the components of a diet drop below the critical level of the body reserves.

The principal feature determining the protein quality for fish nutrition is the level and availability of the essential amino acids (EAA). Generally the principal feature of EAA deficiency is impaired growth. Since fish have a much more limited capacity for carbohydrate metabolism than higher vertebrates, there is little point in increasing carbohydrate levels and only limited information is available on the effects of high-carbohydrate levels. Many reports have indicated that excessive

Table 19.8 Biochemical changes associated with intermediatory metabolism and growth in stressed fish

Determinant	Tissue	Direction of change	Nature of stress
Glucose	Plasma	↑	All forms of stress
Glycogen	Liver, muscle	↓	Physical, respiratory, fright
		↓	Toxicants, starvation
FBPase	Liver	↑	Physical (handling)
G6PDH	Liver	↑	Physical (handling)
AspAT	Liver	↑	Physical (handling)
GDH	Liver	↑	Physical (handling)
Amino acids	Plasma	↑	Physical (handling)
		→	Restraint stress
Proteins	Plasma	↑	Capture stress,
		→	Physical (handling)
LAN	Plasma	↑	Wasting disorders
Collagen	Skeleton	↓	Toxicants
FFA	Plasma	↑	Physical (handling)
		↓	Physical (handling)
Malondialdehyde production	Liver	↑	Toxicants
RNA/DNA ratio	Muscle, whole body	↓	Chronic stress, pollution, starvation

Abbreviations: FBPase = fructose 1, 6-bisphophatase;
G6PDH = glucose 6-phosphate dehydrogenase;
AspAT = aspartate aminotransferase;
GDH = glutamate dehydrogenase;
LAN = leucine aminovaphthylamidase;
FFA = free fatty acids.

Source: Pickering and Pottinger (1995)

Table 19.9 Biochemical changes associated with the osmoregulatory systems of stressed fish

Determinant	Tissue	Direction of change	Nature of stress
Major ions	Plasma	↓	All stresses (freshwater)
	Plasma	↑	All stresses (sea-water)
Osmolarity	Plasma	↓	All stresses (freshwater)
	Plasma	↑	All stresses (sea-water)
ATPase(s)	Gill	↓	Toxicants, low pH, crowding

Source: Pickering and Pottinger (1995)

carbohydrate intake will result in excessive glycogen deposits in the liver, and continued intake results in extensive lipid deposited in the viscera. Dietary disease problems associated with lipid component of the diets appear to be among the most serious and prevalent of all nutritional problems in fish. Deficiency syndromes result due to insufficient certain essential fatty acids. Linolenic acid is particularly important for normal growth. The deficiency signs in essential fatty acid deficiency almost always relate to a swollen, pale liver with fatty infiltration, and there is a consistent anaemia, putatively associated with the lack of secretion of hemopoietin by the

Table 19.10 Biochemical changes in the defence-detoxifying systems of stressed fish

Determinant	Tissue	Direction of change	Nature of stress
Lysozyme	Blood	↓	Physical, elevated ammonia
Haemoglobin	Mucus	↑	Emersion stress
Chemiluminescence	Macrophages	↓	Physical, anoxia
Cytochrome P-450 isoenzymes	Liver	↑	Xenobiotics
[EROD, AHH, T6H FAH, etc.]	Liver	↑	Xenobiotics
Conjugative enzymes	Liver	↑	Xenobiotics
Metallothionein	Liver, other tissues	↑	Cu, Zn, Cd, Hg

Abbreviations: EROD = ethoxyresorufin O-deethylase;
AHH = aryl hydrocarbon hydroxylase;
T6H = testosterone 6β-hydroxylase;
FAH = fatty acid α-hydroxylase.

Source: Pickering and Pottinger (1995)

Table 19.11 Biochemical changes associated with the reproductive system of stressed fish

Determinant	Tissue	Direction of change	Nature of stress
Testosterone	Plasma	↓	Physical, toxicants
II-Ketotestosterone	Plasma	↓	Physical, toxicants
		↑	Chronic Cd exposure
Estradiol	Plasma	↓	Physical, toxicants
Estradiol receptors	Liver	↓	Cortisol implantation, Physical stress
Vitellogenin	Plasma		Physical, toxicants
Total Ca	Plasma	↓	Low pH
Alkalic-labile phosphorus	Plasma	↓	Low pH
Gonadotropin	Pituitary	↓	Cortisol implantation

Source: Pickering and Pottinger (1995)

compromised liver. Mortality is also high, particularly in young, fast-growing fish. The micro-elements (vitamins) and mineral deficiencies have received a lot of attention (see Table 19.12). For details of dietary mineral toxicity, mycotoxins, toxic algae, etc., the reader is referred to Roberts and Bullock (1989).

13. CONTROL OF FISH DISEASES

The principles of control of fish diseases obviously are the same as those for man and domesticated animals. The health of fishes is more affected by their habitat and environment than is that of warm-blooded vertebrates. Therefore, suitable environment is a very important factor for disease prevention in fish culture.

The increase in fish culture over the last decade had led to an increase in the variety and incidence of diseases. Fish may suffer from diseases not normally found in the wild. Under

Table 19.12 Micronutrients

Vitamin	Deficiency Diseases
Fat-soluble vitamins	
A (Retinol)	Impaired growth Exophthalmos, Keratomalacia eyelens displacement Oedema, ascites Depigmentation Deformation of gill arch and operculum
D (Cholecalciferol)	Poor growth Tetany of white skeletal muscles
E (Tocopherols)	Reduced survival and growth Anaemia Ascites
K (Two forms: phylloquinone, farnoquinone)	Prolonged blood clotting Anaemia
Water-soluble vitamins	
Thiamin	Poor growth, anorexia Hyperirritability Convulsions Loss of equilibrium
Riboflavin (B_2)	Impaired growth Anorexia
Pyridoxine	Poor growth Anorexia Epileptiform convulsions Hyperirritability Lowered resistance to handling Erratic, spiral swimming Rapid breathing and gasping Flexing of opercles Rapid onset of rigor mortis
Pantothenic acid	Poor growth and survival Anorexia Clubbed, exudate covered gills
Niacin	Poor growth and feed conversion Anorexia, skin lesions Anaemia
Biotin	Reduced growth and feed conversion Increased mortality Degeneration of gill lamellae; Skin lesions, i.e., blue slime
Folic acid	Slow growth Anorexia Poor feed conversion Anaemia, pale gills
B_{12}	Anaemia

(Contd)

Choline	Impaired growth Poor feed conversion
Inositol	Anorexia, reduced growth Anaemia
C	Anorexia, reduced growth Lordosis, scoliosis Lethargy Haemorrhagic exophthalmia Ascites, anaemia Intramuscular haemorrhage
Mineral Deficiency Diseases	
Iodine	Thyroid tumor in salmonids Hyperemia on the mouth floor and protrusions at the branchial junction; Exophthalmos
Manganese	Poor growth and dwarfism Cataract Reduced food consumption Loss of equilibrium
Zinc	Poor growth Skin and fin lesions Cataract development

crowded conditions, as a result of intensification of fish culture, epidemics may cause severe losses.

Traditional methods of control by chemotherapy are not always effective, may cause pollution, can be costly, and may be subject to prohibitive legislation. In addition, the pathogen involved may develop drug resistance. Newer techniques of control using vaccines are still in an early stage of development and many problems remain to be solved.

An alternative approach, and one which has been recommended by the FAO is to develop disease resistant strains through the genetic manipulation of fish stocks.

In certain selection programmes the level of resistance of pathogens has been increased, but the resistance is never absolute. One reason for this is the multitude of factors likely to be involved in determining whether or not a fish is resistant. Some of these factors are: barrier action of skin and mucous membranes; destruction of foreign organisms by gastric and intestinal secretions; number and activity of leucocytes; number of erythrocytes; ability of macrophages to digest phagocytosed bacteria; bactericidal properties of blood serum; presence of natural antitoxins; ability to form antibodies; pH of blood; detoxying activity of liver; and efficiency in absorption and metabolism of certain vitamins, or other food factors involved in any of these agencies.

Major genes may cause skeletal or other phenotypic deformities and have been implicated in the development of neoplasia. Others may affect viability through pleiotropic effects, or induce stress and therefore predispose to invasion by pathogens. Resistance usually appears, however, to be of a multifactorial nature.

There is significant variation in susceptibility to disease among fish populations. Where possible, specific aspects of infection process should be investigated rather than mortality alone. Difficulties arise because of problems associated with the pathogen. It may be difficult to give standardized doses as well, and to measure infection levels in fish. Besides the variation in fish, the possibility of variation in the pathogen should also be considered. Resistance genes will not necessarily confer resistance to all races of a pathogen.

Evaluation of variation between populations is quicker, and may be a more profitable approach, than selection within a population. In carrying out a selection programme, proper regard should be paid to possible correlated responses to selection. Selection by independent culling or index selection allows other performance characteristics to be considered together with disease resistance. When disease resistant strains are deployed in different geographical areas, or utilized in diverse cultural conditions, genotype-environment interactions should be evaluated (see Price 1985).

13.1 Fish Vaccination

Fish farming has grown significantly during the past few years. Very often fish like trout and salmon are kept at high population densities. This will increase the risk for dramatic disease outbreaks. Although antibiotics can be used for treatment of bacterial diseases, this also has some drawbacks. Repeated use of these drugs can induce antibiotic resistance in micro-organisms. Moreover, harmful residues may be present in the fish brought to the market. In this context it is not surprising that there is increasing interest in prophylaxis of fish diseases by vaccination. Several good reviews or books are available on this subject (see Gudding et al. 1997) (read Kaattari and Piganelli 1996 for vaccination and antibody response, and Manning 1998). In addition to the usual vaccination method by injection, new procedures for bath or immersion methods have been developed, which are imposing less stress on the animals and which are almost as effective as injection. Oral vaccination will usually evoke only minimal immune responses in the host. It is not easy to explain this phenomenon. Oral vaccination studies have yielded variable results. However, oral vaccination has only been studied in fish containing a stomach, which may function as an important immunological barrier. Research has shown that the intestine in almost all fish can be divided into 3 different segments. The first or proximal segment is involved in digestion and absorption of lipids and proteins. The following segment contain epithelial cells with a clear pinocytotic activity, and the third segment or endgut probably plays a role in osmoregulation. In a recent study, the importance of the second gut segment (of carp, stomachless fish) for antigen transport and antigen processing by macrophages was shown. Moreover, numerous lymphocytes (B and T cells) were present in this segment. These cells probably play a role in local (mucosal) responses (refer Kaatari and Piganelli 1996 for account of mucosal immunity. Repeated oral administration of bacterial antigen resulted in antibodies in skin mucus and bile, but not in serum. It is expected that encapsulation of vaccines is needed to ensure that the essential antigenic determinants will reach the second gut segment in an undegraded and immunostimulatory form. This new approach should allow the development of new and effective oral vaccines in the nature (see Davina et al. 1980; Georgpolou and Verrier 1986; Rombout et al. 1986b, 1989a, b, 1993a, b; Rombout and van den Berg 1989; van Muiswinkel 1992; Koumans-van Diepen et al. 1994a, b). According to Joosten et al. (1995), oral vaccination with bioencapsulated vaccines is very promising when applied at the right age. Application at too young an age seems to induce immunosuppression, instead of immunisation.

The development of viral vaccines for aquaculture industry comes under much scrutiny with regard to safety, cost, and effectiveness. The development of recombinant subunit vaccines that avoid the need of attenuated live vaccines for effectiveness fully meet the criteria for safety and effectiveness. The cost of developing viral vaccines will not be as inexpensive as developing bacterial vaccines because the nature of the technology employed to design and implement viral vaccines is far more difficult (Fig. 19.10).

13.2 Genetic Aspects and Disease Resistance

The identification of genes involved in the regulation of defence mechanisms might provide important clues to understand and perhaps to improve disease resistance in fish. Studies with several mammalian species have shown that the products of a particular set of genes, the major histocompatibility complex (MHC), play a key role in the regulation of the immune response. The MHC incorporates a group of genes which are closely linked (refer Manning and Nakanishi 1996 for details; see Secombes 1996; Bernstein et al. 1998). These genes, which show a remarkable

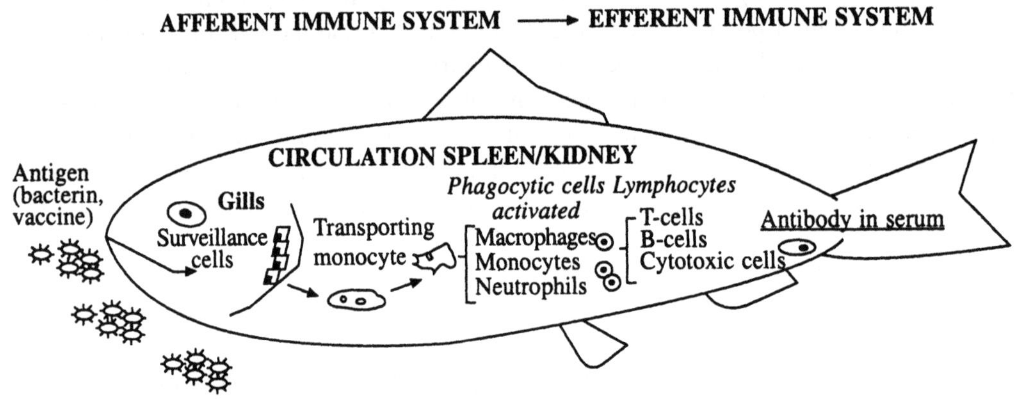

Fig. 19.10 Specific immune response induced in a trout by an antigen bath. The gills absorb the antigens; circulatory monocytes transport the materials to melanomacrophage centres of the immunopoietic organs. The productive response releases humoral antibodies and induces activated cellular responses.
***Source:* Anderson (1990)**

degree of polymorphism, code for membrane glycoproteins which can be divided into 2 classes; I and II (Fig. 19.11). The class I molecules are present on all nucleated cells, whereas class II molecules are more or less restricted to cells of the immune system, i.e., lymphocytes and macrophages. These MHC molecules play an important role in the development of the T cell repertoire (self tolerance) and in the antigen presentation (Fig. 19.12). There is strong evidence that an MHC exists in all vertebrate animals, including fish. Moreover, in the homoiothermic vertebrates like birds and mammals an association has been established between certain MHC alleles and succeptibility to specific diseases. The increasing knowledge of the MHC of fish will certainly be important for our ideas about regulation of the immune response in fish. Several examples of genetic differences in disease resistance in fish have been described, but well-defined genetic markers are lacking. It is tempting to speculate that typing of the MHC will provide us in the future with some of these important markers (van Muiswinkel 1992; see Manning and Nakanishi 1996).

Although there is no need to identify the MHC in all teleostean species, it is likely that it will be investigated in a number of species. The species in which an MHC will be identified are those most used in fish immunological research (e.g., *Cyprinus carpio*, *Oncorhynchus mykiss*, *Salmo salar*, *Oreochromis aureus*, and *Ictalurus punctatus*). In carp, two different symbols are already used to describe major histocompatibility loci, namely K and TLA. In order to reduce confusion

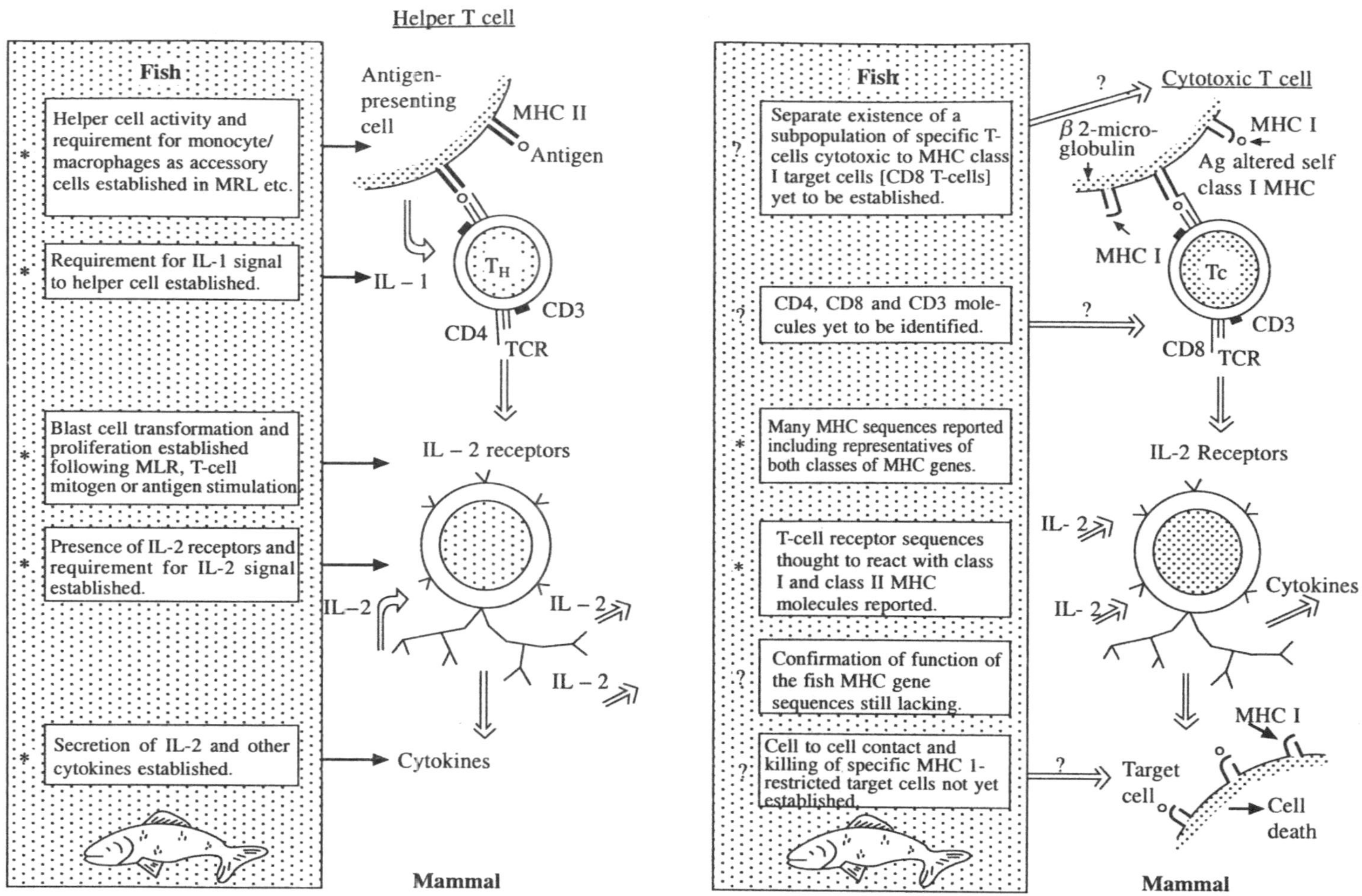

Fig. 19.11 **Present status of research into specific cell-mediated immunity in fish (cf. known responses of mammals). CD3, CD4, CD8, cluster of differentiation (T-cell surface markers, T3, T4 and T8); IL-1, Interleukin 1; IL-2 Interleukin 2, MHC 1, Major histocompatibility complex, class I; MHC II, Major histocompatibility complex class II, MLR, mixed leucocyte reaction; TCR, T-cell receptor.**
Source: **Manning and Nakanishi (1996) (read details)**

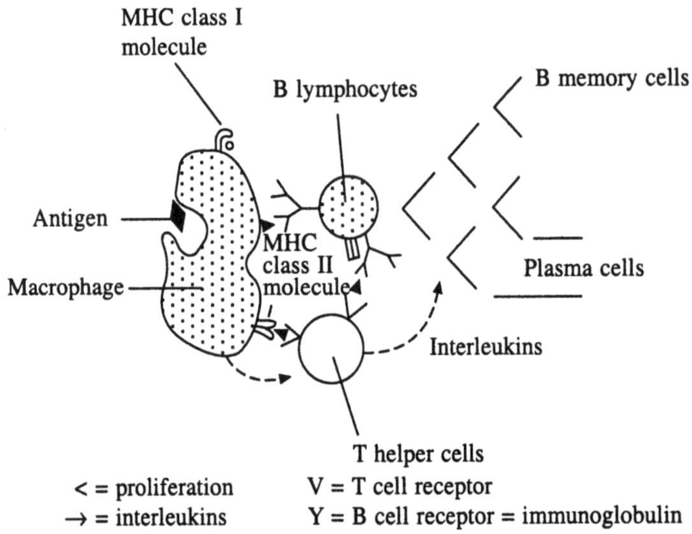

Fig. 19.12 Schematic diagram of cell interactions during the humoral response in vertebrates. Specialized macrophages are able to trap and process foreign molecules or particles, i.e., antigen (♦). These macrophages will present relatively small antigenic determinants (▲) associated with MHC class II molecules to lymphoid cells. Subsequently species T-helper cells (Th) are activated by interaction with the antigenic determinant and factors secreted by the macrophages interleukins. The activated Th cells stimulate the differentiation and proliferation of effector cells as B lymphocytes. B and cytotoxic T cells (not shown in this scheme by secretion of different factors (e.g., interleukins). Depending on the circumstances, B cells will develop into long-live B memory cells or short-lived plasma cells. These plasma cells secrete huge amounts of specific antibodies (immunoglobulins which will bind or kill invading micro-organisms showing the corresponding determinant).
***Source:* Van Muiswinkel (1992)**

in MHC nomenclature, it has been recently proposed to designate the major histocompatibility complexes of the different species according to a standard combination of symbols. Thus, the MHC of carp (*C. carpio*) is proposed to be designated as *Mhc Cyca*, that of the rainbow trout (*O. mykiss*) as *Mhc Onmy*, that of Atlantic salmon (*S. salar*) as *Mhc Sasa*, and that the channel catfish (*I. punctatus*) as *Mhc Icpu* (see review: Stet and Egberts 1991).

13.3 Immunostimulants

The use of immunostimulants in fish culture for the prevention of diseases is a promising new development. At present, substances for treating fish diseases include antibiotics, drugs and chemicals such as those used for sterilizing fish-holding ponds. While each therapeutant is at least partially effective in the treatment of a particular disease, problems arise with accumulation of these substances in the environment as well as the emergence of resistance pathogenic strains when using antibiotics (read Anderson 1996; see Fig. 1, Table VI).

In the past, the immunological approach to preventing fish diseases has been by vaccination against specific pathogens (Anderson 1974; Ellis 1988). Sometimes, the vaccines have contained adjuvants which boost the immune response. Adjuvants are recognized as effective in elevating potency and efficacy of bacterins in fish. Freund's complete and incomplete adjuvants, muramyl dipeptide, FK-565 and newer immunostimulants such as levamisole have been added to vaccines with promising results.

In general, immunostimulants comprise a group of biological and synthetic compounds that enhance defense mechanisms in animals. Immunostimulants such as the beta-glucans, trace mineral and vitamin combinations, and various products derived from many plant and animal sources are effective in preventing diseases when alone, without benefit a vaccine. Several types of beta-glucans seem especially promising for stimulating the non-specific immune response in fish.

When a fish initially encounters a pathogenic micro-organism, the non-specific defense mechanisms are more important than the specific immune response, as the latter requires a longer time for antibody build-up and specific cellular activation. In general, fish have short lifespans and most live in cool water environments which slows development of the specific immune response. These factors may not allow fish to develop the complex physiological pathways that are central to development of an adaptive immune response. The non-specific defense mechanisms, which include phagocytosis and the production of oxidative radicals, are quickly activated by immunostimulants and are rapidly prepared to protect the fish against pathogens. For aging effects on the immune response, see Tatner (1986).

Dietary intake of immunostimulants by fish for the prevention of diseases has definite advantages in large-scale fish culture. In fish farms and hatcheries containing millions of young fish, individual injections are very expensive; immerson, bathing and shower techniques are also costly and involve handling, causing stress. Incorporating an immunostimulant in the feed may be the best way of preventing it to fish. This method could also introduce problems, however, such as how to ensure preservation or prevent denaturation of the immunochemical structure of the immunostimulants from proteases and other antinutritional factors, naturally occurring in many fish feed ingredients. Questions may arise concerning how long feed can be stored without the immunostimulants losing efficacy and potency (i.e., shelf-life), as well as how to mix the substances in the feed for consistency of dose. These problems will be overcome, following examples set by the development in fish nutrition and feed processing technology, which have allowed fish culturists to rely completely on commercially available fish feed.

Dietary intake of immunostimulants by rainbow trout (*Oncorhynchus mykiss*) affects non-specific immunity and protection against furunculosis (Siwicki et al. 1994; see details and literature). Based on the results, the use of immunostimulants for the prevention of diseases in fish culture seems to hold great promise. The fact that we can use specific nutritional supplementation to manipulate the immune response over and above simply providing sufficient vitamins, proteins, and calories in the diet opens a whole new vista for immunology-nutrition collaborative research.

A chapter by Anderson (1996) in respect of this aspect is highly informative.

14. TERATOLOGY

Skeletal anomalies: Environmentally Induced Abnormalities:

Skeletal anomalies, particularly those of spinal column, are commonly observed in fish and are the subject of an extensive literature.

Such anomalies may be genetic, resulting from mutations or recombinations; epigenetic, acquired during embryonic development; or postembryonic, acquired during larval development, at metamorphosis, or during juvenile development. Spinal flexures and compressions, as well as vertebral fusions, have been observed in many teleost species, as have head and fin abnormalities. Evidence exists for a hereditary basis for some skeletal anomalies, but other evidence points to effects of environmental factors such as temperature, salinity, dissolved oxygen, radiation, dietary deficiencies, and toxic chemicals. Recently, increased prevalence of skeletal deformities and anomalies, considered to be pollution-associated, have been recognized in a few fish species for an involvement of various kinds of environmental stress in the occurrence of skeletal anomalies. There is also an appreciable literature concerned with induction of skeletal injuries in fish by exposure to contaminants (Sindermann 1979) (see also Schäperclaus 1992).

The majority of experimental studies that have been conducted on spinal deformities have examined the effects of nutritional deficiencies. Spinal deformities in fish may be induced by deficiencies of vitamin C, vitamin E and the essential amino acid, tryptophan and there are some indications that hypervitaminosis A may cause spinal deformations. Spinal deformities have also been associated will swim bladder dysfunction. Exposure of fish to toxicants has also been reported to result in spinal deformations. (For environmentally induced endocrine abnormalities, read Kime 1999).

15. GENETIC ABNORMALITIES AND EXPERIMENTALLY INDUCED LESIONS

Mutagenicity of marine pollutants has been reported in a number of cases. Sperm cells, eggs and larvae are sensitive to contaminants. It is likely that marine organisms will respond to mutagens in species—specific ways and with different sensitivities.

There is a vast and almost unmanageable literature of published information about the induction of various lesions in fish by experimental exposure to chemical contaminants. Pesticides can, of course, affect fish tissues other than liver. The literature on experimentally induced lesions in fish caused by exposure to heavy metals is unlimited. The contaminant effects on resistance and immune responses have been reported in fish. Environmental stress from contaminants can affect internal resistance to infection in fish by causing a decrease in phagocytic activity or a decrease in antibody synthesis. Both mechanisms have been demonstrated experimentally. Environmental pollution may change the relationships among animal parasites and their fish hosts. There is some evidence for a causal relationship between specific pollutant chemicals and fungus parasitization of fish. Figure 19.13 is a scheme for fish health monitoring.

16. REGENERATION OF TISSUES WITH FOCUS ON INTERNAL ORGANS IN FISH: AN OVERVIEW

Regeneration is a term used to denote the ability of plants, animals and human beings to restore or regrow bodily parts or organs as a result of injury.

It has often been considered that the fish skeletal muscle is incapable of regeneration; this is untrue. Fish muscle is capable of regeneration although to an extremely limited extent. The regenerative processes are similar to those seen in higher vertebrates and include sarcolemmal sprouting and nuclear reduplication.

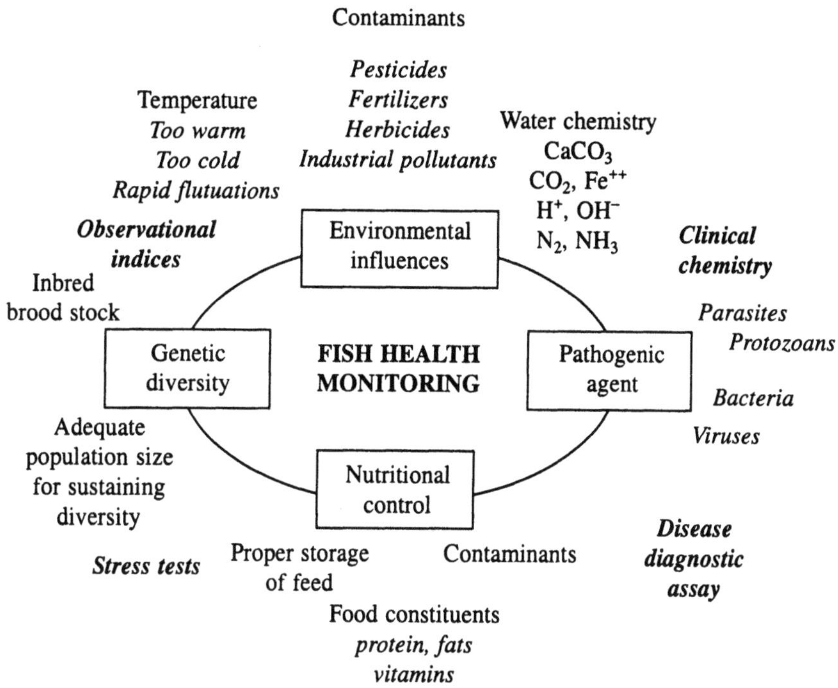

Fig. 19.13 Fish health monitoring involve using shells from various disciplines.
***Source:* Anderson and Barney (1991)**

Interrelations of the individual organs following operations:

The body cavity houses a variety of organs that compete with each other for the available space. If an organ becomes enlarged, this occurs to some extent at the expense of another organ which must contract. It has been reported that the central lobe of kidney becomes enlarged on the side on which the testis or ovary was removed. The spleen becomes enlarged and expands into the space left by the removal of sections of liver. If the swim bladder is removed, the intestine shifts into the space created.

According to investigations and findings so far the carp possesses some ability to regenerate certain inner organs (testis, ovary, spleen) but this depends on whether regenerative tissue in the vicinity of these organs is left in the body after the operation. In the case of other internal organs (liver, kidneys, swim bladder) no regeneration occurs and the wound merely heals.

17. WOUND HEALING

The healing skin wound is affected, like all other metabolic processes, by external temperature. Anderson and Roberts (1975) in a comparison of healing of minor skin wounds in a temperate (*Salmo salar*) and a tropical (*Tanichthys albonubes*) teleost at various temperatures showed that there was a major difference in the rate of wound healing between the highest and the lowest temperatures tested both within and between species. The stress of being held at or beyond the top or bottom of its normal temperature did not per se exert an effect on the healing rate.

The process of wound healing at optimal temperatures, where there is no complication due to the activities of secondary microbial invasion, is similar to that of healing by primary infection in higher animals, viz., there is a pool of tissue fluid and fibrinous exudate, which forms in the wound space and acts a support into which fibroblasts and microphages move to produce an irregular mass of fibrous tissue (read for details: Roberts and Bullock 1976).

Secombes (1996) gave detailed accounts of the (1) cellular events and (2) the control of inflammation. The role of eosinophils in fish inflammatory responses is not clear. The development of inflammatory reactions is controlled by a number of mediators, including cytokines, eicosanoids, complement factors, and other vasoactive compounds released from phagocytes, EGCs/PAs-GL, and thrombocytes.

Cytokines are simple polypeptides or glycoproteins of less than 30 k Da, that act as signalling molecules within the immune system. Conventionally cytokines have been divided into a number of different families, such as interferons (IFN), interleukins (IL), etc. Cytokines play an important role in coordinating leucocyte responses in fish (Secombes et al. 1996; Manning and Nakanishi 1996).

CHAPTER 20
Adaptations to Unique Habitats

1. HYDRODYNAMIC ADAPTATIONS

1.1 Adhesive Organs

Adhesive structures are found in great variety among animals and include various sorts of barbels, hooks, suction cups, and sticky pads. The mechanisms by which these devices operate, be they through use of gluey secretions, suction, surface tension, molecular adhesion, or mechanical grasping, are implicit in their structure. However, tacit assumptions about mechanisms should be treated with caution. In many cases, fine structures and mechanisms of various adhesive devices have not been fully investigated. Many kinds of fishes have adhesive organs formed from modified mouthparts or fins. Such fish can be found in evolutionary diverse families, including the Echeneidae, Sisoridae, Homalopteridae, Gobiidae, Cyclopteridae, and Gobiesocidae.

Suction pressure has been measured in a number of littoral fish, for example *Liparis montagui* is able to vary the pressure with the speed of water flowing over it. Four species of littoral fish (*L. montagui, L. Liparis, Cyclopterus lumpus* and *Apletodon microcephalus*) were able to resist displacement by currents upto 170 cm/s. It has been claimed that the blind side of pleuronectids is capable of acting as a suction disk on smooth surfaces. Pectoral and pelvic fin rays act as 'anti-slip' devices assisting the adhesive organs in many torrential stream fish and possibly also in some littoral species.

A few examples are given here.

The clingfishes: Green and Barber (1988) described the integumental structure and adhesive mechanisms of the ventral adhesive disc of the clingfish, *Gobiesox maeandricus* (Fig. 20.1).

The clingfishes (Gobiesociformes; Gobiesocidae) are a small family with uncertain relationships to other teleost families. All species in the family are characterized by the presence of a concave ventral adhesive disc supported by both pectoral and pelvic girdles. Variations in the structure of the disc are important systematic characters within the group. Gobiesocids are mainly sub-or intertidal species of subtropical and temperate seas but, with their adhesive abilities, some species have invaded torrential tropical mountain streams.

Two distinct types of epidermis are present on the disc surface. Most of the disc is covered with cuboidal epithelium in which four cells types can be distinguished: acidophil cells, mucous cells, epithelial cells, and "alarm substance" cells. The surfaces of the epithelial cells feature whorled microridges typical of teleost epidermis. Near the margin of the disc, on the lateral fin rays, and in the discrete raised patches inside the disc are papillae of stratified columnar epithelium.

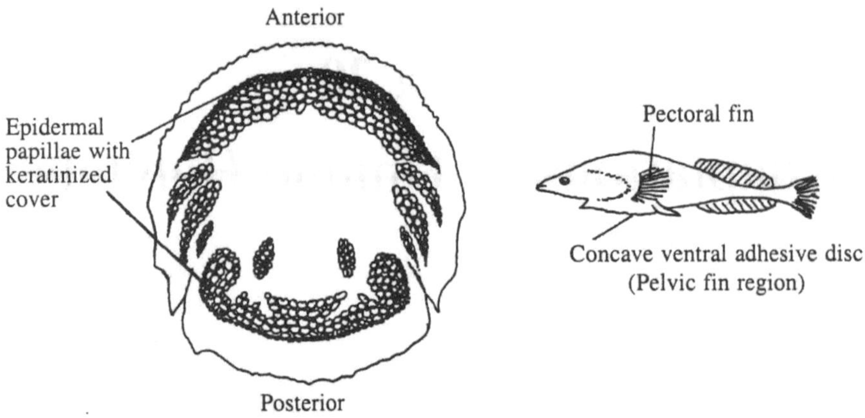

Fig. 20.1 Diagram of the ventral adhesive disc of the northern clingfish, *Gobiesox maeandricus*, showing its general outline in ventral view, the distribution of epidermal papillae on its surface, and its position on the fish.
***Source:* Green and Barber (1988)**

The surface cells of these papillae are microvillous and secrete a non-cellular cuticle which is firmly anchored to the microvillous surface of the papillae. The cuticle is not adhesive. It contains keratin-like proteins but no mucins or mucoproteins are demonstrated histochemically. Connective tissue layers and muscle underlying the epidermis of the disc are arranged to provide structural integrity and shock absorption while allowing some flexibility to the disc and papillae. Negative pressure decays overtime and is periodically renewed by the fish. Thus suction is largely maintained passively and the disc's seal is subject to some leakage. The fish can actively suction pressure when agitated but fatigues and cannot sustain high negative pressures for long. The papillary surface cuticle apparently acts as an abrasion-resistant, frictional surface.

Indian hill stream fishes: Saxena and Chandy (1966; see literature) gave a detailed account on the adhesive apparatus of Indian cyprinid and sisorid fishes (Figs. 20.2–20.4).

The hill stream fishes have developed specialized organs for adhesion in several strategic regions of the body. The adhesive structures are integumentary modifications and serve as highly specialized organs, definite in location, form and size, with their own musculature and nerve supply.

Among the cyprinid hill stream fishes, only the genus *Garra* is known to possess a mental suctorial disc. The general distribution of the mental disc in the genus *Garra* is uniform but the degree of development and area of the disc varies in each species. *G. mullya*, which inhabits comparatively placid streams and lakes, possess a large mental disc, but the frictional devices on the undersurface of its paired fins are feebly developed whereas *G. gotyla*, which is a denizen of more rapidly running streams, has a reduced disc and the frictional devices of the paired fins are better evolved. The epidermal structure of the disc are identical in all species of *Garra* studied, but the differences are more pronounced in the dermal region. As the cornified layer of the epidermis is of no adaptational value to the callous portion, it is very thin but becomes thick and spiny over the border thus resulting in increased friction. The significant part in the organization of the disc is the gradual modification of the dermis in *G. mullya* with increase in body length.

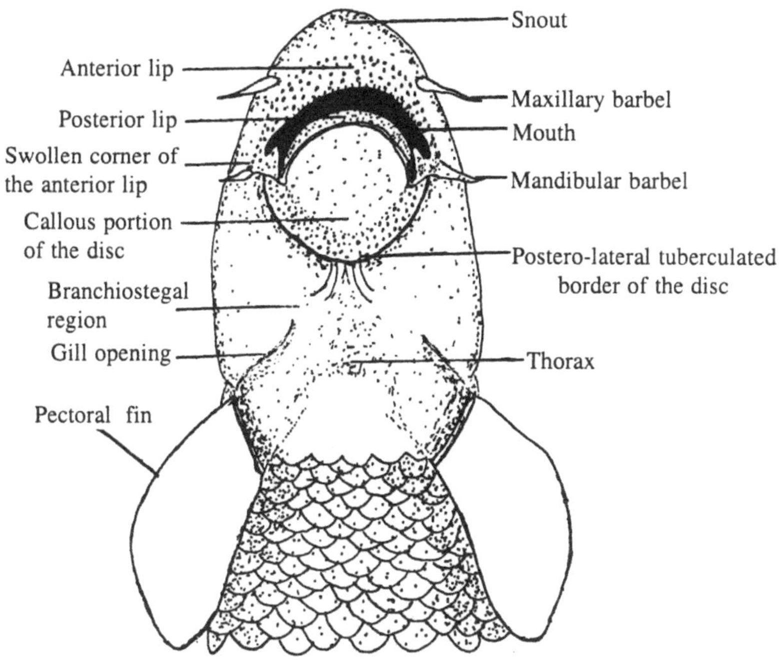

Garra mullya: Mental adhesive apparatus.

Fig. 20.2 ***Garra mullya*. Ventral view of head and trunk.**
***Source:* Saxena (1959)**

For SEM studies on the adhesive apparatus of *G. gotyla* (see Singh et al. 1994). The functional significance of lip-spines and callous pad-mucus has been emphasized.

Increasing complexity in gross anatomy as well as histology is seen in the thoracic apparatus of the sisorid fishes—*Laguvia riberoi*, *Glyptothorax telchitta* and *Pseudecheneis sulcatus*. The simplest condition of the adhesive apparatus is in *Laguvia*. The spines of the stratum corneum, though fully formed, are still slender and less developed than those present on the ridges in *Glyptothorax* (see Singh and Agarwal 1991a; SEM, read Sinha et al. 1990; refer Singh and Agarwal 1993) and *Pseudecheneis* (see Singh and Agarwal 1991b: SEM, 1993). The greatest degree of modification in the structure of spines is met within *Pseudecheneis* (Fig. 20.5). The presence of successive spiny layers and the formation of caps by basal or holdfast cells have not been reported in the spine or its modification in any fish. The varying degree of development and cornification of spines has a close relation with the rapidity of the torrential streams inhabited by the species.

The structure of the thoracic apparatus of sisorid fishes is different from that of the mental disc of *Garra*. What causes the development of two altogether different types of adhesive apparatus in the cyprinid and sisorid fishes which live in a similar ecological niche, is enigmatic. It is all the more so when fishes of the genus *Glyptosternum*, closely related to sisorid genera, are devoid of a thoracic apparatus but possess adhesive ridges on the undersurface of their paired fins. Probably, close study of the habits and habitats of fishes may throw more light on the problem. However, the significance of the absence of a thoracic apparatus in *Glyptosternum* has been utilized in taxonomy.

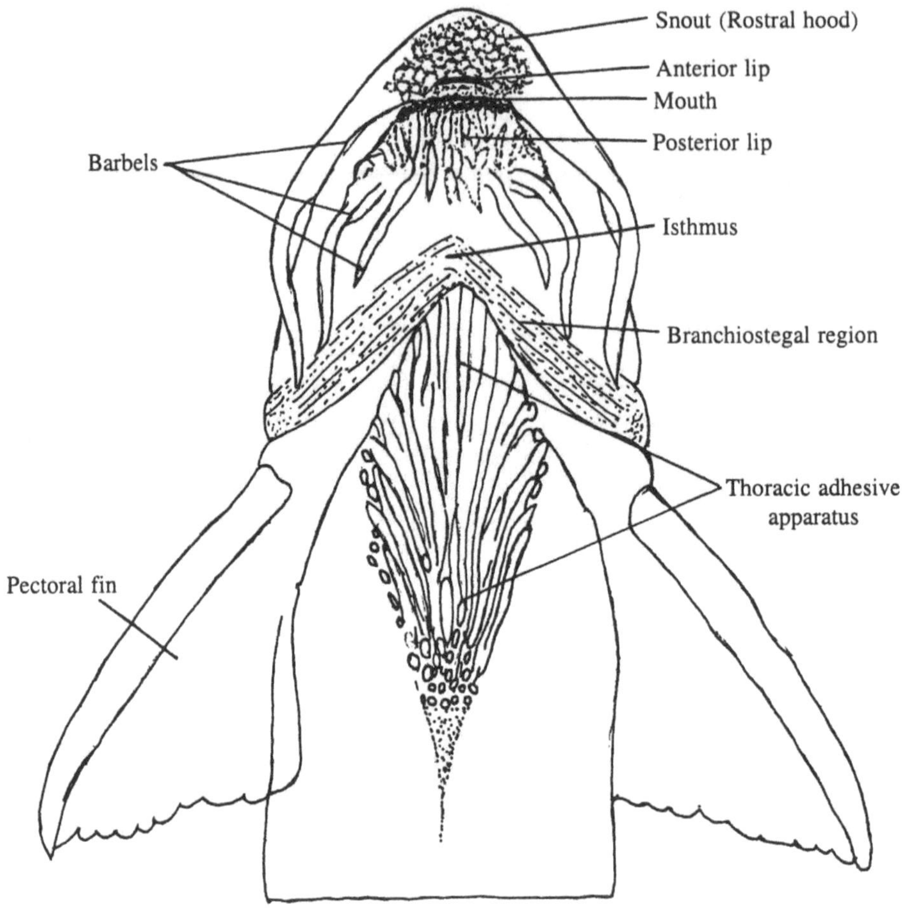

Glyptothorax telchitta: Adhesive apparatus.

Fig. 20.3 ***Glytothorax telchitta*. Ventral view of anterior portion of head and body.**
***Source:* Bhatia (1950)**

Singh (1993), in SEM study of adhesive apparatus of *Crossocheilus latius latius*, pointed out the role of tubercles (with ridges) on labial folds, and the mucus in anti-slip device.

1.2 Streamlining and Boundary Layer

Dorsoventral flattening of the body is characteristic of animals living in fast-flowing streams and it occurs in several families of bottom-living freshwater fish, for example darters (Etheostomatidae), sculpins (Cottidae) and gobies (Eleotridae). In the highly specialized genera *Balitora* and *Glyptosternum* from Indian hill streams, the form is almost leaf-like, but the degree of flattening depends on the strength of the current in the habitats of individual species; there is a complete gradation from the cylindrical form of *Crossocheilus latius*. found in slower water, to the leaf-like form of *Balitora*, found in the fastest water.

Animals of the 'torrential fauna' normally attach themselves to stones and head upstream. They are thus well streamlined and, keeping mostly within the boundary layer, offer minimum resistance to the current. Close contact of the margin fins appear to be important in flatfish,

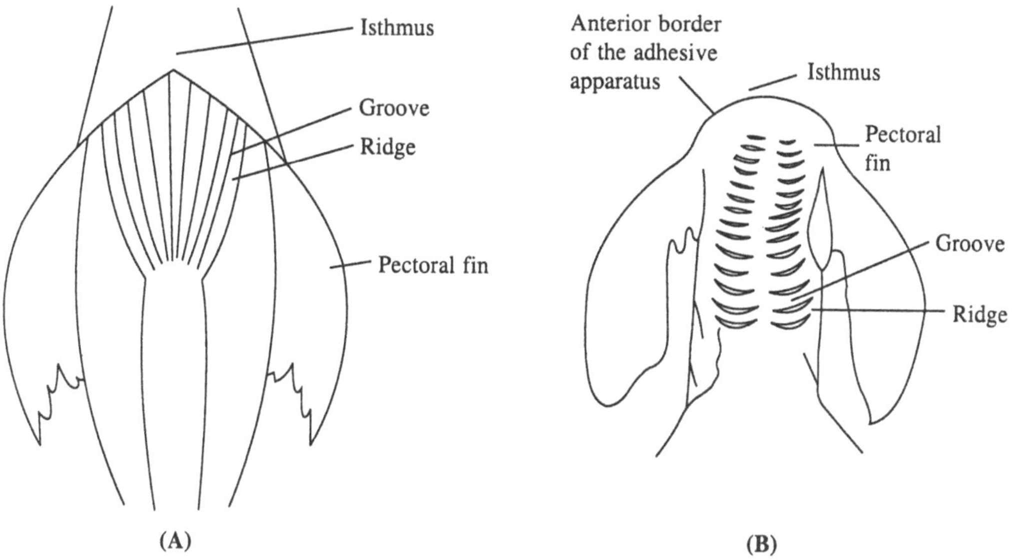

Fig. 20.4 Thoracic adhesive apparatus of (A) *Laguvia riberoi* and (B) *Pseudoecheneis sulcatus*. (Ventral view).

preventing water penetrating beneath the body. Young salmonids, too, which at low current speeds rest with only the edges of the fins touching the bottom, press the entire ventral surface down against the bottom in fast water. Gobies and blennies apparently resist displacement on rocky shores in a similar way (Arnold 1974).

2. DESERT FISHES (WARM SPRINGS) (HEAT RESISTANCE)

Coverage of desert fishes by Deacon and Minckley (1974) gives a comprehensive information about the ecological aspects and the biological responses in deserts.

Desert springs range from swift-flowing origins of creeks (rheocrenes), to quiet, limpid pools (limnocrenes) in caves, in travertine cones, or at ground level (the last perhaps best fitting the classical concept of the oasis), and seepages into raised marshes.

Thermal springs, those which maintain water temperatures somewhat above that of the mean annual air temperature of the area in which they emerge, are sometimes remarkable in their constancy. Chemical constancy of desert spring waters also is a general rule. Springs that are subject to flooding rarely harbour distinctive fish or invertebrate faunas.

Conditions of constant high temperature in springs and of widely fluctuating temperatures in other desert waters may introduce stresses to which only a few fishes have successfully adapted. All fishes have a range of temperature within which they are capable of survival, and they usually have a somewhat narrower range where maximum respiratory efficiency is maintained. One of the warmer records for fishes in North America is that of the spring (44.5°–48.9°C) from which *Lucania browni* (= *Cyprinodon macularis*) was described from northeastern Baja California. Acclimation to warm waters occur much more rapidly than to cold, and thus it follows that heating is likely to have less effect on survival than does rapid cooling.

Some desert species now essentially restricted to constant temperature springs are far more eurythermal that might be expected.

Spring waters are typically circumneutral, buffered by interactions of free carbon dioxide and various ions (principally bicarbonates), but they may sometimes be somewhat acidic. Euryhalinity of some desert fishes seems based upon highly specialized physiological and biochemical mechanisms.

In clear desert pools and in springs, gas production by algae and other photosynthetic organisms in or on bottom sediments may be very high. The spring fishes, *Crenichthys baileyi* and *C. nevadae*, have been examined more thoroughly than other desert fishes with regard to their oxygen relations.

Crenichthys appears to have a markedly enlarged head, which is broad in the opercular region, and this is especially noticeable in larger specimens. Such enlargement accommodates exceptionally well-developed and proportionately large gills, an apparent advantage to a fish living in a low oxygen environment.

Desert habitats, as with most open waters, rarely contain concentrations of free carbon dioxide sufficiently high to create problems for fishes. The highest contents usually are in springs, where in most instances high dissolved solids (especially bicarbonates) tends to buffer any direct effects of the gas.

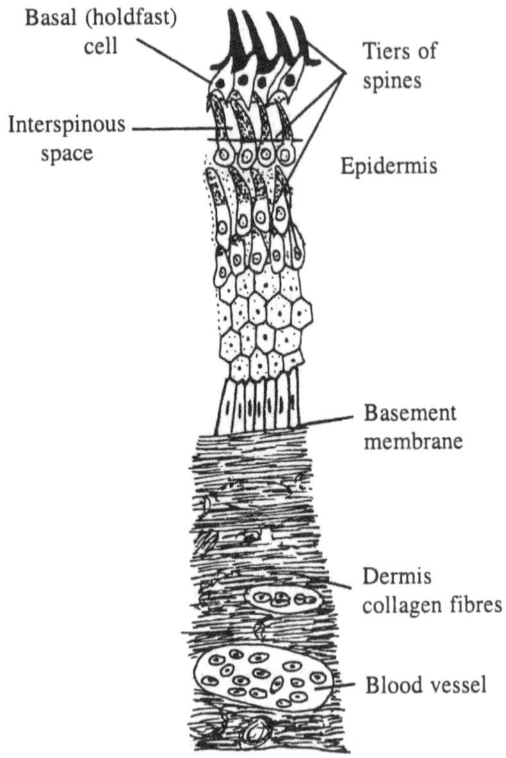

Pseudecheneis sulcatus: L.S. through ridge of thoracic adhesive apparatus.

Fig. 20.5 Magnified portion of the ridge showing caps of first basal cells in the sisorid fish, *Pseudoecheneis sulcatus*.
***Source:* Saxena (1961)**

Fishes native to deserts, almost without exception, have their distribution limited either by geographic barriers, or by a need (or tolerance) for some special environmental feature. Springheads are notoriously depauperate in species, and this holds, generally, for most animal groups. Springheads correspond to the relatively greater constancy of large, low land rivers of temperate zones. Salinities may strongly limit the distribution of species. Temperatures may strongly influence the northern limits of ranges for warm-adapted fishes such as the cyprinodontids, cichlids, and characids. Most desert fishes, particularly those living in springs, seem visually oriented, and are therefore diurnal. The springfishes, *Crenichthys baileyi* and *C. nevadae*, have activity cycles that are light-initiated and dark-inhibited, with some inferred modifying influences by temperature and dissolved oxygen content. Populations of this genus reduce normal activities in response to oxygen levels below 2 mg/liter, temperature above 35°C, or cloud cover. Thermal springs provide a test of sorts for temperature versus day length as primary stimuli for reproduction in fishes, since the constancy of the former may suppress effects of the latter and allow yearround sexual activity. Production of young throughout the year is, however, known from fishes in some thermal springs at higher altitudes. Some fishes living in constant springs exhibit marked seasonality in reproduction that closely resembles that of species of streams and lakes of temperate zones, but the direct relationship of such cycles to physical factors in the environment remains somewhat in doubt.

In warm-spring fishes, the dorsal fin seems to farther back on the body; the head, eye and anterior part of the body is greater in size (perhaps relating to greater gill area); the fins are often more expansive and the body size tend to be smaller. In many respects these features indicate neoteny, and the last characteristics to appear during ontogeny are most labile.

Desert fishes are opportunistic in their feeding, as are most fish species. Foods used by desert fishes provide a pattern of increasing restriction in variety in diet as ichthyofaunal diversity increases. Tolerant forms living alone usually are ubiquitous in feeding. Kinds living in faunistically more saturated habitats exhibit increased specialization of feeding niches.

High water temperatures in deserts may create exceptional nutritional problems for fishes. Metabolic demands at high ambient temperatures must be intense, and a number of authors have noted a maximum in feeding activities at higher environmental temperatures, so long as the temperatures do not rise high enough to suppress activity or to initiate an avoidance reaction. It has been suggested that the system is attuned to a cyclical daily pattern of temnerature, combining maximum intake at high temperature with greater conversion efficiencies at lower thermal levels.

3. PRESSURE ADAPTATION (DEEP-WATER FISHES)

The structure and organic functions of even so highly specialized a vertebrate as a modern spiny-rayed fish show a striking conservatism in their basic plan. But this fundamental pattern is overlain by a palimpset of idiosyncrasies and innovations. Their analysis in terms of the history of the individual organsim and the animal group to which it belongs offers a stimulating challenge.

One inevitable consequence of life in the aquatic environment is an increase in hydrostatic pressure with increasing depth. This physical parameter can have profound consequences for the biology of fishes. Pressure shapes the molecular evolution of deep-living fishes, may be important in the speciation process in the marine environment, and may play a role in determining the bathymetric limits of a species distribution. A variety of data indicate that pressure also influence the behaviour and morphology of deep sea organisms, as well as biochemical and physiological processes (Siebenaller 1987, 1991). Sebert (1997) has given a comprehensive account of pressure effects on shallow-water Fishes.

The deep sea is typically defined as the ocean below the depth of the continental shelf (200 m). The terms epipelagic, mesopelagic, and bathypelagic are used to indicate the vertical distribution of the pelagic species. The epipelagic zone is the photic zone which has a maximum thickness of c. 200 metres, while the mesopelagic zone is the upper parts of the aphotic zone down to a depth of c. 1000 metres. Deeper-living pelagic.species are called bathypelagic in the following, no distinction being made between the bathypelagic and the abysopelagic zones. All species living below the photic zone are regarded as deep-sea fishes. The three terms have been adopted for purely practical reasons. In areas with clear water pelagic deep-sea fishes may possibly be able to perceive daylight down to a depth of c. 900 metres which is roughly coincident with the lower limit of the mesopelagic zone. The residual daylight is thus of some significance for at least some of the species living in the mesopelagic zone, whereas the only source of light available in the bathypelagic zone is bioluminescence.

Studies related to adaptations of animals to light in the sea have been conducted from different angles. The properties of daylight penetrating into the deep-sea are, for depths greater than 200 metres, much simpler than those found above the sea. The light is confined to a narrow wave band in the blue centered around a wavelength of about 475 nm. The angular distribution of light hardly changes with depth, being everywhere symmetrical about the vertical to the surface of the sea. In clear oceanic water, the downward radiance is about 200 times greater than the upward radiance. In very clear oceanic waters, the intensity of light falls rapidly with depth, diminishing about tenfold for every 74 metres increase in depth. In such waters when the sun is high in the sky the downward flux falls from about 10^3 Wm^{-2} at the surface to about 10^{-9} Wm^{-2}, which is close to the threshold of human vision, at about 800 metres.

Deep-water fishes pass their lives in one of the most inaccessible habitats on the planet. They live in a dark cold environment at depths that we can sample in only a fragmentary way. And yet whenever we drop baits into even the deepest ocean, fish soon appear; sometimes very large fish when we consider how little apparent sustenance is available at those depths. How do deep-water fishes spend their lives? Are there well organized, co-adapted communities of fish in the deep or, are there, only random assemblages, brought together perhaps by local changes in the density of food? One approach to this solution lies in the examination of the life history of individual fishes.

Deep-sea fishes occur in all oceans and most major adjacent seas. They are extremely diverse group, including teleosts and cartilaginous fishes. The average depth of the ocean is 3800 metres, with a temperature of 1 to 4°C. Ten metres of depth change in the ocean equals a hydrostatic pressure change of 1 atom.

Evolutionary adaptations to deep-sea life have apparently occurred independently many times in at least some of the 22 orders of fish. Strikingly, the evolution and adaptation of bony fishes to the deep seas, especially to its pelagic environment, have apparently occurred more extensively among taxa that are derived from the relatively primitive group of teleost fishes. Fish of the more derived teleost orders such as the larger order Perciformes, including nearly 150 of the well over 400 families of teleosts have been successful in occupying relatively shallow waters, but have comparatively few deep-sea representatives at the species level (see Weitzman 1997 for systematics of deep-sea fishes).

Physical features

Deep-sea environment is characterized by certain physical features: (1) absence of solar light but presence of biological light, (2) high hydrostatic pressures, (3) low temperatures and (4) oxygen (with increasing depth there appears to be little change in oxygen content except in the "oxygen minimum layers" found at intermediate depths, where the dissolved oxygen in some areas is considerably less than 0.4 m/1, (5) general homogeneity, and (6) a low energy input. Angel (1997) has discussed the nature of water, ocean gradients, morphology of ocean basins, and biophysics and oceanic food webs.

These environmental conditions have long been recognized as posing problems for organisms inhabiting the deep-sea. The adaptations of organisms to the deep-sea environment are manifest in all levels of biological organization from life history and morphology to the biochemical level.

Table 20.1

Features	Mesopelagic, plankton-consuming species	Bathypelagic species
Colour	Many with silvery sides	Black
Photophores	Numerous and well developed in most species	Small or regressed in gonostomatids; a single luminous lure on the females of most certatiods
Jaws	Relatively short	Relatively long
Eyes	Fairly large to very large, with relatively large diopteric parts and sensitive pure-rod retinae	Small of regressed, except in the males of some anglerfishes
Olfactory organs	Moderately developed in both sexes of most species	Regressed in females but large in males of *Cyclothone* spp. and ceratiods (most species)
Cental nervous system	Well developed in all parts	Weakly developed, except in acousticolateralis centres and the forebrain of macrosomatic males
Myotomes	Well developed	Weakly developed
Skeleton	Well ossified, including scales	Weakly ossified; scales usually absent
Swim bladder	Usually present, highly developed	Absent or regressed
Gill system	Gill filaments numerous, bearing very many lamellae	Gill filaments relatively few with a reduced lamellar surface
Kidneys	Relatively large with numerous tubules	Relatively small, with few tubules
Heart	Large	Small

Organization and behaviour of deep-water fishes

The chief morphological differences between mesopelagic and bathypelagic species are given in Table (20.1).

As a general rule there is a decrease in biomass and size of organism with depth. Benthic animals, however are larger and more active than midwater species because of an increase in available food.

Deep-sea animals are mostly predators on plankton or nekton, but since the bathypelagic species have weakly developed skeletons, muscles, eyes and other organs, they must lie in wait tor prey, thus avoiding expenditures of energies in actively seeking and following their food. Those species without a swim bladder must swim much of the time to prevent sinking to the bottom. For a full coverage, the reader is referred to Gartner et al. (1997).

Deep-sea fishes are well adapted to high pressure and low temperatures, and many changes in metabolism would seem to be related to the energy needs of individual animals.

Bathypelagic species have a reduced specific weight and most of the soft tissues assume a gelatinous consistency. The average water content of tissue of the deeper-dwelling species is considerably higher than in shallow-water fishes.

An important behaviour difference is the habit of most mesopelagic species of migrating to the surface or near the surface each night.

The bathypelagic fishes and those mesopelagic species that do not rise to the surface maintain a neutral buoyancy that is facilitated by fat in the liver and swim-bladders, low water temperature, density and viscosity, as well as by the stability of these factors. Fat also functions as an energy reserve. The swim bladder is often absent, thus relieving the fish of the necessity of actively regulating its specific gravity.

A great deal has been learned about the eyes of deep sea animals. There is now a great fund of knowledge about many aspects of bioluminence (refer Chapter 18).

The pineal window is very conspicuous in deep-sea fishes (McNulty 1984). The sensory physiology has been detailed by Montgomery and Pankhurst (1997).

Acanthonus armatus (Ophidiidae)—An example of deep-sea fish:

Fine et al. (1987) reported interesting details of a deep-water (1700–3700 metres) benthopelagic teleost *Acanthonus armatus* (Ophidiidae). The fish has a small tail, most of the body, including the anus, is concentrated in a large head-trunk region, with the massive skull bearing prominent spines on the snout and operculum. The fish lacks a swim bladder and approaches neutral buoyancy by reducing heavy tissues (bone and muscle) and concentrating most of its mass around a huge buoyant cranium filled with low density fluid. Water content as a proportion of flesh mass averages more than 91%. Despite this tissue reduction, *Acanthonus* has been observed swimming slowly just above the sea floor.

Acanthonus has, per unit body weight, the smallest brain and largest semicircular canals of any known teleost and possibly any vertebrate. Pertinent areas of the brainstem and cerebellum are large; this observation suggests that the fish's lateral line and vestibular senses are particularly acute. The huge cranial cavity also contains heavy saccular otoliths, which may indicate that the fish is sensitive to low-frequency sound. Brain size and specialization are consistent with an apparent pattern of low energy requirement, hovering and slow movement over the deep-sea floor, and consumption of small benthic prey in a dark environment.

Similarities have been noted between adaptations of animals to cave and deep-sea environments, and, perhaps not surprisingly, there are a host of similarities between amblyopsid cavefish and *Acanthonus*. Compared with their surface-dwelling relatives, obligate cave-dwellers decrease metabolic rate and the size of the optic system and increase in size and development of the head, cerebellum, semicircular canals, and the lateral line system. Unlike *Acanthonus*, however, cavefish brains are not reduced in size, but do exhibit decreased oxygen utilization.

Brains have high metabolic demand, and numerous studies have shown that other heavy or energetically tissues are reduced or absent in deep-sea fish. The small brains of ophidiid fishes are carried to an extreme in *Acanthonus*. The decreased metabolic demands associated with abyssal existence, coupled with the assumed low metabolism of an animal that is over 90% water, may therefore permit the fish to exist with a small brain. As a result, *Acanthonus* is capable of growing inexpensively to a large size, thereby reducing its chances of being eaten in deep-sea environment.

Anglerfishes:

Deep-sea anglerfishes (Ceratioidei) have the first ray of spinous dorsal located on the skull and modified into an illicium. It is believed to act as a lure for attracting the fish's prey within striking distance of the mouth. In these deep-water forms, which live in perpetual darkness, the terminal bait, or esca, of the illicium is illuminated by a light organ under the control of the animal. The oral and pharyngeal.cavities of the ceratioids are relatively enormous, and many of their other anatomical features are also conditioned by an overall adaptation as "animated fishtrap". The bony skeleton is typically reduced and the musculature, particularly for swimming, is weak. Pelvic fins are lacking and the pectorals, although they retain the basic pediculate structure with elongate radial bones, are reduced in size and scope of movement. Unlike other pediculates, deep-sea anglers are bathypelagic in habit. In keeping with this they have strongly developed black or dark brown pigmentation and the light producing organ of their fishing apparatus. The eyes are small. Certainly the most bizzare characteristic of these fishes is the presence, in certain of them at least, of dwarf parasitic males which become grown fast to the surface of the females. The anatomy of the known parasitic and free-living male ceratioids is quite different from that of the females, which have largely formed the basis for the generalizations about the group. This relationship between the sexes is unique among the vertebrates.

Adaptation as a floating bathypelagic trap has modified many of the *Gigantactis longicirra* structural characteristics: posterior elongation and enlargement of the branchial cavity have displaced the pectorals; specialization of the dorsal branchial elements into a powerful mechanism for swallowing prey has had widespread effects on a number of structures in the preotic region. Special senses of olfaction and vision are poorly developed, facts which are noteworthy in connection with the problem of food getting in general and with angling for prey with a light organ in particular. The presence of the illicium on the skull also has considerable morphogenetic interest because of its widespread effects on skeleton, muscles, nerves, and central nervous system.

The reproductive adaptations of mesopelagic and bathypelagic fishes are correlated with their life history patterns. The eggs and young of bathypelagic fishes run a greater grauntlet of physical changes and predation than those of mesopelagic fishes.

Deep-sea fishes too may have tidal components to their activity patterns. In the Bay of Biscay at depths between 2000 and 4700 metres several species appeared at a trap at low tide but were mostly absent on the ebbing tide. The presence of *Paraliparis bathybius* at a baited trap was only recorded during flood tides, and grenadiers (*Coryphaenoides* spp.) showed faster responses to a bait at a stage in tidal cycle when current speeds were highest, such responses are presumably reflections of the olfactory feeding tactics of the species concerned (Gibson 1992).

Although based on a small sampling of total protein complement of fishes, it appears, at least for enzyme such as dehydrogenases, that fishes colonizing the deep water habitat cope with high hydrostatic pressure by evolving enzymes which are resistant to perturbation. This evolutionary response of insensitivity to hydrostatic pressure additionally solves the problem of the variations in pressure regimes encountered by conspecific populations and individuals. Whether other proteins and membrane-associated systems show other patterns remains to be explored (Siebenaller 1991).

Dehydrogenase enzymes, e.g., lactate and malate dehydrogenases, are especially pressure-sensitive. Species occurring below approximately 500 metres (corresponding to 51 atm pressure) have dehydrogenases with reduced pressure sensitivities relative to shallowerliving species. Thus, selection for pressure-adapted proteins may characterize organisms found in over 80 percent of the biosphere, by volume. The reduced sensitivities to pressure of dehydrogenases from deep-living species are linked to reductions in the catalytic efficiencies (k_{cat} values) of the enzymes, suggesting that adaptation to pressure exacts a price in enzymatic performance. For skeletal muscle actins, pressure adaptations of subunit assembly reactions are observed only in species living below 2,000–3,000 metres. Reductions in volume changes may be achieved by controlling the changes in waters structure, i.e., solvent volume, that accompany catalysis, ligand binding, and protein assembly. These reductions in solvent volume may be effected by reducing the reliance on hydrophobic effects and by closely regulating the shifts in exposure to solvent of water-structuring protein groups over the entire protein surface. Adaptation to pressure may involve amino acid substitutions throughout the protein, not only in the active sites (Somero 1990).

Very few freshwater fishes encounter high pressures. Lake Baikal in Siberia is the only fresh water lake which is 1620 metres deep. Baikal is the deepest known lake, is aerobic throughout its volume, and supports an abundant and diverse fish fauna to its greatest depths.

Baikal presently supports 50 species of fishes representing nine families. One of these families, the cottid Comephoridae, is endemic. The abyssal cottids generally have flabby bodies covered with loose, thin skin. They have reduced eyes and well developed lateral lines. Their bodies are generally whitish or light pale yellow in colour, without spots. Adults of these species generally attain maximum size of 10–20 cm.

The abyssal zone of Baikal has temperatures of 3.4°–3.6° all year round. Light is practically absent. None of the abyssal fishes is known to be luminescent. The oxygen content of deepest water layer is high, usually 75–90% of saturation.

4. POLAR ENVIRONMENT (COLD ADAPTATION: FREEZING RESISTANCE)

Various aspects regarding cold adaptation have been given in reviews (DeVries 1971, 1974, 1978; Hazel 1989; Storey and Storey 1989). To have comprehensive information on different topics, the reader is referred to Biology of Antarctic Fish edited by G. di Prisco, B. Maresca and B. Tota (1991) and Fishes of Antarctica-a biological overview edited by G. di Prisco, E. Pisano and A. Clarke (1998).

In contrast to the temperate regions, the waters of the polar regions are much cooler and have more ice. The Arctic ocean can be thought as a land-locked ocean and, therefore, many of its characteristics are controlled by the climate of the surrounding land masses. The Antarctic in contrast is an ice covered continent surrounded by an unrestricted Southern Ocean. As a consequence, the Arctic region experiences more seasonal changes in temperature than the Antarctic. Thus some regions of the Arctic are ice-free in summer and water temperatures range well above freezing (4° to 7°C). The climate in Antarctica on the other hand is more severe and south of the Antarctic Convergence year-round water temperatures near-2°C are common and ice is abundant. But even here, in some areas small seasonal differences exist in water temperatures and ice cover.

Although the fauna of the polar oceans are not as taxonomically diverse as those of the temperate and tropical seas, they are by no means lacking in numbers of organisms. Around 203 species, 86 genera and 28 families have been recorded from the Antarctic region. The most numerous in terms of species and biomass belong to the suborder Notothenioidei (order Perciformes). They are highly endemic, with 97% of species occurring south of the Antarctic Convergence, Notothenioids are thought to have evolved from a bottom-living ancestral stock which became isolated from temperate waters around 30–40 million years ago following the establishment of

Circum-Antarctic currents. Extant species lack a swim bladder and the vast majority are demersal and relatively sedentary as adults. However, some species have secondarily adapted to pelagic life, enabling them to exploit the abundance of mid-water prey.

The special feature of the environment, and a long period of isolated development, has resulted in the evolution of some unusual forms. The icefishes (family Channichthyidae) are unique among adult vertebrates in that they lack the respiratory pigment haemoglobin, and their muscles contain only minute traces of myoglobin. A common feature of polar fishes is their relatively slow growth rate, delayed sexual maturation, low fecundity, and the production of large yolky eggs. Among notothenioids, oogenesis usually takes two or more years, and there is a lengthened period of embryogenesis which results in the hatching of large advanced larvae with short larval, protracted pelagic juvenile stages. The basic metabolic rates of notothenioids are generally lower than for temperate fish. This is associated with a decrease in protein turnover rates, which in turn is thought to reflect a complex interplay of abiotic (low temperature) and biotic factors (prey availability, reproductive strategy, etc.). In some respects the arctic and antarctic polar marine environments are very similar. Both are characterized by alternating periods of light and darkness, low temperatures and thick ice cover. However, a number of interesting physical distinctions can be made between the two.

The body fluids of most marine teleosts contain only about one third as much salt as does seawater and, as a result, their body fluids freeze between –0.7° and –0.9°C. During winter there are large parts of the polar oceans and coastal north temperate oceans at temperatures of –1.7° to –1.9°C, so that many fishes living in these waters are supercooled by approximately 1°C.

In the Antarctic Ocean a number of fishes exist in a supercooled state throughout life span (e.g., the liparid fish *Liparis* sp. and zoarcid fish *Rhigophila dearborni*). As long as the deep-water Arctic and Antarctic fishes remain in their natural ice-free habitats, their supercooled state is stable enough to permit survival. Tables 20.2–20.4 provide information on certain aspects related to fish fauna in this environment.

Ectotherms generally lack mechanisms for physiological temperature regulation. Consequently, body temperature is determined by ambient temperature and the physical characteristics of the environment. Even the active swimming muscles of teleost fish are greatly within 1°C of ambient water temperature.

The vast majority of ectothermic animals on earth must elude exposure to subzero temperatures to prevent lethal freezing of body fluids. Two strategies occur, freeze tolerance and freeze avoidance.

4.1 Freeze Tolerance and Freeze Avoidance

4.1.1 Biochemical Adaptation for Low Temperature Function

Changes in membrane lipid composition are a primary adaptation for low temperature function. By means of membrane restructure, lipids of appropriate physical properties are matched to the prevailing thermal environment to create membranes with physical properties (fluidity, unit cell dimensions order parameters) largely independent of acclimation temperature and physiological processes (permeability, enzyme and transport activities, receptor and neural functions) which are temperature compensated. The process is termed homeoviscous adaptation. The analogous process of cellular proteins includes: (1) production of new protein isoforms, (2) changes in the

Table 20.2 Freezing and melting points and their differences in Antarctic, Arctic and North-temperate fishes

Species	Water temp (°C)	Freezing point (°C)[1]	Melting point (°C)[2]	Freezing point and melting point differences (°C)
ANTARCTIC				
Trematomus borchgrevinki	−1.9	−2.34	−1.07	1.27
Notothenia gibberifrons	+0.5	−2.24	−1.07	1.17
Rhigophilia dearborni	−1.9	−1.70	−0.94	0.76
Chaenocephalus aceratus	+0.5	−1.45	−0.98	0.47
ARCTIC				
Myoxocephalus scorpioides	−1.7	−2.0	−1.0	1.0
Eleginus gracilus	−1.7	−2.0	−1.2	0.8
NORTH-TEMPERATE				
Myoxocephalus scorpius	−1.2	−1.69	−0.81	0.88
Osmerus mordax	−1.2	−1.22	−0.82	−0.40

1. Freezing point was taken as the temperature of initial ice crystal growth in the presence of seed crystal.
2. Melting point was taken as the temperature at which the last ice crystal melted.

Source: DeVries (1974)

Table 20.3 Seasonal thermal tolerances of some Arctic and Antarctic teleosts

Species	Water temp. and/or thermal history (°C)	Incipient lethal temp. (°C)	
		Lower	Upper
ARCTIC			
Eleginus gracilis (saffron cod)	−1.7°	−1.8°	+15°
Pseudopleuronectes americanus (winter flounder) (winter)	−1.7°	−1.5°	+18°
Pseudopleuronectes americanus (summer)	+18°	−0.7°	+27°
ANTARCTIC			
Rhigophila dearborni	−1.9°	−1.5°	+7°
Rhigophila dearborni (zoarcid fish)	Acclimated to +4°	−1.4°	+8°

Source: DeVries (1978)

Table 20.4 Effect of temperature on electrolyte levels in blood of Temperate, Arctic and Antarctic marine teleosts

Species and season	Water temp. (°C)	Sodium (mequiv/1)	Chloride (mequiv/1)
TEMPERATE			
Fundulus heteroclitus			
warm	+20.0	182	145
cold	−1.5	208	171
ARCTIC			
Myoxocephalus scorpius			
summer	+4.0 to +7.0	–	200
winter	−1.7	216	234
ANTARCTIC			
Trematomus hansomi	−1.9	270	254
Notothenia gibberifrons	0.0 to +1.0	243	220
Trematomus bernacchii			
summer	+0.5	248	231
winter	−1.9	–	254

Source: DeVries (1974)

amounts of enzymes/proteins via alteration of synthesis/degradation rates, and (3) changes in numbers of subcellular organneles (e.g., oxidative potential can be altered by changes in mitochondrial numbers) during acclimation at low temperature.

Homeoviscous Hypothesis of Cold-Adaptation:
An apparently ubiquitous feature of cold-adaptation is an increase in the proportion of chemically unsaturated fatty acids in the lipid component of the cell membrane. This biochemical change serves to maintain an optimum viscosity within the membrane [the Homeoviscous Hypothesis]. Unsaturated lipids tend to be more fluid than their saturated counterparts at any given temperature [contrast polyunsaturated cooking oil with solid (saturated) shortening], thus cold-adapted membrancs rcmain fluid at low tcmperatues, permitting molecular movements within the membrane which are necessary for normal transport of nutrients and ions across the cell membrane. Lipid unsaturation, together with biophysical evidence for increased membrane fluidity, has been reported for cold-adapted representatives of most major groups of living organisms, including bacteria and fishes. Hazel (1995) assessed critically the role of HVA in the thermal adaptation of biological membranes and explored alternative explanations for patterns of thermotropic lipid restructuring.

4.1.2 Freeze Avoidance

Freeze avoidance is the strategy used by polar marine fish to allow survival near the freezing point of seawater (−1.9°C) despite hypoosmotic body fluids. Freezing prevention at subzero temperatures can be achieved in one of two ways: an inhibition of ice nucleation (undercooling) and a colligative depression of freezing point (FP). Adaptation for freeze avoidance exploit both of these mechanisms. Undercooling [maintaining a liquid state below the equilibrium melting point (MP) of body fluids] is the primary mechanism of freezing avoidance in polar marine fish.

Adaptations for freeze avoidance are directed, therefore, at reducing the temperature at which spontaneous nucleation occurs. Two strategies are taken: (1) the elimination of potential ice nucleators including shielding from inoculative freezing by environmental ice and the seasonally active removal or masking of endogenous nucleators, and (2) the addition of specific antifreeze proteins to body fluids which inhibit the growth of embryo ice crystals. These strategies successfully achieve freeze avoidance for polar fish.

4.1.3 Antifreeze Proteins

The blood of cold water fish shows a thermal hysteresis between freezing point and melting point, FPs (the point at which a seed crystal begins to grow) being lower than MPs by about –1.2°C. This hysteresis arises from the actions of specific proteins, called antifreeze peptides which lower the freezing point of body fluids in a non-colligative manner.

4.1.3.1 Fish Antifreeze Peptides

Four structurally different antifreeze proteins—antifreeze glycoprotein (AFGP), and type I, II, III AFPs (antifreeze peptides) from different cold-water fishes-have been well characterized. Glycopeptides are widely distributed in the blood of both Arctic and Antarctic fish; these are composed of repeating sequences of tripeptide (often ala-thr-ala) with a disaccharide residue (galactose and N-acetylglucosamine) attached to the threonine residue (Fig. 20.6). Eight different glycopeptides vary in size from 2600 and 34000 MW. Peptide antifreezes occur in many northern temperate and Arctic fish. Several have a repeating pattern of long sections of nonpolar residues (often alanine accounts for upto 60% of total residues) separated by clusters of polar amino acid residues and with a high alpha-helical content. The protein from the sea raven *Hemitripterus americanus*, by contrast, has a high content of cysteine, little alanine, and a low alpha-helical

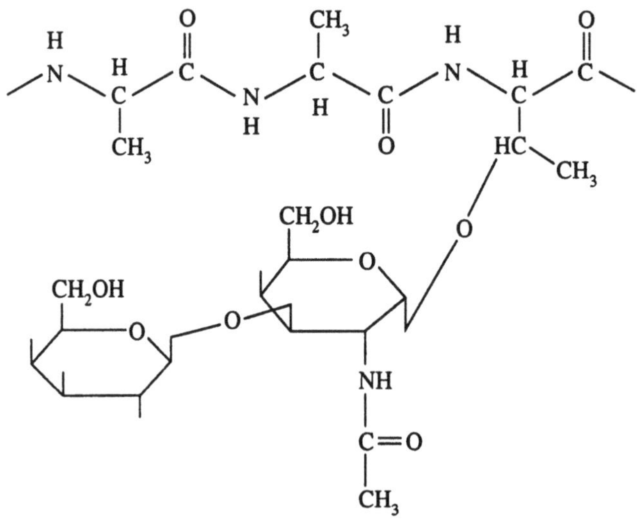

Fig. 20.6 Basic repeating unit of glycopeptide antifreeze molecules. The peptide backbone is composed of two amino acids, alanine and threonine in the sequence of -ala-ala-thr- with a disaccharide of galactose and N-acetylgalactosamine attached to every threonine.
Source: **De Vries (1978)**

content. Cheng (1998) discussed the origin and mechanism of evolution of antifreeze glycoproteins in polar fishes. The evolution of the diverse antifreeze proteins to achieve the same end function qualifies as one of the most interesting model systems for studies of macro-adaptation and molecular evolution.

The mode of action of antifreeze peptides appears to be adsorption to the growing surface of the ice crystal, the secondary structure of the molecules placing the hydroxyl groups of the disaccharide side chains on glycopeptide antifreezers and the aspartate carboxyl and threonine hydroxyl groups on peptide antifreezers in position to hydrogen bond to 0 atoms on the ice surface. Adsorbed antifreeze molecules restrict further ice growth to many small fronts; these have high surface area to volume ratios and growth stops when the ratio exceeds a critical point. Growth can only continue when the surface free energy is lowered, as by lower temperature; thus, FP is effectively lowered.

Antifreeze peptides are present year-round in the blood and intestinal fluid of polar fish but appear only during the winter season in northern temperate fish. Synthesis by the liver in these species is controlled at the level of transcription and responds to both an endogenous rhythms (tied to the reproductive cycle) and to environmental cues [temperature, photoperiod (read Storey and Storey 1989)].

DeVries and Eastman (1981) described the physiology and ecology of notothenioid fishes of the Ross Sea. Perciform fishes of the suborder Notothenioidei are the dominant fishes in Ross Sea and in the Southern Ocean in general. They have evolved physiological adaptations which ensure their survival in ice-laden sea water at a temperature of –1.9°C and extended their habitat range. Among these adaptations are relatively high metabolic rates made possible by enzymes efficient in catalyzing energy-producing reactions at subzero temperature; a restricted range of temperature tolerance; slow growth rates in spite of relatively high metabolic rates; antifreeze glycoproteins which reduce the freezing point of the body fluids below that of sea water and prevent ice crystal growth; aglomerular nephrons which preclude the loss of small molecular weight antifreeze molecules in the urine and mechanisms (especially skeletal reduction and lipid deposition) promoting neutral buoyancy in some species even though swim bladders are absent. Species with different buoyancies are associated with pelagic, cryopelagic, benthopelagic and benthic habitats in the Ross Sea. Physiological adaptations, including buoyancy adaptations in many body systems, and the absence of competition from other fishes allowed this presumably monophyletic group to radiate into several habitats in the Southern Ocean.

5. AN OVERVIEW OF BIOLOGY OF ANTARCTIC FISH

Antarctic fish show many adaptations to their stenothermic cold life histories. The majority of Antarctic fish produce glycoprotein antifreeze compounds which provide resistance to freezing of their hypoosmotic tissues by the subzero environment. J.T. Eastman (1993) published the book "Antarctic fish-biology". The contributions in "Fish of Antarctica" edited by G. di Prisco, E. Pisano and A. Clarke (1998) have updated our knowledge about the Antarctic fish fauna. For special interest in the subject of cold adaptation and stenothermy in Antarctic notothenioid fishes, read Somero et al. (1998).

The blood of Antarctic teleost fishes typically contains fewer erythrocytes and less haemoglobin than that of temperate or tropical species. This feature reaches its extreme in members of the family Channichthyidae, which are the only adult vertebrates known in which haemoglobin and normal erythrocytes are totally absent. Haemoglobinless and naturally cytopenic red-blooded fish species coexist in the Southern Ocean, suggesting that the need for an oxygen-carrying pigment is reduced in both groups.

The haematological peculiarities of Antarctic fishes are considered to be part of a set of characteristics related to a process of cold compensation. The viscosity of corpusculate fluids such as blood is increased substantially by low temperature, thus any reduction in cellular content would tend to reduce the energy required for circulation. At the same time, low temperatures reduce the overall metabolic demand for oxygen, while increasing its solubility in plasma, so that more oxygen can be carried in physical solution, and less needs to be bound to haemoglobin.

Nevertheless, only one of the five families of Antarctic notothenioid fishes has completely forsaken haemoglobin as an oxygen carrier. Why have the Channichthyidae alone taken such a radical course, leaving other families (Nototheniidae, Bathydraconidae, Artedidraconidae, Harpagiferidae) with only partial reductions in haemoglobin? Does haemoglobin remain absolute vital for adequate oxygen transport in the five red-blooded families, or is it a vestigial relict which may be redundant under stress-free living conditions?

The physiological adaptations which enable the family Channichthyidae to prosper without haemoglobin have been discussed in several papers. They include a low metabolic rate; large, well perfused gills; a large blood volume; a larger heart and high stroke volume; large capillary diameter, and cutaneous respiration. The latter is known also to play an important role in gas exchange in several red-blooded Antarctic fishes, but otherwise the red-blooded species lack the morphological features which augment the transport of dissolved oxygen in channichthyids.

It is probably also pertinent that Antarctic teleosts tend to lack the multiplicity of haemoglobins which characterises most other teleosts. All of the Antarctic notothennids except one have only one single major haemoglobin. When present, a minor component (5% of the total) is functionally indistinguishable from the major component. In contrast, blood of most other fishes contain multiple haemoglobins, often showing functional differences which permit oxygen transport even under conditions of extreme acidosis, thereby adapting to environmental changes. In the relatively constant environment of the Antarctic seas, fishes can do well with only a single haemoglobin and under certain conditions, with none at all. For details, articles by di Prisco (1998) and Stam et al. (1998) should be consulted.

Naturally or experimentally-induced anaemia has also been reported in some non-Antarctic fishes, and it has been suggested that many fishes rely on haemoglobin for emergencies, and that oxygen carried in plasma may be sufficient for respiration under minimal working load. Within the spectrum of biological variability, Antarctic fishes represent one extreme in the reliance of vertebrates on haemoglobin-mediated oxygen transport, with the homeothermic mammals and birds lying at the other extreme (see di Prisco et al. 1992).

The majority of Antarctic notothenioid fish also show substantial corporeal accumulations of lipid which are considered to be buoyancy adaptations in this swim bladderless group. For example, lipid has been reported as accounting for up to 37% of the dry weight of oxidative muscle tissues from *Pagothenia borchgrevinki* and *Disstostichus mawsoni* from McMurdo Sound. This feature has led to a prevailing view that lipid is the primary fuel for energy metabolism of Antarctic fish (for discussion, read Sidell 1991; allied references: Hemmingsen 1991; Macdonald and Wells 1991; Bushnell and Jones 1992; Driedzic 1992; Giardina et al. 1998; Wöhrmann 1998).

Notothenioids seem to exhibit the entire range of lateral line morphologies, including free-standing neuromasts, rigid canal systems and membranous canal systems. A study of lateral line function may provide the key to understanding how notothenioids survive the long periods when visual systems are inoperable and that lateral line specializations are a contributing factor to the success of notothenioid fish in the Antarctic ecosystem (Macdonald and Montgomery 1991; see also eye morphology and vestibular labyrinth in this article).

It is generally true that most marine polar fish grow more slowly than related species from warmer water. Explanations have been given by Clarke and North (1991). White (1991) discussed various methods of age determination in Antarctic fish.

In order to capture prey and avoid predators, fish need to swim at very high speeds for short periods. Feeding and escape responses are therefore critical for survival and it seems likely that natural selection would act to maximise burst swimming speed. However, in spite of a long period of evolution to cold conditions, there is evidence that maximum swimming speeds in polar fish are lower than in warm-water species, particularly for larval and juveniles (read Johnston and Johnston 1991).

Antarctic fish have obviously developed mechanisms to compensate for the adverse effects of low temperatures on locomotory performance. Aerobic metabolic capacity and, accordingly, aerobic swimming velocities exhibit clear cold adaptation in Antarctic species (van Dijk et al. 1998).

Loeb et al. (1991) reviewed Antarctic larval fish assemblages. Principal conclusions are: The composition fo Antarctic ichthyoplankton assemblages reflect the highly endemic and predominantly demersal adult fish populations that evolved as a result of environmental cooling, glaciation, and isolation by the Antarctic Polar Front. Seasonal abundance cycles and species succession of the ichthyoplankton assemblages reflect the short, high amplitude production cycles and indicate accommodation to reduce both interspecific competition for food and to enhance predator-prey relations. Distribution patterns within geographical regions reflect: (a) adult fish distributions and spawning sites in shelf and oceanic waters; (b) hydrographic influences (retention versus dispersal zones); and (c) varying degrees of adaptation to pelagic life styles by the early stages. Biogeographical distribution patterns conform to those of the adults and demonstrate similarity of Antarctic continental shelf assemblages relative to peri-Antarctic island assemblages, but also indicate the role of the Circumpolar Current in dispersal of pelagic early stages of some species. The mechanisms by which demersal species maintain their populations over the continental shelves of the isolated peri-Antarctic islands is not understood and requires detailed attention.

Maffia et al. (1993) investigated the adaptation of intestinal cell membrane enzymes to low temperatures in the Antarctic fish *Pagothenia bernacchii*. The results suggested that: (1) the enzymatic activities of the hydrolytic enzymes, alkaline phosphatase, leucine aminopeptidase and maltase, in *P. bernacchii* are comparable to those of a temperate fish *Anguilla anguilla*, despite the very different temperatures of their environment; (2) this characteristic is due to different molecular structures of the enzymes. For further report, see Maffia et al. (1998).

The Antarctic marine environment has a very stable temperature but a pronounced seasonal production cycle. These characteristics have been suggested to be responsible for Antarctic marine organisms having slow but seasonal rates of development. The mechanisms by which low temperature and seasonal variations in food supply may combine to control growth are not clear (Clarke and North 1991).

Nesting behaviour has been reported for an Antarctic fish, the spiny plunder-fish (*Harpagifer antarcticus*), a common shallow-water demersal species. Adult *H. antarcticus* inhabit rubble substrata and spawn 300–1500 eggs into nest during May/July. The nest is guarded by both male and female fish during an incubation period of upto 150 days. After hatching during November/ December the larvae grow at a rate of 0.082 mm d^{-1} (summer) 0.049 mm d^{-1} (winter) and 0.067 mm d^{-1} (annual) calculated from a time-series of samples collected from the field. The rate of growth is slow even among Antarctic species but is similar to the closely related species *H. georgianus*. Adult male and female *H. antarcticus* at the South Orkney Islands attain a maximum standard length of 88 mm and 85 mm respectively.

These is a hypothesis that reproductive strategies among Antarctic poikilotherms are closely linked to the manner and stage at which a species is dependent on the seasonal production cycle (see White and Burren 1992; Christiansen et al. 1998).

Compared with temperate-zone fishes, Antarctic teleost fishes of the family Nototheniidae exhibit a high degree of cold adaptation in the protein secretory system. At water temperatures of –1.5°C, the average liver protein synthetic rate in Antarctic fishes is about twice that predicted and, when extrapolated to 37°C, yields values comparable to mammals. Temperate-zone fishes experience short-term metabolic adjustments during the winter. Antarctic fishes, however, have permanently elevated rates of protein synthesis that probably evolved in response to the constantly cold water.

Lipids play an important role in the biology of Antarctic fishes, especially *Dissostichus* (a swim bladderless, neutrally buoyant fish). The Liveis of *Dissostichus* is responsible for regulating lipid metabolism and for maintaining large peripheral lipid deposits.

Lipid in fish blood is associated with lipoproteins synthesized in the liver. Antarctic fishes exhibit high rates of protein synthesis in most organs including the liver (see Eastman and DeVries 1981).

Sidell et al. (1995) suggested that monounsaturated fatty compounds are the most readily utilized substrates for energy metabolism by oxidative muscle tissues of the Antarctic teleost, *Gobionotothen gibberifrons*.

6. CAVES

Parzefall (1986) gave an account of the behavioural ecology of cave-dwelling fishes. The ecological conditions in caves are characterized by two main factors: nearly all caves have complete darkness and more or less constant temperature.

In the teleostean fish members of about 12 families have colonized caves successfully. They are: Amblyopsidae, Brotulidae (Ophidiidae), Characidae, Clariidae, Cyprinidae, Gobiidae, Ictaluridae, Mastacembalidae. Pimelodidae. Poeciliidae, Trichomycteridae and Synbranchidae. In seven of these families, epigean members are known to be nocturnal and so it is not surprising that those species with a preference for activity in darkness enter caves. But among the ancestors of cave fish, Characidae, Poeciliidae and Cyprinidae include many species with clear diurnal activity. Mainly three of these families (Amblyopsidae, Characidae, Poeciliidae) have been studied for possible behavioural adaptations to cave habitat.

Potential cave-dwellers must have the sense organs and the behaviour necessary to find food and to reproduce in caves.

Suitable food sources and quantity vary from cave to cave. Since there are no green plant producers in the dark, cave fish depend upon food brought in from the outside by the animals or by floods. Compared with the above-ground habitats, most caves do not have an abundance of food. An exception is the sulphur cave where live-bearing tooth carp *Poecilia mexicana* exist at high population density and which has an abundance of food. In various cave populations of *Astyanax mexicanus* the population density is generally higher in the omnivorous fish of epigean habitats. In general the epigean fish get their food visually in open water, whereas cave fish are guided to food by water movements and chemical signals from the food. The cave fish feed at an angle about 45° sustended from the ground whereas the epigean form stands vertically on its head. This difference in behaviour depends on genetic factors, and is inherited independently of eye size and number of melanophores. In addition the taste buds, restricted to the mouth region, are found over the lower jaw and cover the ventral areas of the head in the cave fish. Such an evolutionary improvement of the gustatory equipment needs a lower angle of the body position during feeding, and both these traits could be achieved by small genetic steps during a process of integrative interaction (see also Schemmel 1974, 1980; read Bardach and Atema 1971).

Certain members of Amblyopsidae (e.g., *Chologaster agassizi*, *Typhlichthys subterraneous*, *Amblyopsis speles*), are cave-dwellers. The fish population density in cave is extremely low. Their food, consisting of aquatic invertebrates, is also very scarce. As a consequence cave fish species must swim a great deal to get enough food. Their behavioural changes for food procurement are combined with adaptive changes in the system of free neuromasts and brain anatomy.

Having enough food to reach maturity, the next problem to be solved by cave fish is finding a sexual partner in darkness. Subsequently they need behaviour patterns which provide effective fertilization in the absence of any visual orientation.

For male *Poecilia mexicana*, which actively seeks mates, it is easy to find the female because of the very high population density in the cave. The male checks conspecifics by nipping at the genital region. Females are recognized by a species-specific chemical substance which is produced continuously by mature females. An annual reproductive cycle is known in Amblyopsidae.

Chemical communication is the most important mode of communication in caves. Acoustic and tactile stimuli also exists, but few exact data are available.

It is striking that, despite regressive evolution for agonistic behaviour, schooling, fright reaction and circadian activity, no example exists that shows the gradual reduction of an optical display among cave-dwelling animals. For this reason the absence of visual displays is probably a necessary assumption for the successful colonisation of a lightless habitat.

In some of the caves, food is extremely scarce. Cave-dwelling fish respond to severe food limitation by a decreased metabolic rate and fecundity.

Maxican blind cave fish *Astyanx hubbsi* outperforms visual fish in tasks depending on lateral line output. Its chemical sense is much better developed than that of sighted fish. The auditory capacity in the blind fish is equal to that of eyed ancestor river fish, *A. mexicanus.*

Parzefall (1970) reported that the fleshy genital-pad in females and the number of taste buds in males and females of caveliving *Mollienesia sphenops* show constructive tendencies in comparison with its epigean ancestor and *M. velifera*. The pores of the cephalic canals seem to tend to have incomplete roofing in the cave form. Dijkgraaf (1963) reported that in the blind cave fishes the number of superficial neuromasts with their freely protruding cupulae is sometimes increased so much as to give the whole head a wooly appearance (see Fig. 15.31). The article of

McNulty (1984) covered the pineal organ structure and function and emphasized on some relationships between environmental light and ultrastructural appearance of outer segment. The extent of outer segment degeneration in fishes which live in caves and burrows is variable and appears to depend on the degree to which they are restricted to their underground zones. For instance, the pineal organ of the facultative cave dweller, *Chologaster agassizi*, which is active in springs at night and retreats underground during daylight has photoreceptor cells with normal outer segments showing few irregularities. The eyes of *C. agassizi* are small and probably do not form an image. Progressive morphological adaptations to an underground environment include a reduction in eye size and loss of body pigmentation.

CHAPTER 21

Migration; Schools (Shoals); Habitat Maintenance

1. MIGRATION

1.1 Introduction

Animal migration is a dynamic and diverse process. Running through the diversity is a common thread of adaptation to changing habitats; as seasons change or habitats deteriorate, animals move to new areas, but the distances moved may be long or short, orientation may be simple or complex, and all, part or none of the individuals in a population may migrate. Clearly natural selection has shaped a diversity of patterns and new configurations are still evolving (Dingle 1991).

The spectacular long-distance migration and the mysterious homing ability of fishes have intrigued layman and scientists alike. Research in the past decade has produced major advances in our understanding of the mechanisms by which fishes perform these feats.

There are about 8000 species of fish which live in freshwater and a further 12000 which live in sea; and there are about 120 diadromous species which move regularly between the two environments. Fish are not distributed evenly through the world. About 100 species are found within the Antarctic convergence which suggests that there may not be more than 200 or 300 species in the polar latitudes. There are many more species in temperate waters but the majority-probably at least 70% of the total—are found within the tropics where over 500 species may be found on a single reef (see Harden Jones 1981).

The migrations of diadromous fish species are particularly enigmatic because this behaviour necessitates physiological changes in osmoregulation. Diadromous fishes are found in 28 families and include two distinctly different groups: (i) 87 anadromous species, such as salmon (Salmonidae) and lamprey (Petromyzontidae), which are born in freshwater, migrate to the ocean, and return to freshwater to spawn; and (ii) 41 catadromous species, such as some eels (Anguillidae) and mullets (Mugilidae), which are born in the ocean, migrate to freshwater, and return to the ocean to spawn. The existence of these contrasting directions of migration has long been perplexing. Indeed, it has been described as a paradox in animal migration (Gross et al. 1988).

Many fish are restricted in that they are limited to one geographical area within which they make wintering, spawning, and feeding migrations.

Migration itself is probably an adaptation towards abundance, members of a stock being able to draw upon the resources from a wide area at different stages of their life history and at different seasons. It is perhaps no accident that species of commercial importance are migratory: they are abundant because they are migratory, and they are fished because they are abundant.

Migration is defined as a coming and a going with the seasons on a regular basis and a distinction is made between local and more extensive movements in terms of fish-body lengths (10^3L and 10^5L, respectively), the former being likely to be associated with scalar and the latter with vector quantities. It is suggested that *clue* and *cue* be used, relative to migration, in the sense of information as *where* and *when* respectively. The open ocean is considered as the sea area where oceanic currents are stronger than tidal currents and the importance to migration of the circulatory features of the ocean, at different scales, is discussed. Boundaries (at the surface, between water masses, and near the bottom) are identified as sites where migrants might obtain directional clues (Harden Jones 1984).

Migration is defined as those movements which result in an alternation between two or more separate habitats (i.e., a movement away from one habitat followed eventually by a return to it again) occurring with a regular periodicity (sometimes annual but certainly within the lifespan of an individual) and involving a large fraction of the population (Northcote 1984).

Migration involves elaborate organization within an individual animal and between an animal and its environment so that all aspects are properly integrated in time and place (Meir and Fivizzani 1980).

Migration in fishes logically has its evolutionary origins in the "migration" of required or preferred environments. The extent to which fishes now have come to rely on sign stimuli to guide their migrations must be measured against antecedent responses that are strictly enviroregulatory (Neill 1984).

Most research on migration is concerned primarily with the phenomenon itself and not with effects on related attributes such as body size, age at first reproduction, etc. This is particularly for work on fish in which few attempts have been made to analyze the relationship between migration and life history characteristics (Roff 1991).

A well documented case of diminishing riverine-marine migration involves the Atlantic salmon, *Salmo salar* (Cook 1984; read Brower and Malcolm 1991). We have chosen among a plethora on the subject aspects on fish among the animal taxa though most attention has been accorded to birds and insects. The findings from studies with diadromous species will be drawn upon extensively since they represent a major part of the available data.

1.2 Movements of Fish in Relation to Water Currents

Three-quarters of the Earth's surface is covered by water. Most of this area is occupied by the oceans, which are interconnected and are, on average, about 4000 m deep. The shallower shelf areas, which are less than 200 m deep, occupy a much smaller area but are equally important, if not more so, because of their high biological productivity.

This aspect has been treated in depth. Water currents are thought to provide either a transport system or directional clues for migrating fish. Transport can occur by passive drift, if the fish is a pelagic organism with no external reference points, or by modulated drift, if the fish makes vertical movements in the water column. Orientated movements may occur if the fish can detect the speed and direction of the current.

Passive drift: The pelagic eggs and larvae of both freshwater and marine fish are displaced passively downstream from their spawning sites by the current. Juvenile salmonids, too, probably drift in lakes, reservoirs and rivers and it has been shown that in large lakes they are displaced passively by wind-driven surface currents. Onshore winds apparently also result in the concentration of adult salmonids at the mouths of their spawning rivers. Drift may be important, too, in the migrations of fish in the open sea; it has been suggested for cod, herring, Atlantic salmon and tuna.

Modulated drift: Vertical migration provides the simplest basis for modulated drift, the direction of displacement being determined by the direction of the current when the fish makes it vertical movement.

Orientated movement: Fish whose eggs and larvae are displaced passively from their spawning grounds by the prevailing current must at some point make compensatory movements in the opposite direction to remain within a given geographical area.

Currents: Four distinct current regimes can be recognized (Fig. 21.1). Oceanic currents dominate the high seas while tidal currents are more important on the continental shelves. In estuaries the discharge supplements the ebb tide; in rivers there is only the downstream flow (flow to the sea).

The surface circulation of the main ocean basins is driven largely by the prevailing winds. Its basic pattern is a nearly closed system called a gyre. The northern and southern limits of each gyre are marked by zonal currents flowing east and west. These are joined by meridional boundary currents setting nearly parallel to the continental margins.

The distribution and migratory patterns of many stocks of fishes would appear to be closely related to these gyres, and what might be called the "doctrine of hydrographic containment" now enjoys a measure of respectability. But the ocean currents are neither as simple nor as consistent as the usual charts suggest. The effect of meanders, rings, and eddies (Fig. 21.1) on the abundance, distribution and movements, of fish is something about which little is known. Within the basins occupied by gyres there are water masses with differing temperatures and salinities. At the frontal zones between the water masses horizontal temperature and salinity gradients are more greater than on either side. The zones themselves may vary from a few hundred metres to several thousand metres wide that is from 10^2 to 10^4 times the length of a fish and the temperature gradients may range from greater than 0.01 C/m to 0.001 C/m.

In the open ocean there are features of the vertical structure of the water column that may be important for fish on migration (Fig. 21.1). While the temperature structure is the most obvious-a warm surface layer often lying above a clearly defined thermocline. Current velocities change with depth, and while the transport possibilities offered by undercurrents and countercurrents are generally well known, it is not clear if fish make use of the directional information likely to be available across the rheocline. Furthermore the possibility that fish use the directional information which is probably available from gravity waves at the sea surface, or along internal boundaries, has not yet been seriously considered.

Figure 21.2 shows the relation between oceanic gyrals, production, spawning and fish migration in different water masses.

Tidal currents, which are generated by combined gravitational effects of sun and moon, are

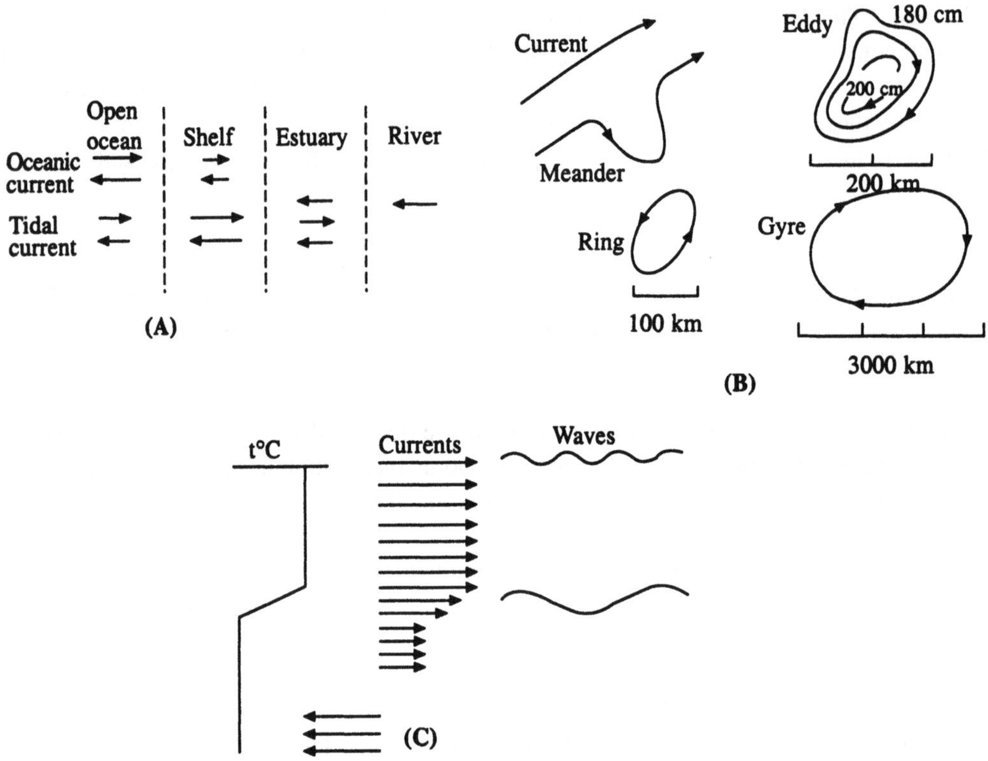

Fig. 21.1 (A) The relative importance of oceanic and tidal currents in the open ocean, continental-shelf waters, estuaries and rivers.
(B) Some horizontal features of the open ocean of importance to the study of fish migration. In the eddy the dynamic height contours are given in centimetres, and the anticyclonic flow is clockwise in the northern hemisphere (as shown) but anticlockwise in the southern hemisphere.
(C) Some vertical features of the open ocean of importance to the study of fish migration: temperature structure; current structure; surface and internal gravity waves.
***Source:* Harden Jones (1984). For (A), see also Arnold (1981)**

present in both oceanic and shelf waters. But their effects are much more pronounced on the continental shelves. Near coasts the rotatory tides of the open sea are replaced by the familiar reversing tides of the seaside. Flood tides, where the water flows towards the coast, are separated by periods of slack water from ebb tides, where water flows away from the coast. Three types of tide can be recognized: diurnal, semidiurnal, and mixed. The daily tidal range is greatest during spring tides, which occur at times of new and full moon, and least during neap tides, which occur near the first and third lunar quarters. The speeds of the tidal currents vary accordingly. An estuary is a semi-closed tidal basin, in which the circulation is a mixing system for fresh and salt water.

The migration pattern or circuits: In general terms the movements of migratory fish are related to those of the water currents. Life histories are geared to regional production cycles and stocks are contained within oceanic gyres. The commercial fisheries are based on migratory fish which winter, spawn and feed in different areas, the distribution of a particular stock often appearing

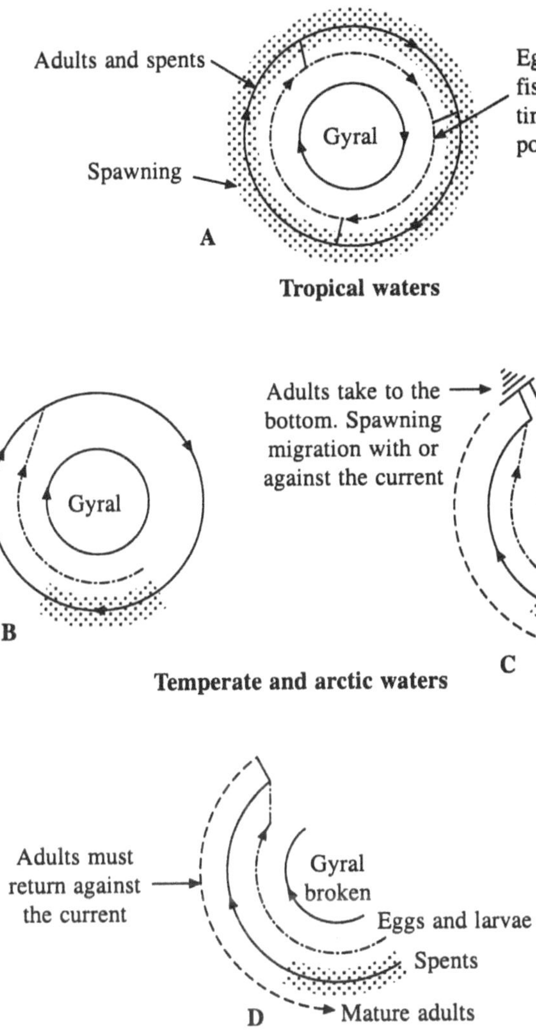

Fig. 21.2 The relation between oceanic gyrals, production, spawning and fish migration in different water masses. A. Gyral restricted to tropical waters; production and spawning continuous. Gyral extends into temperate and arctic waters. B. Production and spawning restricted, gyral complete. C. Production and spawning restricted, gyral complete, but adults may become demersal or there may be an environmental barrier which prevents them completing the circuit. D. Production and spawning restricted, gyral broken.
***Source:* Harden Jones (1965)**

to be contained within a regional hydrographic circulation. The migration patterns of fish can be summarized in the form of simple triangular diagrams as shown in Fig. 21.3. Figure 21.3(a) includes the juvenile and adult components of stock, while Figs. (b) and (c) deal with the mature fish. The mature fish move between their spawning, feeding and wintering areas in a sequence which depends on the season in which they spawn: the spring spawners would follow the sequence F, W, S [(Fig. 21.3 (b)] and autumn spawners the sequence F, S, W [(Fig. 21.3 (c)]. The distance covered during year—the length of the migration circuit—varies between stocks. In temperate and arctic waters there are discrete spawning, nursery and adult feeding grounds and, concomitantly, distinct spawning, feeding and wintering migrations. Eggs and larvae drift downstream with the residual current to the nursery grounds and at a later stage in the life history the adult fish must make a compensatory movement if the population is to remain within the same hydrographic system. These movements are described as denatant and contranatant, respectively. Denatant means swimming, drifting or migrating with the current. In the classic contranatant theory of fish migration, it is inferred that prespawning migrations are active contranatant movements against the prevailing current, while postspawning migrations of spent fish are passive denatant movements with the prevailing current.

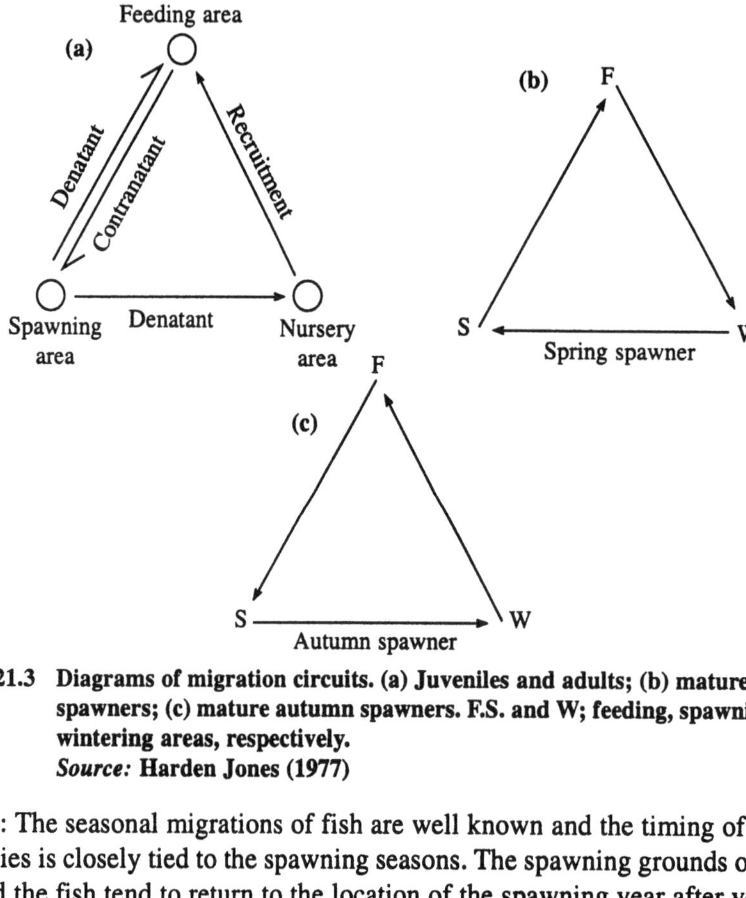

Fig. 21.3 Diagrams of migration circuits. (a) Juveniles and adults; (b) mature spring spawners; (c) mature autumn spawners. F.S. and W; feeding, spawning and wintering areas, respectively.
***Source:* Harden Jones (1977)**

Life histories: The seasonal migrations of fish are well known and the timing of the migrations of many species is closely tied to the spawning seasons. The spawning grounds of many species are fixed, and the fish tend to return to the location of the spawning year after year (Fig. 21.4).

The classical examples of diadromous fish such as salmon (Salmonidae) and eels (Anguillidae) migrate between the sea and fresh water and pass through all four current regimes during their life history. Many more species of truely migratory fish are limited either to the sea or to freshwater. Life histories vary accordingly.

The spawning migration of the salmon is a dramatic event which has stimulated much interest; furthermore salmon are commercially important fish and they are easily and readily available for study during their life in freshwater. Many species of salmon are anadromous spending much of their life history in the sea and only returning to freshwater to spawn. The eggs are laid in gravel beds, or redds, and in most species the young fish migrate to sea as smolts after two or three years in nursery streams or lakes. In two species of Pacific salmon the young migrate to sea as fry shortly after hatching. Eels are catadromous, spawning in the sea and spending most of their adult lives in freshwater. The European eel *Anguilla anguilla,* migrate in the fall from their freshwater habitats and travel to spawning areas in the Sargasso sea. The adults spawn in the early months of the year and then die. The European eel experiences an extensive drift migration across the North Atlantic during its larval stages (leptocephali). Newly metamorphosed elvers arrive in European coastal water two and half years after spawning.

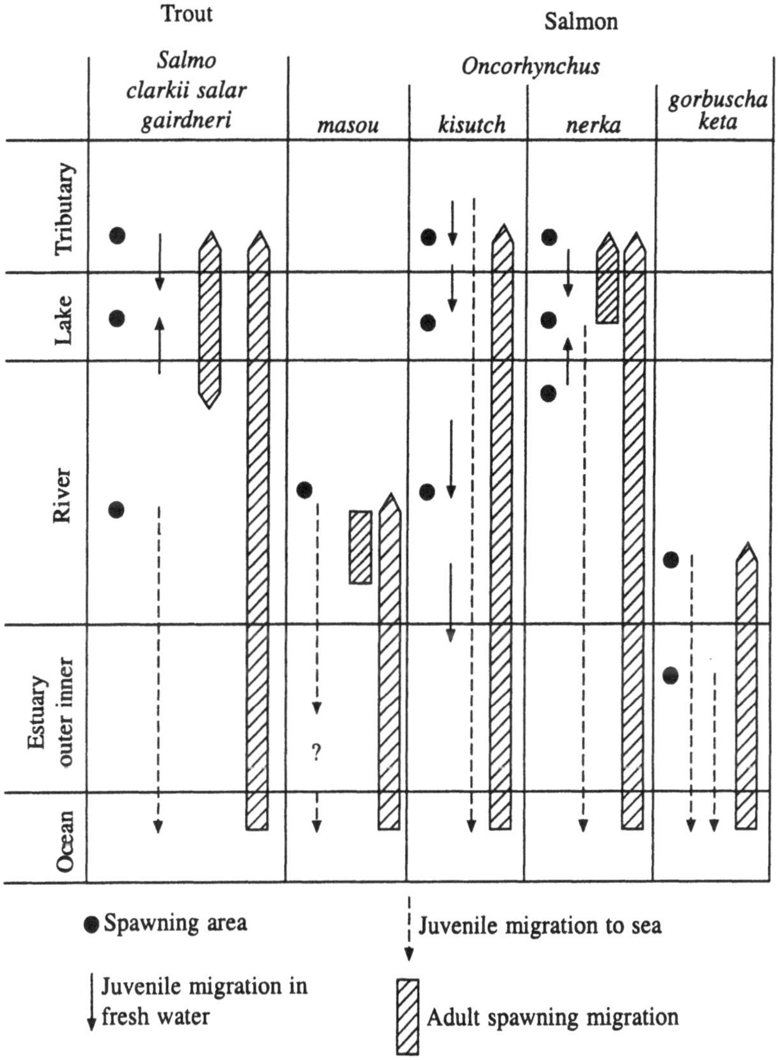

Fig. 21.4 Diagrammatic representation of the spawning areas and the migrations of various species of trout and salmon.
***Source:* Hoar (1965)**

Among pelagic fish there are five species of tuna-bluefin, albacore, bigeye, yellowfin and skipjack which occur more or less continuously across the mojor oceans and which make extensive migrations. In the North Pacific the albacore, *Thunnus alalunga*, spawns in the central subtropical zone. It performs trans-Pacific migrations and supports important Japanese and American Fisheries. There is only one stock contained within the subtropical gyre. On both sides of the Pacific Ocean the distribution of albacore appears to be the fronts between water masses: in the north they feed in the area of divergence towards the edge of the subtropical gyre and in the south in the areas of convergence in the equatorial region.

The plaice, *Pleuronectes platessa*, which is found in the sandy parts of the continental shelf of Europe, is a typical demersal flatfish, whose life history is greatly influenced by both residual and tidal currents. It spawns in winter producing large number of small pelagic eggs. The larvae reach metamorphosis at the end of the planktonic phase. They then become assymetrical and drop to the bottom as miniature flatfish before moving inshore onto intertidal nursery grounds. Tidal transport is a recurrent feature of the plaice's life history from metamorphosis onwards. The plaice in the North Sea migrate by selective tidal stream transport.

Reactions to the current: Flatfish lying on the seabed can detect the current as a result of direct mechanical stimulation. But a fish not in contact with the bottom requires a visual reference point and must be within a few metres of the river bank or seabed to detect its displacement. Fish can detect current speeds of 1–10 cms^{-1}.

There is no evidence that fish out of sight of the bottom can detect the current but fish in midwater in the sea might be able to orientate visually at the rheocline (the velocity gradient occurring vertically across the thermocline) if particles are visible in the water. Alternatively they might be able to detect electric fields generated by the movement of water through the vertical components of the earth's magnetic field. Skates and rays and salmon and eels can detect potential gradients of 0.01–0.1 μ V cm^{-1}; gradients of 0.05–0.5 μV cm^{-1} commonly occur in the ocean.

The basic reaction of fish to currents is rheotaxis. Most fish turn to head into the current and maintain station relative to a fixed reference point. As the speed of the current increases the fish swims faster; a kinesis is superimposed on the taxis. But this behaviour must be modified for migration and at different stages of their life history; diadromous fish must move upstream and downstream. Loss of visual reference points at night may account for the downstream displacement of salmonid fry. But the seaward movements of smolts and adult eels are apparently active orientated movements; the fish head downstream and swim with the current.

There is some evidence that diadromous fish show behavioural changes at the interfaces between the current regimes. A delay at the freshwater/saltwater boundary is common in both adult and juvenile salmonids. Silver eels change from negative to positive rheotaxis on exposure to saltwater. Newly metamorphosed elvers at first drift positively with the tide in coastal waters, progressing only with the residual current. Subsequently, they switch successively to tidal transport and active upstream movement. The second change occurs, in the River Elbe at least, at the limit of tidal incursion.

Oceanic migrations are generally consistent with movements with, rather than against the current. Because they are not well described in relation to oceanographic features, they can be accounted for by a number of different models. These variously invoke passive drift, active downstream migration or some form of independent navigation. Passive drift in the North Equatorial Current and Counter Current can, for example, account for the movements of young skipjack tuna, *Katsuwonus pelamis*, in the Pacific Ocean and jet stream transport can explain those speeds which are significantly faster than average. Two bluefin tuna, *Thunnus thynnus*, which travelled at 4832 km between the Bahamas and Bergen in Norway at an average speed of 40 km/day, probably received substantial assistance from the Gulf Stream and North Atlantic Drift. But in several instances ground speeds estimated from convential tagging experiments suggest that oceanic migrations cannot be accounted for by drift alone. Sockeye salmon, *Oncorhynchus nerka*, in the eastern North Pacific Ocean travel at speeds ranging from 20 to 50 km/day. Two albacore, *Thunnus alalunga*, covered 5239 km from the Central North Pacific to the coast of Washington State in 110 days at an average speed of 48 km/day. Because surface currents in these areas have much lower average speeds it is argued that these are active orientated movements. It has been suggested that salmon follow the ocean circulation in the North Pacific but swim downstream faster then the current, which they detect by means of electric potentials. It has been suggested that some form of biocoordate navigation may be required.

Tidal stream transport: Selective tidal stream is a semi-diurnal rhythm of vertical movement whereby fish are able to use the tidal streams for transport.

Selective tidal stream occurs in plaice *Pleuronectes platessa*, as well as in other species, for example sole (*Solea solea*), American shad (*Alosa sapidissima*), red seabream (*Chrysophrys major*) and dogfish (*Scyliorhinus canicula*). But in all cases the essential depth information is lacking.

It has been shown theoretically that in areas where tides are directional fish may save upto 40% of the energy cost of journey by using tidal stream transport compared with the alternative strategy of swimming continuously in the direction of overall movement.

The movements of adult anadromous fish returning to their home rivers also appear to be related to tide but here the situation is complicated by a period of estuarine delay, during which the fish adapt physiologically to freshwater. Some fish drift passively to and fro with the tide in the vicinity of the saltwater-freshwater interface. But some fish swim against both ebb and flood tides and which make progress upstream into the freshwater do so by holding ground on the ebb tide and drifting or swimming upstream with the flood. Such movements occur in the deeper estuarine channels and salmon are said to hug the bottom following the deepest channel-the thalweg where the current speed is usually least. Once into the freshwater, shad swim much faster than in the brackishwater and appear to vary their speed in response to tide-induced variations in current velocity. This results in an approximately constant ground speed and suggests that the movement is regulated by a visual or optomotor response. Energetically, such behaviour may be less efficient than the alternative strategy of maintaining a constant speed through the water. Similar behaviour is thought to occur in the waters of Long Island Sound before the fish reach the estuary of the Connecticut River but after they have come within the influence of its river. Olfactory stimuli are implicated in this behaviour and also in that of the alewife, *Alosa pseudoharengus*. This clupeid distinguishes its home stream water from that of other streams by olfaction suggesting homeostream odour hypothesis is applicable to anadromous fish generally and not just to salmonids.

The ability to respond differently to ebb and flood tides in estuaries and adjacent coastal waters must be a key factor in the lives of diadromous and other fish which make estuarine migrations. It is not obviously fully understood how such movements are related to the various types of estuarine circulation and further tracking experiments, in which the depth of the fish can be determined, are clearly desirable.

Selective tidal stream transport is probably the optimal strategy for fish migrating in tidal waters. It clearly allows rapid directional movements with a minimal expenditure of energy, very probably without the need for any navigational ability. But it is not the whole story even on the continental shelf and it is unlikely to be applicable in the open ocean.

On the shelf both plaice and cod show contranatant movements, swimming on some occasions against the adverse tide between periods of midwater transport. Such movements invariably occur while the fish is on or close to the bottom and has the necessary visual or tactile stimulus with which to detect the current. But fish out of sight of the bottom may nevertheless swim consistently in one direction; plaice can do so for short periods, salmon for several tidal cycles. Adult pink *Oncorhynchus gorbuscha* and sockeye. *O. nerka* salmon returning to the Fraser River in British Columbia migrate actively through the confined waters of the San Juan Archipelago along the axes of the tidal currents.

Few fish tracks are as yet available from the oceanic regime (Arnold 1974, 1981; Harden Jones 1981, 1984 Arnold and Cook 1984).

1.3 Movement Patterns

1.3.1 Cyclostomes

In all lamprey species there is an upstream migration for spawning. By upstream migration the adult lampreys compensate for downstream drift and/or migration in younger stages. The brook lampreys (*Lampetra planeri*) migrate over comparative short distances, whereas the anadromous species which spend one or several years in the sea may migrate over long distances (Bertmar 1985).

1.3.2 Marine Fishes

The article of Quinn and Brodeur (1991) addresses various aspects of movement patterns in marine fishes.

Migrations are movements of individuals coordinated in space and time. These movements are generally, although not necessarily, accompanied by return movements and are often repeated throughout the life time of the organism. Migrations may occur during one life history stage, examplified by ontogenetic offshore migrations by fishes (e.g., rockfishes *Sebastes* spp.). Planktonic larval and early juvenile recruits settle to the bottom in estuaries or the nearshore environment and subsequently move offshore to deeper water as they grow. This pattern places juveniles in warmer and more productive inshore waters where growth may be faster. Estuarine and inshore nursery grounds are more turbid, often contain higher structural relief and may provide refuge from predation.

Marine migrations are often associated with seasonal variations in temperature and productivity. Mobile species often move towards the equator in fall to avoid cold temperatures and return to higher latitudes in the spring to exploit the seasonal production peaks. There are many examples of seasonal migrations of this kind in northeast Pacific shelf waters (e.g., Pacific hake *Merluccius productus*, dogfish *Squalus acanthias*) and in offshore pelagic waters (e.g., pomfret *Brama japonica* and albacore *Thunnus alalunga*).

The fish communities on coral reefs have distinct diurnal and nocturnal components and many species migrate daily, predominantly at twilight when predators are least efficient. Migratory pattern and feeding period may vary with age or size. The changeover from diurnal to nocturnal communities is less distinct in temperate then tropical reefs. However, diel feeding migrations are also common in temperate nearshore fishes and there are often age-or size-related differences in the tendency to migrate.

Another type of migration is the congregation at spawning areas. Such migrations are often repeated annually but can occur on a daily basis in tropical fishes or only once in a life time in semelparous species. There are many eastern North Pacific fishes, e.g., herring (*Clupea harengus*), Pacific hake that undertake seasonal reproductive migrations. Females which bear the young after fertilization may migrate to release offspring in areas favouring growth and survival (e.g., sharks). Adults may move contranatantly to spawning areas where prevailing currents carry the larvae to favourable nursery areas.

Dispersal: The primary mode of dispersal in many marine organisms is a planktonic larval stage; currents transport young stages to suitable nursery areas. Since the mortality rate associated with drifting into unfavourable areas may be high, many species have developed reproductive patterns to minimize larval dispersal or ensure that the young return to the natal area (e.g., tropical reef fishes). It is argued that retention of larvae in the natal area by oceanographic processes may be more common in many marine populations than dispersal to new areas.

Dispersal is evident in marine fishes by the rapid colonization that occurs after deployment of artificial reefs (for example, rock fishes, in temperate waters). Dispersal is also known in benthic species normally considered to be sedentary (e.g., lingcod *Ophiodon elongatus*). There are many examples of tagged individuals found outside the normal migratory route (e.g., sablefish and spiny dogfish). Tagging studies in the Pacific reveal that most dogfish migrate about 480–800 km north-south along the continental shelf, returning to the same general area, each year. However, some individuals move very long distances including two that crossed the Pacific Ocean (ca. 7,000 km). Dogfish are also occasionally captured on the high seas over deep water. Western North Atlantic dogfish have a similar seasonal north-south migration and trans-Atlantic movements have also been documented.

Home range: The individuals remain within a well-defined area for most of their routine activities. Home range is defined without respect to any other animals, as opposed to a territory which is actively defended against intruders. Home range size varies within species depending on the energetic and reproductive needs of the individual and on the size of the population.

There are numerous examples of territoriality and home range in marine fishes, particularly tropical reef dwellers. In temperate waters of the north-east Pacific, most shallow-water reef rockfishes appear to occupy very limited foraging areas and some show territorial behaviour, even at an early age.

Non-migratory populations: Most fishes are permanent residents of freshwater or marine habitats. However, it has been shown that diadromy is more common and variable than usually believed. In many species, all individuals migrate but diadromy may be facultative as well. Some fishes, notably the Salmonidae in the Atlantic and Pacific, have anadromous and freshwater populations.

Sockeye salmon (*Oncorhynchus nerka*) have an anadromous form that rears in lakes for usually one or two years before migrating to sea and a non-anadromous form ("kokanee") that remains in lakes until maturity. Most populations are anadromous but some sockeye spawn sympatrically with the much smaller kokanee. Assortative mate selection by males apparently minimizes genetic exchange between forms. Among non-salmonids, the threespine stickleback also has an anadromous or *trachurus* form and a freshwater or *leiurus* form. As with the sockeye and kokanee, assortative mating seems to reduce interbreeding between two forms.

Though marine fish populations are more difficult to define than anadromous populations, there are examples of non-migratory populations of marine migrants. Most hake spawn off southern California and migrate north along the coast as far as British Columbia but there is a population in Pudget Sound, Washington which does not migrate.

Partial migration: In some species, not all members of a population migrate. Starry flounder, *Platichthys stellatus*, is apparently such a species. This coastal flatfish, distributed from California around the Pacific rim to Japan, spawns in seawater but juveniles are sometimes found in rivers. There seems to be no information on what proportion of juveniles migrate, how it affects their growth and survival, and whether the tendency represents an alternative life history strategy or an opportunistic tactic. A European congener (*P. flesus*) is also facultatively amphidromous. Individuals may spend upto five years in freshwater and be larger at age than flounder caught at sea.

Lingcod, sablefish and dogfish, are long-lived, broadly distributed fishes that generally reside or migrate within localized areas but some individuals move very long distances. Although the general movement patterns of the species differ, all show occasional dispersal of individuals. For a review on partial migration; niche shift versus sexual maturation in fishes, see Jonsson and Jonsson (1993).

Homing and straying: Many species such as Pacific herring (*Clupea harengus pallasi*) spawn on near-shore substrate at particular sites and times, suggesting that homing may take place. Tagging studies revealed that adult herring tend to spawn in the same area where they previously spawned, though there was less evidence that they spawned in their natal site. Homing seemed to be more precise to certain areas than others (range 6–96%) but there were no data on homing to specific spawning locations. A similar pattern of homing and straying apparently prevails for Atlantic herring (*C. h. harengus*) in Newfoundland.

Variations in home range: There are often intra-specific variations in the size of home ranges and patterns of movement within the range. Rockfish (*Sebastes caurinus* and *S. maliger*) home ranges varied dramatically with the reef type in Puget Sound and corresponded to the seasonal variations in habitat quality.

Rockfishes inhabiting rocky reefs off California displayed three movement patterns. Territorial individuals occupied exclusive or nearly exclusive shelter and feeding ranges. Others, termed "commuters", had discontinuous home ranges with separate shelter and feeding areas. There were also non-territorial "floaters" which did not occupy any single feeding area but were frequently chased from other territories. Many coral reef fishes show a high degree of site tenacity yet may display intraspecific variation in movement patterns. Individual dusky damselfishes

(*Eupomocentrus dorsopunicans*), for example, are territorial but also make frequent forays beyond their defended areas to feed, mate, or assess the availability of better sites.

Distributions associated with oceanographic anomalies: Marine species are normally found within specific geographical areas but there may be exceptions associated with unusual oceanographic conditions. Inter-annual variations in oceanographic conditions have also been associated with variations in the migratory timing and routes of some salmon populations. Oceanographic anomalies not only affect animals over long time scales; some mobile pelagic fishes move in response to day-to-day changes in temperature and productivity.

Range expansions: When a species regularly occurs outside its accepted range, we consider that it might have actually expanded its range. Range expansions of freshwater and anadromous fishes are well known. Expansion can be facilitated by human activities such as canal construction or natural processes such as recession of glaciers.

Range expansions by marine fishes are difficult to demonstrate but it has been suggested that Pacific cod (*Gadus macrocephalus*) may have expanded its range southward along the coast of North America in the 1950s and 1960s while the fine-scaled triggerfish (*Balistes polylepis*) may have expanded its range northward during this century.

1.4 Evolution of Migration and some Life History Parameters in Marine Fishes

Migration is a common phenomenon in marine fishes but the consequences with respect to life history variation have been little explored.

According to Roff (1988), migration is not a behaviour that evolves in isolation from other behavioural or life history components. The probability of migration evolving is enhanced by other factors such as schooling and trade-off between energy used in migration and sequestered for reproduction. These interactions directly influence the optimal age and size at maturity: they may also indirectly affect them by increasing the probability of a species locating unexploited sources (Fig. 21.5). Migratory species are predicted to grow faster, mature later and are larger on average than non-migratory species. It is important to note that theory predicts that large size will be characteristic of migratory species while non-migratory species may be either small or large. Analysis of north temperate marine species confirm these predictions. Figure 21.6 is a summary of the main visual adaptations and their associations with the different types of environmental change. For reasons of clarity, exceptions to this illustrations have been omitted. It should be made clear that there is significant complexity in the ontogenetic trajectories across species.

1.5 Evolution of Diadromy

Gross (1987) and Gross et al. (1988) attempted to clarify several key issues central to understanding this aspect. Ontogenetic changes in osmoregulatory ability will determine the extent and life history pattern of diadromy (see McCormick and Naiman 1985) (Fig. 21.7).

To evolve, migration must maximize fitness or the lifetime product of reproductive success X survivorship. Diadromy, therefore occurs when the gain in fitness from using a second habitat minus the migration costs of moving between habitats exceeds the fitness from staying in only one habitat. The most important biological variable in explaining the presence and direction of diadromous migration is apparently the relative availability of food in sea and freshwater habitats.

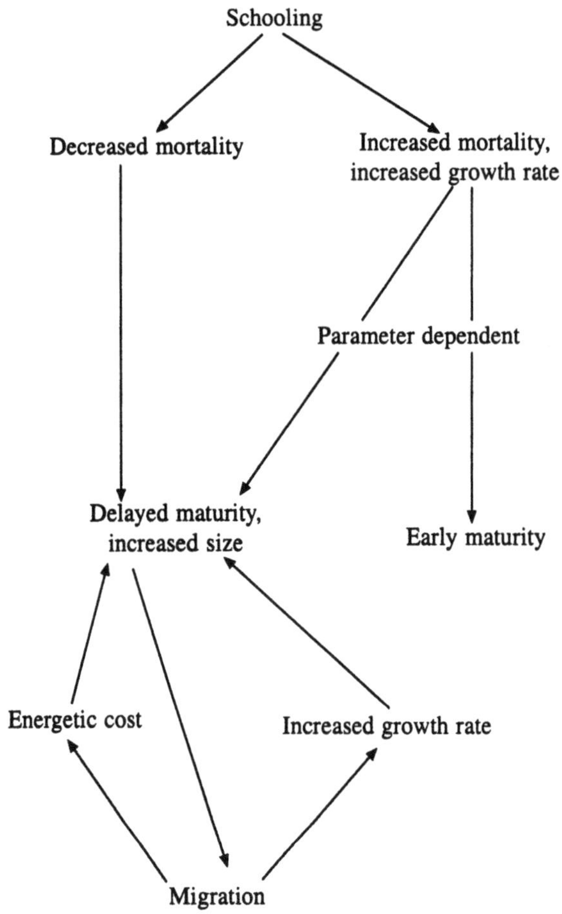

Fig. 21.5 Schematic illustration of how the interaction of behavioural and demographic parameters influence the evolution of migration.
***Source:* Roff (1988)**

Finally, amphidromy is hypothesized to be an ancestral state in the evolution of anadromy and catadromy.

Since the main characteristic of diadromy is alternation between habitats, a single migration model can be developed to explain the evolution of all three diadromous forms: amphidromy, anadromy and catadromy (Fig. 21.8).

Figure 21.9 is an overview of evolution of diadromy. Important points are (1) amphidromous fishes in the tropics are of marine origin, and (2) temperate amphidromous fishes may be of freshwater origin. Evolutionary hypothesis advanced is that of some marine fish species in temperate latitudes are probably of freshwater origin, while some freshwater species in tropical latitudes may be of marine origin.

Diadromous migration, in which some fish species migrate from freshwater and feed in the ocean (anadromous species) and others migrate from the ocean and feed in freshwater, has long been perplexing. However, when the distribution of diadromous species is examined with respect to global patterns in aquatic productivity, this apparent paradox is resolved. The contrasting

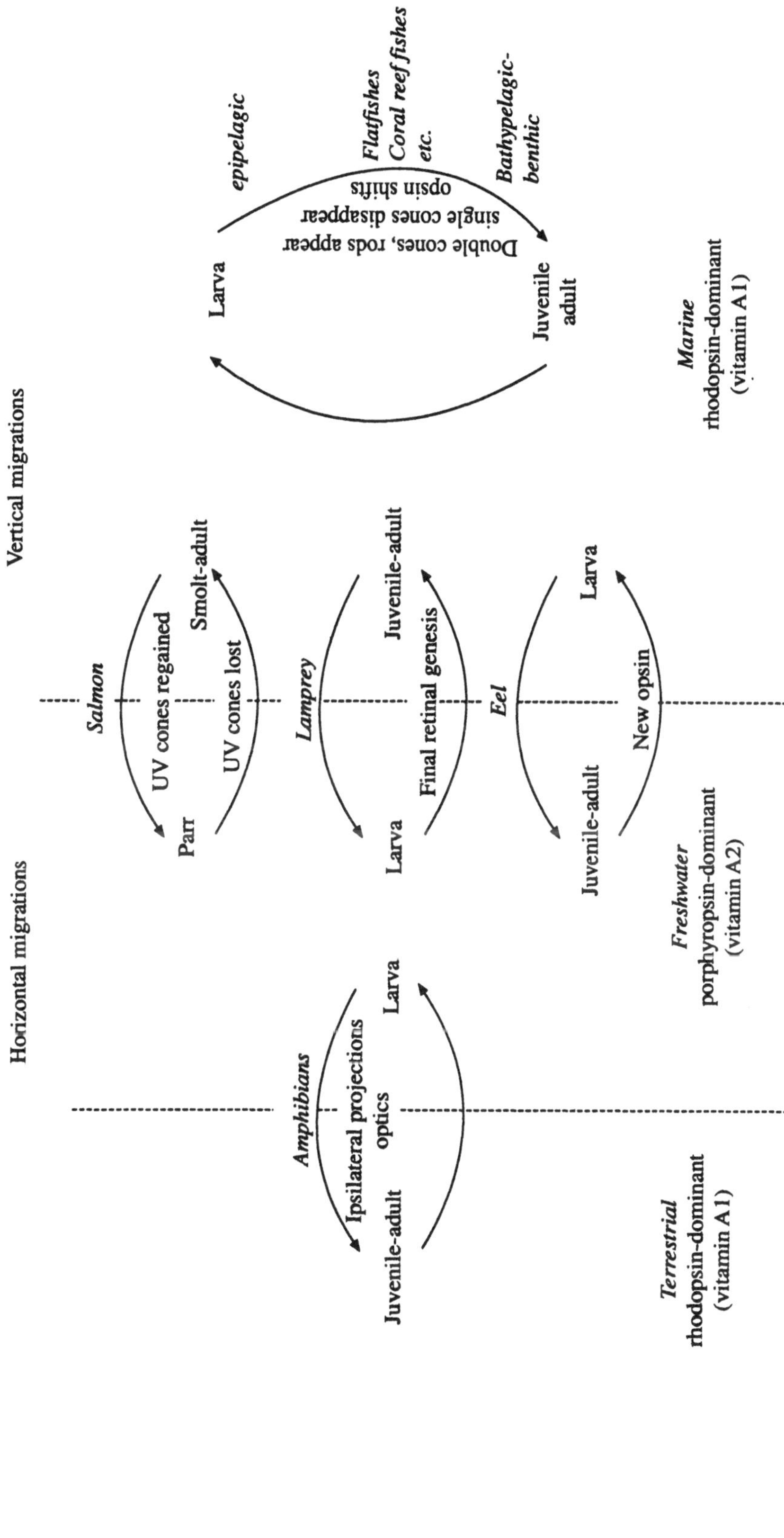

Fig. 21.6
***Source:* Hawryshyn (1998)**

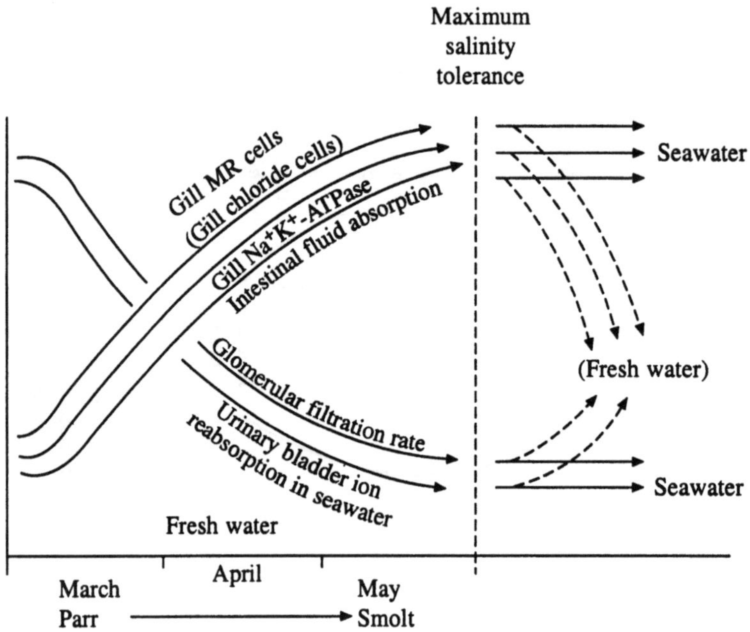

Fig. 21.7 Functional changes in the osmoregulatory organs during the parr-smolt transformation. Functional changes normally associated with osmoregulation in seawater occur in fresh water and result in increased hypoosmoregulatory ability and salinity tolerance. In the absence of exposure to seawater these changes are reversible.
***Source:* McCormick and Saunders (1987)**

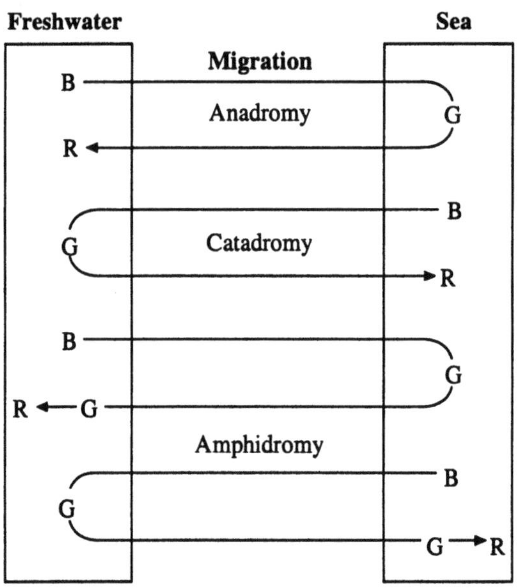

Fig. 21.8 The three life histories of diadromous fishes. Each life history includes occupation of freshwater and sea habitats and migration between them. Anadromous, catadromous and amphidromous life histories are defined by the habitat in which birth (B), growth (G) and reproduction (R) take place. Some growth may occur in the habitat of birth.
***Source:* Gross (1987)**

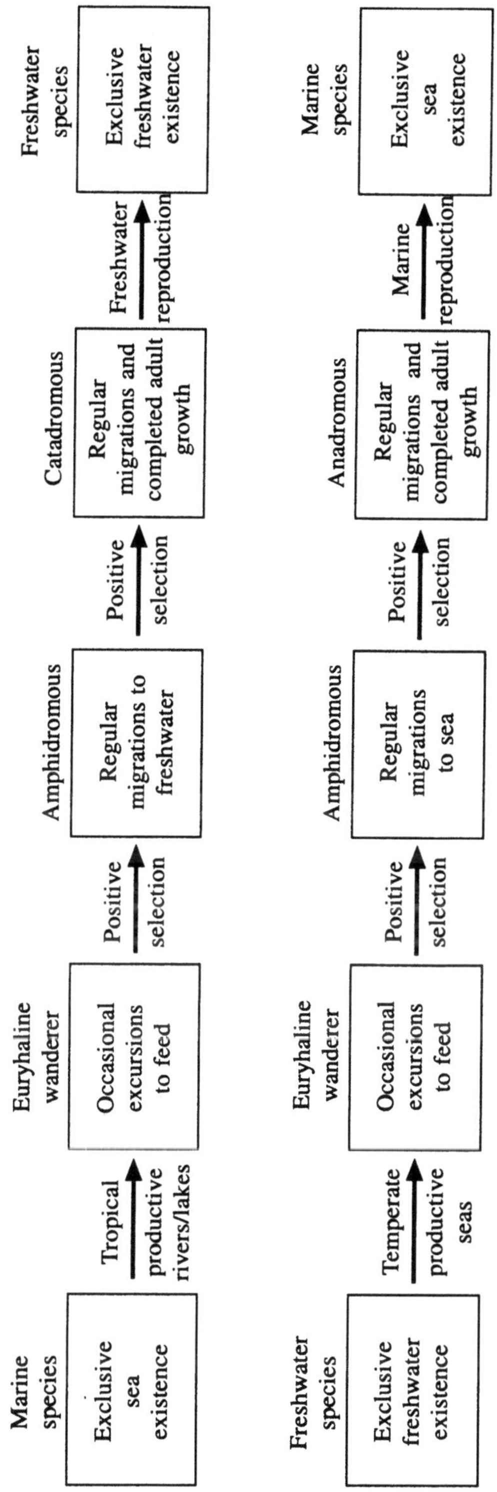

Fig. 21.9 **Overview of the evolution of diadromy in fishes.**
Source: Gross (1987)

directions of migrations can largely be explained by the relative availability of food resources in ocean and freshwater habitats. Oceans are more productive than freshwaters in temperate latitudes, and anadromous species predominate. In contrast, catadromous species generally occur in tropical latitudes where freshwater productivity exceeds that of the ocean. For special attention on anadromous Arctic charr (*Salvelinus alpinus*), see Jorgensen et al. (1997).

2. FISH MIGRATION IN RIVERS

Northcote (1984) focused on the mechanisms of fish migration in rives. Rivers exhibit greater temporal variation in conditions than do coastal or oceanic waters. In rivers migration must ultimately involve both a downstream and an upstream component. The former is generally, but not always, a feature of early life history stages and vice versa. Migration of a given life history stage, however, may involve both downstream and upstream components. Downstream migration may be passive or active, but upstream migration must be active. Downstream migration of juveniles and adults is mainly nocturnal, but sometimes diurnal, especially in turbid waters. Characteristically, maturing adults of riverine and anadromous species must upstream to spawn, but some species or populations may move downstream to spawn. Upstream migration may be nocturnal or diurnal (Fig. 21.10).

3. FEEDING MIGRATIONS

Amongst the intertidal fishes, in order to satisfy their requirements for an adequate food supply many species have to make extensive foraging excursions away from low tide locations. In the great majority of cases these feeding excursions are made at high tide when conditions are more favourable. Although feeding may be possible when immersed, only those forms which have evolved a semiamphibious mode of life can do so efficiently. The extent of these excursions may be related to the abundance of food (Gibson 1982).

Barlow (1986) discussed this aspect from a different angle. Feeding territories have seldom been reported for freshwater fishes in nature. The salmonids, of temperate waters, are well known for holding territories in rivers where they feed on drift. And the algal feeding ayu (*Plecoglossus altivelis*) of Japan is also territorial.

Surely, many freshwater fishes have the behavioural mechanisms necessary for territoriality, site tenacity and aggressiveness which they reveal in captivity.

It is more likely that physical situation mitigates against feeding territories. Algal feeding catfishes in Panama have favoured resting sites, which they might defend. But they move considerable distances up and down stream to follow the outbreak of new food sources. It is also possible that the shallow nature of most streams and rivers makes the territory holder too vulnerable to aerial predators.

Thus a number of factors may be working against feeding territories in freshwater fishes. The most likely factor is rapid and unpredictable change in the riverine environment, such as vast differences in flow rates with precipitation and seasonality. As a consequence, food may be unpredictable in time and space. That would promote the aggregation of many fish at rich patches and work against monogamy. So, too, would long riverine migrations for spawning which occur in many groups. Even without migrations, there is apt to be substantial and unpredictable separation of feeding and breeding sites. All of these would mitigate against stable pair bonds.

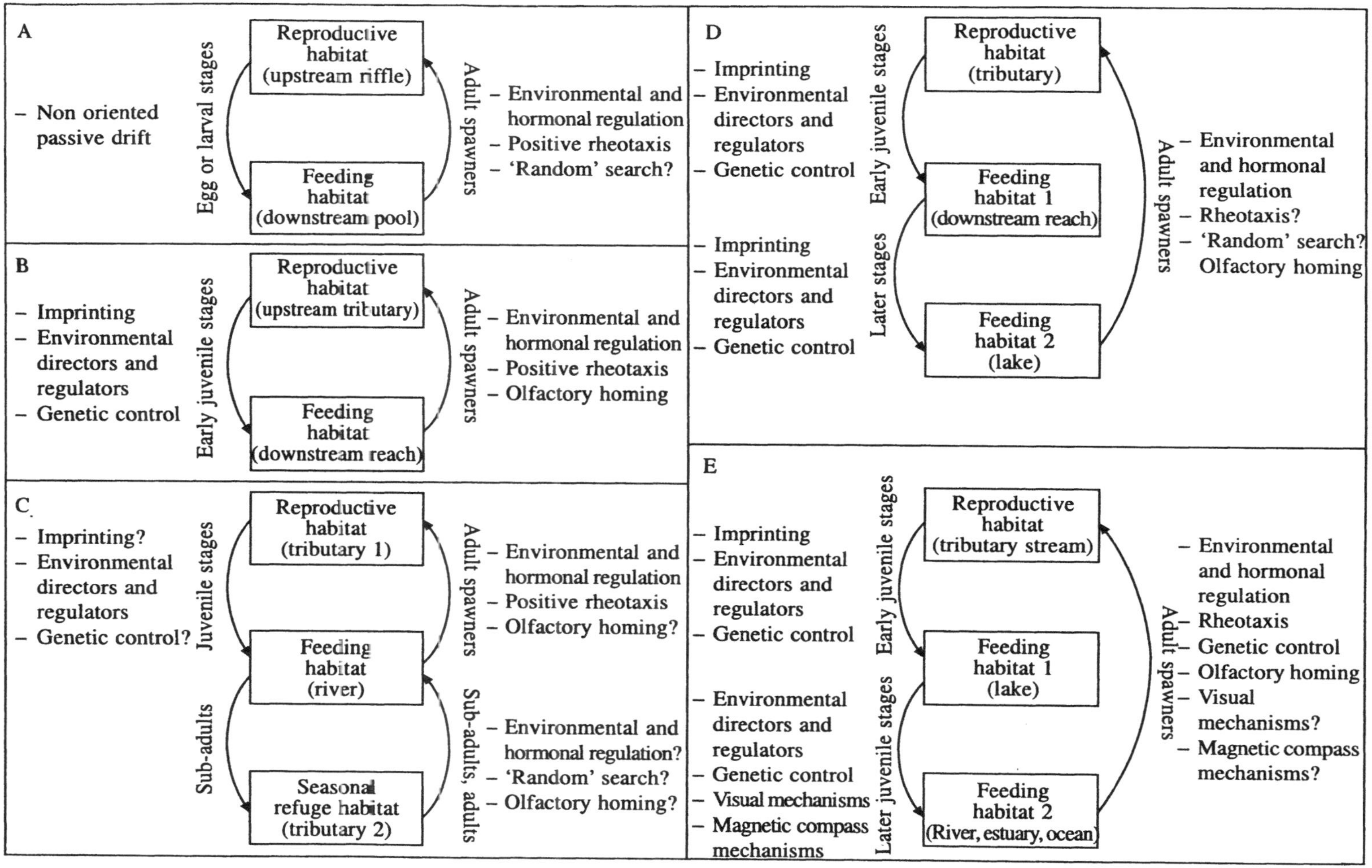

Fig. 21.10 **Hierarchies and sequences in mechanisms of fish migration in rivers.**
Source: **Northcote (1984)**

Whatever the explanation, the common occurrence of feeding territories among coral reef fishes and their apparent absence among freshwater fishes of the tropics is a stunning difference. It appears to be one of the major factors shaping monogamy in the two environments.

4. OVERWINTERING

Most freshwater fishes are in little danger of freezing because the freezing point of freshwater is nearly 0.5°C above the freezing point of their body fluids. The freezing points of serum from freshwater fishes range from –0.50 to –0.65°C. Thus, even when the temperature of a stream or lake drops to its freezing point, there is still a comfortable margin between the freezing point of the fish and that of the water. Only in lakes which freeze to the bottom do fishes encounter conditions where they are likely to freeze. In these environments some fishes such as the Arctic blackfish, *Dallia pectoralis*, and the crucian carp, *Carassius carassius*, overwinter in an inactive state and avoid freezing by burrowing into the warmer mud at the bottom of the frozen lake. In some these fishes may be frozen into the mud or even into a lump of ice yet they remain alive as long as their body fluids do not freeze provided they are capable of anaerobic metabolism (DeVries 1971).

Some shallow-water fishes of the Arctic Ocean avoid freezing migrating to fresh water where the temperature can never fall below 0°C. The Arctic char, *Salvelinus alpinus*, for example, usually leaves the ocean in the autumn and spends the winter in fresh water lakes and streams.

The absence of ice in depths can be explained by the facts that (1) heat is removed from sea water at the surface, (2) ice floats, and (3) increased hydrostatic pressure depresses the freezing point of sea water (0.0075°C for each 10 m increment in depth). Thus, it is not uncommon for supercooled fish to avoid contact with ice by either living in or migrating to deeper water. The long horn sculpin, *Myxocephalus octodecemspinosus*, for example, is common in shallow waters of the Atlantic coast of North America in summer, but during the winter it can only be found in water deeper than 20 m. This fish shows no resistance to freezing in the winter (DeVries 1974).

Many ectotherms simply circumvent the problems associated with cold temperatures by remaining dormant through the winter months. European eels illustrate the winter dormancy pattern. At typical summer temperatures (13–17°C), the eels swim actively and feed regularly, but at temperature below 8°C they cease feeding and burrow in the mud. Similarly, in the brown bullheads, *Ictalurus nebulosus*, cooled to 3°C, spontaneous activity is markedly depressed and the animals enter into a sleeplike state after burying themselves in the mud (Hazel 1989).

The importance of energy stores for overwintering among freshwater fishes depends upon the length and severity of winters, whether or not there is springtime reproduction, body size, latitude and the availability and use of food during overwintering. Fishes can decrease energy demands during winter by reductions in activity, metabolic depression, and entrance in semi-torpidity. Adaptation for coping with hypoxia and anoxia among overwintering freshwater fishes may include metabolic depression, a decrease in blood O_2 affinity, microhabitat selection, air breathing, short-distance migration, biochemical modifications aimed at adjusting glycolytic rates, and alcoholic fermentation (Ultsch 1989; see also Hazel 1989).

Anadromous Arctic charr do not appear to feed whilst overwintering in fresh water, so they must rely on reserves deposited during the summer growth season. Depletion of skeletal lipids is reported (Jorgensen et al. 1997).

5. PHYSIOLOGY OF MIGRATION

In the majority of fishes the spawning migration is a recurring event in the life cycle, albeit with characteristic activities, and with specialized metabolic needs which may temporarily run very high. Many fish undertake extensive breeding migrations often at times when they are not feeding, so that body tissues must be used to provide material for gonad growth and energy for migration. The immediate stimulus is commonly an environmental event. It is known that carriage by water currents forms an integral part of the migrating mechanisms of many adult fish. Diadromous fish have, however, disadvantage that they face the complication of a change in osmotic environment during migration which, quite apart from other problems of migration in itself, can affect endocrine activity.

Although regulation of fish migration is treated as a singular phenomenon for simplicity of presentation, it should be recognized that considerable diversity exists with respect to the taxa studied, the habitats of the fish, and the kinds of migration. Many of the strong migrants (salmon and eels) are members of modern (teleost) bony fishes, Osteichthyes, but there are also migrants that belong to Chondrichthyes (sharks) and Agnatha (lampreys). Differences in evolutionary development may well have produced variations in regulatory mechanisms involved in migration.

The habitats of migratory fish differ considerably with respect to salinity. Some species exist in and are adapted for seawater whereas others are found exclusively in freshwater. Migrators such as salmon and eel which move between fresh- and seawater must be able to adapt to both extremes of salinity (that is, they are amphihaline). In addition, these adaptations to salinity changes must be integrated with other important events during life cycle. The spawning migration of salmon occurs primarily in fresh water whereas that of eels primarily in seawater. This relation is reversed in the trophic migrations of immature fish (see Fig. 21.7).

Because fish are ectothermic their metabolism and behaviour are greatly influenced by variations in water temperature. Seasonal fluctuations in temperature may impose annual cycles in some fish and set the times for migration. Migration may bring the fish into waters of different temperatures. Because temperature also changes with water depth, vertical movements of fish may produce further metabolic alterations. For example, a fish may seek its prey in deep-colder water during the day and migrate at night in warmer surface water where it might use celestial cues for orientation.

Regulatory mechanisms have multiple functions which may be divided into preparatory and adaptive activities. These activities might best be visualized by citing a specific example. The European eel, *Anguilla anguilla*, matures in freshwater ponds and river tributaries where it is known as the yellow eel. Before migration the yellow eel accumulates a store of body fat and undergoes metamorphosis into the silver eel (read Fortaine 1994 for eel "silvering"). The silver eel differs dramatically from the yellow eel in that its osmoregulatory mechanism is prepared for transition to seawater whereas the yellow eel cannot survive a direct transfer to seawater. The heavy fat stores and the osmoregulatory adjustment are distinctly preparatory in that they occur before the time when they are needed most. Adaptive changes may include the mobilization of fat and protein energy in the Sargasso Sea. Further adaptive changes also occur in the osmoregulatory mechanisms in response to increased salinity as the silver eel enters seawater.

While the information on the landlocked sea lamprey (*Petromyzon marinus*) shows a clear correlation between migratory activity and temperature, most observations on other species have tended to emphasize the importance of high water levels. Li et al. (1995) and Li and Sorensen (1997) report that unique conspecific bile acids may function as migratory pheromones in *P. marinus*.

Stimuli and timing: Preparation for migration and the final proximate stimulus depend on an interaction of endogenous and exogenous factors. Preparation includes development of systems responsive to the proximate stimuli. Responsiveness may be periodic so that a fully matured animal may be influenced by proximate environmental cues only at certain times of year. Probably seasonal changes in migratory readiness in fish may be part of an endogenous circannual mechanism. A circannual rhythm of reproductive changes has been reported in a teleost, but the possibility of a circannual mechanism regulating migratory indices has not been examined in fish. Changes in environment cues could be expected to set the phase of such a circannual rhythm. Seasonal changes in day length and especially temperature are involved in regulating animal physiological and behavioural changes.

Change in water temperature is an important environmental stimulus for the initiation of migration in some fish. Relations between temperature decrease and amount of fat stores in stimulation of migration have been reported in several fishes. Other environmental stimuli that initiate migration in responsive fish include flooding (European eel, *Anguilla anguilla*) and weather fronts (Atlantic salmon, *Salmo* sp.).

Studies of daily and seasonal patterns of locomotor activity have been reported in fish. A fascinating interaction of temperature and photoperiod in the regulation of locomotor activity has been reported in juvenile sockeye salmon, *Oncorhynchus nerka*. In addition to entrainment of rhythms, the LD cycle (light-dark) is also necessary in some species, to maintain the expression of a daily rhythm. Nonoptic photoreceptors for photoperiodic entrainment have been reported.

In view of the neuroendocrine disequilibrium which appears to be associated with migratory behaviour, e.g., both with the physiological preparation for migration and for migration itself, it is not astonishing to observe profound changes in several functions (osmoregulation, pigmentation, nutrition). Most studies on hormonal control of fish migration involve movement between areas of different salinity. Timing of migration depends on the physiological state of fish. In primitive vertebrates (lampreys), the contribution of the nervous system in regulating migrations is more important than that of the endocrine system.

Studies of neuroendocrine regulation of fish migration have centered on the thyroid and corticosteroid hormones. An active thyroid appears to be characteristic of many fish during the preparatory period. There is an active hyperplasia of thyroid and of pituitary TSH cells during smoltification of salmonids. Secretion of thyroid hormone T4 also increased at the beginning of smoltification.

Smoltification is an event in salmon development during which the darkly pigmented bottom-dwelling parr is transferred into a pelagic silver smolt. Transformation to smolt just precedes migration to the sea and is associated with a change in salinity preference (see Fig. 21.7) (Table 21.1) (McCormick and Saunders 1987). In coho and chinook salmon (*Oncorhynchus keta* and *O. tschawystchs*) and steelhead trout (*Salmo gairdneri*), a large, 1–2 month "surge" of thyroid hormones during the period of smoltification which may be caused by photoperiod, temperature and phase of moon, has been demonstrated. It should be emphasized that there may be some variability in the role of thyroid hormones in fish migration, related to locomotor activity. Furthermore T4 appears to mobilize lipids in salmonids, an effect which could be important

Table 21.1 Some behavioural and physiological changes coincident with the parr-smolt transformation in salmonids

Behavioural or physiological change
1. Increased deposition of guanine and hypoxanthine in skin and scales (silvering)
2. Increased buoyancy due to increased air volume of swim bladder
3. Alterations in blood haemoglobins (rapid increase in adult forms)
4. Increased schooling behaviour
5. Increased salinity preference
6. Negative rheotaxis
7. Decreased swimming ability

Source: McCormick and Saunders (1987)

metabolically during a long migration. GH may be necessary for proper lipid mobilization during salmonid migration (see Fig. 13.11).

Salmon faces challenges in the life history during transitions. Salmon smolts moving from fresh water into an estuary are prone to extremely high mortality rates. Although the factors responsible for these high death rates are unknown, mortality could result from a variety of factors such as osmoregulatory stress, low food availability, predation, and other forms of competition. Individual survivorship is dependent on the ability of the challenged smolt to out perform inter- and intraspecific opponents. Migration into an estuary requires a dramatic change in osmoregulatory strategy, the success of which is dependent on the physiological status of the smolt.

There is a histological evidence of increased thyroidal activity in the European eel, *Anguilla anguilla*, prior to the catadromous migration of the adult and the anadromous migration (upstream) of the elver. These periods correspond with transformation of the yellow eel to the silver eel and with metamorphosis of the leptocephali larva to the elver, respectively.

Pituitary ACTH and interrenal activity also increase in salmonid smolt and cortisol participates with thyroxine to cause silvering. Similarly plasma cortisol concentration increases during the transformation of the yellow eel to the silver eel. Thyroxine may also influence neural excitability in teleosts as indicated by increased locomotor activity.

Orientation cues (*Visual and nonvisual orientation mechanisms*): Fish use a multitude of environmental cues for orientation during migration. An environmental stimulus, such as water current, temperature or light, may alter fish orientation and act as a "director" of migration, or the stimulus may trigger movement or alter the intensity of movement and act as a "regulator" of migration.

The lunar influence (phase of the moon) may initiate the migration of river lamprey *Lampetra fluviatilis*. Freshwater discharge is the predominant environmental factor responsible for initiating the movement from the sea into the estuary, although temperature may also be a contributory factor. The factors that influence the later stages in upstream migration are not known. In the spawning migration of the brook lamprey *Lampetra planeri*, only temperature showed a significant influence.

There is no conclusive evidence that upstream migrant lampreys tend to home stream as those in which they were borne. It has been suggested that lampreys might find their way into the spawning tributaries by a rheotactic response to the outflow current, since streams with a stronger or more sustained outflow tend to larger spawning populations.

Some fish orient by the sun and have a time-compensated mechanism to take into account the change in the sun's position during the day. The sun may provide useful navigational aids during the seaward migration of the juvenile salmon *Oncorhynchus nerka*; however many other researchers doubt that the sun is an essential navigational cue for fish migration. In addition many migrations occur at night and at great depths where nocturnal celestial cues would have little use. Nevertheless the presence of a time-compensated sun compass is indicative of a circadian neural mechanism that can influence behaviour.

The use of bioluminescence can be effective right upto the surface of the open ocean, and many animals vertically migrate into these areas at night, effectively counterilluminating against moonlight and starlight.

Because the expression of rheotaxis depends on the recognition of visual cues or surface contact as reference for realization of current, rheotaxis is most important for migration in relatively clearly and shallow flowing streams. Other environmental cues may trigger the occurrence or change the sign of rheotaxis. Rainbow trout fry, *Salmo gairdneri*, are negatively rheotactic at low temperatures but exhibit positive rheotaxis with a sharp rise in temperature. Temperature is probably sensed in the brain where warm temperature has been shown to promote the activities of different set of neurons than cold temperature in brook trout, *Salvelinus fontinalis*.

After a long sedentary period of growth, "yellow" eels turn into "silver" eels. Although metamorphosis includes clearcut modifications of the gonadal level, silver eels are still immature and sexual maturity does not proceed further if migration is prevented. This reflects a blockade of the pituitary gonadotropical functions, due to lack of gonadoliberin and to a dopaminergic inhibition of the action of this hormone. It is believed that among the external factors encountered during migration, hydrostatic pressure is an essential signal for relieving the gonadotropic blockade.

Theoretical and experimental studies suggest that the use of inertial clues is not necessarily so unlikely as has been previously thought in ocean migration.

Fishes are highly olfactory animals and olfaction plays a significant role. Odor is an important cue in the anadromous migrations of salmonids and eels. The elver eel reacts to odors, perhaps of conspecifics, from upstream sources never before frequented by the elver whereas the adult reacts to odors imprinted on them in their freshwater habitat before they undertook the catadromous migration. Release of large amounts of MSH and ACTH in smolts may have the effect of making memory permanent (imprinting). This possibility deserves to be examined in greater detail.

Salmon have to navigate in complex water masses. The layered structure of the water is decisive for successful navigation, and that olfaction is crucial in the selection of one of the many layers.

The attractant substances are found in the intestinal content of the smolt and are possibly of bile origin. Physiological experiments have shown that bile is a patent source of odors for fish such as salmon and sea char. The neurons of the olfactory bulb in char react differentially to the odoronts from different char (*Salmo alpinus*) populations. The results of various experiments provide supporting evidence that population-specific odors (pheromones) constitute a trail which the salmon follows on its way home. Atlantic salmon (*Salmo salar*) parr mark the substrate with intraspecific odor. Stream-dwelling brown trout (*Salmo trutta*) can return home when displaced outside their normal range and that olfaction is important for this behaviour.

In non-parasitic and many parasitic lamprey species spawning tends to be performed in groups. The assembly on the spawning grounds may be facilitated by olfactory signals. Joining a group will improve chances of females of finding fit partners as competition among males will be high in a group. Since the females spawn vigorously and usually are spent within a day there is pronounced surplus of males in the nests.

Because salinity gradients are usually very small, the detection of differences probably has limited usefulness in migratory orientation. Notable exceptions might include the layering of seawater by freshwater streams combined with tidal fluctuations. However, some fish can perceive differences as little as 0.0005 M NaCl, and it has been postulated that salmon might use this sense by trial and error searches to move through estuaries. Although it ought not be applied too rigorously the available reports appear to indicate that long daylengths and warm temperatures promote movement into and preference for less saline water in many fish. The differences in the thyroidal activities suggest that thyroxine may not act directly but rather may indirectly influence salinity preference perhaps by augmenting the effects of other regulatory agents. Thyroxine is known to have permissive effects for the activities of other hormones and for the expression of circadian rhythms.

Recent researches have demonstrated that electric and magnetic fields or standing infrasonic waves are used for orientation of migration in several species.

Population-specific, genetically determined factors may influence the responses to an environmental stimulus, e.g., the response to water current in fry of lake inlet and outlet spawning populations. The control mechanisms for riverine migrations are likely organized in sequences or hierachies appropriate to the ontogeny and history of population. The adaptive significance of migration is enigmatic, but probably lies in the fact that optimal habitats for different functions (survival, growth, reproduction are spatially, seasonally and ontogenetically separated (Fig. 21.10).

Social interactions may have an important role in downstream salmonids as well as predator-prey relationships.

Osmoregulation: Fish in freshwater have osmoregulatory problems that are the opposites in several ways from those experienced by fish in seawater. Movements of the great migrators (salmon, eels) from one medium to another during migration demand that the fish be able to osmoregulate in both. The osmoregulatory adjustments, in turn, must be integrated appropriately not only with the physiological requirements of migration itself but also in many instances with reproductive maturation and with differences in water temperature (see Chapter 11).

Transformation of the yellow eel to the silver eel in freshwater is accompanied by osmoregulatory preparation for seawater.

Clearly it is not possible to generalize as to the specific effects of either thyroid hormones, prolactin or cortical steroids on fish migratory behaviour. All of these hormones were probably involved in the pre-adaptations necessary for the evolution of movement between areas of differing salinity. Differences in the specific details for hormonal regulation of the migratory behaviour itself may reflect the separate evolution of such movements in a variety of species and the opportunistic exploitation of osmoregulatory hormones along lines dictated by specific selection pressures and physiological limitations.

Metabolism: A successful migration depends not only upon the appropriate behavioural responses to environmental stimuli, but equally upon closely regulated metabolic changes which enable the fish to mobilize sufficient energy reserves to sustain a movement which may be both prolonged and extensive migration is as much a matter of metabolism as of behaviour. During spawning migrations there are also precise physiological changes controlling gonadal maturation which entail the transfer of metabolic reserves to the gonads. The principal nutritional needs of fish during a spawning migration may be into energy for body maintenance, for gonadal growth, for the performance of migration and, in the case of diadromous species such as salmon, for adaptation to changes in the osmotic environment.

Some species support metabolic demands by increased feeding during migration. Many others migrate in waters that do not consistently support adequate foraging and may cease feeding entirely during migration; the digestive tract may even degenerate extensively (salmon, eel). Large energy stores are necessary for the nonforaging great migrators. Lampreys do not feed during their spawning migration in fresh water. Death often occurs after reproduction in these great migrators.

Of the three types of energy stores (proteins, fats, and carbohydrates), the carbohydrate store (glycogen) is least important in migration. Fat stores are most important but protein may also be metabolized extensively, especially when fat is deficient. Nitrogen excretion in teleost is about $3\frac{1}{2}$ times that in mammals. Fat accumulation occurs during specific intervals that are related to stage of development and time of year in a particular species. However, there is considerable variation in timing among species. Body growth (protein) and fattening are often linked.

Lipids are commonly stored to provide metabolic energy in fish. Fat storages range from about 10 to 50% of the total weight of teleosts at the beginning of migration. The amount varies according to species, but there is also a direct relation between the amount of fat stores and the distance travelled by various populations within a species. Lampreys also accumulate large fat stores (30–50% body weight). Fish store primarily triglycerides (75% of all reserved fat). Most of the remaining body lipid is structural.

Most of the neutral fat stores as well as significant amounts of protein are used up during migration. Relatively few hormones are likely to stimulate an increase in lipogenesis and fat storage. The most potent lipogenic hormone in teleosts appears to be prolactin.

High activity of corticosteroid hormone is an obligatory requirement during migration. Although there is not a direct causative relation between corticosteroid activity and migration, the functions of corticosteroids in fish would seem to be important adjuncts in maintaining metabolic support (see Able 1980, 1991; Meir and Fivizzani 1980; Harden Jones 1981; Hasler 1983; Little 1983; Tucker 1983; Northcote 1984; Bertmar 1985; Fontaine 1989; Døving 1990; Halvorsen and Stabell 1990; Rankin 1991; Arnesesen and Stabell, 1992; Stabell 1992). For a comprehensive account of salmonid smolting, the reader is referred to Boeuf (1993; see also Holmqvist 1993).

6. ENERGETICS OF MIGRATION

Although it is virtually certain that lipids, particularly non-esterified fatty acids provide the major metabolic substrate in fishes, there are few studies of lipid storage in migratory fishes. Major lipid storage sites are in connective tissue or in mesenteries of intestines, skeletal muscle and liver, or in head, skin and tail.

Although many anadromous or catadromous fishes make annual migrations, little attention has been given to patterns of premigratory energy storage. It is known that lipid reserves in the coho salmon, *Oncorhynchus kisutch*, decrease with forced exercise, and that herring deposit large amounts of lipid which are used to move from feeding grounds to breeding sites. However, few complete analyses of storage and utilization of lipids in fishes are available. The most comprehensive model may be found in studies of the sockeye salmon, *O. nerka*. Migratory adult sockeye salmon accumulate considerable fat reserves which are largely used on migratory trips to upstream breeding sites, a distance often over 1200 km. During the migration, the lipid contents of sockeye salmon decreases more than 70% and protein content decreases approximately 40%. Sockeye salmon in other studies consume 91–96% of their fat reserves and 53% of their protein during the course of migration. The major site of lipid storage is not in the flesh, but in head, tail and skin.

Number of factors are considered in finding the energetic costs of migration, e.g., size of fish, speed, current. Fish muscle fibres possess a variety of specialization. Perhaps it is also significant that some of the long-range migrant among fishes, particularly the bluefin tuna demonstrate a remarkable ability to maintain relatively high body temperatures above the ambient temperature of the surrounding water. Undoubtedly, high body temperatures in such fishes (occasionally over 30°C) permit greater activity in cold waters (Blem 1980).

Although recent results do not permit to conclude that plaice use selective tidal stream transport solely to save energy, an energetic analysis indicates that large plaice cannot save energy by using selective tidal stream transport in sea areas with slower tidal stream speeds (mean speed < 0.2 ms^{-1}). They should therefore not exhibit this behaviour in such areas if the mechanism evolved as a means of conserving energy (Metcalfe and Arnold 1990).

In fish the mass-specific energetic costs of active transport decrease with body size, and hence selection would favour large size in migrant species. In fish, migrant species are able to exploit resources unavailable to more sedentary species, and hence should also show an enhanced rate of growth. These predictions are supported by comparisons within populations, between populations, and among species in the group (Roff 1991). For information on reproduction or fecundity, mortality, growth rate, etc., see also Roff (1984), and Jonsson and Jonsson (1993).

7. SCHOOLS (SHOALS)

A fascinating feature of some fish shoals is the highly synchronized movement of all individuals. It is, however, still not clear whether fish shoals make collective decisions or whether they have leaders.

Fish shoals are commonly considered as egalitarian, leaderless groups, suggesting a similar influence of the individuals on movements of the group. There is however, evidence that some individuals have strong position preferences within shoals over prolonged time periods.

Leader-follower relationship have so far been reported only from small 'shoals' consisting just two fish. For larger shoals no leadership has been observed.

Breder (1976) discussed this aspect exhaustively. The closed figure that forms the outline of a school is a remarkably flexible boundary subject to continual transformation. These changes are produced by a large variety of influences both intrinsic and, by a vastly greatly number, extrinsic. Obviously, the most important intrinsic factor in holding a school together is the impulse that causes fishes to assemble, respecting each others necessary swimming room and accepting a common polarization.

The fishes that are outermost along the sides of a school do not form a special boundary layer any more than do those at the front form "leader". Those at the side surfaces differ from the rest only in that they lack fellows on one side. Like those at the front, they are continually changing as their aggregating tendency apparently moves them toward a more central position.

Aside from temporary weakening of bonds by such things as vigorous feeding, reproduction, the coming of a sufficiently dark night, or a particularly violent disturbance, the basic school structure is continuous in obligate schoolers. In facultative schoolers, the school is periodic or of occasional occurrence. True semipermanent intermediates between these two ordinarily distinct modes are not easy to find and are uncertain at best.

Examples of obligate schoolers are clupeids or scombroids, and facultative schoolers are Salmonidae, Cyprinidae and Serranidae.

The intrinsic influences divide naturally into two groups, the first being those of nonorganic elements. Common examples of these are light, water currents, shoreline, sharply mottled bottom patterns, and obstructions. Sharp discontinuities of any of these are especially influential. Organic factors include other schools, large, predatory fishes, fish-catching birds, and rich plankton streaks.

The interaction of a space lattice, vortex, trails, and lubricity of fish surface mucus is shown to be important to the operation and structure of fish schools and significant in terms of locomotor efficiency. This is independent of the various interpretations of possible survival values, protection from predation, and similar ideas—all of which are extremely difficult to prove, even if valid.

A single type of space lattice is shown to approximate the arrangement of fishes in a school on the basis of geometrical reasoning. This is supported by empirical data.

The vortex trails left by each fish, when the fishes are deployed according to the "fish school" lattice, lead each following fish into a series of vortices at a point where the water flow is travelling in the direction in which they are swimming.

The lubricity of the mucus-water mixture that the fish ahead leaves in its vortices decreases the drag on the following fish.

The advantages of the regimented life in a school, as against the freedom of action common to the more or less solitary life, are evidently related to the effectiveness of the drag-reducing mucus in the vortices. The fishes with the least effective mucus appear to take advantage of the schooling life while those with the most effective mucus are more likely to be solitary.

Shaw (1962) estimated that about 2000 or more species of marine fish schooled, especially the planktonophage pelagic species like the clupeids, engraulids and also the mugilids, scombrids and trachurids.

Vision is the predominant sensory system in the schooling of fish although some other stimuli are also involved, and there are close parallels between the bahaviour of fish in schools and in optomotor apparatus. In each situation the fish moves parallel to and in the same direction as the moving field, it changes its speed with that of the field, and it changes its position relative to the field. Furthermore, the optomotor response is shown more strongly when several fish are swimming together. The visual perception of velocity change appears to form the basis of the mechanism by which schools are maintained, suggesting that optomotor response is important in the maintenance of schools. It is not clear if it is important in school formation (Arnold 1974).

Most species investigated school optically with a threshold near 10^{-1} m.c. These is an interesting paradox here in that silverysided schooling fish are very difficult to see when viewed laterally, as by other members of the school (Denton and Nicol 1966). This invisibility depends on a homogeneous distribution of light from all azimuths, hence the school will find it easier to keep together visually at dusk and dawn when the sun is nearest the horizon. There is obviously an adaptive "conflict" between the needs of camouflage and visual contact in the school.

The strategy of fishing may be determined by schooling in terms of net size, design and speed of the tow. Generally speaking the really large catches of fish, such as the Peruvian anchoveta, come from pelagic schooling species. If schools break up at night the density of fish will become reduced and their response to nets may well change (Blaxter 1975).

Magnuson (1978) discussed the energetic aspects of schooling fishes, with emphasis on scombrids. A possible savings of ca 10–20 from schooling has been reported for horse mackerel (*Trachurus mediteraneus*).

Shoal position can have a strong influence on individual fitness. Individuals in front positions of shoals were observed to have higher feeding rates than individuals elsewhere. Individuals rotate their shoal positions according to the tradeoff between energy intake and predation risk. For shoal related details, the reader is referred to a series of papers (Krause et al. 1992, 1996a, b; Bumann and Krause 1993; Krause 1993a, b, c, d, 1994; Krause and Godin 1996).

8. HABITAT MAINTENANCE

An overview of habitat selection has been given by Kramer et al (1997). When habitat selection changes in regular pattern with state of tide or time of day or year, it is considered *migration*. Environmental variables important for habitat selection are given in Table 21.2.

Once established in a favourable zone it is clearly an advantage to maintain position there and the simplest way of doing that is not to move very far. If movement has to be over relatively large distances, to obtain sufficient food for example, then it is advantageous to be able to return to the place formerly occupied. Restricted movement is a common feature of intertidal fishes.

Residence time is regulated by the amount of disturbance caused by wave motion. There is, however, a biological factor which affects residence time or fidelity to a particular pool and that is a fish's age. Young fish are more mobile and they do not stay in the same pool as long as the older ones.

Homing: Restricted patterns of movement cannot be accounted for solely by poor locomotory abilities because many species from different families in widely separated geographical areas return to their original location. The possession of a topographic memory seems to be a common feature of blennies and gobies, and a detailed knowledge of the environment can be rapidly acquired. Vision or olfaction plays a large part in homing. Long-range olfactory homing mechanisms of salmonids is a fine example.

The function of homing and fidelity to a particular area seems to be to enable a fish to utilize and return to a location which provides a suitable refuge over the critical low tide period. Tide pools are the most obvious refuges.

Homing is displayed in Pacific salmon, including pink salmon (*Oncorhynchus gorbuscha*), chum (*O. keta*), coho (*O. kisutch*), sockeye (*O. nerka*), and chinook (*O. tschawytscha*), as well

Table 21.2 Environmental variables associated with fish distribution on a behavioural scale

1. *Physical characteristics of the water*:
   Presence of water
   Permanence of water
   Temperature
   Depth/pressure
   Current speed
   Turbulence
   Suspended material/turbidity
   Light intensity and spectral composition
2. *Chemical characteristics of the water*:
   Dissolved gas (oxygen, carbon dioxide, Hydrogen sulfide)
   pH
   Salinity/conductivity
   Ionic composition
   Other natural dissolved substances (e. g., homing cues, conspecific odours)
   Chemical pollutant
3. *Characteristics of adjacent substratum and structure*:
   Composition of substratum
   Amount, form and composition of other structure (e. g., coral algae, macrophytes, woody debris)
   Availability of caves and holes
4. *Position relative to*:
   Substratum and other structure
   Surface
   Shore and depth of water column
   Permanent refuges for individuals invading temporarily available habitat (e.g., intertidal zone, flood plains, intermittent streams)
5. *Presence or absence of other species*:
   Prey
   Predators and parasites
   Competitors and mutualists

Source: Kramer et al. (1997)

as in Atlantic salmon (*Salmo salar*) and related species, including rainbow trout (*S. gairdneri*), brown trout (*S. trutta*), cutthroat trout (*S. clarki*) and Baltic trout (*Salvelinus alpinus*).

Olfactory mediation of homing, stream odors, olfactory imprinting and role of pheromones in migration are well mentioned.

Information on the spatial and temporal distribution of pacific salmon stocks indicates that marine migratory patterns are not random events. Emigration appears to follow an innately directed pattern of movement, while homing is directed by cues acquired during emigration (Brannon 1984).

Territoriality: One other form of restricted movement which ensures that a favourable location is maintained is territoriality, in which a specific area, usually much smaller than a home range, is actively defended against others of the same and sometimes different species. The possession of a territory guarantees to the holder exclusive rights to feeding, shelter and a spawning site in the defended area. Although food and reproduction are of obvious importance, shelter as a protection against turbulence and predation is equally vital for the small intertidal fishes. Territoriality, dominance hierarchies and the aggressive behaviour that accompanies them are particularly common among gobies, blennies, clingfishes and mudskippers, although the degree to which it is developed varies between species and sometimes even between the sexes of one species. Some species chase others away from their immediate vicinity whereever they happen to be; others are rigidly territorial and defend a well-defined area.

Stream-dwelling fish occupy territories which have usually been regarded as feeding territories, but a fish may move between a number of territories, even within a 24–h period, and accordingly may cover an extended area. Such extended areas are denoted 'home range' or home area'. In the majority of species, however, individual fish will move over relatively short distances and will occupy only a limited area of stream.

A recent coverage of territoriality is given by Grant (1997). Little is known about the behaviour of deep-water fishes in lakes or oceans. Comparative studies of space use in fishes are lacking. Fishes occupy relatively small territories or home ranges compared to birds, mammals or lizards.

CHAPTER 22

Nuclear and Cytogenetic Studies; Atavism

1. INTRODUCTION

Application of molecular genetic techniques in fisheries research has increased dramatically over the past several years, largely due to the increased availability of techniques and an increased awareness of the value of genetic data (see Table 22.1). Today, the boundaries of fisheries related molecular genetic research encompass topics from the identification of markers for stock discrimination to the genetics of pathogenic organisms of commercially important species to the expression of growth factors during maturation.

Buth et al. (1991) presented information on cyprinid allozymic, isozymic, and cytogenetic characteristics and their research applications.

Ozouf-Costaz and Foresti (1992) deal with the most up-to-date methods used for cytogenetic purposes, their feasibility, and underline their deficiencies. Rishi (1989) mentioned the techniques in describing the current status of fish cytogenetics.

The reader may refer the title 'Evolutionary Genetics of Fishes' by B.J. Turner (1984) for a wide coverage. Significant reviews related to fishes are by Utter et al. (1974), Allendorf and Utter (1979), Gold (1979), Bentzlen (1994) Ferguson (1994), Lincoln (1994), Park and Moran (1994), Ward and Grewe (1994), and Wright and Utter (1994). The title: Molecular Genetics in Fisheries (edited by G.R. Carvalho and T.J. Pitcher 1995) has provided a wealth of information.

2. MOLECULAR GENETIC MARKERS

Isozymes: Isozymes are functionally similar but separable forms of enzymes, encoded by one or more loci. Isozymes that are the products of different alleles at the same locus are termed allozymes.

For further information, the reader is referred to Park and Moran (1994) and Verspoor (1998).

A variation of isozyme electrophoresis is the isoelectric focusing (IEF) technique. Although this technique has not achieved as widespread use as isozyme electrophoresis, it has been used in species identification studies as well as population-level studies.

While protein (isozyme) electrophoresis has provided fisheries geneticists with the most genetic data to date, the technique has certain limitations. The resolution of protein electrophoresis is not always adequate for detecting differences between populations or individuals. Because of redundancy in the DNA code that dictates protein sequences, all changes in a gene may not result in a change in the overall charge of the protein expressed, thus many genetic variants are not detected by protein electrophoresis. Furthermore, protein electrophoresis is limited to detecting genetic changes that affect genes that actively express proteins detectable with a histochemical stain. These genes constitute only a small percentage of the whole genome of an animal. The potential amount of genetic variation detectable by DNA methods vastly exceeds the amount detectable by protein methods because DNA sequences are being assayed more directly. Thus, as fisheries geneticists have encountered an increasing number of questions that cannot be resolved with isozymes, DNA methods have generated increasingly more interest. It is worth noting that, although more variability is detectable with DNA methods than with isozyme electrophoresis, the existing data for isozymes in many fish species represent a huge wealth of information that should not be disregarded. Indeed, until a substantial amount of DNA data has been collected for a particular species, an existing isozyme database often represents a more practical source of genetic information (read also Jamieson 1999).

Allozyme electrophoresis is a precise and cost-effective method for comparing populations, species, and closely related genera. Buth (1984) reviewed applications of allozyme variation in studies of cyprinid fishes, and as he predicted, allozyme data are increasingly used, especially for investigations of hybridization and introgression, population structure, biochemical identification of taxa, and phylogenetic systematics. For further information, the reader is referred to Buth et al. (1991). Allozyme applications in cyprinid biogeography are scarce, because few studies employ a broad geographic sampling strategy, or have employed rigorous phylogenetic methods to ascertain the phylogeny of the group. Allozymes are among the least expensive biochemical tools. The reader is also referred to Carvalho and Hauser (1995) for more details.

Isozyme characters, including gene number, tissue-specific expression, and heteropolymer formation have been applied in several studies of fish systematics. Isozyme characters related to gene duplication and silencing may prove especially useful in studies of karyotypic polyploids. Isozyme characters may vary at a number of levels of organization, making it difficult *a priori* to identify an optimal taxonomic level for their application.

2.1 Duplicate Isozyme Loci in Fishes

Genes with a multilocus enzyme system can be related in a number of ways. The products of some loci may be specialists in terms of their intracellular site of expression, i.e., the mitochondrial vs supernatant (= cytosolic) distribution of gene products in systems such as malate hydrogenase, aspartate aminotransferase, and isocitrate dehydrogenase.

The major modes of duplication of such genes include tandem duplication of a single gene and polyploidization involving duplication of an entire genome.

(1) Tandem duplication: Gene duplication involving a single cistron or set of cistrons can be accomplished via errors in mitotic or meiotic processes. The resultant duplicate genes would reside on the same chromosome and, thus, become members of a particular linkage group. While such tandem duplications may yield multitude structural cistrons, the regulatory gene controlling the activity is probably not often duplicated in this process. In fishes, tandem duplication is observed to be rare or, at the very least, uncommon.

The distribution of tandem duplications within lineages depends upon whether their origin was ancient or recent. A particular tandem duplication can be limited to all members of a given species. Tandem duplicates, despite their rarity, can be detected at a number of levels, from the first occurrence in an individual of a species through their postspeciation establishment in a number of phyletic lines.

(2) Polyploidization: Gene duplication involving an entire genome (polyploidy) has been promoted as a major evolutionary phenomenon. With the duplication of the entire chromosome compliment, a duplicate set of unlinked loci under separate regulatory control are created. With one locus to cover the essential previous function, the additional locus if freed to acquire previously forbidden mutations and diverge to ultimately serve new functions.

The relationship between tandem duplication and polyploidy has been summarized as processes that "complement each other as the means of achieving meaningful gene duplication. It would not be suprising if during the course of vertebrate evolution these two means were used alternately". However, it appears that, in fishes, polyploidization has played the greater role.

Polyploid fishes: The polyploid nature of many fishes has usually been suggested on the basis of their karyotype or DNA content. Polyploid fishes are known from six orders: Acipenseriformes, Salmoniformes, Cypriniformes, Siluriformes, Atheriniformes and Perciformes.

Each tetraploid cyprinid species discovered thus far may be the result of an independent polyploidization event. A number of diploid-tetraploid complexes are known among the cobitid fishes. Allotriploids are known in the poeciliid genus *Poeciliopsis*. In Perciformes, only one species, *Lucioperca sandra* provides any suggestion of polyploidy. The search for and analysis of fish polyploids is far from over.

In context of phyletic consequences of gene duplication, views regarding divergence, functional diploidization, retetraploidization have been expressed (read Buth 1983).

The role of polyploidy in the evolution of fishes has been discussed by Schulz (1980). For special attention regarding tetraploidy and the evolution of salmonid fish, the reader is referred to Allendorf and Thorgaard (1984). The details of polyploidy in cyprinid fishes are given by Buth et al. (1991). A revised version of a nomenclature for protein-coding loci in fish is suggested by Shaklee et al. (1990). Morizot (1990) studied the gene maps of poeciliid (*Xiphophorus, Poeciliopsis*), *Fundulus heteroclitus*, centrarchid (*Lepomis*) and salmonid fishes (trouts, salmon, chars). The evidence marshaled amply demonstrated that at least some gene arrangements are conserved among fish orders and may well represent syntenies conserved from the ancestor of teleost fishes (read further for (i) comparisons of fish gene maps with those of other vertebrate, (ii) models for the origins and evolution of vertebrate chromosomes, and (iii) summary and prospects). Further information on *Poeciliopsis* is available in the paper by Morizot et al. (1990).

Analysis of genotypes of 76 polymorphic loci in more than 2600 backcross hybrid individuals derived from intra- and inter-specific genetic crosses of fishes of the genus *Xiphophorus* (Poeciliidae) resulted in the identification of 17 multipoint linkage groups containing 55 protein-coding loci and one sex chromosome-linked pigment pattern gene. Multipoint linkage analyses identified highly probable gene orders for 10 linkage groups. The total genome length was estimated to be ~18 Morgans. Comparisons of the *Xiphophorus* linkage map with those of other fishes, amphibians and mammals suggested that fish gene maps are remarkably similar and probably retain many syntenic groups present in the ancestor of all vertebrates (Morizot et al. 1991).

3. CHROMOSOMES

3.1 Chromosome Numbers and Morphology

The hagfishes *Eptatretus stoutsi* have a chromosomal make up of 46–52 acrocentric chromosomes. An extremely variable number of chromosomes has been reported in *Myxine glutinosa*. Lampreys of the northern hemisphere (*Ichthyomzon gagei*) have a much higher number of chromosomes, ranging from 142 to 174. In some of these, it is extremely difficult to detect the centromere, and they may be true microchromosomes. The rest are all acrocentric. The southern hemisphere lampreys (*Mordacia mordax*) have a chromosome number of 76 and all chromosomes are metacentrics.

In the chimaeras, chromosome numbers of 58 and 86 have been found for *Hydrolagus colliei* and *Chimaera monstrosa* respectively. Both have a mixture of acrocentric chromosomes and microchromosomes.

Amongst the skates and rays, composite chromosome numbers also exist. In the family Rajidae (*Raja clavata*) a range of 98 to 104 chromosomes was found, consisting of a mixture of a few metacentric chromosomes, many acrocentric chromosomes, and many microchromosomes. Stingrays, family Dasyatidae (*Dasyatis sabina*) have 68 to 84 chromosomes. Most of these are metacentrics, the rest being acrocentric, or just barely submetacentric. Electric rays, family Torpedinidae (*Narcine brasiliensis*) have a very low number of 28, all chromosomes being metacentric or submetacentric and most of them are large in size.

Only two species of sharks have been reliably karyotyped, *Squalus acanthias* and *Etmopterus spinax*. The latter has a karyotype of 86 chromosomes, made up of acrocentrics and microchromosomes. *S. acanthias* has a karyotype of 78 chromosomes, comprising a mixture of metacentrics and acrocentrics but no microsomes. In the Chondrostei, consisting of paddlefishes, family Polyodontidae (*Polyodon spathula*), and the sturgeons, family Acipenseridae a high number of chromosomes has been found, ranging from 112 to 240. These are a mixture of metacentrics, submetacentrics and microchromosomes.

Among the gars, family Lepisosteidae, only *Lepisosteus oculatus* (= *productus*) has been karyotyped. It has 68 chromosomes with a mixture of metacentrics, acrocentrics, and microchromosomes. The bowfin, *Amia calva*, has 46 chromosomes, a mixture of submetacentrics to metacentrics and acrocentrics but with no microchromosomes.

In the Teleostei, the most speciose of all fish groups, a diverse array of chromosome numbers has been found. Among the teleosts, no species exhibit microchromosomes. In the osteoglossids (*Osteoglossum bicirrhosum*), there is a variety of chromosomal composition, ranging from 34 to 56, including various mixtures of metacentric and acrocentric chromosomes. Clupeids (*Alosa pseudoharengus*) have 48 acrocentric chromosomes. The Euteleostei, or higher teleosts, have chromosomes ranging from 14 to 140, with various assortments of metacentrics, submetacentrics and acrocentrics. The basic chromosome number has been suggested to be 48.

The Brachiopterygii, comprised solely of the Polypteridae (*Polypterus palmas*) have 36 rather large, submetacentric to metacentric chromosomes.

The lungfishes (*Lepidosiren paradoxa*) are similar to polypterids in having 38 submetacentric to metacentric chromosomes, but they are huge in size. This holds true for *Neoceratodus* and *Protopterus* (read Dingerkus 1979).

Details are available on chromosomes in salmonids (Schulz 1980; Allendorf and Thorgaard 1984), cichlids (Kornfield 1984) and cyprinids (Buth et al. 1991; see for literature). General reviews on fish karyology are of Sola et al. (1981) and Vasiliev (1985). For general information of fish karyology, the reader can consult Rishi (1989). For database of fish chromosomes, read the informative book by M. Klinkhardt, M. Tesche and H. Greven (1995). Recent studies in this direction on notothenioids are from Pisano and Ozouf-Costaz (2000) and Pisano et al. (2001).

3.2 DNA Per Cell Content

It is well established that cells from the same individual at the same state of the cell cycle, and also cells from other individuals of the same species, contain the same amount of DNA. Because this uniformity exists, DNA/cell values have become another basis for comparison between different species and higher systematic categories. When used in conjunction with karyotypes, DNA/cell values can yield information as to whether karyotypic differences are due to minor changes, such as chromosomal rearrangement, or major changes such as tetraploidy. Moreover, since DNA is the basic genetic material, it is believed that these DNA/cell comparisons can give an estimate of the fundamental similarity or difference between groups.

The DNA values given are those per haploid cell. *Eptatretus stouti*, a hagfish, has a rather high value of 2.8 pcg (picograms). The northern hemisphere lampreys exhibit between 1.4 and 2.5 pcg of DNA per cell. The holocephalan *Hydrolagus colliei* has 1.5 to 1.6 pcg. Sharks and rays range between 2.75 and 9.8 pcg. For sturgeons, values of 1.7 to 5.1 pcg have been obtained. The gar, *Lepisosteus oculatus* (= *productus*) has a value of 1.35 pcg, the alligator gar, *Attractosteus spathula* (= *Lepisosteus ferox*), has 1.2 pcg, and the bowfin, *Amia calva*, has 1.23 pcg. Teleosteans exhibit a wide range of DNA values, viz., 0.4 to 4.4 pcg. In *Latimeria*, the value is between 3.465 and 3.8 pcg. In the brachiopterygians the DNA/cell value has been estimated to 4.7 to 5.8 pcg. The species of living dipnoans range between 80.2 and 140 pcg of DNA/cell. The lungfishes probably have the largest amount of DNA/cell of any living animal (Dingerkus 1979). A brief account is also given by Rishi (1989) in this regard.

3.3 Chromosome Banding

3.3.1 C-Banding

C-bands allow heterochromatin, corresponding to highly repeated DNA sequences, to be visualized. They are frequent in the centromere and telomere regions and sometimes close to the NOR location in fish chromosomes. When found interstitially, C-bands can be effective for identifying homologous chromosomes and thus help to characterize and differentiate species. Tandem (centromeric-telemeric fusions) have also been demonstrated by C-banding in a gobiid and in some balistids (Ozouf-Costaz and Foresti 1992; see also Rishi, 1989).

3.3.2 Nuclei and Nuclear Organizer Regions (NORs)

Nucleolar organizer regions (NORs) are regions of certain chromosomes that are composed of highly repetitive ribosomal DNA (rDNA), which code for ribosomal RNA (rDNA) and in the interphase cell forms the nucleolus or nuclei. Studies of nucleoli and NORs in fishes show that diploid species have 2 NORs, usually with two small nucleoli in the interphase nucleus; tetraploid species have four NORs, usually with four nucleoli in the interphase nucleus; and species with high amounts of tandemly duplicated genes have two NORs, usually with two very large nucleoli in a large interphase nucleus. The barb *Barbus tetrazonia*, paddlefish, *Polydon spathula*, and lungfish *Lepidosiren paradoxa*, represent, respectively, the diploid, tetraploid, and tandem gene-duplicated states. The interphase nucleus of *Latimeria* has been shown to contain two small nucleoli, like that of barb, indicating on this basis that *Latimeria* is a typical diploid species (see Dingerkus 1979).

Briefly, NORs are chromosomal sites intimately related to nucleolus formation in interphase and containing the genes for the 185 and 285 ribosomal RNA genes. NOR banding patterns are known for almost 100 cyprinids. Collectively, the interspecific patterns of NOR variation in cyprinids are similar to those in other fishes. Intra-specific variations (heteromorphisms) in NORs have been detected in c. 10–13% of all North American cyprinids examined. Observations indicate that NORs can be taxonomically informative characters (read details for cyprinids, Buth et al. 1991).

NORs can represent excellent specific or population markers and exhibit extensive inter- and intraspecific polymorphism in many fish groups. These variations can affect: (a) the absolute number of NOR sites per genome; (b) the chromosomal location of NOR sites; (c) the relative sizes of individual NOR sites; (d) the number of active NORs (i.e., Ag-NOR sites per cell). Polymorphisms in categories (a) and (b) are usually interspecific; those in (c) and (d) usually intraspecific.

The literature dealing with this subject is rapidly increasing. A spectacular example is the multiple chromosome polymorphism, including 14 different karymorphs, reported in 384 specimens of *Gobius papanellus* (see Ozou-Costaz and Foresti 1992). Intra-individual polymorphism is often due to centric fusion, non-disjunction and hybridization. Examples have been found in the rainbow trout, Atlantic salmon and sockeye salmon. But the significance of the intra-individual polymorphism is far from being understood. Intra-population polymorphism has been observed in rainbow trout, green sunfish, white herring and African toothed carp. The application of banding techniques may further bring to light some of the un-noticed polymorphism in fish species (Rishi 1989).

Supernumerary chromosomes ('B-chromosomes') also often occur in fish. They differ from the regular ones by their generally small size, variable number among specimens or tissues, etc.), non-homology between themselves, and moreover, are usually heterochromatic. The origin and frequency of occurrence of B-chromosomes in fish and their heterochromatic nature are not yet elucidated, but B-chromosomes may well also be involved in karyoevolutionary processes (Ozouf-Costaz and Foresti 1992).

4. MITOCHONDRIAL AND NUCLEAR DNA

Advances in molecular biology (e.g., DNA cloning) have opened frontiers in evolutionary biology. DNA methods have been used in two types of studies: those that examine characteristics of genomes such as genome organization, genome size, or DNA-DNA hybridization, and those that examine the sequence of specific DNA molecules through restriction endonuclease analysis or nucleotide sequencing. Most evolutionary or systematic studies that have focused upon differences at the sequence level have employed mitochondrial DNA (mt DNA). Nuclear genes such as ribosomal DNA (rDNA) or protein-coding genes have been used less frequently. DNA sequences are characterized either by using restriction endonucleases (bacterial enzymes which recognize and cleave DNA molecules at specific 4–6 base sequences), or by directly obtaining the nucleotide sequence of the molecule in question. Restriction endonuclease analysis is used more often than direct sequence analysis because it is quicker and cheaper. The use of DNA technologies for the study of fishes has been limited primarily to mt DNA. Applications of these methods to studies of cyprinids are rare (Buth et al. 1991).

Nuclear genomes: Studies of nuclear genome organization in fishes are limited to a few species. The basic feature or patterns of DNA sequence organization in fishes (including cyprinids), however, appear largely similar to those in other vertebrates if not most eukaryotes. Considerably more has been published on DNA base composition and nucleotide distribution of fish (including cyprinid) nuclear genomes. Genome sizes (nuclear DNA contents) are known for over 100 cyprinid species or subspecies. There is extensive disagreement above the value of DNA-DNA hybridization for systemic inference.

Mitochondrial DNA: Several unique properties of mitochondria DNA (mt DNA) make it particularly useful for molecular investigations: (1) it is relatively small, circular molecule ranging in size from 16.2 to 19.5 kilobases (kb) in fishes, including cyprinids; (2) it is highly compact, with few non-coding regions, and gene content and order appear to be conserved in all vertebrates; (3) it is maternally and clonary inherited without recombination, and usually exists in only one genotype per individual, although there are exceptions, and (4) in vertebrates, it evolves five to ten times as rapidly as scn DNA (single-copy nuclear DNA). This recombination of features permits relatively easy isolation and analysis of mt DNA, particularly for the study of maternal lineages in closely related populations and or species. Few studies have used mt DNA to evaluate cyprinid population structure. The maternal inheritance of mt DNA allows it use as a marker for infering the maternal genome in F_1 hybrids and subsequent backcross individuals, permitting tests of reciprocity of hybridization and directionality of introgression. The lack of paternal contribution of mt DNA, however, makes it necessary to analyze hybrids by using other diagnostic characters (e.g., allozymes, morphology) so that hybrids can be identified. The utility of mt DNA in hybridization studies can be demonstrated by discussing two categories of examples: (1) those in which the maternal component and clonal diversity of unisexual hybrids have been identified, and (2) those in which reciprocity, directionality, and extent of introgression across varying degrees of genetic and taxonomic divergence have been examined.

Systematics and biogeography: Mitochondrial DNA also serves as an excellent marker for inference of hypotheses of phylogenetic relationships, at least among closely related species or species groups, so mt DNA analysis can often be a valuable tool for testing biogeographic hypotheses. Perhaps the most important consideration in phylogenetic analysis using mt DNA is sampling strategy. Variation in rates of evolution among genes should make mt DNA a useful marker for analysing phylogenetic relationships across a variety of taxonomic levels.

Nuclear DNA: As with mt DNA, nuclear genes evolve at different rates, and so can be useful for addressing questions at a variety of hierarchial levels, from individuals within populations to the phylogeny of 'life'. Unlike mt DNA, most nuclear genes (with the exception of sex-lined genes) are diploid, with the degree of complexity varying from single loci to multigene families. Members of multigene families appear to evolve in concert, and hence may be subject to phenomenon of 'molecular drive'.

Two types of nuclear genes have been used for population studies: (1) hypervariable sequences ('DNA fingerprint') or rapidly evolving single copy sequences, and (2) the non-transcribed spacer (NTS) region of the ribosomal DNA (rDNA) gene cluster. These markers have seen limited use, but show promise for the analysis of paternity and maternity testing and for measuring the extent of population structuring.

Nuclear DNA markers have vast potential for the analyses of hybridization and introgression. They can act as unambiguous markers identifying particular taxa and can be used like allozyme markers to study these phenomena. The choice of marker is not constrained by detectability by function (unlike allozymes), providing an immense number of potential markers.

A variety of nuclear DNA markers have employed restriction site maps or DNA sequences of nuclear genes to allow investigators to infer phylogenetic hypotheses at different hierarchical levels, Problems concerning rates of evolution, convergence, and sampling strategies mt DNA are also applicable to analyses of nuclear DNA. Also, as with the analysis of mt DNA, particular genes or portions of genes can be used for phylogenetic analysis at specific levels of divergence. Analyses of nuclear gene variation using recombinant DNA techniques should prove extremely useful as markers for inferring phylogenetic relationships across a wide range of hierarchical categories and organisms (Buth et al. 1991; see also Buth 1984; Ferguson 1994; Park and Moran 1994; Carvalho and Hauser 1995).

Wright and Bentzen (1994) argued the merits of a particular class of nuclear marker, variable number's of tandem repeats (VNTR) loci. VNTRs themselves comprise two (probably related) classes of loci, the minisatellites and microsatellites. The general utility of microsatellites VNTRs to fisheries and aquatic research was highlighted.

5. SEXUALITY AND SEX DETERMINATION

Sexuality: Sexuality in fishes is characterized by its diversity. Hermaphroditism, unisexuality and bisexuality all occur.

Hermaphrodites may be synchronous (simultaneous) when both eggs and sperm mature at the same time, or asynchronous (consecutive) when the fish functions as either male or female at any one time. Among the latter, both protandrous and protogynous forms are known.

Hermaphroditism is usually considered to be mainly a process of sex differentiation rather than sex determination. The involvement of possible genetic factors is not well understood.

Hermaphroditism occur in several orders of fish. This, together with the different forms of the gonad which are observed, indicate that hermaphroditism has evolved several times independently. It is usually considered to have evolved from bisexual or gonochoristic species, although some workers have suggested that, at least in certain groups, hermaphroditism is the primitive type.

The evolutionary ecology of unisexual fishes is distinguished from that of closely related sexual species by two peculiar aspects of their genetics: First, they reproduce, in effect, asexually, and second, their genomic constitutions are invariably identical to those of F_1 hybrids.

Unisexuality in fishes is rarer than hermaphroditism, but examples are known where females produce only female offspring. This may be accomplished by either gynogenesis or hybridogenesis.

In gynogenesis, sperm of a closely related species is used to trigger development of the egg nucleus but does not fuse with it. The male nucleus makes no contribution to the developing embryo. The important genetic consequence of gynogenesis is that daughters are genetic replicates of their mothers. Most of the gynogenetic parthenospecies are also triploid relative to the sexual species from which they were derived. In hybridogenesis, gamete fusion occurs and the paternal genome is expressed, but only the haploid female genome is transmitted to the ovum. Hybridogenetic species are invariably diploid and, indeed, triploidy would seem to preclude hybridogenesis. In hybridogenesis, a reduction division occurs and haploid gametes are produced, but the same genome is always segregated, intact, into functional gametes. Thus, a single genome is inherited clonally.

Poeciliopsis monachlucida is the classic example and serves to illustrate this mode of inheritance. An important consequence of hybridogenesis from the perspective of ecological genetics is that an entire genome is clonally, i.e., asexually inherited.

There is evidence pertaining to the hypothesis that high heterozygosity and heterosis are major components of fitness in unisexual fishes. Parthenogenetic vertebrate species, including unisexual fishes, have the highest levels of heterozygosity ever reported for vertebrates, but it is not certain that heterosis is a direct consequence of heterozygosity.

The Amazon molly *Poecilia (Molliensis) formosa* mates with male *P. latipinna* or *P. mexicana* but reproduces gynogenetically to produce only female offspring. Natural gynogenesis also occurs in certain triploid populations of both *Poeciliopsis* species and *Carassius auratus gibelio*. Hybridogenesis has so far been described only in the genus *Poeciliopsis* where various hybridogenetic forms have been identified.

The majority of fishes reproduce bisexually. Several different mechanisms exist in these fishes and a variety of approaches have been used to investigate the genetic basis of these mechanisms. They include: (a) segregation analysis using marker genes; (b) analysis of sex ratios in intraspecific crosses; (c) analysis of sex ratios in interspecific crosses; (d) inferences from fish produced either by artificial gynogenesis or artificial polyploidy; (e) inferences drawn from so called "sex reversal"; (f) analysis of offspring from fish artificial sex reversal, and (g) cytological analysis of both chromosomes and chromatin (Moore 1984; Price 1984 for details and literature).

Sex determination: Regarding the possible mechanisms of sex determination, fish represent a very heterogeneous group. Most do not exhibit morphologically differentiated sex chromosomes. In some cases it is possible to identify chromosomes carrying sex-determining genes using banding techniques. Morphologically differentiated sex chromosomes are becoming frequently reported in fish, involving male or female heterogamety (XX/XO, XX/XY, ZZ/ZW) and, to a lesser extent, multiple sex chromosome mechanisms ($X_1X_1X_2X_2/X_1X_2Y$, $X_1X_1X_2X_2/X_1X_1X_2$, XX/XY_1Y_2, etc). Heterochromatinization also plays an important role in the differentiation and evolution of sex chromosomes in fish (Ozouf-Costaz and Foresti 1992).

Yamamoto (1969) reviewed sex differentiation. Kallman (1984) reported that sex determination in several species of poeciliid fishes is controlled by a sex chromosome mechanism. A number of species of *Xiphophorus* possess the XX-XY system, and one form, *X. maculatus*, has three sex chromosomes, X, W and Y. In natural populations, females are either WX or WY, XX and males are XY or YY. The sex determining mechanism of rainbow trout is the same as in humans (Fig. 22.1) with the female producing eggs which are all of the X type and the male producing

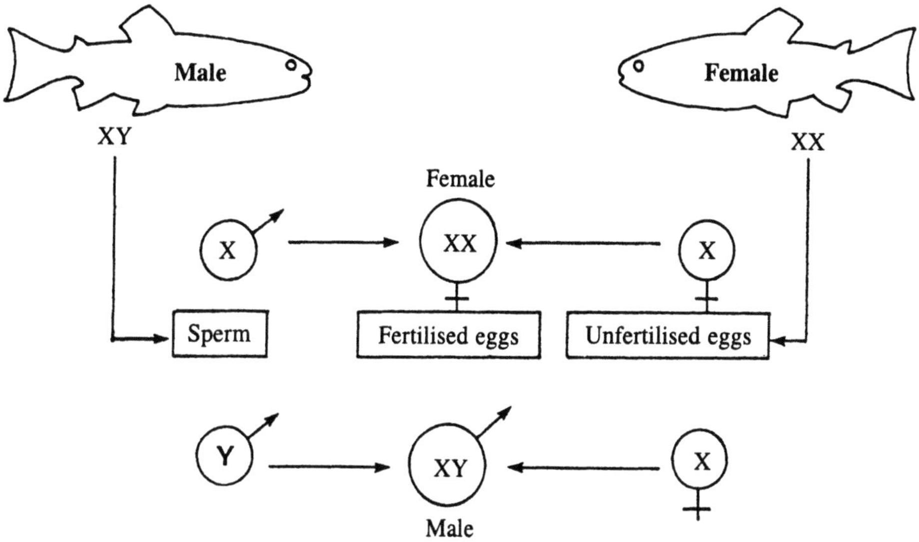

Fig. 22.1 Sex determination in the rainbow trout.

two different types of sperm labelled X and Y. If an egg is fertilized by X sperm it develops into a female but if it is fertilized by a Y sperm a male results. A theory of sex determination has been proposed involving the importance of male-determining gene M (the M gene could be identical with the H-Y locus), autosomal factor T, and suppressor gene S. The reader is also referred to sex markers in cichlids described by Kornfield (1984) and to a review on sexual differentiation by Wachtel et al. (1991; see also Schreibman et al. (1994) (Fig. 22.2). Mouthbrooding cichlid fish *Oreochromis niloticus* sex is determined by genetic factors, temperature level, and genotype-temperature interactions. Temperature-dependent sex differentiation may occur more frequently in fish than previously thought, despite the presence of functional sex chromosomes (Baroiller et al. 1995).

Rishi (1989) gave examples of fishes with different types of sex chromosomes. Tiersch et al. (1991) report that if *SRY* gene (*sex-determining region* Y) is the primary sex determinant in mammals, it is not clear that it has a similar function, or even one that is sex-related, in nonmammals. For the cytogenetics of three populations of the North American cyprinodont fish *Poecilia*

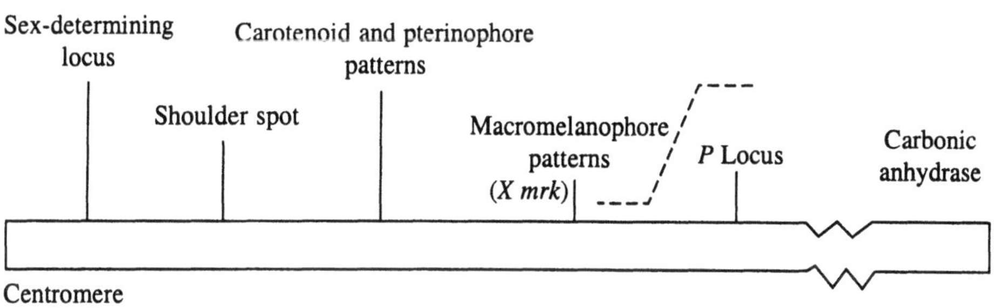

Fig. 22.2 A schematic representation of the *Xiphophorus* sex chromosomes. Broken lines between macromelanophore patterns (*X mrk*) and *P* locus indicates unresolved positional relationship of two loci.
***Source:* Schreibman et al. (1994)**

latipinna, considered to be one of the progenitor species of the genogenetic unisexual *P. formosa*, by C-banding and Ag-staining, refer Sola et al. (1990).

6. ATAVISM

Stiassny (1992) discussed atavisms and evolution. Examples of phylogenetic character reversals abound in the taxonomic literature. An analysis of taxic atavism in the family Cichlidae has been given in respect of (1) neurocranial lateral line foramina; (2) transversus ventralis muscle sections; (3) gill-raker denticulation patterns, and (4) the semi-circular ligament system.

One clade of particular interest is the lamprologine assemblage. This is a highly species-rich, monophyletic radiation of African cichlids found principally in Lake Tanganyika, with a few species distributed throughout the Zaire River. Unfortunately the precise relationships of the lamprologines are not yet solved and their placement within the African radiation is problematical. It is suffice to say that lamprologines are derived African taxa and certainly not immediately basal to the non-tylochromine African radiation.

7. DEVELOPMENTAL GENETICS OF TELEOSTS

The early development of teleost fishes; like that of other vertebrates, consists of an orchestrated progression of both morphological and biochemical events. The biochemical facets of teleost development have been examined by investigating the effects of metabolic inhibitors upon protein levels and the relative times of parental allele activation. In addition, during normal teleost development changing metabolic enzyme activities, isozyme changes and the correlation

Table 22.1 Some uses of molecular genetic information in fisheries research and management

Category	Examples
Systematics	General congruence in phylogenies of *Oncorhynchus* constructed from molecular and morphological data
Intraspecific relationships	Revelation of previously unknown genetic population structure in salmonid species
Legal applications	Invalidation or verification of putative origins or identities of disputed catches
Mixed-stock fisheries	Increasing application in analysing salmonid stock mixtures, and potential value for discriminating any group distinguishable by natural or inducible genetic differences
Fish culture	Problems in hatchery management, including inadequate effective numbers of breeders, and clear distinction between hatchery source populations that had been presumed identical
Stock enhancement	Evaluation of effectiveness of enhancement procedures, quantification of contributions of specific source populations, and genetic marking to identify descendents of enhanced populations
Conservation biology	Detection of distinct native populations and of displacement and introgressions of exogoneous fishes
Population size	Effective population sizes estimated through allele frequency shifts between generations and gamete disequilibrium within generations

Source: Utter (1994)

of specific isozyme changes with cellular differentiation have helped elucidate pattern of gene activation.

7.1 Gene Expression in Teleosts

At least four of the enzyme systems have been examined in various species of teleosts; three of these—lactate dehydrogenase, glucose-6-phosphate dehydrogenase, and phosphogluconate dehydrogenase—exhibit similar patterns of change in various teleosts while the fourth—general estrases-seem quite different.

References

Abad, M.E., Peeze Binkhorst, F.M., Elbal, M.T., Rombout, J.H.W.M. (1987) A comparative immunocytochemical study of the gasteroentero-pancreatic (GEP) endocrine systems in a stomachless and a stomach-containing teleost. Gen. Comp. Endocrinol. **66**: 123–136.

Abdullah, M.A.S., Kime, D.E. (1994) Increased substrate concentration causes a shift from production of 11-oxygenated androgens to 17, 20-dihydroxyprogestogens during the *in vitro* metabolism of 17-hydroxyprogesterone by goldfish testes. Gen. Comp. Endocrinol. **96**: 129–139.

Able, K.P. (1980) Mechanisms of orientation, navigation, and homing. In: Gauthreaux, S.A. Jr. (ed.) Animal migration, orientation and navigation. Academic Press, New York, pp. 283–373.

Able, K.P. (1991) Common themes and variations in animal orientation systems. Am. Zool. **31**: 157–167.

Acierno, R., Agnisola, C., Venzi, R., Tota B. (1990) Performance of the isolated and perfused working heart of the teleost *Conger conger:* study of the inotropic effect of prostacyclin. J. Comp. Physiol. B. **160**: 365–371.

Adriaens, D., Decleyre D., Verraes, W. (1993) Morphology of the pectoral girdle in *Pomatoschistus lozanoi* De Bien, 1923 (Gobiidae) in relation to pectoral fin adduction. Belg. J. Zool. **123**: 135–157.

Adriaensen, D., Scheuermann, D.W., Timmermans J.P., De Groot-Lassel, M.H.A. (1990) Neuroepithelial endocrine cells in the lung of the lungfish *Protopterus aethiopicus:* An electron and fluorescence-microscopical investigation. Acta Anat. **139**: 70–77.

Aerts, P. (1990) Variability of the fast suction feeding process in *Astatotilapia elegans* (Teleostei: Cichlidae): a hypothesis of peripheral feedback control. J. Zool. Lond. **220**: 653–678.

Aerts, P. (1991) Hyoid morphology and movements relative to abducting forces during feeding in *Astatotilapia elegans* (Teleostei: Cichlidae). J. Morphol. **208**: 323–345.

Aerts, P. (1992) Fish biomechanics: purpose or means? Neth. J. Zool. **42**: 430–444.

Aerts, P., De Vree, F. (1993) Feeding performance and muscular constraints in fish. J. Exp. Biol. **177**: 129–147.

Aerts, P., Osse J.W.M., Verraes W. (1987) Model of jaw depression during feeding in *Astatotilapia elegans* (Teleostei: Cichlidae): Mechanisms for energy storage and triggering. J. Morphol. **194**: 85–109.

Afzelius, B.A. (1978) Fine structure of the garfish spermatozoon. J. Ultrastruct. Res. **64**: 309–314.

Agnisola, C., Venzi, R., Houlihan, D.F., Tota, B. (1994) Coronary flow-pressure relationship in the working isolated fish heart: trout (*Oncorhynchus mykiss*) versus torpedo (*Torpedo mamorata*). Phil. Trans. R. Soc. Lond. B **343**: 189–198.

Ahearn, G.A., Behnke, R.D., Zonno, V., Storelli, C. (1992) Kinetic heterogeneity of Na-D-glucose cotransport in teleost gastrointestinal tract. Am. J. Physiol **263**: R1018–1023.

Ahlborn, B., Harper, D.G., Blake, R.W., Ahlborn, D., Cam, M. (1991) Fish without footprints. J. theor. Biol. **148**: 521–533.

Ahne, W. (1979) A survey of viruses occurring in poikilotherm vertebrates. Animal Res. Dev. **9**: 24–34.

Albers, C. (1970) Acid-base balance. In: Hoar, W.S., Randall, D.J. (eds.) Fish physiology, vol. 4. Academic Press, New York, pp. 173–208.

Alexander, R.McN. (1966) Physical aspects of swimbladder function. Biol. Rev. **41**: 141–176.

Alexander, R.McN. (1967) Functional design in fishes. Hutchinson University Library, London.

Alexander, R.McN. (1990) Size, speed and buoyancy adaptations in aquatic animals. Am. Zool. **30**: 189–196.

Alexander, R.McN. (1993) Buoyancy. In: Evans, D.H. (ed.) The physiology of fishes. CRC Press, Boca Raton, pp. 75–97.

Ali, M.A. (ed.) (1975) Vision in fishes. New approaches in research. Plenum Press, New York.

Ali, M.A. (ed.) (1992) Rhythms in fishes. Plenum Press, New York.

Alhossaini, M., Pitcher, T.J. (1988) The relation between daily rings, body growth and environmental factors in plaice, *Pleuronectes platessa* L., juvenile otoliths. J. Fish Biol. **33**: 409–418.

Allendorf, F.W., Thorgaard, G.H. (1984) Tetraploidy and the evolution of salmonid fishes. In: Turner, B.J. (ed.) Evolutionary genetics of fishes. Plenum Press, New York, pp. 1–53.

Allendorf, F.W., Utter, F.M. (1979) Population genetics. In: Hoar, W.S., Randall, D.J., Brett, J.R. (eds.) Fish physiology, vol. 8. Academic Press, New York, pp. 407–454.

Almada, V.C., Santos, R.S. (1995) Parental care in the rocky intertidal: a case study of adaptation and exaptation in Mediterranean and Atlantic blennies. Rev. Fish Biol. Fish. **5**: 23–37.

Altringham, J.D., Johnston, I.A. (1986) Energy cost of contraction in fast and slow muscle fibres isolated from an elasmobranch and an Antarctic teleost fish. J. Exp. Biol. **121**: 239–250.

Altringham, J.D., Johnston, I.A. (1988) Activation of multiply innervated fast and slow myotomal muscle fibres of the teleost *Myoxocephalus scorpius*. J. Exp. Biol. **140**: 313–324.

Alves-Gomes, J.A., Orti, G., Haygood, M., Heiligenberg, W. (1995) Phylogenetic analysis of the South American electric fishes (order Gymnotiformes) and the evolution of their electrogenic system: a synthesis based on morphology, electrophysiology, and mitochondrial sequence data. Mol. Biol. Evol. **12**: 298–318.

Amundsen, P.A., Damsgard, B., Arnesen, A.M., Jobling, M., Jorgensen, E.H. (1995) Experimental evidence of cannibalism and prey specialization in Arctic charr, *Salvelinus alpinus*. Env. Biol. Fish. **43**: 285–293.

Andersen, O., Døving, K.B. (1991) Gonadotropin releasing hormone (GnRH)-novel olfactory stimulant in fish. Neuro Report **2**: 458–460.

Andersen, O., Klungland, H., Kisen, G., Alestrom, P. (1991) GnRH and molecular approaches to control sexual maturation in fish. In: Scott, A.P., Sumpter, J.P., Kime, D.E., Rolfe, M.S. (eds.) Reproductive physiology of fish. Proc. Fourth Int. Symp. Univ. of East Anglia, Norwich, U.K. 7–12 July 1991, pp. 41–42.

Anderson, D.P. (1974) Fish immunology. Snieszko, S.F., Axelrod, H.R. (eds.) T.F.H. Publications, Neptune, New Jersey.

Anderson, P.M. (1995) Urea cycle in fish: molecular and mitochondrial studies. In: Wood, C.M., Shuttleworth, T.J. (eds.) Cellular and molecular approaches to fish ionic regulation, Academic Press, San Diego, pp. 57–83.

Anderson, C.D., Roberts, R.J. (1975) A comparison of the effects of temperature on wound healing in a tropical and a temperate teleost. J. Fish Biol. **7**: 173–182.

Anderson, P.M., Walsh, P.J. (1995) Subcellular localization and biochemical properties of the enzymes of carbamoyl phosphate and urea synthesis in the batrachoidid fishes *Opsanus beta*, *Opsanus tau* and *Porichthys notatus*. J. Exp. Biol. **198**: 755–766.

Andrianov, G.N., Peters, R.C., Bretschneider, F. (1994) Identification of AMPA receptors in catfish electroreceptor organs. Neuro Report **5**: 1056–1058.

Anker, G.C. (1974) Morphology and kinetics of the head of the stickleback, *Gasterosteus aculeatus*. Trans. Zool. Soc. Lond. **32**: 311–416.

Anker, G.C. (1978) Analyses of respiration and feeding movements of the three-spined stickleback, *Gasterosteus aculeatus* L. Neth. J. Zool. **28**: 485–523.

Aota, S., Holmgren, K.D., Gallaugher, P., Randall, D.J. (1990) A possible role for catecholamines in the ventilatory responses associated with internal acidosis or external hypoxia in rainbow trout *Oncorhynchus mykiss*. J. Exp. Biol. **151**: 57–70.

Archer, S.D., Johnston, I.A. (1991) Density of cristae and distribution of mitochondria in the slow muscle fibers of Antarctic fish. Physiol. Zool. **64**: 242–258.

Archer. S.D., Altringham, J.D., Johnston, I.A. (1990) Scaling effects on the neuromuscular system, twitch kinetics and morphometrics of the cod, *Gadus morhua*. Mar. Behav. Physiol. **17**: 137–146.

Arnesen, A.M., Stabell, O.B. (1992) Behaviour of stream-dwelling brown trout towards odours present in homestream water. Chemoecology **3**: 94–100.

Arnesen, A.M., Jorgensen, E.H., Jobling, M. (1993a) Feed intake, growth and osmoregulation in Arctic charr, *Salvelinus alpinus* (L.) following abrupt transfer from freshwater to more saline water. Aquaculture **114**: 327–338.

Arnesen, A.M., Jorgensen, E.H., Jobling, M. (1993b) Feed intake growth and osmoregulation in Arctic charr,

Salvelinus alpinus (L.), transferred from freshwater to saltwater at 8°C during summer and winter. Fish Physiol. Biochem. **12**: 281–292.

Arnesen, A.M., Jorgensen, E.H., Jobling, M. (1994) Feed-growth relationships of Arctic charr transferred from freshwater to saltwater at different seasons. Aquacult. Internet **2**: 114–122.

Arnold, G.P. (1974) Rheotropism in fishes. Biol. Rev. **49**: 515–576.

Arnold, G.P. (1981) Movements of fish in relation to water currents. In: Aidley, D.J. (ed.) Animal migration. Society for experimental biology seminar series 13. Cambridge University Press, Cambridge, pp. 55–79.

Arnold, G.P., Cook, P.H. (1984) Fish migration by selective tidal stream transport: first result by a computer simulation model for the European continental shelf. In: McCleave, J.D., Arnold, G.P., Dodson, J.J., Neill, W.H. (eds.) Mechanisms of migration in fishes. Plenum Press, New York, pp. 227-261.

Atema, J. (1971) Structures and functions of the sense of taste in the catfish (*Ictalurus natalis*). Brain Behav. Evol. **4**: 273–294.

Atkinson, R.J.A., Pelster, B., Bridges, C.R., Taylor, A.C., Morris, S. (1987) Behavioural and physiological adaptations to a burrowing lifestyle in the snakebelly, *Lumpenus lampretaeformis*, and the red band-fish, *Cepola rubescens*. J. Fish Biol. **31**: 639–659.

Avelia, M., Bonancin, M. (1990) Ion fluxes in the gills of freshwater and seawater salmonid fish. In: Truchot, J.P., Lahlou, B (eds.) Animal nutrition and transport processes. 2. Transport, respiration and excretion: Comparative and environment aspects. Comp. Physiol. vol. 6. S. Karger, Basel, pp. 1–13.

Axelsson, M., Driedzic, W.R., Farrell, A.P., Nilsson, S. (1989) Regulation of cardiac output and gut blood flow in the sea raven, *Hemitripterus americanus*. Fish Physiol. Biochem. **6**: 315–326.

Babin, P.J., and Vernier, J.M. (1989). Plasma lipoproteins in fish. J. Lipid Res. **30**: 467–487.

Bailey, J.R., Driedzic, W.R. (1988) Perfusion-independent oxygen extraction in myoglobin-rich hearts. J. Exp. Biol. **135**: 301–315.

Bailey, J.R., Sephton, D.H., Driedzic, W.R. (1990) Oxygen uptake by isolated perfused fish hearts with differing myoglobin concentrations under hypoxic conditions. J. Mol. Cell Cardiol. **22**: 1125–1134.

Bailly, Y., Dunel-Erb, S., Laurent, P. (1992) The neuroepithelial cells of the fish gill filament: Indolamine-immunocytochemistry and innervation. Anat. Rec. **233**: 143–161.

Baird, I.L. (1974a) Anatomical features of the inner ear in submammalian vertebrates. In: Keidel, W.D., Neff, W.D., (eds.) Handbook of sensory physiology, vol. V/1: Auditory system, Springer-Verlag, Berlin, Heidelberg, New York, pp. 159–212.

Baird, I.L. (1974b) Some aspects of the comparative anatomy and evolution of the inner ear in submammalian vertebrates. Brain Behav. Evol. **10**: 11–36.

Baker, B.I. (1991) Melanin-concentrating hormone: a general vertebrate neuropeptide. Int. Rev. Cytol. **126**: 1–47.

Balon, E.K. (1975a) Reproductive guilds of fishes: a proposal and defination. J. Fish Res. Board Can. **32**: 821–864.

Balon, E.K. (1975b) Terminology of intervals in fish development. J. Fish Res. Board Can. **32**: 1663–1670.

Bardach, J.E., Atema, J. (1971) The sense of taste in fishes. In: Beidler, L.M. (ed.) Handbook of sensory physiology, vol. 4, Chemical senses, Part 2 Taste. Springer-Verlag, Berlin, Heidelberg, New York, pp. 293–336.

Barber, M.L., Walsh, P.J. (1993) Interactions of acid-base status and nitrogen excretion and metabolism in the ureogenic teleost *Opsanus beta*. J. Exp. Biol. **185**: 87–105.

Barlow, G.W. (1974) Hexagonal territories. Anim. Behav. **22**: 876–878.

Barlow, G.W. (1981) Patterns of parental investment, dispersal and size among coral reef fishes. Env. Biol. Fish. **6**: 65–85.

Barlow, G.W. (1983) The benefits of being gold: behavioural consequences of polychromatism in the midas cichlid, *Cichlasoma citrinellum*. Env. Biol. Fish. **8**: 235–247.

Barlow, G.W. (1984) Patterns of monogamy among teleost fishes. Arch. Fischwiss. **35**: 75–123.

Barlow, G.W. (1986a) A comparison of monogamy among freshwater and coral-reef fishes. In: Uyeno, T., Arai, R., Taniuchi, T., Matsuura, K. (eds.) Indo-Pacific fish biology: Proceedings of the Second International Conference on Indo-Pacific fishes. Ichthyological Society of Japan, Tokyo, pp. 767–775.

Barlow, G.W. (1986b) Mate choice in the monogamous and polychromatic Midas cichlid, *Cichlasoma citrinellum*. J. Fish Biol. **29** (Suppl. A): 123–133.

Barlow, G.W. (1988) Monogamy in relation to resources. In: Slobodchikoff, C.N. (ed.) The ecology of social behaviour. Academic Press, San Diego, pp. 59–79.

Barlow, G.W. (1989) Choosy fishes: mating and related issues. In: Blanchard, R.J., Brain, P.F., Blanchard, D.C., Parmigiani, S. (eds.) Ethoexperimental approaches to the study of behavior. Kluwer Academic Publishers, Dordrecht, pp. 267–287.

Barlow G.W. (1990) Do the colors of parents, companions and self influence assortative mating in the polychromatic Midas cichlid? Anim. Behav. **40**: 713–722.

Barlow, G.W. (1991) Mating systems among cichlid fishes. In: Keenleyside, M.H.A. (ed.) Cichlid fishes. Behaviour ecology and evolution. Chapman and Hall, London, New York, pp. 173–190.

Barlow, G.W. (1992) Mechanisms of mate choice in monogamy. Workshop (28–30 September 1992) on behavioural mechanisms in evolutionary perspective. pp. 26–29.

Barlow, G.W. (1993a) Fish behavioural ecology: pros, cons and opportunities. Mar. Behav. Physiol. **23**: 7–27.

Barlow, G.W. (1993b) The puzzling paucity of feeding territories among freshwater fishes. Mar. Behav. Physiol **23**: 155–174.

Baroiller, J.F., Chourrout, D., Fostier, A., Jalabert, B. (1995) Temperature and sex chromosomes govern sex ratios of the mouthbrooding cichlid fish *Oreochromis niloticus*. J. Exp. Zool. **273**: 216–223.

Barr, W.A. (1965) The endocrine physiology of fishes. Oceanogr. Mar. Biol. Ann. Rev. **3**: 257–298.

Barry, M.A. (1987a) Afferent and efferent connections of the primary octaval nuclei in the clearance skate, *Raja eglanteria*. J. Comp. Neurol. **266**: 457–477.

Barry, M.A. (1987b) Central connections of the IXth and Xth cranial nerves in the clearnose skate, *Raja eglanteria*. Brain Res. **425**: 159–166.

Barry, M.A., Bennett, M.V.L. (1989) Specialized lateral line receptor system in elasmobranchs: the spiracular organs and vesicles of Savi. In: Coombs, S., Görner P, Münz, H. (eds) The mechanosensory lateral line: neurobiology and evolution. Springer-Verlag, New York, pp. 591–606.

Bass, A.H. (1986) Electric organs revisited: evolution of a vertebrate communication and orientation organ. In: Bullock, T.H., Heiligenberg, W. (eds.) Electroreception, John Wiley, New York, pp. 13–70.

Bass, A.H. (1989) Comparative neurobiology: morphological and neurophysiological correlates of hormonal influences on the electromotor system of fishes. In: Perspectives in neural systems and behavior. Alan R. Liss, Inc., New York, pp. 137–150.

Bass, A.H. (1990) Sounds from the intertidal zone: vocalizing fish. Bioscience **40**: 249–258.

Bass, A.H., Andersen, K. (1991) Inter-and intrasexual dimorphisms in the vocal control system of a teleost fish: motor axon number and size. Brain Behav. Evol. **37**: 204–214.

Bass, A.H., Baker, R. (1990) Sexual dimorphisms in the vocal control system of a teleost fish: morphology of physiologically identified neurons. J. Neurobiol. **21**: 1155–1168.

Bass, A.H., Hopkins, C.D. (1983) Hormonal control of sexual differentiation: changes in electric organ discharge waveform. Science **220**: 971–974.

Bass, A.H., Hopkins, C.D. (1985) Hormonal control of sex differences in the electric organ discharge (EOD) of mormyrid fishes. J. Comp. Physiol. **A156**: 587–604.

Bass, A.H., Marchaterre, M.A. (1989a) Sound-generating (sonic) motor system in a teleost fish (*Porichthys notatus*): sexual polymorphism in the ultrastructure of myofibrils. J. Comp. Neurol. **286**: 141–153.

Bass, A.H., Marchaterre, M.A. (1989b) Sound-generating (sonic) motor system in a teleost fish (*Porichthys notatus*): sexual polymorphisms and general synaptology of sonic motor nucleus. J. Comp. Neurol. **286**: 154–169.

Bass, A.H., Marchaterre, M.A., Baker, R. (1994) Vocal-acoustic pathways in a teleost fish. J. Neurosci. **14**: 4025–4039.

Bastian, J. (1990) Electroreception. In: Stebbins, W.C., Berkley, M.A. (eds.) Comparative signals, vol. 2. John Wiley, New York, pp. 35–88.

Beamish, F.W.H. (1978) Swimming capacity. In: Hoar, W.S., Randall, D.J. (eds.) Fish physiology, vol. 7. Academic Press, New York, pp. 101–187.

Begay, V., Collin, J.P., Falcon, J. (1994a) Calciproteins regulate cyclic AMP content and melatonin secretion in trout pineal photoreceptors. Neuro Report **5**: 2019–2022.

Begay, V., Bois, P., Collin, J.P., Lenfant, J., Falcon, J. (1994b) Calcium and melatonin production in dissociated trout pineal photoreceptor cells in culture. Cell Calcium **16**: 37–46.

Begay, V., Valotaire, Y., Ravault, J.P., Collin, J.P., Falcon, J. (1994c) Detection of estrogen receptor m RNA in trout pineal and retina: estradiol-17β modulates melatonin production by cultured pineal photoreceptor cells. Gen. Comp. Endocrinol. **93**: 61–69.

Bell, J.G., Buddington, R.K., Walton, M.J., Cowey, C.B. (1987) Studies on the putative role of Y-glutamyl transpeptidase in intestinal transport of amino acids in Atlantic salmon. J. Comp. Physiol. **B157**: 161–169.

Belle, N.L., Dufour, S., Marquet, G., Kerdelhue, Fontaine, Y.A. (1988). Etat sexuel et axe gonadotrope hypothalamo-hypophysaire chez les trois especes d'Anguilles de Tahiti (*Anguilla marmorata, A. megastoma et A. obsura*). Bull. Mus. Natn. Hist. Nat. Paris **10A**: 147–160.

Belles-Isles, J.C., FitzGerald, G.J. (1991) Filial cannibalism in sticklebacks: a reproductive management strategy? Ethol. Ecol. Evol. **3**: 49–62.

Bemis, W.E. (1984) Morphology and growth of lepidosirenid lungfish tooth plates (Pisces: Dipnoi). J. Morphol. **179**: 73–93.

Bemis, W.E., Lauder, G.V. (1986) Morphology and function of the feeding apparatus of the lungfish, *Lepidosiren paradoza* (Dipnoi). J. Morphol. **187**: 81–108.

Benmouna, H., Chardon, M., Vandewalle, P. (1989) Feeding movements in a serranid fish: *Serranus scriba* (L.). Fortsch. Zool. **35**: 156–159.

Berenbrink, M., Bridges, C.R. (1994a) Catecholamine-activated sodium/proton exchange in the red blood cells of the marine teleost *Gadus morhua*. J. Exp. Biol. **192**: 253–267.

Berendrink, M., Bridges, C.R. (1994b) Active Na^+-, , Cl^--and HCO_3^--dependent acid extrusion in Atlantic cod red blood cells in winter activated by hypercapnia. J. Exp. Biol. **192**: 239–252.

Bern, H.A. (1969) Urophysis and caudal neurosecretory system. In: Hoar, W.S., Randall, D.J. (eds.) Fish physiology, vol. 2. Academic Press, New York, pp. 399–418.

Bern, H.A. (1990) The caudal neurosecretory system: Quest and bequest. In: Epple, A,. Scanes, C.G., Stetson, M.H. (eds.) Progress in comparative endocrinology. Wiley-Liss, New York, pp. 242–249.

Bern, H.A., Pearson, D., Larson, B.A., Nishioka, R.S. (1985) Neurohormones from fish tails: The caudal neurosecretory system. 1. "Urophysiology" and the caudal neurosecretory system of fishes. Recent Prog. Horm. Res. **41**: 533–552.

Bernstein, J.J. (1970) Anatomy and physiology of the central nervous system. In Hoar, W.S., Randall, D.J. (eds.) Fish physiology, vol. 4. Academic Press, New York, London, pp. 1–90.

Bertmar, G. (1985) General ecology of primitive fishes. In: Foreman, R.E., Gorbman, A., Dodd, J.M., Olsson, R. (eds.) Evolutionary biology of primitive fishes. Plenum Publishing corporation, New York, pp. 11–29.

Bertmar, G., Kapoor, B.G., Miller, R.V. (1969) Epibranchial organs in lower teleostean fishes—an example of structural adaptation. In: Felts, W.J.L., Harrison, R.J. (eds.) International review of general and experimental zoology, vol. 4. Academic Press, New York, London, pp. 1–48.

Best, A.C.G., Bone, Q. (1976) On the integument and photophores of the alepocephalid fishes *Xenodermichthys* and *Photostylus*. J. Mar. biol. Ass. U.K. **56**: 227–236.

Beyenbach, K.W. (1995) Secretory electrolyte transport in renal proximal tubules of fish. In: Wood, C.M., Shuttleworth, T.J. (eds.) Cellular and molecular approahces to fish ionic regulation, Academic Press, San Diego, pp. 85–105.

Beyenbach, K.W., Freire, C.A., Kinne, R.K.H., Kinne-Saffran, E. (1993) Epithelial transport of magnesium in the kidney of fish. Miner. Electrolyte Metab. **19**: 241–249.

Bhat, K.G., Gopal Dutt, N.H. (1990) Cyclic changes in the pituitary gonadotrophs in relation to the ovarian cycle in *Puntius sarana* (Hamilton) (Teleostea). Boll. Zool. **57**: 27–34.

Bicudo, J.E.P.W., Weibel, E.R. (1987) Respiratory exchanges in air breathers versus water breathers: Design contrasts. In: Dejours, P., Bolis, L., Taylor, C.R., Weibel, E.R. (eds.) Comparative physiology: Life in water and on land. Fidia Research Series, vol. 9. Liviana Press, Padova, pp. 15–26.

Bilinski, E. (1974) Biochemical aspects of fish swimming. In: Malins, D.C., Sargent, J.R. (eds.) Biochemical and biophysical perspectives in marine biology, vol. 1. Academic Press, New York, pp. 239–288.

Billard, R., LaGac, F., Loir, M. (1990) Hormonal control of sperm production in teleost fish. In: Epple, A.,

Scanes, C.G., Stetson, M.H. (eds.) Progress in Comparative Endocrinology. Wiley-Liss, New York, pp. 329–335.

Birkeland, C., Neudecker, S. (1981) Foraging behavior of two Caribbean chaetodontids: *Chaetodon capistratus* and *C. aculeatus*. Copeia **1981**: 169–178.

Bjerring, H.C., (1985) Facts and thoughts on piscine phylogeny. In: Foreman, R.E., Gorbman, A., Dodd, J.M., Olsson, R. (eds.) Evolutionary biology of primitive fishes. Plenum Publishing Corporation, New York, pp. 31–57.

Bjerring, H.C. (1989) Apertures of craniate olfactory organs. Acta Zool. Stockholm **70**: 71–85.

Bjerring, H.C. (1993) Yet another interpretation of the coelacanthiform basicranial muscle and its innervation. Acta Zool. Stockholm **74**: 289–299.

Blake, R.W. (1983a) Median and paired fin propulsion. In: Webb, P.W., Weihs, D. (eds.) Fish biomechanics. Praeger, New York, pp. 214–247.

Blake, R.W. (1983b) Fish locomotion. Cambridge University Press, Cambridge, U.K.

Blaxter, J.H.S. (1975) Fish vision and applied research. In: Ali, M.A. (ed.) Vision in fishes. Plenum Publishing Corporation, New York, pp. 757–773.

Blaxter, J.H.S. (1986) Development of sense organs and behaviour of teleost fish larvae with special reference to feeding and predator avoidance. Trans. Am. Fish Soc. **115**: 98–114.

Blaxter, J.H.S. (1988) Pattern and variety in development. In: Hoar, W.S., Randall. D.J. (eds.) Fish physiology, vol. 11A. Academic Press, San Diego, pp. 1–58.

Blaxter, J.H.S. (1992) The effect of temperature on larval fishes. Neth. J. Zool. **42**: 336–357.

Blaxter J.H.S., Denton, E.J., Gray, J.A.B. (1981) Acousticolateralis system in clupeid fishes. In: Tavolga, W.N., Popper, A.N., Fay, R.R. (eds.) Hearing and sound communication in fishes. Springer-Verlag, New York, pp. 39–59.

Bleckmann, H. (1986) Role of lateral line in fish behaviour. In: Pitcher, T.J. (ed.) The behaviour of teleost fishes. Croom Helm, London, pp. 177–202.

Bleckmann, H. (1988) Prey identification and prey localization in surface-feeding fish and fishing spiders. In: Atema, J., Fay, R.R., Popper, A.N., Tavolga, W.N. (eds.) Sensory biology of aquatic animals. Springer-Verlag, Berlin, pp. 619–641.

Bleckmann, H., Bullock, T.H. (1989) Central nervous physiology of the lateral line, with special reference to cartilaginous fishes. In: Coombs, S., Görner, P., Münz, H., (eds.) The mechanosensory lateral line: neurobiology and evolution. Springer-Verlag, New York, pp. 387–408.

Bleckmann, H., Tittel, G., Blubaum-Gronau, E. (1989) The lateral line system of surface-feeding fish: anatomy, physiology and behaviour. In: Coombs, S., Görner, P., Miinz, H. (eds.) The mechanosensory lateral line: neurobiology and evolution. Springer-Verlag, New York, pp. 501–526.

Blem, C.R. (1980) The energetics of migration. In: Gauthreaux, S.A., Jr. (ed.) Animal migration, orientation, and navigation. Academic Press, New York, pp. 175–224.

Block, B.A. (1987) Strategies for regulating brain and eye temperatures: a thermogenic tissue in fish. In: Dejours, P., Bolis, L., Taylor, C.R., Weibel, E.R. (eds.) Comparative physiology: life in water and on land. Fidia Research Series, vol. 9. Liviana Press, Padova pp. 401–420.

Block, B.A., Finnerty, J.R., Stewart, A.F.R., Kidd, J. (1993) Evolution of endothermy in fish: Mapping physiological traits on a molecular phylogeny. Science **260**: 210–214.

Blumer, L.S. (1979) Male parental care in the bony fishes. Quart. Rev. Biol. **54**: 149–161.

Bodznick, D. (1978) Calcium ion: an odorant for natural water discriminations and the migratory behavior of sockeye salmon. J. Comp. Physiol. **A127**: 157–166.

Bodznick, D. (1989) Comparison between electrosensory and mechanosensory lateral line systems. In: Coombs, S., Görner, P., Münz, H. (eds.) The mechanosensory lateral line: neurobiology and evolution. Springer-Verlag, New York, pp. 653–678.

Bodznick, D., Boord, R.L. (1986) Electroreception in Chondrichthyes. In: Bullock, T.H., Heiligenberg, W. (eds.) Electroreception. John Wiley, New York, pp. 225–256.

Bodznick, D., Hjelmstad, G., Bennett, M.V.L. (1993) Accommodation to maintain stimuli in the ampullae of Lorenzini: how an electroreceptive fish achieves sensitivity in a noisy world. Jap. J. Physiol. **43** Suppl.: 231–237.

Boeuf, G. (1993) Salmonid smolting: a pre-adaptation to the oceanic environment. In: Rankin, J.C., Jensen, F.B. (eds.) Fish ecophysiology. Chapman and Hall, London, New York, pp. 105–135.
Boehlert, G.W., Yoklavich, M.M. (1984) Reproduction, embryonic energetics and the maternal-fetal relationship in the viviparous genus *Sebastes* (Pisces: Scorpaenidae). Biol. Bull. **167**: 354–370.
Boehlert, G.W., Yamada, J. (1986) Reproductive mode and energy costs of reproduction in the genus *Sebastes*. Proc. Int Rockfish Symp. Anchorage, Alaska, pp. 143–152.
Boehlert, G.W., Kusakari, M., Shimizu, M., Yamada, J. (1986) Energetics during embryonic development in kurosoi, *Sebastes schlegeli* Hilgendorf. J. Exp. Mar. Biol. Ecol. **101**: 239–256.
Boehlert, G.W., Kusakari, M., Yamada, J. (1991) Oxygen consumption of gestating female *Sebastes schlegeli:* estimating the reproductive costs of livebearing. Env. Biol. Fish **30**: 81–89.
Bolliet, V., Begay, V., Ravault, J.P., Ali, M.A., Collin, J.P., Falcon, J. (1994) Multiple circadian oscillators in the photosensitive pike pineal gland: a study using organ and cell culture. J. Pineal Res. **16**: 77–84.
Bone, Q. (1974) Muscular and energetic aspects of fish swimming. In: Wu, T.W., Brokaw, C.J., Brennan, C. (eds.) Swimming and flying in nature. Plenum Press, New York, pp. 493–528.
Bone, Q. (1978) Locomotor muscle. In: Hoar, W.S., Randall, D.J. (eds.) Fish physiology, vol. 7. Academic Press, New York, pp. 361–424.
Bone, Q., Marshall, N.B., Blaxter, J.H.S. (1995) Biology of fishes. (Second edition) Blackie Academic and Professional, Glasgow.
Boon, J.H., Rombout, J.H.W.M., Booms, G.H.R., Machiels, M.A.M. (1990) Influence of feeding level on the population of C-terminal-gastrin-immunoreactive cells in the digestive tract of young *Clarias gariepinus* (Burchell 1822). Neth. J. Agri. Sci. **38**: 67–76.
Boord, R.L., Campbell, C.B.G. (1977) Structural and functional organization of the lateral line system of sharks. Am. Zool. **17**: 431–441.
Boord, R.L., Montgomery, J.C. (1989) Central mechanosensory lateral line centers and pathways among the elasmobranchs. In: Coombs, S., Görner, P., Münz, H. (eds.) The mechanosensory lateral line: neurobiology and evolution. Springer-Verlag, New York, pp. 323–339.
Borg, B. (1991) Androgens in the three-spined stickleback, *Gasterosteus aculeatus*. In: Scott, A.P., Sumpter, J.P., Kime, D.E., Rolfe, M.S. (eds.) Reproductive physiology of fish. Proc. Fourth Int. Symp. Univ. of East Anglia, Norwich, U.K., 7–12 July 1991, pp. 86–88.
Brainerd, E.L. (1994a) The evolution of lung-gill bimodal breathing and the homology of vertebrate respiratory pumps. Am. Zool. **34**: 289–299.
Brainerd, E.L. (1994b) Pufferfish inflation: functional morphology of postcranial structures in *Diodon holocanthus* (Tetraodontiformes). J. Morphol. **220**: 243–261.
Brand, J.G., Bruch, R.C. (1992) Molecular mechanisms of chemosensory transduction: gustation and olfaction. In: Hara, T.J. (ed.) Fish chemoreception. Chapman and Hall, London, New York, pp. 126–149.
Brannon, E.L. (1984) Influence of stock origin on salmon migratory behaviour. In: McCleave, J.D., Arnold, G.P., Dodson, J.J., Neill, W.H. (eds.) Mechanism of migration in fishes. Plenum Press, New York, pp. 103–111.
Brantley, R.K., Wingfield, J.C. Bass, A.H. (1993a) Sex steroid livels in *Porichthys notatus*, a fish with alternative reproductive tactics, and a review of hormonal bases for male dimorphism among teleost fishes. Horm. Behav. **27**: 332–347.
Brantley, R.K., Marchaterre, M.A., Bass, A.H. (1993b) Androgen effects on vocal muscle structure in a teleost fish with inter-and intrasexual dimorphism. J. Morphol. **216**: 305–318.
Brantley, R.K., Tseng, J., Bass, A.H. (1993c) The ontogeny of inter-and intrasexual vocal muscle dimorphisms in a sound-producing fish. Brain Behav. Evol. **42**: 336–349.
Brantley, R.K., Wingfield, J.C., Bass, A.H. (1993d) Sex steroid levels in *Porichthys notatus*, a fish with alternative reproductive tactics, and a review of the hormonal bases for male dimorphism among teleost fishes. Horm. Behav. **27**: 332–347.
Braun, E.J., Dantzler, W.H. (1987) Mechanisms of hormone actions on renal functions. In: Pang, P.K.T., Schreibman M.P. (eds.) Vertebrate endocrinology: fundamentals and biomedical implications, vol. 2. Academic Press, San Diego, pp. 189–210.
Breder, C.M. Jr. (1976) Fish schools as operational structures. Fish. Bull. **74**: 471–502.

Brennan, J.S., Cailliet, G.M. (1989) Comparative age-determination techniques for white sturgeon in California. Proc. Amer. Fish Soc. **118**: 296–310.
Brett, J.R. (1970) Temperature. Animals. Fishes. In: Kinne, O. (ed.) Marine ecology, vol. 1. Wiley-Interscience, London, pp. 513–560.
Brett, J.R. (1979) Environmental factors and growth. In: Hoar, W.S., Randall, D.J., Brett, J.R. (eds.) Fish Physiology, vol. 8, Bioenergetic and growth. Academic Press, New York, pp. 599–675.
Brett, J.R., Groves, T.D.D. (1979) Physiological energetics. In: Hoar, W.S., Randall. D.J., Brett. J.R. (eds.) Fish Physiology, vol. 8. Bioenergetics and growth. Academic Press, New York, pp. 279–352.
Bridges, C.R. (1988) Respiratory adaptations in intertidal fish. Am. Zool. **28**: 79–96.
Bridges, C.R. (1993) Ecophysiology of intertidal fish. In: Rankin, J.C., Jensen, F.B. (eds.) Fish ecophysiology. Chapman and Hall, London, New York pp. 375–400.
Brown, J.A. (1993) Endocrine responses to environmental pollutants. In: Rankin, J.C., Jensen, F.B. (eds) Fish ecophysiology. Chapman and Hall, London, New York, pp. 276–296.
Brown, G.W. Jr., Brown, S.G. (1985) On urea formation in primitive fishes. In: Foreman, R.E., Gorbman, A., Dodd, J.M., Olsson, R. (eds.) Evolutionary biology of primitive fishes. Plenum Publishing Corporation, New York, pp. 321–337.
Brown, P.S., Brown, S.C. (1987) Osmoregulatory actions of prolactin and other adenohypophysial hormones. In: Pang, P.K.T., Schreibman, M.P. (eds.) Vertebrate endocrinology: fundamentals and biomedical implications, vol. 2. Academic Press, San Diego, pp. 45–84.
Brower, L.P., Malcolm, S.B. (1991) Animal migrations: endangered phenomena. Am. Zool. **31**: 265–276.
Brown, G.E., Smith, R.J.F. (1994) Fathead minnows use chemical cues to discriminate natural shoal mates from unfamiliar conspecifics. J. Chem. Ecol. **20**: 3051–3061.
Brown, G.E., Chivers, D.P., Smith, R.J.F. (1995) Localized defecation by pike: a response to labelling by cyprinid alarm pheromone? Behav. Ecol. Sociobiol. **36**: 105–110.
Brush, R.C., Teeter, J.H. (1990) Cyclic AMP links amino acid chemoreceptors to ion channels in olfactory cilia. Chem. Senses **15**: 419–430.
Bryant, B.P., Harpaz, S., Brand, J.G. (1989) Structure/activity relationships in the arginine taste pathway of the channel catfish. Chem. Senses **14**: 805–815.
Bryant, B.P., Leftheris, K., Quinn, J.V., Brand, J.G. (1993) Molecular structural requirements for binding and activation of L-alanine taste receptors. Amino Acids **4**: 73–88.
Buddington, R.K., Diamond J.M. (1987) Pyloric ceca of fish: a "new" absorptive organ. Am. J. Physiol. **252**: 65–76.
Bullock, A.M. (1980) A comparative study of the epidermal morphology of fishes of the order Anacanthini with special reference to their ecology and distribution. Oceanogr. Mar. Biol. Ann. Rev. **18**: 251–315.
Bullock, A.M., Roberts, R.J., (1975) The dermatology of marine teleost fish. The normal integument. Occanogr. Mar. Biol. Ann. Rev. **13**: 383–411.
Bullock, T.H., Heiligenberg, W. (eds.) (1986) Electroreception. John Wiley, New York.
Bumann, D., Krause, J. (1993) Front individuals lead to shoals of three-spined sticklebacks (*Gasterosteus aculeatus*) and juvenile roach (*Rutilus rutilus*). Behaviour **125**: 189–198.
Burggren, W.W., Johansen K. (1986) Circulation and respiration in lungfishes (Dipnoi). J. Morphol. Suppl. **1**: 217–236.
Burggren, W.W., Bemis, W.E. (1992) Metabolism and ram gill ventilation in juvenile paddlefish, *Polyodon spathula* (Chondrostei: Polydontidae). Physiol. Zool. **65**: 515–539.
Burggren, W.W., Johansen, K., McMahon, B. (1985) Respiration in phyletically ancient fishes. In: Foreman, R.E., Gorbman, A., Dodd, J.M., Olsson, R. (eds.) Evolutionary biology of primitive fishes. Plenum Publishing Corporation, New York, pp. 217–252.
Burleson, M.L. (1995) Oxygen availability: sensory systems . In: Hochackka, P.W., Mommsen, T.P. (eds.). Biochemistry and molecular biology of fishes, vol. 5. Elsevier, Amsterdam, pp. 1–18.
Burleson, M.L., Milsom, W.K. (1993) Sensory receptors in the first gill arch of rainbow trout. Respir. Physiol. **93**: 97–110.
Burleson, M.L., Smatrek, N.J., Milsom, W.K. (1992) Afferent inputs associated with cardioventilatory control in fish. In: Hoar, W.S., Randall, D.J. (eds.) Fish physiology, vol. 12B. Academic Press, New York, pp. 389–426.

Burns, J.R. (1991) Testis and gonopodium development in *Anableps dowi* (Pisces: Anablepidae) correlated with gonadotropic zone area. J. Morphol. **210**: 45–53.
Burns, J.R., Kallman, K.D. (1985) An ovarian regression syndrome in the platyfish, *Xiphophorus maculatus*. J. Exp. Zool. **233**: 301–316.
Burns, J.R., Weitzman, S.H., Grier, H.J., Menzes, N.A. (1995) Internal fertilization, testis and sperm morphology in glandulocaudine fishes (Teleostei: Characidae: Glandulocaudinae). J. Morphol. **224**: 131–145.
Bushnell, P.G., Steffensen, J.F., Schurmann, H., Jones, D.R. (1994) Exercise metabolism in two species of cod in arctic waters. Polar Biol. **14**: 43–48.
Bushmann, P.J., Burns, J.R. (1994) Social control of male sexual maturation in the swordtail characin, *Corynopoma riisei*. J. Fish Biol. **44**: 263–272.
Buth, D.G. (1983) Duplicate isozyme loci in fishes: origins, distribution, phyletic consequences and locus nomenclature. In: Rattazi, M.C., Scandalios, J.G., Whitts, G.S. (eds.) Isozymes: current topics in biological and medical research, vol. 10: genetics and evolution. Allan R. Liss, New York, pp. 381–400.
Buth, D.G. (1984) Allozymes of the cyprinid fishes. In: Turner, B.J. (ed.) Evolutionary genetics of fishes. Plenum Press, New York, pp. 561–590.
Buth, D.G., Dowling, T.E., Gold, J.R. (1991) Molecular and cytological investigations. In: Winfield, I.J., Nelson, J.S. (eds.) Cyprinid fishes. Systematics, biology and exploitation. Chapman and Hall, London, New York, pp. 83–126.
Butler, A.B. (1992) Variation of tectal morphology in teleost fishes. Brain Behav. Evol. **40**: 256–272.
Butler, A.B., Northcutt, R.G. (1992) Retinal projections in the bowfin, *Amia calva:* Cytoarchitectonic and experimental analysis. Brain Behav. Evol. **39**: 169–194.
Butler, A.B., Northcutt, R.G. (1993) The diencephalon of the Pacific herring, *Clupea harengus:* Cytoarchitectonic analysis. J. Comp. Neurol. **328**: 527–546.
Butler, A.B., Saidel, W.M. (1991) Retinal projections in the freshwater butterfly fish, *Pantodon buchholzi* (Osteoglossoidei). Cytoarchitectonic analysis and primary visual pathways. Brain Behav. Evol. **38**: 127–153.
Butler, A.B., Saidel, W.M. (1992) Tectal projection to an unusual nucleus in the diencephalon of a teleost fish, *Pantodon buchholzi*. Neurosci. Lett. **145**: 193–196.
Butler, A.B., Wullimann, M.F., Northcutt, R.G. (1991) Comparative cytoarchitectonic analysis of some visual pretectal nuclei in teleosts. Brain Behav. Evol. **38**: 92–114.
Butler, D.G., Godinouche, M.Z.A. (1995) Functional reabsorption calcium, magnesium and phosphate in the kidneys of freshwater North American eels (*Anguilla rostrata* Le-Sueur) following removal of the corpuscles of Stannius. J. Comp. Physiol. B **165**: 348–358.
Butler, P.J., Day, N. (1993a). The relationship between intracellular pH and swimming performance of brown trout exposed to neutral and sublethal pH. J. Exp. Biol. **176**: 271–284.
Butler, P.J., Day, N. (1993b) The relationship between intracellular pH and seasonal temperature in the brown trout *Salmo trutta*. J. Exp. Biol. **177**: 293–297.
Butler, P.J., Metcalfe, J.D. (1983) Control of respiration and circulation. In: Rankin, J.C., Pitcher, T.J., Duggan, R.T. (eds.) Control processes in fish physiology. Croom Helm, London, pp. 41–65.
Butler, P.J., Metcalfe, J.D. (1988) Cardiovascular and respiratory systems. In: Shuttleworth, T.J. (ed.) Physiology of elasmobranch fish. Springer-Verlag, Berlin, Heidelberg, New York, pp. 1–47.
Bye, V.J. (1984) The role of environmental factors in the timing of reproductive cycles. In: Potts, G.W., Wootton, R.J. (eds.) Fish reproduction: strategies and tactics. Academic Press, London, pp. 187–202.
Cailliet, G.M. (1990) Elasmobranch age determination and verification: an updated review. In: Pratt, H.L., Jr., Gruber, S.H., Taniuchi, T. (eds.) Elasmobranchs as living resources: advances in the biology, ecology, systematics, and the status of fisheries. NOAA Technical Report 90, pp. 157–165.
Cailliet, G.M., Ebeling, A.W. (1990) The vertical distribution and feeding habits of two common midwater fishes (*Leuroglossus stilbius* and *Stenobranchius leucopsarus*) off Santa Barbara. Cal. COFI Rep. **31**: 106–123.
Cailliet, G.M., Radtke, R.L. (1987) A progress report on the electron microsprobe analysis technique for age determination and verification in elasmobranchs. In: Summerfelt, R.C., Hall, G.E. (eds.) The age and growth of fish. Iowa State University Press, Ames, Iowa, pp. 359–369.

Calliet, G.M., Tanaka, S. (1990) Recommendations for research needed to better understand the age and growth of elasmobranchs. In: Pratt, H.L. Jr., Gruber, S.H., Taniuchi, T. (eds.) Elasmobranchs as living resources: Advances in the biology, ecology, systematics, and the status of the fisheries. NOAA Technical Report **90**: 505–507.

Cailliet, G.M., Martin, L.K., Kusher, D., Wolf, P., Welden, B.A. (1983a) Techniques for enhancing vertebral bands in age estimation of California elasmobranchs, Tech. Rep. NMFS, **8**: pp. 157–165.

Cailliet, G.M., Martin, L.K., Harvey, J.T., Kusher, D., Weldon, B.A. (1983b) Preliminary studies on the age and growth of blue, *Prionace glauca*, Common thresher, *Alopias vulpinus*, and shortfin mako, *Isurus* oxyrinchus, sharks from California waters. In: Prince, E., Pulos, L. (eds.) Proc. of the International workshop on age determination of oceanic pelagic fishes: tuna, billfishes, sharks. NOAA Technical Report, NMFS, **8**: 179–188.

Callard, G.V. (1988) Reproductive physiology (B) The male. In: Shuttleworth, T.J. (ed.) Physiology of elasmobranch fishes. Springer-Verlag, Berlin, Heidelberg New York, pp. 292–317.

Callard, G.V. (1991a) Reproduction in male elasmobranch fishes. In: Kinne, R.K.H. (ed.) Oogenesis, spermatogenesis and reproduction. Comp. Physiol. vol. 10. S. Karger, Basel, pp. 104–154.

Callard, G.V. (1991b) Spermatogenesis. In: Pang, P.K.T., Schreibman, M.P. (eds.) Vertebrate endocrinology: fundamentals and biomedical implications, vol. 4A. Academic Press, San Diego, pp. 303–341.

Callard, I.P., Klosterman, L. (1988) Reproductive physiology: (A) The female. In: Shuttleworth, T.J. (ed.) Physiology of elasmobranch fishes. Springer-Verlag, Berlin, Heidelberg, New York, pp. 277–292.

Callard, I.P., Fileti, L.A., Parez, L.E., Sorbera, L.A., Glannoukos, G., Klosterman, L.L., Psang, P., McCracken, J.A. (1992) Role of the corpus luteum and progesterone in the evolution of vertebrate viviparity. Am. Zool. **32**: 264–275.

Cameron, J.N. (1990) Unusual aspects of calcium metabolsim in aquatic animals. Ann. Rev. Physiol. **52**: 77–95.

Cameron, J.N., Heisler, N. (1983) Studies of ammonia in the rainbow trout: physico-chemical parameters, acid-base behaviour and respiratory clearance. J. Exp. Biol. **105**: 107–125.

Cameron, J.N., Heisler, N. (1985) Ammonia transfer across fish gills: a review. In: Giles, R. (ed.) Circulation, respiration and metabolism. Springer-Verlag, Berlin, Heidelberg, New York, pp. 91–108.

Campbell, G. (1970) Autonomic nervous systems. In: Hoar, W.S., Randall, D.J. (eds.) Fish physiology, vol. **4**. Academic Press, New York, pp. 109–132.

Canario, A.V.M. (1991) Sex steroids in marine flatfish. In: Scott, A.P., Sumpter, J.P., Kime, D.E., Rolfe, M.S. (eds.) Reproductive physiology of fish. Proc. Fourth Int. Symp. Univ. of East Anglia, Norwich, U.K., 7–12 July 1991, pp. 71–73.

Cano, T.M., Perry, S.F., Fenwick, J.C. (1994) Cholinergic control of stanniocalcin release in the rainbow trout, *Oncorhynchus mykiss*, and the American eel, *Anguilla rostrata*. Gen. Comp. Endocrinol. **94**: 1–10.

Canty, A.A., Driedzic, W.R. (1987) Evidence that myoglobin does not support heart performance at maximal levels of oxygen demand. J. Exp. Biol. **128**: 469–473.

Capanna, E., Cataudella, S., Monaco, G. (1974) The pharyngeal structure of Mediterranean Mugilidae. Monitore Zool. Ital. (N.S.) **8**: 29–46.

Caprio, J., (1982) High sensitivity and specificity of olfactory and gustatory receptors of catfish to amino acids. In: Hara, T.J. (ed.) Chemoreception in fishes. Elsevier, Amsterdam, pp. 109–134.

Caprio, J. (1984) Olfaction and taste in fish. In: Bolis, L., Keynes, R.D., Maddrell, S.H.P. (eds.) Comparative physiology of sensory systems. Cambridge University Press, Cambridge, pp. 257–283.

Caprio, J. (1988) Peripheral filters and chemoreceptor cells in fishes. In: Atema, J., Fay, R.R., Popper, A.N., Tavolga, W.N. (eds.) Sensory biology of aquatic animals. Springer-Verlag, New York, pp. 313–338.

Caprio, J, Dudek, J., Robinson, J.J. II (1989) Electro-olfactogram and multiunit olfactory receptor responses to binary and tertiary mixtures of amino acids in the channel catfish, *Ictalurus punctatus*. J. Gen. Physiol. **93**: 245–262.

Caprio, J., Brand, J.G., Teeter, J.H., Valentinčič, Kalinoski, D.I., Kohbara, J., Kumazawa, T., Wegert, S. (1993) The taste system of the channel catfish from biophysics to behaviour. TINS (Trends Neurosci.) **16**: 192–197.

Caputi, A., Silva, A., Macadar, O. (1993) Electric organ activation in *Gymnotus carapo:* spinal origin and peripheral mechanisms. J. Comp. Physiol. A **173**: 227–232.

Cardwell, J.R., Liley, N.R. (1990) Current status of behavioural endocrinology in tropical marine teleosts. In: Epple, A., Scannes, C.G., Stetson, M.H. (eds.) Progress in comparative endocrinology. Wiley-Liss, New York, pp. 673–677.

Caseiotta, J.R., Arratia, G. (1993) Jaws and teeth of American cichlids (Pisces: Labroidei). J. Morphol. **217**: 1–36.

Cassano, G., Maffia, M., Vilella, R.S., Storelli, C. (1988) Effects of membrane potential on Na cotransports in eel intestinal brush-border membrane vesicles: Studies with a fluorescent dye. J. Membrane Biol. **101**: 225–236.

Cassano, G., Maffia, M., Ramires, P.A., Vilella, R.S., Storelli, C. (1990) Do neutral amino acids and their N-methylated analogues share the same Na-dependent carrier in brush-border membrane from the eel intestine? Comp. Biochem. Physiol. **96A**: 13–17.

Castejon, O.J., Caraballo, A.J. (1980) Light and scanning electron microscopic study of cerebellar cortex of teleost fishes. Cell Tissue Res. **207**: 211–226.

Cerra, M.C., Canonaco, M., Tota, B. (1992) A quantitative autoradiographic study of ^{125}I atrial natriuretic factor in the heart of a teleost fish (*Conger conger*). J. Exp. Zool. **263**: 215–219.

Chan, S.J., Steiner, D.F. (1994) Structure and expression of insulinlike growth factor genes in fish. In: Sherwood, N., Hew, C. (eds.) Fish physiology, vol. 13. Academic Press, San Diego, pp. 213–224.

Chardon, M., Vandewalle, P. (1991) Acoustico-lateralis system. In: Winfield, I.J., Nelson, J.S. (eds) Cyprinid fishes. Systematics, biology and exploitation. Chapman and Hall, London, New York, pp. 332–352.

Chen, T.T., Marsh, A., Shamblott, M., Chan, K.M., Tang, T.L., Cheng, C.M., Yang, B.Y. (1994) Structure and evolution of fish growth hormone and insulinlike growth factor genes. In: Sherwood, N., Hew, C. (eds.) Fish physiology, vol. 13. Academic Press, San Diego, pp. 179–209.

Chester Jones, I., Phillips, J.G. (1986) The adrenal and interrenal glands. In: Pang, P.K.T., Schreibman, M.P. (eds.) Vertebrate endocrinology: fundamentals and biomedical implications, vol. 1. Morphological considerations. Academic Press, Orlando, pp. 319–350.

Chester Jones, I., Ball, J.N., Henderson, I.W., Sandor, T., Baker, B.I. (1974) Endocrinology of fishes. In: Florkin, M., Scheer, B.T. (eds.) Chemical Zoology. Academic Press, New York, London, pp. 523–593.

Chiba, A., Honma, Y. (1986) Comparative anatomy of the brain-ventricular system of the lamprey and hagfish. In: Uyeno, T., Arai, R., Taniuchi, T., Matsuura, K. (eds.) Indo-Pacific fish, biology: Proc. Second International Conference on Indo-Pacific Fishes. Ichthyological Society of Japan, Tokyo, pp. 76–85.

Chiba, A., Honma, Y. (1989) Innervation and vascularization of the hypothalamo-hypophysial system of the Japanese charr in relation to its reproduction. Physiol. Ecol. Japan Spec. **1**: 455–468.

Chiba, A., Honma, Y. (1991) Immunocytochemical distribution of FMRF amide-like substance in the brain of the cloudy dogfish, *Scyliorhinus torazame*. Cell Tissue Res. **265**: 243–250.

Chiba, A., Honma, Y. (1992) Distribution of neuropeptide Y-like immunoreactivity in the brain and hypophysis of the cloudy dogfish, *Scyliorhinus torazame*. Cell Tissue Res. **268**: 453–461.

Chiba, A., Honma, Y. (1994) Neuropeptide Y-immunoreactive structures in the telencephalon and diencephalon of the white sturgeon, *Acipenser transmontanus*, with special regard to the hypothalamo-hypophyseal system. Arch. Histol. Cytol. **57**: 77–86.

Chiba, A., Torraba, M., Honma, Y., Zapata, A.G. (1988) Occurrence of lymphohaemopoietic tissue in the meninges of the stingray *Dasyatis akajei* (Elasmobranchii, Chondrichthyes). Am. J. Anat. **183**: 268–276.

Chiba, A., Ohnishi, S., Honma, Y. (1989) Immunocytochemical localization of S-100 protein in the hypophysis and saccus vasculosus of the elasmobranchs *Mustelus manazo* and *Scyliorhinus torazame*. Cell Tissue Res. **255**: 255–260.

Chiba, A., Oka, S., Honma, Y. (1994) Ontogenetic development of gonadotropin releasing hormone-like immunoreactive neurons in the brain of the chum salmon, *Oncorhynchus keta*. Neurosci. Lett. **78**: 51–54.

Chiba, A., Honma, Y., Oka, S. (1995) Ontogenetic development of neuropeptide Y-like immunoreactive cells in the gasteroenteropancreatic endocrine system of the dogfish. Cell Tissue Res. **282**: 33–40.

Chiu, J.F., Mack, A.F., Fernald, R.D. (1995) Daily rhythm of cell proliferation in the teleost retina. Brain Res. **673**: 119–125.

Chivers, D.P., Smith, R.J.F. (1993) The role of olfaction in chemosensory-based predator recognition in the fathead minnow, *Pimephales promelas*. J. Chem. Ecol. **19**: 623–633.

Chivers, D.P., Smith, R.J.F. (1994) Fathead minnows, *Pimephales promelas*, acquire predator recognition when alarm substance is associated with the sight of unfamiliar fish. Anim. Behav. **48**: 597–605.

Chivers, D.P., Smith, R.J.F. (1995a) Fathead minnows (*Pimephales promelas*) learn to recognize chemical stimuli from high-risk habitats by the presence of alarm substance. Behav. Ecol. **6**: 155–158.

Chivers, D.P., Smith, R.J.F. (1995b) Chemical recognition of risky habitats is culturally transmitted among fathead minnows, *Pimephales promelas* (Osteichthyes, Cyprinidae). Ethology **99**: 286–296.

Chivers, D.P., Smith, R.J.F. (1995c) Free-living fathead minnows rapidly learn to recognize pike as predators. J. Fish Biol. **46**: 949–954.

Chivers, D.P., Brown, G.E., Smith, R.J.F. (1995a) Acquired recognition of chemical stimuli from pike, *Esox lucius*, by brook sticklebacks, *Culaea inconstans* (Osteichthyes, Gasterosteidae). Ethology **99**: 234–242.

Chivers, D.P., Brown, G.E., Smith, R.J.F. (1995b) Familiarity and shoal cohesion in fathead minnows (*Pimephales promelas*): implications for antipredator behaviour. Can. J. Zool. **73**: 955–960.

Choat, J.H. (1991) The biology of herbivorous fishes on coral reefs. In: Sale, P.F. (ed.) The ecology of fishes on coral reefs. Academic Press, San Diego, pp. 120–155.

Choat, J.H., Bellwood, D.R. (1991) Reef fishes: their history and evolution. In: Sale, P.F. (ed.) The ecology of fishes on coral reefs. Academic Press, San Diego, pp. 39–66.

Christophe, B., Baguet, F. (1983) Luminescence of isolated photocytes from *Porichthys* photophores: adrenergic stimulation. J. Exp. Biol. **104**: 183–192.

Christophe, B., Baguet, F. (1985) The adrenergic control of the photocyte luminescence of the *Porichthys* photophore. Comp. Biochem. Physiol. **81C**: 359–365.

Christiansen, J.S., Martinez, I., Jobling, M., Amin, A.B. (1992) Rapid somatic growth and muscle damage in a salmonid fish. BAM (Basic and Applied Myology), **2**: 235–239.

Cimini, V., Van Noorden, S., Nardini, V. (1989a) Peptides of the gastrointestinal tract of the dogfish (*Scyliorhinus stellaris*). J. Exp. Zool. Suppl. **2**: 146–157.

Cimini, V., Van Noorden, S., Polak, J.M. (1989b) Co-localisation of substance P-, bombesin-and peptide histidine isoluecine (PHI)-like peptides in gut endocrine cells of the dogfish *Scyliorhinus stellaris*. Anat. Embryol. **179**: 615–624.

Cimini, V., Van Noorden, S., Giordano-Lanza, G., Nardini, V., McGregor, G.P., Bloom, S.R., Polak, J.M. (1985) Neuropeptides and 5-HT immunoreactivity in the gastric nerves of the dogfish (*Scyliorhinus stellaris*). Peptides **6**: 373–377.

Claes, G., De Vree, F. (1991a) Kinematics of the pharyngeal jaws during feeding in *Oreochromis niloticus* (Pisces: Perciformes). J. Morphol. **208**: 227–245.

Claes, G., De Vree, F. (1991b) Cineradiographic analysis of the pharyngeal jaw movements during feeding in *Haplochromis burtoni* (Günther, 1893) (Pisces, Cichlidae). Belg. J. Zool. **121**: 227–234.

Claes, G., De Vree, F. (1992) Pharyngeal muscle activities during food processing in *Oreochromis niloticus* (Pisces: Cichlidae). Zool. Jb. Anat. **122**: 173–178.

Claes, G., De Vree, F., Vandewalle, P. (1991) Masticatory operations and the functions of pharyngeal jaw movements in cichlids. Ann. Kon. Mus. Mid. Afr. Zool. Wetensch-Ann. Mus. Roy Afr. Centr. Sc. Zool. **262**: 85–90.

Clements, K.D., Choat J.H. (1995) Fermentation in tropical marine herbivorous fishes. Physiol. Zool. **68**: 355–378.

Clements, K.D., Gleeson, V.P., Slayter, M. (1994) Short-chain fatty acid metabolism in temperate marine herbivorous fish. J. Comp. Physiol. B **164**: 372–377.

Cloutier, R., (1991) Patterns, trends and rates of evolution within the Actinistia. Env. Biol. Fish. **32**: 23–58.

Cochran, R.C., Grier, H.J. (1991) Regulation of sexual succession in the protogynous black sea bass, *Centropristis striatus* (Osteichthyes: Serranidae). Gen. Comp. Endocrinol. **82**: 69–77.

Coetzee, H.L., Nel, M.M., Swanepoel, J.H. (1990) Histological and ultrastructural study of the gastric wall of the freshwater bream, *Oreochromis mossambicus* (Peters) with reference to 'parietal-like' cells. S. Afr. J. Zool. **25**: 1–10.

Coleman, R.M., Gross, M.R., Sargent, R.C. (1985) Parental investment decision rules: a test in bluegill sunfish. Behav. Ecol. Sociobiol. **18**: 59–66.

Coleman, R.M., Gross, M.R. (1991) Parental investment theory: The role of past investment Trends Ecol. Evol. **6**: 404–406.

Colgan, P.W., Brown, J.A. (1988) Dynamics of nest defense by male centrarchid fish. Behav. Proc. **17**: 17–26.

Collie, N.L., Hirano, T. (1987) Mechanisms of hormone actions on intestinal transport. In: Pang, P.K.T., Schreibman, M.P. (eds.) Vertebrate endocrinology: fundamentals and biomedical implications, vol. 2. Regulation of water and electrolytes. Academic Press, San Diego, pp. 239–270.

Conte, F.P., Takano, K., Takemura, A., Boehlert, G.W. (1991) Ontogeny of the sodium pump in embryos of rockfish of the genus *Sebastes*. Env. Biol. Fish. **30**: 127–133.

Coombs, S. (1994) Nearfield detection of dipole sources by the goldfish (*Carassius auratus*) and the mottled sculpin (*Cottus bairdi*). J. Exp. Biol. **190**: 109–129.

Coombs, S., Fay, R.R. (1989) The temporal evolution of masking and frequency selectivity in the goldfish (*Carassius auratus*). J. Acoust. Soc. Am. **86**: 925–933.

Coombs, S., Janssen, J. (1989) Peripheral processing by the lateral line system of the mottled sculpin (*Cottus bairdi*). In: Coombs, S., Görner, P., Münz, H. (eds.) The mechanosensory lateral line: neurobiology and evolution. Springer-Verlag, New York, pp. 299–319.

Coombs, S., Janssen, J. (1990) Water flow detection by the mechanosensory lateral line. In: Stebbins, W.C., Berkley, M.A. (eds.) Comparative perception, vol. **2**: Complex signals. John Wiley, New York, pp. 89–123.

Coombs, S., Montgomery, J. (1992) Fibres innervating different parts of the lateral line system of an Antarctic notothenioid, *Trematomus bernacchii*, have similar frequency responses, despite large variation in the peripheral morphology. Brain Behav. Evol. **40**: 217–233.

Coombs, S., Montgomery, J. (1994) Function and evolution of superficial neuromasts in an Antarctic notothenioid fish. Brain Behav. Evol. **44**: 287–298.

Coombs, S., Janssen, J., Webb, J.F. (1988) Diversity of lateral line systems: evolutionary and functional considerations. In: Atema, J., Fay, R.R., Popper, A.N., Tavolga, W.N. (eds.) Sensory biology of aquatic animals. Springer-Verlag, New York, Berlin, pp. 553–593.

Coombs, S., Görner, P., Münz, H. (eds.) (1989) The mechanosensory lateral line: neurobiology and evolution. Springer-Verlag, New York.

Corbel, M.J. (1975) The immune response in fish: a review. J. Fish Biol. **7**: 539–563.

Corwin, J.T. (1981) Audition in elasmobranchs. In: Tavolga, W.N., Popper, A.N., Fay, R.R. (eds.) Hearing and sound communication in fishes. Springer-Verlag, New York, pp. 81–103.

Cowey, C.B., Sargent, J.R. (1979) Nutrition. In: Hoar, W.S., Randall, D.J. (eds.) Fish physiology, vol. 8. Academic Press, New York, pp. 1–69.

Cowey, C.B., Walton, M.J. (1989) Intermediary metabolism. In Halver, J.E. (ed.) Fish nutrition. (Second edition). Academic Press, San Diego, pp. 259–329.

Craig, J.F. (1985) Aging in fish. Can. J. Zool. **63**: 1–8.

Crawford, J.D. (1992) Individual and sex specificity in the electric organ discharges of breeding mormyrid fish (*Pollimyrus isidori*). J. Exp. Biol. **164**: 79–102.

Crawshaw, L.I. (1979) Responses to rapid temperature changes in vertebrate ectotherms. Am. Zool. **19**: 225–237.

Crim, J.W., Vigna, S.R. (1983) Brain, gut and skin peptide hormones in lower vertebrates. Am. Zool. **23**: 621–638.

Dabrowski, K. (1984) The feeding of fish larvae; present "state of the art" and perspectives. Reprod. Nutr. Dev. **24**: 807–833.

Dabrowski, K. (1986) Ontogenetical aspects of nutritional requirements in fish. Comp. Biochem. Physiol. **85A**: 639–655.

Dabrowski, K., Kock, G. (1989) Absorption of ascorbic acid and ascorbic sulfate and their interaction with minerals in the digestive tract of rainbow trout (*Oncorhynchus mykiss*). Can. J. Fish Aquat. Sci. **46**: 1952–1957.

Dacke, C.G. (1979) Calcium regulation in sub-mammalian vertebrates. Academic Press, London, New York, San Francisco.

Dacke, C.G. (1989) Eicosanoids, steroids and miscellaneous hormones. In: Pang, P.K.T., Schreibman, M.P. (eds.). Vertebrate endocrinology: fundamentals and biomedical implications, vol. 3. Academic Press, San Diego, pp. 171–210.

Danguy, A., Lenglet, G. (1988) Anatomical microscopic study of the gas bladder in *Pangasius sutchi* Fowler (Teleostei, Siluriformes). Annls. Soc. R. Zool. Belg. **118**: 199–209.

Danguy, A., Lenglet, G., Kiss, R.B. (1985) Particularites histologiques de l' oesophage chez *Phractolaemus ansorgei* Boulenger, 1901 (Pisces, Gonorynchiformes). Cybium **9**: 193–201.

Danguy, A.J., Kiss, R., Lenglet, G. (1986) The oesophageal and buccal histology of two freshwater siluriformes (*Schilbe mystus* L. and *Pangasius micronemus* Blkr.) with reference to an unusual cell type. Bull. de la Classee des. Sci. **72**: 62–72.

Danguy, A., Kiss, R., De Launoit, Y., Lenglet, G. (1988) Cell proliferation in the digestive mucosa of fishes as revealed by autoradiography with tritiated thymidine. Z mikrosk-anat. Forsch. Leipzig. **102**: 489–502.

Danguy, A., Genten, F., Gabius, H.J. (1991) Histochemical evolution of application of biotinylated neoglycoproteins for the fish skin. Eur. J. Bas. Appl. Histochem. **35**: 341–357.

Daniel, T.L., Webb, P.W. (1987) Physical determinants of locomotion. In: Dejours, P., Bolis, L., Taylor, C.R., Weibel, E.R. (eds.). Comparative physiology: life in water and on land. Fidia Research Series, Vol. 9, Liviana Press, Padova, pp. 343–369.

Dantzler, W.H. (1982) Studies on nonmammalian nephrons. Kidney Int. **22**: 560–570.

Dantzler, W.H. (1985) Comparative aspects of renal function. In: Seldin, D.W., Giebisch G. (eds.) The kidney: physiology and pathophysiology. Raven Press, New York, pp. 333–364.

Dantzler. W.H. (1995) Renal transport of organic acids and bases in nonmammalian vertebrates. In: Heisler, N. (ed.) Mechanisms of systemic regulation. Advances in Comparative and Environmental Physiology, vol. 22, Springer-Verlag, Berlin, Heidelberg, pp. 27–49.

Danulat, E., Mommsen, T.P. (1990) Norepinephrine: a potent activator of glycogenolysis and gluconeogenesis in rockfish hepatocytes Gen. Comp. Endocrinol. **78**: 12–22.

Davenport, J. (1994) How and why do flying fish fly? Rev. Fish Biol. Fish. **4**: 184–214.

Davie, P.S., Farrell, A.P. (1991) The coronary and luminal circulations of the myocardium of fishes. Can. J. Zool. **69**: 1993–2001.

Davina, J.H.M., Rijkers, G.T., Rombout, J.H.W.M., Timmermans, L.P.M., Van Muiswinkel, W.B. (1980) Lymphoid and non-lymphoid cells in the intestine of cyprinid fish. In: Horton, J.D. (ed.) Development and differentiation of vertebrate lymphocytes. Elsevier/North-Holland Biomedical Press, Amsterdam, pp. 129–140.

Dawson, J. (1969) The keratinised teeth of *Myxine glutinosa*. A histological, histochemical, ultrastructural and experimental study. Acta Zool. Stockholm **50**: 35–68.

Deacon, J.E., Minckley, W.L. (1974) Desert fishes. In: Brown, R.W. (ed.) Desert Biology, vol 2. Academic Press, New York, pp. 385–488.

De Andres, A.V., Muñoz-Chápuli, R., Sans-Coma, V., Garcia-Garrido, L. (1990a) Anatomical studies of the coronary system in elasmobranchs: I. Coronary arteries in lamnoid sharks. Am. J. Anat. **187**: 303–310.

De Andres, A.V., Muñoz-Chápuli, R., Sans-Coma, V., Garcia-Garrido, L. (1990b) Anatomical studies of the coronary system in elasmobranchs: II. Coronary arteries in hexanchoid, squaloid, and carcharhinoid sharks. Anat. Rec. **233**: 429–439.

De Andres, A.V., Muñoz-Chápuli, R., Sans-Coma, V. (1993) Development of the coronary arteries and cardiac veins in the dogfish (*Scyliorhinus canicula*). Anat. Rec. **235**: 436–442.

Defraipont, M., Sovensen, P.W. (1993) Exposure to the pheromone 17a-20B-dihydroxy-4-pregnen-3-one enhances the behavioural spawning success, sperm production and sperm motility of male goldfish. Anim. Behav. **46**: 245–256.

De Graaf, P.J.F. (1989) Control of respiration in the carp (*Cyprinus carpio* L.): Mechanoreceptor input and respiratory rhythms. Chapter 7. Proefschrift Rijksuniversiteit Groningen, pp. 9–124.

De Graaf, P.J.F. (1990) Innervation pattern of the gill arches and gills of the carp (*Cyprinus carpio*). J. Morphol. **206**: 71–78.

Dejours, P. (1994) Environmental factors as determinants in bimodal breathing: an introductory overview. Am. Zool. **34**: 178–183.
Demski, L.S. (1981) Neural control of teleost sound production. In: Tavolga, W.N., Popper, A.N., Fay, R.R. (eds.) Hearing and sound communication in fishes. Springer-Verlag, New York, pp. 427–445.
Demski, L.S. (1992) Chromatophore systems in teleosts and cephalopods: a levels oriented analysis of convergent systems. Brain Behav. Evol. **40**: 141–156.
Demski, L.S., Northcutt, R.G. (1983) The terminal nerve: a new chemosensory system in vertebrates? Science **220**: 435–437.
Dent, J. (1986) The thyroid gland. In: Pang, P.K.T., Schreibman, M.P. (eds.) Vertebrate endocrinology: fundamentals and biomedical implications. Academic Press, Orlando, pp. 175–206.
Denton, E.J., Gray, J.A.B. (1993) Stimulation of the acoustico-lateralis system of clupeid fish by external sources and their own movements. Phil. Trans. R. Soc. Lond. B **341**: 113–127.
Denton, E.J., Rowe, D.M. (1994) Reflective communication between fish, with special reference to the greater sand eel, *Hyperoplus lanceolatus*. Phil. Trans. R. Soc. Lond. B **344**: 221–237.
Devitsina, G.V., Chervova, L.S. (1994) Morphological and physiological aspects of fish chemoreception. In: Singh, H.R. (ed.) Advances in fish biology. Hindustan Publishing Corporation, Delhi, pp. 185–191.
De Vries, A.L. (1971) Freezing resistance in fishes. In: Hoar, W.S., Randall, D.J. (eds.) Fish physiology, vol. 6. Academic Press, New York, London, pp. 157–189.
De Vries, A.L. (1974) Survival at freezing temperatures. In: Sargent, J., Malins, D.C. (eds.) Biochemical and biophysical perspectives in marine biology, vol. 1. Academic Press, London, pp. 289–330.
De Vries, A.L. (1978) The physiology and biochemistry of low temperature adaptations in polar marine ectotherms. In: McWhinnie, M.A. (ed.) Polar research-to the present, and the future. AAAS selected symposium 7, pp. 175–202.
De Vries, A.L., Eastman, J.T. (1981) Physiology and ecology of notothenioid fishes of the Ross Sea. J. Roy. Soc. New Zealand 11, 329–340.
Dewar, H., Brill, R.W., Olson, K.R. (1994) Secondary circulation of the vascular heat exchangers in skipjack tuna, *Katsuwonus pelamis*. J. Exp. Zool. **269**: 566–570.
Dickson, D.H., Graves, D.A. (1979) Fine structure of the lamprey photoreceptors and retinal pigment epithelium (*Petromyzon marinus* L.). Exp. Eye Res. **29**: 45–60.
Dickson, D.H., Graves, D.A. (1981) Giant choroida cells in the lamprey eye (*Petromyzon marinus* L.). Exp. Eye Res. **32**: 301–311.
Dickman, M.C., Schliwa, M., Barlow, G.W. (1988) Melanophore death and disappearance produces color metamorphosis in the polychromatic midas cichlid (*Cichlasoma citrinellum*). Cell Tissue Res. **253**: 9–14.
Dickman, M.C., Annett, C., Barlow, G.W. (1990) Unsuspected cryptic polymorphism in the polychromatic Midas cichlid. Biol. J. Linn. Soc. **30**: 239–249.
Dingle, H. (1991) Evolutionary genetics of animal migration. Am. Zool. **31**: 253–264.
di Prisco, G., Tamburrini, M. (1992) The hemoglobins of marine and freshwater fish: the search for correlations with physiological adaptations. Comp. Biochem. Physiol. **102B**: 661–671.
di Prisco, G., Macdonald, J.A., Brunori, M. (1992) Antarctic fishes survive exposure to carbon monoxide. Experientia **48**: 473–475.
Domenici, P., Blake, R.W. (1991) The kinematics and performance of the escape response in the anglefish (*Pterophyllum eimekei*). J. Exp. Biol. **156**: 187–205.
Domenici, P., Blake, R.W. (1993) Escape trajectories in anglefish (*Pterophyllum eimekei*). J. Exp. Biol. **177**: 253–272.
Donaldson, E.M., Fagerlund, V.H.M., Higgs, D.A., Mc Bride, J.R. (1979) Hormonal enhancement of growth. In: Hoar, W.S., Randall, D.J. (eds.) Fish physiology, vol. 8. Academic Press, New York, pp. 455–597.
Dos Santos, J., Jobling, M. (1991) Gastric emptying in cod, *Gadus morhua* L.: emptying and retention of indigestible solids. J. Fish Biol. **38**: 187–197.
Dos Santos, J., Burkow, I.C., Jobling, M. (1993) Patterns of growth and lipid deposition in cod (*Gadus morhua* L.) fed natural prey and fish-based feeds. Aquaculture **110**: 173–189.

Døving, K.B. (1990) Olfactory cues in salmonid migration. In: Chemical signals in vertebrates, **5**: 573–584.

Døving, K.B., Thommesen, G. (1977) Some properties of the first olfactory system. In: Olfaction and taste **6**, pp. 175–183.

Døving, K.B., Selset, R., Thommesen, G. (1980) Olfactory sensitivity to bile acids in salmonid fishes. Acta Physiol. Scand. **108**: 123–131.

Downhower, J.F., Brown, L. (1975) Superfoetation in fishes and the cost of reproduction. Nature (London), **256**: 345–346.

Downing, A.L., Burns, J.R. (1995) Testis morphology and spermatozeugma formation in three genera of viviparous halfbeaks: *Nomorhamphus*, *Dermogens*, and *Hemirhamphodon* (Teleostei: Hemirhamphidae). J. Morphol. **225**: 329–343.

Dréan, Y.L., Pakdel, F., Valotaire, Y. (1994) Structure and regulation of genes for estrogen receptors. In: Sheerwood, N., Hew, C. (eds.) Fish physiology, vol. 8. Academic Press, San Diego, pp. 331–366.

Driedzic, W.R. (1988) Matching of cardiac oxygen delivery and fuel supply to energy demand in teleosts and cephalopods. Can. J. Zool. **66**: 1078–1083.

Driedzic, W.R. (1992) Cardiac energy metabolism. In: Hoar, W.S., Randall, D.J., Farrell, A.P. (eds.) Fish physiology, vol. 12A. Academic Press, Orlando, pp. 219–266.

Driedzic, W.R., Gesser, H. (1985) Ca^{2+} protection from the negative inotropic effect of contraction frequency on teleost hearts. J. Comp. Physiol. B **156**: 135–142.

Driedzic, W.R., Gesser, H. (1994) Energy metabolism and contractility in ectothermic vertebrate hearts: hypoxia, acidosis and low temperature. Physiol. Rev. **74**: 221–258.

Driedzic, W.R., Hochachka, P.W. (1978) Metabolism in fish during exercise. In: Hoar, W.S., Randall, D.J. (eds.) Fish physiology, vol. 7. Academic Press, New York, pp. 503–543.

Drucker, E.G., Jensen, J.S. (1991) Functional analysis of a specialized prey processing behavior: winnowing by surfperches (Teleostei: Embiotocidae). J. Morphol. **210**: 267–287.

Duguay, S.J., Mommsen, T.P. (1994) Molecular aspects of pancreatic peptides. In: Hoar, W.S., Randall, D.J., Farrell, A.P. (eds.) Fish physiology, vol. 12. Academic Press, San Diego, pp. 225–271.

Dulka, J.G., Stacey, N.E. (1991) Effects of olfactory tract lesions on gonadotropin and milt responses to the female sex pheromone, 17α, 20β-dihydroxy-4 pregnen-3-One, in male goldfish. J. Exp. Zool. **257**: 223–229.

Dulka, J.G., Stacey, N.E., Sorensen, P.W., Van der Kraak (1987) A steroid sex pheromone synchronizes male-female spawning readiness in goldfish. Nature (London) **325**: 251–253.

Dulka, J.G., Sloley, B.D., Stacey, N.E., Peter, R.E. (1992) A reduction in pituitary dopamine turnover is associated with sex pheromone-induced gonadotropin secretion in male goldfish. Gen. Comp. Endocrinol. **86**: 496–505.

Dutta, H.M. (1979) Form and function of jaws during feeding: *Ctenopoma acutirostre*, *Anabas testudineus* and *Macropodus opercularis*. Acta Morphol. Neerl-Scand. **17**: 119–132.

Eales, J.G. (1979) Thyroid functions in cyclostomes and fishes. In: Barrington, E.J.W. (ed.) Hormones and evolution, vol. 1. Academic Press, New York, pp. 341–436.

Eales, J.G. (1985) The peripheral metabolism of thyroid hormones and regulation of thyroidal status in poikilotherms. Can. J. Zool. **63**: 1217–1231.

Eales, J.G. (1988) The influence of nutritional state on thyroid function in various vertebrates. Am. Zool. **28**: 351–362.

Eales, J.G. (1990) Thyroid function in poikilotherms. In: Epple, A., Scannes, C.G., Stetson, M.H. (eds.) Progress in comparative endocrinology, Wiley-Liss, New York, pp. 415–420.

Ebbesson, S.O.E., Northcutt, R.G. (1976) Neurology of anamniotic vertebrates. In: Masterton, R.B., Bitterman, M.E., Campbell, C.B.G., Hotton, N. (eds.) Evolution of brain and behaviour in vertebrates. Lawrence Erlbaum Assoc's., Inc. Hillsdale, N.J., pp. 115–146.

Ecay, T.W., Valentich, I.D. (1991) Chloride secretagogues stimulate inositol phosphate formation in shark rectal gland tubules cultured in suspension. J. Cell Physiol. **146**: 407–416.

Edwards, T.W., Horn, M.H. (1982) Assimilation efficiency of a temperate zone intertidal fish (*Cebidichthys violaceus*) fed diet of microalgae. Mar. Biol. **67**: 247–253.

Eggold, B.T., Motta, P.J. (1992) Ontogenetic dietary shifts and morphological correlates in striped mullet, *Mugil cephalus*. Env. Biol. Fish. **34**: 139–158.

Elbal, M.T., Agulleiro, B. (1986a) A histochemical and ultrastructural study of the gut of *Sparus auratus* (Teleostei). J. Submicroc. Cytol. **18**: 335–347.

Elbal, M.T., Agulleiro, B. (1986b) An immunocytochemical and ultrastructural study of endocrine cells in the gut of a teleost fish, *Sparus auratus* L. Gen. Comp. Endocrinol. **64**: 339–354.

Elbal, M.T., Lozano, M.T., Agulleiro, B. (1988) The endocrine cells in the gut of *Mugil saliens* Risso, 1810 (Teleostei): an immunocytochemical and ultrastructural study. Gen. Comp. Endocrinol. **70**: 231–246.

Ellis, A.E. (ed.) (1988) Fish vaccination. Academic Press, San Diego, U.S.A.

Ellis, T.J., Wagner, G.F. (1995) Post-transcriptional regulation of the stanniocalcin gene by calcium. J. Biol. Chem. **27**: 1960–1965.

Elshoud-Oldenhave, M.J.W. (1979) Prey capture in the pike-perch, *Stizostedion, Lucioperca* (Teleostei, Percidae); a structural and functional analysis. Zoomorphologie **93**: 1–32.

Elshoud-Oldenhave, M.J.W., Osse, J.W.M. (1976) Functional morphology of the feeding system in the ruff-*Gymnocephalus cernua* (L. 1758)-(Teleostei, Percidae). J. Morphol. **150**: 399–422.

Endo, Y., Chiba, A., Honma, Y. (1991) Exocytotic release of neurotransmitters from nerve fibers in the endocrine pancreas (Brockmann body) of the teleost, *Takifugu niphobles*. Neurosc. Lett. **126**: 60–62.

Enger, P.S. (1981) Frequency discrimination in teleosts-central or peripheral? In: Tavolga, W.N., Popper, A.N., Fay, R.R. (eds.) Hearing and sound communication in fishes. Springer-Verlag, New York, pp. 243–255.

Enger, P.S., Kalmijn, A.J., Sand, O. (1989) Behavioural investigations on the functions of the lateral line and inner ear in predation. In: Coombs, S., Görner, P., Münz, H. (eds.) The mechanosensory lateral line: neurobiology and evolution. Springer-Verlag, New York, pp. 575–587.

Epple, A. (1969) The endocrine pancreas. In: Hoar, W.S., Randall, D.J. (eds.) Fish physiology, vol. 2. Academic Press, New York, pp. 275–319.

Epple, A. (1987) Pancreatic islet hormones. In: Pang, P.K.T., Schreibman, M.P. (eds.) Vertebrate endocrinology: fundamentals and biomedical implications, vol. 2. Academic Press, San Diego, pp. 103–119.

Epple, A. (1988) Insulin and other hormones (somatostatin, glucagon and PP) in the neuroendocrine system of some lower vertebrates and that of invertebrates—a mini review. Horm. metabol. Res. **20**: 406–410.

Epple, A., Brinn, J.E. (1986) Pancreatic islets. In: Pang, P.K.T., Schreibman, M.P. (eds.) Vertebrate endocrinology: fundamentals and biomedical implications, vol. 1. Academic Press, Orlando, pp. 279–317.

Esteban, M.A., Meseguer, J. (1994) Phagocytic defence mechanism in sea bass (*Dicentrarchus labrax* L.) an ultrastructural study. Anat. Rec. **240**: 589–597.

Esteban, M.A., Meseguer, J., Garcia-Ayala, L., Agulleiro, B. (1989) Erythropoiesis and thrombopoiesis in the head-kidney of the sea bass (*Dicentrarchus labrax* L.): An ultrastructural study. Arch. Histol. Cytol. **52**: 407–419.

Evans, D.H. (1979) Ionic and osmotic regulation in fish. In: Maloiy, G.M.O. (ed.) Comparative physiology of osmoregulation in animals, vol. 1. Academic Press, London, pp. 305–390.

Evans, D.H. (1980) Kinetic studies of ion transport by fish gill epithelium. Am. J. Physiol. **238**: R224–230.

Evans, D.H. (1986) The role of branchial and dermal epithelia in acid-base regulation in aquatic vertebrates. In: Heisler, N. (ed.) Acid-base regulation in animals. Elsevier, Amsterdam, pp. 139–172.

Evans, D.H. (1987) Acid-base balance and nitrogenous waste excretion in fishes; the aquatic to amphibious transition. In: Dejours, P., Bolis, L., Taylor, C.R., Weibel, E.R. (eds.) Comparative physiology: life in water and on land. Fidia Research Series, vol. 9. Liviana Press, Padova, pp. 121–130.

Evans, D.H. (1990) An emerging role for a cardiac peptide hormone in fish osmoregulation. Annu. Rev. Physiol. **52**: 43–60.

Evans, D.H. (1991) Rat atriopeptin dilates vascular smooth muscle of the ventral aorta from the shark (*Squalus acanthias*) and the hagfish (*Myxine glutinosa*). J. Exp. Biol. **157**: 551–555.

Evans, D.H. (1992) Blood and extracellular fluid volume regulation: role of the renin-angiotensin system, kallikrein-kinn system, and atrial natriuretic peptides. In: Hoar, W.S., Randall, D.J., Farrell, A.P. (eds.) Fish physiology, vol. 12B. Academic Press, Orlando, pp. 135–254.

Evans, D.H. (1993) Osmotic and ionic regulation. In: Evans, D.H. (ed.) The physiology of fishes. CRC Press, Boca Raton, pp. 315–341.

Evans, D.H. (1995) The roles of natriuretic peptide hormones in fish osmoregulation and hemodynamics. In: Heisler, N. (ed.) Mechanisms of systemic regulation. Advances in Comparative and Environmental Physiology, vol. 22. Springer-Verlag, Berlin, Heidelberg, pp. 119–152.

Evans, D.H., Takei, Y. (1992) A putative role for natriuretic peptides in fish osmoregulation. NIPS **7**: 15–19.

Evans, D.H., Toop, T., Donald, J., Forrest, J.N. Jr. (1993) C-type natriuretic peptides are potent dilators of shark vascular smooth muscle. J. Exp. Zool. **265**: 84–87.

Evans, D.L., Gratzek, J.B. (1989) Immune defense mechanisms in fish to protozoan and helminth infections. Am. Zool. **29**: 409–418.

Falcón, J., Thibault, C., Begay, V., Zachmann, A., Collin, J.P. (1992) Regulation of the rhythmic melatonin secretion by fish pineal photoreceptor cells. In: Ali, M.A. (ed.) Rhythms in fishes. Plenum Press, New York, pp. 167–198.

Falcón, J., Begay, V., Goujon, J.M., Voisin, P., Guerlotte, J., Collin, J.P. (1994a) Immunocytochemical localization of hydroxyindole-O-methyltransferase in pineal photoreceptor cells of several fish species. J. Comp. Neurol. **341**: 559–566.

Falcón, J., Bolliet, V., Ravault, J.P., Chesneau, D., Ali, M.A., Collin, J.P. (1994b) Rhythmic secretion of melatonin by the superfused pike pineal organ: thermo-and photoperiod interaction. Neuroendocrinology **60**: 535–543.

Falkmer, S. (1995) Origin of the parenchymal cells of the endocrine pancreas: some phylogenetic and ontogenetic aspects. In: Mignon, M., Jensen, R.T. (eds.) Endocrine tumors of the pancreas. Front. Gastrointest Res. vol. 23. S. Karger, Basel, pp. 2–29.

Fänge, R. (1963) Structure and function of the excretory organs of myxinoids. In: Brodal, A., Fänge, R. (eds.) The biology of *Myxine*. Universitetsforlaget, Oslo, Norway, pp. 516–529.

Fänge, R. (1972) The circulatory system. Hardisty M.W., Potter, I.C. (eds.) The biology of lampreys, vol. 2. Academic Press, London, pp. 241–259.

Fänge, R. (1982a) A comparative study of lymphomyeloid tissue in fish. Devel. Comp. Immunol. Suppl. **2**: 23–33.

Fänge, R. (1982b) Exogenous otoliths of elasmobranchs. J. Mar. Biol. Ass. U.K. **62**: 225.

Fänge, R. (1983) Gas exchange in fish swim bladder. Rev. Physiol. Biochem. Pharmacol. **97**: 111–158.

Fänge, R. (1985) Regulation of blood and body fluids in primitive fish groups. In: Foreman, R.E., Gorbman, A., Dodd, J.M., Olsson, R. (eds.) Evolutionary biology of primitive fishes. Plenum Press, New York, London, pp. 253–272.

Fänge, R. (1986a) Lymphoid organs in sturgeons (Acipenseridae). Veter. Immunol. Immunopathol. **12**: 153–161.

Fänge, R. (1986b) Physiology of haemopoiesis. In: Nilsson, S., Holmgren, S. (eds.) Fish physiology: recent advances. Croom Helm, London, pp. 1–23.

Fange, R. (1987) Lymphomyeloid system and blood cell morphology in elasmobranchs. Arch. Biol. (Bruxelles) **98**: 187–208.

Fänge, R. (1992) Fish blood cells. In: Hoar, W.S., Randall, D.J., Farrell, A.P. (eds.) Fish physiology, vol. 12B. Academic Press, Orlando, pp. 1–54.

Fänge, R., Grove, D. (1979) Digestion. In: Hoar, W.S., Randall, D.J. (eds.) Fish physiology, vol. 8. Academic Press, New York, pp. 161–260.

Fänge, R., Nilsson, S. (1985) The fish spleen: structure and function. Experientia **41**: 152–158.

Faraldi, G., Bonini, E., Farina, L., Tagliafierro, G. (1988) Distribution and ontogeny of glucagon-like cells in the gastrointestinal tract of the cartilaginous fish *Scyliorhinus stellaris* (L.). Acta histochem. **83**: 57–64.

Farrell, A.P. (1984) A review of cardiac performance in the teleost heart: intrinsic and humoral regulation. Can. J. Zool. **62**: 523–536.

Farrell, A.P. (1985) Cardiovascular and hemodynamic energetics of fishes. In: Gilles, R. (ed.) Circulation, respiration and metabolism. Springer-Verlag, Berlin, Heidelberg, pp. 377–385.

Farrell, A.P. (1986) Cardiovascular responses in the sea raven, *Hemitripterus americanus*, elicited by vascular compression. J. Exp. Biol. **122**: 65–80.

Farrell, A.P. (1991a) From hagfish to tuna: a perspective on cardiac function in fish. Physiol. Zool. **64**: 1137–1164.

Farrell, A.P. (1991b) Circulation of body fluids. In: Prosser, C.L. (ed.) Environmental and metabolic animal physiology. Wiley-Liss, New York, pp. 509–558.

Farrell, A.P. (1993) Cardiovascular system. In: Evans, D.H. (ed.) The physiology of fishes. CRC Press, Boca Raton, pp. 219–250.

Farrell, A.P., Jones, D.R. (1992) The heart. In: Hoar, W.S., Randall, D.J., Farrell, A.P. (eds.) Fish physiology, vol. 12A. Academic Press, Orlando, pp. 1–88.

Farrell, A.P., Milligan, C.L. (1986) Myocardial intracellular pH in a perfused rainbow trout heart during extracellular acidosis in the presence and absence of adrenaline. J. Exp. Biol. **125**: 347–359.

Farrell, A.P., Hammons, A.M., Graham, M.S., Tibbits, G.F. (1988) Cardiac growth in rainbow trout, *Salmo gairdneri*. Can. J. Zool. **66**: 2368–2373.

Farrell, A.P., Davie, P.S., Franklin, C.E., Johansen, J.A., Brill, R.W. (1992) Cardiac physiology in tunas. 1. *In vitro* perfused heart preparations from yellowfin and skipjack tunas. Can. J. Zool. **70**: 1200–1210.

Fasano, S., Pierantoni, R., Chieffi, G. (1989) Reproductive biology of elasmobranchs with emphasis on endocrines. J. Exp. Zool. Suppl. **2**: 53–61.

Fasulo, S., Taglifierro, G., Contini, A., Ainis, L., Ricca, M.B., Yanaihara, N., Zaccone, G. (1993) Ectopic expression of bioactive peptides and serotonin in the sacciform gland cells of teleost skin. Arch. Histol. Cytol. **56**: 117–125.

Fay, R.R. (1978) Sound detection and sensory coding by the auditory systems of fishes. In: Mostofsky, D. (ed.) The behaviour of fish and other aquatic animals. Academic Press, New York, pp. 197–235.

Fay, R.R. (1985) Sound intensity processing by the goldfish. J. Acoust. Soc. Am. **78**: 1296–1309.

Fay, R.R. (1992) Analytic listening by the goldfish. Hear. Res. **59**: 101–107.

Fay, R.R. (1994) Perception of temporal acoustic patterns by the goldfish (*Carassius auratus*). Hear. Res. **76**: 158–172.

Fay, R.R., Popper, A.N. (1983) Hearing in fishes: Comparative anatomy of the ear and the neural coding of auditory information. In: Fay, R.R., Gourevitch, G. (eds.) Hearing and other senses: presentations in honour of E.G. Wever. Amphora Press, Groton, CT., pp. 123–148.

Fay, R.R., Ream, T.J. (1986) Acoustic response and tuning in saccular nerve fibers of the goldfish (*Carassius auratus*). J. Acoust. Soc. Am. **79**: 1883–1895.

Fenwick, J.C. (1970) The pineal organ. In: Hoar, W.S., Randall, D.J. (eds.) Fish physiology, vol. 4. Academic Press, New York, pp. 91–108.

Fenwick, J.C. (1989) Calcium exchange across fish gills. In: Pang, P.K.T., Schreibman, M.P. (eds.) Vertebrate endocrinology: fundamentals and biomedical implications, vol. 3. Academic Press, San Diego, pp. 319–342.

Fenwick, J.C., Verbost, P. (1993) A C-terminal fragment of the hormone stanniocalcin is bioactive in eels. Gen. Comp. Endocrinol. **91**: 337–343.

Fenwick, J.C., Vermette, M.G. (1989) Vitamin D_3 and the renal handling of phosphate in American eels. Fish Physiol. Biochem. **7**: 351–358.

Fenwick, J.C., Davison, W., Forster, M.E. (1994) In vivo calcitropic effect of some vitamin D compounds in the marine Antarctic teleost, *Pagothenia bernacehii*. Fish Physiol. Biochem. **12**: 479–484.

Fergusson, M.M., Noakes, D.L.G., Skulason, S., Snorrason, S. (1990) Life-history styles and somatic allocation in iteroparous arctic charr and semelparous pink salmon. Env. Biol. Fish. **28**: 267–272.

Fernald, R.D. (1988) Aquatic adaptations in fish eyes. In: Atema, J., Fay, R.R., Popper, A.N., Tavolga, W.N. (eds.) Sensory biology of aquatic animals. Springer-Verlag, New York, pp. 435–466.

Fernholm, B. (1981) Thread cells from the slime glands of hagfish (Myxinidae). Acta Zool. Stockholm. **62**: 137–145.

Ferrari, M.B., Zakon, H.H. (1989) The medullary pacemaker nucleus is unnecessary for electroreceptor tuning plasticity in *Sternopygus*. J. Neurosci. **9**: 1354–1361.

Fetcho, J.R. (1991) Spinal network of the Mauthner Cell. Brain Behav. Evol. **37**: 298–316.

Fetcho, J.R. (1992) The spinal motor system in early vertebrates and some of its evolutionary changes. Brain Behav. Evol. **40**: 82–97.

Fine, M.L. (1989) Embryonic, larval and adult development of the sonic neuromuscular system in the oyster toadfish. Brain Behav. Evol. **34**: 13–24.

Fine, M.L., Perini, M.A. (1994) Sound production evoked by electrical stimulation of the forebrain in the oyster toadfish. J. Comp. Physiol. A **174**: 173–185.

Fine, M.L., Horn, M.H., Cox, B. (1987) *Acanthonus armatus*, a deep-sea teleost fish with a minute brain and large ears. Proc. R. Soc. Lond. B **230**: 257–265.

Fine, M.L., Bernard, B., Harris, T.M. (1993) Functional morphology of toadfish sonic muscles: relationship to possible fiber division. Can. J. Zool. **71**: 2262–2274.

Finger, T.E. (1982) Somatotopy in the representation of the pectoral fin and free fin rays in the spinal cord of the sea robin, *Prionotus carolinus*. Biol. Bull. **163**: 154–161.

Finger, T.E. (1983) The gustatory system in teleost fish. In: Northcutt, R.G., Davis, R.E. (eds.) Fish neurobiology. vol. 1. Brain stem and sense organs. University of Michigan Press, Ann Arbor, pp. 285–310.

Finger, T.E. (1988a) Organization of chemosensory systems within the brains of bony fishes. In: Atema, J., Fay, R.R., Popper, A.N., Tavolga, W.N. (eds.) Sensory biology of aquatic animals. Springer-Verlag, New York, pp. 339–363.

Finger, T.E., Bullock, T.H. (1982) Thalamic center for the lateral line system in the catfish *Ictalurus nebulosus:* evoked potential evidence. J. Neurobiol. **13**: 39–47.

Finger, T.S., Kanwal, J.S. (1992) Ascending general visceral pathways within the brainstems of two teleost fishes: *Ictalurus punctatus* and *Carassius auratus*. J. Comp. Neurol. **320**: 509–520.

Finger, T.E., Tong, S-L. (1984) Central organization of eighth nerve and mechanosensory lateral line systems in the brainstem of ictalurid catfish. J. Comp. Neurol. **229**: 129–151.

Finger, T.E., Drake, S.K., Kotrschal, K., Womble, M., Dockstader, K.C. (1991) Postlarval growth of the peripheral gustatory system in the channel catfish, *Ictalurus punctatus*. J. Comp. Neurol. **314**: 55–66.

Fishelson, L. (1977) Ultrastructure of the epithelium from the ovary wall of *Dendrochirus brachypterus* (Pteroidae, Teleostei). Cell Tissue Res. **177**: 375–381.

Fishelson, L. (1990) *Rhinomuraena* spp. (Pisces: Muraenidae): the first vertebrate genus with post-anally situated urogenital organs. Mar. Biol. **105**: 253–257.

Fishelson, L. (1995a) Unilateral winged flagellum of sperm in *Badis badis* (Pisces: Teleostei). Copeia **1995**: 241–243.

Fishelson, L. (1995b) Comparative morphology and cytology of the olfactory organs in moray eels with remarks on their foraging behavior. Anat. Rec. **243**: 403–412.

Fishelson, L. (1995c) Cytological and morphological ontogenesis and involution of the thymus in cichlid fishes (Cichlidae, Teleostei). J. Morphol. **223**: 175–190.

Fishelson, L., Montgomery, W.L., Myrberg, A.A., Jr. (1985) A unique symbiosis in the gut of tropical herbivorous surgeonfish (Acanthuridae: Teleostei) from the Red Sea. Science **229**: 49–51.

FitzGerald, G.J. (1992) Egg cannibalism by sticklebacks: spite or selfishness? Behav. Ecol. Sociobiol. **30**: 201–206.

FitzGerald, G.J., Caza, N. (1993) Parental investment in an anadromous population of threespine sticklebacks: an experimental study. Evol. Ecol. **7**: 279–286.

FitzGerald, G.J., Guderley, H., Picard, P. (1989) Hidden reproductive costs in the three-spined stickleback (*Gasterosteus aculeatus*). Exp. Biol. **48**: 295–300.

Fleming, I.A., Gross, M.R. (1989) Evolution of adult female life history and morphology in a Pacific salmon (coho: *Oncorhynchus kisutch*). Evolution **43**: 141–157.

Fleming, I.A., Gross, M.R. (1990) Latitudinal clines: a trade-off between egg number and size in Pacific salmon. Ecology **71**: 1–11.

Fleming, J.R., Crockford, T., Altringham, J.D., Johnson, I.A. (1990) Effects of temperature acclimation on muscle relaxation in the carp: a mechanical, biochemical, and ultrastructural study. J. Exp. Zool. **255**: 286–295.

Flik, G. (1990) Hypocalcin physiology. In: Epple, A., Scanes, C.G., Stetson, M.H. (eds.), Progress in comparative endocrinology. Wiley-Liss, New York, pp. 578–585.

Flik, G., Verbost. P.M. (1993) Calcium transport in fish gills and intestine. J. Exp. Biol. **184**: 17–29.

Flik, G., Labedz, T., Lafeber, F.P.J.G., Wendelaar Bonga, S.E., Pang, P.K.T. (1989) Studies on teleost corpuscles of Stannuis: physiological and biochemical aspects of synthesis and release of hypocalcin in trout, goldfish and eel. Fish Physiol. Biochem. **7**: 343–349.

Flik, G., Atema, W., Fenwick, J.C., Rentier-DeLrue, F., Small, J., Wendelaar Bonga, S.E. (1993) Homologous recombimant growth hormone and calcium metabolism in the tilapia, *Oreochromis mossambicus*, adapted in fresh water. J. Exp. Biol. **185**: 107–119.

Flores, J.A., Burns, J.R. (1993) Ultrastructural study of embryonic and early adult germ cells, and their support cells, in both sexes of *Xiphophorus* (Teleostei: Poeciliidae). Cell Tissue Res. **271**: 263–270.

Fontaine, M. (1975) Physiological mechanisms in migration of marine and amphibaline fish. In: Russell, F.S., Yonge, M. (eds.) Advances in marine biology, vol. 13. Academic Press, London, pp. 241–355.

Fontaine, Y.A. (1975) Hormones in fishes. In: Malins, D.C., Sargent, J.R. (eds.) Biochemical and biophysical perspectives in marine biology, vol. 2. Academic Press, London, pp. 139–212.

Fontaine, Y.A. (1985) Hormonal peptide evolution. In: Foreman, R.E., Gorbman, A., Dodd, J.M., Olsson, R. (eds.) Evolutionary biology of primitive fishes. Plenum Publishing Corporation, New York, pp. 413–432.

Fontaine, Y.A. (1989) Les anguilles: migration et reproduction. Oceanis **15**: 197–206.

Fontaine, Y.A. (1994) L'argenture de l'anguille: metamorphose, anticipation, adaptation. Bull. Fr. Peche. Piscic. **335**: 171–185.

Fontaine, Y.A., Leloup-Hatey, J., Dufour, S. (1991) Endocrine control of reproduction: evolutionary considerations. In: Pang, P.K.T., Schreibman, M.P. (eds.) Vertebrate endocrinology: fundamentals and biomedical implications, vol. 4, Part B. Academic Press, San Diego, pp. 311–331.

Fontaine, Y.A., Pisam, M., Moal, C.L., Rambourg, A. (1995) Silvering and gill "mitochondria-rich" cells in the eel, *Anguilla anguilla*. Cell Tissue Res. **281**: 465–471.

Foran, D. (1991) Evidence of luminous bacterial symbionts in the light organs of myctophid and stomiiform fishes. J. Exp. Zool. **259**: 1–8.

Forey, P.L. (1991) *Latimeria chalumnae* and its pedigree. Env. Biol. Fish. **32**: 75–97.

Forey, P., Janvier, P. (1994) Evolution of the early vertebrates. Amer. Scientist **82**: 554–565.

Forster, M.E., Fenwick, J.C. (1994) Stimulation of calcium efflux from the hagfish, *Eptatretus cirrhatus*, gill pouch by an extract of corpuscles of Stannius from an eel (*Anguilla dieffenbachii*): Teleostei. Gen. Comp. Endocrinol. **94**: 92–102.

Fox, G.Q., Kriebel, M.E., Pappas, G.D. (1990) Morphological, physiological and biochemical observations on skate electric organ. Anat. Embryol. **181**: 305–315.

Francis, R.C., Barlow, G.W. (1993) Social control of primary sex differentiation in the Midas cichlid. Proc. Natl. Acad. Sci. USA **90**: 10673–10675.

Frazetta, T.H. (1994) Feeding mechanisms in sharks and other elasmobranchs. In: Bels, V., Chardon, M., Vandewalle, P. (eds.) Advances in comparative and environmental physiology, vol. 18. Springer-Verlag, Berlin, Heidelberg, pp. 31–57.

Fridberg, G., Bern, H.A. (1968) The urophysis and the caudal neurosecretory system of fishes. Biol. Rev. **43**: 175–199.

Fris, M.B., Horn, M.H. (1993) Effects of diets of different protein content on food consumption, gut retention, protein conversion, and growth of *Cebidichthys violaceus* (Girard), an herbivorous fish of temperate zone marine waters. J. Exp. Mar. Biol. Ecol. **166**: 185–202.

Fry, F.E.J. (1967) Responses of vertebrate poikilotherms to temperature. In: Rose, A.H. (ed.) Thermobiology. Academic Press, London, New York, pp. 375–409.

Fry, F.E.J. (1971) The effect of environmental factors on the physiology of fish: In: Hoar, W.S., Randall, D.J. (eds.) Fish physiology, vol 6. Academic Press, New York, pp. 1–98.

Fry, F.E.J., Hochachka, P.W. (1970) Fish. In: Whittow, G.C. (ed.) Comparative physiology of thermoregulation, vol. 1. Academic Press, New York, London, pp. 79–134.

Fujii, R. (1993a) Cytophysiology of fish chromatophores. Int. Rev. Cytol. **143**: 191–255.

Fujii, R. (1993b) Coloration and chromatophores. In: Evans, D.H. (ed.) The physiology of fishes. CRC Press, Boca Raton, pp. 535–562.

Fujii, R., Wakatabi, H., Oshima, N. (1991) Inositol 1, 4, 5-trisphosphate signals the motile response of fish chromatophores-1. Aggregation of pigment in the tilapia melanophore. J. Exp. Zool. **259**: 9–17.

Fujii, R., Tanaka, Y., Hayashi, H. (1993) Endothelin-1 causes aggregation of pigment in teleostean melanophores. Zool. Sci. **10**: 763–772.

Gannon, B.J., Campbell, G.D., Satchell, G.H. (1972) Monoamine storage in relation to cardiac regulation in the Port Jackson shark *Heterodontus portusjacksoni*. Z. Zellforsch. **131**: 437–450.

Gauldie, R.W. (1990) Phase differences between check ring locations in the orange roughly otolith (*Hoplostethus atlanticus*). Can. J. Fish Aquat. Sci. **47**: 760–765.

Gauldie, R.W. (1991a) The morphology and periodic structures of the otolith of chinook salmon (*Oncorhynchus tshawytscha*), and temperature-dependent variation in otolith microscopic growth increment width. Acta Zool. Stockholm. **72**: 159–179.

Gauldie, R.W. (1991b) Taking stock of genetic concepts in fisheries management. Can. J. Fish Aquat. Sci. **48**: 722–731.

Gauldie, R.W., Nelson, D.G.A. (1990) Otolith growth in fishes. Comp. Biochem. Physiol. **97A**: 119–135.

Gauldie, R.W., Daview, N.M., Coote, G., Vickridge, I. (1990) The relationship between organic material and check rings in fish otoliths. Comp. Biochem. Physiol. **97A**: 461–474.

Gauldie, R.W., Coote, G., Mulligan, K.P., West, I.F., Merrett, N.R. (1991) Otoliths of deep water fishes: structure, chemistry and chemically-coded life histories. Comp. Biochem. Physiol. **100A**: 1–31.

Gee, J.M., Graham, J.B. (1978) Respiratory and hydrostatic functions of the intestine of the catfishes *Hoplosternum thoractum* and *Brochis splendens* (Callichthyidae). J. Exp. Biol. **74**: 1–16.

Genten, F., Danguy, A. (1990) A comparative histochemical analysis of glycoconjugates in secretory cells of fish epidermis by use of biotinylated lectins. Z. mikrosk.-anat Forsch **104**: 835–855.

Georgopoulou, U., Verrier, J.M. (1986) Local immunological response in the posterior intestinal segment of the rainbow trout after oral administration of macromolecules. Dev. Comp. Immunol. **10**: 529–537.

Georgopoulou, U., Sire, M.F., Vernier, J.M. (1985) Macromolecular absorption of proteins by epithelial cells of the posterior intestinal segment and their intracellular digestion in the rainbow trout. Ultrastructural and biochemical study. Biol. Cell **53**: 269–282.

Georgopoulou, U., Dabrowski, K., Sire, M.F., Vernier, J.M. (1988) Absorption of intact proteins by the intestinal epithelium of trout, *Salmo gairdneri*. A luminiscence enzyme immunoassay and cytochemical study. Cell Tissue Res. **251**: 145–152.

Gibb, A.C. (1995) Kinematics of prey capture in a flatfish, *Pleuronichthys verticalis*. J. Exp. Biol. **198**: 1173–1183.

Gibb, A.C., Jayne, B.C., Lauder, G.V. (1994) Kinematics of pectoral fin locomotion in the bluegill sunfish *Lepomis macrochirus*. J. Exp. Biol. **189**: 133–161.

Gibson, R.N. (1982) Recent studies on the biology of intertidal fishes. Oceanogr. Mar. Biol. Ann. Rev. **20**: 363–414.

Gibson, R.N. (1986) Intertidal teleosts: life in a fluctuating environment. In: Pitcher, T.J. (ed.) The behaviour of teleost fishes. Croom Helm, London, pp. 388–408.

Gibson, R.N. (1992) Tidally-synchronized behaviour in marine fishes In: Ali, M.A. (ed.) Rhythms in fishes. Plenum Press, New York, pp. 63–81.

Gilbert, C.R. (1993) Evolution and phylogeny. In: Evans, D.H. (ed.) The physiology of fishes. CRC Press, Boca Raton, pp. 1–45.

Giles, N. (1984) Implications of parental care of offspring for the anti-predator behaviour of adult male and female three-spined sticklebacks, *Gasterosteus aculeatus* L. In: Potts, G.W., Wooton, R. (eds.) Fish reproduction: strategies and tactics. Academic Press, London, pp. 275–289.

Giles, N., Wright, R.M., Nord, M.E. (1986) Cannibalism in pike fry, *Esox lucius* L.: some experiments with fry densities. J. Fish Biol. **29**: 107–113.

Gill, T.S. (1994) The pancreatic islets in teleosts: a brief review. In: Singh, H.R. (ed.) Advances in fish biology. Hindustan Publishing Corporation, Delhi, pp. 98–113.

Goda, H., Toyohara, J., Visconti, M.A., Oshima, N., Fujii, R. (1994) The blue coloration of the common surgeonfish, *Parcanthurus hepatus*-1. Morphological features of chromatophores. Zool. Sci. **11**: 527–535.

Gold, J.R. (1979) Cytogenetics. In: Hoar, W.S., Randall, D.J., Brett, J.R. (eds.) Fish physiology, vol. 8. Academic Press, New York, pp. 353–405.

Goldstein, L., Forster, R.P. (1970) Nitrogen metabolism in fishes. In: Campbell, J.W. (ed.) Comparative biochemistry of nitrogen metabolism, vol. 2. Academic Press, London, pp. 494–518.

Gomahr, A., Palzenberger, M., Kotrschal, K. (1992) Density and distribution of external taste buds in cyprinids. Env. Biol. Fish. **33**: 125–134.

Goolish, E.M. (1991) Aerobic and anaerobic scaling in fish. Biol. Rev. **66**: 33–56.

Goossens, N., Dierickx, K., Vandesande, F. (1977) Immunocytochemical localization of vasotocin and isotocin in the preopticohypophysial neurosecretory system of teleosts. Gen. Comp. Endocrinol. **32**: 371–375.

Gordon, M.S. (1970) Hydrostatic pressure. In: Hoar, W.S., Randall, D.J. (eds.) Fish physiology, vol. 4. Academic Press, New York, pp. 445–464.

Gorlick, D. (1989) Motor innervation of respiratory muscles and an opercular display muscle in Siamese fighting fish *Betta splendens*. J. Comp. Neurol. **290**: 412–422.

Gotceitas, V., Colagan, P. (1989) Predator foraging success and habitat complexity: Quantitative test of the threshold hypothesis. Oecologia **80**: 158–166.

Govoni, J.J., Boehlert, G.W., Watanabe, Y. (1986) The physiology of digestion in fish larvae. Env. Biol. Fish. **16**: 59–77.

Graham, J.B. (1983) Heat transfer. In: Webb, P.W., Weihs, D. (ed.) Fish biomechanics. Praeger, New York, pp. 248–279.

Graham, J.B., Kramer, D.L., Pineda, E. (1978) Comparative respiration of an air-breathing and a non-airbreathing characoid fish and the evolution of aerial respiration in characins. Physiol. Zool. **51**: 279–288.

Grande, L., Bemis, W.E. (1991) Osteology and phylogenetic relationships of fossil and recent paddlefishes (Polyodontidae) with comments on the interrelationships of Acipensiformes. J. Vert. Paleont. 11 Suppl. to No. 1. Soc. Vert. Paleont. Memoir **1**: 1–121.

Grant, J.W.A., Noakes, D.L.G. (1986) A test of a size-selective predation model with juvenile brook charr, *Salvelinus fontinalis*. J. Fish Biol. **29** (Suppl. A): 15–23.

Grant, J.W.A., Noakes, D.L.G., Jonas, K.M. (1989) Spatial distribution of defence and foraging in young-of-the year brook charr, *Salvelinus fontinalis*. J. Anim. Ecol. **58**: 773–784.

Grau, E.G. (1987) Thyroid hormones. In: Pang, P.K.T., Schreibman, M.P. (eds.) Vertebrate endocrinology: fundamentals and biomedical implications, vol. 2. Academic Press, San Diego, pp. 85–102.

Grau, E.G. (1988) Environmental influences on thyroid function in teleost fish. Am. Zool. **28**: 329–335.

Green, D.M., Barber, D.L. (1988) The ventral adhesive disc of the clingfish *Gobiesox maeandricus:* integumental structure and adhesive mechanisms. Can. J. Zool. **66**: 1610–1619.

Greenwood, P.H., Miles, R.S., Patterson, C. (eds.) (1973) Interrelationships of fishes. Suppl. No. 1. Zool. Journ. Linn. Soc. Academic Press, New York.

Grier, H.J. (1992) Chordate testis: the extracellular matrix hypothesis. J. Exp. Zool. **261**: 151–160.

Grier, H.J. (1993) Comparative organization of Sertoli cells including the Sertoli cell barrier. In: Russell, L.D., Griswold, M.D. (eds.) The Sertoli cell. Cache River Press, Clearwater, Florida, pp. 704–739.

Grier, H.J., Fishelson, L. (1995) Colloidal sperm-packaging in mouthbrooding tilapine fishes. Copeia **1995**: 966–970.

Grier, H.J., Parenti, L.R. (1994) Reproductive biology and systematics of phallostethid fishes as revealed by gonad structure. Env. Biol. Fish. **41**: 287–299.

Grier, H.J., Burns, J.R., Flores, J.A. (1981) Testis structure in three species of teleosts with tubular gonopodia. Copeia **1981**: 797–801.

Grier, H.J., Taylor, R.G., Reese, R.O. (1987) The mechanism of tubule elongation during testicular recrudescence in the redfish, *Sciaenops ocellatus* (Perciformes). Proc. V Congr. Europ. Ichthyol. Stockholm. 1985, pp. 285–291.

Grier, H.J., Van den Hurk, Billard, R. (1989) Cytological identification of cell types in the testis of *Esox lucius* and *E. niger*. Cell Tissue Res. **257**: 491–496.

Griffith, R.W. (1991) Guppies, toadfish, lungfish, coelacanths and frogs: a scenario for the evolution of urea retention in fishes. Env. Biol. Fish. **32**: 199–218.

Griffith, R.W., Pang, P.K.T. (1979) Mechanisms of osmoregulation in the coelacanth: evolutionary implications. In: McCosker, J.E., Lagois, M.D. (eds.) The biology and physiology of the living coelacanth. Occas. Pap. Calif. Acad. Sci. No. 34, San Francisco, pp. 79–93.

Gross, M.R. (1982) Sneakers, satellites and parentals: polymorphic mating strategies in North American sunfishes. Z. Tierpsychol. **60**: 1–26.

Gross, M.R. (1987) Evolution of diadromy in fishes. Am. Fish Soc. Symp. **1**: 14–25.

Gross, M.R. (1989) Disruptive selection for alternative life histories in salmon. Nature (London) **313**: 47–48.

Gross, M.R. (1991a) Evolution of alternative reproductive strategies: frequency-dependent sexual selection in male bluegill sunfish. Phil. Trans. R. Soc. Lond. B **332**: 59–66.

Gross, M.R. (1991b) Salmon breeding behaviour and life history evolution in changing environments. Ecology **72**: 1180–1186.

Gross, M.R., Sargent, R.C. (1985) The evolution of male and female parental care in fishes. Am. Zool. **25**: 807–822.

Gross, M.R., Coleman, R.M., McDowall, R.M. (1988) Aquatic productivity and the evolution of diadromous fish migration. Science **239**: 1291–1293.

Grove, B.D., Wourms, J.P. (1991) The follicular placenta of the viviparous fish, *Heterandria formosa*. 1. Ultrastructure and development of the embryonic absorptive surface. J. Morphol. **209**: 265–284.

Grove, B.D., Wourms, J.P. (1994) Follicular placenta of the viviparous fish, *Heterandia formosa:* II. Ultrastructure and development of the follicular epithelium. J. Morphol. **220**: 167–184.

Grove, D.J. (1986). Gastro-intestinal physiology: rates of food processing in fish. In: Nilsson, S., Holmgren, S. (eds.) Fish physiology: recent advances. Croom Helm, London, pp. 140–152.

Gruber, M.S., Myers T.R., Marchaterre, M.A., Bass, A.H., Myer, D.A. (1995) Structure, localization and molecular phylogeny of a GnRH cDNA from a paracanthopterygian fish, the plainfin midshipman (*Porichthys notatus*). Gen. Comp. Endocrinol. **99**: 85–99.

Gruber, S.H. (1977) The visual system of sharks: adaptations and capability. Am. Zool. **17**: 453–469.

Guibbolini, M.E., Henderson, I.W., Mosley, W., Lahlow, B. (1988) Arginine vasotocin binding to isolated branchial cells of the eel: effect of salinity. J. Mol. Endocr. **1**: 125–130.

Guraya, S.S. (1993) Follicular (or oocyte) atresia and its causes and its functional significance in the fish ovary. In: Singh, B.R. (ed.) Advances in fish research, vol. 1. Narendra Publishing House, Delhi, pp. 313–332.

Guraya, S.S. (1994) Nucleolar extrusions and their functional significance in the fish oogenesis. In: Singh, H.R. (ed.) Advances in fish biology. Hindustan Publishing Corporation, Delhi, pp. 155–166.

Habibi, H.R., Peter, R.E. (1991) Gonadotropin releasing hormone (GnRH) receptors in teleosts. In: Scott, A.P., Sumpter, J.P., Kime, D.E., Ralfe, M.S. (eds.) Reproductive physiology of fish. Proc. Fourth Int. Symp. Univ. of East Anglia, Norwich, U.K. 7–12 July 1991, pp. 109–113.

Halvorsen, M., Stabell, O.B. (1990) Homing behaviour of displaced stream-dwelling brown trout. Anim. Behav. **39**: 1089–1097.

Halvorsen, M., Arnesen, A.M., Nilssen, K.J., Jobling, M. (1993) Osmoregulatory ability of anadromous Arctic char, *Salvelinus alpinus* (L.), migrating towards the sea. Aquacult. Fisher. Management **24**: 199–211.

Hamazaki, T.S., Nagahama, Y., Iuchi, I., Yamagami, K. (1989) A glycoprotein from the liver constitutes the inner layer of the envelope (zona pellucida interna) of the fish, *Oryzias latipes*. Dev. Biol. **133**: 101–110.

Hamlett, W.C. (1986) Prenatal nutrient absorptive structures in selachians. In: Uyeno, T., Arai, R., Taniuchi, T., Matsuura, K. (eds.) Indo-Pacific Fish Biology: Proceedings of the Second International Conference on Indo-Pacific fishes. pp. 333–344. Ichthyological Society of Japan, Tokyo.

Hamlett, W.C. (1987) Comparative morphology of the elasmobranch placental barrier. Arch. Biol. (Bruxelles) **98**: 135–162.

Hamlett, W.C. (1988) The elasmobranch liver. A model for chemical carcinogenesis studies in a 'naturally resistant' vertebrate. In: Motta, P. (ed.) Biopathology of the liver: an ultrastructural approach. Kluwer Academic Publishers, Lancaster, U.K., pp. 177–189.

Hamlett, W.C. (1989) Evolution and morphogenesis of the placenta in sharks. J. Exp. Zool. Suppl. **2**: 35–52.

Hamlett, W.C. (1990a) Elasmobranch species as models for studies of placental viviparity and its endocrine regulation. J. Exp. Zool. Suppl. **4**: 129–131.

Hamlett, W.C. (1990b) Subcellular structure and function of cement secreting glands in *Pterophyllum scalare*, a transient larval specialization. J. Submicrosc. Cytol. Pathol. **22**: 27–37.

Hamlett, W.C., Stoner, G.D. (1987) Utilization of the elasmobranch as a model in chemical carcinogenesis studies. Arch. Biol. (Bruxelles) **98**: 243–260.

Hamlett, W.C., Wourms, J.P., Hudson, J.S. (1985a) Ultrastructure of the full-term shark yolk sac placenta. I. Morphology and cellular transport at the fetal attachment site. J Ultrastruct Res. **91**: 192–206.

Hamlett, W.C., Wourms, J.P., Hudson, J.S. (1985b) Ultrastructure of the full-term shark yolk sac placenta. II. The smooth, proximal segment. J. Ultrastruct. Res. **91**: 207–220.

Hamlett, W.C., Wourms, J.P., Hudson, J.S. (1985c) Ultrastructure of the full-term shark yolk sac placenta. III. The maternal attachment site. J. Ultrastruct. Res. **91**: 221–231.

Hamlett, W.C., Wourms, J.P., Smith, J.W. (1985d) Stingray placental analogues: structure of trophonemata in *Rhinoptera bonasus*. J. Submicrosc Cytol. **17**: 541–550.

Hamlett, W.C., Allen, D.J., Stribling, M.D., Schwartz, F.J., Didio, L.J.A. (1985e) Permeability of external gill filaments in the embryonic shark. Electron microscopic observations using horseradish peroxidase as a macromolecular tracer. J. Submicros. Cytol. **17**: 31–40.

Hamlett, W.C., Schwartz, F.J., Didio, L.J.A. (1987) Subcellular organization of the yolk syncytial-endoderm complex in the preimplantation yolk sac of the shark, *Rhizoprionodon terraenovae*. Cell Tissue Res. **247**: 275–285.

Hamlett, W.C., Eulitt, A.M., Jarrell, R.L., Kelly, A.M. (1993) Uterogestation and placentation in elasmobranchs. J. Exp. Zool. **266**: 347–367.

Hanke, W., Kloas, W. (1995) Comparative aspects of regulation and function of the adrenal complex in different groups of vertebrates. Horm. metab. Res. **27**: 389–397.

Hansen, A., Zeiske, E. (1993) Development of the olfactory organ in the zebrafish, *Brachydanio rerio*. J. Comp. Neurol. **333**: 289–300.

Hansen, A., Zeiske, E. (1995) Development of the olfactory organ in the zebrafish: an electron microscopic and immunocytochemical study of early differentiation and growth. Biophysics **40**: 149–170.

Hanssen, R.G.J.M., Aarden, E.M., Van der Venne, W.P.H.G., Pang, P.K.T., Wendelaar Bonga, S.E. (1991) Regulation of secretion of the teleost fish hormone stanniocalcin: effects of extracellular calcium. Gen. Comp. Endocrinol. **84**: 155–163.

Hara, T.J. (1971) Chemoreception. In: Hoar, W.S., Randall, D.J. (eds.) Fish physiology, vol. 5. Academic Press, New York, pp. 79–120.

Hara, T.J. (1975) Olfaction in fish. In: Kerkut, G.A., Phillis, J.W. (eds.) Progress in neurobiology. Pergamon Press, Oxford. pp. 271–355.

Hara, T.J. (1992) Mechanism of olfaction. In: Hara, T.J. (ed.) Fish chemoreception. Chapman and Hall, London, New York, pp. 150–170.

Hara, T.J. (1993a) Roles of olfaction in fish behaviour. In: Pitcher, T.J. (ed.) Behaviour of teleost fishes. (Second edition). Chapman and Hall, London, New York, pp. 171–199.

Hara, T.J. (1993b) Chemoreception. In: Evans, D.H. (ed.) The physiology of fishes. CRC Press, Boca Raton, pp. 191–218.

Hara, T.J. (1994a) The diversity of chemical stimulation in fish olfaction and gustation. In: Singh, H.R. (ed.) Advances in fish biology. Hindustan Publishing Corporation, Delhi, pp. 167–184.

Hara, T.J. (1994b) The diversity of chemical stimulation in fish olfaction and gustation. Rev. Fish. Biol. Fish. **4**: 1–35.

Harden Jones, F.R. (1965) Fish migration and water currents. ICNAF Spec. Publ. No. **6**: 257–266.

Harden Jones, F.R. (1977) Performance and behaviour on migration. In: Steele, J.H. (ed.) Fisheries mathematics. Academic Press, London, New York, pp. 145–170.

Harden Jones, F.R. (1981) Fish migration: strategy and tactics. Society for experimental biology seminar series 13. In: Aidley, D.J. (ed.) Animal migration. Cambridge University Press, Cambridge, pp. 139–165.

Harden Jones, F.R., Scholes, P. (1985) Gas secretion and resorption in the swimbladder of the cod *Gadus morhua*. J. Comp. Physiol. **B 155**: 319–331.

Harmelin-Vivien, M.L. (1978). Comparative feeding behaviours of marine and terrestrial vertebrates (fish and mammals) in a tropical environment. In: Reese, E.S., Lighter, F.J. (eds.) Contrasts in behaviour, adaptations in the aquatic and terrestrial environments. John Wiley, New York, pp. 235–248.

Harmelin-Vivien, M.L. (1989) Implications of feeding specialization on the recruitment processes and community structure of butterflyfishes. Env. Biol. Fish. **25**: 101–110.

Harper, D.G., Blake, R.W. (1991) Prey capture and the fast-start performance of nothern pike *Esox lucius*. J. Exp. Biol. **155**: 175–192.

Hart, P.J.B. (1989) Predicting resource utilization: the utility of optimal foraging models. J. Fish Biol. **35** (Suppl. A): 271–277.

Hart, P.J.B., Hamarin, S.F. (1988) Pike as a selective predator, effects of prey size, availability, cover and pike jaw dimensions. Oikos **51**: 220–226.

Hartschuh, W., Weihe, E., Reinecke, M. (1986) The Merkel Cell In: Bereither-Hahn, J., Matoltsy, A.G., Richards, K.S. (eds.) Biology of the integument, vol. 2. Vertebrates. Springer-Verlag, Berlin, pp. 605–620.

Hasler, A.D. (1983) Synthetic chemicals and pheromones in homing salmon. In: Rankin, J.C., Pitcher, T.J., Duggan, R.T. (eds.) Control processes in fish physiology. Croom Helm, London, pp. 103–116.

Hasselt, M.J.F.M., Van (1978) A kinematic model for the jaw movements in some Labrinae (Pisces, Perciformes). Neth. J. Zool. **28**: 545–558.

Hasselt M.J.F.M., Van (1979) A kinematic jaw model for the rapid wide opening and the closing of the mouth in some Labrinae (Pisces, Perciformes). Neth. J. Zool. **29**: 352–401.

Hawkins, A.D. (1986) Underwater sound and fish behaviour. In: Pitcher, T.J. (ed.) The behaviour of teleost fishes. Croom Helm, London, pp. 114–151.

Hawkins, A.D., Horner, K. (1981) Directional characteristics of primary auditory neurons from the cod ear. In: Tavolga, W.N., Popper, A.N., Fay, R.R. (eds.) Hearing and sound communication in fishes. Springer-Verlag, New York, pp. 311–327.

Hawkins, A.D., Myrberg, A.A., Jr. (1983) Hearing and sound communication underwater. In: Lewis, B. (ed.) Bioacoustics: a comparative approach. Academic Press, New York, pp. 347–405.

Hay, M.E. (1991) Fish-seaweed interactions on coral reefs: effects of herbivorous fishes and adaptations to their prey. In: Sale, P.F. (ed.) The ecology of fishes on coral reefs. Academic Press, New York, pp. 96–119.

Hayama, T., Caprio, J. (1990) Somatotopical organization of the intermediate nucleus of the facial lobe in the channel catfish, *Ictalurus punctatus*. Brain Res. **516**: 137–140.

Hayashi, H., Fujii, R. (1993) Muscarinic cholinoceptors that mediate pigment aggregation are present in the melanophores of cyprinids (*Zacco* spp.). Pigment Cell Res. **6**: 37–44.

Hayashi, H., Fujii, R. (1994) Pharmacological profiles of the subtypes muscarinic cholinoceptors that mediate aggregation of pigment in the melanophores of two species of catfish. Pigment Cell Res. **7**: 175–183.

Hayashi, H., Sugimoto, M., Oshima, N., Fujii, R. (1993) Circadian motile activity of erythrophores in the red abdominal skin of tetra fishes and its possible significance in chromatic adaptation. Pigment Cell Res. **6**: 29–36.

Hazel, J.R. (1989) Cold adaptation in ectotherms: regulation of membrane function and cellular metabolism. In: Wang, L.C.H. (ed.) Advances in comparative and environmental physiology, vol. 4. Springer-Verlag, Berlin, pp. 1–50.

Hazel, J.R. (1993) Thermal biology. In: Evans, D.H. (ed.) The physiology of fishes. CRC Press, Boca Raton, pp. 427–467.

Hazel, J.R. (1995) Thermal adaptation in biological membranes: Is homeoviscous adaptation the explanation? Ann. Rev. Physiol. **57**: 19–42.

Hazon, N., Henderson, I.W. (1985) Factors affecting the secretory dynamics of 1 α-hydroxycorticosterone in the dogfish, *Scyliorhinus canicula*. Gen. Comp. Endocrinol. **59**: 50–55.

Hedrick, M.S., Jones, D.R. (1993) The effects of altered aquatic and aerial respiratory gas concentrations on air-breathing patterns in a primitive fish (*Amia calva*). J. Exp. Biol. **181**: 81–84.

Hedrick, M.S., Burleson, M.L., Jones, D.R., Milsom, W.K. (1991) An examination of central chemosensitivity in an air-breathing fish (*Amia calva*). J. Exp. Biol. **155**: 165–174.

Heijmen, P., Denizot, J.P., Peters, R. (1994) Distribution du glutamate dan l'e'lectrorecepteur ampullaire d'un poisson teleosteen. C.R. Acad. Sci. Paris Sciences de la vie/Neurosciences **317**: 743–747.

Heiligenberg, W. (1990) Electrosensory systems in fish. Synapse **6**: 196–206.

Heiligenberg, W. (1991) The jamming avoidance response of the electric fish, *Eigenmannia:* computational rules and their neuronal implementation. Sem. Neurosci. **3**: 3–18.

Heiligenberg, W. (1993) Electrosensation. In: Evans, D.H. (ed.) The physiology of fishes. CRC Press, Boca Raton, pp. 137–160.

Heiligenberg, W. (1994) The detection and generation of electric communication signals in gymnotiform fish. Fortsch. Zool. **39**: 13–24.

Heiligenberg, W., Rose, G. (1985) Phase and amplitude computations in the midbrain of an electric fish: intracellular studies of neurons participating in the jamming avoidance response of *Eigenmannia*. J. Neurosci. **5**: 515–531.

Heisler, N. (1982) Intracellular and extracellular acid-base regulation in the tropical fresh-water teleost fish *Synbranchus marmoratus* in response to the transition from water breathing to air breathing. J. Exp. Biol. **99**: 9–28.

Heisler, N. (1984) Acid-base regulation in fishes. In: Hoar, W.S., Randall, D.J. (eds.) Fish physiology, vol. 10A. Academic Press, New York, pp. 305–401.

Heisler, N. (1986a) Acid-base regulation in fishes. In: Heisler, N. (ed.) Acid-base regulation in animals. Elsevier, Amsterdam, pp. 309–356.

Heisler, N. (1986b) Mechanisms and limitations of fish acid-base regulation. In: Nilsson, S., Holmgren, S. (eds.) Fish physiology: recent advances. Croom Helm, London, pp. 24–49.

Heisler, N. (1989) Acid-base regulation. In: Shuttleworth, T.J. (ed.) Physiology of elasmobranch fishes. Springer-Verlag, Berlin, pp. 215–252.

Heisler, N. (1990) Mechanisms of ammonia elimination in fishes. In: Truchot, J.P., Lahlou, B. (eds.) Animal nutrition and transport processes. 2. Transport, respiration and excretion: Comparative and environmental aspects. Comp. Physiol. **6**: 137–151.

Heisler, N. (1993) Acid-base regulation. In: Evans, D.H. (ed.) The physiology of fishes. CRC Press, Boca Raton, pp. 343–378.

Helfman, G.S. (1986) Fish behavior by day, night and twilight. In: Pitcher, T.J. (ed.) The behavior of teleost fishes. Croom Helm, London, pp. 366–387.

Hemmingsen, E.A. (1991) Respiratory and cardiovascular adaptations in hemoglobin-free fish: resolved and unresolved problems. In: di Prisco, G., Maresca, B., Tota, B. (eds.) Biology of Antarctic fish. Springer-Verlag, Berlin, pp. 191–203.

Henderson, I.W., Hazon, N., Hughes, K. (1985) Hormones, ionic regulation and kidney function in fishes. In: Laverack, M.S. (ed.) Physiological adaptations of marine animals. Soc. Exp. Biol. **2**: pp. 245–265.

Henderson, R.J., Bell, M.V., Park, M.T., Sargent, J.R., Falcón, J. (1994) Lipid composition of the pineal organ from rainbow trout (*Oncorhynchus mykiss*). Lipids **29**: 311–317.

Henderson, V. (1995) Relationship of the tubular system with other organelles in chloride cells of the channel catfish, *Ictalurus punctatus*. J. Morphol. **224**: 31–45.

Hentschel, H. (1977) The kidney of *Spinachia spinachis* (L.) Flem. (Gasterosteidae, Pisces). I. Investigations of juvenile sticklebacks: anatomy, circulation and fine structure. Z mikrosk-anat. Forsch. Leipzig **91**: 4–21.

Hentschel, H. (1979) The kidney of a teleost, *Spinachia spinachis*. II. Histochemical identification sialic-acid-containing glycoprotein and fine structure of mucus secreting cells. Tissue & Cell **11**: 517–531.

Herman, C.A. (1990) Prostaglandins in lower vertebrates. In: Epple, A., Scannes, C.G., Stetson, M.H. (eds.) Progress in comparative endocrinology. Wiley-Liss, New York, pp. 608–613.

Hermann-Erlee, M.P., Flik, G. (1989) Bone: comparative studies on endocrine involvement in bone metabolism. In: Pang, P.K.T., Schreibman, M.P. (eds.) Vertebrate endocrinology: fundamentals and biomedical implications, vol. 3. Regulation of calcium and phosphate. Academic Press, San Diego, pp. 211–242.

Herring, P.J. (ed.) (1978) Bioluminescence in action. Academic Press, London.

Herring, P.J. (1982) Aspects of the bioluminescence of fishes. Oceanogr. Mar. Biol. Ann. Rev. **20**: 415–470.

Herring, P.J. (1990) Bioluminescent communication in the sea. In: Campbell, P.J., Whitfield, A.K., Maddock, L. (eds.) Light and life in the sea. Cambridge University Press, Cambridge, pp. 245–264.

Heubner, E., Chee, G. (1978) Histological and ultrastructural specialization of the digestive tract of the intestinal air breather *Hoplosternum thoractum* (teleost). J. Morphol. **157**: 301–328.

Hickman, C.P., Jr., Trump, B.F. (1969) The kidney. In: Hoar, W.S., Randall, D.J. (eds.) Fish physiology, vol. 1. Academic Press, New York, pp. 91–239.

Hill, K.T., Cailliet, G.M., Radtke, (1989) A comparative analysis of growth zones in four calcified structures of Pacific blue marlin, *Makaira nigricans*. Fish. Bull. U.S. **87**: 829–843.

Hilliard, R.W., Epple, A., Potter, I.C. (1985) The morphology and histology of the endocrine pancreas of the Southern Hemisphere lamprey, *Geotria australis* Gray. J. Morphol. **184**: 253–261.

Hirano, T. (1986) The spectrum of prolactin action in teleosts. In: C.L. Ralph (ed.) Comparative endocrinology: Developments and directions. pp. 53–74. Alan R. Liss, New York.

Hirano, T. (1989) The corpuscles of Stannius. In: Pang, P.K.T., Schreibman, M.P. (eds.) Vertebrate endocrinology: fundamentals and biomedical implications, vol. 3. Regulation of calcium and phosphate. Academic Press, San Diego, pp. 139–170.

Hirano, T., Ogasawara, T., Bolton, J.P., Collie, N.L., Hasegawa, S., Jwata, T. (1987) Osmoregulatory role of prolactin in lower teleosts. In: Kirsch, R., Lahlou, B. (eds.) Comparative physiology of environmental adaptations vol. 1. 8th ESCP Conf. Strasbourg 1986, S. Karger, Basel, pp. 112–124.

Ho, S.M. (1991) Vitellogenesis. In: Pang, P.K.T., Schreibman, M.P. (eds.) Vertebrate endocrinology: fundamentals and biomedical implications, vol. 4A, Reproduction. Academic Press, San Diego, pp. 91–126.

Hoar, W.S. (1969) Reproduction. In: Hoar, W.S., Randall, D.J. (eds.) Fish physiology, vol. 3. Academic Press, New York, pp. 1–72.

Hoar, W.S. (1988) The physiology of smolting salmonids. In: Hoar, W.S., Randall, D.J. (eds.) Fish physiology, vol. 11B. Academic Press, New York, pp. 275–343.

Hobson, E.S. (1973) Diel feeding migrations in tropical reef fishes. Heigol. wiss. Meer. 24: 361–370.

Hobson, E.S. (1974) Feeding relationships of teleostean fishes on coral reefs in Kona, Hawaii. Fish. Bull. U.S. **72**: 915–1031.

Hobson, E.S. (1978) Aggregating as a defense against predators in aquatic and terrestrial environments. In: Reese, E.S., Lighter, F.J. (eds.) Contrasts in behavior. Adaptations in the aquatic and terrestrial environments. John Wiley, New York, pp. 219–234.

Hobson, E.S. (1994) Ecological relations in the evolution of acanthopterygian fishes in warm-temperate communities of the northeastern Pacific. Env. Biol. Fish. **40**: 49–90.

Hofer, R. (1991) Digestion. In: Winfield, I.J., Nelson, J.S. (eds.) Cyprinid fishes. Systematics, biology and exploitation. Chapman and Hall, London, New York, pp. 413–425.

Holder, J.L., Barlow, G.W., Francis, R.C. (1991) Differences in aggressiveness in the Midas cichlid fish (*Cichlasoma citrinellum*) in relation to sex, reproductive state and the individual. Ethology **88**: 297–306.

Hollis, K.L., Overmier, J.B. (1978) The function of the teleost telencephalon in behavior: a reinforcement mediator. In: The behaviour of fish and other aquatic animals. Academic Press, New York, pp. 137–195.

Holmgren, S., Johnson, A.C., Holstein, B. (1986) Gastro-intestinal peptides in fish. In: Nilsson, S., Holmgren, S. (eds.) Fish physiology: recent advances. Croom Helm, London, pp. 119–139.

Holmqvist, B.I. (1993) Chemoarchitecture of visual and hypophysiotrophic systems in the brain of the Atlantic salmon (*Salmo salar*). Doctoral Thesis. University of Lund, Lund, Sweden (based on 6 original papers).

Holmqvist, B.I., Ekstrom, P. (1995) Hypophysiotrophic systems in the brain of the Atlantic salmon. Neuronal innervation of the pituitary and the origin of pituitary dopamine and neuropeptides identified by means of combined carbocyanine tract tracing and immunocytochemistry. J. Chem. Neuroanat. **8**: 125–145.

Holmqvist, B.I., Ostholm, T, Ekstrom, P. (1992) Retinohypothalamic projections and the suprachiasmatic nucleus of the teleost brain. In: Ali, M.A. (ed.) Rhythms in fishes. Plenum Press, New York, pp. 293–318.

Holmqvist, B.I., Ostholm, T., Alm, P., Ekstrom, P. (1994) Neuroanatomical analysis of the visual and hypophysiotrophic systems in Atlantic salmon (*Salmo salar*) with emphasis on possible mediators of photoperiodic cues during parr-smolt transformation. Aquaculture **121**: 1–12.

Hoogenboezem, W., Van den Boogaart, J.G.M. (1993) Importance of mucus in filter-feeding of bream (*Abramis brama*). Can. J Fish Aquat. Sci. **50**: 472–479.

Hoogenboezem, W., Sibbing, F.A., Osse, J.W.M., Van den Boogaart, J.G.M., Lammens, E.H.R.R., Terlou, A. (1990) X-ray measurements of gill-arch movements in filter-feeding bream, *Abramis brama* (Cyprinidae). J. Fish. Biol. **36**: 47–58.

Hoogenboezem, W., Van den Boogaart, J.G.M., Sibbing, F.A., Lammens, E.H.R.R., Terlou, W.A., Osse, J.W.M. (1991) A new model of particle retention and branchial sieve adustment in filter-feeding bream (*Abramis brama*, Cyprinidae). Can. J. Fish. Aquat. Sci. **48**: 7–18.

Hoogenboezem, W., Lammens, E.H.R.R., Vugt Yu, Osse, J.W.M. (1992) A model for switching between particulate-feeding and filter-feeding in the common bream, *Abramis brama*. Env. Biol. Fish. **33**: 13–21.

Hoogenboezem, W., Lammens, E.H.R.R., MacGillary, P.J., Sibbing, F.A. (1993) Prey retention and sieve adjustment in filter-feeding bream (*Abramis brama*) Cyprinidae. Can. J. Fish. Aquat. Sci. **50**: 465–47.

Hopkins C.D. (1983) Functions and mechanisms in electroreception. In: Northcutt, R.G., Davis, R.E. (eds.) Fish neurobiology, vol. 1. Brain stem and sense organs. University of Michigan Press, Ann Arbor, pp. 215–259.

Hopkins, C.D., Comfort, N.C., Bastian, J., Bass, A.H. (1990) Functional analysis of sexual dimorphism in an electric fish, *Hypopomus pinnecaudatus*, order Gymnotiformes. Brain Behav. Evol. **35**: 350–367.

Horn, M.H. (1970) The swimbladder as a juvenile organ in stromateoid fishes. Breviora No. **359**: 1–9.

Horn, M.H. (1975) Swim bladder state and structure in relation to behavior and mode of life in stromateoid fishes. Fish. Bull. U.S. **73**: 95–109.

Horn, M.H., Riegle, K.C. (1981) Evaporative water loss and intertidal vertical distribution in relation to body size and morphology of stichaeoid fishes from California. J. Exp. Mar. Biol. Ecol. **50**: 273–288.

Horn, M.H., Messer, K.S. (1992) Fish guts as chemical reactors: a model of the alimentary canals of marine herbivorous fishes. Mar. Biol. **113**: 527–535.

Horn, M.H., Murray, S.N., Edwards, T.W. (1982) Dietary selectivity in the field and food preferences in the laboratory for two herbivorous fishes (*Cebidichthys violaceus* and *Xiphister mucosus*) from a temperate intertidal zone. Mar. Biol. **67**: 237–246.

Horn, M.H., Murray, S.N., Fris, M.B., Irelan, C.D. (1990) Diurnal feeding periodicity of an herbivorous blennid fish, *Parablennius sanguinolentus*, in the western Mediterranean. In: Barnes, M., Gibson, R.N. (eds.) Trophic relationships in the marine environment. Proc. 24th Europ. Mar. Biol. Symp. Aberdeen University Press, Aberdeen, pp. 170–182.

Horn, M.H., Mailhiot, K.F., Fris, M.B., MyClanahan (1995) Growth, consumption, assimilation and excretion in the marine herbivorous fish *Cebidichthys violaceus* (Girard) fed natural and high protein diets. J. Exp. Mar. Biol. Ecol. **190**: 97–108.

Horner, K., Hawkins, A.D., Fraser, P.J. (1981) Frequency characteristics of primary auditory neurons from the ear of the cod, *Gadus morhua* L. In: Tavolga, W.N., Popper, A.N., Fay, R.R. (eds.) Hearing and sound communication in fishes. Springer-Verlag, New York, pp. 109–137.

Houlihan, D.F., Mathers, E.M., Foster, A. (1993) Biochemical correlates of growth rate in fish. In: Rankin, J.C., Jensen, F.B. (eds.) Fish ecophysiology. Chapman and Hall, London, New York, pp. 45–71.

Hourigan, T.F., Stanton, F.G., Motta, P.J., Kelly, C.D., Carlson, B. (1989) The feeding ecology of three species of Caribbean anglerfishes (family Pomacanthidae). Env. Biol. Fish. **24**: 105–116.

Hourigan, T.F., Tricas, T.C., Reese, E.S. (1988) Coral reef fishes as indicators of environmental stress in coral reefs. In: Soule, D.F., Kleppel, G.S. (eds.) Marine organism as indicators, Springer-Verlag, New York, pp. 107–135.

Howei, A.J., Lote, C.J., Cunningham, A.A., Zaccone, G., Fasulo, S. (1993) Distribution of immunoreactive Tamm-Horsfall protein in various species in the vertebrate classes. Cell Tissue Res. **274**: 15–19.

Hsiao, S.M., Greeley, M.S., Jr., Wallace, R.A. (1994) Reproductive cycling in female *Fundulus heteroclitus*. Biol. Bull. **186**: 271–284.

Hughes, G.M. (1977a) Dimensions and the respiration of lower vertebrates. In: Pedley, T.J. (ed.) Scale effects in animal locomotion. Academic Press, London, pp. 57–81.

Hugle, D.M., Thuringer, P.L., Smith, R.J.F. (1991) The response of the tidepool sculpin, *Oligocottus maculosus*, to chemical stimuli from injured conspecifics, alarm signalling in the Cottidae (Pisces). Ethology **89**: 322–334.

Huque, T., Brand, J.G., Rabinowitz, J.L. (1992) Metabolism of inositol-1, 4, 5-trisphosphate in the taste organ of the channel catfish, *Ictalurus punctatus*. Comp. Biochem. Physiol. **102B**: 833–839.

Huriaux, F., Vandewalle, P., Philippart, J.C., Focant, B. (1990) Electrophoretic comparison of myosin light chains and parvalbumins of trunk and head muscles from two barbel (*Barbus barbus*) populations. Comp. Biochem. Physiol. **97B**: 547–553.

Huysseune, A., Sire, J.Y., Meunier, F.J. (1994) Comparative study of lower pharyngeal jaw structure in two phenotypes of *Astatoreochromis alluaudi* (Teleostei: Cichlidae). J. Morphol. **221**: 25–43.

Hyatt, K.D. (1979) Feeding strategy. In: Hoar, W.S., Randall, D.J., Brett, J.R. (eds.) Fish physiology, vol. 8. Academic Press, New York, pp. 71–119.

Iles, T.D. (1984) Allocation of resources to gonad and soma in Atlantic herring *Clupea harengus* L. In: Potts, G.W., Wootton, R.J. (eds.) Fish reproduction: strategies and tactics. Academic Press, London, pp. 331–347.

Ince, B.W. (1983) Pancreatic control of metabolism. In: Rankin, J.C., Pitcher, T.J., Duggan, R. (eds.) Control processes in fish physiology. Croom Helm, London, 89–102.

Irelan, C.D., Horn, M.H. (1991) Effects of macrophyte secondary chemicals on food choice and digestive efficiency of *Cebidichthys violaceus* (Girard), an herbivorous fish of temperate marine waters. J. Exp. Mar. Biol. Ecol. **153**: 179–194.

Isaia, J. (1984) Water and nonelectrolyte permeation. In: Hoar, W.S., Randall, D.J. (eds.) Fish physiology, vol. 10B. Academic Press, New York, pp. 1–38.

Ishii, S. (1991) Gonadotropins. In: Pang, P.K.T., Shreibman, M.P. (eds.) Vertebrate endocrinology: fundamentals and biomedical implications, vol. 4B. Reproduction. Academic Press, San Diego, pp. 33–66.

Isokawa, S., Kubota, K., Kosakai, T., Satomura, I., Tsubouchi, M., Sera, A., (1965) Some contributions to study of esophageal sacs and teeth of fishes. J. Nihon Univ. Sch. Dent. **7**: 103–111.

Isokawa, S., Satomura, I., Yamaguchi, K., Yoshie, T., Morimoto, M., (1970). Historadiographic observations on the outer dentin in certain osseous fishes. J. Nihon. Univ. Sch. Dent. **12**: 1–5.

Istin, M. (1975) The structure and formation of calcified tissue In: Malins, D.C., Sargent, J.R. (eds.) Biochemical and biophysical perspectives in marine biology, vol. 2. Academic Press, London, pp. 1–68.

Jackson, K., Abraham, M., Degani, G. (1994) Oocyte maturation triggered by the presence of male in the blue gourami, *Trichogaster trichopterus*. J. Morphol. **220**: 1–9.

Jacob, G.H. (1992) Ultraviolet vision in vertebrates. Am. Zool. **32**: 544–554.

Jakubowski, M. (1987) Ultrastructure of the ampullary electroreceptive organs in *Silurus glanis* (Pisces). Z. mikrosk.-anat. Forsch. Leipzig **101**: 461–472.

Jakubowski, M. (1989) Skin vascularization in fishes compared with that in amphibians. In: Splechtna, H., Hilgers (eds.) Trends in vertebrates morphology. Progress in Zoology, vol. 35. Gustav Fischer, Stuttgart, pp. 542–545.

Jakubowski, M., Whitear, M. (1986) Ultrastructure of taste buds in fishes. Fol. Histochem. Cytobiol. **24**: 310–311.

Jakubowski, M., Whitear, M. (1990) Comparative morphology and cytology of taste buds in teleosts. Z mikrosk-anat. Forsch. Leipzig **104**: 529–560.

Jamieson, I.G., Colgan, P.W. (1989) Eggs in the nests of males and their effect on mate choice in the three-spined stickleback. Anim. Behav. **38**: 859–865.

Janvier, P. (1993) Patterns of diversity in the skull of jawless fishes. In: Hanken, J., Hall, B.K. (eds.) The skull, vol. 2. Patterns of structural and systematic diversity. University of Chicago Press, Chicago, pp. 131–188.

Jayne, B.C., Lauder, G.V. (1993) Red and white muscle activity and kinematics of the escape response of the bluegill sunfish during swimming. J. Comp. Physiol. **A 173**: 495–508.

Jayne, B.C., Lauder, G.V. (1994a) How swimming fish use slow and fast muscle fibres: implications for models of vertebrate muscle requirement. J. Comp. Physiol. **A 175**: 123–131.

Jayne, B.C., Lauder, G.V. (1994b) Comparative morphology of the myomeres and axial skeleton in four genera of centrarchid fishes. J. Morphol. **220**: 185–205.

Jenkins, D.B. (1979) A transmission and scanning electron microscopic study of the saccule in five species of catfishes. Am. J. Anat. **154**: 81–102.

Jensen, F.B., Nikinmaa, M., Weber, R.E. (1983) Effects of exercise stress on acid-base balance and respiratory function of the teleost *Tinca tinca*. Resp. Physiol. **51**: 291–301.

Jensen, F.B., Nikinmaa, M., Weber, R.E. (1993) Environmental perturbations of oxygen transport in teleost fishes: causes, consequences and compensations. In: Rankin, J.C., Jensen, F.B. (eds.) Fish ecophysiology. Chapman and Hall, London, New York, pp. 161–179.

Jeuniaux, C. (1993) Chitinolytic systems in the digestive tract of vertebrates. In: Muzzarelli, R.A.A. (ed.) Chitin enzymology. Eur. Chitin. Soc. Ancona, pp. 233–244.

Jobling, M. (1981). The influences of feeding on the metabolic rate of fishes: a short review. J. Fish Biol. **18**: 385–400.

Jobling, M. (1983a) Growth studies with fish-overcoming the problems of size variation. J. Fish Biol. **22**: 153–157.

Jobling, M. (1983b) A short review and critique of methodology used in fish growth and nutrition studies. J. Fish Biol. **23**: 685–703.

Jobling, M. (1986) Mythical models of gastric emptying and implications for food consumption studies. Env. Biol. Fish. **16**: 35–50.

Jobling, M. (1993) Bioenergetics: feed intake and energy partitioning. In: Rankin, J.C., Jensen, F.B. (eds.) Fish ecophysiology. Chapman and Hall, London, New York, pp. 1–44.

Jobling, M. (1995) Feeding of charr in relation to aquaculture. Nordiac. J. Freshw. Res. **71**: 102–112.

Jobling, M. (1994) Fish bioenergetics. Chapman and Hall, London, New York.

Jobling, M., Daview, P.S. (1980) Effects of feeding on metabolic rate, and the specific dynamic action in plaice, *Pleuronectes platessa* L.J. Fish. Biol. **16**: 629–638.

Jobling, M., Christiansen, J.S., Jorgensen, E.H., Arnesen, A.M. (1993a) The application of X-radiography in feeding and growth studies with fish: a summary of experiments conducted on Arctic charr. Rev. Fish. Sci. **1**: 223–237.

Jobling, M., Baardvik, B.M., Christiansen, J.S., Jorgensen, E.H. (1993b) The effects of prolonged exercise training on growth performance and production parameters in fish. Aquacult. Internat. **1**: 95–111.

Jobling, M., Jorgensen, E.H., Arnesen, A.M., Ringo, E. (1993c) Feeding, growth and environmental requirements of Arctic charr: a review of aquaculture potential. Aquacult. Internat. **1**: 20–46.

Jobling, M., Meloy, O.H., dos Santos, J., Christiansen B. (1994). The compensatory growth response of the Atlantic cod: effects of nutritional history. Aquacult. Internat. **2**: 75–90.

Jobling, M., Johnsen, H.K., Pettersen, G.W., Henderson R.J. (1995). Effect of temperature on reproductive development in Arctic charr, *Salvelinus alpinus* (L.). J. Therm. Biol. **20**: 157–165.

Johansen, K., Burggren, W. (1980) Cardiovascular function in the lower vertebrates. In: Bourne, G.H. (ed.) Hearts and heart-like organs, vol. 1. Academic Press, New York, pp. 61–117.

Johansen, K., Hanson, D. (1968) Functional anatomy of the hearts of lungfishes and amphibians. Am. Zool. **8**: 191–210.

Johnston, I.A. (1981) Structure and function of fish muscles. In: Day, M.H. (ed.) Vertebrate locomotion. Academic Press, London, Symp. Zool. Soc. Lond. No. **48**: 71–113.

Johnston, I.A. (1983a) Dynamic properties of fish muscle. In: Webb, P.W., Weihs, D. (eds.) Fish biomechanics. Praeger, New York, pp. 36–67.

Johnston, I.A. (1983b) On the design of fish myotomal muscles. Mar. Behav. Physiol. **9**: 83–98.

Johnston, I.A. (1989) Antarctic fish muscles-Structure, function and physiology. Antarct. Sci. **1**: 97–108.

Johnston, I.A., Altringham, J.D. (1991) Movement in water: constraints and adaptations. In: Hochachka, P.W., Mommsen, T.P. (eds.) Biochemistry and molecular biology of fishes, vol. 1. Elsevier, Amsterdam, pp. 249–268.

Johnston, I.A., Dunn, J. (1987) Temperature acclimation and metabolism in ectotherms with particular reference to teleost fish. In: Bowler, K., Fuller, B.J. (eds.) Temperature and animal cells. Soc. Exp. Biol. Symp. **4**: 67–93.

Johnson, T.P., Johnson, I.A. (1991) Temperature adaptation and the contractile properties of live muscle fibres from teleost fish. J. Comp. Physiol. B **61**: 27–36.

Johnson, I.A., Camm, J.P., White, M. (1988) Specialisations of swimming muscles in the pelagic antarctic fish *Pleuragramma antarcticum*. Mar. Biol. **100**: 3–12.

Jones, D.R. (1982) Anaerobic exercise in teleost fish. Can. J. Zool. **60**: 1131–1134.

Jones. D.R., Randall. D.J. (1978) The respiratory and circulatory systems during exercise. In: Hoar, W.S., Randall, D.J. (eds.) Fish physiology, vol. 7. Academic Press, New York, pp. 425–501.

Jones, D.R., Brill, R.W., Bushnell, P.G. (1993) Ventricular and arterial dynamics of anaesthetised and swimming tuna. J. Exp. Biol. **182**: 97–112.

Jones, R.E., Baxter, D.C. (1991) Gestation, with emphasis on corpus luteum biology, placentation, and parturition. In: Pang, P.K.T., Schreibman, M.P. (eds.) Vertebrate endocrinology: fundamentals and biomedical implications, vol. 4, Part A. Reproduction. Academic Press, San Diego, pp. 205–302.

Joosten, P.H.M., Aviles-Trigueros M., Sorgeloos, P., Rombout, J.H.W.M. (1995) Oral vaccination of juvenile carp (*Cyprinus carpio*) and gilthead seabream (*Sparus aurata*) with bioencapsulated *Vibrio anguillarum* bacterin. Fish Shellfish Immunol. **5**: 289–299.

Jorgensen, E.H., Jobling, M. (1992) Feeding behaviour and effect of feeding regime on growth of Atlantic salmon, *Salmo salar*. Aquaculture **101**: 135–146.

Jorgensen, E.H., Jobling, M. (1993a) The effects of exercise on growth, food utilisation and osmoregulatory capacity of juvenile Atlantic salmon, *Salmo salar*. Aquaculture **116**: 233–246.

Jorgensen, E.H., Jobling, M. (1993b) Feeding in darkness eliminates density-dependent growth suppression in Arctic charr. Aquacult. Internat. **1**: 90–93.

Jorgensen, E.H., Jobling, M. (1994) Feeding and growth of exercised and unexercised juvenile Atlantic salmon in freshwater, and performance after transfer to seawater. Aquacult. Internat. **2**: 154–164.

Jorgensen, J.M. (1989) Evolution of octavolateralis sensory cells. In: Coombs, S., Görner, P., Münz, H. (eds.) The mechanosensory lateral line: neurobiology and evolution. Springer-Verlag, New York, pp. 115–145.

Jorgensen, J.M., Bullock, T.H. (1987) Organization of the ampullary organs of the African knifefish *Xenomystus nigri* (Teleostei: Notopteridae). J. Neurocytol. **16**: 311–315.

Jurss, K., Bastrop, R. (1995) The function of mitochondria-rich cells (chloride cells) in teleost gills. Rev. Fish Biol. Fish. **5**: 235–255.

Kalinoski, D.L., Huque, T., LaMorte, V.J., Brand, J.G. (1989a) Second-messenger events in taste. In: Brand, J.G., Teeter, J.H., Cagan, R.H., Kare, M.R. (eds.) Chemical Senses, vol. 1. Receptor events and transduction in taste and olfaction. Marcel Dekker, New York, Basel, pp. 85–101.

Kalinoski, D.L., Bryant, B.P., Shaulsky, G., Brand, J.G., Harpaz, S. (1989b) Specific L-arginine taste receptor sites in the catfish, *Ictalurus punctatus:* biochemical and neurophysiological characterization. Brain Res. **488**: 163–173.

Kalinoski, D.L., Johnson, L.C., Bryant, B.P., Brand, J.G. (1992) Selective Interactions of lectins with amino acid taste receptor sites in the channel catfish. Chem. Senses **17**: 381–390.

Kallman, K.D. (1984) A new look at sex determination in poeciliid fishes. In: Turner, B.J. (ed.) Evolutionary genetics of fishes. Plenum Publishing Corporation, New York, pp. 95–171.

Kalmijn, A.J. (1988) Hydrodynamic and acoustic field detection. In: Atema, J., Fay, R.R., Popper, A.N., Tavolga, W.N. (eds.) Sensory biology of aquatic animals. Springer-Verlag, New York, pp. 83–130.

Kalmijn, A.J. (1989) Functional evolution of lateral line and inner ear sensory systems. In: Coombs, S., Görner, P., Münz, H. (eds.) The mechanosensory lateral line: neurobiology and evolution. Springer-Verlag, New York, pp. 187–215.

Kampmeier, O.F. (1969) Evolution and comparative morphology of the lymphatic system. Charles C. Thomas, Springfield, Illinois.

Kaneko, T., Harvey, S., Kline, L.W., Pang, P.KT. (1989) Localization of calcium regulatory hormones in fish. Fish Physiol. Biochem. **7**: 337–342.

Kang, J., Caprio, J. (1991) Electro-olfactogram and multiunit olfactory receptor responses to complex mixtures of amino acids in the channel catfish, *Ictalurus punctatus*. J. Gen. Physiol. **98**: 699–721.

Kang, J., Caprio, J. (1995) In vivo responses of single olfactory receptor neurons in the channel catfish, *Ictalurus punctatus*. J. Neurophysiol. **73**: 172–177.

Kanwal, J.S., Caprio, J. (1983) An electrophysiological investigation of the oropharyngeal (IX–X) taste system in the channel catfish, *Ictalurus punctatus*. J. Comp. Physiol. **150**: 345–357.

Kanwal, J.S., Caprio, J. (1987) Central projections of the glossopharyngeal and vagal nerves in the channel catfish, *Ictalurus punctatus:* clues to differential processing of viscera inputs. J. Comp. Neurol. **264**: 216–230.

Kanwal, J.S., Hidaka, I., Caprio, J. (1987) Taste responses to amino acids from facial nerve branches innervating oral and extraoral taste buds in the channel catfish, *Ictalurus punctatus*. Brain Res. **406**: 105–112.

Kanwal, J.S., Finger, T.E., Caprio, J. (1988) Forebrain connections of the gustatory system in ictalurid catfishes. J. Comp. Neurol. **278**: 353–376.

Kanwal, J.S., Finger, T.E. (1992) Central representation and projections of gustatory systems. In: Hara, T.J. (ed.) Fish chemoreception. Chapman and Hall, London, New York, pp. 79–102.

Kapoor, B.G., Khanna, B. (1993) The potential spectrum of the gut in teleost fishes. In: Singh, B.R. (ed.) Advances in fish research, vol. 1. Narendra Publishing House, Delhi, pp. 221–266.

Kapoor, B.G., Khanna, B. (1994) The alimentary canal of teleosts: a brief survey of structure and function. In: Singh, H.R. (ed.) Advances in fish biology, Hindustan Publishing Corporation, Delhi, pp. 12–24.

Kapoor, B.G., Evans, H.E., Pevzner, R.A. (1975a) The gustatory system in fish. In: Russell, F.S., Yonge, M. (eds.) Advances in marine biology, vol. 13. Academic Press, London, pp. 53–108.

Kapoor, B.G., Smit, H., Verighina, I.A. (1975b) The alimentary canal and digestion in teleosts. In: Russell, F.S., Yonge, M. (eds.) Advances in marine biology, vol. 13. Academic Press, London, pp. 109–239.

Karnaky, K.J., Jr., Valentich, J.D., Currie, M.G., Oehlenscj; Ager, W.F., Kennedy, M.P. (1991) Atriopeptin stimulates chloride secretion in cultured shark rectal gland cells. Am. J. Physiol. **260**: C 1125–C1130.

Kasapi, M.A., Domenici, P., Blake, R.W., Harper, D. (1993) The kinematics and performance of escape responses of the knifefish *Xenomystis nigri*. Can. J. Zool. **71**: 189–195.

Katsuki, Y., Onada, N. (1973) The lateral-line organ of fish as a chemoreceptor. In: Chavin, W. (ed.) Response of fish to environmental changes. Charles C. Thomas, Springfield, Illinois, pp. 389–411.

Kawasaki, M. (1993) Independently evolved jamming avoidance responses employ identical computational algorithms: a behavioural study of the African electric fish, *Gymnarchus niloticus*. J. Comp. Physiol. **A 173**: 9–22.

Kawauchi, H., Itoh, H., Koide, Y. (1991) Additional evidence for duality of fish gonadotropins. In: Scott, A.P., Sumpter, J.P., Kime, D.E., Rolfe, M.S. (eds.) Reproductive physiology of fish. Proc. Fourth Int. Symp. Univ. of East Anglia, Norwich, U.K. 7–12 July 1991, pp. 19–21.

Keenleyside, M.H.A. (1972) The behaviour of *Abudefduf zonatus* (Pisces, Pomacentridae) at Heron Island, Great Barrier Reef. Anim. Behav. **20**: 763–774.

Keenleyside, M.H.A. (1978) Parental care behaviour in fishes and birds. In: Reese, E.S., Lighter, F.J. (eds.) Contrasts in behaviour. Adaptations in the aquatic and terrestrial environments, John Wiley, New York, pp. 3–29.

Keller, C.H., Zakon, H.H., Sanchez, D.Y. (1986) Evidence for a direct effect of androgens upon electroreceptor tuning. J. Comp. Physiol. A **158**: 301–310.

Kemp, N.E. (1984) Organic matrices and mineral crystallites in vertebrate scales, teeth and skeletons. Am. Zool. **24**: 965–976.

Kendall, M.W., Dale, J.E. (1979) Scanning and transmission electron microscopic observations of rainbow trout (*Salmo gairdneri*) gill. J. Fish. Res. Board Can. **36**: 1072–1079.

Kennedy, G., Heiligenberg, W. (1994) Ultrastructural evidence of GABA-ergic inhibition and glutamatergic excitation in the pacemaker nucleus of the gymnotiform electric fish, *Hypopomus*. J. Comp. Physiol. **A 174**: 267–280.

Kershaw, D.R. (1976) A structural and functional interpretation of the cranial anatomy in relation to the feeding of osteoglossoid fishes and a consideration of their phylogeny. Trans. Zool. Soc. Lond. **33**: 173–252.

Kilarski, W., Kozlowska, M. (1987) Comparison of ultrastructural and morphometrical analysis of tonic, white and red muscle fibers in the myotome of teleost fish (*Noemacheilus barbatulus* L.). Z mikrosk-anat. Forsch. Leipzig **101**: 636–648.

Kime, D.E., (1991) Ovarian progestogens in cyprinid fish. In: Scott, A.P., Sumpter, J.P., Kime, D.E., Rolfe, M.S. (eds.) Reproductive physiology of fish. Proc. Fourth Int. Symp. Univ. of East Anglia, Norwich, U.K., 7–12 July 1991, pp 77–79.

Kime, D.E. (1993) 'Classical' and 'non-classical' reproductive steroids in fish. Rev. Fish. Biol. Fish. **3**: 28–48.

Kime, D.E. (1995) The effects of pollution on reproduction in fish. Rev. Fish. Biol. Fish. **5**: 52–96.

Kime, D.E., Abdullah, M.A.S., Sokolowska-Mikolajczyk, M., Epler, P. (1994) Substrate concentration affects *in vitro* of 17-hydroxyprogesterone by ovaries of the carp, *Cyprinus carpio*. Fish Physiol. Biochem. **13**: 317–324.

Kindler. P.M., Philipp, D.P., Gross, M.R., Bahr, J.M. (1989) Serum 11-ketosterone and testosterone concentrations associated with reproduction in male bluegill (*Lepomis macrochirus:* Centrarchidae). Gen. Comp. Endocrinol. **75**: 446–453.

Kindler, P.M., Bahr, J.M., Gross, M.R., Philipp, D.P. (1991) Hormonal regulation of parental care behaviour in nesting male bluegills: Do the effects of bromocriptine suggest a role for prolactin? Physiol. Zool. **64**: 310–322.

King, J.A., Millar, R.P. (1991) Gonadotropin-releasing hormones. In: Pang, P.K.T., Schreibman, M.P. (eds) Vertebrate endocrinology: fundamentals and biomedical implications, vol. 4B. Reproduction. Academic Press, San Diego, pp. 1–31.

King, P.A., Goldstein, L. (1985) Renal excretion of nitrogenous compounds in vertebrates. Renal Physiol. **8**: 261–278.

Kinne, O. (1984) Pisces. In: Kinne, O. (ed.) Diseases of marine animals, Vol. 4. Biologische Anstalt Helgoland, Hamburg, F.R.G., pp. 1–16.

Kloas, W., Reinecke, M., Hanke, W. (1994) Role of the atrial natriuretic peptide for adrenal regulation in the teleost fish *Cyprinus carpio*. Am. J. Physiol. **267**: R 1034–R 1042.

Kohbara, J., Michel, W., Caprio, J. (1992) Responses of single facial taste fibres in the channel catfish, *Ictalurus punctatus*, to amino acids. J. Neurophysiol. **68**: 1012–1026.

Komnick, H. (1986) Chloride cells and salt glands. In: Bereiter-Hahn, J., Matoltsy, A.G., Richards, K.S. (eds) Biology of the integument, vol. 2. Vertebrates. Springer-Verlag, Berlin, pp. 499–516.

Korf, H.W., Oksche, A. (1986) The pineal organ. In: Pang, P.K.T., Schreibman, M.P. (eds.) Vertebrate endocrinology: fundamentals and biomedical implications, vol. 1. Morphological considerations. Academic Press, Orlando, pp. 105–145.

Kormanik, G.A. (1992) Ion and osmoregulation in prenatal elasmobranchs: evolutionary implications. Am. Zool. **32**: 294–302.

Kornfield, I. (1984) Descriptive genetics of cichlid fishes. In: Turner, B.J. (ed.) Evolutionary genetics of fishes. Plenum Publishing Corporation, New York, pp. 591–616.

Korsgaard, B. (1994a) Calcium metabolism in relation to ovarian functions during early and late pregnancy in the viviparous blenny *Zoarces viviparus*. J. Fish Biol. **44**: 661–672.

Korsgaard, B. (1994b) Nitrogen distribution and excretion during embryonic post-yolk sac development in *Zoarces viviparous*. J. Comp. Physiol. B **164**: 42–46.

Korsgaard, B., Weber, R.E. (1989) Maternal-fetal trophic and respiratory relationships in viviparous ectothermic vertebrates. Advances in comparative and environment physiology, vol. 5, Springer-Verlag, Berlin, pp. 209–233.

Kotrschal, K. (1991) Solitary chemosensory cells-taste, common chemical sense or what? Rev. Fish. Biol. Fish. **1**: 3–22.

Kotrschal, K. (1992) Quantitative scanning electron microscopy of solitary chemoreceptor cells in cyprinids and other teleosts. Env. Biol. Fish. **35**: 273–282.

Kotrschal, K. (1995) Ecomorphology of solitary chemosensory cell systems in fish: a review. Env. Biol. Fish. **44**: 143–155.

Kortschal, K. (1996) Solitary chemosensory cells: why do primary aquatic vertebrates need another taste system? Trends Ecol. Evol. **11**: 110–114.

Kotrschal, K. Essler, H. (1995) Goals and approach in the analysis of locomotion in fish, with a focus on laboratory studies. Rev. Fish. Sci. **3**: 171–200.

Kotrschal, K., Whitear, M., Adam, H. (1984) Morphology and histology of the anterior dorsal fin of *Gaidrosparus mediterraneus* (Pisces Teleostei), a specialized sensory organ. Zoomorphology **104**: 365–372.

Kotrschal, K., Peters, R., Atema, J. (1989) A novel chemosensory system in fish: do rocklings (*Ciliata mustela*, Gadidae) use their solitary chemoreceptor cells as fish detectors? Biol. Bull. **117**: 328.

Kotrschal, K., Kinnamon, J.C., Royer, S.M. (1990a) High voltage electron microscopy and 3-D reconstruction of solitary chemosensory cells and Di-I labelling of primary afferent nerves. Proc. XIIth Int. Congr. Elec. Microsc, San Francisco Press, San Francisco, pp. 412–413.

Kotrschal, K., Adam, H., Brandstatter, R., Junger, H., Zaunreiter M., Goldschmid, A (1990b) Larval size constraints determine directional ontogenetic shifts in the visual system of teleosts. Z. Zool. Syst. Evol. Forsch. **28**: 166–182.

Kotrschal, K., Brandstatter, R., Gomahr, A., Junger, H., Patzenberger, M., Zaunreiter. M. (1992) Brain and sensory systems. In: Winfield, I.J., Richards, K.S. (eds.). Cyprinid, fishes. Systematics, biology and exploitation. Chapman and Hall, London, New York, pp. 284–331.

Kotrschal, K., Whitear, M., Finger, T.E. (1993) Spinal and facial innervation of the skin in the gadid fish *Ciliata mustela* (Teleostei). J. Comp. Neurol. **331**: 407–417,

Koumans, J.T.M., Akster, H.A., Osse, J.W.M. (1990) Myosattelite cells of *Cyprinus carpio* (Teleostei) in vitro: isolation, recognition and differentiation. Cell Tissue Res. **261**: 173–181.

Koumans, J.T.M., Akster H.A., Booms, G.H.R., Lemmens, C.J.J., Osse, J.W.M. (1991) Numbers of myosatellite cells in white axial muscle of growing fish: *Cyprinus carpio* L. (Teleostei). Am. J. Anat. **192**: 418–424.

Koumans, J.T.M., Akster, H.A., Booms, G.H.R., Osse, J.W.M. (1993a) Influence of fish size on proliferation and differentiation of cultured myosatellite cells of white axial muscle of carp (*Cyprinus carpio* L.). Differentiation **53**: 1–6.

Koumans, J.T.M., Akster, H.A., Booms, G.H.R., Osse, J.W.M. (1993b) Growth of carp (*Cyprinus carpio*) white axial muscle; hyperplasia and hypertrophy in relation to myonucleus/sarcoplasm ratio and the occurrence of different subclasses of myogenic cells. J. Fish Biol. **43**: 69–80.

Koumans-van Diepen, J.C.E., Van de Lisdonk, M.H.M., Taverne-Thiele, A.J., Verburg-van Kemenade, B. M,L., Rombout, J.H.W.M. (1994a) Characterisation of immunoglobulin-binding leucocytes in carp (*Cyprinus carpio* L.). Dev. Comp. Immunol. **18**: 45–56.

Koumans-van Diepen, J.C.E., Taverne-Thiele, J.J., Van Rens, B.T.T.M., Rombout, J.H.W.M. (1994b) Immunocytochemical and flow cytometric analysis of B cells and plasma cells in carp (*Cyprinus carpio* L.); an ontogenetic study. Fish Shellfish Immunol. **4**: 19–28.

Koya, Y., Munehara, H,, Takano, K., Takahashi, H. (1993a) Effects of extracellular environments on the motility of spermatozoa in several marine sculpins with internal gametic association. Comp. Biochem. Physiol. **106A**: 25–29.

Koya, Y., Takano, K., Takahashi, H. (1993b) Ultrastructural observations on sperm penetration in the egg of elkhorn sculpin, *Alcichthys alcicornis*, showing internal gametic association. Zool. Sci. **10**: 93–101.

Koya, Y., Munehara, H., Takano, K. (1993c) Secretion abilities of epithelia of ovarian wall and ovigerous lamella lining the ovarian cavity of masked greeling, *Hexagrammos octogrammus*. Japan. J. Ichthyol. **40**: 199–208.

Koya, Y., Matsubara, T., Nakagawa, T., (1994a) Efficient artificial fertilization method based on the ovulation cycle in barfin flounder *Verasper moseri*. Fish. Sci. **60**: 537–540.

Koya, Y., Ikeuchi T., Adachi S, Yamauchi, K. (1994b) Embryonic growth during gestation in the viviparous eel pout, *Zoarces elongatus*. Japan. J. Ichthyol. **41**: 338–342.

Koya, Y., Takano, K., Takashashi, H. (1995a) Annual changes in fine structure of inner epithelial lining of the ovary of a marine sculpin, *Alcichthys alcicornis* (Teleostei: Scorpaeniformes), with internal gametic association. J. Morphol. **223**: 85–97.

Koya, Y., Munehara, H., Takano, K. (1995b) Formation of egg adhesive material in masked greenling, *Hexagrammos octogrammus*. Japan. J. Ichthyol. **42**: 45–52.

Kraak, S.B.M., Videler, J.J. (1991) Mate choice in *Aidablennius sphynx* (Teleostei, Blenniidae); females prefer nests containing more eggs. Behaviour **119**: 243–266.

Kramer, B. (1990a) Electrocommunication in teleost fishes. Behavior and experiments. In: Bradshaw, S.D., Burggren, W., Heller, H.C., Ishii, S., Langer, G., Neuweiler, G., Randall, D.J. (eds.) Zoophysiology, vol. 29. Springer-Verlag, Berlin.

Kramer, B. (1990b). Sexual signals in electric fishes. Trends Ecol. Evol. **5**: 247–250.

Kramer, B., Kuhn, B. (1994) Species recognition by the sequence of discharge intervals in weakly electric fishes of the genus *Campylomormyrus* (Mormyridae, Teleostei). Anim. Behav. **48**: 435–445.

Krause, J. (1992) Ideal free distribution and the mechanism of patch profitability assessment in three-spined sticklebacks (*Gasterosteus aculeatus*). Behaviour **123**: 27–37.

Krause, J. (1993a) The relationship between foraging and shoal position in a mixed shoal of roach (*Rutilus rutilus*) and chub (*Leuciscus cephalus*): a field study. Oecologia **93**: 356–359.

Krause, J. (1993b) The effect of 'Schreckstoff' on the shoaling behaviour of the minnow: a test of Hamilton's selfish herd theory. Anim. Behav. **45**: 1019–1024.

Krause, J. (1993c) The influence of hunger on shoal size choice by three-spined sticklebacks, *Gasterosteus aculeatus*. J. Fish. Biol. **43**: 775–780.

Krause, J. (1993d) Positioning behaviour of fish shoals: a cost-benefit analysis. J. Fish Biol. **43** (Suppl. A): 309–314.

Krause, J. (1994a) Differential fitness returns in relation to spatial position in groups. Biol. Rev. **69**: 187–206.

Krause. J. (1994b) The influence of food competition and predation risk on size-assortative shoaling in juvenile chub (*Leuciscus cephalus*). Ethology **96**: 105–116.

Krause, J., Bumann, D., Todt, D. (1992) Relationship between the position preference and nutritional state of individuals in schools of juvenile roach (*Rutilus rutilus*). Behav. Ecol. Sociobiol. **30**: 177–180.

Kroese, A.B.A., Netten, S.M., Van (1989) Sensory transduction in lateral line hair cells. In: Coombs, S., Görner, P., Münz, H. (eds.) The mechanosensory lateral line: neurobiology and evolution Springer-Verlag, New York, pp. 265–284.

Kulczykowska, E. (1995a) Arginine vasotocin melatonin interactions in fish: a hypothesis. Rev. Fish Biol. Fish. **5**: 96–102.

Kundu, R., Mansuri, A.P. (1990) Growth dynamics of myotomal muscle fibres in a carangid, *Caranx malabaricus* (Cuv. et Val.). J. Fish Biol. **36**: 21–27.

Kundu, R., Mansuri, A.P. (1992) Growth of pectoral muscle fibres in relation to somatic growth in some marine fishes. Neth. J. Zool. **42**: 595–606.

Kundu, R., Lakshmi, R., Mansuri, A.P. (1990) Growth dynamics of caudal and pectoral fin muscle fibres in a carangid, *Caranx malabaricus* (Cuv. and Val.), and their possible relation with somatic growth. J. Fish Biol. **37**: 845–852.

Kultz, D., Jurss, K., Jonas, L. (1995) Cellular and epithelial adjustments to altered salinity in the gill and opercular epithelium of a cichlid fish (*Oreochromis mossambicus*). Cell Tissue Res. **279**: 65–73.

Kuwatani, Y., Kafuku, T. (1978) Morphology and function of epibranchial organ studied and inferred on milkfish. Bull. Freshwater Fish Res. Lab. **28**: 221–236.

Lachance, S., FitzGerald, G.J. (1992) Parental care tactics of three-spined sticklebacks living in a harsh environment. Behav. Ecol. **3**: 360–366.

Lacy, E.R., Reale, E. (1985a) The elasmobranch kidney. I. Gross anatomy and general distribution of the nephrons. Anat. Embryol. **173**: 23–34.

Lacy, E.R., Reale, E. (1985b) The elasmobranch kidney. II. Sequence and structure of the nephrons. Anat. Embryol. **173**: 163–186.

Lacy, E.R., Reale, E. (1989) Granulated peripolar epithelial cells in the renal corpuscle of marine elasmobranch fish. Cell Tissue Res. **257**: 61–67.

Lacy, E.R., Reale, E. (1995) Functional morphology of the elasmobranch nephron and retention of urea. In: Wood, C.M., Shuttleworth, T.J. (eds.) Cellular and molecular approaches to fish ionic regulation. Academic Press, San Diego, pp. 107–146.

Lacy, R., Reale, E., Schlusselberg, D.S., Smith, W.K., Woodward, D.J. (1985) A renal countercurrent system in marine elasmobranch fish: a computer-assisted reconstruction. Science **227**: 1351–1354.

Lacy, E.R., Castellucci, M., Reale, E. (1987) The elasmobranch renal capsule: fine structure of Bowman's capsule and the glomerular capillary wall. Anat. Rec. **218**: 294–305.

Lacy, E.R., Luciano, L., Reale, E. (1989) Flagellar cells and ciliary cells in the renal tubule of elasmobranchs. J. Exp. Zool. (Suppl. 2): 186–192.

Ladich, F., Fine, M.L. (1992) Localization of pectoral fin motoneurons (sonic and hovering) in the croaking gourami *Trichopsis vittatus*. Brain Behav. Evol. **39**: 1–7.

Ladich, F., Fine, M.L. (1994) Localization of swimbladder and pectoral motoneurons involved in sound production in pimelodid catfish. Brain Behav. Evol. **44**: 86–100.

Lahnsteiner, F., Patzner, R.A. (1990) Functions of the testicular gland of blennid fish: structural and histochemical investigations. Experientia **46**: 1005–1007.

Lahnsteiner, F., Richtarski, U., Patzner, R.A. (1990) Functions of the testicular gland in two blennid fishes, *Salaria (= Blennius) pavo* and *Lipophrys (= Blennius) dalmatinus* (Blenniidae, Teleostei) as revealed by electron microscopy and enzyme histochemistry. J. Fish Biol. **37**: 85–97.

Lamb, C.F., Caprio, J. (1992) Convergence of oral and extraoral information in the superior secondary gustatory nucleus of the channel catfish. Brain Res. **588**: 201–211.

Lamb, C.F., Caprio, J. (1993) Taste and tactile responsiveness of neurons in the posterior diencephalon of the channel catfish. J. Comp. Neurol. **337**: 419–430.

Lammens, E.H.R.R., Hoogenboezem, W. (1991) Diets and feeding behaviour. In: Winfield, E.J., Nelson, J.S. (eds.) Cyprinid fishes. Systematics, biology and exploitation. Chapman and Hall, London, New York, pp. 353–376.

Land, M.F. (1987) Vision in air and water. In: Dijours, P., Bolis, L., Taylor, C.R., Weibel, E.R. (eds.) Comparative physiology: life in water and on land. Fidia Research Series, vol. 9. Liviana Press, Padova, pp. 289–302.

Langer, M. (1979a) Histologische Untersuchungen an der Teleosteerleber I. Dau Auf ban des Laberparenchyms. Z mikrosk.-anat Forsch. Leipzig **93**: 829–848.

Langer, M. (1979b) Histologische Untersuchungen an der Teleosteerleber II. Das BlutgefaBaystem. Z mikrosk.-anat. Forsch. Leipzig **93**: 849–875.

Langer, M. (1979c) Histologische Untersuchungen an der Teleosteerleber. III. Das galleausfuhrende System. Z mikrosk.-anat. Forsch. Leipzig **93**: 1105–1136.

Langille, B.L., Stevens, E.D., Anantaraman, A. (1983) Cardiovascular and respiratory flow dynamics. In: Webb, P.W., Weihs, D. (eds.) Fish Biomechanics. Praeger, New York, pp. 92–139.

Langille, R.M., Hall, B.K. (1989) Developmental processes, developmental sequences and early vertebrate phylogeny. Biol. Rev. **64**: 73–91.

Larson, B.A., Bern, H.A. (1987) The urophysis and osmoregulation. In: Pang, P.K.T., Schreibman M.P. (eds.) Vertebrate endocrinology: fundamentals and biomedical implications, vol. 2. Regulation of water and electrolytes. Academic Press, San Diego, pp. 143–156.

Lauder, G.V. (1982) Patterns of evolution in the feeding mechanism of actinopterygian fishes. Am. Zool. **22**: 275–285.

Lauder, G.V. (1983a) Food capture. In: Webb, P.W., Weihs, D. (eds.) Fish biomechanics. Praeger, New York, pp. 280–311.

Lauder, G.V. (1983b) Prey capture hydrodynamics in fishes: experimental tests of two models. J. Exp. Biol. **104**: 1–13.

Lauder, G.V. (1984) Pressure and water flow patterns in the respiratory tract of the bass (*Micropterus salmoides*) J. Exp. Biol. **113**: 151–164.

Lauder, G.V. (1985a) Aquatic feeding in lower vertebrates. In: Hildebrand, M., Bramble, D.M., Liem, K.F., Wake, D.B. (eds.) Functional vertebrate morphology. Harvard University Press, Cambridge, pp. 210–229.

Lauder, G.V. (1985b) Functional morphology of the feeding mechanism in lower vertebrates. Fortsch. Zool. **30**: 179–188.

Lauder, G.V. (1989) Caudal fin locomotion in ray-finned fishes: historical and functional analysis. Am. Zool. **29**: 85–102.

Lauder, G.V., Clark, B.D. (1984) Water flow patterns during prey capture by teleost fishes. J. Exp. Biol. **113**: 143–150.

Lauder, G.V., Wainwright, P.C. (1992) Function and history: the pharyngeal jaw apparatus in primitive ray-finned fishes. In: Mayden, R.L. (ed.) Systematics, historical, ecology, and North American freshwater fishes. Stanford University Press, Stanford, pp. 455–471.

Lauder, G.V., Wainwright, P.C., Findeis, E. (1986) Physiological mechanisms of aquatic prey capture in sunfishes: functional determinants of buccal pressure changes. Comp. Biochem. Physiol. **84A**: 729–734.

Laurent, P. (1989) Gill structure and function. Fish. In: Wood, S.C. (ed.) Comparative pulmonary physiology. Current concepts. Marcel Dekker, New York, pp. 69–120.

Laurent, P., Dunel-Erb, S. (1984) The pseudobranch: morphology and function. In: Hoar, W.S., Randall, D.J. (eds.) Fish physiology, vol. 10B. Academic Press, New York, pp. 285–323.

Laurent, P., Perry, S.F. (1991) Environmental effects on fish gill morphology. Physiol. Zool. **64**: 4–25.

Laurent, P., Delaney, R.G., Fishman, A.P. (1978) The vasculature of the gills in the aquatic and aestivating lungfish (*Protopterus aethiopicus*). J. Morphol. **156**: 173–208.

Laverty, G., Clark, N.B. (1989) The kidney. In: Pang, P.K.T., Shreibman, M.P. (eds.) Vertebrate endocrinology: fundamentals and biomedical implications, vol. 3. Academic Press, San Diego, pp. 277–318.

Lechleuthner, A., Schumacher, U., Negele, R.D., Welsch, U. (1989) Lungs of *Polypterus* and *Erpetoichthys*. J. Morphol. **201**: 161–178.

Leatherland, J.F. (1994) Reflections on the thyroidology of fishes: from molecules to humankind. Guelph Ichthyol. Rev. **2**: 1–67.

Le Bail, P.Y., Sire, M.F., Vernier, J.M. (1989) Intestinal transfer of growth hormone into the circulatory system of the rainbow trout, *Salmo gairdneri:* interference by granule cells. J. Exp. Zool. **251**: 101–107.

Leis, J.M. (1991) The pelagic stage of reef fishes: the larval biology of coral reef fishes. In: Sale, P.F. (ed.) The ecology of fishes on coral reefs. Academic Press, New York, pp. 183–230 (references integrated list).

Lemm, C.A., Herman, R.L., Lemarie, D.P., Arzepalo, A. (1993) Effects of diet and environmental salinity on the growth, mortality, and tissue structure of juvenile striped bass. J. Aquat. Anim. Health **5**: 294–305.

Lessard, J., Val, A.L., Aota, S., Randall, D.J. (1995) Why is there no carbonic anhydrase activity available to fish plasma? J. Exp. Biol. **198**: 31–38.

Lie, O., Lambertsen, G. (1985) Digestive lipolytic enzymes in cod (*Gadus morhua*): fatty acid specificity. Comp. Biochem. Physiol. **80B**: 447–450.

Liem, K.F. (1978) Modulatory multiplicity in the functional repertoire of the feeding mechanism in cichlid fishes. 1. Piscivores. J. Morphol. **158**: 323–360.

Liem, K.F. (1979) Modulatory multiplicity in the feeding mechanism in cichlid fishes, as exemplified by the invertebrate pickers of Lake Tanganyika. J. Zool. Lond. **189**: 93–125.

Liem, K.F. (1980) Adaptive significance of intra-and interspecific differences in the feeding reportoires of cichlid fishes. Am. Zool. **20**: 295–314.

Liem, K.F. (1993) Ecomorphology of the teleostean skull. In: Hanken, J., Hall, B.K. (eds.) The skull, vol. 3. Functional and evolutionary mechanisms. University of Chicago Press, Chicago, pp. 422–452.

Liley, N.R (1969) Hormones and reproductive behaviour in fishes. In: Hoar, W.S., Randall, D.J. (eds.) Fish physiology, vol. 3. Academic Press, New York, pp. 73–116.

Liley, N.R., Kroon, F.J. (1995) Male dominance, plasma hormone concentrations, and availability of milt in male rainbow trout (*Oncorhynchus mykiss*). Can. J. Zool. **73**: 826–836.

Liley, N.R., Rouger, Y. (1996) Plasma levels of gonadotropin and 17α, 20β-dihydroxy-4 pregnen-3-one in relation to spawning behaviour of rainbow trout, *Oncorhynchus mykiss* (Walbaum). J. Fish Biol. **37**: 699–711.

Liley, N.R., Stacey, N.E. (1983) Hormones, pheromones and reproductive behavior in fish. In: Hoar, W.S., Randall, D.J., Donaldson, E.M. (eds.) Fish physiology, vol. 9. Academic Press, New York, pp. 1–63.

Liley, N.R., Olsen, K.H., Foote, C.J., van der Kraak, G.J. (1993) Endocrine changes associated with spawning behavior in male kokanee salmon (*Oncorhynchus nerka*) and the effects of anosmia. Horm. Behav. **27:** 470–487.

Lin, H., Randall, D.J. (1990) The effect of varying water pH on the acidification of expired water in rainbow trout. J. Exp. Biol. **149**: 149–160.

Lin, H., Pfeiffer, D.C., Vogl, A.W., Pan, J., Randall, D.J. (1994) Immuno-localization of H^+-ATPase in the gill epithelia of rainbow trout. J. Exp. Biol. **195**: 169–183.

Lindsey, C.C. (1975) Pleomerism, the widespread tendency among related fish species for vertebral number to be correlated with maximum body length. J. Fish Res. Board Can. **32**: 2453–2469.

Lindsey, C.C. (1978) Form, function and locomotry habits in fish. In: Hoar, W.S., Randall, D.J. (eds.) Fish physiology, vol. 8. Academic Press, New York, pp. 1–100.

Little, E.L. (1983) Behavioural function of olfaction and taste in fish. In: Northcutt, R.G., Davis, R.E. (eds.) Fish neurobiology, vol. 1. Brain stem and sense organs. University of Michigan Press, Ann Arbor, pp. 351–376.

Liu, Q., Pitcher, T., Al-Hossaini, M. (1989) Ageing with fisheries length-frequency data, using information about growth. J. Fish Biol. **35**: (Suppl. A): 169–177.

Lobel, P.S. (1989) Spawning behavior of *Chaetodon multicinetus* (Chaetodontidae): pairs and intruders. Env. Biol. Fish. **25**: 125–130.

Loeb, V.J., Kellermann, A.K., Koubbi, P., North, A.W., White, M.G. (1993) Antarctic larval fish assemblages: a review. Bull. Mar. Sci. **53**: 416–449.

Loiselle, P.V., Barlow, G.W. (1978) Do fishes lek like birds? In: Reese, E.S., Lighter, F.J. (eds.) Contrasts in behavior. Adaptations in the aquatic and terrestrial environments. John Wiley, New York, pp. 31–78.

Lomholt, J.P. (1993) Breathing in the aestivating African lungfish, *Protopterus amphibius*. In: Singh, B.R. (ed.) Advances in fish research, vol. 1. Narendra Publishing House, Delhi, pp. 17–34.

Long, S., Giebisch, G. (1979) Comparative physiology of renal tubular transport mechanism. Yale J. Biol. Med. **52**: 525–544.

Lopez-Ruiz, A., Esteban, M.A., Meseguer, J. (1992) Blood cells of the gillhead seabream (*Sparus aurata* L.): Light and electron microscopic studies. Anat. Rec. **234**: 161–171.

Lorenzo, D., Sierra, F., Silva, A., Macadar, O. (1993) Spatial distribution of the medullary command signal within the electric organ of *Gymnotus carapo*. J. Comp. Physiol. A **173**: 221–226.

Losey, G.S., Jr. (1974) *Aspidontus taenicatus:* effects of increased abundance of cleaning symbiosis with notes on pelagic dispersion and *A. filamentosus* (Pisces, Blenniidae). Z Tierpsychol. **34**: 430–435.

Lowenstein, O. (1971) The labyrinth. In: Hoar, W.S., Randall, D.J. (eds.) Fish physiology, vol. 5. Academic Press, New York, pp. 207–240.

Lowenstein, O.E. (1974) Comparative morphology and physiology. In: Kornbhuber, H.H. (ed.) Handbook of sensory physiology, vol. 6, Part 1. Springer-Verlag, Berlin, pp. 75–120.

Lu, M., Wagner, G.F., Renfro, J.L. (1994) Stanniocalcin stimulates phosphate reabsorption by flounder renal proximal tubule in primary culture. Am. J. Physiol. **267**: R1336–R1362.

Lu, Z., Fay, R.R. (1993) Acoustic response properties of single units in the torus semicircularis of the goldfish, *Carassius auratus*. J. Comp. Physiol. **A 173**: 33–48.

Lu, Z., Fay, R.R. (1995) Acoustic response properties of single neurons in the central posterior nucleus of the thalamus of the goldfish, *Carassius auratus*. J. Comp. Physiol. **A 176**: 747–760.

Lund, R. (1986) The diversity and relationships of the Holocephali. In: Uyeno, T., Arai, R., Taniuchi, T., Matsuura, K. (eds.) Indo-Pacific Fish Biology: Proceedings of the Second International Conferences on Indo-Pacific fishes. Ichthyological society of Japan, Tokyo, pp. 97–106.

Lythgoe, J.N., Partridge, J.C. (1991) The modelling of optimal visual pigments of dichromatic teleosts in green coastal water. Vision Res. **31**: 361–371.

Macdonald, J.A., Montgomery, J.C. (1991) The sensory biology of notothenioid fish. In: di Prisco, G., Maresca, B., Tota, B. (eds.) Biology of Antarctic fish. Springer-Verlag, Berlin, Heidelberg, pp. 145–162.

Macdonald, J.A., Wells, R.M.G. (1991) Viscosity of body fluids from Antarctic notothenioid fish. In: di Prisco, G., Maresca, B., Tota, B. (eds.) Biology of Antarctic fish. Springer-Verlag, Berlin, Heidelberg, pp. 163–178.

Macdonald, J.A., Montgomery, J.C., Wells, R.M.G. (1988) The physiology of McMurdo Sound fishes: Current New Zealand research. Comp. Biochem. Physiol. **90B**: 567–578.

MacNichol E.F., Jr., Kunz, Y.W., Levine, J.S., Harosi, F.I., Collins, B.A. (1978) Ellipsosomes: organelles containing a cytochrome-like pigment in the retinal cones of certain fishes. Science **200**: 549–552.

Maddock, R.G., Nicol, J.A.C. (1978) Studies on the eyes of *Hydrolagus* (Pisces: Chimaeridae). Contr. Mar. Sci. **21**: 77–87.

Maetz, J. (1971) Fish gills: mechanisms of salt transfer in fresh water and sea water. Phil. Trans. Roy Soc. Lond. **B262**: 209–249.

Maetz, J. (1974) Aspects of adaptation to hypo-osmotic and hyper-osmotic environments. In: Malins, D.C., Sargent, J.R. (eds.) Biochemical and biophysical perspectives in marine biology, vol. 1. Academic Press, London, pp. 1–167.

Maffia, M., Ahearn, G.A., Vilella, S., Zonno, V., Storelli, C. (1993) Ascorbic acid transport by intestinal brush-border membrane vesicles of the teleost *Anguilla anguilla*. Am. J. Physiol. **264**: R 1248-R 1253.

Magliulo-Cepriano, L., Schreibman, M.P. (1993) The distribution of neuropeptide Y and dynorphin immunoreactivity in the brain and pituitary gland of the plätyfish *Xiphophorus maculatus*, from birth to sexual maturity. Cell Tissue Res. **271**: 87–92.

Magliulo-Cepriano, L., Schreibman, M.P., Blum, V. (1993) The distribution of immunoreactive FMRF-amide neurotensin, in three species of *Xiphophorus* from birth to sexual maturity. Gen. Comp. Endocrinol. **92**: 269–280.

Magliulo-Cepriano, L., Schreibman, M.P., Blum, V. (1994) Distribution of variant forms of immunoreactive gonadotropin-releasing hormone and B-gonadotropins I and II in the platyfish, *Xiphophorus maculatus*, from birth to sexual maturity. Gen. Comp. Endocrinol. **94**: 135–150.

Magnuson, J.J. (1978) Locomotion by scombrid fishes: hydromechanics, morphology, and behavior. In: Hoar, W.S., Randall, D.J. (eds.) Fish physiology, vol. 7. Academic Press, New York, pp. 239–313.

Magurran, A.E., Pitcher, T.J. (1987) Provenance, shoal size and sociobiology of the predator-evasion behaviour in minnow shoals. Proc. R. Soc. Lond. **B229**: 439–465.

Maisey, J.G. (1984) Higher elasmobranch phylogeny and biostratigraphy. Zool. J. Linn. Soc. **82**: 33–54.

Maisey, J.G. (1986) Heads and tails: chordate phylogeny. Cladistics **2**: 201–256.

Maisey, J.G. (1988) Phylogeny of early vertebrate skeletal induction and ossification patterns. In: Hecht, M.K., Wallace, B., Prance, G.T. (eds.) Evolutionary Biology, vol. 22. Plenum Press, New York, pp. 1–36.

Malhotra, Y.R., Jyoti, M.K., Gupta, K. (1989) Reproductive cycles of freshwater fishes. In: Saidapur, S.K. (ed.) Reproductive cycles of Indian vertebrates. Allied Publishers Limited, New Delhi, pp. 58–105.

Mallefet, J., Baguet, F. (1985) Effects of adrenalin on the oxygen consumption and luminescence of the photophores of the mesopelagic fish *Argyropelecus hemigymnus*. J. Exp. Biol. **118**: 341–349.

Mallefet, J., Baguet, F. (1989) Oxygen consumption and luminescence of isolated *Porichthys* photophores in response to adrenergic stimulations. Photochem. Photobiol. **50**: 243–250.

Mallefet, J., Baguet, F. (1991) Metabolic control of *Argyropelecus hemigymnus* photophores: effects of glucose and pyruvate. Can. J. Zool. **69**: 2410–2413.

Mallefet, J., Baguet, F. (1993) Metabolic control of luminescence in isolated photophores of *Porichthys:* effects of glucose on oxygen consumption and luminescence. J. Exp. Biol. **181**: 279–293.

Marshall, W.S. (1995) Transport processes in isolated teleost epithelia: opercular epithelium and urinary bladder. In: Wood, C.M., Shuttleworth, T.J. (eds.) Cellular and molecular approaches to fish ionic regulation. Academic Press, San Diego, pp. 1–23.

Marshall, W.S., Bryson, S.E., Idler, D.R. (1989) Gonadotropin stimulation of K^+ secretion and Na^+ absorption by brook trout (*Salvelinus fontinalis*) sperm duct epithelium. Gen. Comp. Endocrinol. **75**: 118–128.

Marshall, W.S., Bryson, S.E., Wood, C.M. (1992) Calcium transport by isolated skin of rainbow trout. J. Exp. Biol. **166**: 297–316.

Marshall, W.S., Bryson, S.E., Idler, D.R. (1993) Gonadotropin action on brook trout sperm duct epithelium: ion transport stimulation mediated by cAMP and Ca^{2+}. Gen. Comp. Endocrinol. **90**: 232–242.

Marshall, W.S., Bryson, S.E., Burghardt, J.S., Verbost, P.M. (1995) Ca^{2+} transport by opercular epithelium of the fresh water adapted euryhaline teleost, *Fundulus heteroclitus*. J. Comp. Physiol. **B165**: 268–277.

Martin, L.K., Cailliet, G.M. (1988) Age and growth determination of the bat ray, *Myliobatis california* Gill, in central California. Copeia **1988**: 762–773.

Marui, T., Caprio, J. (1992) Teleost gustation. In: Hara, T.J. (ed.) Fish chemoreception. Chapman and Hall, London, New York, pp. 171–198.

Mathis, A., Smith, R.J.F. (1992) Avoidance of areas marked with a chemical alarm substance by fathead minnows (*Pimephales promelas*) in a natural habitat. Can. J. Zool. **70**: 1473–1476.

Mathis, A., Smith, R.J.F. (1993a) Intraspecific and cross-superorder responses to chemical alarm signals by brook stickleback. Ecology **74**: 2395–2404.

Mathis, A., Smith, R.J.F. (1993b) Chemical alarm signals increase the survival time of fathead minnows (*Pimephales promelas*) during encounters wth northern pike (*Esox lucius*). Behav. Ecol. **4**: 260–265.

Mathis, A., Chivers, D.P., Smith, R.J.F. (1993) Population difference in responses of fathead minnows (*Pimephales promelas*) to visual and chemical stimuli from predators. Ethology **93**: 31–40.

Matsubara, T., Hara, A., Takano, K. (1993) Immunohistochemical localization of coelomic fluid-specific protein in the coelomic epithelium and mesovarioum of chun salmon *Oncorchynchus keta*. Nippon Suisan Gakkaishi **59**: 1717–1720 (In Japanese with English summary).

Matsubara, T., Hara, A., Tokano, K. (1994) Transport of circulating protein component into the coelomic cavity during ceolomic fluid production in masu salmon *Oncorhynchus masou*. Nippon Suisan Gakkaishi **60**: 479–483 (In Japanese with English summary).

Matsubara, T., Hara, A., Takano, K. (1995) Cationic constituents, protein concentration and osmolality of coelomic fluid in salmonid fish. Fish. Sci. **61**: 127–130.

Mattisson, A., Fänge, R. (1986) The cellular structure of lymphomyeloid tissues in *Chimaera monstrosa* (Pisces, Holocephali). Biol. Bull. **171**: 660–671.

Mattisson, A., Fänge, R., Zapata, A. (1990) Histology and ultrastructure of the cranial lymphohaemopoietic tissue in *Chimaera monstrosa* (Pisces, Holocephali). Acta Zool. Stockholm. **71**: 97–106.

Matty, A.J. (1966) Endocrine glands in lower vertebrates. In: Felts, W.J.L., Harrison, R.J. (eds.) International review of general and experimental zoology, vol. 2. Academic Press, New York, pp. 43–137.

Matty, A.J. (1985) Fish endocrinology. Chapman and Hall, London, New York.
Mawdesley-Thomas, L.E. (1975) Some diseases of muscle. In: Ribelin, W.E., Migaki, G. (eds.) The pathology of fishes. University of Wisconsin Press, Madison, pp. 343–363.
Mayden, R.L., Rainboth, W.J., Buth, D.J. (1991) Phylogenetic systematics of the cyprinid genera *Mylopharadon* and *Ptychocheilus:* comparative morphometry. Copeia **1991**: 819–834.
McCormick, C.A. (1982) The organization of the octavolateralis area in actinopterygian fishes: a new interpretation. J. Morphol. **171**: 159–181.
McCormick, C.A. (1983) Organization and evolution of the octavolateralis area of fishes. In: Northcutt, R.G., Davis, R.E. (eds.) Fish neurobiology, vol. 1. Brain stem and sense organs. University of Michigan Press, Ann Arbor, pp. 179–214.
McCormick, C.A. (1989) Central lateral line mechanosensory pathways in bony fish. In: Coombs, S., Görner, P., Münz, H. (eds.) The mechanosensory lateral line: neurobiology and evolution. Springer-Verlag, New York, pp. 341–364.
McCormick, C.A. (1992) Evolution of central auditory pathways in anamniotes. In: Webester, D.B., Fay, R.R., Popper, A.N. (eds.) The evolutionary biology of hearing. Springer-Verlag, New York, pp. 323–350.
McCormick, C.A., Bradford M.R., Jr. (1993) The primary octaval nuclei and inner ear afferent projections in the otophysan *Ictalurus punctatus*. Brain Behav. Evol. **42**: 48–68.
McCormick, C.A., Bradford M.R., Jr. (1994) Organization of inner ear endorgan projections in the goldfish, *Carassius auratus*. Brain Behav. Evol. **43**: 189-205.
McCormick, S.D. (1995) Hormonal control of gill Na^+, K^+-ATPase and chloride cell function. In: Wood, C.M., Shuttleworth, T.J. (eds.) Cellular and molecular approaches to fish ionic regulation. Academic Press, San Diego, pp. 285–315.
McCosker, J.E., Lagios, M.D. (eds.) (1979) The biology and physiology of the living coelacanth. Occ Papers Calif. Acad. Sci. **134**: 1–175.
McDonald, D.G., Milligan, C.L. (1992) Chemical properties of the blood. In: Hoar, W.S., Randall, D.J., Farrell, A.P. (eds.) Fish physiology, vol. 12. Academic Press, Orlando, pp. 55–133.
McEnroe, M., Cech, J.J., Jr. (1985) Osmoregulation in juvenile and adult white sturgeon, *Acipenser transmontanus*. Env. Biol. Fish. **14**: 21–30.
McEnroe, M., Cech, J.J., Jr. (1987) Osmoregulation in white sturgeon: life history aspects. Am. Fish Soc. Symp. **1**: 191–196.
McFall-Ngai, M.J. (1983) Adaptations for reflection of bioluminescent light in the gas bladder of *Leiognathus equalus* (Perciformes: Leiognathidae). J. Exp. Zool. **227**: 23–33.
McFall-Ngai, M.J., Dunlap, P.V. (1983) Three new modes of luminescence in the leiognathid fish *Gazza minuta:* discrete projected luminescence, ventral body flash, and buccal luminescence. Mar. Biol. **73**: 227–237.
McFall-Ngai, M.J., Dunlap, P.V. (1984) External and internal sexual dimorphism in leiognathid fishes: morphological evidence for sex-specific bioluminescent signaling. J. Morphol. **182**: 71–83.
McFall-Ngai, M.J., Horwitz, J. (1990) A comparative study of the thermal stability of the vertebrate eye lens: Antarctic ice fish to the desert iguana. Exp. Eye Res. **50**: 703–709.
McFall-Ngai, M.J. Morin, J.G. (1991) Camouflage by disruptive illumination in leiognathids, a family of shallow-water, bioluminescent fishes, J. Exp. Biol. **156**: 119–137.
McFall-Ngai, M., Toller, W. (1991) Frontiers in the study of the biochemistry and molecular biology of vision and luminescence in fishes. In: Hochachka, P.W., Mommsen, T.P. (eds.) Biochemistry and molecular biology of fishes, vol. 1. Elsevier, Amsterdam, pp. 77–107.
McFall-Ngai, M.J., Crescitelli, F., Childress, J., Horwitz, J. (1986) Patterns of pigmentation in the eye lens of the deep-sea hatchetfish, *Argyropelecus affinis* Garman. J. Comp. Physiol. A **159**: 791–800.
McFall-Ngai, M.J., Ding, L., Childress, J., Horwitz, J. (1988) Biochemical characteristics of the pigmentation of mesopelagic fish lenses. Biol. Bull. **175**: 397–402.
McLeese, J.M., Stevens, E.D. (1986) Trypsin from two strains of rainbow trout, *Salmo gairdneri*, is influenced by assay and acclimation temperature. Can. J. Fish Aquat. Sci. **43**: 1664–1667.
McMahon, B.R., Burggren, W.W. (1987) Respiratory physiology of intestinal air breathing in the teleost *Misgurnus anguillicaudatus*. J. Exp. Biol. **133**: 371–393.

McNulty, J.A. (1978) The pineal of the troglophilic fish, *Chologaster agassizi:* an ultrastructural study. J. Neural Transmission **43**: 47–71.

McNulty, J.A. (1979) A comparative light and electron microscopic study of the pineal complex in the deep-sea fishes, *Cyclothone signata* and *C. acclinidens*. J. Morphol. **162**: 1–16.

McNulty, J.A. (1984) Functional morphology of the pineal complex in cyclostomes, elasmobranches, and bony fishes. Pineal Res. Rev. **2**: 1–40.

Meer, H.J. van der, Bowmaker, J.K. (1995) Interspecific variation of photoreceptors in four co-existing haplochromine cichlid fishes. Brain Behav. Evol. **45**: 232–240.

Meier, A.H., Fivizzani, A.J. (1980) Physiology of migration. In: Gauthreaux S.A. Jr. (ed.) Animal migration, orientation, and navigation. Academic Press, New York, pp. 225–282.

Meissl, H., Brandstatter, R. (1992) Photoreceptive functions of the teleost pineal organ and their implications in biological rhythms. In: Ali, M.A. (ed.) Rhythms in fishes. Plenum Press, New York, pp. 235–254.

Meissl, H., Ekstrom, P. (1993) Extrarenal photoreception by pineal systems: a tool for photoperiodic time measurements. Trends in comparat. Biochem. Physiol. **1**: 1233–1240.

Meissl, H., Yáñez, J. (1994) Pineal photosensitivity. A comparison with retinal photoreception. Acta Neurobiol. Exp. **54** (Suppl.): 19–29.

Meseguer, J., Esteban, M.A., Agulleiro, B. (1991) Stromal cells, macrophages and lymphoid cells in the head-kideny of sea bass (*Dicentrarchus labrax* L.). An ultrastructural study. Arch. Histol. Cytol. **54**: 299–309.

Meseguer, M., Esteban, M.A., Garcia-Ayala, L., López-Ruiz, A., Agulleiro, B. (1990) Granulopoiesis in the head-kidney of the sea bass (*Dicentrarchus labrax* L.): An ultrastructural study. Arch. Histol. Cytol. **53**: 287–296.

Meseguer, J., Esteban, M.A., Munoz, J., López-Ruiz, A. (1993) Ultrastructure of the peritoneal exudate cells of seawater teleosts, seabream (*Sparus aurata*) and sea bass (*Dicentrarchus labrax*). Cell Tissue Res. **273**: 301–307.

Meseguer, J., Esteban, M.A., López-Ruiz, A., Bielek, E. (1994a) Ultrastructure of nonspecific cytotoxic cells in teleosts 1. Effector target cell binding in a marine and a freshwater species (seabream: *Sparus aurata* L., and carp: *Cyprinus carpio* L.). Anat. Rec. **239**: 468–474.

Meseguer, J., López-Ruiz, A., Esteban, M.A. (1994b) Melano-macrophages of the seawater teleosts, sea bass *Dicentrarchus labrax*) and gilthead seabream (*Sparus aurata*): morphology, formation and possible function. Cell Tissue Res. **277**: 1–10.

Meseguer, J., López-Ruiz, A., Esteban, M.A. (1994c) Cytochemical characterization of leucocytes from the seawater teleost, gilthead seabream (*Sparus aurata* L.). Histochemistry **102**: 37–44.

Meseguer, J., López-Ruiz, A., Garcia-Ayala, L. (1995) Reticulo-endothelial stroma of the head-kidney from the seawater teleost gilthead seabream (*Sparus aurata* L.): an ultrastructural and cytochemical study. Anat. Rec. **241**: 303–309.

Metzner, W. (1993) The jamming avoidance response in *Eigenmannia* is controlled by two separate motor pathways. J. Neurosci. **13**: 1862–1878.

Meyer, J.H., Zakon, H.H. (1982) Androgens alter the tuning of electroreceptors. Science **217**: 635–637.

Michel, W.C., Caprio, J. (1991) Responses of single facial taste fibers in the sea catfish, *Arius felis*, to amino acids, J. Neurophysiol. **66**: 247–260.

Michel, W.C., Kohbara, J., Caprio, J. (1993) Amino acid receptor sites in the facial taste system of the sea catfish *Arius felis*, J. Comp. Physiol. **A172**: 129–138.

Milinski, M. (1986a) Constraints placed by predators on feeding behaviour. In: Pitcher, T.J. (ed.). The behaviour of teleost fishes. Croom Helm, London, pp. 236–252.

Milinski, M. (1986b) A review of competitive resource sharing under constraints in sticklebacks. J. Fish Biol. **29** (Suppl A): 1–14.

Mills, A., Zakon, H.H. (1987) Coordination of EOD frequency and pulse duration in a weakly electric wave fish: the influence of androgens,. J. Comp. Physiol. **A161**: 417–430.

Miller, P.J. (1979a) A concept of fish phenology. Symp. zool. Soc. Lond. No. **44**: 1–28.

Miller, P.J. (1979b) Adaptiveness and implications of small size in teleosts. Symp. zool. Soc. Lond. No. **44**: 263–306.

Milsom, W.K. (1989) Mechanisms of ventilation in lower vertebrates: adaptations to respiratory and nonrespiratory constraints. Can J. Zool. **67**: 2943–2955.

Milsom, W.K. (1995) Regulation of respiration in lower vertebrates: role of CO_2/pH chemoreceptors. In: Heisler, M. (ed.) Mechanisms of systematic regulation. Advances in comparative and environmental physiology, vol. 21. Springer-Verlag, Berlin, Heidelberg, pp. 61–104.

Milsom, W.K., Brill, R.W. (1986) Oxygen sensitive afferent information arising from the first gill arch of yellowfish tuna. Respir. Physiol. **66**: 193–203.

Miyamoto, T., Restrepo, D., Cragoe, E.J. Jr., Teeter, J.H. (1992) IP_3-and cAMP-induced responses in isolated olfactory receptor neurons from the channel catfish, J. Membrane Biol. **127**: 173–183.

Moller, P., Brown, B. (1990) Meristic characters and electric organ discharge of *Mormyrops curviceps* Roman, (Teleostei: Mormyridae) from the Moa river, Sierra Leone, West Africa. Copeia **1990**: 1031–1040.

Moller, P., Serrier, J., Bowling, D. (1989) Electric organ discharge displays during social encounter in the weakly electric fish *Brienomyrus niger* L. (Mormyridae). Ethology **82**: 177–191.

Mommsen, T.P., Walsh. P.J. (1988) Vitellogenesis and oocyte assembly. In: Hoar. W.S., Randall.D.J. (eds.) Fish. physiology, vol. 11. Academic. Press, New York, pp. 347–404.

Mommsen, T.P., Walsh, P.J. (1989) Evolution of urea synthesis in vertebrates: the piscine connection. Science **243**: 72–75.

Mommsen, T.P., Walsh, P.J. (1992) Biochemical and environmental perspectives on nitrogen metabolism in fishes. Experientia **48**: 583–593.

Montgomery, J.C. (1989) Lateral line detection of planktonic prey. In: Coombs, S., Görner P., Münz H. (eds.) The mechanosensory lateral line: neurobiology and evolution. Springer Verlag, New York, pp. 561–574.

Montgomery, J.C., Macdonald, J.A. (1987) Sensory tuning of lateral line receptors in Antarctic fish to the movements of planktonic prey. Science **235**: 195–196.

Montgomery, J.C., Macdonald, J.A. (1990) Effects of temperature on nervous system: implications for behavioural performance. Am. J. Physiol. **259**: R191–R196.

Montgomery, J.C., Macdonald, J.A. Housley, G.D. (1988) Lateral line function in an Antarctic fish related to the signals produced by planktonic prey. J. Comp. Physiol. **A163**: 827–833.

Montgomery, J.C., Macdonald, J.A., Housley, G.D. (1990) Brain function in Antarctic fish: frequency response analysis of central vestibular units. J. Comp. Physiol. **A166**: 407–410.

Montgomery, J., Coomb, S., Janssen, J. (1994) Form and function relationships in lateral line systems: Comparative data from six species of Antarctic notothenioid fish. Brain Behav. Evol. **44**: 299–306.

Morizot, D.C. (1990) Use of fish gene maps to predict ancestral vertebrate genome organization. In: Ogita, Z.I., Markert, C.L. (eds.) Isozymes: structure, function and use in biology and medicine. Wiley-Liss, New York, pp. 207–234.

Morizot, D.C., Schultz, R.J., Wells, R.S. (1990) Assignment of six enzyme loci to multipoint linkage groups in fishes of the genus *Poeciliopsis* (Poeciliidae): Designation of linkage groups III-V. Biochem. Genet. 83–95.

Morizot, D.C., Slaugenhaupt, S.A., Kallman, K.D., Chakravarti, A. (1991) Genetic linkage map of fishes of the genus *Xlphophorus* (Teleostei: Poeciliidae). Genetics **127**: 399–410.

Morris, S., Bridges, C.R. (1994) Properties of respiratory pigments in bimodal breathing animals: air and water breathing by fish and crustaceans. Am. Zool. **34**: 216–228.

Moss, S.A. (1972a) The feeding mechanism of sharks of the family Carcharhinidae. J. Zool. Lond. **167**: 423–436.

Moss, S.A. (1972b) Tooth replacement and body growth rates in the smooth dogfish, *Mustelus canis* (Mitchill). Copeia **1972**: 808–811.

Moss, S.A. (1977) Feeding mechanisms in sharks. Am. Zool. **17**: 355–364.

Moore, W.S. (1984) Evolutionary ecology of unisexual fishes. In: Turner B.J. (ed.) Evolutionary genetics of fishes. Plenum Publishing Corporation, New York, pp. 329–398.

Motta, P.J. (1982) Functional morphology of the head of the inertial suction feeding butterflyfish, *Chaetodon miliaris* (Perciformes, Chatodontidae). J. Morphol. **174**: 283–312.

Motta, P.J. (1984) Mechanics and functions of jaw protrusions in teleost fishes: a review. Copeia **1984**: 1–18.

Motta, P.J. (1985) Functional morphology of the head of Hawaiian and Mid-Pacific butterflyfishes (Perciformes, Chaetodontidae). Env. Biol. Fish. **13**: 253–276.

Motta, P.J. (1987) A quantitative analysis of ferric iron in butterflyfish teeth (Chaetodontidae, Perciformes) and the relationship to feeding ecology. Can. J. Zool. **65**: 106–112.

Motta, P.J. (1988) Functional morphology of the feeding apparatus of ten species of Pacific butterflyfishes (Perciformes, Chaetodontidae): an ecomorphological approach. Env. Biol. Fish. **22**: 39–67.

Motta, P.J. (1989) Dentition patterns among Pacific and Western Atlantic butterflyfishes (Perciformes, Chaetodontidae): relationship to feeding ecology and evolutionary history. Env. Biol. Fish. **25**: 159–170.

Motta, P.J., Hueter, R.E., Tricas, T.C. (1991) An electromyographic analysis of the biting mechanism of the lemon shark, *Negaprion brevirostris*: functional and evolutionary implications. J. Morphol. **210**: 55–69.

Moura, M.A.F., Farias, I.P., Val A.L. (1994) Effects of temperature on leucocytes of *Colossoma macropomum* and *Hoplosternum littorale* (Pisces). Brazilian J. Med. Biol. Res. **27**: 1589–1598.

Moyle, P.B., Cech, J.J. Jr. (1988) Fishes. An introduction to Ichthyology. (Second edition). Prentice Hall, Englewood Cliffs, New Jersey, U.S.A.

Moy-Thomas, J.A., Miles, R.S. (1971) Palaeozoic Fishes, (Second edition). Chapman and Hall, London, New York.

Muller, M., Osse J.W.M. (1984) Hydrodynamics of suction feeding in fish. Trans. Zool. Soc. Lond. **37**: 51–135.

Muller, M., Osse J.W.M., Verhagen, J.H.G. (1982) A quantitative hydrodynamical model of suction. feeding in fish. J. Theor. Biol. **95**: 49–79.

Muller, M., Van Leeuwen, J.L., Osse J.W.M., Drost, M.R. (1985) Prey capture hydrodynamics in fishes: two approaches. J. Exp. Biol. **119**: 389–394.

Munehara, H., Koya, Y., Takano, K. (1994) Conditions for initiation of fertilization of eggs in the copulating elkhorn sculpin. J. Fish Biol. **45**: 1105–1111.

Munehara, H., Takano, K., Koya, Y. (1991) The little dragon sculpin *Blepsias cirrhosus*, another case of internal gametic association and external fertilization. Japan. J. Ichthyol. **37**: 391–394.

Munk, O. (1968) The eyes of *Amia* and *Lepisosteus* (Pisces, Holostei) compared with the brachiopterygian and teleostean eyes. Vidensk. Meddr. dansk. naturh. Foren. **131**: 109–127.

Munk, O. (1980) Retinal tapetum in the deep-sea teleost *Diretmus argenteus*. Vidensk. Meddr. dansk. naturh. Foren. **142**: 179–186.

Muñoz-Chápuli, R., de Andrés, V. (1995) Anatomy and histology of the cephalic arterial coilings in hammerhead sharks (Genus *Sphyrna*). Acta Zool. Stockholm **76**: 305–307.

Muñoz-Chápuli, R., Garcia-Garrido, L. (1986) Cephalic blood vessels in elasmobranchs: anatomy and phylogenetic considerations. In: Uyeno. T., Arai, R., Taniuchi, T., Matsbura, K. (eds.) Indo-Pacific fish biology. Proc. Second International Conference on Indo-Pacific fishes. Ichthyological Society of Japan, Tokyo, pp. 164–172.

Muñoz-Chápuli, R., de Andrés, V., Ramos, C. (1994a) Tachykinin-like immunoreactivity in the sinus venosus of the dogfish (*Scyliorhinus canicula*). Cell Tissue Res. **278**: 171–175.

Muñoz-Chápuli, R., de Andrés, A.V., Dingerkus, G. (1994b) Coronary artery anatomy and elasmobranch phylogeny. Acta Zool. Stockholm. **75**: 249–254.

Muñoz-Chápuli, R., Macias, D., Ramos, C., de Andrés. V., Gallego A., Navarro, P. (1994c) Cardiac development in the dogfish (*Scyliorhinus canicula*): a model for the study of vertebrate cardiogenesis. Cardioscience **5**: 245–253.

Munshi, J.S.D. (1985) The structure, function and evolution of the accessory respiratory organs of air-breathing fishes of India. Fortsch. Zool. **30**: 353–366.

Munshi, J.S.D. (1990) The bearing of ecological factors on the evolution of air-breathing fishes. In: Manna, G.K., Jana, B.B. (eds.) Impacts of environment on animals and aquaculture, pp. 15–19.

Munshi, J.S.D. (1993) Structure and function of the air-breathing organs of *Heteropneustes fossilis*. In: Singh B.R. (ed.) Advances in fish research, vol. 1. Narendra Publishing House, Delhi, pp. 99–138.

Munshi, J.S.D., Hughes, G.M. (1991) Structure of the respiratory islets of accessory respiratory organs and their relationship with the gills in the climbing perch, *Anabas testudineus* (Teleostei, Perciformes). J. Morphol. **209**: 241–256.

Munshi, J.S.D., Roy, P.K., Ghosh, T.K., Olson, K.R. (1994) Cephalic circulation in the air-breathing snakehead fish, *Channa punctata*, *C. gachua*, and *C. marulius* (Ophiocephalidae, Ophiocephaliformes). Anat. Rec. **238**: 77–91.

Münz, H. (1989) Functional organization of the lateral line periphery. In: Coombs, S., Görner P., Münz, H. (eds.) The mechanosensory lateral line: neurobiology and evolution. Springer-Verlag, New York, 285–297.

Murray, R.W. (1971) Temperature receptors. In: Hoar, W.S., Randall, D.J. (eds.) Fish physiology, vol. 5. Academic Press, New York, London, pp. 121–133.

Musick, J.A., Bruton, M.N., Balon, E.K. (eds.) (1991) The biology of *Latimeria chalumnae* and evolution of coelacanths. Env. Biol. Fish. **32**: 1–436.

Mustafa, T., Agnisola. C., Tota, B. (1992) Myocardial and coronary effects of exogenous arachidonic acid on the isolated and perfused heart preparation and its metabolism in the heart of trout (*Oncorhynchus mykiss*). Comp. Biochem. Physiol. **103C**: 163–167.

Myrberg, A.A. Jr. (1974) Aggression in reef fishes. Am. Zool. **14**: 81–96.

Myrberg, A.A. Jr. (1981) Sound communication and interaction in fishes. In: Tavolga, W.N., Popper. A.N., Fay, R.R. (eds.) Hearing and sound communication in fishes. Springer-Verlag, New York, pp. 395–425.

Nagahama, Y. (1983) The functional morphology of teleost gonads. In: Hoar, W.S., Randall, D.J., Donaldson, E.M. (eds.) Fish physiology, vol. 9A. Academic Press, New York, pp. 223–275.

Nagahama, Y., Yamashita, M. (1989) Mechanisms of synthesis and action of 17α, 20β-dihydroxy-4-pregnen-3-one, a teleost maturation-inducing substance. Fish Physiol. Biochem. **7**: 193–200.

Nagahama, Y., Takemura, A., Takano, K., Adachi, S., Kusakari, M. (1991) Serum steroid hormone levels in the reproductive cycle of *Sebastes taxzanowskii* and *S. schlegeli*. Env. Biol. Fish. **30**: 31–38.

Nagahama, Y., Yoshikuni, M., Yamashita, M., Tanaka, M. (1994) Regulation of oocyte maturation in fish. In: Sherwood, N., Hew, C. (eds.) Fish physiology, vol. 13. Academic Press, San Diego, pp. 393–439.

Nakajima, T. (1979) The development and replacement pattern of the pharyngeal dentition in the Japanese cyprinid fish, *Gnathopogon coerulesceus*. Copeia **1979**: 22–28.

Nakamura, M., Hourigan, T.F., Yamauchi, K., Nagahama, Y., Grau, E.G. (1989) Histological and ultrastructural evidence for the role of gonadal steroid hormones in sex change in the protogynous wrasse *Thallassoma duperrey*. Env. Biol. Fish. **24**: 117–136.

Nayak, M.S., Nadkarni, V.B. (1993) Gonadal development in some teleosts. In: Singh B.R. (ed.) Advances in fish research. vol. 1. Narendra Publishing House, Delhi, pp. 295–312.

Neelakantan, B., Kusuma, N., Bhat, U.G. (1989) Reproductive cycles of marine fishes. In: Saidapur, S.K. (ed.) Reproductive cycles of Indian vertebrates. Allied Publishers Limited, New Delhi, pp. 106–165.

Neighbors, M.A., Horn, M.H. (1991) Nutritional quality of macrophytes eaten and not eaten by two temperate-zone herbivorous fishes: a multivariate comparison. Mar. Biol. **108**: 471–476.

Neill, W.H. (1984) Behavioural enviroregulation's role in fish migration. In: McCleave, J.D., Arnold, G.P., Dodson, J.D., Neill W.H. (eds.) Mechanisms of migration in fishes. Plenum Publishing Corporation, New York, pp. 61–66.

Nel, P.P.C., Nel, M.M., De Wit, L.J.D. (1991) Ultrastructure of the atrioventricular valve and ventricular endo- and myocardium of the tigerfish, *Hydrocynus forskahlii*. J. Morphol. **207**: 129–134.

Nelson, J.S. (1994) Fishes of the world (Third edition). John Wiley, New York.

Netten, S.M. van, (1989) Dynamic behavior and micromechanical properties of the cupula. In: Coombs, S, Görner, P., Münz, H. (eds.) The mechanosensory lateral line: neurobiology and evolution. Springer-Verlag, New York, pp. 247–263.

Netten, S.M. van, Kroese, A.B.A. (1989) Hair cell mechanics control the dynamic behaviour of the lateral line cupula. In: Wilson, J.P., Kept, D.T. (eds.) Cochlear mechanisms. Plenum Publishing Corporation, New York, pp. 47–55.

Neudecker, S. (1979) Effects of grazing and browsing fishes on the zonation of corals in Guam. Ecology **60**: 666–672.

Neudecker, S. (1989) Eye camouflage and false eyespots: chaetodontid responses to predators. Env. Biol. Fish. **25**: 143–157.

Neudecker, S., Lobel, P.S. (1982) Mating systems of chaetodontid and pomacanthid fishes at St. Croix, Z. Tierpsychol. **59**: 299–318.

Nicol, J.A.C. (1975) Studies on the eyes of fishes: Structure and ultrastructure. In: Ali, M.A. (ed.) Vision in fish. Plenum Publishing Corporation, New York, pp. 579–607.

Nicol, J.A.C. (1978) Bioluminescence and vision. In: Herring, P.J. (ed.) Bioluminescence in action. Academic Press, London, pp. 367–398.

Nicol, J.A.C. (1980) Studies on the eyes of toadfishes *Opsanus*. Structure and reflectivity of the stratum argenteum. Can. J. Zool. **58**: 114–121.

Nieuwenhuys, R., Pouwels, E. (1983) The brain stem of actinopterygian fishes. In: Northcutt, R.G., Davis, R.E. (eds.) Fish neurobiology. vol. 1. Brain stem and sense organs. University of Michigan Press, Ann Arbor, pp. 25–88.

Nikinama, M., Boutilier, R.G. (1995) Adrenergic control of red cell pH, organic phosphate concentrations and haemoglobin function in teleost fish. In: Heisler, N. (ed.) Mechanisms of systemic regulation. Advances in comparative and environmental physiology, vol. 21. Springer-Verlag, Berlin, Heidelberg, pp. 107–133.

Nilsson, S. (1986) Control of gill blood flow. In: Nilsson, S., Holmgren, S. (eds.) Fish physiology: recent advances. Croom Helm, London, pp. 86–101.

Nilsson, S. (1994) Evidence for adrenergic nervous control of blood pressure in teleost fish. Physiol. Zool. **67**: 1347–1359.

Nilsson, S., Holmgren, S. (1993) Autonomic nerve functions. In: Evans, D.H. (ed.) The physiology of fishes. CRC Press, Boca Raton, pp. 279–313.

Nishimura, H. (1985a) Evolution of the renin-angiotensin system and its role in control of cardiovascular function in fishes. In: Foreman, R.E., Gorbman, A., Dodd, J.M., Olsson, R. (eds.) Evolutionary biology of primitive fishes. Plenum Publishing Corporation, New York, pp. 275–293.

Nishimura, H. (1985b) Endocrine control of renal handling of solutes and water in vertebrates. Renal Physiol. **8**: 279–300.

Nishimura, H. (1987) Role of the renin-angiotensin system in osmoregulation. In: Pang, P.K.T, Schreibman, M.P. (eds.) Vertebrate endocrinology: fundamentals and biomedical implications, vol. 2. Academic Press, San Diego, pp. 157–187.

Nishimura, H., Madey, M.A. (1989) Signals controlling renin release in aglomerular toadfish. Fish Physiol. Biochem. **7**: 323–329.

Nishimura, H., Ogawa, M. (1973) The renin-angiotensin system in fishes. Am., Zool. **13**: 823–838.

Nishimura, H., Imai, M., Ogawa, M. (1983) Sodium chloride and water transport in the renal distal tubule of the rainbow trout. Am. J. Physiol. **244**: F247–F254.

Noaillac-Depeyre, J. (1985) Localization by immunoelectron microscopy of somatostatin-containing cells in the gastric mucosa of a teleost fish, *Perca fluviatilis*. Biol. Cell **53**: 263–268.

Noaillac-Depeyre, J., Gas, N. (1979) Structure and function of the intestinal epithelial cells in the perch (*Perca fluviatilis* L.). Anat. Rec. **195**: 621–639.

Noaillac-Depeyre, J., Hollande, E. (1981) Evidence for somatostatin, gastrin and pancreatic polypeptide-like substances in the mucosa cells of the gut in fishes with and without stomach. Cell Tissue Res. **216**: 193–203.

Noakes, D.L.G. (1979) Parent-touching behavior by young fishes: incidence, function and causation. Env. Biol. Fish. **4**: 389–400.

Noakes, D.L.G., Godin, J.J. (1988) Ontogeny of behavior and concurrent developmental changes in sensory systems in teleost fishes. In: Hoar, W.S., Randall, D.J. (eds.) Fish physiology, vol. **11B**. Academic Press, New York, pp. 345–395.

Noakes, D.L.G., Skulason, S., Snorrason, S.S. (1989) Alternative life-history styles in salmonine fishes with emphasis on arctic charr, *Salvelinus alpinus*. In: Bruton, M.N. (ed.) Alternative life-history styles of animals. Kluwer Academic Publishers, Dordrecht, pp. 329–346.

Nopanitaya, W., Aghajanian, J., Grisham, J.W., Carson, J.L. (1979) An ultrastructural study of a new type of hepatic perisinusoidal cell in fish. Cell Tissue Res. **198**: 35–42.

Northcote, T.G. (1984) Mechanisms of fish migration in rivers. In: McCleave, J.D., Arnold, G.P., Dodson, J.J., Neill, W.H. (eds.) Mechanisms of migration in fishes. Plenum Publishing Corporation, New York, pp. 317–355.

Northcutt, R.G. (1977a) Elasmobranch central nervous system organization and its possible evolutionary significance. Am. Zool. **17**: 411–429.

Northcutt, R.G. (1978) Brain organization in the cartilaginous fishes. In: Hodgson, E.S., Mathewson R.F. (eds.) Sensory biology of sharks, skates and rays. Office of naval research, Department of the navy, Arlington, Va, pp. 117–193.

Northcutt, R.G. (1981) Audition and the central nervous system of fishes. In: Tavolga, W.N., Popper A.N., Fay, R.R. (eds.) Hearing and sound communication in fishes. Springer-Verlag, New York, pp. 331–355.

Northcutt, R.G. (1986) Electroreception in nonteleost bony fishes. In: Bullock, T.H., Heiligenberg, W. (eds.) Electroreception. John Wiley, New York, pp. 257–285.

Northcutt, R.G. (1989a) Brain variation and phylogenetic trends in elasmobranch fishes. J. Exp. Zool. Suppl. **2**: 83–100.

Northcutt, R.G. (1989b) The phylogenetic distribution and innervation of craniate mechanoreceptive lateral lines. In: Coombs, S., Görner, P., Münz, H. (eds.) The mechanosensory lateral line: neurobiology and evolution. Springer-Verlag, New York, pp. 17–78.

Northcutt, R.G., Bradford, M.R. Jr. (1980) New observations on the organization and evolution of the telencephalon of actinopterygian fishes. In: Ebbesson, S.O.E. (ed.) Comparative neurology of the telencephalon. Plenum Publishing Corporation, New York, pp. 41–98.

Northcutt, R.G., Butler, A.B. (1976) Retinofugal pathways in the longnose gar *Lepisosteus osseus* (Linnaeus). J. Comp. Neurol. **166**: 1–16.

Northcutt, R.G., Butler, A.B. (1991) Retinofugal and retinopetal projections in the green sunfish, *Lepomis cyanellus*. Brain Behav. Evol. **37**: 333–354.

Northcutt, R.G., Butler, A.B. (1993) The diencephalon of the Pacific herring, *Clupea harengus*: retinofugal projections to the diencephalon and optic tectum. J. Comp. Neurol. **328**: 547–561.

Northcutt, R.G., Wulliman, M.F. (1988) The visual system in teleost fishes: morphological patterns and trends. In: Atema, J., Fay, R.R., Popper A.N., Tavolga, W.N. (eds.) Sensory biology of aquatic animals. Springer-Verlag, New York, pp. 515–552.

Nozaki, M. (1985) Tissue distribution of hormonal peptides in primitive fishes. In: Foreman, R.E., Gorbman, A., Dodd, J.M., Olsson, R. (eds.) Evolutionary biology of primitive fishes. Plenum Publishing Corporation, New York, pp. 433–454.

Nozaki, M., Miyata, K., Oota Y., Gorbman, A., Plisetskaya, E.M. (1988) Colocalization of glucagon-like peptide and glucagon immunoreactivities in pancreatic islets and intestine of salmonids. Cell Tissue Res. **253**: 371–375.

Nozaki, M., Naito, N., Swanson, P., Dickhoff, W.W., Nakai, Y., Suzuki, K., Kawauchi, H. (1990) Salmonid pituitary gonadotrophs. II. Ontogeny of GTH I, and GTH II cells in the rainbow trout (*Salmo gairdneri irideus*). Gen. Comp. Endocrinol. **77**: 358–367.

Oguri, M., Omura, Y. (1973) Ultrastructure and functional significance of the pineal organ of teleost. In: Chavin, W. (ed.) Responses of fish to environmental changes. Charles C. Thomas, Springfield, pp. 412–434.

Ojha, J. (1993) Functional organization of teleosteam gills. In: Singh, B.R. (ed.) Advance in fish research, vol. 1. Narendra Publishing House, Delhi, pp. 71–98.

Ojha, J., Misra, A.K. (1993) Interspecific variations in the functional organization, dimensions and scaling of gills in relation to body weight of the Gangetic mullets, *Rhinomugil corsula* (Ham.) and *Sicamugil cascasia* (Ham.) (Mugilidae, Mugiliformes). Proc. Indian natn. Sci. Acad, B. **59**: 567–580.

Ojha, J., Singh, S.K. (1986) Scanning electron microscopy of the gills of a hill-stream fish, *Danio dangilla* (Ham). Arch. Biol. (Bruxelles) **97**: 455–467.

Oka, S., Chiba, A., Honma, Y., Iwanaga, T., Fujita, T. (1993) Development of the caudal neurosecretory system of the chum salmon, *Oncorhynchus keta*, as revealed by immunohistochemistry for urotensins I and II. Cell Tissue Res. **272**: 221–226.

Oka, S., Chiba, A., Honma, Y. (1995) Immunohistochemical distribution of urotensins I and II in the central nervous system of the Senegal bichir, *Polypterus senegalus*. Zool. Sci. **12**: 311–315.

Olsen, K.H., Liley, N.R. (1993) The significance of olfaction and social cues in milt availability, sexual hormone status, and spawning behaviour of male rainbow trout (*Oncorhynchus mykiss*). Gen. Comp. Endocrinol. **89**: 107–118.

Olson, K.R. (1994) Circulatory anatomy in bimodally breathing fish. Am. Zool. **34**: 280–288.

Olson, K.R., Munshi, J.S.D., Ghosh, T.K. and Ojha, J. (1990) Vascular organization of the head and respiratory organs of the air-breathing catfish, *Heteropneustes fossilis*. J. Morphol. **203**: 165–179.

Olson, K.R., Roy, P.K., Ghosh, T.K., Munshi, J.S.D. (1994) Microcirculation of gills and accessory respiratory organs from the air-breathing snakehead fish, *Channa punctata, C. gachua*, and *C. marulius*. Anat. Rec. **238**: 92–107.

Olson, K.R., Chavez, A., Conklin, D.J., Cousins, K.L., Farrell, A.P., Ferlic, R., Keen, J.E., Kne, T., Kowalski, K.A., Veldnan, T. (1994) Localization of angiotensin II responses in the trout cardiovascular system. J. Exp. Biol. **194:** 117–138.

Olsson, R. (1990) Evolution of chordate endocrine organs. In: Apple, A., Scanes, C.G., Stetson, M.H. (eds.) Progress in comparative endocrinology. Wiley-Liss, New York, pp. 272–281.

Omeljaniuk, R.J., Habibi, H.R., Peter, R.E. (1989) Alterations in pituitary GnRH and dopamine receptors associated with the seasonal variation and regulation of gonadotropin release in the goldfish (*Carassius auratus*). Gen. Comp. Endocrinol. **74**: 392–399.

Omura, Y., Korf, H-W, Oguri, M., Oksche, A. (1990) Properties of the blood-brain and blood-cerebrospinal fluid barrier in the circumventricular organs of the diencephalic roof of teleosts. Zool. Sci. **7**: 783–799.

Ono, M., Kawauchi, H. (1994) The somatolactin gene. In: Sherwood, N., Hew, C. (eds.) Fish physiology, vol. 13. Academic Press, San Diego, pp. 159–177.

Onozato, H., Watabe, N. (1979) Studies on fish scale formation and resorption. III. Fine structure and calcification of the fibrillary plates of the scales in *Carassius auratus* (Cypriniformes: Cyprinidae). Cell Tissue Res. **201**: 409–422.

Orvig, T. (ed.) (1968) Current problems of lower vertebrate phylogeny. Nobel Symposium 4. Almqvist and Wiksell, Stockholm.

Orvig, T. (1977) A study of odontodes ('dermal teeth') from developmental, structural, functional and phyletic points of view. In: Andrews, S.M., Miles, R.S., Walker, A.D. (eds.) Problems in vertebrate evolution. Linnean Soc. Symp. Ser. **4**: 53–75.

Orvig, T. (1980) Histologic studies of ostracoderms, placoderms and fossil elasmobranchs. 4. Ptyctodontid tooth plate and their bearing on holocephalan ancestry: the condition of *Ctenurella* and *Ptyctodus*. Zool. Scr. **9**: 219–239.

Orvig, T. (1985) Histologic studies of ostracoderms, placoderms and fossil elasmobranchs. 5. Ptyctodontid tooth plates and their bearing on holocephalan ancestry: the condition of chimaerids. Zool. Scr. **14**: 55–79.

Osse, J.W.M. (1985) Jaw protrusion, an optimization of the feeding apparatus of teleosts ? Acta Biotheor. **34**: 219–232.

Osse, J.W.M. (1990) Form changes in fish larvae in relation to changing demands of function. Neth, J. Zool. **40**: 262–385.

O'Toole, L.B., Armour, K.J., Decourt, C., Hazon, N., Lahlou, B., Henderson, I.W. (1990) Secretory patterns of 1 α-hydroxycorticosterone in the isolated perifused interrenal gland of the dogfish, *Scyliorhinus canicula*. J. Mol. Endocr. **5**: 55–60.

Pang, P.K.T., Griffith, R.W., Atz, J.W. (1977) Osmoregulation in elasmobranchs. Am. Zool. **17**: 365–377.

Pang, P.K.T., Pang, R.K., Wendelaar Bonga, S.E. (1989) Hormones and calcium regulation in vertebrates: an evolutionary and overall consideration. In: Pang, P.K.T., Schreibman, M.P. (eds.) Vertebrate endocrinology: fundamentals and biomedical implications, vol. 3. Academic Press, San Diego, pp. 343–352.

Pankhurst, N.W., Carragher, J.F. (1991) Seasonal endocrine cycles in marine teleosts. In: Scott, A.P., Sumpter, J.P., Kime, D.E., Rolfe, M.S. (eds.) Reproductive physiology of fish. Proc. Fourth Int. Symp. Univ. of East Anglia, Norwich, U.K. 7–12 July 1991, pp. 131–135.

Parsons, G.R., Grier, H.J. (1992) Seasonal changes in shark testicular structure and spermatogenesis. J. Exp. Zool. **261**: 173–174.

Partridge, J.C. (1990) The colour sensitivity and vision in fishes. In: Herring, P.J., Campell, A.K., Whitfield, M., Maddock, L. (eds.) Light and life in the sea. Cambridge University Press, Cambridge, pp. 167–184.

Partridge, J.C., Archer, S.N. (1992) Single and multiple visual pigments in deep-sea fishes. J. mar. Biol. Ass. U.K. **72**: 113–130.

Partridge, J.C., Archer, S.N., Lythgoe, J.N. (1988) Visual pigments in the individual rods of deep-sea fishes. J. Comp. Physiol. A **162**: 543–550.

Partridge, J.C., Shand, J., Archer, S.N., Lythgoe, J.N., Groningen-Luyben, W.A.H.M. van. (1989) Interspecific variation in the visual pigments of deep-sea fishes. J. Comp. Physiol. **A164**: 513–529.

Parzefall, J. (1986) Behavioural ecology of cave-dwelling fishes. In: Pitcher T.J. (ed.) The behaviour of teleost fishes. Croom Helm, London, pp. 433–458.

Pelster, B. (1994) Adrenergic control of swimbladder perfusion in the European eel *Anguilla anguilla*. J. Exp. Biol. **189**: 237–250.

Pelster, B., Scheid, P. (1992) Countercurrent concentration and gas secretion in the fish swim bladder. Physiol. Zool. **65**: 1–16.

Pelster, B., Scheid, P. (1993a) Metabolism of the swimbladder epithelium and the single concentrating effect. Comp, Biochem. Physiol. **105A:** 383–388.

Pelster, B., Scheid, P. (1993b) Glucose metabolism of the swimbladder tissue of the European eel *Anguilla anguilla*. J. Exp. Biol. **185**: 169–178.

Pelster, B., Scheid, P. (1994) Gegenstromkonzentrierung in der Schwimmblasse. Die Erzeugung hoher Gaspartialdriicke durch das Zusammenspiel von. Zellularem stoffwechsel and Organdurchblutung. Naturwissenschaften **81**: 473–480.

Pelster, B., Bridges, C.R., Grieshaber, M.R. (1988) Respiratory adaptations of the burrowing marine teleost *Lumpenus lampretaeformis* (Walboum). II. Metabolic adaptations. J. Exp. Mar. Biol. Ecol. **124**: 43–55.

Pelster, B., Hicks, J., Driedzic., W.R. (1994) Contribution of the pentose phosphate shunt to the formation of CO_2 in swimbladder tissue of the eel. J. Exp. Biol. **197**: 119–128.

Perks, A.M. (1987) The neurohypophysis: the place of comparative studies in biology and medicine. In: Pang P.K.T., Schreibman, M.P. (eds.) Vertebrate endocrinology: fundamentals and biomedical implications. vol. 2. Academic Press, San Diego, pp. 9–44.

Perrone, M. Jr., Zaret, T.M. (1979) Parental care patterns of fishes. Amer. Nat. **113**: 351–361.

Perry, S.F., Laurent, P. (1990) The role of carbonic anhydrase in carbon dioxide excretion, acid-base balance and ionic regulation in aquatic gill breathers. In: Truchot J-P, Lahlou, B. (eds.) 1. Animal nutrition and transport processes. 2. Transport, respiration and excretion: Comparative and environmental aspects. Comp. Physiol. **6**: 39–57.

Perry, S.F., Wood, C.M. (1989) Control and coordination of gas transfer in fishes. Can. J. Zool. **67**: 2961–2970.

Persson, L. (1991) Interspecific interactions. In: Winfield, I.J., Nelson, J.S. (eds.) Cyprinid fishes. Systematics, biology and exploitation. Chapman and Hall, London, New York, pp. 530–551.

Peter, R.E. (1983) The brain and neurohormones in teleost reproduction. In: Hoar, W.S., Randall, D.J., Donaldson, E.M. (eds.) Fish physiology, vol. 9 A. Academic Press, New York, pp. 97–135.

Peter, R.E., Fryer, J.N. (1983) Endocrine functions of the hypothalamus of actinopterygians. In: Davis, R.E., Northcutt, R.G. (eds.) Fish neurobiology, vol. 2: Higher brain areas and functions. University of Michigan Press, Ann Arbor, pp. 165–201.

Peter, R.E., Habibi, H.R., Marchant, T.A., Nahorniak, C.S. (1987) Vertebrate gonadotropin-releasing hormones: phylogeny and structure function relationships. Ann. New York Acad. Sci. **519**: 299–309.

Peter, R.E., Habibi, H.R., Chang, J.P., Nahorniak, C.S., Y.U. K.L., Huang, Y.P., Merchant, T.A. (1990) Actions of gonadotropin-releasing hormone (GnRH) in the goldfish. In: Apple, A., Scanes, C.G., Stetson, M.H. (eds.) Progress in comparative endocrinology. Wiley-Liss, Inc, New York, pp. 393–398.

Peter, R.E., Trudeau, V.L., Sloley, B.D., Peng, C., Nahorniak, C.S. (1991) Actiona of catecholamines, peptides and sex steroids in regulation of gonadotropin-II in the goldfish. In: Scott, A.P., Sumpter, J.P., Kime, D.E., Rolfe, M.S. (eds.) Reproductive physiology of fish. Proc. Fourth Int. Symp. Univ. of East Anglia., Norwich, U.K., 7–12 July 1991, pp. 30–34.

Peters, R.C. (1993) Orientation in catfish and dogfish. Paper No. 33. Paper presented at R.I.N. 93. Orientation and Navigation. Birds, humans and other animals. The 1993 Conference of the Royal Institute of Navigation, 15–17 April 1993. Oxford.

Peters, R.C., Kotrschal, K., Atema, J. (1989) A novel chemosensory system in fish: electrophysiological evidence for mucus detection by solitary chemosensory cells in rocklings (*Ciliata mustela*, Gadidae). Biol. Bull. **177**: 329.

Peters, R.C., Kotrschal, K., Krautgartner, W.D. (1991) Solitary chemoreceptor cells of *Ciliata mustela* (Gadidae, Teleostei) are tuned to mucoid stimuli. Chem. Senses **16**: 31–42.

Phleger, C.F. (1987) Bone lipids in tropical reef fishes. Comp. Biochem. Physiol. **86B**: 509–512.

Phleger, C.F. (1988a) Bone lipids of Jamaican reef fishes. Comp. Biochem. Physiol. **90B**: 279–283.

Phleger, C.F. (1988b) The importance of skull lipid as an energy reserve during starvation in the ocean surgeon, *Acanthurus bahianus*. Comp. Biochem. Physiol. **91A**: 97–100.

Phleger, C.F. (1991) Biochemical aspects of buoyancy in fishes. In: Hochachka, P.W., Mommsen, T.P. (eds.) Biochemistry and molecular biology of fishes, vol. 1. Elsevier, Amsterdam, pp. 209–247.

Phleger, C.F., Laub, R.I. (1989) Skeletal fatty acids in fish from different depths off Jamaica. Comp. Biochem. Physiol. **94B**: 339–334.

Phleger, C.F., Wambeke, S.R. (1994) Bone lipids and fatty acids of Peru fish. Comp. Biochem. Physiol. **109B**: 145–152.

Pitcher, T.J. (1986) Predators and food are the keys to understanding fish shoals: a review of recent experiments. Naturaliste Can. (Rev. Ecol. syst.) **113**: 225–233.

Pitcher, T.J., House, A.C. (1987) Foraging rules for group feeders: area copying depends upon food density in shoaling goldfish. Ethology **76**: 161–167.

Pitcher, T.J., Turner, J.R. (1986) Danger at dawn: experimental support for the twilight hypothesis in shoaling minnows. J. Fish Biol. **29** (Suppl. A): 59–70.

Pitcher, T.J., Lang, S.H., Turner, J.A. (1988) A risk-balancing trade off between foraging rewards and predation hazard in a shoaling fish. Behav. Ecol. Sociobiol. **22**: 225–228.

Piiper, J. (1989) Gas-exchange efficiency of fish gills and bird lung. In: Paganelli, C.V., Farhi, L.E. (eds.) Physiological function in special environments. Springer-Verlag, New York, pp. 159–171.

Piiper, J., Scheid, P. (1989) Gas exchange. Theory, models, and experimental data. In: Wood, S.C., (ed.) Comparative pulmonary physiology. Current concepts. Marcel Dekker, New York, pp. 369–416.

Piomelli, D., Pinto. A., Tota, B. (1985) Divergence in vascular actions of prostacyclin during vertebrate evolution. J. Exp. Zool. **233**: 127–131.

Platt, C. (1983) The peripheral vestibular system of fishes. In: Northcutt, R.G., Davis, R.E. (eds.) Fish neurobiology, vol. 1. Brain stem and sense organs. University of Michigan Press, Ann Arbor, pp. 89–124.

Plisetskaya, E. (1985) Some aspects of hormonal regulation of metabolism in agnathans. In: Foreman, R.E., Gorbman, A., Dodd, J.M., Olsson, R. (eds.) Evolutionary biology of primitive fishes. Plenum Publishing Corporation, New York, pp. 339–361.

Plisetskaya, E.M. (1989) Physiology of fish endocrine pancreas. Fish Physiol. Biochem. **7**: 39–48.

Plisetskaya, E.M. (1990) Recent studies of fish pancreatic hormones: selected topics. Zool. Sci. **7**: 335–353.

Pool, C.W., Raamsdonk, W. van. Diegenbach, P.C., Mijzen, P., Schenkkan, E.J., Stelt, A. van der (1976) Muscle fibre typing with sera against myosin and actin. Acta Histochem. **57**: 20–33.

Popper, A.N. (1983) Organization of the inner ear and auditory processing. In: Northcutt, R.G., Davis, R.E. (eds.) Fish neurobiology, vol. 1. Brain stem and sense organs. University of Michigan Press, Ann Arbor, pp. 125–178.

Popper, A.N., Fay, R.R. (1973) Sound detection and processing by teleost fishes: a critical review. J. Acoust. Soc. Am. **53**: 1515–1529.

Popper, A.N., Fay, R.R. (1977) Structure and function of the elasmobranch auditory system. Am. Zool. **17**: 443–452.

Popper, A.N., Fay, R.R. (1984) Sound detection and processing by teleost fish: a selective review. In: Bolis, L., Keynes, R.D., Maddrell, S.H.P. (eds.) Comparative physiology of sensory systems. Cambridge University Press, Cambridge, pp. 67–101.

Popper, A.N., Fay, R.R. (1993) Sound detection and processing by fish: critical review and major research questions. Brain Behav. Evol. **41**: 14–38.

Popper, A.N., Platt, C. (1993) Inner ear and lateral line. In: Evans, D.H. (ed.) The physiology of fishes. CRC Press, Boca Raton, pp. 99–136.

Popper, A.N., Platt, C., Saidel, W.M. (1982) Acoustic functions in the fish ear. Trends Neurosci. **5**: 276–280.

Potts, G.W. (1985) The nest structure of the corkwing wrasse, *Crenilabrus melops* (Labridae: Teleostei), J. Mar. Biol. Ass. U.K. **65**: 531–546.

Potts, G.W., Wootton, R.J. (eds.)(1984) Fish reproduction: Strategies and tactics. Academic Press, London.

Potts, W.T.W., Hedges, A.J. (1991) Gill potentials in marine teleosts. J. Comp. Physiol, B **161**: 401–405.

Potts, W.T.W., McWilliams, P.G. (1989) The effects of hydrogen and aluminium ions on fish gills. In: Morris, R., Taylor, E.W., Brown, D.J.A., Brown, J.A. (eds.) Acid toxicity and aquatic animals. Cambridge University Press, Cambridge, pp. 201–220.

Poulin, R., FitzGerald, G.J. (1989) Risk of parasitism and microhabitat selection in juvenile sticklebacks. Can. J. Zool. **67**: 14–18.

Prasad Rao, P.D. (1994) Neuroendocrine control of pituitary gland in teleosts. In: Singh H.R. (ed.) Advances in fish biology. Hindustan Publishing Corporation, Delhi, pp. 114–120.

Prased Rao, P.D., Finger, T.E. (1984) Asymmetry of the olfactory system in the brain of the winter flounder, *Pseudopleurorectes americanus*. J. Comp. Neurol. **225**: 492–510.

Prasad Rao, P.D., Sharma, S.C. (1982) Retinofugal pathways in juvenile and adult channel catfish, *Ictalurus (Ameiurus) punctatus*: an HRP and autoradiographic study. J. Comp. Neurol. **210**: 37–48

Prasad Rao, P.D., Job, T.J., Schreibman, M.P. (1993) Hypophysiotropic neurons in the hypothalamus of the catfish *Clarias batrachus*: A cobaltous lysine and HRP study. Brain Behav. Evol. **42**: 24–38.

Price, D.J. (1984) Genetics of sex determination in fishes—a brief review. In: Potts G.W., Wooton, R.J. (eds.) Fish reproduction: strategies and tactics, Academic Press, London, pp. 77–89.

Price, D.J. (1985) Genetics of susceptibility and resistance to disease in fishes. J. Fish Biol. **26**: 509–519.

Prunet, P., Auperin, B. (1994) Prolactin receptors. In: Sherwood, N., Hew, C. (eds.) Fish physiology. vol. 13. Academic Press, San Diego, pp. 367–391.

Purdom, C.E. (1979) Genetics of growth and reproduction in teleosts. Symp. Zool. Soc. Lond. No. **44**: 207–217.

Quesada, J., Villena, M.I., Agulleiro, B. (1990) Structure of the spleen of the sea bass (*Dicentrarchus labrax*): a light and electron microscopic study. J. Morphol. **206**: 273–281.

Quesada, J., Villena, M.I., Navarro, V. (1994) Ontogeny of the sea bass spleen (*Dicentrarchus labrax*): a light and electron microscopic study. J. Morphol. **221:** 161–176.

Quinn, T.P., Brodeur, R.D. (1991) Intra-specific variations in the movement patterns of marine animals. Am. Zool. **31**: 231–241.

Raamsdonk, W., van, Pool, C.W., Kronnie, G.T. (In press) Differentiation of muscle fibre types in the teleost *Brachydamo rerio* (cited from Raamsdonk, W. van. 1978 Thesis). De Ontwikkeling van de Rompmusculatuur van *Brachydamnio rerio* (de zebravis). Zoologisch Laboratorium van de Universiteit van Amsterdam.

Raguse-Degener, G., Pietschmann, M., Walvig, F., Stolte, H. (1980) Excretory systems in the hagfish *Myxine glutinosa*. Contr. Nephrol. **19**: 1–8.

Rajjo, I.M., Vigna, S.R., Crim, J.W. (1988) Actions of cholecytokinin-related peptides on the gallbladder of bony fishes *in vitro*. Comp. Biochem. Physiol. **90C**: 267–273.

Rajjo, I.M., Vigna, S.R., Crim, J.W. (1989) Immunohistochemical localization of bombesin-like peptides in the digestive tract of the bowfin, *Amia calva*. Comp. Biochem. Physiol. **94C**: 405–409.

Randall, D.J. (1968) Functional morphology of the heart in fishes. Am. Zool. **8**: 179–189.

Randall, D.J. (1970) The circulatory system. In: Hoar, W.S., Randall, D.J. (eds.) Fish physiology, vol. 4. Academic Press, New York, pp. 133–172.

Randall, D.J. (1985) Shunts in fish gills. In: Johansen, K., Burggren, W. (eds.) Cardiovascular shunts: phylogenetic, ontogenetic and clinical aspects. Alfred Benzon Symposium No. 21. Munskgaard, Copenhagen, pp. 71–87.

Randall, D.J. (1990) Control and coordination of gas exchange in water breathers. In: Boutlier, R.G. (ed.) Advances in comparative and environmental physiology, vol. 6. Springer-Verlag, Berlin, pp. 253–278.

Randall, D.J. (1994) Cardio-respiratory modeling in fishes and the consequences of the evolution of air-breathing. Cardioscience **5**: 165–171.

Randall, D.J., Lin, H. (1993) Effects of water pH on gas and ion transfer across fish gills. In: Rankin J.C., Jensen, F.B. (eds.) Fish ecophysiology. Chapman and Hall, London, New York, pp. 265–275.

Randall, D.J., Wright, P.A. (1987) Ammonia distribution and excretion in fish. Fish Physiol. Biochem. **3**: 107–120.

Randall, D.J., Wright, P.A. (1989) The interaction between carbon dioxide and ammonia excretion and water pH in fish. Can. J. Zool. **67**: 2936–2942.

Randall, D.J., Lin, H., Wright, P.A. (1991) Gill flow and the chemistry of the boundary layer. Physiol. Zool. **64**: 26–38.

Randall, D.J., Mense, D., Boutlier, R.G. (1987) The effects of burst swimming on aerobic swimming in chinook salmon (*Oncorhynchus tshawytscha*). Mar. Behav. Physiol. **13**: 77–88.

Randall, D.J., Wood, C.M., Perry, S.F., Bergman, H., Maloiy, G.M.O., Mommsen. T.P., Wright, P.A. (1989) Urea excretion as a strategy for survival in fish living in a very alkaline environment. Nature (London) **337**: 165–166.

Randall, C.F., Bromage, N.R., Thorpe, J.E., Miles, M.S. (1994) Photoperiod, melatonin and the timing of smoltification in salmonid fish. Aquaculture **121**: 289–300.

Rand-Weaver, M., Kawauchi, H. (1993) Growth hormone, prolactin and somatolactin: a structural overview. In: Hochachka, P.W., Mommsen, T.P. (eds.) Biochemistry and molecular biology of fishes, vol. 2. Elsevier, Amsterdam, pp. 39–56.

Rankin, M.A. (1991) Endocrine effects on migration. Am. Zool. **31**: 217–230.

Raschi, W. (1986) A morphological analysis of the ampullae of Lorenzini in selected skates (Pisces, Rajoidei). J. Morphol. **189**: 225–247.

Raschi, W., Adams, W.H. (1988) Depth-related modifications in the electroreceptive system of the eurybathic skate, *Raja radiata* (Chondrichthyes: Rajidae). Copeia **1988**: 116–123.

Raschi, W., Mackanos, L.A. (1987) Evolutionary trends in the peripheral component of the chondrichthyan electroreceptors: a morphological reivew. Arch. Biol. (Bruxelles) **98**: 163–186.

Raschi, W., Mackanos, L.A. (1989) The structure of the ampullae of Lorenzini in *Dasyatis garouaensis* and its implications on the evolution of freshwater electroreceptive systems. J. Exp. Zool. Suppl. **2**: 101–111.

Reale, E., Lacy, E.R. (1987) Some observations on the elasmobranch kidney. Arch. Biol. (Bruxelles) **98**: 231–241.

Reichenbach-Klinke, H.H. (1966) Krankheiten und Schädigungen der Fische. Gustav Fischer Verlag, Stuttgart.

Reebs, S. (1992) Sleep, inactivity and circadian rhythms in fish. In: Ali, M.A. (ed.) Rhythms in fishes. Plenum Press, New York, pp. 127–135.

Rees, J.F., Baguet, F. (1988) Metabolic control of spontaneous glowing in isolated photophores of *Porichthys*. J. Exp. Biol. **135**: 289–299.

Rees, J.F., Kacuass, M., Baguet, F. (1991) Purinergic receptors as a regulatory process of luminescence in *Porichthys* luminous organs,. J. Exp. Zool. **258**: 181–188.

Rees, J.F., Thompson, E.M., Baguet, F., Tsuji, F.I. (1992) Evidence for the utilization of coelenterazine as the luminescent substrate in *Argyropelecus* photophores. Mol. Mar. Biol. Biotech. **1**: 219–225.

Reese, E.S. (1989) Orientation behavior of butterflyfishes (family Chaetodontidae) on coral reefs: spatial learning of route specific landmarks and cognitive maps. Env. Biol. Fish. **25**: 79–86.

Reese, E.S. (1991) How behavior influences community structure of butterflyfishes (Family Chaetodontidae) on Pacific coral reefs. Ecol. Int. Bull. **19**: 29–41.

Reese, E.S. (1993) Reef fishes as indicators of conditions on coral reefs. Proc. Collon. Global Aspects of coral reefs: health, hazards, and history. University of Miami, Florida, pp. 59–65.

Reif, W.E. (1982) Evolution of dermal skeleton and dentition in biology. In: Hecht, M.K., Wallace, B., Prauce, G.T. (eds.) Evolutionary vertebrates/The odontode regulation theory. Evol. Biol. **15**: 287–368.

Reifel, C.W. (1988) Endocrine cells in the gastrointestinal tract of a stomachless teleostean fish. Anat. Anz. Jena **167**: 259–263.

Reifel, C.W., Linden, R.D. (1983) Endocrine cells in the gastrointestinal tracts of 3 teleostean species. Anat. Anz. Jena **154**: 413–418.

Reifel, C.W., Martin-Sorensen, M., Samloff, I.M. (1983) Gastrin immunoreactive cells in the gastrointestinal tracts from four species of fish. Can. J. Zool. **61**: 1464–1468.

Reshkin, S.J., Cassano, G., Womersley, C., Ahearn, G.A. (1988) Preservation of glucose transport and enzyme activity in fish intestinal brush-border and basolateral membrane vesicles. J. Exp. Biol. **140**: 123–135.

Reshkin, S.J., Vilella, S., Ahearn, G.A., Storelli, C. (1989) Basolateral inositol transport by intestines of carnivorous and herbivorous teleosts. Am. J. Physiol. **256**: G 509-G516.

Resink, J.W., Voorthius, P.K., Van den Hurk, R., Vullings, H.G.B., Van Oordt, P.G.W.J. (1989) Pheromone detection and olfactory pathways in the brain of the female African catfish, *Clarias gariepinus*. Cell Tissue Res. **256**: 337–345.

Restrepo, D., Boyle, A.G. (1991) Stimulation of olfactory receptors alters regulation of (Cai) in olfactory neurons of the catfish (*Ictalurus punctatus*). J. Membrane Biol. **120**: 223–232.
Restrepo, D., Miyamoto, T., Bryant, B.P., Teeter, J.H. (1990) Odor stimuli trigger influx of calcium into olfactory neurons of the channel catfish. Science **249**: 1166–1168.
Reutter, K. (1978) Taste organ in the bullhead (Teleostei). Adv. Anat. Embryol. Cell Biol. **55**: 1–98.
Reutter, K. (1982) Taste organ in the barbel of the bullhead. In: Hara, T.J. (ed.) Chemoreception in fishes. Elsevier, Amsterdam, pp. 77–91.
Reutter, K. (1986) Chemoreceptors. In: Bereiter-Hahn, J., Matoltsy, A.G., Richards, K.S. (eds.) Biology of the integument, vol. 2. Vertebrates. Springer-Verlag, Berlin, pp. 586–604.
Reutter, K. (1992) Structure of the peripheral gustatory organ, represented by the siluroid fish *Plotosus lineatus* (Thünberg) In: Hara, T.J. (ed.) Fish chemoreception. Chapman and Hall, London, New York, pp. 60–78.
Reutter, K., Witt, M. (1993) Morphology of vertebrate taste organs and their nerve supply. In: Simon, S.A., Roper, S.D. (eds.) Mechanisms of taste transduction. CRC Press, Boca Raton, pp. 29–82.
Reynolds, J.D., Gross, M.R. (1992) Female mate preference enhances offspring growth and reproduction in a fish, *Poecilia reticulata*. Proc, R. Soc. Lond. **B. 250**: 57–62.
Reynolds, J.D., Gross, M.R., Coombs, M.J. (1993) Environmental conditions and male morphology determine alternative mating behaviour in Trinidadian guppies. Anim. Behav. **45**: 145–152.
Ribelin, W.E., Migaki, G. (1975) The pathology of fishes. University of Wisconsin Press, Madison.
Richards, B.A., Wainwright, P.C. (1995) Scaling the feeding mechanism of largemouth bass (*Micropterus salmoides*): kinematics of prey capture. J. Exp. Zool. **198**: 419–433.
Richman, N.J., Ford, C.A., Helms, L.M.H., Cooke, I.M., Pang, P.K.T., Grau, E.G. (1991) The loss of $^{45}CA^{2+}$ associated with prolactin release from the tilapia (*Oreochromis mossambicus*) rostral pars distalis. Gen Comp. Endocrinol. **83**: 56–67.
Ricker W.E. (1979) Growth rates and models. In: Hoar, W.S., Randall, D.J. (eds.) Fish physiology, vol. 8. Academic Press, New York, pp. 677–742.
Rishi, K.K. (1989) Current status of fish cytogenetics. In: Das, P., Jhingaran, A.G. (eds.) Fish genetics in India. Today and Tomorrow's Printers and Publishers, New Delhi, pp. 1–20.
Rivers, R., Lacy, E.R., Grabowski G.M., Luciano, L., Reale, E. (1995) Optical analysis of the flagellar ribbon beat cycle in the fish nephron., J. Submicrosc Cytol. Pathol. **27**: 407–416.
Roberts, J.L. (1994) Temperature acclimatation and behavioural thermoregulation in cold-blooded animals. Federation Proc. **33**: 2155–2161.
Roberts, R.J., Bullock, A.M. (1976) The dermatology of marine teleost fish. II. Dermatopathology of the integument. Oceanogr. Mar. Biol. Ann. Rev. **14**: 227–246.
Roberts, R.J., Bullock, A.M. (1989) Nutritional pathology. In: Halver, J.E. (ed.) Fish nutrition, (Second edition). Academic Press, New York, pp. 424–475.
Robertson, D.R. (1986) The ultimobranchial body. In: Pang, P.K.T., Schreibman, M.P., Vertebrate endocrinology: fundamentals and biomedical implications, vol. 1. Academic Press, Orlando, pp. 235–259.
Robertson, D.R. (1989) Regulation of calcium and phosphate by intestine. In: Pang P.K.T., Schreibman, M.P. (eds.) Vertebrate endocrinology: fundamentals and biomedical implications, vol. 3. Academic Press, San Diego, pp. 243–275.
Robinson, C.J., Pitcher, T.J. (1989a) The influence of hunger and ration level on shoal density, polarization and swimming speed of herring, *Clupea harengus* L. J. Fish Biol. **34**: 631–633.
Robinson, C.J., Pitcher, T.J. (1989b) Hunger motivation as a promotor of different behaviours within a shoal of herring: selection for homogeneity in fish shoal? J. Fish Biol. **35**: 459–460.
Rocha, E., Monteiro, R.A.F., Pareira, C.A. (1994a) The liver of the brown trout, *Salmo trutta fario*: a light and electron microscope study. J. Anat. **185**: 241–249.
Rocha, E., Monteiro, R.A.F., Pareira, C.A. (1994b) Presence of rodlet cells in the intrahepatic biliary ducts of the brown trout, *Salmo trutta fario* Linnaeus, 1758 (Teleostei, Salmonidae). Can. J. Zool. **72**: 1683–1687.
Rocha, E., Monteiro, R.A.F., Pareira, C.A. (1995) Microanatomical organization of hepatic stroma of the brown trout, *Salmo trutta fario* (Teleostei, Salmonidae): qualitative and quantitative approach. J. Morphol. **223**: 1–11.

Roff, D.A. (1984) The evolution of life history parameters in teleosts. Can. J. Fish Aquat. Sci. **41**: 989–1000.

Roff, D.A. (1988) The evolution of migration and some life history parameters in marine fishes. Env. Biol. Fish. **22**: 133–146.

Roff, D.A. (1991) Life history consequences of bioenergetic and biomechanical constraints on migration. Am. Zool. **31**: 205–215.

Rogers, P.H., Popper, A.N., Hastings, M.C., Saidel, W.M. (1988) Processing of acoustic signals in the auditory system of bony fish. J. Acoust Soc. Am. **83**: 338–349.

Rogers, W., Barlow, G.W. (1991) Sex differences in mate choice in a monogamous biparental fish, the Midas cichlid (*Cichlasoma citrinellum*). Ethology **87**: 249–261.

Romano, P.M., Ahearn, G.A., Storelli, C. (1989) Na-dependent L-glutamate transport by eel intestinal, BBMV: role of K^+ and CL^-. Am. J. Physiol. **257**: R180.

Rombough, P.J., Moroz, B.M. (1990) The scaling and potential importance of cutaneous and branchial surfaces in respiratory gas exchange in young chinook salmon (*Oncorhynchus tshawytscha*) J. Exp. Biol. **154**: 1–12.

Rombout, J.H.W.M. (1977) Enteroendocrine cells in the digestive tract of *Barbus conchonius* (Teleostei, Cyprinidae). Cell Tissue Res. **185**: 435–450.

Rombout, J.H.W.M., van den Berg, A.A. (1989) Immunological importance of the second gut segment of carp. I. Uptake and processing of antigens by epithelial cells and macrophages. J. Fish Biol. **35**: 13–22.

Rombout, J.H.W.M., Grinten, C.P.M. van der., Peeze Binkhorst, F.M., Tavernl-Thiele, J.J., Schooneveld, H. (1986a) Immunocytochemical identification and localization of peptide hormones in the gastro-entero-pancreatic (GEP) endocrine system of the mouse and a stomachless fish, *Barbus conchonius*. Histochemistry **84**: 471–483.

Rombout, J.H.W.M., Block, L.J., Lamers, C.H.J., Egberts, E. (1986b) Immunization of carp. (*Cyprinus carpio*) with *Vibrio anguillarum* bacterin: indications for a common mucosal immune system. Dev. Comp. Immunol. **10**: 341–351.

Rombout, J.H.W.M., Bot, H.E., Taverne-Thiele, J.J. (1989a) Immunological importance of the second gut segment of carp. II. Characterization of mucosal leucocytes. J. Fish Biol. **35**: 167–178.

Rombout, J.H.W.M., van den Berg, A.A., van dan Berg, C.T.G.A., Witte, P., Egberts, E. (1989b) Immunological importance of the second gut segment of carp. III. Systemic and/or mucosal immune responses after immunization with soluble or particulate antigen. J. Fish Biol. **35**: 179–186.

Rombout, J.H.W.M., Taverne-Thiele, A.J., Villena, M.I. (1993a) The gut-associated lymphoid tissue (GALT) of carp. (*Cyprinus carpio* L.): an immunocytochemical analysis. Dev. Comp. Immunol. **17**: 55–66.

Rombout, J.H.W.M., Taverne, N., Kamp Mvande, Taverne-Thiele, A.J. (1993b) Differences in mucus and serum immunoglobin of carp. (*Cyprinus carpio* L.). Dev. Comp. Immunol. **17**: 309–317.

Roper, S.D. (1989) The cell biology of vertebrate taste receptors. Ann. Rev. Neurosci. **12**: 329–353.

Rowland, W.J. (1974) Reproductive behavior of the fourspine stickleback, *Apeltes quadracus*. Copeia **1974**: 183–194.

Ryer, C.H., Olla, B.L. (1995) Influences of food distribution on fish foraging behaviour. Anim. Behav. **49**: 411–418.

Saha, N., Ratha, B.K. (1986) Effects of ammonia stress on ureogenesis in a freshwater air-breathing teleost, *Heteropneustes fossilis*. In: Kon, O.L., Chung, M.C.M., Hwang, P.L.H., Leong, S.F., Loke, K.H., Thiyagaraj, Wong P.T.H. (eds.) Contemporary themes in biochemistry, vol. 6. Cambridge University Press, Cambridge, pp. 342–343.

Saha, N., Ratha, B.K. (1987) Active ureogenesis in a freshwater air-breathing teleost, *Heteropneustes fossilis*. J. Exp. Zool. **241**: 137–141.

Saha, N., Ratha, B.K. (1989) Comparative study of ureogenesis in freshwater air-breathing teleosts. J. Exp. Zool. **252**: 1–8.

Saidapur, S.K. (1978) Follicular atresia in the ovaries of nonmammalian vertebrates. Int. Rev. Cytol. **54**: 225–244.

Saidapur, S.K. (1982) Structure and function of postovulatory follicles (corpora lutea) in the ovaries of nonmammalian vertebrates. Int. Rev. Cytol. **75**: 243–285.

Saidapur, S.K. (1989) Reproductive cycles: an overview. In: Saidapur, S.K. (ed.) Reproductive cycles of Indian vertebrates. Allied Publishers Limited, New Delhi, pp. 427–437.

Saidel, W.M., Butler, A.B. (1991) Retinal projections in the freshwater butterfly fish, *Pantodon buchholzi* (Osteoglossoidei) II. Differential projections of the dorsal and ventral hemiretinas. Brain Behav. Evol. **38**: 154–168.

Samejima, M., Happe, H.K., Murrin, L.C., Pfeiffer, R.E., Ebadi, M. (1994) Distribution of cholinergic and dopaminergic receptors in rainbow trout pineal gland. J. Pineal Res. **16**: 37–43.

Sampath Kumar, R. (1994) Existence of adrenodoxin-like peptide in teleost fishes: immunohistochemical evidence. J. Exp. Zool. **270**: 557–561.

Sanderson, S.L., Wassersug, R. (1993) Convergent and alternative designs for vertebrate suspension feeding. In: Hanken, J., Hall, B.K. (eds.) The skull, vol. 3: functional and evolutionary mechanisms. University of Chicago Press, Chicago, pp. 37–112.

Sanderson, S.L., Cech, J.J. Jr, Cheer, A.Y. (1994) Paddlefish buccal flow velocity during ram suspension feeding and ram ventilation. J. Exp. Biol. **186**: 145–156.

Sanford, C.P., Lauder, G.V. (1989) Functional morphology of the "tonguebite" in the osteoglossomorph fish *Notopterus*. J. Morphol. **202**: 379–408.

Sans-Coma V., Gallego, A., Muñoz-Chápuli, R., Andrés, A.V. de, Duran, A.C., Fernandez, B. (1995) Anatomy and histology of the cardiac conal valves of the adult dogfish (*Scyliorhinus canicula*). Anat. Rec. **241**: 496–504.

Santer, R.M. (1985) Morphology and innervation of the fish heart. Adv. Anat. Embryol. Cell Biol. **89**: 1–102.

Santer, R.M., Green Walker, M. (1980) Morphological studies on the ventricle of teleost and elasmobranch hearts. J. Zool. Lond. **190**: 259–272.

Santos, R.S. (1995) Allopaternal care in the redlip blenny. J. Fish Biol. **47**: 350–353.

Sargent, J.R., Henderson, R.J., Tocher, D.R. (1989) The lipids. In: Halver, J.E. (ed.). Fish Nutrition. (Second edition). Academic Press, San Diego, pp. 153–218.

Sargent, R.C., Taylor, P.D., Gross, M.R. (1987) Parental care and the evolution of egg sizes in fishes. Am. Nat. **129**: 32–46.

Sasse, D., Pfeiffer, W., Aronold, M. (1970) Epidermale organe am. kopf von *Morlius charysophakedion* (Cyprinidae, Ostariophysi, Pisces). Z. Zellforsch. **103**: 218–231.

Satchell, G.H. (1976) The circulatory system of air-breathing fish. In: Hughes, G.M. (ed.) Respiration of amphibious vertebrates. Academic Press, London, New York, pp. 105–124.

Satchell, G.H. (1978) Microcirculation in fishes. In: Kaley G, Altura, B.M. (eds.) Microcirculation, vol. 2. University Park Press, Baltimore, pp. 619–647.

Satchell, G.H. (1992) The venous system. In: Hoar, W.S., Randall, D.J., Farrell, A.P. (eds.) Fish physiology, vol. 12A. Academic Press, Orlando, pp. 141–183.

Sathyanesan, A.G. (1994) Neuroendocrinology of reproduction in the teleostean fishes. In: Singh, H.R. (ed.) Advances in fish biology. Hindustan Publishing Corporation, Delhi, pp. 132–146.

Saxena, S.C., Chandy, M. (1966) Adhesive apparatus in certain Indian hill stream fishes. J. Zool. Lond. **148**: 315–340.

Sayer, M.D.J., Daverport, J. (1991) Amphibious fish: why do they leave water? Rev. Fish Biol. Fish. **1**: 159–181.

Schaffer, W.M. (1979) Equivalence of maximizing reproductive value and fitness in the case of reproductive strategies. Proc. Natl. Acad. Sci. U.S.A. **76**: 3567–3569.

Schellart, N.A.M. (1992) Interrelations between the auditory, the visual and the lateral line systems of teleosts: a miniview of modelling sensory capabilities. Neth. J. Zool. **42**: 459–477.

Schemmel, C. (1980) Studies on the genetics of feeding behaviour in the cavefish *Astyanax mexicanus f. anoptichthys*. An example of apparent monofactorial inheritance by polygenes. Z. Tierpsychol, **53**: 9–22.

Scheid, P., Pelster, B. (1995) Gas exchange in the fish swimbladder. In: Heisler, N. (ed.) Mechanisms of systemic regulation. Advances in comparative and environmental physiology, vol. 21. Springer-Verlag, Heidelberg, pp. 41–59.

Scheid, P., Pelster, B., Kobayashi, H. (1990) Gas exchange in the fish swimbladder. In: Piiper. J., Goldstick, T.K., Meyer, M. (eds.) Oxygen transport to tissue, vol. 12. Plenum Press, New York, pp. 735–742.

Schindler, J.F. (1990) Retrograde trafficking of tracer protein by the internal ovarian epithelium in gravid goodeid teleosts. Anat. Rec. **226**: 177–186.

Schindler, J.F., de Vries, U. (1986a) Ultrastructure of embryonic anal processes in *Girardinichthys viviparus* (Cyprinodontiformes, Osteichthyes). J. Morphol. **188**: 203–224.
Schindler, J.F., de Vries, U. (1986b) Surface ultrastructure of trophotaeniae, placental analogues appending to goodeid embryos. Z. mikrosk.- anat. Forsch. Leipzig **100**: 939–949.
Schindler, J.F., de Vries, U. (1987a) Maternal-embryonic relationships in the goodeid teleost, *Xenoophorus captivus*. Embryonic structural adaptations to viviparity. Cell Tissue Res. **247**: 325–338.
Schindler, J.F., de Vries, U. (1987b) Protein uptake and transport by trophotaenial absorptive cells in two species of goodeid embryos. J. Exp. Zool. **241**: 17–29.
Schindler, J.F., de Vries, U. (1988a) Ovarian structural specializations facilitate aplacental matrotrophy in *Jenynsia lineata* (Cyprinodontiformes, Osteichthyes). J. Morphol **198**: 331–339.
Schindler, J.F., de Vries, U. (1988b) Maternal-embryonic relationships in the goodeid teleost, *Xenoophorus captivus*. The vacuolar apparatus in trophotaenial absorptive cells and its role in micromolecular transport. Cell Tissue Res. **253**: 115–128.
Schindler, J.F., de Vries, U. (1988c) Endocytosis at 0°C, 5°C and 10°C in trophotaenial absorptive cells of goodeid embryos (Teleostei). Cell Tissue Res. **254**: 399–402.
Schindler, J.F., de Vries, U. (1990) Membrane differentiations of an absorptive epithelium covering embryonic trophotaeniae in a goodeid teleost. Cell Tissue Res. **259**: 321–330.
Schindler, J.F., Greven, H. (1992) Protein-gold transport in the endocytic complex of trophotaenial absorptive cells in the embryos of a goodied teleost. Anat. Rec. **233**: 387–398.
Schindler, J.F., Hamlett, W.C. (1993) Maternal-embryonic relations in viviparous teleosts. J. Exp. Zool. **266**: 378–393.
Schindler, J.F., Kujat, R. (1990) Structure and function of the trophotaeniae placenta in the goodeid teleost, *Skiffia bilineata*. Z. mikros. -anat. Forsch. Leipzig. **104**: 241–257.
Schindler, J.F., Radda, A.C. (1990) Die Trophotanien-Placenta des Genus *Allodontichthys* (Cyprinodontiformes: Goodeidae). Acta Biol. Benrodis. **2**: 57–69.
Schindler, J.F., Kujat, R., de Vries, U. (1988) Maternal-embryonic relationships in the goodeid teleost, *Xenoophorus captivus*. The internal ovarian epithelium and the embryotrophic fluid. Cell Tissue Res. **254**: 177–182.
Schikorski, T., Braun, N., Zimmerman, H. (1994) Pojection of brain stem neurons to the giant electromotoneurons in the cervical spinal cord of the electric catfish *Malapterurus*. Brain Behav. Evol. **43**: 306–318.
Scholinck, D.A., Mangum, C.P. (1991) Sensitivity of hemoglobins to intracellular effectors: primitive and derived features. J. Exp. Zool. **259**: 32–42.
Schreibman, M.P. (1986) Pituitary gland. In: Pang, P.K.T., Schreibman, M.P. (eds.) Vertebrate endocrinology: fundamentals and biomedical implications, vol. 1. Academic Press, Orlando, pp. 11–55
Schreibman, M.P. (1987) Aging of the neuroendocrine system. In: Norris, D.O., Jones, R.E. (eds.) Hormones and reproduction in fishes, amphibians and reptiles. Plenum Publishing Corporation, New York, pp. 563–584.
Schreibman, M.P., Holtzman, S., Cepriano, L. (1990) The life cycle of the brain-pituitary-gonad-axis in teleosts. In: Epple, A., Scanes, C.G., Stetson, M.H. (eds.). Progress in comparative endocrinology. Wiley-Liss, Inc., New York, pp. 399–408
Schultze, H.P. (1986) Dipnoans as sarcopterygians, J. Morphol. (Suppl.) **1**: 39–74.
Schultze, H.P. (1991) CT scan reconstruction of the palate region. Env. Biol. Fish. **32**: 183–192.
Schultze, H.P. (1993) Patterns of diversity in the skulls of jawed fishes. In: Hanken, J., Hall. B.K. (eds.) The skull, vol 2: Patterns of structural and systematic diversity. University of Chicago Press, Chicago, pp. 189–254.
Schultze, H.P., Campbell, K.S.W. (1986) Characterization of the Dipnoi, a monophyletic group,. J. Morphol. (Suppl.) **1**: 25–37.
Schulz, R.J. (1980) Role of polyploidy in the evolution of fishes. In: Lewis, W.H. (ed.) Polyploidy: Biological relevance. Plenum Publishing Corporation, New York, pp. 313–340.
Schwassmann, H.O. (1971) Biological rhythms. In: Hoar, W.S., Randall, D.J. (eds.) Fish physiology, vol. 6. Academic Press, New York, London, pp. 371–428.
Scott, A.P., Sumpter, J.P., Kime, D.E., Rolfe, M.S. (eds.) (1991) Reproductive physiology of fish. Proc. Fourth Int. Symp. University of East Anglia, Norwich, U.K., 7–12 July 1991.
Senger, H., Storch, V., Reinecke, M., Kloas, W., Hanke, W. (1994) The development of functional digestive and metabolic organs in turbot, *Scophthalmus maximus*. Mar. Biol. **119**: 471–486.

Selcer K.W., Leavitt, W.W. (1991) Estrogens and progestins. In: Pang P.K.T., Schreibman, M.P. (eds.) Vertebrate, endocrinology: fundamentals and biomedical implications, vol. 4B. Academic Press, San Diego, pp. 67–114.

Selset, R. (1980) Chemical methods for fractionation of odorants produced by char smolts and tentative suggestions for pheromone origins. Acta Physiol. Scand. **108**: 97–103.

Selset, R., Døving, K.B. (1980) Behaviour of mature anadromous char (*Salmo alpinus* L.) towards odorants produced by smolts of their own population. Acta Physiol. Scand. **108**: 113–122.

Sephton, D.H., Driedzic, W.R. (1994) Adrenergic stimulation of glycolysis without change in glycolytic enzyme binding in rainbow trout (*Oncorhynchus mykiss*) erythrocytes. Can. J. Zool. **72**: 950–953.

Sephton, D.H., MacPhee, W.L., Driedzic, W.R. (1991) Metabolic enzyme activities, oxygen consumption and glucose utilization in sea raven (*Hemitripterus americanus*) erythrocytes. J. Exp. Biol. **159**: 407–418.

Serrier, J., Moller, P. (1989) Patterns of electric organ discharge activity in the weakly electric fish *Brienomyrus niger* L. (Mormyridae). Exp. Biol. **48**: 235–244.

Shaklee, J.B., Allendorf, F.W., Morizot, D.C., Whitt, G.S. (1990) Gene nomenclature for protein-coding loci in fish. Trans. Amer. Fish. Soc. **119**: 2–15.

Shapiro, D.Y. (1979) Social behavior, group structure, and the control of sex reversal in hermaphroditic fish. Adv. Study Behav. **10**: 43–102.

Shapiro, D.Y. (1980) Serial female sex changes after simultaneous removal of males from social groups of a coral reef fish. Science **209**: 1136–1137.

Shapiro, D.Y. (1981) Size, maturation and the social control of sex reversal in a coral reef fish *Acanthias squamipinnis*. J. Zool. Lond. **193**: 105–128.

Shapiro, D.Y. (1986a) Subgroup and group development in a sex-changing fish. Anim. Behav. **34**: 716–726.

Shapiro, D.Y. (1986b) Intra-group home ranges in a female-biased group of sex-changing fish. Anim. Behav. **34**: 865–870.

Shapiro, D.Y. (1987a) Reproduction in groupers. In: Polovina, J.J., Ralston S. (eds.) Tropical snappers and groupers: Biology and fisheries management. Westview Press, Boulder, Colorado, pp. 295–327.

Shapiro, D.Y. (1987b) Differentiation and evolution of sex change in fishes. Bioscience **37**: 490–497.

Shapiro, D.Y. (1988a) Behavioural influences on gene structure and other new ideas concerning sex change in fishes. Env. Biol. Fish. **23**: 283–297.

Shapiro, D.Y. (1988b) Variation of group composition and spatial structure with group size in a sex-changing fish. Anim. Behav. **36**: 140–149.

Shapiro, D.Y. (1989) Sex change as an alternative life-history style. In: Bruton M.N. (ed.) Alternative life-history styles of animals. Kluwer Academic Publishers, Dordrecht, pp. 177–195.

Shapiro, D.Y. (1991) Intraspecific variability in a social systems of coral reef fishes. In: Sale, P.F. (ed.) The ecology of fishes on coral reefs. Academic Press, London, pp. 331–335.

Shapiro, D.Y. (1992) Plasticity of gonadal development and protandry in fishes. J. Exp. Zool. **261**: 194–203.

Shapiro, D.Y., Boulon, R.H. Jr. (1987) Evenly dispersed social groups and intergroup competition for juveniles in a coral-reef fish. Behav. Ecol. Sociobiol. **21**: 343–350.

Shapiro, D.Y., Lubbock, R. (1980) Group sex ratio and sex reversal. J. Theor. Biol. **82**: 411–426.

Shelton, G., Jones, D.R., Milsom, W.K. (1986) Control of breathing in ectothermic vertebrates. In: Fishman, A.P., Cherniak, N.S., Widdicomby, J.G., Greiger, S.R. (eds.) Handbook of physiology, Section, 3. The respiratory system, vol. 2. Control of breathing, Part 2. Am. Physiol. Soc, Bethesda, pp. 857–909.

Sheridan, M.A. (1988) Lipid dynamics in fish; aspects of absorption, transportation, deposition and mobilization. Comp. Biochem. Physiol. **90B**: 679–690.

Sherwood, N.M., Coe, I.R. (1991) Neuropeptides and their genes in fish. In: Scott, A.P., Sumpter J.P., Kime, D.E., Rolfe, M.S. (eds.) Reproductive physiology of fish. Proc. Fourth Int. Symp. University of East Anglia, Norwich, U.K. 7-12 July 1991, pp. 38–40.

Shimuzu, M., Kusakari, M., Yoklavich, M.M., Boehlert, G.W., Yamada, J. (1991) Ultrastructure of the epidermis and digestive tract in *Sebastes* embryos, with special reference to the uptake of exogenous nutrients. Env. Biol. Fish. **30**: 155–163.

Shumway, C.A. (1989) Multiple electrosensory maps in the medulla of weakly electric gymnotiform fish. 1. Physiological differences. J. Neurosci. **9**: 4388–4399.

Shumway, C.A., Zelick, R.D. (1988) Sex recognition and neuronal coding of electric organ discharge waveform in the pulse-type weakly electric fish, *Hypopomus occidentalis*. J. Comp. Physiol. A **163**: 465–478.
Shuttleworth, T.J. (1987) Salt gland function in osmoregulation in terrestrial and aquatic environments. In: Dejours, P., Bolis, L., Taylor, C.R., Weibel, E.R. (eds.) Comparative physiology: life in water and on land. Fidia Research Series, vol. 9. Liviana Press, Padova, pp. 537–548.
Sibbing, F.A. (1982) Pharyngeal mastication and food transport in the carp. (*Cyprinus carpio* L.): a cineradiographic and electromyographic study. J. Morphol. **172**: 223–258.
Sibbing, F.A. (1988) Specializations and limitations in the utilization of food resources by the carp, *Cyprinus carpio*: a study of oral food processing. Env. Biol. Fish. **22**: 161–178.
Sibbing, F.A. (1991a) Food capture and oral processing. In: Winfield, I.J., Nelson, J.S. (eds.) Cyprinid fishes. Systematics, biology and exploitation. Chapman and Hall, London, New York, pp. 377–425.
Sibbing, F.A. (1991b) Food processing by mastication in cyprinid fish. In: Vincent, J.F.V., Lillford, P.J. (eds.) Feeding and the texture of food. Society for Experimental Biology, Seminar series. n. 44. Cambridge University Press, Cambridge, pp. 57–92.
Sibbing, F.A., Uribe, R. (1986) Regional specializations in the oro-pharyngeal wall and food processing in the carp. (*Cyprinus carpio* L.). Neth. J. Zool. **35**: 377–422.
Sibbing, F.A., Osse, J.W.M., Terlou, W.A. (1986) Food handling in the carp (*Cyprinus carpio*): its movement patterns, mechanisms and limitations, J. Zool. Lond. **210**: 161–203.
Sidell, B.D. (1991) Physiological roles of high lipid content in tissues of Antarctic fish species. In: di Prisco, G., Maresca, B., Tota, B. (eds.) Biology of Antarctic fish. Springer-Verlag, Berlin, Heidelberg, 220–231.
Sidell, B.D., Driedzic, W.R. (1985) Relationship between cardiac energy metabolism and cardiac work demand in fishes. In: Gilles, R. (ed.) Circulation, respiration and metabolism. Springer-Verlag, Berlin, Heidelberg, pp. 386–401.
Sidell, B.D., Crockett, E.L., Driedzic, W.R. (1995) Antarctic fish tissues preferentially catabolize monoenoic fatty acids. J. Exp. Biol. **271**: 73–81.
Siebenaller, J.F. (1987) Biochemical adaptation in deep-sea animals. In: Jannasch, H.W., Zimmerman A.M., Marquis, R.E. (eds.) Current perspectives in high pressure biology. Academic Press, London, pp. 33–48.
Siebenaller, J.F. (1991) Pressure as an environmental variable: magnitude and mechanisms of perturbation. In: Hochachka, P.W., Mommsen, T.P. (eds.) Biochemistry and molecular biology of fishes, vol. 1. Elsevier, Amsterdam, pp. 323–343.
Silver, W.L., Finger, T.E. (1984) Electrophysiological examination of non-olfactory, non-gustatory chemosense in the searobin, *Prionotus carolinus*. J. Comp. Physiol. A **154:** 167–174.
Simpson, B.R.C. (1979) The phenology of annual killifishes. Symp. Zool. Soc. Lond. No. **44**: 243–261.
Sindermann, C.J. (1966) Diseases of marine fishes. Adv. mar. Biol. **4**: 1–89.
Sindermann, C.J. (1979) Pollution-associated diseases and abnormalities of fish and shellfish: a review. Fish. Bull. U.S. **76**: 717–749.
Singh, B.R. (1993) Development, origin and evolution of the neomorphic air breathing organs in teleosts. In: Singh B.R. (ed.) Advances in fish research, vol. 1. Narendra Publishing House, Delhi, pp. 1–16.
Singh, N. (1993) The scanning electron microscopic structure of the 'adhesive apparatus' and the mechanism of adhesion in hillstream teleost, *Crossocheilus latius latius* (Family-Cyprinidae) from Garhwal Himalaya. J. Anim. Morphol. Physiol. **40**: 109–114.
Singh, N. (1994) Scanning electron microscopic study of the olfactory epithelium of four coldwater hillstream teleosts from Garhwall hills (India). J. Biosci. **19**: 91–102.
Singh, N., Agarwal, N.K. (1991a) SEM surface structure of the adhesive organ of the hillstream fish *Glyptothorax pectinopterus* (Teleostei: Sisoridae) from the Garhwall Hills. Funct. Develop Morphol. **1**: 11–13.
Singh, N., Agarwal, N.K. (1991b) The SEM surface structure of the adhesive organ of *Pseudoecheneis sulcatus* McClelland (Teleostei: Sisoridae), from Garhwal Himalayan hillstreams. Acta Ichthyol. et. Piscat. **21**: 29–35.
Singh, N., Agarwal, N.K. (1993) Organs of adhesion in four hillstream fishes, a comparative morphological studies. In: Singh, H.R. (ed.) Advances in Limnology. Narendra Publishing House, Delhi, pp. 311–316.
Singh, N., Agarwal, N.K., Singh, H.R. (1994) SEM investigations on the 'adhesive apparatus' of *Garra gotyla gotyla* (Family Cyprinidae) from Garhwal Himalaya. In: Singh, H.R. (ed.) Advances in fish biology, Hindustan Publishing Corporation, Delhi, pp. 281–291.

Singh, N., Bhatt, K.C., Bahuguna, M.K., Kumar, D. (1995) Fine structure of olfactory epithelium in *Schizothoraichthys progastus* McClelland and *Schizothorox richardsonii* Gray (Cyprinidae: Teleostei) from Garhwal Himalaya (India). J. Biosci. **20**: 385–396.

Singh, T.P., Lal, B. (1994) Endocrine physiology of reproduction in Indian catfishes. In: Singh, H.R. (ed.) Advances in fish biology. Hindustan Publishing Corporation, Delhi, pp. 147–154.

Sire. J.Y. (1982) Regeneration des ecailles d'un Cichlidae, *Hemichromis bimaculatus* (Gill) (Teleostei, Perciforme). I. Morphogense, structure et mineralisation. Ann. Sci. Nat. Zool. Paris **4**: 153–169.

Sire, J.Y. (1989a) The same cell lineage is involved in scale formation and regeneration in the teleost fish *Hemichromis bimaculatus*. Tissue Cell **21**: 447–462.

Sire, J.Y. (1989b) Scales in young *Polypterus senegalus* are elasmoid: new phylogenetic implications. Am. J. Anat. **186**: 315–323.

Sire, J.Y. (1990) From ganoid to elasmoid scales in the actinopterygian fishes. Neth. J. Zool. **40**: 75–92.

Sire, J.Y. (1993) Development and fine structure of the bony scutes in *Corydoras arcuatus* (Siluriformes, Callichthyidae). J. Morphol. **215**: 225–244.

Sire, J.Y., Arnulf, I. (1990) The development of squamation in four teleostean fishes with a survey of the literature. Japan. J..Ichthyol. **37**: 133–143.

Sire, J.Y., Huysseune, A. (1993) Fine structure of the developing frontal bones and scales of the cranial vault in the cichlid fish *Hemichromis bimaculatus* (Teleostei, Perciformes). Cell Tissue Res. **273**: 511–524.

Sire, J.Y., Meunier, F.J. (1993) Ornamentation superficielle et. structure des plaques osseuses dermiques de quelqries Siluriformes cuirasses (Loricariidae, Callichthyidae, Doradidae). Ann. Sci. Nat. Zool. Paris **14**: 101–123.

Sire, M.F., Vernier, J.M. (1981) Etude ultrastructurale de la synthese de chylomicrons au cours de l'absorption intestinale des. lipides chez. la Truite. Influence de la nature des. acides gras ingere's. Biol. Cell **40**: 47–62.

Sire, M.F., Lutton, C., Vernier, J.M. (1981) New views on intestinal absorption of lipids in teleostean fishes: an ultrastructural and biochemical study in the rainbow trout. J. Lipid Res. 22: 81–94.

Sire, J.Y., Boulekbache, H., Joly, C. (1990) Epidermal-dermal and fibronectin cell-interactions during fish scale regeneration: immunofluorescence and TEM studies. Biol. Cell **68**: 147–158.

Sivak, J.G. (1978) A survey of vertebrate strategies for vision in air and water. In: Ali, M.A. (ed.) Sensory ecology. Plenum Publishing Corporation, New York, pp. 503–519.

Sivak, J.G. (1990) Optical variability of the fish lens. In: Douglas, R.H., Diamgoz, M.B.A. (eds.) The visual system of fish. Chapman and Hall, London, New York, pp. 63–80.

Siwicki, A.K., Anderson, D.P., Rumsey, G.L. (1994) Dietary intake of immunostimulants by rainbow trout affects non-specific immunity and protection against furunculosis. Vet. Immunol. Immunopath. **41**: 125–139.

Skulason, S., Snorrason, S.S., Noakes, D.L.G., Ferguson, M.M., Malmquist, H.J. (1989) Segregation in spawning and early life history among polymorphic Arctic charr, *Salvelinus alpinus*, in Thingvallavatn, Iceland. J. Fish Biol. 35 (Suppl. A): 225–232.

Smatresk, N.J. (1988) Control of the respiratory mode in air-breathing fishes. Can. J. Zool. **66**: 144–151.

Smatresk, N.J. (1994) Respiratory control in the transition from water to air breathing in vertebrates. Am. Zool. **34**: 264–279.

Smatresk, N.J., Aziz, S.Q. (1987) Characteristics of lung mechanoreceptors in spotted gar, *Lepisosteus oculatus*. Am. J. Physiol. **252**: R 1066–R 1072.

Smit, H. (1980) Some aspects of environmental (phenotypic) adaptations in fishes. Neth. J. Zool. **30**: 179–207.

Smith, L.S. (1989) Digestive functions in teleost fishes. In: Halver, J.E. (ed.) Fish nutrition. (Second edition). Academic Press, San Diego, pp. 331–421.

Smith, R.J.F. (1982) The adaptive significance of the alarm substance-fright reaction system. In: Hara, T.J. (ed.) Chemoreception in fishes. Elsevier, Amsterdam, pp. 327–342.

Smith, R.J.F (1986) The evolution of chemical alarm signals in fishes. In: Duvall, D., Muller-Schwarze, D., Silverstein, R.M. (eds.) Chemical signals in vertebrates, vol. 4. Plenum Publishing Corporation, New York, pp. 99–115.

Smith, R.J.F. (1992) Alarm signals in fishes. Rev. Fish. Biol. Fish. **2**: 33–63.
Smith, R.R. (1989) Nutritional energetics. In: Halver, J.E. (ed.) Fish nutrition. (Second edition). Academic Press, San Diego, pp, 1–29.
Smith C, Reay, P. (1991) Cannibalism in teleost fish. Rev. Fish. Biol. Fish. **1**: 41–64.
Smith-Grayton, P.K., Keenleyside, M.H.A. (1978) Male-female parental roles in *Herotilapia multispinosa* (Pisces: Cichlidae). Anim. Behav. **26**: 520–526.
Smolka, A.J., Lacy, E.R., Luciano L, Reale, E. (1994) Identification of gastric, H., K-ATPase in an early vertebrate, the Atlantic stingray *Dasyatis sabina*. J. Histochem. Cytochem. **42**: 1323–1332.
Snieszko, S.F. (1972) Nutritional fish diseases. In: Halver, J.E. (ed.) Fish nutrition. Academic Press, New York, London, pp. 403–437.
Snieszko, S.F. (1973) Recent advances in scientific knowledge and developments pertaining to diseases of fishes. In: Advances in veterinary science and comparative medicine. **17**: 291–314.
Sokabe, H., Ogawa, M. (1974) Comparative studies of the juxtaglomerular apparatus. Int. Rev. Cytol. **37**: 271–327.
Sola, L., Monaco, P.J., Rasch, E.M. (1990) Cytogenetics of bisexual/unisexual species of *Poecilia*. I.C.-bands, Ag-NOR polymorphisms, and sex chromosomes of *Poecilia latipinna*. Cytogenet. Cell Genet. **53**: 148–154.
Somero, G.N. (1990) Life at low volume change: hydrostatic pressure as a selective factor in the aquatic environment. Am. Zool. **30**: 123–135.
Sorensen, P.W. (1992) Hormones, pheromones and chemoreception. In: Hara, T.J. (ed.) Fish chemoreception. Chapman and Hall, London, New York, pp. 199–228.
Sorensen, P.W., Stacey, N.E. (1990) Identified hormonal pheromones in the goldfish: the basis for a model of sex pheromone function in teleost fish. In: Macdonald, D.W., Muller-Schwarze, D., Natynczuk, S.E. (eds.) Chemical signals in vertebrates, vol. 5. Oxford University Press, Oxford, pp. 302–311.
Sorensen, P.W., Hara, T.J., Stacey, N.E., Goetz, F.W.M. (1988) F prostaglandins function as potent olfactory stimulants that comprise the postovulatory female sex pheromone in goldfish. Biol. Reprod. **39**: 1039–1050.
Sower, S.A. (1990) Gonadotropin-releasing hormone in primitive fishes. In: Epple, A., Scanes, C.G., Stetson, M.H. (eds.) Progress in comparative endocrinology. Wiley-Liss, New York, pp. 73–78.
Sparks, N.H.C., Motta, P.J., Shellis, R.P., Wade, V.J., Mann, S. (1990) An analytical electron microscopy study of iron-rich teeth from the butterflyfish (*Chaetodon ornatissimus*). J. Exp. Biol. **151**: 371–385.
Sriwastwa, V.M.S., Srivastav, S.R. (1994) Structure and function of fish pineal. In: Singh, H.R. (ed.) Advances in fish biology. Hindustan Publishing Corporation, Delhi, pp. 121–131.
Stabell, O.B. (1992) Olfactory control of homing behaviour in salmonids. In: Hara, T.J. (ed.) Fish chemoreception. Chapman and Hall, London, New York, pp. 249–270.
Stabell, O.B., Selset, R. (1980) Comparison of mucus collecting methods in fish olfaction. Acta Physiol. Scand. **108**: 91–96.
Stacey, N.E. (1983) Hormones and reproductive behaviour in teleosts. In: Rankin, J.C., Pitcher, T.J., Duggan, R.T. (eds.) Control processes in fish physiology. Croom Helm, London, pp. 117–129.
Stacey, N.E. (1987) Roles of hormones and pheromones in fish reproductive behaviour. In: Crews, D. (ed.) Psychobiology of reproductive behaviour: an evolutionary perspective. Prentice-Hall, Englewood Cliffs, New Jersey, pp. 28–60.
Stacey, N.E., Sorensen, P.W. (1986) $17\alpha 20\beta$-dihydroxy-4-pregnen-3-one: a steroidal primer pheromone increasing milt volume in the goldfish, *Carassius auratus*. Can. J. Zool. **64**: 2412–2417.
Stacey, N.E., Sorensen, P.W. (1991) Function and evolution of fish hormonal pheromones. In: Hochachka, P.W., Mommsen. T.P. (eds.) The biochemistry and molecular biology of fishes, vol. 1. Elsevier, Amsterdam, pp. 109–135.
Stacey, N.E., Kyle, A.L., Liley, N.R. (1986) Fish reproductive pheromones. In: Duvall, D., Muller-Schwarze, D., Silverstein, R.M. (eds.) Chemical signals in vertebrates 4. Plenum Publishing Corporation, New York, pp. 117–133.
Stacey, N.E., Sorensen, P.W., Dulka, J.G., Cardwell, J.R., Irvine A.S. (1991) Fish sex pheromones: Current status and potential applications. Bull. Inst. Zool. Academia. Sinica Monograph **16**: 189–227.
Stacey, N.E., Cardwell, J.R., Liley, N.R., Scott, A.P., Sorensen, P.W. (1993) Hormones as sex pheromones in fish. Twelfth International Congress in Comparative Endocrinology, Toronto, Canada, May 1993, pp. 1–23.

Sterba, T., Wagner, G.F., Schroedter, I.C., Friesen, H.G. (1993) In situ detection and distribution of stanniocalcin mRNA in the corpuscles of Stannius of sockeye salmon, *Onocorhynchus nerka*. Mol. Cell Endocrinol. **90**: 179–185.

Stet, R.J.M., Egberts, E. (1991) The histocompatibility system in teleostean fishes: from multiple histocompatibility loci to a major histocompatibility complex. Fish Shellfish Immunol. **1**: 1–16.

Stevens, E.D., Neill, W.H. (1978) Body temperature relations of tunas, especially skipjack. In: Hoar, W.S., Randall, D.J. (eds.) Fish physiology, vol. 7. Academic Press, New York, pp. 315–359.

Steffensen, J.F., Lomholt, J.P. (1992) The secondary vascular system. In: Hoar. W.S., Randall, D.J., Farrell, A.P. (eds.) Fish physiology, vol. 12A. Academic Press, Orlando, pp. 185–217.

Stiassny, M.L.J., Gerstner, C.L. (1992) The parental care behaviour of *Paratilapia polleni* (Perciformes, Labroidei), a phylogenetically primitive cichlid from Madagascar, with a discussion of the evolution of maternal care in the family Cichlidae. Env. Biol. Fish. **24**: 219–233.

Stiassny, M.L.J., Mezey, J.G. (1993) Egg attachment systems in the family Cichlidae (Perciformes: Labroidei), with some comments on their significance for phylogenetic studies. Amer. Mus. Novit. No. **3058**: 1–11.

Stipp, A.C.M., Ferri, S., Sesso, A. (1980) Fine structural analysis of a teleost exocrine pancreas cellular components—A freeze-fracture and transmission electron microscopic study. Anat. Anz. **147**: 60–75.

Storelli, C, Vilella, S., Paola Romano, M., Maffia, M., Cassano, G. (1989) Brush-border amino acid transport mechanisms in carnivorous eel intestine. Am. J. Physiol **257**: R506–R510.

Stroband, H.W.J., Debets, F.M.H. (1978) The ultrastructure and renewal of the intestinal epithelium of the juvenile grasscarp *Ctenopharyngodon idella* (Val.). Cell Tissue Res. **187**: 181–200.

Stroband, H.W.J., Kroon, A.G. (1981) The development of the stomach in *Clarias lazera* and the intestinal absorption of protein macromolecules. Cell Tissue Res. **215**: 397–415.

Stroband, H.W.J., Meer, H.V.D., Timmermans, L.P.M. (1979) Regional functional differentiation in the gut of the grasscarp, *Ctenopharyngodon idella* (Val.). Histochemistry **64**: 235–249.

Storey, K.B., Storey, J.M. (1989) Freeze tolerance and freeze avoidance in ectotherms. In: Wang, L.C.H. (ed.) Advances in comparative and environmental physiology, vol. 4. Springer-Verlag, Berlin, Heidelberg, pp. 51–82.

Sullivan, K.M. (1986) Physiology of feeding and starvation in overwintering freshwater fishes. In: Simenstad, C.A., Cailliet. G.M. (eds.) Contemporary studies on fish feeding. Dr. W. Junk Publishers, Dordrecht, pp. 259–268.

Suzuki, M., Kubokawa, K., Nagasawa, H., Urano, A. (1994) Sequence analysis of vasotocin cDNAs of the lamprey, *Lampetra japonica*, and the hagfish, *Eptatretus burgeri*: evolution of cyclostome vasotocin precursors. J. Endocrinol. 144 (in press).

Swanson, P. (1991) Salmon gonadotropins: reconciling old and new ideas. In: Scott, A.P., Sumpter J.P., Kime, D.E., Rolfe, M.S. (eds.) Reproductive physiology of fish. Proc. Fourth Int. Symp. Univ. of East Anglia, Norwich, U.K., 7–12 July 1991, pp. 2–7.

Symmons, S. (1979) Notochordal and elastic components of the axial skeleton of fishes and their functions in locomotion. J. Zool. Lond. **189**: 157–206.

Szabo, T., Denizot, J.P. (1993) Long ascending spinal pathways similar to those of higher vertebrates occur in certain teleost fish: a possible proprioceptive control system of electrosensory processing. In: Singh, H.R. (ed.) Advances in fish biology. Hindustan Publishing Corporation, Delhi, pp. 261–269.

Tagliafierro G., Faraldi, G., Pozzi, M.G. (1984) Ultrastructure of the pyloric endocrine cells in the cartilaginous fish *Scyliorhinus stellaris*. Boll. Zool. **51**: 295–311.

Tagliafierro, G., Faraldi, G., Bandelloni, R. (1985) Distribution, histochemistry and ultrastructure of somatostatin-like immunoreactive cells in the gastroenteric tract of the cartilaginous fish *Scyliorhinus stellaris* (L.). Histochem. J. **17**: 1033–1041.

Tagliafierro, G., Farina, L., Faraldi, G., Rossi, G.G., Vacchi, M. (1989a) Distribution of somatostatin and glucagon immunoreactive cells in the gastric mucosa of some cartilaginous fishes. Gen. Comp. Endocrinol. **75**: 1–9.

Tagliafierro, G., Rossi, G.G., Bonini, E., Faraldi, G., Farina, L. (1989b) Ontogeny and differentiation of regulatory peptide- and serotonin-immunoreactivity in the gastrointestinal tract of an elasmobranch. J. Exp. Zool. **2**: 165–174.

Tagliafierro, G., Bonini, E., Feraldi, G., Farina, L., Rossi, G.G. (1998) Distribution and ontogeny of VIP-like immunoreactivity in the gastro-entero pancreatic system of a cartilaginous fish, *Scyliorhinus stellaris*. Cell Tissue Res. **253**: 23–28.

Takei, Y. (1993) Role of peptide hormones in fish osmoregulation. In: Rankin, J.C., Jensen, F.B. (eds.) Fish ecophysiology. Chapman and Hall, London, New York, pp. 136–160.

Takemura. A., Hara, A., Takano, K. (1991a) Immunochemical identification and partial characterization of female-specific serum proteins in white-edged rockfish, *Sebastes taczanowskii*. Env. Biol. Fish. **30**: 49–65.

Takano, K., Takemura, A., Furihota, M., Nakanishi, T., Hara, A. (1991b) Annual reproductive and spawning cycles of female *Sebastiscus marmoratus*. Env. Biol. Fish. **30**: 39–48.

Takemura, A., Nunomura, W., Takano, K., Hirai, H. (1993) Novel female protein with biological activities of C-reactive protein and lectin in a viviparous rockfish, *Sebastes taczanowskii*. J. Exp. Zool. **266**: 188–194.

Takemura, A., Takano, K., Takahashi, H. (1995) The uptake of macromolecular materials in the hindgut of viviparous rockfish embryos. J. Fish Biol. **46**: 485–493.

Tamura, E., Honma, Y. (1973) Histological changes in the organs and tissues of gobiid fishes throughout its life-span-V. Seasonal changes in the branchial organs of the flat-head goby in relation to sexual maturity. Bull Jap. Soc. Sci. Fish. **39**: 1003–1011.

Tavolga, W.N. (1971) Sound production and detection. In: Hoar, W.S., Randall, D.J. (eds.) Fish physiology, vol. 5. Academic Press, New York, London, pp. 135–205.

Teeter, J.H., Sugimoto, K., Brand, J.G. (1989) Ionic currents in taste cells and reconstituted taste epithelial membranes. In: Brand J.G., Teeter, J.H., Cagan, R.H., Kare, M.R. (eds.) Chemical senses, vol. 1: Receptor events and transduction in taste and olfaction. Marcel Dekker, New York, Basel, pp. 151–170.

Teeter, J.H., Brand, J.G., Kumazawa, T. (1990) A stimulus-activated conductance in isolated taste epithelial membranes. Biophys. J. **58**: 253–259.

Terakado, K., Ogawa, M., Hashimoto, Y., Matsuzaki, H. (1975). Ultrastructure of the thread cells in the slime gland of Japanese hagfishes, *Paramyxine atami* and *Eptatretus burgeri*. Cell Tissue Res. **159**: 311–323.

Terlou, M., Gielen, J.T., Notenboom, C.D. (1985) Nucleus lateralis tuberis pars lateralis and background adaptation in the rainbow trout, *Salmo gairdneri*. Cell Tissue Res. **240**: 185–190.

Thibault, C., Collin, J.P., Falcón, J. (1993a) Intrapineal circadian oscillator(s), cyclic nucleotides and melatonin production in pineal photoreceptor cells. In: Touitou, Y., Arendt, J., Pevet, P. (eds.) Melatonin and the pineal gland—From basic science to clinical application. Elsevier, Amsterdam, pp. 11–18.

Thibault, C., Falcón, J., Greenhouse, S.S., Lowery, C.A., Gern,W.A., Collin, J.P. (1993b) Regulation of melatonin production by pineal photoreceptor cells: role of cyclic nucleotides in the trout (*Oncorhynchus mykiss*). J. Neurochem. **61**: 332–339.

Thibault, R.E., Schultz, R.J. (1978) Reproductive adaptations among viviparous fishes (Cyprinodontiformes: Poecillidae). Evolution **32**: 320–333.

Thomson, K.S. (1969) The biology of the lobe-finned fishes. Biol. Rev. **44**: 91–154.

Thomson, K.S. (1970) Intracranial movement in the coelacanth *Latimeria chalumnae* Smith (Osteichthyes, Crossopterygii). Postilla **149**: 1–12.

Thorpe, J. (ed.) (1978) Rhythmic activity of fishes. Academic Press, London, New York.

Tibbits, G.F., Moyes, C.D. (1992) Excitation-contraction coupling in the teleost heart. In: Hoar, W.S., Randall, D.J., Farrell, A.P. (eds.) Fish physiology, vol. 12A. Academic Press, Orlando, pp. 267–304.

Tiersch, T.R., Mitchell, M.J., Wachtel, S.S. (1991) Studies on the phylogenetic conservation of the *SRY* gene. Hum. Genet. **87**: 571–573.

Timms, A.M., Keenleyside, M.H.A. (1975) The reproductive behaviour of *Aequidens paraguayensis* (Pisces, Cichlidae). Z. Tierpsychol. **39**: 8–23.

Todd, E.S. (1973) A preliminary report of the respiratory pump in the Dactyloscopidae. Copeia **1973**: 115–119.

Tota, B. (1983) Vascular and metabolic zonation in the ventricular myocardium of mammals and fishes. Comp. Biochem. Physiol. **76A**: 423–437.

Tota, B., Cimini, V., Salvatore, G., Zummo, G. (1983) Comparative study of the arterial and lacunary systems of the ventricular myocardium of elasmobranch and teleost fishes. Am. J. Anat. **167**: 15–32.

Tota, B., Acierno, R., Agnisola, C. (1991a) Mechanical performance of the isolated and perfused heart of the haemoglobinless Antarctic icefish *Chionodraco hamatus* (Lönnberg): effects of loading conditions and temperature. Phil. Trans. R. Soc, Lond, B **332**: 191–198.

Tota, B., Agnisola C., Schioppa, M., Acierno, R., Harrison, P., Zummo, G. (1991b) Structural and mechanical characteristics of the heart of the icefish *Chionodraco hamatus* (Lönnberg). In: di Prisco. G., Maresca, B., Tota, B. (eds.) Biology of Antarctic fish, Springer-Verlag, Berlin, Heidelberg, pp. 204–219.

Tricas, T.C. (1979) Relationships of the blueshark, *Prionace glauca*, and its prey species near Santa Catalina Island, California. Fish. Bull. **77**: 175–182.

Tricas, T.C. (1982) Bioelectric-mediated predation by swell sharks, *Cephaloscyllium ventriosum*. Copeia **1982**: 948–952.

Tricas, T.C. (1985) Feeding ethology of the white shark, *Carcharodon carcharias*. Mem. Southern. Calif Acad. Sci. **9**: 81–91.

Tricase, T.C. (1989a) Determinants of feeding territory size in the corallivorous butterflyfish, *Chaetodon multicinetus*. Anim. Behav. **37**: 830–841.

Tricas, T.C. (1989b) Prey selection by coral-feeding butterflyfishes: strategies to maximize the profit. Env. Biol. Fish. **25**: 171–185.

Tricas. T.C., Highstein, S.M. (1991) Action of the octavolateralis efferent system upon the lateral line of free-swimming toadfish, *Opsanus tau*. J. Comp. Physiol. A **169**: 25–37.

Tricas, T.C., McCosker, J.E. (1984) Predatory behavior of the white shark (*Carcharodon carcharias*), with notes on its biology. Proc. Calif. Acad. Sci. **14**: 221–238.

Tricas, T.C., Michael, S.W. (1996) Electrosensory optimization to conspecific phasic signals for mating. Neurosci. Lett. **202**: 1–4.

Tricas, T.C., Taylor, L.R., Neftel, G. (1981) Diel behavior of the tiger shark, *Galeocerdo cuvier*, at French Frigate Shoals, Hawaiian Islands. Copeia **1981**: 904–908.

Tucker, D. (1983) Fish chemoreception: peripheral anatomy and physiology. In: Northcutt, R.G., Davis, R.E. (eds.) Fish neurobiology, Brain stem and sense organs, vol. 1. Univrsity of Michigan Press, Ann. Arbor, pp. 311–349.

Turingan, R.G., Wainwright, P.C. (1993) Morphological and functional bases of durophagy in the queen triggerfish, *Balistes vetula* (Pisces, Tetraodontiformes). J. Morphol. **215**: 101–118.

Turingan, R.G., Wainwright, P.C., Hensley, D.A. (1995) Interpopulation variation in prey use and feeding biomechanics in Caribbean triggerfishes. Oecologia **102**: 296–304.

Uehara, K., Miyoshi, S. (1987) Structure of the upper teeth of the filefish, *Stephanolepis cirrhifer*. Anat. Rec. **217**: 16–22.

Uehara, K., Miyoshi, S. (1993) Structure of comblike teeth of the ayu sweetfish, *Plecoglossus altivelis* (Teleostei: Isospondyli): 1. Denticles and tooth attachment. J. Morphol. **217**: 229–238.

Uehara, K., Miyoshi, S., Toh, H. (1983) Fine structure of the horny teeth of the lamprey, *Entosphenus japonicus*. Cell Tissue Res. **231**: 1–15.

Uliniski, P.S. (1980) Functional morphology of the vertebrate visual system: an essay on the evolution of complex systems. Am. Zool. **20**: 229–246.

Ultsch, G.R. (1989) Ecology and physiology of hibernation and overwintering among freshwater fishes, turtles, and snakes. Biol. Rev. **64**: 435–516.

Underwood, H. (1990) The pineal and melatonin: regulators of circadian function in lower vertebrates. Experientia **46**: 120–128.

Underwood, H., Groos, G. (1982) Vertebrate circadian rhythms: retinal and extraretinal photoreception. Experientia **38**: 1013–1021.

Urano, A., Kubokawa, K., Hiraoka, S. (1994) Expression of the vasotocin and isotocin gene family in fish. In: Sherwood N., Hew, C. (eds.) Fish physiology, vol. 13. Academic Press, San Diego, pp. 101–132.

Urasa, F.M., Wendelaar Bonga, S.E. (1987) Effects of calcium and phosphate on the corpuscles of Stannius of the teleost fish *Oreochromis mossambicus*. Cell Tissue Res. **249**: 681–690.

Ursin, E. (1979) Principles of growth in fishes, Symp. Zool. Soc. Lond. No. **44**: 63–87.

Utter, F.M., Hodgins, H.O., Allendorf, F.W. (1974) Biochemical genetics of fishes. In: Malins D.C., Sargent, J.R. (eds.) Biochemical and biophysical perspectives in marine biology, vol. 1. Academic Press, London, New York, San Francisco, pp. 213–238.

Val, A.L. (1993) Adaptations of fishes to extreme conditions in fresh waters. In: Bicudo, J.E. (ed.) The vertebrate gas transport cascade: adaptations to environment and mode of life. CRC Press, Boca Raton, pp. 43–53.

Val, A.L., Affonso, E.G., Souza, R.H.S., Almeida-Val, V.M.F., Moura, M.A.F. (1992) Inositol pentaphosphate in the erythrocytes of an Amazonian fish, the pirarucu (*Arapaima gigas*). Can. J. Zool. **70**: 852–855.

Val, A.L., Lessard, J., Randall, D.J. (1995) Effects of hypoxia on rainbow trout (*Oncorhynchus mykiss*): intraerythrocytic phosphates. J. Exp. Biol. **198**: 305–310.

Valentinčič, T.B., Caprio, J. (1994) Chemical and visual control of feeding and escape behaviors in the channel catfish *Ictalurus punctatus*. Physiol. Behav. **55**: 845–855.

Valentinčič, T., Wegert, S., Caprio, J. (1994) Learned olfactory discrimination versus innate taste responses to amino acids in channel catfish (*Ictalurus punctatus*). Physiol. Behav. **55**: 865–873.

Valentich, J.D., Karnaky, K.J. Jr., Moran, W.M. (1995) Phenotypic expression and natriuretic peptide-activated chloride secretion in cultured shark (*Squalus acanthias*) rectal gland epithelial cells. In: Wood, C.M., Shuttleworth T.J. (eds.) Cellular and molecular approaches to fish ionic regulation. Academic Press, San Diego, pp. 173–205.

Valentich, J.D., Karnaky, K.J. Jr., Ecay, T.W. Ultrastructural and cytochemical characterization of cultured dogfish rectal gland cells. Am. J. Physiol. (in press).

Valerio, M., Barlow, G.W. (1986) Ontogeny of young Midas cichlids: A study of feeding, filial cannibalism and agonism in relation to differences in size. Biol. Behav. **11**: 16–35.

Van den Berg C, Sibbing, F.A., Osse, J.W.M., Hoogenboezem, W. (1992) Structure, development and function of the branchial sieve of the common bream, *Abramis brama*, white bream, *Blicca bjoerkna* and roach, *Rutilus rutilus*. Env. Biol. Fish. **33**: 105–124.

Van Leeuwen, J.L. (1984) A quantitative study of flow in prey capture by rainbow trout, *Salmo gairdneri* with general consideration of actinopterygian feeding mechanism. Trans. Zool. Soc. Lond. **37**: 171–227.

Van Leeuwen, J.L., Muller, M. (1983) The recording and interpretation of pressures in prey-sucking fish. Neth. J. Zool. **33**: 425–475.

Van Leeuwen, J.L., Muller, M. (1984) Optimum sucking technique for predatory fish. Trans. Zool. Soc. Lond. **37**: 137–169.

Van den Berg, C., van den Boogart, J.G.M., Sibbing, F.A., Lammens, E.H.R.R., Osse, J.W.M. (1993) Shape of zooplankton and retention in filter-feeding: quantitative comparison between industrial sieves and the branchial sieves of common bream (*Abramis brama*) and white bream (*Blicca bjoerkna*). Can. J. Fish Aquat. Sci. **50**: 716–724.

Van den Berghe, E.P., Gross, M.R. (1989) Natural selection resulting from female breeding competition in a pacific salmon (coho-*Oncorhynchus kisutch*). Evolution **43**: 125–140.

Van den Hurk, R., Lambert, J.G.D. (1983) Ovarian steroid glucuronides function as sex pheromones for male zebrafish, *Brachydanio rerio*. Can. J. Zool. **61**: 2381–2387.

Van den Hurk, R., Resink, J.W. (1992) Male reproductive system as sex pheromone producer in teleost fish. J. Exp. Zool. **261**: 204–213.

Van den Hurk, R., Resink, J.W., Pente, J. (1987) The seminal vesicle of the African catfish, *Clarias gariepinus*. A histological, histochemical, enzyme-histochemical, ultrastructural and physiological study. Cell Tissue Res. **247**: 573–582.

Van der, Meer H.J. (1989) Ecological significance of retinal receptor patterns in 4 sympatric haplochromine cichlids. Ann. Mus. Roy. Afr. Centr. Sc. Zool. **257**: 51–56.

Van der, Meer, H.J. (1991) Determination of photic thresholds in two sympatric cichlids using optomotor response. Ann. Mus. Roy. Afr. Centr. Sc. Zool. **262**: 91–96.

Van der, Meer, H.J. (1992) Constructional morphology of photoreceptor patterns in percomorph fish. Acta. Biotheor, **40**: 51–85.

Van der, Meer, H.J., (1993) Light-induced modulation of retinal development in the cichlid fish *Haplochromis sauvagei* (Pfeffer 1896). Zool. J. Linn Soc. **108**: 271–285.

Van der, Meer, H.J. (1994) Ontogenetic change of visual thresholds in the cichlid fish *Haplochromis sauvagei*. Brain Behav. Evol. **44**: 40–49.

Van der, Meer, H.J. (1995) Visual resolution during growth in a cichlid fish: a morphological and behavioural case study. Brain Behav. Evol. **45**: 25–33.

Van der, Meer, H.J., Anker, G.C.H. (1984) Retinal resolving power and sensitivity of the photopic system in seven haplochromine species (Teleostei, Cichlidae) Neth. J. Zool. **34:** 197–209.
Van der, Meer, H.J., Anker, G.C.H. (1986) The influence of light deprivation on the development of the eye and retina in the cichlid *Sarotherodon mossambicus* (Teleostei). Neth, J. Zool. **36**: 480–498.
Vandewalle, P., Chardon, M. (1991) A new hypothesis on the air flow in air breathing in *Clarias gariepinus* (Teleostei, Siluriformes). Belg. J. Zool. **121**: 73–80.
Vandewalle, P., Huyssene, A., Aerts, P., Verraes, W. (1994) The pharyngeal apparatus in teleost feeding. In: Bels, V., Chardon, M., Vandewalle, P. (eds.) Advances in comparative and environmental physiology, vol. 18. Springer-Verlag, Berlin, Heidelberg, pp. 59–92.
Vandewalle, P., Saintin, P., Chardon, M. (1995) Structures and movements of the buccal and pharyngeal jaws in relation to feeding in *Diplodus sargus*. J. Fish Biol. **46**: 623–656.
Vandewalle, P., Radermaker, F., Surlemont, C., Chardon, M. (1990) Apparition of the Weberian characters in *Barbus barbus* (Linne 1758) (Pisces Cyprinidae). Zool. Anz. **225**: 362–376.
Van der Velden, J.A., Spanings, F.A.T., Flik G.,, Wendelaar Bonga, S.E. (1991) Early life stages of carp. (*Cyprinus carpio* L.) depend on ambient magnesium for their development. J. Exp. Biol. **158**: 431–438.
Van der Velden, J.A., Groot, J.A., Flik, G., Polak, P., Kolar, Z.I. (1990) Magnesium transport in fish intestine. J. Exp. Biol. **152**: 587–592.
Van der Velden, J.A., Spanings, F.A.T., Flik, G., Zegers, C., Kolar, Z.I., Wendelaar Bonga, S.E. (1991). Growth rate and tissue magnesium concentration in adult freshwater tilapia, *Oreochromis mossambicus* (Peters), fed diets differing in magnesium content. J. Fish Biol. **39**: 83–91.
Van Muiswinkel, W.B. (1992) Fish immunology and fish health. Neth, J. Zool. **42**: 494–499.
Venegas, H. (1983) Organization and physiology of the teleostean optic tectum. In: Davis, R.E., Northcutt, R.G. (eds.) Fish neurobiology, vol. 2. Higher brain areas and functions. University of Michigan Press, Ann Arbor, pp. 43-90.
Van Noorden, S. (1990) Gut hormones in cyclostomes. Fish Physiol. Biochem. **8**: 399–408.
Verbost, P.M., Flik, G., Fenwick, J.C., Greco, A.M., Pang, P.K.T., Wendelaar Bonga, S.E. (1993) Branchial calcium uptake: possible mechanisms of control by stanniocalcin. Fish Physiol. Biochem. **11**: 205–215.
Vernier, J.M. (1990) Intestine ultrastructure in relation to lipid and protein absorption in teleost fish. In: Mellinger J. (eds.) Animal nutrition and transport processes. 1. Nutrition in wildlife and domestic animals, Comp. Physiol. vol. 5, S. Karger, Basel, pp. 166–175.
Verri, T., Maffia, M., Storelli, C. (1992) H^+/glycyl-glycine cotransport in eel. intestinal brush-border membrane vesicles: studies with the pH-sensitive dye Acridine organe. Biocheim. Biophys. Acta **1110**: 123–126.
Videler, J. (1993) Fish swimming. Chapman and Hall. London, New York.
Videler, J.J., Wardle, C.S. (1991) Fish swimming stride by stride: Speed limits and endurance. Rev. Fish Biol. Fish. **1**: 23–40.
Viete, S., Heiligenberg, W. (1991) The development of the jamming avoidance response (JAR) in *Eigenmannia*: an innate behavior indeed. J. Comp. Physiol. A **169**: 15–23.
Vigna, S.R. (1983) Evolution of endocrine regulation of gastrointestinal function in lower vertebrates. Am. Zool. **23**: 729–738.
Vigna, S.R. (1985) Functional evolution of gastrointestinal hormones. In: Foreman R.E., Gorbman A., Dodd, J.M., Olsson, R. (eds.) Evolutionary biology of primitive fishes. Plenum Publishing Corporation, New York, pp. 401–412.
Vigna, S.R. (1986) Gastrointestinal tract. In: Pang, P.K.T., Schreibman, M.P. (eds.) Vertebrate endocrinology: fundamentals and biomedical implications, vol. 1. Academic Press, Orlando, pp. 261–278.
Vigna, S.R., Thorndyke, M.C., Williams, J.A. (1986) Evidence for a common evolutionary origin of brain and pancreas cholecytokinin receptors. Proc. Nat. Acad. Sci. U.S.A **83**: 4355–4359.
Vilella. S., Ahearn, G.A., Cassano, G., Storelli, C. (1988) Na-dependent L-proline transport by eel intestinal brush-border membrane vesicles. Am. J. Physiol. **255**: R648–R653.
Vilella, S., Cassano, G., Storelli, C. (1989a) How many Na^+-dependent carriers for L-alanine and L-proline in the eel intestine? Studies with brush-border membrane vesicles. Biochim. Biophys. Acta **984**: 188–192.
Vilella, S., Reshkin, S.J., Storelli, C., Ahearn, C.A. (1989b) Brush-border inositol transport by intestines of carnivorous and herbivorous teleosts. Am. J. Physiol. **256**: G501–G508.

Vilella. S., Ahearn, G.A., Cassano, G., Maffia M., Storelli, C. (1990) Lysine transport by brush-border membrane vesicles of eel intestine: interaction with neutral amino acids. Am. J. Physiol. **259**: R 1181–R 1188.

Vincent, A.C.J. (1992) Prospects for sex reversal in teleost fishes. Neth. J. Zool. **42**: 392–399.

Vincent, A.C.J. (1994a) Seahorses exhibit conventional sex roles in mating competition, despite male pregnancy. Behaviour **128**: 135–151.

Vincent, A.C.J. (1994b) Operational sex ratios in seahorses. Behaviour **128**: 153–167.

Vincent, A.C.J. (1995) A role for daily greetings in seahorse pair bonds. Anim. Behav. **49**: 258–260.

Vincent, A.C.J., Sadler, L.M. (1995) Faithful pair bonds in wild seahorses, *Hippocampus whitei*. Anim. Behav. **50**: (In press).

Vincent, A., Ahnesjo, I., Berglund, A, Roseqvist, G. (1992) Pipefishes and seahorses: are they all sex role reversed ? Trend. Ecol. Evol. **7**: 237–241.

Vincent, A.C.J., Berglund, A., Ahnesjo, I (1995) Reproductive ecology of five pipefish species in one eelgrass meadow. Env. Biol. Fish. (In press).

Vischer, H.A. (1989) The development of lateral-line receptors in *Eigenmannia* (Teleostei, Gymnotiformes). I. The mechanoreceptive lateral-line system. Brain Behav. Evol. **33**: 205–222.

Vischer, H.A. (1990) The morphology of the lateral line system in 3 species of Pacific cottoid fishes occupying disparate habitats. Experientia **46**: 244–250.

Vischer, H.A., Lannoo, M.J., Heiligenberg, W. (1989) Development of the electrosensory system in *Eigenmannia* (Gymnotiformes): I. The peripheral nervous system. J. Comp. Neurol. **290**: 16–40.

Vogel, W.O.P. (1978) Arteriovenous anastomes in the afferent region of trout gill filaments *Salmo gairdneri* Richardson, Teleostei). Zoomorphologie **90**: 205–212.

Vogel, W.O.P. (1985) Systemic vasculature anastomoses, primary and secondary vessels in fish, and the phylogeny of lymphatics. In: Johansen K., Burggren W.W. (eds.) Cardiovascular shunts, A. Benzon Symposium 21. Munksgaard, Copenhagen, Denmark, pp. 143–159.

Vogel, W., Vogel, V., Kremers, H. (1973) New aspects of the intrafilamental vascular system in gills of a euryhaline teleost, *Tilapia mossambica*. Z. Zellforsch. **144**: 573–683.

Vogel, W., Vogel, V., Schlote, W. (1974) Ultrastructural study of arteriovenous anastomoses in gill filaments of *Tilapia mossambica*. Cell Tissue Res. **155**: 491–512.

Voneida, T.J., Fish, S.E. (1984) Central nervous system changes related to the reduction of visual input in a naturally blind fish (*Astyanax hubbsi*). Am. Zool. **24**: 775–782.

Vonk, H.J., Western, J.R.H. (1984) Comparative biochemistry and physiology of enzymatic digestion. Academic Press, New York.

Wachtel, S.S., Wachtel, G., Nakamura, D. (1991a) Sexual differentiation. In: Pang, P.K.T., Schreibman, M.P. (eds.) Vertebrate endocrinology: fundamentals and biomedical implications, vol. 4B. Academic Press, San Diego, pp. 149–180.

Wachtel, S.S., Demas, S., Tiersch, T., Pechan, P., Shapiro, D. (1991b) Bkm satellite, DNA and *ZFY* in the coral reef fish *Acanthias squamipinnis*. Genome **34**: 612–617.

Wagner, G.F. (1994) The molecular biology of the corpuscles of Stannius and regulation of stanniocalcin gene expression. In: Sherwood, N., Hew, C. (eds.) Fish physiology, vol. 13. Academic Press, San Diego, pp. 273–306.

Wagner, G.F., Jaworski, E. (1994) Calcium regulates stanniocalcin mRNA levels in primary cultured rainbow trout corpuscles of Stannius. Mol. Cell Endocrinol. **99**: 315–322.

Wagner, G.F., Fenwick, J.C., Park, C.M., Milliken, C., Copp, D.H., Friesen, H.G. (1988) Comparative biochemistry and physiology of teleocalcin from sockeye and coho. salmon. Gen. Comp. Endocrinol. **72**: 237–246.

Wagner G.F., Fargher, R.C., Milliken, C., Mckeown, B.A., Copp, D.H. (1993) The gill calcium transport cycle in rainbow trout is correlated with plasma levels of bioactive, not immunoreactive stanniocalcin. Mol Cell Endocrinol. **93**: 185–191.

Wainwright, P.C. (1987) Biomechanical limits to ecological performance: mollusc-crushing by the Caribbean hogfish, *Lachnolaimus maximus* (Labridae). J. Zool. Lond. **213**: 283–297.

Wainwright, P.C. (1988) Morphology and ecology: functional basis of feeding constraints in Caribbean labrid fishes. Ecology **69**: 635–645.

Wainwright, P.C. (1989a) Prey processing in haemulid fishes: patterns of variation in pharyngeal jaw muscle activity. J. Exp. Biol. **141**: 359–375.
Wainwright, P.C. (1989b) Functional morphology of the pharyngeal jaw apparatus in perciform fishes: an experimental analysis of the Hacmulidae. J. Morphol. **200**: 231–245.
Wainwright, P.C., Richard, B.A. (1995) Scaling the feeding mechanism of the largemouth bass (*Micropterus salmoides*): motor pattern. J. Exp. Biol. **198**: 1161–1171.
Wainwright, P.C., Turingan, R.G. (1993) Coupled *versus* uncoupled functional systems: motor plasticity in the queen triggerfish *Balistes vetula*. J. Exp. Biol. **180**: 209–227.
Wainwright, P.C., Sanford, C.P., Reilly, S.M., Lauder, G.V. (1989) Evolution of motor patterns: Aquatic feeding in salamander and ray-finned fishes. Brain Behav. Evol. **34**: 329–341.
Wainwright, P.C., Turingan, R.G., Brainerd, E.L. (1995) Functional morphology of pufferfish inflation; mechanism of the buccal pump. Copeia **1995**: 614–625.
Wake, M.H. (1986) Urogenital morphology of dipnoans, with comparisons to other fishes and to amphibians. J. Morph. Suppl. **1**: 199–216.
Wake, M.H. (1992) Evolutionary scenarios, homology and convergence of structural specializations for vertebrate viviparity. Am. Zool. **32**: 256–263.
Walsh, P.J., Henry, R.P. (1991) Carbon dioxide and ammonia metabolism and exchange. In: Hochachka, P.W., Mommsen, T.P. (eds.) Biochemistry and molecular biology of fishes, vol. 1. Elsevier, Amsterdam, pp. 101–207.
Walsh, P.J., Milligan, C.L. (1995) Effects of feeding and confinement on nitrogen metabolism and excretion in the gulf toadfish *Opsanus beta*. J. Exp. Biol. **198**: 1559–1566.
Walsh, P.J., Tucker, B.C., Hopkins, T.E. (1994a) Effects of confinement/crowding on ureogenesis in the gulf toadfish *Opsanus beta*. J. Exp. Biol. **191**: 195–206.
Walsh, P.J., Wood. C.M., Perry, S.F., Thomas, S. (1994b) Urea transport by hepatocytes and red blood cells of selected elasmobranch and teleost fishes. J. Exp. Biol. **193**: 321–335.
Walsh, P.J., Mommsen, T.P., Bass, A.H. (1995) Biochemical and molecular aspects of singing in batrachoidid fishes. In: Hochachka, P.W., Mommsen, T.P. (eds.) Biochemistry and molecular biology of fishes, vol. 4. Elsevier, Amsterdam, pp. 279–289.
Walsh, P.J., Bergman, H.L., Narahara, A., Wood, C.M., Wright, P.A., Randall, D.J., Maina, J.N., Laurent, P. (1993) Effects of ammonia on survival, swimming and activities of enzymes of nitrogen metabolism in the lake Magadi tilapia *Oreochromis alcalicus* Grahami. J. Exp. Biol. **180**: 323–387.
Wardle, C.S. (1971) New observations on the lymph system of the plaice *Pleuronectes platessa* and other teleosts. J. mar. biol. Ass. U.K. **51**: 977–990.
Ware, D.M. (1984) Fitness of different reproductive strategies in teleost fishes. In: Wooton R., Potts G. (eds.) Fish reproduction: strategies and tactics. Academic Press, London, pp. 349–365.
Wassersug, R.J., Johnson, R.K. (1976) A remarkable pyloric caecum in the evermannellid genus *Coccorella* with notes on gut structure and function in alepisauroid fishes (Pisces, Myctophiformes). J. Zool. Lond. 179–289.
Watts, R.L., Watts, D.C. (1974) Nitrogen metabolism in fishes. In: Florkin, M., Scheer, B.T. (eds.) Chemical zoology, vol. 8. Deuterostomians, cyclostomes and fishes. Academic Press, New York, pp. 369–446.
Webb, J.F. (1989a) Gross morphology and evolution of the mechanoreceptive lateral-line system in teleost fishes. Brain Behav. Evol. **33**: 34–53.
Webb, J.F. (1989b) Developmental constraints and evolution of the lateral line system in teleost fishes. In: Coombs, S., Görner, P., Münz, H. (eds.) The mechanosensory lateral line: neurobiology and evolution. Springer-Verlag, New York, pp. 79–97.
Webb, J.F. (1989c) Neuromast morphology and lateral line trunk canal ontogeny in two species of cichlids: an SEM study. J. Morphol. **202**: 53–68.
Webb, J.F. (1990) Comparative morphology and evolution of the lateral line system in the Labridae (Perciformes: Labroidei). Copeia **1990**: 137–146.
Webb, P.W. (1978) Hydrodynamics: nonscombroid fish. In: Hoar, W.S., Randall, D.J. (eds.) Fish physiology, vol. 7. Academic Press, New York, pp. 189–237.
Webb, P.W. (1984a) Body form, locomotion and foraging in aquatic vertebrates. Am. Zool. **24**: 107–120.

Webb, P.W. (1984b) Chase response latencies of some teleostean piscivores. Comp. Biochem. Physiol. **79A**: 45–48.

Webb, P.W. (1986a) Effect of body form and response threshold on the vulnerability of four species of teleost prey attached by largemouth bass (*Micropterus salmoides*). Can. J. Fish Aquat Sci. **43**: 763–771.

Webb, P.W. (1986b) Locomotion and predator-prey relationships. In: Feder, M.E., Lauder, G.V. (eds.) Predator-prey relationships. Chicago University Press, Chicago, pp. 24–41.

Webb, P.W. (1988) Simple physical principles and vertebrate aquatic locomotion. Am. Zool. **28**: 709–725.

Webb, P.W. (1990) How does benthic living affect body volume, tissue composition, and density of fishes? Can. J. Zool. **68**: 1250–1255.

Webb, P.W. (1992) Is the high cost of body/caudal fin undulatory swimming due to increased friction drag or inertial recoil? J. Exp. Biol. **162**: 157–166.

Webb, P.W. (1993a) Swimming. In: Evans, D.H. (ed.) The physiology of fishes. CRC Press, Boca Raton, pp. 47–73.

Webb. P.W. (1993b) Is the tilting behaviour at low swimming speeds unique to negatively buoyant fish? Observations on steelhead trout, *Oncorhynchus mykiss* and blue gill, *Lepomis macrochirus*. J. Fish Biol. **43**: 687–694.

Webb, P.W. (1993c) The effect of solid and porous channel walls on steady swimming of steelhead trout *Oncorhynchus mykiss*. J. Exp. Biol. **178**: 97–108.

Webb, P.W. (1994) Exercise performance of fish. In: Jones, J.H. (ed.) Advances in veterinary sciences and comparative medicine, vol. 38B. Academic Press, San Diego, pp. 1–49.

Webb, P.W., Weihs D. (1994) Hydrostatic stability of fish with swim bladders: not all fish are unstable. Can. J. Zool. **72**: 1149–1154.

Webb, P.W., Sims, D., Schultz, W.W. (1991) The effects of an air/water surface on the fast-start performance of rainbow trout (*Oncorhynchus mykiss*). J. Exp. Biol. **155**: 219–226.

Webb, P.W., Hardy, D.H., Hehl, V.L. (1992) The effect of armored skin on the swimming of longonse gar, *Lepisosteus osseus*. Can. J. Zool. **70**: 1173–1179.

Webb, R.E., Jensen, F.B. (1988) Functional adaptations in hemoglobins from ectothermic vertebrates. Ann, Rev. Physiol. **50**: 161–179.

Weber, R.E. (1990) Functional significance and structural basis of multiple hemoglobins with special reference to ectothermic vertebrates. In: Truchot J-P, Lahlou B (ed.) Animal nutrition and transport processes. 2. Transport, respiration and excretion: Comparative and environmental aspects. Comp. Physiol. vol. 6. S. Karger, Basel, pp. 58–75.

Weber, R.E. (1992) Molecular strategies in the adaptation of vertebrate hemoglobin function. In: Wood, S.C., Weber, R.E., Hargens, A.R., Millard R.W. (eds.) Physiological adaptations in vertebrates: respiration, circulation and metabolism. Marcel Dekker, New York, pp. 257–277.

Weber, R.E., Wells, R.M.G. (1989) Hemoglobin structure and function. In: Wood, S.C. (ed.) Comparative pulmonay physiology-Current concepts. Lung biology in health and disease, vol. 39. Marcel Dekker, New York, pp. 279–310.

Wells, R.M.G., Weber, R.E. (1983) Oxygenational properties and phosphorylated metabolic intermediates in blood and erythrocytes of the dogfish, *Squalus acanthias*. J. Exp. Biol. **103**: 95–108.

Wells, R.M.G., Weber, R.E. (1990) The spleen in hypoxic and exercised reinbow trout. J. Exp. Biol. **150**: 461–466.

Wells, R.M.G., Weber, R.E. (1991) Is there an optimal haemotocrit for rainbow trout, *Oncorhynchus mykiss* (Walbaum)? An interpretation of recent data based on blood viscosity measurements. J. Fish Biol. **38**: 53–65.

Wells, R.M.G., Macdonald, J.A., di Prisco, G. (1990) Thin-blooded Antarctic fishes: a rheological comparison of the haemoglobin-free icefishes (*Chionodraco kathleenae* and *Cryodraco antarcticus* with a red-blooded nototheniid, *Pagothenia bernacchii*. J. Fish Biol. **36**: 595–609.

Wendelaar Bonga, S.E. (1993) Endocrinology. In: Evans, D.H. (ed.) The physiology of fishes. CRC Press, Boca Raton, pp. 469–502.

Wendelaar Bonga, S.E., Pang, P.K.T. (1986) Stannius corpuscles. In: Pant, P.K.T., Schreibman, M.P. (eds.) Vertebrate endocrinology: fundamentals and biomedical implications, vol. 1. Academic Press, New York, pp. 439–464.

Wendelaar Bonga, S.E., Pang, P.K.T. (1989) Pituitary hormones. In: Pang, P.K.T., Schreibman, M.P. (eds.) Vertebrate endocrinology: fundamentals and biomedical implications, vol. 3. Academic Press, New York, pp. 105–137.

Wendelaar Bonga, S.E., Smits, P.W.J.M., Flik, G., Kaneko, T., Pang, P.K.T. (1989) Immunocytochemical localization of hypocalcin in the endocrine cells of the corpuscles of Stannius in three teleost species (trout, flounder and goldfish). Cell Tissue Res. **255**: 651–656.

Wendelaar Bonga, S.E., Flik, G., van der Velden, J.A., Kolar, Z. (1990) Prolactin cell activity and sodium balance in the acid tolerant mudminnow *Umbra pygmaea* in acid and neutral water. Gen. Comp. Endocrinol. **73**: 421–432.

Wendelaar Bonga, S.E., Flik, G., Balm, P.H.M., Van der Meij J.C.A. (1995) The ultrastructure of chloride cells in the teleost *Oreochromis mossambicus* during exposure to acidified water. Cell Tissue Res. **259**: 575–585.

Wersall, J., Bagger-Sjoback. D. (1974) Morphology of the vestibular sense organ. In: Kornhuber, H.H. (ed.) Handbook of sensory physiology, vol. 6. part 1. Springer-Verlag, Berlin, pp. 123–170.

Westneat, M.W., Wainwright, P.C. (1989) Feeding mechanism of *Epibulus insidator* (Labridae: Teleostei): evolution of a novel functional system. J. Morphol. **202**: 129–150.

White, M.G. (1991) Age determination in Antarctic fish. In: di Prisco, G., Maresca, B., Tota, B. (eds.) Biology of Antarctic fish. Springer-Verlag, Berlin, Heidelberg, pp. 87–100.

White, M.G., Burren, P.J. (1992) Reproduction and larval growth of *Harpagifer antarcticus* Nybelin (Pisces, Notothenioidei). Ant. Science **4**: 421–430.

Whitear, M. (1986a) Epidermis. In: Bereiter-Hahn, J., Matoltsy, A.G., Richards, K.S. (eds.) Biology of the integument, vol. 2: Vertebrates. Springer-Verlag, Berlin, Heidelberg, New York, pp. 8–38.

Whitear, M. (1986b) Dermis. In: Bereiter-Hahn, J., Matoltsy, A.G., Richards, K.S. (eds.) Biology of the integument, vol. 2. Vertebrates. Springer-Verlag, Berlin, Heidelberg, New York, pp. 39–64.

Whitear, M. (1992) Solitary chemosensory cells. In: Hara, T.J. (ed.) Fish chemoreception, Chapman and Hall, London, New York, pp. 103–125.

Whitear, M., Zaccone, G., Fasulo, S., Licata, A. (1991a) Fine structure of the axillary gland in the brown bullhead (*Ictalurus nebulosus*). J. Zool. Lond. **224**: 669–676.

Whitear, M., Zaccone, G., Ricca, M.B., LoCassio, P. (1991b) Fine structure and histochemistry of the venom gland in the Indian stinging catfish (*Heteropneustes fossilis*). J. Zool. Lond. **224**: 429–489.

Whittear, J.M. (1991) Artificial regulation of reproductive cycles. In: Pang, P.K.T., Schreibman, M.P. (eds.) Vertebrate endocrinology: fundamentals and biomedical implications. vol. 4B. Academic Press, San Diego, pp. 243–268.

Wisenden, B.D., Chivers, D.P., Brown, G.E., Smith, R.J.F. (1995) The role of experience in risk assessment: avoidance of areas chemically labelled with fathead minnow alarm pheromone by conspecifics and heterospecifics. Ecoscience **2**: 116–122.

Witt, M., Reutter, K. (1988) Lectin histochemistry on mucous substances of the taste buds and adjacent epithelia of different vertebrates. Histochemistry **88**: 453–461.

Witt, M., Reutter, K. (1990) Electron microscopic demonstration of lectin binding sites in the taste buds of the European catfish *Silurus glanis* (Teleostei). Histochemistry **94**: 617–628.

Wood, C.M. (1989) The physiological problems of fish in acid waters. In: Morris, R., Taylor, E.W., Brown, D.J.A., Brown, J.A. (eds.) Acid toxicity and aquatic animals. Cambridge University Press, Cambridge, pp. 125–152.

Wood, C.M. (1991) Branchial ion and acid-base transfer in freshwater teleost fish: environmental hyperoxia as a probe. Physiol. Zool. **64**: 68–102.

Wood, C.M. (1993) Ammonia and urea metabolism and excretion. In: Evans, D.H. (ed.) The physiology of fishes. CRC Press, Boca Raton, pp. 379–425.

Wood, C.M., Goss, G.G. (1990) Kinetic analysis of the relationships between ion exchange and acid base regulation at the gills of freshwater fish. In: Truchot, J.P., Lahlou, B. (eds.) Animal nutrition and transport processes. 2. Transport, respiration and excretion: Comparative and environmental aspects. Comp. Physiol. **6**: 119–136.

Wood, C.M., Perry, S.F., Wright, P.A., Bergman, H.L., Randall, D.J, (1989) Ammonia and urea dynamics in the lake Magadi tilapia, a ureotelic teleost fish adapted to an extermely alkaline environment. Respir. Physiol. **77**: 1–20.

Wood, C.M., Bergman, H.L., Laurent, P., Maina, J.N., Narahara, A., Walsh, P.J. (1994) Urea production, acid-base regulation and their interactions in the Lake Magadi tilapia, a unique teleost adapted to highly alkaline environment. J. Exp. Biol. **189**: 13–36.

Wood, C.M., Hopkins, T.E., Hogstrand, C., Walsh, P.J. (1995) Pulsatile urea excretion in the ureagenic toadfish *Opsanus beta*: an analysis of rates and routes. J. Exp. Biol. **198**: 1729–1741.

Woodhead, A.D. (1975) Endocrine physiology of fish migration. Oceanogr. Mar. Biol. Ann. Rev. **13**: 287–383.

Woodhead, A.D. (1979) Senescence in fishes. Symp. Zool. Soc. Lond. No. **44**: 179–205.

Wooton, R.J. (1990) Ecology of teleost fishes. Chapman and Hall, London, New York.

Wooton, R.J. (1992) Constraints in the evolution of fish life histories. Neth. J. Zool. **42**: 291–303.

Wooton, R.J., Potts, G.W. (eds.) (1984) Fish reproduction: strategies and tactics. Academic Press, London.

Wourms, J.P., Atz, J.W., Stribling, M.D. (1991) Viviparity and the maternal-embryonic relationship in the coelacanth *Latimeria chalumnae*. Env. Biol. Fish. **32**: 225–248.

Wright, P.A., Randall, D.J. (1987) The interaction between ammonia and carbon dioxide stores and excretion rates in fish. Annls. Soc. R. Zool. Belg. 117 Suppl. 1 Ecophysiology of acid stress in aquatic organisms, pp. 321–329.

Wright, P.A., Randall, D.J., Perry, S.F. (1989) Fish gill water boundary layer: a site of linkage between carbon dioxide and ammonia excretion. J. Comp. Physiol,. B **158**: 627–635.

Wright, S.H. (1991) The interface of animal and aqueous environment: strategies and constraints on the maintenance of solute balance. In: Hochachka, P.W., Mommsen, T.P. (eds.) Biochemistry and molecular biology of fishes, vol. 1. Elsevier, Amsterdam, pp. 165–180.

Wu, E.H. (1994) Kinematic analysis of jaw protrusion in orectolobiform sharks: a new mechanism for jaw protrusion in elasmobranchs. J. Morphol. **222**: 175–190.

Xiang, H., Taylor, E.W., Whitely, N., Randall, D.J. (1994) Modulation of noradrenergic action by neuropeptide, Y in dogfish (*Squalus acanthias*) hearts. Physiol. Zool. **67**: 204–215.

Yamada, J., Watabe, N. (1979) Studies on fish scale formation and resorption. I. Fine structure and calcification of the scales in *Fundulus heteroclitus*. (Atheriniformes: Cyprinodontidae). J. Morphol. **159**: 49–66.

Yamamoto, M. (1978) Morphological changes in the esophageal epithelium of the eel, *Anguilla japonica*, during adaptation to seawater. Cell Tissue Res. **193**: 25–38.

Yamamoto, M. (1982) Comparative morphology of the peripheral olfactory organ in teleosts. In: Hara, T.J. (ed.) Chemoreception in fishes. Elsevier, Amsterdam, pp. 39–59.

Yamamoto, T. (1961) The physiology of fertilization in fish egg. Int. Rev. Cytol. **12**: 361–405.

Yamamoto, T. (1969) Sex differentiation. In: Hoar, W.S., Randall, D.J. (eds.) Fish physiology, vol. 3. Academic Press, New York, pp. 117–175.

Yamauchi, A (1980) Fine structure of the fish heart. In: Bourne, G.H. (ed.) Hearts and heart-like organs, vol. 1. Academic Press, New York, pp. 119–148.

Yates, G.T. (1983) Hydromechanics of body and caudal fin propulsion. In: Webb. P.W., Weihs, D. (eds.) Fish biomechanics. Praeger, New York, pp. 177–213.

Ye, X., Randall, D.J., He, X. (1991) The effect of acid water on oxygen consumption, circulating catecholamines and blood ionic and acid-base status on rainbow trout (*Salmo gairdneri*, Richardson). Fish Physiol. Biochem. **9**: 23–30.

Yew, D.T., Yoshihara, H.M. (1977) An ultrastructural study on the retina of the blind cave fish (*Astyanax hubbusi*). Cytologia **42**: 175–180.

Yodin, K.G., Cailliet, G.M. (1990) Age and growth of the gray smoothhound, *Mustelus californicus*, and the brown smoothhound, *M. henlei*, sharks from central California. Copeia **1990**: 191–204.

Yoklavich, M.M., Boehlert, G.W. (1991) Uptake and utilisation of ^{14}C– glycine by embryos of *Sebastes melanops*. Env. Biol. Fish. **30**: 147–153.

Yokota, S.D., Benyajati, S., Dantzler, W.H. (1985) Comparative aspects of glomerular filtration in vertebrates. Renal. Physiol. Basel **8**: 193–221.

Young, J.Z. (1981) The life of Vertebrates, (Third edition). Clarendon Press, Oxford.

Zaccone, G. (1986) Neuron-specific enolase and serotonin in the Merkel cells of conger-eel. (*Conger conger*) epidermis: An immunohistochemical study. Histochemistry **85**: 29–34.

Zaccone, G., Fasulo, S., Cascio, P.L., Licata, A., Ainis, L. (1987) Occurrence of gastrin-releasing peptide-like and bombesin-like immunoreactivities in the epidermal sacciform gland cells of the clingfish *Lepadogaster candollei*. Bas. Appl. Histochem. **31**: 455–464.

Zacoone, G., Fasulo, S., Ainis, L., Contini, A. (1989a) Localization of calmodulin positive immunoreactivity in the surface epidermis of the brown trout, *Salmo trutta*. Histochemistry **91**: 13–16.

Zaccone, G., Goniakowska-Witalinska, L., Lauweryas, J.M., Fasulo, S., Tagliafierro, G. (1989b) Fine structure and serotonin immunohistochemistry of the neuroendocrine cells in the lungs of the bichirs *Polypterus delhezi* and *P. ornatipinnis*. Bas. Appl. Histochem. **33**: 277–287.

Zaccone, G., Tagliafierro, G., Goniakowska-Witalinska, L., Fasulo, S., Ainis, L., Mauceri, A. (1989c) Serotonin-like immunoreactive cells in the pulmonary epithelium of ancient fish species. Histochemistry **92**: 61–63.

Zaccone, G., Ainis L., Fasulo, S., Cascio, P.L., Mauceri, A., Licata, A. (1989d) Merkel cells detected by immunocytochemistry in the epidermis of the freshwater catfish *Heteropneustes fossilis* (Bloch). Zool. Jb. Anat. **119**: 129–134.

Zaccone, G., Lauweryns, J.M., Fasulo, S., Tagliafierro, G., Ainis, L., Licata, A. (1992a) Immunocytochemical localization of serotonin and neuropeptides in the neuroendocrine paraneurons of teleost and lungfish gills. Acta Zool. Stockholm. **73**: 177–183.

Zaccone, G., Wendelaar Bonga, S.E., Flik. G., Fasulo, S., Licata, A., Cascio, P.L., Mauceri, A., Lauriano, E.R. (1992b) Localization of calbindin D28K-like immunoreactivity in fish gill: a light microscopic and immunoelectron histochemical study. Regulatory Peptides **41**: 198–208.

Zaccone, G., Fasulo, S., Ainis, L. (1994) Distribution patterns of the paraneural endocrine cells in the skin, gills and the airways of fishes as determined by immunohistochemical and histological methods. Histochemical Journal **26**: 606–629.

Zaccone, G., Fasulo, S., Ainis, L. (1995a) Neuroendocrine epithelial cell system in respiratory organs of air-breathing and teleost fishes. Int. Rev. Cytol. **157**: 277–314.

Zaccone, G., Howie, A.J., Maucere, A., Fasulo, S., Lo Casio, P., Youson, J.H. (1995b) Distribution patterns of cytokeratins in epidermis and horny teeth of the adult sea lamprey, *Petromyzon marinus*. Folia Histochem. et Cytobiol. **33**: 69–75.

Zander, C.D. (1979) Morphology of skin structures in the head region of *Paragobiodon echinocephalus* (Ruppell) (Gobiidae, Pisces). Zool. Anz. Jena. **202**: 226–234.

Zakon, H.H. (1984) Postembryonic changes in the peripheral electrosensory system of a weakly electric fish: addition of receptor organs with age. J. Comp. Neurol. **228**: 557–570.

Zakon, H.H. (1987) Hormone-mediated plasticity in the electrosensory system of weakly electric fish. Trends Neurosi. **10**: 416–421.

Zakon, H.H. (1988) The electroreceptors: diversity in structure and function. In: Atema. J., Fay R.R., Popper, A.N., Tavolga, W. (eds.) Sensory biology of aquatic animals. Springer-Verlag, Berlin, pp. 813–850.

Zakon, H.H. (1991) Regeneration of electroreceptors in weakly electric fish. In: Regeneration of vertebrate sensory receptor cells. Wiley, Chichester (Ciba Foundation Symposium 160), pp. 294–313.

Zapata, A.G. (1980) Ultrastructure of elasmobranch lymphoid tissue. 1. Thymus and spleen. Develop. Comp. Immun. **4**: 459–472.

Zapata, A.G. (1981) Ultrastructure of elasmobranch lymphoid tissue. 2. Leydig's and epigonal organs. Develop. Comp. Immun. **5**: 43–52.

Zapata, A.G. (1983) Phylogeny of the fish immune system. Bull. Inst. Pasteur **81**: 165–186.

Zapata, A.G., Fänge, R., Mattisson, A., Villena, A. (1984) Plasma cells in adult Atlantic hagfish, *Myxine glutinosa*. Cell Tissue Res. **235**: 691–693.

Zeiske, E., Theisen, B., Breucker, H. (1992) Structure, development, and evolutionary aspects of the peripheral olfactory system. In: Hara, T.J. (ed.) Fish chemoreception. Chapman and Hall, London, New York, pp. 13–39.

Zeiske, E., Thiesen, B., Breucker, H. (1994) The olfactory organ of the hardhead sea catfish, *Arius felis* (L.): gross morphology and fine structure. Acta Zool. Stockholm. **75**: 115–123.

Zottoli, S.J., Danielson, P.D. (1989) Lateral line afferent and efferent systems of the goldfish with special reference to the Mauthner cell. In: Coombs, S., Görner, P., Münz, H. (eds.) The mechanosensory lateral line: neurobiology and evolution. Springer-Verlag, New York, pp. 461–478.

Zottoli, S.J., Hangen, D.H., Faber, D.S. (1984) The axon reaction of the goldfish Mauthner cell and factors that influence its morphological variability. J. Comp. Neurol. **230**: 497–516.

Zottoli, S.J., Marek, L.E., Agostini, M.A., Strittmatter, S.L. (1987) Morphological and physiological survival of goldfish Mauthner axon isolated from their somata by spinal cord crush. J. Comp. Neurol. **255**: 272–282.

Zupane, G.K.H., Heiligenberg, W. (1989) Sexual maturity-dependent changes in neuronal morphology in the prepacemaker nucleus of adult weakly electric knifefish, *Eigenmannia*. J. Neurosci. **9**: 3816–3827.

Zyznar, E.S., Cross, F.B., Nicol, J.A.C. (1978) Uric acid in the tapetum lucidum of mooneyes *Hiodon*. (Hiodontidae Teleostei). Proc. R. Soc. Lond. B. **201**: 1–16.

Appendix to References-1

Abraham, M., Hilge, V., Riehl, R., Iger, Y. (1993) Muco-follicle cells of the jelly coat in the oocyte envelope of the sheathfish (*Silurus glanis* L.). J. Morphol. **217**: 37–43

Akoev, G.N., Andrianov, G.N. (1989) Synaptic transmission in the mechano- and electroreceptors of the acousticolateral system. Progr. Sensory Physiol. **9:** 53–95.

Alexander, R.McN (1973) Jaw mechanisms of the coelacanth *Latimeria*. Copeia 1973: 156–158.

Ali, M.A., Wagner, H.J., (1976) Scanning electron microscopy of four teleostean retinas. Rev. Can. Biol. **35**: 199–210.

Alves-Gomes, J., Hopkins, C.D. (1997) Molecular insights into the phylogeny of mormyriform fishes and the evolution of their electric organs. Brain Behav. Evol. **49**: 324–351.

Amagai, S. (1998) Time coding in the midbrain of mormyrid electric fish. II. Stimulus selectivity in the nucleus exterolateralis pars posterior. J. Comp. Physiol. **A182:** 131–143.

Amagai, S., Friedman, M.A., Hopkins, C.D. (1998) Time coding in the midbrain of mormyrid electric fish. I. Physiology and anatomy of cells in the nucleus exterolateralis pars anterior. J. Comp. Physiol. **A182:** 115–130.

Andrianov, G.N., Bretschneider, F., Peters, R.C. (1992) The acions of L-glutamate and its agonists on the ampullary electroreceptor organs of the catfish *Ictalurus nebulosus*. Comp. Biochem. Physiol. **103C**: 65–71.

Andrianov, G.N., Peters, R.C., Bretschneider, F. (1994) Identification of AMPA receptors in catfish electroreceptor organs. NeuroReport **5:** 1056–1058.

Andrianov, G.N., Bretschneider, F., Peters, R.C. (1996) Mode of operation of ampullary electroreceptor organs of freshwater catfish. Prim. Sensory Neuron **1:** 231–242.

Andrianov, G.N., Bretschneider, F., Peters, R.C. (1997) Electrophysiological demonstration of N-methyl-D-aspartate receptors at the afferent synapse of catfish electroreceptor organs. Neuroscience **79**: 1231–1237.

Angel, M.V. (1997) What is the deep sea? In Hoar, W.S., Randall, D.J., Farrell, A.P. (eds.) Fish physiology, vol. 16. Deep-sea fishes. Academic Press, San Diego, pp. 1–41.

Arnott, H.J., Maciolek, H.J., Nicol, J.A.C. (1970) Retinal tapetum lucidum: a novel reflecting system in the eye of teleosts. Science **169:** 478–480.

Baguet, F. (1975) Excitation and control of isolated photophores of luminous fishes. Progr. Neurobiol. **5**: 97–125.

Bakker, T.C.M., Milinski, M. (1993) The advantages of being red: sexual selection in the stickleback. Mar. Behav. Physiol. **23**: 287–300.

Barber, V.C., Emerson, C.J. (1980) Scanning electron microscopic observations on the ear of the skate, *Raja ocellata*. Cell Tissue Res. **205:** 199–215.

Bass, A.H. (1989) Evolution of vertebrate motor systems for acoustic and electrical communication: peripheral and central elements. Brain Behav. Evol. **33**: 237–247.

Barnett, C.H. (1951) The structure and function of the choroidal glands of teleostean fish. J. Anat. **85**: 113–119.

Baustian, M.D., Beyenbach, K.W. (1993) Isolated renal tubules: studies of transepithelial secretory mechanism in teleost and elasmobranch fish. In: Brown, J.A., Balment, R.J., Rankin, J.C. (eds.) New insights in vertebrate kidney function. Cambridge University Press, Cambridge, pp. 45–62.

Baustian, M.D., Wang, S.Q., Beyenbach, K.W. (1997) Adaptive responses of aglomerular toadfish to dilute sea water. J. Comp. Physiol. **B167:** 61–70.

Bennett, M.V.L. (1970) Comparative physiology: electric organs. Ann. Rev. Physiol. **32**: 471–528.

Bennett, M.V.L. (1971a) Electric organs. In: Hoar, W.S., Randall, D.J. (eds.) Fish physiology, vol. 5. Academic Press, New York, London, pp. 347–491.

Bennett, M.V.L. (1971b) Electroreception. In: Hoar, W.S., Randall, D.J. (eds.) Fish physiology, vol. 5. Academic Press, New York, London, pp. 493–575.

Benjamin, M. (1988) Mucochondroid (mucous connective) tissues in the heads of teleosts. Anat. Embryol. **178:** 461–474.

Benjamin, M. (1989) Hyaline-cell cartilage (chondroid) in the heads of teleosts. Anat. Embryol. **179:** 285–303.

Benjamin, M., Ralphs, J.R. (1991) Extracellular matrix of connective tissues in the heads of teleosts. J. Anat. **179:** 137–148.

Berget, B.A., Wainwright, P.C. (1997) Morphology and kinematics of prey capture in the syngnathid fishes *Hippocampus erectus* and *Syngnathus floridae*. Mar, Biol. **127:** 563–570.

Berglund, A., Rosenqvist, G., Svensson, I. (1986a) Reversed sex roles and parental energy investment in zygotes of two pipefish (Syngnathidae) species. Mar. Ecol. Prog. Ser. **29:** 209–215.

Berglund, A., Rosenqvist, G., Svensson, I. (1986b) Mate choice and sexual dimorphism in two pipefish species (Syngnathidae). Behav. Ecol. Sociobiol. **19**: 301–307.

Bernier, N.J., Harris, J., Lessard, J., Randall, D.J. (1996a) Adenosine receptor blockade and hypoxia-tolerance in rainbow trout and Pacific hagfish. 1. Effects on anaerobic metabolism. J. Exp. Biol. **199**: 485–495.

Bernier, N.J., Fuentes, J., Randall, D.J. (1996b) Adenosine receptor blockade and hypoxia-tolerance in rainbow trout and Pacific hagfish. II. Effects on plasma catecholamines and erythrocytes. J. Exp. Biol. **199**: 497–507.

Bernstein, R.M., Schlutor, S.F., Marchalonis, J.J. (1998) Immunity. In: Evans, D.H. (ed.) The physiology of fishes. (Second edition). CRC Press, Boca Raton, pp. 215–242.

Bertmar, G. (1972) Secondary folding of olfactory organ in young and adult sea trout. Acta Zool. Stockholm **33**: 113–120.

Berwert, L., Segner, H., Reincke, M. (1995) Ontogeny of IGF-1 and the classical islet hormones in the turbot, *Scophthalmus maximus*. Peptides **16**: 113–122.

Best, A.C.G., Nicol, J.A.C. (1978) Notes on the retina and tapetum lucidum of *Howella* (Teleostei: Cheiloolipteridae). J. mar. Biol. Ass. U.K. **58:** 735–738.

Best, A.C.G., Nicol, J.A.C. (1979) On the eye of the goldeye *Hiodon alosoides* (Teleostei: Hiodontidae). J. Zool. Lond. **188**: 309–332.

Beyenbach, K.W., Dantzler, W.H. (1990) Comparative kidney tubule sources, isolation, perfusion, and function. In: Fleischer, S., Fleischer, B. (eds.) Methods in enzymology, vol. 191. Academic Press, San. Diego, pp. 167–226.

Beyenbach, K.W., Liu, P.L.F. (1996) Mechanism of fluid secretion common to aglomerular and glomerular kidneys. Kidney Int. **49:** 1543–1548.

Bowmaker, J.K., Kunz, Y.W. (1987) Ultraviolet receptors, tetrachromatic colour vision and retinal mosaic in the brown trout (*Salmo trutta*): age-dependent changes. Vision Res. **27**: 2101–2108.

Bowmaker, J.K., Thorpe, A., Douglas, R.H. (1991) Ultraviolet-sensitive cones in the goldfish. Vision Res. **31**: 349–352.

Brackevelt, C.R. (1980a) The fine structure of the retinal epithelium in the scissortail (*Rasbora trilineata*) (Teleost). Anat. Anz. Jena **148**: 225–235.

Brackevelt, C.R. (1980b) Fine structure of the retinal epithelium and tapetum lucidum in the giant danio (*Danio malabaricus*) (Teleost). Anat. Embryol. **158**: 317–328.

Brill, R.W. (1996) Selective advantages conferred by the high performance physiology of tunas, billfishes, and dolphin fish. Comp. Biochem. Physiol. **113A**: 3–15.

Britz, R., Kokoscha, M., Riehl, R. (1995) The anabantoid genera *Ctenops, Luciocephalus, Parasphaerichthys,* and *Sphaerichthys* (Teleostei: Perciformes) as a monophyletic group: evidence from egg surface structure and reproductive behaviour. Japan, J. Ichthyol. **42**: 71–79.

Browman, H.I., Hawryshyn, C.W. (1992) Thyroxine induces a precocial loss of ultraviolet photosensitivity in rainbow trout (*Oncorhynchus mykiss*, Teleostei). Vision Res. **32**: 2303–2312.

Buchholz, H., Schonrock, C., Fehr, S., Richter, D. (1995) Sequence analysis of a cDNA encoding an isotocin precursor and localization of the corresponding mRNA in the brain of the cartilaginous fish *Torpedo marmorata*. Mol. Mar. Biol. Biotech **4**: 179–184.

Burggren, W.W. (1979) Bimodal gas exchange during variation in environmental oxygen and carbon dioxide in the air breathing fish *Trichogaster trichopterus*. J. Exp Biol. **82**: 197–213.

Burggren, W., Dunn, J., Barnard, K. (1979) Branchial circulation and gill morphometrics in the sturgeon *Acipenser transmontanus*, an ancient chondrostean fish. Can. J. Zool. **57**: 2160–2170.

Burleson, M.L., Shipman, B.N., Smatresk, N.J. (1998) Ventilation and acid-base recovery following exhaustive activity in an air-breathing fish. J. Exp. Biol. **201**: 1359–1368.

Burns, J.R., Weitzman, S.H. (1996) Novel gill-derived gland in the male swordtail characin, *Corynopoma riisei* (Teleostei: Characidae: Glandulocaudinae). Copeia **1996:** 627–633.

Burns, J.R., Weitzman, S.H., Malabarba, L.R. (1997) Insemination in eight species of cheirodontine fishes (Teleostei: Charcidae: Cheirodontine). Copeia **1997**: 433–438.

Bushnell, P.G., Jones, D.R., Farrell, A.P. (1992) The arterial system. In: Hoar, W.S., Randall, D.J., Farrell, A.P. (eds.) Fish physiology, vol. 12A. Academic Press, Orlando, pp. 89–139.

Bushnell, P.J., Conklin, D.J., Duff, D.W., Olson, K.R. (1998) Tissue and whole-body extracellular, red blood cell and albumin spaces in the rainbow trout as a function of time: a reappraisal of the volume of the secondary circulation. J. Exp. Biol. **201**: 1381–1391.

Carvalho, G.R., Hauser, L. (1995) Molecular genetics and the stock concept in fisheries. In: Carvalho, G.R., Pitcher, T.J. (eds.) Molecular genetics in fisheries. Chapman and Hall, London, New York, pp. 55–79.

Chagas, C., Neto, V.M. (1989) *Electrophorus electricus*: an overview. Melánges de neurophysiologie Monnier: 63–76.

Chakrabarti, P., Mandal, D.K., Ganguli, S. (1992) A scanning electron microscope study of the mucosal epithelium of the alimentary canal of stomach bearing herbivorous fish *Oreochromis mossambicus* (Peters) Eur. Arch. Biol. **103**: 265–270.

Chambers, R.C. (1993) Phenotypic variability in fish populations and its representations in individual-based models. Trans. Am. Fish. Soc. **122**: 404–414.

Chambers, R.C., Leggett, W.C. (1992) Possible causes and consequences of variation in age and size at metamorphosis in flatfishes (Pleuronectiformes): an analysis at the individual, population, and species levels. Neth. J. Sea. Res. **29:** 7–24.

Chambers, R.C., Leggett, W.C. (1996) Maternal influences on variation in egg sizes in temperate marine fishes. Am. Zool. **36:** 180–196.

Chandra, S., Morrison, G.H., Beyenbach, K.W. (1997) Identification of Mg-transporting renal tubules and cells by ion microscopy imaging of stable isotopes. Am. J. Physiol. **273:** F 939-F948.

Chardon, M, Vandewalle, P. (1997) Evolutionary trends and possible origin of the Weberian apparatus. Neth. J. Zool. **47**: 383–403.

Chiba, A. (1990) Structure of the corpuscles of Stannius in the euryhaline teleost *Takifugu niphobles* (Tetraodontiformes, Teleostei), with special regard to vascularization and extravascular channel system. Anat. Embryol. **182**: 161–169.

Chopin, L.K., Bennett, M.B. (1995) The regulation of branchial blood flow in the blacktip reef shark, *Carcharhinus melanopterus* (Carcharhinidae: Elasmobranchii). Comp. Biochem. Physiol. **112A:** 35–41.

Chopin, L.K., Bennett, M.B. (1996a) Morphology and tyrosine-hydroxylase immunohistochemistry of the systemic secondary vessel system of the blue catfish, *Arius graeffei*. J. Morphol. **229**: 347–356.

Chopin, L.K., Bennett, M.B. (1996b) Cellular ultrastructure and catecholamine histofluorescence of the heart of the Australian lungfish, *Neoceratodus forsteri*. J. Morphol. **223**: 191–201.

Claiborne, J.B. (1998) Acid-base regulation. In: Evans, D.H. (ed.) The physiology of fishes. (Second edition). CRC Press, Boca Raton, pp. 177–198.

Claiborne, J.B., Perry, E., Bellows, S, Campbell, J. (1997) Mechanisms of acid-base excretion across the gills of a marine fish. J. Exp. Zool. **279**: 509–520.

Collin, H.B., Collin, S.P. (1995) Ultrastructure and organisation of the cornea, lens and iris in the pipefish, *Corythoichthyes paxtoni* (Syngnathidae, Teleostei). Histol. Histopathol. **10**: 313–323.

Collin, H.B., Collin, S.P. (1996) The fine structure of the cornea of the salamanderfish, *Lepidogalaxias salamandroides* (Lepidogalaxiidae, Teleostei). Cornea **15:** 414–426.

Collin, S.P. (1988) The retina of the shovel-nosed ray, *Rhinobatos batillum* (Rhinobatidae): morphology and quantitative analysis of the ganglion, amacrine and bipolar cell populations. Exp. Biol. **47**: 195–207.

Collin, S.P. (1997) Specialisations of the teleost visual system: adaptive diversity from shallow-water to deep-sea. Acta Physiol. Scand. **161:** (Suppl. 638): 5–24.

Collin, S.P., Ali, M. (1994) Multiple areas of acute vision in two freshwater teleosts, the creek chub, *Semotilus atromaculatus* (Mitchill) and the cutlips minnow, *Exoglossium maxillingua* (Lesuer). Can. J. Zool. **72**: 721–730.

Collin, S.P., Collin, H.B. (1988) The morphology of the retina and lens of the sandlance, *Limnichthyes fasciatus* (Creeiidae). Exp. Biol. **47**: 209–218.

Collin, S.P., Fritzch, B. (1993) Observations on the shape of the lens in the eye of the silver lamprey, *Ichthyomyzon unicuspis*. Can. J. Zool. **71**: 34–41.

Collin, S.P., Pettigrew (1988a) Retinal topography in reef teleosts. I. Some species with well-developed areae but poorly developed streaks. Brain Behav. Evol. **31**: 269–282.

Collin, S.P., Pettigrew, J.D. (1988b) Retinal topography in reef teleosts. II. Some species with prominent horizontal streaks and high-density areae. Brain Behav. Evol. **31**: 283–295.

Collin, S.P., Northcutt, R.G. (1993) The visual system of the Florida garfish, *Lepisosteus platyrhincus* (Ginglymodi). Retinal ganglion cells. Brain Behav. Evol. **42:** 295–320.

Collin, S.P., Partridge, J.C. (1996) Retinal specializations in the eyes of deep-sea teleosts. J. Fish Biol. 49 (Suppl. A): 157–174.

Collin, S.P., Collin, H.B., Ali, M.A. (1996a) Fine structure of the retina and pigment epithelium in the creek, chub, *Semotilus atromaculatus* (Cyprinidae, Teleostei). Histol. Histopathol. **11**: 41–53.

Collin, S.P., Collin, H.B., Ali, M.A. (1996b) Ultrastructure and organization of the retina and pigment epithelium in the cutlips minnow, *Exoglossum maxillingua* (Cyprinidae, Teleostei). Histol. Histopathol. **11**: 55–69.

Collin, S.P., Hoskin, R.V., Partridge, J.C. (1997) Tubular eyes of deep-sea fishes: a comparative study of retinal topography. Brain Behav. Evol. **50**: 335–357.

Conklin, D.J., Chavas, A., Duff, D.W., Weaver, L, Jr., Zhang, Y., Olson, K.R. (1997) Cardiovascular effects of arginine vasotocin in the rainbow trout *Oncorhynchus mykiss*. J. Exp. Biol. **200**: 2821–2832.

Coombs, S. (1995) Natural orienting behaviors for measuring lateral line function. In: Klump, G.M., Dooling, R.J., Fay, R.R., Stabbins, W.C. (eds.) Methods in comparative psychoacoustics. Birkhauser Verlag, Basel, pp. 237–248.

Coombs, S., Conley, R.A. (1997) Dipole source localization by the mottled sculpin II. The role of lateral line excitation patterns. J. Comp. Physiol. **A 180:** 401–415.

Coombs, S., Hastings, M., Finneran, J. (1996) Modeling and measuring lateral line excitation patterns to changing dipole source locations. J. Comp. Physiol. **A 178**: 359–371.

Cordeiro, M.C.R., Neto, V.M., Benchimol, M., Faria, M.V.C., Chagas, C. (1995) Microheterogeneity of desmin in the electric organ and dorsal muscle of th electric eel *Electrophorus electricus*. Comp. Biochem. Physiol. **111A:** 345–350.

Cordeiro, M.C.R., Benchimol, M., Mermelstein, C.S., Sa' L.A., Faria, M.V.C., Chagas, C., Neto. V.M. (1996) Desmin filaments in the electrocytes of the electric organ of the electric eel. *Electrophorus electricus*. Cell Tissue Res. **285:** 387–393.

Costa, M.L., Neto. V.M., Chagas, C. (1988) Desmin heterogeneity in the main electric organ of *Electrophorus electricus*. Biochimie **70**: 783–789.

Coucelo, J., Coucelo, J., Azevedo, J. (1996) Ultrasonography characterization of heart morphology and blood flow of lower vertebrates. J. Exp. Zool. **275:** 73–82.

Coughlin, D.J., Hawryshyn, C.W. (1994) The contribution of ultraviolet and short-wavelength sensitive cone mechanisms to color vision in rainbow trout. Brain Behav. Evol. **43**: 219–232.

Coughlin, D.J., Hawryshyn, C.W. (1995) A cellular basis for polarized-light vision in rainbow trout. J. Comp. Physiol. A **176**: 261–272.

Coughlin, D.J., Rome, L.C. (1996) The roles of pink and red muscle in powering steady swimming in scup, *Stemotomus chrysops*. Am. Zool. **36**: 666–677.

Coughlin, D.J., Strickler, J.R., Sanderson, B. (1992) Swimming and search behaviour in clownfish, *Amphiprion perideraion*, larvae. Anim. Behav. **44:** 427–440.

Coughlin, D.J., Zhang, G, Rome, L.C. (1996) Contraction dynamics and power production of pink muscle of the scup. (*Stenotomus chrysops*). J. Exp. Biol. **199:** 2703–2712.

Csányi, V., Dóka, A. (1993) Learning interactions between prey and predator fish. Mar. Behav. Physiol. **23**: 63–78.

Demski, L.S. (1987) Diversiity in reproductive patterns and behavior in teleost fishes. In: D. Crews (ed.) Physchobiology of reproductive behavior: an evolutionary perspective. Prentice-Hall, Inc. Englewood Cliffs, New Jersy pp. 1–27.

di Prisco, G., Maresca, B., Tota, B. (eds.) (1991) Biology of Antarctic fish. Springer-Verlag, Berlin, Heidelberg, New York.

Dickinson, K.A. (1996) Locomotor muscle of high-performance fishes: what do comparisons of tunas with ectothermic sister taxa reveal? Comp. Biochem. Physiol. **113A**: 39–49.

Dickinson, M.H. (1996a) Unsteady mechanisms of force generation in aquatic and aerial locomotion. Am. Zool. **36**: 537–554.

Donald, J.A. (1998) Autonomic nervous system. In: Evans, D.H. (ed.) Physiology of fishes. (Second edition). CRC Press, Boca. Raton, pp. 407–439.

Donnelly, J, Torres, J.J., Crabtree, R.E. (1995) Proximate compositions and nucleic acid content of premetamorphic leptocephalus larvae of the congrid eel. *Ariosoma balearicum*. Mar. Biol. **123**: 851–858.

Douglas, R.H., Bowmaker, J.K., Kunz, Y.W. (1989) Ultraviolet vision in fish. In: Kulikowski, J.J., Dickinson, C.M., Murray, I.J. (eds.) Seeing contour and colour. Pergamon Press, Oxford, pp. 601–616.

Douglas, R.H., Hawryshyn, C.W. (1990) Behavioural studies of fish vision: an analysis of visual capabilities. In: Douglas, R.H., Diamgoz, M.B.A. (eds.) The visual system of fish. Chapman and Hall, London, New York, pp. 373–418.

Driedzic, W.R., Almeida-Val, V.M.F. (1996) Enzymes of cardiac energy metabolism in Amazonian teleosts and the fresh-water stingray (*Potamotrygon hystrox*). J. Exp. Zool. **274**: 327–333.

Driedzic, W.R., Bailey, J.R., Sephton, D.H. (1998) Cardiac adaptations to low temperature in non-polar teleost fish. J. Exp. Zool. 275: 186–195.

Drucker, E.G. (1996) The use of gait transition speed in comparative studies of fish locomotion. Am. Zool. **36:** 555–566.

Drucker, E.G., Jensen, J.S. (1996a) Pectoral fin locomotion in the striped surfperch, I. Kinematic effects of swimming speed and body size. J. Exp. Biol. **199**: 2235–2242.

Drucker, E.G., Jensen, J.S. (1996b) Pectoral fin locomotion in the striped surfperch, II. Scaling swimming kinematics and performance at a gait transition. J. Exp. Biol. **199**: 2243–2252.

Drucker, E.G., Jensen, J.S. (1997) Kinematic and electromyographic analysis of steady pectoral fin swimming in the surfperches. J. Exp. Biol. **200**: 1709–1723.

Duan, C., Plisetskaya, E.M., Dickhoff, W.W. (1995) Expression of insulin-like growth factor in normally and abnormally developing coho salmon (*Oncorhynchus kisutch*). Endocrinology. **136**: 446–452.

Duff, D.W., Concklin, D.J., Olson, K.R. (1997) Effect of atrial natriuretic peptide on fluid volume and glomerular filteration in the rainbow trout. J. Exp. Zool. **278**: 215–220.

Dugatkin, L.A. Godin, J.G.J. (1993) Female mate copying in the guppy (*Poecilia reticulata*): age-dependent effects. Behav. Ecol. **4:** 289–292.

Duguay, S.T., Mommsen, T.P. (1994) Molecular aspects of pancreatic peptides. In: Sherwood, N.M., Hew, C.L. (eds.) Fish physiology, vol. 13. Academic Press, San Diego, pp. 225–271.

Eales, J.G., Brown, S.B. (1993) Measurement and regulation of thyroidal status in teleost fish. Rev. Fish Biol. Fish. **3**: 299–347.

Eastman, J.T., DeVries, E. (1997) Morphology of the digestive system of Antarctic nototheniid fishes, Polar. Biol. **17:** 1–13.

Emde, G. von der. (1998a) Capacitance detection in the wave-type electric fish *Eigenmannia* during active electrolocation. J. Comp. Physiol. **A 1821**: 217–224.

Emde, G. von der. (1998b) Electroreception. In: Evans, D.H. (ed.) The physiology of fishes. (Second edition). CRC Press, Boca Raton, pp. 313–343.

Emde, G. von der., Bleckmann, H. (1997) Waveform tuning of electroreceptor cells in the weakly electric fish, *Gnathonemus petersii*. J. Comp. Physiol. **A 181**: 511–524.

Epple, A., Scanes, C.G., Stetson, M.H. (eds.) (1990) Progress in comparative endocrinology. Proc. XIth. Int. Symp. Comp. Endocrinol., Malaga, Spain, May 14–20, 1989. Wiley-Liss, New York.

Evans, R.R. (1973) The swimbladder and associated structures in Western Atlantic sea robins (Triglidae). Copeia **1973**: 315–321.

Evelyn, T.P.T. (1996) Infection and disease. In: Iwama, G., Nakanishi, T. (eds.) Fish physiology, vol. 15. The fish immune system: organism, pathogen, and environment. Academic Press, San Diego, pp. 339–366.

Fago, A., Weber, R.E. (1995) The hemoglobin system of the hagfish *Myxine glutinosa*: aggregation state and functional properties. Biochim. Biophys. Acta **1249:** 109–115.

Farrell, A.P.(1996) Features heightening cardiovascular performance in fishes, with special references to tunas. Comp. Biochem. Physiol. **113A:** 61–67.

Fedrigo, O., Meunier, F.J., Duhamel, G., Sire, J-Y. (1996) Etude morphologique et histologique des scutes Macroramphosidae (Teleostei, Gasterosteiformes) Cah. Biol. Mar. **37**: 189–203.

Ferguson, M. (1994) The role of molecular genetic markers in the management of cultured fishes. Rev. Fish Biol. Fish. **4:** 351–373.

Feverlein, R.J., Weber, R.E. (1994) Rapid and simultaneous measurement of anodic and cathodic haemoglobins and ATP and GTP concentrations in minute quantities of fish blood. J. Exp. Biol. 189: 272–277.

Finger, T.E. (1997) Feeding patterns and brain evolution in ostariophysian fishes. Acta Physiol. Scand. 161 (Suppl. 638): 59–66.

Finger, T.E. (1988b) Sensorimotor mapping and oropharyngeal reflexes in goldfish, *Carassius auratus*. Brain Behav. Evol. **31**: 17–24.

Fishelson, L. (1994) Comparative internal morphology of deep-sea eel, with particular emphasis on gonads and gut structure. J. Fish. Biol. **44:** 75–101.

Fishelson, L. (1996) Skin morphology and cytology in marine eels adapted to different lifestyles. Anat. Rec. **246**: 15–29.

Fishelson, L. (1997) Comparative ontogenesis and cytomorphology of the nasal organs in some species of cichlid fish (Cichlidae, Teleostei). J. Zool. Lond. **243**: 281–294.

Fishelson, L., Baranes, A. (1997) Ontogenesis and cytomorphology of the nasal olfactory organs in the Oman Shark, *Lago omanensis* (Triakidae) in the Gulf of Aqaba, Red Sea. Anat. Rec. **249**: 409–421.

Fishelson, L., Sharon, O. (1997) Spatial and foraging behaviour, diet and morphogenesis of postlarvae and juveniles of *Pempheris vanicolensis* in the Gulf of Aqaba, Red Sea. J. Fish Biol. **51**: 251–265.

Flock, Å. (1965) Electron microscopic and electrophysiological studies on the lateral line canal organ, Acta Oto-Laryngol. Suppl. **199**: 7–90.

Focant, B., Melot, F., Collin, S., Vandewalle, P., Huriaux, F. (1996) Distribution of myosin and parvalbumin isoforms during the development of the catfish. Arch. Physiol. Biochem. 104, B18.

Freire, C.A., Kinne, R.K.H., Kinne-Saffran, E., Beyenbach, K.W. (1996) Electrodiffusive transport of Mg across renal membrane vesicles of the rainbow trout. *Oncorhynchus mykiss*. Am. J. Physiol. **270:** F739–F748.

Friedman, M.A., Hopkins, C.D. (1996) Tracking individual mormyrid electric fish in the field using electric organ discharge waveforms. Anim. Behav. **51**: 391–407.

Friedman, M.A., Hopkins, C.D. (1998) Neural substrates for species recognition in the time-coding electrosensory pathway of mormyrid electric fish. J. Neurosci. **18**: 1171–1185.

Fritzsch, B., Collin, S.P. (1990) Dendritic distribution of two populations of ganglion cells and the retinopetal fibers in the retina of the silver lamprey (*Ichthyomyzon unicuspis*). Visual Neurosci. **4**: 533–545.

Fuentes, J., Eddy, F.B. (1997) Drinking in marine, euryhaline and freshwater teleost fish. In: Hazon, N., Eddy, F.B., Flik, G. (eds.) Ionic regulation in animals. A tribute to Professor W.T.W. Potts, Springer-Verlag, Berlin, pp. 135–149.

Fujii, R., Oshima, N. (1994) Factors influencing activities of fish chromatophores. In: Gilles, R. (ed.) Advances in Comparative and Environmental Physiology, vol. 20. Springer-Verlag, Berlin, Heidelberg, pp. 1–54.

Fujii, R., Yamada, T., Fujita, T., Oshima, N. (1997) An improved photoelectric method for recording motile responses of individual leucophores. Zool. Sci. **14:** 549–558.

Fujita, T., Fujii, R. (1997) Endothelins dispose light-scattering organelles in leucophores of the Medaka, *Oryzias latipes*. Zool. Sci. **14:** 559–569.

Galis, F. (1993) Morphological constraints on behaviour through ontogeny: the importance of developmental constraints. Mar. Behav. Physiol. **23**: 119–135.

Gartner, J.V. Jr., Crabtree, R.E. Sulah, K.J. (1997) Feeding at depth. In: Hoar, W.S., Randall D.J., Farrell A.P. (eds.) Fish physiology, vol. 16. Deep-sea fishes. Academic Press, San Diego, pp. 115–193.

Geerlink, P.J. (1987) The role of the pectoral fins in braking of mackerel, cod and saithe. Neth. J. Zool. **37:** 81–104.

Geerlink, P.J. (1989) Pectoral fin morphology: a simple relation with movement pattern? Neth. J. Zool. **39:** 166–193.

Geerlink, P.J. (1993) Pectoral fin kinematics of *Coris formosa* (Teleostei, Labridae). Neth. J. Zool. **33:** 515–531.

Geerlink, P.J., Videler, J.J. (1987) The relation between structure and bending properties of teleost fin rays. Neth. J. Zool. **17:** 59–80.

Gibb, A.C. (1995) Kinematics of prey capture in a flatfish, *Pleuronichthys verticalis*. J. Exp. Biol. 1173–1183.

Gibbs, A.G. (1997) Biochemistry at depth. In: Hoar, W.S., Randall, D.J., Farrell, A.P. (eds.) Fish physiology, vol. 16. Deep-sea fishes. Academic Press, San Diego, pp. 239–277.

Gillis, G.B. (1996) Undulatory locomotion in elongate aquatic vertebrates: anguilliform swimming since Sir. James Gray. Am. Zool. **36:** 656–665.

Gilmour, K.M. (1998) Gas exchange. In: Evans, D.H. (ed.) The physiology of fishes. (Second edition). CRC Press, Boca Raton, pp. 101–127.

Goda, M., Fujii, R. (1995) Blue chromatophores in two species of callionymid fish. Zool. Sci. **12:** 811–813.

Goda, M., Fujii, R. (1996) Biology of the chromatophores of the ice goby, *Leucopsarion petersii.* Zool. Sci. **13:** 783–793.

Godin, J.G.J. (1995) Predation risk and alternative mating tactics in male Trinidadian guppies (*Poecilia reticulata*). Oecologia **103:** 224–229.

Godin, J.G.J., Briggs, S.E. (1996) Female mate choice under predation risk in the guppy. Anim. Behav. **51**: 117–130.

Godin, J.G.J., Crossman S.L. (1994) Hunger-dependent predator inspection and foraging behaviours in the threespine stickleback (*Gasterosteus aculeatus*) under predation risk, Behav. Ecol. Sociobiol. **34:** 359–366.

Godin, J.G.J., Davis, S.A. (1995) Who dares, benefits: predator approach behaviour in the guppy (*Poecilia reticulata*) deters predator pursuit. Proc. R. Soc. Lond. B259: 193–200.

Godin, J.G.J., Dugatkin, L.A. (1995) Variability and repeatability of female mating preference in the guppy. Anim. Behav. **49:** 1427–1433.

Godin, J.G.J., Dugatkin, L.A. (1996) Female mating preference for bold males in the guppy, *Poecilia reticulata*. Proc. Natl. Acad. Sci. U.S.A., **93**: 10262–10267.

Goehler, L.E., Finger, T.E. (1992) Functional organization of vagal reflex systems in the brain stem of the goldfish, *Carassius auratus.* J. Comp. Neurol. **319**: 463–478.

Gonzales, M.E., Blanquez, M.J., Rojo, C. (1996) Early gill development in the rainbow trout, *Oncorhynchus mykiss*. J. Morphol. **229**: 201–217.

Grant, J.W.A. (1993) Whether or not to defend? The influence of resource distribution. Mar. Behav. Physiol. **23:** 137–153.

Grant. J.W.A. (1997) Territoriality. In: Godin J.G.J. (ed.) Behavioural ecology of teleost fishes. Oxford University Press, Oxford, pp. 81–103.

Grant, K, Meek, J., Sugawara, Y., Vernon, M., Denizot, J.P., Hafmans, T.G.H., Serrier, J., Szabo, T. (1996) Projection neurons of the mormyrid electrosensory lateral line lobe: morphology, immunohistochemistry and synaptology, J. Comp. Neurol. **375**: 18–42.

Grondona, J.M., Perez, J., Lopez-Avalos, M.D., Nuolart, F.J., Feruzzo, B., Fernandez-Llebrez, P, Rodriquez, E.M. (1994) Analysis of the secretory glycoproteins of the subcommissural organ of the dogfish (*Scyliorhinus canicula*). Mol. Brain Res. **26:** 299–308.

Grubich, J.R., Wainwright, P.C. (1997) Motor basis of suction feeding performance in large mouth bass, *Micropterus salmoides*. J. Exp. Zool. **277**: 1–13.

Grundfest, H. (1967) Comparative physiology of electric organs of elasmobranch fishes. In: Gilbert, P.W., Mathewson, R.F., Rall, D.P. (eds.) Sharks, skates, and rays. John Hopkins Press, Baltimore, Maryland, pp. 399–432.

Gudding, R., Lillehaug, A., Midtlynrg, P., Brown, F. (eds.) (1997) Fish Vaccinology. S. Karger, Basel.
Guraya, S.S. (1993) Follicular (or oocyte) atresia and its causes and functional significance in the fish ovary. In: Singh, B.R. (ed.) Advances in fish research, vol. 1. Narendra Publishing House, Delhi, pp. 313–332.
Guraya, S.S. (1996) Recent advances in functional morphology of follicular wall, egg-surface components, and micropyle in the fish ovary. In: Munshi, J.S.D., Dutta, H.M. (eds.) Fish morphology-horizon of new research. Oxford and IBH Publishing Co. Pvt. Ltd. New Delhi, pp. 111–145.
Gutierrez, J., Parrizas., M, Maestro, M.A., Navarro, I., Plisetskaya, E.M. (1995) Insulin and IGF-1 binding and tyrosine kinase activity in fish heart. J. Endocrinol. **146:** 35–44.
Haedrich, R.L., Wittenberg, J. Nelson, J.G. (1973) A septum in the eye of osteoglossoid fishes. Copeia **1973:** 594–595.
Hamlett, W.C., Schwartz, F.J., Schmeinda, R., Cuevas, E. (1996) Anatomy, histology and development of the cardiac vascular system in elasmobranchs. J. Exp. Zool. **275:** 83–94.
Hara, T.J., Zhang, C. (1996) Spatial projections to the olfactory bulb of functionally distinct and randomly distributed primary neurons in salmonid fishes. Neurosci. Res. **26:** 65–74.
Hara, T.J., Zhang, C. (1998) Topographic bulbar projections and dual neural pathways of the primary olfactory neurons in salmonid fishes. Neuroscience **82:** 301–313.
Hawryshyn, C.W. (1992) Polarization vision in fish. Amer. Scientist **80:** 164–175.
Hawryshyn, C.W., McFarland, W.N., (1987) Cone photoreceptor mechanism and the detection of polarized light in fish. J. Comp. Physiol. **A 160**: 459–465.
Hawryshyn, C.W., Arnold, M.G., Chaisson, D.J., Martin, P.C. (1989) The ontogeny of ultraviolet photosensitivity in rainbow trout (*Salmo gairdneri*). Visual Neurosci. **2:** 247–254.
Hayashi, H., Nakamura, S., Fujii, R. (1996) The endothelin receptors that mediate aggregation of pigment in fish melanophores. Comp. Biochem. Physiol. **115B**: 143–152.
Hazon, N., Balment, R.J. (1998) Endocrinology. In: Evans, D.H. (ed.) Physiology of fishes. (Second edition). CRC Press, Boca. Raton., pp. 441–463.
Hazon, N., Tierney, M.L., Anderson, G., Mackenzie, S., Cutler, C., Cramb, G. (1997). Ion and water balance in elasmobranch fish. In: Hazon, N., Eddy, F.B., Flik, G. (eds.) Ionic regulations in animals: a tribute to Professor W.T.W. Potts. Springer-Verlag, Berlin, Heidelberg, pp. 70–86.
Hebrank, J.H, Hebrank, M.R. Long, J.H. Jr., Block, B.A., Wainwright S.A. (1990) Backbone mechanics of the blue marlin *Makaira nigricans* (Pisces, Istiophoridae). J. Exp. Biol. **148:** 449–459.
Heierhorst, J., Morley, S.D., Figureroa, J., Krentler, C., Lederis, K. (1989) Vasotocin and isotocin precursors from the white sucker, *Catastomus commersoni*: cloning and sequence analysis of the cDNAs. Proc. Natl. Acad. Sci. U.S.A., **86**: 5242–5246.
Heierhorst, J., Mahlmann, S., Morley, S.D., Coe, I.R., Sherwood, N.M., Richter, D. (1990) Molecular cloning of two distinct vasotocin precursor cDNAs from chum. salmon (*Oncorhynchus keta*) suggests an ancient gene duplication. F.E.B.S. Lett. **260**: 301–304.
Hills, B.A. (1996) Effects of gill dimensions on respiration. In: Munshi, J.S.D., Dutta, H.M. (eds.) Fish morphology-horizon of new research. Oxford & IBH Publishing Co. Pvt. Ltd., New Delhi, pp. 235–247.
Hopkins, C.D. (1995) Convergent designs for electrogenesis and electroreception. Curr. Opin. Neurobiol. **5:** 769–777.
Hughes, R.N. (1997) Diet selection. In: Godin, J.G.J. (ed.) Behavioural ecology of teleost fishes. Oxford University Press, Oxford, pp. 134–159.
Huntingford, F.A., Wright, P.J. (1993) The development of adaptive variation in predator avoidance in freshwater fishes. Mar. Behav. Physiol. **23:** 45–61.
Huriaux, F., Melot, F., Vandewalle, P., Collin, S., Focant, B. (1996) Parvalbumin isotypes in white muscle from three teleost fish: characterization and their expression during development. Comp. Biochem. Physiol. **113B**, 475–484.
Huriaux, F., Collin, S., Vandewalle, P., Philippart, J.C., Focant, B. (1997) Characterization of parvalbumin isotypes in white muscle from the barbel and expression during development. J. Fish Biol. **50:** 821–836.
Hutchings, J.A. (1993) Behavioural implications of intraspecific life history variation. In: Huntingford, F.A., Torricelli, P. (eds.) Behavioural ecology of fishes. Harwood Academic Publishers, Chur, Switzerland, pp. 187–203.

Hutchings, J.A. (1996) Adaptive phenotypic plasticity in brook trout, *Salvelinus fontinalis*, life histories. Ecoscience **3**: 25–32.

Irvine, I.A.S., Sorensen, P.W. (1993) Acute olfactory sensitivity of wild common carp, *Cyprinus carpio*, to goldfish hormonal sex pheromones is influenced by gonadal maturity. Can. J. Zool. **71**: 2199–2210.

Iwama, G., Nakanishi, T. (eds.) (1996) Fish physiology, vol. 15. The fish immune system: organism, pathogen, and environment. Academic Press, San Diego.

Jalabert, B., Fostier, A., Breton, B., Weil, C. (1991) Oocyte maturation in vertebrates. In: Pang, P.K.T., Schreibman, M.P. (eds.) Vertebrate endocrinology: fundamentals and biomedical implications, vol. 4. Academic Press, San Diego. pp. 23–90.

Jayne, B.C., Lauder, G.V. (1996) New data on axial locomotion in fishes: how speed affects diversity of kinematics and motor patterns. Am. Zool. **36:** 642–655.

Jayne, B.C., Lozada, A.F., Lauder, G.V. (1996) Function of the dorsal fin in bluegill sunfish: motor patterns during four distinct locomotor behaviors. J. Morphol. **228**: 307–326.

Jobling, M., Koskela, J. (1996) Interindividual variations in feeding and growth in rainbow trout during restricted feeding and in a subsequent period of compensatory growth. J. Fish Biol. **49**: 658–667.

Jorgensen, E.H., Johansen, S.J.S., Jobling, M. (1997) Seasonal patterns of growth, lipid deposition and lipid depletion in anadromous Arctic charr. J. Fish Biol. **51:** 312–326.

Kaattari, S.L., Piganelli, J.D. (1996) The specific immune system: humoral defense. In: Iwama, G., Nakanishi, T. (eds.) Fish physiology, vol. 15. The fish immune system: organism, pathogen, and environment. Academic Press, San Diego, pp. 297–254.

Kalmijn, A.J. (1997) Electric and near-field acoustic detection, a comparative study. Acta Physiol. Scand. 161 (Suppl. 638): 25–38.

Kang, J., Caprio, J. (1997) In vivo responses of single olfactory neurons of channel catfish to binary mixtures of amino acids. J. Neurophysiol. **77**: 1–8.

Kanwal, J.S., Finger, T.E. (1997) Parallel medullary gustatospinal pathways in a catfish: possible neural substrates for taste-mediated food search. J. Neurosci. **17**: 4873–4885.

Karnaky, K.J. Jr. (1998) Osmotic and ionic regulation. In: Evans D.H. (ed.) The physiology of fishes. (Second edition). CRC Press, Boca Raton, pp. 157–176.

Kirschbaum, F. (1980) Ontogeny of both larval electric organ and electromotoneurones in *Pollimyrus isidori* (Mormyridae, Teleostei) In: Szabó, T., Czéh, G. (eds.) Advances in physiological sciences, vol. 31. Sensory biology of aquatic lower vertebrates. Pergamon Press, Budapest, Hungary, pp. 129–157.

Kemp, A. (1995) Marginal tooth-bearing bones in the lower jaw of the recent Australian lungfish, *Neoceratodus forsteri* (Osteichthyes, Dipnoi). J. Morphol. **225**: 345–355.

Kemp, A. (1996) Role of epidermal cilia in development of the Australian lungfish, *Neoceratodus forsteri* (Osteichthyes: Dipnoi). J. Morphol. **228**: 203–221.

Kestern, R.E. van, Tensen, C.P., Smit, A.B., Minnen, J. van., Kolakowski, L.F., Meyerhof, W., Richter, D., Heerikhuizen, H. van., Vreugdenhil, E., Geraerts, W.P.M. (1996) Co-evolution of ligand-receptor pairs in the vasopressin/oxytocin superfamily at bioactive peptides. J. biol. Chem. **271**: 3619–3626.

Kisia, S.M. (1996) Structure of fish locomotory muscle. In: Munshi, J.S.D., Datta, H.M. (eds.) Fish morphology-horizon of new research. Oxford & IBH Publishing Co. Pvt. Ltd., New Delhi, pp. 169–178.

Klitgaard, T. (1978) Morphology and histology of the heart of the Australian lungfish, *Neoceratodus forsteri* (Krefft). Acta Zool. Stockholm **59**: 187–198.

Kohbara, J., Caprio, J. (1996) Taste responses to binary mixtures of amino acids in the sea catfish, *Arius felis*. Chem. Senses **21:** 45–53.

Korf, H.W. (1996) Innervation of the pineal gland. In: Burnstock, G. (ed.) The autonomic nervous system, vol. 10. Autonomic-endocrine interactions. Harwood Academic Publishers, Amsterdam, pp. 129–180.

Korf, H.W., Kroeber, S., Schomerus, C. (1997) Regulation of the intracellular concentration of free calcium ions in pinealocytes of the rainbow trout and the rat. Biol. Signals **6**: 201–211.

Korsmeyer, K.E., Dewer, H., Lai, N.C., Graham, J.B. (1996) The aerobic capacity of tunas: adaptation for multiple metabolic demands. Comp. Biochem. Physiol. **113A**: 17–24.

Kramer, B., Teubl, H. (1993) Spectral phase sensitivity in the communication of wave-type electric fish. Naturwissenschaften **80:** 575–578.

Kramer, D.L., Rangeley, R.W., Chapman, L.J. (1997) Habitat selection: patterns of spatial distribution from behavioural decisions. In: Godin J.G.J., (ed.) Behavioural ecology of teleost fishes. Oxford University Press, Oxford, pp. 37–80.

Krause, J., Godin, J.G.J. (1994) Shoal choice in the banded killifish (*Fundulus diaphanus*, Teleostei, Cyprinodontidae): effects of predation risk, fish size, species composition and size of shoals. Ethology **98**: 128–136.

Krause, J., Godin, J.G.J. (1995) Predator preferences for attacking particular prey group sizes: consequences for predator hunting success and prey predation risk. Anim. Behav. **50:** 465–473.

Krause, J., Godin, J.G.J. (1996a) Influence of prey foraging posture on flight behaviour and predation risk: predators take advantage of unwary prey. Behav. Ecol. **71**: 264–271.

Krause, J., Godin, J.G.J. (1996b) Influence of parasitism on shoal choice in the banded killifish (*Fundulus diaphanus*, Teleostei, Cyprinodontidae). Ethology **102**: 40–49.

Krause, J., Godin, J.G.J., Brown, D. (1996a) Phenotypic variability within and between fish shoals. Ecology **77:** 1586–1591.

Krause, J., Godin, J.G.J., Brown, D. (1996b) Size-assortativeness in multiple-species fish shoals. J. Fish Biol. **49:** 221–225.

Kroeber, S., Schomerus, C., Korf, H.W. (1997) Calcium oscillations in a subpopulation of S-antigen-immunoreactive pinealocytes of the rainbow trout (*Onorhynchus mykiss*). Brain Res. **744**: 68–76.

Kulczykowska, E. (1997) Response of circulating arginine vasotocin and isotocin to rapid osmotic challenge in rainbow trout. Comp. Biochem. Physiol. **118 A:** 773–778.

Kulczykowska, E., Iuvone, P.M. (1998) Highly sensitive and specific assay of plasma melatonin using high-performance liquid chromatography with fluorescence detection preceded by solid-phase extraction. J. Chromatogr. **36**: 175–178.

Kulczykowska, E., Stolarski, J. (1996) Diurnal changes in plasma arginine vasotocin and isotocin in rainbow trout adapted to fresh water and brackish Baltic water. Gen. Comp. Endocrinol. **104**: 197–202.

Kunz, Y.W. (1987) Tracts of putative ultraviolet receptors in the retina of the two-year-old brown trout (*Salmo trutta*) and the Atlantic salmon (*Salmo salar*). Experentia **47**: 1202–1204.

Kunz, Y.W., Callaghan, E. (1989) Embryonic fissures in teleost eyes and their possible role in detection of polarized light. Trans. Amer. Fish Soc. **118:** 195–202.

Kunz, Y.W., Shuilleabhain, M.N., Callaghan, E. (1985) The eye of the venomous marine teleost *Trachinus vipera* with special reference to the structure and ultrastructure of visual cells and pigment epithelium. Exp. Biol. **43**: 161–178.

Kunz, Y.W., Wildenburg, G., Goodrich, L., Callaghan, E. (1994) The fate of ultraviolet receptors in the retina of the Atlantic salmon (*Salmo salar*). Vision Res. **34**: 1375–1383.

Kvarnemo, C., Forsgren, E., Magnhagen, C. (1995) Effects of sex ratio on intra-and intersexual behaviour in sand gobies. Anim. Behav. **50:** 1455–1461.

Ladich, F. (1985) Ontogenetische Entwicklung der Pseudobranchie von *Macropodus opercularis* (Anabantoidea, Teleostei). Zool. Jb. Anat, **113:** 119–132.

Ladich, F. (1987) Anatomie und Feinstruktur der Pseudobranchie von *Macropodus opercularis* (Anabantoidei, Teleostei). Zool. Jb. Anat. **115**: 489–505.

Lam, T.J. (1983) Environmental influences on gonadal activity in fish. In: Hoar, W.S., Randall, D.J., Donaldson, E.M. (eds.) Fish physiology, vol. 9B. Academic Press, London, pp. 65–116.

Lanzing, W.J.R. (1974) Sound production in the cichlid *Tilapia mossambica* Peters. J. Fish Biol. **6**: 341–347.

Lauff, R.F., Wood, C.M. (1996) Respiratory gas exchange, nitrogenous waste excretion, and fuel usage during starvation in juvenile rainbow trout, *Oncorhynchus mykiss*. J. Comp. Physiol. **B165**: 542–551.

Laurent, P. (1996) Vascular organization of lungfish, a landmark in ontogeny and phylogeny of air-breathers. In: Munshi, J.S.D., Dutta, H.M. (eds.) Fish morphology-horizon of new research. Oxford & IBH Publishing Co. Pvt. Ltd. New Delhi, pp. 47–58.

Lessard, J., Val, A.L., Aota, S., Randall, D.J. (1995) Why is there no carbonic anhydrase activity available to fish plasma? J. Exp. Biol. **198:** 31–38.

Li, W., Sorensen, P.W. (1997) Highly independent olfactory receptor sites for naturally occurring bile acids in the sea lamprey, *Petromyzon marinus*. J. Comp. Physiol. **A 180**: 429–438.

Li, W., Sorensen, P.W., Gallaher, D.D. (1995) The olfactory system of migratory adult sea lamprey (*Petromyzon marinus*) is specifically and acutely sensitive to unique bile acids released by conspecific larvae. J. Gen. Physiol. **105**: 569–587.

Lincoln, R. (1994) Molecular genetics applications in fisheries: snake oil or restorative? Rev. Fish Biol. Fish. **4:** 389–392.

Lomholt, J.P., Johansen, K. (1976) Gas exchange in the amphibious fish, *Amphipnous cuchia.* J. Comp. Physiol. **107**: 141–157.

Long, J.A. (1995) The rise of fishes. John Hopkin University Press, Baltimore.

Long, J.H. Jr., (1992) Stiffness and damping forces in the intervertebral joints of blue marlin (*Makaira nigricans*). J. Exp. Biol. **162**: 131–155.

Long, J.H. Jr, (1995) Morphology, mechanics and locomotion: the relation between notochord and swimming motions in sturgeon. Env. Biol. Fish. **44**: 199–211.

Long, J.H. Jr. Nipper, K.S. (1996) The importance of body stiffness in undulatory propulsion. Am. Zool. **36**: 678–694.

Long, J.H. Jr., McHenry, M.J., Boetticher, N.C. (1994) Undulatory swimming: how travelling waves are produced and modulated in sunfish (*Lepomis gibbosus*). J. Exp. Biol. **192**: 129–145.

Long, J.H. Jr., Hale, M.E., McHenry, M.J., Westneat, M.W. (1996) Functions of fish skin: flexural stiffness and steady swimming of longnose gar *Lepisosteus osseus*. J. Exp. Biol. **199**: 2139–2151.

Losey. G.S. Jr. (1993) Knowledge of proximate causes aids our understanding of function and evolutionary history. Mar. Behav. Physiol. **23:** 175–186.

Maddock, R.G., Nicol, J.A.C. (1978) Studies on the eyes of *Hydrolagus* (Pisces: Chimaeridae). Cont. Mar. Sci. **21**: 77–87.

Magnhagen, C. (1991) Predation risk as a cost of reproduction. Trends Ecol. Evol. **6:** 183–186.

Magnhagen, C. (1992) Parental care and predation risk in fish. Ann. Zool. Fennici. **29**: 227–232.

Magnhagen, C. (1993) Conflicting demands in gobies: when to eat, reproduce and avoid predators. Mar. Behav. Physiol. **23:** 79–90.

Magnhagen, C. (1995) Sneaking behaviour and nest defence are affected by predation risk in the common goby. Anim Behav. **50**: 1123–1128.

Magnhagen, C., Vestergaard, K. (1993) Brood size and offspring age affect risk-taking and aggression in nest-guarding common gobies. Behaviour **125**: 233–243.

Magurran, A.E., Seghers, B.H., Carvalho, G.R., Shaw, P.W. (1993) Evolution of adaptive variation in antipredator behaviour. Mar. Behav. Physiol. **23:** 29–44.

Maina, J.N., Wood, C.M., Norahara, A., Bergman, H.L., Laurent, P., Walsh, P. (1996) Morphology of the swim (air) bladder of a cichlid teleost: *Oreochromis grahami* (Trewavas, 1983), a fish adapted to a hyperosmotic, alkaline, and hypoxic environment: a brief outline of the structure and function of the swimbladder. In Munshi, J.S.D., Dutta, H.M. (eds). Fish morphology-horizon of new research. Oxford & IBH Publishing Co. Pvt. Ltd. New Delhi, pp. 179–192.

Mallefet, J., Shimomura, O. (1995) Presence of coelenterazine in mesopelagic fishes from the Strait of Messina. Mar. Biol. **124:** 381–385.

Manning, M.J. (1998) Immune defence systems. In: Black, K.D., Pickering, A.D. (eds.). Biology of farmed fish, Sheffield Academic Press, Sheffield, pp. 180–221.

Manning, M.J., Nakanishi, T. (1996) The specific immune system: cellular defenses. In: Iwama, G., Nakanishi, T. (eds.) Fish physiology, vol. 15. The fish immune system: organism, pathogen, and environment. Academic Press, San Diego, pp. 159–205.

Marconato, A., Shapiro, D.Y. (1996) Sperm allocation, sperm production and fertilization rates in the bucktooth parrotfish. Anim. Behav. **52:** 971–980.

Marshall, N.B. (1967) The olfactory organs of bathypelagic fishes. Symp. zool. Soc. Lond. No. **19**: 57–70.

Marshall, W.S., Bryson, S.E., Darling, P., Whitten, C., Patrick, M., Wilkie, M., Wood, C.M., Buckland-Nicks, J. (1997) NaCl transport and ultrastructure of opercular epithelium from a freshwater-adapted euryhaline teleost, *Fundulus heteroclitus*. J. Exp. Zool. **277**: 23–37.

Martin, K.L.M. (1991) Facultative aerial respiration in an intertidal sculpin, *Clinocottus analis.* (Scorpaeniformes: Cottidae). Physiol. Zool. **64**: 1341–1355.

Martin, K.L.M. (1993) Aerial release of CO_2 and respiratory exchange rates in intertidal fishes out of water. Env. Biol. Fish. **37**: 189–196.

Martin, K.L.M. (1996) An ecological gradient in air-breathing ability among marine cottid fishes. Physiol. Zool. **69:** 1096–1113.

Martin, K.L.M., Lighton, J.R.B. (1989) Aerial CO_2 and O_2 exchange during terrestrial activity in an amphibious fish, *Alticus kirki* (Blenniidae). Copeia **1989:** 723–727.

Mathieu-Costello, O., Brill, R.W., Hochachka, P.W. (1996) Structural basis for oxygen delivery: muscle capillaries and manifolds in tuna red muscle. Comp. Biochem. Physiol. **113A**: 25–31.

McClelland, G., Zwingelstein, G., Weber, J.M., Brichon, G. (1995) Lipid composition of tissue and plasma in two Mediterranean fishes, the gilt-head sea bream (*Chrysophyrys auratus*) and European seabass (*Dicentrarchus labrax*). Can. J. Fish Aquat. Sci. **52**: 161–170.

McCormick, C.A. (1997) Organization and connection of octaval and lateral line centers in the medulla of a clupeid, *Dorosoma cepedianum*. Hearing Res. **110**: 39–60.

McCormack, C.A., Hayden, T.J., Kunz, Y. W. (1989) Ontogenesis of diurnal rhythms of cAMP concentration, outer segment disc shedding and retinomotor movements in the eye of the brown trout. Brain Behav. Evol. **34**: 65–72.

McCormick, S.D., Naiman, R.J. (1985) Hypoosomoregulation in an anadromous teleost: influence of sex and maturation. J. Exp. Zool. **234:** 193–198.

McCormick, S.D., Saunders, R.L. (1987) Preparatory physiological adaptations for marine life of salmonids: osmoregulation, growth, and metabolism. Am, Fish Soc. Symp. **1**: 211–229.

McFall-Ngai, M.J. (1990) Crypsis in the pelagic environment. Am. Zool. **30**: 175–188.

McHenry, M.J., Pell, C.A., Long, J.H. Jr. (1995) Mechanical control of swimming speed: stiffness and axial waveform in undulating fish models. J. Exp. Biol. **198:** 2293–2305.

Meek, J., Grant, K., Sugawara, Y., Hafmans, T.G.M., Veron, M., Denizot J.P. (1996) Interneurons of the ganglionic layer in the mormyrid electrosensory lateral line lobe: morphology, immunohistochemistry and synaptology. J. Comp. Neurol. **375**: 43–65.

Meunier, F.J. (1994) Structure et mineralisation des ecailles de Quelques Characiformes de Guyane (Osteichthyes, Teleostei). Rev. Hydrobiol. trop. **27:** 407–422.

Meunier, F.J., Huysseune, A. (1992) The concept of bone tissues in Osteichthyes. Neth. J. Zool. **42**: 445–458.

Meunier, F.J., Poplin, C. (1995) Paleohistological study of the scales of *Amia robusta* Priem, 1901, Amiidae from the Thanetian (Paleocene) of Cernay (France). Geobios M.S. No. **19**: 39–43.

Meunier, F.J., Rojas-Beltran, R., Boujard, T. Lecomte, F. (1994) Rythmes saisonniers de la croissance chez. qualques Teleosteens de Guyane francaise. Rev. Hydriobiol. trop. **27**: 423–440.

Meurling, P., Rodriguez, E.M., Pena, P., Grondona, J.M., Perez, J. (1996) Hypophysial and extrahypophysial projections of the neurosecretory system of cartilaginous fishes: an immunocytochemical study using a polyclonal antibody against dogfish neurophysin. J. Comp. Neurol. **373**: 400–421.

Meyer-Rochow, V.B., Devine, C.E. (1986) Ultrastructural observations and pH-measurements on red and white muscle from Antarctic fish. Polar Biol. **6:** 241–246.

Meyer-Rochow, V.B., Ingram, J.R. (1993) Red-white muscle distribution and fibre growth dynamics: a comparison between lacustrine and riverine populations of the Southern smelt *Retropinna retropinna* Richardson. Proc. R. Soc. Lond. **B. 252**: 85–92.

Meyer-Rochow, V.B., Klyne, M:A. (1982) Retinal organization of the eyes of three nototheniid fishes from the Ross. Sea. (Antarctica). Genebaurs morph. Jahrb. Leipzig **5**: 762–777.

Meyer-Rochow, V.B., Pyle, C.A. (1980) Fatty acid analysis of lens and retina of two Antarctic fish and of the head and body of the Antarctic amphipod *Orchomene plebs*. Comp. Biochem. Physiol. **65B**: 395–398.

Meyer-Rochow, V.B., Tiang, M.K. (1978) Visual behaviour, eye and retina of the parasitic fish *Carapus mourlani*. Biol. Bull. **155:** 576–585.

Meyer-Rochow, V.B., Ishihara, Y., Ingram, J.R. (1994) Cytochemical and histochemical details of muscle fibres in the Southern smelt *Retropinna retropinna* (Pisces; Galaxioidei). Zool. Sci. **11**: 55–62.

Meyer-Rochow, V.B., Morita, Y., Tamotsu S. (1999) Immuno-cytochemical observations on pineal organ and retina of the Antarctic teleosts *Pagothenia borchgrevinki* and *Trematomus bernachii*. J. Neurocytol. **28**: 125–130.

Miles, A.E.W. and Poole, D.F.G. (1967) The history and general organization of dentitious. In: Miles, A.E.W. (ed.) Structural and chemical organization of teeth, vol. 1. Chapter 1. Academic Press, New York, pp. 3–44.

Milinski, M. (1990) Parasites and host decision-making. In: Barnard, C.J., Behnke, J.M. (eds.) Parasitism and host behaviour. Taylor and Francis, London, pp. 95–116.

Moller, H. (1985) A critical review on the role of pollution as a cause of fish diseases. In: Ellis, A.E. (ed.) Fish and shellfish pathology. Academic Press, London, pp. 169–182.

Moller, P. (ed.) (1995) Electric fishes. History and behavior. Chapman and Hall, London, New York.

Mommsen, T.P. (1998) Growth and metabolism. In: Evans, D.H. (ed.) Physiology of fishes. (Second edition). CRC Press, Boca. Raton, pp. 65–97.

Montgomery, J., Pankhurst, N. (1997) Sensory physiology. In: Hoar, W.S., Randall, D.J., Farrell, A.P (eds.) Fish physiology, vol. 16. Deep-sea fishes. Academic Press, San Diego, pp. 325–349.

Montgomery, J., Coombs, S., Halstead, M.W. (1995) Biology of the mechanosensory lateral line in fishes. Rev. Fish Biol. Fish. **5**: 399–416.

Moon, T.W., Castejon, C., Banos, N., Maestro, M.A., Plisetskaya, E.M., Gutierrez, J., Navarro, I. (1996) Insulin and IGF-1 binding in isolated trout cardiocytes. Gen. Comp. Endocrinol. **103**: 264–272.

Moriyama, S., Dickhoff, W.W., Plisetskaya, E.M. (1995) Isolation and characterization of insulin-like growth factor-1 from rainbow trout, *Oncorhynchus mykiss*. Gen. Comp. Endocrinol. **99**: 221–229.

Moriyama, S., Kawaga, H., Duan, C., Dickhoff, W.W., Plisetskaya, E.M. (1997) Characterization of two forms of recombinant salmon insulin-like growth factor-1: activities and complexing with insulin-like growth factor-1 binding proteins. Comp. Biochem. Physiol. **117C**: 201–206.

Motta, P.J., Tricas, T.C., Hueter, R.E., Summers, A.P. (1997) Feeding mechanism and functional morpholory of the jaws of the lemon shark *Negaprion brevirostris* (Chondrichthyes, Carcharhinidae). J. Exp. Biol RoM 2765–2780.

Nakao, T., Uchinomiya, K. (1978) A study of the blood vascular system of lamprey gill filament. Am. J. Korf **151**: 239–264.

Nikinmaa, M., Weber, R.E. (1993) Gas transport in lamprey erythrocytes. In: Bicudo J.E.P.W. (ed.) The vertebrate gas transport cascade: adaptations to environment and mode of life. CRC Press, Boca Raton, pp. 179–189.

Nilsson, S. (1997) The autonomic nervous system of the dog and the dogfish: a comparative approach. Acta Physiol. Scand. **161**: (Suppl. 638): 39–46.

Northcutt, R.G., Neary, T.J., Senn, D.G. (1979) Observations on the brain of the coelacanth *Latimeria chalumnae*: external anatomy and quantitative analysis. J. Morphol. **155**: 181–192.

Nurnberger, F., Korf, H.W., Ali, M.A., Noakes, D.L.G. (1996) Salinity and vasotocin immunoreactivity in the brain of *Rivulus marmoratus* (Teleostei). Naturwissenschaften **83**: 326–328.

Obika, M., Meyer-Rochow, V.B. (1986) Ultrastructure of microtubules in dermal melanophores and spinal nerve of the Antarctic teleost *Pagothenia borchgrevinki*. Cell Tissue Res. **244**: 339–343.

Obika, M., Meyer-Rochow, V.B. (1990) Dermal and epidermal chromatophores of the Antarctic teleost *Trematomus bernacchii*. Pigment Cell Res. **3:** 33–37.

Ogawa, K., Marui, T., Caprio, J. (1997a) Bimodal (taste/tactile) fibers innervate the maxillary barbel in the channel catfish. Chem. Senses **22**: 477–482.

Ogawa, K., Marui, T., Caprio, J. (1997b) Quinine suppression of single facial taste fiber responses in the channel catfish. Brain Res. **769**: 263–272.

Ohta, T. (1985) Electron microscopic observations on sperm entry and pronuclear formation in naked eggs of the rose bitterling in polyspermic fertilization. J. Exp. Zool. **234:** 273–281.

Ohta, T. (1991) Initial stages of sperm-egg fusion in the freshwater teleost, *Rhodeus ocellatus ocellatus*. Anat. Rec. **229:** 195–202.

Ohta, T., Nashirozawa, C. (1996) Sperm. penetration and transformation of sperm entry site in eggs of the freshwater teleost *Rhodeus ocellatus ocellatus*. J. Morphol. **229**: 191–200.

Ohta, T., Iwamatsu, T., Tanaka, M., Yoshimoto, Y. (1990) Cortical alveolus breakdown in the eggs of the freshwater teleost *Rhodeus ocellatus ocellatus*. Anat. Rec. 486–496.

Olson, K.R. (1991) Vasculature of the fish gill: anatomical correlates of physiological functions. J. Elect. Microsc. Tech. **19:** 389–405.

Olson, K.R. (1992) Blood and extracellular fluid volume regulation: role of renin–angiotensin system, kallikrein-kinin system. In: Hoar, W.S., Randall, D.J., Farrell, A.P. (eds.) Fish physiology, vol. 12B. Academic Press, Orlando, pp. 135–254.

Olson, K.R. (1996a) Scanning electron microscopy of the fish gill. In: Munshi, J.S.D., Dutta, H.M. (eds.) Fish morphology-horizon of new research. Oxford & IBH Publishing Co. Pvt. Ltd., New Delhi, pp. 31–45.

Olson, K.R. (1996b) Secondary circulation in fish: anatomical organization and physiological significance. J. Exp. Zool. **275:** 172–185.

Olson, K.R. (1998a) Hormone metabolism by the fish gill. Comp. Biochem. Physiol. **119A**: 55–65.

Olson, K.R. (1998b) The cardiovascular system. In: Evans, D.H. (ed.) The physiology of fishes. (Second edition). CRC Press, Boca Raton, pp. 129–154.

Olson, K.R., Conklin, D.J., Weaver, L. Jr., Duff, D.W., Herman, C.A., Wang, X., Conlon, J.M. (1997a) Cardiovascular effects of homologous bradykinin in rainbow trout. Am. J. Physiol. **272**: R1112–R1120.

Olson, K.R., Conklin, D.J., Conlon, J.M., Kellog, M., Smith, M.P., Weaver, L. Jr., Bushnell, P.G., Duff, D.W. (1997b) Physiological inactivation of vasoactive hormones in rainbow trout. J. Exp. Zool. **279**: 254–264.

Omura, Y., Ali, M.A. (1980) Responses of pineal photoreceptors in the brook and rainbow trout. Cell Tissue Res. **208**: 111–122.

Oshima, N., Wannitikul, P. (1996) Signal transduction of MCH in melanophores of the tilapia, *Oreochromis niloticus*. Zool. Sci. **13:** 351–356.

Osse, J.W.M., Sibbing, F.A., Boogaart, J.G.M. van den. (1997) Intra-oral food manipulation of carp and other cyprinids: adaptations and limitations. Acta Physiol. Scand. (Suppl. 638): 47–57.

Par khurst, N.W. (1998) Reproduction. In: Black, K.D., Pickering, A.D. (eds.) Biology of farmed fish. Sheffield Academic Press, Sheffield, pp. 1–26.

B.W.K., Moran, P. (1994) Developments in molecular genetic techniques in fisheries. Rev. Fish Biol. Fish. rtn/2f2–299.

Parrizas, M., Plisetskaya, E.M., Planas, J., Gutierrez, J. (1995) Abundant insulin-like growth factor-1 (IGF-1) receptor binding in fish skeletal muscle. Gen. Comp. Endocrinol. **98**: 16–25.

Patey, C.P., Driedzic, W.R. (1997) Cold acclimation increases activities of mitochondrial long-chain acyl-CoA synthetase and carnitine acyl-CoA transferase I in heart of rainbow trout (*Oncorhynchus mykiss*). Can. J. Zool. **75**: 324–331.

Pelster, B. (1995a) Mechanism of acid release in isolated gas gland cells of the European eel. *Anguilla anguilla*. Am. J. Physiol. **269**: R 793–R799.

Pelster, B. (1995b) Metabolism of the swimbladder tissue. In: Hochachka, P.W., Mommsen, T.P. (eds.) Biochemistry and molecular biology of fishes, vol. 4. Elsevier, Amsterdam, pp. 101–118.

Pelster, B. (1997) Buoyancy at depth. In: Hoar. W.S., Randall, D.J., Farrell, A.P. (eds.) Fish physiology, vol. 16. Deep-sea fishes. Academic Press, San Diego, pp. 195–237.

Pelster, B. (1998) Buoyancy. In: Evans, D.H. (ed.) The physiology of fishes. (Second edition). CRC Press, Boca Raton, pp. 25–42.

Pelster, B., Pott, L. (1996) Control of acid secretion in cultured gas gland cells of the European eel. *Anguilla anguilla*. Am. J. Physiol. **270**: R578–R584.

Pelster, B., Bailey, J.R., Driedzic, W.R. (1997) The fate of ^{14}C infused into the swim bladder of the American eel, *Anguilla rostrata*. Can. J. Zool. **75**: 116–121.

Persson, L., Diehl, S., Eklov, P., Christensen, B. (1997) Flexibility in fish behaviour: consequences at the population and community levels. In: Godin, J.G.J. (ed.) Behavioural ecology of teleost fishes. Oxford University Press, Oxford, pp. 316–343.

Pettigrew, J.D., Collin, S.P. (1995) Terrestrial optics in an aquatic eye: the sandlance, *Limnichthytes fasciatus* (Creediidae, Teleostei). J. Comp. Physiol. **A 177:** 397–408.

Phleger, C.F., Nichols, P.D., Virtue, P. (1997) The lipid, fatty acid and fatty alcohol composition of the myctophid fish *Electrona antarctica*: high level of wax esters and food-chain implications. Antarctic Sci. **9:** 258–265.

Pickering, A.D. (1998) Stress responses of farmed fish. In: Black, K.D., Pickering, A.D. (eds.) Biology of farmed fish. Sheffield Academic Press, Sheffield, pp. 222–255.

Piotrowski, T., Northcutt, R.G. (1996) The cranial nerves of the Senegal bichir, *Polypterus senegalus* (Osteichthyes: Actinopterygii: Cladistia). Brain Behav. Evol. **47:** 55–102.

Plisetskaya, E.M. (1995) Peptides of insulin and glucagon superfamilies in fish. Neth. J. Zool. **45:** 181–188.

Plisetskaya, E.M., Duan, C. (1994) Insulin and insulin-like growth factor 1 in coho. salmon *Oncorhynchus kisutch* injected with streptozotocin. Am. J. Physiol. **267**: R1408–R1412.

Plisetskaya, E.M., Mommsen, T.P. (1996) Glucagon and glucagon-like peptides in fishes. Int. Rev. Cytol. **168:** 187–257.

Polinov, A.L., Kuzik, V.V̇., Danilova, O.A. (1997) The hypothalamo-hypophyseal system in Acipenseridae. XI. Morphological and immunohistochemical analysis of nonapeptidergic and corticocoliberin-immunoreactive elements in hypophysectomized sterlet (*Acipenser ruthenus* L.). Gen. Comp. Endocrinol. **105**: 314–322.

Ralston, K.R., Wainwright, P.C. (1997) Functional consequences of trophic specialization in pufferfishes. Funct. Ecol. **11**: 43–52.

Randall, D.J., Val, A.L. (1995) The role of carbonic anhydrase in aquatic gas exchage. In: Heisler, N. (ed.) Mechanisms of systemic regulation. Respiration and circulation. Advances in comparative and environmental physiology, vol. 21. Springer-Verlag, Berlin, Heidelberg, pp. 25–39.

Randall, D.J., Brauner, C. (1998) Interactions between ion and gas transfer in freshwater teleost fish. Comp. Biochem. Physiol. **119A**: 3–8.

Rankin, J.C. (1997) Osmotic and ionic regulation in cyclostomes. In: Hazon, N, Eddy, F.B., Flik, G. (Eds.) Ionic regulation in animals: A tribute to Professor, W.T.W. Potts. Springer-Verlag, Berlin, Heidelberg, pp. 50–69.

Raschi, W., Tabit, C. (1992) Functional aspects of placoid scales: a review and update. Aus. J. Mar. Freshw. Res. **43**: 123–147.

Rasotto, M.B., Manni, L. (1995) The juxtatesticular body of jawfishes (Teleostei, Opisthognathidae): comparative morphology and fine structure. J. Morphol. **226**: 237–246.

Reinecke, M. (1995) Peptides related to insulin-like growth factor 1 and 2 in the gastro-entero-pancreatic system of lower vertebrates. Neth. J. Zool. **45**: 192–194.

Reincke, M, Müller, C., Segner, H. (1997a) An immunohistochemical analysis of the ontogeny, distribution and coexistence of 12 regulatory peptides and serotonin in endocrine cells and nerve fibers of the digestive tract of the turbot, *Scophthalmus maximus* (Teleostei). Anat. Embryol. **195**: 87–102.

Reinecke, M., Schmid, A., Ermatinger, R., Loffing-Cueni, D. (1997b) Insulin-like growth factor 1 in the teleost *Oreochromis mossambicus*, the tilapia: gene sequence, tissue expression, and cellular localization. Endocrinology **138**: 3613–3619.

Reznick, D., Callahan, H., Llauredo, R. (1996) Maternal effects on offspring quality in poeciliid fishes. Am. Zool. **36:** 147–156.

Riedel, G. (1997) The forebrain of the blind cave fish *Astyanax hubbsi* (Characidae). I. General anatomy of the telencephalon. Brain Behav. Evol. **49**: 20–38.

Riehl, R., (1993) Surface morphology and micropyle as a tool for identifying fish eggs by scanning electron microscopy. Microscopy and Analysis: 29–31.

Riehl, R. (1996) The ecological significance of the egg envelope in teleosts with special reference to limnic species. Limnologica **26**: 183–189.

Riehl, R., Gotting, K.J. (1974) Zu struktur und Vorkommen der Mikropyle an Eizellen and Eiern von. Knochenfischen (Teleostei). Arch. Hydrobiol. **74**: 393–402.

Riehl, R., Kokoscha, M. (1993) A unique surface pattern and micropylar apparatus in the eggs of *Luciocephalus* sp. (Perciformes, Luciocephalidae). J. Fish Biol. **43**: 617–620.

Riehl, R., Meinel, W. (1994) Die Eier hemischer Fische 8. Kaulbarsch-*Gymnocephalus cernuus* (Linnaeus, 1758) mit Anmerkungen zum taxonomischen Status von. *Gymnocephalus baloni* (Holcik and Hensel, 1974). Fischokologie. **7:** 25–33.

Rink, E., Wullimann, M.F. (1998) Some forebrain connections of the gustatory system in the goldfish *Carassius auratus* visualized by separate DiI application to the hypothalamic inferior lobe and the torus lateralis. J. Comp. Neurol. **394**: 152–170.

Roberts, B.L. (1978) Mechanoreceptors and the behavior of elasmobranch fishes with special reference to the acoustico-laterals system. In: Hodgson, E.S., Mathewson, R.F. (eds.) Sensory biology of sharks, skates and rays. Office of Naval Research, Department of the Navy, Arlington, Va, pp. 331–390.

Roberts, B.L., Meredith, G.E. (1989) The efferent system. In: Coombs, S., Görner, P., Münz, H. (eds.) The mechanosensory lateral line: neurobiology and evolution. Springer-Verlag, New York, pp. 445–459.

Roberts, J.L. (1975a) Active branchial and ram gill ventilation in fishes. Biol. Bull. **148**: 85–105.

Roberts, J.L. (1975b) Respiratory adaptations of aquatic animals. In: Verneberg, F.J. (ed.) Physiological adaptation of aquatic animals. Intext Educational Publication, New York, pp. 395–414.

Rocha, E., Monteiro, R.A.F., Pareira, C.A. (1996) The pale-grey interhepatocytic cells of brown trout (*Salmo trutta*) are a subpopulation of liver resident macrophages or do they establish a different cellular type? J. Submicrosc Cytol. Pathol. **28:** 357–368.

Rocha, E., Monteiro, R.A.F., Pereira, C.A. (1997) Liver of the brown trout, *Salmo trutta* (Teleostei, Salmonidae); a stereological study at light and electron microscopic levels. Anat. Rec. **247**: 317–328.

Rosenqvist, G. (1990) Male mate choice and female-female composition for mates in the pipefish *Nerophis ophidion*. Anim. Behav. Physiol. **39**: 1110–1115.

Rosenqvist, G. (1993) Sex role reversal in a pipefish. Mar. Behav. Physiol. **23**: 219–230.

Rosenqvist, G., Houde, A. (1997) Prior exposure to male phenotypes influences mate choice in the guppy, *Poecilia reticulata*. Behav. Ecol. **8**: 194–198.

Roy, P.K., Munshi, J.S.D. (1996) Morphometrics of the respiratory system of air-breathing fishes of India. In: Munshi, J.S.D., Dutta, H.M. (eds.) Fish Morphology-horizon of new research. Oxford & IBH Publishing Co. Pvt. Ltd., New Delhi, pp. 203–234.

Sa, L.A., Neto, V.M., De. Oliveira, M.M., Chagas, C. (1991) Heterogeneity of purified actin in the electric organ of the electric eel. *Electrophorus electricus*. J. Exp. Zool. **257**: 43–50.

Saether, B.S., Johnson, H.K., Jobling, M. (1996) Seasonal changes in food consumption and growth of Arctic charr exposed either simulated natural or a **12**: 12 L.D. photoperiod at constant water temperature. J. Fish Biol. **48**: 1113–1122.

Sakamoto, T., McCormick, S.D., Hirano, T. (1993) Osmoregulatory actions of growth hormone and its mode of action in salmonids: a review. Fish Physiol. Biochem. **11**: 155–164.

Schaperclaus, W. (1992) Fish diseases, vol. 2. Balkema, Rotterdam.

Schmitt, E., Kunz, Y.W. (1989) Retinal morphogenesis in the rainbow trout, *Salmo gairdneri*, Brain Behav. Evol. **34**: 48–64.

Schreck, C.B. (1996) Immunomodulation: endogenous factors. In: Iwama, G., Nakanishi, T. (eds.) Fish physiology, vol. 15. The fish immune system: organism, pathogen, and environment. Academic Press, San Diego, pp. 311–337.

Secombes, C.J. (1994) Enhancement of fish phagocyte activity. Fish Shellfish Immunol. **4:** 421–436.

Secombes, C.J. (1996) The nonspecific immune system: cellular defenses. In: Iwama, G., Nakanishi, (eds.) Fish physiology, vol. 15. The fish immune system: organism, pathogen, and environment. Academic Press, San Diego, pp. 63–103.

Secombes, C.J., Chappel, L.H. (1996) Fish immune responses to experimental and natural infection with helminth parasites. Ann. Rev. Fish Dis. **6**: 167–177.

Secombes, C.J., Hardie, L.J., Daniels, G. (1996) Cytokines in fish: an update. Fish Shellfish Immunol. **6:** 291–304.

Sebert, P. (1997) Pressure effects on shallow-water fishes. In: Hoar, W.S., Randall, D.J., Farrell, A.P. (eds.) Fish physiology vol. 16. Deep-Sea fishes. Academic Press, San Diego, pp. 279–323.

Segner, H., Storch, V., Reinecke, M., Kloas, W., Hanke, W. (1994) The development of functional digestive and metabolic organs in turbot, *Scophthalamus maximus*. Mar. Biol. **119**: 471–486.

Segner, H, Storch, V., Reinecke, M., Kloas, W., Hanke, W. (1995) A tabular overview of organogenesis in larval turbot (*Scophthalmus maximus* L.). I.C.E.S. mar. Sci. Symp. **201:** 35–39.

Shanbhag, B.A., Saidapur, S.K. (1996) Atretic follicles and corpora lutea in the ovaries of fishes: structure-function correlations and significance. In: Munshi, J.S.D., Dutta, H.M. (eds.) Fish morphology-horizon of new research. Oxford & IBH Publishing Co. Pvt. Ltd., New Delhi, pp. 147–168.

Shand, J. (1997) Ontogenetic changes in retinal structure and visual acuity: a comparative study of coral-reef teleosts with differing post-settlement lifestyles. Env. Biol. Fish. **49**: 307–322.

Shapiro, D.Y., Appeldoorn, R.S., Hensley, D.A. Ray, M. (1997) Water flow and spawning time in a coral reef fish. Proc. 8th Int. Coral. Reef Sym. **2:** 1121–1126.

Shearer, K.D., Silverstein, J.T., Plisetskaya, E.M., (1997) Role of adiposity in food intake control of juvenile chinook salmon (*Oncorhynchus tshawytscha*). Comp. Biochem. Physiol. **118A:** 1209–1215.

Shedpure, M., Pati, A.K. (1995) The pineal gland: structural and functional diversity. Ind. J. Exp. Biol. **33**: 625–640.

Shelton, G. (1970) The regulation of breathing. In: Hoar, W.S., Randall, D.J. (eds.) Fish physiology, vol. 4. Academic Press, New York, pp. 293–359.

Shieh, K.T., Wilson, W., Winslow, M., McBride, D.W. Jr., Hopkins, C.D. (1996) Short-range orientation in electric fish: an experimental study of passive electrolocation. J. Exp. Biol. **199**: 2383–2393.

Singh, H.R. (1994) The structure and function of the saccus vasculosus. In: Singh, H.R. (ed.) Advances in fish biology. Hindustan Publishing Corporation, New Delhi, pp. 246–260.

Sinha, G.M., Chakrabarti, P. (1989) Micro-architecture of the gut epithelium in *Labeo calbasu* (Hamilton): a scanning electron microscopic study. Arch. Biol. (Bruxelles) **100**: 283–294.

Sire, J.Y., Huysseune, A (1996) Structure and development of the odontodes in an armoured catfish, *Corydoras aeneus* (Siluriformes, Callichthyidae). Acta Zool. Stockholm **77:** 51–72.

Sire, J.Y., Quilhac, A., Bourguignon, A., Allizard, F. (1997) Evidence for participation of the epidermis in the deposition of superficial layer of scales in zebrafish (*Damio rerio*): A SEM and TEM study. J. Morphol. **231**: 161–174.

Smith, C., Wootton, R.J. (1995) The costs of parental care in teleost fishes. Rev. Fish Biol. Fish. **5:** 7–22.

Smith, R.J.F. (1997) Avoiding and deterring predators. In: Godin, J.G.J. (1997) Behavioural ecology of teleost fishes. Oxford University Press, Oxford, pp. 163–190.

Sorensen, B., Weber, R.E. (1995) Effects of oxygenation and the stress hormones adrenaline and cortisol on the viscosity of blood from the trout. J. Exp. Biol. **198:** 953–959.

Sorensen P.W. (1996) Biological responsiveness to pheromones provides fundamental and unique insight into olfactory function. Chem. Senses **21:** 245–256.

Sorensen, P.W., Scott, A.P. (1994) The evolution of hormonal sex pheromones in teleost fish: poor correlation between the pattern of steroid release by goldfish and olfactory sensitivity suggests that these cues evolved as a result of chemical spying rather than signal specialization. Acta Physiol. Scand. **152**: 191–205.

Srivastava, G.B.L., Gopesh, A., Singh, M. (1981) A new neurosecretory system in fish, located in the gill region. Experientia. **37**: 850–851.

Stevens, E.D., Lam, H.M., Kendall, J. (1974) Vascular anatomy of the counter-current heat exchanger of skipjack tuna. J. Exp. Biol. **61**: 145–153.

Stoiber, W., Sänger, A.M. (1996) An electron microscopic investigation into the possible source of new muscle fibres in teleost fish. Anat. Embryol. **194**: 569–579.

Sugimoto, M., Kawamura, T., Fujii, R., Oshima, N. (1994) Changes in the responsiveness of melanophores to electrical nervous stimulation after prolonged background adaptation in the medaka, *Oryzias latipes*. Zool. Sci. **11:** 39–44.

Tibbits, G.F. (1996) Towards a molecular explanation of the high performance of the tuna heart. Comp. Biochem. Physiol. **113A:** 77–82.

Tong, S.L. (1982) The nucleus praeeminentialis: an electro-and mechanoreceptor center in the brainstem of the catfish. J. Comp. Physiol. **145**: 299–309.

Tong, S.L., Bullock, T.H. (1982) Electroreceptive representation and its dynamics in the cerebellum of the catfish, *Ictalurus nebulosus* (Ictaluridae Siluriformes). J. Comp. Physiol. **145**: 289–298.

Tong, S.L., Finger, T.E. (1983) Central organization of the electrosensory lateral line system in bullhead catfish *Ictalurus nebulosus*. J. Comp. Neurol. **217**: 1–16.

Torricelli, P., Lugli, M., Bobbio, L. (1993) The importance of male-male competition and sexually selected dimorphic traits for male reproductive success in site-attached fishes with paternal care: the case of the freshwater goby *Padogobius martensi*, Mar. Behav. Physiol. **23**: 231–256.

Tota, B., Gattuso, A (1996) Heart ventricle pumps in teleosts and elasmobranchs: a morphodynamic approach. J. Exp. Zool. **275**: 162–171.

Tveitten, H., Johnsen, H.K., Jobling, M. (1996) Influence of maturity status on the annual cycles of feeding and growth in Arctic charr. reared at constant temperature. J. Fish Biol. **48:** 910–924.

Utter, F.M. (1994) Perspectives of molecular genetics and fisheries into the 21st century. Rev. Fish Biol. Fish. **4**: 374–378.

Valentinčič, T., Caprio, J. (1997) Visual and chemical release of feeding behavior in adult rainbow trout. Chem. Senses **22**: 375–382.

Van den Berg, C., Van Snik, G.J.M., van der Boogart, J.G.M., Sibbing, F.A., Osse, J.W.M. (1994) Comparative microanatomy of the branchial sieve in three sympatric cyprinid species, related to filter-feeding mechanisms. J. Morphol. **219**: 73–87.

Van der Kraak, G., Chang, J.P., Janz. D.M. (1998) Reproduction. In: Evans, D.H. (ed.) Physiology of fishes. (Second edition). CRC Press, Boca Raton, pp. 465–488.

Van Rooij, J.M., Videler, J.J. (1996) Estimating oxygen uptake rate from ventilation frequency in the reef fish *Sparisoma viride.* Mar. Ecol. Prog. Ser. **132:** 31–41.

Van Rooij, J.M., Bruggemann, J.H., Videler, J.J., Breeman, A.M. (1995a) Ontogenetic, social, spatial and seasonal variations in condition of the reef herbivore *Sparisoma viride*. Mar. Biol. **123:** 269–275.

Van Rooij, J.M., Bruggemann, J.H., Videler, J.J., Breeman, A.M. (1995b) Plastic growth of the herbivorous reef fish *Sparisoma viride:* field evidence for a trade-off between growth and reproduction. Mar. Ecol. Prog. Ser. **122:** 93–105.

Van Rooij, J.M., Jong, E. de, Vaandrager, F., Videler, J.J. (1996a). Resource and habitat sharing by the stoplight parrotfish, *Sparisoma viride*, a Caribbean reef herbivore. Env. Biol. Fish. **47**: 81–91.

Van Rooij, J.M., Kroom, F.J., Videler, J.J. (1996b) The social and mating system of the herbivorous reef fish *Sparisoma viride:* one-male versus multi-male groups. Env. Biol. Fish. **47**: 353–378.

Van Weerd, J.H., Richter, C.J.J. (1991) Sex pheromones and ovarian development in teleost fish. Comp. Biochem. Physiol. **100A**: 517-527.

Verspoor, E. (1998) Molecular markers and the genetic management of farmed fish. In: Black, K.D., Pickering, A.D. (eds.) Biology of farmed fish. Sheffield Academic Press, Sheffield, pp. 355–382.

Videler, J.J. (1995) Body surface adaptations to boundary-layer dynamics. Soc. Exp. Biol. pp. 1–20.

Videler, J.J., Nolet, B.A. (1990) Costs of swimming measured at optimum speed: scale effect, differences between swimming styles, taxonomic groups and submerged and surface swimming. Comp. Biochem. Physiol. **97A**: 91–99.

Wagemans, F., Focant, B., Vandewalle, P. (1998) Early development of the cephalic skeleton in the turbot. J. Fish Biol. **52**: 166–204.

Wagner, H.J., Menzes, N.A., Ali, M.A. (1976) Retinal adaptations in some Brazilian tide pool fishes (Teleostei). Zoomorphologie **83**: 209–226.

Wainwright, P.C. (1996) Ecological explanation through functional morphology: the feeding biology of sunfishes. Ecology **77**: 1336–1343.

Wainwright, P.C., Turingan, R.G. (1997) Evolution of pufferfish inflation behavior. Evolution **51:** 506–518.

Walsh, P.J. (1997) Evolution and regulation of urea synthesis and ureotely in (batrachoidid) fishes. Annu. Rev. Physiol. **59**: 299–323.

Walsh, P.J. (1998) Nitrogen excretion and metabolism. In: Evans, D.H. (ed.) The physiology of fishes. (Second edition). CRC Press, Boca Raton, pp. 199–214.

Ward, R.D., Grewe, P.M. (1994) Appraisal of molecular genetic techniques in fisheries. Rev. Fish Biol. Fish **4:** 300–325.

Webb, P.W., Weihs D. (1986) Functional locomotor morphology of early life history stages of fishes. Trans. Amer. Fish. Soc. **115**: 115–127.

Weatherley, A.H., Gill, H.S. (1987) The biology of fish growth. Academic Press, London.

Weber, J.-M. (1991) Effect of endurance swimming on the lactate kinetics of rainbow trout. J. Exp. Biol. **158**: 463–476.

Weber, J.-M., Haman, F. (1996) Pathways for metabolic fuels and oxygen in high performance fish. Comp. Biochem. Physiol. **113A**: 33–38.

Weber, J.-M., Zwingelstein, G. (1995) Circulatory substrate fluxes and their regulation. In: Hochachka, P.W., Mommsen, T.P. (eds.) Biochemistry and molecular biology of fishes, vol. 4. Elsevier, Amsterdam, pp. 15–32.

Weitzman, S.H. (1997) Systematics of deep-sea fishes. In: Hoar, W.S., Randall, D.J., Farrell, A.P. (eds.) Fish physiology, vol. 16. Deep-sea fishes. Academic Press, San Diego, pp. 43–77.

Wilga, C.D., Motta, P.J. (1998) Conservation and variation in feeding mechanism of the spiny dogfish *Squalus acanthias.* J. Exp. Biol. **201**: 1345–1358.

Witten, P.E. (1997) Enzyme histochemical characteristics of osteoblasts and mononucleated osteoclasts in a teleost fish with acellular bone (*Oreochromis niloticus*, Cichlidae). Cell Tissue Res. **287**: 591–599.

Witten, P.E., Willwock, W. (1997) Growth requires bone resorption at particular skeletal elements in a teleost fish with acellular bone (*Oreochromis niloticus*, Teleostei: Cichlidae). J. Appl. Ichthyol **13**: 149–158.

Wood, C.M., Hopkin, T.E., Walsh, P.J. (1997) Pulsatile urea excretion in the toadfish (*Opsanus beta*) is due to a pulsatile excretion mechanism, not a pulsatile production mechanism. J. Exp. Biol. **200**: 1039–1046.

Wright, J.M., Bentzen, P. (1994) Microsatellites: genetic markers for the future. Rev. Fish Biol. Fish. **4**: 384–388.

Wright, P.J., Huntingford, F.A. (1993) Daily growth increments in the otolith of the three-spined stickleback, *Gasterosteus aculeatus*. J. Fish Biol. **42**: 65–77.

Wright, P.J., Talbot, C., Thorpe, J.E. (1992) Otolith calcification in Atlantic salmon parr, *Salmo salar* L. and its relation to photoperiod and calcium metabolism. J. Fish Biol. **40**: 779–790.

Wubbels, R.J., Schellart, N.A.M. (1997) Neuronal coding of sound detection in the auditory midbrain of the rainbow trout. J. Neurophysiol. **77**: 3060–3074.

Wullimann, M.F. (1997) Major patterns of visual brain organization in teleosts and their relation to prehistoric events and the paleontological record. Peleobiology **23**: 101–114.

Yacob, A., Kunz, Y.W. (1977) 'Disc shedding' in the cone outer segments of the teleost, *Poecilia reticulata*, P. Cell Tissue Res. **181**: 487–492.

Yano, T. (1996) The nonspecific immune system: humoral defense. In: Iwama, G., Nakanishi, T. (eds.) Fish physiological, vol. 15. The fish immune system: organism, pathogen, and environment. Academic Press, San Diego, pp. 106–159.

Yoshiyama, R.M., Cech, J.J. Jr. (1994) Aerial respiration by rocky intertidal fishes of California and Oregon. Copeia **1994**: 153–158.

Zaccone, G., Fasulo, S., Ainis, L., Licata, A. (1997) Paraneurons in the gills and airways of fishes. Microsc. Res. Techn. **37**: 4–12.

Zapata, A.G. Torroba, M., Sacedon, R., Varas, A., Vicente, A. (1996b) Structure of the lymphoid organs of elasmobranchs. J. Exp. Zool. **275**: 125–143.

Zhang, Y., Weaver, L. Jr., Ibeawuchi, A., Olson, K.R. (1997) Catecholaminergic regulation of venous function in the trout. Am. J. Physiol. **274**: R1195–R1202.

Zippel , H.P. (1993) Regeneration in the peripheral and the central olfactory system: a review of morphological, physiological and behavioural aspects. J. Hirnforsch. **34**: 207–229.

Zippel, H.P., Sorensen, P.W., Hansen, A. (1997) High correlation between microvillous olfactory receptor cell abundance and sensitivity to pheromones in olfactory nerve-sectioned goldfish. J. Comp. Physiol. **A 180**: 39–52.

Zylberberg, L., Meunier, F.J. (1996) Ultrastructural data on the melanophores associated with the cellular elasmoid scales in *Leporinus friderici* (Teleostei: Ostariophysi, Anastomidae): their putative participation in scale matrix formation. J. Morphol. **228**: 155–164.

Zyznar, E.S., Ali, M.A. (1975) An interpretative study of the organization of the visual cells and tapetum lucidum of *Stizostedion*. Can. J. Zool. **53**: 180–196.

Appendix to References-2

Adam, H. (1959) Hypophyse und hypothalamo-neurohypophysäres neurosekret system bei cyclostomen *Myxine glutinosa* und *Bdellostoma stouti*. Verh. Deutsch. Zoologi. Gesselschaft Münster/Westf.: 157–171.

Ali, M.A., Anctil, M., Cervetto, L. (1978) Photoreception. In: Ali, M.A. (ed.) Sensory ecology. Review and perspectives. Plenum Press, New York, pp. 467–502.

Anderson, D.P. (1990) Immunological indicators: effects of environmental stress on immune protection and disease outbreaks. Am. Fish. Soc. Symp. **8**: 38–50.

Anderson, D.P. (1996) Environmental factors in fish health: immunological aspects. In: Iwama, G., Nakanishi, T. (ed.) Fish physiology, Vol. 15. The fish immune system: organism, pathogen, and environment. Academic Press, San Diego, pp. 289–310.

Anderson, D.P., Barney, P.J. (1991) the role of the diagnostic laboratory in fish disease control. Annual Rev. Fish Dis. pp. 41–62.

Anderson, P.M. (1995) Molecular aspects of carbamoly phosphate synthesis. In: Walsh, P.J., Wright, P. (eds.) Nitrogen metabolism and excretion. CRC Press, Boca Raton, pp. 33–49.

Atz, J.W. (1952) Internal nares in the teleost, *Astroscopus*. Anat. Rec. **113**: 105–115.

Ariëns Kappers, C.U., Huber, G.C., Crosby, E.C. (1936) The comparative anatomy of the nervous system of vertebrates, including man. Macmillan, New York.

Arnott, H.J., Nicol, J.A.C. (1970) Reflection of ratfish skin (*Hydrolagus colliei*). Can. J. Zool. **48**: 137–151.

Aronson, L.R. (1963) The central nervous system of sharks and bony fishes with special reference to sensory and integrative mechanisms. In: Gilbert, P.W. (ed.) Sharks and survival. D.C. Heath, Boston, Mass. pp. 165–241.

Augustinsson, K.B., Fänge, R., Johnels, A., Östlund, E. (1955) Histological, physiological and biochemical studies on the heart of two cyclostomes, hagfish (*Myxine*) and lamprey (*Lampetra*). J. Physiol. **131**: 257–276.

Barlow, G.W., Siri, P. (1994) Polychromatic midas cichlids respond to dummy opponents: colour, contrast and context. Behaviour 130: 77–112.

Barry, K.L., Hawryshn, C.W. (1999) Effects of incident light and background conditions of Hawaiian coral reef fish. J. Mar. Biol. Ass. U.K. **79**: 495–508.

Becker, K., Fishelson, L., Amseigruber, W.M. (2001) Cytological outogenesis and involution of the thymus and head-kidney in juvenile and old domestic carp: Is ageing in fish a chronological or growth-related phenomenon? J. Appl. Ichthyol. **17**: 1–13.

Bell, C.C. (1979) Central nervous system physiology of electroreception, a review. J. Physiol. Paris **75**: 361–379.

Bhatia, B. (1950) Adaptive modifications in the hill-stream catfish, *Glyptothorax telchitta* (Hamilton). Proc. Natl. Inst. Sci. India **16**: 271–285.

Blaxter, J.H.S. (1975) Fish vision and applied research. In: Ali, M. (ed.) Vision in fishes. Plenum Publishing Corporation, New York, pp. 757–773.

Blaxter, J.H.S., Tyler, P. (1978) Physiology and function of the swimbladder. Adv. Comp. Physiol. Biochem. **7**: 311–367.

Bleckmann, H., Hofmann, M.H. (1999) Special senses. In: Hamlett, W.C. (ed.) Sharks, skates, and rays. John Hopkins University Press, Baltimore, London, pp. 300–328.

Block, B.A., Stevens, E.D. (eds.) (2001) Tuna. Fish physiology, vol. 19. Academic Press, San Diego.

Boeuf, G., Le Bail P.Y. (1999) Does light have an influence on fish growth? Aquaculture **177**: 129–152.

Boeuf, G., Boujard, D., Person-Le Ruyet, J. (1999) Control of the somatic growth in turbot. J. Fish Biol. **55** (Suppl. A): 128–147.

Bone, Q. (1972) Buoyancy and hydrodynamic functions of integument in the castor oil fish, *Ruvettus pretiosus* (Pisces: Gemptylidae). Copeia **1972**: 78–87.

Bone Q., Brook, C.E.R. (1973) On *Schedophilus medusophagus:* (Stromateoidei) J. Mar. biol. Ass. U.K. **53**: 752–761.

Bozzano, A., Collin, S.P. (2000) Retinal ganglion cell topography in elasmobranchs. Brain Behav. Evol. **55**: 191–208.

Braun, C.B. (1998) Schreiner organs: a new craniate chemosensory modality in hagfishes. J. Comp. Neurol. **392**: 135–163.

Braun, C.B., Northcutt, R.G. (1997) The lateral line system of hagfishes (Craniata: Myxinoidea). Acta Zool. Stockholm **78**: 247–268.

Braun, C.B., Northcutt, R.G. (1998) Cutaneous exteroreceptors and their innervation in hagfishes. In: Jorgensen. J.M., Lomholt, J.P., Weber, R.E., Malte, H. (eds.) The biology of hagfishes. Chapman and Hall, London, New York, pp. 512–532.

Brodal, A., Fänge, R. (eds.) (1963) The biology of *Myxine*. Universitetsforlaget, Oslo, Norway.

Brauner, C.J., Jensen, F.B. (1999) O_2 and CO_2 exchange in fish: the non-linear release of Bohr/Haldane protons with oxygenation. In: Val, A.L., Almeida-Val, V.M.F. (eds.) Biology of tropical fishes. INPA, Manaus, pp. 393–400.

Brauner, C.J., Randall, D.J. (1998) The linkage between oxygen and carbon dioxide transport. In: Perry, S.F., Tufts B.L. (eds.) Fish physiology, vol. 17. Fish respiration. Academic Press, New York, pp. 283–319.

Bretschneider, F., van den Berg, A.V., Peters, R.C. (2001) Mechanoreception: hearing and lateral line. In: Kapoor, B.G., Hara, T.J. (eds.) Sensory biology of jawed fishes: new insights. Oxford & IBH Publishing Co. Pvt. Ltd., New Delhi. pp. 215–253.

Bridges, C.R. (1999) Ecophysiological adaptationes in intertidal rockpool fishes. In: Mittal, A.K. Eddy, F.B., Munshi, J.S.D. (eds.) Water/air transition in biology. Oxford & IBH Publishing Co. Pvt. Ltd., New Delhi, pp. 67–92.

Buslé, J., Anadon, G. (1996) The structure and function of fish liver. In: Munshi, J.S.D., Dutta, H.M. (eds.) Fish morphology-horizon of new research. Oxford & IBH Publishing Co. Pvt. Ltd., New Delhi, pp. 77–93.

Bullock, T.H. (1981) Comparisons of the electric and acoustic senses and their central processing. In: Tavolga, W.N., Popper, A.N., Fay, R.R. (eds.) Hearing and sound communication in fishes. Springer-Verlag, New York, pp. 125–171.

Burnstock, G. (1959) The innervation of the gut of the brown trout (*Salmo trutta*). Quart. J. Micros. Sci. **100**: 199–220.

Burton, D. (1999) Environmentally related morphological plasticity in the teleost epidermis. In: Mittal, A.K., Eddy, F.B., Munshi, J.S.D. (eds.) Water/air transition in biology. Oxford & IBH Publishing Co. Pvt. Ltd., New Delhi, pp. 117–129.

Cahn, P.H., Shaw, E. (1965) A method for studying lateral line cupular bending in juvenile fishes. Bull. Mar. Sci. **15**: 1060–1071.

Cerra, M.C., Mazza, R., Pellegrino, D., Tota, B. (1998) Natriuretic peptides in Antartic teleosts: cardic receptors in *Chionodraco hamatus* and *Trematomus bernacchii*. In: di Prisco, G., Pisano, E., Clarke A. (eds.) Fishes of Antarctica. A biological overview. Springer-Verlag Italia, Milano, pp. 205–213.

Chadwich, T.D., Wright, P.A. (1999) Nitrogen excretion and expression of urea cycle enzymes in the Atlantic cod (*Gadus morhua* L.): a comparison of early life stages with adults. J. Exp. Biol. **202**: 2653–2662.

Chardon, M., Vandewalle, P. (1989) About identification of swimbladder sheets in Ostariophysi. Fortsch. Zool. **35**: 264–268.

Cheng C-H.C. (1998) Origin and mechanism of evolution of antifreeze glycoproteins in polar fish. In: di Prisco G., Pisano, E., Clarke A., (eds.) Fishes of Antarctica. A biological overview. Springer-Verlag Italia, Milano, pp. 311–328.

Chiba, A., Sohn, Y.C., Honma, Y. (1996a) Immunohistochemical and ultrastructural characterization of the terminal nerve ganglion cells of the ayu, *Placoglossas altevelis* (Salmoniformes, Teleostei). Anat. Rec. **246**: 549–556.

Chiba, A., Sohn, Y.C., Oka, S. Honma, Y. (1996b) Localization of neuropeptide Y-like immunoreactivity in the saccus vasculosus of cartilagimous and bony fishes. Zool. Sci. **13**: 347–350.

Christiansen, J.S., Fevolden, S.E., Karamushko, O.V., Karamuschko, L.I. (1998) Maternal output in polar fish reproduction. In: di Prisco, G., Pisano, E., Clarke, A. (eds.) Fishes of Antarctica. A biological overview. Springer-Verlag Italia, Milano, pp. 41–52.

Clifton, K.B., Motta, P.J. (1998) Feeding morphology, diet, and ecomorphological relationships among five Caribbean labrids (Teleostei, Labridae). Copeia **1998**: 953–966.

Collette, B.B. (1977) Epidermal breeding tubercles and bony contact organs in fisheo. Symp. zool. Soc. Lond. No. **39**: 225–268.

Collin, H.B., Collin, S.P. (2000) The corneal surface of aquatic vertebrates: microstructures with optical and nutritional function? Phil. Trans. R. Soc. Lond. **B 355**: 1171–1176.

Collin, S.P., Collin, H.B. (1998) The deep-sea teleost cornea: a comparative study of gadiform fishes. Histol. Histopathol. **13**: 325–336.

Collin, S.P., Collin, H.B. (2001) The fish cornea: Adaptations for different aquatic environments. In: Kapoor, B.G., Hara, T.J. (eds.) Sensory biology of jawed fishes: new insights. Oxford & IBH Publishing Co. Pvt. Ltd., New Delhi, pp. 57–96.

Collin, S.P., Potter, I.C. (2000) The ocular morphology of the southern hemisphere lamprey *Mordacia mordax* Richardson with special reference to a single class of photoreceptor and a retinal tapetum. Brain Behav. Evol. **55**: 120–138.

Collin, S.P., Potter, I.C., Braeckevelt, C.R. (1999) The ocular morphology of the southern hemisphere lamprey *Geotria australis* Gray, with special reference to optical specialisations and the characterisation and phylogeny of photoreceptor types. Brain Behav. Evol. **54**: 96–118.

Collin, S.P., Llyoid, D.J., Wagner H.J. (2000) Foveate vision in deep-sea teleosts: a comparison of primary visual and olfactory inputs. Phil. Trans. R. Soc. Lond. **B 355**: 1315–1320.

Coombs, S., (1999) Signal detection theory, lateral line excitation patterns and prey capture behaviour of mottled sculpin. Anim. Behav. **58**: 421–430.

Coombs, S., Montogomery, J.C. (1999) The enigmatic lateral line system. In: Fay, R.R., Popper, A.N. (eds.) Comparative hearing. Springer-Verlag, New York, pp. 319–362.

Cosica, M.R., Oreste, U. (1998) Antartic fish immunoglobulins: preliminary data on structure and antibody specificity. In: di Prisco, G., Pisano, E., Clarke A. (eds.) Fishes of Antarctica. A biological overview. Springer-Verlag Italia, Milano, pp. 175–184.

Coucelo, J., Coucelo, J., Azevedo, J. (1996) Ultrasonography characterization of heart morphology and blood flow of lower vertebrates. J. Exp. Zool. **275**: 73–82.

Coughlin, D.J. (2000) Power production during steady swimming in large mouthbass and rainbow trout. J. Exp. Biol. 203: 617–629.

Coughlin, D.J., Rome, L.C. (1999) Muscle activity in steady swimming scup, *Stenotomus chrysops*, varies with fiber types and body position Biol. Bull. **196**: 145–152.

Davis, R.E., Northcutt, R.G. (eds.) (1983) Fish neurobiology, vol. 2 Higher brain areas and functions. University of Michigan Press, Ann Arbor.

Dawson, J. (1969) The keratinised teeth of *Myxine glutinosa*. A histological, histochemical, ultrastructural and experimental study. Acta zool. Stockholm **50**: 35–68.

di Prisco, G. (1998) Molecular adaptations of Antarctic fish hemoglobins. In: di Prisco, G. Pisano, E., Clarke, A. (eds.) Fishes of Antarctica. A biological overview. Springer-Verlag Italia, Milano, pp. 339–353.

di Prisco, G., D' Avino, R., Caruso, C., Tamburini, M., Camardella, L., Rutingliano, B., Carratore, V., Romano, M. (1991) The biochemistry of oxygen transport in red-blooded antarctic fish. In: di Prisco, G., Maresca, B., Tota, B. (eds.) Biology of Antarctic fish. Springer-Verlag, Berlin, pp. 263–281.

di Prisco, G., Pisano, E., Clarke, A. (eds.) (1998) Fishes of Antarctica. A biological overview. Springer-Verlag Italia, Milano.

Diogo, R., Chardon, M. (2000) Anatomie et fonction des structures cephaliques associee a la prise de nourriture cheiz le genre *Chrysichthys* (Teleostei: Siluriformes). Belg. J. Zool. **130**: 21–37.

Dodd, J.M., Kerr, T. (1963) Comparative morphology and histology of the hypothalamo-neurohypophysial system. Proc. Zool. Soc. Lond. No. **9**: 5–27.

Douglas, R.H., Diamgoz, M.B.A. (eds.) (1990) The visual system of fish. Chapman and Hall, London, New York.

Døving, K.B. (1998) The olfactory system of hagfishes. In: Jorgensen, J.M., Lomholt, J.P., Weber, R.E., Malte, H. (eds.) The biology of hagfishes. Chapman and Hall, London, New York, pp. 533–540.

Döving, K.B., Gemne, G. (1965) Electrophysiological and histological properties of the olfactory tract of the burbot (*Lota lota L.*). J. Neurophysiol. **28**: 139–153.

Donoghue, P.C.J., Forey, P.L., Aldridge R.J. (2000) Conodont affinity and chordate phylogeny. Biol. Rev. **75**: 191–251.

Drucker, E.G., Lauder, G.V. (2000) A hydrodynamic analysis of fish swimming speed: wake structure and locomotor force in slow and fast labriform swimmers. J. Exp. Biol. **203**: 2379–2393.

Drucker, E.G., Lauder, G.V. (2001) Wake dynamics and fluid forces of turning maneuvers in sunfish. J. Exp. Biol. **204**: 431–442.

Eastman, J.T. (1993) Antartic fish biology: evolution in a unique environment. Academic Press, San Diego.

Edmonds M.A., Motta, P.J., Hueter R.E. (2001). Food capture kinematics of the suction feeding horn shark, *Heterodontus francisci*. Env. Biol. fish. **62**: 415–427.

Edwards A.S., Herring, P.J. (1977) Observations on the comparative morphology and operation of the photogenic tissues of myctoplid fishes. Mar. Biol. **41**: 59–70.

Egginton, S., Rankin, J.C. (1998) Vascular adaptations for a low pressure/high flow blood supply to locomotory muscles of Antarctic icefish. In: di Prisco, G., Pisano, E., Clarke A. (eds.) Fishes of Antarctica. A biological overview. Springer-Verlag Italia, Milano, pp. 185–195.

Ekstom, P., Meissl, H. (1997) The pineal organ of teleost fishes. Rev. Fish Biol. Fish. **7**: 199–284.

Elies, G., Duval, H., Bonnec, G., Wolff, J., Boeuff, G., Boujard, D. (1999) Insulin and insulin-like growth factor-1 receptors in an evoluted fish, the turbot: cDNA cloning and mRNA expression. Mol. Cell. Endocrinal. **158**: 173–185.

Emde, G. von der., Schwarz, S. (2001) Detection of electric signals in jawed fishes. In: Kapoor, B.G., Hara, T.J. (eds.) Sensory biology of jawed fishes: new insights. Oxford & IBH Publishing Co. Pvt. Ltd., New Delhi, pp. 161–180.

Endean, R. (1961) A study of the distribution, habitat, behaviour, venom apparatus, and the venom of the stone-fish. Aust. J. Mar. Freshw. Res. **12**: 177–190.

Epple, A., Brinn, J.E. (1987) The comparative physiology of pancreatic islets. Springer-Verlag, Berlin.

Esteban, M.A., Muñoz, J., Meseguer, J. (2000) Blood cells of sea bass (*Dicentrarchus labrax* L.). Flow cytometric and microscopic studies. Anat. Rec. **258**: 80–89.

Fago, A., Weber, R.E. (1998) Hagfish haemoblobins. In: Jorgensen, J.M., Lomholt, J.P., Weber, R.E., Malte, H. (eds.) The biology of hagfishes. Chapman and Hall, London, New York, pp. 321–333.

Fänge, R. (1953) The mechanisms of gas transport in the euphysoclist swimbladder. Acta Physiol. Scand. **30** (Suppl. 110): 1–133.

Fänge, R. Holmgren, S., Nilsson, S. (1976) Autonomic nerve control of the swimbladder of the goldsinny wrasse, *Ctenolabrus rupestris*. Acta Physiol. Scand. **97**: 292–303.

Fernholm, B. (1998) Hagfish systematics. In: Jorgensen, J.M., Lomholt, J.P., Weber, R.E., Malte, H. (eds.) The biology of hagfishes. Chapman and Hall, London, New York, pp. 33–44.

Fines, G.A., Ballantyne, J.S., Wright, P.A. (2001) Active urea transport and an unusual basolateral membrane composition in the gills of a marine elasmobranch. Am J. Physiol. Regulatory Integrative Comp. Physiol. **280**: R16–R24.

Fishelson, L. (1968) Structure of the vertebral column in *Lepadichthys lineatus*. A clingfish associated with crinoids. Copeia **1968**: 859–861.

Fujii, R. (1973) Responses of fish pigment cells to environmental changes. In: Chavin, W. (ed.) Responses of fish to environmental changes. Charles C. Thomas, Springfield, Illinois, pp. 342–362.

Fuiman, L.A., Magurran, A.E. (1994) Development of predator defences in fishes. Rev. Fish Biol. Fish. **4**: 145–183.

Giardina, B., Mordente, A., Zappacosta, B., Calla, C., Colacicco, L. Gozzo, M.L. Lippa, S. (1998) The oxidative metabolism of Antartic fish: some peculiar aspects of cold adaptation. In: di Prisco, G., Pisano, E., Clarke, A. (eds.) Fishes of Antarctica. A biological overview. Springer-Verlag Italia, Milano. pp. 129–138.

Gilloteaux, J., Oldham, C.K., Biaglanti-Risbourg, S. (1996) Ultrastructural diversity of the biliary tract and the gallbladder in fish. In: Munshi, J.S.D., Dutta, H.M. (eds.) Fish morphology-horizon of new research. Oxford & IBH Publishing Co. Pvt. Ltd., New Delhi, pp. 95–110.

Glass, M.L., Soncini, R, (1997) Physiological shunts in gills of teleost fish: assessment of the evidence. In: Val., A.L., Almeida-Val., V.M.F. (eds) Biology of tropical fishes. INPA, Manaus, pp. 333–340.

Gluckmann, I., Vandewalle, P. (1998) Morphofunctional analysis of the feeding apparatus in four Pomacentridae species: *Dascyllus aruanus, Chromis retrofasciata, Chrysiptera biocellata* and *C. unimaculata*. Ital. J. Zool. **65** Suppl.: 421–424.

Gopesh, A., Srivastava, C.B.L. (1997) Paraneuronic nature of fish pseudobranchial neurosecretory cells. J. Freshwater Biol. **9**: 78–81.

Graham, J.B. (1976) Respiratory adaptations of marine air-breathing fishes. In: Hughes, G.M. (ed.) Respiration of amphibious vertebrates. Academic Press, London, pp. 165–187.

Greer Walker, M. (1970) Growth and development of the skeletal muscle fibres of the cod (*Gadus morhua* L.). J. Cons. int. Explor. Mer. **33**: 228–244.

Greer Walker, M., Horwood, J., Emerson, L. (1980) On the morphology and function of red and white skeletal muscle in the *Anchovies engraulis encrasicolus* L. and *E. mordax* Girard. J. mar. biol. Ass. U.K. **60**: 31–37.

Grosell, M., Jensen, F.B. (1999) NO_2 uptake and HCO_3 excretion in the intestine of the European flounder (*Platichthys flesus*). J. Exp. Biol. **202**: 2103–2110.

Guraya, S.S. (1999) Recent advances in gonadal development, sex differentiation and maturation, and sex reversal in fish. In: Saksena, D.N. (ed.) Ichthyology: recent research advances. Oxford & IBH Publishing Co. Pvt. Ltd., New Delhi, pp. 31–45.

Halstead, B.W. (1971) Venomous fishes. In: Bucherl, W., Buckley, E., Deulof, V. (eds.) Venomous animals and their venoms, vol. 2. Academic Press, New York, London, pp. 587–626.

Halstead, B.W., Bunker, N.C. (1952) The venom apparatus of the ratfish *Hydrolagus* colliei. Copeia **1952**: 128–133.

Halstead, B.W., Kuninobu, L.S., Hebard, H.G. (1953) Catfish, *Galeichthys felis* (Linnaeus). Trans. Amer. Fish. Soc. **72**: 297–314.

Halstead, B.W., Chitwood, M.J., Modglin, F.R. (1955) The anatomy of the venom apparatus of the zebrafish, *Pterois volitans* (Linnaeus). Anat. Rec. **122**: 317–333.

Hama, K. (1978) A study on the fine structure of the pit organ of the common Japanese sea eel *Conger myriaster*. Cell Tissue Res. **189**: 375–388.

Hamdani, El-Hassan., Stabell, O.B., Alexander, G., Døving, K.B. (2000) Alarm reaction in the crucian carp is mediated by the medial bundle of the medial olfactory tract. Chem. Senses **25**: 103–109.

Hamlett, W.C. (ed.) (1999) Sharks, skates and rays. The biology of elasmobranch fishes. The Johns Hopkins University Press, Baltimore, London.

Haneda, Y. (1955) Luminous organisms of Japan and the For East. In: Johnson, F.H. (ed.) The luminescence of biological systems. Amer. Assoc. Adv. Sci. Washington, D.C., pp. 335–385.

Haneda, Y., Johnson, F.H. (1962) The comparative anatomy of the indirect type of photogenic system of luminescent fishes, with special reference in *Paraprlacanthus beryciformes*. *Science Rep. Yokosuka city Museum No*. **7**: 1–10.

Haneda, Y., Tsuji, F.I., Sujiyama, N. (1969) Luminescent systems in apogonid fishes from the Philippines. Science **165**: 188–190.

Hansen, A., Zippel, H.P., Sorensen, P.W., Caprio, J. (1999) Ultrastructure of the olfactory epithelium in intact, axotomized, and bulbectomized goldfish, *Carassus auratus*. Microsc. Res. Tech. **45**: 325–338.

Hara, T.J. (ed.) (1982) Chemoreception in fishes. Chapman and Hall, London, New York.

Hara, T.J. (ed.) (1992) Fish chemoreception. Elsevier, Amsterdam.

Harden Jones, F.R. (1984) A view from the ocean. In: McCleave, J.D., Arnold, G.P., Dodson, J.J., Neill, W.H. (eds.) Mechanisms of migration in fishes. Plenum Publishing Corporation, New York, pp. 1–26.

Harder, V., Souza, R.H.S., Severi, W., Rantin, F.T., Bridges, C.R. (1999) The South American lungfish-Adaptations to an extreme habitat. In: Val, A.L., Almeida-Val, V.M.F. (eds.) Biology of tropical fishes. INPA, Manaus, pp. 87–98.

Hart, P.J.B. (1986) Foraging in teleost fishes. In: Pitcher, T.J. (ed.) The behaviour of teleost fishes. Croom Helm, London, pp. 211–235.

Hawkins, A.D., Amorim, M.C.P. (2000) Spawning sounds of the male haddock *Melanogrammu) alglefirus*. Env. Biol. Fish. **59**: 29–41.

Hawryshyn, C.W. (1998) Vision. In: Evans, D.H. (ed.) The physiology of fishes. (Second edition). CRC Press, Boca Raton, pp. 345–374.

Heilingenberg, W. (1997) Principles of electrolocation and jamming avoidance in electric fish. A meuroethological approach. Springer-Verlag, Berlin.

Henglmüller, S.M., Ladich, F. (1999) Development of agonistic behaviour and vocalization in croaking gouramis. J. Fish Biol. **54**: 380–395.

Herring, P.J. (2000) The biology of the deep ocean. Oxford University Press, Oxford.

Heisler, N. (1999) Passive ion permeability as a limiting factor for acid-base regulation in fish. In: Val, A.L., Almeida-Val, V.M.F. (eds.) Biology of tropical fish. INPA, Manaus, pp. 375–392.

Hinton, D.E., Lauren, D.J. (1990) Integrative histopathological approaches to detecting effects of environmental stressors on fishes. Trans. Fish. Soc. Symp. **8**: 51–66.

Hoar, W.S. (1965) The endocrine system as a chemical link between the organism and its environment. Trans. R. Soc. Can. **3**: 175–200.

Hofmann, M.H. (2001) The role of the fish telencephalon in sensory information processing. In: Kapoor, B.G., Hara, T.J. (eds.) Sensory biology of jawed fishes: new insights. Oxford & IBH Publishing Co. Pvt. Ltd., New Delhi, pp. 255–274.

Hopkins, T.E., Wood, C.M., Walsh, P.J. (1999) Nitrogen metabolism and excretion in an intertidal population of the gulf toadfish (*Opsanus beta*). Mar. Fresh. Behav. Physiol. **33**: 21–34.

Horn, M., Martin, K., Chotkowski, M. (eds.) (1998) Intertidal fish. Academic Press, San Diego.

Huet, L., Goosse, V., Parmentier, E., Vandewalle, P. (1999) About some skeletal particularities of the first vertebrae related to the mode of prey capture in *Uranoscopus scaber* (Uranoscopidae), Cybium **23**: 161–167.

Hughes, G.M. (1977) Ultrastructure of blood capillaries at the respiratory surfaces of fishes. Problémes de morphologie et d' embryologie expérimentales et compares. pp. 366–370.

Huysseune, A., Sire, J.Y. (1998) Evolution of patterns and processes in teeth and tooth-related tissues in non-mammalian vertebrates. Eur. J. Oral Sci. 106 (Suppl. 1): 437–481.

Inohaya, K., Yasumasu. S., Ishimaru, M., Ohyama, A., Iuchi. I, Yamagami, K. (1995) Temporal and spatial patterns of gene expression for the hatching enzyme in the teleost embryo, *Oryzias latipes*, Dev. Biol. **171**: 374–385.

Iwai, T. (1960) Notes on the luminous organ of the apogonid fish, *Siphamia majimai*. Ann. and Mag. Nat. Hist. Ser. 13, **2**: 545–550.

Iwama, G.K., Vijayan, M.M., Morgan, J.D. (1999) The stress response in fish. In: Saksena, D.N. (ed.) Ichthyology: recent research advances. Oxford & IBH Publishing Co. Pvt. Ltd., New Delhi, pp. 47–57.

Jakubowski, M., Zuwala, K. (2000) Taste organs in lower vertebrates: morphology of the gustatory organs in fishes. In: Dutta, H.M., Munshi, J.S.D. (eds.) Vertebrate functional morphology-horizon of research in the 21st century. Science Publishers, Inc. Enfield (NH), USA, pp. 159–172.

Jamieson, A. (1999) The necessity of applying genetic markers in fish stock management. In: Val, A.L., Almeida-Val, V.M.F. (eds.) Biology of tropical fishes. INPA, Manaus, pp. 293–302.

Jensen, D. (1965) The aneural heart of the hagfish. Ann. N.Y. Acad. Sci. **127**: 443–458.

Jensen, F.B., Fago, A., Weber, R.E. (1998) Hemoglobin structure and function. In: Perry, S.F., Tufts, B.L. (eds.) Fish physiology, vol. 17. Fish respiration. Academic Press, San Diego, pp. 1–40.

Jha, J., Haque, Z. (1999) Functional organisation and scaling of bimodal gas exchangers in relation to body weight of an estuarine amphibious goby, *Pseudapocryptes lanceolaas*. In: Saksena, D.N. (ed.) Ichthyology: recent research advances. Oxford & IBH Publishing Co. Pvt. Ltd., New Delhi, pp. 261–280.

Jobling, M. (1999) Methols for the study of fish feeding and growth. In: Saksena, D.N. (ed.) Ichthyology: recent research advances. Oxford & IBH Publishing Co. Pvt. Ltd., New Delhi, pp. 75–109.

Johnston, I.A. (ed.) (2000) Muscle development and growth, Fish physiology, vol. 18. Academic Press, San Diego.

Jones, D.R. (1999) The teleost bulbus arteriosus: form and functions. In: Saksena, D.N. (ed.) Ichthyology: recent research advances. Oxford & IBH Publishing Co. Pvt. Ltd., New Delhi, pp. 111–116.

Jorgensen, C.B., Larsen, L.O. (1960) Comparative aspects of hypothalamic-hypophyseal relationships. Ergeb. der Biol. **22**: 1–29.

Jorgensen, J.H., Lomholt, J.P., Weber, R.E., Malte, H. (eds.) (1998) The biology of hagfishes. Chapman and Hall, London, New York.

Kasumyan, A.O., Malyukina, G.A., Devitsina, G.V., Marusov, E.A., Lebedeva, N.E., Chervova, L.S., Golovkina, T.V., Sidorov, S.S., Golubev, A.V. (1999) Chemosensory systems in fishes: morphology, physiology and role in behaviour. In: Saksena, D.N. (ed.) Ichthyology: recent research advances. Oxford & IBH Publishing Co. Pvt. Ltd., New Delhi, pp. 117–144.

Kerr, T. (1942) A comparative study of some teleostean pituitaries. Proc. Zool. Soc. Lond. **112**: 37–56.

Kerr, T. (1952) The scales of primitive living actinopterygians. Proc. Zool. Soc. Lond. **122**: 55–78.

Kerr, T. (1955) The scales of modern lungfish. Proc. Zool. Soc. Lond. **125**: 335–345.

Kime, D.E. (1999) Environmentally induced endocrine abnormalities in fish. In: Hester, R.E., Harrison, R.M. (eds.) Endocrine disrupting chemicals. Issues in environmental science and technology, No. 12. The Royal Society of Chemistry, Cambridge, UK., pp. 27–48.

Kawamoto, N., Higashi, T. (1965) Studies on the structure and function of the pharyngeal organ in the mullet, *Mugil cephalus* Linne. Bull. Coll. Agr. and Vet. Med., Nihon Univ. No. **21**: 1–8.

Kisia, S.M. (1999) Oxygen diffusion in gills and muscle of fish in relation to changes in some environmental and physiological conditions. In: Mittal, A.K., Eddy, F.B., Munishi, J.S.D. (eds.) Water/air transiton in biology. Oxford & IBH Publishing Co. Pvt. Ltd., New Delhi, pp. 249–256.

Klinkhardt, M. Tesche, M., Greven, H. (1995) Database of fish chromosomes. Westarp Wissenschaften, Magdeburg.

Knowles, F., Vollrath, L., Meurling, P. (1975) Cytology and neuroendocrine relations of the pituitary of the dogfish, *Scyliorhinus canicula*. Proc. R. Soc. Lond. B. **191**: 507–523.

Kobayasi, H. (1953) Comparative studies of the scale in Japanese freshwater fishes, with special reference to phylogeny and evolution. Jap. Jour. Ichthyol. **2**: 246–260.

Kobayashi, H., Okada, K., Yamada, C., Okawara, Y. (1986) The caudal neurosecretory system in fishes. In: Pang, P.K.T., Schreibman, M.P. (eds.) Vertebrate endocrinology: fundamentals and biomedical implications, vol. 1. Academic Press, Orlando, pp. 147–174.

Kong, H., Kahatapitiya., Kingsley, K., Salo, W.L., Anderson, P.M., Wang, Y.S., Walsh, P.J. (2000) Induction of carbomoyl phosphate synthetase III and glutamine synthetase mRNA confinement stress in gulf toadfish (*Opsanus beta*). J. Exp. Biol. **203**: 311–320.

Kramer, B. (1994) Communication hehaviour and sensory mechanisms in weakly electric fishes. In: Slater, P.J.B., Rosenblatt, J.S., Snowdom, C.T. Milinski, M. (eds.) Advances in the study of behavior, vol. 23. Academic Press, San Diego, pp. 233–270.

Kramer, B. (1997) Electric organ discharges and their relation to sex in mormyrid fishes. Naturwissenschaften **84**: 119–121.

Kramer, B. (1999) Mechanisms of signal analysis in *Eigenmannia* (Gymnotiformes): the jamming avoidance response and communication. In: Val, A.L., Almeida-Val, V.M.F. (eds.) Biology of tropical fishes. INPA, Manaus, pp. 41–61.

Kramer, B. (2001) Signal analysis in the communication of a weakly electric fish, *Eigenmanna virescens* (Gymnotiformes). In: Backhaus, W. (ed.) Neuronal coding of perceptual systems. Series in bisphysics and biocybcrnctics, vol. 9. World Scientific, New Jersey, pp. 314–333.

Kroese, A.B.A., Netten, S.M. van. (1986) The application of incident light polarization microscopy for the visualization of vertebrate sensory hair cells *in vivo*. J. Microscopy **145**: 309–317.

Kunz, Y.W. (1990) Ontogeny of retinal pigment epithelium photoreceptor complex and development of rhythmic metabolism under ambient light conditions. In: Osborne, N., Chader, J. (eds.) Progress in retinal research, vol. 9. Pergamon Press, Oxford, pp. 135–196.

Kunz, Y.W. (2001) Development of the eye of teleosts. In: Kapoor, B.G., Hara, T.J. (eds.) Sensory biology of jawed fishes: new insights. Oxford & IBH Publishing Co. Pvt. Ltd., New Delhi, pp. 1–18.

Ladich, F. (2000) Acoustic communication and the evolution of hearing in fishes. Phil. Trans. R. Soc. Lond. B. **355**: 1285–1288.

Ladich, F., Bass, A.H. (1996) Sonic/Vocal acousticolateralis pathways in teleost fishes: a transneuronal biocytin study in mochokid catfish. J. Comp. Neurol. **374**: 493–505.

Laurent, P., Rouzeau, J-D. (1972) Afferent neural activity from pseudobranch of teleosts. Effects of P_{O_2}, pH, osmotic pressure and Na^+ ions. Respir. Physiol. **14**: 307–331.

Lebedeva, N.Y. (1999) Skin and superficial mucus of fish: biochemical structure and functional role. In: Saksena, D.N. (ed.) Ichthyology: recent research advances. Oxford & IBH Publishing Co. Pvt. Ltd., New Delhi, pp. 177–193.

Lesiuk, T.P., Lindsey, C.C. (1978) Morphological peculiarities in the neck_bending Amazonian characoid fish. *Rhaphiodon vulpinus*. Can. J. Zool. **56**: 991–997.

Lindley, T.E., Scheiderer, C.L., Walsh, P.J., Wood, C.M., Bergman, H.L., Bergman, A.L., Laurent, P., Wilson, P., Anderson, P.M. (1999) Muscle as the primary site of urea cycle enzyme activity in an alkaline lake-adapted tilapia (*Oreochromis alcalicus grahami*). J. Biol. Chem. **274**: 29858–29861.

Lorenzo, D., Silva, A., Caputi, A.A., Borde, M., Macadar, O. (2001) Electrogeneration in South American weakly electric fish. In: Kapoor, B.G., Hara, T.J. (eds.) Sensory biology of jawed fishes: new insights. Oxford & IBH Publishing Co. Pvt. Ltd., New Delhi, pp. 121–159.

Maffia, M., Acierno, R., Rollo, M., Stoelli, C. (1998) Ion and metabolite transport through the intestinal membranes of the Antartic fish *Trematomus* bernacchii. In: di Prisco, G., Pisano, E., Clarke A. (eds.) Fishes of Antarctica. A biological overview. Springer-Verlag Italia, Milano. pp. 237–246.

Masagaki, A., Fujii, R. (1999) Differential actions of melatonin on melanophores of the threeline pencilfish, *Nannostomus trifasciatus*. Zool. Sci. **16**: 35–42.

Majumdar, N.N. (1952) On the anatomy and histology of the palate of *Cirrhina mrigala* (Hamilton), with special reference to the papillae present on it tagether with comments on their possible respiratory function. J. Zool. Soc. India **3**: 255–265.

McAllister, D.E. (1967) The significance of ventral bioluminescence in fishes. Journ. Fish. Res. Bd. Canada **24**: 537–554.

Meek, J., Nieuwenhuys, R. (1998) Holosteans and teleosts. In: Nieuwenhuys, R., Donkelaar, H.J. ten., Nicholson, C. The central nervous system of vertebrates, vol. 2. Springer-Verlag, Berlin, Heidelberg, New York, pp. 759–937.

Meissl, H. (1997) Photic regulation of pineal function. Biol. Cell **89**: 549–554.

Meseguer, J., Esteban, M.A., Mulero, V. (1999) Natural cytotoxic response in fish: a review. In: Saksena, D.N. (ed.) Ichthyology: recent research advances. Oxford IBH Publishing Co. Pvt. Ltd., New Delhi, pp. 195–208.

Metcalfe, J.D., Butler, P.J. (1986) The functional anatomy of the gills of the dogfish (*Scyliorhinus canicula*). J. Zool. Lond. A **208**: 519–530.

Metcalfe, J.D., Butler, P.J. (1988). The effects of alpha—and beta—adrenergic receptor blocade on gas exchange in the dogfish. (*Scyliorhinus canicula L.*) during normoxia and hypoxia. J. Comp. Physiol. B **158**: 39–44.

Milsom, W.K. (1999) Control of breathing in fish: role of chemoreceptor inputs. In: Saksena, D.N. (ed.) Ichthyology: recent research advances. Oxford and IBH Publishing Co. Pvt. Ltd., New Delhi, pp. 203–223.

Milsom, W.K., Sundin, L., Reid, S., Kalinim, A., Rantin, F.T. (1999) Chemoreceptor control of cardiovascular reflexes. In: Val, A.L., Almeida-Val, V.M.F. (eds.) Biology of tropical fishes. INPA, Manaus, pp. 363–374.

Mogdans, J., Bleckmann, H. (2001) The mechanosensory lateral line of jawed fishes. In: Kapoor, B.G., Hara, T.J. (eds.) Sensory biology of jawed fishes: new insights. Oxford & IBH Publishing Co. Pvt. Ltd., New Delhi, pp. 181–213.

Montgomery J.C., Macdonald, J.A. (1998) Evolution of sensory systems: a comparison of Antarctic and deep-sea ichthyofauna. In: di Prisco, G., Pisano, E., Clarke A. (eds.) Fishes of Antarctica. A biological overview. Springer-Verlag Italia, Milano, pp. 329–338.

Motta, P.J., Wilga, C.D. (1999) Anatomy of the feeding apparatus of the nurse shark, *Ginglymostoma cirratum*. J. Morphol. **241**: 33–60.

Motta, P.J., Wilga, C.D. (2001) Advances in the study of feeding behaviors, mechanisms, and mechanics of sharks. Env. Biol. Fish. **60**: 131–156.

Munoz, R.C., Motta, P.J. (2000) Interspecific aggression between two parrotfishes (*Sparisoma*, Scaridae) in the Florida Keys. Copeia **2000**: 674–683.

Munk, O. (1966) Ocular anatomy of some deep-sea teleosts. Dana-Report No. 70, pp. 5–71.

Munk, O., Bartelsen, E. (1980) On the esca light organ and its associated light-guiding structures in the deep-sea anglerfish *Chaenophyrne draco* (Pisces, Ceratioidei). Vidensk. Meddr. dansk naturh. Foren. **142**: 103–129.

Munshi, J.S.D. (1962) On the accessory respiratory organs of *Heteropneustes fossilis* Bloch. Proc. Roy. Soc. Edn. **B68**: 128–146.

Munshi, J.S.D. (1976) Gross and fine structure of the respiratory organs of air-breathing fishes. In: Hughes, G.M. (ed.) Respiration of amphibious vertebrates, Academic Press, London, New York, pp. 73–104.

Munshi, J.S.D., Olson, K.R., Ojha, J., Ghosh, T.K. (1986) Morphology and vascular anatomy of the accessory respiratory organs of the air-breathing climbing perch, *Anabas testudineus* (Bloch). Am. J. Anat. **176**: 321–331.

Munshi, J.S.D., Olson, K.R., Ghosh, T.K., Ojha, J. (1990) Vasculature of the head and respiratory argans in an obligate air-breathing fish, the swamp eel *Monopterus* (=*Amphipnous*) cuchia. J. Morphol. **203**: 181–201.

Murata, N., Fujii, R. (2000) Pigment—aggregating action of endothelins on medaka xanthophores. Zool. Sci. **17**: 853–862.

Nagahama, Y. (1986) Testis In: Pang, P.K.T., Schreibman, M.P. (eds.) Vertebrate endocrinology: fundamentals and biomedical implications, vol. 1. Academic Press, San Diego, pp. 399–437.

Nandi, J. (1962) The structure of the interrenal gland in teleost fishes. Univ. California Publ. Zoöl. **65**: 129–212.

Nath, P. (1999) Some aspects of teleost vitellogenesis. In: Saksena, D.N. (ed.) Ichthyology: recent research advances. Oxford & IBH Publishing Co. Pvt. Ltd., New Delhi, pp. 249–259.

Nauen, J.C., Lauder, G.V. (2000) Locomotion in scombrid fishes: morphology and kinematics of the finlets of the chub mackerel *Scomber japonicus*. J. Exp. Biol. **203**: 2247–2259.

Nelson, E.M. (1955) The morphology of the swim bladder and auditory bulla in the Holocentridae. Fieldiana: Zool. **37**: 121–130.

Nicol, J.A.C. (1967) The luminescence of fishes. Symp. Zool. Soc. Lond No. **19**: 27–55.

Nicol, J.A.C. (1989) The eyes of fishes. Clarendon Press, Oxford.

Nieuwenhuys, R. (1998a) Brachiopterygian fishes. In: Nieuwenhuys, R., Donkelaar, H.J. ten., Nicholson, C. The central nervous system of vertebrates, vol. 1. Springer-Verlag, Berlin, Heidelberg, New York, pp. 655–699.

Nieuwenhuys, R. (1998b) Chondrostean fishes. In: Nieuwenhuys, R., Donkelaar, H.J. ten., Nicholson, C. The central nervous system of vertebrates, vol. 1 Springer-Verlag, Berlin, Heidelberg, New York, pp. 701–757.

Nieuwenhuys, R. (1998c) Lungfishes. In: Nieuwenhuys, R., Donkelaar, H.J. ten., Nicholson, C. The central nervous system of vertebrates, vol. 2. Springer-Verlag, Berlin, Heidelberg, New York, pp. 939–1006.

Nieuwenhuys, R. (1998d) The coelacanth *Latimeria chalumnae*. In: Nieuwenhuys, R., Donkelaar, H.J. ten., Nicholson, C. The central nervous system of vertebrates, vol. 2. Springer-Verlag, Berlin, Heidelberg, New York, pp. 1007–1045.

Nieuwenhuys, R., Donkelaar, H.J. ten., Nicholson, C. (1998) The central nervous system of vertebrates. Vols. 1, 2. Springer-Verlag, Berlin, Heidelberg, New York.

Nikinmaa, M., Salama, A. (1998) Oxygen transport in fish. In: Perry, S.F., Tufts, B.L. (eds.) Fish physiology, vol. 17. Fish respiration. Academic Press, New York, pp. 141–184.

Nilsson, S. (1971) Adrenergic innervation and drug responses of the oval sphincter in the swimbladder of the cod (*Gadus morhua*). Acta Physiol. Scand. **83**: 446–453.

Nilsson, S. (1972) Autonomic vasomotor innervation in the gas gland of the swimbladder of a teleost (*Gadus morhua*). Comp. gen. Pharmac. **3**: 371–375.

Northcutt, R.G. (1985) The brain and sense organs of the earliest vertebrates: reconstruction of a morphotype. In: Foreman, R.E., Gorbman, A., Dodd, J.M., Olsson, R. (eds.) Evolutionary biology of primitive fishes. Plenum Publishing Corporation, New York, pp. 81–112.

Northcutt. R.G., Davis, R.E. (eds.) (1983a) Fish neurobiology, vol. 1. Brain stem and sense organs. University of Michigan Press, Ann Arbor.

Northcutt, R.G., Davis, R.E. (1983b) Telencephalic organization in ray-finned fishes. In: Davis, R.E., Northcutt, R.G. (eds.) Fish neurobiology, vol. 2. Higher brain areas and functions. University of Michigan Press, Ann Arbor, pp. 203–236.

Novales Flamarique, I., Hawryshyn, C.W. (1998) Photoreceptor types and their relation to the spectral and polarization sensitivities of clupeid fishes. J. Comp. Physiol. **A 182**: 793–803.

Osse, J.W.M., van den Boogaart, J.G.M. (1995) Fish larvae, development, allometric growth and the aquatic environment. ICES mar. Sci. Symp. **201**: 21–34.

Overholtzer, K.L., Motta, P.J. (1999) Comparative resource use by juvenile parrotfishes in the Florida Keys. Mar. Ecol. Prog. Ser. **177**: 177–187.

Pang, P.K.T., Schreibman, M.P. (eds.) (1986–1991) Vertebrate endocrinology: fundamentals and biomedical implications, vols. 1–4. Academic Press, Orlando, San Diego, New York.

Parenti, L.R., Thomas, K.R. (1998) Pharyngeal jaw morphology and homology in sicydiine gobies (Teleostei: Gobiidae) and allies. J. Morphol. **237**: 257–274.

Parkyn, D.C., Hawryshyn, C.W. (2000) Spectral and ultraviolet-polarisation sensitivity in juvenile salmonids: a comparative analysis using electrophysiology. J. Exp. Biol. **203**: 1173–1191.

Parmentier, E., Chardon, M., Poulicek, M., Busseres, J.C., Vandewalle, P. (1998) Morphology of the buccal apparatus and related structures in four species of Carapidae. Aust. J. Zool. **46**: 391–404.

Part, P., Wood, C.M., Gilmour, K.M., Perry, S.F., Laurent, P., Zadunaisky, J., Walsh, P.J. (1999) Urea and water permeability in the ureotelic gulf toadfish (*Opsanus beta*). J. Exp. Zool. **283**: 1–12.

Patzner, R.A. (1998) Gonads and reproduction in hagfishes. In: Jorgensen, J.M., Lomholt, J.P., Weber, R.E., Malte, H. (eds.) The biology of hagfishes. Chapman and Hall, London, New York, pp. 378–395.

Perks, A.M. (1999) Neurohypophysical principles at the base of the vertebrate tree: a personal and historical view of the elasmobranchs and cyclostomes. In: Saksena, D.N. (ed.) Ichthyology: recent research advances. Oxford & IBH Publishing Co. Pvt. Ltd., New Delhi, pp. 281–282.

Perry, S.F., McDonald, G. (1993) Gas exchange. In: Evans, D.H. (ed.) The physiology of fishes. CRC Press, Boca Raton, pp. 251–278.

Peters, R.C., Struik, M.L., Bretschneider, F. (2001) Electroreception in freshwater catfish: the biologically adequate stimulus. In: Kapoor, B.G., Hara, T.J. (eds.) Sensory biology of jawed fishes: new insights. Oxford & IBH Publishing Co. Pvt. Ltd., New Delhi, pp. 275–296.

Pettigrew, J.D., Collin, S.P., Fritsches, K. (2000) Prey capture and accommodation in the sandlance, *Limnichthyes fasciatus* (Creediidae; Telestei). J. Comp. Physiol. **A186**: 247–260.

Phleger, C.F. (1998) Buoyancy in marine fishes: direct and indirect role of lipids. Am. Zool. **38**: 321–330.

Pickering, A.D. (1997) Husbandry and stress. In: Bernoth, E.M., Ellis, A.E., Midtlyng, P.J., Olivier, G., Smith, P.R. (eds.) Furunculosis. Furunculosis. Academic Press London, pp. 178–202.

Pickering, A.D., Pottinger, T.G. (1995) Biochemical effects of stress. In: Hochachka, P.W., Mommsen, T.P. (eds.) Biochemistry and molecular biology of fishes, vol. 5. Elsevier, Amsterdam, pp. 349–379.

Pisano, E., Mazzei, F., Derome, N., Ozouf-Costaz, C., Hurea, J.-C., di Prisco, G. (2001) Cytogenetics of the bathydraconid fish *Gymnodraco acuticeps* (Perciformes, Notothenioidei) from Terra Nova Bay, Ross Sea. Polar Biol. **24**: 846–852.

Popper, A.N., Fay, R.R. (1999) The auditory periphery in fishes. In: Fay, R.R., Popper, A.N. (eds.) Comparative hearing. Springer-Verlag, New York, pp. 43–100.

Raison, R.L., dos Remedios, N.J. (1998) The hagfish immune system. In: Jorgensen, J.M., Lomholt, J.P., Weber, R.E., Malte, H. (eds.) The biology of hagfishes. Chapman and Hall, London, New York, pp. 334–344.

Randall, D.J., Brauner, C.J. (1998) Interactions between ion and gas transfer in freshwater teleost fish. Comp. Biochem. Physiol. **119** A: 3–8.

Randall, D.J., Brauner, C.J. (1999) Factors influencing the optimization of haemoglobin—oxygen transport in fish. In: Saksena, D.N. (ed.) Ichthyology: recent research advances. Oxford & IBH Publishing Co. Pvt. Ltd., New Delhi, pp. 293–298.

Raschi, W.G., Aadlond, C. Keithar, E.D. (2001) A morphological and functional analysis of the ampullae of Lorenzini in selected galeoid sharks. In: Kapoor, B.G., Hara, T.J. (eds.) Sensory biology of jawed fishes: new insights. Oxford & IBH Publishing Co Pvt. Ltd., New Delhi. pp. 297–316.

Reese, E.S. (1991) How behaviour influences community structure of butterflyfishes (Family Chaetodontidae) on Pacific coral reefs. Ecol. Int. Bull. **19**: 29–41.

Reese, E.S. (1993) Reef fishes as indicators of conditions on coral reefs. Proc. Coll. Global aspectls of coral reef: health, hazards, and history. University of Miami, Florida, pp. 59–65.

Reid, S.G. (1999) Control of catecholamine secretion from fish chromaffin cells. In: Val, A.L., Almeida-Val, V.M.F. (eds.) Biology of tropical fishes. INPA, Manaus, pp. 401–411.

Rishi, S. Rishi, K.K., Gill, P.K. (1999) Sex chromosomes in fishes. In: Saksena, D.N. (ed.) Ichthyology: recent research advances. Oxford & IBH Publishing Co. Pvt. Ltd., New Delhi, pp. 309–320.

Rocha, E., Monteiro, R.A.F. (1999) Histology and cytology of fish liver: a review. In: Saksena, D.N. (ed.) Ichthyology: recent research advances. Oxford & IBH Publishing Co. Pvt. Ltd., New Delhi, pp. 321–344.
Roth, A., Tscharntke, H. (1976) Ultrastructue of the ampullary elctroreceptors in lungfish and Branchiopterygii Cell Tissue Res. 173: 95–108.
Sadovy, Y., Shapiro, D.Y. (1987) Criteria for the diagnosis of hermaphroditism in fishes. Copeia **1987**: 136–156.
Satchell, G.H. (1991) Physiology and form of fish circulation. Cambridge University Press, Cambridge.
Saxena, S.C. (1959) Adhesive apparatus of a hill-stream cyprinid fish *Garra mullya* (Sykes). Proc. Natl. Inst Sci. India **25**: 205–214.
Schreibman, M.P., Schartl, M., Kallman, K.P., Magliulo–Capriano, L. (1994) Molecular approaches to study the genetic regulation of the fish reproductive system. In: Perspectives in comparative endocrinology, pp. 343–351.
Schwassmann, H.O. (1978) Ecological aspects of electroreception. In: Ali, M.A. (ed.) Sensory ecology. Review and perspectives. Plenum Press New York, pp. 521–533.
Shand, J., Foster, R.G. (1999) The extraretinal photoreceptors of non-mammalian vertebrates. In: Archer, S.N., Djamgoz, M.B.A, Loew, E.R., Partridge, J.C., Vallerga, S. (eds.) Adaptive mechanism in the ecology of vision. Kluwer Academic Publishers, Dordrecht, pp. 197–222.
Shand, J., Archer, M.A., Collin, S.P. (1999) Ontogenetic changes in the retinal photoreceptor mosaic in a fish, the black bream, *Acanthopagrus butcheri*, J. Comp. Neurol. **412**: 203–217.
Shand, J., Chin, S.M., Harman, A.M., Collin, S.P. (2000a) The relationship between the position of the retinal area centralis and feeding behaviour in juvenile black bream *Acanthopagrus butcheri* (Sparidae: Teleostei). Phil. Trans. R. Soc. Lond. B **355**: 1183–1186.
Shand, J., Chin, S.M., Harman, A.M., Moore, S., Collin, S.P. (2001b) Variability in the location if the retinal ganglion cell area centralis is correlated with ontogentic changes in feeding behavior in the black bream, *Acanthopagrus butcheri* (Sparidae, Teleostei). Brain Behav. Evol. **55**: 176–190.
Shephard, K.L. (1994) Functions for fish mucus. Rev. Fish Biol. Fish. **4**: 401–429.
Sheridan, M.A. (1994) Regulation of lipid metabolism in poikilothermic vertebrates. Comp. Biochem. Physiol. **107B**: 495–508.
Sindermann, C.J. (1966) Diseases of marine fishes. Adv. mar. Biol. **4**: 1–89.
Sire, J.Y. Huysseune, A., Meunier, F.J. (1990) Osteoclass in teleost fish: light and electro-microscopical observations. Cell Tissue Res. 260: 85–94.
Sivak, J.G., Andison, M.E., Pardue, M.T. (1999) Vertebrate optical structure. In: Archer, S.N., Djamgoz, M.B.A., Loew, E.R., Partridge, J.C., Vallerga, S. (eds.) Adaptive mechanisms in the ecology of vision. Kluwer Academic Publishers, Dordrecht, pp. 73–94.
Smeets, W.J.A.J. (1998) Cartilaginous fishes. In: Nieuwenhuys, R., Donkelaar, H.J. ten., Nicholson, C. The central nervous system of vertebrates, vol. 1. Springer-Verlag, Berlin, Heidelberg, New York, pp. 551–654.
Smith, C.P., Wright, P.A. (1999) Molecular characterization in an elasmobranch urea transporter. Amer. J. Physical. Regulatory Integrative Comp. Physical. **276**. R622–R626.
Somero, G.N., Fields, P.A., Hofmann, G.E., Weinstein, R.B., Kawall, H. (1998) Cold adaptation and stenothermy in Antarctic notothenioid fishes: What has been gained and what has been lost? In: di Prisco, G., Pisano, E., Clarke A. (eds.) Fishes of Antarctica. A biological overview. Springer-Verlag Italia, Milano. pp. 97–109.
Srivastava, C.B.L. (1996) Catfish morphology: a reappraisal. In: Munshi, J.S.D., Dutta, H.M. (eds.) Fish morphology-horizon of new research. Oxford and IBH Publishing Co. Pvt. Ltd., New Delhi, pp. 21–30.
Srivastava, C.B.L. Gopesh, A. (1987) A third system of neurosecretion in fish-the pseudobranchial neurosecretory system. In: Bhatnagar, V.S. (ed.) Advances in cytology and genetics, pp. 122–128. (Publisher: Editor).
Srivastava, C.B.L., Singh, M. (1980) Occurrence of carotid labyrinth in the catfish group of teleost fishes. Experientia **36**: 651–652.
Srivastava, C.B.L. Gopesh, A., Singh, M. (1981) A new neurosecretory system in fish, located in the gill region. Experientia **37**: 850–851.
Srivastava, C.B.L., Tripathi, S. Pandey, K.N., Singh, M. (1988). Fish carotid labyrinth—a transformed pseudobranch. Arch Biol. (Bruxelles) **99**: 33–50.

Srivastava, C.B.L., Tripathi, S., Pandey, K.N. (1994) Fish carotid labyrinth. In: Singh, H.R. (ed.) Advances in fish biology and fisheries, vol. 1. Hindustan Publishing Corporation, Delhi, pp. 84–97.

Stam, W.T., Beintema, J.J., D' Avino, R.D., Tamburrini, M., Cocca, E., di Prisco, G.D. (1998) Evolutionary studies on teleost hemoglobin sequences. In: di Prisco, G., Pisano, E., Clarke, A. (eds.) Fishes of Antarctica. A biological overview. Springer-Verlag Italia, Milano. pp. 355–359.

Stiassny, M.L.J., Parenti, L.R., Davidson, G.D. (eds.) (1996) Interrelationships of fishes. Academic Press, San Diego.

Tagliafierro, G. (1999) The neuroendocrine system in the gastrointestinal tract and pancreas of fishes. In: Saksena, D.N. (ed.) Ichthyology: recent research advances. Oxford & IBH Publishing Co. Pvt. Ltd., New Delhi, pp. 369–388.

Tagliafierro, G., Faraldi, G. (1984) Immunohistochemical localization of peptide hormone producing cells in the gastro-entero-pancreatic system of an elasmobranchian cartilaginous fish. 7th Int. Cong. Histochemistry and Cytochemistry, Helsinki, Finland, 5–11 August, 1984, p. 424.

Tagliafierro, G., Zaccone, G. (2001) Morphology and immunohistochemistry of taste buds in bony fishes. In: Kapoor, B.G., Hara, T.J. (eds.) Sensory biology of jawed fishes: new insights. Oxford & IBH Publishing Co. Pvt. Ltd., New Delhi, pp. 335–345.

Tagliafierro, G., Carline, M., Faraldi, G., Gallus, L. (1998) The neuroendocrine system in the intestinal tract and pancreas of Antartic fish. In: di Prisco, G., Pisano, E., Clarke A. (eds.) Fishes of Antarctica. A biological overview. Springer-Verlag Italia, Milano. pp. 247–256.

Tester, A.L. (1963) Olfaction, gustation, and the common chemical senses in sharks. In: Gilbert, P.W. (ed.) Sharks and survival. D.C. Heath and Co., Boston, pp. 255–282.

Theisen, B. (1972) Ultrastructure of the olfactory epithelium in the Australian lungfish *Neoceratodus forsteri*. Acta Zool. Stockholm **53**: 205–218.

Theisen, B. (1973) The olfactory system in the hagfish *Myxine glutinosa*. I. Fine structure of the apical part system of the olfactory epithelium. Acta Zool. Stockholm **54**: 271–284.

Todgham, A.E., Anderson, P.M., Wright, P.A. (2001) Effects of exercise on nitrogen excretion, carbomoyl phosphate synthetase III activity and related urea cycle enzymes in muscle and liver tissues of juvenile rainbow trout (*Oncorhynchus mykiss*). Comp. Biochem. Physical. A **129**: 527–539.

Val, A.L., Marcon, J.L., da Costa, O.T.F., Barcellos, J.F.M., Garcia, J.T.M., Almeida-Val, V.M.F. (1999) Fishes of the Amazon: surviving environmental changes. In: Saksena, D.N. (ed.) Ichthyology: recent research advances. Oxford & IBH Publishing Co. Pvt. Ltd., New Delhi, pp. 389–402.

van Dijk, P.L.M., Hardewing, I., Pörtner, H.O. (1998) Exercise in the cold: high energy turnover in Antarctic fish. In: di Prisco, G., Pisano, E., Clarke, A. (eds.) Fishes of Antarctica. A biological overview. Springer-Verlag Italia, Milano. pp. 225–236.

Vandewalle, P., Parmentier, E., Poulicek, M., Bussers J.C., Chardon, M. (1998) Distinctive anatomical features of the branchial basket in four Carapidae species (Ophidiiformes, Paracanthopterygii. Eur. J. Morph. **36**: 153–164.

Vanegas, H., Ito, H. (1983) Morphological aspects of teleostean visual system: a review. Brain Res. Rev. **6**: 117–137.

Viladiu, C., Vandewalle, P., Osse, J.W.M., Casinus, A. (1999) Suction feeding strategies of two species of Mediterranean Serranidac (*Serranus cabrilla* and *Serranus seriba*). Neth. J. Zool. **49**: 81–95.

Wagner, H.J. Fröhlich, E., Negishi, K., Collin, S.P. (1998) The eyes of deep-sea fishes II. Functional morphology of the retina. Prog. Retinal Eye Res. **17**: 637–685.

Wainwright, P.C. (1999) Ecomorphology of prey capture in fishes. In: Saksena, D.N. (ed.) Ichthyology: recent research advances. Oxford & IBH Publishing Co. Pvt. Ltd., New Delhi, pp. 403–415.

Walsh, P.J., Wood, C.M. Moon, T.W. (1998) Red blood cell metabolism. In: Perry, S.F., Tufts, B.L. (eds) Fish Physiology, vol. 17. Fish respiration, Academic Press, New York, pp. 41–73.

Walsh, P.J., Handel-Fernandez, M.E., Vincek, V. (1999) characterization and sequencing of glutamine synthetase cDNA from liver of the ureotelic gulf toadfish (*Opsanus beta*). Comp. Biochem. Physiol. **B124**: 215–259.

Weber, R.E. (1996) Haemoglobin adaptationss in Amazonian and temperate fish with special reference to hypoxia, allasteric effectors and functional heterogeneity. In: Val. A.L., Almeida-Val, V.M.F., Randall, D.J. (eds.) Physiology and biochemistry of the fishes of the Amazon. INPA Manaus, pp. 75–90.

Weisel, G.F. (1975) The integument of the paddlefish. *Polyodon spathula*. J. Morphol. **145**: 143–150.

Weisel, G.F. (1978) The integument and caudal filament of the shovelnose sturgeon, *Scaphirhynchus platorynchus*. Amer. Midland Nat. **100**: 179–189.

Weisel, G.F. (1979) Histology of the feeding and digestive organs of the shovelnose sturgeon, *Scaphirhynchus platorynchus*. Copeia **1979**: 518–525.

Whitear, M. (1993) Epithelial sensory cells in fish. In: Singh, B.R. (ed.) Advances in fish research, vol. 1. Narendra Publishing House, Delhi, pp. 169–184.

Wicht, H., Nieuwenhuys, R. (1998) Hagfishes (Myxinoidea). In: Nieuwenhuys, R., Donkelaar, H.J. ten., Nicholson, C. The central nervous system of vertebrates, vol. 1. Springer-Verlag, Berlin, Heidelberg, New York, pp. 497–549.

Widder, E.S. (1999) Biolumineocence. In: Archer, S.N., Djamgoz, M.B.A., Loew, E.R., Partridge, J.C., Vallerga, S. (eds.) Adaptive mechanisms in the ecology of vision. Kluwer Academic Publishers, Dordrecht, pp. 555–581.

Wiley, M.L., Collette, B.B. (1970) Breeding tubercles and contact organs in fishes: their occurrence, structure, and significance. Bull. Amer. Mus. Nat. Hist. **143**: 143–216.

Wilga, C.D., Lauder, G.V. (1999) Locomotion in sturgeon: function of the pectoral fins. J. Exp. Biol. **202**: 2413–2432.

Wilga, C.D, Motta, P.J. (1998) Feeding mechanism of the Atlantic guitarfish *Rhinobatos lentiginosus*: modulation of kinematic and motor activity. J. Exp. Biol. **201**. 3167–3184.

Wilga, C.D., Motta, P.J. (2000) Durophagy in sharks: feeding mechanics of the hammerhead *Sphyrna tiburo*. J. Exp. Biol. **203**: 2781–2796.

Wilga, C.D., Wainright, P.C., Motta, P.J. (2000) Evolution of jaw depression mechanics in aquatic vertebrates: insights from Chondrichthyes. Biol. J. Linn. Soc. **71**: 165–185.

Witten, P.E., Villwock, W., Watermann B. (1999a) About the use of radiological, histochemical procedures to visualize bone resorpation at developing and at pathological altered skeletons of advanced teleosts. In: Wedekhind, H. (ed.) Krankheiten der Aquatischen Organismen EAFP-Schrift Zur Tagung der deutschen Sketion der European Association of Fish Pathologists (EAFB) am 23–25, September 1998. Institut fur Binnenfischerei Postdam Sacrow e.v., GroB Glinicke: 188–200.

Witten, P.E., Holliday, L.S., Delling, G., Hall, B.K. (1999b) Immunohistochemical identification of a vascular proton pump (V-ATPase) in bone-resorbing cells of an advanced teleost species, *Oreochromis niloticus*. J. Fish Biol. **55**: 1258–1272.

Wohrmann, A.P.A. (1998) Aspects of eco-physiological adaptations in Antarctic fish. In: di Prisco, G., Pisano, E., Clarke, A. (eds.) Fishes of Antarctica. A biological overview. Springer-Verlag Italia, Milano. pp. 119–128.

Wright, G.M., Keeley, F.W., DeMont, M.E. (1998) Hagfish cartilage. In: Jorgensen, J.M., Lomoholt, J.P., Weber, R.E., Malte, H. (eds.) The biology of hagfishes. Chapaman and Hall, London, New York, pp. 160–170.

Wright, P.A., Anderson P.M. (eds.) (2001) Nitrogen excretion. Fish physiology, vol. 20. Academic Press, San Diego.

Yamagami, K. (1996) Studies on the hatching enzyme (choriolysin) and its substrate, egg envelope, constructed of all precursors (chriogenins) in *Oryzias latipes*: a sequel to the information in 1991/1992. Zool. Sci. **13**: 331–340.

Yata, H., Tsuji, F.I. (1976) The luminescent systems of pony-fishes. J. Morphol. **150**: 539–552.

Yew, D.T., Cho, E.Y.P., Kwong, W.H., Lai, H.W.L., Li, W.W.Y., Lam, K.L. (2001) The eyes of jawed fishes. In: Kapoor, B.G., Hara, T.J. (eds.) Sensory biology of jawed fishes: new insights. Oxford & IBH Publishing Co. Pvt. Ltd., New Delhi, pp. 19–55.

Youson, J.H. (1985) Organ development and specialization in lamprey species. In: Foreman, R.E., Gorbman, A., Dodd, J.M., Olsson, R. (eds.) Evolutionary biology of primitive fishes. Plenum Publishing Corporation, New York, pp. 141–164.

Yazdani, G.M., Alexander, R. McN. (1967) Respiratory currents of flatfish. Nature (London) **213**: 96–97.

Zaccone, G., Ainis, L., Fasulo, S., Mauceri, A. Lauriano, E.R. Licata, A. Lo Cascio, P. (1999a) Paraneurons in the skin epithelium of the fishes. Mittal, A.K., Eddy, F.B., Munshi, J.S.D. (eds.) Water/air transtion in biology Oxford & IBH Publishing Co. Pvt. Ltd., New Delhi pp. 187–213.

Zaccone, G., Mauceri, A. Ainis, L., Fasulo, S., Licata, A. (1999b) Paraneurons in the skin and gills of fishes. In: Saksena, D.N. (ed.) Ichthyology: recent research advances. Oxford and IBH Publishing Co. Pvt. Ltd., New Delhi, pp. 417–447.

Zaccone, G. Kapoor, B.G., Fasulo, S., Ainis, L. (2001) Structural, histochemical and functional aspects of the epidermis of fishes. In: Southward, A.J., Tyler, P.A., Young, C.M., Fuiman, L.A. (eds.) Advances in marine biology, vol. 40. Academic Press, London, pp. 253–348.

Zapata, A.G., Chiba, A. Varas, A. (1996a) Cells and tissues of the immune system of fish. In: Iwama, G., Nakanishi, T. (eds.) Fish physiology, vol. 15. The fish immune system: organisms, pathogen and environment. Academic Press, San Diego, pp. 1–62.

Zieliniski, B.S., Hara, T.J. (2001) The neurobiology of fish olfaction. In: Kapoor, B.G., Hara, T.J. (eds.) Sensory biology of jawed fishes: new insights. Oxford & IBH Publishing Co. Pvt. Ltd., New Delhi, pp. 347–366.

Subject Index

MIX
Papier aus verantwortungsvollen Quellen
Paper from responsible sources
FSC® C105338

If you have any concerns about our products,
you can contact us on
ProductSafety@springernature.com

In case Publisher is established outside the EU,
the EU authorized representative is:
Springer Nature Customer Service Center GmbH
Europaplatz 3, 69115 Heidelberg, Germany

Printed by Libri Plureos GmbH
in Hamburg, Germany